Conference Record of

The Twenty-Fifth Asilomar Conference on Signals, Systems & Computers

Volume 1 of 2

1951-1991
40 YEARS OF SERVICE

IEEE COMPUTER SOCIETY
A member society of the
Institute of Electrical and Electronics Engineers, Inc.

Conference Record

Twenty-Fifth Asilomar Conference on Signals, Systems & Computers

NOVEMBER 4–6, 1991
PACIFIC GROVE, CALIFORNIA

Volume 1 of 2

1951 · 1991
IEEE Computer Society Press

The Institute of Electrical and Electronics Engineers, Inc.

Conference Record of

The Twenty-Fifth Asilomar Conference on Signals, Systems & Computers

Volume 1 of 2

November 4–6, 1991 *Pacific Grove, California*

Edited by
Ray R. Chen
San Jose State University

Sponsored by
Naval Postgraduate School
Monterey, California
and
San Jose State University
San Jose, California

In Cooperation with
IEEE Computer Society

1951–1991
40 YEARS OF SERVICE

IEEE Computer Society Press
Los Alamitos, California
Washington • Brussels • Tokyo

The papers in this book comprise the proceedings of the meeting mentioned on the cover and title page. They reflect the authors' opinions and, in the interests of timely dissemination, are published as presented and without change. Their inclusion in this publication does not necessarily constitute endorsement by the editors, the IEEE Computer Society Press, or the Institute of Electrical and Electronics Engineers, Inc.

Published by the
IEEE Computer Society Press
10662 Los Vaqueros Circle
PO Box 3014
Los Alamitos, CA 90720-1264

IEEE Computer Society Press Order Number 2470
Library of Congress Number 89-640232
ISSN 1058-6393
IEEE Catalog Number 91CH3112-0
ISBN 0-8186-2470-1 (paper)
ISBN 0-8186-2471-X (microfiche)
ISBN 0-8186-2472-8 (case)

Additional copies can be ordered from

IEEE Computer Society Press
Customer Service Center
10662 Los Vaqueros Circle
PO Box 3014
Los Alamitos, CA 90720-1264

IEEE Service Center
445 Hoes Lane
PO Box 1331
Piscataway, NJ 08855-1331

IEEE Computer Society
13, avenue de l'Aquilon
B-1200 Brussels
BELGIUM

IEEE Computer Society
Ooshima Building
2-19-1 Minami-Aoyama
Minato-ku, Tokyo 107
JAPAN

Cover designed by Jack Ballestero

Editorial Production: Edna Straub, Penny Storms, and Lisa O'Conner
Printed in the United States of America by McNaughton & Gunn, Inc.

THE INSTITUTE OF ELECTRICAL AND ELECTRONICS ENGINEERS, INC.

Foreword

The Twenty-Fifth Annual Asilomar Conference on Signals, Systems & Computers was held November 4–6, 1991, at the Asilomar Hotel and Conference Grounds, Pacific Grove, California. The conference was sponsored by the Naval Postgraduate School and San Jose State University, in cooperation with the IEEE Computer Society.

The conference was devoted to the presentation of new ideas and preliminary research results in the areas of theoretical and applied signal processing, computer algorithms, circuits, systems, communications, and image processing.

The papers contained in this record were selected after careful screening from many excellent submissions. The 1991 Conference Committee consists of

General Chair:	Frederic J. Harris *San Diego State University*
Technical Program Chair:	Neil Jablon *AT&T Bell Labs*
Technical Program Committee:	John J. Shynk *University of California, Santa Barbara*
	Hari Krishna *Syracuse University*
	James Ritcey *University of Washington*
	Robert G. Redinbo *University of California, Davis*
	Huseyin Abut *San Diego State University*
	Rabinder N. Madan *Office of Naval Research*
Publications Chair:	Ray R. Chen *San Jose State University*
Publicity Chair:	Avtar Singh *San Jose State University*
NPS Coordinator:	Monique Fargues *Naval Postgraduate School*
Finance Chair:	Ralph Hippenstiel *Naval Postgraduate School*

Gratitude is expressed to the attendees, authors, session chairs, organizers, and reviewers whose interest and efforts contributed much to the success of the conference.

The Twenty-Sixth Annual Asilomar Conference on Signals, Systems & Computers is scheduled for October 26–28, 1992 at the Asilomar Hotel and Conference Grounds, Pacific Grove, California. Further information about this conference may be obtained by contacting the Technical Program Chair, James A. Ritcey, Department of Electrical Engineering FT-10, University of Washington, Seattle, WA 98195.

Ray R. Chen
Editor

Table of Contents

Volume 1

Session MP4: Radar and Sonar Signal Processing
Chair: *J. Ritcey*, University of Washington, Seattle, WA

Session MP5: Knowledge-Based Signal Processing
Chair: *D.D. Sworder*, University of California, San Diego, La Jolla, CA

Session MP6: Adaptive Filtering I
Chair: *N.J. Bershad*, University of California, Irvine, CA

Session MP7: Matrix-Based Signal Processing (Poster Session)
Chair: *M.P. Fargues*, Naval Postgraduate School, Monterey, CA

Session MP8: Speech Processing II
Chair: *H. Abut*, San Diego State University, San Diego, CA

Session TA1: Spectral Estimation
Chair: *R. Hippenstiel*, Naval Postgraduate School, Monterey, CA

Session TA2a: Arithmetic for DSP

Chair: *L. Ciminiera*, Politecnico di Torino, Torino, Italy

Session TA2b: DSP Implementations

Chair: *B.P.M. Tao*, Rockwell International Corporation, Newport Beach, CA

Session TA3: Image Coding

Chair: *K.S. Thyagarajan*, San Diego State University, San Diego, CA

Session TA4: Neural Networks for Signal Processing

Chair: *J.-N. Hwang*, University of Washington, Seattle, WA

Volume 2

Session TP1: Blind Adaptive Processing
Chair: *B.G. Agee*, Radix Technologies, Inc., San Jose, CA

Session TP2: Quadrature-Mirror Filters, Wavelets, and Applications
Chair: *B.C. Levy*, University of California, Davis, CA

Session TP3: Biologically Motivated Signal Processing
Chair: *R.O. Duda*, San Jose State University, San Jose, CA

Session TP4: Military Applications
Chairs: *J.T. Rickard*, Orincon Corporation, San Diego, CA, and
G.M. Dillard, Naval Ocean Systems Center, San Diego, CA

Session TP5: Communication and Estimation
Chair: *J. Burl*, Naval Postgraduate School, Monterey, CA

Session TP6: Signal Subspace Methods
Chair: *A. Nehorai*, Yale University, New Haven, CT

Session TP7: Speech Processing I
Chair: *H. Abut*, San Diego State University, San Diego, CA

Session WA6: Communication Systems
Chairs: *L.J. Griffiths and M.J. Rude*, University of Southern California, Los Angeles, CA

Session WA7a: Computer Vision (Poster Session)
Chair: *K.S. Arun*, University of Illinois at Urbana-Champaign, Urbana, IL

Session WA7b: Testing and Fault Tolerance (Poster Session)
Chair: *C.-H. Lee*, Naval Postgraduate School, Monterey, CA

Preamble Recognition and Synchronization Detection in Partial-Response Systems

F. Dolivo and W. Schott
IBM Research Division, Zurich Research Laboratory, 8803 Rüschlikon, Switzerland

Abstract

To achieve fast initial gain and timing-phase adjustment in receivers for communication or storage systems using partial-response signaling, a synchronization preamble preceding user data is transmitted. A method for recognizing that synchronization has been achieved and that the preamble is being received, and for detecting the end of preamble is presented. The method is based on a maximum-likelihood test. Its efficiency is demonstrated by computing the probability of not recognizing the preamble when it is actually received and the probability of falsely recognizing the preamble when in fact user data are received.

1 Introduction

In receivers for communication or storage systems [1-3], the output signal of the receive filter is scaled and sampled prior to performing symbol-by-symbol or symbol sequence detection. A gain-control and a timing-recovery scheme [3,4] adjust the receiver gain and sample-timing phase to minimize sample-amplitude errors and undesired intersymbol interference. For fast initial adjustment of the gain and timing phase, a special synchronization preamble is sent ahead of user data. This paper describes a method to reliably recognize in binary partial-response (PR) class-IV (PR-IV) systems [2] that synchronization has been achieved and the preamble is being received, and to detect the end of the preamble. Recognition of these events is required to control operations in the receiver during the start-up phase and to signal the start of user data to the system.

The synchronization preamble consists of a timing sequence followed by a known sequence of bits marking its end. The timing sequence does not satisfy the constraints of the rate 8/9 code [5] used in PR systems and serves as a unique synchronization mark. The method for detecting preamble reception is based on a maximum-likelihood (ML) test performed on the samples received after synchronization has been achieved. This test determines whether the received samples are more likely to be caused by the timing sequence or by a rate 8/9 encoded data sequence. Once the timing sequence has been recognized, the ML test is used to detect the end of the preamble. Two kinds of errors, namely failing to recognize the preamble when it has indeed been received and falsely recognizing it when in fact user data have been received, will affect the reliability of the receiver start-up. The efficiency of the method to detect reception of the preamble was assessed by computing the miss and false-alarm probabilities associated with these two events.

Section 2 describes a PR system and its receiver start-up procedure. Section 3 describes the preamble-recognition method, and the resulting miss and false-alarm probabilities are derived in Section 4.

2 Partial-Response Signaling System

A PR signaling system is shown in Fig. 1. The binary sequence $\{b_n\}$, $b_n \in (0,1)$, consisting of either user data encoded by the rate 8/9 code or the preamble sequence employed for fast receiver synchronization, is precoded into the binary sequence $\{b_n'\} = \{b_n \oplus b_{n-2}'\}$, where $\oplus$ denotes modulo 2 addition. The precoded sequence is then converted into the data sequence $\{a_n\}$ by translating $b_n' = 0$ into $a_n = -1$ and $b_n' = 1$ into $a_n = +1$. The sequence $\{a_n = \pm 1\}$ is sent at the signaling rate $1/T$ through a communication (transmit filter and transmission medium) or storage channel and a band-limiting receive filter. The gain γ of the variable gain amplifier (VGA) is adjusted to compensate the unknown gain ζ of the channel and receive filter so that the total gain $g = \zeta\,\gamma$ of the overall system becomes unity. For PR-IV signaling, the response of the overall system with $g = 1$ to an isolated input signal $a_0 = +1$ is given by $h(t) = p(t) - p(t-2T)$ where $p(t)$ is a Nyquist signal element [3]. Sampling the received signal $y(t)$ (see Fig. 1) at times $nT + \tau$ yields

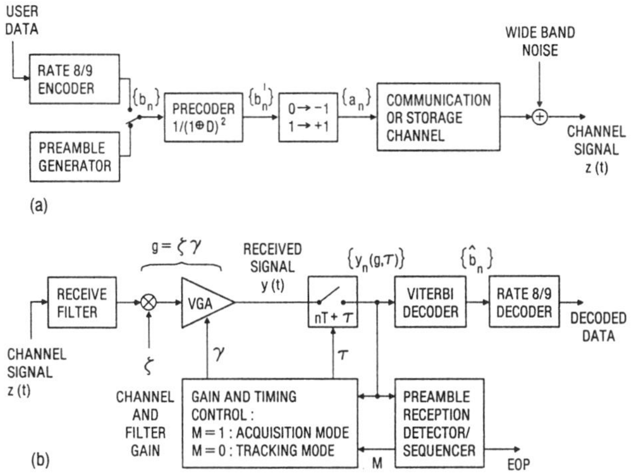

Figure 1. Partial-response signaling system. (a) Transmission and (b) reception.

$$y_n(g, \tau) = y(nT + \tau)$$
$$= g\left(\sum_m a_m\, h[(n-m)T+\tau] + r(nT+\tau)\right), \tag{1}$$

where $r(\bullet)$ denotes noise band-limited by the receive filter. For the correct gain $g=1$ and timing phase $\tau=0$, at ideal sampling instants nT, the received samples are given by

$$y_n(g=1, \tau=0) = x_n + r_n\,, \tag{2}$$

where, since $h(0)=1$, $h(2T)=-1$, and $h(nT)=0$, $n \neq 0,\ 2$, the data-signal samples are given by

$$x_n = a_n - a_{n-2}\,, \qquad x_n \in \{-2, 0, +2\}\,. \tag{3}$$

The noise samples $\{r_n\}$ are assumed to be uncorrelated and Gaussian with zero mean and standard deviation σ. The samples $\{x_n\}$ are the outputs of a discrete channel defined by the polynomial $1-D^2$, where D refers to a delay T, and depend for n even (odd) on the inputs $\{a_n,\ n\,\text{even (odd)}\}$. Thus PR-IV sequences can be viewed as two independent interleaved dicode PR sequences with polynomial $1-D'$, where D' refers to a delay $2T$ [1,3]. The PR dicode system can be described by the two-state trellis diagram shown in Fig. 2. The Viterbi decoder performs maximum-likelihood sequence detection by operating in an interleaved fashion on the two dicode sequences [1,3]. The probability P_e of an error event occurring is well-approximated at moderate and high signal-to-noise ratios by [2]

$$P_e \simeq 2Q(d_{\text{free}}/2\sigma) = 2Q(\sqrt{2}/\sigma)\,, \tag{4}$$

where $d_{\text{free}} = \sqrt{8}$ is the minimum Euclidian distance between any paths that separate and remerge in the two-state trellis diagram and

$$Q(x) = (1/\sqrt{2\pi})\int_x^\infty \exp(-y^2/2)\,dy\,. \tag{5}$$

The VGA gain γ and timing phase τ are adjusted by control schemes. For initial acquisition of gain and timing phase [4], the binary timing sequence

$$\{b_n\} = \{\ldots 1, 1, 1, 1, 1, 1, 1, 1 \ldots\} \tag{6}$$

is sent by the preamble generator, resulting in the timing data sequence

$$\{a_n\} = \{\ldots +1, +1, -1, -1, +1, +1, \ldots\}\,. \tag{7}$$

Special algorithms exploiting knowledge of the timing data sequence are employed in the acquisition mode.

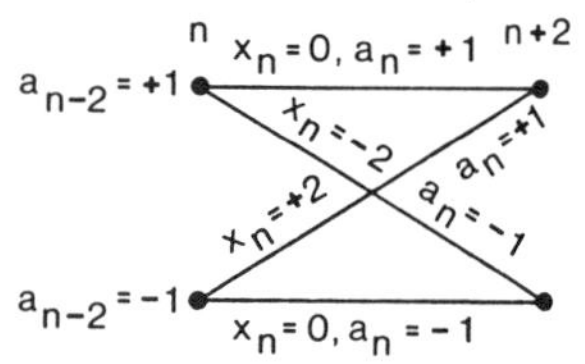

Figure 2. Trellis diagram for a dicode system.

Different algorithms and lower control-loop gains are used for gain and phase tracking during reception of random data. The operation mode of the control schemes is determined by a preamble-reception detector and a sequencer (see Fig. 1).

Codes for systems using PR-IV signaling with maximum-likelihood sequence detection [5] are described by the parameters (M, N, K), where M and K are the maximum run-length of 0's and 1's in the encoded binary sequence and N is the maximum number of consecutive 0's in the two subsequences with even or odd time indices. The constraint on M limits the run-length of zero-valued data-signal samples and guarantees satisfactory operation of gain and timing control. The constraint on N assures proper operation of the Viterbi decoder with a path-memory length of $2N$ bits. The constraint on K ensures that the all-1 binary timing sequence (6) can be employed as a unique synchronization mark.

3 Preamble Recognition

For fast startup of the receiver, a synchronization preamble is sent ahead of rate 8/9 encoded user data. The synchronization preamble comprises the timing sequence for fast initial acquisition of gain and timing phase, and a flag to mark the end of the preamble. The timing sequence is a unique synchronization mark that will be discriminated from rate 8/9 encoded data by a ML test. This test assumes that the received samples exhibit the PR-IV signaling format (see Eqs. (2) and (3)) and must therefore be performed when gain and timing phase are correctly adjusted, i.e., after synchronization.

3.1 Synchronization Detection

To achieve initial acquisition of the gain and timing phase, algorithms exploiting knowledge of the timing data sequence are activated in the receiver. Since $a_n = -a_{n-2}$ in the timing data sequence (7), the data signals (3) result in the two-level sequence

$$\{x_n = 2a_n\} = \{\ldots +2, +2, -2, -2, +2, +2, \ldots\}\,. \tag{8}$$

The gain control and timing recovery techniques are based on the computation of gain and timing gradients from received signal samples and estimates of the data signals (8). For timing-phase acquisition, it has been shown in [4] that, to eliminate hangup situations at the unstable equilibrium point halfway between the desired sampling instants, data-signal estimates should be reconstructed with a variable decision threshold

$$\hat{x}_n = \begin{cases} +2 & \text{for } y_n(g,\tau) \geq \eta_n \\ -2 & \text{for } y_n(g,\tau) < \eta_n\,, \end{cases} \tag{9}$$

where the decision threshold η_n is given by

$$\eta_n = \varepsilon\,\text{sign}\,(\hat{x}_{n-2})\,, \qquad \varepsilon \geq 0\,. \tag{10}$$

The error $e_n = y_n(g, \tau) - \hat{x}_n$ reduces to the noise r_n as the gain g and timing phase τ converge to their correct settings $g=1$ and $\tau=0$ (see Eq. (2)). Thus, correct adjustment of the gain and timing phase can be inferred from the frequency of occurrence of errors of magnitude smaller than δ and will be determined from the running sum

$$S_n = \begin{cases} S_{n-1} + 1 & \text{if } |e_n| < \delta \\ \max\,(S_{n-1} - 1,\ 0) & \text{otherwise}. \end{cases} \qquad (11)$$

This running sum is initialized to 0 at the time the receiver is put into the acquisition mode and remains non-negative. Synchronization is detected when the running sum reaches the value J, i.e., at the time index n_S for which $S_{n_S} = J$ for the first time. Computer simulations have shown the effectiveness of this method for $\delta = 0.5$ and $J = 5$.

3.2 Timing-Sequence Recognition

Synchronization can be achieved not only when the preamble is encountered but might also happen, given sufficient time and favorable initial conditions, while receiving rate 8/9 encoded data. These two cases will be distinguished by a test that determines whether the received samples are more likely to be caused by the rate 8/9 encoded data or by a replica of the timing data sequence (7) reconstructed in the receiver,

$$a_n^T = \begin{cases} \hat{x}_n/2 & \text{for } n = n_S,\ n_S + 1 \\ -a_{n-2}^T & \text{for } n \geq n_S + 2, \end{cases} \qquad (12)$$

where n_S is the time index at which synchronization is detected. The values of a_n^T at times n_S and $n_S + 1$ determine the phase of the reconstructed timing data sequence and are derived from Eq. (8), with $\hat{x}_n$ (see Eq. (9)) used in place of x_n. The non-zero variable threshold (10) greatly increases the likelihood that the sequence $\{\hat{x}_n\}$ will satisfy the condition $\hat{x}_n = -\hat{x}_{n-2}$, as should be the case according to Eq. (8), and makes the reconstruction method (9) very reliable. The probability of incorrectly reconstructing a_n^T at $n = n_S$ or $n_S + 1$, given by $P(\{a_n^T \neq a_n\}) = Q((2 + \varepsilon)/\sigma)$, is much smaller than the error-event probability P_e (see Eq. (4)). For $\varepsilon = 2$ and $\sigma = 0.32$, we have $P_e = 10^{-5}$ and $P(\{a_n^T \neq a_n\}) = 4 \times 10^{-36}$. It is thus extremely improbable that the phase of the replica (12) will be reconstructed incorrectly when the timing data sequence (7) is actually received.

The path of the timing data sequence replica $\{a_n^T\}$ in the trellis diagram for one of the two independent dicode schemes (n even or odd, see Fig. 2) is shown in Fig. 3. A data sequence that separates from $\{a_n^T\}$ at time $n = k - 4$ must remerge with it at time k to be at the minimum Euclidean distance $d_{\text{free}} = \sqrt{8}$. This single violation of $\{a_n^T\}$ at time $k - 2$,

$$\{a_n^V\} = \{\ldots\, a_{k-6}^V = -a_k^T,\ a_{k-4}^V = +a_k^T,\ a_{k-2}^V = +a_k^T,\ a_k^V = +a_k^T,\ a_{k+2}^V = -a_k^T \ldots\}, \qquad (13)$$

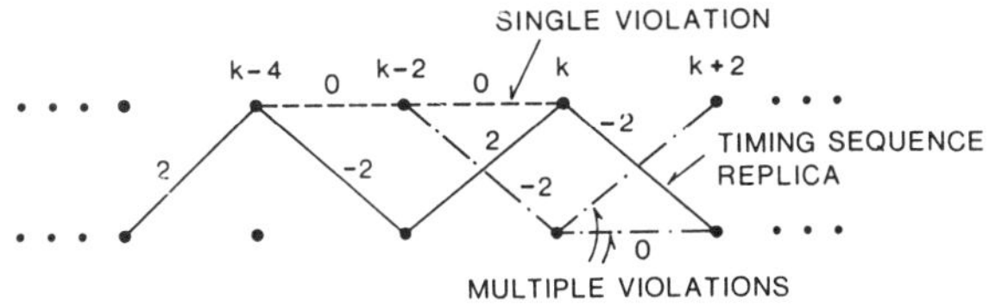

Figure 3. Timing-sequence replica and violations.

can be detected with a simple test derived from the ML principle. A single violation has occurred if the received samples $\{y_n(g, \tau)\}$ are closer in the mean-squared sense to the data samples $\{x_n^V = a_n^V - a_{n-2}^V\}$ than to $\{x_n^T = a_n^T - a_{n-2}^T\}$ (see (3)), i.e., if

$$\sum_n \left[y_n(g, \tau) - x_n^V\right]^2 < \sum_n \left[y_n(g, \tau) - x_n^T\right]^2. \qquad (14)$$

Using (13), this condition results in deciding at time k that a single violation occurred at time $k - 2$ if

$$a_k^T\,[y_k(g, \tau) - y_{k-2}(g, \tau)] < \rho\,, \qquad (15)$$

where $\rho = 2$, and that a timing pattern was present otherwise. Selecting $\rho \neq 2$ allows the decision to be biased in favor of a violation for $\rho > 2$ or the timing pattern for $\rho < 2$. When the gain and timing phase are correctly adjusted and for a given data sequence $\{a_n\}$, the probability that inequality (15) is not satisfied can be derived using Eqs. (2) and (3) and is given by:

$$Q\left(\left[\rho - a_k^T\,(a_k - 2a_{k-2} + a_{k-4})\right] / \sqrt{2}\,\sigma\right). \qquad (16)$$

The probability for a timing data sequence to be detected at time k when actually the single violation (13) is present is obtained from (16) with $a_n = a_n^V = a_k^T$ for $n = k,\ k - 2$, and $k - 4$:

$$P_{T/V} = Q(\rho / \sqrt{2}\,\sigma)\,. \qquad (17)$$

The probability of the other type of error to occur, which detects a violation when a timing pattern is present, is derived from (16) using $a_{k-4} = -a_{k-2} = a_k = a_k^T$:

$$P_{V/T} = 1 - Q\left((\rho - 4) / \sqrt{2}\,\sigma\right) = Q\left((4 - \rho) / \sqrt{2}\,\sigma\right). \qquad (18)$$

For $\rho = 2$, the two types of errors occur with an equal probability of $Q(\sqrt{2}/\sigma)$ which is half the error-event probability P_e (see Eq. (4)).

Violations to the timing data sequence replica other than the single violation (13) do not remerge at time k (see Fig. 3). The probability of failing to detect these violations with the ML test (15), obtained from (16) with $a_{k-4} = a_{k-2} = -a_k = a_k^T$, is given by $Q\left((\rho + 2) / \sqrt{2}\,\sigma\right)$ and is much smaller than $P_{T/V}$ (see (17)). Such violations are therefore detected even more reliably than single violations.

The (M, N, K) codes limit the run-length of 1's in the encoded sequence $\{b_n\}$ to K and thus enforce the condition that violations to the timing data sequence $\{a_n\}$ occur at least every $K + 3$ symbols. The timing data sequence will be recognized when, after synchronization has been detected with test (11), no more than n_V violations in a sequence of n_ℓ symbols are detected with the ML test (15). The probability P_{false}^{TR} of falsely recognizing the timing data sequence while receiving rate 8/9 encoded data and the probability P_{miss}^{TR} of failing to recognize the timing data sequence while receiving it will depend on the code used and the parameters n_V and n_ℓ. These probabilities will be evaluated in the next section for the (4,6,4) code.

3.3 End-Of-Preamble Detection

A data pattern of opposite phase to the timing data sequence is farthest apart in terms of Euclidean distance from the timing-data sequence replica (12) and can be more reliably recognized by the ML test (15) than any other pattern can. Therefore, the end-of-preamble (EOP) flag is selected as shown in Fig. 4,

$$a_1^F = +c, \quad a_2^F = +d, \quad a_3^F = -c, \quad a_4^F = -d \ldots , \qquad (19)$$

where $c = \pm 1$ and $d = \pm 1$ are the values assumed by the last two data symbols a_{L-1} and a_L of the timing data sequence. This choice provides the desired phase change with respect to the timing sequence replica (12). The probability that the timing data sequence violation generated by the ith data signal in flag (19) will not be recognized by the ML test (15) is obtained from (16) with $k = L + i$ using the sequence $\{a_n\}$ shown in Fig. 4 and noting that $a_{L+i}^T = -a_{L+i}$:

$$P_{\text{fail}}^F(i) = \begin{cases} Q\left[(\rho - 2)/\sqrt{2}\,\sigma\right] & \text{for } i = 1,\ 2 \\ Q\left[(\rho + 2)/\sqrt{2}\,\sigma\right] & \text{for } i = 3,\ 4 \\ Q\left[(\rho + 4)/\sqrt{2}\,\sigma\right] & \text{for } i \geq 5\,. \end{cases} \qquad (20)$$

The EOP flag will be recognized when n_F violations have been detected after the timing data sequence has been recognized. The probability of failing to detect the end of the preamble depends on n_F and the probabilities (18) and (20). This probability is computed in the next section.

TIMING SEQUENCE					EOP FLAG				
... a_{L-3}	a_{L-2}	a_{L-1}	a_L		a_{L+1}	a_{L+2}	a_{L+3}	a_{L+4}	...
... $-c$	$-d$	$+c$	$+d$		$+c$	$+d$	$-c$	$-d$	...
					a_1^F	a_2^F	a_3^F	a_4^F	...

TIMING SEQUENCE REPLICA								
... a_{L-3}^T	a_{L-2}^T	a_{L-1}^T	a_L^T	a_{L+1}^T	a_{L+2}^T	a_{L+3}^T	a_{L+4}^T	...
... $-c$	$-d$	$+c$	$+d$	$-c$	$-d$	$+c$	$+d$	...

Figure 4. End-of-preamble (EOP) flag and timing sequence replica.

4 Miss and False-Alarm Probabilities

4.1 Timing-Sequence Recognition

After detecting synchronization, the reception of the timing-data sequence will be recognized when the ML test (15) detects no more than n_V times the presence of a violation in a sequence of n_ℓ received samples $y_n(g, \tau)$. The probability P_{miss}^{TR} of not recognizing the timing data sequence when it is in fact received is given by the probability that the ML test (15) will detect at least $n_V + 1$ times the reception of a violation in a sequence of n_ℓ samples (2) with data signals given by (8). At moderate and high signal-to-noise ratios, the n_ℓ successive tests can be assumed to be statistically independent and the probability that the ML test (15) will decide for a violation more than $n_V + 1$ times can be neglected. The probability P_{miss}^{TR} can thus be approximated by

$$P_{\text{miss}}^{TR} \simeq \binom{n_\ell}{n_V+1} P_{V/T}^{\,n_V+1} (1 - P_{V/T})^{n_\ell - n_V - 1} , \qquad (21)$$

where $P_{V/T}$ is given by (18).

The probability of falsely recognizing the timing data sequence while actually receiving data encoded with the rate 8/9 (4,6,4) code is evaluated under the assumption that the binary encoded sequence

$$\{ \ldots\ 11110101111010\ \ldots \} \qquad (22)$$

is received after detecting synchronization at time $n = n_s$. This is a worst-case situation since this sequence results in the least number of violations in the sequence of received samples. For the worst-case sequence (22), obtaining n_p violations requires that at least $n_\ell = 7n_p$ samples are received. In order to distinguish the timing data sequence from an encoded data sequence, n_p must be greater than the parameter n_V. For recognizing reception of the timing sequence, the decision threshold in the ML test (15) is normally set to $\rho = 2$. In this case, the probability P_{false}^{TR} of falsely recognizing the timing data sequence while receiving the worst-case data sequence (22) can be approximated at moderate and high signal-to-noise ratios by

$$P_{\text{false}}^{TR} \simeq (1/4)^{n_p} \binom{n_p}{n_V} P_{T/V}^{\,n_p - n_V} \times (1 - P_{T/V})^{n_V + n_\ell - 3n_p} , \qquad (23)$$

where $P_{T/V}$ is given in (17).

In Table 1, the miss and false-alarm probabilities P_{miss}^{TR} and P_{false}^{TR} for timing-sequence recognition are given for various parameters n_p and n_V. The probabilities were calculated for an error-event probability of $P_e = 10^{-5}$. The decision threshold ρ of the ML test (15) was set to 2. The results show that deciding that the timing sequence is being received when no more than $n_V = 1$ violation is detected in $n_\ell = 14$ or 21 received samples results in miss and false-alarm probabilities that are significantly smaller than the error-event probability.

4.2 End-of-Preamble Flag Recognition

The EOP flag (see Eq. (19)) will be recognized as soon as the ML test (15) detects n_F violations in the received sample sequence after the timing data sequence has been recognized at time $n = n_T$. The probability P_{false}^{FR} of falsely accepting flag reception while still receiving the timing data sequence can be approximated at moderate to high signal-to-noise ratios by the probability that the ML test (15) will detect a violation exactly n_F times while receiving the $L - n_T$ timing data symbols preceding the EOP flag:

$$P_{\text{false}}^{FR} \simeq \binom{L - n_T}{n_F} P_{V/T}^{\,n_F} (1 - P_{V/T})^{L - n_T - n_F} , \qquad (24)$$

where $P_{V/T}$ is given by (18).

The probability $P^{FR}_{miss}(I)$ of not recognizing the EOP flag before or while receiving its Ith data signal is calculated under the assumption that the flag has not been falsely detected during timing data sequence reception. When the parameter n_F is equal to 1 this simply means that the ML test (15) fails to recognize the reception of the first I data signals of the flag and, at moderate and high signal-to-noise ratios,

$$P^{FR}_{miss}(I) \simeq \prod_{i=1}^{I} P^{F}_{fail}(i) \qquad \text{for } n_F = 1\,, \tag{25}$$

where $P^F_{fail}(i)$, $i = 1, 2, 3, \ldots$, are given by (20). The formulas for n_F greater than 1 are more complex and will not be given here.

In Table 2, the false-alarm and miss probabilities P^{FR}_{false} and $P^{FR}_{miss}(I)$ for EOP-flag recognition are given as a function of the parameters I and n_F. The probabilities were calculated for an error-event probability of $P_e = 10^{-5}$ and it was assumed that $L - n_T = 27$ timing data sequence samples are received after timing-sequence recognition. The decision threshold ρ of the ML test (15) was set to 2 or 0, respectively. The results show that to decrease the probability of false-alarm it is advantageous to bias the ML test (15) selecting $\rho = 0$ to reduce the probability (18) of detecting a violation while still receiving the timing sequence. For $\rho = 2$, $n_F = 1$ or 2 or $\rho = 0$, $n_F = 1$ the probability of not recognizing the EOP flag before or while receiving its fourth data signal is much lower than the error-event probability.

References

[1] H. Kobayashi, "Application of probabilistic decoding to digital magnetic recording systems," *IBM J. Res. Develop.*, Vol. 15, pp. 65-74, Jan. 1971.

[2] G. D. Forney, "Maximum-likelihood sequence estimation of digital sequences in the presence of intersymbol interference," *IEEE Trans. Inform. Theory*, Vol. IT-18, No. 3, pp. 363-378, May 1972.

[3] F. Dolivo, "Signal processing for high-density magnetic recording," *COMPEURO'89*, Hamburg, W. Germany, May 1989, in *Proc. VLSI and Computer Peripherals*, edited by W.E. Pröbster and H. Reiner, pp. 1.91-1.96. Computer Society Press of the IEEE, 1989.

[4] F. Dolivo, W. Schott and G. Ungerboeck, "Fast timing recovery for partial-response signaling systems," *International Conference on Communications '89*, Boston, MA, June 1989, *ICC'89 Conference Record*, Vol. 1, pp. 18.5.1-18.5.5.

[5] R. Cideciyan and F. Dolivo, "Data Coding for Fast Start-Up of PRML Receivers," European Patent Application EP 0 413 976.

Table 1. Miss and false-alarm probabilities for timing sequence recognition. Timing sequence detected if no more than n_V violations are recognized in a sequence of n_e received samples; false-alarm probabilities are computed for the worst-case data sequence encoded with the (4,6,4) code. The corresponding error-event probabilitity is $P_e = 10^{-5}$ and the ML-test threshold is $\rho = 2$.

n_p	n_e	$n_V = 0$: P^{TR}_{miss}	P^{TR}_{false}	$n_V = 1$: P^{TR}_{miss}	P^{TR}_{false}	$n_V = 2$: P^{TR}_{miss}	P^{TR}_{false}
1	7	3.5×10^{-5}	1.2×10^{-6}				
2	14	7.0×10^{-5}	1.6×10^{-12}	2.3×10^{-9}	6.2×10^{-7}		
3	21	1.0×10^{-4}	2.0×10^{-18}	5.2×10^{-9}	1.2×10^{-12}	1.7×10^{-13}	2.3×10^{-7}

Table 2. False-alarm and miss probabilities for end-of-preamble (EOP) flag recognition. EOP flag detected as soon as n_F violations are recognized after timing-sequence recognition. For the miss probability, it is assumed that $L - n_T = 27$ timing sequence samples are still received before encountering the EOP flag. The corresponding error-event probabilitity is $P_e = 10^{-5}$. The ML-test threshold was set to $\rho = 2$ and 0, respectively.

	n_F	P^{FR}_{false}	$P^{FR}_{miss}(1)$	$P^{FR}_{miss}(2)$	$P^{FR}_{miss}(3)$	$P^{FR}_{miss}(4)$	$P^{FR}_{miss}(5)$	$P^{FR}_{miss}(6)$
	1	1.3×10^{-4}	0.5	2.5×10^{-1}	1.2×10^{-19}	5.9×10^{-38}	1.2×10^{-77}	
$\rho = 2$	2	8.8×10^{-9}	1.0	7.5×10^{-1}	2.5×10^{-1}	2.4×10^{-19}	5.9×10^{-38}	2.4×10^{-77}
	3	3.7×10^{-13}	1.0	1.0	7.5×10^{-1}	2.5×10^{-1}	2.4×10^{-19}	5.9×10^{-38}

	n_F	P^{FR}_{false}	$P^{FR}_{miss}(1)$	$P^{FR}_{miss}(2)$	$P^{FR}_{miss}(3)$	$P^{FR}_{miss}(4)$	$P^{FR}_{miss}(5)$	$P^{FR}_{miss}(6)$
	1	1.4×10^{-17}	1.0	1.0	5.0×10^{-6}	2.5×10^{-11}	1.2×10^{-29}	5.7×10^{-48}
$\rho = 0$	2	8.9×10^{-35}	1.0	1.0	1.0	9.9×10^{-6}	2.5×10^{-11}	2.4×10^{-29}
	3	3.7×10^{-52}	1.0	1.0	1.0	1.0	9.9×10^{-6}	2.5×10^{-11}

Non-Linear Identification on the Digital Magnetic Recording Channel

Nicholas P. Sands* and John M. Cioffi

Information Systems Laboratory
Stanford University
Stanford, CA 94305-4055

Abstract

We describe a method for identifying the digital magnetic recording channel in which data dependent nonlinear effects are present. The method can be extended to any discrete-input, discrete-time channel with finite memory. The channel will be considered as a finite state machine or lookup table, in which the number of states is related to the memory length of the channel. The model parameters can be derived using linear least-squares estimation or by adaptation of the model to minimize mean-squared error. A simple linear transformation (Hadamard transform) will be used to convert the finite state machine outputs to the coefficients of a Volterra Series model.

Prototype channels with various recording densities are studied with this technique.

The finite state machine description can be used to construct a trellis, from which receiver performance bounds can be derived.

1 Introduction

In discussions of signal detection schemes for digital magnetic recording, the channel has typically been considered as linear in the limited sense that superposition holds. As recording density is increased, this assumption breaks down, and nonlinear intersymbol interference is observed [1, 2]. The physical sources for this breakdown have been attributed to transition shifting during the write process and the prescence of previous magnetization in the medium [2, 3, 4].

Various models have been suggested to account for nonlinearities in this channel. The Volterra Series approach has been reported [5, 6] and the use of a finite state machine has also been discussed in [5] and [7].

In this paper, we present an identification procedure based on the finite state machine model. This model is general for a discrete-time channel with binary inputs and finite time response. The model parameters are derived using a least-squares technique, although adaptive methods could also be used. By appropriate arrangement of the model parameters, we show that a convenient linear transformation - the Hadamard transform[6] - yields the Volterra Series model for the channel. We then extract those Volterra

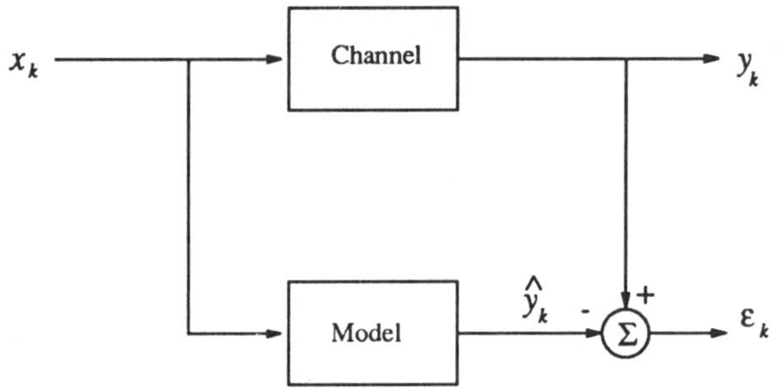

Figure 1: Channel Identification

Series kernels that have appreciable amplitude. The finite state machine(FSM) model is equivalent to a trellis description, and as such, the receiver performance bound for the channel under maximum likelihood sequence detection can be found.

2 Theory

2.1 Chip Decomposition

The channel output $y(t)$ is decomposed into signal chips [8], $y_k(t)$ each of duration T, defined by

$$y_k(t) = y(t + kT) \quad 0 \leq t < T. \tag{1}$$

We can reconstruct $y(t)$ using

$$y(t) = \sum_k y_k(t - kT). \tag{2}$$

For practical purposes, we will deal with a sampled version of $y(t)$ at some arbitrary multiple ρ of the symbol rate $1/T$. The ith sample in the kth symbol period is then

$$y\,(t = (k + i/\rho)T) = y_{k,i}.$$

so that the chips are now vectors of length ρ, which are concatenated to form the sequence of channel output samples. We call this a *fractionally spaced* channel model. Note that we can let ρ grow to be arbitrarily large, in which case we have the continuous time function again.

*Supported by IBM Resident Study Program

1058-6393/91 $01.00 © 1991 IEEE

2.2 Volterra Series

Consider a $1 \times 2^\nu$ row vector $\mathbf{x}_k^{(\nu)}$, a function of the channel input sequence $\{x_k\}$, defined recursively as follows

$$\begin{aligned} \mathbf{x}_k^{(0)} &= [1] \\ \mathbf{x}_k^{(1)} &= [1 \quad x_k] \\ \mathbf{x}_k^{(2)} &= [1 \quad x_k \quad x_{k-1} \quad x_k x_{k-1}] \\ &\vdots \\ \mathbf{x}_k^{(n+1)} &= \left[\mathbf{x}_k^{(n)} \mid x_{k-n}\mathbf{x}_k^{(n)}\right] \end{aligned}$$

for n up to $\nu - 1$. Also $\chi_{k,i}$ is the i-th element of $\mathbf{x}_k^{(\nu)}$. We will use the chip decomposition of (1) to express the model output as a sequence of chips $\hat{y}_k(t)$. We define the ν-order Volterra Series expansion for a channel of memory ν such that the kth chip is given by

$$\hat{y}_k(t) = \sum_{i=1}^{2^\nu} \chi_{k,i} v_i(t), \qquad (3)$$

where $v_i(t)$ is a chip decomposition of the Volterra kernels of the channel.

Further, in the fractionally spaced channel model, the kth chip can be expressed as a vector $\hat{\mathbf{y}}_k$ of length ρ, given by

$$\hat{\mathbf{y}}_k = \mathbf{x}_k^{(\nu)}\mathbf{V}, \qquad (4)$$

where the $2^\nu \times \rho$ matrix $\mathbf{V}$ consists of the chips of the 2^ν Volterra kernels, sampled at ρ/T.

2.3 Hadamard Transform

The Hadamard transformation can be used to map the FSM model parameters to the coefficients of the Volterra series model. There is a computationally efficient recursive method to do the transformation. The matrix is defined recursively

$$\mathbf{H}_0 = [1] \qquad (5)$$

$$\mathbf{H}_{n+1} = \begin{bmatrix} \mathbf{H}_n & \mathbf{H}_n \\ \mathbf{H}_n & -\mathbf{H}_n \end{bmatrix} \qquad (6)$$

The matrix has some useful properties: it is symmetric, as can be seen from the definition; and its inverse is just a scaled version of the matrix itself. Compute the product

$$\mathbf{H}_{n+1}\mathbf{H}_{n+1} = 2\begin{bmatrix} \mathbf{H}_n\mathbf{H}_n & 0 \\ 0 & \mathbf{H}_n\mathbf{H}_n \end{bmatrix} \qquad (7)$$

repeating the recursion for n down to 0 for which

$$\mathbf{H}_0\mathbf{H}_0 = [1] \qquad (8)$$

Thus,

$$\mathbf{H}_n\mathbf{H}_n = 2^n\mathbf{I} \qquad (9)$$

and we can write the inverse of $\mathbf{H}$

$$\mathbf{H}_n^{-1} = \mathbf{H}_n/2^n \qquad (10)$$

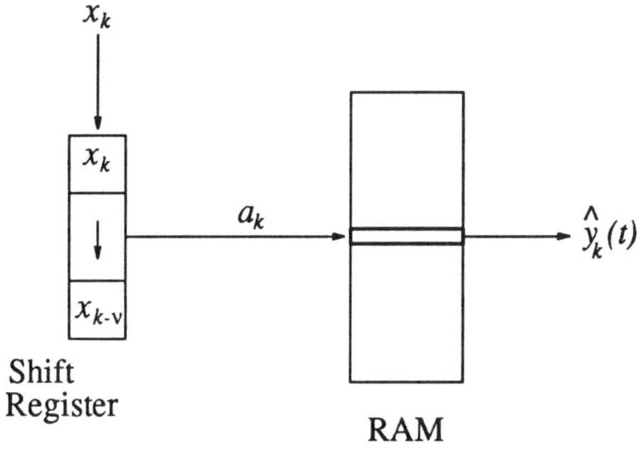

Figure 2: Finite state machine

2.4 Finite State Machine Model

In the case of saturation recording, the channel is constrained to have binary inputs, that is the field that is used to magnetize the medium takes one of two values. If we further assume that the memory of the channel is limited to at most ν symbols, we can ascribe the memory of past inputs to one of a finite number $2^{\nu-1}$ states. Then the channel has the Markov property, namely the next output depends on the current input and the current state. This can also be construed as a table in which the state and current input are combined into a binary address. Then this can be encoded into a integer that forms a pointer to a table entry. The channel outputs are the entries in the table.

The address pointer at the kth symbol will be denoted by $a_k^{(\nu)} \in \{0, \ldots, 2^\nu - 1\}$ and is computed using

$$a_k^{(\nu)} = \sum_{i=0}^{\nu-1} (1 + x_{k-i})\, 2^{i-1} \qquad (11)$$

we can also establish the recursion

$$a_k^{(n+1)} = a_k^{(n)} + (1 + x_{k-n})\, 2^n \qquad (12)$$

for $n = 0, 1, \ldots \nu - 1$. We express the FSM channel model as a $2^\nu \times \rho$ matrix $\mathbf{R}$ with each row representing the ρ samples of the 2^ν FSM model chips.

Thus the kth output chip is given by

$$\hat{\mathbf{y}}_k = \mathbf{s}_k^{(\nu)}\mathbf{R} \qquad (13)$$

where the vector $\mathbf{s}_k^{(\nu)}$ is a *selector* such that only one of its elements, namely the $a_k^{(\nu)}$th, is one and all other elements are zero.

Using the recursion in (12), we observe that

$$\begin{aligned} \mathbf{s}_k^{(n+1)} &= 2\left[\mathbf{s}_k^{(n)} \,|0\ldots0\right] \qquad x_{k-n} = +1 \\ &= 2\left[0\ldots0|\, \mathbf{s}_k^{(n)}\right] \qquad x_{k-n} = -1. \end{aligned} \qquad (14)$$

Given these relations, we will use the Hadamard matrix to compute

$$\mathbf{x}_k^{(n+1)}\mathbf{H}_{n+1}$$

$$= \left[\mathbf{x}_k^{(n)}\mathbf{H}_n(1+x_{k-n}) \;\middle|\; \mathbf{x}_k^{(n)}\mathbf{H}_n(1-x_{k-n})\right] \quad (15)$$

Thus

$$\begin{aligned} \mathbf{x}_k^{(n+1)}\mathbf{H}_{n+1} &= 2\left[\mathbf{x}_k^{(n)}\mathbf{H}_n \; |0\ldots0\right] & x_{k-n} = +1 \\ &= 2\left[0\ldots0|\;\mathbf{x}_k^{(n)}\mathbf{H}_n\right] & x_{k-n} = -1 \end{aligned}$$

We can see from these recursive relations that if $\mathbf{x}_k^{(n)}\mathbf{H}_n = \mathbf{s}_k^{(n)}$ then $\mathbf{x}_k^{(n+1)}\mathbf{H}_{n+1} = \mathbf{s}_k^{(n+1)}$. Then to complete the proof, set $n = 0$

$$\mathbf{x}_k^{(0)}\mathbf{H}_0 = [1][1] = [1] = \mathbf{s}_k^{(0)} \quad (16)$$

Thus we have an inductive proof that

$$\mathbf{x}_k^{(n)}\mathbf{H}_n = \mathbf{s}_k^{(n)} \quad (17)$$

for $n = 0, 1, \ldots, \nu$. Having established this relation we proceed to show the relation between the Volterra Series expansion and the FSM matrix $\mathbf{R}$.

Recall the Volterra Series expansion

$$\begin{aligned} \hat{\mathbf{y}}_k &= \mathbf{x}_k^{(\nu)}\mathbf{V} & (18) \\ &= \mathbf{x}_k^{(\nu)}\mathbf{H}\mathbf{H}^{-1}\mathbf{V}. & (19) \end{aligned}$$

Replacing $\mathbf{x}_k^{(\nu)}\mathbf{H}_\nu = \mathbf{s}_k^{(\nu)}$ and $\mathbf{H}^{-1} = \mathbf{H}/2^\nu$ we obtain

$$\hat{\mathbf{y}}_k = \mathbf{s}_k^{(\nu)}\mathbf{H}\mathbf{V}/2^\nu. \quad (20)$$

By equality with (13)

$$\begin{aligned} \mathbf{R} &= \mathbf{H}\mathbf{V}/2^\nu & (21) \\ \mathbf{V} &= \mathbf{H}\mathbf{R} & (22) \end{aligned}$$

Thus the Hadamard matrix is established as the relation between the Volterra kernels and the FSM model parameters.

2.5 Least-squares Identification

We consider the fractionally spaced samples of $y(t)$ as a set of ρ *subchannels* in the terminology of [9]. In this way, the row vector $\mathbf{y}_k$ consists of

$$\mathbf{y}_k = [y_{k,1} \quad y_{k,2} \quad \cdots \quad y_{k,\rho}] \quad (23)$$

and the columns of matrix $\mathbf{R}$ are formed by subchannel models $\mathbf{r}_i$

$$\mathbf{R} = [\mathbf{r}_1 \quad \mathbf{r}_2 \quad \ldots \quad \mathbf{r}_\rho] \quad (24)$$

We now assume a fixed value for ν in (13). Drop the superscript, and consider the estimate of the i-th such subchannel

$$\hat{y}_{k,i} = \mathbf{s}_k\mathbf{r}_i \quad (25)$$

We compute the error between this and the observed channel output $y_{k,i}$

$$e_{k,i} = y_{k,i} - \mathbf{s}_k\mathbf{r}_i \quad (26)$$

Although our model for the system is non-linear with respect to the original input sequence $\{x_k\}$, the error is linear with respect to the model parameters. Therefore we can use the results from linear estimation theory and find the set of model parameters $\mathbf{R}$ that minimizes the error sequence in the least squares sense.

Notice that we have in effect expanded our definition of *input*; from the sequence x_k to a nonlinear but completely defined function of the sequence $\mathbf{s}_k$ in accordance with (14), and in the process, retained linearity in the model parameters.

From linear estimation theory we invoke the Orthogonality Principle, which states that the error sequence for the i-th subchannel $e_{k,i}$ is orthogonal to the available data $\mathbf{s}_k$.

$$E\left[\mathbf{s}_k e_{k,i}\right] = 0 \quad (27)$$

In practice we can observe the channel's response to a finite input sequence, say of length N, so we estimate the expectation based on a finite amount of data. Thus we have a $N \times 2^\nu$ matrix $\mathbf{S}$ expressing the selector corresponding to each ν-tuple of the input sequence,

$$\mathbf{S} = \begin{bmatrix} \mathbf{s}_1 \\ \mathbf{s}_2 \\ \vdots \\ \mathbf{s}_N \end{bmatrix}, \quad (28)$$

and a $N \times 1$ column vector $\mathbf{y}_i$ containing the samples of the channel output. The estimate of the expectation then becomes a sum over the N sets of data, and can be expressed as the matrix product

$$\mathbf{S}'\mathbf{e}_i = 0 \quad . \quad (29)$$

Then since

$$\mathbf{e}_i = \mathbf{y}_i - \mathbf{S}\mathbf{r}_i \quad , \quad (30)$$

we obtain the minimum sample square error (MSSE) solution

$$\hat{\mathbf{r}}_{i,opt} = (\mathbf{S}'\mathbf{S})^{-1}\mathbf{S}'\mathbf{y}_i \quad (31)$$

and the MSSE is given by

$$\xi_{min,i} = (\mathbf{y}_i'\mathbf{y}_i - \mathbf{y}_i'\mathbf{S}\hat{\mathbf{r}}_{i,opt})/N \quad (32)$$

We can assemble the ρ subchannels together, and write

$$\begin{aligned} \mathbf{Y} &= [\mathbf{y}_1 \quad \mathbf{y}_2 \quad \ldots \quad \mathbf{y}_\rho] & (33) \\ \mathbf{E} &= [\mathbf{e}_1 \quad \mathbf{e}_2 \quad \ldots \quad \mathbf{e}_\rho]\,. & (34) \end{aligned}$$

The complete FSM model is

$$\hat{\mathbf{R}}_{opt} = (\mathbf{S}'\mathbf{S})^{-1}\mathbf{S}'\mathbf{Y} \quad . \quad (35)$$

The minimum mean sample squared error over all the ρ subchannels is given by $\sum_i \xi_{min,i}/\rho$, that is, the mean of the diagonal elements of the matrix $\mathbf{E}'\mathbf{E}$. This can be expressed as

$$\hat{\xi}_{min} = trace\left(\mathbf{Y}'\mathbf{Y} - \mathbf{Y}'\mathbf{S}\hat{\mathbf{R}}_{opt}\right)/\rho N \qquad (36)$$

Note that in practice to find $\hat{\mathbf{R}}_{opt}$ it will not be necessary to compute the pseudo-inverse of $\mathbf{S}$, as implied by (35). Due to the sparse structure of $\mathbf{S}$, a special algorithm has been devised to compute $\hat{\mathbf{R}}_{opt}$ without matrix inversion.

2.6 Adaptive Identification

The use of an adaptive procedure for channel identification is suggested in [10]. Even though our model is nonlinear in nature, the error is linear with respect to the model parameters. The squared error will be quadratic, suggesting the use of a gradient search method, such as the LMS algorithm [10]. For the case of a linear model (FIR), the model parameters $\mathbf{w}_k$ are updated according to

$$\mathbf{w}_{k+1} = \mathbf{w}_k + 2\mu e_k \mathbf{x}_k \qquad (37)$$

Notice that at each iteration, all the parameters get updated.

For the FSM model, we have to modify the algorithm slightly [11] since the error is a nonlinear function of the channel input sequence x_k

$$r_{k,i} = r_{k,i} + 2\mu e_{k,i} \qquad (38)$$

In this case, only one of the model parameters receives an update. This leads to a slower convergence process, by a factor of $2^\nu/\nu$. This algorithm approaches the MMSE solution of (35), but with some excess error due to the gradient estimate.

3 Identification Procedure

3.1 Data Acquisition

The channel input sequence $\{x_k\}$ was obtained from a pseudo-random bit sequence (PRBS) rate $1/T$. The NRZI precode was applied - equivalent to $\frac{1}{1\oplus D}$, where $\oplus$ denotes modulo-2 addition. This is a standard technique in digital magnetic recording and is used to cancel an inherent $1-D$ factor in the read-back process. It also has the desirable effect of making the output insensitive to signal inversions in the channel.

Next, the precoded binary sequence is modulated. The transmitter basis functions $\varphi(t)$ are defined as

$$\begin{aligned}\varphi(t) &= 1 \quad 0 \le t < T \\ &= 0 \quad otherwise \;.\end{aligned}$$

The channel input is then

$$x(t) = \sum_k x_k \varphi(t - kT) \quad . \qquad (39)$$

where the sequence $\{x_k\} \in \{+1, -1\}$ is the pseudo-random sequence of period N.

normalized density	FIR $\nu = 8$	FIR $\nu = 256$	FSM $\nu = 8$
1.1	14.24	14.76	18.39
1.3	14.40	14.98	22.48
1.5	13.85	14.17	20.56
1.7	13.29	14.32	21.39
1.9	12.87	15.29	19.88

Table 1: SDR Results Summary for FIR and FSM models

The magnetic recording system used in this work consisted of a thin film disk and magnetoresistive(MR) head.

The PRBS was written at transition densities exceeding the nominal inner-diameter design density for this head and disk by factors up to 1.9.

Linear recording density is often characterized by the width of the step response at half its maximum amplitude, scaled to the symbol period - pw_{50}/T. The step responses on these channels were estimated and the pw_{50}/T ranged from 1.73 to 2.58.

A digital sampling oscilloscope was used to record the channel output. This instrument can digitize at a rate of 200 x 10^6 samples per second at 8 bits resolution and has a waveform memory of 40000 samples. This gives an oversampling ratio relative to the symbol rate of approximately 4, and with a sequence of length 1023, allows the response to nearly 10 repetitions of the input sequence to be digitized at once.

3.2 Post Processing

The digitized data sets undergo a number of processes to condition them for use in channel identification. A low-pass pre-filter was applied to filter some spurious high frequency noise. Resampling is necessary to re-create the frequency and phase of the write clock, since the digitizer was clocked asynchronously. Then, a cross-correlation technique was used to synchronize the read data with the written sequence. Noise averaging was done by averaging each repetition of the response to the sequence.

4 Results

The performance measure for modeling accuracy is residual mean-squared-error $E[(y_k - \hat{y}_k)^2]$. This is converted to ratio of signal to distortion (SDR), which is defined as

$$SDR = 10\log_{10}\left[\frac{E[y_k^2]}{E[(y_k - \hat{y}_k)^2]}\right] \qquad (40)$$

This is approximated from the results of (35) and (36) and using $E[y_k^2] = trace(\mathbf{Y}'\mathbf{Y})/\rho N$ we have

$$SDR = -10\log_{10}\left[1 - trace(\mathbf{Y}'\mathbf{S}\hat{\mathbf{R}}_{opt})/trace(\mathbf{Y}'\mathbf{Y})\right] \qquad (41)$$

The results of the channel identification process are shown in Table 1. Note that the FSM models give a clear 4-5dB gain over the linear models.

Having obtained the FSM model $\mathbf{R}_{opt}$, we now transform to a Volterra Series representation $\mathbf{V}$, via

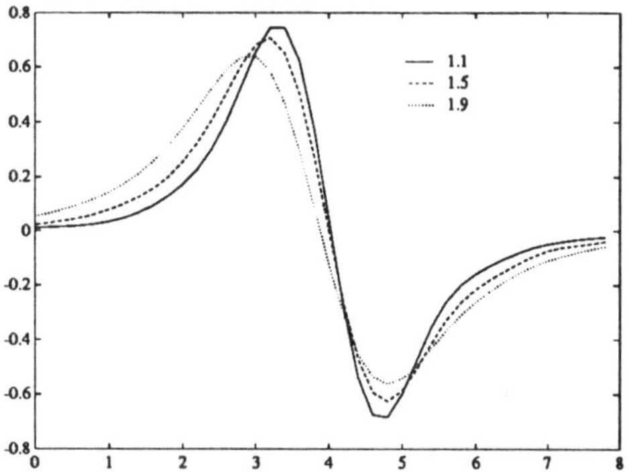

Figure 3: First order (linear) kernels.

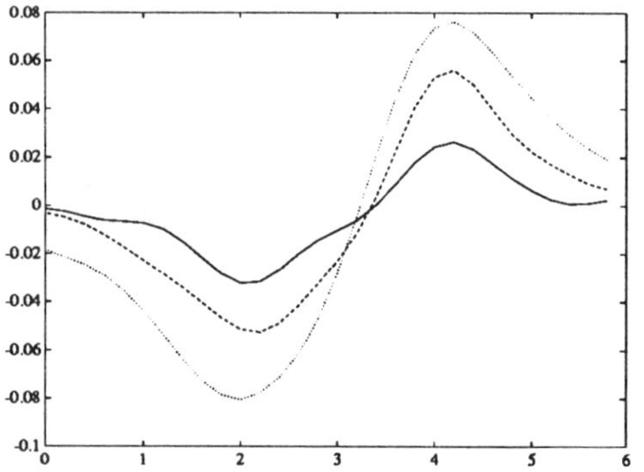

Figure 4: Kernels corresponding to $x_k x_{k-1} x_{k-2}$

(22). Since in most practical channels the types and extent of nonlinearity will be limited, many of the rows of **V** will be negligible, and we can focus on some specific types of nonlinearity. The first order kernel - the linear pulse response - is obtained by concatenating [8] those rows of **V** with row indices which are integral powers of 2. Other kernels are obtained in a similar way. Figure 3 shows the linear kernels, demonstrating increased dispersion as the recording density increases. The largest nonlinear kernel was the third-order nearest-neighbor corresponding to $x_k x_{k-1} x_{k-2}$, as shown in Figure 4. We interpret this interaction as evidence of the proximity induced transition shift [2, 4].

5 Conclusion

We have observed that a purely linear model has limited accuracy on the magnetic recording channel at high density. Only 15dB of SDR was possible, even with a very long FIR model. The finite state machine model is equivalent to a complete Volterra Series expansion for these channels, given the discrete time, binary input sequence. In this sense the FSM model is general, if the channel is not time-varying. Using a 128 state model, we obtained SDR of about 20dB, which is a significant improvement over the FIR model. Another advantage of this representation is that it can be used directly to assess the receiver performance bound under maximum likelihood sequence detection (MLSD).

The disadvantages of such a model are involved with the number of parameters to be estimated(2^ν) and the fact that inspection of the parameter vector does not give any special insight into physical effects in the channel. We have therefore used the Hadamard matrix to transform the FSM parameter vector to the equivalent VS kernels. In this way, the higher order kernels can be inspected and the interactions from which they originate can be deduced.

References

[1] Paul Newby and Roger Wood. "The Effects of Nonlinear Distortion on Class IV Partial Response". *IEEE Transactions on Magnetics*, MAG-22:1203–1205, September 1986.

[2] D. Palmer, P. Ziperovich, R. Wood, and T. Howell. "Identification of Non-linear write effects using pseudorandom sequence". *IEEE Transactions on Magnetics*, MAG-23:2377–2379, September 1987.

[3] C. Michael Melas, Patrick Arnett, Irene Beardsley, and Dean Palmer. "Nonlinear Superposition in Saturation Recording of Disk Media". *IEEE Transactions on Magnetics*, MAG-23:2019–2021, September 1987.

[4] C. Tsang and Y-S. Tang. "Time-Domain Study of Proximity-Effect Induced Transition Shifts". *IEEE Transactions on Magnetics*, MAG-27:795–802, March 1991.

[5] O. Agazzi and D. G. Messerschmitt. "Nonlinear Echo Cancellation of Data Signals". *IEEE Transactions on Communications*, COM-30:2421–2433, November 1982.

[6] R. Hermann. "Volterra Modeling of Digital Magnetic Saturation Recording Channels". *IEEE Transactions on Magnetics*, MAG-26:2125–2127, September 1990.

[7] K.D. Fisher, J.M.Cioffi, and H.Thapar. "Modeling Thin-Film Storage Channels". *IEEE Transactions on Magnetics*, MAG-25(5), September 1989.

[8] Roger Wood. "Maximum Likelihood Reception on Nonlinear Dispersive Data Channels". *Electronics Letters*, 14:669–671, September 1978.

[9] J. M. Cioffi. "Least-squares Storage Channel Identification". *IBM Journal of Research and Development*, 30:310–320, May 1986.

[10] Bernard Widrow and Samuel D. Stearns. *"Adaptive Signal Processing"*. Prentice-Hall, Englewood Cliffs, NJ, 1985.

[11] K. D. Fisher, J. M. Cioffi, W. L. Abbott, P. S. Bednarz, and C. M. Melas. "An Adaptive RAM-DFE for Storage Channels". To be published, 1991.

Equalizer Performance in a Magnetic Recording Channel with Asymmetric Response

Dennis J. Tyner and John G. Proakis
CDSP Center for Research and Graduate Studies
Department of Electrical and Computer Engineering
Northeastern University
Boston, MA 02115

Abstract

The performance of linear and decision-feedback equalization and maximum-likelihood sequence estimation techniques is evaluated for a digital magnetic recording channel having an asymmetric pulse response. In particular, we evaluate the performance of linear and decision-feedback equalizers which shape the signal step response at the output of the equalizer into responses of the form $(1+D)^n$, *where* n *is a non-negative integer. Alternatively, partial responses of length* L *that minimize the mean-square-error at the output of a linear equalizer are derived and their performance is compared to the* $(1+D)^n$ *equalizers. Comparisons are made to the performance of equalizers in a magnetic recording channel having a symmetric pulse response.*

1 Introduction

In magnetic recording, magneto-resistive (MR) read heads are being used in high density environments because of their high read sensitivity and their ability to suppress head and electronic noise effects [1]. The use of such heads, however, can result in positive-negative peak '*asymmetry*' in the data signal at the output of the read head [1, 2, 3]. That is, the magnitude of the response due to a positive magnetic transition differs from that of a negative magnetic transition. This peak asymmetry, as well as the channel intersymbol interference (ISI) and noise, must be considered when designing a receiver/detector that estimates the stored symbols with low error probability.

In section 2, a channel model that accounts for the asymmetry is presented and an equivalent discrete-time white noise model is derived. Bounds on the output signal-to-noise ratio (SNR) of the optimal receiver in an asymmetric environment are given in section 3. In section 4 we consider suboptimal receiver/detection schemes, such as the zero-forcing (ZF) linear and decision-feedback equalizers which shorten the overall response to a family of responses (including full and partial) of the form $(1+D)^n$, for $n = 0, 1, 2$. Comparisons of their output SNR's are made relative to the full response equalizer and to their corresponding symmetric channel equalizers. In section 5 we simultaneously solve for the optimal linear equalizer and partial response (PR) of desired length that correspond to the minimum mean-square-error at the equalizer output. Using an alternate measure of goodness, in section 6 we employ Monte Carlo simulation to obtain the error rate performance of these equalizers followed with either a symbol-by-symbol (SBS) detector or a maximum-likelihood sequence estimator implemented in the form of the Viterbi Algorithm (VA). Finally, the concluding remarks and discussion are found in section 7.

2 Channel Model

A model for the asymmetric read/write process is given in figure 1. The input to the channel is the binary sequence $\{a_k\}$ having T-spaced discrete elements that take on values of 0 or 1 independently. This binary sequence is written onto the magnetic media using non-return-to-zero (NRZ) modulation that creates a two-level magnetization waveform, $m(t)$, containing transitions (alternating in direction) wherever adjacent input elements differ from each other ($a_k \neq a_{k-1}$). The response of the read head to a magnetic transition is modeled with a scaled Lorentzian pulse, where the scale factor depends upon whether the transition is from a negative level to a positive level (positive transition) or from a positive level to

1058-6393/91 $01.00

a negative level (negative transition). The Lorentzian pulse is described mathematically as

$$s(t) = \frac{1}{1 + (\frac{2t}{pw_{50}})^2} , \quad (1)$$

where pw_{50} is called the '*half-amplitude width*' and is equivalent to the amount of time that $s(t)$ is greater than or equal to half of its peak value. (The peak value of $s(t)$ is 1 and occurs at $t = 0$.)

Positive-negative peak asymmetry in the digital magnetic recording channel is defined to be the absolute difference between the peak magnitudes of the positive and negative transition responses divided by the magnitude of the larger of the two [1, 3]. For our model, a positive transition in $m(t)$ results in an output response of $2s(t)$, while a negative transition leads to an output response of $-2\alpha s(t)$ for $0.5 < \alpha < 1.0$, which implies an asymmetry of $(1-\alpha)$. Here, we arbitrarily chose the positive transition response to have a larger magnitude than the response to a negative transition, but we could have easily inverted this in our model and the equalizer performance, as discussed later, would remain unchanged. The change, however, would have been reflected in the value of the equalizer output signal levels.

The read response, $y(t)$, is nothing more than the sum of all the transition responses and the channel output, $r(t)$, is the sum of $y(t)$ and $n(t)$, where $n(t)$ is a white Gaussian noise process, which is independent of a_k, having power spectral density N_0. To express $r(t)$ in a closed form expression, we define a new three-level sequence, $\{I_k\}$, having elements defined in terms of the input sequence as

$$I_k = \begin{cases} 2, & a_k = 1,\ a_{k-1} = 0 \\ 0, & a_k = a_{k-1} \\ -2\alpha, & a_k = 0,\ a_{k-1} = 1 \end{cases} , \quad (2)$$

where adjacent non-zero elements of $\{I_k\}$ are restricted to sign alternations and occur wherever there are steps in $m(t)$. For this reason, we call the Lorentzian signal $s(t)$ the unit step response of the magnetic channel. Now, the signal $r(t)$ can be expressed in terms of the elements of the ternary sequence $\{I_k\}$ in a more desirable form as

$$r(t) = \sum_{k=-\infty}^{\infty} I_k s(t - kT) + n(t) , \quad (3)$$

where the asymmetry is present in the symbol sequence $\{I_k\}$, the noise is represented by $n(t)$, and the ISI is a consequence of the infinite duration of the signal $s(t)$.

It is well known that the channel model described in (3) can be represented with the equivalent discrete-time white noise filter model shown in figure 2 [4]. Here, $\{I_k\}$ is the same ternary sequence described in (2), n_k is the discrete additive white Gaussian noise with variance N_0, $\{g_k\}$ are the discrete filter coefficients, and r_k is related to these three by

$$r_k = (I * g)_k + n_k , \quad (4)$$

where $*$ indicates convolution. Given that x_k is the sampled autocorrelation function of $s(t)$ described by

$$x_k = \int_{-\infty}^{\infty} s(t)s(t + kT)\, dt \quad (5)$$

and

$$X(D) = \sum_{k=-\infty}^{\infty} x_k D^k \quad (6)$$

is the D-transform of x_k where D is the unit delay operator, then the filter coefficients, $\{g_k\}$, can be found by performing a spectral factorization of $X(D)$, where $X(D) = G(D)G(D^{-1})$ [5]. For the Lorentzian channel model it has been shown [6] that the spectral characteristics of X can be written in closed form as

$$X(f) = \frac{\pi^2 pw_{50} S}{4\sinh(\pi S)} \cosh(2\pi S(\frac{1}{2} - f)) , \ 0 \le f \le 1 . \quad (7)$$

In (7), f is the normalized frequency defined as

$$f = \frac{\omega T}{2\pi} \quad (8)$$

and S is the normalized information density expressed in terms of the bit rate as

$$S = \frac{pw_{50}}{T} . \quad (9)$$

One possible factorization of $X(D)$ leads to the following representation of the filter coefficients [6];

$$g_k = \left[\frac{pw_{50}S}{8} \tanh\left(\frac{\pi S}{2}\right)\right]^{\frac{1}{2}} \frac{k + \frac{S}{2}}{k^2 + (\frac{S}{2})^2} . \quad (10)$$

As in the continuous-time channel model with $s(t)$, g_k is also seen to have infinite support, which leads to ISI over infinitely many symbols.

3 The Optimal Receiver

It has been shown [5] that the optimal receiver/detector structure for the ISI white noise channel described in (3) consists of a filter matched to $s(t)$ followed by a symbol rate sampler and a maximum-likelihood sequence estimator that processes the matched filter output samples. If these samples are denoted by z_k, then

$$z_k = \sum_{i=-\infty}^{\infty} I_i x_{k-i} + v_k \ . \tag{11}$$

In (11), v_k is the sampled noise at the output of the matched filter and is related to its continuous-time counterpart by

$$v_k = \int_{-\infty}^{\infty} n(t)s(t-kt)\,dt \ . \tag{12}$$

In evaluating the performance of the matched filter we calculate an upper bound on the output SNR that is achieved by the filter matched to an isolated transition, which operates in the absence of ISI. To find this output SNR bound, we calculate the signal power of the sampled output of the MF by computing the average over all possible unitary signal inputs (2,0, or -2α) and dividing it by the output noise power. In this manner, we find the the signal power to be

$$P = E\{(I_0 x_0)^2\} = \frac{1+\alpha^2}{2} x_0^2 \tag{13}$$

and the noise power is

$$P_N = E\{v_0^2\} = N_0 x_0 \ . \tag{14}$$

From (5) we have

$$x_0 = \frac{\pi p w_{50}}{4} \tag{15}$$

and, hence, the average output SNR for the MF operating in the asymmetric channel on an isolated transition is

$$SNR_{MF}^a = (1+\alpha^2)\frac{\pi p w_{50}}{4N_0} \ . \tag{16}$$

In the symmetric channel, $\alpha = 1$ and the symmetric MF bound is

$$SNR_{MF}^s = \frac{\pi p w_{50}}{2N_0} \ . \tag{17}$$

With the results of (16) and (17) we define a loss due the asymmetry as

$$SNR_{MF}^{loss} = \frac{SNR_{MF}^s}{SNR_{MF}^a} = \frac{2}{(1+\alpha^2)} \ . \tag{18}$$

It is interesting to note that (16), (17) and consequently (18) are independent of the normalized information density S. The reason for this non-dependence is that S only affects the minimum spacing between transitions in $m(t)$, which determines the severity of the ISI, while the MF bound is only dependent upon the response to a single input symbol which, at most, depends upon one transition and consequently no ISI.

4 Performance of Equalizers for $(1+D)^n$ Response

Since the computational complexity of the VA-based MLSE increases exponentially with the length of the channel impulse response, it becomes evident that with the infinite length of g_k as described in (10), the optimal detector becomes unrealizable. For this reason we consider ZF equalizers that shorten the total impulse response (the channel in cascade with the equalizer) to a response of the form $(1+D)^n$, for $n = 0,1,2$. Unlike the symmetric channel, the signal levels at the output of the equalizer, shown as solid circles in figure 3, are not equi-spaced. Furthermore, knowing that the symmetric channel equalizer signal outputs are $\{-2,0,2\}$ for $n = 0,1$ and $\{-4,-2,0,2,4\}$ for $n = 2$, then from figure 3 we notice that except for the full response of $n = 0$, there is an increase in the number of possible output levels in the asymmetric channel with respect to the symmetric case, where the difference is more significant for larger n. This increase in the number of equalizer output signal levels leads to a degradation in the SNR achieved at the output of the ZF equalizers.

In describing the performance of these ZF equalizers, we begin first by discussing the equalizer output signal power. Assuming that an appropriate precoder [7] is used, each of the output signal levels will map into a unique input I_k, or equivalently, a_k. With the precoding, SBS detection can be used while catastrophic error propagation is eliminated. In figure 3, adjacent signal levels that map into the same input element a_k are grouped together in shaded blocks and considered as a single output symbol. From this, we define the threshold levels for the decision regions as being one-half of the distance between adjacent output symbols. In this way, any noise-corrupted equalizer output sample that falls into a particular decision region will be mapped into its appropriate input symbol which corresponds to the output symbol that is closest in distance to the output sample. We then define the output signal power, $P(n,\alpha)$, to be the average of the square of the half distances, leading to

$$\begin{aligned} P(0,\alpha) &= \frac{1+\alpha^2}{2}, \\ P(1,\alpha) &= \alpha^2, \\ P(2,\alpha) &= \frac{5\alpha^2-4\alpha+1}{2}. \end{aligned} \tag{19}$$

Here, besides the obvious dependence on α, we also see that the output signal power depends upon the particular response that is chosen, whereas in the symmetric case, the output signal power is unity for all n.

The noise power at the output of the ZF linear equalizer (LE) is [5]

$$P_N^{ZF-LE} = \int_0^1 N_0 \frac{|P_n(f)|^2}{X(f)} \, df \,, \qquad (20)$$

where $P_n(f)$ is the frequency response of the $(1+D)^n$ signal which can be expressed as

$$P_n(f) = 2^n e^{-j2\pi f n} \cos^n(\pi f) \,. \qquad (21)$$

With the aid of (7) and (19)-(21), we find the output SNR for the asymmetric ZF LE to be

$$SNR^a_{ZF-LE} = \frac{P(n,\alpha)}{P_N^{ZF-LE}} = \frac{\pi p w_{50}}{4N_0} \frac{P_S(n,\alpha)}{K_L(n,S)} \,, \qquad (22)$$

where

$$K_L(n,S) \doteq \frac{4^n \sinh(\pi S)}{\pi S} \int_0^1 \frac{\cos^{2n}(\pi f)}{\cosh(2\pi S(\frac{1}{2} - f))} \, df \,. \qquad (23)$$

We use the output SNR of the full response equalizer as a performance benchmark and define a '*response gain*', RG, as a ratio of the output SNR of the type n PR equalizer to the output SNR of the full response equalizer. For the LE in the asymmetric channel we have

$$RG_{ZF-LE}(n) = \frac{2P(n,\alpha)}{1+\alpha^2} \frac{K_L(0,S)}{K_L(n,S)} \,. \qquad (24)$$

As seen in figure 4, for $S = 3$ the partial response equalizers having $n = 1$ and 2 perform significantly better (about 6 dB) than the full response equalizer at large α. However, as the asymmetry increases (α decreases), the performance of the $n = 0$ equalizer more closely resembles that of the other two.

We now turn our attention to the ZF decision-feedback equalizer (DFE) and its output noise power, which, under the assumption of correct feedback symbols, can be expressed as [5]

$$P_N^{ZF-DFE} = e^{\left(\int_0^1 \ln \frac{N_0 |P_n(f)|^2}{X(f)} \, df\right)} \,. \qquad (25)$$

Using properties of the natural logarithm we can expand the integral in (25) to isolate its dependence on n. In doing so we have

$$P_N^{ZF-DFE} = e^{\left(\int_0^1 \ln \frac{N_0}{X(f)} \, df\right)} e^{\left(\int_0^1 \ln |P_n(f)|^2 \, df\right)} \,, \qquad (26)$$

where only the second exponential term in (26) is a function of n. After replacing $P_n(f)$ by (21) and with the help of [8], this term is found to be exactly equal to 1 for all n, and therefore

$$P_N^{ZF-DFE} = e^{\left(\int_0^1 \ln \frac{N_0}{X(f)} \, df\right)} \,. \qquad (27)$$

Hence, unlike the ZF LE, the noise power at the output of the ZF DFE is independent of the choice of n. Its output SNR, however, is still dependent on n and consequently, the output SNR of the ZF DFE in the asymmetric channel remains dependent upon the response selection, unlike the output SNR of the symmetric channel ZF DFE which was shown to be independent of n [9]. Using (7), (19), and (27), we find this asymmetric channel output SNR to be

$$SNR^A_{ZF-DFE} = \frac{P_S(n,\alpha)}{P_N^{ZF-DFE}} = \frac{\pi p w_{50}}{4N_0} \frac{P(n,\alpha)}{K_D(S)} \,, \qquad (28)$$

where

$$K_D(S) \doteq \frac{\sinh(\pi S)}{\pi S} \, e^{-\left(\int_0^1 \ln[\cosh(2\pi S(\frac{1}{2}-f))] \, df\right)} \,. \qquad (29)$$

As with the ZF LE, we define the RG for the ZF DFE as the ratio of the output SNR of the type n PR DFE to the output SNR of the full response DFE. That is,

$$RG_{ZF-DFE}(n) = \frac{2P(n,\alpha)}{1+\alpha^2} \,. \qquad (30)$$

Figure 5 shows the RG_{ZF-DFE} as a function of α for $S = 3$. As noted earlier, at $\alpha = 1$ (symmetric channel) all three equalizers perform equally well and therefore there is no gain noted for any of the PR DFE's. As α decreases, the PR's show a loss in output SNR with respect to the full response equalizer. Therefore, the full response ZF-DFE is better than the PR DFE's in an asymmetric channel. Also, the smaller the value of n, the better the performance of the ZF DFE.

To show the advantage of the DFE over the LE, we compare their output SNR's and define the gain of the DFE to be

$$\text{gain}_{DFE} \doteq \frac{SNR^a_{ZF-DFE}}{SNR^a_{ZF-LE}} = \frac{K_L(n,S)}{K_D(S)} \,. \qquad (31)$$

Since the corresponding linear and decision-feedback equalizers have the same dependence on α, the DFE gain is shown in figure 6 as a function of S. As expected, when the normalized density is large, the ISI becomes more severe and the DFE's perform much better than the LE's.

5 Generalized Partial Response Shapes

Partial responses of the form $(1+D)^n$ have been used in magnetic recording because of their close spectral match to the symmetric recording channel and for their equi-spaced signal levels that are beneficial when using SBS detection. When asymmetry is included in the magnetic channel model, these PR signals no longer maintain equal spacing between adjacent levels, as seen in figure 3, which leads to a loss in the output SNR that is achieved with the use of a SBS detector. Because of this degrading performance we search for alternate PR's which reduce the noise enhancement of the equalizer. In particular, we find PR's that minimize the mean-square-error (MSE), similar to the responses described by Falconer and Magee [11], and employ a sequence estimator rather than a SBS detector to estimate the input symbols. Here, we define generalized PR signals as those signals that contain non-integer valued controlled ISI.

We begin by defining the desired length of the PR. In this manner, we have the desired output at time k as

$$d_k = \sum_{i=-m_1}^{m_2} p_i I_{k-i} \, , \tag{32}$$

where the p_i's are the discrete time partial response values and the desired length of the response is $L = m_1 + m_2 + 1$. If z_k is the output of the equalizer at time k, then the MSE at that time instant is

$$\varepsilon_k^2 = (z_k - d_k)^2 \, . \tag{33}$$

The values of the $2K+1$ linear equalizer tap weights, $\{c_i\}_{i=-K}^{K}$, and the values of the PR are then calculated simultaneously by solving the following sets of linear equations;

$$\begin{aligned} E\left\{\frac{\partial \varepsilon_k^2}{\partial c_i}\right\} &= 0, \quad -K \le i \le K \\ E\left\{\frac{\partial \varepsilon_k^2}{\partial p_i}\right\} &= 0, \quad -m_1 \le i \le m_2 \, . \end{aligned} \tag{34}$$

To ensure that we avoid the trivial solution, we also impose the constraint that the desired response of the '*negative unit pulse*' has a value of p_0 that is equal to minus one. This restriction leads to

$$p_0 = \frac{1}{\alpha}(1 + p_{-1}) \, . \tag{35}$$

Substitutions of (35) into (34) will eliminate p_0 and reduce the number of equations to $2K+L$. For efficient use of the Viterbi algorithm a small L is more desirable since the computational complexity of the algorithm is exponentially dependent on L.

6 Evaluation and Comparison of Performance

In this section we obtain the performance of the equalizers in terms of their bit-error rate (BER) by using computer simulation of the continuous-time channel described in (3). The channel response, $s(t)$, was implemented as a discrete-time tapped delay line (TDL) with coefficients spaced by $T/4$. The samples of the additive white Gaussian noise, $n(t)$ and the input symbols, a_k, were developed using a uniform random number generator and in the case of the noise, a uniform-to-Gaussian conversion as described in [10]. A precoder was used on the input symbols prior to their conversion to the ternary sequence described by (2) when the equalizers shaped the signal to a partial response and were followed with a SBS detector. The samples of $r(t)$ were then passed through a unity gain low-pass filter (LPF), having an approximate normalized bandwidth of $1/T$ Hz and a maximum stopband amplitude of -42.5 dB, which was used to eliminate aliasing and reduce the high frequency noise content. Calculation of the filter coefficients is described in detail in chapter 5 of [10]. The LPF outputs were down-sampled and then processed by $T/2$ fractionally-spaced linear and decision-feedback equalizers. Both equalizers operated on forty-one feed-forward taps and the DFE was supplemented with an additional ten feedback taps. The equalizer tap weights were adjusted based on a steepest-descent recursive algorithm while using a known pseudo-random training sequence.

Estimation of the input symbols is done with either a SBS detector or a sequence estimator implemented via the VA. In the case of the DFE followed by the VA, temporary decisions used in the feedback path of the equalizer were estimated with a SBS detector.

Because the magnetic recording channel is an amplitude-limited channel, we determine the BER as a function of the equalizer input SNR defined to be

$$(SNR)_I \doteq \frac{(\text{peak of positive transition response})^2}{\text{noise power}} \, . \tag{36}$$

The quantities listed on the left hand side of (32) are evaluated at the input to the ZF equalizers.

Figure 7 illustrates the full and partial response equalizer performance with both the SBS detector and the VA. The results, given for a relatively large normalized density ratio of $S = 3$, compare the symmetric channel performance with a channel having 20% asymmetry. As a reference, a lower bound is shown which results from a matched filter receiver followed by an MLSE. This bound corresponds to the occurrence of a minimum distance error event [4] and the minimum distance was obtained via an exhaustive computer search. The full response equalizers using SBS detection show a loss of about 1 dB for the asymmetric channel relative to the symmetric channel. This agrees with the loss in signal power that is experienced at the output of the $n = 0$ equalizer as given by (19). Here, for a BER of 10^{-4}, the DFE shows an overwhelming improvement over the LE of close to 8 dB for both the symmetric and asymmetric channel. Since the full response LE result is so poor, we only consider the VA with the DFE, where we see a modest improvement of about 1/2 dB with respect to the SBS detector.

Turning our attention to the $n = 1$ PR equalizers, we see a loss due to the asymmetry of approximately 2 dB for the LE and close to 3 dB for the DFE for SBS detection. The LE loss is expected from (19). However, the additional loss of the DFE can be attributed to the propagation of feedback errors which become more severe when the detector must distinguish between the increased number of output levels for increasing n. Unlike the full response equalizers, the advantage of the DFE is only 1/2 dB. When the equalizers are followed with a Viterbi detector, the DFE picks up an additional 1 dB while the LE gains about 2 dB, leading to a slight overall benefit for the LE.

Finally, for the $n = 2$ PR equalizers we see that in an asymmetric channel the two equalizers lead to the same BER when they are used with the SBS detector. Again, the LE shows the expected asymmetric loss of 3 dB as predicted by (19) while the increased DFE loss of 5 dB is a result of the increasing number of levels that must be distinguished by the detector for use in the feedback path. The LE with the Viterbi detector shows a 5 dB gain over the SBS equivalent while the DFE gain is about 3 dB. This gives the LE with the VA an approximate 2 dB advantage over the DFE with the VA.

The advantage of the DFE over the LE with SBS detection, which was anticipated by the results shown in figure 6, relied on the fact that correct feedback symbols were used. In the more realistic environment, where that is not the case, we note that feedback errors can degrade the performance of the DFE significantly. Again, as n gets larger, the effect of feedback errors becomes more substantial and linear equalization eventually becomes more beneficial than decision-feedback equalization.

Now, we consider a linear equalizer which shapes the output to the optimal signal response of length 5 subject to the constraints of (34) and (35). For $S = 3$, an input SNR of 25 dB, a 21-tap T/2 fractionally-spaced equalizer and a channel having 20% asymmetry, we find the optimal responses for $l = 3, 5, 7, 9$ as shown in figure 8 and the corresponding optimal equalizer having spectral characteristics shown in figure 9. If we define the noise enhancement (NE) of the linear equalizer to be the sum of the squares of the tap weights, we find the NE is reduced as the desired PR increases in length. For $L = 3$ we have a NE of 6.83 dB while for $L = 9$ the NE is reduced to .88 dB. In comparison, the NE of the equivalent $(1 + D)^n$ MSE linear equalizers is 8.3 dB for $n = 0$, 5.4 dB for $n = 1$ and 6.5 dB for $n = 2$, showing that the noise power at the output of the optimal equalizer is smaller than that of the $(1 + D)^n$ equalizers for $L \geq 5$. To find the SNR improvement, however, we must also consider the gain (i.e. d_{min}) from the signal response when using the VA. Doing so, as calculated in table 1, we define the gain in the output SNR for the equalizers using the full response equalizer as a reference. Comparing the gains, we find that the $n = 2$ PR equalizer outperforms any of the optimal equalizers listed in the table, despite the significant reduction in NE that is achievable.

7 Summary and Conclusions

A model for the asymmetric recording channel was presented, where the asymmetry was included in the channel input symbols. Equalizers were used to shape the signal into an output response of the form $(1+D)^n$. We found that the signal loss due to asymmetry at the output of the equalizers using SBS detection was the same for both the LE and the DFE, dependent upon the target response chosen. We have also seen that as n increases, this signal loss increases because the number of possible signal levels at the equalizer output grows as the output response increases in length. As was demonstrated with the computer simulation, with more signal levels, the DFE is more susceptible to feedback errors because unlike the LE, the SBS detector for the DFE must distinguish among each of these output signals. While some adjacent signal levels are close to each other when n increases, the SBS detector for the LE must only decide on output symbols. For the DFE/SBS detector, it was also noted that although the noise power is independent of the choice of n, because of the output signal power, the output SNR is dependent on the desired output response.

Generalized PR shapes that minimize the MSE were studied. Although the NE of these equalizers was reduced for large enough L, the value of d_{min} was also reduced, leading to a better performance for the $n = 2$ LE.

Finally, for our computer simulation, we find the best results for $S = 3$ are obtained when using a linear equalizer followed with a Viterbi detector and using an $n = 2$ partial response. The best performance using a DFE was found to include a Viterbi detector and a partial response of either $n = 1$ or 2.

References

[1] C. Tsang, M. Chen, T. Yogi, and K. Ju, "Gigabit Density Recording Using Dual Element MR/Inductive Heads on Thin-Film Disks," *IEEE Transactions on Magnetics*, Vol. MAG-26, pp 1689–1693, September 1990.

[2] R. Hermann, "Volterra Modeling of Digital Magnetic Saturation Recording Channels," *IEEE Transactions on Magnetics*, Vol. MAG-26, pp 2125–2127, September 1990.

[3] T. D. Howell, D.P. McCown, T. A. Diola, Y. Tang, K. R. Hense and R. L. Gee, "Error Rate Performance of Experimental Gigabit Per Square Inch Recording Components," *IEEE Transactions on Magnetics*, Vol. MAG-26, pp 2298–2302, September 1990.

[4] G. D. Forney, "Maximum-Likelihood Sequence Estimation of Digital Sequences in the Presence of Intersymbol Interference," *IEEE Transactions on Information Theory*, Vol. IT-18, pp 363–378, May 1972.

[5] J. G. Proakis, *Digital Communications*, McGraw-Hill, New York, 1989.

[6] J. W. M. Bergmans, "Discrete-Time Models for Digital Magnetic Recording," *Phillips Journal of Research*, Vol. 41 No. 6, pp 531–558, 1986.

[7] H. Kobayashi and D. T. Tang, "Application of Partial-Response Channel Coding to Magnetic Recording Systems," *IBM Journal of Research and Development*, Vol. 15, pp 368–375, July 1970.

[8] I. S. Gradshteyn and I. M. Ryzhik, *Table of Integrals, Series and Products*, Academic Press, New York, 1980.

[9] J. W. M. Bergmans, "Partial Response Equalization," *Phillips Journal of Research*, Vol. 42 No. 2, pp 209–245, 1987.

[10] J. G. Proakis and D. G. Manolakis, *Introduction to Digital Signal Processing*, Macmillan, New York, 1988.

[11] D. D. Falconer and F. R. Magee Jr., "Adaptive Channel Memory Truncation for Maximum Likelihood Sequence Estimation," *The Bell System Technical Journal*, Vol. 52, pp 1541–1562, November 1973.

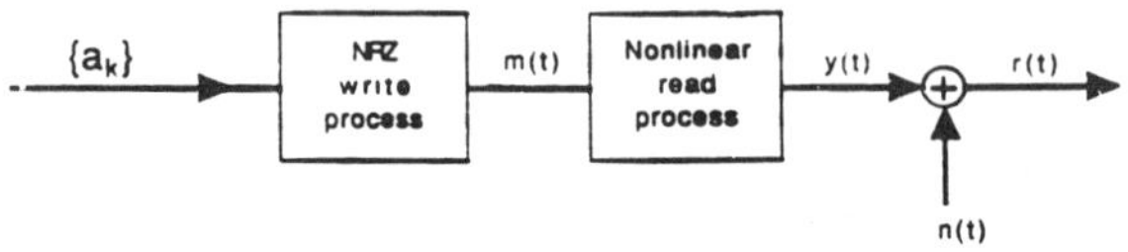

Fig. 1: Channel model for the asymmetric read/write process.

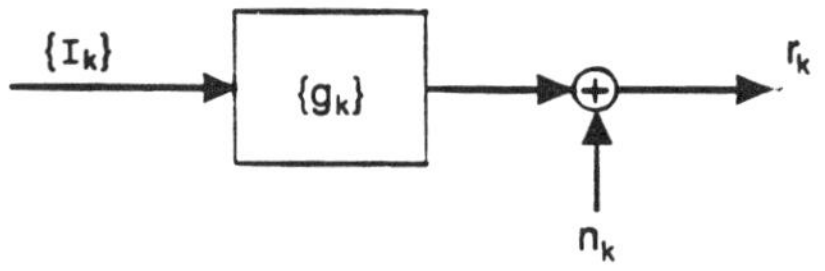

Fig. 2: Equivalent discrete-time asymmetric channel model.

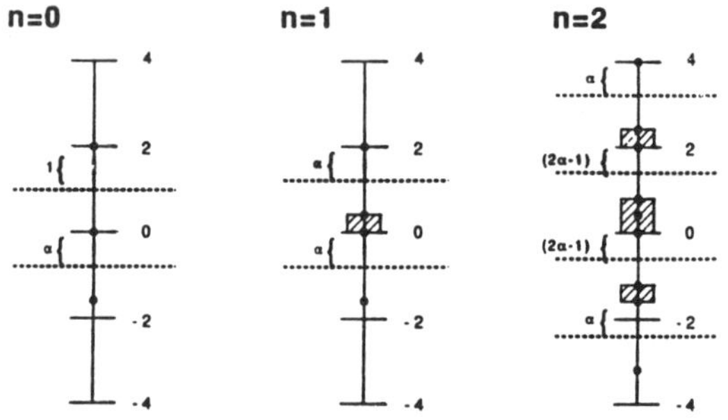

Fig. 3: Signal levels at the output of the equalizers. Dotted lines show the thresholds between decision regions.

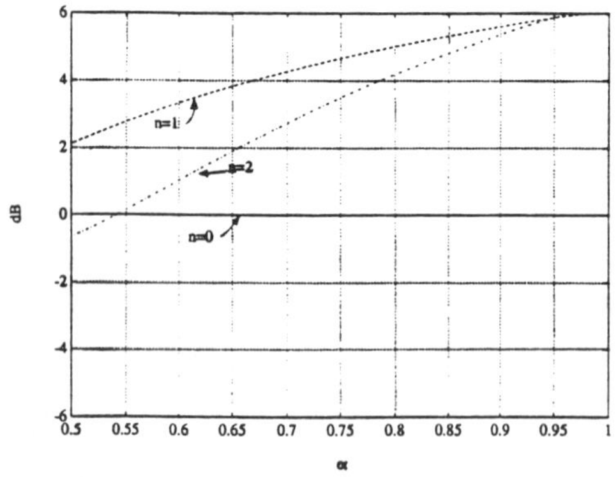

Fig. 4: RG for the ZF LE's with $S = 3$.

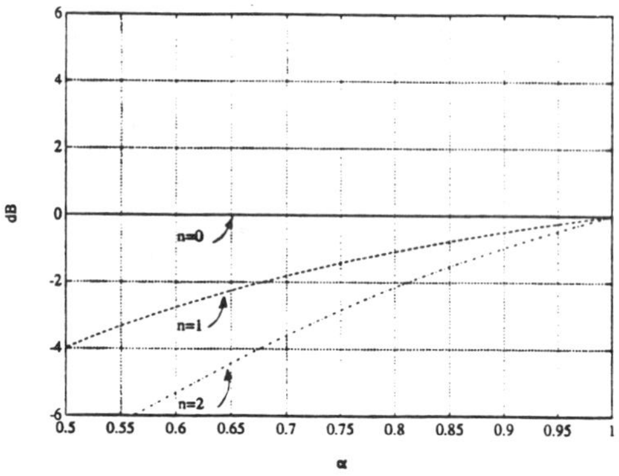

Fig. 5: RG for the ZF DFE's with $S = 3$.

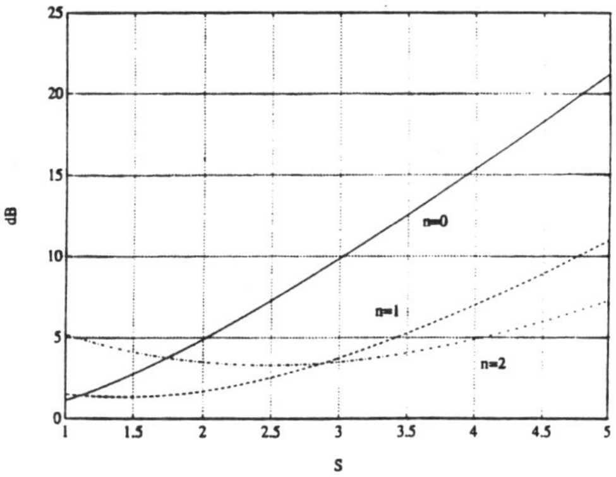

Fig. 6: Gain of the DFE over the LE.

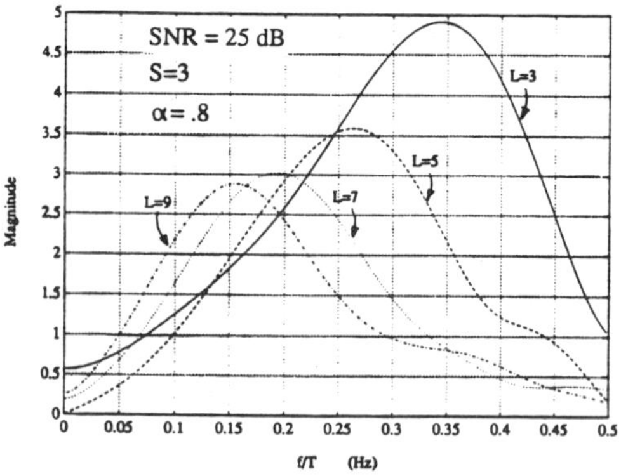

Fig. 9: Frequency response of the optimal PR LE's.

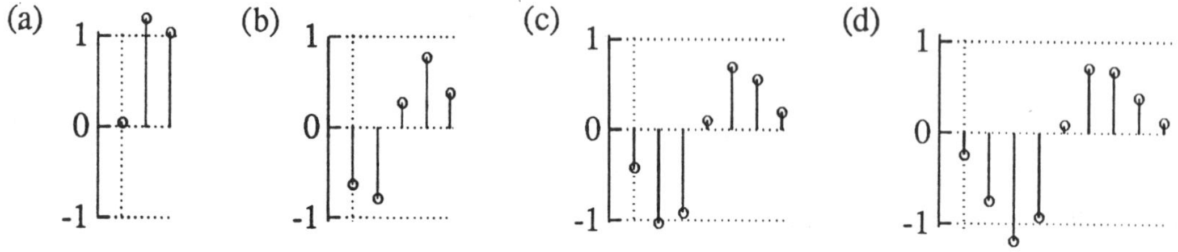

Fig. 8: Optimal PR's for $SNR = 25dB$, $\alpha = .8$ and $S = 3$.

PR	NE (dB)	d_{min} (dB)	gain (dB)
n = 0	8.28	7.09	—
n = 1	5.41	7.09	2.77
n = 2	6.47	9.82	4.54
optimal (L=3)	6.83	7.92	2.28
optimal (L=5)	3.49	5.25	2.95
optimal (L=7)	1.57	4.24	3.86
optimal (L=9)	0.88	2.12	2.43

Table 1: PR gain for $SNR = 25dB$, $\alpha = .8$ and $S = 3$.

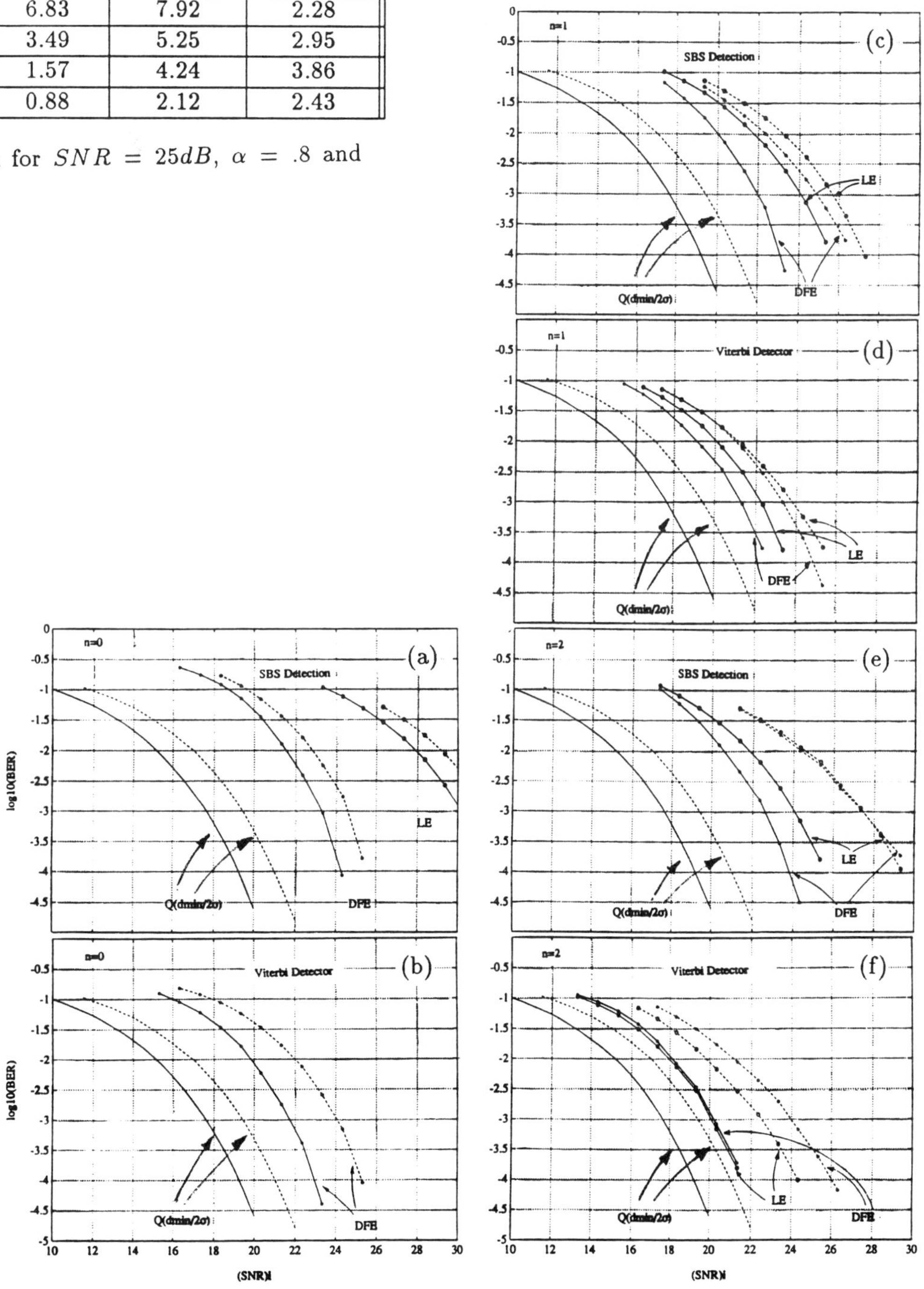

Fig. 7: BER performance for $\alpha = 1$ (solid lines) and $\alpha = .8$ (dashed lines).

Source-Channel Coding for Digital Magnetic Recording

Steven W. McLaughlin and David L. Neuhoff

Department of Electrical Engineering and Computer Science
University of Michigan
Ann Arbor, MI 48109-2122

Abstract

Digital magnetic recording systems are typically designed to store bits with great reliability (e.g. error probability 10^{-10}). Although such reliability is essential for many types of data, it is unnecessary for storing the binary representation of analog data. This paper explores the possibility of increasing the storage of analog data samples on magnetic media by increasing the density with which bits are stored and using source-channel codes to both compress the data and protect it from the increased error rate that results from higher bit density.

To determine what gains are possible using this approach, we simulate a system consisting of an IID Gaussian data source, a tree structured vector quantizer for source-channel coding, a channel model similar to that of French and Wolf, and two kinds of detector/demodulators (gated peak detection, and $1\text{-}D^2$ partial response maximum likelihood). For IID Gaussian data, for which only modest compression is possible, it is found that gains on the order of 3 to 4 in data density are achievable with our approach. Similar results are expected for real sources such as speech, images or music.

1. Introduction

In digital magnetic recording one is concerned with storing as many bits as possible on a disk or other magnetic medium with high reliability. The performance of a magnetic recording system is measured in terms of its *bit density* (in bits per unit area) and its *bit error probability* (usually on the order of 10^{-10}). This paper explores the possibility of dedicating a disk to store the binary representation of analog data samples (e.g. sampled speech, pixels from an image). Here it is unnecessary to maintain the reliability of conventional magnetic recording systems, in fact error probabilities on the order of 10^{-3} may go unnoticed. Since the error probability increases with the bit density, one might increase the bit density over a conventional recording system and then use a *source-channel code* to both compress the data samples (a source code) and protect against the increased error rate (a channel code).

A digital magnetic recording system that stores samples from a data source is illustrated in Figure 1. We consider such a system to be composed of an encoder, recording channel, demodulator and decoder. In contrast to the usual measure of performance, this system's performance is determined by its *data density* (in data samples per unit area) and its average *distortion.* We have adopted mean squared error as the distortion measure because of its tractability and because we wish to focus on source-channel coding issues rather than perceptual issues.

The data density is a function of the *rate* R of the source-channel code (in bits per data sample), and the media area that one bit occupies. Each channel use occupies an area vTW where W is the track width, v is the velocity of the media and T is the duration of each channel bit. Specifically, the data density r (in data samples per unit area) of the system is given by

$$r = \frac{1}{R}\frac{1}{vTW}\left(\frac{\text{data samples}}{\text{bit}}\frac{\text{bit}}{\text{unit area}}\right)$$

The goal of this paper is to determine how large r can be made by suitable choices of the source-channel code (and rate R), T, W, and demodulator, subject to some distortion constraint on the reproduced data samples. To accomplish this we need a channel model that accurately captures changes in T and W and candidate source-channel codes that yield low distortion on a noisy channel. These are addressed in Sections 2, 3 and 4. Section 5 gives the resulting data densities of these codes and compares them with the maximum achievable data density that can be found from information theory.

2. Channel Model

This section describes a model which is similar to that of French and Wolf [1]. As illustrated in Figure 2, the input to the model is the sequence of bits $\{X_i\}$ ($X_i = \pm 1$) and its output is the analog signal

$$R(t) = \left(\sum_i X_i \, A \, p_T(t - iT) + n_m(t) \right) * h_c(t) + n_e(t)$$

where the model is characterized by a linear filter $h_c(t)$, write current pulse $p_T(t)$, constant A, media white noise $n_m(t)$, and read circuit electronic noise $n_e(t)$. Our model differs from the French-Wolf model in that we have added an additional constraint on the input, namely, data is written every T seconds in our model, whereas the input to French-Wolf model is any two-level waveform.

The channel filter impulse response $h_c(t)$ is chosen to correspond to the Lorentzian step response L(t) of the form [2]

$$L(t) = \frac{1}{1+(\frac{2t}{\beta})^2}$$

which gives $h_c(t) = d/dt\, L(t)$. Using this one may rewrite R(t) as

$$R(t) = \sum_i (X_i - X_{i-1}) \, A \, L(t-iT) + n(t)$$

1058-6393/91 $01.00 © 1991 IEEE

where now it is clear that $R(t)$ is composed of Lorentzian pulses that occur only when successive X_i's differ.

Two noises are accounted for in the model: electronic noise $n_e(t)$ in the read circuitry and media noise, $n_m(t) * h_c(t)$ which is due to the particulate nature of the magnetic medium. Following [1] we model $n_e(t)$ as a white Gaussian noise with power spectral density $N_e/2$ and media white noise $n_m(t)$ as a Gaussian noise whose spectrum is shaped by $|H_c(\omega)|^2$.

To model the effects of changing T and W we introduce the notions of *lineal* and *track densities*. Recall that our initial goal is to increase the bit density over a system that has very small error probability (e.g. 10^{-10}). A system with this high reliability is assumed to be characterized by some nominal values of track width W_o, bit spacing T_o and Lorentzian pulse height A_o. Since our idea is to decrease both T and W to achieve a larger data density we define the (relative) *lineal density* to be

$$\delta = \frac{T_o}{T}$$

and the (relative) *track density* to be

$$\gamma = \frac{W_o}{W}$$

where δ and γ measure the increase in bit density in the lineal and and radial directions, respectively. We first note that decreasing T (increasing δ) increases the intersymbol interference in the readback signal R(t) that occurs between adjacent Lorentzian pulses. As in [1] we assume that the effects of changing γ are

$$A = \frac{A_o}{\gamma}, \quad N_m = \frac{N_{m,o}}{\gamma}$$

In summary the independent "physical" parameters of the recording channel model are W_o, A_o, $N_{m,o}$, N_e, T_o and v. The parameters T and W (equivalently, δ and γ) are presumed to be subject to our choice.

For our work later in finding data densities we must find values for the channel parameters. Actually it is sufficient to find values for the parameters W_o, T_o, $A_o^2/N_{m,o}$ and A_o^2/N_e. To find these we consider the "typical" performance of a "typical" recording system, and reverse engineer to find the parameter values. A typical system uses gated peak detection [3] and achieves error probability 10^{-10}. Also in a typical system about 70% of the total noise power is attributable to media noise (we also later consider 90%). We find the parameters such that a typical system, that uses our channel model with nominal values $T = T_o$, $W = W_o$, $A = A_o$ and $N_m = N_{m,o}$, achieves the typical performance. We choose $T_o = \beta = 10^{-7}$ which is also typical of many recording systems [3].

Although not derived here, it is straightforward (albeit tedious) to find the remaining parameters [9]. In particular we find an approximate expression for error probability in terms of $A_o^2/N_{m,o}$ and A_o^2/N_e

$$P_E = f(A_o^2/N_{m,o}, A_o^2/N_e) = 10^{-10}.$$

and an exact expression for the media noise fraction,

$$\frac{\text{media noise power}}{\text{total noise power}} = g(A_o^2/N_{m,o}, A_o^2/N_e) = .7 \text{ (or .9)}$$

From these two equations one can compute the channel parameters $A_o^2/N_{m,o}$ and A_o^2/N_e. For the case of 70% media noise, $A_o^2/N_{m,o} = 2.12\text{x}10^9$ and $A_o^2/N_e = 5.51\text{x}10^9$ and for 90% media noise, $A_o^2/N_{m,o} = 1.65\text{x}10^9$ and $A_o^2/N_e = 1.65\text{x}10^{10}$. This completes the description of the channel model.

3. Source Channel Codes

We consider a class of combined source-channel codes called *noisy channel vector quantizers* (NCVQ's), which are, essentially, vector quantizers designed with attention to the fact that they are used on a noisy channel [5]. We, in fact, consider the special case of *nonredundant* NCVQ's, meaning no redundant error protection bits are added to the usual vector quantizer designed for the noiseless channel.

Suppose we must encode the samples from a real-valued data source (e.g. speech) for transmission over a discrete-time, discrete-alphabet noisy channel. We model the data samples as a continuous-valued random process $\{U_i\}$ with $p_k(\underline{u})$ denoting the joint probability density of random variables $U_1,...,U_k$. We assume the channel is binary and that for any binary L-tuples v and w we let $q_L(\underline{w}|\underline{v})$ denote the probability that the channel outputs $\underline{w}$ given that its input is $\underline{v}$.

The NCVQ operates as follows: Given a k-dimensional data sample $\underline{u} = (u_1,...,u_k)$, an *encoder* determines the *cell* S_i of a *partition* $S = \{S_1,...,S_N\}$ in which $\underline{u}$ lies and produces an L-bit codeword $\underline{e}(i)$, which is stored in the media ($\underline{e}$ is an assignment rule). The decoder receives the L-bit channel output $\underline{c}$ (which happens with probability $q_L(\underline{c}|\underline{e}(i))$), and produces $\underline{y}_{d(\underline{c})}$ from a *codebook* $C = \{\underline{y}_i : 1 \le i \le 2^L\}$ as the reproduction of $\underline{u}$ (d is a decoding rule). From here on we restrict ourselves to case where $N=2^L$, meaning the number of channel inputs is equal to the number of channel outputs. In this case it is common to choose $e(\bullet)$ to be invertible and $d(\bullet)$ to be its inverse, namely $d = e^{-1}$.

The performance of an NCVQ is measured by its *rate* R and *distortion* D. The rate R is defined to be L/k, the number of channel bits per data sample. The distortion D is defined to be the mean squared error

$$D = \frac{1}{k} E\|\underline{U}-\underline{Y}\|^2$$

$$= \frac{1}{k} \sum_{i=1}^{N} \sum_{j=1}^{N} q_L(\underline{e}(j)|\underline{e}(i)) \int_{S_i} \|\underline{u}-\underline{y}_j\|^2 p_k(\underline{u})\, d\underline{u}$$

where $\underline{U}$ denotes a data sample vector and $\underline{Y}$ denotes the reproduction produced by the decoder, and where the expectation is with respect to the randomness in both the source and channel.

We assume that the codebook consists of centroids [6] and therefore one may straightforwardly decompose the distortion into $D = D_S + D_C$,
where D_S

$$D_S = \frac{1}{k} \sum_{i=1}^{N} \int_{S_i} \|\underline{u}-\underline{y}_i\|^2 \, p_k(\underline{u}) \, d\underline{u}$$

$$D_C = \frac{1}{k} \sum_{i=1}^{N} P(\underline{U} \in S_i) \sum_{j=1}^{N} q_L(\underline{e}(j)|\underline{e}(i)) \, \|\underline{y}_i-\underline{y}_j\|^2$$

The term D_S is called the *source distortion*, for it is the distortion that would occur if the source vectors were quantized with the partition S and reconstructed with no channel noise. The term D_C is called the *channel distortion*. It is the component of the distortion that is influenced by the noisy channel.

Algorithms have been developed for finding optimum NCVQ's [4,5]. While the NCVQ's produced by such algorithms are good, they are, unfortunately, completely unstructured and have large complexity when N is large. One way to obtain low complexity NCVQ's is to adapt low complexity, structured vector quantizers, designed for operation with a *noiseless* channel, to operation over a noisy channel. Indeed, this is a common way to design an NCVQ . Specifically we need to choose the channel assignment rule e(•) to make the channel distortion small In fact, the channel assignment may be viewed as a channel code, except that since $N = 2^L$ there is no "redundancy" to enable error correction.

A number of methods for designing good assignment rules have been proposed [6]. Unfortunately, these methods are complex and produce unstructured assignments that can be implemented only by table lookup. While such table lookups are conceptually trivial, they do require considerable storage when N is large. Consequently, there is need to find low complexity, structured assignment rules to go with the low complexity, structured vector quantizers.

In the next section we propose and evaluate the merits of tree-structured vector quantization (TSVQ) as the quantizer in such a scheme. TSVQ has low partitioning complexity, and there is a simple assignment rule that performs well.

4. Tree-Structured NCVQ

Tree-structured vector quantization (TSVQ) [7] is a low complexity method of vector quantization, whose performance on a noiseless channel nearly equals that of the best unstructured vector quantizers [7,8]. Very briefly, the partitioning associated with a k-dimensional TSVQ of size $N = 2^M$ is accomplished with the aid of a complete binary tree of depth M , with a *testvector* of dimension k associated with each node except the root. Given a source vector $\underline{u}$, the search for the cell that contains it proceeds as follows: First $\underline{u}$ is compared with the two test vectors at depth 1, designated nodes 1 and 2 . Binary comparisons continue until the terminal node of the tree, where the output of the final comparison gives the quantization cell S_i (S_i is the set of all $\underline{u}'$ that would also lead to the same terminal node). When used with a noiseless channel, it is customary for the encoder to output the sequence of 0's and 1's corresponding to the binary decisions at each depth. This sequence is called a *pathmap* since it directs the way to the terminal node corresponding to the cell in which the source vector lies. We have found that this pathmap assignment rule works well for the noisy channel. In this case a "good" rule is, essentially, one where channel inputs $\underline{e}(i)$ that are close in Hamming distance are assigned to partition cells S_i that are close in Euclidean distance. Note that the pathmap rule makes assignments that are roughly consistent with the notion that channel inputs that are close in Hamming distance be assigned to cells that are close in Euclidean distance. For instance two nodes stemming from the same parent node correspond to adjacent cells (which are are close in Euclidean distance), and have assignments that are close in Hamming distance, because their pathmaps differ only in the last branches. An additional advantage of the pathmap assignment is that it comes for free from the usual way of partitioning. It is possible to show [9] that the TSVQ with pathmap assignment performs comparably to the optimum NCVQ's given in [5], and in fact comes with in 1dB of the performance of the best possible NCVQ's.

5. Achievable Data Density and Conclusions

We are interested in finding what data densities and distortions can be achieved with our approach of increasing bit density and using source-channel codes. To do this we have simulated a system, shown in Figure 3, consisting of an IID Gaussian source (for tractability), TSVQ with pathmap assignment, our channel model (with parameters found earlier), various choices of γ and δ , and two kinds of demodulators (peak detection and $1-D^2$ PRML). To reduce the number of variables in the simulations we chose a target distortion and consider only those systems with $D \le .01$, which, for the unit variance IID Gaussian data, means that the distortion can be at most 1 percent. Therefore for a given TSVQ and demodulator, we are interested in those (γ, δ) pairs that give distortion $D \le .01$. To do this, in the simulations we first fix γ and then find the largest δ , denoted δ_γ , such that $D = D_S + D_C = .01$, where it is only D_C that increases with γ and δ. The resulting data density is the system data density at distortion D , denoted $r(\delta_\gamma, \gamma, D)$ whose value is

$$\frac{1}{R} \frac{\delta_\gamma \gamma}{v W_o T_o}$$

We compare this data density to maximum achievable data density (see [9,10]) of any code and the given channel model with $\gamma = \delta = 1$. Information theory says that the maximum data density achievable by any code that achieves distortion D for the "typical" system ($\delta = \gamma = 1$) is

$$r_o(D) = \frac{C'}{R(D) v W_o T_o}$$

where C' is the channel capacity (in bits per channel use) and R(D) (in bits per data sample) is the rate distortion function . The channel capacity is the maximum rate at which reliable communications is possible, and the rate-distortion function gives the minimum rate at which encoding with distortion D is possible.

Instead of plotting the data density $r(\delta_\gamma, \gamma, D)$ we define and plot

$$\underline{r}(\delta_\gamma, \gamma, D) = \frac{r(\delta_\gamma, \gamma, D)}{r_0(D)}$$

which is the increase in data density over the best possible data density achievable for the "typical" magnetic recording system that achieves distortion D. Specifically we plot a point representing $(\delta_\gamma, \underline{r}(\delta_\gamma, \gamma, D))$. Such a point can also be compared to the curve $\underline{r}^*(\delta, \gamma, .01)$ which is the maximum data density achievable for any given γ and δ.

Figures 4 and 5 show the simulation results for $\rho = .7$ and $.9$, respectively. Each plots show the performance for several different TSVQ codes for both the peak detection and $1\text{-}D^2$ decoders. The plots shows the increase in data density, $\underline{r}(\delta_\gamma, \gamma, D)$. There are a number of curves on each plot and each curve represents the data density achieved for a certain TSVQ with dimension k and rate R (labelled (k,R)), for different γ's. The point farthest to the right on each curve is for $\gamma = 1$ and the remainder of the points are for increasing integer values of γ. The straight line plots are the maximum achievable data density $\underline{r}^*(\delta, \gamma, .01)$ for any γ and δ.

These plots show that data density gains are different for peak detection and $1\text{-}D^2$ equalization. For peak detection increases in data density are due primarily to increases in γ and multidimensional TSVQ's. It is interesting to note that there is virtually no gain attributable to increases in δ. This is because as δ is increased the intersymbol interference (ISI) increases rapidly and the effective signal to noise ratio in peak detection drops sharply. Gains for peak detection are on the order of 3 - 4 and require γ between 4 and 6 to achieve this. For $1\text{-}D^2$ equalization the data density gains are due to δ and multidimensional TSVQ's. This shows the inherent advantage of controlling the ISI through equalization. Here increases in δ do not increase ISI, and even though high frequency noise is amplified by the equalizer, the net decrease in signal to noise ratio is not as sharp as in the peak detection case. Therefore one does not have to increase γ as much to achieve comparable increases in data density. Gains in this case are on the order of 2 - 3.5 with δ of about 1.5 - 3 and γ of about 2.

References

[1] C.A. French and J.K. Wolf, "Bounds on the Capacity of a Peak Power Constrained Gaussian Channel," *IEEE Trans. Magnetics*, vol MAG-24, No.5, Sept 1988.

[2] H.K. Thapar and A.M. Patel, "A Class of Partial Response Systems for Increasing Storage Density in Magnetic Recording," *IEEE Transactions on Magnetics*, Vol. MAG-23, No.5, September 1987.

[3] J.M. Cioffi, W.L. Abbott, H.K. Thapar, C.M. Melas, K.D. Fisher, "Adaptive Equalization in Magnetic Disk Storage Systems," *IEEE Communications Magazine*, February, 1990.

[4] H. Kumazawa, M. Kasahara, and T. Namekawa, "A Construction of Vector Quantizers for Noisy Channels," *Electronics and Engineering in Japan*, vol 67-B, No.4, 1984.

[5] N. Farvardin and V. Vaishampayan, "Optimal Quantizer Design for Noisy Channels: An Approach to Combined Source-Channel Coding," *IEEE Trans. Info. Theory*, vol IT-33, pp.827-838, Nov. 1987.

[6] N. Farvardin, "A Study of Vector Quantizers for Noisy Channels," *IEEE Trans. Info. Theory*, vol IT-36, No.4, pp.799-809, July 1990.

[7] R.M. Gray and Y. Linde, "Vector quantizers and predictive quantizers for Gauss-Markov sources," *IEEE Trans. Commun.*, Vol. COM-30, pp. 381-389, Feb. 1982.

[8] D.L. Neuhoff and D.H. Lee, "On the Performance of Tree-Structured Vector Quantization," *ICASSP91*, Vol. 4, pp. 2277-2280, Toronto, May 1991.

[9] D.L. Neuhoff and S.W. McLaughlin, "Source-Channel Coding for Digital Magnetic Recording," in preparation.

[10] D.L. Neuhoff and S.W. McLaughlin, "Upper Bounds to the Capacity of the Magnetic Recording Channel" in preparation.

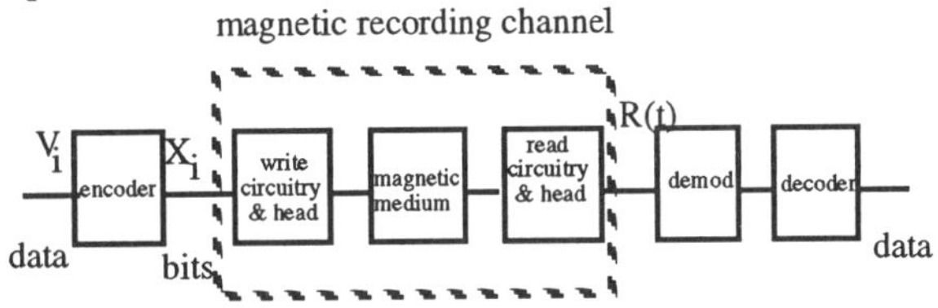

Figure 1: A digital magnetic recording system

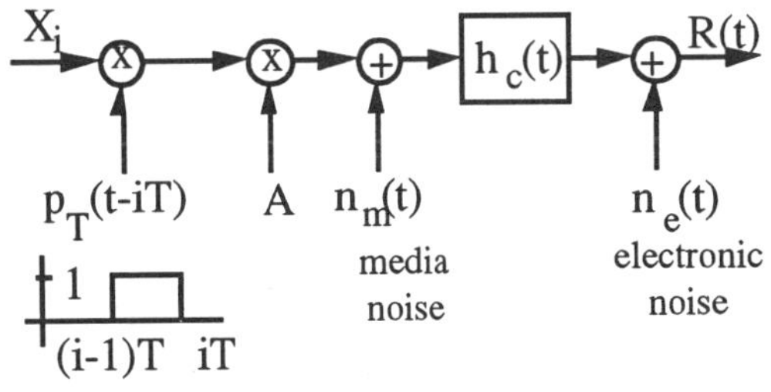

Figure 2: The model of the magnetic recording channel

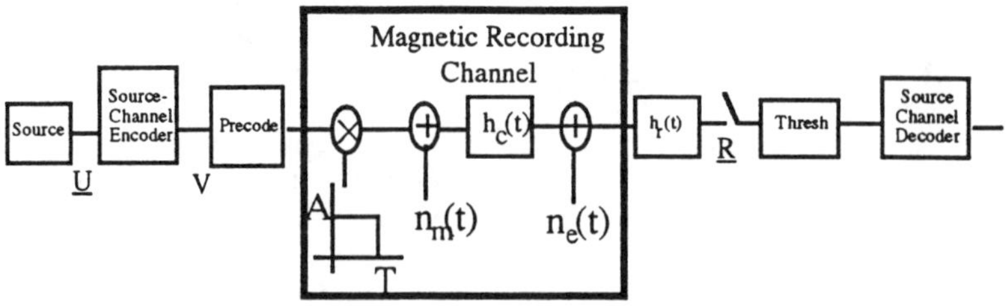

Figure 3: The System to be Simulated

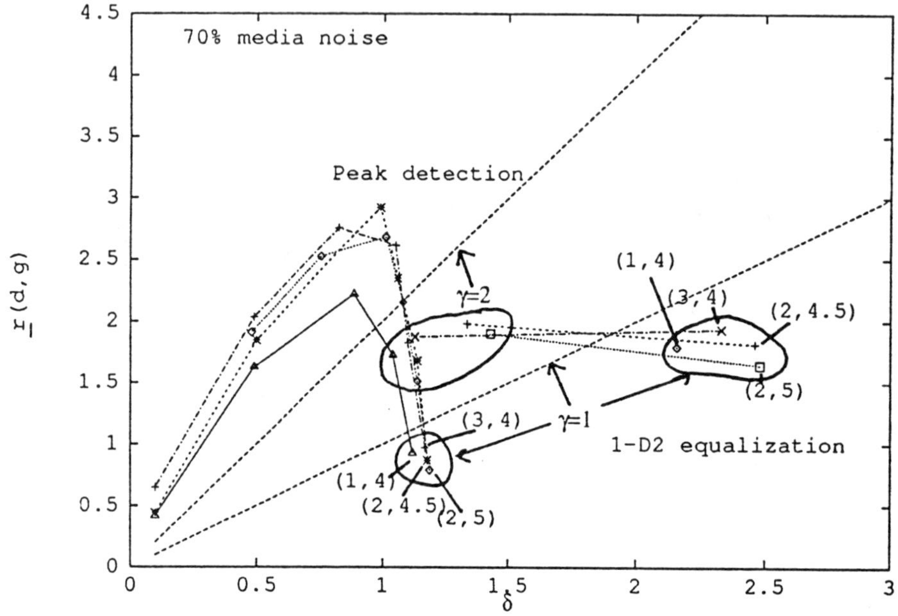

Figure 4: Data Density vs. δ for 70% media noise, various γ's, and TSVQ's (k,R).

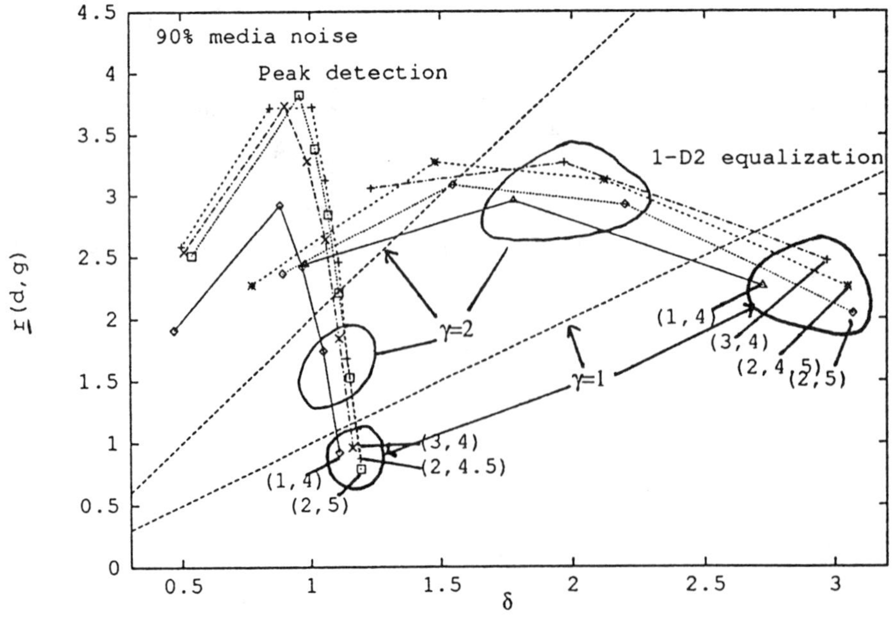

Figure 5: Data Density vs. δ for 90% media noise, various γ's, and TSVQ's (k,R).

Detection in Magnetic Recording with Cross-Track Interference *

Mathew P. Vea José M.F. Moura

Data Storage Systems Center, Carnegie Mellon University

Abstract–We have analyzed the simple case of an isolated pulse with a single interfering pulse to get insight into the practial cross-track interference problem. We treat the problem in signal space to gain an intuitive geometric interpretation. We derive the Maximum Likelihood detector, the Linear Minimum Mean Squared Error (LMMSE) detector, and the minimum probability of error (mpe) linear detector. The mpe detector was found to closely approach the performance of the optimal ML nonlinear detector, and to significantly improve on the performance of the LMMSE detector as well a detector naively designed to match a channel without cross-track interference. These results provide insight and motivation to design equalizers to improve detector performance in the presence of such interference.

1 Introduction

Interference from information recorded off the track of interest is a significant factor limiting the performance of magnetic recording channels. This is especially true of hard disk storage systems, where the interference comes from adjacent tracks as well as "old information" in the guard band. This interference is not Gaussian, and so requires separate treatment from that of other noise sources in the magnetic recording channel. Cross-track or intertrack interference has been treated before, but often is assumed to have the same channel response as the on-track response[1]. In fact, the response of the channel to adjacent tracks will be different from its on-track response due to the side reading properties of the head [4]. This is due to a broader pulse response of the head when side reading.

The work reported here is an initial study of the cross-track interference problem. We study the case of an isolated pulse and a single interfering pulse with distinct pulse shapes. We have derived three detection schemes, the Maximum Likelihood detector and two linear schemes. These schemes are compared, at different interference levels and pulse shapes, to the performance of a detector which lacks knowledge of the interference.

*This material is based on work supported by the National Science Foundation under Grant No. ECD-8907068. The government has certain rights in this material

2 Problem Formulation

The isolated pulse problem involves the detection of a desired data symbol α from the observation $x(t)$, which includes the interfering symbol β and additive white Gaussian noise $n(t)$:

$$x(t) = \alpha p_1(t) + \beta p_2(t) + n(t). \qquad (1)$$

The noise is white with spectral density σ^2. The desired symbol α takes on values $\{-1, 1\}$ with equal probability, β takes on equally likely values $\{-\rho, \rho\}$ and α and β are statistically independent. Since α and β are binary, the pulses $p_1(t)$ and $p_2(t)$ represent the pulse, or dibit, response of the magnetic recording channel. These pulses are characterized by their inner product:

$$\gamma = < p_1(t), p_2(t) > . \qquad (2)$$

In this problem all symbols are real-valued, so the inner product is defined as:

$$< x(t), y(t) > = \int_{-\infty}^{\infty} x(t)y(t)dt.$$

The pulses are of unit energy, i.e. $< p_i(t), p_i(t) > = 1$. The isolated pulse problem as stated is completely characterized by three parameters: interference correlation γ, interference amplitude ρ, and noise power σ^2.

The continuous time problem of equation (1) can be transformed to a two-dimensional detection problem by passing the observed signal $x(t)$ through sampled filters which are matched to the pulse shapes $p_1(t)$ and $p_2(t)$. This filtering operation is equivalent to taking the inner product of $x(t)$ with $p_1(t)$ and $p_2(t)$. The familiar Gram-Schmidt procedure can be used to create a vector of orthonormal basis functions $\mathbf{\Phi}(t)$ from $p_1(t)$ and $p_2(t)$:

$$\mathbf{\Phi}(\mathrm{t}) = \begin{bmatrix} \phi_1(t) \\ \phi_2(t) \end{bmatrix} = \begin{bmatrix} 1 & 0 \\ \frac{-\gamma}{\sqrt{1-\gamma^2}} & \frac{1}{\sqrt{1-\gamma^2}} \end{bmatrix} \begin{bmatrix} p_1(t) \\ p_2(t) \end{bmatrix}$$

1058-6393/91 $01.00 © 1991 IEEE

$$= \mathbf{A}^{-1}\mathbf{P}(\mathbf{t}). \quad (3)$$

$\mathbf{P}(t)$ is the vector of pulse functions and $\mathbf{A}^{-1}$ is the orthonormalizing transformation such that $\mathbf{P}(t) = \mathbf{A}\mathbf{\Phi}(t)$. The statistic $\mathbf{x}$ is formed as the result of the inner product of $x(t)$ and $\mathbf{\Phi}(t)$:

$$\mathbf{x} = < \mathbf{\Phi}(t), x(t) > . \quad (4)$$

The notation here is somewhat unusual: each element of $\mathbf{x}$ is equal to the inner product of the corresponding element time function of $\mathbf{\Phi}(t)$ with $x(t)$.

The constituents of $\mathbf{x}$, x_1 and x_2, together form a sufficient statistic for the detection of α. This is easily shown to be true by contructing a complete set of orthonormal basis functions [3, p. 171] $\{\phi_i(t)\}$ whose first two members are $\phi_1(t)$ and $\phi_2(t)$. The statistics formed by taking the inner product with the higher order basis functions ($x_j =< \phi_j(t), x(t) >$ for $j > 2$) are independent of α, and therefore give no additional information for the detection of α. Thus, the continuous time observation $x(t)$ has been reduced to two dimensions without loss of information. The resulting statistic $\mathbf{x}$ is a function of the symbol vector $\mathbf{\Theta} = [\alpha\ \beta]^T$ and the noise vector $\mathbf{n}$: [1]

$$\begin{aligned} \mathbf{x} &= < \mathbf{\Phi}(t), \mathbf{P}^T(t)\Theta + n(t) > \\ &= < \mathbf{\Phi}(t), \mathbf{\Phi}^T(t)\mathbf{A}^T\Theta > + < \mathbf{\Phi}(t), n(t) > \\ &= \mathbf{A}^T\Theta + \mathbf{n}. \end{aligned} \quad (5)$$

The noise $\mathbf{n}$ is zero-mean and Gaussian with variance $\sigma^2\mathbf{I}$. The two noise components n_1 and n_2 are independent as a result of the choice of an orthonormal basis.

It is evident from equation (5) that the probability distribution of $\mathbf{x}$ conditioned on Θ, $p(\mathbf{x}|\Theta)$, is Gaussian with mean $\mathbf{A}^T\Theta$ and variance $\sigma^2 I$. The detection problem can be visualized in signal space, as shown in Figure 1, with the conditional mean of $\mathbf{x}$ plotted for all values of α and β. The four different symbol pairs $\{\Theta_i\}$ shown in Figure 1 are defined to take on the following values of α and β:

$$\theta_1 = \begin{bmatrix} 1 \\ \rho \end{bmatrix}, \theta_2 = \begin{bmatrix} 1 \\ -\rho \end{bmatrix}, \theta_3 = \begin{bmatrix} -1 \\ \rho \end{bmatrix}, \theta_4 = \begin{bmatrix} -1 \\ -\rho \end{bmatrix} \quad (6)$$

The problem of detecting α is reduced to drawing a boundary in signal space to distinguish the region where $\alpha = 1$ is decided from the decision region for $\alpha = -1$. The detection schemes derived in the following section differ only in the criterion for which they draw this boundary.

[1] Again, the inner product here is applied to each element time function in the vector of time functions. Thus, $< \mathbf{\Phi}(\mathbf{t}), \mathbf{\Phi}(t)^T(t) >= I$, the identity matrix

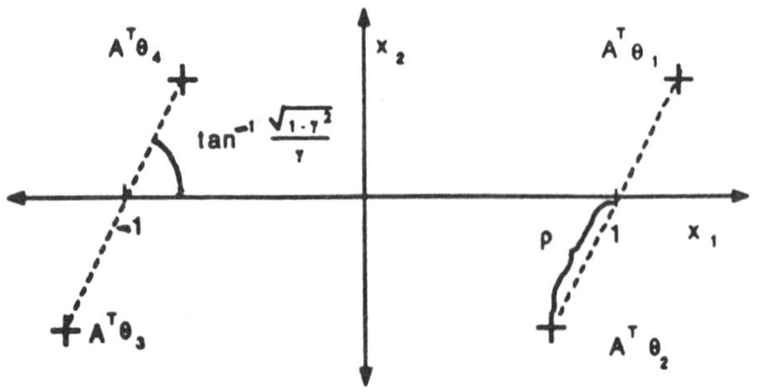

Figure 1: Signal space representation of isolated pulse detection

3 Detector design

The ideal detector has the minimum probability of error among all detection schemes. For the case of equally likely symbols, this detector is the Maximum Likelihood (ML) detector. The ML detector provides an upper bound on detector performance.

The decision boundary for the ML detector is drawn by comparing the likelihood ratio $\Lambda(\mathbf{x})$ to zero:

$$\Lambda(\mathbf{x}) = \frac{p(\mathbf{x}|\alpha = 1)}{p(\mathbf{x}|\alpha = -1)} \lessgtr 1. \quad (7)$$

Since this is a binary detection problem, there is only one decision threshold and the problem can be reduced to a single dimension. This problem is a composite problem because the interfering symbol β is an unwanted parameter. In composite problems, the effect of the unwanted parameter is averaged out by summing over all values of the parameter. In this case, there are two equiprobable values for β. The likelihood function is the ratio of two sums of two Gaussian functions:

$$\Lambda(\mathbf{x}) = \frac{p(\mathbf{x}|\Theta_1) + p(\mathbf{x}|\Theta_2)}{p(\mathbf{x}|\Theta_3) + p(\mathbf{x}|\Theta_4)}. \quad (8)$$

The likelihood ratio is clearly a nonlinear function of $\mathbf{x}$. The ML detector is impractical for this reason, but its probability of error is important to bound the performance of other detection schemes. The probability of error can be computed numerically by integrating the likelihood ratio over the decision region in the two-dimensional signal space:

$$P(e) = \int_{\Lambda(\mathbf{X})>0} p(\mathbf{x}|\alpha = -1)d\mathbf{x}. \quad (9)$$

3.1 Linear detectors

Although the signal space of Figure 1 is two dimensional, only one dimension is required to detect α. If

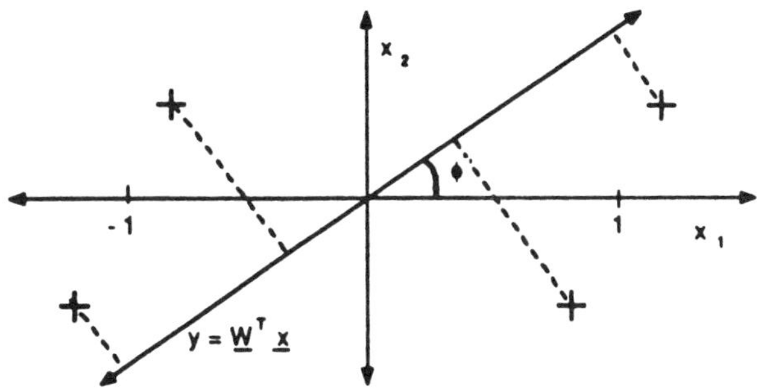

Figure 2: Linear detector signal space

the resulting scalar signal is a *linear* function of the statistic $\mathbf{x}$, then the detection scheme is linear. Specifically, the statistic $\mathbf{x}$ is reduced to the statistic y by the weight vector w:

$$y = \mathbf{w}^T\mathbf{x} \lessgtr T \tag{10}$$

and then y is compared to a threshold T to decide α. The decision boundary of a linear detector is the straight line whose equation is $y = \mathbf{w}^T\mathbf{x} = T$. Another interpretation of the linear detector is that it projects each point in the two-dimensional signal space onto a line perpendicular to the decision threshold, as shown in Figure 2.

A major advantage of linear detectors is that they require just one inner product, while nonlinear detectors require two. This can be shown by interchanging the order of the weighting and the inner product operators:

$$\begin{aligned} y = \mathbf{w}^T\mathbf{x} &= \mathbf{w}^T < \mathbf{\Phi}(t), x(t) > \\ &= < \mathbf{w}^T\mathbf{\Phi}(t), x(t) > . \end{aligned} \tag{11}$$

Thus two filters with impulse responses $\phi_1(t)$ and $\phi_2(t)$ can be replaced by a single filter whose impulse response is a weighted function of the matched filters: $\mathbf{w}^T\mathbf{\Phi}(t)$.

The probability of error of a linear detector is a function of the weight vector $\mathbf{w}$. The derivation is straightforward and relies on the fact that y is a linear function of x, so y given symbols $\mathbf{\Theta}$ is Gaussian. Specifically, $y|\mathbf{\Theta}$ has the distribution $N(\mathbf{w}^T A^T \mathbf{\Theta}, \sigma^2\mathbf{w}^T\mathbf{w})$. The problem was structured such that there are two symmetric symbol pairs: $\Theta_1 = -\Theta_4$ and $\Theta_2 = -\Theta_3$. The signal space for y is just a line, with the four conditional means representing the four possible symbol pairs. For the linear detector, the conditional means $\{\mathbf{w}^T A^T \Theta_i\}$ are correspondingly grouped into symmetric pairs, as shown in Figure 2. For the purpose of this derivation, we assume that $\mathbf{w}^T A^T \Theta_1 > 0$ and $\mathbf{w}^T A^T \Theta_2 > 0$. [2] In this case the optimum (ML) decision threshold will be zero. This threshold can be derived by constructing the likelihood ratio and noting that $p(y|\alpha = 1) = p(-y|\alpha = -1)$. The probability of error for a linear detector is therefore a simple function of the conditional means. Formally, the probability of error is equal to:

$$\begin{aligned} P(e) &= \frac{1}{2}P(e|\Theta_1) + \frac{1}{2}P(e|\Theta_2) \\ &= \frac{1}{2}Q(\frac{\mathbf{w}^T A^T \Theta_1}{\sigma\sqrt{\mathbf{w}^T\mathbf{w}}}) + \frac{1}{2}Q(\frac{\mathbf{w}^T A^T \Theta_2}{\sigma\sqrt{\mathbf{w}^T\mathbf{w}}}), \end{aligned} \tag{12}$$

where $Q(x)$ is an error function:

$$Q(x) = \frac{1}{\sqrt{2\pi}}\int_x^\infty e^{-y^2/2}dy. \tag{13}$$

Equation (12) follows from the assumption of equiprobable symbols and the observation that $P(e|\Theta_1) = P(e|\Theta_4)$ and $P(e|\Theta_2) = P(e|\Theta_3)$ due to symmetry.

3.2 LMMSE detector

The derivation of the LMMSE weight vector follows from the Orthogonality Principle, which states that the error signal is orthogonal to the observations for the optimal $\mathbf{w}$. For the linear detector of equation (11), y functions as the estimate of α, and therefore the error signal is equal to $y - \alpha$. The Orthogonality Principle states that:

$$\begin{aligned} E\{(y-\alpha)\mathbf{x}\} &= 0 \\ \text{or} \qquad E\{y\mathbf{x}\} &= E\{\alpha\mathbf{x}\} \end{aligned} \tag{14}$$

The optimal weight vector $\mathbf{w}$ is derived by substituting the definitions for y (11) and $\mathbf{x}$ (5) into equation (14):

$$(A^T E\{\mathbf{\Theta}\mathbf{\Theta}^T\} + \sigma^2 I)\mathbf{w} = A^T E\{\alpha\mathbf{\Theta}\} \tag{15}$$

This equation follows from the assumptions that all symbols and noise are independent with zero mean. The final expression for $\mathbf{w}$ is found by substituting the statistics for α and β and solving the linear equations:

$$\mathbf{w} = \frac{1}{\sigma^4 + (1+\rho^2)\sigma^2 + \rho^2(1-\gamma^2)}\begin{bmatrix} \sigma^2 + (1+\gamma^2)\rho^2 \\ -\gamma\rho^2\sqrt{1-\gamma^2} \end{bmatrix} \tag{16}$$

[2] This assumption does not contrain our solution unduly. If one of $\mathbf{w}^T A^T \Theta_1$ and $\mathbf{w}^T A^T \Theta_2$ is negative and the other positive, the detector will have poor performance; this case is not of interest. If they are both negative for a given $\mathbf{w}$, the detector performance is identical to the case with weight vector $-\mathbf{w}$ where they are both positive.

3.3 Mpe linear detector

The LMMSE detector minimizes the probability of error if the only interference is Gaussian noise. It does not minimize probability of error for this case, because the interfering symbol β is not Gaussian. The linear detector which minimizes probability of error can be derived by making a few simple observations about equation (12). The probability of error is a sum of two error functions. For most cases, $P(e)$ will be dominated by the error function with the smaller (less negative) argument. Therefore a simple approximation to the minimum probability of error receiver is to maximize the minimum amplitude of the two arguments. That is, the weight vector $\mathbf{w}$ is chosen according to:

$$\max_{\mathbf{w}}\{\min(\mathbf{w}^T A^T \Theta_1, \mathbf{w}^T A^T \Theta_2)\}. \qquad (17)$$

A second observation about the arguments is that they are both of the form $\frac{\mathbf{w}^T A^T \Theta_i}{\sigma\sqrt{\mathbf{w}^T\mathbf{w}}}$, which is a simply a projection of $A^T\Theta_i$ onto the line $\mathbf{w}^T\mathbf{x} = 0$. These arguments are therefore independent of the magnitude of $\mathbf{w}$, and so also is the probability of error. Therefore we can assume that the magnitude of $\mathbf{w}$ is one and that it takes on the form:

$$\mathbf{w} = \begin{bmatrix} \cos\phi \\ \sin\phi \end{bmatrix} \qquad (18)$$

where ϕ is the angle which the projection line makes with the x-axis. Therefore the optimization problem is reduced to one dimension: choose ϕ to maximize equation (17). In fact, an exact value for ϕ can be computed by taking the derivative of the probability of error equation 12 with respect to ϕ, and setting it equal to zero. Equation (17) provides a more intuitive solution, which we will follow.

We have restricted ourselves to the case where both error function arguments are positive. When the two arguments are plotted as functions of ϕ, as shown in Figures 3 and 4, it is apparent that there are two regions of the plot:

$$\begin{aligned} \text{Region I}: \quad \frac{-(1+\rho\gamma)}{\rho\sqrt{1-\gamma^2}} &< \tan\phi < \frac{-\gamma}{\sqrt{1-\gamma^2}} \\ \text{where } \mathbf{w}^T A^T \Theta_2 &> \mathbf{w}^T A^T \Theta_1 > 0 \\ \text{Region II}: \quad \frac{-\gamma}{\sqrt{1-\gamma^2}} &< \tan\phi < \frac{1-\rho\gamma}{\rho\sqrt{1-\gamma^2}} \\ \text{where } \mathbf{w}^T A^T \Theta_1 &> \mathbf{w}^T A^T \Theta_2 > 0 \qquad (19) \end{aligned}$$

The value of ϕ that solves equation (17) will either occur at the intersection of the two argument functions, or at the maximum of one of the two functions. Since

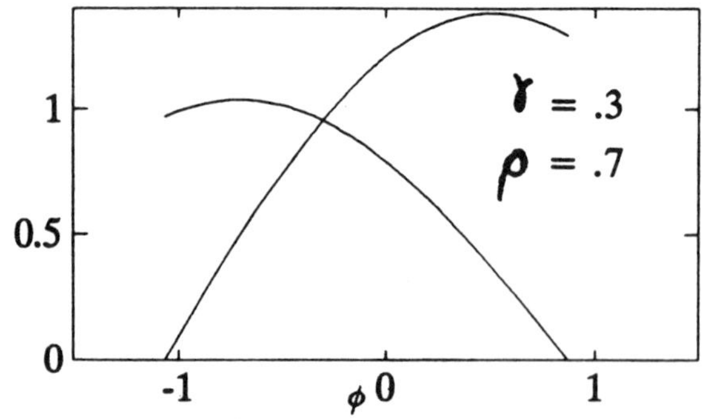

Figure 3: Case 1: distance metrics vs. ϕ

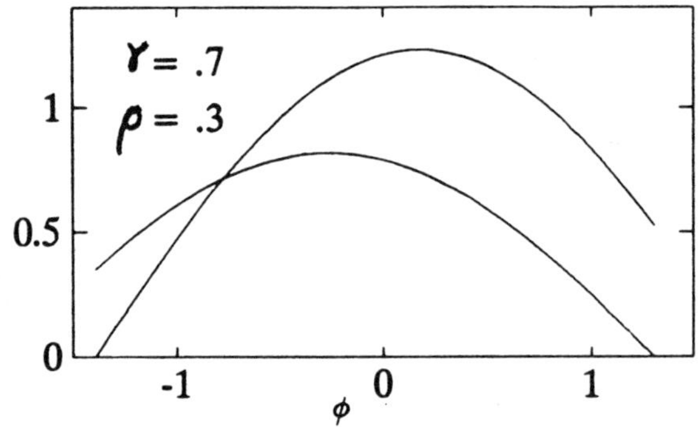

Figure 4: Case 2 distance metrics vs. ϕ

the arguments are projection funcitions, each is maximized when $\mathbf{w}$ is aligned with $A^T\Theta_i$. A little bit of analysis shows that the maximum of $\mathbf{w}^T A^T \Theta_1$ always occurs in region II, and therefore is never the minimum of the two arguments. The maximum of the second argument, $\mathbf{w}^T A^T \Theta_2$, will occur in region II if $\rho < \gamma$. So the design of the minimum probability linear detector follows one of two cases. In case 1, the optimal weight vector maximizes $\mathbf{w}^T A^T \Theta_2$. In case 2, the optimal weight vector occurs where the two error functions are equal. Specifically,:

$$\begin{aligned} \rho > \gamma: \ \mathbf{w} &= \begin{bmatrix} \sqrt{1-\gamma^2} \\ -\gamma \end{bmatrix} \\ \rho < \gamma: \ \mathbf{w} &= \frac{1}{\sqrt{1-2\rho\gamma+\rho^2}} \begin{bmatrix} 1-\rho\gamma \\ \sqrt{1-\gamma^2} \end{bmatrix} \qquad (20) \end{aligned}$$

The distance metrics from equation (12) are plotted for these two cases in Figures 3 and 4.

4 Performance Comparison

The probability of error for each of the three detection schemes, the ML, the LMMSE, and the minimum probability of error (mpe) detectors, were computed

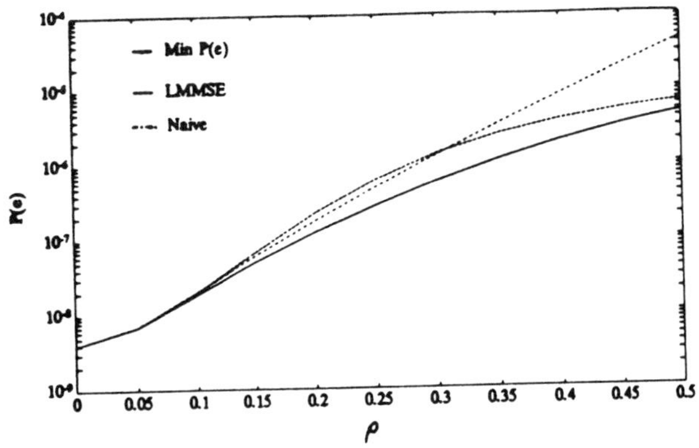

Figure 5: Detector performance vs. ρ

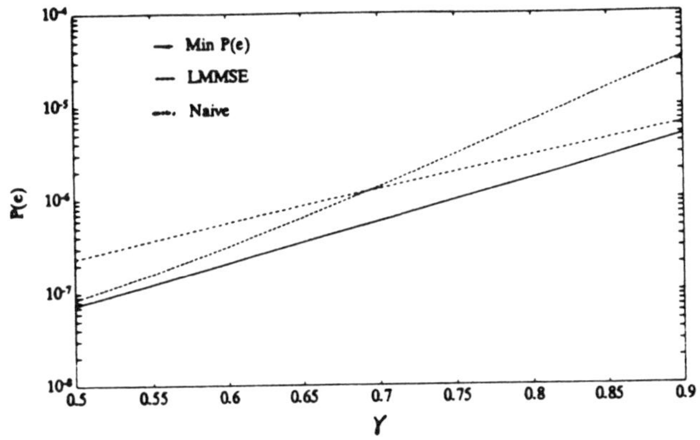

Figure 6: Detector performance vs. γ

for varying values of γ and ρ. In addition, the performance of a fourth detector was computed for comparison. This detector, referred to in figures as the "naive" detector, consists of an input filter with response $p_1(t)$ and an optimal slicer. Thus, it lacks knowledge of the cross-track response.

The performance of the mpe detector as plotted on the scale of Figures 5 and 6 is indistinguishable from that of the ML detector, so the ML detector was not included in the plots. This signifies that a linear detector is sufficient to achieve near-optimal performance for cases of interest to magnetic recording. Only when correlation γ approaches zero does the mpe performance diverge from the ML detector. The performance of the three schemes is plotted as a function of the interference correlation coefficient ρ in Figure 5, where $\gamma = .7$ and $\sigma^2 = .03$. Note the surprisingly poor performance of the LMMSE detector, which is outperformed by the naive detector at moderate levels of interference ($\rho < .3$). Figure 6 compares the detector performance as a function of γ, with $\rho = .3$. Again, note the poor performance of the LMMSE detector, which is outperformed by the naive detector at high interference correlation ($\gamma > .7$).

5 Summary

This study is a simplification of the cross-track interference problem. There may be more than one interfering track, for example when the head reads information form both the guard band and an adjacent track. The level of interference and its pulse shape will be neither known nor constant. Also, we have ignored intersymbol interference to focus on intertrack interference. Nonetheless, in reducing the problem to its essentials we have gained insight. We demonstrated that the LMMSE equalizer is suboptimal in the presence of intertrack interference. This is significant, since the MMSE criterion is commonly used to design adaptive filters and equalizers[2]. We also showed that significant improvement can result from incorporating knowledge of the cross-track response into the equalizer design. And finally, we characterized the interference problem concisely in terms of pulse shape and amplitude and gained an intuitive geometric description of intertrack interference in signal space.

References

[1] W.L. Abbott, J.M. Cioffi, and H.M. Thapar, "Performance of digital magnetic recording with equalization and offtrack interference", *IEEE Trans. on Magnetics*, vol. MAG-27, No.1, pp. 705-716, Jan. 1991.

[2] D.G. Messerschmitt, L.C. Barbosa, and T.D. Howell, "A study of sampling detectors for magnetic recording", Tech. Report RJ 4081, IBM Almaden Research Center, San Jose, Ca., July 1984.

[3] H.L. Van Trees, *Detection, Estimation, and Modulation Theory*, Wiley, 1968.

[4] M.P. Vea and J.M.F. Moura, "Magnetic recording channel model with intertrack interference", *IEEE Trans. on Magnetics*, vol. MAG-27, No. 6, Nov. 1991.

New Equalizer Targets for Sampled Magnetic Recording Systems*

James Fitzpatrick, Jack Keil Wolf and Lineu Barbosa
Center for Magnetic Recording Research
Mail Code 0401
University of California - San Diego
9500 Gilman Dr.
La Jolla, CA 92093-0401

Abstract

In high density magnetic recording, signals are often equalized to a partial response target of the form $(1-D)(1+D)^n$ and then detected by a Viterbi detector using Euclidean distance. When the recorded signals are in the presence of additive white Gaussian noise, the above detector is not a true maximum likelihood detector [1], for it ignores the correlation in the noise introduced by the equalizer. In this work we use the ideas presented in [1] to evaluate the performance of recording systems using traditional equalizer targets and to show methods of finding improved channel targets for a fixed complexity. The target $1+.25D-.75D^2-.5D^3$ is shown to outperform the EPR4 target over a large range of densities. Furthermore it is shown that in the absence of coding there exists little opportunity for performance improvements beyond those found with the above target.

1 Introduction

In 1972 Forney [2] developed a method for finding the maximum likelihood sequence estimate (MLSE) of a recorded sequence over a channel with intersymbol interference. He showed that for a channel with impulse response spanning (L+1) bit periods in additive white Gaussian noise (AWGN), MLSE could be performed by applying the Viterbi algorithm to a trellis with 2^L states. In subsequent works [1, 3] it was shown that MLSE can be implemented many different ways, but that the number of trellis states necessary is always 2^L. For channels whose impulse response spans large numbers of bit periods, the complexity of attaining true MLSE can be prohibitive. As a consequence many lower performance techniques to approximate MLSE have been proposed.

In this paper we consider the following system. A linear channel with pulse response of duration (L+1)•T seconds is assumed, where T is one bit interval and L is large, possibly infinite. The received signal is corrupted by AWGN and filtered with a matched filter sampled every T seconds. The output is then equalized to have a duration (M+1) nonzero samples with M<<L. The equalization introduces correlation to the noise which is subsequently ignored in the 2^M state Viterbi detector following the equalizer. The Viterbi detector uses a squared Euclidean metric on a trellis based upon the target of the equalizer. Figure 1 shows the channel and approximate MLSE considered here. Note that block labeled Viterbi detector represents a device that implements the Viterbi algorithm. If it were based upon the full response of the channel (i.e. no equalization), then the Viterbi detector would be a maximum likelihood detector.

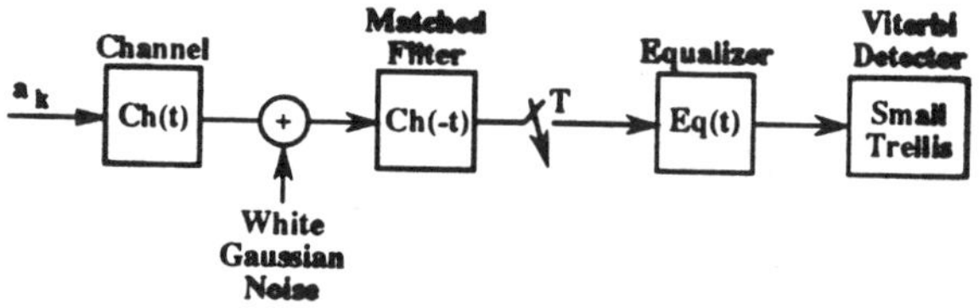

Figure 1. Channel model and approximate MLSE.

In [1] an algebraic method for determining the probability of mistaking one sequence for another was developed for the system just described. We use the techniques of [1] to evaluate the loss in performance when classical partial response targets, such as $1-D^2$ (PR4) or $1+D-D^2-D^3$ (EPR4) are used on: an ideal Lorentzian (step response) channel; a helical scan tape channel; and an experimental thin film disk channel. We investigate the use of alternative equalizer targets and find that for a fixed complexity (i.e. number of trellis states) and for a large range of densities, the PR4 target is the best possible on all three channels. However, for eight state trellises approximately .5dB performance improvement can be attained by equalizing to targets other than EPR4.

In section 2 an important result from [1] is reviewed. In section 3 the theory is applied to three magnetic recording channels and simulation results are presented.

* This research is partially supported by the Center for Magnetic Recording Research, UCSD and the National Science Foundation under grant NCR-9105639.

2 Background

We assume the channel model of Figure 1 has been converted to its discrete time equivalent (DTE) by whitened matched filtering the received signal [2]. The equalizer converts the channel response to a shortened discrete target, which acts as the basis of the trellis for the Viterbi detector. The system can be described in terms of D transforms, with channel $f(D)$, equalizer $g(D)$, target response $h(D)$, and independent, identically distributed (i.i.d.) Gaussian noise samples $n(D)$. The variance of the noise samples is σ^2. Figure 2 shows a block diagram of the channel and detector. We refer to Viterbi detectors that act on an equalized version of the channel response as approximate maximum likelihood (AML) detectors.

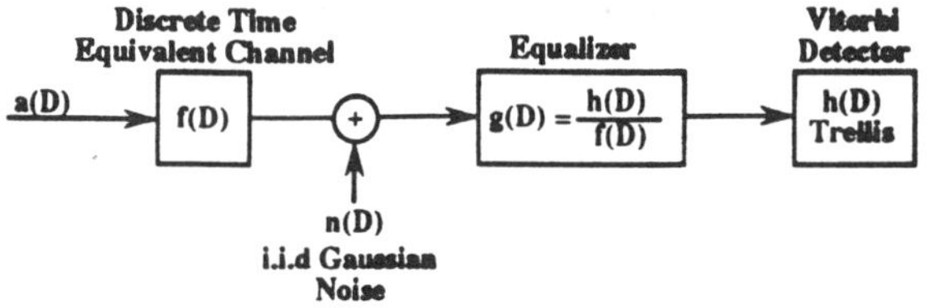

Figure 2. Discrete time model of approximate MLSE.

For a channel input sequence $a_0(D)$, the output of the equalizer is $a_0(D)h(D) + n_c(D)$ where $n_c(D)$ is correlated Gaussian noise. The probability of the Viterbi detector mistaking the received sequence for a sequence resulting from transmitting $a_1(D)$ is the probability that the noise in the direction of $(a_0(D)-a_1(D))h(D) = e(D)h(D)$ exceeds half the distance between $a_0(D)h(D)$ and $a_1(D)h(D)$. We refer to the probability of mistaking any two sequences that differ by $e(D)$ as the probability of the error event $e(D)$, or simply, $\Pr(e(D))$. Since the noise at the detector is Gaussian, this probability can be expressed in terms of the ratio of the energy of the error sequence (properly defined) at the output of the equalizer to the variance of the noise (again properly defined) at the same point. If we define the squared apparent distance, $(d_{app})^2$, of $e(D)$ as the energy in $e(D)h(D)$, then the probability of error can be written as

$$\Pr(e(D)) = Q\left(\frac{d_{app}}{2\sigma'}\right)$$

where $(\sigma')^2$ is the variance of the noise in the direction of $e(D)h(D)$ at the output of the equalizer and

$$Q(x) = \frac{1}{\sqrt{2\pi}}\int_x^{\infty} \exp(-y^2/2)dy.$$

The apparent distance of an error sequence represents the distance an error event appears to have at the output of an equalizer when noise correlation is ignored. For example, if a channel is equalized to PR4 (i.e. $f(D)g(D) = h(D) = 1-D^2$), the squared apparent distance between two sequences, $a_0(D)$ and $a_1(D)$ that differ by $e(D) = 1+D$, is normally calculated by taking the squared distance between equalized channel's responses to each sequence. Equivalently, we take the sum of the squares of the response to the error sequence. In this example, $d_{app} = |e(D)h(D)| = \sqrt{[(1)^2+(1)^2+(-1)^2+(-1)^2]}$. The apparent distance is misleading, for it is independent of the original channel response and thus ignores the noise enhancement and colorization introduced by the equalizer. The goal of this paper is to characterize the performance degradation that occurs when the correlation of the noise is ignored.

The variance of the noise in the direction of an error event, $(\sigma')^2$, can be expressed as $(\gamma\sigma)^2$ where σ^2 is the variance of the noise at the input to the equalizer and γ^2 is the increase in noise power that arises through the detection process described above. We define the effective distance, d_{eff}, of an error event as

$$d_{eff} = \frac{d_{app}}{\gamma}.$$

The effective distance represents the distance one actually achieves with AML detection. Using d_{eff} we can express the probability of an error sequence as

$$\Pr(e(D)) = Q\left(\frac{d_{eff}}{2\sigma}\right). \tag{1}$$

In [1] an expression was derived for the probability of an error sequence in terms of the original channel and the equalizer target. By phrasing the derived probability of error in terms of d_{eff}, we find that

$$d_{eff} = \frac{\langle e, Me\rangle}{\sqrt{\langle e, MM_0^{-1}Me\rangle}} \tag{2}$$

where $\langle \bullet,\bullet \rangle$ is used to denote the inner product between two vectors, M is the kernel of the quadratic associated with the equalizer target, and M_0 is the kernel of the quadratic associated with the real channel. In terms of the notation of Figure 2, $M(D) = h(D)h(D^{-1})$, $M_0(D) = f(D)f(D^{-1})$ and the inner product of two polynomials is just the constant term resulting after taking their product. To further clarify Eq. 2 we work out an example.

Let the nominal DTE channel have D transform $f(D) = 1+.5D+.1D^2$ and let the simplified channel target be $h(D) = 1+D$. Then we find $M(D) = D^{-1} + 2 + D$ and $M_0(D) = .1D^{-2} + .55D^{-1} + 1.26 + .55D + .1D^2$. For the error pattern $e(D) = 1 - D$, the squared apparent distance is

$$d_{app}^2 = \langle e, Me\rangle = \langle e(D)h(D), e(D)h(D)\rangle$$

which equals 2, the constant term of $D^{-2} + 2 + D^2$. Temporarily dropping the D argument, the denominator of Eq. 2 can be written as the constant term in the polynomial $(ehh/f)(ehh/f)^*$, where $()^*$ represents complex conjugation. For the present example, $ehh/f = 1 + 1.5D + .15D^2 - .225D^3 + .0975D^4 + \ldots$, and the constant term of $(ehh/f)(ehh/f)^*$ can be seen to be the sum of the coefficients squared, or 3.33. Thus the effective distance of the error event $e(D) = 1 - D$ is $d_{eff} = 2/\sqrt{3.33} \approx 1.096$ which is less than the apparent distance, $d_{app} = \sqrt{2}$. The true distance, d_{true}, of an error event is the distance the event achieves when all noise correlation is properly considered. It determines the performance of the true, optimal MLSE. The squared true distance of an error event is most easily computed as the energy in the channel's response to the error event: that is, $d_{true} = |e(D)f(D)| = \sqrt{1.42} \approx 1.192$, for $e(D) = 1 - D$.

An AML detector always performs somewhat worse than a true MLSE detector. We measure this loss as

$$Loss = 10 \cdot \log\left(\frac{d_{eff}^2}{d_{true}^2}\right) dB. \tag{3}$$

In the example above, the loss associated with $e(D) = 1-D$ when the channel is equalized to $1+D$ is -.73dB.

3 Application to Magnetic Recording

3.1 The Channels

We studied the performance of AML detectors for three recording channels. The first was the classical channel with the Lorentzian step response $s(t)$,

$$s(t) = \frac{K}{1 + \left(2t/pw_{50}\right)^2}$$

where pw_{50} is the pulse width at half the peak amplitude of the pulse and K is an energy normalization constant. The second channel was a helical scan tape recorder with a ferrite head on thick particulate media and the final channel was a thin film disk system. Figure 3 shows the Lorentzian and the two, measured and averaged step responses.

For each response we determined a pulse width at half the peak amplitude. We defined the recording density, a, as the ratio of pulse width at half amplitude to the channel bit period, T. Analyses were performed for a in the range $1 \leq a \leq 3$. As in the above we assumed that linear superposition holds.

For the measured channels, the DTE channel was calculated by matched filtering the step response, sampling the result, and spectrally factoring the polynomial created by sampling. Unfortunately, this factorization cannot be performed exactly for the Lorentzian pulse, so the matched filter response of the Lorentzian pulse was truncated to 101 terms and then factored. After finding the DTE channel for each system, we normalized the step response energy and convolved it with $(1-D)$, so that error sequences could be expressed using NRZ notation (i.e. we used the channel's pulse response rather than its step response).

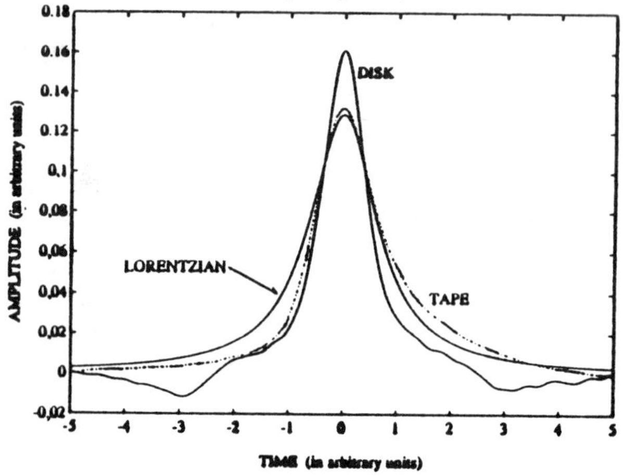

Figure 3. Step response of the Lorentzian, the tape, and the disk channels.

3.2 Distance Calculations

In order to estimate the performance of approximate ML (AML) detector systems we made two assumptions. First, we assumed that the probability of bit error is dominated by the minimum distance error event. That is, $P_e \approx Q(d_{min}/(2\sigma))$. Second, the set of error patterns having smallest distance on the true channel (i.e. at the input to the equalizer) are assumed to be the set of error patterns having the smallest effective distance in the AML system. Short of an exhaustive search, it is very difficult to determine the validity of this assumption. Later in this section we verify our assumption through simulation.

For the each of the 3 channels, the 9 error patterns with smallest true distance were found for densities between 1. and 3. For all channels below density 2.25, the minimum distance error event was the isolated dibit, $e(D) = 1$. At density 2.25 and above, the error event $e(D) = 1-D+D^2$ had the minimum distance. Figure 4 shows a plot of the distance of four error events as a function of recording density for the Lorentzian channel. The distance of the same error events on the other channels followed very closely the results shown in Figure 4. Notice that as recording densities increase, a severe penalty in performance is inevitable for a constant noise power. Not only is the minimum distance error event becoming small rapidly, but the number and length of error events with near minimum distance also is growing quickly. It is important to keep in mind that the true distances plotted in Figure 4 upper bound the corresponding effective distances when suboptimal approaches are used.

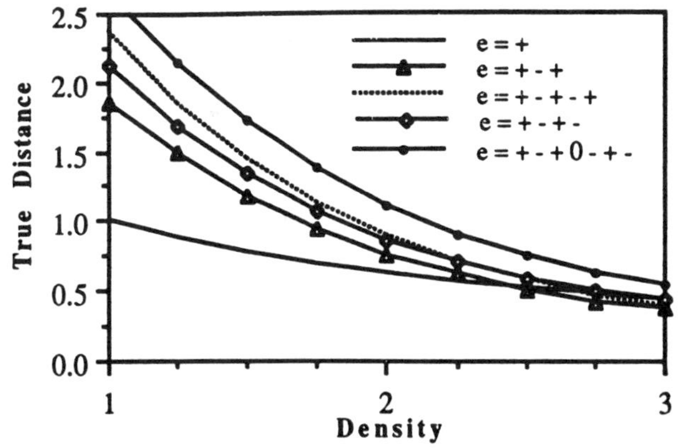

Figure 4 True distance on a Lorentzian channel.

We compare detectors of a fixed complexity, where complexity is defined in terms of the number of states used in the Viterbi decoder. For example, PR4 is compared to other channels of the form $(1-D)(1+\alpha D)$ and the EPR4 target is compared with target channels of the form $(1-D)(1-(\mu+\upsilon i)D)(1-(\mu-\upsilon i)D)$, where $i = \sqrt{-1}$ and infinite precision values of α, μ and υ are assumed.

Starting with the 4 state trellis targets, we calculated the effective distance when the target channel was equalized to $(1-D)(1+\alpha D)$ for each error event and each density. The value of α was varied from -1. to 2.5 and the best target as well as the effective distance of the worst case error pattern for that target was determined at each density. For $a > 1.75$ on the disk channel and for $a > 2$ on the tape and Lorentzian channels, PR4 was the best possible target. The tape and Lorentzian channels exhibited similar amounts of loss when equalized to PR4, while loss of the disk channel increased rapidly at high densities. At density 2.75 the Lorentzian and tape channels lose approximately 3 dB compared to ML detection, while the disk system loses over 5.5dB in performance. In Figure 5 we plot loss of the minimum distance error events versus density for the PR4 target as well as for the best 4 state target at each density for both the disk and Lorentzian channels.

Analogous computations were carried out for 8 state targets of the form $(1-D)(1-(\mu+\upsilon i)D)(1-(\mu-\upsilon i)D)$ for $-2.75 \leq \mu \leq 0.0$ and $0.0 \leq \upsilon \leq 1.35$. Unlike the results for 4 state targets, the minimum effective distance for the EPR4 channel diverges from the best possible 8 state target for a range of densities. We consider first the Lorentzian and the helical scan tape channels since their performances were very similar.

After carefully examining the data an alternative equalizer target, that achieves better effective than EPR4

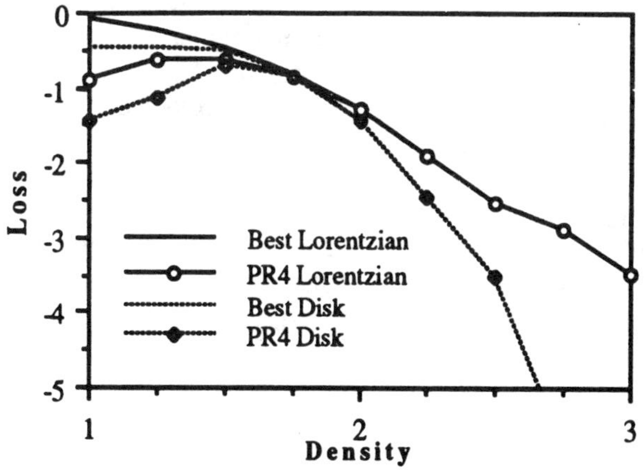

Figure 5. Loss vs. density for 4 state targets on disk and Lorentzian channels.

for $a \leq 2.75$, was found. The target shows very little loss compared to the best possible 8 state target at densities between 2 and 2.75. Figure 6 shows a plot of loss vs density on the Lorentzian channel for: the best possible 8 state target; the EPR4 target; and for $(1-D)(1+1.25D+.5D^2)$. The discontinuous jump in the EPR4 loss curve occurs when the true minimum distance error event and the effective minimum distance error events simultaneously change.

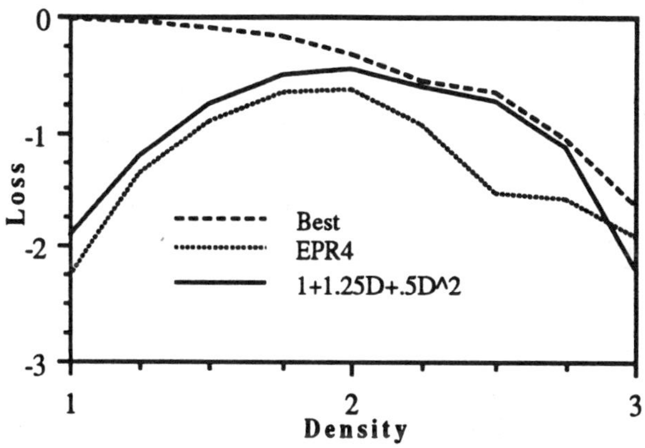

Figure 6. Loss versus density for 8 state targets on the Lorentzian channel.

These results thus far can be interpreted qualitatively in the frequency domain. The frequency response of the EPR4 target and $(1-D)(1+1.25D+.5D^2)$ are very similar except at frequencies near π/T, the Nyquist frequency. EPR4 has a second order null at π/T while the alternative target has only a 'dip' in its response. At very low recording densities there is significant response in the Lorentzian channel at the Nyquist frequency. However, as densities increase, the power shifts towards lower frequencies and less energy is found at π/T. Since there is always power in the real channel at π/T, the new target outperforms EPR4 at low and intermediate densities until that power in the channel at the Nyquist frequency is better approximated by a null than by a 'dip'.

The results of Figure 6 must be considered in light of the fact that the probability of bit error is not solely a function of the minimum distance error event. As was shown in Figure 4 there are many error events with nearly the minimum distance on the true channel at very high densities. Ultimately, a distance profile is necessary to compare different equalizer targets. In Figure 7 we plot the simulated probability of bit error versus signal to noise ratio for a Lorentzian channel equalized to either EPR4 or $(1+.25D-.75D^2-.5D^3)$ at 3 different densities. At densities $a=2.0$ and $a=2.5$, the new target outperforms EPR4 by about .5dB. At density $a=3$ the EPR4 outperforms the new target as is predicted in Figure 6. Similar results were found for the tape channel.

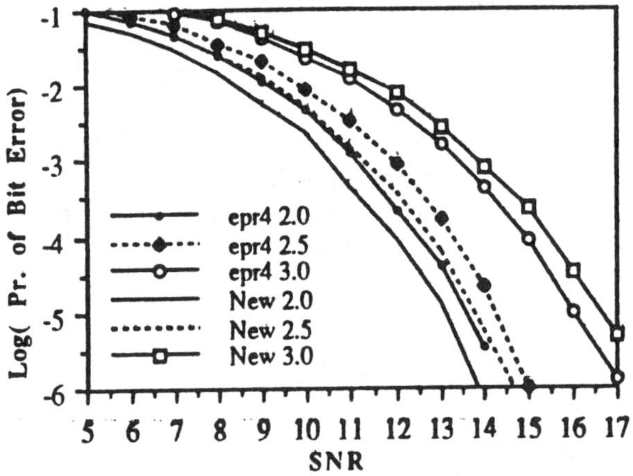

Figure 7. Probability of bit error vs signal to noise ratio for Lorentzian channel

The thin film disk channel has a different behavior than the previous channels. Undershoots in the step response and corresponding 'wiggles' in the response's frequency domain representation made it difficult to find equalizer targets that performed well over large ranges of densities. The $(1-D)(1+1.25D+.5D^2)$ target averaged only a few tenths of a dB better than EPR4 for $a \leq 2.25$, but became much worse at higher densities. Other targets such as $(1-D)(1+3D+2.5D^2)$ and $(1-D)(1+2D+1.25D^2)$ matched the best possible 8 state detector performance for certain densities, but fell off rapidly outside of these regions. In Figure 8 we show the loss versus density curves for the best possible 8 state target, the EPR4 target and for $(1-D)(1+3D+2.5D^2)$. It is important to keep in mind that loss is measured with respect to the performance attainable for a true ML detector. The fact that the loss for the EPR4 channel is less at $a=2.5$ than at $a=2.25$ means only that the target approximates ML detection better at $a=2.5$ than at $a=2.25$. The absolute performance of the AML channel falls off with density as is implied by Figure 4.

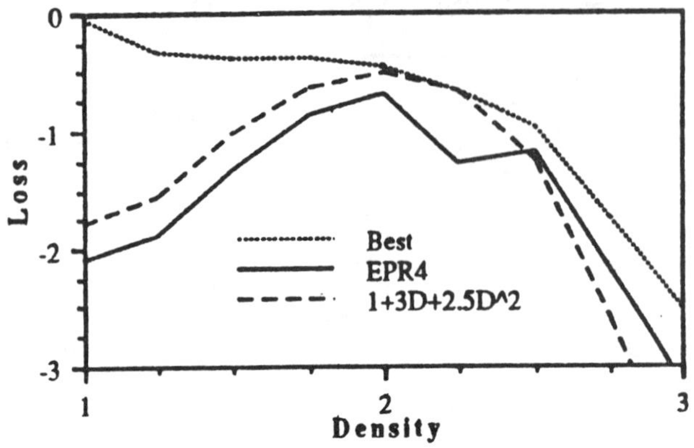

Figure 8. Loss versus density thin film disk channel.

4 Conclusions

We have used the techniques of [1] to estimate the performance degradation of several different equalizer targets for a tape channel, a thin film disk channel and for an ideal Lorentzian channel at a range of densities. Virtually no improvement over the PR4 target can be found for targets decodable with a 4 state Viterbi detector, while approximately .5dB gain can be achieved over the EPR4 channel. The same techniques can be applied to other channels to find targets that optimize performance for a fixed complexity.

In the future this work will be extended to include the effects of coding on equalizer target choice.

Acknowledgements

The authors would like to thank Alan Armstrong for providing us with the measured disk and tape responses.

References

[1] L. C. Barbosa, "Maximum Likelihood Sequence Estimators: A Geometric View", *IEEE Trans. on Info. Th.*, Vol. 35, No. 2, March 1989, pp. 419-27.

[2] G. D. Forney, "Maximum-Likelihood Sequence Estimation of Digital Sequences in the Presence of Intersymbol Interference", *IEEE Trans. on Info. Th.*, Vol. 18, No. 3, May 1972, pp. 363-77.

[3] G. Ungerboeck, "Adaptive Maximum-Likelihood Receiver for Carrier-Modulated Data-Transmission Systems", *IEEE Trans. on Comm. Th.*, Vol. 22, No. 5, May 1974, pp. 624-36.

Parallel Error-Trapping and Error-Detection Decoding

Patrick Lee and Stanley Chang

Western Digital Corp., San Jose CA, 95119

Abstract

In this paper, we present a parallel decoding method for a concatenated outer (nearest the channel) error-correcting code (ECC) and inner (nearest the user) error-detecting code (EDC). This method can be used if (1) both the ECC and EDC are linear codes, (2) the ECC decoder can be implemented as an error-trapping decoder and (3) the EDC decoder is realized by a polynomial division circuit. Compared to a serial implementation, the parallel decoding has less decoding delay and requires very minimal additional hardware.

1. Introduction.

One method of improving the integrity of data transmitted over a noisy channel is to use a concatenated code consisting of an inner (nearest the channel) error-correcting code (ECC) and an outer (nearest the user) error-detecting (EDC) code. With this approach, the user data is first encoded by EDC encoder and then by the ECC encoder. As illustrated in Fig. 1, the data can be decoded serially. That is, the received data is first decoded by the ECC decoder, which hopefully corrects all the channel induced errors, and subsequently by the EDC decoder, which then checks for any undetected errors in the user data. In particular, note that the coding delay (i.e. time required by the decoder to process the data) for the serial approach is the sum of the delay for the ECC decoder and the EDC decoder.

Because of the structure of the ECC and EDC, the serial decoding approach is often the only means of recovering the data (for example, see [1]-[4]). However, as will be shown, if

(1) the ECC and EDC are both linear codes,
(2) the ECC decoder can be implemented as an error-trapping decoder, and
(3) EDC decoder can be implemented by polynomial division circuit defined over some finite field, GF(q),

then decoding operations can be implemented using the parallel decoding structure shown in Fig. 2.

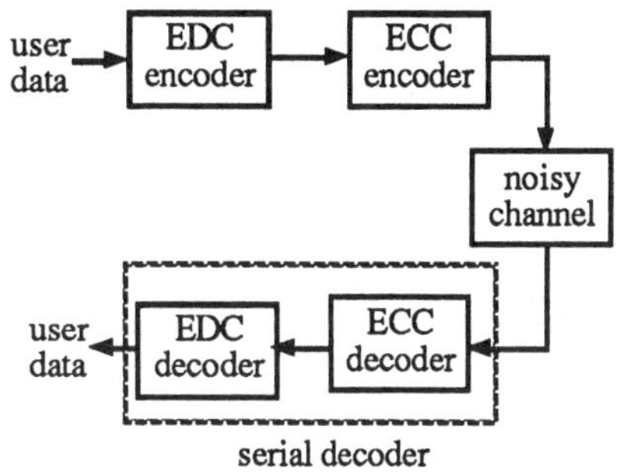

Figure 1. Serial Decoding

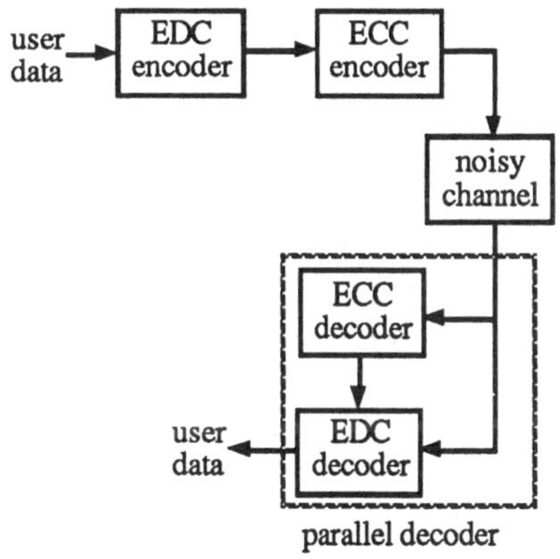

Figure 2. Parallel Decoding

The parallel approach is attractive because it involves less coding delay than the serial approach with very minimal increase in additional hardware. Specifically, the worst case coding delay for the parallel approach is only slightly longer than the worst case coding delay for the ECC decoder. Further, it can be used with any ECC

1058-6393/91 $01.00 © 1991 IEEE

and EDC which belong to the class of cyclic codes (including CRC's, BCH and Reed-Solomon codes) as well as some burst error-correcting codes.

In Section 2, we describe the parallel decoding algorithm when the received codeword is corrected by both a forward and a backwards error-trapping ECC decoder, and in Section 3, we present a mathematical justification for the algorithm.

2. The Parallel Decoding Algorithm.

The basic idea of the algorithm is as follows. First, we compute the syndrome for the EDC from the (uncorrected) received codeword, $v(x)= c(x) + e(x)$, where $c(x)$ is the transmitted codeword polynomial and $e(x)$ is the error polynomial, using the usual division circuit. Specifically, we compute

$$s(x)=R_{g(x)}[v(x)]= R_{g(x)}[c(x)] + R_{g(x)}[e(x)] = R_{g(x)}[e(x)],$$

where $R_{g(x)}[e(x)]$ denotes the remainder when $e(x)$ is divided by the EDC generator polynomial,

$$g(x)= g_0 + g_1x + ... + g_mx^m.$$

If

$$s(x)= s_0 + s_1x + ... + s_{m-1}x^{m-1} \neq 0,$$

then the received codeword must be corrected by the ECC decoder. When corrections are necessary, we can avoid having to recompute the EDC syndrome *after* the corrections by simply computing a new syndrome, $\sigma(x)$, from the "correction" polynomial, $\varepsilon(x)$, generated by the ECC decoder. Explicitly, we compute

$$\sigma(x)= R_{g(x)}[\ \varepsilon(x)\ ]= \sigma_0 + \sigma_1x + \sigma_{m-1}x^{m-1},$$

where

$$\varepsilon(x)= \varepsilon_0 + \varepsilon_1x + ... + \varepsilon_nx^n,$$

and ε_k is the error magnitude for the kth symbol generated by the ECC decoder. Next, we compare $s(x)$ and $\sigma(x)$. If $s(x)= \sigma(x)$, then we assume that all errors have been correctly decoded by the ECC decoder, otherwise we assume that the ECC decoder has miscorrected the data. Further, since $s(x)= \sigma(x)$ if and only if $s(x)+\sigma(x)= 0$ (modulo q), it follows that we can simplify the implementation by computing

$$\sigma(x)= R_{g(x)}[\ \varepsilon(x)+s(x)\ ].$$

In this case, if $\sigma(x)= 0$, we assume that the received data was correctly decoded, otherwise ($\sigma(x)\neq 0$), we assume that the received data was miscorrected. Thus, since this amounts to just adding the syndrome symbols (computed from the received codeword) to the appropriate correction symbols generated by the ECC decoder, the parallel decoding algorithm can be implemented as shown in Fig. 3.

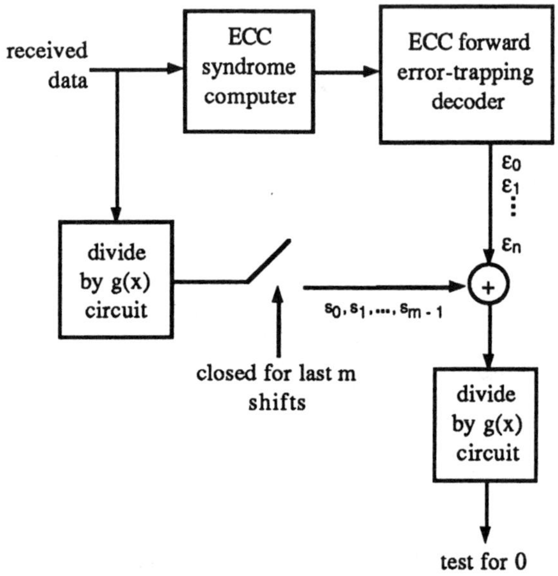

Figure 3. Parallel Decoding With Forward Error-Trapping ECC Decoder

However, for shortened codes, it is sometimes more efficient to perform "backwards" error-trapping, in which the ECC decoder outputs the reciprocal correction polynomial,

$$\varepsilon^*(x)= x^n\varepsilon(1/x)= \varepsilon_n + \varepsilon_{n-1}x + ... + \varepsilon_0x^n.$$

As a consequence, the EDC decoder must compute the syndrome given the sequence of ECC corrections in reverse order (i.e. instead of processing the sequence ε_n, $\varepsilon_{n-1},...,\varepsilon_0$, the EDC decoder must process the sequence $\varepsilon_0,\varepsilon_1,...,\varepsilon_n$). In this case, parallel decoding can be implemented as shown in Fig. 2. In particular, note for the backwards case, $\sigma(x)$ is given by

$$\sigma(x)= R_{g^*(x)}[\ x^{n-m+1}s^*(x) + \varepsilon^*(x)\ ],$$

where

$$s^*(x)= x^{m-1}s(1/x)= s_{m-1} + s_{m-2}x + s_0x^{m-1}$$

is the reciprocal polynomial of the (forward) syndrome, $s(x)$, and

$$g^*(x)= x^m g(1/x)= g_m + \quad +g_1x^{m-1} + g_0x^m$$

is the reciprocal of the generator polynomial, $g(x)$. That is, for backwards decoding, the syndrome bits are delivered in reverse order, where the first output bit of the EDC syndrome computer is s_0 (for forward decoding, the first output bit is s_{m-1}).

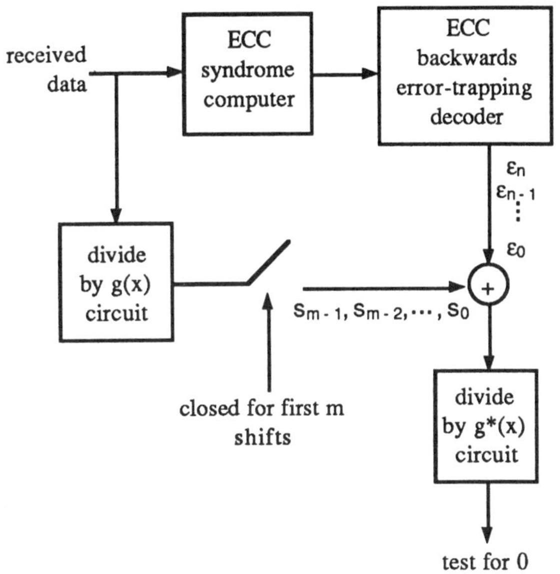

Figure 4. Parallel Decoding With Backward Error-Trapping ECC Decoder

3. Mathematical Justification.

In this section, we provide the mathematical justification for the algorithm. Specifically, we first show that the parallel decoding algorithm is equivalent to the serial decoding method in the sense that any error pattern which is detectable using the serial method is also detectable using the parallel method.

We begin by first observing that $R_{g(x)}[v(x)+\varepsilon(x)]$, the EDC syndrome computed using the serial method is equal to $\sigma(x)$, the EDC syndrome computed for the parallel method using the forward ECC error-trapping decoder. Specifically,

$$\begin{aligned} R_{g(x)}[\ v(x)+\varepsilon(x)\ ] &= R_{g(x)}[\ v(x)\ ] + R_{g(x)}[\ \varepsilon(x)\ ] \\ &= s(x) + R_{g(x)}[\ \varepsilon(x)\ ] \\ &= R_{g(x)}[\ s(x) + \varepsilon(x)\ ]= \sigma(x). \end{aligned}$$

Thus, it follows immediately that parallel decoding is equivalent to serial decoding when the ECC decoder uses a forward error-trapping algorithm.

To show that the parallel algorithm also works when used in conjunction with a backwards ECC error-trapping decoder, it is sufficient to show that

$$R_{g(x)}[\ v(x) + \varepsilon(x)\ ]= R_{g(x)}[\ e(x) + \varepsilon(x)\ ]= 0$$

if and only if

$$R_{g^*(x)}[\ x^{n-m+1}s^*(x) + \varepsilon^*(x)\ ]= 0.$$

To see that this is true, first note that the error polynomial, $e(x)$, can be written as

$$e(x)= Q(x)g(x) + s(x),$$

where $Q(x)$ is a polynomial of degree $n-m$, $g(x)$ has degree m, and $s(x)$ has degree $m-1$. Further, since $e^*(x)$, the reciprocal polynomial of $e(x)$, can be written as

$$\begin{aligned} e^*(x) &= x^n e(1/x) \\ &= x^n\left[\ Q(1/x)g(1/x) + s(1/x)\ \right] \\ &= \left[x^{n-m}Q(1/x)\right]\left[x^m g(1/x)\right] + x^{n-m+1}\left[x^{m-1}s(1/x)\right] \\ &= Q^*(x)g^*(x) + x^{n-m+1}s^*(x), \end{aligned}$$

it follows that

$$R_{g^*(x)}[\ e^*(x)\ ]= R_{g^*(x)}[\ x^{n-m+1}s^*(x)\ ].$$

In particular,

$$R_{g^*(x)}\left[x^{n-m+1}s^*(x) + \varepsilon^*(x)\right]= R_{g^*(x)}\left[\ e^*(x) + \varepsilon^*(x)\ \right].$$

Thus, we only need to show that $R_{g^*(x)}[e^*(x)+ \varepsilon^*(x)]=$ 0 if and only if $R_{g(x)}[\ e(x) + \varepsilon(x)\ ]= 0$. Now assume $R_{g^*(x)}[e^*(x)+\varepsilon^*(x)]= 0$. In this case, $e^*(x)+\varepsilon^*(x)$ must be divisible by $g^*(x)$. That is, there is a polynomial, q(x) such that

$$e^*(x) + \varepsilon^*(x)= q(x)g^*(x).$$

However, since for any two polynomials, $a(x)$ and $b(x)$, of finite degree,

$$[\,a^*(x)\,]^* = a(x) \text{ and } [\,a(x)b(x)\,]^* = a^*(x)b^*(x),$$

where $[\,f(x)\,]^*$ is the reciprocal polynomial of $f(x)$, it follows that

$$e(x)+\varepsilon(x) = [\,e^*(x)+\varepsilon^*(x)\,]^* = [\,q(x)g^*(x)\,]^* = q^*(x)g(x).$$

Thus, $e(x)+\varepsilon(x)$ is divisible by $g(x)$ and $R_{g(x)}[e(x)+e(x)]=0$. Since using a similar argument shows the reverse is true, it then follows that parallel decoding in conjunction with a backwards error-trapping ECC decoder is equivalent to serial decoding.

References.

[1] E.R. Berlekamp, *Algerbraic Coding Theory*, Aegean Park Press, 1984 (rev)

[2] R.E. Blahut, *Theory and Practice of Error Control Codes*, Addison-Wesley, 1984

[3] W.W. Peterson and E. J. Weldon, Jr., *Error-Correcting Codes*, MIT Press, 1988

[4] S. Lin and D.J. Cosetello, Jr. *Error Control Coding: Fundementals and Applications*, Prentice-Hall, 1983

EFFECTS OF RESOURCE SATURATION IN REAL-TIME COMPUTING ON DATA FLOW ARCHITECTURES

S. Som, R.R. Mielke, and J.W. Stoughton
Department of Electrical and Computer Engineering
Old Dominion University
Norfolk, Virginia 23529-0246

Abstract

This paper is concerned with resource saturation in dataflow architectures operating in real-time. Resource saturation is created by excessive concurrency in the computational problem and can make real-time system performance unpredictable. A computational model for dataflow architectures, called ATAMM, is described and used for achieving the best possible time performance for a given number of resources while retaining predictability and periodic steady state operation. Concurrency control measures are developed. Simulation results are presented to validate the ATAMM model and to illustrate that an uncontrolled execution of a dataflow graph can easily lead to unpredictable and degraded performance.

1. Introduction

The dataflow concept has already gained wide acceptance as an excellent computational model for multiprocessor systems. An emerging area of interest is to use dataflow computers for real-time computing [1, 2, 3]. The main objective is to improve performance by exploiting concurrency in the computational algorithm. A few example of dataflow architectures developed for large-grained real-time algorithms are the Advanced Development Model (ADM) avionics computer [4], the Generic VHSIC Spaceborne Computer (GVSC) [4], and the MAX dataflow machine [2].

Performance of real-time systems has been greatly improved by concurrent processing. However, a little known but tremendously significant problem in increasing computational concurrency with dataflow model or concurrent processing in general, is that excessive concurrency can saturate or even deadlock any architecture having finite resources [5, 6]. The effects of resource saturation (a phenomenon of creating more computational tasks than the available resources) are that data packets experience unnecessary waiting times and require extra storage space, and system performance becomes unpredictable. The execution of algorithms must be controlled so that resource need does not exceed resource availability.

Resource saturation is totally unacceptable in real-time computing because of the usual requirement of high degree of predictability. The problem is even more acute for large-grained multicomputer systems as described in [4] and [2] because of relatively fewer processors. Algorithm To Architecture Mapping Model, or ATAMM, is a marked graph dataflow model which can control concurrency in real-time computing based on the number of available resources [4]. The ATAMM model is described briefly in Section II. In Section III, the ATAMM model is used to develop concurrency control strategies for predictable performance. Then in Section IV, concurrency control measures are elaborated and examples are presented for resource saturation by uncontrolled execution of the dataflow graph. All results and claims are supported by simulation of a space surveillance algorithm.

2. The ATAMM Model

ATAMM describes algorithm execution on a dataflow architecture by three Petri Net marked graphs: the algorithm marked graph (AMG), the node marked graph (NMG), and the computational marked graph (CMG). An algorithm marked graph is a marked graph version of a directed or a dataflow graph which represents a specific algorithm decomposition. Transitions (graph nodes) and places (graph edges) represent algorithm operations and algorithm operands, respectively. The transition times represent the computational times required for the algorithm operations. All edges have a queue of buffers and can accommodate more than one token at a time. As an example to illustrate the ATAMM model, consider a space surveillance algorithm having the AMG shown in Figure 1. Node labels describe algorithm operations. Signals which are transferred from one node to another are shown as directed edges.

1058-6393/91 $01.00

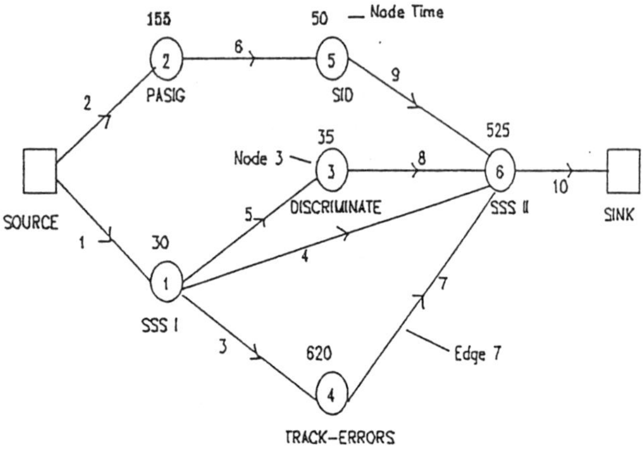

Figure 1. Algorithm Marked Graph for a space surveillance algorithm.

The node marked graph (NMG) is a representation of the execution of an AMG node by a functional unit which is also termed as a resource or a processor. The NMG for an AMG node is shown in Figure 2. A functional unit must be available for assignment to the algorithm operation before the read transition can fire. Once assigned, the functional unit is used to implement the read, process, and write operations before being returned to a queue of available functional units. The computational marked graph (CMG) is constructed from the AMG by replacing every AMG node by the corresponding NMG. AMG places are replaced by place pairs, a forward directed place representing data-flow and a backward directed place representing control-flow. The CMG corresponding to the algorithm marked graph of Figure 1 is shown in Figure 3.

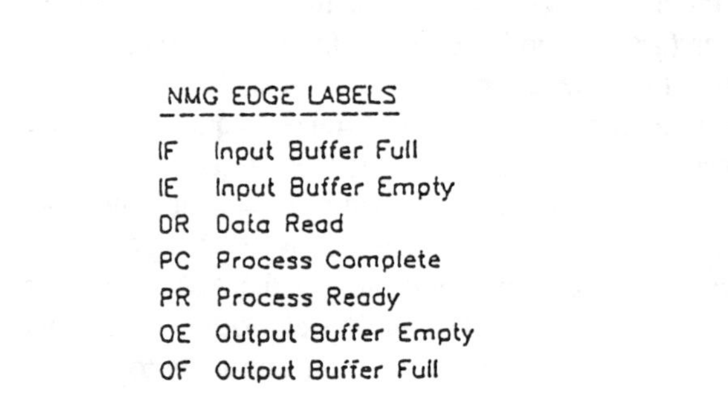

Figure 2. Node Marked Graph.

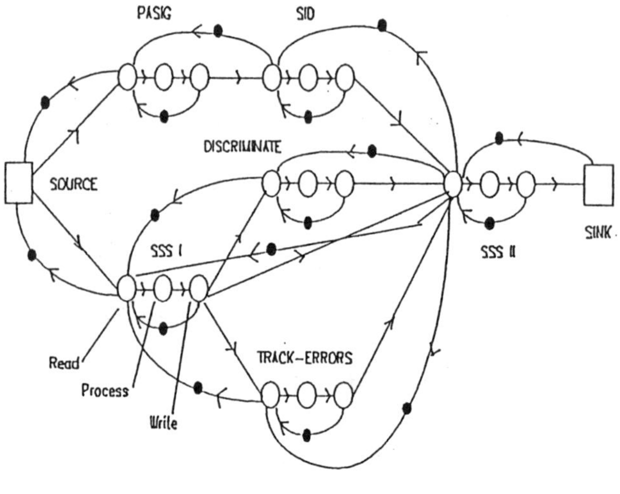

Figure 3. The Computational Marked Graph for the space surveillance algorithm.

Two measures of time performance, TBIO and TBO, are defined next. The performance measure TBIO (time between input and output) is the elapsed computing time between an algorithm input and the corresponding algorithm output. Therefore, TBIO is an indicator of computing time. It is shown in [6, 7] that the algorithm imposed lower bound for TBIO, denoted $TBIO_{LB}$, is given by the sum of transition times for nodes contained in the longest directed path from the input source to the output sink in the AMG. The performance measure TBO (time between outputs) is the elapsed computing time between successive algorithm outputs. Therefore, the inverse of average TBO is an indicator of output per unit time or throughput. It is shown in [6] that the algorithm imposed lower bound for average TBO (TBO_{LB}) is given by the largest time per token of all directed circuits in the CMG. The number of resources of the architecture is denoted by R.

3. Performance Prediction

The goal of this section is to briefly describe a design procedure to achieve predictable performance by preventing resource saturation. For a given algorithm decomposition, the parameters TBIO, TBO and R define an operating point for ATAMM. The display of all operating points on a graph of TBO versus TBIO with R indicated as a parameter is called the ATAMM performance plane. The ATAMM performance plane, illustrated in Figure 4, is extremely useful for selecting system operating strategies. The best system performance is achieved by operating at point B where $TBIO = TBIO_{LB}$ and $TBO = TBO_{LB}$. For this operating point, input data packets are injected at the interval of TBO_{LB} by adjusting the source transition time. The maximum rate of input injection for any operating point should not exceed 1/TBO to avoid congestion in the AMG with extra data packets.

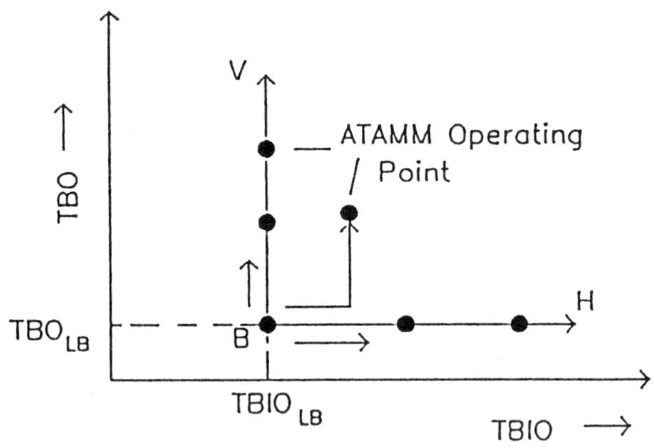

Figure 4. ATAMM Performance Plane

Two key ideas for shifting the operating point for lowering resource requirements are to increase the input injection interval above TBO_{LB} and to modify the algorithm marked graph to obtain a predictable performance on a given hardware capability. Modifications are achieved by adding control edges and buffers as described below. A control edge is an AMG edge which imposes a precedence relation among two AMG nodes but does not imply data dependency. This kind of edge is used to delay the firing of an AMG node. The buffers are inserted on AMG edges to provide additional storage space for the output of an AMG node.

The operating points for the space surveillance algorithm of Figure 1 are shown in Figure 5. Read and write times are assumed to be 11 and 14 time units, respectively. These operating points are achieved by algorithm modification using control edges and adjusting the required number of buffers, and by controlling injection rate. Operating point B is the operating point which achieves the lower bounds for TBIO and TBO with three resources or more. This point is achieved for three resources by adding a control edge from node 3 to node 2 and an extra buffer on edge 4. From the operating point diagram of Figure 5, there is more than one possible operating point for a given value of resources. The selection of which point is actually used is made off-line by the system user prior to algorithm execution.

In order to validate the prediction of performance and resource requirements by the ATAMM model, a number of software support tools have been developed [8, 9]. The support tools include a simulator and an analyzer which can simulate and analyze the execution of a dataflow graph on a dataflow architecture. These software tools are developed in C language for Windows 2.1/3.0 running under MS-DOS on the IBM PC or compatibles. The operating point (TBIO = 1250, TBO = 656) of Figure 5 is simulated for fifteen data packets.

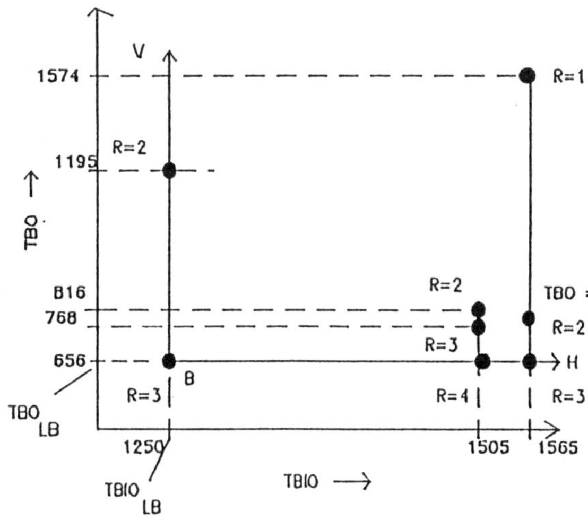

Figure 5. ATAMM operating points for the space surveillance algorithm.

The minimum input injection interval at the source is set to 725 time units. This value of input injection interval is about 11% higher than the TBO shown in Figure 5 to account for overheads such as bus contention. The number of resources is set to 3. The time performance TBIO and TBO for data packets are displayed in Figure 6. TBI is the measured time between injection of successive input data packets. From the display, the values of TBIO and TBO at steady state are 1401 and 730, respectively. The time performance reaches steady state beginning with the fourth data packet. Another display called total graph play (TGP) is shown in Figure 7 which depicts execution of all graph nodes. The execution of AMG nodes are shown by hatched bars on separate rows of the display in numerical order with node 1 on the bottom of the screen. The data packet number is indicated in the middle of the node bar. It can be observed that all graph nodes are executed periodically resulting in periodic outputs and periodic resource requirements.

4. Resource Saturation

A dataflow graph for real-time computing can experience resource saturation due to excessive concurrency in the algorithm resulting in unpredictable performance. Computational concurrency occurs in two ways. First, several nodes of the dataflow graph may be performed simultaneously for a single data packet. This type of concurrency is referred to as parallel concurrency

PERFORMANCE

Select

PACKET	TBI	TBO	TBIO	R
1	34	1394	1360	3
2	730	755	1385	3
3	730	746	1401	3
4	730	730	1401	3
5	730	730	1401	3
6	730	730	1401	3
7	730	730	1401	3
8	730	730	1401	3
9	730	730	1401	3
10	730	730	1401	3
11	730	730	1401	3
12	730	730	1401	3
13	730	730	1401	3
14	730	730	1401	3
15	730	666	1337	3

Figure 6. Performance display for operating point B.

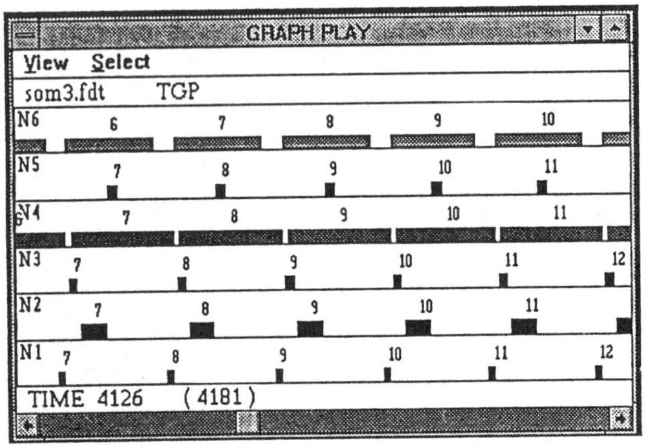

Figure 7. Total Graph Play for operating point B.

because it is the result of inherent parallelism in the dataflow graph. Second, nodes of the dataflow graph belonging to different data packets may be performed simultaneously in the computing system. This is an unique feature of real-time computing as the same set of computational algorithms are usually executed repetitively for successive data packets. This type of concurrency is called pipeline concurrency because the graph nodes are repeated like stages of a pipeline.

Improper scheduling of the dataflow graph can create excessive parallel concurrency of algorithm nodes and can easily lead to resource saturation. In the ATAMM model, control edges are used to establish precedence relations to control parallel concurrency. The level of parallel concurrency in a dataflow graph should be proportional to the parallelism in the architecture. Hence, the selection of an operating point must be reevaluated when the number of resources is decreased due to fault conditions. Otherwise, the system may become saturated with enabled AMG nodes.

Pipeline concurrency is controlled by adjusting the minimum injection interval between input data packets. If the input data injection rate is higher than the ATAMM specified TBO, it can cause resource saturation by generating excessive pipeline concurrency. This leads to operation at a steady-state where TBIO is much greater than the longest path time between input and output. This occurs because the pipeline from input to output becomes congested with extra data packets which must wait for successor nodes to complete. Also, a resource must be free to process an enabled AMG node. In dynamic memory allocation schemes, this will tremendously increase memory requirements to accommodate these additional data tokens. Another effect of driving the dataflow architecture above its capability, is that the AMG outputs may not be generated at regular intervals which is desirable for control and signal processing applications. Therefore, the ATAMM model suggests that the minimum input data injection interval be controlled to the best possible TBO value for a given R which the model predicts. This injection control scheme eliminates data packet congestion and allows periodic steady state operation where TBIO and TBO remain same for all data packets.

In order to illustrate resource saturation caused by excessive parallel concurrency, the space surveillance algorithm is simulated for operating point B but without the control edge and extra buffer as described in the previous section. The time performance TBIO and TBO for data packets are displayed in Figure 8. From the display, the values of TBIO and TBO are much higher than that in Figure 6 and the system does not reach steady state. This shows the effectiveness of the ATAMM model in predicting the need for control edges and extra storage spaces. Also, the AMG of Figure 1 is simulated for operating point B but with one less resource. The performance TBIO and TBO are displayed in Figure 9 which again shows the effects of resource saturation in degrading performance. This shows that if R goes down due to a fault or routine maintenance, the AMG has to be rescheduled.

PERFORMANCE

Select

PACKET	TBI	TBO	TBIO	R
1	34	1405	1371	3
2	730	825	1466	3
3	730	791	1527	3
4	746	812	1593	3
5	730	798	1661	3
6	730	800	1731	3
7	730	795	1796	3
8	730	795	1861	3
9	730	795	1926	3
10	730	795	1991	3
11	730	854	2115	3
12	730	862	2247	3
13	800	862	2309	3
14	862	846	2293	3
15	862	739	2170	3

Figure 8. Performance display without control edges and buffers.

PERFORMANCE

Select

PACKET	TBI	TBO	TBIO	R
1	27	1388	1361	2
2	730	992	1623	2
3	730	959	1852	2
4	730	959	2081	2
5	730	959	2310	2
6	862	959	2407	2
7	959	959	2407	2
8	959	959	2407	2
9	959	959	2407	2
10	959	959	2407	2
11	959	959	2407	2
12	959	959	2407	2
13	959	959	2407	2
14	959	955	2403	2
15	959	915	2359	2

Figure 9. Performance display with R=2.

For illustrating resource saturation by excessive pipeline concurrency, the space surveillance graph is simulated for operating point B but with the minimum input injection interval equal to 300 time units. From the performance display of Figure 10, the steady state values of TBIO and TBO are 2309 and 862, respectively, which are much larger than the corresponding values in Figure 6. The reason is that the lower minimum injection interval allows more input data tokens to enter the graph, but dataflow graph cannot be executed at a rate faster than $1/TBO_{LB}$. The result is that data tokens simply wait on the algorithm edges and TBIO is increased sharply. Also this increase in computing time is dependent on token positions in the graph and hence is very difficult to predict.

PERFORMANCE

Select

PACKET	TBI	TBO	TBIO	R
1	34	1411	1377	3
2	305	862	1934	3
3	487	862	2309	3
4	862	862	2309	3
5	862	862	2309	3
6	862	862	2309	3
7	862	862	2309	3
8	862	862	2309	3
9	862	862	2309	3
10	862	862	2309	3
11	862	862	2309	3
12	862	862	2309	3
13	862	862	2309	3
14	862	846	2293	3
15	862	739	2170	3

Figure 10. Performance display with minimum injection interval = 300 time units.

Conclusions

The key idea of the ATAMM model is to prevent resource saturation and never allow the system to go into a resource limited mode. Resource saturation is the direct result of excessive concurrency in the dataflow graph which the architecture and algorithm cannot perform, resulting in unpredictable token movement along the dataflow graph. The ATAMM model is being implemented on two VHSIC architectures, the ADM system of Westinghouse and the GVSC system of IBM [4]. Current research topics include modeling of algorithms with variable node times, fault tolerance, and overhead estimation.

Acknowledgment

The work reported here was supported in part by the NASA Langley Research Center under Grants NCC1-136 and NAG-1-683.

References

[1] E.A. Lee and D.G. Messerschmitt, "Static Scheduling of Synchronous Data Flow Programs for Digital Signal Processing," IEEE Transactions on Computers, vol. 36, pp. 24-35, January 1987.

[2] R.D. Rasmussen, G.S. Bolotin, N.J. Dimopoulos, B.F. Lewis, and R.M. Manning, "Advanced General Purpose Multicomputer for Space Applications," Proc. of the 1987 International Conference on Parallel Processing, University Park, PA, August 17-21 1987.

[3] Kesab. K. Parhi and David G. Messerschmitt, "Static Rate-Optimal Scheduling of Iterative Data-Flow Programs via Optimum Unfolding," IEEE Transactions on Computers, pp. 178-195, vol. 40, No. 2, February 1991.

[4] R. Mielke, J. Stoughton, S. Som, R. Obando, M. Malekpour, and B. Mandala "ATAMM Multicomputer Operating System Functional Specification," NASA Contractor Report 4339, Grant NCC1-136, August 1990.

[5] D.E. Culler and Arvind, "Resource Requirements of Dataflow Programs," Proc. of the 15th Annual International Symposium on Computer Architecture, Honolulu, Hawaii, May 30 - June 2, 1988.

[6] S. Som, "Performance Modeling and Enhancement for the ATAMM Data Flow Architecture," Ph.D. Dissertation, Old Dominion University, Norfolk, VA, May 1989.

[7] S. Som, R.R. Mielke, and J.W. Stoughton, "Strategies for Predictability in Real-Time Data-Flow Architectures," Proc. of the 11th IEEE Real-Time System Symposium, pp. 226-235, Orlando, Florida, December 5-7, 1990.

[8] R.L. Jones, III, "Diagnostic Software for Concurrent Processing Computer Systems," Master's Thesis, Old Dominion University, April 1990.

[9] M.R. Malekpour, "Simulator for Concurrent Processing Data Flow Architectures," Master's Thesis, Old Dominion University, March 1991.

FAST STRUCTURED SPARSE UNARY MATRIX OPERATIONS ON DENSE SYSTOLIC ARRAYS[1]

*A. O. Barbir * , J. L. Aravena ** , and K. Arfaoui**

* Computer Science, Western Carolina University, Cullowhee, NC 28723

** Electrical & Comp. Engr., Louisiana State University, Baton Rouge, LA 70803

Abstract

The paper presents a systematic method for increasing the efficiency of a class of systolic arrays implementing unary matrix operations. The method reported here uses the concept of timing graph to display the evolution of computations as function of the compute cycle. In an extension of previous results this paper introduces the concept of timing graph decoupling as a procedure to determine structured sparsity patterns that can be executed more efficiently in a dense array. The technique is applied here to **LU** and **QR** decomposition arrays.

I. Introduction

Recent advances in VLSI technology have made it possible to design and fabricate specialized parallel architecture for numerous applications. Parallel structures such as systolic arrays offer attractive improvement in the computational speed over the general purpose sequential computer.

Conventional methods for designing systolic arrays are based on the generic form of an algorithm. This design methodology results in arrays that are incapable of adapting to the changing nature of real-time applications. For example, an array designed to compute the **LU** decomposition of a dense matrix will take the same time to implement the decomposition of a banded sparse matrix.

In general systolic arrays have very limited reconfiguration capabilities. This fact prompted researchers to consider the addition of switching networks as a mean of increasing the flexibility of an array [1-4]. This concept tries to reconfigure the array connectivity such that it matches the problem being solved. Although this technique increases the utility of an array, but it is rather limited and expensive to implement.

In this study we consider an alternative method for increasing the class of algorithms that can be implemented in a given array. Here, we assume that the resulting connectivity of the array is fixed. However, we look at the compute cycle number and the way data elements are fed into the array as extra degrees of freedom. This strategy allows the derivation of procedures that identifies the collection of structured sparsity patterns that can be executed in an array in a reduced number of cycles.

The paper is organized as follows: Section 2 summarizes previous work and basic notations including the Timing Graph (TG) representation of an array operation as a function of the compute cycle. Section 3 introduces procedures to decouple the timing graphs in order to determine structured sparsity patterns. Section 4 applies the timing graph decoupling to matrix decomposition arrays.

2. Background and Basic Notations

A large class of algorithms that are suitable for systolic array implementation can be represented in a high level language (HLL) as a set of nested loops [2] characterized by a set of assignment statements and possibly some conditional branch statements. To differentiate between the states of the variables, a variable is defined as *used* if it appears on the right hand side of an assignment statement, and is *generated* if it appears on the left hand side of the statement.

Between variables generated at different index points in the HLL description of the algorithm, there are some dependencies that describe the algorithm's internal structure and dictate its communication requirements. These dependencies are described by what is known as data dependence vectors [2]. A data dependence vector is defined as *fixed* if all its components are constant, otherwise, it is defined as *variable*. Algorithms characterized by variable data dependence vectors are in general not suitable for systolic arrays implementation. However, in [2] it was shown that a dependence vector with l variables can be replaced by a set of $2l+1$ fixed vectors. This leads to the regularization of data flow in

[1] Work was funded in part by NSF Grant USE-9152384.

1058-6393/91 $01.00 © 1991 IEEE

the algorithm and allows its mapping into an array.

An algorithm described in HLL is suitable for systolic array implementation if its variables are pipelined [3] and has all of its data dependencies transformed into a set of fixed [3] data dependence vectors. In this case, it is referred to as a *systolic* algorithm.

In matrix algorithms, it is possible to distinguish between two types of *fixed* data dependence vectors, mainly *flexible* and *non-flexible* vectors. Matrix-matrix (vector) algorithms are characterized by flexible data dependence vectors. This is the case because it is possible to re-arrange the order of computations without effecting the overall results. Other matrix algorithms such as the **LU** and **QR** algorithms are characterized by *non-flexible* data dependence vectors. This means that the order of the computations is important.

The HLL description of a systolic algorithm is static, meaning that it does not provide any information about the execution time of the computations. An ordered execution of a systolic algorithm generates the following 6-tuple $A\ (J^m, JI, JO, FI, D, V)$ (1) where, J^m is the algorithm domain; JI is the set of the algorithm *used* data elements; JO is the set of the *generated* data elements; $FI = \{f_1, ..., f_r\}$; set of memoryless functions that map the input data to the output data; $D = \{d_1, ..., d_v\}$ is the set of data dependence vectors; $V = \{a_1, ..., a_l\}$ set of variables defined in J^m.

This algorithm model is independent of any hardware implementation. To establish the implementation of an algorithm in the array, it is necessary to establish a correspondence between algorithm variables and array variables. In this regard, a systolic array implementing the algorithm must produce the same tuples described in (1), with the exception, that the sets are generated at different locations in space and are distributed in time. The array imposes an execution ordering, on the algorithm variables, characterized by a time parameter and satisfies: i)If $a(j)\ \varepsilon\ JI$, $j\ \varepsilon\ J^m$ is an algorithm used element then , it must be an input element to an appropriate PE at a unique cycle k, ii) If $a(j)\ \varepsilon\ JO$, $j\ \varepsilon\ J^m$ is an algorithm generated element then, it must be an output element from an appropriate PE in the array at a unique cycle k, iii) To any operation in the original algorithm such as addition, multiplication, etc., an identical operation must be performed by the array processing elements, iv) Data dependencies of the systolic algorithm are preserved in the array.

The array operations as a function of the compute cycle,is modeled as a directed weighted computing graph (DWCG) [3], where, the vertices (nodes) correspond to the PEs, the directed arcs corresponds to the links, the graph is capable of transferring data elements between the nodes, such that at each cycle processed data elements on node i are conveyed to node j on arc (i,j). At each cycle, the graph enters a new state typified by the new data elements in all the vertices. The weight of an arc in the DWCG, is a unique integer, greater than zero, identifying the corresponding path. Details on the assignment of these weights are given in [3].

The array connectivity is defined by the DWCG weighted adjacency matrix denoted by $\mathbf{P}\ \varepsilon\ \mathbf{R}^{NxN}$, where N is the total number of PEs in the array. The completion of the degree of the boundary nodes in the DWCG requires the assumption that external data elements are supplied to those nodes by user induced input/output nodes. The complete graph will be referred to as the array graph (AG). In the AG, the user induced input nodes supply their data elements to the array, in the form of a timing function denoted by $\pi_{kw}(n)$, where, k is the cycle number, w is the weight of the arc corresponding to data elements $a(.)$, and n is the input node number.

2.1. Array Timing Graph

The operations of the nodes in the AG at each compute cycle are described by the timing graph (TG). The TG is an acyclic graph that consists of a total of T time levels (number of compute cycles). All the information needed to define the state of a node is included in the representation. Only active nodes are represented at each level in the TG. The construction of the TG retains the structure of the array operations, its connectivity and the integrity of the flow of data elements between the PEs of the systolic array.

In the TG , a time axis must be defined that indicates the state of each PE in the array at every compute cycle. This time axis will be referred to as the node direction vector and for node j is denoted by $\mathbf{d_j}$. Furthermore, all elements $a(.)$ belonging to the same algorithm variable, follows lines that are parallel to a direction vector denoted by $\mathbf{V_a}$. The details for constructing the TG are given in [3].

Consider a node g, represented in the TG at time levels $i, i+1, ..., j$. The collection of elements available on the node's input arc q, that carry elements belonging to the same data path (specified by the direction vector $\mathbf{V_a}$) for those time levels, define a timing sequence, denoted by $\mathbf{X}(\mathbf{V_a}, g, j-i+1)$ [3]. Timing sequences on all lines must be of length T (designed execution time of the array). If one sequence is shorter than the other, zero entries must be appended to the sequence in the proper time slots , such that the resulting sequence is of length T.

It is important at this point of the discussion to notice that timing sequences can be manipulated [3]. It is possible to slide the sequence up or down on the time axis. To slide the sequence up n times , one can write $U^n(\mathbf{X}(.))=U(U^{n-1}(\mathrm{X}(.)))$. Sliding up a sequence one cycle is equivalent to deleting the first element in the sequence and appending a zero to its last entry. Timing sequences can also be slid down. Sliding a sequence down by one cycle is equivalent to inserting a new element in the first time slot, and deleting the last element in the sequence. To slide a sequence down n times, one can write $D^n(\mathbf{X}(.))=D(D^{n-1}(\mathbf{X}(.)))$.

The sliding operations can be used effectively to derive the modified timing functions of the user induced input nodes, when the TG of the array is modified. For example, consider the **LU** array for the *4x4*, **A** matrix shown in Figure 1. The TG is depicted in Figure 2. The timing sequence on node *1* starting at time level k_o and ending at time level k_9 for the $a(.)$ data elements is $\mathbf{X}(\mathbf{V_a},1,9)=\{a_{11}, a_{21}, a_{31}, a_{41}, o, o,o,o,o,o\}$. Sliding the sequence up by one time slot leads to the new sequence $\mathbf{X1}(.)=\{a_{21}, a_{31}, a_{41}, o, o,o,o,o,o\}$. At each compute cycle the array executes a set of computations so that after each cycle it is possible to examine the partial results and establish their functional dependence. In this regard, let Ψ be the set of all operations executed in the array and depicted in the TG. Given the time axis and a corresponding array this set of operations is evaluated in T cycles. The set Ψ_k, on the other hand, contains all operations scheduled up to the k-th compute cycle. Thus for $k < T$ this set is a proper subset of Ψ. For specific realizations of the algorithm some operations are redundant; typically because they involve additions of, or multiplications by, zero. These operations can be eliminated without affecting the final outcome. Thus, it is possible to define the set Ψ_r of required operations. A TG displaying only the set Ψ_r, is termed a decoupled timing graph (DTG). If the set of required operations is a proper subset of some Ψ_r, then the complete set of required operations could be executed in a reduced number of cycles. Such specific realizations are termed *fast realizations* of the algorithm. The result will be families of structured sparse matrices executable in a single array more efficiently than the generic form.

To find a fast realization one must determine the collection of operations that, for a given set of parameters, can be omitted without affecting the final numerical value. The restriction of operating an array for k cycles requires that the total number of levels in the TG be reduced to k. TG decoupling will be used as the vehicle for determining such forms.

3. Timing Graph Decoupling

Timing graph decoupling is defined as the process of determining those entries in the input data that can be omitted or forced to a particular pre-set value, that lead to decomposing the array operations into sub-arrays implementing smaller size problems. Array decomposition speeds up the response time of the generic array because, it skips those entries from the input that are irrelevant to the final results and it trads a delay element in the input with an entry from the input elements to be processed.

Due to the nature of the algorithms at hand, it is important to note that if an entry is omitted or forced to a particular value, such as zero, care must be taken to insure the numerical correctness of all elements that are dependent on it. Next, a procedure is introduced that marks all the nodes of the array that can skip their input data. Once the collection of those elements are determined, another procedure is presented that decomposes the array into sub-arrays implementing a problem of smaller dimension.

Procedure D

This procedure determines the collection of entries in a unary array that can be assigned zero value, and skipped from the array input data.

1- For $i = 0$ to k do
   Construct the set $M(i) = \{n_j\}$, where n_j is a boundary node at level i in the TG.
   $b(i)$ holds the number of nodes at time i
2- Let $\mathbf{V_a}$ be the direction vector for $a(.)$.
3- For $kk = 0$ to k do
   For $j = 1$ to $b(kk)$ do
   i) For each $n_j \in M(kk)$ do
      Evaluate the node functions at cycle kk for the modified numerical value of the element.
      If the weight of the node is not defined, mark the node as ineligible for omission.
   ii) Let m be the total number of nodes that lie on a line parallel to $\mathbf{V_a}$ and starting at node n_j.
      For $i = 1$ to m do
      Evaluate the functions of node n_i.
      If the weight of the node is not defined, mark the node as ineligible for skipping.

End (Procedure D)

Procedure NSA

This procedure determines the structure of the modified sub-arrays and the new input timing functions.

1- For each boundary node marked as eligible for skipping in the TG identify the vector $\mathbf{V_a}$ that carries the input data elements.
2- If all nodes on $\mathbf{V_a}$ are eligible for omission, find i and j the first and last time levels of the nodes.
 i) For $k = i+1$ to j do
   a) Move the boundary node and all the nodes that lie on $\mathbf{V_a}$ to level k-1 if and only if all the nodes along the lines stay properly connected as specified by **P**.
   b) The modified input timing function of the

boundary node is $\pi'(.) = U^{k-1}(\pi(.))$.

3- Repeat step 2, for all boundary nodes that are eligible for omission.

End (* Procedure NSA *).

The collection of fast realizations executable on unary arrays can be enhanced by the TG recursive decoupling. Next, a procedure is presented that determines the collection of fast forms implementable in a given array.

Procedure TGD

1- Invoke procedure **D** for the full size array.

2- Invoke procedure **NSA** for the full size array.

3- Re-label the nodes in the TG into separate sets, here each set corresponds to the nodes of one sub array. Let l be the number of the sets.

4- For $ll = 1$ to l do

i) For $i = 0$ to k do

Construct the set $M(i) = \{ n_j \}$, where n_j is a boundary node at level i in the TG, of the set ll.

$b(i)$ holds the number of nodes at time level i.

ii) For each boundary node, identify the vector $\mathbf{V_a}$.

iii) For $kk = 0$ to k do

For $j = 1$ to $b(kk)$ do

For each $n_j \in M(kk)$ do

Evaluate the node functions for cycle kk for the modified numerical value of the element. If the weight of the node is not defined, mark the node as ineligible for skipping.

iv) Let m be the total number of nodes that lie on a line parallel to $\mathbf{V_a}$ and starting at node n_j.

For $i = 1$ to m

Evaluate the mapping functions for node n_i.

If the weight of the node is not defined, mark it as ineligible for skipping.

v) If all nodes belonging to $\mathbf{V_a}$ are eligible for skipping, find i and j, the first and last levels.

For $k = i+1$ to j do

a) Move the boundary node and all the nodes that lie on $\mathbf{V_a}$ to level $k-1$ if and only if all the nodes stay properly connected as specified by **P**.

b) New node timing function is $\pi'(.) = U^{k-1}(\pi(.))$.

c) Repeat step 2 for all eligible boundary.

End (* Procedure TGD *).

4. Fast LU and QR Decompositions

Ttwo examples of unary arrays, namely, the **LU** and the **QR** factorization arrays are studied. The developed concepts are applied to decompose these arrays into sub-arrays implementing problems of smaller dimension. The collection of all fast forms executable on the arrays are also determined.

4.1. LU Decomposition Array

Consider the **LU** array of Figure 1. Procedure **D** produces the nodes that can skip a zero entry in their input data. The collection of nodes that permit the omission of their input entries are shown in the dotted boxes in Figure 3. Procedure **NSA** produces the modified array of Figure 3. The array is decomposed into two sub-arrays. The modified array produces results in $2n$ cycles, while the generic array requires $3n$ cycles. The fast form of **A** is given in Figure 4. Recursive decoupling produces the collection of fast forms for the first sub-array given in Figure 5. The collection of fast forms for the second sub-array is given in Figure 6.

4.2. QR Triangularization Array

Consider the **QR** array for $n = 4$ depicted in Figure 7. The TG of the array is given in Figure 8. Procedure **D** results in the collection of nodes that can skip a zero entry. The nodes are identified in Figure 9 in dotted boxes. The decomposed array is depicted in Figure 9 with the modified input timing functions. The array consists of two sub-arrays. The fast form of **A** is given in Figure 10. The fast **QR** array execution time is $2n-1$ cycles as opposed to the generic dense array that performs the computations in $3n-1$ cycles. The boxed PEs in Figure 9 will act as delay elements. Recursive decoupling produces the collection of fast forms for the first sub-array given in Figure 11. For the second sub-array the collection of fast forms are in Figure 12.

5. Conclusion

The paper presented a new technique that determines the family of structured sparsity patterns that can be executed on systolic arrays designed for dense unary matrix operations in a reduced number of cycles. The proposed technique is primarily an analysis tool that has important applications in the design of computing structures with reduced computational complexity. It provides a means of increasing the class of implementable algorithms on a single array.

6. References

1- H.T. Kung, " *Special-purpose devices for signal and image processing : an opportunity in VLSI* ", Proc. SPIE, vol. 241, July 1980, pp. 76-84.

2- H. R. Barada, " *A comprehensive methodology for algorithmcharacterization, regularization and mapping into optimal VLSI arrays* ", Ph.D. dissertation, Louisiana State University, Baton Rouge, Louisiana, Aug. 1989.

3- A. O. Barbir, " *Systolic Array Implementations with Reduced Compute Time* ", Ph.D. Dissertation, Louisiana State Uni., Dec. 1989.

4- J. L. Aravena and A. O. Barbir, " *A Class of Complexity High Concurrence Algorithms* ", IEEE Trans. on Parallel & Distributed Systems, Oct. 1991.

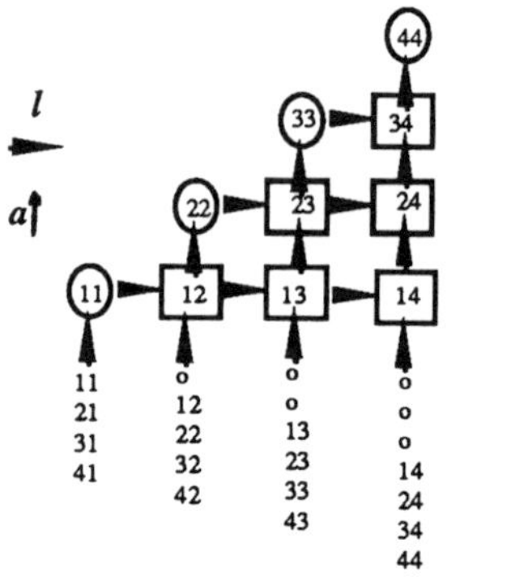

Figure 1. LU array, n = 4.

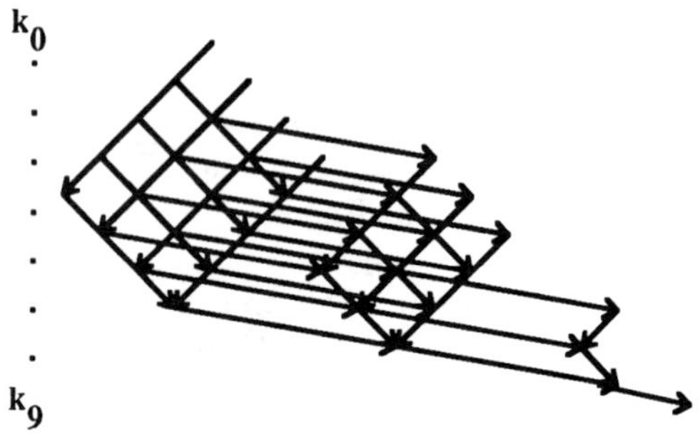

Figure 2. TG LU array.

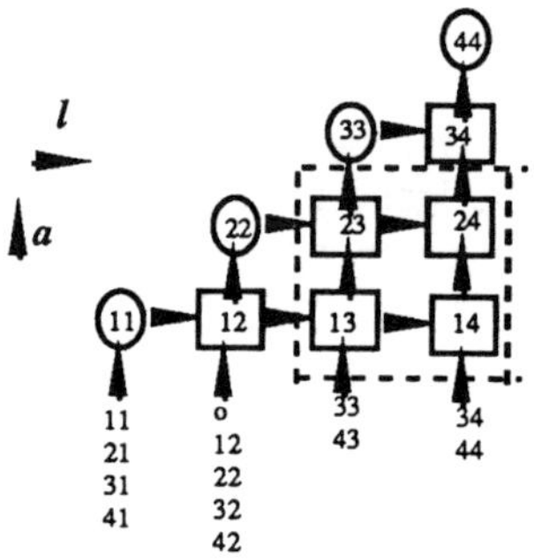

Figure 3. Fast LU array, n = 4.

$$\begin{bmatrix} 11 & 12 & o & o \\ 21 & 22 & o & o \\ 31 & 32 & 33 & 34 \\ 41 & 42 & 43 & 44 \end{bmatrix}$$

Figure 4. Fast Form of A for LU.

$$\begin{bmatrix} 11 & 12 \\ 21 & 22 \\ 31 & 32 \\ o & o \end{bmatrix}, \begin{bmatrix} 11 & 12 \\ 21 & 22 \\ o & o \\ o & o \end{bmatrix}, \begin{bmatrix} 11 & o \\ 21 & 22 \\ o & o \\ o & o \end{bmatrix}, \begin{bmatrix} 11 & o \\ o & 22 \\ o & o \\ o & o \end{bmatrix}.$$

Figure 5. Fast forms for LU first sub-array.

$$\begin{bmatrix} 33 & 34 \\ o & 44 \end{bmatrix}, \begin{bmatrix} 33 & o \\ o & 44 \end{bmatrix}.$$

Figure 6. Fast forms for LU 2nd sub-array.

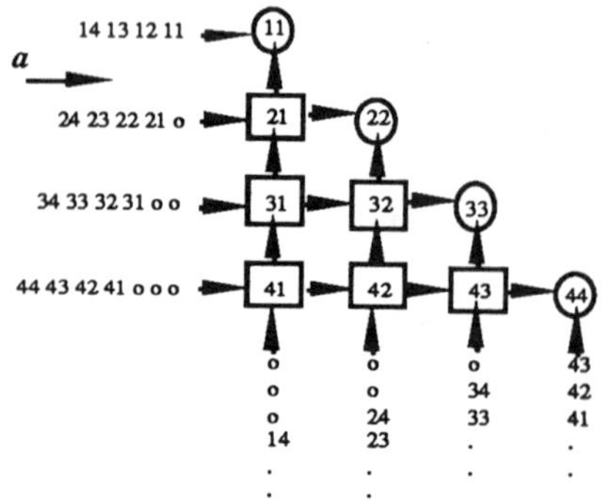

Figure 7. QR array, n = 4.

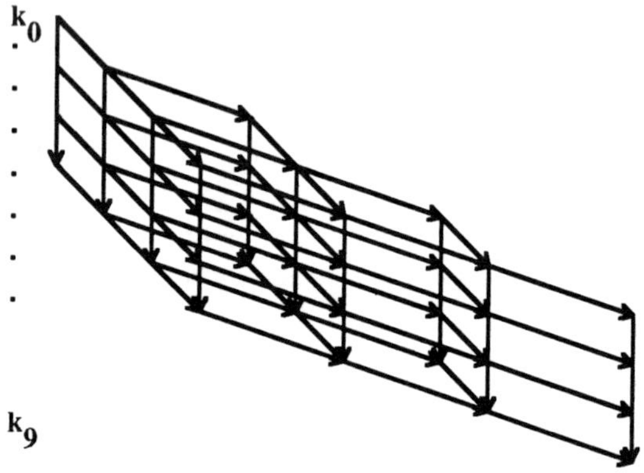

Figure 8. TG QR array.

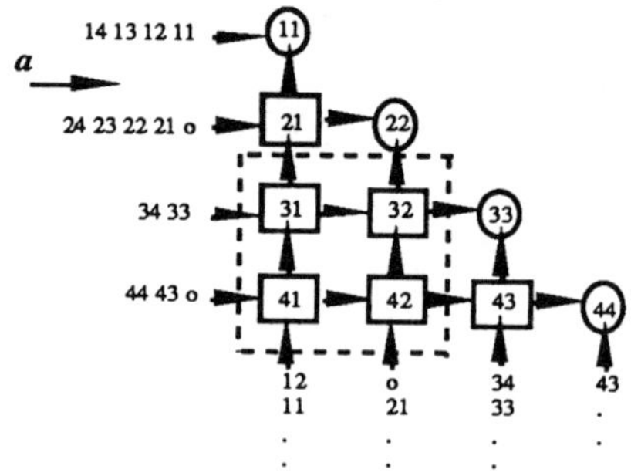

Figure 9. Fast QR array, n = 4.

$$\begin{bmatrix} 11 & 12 & 13 & 14 \\ 21 & 22 & 23 & 24 \\ o & o & 33 & 34 \\ o & o & 43 & 44 \end{bmatrix}$$

Figure 10. Fast Form of A for QR array.

$$\begin{bmatrix} 11 & 12 & 13 & 14 \\ 21 & 22 & o & o \end{bmatrix}, \begin{bmatrix} 11 & 12 & o & o \\ 21 & 22 & o & o \end{bmatrix}$$

$$\begin{bmatrix} 11 & o & o & o \\ 21 & 22 & o & o \end{bmatrix}, \begin{bmatrix} 11 & o & o & o \\ o & 22 & o & o \end{bmatrix}$$

Figure 11. Fast forms for QR first sub-array.

$$\begin{bmatrix} 33 & o \\ 43 & 44 \end{bmatrix}, \begin{bmatrix} 33 & o \\ o & 44 \end{bmatrix}.$$

Figure 12. Fast forms for QR 2nd sub-array.

On Reducing Hotspot Contention in Multistage Networks

Md. Najmul Karim and Ahmed El-Amawy

Louisiana State University, Baton Rouge, Louisiana

Abstract

To prevent tree saturation in multistage networks a simple combining technique is proposed. Instead of considering each queue, we consider two queues at a time for comparison and combining of requests. In the new scheme, the queues which are traditionally located at the output of a switch are moved to the input of the switch in the next stage to which the queue leads. This facilitates combining of requests which are headed to the same switch by using both IN queues for combining. The new scheme uses simpler queueing structure compared to those proposed for conventional pairwise combining. The proposed technique eliminates tree saturation even in very large networks. It has been shown through extensive simulation that the proposed scheme outperforms most if not all previously suggested schemes.

1. Introduction

Congestion in buffered multistage interconnection networks is typically minimum if the requests are perfectly distributed over all the memory modules. However, if the traffic pattern is not well distributed, congestion is created in the network and performance is seriously degraded [6]. Non-uniform distribution of the traffic pattern usually occurs because some of the memory locations are referenced more frequently than others. Such locations are often referred to as *hotspots.* The basic causes of hotspots are synchronizing requests, requests for resource management, peculiar array access pattern or transient heavy load to a particular memory module.

Over the last few years the hotspot (tree saturation) problem has drawn the attention of many researchers [1]-[10]. Several approaches have been proposed for alleviating the problem. The approaches can be classified into two general categories. Software-based and hardware-based. Software techniques rely on identifying program segments that could cause hotspots. Those segments are then reprogrammed such that hotspot contention does not arise [10].

Hardware-based approaches can be subdivided into combining [3]-[5] and non-combining techniques [6]-[8]. In combining techniques the rate of hotspot requests reaching the hot memory module is reduced by combining requests destined to the hotspot. In non-combining techniques the flow of hotspot requests through the network is kept under control.

This paper proposes a new hardware combining approach for reducing the effects of hotspots on performance. It emphasizes eliminating the deficiency of the pairwise combining technique where combinations take place at lower rates, especially at the earlier stages. Our new approach provides some of the smallest message delays reported to date but may incur some extra hardware cost.

Section 2 provides some preliminaries. Section 3 introduces the model and summarizes simulation results based on the proposed scheme. Section 4 is a conclusion.

2. Preliminaries

A hotspot creates congestion in the network that blocks not only hotspot requests, but also non -hotspot requests. These requests pass through a tree of buffers which originates at the processors and converge to the hot memory module. As a result, the other requests that need to use part of this path are also blocked. The philosophy behind a combining scheme is to combine these hotspot requests at successive stages so that the buffers do not get filled up with many requests. However if the combining rate is not sufficiently high, the problem of tree saturation will persist. This is the main limitation of pairwise combining.

Since our new scheme attempts to overcome the limitations of conventional pairwise combining, it seems appropriate to briefly, review pairwise combining. Then we will present our scheme and show how the limitations of pairwise combining are overcome.

1058-6393/91 $01.00

The combining hardware for pairwise combining is shown in Figure 1. The requests enter through the IN column of a particular output queue, and leave thorough the OUT column of the same queue. While moving through the IN column, a request checks the adjacent position in the OUT column, and moves to that position if it is empty. If the adjacent position on the other hand contains a request to the same address, the request from the IN column is moved to the CHUTE column. The request in the CHUTE column and the matched request in the OUT column proceed downwards synchronously. These requests leave the corresponding columns simultaneously and enter a combining logic where they combine. A record of the combining is kept in the wait buffer to allow individual responses in the return path.

The problem here is, however, that at the early stages, hotspot requests do not meet in sufficient numbers in order to make combining effective. Though the hardware facility for combining is available at the output buffers of every stage, many hotspot requests propagate to the succeeding stages without being combined. Since they follow a converging tree, they accumulate at the succeeding stages at much higher rates. In particular if r is the probability that a processor makes a request in each cycle and if h is the probability that the issued request is directed towards a hotspot, then the hotspot rate at a switch in the first stage (stage 0) = $r \times h$

If there is no attempt to control the rate of hotspot requests (e.g. through combining), then at each succeeding stage this rate doubles.

If f_0 is the (failure) probability that a request will go to the next stage from stage 0 without being combined and $f_1, f_2, \ldots, f_i$ are the corresponding failure probabilities at stages 1,2, ..., i respectively, then considering only switches in the saturation tree

hotspot request rate at stage 1 is

$$(1-f_0)rh + 2f_0 rh = (1+f_o)rh$$

hotspot request rate at stage 2 is

$$(1-f_1)(1+f_0)rh + 2f_1(1+f_0)rh = rh(1+f_0)(1+f_1)$$

Similarly, hotspot request rate at stage i is

$$rh(1+f_0)(1+f_1)....(1+f_{i-1})$$

So even with combining, the hotspot request rate keeps on increasing at successive stages. This fact is the motive behind the work presented in this paper.

3. The new combining technique

3.1 Network and switch Structure

Our proposed combining technique called *input queue combining*, tries to increase the availability of hotspot requests for combining at different stages. Instead of combining enqueued hotspot request in the two columns (IN and OUT) of an output queue, a pair of queues are combined. The two output queues of the same switching element however will not contain requests directed to the same hotspot. So it is logical to choose two output queues from two different switching elements. The pairs of output queues at any stage are chosen in such a way that each pair feeds the same switching element in the next stage.

The pair of queues at the previous stage, however may not be physically close to make combining easy. We therefore suggest a modification in the structure of the network and the switching element, so that instead of having queues at switch outputs, the queues are moved to the next stage inputs. In other words, an output queue at stage i is moved to the switch it feeds in stage $i+1$ and treated as input queue. This suggested modification in the structure is illustrated in Figure 2.

Comparison and combination of enqueued hotspot requests in the queue pair of a switching element can now be performed easily. If the requests in a corresponding position of the queue pair match they are combined. Choosing a queue to put the combined hotspot request in is done at random. A record of the combination is kept in the wait buffer for a pair of queues. The overall combining structure is therefore similar to that shown in Figure 3.

3.2. The Model

We now state the model on which our simulations are based. We assume a non-uniform traffic model, where each request has a finite probability of being directed to the hotspot. If a request is not directed to that location, it has equal probability of being directed to all the memory modules. For the purpose of analysis an n stage buffered omega network consisting of 2×2 crossbar switches is assumed. The network is considered packet switched and synchronous. The routing scheme is based on the destination tag, and the source address is sent along with the memory address. The following further assumptions are also made.

1. Each request is a single packet.
2. The input queue is capable of accepting two packets simultaneously. If at some cycle a queue has only one free location and two requests are directed to it, the queue randomly accepts one of the two.
3. A request is put in a queue at the first empty location nearest to the head.
4. The service time of a request in a queue is the same as the network cycle time.
5. The enqueuing and dequeueing processes are overlapped.
6. Each processor has an infinite queue.

In the model that we have just described, we stated that an input queue may accept two packets per network cycle (if there is enough space for them). Although similar assumptions have been adopted with other schemes [3]-[8], the situation is slightly more involved in our case. For an input queue to accept two requests in a cycle, either :

i) Two links are utilized two connect an output of a switch at stage i to the input queue at stage $i+1$; or,

ii) A single link is utilized between two switches in successive stages but two packets may be time multiplexed in a single network cycle.

The first choice incurs high (virtually doubled) link costs compared to that of a standard network. The alternative choice may be reasonable to assume since a network cycle is assumed to be long enough for a read-modify-write cycle. Which choice to take will depend on several factors such as actual network cycle, desired hardware cost, and integration level.

3.3. Simulation Results

An omega network of size 256×256 has been simulated. The simulation has been performed for both the forward and return paths. Since delay in either path depends on the length of the buffer in the respective path, we have simulated both the new scheme and conventional pairwise combining assuming the same size, buffers for the forward, return wait buffers. Simulation results for different network size with a fixed percentage of hotspot request (10%) are shown in Figure 4. The results clearly show the improvement achieved by using the new scheme, over pairwise combining.

To verify that the lower delay achieved by the proposed technique is mainly due to increased availability of hotspot requests for combining, we have counted the number of combinations that took place with both pairwise combining and the new scheme for fixed number of cycles. We have found that the number of combined requests is consistently higher with the new scheme at every stage, for different request and hotspot rates, and network sizes. This supports our prediction that the new scheme overcomes the main limitation of pairwise combining namely, low rate of combining.

Although conventional pairwise combining cannot solve the problem in larger sized networks [6] our scheme works well for such networks as well. Figure 4 also shows the performance of our scheme and that of pairwise combining for larger networks like 512×512 and 1024×1024.

Finally it is worth pointing out that the message delay offered by our scheme is lower then that reported in most (if not all) previous schemes.

4. Conclusion

We have proposed a new technique for reducing hotspot contention in multistage interconnection networks. The new technique tries to increase the rate at which the requests are combined. Instead of considering a single queue for request combination, we consider the two queues leading to the same switching element and the network. This has been found to increase combining at any switching node in the saturation tree. For combining requests in two queues, we have proposed a modified structure for the switching element. The proposed technique has been found very effective for eliminating the tree saturation problem. The queue structure becomes much simpler compared to pairwise combining and the component count in the queue is also reduced. The delays reported in our study are among the lowest reported to date even for very large networks.

References

[1] LEE R. : "On Hotspot Contention", *ACM SIG Comput.*, vol 13, pp. 15-20, Dec.1985.

[2] KUMAR M. and PFISTER G. : "The Onset of Hot Spot Contention", *Proc. Int'l Conf. Parallel Processing*, Aug 1986, pp. 46-50.

[3] EAGER D. and HO S.W. : "A Novel Strategy For Controlling Hot Spot Congestion", *Proc. Int'l Conf. Parallel Processing*, Aug. 1989, pp. I14-18.

[4] DIAS M.D. and KUMAR M. : "Preventing Congestion in Multistage Networks In the Pres-

ence of Hotspots", *Proc. Int'l Conf. Parallel Proc.*, Aug 1989 pp. 19-13.

[5] SCOTT, S.L. and SOHI G.S. : " The use of Feedback in Multiprocessors and Its Application to Tree Saturation Control",*EEE Trans Computers*, Vol.j1, No. 4, October 1990, pp. 385-398.

[6] PFISTER G.F and NORTON V.A : "HotSpot Contention and Combining in Multistage Interconnection Networks", *IEEE Trans. Computers*, vol C-34, No. 10, Oct. 1985 pp. 943 - 848.

[7] LEE G., KRUSKAL C.P. and KUCK D.J : "The Effectiveness of Combining in Shared Memory Parallel Computers in the Presence of Hot Spots", *Proc. Int'l Conf. Parallel Processing*, Aug. 1986 pp. 35-41.

[8] TZENG N. : "Design of a Novel Combining Structure for Shared Memory Multiprocessor", *Proc. Int'l Conf. Parallel Processing*, 1989, pp. I1-8.

[9] TANG P. and YEW P.C. : "Algorithms for Distributing Hotspot Addressing in Large Multiprocessor System", *Center for Supercomputing R.&D.*, Univ. Illinois, Urbana, Cedar doc. 617, Dec. 1986.

[10] YEW P., TZENG N. and LAWRIE D. : "Distributing Hot Spot Addressing in Large Scale Multiprocessors", *IEEE Trans Computers*, Vol C-36, No. 4, April 1987 pp. 388-395.

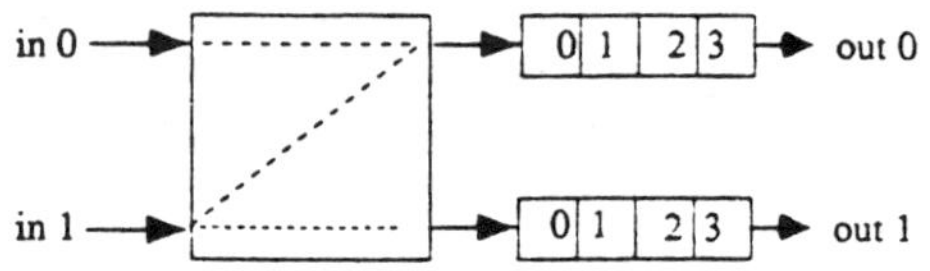

Figure 1(a) : The convergence of hotspot requests in one output queue.

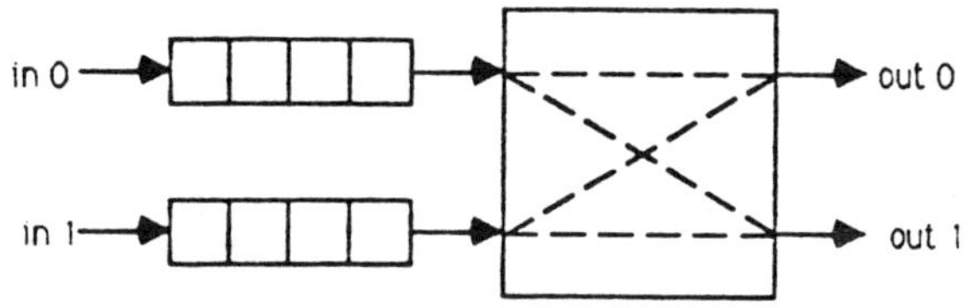

Figure 2 : Modification in the switching element.

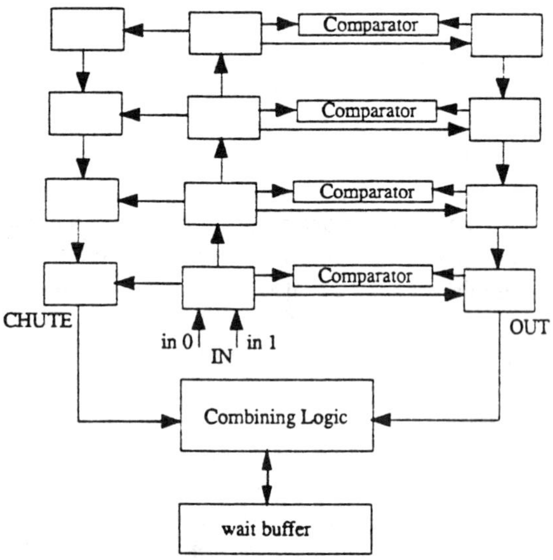

Figure 1(b) : The combining structure in an output queue in pairwise combining technique.

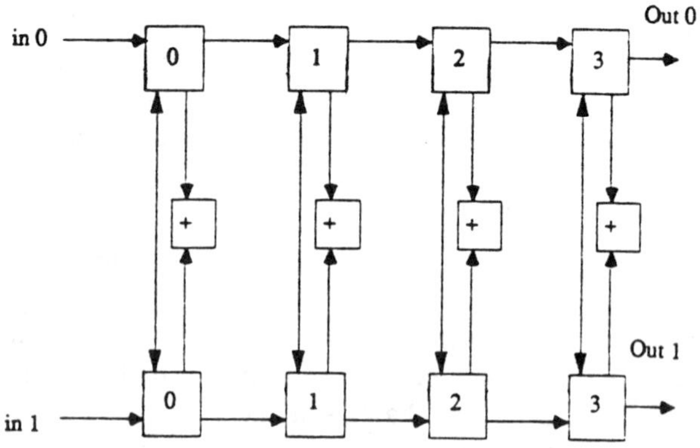

Figure 3 : The overall combining structure of the proposed technique.

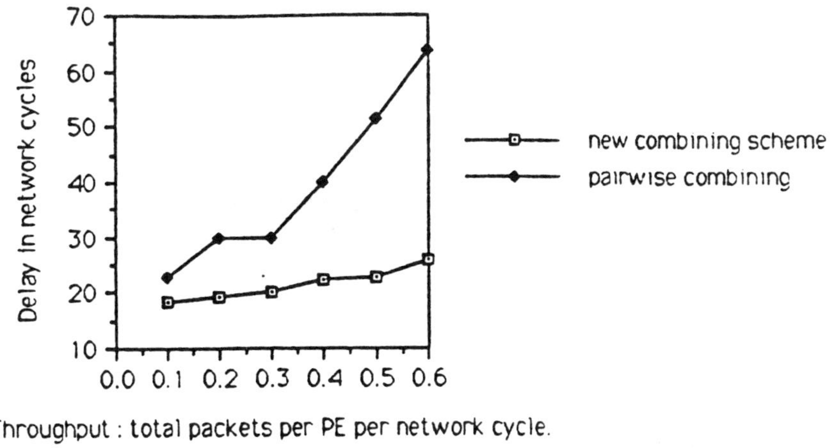

(a)

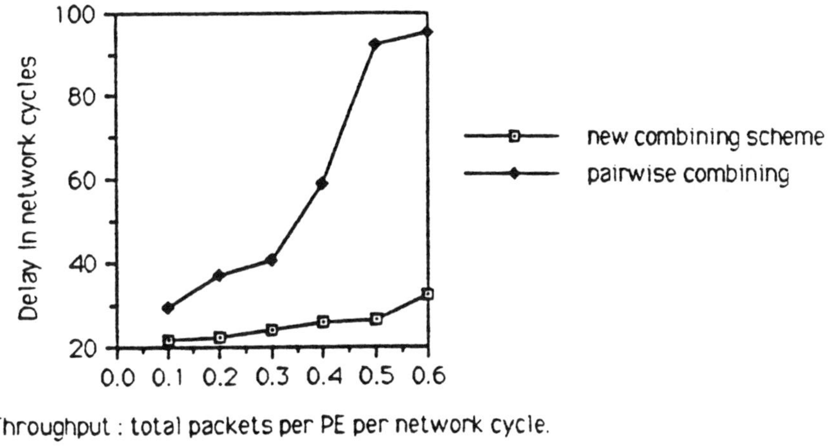

(b)

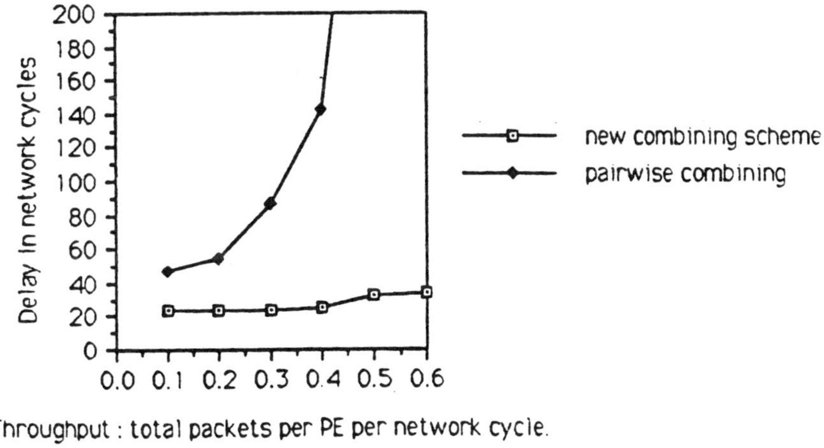

(c)

Figure4 : The delay in network cycles versus throughput for 10 percent hotspot requests.

a : 256 X 256 network size.
b : 512 X 512 network size.
c : 1024 X 1024 network size.

Memory Management in Real-time Multiprocessors

Kasi Anantha, David Tangman and Daniel Gold
Department of Mathematical Sciences
San Diego State University San Diego, CA 92182

Abstract

In multiprocessor systems, a common programming error occurs when one of the processors accidently modifies the instructions or data of another processor. These errors can be extremely difficult to detect, but can be eliminated through the use of memory management hardware. The KUMARAN II architecture is a multiprocessor architecture (based on seven MC68030 processors) targeted for real-time applications. The basic philosophy of the KUMARAN II architecture is that the real-time system developer should be primarily concerned with the application and not with the internal hardware. Thus, the architecture has been designed to automatically implement and enforce the user-entered, high-level memory definitions and specifications. This paper describes the specification and the implementation of user-interface to KUMARAN II memory management.

1 Introduction

Real-time programmers usually avoid memory management; the overhead of translating virtual addresses to physical addresses, as well as the penalty for accessing secondary storage when servicing a page fault, makes memory management unattractive for many real-time applications. Thus, real-time applications are generally memory-resident, and it is up to the programmer to carefully design and manage memory resources. However, in a multiprocessor system with shared memory space, this technique can induce extremely subtle programming errors, as illegal accesses by an errant processor are often difficult to detect. Memory management in Kumaran II has been designed specifically to solve these problems.

Kumaran II contains seven MC68030 processors and 32 Megabytes of physical memory. Kumaran II enforces a strict memory management discipline which partitions the 32-Megabyte memory into seven, 4-Megabyte local memories (one for each processor) and one, 4-Megabyte shared memory area. Furthermore, Kumaran II prevents processes from writing to local and shared memory pages which are marked as read-only.

1.1 Previous Work

CHIMERA[1] is a real time computing environment developed at Carnegie-Mellon University for the Reconfigurable Modular Manipulator System (RMMS) project. CHIMERA consists of one or more Ironics 68020 CPU boards, acting as real-time computing engines, residing on the VME bus of a Sun 3 workstation. This computing environment stressed the following two aspects:

1. 'High-level computing, performed on a general purpose processor, for real-time program development and offline storage and analysis of experimental results.'

2. High performance numerical computing on dedicated real-time processors to implement the control law, and sample and buffer data for later storage on the general purpose system.

The CONDOR[2] system on the other hand was developed at MIT for controlling the Utah-MIT hand and other complex robots. In the second version of the CONDOR hardware a Sun-3 forms the main development machine while the real time controller utilizes Ironics Corporation 16 MHz Motorola 68020 based single board computers equipped with Motorola 68881 floating point processors. The control processors are linked using a VME bus and are coupled to the development host through a VME bus to VME bus adaptor.

The Kumaran[3,4,5] multiprocessor system used Motorola MVME136A boards based on MC68020 and MC68881 processors, and AT&T System V on a Motorola MVME147 board for program development and data storage and analysis. The real time CPU boards (MVME136A) and the development CPU (MVME147) were kept on the same VME bus chassis. Real time CPUs communicated with the Unix CPU via the VME bus using special device drivers.

1058-6393/91 $01.00

The Kumaran II[6] multiprocessor system replaced the Motorola MVME136A boards with faster MVME143-2 boards containing an MC68030 at 33 MHz and MC68882 floating point units. Later the Unix CPU was moved to its own chassis and connected to the real time CPUs using a VME bus to VME bus adaptor.

It was noticed during the software development and experimentation that the multiprocessors occasionally overwrote the instructions and data space of another processor resulting in a program crash. The system was left in a state from which debugging was extremely difficult. The destructive interference of these processors could be alleviated by the use of the virtual memory system provided by the MC68030. In addition, the use of virtual memory further improved the development of application programs by eliminating the need for individual process to know the physical address used by shared variables. Each process can access a shared variable using its own virtual address and the MMU will map the virtual address to the physical address automatically.

1.2 System Design

Kumaran II provides the application designer with a distributed, real-time memory management environment which partitions the VME bus address space into eight sections: an I/O space (low memory), a separate 4MB processor-local memory space for each of the 6 real-time Kumaran II multiprocessors, and a 4MB common memory space. By default, every Kumaran II process has access to the I/O space and its own processor-local RAM, but must use the facilities provided by the memory management executive calls to establish and manipulate access to the common memory partition. This result is achieved through a built-in Kumaran II process construction procedure, a set of callable executive routines, and the internal MMU of each of the MC68030 processors within Kumaran II.

The built-in Kumaran II process construction procedure is implemented as a Korn-shell script called "cmmake" which accepts a user-specified "buildfile" as input, and produces a UNIX "makefile" as output. Example 2(a) describes the general format for a cmmake buildfile. Note that white space characters may be freely used within a buildfile, and that the ".ext" file extension for object files within a "OBJECTS/ENDOBJECTS" block must either be ".sa" (MC68030 assembly language sources) or ".c" (standard C source files). The *program name* parameter is the name to assign to the executable image (COFF format) that will be produced by the make utility. For example, suppose an application (to be named exampleprog) consists of 5 source files: top.c, sub1.c, sub2.c, sub3.c, and drivers.sa (MC68030 assembly language source). The application designer would create a buildfile (named "mybuild)" similar to that of example 2(b), and invoke the cmmake script through the Korn-shell.

If cmmake detects no errors either in the format of the buildfile or in the sources that are specified, a file named "makefile", acceptable as input to the UNIX make utility, will be created in the current directory path. example 3 displays the makefile that would be produced from the buildfile for exampleprog. The programmer can further customize the makefile (e.g. add dependencies) if so desired, but unless the *program name* or *objectX.ext* files which occur in the buildfile are changed, there is no reason to re-run cmmake.

Notice that in example 3, the reference to "$(CM)/header.o" and "$(CM)/cm.o" objects. The header.o object is the first object linked into the executable image, and contains the memory-management initialization code. After header.o has initialized the run-time memory environment for the MC68030 processor, it calls the "start" entry point into the application, as Kumaran II requires the programmer to name the top-level routine of the application "start" instead of "main". The cm.o object contains the common memory, memory management executive calls which Kumaran II provides to its applications.

Kumaran II memory management segments the 4MB common-memory partition into 128, 32KB blocks. The Kumaran II memory management executive calls (provided in the cm.o object) allow a process to dynamically manipulate access and rights to these 32KB blocks of common memory. A brief description of these calls is given below.

```
int  address; /* VME bus address */
int  pagnum; /* Common memory page
 number 0..127 */
int  rights; /* 0=none; 1=read only;
 2=read/write */
int  size; /* # of bytes requested */


(1) int cmset( pagnum, rights, size )

   This function sets the access ''rights''
   to the common memory page(s) required for
   a block of ''size'' bytes.  This is accom-
```

```
plished by identifying and modifying the
common memory page descriptors according
to the requested access rights.  If the
''size'' parameter is greater than 32KB,
then the access rights for multiple, cont-
iguous, common memory pages will be set.

RETURNS:
   0: ok
  -1: illegal page number
  -2: illegal rights
  -3: illegal size

(2) char *cmlink( pagnum )

    This function will return a pointer to the
    requested common memory block.  It is up
    to the user to type cast the returned
    pointer to the appropriate data structure
    type.

    RETURNS:
     -1: illegal page number
      0: no access rights
    > 0: pointer to first byte of the
         common page.

(3) int cmtest( pagnum )

    This function will return the access
    rights that have been established for a
    common memory block.

    RETURNS:
     -1: illegal page number
      0: no access rights
      1: read-only access
      2: read/write access
```

Example 1. memory management calls

The header.o object file implements the Kumaran II virtual memory model directly in the MC68030 MMU. The transparent translation (TT0) registers of the MC68030 MMU are initialized to allow for transparent translation of all I/O space addresses, and the address translation tree entirely in processor-local memory space. The entire VME bus address space is partitioned into 32KB pages, and the high-order 6 bits all virtual addresses will be ignored (thus, physical memory is limited to 64MB).

The CPU root pointer (CRP) points to the first-level of the address translation tree. The TIA field of the translation control register is set to 4 bits, and the TIB field is set to 7 bits. Thus, the highest-order 4 bits of a virtual address (bits 22-25) partition the virtual address space into sixteen segments. The 7-bit TIB field further divides each of the 16 4MB segments into 128, 32KB pages. Since the 4MB of local-memory addresses range from 0 to 400000(hex) and the 4MB of common-memory occupy the highest 4MB of the virtual address space, 2800000(hex) to 3200000(hex), all of the descriptors in the first-level translation table are invalid, except for the first and the tenth (which contain pointers to the second-level page descriptor tables for the local-space and common-space). If a Kumaran II process attempts to access an address that is described by one of the invalid descriptors in the first-level table, a bus-error exception is raised and the access is disallowed. The second-level tables for the first and the tenth entries in the first-level table (the local-memory and common-memory segments respectively) each contain 128 page descriptors which directly map to the 128, 32KB pages of physical memory at those locations. Once the internal tree traversal algorithm of the MC68030 MMU locates a page descriptor in the second-level table, the translation is immediately cached in the internal Address Translation Cache (ATC) of the MC68030.

The Kumaran II programmer constructs the application around a distributed processing model that consists of six MC68030 processors (each of which have exclusive access to a local 4MB RAM and shared access to a 4MB RAM) and the set of Kumaran II common memory executive calls. Next, the programmer creates a buildfile which lists each source file in the application, as well as assigns a name for the process, and runs the cmmake Korn-shell script to construct the makefile for the target Kumaran II process. Finally the programmer invokes the UNIX make utility, and if no compilation or linking errors are detected, an executable Kumaran II image (COFF format) is produced. Using the "loadunix" utility described in [6], the programmer can then load the executable image onto a selected MC68030 processing node.

```
(a)  PROGRAM exampleprog
     OBJECTS
          top.c
          sub1.c
          sub2.c
          sub3.c
          drivers.sa
     ENDOBJECTS

(b)  PROGRAM <program_name>
```

```
OBJECTS
    <object1.ext>
    <object2.ext>
    ...
    <objectN.ext>
ENDOBJECTS
```

Example 2. buildfile format and example

```
CMOBJS = $(CM)/header.o $(CM)/cm.o
OBJS = top.o sub1.o sub2.o sub3.o drivers.o

exampleprog: $(OBJS)
   ld -o exampleprog $(CMOBJS) $(OBJS) \
   $(CM)/mv143.ld -m -lc > exampleprog.map

top.o: top.c
       gcc -c -w -I\$(CM) top.c
sub1.o: sub1.c
       gcc -c -w -I\$(CM) sub1.c
sub2.o: sub2.c
       gcc -c -w -I\$(CM) sub2.c
sub3.o: sub3.c
       gcc -c -w -I\$(CM) sub3.c
drivers.o: drivers.ro
       cvtu -t0 drivers.ro
       asm drivers.sa -w +p=68030
```

Example 3. makefile for exampleprog

1.3 Results

The virtual memory implementation was applied to three application programs. We briefly describe those applications here.

The first application studies the internal cache architecture of the MC68030 with different cache sizes for instruction and data caches. The MC68030 uses 256 bytes each for instruction and data caches. The simulation varied the cache sizes from 256 bytes to 16K bytes and estimated the cache hit rate by running the simulation on 2.5 MB of instruction and data traces obtained from programs which ran on a MC68030 processor.

First the simulation was run using one processor and then on multiple processors after loading the trace code in common global memory. The time taken to run the simulation was reduced by approximately 1/N where N was the number of processors used. Contrary to initial expectation the VME bus contention did not affect the results.

The results are presented in table 1 and table 2.

The second application used Kumaran II for fast parallel training of a neural network[8]. The most time consuming part of any neural network model is in the training of the network and finding the optimal parameters for the weights. We reduced the search time by running the model on all six processors of Kumaran II. Each processor was started with a slightly different normal perturbations of the weights with minimum squared error obtained from the previous runs. The minimum squared error after 10,000 iterations is plotted in Figure 1 as a function of number of iterations. While slow convergence is evident, the results are better than the uniprocessor case plotted in Figure 2.

For the third application the sweptsine control system[5] was recoded to implement the virtual memory[8].

1.4 Conclusions

Recent trend in embedded control is to avoid virtual memory. Our experience has found that using virtual memory eliminates the chance of processors modifying the instructions and data of another processor and their own code. This in turn has reduced the amount of time spent in debugging multiprocessor applications. By avoiding paging to a secondary storage device and multi-level page tables, the normal overhead found in standard virtual memory implementations has been eliminated.

Acknowledgements

The authors would like to thank Bob Hilchey for hardware support in maintaining the Kumaran II multiprocessor system and Arul Anantha for system software support.

References

[1] Donald Schmitz, Pradeep Khosla, Regis Hoffman and Takeo Kanade, "CHIMERA: A Real-time Programming Environment", Proceedings of the IEEE International Conference on "Robotics and Automation", IEEE, 1989

[2] Sundar Narasimhan, David M. Siegel, John M. Hollerbach, "Condor: A Revised Architecture for Controlling the Utah-MIT hand", Proceedings of the IEEE International Conference on "Robotics and Automation", IEEE, 1988

[3] K. Anantha, M. Vuskovic and T. Liang, "A Multiprocessor Implementation of Inverse Robot Dynamics", International Journal of Mini and Microcomputers, Vol. 11, No. 3, 1989.

[4] K. Anantha, M. Vuskovic and T. Liang,"A Multiprocessor Test bed for Robotics Applications", Mathematical and Computer Modelling, Vol. 14, pp 444-449,1990

[5] K. Anantha and D. Gold, "Kumaran - A Multiprocessor System for Swept Sine Vibration Control", Proceedings of the Twenty Third Asilomar Conference on Signals, Systems and Computers, October, 1989.

[6] K. Anantha, D. Gold and Arul Anantha,"A Multiprocessor System for Real-time Applications", Proceedings of the Twenty Fourth Asilomar Conference on Signals, Systems and Computers, November, 1990.

[7] R. Swiniarski, K. Anantha and D. Gold, "Kumaran - A Multiprocessor System for Fast Parallel Training and Real-time Fuel-Time Optimal Control of Industrial Robots Using A Neural Network Approach", Special Session, "Neural Networks in System and Control", 8th International Congress of Cybernetics and Systems", New York, Jun 11-15, 1990

[8] D. Gold and K. Anantha, "Software Development and Real-time Target Systems on a Common Backplane", Proceedings of the Twenty Fifth Asilomar Conference on Signals, Systems and Computers, November, 1991.

Cache size	Instruction Cache Hit Rate	Data Cache Hit Rate
256 B	89.9	55.9
1 KB	92.7	66.0
4 KB	96.3	74.4
16 KB	98.4	75.9

Table 1
Cache Hit Rate versus Cache Size

Number of CPU	Execution time
1	27 secs
2	14 secs
4	7 secs
6	4.5 secs

Table 2
Execution time versus number of processors

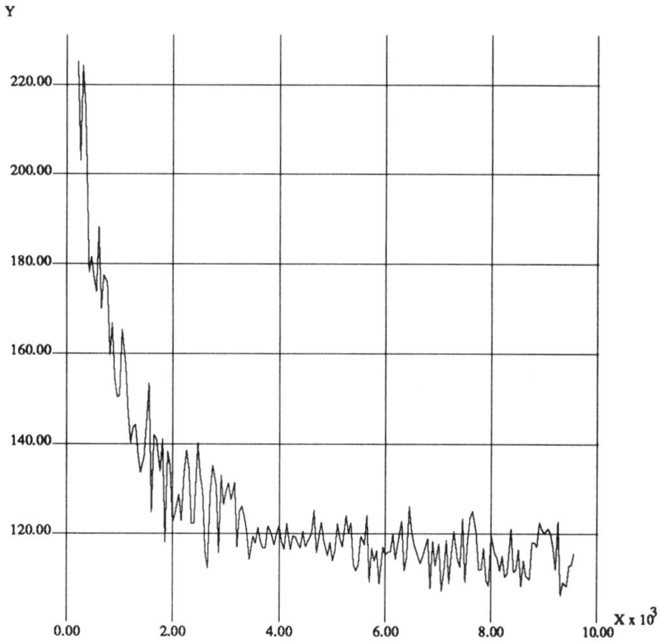

Figure 1. Squared Error for Kumaran2 vs . number of iterations

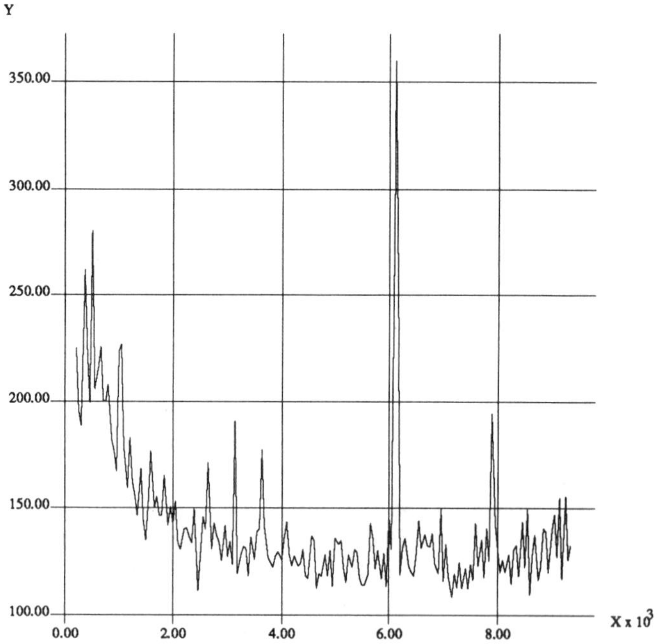

Figure 2. Squared Error for a Uniprocessor system vs. number of itratons

A Scalable Orthogonal Multiprocessor Architecture

Cheng-Hsien Tung
Computer Science and Engineering Department
The University of Connecticut
Storrs, CT 06269-3155

Abstract

The demand for large scale parallel computer systems has been intentified recently. Since shared memory machines can support many different programming models, they are candidates for parallel systems. Shared bus architectures are attractive only for small scale systems. One reason for this is that the number of processors which can be put on the bus is limited. This paper proposes a multiple bus architecture capable of being scaled to massive dimension to handle large amounts of data. This architecture has low latency compared with other multiprocessor architectures which can support the same number of processors.

1 Introduction

Many different parallel architectures have been proposed to meet the growing computational demands. Both SIMD and MIMD architectures have been designed. Among MIMD architectures, both shared memory and message passing configurations exist. Because shared memory machines can provide a globally shared physical address space which leaves the choice of programming model to the programmers, it is always the first preference of the users.

Shared memory MIMD designs have been studied extensively in the past. Three different kind of interconnection networks (crossbar, multistage switching networks and bus) have been chosen to build various shared memory parallel machines. The cost associated with crossbar switch is extremely high. That is the reason why crossbar can only be used in a machine with few processors and memory modules.

Many multiprocessors with hundreds of processors have been designed based on multistage interconnection networks. NYU Ultracomputer, BBN Butterfly and IBM RP3 [1] all fall in this category. While the multistage switch network can connect a large number of processors and memory modules together, this approach has several drawbacks.

First, the memory accessing latency may become a major problem in these systems. Second, to enforce cache coherence is difficult. Third, the data path width is restricted because of the physical pin limitations on the interconnection network.

The third choice to build a multiprocessor is to connect many processors to a single bus. many multiprocessors are built around a single system bus, eg. Encore's Multimax and Sequent's Symmetry. While these multiprocessors have been popular in small scale parallel systems, they are not appropriate for massively parallel systems.

The major drawback to implement a very large scale single bus computer architecture is the problem of bus contention. The bottleneck created at the bus can slow the system substantially as processors are forced to encounter lengthy waits for bus time. To avoid saturating the bus, only a few processors can be attached to this system bus. Private cache associated with each individual processor can significantly relieve the contention problem, but the maximum number of processors allowed to be attached to the bus is still very limited.

However, the advantages of a bus based multiprocessor are attractive and deserve further exploration. Recently, many researchers are working hard to design bus based large scale multiprocessor systems which can contain hundreds of processors, not just tens like Symmetry or Multimax. Fully connected multiple bus system, hierarchical bus architecture, multi level cache/bus structure, cluster shared bus system and several orthogonal shared bus architectures are examples demonstrating the efforts to design large scale multiprocessor systems to circumvent the contention problem associated with a single bus.

In this paper, we proposed a multiple bus architecture capable of supporting a large number of processors. We will mention several previously proposed large scale bus based multiprocessors in the next section. In the third section, we will present the partially orthogonal multibus multiprocessor system which can be scaled up to support a large number of processors. The summary is given in the last section.

2 Previously proposed architectures

While shared memory multiprocessors are appealing in the sense that they are conceptually simple, and relatively easy to program, they have several drawbacks. Perhaps the most pressing drawback of these has been the bottlenecks encountered at both the connection mechanisms and memory module(s) [2].

To combat these problems, proponents of shared buses have proposed multiple bus, multiple memory module architectures to increase bandwidth at these bottleneck points. Theoretically, adding additional busses to the system permits a higher bandwidth channel to memory. In some schemes [3], memory and processors are fully connected by a set of busses. In other schemes, a hierarchical cache/bus architecture is proposed [4,5]. Bus based orthogonal architectures where the data paths within the machine are not fully connected have also been proposed [6,7].

In this section, we will mention several large scale bus based multiprocessor systems proposed in the literature in the following paragraphs.

First, a fully connected multiple bus architecture [3] is shown in Figure 1. As can be seen in the figure, each processor(with private cache) as well as each memory module is connected to each bus. This architecture introduces several problems. First of all, two stages of arbitration are required, one to delegate bus resources to processors, and a second to delegate memory resources to busses. Another problem of fully connected multiple bus systems is that - winning arbitration for a bus cycle does not necessarily guarantee a memory access. Also, it is difficult to maintain cache coherence on this system.

The second proposed architecture is a hierarchical multiprocessor with two or more levels of private caches and shared buses [4,5] as shown in Figure 2. The level two cache is used as a filter between two level of buses. Most cache misses in the first level cache can be satisfied in the second level cache which can substantially reduce the contention problem with level two bus and memory modules. As shown in the simulations [4], this architecture can support more than one hundred processors. Also extended versions of shared bus multicache coherence protocols are used to maintain coherence among all the caches in the system.

Third, one architecture aimed at achieving parallelism on the scale of one thousand or more processors is the Wisconsin Multicube [6]. The Wisconsin Multicube is a 2 dimensional grid of buses with a processor situated at every bus intersection as shown in Figure 3. Each processor has a cache which is capable of snooping both of the attached buses. A cache miss can be satisfied by at most two bus transactions. However, the problem with the Wisconsin Multicube involves the effect of large scale data sharing. Because any cache coherence operation results in a broadcast over N+1 of the system buses for an NxN mesh, this is likely to be difficult in situations where large amounts of data sharing takes place.

Finally, a multiple bus shared memory multiprocessor architecture with multiple caches associated with each processor has been studie [8,9]. In [8], this architecture is called the Partially Orthogonal Multibus Multiprocessor(POMM). This architecture enjoys several advantages which we will briefly mention below. The POMM can consist of tens of processors with current technology. Also, the POMM can be used as a basic building block to build a large scale multiprocessor system. There are two approaches to integrate many POMMs to form a larger system. We will discuss them in the following section.

3 The Partially Orthogonal Multibus Multiprocessor

The partially orthogonal multibus multiprocessor(POMM) is an architecture which tries to combine several single bus multiprocessor systems together to form a larger multiprocessor system. As shown in Figure 4, the POMM can be viewed as B(where B=4 in this case) copies of single bus multiprocessor system combined together. This architecture is scalable and capable of supporting a large number of independent processors. Further, applications which run on the machine would be able to take advantage of various degrees of data sharing without substantially affecting the performance of the system.

The POMM consists of P processors each connected to the same set of B busses. Each bus is connected to one memory module which it does not share with any other bus. Each processor can access all areas of memory. The main memory address determines which bus a processor will use for a given access.

Each processor is connected to each bus through a bus snooping cache controller. This results in B separate caches and cache controllers at each processor.

Although this appears complex, the result is that each processor can maintain coherence in all caches independently without incurring the additional penalty of using a global directory.

Due to the connection strategy, bottlenecks will not appear at memory modules. Any conflicts will be resolved by delegation of the bus. A set of B arbiters is required to delegate the bus resources. Further, these arbiters are very simple in design, as each is only a one of P arbiter which can work on a simple protocol.

So the POMM architecture enjoys the following advantages:

1. The B of P arbiter required by the fully connected multiple bus system is replaced by a set of B one of P arbiters, and a second arbitration stage is not required. The resulting arbitration cycle is shorter.

2. Cache coherence can be maintained through bus snooping techniques, and does not require a complex directory mangement strategy.

3. Since arbitration for the bus and memory access is done in a single stage, the problems of memory contention and bus contention have been united and largely eliminated.

Each processor can access any memory location through only one bus in the POMM. While in hierarchical or cluster architectures, a memory access to a remote memory module, at least two level of buses are required in the memory access, the POMM provides a better structure for data sharing while the latency and cache coherence overhead can be minimized.

A recent study [10] showed the comparison between wide single bus architecture and multibus architecture similar to the POMM. It was found that in all cases tested, the multiple bus architecture outperformed the wide bus architecture.

We also are currently studying the performance of the POMM through discrete event simulations. The preliminary results showed that the POMM had better processor and bus utilization. The bus capacity used in maintaining cache coherence is alleviated. A more detailed description about the simulation model and some results is in [11].

As in a hierarchical system or a cluster architecture, a single bus multiprocessor can be used as the building block to construct larger system, the POMM can also be chosen as the basic unit to build large scale multiprocessor systems with hundreds even thousands of processors. Those systems with POMM as building blocks should still have similar advantages possessed by the POMM, such as low memory access latency, snooping cache coherence mechanisms and high bus bandwidth.

Here we propose two forms of architectures which both use POMMs as the basic building blocks. The first one is named as Connected POMM(CPOMM) and the second one is called Hierarchical POMM(HPOMM).

3.1 CPOMM

The CPOMM is simply to connect several POMMs using a second level system bus as shown in Figure 5. So the original POMM memory modules can be called intermediate memory or can be considered as the second level caches. Assuming the second level cache is large enough, that most memory accesses will be satisfied in this level without having to request a system bus transaction.

Within a POMM, snooping cache coherence protocol can be easily enforced. However, in a system like CPOMM with two levels of caches, to implement a system wide snooping cache coherence protocol, the well-known snooping cache coherence protocols for single bus case must be extended. The second level cache must snoop bus activities on both sides and the invalidation requests must be carefully handled. Especially when shared data are cached in different POMMs, the invalidation must be broadcast on the system bus also. In [5,6], the required extensions to the single bus cache coherence protocols for multi level cache architectures have been studied carefully that we will not repeat here. Through the extended cache coherence protocol, a CPOMM can enforce a system wide cache coherence based on a snooping mechanism. This is certainly one of the advantages selecting the bus structure as the interconnection network. Otherwise a directory based protocol must be adopted.

If we consider that a single bus can support 16 master devices(e.g. processors) without severe performance degradation from contention, then a POMM with 4 buses can accommodate 64 processors. Assume that the system bus can also support 16 intermediate memory modules(second level caches), then a CPOMM can consist of 256 processors. To construct a system of this size using crossbar or multistage interconnection network can be very expensive. A bus based system like CPOMM is certainly cheaper and easier to build.

3.2 HPOMM

The second architecture to build an even larger system using POMMs as building blocks is called the Hierarchical Partially Orthogonal Multibus Multiprocessor (HPOMM). The idea is to apply the original POMM construction principle recursively. As shown in the Figure 6, a second level of POMM is built. Similarly, the cache coherence protocol used for hierarchical cache/bus architectures [5] can be also utilized here.

The size of the system is significantly extended. Again we assume each POMM consists of 64 processors. Then a HPOMM with 4 second level buses can have 64x64 = 4096 processors which is a massively parallel computer. Unlike many massively parallel systems based on grid interconnection network, HPOMM can support much larger amount of data sharing and is general enough for many different algorithms.

4 Summary

A very interesting and promising multiprocessor architecture is proposed. This architecture has an orthogonal bus structure which maximizes bus bandwidth while reducing contention and arbitration overhead. The processors are each fully connected to memory through isolated busses. It has a low latency compared with other multiprocessor architectures which can support the same number of processors. The cache coherence is maintained easily on each bus. Other than the advantages mentioned above, the architecture also enjoys the followings:

1. This architecture is shared memory multiprocessor system. So it supports the shared memory programming model. It makes the programming tasks easier to accomplish. Many hardware supported features are transparent to users. Also the software can be designed to be independent of the exact machine dimensions.

2. It is modular and scalable. The basic POMM can be used as building block to build large scale multiprocessor systems. Due to this, we believe that this architecture is capable of supporting hundreds even thousands of processors under current bus technology. In addition, we see no impediment to scaling it further as bus technology improves.

The relative simplicity of this architecture, when compared to other multiprocessor systems with the same number of processors, makes it a good candidate for parallel processing economically as well as operationally. Since it is possible to support a large number of processors, it can also be successful in the massively parallel computing arena.

References

[1] J. J. Dongarra, "Experimental Parallel Computing Architectures," North Holland, 1987.

[2] P. Stenstrom, "Reducing Contention in Shared-Memory Multiprocessors," IEEE Computer, Nov. 1988, pp. 26-37.

[3] T. N. Mudge, J. P. Hayes, and D. C. Winsor, "Multiple Bus Architectures," IEEE Computer, June 1987, pp. 42-48.

[4] A. W. Wilson Jr., "Hierarchical Cache/Bus Architecture for Shared Memory Multiprocessors," ISCA, 1987, pp. 244-252.

[5] J-L. Baer and W-H. Wang, "On the Inclusion properties for Multi-Level cache Hierarchies," ISCA, 1988, pp. 73-80.

[6] J.R. Goodman and P.J. Woest, "The Wisconsin Multicube: A New Large-Scale Cache Coherent Multiprocessor," ISCA, 1988, pp. 422-431.

[7] K. Hwang, P. S. Tseng and D. Kim, "An Orthogonal Multiprocessor for Parallel Scientific Computation," IEEETC, Jan. 1989, pp. 47-61.

[8] C.H. Tung and C.W. McCarron, "A High Performance Multiprocessor with Partially Orthogonal Multibus and Memory," ISMM Int. conf. on PDCS, Oct. 1990, pp. 62-66.

[9] D. C. Winsor and T. N. Mudge, "Crosspoint Cache Architectures," Int. Conf. on Parallel Processing, 1987, pp. 266-269.

[10] A. Hopper, A. Jones and D. Lioupis, "Multiple vs. Wide Bus Multiprocessor," ISCA, 1989, pp. 300-306.

[11] C. W. McCarron, "Advantages of a Partially Orthogonal Multibus Multiprocessor Architecture," Master Thesis(in preparation), CSE, UCONN, 1991.

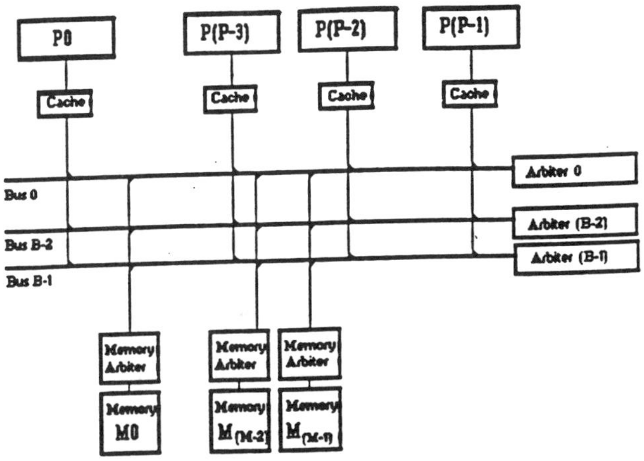

Figure 1 The Fully Connected Multibus

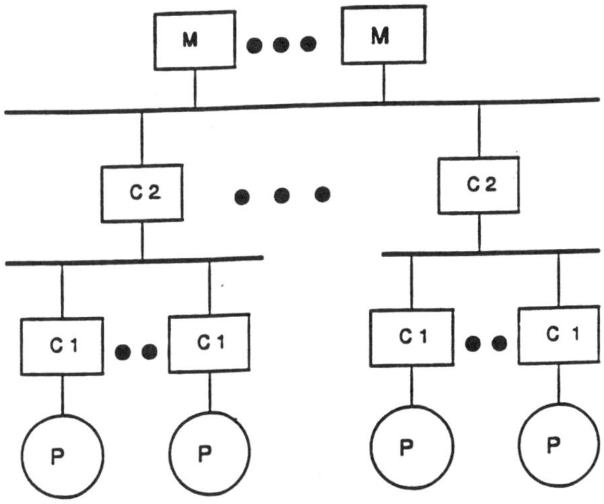

Figure 2 The Hierarchical Cache/Bus Architecture

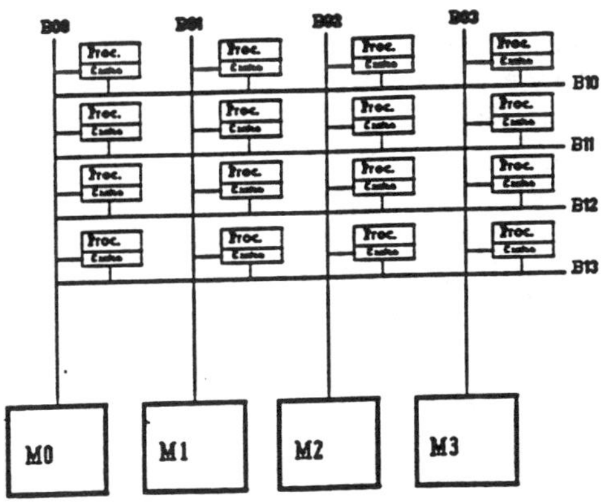

Figure 3 The Wisconsin Multicube

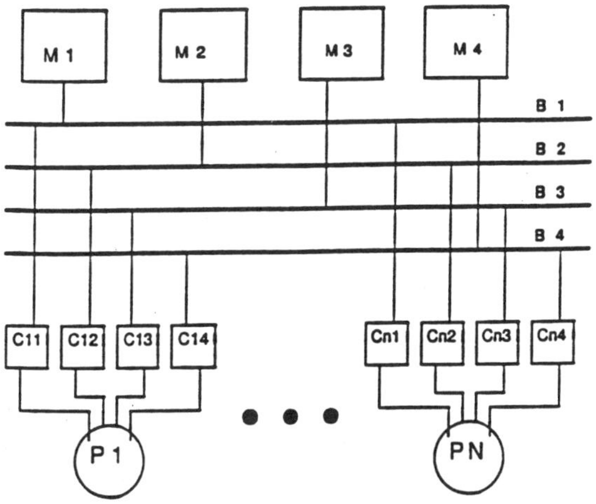

Figure 4 The POMM Architecture

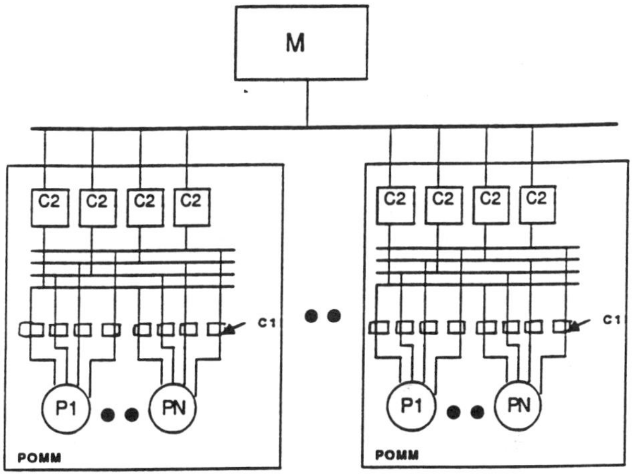

Figure 5 The Connected POMM (CPOMM)

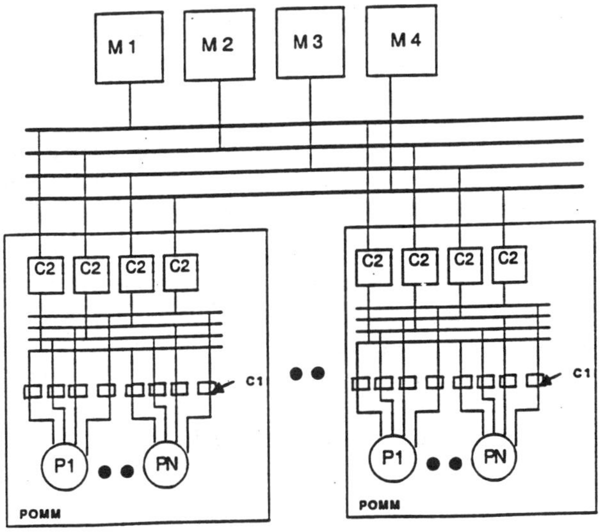

Figure 6 The Hierarchical POMM (HPOMM)

High-Speed Huffman Decoder Architectures[1]

Keshab K. Parhi
Department of Electrical Engineering
University of Minnesota, Minneapolis, MN, 55455

ABSTRACT

This paper presents pipelined and parallel architectures for high-speed implementation of Huffman decoders (which belong to the class of tree based decoders) using *look-ahead* computation techniques. Huffman decoders are used in high-definition television (HDTV) video and other data compression systems. The achievable speed in these decoders is inherently limited due to their sequential nature of computation. The unequal code word length of the Huffman code words makes it difficult to apply look-ahead. This problem is overcome by representing Huffman decoders as finite state machines (which can exploit look-ahead). Pipelined and parallel (or *incremental block*) architectures for high-speed implementation of Huffman decoders are presented. The proposed approach is useful for high-speed Huffman decoder implementations where the number of symbols of the decoder is low.

1. Introduction

The state update operation associated with feedback loops limits the opportunities for pipelining and parallel processing in recursive signal processing algorithms. The computation associated with a feedback loop imposes an inherent upper bound on the achievable iteration rate, referred to as the *iteration bound* [1-2]. High-speed implementation of these recursive signal processing algorithms was considered a major challenge a few years ago. In the past few years, we proposed *look-ahead techniques* and showed that appropriate *algorithm transformations* can preserve the input-output characteristics of the algorithm and can lead to new structures suitable for high-speed implementations [3].

In the past, we proposed *look-ahead computation* and *decomposition* techniques to pipeline a few recursive algorithms with *logarithmic* increase in hardware complexity with respect to increase in speed [4-5]. We also proposed *incremental computation* techniques to obtain block structures with linear hardware increase with respect to increase in speed [6-7]. Furthermore, we combined pipelining with block processing in recursive systems [6]. We considered pipelining in different recursive algorithms, including one-dimensional recursive filters [4-7], two-dimensional recursive filters [8], adaptive digital filters [9-10], algorithms containing quantizer loops [11], piecewise linear recursive loops [11], add-compare-select loops [12-16], and finite state machines [3,17]. Look-ahead, referred to as *recursive doubling*, has also been used to obtain parallel implementation of linear recurrence systems [18-19].

This paper considers high-speed architecture design of Huffman decoders [20-22]. These decoders are used in high-definition television (HDTV) video and other data compression systems. The unequal code word length in Huffman codes makes it difficult to apply look-ahead. However, we can represent a Huffman decoder using equivalent finite state machines, which can exploit look-ahead. In this paper, we exploit look-ahead and incremental computation techniques to design pipelined and parallel architectures for Huffman decoders.

Although applicability of look-ahead in finite state machines was observed in [3,17], completely pipelined high-speed Huffman decoder architectures have not been presented in a systematic way. The unequal code wordlength in Huffman decoders makes it difficult to apply look-ahead in a traditional way. However, by transforming the Huffman decoder to an equivalent finite state machine, we can apply look-ahead. The finite state machine is designed such that each code word boundary resets to a unique state. In section 2 of this paper, we present completely pipelined and parallel Huffman decoder implementations. It is observed that the *decomposition* technique (which relied upon *reuse* of intermediate results) is not always useful in Huffman decoder implementations. This is because the complexity of a storage unit or a latch is more than the complexity of a computation module used in Huffman decoders (such as a NAND or NOR gate). Furthermore, logic synthesis tools can be used to reduce the *precomputation overhead* associated with look-ahead implementation of Huffman decoders.

2. Architectures For Huffman Decoders

Huffman decoders [20] are used in HDTV video compression systems. In this decoder, probability of occurrence of different symbols are assumed to be known. More probable symbols are assigned shorter

[1] This research was supported by the the Army Research Office under contract number DAAL03-90-G-0063.

1058-6393/91 $01.00 © 1991 IEEE

code words; this leads to an overall reduction in the number of bits to be transmitted.

Consider an example of Huffman decoder which needs to decode one of five symbols, denoted as a, b, c, d, and e. Let the probability of occurrence of these symbols be respectively 0.5, 0.25, 0.1, 0.1, and 0.05. The Huffman code for this symbol set is constructed using the Huffman tree shown in Fig. 1. The probabilities of the two lowest probability symbols are added in each step. At the end, we assign code words to symbols starting from the top of the tree such that the higher probability occurrence is assigned a code 1 and lower a 0 (ties are broken arbitrarily). For the example symbol set, the chosen code words are a:0, b:10, c:110, d:1111, and e:1110. The average code word length is 1.9 bits (a simple binary encoding would have required three bits per symbol).

The use of unequal length code words for different symbols makes it difficult to use *look-ahead* to obtain faster implementation of Huffman decoders. But we can obtain a four-state finite state machine representation of the Huffman decoder as shown in Fig. 2. At the end of each codeword, the finite state machine returns to the initial state S_1. The finite state machine processes one input bit, represented as $x(n)$, and has five outputs represented as $a(n)$, $b(n)$, $c(n)$, $d(n)$, and $e(n)$. Each output signal wire represents the presence or absence of the corresponding symbol in that cycle. For example, $c(n)$ is 1 if the decoder detects the symbol c in cycle n, and 0 otherwise.

The finite state machine can be represented by the state update representation

$$\mathbf{s}(n+1) = \mathbf{s}(n)\mathbf{T}(n) \tag{2.1}$$

where

$$\mathbf{s}(n) = \begin{bmatrix} s_1(n) \; s_2(n) \; s_3(n) \; s_4(n) \end{bmatrix},$$

and

$$\mathbf{T}(n) = \begin{bmatrix} \overline{x(n)} & x(n) & 0 & 0 \\ \overline{x(n)} & 0 & x(n) & 0 \\ \overline{x(n)} & 0 & 0 & x(n) \\ 1 & 0 & 0 & 0 \end{bmatrix},$$

and the output representation

$$a(n) = s_1(n)\,\overline{x(n)}\,,\; b(n) = s_2(n)\,\overline{x(n)}\,,$$

$$c(n) = s_3(n)\,\overline{x(n)}\,,\; d(n) = s_4(n)\,x(n)\,,\; e(n) = s_4(n)\,\overline{x(n)}\,.$$

A hardware implementation of the sequential decoder described by (2.1) is shown in Fig. 3. The recursive loop in this architecture limits the maximum achievable decoding rate. But we can rewrite (2.1) as

$$\mathbf{s}(n+2) = \mathbf{s}(n)\mathbf{T}(n)\mathbf{T}(n+1)\,, \tag{2.2}$$

which is equivalent to two steps of look-ahead (in general, we can express $\mathbf{s}(n+M)$ in terms of $\mathbf{s}(n)$ to obtain M steps of look-ahead). The expression $\mathbf{T}(n)\mathbf{T}(n+1)$ can be computed outside the loop, and (2.2) can then be used to pipeline the decoder by two stages thus leading to a two times speed improvement. Furthermore, we can use logic minimization to reduce the implementation complexity of $\mathbf{T}(n)\mathbf{T}(n+1)$. After some minimization, we get

$$\mathbf{T}(n)\mathbf{T}(n+1) =$$

$$\begin{bmatrix} \overline{x(n+1)} & \overline{x(n)}x(n+1) & x(n)x(n+1) & 0 \\ \overline{x(n+1)} & \overline{x(n)}x(n+1) & 0 & x(n)x(n+1) \\ x(n)+\overline{x(n+1)} & \overline{x(n)}x(n+1) & 0 & 0 \\ \overline{x(n+1)} & x(n+1) & 0 & 0 \end{bmatrix}$$

The concurrency available in finite state machine representation in (2.2) can be exploited either in the form of pipelining or parallel (block) processing. Fig. 4 shows the two-stage pipelined realization corresponding to (2.2) and Fig. 5 shows the parallel realization corresponding to (2.2) with block size two. Both the architectures in Fig. 4 and Fig. 5 can decode at twice faster rate compared to the architecture in Fig. 3. The actual improvement in speed in the pipelined architecture can be obtained after redistributing the latches and after pipelining the non-recursive portion [4-8]. The parallel architecture is based on *block state update* principle, where we update the states $s_i(2k+2)$ using the states $s_i(2k)$ and each latch is 2–*slow* [6]. Furthermore, the *incremental computation* technique is used to compute the *missed* states $s_i(2k+1)$ using the available states $s_i(2k)$ using (2.1). The advantage of the incremental state computation is that the complexity increases linearly with respect to speed.

We can also combine pipelining with parallel processing. Pipelined and parallel architectures are designed using the look-ahead and incremental output computation technique. To design a M stage loop pipelined L-input decoder, we need to update the states $s_i(kL+ML)$ using the states $s_i(kL)$. For a two stage pipelined two input decoder, the state update operation is described by

$$\mathbf{s}(2k+4) = \mathbf{s}(2k)\mathbf{T}(2k)\mathbf{T}(2k+1)\mathbf{T}(2k+2)\mathbf{T}(2k+3)$$

$$= \mathbf{s}(2k)\mathbf{T}'(2k+3)\,, \tag{2.3}$$

where $\mathbf{T}'(2k+3)$ is

$$\begin{bmatrix} \overline{x(2k+3)}+x(2k+2)x(2k+1)x(2k) & \overline{x(2k+2)}x(2k+3) & \overline{x(2k+1)}x(2k+2)x(2k+3) & \overline{x(2k)}x(2k+1)x(2k+2)x(2k+3) \\ \overline{x(2k+3)} & x(2k+3)[\overline{x(2k+2)}+x(2k)x(2k+1)] & \overline{x(2k+1)}x(2k+2)x(2k+3) & \overline{x(2k)}x(2k+1)x(2k+2)x(2k+3) \\ \overline{x(2k+3)} & \overline{x(2k+2)}x(2k+3) & x(2k+2)x(2k+3)[x(2k)+\overline{x(2k+1)}] & \overline{x(2k)}x(2k+1)x(2k+2)x(2k+3) \\ \overline{x(2k+3)} & \overline{x(2k+2)}x(2k+3) & \overline{x(2k+1)}x(2k+2)x(2k+3) & x(2k+1)x(2k+2)x(2k+3) \end{bmatrix}.$$

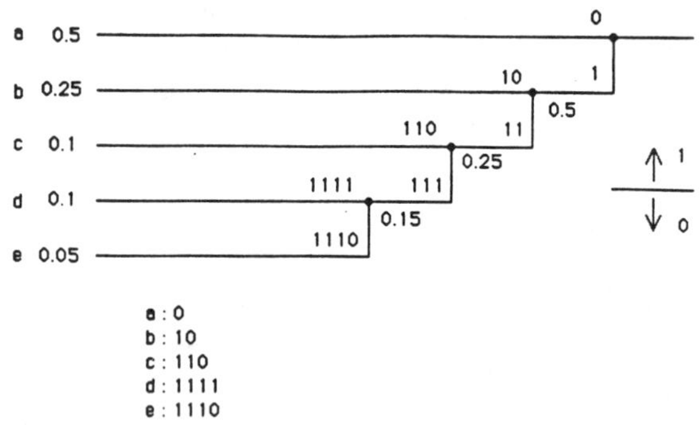

Fig. 1: A Huffman tree.

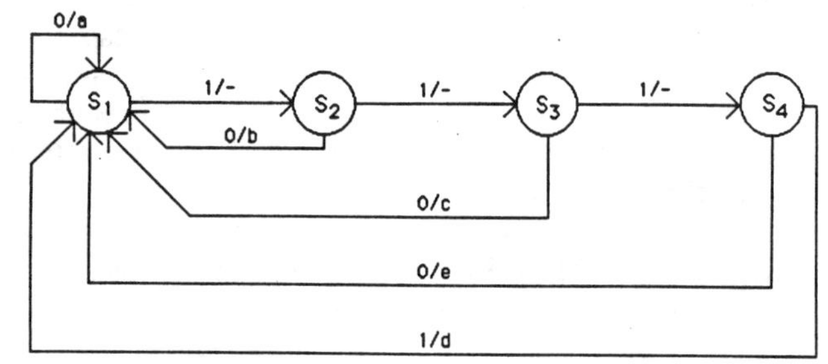

(a)

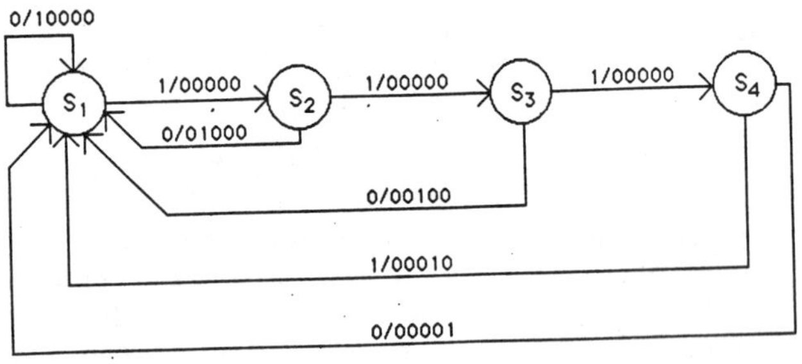

(b)

Fig. 2: (a) A finite state machine representation of the Huffman decoder, (b) The equivalent decoder finite state machine with five outputs.

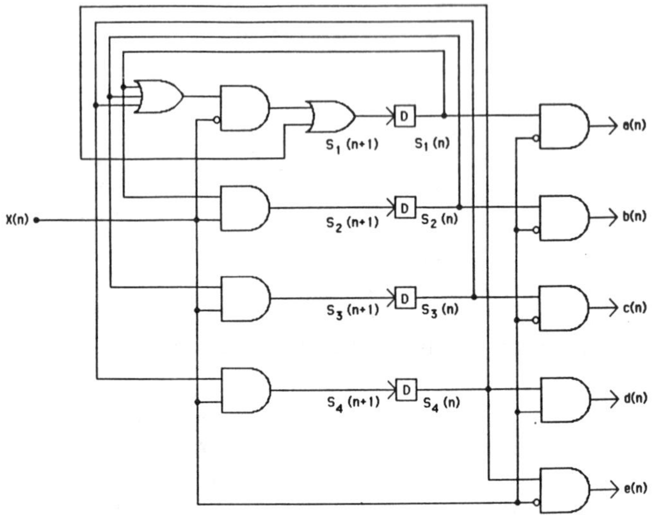

Fig. 3: Hardware implementation of the finite state machine decoder of Fig. 2.

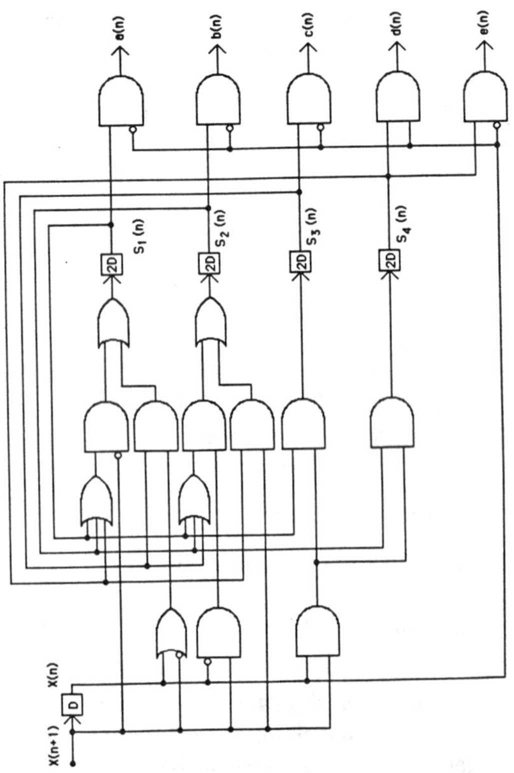

Fig. 4: Two-stage pipelined decoder obtained using look-ahead transformation on the system in Fig. 3.

The two-stage pipelined two-input decoder is shown in Fig. 6. The decoder processes two input bits $x(2k+3)$ and $x(2k+2)$. These two inputs are delayed by one cycle to obtain $x(2k+1)$ and $x(2k)$ respectively. The outputs are obtained using incremental computation, and the actual improvement due to pipelining is achieved after redistribution of latches. This architecture is four times faster than that in Fig. 3. We have not reused the intermediate results, since such reuse requires more hardware. For example, we could delay $x(2k+2)x(2k+3)$ by one cycle to obtain $x(2k)x(2k+1)$, but such reuse would require one latch, which is more expensive than an "and" gate required to compute the expression $x(2k)x(2k+1)$. The reuse of intermediate results led to significant reduction in hardware in linear recursive systems [4-6], but is not much useful in high-speed Huffman decoder design. Furthermore, logic minimization can reduce the overall implementation complexity of the hardware overhead due to look-ahead computation.

3. Conclusion

The look-ahead technique has been applied to design high-speed Huffman decoders using pipelining or parallel processing. This approach is well suited for high-speed Huffman decoder design, where the number of symbols to be decoded is low. For decoders with large symbol set, this approach may not be very practical. Future work needs to be directed towards design of new decoder algorithms which are inherently pipelinable.

4. References

[1] M. Renfors, and Y. Neuvo, "The Maximum Sampling Rate of Digital Filters under Hardware Speed Constraints", *IEEE Trans. on Circuits and Systems*, **28**(3), pp. 196-202, March 1981

[2] K.K. Parhi, and D.G. Messerschmitt, "Static Rate-Optimal Scheduling of Iterative Data Flow Programs via Optimum Unfolding", *IEEE Trans. on Computers*, **40**(2), February 1991, pp. 178-195

[3] K.K. Parhi, "Algorithm Transformation Techniques for Concurrent Processors", *Proceedings of the IEEE*, (Special issue on supercomputer technology), **77**(12), December 1989, pp. 1879-1895

[4] K.K. Parhi, and D.G. Messerschmitt, "Pipeline Interleaving and Parallelism in recursive Digital Filters, Part I: Pipelining using Scattered Look-Ahead and Decomposition", *IEEE Transactions on Acoustics, Speech, and Signal Processing*, July 1989, pp. 1099-1117

[5] K.K. Parhi, and M. Hatamian, "A High Sample Rate Recursive Filter Chip", *in VLSI Signal Processing III*, Chapter 1, IEEE Press, (Proc. of the Third IEEE VLSI Signal Processing Workshop), Monterey, CA, Nov. 1988 (see also IEEE Journal of Solid State Circuits, February 1992)

[6] K.K. Parhi, and D.G. Messerschmitt, "Pipeline Interleaving and Parallelism in recursive Digital Filters, Part II: Pipelined Incremental Block Filtering", *IEEE Transactions on Acoustics, Speech, and Signal Processing*, July 1989

[7] C.W. Wu, and P.R. Cappello, "Application Specific CAD of VLSI Second Order Sections", *IEEE Trans. on Acoustics, Speech, and Signal Processing*, **Vol. 36**, May 1988, pp. 813-825

[8] K.K. Parhi, and D.G. Messerschmitt, "Concurrent Architectures for Two-Dimensional Recursive Digital Filtering", *IEEE Trans. on Circuits and Systems*, **36**(6), June 1989, pp. 813-829

[9] K.K. Parhi and D.G. Messerschmitt, "Concurrent Cellular VLSI Adaptive Filter Architectures", *IEEE Transactions on Circuits and Systems*, **34**(10), October 1987, pp. 1141-1151

[10] T. Meng, and D.G. Messerschmitt, "Arbitrarily High Sampling Rate Adaptive Filters", *IEEE Trans. on Acoustics, Speech, and Signal Processing*, April 1987

[11] K.K. Parhi, "Pipelining In Algorithms with Quantizer Loops", *IEEE Trans. on Circuits and Systems*, **38**(7), July 1991

[12] G. Fettweis, and H. Meyr, "Parallel Viterbi Decoding by Breaking the Compare-Select Feedback Bottleneck", *Proc. of the IEEE International Conference on Communications*, Philadelphia, pp. 719-723, June 1988 (See also *IEEE Trans. on Communications*, August 1989)

[13] K.K. Parhi, "Look-Ahead in Dynamic Programming and Quantizer Loops", *IEEE International Symposium on Circuits and Systems*, Oregon, May 1989, pp. 1382-1387

[14] K.K. Parhi, "Pipelining in Dynamic Programming Architectures", *IEEE Trans. on Signal Processing*, June 1991

[15] C.Y. Chang, and K. Yao, "Viterbi Decoding by Systolic Array", *Ann. Allerton Conference on Communications, Controls, and Computing*, Oct. 1985, pp. 430-439

[16] H. Thapar, and J. Cioffi, "A Block Processing Method for Designing High-Speed Viterbi Decoders", *Proceedings of the IEEE Int. Conf. on Communications*, 1989, pp. 836-840

[17] H.D. Lin, and D.G. Messerschmitt, "Improving the Iteration Bound of Finite State Machines", *in Proc. of IEEE International Symposium of Circuits and Systems*, may 1989, Portland (Oregon), pp. 1328-1331

[18] P.M. Kogge, and H.S. Stone, "A Parallel Algorithm for the Efficient Solution of a General Class of Recurrence Equations", *IEEE Trans. on Computers*, **22**(8), pp. 786-793, 1973

[19] P.M. Kogge, "Parallel Solution of Recurrence Problems", *IBM J. Res. and Development*, pp. 138-148, March 1974

[20] D.A. Huffman, "A Method for the Construction of Minimum Redundancy Codes", *Proc. of the IRE*, **68**(7), Sept. 1952, pp. 1098-1101

[21] S.-M. Lei, M.-T. Sun, K. Ramachandran, and S. Palaniraj, "VLSI Implementation of An Entropy Coder and Decoder For Advanced TV Applications", *in Proc. of IEEE Int. Symp. of Circuits and Systems*, May 1990, New Orleans, pp. 3030-3033

[22] H.D. Lin, and D.G. Messerschmitt, "High Throughput Reconstruction of Huffman-Coded Images", *in Proc. of IEEE Conf. on Computer Design*, Oct. 1989

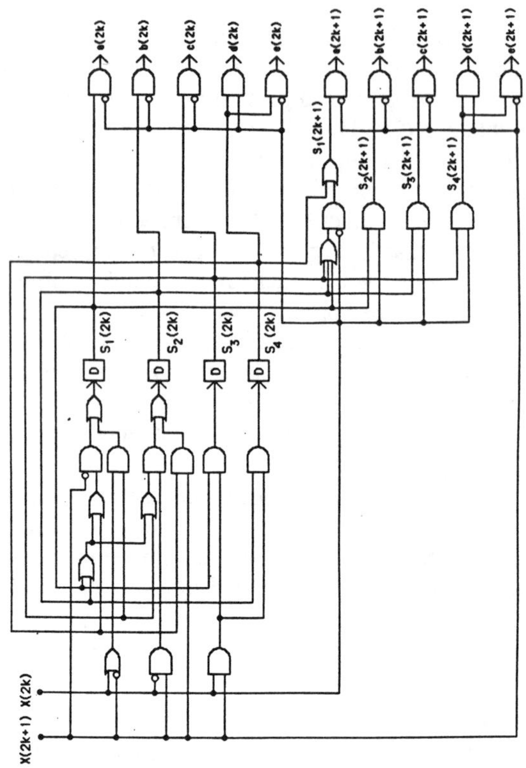

Fig. 5: Incremental block decoder obtained using look-ahead computation and incremental output computation for block size two.

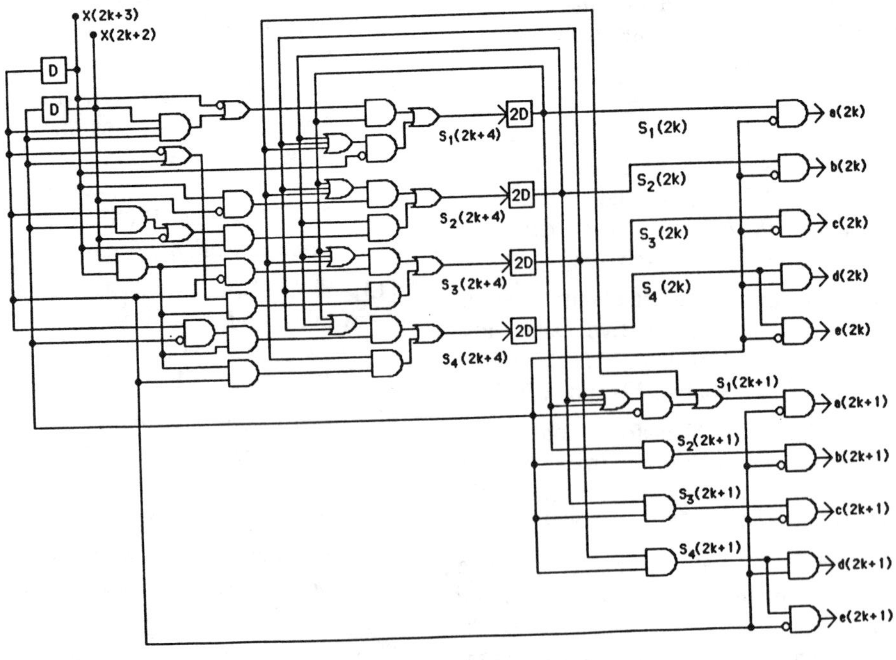

Fig. 6: Two-stage pipelined incremental block decoder with block size two.

Software Development and Real-time Target Systems on a Common Backplane

Daniel Gold and Kasi Anantha
Department of Mathematical Sciences
San Diego State University San Diego, CA 92182

Abstract

An architecture where the development and real-time target platforms share a common backplane is highly desirable. This approach has many advantages yet can also cause serious problems. Problems include resetting the developing system while trying to reset the real-time targets, host memory not protected from errant real-time processes and processors, and real-time interrupts affecting the host system. These problems are alleviated by connecting host and target system backplanes via a VMEbus-to-VMEbus adapter, which links together the address, data and control lines of both systems. This paper discusses the advantages of a backplane connected system including host or remote (via network or telephone) access and monitoring for development, testing and debugging of real-time applications.

1 Introduction

A real time programming environment requires a real-time target system for testing the application and a host system for software development. The hardware and software requirements for these systems vary widely. A host development system should have good time sharing and network capability whereas the real-time system should have a simple kernel that provides fast real time response to interrupts without the overhead normally associated with time shared system. However if the two systems are not connected by a fast data path much time can be lost in loading the real-time programs on the target and collecting the real-time data for postprocessing.

Real-time target software development is typically performed on a host system (e.g. workstation or PC). The software must then be transferred to the target system. Earlier methods included burning PROMS and RS-232, and more recently via networks (e.g. Ethernet via TCP/IP). While PROMS may be the ultimate goal, the process of burning and replacing PROMS quickly becomes tedious, especially for small changes. Although less work than burning PROMS, RS-232 is still slow compared to Ethernet or backplane speed. Ethernet downloading, however, requires network software to be resident in the target system, which may not always be true. The network software may also slow down the target CPU, as well as using extra memory for network buffers. In order to provide a fast data path between the realtime target system and the host development system, both can reside on the same VME bus backplane as was done in Kumaran2 architecture.

Host and target systems on the same backplane, however, created new problems. Trying to reset the target processors also resulted in resetting (and crashing) of the host system. Additionally, interrupts meant for real time processors interrupt the host system, and target processors modify the Unix system memory.

The VMEbus-to-VMEbus adapter provides a fast data path between the target and host systems, while protecting the host system by masking off interrupts and the VMEbus SYSRESET* line, and preventing the target system from accessing host system memory.

2 Previous Work

CHIMERA[1] is a real time computing environment developed at Carnegie-Mellon University for Reconfigurable Modular Manipulator System (RMMS) project. CHIMERA consists of one or more Ironics 68020 CPU boards, acting as real-time computing engines, residing on the VME bus of a Sun 3 workstation. This computing environment stressed the following two aspects:

1. 'High-level computing, performed on a general purpose processor, for real-time program development and offline storage and analysis of experimental results.'

1058-6393/91 $01.00 © 1991 IEEE

2. High performance numerical computing on dedicated real-time processors to implement the control law, and sample and buffer data for later storage on the general purpose system.

CONDOR[2] system on the other hand was developed at MIT for controlling the Utah-MIT hand and other complex robots. In the second version of the CONDOR hardware a Sun-3 forms the main development machine while the real time controller utilizes Ironics Corporation 16 MHz Motorola 68020 based single board computers equipped with Motorola 68881 floating point unit. The control processors are linked using a VME bus and are coupled to the development host through a VME bus to VME bus adapter.

Kumaran[3,4,5] multiprocessor system used Motorola MVME136A boards based on MC68020 and MC68881 processors and AT&T System V on an MVME147 board for program development, data storage and analysis. The real time CPU boards (MVME136A) and the development CPU (MVME147) were kept on the same VME bus chassis. Real time CPUs communicated with the Unix CPU via the VME bus using special device drivers.

Kumaran II[6] multiprocessor system replaced the Motorola MVME136A boards with faster MVME143-2 boards with MC68030 at 33 MHz and MC68882 floating point units. Later the Unix CPU was moved to its own chassis and connected to the real time CPUs using a VMEbus to VMEbus adapter.

3 System Design

Connecting a target and host system via a bus-to-bus adapter affords several advantages including:

1. Much faster downloading, via the VMEbus. A one megabyte file can be loaded into a single target CPU in under 9 seconds, and 6 targets (in parallel) in under 28 seconds.

2. Communication between host and target system, since the bus-to-bus adapter may be jumpered to pass VMEbus interrupts between host and target. Thus processes on the host may start a target processor executing, and wait for a VMEbus interrupt to signify completion. Host processors may also send VMEbus interrupts to the target system.

3. The target system memory appears as an extension to the host system memory. Thus target system memory locations may be examined/modified via commands on the host. The target system, however, is prevented from accessing the host system memory by hardware jumpers on the adapter card. This feature is very powerful for remote debugging. Values may be initialized in the target system memory, remote programs executed, and data saved/retrieved from target memory, all from the host system. Additionally, host-target synchronization and semaphores may be implemented in target system memory.

4. Remote reset of the target system. The bus-to-bus adapter is capable of driving the VMEbus SYSRESET* control line via a software command. Thus the host system (via software) can reset the target processors when necessary. The target system is prevented from driving the host system SYSRESET* line via hardware jumpers.

Figure 1 shows the connection of a Unix based host for software development and debugging, and a real-time target system. The target system consists of multiple target CPU boards, memory, and other I/O cards. Each CPU card may be individually reset, loaded, started, stopped , or debugged remotely. CPU boards may also be reset all at once (via SYSRESET*) and facilities exist for parallel loading of identical code to multiple target processors.

4 System Software

The Unix-based host system provides a development environment and access to the real-time processors under Unix. This includes the ability to download programs to the target processors, start their execution and retrieve data from each CPU's local memory after program completion. Other utilities provide the capability to reset, halt, and interrupt any of the target processors from Unix, as well as receive interrupts from the target processors.. Since Kumaran also supports an Ethernet interface, programs can be developed, loaded, executed, and debugged from remote machines.

In order to access the target processors under Unix, special device drivers (for both the target boards and the bus-to-bus adapter) and related utilities were developed for Kumaran. The target CPU device driver allows Unix processes to communicate with the target boards via VMEbus interrupts, since the MVME143 is capable of generating VMEbus interrupts. The driver also prevents anyone else from accessing a target board while it is currently in use. When a Unix process wishes to communicate with a target board, the device driver blocks the process. When the target board generates a VMEbus level 1 interrupt, the driver determines which board generated the interrupt, and unblocks any processes waiting for that board. The device driver for the bus-to-bus adapter allows Unix processes to reset the target system via the SYSRESET* line, as well as vectoring target system VMEbus interrupts to the correct host process. Special utilities developed give Unix processes the capability to reset, halt, interrupt, start, stop, and load each target processor. In addition, other utilities were developed to access the local RAM on each CPU board. This is possible because the local RAM on each CPU is seen as an extension to the Unix memory. For example, the first 8 MB of physical memory is reserved for Unix, the next 4 MB is reserved for the color graphics card, the next 4 MB is local RAM of CPU 1 and so on. The local RAM on each CPU is private in that it is not available under Unix for user processes, but Unix knows it exists and allows access via the bus-to-bus adapter and associated device driver. Memory access utilities include the ability to write local CPU memory (for downloading programs to a target board), read local CPU memory (for saving data from a particular CPU), and modifying common memory (i.e. to clear/set synchronization lock).

The executable program files are in COFF (Common Object File Format) format in the Unix file system. Since each board's local memory is accessible over the VME bus we can download COFF format programs by writing directly to a particular CPU's local memory under Unix.

Data generated by the six MVME143 boards are saved in the target processors memory. The MVME143 target boards communicate with the Unix host via message passing in common memory or interrupts. Upon completion of the real time application, the host system is notified and then communicates with the target processors to transfer data to the Unix side, saved as multiple Unix files. The Unix files are more convenient for postprocessing of the data and generating graphs.

5 Sweptsine Feedback Control System

Although other test methods such as random testing and shock testing are now widely used for structural analysis, swept sine testing is best suited for getting precise details of resonances of structures. The control channel signal is compared to the control reference. The difference is used to change the system gain so that the drive signal to an external power amplifier is properly modified. The signal from the power amplifier is used to vibrate a shaker table on which the structure is mounted. Transducers mounted on the structures provide response signals for resonance analysis. An earlier paper described in detail the swept sine control system built using Kumaran[5]. We will briefly describe the control system.

The signal from the transducer is filtered using an analog filter with cut off frequency set to 2 KHz. The signal is sampled at 5 KHz using an A/D with 12 bit resolution. In order to provide sufficient precision auto ranging is implemented using an input attenuator and an rms detector. The rms detector is implemented with a digital Butterworth filter whose parameters are varied as a function of frequency of the drive signal.

A servo loop is implemented in software to control the output attenuator so that the input signal follows a reference profile. The detector provides rms values at a constant 5 KHz.

The software is implemented by five processes namely *sinegen, analyze, control, freqdata* and *sweep* running on five processors. They use local memories of the processors for instruction and local data and global memory for shared variables. The process sinegen generates the the digital wave form at 10 KHz sample rate and loads the input and output attenuators and collect response data at 5 KHz. The *analyze* implements the rms detector, whereas *control* uses the rms value to implement a proportional controller for the log of the amplitude. The process *sweep* implements a log sweep rate and the process *freqdata* collects the frequency, drive output and rms value of the reference signal, and estimated transfer function. When the data for single run has been taken, *freqdata* interrupts the Unix process so that the data can be saved in the Unix file system.

Both sweptsine control system[5] and host system software were modified to make use of the features provided by the bus-to-bus adapter. Both host and target software were also modified to take advantage of the virtual memory implementation[7]. The real time software is generated on the host system, and downloaded over the backplane to the target system (via the bus-to-bus adapter). The target software is then remotely started by the host, and communicates with the host via interrupts through the bus-to-bus adapter. Interrupts flag such events as data collection complete, and problems during the experiment requiring premature termination of the experiment. When the real time software has collected a pre-determined amount of data, the host is interrupted and reads data from the target board into the Unix file system. The real time processors implement a double buffering scheme for continuous data recording while the Unix processor reads the data from the previous experiment. Since the data path between the host and target systems is the backplane (via the bus-to-bus adapter), only double buffering is needed to insure continuous data collection. If a slower data path was used, multiple buffers would be necessary, or the host would be forced to wait until all data collection had completed to transfer data from the target system to the host system. Should the host receive an interrupt signifying a problem, all target boards are reset via the bus-to-bus adapter, and leftover data saved to the host.

6 Results

Earlier sweptsine control system[5] software used an external circuit to emulate a peak notch filter. Since the emphasis in this paper is on the host/target system architecture and real time memory management, the external circuitry was removed and the system run closed loop. Figure 2 displays drive signal, control signal and transfer function as a function of frequency. The spikes are due to the noise from the switching power supply.

6.1 Conclusions

The use of bus-to-bus adaptor enabled remote reset of the Kumaran II target processors before the start of the tests, and isolate the Unix system from illegal access by the target processors. The virtual memory implementation protected each processor from destructive interference by other processors. The new software is being used by other researchers using Kumaran II. Future plans include debugging tools on the host system so that tests can be debugged remotely. The debugging tools will include an individual window for each processor so they can display results concurrently.

Acknowledgements

The authors would like to thank Bob Hilchey for hardware support in maintaining the Kumaran II multiprocessor system and Arul Anantha for system software support.

References

[1] Donald Schmitz, Pradeep Khosla, Regis Hoffman and Takeo Kanade, "CHIMERA: A Real-time Programming Environment", Proceedings of the IEEE International Conference on "Robotics and Automation", IEEE, 1989

[2] Sundar Narasimhan, David M. Siegel, John M. Hollerbach, "Condor: A Revised Architecture for Controlling the Utah-MIT hand", Proceedings of the IEEE International Conference on "Robotics and Automation", IEEE, 1988

[3] K. Anantha, Marko Vuskovic and Ting Liang, "A Multiprocessor Implementation of Inverse Robot Dynamics", International Journal of Mini and Microcomputers, Vol. 11, No. 3, 1989.

[4] K. Anantha, M. Vuskovic and T. Liang,"A Multiprocessor Test bed for Robotics Applications", Mathematical and Computer Modelling, Vol. 14, pp 444-449,1990

[5] K. Anantha and Daniel Gold, "Kumaran- A Multiprocessor System for Swept Sine Vibration Control", Proceedings of the Twenty Third Asilomar Conference on Signals, Systems and Computers, October, 1989.

[6] K. Anantha, Daniel Gold and Arul Anantha,"A Multiprocessor System for Realtime Applications", Proceedings of the Twenty Fourth Asilomar Conference on Signals, Systems and Computers, November, 1990.

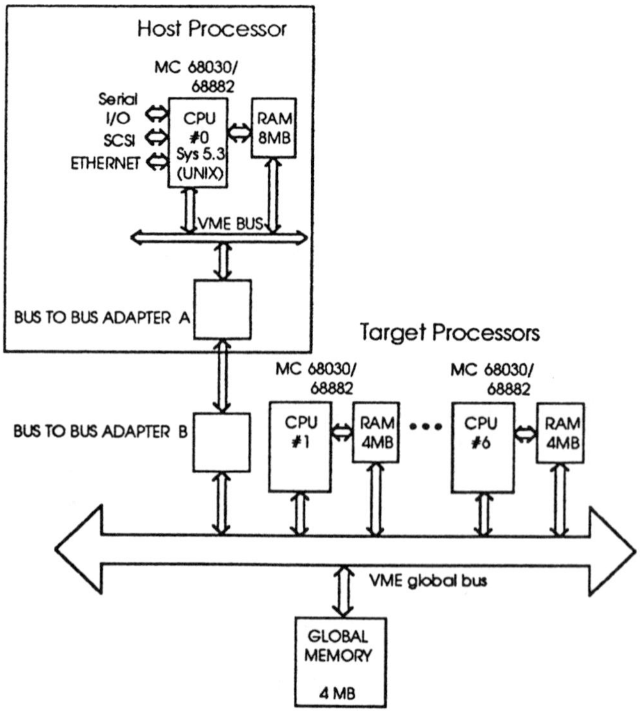

Fig. 1 Configuration of the KUMARAN II multiprocessor

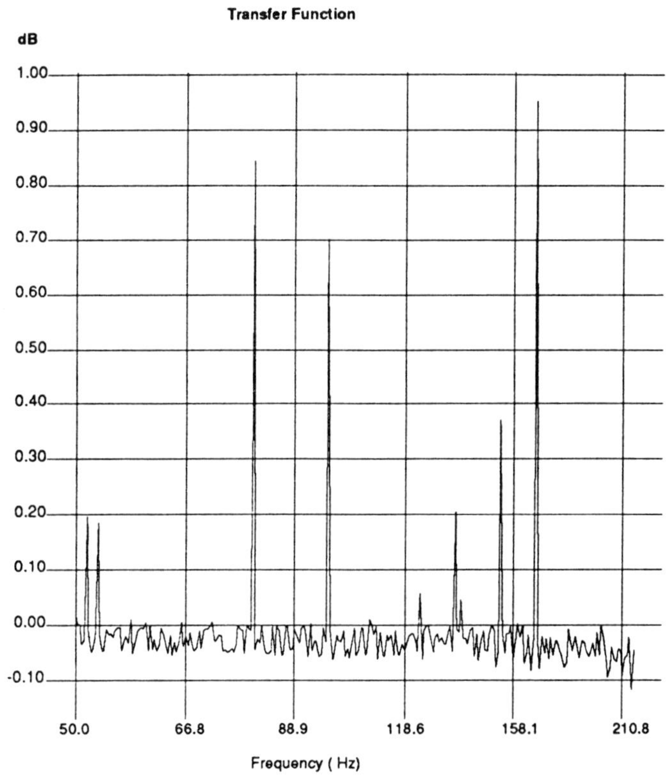

Figure 2. Transfer function vs. Frequency

VLSI Chip Architecture for Real-Time Ambiguity Function Computation

Nenad Marinovich
Department of Electrical Engineering
City College of CUNY
New York, NY 10031

Vojin G. Oklobdzija
Department of Electrical Engineering and
Computer Science
University of California
Davis, CA 95616

Abstract

A chip architecture is developed that evaluates the cross-ambiguity function samples on an arbitrary sampling grid at real-time radar rates. It can be scaled to match different date lengths and sampling grid sizes. It is shown to be potentially quite useful for a wide range of typical radar system.

1. Introduction

It is well known that, under ideal conditions, the optimal procedure for joint estimation of the delay and Doppler shift of a radar echo requires computation of the complex cross-ambiguity function between the transmitted and the received signals. Even under non-ideal conditions, in a real world operating environment, this procedure is considered to be the best suboptimal one. However, prohibitively high computational complexity and the lack of efficient algorithms and appropriate hardware have been preventing its implementation in practice. Recently, a new algorithm for efficient computation of the decimated cross-ambiguity function on arbitrary sampling grids have been developped by Auslander, Gertner, and Tolimieri (AGT algorithm)[1,2]. Its important feature is that the computations on complementary, decimated sampling grids can be performed independently, in parallel, in the same time it takes to compute the FFT of the received data vector. Combined with the progress in VLSI technology, this makes the implementation of the cross-ambiguity function based receiver feasible at present time. In this paper, based on a simple modification of the AGT algorithm, we develop a VLSI chip architecture for real-time computation of the cross-ambiguity function magnitude. Using actual radar examples, we discuss the design trade-offs, and demonstrate the scalability of the architecture needed to accommodate different data lengths and sampling grid sizes.

2.Background

Let x(t) and y(t) denote complex envelopes of the transmitted and received signals, respectively, and x(n) and y(n) sequences of their samples. The complex cross-ambiguity function between the received and transmitted signals is defined as:

$$(1) \qquad A_{yx}(\nu,\tau) = \int_{-\infty}^{\infty} y(t)x^{*}(t-\tau)e^{j2\pi\nu t}dt$$

Regularly sampled CAF is defined on an L x L sampling grid as

$$(2) \qquad A_{yx}(k\Delta\nu, l\Delta\tau) = \sum_{i=0}^{L} y(i)x^{*}(i-l)e^{j\frac{2\pi}{L}ik} di$$

$$0 \le l < L \qquad -\frac{L}{2} \le k < \frac{L}{2}$$

where L>C+D-1, with C and D being the transmited and received sequence lengths, respectively. Given that data volume L is very large, typically in thousands, the high computational cost in Eq. (2) becomes readily apparent even when FFT is used. In addition, what makes Eq. (2) impractical to use is the requirement, typically encountered in practice, that the samples of the CAF need to be computed on a considerably smaller sampling grid corresponding to a desired quantization of delay and Doppler shifts. This is indicated in Fig. 1 where P and R are the numbers of Doppler and range measurement bins, respectively. However, the efficient computation of the CAF samples on a desired grid is possible with the help of the new AGT algorithm [1,2] that is outlined below.

Initial step requires the computation of the discrete Zak transform (DZT)[1] of the received data vector. If L=MN, the DZT is defined as an M x N matrix

$$(3) \qquad Z_y(r,s) = \sum_{i=0}^{N-1} y(s+iM)e^{-j\frac{2\pi}{N}ir}$$

This computation can be visualized in the following way: first, rearange the data vector into an M x N matrix by filling the rows with successive blocks of M data samples; then, replace each column by its discrete Fourier transform (DFT) In the next step, pointwise multiplication of the received data DZT with the precomputed DZT of the transmitted data is performed, followed by the 2-D inverse DFT of the result. This gives the L samples of the CAF on a decimated M x N sampling grid, with sample spacing along the Doppler and delay axes being N $\Delta\nu$ and M $\Delta\tau$ respectively:

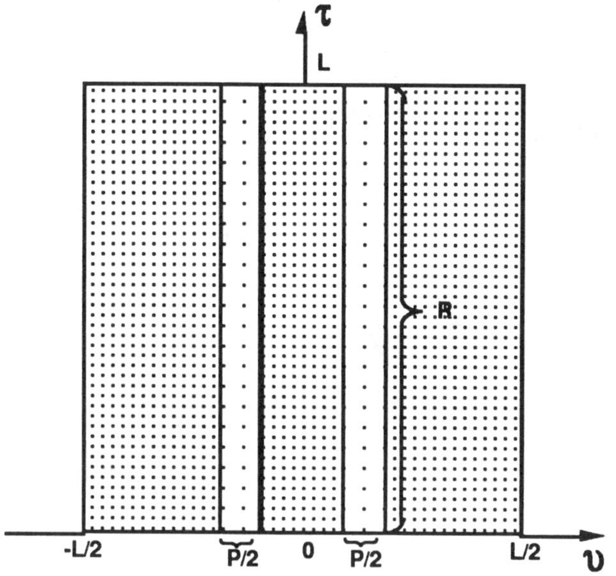

Fig. 1 Regular (fine) and desired (coarse) sampling grids

(4) $A_{yx}(mN\Delta\nu, nM\Delta\tau) = 2D-DFT^{-1}\{Z_y(r,s)Z^*$

$$= \frac{1}{MN}\sum_{r=0}^{N-1}\sum_{s=0}^{M-1} Z_y(r,s)Z^*_x(r,s)e^{j2\pi(\frac{rn}{N}+\frac{ms}{M})}$$

When FFT algorithms are used for DFT computation, this procedure takes about the same amount of computation as an L-point FFT of the raw data. An important property of the CAF that permits its computation on arbitrary domains [2] is the time and frequency shift property:

(5) $x_{\phi\theta}(t) = x(t-\theta)e^{-j2\pi\phi t} \Rightarrow$

$$A_{yx_{\phi\theta}}(\nu,\tau) = e^{j2\pi\nu\theta}A_{yx}(\nu+\phi,\tau+\theta) \Rightarrow$$

$$|A_{yx}(\nu+\phi,\tau+\theta)| = |A_{yx_{\phi\theta}}(\nu,\tau)|$$

In particular, to compute the magnitude of the CAF on another sampling grid with arbitrary offsets ϕ and θ, we use the AGT algorithm with precomputed DZT for those offsets:

(6) $|A_{yx}(mN\Delta\nu+\phi, nM\Delta\tau+\theta)| =$

$$|A_{yx_{\phi\theta}}(mN\Delta\nu, nM\Delta\tau)|$$

For each desired offset pair (ϕ, θ), we repeat the computation of (6) using the AGT algorithm, Eqs. (3) and (4), to obtain the samples of the CAF magnitude on the desired uniform sampling grid. Note that all the computations performed for different ϕ, θ pairs after the DZT of the received data is determined, are independent and can be performed in parallel.

3.Algorithm

The computation outlined above is too wasteful for the non-uniform sampling grids such as the one displayed in Fig. 1. However, a simple modification permits its use in such a case, too. Namely, instead of computing all M x N values in (6) for each offset pair, we should only compute N samples along the m=0 axis

(7) $|A_{yx}(0+\phi, nM\Delta\tau+\theta)| =$

$$= |2D-DFT^{-1}\{Z_y(r,s)Z^*_{x_{\phi\theta}}(r,s)\}|_{m=0}$$

$$= |\frac{1}{MN}\sum_{r=0}^{N-1} e^{j\frac{2\pi}{N}rn}\sum_{s=0}^{M-1} Z_y(r,s)Z^*_{x_{\phi\theta}}(r,s)|$$

for each pair of offsets

(8) $\phi = \pm(\nu_0 + p\alpha) \qquad p = 0,1,\ldots\ldots,\frac{P}{2}-1$,

$$\theta = q\beta \quad \beta = \frac{M\Delta\tau}{Q} \quad q = 0,1,\ldots\ldots,Q-1 \quad Q = \frac{R}{N}$$

Note that the 2-D M x N inverse FFT in (4) has been replaced by an N-point inverse FFT and N x (M-1) additions. The total number of multiplications required if radix-2 Cooley-Tukey FFT is used is

(9) PQN(log N + 1)/2 + MN(log N)/2 =
=(PR + L)(log N)/2

Our goal is to develop a modular architecture that can be integrated on a chip while allowing the independent computations to proceed in parallel. With that in mind we organize the computation (7) in the following reduced grid CAF algorithm (RGCAF) that minimizes necessary RAM storage:

BEGIN RGCAF
FOR *each column of the input data matrix* DO
- *compute its* FFT
FOR *each* ϕ, θ *pair* DO IN PARALLEL
- *pointwise multiply with respective column of the precomputed reference* DZT *of* x $_{\phi,\theta}$
- *accumulate the resulting column vector over all input columns*
DONE
DONE
FOR *each* ϕ,θ *pair* DO IN PARALLEL
- *compute the inverse* FFT *of the result*
- *output the magnitude of the result*
DONE
END RGCAF

Visual illustration of this computation is given in Fig. 2.

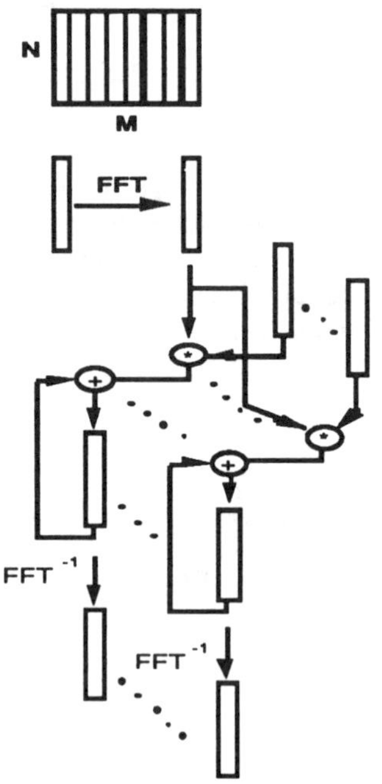

Fig.2 Computation Flow

4.Architecture

Examination of the RGCAF algorithm indicates that two processing modules are required: one to compute the N-point FFT of the input data matrix column - input module (IPM), and the other to perform the pointwise vector multiplication/accumulation and the inverse N-point FFT - parallel processing module (PPM). Data-path that computes the RGCAF algorithm is shown in Fig 3. The core computation in both modules is that of an N-point FFT. Least complex hardware implementation of the FFT for our purposes is based on repetitive use of a basic radix-2 Cooley-Tukey FFT butterfly processing element (PE) and requires two N-word RAM buffers to alternate the storage of input and output vectors during successive stages of the radix-2 Cooley-Tukey FFT algorithm. In addition, an N-word ROM of FFT coeficients is required in each processing module, and different PPM's have an L-word ROM to store the precomputed DZT of the delay and Doppler shifted transmitted signal for different ϕ,θ shifts. Finally, a magnitude estimator is a combinational network that generates an approximation of a complex number magnitude according to the following simple rule:

(10) $A = R + jI$

$$|A| = \max \begin{Bmatrix} |R| \\ \frac{7}{8}|R| + \frac{1}{2}|I| \\ \frac{1}{2}|R| + \frac{7}{8}|I| \\ |I| \end{Bmatrix}$$

This magnitude estimator has maximum error of 3% [3].

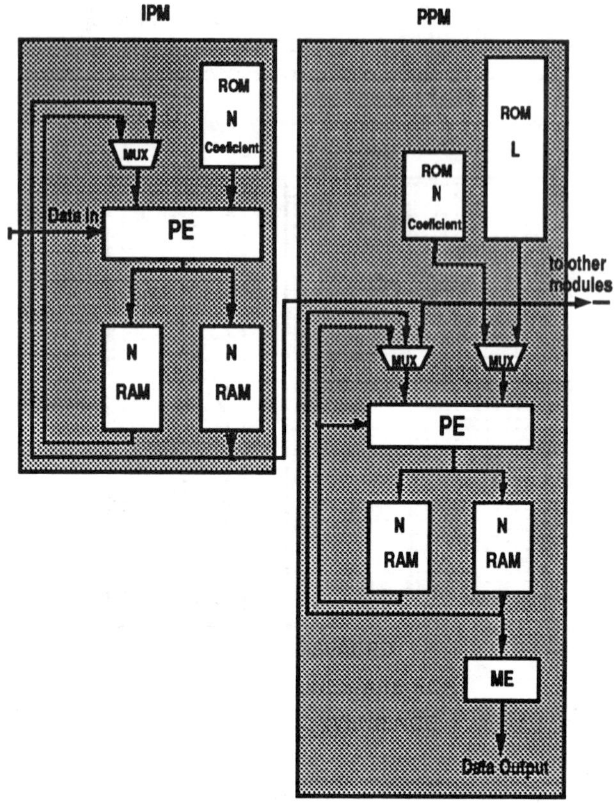

Fig. 3. Organization of the Data-Path

To accommodate the large dynamic range typical in radar applications and the fast growth of the signal through the processing chain, the complex data is internally represented using a following hybrid floating-point format[3]:

$$(11) \qquad (m_r + jm_i)2^e$$

where m_r and m_i are real and imaginary component mantissas, represented as 2's complement fractions, and e is the positive exponent whose initial value is set to zero for the input signal. Since the data amplitude will only grow during the processing, exponent is limited to be a positive integer. Rather broad range of radar applications can be accommodated with 11 bit mantissa and 5 bit exponent which requires a 27-bit complex word size[3]. The floating-point butterfly processing element which may pass through the data from its left side input, and permits pointwise vector multiply/accumulate operation is shown in Fig. 4.

It is implemented as a four-stage pipeline that computes the inverse FFT in the follwing manner:

1) - first butterfly input is loaded from the input RAM buffer into SR1;
2) - SR2 is loaded with the product of SR1 and 1/2 of an FFT coefficient;

- SR3 is loaded from the input RAM buffer with 1/2 of the second butterfly operand;
3) - difference SR3-SR2 is placed in SR5;
- sum SR3+SR2 is placed in SR4;
- SR1 is loaded from the input RAM buffer with the third input operand;
4) - first butterfly output from SR5 is stored into the other RAM buffer;
- SR5 is loaded with SR4;
- SR2 is loaded with the product of SR1 and 1/2 of an FFT coefficient;
- SR3 is loaded from the input RAM buffer with 1/2 of the fourth input operand;
5) - second butterfly output from SR5 is stored into the other RAM buffer;
- difference SR3-SR2 is placed in SR5;
- sum SR3+SR2 is placed in SR4;
- SR1 is loaded from the input RAM buffer with the fifth input operand;
6) - third butterfly output from SR5 is stored into the other RAM buffer;
- SR5 is loaded with SR4;
- SR2 is loaded with the product of SR1 and 1/2 of an FFT coefficient;
- SR3 is loaded from the input RAM buffer with 1/2 of the sixth input operand;
7) - fourth butterfly output from SR5 is stored into the other RAM buffer;
- difference SR3-SR2 is placed in SR5;
- sum SR3+SR2 is placed in SR4;
- SR1 is loaded from the input RAM buffer with the seventh input operand;
.......continue.......

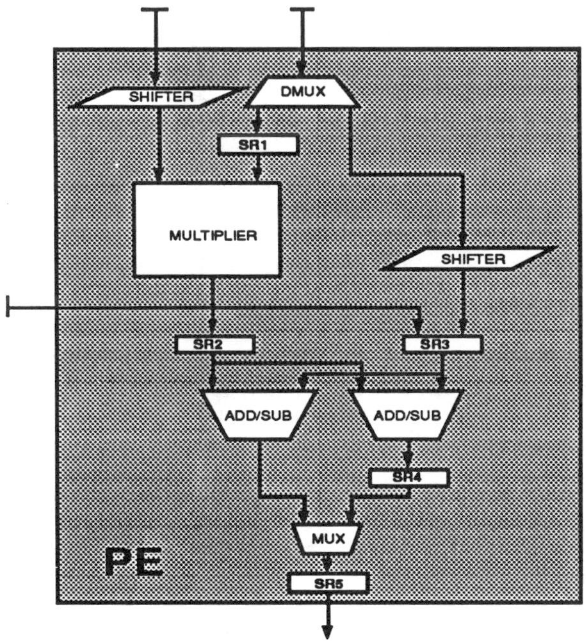

Fig.4. Organization of the Processing Element

This process is continued until all the data from the input RAM buffer has been processed. Then the role of the input and output buffer is exchanged and the whole process is repeated for the next stage of the FFT computation. In computation of the forward FFT, division of operands by 2 is not performed.
The whole FFT computation takes approximately N(log N) clock cycles (neglecting the pipeline latency). We can now estimate the total execution time to be approximately equal to

(12) $Tx=[L(\log N + 1)]Tc$

This is less than the time needed for the computation of a single L-point FFT of the raw data with circuits of comparable complexity. Regularity and modularity of the proposed architecture make it easy to integrate. It can be scaled to accomodate different data lengths L. On one extreme, for huge data lengths, an IPM and a single PPM would be put on a single chip and P x R/N chips will be required to generate P x R CAF samples. On the other, for short data lengths, an IPM and many PPM's would fit on a single chip. How many chips would be required depends on the desired number of Doppler bins P, and on how many of the R range bins are computed per PPM. The trade-offs involved are discussed in the next section.

5.Radar Examples

Practical implications of the proposed chip architecture can be assessed by considering how could it be used in actual radar systems. With this goal, we consider several typical pulsed Doppler radar systems [4]. We estimate the equivalent gate count of the required processing modules in each case, and estimate the chip count knowing that, with the current state of the art, up to about 500,000 gates are available on a single chip. With the present state of the art, it is estimated that the clock cycle of 10nsec is sufficient for proper operation of the processing element in Fig. 4 and for memory access[5]. This value is used when estimating the execution times.

A. Airborne early warning radar (AEWR)

This radar operates in the UHF band, with the range 300 nautical miles = 1.8msec, the pulse repetition frequency PRF=300Hz, data block L=32K samples, R=4K range bins, and P=8 Doppler bins. Using N=1K and M=32, we estimate the gate count to be 120K for IPM and 220K for PPM.

Consequently, only one PPM together with an IPM can be put on single chip that will compute 1K range samples. Therefore, four chips are required per Doppler bin, and total of 4P=32 chips are required.

Total execution time is approximately T=3.2μsec. The input data block rate is determined by the longer of the maximum time-of-flight (2 x 1.8msec) and the pulse repetition interval (3.3msec). It equals one L-point data block per 3.6 msec in this case. Therefore, to keep up with the incoming data rate, one 32-chip set is sufficient.

B. 2-D Air surveillance radar (ASR)

This is an S-band radar, with range 60 nmi = 0.36msec, PRF=1.2KHz, data block L=4K, R=1K range bins, and P=8 Doppler bins. Using N=1K and M=4, the estimated gate count is 20K for IPM and 125K for PPM.

With room to spare, three PPM's can fit on a single chip together with an IPM and all the range samples for three Doppler bins can be computed with a single chip. Three chips are required for all the desired Doppler bins.

Each CAF computation can be completed in approximately 0.44msec. The input data block rate is one block per 0.83msec. Again, one set of three chips is sufficient to keep up with the incoming data.

C. Portable battlefield reconnaissance radar (BRR)

This is a Ku-band radar, with range 6 nmi = 36μsec, PRF=12KHz, data block L=1K, R=256 range bins, and P=10 Doppler bins. Using N=256 and M=4, the estimated gate count is 6K for IPM and 32K for PPM.

Twelve PPM's can fit on a single chip which is sufficient for the computation on all the range and Doppler bins.

Execution time is approximately 92μsec, which can keep up with the data coming in at the rate of one block per 100μsec. In this case one chip is sufficient to implement the radar's Doppler processor.

D. Airborne intercept radar (AIR)

Operating in X-band this nose-mounted radar has range of 30nmi = 180usec, PRF=150KHz, data block L=1024, R=256 range bins, and P=50 Doppler bins. Using N=256 and M=4, the estimated gate count is 6K for IPM and 32K for PPM.

Fifteen PPM's can fit on a single chip and it can compute all the range samples for fifteen Doppler bins. Four chips are needed for all the required bin computations.

It should take approximately 92μsec to complete the computation. To keep up with the incoming data rate of one block per 360μsec only one set of four chips is needed.

6. Summary

A VLSI chip architecture for real-time cross-ambiguity function computation in radar systems was developed and its practical implications were assessed. It is based on a modification of a recent algorithm for decimated ambiguity function computation[1,2].

Bulk of the chip is memory, both RAM and ROM. FFT is performed by repetitive use of a single butterfly computing element. With the present state of the art, the resulting low circuit complexity and simple control allow integration of the processing modules on one or more chips, depending on the radar system requirements, avoiding the chip-crossing penalties and improving speed.

The architecture is regular, easy to integrate, and modular. It can be scaled to match different data lengths. To accomodate various sampling grid sizes and input data rates, chips can be parallelized and pipelined with virtualy no overhead. Potential utility of this architecture for a broad range of radar applications was illustrated on several examples of typical systems.

7. References

[1] L. Auslander, I.G. Gertner, and R. Tolimieri, "*The discrete Zak transform application to time-frequency analysis and synthesis of nonstationary signals*", IEEE Trans. Sig. Proc., vol 39, no. 4, pp. 825-835, April 1991.

[2] L. Auslander and I.G. Gertner, "*Computing the ambiguity function on various domains*", submitted for publication.

[3] J.H. McClellan and R.J. Purdy, "*Applications of digital signal processing to radar*", Chapter 5 in "*Applications of Digital Signal Processing*", A.V. Oppenheim (ed.), Prentice-Hall, 1978.

[4] D.C. Schleher, "*MTI and Pulsed Doppler Radar*", Artech House, 1991.

[5] R.K. Montoye, E. Hokenek, and S.L. Runyon, "*Design of the IBM RISC System/6000 floating-point unit*", IBM Jour. Res. and Devel., vol 34, no. 1, pp. 59-70, Jan. 1991.

An Implementation of a High-Level Language for a Data-Driven Processor

Tetsuo Yamasaki *, Koichi Munakata *, Kenji Shima *,
Shin'ichi Yoshida **, and
Hiroaki Terada ***
Mitsubishi Electric Corporation *
Sharp Corporation **
Osaka University ***

Abstract

Data-driven processors are suitable for multi-processor systems, whose machine codes (dataflow graphs) are readily subdivided for allocation to multiple processors.

We have been developing a superpipelined data-driven processor and its development environment, a high-level language (FGCL) and its language processing system.

FGCL has several features particularly suitable for signal processing and parallel processing such as parallel execution statements, a data transfer mechanism between executions of a program with multiple input data-sets, and delay and history classes.

1. Introduction

The computing performance of processors has been improving at a consistently rapid pace in recent years, but more computing power is desired for massive calculations in fields which include signal processing, image processing, computer graphics and simulations of natural phenomena. In order to satisfy these demands, multiprocessor systems have been implemented for practical use, as well as parallel and distributed processing. But designing a software that fully utilizes these systems' potentials is much more difficult than for single-processor systems. Data-driven processors are suitable for parallel processing on a multiprocessor configuration. This is because data are processed in an order that is dependent only upon the data's requirements, and multiprocessor configuration is possible because dataflow graphs, which correspond to the machine codes, are readily subdivided for allocation to processors.

We have been developing a superpipelined data-driven processor, in addition to its development environment [1]~[9]. A high-level language for our data-driven processor and its language processing system, implemented on an engineering work station, are described in this paper.

There are two alternatives in selecting a language for a new data-driven processor architecture. One is to select a language designed especially for the data-driven processor. The other is to adopt a conventional language for the data-driven processor. We chose the latter because most programmers are not willing to use an unfamiliar language, and would prefer the most familiar languages, for example, the C language, to solve their problems. Thus, we intend to maintain compatibility with the C language. This makes it possible to utilize the ever accumulating software resource, and makes it easier for users to utilize the new processor, which features a unique architecture. Our language, which we have coined as "Flow Graph C Language" (FGCL), has several features suitable for parallel processing and signal processing other than compatibility with C.

2. Features of FGCL

2.1 Subset of C language

Barring a few exceptions [10], languages for data-driven processors must observe the single-assignment rule, a rule that is unfamiliar to many users. With FGCL, users do not have to follow the rule, and can write programs just like those written in C language.

In FGCL, some features of C are not implemented, such as the goto-statement, label, and "a pointer to a function", because according to the data-driven principle, the order of execution should be determined only by data dependency and not by control. It would take more effort to implement these features in data-driven architecture than in conventional control-driven architecture, moreover it would depart significantly from the data-driven principle.

2.2 Extended features

a)Delay-class

Delay-class is introduced in order to pass values from one generation to the next in the execution of a program with multiple data-sets. For example, delay-class would make it easier to describe signalflow graphs as shown in figure 1. A generation-identifier is one of the identifiers stored in packets on data-driven processors for the identification of execution environments, such as the order of the input data-sets. Each data-set is a set of data which is needed to execute a program once.

b)History-class

History-class is introduced in order to transfer values from one generation to another. There can be one or more differences between those generations and the differences are specified by the declaration shown in

1058-6393/91 $01.00 © 1991 IEEE

figure 2.

c)Fstruct

Fstruct is introduced in order to enhance describability of function arguments. Because FGCL permits users to declare output parameters of functions and requires parameters to be written explicitly, parameter-lists are likely to become long by listing all the interfacing variables instead of concealing them behind a pointer. Figure 3 is an example, which shows that expression-A is equivalent to expression-B.

```
example_delay(u;y)
float     u, y;
{
        delay float d1=0.0, d2=0.0;
        float     c;
        c = u-B1*d1-B2*d2;
        y = (c+A1*d1+A2*d2)*A0;
        d1 = c;  -----------------  *
        d2 = d1;  ----------------  *
        return(y);
}
```

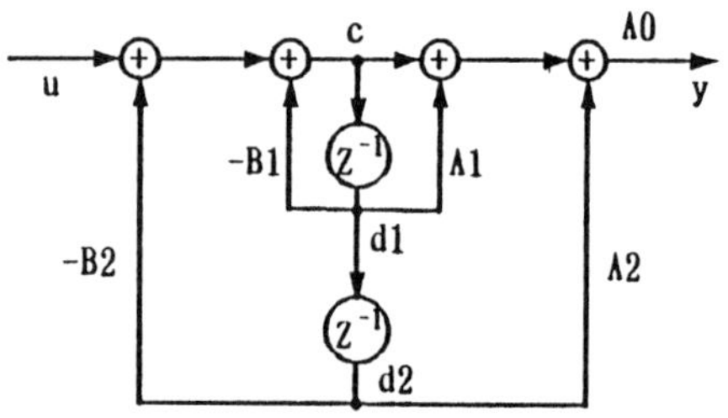

*)Values are passed to delay-class variables in the next generation.

Fig. 1. Signalflow graph represented by equations with delay-class variables, d1 and d2.

```
history int A[2][3]#3# = {{{1,2,3},{4,5,6}},         /* 1st generation */
                          {{7,8,9},{10,11,12}},      /* 2nd generation */
                          {{13,14,15},{16,17,18}}}   /* 3rd generation */
    A[0][0] = A[0][0]+x;
```

[execution manner]

generation	x	A[0][0](before updating)	A[0][0](after updating)
1	1	1	2
2	2	7	9
3	3	13	16
4	4	2	6
5	3	9	12
6	2	16	18
7	1	6	7
8	0	12	12

Arrows represent the transfer of values between generations whose gap is specified as 3 in the declaration.

Fig. 2. Execution manner of a equation containing an array A, a history-class variable.

d)Sequential-execution construct

An example of a sequential-statement is shown in figure 4. statement1 through statement3 are designed to be executed sequentially. Statements 0, 4, and 5 are executed in parallel, where its order of execution is undeterminable if there is no data dependency among them.

e)Parallel-execution construct

The para-statement is introduced to explicitly declare the program's parallel processes.

3. FGCL language processing system

3.1 Implementation of FGCL

a)Assignment of variables

Auto-type variables are treated as "flow variables" (data in pipeline stages, matching memory and queues in a data-driven processor, namely active data), those corresponding to register variables in conventional systems, and other variables are allocated in memory.

b)Memory synchronization

One of the biggest problem in data-driven processors, as in parallel processing systems, is to avoid the side effects of memory access procedures. In this implementation, the sequence of memory access described in the program is assured by the language processing system. But here, memory synchronization is performed on the condition that memory pointers have the same identifier or synchronization is performed between variables which have the same identifier. Thus if a memory pointer is assigned to another memory pointer, synchronization of the two is not performed even if both pointers are pointing to the same memory address.

Synchronization is performed as follows, with each case shown in figure 5. The underlying principle of the method is that nodes which only refer to memory can be executed in parallel as shown in figure 5(a). And if there is a node that updates memory which precedes

```
/* Expression-A */
fstruct variable_set {
                    int     b;
                    short   c;
                    int     *d;
                  } a;
int     w, x;
func(a, w;x);

/* Expression-B */
int     b;
short   c;
int     *d;
int     w, x;
func(b, c, d, w;x);
```

Fig. 3. Example of fstruct used in parameter description of function call.

memory reference nodes, the updating node should be executed only after the execution of all the nodes which refer to memory as shown in figure 5(b). A node referring to memory should receive a pointer to memory from the nearest preceding node which updates memory as shown in figure 5(c).

An asynchronous memory pointer variable can be used by declaring the pointer as "asynchro". This variable system does not perform synchronization of memory access. In this case, the user is responsible for ensuring the consistency of memory access.

c)Switch-statement

Each "case" part in the switch-statement is evaluated in parallel. Processing time necessary to determine which case is to be executed is independent of the number of cases and thus is a constant.

d)Para-statement

If no data dependency exists for each repetition of a for-statement and the number of times repeated is statically determined, the para-statement can be applied so that each repetition can be executed in parallel. This makes the process's execution faster than that for the for-statement because in this implementation, loop-constructs are synchronously executed in order to prohibit shortage of color-identifier which is used to identify packets in repetition stages in parallel execution of loop-constructs.

Example-A shown in figure 6 expresses the following calculation.

square = 1*1 + 2*2 + ······ + 9*9;

Each multiplication in it is executed in parallel. Example-B shown in figure 6 expresses the following calculations:

square[1] = 1*1; ······ square[9] = 9*9;

They are executed in parallel. After all of the calculations are completed, a pointer to the variable "square" is output from the para-statement block with synchronization.

e)Fstruct class

Fstruct is implemented as a structure whose members are not allocated to memory. In conventional system terms, this is same as when all members are in registers.

f)Loop-constructs

In loop-constructs, dataflow is synchronously controlled as shown in figure 7. The for-body block corresponds to statements executed repeatedly in a for-statement and the for-gating block controls whether to repeat or end the loop depending upon the result from the for-condition block. Each repetition of a for-statement is synchronized with a for-gating block.

g)Control constructs

Break-statements, return-statements, and continue-statements are not desirable to the dataflow execution mechanism, but in C programs, they are frequently used. Thus for compatibility with C language, we have allowed the use of these expressions.

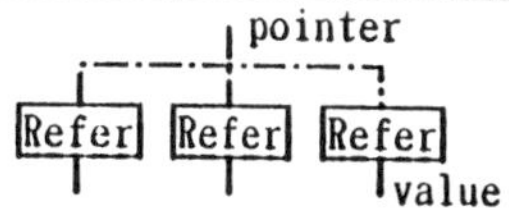

(a) Referring nodes are executable in parallel.

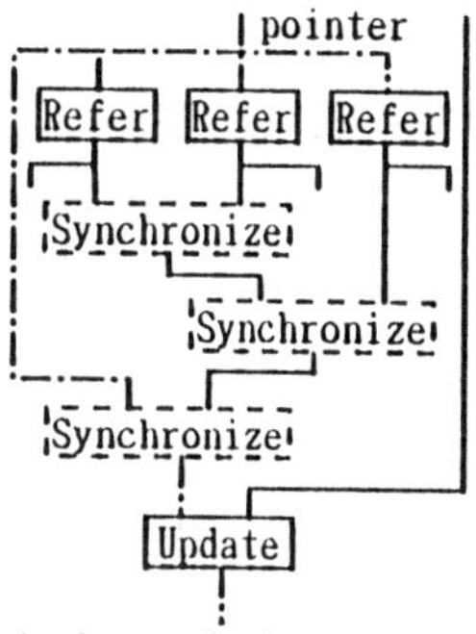

(b) Updating node is executable after execution of all the preceding nodes have completed their reference to memory.

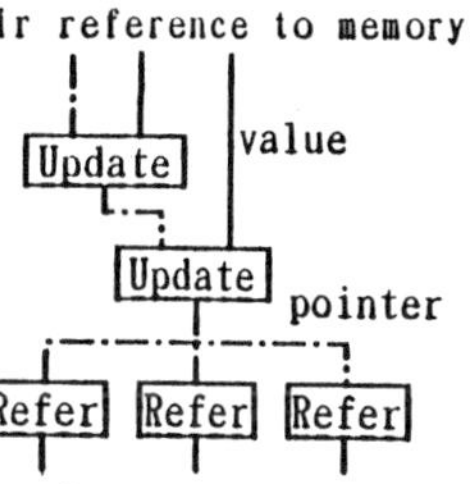

(c) Referring node receives a pointer from the nearest updating node.

Fig. 5. Synchronization of memory access.

```
statement0;                              Execution order
sequential{
          statement1;  statement0  statement1  statement4  statement5
          statement2;                   ↓
          statement3;              statement2
}                                       ↓
statement4;                        statement3
statement5;
```

Fig. 4. Execution manner of a program including a sequential-statement.

```
/* Example-A */
para(i = 1; i < 10; i++) {
    square = i*i;
} sum(square);

/* Example-B */
int    square[10];
para(i = 1; i < 10; i++) {
    square[i] = i*i;
} nosum(square);
```

Fig. 6. Examples of para-statement.

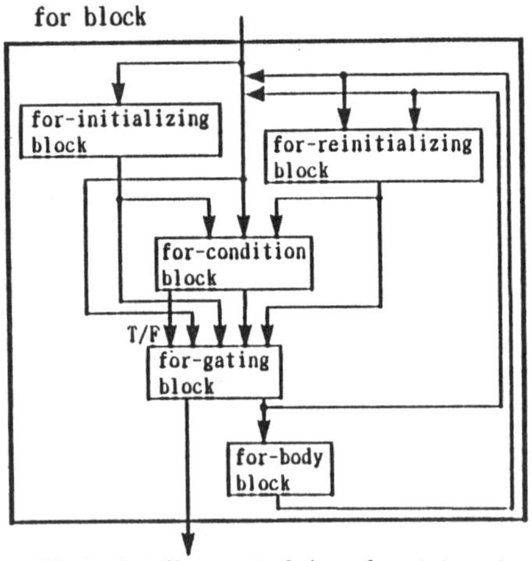

Fig. 7. Dataflow control in a for-statement.

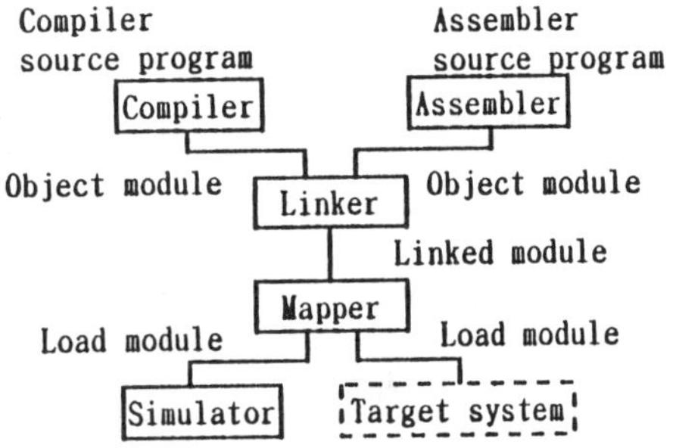

Fig. 8. Configuration of FGCL language processing system.

```
main(x, a1, a2, a3, a4;y, z)
int      x, a1, a2, a3, a4, y, z;
{
        y = a1*x*x*x + a2*x*x + a3*x + a4;
        z = a1*x*x*x - a2*x*x - a3*x - a4;
        return(y, z);
}
```

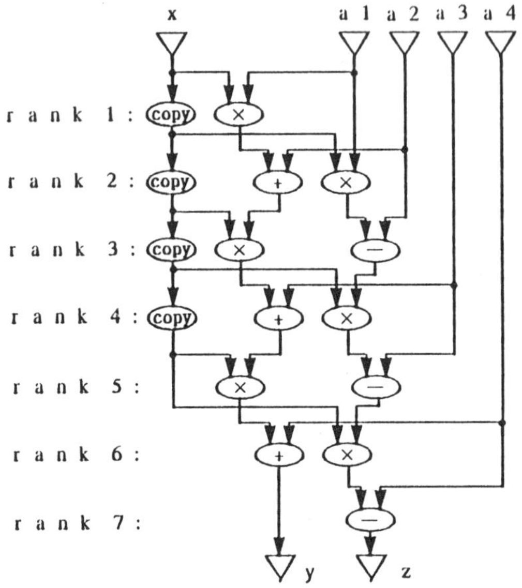

Fig. 9. Example of dataflow graph with the calculated rank.

3.2 Configuration of the language processing system

The flow of the FGCL language processing system is shown in figure 8 in which a load module is produced from a combination of FGCL source programs and assembly-language representations.

a) Compiler

The compiler extracts the parallel executable parts from programs, which sequentially and procedurally describe algorithms to solve target subjects, according to data dependencies. At first, syntax analysis and dataflow analysis are performed, followed by organization of control structures by dataflow and by operations needed for control. Then memory access synchronization and rank analysis of each function module for node numbering procedure are performed.

b) Linker

The linker links dataflow graphs and calculates the rank of the linked dataflow graph from partially pre-calculated ranks of each dataflow graph to prepare it for assigning node number in the mapper. An example of dataflow graph with the calculated rank is shown in figure 9.

c) Mapper

The mapper divides and allocates program nodes and data in memory to multiple processors according to user's specifications such as node-level specification and function-level specification. Allocation can also be done automatically by specified algorithms, such as vertical and horizontal division of nodes, depending on the configuration information of the processor system. It is easy to divide a dataflow graph into smaller pieces, but some attention must be paid in allocating them to memory. In this implementation, nodes accessing the same data in memory are allocated to one processor. This is due to the way memory access in the processor is handled where packets (in the form of nodes in the execution on data-driven processors) which accessed a memory unit are not allowed to receive information enabling them to go to any other processor except the processor with the memory unit. After the division and allocation of nodes and memory to multiple processors, the mapper numbers the nodes according to the rank determined in the linker. The mapper also generates information of distribution of nodes with respect to ranks. An example is shown in figure 10. Users can determine how to map nodes to multiple processors by referring to this information.

d) Simulator

The simulator executes the load modules. Simulation is performed exactly on the basis of the data-driven processor's architecture shown in figure 11. It also has program debugging functions, such as break-point execution, step-by-step execution, execution trace, in order to debug parts of programs.

e)Assembler

Assembly-language is also available in programming. Users can directly describe the dataflow graph with it, and can describe critical portions of programs.

3.3 Procedures in Compilation

The compiler is composed of three sections, pass1 and pass2.

3.3.1 Pass1

1)Extraction of constructs

The control structures in programs, such as branch-constructs, loop-constructs, and formula are extracted and put into the corresponding templates, which are called blocks. Basically, a block is composed of an input interface part, an output interface part, and a body part. A formula block is the simplest of all blocks. The others have inner nested blocks, which are called sub blocks.

2)Transformation of formula into dataflow graph

Statements in formula blocks are transformed into dataflow graphs. If only a formula is included inside the body of a block, that becomes a formula block. Statements in the body are analyzed into dataflow-graph-style information, in which operations are represented by nodes and data dependencies are represented by arcs between nodes, variables referred in the statements are listed on the input interface part of the block, and variables updated in the statements are listed on the output interface part.

3)Connection of dataflow between blocks

Dataflow between blocks is connected as follows. At first, in the deepest part of nested blocks, variables on the input interface part are added to the input interface part of the nesting block, and variables on the output interface part are added to the output interface part of the nesting block. The procedure is iterated until it reaches the top block. After that, from the lowest data stream of the dataflow graph to the upper data stream, dataflow is made by connecting data on the same type interface.

3.3.2 Pass2

1)Synchronization of memory access

Synchronization of variables allocated to memory, or variables pointing to memory are performed as described in section 3.1(b).

2)Rank analysis of nodes for nodes numbering

In object modules, each node is numbered according to data dependencies. The nodes in the upper stream of the dataflow graph should have smaller numbers than those in the lower stream of the graph in our architecture in order to avoid hash conflicts.

3.4 Remained subjects

Some restrictions on the C language remain mainly because of differences between data-driven execution mechanism and control-driven execution mechanism found in traditional architecture.

Global variables concealed in a function, which are called from another function, cannot be recognized when those functions are separately compiled. In this case, data paths for global variables is necessary, which are allocated on memory, and perform memory synchronization in the linking procedure. And furthermore, if function calls are made in constructs such as loops and branches, data paths have to be controlled according to the construct. In the present version of our compiler and linker, global variables are allowed, but those requiring control of data paths for program constructs are not supported. Full support of global variables with less efforts in linker has to be considered.

4. Conclusions

Implemented language processing system cannot fully extract parallel-executable portions hidden in source programs. Therefore users have to write programs being conscious of parallel procedure to utilize the maximum ability of a data-driven processor. However, besides compatibility with the conventional language, FGCL is extended with several specific features for parallel processing and signal processing to exploit the potential ability of a data-driven processor.

```
-------------------------------------------------------------------
total ranks =        7
total nodes =       14
-------------------------------------------------------------------
execution | total | 0   10   20   30   40   50   60   70   80   90   100(%)
  rank    | nodes |
    1     |   1   | *********
    2     |   5   | ******************************************
    3     |   2   | *****************
    4     |   1   | *********
    5     |   2   | *****************
    6     |   2   | *****************
    7     |   1   | *********
-------------------------------------------------------------------
```

Fig. 10. Distribution of nodes against ranks.

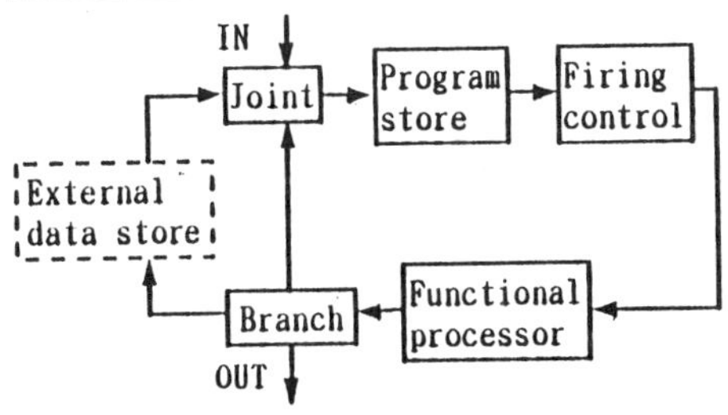

Fig. 11. Block diagram of the data-driven processor.

Data-driven processors and FGCL will serve as a useful platform for building applications for multiprocessor systems. This is because dataflow graphs are readily divided for processors, execution of dataflow graphs with multiprocessors are synchronized automatically by data's flow across them, and FGCL is familiar to use for C programmers.

We have implemented the language processing system, and are now working on its test and evaluation. Evaluations are performed upon an emulation system, and will soon be realized on a real system composed with superpipelined data-driven processors [9]. Other subjects to be taken up in the future are the optimization of loop constructs, extraction of more parallel-processing portions hidden in sequentially described programs, and improvement in the synchronization procedure of memory access. Mapping method is also an important subject for multiprocessor systems to utilize their potential ability.

Acknowledgments

The authors are grateful to T. Okamoto, K. Yoda, and Dr. S. Miyata (Sharp Corporation) for their participation in this work. The authors are also grateful to Dr. I. Nakahori (Mitsubishi Electric Corporation) for his encouragements on this study.

References

[1]T. Okamoto, S. Yoshida, T. Yamasaki, K. Shima, H. Terada, et al., "An implementation method of a high level language for data-driven processor (in Japanese)," in Proc. JSPP'89, pp. 227-233, Feb. 1989.

[2]H. Nishikawa, H. Terada, K. Shima, et al., "Architecture of a one-chip data-driven processor: Q-p," in Proc. 16th Int'l Conf. Parallel Processing, pp. 319-326, Aug. 1987.

[3]K. Asada, H. Terada, K. Shima, et al., "Hardware structure of a one-chip data-driven processor:Q-p," in Proc. 16th Int'l Conf. Parallel Processing, pp. 327-329, Aug. 1987.

[4]H. Terada, K. Shima, et al., "VLSI design of a one-chip data-driven processor:Q-v1," in Proc. Fall Joint Comput. Conf., pp. 594-601, Oct. 1987.

[5]S. Komori, T. Yamasaki, K. Shima, H. Terada, et al., "An elastic pipeline mechanism by self-timed circuits," IEEE J. Solid-State Circuits, Vol. 23, No. 1, pp. 111-117, Feb. 1988.

[6]T. Yamasaki, K. Shima, H. Terada, et al., "VLSI implementation of a variable-length pipeline scheme for data-driven processors," IEEE J. Solid-State Circuits, vol. 24, no. 4, pp. 933-937, Aug. 1989.

[7]S. Komori, T. Yamasaki, K. Shima, H. Terada, et al., "A 40-Mflops 32-bit floating-point processor," in Dig. Tech. Papers IEEE ISSCC, pp. 46-47, Feb. 1989.

[8]S. Komori, K. Shima, H. Terada, et al., "The data-driven microprocessor," IEEE MICRO, pp. 45-59, June 1989.

[9]S. Komori, H. Tarada, et al., "A 50 MFLOPS Superpipelined Data-Driven Microprocessor," in Dig. Tech. Papers IEEE ISSCC, pp. 92-93, Feb. 1991.

[10]S. Sekiguchi, et al., "Implementation of a dataflow parallel language DFC II (in Japanese)," in Proc. JSPP'90, pp. 297-304, May 1990.

Fixed Size Array Architectures for Computing Arithmetic Fourier Transform*

Heonchul Park

Department of EE-Systems, EEB-243
University of Southern California
Los Angeles, CA 90089-2562
hpark@halcyon.usc.edu

Viktor K. Prasanna

Department of EE-Systems, EEB-244
University of Southern California
Los Angeles, CA 90089-2562
prasanna@halcyon.usc.edu

Abstract

We propose modular and area efficient VLSI architectures for computing the Arithmetic Fourier Transform (AFT). The proposed design employs $2p + 1$ PEs and external memory of size $O(N)$ to compute $2N + 1$ Fourier coefficients, $1 \leq p \leq N$. Each PE has an adder and a fixed amount of local storage and one PE has a multiplier. I/O with the host is performed using a fixed number of channels. By suitable choice of I/O schedule and activation of PEs, non-uniform data dependencies in the AFT computation which require non-equidistant inputs and assignment of Möbius function values are resolved. The design achieves linear speed-up and can also be used to compute the Discrete Cosine Transform.

1 Introduction

Recently, Tufts and Sadasiv independently discovered the Arithmetic Fourier Transform (AFT) [8] which had been developed by Bruns [1] to compute the coefficients of the Fourier series of a periodic function using the Möbius inversion of series. Reed *et al.* [5] extended the AFT algorithm to compute the Fourier coefficients of both even and odd components of a periodic function. In [5], they employed zero-order interpolation to improve the efficiency of the computation and to reduce the number of inputs. The AFT algorithm was extended in [6] by employing symmetrical computing sequences to obtain the real and imaginary parts of the Fourier coefficients. To compute the Fourier coefficients using the algorithm in [6], $2N$ alternating averages are computed by applying zero-order interpolation to the inputs. The Fourier coefficients are computed by applying the Möbius inversion formula to the alternating averages. The AFT algorithm using the zero-order interpolation requires $O(N^2)$ additions for computing the alternating averages and the Fourier coefficients, and $O(N)$ multiplications for scaling the alternating averages.

A VLSI design for computing the AFT is proposed in [6]. The design is based on a tree architecture and each node has butterfly structure. To compute $2N + 1$ point Fourier coefficients, $O(N^2)$ PEs are required with $O(N)$ PEs having a multiplier in them. Also, broadcast of input data is performed.

In this paper, we propose modular and area efficient VLSI design for implementing the AFT algorithm based on the algorithm in [6]. By careful design, non-regular data dependencies which require non-equidistant inputs and Möbius function value assignments in the computation are resolved. Our design leads to modular linear arrays with constant hardware in each PE, and few control lines. It consists of two sets of p PEs and one multiplier with fixed amount of local storage in each PE. Also, it uses $O(N)$ external memory for storing the scaling factors, Möbius function values, and intermediate results. All the data and control signals flow from left to right. Our design requires fixed I/O bandwidth with the host and takes $2N^2/p + 3p + \lceil p/2 \rceil$ time units to finish the computation. Compared with conventional FFT/DFT architectures, the VLSI area occupied by a PE is significantly reduced, since no multiplier is employed in the PE. The DCT algorithm adopted from the AFT has similar complexity as the AFT. The proposed design can compute the DCT with minor modifications.

This paper is organized as follows. In the next Section, we briefly introduce the AFT algorithm. De-

*This research was supported in part by NSF under grant IRI-9145810.

tailed design for computing the AFT is presented in Section 3. Concluding remarks are made in Section 4.

2 Arithmetic Fourier Transform

In this section, we briefly introduce the AFT and an algorithm to compute it which is used in our design. Details can be found in [5] and [6].

To compute the AFT for given $2N$ input samples, $2N$ alternating averages are computed using zero-order interpolation, and an average over $2N$ input samples is computed. These averages are scaled by multiplying with certain scaling factors. $2N+1$ Fourier coefficients are computed by applying the Möbius inversion formula to the scaled alternating averages. The AFT algorithm is derived by the following definitions and theorem. The Möbius function $\mu(n)$ is defined by

$$\begin{aligned} \mu(n) &= 1 && \text{if } n=1 \\ &= (-1)^r && \text{if } n = p_1p_2\cdots p_r \\ & && \text{where } p_i\text{'s are distinct primes} \\ &= 0 && \text{if } p^2|n \text{ for some prime } p \end{aligned}$$

where $p^2|n$ means integer p^2 divides integer n.

The following Möbius inversion formula can invert certain series of functions under given conditions. It plays a key role in the computation of the Fourier coefficients.

Möbius inversion formula [2]: Assume that n is a positive integer, $f(n)$ is a nonvanishing function in the interval, $1 \le n \le N$, and $f(n) = 0$ for $n > N$. Also, if $g(n) = \sum_{m=1}^{\lfloor N/n \rfloor} f(mn)$ then $f(n) = \sum_{k=1}^{\lfloor N/n \rfloor} \mu(k)g(kn)$ where $\lfloor x \rfloor$ denotes the integer part of real number x.

Denote $A(t)$ as a finite Fourier series of period T. $A(t)$ is as follows. $A(t) = \sum_{n=-N}^{N} \alpha_n e^{i2\pi n f_0 t}$ where $f_0 = 1/T$ and α_n is a coefficient of nonvanishing function in the interval, $-N \le n \le N$, and $\alpha_n = 0$ for $n > N$ and $n < -N$.

The coefficients α_n are given by:

$$\alpha_n = \frac{1}{T}\int_0^T A(t)e^{-i2\pi n f_0 t}dt$$

for $n = 0, \pm1, \pm2, \cdots, \pm N$. If the mean α_0 is removed from $A(t)$, then

$$\overline{A}(t) = \sum_{n=1}^{N} a_n \cos 2\pi n f_0 t + \sum_{n=1}^{N} b_n \sin 2\pi n f_0 t$$

where a_n and b_n are real and imaginary parts of the Fourier coefficients, respectively. If the periodic function $\overline{A}(t)$ is shifted in time by αT, $-1 < \alpha < 1$, then

$$\overline{A}(t+\alpha T) = \sum_{n=1}^{N} c_n(\alpha)\cos 2\pi n f_0 t + \sum_{n=1}^{N} d_n(\alpha)\sin 2\pi n f_0 t$$

where $c_n(\alpha) = a_n \cos 2\pi n\alpha + b_n \sin 2\pi n\alpha$ and $d_n(\alpha) = -a_n \sin 2\pi n\alpha + b_n \cos 2\pi n\alpha$. The Fourier coefficients a_n and b_n can be derived from $c_n(\alpha)$. If an alternating average $B(2n,\alpha)$ is defined using input sequence $A(t)$, then $c_n(\alpha)$ can be represented by the series $B(2n,\alpha)$. $B(2n,\alpha)$ denotes the $2n$th average of the $2n$ values $A(mT/2n + \alpha T), 0 \le m \le 2n-1:$

$$B(2n,\alpha) = \frac{1}{2n}\sum_{m=0}^{2n-1}(-1)^m A(mT/2n + \alpha T)$$

Using the Möbius inversion formula for finite series $B(2n,\alpha)$, the coefficients $c_n(\alpha)$ are given by

$$c_n(\alpha) = \sum_{l=1,3,\cdots}^{\lfloor N/n \rfloor} \mu(l)B(2nl,\alpha).$$

The following Fourier coefficients a_n and b_n of the periodic input function $A(t)$ are derived from $c_n(\alpha)$ by setting α equal to 0 and $1/4n$, respectively.

$$\begin{aligned} a_0 &= \frac{1}{2N}\sum_{m=1}^{2N} A(m) \\ a_n &= \sum_{l=1,3,}^{\lfloor \frac{N}{n} \rfloor} \mu(l)B(2nl,0), n = 1,\cdots,N \\ b_n &= \sum_{l=1,3,}^{\lfloor \frac{N}{n} \rfloor} \mu(l)(-1)^{\frac{l-1}{2}}B(2nl,\frac{1}{4nl}), n = 1,\cdots,N \end{aligned}$$

Similarly, the coefficients of Discrete Cosine Transform (DCT) can be obtained by multiplying a_n with $cos(n\pi/2N) - sin(n\pi/2N)$.

The alternating averages $B(2n,\alpha)$ are computed by applying zero-order interpolation to the equidistant input samples of $A(t)$. The Fourier coefficients, a_n and b_n, are computed by applying the Möbius inversion formula to the alternating averages $B(2n,\alpha)$ after multiplying with the scaling factors. Zero-order interpolation is used for computational efficiency [6]. The error which results from zero-order interpolation is tolerable (see [5] and [6]). The AFT algorithm does not require complex multiplications which are needed in the conventional FFT algorithms. The algorithm has $O(N^2)$ time complexity on a serial machine with $2N$ inputs.

Notice that the AFT can be computed using matrix-vector multiplications like the DFT. However, if N PEs are employed to obtain $O(N)$ speed-up, it will lead to an inefficient design since $O(N^2)$ memory is required to store two $N \times 2N$ coefficient matrices for zero-order interpolation and two $N \times N$ coefficient matrices for storing the Möbius function values. Also, $O(N)$ I/O bandwidth with the host is needed to achieve $O(N)$ speed-up [3].

3 Fixed Size Linear Arrays

In this section, fixed size arrays for computing the AFT are designed which lead to linear speed-up with $2p+1$ PEs, $1 \le p \le N$ are presented. For sake of simplicity, assume that N is divisible by p.

A block diagram of fixed size arrays is shown in figure 1. The alternating average $B(2n, \alpha)$, $1 \le n \le N$, is computed in $PE_{1,n \bmod p}$ during the $\lceil n/p \rceil$th iteration. PE_{mul} computes the Fourier coefficient a_0 and performs multiplication with the scaling factors. The Fourier coefficients are computed in $PE_{2,i}$, $1 \le i \le p$. The external memory for storing the scaling factors is an $N \times 1$ array with the ith location having $\frac{1}{2i}$, $1 \le i \le N$. The external memory for storing the Möbius function values is an $\lfloor N/2 \rfloor \times 1$ array with the ith location having $\mu(2i+1)$, $1 \le i \le \lfloor N/2 \rfloor$. Thus, $O(N)$ memory is required to store the coefficients. B_I is used to store the input window, B_F is used to keep the intermediate results of the computation and the Möbius function values fed back from the rightmost PE in $group_2$. Since p alternating averages are computed every $2N$ cycles and the time for computing the partial sums of the Fourier coefficients in $group_2$ is less than $2N$ units, the PEs in $group_2$ should wait until the next p alternating averages become available.

The data in B_I is fed into the leftmost PE in $group_1$ N/p times. Every $2N$ cycles, p alternating averages are computed. These outputs are scaled in PE_{mul} and fed into the leftmost PE in $group_2$. When the ith block of p alternating averages, $B(2[(i-1)p+1], \alpha), \cdots, B(2ip, \alpha)$, is available in $group_2$, the data in B_F is fed into $group_2$. The intermediate data from the $PE_{2,p}$ is fedback into B_F. New Möbius function values in the external memory are fed into the array, after the last Möbius function value in B_F is input to the array. Notice that $B(2n, \alpha)$ corresponds to the Fourier coefficient a_n and b_n, if $n > N/3$. The p PEs in $group_2$ compute the Fourier coefficients $a_1, b_1, \cdots, a_{N/3}, b_{N/3}$ using $2N$ alternating averages.

Figures 2 and 3 show the internal structure of the PEs in $group_1$ and $group_2$, respectively. Data channels with different speeds are employed to transport the data between adjacent PEs. These data channels can be implemented by delay elements within the PEs. Data in the slow channels takes two or three time units to pass through a PE, while the data in the fast channel takes one time unit to pass through a PE. F_* and S_* denote the fast and slow channels, respectively. In $group_1$, input data is fed into the leftmost PE through the data channel F_d. The data channel F_x carries the index corresponding to the input data. Alternating averages are transferred to PE_{mul} through F_a and F_b. In $group_2$, the Fourier coefficients a_0, $a_i(b_i)$, $N/2+1 \le i \le N$, are obtained through $F_{a1}(F_{b1})$. The rest of coefficients $a_i(b_i)$, $1 \le i \le N/2$, are obtained form $S_a(S_b)$. The slow channels S_{xa} and S_{xb} carry the index corresponding to the data in S_a and S_b, respectively. The Möbius function values used for computing the real(imaginary) coefficients travel through two bits channels $F_{ma}(F_{mb}), S_{ma}(S_{mb})$.

Notice that the registers in the slow channels can be overwritten, if the data moves at a fixed speed in the slow channel and the number of registers in the slow channels is fixed. This imbalance in data flow occurs since the PEs in the left side in $group_2$ send more number of Möbius function values to the slow channel than those on the right side. To avoid this situation, the distance between two consecutive partial sums (the number of registers between two consecutive partial sums in the slow channel) should increase, as the computation proceeds. The control of the distance between two consecutive partial sums is achieved by an extra delay element in the slow channel.

The operations of the PEs in $group_1$ and $group_2$ are shown in figures 4 and 5, respectively. Additional details can be found in [4].

The time to compute $2N$ alternating averages in the first group of p PEs is $2N^2/p+2p-1$ units. The total time for computing the $2N+1$ Fourier coefficients is $2N^2/p+3p+\lceil p/2 \rceil$ units.

Let the number of data arrivals per unit time be R_I. Suppose the number of data processed in the array per unit time is R_C. Since the computation performed by the PEs in $group_1$ takes more time than that performed by the PEs in $group_2$, the number of PEs needed for a given real-time application is determined by the execution speed of the first group of p PEs. The number of PEs, p, should be chosen to satisfy:

$$(p/N)R_C \ge R_I$$

The above design can be extended to compute the DCT by adding a multiplier at the end of PEs in $group_2$. The performance is similar to the computation of AFT. The hardware overhead is one additional multiplier and storage for the coefficients needed for scaling.

4 Conclusions

We have derived linear arrays for computing the AFT and the DCT. The proposed design for computing the AFT has $2p$ PEs and one multiplier, $1 \leq p \leq N$. The design uses simple PEs (without multipliers) and a small external memory for storing the coefficients, compared with the known DFT/FFT VLSI designs. The fixed size array can compute the AFT with less than $2N$ PEs as long as sufficient external memory is available. Also, few control signals are needed. To process $2N$ inputs, the array takes $O(N^2/p)$ time units. The designs are modular; extending the design to solve larger size problem does not affect the internal structure of the PEs.

References

[1] H. Bruns, *Grundlinien des Wissenschaftlichichnen Rechnens.* Leipzig: B. G. Teubner Velag, 1903.

[2] L. K. Hua, *Introduction to Number Theory.* Springer-Verlag: New York, 1982.

[3] S. Y. Kung, *VLSI array processors,* Prentice-Hall, 1988.

[4] H. Park and V. K. Prasanna, "Modular VLSI architectures for computing arithmetic Fourier transform," *IRIS # 273, Tech. Rep., USC EE-Systems Dept.,* Aug. 1991.

[5] I. S. Reed, D. W. Tufts, X. Yu, T. K. Truong, M. Shih, and X. Yin, "Fourier Analysis and Signal Processing by Use of the Möbius Inversion Formula," *IEEE Trans. ASSP.,* Vol. 38, No. 3, pp. 458-470, Mar. 1990.

[6] I. S. Reed, M. T. Shih, T. K. Truong, E. Hendon, and D. W. Tufts, "A VLSI Architecture for Simplified Arithmetic Fourier Transform Algorithm," *Proc. Int. Conf. Application Specific Array Processors,* 1990.

[7] D. W. Tufts and G. Sadasiv, "The Arithmetic Fourier Transform," *IEEE ASSP Mag.,* pp. 13-17, Jan. 1988.

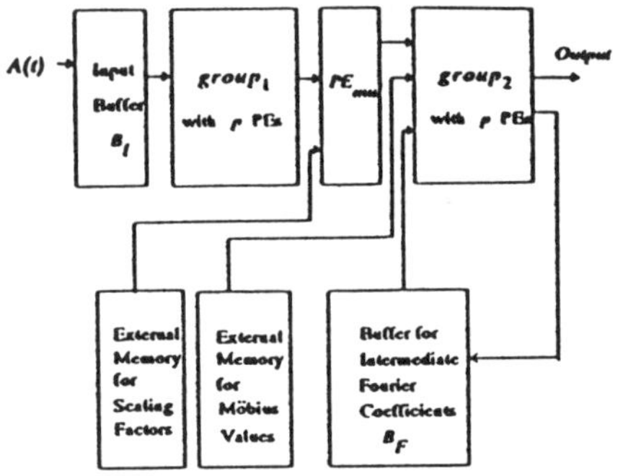

Figure 1: Overall block diagram of fixed size array

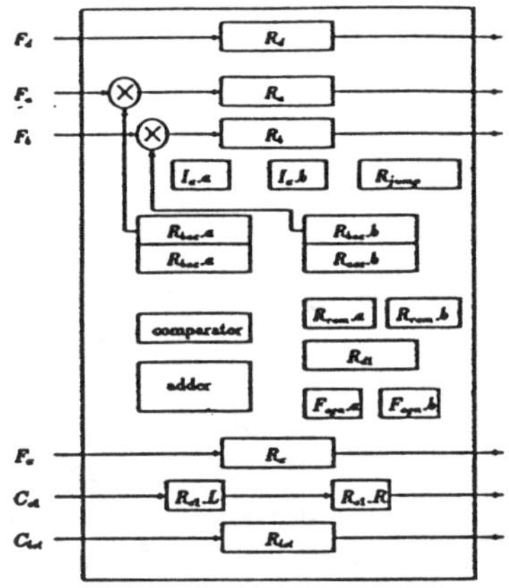

Figure 2: Structure of PE in $group_1$

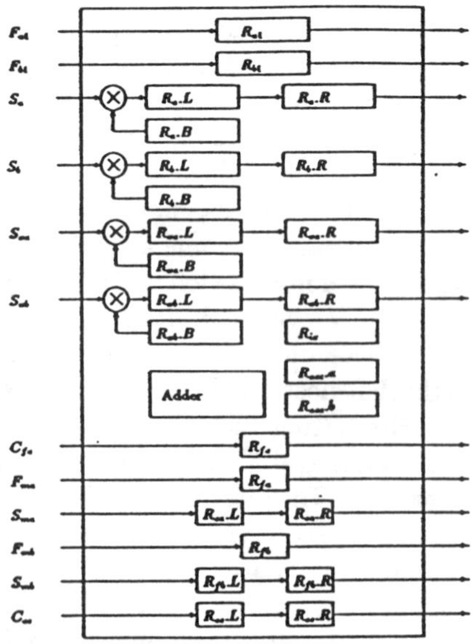

Figure 3: Structure of PE in $group_2$

//Computing $B(2n, 0)$
if $R_x - I_x.a \geq 0$ then
begin
if $|R_{rem}.a| > |R_x - I_x.a|$ then
$R_{acc}.a \leftarrow R_{acc}.a + F_{sgn}.a \times R_d$
else
$R_{acc}.a \leftarrow R_{acc}.a + F_{sgn}.a \times R_{d1}$
$I_x.a \leftarrow I_x.a + R_{jump}$;
if $F_{sgn}.a = 1$ then $F_{sgn}.a \leftarrow -1$
else $F_{sgn}.a \leftarrow -1$
end
else $R_{rem}.a \leftarrow R_x - I_x.a$
//Computing $B(2n, 1/4n)$
if $R_x - I_x.b \geq 0$ then
begin
if $|R_{rem}.b| > |R_x - I_x.b|$ then
$R_{acc}.b \leftarrow R_{acc}.b + F_{sgn}.b \times R_d$
else
$R_{acc}.b \leftarrow R_{acc}.b + F_{sgn}.b \times R_{d1}$
$I_x.b \leftarrow I_x.b + R_{jump}$;
if $F_{sgn}.b = 1$ then $F_{sgn}.b \leftarrow -1$
else $F_{sgn}.b \leftarrow -1$
end
else
begin
$R_{rem}.a \leftarrow R_x - I_x.a$;
$R_{d1} \leftarrow R_d$
end
//Transporting the computed data to PE_{mul} and initializing the PE
if $R_{lst} = 1$ then
begin
$R_{bac}.a \leftarrow R_{acc}.a$;
$R_{bac}.b \leftarrow R_{acc}.b$;
$R_{acc}.a \leftarrow 0$;
$R_{acc}.b \leftarrow 0$;
$I_x.a \leftarrow 0$;
$I_x.b \leftarrow \frac{1}{2}R_{jump}$;
$F_{sgn}.a \leftarrow 1$;
$F_{sgn}.b \leftarrow 1$
end
if $R_{s1}.L = 1$ then
begin
$R_a \leftarrow R_{bac}.a$;
$R_b \leftarrow R_{bac}.b$
end

Figure 4: Operation of PEs in $group_1$

//Loading data into accumulators
if $R_{ss}.L = 1$ then
begin
$R_{acc}.a \leftarrow R_{a1}$;
$R_{acc}.b \leftarrow R_{b1}$
end
//Initializing the data channel S_a and S_b
if $R_{fs} = 1$ then
begin
$R_a.L \leftarrow R_{acc}.a$;
$R_b.L \leftarrow R_{acc}.b$
end
//Computing the real part of the Fourier coefficients
if $R_{ix} = R_{xa}.L$ then
begin
$R_a.B \leftarrow R_a.L + R_{sa}.L \times R_{acc}.a$;
$R_a.L \leftarrow R_a.R$;
$R_{fa} \leftarrow R_{sa}.L$;
$R_{xa}.B \leftarrow R_{xa}.L + R_{xa}.R$;
$R_{xa}.L \leftarrow R_{xa}.R$
end
//Computing the imaginary part of the Fourier coefficients
if $R_{ix} = R_{xb}.L$ then
begin
$R_b.B \leftarrow R_b.L + R_{sb}.L \times R_{acc}.b$;
$R_b.L \leftarrow R_b.R$;
$R_{fb} \leftarrow R_{sb}.L$;
$R_{xb}.B \leftarrow R_{xb}.L + R_{xb}.R$;
$R_{xb}.L \leftarrow R_{xb}.R$
end
//Control the movement of the Möbius function values
if $R_{fa} \neq 2$ and $R_{sa}.R \neq 2$ and $R_{sa}.L = 2$ then
$R_{sa}.L \leftarrow R_{fa}$
if $R_{fb} \neq 2$ and $R_{sb}.R \neq 2$ and $R_{sb}.L = 2$ then
$R_{sb}.L \leftarrow R_{fb}$

Figure 5: Operation of PEs in $group_2$

Implementation Aspects of Convolvers Using Non-Binary Arithmetic

Subbu Turimella and Alexander Skavantzos

Louisiana State University, Baton Rouge, Louisiana

Abstract

The Mersenne and Fermat Number Transforms have been considered attractive tools for computing convolutions. They result in minimum number of multiplications, rely on simple scalings by powers of two and they use arithmetic modulo Mersenne or Fermat numbers. These transforms, however, suffer from some disadvantages. The transformation length is limited by the word length of the modulus used, while the choices of suitable moduli are limited too. In this paper, we consider transforms using arithmetic modulo M *with* $M=r^t \pm c$. *This way, higher transformation lengths are obtained and higher flexibility in choosing the modulus* M *is achieved. The new transforms rely on scalings by powers of* r *and in order to efficiently perform these scalings we use radix-r arithmetic (with* $r>2$*). Then, such scalings become simple shifts and rotates in the radix-r system.*

I. Introduction

"Cyclic and skew cyclic convolutions are the basic building blocks in the implementation of Finite-Impulse-Response (FIR) filters, computation of linear convolutions, auto correlation and cross correlation functions [1]. The straight forward approach for computing the cyclic convolution of two n-point sequences relies on n^2 multiplications followed by n(n-1) additions. However, the convolution property of certain transforms can be used in order to compute such cyclic convolutions using only n multiplications, scaling by constants operations and additions. The Discrete Fourier Transform (DFT) is one such transform. This transform, however, relies on complex multiplications and introduces roundoff error [2].

On the other hand, the Number Theoretic Transforms (NTTs) are transforms defined in rings of integers modulo an integer M, have the convolution property and while introducing no roundoff error at all, they still call for only n multiplications, scalings by constants and additions, when used in computing n-point cyclic convolutions [3]-[4].

Considering two n-point sequences a_i and b_i, their cyclic convolution c_i, $i = 0, 1, .., n-1$, can be computed using NTTs as follows:

1. Compute the NTTs of the sequences a_i and b_i. These NTTs are A_i and B_i with

$$A_i = < \sum_{k=0}^{n-1} a_k \alpha^{<k.i>_n} >_M$$
$$B_i = < \sum_{k=0}^{n-1} b_k \alpha^{<k.i>_n} >_M \qquad i = 0,1,...,n-1 \quad (1)$$

   where $<x>_y$ denotes the operation x modulo y and α is the nth root of unity in the ring of integers modulo M.

2. Compute the sequence C_i, $i = 0, 1, \cdots, n-1$, where

$$C_i = < A_i \times B_i >_M, \qquad i = 0, 1, \cdots, n-1 \quad (2)$$

3. Compute the Inverse NTT of C_i, which is

$$c_i = < n^{-1} \sum_{k=0}^{n-1} C_k \alpha^{-<ki>_n} >_M \quad (3)$$

The sequence c_i of equation (3) is the cyclic convolution of the sequences a_i and b_i. Equations (1)-(3) dictate that the computation of the cyclic convolution of two n-point sequences can be computed using only n real multiplications plus overhead imposed by the NTT and NTT^{-1}. This overhead consists of additions and scaling by powers of the constant α operations. However, although the operations of addition and scaling by constant are much simpler than the operation of the two-operand multiplication, the number of such operations is much larger than the number of two-operand multiplications, when using the NTT approach. As a result, efficient scaling policies should be investigated when implementing convolutions using Number Theoretic Transforms.

1058-6393/91 $01.00 © 1991 IEEE

So far, some interesting NTTs have been explored and implemented. They are the Mersenne Number Transform (MNT) and the Fermat Number Transform (FNT) [3]-[4]. The MNTs and FNTs have the advantages that the moduli M are of the form $2^t - 1$ or $2^t + 1$ and the roots of unity α are powers of two. Such advantages imply attractive implementations of the above transforms, since scaling by powers of two can be implemented with simple binary shift or rotate operations. However, these transforms suffer from some disadvantages; the transformation length is limited by the word length of the modulus used, while the choices of the suitable moduli are limited too. To solve this problem, several approaches have been suggested, with one approach suggested by Balla and Antoniou [5] who have proposed a transformation with α, (the root of unity), as three and computed the transform by using ternary modular arithmetic.

In this paper, we relax the restrictions imposed by the MNTs and FNTs further. We extend the Number Theoretic Transforms so that the moduli used are of the form

$$M = \alpha^t \pm c \qquad (4)$$

where α is the nth root of unity modulo M. This way, higher transformation lengths are obtained and higher flexibility in choosing the modulus M is achieved. Some choices of n,M, α, t and c so that the NTT transformations exist are given in [6]. These new NTTs would rely on scaling by powers of α operations rather than by powers of two. In order to efficiently perform the scalings by the powers of α, we represent our data as radix-α data instead of binary. The radix-α arithmetic is performed in the Non-Redundant Signed Digit System (NRSD). Since our data are radix-α data, scalings by powers of α will be implemented as simple shifts and rotates in our radix-α system. A brief discussion on the NRSD is offered in Section II, while the heart of our new NTTs, which is the radix-r scaler, is described in Section III. Due to restrictions in the size of the paper, the remaining components necessary for implementing convolutions based on the NTT equations (1)-(3) are not given here. These components are the binary to radix-r converter, the radix-r adder and the radix-r multiplier. Their designs are given in [6].

II. Non-Redundant Signed Digit System (NRSD)

In a number system, cardinality is the number of digit symbols in the digit set. When the cardinality of the system is higher than the radix, then the resulting number system is redundant and a number in this system can have more than one representations. The redundancy in a system can be easily achieved by including signed digits in the digit set. For example, consider a radix three system with the digit set $\{\bar{2},\bar{1},0,1,2\}$ where $\bar{d}$ indicates $-d$. Here the cardinality of the system is five and the radix is there. It can be observed that (101) and $(11\bar{2})$ both represent the decimal value 10 in such a system. When signed digits are included in the system, the system is called signed digit number system. The main advantage of the redundant signed digit number system is that it supports carry free addition. So far, in the technical literature, signed digits were only used to obtain redundancy in number systems for exploiting the above mentioned advantage. Several signed digit number systems are summarized in [7].

Alternatively, we can define a signed digit number system without redundancy, by simple taking the digit set with the symbols from $\lfloor -(r-1)/2 \rfloor$ to $\lceil +(r-1)/2 \rceil$. Although this type of system, which is called non-redundant sign digit number system (NRSD), does not support the carry free addition, it does have an interesting property; that is, the maximum possible product of two digits in this system is at most $((r-1)/2)^2$, while in the case of conventional radix-r system, that is when the symbols range from 0 to $r-1$, the product can grow up to $(r-1)^2$. Clearly, this system reduces the result by a factor of four and the difference grows polynomially with r. The use of the NRSD system is more advantageous in the cases where r is an odd integer, because the digit set is symmetric. Hereafter, whenever we consider radix-r NRSD system, we assume that r is an odd integer. Regarding the notation of the digits, since the digit set is symmetric about zero, a simple encoding would be done by using ordered pairs (s,v) for every digit in the number. Here s is a single bit for indicating the sign of the digit (1 for a negative digit and 0 for a positive) and v indicates the number of binary bits required to represent the magnitude of the digit. Notice that

$$v = \log_2 \lceil r/2 \rceil \qquad (5)$$

and that we require $v+1$ bits to represent every signed digit. A negative zero is not included in the digit set and is used as a don't care condition. The addition and multiplication rules of the NRSD system are given in [6].

III. Design of Scaler

The function of the scaler is to perform the

scaling operations $< a_i\ r^s >_M$ (dictated by (1) and (3)), where

$$M = r^t - c \qquad (6)$$

and

$$r^n \equiv <1>_M$$
$$r^{n/2} \equiv <-1>_M \text{ if } n \text{ is even} \qquad (7)$$
$$<r^t>_M = c$$

The data to be scaled, (a_i's), are assumed to be radix-r data while our arithmetic is performed in the NRSD. The scaling takes place according to the following equation

$$r^{s+p} = \qquad (8)$$

$$\begin{cases} r^{p+s-n} & \text{if } p+s \ge n & \text{case (i)} \\ -r^{p+s-n/2} & \text{if } n/2 \le p+s < n \text{ and } n \text{ even} & \text{case (ii)} \\ c \times r^{p+s-t} & \text{if } t \le p+s < n/2 & \text{case (iii)} \end{cases}$$

The scaler first calculates the new digit position $s+p$ by using (8). It then adds all the digits that map to the same position while propagating the carries appropriately. The ranges of the mapped digits are as follows:

Range 1. If either case (i) or (ii) of equation (8) applies, then the digit can range from $-(r-1)/2$ to $(r-1)/2$.

Range 2. If case (iii) of equation (8) applies, then the range of the digit is from $-(r-1)^2/4$ to $(r-1)^2/4$ due to the fact that the digit is now a product of two NRSD digits.

Since each PE has a two-digit output, (E = sum and F = carry), the maximum sum of the inputs that can be accommodated is up to:

$$r \times max(F) + max(E) = \qquad (9)$$

$$r \times \frac{(r-1)}{2} + \frac{(r-1)}{2} = \frac{r^2-1}{2}$$

while it can be observed that

$$\frac{r^2-1}{2} = 2 \cdot \left[\frac{r-1}{2}\right]^2 + 2.\frac{r-1}{2} \qquad (10)$$

$$\frac{r^2-1}{2} \ge \left[\frac{r-1}{2}\right]^2 + 3.\frac{r-1}{2} \text{ for } r > 2 \qquad (11)$$

$$\frac{r^2-1}{2} < 3 \cdot \left[\frac{r-1}{2}\right]^2 \quad \text{for } r > 5 \qquad (12)$$

It can be seen, from equations (9) and (10), that each PE in the scaler can accept two inputs of Range 1 and two of Range 2.

The scaling unit consists of n sub-scalers, each performing the scaling operations by $r^0, \ldots, r^{n-1}$. The sub-scaler that scales by a factor of r^0 is trivial. Figure 1 shows a block diagram of the sub-scaler. Each sub-scaler has four blocks of PEs. In blocks one and three, each PE has four inputs, while in the blocks two and four each PE has only two inputs. The function of the PEs in all the blocks is the same; that is, each PE adds all the inputs and produces a two digit output.

Block 1 has (q-1) layers, where q is the number of digits in c. The function of this block is to add all the digits that map to same position by equation (8) and express the result in t-digits. The output of the block is the result of the addition in two components, sum and carry. If r^s is the scaling factor, then for the first layer, input A consists of the digits in the input number for which equation (8) does not apply, shifted to the left by s positions, while input B consists of the digits for which either case (i) or case (ii) of equation (8) applies. From the second level onwards, these inputs are the outputs of the previous layer, i.e., input A is the output E (sum component) and the input B is the F (carry component). Notice that, due to the shift in F, the most significant digit, say F_{out}, overflows. This F_{out} is multiplied by c while $c \times F_{out}$ becomes the third input (input C) of the next layer. The third input (input C) for the first layer, as well as the fourth input (input D) through out the block are derived from the digits of the input number subject to the case (iii) of equation (8). These digits are organized as follows:

> Let the scaling factor of the unit be r^s and let $(a_{i+j}, a_{i+j-1}, \ldots, a_i)$ be the digit sequence in the input number such that
>
> $$t \le (i+s) \quad \text{and} \quad n/2 > i+j+s$$
>
> so that case (iii) of equation (8) is applicable to these digits. If $c_{q-1}, c_{q-2}, \ldots, c_0$ are the digits of c, then c_0 times $(a_{i+j}, a_{i+j-1}, \ldots, a_i)$ is the third input in the first layer while the fourth input (input D) throughout the block 1 is obtained as $c_u \times (a_{i+j}, a_{i+j-1}, \ldots, a_i) \times r^u$ where u is the layer number and $1 \le u \le q-1$.

Block 2 has t layers, but t+2 positions. The function of this block is to suppress the carry component in the input from the block 1 up to t positions. Here we do not apply the equation (8), to avoid the infinite loops due to limitations of modular additions in NRSD system, as explained in reference [6]. Since the component in the result can grow up to t+2 digits, we have t+2 positions in this block. Observe that F_{t+1} would always be zero. In the blocks 2 and 4, inputs to the PEs in each layer are the 2 outputs of the previous layer, E and F. Since F_{out}, (the carry over flow of a layer), is always zero, it is neglected.

The purpose of the block 3 is to simply compress the t+2 digit result of the block 2 to t+1 digits with the help of the equation (8). Here, input A and B are the sum and the carry shifted by one digit to the left from the previous level, while, the inputs C and D are $c \times E_{t+1}$, $c \times F_t$ respectively, where E_{t+1} is the sum output of $t+1^{th}$ position and F_t is the carryout of the tth position of the previous layer. In the block 4, which is t+1 digits long, we compress the carry completely and output the result in t+1 digits. Again we have t+1 positions to avoid the infinite loops in addition.

It can be easily observed that, the combined function of the later three blocks is to simply compress the two component result, (sum and carry), into one. Therefore, these 3 blocks together can be called Carry Propagate Adder (CPA).

IV. Conclusion

In this paper we considered implementation aspects of convolutions using Number Theoretic Transform Techniques and non-binary arithmetic. The main motivation was to relax the restrictions imposed by the well known Mersenne and Fermat Number Transforms. These restrictions were the limited transformation length and lack of flexibility in choosing the modulus used for the arithmetic.

Although the MNTs and FNTs use moduli of the form $2^t - 1$ or $2^t + 1$, our transforms use the moduli of the form $M = r^t \pm c$. Such transforms rely on scalings by powers of r and result in minimum number of multiplications whenever used in implementing convolutions. In order to efficiently perform the scalings by powers of r, we have encoded our data as radix-r data and performed all the operations using the radix-r Non-Redundant Signed Digit System. By doing so, the dominating operations of scaling by powers of r have been translated into simple shifts or rotates in the radix-r system.

Acknowledgment

This work has been supported in part by NSF Grant No. CCR8910187.

References

[1] J. H. McClellan and C. M. Rader, *Number Theory in Digital Signal Processing,* Prentice-Hall, Englewood Cliffs, NJ, 1979.

[2] A. V. Oppenheim and C. Weinstein, "Effects of finite register length in digital filtering and the fast Fourier transform," *Proc. IEEE,* vol. 60, pp. 957-976, August 1972.

[3] C. M. Rader, "Discrete Convolutions via Mersenne Transforms," *IEEE Trans. on Computers,* vol. C-21, pp. 1269-1273, December 1972.

[4] R. C. Agarwal and C. S. Burrus, "Fast Convolution Using Fermat Number Transforms with Applications to Digital Filtering," *IEEE Trans. on Acoustics, Speech and Signal Processing,* vol. ASSP-22, pp. 87-97, April 1974.

[5] P. C. Balla and A. Antoniou, "Number Theoretic Transform Based on Ternary Arithmetic and Its Application to Cyclic Convolution," *IEEE Trans. on Circuits and Systems,* vol. CAS-30, pp. 504-505, July 1983.

[6] S. Turimella, "Implementation Aspects of Convolvers Using Non-Redundant Signed Digit Number System," MS thesis, Louisiana State University, 1991.

[7] B. Parhami, "Generalized Signed-Digit Number Systems: A Unifying Framework for Redundant Number Representations," *IEEE Trans. on Computers,* vol. C-39, No. 1, pp. 89-98, January 1990.

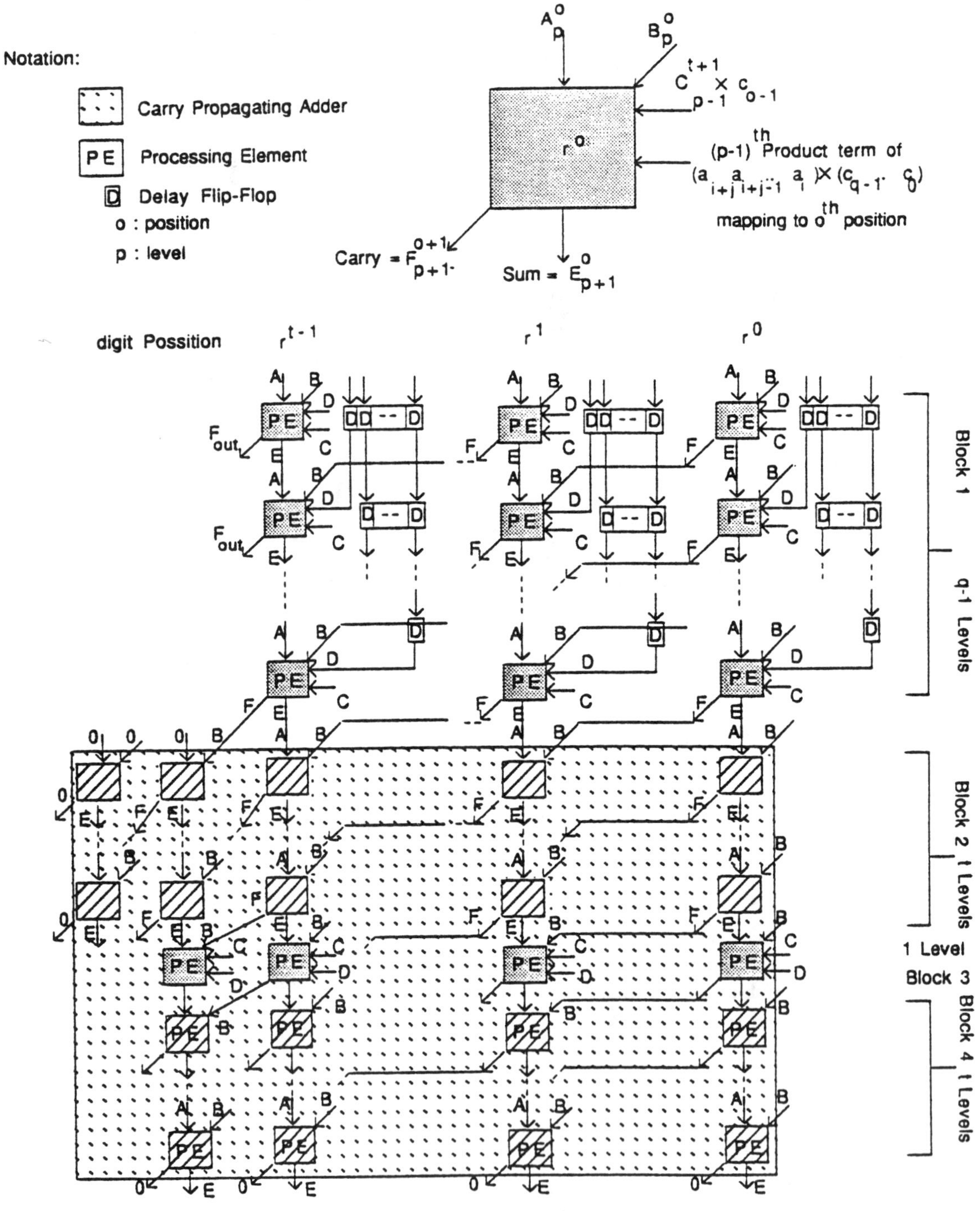

Figure 1: Radix-r Sub-scaler Performing Arithmetic Modulo M.

SUBBAND DFT - INTERPRETATION, ACCURACY, AND COMPUTATIONAL COMPLEXITY

OGNJAN V. SHENTOV†, ABDUL N. HOSSEN§, SANJIT K. MITRA†, ULRICH HEUTE§

†Dept. of Electrical & Computer Engg., University of California, Santa Barbara, CA 93106 USA
§AG Digitale Signalverarbeitung, Ruhr-Universitaet, D-4630 Bochum, Germany

Abstract

A recent DFT computation method based on a sub-band decomposition is investigated [1],[2]. The two distinct parts of the algorithm, a preprocessing Hadamard-transform stage and a "correction" stage, are interpreted as a filter-bank plus a recombination network, and are compared to the corresponding filter-bank interpretation of the direct DFT and classical FFT algorithms. The comparison provides a better understanding of the nature of the errors due to dropping of sub-bands in the approximate DFT computation of narrow-band signals. Formulae for the approximation errors due to linear distortion and to aliasing are further derived. The computational complexity of the algorithm is analyzed both theoretically and in terms of running-time measurements. With these insights, the novel algorithm is compared to existing methods, especially for the computation of a limited number of frequency points. Previously reported complexity estimates are refined and extended.

1. Introduction

The Discrete Fourier Transform (DFT) of a length-N sequence $\{x(n)\}$ is defined by

$$X(k) = \sum_{n=0}^{N-1} x(n) W_N^{nk}, \qquad k = 0,1,\ldots,N-1, \tag{1}$$

with $W_N = e^{-j2\pi/N}$. The DFT can be computed efficiently by one of the well-known Fast Fourier Transform (FFT) algorithms. A classical and widely used example is the radix-2, decimation-in-time algorithm often referred to as "Cooley-Tukey (CT-) FFT", based on a repeated decomposition of the input sequence into even- and odd-indexed subsequences [3].

The repeated decomposition of a larger size problem into a number of easier to solve, smaller-size ones is typical for all FFT-type algorithms. It is also the basis of the newly proposed algorithm [1],[2], referred to as "Sub-Band (SB-) DFT". Here, however, the decomposition has a physical meaning: The shorter-length subsequences correspond to roughly separated frequency bands of the input signal.

Due to the frequency separation of the subsequences in the SB-DFT, the algorithm can provide computation of all frequency points, leading to a "full-band analysis", or only of a limited number of (adjacent) frequency points, termed "partial-band analysis". If portions of the spectrum are known to have negligible energy contribution, dropping out the processing of the corresponding subsequences can result in significant computational savings in the partial-band analysis. This is the case, when narrow-band signals are to be processed, where a-priori knowledge of the underlying physical process or some estimation mechanism indicates which band(s) should be retained. The computation of the short-size transforms at each stage can itself be done via an ordinary FFT, or be further decomposed using the proposed algorithm. This feature provides additional user-flexibility to conduct the computations in the required bands only and to use available hardware and/or software.

2. Full-Band Analysis

2.1 Signal-Flow Graphs

In the first step of the algorithm, the input sequence $\{x(n)\}$ is decomposed into a low-pass filtered and a high-pass filtered subsequence, termed $\{g_L(n)\}$ and $\{g_H(n)\}$ respectively, of length N/2 each:

$$g_L(n) = \tfrac{1}{2}[x(2n) + x(2n+1)], \tag{2a}$$

$$g_H(n) = \tfrac{1}{2}[x(2n) - x(2n+1)]. \tag{2b}$$

The desired size-N DFT $\{X(k)\}$ can be obtained from the two size-N/2 DFTs $\{G_L(k)\} = \mathrm{DFT}\,\{g_L(n)\}$ and $\{G_H(k)\} = \mathrm{DFT}\{g_H(n)\}$ according to [2]:

$$X(k) = (1 + W_N^k)G_L(k) + (1 - W_N^k)G_H(k). \tag{3}$$

This stage of the algorithm is shown in Fig. 1. The frequency-dependent factors multiplying the transforms $G_L(k)$ and $G_H(k)$ are incorporated in the block denoted "correction factors". As indicated, the two half-size DFTs can be calculated by two (standard) FFTs.

The filtering decomposition in Eq. (2) can, of course, be applied for the low- and high-pass filtered subsequences $\{g_L(n)\}$ and $\{g_H(n)\}$ again, and it can be further repeated to their respective filtered subsequences as well. All filtering stages can be combined into one block, and all down-sampling steps can be realized together by carrying out a block transform every N samples. This part of the algorithm can be implemented by a chain of delays and an addition/subtraction network. The remaining part consists of modified versions of the "correction" factors in Eq. (3), shown in Fig. 1. The SFG in Fig. 2 illustrates the result of apply-

1058-6393/91 $01.00 © 1991 IEEE

ing a complete decomposition of an input sequence of length N = 16, that is, until sequences of length 2 remain.

In Fig. 2, the left-hand part is a realization of the preprocessing stage. It has the structure of a radix-2, decimation-in-frequency FFT, but there are no multiplications. An examination of the network reveals that it is a realization of a fast Hadamard transform.

The right half of the SFG is the generalized "correction factors" block in Fig. 1. It contains all complex factors $W_N{}^i$ typical to any FFT algorithm, though in a "double-butterfly" structure. Simplifications of the graph are possible whenever trivial multipliers such as $W_N{}^0$ are used.

2.2 Filter-Bank Interpretation

The DFT definition in Eq. (1) can be rewritten as

$$X(k) = \sum_{n=0}^{N-1} x_k(n), \qquad (4)$$

where $x_k(n) = x(n)W_N{}^{nk}$. The DFT computation in Eq. (4) can be interpreted as a combination of N complex modulators with frequencies $2k\pi/N$, and N identical lowpass filters [4], sampled at time instants $n = rN-1$, with r being integer. These lowpass filters have impulse responses

$$h(n) = \begin{cases} 1, & \text{for } n = 0,1,\dots,N-1, \\ 0, & \text{otherwise,} \end{cases} \qquad (5)$$

and transfer functions

$$H(z) = (1-z^{-N})/(1-z^{-1}). \qquad (6)$$

The corresponding magnitude response for N = 16 is shown in Fig. 3.

A similar filter-bank interpretation can be given to the CT-type FFT. The only difference in this case is in the representation of the averaging filter: While the DFT contains this summation in a direct form (see Eq. (4)), the classical FFTs can be shown to contain a cascade structure [4]; for the decimation-in-time, radix-2 FFT ($N = 2^\mu$), for instance, the transfer function is found as

$$H(z) = z^{-(N-1)} \prod_{i=0}^{\mu-1} (1 + z^{2^i}). \qquad (7)$$

It should be stressed that, in both the DFT and FFT cases, the filters are identical for all channels k. The differences between them are only based on the modulation terms $e^{-j2\pi kn/N}$, i.e. on the angular shift by a factor of $-2\pi k/N$ of the transfer function zeros of the k-th channel.

In the case of the SB-DFT computation, a filter-bank is an intrinsic part of the algorithm. The filters in this case, even though symmetric with respect to N/2, are different at each output node of the Hadamard block. This difference stems from the fact that, due to the addition/subtraction operations in this network, some parts of the modulations are already realized here, leading to filters with differing center frequencies. As a result, the filters are band-pass instead of all being lowpass, and they are no longer shifted versions of just one pattern. Figure 3 shows the magnitude responses of two of the Hadamard filters (e.g., k = 1 and k = 2, for N = 16).

These filters have clearly poorer characteristics compared to the corresponding, identical filters of the standard DFT. Thus, the task of the correction network in Fig. 2 is to combine the Hadamard filters such that, at the output of the correction-factor block, shifted versions of Fig. 3 and of H(z) as in Eq. (6) result.

2.3 Algorithmic Complexity

This section gives a comparison of the computational complexities of the standard CT-FFT and the new algorithm, as presented in a generalized form in Fig. 2. The comparison is based on the total operational count (additions and multiplications) of both algorithms. A real-valued input signal is assumed.

The SFG in Fig. 2 contains ($N \log_2 N$) real additions in the Hadamard network, plus ($5N \log_2 N - 5N$) real additions and ($2N \log_2 N - 2N$) real multiplications in the recombination part; so, a total of ($6N \log_2 N - 5N$) real additions and ($2N \log_2 N - 2N$) real multiplications are required. Since the CT-FFT needs only ($3N \log_2 N$) real additions and ($2N \log_2 N$) real multiplications, the CT-FFT is of course superior for a full-band analysis. Indeed, from run-time experiments (using an AT 286/10 MHz with a co-processor 287), the execution time of the full-band SB-DFT is found to be 20 to 70 percent higher than that of the CT-FFT, when N varies from 8 to 1024.

3. Partial-Band Analysis

3.1 Band Selection

As pointed out above, computational savings are possible, if the signal spectrum is concentrated in a certain frequency band and the remaining bands have negligible energy contributions. The selection of the primary band can be done from a-priori knowledge of the underlying physical process or by using some estimation algorithm. We briefly present next the algorithm for primary band estimation, introduced first in [5].

The proposed estimation technique is closely related to the SB-DFT algorithm and can be implemented as a modification of it. The main idea is to obtain an estimate of the first autocorrelation coefficient of the input sequence at each stage and thus decide which of the filtered subsequences has a higher energy content. This is done by comparing the absolute magnitudes of corresponding samples of the lowpass/highpass filtered subsequences and storing the number of positive comparisons in counters. A simple interpretation of the output of the counters at different stages allows to dynamically determine estimates of the power spectrum density of the input signal in each band. The process requires no additional operations, runs in parallel to the main algorithm, and can be used on-line [5].

3.2 Half-Band Lowpass Analysis

As an introduction to the general case, we shall first deal with the case of non-negligible energy only in the frequency range $f \in [0, f_s/4]$ (with f_s denoting the sampling frequency) or, equivalently, $k \in \{0, 1, ..., N/4 - 1\}$. In this case, Eq. (3) reduces to [2]:

$$X(k) \cong \hat{X}(k) = (1 + W_N^k) G_L(k), \quad (k = 0,1,\ldots,\frac{N}{4}-1). \qquad (8)$$

This approximation is equivalent to neglecting the high-frequency branch in Fig. 1 and cancelling the corresponding lines of the SFG in Fig. 2. Only N/4 frequency points need to be calculated for real signals. These simplifications result in a half-band SFG.

3.2.1 Filter-bank Interpretation

While in the full-band analysis the weighting network recombines the differing Hadamard bandpasses to equispaced filters of the same form as known from the DFT/ FFT, the recombination now is incomplete, leading to linear distortions, leakage, and, thereby, to aliasing. Figure 3 shows the magnitude response for the output X(1) of a half-band lowpass analysis for the N = 16 case.

3.2.2 Algorithmic Complexity

The simplified SFG of the half-band lowpass analysis and a similar SFG, but without simplification for the case of trivial multiplications are compared in their complexities (number of operations) with the full-band SFG of Fig. 2.

Using the non-simplified half-band SFG, we need ($3N \log_2 N - N$) real additions and ($N \log_2 N - N$) real multiplications less than with the full-band SFG in Fig. 2; the simplified SFG of the half-band lowpass analysis scheme will save ($3N/2 - 4$) real additions and N real multiplications more. This means that now the new algorithm needs only slightly more additions, but considerably less multiplications than the CT-FFT: Already for a halfband transform, it becomes preferable, as verified by the computation times reported in the following section.

3.3 General Narrow-Band Analysis

3.3.1 Approximation Errors

In the general case of partial-band analysis, we consider the computation of P = N/2M adjacent frequency points, where both N and M are assumed to be powers of 2 ($M = 2^m$). For simplicity, it is also assumed that the computational algorithm is organized in a tree structure which allows us to reduce the band of interest in half at each stage by discarding the unwanted subsequences. Two main types of approximation errors are involved in this process: linear distortions corresponding to non-constant frequency responses in the band of interest, and aliasing due to non-ideal attenuation in the zeros of the original DFT filters.

Linear Distortion. The one-stage approximation in Eq. (8) can be expressed as [2]

$$\hat{X}(k) = \frac{1}{2}[1 + \cos(2\pi k / N)] X(k) - j\frac{1}{2}[\sin(2\pi k / N)] X(k + \frac{N}{2}), \qquad (9)$$

where $\hat{X}(k)$ is the approximate, and X(k) is the true signal spectrum. If the signal has no energy content above k = N/4, the second term in Eq. (9) can be dropped. The result is a value $\hat{X}(k)$ which is diminished by the factor $^1/_2$ [1+cos(2πk/N)] - see the maximum of the magnitude response in Fig. 3, which should be equal to N = 16, ideally. From this, the exact transform X(k) is simply obtained by the "equalization"

$$X(k) = 2\hat{X}(k) / [1 + \cos(2\pi k / N)]. \qquad (10)$$

In general, for an m-stage reduction (to N/2M DFT values) with arbitrary band selection (like low-low-high-low-high etc.), the generalization of Eq. (10) can be shown to be:

$$X(k) = \frac{2^m \hat{X}(k)}{\prod_{i=1}^{m} [1 + s(i) \cos(2^i \pi k / N)]}, \qquad (11)$$

with s(i) = 1 for a low-pass choice, and s(i) = –1 for a high-pass choice in stage i, respectively. Eq. (11) is a general concise expression that allows for a complete compensation of the linear distortion in the band of interest.

Aliasing Errors. Unlike the linear distortions, the aliasing errors induced by the decimation process are irreversible. It can be shown that for an m-stage reduction, the approximate transform will contain (M –1) aliasing terms, uniformly distributed on the frequency axis. $\hat{X}(k)$ can be expressed as follows [2]:

$$\hat{X}(k) = A_0 X(k) + A_1 X(k + \frac{N}{M}) + A_2 X(k + \frac{2N}{M}) + \cdots + A_{M-1} X(k + \frac{N(M-1)}{M}). \qquad (12)$$

It can be seen from Eq. (12) that for N = 16, M = 2, and if $X(9) \neq 0$, there will be an aliasing error in calculating X(1).

The coefficients A_i depend on the frequency index k. For instance, if M = 4 and m = 2, in the case of low-low reduction stages, they are expressed as

$$\begin{aligned} A_0 &= \frac{1}{4}[1 + \cos(\frac{2\pi k}{N})][1 + \cos(\frac{4\pi k}{N})], \\ A_1 &= \frac{1}{8}(1 - j)\sin(\frac{4\pi k}{N})[1 + \cos(\frac{2\pi k}{N}) - \sin(\frac{2\pi k}{N})], \\ A_2 &= -j\frac{1}{4}\sin(\frac{2\pi k}{N})[1 + \cos(\frac{4\pi k}{N})], \\ A_3 &= -\frac{1}{8}(1 + j)\sin(\frac{4\pi k}{N})[1 + \cos(\frac{2\pi k}{N}) + \sin(\frac{2\pi k}{N})]. \end{aligned} \qquad (13)$$

A general expression for the aliasing error AE(k), after compensation of the linear distortion and under the assumption of small components with equal amplitudes λ outside the band of interest, is found to be

$$\frac{AE(k)}{\lambda} = 1 - \prod_{i=1}^{m} \frac{[1 + s(i) W_N^{(2^{i-1})k}]}{[1 + s(i)\cos(2^i \pi k / N)]},$$

where s(i) is defined as in Eq. (11).

Figure 4 illustrates the normalized aliasing error, for $N = 128$ and differing values of $M = 2^m$, in the case where in all m reduction stages the low-pass branch is followed. As can be expected, for a fixed frequency index k, the aliasing error increases with increasing m.

3.3.2 Computational Complexity

The computational complexity of the proposed algorithm is first examined from a theoretical point-of-view, i.e. in terms of the total operational count (additions, multiplications) and compared to existing methods which calculate the DFT in $P < N$ points (direct DFT [3], Goertzel algorithm [6], pruned FFT [7], Sorensen-Burrus method [8]). Our comparison indicates that the SB-DFT approach is the fastest method in all partial analysis cases, with the following exceptions:

1. For P =1, the direct DFT is (trivially!) the fastest method .
2. For $P = 2$, $N \leq 16$, the Goertzel algorithm is the fastest method.
3. For $P \geq N/2$,the three-butterfly split-radix FFT [9] is the simplest and fastest algorithm.

The number of real operations required in the three different forms of the SB-DFT are found to be :

1. Form of generalized Fig.1 :

$$\# MUL = 2N\mu / M, \quad (15a)$$
$$\# ADD = [N(M-1) + 3N\mu - Nm] / M, \quad (15b)$$

2. Form of generalized Fig.2 :

$$\# MUL = 2N(\mu - 1) / M, \quad (16a)$$
$$\# ADD = [N(M-2) + 6N\mu - 4N(m+1)] / M, \quad (16b)$$

3. Form of generalized Fig.2 (with simplification in case of multiplications with W_N^0):

$$\# MUL = [2N(\mu - 2) / M] - 4(m-1), \quad (17a)$$
$$\# ADD = [N(M-2) + 6N\mu - 4Nm - 7N] / M - 2m + 6. \quad (17b)$$

Figure 5 shows the total number of operations (multiplications + additions) for these three forms and three other methods: A pruned FFT [7], the 3-butterfly split-radix FFT [9], and the method due to Sorensen and Burrus [8], for the special case $N = 128$.

In addition to the operational count, an important consideration in evaluating the relative merits of different computational algorithms is the programming complexity and various other factors like memory access time, addressing etc.. Figure 6 shows a comparison of the computation times of all algorithms for a different number of frequency points and normal processing conditions (the SB-DFT programs include linear-distortion corrections, and a 2-butterfly split-radix FFT is used instead of the 3-butterfly split-radix FFT). The different forms of the subband reduction technique of Figs. 1 and 2 are programmed for the general case $M = 2^m$ to calculate the DFT in $P = N/2M = 2^r$ points with $P < N = 2^\mu$.

Both the operational count and the running-time comparison indicate that the SB-DFT algorithm is to be preferred in cases when the speed of computation is the most important factor and some approximation error can be tolerated.

4. Conclusions

A detailed analysis of a newly proposed FFT-type algorithm has been presented. Several variants were introduced in the form of SFG descriptions. The main characteristic of the approach is the frequency-separation property of the subsequences involved in the decomposition process. A novel filter-bank interpretation of the procedure was presented, allowing to understand the errors occuring in the method's use for an approximate partial-band transform. These errors were studied in depth to obtain general formulae describing their nature, whatever the number and type of decomposition stages might be. Also previously presented results on the computational complexity were refined and extended to be comparable with other methods computing a limited number of frequency points. As a result, the SB-DFT algorithm was shown to be the fastest method under relatively wide conditions.

5. Acknowledgements

This work was supported in part by a University of California MICRO grant with matching support from the Rockwell Corp. and ESL Inc., and in part by a grant from DAAD (German Academic Exchange Service).

6. References

[1] S. K. Mitra et. al., "A Method For Fast Approximate Computation Of Discrete Time Transforms," *Proc. ICASSP*, Albuquerque, New Mexico, April 1990, pp.2025-2028.

[2] S. K. Mitra et. al., "DFT Calculation via Subband Decomposition," *Proc. EUSIPCO'90*, Barcelona, Spain, Sept. 1990, pp. 501-504.

[3] A. V. Oppenheim, and R. W. Schafer, *Digital Signal Processing*, Prentice Hall, Englewood Cliffs, N.J., 1975.

[4] U. Heute, "Results of a Deterministic Analysis of FFT Coefficient Errors," *Signal Processing*, Vol. 3 (1981), pp. 321-331.

[5] O.V. Shentov and S. K. Mitra, "A Simple Method for Power Spectral Estimation Using Subband Decomposition," *Proc. ICASSP*, Toronto, Canada, May 1991, pp. 3153-3156 .

[6] C. S. Burrus and T. W. Parks, *DFT/FFT and Convolution Algorithms*, John Wiley and Sons, New York, NY, 1985.

[7] K. Nagai, "Pruning the Decimation-in-Time FFT Algorithm with Frequency Shift," *IEEE Trans. Acoust., Speech, Signal Processing*, Vol. ASSP-34, no. 4, Aug. 1986, pp. 1008-1010.

[8] H. V. Sorensen et. al., "A New Efficient Algorithm for Computing a Few DFT Points," *Proc. IEEE Int. Symp. on Circuits and Systems*, Espoo, Finland, June 1988, pp. 1915-1918.
[9] H. V. Sorensen et. al., "On Computing the Split-Radix FFT," *IEEE Trans. Acoust., Speech, Signal Processing*, Vol. ASSP-34, no.1, Feb. 1986, pp. 152-156.

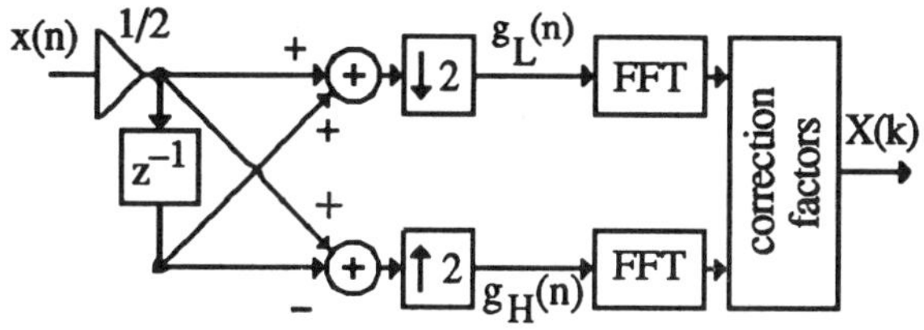

Figure 1. The two-band decomposition scheme.

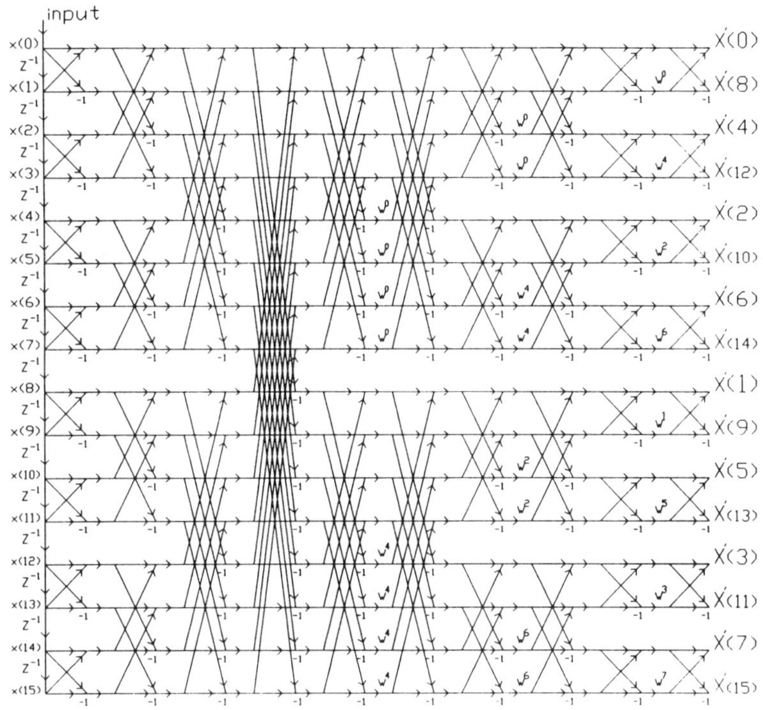

Figure 2. Signal-flow graph of a 16-point full-band SB-DFT with the factors 1/2 of Eq. (2) omitted.

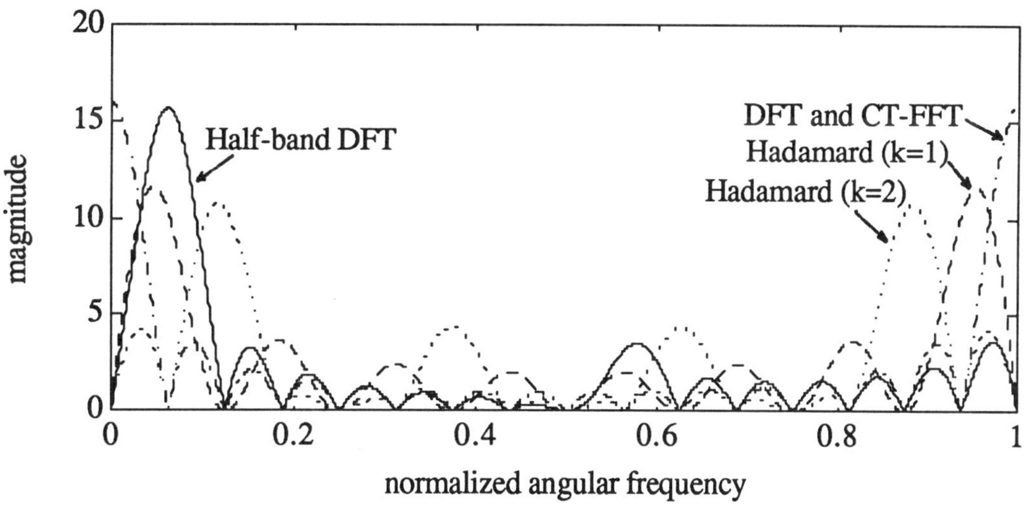

Figure 3. Magnitude responses.

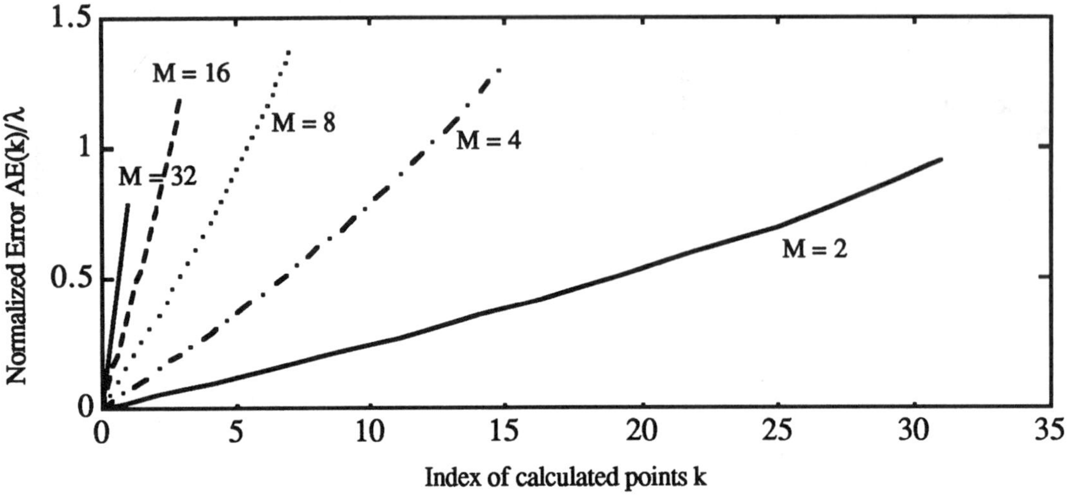

Figure 4. Normalized aliasing error (N = 128).

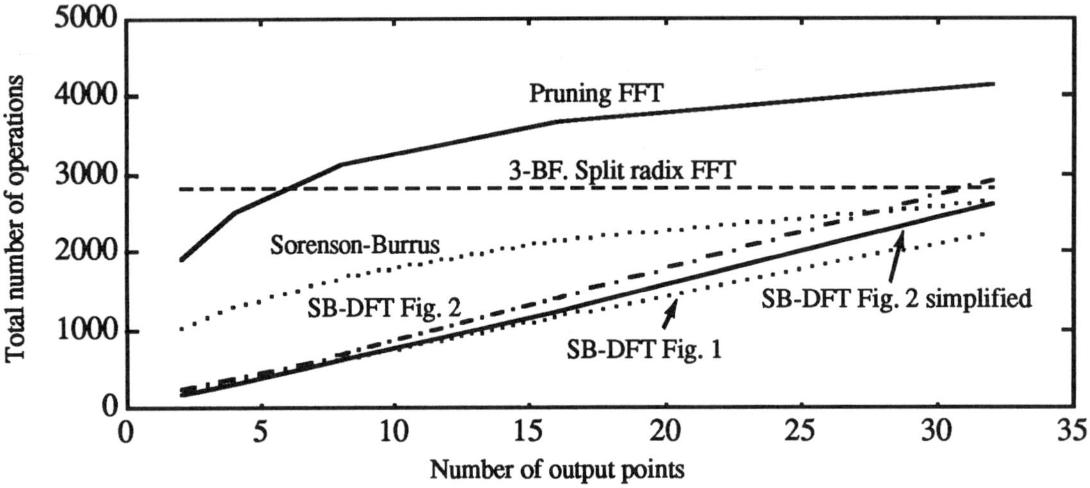

Figure 5. Computational complexity comparison.

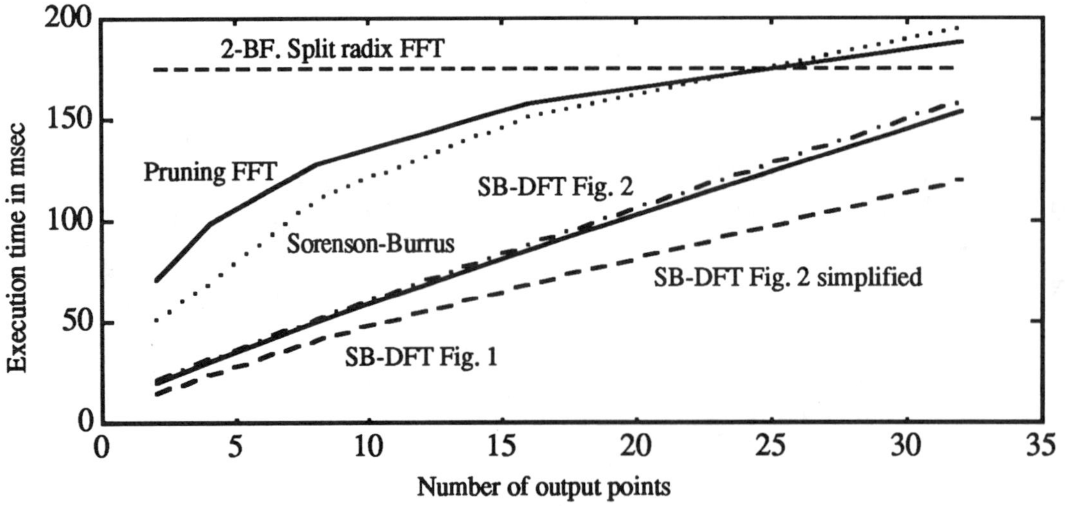

Figure 6. Running-time comparison.

Multimegapoint FFT's Running on Workstation Computers

Michael J. Grimm and George A. Zimmerman

Jet Propulsion Laboratory
California Institute of Technology
4800 Oak Grove Drive, MS 238-343
Pasadena, California 91109

Abstract

This paper describes a method for performing multimegapoint Fast Fourier Transforms (FFT's) on workstation-class computers. It demonstrates that although the workstation's compute speed may be adequate to perform large FFT's; if care is not given to memory array access, thrashing will severely degrade performance. The matrix FFT, combined with a file oriented matrix transpose are shown to be an easy way to prevent thrashing. This method was applied to perform simulations of a 2^{23} *(8,388,608) point real-only FFT processor which will be used in NASA's Search for Extraterrestrial Intelligence program.*

1. Introduction

Increases in the speed and storage capacity of commonly available workstation computers have made it possible to perform high-resolution spectral analysis of multimegapoint data sets on these machines. Given that current workstation computers easily achieve throughputs exceeding one million computes per second on compute limited problems, and a 2^n point radix-2 Fast Fourier Transform (FFT) requires approximately $5 \times n \times 2^n$ computes, a workstation might compute a 4,194,304 point FFT in under 8 minutes, and a 16 megapoint FFT in 32 minutes. However, even though virtual memory machines can handle the large array sizes required for multimegapoint FFT's, unless the FFT routine is designed with actual memory size constraints in mind; these compute speeds for larger size data sets will not be achieved.

This paper shows that the data addressing required for an in-place FFT whose memory requirement significantly exceeds the physical memory size of a machine causes an unacceptable level of virtual memory disk access, commonly known as "thrashing". In addition, a matrix-decomposition FFT is shown to replace the reordering the large data set with matrix transpositions, using a two pass memory-efficient matrix transposition technique. The resulting FFT technique has been used to perform simulations of special purpose hardware for an 80 MHz real time 8 megapoint real-only FFT being developed for use in NASA's Search for Extraterrestrial Intelligence [1].

2. FFT Data Ordering and Virtual Memory Machines

In-place FFT algorithms are appropriate to use when the data array of points being transformed fit into the physical memory of the workstation. However, even though in-place FFT algorithms are memory efficient, they still require a single memory array the size of the transform. If intermediate results are stored as complex valued pairs of 32 bit floating point numbers, the memory array requires 8 bytes times the number of points in the data set. For a 16 million point data set, the memory array size is 128 megabytes. On workstation computers, the usual answer to such large data sets is to use virtual memory, shuttling pages between disk and physical memory as required. Most in-place FFT algorithms require radix-reversal of the data in one form or another during the transform computation. The difficulty of performing radix-2-reversal in a virtual address machine where the size of the data array exceeds the size of available physical memory will now be analyzed.

Assume that a data page will hold a power-of-two data points, $S = 2^{SPOW}$, and that the data array to be radix-2 reversed contains $N = 2^n$ data points. Then, the array requires $P = 2^{n-SPOW}$ virtual pages. Radix-2 reversal interchanges data point i with the data point whose index is the radix-2 reversal of i. A simple one-pass sequence of address pairs to be interchanged is:

$$(0 \rightleftarrows 0),(1 \rightleftarrows N/2),(2 \rightleftarrows N/4),(3 \rightleftarrows 3N/4) \cdots$$

Notice that the first P entries of the destination

1058-6393/91 $01.00

addresses access a single point in each of the P virtual pages of the data array. It is easy to see that if the total number of points is at least S^2, i.e., $P \geq S$, radix-2 bit reversal guarantees that interchanging all the points in any one page touches at least one single point in each of S other pages. Because the number of interchanges to be performed decreases as groups of P pages are traversed, the total number of page boundaries that must be crossed is given by:

$$\frac{\max(S,P)}{S} \sum_{i=1}^{\min(S,P)} (\min(P,S)-i) = \frac{P}{2}(\min(S,P) - 1).$$

Not all of these pages will have to be transferred to and from disk, especially if the physical memory available is close to the array size. However, for large arrays the procedure must repeatedly interchange points in different memory pages causing the virtual memory system to constantly swap data between physical memory and the disk. Since the individual pages must be both read from and written out to the disk, this can require almost $(\min(S,P) - 1)P \times$ *bytes per page* of disk I/O. With a typical page size of 4 kbytes, and 8 byte complex words, S would be 512, and requiring $P = 8192$ pages for a 4 megapoint FFT. If most of the pages require swapping from disk, this can require almost 16 Gbytes of disk I/O, which, at 1 Mbyte per second disk rates, requires about 4.5 hours of disk transfer time, ignoring the seek time per page which can be as large or larger than the page transfer time. This time might appear reasonable on the surface, as the run can finish overnight; however, remember that the computation of the 4 megapoint FFT requires less than 8 minutes. The problem rapidly grows intolerable as the number of points in the FFT increases. For example, a 16 megapoint FFT that requires under one hour's computation can require almost 64 Gbytes of disk I/O, adding about 18 hours of transfer time to execution time.

One solution to the thrashing problem is to install more DRAM in the workstation; however, it is often either technically or economically impossible to sufficiently upgrade the workstation. The authors have adopted a software solution to implement a simulator for an 8 megapoint real-input FFT. By using a matrix-decomposition FFT algorithm the data array sizes have been reduced and isolated disk I/O so that it can be explicitly managed in the FFT software.

3. Matrix FFT Algorithm

The matrix Discrete Fourier Transform (DFT) algorithm, derived in detail in (L.R Rabiner & Gold, pp. 371-377), is summarized below.
Consider an $N = KL$ point sequence to be transformed. The DFT of the N point sequence $x(n)$, $n = 0,1,...N\text{-}1$ is

$$X(k) = \sum_{n=0}^{N-1} x(n) e^{-j2\pi\frac{nk}{N}} \quad k = 0,1,...N\text{-}1.$$

Now, let $n = Ml + m$ and $k = Lr + s$, then:

$$X(k) = X(s,r) = \sum_{m=0}^{M-1} \sum_{l=0}^{L-1} x(l,m) e^{-j2\pi\frac{(Ml+m)(Lr+s)}{N}}$$

$$k = 0,1, \cdots N\text{-}1 \;\; (r = 0,1, \cdots M\text{-}1;\, s = 0,1, \cdots L\text{-}1).$$

By factoring the exponential term and rearranging the summation, the following is obtained:

$$X(s,r) = \sum_{m=0}^{M-1} e^{-j2\pi\frac{mr}{M}} \left(e^{-j2\pi\frac{ms}{ML}} \left(\sum_{l=0}^{L-1} x(l,m) e^{-j2\pi\frac{ls}{L}} \right)\right)$$

which, recognizing the individual DFT's in the sums, is:

$$X(s,r) = DFT_m[\, e^{-j2\pi\frac{ms}{ML}} DFT_l[x(l,m)]\,],$$

where DFT_m denotes the DFT taken over the index m, and DFT_l denotes the DFT taken over the index l. With this re-indexing, if the data is considered as an L by M matrix, the DFT may be broken into separate DFT's of the rows and columns with a phase correction (twiddle factor) between the row and column DFT's. The sequence of operations is enumerated in the next section.

The matrix FFT algorithm replaces the requirement for radix reversal of the full data array with the requirement to perform a matrix transpose of the full data array. Radix reversal is just a transpose of a multidimensional array, so efficient methods of transposing a large array can easily be applied to radix reversing the large array. However, since the matrix FFT algorithm localizes data accesses into independent row and column FFT's, it is preferable to the straight radix-2 or radix-4 algorithms.

4. Implementation of Large FFT's on Small Memory Machines

The matrix algorithm can be built for a small memory machine, and hence is ideal for multimegapoint FFT's running on ordinary workstation-class computers.
This note briefly describes the matrix FFT implementation and a method of performing the required matrix transposes without thrashing.

Suppose there is an $N = K \times L$ file (sequence) of data points for which the FFT is desired.

1. Logically form the data points into a K by L point matrix and actually transpose the file into a L by K point file .
2. Now read the transposed file in K point blocks, perform K point FFT's on the blocks, and write the FFT'ed blocks to a file.
3. Transpose the L by K point file to a K by L point file.
4. Now read the transposed file in L point blocks, multiply the blocks by a row of the twiddle matrix, perform a L point FFT on the result, and write the FFT'ed results to file.
5. Now transpose the K by L point file to an L by K point result file which when treated serially, contains frequency points in sequential order.

The short FFT's can be performed with a conventional in-place algorithm; however, the matrix FFT algorithm can provide some significant advantages if the FFT routine is tailored to take advantage of them. These advantages of the matrix algorithm have to do with the generation of radix-reversed addresses and with the generation of twiddle values. In the matrix FFT algorithm the only radix-reversed addresses are for K or L point vectors. The values of K and L are in the thousands; so the radix-reversed addresses can be calculated once and then reused thousands of times. Specifically, a K point address vector can be calculated once and reused L times; then a L point address vector can be calculated once and reused K times. The twiddle vectors within the K and L point FFT's can be calculated once and then reused more than $K+L$ times. However, a single multimegapoint twiddle vector must be used between the K and L point FFT's. This "twiddle in the middle" is the only disadvantage of using the matrix FFT algorithm rather than using the straight radix 2^n algorithm.

Even with computational advantages to be gained in the row and column FFT's, the heart of the matrix FFT process is the matrix transposition stages. The file oriented transpose that will be described is a two pass operation. Sequential blocks of the input file are read into DRAM. The blocks are partially transposed, and sub-blocks are scatter written to the output file. Next, sequential blocks of the output file are read to DRAM, the transpose is completed, and the block is written back to the output file. The output file now contains the transpose of the input file. The sub-blocks that are written during the first pass are several kilobytes long, and the DRAM block size is in the low megabytes. These sizes must be chosen to be large enough to assure reasonable disk I/O speed, but not so large as to be impractical on the target workstation.

During the second pass of the transpose operation, data is available in DRAM in transpose order. Hence the transpose operation can be combined with the next arithmetic pass. For instance, transposing the input data can be combined with performing the K point initial FFT's. The FFT algorithm can be restated as follows:

1. Transpose the input file, perform K point FFT's on the transposed input, and put the result in the output file.
2. Transpose the output file, perform the twiddle multiplies followed by L point FFT's, and put the result in the scratch file.
3. Transpose the scratch file to the output file.

Each of the FFT algorithm's three transpose operations is done with four disk I/O operations: read, write, read, write. Thus the total amount of disk I/O incurred during the FFT algorithm is twelve times the file size.

As a result, if the workstation can perform disk I/O at 1 megabyte per second; the total disk I/O time for a 16 megapoint FFT is 26 minutes. If the machine achieves 1 million computations per second, the compute time should be 32 minutes, and the 16 megapoint FFT will complete in about one hour.

5. Matrix Transpose Without Thrashing

The file oriented transpose algorithm dates back to before the advent of virtual address computers. A K by L matrix is transposed using a buffer that can hold B points. If B is much greater than K and L, but smaller than $K \times L$; then factor K into $K_1 \times K_2$, and L

into $L_1 \times L_2$. K_2 should be chosen to be as large as possible with the restriction that $K_2 \times L \le B$. Likewise L_2 should be chosen to be as large as possible with the restriction that $L_2 \times K \le B$. Treat the matrix as if it were four dimensional: K_1 by K_2 by L_1 by L_2 where K_1 is the slowest moving dimension.

The transpose is performed in two passes. First each of the K_1 three dimensional sub-matrices is sequentially processed. Read a K_2 by L_1 by L_2 sub-matrix into the buffer, and then transpose it to a L_1 by L_2 by K_2 matrix. Now write each of the L_1 two dimensional L_2 by K_2 sub-matrices to the output file. The disk addresses for the two dimensional sub-matrices are calculated so that on completion of the first pass, the output file contains a L_1 by K_1 by L_2 by K_2 transposed version of the input file.

The second pass of the transpose sequentially processes each of the L_1 three dimensional sub-matrices. The K_1 by L_2 by K_2 sub-matrix is read into the buffer and then transposed into a L_2 by K_1 by K_2 sub-matrix. The transposed sub-matrix is written back to the output matrix where it was read from. The result of this pass is that the output file contains the two dimensional L by K transpose of the K by L input file.

6. Simulation Notes

The initial application of the techniques described above was in simulating a 24-bit fixed-point arithmetic implementation of an 8 megapoint real FFT as part of a polyphase FFT filter bank [1]. Because the purpose of the simulation was to bit-for-bit functionally simulate the arithmetic in an application-specific integrated circuit (ASIC), the full computational power of the workstation's CPU was utilized. The simulations were run overnight on a SUN4/360. The input functions for the simulations were sinusoids, chirps and noise. As a result, the input vectors could be generated on the fly in the transpose order required by the matrix FFT algorithm. Under these conditions, the eight megapoint FFT required approximately one hour to complete.

The simulator for the fixed-point FFT system was converted to a file-oriented floating-point FFT program. This program has been used to transform run-length compressed files of 12-bit quantized filter coefficients. The data was run length compressed so that input points could be computed on the fly in transpose order in a manner similar to that described for coefficient storage in [1]. Table I gives wall clock execution times for FFT's running on a SPARCstation 2 with 32 megabytes of memory. The "User time" measurement in Table I is as defined for the UNIX operating system, i.e., time spent on the process excluding disk swaps and I/O waits.

Table I: Example Execution Times for FFT's on a SPARCstation 2

FFT Size (complex points)	Wall Clock Time (sec)	User Time (sec)
16 M	2166.0	1246.8
8 M	975.6	612.9
4 M	480.9	308.6
2 M	201.2	149.4
1 M	97.9	74.7
256 k	25.4	20.1
64 k	6.2	5.6

Wall clock execution time for a sixteen megapoint FFT was thirty-six minutes. However, for thirteen of the thirty-six minutes, the CPU was idle, presumably waiting on disk I/O. Separate timings of the matrix transposition routine confirmed this hypothesis. This indicates that in applications where execution speed is critical, one might double buffer the data blocks and overlap compute and disk wait time, thus gaining a 35% decrease in wall clock execution time for the sixteen megapoint FFT.

With the quite reasonable compute times reported above, storing the large files necessary for intermediate storage and output became the major problem. A disk-efficient version of the program was formed to compute either amplitude or phase of the complex output. By only computing amplitude or phase, the data volume is haved. As a result, the final matrix transposition can put the output back into the intermediate storage file, dropping the full-FFT length storage requirement from two files to one.

7. Conclusions

Computation rates for commonly available workstations and personal computers have improved dramatically in recent years. Multimegapoint FFT computations, once compute-limited, can now be computed in less than 1 hour's computation. However, the memory capacity of commonly available computers has yet to increase to the level required for in-memory computation of these transforms. Demand-paged virtual memory can be defeated by the complicated data ordering in algorithms such as the FFT. As a result, careful thought must be given to mass storage (disk) I/O in the choice and implementation of the algorithm.

The matrix decomposition FFT is an algorithm well-suited to memory-limited systems. Use of this algorithm, together with an I/O efficient matrix transposition routine, allows very large transforms to be performed with reasonable execution times on common workstation computers or even high-end personal computers. This application of commonly available computers enables previously prohibitive simulations of large real-time signal processing systems, multimegapoint high-resolution data analysis, and many other applications requiring spectral analysis of large vectors. This ability to analyze vast amounts of data leaves the user with a familiar challenge: how to best reduce it to a useable, understandable summary form.

8. Acknowledgement

The research described in this paper was carried out by the Jet Propulsion Laboratory, California Institute of Technology, under a contract with the National Aeronautics and Space Administration.

9. References

1. G.A. Zimmerman, M.F. Garyantes, and M.J. Grimm, "A 640 MHz, 32 Megachannel Polyphase-FFT Spectrum Analyzer," *Presented at the 25th Annual Asilomar Conference on Signals, Systems and Computers* ,Pacific Grove, California, November 4-6, 1991.

A 640 MHz 32 Megachannel Real-Time Polyphase-FFT Spectrum Analyzer

George A. Zimmerman, Michael F. Garyantes, and Michael J. Grimm

Jet Propulsion Laboratory
California Institute of Technology
4800 Oak Grove Drive, M/S 238-343
Pasadena, CA 91109

Abstract

This paper describes a polyphase-FFT spectrum analyzer being designed for NASA's Search for Extraterrestrial Intelligence (SETI) Sky Survey at the Jet Propulsion Laboratory. By replacing the time-domain multiplicative window preprocessing with polyphase filter processing, much of the processing loss of windowed FFT's can be eliminated. Polyphase coefficient memory costs are minimized by effective use of run-length compression. Finite word length effects are analyzed, producing a balanced system with 8-bit inputs, 16-bit fixed-point polyphase arithmetic and 24-bit fixed-point FFT arithmetic. Fixed-point renormalization midway through the computation is seen to be naturally accommodated by the matrix FFT algorithm proposed.

1. Introduction

A 2^{25} (33,554,432) channel, 640 MHz wide polyphase-Fast Fourier Transform (FFT) spectrum analyzer is being designed at the Jet Propulsion Laboratory for the Search for Extraterrestrial Intelligence (SETI) Sky Survey. This spectrum analyzer will be used to separate two 320 MHz wide polarizations into channels approximately 20 Hz wide for input to SETI signal detection algorithms. Construction of a prototype windowed-FFT spectrum analyzer [1] with 40 MHz of bandwidth and 2^{21} (2,097,152) channels has recently been completed. The new spectrum analyzer design, similar to the prototype machine in many respects, is functionally divided into eight identical 80 MHz, 4 Megachannel, real-input polyphase-FFT filter banks, each implemented as a pipelined special purpose hardware signal processor. The spectrum analyzer functions consist of polyphase preprocessing, a 4 Megapoint matrix-algorithm FFT and trigonometric recombination ("real-adjust") to compute the positive half of an 8 Megapoint real FFT.

Other than the increases in bandwidth and number of channels, the main architectural difference from the prototype spectrum analyzer system is that the new design is a polyphase-FFT spectrum analyzer rather than a windowed-FFT spectrum analyzer. The advantages of polyphase-FFT spectrum analysis, as well as a review of the supporting theory are presented in [2], and are only touched on here. Similarly, features that the spectrum analyzer has in common with the prototype system can be found in [1], and are not described in detail here.

The remainder of this paper will be divided into the following three main sections: a general description of the spectrum analyzer, a description of the polyphase-FFT filter bank implementation, and a discussion of finite word length effects in the fixed-point arithmetic implementation.

2. General Description

The 320 MHz, dual polarization system is divided into four 80 MHz sub-bands per polarization. Each of the 80 MHz sub-bands, or "slices" is identical to the others, and each can be operated independently as an 80 MHz spectrum analyzer.

Like the prototype system, each 80 MHz slice of the spectrum analyzer is a pipelined architecture, allowing all stages of the polyphase-FFT algorithm to execute concurrently. As in the prototype system, a stage "bypass" capability together with stimulus and response buffers provides built-in testability. Each 80 MHz slice of the system will have an 8 bit A/D converter as its input. The FFT will use 24-bit fixed-point arithmetic implemented as an application-specific integrated circuit (ASIC). All of the FFT arithmetic is performed in 24-bit fixed-point, allowing the transforms before and after the matrix transposition to use identical board designs. In contrast, the prototype performs the column FFT's in 16-bit fixed-point arithmetic and the row FFT's in 32-bit floating point arithmetic requiring two unique board designs.

Architectural improvements in the FFT portion of the new spectrum analyzer include the replacement of two

1058-6393/91 $01.00 © 1991 IEEE

spectrum length (4 complex megapoint) double buffers with single buffers and the removal of a third spectrum length double buffer. This improvement significantly reduces the amount of memory in the spectrum analyzer, reducing both size and cost.

The user-loadable window function in the prototype system has been replaced by polyphase filter preprocessing. The polyphase-FFT filter bank approach to spectrum analysis provides greater interference isolation and significantly flatter transfer functions in the output spectral channels. The polyphase preprocessor is capable of operating as a window function, because a windowed FFT is a degenerate case of a polyphase-FFT filter bank.

As in the prototype system, the FFT is implemented using a matrix-style FFT pipeline. By decomposing the transform into shorter row and column transforms, the long delay memories and coefficient ("twiddle factor") storage required for the 4 Megapoint FFT are concentrated in a single matrix transposition and complex multiplication stage. The spectrum analyzer performs a 4 Megapoint (2^{22}) FFT as 1024 4096-point column FFT's followed by 4096 1024-point row FFT's with multiplication of the matrix entries by complex rotation factors between the row and column transforms. The row and column FFT's are each performed as pipelined radix-4 FFT's.

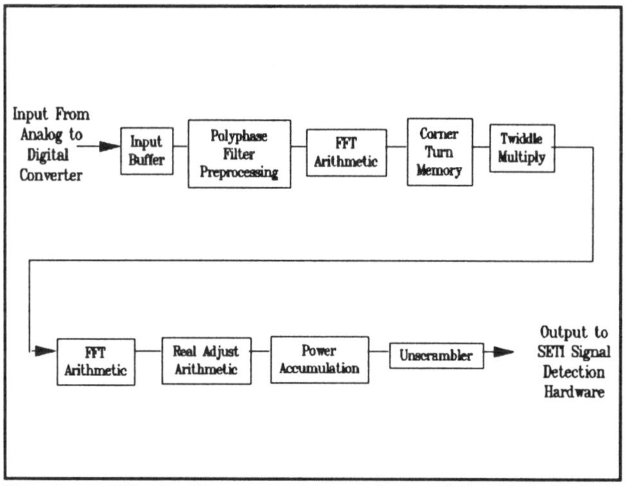

Figure 1: Block diagram of one of the eight identical 80 MHz, single polarization slices in the digital spectrum analyzer.

While the prototype system was implemented in wire-wrap technology, the greater bandwidth and 80 MHz modular nature of the new system naturally lead to implementation as printed circuit boards. While it is functionally quite similar to the prototype, the detailed hardware designs are all new so that the new system can take advantage of improvements in technology. A functional block diagram is shown in Figure 1.

3. Polyphase-FFT Structure

Because the spectrum analyzer channels are considered independently by the signal detection algorithms, the worst case processing loss [3], that is, the maximum attenuation of a sinusoidal signal in a channel's passband (in dB) plus the ratio of a channel's equivalent noise bandwidth to the channel spacing (in dB), is crucial to the system's sensitivity to weak signals. This is rarely less than 3 dB for windowed FFT's [3]. This is a result of the fact that for a windowed FFT, the time aperture over which the signal is considered (in seconds) is exactly equal to the reciprocal of the FFT channel spacing (in Hz).

Polyphase-FFT spectrum analysis is related to windowed-FFT spectrum analysis in the following manner. Consider an FFT with channel spacing N_t times finer than the desired channel spacing. This finer FFT operates on a time aperture N_t times longer than a minimum-length (nominal) FFT providing the desired channel spacing. If the nominal FFT time aperture is N_1 samples, the longer FFT's is $N_1 x N_t$ samples. Apply a time-domain multiplicative window with the desired low pass filter transfer function to the longer FFT, compute the FFT, and discard all but every N_t *th* FFT channel. That is, retain only those FFT channels whose center frequencies correspond to a center frequency in the nominal length N_1 transform. Now shift the $N_1 x N_t$ sample input vector by N_1 samples and repeat the procedure. This process is equivalent to a polyphase-FFT filter bank, implementing a bank of identically shaped FIR band pass filters each centered on the channel center frequencies of the nominal, N_1-long FFT.

In summary, the polyphase-FFT filter bank operates on a time aperture larger than its resolution, allowing channel shapes to encompass any transfer function that can be implemented as an $N_1 x N_t$ tap FIR filter. With as few as 8 taps per polyphase branch it is possible to reduce the worst case processing loss to less than 1 dB. In fact, by including polyphase-filter preprocessing step prior to computing the FFT, as much as 2.6 to 2.9 dB can be gained in worst-case processing loss over Hanning or Blackman windowed FFT's. In large FFT systems such as this one, the polyphase preprocessing has a small computational cost when compared to the FFT.

However, substantial gains over windowed-FFT techniques are made even for systems with 2 or 4 tap polyphase branches, making polyphase preprocessing more than appropriate for smaller spectrum analyzer systems. For more discussion of polyphase-FFT filter banks, see [2].

The main cost incurred in the polyphase preprocessing is in memory. The polyphase preprocessing requires that N_1N_t input points must be stored at any one time, whereas in the windowed FFT system only N_1 points of storage were required. This storage requirement is mitigated somewhat by the fact that the points to be stored are of the input word length which is typically much shorter than the FFT arithmetic word length. In addition, the polyphase coefficients may be quantized to only slightly longer than the input word length without significant loss. In our implementation, the input word length is 8 bits, and 12 bits is sufficient for the polyphase coefficients. Due to the fact that our desired channel transfer functions are based on ideal band pass filters, the prototype low pass polyphase transfer function does not deviate significantly from an ideal low pass filter. Correspondingly, the polyphase coefficients do not deviate significantly from the Fourier transform of an ideal low pass filter, the Sinc function. As a result, the maximum rate of change of coefficients is limited, and a minimum run length can be found for quantized coefficients. For 12 bit coefficient quantization, the minimum run length is slightly more than 2048 coefficients, allowing for effective use of run-length compression to minimize the coefficient memory cost.

The minimum run length is greater than the row (second) FFT length, guaranteeing that there will be at most one transition per row. As a result, the rows may be separately run length compressed, simply by identifying two coefficient values and the location of the transition. This is advantageous in a pipelined signal processor, where the order of data points into the board may be scrambled both within columns and by columns, because it allows run length coding to be efficiently implemented as shown in Figure 2.

Hardware simplifications result by setting the real and imaginary polyphase coefficients equal. Since the FFT is to be performed with the real-adjust algorithm, the real samples represent even time indices while the imaginary samples represent the odd time indices. Constraining real and imaginary polyphase coefficients to be identical means that the impulse response of the prototype low pass FIR filter is constrained to change values only on even samples. Such a constraint will alter the transfer

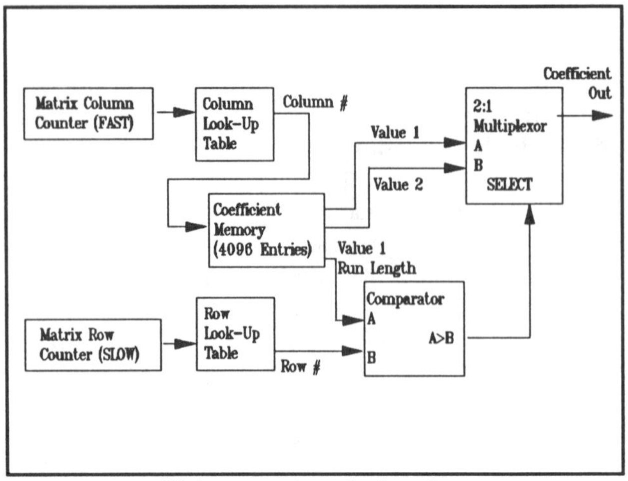

Figure 2: Coefficient run length decoder

function of the resulting filter. One can use a matrix formulation of the FFT, ($N_1 \times N_t$ by 2 real-only points) to examine the transfer function of the resulting filter. Since the impulse response only changes value on even samples, the initial two column FFT's ($N_1 \times N_t$) are identical, and, in fact, both are the desired low pass prototype transfer functions centered on the zero frequency channel. The relevant portions of the entire transfer function are computed by taking the two-point FFT's of the "twiddle-in-the-middle" complex rotation factors $\exp(-2\pi j \times (row\# \times col\#)/N_1)$ in the passband and near transition band regions of the low pass initial FFT results. The resulting transfer function is mainly low pass, with a replicant passband, attenuated by 135 dB, centered at the Nyquist frequency (π). For a system with a 41 dB INR sinusoidal input, this results in a spur 28 dB below the noise floor in the output, requiring about 360,000 accumulations, or about 5.2 hours to reach a 0 dB channel output Signal-to-Noise Ratio (SNR). It is also important to note that this spur is at the limit of the 24-bit representation, since the power of the smallest 24-bit number ($2^{-23} + 0j$) is 141.5 dB down from the largest power value ($-1 - j$), and, as such, is at about the same level as output arithmetic and quantization noise.

4. Finite Word Length Effects

The spectrum analyzer has been designed with 8 bit input quantization based on considerations of A/D converter technology and the maximum expected RFI levels. Noise due to finite word lengths can be divided between the A/D converter and the spectrum analyzer arithmetic. A balanced system would have roughly equal contributions from each source. The purpose of the SETI system is to detect small signals in the presence of interference and noise. Arithmetic and quantization noise will degrade the system performance by effectively increasing the input noise level, decreasing the SNR. The resulting noise power out will be:

$$\sigma^2_{out} = \sigma^2_{in} + \sigma^2_{NQ} + \sigma^2_{Arith} = \sigma^2_{in}\left(1 + \frac{\sigma^2_{NQ}}{\sigma^2_{in}}\left(1+\frac{\sigma^2_{Arith}}{\sigma^2_{NQ}}\right)\right)$$

where σ^2_{in} and σ^2_{out} are the input and output noise power, and σ^2_{NQ} and σ^2_{Arith} are the quantization and arithmetic noise respectively. It is apparent that the sensitivity loss is:

$$\frac{\sigma^2_{out}}{\sigma^2_{in}} = 1 + \frac{\sigma^2_{NQ}}{\sigma^2_{in}}\left(1+\frac{\sigma^2_{Arith}}{\sigma^2_{NQ}}\right),$$

In order to put the input noise and the two digital noise factors into the same units, a translation from RMS noise (volts) to quantization levels must be used; however, given a fixed input quantizer word length, it is apparent that the loss will be controlled by the ratio of the arithmetic to quantization noise.

Using the techniques in [4], the noise due to fixed-point multiplication roundoffs present in an output can be computed. This computation assumes that all roundoff noise sources are white, mutually uncorrelated, zero mean and uniformly distributed with a maximum value of 1/2 one least significant bit. A radix-4 decimation-in-frequency FFT implementation was assumed. Each radix-4 FFT stage scales the data by 1/4 to ensure no overflows occur. Blind overflow-protection scaling of the FFT stages would result in few significant bits to represent uncorrupted, noise-only or weak signal outputs. so the 24 bit words must be automatically renormalized at some point in the computation.

Note that up-shifting the data at an intermediate processing stage to renormalize it amplifies all sources of noise prior to the renormalization. As a result, the ratio of the arithmetic noise to input quantization noise will depend on the renormalization shift. When the renormalization shift is small, the majority of the arithmetic noise is due to the final stages of arithmetic, after the renormalization, and the renormalizing reduces the relative effect of the fixed-point arithmetic. Note that the amount of renormalization shift depends on the maximum peak in the spectrum. In most cases, strong interference will be wideband relative to a spectrum analyzer channel (19 Hz), allowing a significant (> 1) renormalization shift to be used.

Because renormalization amplifies noise sources before it, the point in processing at which data renormalization occurs will be a factor in the overall arithmetic noise. In a pipelined processor implementation, renormalization requires a buffer capable of holding the entire vector. The matrix FFT algorithm was chosen for the FFT implementation based on experience with the SETI prototype [1]. This initial reason for this choice was because the algorithm concentrates pipeline delay memories and coefficient memory in a central matrix transposition and vector multiplication stage. In fixed-point FFT's, another advantage becomes apparent: the matrix transposition buffer provides a natural location for the vector renormalization. Renormalization can therefore be performed at the corner-turn memory, prior to the twiddle multiplication stage, without the cost of an additional buffer.

Making the usual assumption of white, zero mean, uniformly distributed input quantizer noise, noise due to the input quantizer was computed. The resulting ratio of the Arithmetic Noise to input quantization noise is given in TABLE I. These results can then be used to calculate arithmetic losses at a given input SNR.

TABLE I: Arithmetic Noise / Input Quantization Noise

Renormalization Shift	Input 8 bits	Quantization 10 bits
0	-0.6 dB	11.4 dB
1	-6.7 dB	5.4 dB
2	-12.7 dB	-0.6 dB

Losses due to finite word lengths were predicted by comparing the predicted SNR with the SNR that would be output if the input quantization and arithmetic were infinitely precise. The ideal, infinite precision SNR is given by:

$$SNR_{ideal\ out}(dB) = SNR_{in}(dB) + Gain_{FFT}(dB) + Gain_{PP}(dB) - EQNB_{PP}(dB)$$

where the $Gain_{FFT}$ is the gain due to the number of

channels in the FFT (2^{22} channels = 66.23 dB), $Gain_{PP}$ is the gain of the polyphase-FFT filter transfer function, dependent on the signal's frequency within the resolved FFT channel, and $EQNB_{PP}$ is the equivalent noise bandwidth of the polyphase-FFT filter transfer function relative to an ideal bandpass filter. For our analysis cases, we will consider a strong signal positioned within an FFT channel such that the processing loss, $Gain_{PP}+EQNB_{PP}$, is 1 dB, an achievable worst case processing loss for 8-tap polyphase filters.

Losses at various input SNR's are shown in TABLE II. It is important to remember that at input SNR's above 40 dB (output SNR's above 106 dB), quantization spurs from an 8 bit input quantizer will become noticeable in the output, defining the upper limit of the spectrum analyzer's useful dynamic range in the presence of strong sinusoidal signals [5].

TABLE II: Loss Due to Finite Word Lengths

Input Quantizer	SNR_{in}	Ideal SNR_{out}	Predicted Loss
8 bits	30 dB	95.2 dB	0.1 dB
8 bits	35 dB	100.2 dB	0.2 dB
8 bits	38 dB	103.2 dB	0.3 dB
8 bits	41 dB	106.2 dB	0.6 dB
8 bits	47 dB	112.2 dB	1.9 dB
10 bits	47 dB	112.2 dB	0.5 dB

6. Conclusions

This paper has described the latest results in a continuing effort of building more sensitive and broader bandwidth multichannel spectrum analyzers [1,6]. The thrusts of the current effort, described in this paper, have been in the areas of signal processing, machine architecture, and technology utilization. The introduction of the polyphase-FFT provides signal processing gain superior to windowed FFT's, nearly independent of frequency location within a channel. Improvements in machine architecture have allowed the use of fixed-point arithmetic, and have saved cost in both memory and arithmetic parts. Through the effective utilization of advances in technology, the bandwidth and number of channels of a single processor unit have been each been doubled. The signal processor will be capable of operation at the limits of its 8 bit A/D converter, allowing up to 41 dB input SNR with less than 1 dB loss in sensitivity. The system will be constructed at the Jet Propulsion Laboratory over the next four years, with the first 80 MHz slice being completed in 1994.

Acknowledgement

The research described in this paper was carried out by the Jet Propulsion Laboratory, California Institute of Technology, under a contract with the National Aeronautics and Space Administration.

References

1. M.P. Quirk, H.C. Wilck, M.F. Garyantes, and M.J. Grimm, "A Wideband, High-Resolution Spectrum Analyzer," *Transactions of the ASSP*, Vol. 36, p. 1858, 1988.

2. G.A. Zimmerman and S. Gulkis, "Polyphase DFT Spectrum Analysis for the Search for Extraterrestrial Intelligence Sky Survey," *TDA Progress Report 42-107*, Vol. July-September 1991, Jet Propulsion Laboratory, Pasadena, California, 1991.

3. F.J. Harris, "On the Use of Windows for Harmonic Analysis with the Discrete Fourier Transform," *Proceedings of the IEEE*, Vol. 66 No. 1, p. 51, 1978.

4. A.V. Oppenheim and R.W. Schafer, *Digital Signal Processing*, Prentice-Hall, 1975.

5. M. Flanagan, "The Behavior of Quantization Spectra as a Function of Signal-to-Noise Ratio," *TDA Progress Report 42-107*, Vol. July-September 1991, Jet Propulsion Laboratory, Pasadena, California, 1991.

6. G.A. Morris and H.C. Wilck, "JPL 2^{20} Channel 300 MHz Bandwidth Digital Spectrum Analyzer," presented at the IEEE International Conference on Acoustics, Speech and Signal Processing, Tulsa, Oklahoma, 1978.

Algorithms for Length 15 and 30 Discrete Cosine Transform

Zhongde Wang, G.A. Jullien, W. C. Miller

VLSI Research Group, Department of Electrical Engineering
University of Windsor, Windsor, Ontario, Canada N9B 3P4

Abstract

Efficient algorithms for computing the DCT abound in the literature. In this paper we show that particular efficiencies can be realized if we use a sample length radix of 15. Using sparse matrix factorization, algorithms for length 15 and 30 discrete cosine transforms (DCT) are developed that require only a single cascaded multiplier in the centre of the flow graph. We show that these algorithms are more efficient than the commonly used radix-2 algorithms for the DCT, especially in terms of the number of required multiplications.

1 Introduction

Since the discovery of the discrete cosine transform (DCT) in 1974 [1], it has become the major tool for image coding [2]. During the last decade, many algorithms for the DCT have been developed [3-8]. Most algorithms for the DCT are radix-2 based [3-7]. However, if algorithms are evaluated with counts of operations, one may find that the radix-2 algorithms are not the most efficient. Recently, Lee has given an index mapping scheme for computing the DCT for lengths which can be factorized into mutually prime factors [8]. The index mapping procedure given by Lee, however, is rather complicated. Since for most DCT applications, image compression for example, the transform length is not very large ($n \leq 32$), it is more desirable to develop special algorithms to compute the DCT for such small lengths. In this paper, we develop algorithms for length 15 and 30 DCTs. They can be used as basic blocks to build algorithms for DCT of length $15*2^k$.

2 Decomposition of the DCT matrix

The commonly used DCT (DCT-II [9]) matrix is given by:

$$C_N^{II} = \left[k_m \cos m(n+\frac{1}{2})\frac{p}{N} \right] \qquad (1)$$

$$m, n = (0, 1, \ldots\ldots, N-1)$$

$$k_m = \begin{cases} \frac{\sqrt{2}}{2} & m=0 \\ 1 & \text{otherwise} \end{cases}$$

Since k_m does not affect the analysis, it will be omitted in the following discussion.

It has been shown that for N even (N=2M), C_N^{II} can be factorized into matrices with lower order without any multiplications [9]

$$C_{2M}^{II} = P \begin{bmatrix} C_M^{II} & 0 \\ 0 & C_M^{IV} \end{bmatrix} A_{2M} \qquad (2)$$

where

$$P = \begin{bmatrix} 10\ldots\ldots\ldots\ldots0 \\ 0\ldots\ldots\ldots\ldots01 \\ 010\ldots\ldots\ldots0 \\ 0\ldots\ldots\ldots010 \\ \ldots\ldots\ldots\ldots\ldots \\ \ldots\ldots\ldots\ldots\ldots \\ \ldots\ldots\ldots\ldots\ldots \\ 0\ldots010\ldots\ldots \\ 0\ldots\ldots010\ldots. \end{bmatrix} \qquad (3)$$

is a permutation matrix,

$$C_N^{IV} = \left[\cos (m+\frac{1}{2})(n+\frac{1}{2})\frac{\pi}{N} \right] \qquad (4)$$

$$m,n = 0, 1, \ldots\ldots, N-1$$

is the even type IV DCT matrix, and

$$A_{2M} = \begin{bmatrix} I_M & \hat{I}_M \\ \hat{I}_M & -I_M \end{bmatrix} \qquad (5)$$

is the even-odd decomposition matrix which involves additions only. I is the identity matrix; $\hat{I}$ is the antidiagonal identity matrix.

We now introduce a new factorization that will ultimately lead to the single cascaded multiplier implementation. For N odd (N=2M+1), the DCT matrix can be factorized in the same way as that given in (2):

$$C^{II}_{2M+1} = \mathbf{P}\begin{bmatrix} C^{IIO}_{M+1} & 0 \\ 0 & C^{IVO}_{M} \end{bmatrix} A_{2M+1} \quad (6)$$

where **P** is the same permutation matrix as given by (3),

$$A_{2M+1} = \begin{bmatrix} I_M & & \hat{I}_M \\ 0 & 1 & 0 \\ \hat{I}_M & & -I_M \end{bmatrix} \quad (7)$$

and both

$$C^{IIO}_{M+1} = \left[\cos\left(m(n+\tfrac{1}{2})\,\frac{2\pi}{2M+1}\right)\right] \quad (8)$$

$$m,n = 0,\ 1,\ \ldots\ldots,\ M$$

and

$$C^{IVO}_{M+1} = \left[\cos\left((m+\tfrac{1}{2})(n+\tfrac{1}{2})\,\frac{2\pi}{2M+1}\right)\right] \quad (9)$$

$$m,n = 0,\ 1,\ \ldots\ldots,\ M-1$$

are odd type DCT [3] (ODCT-II and ODCT-IV) matrices.

C^{IV}_{2M+1} can be factorized into lower order matrices as

$$C^{IV}_{2M+1} =$$

$$\begin{bmatrix} I_{M+1} & 0 \\ 0 & D_M \end{bmatrix} A_{2M+1} \begin{bmatrix} R_{M+1} & 0 \\ 0 & Q_M \end{bmatrix} A_{2M+1} \begin{bmatrix} I_{M+1} & 0 \\ 0 & D_M \end{bmatrix} \quad (10)$$

where A_{2M+1} is given by (7),

$D_M = \mathrm{diag}[\ 1,-1,1,-1,\ \ldots\ldots,1,\ -1]$ for M even

$= \mathrm{diag}[-1,1,-1,1,\ \ldots\ldots,1,-1]$ for M odd

The elements of R_{M+1} and Q_M at the ith row and jth column are given by

$$r(i,j) = \frac{1}{2}\{\cos[(i+\tfrac{1}{2})(j+\tfrac{1}{2})\tfrac{\pi}{2M+1}] + (-1)^{M-i}\cos[(2M-i+\tfrac{1}{2})(j+\tfrac{1}{2})\tfrac{\pi}{2M+1}]\}$$

and

$$q(i,j) = \frac{1}{2}\{\cos[(i+\tfrac{1}{2})(j+\tfrac{1}{2})\tfrac{\pi}{2M+1}] - (-1)^{M-i}\cos[(2M-i+\tfrac{1}{2})(j+\tfrac{1}{2})\tfrac{\pi}{2M+1}]\}$$

Using trigonometric identities

$$\cos a + \cos b = 2\cos\frac{a+b}{2}\cos\frac{a-b}{2}$$
$$\cos a - \cos b = 2\sin\frac{a+b}{2}\sin\frac{b-a}{2} \quad (11)$$

r(i,j) and q(i,j) can be simplified. For M-i even, let $M\text{-}i=2p_1=2M-2p_2$.

$$p_1 \in \left[0,\ \frac{M-1}{2}\right] \qquad p_2 \in \left[\frac{M+1}{2},\ M-1\right]$$

Then r(i,j) and q(i,j) become:

$$r(i,j) = \cos[(i+\tfrac{1}{2})\tfrac{\pi}{2}]\cos[(M-i)(j+\tfrac{1}{2})\tfrac{\pi}{2M+1}] = \frac{\sqrt{2}}{2}(-1)^{\lfloor\frac{M+1}{2}\rfloor}\cos[p_1(j+\tfrac{1}{2})\tfrac{2\pi}{2M+1}] \quad (12a)$$

and

$$q(i,j) = \sin[(i+\tfrac{1}{2})\tfrac{\pi}{2}]\sin[(M-i)(j+\tfrac{1}{2})\tfrac{\pi}{2M+1}] = \sin[(i+\tfrac{1}{2})\tfrac{\pi}{2}]\sin[(2M+1-2p_2-1)(j+\tfrac{1}{2})\tfrac{\pi}{2M+1}]$$

$$= \frac{\sqrt{2}}{2}(-1)^{\lfloor\frac{M}{2}\rfloor+j}\cos[(p_2+\tfrac{1}{2})(j+\tfrac{1}{2})\tfrac{2\pi}{2M+1}] \quad (12b)$$

For M-i odd, let $M\text{-}i=2p_2+1=2M-2p_1+1$.

$$p_1 \in \left[\frac{M+1}{2},\ M\right] \qquad p_2 \in \left[0,\ \frac{M-1}{2}\right]$$

Then r(i,j) and q(i,j) become

$$r(i,j) = \sin[(i+\tfrac{1}{2})\tfrac{\pi}{2}]\sin[(M-i)(j+\tfrac{1}{2})\tfrac{\pi}{2M+1}] = \sin[(i+\tfrac{1}{2})\tfrac{\pi}{2}]\sin[(2M+1-2p_1-1)(j+\tfrac{1}{2})\tfrac{\pi}{2M+1}]$$

$$= \frac{\sqrt{2}}{2}(-1)^{\lfloor\frac{M}{2}\rfloor+j}\cos[p_1(j+\tfrac{1}{2})\tfrac{2\pi}{2M+1}] \quad (13a)$$

and

$$q(i,j) = \cos[(i+\tfrac{1}{2})\tfrac{\pi}{2}]\cos[(M-i)(j+\tfrac{1}{2})\tfrac{\pi}{2M+1}] = \frac{\sqrt{2}}{2}(-1)^{\lfloor\frac{M+1}{2}\rfloor}\cos[(p_2+\tfrac{1}{2})(j+\tfrac{1}{2})\tfrac{2\pi}{2M+1}] \quad (13b)$$

where $\lfloor . \rfloor$ represents the integer part of the argument. Therefore, except for some permutation and sign changes, R_{M+1} and Q_M are essentially equivalent to C^{IIO}_{M+1} and C^{IVO}_{M}, respectively. Thus, for N=30, *and without any multiplications*, we can decompose C^{II}_{30} into C^{IIO}_{8},

C_7^{IVO}, R_8, and Q_7. It is sufficient for our algorithm to discuss the further decomposition of C_8^{IIO} and C_7^{IVO} matrices only.

3 Decomposition of C_8^{IIO} and C_7^{IVO}

The explicit form of C_8^{IIO} and C_7^{IVO} are given by

$$C_8^{IIO} = \begin{bmatrix} 1 & 1 & 1 & 1 & 1 & 1 & 1 & 1 \\ c_2 & c_6 & 0.5 & c_{14} & -c_{12} & -c_8 & -c_4 & -1 \\ c_4 & c_{12} & -0.5 & -c_2 & -c_6 & -c_{14} & -c_{12} & 1 \\ c_6 & -c_{12} & -1 & -c_{12} & c_6 & c_6 & -c_{12} & -1 \\ c_8 & -c_6 & -0.5 & c_4 & c_{12} & -c_2 & -c_{14} & 1 \\ 0.5 & -1 & 0.5 & 0.5 & -1 & 0.5 & 0.5 & -1 \\ c_{12} & -c_6 & 1 & -c_6 & c_{12} & c_{12} & -c_6 & 1 \\ c_{14} & -c_{12} & 0.5 & -c_8 & c_6 & -c_4 & c_2 & -1 \end{bmatrix} \tag{14}$$

and

$$C_7^{IVO} = \begin{bmatrix} c_1 & c_3 & c_5 & c_7 & c_9 & c_{11} & c_{13} \\ c_3 & c_9 & 0 & -c_9 & -c_3 & -c_3 & -c_9 \\ c_5 & 0 & -c_5 & -c_5 & 0 & c_5 & c_5 \\ c_7 & -c_9 & -c_5 & c_{11} & c_3 & -c_{13} & -c_1 \\ c_9 & -c_3 & 0 & c_3 & -c_9 & -c_9 & c_3 \\ c_{11} & -c_3 & c_5 & -c_{13} & -c_9 & c_1 & -c_7 \\ c_{13} & -c_9 & c_5 & -c_1 & c_3 & c_7 & c_{11} \end{bmatrix} \tag{15}$$

where $c_i=\cos(i\pi/30)$. With some manipulation and using the following trigonometric identities

$c_2 - c_8 = c_{12}$; $c_4 - c_{14} = c_6$; $c_6 - c_{12} = 0.5$;
$c_7 + c_{13} = c_3$; $c_1 - c_{11} = c_9$; $c_1 + c_{11} = 2c_5 c_6$;
$c_{13} - c_7 = 2c_5 c_{12}$; $c_1 + c_{11} + c_{13} - c_7 = c_5$,

C_8^{IIO} and C_7^{IVO} can be factorized in a similar way to the Winograd small N DFT algorithm [11] as:

$$C_8^{IIO} = A F^T U F B \tag{16}$$

and

$$C_7^{IVO} = G^T T G \tag{17}$$

where $U = \mathrm{diag}\{m_1, m_2, \ldots\ldots m_9\}$ and $T = \mathrm{diag}\{t_1, t_2, \ldots\ldots t_9\}$ are diagonal matrices which contain all necessary multiplications.

A, B, F, and G are so called incidence matrices which contain no multiplications. Their elements are 1, -1, or 0. The multiplication coefficients are given by Table 1.

Table 1 Multiplier Coefficients for 15 DCT-II

$m_1 = 1$	$m_2 = 1.875$	$m_3 = 0.75(c_6+c_{12})$
$m_4 = -1.25$	$m_5 = 0.5(c_4+c_{14}-c_2-c_8)$	$m_6 = -1.5$
$m_7 = -0.5(c_6+c_{12})$	$m_8 = -\frac{1}{2}(c_4+c_{14}+c_2+c_8)$	$m_9 = 0.5(c_2+c_8)$
$t_1 = -c_3$	$t_2 = \frac{1}{4}(c_1+c_{11}+c_7-c_{13})$	$t_3 = c_3 + c_9$
$t_4 = -c_5$	$t_5 = -1.5(c_3+c_9)$	$t_6 = c_3 - c_9$
$t_7 = 1.5(c_9 - c_3)$	$t_8 = 1.25c_5$	$t_9 = 1.5c_3$

The explicit forms of A, B, F, and G are given by:

$$A=\begin{bmatrix} 1&0&0&0&0&0&0&0\\ 0&-1&0&0&-1&0&0&0\\ 0&0&1&0&0&0&0&1\\ 0&0&0&-1&0&0&-1&0\\ 0&1&0&0&-1&0&0&0\\ 0&0&0&0&0&-1&0&0\\ 0&0&0&1&0&0&-1&0\\ 0&0&-1&0&0&0&0&1 \end{bmatrix} \tag{18}$$

$$B=\begin{bmatrix} 0&0&0&0&0&0&0&1\\ 0&0&0&1&0&0&1&0\\ 1&0&0&0&0&1&0&0\\ 0&1&0&0&1&0&0&0\\ 0&0&0&-1&0&0&1&0\\ 0&0&1&0&0&0&0&0\\ 0&-1&0&0&1&0&0&0\\ -1&0&0&0&0&1&0&0 \end{bmatrix} \tag{19}$$

$$F=\begin{bmatrix} 1&0&0&1&0&0&0&0\\ 0&1&0&0&0&0&0&0\\ 0&0&1&0&0&0&0&0\\ 0&0&0&1&0&0&0&0\\ 0&0&0&0&1&0&0&0\\ 0&1&0&0&0&1&0&0\\ 0&0&-1&0&0&0&1&0\\ 0&0&0&0&0&0&0&1\\ 0&0&0&1&0&0&0&0 \end{bmatrix}\begin{bmatrix} 1&0&0&0&0&1&0&0\\ 0&1&0&0&0&0&0&0\\ 0&0&1&0&0&0&0&0\\ 0&1&0&1&0&0&0&0\\ 0&0&0&0&1&0&0&0\\ 0&0&0&0&0&1&0&0\\ 0&0&0&0&0&0&1&0\\ 0&0&0&0&0&0&0&1 \end{bmatrix}\begin{bmatrix} 1&&0\\ &\begin{matrix}1&1\\1&-1\end{matrix}&\\ 0&&I_5 \end{bmatrix} \tag{20}$$

$$G=\begin{bmatrix} 0&1&0&0&1&0&0\\ 1&0&0&0&0&0&0\\ 0&1&0&0&0&0&0\\ 0&0&1&0&0&0&-1\\ 0&0&0&1&0&0&0\\ 0&0&0&0&1&0&0\\ 0&0&0&0&0&1&0\\ 0&0&0&0&0&0&1\\ 0&0&0&1&0&1&0 \end{bmatrix}\begin{bmatrix} 1&0&0&0&0&0&1\\ 0&1&0&-1&0&0&0\\ 0&0&1&0&0&0&0\\ 0&0&0&1&0&0&0\\ 0&0&0&0&1&-1&0\\ 0&0&0&0&0&1&0\\ 1&0&0&0&0&0&-1 \end{bmatrix}\begin{bmatrix} 1&0&0&0&0&1&0\\ 0&1&0&0&0&0&0\\ 0&0&1&0&0&0&0\\ 0&0&0&1&0&0&1\\ 0&0&0&0&1&0&0\\ 1&0&0&0&0&-1&0\\ 0&0&0&1&0&0&-1 \end{bmatrix} \tag{21}$$

Based on these matrix factorizations, we can draw the signal flowgraph for the length-15 DCT-II as shown in Fig. 1.

4 Computational Complexity

Direct counting from Fig. 1 shows that the length-15 DCT-II requires 67 additions and 18 multiplications. Equations (10), (12) and (13) show that the length 15 DCT-IV requires the same number of multiplications as the length 15 DCT-II, but 14 more additions. Therefore, the length-30 DCT-II requires 36 multiplications and 2x67+14+30=178 additions. Table 2 summarizes the computational complexity of the proposed algorithms, where M and A represent the number of multiplications and additions, respectively.

An important point is the single cascaded multiplier, this will reduce the scaling requirement for fixed point computation of the algorithm.

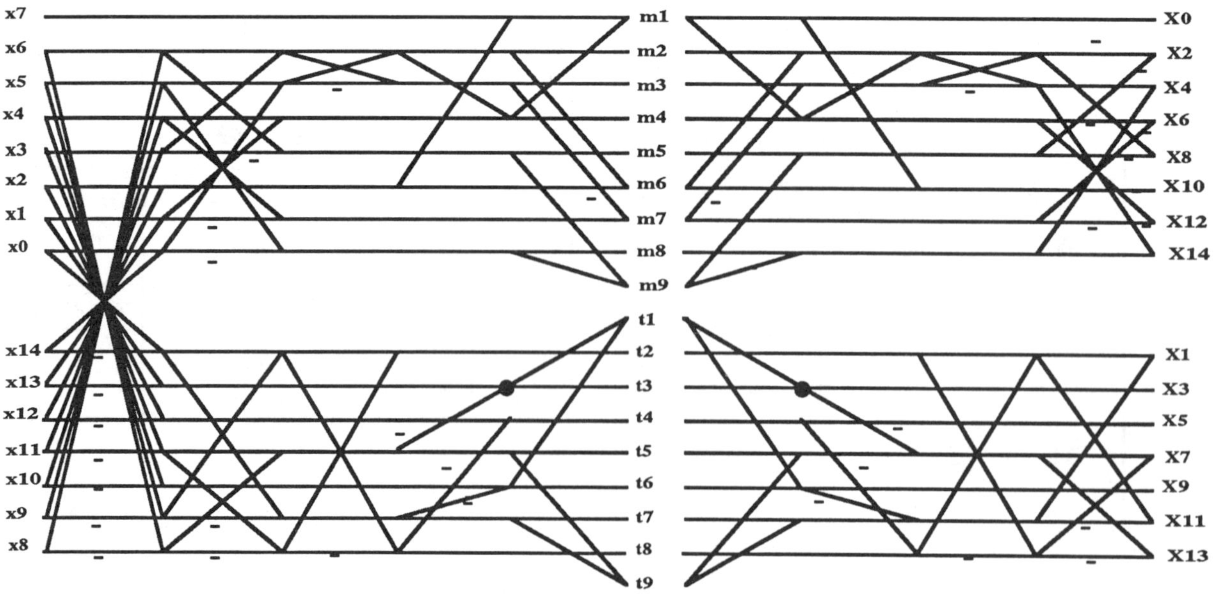

Fig. 1 Signal Flowgraph for length 15 DCT-II

Table 2 Comparison of Computational Complexity

	New Algorithm				Lee's Index Mapping				Radix-2 Algorithm			
N	M	$\frac{M}{N}$	A	$\frac{A}{N}$	M	$\frac{M}{N}$	A	$\frac{A}{N}$	M	$\frac{M}{N}$	A	$\frac{A}{N}$
15	18	1.2	67	4.47	20	1.33	77	5.13				
16									32	2.0	81	5.06
30	36	1.2	178	5.67	55	1.83	198	6.6				
32									80	2.5	209	6.53

For comparison, Table 2 also lists the computational complexity of two other algorithms: the efficient radix-2 algorithm; and Lee's index mapping algorithm. From table 2 one may conclude that the proposed algorithm is more efficient than either the radix-2 algorithm or Lee's index mapping algorithm.

The proposed algorithms for computing length 15 and 30 DCT-II can be used as bases for developing algorithms for computing length-$15x2^k$ DCT-II by using the decomposition procedure developed in [6]. We would like to point out that extended algorithms developed in this way are also more efficient than the radix-2 FCT and Lee's index mapping algorithms.

5 Conclusions

In this paper we have developed special algorithms for computing length 15 and 30 DCT-II, based on sparse matrix factorization. Complete matrix factorizations required to compute the algorithms are presented and we have shown, by computational complexity comparisons, that these algorithms are more efficient than either the radix-2 algorithm or Lee's index mapping algorithm. Since length 15 and 30 are suitable lengths for image processing, we expect that there would be useful hardware implementations based on these flowgraphs. The single cascaded multiplication is a particularly interesting feature for fixed point implementation.

6 References

[1] N. Ahmed, T. Natarajan, and K. R. Rao (1974): Discrete cosine transform, *IEEE Trans. Comput.*, vol. C-23, pp. 90-93.

[2] P. H. Ang, P. A. Ruetz and D. Auld (1991): Video compression makes big gains, *IEEE Spectrum*, Oct., pp. 16-19.

[3] B. G. Lee (1984): A new algorithm for the discrete cosine transform, *IEEE Trans. Acoust. Speech, Sig. Process.* vol. ASSP-32, pp.1243-45.

[4] Y. Morikawa, H. Hamada, and H. Yamane (1985): A fast algorithm for the cosine transform based on successive order reduction of the Tchebycheff polynomial, *Trans. Inst. Elec. Com. Eng. Japan*, J68-A, 173-80.

[5] M. Vetterlie and H. Nussbaumer (1984): Simple FFT and DCT algorithms with reduced number of operation, *Sig. Proc.* 6, pp.256-78.

[6] Z. Wang (1985): On computing the discrete Fourier and cosine transforms, *IEEE Trans. Acoust. Speech, Sig. Process.* vol. ASSP-33, pp.1141-44.

[7] H. S. Hou (1987): A fast recursive algorithm for computing the discrete cosine transform, ibid, ASSP-35, pp.1455-1461.

[8] B. G. Lee (1987): Input and output mapping for a prime-factor-decomposited computation of discrete cosine transform, *IEEE Trans. Acoust. Speech, Signal Process.* vol ASSP-37, pp.237-244.

[9] Z. Wang, (1984): Fast algorithms for the discrete W transform and for the discrete Fourier transform", *IEEE Trans. Acoust., Speech, Signal Process.* vol. ASSP-32, pp.803-816.

[10] Z. Wang and B. R. Hunt, (1985): The discrete W transform, *Appl. Math. Comput.* vol.16, pp. 19-48.

[11] S. Winograd (1978): On computing the discrete Fourier transform, *Math. Comput.* vol. 32, pp.178-199.

The Discrete Polynomial Transform (DPT), Its Properties and Applications*

Shimon Peleg
Elec. Eng. & Comp. Sc.
University of Cal., Davis
Davis, CA 95616

Boaz Porat
Electrical Engineering
Technion-I.I.T.
Haifa, Israel 32000

Benjamin Friedlander
Elec.Eng. & Comp. Sc.
University of Cal., Davis
Davis, CA 95616

Abstract

The Discrete Polynomial Transform (DPT) is a new tool for analyzing constant-amplitude continuous-phase signals. It is based on modeling the signal phase by a polynomial function of time (on a finite interval). The main properties of the DPT are its ability to identify the degree of the phase polynomial and to estimate its coefficients. The transform is robust to deviations from the ideal signal model, such as slowly-varying amplitude, additive noise and non-polynomial (but continuous) phase. The paper motivates the transform by showing its effect on polynomial-phase signals. Then its main mathematical properties are listed and proofs are provided for some. It is shown how to use those properties to derive estimation and classification algorithms for continuous-phase signals. Some recent results on the statistical accuracy of these algorithms are stated. Finally, applications of the DPT to certain radar and communication problems are discussed.

1 Introduction

This paper introduces a new nonlinear transform for discrete-time complex signals. The transform is aimed at analyzing signals with constant amplitude and continuous-phase. Signals like that are common in radar and communication applications. For example, radar return echoes from maneuvering target have continuous-phase characteristics, depending on the target maneuvers.

A recent paper [3] has introduced the PT as a new nonlinear transform for continuous-time complex signals. The current paper deals exclusively with the discrete version of the PT - the DPT.

In the next section we define the operator $DP_M[s_n,\tau]$, the transform $DPT_M[s_n,\omega,\tau]$ and explore their relation to discrete-time complex signals having constant amplitude and polynomial-phase.

In the third section, motivated by the properties of the transform, we introduce algorithms for estimating the parameters of polynomial-phase signals. Then we show that we can classify such signals according to their degree. Using the Weierstrass approximation theorem we consider signals whose phases are continuous, but not necessarily polynomial.

[1]This work was supported in part by the University of California MICRO program and Applied Signal Technology, Inc.

In the final section we discuss the performance of the algorithms in presence of additive white and when the signal's amplitude is not constant. We give a brief survey of applications followed by detailed example.

2 The operator $\mathcal{DP}_M[s_n,\tau]$ and the transform $DPT_M[s_n,\omega,\tau]$

This section introduces the operator $\mathcal{DP}_M[s_n,\tau]$, the transform $DPT_M[s_n,\omega,\tau]$ and explore their relation to discrete-time complex signals having constant amplitude and polynomial phase.

Consider a discrete-time complex signal s_n and let M and τ be positive integers. Let $L+(M-1)\tau$ be the number of signal samples (where L is nonnegative). The operator $\mathcal{DP}_M[s_n,\tau]$ is defined by

$$\mathcal{DP}_M[s_n,\tau] = \prod_{q=0}^{M-1} [s_{n+q\tau}^{\$_q}]^{\binom{M-1}{q}}, \qquad n = 1,...,L, \quad (1)$$

where $\$_q$ is defined by

$$s_n^{\$_q} = \begin{cases} s_n, & \text{when q is even,} \\ s_n^*, & \text{when q is odd.} \end{cases} \quad (2)$$

M is called the order of the operator, and τ is called the delay parameter.

The following special cases illustrate this definition.

$$\mathcal{DP}_1[s_n,\tau] = s_n. \quad (3)$$

$$\mathcal{DP}_2[s_n,\tau] = s_n s_{n+\tau}^*. \quad (4)$$

$$\mathcal{DP}_3[s_n,\tau] = s_n [s_{n+\tau}^*]^2 s_{n+2\tau}. \quad (5)$$

The motivation for introducing the $\mathcal{DP}_M[s_n,\tau]$ is in the following two theorems.

Theorem 1 *(The Main Theorem)*
Let s_n be a discrete-time complex signal with constant amplitude and polynomial phase, and let M be the degree of the phase polynomial, i.e.,

$$s_n = \exp\{j \sum_{m=0}^{M} a_m (n\Delta)^m\}. \quad (6)$$

1058-6393/91 $01.00 © 1991 IEEE

Then, for all positive τ

$$\mathcal{DP}_M[s_n,\tau] = \exp\{j[\omega_0 n\Delta + \varphi_0]\}, \qquad (7)$$

where

$$\omega_0 = (-1)^{M-1}M!(\tau\Delta)^{M-1}a_M, \qquad (8)$$

and

$$\begin{aligned}\varphi_0 &= (-1)^{M-1}[(M-1)!(\tau\Delta)^{M-1}a_{M-1}\\ &+ 0.5(M-1)M!(\tau\Delta)^{M-1}a_M]. \qquad (9)\end{aligned}$$

Proof: The proof is based on the theorem in the Appendix.

Thus, we can conclude that for all positive τ, $\mathcal{DP}_M[s_n,\tau]$ is a single tone with frequency ω_0.

Theorem 2 *If* s_n *is a complex constant-amplitude signal, i.e.,*

$$s_n = \exp\{j\varphi_n\}, \qquad (10)$$

and

$$\mathcal{DP}_M[s_n,1] = \exp\{j[\omega_0 n\Delta + \varphi_0]\}, \qquad (11)$$

with $\omega_0 \neq 0$*, then* s_n *has polynomial-phase of degree* M *and* a_M *is equal to*

$$a_M = (-1)^{M-1}\frac{\omega_0}{M!\Delta^{M-1}}. \qquad (12)$$

Remark: The theorem is true only for $\tau = 1$.

Proof: Since

$$\mathcal{DP}_M[s_n,1] = \exp\left\{j\sum_{q=0}^{M-1}(-1)^q\binom{M-1}{q}\varphi_{n+q}\right\} \qquad (13)$$

we get

$$\sum_{q=0}^{M-1}(-1)^q\binom{M-1}{q}\varphi_{n+q} = \omega_0 n\Delta + \varphi_0. \qquad (14)$$

Using a $\mathcal{Z}$ transformation on both sides we get

$$\sum_{q=0}^{M-1}(-1)^q\binom{M-1}{q}z^q\Phi(z) = \omega_0\frac{z}{(z-1)^2}\Delta + \varphi_0\frac{z}{z-1}, \qquad (15)$$

where

$$\Phi(z) = \mathcal{Z}\{\varphi_n\}. \qquad (16)$$

But,

$$\sum_{q=0}^{M-1}(-1)^q\binom{M-1}{q}z^q = (-1)^{M-1}(z-1)^{M-1}. \qquad (17)$$

So,

$$\Phi(z) = (-1)^{M-1}\left[\omega_0\frac{z}{(z-1)^{M+1}}\Delta + \varphi_0\frac{z}{(z-1)^M}\right]. \qquad (18)$$

Since

$$\frac{z}{(z-1)^m} = \mathcal{Z}\left\{\frac{1}{(m-1)!}\prod_{k=1}^{m-1}(n-k+1)\right\}, \qquad (19)$$

we get that

$$\begin{aligned}\varphi_n &= (-1)^{M-1}\omega_0\frac{\Delta}{M!}\prod_{k=1}^{M}(n-k+1)\\ &+ (-1)^{M-1}\varphi_0\frac{1}{(M-1)!}\prod_{k=1}^{M-1}(n-k+1) \qquad (20)\end{aligned}$$

This is a polynomial of degree M and a_M as shown in equation (12). □

By the above two theorems, the $\mathcal{DP}_M[z_n,\tau]$ can be used to identify polynomial-phase signals, find their degree and estimate their parameters.

The operator $\mathcal{DP}_M[s_n,\tau]$ has the following properties (given without proofs):

Property 1:

$$\mathcal{DP}_M[s_1 s_2,\tau] = \mathcal{DP}_M[s_1,\tau]\mathcal{DP}_M[s_2,\tau]. \qquad (21)$$

Property 2:

$$\mathcal{DP}_M[b_0,\tau] = b_0^{2^{M-1}}. \qquad (22)$$

Property 3: For $k = 0, 1, ..., M-2$

$$\mathcal{DP}_M\left[\exp\left\{j\sum_{m=0}^{k}a_m(n\Delta)^m\right\},\tau\right] = 1. \qquad (23)$$

Property 4:

$$\mathcal{DP}_M\left[\exp\left\{j\sum_{m=0}^{M-1}a_m(n\Delta)^m\right\},\tau\right] = \exp\{j\varphi_0\}, \qquad (24)$$

where φ_0 is shown in equation (9).

Property 5:

$$\mathcal{DP}_M[s_n,\tau] = \mathcal{DP}_{M-1}[s_n,\tau]\{\mathcal{DP}_{M-1}[s_{n+\tau},\tau]\}^*. \qquad (25)$$

The DPT of order M of s_n is defined as the Discrete Fourier Transform of $\mathcal{DP}_M[s_n,\tau]$, i.e.,

$$DPT_M[s_n,\omega,\tau] = \sum_{n=1}^{L}\mathcal{DP}_M[s_n,\tau]\exp\{-j\omega n\Delta\}. \qquad (26)$$

The special case of $M = 2$ yields the Discrete Ambiguity Function. The higher-order transforms are new.

A conclusion from Theorem 1 and the definition of DPT of order M is:

Let s_n be a polynomial phase signal of M degree then $|DPT_M[s_n, \omega, \tau]|$ has a global maximum in $\omega = \omega_0$ (see equation (8)).

Since

$$DPT_M[s_n, \omega, \tau] = DFT\{DP_M[s_n, \tau], \omega\}, \qquad (27)$$

all the properties of the discrete Fourier transform apply to this case. For example

$$DPT_M[s_1 s_2, \omega, \tau] = DPT_M[s_1, \omega, \tau] \otimes DPT_M[s_2, \omega, \tau]. \qquad (28)$$

The $\otimes$ is the symbol for circular convolution. So, we get that the DPT_M of multiplication of two signals is the convolution of the DPT_M of each signal.

3 Estimating the parameters of constant amplitude continuous-phase signals

This section describes the application of the DPT_M transform for estimating the parameters of continuous-phase signals. Motivated by Theorem 1 and Theorem 2 we introduce an algorithm for estimating the parameters of polynomial-phase signals and classifying such signals according to their degree. Then, we consider signals whose phase is continuous, but not necessarily polynomial.

Let us assume that the signal is known to have polynomial phase of degree M at most, and we wish to estimate the phase coefficients. The algorithm we introduce estimates the coefficients $\{a_m\}$ sequentially, starting at the highest order coefficient a_M. At each step, (except for the first one), the effect of the phase term of the higher order is removed. The Algorithm is as follows.

1) Substitute $m = M$ and $s_n^{(m)} = s_n, \quad 1 \le n \le N$.

2) Compute $\hat{a}_m$ by

$$\hat{a}_m = \frac{(-1)^{m-1}}{m!(\Delta\tau)^{m-1}} \arg\max_{\omega}\{|DPT_m[s_n^{(m)}, \omega, \tau]|\}. \qquad (29)$$

3) Let

$$s_n^{(m-1)} = s_n^{(m)} \exp\{-j\hat{a}_m(n\Delta)^m\}, \quad 1 \le n \le N. \qquad (30)$$

4) Substitute $m = m - 1$. If $m \ge 1$, go to step 2).

The choice of τ in step 2) will affect the accuracy of the estimated coefficients. Based on statistical analysis, we have found that, when the number of measurements and the signal-to-noise ratio tend to infinity, the optimal value of τ is $\lfloor N/(m+2) \rfloor$ when $m > 3$, and $\lfloor N/m \rfloor$ when $m \le 3$.

To compute step 2), it is recommended to use the FFT. The frequency that brings to maximum can be found as in the standard frequency estimation from FFT.

If the signal is a signal with polynomial-phase of degree m less then M, the algorithm gives in step 2) that $\hat{a}_M \approx 0$. The decision $\hat{a}_M = 0$ can be based on comparison to the Cramer-Rao lower bound. (See [5] for approximation for the bound.) Using this fact, we can change the algorithm such that in step 3) there is no phase removal as long as $\hat{a}_m = 0$. When m is reached such that $\hat{a}_m \neq 0$, the algorithm includes the phase removal.

Using the above property, the algorithm can be used to classify polynomial-phase signals according to their degree. So, for signals that are known to have a phase degree of M at most, we can simultaneously find the true degree and estimate the coefficients of the polynomial.

Now, consider signals with continuous-phase, but not necessarily polynomial. By the Weierstrass theorem [2], every continuous function on a closed interval can be uniformly approximated by polynomials. Thus, for sufficiently large M, we can approximate a continuous phase by a polynomial such that the approximation is satisfactory. Hence the algorithm can be used to obtain polynomial phase approximations for signals having a continuous phase.

4 Extensions and applications

In this section we briefly discuss the cases of noisy signal and of non-constant amplitude. We then mention some applications and provide an example.

Let us assume we are given a discretized noisy signal, that is

$$z_n = s_n + w_n = b_0 \exp\left\{j \sum_{m=0}^{M} a_m (n\Delta)^m\right\} + w_n. \qquad (31)$$

The statistical analysis of the case $M = 2$ is given in [4]. Through an approximate statistical analysis, we proved that the value $\tau = N/2$ minimizes the variance of the estimate. For this value of τ, the performance of the algorithm was found to be only several percent worse than the CRB, at moderate and high SNRs.

The statistical analysis of the $M > 2$ will be given in a future publication.

In [3] we developed a simple criterion for determining the operating range of the transform. When the SNR is sufficiently large, the following relation must hold

$$N \ge \frac{25}{\mathrm{SNR}} \binom{2M-2}{M-1}. \qquad (32)$$

Next we consider the case of amplitude variations. Let assume a signal of the form:

$$s_n = b_n \exp\left\{j \sum_{m=0}^{M} a_m (n\Delta)^m\right\}, \qquad (33)$$

where b_n is a real discrete function. So,

$$DPT_M[s_n, \omega, \tau] = DPT_M[b_n, \omega, \tau] \otimes DFT\{\exp\{j[w_0 n\Delta + \varphi_0]\}\}, \qquad (34)$$

where ω_0 and φ_0 are given in equations (8) and (9). Hence

$$|DPT_M[s_n, \omega, \tau]| = |DPT_M[b_n, \omega - \omega_0, \tau]|. \quad (35)$$

As can be seen, as long as the maximum of the absolute value of $DPT_M[b_n, \omega, \tau]$ is at $\omega = 0$, the algorithm would yield the correct coefficient a_M. This is the case if the amplitude variations of the signal are slow compared to the phase variations. The larger the bandwidth of $DPT_M[b_n, \omega, \tau]$, the less are the effective SNR and the resolution.

The transform and algorithms are potentially useful in various areas. As mentioned in section 3 the transform can be used for signal classification. The second algorithm can be used to classify polynomial phase signals according to their degree, while simultaneously estimating their coefficients. Another application is instantaneous phase and frequency estimation. Two examples of this application are shown in [3]. A possible radar application is the estimation of dynamic target parameters in the context of pulse-Doppler radar. In [4] we have demonstrated the use of the transform to the problem of estimating the kinematic parameters of an accelerating target. A possible communication application is modulation classification of communication signals from passive measurements.

Finally we illustrate the performance of the transform in a typical application. Signals with high-order polynomial-phase are typically broadband. As such, we expect them, in practical situations, to be received with the interference of low-order polynomial-phase signals, such as single tones. In the following example we examine a QFM (quadratic frequency modulation, ie, third-order polynomial-phase) signal jammed by a chirp (second-order polynomial-phase) signal and four single tones. For the QFM signal the phase coefficients are: $a_3 = 1.25 \cdot 10^{17}, a_2 = -10^{12}$ and $a_1 = 2\pi 5 \cdot 10^6$. The $b_0 = 1$. For the chirp signal the phase coefficients are: $a_2 = 5 \cdot 10^{11}$ and $a_1 = 2\pi 10^6$. The $b_0 = 0.5$. Finally, for four single tones the coefficients are $a_1 = 2\pi 20 \cdot 10^6$ for the first, $a_1 = 2\pi 40 \cdot 10^6$ for the second, $a_1 = 2\pi 60 \cdot 10^6$ for the third and $a_1 = 2\pi 80 \cdot 10^6$ for the fourth. The amplitude of all four single tones is 0.25. The sampling frequency is 100MHz, the SNR is 3dB over the sampling frequency, and the number of samples is 4096.

Fig. 1 illustrates the magnitude of the FFT of the received signals. Fig. 2 shows the magnitude of DPT_3 of the sampled data, computed with $\tau = \lfloor N/3 \rfloor$. The spectral line corresponding to the $a_3(n\Delta)^3$ term of the phase is clearly visible. The estimated value of a_3 is $1.2501 \cdot 10^{17}$. Fig. 3 shows the magnitude of DPT_2 of the received data after the effect of the phase term of the third order was removed, computed with $\tau = N/2$. The spectral line corresponding to the $a_2(n\Delta)^2$ term of the phase is clearly visible. The estimated value of a_2 is $-1.0008 \cdot 10^{12}$. Finally, Fig. 4 shows the magnitude of the FFT of the received data after the effect of the phase terms of the third and second order were removed. The spectral line corresponding to the $a_1(n\Delta)$ term of the phase is clear. The estimated value of a_1

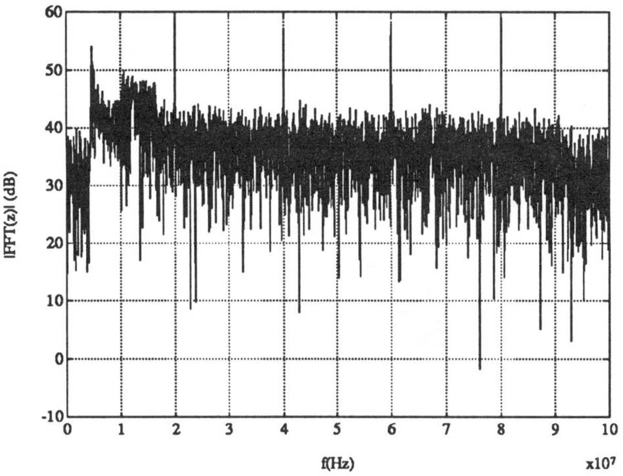

Figure 1: Spectrum of the received signals

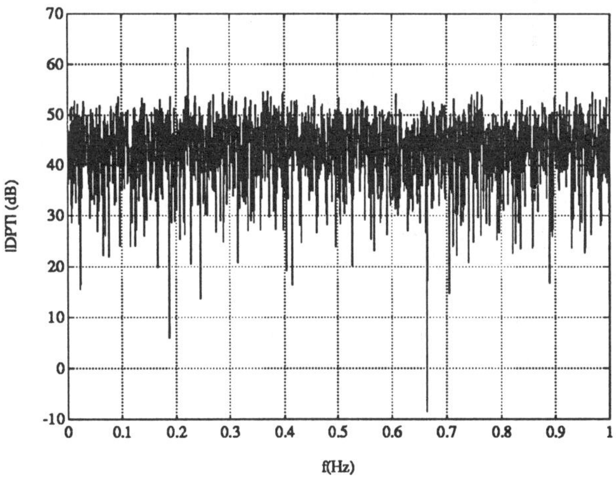

Figure 2: Magnitude of DPT_3

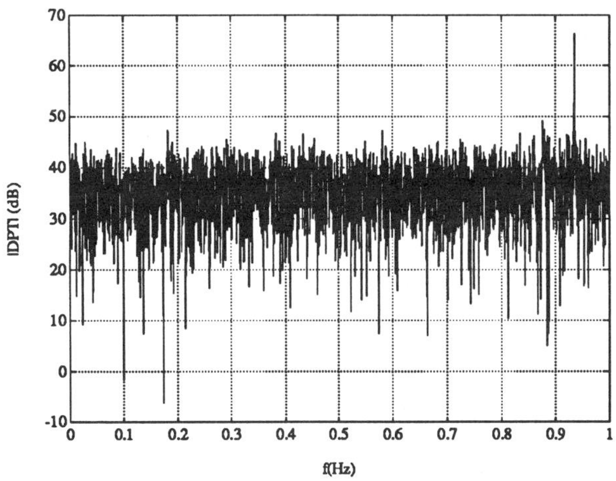

Figure 3: Magnitude of DPT_2

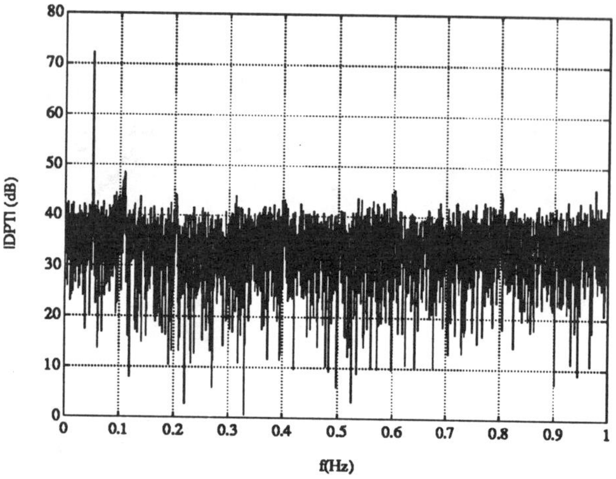

Figure 4: Magnitude of FFT after removing the high order terms

is $2\pi 5.0026 \cdot 10^6$. As we notice the interfering signals are spread all over the spectrum and do not hamper the accurate estimation of the phase coefficients.

Appendix A

Let $A(r,l)$ be the number of distributions of r balls into l boxes, in such away that no box remains empty. According to [1, p.58] we have that

$$(-1)^l A(r,l) \stackrel{def}{=} \sum_{q=0}^{l} (-1)^q \binom{l}{q} q^r. \tag{36}$$

Conclusion A1: For every $r < l$,

$$\sum_{q=0}^{l} (-1)^q \binom{l}{q} q^r = (-1)^l A(r,l) \equiv 0. \tag{37}$$

Conclusion A2: For $r = l$,

$$\sum_{q=0}^{l} (-1)^q \binom{l}{q} q^l = (-1)^l A(l,l) = (-1)^l l!. \tag{38}$$

Conclusion A3: For $r = l+1$,

$$\sum_{q=0}^{l} (-1)^q \binom{l}{q} q^{l+1} = (-1)^l A(l+1,l) = (-1)^l \frac{l}{2}(l+1)!. \tag{39}$$

Theorem 3 *We have that*

$$\sum_{m=0}^{M} a_m \Delta^m \sum_{q=0}^{M-1} (-1)^q \binom{M-1}{q} (n+q\tau)^m = \omega_0 n\Delta + \varphi_0, \tag{40}$$

where

$$\omega_0 = (-1)^{M-1} M! (\tau\Delta)^{M-1} a_M, \tag{41}$$

and

$$\begin{aligned} \varphi_0 &= (-1)^{M-1}(M-1)!(\tau\Delta)^{M-1} a_{M-1} \\ &\quad + (-1)^{M-1}\frac{M-1}{2} M!(\tau\Delta)^M a_M]. \end{aligned} \tag{42}$$

Proof:

$$\begin{aligned} &\sum_{m=0}^{M} a_m \Delta^m \sum_{q=0}^{M-1} (-1)^q \binom{M-1}{q} (n+q\tau)^m \\ &= \sum_{m=0}^{M} a_m \Delta^m \sum_{q=0}^{M-1} (-1)^q \binom{M-1}{q} \sum_{k=0}^{m} \binom{n}{k} n^{m-k} (q\tau)^k \\ &= \sum_{m=0}^{M} a_m \Delta^m \sum_{k=0}^{m} \binom{n}{k} n^{m-k} \tau^k \sum_{q=0}^{M-1} (-1)^q \binom{M-1}{q} q^k \\ &= (-1)^{M-1} \sum_{m=0}^{M} a_m \Delta^m \sum_{k=0}^{m} \binom{n}{k} n^{m-k} \tau^k A(k, M-1) \\ &= \omega_0 n\Delta + \varphi_0. \end{aligned} \tag{43}$$

□

References

[1] W. Feller, *An Introduction to Probability Theory and its Applications.* New-York: John Wiley, 1968.

[2] W. Rudin, *Principles of Mathematical Analysis*, 3rd ed. New-York: McGraw-Hill, 1976.

[3] S. Peleg and B. Porat, "Estimation and Classification of Polynomial-Phase Signals," *IEEE Trans. Information Theory*, Vol. 37, pp. 422-430, 1991.

[4] S. Peleg and B. Porat, "Linear FM Parameter Estimation from Discrete Time Observations," *IEEE Trans. Aerospace and Electronic Systems*, Vol. 27, pp. 607-616, 1991.

[5] S. Peleg and B. Porat, "The Cramer-Rao Lower Bound for Signals with Constant Amplitude and Polynomial Phase," *IEEE Trans. on Signal Processing*, Vol. 39, pp. 749-752, 1991.

The Systematic Design of Area Efficient VLSI Architectures for the Discrete Fourier Transform

Archie W. Julien * & William G. Bliss †
Department of Electrical Engineering
Texas A&M University
College Station, TX, 77843-3128

Abstract

Systematic application of linear projection and scheduling to Dependence Graph (DG) algorithm representations is used to generate asymptotically optimal DFT layouts in a VLSI computing model. Two-stage Prime Factor Maps are used to generate regular and minimum Area Period (AP) architectures. Offset parallel multi-projections generate overlapped twin-PE arrays. Systematic multi-projection of six-dimensional 3-factor DGs and eight-dimensional 4-factor DGs generate successively faster AT^2 optimal and regular architectures, but with increasing wire area.

1 Introduction

We consider the systematic design of efficient, real time, application-specific, and regular VLSI arrays for the DFT. Minimum AP architectures achieve the maximum throughput per unit area. Regular arrays composed of many simple cells are highly amenable to efficient fault-tolerant techniques, such as spare row/column switching.

We adopt the VLSI word-level grid model of Thompson [1] as a technology independent figure of merit which penalizes wire area and memory. The VLSI bound for low latency, T is $AT^2 = \Omega(N^2)$, while the bound for low period is $AP = \Omega(N^{3/2})$

A minimum AP 2-D mesh with N PEs in [1] executed the power of two FFT by emulating the butterfly network, but, control was complicated by time varying PE function. Faster architectures are VLSI model analyzed in [2]. A class of minimum AT^2 architectures uses two-stage factorizations [3], but requires orthogonal trees and some long wires. Methods that offload data for storage and permutation [4] [5], underestimate complexity. An isolated linear systolic array [6]

*This research was supported by a grant from the Texas Advanced Research Program

†email: bliss@ee.tamu.edu

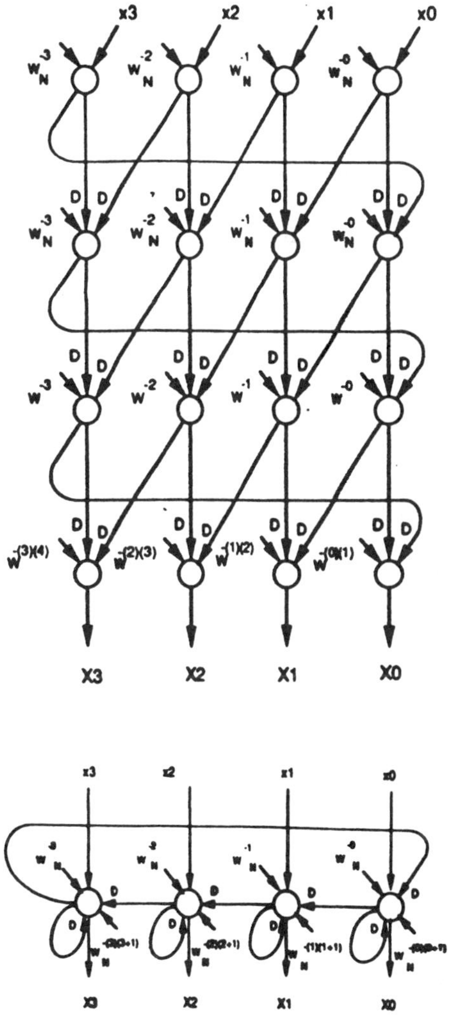

Figure 1: Spiral Mesh DG for modified Goertzel's DFT, and Ring Array SFG for in place calculation.

1058-6393/91 $01.00 © 1991 IEEE

is suboptimal and a triangular systolic array [7] are suboptimal with $AP = O(N^2)$. This paper extends [8]. Details are in [9].

Define linear projection and scheduling [10]: a DG node $\vec{n}$ is mapped to a Signal Flow Graph (SFG) Processing Element (PE) $\vec{p}$ by projection along $\vec{d}$, $\vec{p} = \mathbf{P}^T \vec{n}$, where $\mathbf{P}$ is a processor basis orthogonal to $\vec{d}$. The computation of node $\vec{n}$ is linearly scheduled by the $\vec{s}$ to execute at time $t(\vec{n}) = \vec{s}^T \vec{n}$.

2 Kernel Algorithms

The Vector-Matrix (VM) DFT with N^2 Multiply Accumulates (MAs) is

$$X(k) = \sum_{n=0}^{N-1} x(n) W_N^{-nk}, \quad W_N \equiv e^{j\frac{2\pi}{N}}. \qquad (1)$$

so its AP complexity will be at least $O(N^2)$. The minimum AP systolic mesh in [8] was based on flexible choice of orthogonal row and column kernels, so we digress and derive some $O(N^2)$ kernels.

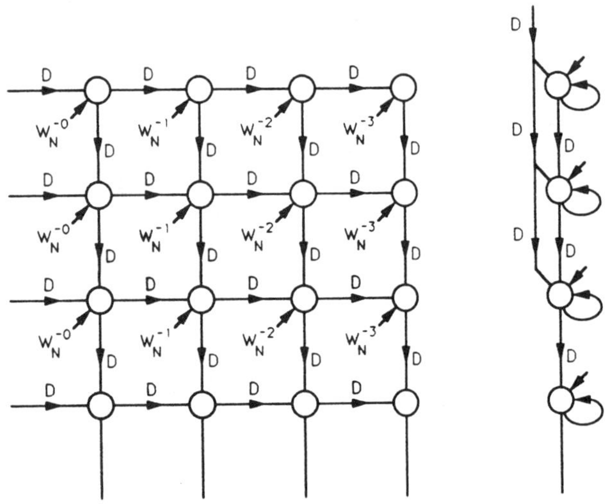

Figure 2: Orthogonal Mesh DG for Goertzel's DFT, and Linear Array SFG for parallel-in serial-out.

The VM spiral mesh DG and associated ring SFG are well known, but distribution or storage of W_N terms is problematic. This can be overcome by modifying Goertzel's algorithm as in the spiral DG of Fig.1. The DG is similar to VM, but Goertzel's implements nested multiplications. This DG is easily derived by using circular shifts of the DFT as a final correction factor to compensate for shifted data. The systolic ring SFG is created by projecting and scheduling down. A DG for Goertzel's algorithm can have orthogonal I/O, similar to V-M products, as in Fig.2. Projection to the East and a NW to SE schedule give the linear (vertical) systolic SFG with parallel-in serial-out I/O. Similarly, if the DG of Fig.2 is projected down with the same schedule, then a linear (horizontal) systolic SFG with serial-in parallel-out I/O is created. For all of these SFGs the W_N coefficients are easy to distribute or store.

It was shown in [8] that it is useful to hierarchically organize DFTs, using various kernel functions. The square root FFT butterfly was shown to be AP optimal when used with a pipelined matrix transpose.

3 Two-Stage DFT Algorithm Mapping

Two factor index maps where $N = LM$ are described by

$$n = (K_1 l + K_2 m) \ mod \ N, \quad k = (K_3 q + K_4 p) \ mod \ N$$

If $GCD(L, M) = 1$, then the input map is a Prime Factor Map (PFM) only if [11] $\{K_1 = c_1 M$ and $K_2 = c_2 L\}$ and $\{GCD(K_1, L) = GCD(K_2, M) = 1\}$. Other one-to-one maps are Common Factor Maps (CFM). Derivation of two-factor CFM architectures are in [8] [9]. CFM maps always require Twiddle Factors (TF) between stages, and conventional row-major column-major mappings are possible. PFMs don't require TFs, but don't allow these forms. Substituting the PFM into (1)

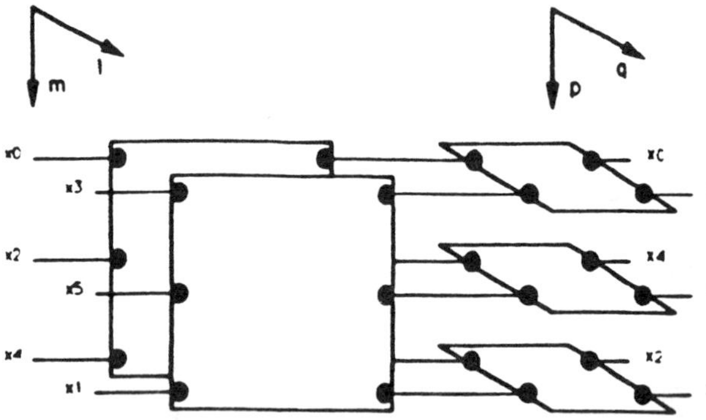

Figure 3: PFM Kernel DG: $N = 6,\ L = 2,\ M = 3$.

$$Y(q,p) = \sum_{l=0}^{L-1} \sum_{m=0}^{M-1} y(l,m) W_N^{-K_1 K_3 l q} W_N^{-K_1 K_4 l p}$$

$$W_N^{-K_2 K_3 m q} W_N^{-K_2 K_4 m p}. \qquad (2)$$

Using $K_2 = c_2 L,\ K_3 = c_3 M$ and $K_1 = c_1 M,\ K_4 = c_4 L$, gives

$$Y(q,p) = \sum_{l=0}^{L-1} \quad \sum_{m=0}^{M-1} y(l,m) W_N^{-(c_1 M)(c_3 M) l q}$$

$$W_N^{-(c_1 M)(c_4 L)lp} W_N^{-(c_2 L)(c_3 M)mq} W_N^{-(c_2 L)(c_4 L)mp}. \qquad (3)$$

But $W_N^{-(c_1 M)(c_4 L)lp} = W_N^{-(c_2 L)(c_3 M)mq} = 1$, so

$$Y(q,p) = \sum_{l=0}^{L-1} \sum_{m=0}^{M-1} y(l,m) W_L^{-c_1 c_3 M l q} W_M^{-c_2 c_4 L m p},$$

where the nesting may be calculated in either order. Using Good's [12] coefficients; $c_1 = c_2 = 1$, $c_3 = M^{-1} mod\ L$, $c_4 = L^{-1}\ mod\ M$, e.g, if $N = 12$, $L = 4$, $M = 3$, then $K_3 = 3M$ and $K_4 = L$.

$$Y(q,p) = \sum_l [\sum_m y(l,m) W_M^{-mp}] W_L^{-lq} \qquad (4)$$

The inner sums are M-point *column*-DFTs on $y(l,m)$, and call the result $y1(l,p)$. The outer sums are then M-point *row*-DFTs on $y1(l,p)$, yielding the final DFT.

$$y1(l,p) \equiv \sum_{m=0}^{M-1} y(l,m) W_M^{-mp} \qquad (5)$$

$$Y(q,p) = \sum_{l=0}^{L-1} y1(l,p) W_L^{-lq} \qquad (6)$$

The number of MAs for PFM DFT is LM^2 for the column DFTs and ML^2 for the row DFTs, so the total is $N(M+L)$. Choosing $M \approx L \approx N^{1/2}$ means approximately $2N^{3/2}$ MAs are required.

A generic PFM hierarchical DG is shown in Fig. 3. Note that a DFT plane can be performed by any kernel. If a factor L (or M) is not prime, that DFT may be further decomposed. We can write (5) and (6) in single assignment form, creating the DG of Fig.4. Taking m and q as collinear and projecting and scheduling along m, q we have

$$\vec{d}(l,m,q,p) = \vec{s}(l,m,q,p) = (0,1,1,0)^T \qquad (7)$$

So the SFG space, $\mathbf{P}$, lies in the l,p plane. A DG node at (l,m,q,p) simply projects to the PE at $(l,0,0,p)$. Vertical edges in the DG become wires of length 0, i.e., self loops

$$\vec{e}_{SFG} = \mathbf{P}^T \vec{e}_{DG} = \vec{0} \qquad (8)$$

where the delay in the self-loop is

$$D(\vec{e}_{SFG}) = \vec{s}^{\,T} \vec{e}_{DG} = 1 \qquad (9)$$

The resultant SFG is shown in Fig.5. This SFG can be made systolic by choosing the alternate linear schedule $\vec{s}^T = (1,1,1)$, where m and q are collinear, i.e., $q = m + M$.

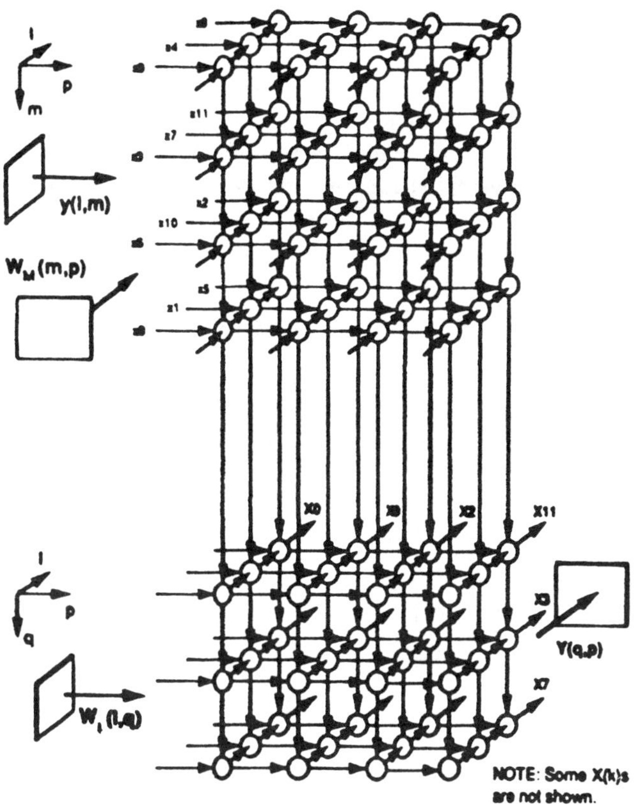

Figure 4: DG for two-stage PFM: $N = 12$, $L = 3$, $M = 4$.

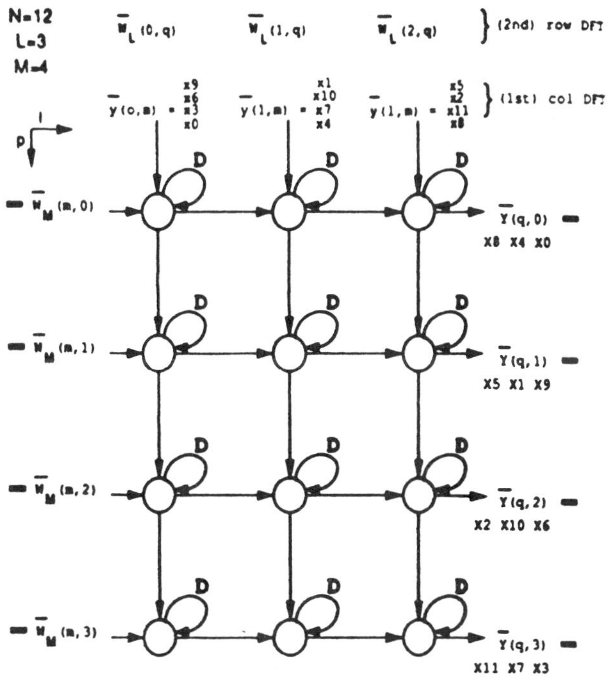

Figure 5: SFG for two-stage PFM: $N = 12$, $L = 3$, $M = 4$.

A twin PE array with approximately double throughput can be derived by using offset parallel projections in the m and q solids. Projecting along m in the top solid, offsetting a projection along q in the bottom solid, and overlapping the default schedules gives Fig. 6, illustrated for a CFM. Because the period is limited by $max(L, M)$, a free time slot is always available to perform the TF multiplies if $GCD(L, M) = 1$.

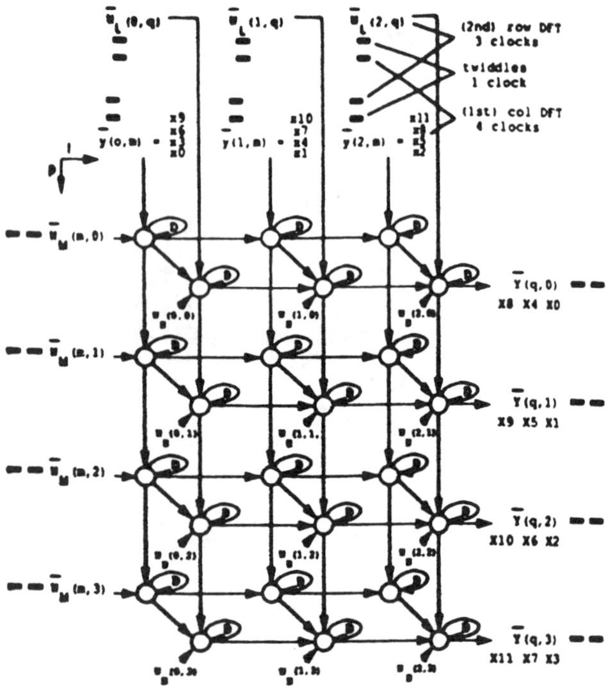

Figure 6: Twin PE two-stage CFM SFG: $N = 12$, $L = 3$, $M = 4$.

4 Three-Stage DFT Index Maps

A 2-stage DFT index map can be extended to 3-stages, $N = STU$, of the form

$$n = (K_1 s + K_2 t + K_3 u) \; mod \; N,$$

$$k = (K_4 r + K_5 q + K_6 p) \; mod \; N$$

With constraints on K_i to assure one-to-one mapping, substitute n and k into (1), obtaining

$$\begin{aligned} Y(r,q,p) = \; & \textstyle\sum_{s=0}^{S-1} \sum_{t=0}^{T-1} \sum_{u=0}^{U-1} y(s,t,u) W_N^{-K_1K_4sr} \\ & W_N^{-K_1K_5sq} W_N^{-K_1K_6sp} \\ & W_N^{-K_2K_4tr} W_N^{-K_2K_5tq} \\ & W_N^{-K_2K_6tp} W_N^{-K_3K_4ur} \\ & W_N^{-K_3K_5uq} W_N^{-K_3K_6up}. \qquad (10) \end{aligned}$$

If S, T, and U are relatively prime, choose PFM $K_1 = c_1TU$, $K_2 = c_2SU$, $K_3 = c_3ST$, $K_4 = c_4TU$, $K_5 = c_5SU$, $K_6 = c_6ST$,so (10) becomes,

$$\begin{aligned} Y(r,q,p) = & \sum_{s=0}^{S-1}\sum_{t=0}^{T-1}\sum_{u=0}^{U-1} y(s,t,u) \\ & W_N^{-c_1TUc_4TUsr} W_N^{-c_1TUc_5SUsq} W_N^{-c_1TUc_6STsp} \\ & W_N^{-c_2SUc_4TUtr} W_N^{-c_2SUc_5SUtq} W_N^{-c_2SUc_6STtp} \\ & W_N^{-c_3STc_4TUur} W_N^{-c_3STc_5SUuq} W_N^{-c_3STc_6STup}. \qquad (11) \end{aligned}$$

Since $STU = N$, (11) reduces to

$$Y(r,q,p) = \sum_{s=0}^{S-1}\{\sum_{t=0}^{T-1} [\textstyle\sum_{u=0}^{U-1} y(s,t,u) W_U^{-c_3c_6STup}] \; W_T^{-c_2c_5SUtq}\} W_S^{-c_1c_4TUsr}.$$

Applying the technique with Good's coefficients:

$$c_1 = c_2 = c_3 = 1, \; c_4 = (TU)^{-1} mod \; S,$$
$$c_5 = (SU)^{-1} mod \; T, \text{ and } c_6 = (ST)^{-1} mod \; U,$$

$$Y(r,q,p) = \sum_{s=0}^{S-1}\{\sum_{t=0}^{T-1}[\sum_{u=0}^{U-1} y(s,t,u) W_U^{-up}] \; W_T^{-tq}\} W_S^{-sr}. \qquad (12)$$

The summations perform ST U-point DFTs, SU T-point DFTs, and finally, TU S-point DFTs, and the nesting may be calculated in any order. The number of MAs is therefore $N(S+T+U)$. If $S \approx T \approx U \approx N^{1/3}$, then $3N^{4/3}$ MAs are required. The sums in 12 create a 6-D DG. By three projections we arrive at a 3-D (near) cube. By using a skew projection we create the SFG of Fig.7. This SFG has $A = O(N^{4/3})$ and $T = O(N^{1/3})$, so it is AT^2 optimal. The input phase DFT uses ser-par Goertzel's, the middle phase DFT uses par-par Goertzel's ring, and the output phase DFT uses par-ser Goertzel's.

Extending this method, let each factor of the 2-stage PFM DFT, be composed of two relatively prime, nearly equal factors; $N_i \approx \sqrt{L} \approx \sqrt{M} \approx N^{1/4}$. Then each L- and M-point DFT within the two smaller $N^{3/2}$ DFT stages may be reduced to $2L^{3/2} \approx 2M^{3/2} \approx 2N^{3/4}$ MAs. This results in $L2M^{3/2} + M2L^{3/2} \approx 4N^{5/4}$ MAs for the 4-stage PFM. The mapping of the input and output indices can be

$$\begin{aligned} n &= K_1n_1 + K_2n_2 + K_3n_3 + K_4n_4 \\ k &= K_5k_1 + K_6k_2 + K_7k_3 + K_8k_4. \qquad (13) \end{aligned}$$

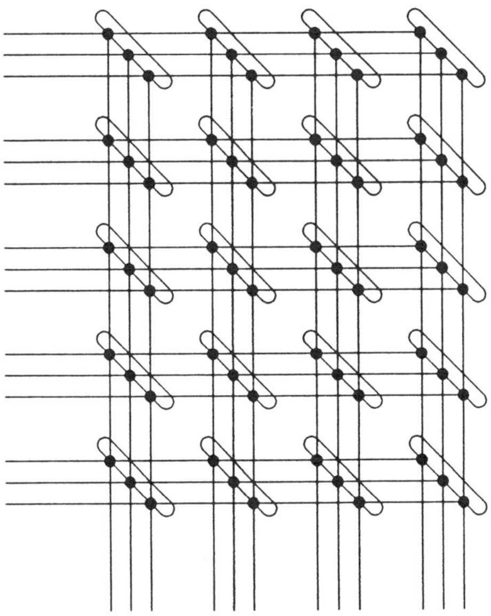

Figure 7: Cube embedding SFG 3-stage PFM.

Substituting into (1) and using Good's coefficients

$$Y(k_4, k_3, k_2, k_1) = \sum_{n_4=0}^{N_4-1} \left\{ \sum_{n_3=0}^{N_3-1} \left[\sum_{n_2=0}^{N_2-1} \left(\sum_{n_1=0}^{N_1-1} y(n_4, n_3, n_2, n_1) W_U^{-n_1 k_1} \right) W_U^{-n_2 k_2} \right] W_T^{-n_3 k_3} \right\} W_S^{-n_4 k_4}.$$

These sums create a 8-D DG. With four projections we arrive at a (near) 4-D hyper-cube. By using two skew projections we create a 'mesh' SFG of $N^{1/2}$ cells, where each cell has $N^{1/2}$ PEs connected as in Fig.8. This SFG has $A = O(N^{3/2})$ and $T = O(N^{1/4})$, so it is AT^2 optimal. The input phase DFT uses ser-par Goertzel's, two middle phase DFTs use par-par Goertzel's ring, and the output phase DFT uses par-ser Goertzel's.

5 Conclusions

We have demonstrated that PFM algorithms can be systematically mapped to regular systolic VLSI arrays that are AP efficient in the VLSI model. An offset piecewise parallel projection generates a twin-PE array. Three and four factor algorithms are systematically mapped to two dimensional hyper-cube embeddings and are AT^2 optimal.

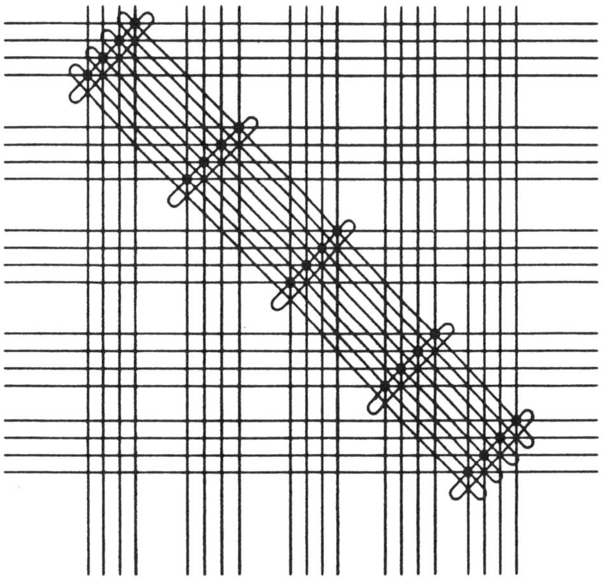

Figure 8: One cell in embedding of 4-cube.

References

[1] C. Thompson, "Area-time complexity for VLSI," in *Caltech Conference on VLSI*, pp. 495–508, Jan. 1979.

[2] C. Thompson, "Fourier Transforms in VLSI," *IEEE Transactions on Computers*, vol. 32, pp. 1047–1057, Nov. 1983.

[3] C. Bilardi and M. Sarrafzadeh, "Optimal Discrete Fourier Transform in VLSI," *VLSI: Algorithms and Architectures*, pp. 79–89, 1985.

[4] C. Jesshope, "The implementation of Fast Radix 2 Transforms on array processors," *IEEE Transactions on Computers*, pp. 20–27, 1980.

[5] M. J. Irwin, R. M. Owens, T. P. Kelliher, K.-K. Leung, and M. Vishwanath, "The Arithmetic Cube II: A second generation VLSI DSP processor," in *Proc. Int. Conf. ASSP*, pp. 1125–1128, 1991.

[6] J. Beraldin, T. Aboulnasr, and W. Steenaart, "Efficient one-dimensional systolic array realization of the Discrete Fourier Transform," *IEEE Transactions on Circuits and Systems*, vol. 36, pp. 95–100, Jan. 1989.

[7] P. Gader, "Bidiagonal factorization of Fourier Matrices and systolic algorithms for computing Discrete Fourier Transforms," *IEEE Transactions on Acoustics, Speech, and Signal Processing*, vol. 37, pp. 1280–1283, Aug. 1989.

[8] W. G. Bliss and A. W. Julien, "Efficient and reliable VLSI algorithms and architectures for the Discrete Fourier Transform," in *Proc. Int. Conf. ASSP*, 1990.

[9] A. Julien, "The systematic design of area efficient VLSI architectures,". Ph.D. dissertation, Texas A&M Univ., Dept EE, 1991.

[10] S. Kung, *VLSI Array Processors*. Englewood Cliffs, NJ: Prentice Hall, Inc., 1988.

[11] C. Burrus and T. Parks, *DFT/FFT and Convolution Algorithms*. John Wiley & Sons, 1985.

[12] I. Good, "The relationship between two Fast Fourier Transforms," *IEEE Trans. Comput.*, vol. C-20, pp. 310–317, Mar. 1971.

A Comparison of Resource Constrained and Classical Detection Theory

J.S. Bird and M.M. Goulding*
School of Engineering Science
Simon Fraser University
Burnaby, B.C., CANADA
V5A 1S6

Abstract

Real world detection systems are designed to allocate limited resources. This paper describes a technique to include the effect of these limitations in the system design. The resulting detection criterion is compared with classical detection criteria and seen to provide better performance.

1 Introduction

Signal detection theory has come a long way since its early beginnings in the late 1940's. Mathematical advances in the field of hypothesis testing and engineering advances in radar and sonar have resulted in systems that can detect and track targets at farther and farther ranges. Statistical sensor deployments, which necessarily involve decision theory, are still plagued, however, by the same three difficulties outlined by Middleton in 1960 [1], namely,

1. the apparent arbitrariness of cost assignments,
2. the inadequacy of the a-priori information, and
3. the uncertainty in the choice of an optimality criterion.

These difficulties are a direct result of working in the real world where we cannot avoid paying the price for ignorance. The resource constrained detection theory outlined in this paper is a unique approach to the fusion of classical decision theory and real world problems.

Resource constrained detection is best described by example. Consider a very simple system consisting of a gun control radar and a gun designed to protect a ship. Assume the gun is capable of firing one round per second and is assigned the task of destroying all incoming targets. We would like to design a system that has the highest probability of intercepting the targets. A typical strategy to accomplish this is to set the radar threshold to yield a given false alarm probability and shoot at any target that crosses that threshold. This strategy results in a firing rate that is acceptable provided no targets are present and the false alarm rate is less than the maximum firing rate. As targets appear the firing rate increases until, if the number of targets is large enough, targets are missed because the gun is busy. This is a classical overload condition that puts the ship in jeopardy. To reduce the potential for overload the radar false alarm rate is lowered. But this reduces the radar's ability to detect small targets even when the number of targets is small and there is no danger of an overload; so again the ship is vulnerable.

It is clear that in this simple example the degree of protection is determined by the firing rate of the gun and therefore the gun should be fully utilized. The radar should simply direct the gun to the radar cells most likely to contain targets, subject to the rate of the gun. This is a resource constrained detection system where the resource is the gun's firing rate.

The threshold for the radar is easily set to keep the gun busy but not overloaded. There need be no appeal to probability distributions etc. to calculate false alarm rates. If no targets are present the threshold drops to keep the gun firing, albeit at nothing. But the radar is very sensitive to small targets and therefore the ship is optimally protected. As the number of radar cells crossing the threshold increases so does the threshold and therefore the cells most likely to contain targets are passed to the gun. As discussed in section 3, the detection probability for this strategy is always greater than or equal to that of the strategy first described.

While it seems that the resource constrained criterion is wasteful it does provide the ship with the greater degree of protection. After all, is that not the reason for the system in the first place? If the criterion is held in disfavour then there is a hidden resource that is actually driving the design. If that is the case this hidden resource should be brought into the open and the control criterion built around it. Such a hidden resource might be the total number of shells fired; either the supply is limited or the ship is only worth so many shells.

Whatever the case may be, real world detection systems are usually governed by a variation on the resource constraint principle. Therefore it is beneficial

*This work supported in part by the Natural Sciences and Engineering Research Council of Canada under operating grant OGPIN 011, and by the Department of National Defence under contract W7714-9-9180/01-SZ.

1058-6393/91 $01.00 © 1991 IEEE

to develop a detection theory around this principle to guide future designs and to assess the performance of the present.

2 The Classical Approach

In the classical approach to detection theory the likelihood function plays a central role. It is beyond the scope of this paper to develop the arguments supporting the use of this function and therefore the reader is referred to the excellent books by Middleton [1] or Poor [2]. The essence of the detection problem is that a signal is received and a test statistic developed. (A test statistic is a random variable that tends to be large if a target is present and small if it is not.) If the test statistic is above an assigned threshold a target is declared present; otherwise no target is declared. Classical detection theory assigns a value to the threshold that depends on the probability distributions of the test statistic and the prevailing optimization criterion.

There are a number of criteria that are employed to set the threshold in detection systems. We will discuss three of them: the Neyman-Pearson (the most common for radar applications), the Bayes (a cost function approach) and the minimum equivocation (an information theoretic approach). Table 1 lists the mathematical form of these criteria and defines the associated terms.

The Neyman-Pearson criterion is the one that is employed in the radar/gun example given above. After adopting an acceptable false alarm probability (pf), one uses the probability distribution of the test statistic under the assumption of no targets present to calculate the corresponding threshold. There are a number of difficulties with this approach. First, the definition of an acceptable false alarm probability is likely to prove elusive. As we saw in the example, setting pf high will cause the system to overload with the presence of only a few targets. On the other hand, setting it low lowers the sensitivity of the system to small targets even when the system is not busy with large ones.

On the positive side, the Neyman-Pearson criterion does provide an objective means of setting the threshold, without regard for probability distributions, assuming one can actually measure pf. Unfortunately, observing (and therefore measuring) pf is not always possible. One has to make sure there are no targets present and typically wait for a long time if pf is to be set to a low value.

The Bayes criterion provides a method by which costs can be incorporated into the decision process. Typically cost assignments are made for errors and correct decisions and then the threshold is set to minimize the total risk (cost). This criterion results in a likelihood test like the Neyman-Pearson, but with a different threshold. The threshold is a function of the costs and the a priori statistics of a target being present. Unlike the Neyman-Pearson test, however, the threshold can not be determined objectively because there is no objective parameter to measure. Since the threshold in a likelihood test is the slope of the ROC curve [3] a Bayes test assigns a value to this slope. In order to set the threshold in practice one has to know the ROC curve and hence is required to have complete knowledge of the decision statistics.

A variation on the Bayes test is obtained when the cost of making a correct decision is zero and the error costs are equal. For this situation the threshold is the ratio of the probability of a target not being present to the probability of a target being present. This assignment of costs minimizes the probability of error and is called an "ideal observer" by Middleton [1].

There are numerous problems associated with applying the Bayes criterion to radar or sonar detection systems. First, the cost assignments are clearly subjective. Second, the a priori target population statistics must be known. Finally, the Bayes criterion defines the threshold in terms of the slope of the ROC curve and therefore the probability distributions of the test statistic under the target present and absent hypotheses must be known explicitly.

Various attempts have been made to mitigate the defects listed above by bounding costs and averaging over possible target probability distributions. This results in the minimax rules which are essentially compromises resulting from a less than ideal decision criterion and from our ignorance.

The minimum equivocation (ME) criterion is similar to the Bayes test in that it can be viewed as an assignment of cost functions to the various outcomes of the decision. ([1],[3]) In this situation, however, the costs are well defined (in terms of the distribution of the test statistic) and relate to the uncertainty in the decision. This criterion minimizes the uncertainty, as defined by Shannon's information measure. When the probability of making each kind of error is equal, the ME criterion reduces to the Bayes ideal observer. The ME criterion suffers from the same problems as the Bayes criterion when applied in practice. As it sets the slope of the ROC curve at the required operating point the ME criterion requires knowledge of the ROC curve.

Nonparametric hypothesis testing [2] is one approach to detection systems that helps alleviate some of the above problems. Nonparametric detectors seek to make the decision process less sensitive to prevailing statistics by creating a test statistic that is more robust in the sense that it depends on a general characteristic of the distributions. For example, the distribution of the test statistic might be required to be an even function under the target present and absent hypotheses. The essence of this approach is that having admitted our ignorance, we proceed to develop test statistics for which we know the distributions, and then develop optimum tests based on these known distributions. These tests are seldom optimum in the sense that if we knew the distributions of the original data we could do better. But we don't and therefore must be satisfied with the resulting suboptimum performance, which often is adequate.

Even with nonparametric tests we have a problem, however. Once the statistics of the test statistic have been decided upon we still have to use a Neyman-Pearson, Bayes or other such criterion to set the threshold. The Bayes and ME criterion are quickly discarded because they require knowledge of the target concentration (a priori probability of a target). The Neyman-Pearson criterion on the other hand is very useful. This approach has resulted in the many non-parametric CFAR (constant false alarm) techniques that are available to detection systems designers today [4].

Nonparametric tests do not, however, remove the basic problem with the Neyman-Pearson test, namely the setting of the false alarm probability. We no longer have to measure it because we know what it is, but we still have to set it to some value. For example, in the radar/gun illustration given above we could set the false alarm rate but we still did not have optimum protection for the ship.

The rate constraint test [5] is a special form of resource constrained tests where the number of threshold crossings that can be handled determines the constraint on the resource. Unlike the Bayes and ME tests implementation of the rate constraint test does not require knowledge of the statistics. Also, because the hit rate is observable and this observability is independent of the presence or absence of a target, practical realization of the rate constraint test is possible. We will now discuss rate constraint tests in more detail.

3 Rate Constrained Tests

From the radar/gun-control problem, we see that there exists at least one example where optimum performance from the detection system results when the detection probability, *pd* is maximized without respect to the cost of false alarms. The Rate-Constraint Criterion [5] is the solution to the problem of maximizing *pd* subject to a limitation on the total number of declared hits. (If there were no constraint, the optimal threshold would be zero - yielding $pd = pf = 1$.)

For example, consider the detection system of figure 1. Sensory data arriving from the radar at a rate r_s cells/unit time is presented to a two stage system. The first stage of the system is a likelihood ratio test with threshold ϕ; those cells exceeding ϕ are passed on to the rate constrained verifier. In general, the verifier could be any rate constrained resource; for example it could be a band-limited communications channel, higher level processing such as a tracker, or even a control system for an autonomous vehicle that must act based on the detection decisions of the first stage [6]. We assume that the verifier is capable of processing cells at a rate r_v, where $r_v < r_s$, and that the performance of the verifier is independent of the its actual processing load. If the number of cells presented to the verifier exceeds r_v, we assume that uniform randomization occurs. Thus

$$pd_{1,eff} = \begin{cases} pd_1(\phi) & \text{if } r_{o,1}(\phi) \le r_v \\ pd_1(\phi)\frac{r_v}{r_{o,1}(\phi)} & \text{if } r_{o,1}(\phi) > r_v \end{cases} \quad (1)$$

where

$$r_{o,1}(\phi) = r_s[\pi_1 pd_1(\phi) + \pi_0 pf_1(\phi)] \quad (2)$$

is the output rate of the first stage, π_1 is the a priori probability of a target being present in any given cell, π_0 is the a priori probability of a given cell being empty, and $pd_1(\phi)$ and $pf_1(\phi)$ are respectively, the detection and false alarm probabilities of the first stage. It can be shown [5] that in order to maximize $pd_{1,eff}$, ϕ should be set such that $r_{o,1} = r_v$. That is, the output rate from stage 1 should be set to match the verifier's maximum processing rate.

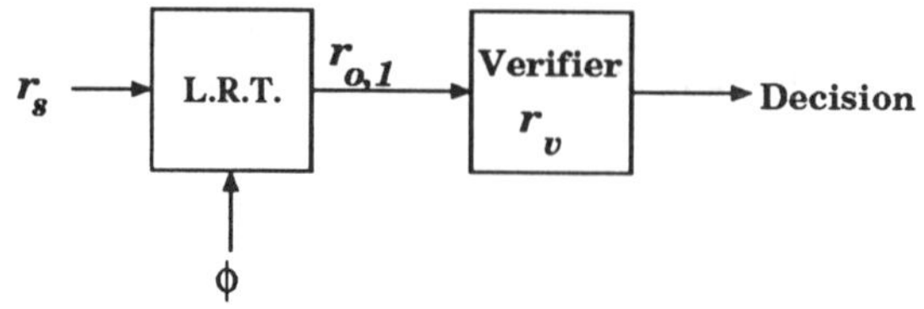

Figure 1: A simple two stage detection system consisting of a likelihood ratio test (L.R.T.) and a rate-constrained verifier.

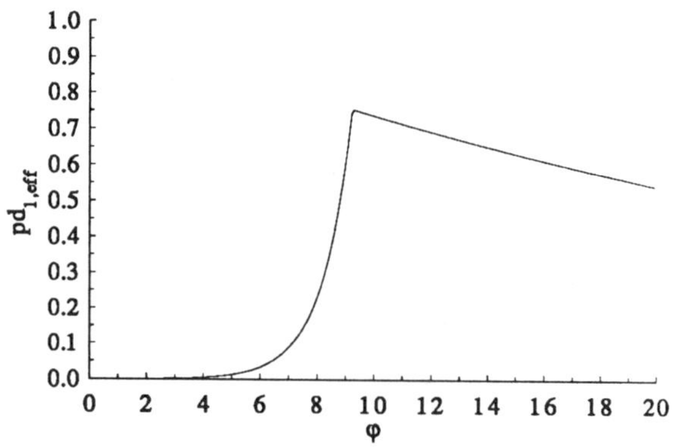

Figure 2: Effective Detection Probability Versus Threshold for a rate-constrained verifier.

For example, consider figure 2 which is a plot of $pd_{1,eff}$ versus ϕ when $r_v = 10^{-4} r_s$, and $\pi_1 = 10^{-6}$. The targets are assumed to be Rayleigh distributed in Gaussian noise, so that [7]

$$pd_1(\phi) = \exp\left(-\frac{\phi}{1+\sigma}\right) \text{ and } pf_1(\phi) = \exp(-\phi) \quad (3)$$

where σ is the signal-to-noise ratio, assumed to be 15dB for this example.

The effect of setting ϕ too low and thus causing randomization is clearly evident. Less drastic, but also obvious is the effect of setting ϕ higher than the rate-constrained optimal value.

Thus far we have assumed that the verifier is not involved in the overall system detection performance. If the verifier is a form of decision device, its detection and false alarm probabilities must be considered in the system performance. It can be shown [5] that if

$$\frac{\partial \log pd_v}{\partial \log pf_v} < \frac{\partial \log pd_1}{\partial \log pf_1} \quad (4)$$

maximum detection performance for the system will result from maximizing $pd_{1,eff}$. Condition 4 means that the verifier does a better job of discriminating between targets and empty cells - this is to be expected since it operates at a slower rate than the first stage.

Consider the addition of an independent intermediate stage between stage 1 and the verifier. It can be shown [5] that if the intermediate stage has a maximum operating rate of r_2 cells/unit time where $r_s > r_2 > r_v$ and if

$$\frac{\partial \log pd_v}{\partial \log pf_v} < \frac{\partial \log pd_2}{\partial \log pf_2} < \frac{\partial \log pd_1}{\partial \log pf_1}$$

then maximum overall detection probability ($pd_t = pd_1 pd_2 pd_v$) will result when the thresholds for the first two tests are such that their output rate matches the input rate of the next stage. This is a general result that may be extended to as many stages as required. Since the optimal threshold for the intermediate stages will be non-zero, it follows that each additional stage must increase the overall detection probability, provided that the above condition holds. Note that the price for increased performance is increased hardware and increased latency in the decision making process.

4 Further Discussion

Resource constraint tests implicitly imply a resource that must be managed and an objective for the detection system. For the radar/gun example the resource was the gun firing rate and the objective was to protect the ship. In designing a detection system the question of application or objective must be addressed first. Why do we want this system, what is it supposed to do, and how much are we willing to spend to get it done? Again in the example, somebody decided that a gun with a one second firing rate was all that was necessary or all they were willing to spend to protect the ship. Two guns would have been better, doubling the firing rate.

Other factors could have been brought into the example such as target identification. It might not be desirable to shoot at just anything if one's own aircraft are in the area. Therefore a means of identifying the target has to be instituted. If the process of identification is slower than the firing rate of the gun then identification becomes the limiting resource.

The essence of resource constrained detection is that the radar, sonar, etc. is viewed as a device that identifies the k most likely cells to contain targets whether there are targets present or not. It is up to the rest of the system to decide what to do with those cells. For example, suppose a radar always listed the five cells most likely to contain targets. This would not seem very useful except perhaps in the radar/gun example. But an operator could easily use coincidence of cells in adjacent lists to build significance. This view of a radar radically effects its design. Throughput and the building of coincidence become the design parameters of interest and probability distributions and thresholding become secondary concerns. This is not to say that optimum tests, nonparametric techniques, and normalization for noise and clutter are not important because they are. The better the test, the better the resource constraint procedure is going to work. Resource constraint detection is a means by which the detection process can be controlled in practical applications because it reflects the real objectives of the system and the constraints placed upon it.

Constraints on the detection system come from the environment as well as the resource and this must be recognized. Examples are: the time constant of target evolution and the nature of the support for coincidence. Targets are detectable for a finite period of time after which they may be lost and a new detection sequence required to detect them. If the radar does not process cells within this time period performance drops.

Support for coincidence generally comes from two sources: time or space. Coincidence over time is employed to support the presence of a target when the target is of small extent when compared to a test cell. For extended targets such as those that would be the object of an obstacle avoidance sonar, coincidence could be required over space (across cells). In either case resources are consumed to establish the coincidence. One of these resources might well be the constraining resource on the system.

References

[1] D. Middleton, *An Introduction to Statistical Communications Theory*. McGraw-Hill, 1960.

[2] H. Poor, *An Introduction to Signal Detection and Estimation*. Springer-Verlag, 1989.

[3] I. Hoballah and P. Varshney, "An information theoretic approach to the distributed detection problem," *IEEE Trans. Info. Theory*, vol. IT-35, pp. 988–994, Sept. 1989.

[4] N. Atkins and A. Morley, "Maintenance of constant false alarm rate," in *Proceedings of IEE Conference on Radar 87*, pp. 499–503, Oct. 1987.

[5] J. Bird and M. Goulding, "Rate constrained target detection," *accepted for publication in IEEE Trans. Aerospace and Electronic Systems.*

[6] J. Bird and M. Goulding, "Obstacle avoidance with resource constrained detection," *7th International Symposium on Unmanned Untethered Submersible Technology*, pp. 441–452, 1991.

[7] J. Di Franco and W. Rubin, *Radar Detection*. Prentice-Hall, 1968.

Table 1: Comparison of Detection Criteria

Name	Detection Criteria	Notes for Practical Implementation
Neyman-Pearson	$L \underset{H_0}{\overset{H_1}{\gtrless}} \lambda$ where λ is such that pf is constant.	pf is unobservable unless sure target is absent. Takes a long time to estimate if small or we need the distribution of the test statistic under H_0 to calculate
Bayes	$L \underset{H_0}{\overset{H_1}{\gtrless}} \frac{\pi_0(C_{10}-C_{00})}{\pi_1(C_{01}-C_{11})}$ If $C_{00} = C_{11} = 0$ and $C_{01} = C_{10}$ we have the Ideal Observer: $L \underset{H_0}{\overset{H_1}{\gtrless}} \frac{\pi_0}{\pi_1}$ (gives minimum error)	C_{01}, C_{10}, C_{00} and C_{11} are subjective and arbitrary. π_0 and π_1 are unknown.
Minimum Equivocation (M.E.)	$L \underset{H_0}{\overset{H_1}{\gtrless}} \frac{\pi_0 \ln(\frac{1-pf}{pf}(\frac{r}{1-r}))}{\pi_1 \ln(\frac{pd}{1-pd}(\frac{1-r}{r}))}$	π_0 and π_1 are unknown. pf and pd require probability distributions under H_0 and H_1.
Rate Constraint	$L \underset{H_0}{\overset{H_1}{\gtrless}} \phi$ where ϕ is such that $r =$ constant determined by resources being controlled by the test.	r is observable (doesn't require distributions - although may use distributions if available) Needs a follow-on resource to establish significance.

Key

H_0	Target Absent Hypothesis (null hypotheses)
H_1	Target Present Hypothesis (alternative hypothesis)
π_0	A priori probability of no target being present
π_1	A priori probability of a target being present ($\pi_0 + \pi_1 = 1$)
C_{ij}	Cost of declaring H_i when H_j is true
pf	probability of false alarm
pd	probability of detection
r	hit rate ($r = \pi_1 pd + \pi_0 pf$)
L	likelihood ratio, $L = \frac{p(x\|H_1)}{p(x\|H_0)}$ where $p(x\|H_i)$ is the probability distribution of the test statistic conditioned on H_i.
$L \underset{H_0}{\overset{H_1}{\gtrless}} k$	Test that accepts H_0 if $L < k$, otherwise accepts H_1.

Nonlinear Kalman Filtering of Long-Baseline, Short-Baseline, GPS, and Depth Measurements

B.M. Bell, B.M. Howe, J.A. Mercer, and R. C. Spindel

Applied Physics Laboratory, College of Ocean and Fishery Sciences,
University of Washington, Seattle, Washington 98105

Abstract

In a recent Moving Ship Tomography experiment we deployed a vertical array of receiving hydrophones 300 times around a 1000 km diameter circle south of Bermuda; six acoustic sources within the circle transmitted every 3 hours. We plan to use the travel times measured along the source/receiver paths to reconstruct the three-dimensional sound speed field in the interior of the circle. Accurate location of the hydrophone receivers is crucial to this plan. A long-baseline tracking system, consisting of floating buoys and the ship, acoustically measured the slant range to each hydrophone (~1 km in depth and ~1 km in horizontal range). The range to orbiting satellites was measured with GPS. A short-baseline acoustic system measured range and direction from the ship to a beacon on the cable, and to each buoy. Depths at various points near the hydrophones were measured using three different pressure sensors. Here we describe a Kalman filter that combines these primary measurements and some supporting measurements to form an estimate of the buoy and hydrophone locations. We use an iterated correction step to adjust for the nonlinearities and an outlier detection scheme to make the filter robust.

Introduction

Moving Ship Tomography measures acoustic travel times between transmitters moored above the ocean floor and hydrophones on a cable tethered to a ship. The purpose is to determine the three-dimensional sound speed structure in the water between several transmitters and many reception points. Knowing the precise location of the hydrophones when they receive a signal is crucial to this determination. In this paper we consider a Kalman filter that determines the location of a subarray of the hydrophones near the end of the cable (see Figure 1).

In the Bermuda MST experiment, a measurement sequence was repeated every 3 hours. One or two floating buoys were first deployed, loosely tethered to the ship by a line. Then the ship was moved into a position where it could drift and maintain a good tracking geometry for itself, the buoys, and the array cable which was deployed next. The cable was lowered about 1 km, and a flexible floating spar was attached. The ship's screw was then "declutched" so that the ship drifted freely with the wind and current, and the cable was payed out so as to preclude tension. Thus, a constant velocity model is an accurate approximation for the

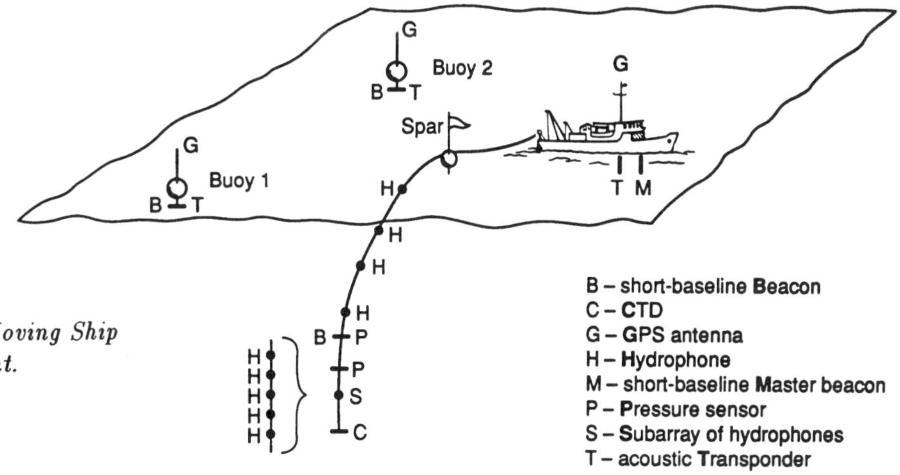

Figure 1. Tracking system for Moving Ship Tomography experiment.

1058-6393/91 $01.00 © 1991 IEEE

motion of the buoys and the hydrophone subarray relative to the ship.

Four types of data were collected every second to facilitate determination of the hydrophone locations: Global Positioning System (GPS) data consisting of pseudorange measurements from orbiting satellites to the ship and to the buoys; acoustic ranges from the buoys to the hydrophones and from the ship to the hydrophones; range and direction between the ship and beacons on the cable and buoys; and depth data consisting of pressure sensor measurements at points near the subarray. The Kalman filter provides a convenient format in which to combine these data to get an estimate of the hydrophone subarray's location. (Attitude data, i.e., heave, roll, and pitch, were also collected. They are used to convert the other data to east, north, and height coordinates with one origin.)

In sections 1–4 we discuss each of the data types. In sections 5 and 6 we present the Kalman filter model. Section 7 gives the results of applying the filter to the data collected during one particular stop. It also includes the iterated correction step and outlier detection schemes. In the final section we discuss possible improvements to the method.

1. Differential GPS Data

At each ship stop, one or two floating buoys were deployed. Each buoy contained a four-channel GPS receiver, and the ship had an eight-channel GPS receiver. (For a description of these receivers, see Ref. [1].) The GPS receivers measured pseudorange to the orbiting GPS satellites. The positions of these satellites can be derived from their ephemerides which were also recorded. We also know approximately how high the ship and the buoys were above the Earth's surface. From these measurements and facts we can infer the location of the buoys and the ship. There are biases of up to 150 m in the GPS information stream, but they do not affect relative accuracies if pseudoranges to the same satellites are used to infer the location of the buoys and the ship [2,3].

Let m be the number of satellites that are in common between the ship and the buoy, and let $p \in \Re^m$ be a vector of pseudoranges to the satellites. Let h be the height of the object being tracked, $x \in \Re^3$ be the current position of the buoy or ship in Cartesian coordinates, and b the pseudorange clock bias in the GPS receiver. Assume that the errors in pseudorange and in height are mutually independent. If the data errors are normally distributed, the negative log of the likelihood function is

$$l(p,h,x,b) = \frac{[e(x)-h]^2}{\sigma_H^2} + \sum_{i=1}^{m} \frac{[|x-s_i| - (p_i - b)]^2}{\sigma_r^2},$$

where $e(x)$ is the elevation as a function of position in Cartesian coordinates, $|\cdot|$ is the distance function in $\Re^3$, s_i is the location of the ith satellite, σ_H^2 is the variance in the height measurement, and σ_r^2 is the error in the pseudorange measurement. (We actually used the differential variances in height and pseudorange for these values; see discussion in the next paragraph.) Given the pseudorange data and height estimate, we compute a position estimate by minimizing $l(p,h,x,b)$ over (x,b). This can be thought of as an implicit function f mapping the pseudoranges and height to the position and time bias estimates.

The processing above is performed for both the ship and the buoys. Let (p^s, h^s) and (p^b, h^b) be the pseudoranges and height corresponding to the ship and one of the buoys. Let x^s and x^b be the corresponding estimates of the ship and buoy locations. We make a local affine approximation to the function f and obtain the relation $\nabla f(p,h)(p^b - p^s, h^b - h^s) = x^b - x^s$. Note that the differences have much smaller errors than the values being differenced [2,3]. The variance in the relative position estimate $x^b - x^s$ is approximately $\nabla f V \nabla f^{\mathrm{T}}$ where V is the variance of $(p^b - p^s, h^b - h^s)$, which we assume is diagonal with entries σ_r^2 and σ_H^2. The relative GPS position $x^b - x^s$ and the variances V are used by the Kalman filter. The north component of the GPS relative positions of Buoy #2 during the sample ship stop is plotted in Figure 2a. The corresponding variance is plotted in 2b.

2. Acoustic Ranging

Each floating buoy and the ship had an acoustic pinger synchronized to the common GPS time base with accuracy better than 1 μs. The ship and buoys each had their own acoustic frequency of transmission. Acoustic pulses arrived at the hydrophones on the cable, were converted to electric signals, and then transmitted via the cable. The arrival time of an acoustic pulse was determined in the following manner [4]: "... each receiver consists of a digitally tunable pre-filter with gain control, followed by a clipper, then followed by a digitally tunable quadrature demodulator. The quadrature outputs of the demodulator are low-pass filtered, and the output absolute values are summed to provide the signal envelope." The envelope output is compared against a threshold value every 100 μs. If the output rises above the threshold, a pulse reception time is stored. The arrival time errors are assumed

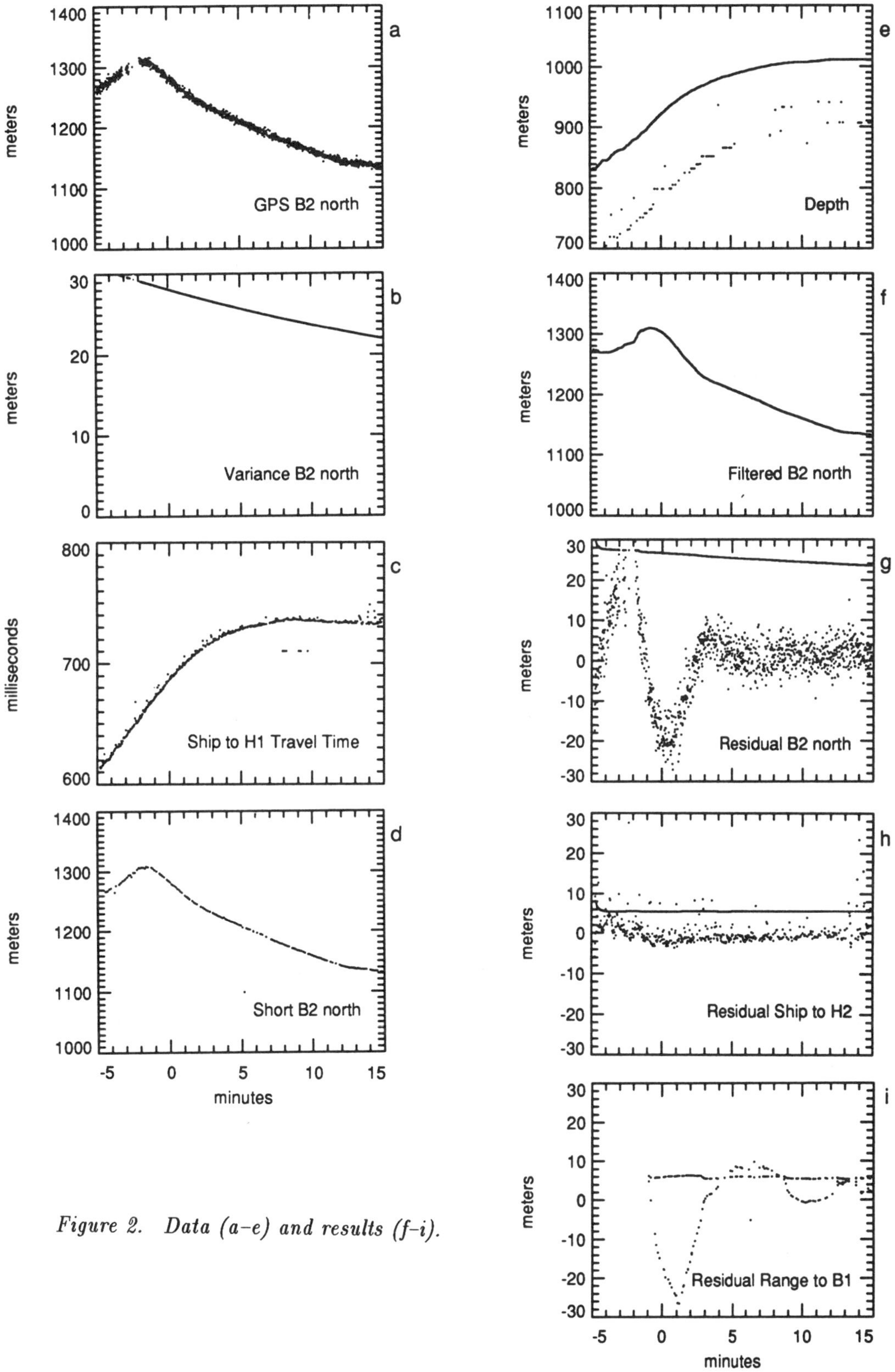

Figure 2. Data (a–e) and results (f–i).

to be independent. The travel times from the ship to hydrophone #1 for our sample ship stop are plotted in Figure 2c.

3. Short-Baseline Tracking System

A Nautronix short-baseline tracking system was also used. Its coded beacons operated remotely in the transponder mode; i.e., a ship transducer pinged, and the beacons heard this and replied. The replies were heard by a short-baseline array of receivers, resulting in range and direction measurement between the ship and the beacon [5]. The roll, pitch, and heading of the ship were measured by another set of sensors. These values were used to convert the range and direction measurements to east, north, and height coordinates. The short-baseline data from the beacons on the buoys are two-dimensional because we know the height of the buoys relative to the ship. Figure 2d shows the short-baseline result for the north component of the position of Buoy #2 for our sample ship stop.

The values obtained from the short-baseline system are in east, north, and height coordinates. But errors in these data are closer to being independent and Gaussian in range and direction coordinates, so the data vector is converted from east, north, and height to range and direction. This is done in two dimensions for the buoy location and three dimensions for the location of the cable beacon. (See the variables d_k, e_k, s_k, and t_k in Section 5.)

Sometimes the short-baseline system gave a direction that was either 90 or 180 degrees from the correct one. It is necessary to filter out the direction data when this happens. Our Kalman filter handles this by rejecting data based on the residual magnitude.

4. Depth Data

Three pressure sensors were used to measure depth. Two were Paroscientific gauges: one was part of the Sea-Bird conductivity, temperature, and pressure package (CTD), and the other was packaged with a tilter and compass (TCPP). The third was a crude pressure sensor in one of the short-baseline beacons. The pressure sensor in the TCPP was not functioning properly during the sample ship stop. This does not create a problem for our Kalman filter because the filter can delete any element from the data set and run on less information. This reduced model is constructed at each time point to handle the case in which data are missing for some time and then available again. Figure 2e is a plot of the depth data for our sample ship stop. The solid curve corresponds to the CTD data, and the sparse points correspond to the crude pressure sensor. Note that the array was descending during the sample stop.

5. The Data Model

We use $z_k \in \Re^m$ to represent the data vector at the kth time step, $x_k \in \Re^n$ to represent the state vector, $v_k \in \Re^m$ to represent the measurement error, and $h : \Re^n \longrightarrow \Re^m$ to represent a noiseless measurement as a function of state value. The measurement equation is

$$z_k = h(x_k) + v_k .$$

The state vector is composed of the positions and velocities for the buoys and the CTD located at the end of the cable. Let b_k be the positions of the buoys at the kth time step relative to the ship in east and north coordinates (buoy height is taken to be constant). Let c_k be the position of the CTD relative to the ship in east, north, and height coordinates. The state vector is given by

$$x_k = \begin{pmatrix} b_k \\ c_k \\ \dot{b}_k \\ \dot{c}_k \end{pmatrix} = \begin{pmatrix} \text{Horizontal locations of buoys} \\ \text{Location of CTD} \\ \text{Horizontal velocities of buoys} \\ \text{Velocity of CTD} \end{pmatrix} .$$

Note that b_k may be the location of more than one buoy; we use the notation $b_{k,i}$ for the location of the ith buoy.

The data vector is composed of the GPS solution for the relative buoy locations in east and north coordinates, the short-baseline ranges from the ship to the buoy and cable beacons, the short-baseline directions from the ship to the buoy and cable beacons, the acoustic ranges from the ship to each of the hydrophones, and three depth measurements. Let M be the number of buoys and L be the number of hydrophones. Let $g_k \in \Re^{2M}$ be the GPS location of the buoys, $s_k \in \Re^M$ be the short-baseline ranges from the ship to the buoys, $t_k \in \Re$ be the short-baseline range from the ship to the cable beacon, $d_k \in \Re^{2M}$ be the short-baseline directions from the ship to the buoys, $e_k \in \Re^3$ be the short-baseline direction from the ship to the cable beacon, $q_k \in \Re^{ML}$ be the acoustic ranges from the buoys to the hydrophones, $r_k \in \Re^L$ be the acoustic ranges from the ship to the hydrophones, and $p_k \in \Re^3$ be the depth measurements. Let SBL denote the short-baseline system and AR the acoustic range. The data vector is given by

$$z_k = \begin{pmatrix} g_k \\ s_k \\ t_k \\ d_k \\ e_k \\ q_k \\ r_k \\ p_k \end{pmatrix} = \begin{pmatrix} \text{GPS buoy locations, horizontal} \\ \text{SBL ranges, ship} \rightarrow \text{buoys} \\ \text{SBL range, ship} \rightarrow \text{cable} \\ \text{SBL directions, ship} \rightarrow \text{buoys} \\ \text{SBL direction, ship} \rightarrow \text{cable} \\ \text{AR, buoys} \rightarrow \text{hydrophones} \\ \text{AR, ship} \rightarrow \text{hydrophones} \\ \text{Pressure sensor depths} \end{pmatrix} .$$

The function h maps the state x_k to the corresponding noiseless data vector. Let h_g be the components of h that correspond to g_k. Define h_s, h_t, h_d, h_e, h_q, h_r, and h_p in a similar fashion. We model the end of the cable as a rigid vertical stick. The location of the jth hydrophone is $H_k = c_k + (0, 0, d_j)^{\text{T}}$, where d_j is the distance that the jth hydrophone is above the CTD. Let $B_{k,i} \in \Re^3$ be the location of the ith buoy in east, north, and height coordinates. Let $C_k \in \Re$ be the height component of the vector c_k, and let $D_j \in \Re$ be the height that the jth pressure sensor is above the CTD. Our model for a noiseless data vector is

$$h(x_k) = \begin{pmatrix} h_g(x_k) \\ h_s(x_k) \\ h_t(x_k) \\ h_d(x_k) \\ h_e(x_k) \\ h_q(x_k) \\ h_r(x_k) \\ h_p(x_k) \end{pmatrix} = \begin{pmatrix} b_k \\ |b_{k,i}| \text{ for } i = 1, M \\ |c_k| \\ b_{k,i}|b_{k,i}|^{-1} \text{ for } i = 1, M \\ c_k|c_k|^{-1} \\ |B_{k,i} - H_k| \text{ for } j = 1, L; i = 1, M \\ |H_k| \text{ for } j = 1, L \\ -(C_k + D_1, C_k + D_2, C_k + D_3)^{\text{T}} \end{pmatrix} .$$

Let G_k be the covariance matrix for the buoy locations (g_k), σ_s be the standard error for short-baseline ranges (s_k and t_k), σ_d be the standard error for short-baseline directions (d_k and e_k), σ_q be the standard error for acoustic travel times (q_k and r_k), and S_p be the covariance matrix for the depth measurements (p_k). The error in the depth measurements is assumed independent so S_p is diagonal. The error in the buoy GPS locations is assumed independent from one buoy to the next. Thus the matrix G_k is block diagonal with as many blocks as there are buoys. The total covariance matrix is

$$R_k = E[v_k v_k^{\text{T}}] = \begin{pmatrix} G_k & 0 & 0 & 0 & 0 & 0 & 0 & 0 \\ 0 & \sigma_s^2\text{I} & 0 & 0 & 0 & 0 & 0 & 0 \\ 0 & 0 & \sigma_s^2\text{I} & 0 & 0 & 0 & 0 & 0 \\ 0 & 0 & 0 & \sigma_d^2\text{I} & 0 & 0 & 0 & 0 \\ 0 & 0 & 0 & 0 & \sigma_d^2\text{I} & 0 & 0 & 0 \\ 0 & 0 & 0 & 0 & 0 & \sigma_q^2\text{I} & 0 & 0 \\ 0 & 0 & 0 & 0 & 0 & 0 & \sigma_q^2\text{I} & 0 \\ 0 & 0 & 0 & 0 & 0 & 0 & 0 & S_p \end{pmatrix} .$$

The division of R_k into sub-blocks above corresponds to the eight components in the definition of z_k.

6. The Transition Model

We use $g : \Re^n \rightarrow \Re^n$ to represent a noiseless transition and $w_k \in \Re^n$ to represent the transition noise. The transition equation is

$$x_{k+1} = g(x_k) + w_k .$$

Let α be the time between data values. The transition function g in our model is given by a linear approximation for position and a constant approximation for velocity; i.e.,

$$g(x_k) = \begin{pmatrix} \text{I} & 0 & \alpha & 0 \\ 0 & \text{I} & 0 & \alpha \\ 0 & 0 & \text{I} & 0 \\ 0 & 0 & 0 & \text{I} \end{pmatrix} x_k = \begin{pmatrix} b_k + \alpha\dot{b}_k \\ c_k + \alpha\dot{c}_k \\ \dot{b}_k \\ \dot{c}_k \end{pmatrix} .$$

We model the noise in the transition equation as component-wise-independent with a different level for the positions and the velocities. Let σ_l be the standard error of each of the location components of w_k, and let σ_v be the standard error of each of the velocity components of w_k. Our model for the noise w_k is a Gaussian random variable with variance given by

$$Q_k = E[w_k w_k^{\text{T}}] = \begin{pmatrix} \sigma_l^2\text{I} & 0 & 0 & 0 \\ 0 & \sigma_l^2\text{I} & 0 & 0 \\ 0 & 0 & \sigma_v^2\text{I} & 0 \\ 0 & 0 & 0 & \sigma_v^2\text{I} \end{pmatrix} .$$

7. Sample Ship Stop

In this section we discuss our Kalman filter by looking at the results for a sample ship stop. We use a forward filter with an iterated correction step to estimate the state variable values. A discussion of this method for nonlinear measurement functions can be found in [6]. Time in our sample stop is relative to 6:00 a.m. on day 193 of 1991.

Sometimes the data have outliers; see for example the travel time plot in Figure 2c or the short-baseline depth measurements in 2e. The forward Kalman filter provides a predicted estimate of the state variable from which we compute a predicted data estimate by applying h. We form a bound for each component of the data that is five times the standard error of the data plus the standard error of the predicted data value. We also form a residual for each component by subtracting the predicted data from the actual data. If the absolute value of the residual is greater than the bound, that component is considered an outlier and ignored.

Table I lists the standard error values that correspond to the sample ship stop. The filtered position estimate for the north component of the location of Buoy #2 is plotted in Figure 2f. The corresponding GPS residuals and bounds are plotted in 2g. (The solid line represents the bound for the residuals. Our filter ignores data if its residual is outside the corresponding bound.) Note that the residuals in Figure 2g have structure between −5 and 5 minutes. The ship's screw was not "declutched" until 6 minutes before the hour, and the end of the cable was accelerating in the vertical direction during the first 10 minutes of this ship stop (see Figure 2e). Once the ship was allowed to drift for 10 minutes, and the cable end came to its final depth, the relative accelerations died out and our linear transition model became a good approximation of the system's motion. Thus we have disagreement between the data and filtered values at the beginning and good agreement at the end (minutes 5–15).

Table I. Standard errors for sample ship stop.

σ_H	height estimates	1 m
σ_r	pseudoranges	5 m
σ_s	short-baseline ranges	1 m
σ_d	short-baseline direction	0.01
σ_q	acoustic ranges	1 m
σ_1	TCPP depth	2 m
σ_2	CTD depth	2 m
σ_3	beacon pressure sensor	10 m
σ_l	transition position	0 m
σ_v	transition velocity	0.001 m/s

The travel times had quite a few points that did not follow the general shape of the data curves (see Figure 2c). These points are more pronounced in the acoustic range residual plot in Figure 2h. The outliers that are farther from zero than the solid line in these plots are ignored by our filter.

8. Future Work

The filter presented is forward looking; i.e., only previous information is used for estimating the position at any point in time. We are currently implementing a smoothing version of the nonlinear Kalman filter that does not have this defect.

The vertical stiff array model presented is inappropriate for estimating the position of the hydrophones that are far from the end of the cable. To make such an estimate, we need to allow some variation in the separation between hydrophones but at the same time constrain the positions to reflect the fact that the shape of the cable is not arbitrary.

Most of the residuals appear to be random and within their error bounds between the 5 minute and 15 minute marks. The one exception is the short-baseline range from the ship to Buoy #1 (see Figure 2i). The residuals for these data have a structure that is unresolved at this time.

References

[1] Motorola Inc., "Eagle Mini-Ranger Global Positioning System Receiver, User's Manual," Doc. No. 68-P29027U, Tempe, Arizona, March 1987.

[2] D. Wells (Ed.) et al., "Guide to GPS Positioning," Solutions 10.14, Canadian GPS Associates, Ottawa, May 1987.

[3] R. P. Denaro and R. M. Kalafus, "Differential operation of NAVSTAR GPS for enhanced accuracy," in *NATO AG RD Lecture Series, the NAVSTAR GPS System,* AGARD-LS-161, 1988.

[4] Benthos Undersea Systems Technology, "DS 7000 Acoustic Deck Unit, Preliminary Manual," p. 9, Doc. No. M-500-004, Falmouth, MA, 17 December 1990.

[5] Nautronix Ltd., "Acutrak Introduction Manual for the ATS-SO2, ATS-SO4 and ATS-LR08 Models," pp. 1–8, Perth, Australia, 21 January 1988.

[6] B. M. Bell and F. W. Cathey, "The iterated Kalman filter update as a Gauss-Newton method" (submitted to *IEEE Transactions on Automatic Control*), June 1991.

Chernoff Bounds for Class A Noise

Pamela A. Nielsen*
Dept. of ECE
University of Arizona
Tucson, Az. 85721

Abstract

The goal is, using a very large passive array, to determine the performance limits of a detector. The signal of interest is narrowband with a Gaussian envelope, and the contaminating noise is multivariate Class-A. Two different multivariate models for the Class A family are presented. One of the models is appropriate for array processing applications. The data is spatially dependent and temporally independent. It is shown, in the spatially independent case, that the Chernoff approximation does closely approximate the performance of the optimal detector. It is shown that the approximation improves as the number of samples increases. Unfortunately, it is also shown that the Chernoff approximation requires numerical evaluation of a M-dimensional integral. For the application here, M may be as large as 150, ruling out this approach. Two alternative approaches are examined. First, approximating the Class A model by a Gaussian model is shown to result in a poor approximation. Second, the exact likelihood ratio is approximated by a piecewise function. While the approximation can be done with very good accuracy, the bound must be evaluated numerically.

1 Introduction

The problem of interest is the passive detection of a subsurface moving source using a stationary, very large array. The source will radiate a wide band of low frequency components and is assumed to be in the far-field of the array. Since the source is in the far-field, the signal strengths will be small and the large array is used for coherent processing of the signal, e.g. some type of spatial processing or beamforming will be used. The goal is to evaluate different detection methods.

This problem is quite complex, and presumably it will take several years for the above questions to be solved to the accuracy desired. Unlike radar, the frequencies of interest are low, and the transmission medium, the ocean, will distort the transmitted signal in a more complicated manner. Hence, assumptions which are often acceptable in radar applications are not necessarily acceptable in this application.

The results presented in this paper are an outgrowth of the problem stated above. For an assumed noise model, the Chernoff bound was investigated and compared to simulated and theoretical performance measures. The goal of the investigation was to determine the usefulness of using the Chernoff bounds on the probability of false alarm and the probability of a miss, for a very large passive array.

*Thanks to Lawrence Livermore National Laboratory and the Engineering Research Division

2 Signal Model

The ultimate interest is in solutions for a wideband source. Since a wideband source can be modeled as a sum of narrowband sources, we constrain the analysis to the case of a narrowband source. The results can be generalized to the wideband case simply by processing in parallel different frequency bands. The narrowband radiated signal can be represented as

$$s_t(t) = \alpha(t)cos(\omega_c t + \phi(t)) \qquad (1)$$

where the carrier frequency ω_c is known, and the envelope and phase functions, α and ϕ are random and slowly varying. The functions α and ϕ are approximately constant as the wave sweeps across the array. The phase is assumed to be uniformly distributed on the interval $[0, 2\pi)$ and the envelope is a zero mean Gaussian random variable. The variance of α is dependent on the carrier frequency, and is given as $\sigma^2(\omega_c) = \sigma^2$. For convenience, the complex analytic representation of the signal will be used, $\tilde{s}_t(t) = \alpha(t)e^{j\phi(t)}e^{i\omega_c t}$.

Since the signal is narrowband and the ocean, in deep water, can be accurately modeled as a waveguide, the normal model approximation for the solution to the full wave equation can be used. Denoting the source position as $\underline{\theta}$ (this is a three-dimensional vector, either in rectangular or spherical coordinates) and the position of the ith sensor as $\underline{\chi}_i$, then, from Clay and Medwin [1] the received acoustic pressure wave can be written as $\tilde{s}(t) = \alpha(t)e^{j\phi(t)}e^{i\omega_c t}e(\omega_c, \theta)$ where e is a function of: the ambient density at position $\underline{\chi}_i$; the mode numbers m (the values of m are dependent on the waveguide dimensions and the source frequency); the source excitation coefficient for the mth mode; the horizontal wave number for the mth mode (the wave number is range dependent only); the range between the source position and the ith sensor; and the mth mode eigenfunction. It should be understood, that if the source frequency is changed, then the coefficient values in e will change.

Rewriting the received analytic signal in vector notation, yields

$$\underline{\tilde{s}}(t) = \alpha(t)e^{j\phi(t)}e^{i\omega_c t}\underline{e}(\omega_c,\theta) \quad (2)$$

where the dimension of the replicant vector $\underline{e}$ is equal to the number of sensors M. Note that the vector $\underline{e}$ can be viewed as a steering vector for a beamforming operation. The mean of the received signal, $\underline{\tilde{s}}(t)$, is zero, and the signal covariance is

$$K_s = E\left(\underline{\tilde{s}}(t)\underline{\tilde{s}}(t)^H\right) = \sigma^2(\omega_c)\underline{e}(\omega_c,\theta)\underline{e}(\omega_c,\theta)^H \quad (3)$$

The received signal covariance is dependent on the signal energy through σ^2 and on the spatial covariance, through $\underline{ee}^H$. Note the familiar decoupling of the temporal and spatial components, which occurs in narrowband beamforming.

Since the envelope is Gaussian, the complex analytic signal is Gaussian [5] with a density function

$$p_{\underline{\tilde{s}}(t)}(\underline{x}) = \frac{1}{\pi^M |det K_s|}e^{-\underline{x}^H K_s^{-1}\underline{x}} \quad (4)$$

3 Noise Model

Ocean noise has been well documented as being non-Gaussian. A few recent papers demonstrating this are [7], [8], [9], [10]. It has been suggested that an appropriate model for oceanic noise is Middleton's Class-A model [8]. The Class-A model is a three-parameter family of density functions, mathematically described as [3], [4]

$$p(x) = e^{-A}\sum_{m=0}^{\infty}\frac{A^m}{m!}\frac{1}{\sqrt{2\pi\sigma_m^2}}e^{-x^2/2\sigma_m^2} \quad (5)$$

Clearly, the Class-A family is a Poisson weighted infinite sum of Gaussian density functions. The parameter $A > 0$ is the " Overlap Index". The overlap index "measures the mean deviation of a typical noise source from the average.." [4]. For the Class-A family, the parameter A is small, usually in the range 10^{-4} to 0.5.

The second parameter of the Class-A model is the "Gaussian Factor" Γ. This parameter is defined as the ratio of the Gaussian variance σ_g^2 to that of the variance or intensity of the non-Gaussian component. Γ relates the overlap index A to the variance of the mth term through

$$\sigma_m^2 = \frac{m/A + \Gamma}{1+\Gamma} \quad (6)$$

$\Gamma > 0$ and is typically 10^{-5} to 0.1.

The final parameter of the Class-A model is the intensity of the non-Gaussian component. The Class A family was originally derived to model the electromagnetic interference caused by man-made sources. In the development, the noise was modeled as having two components, a Gaussian component with variance σ_g^2, and a non-Gaussian component (Class A). The third parameter measures the relative strengths of the Gaussian and the non-Gaussian component. This parameter is implicit and is important when fitting experimental data to the family.

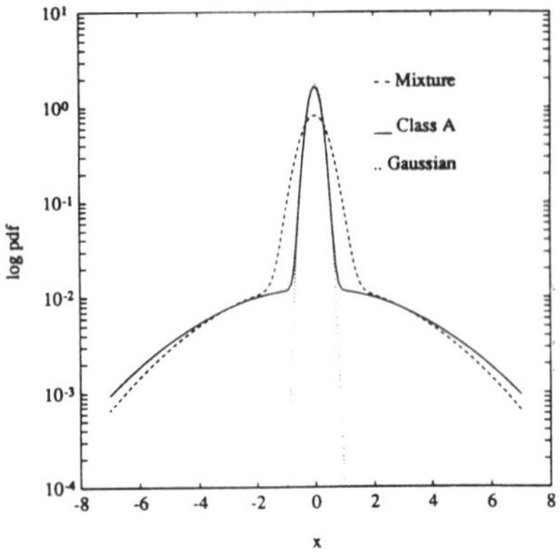

Figure 1: Comparison of Class A and Mixture Density Functions

The Class A density function is symmetric about its mean of zero. As the parameter A tends to infinity, the model approaches a Gaussian density. The variance is a monotonic increasing function of the index m. It can be easily shown that $\sigma_{m+1}^2/\sigma_m^2 = 1 + 1/(A\Gamma)$.

Working with an infinite series it typically not appealing or convenient. We therefore wish to truncate the infinite series to two terms. Spaulding [6], has shown that the series, for small A and Γ, can be truncated to two terms with less than a 2% error as

$$p_T(x) = \left\{\frac{e^{-A}}{\sqrt{2\pi\sigma_0^2}}e^{-x^2/2\sigma_0^2} + \frac{1-e^{-A}}{\sqrt{2\pi\sigma_1^2}}e^{-x^2/2\sigma_1^2}\right\} \quad (7)$$

where σ_0 and σ_1 are defined from (6). The standard Gauss-Gauss mixture model can be written as

$$p_m(x) = \frac{\epsilon}{\sqrt{2\pi\sigma_{m0}^2}}e^{-x^2/2\sigma_{m0}^2} + \frac{1-\epsilon}{\sqrt{2\pi\sigma_{m1}^2}}e^{-x^2/2\sigma_{m1}^2} \quad (8)$$

which is often used in the literature, and which some claim is a truncation of the Class-A model. Typically, the mixture parameter $\epsilon \in (0,1)$, $\sigma_0^2 < 1$, and $\sigma_1^2 > 1$. The total variance is $\sigma^2 = \epsilon\sigma_{m0}^2 + (1-\epsilon)\sigma_{m1}^2$. The difference between p_m and p_T is illustrated in Figure 1. The Gaussian curve has a variance of 0.04762. The Class A noise has parameters $A = 0.1$ and $\Gamma = 0.05$. Therefore, $e^{-A} = 0.905; \sigma_0^2 = 0.04762; \sigma_1^2 = 9.5714$, and the variance is 0.9523. For the mixture model, the mixture parameter $\epsilon = 0.905$. There are many choices of variances, σ_{m0} and σ_{m1} which will result in the total variance being 0.9523, the same as the Class-A truncated model. The plotted curve has $\sigma_{m0} = 0.2$ and $\sigma_{m1} = 8.11$. The mixture model and the Class-A models may be the same, but there are more constraints on the parameter choices for the Class-A model.

3.1 Multivariate Model

The data received by the array will be spatially and possibly temporally correlated. For one snap-shot in time, the data will have a spatial multivariate density function. There are at least two ways that the scalar version of the Class-A model can be modified to handle dependent data.

Case I: Set $A > 0$ but set Γ to be a $M \times M$ matrix. Then the mth covariance matrix can be written as

$$K_m = (I + \Gamma)^{-1}(\frac{m}{A}I + \Gamma) \quad (9)$$

and the multivariate density written as

$$p(\underline{x}) = \sum_{m=0}^{\infty} \frac{A^m}{m!} \frac{e^{-A}}{\sqrt{(2\pi)^{M/2}|detK_m|}} e^{-\underline{x}^T K_m^{-1} \underline{x}/2} \quad (10)$$

The multivariate density has a mean of zero and a covariance given by $E\underline{x}\underline{x}^T = e^{-A}\sum_{m=0}^{\infty} \frac{A^m}{m!} K_m$. The problem with this approach is the interpretation of Γ.

Case II: An alternative to the above approach is to let both A and Γ be scalars, as before. Hence, the covariance of the mth term is $K_m = \sigma_m^2 K_0$ with the scalar's σ_m^2 defined as in (6). K_0 could be interpreted as the spatial covariance matrix, e.g. the covariance of the inputs to the different sensors *at the same time*. The multivariate density can be written as

$$p(\underline{x}) = \sum_{m=0}^{\infty} \frac{A^m}{m!} \frac{e^{-A}}{\sqrt{(2\pi)^{M/2}|detK_0|\sigma_m}} e^{-\underline{x}^T K_m^{-1} \underline{x}/2} \quad (11)$$

The multivariate density has a zero mean and a covariance of

$$E\underline{x}\underline{x}^T = K_0 e^{-A} \sum_{m=0}^{\infty} \frac{A^m}{m!} \sigma_m^2 \quad (12)$$

This model can be truncated to two terms, in the manner of Spaulding [6] yielding the following Class-A model

$$p_T(\underline{x}) = \left\{ \frac{e^{-A}}{\sqrt{(2\pi)^{M/2}|detK_0|\sigma_0}} e^{-\underline{x}^T K_0^{-1} \underline{x}/(2\sigma_0^2)} + \frac{1 - e^{-A}}{\sqrt{(2\pi)^{M/2}|detK_0|\sigma_1}} e^{-\underline{x}^T K_0^{-1} \underline{x}/(2\sigma_1^2)} \right\} \quad (13)$$

with σ_0^2, σ_1^2 defined as before, and K_0 the spatial covariance.

4 Hypothesis Problem

The detection problem can be modeled as the following simple binary hypothesis test [2].

$$H_0 : \underline{X} = \underline{N} \qquad H_1 : \underline{X} = \underline{N} + \underline{S} \quad (14)$$

$\underline{N}$ contains the M-samples of the noise, and $\underline{S}$ contains the M-samples of the signal due to the source, received at the array at some time t.

The noise is multivariate Class-A. Using the complex analytic representation,

$$p_T(\underline{x}|H_0) = e^{-A} \left\{ \frac{1}{\pi^M |detK_0|\sigma_0^2} e^{-\underline{x}^H K_0^{-1} \underline{x}/(\sigma_0^2)} + \frac{e^A - 1}{\pi^M |detK_0|\sigma_1^2} e^{-\underline{x}^H K_0^{-1} \underline{x}/(\sigma_1^2)} \right\} \quad (15)$$

For the alternative hypothesis, the conditional density function is the convolution of the density of the signal with that of the noise. On recognizing that the convolution of two Gaussian densities yields a Gaussian density, the conditional density given that the signal is present can be quickly written as above by substituting $K_0\sigma_0^2 + K_s$ for $K_0\sigma_0^2$ and by substituting $K_0\sigma_1^2 + K_s$ for $K_0\sigma_1^2$. This density is not Class-A because the variances do not follow the update rule defined by (6).

These expressions for the conditional density functions are, of course, for a single snap-shot in time. In practice data will be collected over a time interval, hence the exact density representation will be multivariate spatially and temporally. If the samples can be taken far enough apart, then the samples will be uncorrelated and possibly can be considered temporally independent. Let N_t be the number of time snapshots. Assuming independence, the conditional density under each hypothesis is the product of the conditional densities above, $p_T(\mathbf{x}|H_0) = \prod_{l=1}^{N_t} p_T(\underline{x}_l|H_0)$. Here, $\mathbf{x}$ is the $M \times N_t$ matrix of data, and $\underline{x}_l$ is the lth column which corresponds to a particular time snapshot. For the alternative hypothesis, the conditional density function can be similarly written.

5 Chernoff Bounds

The usefulness of the Chernoff bound to determine upper bounds on the false alarm and miss probabilities was investigated for the problem described. The Chernoff bound [2] hinges on evaluating $\mu(s)$, where

$$\mu(s) = log \left\{ \int_{-\infty}^{\infty} \left(\frac{p(\underline{x}|H_1)}{p(\underline{x}|H_0)} \right)^s p(\underline{x}|H_0) d\underline{x} \right\} \quad (16)$$

If μ exists, then the false alarm and miss probabilities can be bounded above by

$$P_{fa} \leq e^{\mu(s) - s\gamma} \qquad P_m \leq e^{\mu(s) + (1-s)\gamma} \quad (17)$$

for $0 < s < 1$. The best bound is found by optimizing over s, hence s is chosen such that $\gamma = \dot{\mu}(s)$ [2].

If the central limit theorem can be applied to yield the following approximate expressions for the error probabilities

$$\begin{aligned} P_{fa} &\approx erfc(s\sqrt{\ddot{\mu}(s)}) e^{\mu(s) - s\dot{\mu}(s) + s^2\ddot{\mu}(s)/2} \quad (18) \\ P_m &\approx erfc((1-s)\sqrt{\ddot{\mu}(s)}) e^{\mu(s) + (1-s)\dot{\mu}(s) + (1-s)^2\ddot{\mu}(s)/2} \end{aligned}$$

where $0 < s < 1$ and $erfc$ is the complimentary error function. Generally, the Chernoff bounds are loose for N_t small; however, as N_t grows then the bounds become tighter.

5.1 Bound Evaluation

For the problem stated above, the first step to evaluating the Chernoff bound is to evaluate μ. (Let $N_t = 1$ but $M > 1$). Referring to (16) and the descriptions of the density of the data under the two different hypotheses, e.g. see (15) and the discussion following, it is clear that the necessary integral cannot be evaluated analytically and must be evaluated numerically for arbitrary signal and noise covariance matrices. Accurate numerical evaluation of integrals for $M > 2$ dimensions is very difficult. For our case, M could be quite large, say 100 to 150, and numerical integration is not feasible.

As an alternative, we evaluate the univariate case. The intention is to determine if the bounds are accurate and if an alternative method for evaluating the μ function can be derived. Perhaps, some conclusions can be drawn about the bound for the multivariate case from the univariate case results. For $M = 1$ the scalar density functions will be used. Hence, $p(x|H_0)$ is defined by (7) and $p(x|H_1)$ is defined as by (7) with the substitutions $\sigma_0^2 + \sigma_s^2$ for σ_0^2 and $\sigma_1^2 + \sigma_s^2$ for σ_1^2. Numerical techniques are needed to evaluate μ and the first two derivatives. The extension to the situation when $N_t > 1$ and the data are temporally independent is straight forward on recognizing that the multivariate density function will be a product of univariate densities, and that the logarithm will convert the product to a sum [2].

6 Results

The Chernoff bound was evaluated numerically for the case when the noise was Class-A with parameters $A = 0.1$ and $\Gamma = 0.05$. The number of snap shots was taken to be $N_t = 1, 2, 5$, and the sample by sample SNR was either one or two. Using the expressions above bounds on the probability of false alarm and detection were obtained. These bounds were compared to the theoretical performance and to performance determined via simulation. For theoretical performance, the optimal log-likelihood ratio detector was determined and the decision regions for the null and alternative hypothesis were determined. Then, the conditional density functions were numerically integrated over these regions to determine the error probabilities.

The performance was simulated by generating data belonging to the Class-A family and the the conditional density functions under the alternative hypothesis. The 4000 samples were generated using the rejection method. Comparison of empirical density functions of the data with theoretical density functions shows very good agreement.

The ROC comparing the simulated and the theoretical probabilities is shown in Figure 2. The detection probability is simulated very accurately, while there is a small error in the false alarm probability. The effect of the small error in the false alarm probability, is that for a fixed probability of detection, the simulation slightly over estimates the probability of false alarm.

Figures 3 plots the simulated performance as compared to the Chernoff bound. The results in the figures

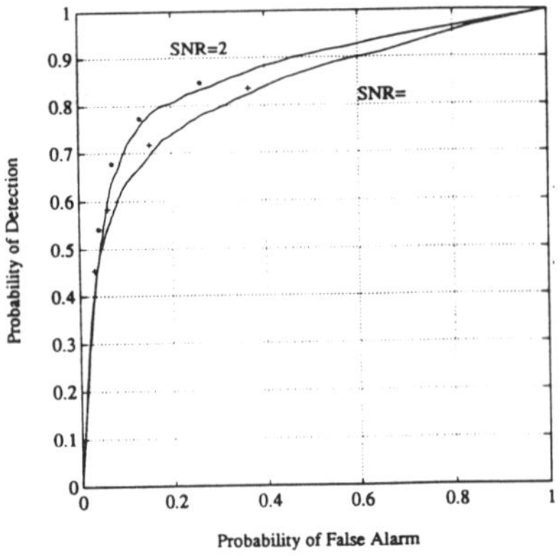

Figure 2: ROC of the simulated and theoretical performance.

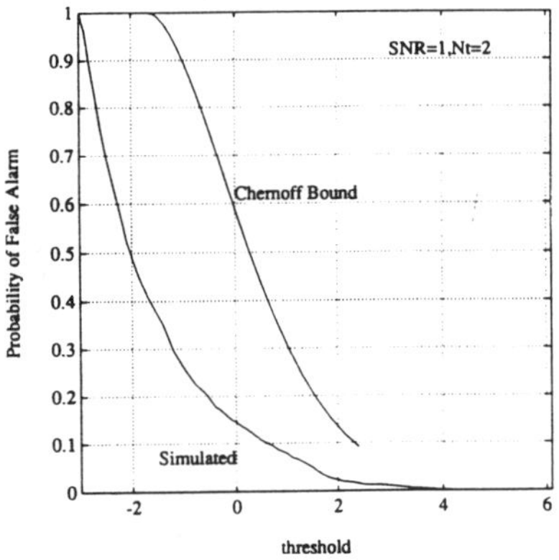

Figure 3: Comparison of Chernoff bounds and simulated performance.

are typical. The Chernoff bound on the false alarm probability is much larger than the actual false alarm probability, and the lower bound on the probability of detection is much lower than the actual probability. The bounds are so loose that they are not useful and they are not a good indicator of the detectors performance. As N_t is increased, the bounds become tighter, but they are still too loose to be useful.

Since the bound was so loose, and since we will use multiple snap shots ultimately, the approximate expressions for the error probabilities were evaluated. The results are shown in Figure 4. There is a surprisingly good agreement between the theoretical, simulated, and approximate expressions. As expected, the approximate expression becomes more accurate as the number of snap shots increases.

To summarize, we can evaluate the Chernoff bounds. In fact, the approximate expressions are reasonably good even for N_t small, but the extension to the case when the data is spatially dependent is not clear. Hence, the original problem has not been solved.

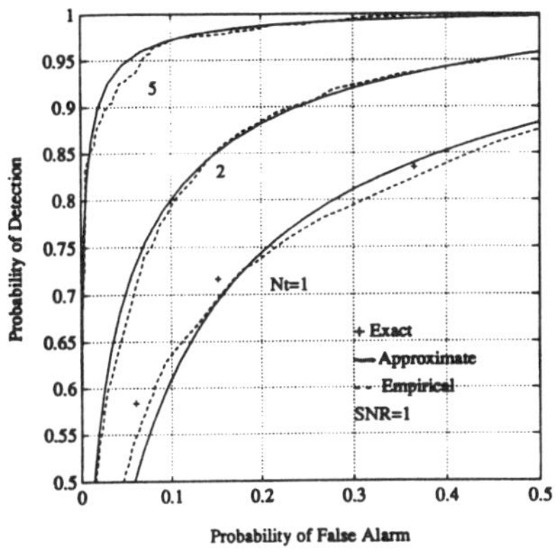

Figure 4: Comparison of Chernoff approximations and simulated performance.

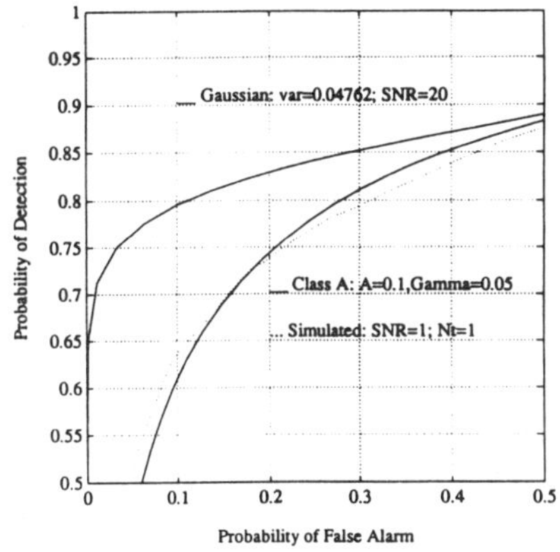

Figure 5: Comparison of performance for Gaussian and Class A noise.

6.1 Gaussian Approximation

Suppose the Class A noise is approximated as Gaussian. Will this cause a significant change in the Chernoff bounds and/or approximations? Unfortunately, it will. If the noise has a variance of σ_n^2 and the signal has a variance of σ_s^2, then for real data the μ function can be evaluated to be

$$\mu(s) = \left\{ \log((1+SNR)^{(1-s)}/(1+(1-s)SNR)) \right\} /2$$

where $SNR = \sigma_s^2/\sigma_0^2$. For the complex case, the function is the same except the $SNR = \sigma_s^2 \underline{e}^H K_n^{-1} \underline{e}$ and the factor of one-half disappears because the samples are complex. The SNR is essentially the output of a beamformer. The expressions above can be easily evaluated without any integration. The question is really whether the above μ function is a good approximation of the μ function for Class-A noise.

Figure 5 compares the approximate expressions for the probability of false alarm and detection for the Class A and the Gaussian cases. Note the difference in the SNR. The Gaussian noise was modeled as having a variance $\sigma_n^2 = \sigma_0^2$. The signal variance was chosen such that the sample-by-sample SNR for the Class A noise was unity. It is important to notice that the *shapes* of the two curves are different. This suggests that assuming the noise is Gaussian will not provide an accurate estimate of the error probabilities. The graph shown is typical, though comparisons were made for different SNR and different temporal sample lengths N_t.

An explanation for the difference can be found by examining the shape of the likelihood ratio. The likelihood ratio for Class A noise censors some of the large values (it does increase towards infinity as the amplitude increases towards infinity). To conclude, using a simple Gaussian approximation is not appropriate and will provide results which are misleading.

6.2 Piece-wise Approximation

A better approach than using the Gaussian approximation, might be to approximate the likelihood ratio, $L(x) = p(x|H_1)/p(x|H_0)$ in the calculation of the μ function, with a simpler function. Several different functions were considered, but it was decided to break the log-likelihood ratio, $l(x) = \log(L(x))$, into three regions. The log-likelihood ratio was approximated as follows. For small x, set $\hat{l}(x)$ equal to the log-likelihood ratio of two Gaussian random variables, each with a mean of zero, and variances of $\sigma_s^2 + \sigma_0^2$ and σ_0^2 respectively. Therefore for small x, $\hat{l}(x) = a_1 x^2 + a_2$, with the constants a_1 and a_2 determined by the variances. The break point is the value of x such that $\hat{l}(x) = \max(l(x))$. For moderate x the log-likelihood ratio is approximated by the line $\hat{l}(x) = mx + b$. For large x, $\hat{l}(x) = 0$. The break point between the last two regions determines the values of the slope and intercept. These two parameters were chosen such that the line starting at the end point of region one approximately passed through the inflection point of $l(x)$. Finally, the we have $\hat{L} = \exp(\hat{l})$ completing the approximation. For the case when the $SNR = 1$, $A = 0.1$, and $\Gamma = 0.05$, an example of the piece-wise approximation is shown in Figure 6. The break points between the regions are ± 0.6755 and ± 3.755.

The μ function using this approximation can be evaluated. Note that while the approximation is not differentiable in x, it does not imply that μ is not differentiable with respect to s. A comparison of the μ functions using the exact and the piece-wise approximation is shown in Figure 7. The curves are indistinguishable.

To conclude, we *can* accurately approximate the likelihood ratio and evaluate μ. There is no reason to believe that the first and second derivatives cannot be easily evaluated also. However, the original problem has not really been solved. That is, the solution method does not remove the need for evaluating a M-dimensional integral numerically. In the Gaussian case the integral limits are $\pm\infty$, which allows the integral to be evaluated by a proper choice of constants. For a piecewise approximation, the integral is broken into a sum where at least one limit is finite. Hence, the integral must still be evaluated numerically and the original problem of evaluating the Chernoff bound for

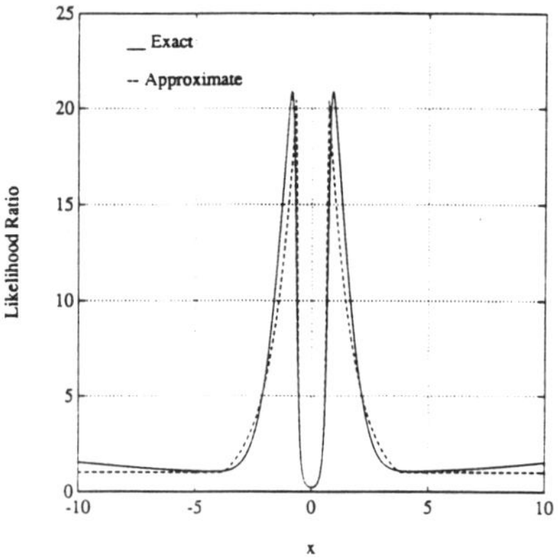

Figure 6: Piece-wise approximation of the likelihood ratio.

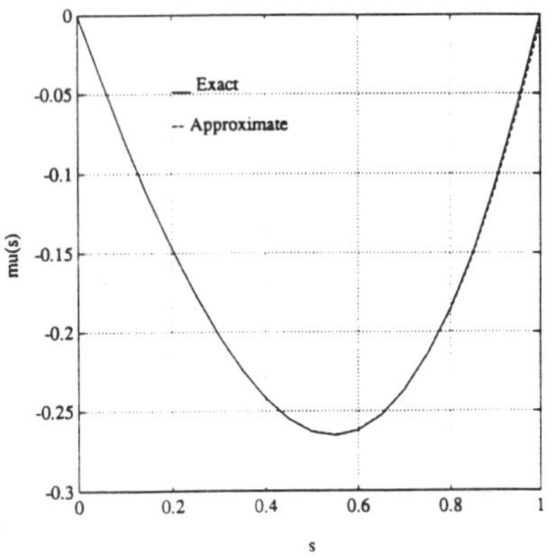

Figure 7: Piece-wise approximation of the likelihood ratio.

data which is temporally independent, but is spatially dependent has not been solved.

7 Conclusions

The original problem is the determination of the performance of a very large array. The signal was assumed to be narrowband with a Gaussian envelope, and the noise was modeled as being a member of the multivariate Class-A family. Two different multivariate models for the Class A family were presented. One of the models is appropriate for array processing applications. The data was assumed to be spatially dependent and temporally independent. It was found that the Chernoff approximation does closely approximate the performance for this model if the array dimension is one, that is, if the data is spatially independent. It was found that the approximation improves as the number of samples increases. Unfortunately, it was also found that direct calculation of the Chernoff approximation must be numerical, involving a M-dimensional numerical integral. For the application here, M may be as large as 150, ruling out this possibility. The bounds were compared to theoretical and simulated performance where possible.

Two alternative approaches were examined. First, approximating the Class A model by a Gaussian model was shown to result in a poor approximation. Second, the exact likelihood ratio was approximated by a piece-wise function. The approximation can be done with very good accuracy, though the bound must be evaluated numerically. Therefore, the original problem has not been solved. The author would like to thank D. Harris and J. Candy. Additionally, the author is interested in hearing about alternative approaches.

References

[1] C.S. Clay and H. Medwin, *Acoustical Oceanography: Principles and Applications*, John Wiley and Sons, New York, 1977.

[2] H.L. Van Trees, *Detection, Estimation, and Modulation Theory: Part I*, John Wiley and Sons, New York, 1968.

[3] D. Middleton, *Introduction to Statistical Communication Theory*, Peninsula Press, Los Altos, CA, 1987.

[4] D. Middleton, "Procedures for Determining the Parameters of the First-Order Canonical Models of Class A and Class B Electromagnetic Interference," *IEEE Trans. on EMC*, EMC-21(3), 190-208, 1979.

[5] K. Miller, "Complex Gaussian Processes", *SIAM Review*, 11(4), 544-567, 1969.

[6] A.D. Spaulding D. Middleton, "Optimum Reception in Impulsive Interference Environment," *OT Report 75-67*, June 1975. (Office of Telecom., U.S. Dept. of Commerce, 325 Broadway, Boulder, CO. 80303), see 193-195.

[7] P.L. Brockett, M. Hinich, and G.R. Wilson, "Bispectral Characterization of Ocean Acoustic Time Series: Nonlinearity and Non-Gaussianity", in *Topics in Non-Gaussian Signal Processing*, Eds. E.J. Wegman, S.C. Schwartz, and J.B. Thomas, Springer-Verlag, New York, 2-16, 1989.

[8] D.R. Powell and G.R. Wilson, "Class A Modeling of Ocean Noise Processes", in *Topics in Non-Gaussian Signal Processing*, Eds. E.J. Wegman, S.C. Schwartz, and J.B. Thomas, Springer-Verlag, New York, 17-28, 1989.

[9] F.W. Machell, C.S. Penrod, and G.E. Ellis, "Statistical Characteristics of Ocean Acoustic Noise Processes," in *Topics in Non-Gaussian Signal Processing*, Eds. E.J. Wegman, S.C. Schwartz, and J.B. Thomas, Springer-Verlag, New York, 29-57, 1989.

[10] P.A. Nielsen and J.B. Thomas, "A Comparison of Parametric and Non-Parametric Performance Levels for Underwater Noise", *J. Acous. Soc. Am.*, 225-236, 1990.

Spread Spectrum Code, Doppler, and Multipath Tracking Using Gaussian Sum Filtering

Alfred W. Fuxjaeger

CIPR
Department of ECE
University of California
Santa Barbara, CA 93106

Ronald A. Iltis

CIPR
Department of ECE
University of California
Santa Barbara, CA 93106

Abstract

We consider the problem of obtaining joint estimates of code delay, τ, multipath coefficient vector $\mathbf{f}$, radial Doppler velocity, v_r, and carrier phase due to Doppler shift, ϕ distorting a direct-sequence spread-spectrum signal during propagation. Here we apply parallel Kalman filtering in the form of the Gaussian Sum algorithm. The Gaussian Sum filter is capable of updating the joint density of $\tau, \mathbf{f}, v_r, \phi$ given the measurements, Z^k, (i.e. $p\left(\tau, \mathbf{f}, v_r, \phi | Z^k\right)$). In particular, we are interested in obtaining an estimate of the radial Doppler velocity, v_r in the presence of a large initial uncertainty in that variable. A Gaussian Sum filter that uses enough terms to approximate the joint density will be less susceptible to divergence than an estimator consisting of one Kalman filter. Bit error rate results are given to demonstrate the usefulness of the parameter estimates in a RAKE receiver for DPSK encoded binary data.

1 Introduction

Estimates of the unknown code timing, τ, multipath channel coefficients, $\mathbf{f}$ and relative platform velocity v_r, distorting a transmitted DS spread spectrum signal must be computed if a reasonable bit error rate (BER) performance is to be obtained in a coherent receiver. Although the problem of timing-only estimation has been thoroughly studied, [1],[2], the problem of synchronization in the presence of multipath and code Doppler has received less attention. A scheme for synchronization in the presence of unknown timing, and Doppler spreading (fading) given a-priori knowledge of the scattering function of the transmission medium was considered in [3]. Estimation of v_r, varying jointly with $\tau, \mathbf{f}$ was the subject of [4]. An approach using a set of parallel Kalman filters to track $\tau, \mathbf{f}$ varying in time while obtaining an estimate of the unknown parameter, v_r was used in [5]. However, divergence limited the usefulness of the estimators in [4],[5].

Herein we consider the use of a Gaussian Sum filter to estimate τ, the DS spread spectrum signal time delay, $\mathbf{f}$ the complex-valued fading multipath coefficients distorting the received signal, v_r the radial velocity between the transmitter and receiver causing the Doppler shift, and ϕ, the instantaneous phase of the Doppler shift induced sinusoid modulating the received signal. The estimator will take the form of a bank of Kalman filters running in parallel, each updating the mean vector and the covariance matrix of a Gaussian density that is combined in a weighted sum to approximate $p\left(\tau, \mathbf{f}, v_r, \phi | Z^k\right)$. The weights, α_i used in the Gaussian Sum will be obtained by a recursion on the Extended Kalman filter (EKF) innovations. The estimates of $\tau, \mathbf{f}, v_r, \phi$ together with α_i will then be used in a simulated DS spread spectrum receiver.

2 Signal and Channel Models

The signal model we consider in this paper is as described in [7], and corresponds to an underwater acoustic communication system.

$$\begin{array}{l} \text{Spread spectrum carrier frequency - } f_c = 10\text{kHz} . \\ \text{PN code bandwidth} = 1/T_c = 5\text{kHz} . \\ \text{Bit rate} = 1/T_b = 333 \text{ bits/sec} . \\ \text{Processing gain} = 15 \text{ chips/bit.} \\ T_s = T_c/2 . \\ \text{Signal propagation velocity} = C = 4900 \text{ ft./sec} \end{array} \tag{1}$$

The received signal is assumed to be down converted from the carrier frequency to the baseband and is thus described by a complex-valued baseband equivalent signal. The signal, $s_d\left(k, \tau(k), \mathbf{f}(k), \phi(k)\right)$ represents the noiseless received waveform distorted by multipath and Doppler shift and is given as follows,

$$s_d\left(k, \tau(k), \mathbf{f}(k), \phi(k)\right) \triangleq \sum_{l=0}^{N_f-1} f_l\left(kT_s\right) e^{-2\pi j \phi(k)} s\left((k-l)T_s + \tau(k)\right). \tag{2}$$

The $f_l\left(kT_s\right)$ are the complex-valued multipath coefficients. Here $\phi(k)$ represents the accumulated phase shift of the carrier modulating the received signal due to Doppler shift velocity $v_r(k)$. State variable, $\tau(k)$ represents the accumulated timing offset due to timing jitter and code Doppler. Following [4], $s(t)$ of equation (2) is a low-pass filtered DS spread spectrum waveform. When modulated by binary bits, d_n it is as follows,

$$s(t) = \sum_{n=-\infty}^{\infty} d_n \sqrt{\frac{2E_b}{T_b}} PN_{lp}(t - nT_b) \quad (3)$$

The actual received signal $\mathbf{z}(k)$ in additive noise is,

$$\mathbf{z}(k) = h(\mathbf{x}(k), k) + \begin{bmatrix} Re\{\mathbf{n}(k)\} \\ Im\{\mathbf{n}(k)\} \end{bmatrix} \quad (4)$$

where,

$$h(\mathbf{x}(k), k) \triangleq \begin{bmatrix} Re\{s_d(k, \tau(k), \mathbf{f}(k), \phi(k))\} \\ Im\{s_d(k, \tau(k), \mathbf{f}(k), \phi(k))\} \end{bmatrix} \quad (5)$$

The noise process, $\mathbf{n}(\mathbf{k})$, is a circular white Gaussian (complex-valued) sequence (AWGN) such that,

$$E\{\mathbf{n}(k)\mathbf{n}^*(l)\} = \frac{4N_0}{T_c}\delta_{k,l}. \quad (6)$$

Define a composite state vector, $\mathbf{x}(k) \in R^{2N_f+3}$, containing the signal parameters.

$$\mathbf{x}(k) = [\tau(k), Re\{f_0(k)\}, Re\{f_1(k)\}, ..., Re\{f_{N_f-1}(k)\}, Im\{f_0(k)\}, Im\{f_1(k)\}, ..., Im\{f_{N_f-1}(k)\}, v_r(k), \phi(k)]^T. \quad (7)$$

Note that the multipath coefficients, $f_l(k)$ have been split into real and imaginary components where,

$$f_l(k) \triangleq Ref_l(k) + jImf_l(k). \quad (8)$$

The accumulated timing, $\tau(k)$ varies according to the autoregressive, (AR) process equation,

$$\tau(k) = g_\tau \tau(k-1) - g_{\tau,v_r} v_r(k) + w_\tau(k). \quad (9)$$

The variable, g_{τ,v_r} is a proportionality constant relating the radial Doppler velocity, $v_r(k)$ to the resulting accumulated timing offset (code Doppler).

$$g_{\tau,v_r} \triangleq \frac{T_s}{C}. \quad (10)$$

The channel coefficients also vary according to an AR process,

$$Ref_l(k+1) = g_f Ref_l(k) + w_{r,f}(k), \quad (11)$$

$$Imf_l(k+1) = g_f Imf_l(k) + w_{q,f}(k), \quad (12)$$

for $l = 0, ..., N_f - 1$.

The relative platform velocity $v_r(k)$ is given by,

$$v_r(k+1) = g_{v_r} v_r(k) + w_{v_r}(k). \quad (13)$$

The phase $\phi(k)$ varies according to,

$$\phi(k) \triangleq \phi(k-1) + g_{\phi,v_r} v_r(k) \quad (14)$$

$$g_{\phi,v_r} \triangleq \frac{f_c T_s}{C} \quad (15)$$

Where f_c is the frequency of the transmitted carrier. Variable, $g_{v_r,\phi}$ is a proportionality constant that accounts for the contribution of $v_r(k)$ to $\phi(k)$, the carrier Doppler effect.

In state variable form this becomes,

$$\mathbf{x}(k+1) = \mathbf{F}\mathbf{x}(k) + \mathbf{G}\mathbf{w}(k). \quad (16)$$

where $\mathbf{F}$ is defined in terms of the constant parameters of equations (9-16). The $w_f(k), w_\tau(k), w_{v_r}(k)$ are AWGN processes with variances, $\sigma_f^2, \sigma_\tau^2, \sigma_{v_r}^2$ respectively. Matrix, $\mathbf{G} \in R^{2N_f+3, 2N_f+3}$ is taken to be identity. The constant, g_f determines the channel Doppler spread. The constant g_τ, determines the timing jitter bandwidth. Parameter g_{v_r}, is taken very close to unity, because we will not expect the radial velocity, v_r to vary greatly over the time duration of the simulation.

2.1 Modification of the AR Model

We find it useful to depart slightly from the AR model of equation (16). From equations (9), and (14) we see that $\tau(k), \phi(k)$ tend to increase in time. If $\tau(k), \phi(k)$ were to become large enough, arithmetic precision problems would eventually preclude the accurate operation of an EKF based estimator. Hence we propose to model $\phi(k), \tau(k)$ as varying modulo 2π and modulo $L_{PN}T_c$ respectively. Here x modulo y is defined,

$$\text{mod}(x, y) \triangleq x - \left\lfloor \frac{x}{y} \right\rfloor y. \quad (17)$$

As the state vector of equation (16) propagates, $\tau(k)$ is continually reset to $\text{mod}(\tau(k), L_{PN}T_c)$ and $\phi(k)$ to $\text{mod}(\phi, 2\pi)$. Note that the signal and signal derivatives do not change in value when they are evaluated at $\phi(k), \tau(k)$ truncated modulo 2π and L_{PN} as in equation (17). Hence Kalman filter measurement updates (unlike the Kalman filter time updates) are not affected. Thus as long as the Kalman gain remains large enough we may not experience problems when truncating, $\phi(k), \tau(k)$.

3 Gaussian Sum Algorithm

To circumvent divergence problems arising from joint estimation involving v_r, we propose to use the Gaussian Sum algorithm,[8]. This involves initial state estimates, $\widehat{x}_i(0|0)$ for $i = 1, ..., N_{filt}$ spaced on a 1-dimensional grid in the v_r dimension. Thus the $\widehat{x}_i(0|0)$ differ only in their $2N_f + 2$-th components, $\widehat{v}_{r,i}(0|0)$. We choose the $\widehat{v}_{r,i}(0|0)$ to cover the entire range of uncertainty in $v_r(0)$ and with equal spacing. Intuitively, this can ensure that at least one of the EKF's comprising the Gaussian Sum estimator is initially conditioned on a $\widehat{v}_{r,i}(0|0)$ close to $v_r(0)$, and that an acceptable estimate of $\tau(k)$ can be obtained, even if the estimate of $v_r(k)$ is not accurate. This approach is similar to that in [5]. We assume that timing acquisition has given us an estimate $\widehat{\tau}(0|0)$ accurate to within half a chip of $\tau(0)$ before the Gaussian Sum algorithm is implemented.

The equations for Gaussian Sum filtering [8] are reviewed as follows. The Gaussian Sum algorithm for observation nonlinearity assumes the state dynamic model of equations, (4),(16). In the application considered here, $\mathbf{z}(k) \in R^2$.

The cumulative measurement sequence is defined,

$$Z^k \triangleq \{\mathbf{z}(1), \mathbf{z}(2), ..., \mathbf{z}(k)\}. \quad (18)$$

We will use the following notation for a multivariate Gaussian density,

$$\mathcal{N}(\mathbf{x}, \mathbf{R}) \triangleq 2\pi^{-n/2} |\mathbf{R}|^{-1/2} \exp\left(-\frac{1}{2}\mathbf{x}^T\mathbf{R}^{-1}\mathbf{x}\right) \quad (19)$$

In Gaussian Sum filtering, the conditional density, $p\left(\mathbf{x}(k)|Z^{k-1}\right)$ is assumed to be given as a weighted sum of Gaussian densities,

$$p\left(\mathbf{x}(k)|Z^{k-1}\right) = \sum_{i=1}^{N_{filt}} \alpha_i(k-1) \\ \mathcal{N}\left(\mathbf{x}(k) - \widehat{\mathbf{x}}_i(k|k-1), \mathbf{P}_i(k|k-1)\right) \quad (20)$$

It can be shown that [8]

$$p\left(\mathbf{x}(k)|Z^k\right) \\ \approx \frac{1}{c}\sum_{i=1}^{N_{filt}} \alpha_i(k-1)\beta_i(k)\mathcal{N}\left(\mathbf{x}(k) - \widehat{\mathbf{x}}_i(k|k), \mathbf{P}_i(k|k)\right), \quad (21)$$

where $1/c$ is a normalizing constant. Here $\beta_i(k)$ of equation (21) is,

$$\beta_i(k) \triangleq \mathcal{N}\left(\mathbf{z}(k) - h\left(\widehat{\mathbf{x}}_i(k|k-1), k\right), \mathbf{\Omega}_i(k)\right) \quad (22)$$

and $\mathbf{\Omega}_i(k)$ is the innovations covariance matrix.

Computation of $\widehat{\mathbf{x}}_i(k|k), \mathbf{P}_i(k|k)$ in equation, (21) is by the ordinary EKF measurement updates as follows.

$$\mathbf{P}_i(k|k) = \\ \mathbf{P}_i(k|k-1) \\ -\mathbf{P}_i(k|k-1)\mathbf{H}_i(k)\mathbf{\Omega}_i(k)^{-1}\mathbf{H}_i^T(k)\mathbf{P}_i(k|k-1) \quad (23)$$

$$\widehat{\mathbf{x}}_i(k|k) = \\ \widehat{\mathbf{x}}_i(k|k-1) + \mathbf{P}_i(k|k-1)\mathbf{H}_i(k)\mathbf{\Omega}_i(k)^{-1} \\ \left(\mathbf{z}(k) - h\left(\widehat{\mathbf{x}}_i(k|k-1), k\right)\right) \quad (24)$$

$$\mathbf{\Omega}_i(k) \triangleq \mathbf{H}_i^T(k)\mathbf{P}_i(k|k-1)\mathbf{H}_i(k) + \mathbf{R} \quad (25)$$

where the gradient $\mathbf{H}_i(k)$ is defined by,

$$\mathbf{H}_i(k) \triangleq \left.\frac{\partial h(\mathbf{x}(k), k)}{\partial \mathbf{x}(k)}\right|_{\mathbf{x}(k) = \widehat{\mathbf{x}}_i(k|k-1)} \quad (26)$$

This results in a recursive update for $\alpha_i(k)$, the weighting coefficient associated with the i-th EKF.

$$\alpha_i(k) = \\ \frac{1}{c}\alpha_i(k-1)\mathcal{N}\left(\mathbf{z}(k) - h\left(\widehat{\mathbf{x}}_i(k|k-1), k\right), \mathbf{\Omega}_i(k)\right) \quad (27)$$

It can be shown [8] that,

$$p\left(\mathbf{x}(k+1)|Z^k\right) = \\ \frac{1}{c}\sum_{i=1}^{N_{filt}} \alpha_i(k)\mathcal{N}\left(\mathbf{x}(k+1) - \widehat{\mathbf{x}}_i(k+1|k), \mathbf{P}_i(k+1|k)\right) \quad (28)$$

The time updates for the covariances, $\mathbf{P}_i(k+1|k)$ and means, $\widehat{\mathbf{x}}_i(k+1|k)$ are,

$$\mathbf{P}_i(k+1|k) = \mathbf{F}^T\mathbf{P}_i(k|k)\mathbf{F} + \mathbf{G}^T\mathbf{Q}\mathbf{G}. \quad (29)$$

and,

$$\widehat{\mathbf{x}}_i(k+1|k) = \mathbf{F}\widehat{\mathbf{x}}_i(k|k) \quad (30)$$

respectively. The elements of the diagonal matrix, $\mathbf{Q}$ are the process noise variances.

4 RAKE Receiver

We simulate a RAKE receiver that uses the signal parameter estimates obtained from the Gaussian Sum filter. Define a RAKE correlator output,

$$X(n) \triangleq \sum_{k=k(n)}^{k(n+1)} r(k)\, s_d^*\left(k, \widehat{\tau}(k), \widehat{\mathbf{f}}(k), \widehat{\phi}(k)\right) \quad (31)$$

with $r(k)$ defined in terms of the components of $\mathbf{z}(k)$ of equation,(4) as,

$$r(k) \triangleq z_1(k) + jz_2(k) \quad (32)$$

The estimates of $\tau, \mathbf{f}, v_r, \phi$ appearing in equation (31) are obtained from the Gaussian Sum filter as the appropriate components (see equation (7)) of the overall state estimate,

$$E\left(\mathbf{x}(k)|Z^k\right) = \sum_{i=1}^{N_{filt}} \alpha_i(k)\widehat{\mathbf{x}}_i(k|k) \quad (33)$$

In the presence of Doppler shift we must carefully consider the location of the bit boundaries. The sample number for the start of the n-th bit is defined to be $k(n)$. It is reasonable for the receiver to consider sample k as part of the n-th received bit if,

$$n = \left\lfloor \frac{kT_s + \widehat{\tau}(k)}{L_{PN}T_c} \right\rfloor \quad (34)$$

where $\widehat{\tau}(k)$ is an overall estimate of $\tau(k)$ given by the first component of the left hand side of the vector of equation, (33). The sample number, $k(n)$ for the start of the n-th received data bit is given by,

$$k(n) = \left\{ k : \left((kT_s + \hat{\tau}(k)) - \left\lfloor \frac{(kT_s + \hat{\tau}(k))}{L_{PN}T_c} \right\rfloor L_{PN}T_c < .5 \right) \bigcap \left(n = \left\lfloor \frac{kT_s + \hat{\tau}(k)}{L_{PN}T_c} \right\rfloor \right) \right\} \quad (35)$$

Sample number, $k(n)$ is the unique k in the n-th bit for which the total timing excess over nL_{PN} chips (i.e. n bits) is less than 1/2 chip.

The DPSK RAKE receiver is shown in Figure 1. The decision variable is given by $U(n)$, where,

$$U(n) = Re\{X(n)X^*(n-1)\} \ . \quad (36)$$

The receiver decides for binary, 1 if $U(n) > 0$ and for binary 0 otherwise. The instantaneous bit error rate (BER) for the receiver is computed by first obtaining the means and covariances of $X(n)$ as a function of parameters and parameter estimates as in [7]. These quantities are then substituted into the analytic formula for BER in DPSK signaling,[6].

4.1 Decision Feedback

In actual decision feedback simulation, we choose to compare the number of correctly received DPSK data bits to the transmitted data bits, and arrive at a single overall number for the BER, as well as plot an instantaneous BER. Recall that the received signal is given by equations (3), and (2). Data is DPSK encoded, and the n-th transmitted data bit is d'_n where,

$$d'_n = d_n d_{n-1} \quad (37)$$

Where d_n is the binary signal bit of equation (3). The Gaussian Sum filter processes the received signal samples from the previous bit, (i.e. from $k(n) - k(n-1)$ samples ago,) demodulated with the detected estimate of d_{n-1} namely, $\hat{d}_{n-1}$. The DPSK decision variable is still given by $U(n)$ of equation (36). However the receiver uses the state estimate available at the end of the last completely received data bit. Thus,

$$X(n) = \sum_{k=k(n)}^{k(n+1)} r(k)\, s_d^*\Big(k,\ \hat{\tau}(k(n)),\ \hat{\mathbf{f}}(k(n)),\ \hat{\phi}(k(n))\Big). \quad (38)$$

The decoded data bit $\hat{d'}_n$ is given by,

$$\hat{d'}_n = \mathrm{sgn}\,[U(n)] \quad (39)$$

and the signal bit decision, $\hat{d}_n$ by,

$$\hat{d}_n = \hat{d}'_n \hat{d}_{n-1}. \quad (40)$$

Direct comparison of $\hat{d'}$ and d' gives the BER determined by Monte Carlo simulation.

5 Results

The simulated received signal was DPSK data modulated, and the receiver operated in the decision feedback mode. The estimator began operation with $N_{filt} = 5$ EKF's, and EKF i was omitted from the estimator when $\alpha_i(k) < t_\alpha = 10^{-2}$. Here t_α is a threshold. The true radial velocity $v_r = 7$ kts. (Here, kts denotes knots, a unit of velocity. 1 knot = 1.6878 ft./sec) The $2N_f + 2$-th component of the state vector of the i-th EKF was conditioned on $\hat{v}_{r,i}(0|0)$ where, $\hat{v}_{r,1}(0|0) = 5.5, \hat{v}_{r,2}(0|0) = 8.5, \hat{v}_{r,3}(0|0) = 11.5, \hat{v}_{r,4}(0|0) = 14.5, \hat{v}_{r,5}(0|0) = 17.5$ kts. In this simulation, the emphasis was on operating the receiver with a large initial uncertainty range in v_r.

Results for $E_b/N_0 = 13$dB are shown in Figure 2. Parameter, $g_f = .9999$, and thus the Doppler spread is .2 Hz. Since, $g_{v_r} = 1$ and $\sigma^2_{v_r} = 10^{-10}$, the radial Doppler, $v_r(k)$ was practically constant throughout. Timing jitter was determined by AR parameters, $g_\tau = .9999, \sigma^2_\tau = 10^{-5}, \sigma^2_{\mathbf{f}} = 6.66 \times 10^{-5}$. The time-averaged analytic BER in Figure 2 is 2.27×10^{-2}, and the time averaged ideal BER is 5.73×10^{-3}. The actual BER obtained from Monte Carlo simulation was $386/9960 = 3.87 \times 10^{-2}$, a quantity in close agreement with the time average of the actual BER. The long term estimation error in $v_r(k)$ shown in Figure 3 is .35 kts.

6 Conclusions

We have demonstrated that the Gaussian Sum algorithm could be applied to the problem of joint estimation of timing, multipath coefficients and radial Doppler velocity (i.e. $\tau, \mathbf{f}, v_r$) for a DS spread spectrum signal. Beginning with a large initial uncertainty in v_r, the N_{filt} EKF's comprising the estimator acquired a coarse estimate of v_r needed to ensure that the timing offset, $\tau(k)$ could be tracked continuously. For data-modulated received signals, bit error rate simulations showed that the receiver was capable of coarse Doppler estimation and could solve the crucial problem of tracking the code delay, τ to within an accuracy of 1/2 of a PN sequence chip duration.

References

[1] J.K. Holmes and C.C. Chen, "Acquisition time performance of PN spread spectrum systems," *IEEE Transactions on Communications*, vol. COM-25, pp. 778–783, Aug. 1977.

[2] R.E. Ziemer and R.L. Peterson, *Digital Communications and Spread Spectrum Systems.* N.Y., N.Y. MacMillan Publishing Co., 1985.

[3] S.S. Soliman, "Tracking loop for fading dispersive channels," *IEEE Transactions on Communications*, vol. 38, pp. 292–299, Mar. 1990.

[4] R.A. Iltis and A.W. Fuxjaeger, "A digital DS spread spectrum receiver with joint channel and Doppler shift estimation," *IEEE Transactions on Communications*, vol. 39, pp. 1255–1267, Aug. 1991.

[5] A.W. Fuxjaeger and R.A. Iltis, "Adaptive parameter estimation using parallel Kalman filtering for spread spectrum code and Doppler tracking," Tech. Rep. CIPR 91-05, University of California Santa Barbara, 1991.

[6] J.G. Proakis, *Digital Communications Systems*. N.Y., N.Y. McGraw-Hill, 1983.

[7] R.A. Iltis, "A digital DS spread spectrum receiver with joint channel and Doppler shift estimation," in *Twenty-third Asilomar Conference on Signals, Systems and Computers, Vol.I*, pp. 206–210, Oct. 1989.

[8] D.L. Alspach and H.W. Sorenson, "Nonlinear Bayesian estimation using Gaussian sum approximations," *IEEE Transactions on Automatic Control*, vol. 17, pp. 439–448, Aug. 1972.

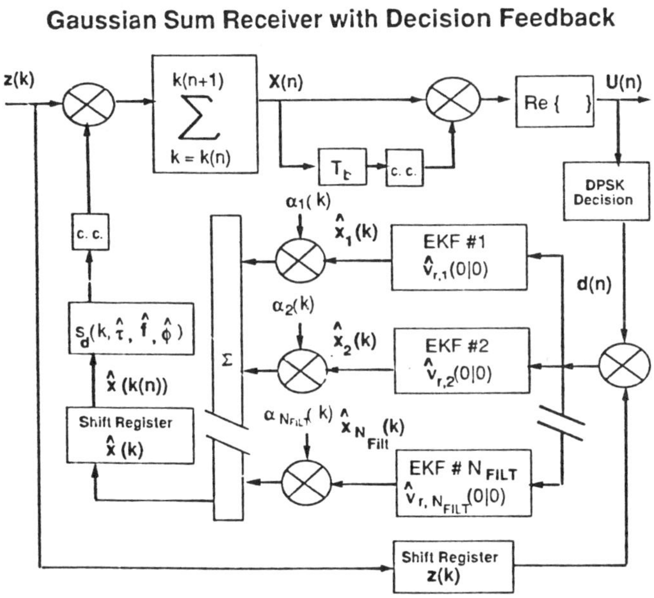

Figure 1: RAKE receiver

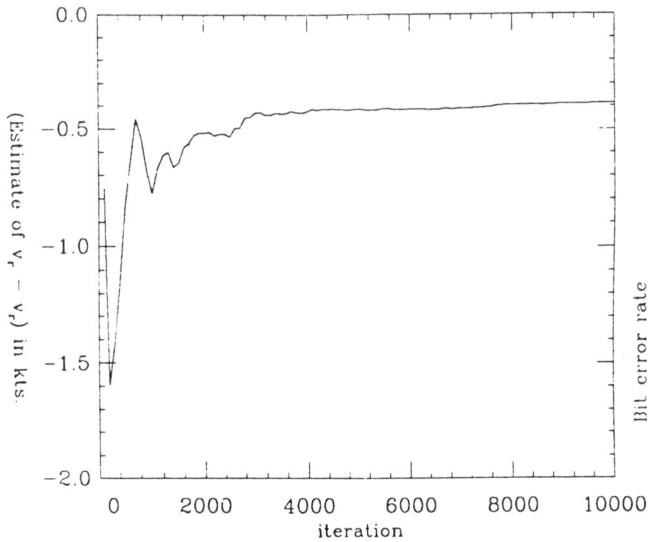

Figure 3: Error in estimation of v_r, for E_b/N_0 = 13dB,$g_f = g_r = .9999$

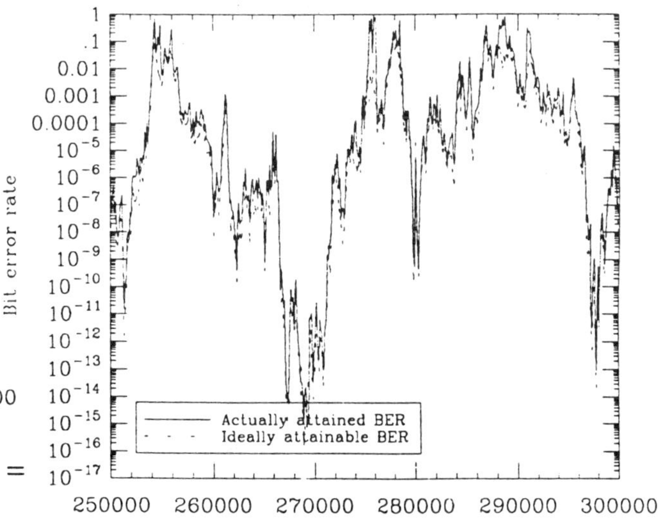

Figure 2: Instantaneous BER vs. iteration, for E_b/N_0 = 13dB, $g_f = g_r = .9999$

Simulation of Correlated Non-Gaussian Interference for Radar Signal Detection

M. Rangaswamy, D.D. Weiner
Department of Electrical
and Computer Engineering
Syracuse University
Syracuse, NY-13244-1240

A. Ozturk
Department of Computer
Engineering
Ege University
Izmir, Turkey

Abstract

This paper presents new canonical computer simulation procedures for generating correlated non-Gaussian radar clutter which can be characterized by spherically invariant random processes (SIRP). The probability density function (PDF) of every random vector obtained by sampling an SIRP is a function of a non-negative quadratic form. The PDF of this quadratic form is used for assessing the performance of the simulation procedures. The simplicity of the proposed procedures make them practical for use in a variety of applications. [1]

1 Introduction

This investigation is motivated by a desire to simulate correlated non-Gaussian radar clutter. Various investigators have reported experimental results where non-Gaussian marginal probability density functions (PDF) have been used to model the clutter. Usually, radars process N samples at a time. Statistical characterization of the clutter requires the specification of the joint PDF of the N samples. In addition, the clutter may be highly correlated. Hence, the joint PDF must take into account the correlation between samples. The theory of spherically invariant random processes (SIRP) provides a powerful mechanism for obtaining the joint PDF of the N correlated non-Gaussian clutter samples. Every random vector obtained by sampling an SIRP is called a spherically invariant random vector (SIRV). This paper presents algorithms for the simulation of K-distributed and Weibull distributed SIRVs. Another issue addressed in this paper is the performance assessment of the simulation procedures. It has been pointed out in [9] that the quadratic form appearing in the PDF of the SIRV contains all the information necessary to identify the PDF of the underlying SIRV. We make use of this result in order to assess the performance of the simulation procedures. Some interesting simulation schemes for SIRVs have been proposed in [2] and [3].

In Section 2, we present some definitions and preliminaries pertaining to the theory of spherically invariant random processes. Section 3 deals with the proposed simulation procedures. Finally, performance assessment of the simulation procedures is discussed in Section 4.

2 Preliminaries

We begin by defining the terms spherically invariant random vector and spherically invariant random process. A spherically invariant random vector (SIRV) is a random vector (real or complex) whose PDF is uniquely determined by the specification of a mean vector, a covariance matrix and a characteristic first order PDF. A spherically invariant random process (SIRP) is a random process (real or complex) such that every random vector obtained by sampling this process is an SIRV. The work of Yao [10] gave rise to a representation theorem which can be stated as follows.

Theorem 1 *If a random vector is an SIRV, then there exists a non-negative random variable S such that the PDF of the random vector conditioned on S is a multivariate Gaussian PDF.*

For the problem of radar clutter modeling and simulation, it is desirable to control the non-Gaussian PDF and its correlation properties independently. Therefore, we consider the product given by $\mathbf{X} = \mathbf{Z}S$, where $\mathbf{X} = [X_1 \dots X_N]^T$ denotes the SIRV, $\mathbf{Z} = [Z_1 \dots Z_N]^T$ is a Gaussian random vector with zero mean and covariance matrix $\mathbf{M}$ and S is a non-negative random variable with PDF $f_S(s)$. For convenience of analysis, $\mathbf{Z}$ and S are assumed to be statistically independent. The PDF of $\mathbf{X}$ conditioned on S is

$$f_{\mathbf{X}|S}(\mathbf{x}|s) = (2\pi)^{-\frac{N}{2}} |\mathbf{M}|^{-\frac{1}{2}} s^{-N} exp(-\frac{q}{2s^2}) \qquad (1)$$

where q is a non-negative quadratic form given by $q = \mathbf{x}^{\mathbf{T}}\mathbf{M}^{-\mathbf{1}}\mathbf{x}$ and $|\mathbf{M}|$ denotes the determinant of the covariance matrix $\mathbf{M}$. The PDF of $\mathbf{X}$ is given by

$$f_{\mathbf{X}}(\mathbf{x}) = (2\pi)^{-\frac{N}{2}} |\mathbf{M}|^{-\frac{1}{2}} h_N(q) \qquad (2)$$

[1] This work was supported by the Rome Laboratory, U.S. Air Force, through a subcontract with Kaman Science Corporation under contract number SCEE-DACS-P35498-2. Dr. Weiner participated on this effort as a member of Rome Laboratory through the AFOSR University Resident Research Program.

1058-6393/91 $01.00 © 1991 IEEE

where

$$h_N(q) = \int_0^\infty s^{-N} exp(-\frac{q}{2s^2}) f_S(s) ds. \quad (3)$$

The PDF of the random variable S is called the characteristic PDF of the SIRV. We point out that the covariance matrix of the SIRV is given by $\mathbf{\Sigma} = \mathbf{M}E(S^2)$ where $E(S^2)$ is the mean square value of the random variable S. It is seen that the covariance matrix of the SIRV normalized by the mean square value of S is the covariance matrix of the Gaussian random vector. Note that it is possible to set the covariance matrix of the SIRV equal to that of the Gaussian random vector by requiring that $E(S^2)$ be equal to unity. Several techniques are available in [4], and [9] for obtaining $h_N(q)$. Note that the Gaussian random vector is a special case of an SIRV and is obtained when $f_S(s) = \delta(s-1)$ where $\delta(t)$ is the unit impulse function.

Many of the attractive properties of Gaussian random vectors also apply to SIRVs. The most relevant property of SIRVs for the purpose of computer simulation is the closure property under linear transformation [10] stated below.

Theorem 2 *If* $\mathbf{X}$ *is an SIRV with characteristic PDF* $f_S(s)$, *then*

$$\mathbf{Y} = \mathbf{AX} + \mathbf{b} \quad (4)$$

is also an SIRV with the same characteristic PDF. It is assumed that $\mathbf{A}$ *is a nonsingular matrix and* $\mathbf{b}$ *is a known vector having the same dimension as* $\mathbf{X}$.

Theorem 2 provides us with a powerful technique for simulating SIRVs. A white SIRV is defined as one that has a diagonal covariance matrix. In other words, the components of the SIRV are uncorrelated but not necessarily independent. We can start with a zero mean white SIRV $\mathbf{X}$ having identity covariance matrix and perform the linear transformation given by eq (4) to obtain an SIRV $\mathbf{Y}$ having a non-zero mean and desired covariance matrix $\mathbf{\Sigma}$. The matrix $\mathbf{A}$ and the vector $\mathbf{b}$ are given by

$$\mathbf{A} = \mathbf{ED}^{\frac{1}{2}} \quad (5)$$

$$\mathbf{b} = \mu_{\mathbf{y}} \quad (6)$$

where $\mathbf{E}$ is the matrix of normalized eigen vectors of the covariance matrix $\mathbf{\Sigma}$, $\mathbf{D}$ is the diagonal matrix of eigen values of $\mathbf{\Sigma}$ and $\mu_{\mathbf{y}}$ is the desired non-zero mean vector.

In many instances it is not possible to obtain $f_S(s)$ for an SIRV in closed form, even though its existence is guaranteed. In such cases, an alternate approach must be used in order to characterize the SIRV. The following theorem provided in [4] can be used to completely characterize a white SIRV having zero mean and identity covariance matrix.

Theorem 3 *A random vector* $\mathbf{X} = [X_1 \dots X_N]^T$ *is a zero mean white SIRV having identity covariance matrix if and only if there exist random variables* $R \in (0,\infty)$, $\Theta \in (0, 2\pi)$ *and* $\Phi_k \in (0,\pi)$, $(k = 1, \dots N-2)$ *such that when the components of* $\mathbf{X}$ *are expressed in the generalized spherical coordinates*

$$\begin{gathered} X_1 = R\cos(\Phi_1) \\ X_k = R\cos(\Phi_k)\prod_{i=1}^{k-1}\sin(\Phi_i) \quad (1 < k \leq N-2) \\ X_{N-1} = R\cos(\Theta)\prod_{i=1}^{N-2}\sin(\Phi_i) \\ X_N = R\sin(\Theta)\prod_{i=1}^{N-2}\sin(\Phi_i) \end{gathered} \quad (7)$$

then the random variables R, Θ *and* Φ_k *are mutually and statistically independent having PDFs of the form*

$$\begin{gathered} f_R(r) = \frac{r^{N-1}}{2^{\frac{N}{2}-1}\Gamma(\frac{N}{2})} h_N(r^2) u(r) \\ f_{\Phi_k}(\phi_k) = \frac{\Gamma(\frac{N-k+1}{2})}{\sqrt{\pi}\Gamma(\frac{N-k}{2})} sin^{N-1-k}(\phi_k)[u(\phi_k) - u(\phi_k - \pi)] \\ f_\Theta(\theta) = (2\pi)^{-1}[u(\theta) - u(\theta - 2\pi)] \end{gathered} \quad (8)$$

where $\Gamma(\nu)$ *is the Eulero Gamma function and* $u(t)$ *is the unit step function.*

As a consequence of Theorem 3, any SIRV with zero mean and identity covariance matrix can be represented in generalized spherical coordinates which are mutually and statistically independent regardless of the SIRV considered. Also, note that the PDFs of Θ and Φ_k, $(k = 1, \dots N-2)$ are functionally independent of the white SIRV considered. Only the PDF of R changes from one white SIRV to another. Note that $R^2 = \sum_{k=1}^{N} X_k^2 = \mathbf{X}^\mathbf{T}\mathbf{X}$. Hence R is the norm of the SIRV.

Another important feature of the SIRV is that the quadratic form appearing in its PDF contains all the information necessary to identify the PDF. It follows that knowledge of the PDF of the quadratic form of the SIRV is sufficient to identify the PDF of the corresponding SIRV [9].

Theorem 4 *The PDF of the quadratic form appearing in eq (2) is given by*

$$f_Q(q) = \frac{1}{2^{\frac{N}{2}}\Gamma(\frac{N}{2})} q^{\frac{N}{2}-1} h_N(q) \quad (0 \leq q \leq \infty) \quad (9)$$

and remains unchanged regardless of whether or not the SIRV is white.

The theorems provided in this section will be made use of in the proposed simulation approach, discussed in Section 3, and in assessing the performance of the simulation procedure, discussed in Section 4.

3 Proposed Simulation Procedures

In this section we present algorithms for the simulation of the K-distributed and Weibull distributed SIRVs. Let $\mathbf{Y} = [Y_{c1}, Y_{s1} \dots, Y_{cN}, Y_{sN}]^T$ where the subscripts c and s denote the in phase and out of phase quadrature components of the N complex radar samples. The K-distributed envelope PDF is expressed as

$$f_R(r) = \frac{2b}{\Gamma(\alpha)}(\frac{br}{2})^\alpha K_{\alpha-1}(br) \quad (0 \leq r \leq \infty) \quad (10)$$

where α is the shape parameter of the distribution, b denotes the scale parameter of the distribution and $K_N(t)$ is the Nth order modified Bessel function of the second kind. The corresponding SIRV PDF of the $2N$ quadrature components is given by [9]

$$f_{\mathbf{Y}}(\mathbf{y}) = (2\pi)^{-N}|\mathbf{M}|^{-\frac{1}{2}}h_{2N}(q) \tag{11}$$

where

$$h_{2N}(q) = \frac{b^{2N}}{\Gamma(\alpha)}\frac{(b\sqrt{q})^{\alpha-N}}{2^{\alpha-1}}K_{N-\alpha}(b\sqrt{q}). \tag{12}$$

The characteristic PDF for the K-distributed SIRV is the generalized Chi PDF. In order to obtain the random variable S with PDF $f_S(s)$ and having unit mean square value, we start with a related PDF given by

$$f_V(v) = \frac{2b}{\Gamma(\alpha)2^{\alpha}}(bv)^{2\alpha-1}exp(-\frac{b^2v^2}{2})u(v) \tag{13}$$

where b is the scale parameter, α is the shape parameter and $u(v)$ is the unit step function. The mean square value of V denoted by a^2 is equal to $\frac{2\alpha}{b^2}$. The random variable S is then given by $S = \frac{V}{a}$. The simulation procedure for this SIRV PDF is fairly simple and is stated below.

(1) Generate a white zero mean Gaussian random vector $\mathbf{Z}$, having identity covariance matrix.

(2) Then generate a random variable V from the PDF $f_V(v)$. Denote the mean square value of V by a^2.

(3) Normalize the random variable V by a to obtain S. In other words generate $S = \frac{V}{a}$.

(4) Generate the product given by $\mathbf{X} = \mathbf{Z}S$. At this step, we have a white SIRV having zero mean and identity covariance matrix.

(5) Finally, perform the linear transformation given by eq (4) to obtain the SIRV $\mathbf{Y}$ with desired mean and covariance matrix.

Fig 1 shows the simulation procedure presented above.

The Weibull distributed envelope PDF is given by

$$f_R(r) = abr^{b-1}exp(-ar^b) \quad (0 \leq r \leq \infty) \tag{14}$$

where a is the scale parameter and b is the shape parameter. The corresponding SIRV PDF from [9] is

$$f_{\mathbf{Y}}(\mathbf{y}) = (2\pi)^{-N}|\mathbf{M}|^{-\frac{1}{2}}h_{2N}(q) \tag{15}$$

where

$$h_{2N}(q) = \sum_{k=1}^{N} C_k q^{\frac{kb}{2}-N}exp(-aq^{\frac{b}{2}}), \tag{16}$$

and

$$C_k = \sum_{m=1}^{k}(-1)^{m+N}2^N\frac{a^k}{k!}\begin{pmatrix} k \\ m \end{pmatrix}\frac{\Gamma(1+\frac{mb}{2})}{\Gamma(1+\frac{mb}{2}-N)}. \tag{17}$$

We recognize for the Weibull SIRV that the characteristic PDF $f_S(s)$ is not known in closed form. Therefore, an alternate simulation procedure must be developed. This simulation procedure requires the capability to generate the independent random variables R, Θ and Φ_k $(k = 1,\ldots N-2)$ having PDFs of the form of eq(8). Generation of the variates Φ_k $(k = 1,\ldots N-2)$ is extremely difficult from a computational standpoint. This problem is overcome as follows. Since the Gaussian random vector belongs to the family of SIRVs, a zero mean white Gaussian random vector with identity covariance matrix admits a representation of the form of eq(7). Let R_G denote the norm of the white Gaussian random vector. In order to generate the desired SIRV, we then make use of the fact that the PDFs of Θ and Φ_k are invariant to the choice of the SIRV. Note that the random variable S is nothing but the ratio of the norm of the white SIRV to the norm of the white Gaussian random vector. The simulation procedure is stated below:

(1) Generate a white, zero mean Gaussian random vector $\mathbf{Z}$ having identity covariance matrix.

(2) Compute the norm $R_G = ||\mathbf{Z}||$ of the white Gaussian random vector, noting that $R_G^2 = \mathbf{Z}^{\mathbf{T}}\mathbf{Z}$.

(3) Generate the norm $R = ||\mathbf{X}||$ of the white Weibull SIRV $\mathbf{X}$, from the PDF of R given by eq(8).

(4) Obtain the random variable S given by $S = \frac{R}{R_G}$.

(5) Generate the white SIRV $\mathbf{X}$ by taking the product $\mathbf{X} = \mathbf{Z}S$.

(6) Finally, perform the linear transformation given by eq (4) to obtain the SIRV $\mathbf{Y}$ with desired mean and covariance matrix.

The simulation procedure is shown schematically in Fig 2. Note that the simulation procedure avoids the generation of the variables Θ and Φ_k $(k = 1,\ldots N-2)$. The generation procedure for a white Gaussian random vector is well known. Therefore, we need to concern ourselves only with the development of a suitable generation scheme for the norm R of the white SIRV $\mathbf{X}$. In this paper we generate R by making use of the approach called the 'Rejection Method'. The rejection method can be used to generate random variables whose cumulative distribution functions are not known, but whose PDFs are explicitly known [1]. The rejection procedure assumes knowledge of the maximum value of the PDF of R for a given SIRV PDF and a finite range of the PDF of R so that the area under the PDF curve is close to unity. These quantities are denoted by c and b, respectively. The Rejection method is summarized below:

(1) Generate a uniform random variate U_1 on the interval $(0, b)$.

(2) Generate another uniform variate U_2 on the interval $(0, c)$.

(3) If $U_2 \leq f_R(U_1)$, then $R = U_1$. Otherwise, reject U_1 and return to step 1.

A major drawback of this approach is that it becomes inefficient when U_1 is frequently rejected in step 3. In order to improve the efficiency of the algorithm, the generalized rejection method is used. The generalized rejection method requires the use of a PDF called the majorizing PDF [1], [6] that bounds the PDF of the variable being generated. A variable from the majorizing PDF is then used in step 1 above instead of the uniform random variable.

Note that the simulation procedures of Fig 1 and Fig 2 are canonical in the sense that only V and R, respectively change from the simulation of one SIRV to another. Note that the scheme of Fig 2 can be used even when $f_S(s)$ is known. However, the scheme of Fig 1 is preferred when S can be generated easily. In both schemes, the order of the product device and the linear transformation can be interchanged without affecting the output $\mathbf{Y}$. Finally, we point out that both of the simulation procedures presented in this section are consistent with the representation theorem for SIRVs.

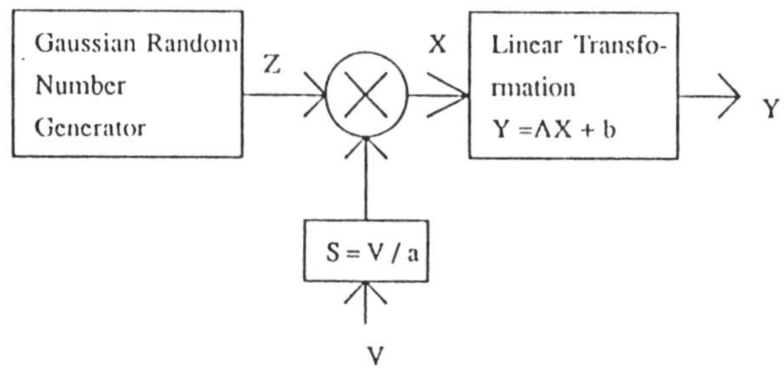

Figure 1: Simulation Scheme for SIRVs with Known Characteristic PDF

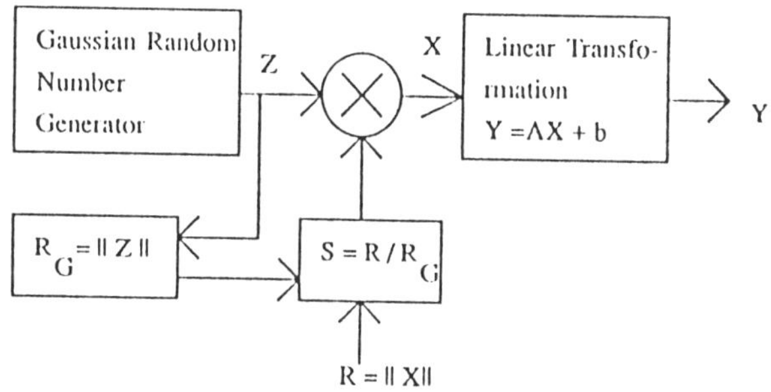

Figure 2: Simulation Scheme for SIRVs with Unknown Characteristic PDF

4 Performance Assessment of the Simulation Schemes

In this section we concern ourselves with the performance assessment of the simulation procedures developed in section 3. We point out that the simulation procedures developed in section 3 are exact in the sense that they are based upon the representation theorem. Hence, departures from the exact SIRVs will depend for the most part on the nonideality of the uniform random number generators used. Empirical assessment of the simulation procedures is necessary for practical applications.

An attractive feature of SIRVs is that the quadratic form q appearing in the SIRV PDF contains all the information necessary for identifying the PDF of the SIRV. In other words, knowledge of the PDF of the quadratic form is sufficient to determine the underlying SIRV PDF. Furthermore, the quadratic form PDF remains unchanged regardless of whether the SIRV is white or colored. The PDF of the quadratic form appearing in the SIRV PDF is given by eq (9). For the radar problem where we deal with N complex samples or $m = 2N$ quadrature components, note that we make use of eq (9) with N replaced by $2N$. Hence, we base our goodness of fit test procedure for the generated SIRVs on the PDF of the quadratic form q. Note that we have now reduced the multivariate problem to an equivalent univariate problem involving the goodness of fit test for the PDF of the quadratic form.

In the examples presented in this section, we generated 1000 realizations of the random vector $\mathbf{Y}$ with 2 complex samples and obtained one thousand samples of the quadratic form Q for the K-distributed and Weibull SIRVs. In each case, we used the corresponding theoretical PDF of the quadratic form given by eq (9) to test for the distribution of the generated quadratic form. The frequency histograms for the generated data and the corresponding theoretical PDFs are shown in figures 3-4. In addition, a Chi-Square test was performed on the generated data with the Type-1 error fixed at 0.05 and the null hypothesis was not rejected in each case. The frequency histograms provide a good idea about the true distributions for large sample sizes. Observe that the empirical PDFs are very close to the theoretical PDFs. Note that the procedure used in this section to assess the distributional assumptions of the random samples from the SIRV PDFs is a formal goodness of fit test. Similar procedures have been proposed to test for multivariate normality in [8] and [7].

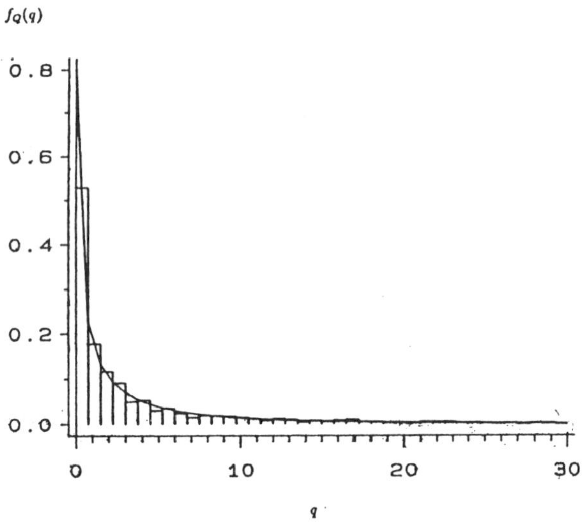

Figure 3: Theoretical and Empirical Quadratic form PDFs for K-distributed SIRV: $\alpha = 0.5$, $b = 1$

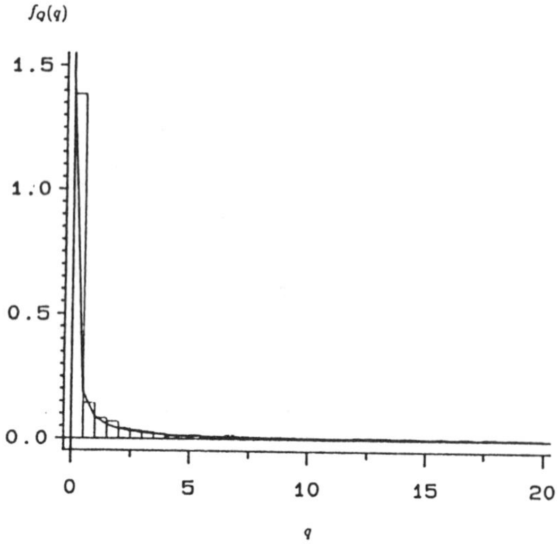

Figure 4: Theoretical and Empirical Quadratic form PDFs for Weibull SIRV: $b = 0.5$, $a = 1.86$

5 Conclusion

In this paper, we have presented two schemes that can be used in practice to simulate correlated non-gaussian radar clutter, when the clutter can be modeled as a spherically invariant random process. We pointed out that the simulation schemes developed are canonical and do not change form from the simulation of one SIRV to another. A new approach, based on the PDF of the quadratic form appearing in the SIRV PDF, was used to perform a formal goodness of fit test in order to assess performance of the proposed simulation schemes. Performance assessment based on this scheme showed excellent agreement between the theoretical and empirical PDFs of the quadratic form. Finally, it was pointed out that use of this technique reduced the goodness of fit test from a multivariate testing procedure to a univariate testing procedure resulting in tremendous processing simplicity. Therefore, this procedure lends itself very well to practical applications.

References

[1] Bratley, P., Fox, B.L. and Schrage, L.E."A guide to Simulation", Springer-Verlag, New York, 1987

[2] Brehm, H. and Stammler, W."Description and Generation of Spherically Invariant Speech-Model Signals", *Signal Process.*, 1987, **12**, (2), pp. 119-141

[3] Conte, E., Longo, M. and Lops, M."Modelling and Simulation of non-Rayleigh Radar Clutter", *IEE Proc.F, Commun., Radar, & Signal Process.*, 1991, **138**, (2), pp. 121-130

[4] Goldman, J. "Detection in the Presence of Spherically Symmetric Random Processes", *ibid.*, 1976, **IT-22**, pp. 52-58

[5] Gradshteyn, I.S. and Ryzhik, I.M."Table of Integrals, Series and Products", Academic Press Inc., New York, 1980

[6] Johnson, M.E."Multivariate Statistical Simulation", John Wiley and sons, New York, 1987

[7] Koziol, J.A."A class of Invariant Procedures for Assessing Multivariate Normality", Biometrika 69, 2, 1982, pp. 423-427

[8] Ozturk, A. "A New Method For Univariate and Multivariate Distribution Identification", Submitted for publication.

[9] Rangaswamy, M., Weiner, D. and Ozturk, A."Spherically Invariant Random Processes for Modeling and Distribution Identification of Non-Gaussian Random Vectors", Accepted for publication in **IEEE-AES** trans.

[10] Yao, K. "A Representation Theorem and Its Applications to Spherically Invariant Random Processes", *IEEE Trans.*, 1973, **IT-19**, pp. 600-608

Experimental Results for Adaptive Sidelobe Cancellation Techniques Applied to an HF Array

Richard A. Games, Stephen A. Townes, and Ronald T. Williams

The MITRE Corporation
Burlington Road
Bedford, MA 01730

Abstract

This paper describes an experimental analysis of adaptive array processing techniques conducted using MITRE's 16-element high frequency (HF) array located in Bedford, Massachusetts. Our results demonstrate the ability of these techniques to cancel both narrow-bandwidth interference and wide-bandwidth noise. We also discuss the effects of calibration errors and the nonstationary HF channel on beamforming and constrained adaptive processing. We demonstrate that the HF channel will support both beamforming and constrained processing over short time intervals. However, for longer periods robust adaptive array processing techniques may be required. This paper considers two robust methods, namely, the diagonal loading algorithm and a principal components projection method. Finally, we show that adaptive array processing algorithms can be used in conjunction with a matched filter to extract a spread-spectrum signal from interference.

1 Introduction

Most communication, direction finding, and radar applications involve detecting and estimating a desired signal that may be corrupted by additive noise and interference. Signal processing is often applied to filter out those interference and noise components that can be separated in some dimension from the signal of interest. This paper considers techniques that take advantage of the directivity derived from using an array of sensors to remove interference and noise that arrive at the antenna array from a different direction than the desired signal. We describe interference and noise cancellation results obtained by applying adaptive digital beamforming techniques to data recorded from a 16-element linear array located at the MITRE Corporation in Bedford, Massachusetts. This array operates with up to one megahertz (MHz) of bandwidth in the high frequency (HF) radio band (3 to 30 MHz) and is part of a wide-bandwidth HF experimental program designed to support the development of advanced HF communication, direction finding, and over-the-horizon radar systems.

1.1 Background

The HF band tends to be densely populated with strong narrow-bandwidth interferers. In applications where the desired signal has a wide bandwidth, it may not be possible (or even desirable) to transmit in bands that are free from such interference. Adaptive array processing offers the possibility of removing narrow-bandwidth interferers without affecting the desired signal. To increase the available degrees of freedom, subbanding can also be incorporated in an adaptive beamforming scheme to cancel a large number of interferers across a wide bandwidth.

Cancellation of HF wide-bandwidth background noise also provides a significant opportunity for effectively applying adaptive array processing techniques. In the lower range of the HF band, particularly at night, the background noise tends to be dominated by worldwide thunderstorm activity and is composed of many spatially confined impulses. Adaptive techniques can null lightning bursts and can potentially achieve significant improvements in output signal-to-interference-plus-noise ratio (SINR). Griffiths [6] has used the Wide Aperture HF Radio Research Facility (WARF) to show that for a narrow-bandwidth HF array, both interference and background noise can be cancelled using adaptive array processing techniques. Our goal has been to develop adaptive algorithms that can cancel directional atmospheric noise and narrow-bandwidth interference over a wide bandwidth (nominally, 1.024 MHz).

To prevent cancellation of the desired signal, constraints can be applied to the adaptive array. To form these constraints the response of the array to the signal

1058-6393/91 $01.00 © 1991 IEEE

must be known. Usually both the signal and the interference propagate to the receiving array through the ionosphere. The ionosphere, however, is nonstationary and often induces significant changes in important signal and interference parameters, including directions of arrival and signal strength. The variability of the propagation environment makes it difficult to properly constrain the adaptive array processor to prevent signal cancellation. Moreover, the large apertures needed to operate effectively at HF tend to make accurate calibration of the array difficult. Robust methods designed to overcome the effects of both the HF channel and calibration errors must be employed.

1.2 Hardware

The experimental facility consists of two linear 16-element arrays whose boresights point northeast and southwest. The arrays are separated by a 20-foot high backscreen. Both arrays are composed of 16 vertical dipoles spaced 25 feet apart with a wide-bandwidth preamplifier built into each element. Due to the high input impedance of the preamplifiers, the antenna elements are essentially configured as field probes with little or no mutual coupling. The outputs of the elements are fed from the preamplifiers to a bank of radio receivers through sets of coaxial cables. To minimize phase offset errors, the cables are cut to equal lengths. Initially, the cables lead to a small structure located a short distance from the array. At this point the signals are passed through an additional set of preamplifiers and transmitted to the receivers located in the main laboratory complex through another set of coaxial cables cut to identical lengths of approximately 1200 feet.

On input to the digitizing receivers, the array elemental signals are first passed through a series of mixer-amplifier stages and bandpass filters that reduce their bandwidths to approximately 1.2 MHz and translate their center frequencies to 10.3 MHz. The output of the analog filters are then passed through 12-bit analog-to-digital converters having sampling rates of 8.192 million samples per second. The signals are then translated to baseband and processed using 128-tap FIR filters that (1) perform a 4-to-1 decimation to achieve a complex sampling rate of 2.048 MHz, (2) reduce the signal bandwidth to 1.024 MHz, and (3) give the in-phase and quadrature components as output. The signal samples are then stored using a high-speed RAM that can hold up to 1 second of data. After recording has been completed, the data set is transferred to high-density magnetic tapes. The resulting data files contain 16-bit inphase and quadrature samples. The recording system limits the amount of contiguous data that can be measured to 8 megabytes per channel, or roughly one second of data every 6 minutes.

The array is calibrated by injecting tones into the preamplifiers and subsequently measuring the amplitude and phase response at the output of the receivers. The calibration results are recorded and can be used to eliminate phase offsets and amplitude variations in data files recorded at approximately the same time. It should be noted that the array elements are not included in the calibration loop.

1.3 Sidelobe Cancellation

The bulk of our analysis work has focused on adaptive sidelobe cancellers. Sidelobe cancellation represents a convenient framework for implementing partially adaptive array processing techniques. They can be decomposed into two essential components: the primary element and the auxiliary array. The primary element corresponds to either a single element or the beamformed output of an antenna array. It is generally designed to have relatively high gain in the direction of the desired signal while it attenuates signals that arrive in its sidelobes. The adaptively weighted outputs of the auxiliary array elements are summed and the result is subtracted from the output of the primary element to cancel interference and noise. To prevent the signal from being cancelled, either the signal in the auxiliary array must be weak relative to the background noise or constraints must be applied to the auxiliary array weights to insure that the adapted beampattern has unit gain in the direction of the desired signal. If we assume that the signal is narrowband, then the desired array response can be achieved by requiring that the inner product between the steering vector of the desired signal and the vector of auxiliary array weights is zero.

In our analysis we have concentrated our efforts on block processing based least-squares sidelobe cancellation techniques, since they appear to be a practical approach to cancelling interference and noise for the HF radar and communications applications that motivate our work. Let $\mathbf{x}$ denote a length-N vector containing N samples measured by the primary element and let $\mathbf{Y}$ be an $N \times L$ matrix with its ith column containing N samples from element i. We let $\mathbf{w}$ denote a vector of auxiliary array weights. The least-squares approach requires that the Euclidean norm of the residual vector, $\mathbf{z} = \mathbf{x} - \mathbf{Y}\mathbf{w}$, be minimized subject to the applied constraints. That is,

$$\min_{\substack{\mathbf{w} \\ \mathbf{C}^H\mathbf{w}=0}} \|\mathbf{x} - \mathbf{Y}\mathbf{w}\|_2^2 \quad (1)$$

where $\mathbf{C}$ is an $L \times K$ matrix whose columns correspond to the K constraints applied to the auxiliary array weight vector and the superscript "H" denotes the Hermitian transpose operator. These constraints are designed to prevent cancellation of the desired signal and in some cases to shape the beampattern of the array to achieve some desired response. Equation (1) corresponds to a constrained least-squares problem whose solution can be obtained using standard techniques ([5], pages 410–411, and [7]).

To apply adaptive sidelobe cancellation to a wideband system, a sub-banding approach is used [4]. This technique passes the samples measured at each sensor through a bank of narrowband filters to divide the wideband data into a set of narrowband "frequency bins." The output of each filter is then processed independently using either the sidelobe cancellation algorithm described above or some alternative adaptive processing algorithm. The residuals associated with the various frequency bins can be appropriately combined to reconstruct the wideband signal. Typically, the sub-banding is performed using FFTs. The data is divided into N blocks of length M, and an M-point FFT is then applied to each block. Windowing may be applied to the data to improve the accuracy of the filter and to reduce anomalous effects associated with the blocking procedure.

2 Interference and Noise Cancellation

Adaptive array processing techniques are designed to contend with a variety of interference sources in the HF band, including narrow-bandwidth man-made interference and wide-bandwidth atmospheric noise. In this section we describe experimental results designed to assess the amount of interference and noise cancellation that can be achieved across a 1.024 MHz band using a constrained least-squares sidelobe canceller. We also assess the sensitivity of the constraints to calibration errors and mismatch.

2.1 Interference Cancellation

The data file b0715005 was recorded using the southwest array on 15 July 1991 at 21:06 Greenwich mean time (GMT). The center frequency was set at 17.6125 MHz. The signal located at the center of the band was an amplitude modulated shortwave broadcast from radio station WYFR located at Okeechobee, Florida. In a typical experiment, 16 milliseconds (ms) of data was processed. At the 2.048 million samples per second data rate, this corresponds to 2^{15} complex samples. The data file was divided into 32 blocks of 1024 samples, and an FFT was applied to each block. The width of each respective sub-band was 2 kHz. Block least-squares sidelobe cancellation techniques were then independently applied to each of the 1024 sub-bands, using a block of 32 samples. An inverse FFT was taken to obtain the time-domain version of the output residual.

Figures 1 and 2 depict the periodograms of the beamformed primary element and the output residual after applying the sub-banded sidelobe canceller described above to data from the fourth 16 ms block of file b0715005. In these figures the normalized frequency on the x-axis corresponds to the true base-band frequency divided by 1.024 MHz. Elements 1 through 6, 10, and 13 were used to perform beamforming with a uniform taper across the array. These same elements were later used as auxiliary array sensors. The carrier of the signal was used to determine the response of the array and to subsequently determine the constraint. The steering vector associated with the signal was measured by determining the amplitude and phase response of the antenna elements to the carrier using FFTs. After applying constrained sidelobe cancellation (constraint derived from the first 16 ms block and applied to the fourth 16 ms block) to the data file, both the noise level and the narrow-bandwidth interference were significantly reduced, while the signal level remained essentially unchanged. The reduction in background noise level found when figures 1 and 2 are compared is due to the correlation between the beamformed primary-element noise samples and the receiver noise in the auxiliary array sensors; it does not represent cancellation of HF atmospheric noise. This correlation is, of course, due to the fact that many of the elements used to form the primary element were later employed as auxiliary array elements. The noise level shown in figure 2 is identical to that measured in a single element when no adaptive processing is applied.

We also observed that not all of the interferers are completely reduced in figure 2. For example, there exist a number of interferers between 0 and 0.1 normalized frequency that are well above the background noise level. We believe that the primary cause of such incompletely cancelled interference is the leakage of the FFT filter used in the sub-banding. Interferers that are

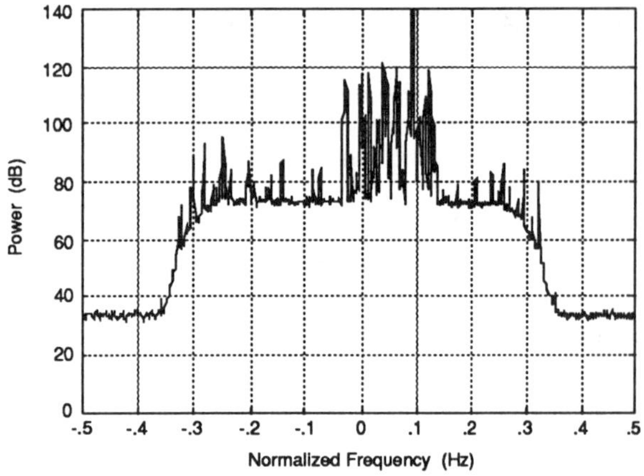

Figure 1. 1024-Point Periodogram of the (Beamformed) Primary Element

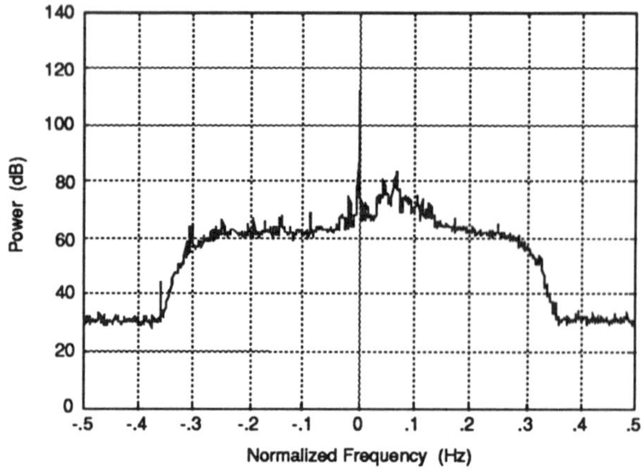

Figure 2. Periodogram of the Residual Obtained Using Eight Auxiliaries

not exactly on an FFT bin are spread into all the remaining bins. In areas of the band with a high density of narrow-bandwidth interferers, each bin could contain significant components of interferers from many different directions, overwhelming the number of degrees of freedom available for cancellation.

Finally in analyzing the response of the array to the signal transmitted by WYFR and other stations, including a tone broadcast from Homestead, Florida, we found that array response remained relatively stable over short time intervals of 64 ms or less. However, for intervals greater than 1 minute, significant changes in the array response occurred that resulted in significant signal cancellation when constrained array processing techniques were applied. These variations were most likely due to fluctuations in the HF channel, and to variations in the local environment. The results suggest the need for robust methods for performing constrained processing (see section 2.3).

2.2 Cancellation of HF Background Noise

At times (particularly during summer nights in the lower parts of the HF band) impulsive atmospheric events due to distant lightning storms combine to become the dominant component of the background noise received by the array. It is conceivable that these distant storms produce noise bursts of sufficient directivity that they can be cancelled using spatial filtering. If this is true, then lowering the background noise level would benefit a variety of HF applications. It is important to determine whether the HF background noise can be reduced, how reliably this can be accomplished, and what levels of background noise attenuation can be achieved. An experiment demonstrating that up to 10 dB of HF background noise cancellation can be achieved is described below. We have also conducted experiments not reported here, in which locally-generated powerful noise bursts due to thunderstorm activity or urban noise were routinely cancelled using a least-squares sidelobe canceller.

File b0501004 was collected using the southwest array on 1 May 1991 at 3:02 GMT. The center frequency was at 9.244 MHz. Figure 3 shows the 1024-point periodogram of the first sensor. A total of 2^{17} complex samples (64 ms) were processed, corresponding to 128 blocks of size 1024. The region of the band between normalized frequencies -0.2 to -0.1 can be used to establish a background noise level of approximately 70 dB. Figure 4 shows the results of scanning a beam from $-60°$ to $60°$ over the frequency range of 8.8315 to 8.9565 MHz. These results suggest the presence of a strong noise source emanating from the southeast at $-15°$. Furthermore, most of the interference measured by the array also arrived from this sector. These results suggest significant propagation from the southeastern United States and the Caribbean, where thunderstorm activity would be expected.

In analyzing the data the first array element was selected as the primary element and a sub-banded unconstrained sidelobe canceller was applied to the 128 blocks using up to 15 auxiliaries. Figure 3 also shows the periodogram of the residual when 15 auxiliaries were used in the cancellation. The background noise level dropped 12 dB to about 58 dB and the narrow-bandwidth interferers were cancelled across the band. Note the rise in the out-of-band noise floor due to FFT leakage between sub-bands of the strong interferers.

We conclude that the background noise in this case had a significant directional component, and that this component was reduced by the least-squares sidelobe canceller.

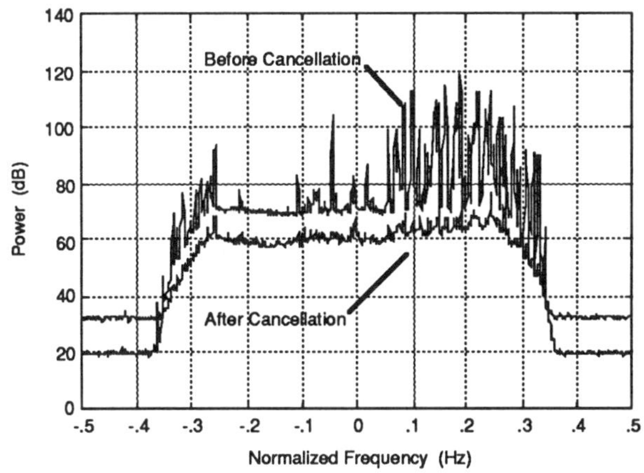

Figure 3. Cancellation of HF Background Noise Using 15 Auxiliaries

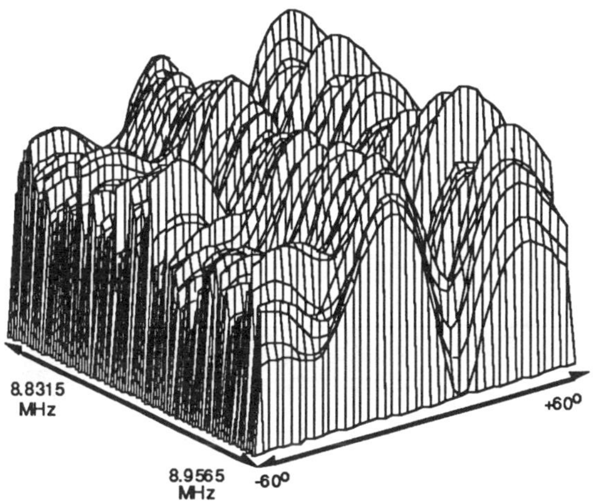

Figure 4. Beamscan Showing Wide-bandwidth Noise Source at $-15°$

2.3 Robust Constrained Techniques

Two methods for reducing the sensitivity of adaptive array processing techniques to calibration errors that provide improved performance over more conventional techniques were applied to data from the local HF array: the diagonal loading algorithm [2] and a signal subspace projection-based approach [3].

To assess the performance gains that could be achieved, we applied diagonal loading to a number of data files containing known narrow-bandwidth signals. In general we found that the output SINR of a fully adaptive array processing system with look direction constraints could be increased by between 2 dB and 15 dB over more conventional techniques, provided the diagonal loading factor was properly selected.

An alternative to the diagonal loading algorithm has recently been proposed by Feldman and Griffiths [3]. In this approach the constraints used to preserve the main beam of the array pattern are projected onto the signal subspace of the auxiliary array data matrix, and the resulting projection is then used to constrain the auxiliary array weights. In applying this technique to a sub-band containing a strong narrow-bandwidth signal, we generally found that this projection based approach was an effective method for applying constraints when calibration errors existed. However, we have so far only considered the simple case in which the signal subspace ws one-dimensional. In the future it will be useful to apply this technique to more complicated signal and interference environments to better assess its usefulness.

3 Enhancing PN Sequence Detection

In this section we describe data analysis results that use an adaptive sidelobe canceller to improve the detection of spread-spectrum signals. We focused on issues affecting the performance of the PN matched filter including finite sample effects, the impact of constraints, and the blocking effects due to the FFT sub-banding block. One concern going into the study was that the independent cancellations in each of the sub-bands might corrupt the underlying signal structure so as to make PN matched filter detection impossible. Our results show that this is not the case, although they also indicate that some correlation degradation may result due to signal cancellation caused by main-beam interference and finite sample effects. For a wide-bandwidth signal, the presence of a similarly directed wide-bandwidth interfering source could result in complete cancellation. Similarly directed narrow-bandwidth interferers pose more or less of a problem depending on the percentage of the band affected. This suggests that constraints may be required and that large cancellation blocks be used (say, four to eight times the number of auxiliaries).

3.1 Interference Cancellation

A number of data files were collected in late May

1991 that contained repetitions of a 4095-chip PN sequence transmitted from Homestead, Florida. Each chip of the PN sequence was sampled four times, and the power level of each sample was varied between files. Matched filtering alone was unable to detect the presence of the wide-bandwidth signal irrespective of the signal power used at the transmitter. The matched filtering was performed by forming the cyclic crosscorrelation between a 64 ms received file (2^{17} complex points) and the PN reference sequence. Adaptive spatial processing was successfully applied to a number of these files, and the result we describe is representative.

The data file b0522007 was recorded on the southwest array on 22 May 1991 at 18:07:05 GMT. The center frequency was 15.848 MHz, and the power level of the wide-bandwidth PN sequence was set so that it was below the receiver noise. In these experiments FFT blocks of size up to 16,384 were considered, limiting the number of frequency-domain snapshots to 8. To maintain a sufficient ratio of auxiliaries to snapshots, only four auxiliaries were chosen (sensors 4, 5, 6, and 7). Sensor 3 was selected as the primary element. The 16,384-point periodogram of the primary element is shown in figure 5. Note the narrow-bandwidth interference found over the 1.024 MHz band of interest.

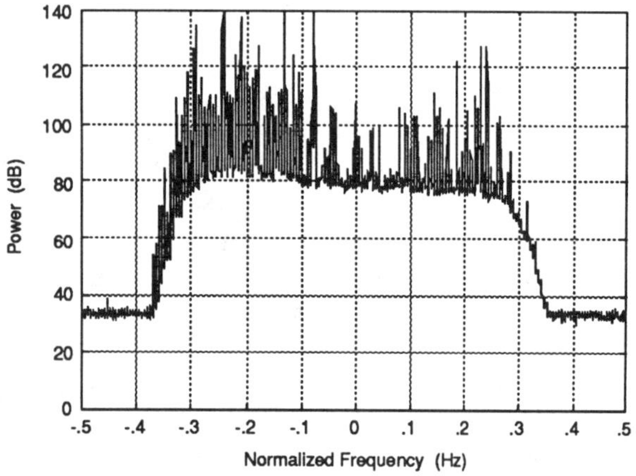

Figure 5. Periodogram of the Primary Element Showing the Presence of Narrowband Interference

Sub-banding was performed on the array data and constrained least-squares sidelobe cancellers were then applied to each frequency bin to cancel the narrow-bandwidth interference. The number of auxiliaries was varied from one to four. An inverse FFT was used to transform the spatially filtered data back to the time-domain, and a matched filter was subsequently applied to the residual vector.

To derive the constraint, approximate amplitude and phase relationships for the wide-bandwidth signal were determined by considering the correlation peaks produced by a PN matched filter at each array element. The PN matched filter output was computed using the received data from each auxiliary element after it had been subjected to 10% frequency-domain excision. The amplitude and phase of the third peak (out of the eight peaks) was used to derive a single constraint, which was applied to all sub-bands in a constrained sidelobe cancellation procedure. The output of the matched filter for 2^{14} sub-bands is shown in figure 6 for four auxiliaries, corresponding to three degrees of freedom. The correlation spikes associated with the signal are apparent, and the multipath structure of the correlation peaks is evident. The appearance of two returns in each cycle was due to multiple reflections off different layers of the ionosphere. From the periodogram of the constrained residual (not shown here), we see that the single constraint used to preserve the signal also prevented considerable narrow-bandwidth interference cancellation, most likely because the poor resolution of the primary element meant that the single constraint decreased the null depths produced by the sidelobe canceller for a considerable number of narrow-bandwidth interferers.

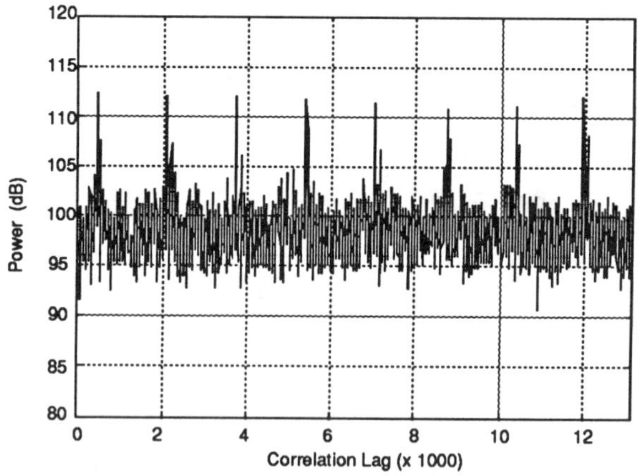

Figure 6. Matched Filter Output for the Residual of a Four-Element Constrained Sidelobe Canceller

3.2 FFT Blocking Effects

In the simplest form of the sub-banded concurrent-block sidelobe cancellation system, the input data from the primary and auxiliary elements are divided into contiguous blocks, called *FFT blocks*, of size B, and a B-point FFT is applied to each of these blocks. When the final residual is obtained by applying an inverse FFT to the cancelled frequency-domain data, periodic spikes appear near the boundary of the FFT blocks. This same phenomenon has been observed previously for frequency-domain excision [1] and it is well understood: a narrow-bandwidth interferer whose frequency is not exactly "on-bin", i.e., not an integral multiple of $1/B$, contributes to every FFT coefficient; the frequency-domain filter, be it excision or sidelobe cancellation, removes or attenuates only the large components and leaves essentially unaffected the components that are below some threshold; and it is these many small components that the inverse FFT uses to reconstruct the narrow-bandwidth signal near the boundary of the FFT block. The result is that the filtered narrow-bandwidth interferer has most of its remaining power concentrated at the edges of the FFT block. The incompletely cancelled interference becomes more localized at the boundary due to increased FFT resolution, and these boundaries occur less frequently as the length of the FFT block is increased, implying that this effect becomes less important for longer FFT blocks.

A number of remedies for the FFT blocking effects have been suggested, including the application of a second time-domain excision to the spikes in the residual, windowing the initial FFTs, and overlap-and-add techniques. There are trade-offs involved when these techniques are applied, however, since the desired signal is also being affected.

4 Conclusion

This paper described results obtained from off-line processing of HF array data using a variety of adaptive array processing methods. We demonstrated that both interference and noise can be cancelled across a 1 MHz band using adaptive array processing methods. Strong wide-bandwidth noise bursts can be routinely cancelled in the time domain using simple, computationally efficient least-squares methods, and the results suggest that at times the HF background noise can be substantially suppressed. We found that while the array hardware and the HF environment are essentially stable over short time intervals (64 ms), significant variations may occur over longer periods (1 minute), and these variations may significantly impact the performance of conventional beamformers and constrained least-squares adaptive array algorithms. Robust techniques, including diagonal loading and a signal subspace projection based approach, were successfully applied to reduce the sensitivity of the adaptive array to both calibration errors and fluctuations in the HF channel. Finally, adaptive sidelobe cancellation techniques were combined with matched filtering to detect spread-spectrum signals. The underlying structure of the PN sequence was preserved while the interference was cancelled.

Acknowledgments

This work was supported by the United States Air Force Electronic Systems Division under contract F19628-89-C-0001. The authors would like to thank Pat L. Durante for developing the computer code used in this analysis. The hardware used in this experimental effort was developed as part of MITRE's wide-bandwidth HF experimental program. We are particularly indebted to Charles J. Beanland, Glenn L. Glatfelter, and Jeffrey Manosh, who designed and developed the array hardware described in section 1.2.

References

1. Abraham, L. G., M. I. Clune, D. C. Rogers, and B. G Thornton, 1989, Stationarity of Wideband HF Noise after Interference Excision, *Proc. 1989 IEEE Mil. Comm. Conf.*, Boston, MA, pp. 48.3.1–48.3.6.

2. Cox, H., R. M. Zeskind, and M. M. Owen, 1987, Robust Adaptive Beamforming *IEEE Trans. Acous., Speech, Signal Processing,* Vol. ASSP-35, pp. 1365–1375.

3. Feldman, D. D., and L. J. Griffiths, 1991, A Constraint Projection Approach for Robust Adaptive Beamforming, *Proc. IEEE ICASSP,* Toronto, Canada, pp. 1381–1384.

4. Gabriel, W. F., 1986, Adaptive Digital Processing Investigation of DFT Subbanding vs Transversal Filter Canceler, NRL Report 8981, Naval Res. Lab., Washington, DC.

5. Golub, G. H., and C. F. Van Loan, 1983, Matrix Computations, The Johns Hopkins University Press, Baltimore, Maryland.

6. Griffiths L. J., 1976, Time-Domain Adaptive Beamforming of HF Backscatter Radar Signals, *IEEE Trans. Antennas Propagat.,* Vol. AP–24, pp. 707–720.

7. Griffiths, L. J., and C. W. Jim, 1982, An Alternative Approach to Linearly Constrained Adaptive Beamforming, *IEEE Trans. Antennas Propagat.,* Vol. AP–30, pp. 27–34.

A Method of Cepstral Enhancement For Detection and Parameter Estimation of Echoes

B. Armstrong and N. Ahmed

Department of EECE
University of New Mexico
Albuquerque, NM 87131

Abstract

Transmission channels such as that of seismic paths, radar and sonar etc.., are often encumbered by multipath reflections which can be reduced using cepstrum analysis. The cosinusoidal ripple representing the echo delay and gain in the power cepstrum have typically been estimated using DFT methods. When additive noise is present in the input composite signal, the cosinusoidal ripple is corrupted by multiplicative noise. In this paper adaptive FIR and IIR frequency estimation methods are utilized in a detection scheme to identify quefrency in the log power spectrum. In addition, an adjustable window based on a Butterworth design is suggested for use in isolating the cosinusoidal frequency. The advantages and disadvantages of windowing are discussed. Also the problems associated with complex cepstrum processing are discussed as they relate to echo detection and delay estimation. Comparisons are made of the proposed processing methods using a standard broadband rectangular function with random noise added in addition to actual echoic data.

I. Introduction

The propagation of many types of signals inherently contain the effects of reverberation, multipath or additive echo reflections. In applications where multiple sensors are allowed, beamforming, interferometer approximations or the bicepstrum can be applied [1]. Cepstrum analysis has been useful for applications where only one sensor is available, in reducing reflection interference or estimating multichannel gains or delays [2]. It is assumed when applying cepstrum processing, that the basic direct path signal is unknown a priori. The application areas encompass radar and sonar, speech, marine and earth seismology, and many others. The area of interest in this paper is that of broadband signals arising from an ionospheric dispersive channel which have been embedded in background random noise.

The applicability and effectiveness of cepstrum processing is highly data dependent where echo size, additive random noise, and relative signal-noise bandwidth are of primary concern [2,3]. The areas related to the overall task of echo removal in a multipath envirmnent, of interest in this paper, are those of echo detection and parameter estimation which are most easily accomplished using the power cepstrum, rather than the complex cepstrum. The complex cepstrum, used for deconvolution, is not stressed since it is more susceptible to background noise and phase wrapping error than the power cepstrum [2].

This paper has two objectives: first, to briefly review the techniques of cepstrum signal processing noting some limitations in detection and parameter estimation; and second, to present new procedures which offer significant improvement to cepstrum processing in both detection and parameter estimation. Improvement in detectability has been achieved by the use of window functions applied to the log spectrum including, the rectangular window, and Hamming and modified Hamming windows [3]. These windows are limited in design of the signal to noise bandwidth and thus a modified Butterworth window which is suggested for its signal to noise bandwidth selectivity. Window functions are are of limited use for broadband signals.

To further improve echo detectability various methods of spectral estimation are suggested. Since random noise erodes the cosinusoidal ripple representing an echo in the log spectrum, the adaptive finite impulse response (FIR) and infinite impulse response (IIR) methods of line enhancement [4] are suggested for application to the log spectrum. Results of both adaptive line enhancement (ALE) procedures applied to broadband signals in background random noise, are provided for comparison.

II. Power Cepstrum: Review

In applications such as seismic processing time delays and amplitudes of reflections are of primary interest. However, basic wavelet reconstruction is not necessary and the power cepstrum is preferred over other techniques including the complex cepstrum. When extraction of the basic wavelet is desired as in sonar or radar applications, the complex cepstrum can be of great value since the inverse processing can, in many instances, be achieved. However, if phase wrapping occurs, deconvolution or even detection can become impossible with the complex cepstrum.

It has been shown, as will be repeated below, that the effect of a delayed echo will appear as a cosinusoidal ripple in the log spectrum [2]. The frequency of the ripple can be seen by calculating the spectrum of the log spectrum, which now has the units

1058-6393/91 $01.00 © 1991 IEEE

of time representing the amount of delay of the echo (termed "ephoc").

Using the definition of the power cepstrum of a data sequence in [2], as the square of the inverse z-transform of the logarithm of the magnitude squared of the z-transform of the data sequence, we can write

$$x_{pc}(nT) = \left[Z^{-1}(\log|X(z)|^2)\right]^2$$

$$= \left\{\frac{1}{2\pi j}\oint_c \log|X(z)|^2 z^{n-1}dz\right\}^2 \quad (1)$$

where $X(z)$ is the z-transform of the data sequence $x(nT)$. A block diagram illustrating the algorithm for forming the power cepstrum is shown in Fig. 1.

The convolution of two sequences can be expressed as

$$x(nT) = f(nT) * g(nT) \quad (2)$$

or finding the magnitude squared of the z-transform gives

$$|X(z)|^2 = |F(z)|^2|G(z)|^2 \quad (3)$$

and taking logarithms yields

$$\log|X(z)|^2 = \log|F(z)|^2 + \log|G(z)|^2. \quad (4)$$

If we apply (1) then

$$x_{pc}(nT) = f_{pc}(nT) + g_{pc}(nT) + cross\ terms \quad (5)$$

If the power cepstra occupy different quefrency ranges or the final squaring operation in (1) is omited, then (5) can be reduced to

$$x_{pc}(nT) = f_{pc}(nT) + g_{pc}(nT) \quad (6)$$

We note that liftering (filtering) in the quefrency (frequency) domain can be used to separate the individual contributions of the respective cepstra.

Consider the case of a composite signal consisting of a basic wavelet and a single echo. In this case the discrete time impulse train function is

$$g(nT) = \delta(nT) + a\,\delta(nT - n_o T) \quad (7)$$

where $\delta(nT)$ denotes the *unit sample sequence*. If we evaluate (4) on the unit circle ($z = e^{j\omega T}$) the log of the power spectrum becomes

$$\log|X(e^{j\omega T})|^2 = \log|F(e^{j\omega T})|^2$$
$$+ \log[1+a^2+2a\cos(\omega n_o T)]$$
$$= \log|F(e^{j\omega T})|^2 + \log(1+a^2)$$
$$+ \log\left[1 + \frac{2a}{1+a^2}\cos(\omega n_o T)\right] \quad (8)$$

By expanding the third term on the right of (8) in a power series (excluding the point values $a = \pm 1$ and $\cos(\omega n_o T) = \pm 1$ [2]) we obtain

$$\log\left[1 + a_o\cos(\omega n_o T)\right] =$$
$$\sum_{m=1}^{\infty}\frac{(-1)^{m+1}}{m}\left[a_o\cos(\omega n_o T)\right]^m \quad (9)$$

where $a_o = \frac{2a}{(1+a^{2)}}$. From the expression in (9) it can be seen that the composite signal will contain cosinusoidal ripples in the corresponding log magnitude squared of the z-transform whose gamnitude (amplitude) and quefrency (ripple frequency) are related to the echo amplitude (a) and delay $(n_o T)$ respectively. Substituting (9) into (8) and applying (1) will result in the power cepstrum which will contain peaks at quefrencies of $(n_o T)$ and multiples thereof. These peaks can be detected providing the quefrency content of $|F(e^{j\omega T})|^2$, or effects of random noise, does not corrupt the impulses at positive integer multiples of $(n_o T)$. The echo arrival times can be estimated by noting the time of occurrence of the first peak or time difference between periodic peaks in the power cepstrum.

The amplitude of an echo can be estimated from the power cepstrum or the complex cepstrum since the peaks representing impulse functions correspond to the cosinusoidal ripples in (9). Random noise present in the input however, can erode the the gain of the impulse functions. In addition, aliasing of cepstral peaks from each echo in (9) can cause problems in detection or parameter estimation. Zero padding the input alleviates the problem of aliasing. Either of these problems can produce extraneous peaks as will similar cross product terms arising from multiecho interference.

The most limiting problem in cepstrum analysis is that it is severely distorted by the presence of ambient random noise. A composite signal with a single echo when random noise is added can be modeled as

$$x(nT) = f(nT) + af(nT - n_o T) + w(nT) \quad (10)$$

where $w(nT)$ is a time random variable. Substituting $A = |F(\omega)|$ and $B = |W(\omega)|$ and assuming $|a| < 1$, it can be shown that the real part of the log of the Fourier transform of the sampled sequence in (10) is given as

$$\mathrm{Re}[\log|X(\omega)|e^{j\theta_x}] = \frac{1}{2}\log[2AB\cos(\theta_f - \theta_w)]$$
$$+ \frac{1}{2}\log\left[1 + \left[\frac{B}{2A} + \frac{a^2 A}{2B} + \frac{A}{2B}\right]\sec(\theta_f - \theta_w)\right.$$
$$+ \frac{aA}{B}\cos(\omega n_o T)\sec(\theta_f - \theta_w)$$
$$\left. + \frac{a\cos(\theta_f - \theta_w - \omega n_o T)}{\cos(\theta_f - \theta_w)}\right] \quad (11)$$

where ω is radian frequency and θ_f and θ_w are the phase of the primary signal and noise respectively. The power cepstrum is computed from (11). We see that the term involving $\cos(\omega n_o T)$ in the power cepstrum, computed for the non-noise case in (8), when compared to (11), is corrupted by the multiplicative noise term $\sec(\theta_f - \theta_w)/B$. In addition, additive random noise is apparent in other terms in (11).

III. Echo Detection

Although the power cepstrum is not invertible for wavelet recovery, it has found advantage over the complex cepstrum in echo detection and parameter estimation [5,8]. As was noted in (11) however, the effects of additive random noise in the input composite wavelet, represents one of the most severe restrictions to echo detection, gain estimation and signal recovery in cepstrum processing.

III.1 Windowing the Log Spectrum

In the case of a single echo, detection can be achieved in a background of Gaussian white random noise with an SNR of $20dB$ without additional processing according to Kemerait [2]. Using Hanning smoothing of the log magnitude, however, Kemerait reports detection with a SNR as low as $2\ dB$ for the same example. However, Hassab has determined that those results have not been reproduced [3]. Hanning smoothing of the log magnitude is equivalent to applying a Hanning window to the transform of the log magnitude.

Windowing of the log magnitude spectrum has demonstrated improvement in detectability by minimizing leakage and more importantly, selecting frequency regions of higher SNR in the log spectrum. Thus the ratio of the signal and noise bandwidths is important in window design [3].

A modified Hamming window has been utilized by Hassab [3] which enhances the effective measures of signal bandwidth relative to noise bandwidth and the SNR. If the window is not designed with this in mind an increase in the detection threshold will occur. This window function is given as

$$W_r = 0.56 - 0.44\cos(2\pi m/L) \tag{12}$$

Hassab has obtained echo detection at an SNR of $5\ dB$ (and a biased detection at $0\ dB$). For this signal traditional windows have produced detectable results only to within an SNR of $20\ dB$ [9].

Window width and and frequency selectivity can be specified using an inverted Butterworth magnitude response for windowing the log power spectrum. In addition, the filter order of the Butterworth response can be selected to minimize signal truncation effects in the log power spectrum. This window is far more adjustable for signal-to-noise bandwidth improvement than any of the other windows reviewed. For example, the modified Butterworth window of order $n=2$ is given as

$$|H(j\omega)|^2 = 1 - \frac{1}{1+(\omega/\omega_c)^4} \tag{13}$$

where ω_c is the cutoff frequency.

III.2 Detection Using Adaptive Line Enhancement

Random noise in the input wavelet causes many spurious peaks to appear in the cepstrum [2,3]. Since these peaks adversely affect all cepstrum processing we consider a method of discriminating between cepstral peaks due to deterministic sources and those due to random origin in the log spectrum. This method applied to the log spectrum is called *Adaptive Line Enhancement* (ALE). The problem of identifying cosinusoidal ripples with randomly varying amplitude is similar to identifying pure sinusoids in broadband noise.

III.2.1 The Nonrecursive Adaptive Line Enhancer

The Block diagram of Fig. 2. represents a Δ-step predictor where $\hat{x}_k$ is the predicted value of the input x_k and ε_k where L is the number of filter weights which are denoted w_i.

The weights of the ADP can be updated using the LMS (least-mean-square) algorithm [4], which is as follows:

$$w_l(k+1) = w_l(k) + \hat{\mu}e(k)x(k-l-\Delta) \quad 1 \le l \le L \tag{14}$$

where $e(k) = x(k) - y(k)$, and $\hat{\mu}$ is the normalized convergence parameter, which is defined as

$$\hat{\mu} = \frac{(1-\beta)}{\sigma_x^2(k)} \tag{15}$$

where

$$\sigma_x^2(k) = \beta\sigma_x^2(k) + (1-\beta)x^2(k) \tag{16}$$

with $0 < \beta < 1$ being a smoothing parameter. The quantity $\sigma_x^2(k)$ is simply an estimate of the input signal power at time k.

The linear predictor acts as a decorrelator filter and strives to remove the uncorrelated components in the input x_k, producing an estimate $\hat{x}_k$ that is virtually uncorrelated to the error beyond $\Delta-1$ samples.

III.2.2 The Recursive Adaptive Line Enhancer

A recursive adaptive LMS type algorithm replaces the linear filter $H(z)$ with a fixed time delay of Δ of Fig. 2.

Several properties are desirable for the transfer function $z^{-\Delta}H_k(z)$; it should have unit gain and zero phase at other frequencies to suppress noise in y_k. The simplest transfer function shown to have this form is [4]

$$H_{\theta k}(z) = \frac{(1-r^2)(\cos\theta_k - z^{-1})}{1-(1+r^2)z^{-1}\cos\theta_k + r^2 z^{-2}} \tag{17}$$

with $\Delta = 1$, poles of radius r and unity gain at frequency θ_k. By assuming that $H_k(z)$ is stable with complex poles of radius r, we note the additional constraints

$$0 < r < 1$$

$$|(1+r^2)\cos\theta_k| < 2r \tag{18}$$

At a given radius r the ALE is searching for a minimum error only by adjusting the frequency θ_k. However, the constraints limit the range of θ_k to a range of frequencies . For example, with $r = 0.95$, frequencies from $\theta_k = 0.05\pi$ to 0.95π are covered.

We assume that r is constant during adaptation, thus fixing the bandwidth of the adaptive filter.

A nonrecursive relationship has been derived for the estimate of the gradient, based on the output y_k, denoted α_k [4]:

$$\alpha_k = \left[\frac{1-r^2}{1+r^2}\right]x_{k-1} + y_{k-1} \tag{19}$$

which is used in the weight update equation know as the modified recursive least mean square algorithm

(MRLMS). The MRLMS algorithm is given as

$$w_{k+1} = w_k + \rho \frac{\varepsilon_k \alpha_k}{\Psi_k} \qquad MRLMS \qquad (20)$$

where ρ is a convergence parameter and

$$\Psi_k = \beta \Psi_{k-1} + (1-\beta)\alpha_k^2 \qquad (21)$$

The memory factor β is chosen in the range $0<\beta<1$.

For a detailed study of the convergence properties of these algorithms see reference [4].

IV. Applications and Results

The examples which follow illustrate detection with improvements shown using the suggested modified Butterworth window, and spectral estimation of the cosinusoidal ripples arising from the presence of echoes using the FIR-ALE and IIR-ALE algorithms. Further, estimation of echo parameters is performed using the power cepstrum or the power cepstrum without the final squaring operation in (5).

The waveforms used to illustrate the use of the complex and power cepstrum are 1) a damped sinusoidal model and 2) a broadband rectangular shaped waveform passed through an ionospheric model. All time and cepstral waveforms are labeled in samples rather than seconds for ease in determining echo delays.

The damped sinusoidal model is defined by the waveshape of $e^{-at} \sin bt$ where the damping coefficient is $a = 19.84 x\ 10^6$ and the radian frequency is $b = 4.496 x\ 10^6$. The sampling frequency was 400 MHz.

Fig. 3a represents the composite of the basic and echo wavelet which is scaled by the normalized gain A = 0.3 and delayed by 42 samples with Gaussian random noise of zero mean and variance 0.00003 added for an SNR of $20dB$. The real part of the IDFT of the power cepstrum for this waveform is shown in Fig. 3b.

Improvement in detection of echoic peaks by windowing of the log power spectrum, using a modified Hamming window [3], is observed in the real part of the IDFT of the power cepstrum of Fig. 3c. The cepstral peak at sample 42 stands out more clearly.

To allow greater control over the window width, the modified Butterworth window was applied to the log spectrum of this input composite resulting in a peak with a larger threshold as shown in the cepstral waveform of Fig. 3d. Adjusting the width and filter order to the values $fc = 0.6$ and $n = 2$, the signal strength relative to the background noise is optimized. The peak at sample 42 in the cepstrum using the modified Butterworth is 80% above the other cepstral peaks, excluding the low quefrency cepstra of the basic wavelet. The cepstral peak of the same quefrency is 11.4% for modified Hamming window.

A broadband signal example representative of the midrange frequency of an electro magnitic pulse (EMP) waveform is that of the rectangular wave passed through the ionospheric model and deconvolved. This waveform is that of an 11 sample rectangular waveform sampled at 400 Mz and passed through an ionospheric model which effectively filters the signal leaving frequency and phase preserved from 50 to 200 MHz.

With an echo of gain A = 0.4 delayed by 70 samples and Gaussian random noise with zero mean and variance 0.00265 added, the time waveform is shown in Fig. 4a. The SNR for this composite waveform is 5 dB. The log spectrum was high pass filtered to suppress the low frequency of the basic wavelet. The high pass filter applied was that of a Butterworth response with normalized cutoff frequency $f_c = 0.05$ and order $n = 32$ chosen arbitrarily. The corresponding real part of the IDFT of the power cepstrum is presented in Fig. 4b. Note that several peaks in this cepstrum could be potential echoes and the peak at sample 70 is not exceptional. Zero padding of the input was done to minimize aliasing of the power cepstrum.

To improve echo detection adaptive line enhancement was performed on the log power spectrum of this rectangular waveform example. Both the FIR-ALE and the IIR-ALE were applied for effectiveness comparison. The measure of effectiveness of the method is the amount the detection threshold or the background cepstra was lowered. The IIR-ALE cannot be evaluated using these criteria since the shape of its transfer function is fixed and correct frequency tracking is a more accurate effectiveness measure.

The FIR-ALE was applied to the Butterworth filtered log power spectrum resulting in the real part of the IDFT of the power cepstrum of Fig. 4.c. The adaptive parameters used were $L = 20$ weights and a convergence parameter of $\mu = 0.0005$. The most notable effect of the FIR-ALE filter is that all components have been suppressed except the one at sample 70, the cepstral peak of the added echo. The detection threshold in the cepstrum of Fig. 4b is 0.18293 and following application of the FIR-ALE becomes 0.09156, which is 50% lower.

For comparison, the IIR-ALE was studied in this application for possible advantages. The form of the adaptive filter used is that of the MRLMS algorithm. The data was passed through the adaptive filter several times to achieve convergence. The value of convergence was $\mu_4 = 0.0005$. Other parameter values used were the normalization smoothing parameter $\beta = 0.99$. The initial estimate of power in the gradient for normalization was $\rho = 0.001$ and pole radius $r^2 = 0.95$. The algorithm was started from various initial weight values ω_o and always converged to the desired cosinusoidal ripple of $\omega_f = 1.26$. The resulting cepstra is shown for comparison in Fig. 4d where a peak is exceedingly clear at sample 70.

V. Summary and Conclusion

The methods designed to improve the power cepstrum's ability to detect echoes include windowing the log power spectrum among others. It is evident that these methods can also be applied to the complex cepstrum. The modified Butterworth window demonstrates improved detection thresholds over the other window functions studied, including the modified Hamming window. The adaptive line enhancement filters are shown to significantly reduce detection thresholds for a simulated ionospheric model rectangular waveform with added noise and an SNR of $5dB$. The detection threshold using the FIR-ALE was reduced by 50% for the unknown cosinusoidal frequency of this waveform. The reduction of the detection threshold using the IIR-ALE was an even larger 80%. In addition, the frequency tracking capability of the IIR-ALE can yield more exact estimates of echo delay. Then, identifying more accurately where the echo is in the power cepstrum the echo gain can be estimated more acurately. Areas of future research are in finding methods to isolate the randomly varying cosinusoidal ripple using possibly block processing of a combination of the IIR and FIR ALE methods. Finally, the techniques presented in this paper demonstrate an effective method for detecting echoes and estimating their parameters.

References

[1] C.L. Nikias and F. Liu, "Bicepstrum Computation Based on Second- and Third-order Statistics with Application", Proc. IEEE Int. Conf. Acoustic Speech and Signal Proc., Vol.5, No.EU, pp.2381-2385, April 1990.

[2] Donald G. Childers, David P. Skinner, and Robert C. Kemerait, "The Cepstrum: A guide to Processing," Proc. IEEE, vol. 65, no 10, pp. 1428- 1443, Oct. 1977.

[3] J. C. Hassab and R. E. Boucher, "Improved cepstrum performance through windowing of log spectrum," J. of Sound and Vibration, vol. 58, No. 4, pp. 597-598, 1978.

[4] N. Ahmed, *Adaptive Filtering: Handbook of Digital Signal Processing* , Academic Press. Inc., San Diego, Calif., 1987, pp. 857-897.

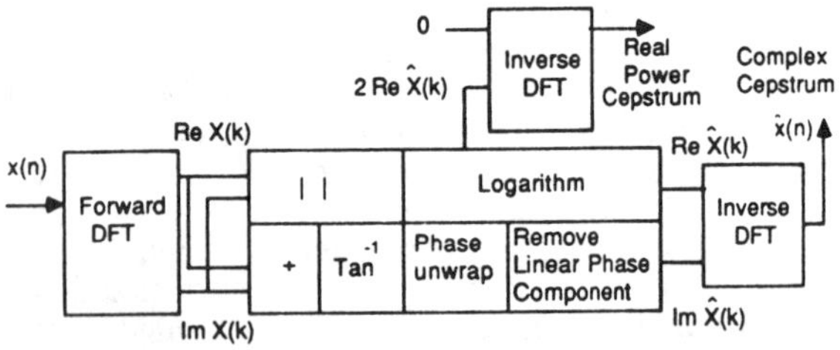

Fig. 1. Block processing for the complex and real power cepstrum.

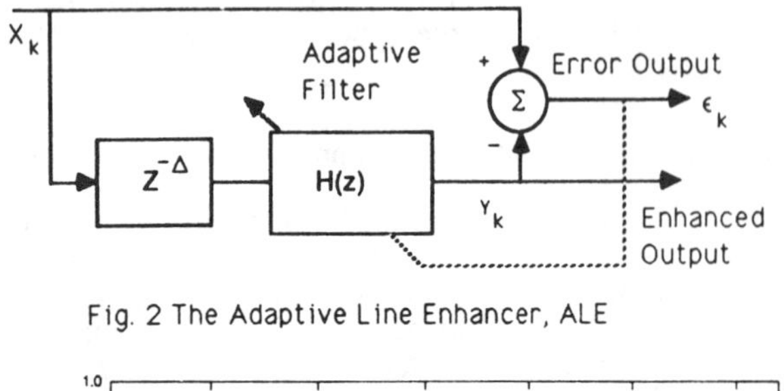

Fig. 2 The Adaptive Line Enhancer, ALE

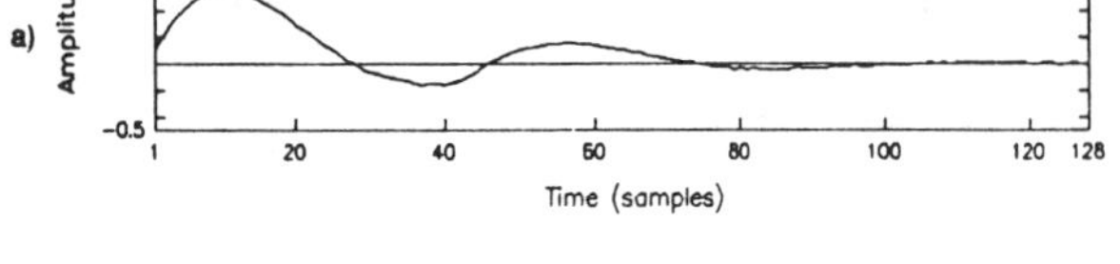

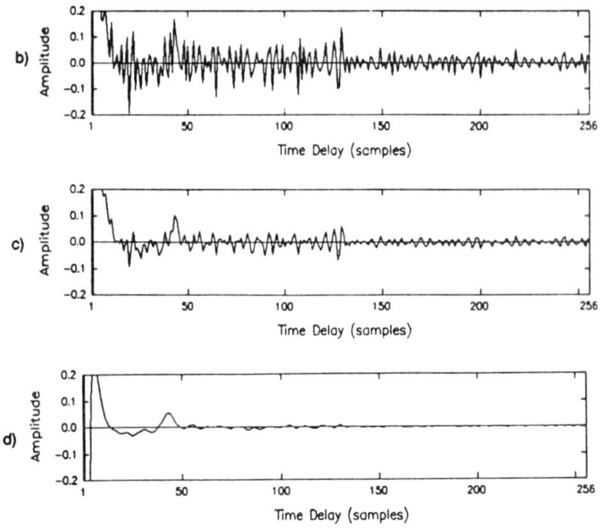

Fig. 3 Damped sinusoidal time waveform with additive Gaussian noise with a SNR =20 dB plus scaled echo (A=0.3) delay 32 samples; a) time waveform, b) real part of the final IDFT of the power cepstrum c) processed with the modified Hamming window d) and processed with the modified Butterworth window.

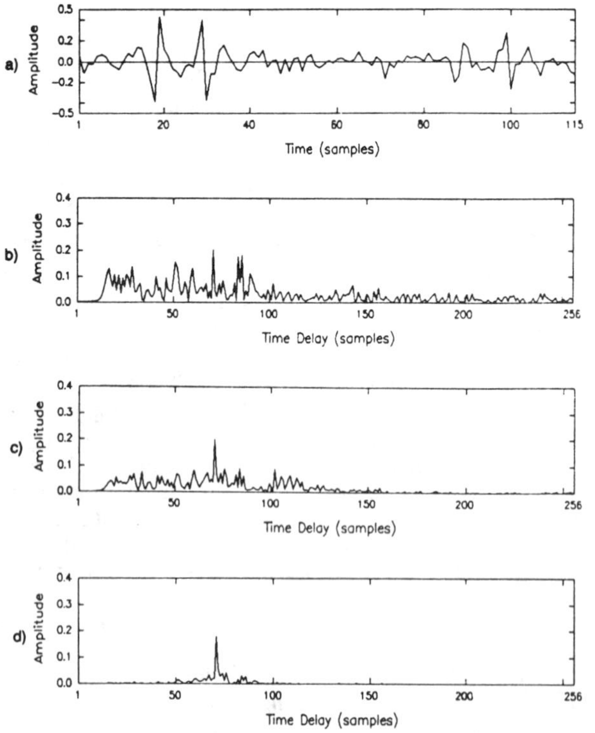

Fig. 4 Broadband ionospheric model example with scaled (A=0.4) single echo and zero mean Gaussian random noise added; SNR=5 dB. a) time waveform b) high pass filtered power cepstrum c) power cepstrum of high pass and FIR-ALE filtered log spectrum d) power cepstrum of high pass and IIR-ALE filtered log spectrum.

Incorporating Human Expertise in Computer Algorithms for Biomedical Applications

A.V. Sebald

University of California, San Diego,La Jolla CA 92093
sebald@ucsd.edu

Abstract

Since humans (M.D.s, RN's) regularly deliver care to patients with a great deal of success, it would seem logical that computer algorithms designed to assist in patient care could profit from the inclusion of human expertise. This paper discusses practical constraints arising in such an approach.

I. Introduction

Human experts seem to have great contributions to make in the movement toward ever more intelligent computer systems. In fact, they often are the only examples of intelligent controllers able to successfully solve the control problems of interest. The motivation to incorporate such expertise in computer system is very strong. This paper discusses the issue in fairly broad terms, with the purpose of sharing personal experiences from the medical field.

In this paper, we emphasize the paradigm of humans solving very complex control problems in real time. Good examples include trauma care or anesthesiology. The goal is to gain information about how humans solve these problems and to develop concepts applicable to autonomous machine prostheses for problems like orthostatic hypotension (subnormal blood pressure upon standing.)

There are basically two main issues here. First, what kinds of human expertise can be brought to bear on the design of intelligent computer systems? Second, what kinds of problems arise in successfully gleaning the desired expertise in a form which can be useful?

II How can Human Expertise be Useful?

Rules: Perhaps the first type of human expertise which comes to mind is that associated with rules of analysis common to expert systems. While this is indeed very useful, there are many equally important forms. Let us, however, first consider this type of human information.

Human decision making is often based on classical rules: "If it is raining, take an umbrella". Given the list of rules, the technology exists in PROLOG and LISP to make inferences and most other required operations on such rules[1]. Much human expertise of this type is best described in linguistic variables: "If the pressure is high and rising rapidly, then give an aggressive bolus of the drug." This of, course can be easily handled with fuzzy techniques[2]. In fact, there now exist fuzzy PROLOG implementations, so there is no real lack of flexibility for incorporating information of either of these types.

Performance indices: A second type of human expertise we have found to be very important is specification of appropriate performance indices. To take the blood pressure control example for a moment, it turns out that desired performance changes with both the type of patient and the target pressure at any given time. For example, in most cases, low targets coincide with tight tolerances on the down side and relatively looser tolerances on the up side. The asymmetry is reversed for high targets. The type of patient (e.g. elderly male cardiac patient vs healthy young athlete with a shoulder injury) will also influence the strength of penalties on either side of the target. In fact, it is common that staying in some desired range is good enough and penalties for being outside the range are somewhat artificial in that being outside the range is essentially to be avoided.

Another common humanly specified performance goal is that the controlled variable should not stay outside the desired range for more than a certain maximum time (which time is also condition and patient dependent). Furthermore, typically, the controlled variable should be precluded at all costs from exceeding some absolute upper and lower bounds (which bounds are also patient and condition dependent).

Finally, desired response speed also depends on circumstances and patients. Sometimes, the required

1058-6393/91 $01.00

speed of response is actually determined in consultation with the perfusionist or surgeon as a case proceeds.

While these humanly specified performance criteria can easily be encoded with standard or fuzzy approaches, it is clear that standard quadratic performance indices have very little validity, that non trivial amounts of code need to be associated with computation of the true objective functions, and that this complexity makes design and testing of controller schemes based on analytical results much less attractive.

Target values: Finally, it sometimes happens that human experts do not know what performance targets the body actually uses. For example, it is apparently not known what the body's control systems use as a target for blood glucose levels. Furthermore, there is great disagreement over whether the body actually has a target for arterial pressure to which the body's control systems respond. Some say that the resulting controlled pressure is simply the equilibrium point of the current configurations of the body's systems.

How well is the intelligent system working? Humans who know how to perform complex tasks might also provide information on how well a computerized system is performing the same task. By watching it attack the problem, the human may be able to provide embellishments to previous information and even new rules etc. As discussed below, however, this human input can be difficult to interpret and may, in fact be irrelevant.

III. How can Human Expertise be Reliably Obtained?

Why not ask? The most obvious way to obtain information of the above types is to simply ask for it. For example, one could simply watch an expert performing some task and simply ask why she did what she did when she did it. Unfortunately, such a query is not simple. First, there are pros and cons associated with whether one should ask hypothetically (in the quiet of the expert's office) or whether one should monitor what the person is actually doing. Of course, the goal is to delve into the actual decision making process without hindering performance.

One is actually trying to glean a great deal of information here. How does the expert know that something needs to be done? Was it a change in the sound of the process, a change in an instrument reading or what? Second, how does the expert decide what to do? Third, how does the expert know whether the proper thing was done or whether some further correction is needed?

We have found that this process is non trivial. Conversation with colleagues confirms that we are not alone in this view.

Let us examine some common problems. When warranted, the discussion of each issue is organized in two parts. Part 1 examines the process of discussion with the expert regarding some abstract question (e.g. a discussion in the expert's office about a hypothetical problem.) Part two examines the process of discussion during or after some decision process has actually been applied to a real world problem.

Problem 1: The expert does not know the answer

This problem occurs in several forms. First, One asks what should be done in a certain case and the expert responds with "I don't know". This is not a problem unless the expert feels such an answer is improper. Some professions seem to emphasize confidence with the consequent, albeit unintended, result that the expert will hedge rather than say "I don't know". The hedging sometimes results in a defensive posture and further acquisition of information may be difficult.

Alternatively, we have found that sometimes the expert simply does not know how he arrived at some decision. A common reason is that, after much practice, the human expert does things somewhat automatically and they may be unconscious of the actual detailed reasoning process.

a) discussion in the abstract:

This tends to be less intense since contradictions do not arise.

b) discussion of an actual case:

Here, things are more difficult. Sometimes the expert simply cannot express why an action was taken. The action just seemed like the right thing to do. Psychologically, people do not like to be put on the spot trying to explain, even to themselves that they didn't do something capricious or even lucky. Some colleagues have found that one should think of this whole affair as a two step process in which one discusses the issues first in the abstract and then after the fact. Some have found it important to record the early session because of discrepancies in memory over what transpired in the first session.

Problem 2: The expert may remember what happened differently than what actually happened.

Recall of specific events in a complex procedure is not perfect.

Sometimes, the expert remembers what reasoning procedure he consciously used but external information may indicate that the expert may have unconsciously used a different approach from the one he thought he had.

Problem 3: The answers may be of no use if their implementation requires information (e.g. smells or psychological atmosphere in the room) not available to the computer.

Problem 4: The expert watches a computer algorithm perform a task and expresses concern over its intelligence.

This seems to arise from two main sources. First, if the expert and the engineer have very different backgrounds, the expert may inquire why the computer is doing something and be met with an explanation involving poles and zeros or even Ricatti equations. The computer system may be doing something perfectly proper, but the expert may remain unsatisfied by the explanation. This can have very serious consequences if the expert needs to understand how the machine "thinks" prior to committing it to actual patient care.

The second source of difficulty here is that experts seem to generate their own styles of action. These styles are based on much experience and they are trusted and used regularly by the individual. Often, such styles are the norm for groups of experts. The problem occurs when computers manifest a "style" grossly different from the one used by the expert. Given that the computer's approach is a hybrid of the expert and the engineer, such differences are not unlikely. Once again, the concern here is expert confidence in the machine's ability to perform.

Why not measure everything and perform system identification? One is tempted to propose that the ideal approach is to record the entire session in which the human actually performs the task in question. One would need to record all actions and measure and record all relevant variables. One may even wish to monitor such things as eye/head movement to get information on what is being observed at any given time.

One should probably also videotape the session to get information on the sounds etc occuring during the process.

The general idea would be to use sophisticated system identification techniques to build an empirically based phenomenological model of the human expert's technique.

Clearly, there are also problems with this approach. The computer interface may substantially alter the normal intimacy between the expert and the problem being solved. At a minimum, extensive familiarization may be needed.

While reliable information may result, measurements like these are quite intrusive and may simply be infeasible.

Experts are only human: The obvious point in all of this is that when gathering information on human expertise, one must take a great deal of care with the humanity of the expert. Key points include:
1) the desire for professional peer approval and respect,
2) the fear of the appearance of acting capriciously,
3) the fear of replacement by machines or other humans (unwillingness to share tricks of the trade),
4) a defensive response to pressure for additional justification, and
5) the fear of having a serious mistake captured in great detail on computer records or videotape.

Can we hope to put a human's capability in a computer? It is very possible that the problems we have been discussing stem from a bad match between the hardware and software architectures of a human and the best computer systems. The human's gestalt may be crucial.

The bottom line:
1) People like to feel uniquely qualified in their fields of expertise and generally do not like to be pressed heavily on why they do what they do.

2) People are reluctant to have someone examine every detail of their actions. They tend to want to focus on the quality of the outcome.

IV. Suggestions

1) One may need fairly complicated system to glean reliable info while not raising defensive barriers. Such a system would be designed expressly to build confidence between the expert and the engineer in order to construct an intellectual common ground and specifically elicit information in the least threatening manner. This can be extremely time consuming and is very difficult to achieve.

2) In spite of the above comments on avoiding situations which put the expert in a defensive posture, there already are mechanisms for studying failures (grand rounds, M&Ms (mortality and morbidity conferences) etc). These mechanisms often provide lots of information and often provide a positive atmosphere involving the common search for a better solution.

3) Most of the problems outlined in this discussion stem from a lack of common language or culture between the expert and the engineer. Thus, the best way to solve problem 1) above might be to teach computing to the experts (e.g. MDs) and let them develop their own intelligent systems. A second approach is to provide a kind of erector set which would permit the experts to easily build systems of their own design. People with MDs and PhDs or advanced degrees in engineering or computer science are becoming very common.

Caveat Emptor: This piece is intended to encourage discussion on this important topic. Most things discussed here are merely personal opinions based on experience. Perhaps the most important result of this exercise would be an indication from others of the degree of universality of these experiences.

References

[1] Bratko, Ivan, **Prolog Programming for Artificial Intelligence**, Addison Wesley, 1990.

[2] Isaka, Satoru, **On the Design and Architecture of Adaptive Fuzzy Controllers and Their Application to a Biomedical Control Problem,** PhD Thesis, University of CA San Diego, 1989.

Image Enhancement and Filtering Using Wavelets

Kenneth E. Prager and Paul F. Singer

Hughes Aircraft Company
Electro-Optical and Data Systems Group
P.O. Box 902
El Segundo, CA 90245

Abstract

We report on the results of our analysis of the wavelet transform and its use as a tool for image enhancement and filtering in knowledge-based systems. We start by reviewing the concept of the wavelet transform. We then present a convolution theorem for the wavelet transform and describe a filter model which uses a wavelet orthonormal basis set. This model is compared to the linear filter model for image enhancement and filtering. We then apply the wavelet filter to a set of degraded sample imagery. Applied operations include edge detection and sharpening.

I. Introduction

The topic of multiresolution and wavelet transforms has received a great deal of attention in recent years. Wavelet transforms are important because they provide a means for localization of signals in both frequency and time [2, 9]. This leads to many applications such as signal analysis [7], image compression [3,5,6] and applied mathematics [9].

In this paper we demonstrate how the traditional signal and image processing operation of *Finite Impulse Response* (FIR) filtering can be encompassed within a discrete wavelet transform. Examples of FIR filter applications include temporal filtering for the reduction of time-random noise, spatial filtering for the reduction of correlated noise—using either fixed or adaptive coefficients, edge sharpening and edge detection.

To appreciate the benefits of applying FIR filtering within the wavelet domain, consider the application of image compression. Suppose that video imagery from a sensor in a remote location is to be compressed, using wavelet compression, for transmission to a central location. Further, suppose that the sensor is subject to noise. It would then be desirable to filter this signal at some point. This filtering operation may be performed in the transform domain or combined with the wavelet transform filter bank and applied as part of the compression or decompression process.

Note that the more powerful tool is the case where filtering is performed within the wavelet domain. In this implementation, simple filters with relatively few coefficients, applied in the wavelet domain, can represent complex filters in the temporal or spatial domain. This is due to the fact that the wavelet decomposition of a signal separates the the signal into bands of different frequency or scale.

This paper is constructed as follows. The important results of Mallat's wavelet algorithm [6] are reviewed in Section II. We concentrate on applications of Mallat's algorithm because it represents an efficient method for performing wavelet decomposition. A shift theorem and a convolution theorem are developed in Section III. In the wavelet domain, the convolution of two functions in the temporal or spatial domain is shown to be equal to the wavelet transform of shifted versions of one of the functions, convolved with the other function. The shift theorem is used to show that convolution in the wavelet domain can be thought of in other ways. The theorems developed in Section III are extended to two-dimensions in Section IV, for application to imagery. Several examples of filtering in the wavelet domain are presented in Section V.

II. Wavelet Representation of Signals

In this section we review the important results of multiresolution analysis and wavelet representation of signals. The notation is similar to [6], where the scaling function $\phi(x)$ is defined as $\phi_{2^j}(x) = 2^j\phi(2^j x)$. Then, the family of functions $\left(\sqrt{2^{-j}}\,\phi_{2^j}(x-2^{-j}n)\right)_{n\in Z}$, represents an orthonormal basis of the vector space V_{2^j}. V_{2^j} can be thought of as the set of all approximations, at resolution 2^j, of the $L^2(R)$ functions. Similarly, the family of functions $\left(\sqrt{2^{-j}}\,\psi_{2^j}(x-2^{-j}n)\right)_{n\in Z}$, represents an orthonormal basis of the vector space O_{2^j}, which is the set of all differences, at resolution 2^j, of the $L^2(R)$ functions. That is, $V_{2^{j+1}} = O_{2^j} \oplus V_{2^j}$. The orthogonal wavelet $\psi(x)$ is defined as $\psi_{2^j}(x) = 2^j\psi(2^j x)$.

A. Notation

For two functions, $f(x)\in L^2(R)$ and $g(x)\in L^2(R)$, the inner product is written as

1058-6393/91 $01.00 © 1991 IEEE

$$\langle f(x), g(x)\rangle = \int_{-\infty}^{+\infty} f(x) g(x)\, dx \quad , \tag{1}$$

where R is the set of real numbers and $L^2(R)$ is the vector space of measurable, square-integrable one-dimensional functions.

For two sequences, $f(n) \in l^2(Z)$ and $g(n) \in l^2(Z)$, the discrete inner product is written as

$$\langle f(n), g(n)\rangle_n = \sum_{n=-\infty}^{+\infty} f(n) g(n) \quad , \tag{2}$$

where Z is the set of integers and $l^2(Z)$ is the vector space of square-summable one-dimensional sequences.

A. The Multiresolution Transform

The continuous approximation [6] of a signal $f(x)$ at resolution 2^j is given by

$$A_{2^j} f(x) = 2^{-j} \langle \langle f(u), \phi_{2^j}(u - 2^{-j}n)\rangle \phi_{2^j}(x - 2^{-j}n)\rangle_n . \tag{3}$$

The discrete approximation is thus

$$A^d_{2^j} f = (\langle f(u), \phi_{2^j}(u - 2^{-j}n)\rangle)_{n \in Z} . \tag{4}$$

The discrete approximation can also be expanded and written as

$$\begin{aligned} A^d_{2^j} f &= \langle f(u), \phi_{2^j}(u - 2^{-j}n)\rangle \\ &= \langle \tilde{h}(2n-k) \langle f(u), \phi_{2^{j+1}}(u - 2^{-j-1}k)\rangle\rangle_k \\ &= \langle \tilde{h}(2n-k) A^d_{2^{j+1}} f\rangle_k \end{aligned} \quad , \tag{5}$$

where

$$\tilde{h}(-n) = h(n) = \langle \phi_{2^{-1}}(u), \phi(u-n)\rangle . \tag{6}$$

B. The Wavelet Transform

In a manner similar to the multiresolution transform, we can define the orthogonal projection of the signal $f(x)$ onto O_{2^j} as

$$W_{2^j} f(x) = 2^{-j} \langle \langle f(u), \psi_{2^j}(u - 2^{-j}n)\rangle \psi_{2^j}(x - 2^{-j}n)\rangle_n . \tag{7}$$

This operator represents the continuous detail signal [6] of $f(x)$ at resolution 2^j. The discrete approximation is thus

$$W^d_{2^j} f = (\langle f(u), \psi_{2^j}(u - 2^{-j}n)\rangle)_{n \in Z} . \tag{8}$$

Since O_{2^j} is orthogonal to V_{2^j} and $O_{2^j} \oplus V_{2^j} = V_{2^{j+1}}$, $W^d_{2^j} f$ can be thought of as the difference of information between $A^d_{2^{j+1}} f$ and $A^d_{2^j} f$, Equations (4) and (8) lead to a filter bank implementation [8] for the wavelet decomposition of $f(x)$, as shown in Fig. 1(a). In a manner analogous to (5), (8) can be expanded and written as

$$\begin{aligned} W^d_{2^j} f &= \langle f(u), \psi_{2^j}(u - 2^{-j}n)\rangle \\ &= \langle \tilde{g}(2n-k) \langle f(u), \phi_{2^{j+1}}(u - 2^{-j-1}k)\rangle\rangle_k \\ &= \langle \tilde{g}(2n-k) A^d_{2^{j+1}} f\rangle_k \end{aligned} \quad , \tag{9}$$

where

$$\tilde{g}(-n) = g(n) = \langle \psi_{2^{-1}}(u), \phi(u-n)\rangle . \tag{10}$$

Equation (5) combined with (9) leads to a binary tree or pyramid implementation [1,6,8], as shown in Fig. 2(a). Note that the scaling filter, H, and the wavelet filter, G, are quadrature mirror filters [6,8]. Thus, their impulse responses are related by

$$g(n) = (-1)^{1-n} h(1-n) \tag{11}$$

The block diagrams shown in figures 1(a) and 2(a) demonstrate how a continuous signal, $f(x)$, can be decomposed into a set of discrete signals $(A_{2^{-J}} f, (D_{2^j} f)_{-J \le j \le -1})$. Because of sub-sampling at each stage, the wavelet representation contains the same number of samples as the original signal [6].

C. Signal Reconstruction

The vector space O_{2^j} is the orthogonal complement of V_{2^j} in $V_{2^{j+1}}$. Thus, $(\sqrt{2^{-j}}\, \phi_{2^j}(x - 2^{-j}n), \sqrt{2^{-j}}\, \psi_{2^j}(x - 2^{-j}n))_{n \in Z}$, is an orthonormal basis of $V_{2^{j+1}}$. Therefore, the discrete approximation [6] at resolution 2^{j+1} can be represented by

$$\begin{aligned} A^d_{2^{j+1}} f &= \langle f(u), \phi_{2^{j+1}}(u - 2^{-j-1}n)\rangle \\ &= 2\langle h(n-2k) A^d_{2^j} f\rangle_k \\ &\quad + 2\langle g(n-2k) W^d_{2^j} f\rangle_k \end{aligned} \quad . \tag{12}$$

Fig. 2(b) demonstrates how the signal $f(x)$ can be reconstructed from a wavelet representation.

III. Filtering in the Wavelet Domain

In this section we develop a set of theorems which can be used to design and analyze filters which operate in the wavelet domain. Intuitively, if one were to amplify all of the coefficients of, for example $W^d_{1/2} f$, we would expect the reconstructed signal to exhibit stronger high frequency features. This is due to the fact that $W^d_{1/2} f$ represents the high-pass portion of the signal. The utility of these theorems lies in the fact that filtering can be performed as part of the transform process or within the transform domain. In the latter case, simple filters in the wavelet domain can represent complex filters in the spatial domain.

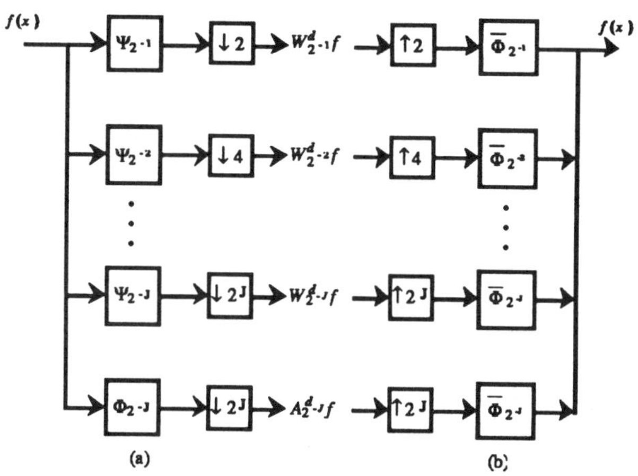

Fig. 1. Filter bank implementation. (a) Wavelet decomposition using (4) and (8). (b) Signal reconstruction.

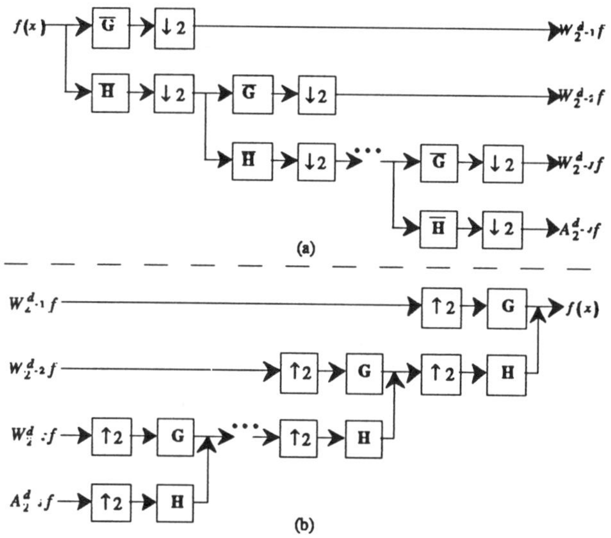

Fig. 2. Binary tree implementation. (a) Wavelet decomposition using (5) and (9). (b) Signal reconstruction.

In what follows, we consider continuous signals $f(x)$. However, for implementations involving digital computers, the extension to the discrete-time case is relatively straightforward.

A. Convolution in the Wavelet Domain

Theorem 1a (Shift Theorem): The discrete approximation $A_{2^j}^d f_v$ of the signal $f(u-v)$ is equivalent to the continuous approximation at resolution 2^j, $A_{2^j} f(x)$, evaluated at $x = 2^{-j}k - v$, where

$$A_{2^j} f(x) = 2^{-j} \left\langle A_{2^j}^d f, \phi_{2^j}\left(x - 2^{-j} n\right)\right\rangle_n . \quad (13)$$

That is

$$\begin{aligned} A_{2^j}^d f_v &= \left\langle f(u-v), \phi_{2^j}\left(u - 2^{-j} n\right)\right\rangle \\ &= A_{2^j} f(x)\big|_{x = 2^{-j}k - v} \end{aligned} \quad . \quad (14)$$

Theorem 1b (Shift Theorem): The discrete detail signal $W_{2^j}^d f_v$ of the signal $f(u-v)$ is equivalent to the continuous detail signal at resolution 2^j, $W_{2^j} f(x)$, evaluated at $x = 2^{-j}k - v$, where

$$W_{2^j} f(x) = 2^{-j} \left\langle W_{2^j}^d f, \psi_{2^j}\left(x - 2^{-j} n\right)\right\rangle_n . \quad (15)$$

That is

$$\begin{aligned} W_{2^j}^d f_v &= \left\langle f(u-v), \psi_{2^j}\left(u - 2^{-j} n\right)\right\rangle \\ &= W_{2^j} f(x)\big|_{x = 2^{-j}k - v} \end{aligned} \quad . \quad (16)$$

Proof: We prove Theorem 1a. The proof for Theorem 1b follows directly. Using (3), (13) can be written as

$$\begin{aligned} A_{2^j} f(x) = 2^{-j} \Big\langle \left\langle f(u), \phi_{2^j}\left(u - 2^{-j} n\right)\right\rangle, \\ \phi_{2^j}\left(x - 2^{-j} n\right) \Big\rangle_n \end{aligned} \quad , \quad (17)$$

which is equivalent to

$$A_{2^j} f(x) = \left\langle f(u), \phi_{2^j}(u - x)\right\rangle . \quad (18)$$

Inserting $x = 2^{-j}k - v$ into (18) yields

$$A_{2^j} f(x) = \left\langle f(u), \phi_{2^j}\left((u + v) - 2^{-j} k\right)\right\rangle . \quad (19)$$

Then, by a change of variables, we have

$$A_{2^j} f(x) = \left\langle f(u-v), \phi_{2^j}\left(u - 2^{-j} n\right)\right\rangle . \quad \blacklozenge \quad (20)$$

The utility of this theorem lies in the fact that it is generally easier to implement an algorithm for reconstruction via (14) (or (16)) than it is to compute shifted versions of the discrete approximation (or detail) signal directly.

We now consider the wavelet representation of the signal $f'(x)$, where

$$f'(x) = \int_{-\infty}^{+\infty} f(x-u)\, b(u)\, du = \left\langle f(x-u), b(u)\right\rangle \quad . \quad (21)$$

Theorem 2a (Convolution Theorem): The discrete approximation of the signal $f'(x)$ at resolution 2^j is equal to the inner product of the discrete approximation of shifted versions of $f(u-v)$ and the filter kernel $b(v)$.

Proof:

$$\begin{aligned} A_{2^j}^d f' &= \left\langle f'(u), \phi_{2^j}\left(u - 2^{-j} n\right)\right\rangle \\ &= \left\langle \left\langle f(u-v), b(v)\right\rangle, \phi_{2^j}\left(u - 2^{-j} n\right)\right\rangle \\ &= \left\langle \left\langle f(u-v), \phi_{2^j}\left(u - 2^{-j} n\right)\right\rangle, b(v)\right\rangle . \end{aligned} \quad \blacklozenge \quad (22)$$

This can be equivalently stated as

$$A_{2^j}^d f' = \left\langle \left\langle b(u-v), \phi_{2^j}\left(u - 2^{-j} n\right)\right\rangle, f(v)\right\rangle . \quad (23)$$

Theorem 2b (Convolution Theorem): Given a signal $f'(x)$ as in (21), the discrete detail signal at resolution 2^j is equal to the inner product of the discrete detail signal of shifted versions of $f(u-v)$ and the filter kernel $b(v)$.

Proof:

$$\begin{aligned} W_{2^j}^d f' &= \langle f'(u), \psi_{2^j}(u-2^{-j}n)\rangle \\ &= \langle\langle f(u-v), b(v)\rangle, \psi_{2^j}(u-2^{-j}n)\rangle \\ &= \langle\langle f(u-v), \psi_{2^j}(u-2^{-j}n)\rangle, b(v)\rangle. \quad \blacklozenge \end{aligned} \tag{24}$$

This can be equivalently stated as

$$W_{2^j}^d f' = \langle\langle b(u-v), \psi_{2^j}(u-2^{-j}n)\rangle, f(v)\rangle. \tag{25}$$

B. Implementation of Wavelet-Based Filters

There a several ways in which Theorem 2 can be applied: as part of a filter bank; as part of a binary tree; or within the transform domain.

For a filter bank implementation, Theorem 1 can be applied directly to (23) and (25), yielding

$$\begin{aligned} A_{2^j}^d f' &= \langle\langle A_{2^j}^d b, \phi(n-k-2^j v)\rangle_k, f(v)\rangle \\ W_{2^j}^d f' &= \langle\langle W_{2^j}^d b, \psi(n-k-2^j v)\rangle_k, f(v)\rangle. \end{aligned} \tag{26}$$

Equation (26) can also be thought of in terms of matrix-vector notation as

$$\begin{aligned} \underline{\mathbf{Af}}' &= \underline{\mathbf{Ab}}^T \Phi\, \underline{\mathbf{f}} \\ \underline{\mathbf{Wf}}' &= \underline{\mathbf{Wb}}^T \Psi\, \underline{\mathbf{f}}. \end{aligned} \tag{27}$$

This also leads to the following corollary.

Corollary: The transformed scaling and wavelet bases, ϕ' and ψ', defined as

$$\begin{aligned} \phi'_{2^j}(x) &= \langle b(u), \phi_{2^j}(u-x)\rangle \\ \psi'_{2^j}(x) &= \langle b(u), \psi_{2^j}(u-x)\rangle \end{aligned} \tag{28}$$

are orthogonal bases.

For a binary tree implementation, (23) and (25) can be expanded in a manner similar to (5) and (9) (see [6]), yielding

$$\begin{aligned} A_{2^j}^d f' &= \langle \tilde{h}'_j(2n-k) A_{2^{j+1}}^d f\rangle_k \\ W_{2^j}^d f' &= \langle \tilde{g}'_j(2n-k) A_{2^{j+1}}^d f\rangle_k, \end{aligned} \tag{29}$$

where v is confined to integral shifts, l, and

$$\begin{aligned} \tilde{h}'_j(-n) &= h'_j(n) \\ &= \langle \phi_{2^{-1}}(u), \langle \phi(u-n-2^{j+1}l), b(l)\rangle_l\rangle \\ \tilde{g}'_j(-n) &= g'_j(n) \\ &= \langle \psi_{2^{-1}}(u), \langle \phi(u-n-2^{j+1}l), b(l)\rangle_l\rangle. \end{aligned} \tag{30}$$

Finally, with a change of variables and setting $v = 2^{-j}l$, equations (23) and (25) can be written as

$$\begin{aligned} A_{2^j}^d f' &= \langle\langle f(u), \phi_{2^j}(u-2^{-j}l)\rangle, b(2^{-j}(n-l))\rangle_l \\ &= \langle A_{2^j}^d f, b(2^{-j}(n-l))\rangle_l \\ W_{2^j}^d f' &= \langle\langle f(u), \psi_{2^j}(u-2^{-j}l)\rangle, b(2^{-j}(n-l))\rangle_l \\ &= \langle W_{2^j}^d f, b(2^{-j}(n-l))\rangle_l \end{aligned} \tag{31}$$

for filtering directly within the transform domain.

Equations (26), (29) and (31) provide equivalent methods for the synthesis of filters. In the case of (26) and (29), filtering is performed as part of the signal decomposition process, while (31) allows for filtering directly within the wavelet domain. Figures 3 and 4 demonstrate how filtering can be applied during the decomposition process. Note that in Fig. 4(a), the filters must be recomputed at each stage.

The results of Theorem 2 can also be used for the analysis of filtering and other operations (such as level quantization) performed in the wavelet domain.

IV. Extension of the Filter Model to Images

A. Wavelet Representation of Images

Here we only consider the case of separable multiresolution approximations of $L^2(R^2)$. If $(V_{2^j})_{j\in Z}$ is such a multiresolution approximation then $(2^{-j}\Phi_{2^j}(x-2^{-j}n, y-2^{-j}m))_{(n,m)\in Z^2}$ forms an orthonormal basis of V_{2^j}. If V_{2^j} can be decomposed as $V_{2^j} = V_{2^j}^1 \otimes V_{2^j}^1$ then $\Phi(x,y) = \phi(x)\phi(y)$ [6].

The discrete approximation of the signal $f(x,y)$, at resolution 2^j, is given by

$$A_{2^j}^d f = \langle f(u,v), \phi_{2^j}(u-2^{-j}n)\phi_{2^j}(v-2^{-j}m)\rangle. \tag{32}$$

According to Mallat, the discrete detail signal at resolution 2^j is equal to the orthogonal projection of the signal on the orthogonal complement of V_{2^j} in $V_{2^{j+1}}$. If O_{2^j} is this orthogonal complement then

$$\begin{aligned} &(2^{-j}\Psi_{2^j}^1(x-2^{-j}n, y-2^{-j}m), \\ &2^{-j}\Psi_{2^j}^2(x-2^{-j}n, y-2^{-j}m), \\ &2^{-j}\Psi_{2^j}^3(x-2^{-j}n, y-2^{-j}m))_{(n,m)\in Z^2} \end{aligned} \tag{33}$$

is an orthonormal basis of O_{2^j}, where

$$\begin{aligned} \Psi^1(x,y) &= \phi(x)\psi(y) \\ \Psi^2(x,y) &= \psi(x)\phi(y) \\ \Psi^3(x,y) &= \psi(x)\psi(y). \end{aligned} \tag{34}$$

The discrete detail signal of $f(x,y)$ thus consists of three components:

$$\begin{aligned} W_{2^j}^{d1} f &= \langle f(u,v), \phi_{2^j}(u-2^{-j}n)\psi_{2^j}(v-2^{-j}m)\rangle \\ W_{2^j}^{d2} f &= \langle f(u,v), \psi_{2^j}(u-2^{-j}n)\phi_{2^j}(v-2^{-j}m)\rangle \\ W_{2^j}^{d3} f &= \langle f(u,v), \psi_{2^j}(u-2^{-j}n)\psi_{2^j}(v-2^{-j}m)\rangle, \end{aligned} \tag{35}$$

where $W_{2^j}^{d1} f$ represents the vertical high frequencies and horizontal low frequencies, $W_{2^j}^{d2} f$ represents the horizontal high frequencies and vertical low frequencies, and $W_{2^j}^{d3} f$ represents the vertical and horizontal high frequencies [6].

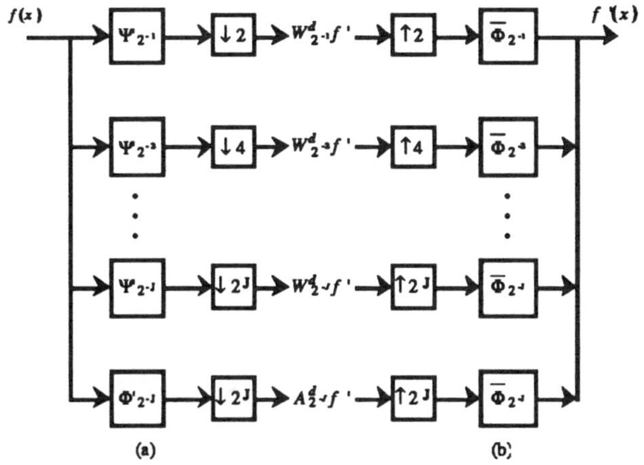

Fig. 3. (a) Filter/wavlet decomposition. Φ' and Ψ' are obtained from (28). (b) Signal reconstruction using the same filter set as in Fig. 1(b). The output f '(x) is a filtered version of the input.

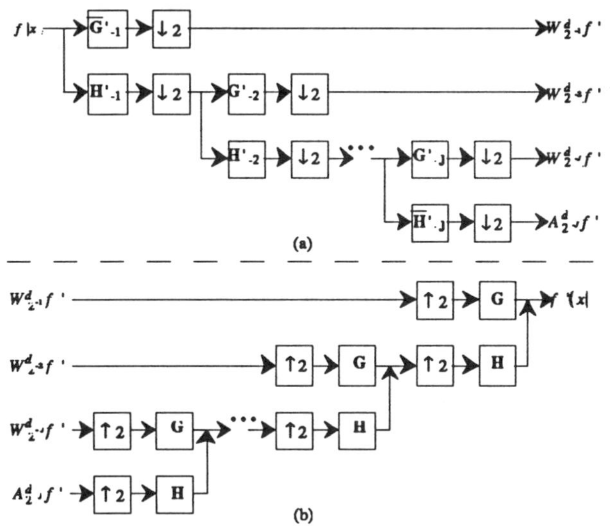

Fig. 4. (a) Filter/wavlet decomposition. $\tilde{G}'$ and $\tilde{H}'$ are obtained from (30). (b) Signal reconstruction using the same filter set as in Fig. 2(b). The output f '(x) is a filtered version of the input.

Note that when $\Phi(x,y)$ and $(\Psi^i(x,y),\ i=1,\ 2,\ 3)$ are separable, the two-dimensional transform can also be implemented using a pyramid algorithm. This is accomplished by first transforming the rows and then the columns of the image [6].

B. Filter Model

In this case, we consider the signal

$$f'(x,y) = \iint_{-\infty}^{+\infty} f(x-u, y-v)\, b(u,v)\, du\, dv$$
$$= \langle f(x-u, y-v),\ b(u,v)\rangle \quad . \quad (36)$$

The multiresolution approximation of $f'(x,y)$, at resolution 2^j, is given by

$$A^d_{2^j} f' = \langle\langle A^d_{2^j} b,\ \phi(n-k-2^j u)\ \phi(m-l-2^j v)\rangle_{k,l},\ f(u,v)\rangle \quad . \quad (37)$$

The discrete detail signal is again made up of three components, which can be written as:

$$W^{d1}_{2^j} f' = \langle\langle W^{d1}_{2^j} b,\ \phi(n-k-2^j u)\ \phi(m-l-2^j v)\rangle_{k,l},\ f(u,v)\rangle$$
$$W^{d2}_{2^j} f' = \langle\langle W^{d2}_{2^j} b,\ \phi(n-k-2^j u)\ \phi(m-l-2^j v)\rangle_{k,l},\ f(u,v)\rangle$$
$$W^{d3}_{2^j} f' = \langle\langle W^{d3}_{2^j} b,\ \phi(n-k-2^j u)\ \phi(m-l-2^j v)\rangle_{k,l},\ f(u,v)\rangle \quad . \quad (38)$$

Equations (37) and (38) are written in a form which leads one to consider the operation of filtering as part of the transform operation. They can also be written in a manner similar to (31), in which case filtering is performed within the wavelet domain.

V. Applications of the Filter Model

In this section we demonstrate the application of image enhancement within the wavelet domain. For the two examples shown here we extended (31) to two-dimensions and performed the operations directly within the transform domain.

A. Edge Enhancement

We transformed the original image, shown in Fig. 5, into a 3x3 level wavelet image using Daubechies' D3 coefficients [1]. The $(W^{d\,i}_{2^{-1}} f,\ i=1, 2, 3)$ portions were then filtered using a (3x3) unsharp masking kernel [4]. This technique is used to increase the sharpness of edges. The inverse transform was computed and the result is shown in Fig. 6.

Note that only the high frequency portions of the wavelet image were operated on, accounting for three quarters of the total number of pixels. A (6x6) kernel operating over the entire image would be required to obtain the equivalent result in the spatial domain. This represents a computational savings of five if the transform overhead is ignored.

B. Edge Detection

In this case, the original image was transformed into a 4x4 level wavelet image using the D3 coefficients. The $(W^{d\,i}_{2^{-1}} f,\ W^{d\,i}_{2^{-2}} f,\ i=1, 2, 3)$ portions of the image, representing the high frequencies, were retained and the low frequencies were discarded. We then performed the reconstruction and thresholding. The result is shown in Fig. 7.

Fig. 5. Original image.

Fig. 6. Edge enhanced image obtained using unsharp masking technique within the transform domain.

VI. Conclusion

We have described how FIR filtering can be combined with wavelet transforms to enhance signals and imagery. The Shift and Convolution Theorems can be used for both the analysis of operations applied within the wavelet domain and filter synthesis.

In the case of filter synthesis, filters can be applied during wavelet decomposition or reconstruction in either a filter bank or binary tree implementation. They can also be applied directly within the transform domain.

Fig. 7. Binary version of high pass portion of Fig. 5, obtained by suppressing low frequency components in the transform domain. Threshold was set at $z=79$.

References

[1] I. Daubechies, "Orthonormal Basis of Compactly Supported Wavelets," *Comm. on Pure and Applied Math.*, Vol.. 41, pp. 909-996, 1988.

[2] I. Daubechies, "The Wavelet Transform, Time-Frequency Localization and Signal Analysis," *IEEE Trans. Inform. Theory.*, Vol. 36, pp. 961-1005, 1990.

[3] J. Daugman, "Complete Discrete 2D Gabor Transforms by Neural Networks for Image Analysis and Compression," *IEEE Trans. Acoust., Speech, Signal Proc.*, Vol. 36, pp. 1169-1179, 1988.

[4] A. K. Jain, *Fundamentals of Digital Image Processing*, Prentice-Hall, 1989.

[5] A. K. Lewis and G. Knowles, "Video Compression Using 3D Wavelet Transforms," *Electronics Letters*, Vol. 26, pp. 396-398, 1990.

[6] S. Mallat, "A Theory For Multiresolution Signal Decomposition: The Wavelet Transform," *IEEE Trans. Pattern Anal. Machine Intell.*, Vol. 11, pp. 674-693, 1989.

[7] F. B. Tuteur, "Wavelet Transformations in Signal Detection," in *Proc. 1988 IEEE Int. Conf. Acoust., Speech, Signal Proc.*, New York, NY, Apr. 11-14, 1988, pp. 1435-1438.

[8] P. P. Vaidyanathan, "Lossless Systems in Wavelet Transforms," in *ISCAS*, Singapore, 1991.

[9] O. Rioul and M. Vetterli, "Wavelets and Signal Processing," *IEEE Signal Proc. Mag.*, Vol. 8, pp. 14-38, 1991.

Knowledge-Based Estimation Methods

D.D. Sworder
Dept. of ECE
Univ. of California, San Diego
La Jolla, CA 92093-0407

P.F. Singer and D. Doria
Hughes Aircraft Co.
El Segundo, CA 90245

R.G. Hutchins
ORINCON Corp.
San Diego, CA 92121

Abstract

At its most basic level, an estimation algorithm is a causal mapping from a spatiotemporal observation to an approximation of a primary process. To gain insight into the idiosyncracies of knowledge-based estimators, and to contrast them with those of their more established brethren, this paper considers the performance of an image-enhanced estimator in a specific scenario; an agile target moving in the plane with intermittent maneuvers is tracked from a fixed location.

1. Introduction

At its most basic level, an estimation algorithm is a causal mapping from a spatiotemporal observation (the measurement) to an approximation of a primary process (the signal or state depending on the context). Estimators have become increasingly sophisticated as more versatile online processors have become available. If the signal has a suitable structure, improved estimation and prediction is achieved through model-based synthesis procedures. These approaches, as their name suggests, use a comprehensive model to delineate both the signal dynamics and the precise relationship between the signal and its measurement. Through the model, the estimator is tuned to subtle patterns in the measurement, thus enabling good performance to be achieved in the presence of significant measurement ambiguity.

Perhaps the most widely studied model-based algorithm is the Kalman filter and its lineal variants. In a tracking problem, the dynamic features of the target would be represented by a state process $\{x_t\}$ having as components; positions, velocities and perhaps accelerations in an appropriate reference frame. The observation process $\{y_t\}$ would be derived from a radar or a similar device that provides a measurement---albeit a noisy one---of the center-of-reflection location of the target; e.g., range and bearing. To illustrate these ideas in the context of a specific example, consider the problem of tracking the motion of an agile object in the X-Y plane. A simple model for an evasive target moving at essentially constant speed is;

$$d\begin{pmatrix} x \\ y \\ V_x \\ V_y \end{pmatrix} = \begin{bmatrix} 0 & 0 & 1 & 0 \\ 0 & 0 & 0 & 1 \\ 0 & 0 & 0 & -\omega_\psi \\ 0 & 0 & \omega_\psi & 0 \end{bmatrix} \begin{pmatrix} x \\ y \\ V_x \\ V_y \end{pmatrix} dt + \begin{bmatrix} 0 & 0 \\ 0 & 0 \\ 1 & 0 \\ 0 & 1 \end{bmatrix} d\begin{pmatrix} w_x \\ w_y \end{pmatrix}$$

where $\{x_t, y_t\}$ are position coordinates, and $\{V_x, V_y\}$ are associated velocities. The target is subject to two types of acceleration; a wide band omnidirectional acceleration described by the Brownian motion $\{w_x, w_y\}$ of intensity W, and a maneuver acceleration represented by the turn rate process $\{\omega_\psi\}$. The former fits well within the conventional linear-Gauss-Markov (LGM) framework, but the latter is troublesome in two respects. First, because of their temporal structure, the sample paths of the maneuver accelerations are better described by a discontinuous rather than a continuous process, and are clearly nongaussian. Secondly, the turn enters into the target dynamics as a state multiplier, and consequently, renders the dynamic equation nonlinear in the augmented state space.

In many applications, the measured relations between the indicated variables are nonlinear. With a smooth nonlinearity, it is possible to linearize the measurement equation about the estimated state. If the linear (or linearized) equations are adequate, there is a well known solution to the mean-square inference problem; the (extended) Kalman filter (EKF). Let the measurement equation be given by

$$dy_t = Dx_t dt + dn_t$$

and denote the information pattern generated by the sensor measurements by $\{Y_t\}$. The best estimate of the state ($\hat{x}_t = E\{x_t | Y_t\}$) satisfies an equation of the form:

$$d\hat{x}_t = A\hat{x}_t dt + P_{xx} D' R_x^{-1} d\nu_t$$

with $d\nu_t$, the increment of the innovations process, and P_{xx}, the error covariance matrix:

$$(d/dt)P_{xx} = AP_{xx} + P_{xx}A' + W - P_{xx}D'R_x^{-1}DP_{xx}$$

subject to appropriate initial conditions. The error covariance is contingent upon the intensity of the exogenous processes in both state and observation; e.g., as W increases, the increment in $\{P_{xx}\}$ increases proportionately. This has an intuitive justification. As the state process becomes more volatile, P_{xx} increases, and through this intermediary, the EKF becomes "faster" and more responsive to state changes. Of course, as a corollary to this, the same filter will amplify the measurement noise $\{n_t\}$.

Recently, alternative tracking architectures have become feasible as novel sensors have been developed. Video and FLIR sensors create a sequence of pictures of a scene containing the target. No longer is the observation restricted to a point-equivalent target, but instead, features of spatial extent can be extracted from the image. The raw image data is received as a sequence of matrices of grey levels, generated in packets or frames at a fixed frame rate λ frames/sec. An image processor recasts the data, and extracts relevant features from it. The processor acts as a pattern classifier, placing each image in one of a prespecified set of topical bins. The information obtained from the image is often complementary to that obtained from a point-location sensor; e.g., target type and orientation can be measured from shape data rather than being inferred from target motion. However, the errors inherent in the image link are quite distinct from those found in the conventional observation links. The relevant errors are misclassifications as occur when the target is placed in feature bin i despite the fact that the correct choice is bin j. Misclassifications depend upon the fidelity of the image and the sophistication of the processing, and these in turn depend upon depend on range and geometry, the angular bin size, the sensitivity of the image elements and the processing algorithm, etc.

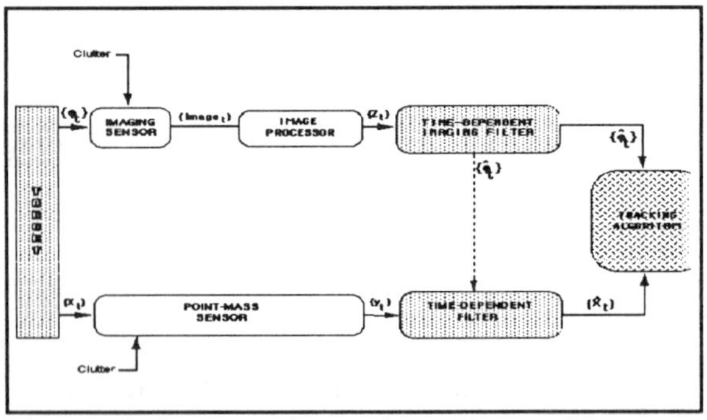

Figure 1: Image-enhanced tracking architecture

The maneuvering target problem has received considerable attention in the literature where it is usually assumed that estimation and prediction is accomplished using bearing and perhaps range data. The simplest estimators use an EKF derived by replacing the turn rate process with additive white noise, with additive colored noise, or by augmenting W with pseudonoise. It will be assumed here that image augmentation is exploited. Figure 1 shows a tracking architecture proposed in several references which is useful for integrating the information derived from an imager with that from conventional sensors. The lower path shows the usual mapping from center-of-reflection observations to an estimate of target location. The upper path is image-specific, and uses observations of target shape to augment the point-location data in the lower path.

To gain insight into the idiosyncracies of image-enhanced estimators, and to contrast them with those of their more established brethren, this paper considers the performance of an image-enhanced estimator in a specific scenario; an agile target moving in the plane with intermittent maneuvers is tracked from a fixed location. This application is fundamentally nonlinear and nongaussian, and illustrates both the promise and the limitations associated with image augmentation. The next section introduces some of the issues that arise in image generation and interpretation in a tactical environment.

2. Knowledge-based Image interpretation

To implement to upper path in Figure 1, the orientation of a target in R^3 must be inferred from the image. A 2-D projection provides the only data upon which this inference can be made. Unfortunately, distortions in both the visible and the infrared bands can cause changes in the apparent shape and internal structure of the target. For example, smoke and contrails in the visible band and plumes in the infrared band can cause the ostensible contour of an aircraft to differ from that of a prestored model of the target. Shadows or internal reflections of sunlight can cause local edges in the interior of a visible image. Internal heating or external heat sources can likewise change the internal distribution of thermal signatures. Especially for ground vehicles, the likelihood of partial occlusion is significant in some situations. Pattern recognition approaches that involve global features of the object (e.g., moment invariants or Fourier descriptors) degrade when confronted with even local changes in perceived shape.

To achieve robust performance in the face of these disturbances, a model based approach to target matching that utilizes local features of the observed target signature has been adopted. This approach employs the Generalized Hough Transform (GHT). The GHT algorithm encodes the target contour into a prestored table (the R-table) with distance and the angle to a selected reference point used as indices. For each sample image, the R-table features of a measured contour are computed. The sorting process proceeds by matching computed R-features with prestored points associated with each view. The data-to-model matching calculation yields an estimate of the model reference point position and the rotation. The algorithm accomplishes the match by "voting" for a particular value of the position and orientation of the target. The votes are accumulated as a three dimensional histogram in what is called P-space. Measured range is used to scale the match.

The votes resulting from ambiguous data tend to spread out in an amorphous region of P-space. The correct data-to-model matchings tend be more concentrated. From the histogram the position and orientation of the target reference point can be estimated.

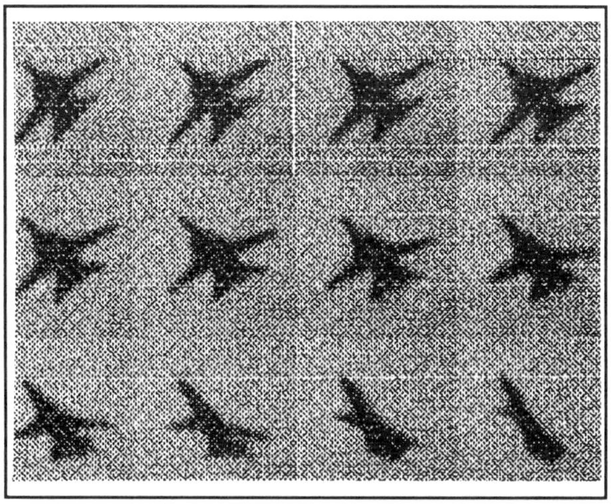

To evaluate the feasibility of estimating the orientation of a moving target, an experiment on actual data set was taken from the IMPRINT (Image Processing for the Identification of Non-Cooperative Targets) data base. Hughes Aircraft Company under the sponsorship of the Naval Air Development Center in Warminster, Pa., collected video imagery of military aircraft and evaluated the usefulness of a variety of pattern recognition techniques for target identification. For a subset of the IMPRINT imagery, target orientation coordinates were recorded from an on-board inertial navigation system (INS). These were used to evaluate the errors in the estimated target orientation obtained with the GHT. Figure 2 shows the sequence of imagery used in this experiment, in which an F-14 is undergoing a rolling maneuver.

The "truth" model was obtained using plastic scale models of the anticipated targets mounted on a mechanism that included computer controlled gimbals such that the aircraft was able to be rotated about its yaw and roll axes. The quantization of both the yaw and roll axes of the model imagery was 30 degrees in this experiment. Using the best fit to the model in yaw and roll the orientation of the longitudinal axis of the aircraft can be estimated along with aircraft type. To obtain an estimate of the accuracy of the classification procedure, the angle between the unit vectors obtained from the GHT and from the INS ground truth was calculated for each video frame. It was found that the mean difference in the angles was 16.6 degrees with a standard deviation of 8.7 degrees.

3. Image-Based Models

To integrate the image information into an estimation algorithm like that of the EKF, the encounter model must be completed. There are two accelerations influencing the target path. Because the turn rate tends to be nearly constant over intervals with sudden changes at unpredictable times, a useful model is created by partitioning the range of turn rates into discrete levels. Let $\{\alpha_t\}$ be the maneuver indicator process; $\alpha_t=e_i$ if $\omega_\psi=a_i$ at time t. The target is described by a family of K stochastic differential equations indexed by the maneuver acceleration. The most common way to quantify the dynamics of the maneuver process is to suppose the successive maneuver modes are represented by a Markov chain with transition rate matrix q.

Before analyzing a tracker with the full sensor suite, consider an architecture which has no image path; i.e., suppose the tracker avails itself of just the lower path in Figure 1 using a range-bearing measurement. These measurements will be taken at a fixed rate (10 samples/sec.) with independent errors. A 4-dimensional filter is sufficient to generate the state estimate (position and velocity). The filter gain depends parametrically on the intensity of the exogenous disturbance, and the filter will be denoted by EKF. A slight increase in model complexity gives a more realistic motion equation. Rather than neglecting the turn rate, it can be replaced with a Gaussian surrogate. The turn rate shaping filter is a two parameter family, with τ the time constant, and $R(0)=E\{(\omega_\psi)^2\}$. The resulting estimator will be called EKF_S.

In the dual-path tracking architecture shown in Figure 1, the sensor suite contains an imager in addition to the conventional range-bearing sensors. As pointed out earlier, the imager is a feature matching block, and its errors are not well described by the additive Gaussian noise. To create a more realistic model of the orientation classifier, suppose that the imager collects data at a rate of λ frames/sec. and places each target image into one of L equally spaced orientation bins. The output of the processor can be written as an L-dimensional counting process $\{\sigma_t\}$, the ith component of which is the number of times the target has been placed in bin i on the interval [0,t]. This sequence of counts (or symbols) can be interpreted by a temporal processor to give the relative likelihoods of the various turn rate hypotheses. This development is carried out in [1]. The quality of the imager is determined by the L×L-discernibility matrix $Þ=[Þ_{ij}]$ where $Þ_{ij}$ is the probability that bin i will be selected by the processor if bin j contains the true target orientation at time of image creation; i.e., (see [2])

The fidelity of image interpretation is a function of three things: the frame rate (λ), the ability of the image to

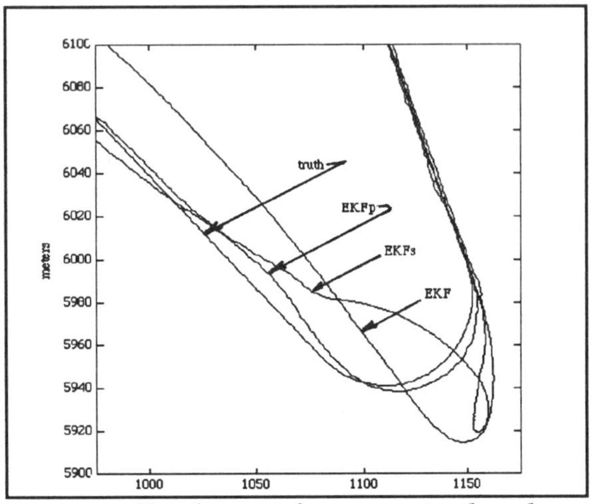

Figure 3: Tracker performance in the plane.

correctly classify a single image (Þ), and the tempo of the encounter (q). In [3], an equation for the conditional probabilities of the turn rate was derived, and used to modify an EKF. This filter will be called EKF_P. To illustrate the contrasts in performance, consider a specific vignette with a target initially located at (x_0,y_0)= (6.4, 1.0)KM, and initial velocity (v_{x0},v_{y0})= (5.0,-13.3) meters/-sec. Moving at constant velocity for $t\in[0, 30)$ sec., a 0.5 g turn is made for $t\in[30,38)$, after which the target returns to constant velocity motion. The omnidirectional accelerations will be assumed to be slight; $(dw_x)^2=(dw_x)^2$ =0.01dt $(m/sec^2)^2$. Figure 3 shows the target path (truth) without the wide band acceleration. Tracking will be accomplished with a 10 samples/sec. range-bearing sensor located at the origin with Gaussian errors of standard deviation 5.0 meters and 0.25° respectively, independent in time and type. If an imager is available, it will be collocated, and make measurements at the same rate, placing the target in angular bins of width 22.5°. It will be supposed that the lateral accelerations in the maneuver model are given by {a_1=20°/sec. turn right, a_2=no turn, a_3=20°/sec. turn left}, and that the chain {α_t} is symmetric about the coasting mode; i.e., the probabilistic character of a_t= 1 and 3 are identical. The elements of the q-matrix are determined jointly by the mean sojourn times in each acceleration mode, and the likelihood of alternative modal transitions.

Both the motion model and the observation equation are nonlinear. Furthermore, if the vignette is partitioned into pre-, post-, and maneuver coincident segments, the relative performance of the trackers differs as a function of the portion of the path that is being reviewed. A good indication of performance can, however, be determined by simulation. Figure 3 shows the position response of selected estimators. To display tracker biases, a 20-trial mean sample path is shown rather than a (noisier) single sample path. All of the filters perform well in the premaneuver phase, but both EKF and EKF_S have difficulty following a turn; the latter making a pirouette at apogee. Only EKF_P performs satisfactorily in the postmaneuver initiation phase of the engagement.

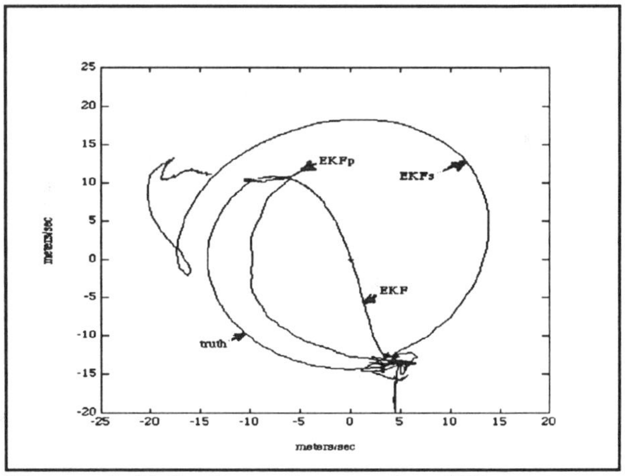

Figure 4: Velocity estimates for different trackers.

Figure 4 shows sample-mean velocity profiles, and the nonimaging trackers are not good. The reason for the clearly degraded velocity estimates is the fact that velocity is a slack variable in the estimation algorithm. It is included in the model, and the EKFs estimate its value concurrently with estimates of the location variables, but there is no direct velocity measurement. The filter tends to assign observation residuals to slack variables to a much greater degree than one might expect. This is true to some extent during the nonmaneuvering phase, (ω_ψ=0), where the model underlying the EKF is valid, but the effect is magnified during a maneuver. In this phase of the scenario, there is conspicuous misidentification of the velocity by both EKF and EKF_S, with neither recognizing the correct sense of rotation; the pirouette yields a velocity profile that is the mirror image of the actual path. This poor performance in velocity is magnified by prediction algorithms, the simplest of which extrapolate to the predicted position along the direction of the velocity estimate. The small velocity errors in EKF_P portend much improved prediction fidelity.

4. Performance Sensitivities

As noted earlier, the fidelity of the upper link in Figure 1 depends on the number of bins, the processor fidelity and the frame rate. Although L and λ are fixed by design, Þ changes during an encounter. The discernibility matrix is clearly dependant on range because target extent changes as a function of the square of range. A target which subtends few pixels offers little spatial structure to the imager, and the imager will tend to classify orientation arbitrarily; i.e., σ_t will tend to be uniformly distributed.

As the target approaches the imager, and as more target pixels become available, silhouette projection errors come to the fore. The target silhouette is compatible with two distinct orientation angles, thus creating an equivocal situation for the processor. When the number of pixels is large enough, this projection ambiguity can be resolved from an analysis of the internal features of the target. For the encounter discussed in the previous section, such errors had little influence on performance. This was due to two special aspects of the target path. The initial estimate of orientation was correct, and the kinematics were sufficiently simple to prevent subsequent confusion regarding the projection of the image. Further, the geometry of the path was such as to avoid having the target change its turn while broadside to the sensor. Such situations create maximum projection effect.

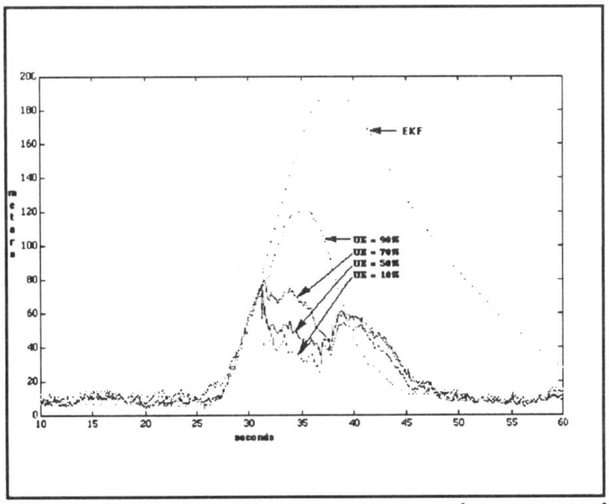

Figure 5: Estimation error as a function of discernibility.

Nearest neighbor errors are important in certain circumstances, but were not studied in detail here. At long ranges, uniform errors are dominant, while at short ranges, projection errors become so. The region in P-space most closely associated with nearest neighbor errors is in an intermediate zone. Note however that nearest neighbor errors are very important since a move to a adjacent bin is the initial signature of a maneuver. False alarm rates are sensitive to the probability of landing in the neighboring bin.

Figure 5 shows the mean 5-second prediction errors on this scenario for different levels of uniform error. For a sample of size ten, the degradation of performance is apparent, but even when the probability of error is 70% the predictor performs well. Above this level, performance falls off quickly, and the hybrid filter/predictor reduces to the EKF (also shown) when the uniform error reaches 100%.

5. Conclusions

New sensors require novel architectures to capture their potential for performance improvement. It has been shown that the response of an image-enhanced tracking loop is superior to that attainable from conventional trackers without imaging capability. From this study, it is evident that an image enhanced tracker can achieve good quiescent performance without a severe penalty during turn. This exchange is difficult to make with conventional algorithms because they tend to balance improvements in one part of the encounter with degradation elsewhere.

Acknowledgements

This research was partially supported by a grant from the Hughes Aircraft Co., and by the MICRO Program of the State of California under Project No. 91-156.

References

1. D.D. Sworder and R.G. Hutchins, "Maneuver Estimation Using Measurements of Orientation," *IEEE Trans. on Aerospace and Electronic Systems*, Vol. AES-26, 5 (1990) 625-638.

2. D.D Sworder and R.G. Hutchins, "Image-Enhanced Tracking," *IEEE Trans.on Aerospace and Electronic Systems*, Vol. AES-25, 5, September 1989, 701-710.

3. R.G. Hutchins and D.D. Sworder, "Image Fusion Algorithms for Tracking Maneuvering Targets," *Journal of Guidance, Control and Dynamics*, 1991 (to appear).

A Non-Markov Finite Dimensional Filter

Robert J. Elliott*
Department of Statistics
and Applied Probability
University of Alberta
Edmonton, Alberta T6G 2G1

David D. Sworder
Department of Applied Mechanics
and Engineering Science
University of California, San Diego
La Jolla, California 92093-0411

Thomas J. Taylor
Department of Mathematics
Arizona State University
Tempe, Arizona 85287

Abstract

The state space of a general, not necessarily Markov, finite state space process is identified with the set of unit vectors in a Euclidean space. A filtering problem is considered where the observation process records only jumps between certain subsets of the states. A recursive equation for the filtered estimate of the state is obtained. If the transition rates are measurable with respect to the observations this filter is finite dimensional. The corresponding Zakai equation is also derived.

1 Introduction

The situation and results presented below generalize those in [1]. Consider a process S_t, $t \geq 0$, with right-constant sample paths defined on a probability space (Ω, F, P). Its state space is an arbitrary finite set $\widehat{S} = \{s_1, \ldots, s_N\}$. By considering the functions $\phi_k(s_i)$, defined so that $\phi_k(s_i) = 0$ if $i \neq k$ and $\phi_k(s_k) = 1$, and writing $X_t := (\phi_1(S_t), \ldots, \phi_N(S_t))$ we see that equivalently we can consider a process X_t, $t \geq 0$, whose state space is the set $S = \{e_1, \ldots, e_N\}$ of unit (column) vectors $e_i = (0, \ldots, 1, 0, \ldots, 0)^*$ of R^N. (Here * denotes transpose.) $\{F_t\}$ will be the right continuous complete filtration generated by X.

Suppose that in any finite time interval X_t has, almost surely, only finitely many jumps, (this is implied by the boundedness of a_{ii} defined below), and write $T_k(\omega)$, $k \in N$, for the k-th jump time. $\delta_{T_k(\omega)}(dr)$ will denote the unit mass at time $T_k(\omega)$ and, if $X_{T_k(\omega)} = e_{i_k(\omega)}$, write $\delta_{i_k(\omega)}(i)$ for the function which is 1 if $i = i_k(\omega)$ and 0 otherwise. X_t is a multivariate jump process and we can write

$$\begin{aligned} X_t &= X_0 + \sum_{0<r\leq t} \Delta X_r, \text{ where } \Delta X_r = X_r - X_{r-} \\ &= X_0 + \int_0^t \sum_i (e_i - X_{r-}) \\ &\quad \times \Big(\sum_k \delta_{T_k(\omega)}(dr) \delta_{i_k(\omega)}(i) \Big). \end{aligned} \tag{1.1}$$

*Research partially supported by NSERC Grant A-7964.

The random measure $\sum_k \delta_{T_k(\omega)}(dr)\delta_{i_k(\omega)}(i)$ picks out the jump size and the jump times and, following Jacod [2], has a predictable compensator $\nu^p(dr, e_i)$. In turn, ν^p factors into its Lévy system:

$$\nu^p(dr, e_i) = \lambda(e_i, X_{r-}, r, \omega) d\Lambda(r, X_{r-}, \omega).$$

Roughly speaking, $d\Lambda(r, X_{r-}, \omega)$ is the conditional probability that the next jump occurs at time r given the previous history of the process. (In the Markov case the conditioning is only on the event that the process is in state X_{r-} just before r.) $\lambda(e_i, X_{r-}, r, \omega)$, defined for $e_i \neq X_{r-}$, is the conditional probability that the process jumps from X_{r-} to $X_r = e_i$, given there is a jump at time r and given the previous history of the process.

$d\Lambda$ is a non-negative random measure on R^+. For simplicity we shall suppose it is absolutely continuous with respect to Lebesgue measure, so that there is a predictable Radon–Nikodym derivative $a(r, X_{r-}, \omega)$ such that

$$d\Lambda(r, X_{r-}, \omega) = a(r, X_{r-}, \omega) dr.$$

Writing $a(e_i, X_{r-}, r, \omega) = \lambda(e_i, X_{r-}, r, \omega) a(r, X_{r-}, \omega)$ for $e_i \neq X_{r-}$ we see

$$\sum_{e_i \neq X_{r-}} a(e_i, X_{r-}, r, \omega) = a(r, X_{r-}, \omega).$$

If we define a matrix $A(r, \omega) = (a_{ij}(r, \omega))$ by putting

$$\begin{aligned} a_{ij}(r,\omega) &= a(e_j, e_i, r, \omega) \quad \text{if } i \neq j \\ \text{and} \quad a_{ii}(r,\omega) &= -a(r, e_i, \omega), \quad 1 \leq i \leq N, \end{aligned}$$

then, (with obvious notation),

$$\sum_i (e_i - X_{r-}) a_{iX_{r-}} = A(r,\omega) X_{r-}$$

and the representation (1.1) can be written

$$\begin{aligned} X_t &= X_0 = \int_0^t \sum_i (e_i - X_{r-}) \\ &\quad \times \Big(\sum_k \delta_{T_k(\omega)}(dr) \delta_{i_k(\omega)}(i) - a_{iX_{r-}} dr \Big) \\ &\quad + \int_0^t A(r,\omega) X_{r-} dr. \qquad (1.2) \end{aligned}$$

Note X is a Markov process if and only if the elements of A are deterministic.

The decomposition (1.2) expresses X as the sum of a martingale M and a predictable process of integrable variation. The martingale M is

$$\begin{aligned} M_t &= \int_0^t \sum_i (e_i - X_{r-}) \\ &\quad \times \Big(\sum_k \delta_{T_k(\omega)}(dr) \delta_{i_k(\omega)}(i) - a_{iX_{r-}} dr \Big) \\ &= X_t - X_0 - \int_0^t A(r,\omega) X_{r-} dr \qquad (1.3) \\ &= X_t - X_0 - \int_0^t A(r,\omega) X_r dr, \qquad (1.4) \end{aligned}$$

because for almost all ω $X_r(\omega) = X_{r-}(\omega)$ except for countably many r. (This observation will be used to equate similar integrals below.)

Suppose $x, y \in R^N$; the scalar product will be denoted by $x \cdot y$.

For unit vectors $e_i, e_j \in S$, $i \neq j$, consider the stochastic integral

$$M_t^{ij} = \int_0^t (e_i \cdot X_{r-})(e_j \cdot dM_r). \qquad (1.5)$$

Note the integrand is predictable, so M^{ij} is a martingale. Now

$$\begin{aligned} (e_i \cdot X_{r-})(e_j \cdot dX_r) &= (e_i \cdot X_{r-})(e_j \cdot \Delta X_r) \\ &= (e_i \cdot X_{r-})(e_j \cdot (X_r - X_{r-})) \\ &= I(X_{r-} = e_i,\ X_r = e_j) \ (1.6) \end{aligned}$$

Substituting (1.3) in (1.5)

$$\begin{aligned} M_t^{ij} &= \int_0^t (e_i \cdot X_{r-})(e_j \cdot dX_r) \\ &\quad - \int_0^t (e_i \cdot X_{r-})(e_j \cdot A(r,\omega) X_{r-}) dr. \end{aligned}$$

Write $N_t(i,j)$ for the number of jumps of process X from e_i to e_j up to time t. Then using (1.6)

$$\begin{aligned} M_t^{ij} &= N_t(i,j) - \int_0^t I(X_{r-} = e_i) a_{ji}(r,\omega) dr \\ &= N_t(i,j) - \int_0^t I(X_r = e_i) a_{ji}(r,\omega) dr, \end{aligned}$$

(again because $X_r = X_{r-}$ a.s. except for countably many r). Therefore we can write

$$N_t(i,j) = \int_0^t I(X_r = e_i) a_{ji}(r,\omega) dr + M_t^{ij}. \qquad (1.7)$$

2 The Observation Process

Suppose $\{1, 2, \ldots, N\} = A(1) \cup A(2) \cup \ldots \cup A(d)$ where $A(k) \cap A(\ell) = \emptyset$ if $k \neq \ell$. For set $A(k)$ write $I(A(k))$ for the vector $\sum_{i \in A(k)} e_i$. For $k, \ell \in \{1, \ldots, d\}$, $k \neq \ell$, define

$$Y_t^{k\ell} = \sum_{(i,j) \in A(k) \times A(\ell)} N_t^{ij}$$

so from (1.6)

$$Y_t^{k\ell} = \int_0^t (I(A(k)) \cdot X_{r-})(I(A(\ell)) \cdot dX_r).$$

Substituting from (1.3) we obtain, similarly to (1.7),

$$Y_t^{k\ell} = \int_0^t h^{k\ell}(r) dr + Q_t^{k\ell}$$

where

$$\begin{aligned} h^{k\ell}(r) &= \sum_{(i,j) \in A(k) \times A(\ell)} I(X_r = e_i) a_{ji}(r,\omega) \\ \text{and} \quad Q_t^{k\ell} &= \sum_{(i,j) \in A(k) \times A(\ell)} M_t^{ij}. \end{aligned}$$

The *Observation Process* will be a set of processes of the form $Y_t^{k\ell}$. That is, we suppose there is a collection of pairs (k_1, ℓ_1), (k_2, ℓ_2), $\ldots$, (k_p, ℓ_p) with k_i, $\ell_i \in \{1, \ldots, d\}$, $k_i \neq \ell_i$, and $(k_i, \ell_i) \neq (k_j, \ell_j)$ for $i \neq j$. For simplicity we shall write Y^j for $Y^{k_j \ell_j}$, etc., so the observation process is

$$\begin{aligned} Y_t &:= (Y_t^1, Y_t^2, \ldots, Y_t^p)^*, \\ \text{where} \quad Y_t^j &= \int_0^t h^j(r) dr + Q_t^j. \qquad (2.1) \end{aligned}$$

Note that if $i \neq j$ Y^i jumps at different times to Y^j, so the martingales Q^i and Q^j are orthogonal.

Write $\{\mathcal{Y}_t\}$ for the right continuous, complete filtration generated by Y.

If $\{\phi_t\}$, $t \geq 0$, is a process the $\mathcal{Y}$-optional projection of ϕ will be denoted by $\hat{\phi}$, so that for all $t \geq 0$, $\hat{\phi}_t = E[\phi_t \mid \mathcal{Y}_t]$ a.s. Our objective is to obtain a recursive equation for $\widehat{X}_t = E[X_t \mid \mathcal{Y}_t]$.

Definition 2.1 *The* Innovation Process *is*

$$\widetilde{Q}_t := (\widetilde{Q}_t^1, \ldots, \widetilde{Q}_t^p)^*$$

where

$$\widetilde{Q}_t^j = Y_t^j - \int_0^t \hat{h}^j(r)dr.$$

Simple arguments, using Fubini's theorem, show that $\widetilde{Q}^j$ is a $\mathcal{Y}$- martingale. We can, therefore, write

$$Y_t^j = \int_0^t \hat{h}^j(r)dr + \widetilde{Q}_t^j. \tag{2.2}$$

Similar calculations, again using Fubini's theorem and (1.4), show that the process

$$\widetilde{M}_t := \widehat{X}_t - \widehat{X}_0 - \int_0^t \widehat{A(r,\omega)X_r}dr$$

is an R^N-valued martingale with respect to the $\mathcal{Y}$-filtration. Because $\mathcal{Y}_0$ is the trivial σ-field, $\{\Omega, \phi\}$, $\widehat{X}_0$ is a constant vector, the initial distribution of X.

From the representation result for martingales with respect to a multivariate jump process $\widetilde{M}$ can be represented as

$$\begin{aligned}\widetilde{M}_t &= \int_0^t \gamma_r d\widetilde{Q}_r \\ &= \sum_{j=1}^p \int_0^t \gamma_r^j d\widetilde{Q}_r^j.\end{aligned}$$

Here γ is a $\mathcal{Y}$-predictable $N \times p$ matrix valued process with columns

$$\gamma_r^j = (\gamma_r^{1j}, \ldots, \gamma_r^{Nj})^*, \quad j = 1, \ldots, p.$$

We, therefore, have

$$\widehat{X}_t = \widehat{X}_0 + \int_0^t \widehat{A(r,\omega)X_r}dr + \int_0^t \gamma_r d\widetilde{Q}_r. \tag{2.3}$$

Theorem 2.2

$$\begin{aligned}\gamma_r^j \;=\; & \hat{h}^j(r)^{-1} I_{\hat{h}^j(r)\neq 0}\Big\{ \sum_{(\sigma,\rho)\in A(k_j)\times A(\ell_j)} \\ & \times \widehat{(e_\sigma \cdot X_r)a_{\rho\sigma}(r,\omega)} \cdot e_\rho - \hat{h}^j(r) \cdot \widehat{X}_r\Big\}.\end{aligned}$$

Proof. The filtering problem will be solved if we determine the elements of γ. This we do by calculating $\widehat{X}_t \otimes Y_t$ two ways; in fact we calculate the j-th column $\widehat{X}_t \cdot Y_t^j$ in two ways. (For $x = (x_1, \ldots, x_N)^* \in R^N$ and $y = (y_1, \ldots, y_p)^* \in R^p$ the tensor product $x \otimes y$ can be thought of as the $N \times p$ matrix with elements $x_i y_j$; consequently the j-th column of $x \otimes y$ is $(x_1 y_j, x_2 y_j, \ldots, x_N y_j)^* = x \cdot y_j$.) Now

$$\begin{aligned}X_t \cdot Y_t^j &= \int_0^t X_{r-} \cdot dY_r^j \\ &\quad + \int_0^t dX_r \cdot Y_{r-}^j + \sum_{0<r\leq t} \Delta X_r \cdot \Delta Y_r^j\end{aligned}$$

and from (1.4) this is

$$\begin{aligned}&= \int_0^t X_{r-} \cdot dY_r^j + \int_0^t A(r,\omega)X_r \cdot Y_{r-}dr \\ &\quad + \int_0^t dM_r \cdot Y_{r-} + \sum_{0<r\leq t} (X_r - X_{r-})\Delta Y_r^j \quad (2.4)\end{aligned}$$

Note $\sum_{0<r\leq t} X_{r-}\Delta Y_r^j = \int_0^t X_{r-}dY_r^j$. Also $\Delta Y_r^j = (I(A(k_j)) \cdot X_{r-})(I(A(\ell_j)) \cdot X_r)$ and $X_r \cdot (e_\rho \cdot X_r) = e_\rho(e_\rho \cdot X_r)$, $1 \leq \rho \leq N$, so $X_r \cdot \Delta Y_r^j = \sum_{\rho\in A(\ell_j)}(I(A(k_j)) \cdot X_{r-})(e_\rho \cdot X_r)e_\rho$. Therefore,

$$\begin{aligned}&\sum_{0<r\leq t} \Delta X_r \cdot \Delta Y_r^j \\ &= \sum_{0<r\leq t} (X_r - X_{r-}) \cdot \Delta Y_r^j \\ &= \sum_{\rho\in A(\ell_j)} (I(A(k_j)) \cdot X_{r-})(e_\rho \cdot X_r)e_\rho \\ &\quad - \int_0^t X_{r-} \cdot dY_r^j \\ &= \sum_{(\sigma,\rho)\in A(k_j)\times A(\ell_j)} \int_0^t (e_\sigma \cdot X_r)a_{\rho\sigma}(r,\omega)dr \cdot e_\rho \\ &\quad + \sum_{(\sigma,\rho)\in A(k_j)\times A(\ell_j)} Q_t^{\sigma\rho} \cdot e_\rho - \int_0^t X_{r-} \cdot dY_r^j.\end{aligned}$$

Substituting in (2.4) we see

$$\begin{aligned}X_t \cdot Y_t^j &= \int_0^t A(r,\omega)X_r \cdot Y_{r-}^j dr + \int_0^t dM_r \cdot Y_{r-} \\ &+ \sum_{(\sigma,\rho)\in A(k_j)\times A(e_j)} \int_0^t (e_\sigma \cdot X_r)a_{\rho\sigma}(r,\omega)dr \cdot e_\rho \\ &+ \sum_{(\sigma,\rho)\in A(k_j)\times A(\ell_j)} Q_t^{\sigma\rho} \cdot e_\rho. \end{aligned} \tag{2.5}$$

Taking the $\mathcal{Y}$-optional projection of each side of (2.5) and using the fact that

$$\widehat{A(r,\omega)X_r \cdot Y_{r-}^j} = \widehat{A(r,\omega)X_r} \cdot Y_{r-}^j$$

we have

$$\widehat{X}_t \cdot Y_t^j = \int_0^t \widehat{A(r,\omega)X_r \cdot Y_{r-}^j}\, dr + \sum_{(\sigma,\rho)\in A(k_j)\times A(\ell_j)}$$
$$\times \int_0^t \widehat{(e_\sigma \cdot X_r) a_{\rho\sigma}(r,\omega)} \cdot dr e_\rho + H_t^1 \qquad (2.6)$$

where H^1 is a square integrable $\mathcal{Y}$-martingale. However, from (2.2) and (2.3) we also have

$$\begin{aligned}\widehat{X}_t \cdot Y_t^j &= \int_0^t \widehat{X}_{r-} dY_r^j + \int_0^t d\widehat{X}_r \cdot Y_{r-}^j \\ &+ \sum_{0<r\le t} \Delta\widehat{X}_r \cdot \Delta Y_r^j \\ &= \int_0^t \widehat{X}_{r-}\hat{h}^j(r)dr + \int_0^t \widehat{X}_{r-} d\widetilde{Q}_r^j \\ &+ \int_0^t \widehat{A(r,\omega)X_r} \cdot Y_{r-}^j dr + \int_0^t \gamma_r Y_{r-}^j d\widetilde{Q}_r \\ &+ \sum_{0<r\le t} \gamma_r^j \cdot \Delta Y_r^j\end{aligned}$$

because the martingales $\widetilde{Q}^i$ and $\widetilde{Q}^j$ jump at different times if $i \neq j$. That is

$$\begin{aligned}\widehat{X}_t \cdot Y_t^j &= \int_0^t \widehat{X}_r \cdot \hat{h}^j(r) dr + \int_0^t \widehat{A(r,\omega)X_r} \cdot Y_{r-}^j dr \\ &+ \int_0^t \gamma_r^j \hat{h}^j(r) dr + H_t^2, \qquad (2.7)\end{aligned}$$

where H^2 is a square integrable $\mathcal{Y}$-martingale. Now $\widehat{X}_t \cdot Y_t^j$ is a special semimartingale so the decompositions (2.6) and (2.7), into the sum of a martingale and a predictable bounded variation process, must be the same. Therefore, equating the bounded variation processes we see

$$\gamma_r^j = \hat{h}^j(r)^{-1} \cdot I_{\hat{h}^j(r)\neq 0}\Big\{ \sum_{(\sigma,\ell)\in A(k_j)\times A(\ell_j)}$$
$$\times \widehat{(e_\sigma \cdot X_r) a_{\rho\sigma}(r,\omega)} e_\rho - \hat{h}^j(r)\widehat{X}_r \Big\}.$$

Corollary 2.3 *If the entries $a_{ij}(t,\omega)$ of the matrix $A(t,\omega)$ are adapted to the $\mathcal{Y}$-filtration then*

$$\widehat{(e_\sigma \cdot X_r) a_{\rho\sigma}(r,\omega)} = (e_\sigma \cdot \widehat{X}_r) a_{\rho\sigma}(r,\omega)$$

so

$$\hat{h}^j(r) = \sum_{(\sigma,\rho)\in A(k_j)\times A(\ell_j)} (e_\sigma \cdot \widehat{X}_r) a_{\rho\sigma}(r,\omega)$$

and $\widehat{X}_r$ is given by a finite dimensional filter. That is

$$\widehat{X}_t = \widehat{X}_0 + \int_0^t A(r,\omega)\widehat{X}_r dr$$
$$+ \int_0^t \sum_{j=1}^p \hat{h}^j(r)^{-1} I_{\hat{h}^j(r)\neq 0} \Big\{ \sum_{(\sigma,\rho)\in A(k_j)\times A(\ell_j)}$$
$$\times (e_\sigma \cdot \widehat{X}_r) a_{\rho\sigma}(r,\omega) e_\rho - \hat{h}^j(r)\widehat{X}_r \Big\} d\widetilde{Q}_r^j. \qquad (2.8)$$

3 An Application

To illustrate the utility of Corollary 2.3, consider a problem arising when tracking an agile target. The details of the tracker design are discussed in [3], but can be summarized as follows. An image-enhanced tracker uses a sequence of "pictures" of the target to augment center-of-reflection measurements (say range-bearing) in an estimation algorithm. An imager is particularly useful because target orientation changes before there is a discernable path deviation, and target orientation is directly obtainable from image analysis. An estimate of the magnitude and direction of the turn can be used to adjust the parameters of an extended Kalman filter (EKF), and the direction of the target's velocity vector can be used for predicting future location.

To see how this is done, suppose the set of possible maneuvers is partitioned into K levels; the turn rate process $\{\omega_t\}$ satisfies $\omega_t \in \{a_1, \ldots, a_K\}$. For notational convenience, let $\{\alpha_t\}$ be the maneuver indicator process, $\alpha_t = e_i \in \mathcal{R}^K$ if the turn rate at time t is in the i-th bin, and let $\{1, \ldots, K\} = K$.[1] The target motion model is given by a family of stochastic differential equations indexed by the current maneuver acceleration. The simplest way to delineate the maneuver dynamics is to suppose the successive maneuver modes are represented by an F-Markov chain with transition rate matrix q;

$$\mathcal{P}(\alpha_{t+\varepsilon} = e_j \mid \alpha_t = e_i) = \delta_{ij} + q_{ij}\varepsilon + o(\varepsilon) \qquad i,j,\in K \qquad (3.1)$$

This is a tractable and oft used model, the parameters of which can be expressed in terms of the mean duration of each maneuver mode, and the transition probabilities between modes. Equation (3.1) has been proposed as a description the temporal dynamics of pilot induced acceleration [4].

The image processor classifies a given image in one of L angular bins. The output of the image processor can be written as an L-dimensional counting process $\{\sigma_t\}$, the i-th component of which is the number of times on $[0,t]$ that the target has been placed in bin i; i.e., $\{\sigma_{it}\}$ simply counts the number of times the image processor has generated the symbol corresponding to the i-th angular orientation. The quality of the imager is determined by the frame rate λ and the $L \times L$-discernibility matrix $\Phi = [\Phi_{ij}]$ where Φ_{ij} is the probability that symbol i will be generated by the processor if j is the target orientation at time of image creation. Let the orientation indicator process be given by $\{\rho_t\}$; $\rho_t = e_i \in \mathcal{R}^L$ if the velocity vector is in the i-th angular bin. Then Φ can be written:

$$\Phi_{ij} = \mathcal{P}(\Delta\sigma_i \neq 0 \mid \rho = e_j \text{ and } \Delta\sigma \neq 0). \qquad (3.2)$$

[1] The dimension of the unit vector will correspond to the applications.

Equation (3.2) yields the likelihood of correctly inferring the true orientation from a single image frame. The tracker actually uses a sequence of counts (or symbols) in a temporal processor to give the conditional probabilities of the various turn rate hypotheses. To develop the estimator note that the composite status of the target is given by $X_t = \{\alpha_t \otimes \rho_t\}$, the Kronecker product of maneuver and orientation. The components of the vector X_t can be written as a pair if it is important to distinguish its progenitors; i.e., $X_{(ij)} = \alpha_i \rho_j$. The aggregation implicit in $\{\alpha_t\}$ leads to uncertainty regarding the evolution of the angular process even if the turn rate were known. To account for this, assume that if α_t were e_i, $\{\rho_t\}$ would be a Markov process with transition matrix $\{Q^i, i \in K\}$. Define the $KL \times KL$ matrix $A' = (q \otimes \mathbf{I}_L) + Q^*$ where the block diagonal matrix $Q^* = \text{diag}(Q^1, \ldots, Q^K)$. Then $\{X_t\}$ is a Markov process with generator A.

From X_t, both $\alpha_t = (\mathbf{I}_K \otimes \mathbf{1}'_L)X_t$ (or $\sum_{j \in L} X_{(ij)}$) and $\rho_t = (\mathbf{1}'_K \otimes \mathbf{I}_L)X_t$ (or $\sum_{i \in K} X_{(ij)}$) can be determined where $\mathbf{I}_K$ is a K dimensional identity matrix and $\mathbf{1}_K$ is a K-vector of ones. Let $\{\mathbf{Z}_t\}$ be the filtration generated by $\{\sigma_t\}$, and let $\{\hat{\alpha}_t\}$ be the vector of conditional probabilities of individual maneuver accelerations. It is shown in [5] that $\{\hat{\alpha}_t\}$ and $\{\hat{\rho}_t\}$ can be deduced from the solution to a nonlinear stochastic differential equation:

$$\hat{\alpha}_t = (\mathbf{I}_K \otimes \mathbf{1}'_L)\widehat{X}_t \tag{3.3}$$

$$\hat{\rho}_t = (\mathbf{1}'_K \otimes \mathbf{I}_L)\widehat{X}_t \tag{3.4}$$

subject to:

$$d\widehat{X}_t = A'\widehat{X}_t dt + (\text{diag}(\widehat{X}_t) - \widehat{X}_t\widehat{X}'_t)(\mathbf{1}'_K \otimes \mathbf{I}_L) \times \Lambda' \text{diag}(\Lambda(\mathbf{1}'_K \otimes \mathbf{I}'_L)\widehat{X}_t)^{-1} d\sigma_t \tag{3.5}$$

where the quality of the imager is delineated by the $L \times L$ matrix $\Lambda = \lambda\Phi$, and the maneuver dynamics are rendered in terms of the $KL \times KL$ matrix A.

The estimates generated by (3.3) and (3.4) are used to adjust the gain of the EKF, and also to give the magnitude and direction of the turn rate. There are, however, difficulties in implementing the estimator as written. First, the algorithm operates at the frame rate of the imager, i.e., (3.5) is updated λ times/sec., and the adjustments to the maneuver adaptive EKF must be made at this same rate. Secondly, the number of orientation bins influences the fidelity of the estimate of the direction of the velocity; a finer angular grid is potentially reflected in a sharper velocity estimate. Unfortunately, increasing L makes the algorithm more complicated (of higher dimension), and it increases the likelihood of imager errors ($\|\Phi - \mathbf{I}_L\|$ is larger). Broadening the bins would make image classification more accurate, albeit at a coarser level.

Simulation suggests that the most important information is produced when the processor detects a change in angular orientation; although the complete $\{\sigma_t\}$ process is used in (3.5), and it contributes to the detailed structure of $\{\widehat{X}_t\}$, the biggest increments in the location estimate occur when the observation passes from bin to bin. Suppose that a subpartition of the orientation grid is created by aggregating contiguous angular bins in groups of r; i.e., let $L = rd$ where r is a positive integer. Then d is the (coarse) dimension of the angular grid as seen by the imager, and L the (fine) dimension of the angular grid of the estimate. In the notation of Section 2, $KL = N$, the dimension of X_t. The partitioning of the observation space indicated in Section 2 is easily accomplished by summing the appropriate elements of X_t. For example, the first distinguishable orientation bin can be written as $A(1) = \sum_{i \in K, j \in R} X_{(ij)}$. The set $\mathbf{KL}$ is partitioned in $A(1) \cup A(2) \cup \ldots \cup A(d)$ where each of the $A(i)$ is a Kr-grouping associated with a contiguous angular states and an arbitrary maneuver.

Suppose the observation is restricted to transitions $\{A(i) \rightarrow A(j); i,j \in \mathbf{d}\}$; i.e., $\{Y_t\}$. Because of the coarseness of the classification, the errors in interpretation are smaller, and in some instances can be neglected. Further, the observation rate, and hence the frequency of adapting the EKF is much slower. The requisite estimation algorithm is given in (2.8). It is interesting to note that one simplification used in (3.5) cannot be carried forward to (2.8). Equation (3.5) is linear between observations with a nonlinear update, and this fact simplifies its solution. This reduced form is a result of the fact that the frame rate is constant, and as a consequence, the absence of a count in $\{\sigma_t\}$ conveys no relevant information. In contrast, absence of a transition is meaningful in (2.8), and the innovations process is the appropriate excitation.

References

[1] R.J. Elliott. "A partially observed control problem for Markov chains," *Applied Math. and Optimization*, (to appear).

[2] J. Jacod. *Calcul stochastique et problèmes de martingales*, Lecture Notes in Math 794. Springer–Verlag, Berlin-Heidelberg-New York, 1979.

[3] D.D. Sworder and R.G. Hutchins, "Improved Tracking of an Agile Target," *Journal of Guidance, Control and Dynamics*, 1991 (to appear).

[4] J.R. Cloutier, J.H. Evers and J.J. Feeley, "Assessment of Air-to-Air Missile Guidance and Control Technology," *IEEE Control Systems Magazine*, Oct. 1989, 27–34.

[5] D.D. Sworder and R.G. Hutchins, "Maneuver Estimation Using Measurements of Orientation," *IEEE Trans. on Aerospace and Electronic Systems*, Vol. AES-26, 5 (1990) 625–638.

Model Variety Analysis: Determining the Need for Knowledge-Based Signal Processing

Avi Weiss, Erkan Dorken, Hamid Nawab
ECS Department
Boston University

Abstract

To utilize a mathematically formulated signal processing algorithm such as the short-time Fourier transform (STFT) for signal understanding applications, it is important to establish an analytic framework for determining whether or not adjustment of the algorithm control parameters (window length, FFT length, etc.) is necessary to guarantee "undistorted" output for all signals in the application domain. If control parameter adjustment is found to be necessary, a signal processing algorithm is said to have a model variety problem with respect to the class of input signals in the application domain. The necessity for such parameter adjustments implies that knowledge-based techniques must be used to adjust the signal processing parameters. In this paper, we present a systematic framework for analytically determining whether a given signal processing algorithm has a model variety problem with respect to a particular application domain. The framework is illustrated for the STFT algorithm as applied to the problem of non-speech sound analysis. [1]

1 Introduction

The signal processing problems addressed in this paper arise in the context of computer-based interpretation of signals, an area often referred to as *signal understanding*. Systems for signal understanding take signals as input and typically provide assertions about the signal generating environment as their output. For example, a sound understanding system may take an acoustic signal as input and produce at its output an assertion such as "the telephone is ringing". The research reported in this paper is concerned with the integration of mathematically formulated signal processing algorithms into signal understanding systems.

It is generally desirable to design signal processing algorithms on the basis of a mathematically tractable solution to a mathematically formulated problem. The need for mathematical tractability leads to restrictions being placed on the mathematical structure assumed for the input signals and the signal processing output. For example, the Weiner filter for additive noise removal is mathematically derived by assuming statistical models for the desired signal and the additive noise, and the output is required to be the optimum least-squares estimate of the desired signal. Thus, the Weiner filter can be used for noise removal provided that its inputs satisfy the assumed statistical models and the optimum least-squares estimate of the desired signal is considered to be an adequate criterion for noise "removal".

[1] This work was partially supported by ROME Laboratory under contract F30602-91-C-0038

In some signal understanding applications, the following two requirements have a direct bearing on the design of signal processing algorithms:

Requirement 1: The need to process a tremendous variety of signal types due to the situation-dependent nature of the input. For example, a sound understanding system has to deal with inputs including narrowband, harmonic, "noise-like", impulsive and frequency modulated signals and combinations thereof.

Requirement 2: The need to change processing goals in a context-dependent way. For example the goal of a sound understanding system might be to respond to either the sounds of an infant or a ringing telephone and to ignore other sound sources. If an infant sound is detected, the system's goal might then switch to determining whether the infant is crying or choking.

For signal understanding systems with the above requirements, it is unrealistic to design a single mathematically derived signal processing algorithm that can be applied to all possible input signals to produce the desired output information for each input. There are too many types of input signals, each type best suited to a different kind of mathematical formulation. Additionally, the form of the desired output information has different mathematical formulations in accordance with the assumed nature of the input signal and the current goals of the signal understanding system. If mathematically derived signal processing is to be utilized, one possible approach is for the signal understanding system to have access to a "database" of mathematically derived algorithms. This "database" is indexed by the types of assumptions made about the input signal and the type of output information desired in accordance with the current goals of the signal understanding system. It therefore becomes necessary for the signal understanding system to be able to carry out a context-dependent *search* through the algorithm database.

1058-6393/91 $01.00 © 1991 IEEE

ample, algorithms for time-frequency analysis of signals differ from each other because of the assumptions they make about the spectral characteristics of the input signal and the manner in which these characteristics vary as a function of time. Therefore, from a high-level perspective, the structure of the different algorithms is the same, but at more detailed levels, there are differences. These differences, however, can often be represented in terms of the change of values of control parameters.[1] The algorithm search is thus transformed into a search for appropriate control parameter values for the "packaged" algorithm.

Suppose a particular input signal is to be processed by a signal processing algorithm with adjustable control parameters. In order to select appropriate values for the control parameters, the system must consider the current system goals as well as knowledge about certain characteristics of the particular input signal. This leads to the dilemma that choosing the appropriate control parameter values requires knowledge about the signal, but this knowledge can only be obtained by first processing the signal with an algorithm with appropriate control parameter settings. An iterative procedure [1] for converging to the appropriate control parameter values has been proposed on an empirical basis. The iterative procedure begins by using the best available guess for the control parameter values (in the worst case, the system may resort to an arbitrary assignment for the control parameter values). The input signal is then processed using the chosen control parameter values. The processing output is analyzed for evidence of inappropriate control parameter values. This analysis is based on the idea that if the chosen control parameters are not appropriate, the output does not produce meaningful assertions about the signal generating environment. For example, the processing output data may have characteristics incompatible with the physics of sound production. Alternatively, the output may exhibit characteristics that are inconsistent with interpretations from previous time frames. In such cases, the processing output data is said to be "distorted". A knowledge-based system then analyzes the nature of those distortions and hypothesizes new control parameters for the processing algorithm. The input signal is then re-processed using the newly hypothesized control parameter values. This iterative process continues until output data is obtained that is considered to be "undistorted" by the signal understanding system. The concepts of "distorted" and "undistorted" output, are formally defined and illustrated in this paper.

Without a rigorous definition for distorted output, it would not be possible to *analytically* determine whether or not an algorithm requires control parameter adjustments if the inputs are restricted to a particular class of signals. Clearly, if there exists a single assignment for the control parameters that ensures undistorted output for *all* possible input signals in an application domain, there would be no need for control parameter adjustment. If this is not the case, the algorithm is said to have a "model variety problem" with respect to that application domain. This terminology is based on the fact that varying the values of the control parameters of an algorithm corresponds conceptually to varying the mathematical model used to derive the algorithm. It is interesting to note that if an SPA has a model variety problem with respect to a class of signals S, then it also has a model variety problem with respect to any class of signals that contains the set S. A corollary of this observation is that in a practical application with a possibly open-ended class of signals, one can simply choose a mathematically tractable subset of possible input signals and show that the SPA has a model variety problem with respect to that subset. In this paper, a precise mathematical formulation of the model variety problem is constructed and illustrated through examples. This formulation permits the use of *analytical methods* in determining the need for knowledge-based signal processing, or alternatively, whether or not algorithm search is necessary to utilize mathematically derived signal processing algorithms in a signal understanding system for an application with a specified class of input signals.

[1] The FFT-length is an example of a control parameter value that distinguishes different FFT algorithms.

2 The "Undistorted Output" Concept

To gain insight into the notion of "undistorted output" of a signal processing algorithm (SPA), we present an illustrative example which involves short-time Fourier transform (STFT) processing of signals. The resulting insight will be used to develop an objective definition of "undistorted SPA output". The signal processing algorithm in this application is the short-time Fourier transform (STFT) [5]. The STFT algorithm is controlled by three parameters; the analysis window length, N_w, the number of FFT points to calculate, N_{fft}, and the decimation factor, L. The output of the STFT is used to determine the change in frequency content over time of a signal $x(n)$. In order for the STFT output to be "undistorted", locations of the maxima in output must correspond (within a given tolerance) to the frequencies present in the signal $x(n)$. The signal $x(n)$ may, for example, be generated from a linear combination of a sinusoid and a chirp,

$$x(n) = \sin(\phi_1(n)) + \sin(\phi_2(n))$$

where $\phi_1(n) = \omega_1 n^2/2$ and $\phi_2(n) = \omega_2 n$. Whether or not the STFT output is considered "undistorted" depends upon how well $\omega_1(n) = d(\phi_1(n))/dn$ and $\omega_2(n) = d(\phi_2(n))/dn$ can be estimated from the STFT output.

The STFT example presented above can be generalized in the following representation. The "undistorted output" concept can then be stated in terms of this representation.

Signal generation process: The signal $x(n)$ is the output of a hypothetical signal generating process (SGP), which has m parameters $q_1 \ldots q_m$.

Signal processing algorithm: A signal processing algorithm with algorithm control parameters

$p_1 \ldots p_r$ which processes $x(n)$ and produces an output.

Compatibility criteria: The SPA output is considered to be an "undistorted output" if the parameters $q_1 \ldots q_s$ (where $s \leq m$) of the SGP can be estimated to within a specified tolerance. Parameter-estimation procedure $f_i\{output[\xi_1 \ldots \xi_k]\}$ is defined to form the parameter estimate $\hat{q}_i$. Compatibility criteria are derived from the application domain and stipulate the tolerance which $(\hat{q}_i - q_i)^2$ must satisfy.

In the next section, the formalized concept of "undistorted output" is used to build an objective definition for the *partitioning* of the set S consisting of all possible input signals into two complementary sets, U and D, which respectively contain the signals for which a signal processing algorithm produces undistorted and distorted outputs.

3 Model Variety Analysis

As discussed in previous sections, to utilize mathematically formulated signal processing algorithms for signal understanding applications, it is important to establish an analytic procedure for determining whether or not adjustment of the algorithm control parameters is necessary to guarantee undistorted output for all signals in the application domain. In this section, the formal definition of "undistorted output" (as presented in the previous section) is used as the basis of analyzing mathematically formulated signal processing algorithms in order to determine whether they give rise to the model variety problem with respect to a specified class of input signals. The analysis framework is illustrated for the STFT algorithm as applied to the problem of non-speech sound analysis.

3.1 The Model Space

The objective definition of "undistorted output" from the previous section can now be used to formulate a set of conditions that, if satisfied by $input[n]$, guarantee that the SPA output will be "undistorted". The algorithm control parameters are variables in the condition statements (see section 3.3 for examples) and are the means through which the conditions can be varied. The condition statements and their variables comprise the *model-space* of the SPA. Setting the control parameters to specific values selects one SPA *model* from the model-space. The model selected from the model-space *partitions* the set S of all possible input signals in the application domain into two complementary subsets. One subset, U, contains the signals that satisfy the model conditions, for which the SPA produces "undistorted" output. The other subset, D, contains the signals that do not satisfy the model conditions, for which the SPA produces "distorted" output.

3.2 Model Variety Problem

The model variety problem can now be stated in terms of the model space of an SPA:

> *The model variety problem arises when there is no model in the model-space that* **is optimal in some sense** *for all possible SPA inputs in a particular application domain.*

In many SPA's, any particular control parameter often occurs as a variable in *more* than one condition in the SPA model (see section 3.3). In such cases, it is possible for the conditions to place conflicting requirements on the control parameter. Such conflicting requirements may give rise to the inherent model variety problem:

> *The inherent model variety problem arises when there is no model in the model-space* **that gives undistorted output** *for all possible inputs in a particular application domain.*

That is, there is no model in the model-space for which D, the set of input signals for which the algorithm produces distorted output, is the null set. In addition to ensuring "undistorted output" from the SPA, sometimes there is the concern of minimizing the computational cost in obtaining undistorted output. For example, consider a situation where undistorted output is attainable with a 512 point FFT as well as a 1024 point FFT. The 512 point FFT would be the preferred algorithm if computational efficiency is of concern, as it requires fewer arithmetic operations. In such applications, in addition to guaranteeing undistorted output, it is desired to achieve this undistorted output at minimum computational cost. For such applications, the efficiency model variety problem can be defined:

> *The efficiency model variety problem arises when there is no model in the model-space* **that gives undistorted output at minimum computational cost** *for all possible inputs in a particular application domain.*

That is, although there may exist a model in the model-space for which D is the null set for *all* possible input signals, that model is not the most efficient one for *each* input signal.

3.3 Sound Analysis Example

Signal processing techniques that have been incorporated in sound understanding applications are largely guided by the results from studies of human auditory processing. These studies indicate that human auditory processing involves the decomposition of the input sound signal into time-dependent spectral components [7]. In our example, the STFT is used to capture the time-dependent frequency content of the sound signals. A peak-picking algorithm is then applied to the STFT output to estimate the frequency locations of significant spectral components. A mathematically tractable subset (S) of possible sound signals is choosen to show that the STFT has a model variety problem with respect to the sound signals involved in the analysis. The signals in S consist of linear combinations of sinusoids and chirps. The chosen type of signals reflects the frequency-modulated narrowband characteristics of many sound signals. The frequency estimates are required to be within 10%

of their true values. Additionally, chirp frequency changes of $\frac{2\pi}{1000}$ radians or higher must be detected.

Signal Generation Process: The signal generation process generates the signal

$$x(n) = \sin \phi_1(n) + \sin \omega_2 n$$

where $\phi_1(n) = \omega_1 n^2/2$. The signal generation process has the parameters: $q_1(n) = d(\phi_1(n))/dn = \omega_1 n$ and $q_2 = \omega_2$.

Signal processing algorithm: The signal processing algorithm (SPA) is the STFT. Two parameter estimation procedures are defined: $f_1\{output[n,k]\}$ and $f_2\{output[n,k]\}$, which form parameter estimates $\hat{q}_1(n)$ and $\hat{q}_2$ respectively. The procedure f_1 determines the frequency location values of the two largest maxima in the STFT output and selects the lower frequency location value. The procedure f_2 determines the frequency location values of the two largest maxima in the STFT output and selects the higher frequency location value.

Compatibility criteria: The criteria for compatibility between the input and output are restated in terms of the parameters q_1, q_2, $\hat{q}_1$ and $\hat{q}_2$:

$$\left|\frac{\hat{q}_1(n) - q_1(n)}{q_1(n)}\right| \leq 0.1 \tag{1}$$

$$\left|\frac{\hat{q}_2 - q_2}{q_2}\right| \leq 0.1 \tag{2}$$

$$\textit{if } q_1(n) \leq q_2, \textit{ then } \hat{q}_1(n) \leq \hat{q}_2 \tag{3}$$

$$||\hat{q}_1(n) - \hat{q}_1(m)| - |q_1(n) - q_1(m)|| < \frac{2\pi}{1000} \tag{4}$$

Criteria 1 and 2 stipulate the requirement that the estimate of each sinusoid's frequency be within 10% of its true value. Criterion 3 stipulates the requirement that the estimate tolerances be chosen to ensure that the estimated frequencies of the sinusoid and the chirp maintain the order relationship of their actual frequencies. Criterion 4 stipulates the requirement that chirp frequency changes of $\frac{2\pi}{1000}$ radians or higher must be detected.

Derivation of model-space from compatibility criteria and SPA properties: The conditions which must be satisfied by the SPA input in order to give "undistorted output" can be derived from the STFT algorithm properties and the compatibility criteria listed above. These conditions and their sources are presented in table 1.

1. *Sufficient-data condition:* This condition imposes a lower bound on the window length, N_w. It is derived on the basis of ensuring that the two spectral peaks due to a single sinusoid may be unambiguously detected from the Fourier transform of a short-time section. A sinusoid of frequency $2\pi/N_s$ has one frequency component at $2\pi/N_s$ and the other at $-2\pi/N_s$. If the window length has the value N_w, each of these components can be assumed to have significant contributions up to $2\pi/N_w$ away from the actual frequency of the sinusoid. To ensure that the positive frequency component does not significantly interact with the negative frequency component, it is required that

$$\frac{2\pi}{N_w} < \frac{2\pi}{N_s} \implies N_w > N_s$$

and $N_s = 2\pi/\omega_s$, where ω_s is the frequency of the sinusoid. Clearly, the greatest lower bound for N_w is obtained for the smallest ω_s.

1.	$N_w > \min(\omega_1, \omega_2) > \frac{2\pi}{N_w}$
2.	$\|n - m\| \geq N_w$ where n and m are chosen to maximize $\|n - m\|$ and $\|\omega_1(n) - \omega_1(m)\| \leq \frac{2\pi}{1000}$
3.	$\|\omega_1 - \omega_2\| > \frac{2\pi}{N_w}$
4.	$(0.1)\min(\omega_1, \omega_2) > \frac{2\pi}{N_{fft}}$
5.	$\|\omega_1 - \omega_2\| > \frac{2\pi}{N_{fft}}$

Table 1: STFT model-space conditions. The parameters ω_1 and ω_2 respectively represent the frequencies of the chirp and the constant-frequency sinusoid in each input signal.

2. *Time-resolution condition:* This condition imposes an upper bound on the window length, N_w. It is derived on the basis of ensuring that the chirp frequency does not have too great a change within a given short-time section. Note that the STFT assigns only one frequency to the chirp over any short-time section of the signal. Compatibility criterion 4 requires that if the chirp frequency changes by more than $\frac{2\pi}{1000}$ radians, then the estimate of the chirp frequency must also change. Therefore, the window length N_w must be chosen to ensure that the chirp frequency does not change by more than $\frac{2\pi}{1000}$ radians over any short-time section of length N_w.

3. *Frequency-resolution condition:* This condition places a lower bound on the window length N_w in order to ensure that the two sinusoids in a short-time section can be unambiguously detected. If the window length has value N_w, each sinusoid has spectral peaks whose width may be approximated by $2\pi/N_w$. In order to ensure that aliasing does not occur due to interaction between the main lobes of the two sinusoids, the frequencies of the two sinusoids are required to be greater than $2\pi/N_w$ apart from each other.

4. *Frequency-precision condition:* This condition imposes a lower bound on N_{fft} in order to ensure that the frequency of a sinusoid can be estimated to within

the precision required by compatibility criteria 1 and 2. The binwidth of the FFT is given by $2\pi/N_{fft}$. This means that the FFT estimate cannot be more precise than $2\pi/N_{fft}$. Compatibility criteria 1 and 2 require that frequency estimates be within 10% of the actual values. This can be accomplished by requiring the FFT binwidth to be, at most, 10% of the smallest frequency sinusoid.

5. *Binwidth-resolution condition:* This condition imposes a lower bound on N_{fft} in order to satisfy the requirements of compatibility criterion 4, which stipulates that if the frequency spacing between the two sinusoids is less than the 10% frequency precision requirement, then their frequencies are to be estimated with greater precision. Clearly, this more precise estimation can be accomplished by requiring the FFT binwidth to be less than the frequency distance between the two sinusoids.

Model variety analysis: The inherent model variety problem arises in this application because for some input signals, the time resolution condition and the frequency resolution condition place *conflicting requirements* on the window length N_w. To illustrate this, consider an $x - y$ space where $x \geq 0$ and $y \geq 0$. Furthermore, let any point (x_o, y_o) in this space correspond to those signals produced by the SGP that have the property $y_o = |\omega_1 - \omega_2|/2\pi$ and $x_o = |n - m|$ where n and m are chosen to maximize $|n - m|$ while ensuring $|\omega_1(n) - \omega_1(m)| \leq \frac{2\pi}{1000}$. Setting the window length to the value N_w will set the conditions 2 and 3 shown in table 1 and partitions the set S of possible input signals into two subsets, U and D, which respectively contain the signals for which the STFT produces undistorted and distorted outputs. Subsequently, any signal produced by the SGP with $x \geq N_w$ and $y \geq 2\pi/N_w$ would then satisfy these two conditions. These signals are shown in region U of figure 1. Clearly, re-setting the window length to $N_w^{'}$ where $N_w^{'} > N_w$ will re-partition the set of all possible input signals, altering the subset U which contains the signals that could satisfy conditions 2 and 3 in table 1. If the set S of all possible input signals includes all the signals in the region $\{(x, y)(x > 0, y > 0)\}$, it is evident that there is no value for the window length that could reduce the set D, containing the input signals for which the STFT will produce distorted output, to the null set. This application is said to have an *inherent* model variety problem.

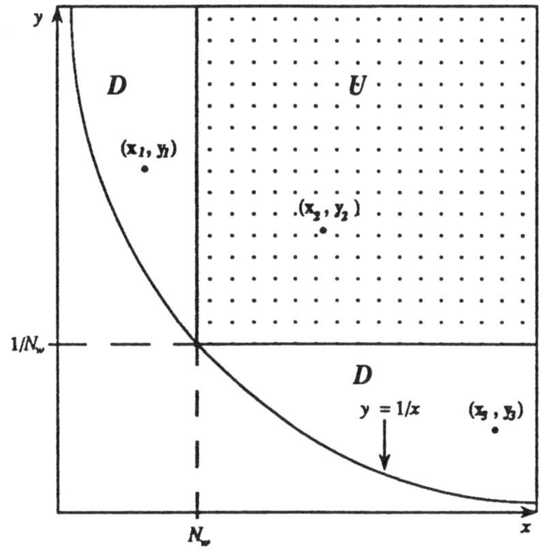

Figure 1: Illustration of the partitioning of the set S of possible input signal into subsets U and D, which respectively contain the signals for which the STFT produces undistorted and distorted outputs: Any point (x, y) represents signals for which $x = |n - m|$ which maximizes $|n - m|$ while ensuring $|\omega_1(n) - \omega_1(m)| \leq \frac{2\pi}{1000}$ and $y = |\omega_1 - \omega_2|/2\pi$. Region U: set of input signals that satisfy model-space conditions 2 & 3 (from table 1) for a window length of N_w, for which the STFT will produce undistorted output. Region D: signals that do not satisfy model-space conditions 2 & 3 for a window length of N_w, for which the STFT will produce distorted output.

References

[1] H. Nawab and V. Lesser, "High-Level Adaptive Signal Processing", *NAIC Final Report*, vol 17, October 1989.

[2] H. Nawab, V. Lesser, E. Miliois, "Diagnosis Using the Underlying Theory of a Signal Processing System", *IEEE Transactions on Systems, Man and Cybernetics. Special Issue on Diagnostic Reasoning.* May/June 1987.

[3] A. Oppenheim and A. Willsky, *Signals and Systems*, Prentice Hall, New Jersey, 1983.

[4] A. Oppenheim and R. Schafer, *Digital Signal Processing*, Prentice Hall, New Jersey, 1975.

[5] H. Nawab and T. Quatieri, "Short-Time Fourier Transform", *Advanced Topics in Signal Processing*, Prentice Hall, New Jersey, 1988.

[6] J. McClellan, "Parameteric Signal Modeling", *Advanced Topics in Signal Processing*, Prentice Hall, New Jersey, 1988.

[7] Bregman S. Albert, *Auditory Scene Analysis*, MIT Press, 1990.

Evidence Generation and Representation for Model Uncertainty Management In Nonlinear State Estimation

David Ferkinhoff, Kai Gong, Kathleen Keay, John MacDonald, Steven Nardone*

Organization: Naval Underwater Systems Center
Address: Bldg 1171-1, Code 2211
Newport, RI 02841-5047
Phone # (401) 841-3354
Fax # (401) 841-4749

*University of Massachusetts Dartmouth
Department of Electrical Engineering
North Dartmouth, MA 02747
(508) 999-8466

Abstract

This paper addresses the problem of managing modeling uncertainty in nonlinear state estimation. Oftentimes, modeling errors in the state estimation process produce deterministic features in the predicted residual sequence. A method for assessing model accuracy and providing alternate model selections is developed that involves the application of evidence generation and evidential reasoning. The evidence is the existence of non-zero coefficients of a regression fit to the predicted residuals. A belief function, developed from a classical multiple hypothesis likelihood ratio test, is used to map the evidence into the basic probability assignments for a feature frame of discernment (Θ = {no feature, feature}). Dempster-Shafer (D-S) theory of evidential reasoning is used to map the belief in the existence of a feature into belief in model hypotheses and the subsequent measures of support and plausibility. This approach provides a robust and flexible means for assessing model accuracy and suggests an interpretation in the sense of D-S theory of Bayesian hypothesis testing. Tracking problem results are presented that compare the performance of the D-S method to that of a traditional model assessment approach.

1: Introduction

Parameter estimation in nonlinear dynamic systems requires the processing of data under various kinematic, environmental, and sensor models [1,2,3]. However, nonlinear models of sufficient accuracy to describe the underlying principles and assumptions associated with the process increase the possibility of multiple solutions, as well as the potential for incurring system instability. Thus, the judicious selection of appropriate models is an important consideration and is critical if accurate state estimates are to be achieved. The model selection problem becomes even more difficult when the models can change dynamically. Traditionally, modeling hypotheses are examined using a combination of either sequential or batch state estimation techniques with multiple hypothesis testing procedures to identify models that produce consistent results [4]. Because these methods of model assessment apply all models simultaneously, large amounts of computational resources are required to achieve real-time performance.

It is well known that modeling errors can induce identifiable features into the predicted residual sequence. A nontraditional processor for model assessment was proposed previously [1] to exploit this relationship and is illustrated in figure 1. That approach used state estimation and hypothesis testing coupled with Dempster-Shafer (DS) evidential reasoning [5,6] to resolve modeling uncertainty. The detection of features in the predicted residuals for multiple measurement types was used as evidence of modeling error. The belief corresponding to the existence of features was expressed with fixed values, determined empirically, for the basic probability assignment (BPA) in the DS measurement frame of discernment. The measurement frames were related to the model frame by compatibility mappings [1,6]. Dempster's rule of combination was used to merge the model frame from each measurement type into a single soultion frame, and DS theory of evidential reasoning was used to generate support and plausibility for each model hypothesis. Subsequent processing with the selected model hypotheses generated additional evidence that allowed for further pruning. Monte Carlo experiments indicated that accumulation of evidence with time could resolve model ambiguity, provided that the effects of the mismodeling were uniquely observable. Because only those models deemed plausible of having produced the observed features are selected for subsequent processing, this approach can offer a potentially significant cost savings over traditional techniques. Subsequent studies [7,8] addressed multiple feature detection using classical approaches and an artificial neural network. That work provided a basis for the development of a belief function that could continuously map the feature test statistics into belief in the existence of the features.

This paper presents two interrelated extensions of the previously proposed configuration. First, a fundamentally different gallery is used that incorporates an individual feature frame in the Dempster-Shafer evidential reasoner (DS-ER), shown in figure 2. Second, and perhaps the more notable contribution, is the replacement of fixed BPA's with a continuous belief function. The function uses the result of the hypothesis test to distribute belief to the elements of the individual feature frames of discernment. As such, this functional relationship represents a departure from a classical Bayesian probablistic interpretation of the test statistic to the Dempster-Shafer theory of evidence.

The specific application considered is the tracking of a single dynamic contact by an observer with data from two spatially diverse passive acoustic sensors in an ocean environment. One sensor provides measurements of azimuth bearing (β_A), depression/elevation angle (DE) and frequency (F_A). The second sensor is a line array that provides measurements of conical angle (β_C), and frequency (F_C). The conical angle is the angle of signal arrival relative to the array axis and is therefore a function of both DE and β_A [9]. Initially, data prior to the time of a model anomaly are processed with the correct model to obtain a maximum likelihood estimate. The subsequent segment of data is assessed for consistent models. Of course, this tacitly assumes that the data are always correctly

1058-6393/91 $01.00 © 1991 IEEE

partitioned into separate segments, which is convenient for this investigation. In more realistic settings, partitioning the data is required [7,8] and would affect performance. However, the goal here is to isolate the performance of the DS-ER in conjunction with the belief functions.

Four modeling hypotheses are considered. The baseline model, or null hypotheses (H_o), is a constant-velocity contact with direct-path signal arrival and zero-mean measurement noise. This model serves as the correct model prior to the occurrence of an anomaly. A change in propagation path (PP) to a bottom bounce signal represents one type of modeling anomaly. Another is a change in contact velocity or a contact maneuver (TM). The last hypothesis is the unknown or unidentified model anomaly (UK), which has been included to capture model anomalies that cannot be classified presently as a PP or TM.

Monte Carlo results are presented that compare likelihood values of a classical hypothesis test for model assessment with the support and plausibility measures computed by the DS-ER. In principle, the support and plausibility measures obtained provide lower and upper bounds, respectively, on the belief in the hypotheses [5,6]. For the experiment conducted here, the DS-ER can identify the correct model under favorable conditions. As the conditions deteriorate, the plausibility of incorrect models increases, while support for the correct model decreases. Under the worst conditions all models become equally plausible, with relatively little support, which is an expression of a state of total ignorance. Of particular interest is the ability of the DS-ER process to preselect models for subsequent processing, which offers a potentially significant computational cost savings over more traditional approaches. Although limited, this application is sufficient to demonstrate both the feasibility of the DS-ER approach and the utility and expandability of the specific design.

2: A classical model assessment approach

For the problem considered here, a state estimate is obtained for each model hypothesis. H_o is used prior to the time of model anomaly, while the H_i's are used for subsequent processing. Consider the case where the set of L modeling hypotheses $\{H_i\}$ represents all the possible models. The measurements Z are functions of the state X and are corrupted by additive Gaussian white noise N such that

$$Z = Z_T(X) + N . \qquad (1)$$

Here $Z_T(X)$ is the correct composite function that models both the measurement process and the state dynamics. The probability density function for k measurements conditioned on the state and the model hypothesis H_i is used as the likelihood function; i.e.,

$$P(Z|X,H_i) = (2\pi\sigma)^{-k/2} \exp[\frac{-1}{2\sigma^2} \|Z - Z_i(X)\|^2], \qquad (2)$$

where σ is the measurement noise standard deviation. The maximum likelihood estimate (MLE) is the value of the state that maximizes the likelihood function, and it can be obtained by solving the log-likelihood equations [10]:

$$\frac{\partial \ln[P(Z|X,H_i)]}{\partial X} \Big|_{X=\hat{X}_i} = 0. \qquad (3)$$

While many techniques for solving systems of nonlinear equations may be employed, here a Gauss-Newton iterative method is used. A separate MLE is obtained for each of the L model hypotheses. The individual estimates are ranked according to how well they fit the data. A quadratic performance index,

$$J_i = \frac{1}{k} \|Z - Zi(\hat{X}_i)\|^2{}_{\sigma^{-2}}, \qquad (4)$$

is used to measure the fit. Pruning of the hypotheses is accomplished by comparing all performance indices to a threshold. Noting that the performance index is a χ^2 (chi-squared) random variable, one can set the threshold to achieve a desired probability of false alarm. Only those hypotheses likely to have been involved in the measurement process are retained and may be ranked using the weights

$$w_i = J_i^{-1} \Big/ \sum_{i=0}^{L-1} J_i^{-1} . \qquad (5)$$

Any further reduction or resolution in the model hypotheses requires additional information. Because a state estimate is obtained for each hypothesis, this approach is computationally intensive.

3: Dempster-Shafer evidential reasoning

The Dempster-Shafer theory of evidential reasoning is one of several approaches for producing inferences from uncertain information [11]. Its appeal for application to model assessment is that it intrinsically accommodates the expression of ignorance. The basic structure for DS evidential reasoning is the frame of discernment. The frame of discernment Θ is a set of mutually exclusive and exhaustive hypotheses [6]:

$$\Theta = \{H_o \ldots H_{L-1}\} \qquad (6)$$

The power set is formed as the collection of all subsets (A_i) of Θ, excluding the null set, and is denoted by 2^Θ. A BPA, denoted as $m(A_i)$, represents the belief that the hypotheses of A_i are true. BPA's are assigned to all the elements of the power set subject to the constraints that the individual BPA's are bounded between zero and one and their sum is unity. The additional measures of plausibility and support for each element can also be computed. The support for A_i, denoted by $S(A_i)$, is the sum of the BPA's over all the subsets of A_i, i.e.,

$$S(A_i) = \sum_{a \in A_i} m(a) , \qquad (7)$$

and represents belief that directly supports A_i. The plausibility of A_i is a measure of the lack of disconfirming belief and is one minus the belief in the complement of A_i; i.e.,

$$Pl(A_i) = 1.0 - S(A_i) = 1.0 - \sum_{a \in A_i} m(a). \qquad (8)$$

Note that, in general, support is less than or equal to plausibility and both are bounded by zero and one. The uncertainty in assigning a belief to the hypotheses contained in element A_i is the difference between the plausibility and support. Dempster-Shafer theory includes Bayesian probability as a special case when all belief is distributed among the singleton sets $\{H_o\}, \{H_1\}, \ldots, \{H_{L-1}\}$.

If several bodies of evidence exist for a frame, the resulting beliefs can be combined using Dempster's combination rule:

$$m_c(A_i) = \frac{1}{\alpha} \sum_{A_j \cap A_k = A_i} m_1(A_j) m_2(A_k), \quad (9a)$$

where $m_1(A_j)$ and $m_2(A_k)$ are the belief in A_j and A_k generated from the bodies of evidence E1 and E2. The term α is the renormalization constant necessary to account for belief being placed into the null intersection ϕ and is

$$\alpha = 1 - \sum_{A_j \cap A_k = \phi} m_1(A_j) m_2(A_k) \ . \quad (9b)$$

In both expressions, the indices j and k range over all the elements of the power set. Dempster's combination rule is both commutative and associative.

Compatibility maps, which are used to map belief in one frame of discernment to a different frame, are useful for combining belief in dissimilar frames into belief in a common frame [6,12]. The compatibility map defines the relationship between the elements of two frames $\Theta_{A,B}$ where

$$\Theta_{A,B} \subseteq \Theta_A \times \Theta_B, \quad (10a)$$

and the compatibility mapping is

$$C_{A \to B}(A_k) = \{b_j \mid (a_i, b_j) \in \Theta_{A,B}, a_i \in A_k\}, \quad (10b)$$

where at least one pair (a_i, b_j) is specified for each A_k in Θ_A. With a network of compatibility relations, different frames of reasoning can be linked together. The collection of frames and compatibility relations is called a gallery.

4: Application of Dempster-Shafer evidential reasoning

4.1: Gallery description

The first level of reasoning is at the individual feature frame of discernment θ = {no feature ($\overline{F}$), feature (F)} as shown in figure 2. For each measurement type, a traditional likelihood ratio test is used to generate evidence pertaining to the existence of the features of a step (jump) and/or a ramp (drift). The belief function, to be described shortly, maps the evidence of a feature into belief in the existence of the feature within an individual feature frame. Separate individual feature frames θ_j and θ_d are used for jump and drift, respectively .

The second level is composed of composite feature frames of discernment θ_F = {no features ($\overline{F}$), jump-only (J), drift-only (D), jump and drift (JD)}, again one for each measurement type. Belief in the composite feature frame is obtained from belief in frames θ_j and θ_d through the compatibility relationship defined in table I and Dempster's rule of combination illustrated in table II. Using the compatibility map of table I, two composite feature frames are generated, one for the jump feature (θ_{Fj}) and one for the drift feature (θ_{Fd}). From table I it is seen that the existence of a feature in θ_j is compatible with the hypotheses of jump-only or jump and drift (J,JD) in θ_F. Consequently, the belief in jump from θ_j is mapped into belief in the element (J,JD) of θ_{Fj}. Similar reasoning applies to the remaining terms in the compatibility relationship. Note that belief in the elements of the individual feature frames is mapped to belief in subsets of the composite feature frame, so belief can be assigned directly into this frame. Although the system beliefs were implemented directly in the composite feature frame, the above description is useful in developing the belief functions.

Beliefs in the two composite feature frames θ_{Fj} and θ_{Fd} are merged using Dempster's combination rule to form a single composite feature frame for each measurement type as illustrated in table II. The belief assigned to the intersection of frame elements is the renormalized product of the belief in the header subsets as defined by equation (9). Thus, for example, the belief in $\overline{F}$ of the combined frame is

$$m_c(\overline{F}) = \frac{1}{\alpha} m_d(\overline{F},J) m_j(\overline{F},D). \quad (11)$$

For these frames there are no intersections that result in the null set; hence, α equals one. Note that although the constituent composite feature frames have no belief in the singleton hypotheses, the combined frame may have belief in the singletons as shown in table II.

The compatibility relations, defined in table III, map the belief in the separate composite feature frames into beliefs in the various possible modeling hypotheses θ_M = {null, maneuver, propagation path, unknown}. The compatibility relationships were established by a perturbation analysis [1]. In this approach, the model change introduces a perturbation into the system equations, and a Taylor series expansion is used to determine the effects on the predicted residuals. Thus, for example, a change in the platform velocity can introduce either jump or jump and drift in the frequency residuals and a drift in azimuth bearing. A separate compatibility relationship is developed for each composite feature frame resulting in one model frame for each measurement type. Dempster's combination rule is used to combine all the model frames from each measurement type into a single solution frame that represents the composite belief in model hypotheses from all the evidence. Although limited, the framework of this gallery is readily extendable to include additional features in the individual feature frame, such as curvature, or other evidence through suitable compatibility relations, as well as additional model hypotheses through extensions of the model frame of discernment.

4.2: Belief function

The belief functions map the evidence pertaining to the existence of features in the predicted residuals into a belief in the existence of a particular feature. Thus, it must distribute belief over the sets of hypotheses of ({$\overline{F}$},{F},{$\overline{F}$,F}). The bases for distributing belief are the feature detection (P_D) and false alarm (P_F) probabilities obtained from a multiple hypothesis likelihood ratio test. The features are extracted from a regression fit of a second-order polynomial to the residuals:

$$\hat{m} = [B^T B]^{-1} B^T [Z - Z_o(\hat{X})] \ , \quad (12)$$

where $Z_o(\hat{X})$ are the predicted measurements associated with hypothesis H_o, and B is the matrix

$$B = \begin{bmatrix} & \cdot & \\ 1 & t_k & \frac{1}{2} t_k^2 \\ & \cdot & \end{bmatrix}, \quad (13)$$

and $\hat{m} = [\hat{a}_o, \hat{a}_1, \hat{a}_2]^T$ is the vector of estimated regression coefficients identified with the specific features of jump (a_o), drift (a_1), and curvature (a_2). Although curvature may be present in the residual sequence, it is not used presently as evidence in the DS-ER because it is not a critical factor in assessing the given modeling anomaly types. However, its presence, or even higher coefficients, may be necessary in the regression fit to obtain unbiased estimates of the jump and drift coefficients. Here, the significance of the curvature coefficient is rejected if its test statistic (discussed below) falls below a threshold set to achieve a 10-percent false alarm rate. Other techniques are also available to determine the appropriate order of the regression fit [13].

A multiple hypothesis test for these features, in the case of negligible state estimation error [8], is the basis on which the belief functions are developed. The test statistic ℓ is compared to a threshold ε for the multiple hypothesis test as

$$\ell \underset{\bar{H}_j}{\overset{\bar{H}_i}{\gtrless}} \varepsilon, \tag{14}$$

where H_j is the hypothesis that all the features exist, H_i is the hypothesis that the ith feature is not significant, and

$$\ell = \frac{\|\hat{M}_i - \hat{M}_j\|^2}{\sigma^2}. \tag{15}$$

Here, $\hat{M}_k = B\hat{m}_k$, where the $\hat{m}_k$ are separate estimates with the appropriate coefficient set to zero, and the test statistic ℓ is an estimate of signal (feature)-to-noise ratio (SNR) [8]. Traditionally, the threshold is set to satisfy a specified probability of false alarm criterion.

The test statistic is the squared difference between the two regression fits, $\hat{M}_j$ with all the coefficients and $\hat{M}_i$ with some coefficients set to zero. The test statistic is a χ^2 random variable with the number of degrees of freedom equal to the number of features being tested, one in this case. Since the ℓ is χ^2 with one degree of freedom, it may be alternately viewed as the square of a univariant Guassian random variable [10]. Under the hypothesis that a feature is not present, the corresponding Gaussian distribution is zero mean. Under the hypothesis that a feature is present, the Gaussian distribution has a nonzero mean equal to the square root of the test statistic. In the absence of any prior knowledge about the feature, the threshold is taken at the point of equal likelihood of the two densities, which is $\varepsilon = \sqrt{\ell/2}$. Note that the threshold varies with the estimated SNR and therefore responds dynamically to changing signal levels.

The P_F is associated with the belief that the estimated SNR value belongs to the "no feature" distribution (true SNR = 0), while P_D is associated with the belief that the estimated SNR value belongs to the "feature present" distribution (true SNR $\neq$ 0). The belief that a feature is present is directly proportional to the estimated SNR, which is analogous to a signal excess (excess beyond SNR = 0). Consequently, the belief that a feature is present is set to the difference between P_D and P_F:

$$B_F = P_D - P_F. \tag{16}$$

This definition of belief is similar to that given in [14].

The belief in no feature B_C is more complicated. Two separate conditions lead to zero SNR. In one case, noise may approach infinity; in the other, the signal level may approach zero. In the large noise case, although SNR is approaching zero, the placing of all belief into $\bar{F}$ results in low plausibility of a feature being present. This is clearly undesirable because, although high-noise inhibits a feature detection, one could be present; thus, the plausibility of the existence of a feature should be high. For the simple structure of the individual feature frames, a high plausibility for the feature implies a low belief in $\bar{F}$. Thus, B_C should be a function of both the estimated SNR as well as a measure of the significance of the noise level. Following this line of reasoning, the B_C is established as the joint probability of two (assumed) independent events: the probability of false alarm P_F given the estimated SNR, and the belief in the ability to detect a reference level feature in the given noise level. Consequently,

$$B_C = (P_{DR} - P_{FR})\ P_F, \tag{17}$$

where P_{DR} and P_{FR} are, respectively, the detection and false alarm probabilities of a reference signal in the given noise level. The reference signals can be set at the minimum level of bias that can be reasonably tolerated. The importance of the reference signal level is discussed in the experimental results section. Because $B_C \leq P_F$, it follows that $B_C + B_F \leq 1$, and so the remaining belief is assigned to the frame and represents ignorance as

$$B_\theta = 1 - P_D + [\ 1 - (P_{DR} - P_{FR})]P_F\ . \tag{18}$$

The belief functions for no feature, feature, and uncertainty vs SNR are shown in figures 3a and b for a high-noise and low-noise case, respectively. Here, the reference is a constant bias level of 0.3 and noise standard deviation as noted. Note that in high-noise and low-SNR conditions, B_F and B_C are small while B_θ is high, indicating that not much is known about the presence or absence of a feature. For low-noise and low-SNR most of the belief goes into $\bar{F}$. For high-SNR most of the belief is in F regardless of the noise level.

5: Experimental results and discussion

Experiments were conducted using computer-generated data. Results were obtained for both the classical approach and the proposed evidential reasoner. In the scenario depicted in figure 4 the observer executes three legs, each 900 seconds long, at a speed of 10 m/s. The course on the first, second, and third legs is, respectively, 45°, 315°, and 45°. The relative depth to the ocean bottom is 4800 m for both the contact and observer. The initial contact bearing is 0° at 15 km, course is 0°, and speed is 10 m/s . The available measurements are sampled every 20 seconds. For the purposes of these experiments, a single modeling anomaly, when present, always occurs in the middle of the third observer leg. All data are processed up to the time of the modeling anomaly with the correct model to obtain a maximum likelihood estimate. The data subsequent to the time of anomaly are then examined for model assessment. The TM model is a slight maneuver from a course of 0° to 5°, while the UK is a 1° bias on β_A. The goal of the classical technique is to rank the hypotheses with respect to their measures of likelihood given by equation (5), while the goal of the DS-ER is to preselect models for further processing by classical techniques. Results for the UK were not obtained for the classical approach.

Figure 5 shows an example of conical angle and frequency residuals from a single experiment for a contact maneuver in a low-noise environment. Although the contact maneuver is slight, its effect on the predicted residuals is clearly seen as a jump in frequency and a drift in conical angle. For this experiment the features were easily detected and good results were obtained. As noise levels increase, the detectability of the features decreases, and a corresponding degradation occurs in the performance of the model assessment algorithms.

Tables IV and V present the results of a Monte Carlo experiment with 30 trials. All table entries are average values in percent. The left and right halves of table IV contain results, using data from sensor 2 only, for a low noise level (twice that used in figure 5) and a high noise level (eight times that used in figure 5), respectively, while table V contains results using only angle data from both sensors. The tables are confusion matrices where the horizontal headers indicate the actual model anomaly and the vertical headers indicate the hypothesized model. Thus, the first column of table IV contains the low-noise results for no modeling anomaly, the second column contains low-noise results for a change in propagation path, etc. Each cell contains five entries organized into three rows. The first entry is the weight from the classical model assessment approach. The second and third rows are, respectively, the support and plausibility for the hypothesis from the DS-ER. The left column for support and plausibility in each cell was obtained using high reference signal levels (maximum value of 3 for jump or drift), whereas the right column was obtained using low reference signal levels (maximum value of 0.5).

From table IV for the low-noise condition, it is seen that when no model anomaly occurs the classical assessment will show that TM fits the data slightly better than H_o. This will always be the case when H_o is true because a better fit to the data is always obtained with two contact legs vs a single contact leg. Although this is an annoyance, it is easily compensated for by a cluster test on the contact velocities [1]. However, because the focus of this paper is on the DS-ER and its belief functions, no correction was deemed necessary and the problem is largely ignored here and elsewhere in the experimental results. Thus, for the low-noise case illustrated in the table IV, the correct model is readily identified by either the classical method or the DS-ER. For the TM anomaly, the support and plausibility of TM are, respectively, 63.9 percent and 77.2 percent under high reference signal levels, and 60.4 percent and 71.3 percent under low reference signal levels. Note that the correct hypothesis is well supported with a relatively low level of uncertainty (~14 percent). Additionally, there is negligible support or plausibility for H_o under both high and low reference signals, which indicates that there is essentially no evidence to support belief in "no feature." This is precisely the conclusion that should be drawn from the residual plots of figure 5. Finally, because the unknown anomaly is a bias in the azimuth bearing of sensor 1, it has no effect on the measurements of sensor 2. Thus, the UK anomaly of column 4 produces results that are nearly identical to those of H_o in column 1.

The results for high noise were obtained under similar conditions except that the noise levels were increased by a factor of four. Comparison of the results shows the anticipated degradation in performance as the features become masked by the effects of the noise. In particular, note the results for a TM anomaly. As the noise level increased it masked the features; therefore, detectability (P_D-P_F) was diminished, resulting in an erosion of support for TM. Because the reference signal level primarily affects the belief in "no feature," and hence the plausibility of a feature, changing the reference level has a much more dramatic effect on plausibility and support under high-noise conditions than under low-noise conditions. This is seen by comparing the change in plausibilities with reference level for low- and high-noise conditions. Furthermore, note that the plausibility of PP and UK increased dramatically in the high-noise environment, while there was little support for any of the hypotheses. This indicates that while no features were detected, there was also little belief in $\overline{F}$, thereby expressing ignorance. These results are consistent with the conclusions that should be drawn from the data. Although the reference levels are somewhat subjective, it is seen that the sensitivity of a hypothesis' plausibility to the variability in the reference signal level, and not necessarily the establishment of an absolute value, is an important consideration.

Table V shows similar results for low- and high-noise conditions, respectively, for the case where the angle measurements are available from both sensors. Note that for the low-noise case the loss of evidence in the frequency residuals is adequately compensated for by the information in the azimuth bearing and DE angle. In particular, note that the UK hypothesis is easily identified in the low-noise case.

Although limited results are presented to make the specific points stated above, additional results not presented here yielded similar performance as noise and reference signal levels were increased or decreased and various combinations of measurements were used.

6: Summary and conclusions

This paper, which is an extension of previous work [1], has presented an approach to the management of model uncertainty in state estimation. Although the specific application of contact tracking with passive acoustic data in the ocean environment is considered, the method can be applied to other problems. The approach uses the Dempster-Shafer theory of evidential reasoning (DS-ER) to assess plausible models for processing the data. The gallery, a collection of frames and compatibility relationships, was developed with unique features that allowed for expansion at many levels. Evidence pertaining to the existence of features in predicted residuals is extracted from the coefficients of a regression fit and used for model assessment. A belief function that maps the evidence into the basic probability assignments of the DS-ER frame of discernment $\theta = \{\text{no feature, feature}\}$ was developed from a classical multiple hypothesis likelihood ratio test. The resulting beliefs were based on the detection and false alarm probabilities established by a dynamic threshold and the detectability of a reference signal in noise. As such, this approach represents an interpretation of Bayesian hypothesis testing in the sense of DS-ER. The relative distribution of belief in the existence or nonexistence of features was observed to be consistent with the evidence available in the predicted residual sequence.

Monte Carlo results were obtained and compared for both DS-ER and a classical model assessment approach. The classical method employed a ranking scheme based on weights derived from the mean squared value of the residual error. Because classical methods typically involve the generation of a state estimate for every model hypothesis, the DS-ER offers a potential for substantial computational cost savings. This is an important consideration for many applications. Although limited results were presented, the experiments demonstrated the feasibility and robustness of the DS-ER approach to preselect plausible hypotheses, and thereby established the suitability of the belief functions and associated gallery.

Acknowledgements

This study was conducted under the combat control program sponsored by the Chief of Naval Research, Office of Naval Technology, program manager D. Houser (OCNR-232).

References

1. J. G. Baylog, A. A. Magliaro, S. M. Zile, and K. F. Gong, "Underwater Tracking in the Presence of Modeling Uncertainty," Proceedings of the Twenty-First Asilomar Conference on Signals, Systems and Computers, November 1987.

2. D. Kolb and F. Holister, "Bearings-Only Contact Parameter Estimation," Proceedings of the First Asilomar Conference on Circuits and Systems, 1967.

3. K.F. Gong, S. E. Hammel, S. C. Nardone, and A. G. Lindgren, "Three Dimensional Contact Parameter Estimation," Proceedings of the Sixteenth Asilomar Conference on Circuits, Systems and Computers, November 1982.

4. K.F. Gong, J. G. Baylog, and A. A. Magliaro "A Decision-Directed Approach to Solution Integration for Tracking in an Underwater Environment," Proceedings of the Eighteenth Asilomar Conference on Circuits, Systems and Computers, November 1984.

5. A. P. Dempster, "A Generalization of Bayesian Inference," Journal of the Royal Statistical Society, Series B, 30, pp205-247, 1968.

6. G. Shafer, A Mathematical Theory of Evidence, Princeton University Press, Princeton, NJ, 1976.

7. D. Ferkinhoff, J. Baylog, K. Gong, and S. Nardone, "Feature Extraction and Interpretation for Dynamic System Model Resolution," Proceedings of the Twenty-Fourth Asilomar Conference on Circuits, Systems and Computers, November 1990.

8. D. Ferkinhoff, J. Baylog, K. Gong, and S. Nardone, "Feature Detection for Model Assessment in State Estimation," Technical Report 7064, Naval Underwater Systems Center, Newport, RI, 1991 (in publication).

9. "Underwater Target Tracking Using Measurements From a Linear Array of Omnidirectional Acoustic Elements," Technical Report 6010, Naval Underwater Systems Center, Newport, RI, 15 July 1981 .

10. H. L. Van Trees, Detection Estimation and Modulation Theory, John Wiley & Sons, New York, 1968.

11. E. J. Horvitz, D. E. Heckerman, and C. P. Langlota, "A Framework for Comparing Alternative Formulations for Plausible Reasoning," Proceedings AAAI-86 Fifth National Conference on Artificial Intelligence, 11-15 August 1986, Philadelphia, PA, Vol. I.

12. J. D. Lowrance, T. D. Garvey, and T. M. Strat, "A Framework for Evidential Reasoning Systems," Proceedings AAAI-86 Fifth National Conference on Artificial Intelligence, 11-15 August 1986, Philadelphia, PA, Vol. I.

13. M. L. Graham and F. O'Brien, "Rank Analysis Module for Target Motion Analysis," Patent Application, Navy Case 72911, 20 June 1990.

14. G. Shafer and J. Pearl, "Numerical Uncertainty in Expert Systems," Readings in Uncertain Reasoning, Morgan Kaufmann Publishers Inc., San Mateo, CA, 1990.

Table I: Compatibility relationship for mapping from the individual frame of discernment to the composite frame of discernment

		$\bar{F}$	J	D	JD	Result
θ_j	$\bar{F}j$	✓		✓		$\bar{F}$,D
	Fj		✓		✓	J,JD
	θ_j	✓	✓	✓	✓	θ_F
θ_d	$\bar{F}d$	✓	✓			$\bar{F}$,J
	Fd			✓	✓	D,JD
	θ_d	✓	✓	✓	✓	θ_F

$\theta_j = \{ \bar{F}j , Fj \}$

$\theta_d = \{ \overline{Fd} , Fd \}$

$\theta_F = \{ \bar{F} , J , D , JD \}$

Table II: Tabular form of Dempster's combination rule

θ_{F_d} \ θ_{F_j}	$m_j(\bar{F},D)$	$m_j(J,JD)$	$m_j(\theta_F)$
$m_d(\bar{F},J)$	$m_c(\bar{F})$	$m_c(J)$	$m_c(\bar{F},J)$
$m_d(D,JD)$	$m_c(D)$	$m_c(JD)$	$m_c(D,JD)$
$m_d(\theta_F)$	$m_c(\bar{F},D)$	$m_c(J,JD)$	$m_c(\theta_F)$

Table III: Compatibility relationship for mapping from the composite feature frame of discernment to the model frame of discernment

Composite Feature Frame θ_F																	
β_C				Fa & Fc				DE				β_A					
JD	D	J	$\bar{F}$	JD	D	J	$\bar{F}$	JD	D	J	$\bar{F}$	JD	D	J	$\bar{F}$		Model Frame θ_M
			✓				✓				✓				✓	H0	
✓		✓		✓				✓							✓	PP	
	✓			✓		✓					✓		✓			TM	
					✓				✓	✓		✓		✓		UK	

Table IV: Experimental results for sensor 2 only

	Low Noise								High Noise							
	H0		PP		TM		UK		H0		PP		TM		UK	
	HREF	LREF	HREF	LREF	HREF	LREF	HREF	LREF	HREF	LREF	HREF	LREF	HREF	LREF	HREF	LREF
H0 W	47.4		18.4		20.9		47.3		40.8		03.8		34.0		47.3	
H0 S	78.5	68.2	01.5	00.0	00.0	02.2	78.1	67.6	76.9	03.1	00.0	00.0	49.7	01.2	76.9	03.1
H0 P	82.6	77.4	01.7	00.5	00.0	02.8	82.1	76.9	82.0	59.1	00.0	00.0	54.6	25.7	82.0	59.1
PP W	02.1		52.4		04.0		02.2		15.2		97.5		02.6		03.0	
PP S	06.2	08.0	96.5	97.8	22.8	25.1	06.4	07.8	05.9	15.4	99.9	99.9	14.4	17.1	05.9	15.4
PP P	08.1	11.7	96.8	99.5	36.1	35.8	08.2	11.6	08.0	73.8	99.9	99.9	17.9	78.3	08.0	73.7
TM W	50.5		29.2		75.1		50.5		44.0		05.2		44.7		50.7	
TM S	09.8	11.8	01.5	00.3	63.9	60.4	10.1	12.4	10.4	04.6	00.0	00.0	26.2	10.4	10.4	04.6
TM P	13.9	20.3	01.8	02.0	77.2	71.3	14.1	20.9	15.4	71.1	00.0	00.0	32.5	75.1	15.4	71.1
UK W	NA		NA		NA		NA		NA		NA		NA		NA	
UK S	00.3	01.1	00.0	00.0	00.0	01.2	00.3	00.7	00.4	01.0	00.0	00.0	02.2	02.8	00.3	01.0
UK P	00.6	03.5	00.1	00.5	00.0	01.8	01.1	03.6	01.5	50.7	00.0	00.0	03.5	36.8	01.5	50.7
Using Measurements β_C F_C																

Table V: Experimental results for angle data only from both sensors

	Low Noise								High Noise							
	H0		PP		TM		UK		H0		PP		TM		UK	
	HREF	LREF	HREF	LREF	HREF	LREF	HREF	LREF	HREF	LREF	HREF	LREF	HREF	LREF	HREF	LREF
H0 W	48.7		00.1		20.9		48.4		47.6		01.2		47.4		47.5	
H0 S	70.4	54.5	00.0	00.0	10.3	07.4	00.2	00.0	67.2	01.0	00.0	00.0	63.3	00.9	53.8	00.3
H0 P	74.9	66.9	00.0	00.0	11.7	10.0	00.2	00.0	73.0	38.4	00.0	00.0	68.5	36.7	58.3	13.0
PP W	00.2		99.8		04.0		00.3		02.5		97.5		02.6		03.0	
PP S	10.1	11.5	99.9	99.9	02.1	02.7	00.0	00.0	10.2	18.4	99.6	98.3	09.2	17.0	05.7	07.7
PP P	11.0	13.3	99.9	99.9	02.3	03.1	00.0	00.0	11.3	55.7	99.6	99.1	10.2	51.4	06.8	21.0
TM W	51.1		00.1		75.1		51.3		49.9		01.3		50.0		48.7	
TM S	12.8	16.4	00.0	00.0	83.9	83.7	00.0	00.0	14.0	07.1	00.0	00.0	19.6	09.8	10.7	02.4
TM P	16.8	28.1	00.0	00.0	85.4	86.7	00.0	00.0	19.0	52.7	00.0	00.0	24.2	54.9	14.5	18.2
UK W	NA		NA		NA		NA		NA		NA		NA		NA	
UK S	02.1	02.5	00.0	00.0	02.0	02.9	99.8	99.9	02.4	18.1	00.4	00.9	02.2	18.9	23.3	69.9
UK P	04.1	05.3	00.0	00.0	02.2	03.5	99.8	99.9	03.0	56.1	00.4	01.6	02.3	55.8	23.8	83.9
Using Measurements β_A DE β_C																

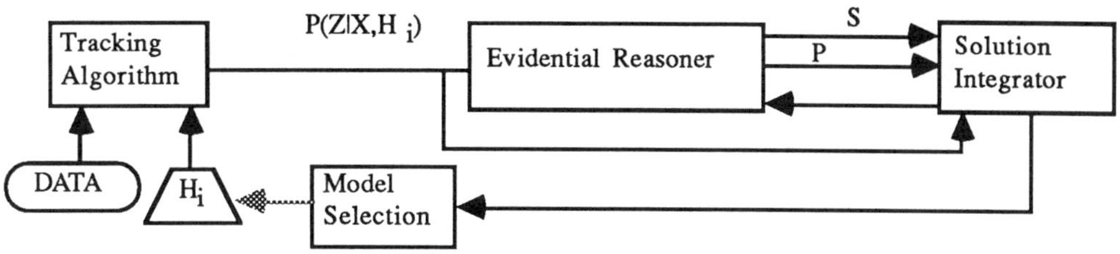

Figure 1: Block diagram of a model assessment system

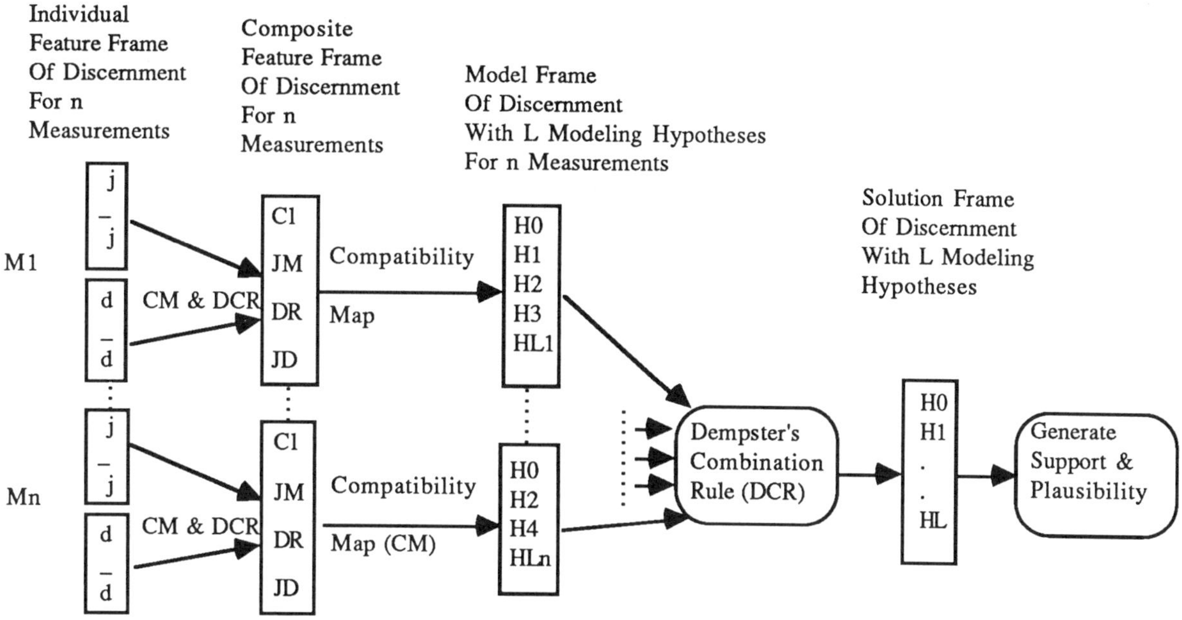

Figure 2: Block diagram of the gallery

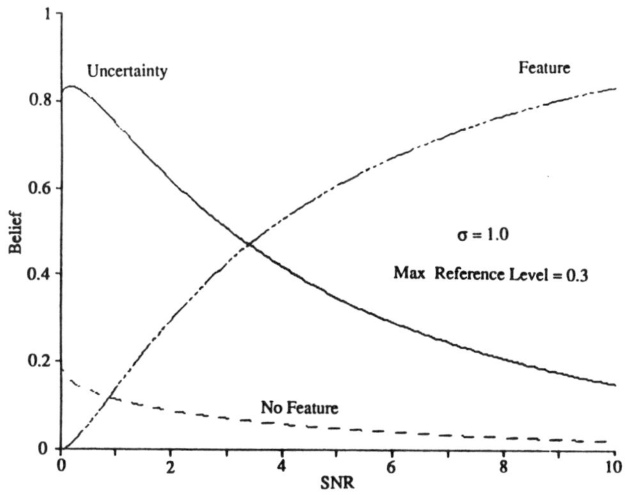

Figure 3a: Belief function - high noise

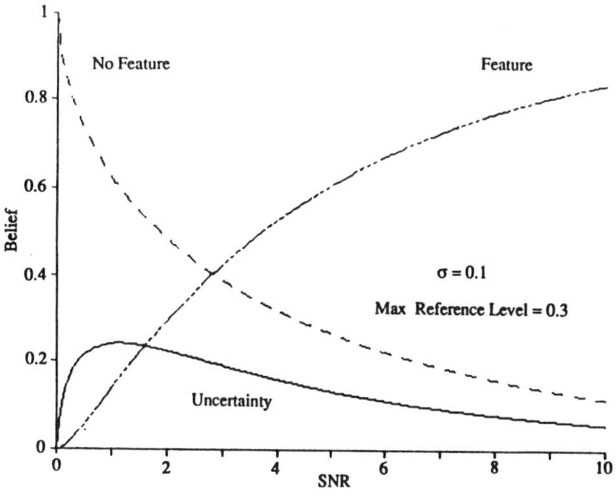

Figure 3b: Belief function - low noise

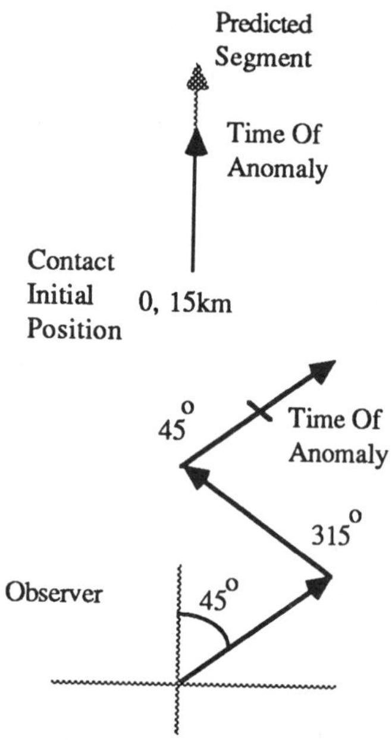

Figure 4: Geographic plot of contact and observer scenario

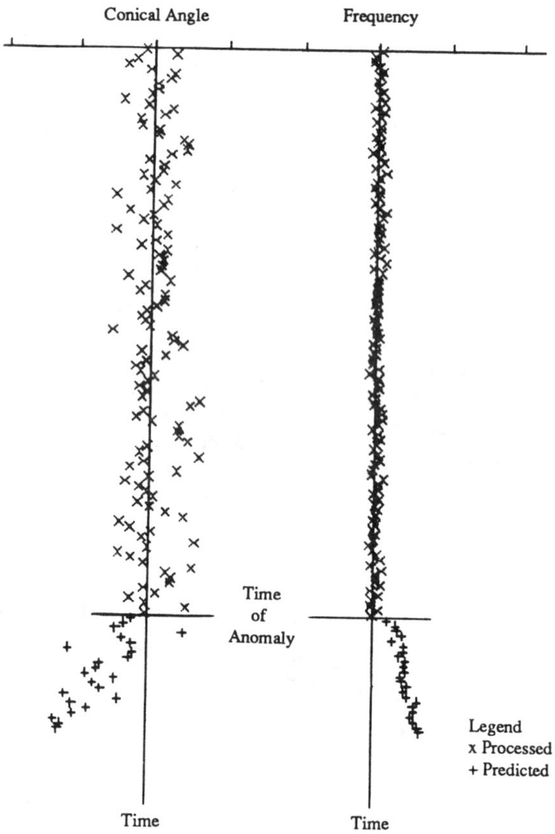

Figure 5: Sensor 2 residuals

Communications Equalization Using Non-Linear Adaptive Networks

Colin F.N. Cowan
Dept. of Electronic & Electrical Engineering
Loughborough University of Technology
Leicestershire, LE11 3TU, England

Abstract

The process of detection and equalization in data communications systems has long been seen as the realm of maximum likelihood sequence estimation on the one hand and linear adaptive filters on the other. This paper sets out to reexamine the equalization process and first demonstrates that linear equalizers are in fact sub-optimum when noise is present in the communications channel. A non-linear structure, on the other hand, can provide a near optimum decision boundary but one which still does not approach the performance of MLSE.

The paper examines the various performance and complexity issues raised in the deployment of such non-linear adaptive systems. Further the development of a multichannel adaptive solution (in this single channel temporal case) provides a possible parallel to the MLSE solution.

1 Introduction

The subject of adaptive equalization [1] has been commonly viewed as an inverse filtering problem where the objective of the filter is to restore the signal present at the input to a communications channel given that only observations of the channel output are available to the processor. It is normally assumed that the channel transfer characteristic is linear and that additive noise is not a major effect. The fact that the channel is a-priori unknown and varies slowly with time is the reason for the use of filters which are adaptive in nature.

In the situation where there is no additive noise in the channel and the channel is minimum phase in nature it is possible to form this type of equalization perfectly by deploying a decision feedback equalizer (d.f.e.) [1]. However, when the channel is non-minimum phase or there is significant noise present then only an approximation to the true inverse may be made. It is, of course, well known that in this case an optimum detection strategy may be adopted by using maximum likelihood sequence estimation (MLSE) [2]. However, this is a solution which involves an exponential increase in complexity dependent on the channel intersymbol interference (ISI) spread.

This paper sets out to describe a sub-optimal equalization/detection process which depends on the use of non-linear adaptive processing of the observed channel output [3]. This concept is developed to encompass the idea of sequence estimation as a logical extension to this classification process [4]. The form of sequence estimation is equivalent to MLSE detection in the limit but significant savings in computational complexity are attained by the use of the approximations suggested in this paper.

The paper first describes how the equalization process may be viewed as a classification problem and describes how an optimum (non-linear) classification strategy is formed. One solution to this classification problem is provided which allows classification on a symbol by symbol basis with an improved bit error rate (ber) performance relative to the linear system. The extension of this idea to encompass sequence estimation is then described by the use of Mahalanobis distance classifiers [4], which are essentially generalisations of radial basis function (RBF) artificial neural networks. By, once again using a symbol by symbol detection strategy it is shown how this approach compares with the optimal Euclidean classification approach when the ISI is known perfectly (the MLSE case).

2 Equalization as a Classification Problem

The problem of reconfiguring the process of equalization as a classification problem can be done most easily by considering a very simple example. If we consider the situation where the transmitted data sequence is randomly chosen from the binary set {1,-1} and the channel has an ISI spread of only 2 symbols, given by:

$$H(z) = 1 + 0.5z^{-1}$$

1058-6393/91 $01.00 © 1991 IEEE

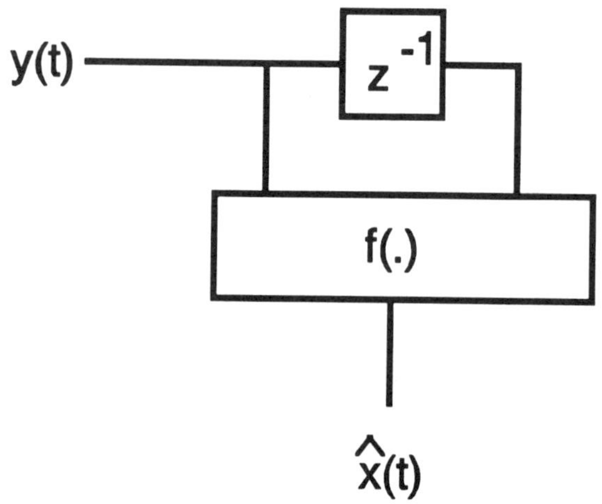

Figure 1 : Structure of the general 2 tap equalizer

The postulated equalizer is then taken as one which uses 2 successive received samples to form its estimate, see figure 1. The number of possible received states is the 8, and these are illustrated in figure 2. In this diagram the symbol, o, corresponds to a state where the zero lag received symbol is -1 and the, x, corresponds to +1. Obviously it is possible to correctly classify these states by a linear system since we can draw a straight line (see figure 2) which has -1 states to the left and +1 states to the right. This situation corresponds to the linear equalizer, where the function f{•} in figure 1 is a linear weighted sum followed by the signum function.

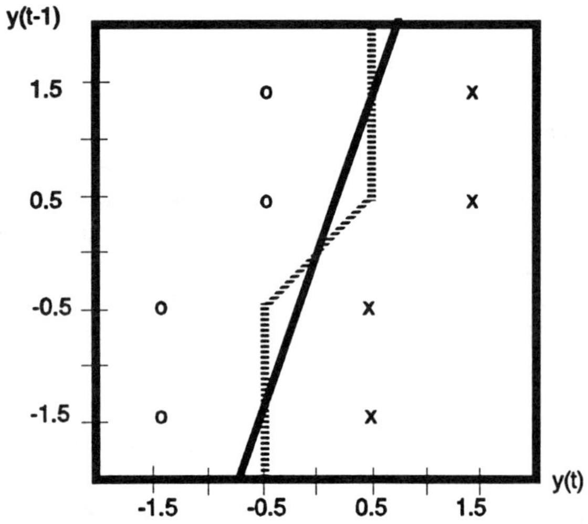

Figure 2 : Received signal states for a 2 tap equalizer given a channel of $1 + 0.5z^{-1}$.

However, if we now add noise to the observations and apply a minimum a-posteriori error criterion (MAP) then the optimum decision boundary is that shown as a broken line in figure 2. This non-linear boundary may not be realised using a standard equalizer and requires f{•} to be some non-linear functional. It has already been shown that this boundary may be formed by using a number of non-linear combiner techniques including the multi-layer perceptron (MLP) [3], the Volterra series [5] or radial basis functions [6].

Any of these techniques allow an improvement in ber relative to the equivalent linear equalizer along the lines of that indicated in figure 3.

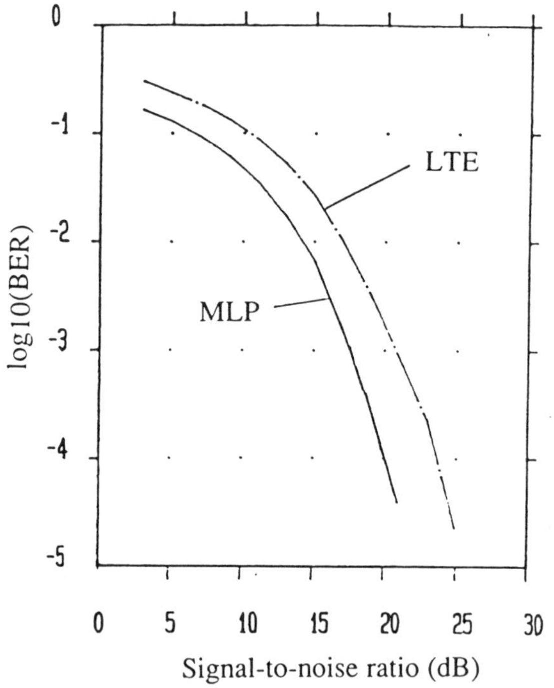

Figure 3 : Bit error rate curves for optimum (MLP) and linear equalizers (LTE).

3 Sequence Estimation

However, if we examine the simple problem postulated in figure 2 it may be noted that simply classifying the current symbol as a +1 or -1 is not truly making use of all the available information. We can in fact determine the sequence of the last 2 transmitted symbols from this information (even when there is no ISI), figure 4. By doing this we have available additional information due to the time shifted nature of successive samples. That is, there are only certain permitted transitions between states [4] and this information may be used, in conjunction with the Viterbi algorithm, to further improve the ber.

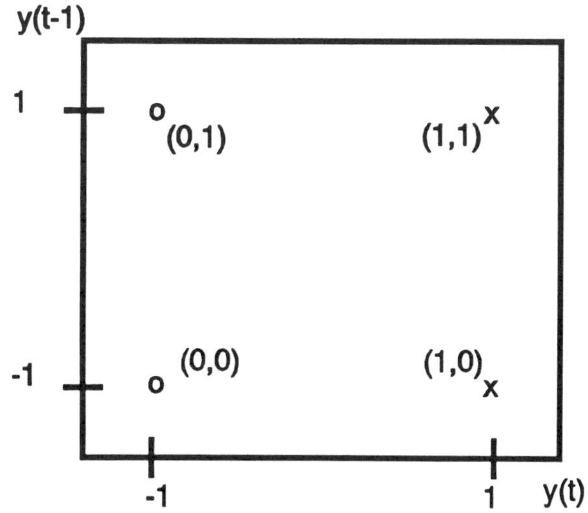

Figure 4 : Received signal states for zero ISI.

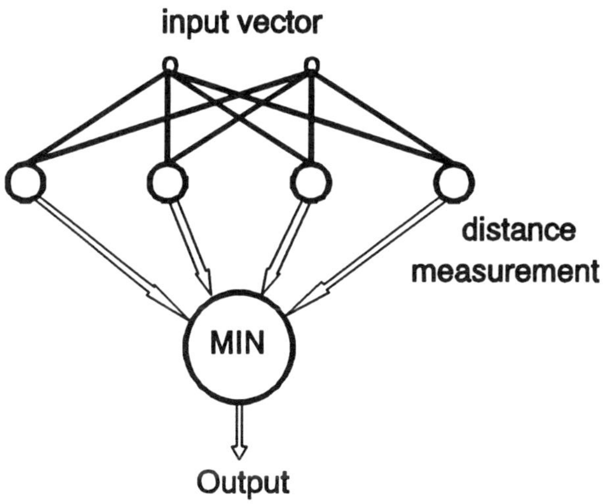

Figure 5 : Structure of the non-linear classifier.

In the case where there is no ISI this is a simple procedure which involves the measurement of the statistical distance of an observed state from one of the known transmitted states. For the case of additive Gaussian noise this may be done (optimally) by the use of a Euclidean distance measure, see figure 5. This indeed corresponds to the situation which exists with MLSE. However, if we return to the situation in figure 2 we note that the number of states has doubled (given only one sample spread of ISI). This means that twice the number of classifier centres must be used to estimate the sequence and an accurate channel model must exist to locate those centres. This is precisely the situation which exists with the MLSE detector and shows us how the explosive growth in complexity occurs with this structure.

However, we may sacrifice the optimality of the classifier, to some degree, by allowing one cluster to encompass more than one observed state. The Euclidean distance measure would be no longer applicable in this case since the point spread in the data obviates the Gaussian assumption, figure 2. If we instead allow the clusters to have a hyper-elliptic shape then our distance measure will more closely approximate the actual (bi-modal) cluster. Such a modification is possible by the use of a Mahalanobis classifier [4] rather than the Euclidean. The result of applying such a technique to the problem in figure 2 is illustrated in figure 6 where the ellipses correspond to contours of equal distance from the statistical centre. This solution to the problem involves a lesser burden in computational terms than the MLSE case but is clearly sub-optimum. It does have the additional advantage that no channel model is required.

The final result shown in figure 7 illustrates the comparative symbol error rate comparison for the case where the centres are known and the Euclidean classifier is used (KCD) and for the Mahalanobis distance (MD) classifier. The channel used here was:

$$H(z) = 0.3482 + 0.8704z^{-1} + 0.3482z^{-2}$$

The multiple MD curves correspond to different numbers of states contained within a single classifier ellipse. Where this number is 1 then the KCD results. The number, g, describes the number of groups, or classifiers, used to describe each detected state. Specifically, the number of clusters used in each case is 2^g. So with an ISI spread of 3 and an order 3 equalizer, used in figure 7, the KCD uses 32 centres to do its classification. The nearest MD curve is for g=3, that is it uses 8 centres. This comparison clearly shows a highly advantageous trade-off in favour of the MD classifier.

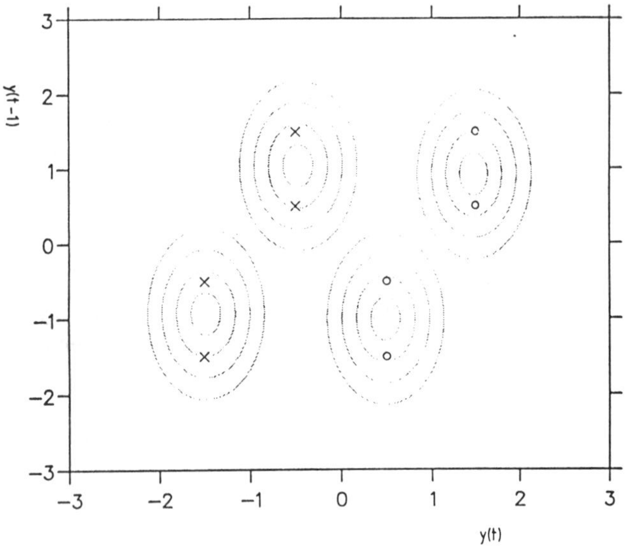

Figure 6 : Contours of equal statistical distance using the Mahalanobis distance classifier.

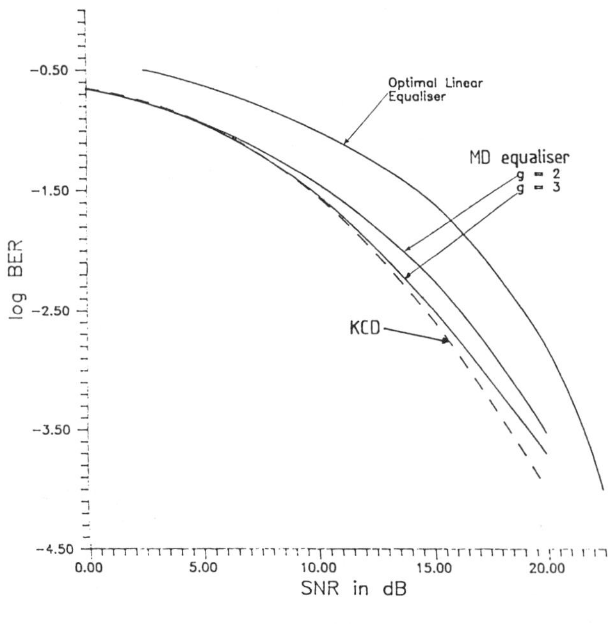

Figure 7 : Bit error rate comparisons for equalizer order 3 and ISI spread of 3 symbols.

4 Conclusions

This paper has described the use of shaped clustering techniques for the data equalization problem. The technique provides a sub-optimum approximation to MLSE detection with, potentially, much reduced complexity. Future work will aim to implement the full system including the Viterbi processor and also to examine some aspects of training which are, currently, adversely affected by the presence of multiple states in a single cluster.

Acknowledgements

This work was undertaken with sponsorship from the UK Science and Engineering Research Council. The author would like to acknowledge the valuable contributions made by Samson See.

References

1. Quereshi, S.U.H., "Adaptive Equalization", Proc. IEEE, vol.73, no.9, Sept.1985, pp.1349-1387.

2. Forney, G.D., "Maximum-likelihood sequence estimation of digital sequences in the presence of intersymbol interference", IEEE Trans.Inf.Th., vol.18, no.3, May 1972, pp.363-377.

3. Gibson, G.J., Siu, S. and Cowan, C.F.N., "The application of nonlinear structures to the reconstruction of binary signals", IEEE Trans. SP, vol.39, (8), Aug.1991, pp.1877-1884.

4. Cowan, C.F.N., "Non-linear adaptive equalisation", Proc. IEE Conf. Pub. no. 340, Sept. 1991, pp.1-5.

5. Chen, S., Gibson, G.J. and Cowan, C.F.N., "Adaptive channel equalisation using a polynominal-perceptron structure", IEE Proc. Pt. I, vol.137, no.5, Oct. 1990, pp.257-264.

6. Chen, S., Gibson, G.J., Cowan, C.F.N. and Grant, P.M., "Reconstruction of binary signals using an adaptive radial basis function equaliser", Signal Processing, vol.22, no. 1, Jan.1991, pp.77-93.

IIR Filtering for Acoustic Echo Cancellation

Mamadou MBOUP[1], Madeleine BONNET[1,2]

1. Laboratoire des Signaux et Systèmes et Groupement de Recherche TDSI, CNRS-ESE, Plateau de Moulon, 91192 Gif-sur-Yvette Cedex, France.

2. Université Paris V–Sorbonne, UFR MSI, 12 rue Cujas, 75005 Paris, France.

Abstract

Acoustic echo cancellation is based on the identification of an acoustic coupling channel by an adaptive filter. The response length of the channel induces a very high complexity while its rapid variations require a fast processing. The actual structure of the echo path is of course unknown. It is usually modelized by a FIR filter which behavior is well understood. However such filters may require up to thousands adaptive weights to correctly identify the corresponding impulse response. The few attempts found in the literature, using IIR filters to reduce the complexity, show poor performance as compared to those of the FIR filters. The fundamental question of whether the poor performance stem from a sub-optimality of the updating procedures or from an irrelevancy of the pole-zero structure itself, is investigated in this paper.

1 Introduction

The acoustic coupling between loudspeaker and microphone is a severe problem for modern audio terminals such as teleconference rooms, hands-free and radio-mobile telephones. Acoustic echo cancellation is based on the simulation of the echo path by an adaptive filter. The estimated echo signal is then subtracted from the real echo to give the residual echo.

The echo path is characterized by a very long and time varying impulse response. Traditionally, the adaptive filter is of finite impulse response (FIR) type [1], with the advantages of guaranteed stability during adaptation and unimodal mean square error (MSE) surface.

However, the FIR structure is computationally rather expensive, and the adaptive filter may require up to thousands of coefficients to match the impulse response. Consequently, there has been a growth of interest in adaptive infinite impulse response (IIR) filters for acoustic echo cancellation, with the hope of a significant reduction of complexity.

The theory of adaptive recursive filters is still incomplete. Indeed, the analysis of the algorithms is much more complex than for FIR filters because of the nonlinear property of the output error [2]. The resulting multimodal nature of the MSE surface could then lead to sub-optimal performance. Another problem is that the filter can become unstable if one or more poles update outside the unit circle for a significant amount of time. Stability monitoring is then required.

A first attempt consisted in comparing the performances of FIR and IIR adaptive filters to identify an actual echo path. Several simulations have shown that for the same number of parameters, the FIR adaptive filter behaved better than the IIR one. Similar attempts have been made concerning the same application by other authors [3]. The relatively poor performance obtained with recursive adaptive filters leads some authors conclude for the inefficiency of IIR filters in this application, while others still believe the contrary [4].

The fundamental question of whether the above-mentioned poor performance stem from a sub-optimality of the updating procedures or from an irrelevancy of the pole-zero structure itself, remains so far unanswered. In this paper, we investigate this question from two points of view, i.e.:

- *does it make any physical sense to fit a pole-zero model to the phenomenon underlying the echo?*
- the echo path can be entirely characterized by its impulse response (which could be rigorously infinite). A FIR transfer function model can exactly match the useful part of it. Now, *is it possible to approximate with a small error bound, this all-zero transfer function by a pole-zero transfer function with fewer parameters ?*

These two points are treated respectively in terms of wave equation in a closed space and Hankel-norm approximation theory.

2 Physical Interpretation

The main idea in this section is to understand and interprete what would be the meaning of the presence of poles, in a pole/zero transfer function model, for acoustic echo path modelling. We know *a priori* that if there is any physical meaning then, it is necessarely related to the vibrational modes of the sound wave

1058-6393/91 $01.00

in the room. By this way, Zitzewitz and al. [3] suggested to assign two complex conjugate poles to each maximum of the echo path frequency response. This method is however not fruitful since, as shown in figure 1, there are no emphasized maxima.

Another method is to start with the wave equation in a rigid room i.e.:

$$\Delta p + k^2 p = 0, \qquad k = \frac{\omega}{c} \tag{1}$$

where p is the sound pressure, ω is an angular frequency and c is the sound velocity. The eigenfunctions of this equation correspond to the vibrational modes of the sound pressure. In the case of a rectangular room, the number of these vibrational modes in the frequency range from 0 to f, is given by [5]:

$$N_f = \frac{4\pi}{3} V \left(\frac{f}{c}\right)^3 + \frac{\pi}{4} S \left(\frac{f}{c}\right)^2 + \frac{L}{8}\frac{f}{c}. \tag{2}$$

In this expression, V is the volume, S the area of all walls and L the sum of all edge lengths of the walls of the rectangular room.

To give an example: for a room of dimensions $L_x = 5m$, $L_y = 4m$ and $L_z = 2.8m$, the number of eigenfrequencies in the range of [0 , 3.4kHz] is: $N_f = 241\ 730$. Since this astronomic value represents half the number of poles that would be required to exactly model the physical phenomenon, it is attractive, at first glance, to argue for the irrelevance of a pole-zero model. However, the acoustical phenomenon is so complicated that it can be investigated only under statistical considerations. Indeed, the eigenfrequencies are highly overlapping and, therefore, they can be reduced to some averages which provide a much more parsimonious number of vibrational modes.

However, this method provides only a qualitative answer, explaining how a pole-zero model would relate to the physical phenomenon of acoustic echo. As for the method of [3], it is not explicit wrt the order of the model.

3 Transfer function approximation

In this section, a method is presented to approximate a finite impulse response transfer function model $H(z)$, by a pole-zero transfer function model. This method was proposed by Glover [6] and provides an approximation error bound. When applied to the case where $H(z)$ represents the useful part of the acoustic echo path, we show that the corresponding error bound is small.

Let $H(z)$ be given by:

$$H(z) = h_0 + h_1 z^{-1} + \cdots + h_L z^{-L} \tag{3}$$

3.1 Balanced Realization

Consider the dynamical representation of the input-output system of $H(z)$:

$$\begin{cases} X(n+1) &= \mathbf{A}X(n) + Bx(n) \\ y(n) &= \mathbf{C}X(n) + Dx(n) \end{cases} \tag{4}$$

where

$$\mathbf{A} = \begin{bmatrix} 0 & 0 & & & \\ 1 & 0 & & \mathbf{0} & \\ & \ddots & \ddots & & \\ & & 1 & 0 & \\ & & & 1 & 0 \end{bmatrix}; \; B = [\ 1, 0, \ \cdots \ , 0\ ]^T;$$

$$D = h_0; \; \mathbf{C} = [\ h_1, \ \cdots \ , h_L\ ];$$

$$X(n) = [\ x(n-1), \ \cdots \ , x(n-L)\ ]^T.$$

It is straightforward to verify that :

$$H(z) = \mathbf{C}(z\mathbf{I} - \mathbf{A})^{-1}B + D \tag{5}$$

Let $\mathbf{P}$ and $\mathbf{Q}$ be the solutions of the following Lyapunov equations:

$$\begin{cases} \mathbf{APA}^T + BB^T &= \mathbf{P} \\ \mathbf{AQA}^T + C^TC &= \mathbf{Q} \end{cases} \tag{6}$$

The matrices $\mathbf{P}$ and $\mathbf{Q}$ are symmetric, they are respectively positively defined iff the system is respectively completely controllable and completely observable i.e.:

$$\begin{cases} rg\,[\ B, \ \mathbf{A}B, \ \cdots \ , \mathbf{A}^{L-1}B\ ] &= L \\ rg\,[\ \mathbf{C}^T, \ \mathbf{A}^T\mathbf{C}^T, \ \cdots \ , \mathbf{A}^{L-1^T}\mathbf{C}^T\ ] &= L \end{cases} \tag{7}$$

These matrices are respectively the controllability and observability gramiams. The transfer function $H(z)$ given by eq(3) is invariant under regular state space transformation. Furthermore, $\mathbf{P} = \mathbf{I}$ is a solution of eq(6), and $\mathbf{Q}$ can be diagonalized as:

$$\mathbf{Q} = \mathbf{U}\mathbf{\Sigma}^2\mathbf{U}^T \tag{8}$$

with

$$\mathbf{U}\mathbf{U}^T = \mathbf{I} \ and \ \mathbf{\Sigma} = \mathrm{diag}[\sigma_1, \ldots, \sigma_L]$$

Consider now the state space transformation $\mathbf{T}$ defined as follow:

$$\mathbf{T} = \mathbf{U}\mathbf{\Sigma}^{-\frac{1}{2}} \tag{9}$$

$$\mathbf{A}_e = \mathbf{T}^{-1}\mathbf{AT}\ ;\ B_e = \mathbf{T}^{-1}B\ ;\ \mathbf{C}_e = \mathbf{CT}^{-1} \tag{10}$$

It is straightforward that with this transformation, the controllability and observability gramians are equal and diagonal:

$$\mathbf{P}_e = \mathbf{Q}_e = \mathbf{\Sigma}. \tag{11}$$

The corresponding state space representation is therefore said to be *balanced*.

3.2 Hankel-Norm Approximation

Consider in the sequel the balanced realization ($\mathbf{A}_e$, B_e, $\mathbf{C}_e$) with the partition :

$$\mathbf{A}_e = \begin{bmatrix} \mathbf{A}^e{}_{11} & \vdots & \mathbf{A}^e{}_{12} \\ \cdots & & \cdots \\ \mathbf{A}^e{}_{21} & \vdots & \mathbf{A}^e{}_{22} \end{bmatrix} \; ; \; B_e = \begin{bmatrix} B^e{}_1 \\ \cdots \\ B^e{}_2 \end{bmatrix};$$

$$\mathbf{C}_e = \begin{bmatrix} \mathbf{C}^e{}_1 & \vdots & \mathbf{C}^e{}_2 \end{bmatrix} \; ; \boldsymbol{\Sigma} = \begin{bmatrix} \boldsymbol{\Sigma}_1 & 0 \\ 0 & \boldsymbol{\Sigma}_2 \end{bmatrix}$$

with

$$\begin{cases} \boldsymbol{\Sigma}_1 &= \mathrm{diag}[\sigma_1, \ldots, \sigma_K] \\ \boldsymbol{\Sigma}_2 &= \mathrm{diag}[\sigma_{K+1}, \ldots, \sigma_L] \end{cases}$$

where, by convention,

$$\sigma_1 \geq \sigma_2 \geq \ldots \geq \sigma_L \geq 0 \; . \tag{12}$$

If the pole-zero transfer function $\widehat{H}(z) = \frac{P(z)}{Q(z)}$ is the Hankel-norm approximation of $H(z)$, it is shown [6] that the error bound is:

$$||H(z) - \widehat{H}(z)||_{\mathrm{H}} \leq 2 \; \mathrm{Tr}(\mathrm{diag}[\sigma_{1+K}, ., \sigma_K]), \tag{13}$$

where

- $||H(z)||_{\mathrm{H}} \stackrel{def}{=} \lambda^{\frac{1}{2}}{}_{max}(\mathbf{PQ})$
- $K = N - M$, *N being the number of coefficients of* $H(z)$
- $M = M_p + M_q$, M_p and M_q *being, respectively, the number of coefficients of* $P(z)$ and $Q(z)$.

3.3 Application

Let $H(z)$ represent the useful part of an acoustic echo path impulse response and $\mathcal{H}$ be the Hankel matrix of the coefficients of $H(z)$. Then, we have:

$$\mathcal{H} = \begin{bmatrix} h_1 & h_2 & \cdots & h_L \\ h_2 & \cdots & h_L & 0 \\ & & \mathbf{0} & \\ h_L & & & \end{bmatrix} \tag{14}$$

Since $\mathcal{H}$ is symmetric, it can be diagonalised as:

$$\mathcal{H} = \mathbf{V} \boldsymbol{\Lambda} \mathbf{V}^T \tag{15}$$

where $\boldsymbol{\Lambda}$ is diagonal. Furthermore, it is shown [7] that the eigenvalues of this Hankel matrix are related to the elements of $\boldsymbol{\Sigma}$ through:

$$\boldsymbol{\Sigma} = |\boldsymbol{\Lambda}| \tag{16}$$

Therefore, from equations (14) and (16), we have:

$$\det(\boldsymbol{\Sigma}) = |\det(\mathcal{H})| \tag{17}$$

$$\Leftrightarrow \quad \prod_{i=1}^{L} \sigma_i = |h_L|^L \tag{18}$$

In the case of room acoustic, the above product is relatively very small because of the exponential decrease of the impulse response (see figure 2). It results an important number of insignificant value of σ_i and then a small approximation error bound is obtained with few parameters.

In figure 3, a normalized Hankel-norm approximation error bound is ploted vs the order of the pole-zero model. The all-zero transfer function to be approximated was the first 512 coefficients of an actual radio-mobile echo path. In this figure, a 128 order of a pole-zero model yields a normalized error bound of −30 dB.

4 Conclusion

The encouraging results suggest that IIR adaptive filtering must be considered as a possible alternative to FIR adaptive filtering for the acoustic echo cancellation problem. Further investigation is still necessary, mainly in the direction of adaptive algorithm design.

References

[1] A. Gilloire, "State of the Art in Acoustic Echo Cancellation," Adaptive Algorithms: Applications and Non Classical Schemes, *Publicacions da Universidade de Vigo*, 1991, pp 20–31.

[2] C. R. Johnson, Jr., "Adaptive IIR Filtering: Current Results and Open Issues," *IEEE Trans. Inform. Theory*, vol. 30, n° 2, March 1984, pp. 237–250.

[3] A. V. Zitzewitz, "Considerations on Acoustic Echo Cancellation Based on Real Time Experiments," in *Proc. EUSIPCO-90*, Barcelona, Spain, 1990, pp. 1987–1990.

[4] S. J. Flockton, S. Gudvangen, "Pole–Zero Modelling of Room Transfer Functions," 2nd *Intern. Workshop on Acous. Echo Cancellation*, L'Aquila, Italy, Sept 1991.

[5] H. Kuttruff, *Room Acoustics*, Applied Science Publishers, LTD, London 1979.

[6] K. Glover, "All Optimal Hankel-norm Approximations of Linear Multivariable Systems and their L^∞-error Bounds," *Int. Journ. of Cont.* vol. 39, n° 6, 1984, pp. 1115–1193.

[7] B. Beliczynski, G. D. Cain and I. Kale, "Conversion of FIR Prototype Filters to IIR Filters by Means of Balanced Model Reduction," *ISSPA-90*, Gold-Coast, Australia, Aug. 1990, pp. 85–90.

Figure 1: Frequency response of an actual teleconference room

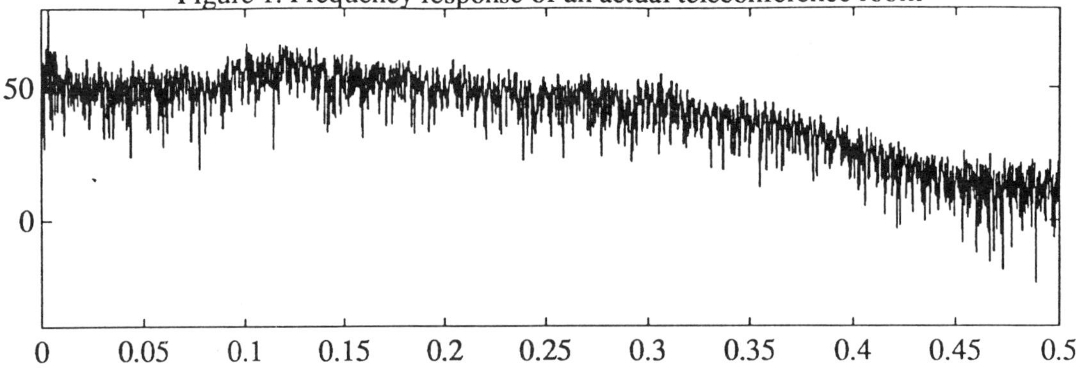

Figure 2: Impulse response of a teleconference room

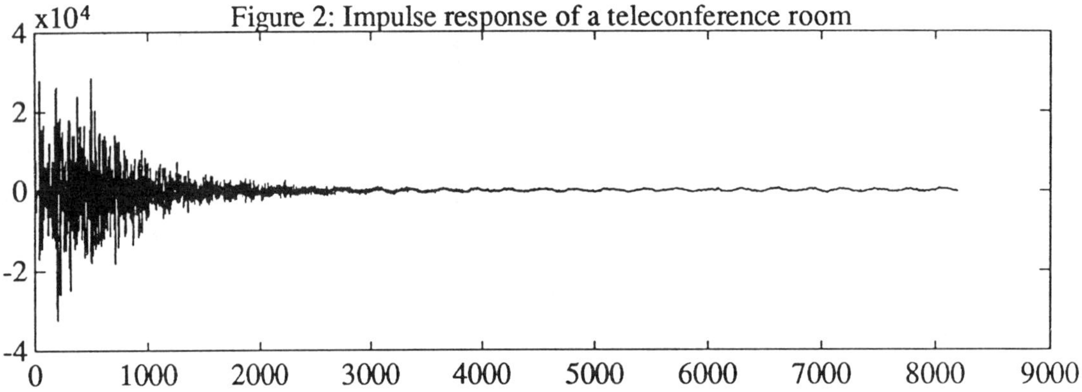

Figure 3: Approximation error bound (dB) vs order of the IIR filter

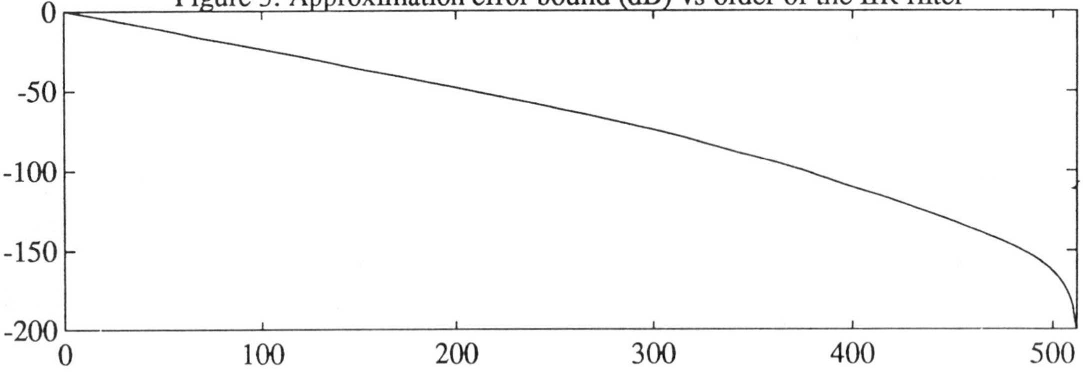

Output-Error Adaptive Bilinear Filters

Junghsi Lee and V. John Mathews

Department of Electrical Engineering
University of Utah
Salt Lake City, UTAH 84112

Abstract

This paper presents an overview of several gradient-type recursive algorithms for adaptive nonlinear filters equipped with bilinear system models. Bilinear models are attractive because they can approximate a large class of nonlinear systems with great parsimony in the use of coefficients. Two algorithms of complexity $O(N^3)$ (N is the memory span of the bilinear system model used in the adaptive filter) multiplications per sample and two other algorithms of complexity $O(N^2)$ multiplications per sample are presented in this paper. The results of several computer experiments show that at least one of the $O(N^2)$ complexity algorithms works almost as well as the more complex algorithms.

1 Introduction

This paper introduces and compares several adaptive nonlinear filters that employ the bilinear system model. The input-output relationship for a bilinear system is given by

$$y(n) = \sum_{i=0}^{N-1} a_i x(n-i) + \sum_{i=1}^{N-1} b_i y(n-i) + \sum_{i=0}^{N-1}\sum_{j=1}^{N-1} c_{i,j} x(n-i) y(n-j) \qquad (1)$$

where x(n) and y(n) are the input and output signals, respectively, of the system. This model is attractive because it can adequately approximate a large class of nonlinear systems with good parsimony in the use of coefficients, much like linear infinite-impulse-response (IIR) filters can approximate a large class of linear systems with good parsimony. Bilinear system models have been used in a wide class of problems in engineering, biology, socio-economics, and ecology [2], [7].

In spite of the potential benefits of bilinear system models, the theory of adaptive nonlinear filters employing nonlinear feedback structures is still very much in its infancy. Some very recent work in adaptive bilinear filtering employ recursive least-squares (RLS) adaptation algorithms or its variations [1], [5]. Among the techniques that are available, even the most computationally efficient schemes require $O(N^3)$ multiplications per sample. It is the purpose of this paper to introduce several LMS type adaptive bilinear filters. Among these filters are two simplified structures which need $O(N^2)$ multiplications per sample and still performs reasonably well.

The remainder of this paper is organized as follows. The following section describes several LMS-type adaptive bilinear filtering algorithms. Section III contains simulation results that demonstrate and compare the performances of the algorithms. Finally, the concluding remarks are made in Section IV.

2 Derivations of the Adaptive Bilinear Filters

We will first derive an output-error LMS-type adaptive bilinear filtering algorithm. For convenience, let us denote it as ABF1. This algorithm has a computational complexity of N^3+3N^2-N. We also present a simplified version (ABF2) of the ABF1. ABF2 requires $5N^2-N$ computations per sample. A further simplified version (AFB3) that needs $2N^2+2N$ computations per sample only will then be introduced. Finally, we will present bilinear filtering algorithm (ABF4) which is the extension to Fan's work [3]. The computational complexity of ABF4 is the same as that of ABF1. It must be noted that all the algorithms are extensions of adaptive IIR filters available for linear system models to the nonlinear case. Our objective is to study the usefulness of such techniques in adaptive nonlinear filtering problems.

2.1 Derivation of ABF1

Let d(n) and x(n) represent the desired response signal and the input signal, respectively, to the adaptive filter. Let $\hat{d}(n)$ denote the *a priori* output of the adaptive filter at time n, i.e.,

1058-6393/91 $01.00 © 1991 IEEE

$$\hat{d}(n) = \sum_{i=0}^{N-1} \hat{a}_i(n-1)x(n-i) + \sum_{i=1}^{N-1} \hat{b}_i(n-1)\hat{d}(n-i)$$

$$+ \sum_{i=0}^{N-1}\sum_{j=1}^{N-1} \hat{c}_{i,j}(n-1)x(n-i)\hat{d}(n-j) \quad (2)$$

In (2), $\hat{a}_i(n-1)$, $\hat{b}_i(n-1)$ and $\hat{c}_{i,j}(n-1)$ are the coefficients at time n-1 of the adaptive bilinear filter.

For simplicity, vector representation will be employed. The input vector $\hat{X}(n)$ at time n, which has N^2+N-1 elements, is defined as

$$\begin{aligned}\hat{X}(n) = [&x(n), x(n)\hat{d}(n-1), ..., x(n)\hat{d}(n-N+1), \hat{d}(n-1), \\ &x(n-1)\hat{d}(n-1), ..., x(n-N+1)\hat{d}(n-1), ..., \\ &x(n-N+2), x(n-N+2)\hat{d}(n-N+1), \hat{d}(n-N+1), \\ &x(n-N+1)\hat{d}(n-N+1), x(n-N+1)]^T, \end{aligned} \quad (3)$$

where the superscript T denotes transposition. Also, the N^2+2N-1 coefficient vector W_n at time n is defined as

$$\begin{aligned}W_n = [&\hat{a}_0(n), \hat{c}_{0,1}(n), ..., \hat{c}_{0,N-1}(n), \hat{b}_1(n), \hat{c}_{1,1}(n), ..., \hat{c}_{N-1,1}(n) \\ &, ..., \hat{a}_{N-2}(n), \hat{c}_{N-2,N-1}(n), \hat{b}_{N-1}(n), \hat{c}_{N-1,N-1}(n), \hat{a}_{N-1}(n)]^T\end{aligned} \quad (4)$$

The problem is then to find a gradient descent solution for the coefficients of the adaptive filter which attempts to minimize the cost function

$$\begin{aligned}J(n) &= (d(n) - \hat{d}(n))^2 \\ &= (d(n) - W_{n-1}^T\hat{X}(n))^2\end{aligned} \quad (5)$$

at each time. The gradient descent solution is given by

$$W_n = W_{n-1} - \frac{\mu}{2}\frac{\partial(d(n) - W_{n-1}^T\hat{X}(n))^2}{\partial W_{n-1}} \quad (6)$$

where μ is a small positive constant that controls the rate at which the adaptive filter converges. Note that $\hat{d}(n)$ and therefore $\hat{X}(n)$ are functions of W_{n-1}. The gradient in (6) may be obtained from

$$-\frac{\partial(d(n) - W_{n-1}^T\hat{X}(n))}{\partial W_{n-1}} = \hat{X}(n) + \sum_{i=1}^{N-1}\hat{b}_i(n-1)\frac{\partial\hat{d}(n-i)}{\partial W_{n-1}}$$

$$+ \sum_{i=0}^{N-1}\sum_{j=1}^{N-1}\hat{c}_{i,j}(n-1)x(n-i)\frac{\partial\hat{d}(n-j)}{\partial W_{n-1}} . \quad (7)$$

Note that (7) indicates the necessity of reevaluation of the derivatives of past values of $\hat{d}$ with respect to W_{n-1}. An assumption commonly employed in the adaptive IIR filtering literature is that μ is sufficiently small such that $W_{n-1} \approx W_{n-2} \approx ... \approx W_{n-N}$ [6]. Using this approximation and the fact that d(n) is not a function of W_{n-1}, we may write (7) as

$$\frac{\partial\hat{d}(n)}{\partial W_{n-1}} = \hat{X}(n) + \sum_{i=1}^{N-1}\hat{b}_i(n-1)\frac{\partial\hat{d}(n-i)}{\partial W_{n-1-i}} +$$

$$\sum_{i=0}^{N-1}\sum_{j=1}^{N-1}\hat{c}_{i,j}(n-1)x(n-i)\frac{\partial\hat{d}(n-j)}{\partial W_{n-1-j}} . \quad (8)$$

Let

$$\psi(n) = \frac{\partial\hat{d}(n)}{\partial W_{n-1}} \quad (9)$$

Then, (8) simplifies to

$$\psi(n) = \hat{X}(n) + \sum_{i=1}^{N-1}\hat{b}_i(n-1)\psi(n-i)$$

$$+ \sum_{i=0}^{N-1}\sum_{j=1}^{N-1}\hat{c}_{i,j}(n-1)x(n-i)\psi(n-j) . \quad (10)$$

Finally, we summarizes ABF1 as follows.

$$\varepsilon(n) = d(n) - W_{n-1}^T\hat{X}(n) \quad (11)$$

$$\psi(n) = \hat{X}(n) + \sum_{j=1}^{N-1}\psi(n-j)\{\hat{b}_j(n-1) + \sum_{i=0}^{N-1}\hat{c}_{i,j}(n-1)x(n-i)\} \quad (12)$$

$$W_n = W_{n-1} + \mu\varepsilon(n)\psi(n) \quad (13)$$

The above algorithm is an adaptation of [6] to the bilinear case. As discussed earlier, the computational complexity of ABF1 is N^3+3N^2-N multiplications per sample and it needs to store N-1 past values of the N^2+N-1 element vector ψ. By employing arguments similar to those used in [6], the computational burden of the adaptive filter can be reduced. This will result in algorithm ABF2.

2.2 Derivation of ABF2

The key to the derivation of ABF2 is the fact that the last N^2-N-1 elements of $\hat{X}(n)$ in (3) are delayed versions of the first 2N elements of $\hat{X}(n)$. This motivates the approximation of replacing the last N^2-N-1 elements of ψ with appropriate delayed versions of the first 2N elements of ψ. Instead of (12), ABF2 uses

$$\psi_{2N}(n) = \hat{X}_{2N}(n) + \sum_{j=1}^{N-1}\psi_{2N}(n-j)\,\{\,\hat{b}_j(n-1) + \sum_{i=0}^{N-1}\hat{c}_{i,j}(n-1)x(n-i)\,\}\,, \quad (14)$$

where ψ_{2N} and $\hat{X}_{2N}$ denote the vectors that contain the first 2N entries of ψ and $\hat{X}$, respectively. The N^2+N-1 element vector ψ, which is needed for (13), is constructed from the current and past values of ψ_{2N}. Note that ABF2 needs to store N-1 past values of the 2N element vector ψ_{2N} and requires only $5N^2$–N multiplications per sample.

2.3 Derivation of ABF3

In 1976, Feintuch proposed a highly simplified linear adaptive recursive LMS filtering algorithm [4]. He proposed to use

$$\psi(n) = \hat{X}(n) \quad (15)$$

for the gradient vector. This can be viewed as a simplified version obtained by neglecting the second term on the right hand side of (12). Adaptation of this idea will result in an adaptive bilinear filter that needs only $2N^2$+2N multiplications per sample.

2.4 Derivation of ABF4

Recently, Fan proposed some adaptive IIR filters [3] which are based on the Steiglitz-McBride scheme. We have also extended Fan's filters to adaptive bilinear filters. The resulting algorithms are similar to ABF1 and ABF2, depending on whether ψ or ψ_{2N} is employed. We will discuss the scheme (ABF4) involving ψ here. The only difference be-tween ABF1 and ABF4 is that the latter uses d, in comparison with $\hat{d}$ used by ABF1, in (12). That is

$$\psi(n) = X(n) + \sum_{j=1}^{N-1}\psi(n-j)\{\hat{b}_j(n-1) + \sum_{i=0}^{N-1}\hat{c}_{i,j}(n-1)x(n-i)\}\,, \quad (16)$$

where X(n) is defined in a similar way as $\hat{X}(n)$ with the exception that d is used in place of $\hat{d}$ in the definition.

3 Simulation Results

In this section, we present the results of several experiments. The problem considered here is that of identifying a bilinear system with N=3. The coefficients of the unknown bilinear system can be found in Table I. The input signal x(n) was a white, zero-mean and pseudorandom Gaussian noise. The variance of x(n) was 0.18, and the power of y(n) was about unity. We first considered a white, Gaussian, zero-mean measurement noise with variance ξ_o. Two different noise levels were studied: ξ_o=0.001 and ξ_o=0.1. We also conducted the experiments where the measurement noise was colored. The colored noise was obtained through the equation

$$v(n) = \eta(n) - 0.1v(n) + 0.8v(n-2), \quad (17)$$

where $\eta(n)$ belonged to a zero-mean white Gaussian process. The variance of $\eta(n)$ was adjusted such that the variance of v(n) (for convenience, denote it as ξ_o again) was either 0.1 or 0.001 approximately. The adaptive filter was implemented with the same structure and the same number of coefficients as that of the unknown system. The step-size μ was set to 0.009. All of the results presented were obtained from ensemble averages over fifty independent runs.

Tables II and III contain the mean and the variance values of selected filter coefficients when the measurement noises were white and colored, respectively. These values were obtained by time averaging the corresponding ensemble averaged sequences in the range [9000,10000]. The steady-state *a priori* mean-squared error values obtained in a similar manner are given in Table IV. All the algorithms seem to work reasonably well. It appears that the performances of ABF1 and ABF2 are very similar even though ABF2 is an order of magnitude less complex than ABF1. ABF3 also works quite well when the measurement noise is white. The coefficient trajectories, which are not presented here, indicated that AFB4 has a slower convergence rate than the other algorithms. Also, AFB4 does not work well when the measurement noise is colored and the noise level is relatively high.

4 Concluding Remarks

This paper presented an overview of several LMS type adaptive bilinear filtering algorithms. The results of a large number of experiments showed that the algorithms work well in several situations. While it is too early to draw definitive conclusions, it is possible that the more efficient algorithms perform as well as the others in adaptive identification of bilinear systems. On the basis of the experiments presented, it appears that ABF2, because of its simplicity and good performance, may be a compromise choice in many applications. However, much more analytical and empirical works need to be done to further evaluate the performance of these algorithms.

One significant problem that we did not address in this paper is that of the stability of the identified systems. In general, the stability conditions for bilinear systems depends on the characteristics of the input signal also, and this complicates the problem. While there are several works on the stability of bilinear systems [8], [9] available in the literature, the on-line implementation of such schemes to monitor the stability of the systems is very computationally intense. We are currently investigating simplified algorithms for real-time stability monitoring.

Acknowledgement

This work was supported in part by NSF Grant No. MIP 8922146.

References

[1] H. K. Baik and V. J. Mathews, "Adaptive Lattice Bilinear Filters," submitted to *IEEE Transactions on Acoustics, Speech, and Signal Processing*, December 1990.

[2] C. Bruni, G. Dipillo and G. Koch, "Bilinear Systems: An Appealing Class of 'Nearly Linear' Systems in Theory and Applications,"*IEEE Transactions on Automatic Control*, Vol. AC-19, No. 4, pp. 334-348, August 1974.

[3] H. Fan and W. K. Jenkins, "A New Adaptive IIR Filter," *IEEE Transactions on Circuits and Systems*, Vol. CAS-33, No. 10, pp. 939-947, October 1986.

[4] P. L. Feintuch, "An Adaptive Recursive LMS Filter," *Proceedings of the IEEE*, Vol. 64, pp. 1622-1624, November 1976.

[5] F. Fnaiech and L. Ljung," Recursive Identification of Bilinear Systems," *International Journal of Control*, Vol. 45, No. 2, pp. 453-470, 1987.

[6] C. R. Johnson, Jr., "Adaptive IIR Filtering: Current Results and Open Issues," *IEEE Transactions on Information Theory*, Vol. IT-30, No. 2, pp. 237-250, March 1984.

[7] R. R. Mohler, *Bilinear Control Processes*, Academic Press, New York, 1973.

[8] C. S. Kubrusly and O. L. V. Costa, "Mean Square Stability Conditions for Discrete Stochastic Bilinear Systems," *IEEE Transactions on Automatic Control*, Vol. AC-30, No. 11, pp. 1082-1087, November 1985.

[9] J. Liu, "On the Existence of a General Multiple Bilinear Time Series," *Journal of Time Series Analysis*, Vol. 10, No. 4, pp. 341-355, 1989.

Table I
Coefficients of the Unknown Bilinear System
Used in the Experiments

a_0= 1.0	a_1= 1.0	a_2= 1.0	b_1= 0.5	b_2= - 0.5	
$c_{0,1}$= 0.3	$c_{0,2}$= 0.1	$c_{1,1}$= - 0.2	$c_{1,2}$= - 0.2	$c_{2,1}$= 0.1	$c_{2,2}$= 0.3

Table IV
Mean-Squared Error

	ABF1		ABF2		ABF3		ABF4	
	ξ_o=0.001	ξ_o=0.1	ξ_o=0.001	ξ_o=0.1	ξ_o=0.001	ξ_o=0.1	ξ_o=0.001	ξ_o=0.1
White Noise	1.029×10^{-3}	0.1050	1.029×10^{-3}	0.1051	1.014×10^{-3}	0.1036	1.032×10^{-3}	0.1054
Colored Noise	1.056×10^{-3}	0.1031	1.056×10^{-3}	0.1030	1.059×10^{-3}	0.1034	1.057×10^{-3}	0.1109

Table II
Mean and Variance of Filter Coefficients
(Measurement Noise is White)

		ABF1		ABF2		ABF3		ABF4	
		ξ_o=0.001	ξ_o=0.1	ξ_o=0.001	ξ_o=0.1	ξ_o=0.001	ξ_o=0.1	ξ_o=0.001	ξ_o=0.1
a_1	Mean	0.9999	1.0057	1.0001	1.0089	0.9994	1.0077	0.9999	1.0074
	Variance	4.78×10^{-6}	3.80×10^{-4}	5.12×10^{-6}	4.64×10^{-4}	4.85×10^{-6}	4.77×10^{-4}	4.73×10^{-6}	4.77×10^{-4}
b_1	Mean	0.5004	0.4943	0.5003	0.4914	0.5001	0.4957	0.5016	0.4923
	Variance	4.79×10^{-6}	5.93×10^{-4}	5.02×10^{-6}	5.97×10^{-4}	3.23×10^{-6}	3.98×10^{-4}	4.86×10^{-6}	6.48×10^{-4}
$c_{2,2}$	Mean	0.3001	0.2941	0.2996	0.2927	0.2999	0.2949	0.3015	0.2956
	Variance	5.15×10^{-6}	5.77×10^{-4}	5.74×10^{-6}	7.63×10^{-4}	6.57×10^{-6}	4.80×10^{-4}	5.22×10^{-6}	9.18×10^{-4}

Table III
Mean and Variance of Filter Coefficients
(Measurement Noise is Colored)

		ABF1		ABF2		ABF3		ABF4	
		ξ_o=0.001	ξ_o=0.1	ξ_o=0.001	ξ_o=0.1	ξ_o=0.001	ξ_o=0.1	ξ_o=0.001	ξ_o=0.1
a_1	Mean	1.0002	1.0027	1.0002	1.0047	0.9995	1.0044	1.0011	1.1000
	Variance	1.78×10^{-6}	2.09×10^{-4}	2.23×10^{-6}	2.28×10^{-4}	4.53×10^{-6}	5.31×10^{-4}	1.86×10^{-6}	5.07×10^{-4}
b_1	Mean	0.5006	0.4989	0.5006	0.4975	0.5005	0.4988	0.5010	0.4027
	Variance	1.59×10^{-6}	1.53×10^{-4}	1.89×10^{-6}	1.84×10^{-4}	2.43×10^{-6}	2.12×10^{-4}	1.65×10^{-6}	9.74×10^{-4}
$c_{2,2}$	Mean	0.3007	0.3018	0.3004	0.3006	0.3004	0.3032	0.3014	0.2220
	Variance	2.56×10^{-6}	2.33×10^{-4}	3.43×10^{-6}	2.85×10^{-4}	7.67×10^{-6}	5.50×10^{-4}	2.58×10^{-6}	6.09×10^{-4}

Stochastic Convergence Analysis of a Two-Layer Perceptron for a System Identification Model[1]

Neil J. Bershad[2], Colin F.N. Cowan[3] and John J. Shynk[4]

1. This work was partially supported by the Science and Engineering Council of Great Britain, Lawrence Livermore National Laboratory, Rockwell International Corporation, and the University of California MICRO Program.

2. Department of Electrical and Computer Engineering, University of California, Irvine, CA 92717.

3. Department of Electronic and Electrical Engineering, University of Technology, Loughborough, Leicestershire, LE11 3TU, United Kingdom.

4. Center for Information Processing Research, Department of Electrical and Computer Engineering, University of California, Santa Barbara, CA 93106

Abstract

This paper studies the stationary points of a two-layer perceptron which attempts to identify the parameters of a specific nonlinear system. The training sequence is modelled as the binary output of the nonlinear system when the input is comprised of an independent sequence of zero mean Gaussian vectors with independent components. The training rule is a modified version of Rosenblatt's algorithm. It is shown that the two-layer perceptron correctly identifies all parameters of the unknown nonlinear system.

I. Introduction

The backpropagation (BP) algorithm is a widely used training procedure that adjusts the connection weights of a multilayer perceptron [1], [2]. It is a gradient-descent method that minimizes the mean-square error (MSE) between the perceptron output signals and a set of training sequences or desired response signals [3]. The configuration of a perceptron with two-layers (one hidden layer) is shown in Fig. 1. The BP algorithm requires that the threshold elements at the summing nodes be differentiable nonlinearities.

In this paper, we study the stationary points of a two-layer perceptron structure which attempts to identify the parameters of a specific nonlinear system as shown in Fig. 2. The nonlinear system has the same type of structure as would exist for a two-layer perceptron. The training sequence is generated by passing a Gaussian data vector X(n) through two linear systems F_1 and F_2. The output of each linear system is passed through separate signum functions (hard nonlinearities). The outputs of the signum functions are weighted by h_1, h_2, respectively, and summed and processed by a second differentiable nonlinearity. Depending on F_1, F_2, h_1, h_2, and $g(\cdot)$, this structure defines a wedge-shaped decision region in N-space (the intersection of two hyperplanes, each of which passes through the origin). For N = 2, this corresponds to the region between two straight lines which intersect at the origin. The adaptive portion of Fig. 2, corresponding to a two-layer perceptron, consists of the same nonlinear structure but with unknown weights W_1, W_2, q_1, and q_2. Note that the nonlinearities following W_1 and W_2 are not differentiable. This implies that W_1, W_2, q_1, and q_2 cannot be adapted using the BP algorithm. The output nonlinearity $g(\cdot)$ in the system model is assumed to be an arbitrary odd symmetric differentiable function.

II. System Identification Model

A. Gaussian Model for the Training Data

The input vector $X^T(n) = [x_1(n), x_2(n), \ldots, x_N(n)]$ consists of independent, identically distributed (i.i.d.) Gaussian variates with zero means and common variance σ_x^2, i.e., $E[x_i(n)] = 0$ all i and n

$$E[x_i(n)\, x_j(m)] = \begin{cases} 0, & i \neq j \text{ or } n \neq m \\ \sigma_x^2, & i = j \text{ and } n = m \end{cases} \tag{2.1}$$

Hence, the covariance matrix for X(n) is modeled as,

$$E[X(n)\, X^T(n)] = \sigma_x^2\, I. \tag{2.2}$$

The network is trained as follows. A vector X(n) is randomly selected. Two intermediary training signals with values ±1 are generated according to $\mathrm{sgn}[d_i(n)]$, i = 1, 2,

where $$d_1(n) = F_1^T X(n), \quad d_2(n) = F_2^T X(n). \tag{2.3}$$

The final training sequence is given by

$$d(n)=g[h_1\, \mathrm{sgn}(d_1) + h_2\, \mathrm{sgn}(d_2)] \tag{2.4}$$

1058-6393/91 $01.00 © 1991 IEEE

$$\text{where} \qquad \mathrm{sgn}(x) = \begin{cases} 1, & x \geq 0 \\ -1, & x < 0 \end{cases} \qquad (2.5)$$

If $g(x) = \mathrm{sgn}(x)$, then $d(n) = \pm 1$. Otherwise, for a soft nonlinearity, $d(n)$ lies in the interval (-1, 1). The output $y(n)$ of the two-layer neural net is given by

$$y(n) = g[(q_1 \,\mathrm{sgn}(W_1^T X(n)) + q_2 \,\mathrm{sgn}\,(W_2^T X(n))]. \qquad (2.6)$$

The error signal (which drives the learning algorithm) is

$$e(n) = d(n) - y(n). \qquad (2.7)$$

Clearly, $e(n) = 0$ when the parameters of the two-layer perceptron are equal to the parameters of the nonlinear system, i.e., $W_1 = F_1$, $W_2 = F_2$, $q_1 = h_1$, and $q_2 = h_2$. Is this solution unique or are there other possible solutions? Obviously, if $W_1 = k_1 F_1$ and $W_2 = k_2 F_2$ (k_1, k_2 are positive scalars) and $q_1 = h_1$, $q_2 = h_2$, then $e(n) = 0$ because the signum functions are amplitude insensitive. If $e(n) \neq 0$, then the parameters of the neural net must be adjusted so as to reduce the error for subsequent inputs.

B. Training Algorithm

There are (2N + 2) weights that need to be adjusted in the perceptron. Because of the signum function nonlinearities in the hidden layer, one cannot use the multilayer backpropagation algorithm. Instead, consider a two stage process which first adjusts the outer layer for any given $W_1(n)$, $W_2(n)$. Then, using Rosenblatt's algorithm [4], [5] for the output layer,

$$q_1(n+1) = q_1(n) + \mu_{q_1} e(n)\,\mathrm{sgn}[W_1^T(n)\, X(n)] \qquad (2.8)$$

$$q_2(n+1) = q_2(n) + \mu_{q_2} e(n)\,\mathrm{sgn}[W_2^T(n)\, X(n)] \qquad (2.9)$$

These recursions essentially view $\mathrm{sgn}\,[W_1^T(n)\, X(n)]$ and $\mathrm{sgn}\,[W_2^T X(n)]$ as binary input data to a single-layer perceptron. Thus, their stationary points are defined by the orthogonality principle, i.e.,

$$E\Big[e(n)\,\mathrm{sgn}[W_i^T(n)\, X(n)]\Big] = 0, \; i = 1, 2. \qquad (2.10)$$

Since (2.10) depends on $W_i(n)$, one should also specify a set of equations for updating $W_i(n)$. However, we shall subsequently show that (2.10) is sufficient to define the stationary points for the system identification problem studied here.

III. Analysis of the Stationary Points

To find the stationary points defined by Eq. (2.10), one needs to evaluate expectations of the form

$$E\{g[h_1 \,\mathrm{sgn}\, d_1 + h_2 \,\mathrm{sgn}\, d_2]\,\mathrm{sgn}\, z_i\}, \quad i = 1, 2 \qquad (3.1)$$

$$E\,\{g[q_1 \,\mathrm{sgn}\, z_1 + q_2 \,\mathrm{sgn}\, z_2]\,\mathrm{sgn}\, z_i\}, \quad i = 1, 2 \qquad (3.2)$$

where the time index, n, is deleted for convenience. Now, given W_i, i = 1, 2, d_1, d_2, z_1, and z_2 are jointly Gaussian variates with zero means, and variances and covariances given by $E[d_1^2] = \sigma_x^2 F_1^T F_1$, $E[d_2^2] = \sigma_x^2 F_2^T F_2$,

$E[z_1^2] = \sigma_x^2 W_1^T W_1$, $E[d_2^2] = \sigma_x^2 W_2^T W_2$, $E[d_1 z_1] = \sigma_x^2 F_1^T W_1$,

$E[d_1 z_2] = \sigma_x^2 F_1^T W_2$, $E[d_2 z_1] = \sigma_x^2 F_2^T W_1$, $E[d_2 z_2] = \sigma_x^2 F_2^T W_2$,

$E[z_1 z_2] = \sigma_x^2 W_1^T W_2$, (3.3). It is shown in [6] that

$$
\begin{aligned}
E\{g[h_1 \,\mathrm{sgn}\, d_1 + h_2 \,\mathrm{sgn}\, d_2]\,\mathrm{sgn}\, z_i\} = \\
\frac{g(h_1+h_2)}{\pi}\left[\sin^{-1}\left\{\frac{F_1^T W_i}{\sqrt{F_1^T F_1 W_i^T W_i}}\right\} + \sin^{-1}\left\{\frac{F_2^T W_i}{\sqrt{F_2^T F_2 W_i^T W_i}}\right\}\right] \\
+\frac{g(h_1-h_2)}{\pi}\left[\sin^{-1}\left\{\frac{F_1^T W_i}{\sqrt{F_1^T F_1 W_i^T W_i}}\right\} - \sin^{-1}\left\{\frac{F_2^T W_i}{\sqrt{F_2^T F_2 W_i^T W_i}}\right\}\right] \\
+\frac{g(-h_1-h_2)}{\pi}\left[-\sin^{-1}\left\{\frac{F_1^T W_i}{\sqrt{F_1^T F_1 W_i^T W_i}}\right\} - \sin^{-1}\left\{\frac{F_2^T W_i}{\sqrt{F_2^T F_2 W_i^T W_i}}\right\}\right] \\
+\frac{g(-h_1+h_2)}{\pi}\left[-\sin^{-1}\left\{\frac{F_1^T W_i}{\sqrt{F_1^T F_1 W_i^T W_i}}\right\} + \sin^{-1}\left\{\frac{F_2^T W_i}{\sqrt{F_2^T F_2 W_i^T W_i}}\right\}\right]
\end{aligned}
$$

and $\qquad i = 1, 2 \qquad$ (3.4)

$$
\begin{aligned}
2\; E\{g[q_1 \,\mathrm{sgn}\, z_1 + q_2 \,\mathrm{sgn}\, z_2]\,\mathrm{sgn}\, z_1\} = \\
g(q_1+q_2)\left[\frac{1}{2} + \frac{1}{\pi}\sin^{-1}\left\{\frac{W_1^T W_2}{\sqrt{W_1^T W_1 W_2^T W_2}}\right\}\right] \\
+\, g(q_1-q_2)\left[\frac{1}{2} - \frac{1}{\pi}\sin^{-1}\left\{\frac{W_1^T W_2}{\sqrt{W_1^T W_1 W_2^T W_2}}\right\}\right] \\
-g(-q_1-q_2)\left[\frac{1}{2} + \frac{1}{\pi}\sin^{-1}\left\{\frac{W_1^T W_2}{\sqrt{W_1^T W_1 W_2^T W_2}}\right\}\right] \\
-g(-q_1-q_2)\left[\frac{1}{2} + \frac{1}{\pi}\sin^{-1}\left\{\frac{W_1^T W_2}{\sqrt{W_1^T W_1 W_2^T W_2}}\right\}\right] \\
-\, g(-q_1+q_2)\left[\frac{1}{2} - \frac{1}{\pi}\sin^{-1}\left\{\frac{W_1^T W_2}{\sqrt{W_1^T W_1 W_2^T W_2}}\right\}\right] \\
= V(q_1, q_2, W_1^T W_2) \qquad (3.5)
\end{aligned}
$$

By permuting subscripts, a similar result is obtained for

$$2\; E\{g[q_1 \,\mathrm{sgn}\, z_1 + q_2 \,\mathrm{sgn}\, z_2]\,\mathrm{sgn}\, z_2\} = V(q_2, q_1, W_1^T W_2) \qquad (3.6)$$

Eqs. (3.4)-(3.6) hold for any h_1, h_2, F_1, F_2 and any function g(.). The solution to these equations for arbitrary g(.) is not obvious. However, if we choose g(x) = sgn (x) so the training sequence d(n)=±1 and the output sequence is y(n)=±1, then considerable simplification occurs in the solution. However, first it is useful to consider the shape of the decision regions in the observation space of X(n) for g(x)=sgn x. From (2.4),

$$d(n) = \text{sgn} \{ h_1 \,\text{sgn}[F_1^T X(n)] + h_2 \,\text{sgn}[F_2^T X(n)] \}. \quad (3.7)$$

Without loss of generality, assume h_1 and h_2 are nonnegative.

Then if $h_1 > h_2 > 0$, $d(n)=\begin{cases} 1 & \text{iff } F_1^T X(n) \ge 0 \\ -1 & \text{iff } F_1^T X(n) < 0 \end{cases}$

independent of F_2. (3.8)

Alternatively, if $0 < h_1 < h_2$, $d(n)=\begin{cases} 1 & \text{iff } F_2^T X(n) \ge 0 \\ -1 & \text{iff } F_2^T X(n) < 0 \end{cases}$

independent of F_1. (3.9)

Hence, the only case when d(n) depends on both F_1 and F_2 is for h_1=h_2. Thus

$$d(n)=\begin{cases} -1, & \text{iff } F_1^T X(n) < 0 \text{ and } F_2^T X(n) < 0 \\ 1, & \text{otherwise} \end{cases} \quad (3.10)$$

For this case of interest, . (3.4)-(3.6) reduce to

$$2\sin^{-1}\left\{\frac{F_1^T W_1}{\sqrt{F_1^T F_1 W_2^T W_2}}\right\} + 2\sin^{-1}\left\{\frac{F_2^T W_1}{\sqrt{F_2^T F_2 W_2^T W_2}}\right\} =$$
$$\text{sgn}(q_1+q_2)\left[\frac{\pi}{2} + \sin^{-1}\left\{\frac{W_1^T W_2}{\sqrt{W_1^T W_1 W_2^T W_2}}\right\}\right]$$
$$+\text{sgn}(q_1-q_2)\left[\frac{\pi}{2} - \sin^{-1}\left\{\frac{W_1^T W_2}{\sqrt{W_1^T W_1 W_2^T W_2}}\right\}\right]$$
$$-\text{sgn}(-q_1-q_2)\left[\frac{\pi}{2} + \sin^{-1}\left\{\frac{W_1^T W_2}{\sqrt{W_1^T W_1 W_2^T W_2}}\right\}\right]$$
$$-\text{sgn}(-q_1+q_2)\left[\frac{\pi}{2} - \sin^{-1}\left\{\frac{W_1^T W_2}{\sqrt{W_1^T W_1 W_2^T W_2}}\right\}\right] \quad (3.11)$$

$$2\sin^{-1}\left\{\frac{F_1^T W_2}{\sqrt{F_1^T F_1 W_2^T W_2}}\right\} + 2\sin^{-1}\left\{\frac{F_2^T W_2}{\sqrt{F_2^T F_2 W_2^T W_2}}\right\} =$$
$$\text{sgn}(q_1+q_2)\left[\frac{\pi}{2} + \sin^{-1}\left\{\frac{W_1^T W_2}{\sqrt{W_1^T W_1 W_2^T W_2}}\right\}\right]$$
$$+\text{sgn}(q_1-q_2)\left[\frac{\pi}{2} - \sin^{-1}\left\{\frac{W_1^T W_2}{\sqrt{W_1^T W_1 W_2^T W_2}}\right\}\right]$$
$$-\text{sgn}(-q_1-q_2)\left[\frac{\pi}{2} + \sin^{-1}\left\{\frac{W_1^T W_2}{\sqrt{W_1^T W_1 W_2^T W_2}}\right\}\right]$$
$$-\text{sgn}(-q_1+q_2)\left[\frac{\pi}{2} - \sin^{-1}\left\{\frac{W_1^T W_2}{\sqrt{W_1^T W_1 W_2^T W_2}}\right\}\right] \quad (3.12)$$

What are the possible solutions to (3.11) and (3.12)? Suppose $q_1 > q_2 > 0$. Then

$$\sin^{-1}\left\{\frac{F_1^T W_1}{\sqrt{F_1^T F_1 W_1^T W_1}}\right\} + \sin^{-1}\left\{\frac{F_2^T W_1}{\sqrt{F_2^T F_2 W_1^T W_1}}\right\} = \pi \quad (3.13)$$

$$\sin^{-1}\left\{\frac{F_1^T W_2}{\sqrt{F_1^T F_1 W_2^T W_2}}\right\} + \sin^{-1}\left\{\frac{F_2^T W_2}{\sqrt{F_2^T F_2 W_2^T W_2}}\right\} =$$
$$2\sin^{-1}\left\{\frac{W_1^T W_2}{\sqrt{W_1^T W_1 W_2^T W_2}}\right\} \quad (3.14)$$

The first equation has no solution unless F_1=k F_2 (k is a positive scalar) since both $\sin^{-1}(\cdot)$ functions cannot be $\frac{\pi}{2}$ for $F_1 \neq k\, F_2$. This is a trivial case. Similar arguments hold for $q_2 > q_1 > 0$. Hence, the only solution to Eq. (3.11) is for q_1=q_2 > 0. For this case, (3.11) and (3.12) become

$$\sin^{-1}\left\{\frac{F_1^T W_1}{\sqrt{F_1^T F_1 W_1^T W_1}}\right\} + \sin^{-1}\left\{\frac{F_2^T W_1}{\sqrt{F_2^T F_2 W_1^T W_1}}\right\} =$$
$$\frac{\pi}{2} + \sin^{-1}\left\{\frac{W_1^T W_2}{\sqrt{W_1^T W_1 W_2^T W_2}}\right\} \quad (3.15)$$

$$\sin^{-1}\left\{\frac{F_1^T W_2}{\sqrt{F_1^T F_1 W_2^T W_2}}\right\} + \sin^{-1}\left\{\frac{F_2^T W_2}{\sqrt{F_2^T F_2 W_2^T W_2}}\right\} =$$

$$\frac{\pi}{2} + \sin^{-1}\left\{\frac{W_1^T W_2}{\sqrt{W_1^T W_1 W_2^T W_2}}\right\} \quad (3.16)$$

Note that although (3.15) and (3.16) were derived from conditions for the stationary points of the output layer, these equations describe conditions on the weights in the hidden layer. We have shown that $h_1=h_2 > 0$ requires $q_1=q_2 > 0$ for any value of h_1 or q_1. This condition occurs because the signum function is amplitude independent. Are the solutions to (3.15) and (3.16) unique or are there multiple stationary points? There are two cases.

A. Orthogonal Hyperplanes ($F_1^T F_2 = 0$)

Assume a solution of the form

$$W_1 = \alpha F_1 + \sqrt{1 - \alpha^2}\ F_2 ,$$

$$W_2 = \beta F_1 + \sqrt{1 - \beta^2}\ F_2 \quad (3.17)$$

where α and β are constrained to be non negative (i.e. one would expect that any stationary points would require that projections of W_1 and W_2 on F_1 and F_2 be non negative).

Then $F_1^T W_1 = \alpha$, $F_2^T W_1 = \sqrt{1-\alpha^2}$, $F_1^T W_2 = \beta$, $F_2^T W_2 = \sqrt{1-\beta^2}$, $W_1^T W_2 = \alpha\beta + \sqrt{1-\alpha^2}\sqrt{1-\beta^2}$ (3.18)

Inserting (3.18) into (3.15) and (3.16) yields

$$\alpha = 1,\ \ \beta = 0 \text{ or } \alpha = 0,\ \ \beta = 1 \quad (3.19)$$

Thus, W_1 is collinear with F_1 and W_2 is collinear with F_2 (i.e. $W_1=k_1F_1$, $W_2=k_2 F_2$ where k_1 and k_2 are arbitrary positive constants) or vice-versa.

B. Non-Orthogonal Hyperplanes ($F_1^T F_2 \neq 0$)

Suppose that $F_2 = a F_1 + \sqrt{1 - a^2}\ G$ (3.20)

with $F_1^T G = 0$, $G^T G = 1$. Hence $F_2^T F_1 = a$.

Let $W_1=\alpha F_1 + \sqrt{1 - \alpha^2}\ G$,

$W_2=\beta F_1 + \sqrt{1 - \beta^2}\ G$, (3.21)

$F_1^T W_1=\alpha$, $F_2^T W_1=a\alpha + \sqrt{1 - a^2}\ \sqrt{1 - \alpha^2}$,

$F_1^T W_2=\beta$, $F_2^T W_2 = a\beta + \sqrt{1 - a^2}\ \sqrt{1 - \beta^2}$ (3.22)

Obvious solutions to (3.15) and (3.16) are: 1) $\alpha = 1, \beta = a$, 2) $\beta = 1, \alpha = a$. This result corresponds to the desired solutions $W_1=F_1$, $W_2=F_2$ or $W_1=F_2$, $W_2=F_1$. Thus, as for the orthogonal case, W_1 is collinear with F_1 and W_2 is collinear with F_2 (i.e., $W_1=k_1F_1$, $W_2=k_2 F_2$ where k_1 and k_2 are arbitrary positive constants) or vice-versa.

Since (3.15) and (3.16) are independent of amplitude, any solution of the form $W_1=k_1F_1$, $W_2=k_2F_2$ or $W_1=k_1F_2$, $W_2=k_2 F_1$ for arbitrary $k_1, k_2 > 0$ is a stationary point for this two-layer perceptron. Other solutions to (3.15) and (3.16) for the non-orthogonal case have not been found.

IV. Conclusions

The stationary points of a two-layer perceptron have been evaluated for a system identification problem with Gaussian inputs. Ideally, the perceptron will make no errors when the various linear combiners generate hyperplanes which exactly match the system to be identified. It is shown that, on an ensemble average basis (i.e., averaged over the random Gaussian training data), the stationary points of the perceptron correspond to the same zero error points. No non-optimum stationary points were obtained. Interestingly, the stationary points were shown to depend uniquely on the output layer weights. There was no need to study the weight update algorithm for the hidden layer.

The uniqueness of these solutions may be a direct result of the significant constraints on the model for the training sequence. Removal of some of these constraints (e.g. by introducing bias terms) makes the problem more realistic but at the same time makes the analysis increasingly difficult. Future work in this area is being directed along these lines.

References

1. D.E. Rumelhart and J.L. McClelland, *Parallel Distributed Processing: Exploration in the Microstructure of Cognition*, Vol. I, MIT Press, Cambridge, MA, 1986.
2. R.P. Lippmann, "An Introduction to Computing with Neural Nets," *IEEE ASSP Magazine*, Vol. 4, pp. 4-22, Apr. 1987.
3. B. Widrow and M.A. Lehr, "30 Years of Adaptive Neural Networks: Perceptron, Madaline and Backpropagation," *Proc. of IEEE*, Vol. 78, No. 9, pp. 1415-1441, Sept. 1990.
4. J.J. Shynk and S. Roy, "Convergence Properties and Stationary Points of a Perceptron Learning Algorithm," *Proc. of IEEE*, Vol. 78, pp. 1599-1604, Oct. 1990.
5. J.J. Shynk and N.J. Bershad, "Steady-State Analysis of a Single-Layer Perceptron Based on a System Identification Model with Bias Terms," *IEEE Trans. on Circuits and Systems*, Vol. 38, No. 9, pp. 1030-1042, Sept. 1991.
6. N. J. Bershad, C.F.N.Cowan and J.J. Shynk, "Stochastic Convergence Analysis of a Two-Layer Perceptron Learning Algorithm for a System Identification Model," submitted.

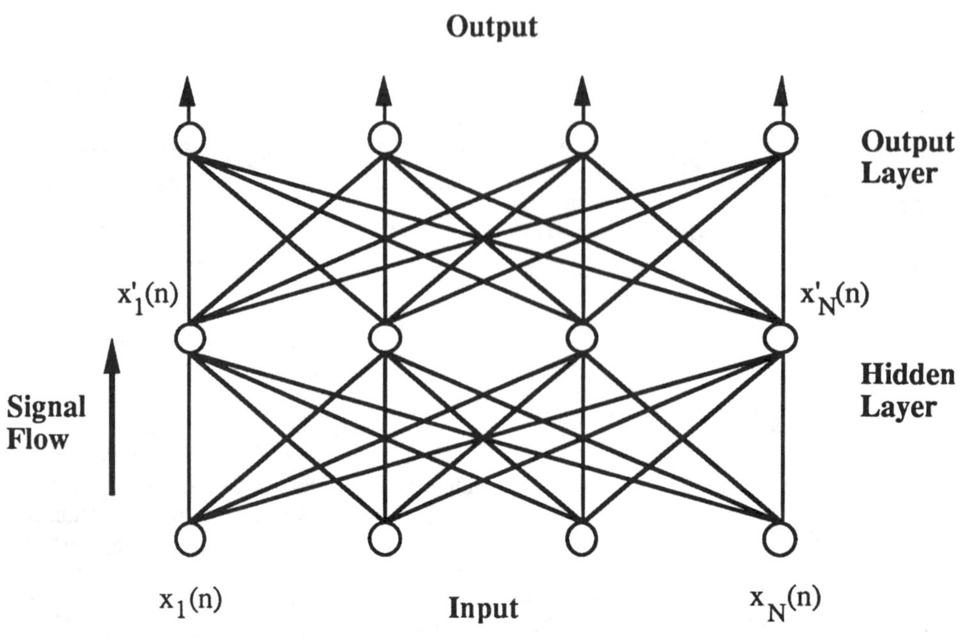

Figure 1. Two-Layer Perceptron.

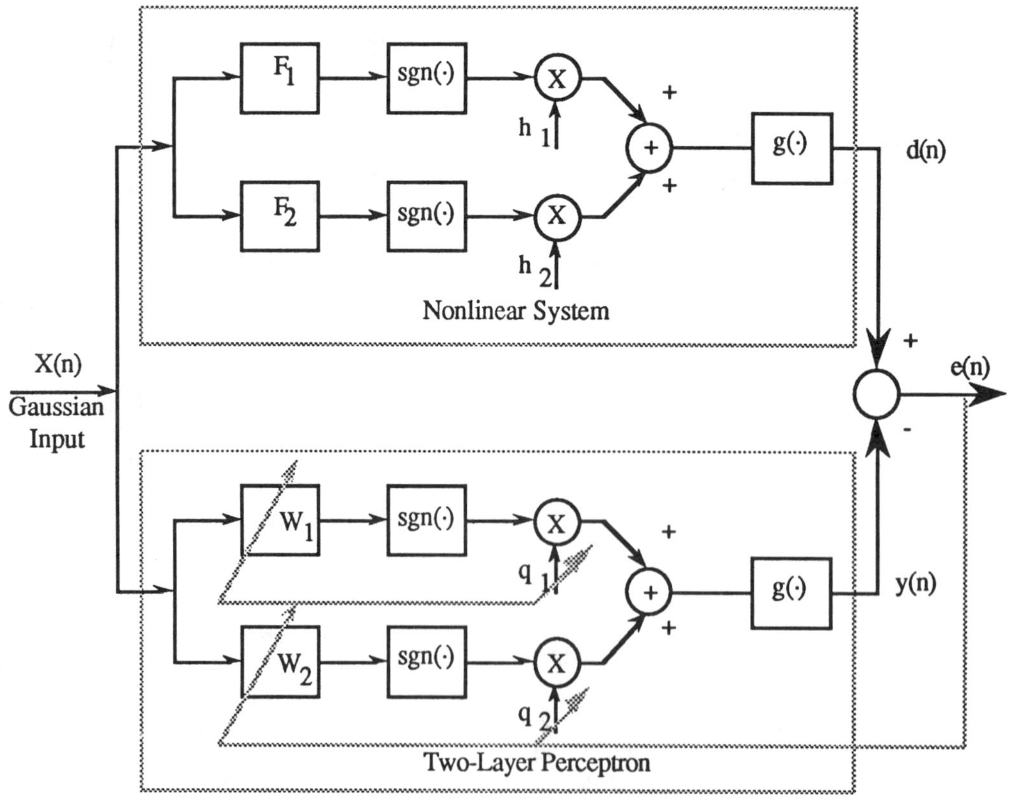

Figure 2. Nonlinear System Identification Using a Two-Layer Perceptron.

Performance Analysis of a McClellan Transformation-Based 2-D Adaptive Digital Filter

Jeffrey. C. Strait and W. Kenneth Jenkins
Department of Electrical and Computer Engineering and
The Coordinated Science Laboratory
University of Illinois
Urbana, Illinois 61801

Abstract

A 2-D adaptive filter structure based on the McClellan transformation design technique for 2-D FIR filters is capable achieving improved convergence rates and reduced computational efficiency in 2-D adaptive filters. It been shown experimentally that if the transformation structure is constrained by apriori knowledge of contour shapes in the frequency domain, the 2-D adaptive algorithm greatly reduces computational requirements and produces more rapid learning characteristics, as compared to the 2-D direct form. This paper presents an analysis of the McClellan transformation 2-D adaptive filter to illustrate that the learning characteristics are similar to those of a 1-D LMS adaptive filter.

1 Introduction

In recent years there has been increased interest in applying 2-D adaptive digital filters to the problems of image processing, computer vision, and remote sensing. However, due to the large number of adaptive parameters needed for 2-D system identification, 2-D applications often require too much computation and are too slow to converge in many real-time nonstationary environments. In 1988 Hadhoud and Thomas [1] published a 2-D direct form LMS algorithm for the estimation of nonstationary images. Their work is one of only a few published studies in which 1-D adaptive filtering theory has been extended to two dimensions in an effort to address modern problems in image processing.

The McClellan transformation [2] is a 2-D digital filter design technique that uses a 1-D FIR digital filter prototype as the basis for obtaining a near optimal 2-D design. The impulse response (as well as the frequency response) of the 1-D prototype is mapped by the "McClellan transformation" into a 2D digital filter that inherits its frequency characteristics from the prototype. The frequency response of the 2-D FIR digital filter is controlled by the impulse response of the 1-D prototype, h(n), n = 0, . . . , M-1, and a set of four "contour" parameters A, B, C, and D which define the mapping. Once the contour parameters are fixed, the class of filters that can be obtained becomes constrained, i.e., only limited types of frequency responses can be obtained by varying h(n) once the A, B, C, and D's are fixed. In particular, contour parameters have been found to produce frequency responses that are circular, elliptical, and fan symmetric. McClellan and Chan [3] later proposed the filter structure shown in Fig. 1 (without the orthogonal transformation) for realizing the transformation filter. This structure, which uses the h(n)'s and the contour parameters A, B, C, and D as explicit multiplier coefficients, is the basis for the 2-D adaptive filter that is further studied in the sections to follow.

An adaptive 2-D transformation filter is obtained by applying an LMS update algorithm to the parameters of the McClellan-Chan filter structure. When the A, B, C, and D's are fixed and the h(n)'s are adaptively adjusted, a constrained 2-D adaptive filter results that requires the update of only N+1 parameters at each iteration, where M = 2N+1 is the length of the 1-D prototype. The classes of adaptive systems that can be obtained include many important types of filters that are useful in real-time image processing [4,5].

More recently Shapiro and Staelin [6] reported a least squares (LS) algorithm for adapting the parameters of the 1-D prototype in the McClellan transformation structure. They showed how the LS adaptive algorithm can be efficiently implemented in a systolic array architecture.

2 McClellan Transformation 2-D Filters

The McClellan transformation technique can be used to design 2-D FIR filters with either zero or linear phase from 1-D FIR filters of either zero or linear phase. This is achieved by applying an algebraic mapping from the 1-D frequency domain to the 2-D frequency domain. Hence, a single frequency in the 1-D is mapped to a contour of constant value in the 2-D plane. The value that the frequency response of the 2-D FIR filter takes on this contour will be the value of the 1-D frequency response at the value of ω associated with the contour. The shape of the transition band in the two dimensional frequency response is determined by the nature of the transformation contours, while the location of the transition band of the 2-D filter depends on where the 1-D prototype filter has its transition band. Thus, designing filters using the McClellan transformation technique can be broken down into two distinct steps: 1) a set of contours is defined such that the contour corresponding to the passband edge of the final 2-D filter has a desired shape, and 2) a 1-D prototype filter is designed with an appropriate transition band.

3 Adaptive Filtering with Constrained Contours

The 2-D adaptive transformation filter will be called a 2-D constrained transformation filter (CTF) when the contour parameters A, B, C, and D are held constant, and the a_k's (or equivalently, the h(n)'s) are adaptively adjusted to minimize the mean-square error $e(n_1, n_2)$ between the output of the adaptive system and a training signal $d(n_1, n_2)$. In system identification problems the training signal $d(n_1, n_2)$ is sometimes obtained by passing white noise through an unknown system to be identified. The adaptive algorithm updates the a_k coefficients at each iteration with a term that is approximately proportional to the gradient of the mean-squared error (MSE) with respect to the coefficient vector. Thus for a horizontal update direction (n_1 direction) the steepest descent algorithm takes the form

$$a_k(n_1+1, n_2) = a_k(n_1, n_2) - \mu \nabla_{e2}(n_1, n_2), \qquad (1)$$

where μ is the step size. In a manner analogous to the conventional LMS algorithm, the gradient of the mean-square error is approximated by the gradient of the instantaneous squared error, i.e.

$$\nabla_{e2}(n_1, n_2) \approx \partial e^2(n_1, n_2)/\partial a_k = 2e(n_1, n_2)\partial e(n_1, n_2)/\partial a_k, \qquad (2)$$

where $e(n_1, n_2) = d(n_1, n_2) - y(n_1, n_2)$. The output of the adaptive structure can be expressed as the inverse Fourier transform of the Fourier transform of the input signal $x(n_1, n_2)$ multiplied by the frequency response of the 2-D FIR structure, i.e.,

$$y(n_1, n_2) = F^{-1}\{X(\omega_1, \omega_2)\, H(\omega_1, \omega_2)\}, \qquad (3)$$

where $H(\omega_1, \omega_2)$ is the frequency response of the McClellan FIR structure. The error error can be expressed as

$$e(n_1, n_2) = d(n_1, n_2) - F^{-1}\{\sum_{n=0}^{N} a_n T_n[F(\omega_1, \omega_2)]X(\omega_1, \omega_2)\}. \qquad (4)$$

Next Eq.(4) is differentiated with respect to a_n and substituted into Eqs. (1) and (2), resulting in the final update equation

$$a_k(n_1+1, n_2) = a_k(n_1, n_2) + \mu e(n_1, n_2) s_k(n_1, n_2), \qquad (5)$$

where $s_k(n_1, n_2) = F^{-1}\{T_k[F(\omega_1, \omega_2)]X(\omega_1, \omega_2)\}$, k = 0, . . . , N, is simply the signal at the node in the 2-D network of Fig. 1 just before the k-th down-feed multiplier. Similar update equations can be written for stepping in other directions in the (n_1, n_2) - plane.

For a CTF of size (2N+1) x (2N+1) there are N+1 coefficients to be updated at each time step. Thus, as the filter order is increased, the computational complexity of the update algorithm increases proportionately to N rather than N^2, as is the case with a direct form LMS adaptive filter.

Figures 2 and 3 show learning curves for an experiment where the adaptive behavior of the constrained transformation filter is compared to a conventional 2-D direct form LMS filter for the identification of a system with a circularly symmetric frequency response. Note from Fig. 2 that after 500 iterations the direct form LMS filter has reduced the MSE to only about -25 dB., whereas in Fig. 3 it is seen that the CTF has reduced the error to approximately -130 dB. This demonstrates the remarkably faster convergence rate of the CTF structure, which has been confirmed in scores of experiments conducted with circular, elliptical, and fan shaped characteristics. In all of these experiments it was observed that the CTF converges much faster than the direct 2-D LMS structure, and in many cases it appears that the learning curve for the CTF is somewhat less noisy than the corresponding one for the direct form.

The following list summarizes the important properties of the CTF adaptive 2-D architecture:

1) The computational complexity of adaptive update mechanism for the the CTF is of order $O\{N\}$, as compared to order $O\{N^2\}$ for the LMS direct form. The price paid for this improved performance is that the adaptive filter is "pre-constrained" according to the contours which are established by fixing the mapping parameters of the McClellan transformation.

2) The CTF learns much more rapidly than an LMS direct form filter. The learning rate falls between the extremes of the direct form 2-D in the slow extreme, and the rate of a 1-D filter characterized by the prototype in the fast extreme.

3) Experiments show that the update direction in the (n_1, n_2) - plane does not affect the rate of convergence very much as long as the input is white noise. It was found that horizontal, vertical, and 45° update directions resulted in very similar learning curves.

4) As in 1-D adaptive filters, the rate of convergence of 2-D adaptive filters is affected by coloring on the input

noise. However, in the rate of convergence for the CTF architecture appears to be dictated by the widest bandwidth in the 2-D frequency space, rather than by the narrowest. This means that if the input noise is wide-band along one frequency axis (ω_1) and narrow-band along the other (ω_2), the CTF tends to adapt rapidly, as if the whole input spectrum had the widest bandwidth. In this sense the symmetry constraints imposed by the McClellan design procedure create a robust structure that selects the "best case" performance.

4 Improvements with an Orthogonal Transform

When the transformation is fixed, i.e., when A, B, C, and D are held constant, the transformation filters will shape the power spectral density of the input noise in the process of generating the signal $\mathbf{s} = [s_0(n_1, n_2), \ldots, s_N(n_1, n_2)]^T$. This implies that the magnitude of the spectrum of $\mathbf{s}$ will not in general be flat, and hence the components of $\mathbf{s}$ will be correlated. This coloring, inadvertently imposed on the signal $\mathbf{s}$ by the transformation filters, slows the convergence rate of the adaptation process. This undesirable coloring of $\mathbf{s}$ can be removed by using a 1-D orthogonal transform of length N + 1 and a power normalization algorithm to equalize the power in the spectral components of $\mathbf{s}$ along the spatial coordinate. This modified structure incorporating an orthogonal transform to produce $\mathbf{S} = T\{\mathbf{s}\}$ is depicted in Fig. 1 when the orthogonal transform $T\{\cdot\}$ is included in the structure. In this work the DFT (or FFT) is used to orthogonalize the signal $\mathbf{S}$, although other orthogonal transform are also candidates (e.g., the discrete cosine transform, the discrete Hartley transform, the Walsh-Hadamard transform, etc.).

In the following discussion it is assumed that the input to the system is 2-D white noise, so the objective of the orthogonalization and power normalization will be to remove the coloring imposed by the transformation filters. (Coloring on the input signal can also be removed by this procedure, but this is not considered at this time). Figure 4 shows the average magnitude squared spectrum of $\mathbf{S}$ for transformation filters with fixed circular contours. Samples from these spectra will be used to normalize the spectral components of $\mathbf{S}$ so that the normalized magnitude spectrum of $\mathbf{S}$ is flat (white), thereby removing the correlation between the components of $\mathbf{S}$ and improving the convergence rate of the 2-D CTF structure. If σ_k^2 is the power of the k-th component $S_k(n_1, n_2)$, where $\mathbf{S} = T\{\mathbf{s}\}$, then the normalized signal can be denoted by $\mathbf{S}_{\sigma^2} = [S_0(n_1, n_2)/\sigma_0^2, \ldots, S_N(n_1, n_2)/\sigma_N^2]^T$. Using this notation, the update equation for the power normalized CTF structure becomes

$$\mathbf{a}_{j+1} = \mathbf{a}_j + 2\mu e(n_1, n_2)\, \mathbf{S}_{\sigma^2}^*, \tag{6}$$

where $\mathbf{a} = [a_1, a_2, \ldots, a_N]^T$, the subscript "j" is used to denote an increment in either one of the n_1 or n_2 indices, and the * denotes complex conjugate.

Figures 5 and 6 present learning curves for experiments comparing an 11 x 11 circularly symmetric CTF both with and without orthogonalization (FFT) and power normalization. Without orthogonalization and power normalization the CTF converged to -140 dB. in approximately 500 iterations. However, with the orthogonalization and power normalization (whitening) a level of -140 dB. was reached in approximately 200 iterations. Note that as long as the contours are fixed with prescribed contour parameters and the input signal is white, then the power normalization factors can be calculated apriori and the whitening can be achieved as a deterministic power normalization using pre-calculated constants. If the input signal is colored with an unknown spectral shape, then the power normalization factors must be estimated on-line and the procedure becomes more computationally expensive and less accurate.

5 Mathematical Analysis of Performance

In previous work the performance characteristics of the CTF adaptive filter were studied through experiments and computer simulations. At the present time, little is known about the theoretical limits of the algorithm, or how it compares to the 2-D direct form LMS algorithm. This section develops an analysis of the CTF 2-D adaptive filter which characterizes the learning characteristics of the inherent adaptive algorithm. If

$$\mathbf{s} = [s_0(n_1, n_2), \ldots, s_N(n_1, n_2)]^T \tag{7}$$

and

$$\mathbf{a} = [a_0, a_2, \ldots a_N]^T, \tag{8}$$

then Eq. (4) can be expressed as

$$y(n_1, n_2) = \mathbf{a}^T\mathbf{s} = \mathbf{s}^T\mathbf{a}. \tag{9}$$

Then the MSE (learning curve) can be expressed as

$$\text{MSE} = E\{e_2(n_1, n_2)\} = E\{d(n_1, n_2)\} - 2E\{d(n_1, n_2)\mathbf{s}^T\}\mathbf{a} + \mathbf{a}^T E\{\mathbf{s}\mathbf{s}^T\}\mathbf{a}. \tag{10}$$

If $\mathbf{R}_{ss} = E\{\mathbf{s}\mathbf{s}^T\}$ is defined to be the "intermediate autocorrelation matrix", and $\mathbf{p} = E\{d(n_1, n_2)\mathbf{s}^T$ is the "cross correlation vector" between the intermediate signals and the desired signal, then the MSE can be rewritten as

$$\text{MSE} = E\{e_2(n_1, n_2)\} = E\{d(n_1, n_2)\} - 2\mathbf{pa} + \mathbf{a}^T\mathbf{R}_{ss}\mathbf{a}. \tag{11}$$

The minimum MSE solution to Eq. (11) is found to be

$$a^* = R_{ss}^{-1} p. \tag{12}$$

The solution $\mathbf{a}^*$ in Eq. (12) is the Wiener solution expressed in terms of the intermediate correlation quantities $\mathbf{p}$ and $\mathbf{R}_{ss}$. Note that each $s_i(n_1, n_2)$, $i = 0, \ldots, N$ is obtained by passing the input signal through a linear time invariant filter as long as the contour parameters A, B, C, and D are fixed at constant values. Let $h_i(n_1, n_2)$ denote the impulse response from the input to the point where $s_i(n_1, n_2)$ is produced. The support of $h_i(n_1, n_2)$ is a (2i+1) x (2i+1) square, the size of which increases with increasing i. Next the columns of $h_i(n_1, n_2)$ are concatenated into a single column vector of length $(2i+1)^2$, denoted $\mathbf{t}_i = [h_i(2i+1,2i+1), \ldots, h_i(0, 2i+1), h_i(2i+1, 2i), \ldots\ldots, h_i(0, 0)]^T$. Then it can be shown that $s_i = \mathbf{x}^T\mathbf{t}_i$, $s_j = \mathbf{x}^T\mathbf{t}_j$, $s_i^T s_j = \mathbf{t}_i^T\mathbf{X}\mathbf{X}^T\mathbf{t}_j$, and $\mathbf{R}_{ss} = \mathbf{t}\mathbf{R}_{xx}\mathbf{t}^T$, where $\mathbf{t} = [t_0, t_1, \ldots, t_N]^T$, $\mathbf{R}_{xx}$ is the input autocorrelation, and $\mathbf{X}$ is the column concatenation of $\mathbf{X}(n_1, n_2)$, an (N+1) x (N+1) block of input samples of $x(n_1, n_2)$. Using these conventions, the following expression is obtained for the expected value of $\mathbf{a}$:

$$E\{a_{j+1}\} = [\mathbf{I} - 2\mu\mathbf{t}\mathbf{R}_{xx}\mathbf{t}^T]E\{a_j\} + 2\mu\mathbf{R}_{ss}\mathbf{a}^* \tag{13}$$

If $\mathbf{c}$ denotes the $\mathbf{a}$ vector after it has been translated and rotated by a diagonalization of Eq. (13), the result becomes

$$E\{\mathbf{c}_{j+1}\} = [\mathbf{I} - 2\mu\lambda]E\{\mathbf{c}_j\}, \tag{14}$$

where λ is a diagonal matrix which contains the eigenvalues of $\mathbf{R}_{ss}$ on the main diagonal. The coloring that is imposed on the input signal by the transformation filters causes the eigenvalues to take on different values; the orthogonalization and power normalization reconditions the intermediate autocorrelation matrix, causing the final eigenvalues to become equal .

Finally, let $N\sigma^2_{ss}$ be the variance of the input, where σ^2_{ss} is the value of each entry (eigenvalue) on the main diagonal of the whitened intermediate autocorrelation matrix. It can then be shown that for small values of the iteration index j in Eq. (14), the time constant of the "pth mode" of the weight vector is approximately given by

$$T_p = \frac{1}{2\mu\,\sigma^2_{ss}} = \frac{(N+1)}{2\mu\,\sigma^2_{xx}} \tag{15}$$

This result shows that the time constants which govern the convergence of the vector $\mathbf{a}$ (or $\mathbf{c}$) depend approximately on (N+1) , where N+1 is the length of the 1-D prototype FIR in the CTF structure. When a similar analysis is done for an equivalent 2-D LMS adaptive filter, the result obtained is similar except that the time constants for that case depend on $(2N+1)^2$, where (2N+1) x (2N+1) is the size of the equivalent 2-D LMS adaptive filter. This result illustrates that the CTF should indeed converge much faster than the conventional 2-D LMS structure, a behavioral characteristic that has been observed in many experiments previously published [4,5].

Acknowledgements

This work is supported in part by the Joint Services Electronics Program under contract N00014-84-C-0149, and in part by SDIO/IST, managed by the Army Research Office under contract DAAL03-86-K-0111.

References

[1] M. M. Hadhoud and D. W. Thomas, "The two-dimensional adaptive LMS (TDLMS) algorithm," *IEEE Transactions on Circuits and Systems*, vol. 35, no. 5, pp 485-494, May 1988.

[2] Mersereau, R. M., Mecklenbrauker, W. F. G. and Quatieri Jr., T. F., "McClellan transformations for two-dimensional digital filtering: I - Design," *IEEE Transactions on Circuits and Systems*, vol. CAS-23, no. 7, pp 405-414, 1976.

[3] McClellan, J. H. and Chan, D. S. K., "A 2-D FIR filter structure derived from the Chebyshev recursion," *IEEE Transactions on Circuits and Systems*, vol. CAS-24, no. 7, pp 372-378, 1977.

[4] W. K. Jenkins and R. P. Faust, "A constrained two-dimensional adaptive digital filter with reduced computational complexity," *Proc. of the 1988 International Symposium on Circuits and Systems*, Espoo, Finland, pp 2655-2658, June 1988.

[5] J. C. Strait and W. K. Jenkins, "Filter Architectures and Adaptive Algorithms for 2-D Adaptive Digital Signal Processing," *Proceedings of the 1989 International Conference on Acoustics, Speech, and signal Processing*, Glasgow, Scotland, pp 932-935, June 1989.

[6] J. M. Shapiro and D. H. Staelin, "Multidimensional Adaptive Filtering via McClellan Transformations," *Proceedings of the 1990 International Conference on Acoustics, Speech, and Signal Processing*, Albuquerque, NM, pp 2009-2012, April 1990.

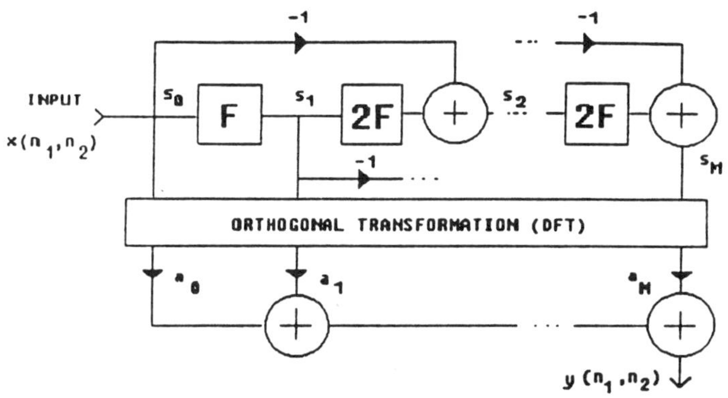

Figure 1. Realization of a 2-D FIR digital filter by means of the Chebyshev recursion (including orthogonalization).

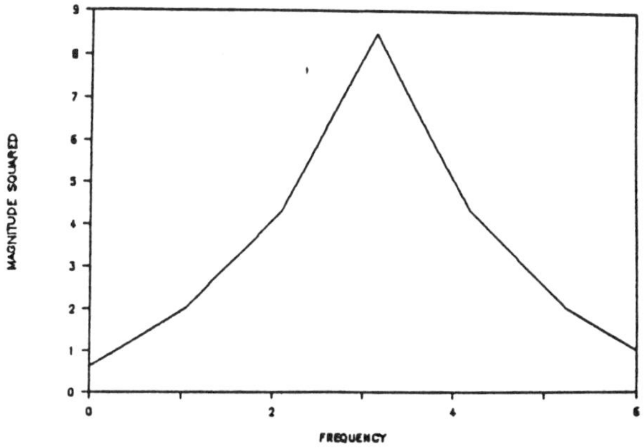

Figure 4. Average squared magnitude spectrum of the signal S for circularly symmetric transformation filters.

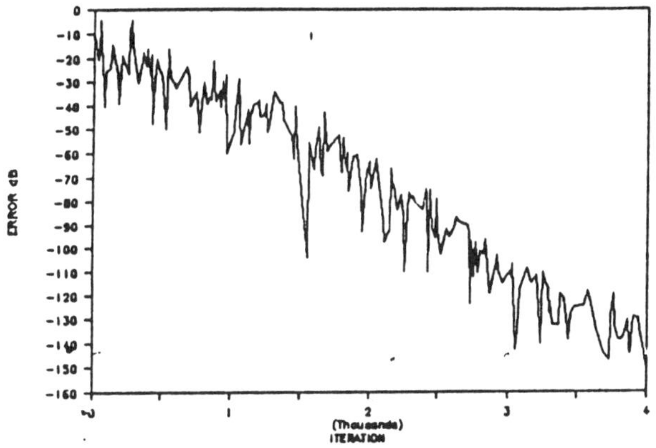

Figure 2. Learning curve for an 11 x 11 system identification example using a (4-quadrant) symmetric direct form 2-D LMS FIR filter with white noise input and $\mu = 0.005$. (circular symmetry)

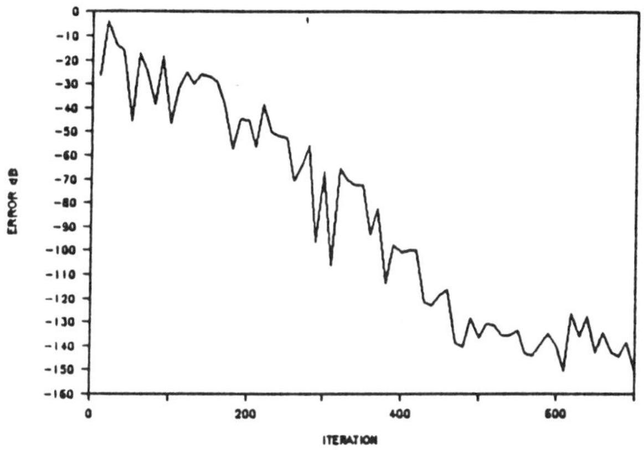

Figure 5. Learning curve for an 11 x 11 constrained transformation filter with circular contours, no orthogonal transform, and $\mu = 0.025$.

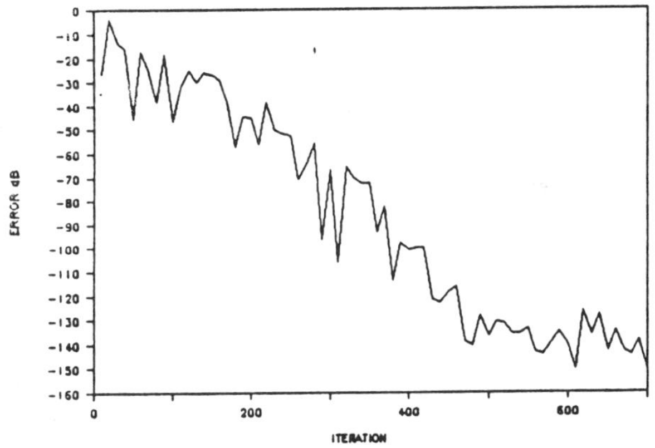

Figure 3. Learning curve for an 11 x 11 system identification example using a 2-D transformation adaptive filter with white noise input and $\mu = 0.025$. (circular symmetry)

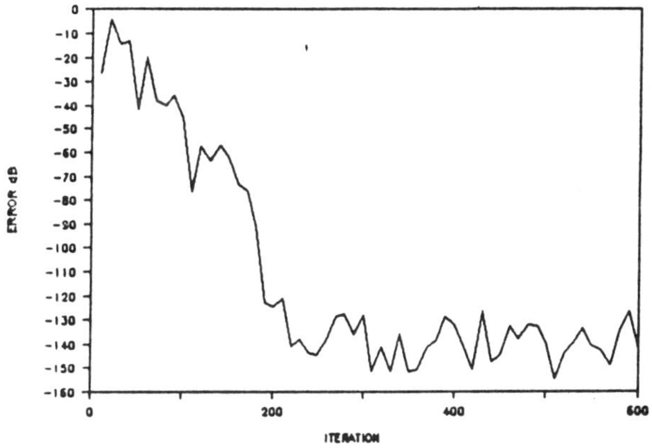

Figure 6. Learning curve for an 11 x 11 constrained transformation filter with circular contours, an orthogonal transform, and $\mu = 0.080$.

Iterative Methods for Estimation of 2-D AR Parameters using a Data-Adaptive Toeplitz Approximation Algorithm

John C. Eremic and Murali Tummala
Naval Postgraduate School
Monterey, California 93943

Abstract

A new two-dimensional data-adaptive algorithm utilizing the iterative Toeplitz approximation method is presented. This algorithm provides a means for efficient estimation of 2–D AR parameters representing spatially variant data arrays. Its convergence performance is comparable to that of the 2–D Recursive Least Squares (RLS) algorithm but is computationally more efficient. The savings in computation is realized by reducing the size of the matrix to be inverted when solving the AR model normal equations. The ability of the algorithm to accurately estimate the model parameters using very small data sets and various windowing schemes are evaluated. Single quadrant and combined quadrant (CQ) spectral estimates are compared for quarter-plane (QP), and nonsymmetric half-plane (NSHP) regions of support.

1 Introduction

The traditional least squares or Wiener filtering based 2–D AR parameter estimation methods [1] provide accurate spectral estimates but require the the inversion of a relatively large autocorrelation matrix. This is undesirable since matrix inversion is computationally expensive. Iterative methods, using a Toeplitz approximation algorithm, have been suggested [2, 3] as an alternative to the direct Wiener filter solution. The iterative Toeplitz approximation method has proven to be an effective means to estimate the parameters of fixed data arrays without having to invert the correlation matrix. This method is particularly useful when data arrays of limited size must be processed.

The results obtained using iterative methods to obtain the AR parameters of fixed data arrays suggest that these methods may be used to some advantage when estimating parameters adaptively. Adaptive parameter estimation is necessary in many digital signal processing applications where the data is nonstationary. In these applications, real-time or near real-time processing is often desirable. This requires that any algorithm used must be computationally efficient and be capable of providing accurate estimates obtained from a minimum of data. This paper addresses the implementation of the iterative Toeplitz approximation method for 2–D parameter estimation of non-stationary data. It is shown that this algorithm can be efficiently implemented to obtain accurate estimates from very small data arrays.

2 Solving The Normal Equations by Direct Inversion

In the 2–D least squares problem, the error between the random process $x[n_1, n_2]$ and its estimate $\hat{x}[n_1, n_2]$ is given by

$$e[n_1, n_2] = x[n_1, n_2] - \hat{x}[n_1, n_2] \tag{1}$$

or in expanded form as

$$e[n_1, n_2] = x[n_1, n_2] + \sum_{(i,j)\,\in A}\sum a(i,j)x[n_1 - i, n_2 - j]\,, \tag{2}$$

for $(i,j) \neq (0,0)$. The objective of the least squares method is to minimize the squared errors from a particular set of these terms [4]. If we let $a(0,0) = 1$, we can express (2) as

$$e[n_1, n_2] = \sum_{(i,j)\,\in A}\sum a(i,j)x[n_1 - i, n_2 - j] \tag{3}$$

which is compactly represented in vector notation as $\mathbf{e} = \mathbf{Xa}$. The error $e[n_1, n_2]$ is computed for each position of the filter, then normal equations are formed by multiplying both sides of (3) by $\mathbf{X}^T$ and applying the orthogonality principle [5]. This results in

$$\mathbf{Ra} = \mathcal{E} \tag{4}$$

where $\mathbf{R} = \mathbf{X}^T\mathbf{X}$ and the error variance vector $\mathcal{E} = \mathbf{X}^T\mathbf{e} = [\mathcal{E}^{(0)T}, 0, \ldots, 0]^T$ with $\mathcal{E}^{(0)} = [\sigma_\epsilon^2, 0, \ldots, 0]^T$. The parameter estimates are obtained from (4) by muliplying both sides of the equation by the inverse of $\mathbf{R}$. This is referred to as the *direct inversion method.*

3 Solving The Normal Equations using Iterative Methods

An alternative to direct inversion is to take advantage of the near Toeplitz-block-Toeplitz structure of the autocorrelation matrix and successively partition the normal equation [2]. This partitioning permits the estimation of parameters while requiring the inversion of a matrix that is significantly smaller than the original autocorrelation matrix. The following discussion develops the iterative solution for QP region of support.

In order to iteratively solve the QP normal equations we must begin by dividing both sides of (4) by σ^2. This results in a modified normal equation given by

$$\begin{bmatrix} \mathbf{R}_0 & \mathbf{R}_{-1} & \cdots & \mathbf{R}_{-P_1+1} \\ \mathbf{R}_1 & \mathbf{R}_0 & \cdots & \mathbf{R}_{-P_1+2} \\ \vdots & \vdots & \ddots & \vdots \\ \mathbf{R}_{P_1-1} & \mathbf{R}_{P_1-2} & \cdots & \mathbf{R}_0 \end{bmatrix} \begin{bmatrix} \mathbf{a}_0'' \\ \mathbf{a}_1'' \\ \vdots \\ \mathbf{a}_{P_1-1}'' \end{bmatrix} = \mathcal{E} \, . \tag{5}$$

where $\mathbf{R}_k$ are $P_1 \times P_2$ Toeplitz blocks of correlation terms, $\mathcal{E} = [1\,0\,0, \ldots, 0]^T$ and $\mathbf{a}_i'' = \frac{1}{\sigma^2}\mathbf{a}_i$ for $i = 1, 2, \ldots, P_1-1$. We now partition the normal equation (5) to form an iterative solution. We can combine the resulting iterative solutions from the successive levels of partitioning into a set of compact iterative summations

$$\mathbf{a}_0^{\prime\prime(k)} = \mathbf{a}_0^{\prime\prime(0)} - \mathbf{R}_0^{-1} \sum_{i=1}^{P_1-1} \mathbf{R}_{-i}\mathbf{a}_i^{\prime\prime(k-1)} \, , \tag{6}$$

$$\mathbf{a}_j^{\prime\prime(k)} = -\mathbf{R}_0^{-1} \sum_{i=0}^{P_1-1} \mathbf{R}_{j-i}\mathbf{a}_i^{\prime\prime(k-1)} \, , (i \neq j) \, , \tag{7}$$

where $\mathbf{a}_0^{\prime\prime(0)} = \mathbf{R}_0^{-1}\mathcal{E}^{(0)}$, $\mathbf{a}_j^{\prime\prime(0)} = -\mathbf{R}_0^{-1}\mathbf{R}_j\mathbf{a}_0^{\prime\prime(0)}$, and $j = 1, 2, \ldots, P_1 - 1$. The index of iteration is k and $\mathbf{a}_i''$ are the $P_2 \times 1$ parameter vectors. The solution of (6)-(7) requires $\mathcal{O}(P_1^2P_2^2K)$ multiplications, where K represents the number of iterations to converge to the true parameter values. An additional $P_1P_2^2$ multiplications are required to compute the initial parameter estimates $\mathbf{a}_j^{\prime\prime(0)}$ for $j = 1, 2, \ldots, P_2 - 1$. This results in total multplications of $\mathcal{O}(P_1^2P_2^2K + P_1P_2^2)$. Depending on the number of iterations required to converge to the true parameters, this total compares with the algorithms of Wax and Kalaith [6] and Akaike [7] which require $\mathcal{O}(P_1^3P_2^2)$ operations. When large arrays are used this represents a noticeable savings over direct inversion which requires $\mathcal{O}(P_1^3P_2^3)$ multiplications.

The derivation of the iterative NSHP solution follows a procedure of succesive partitioning similar to that of QP support.

4 The Iterative Solution Using Toeplitz Approximation

The discussion of the previous section assumes an autocorrelation matrix with Toeplitz-block-Toeplitz structure. In most cases the autocorrelation matrix must be formed from a limited, and possibly very small, amount of sampled data. In this situation, it is best to use the covariance method to estimate the correlation matrix to reduce the bias in the estimate. However, since this method does not have Toeplitz-block-Toeplitz structure, it is necessary to modify the iterative algorithm described above.

For first quadrant parameter estimates, the covariance estimate $\mathbf{R}$ has the form

$$\begin{bmatrix} \mathbf{R}_{0,0} & \mathbf{R}_{0,1} & \cdots & \mathbf{R}_{0,P_1-1} \\ \mathbf{R}_{1,0} & \mathbf{R}_{1,1} & \cdots & \mathbf{R}_{1,P_1-1} \\ \vdots & \vdots & \ddots & \vdots \\ \mathbf{R}_{P_1-1,0} & \mathbf{R}_{P_1-1,1} & \cdots & \mathbf{R}_{P_1-1,P_1-1} \end{bmatrix} . \tag{8}$$

A proven procedure [2] that can be used to form the Toeplitz approximation is to first average the diagonal blocks $\mathbf{R}_{j,j}$, $i = j$ and then form a Toeplitz $\mathbf{T}$ matrix from this averaged block $\mathbf{R}_{avg}$ by averaging its diagonal elements. The diagonal elements of $\mathbf{T}$ can be obtained as follows

$$t(i) = \frac{1}{P_2 - i} \sum_{j=0}^{P_2-i-1} \mathbf{R}_{avg}(i + j, j) \tag{9}$$

where $i = 0, 1, \ldots, P_2 - 1$.

This done, each diagonal block may be written as

$$\mathbf{R}_{j,j} = \mathbf{T} + \Delta_{j,j} \tag{10}$$

where $\Delta_{j,j}$ is the difference between $\mathbf{R}_{j,j}$ and the Toeplitz approximation $\mathbf{T}$. Combining the successive iterative equations, using the methods described above, results in our solution to the normal equations using the *forward* iterative algorithm for the QP case, which takes the form

$$\mathbf{a}_0''^{(k)} = \mathbf{a}_0''^{(0)} - \mathbf{T}^{-1}\Delta_{0,0}\mathbf{a}_0''^{(k-1)} - \mathbf{T}^{-1}\sum_{i=1}^{P_1-1}\mathbf{R}_{0,i}\mathbf{a}_i''^{(k-1)} \tag{11}$$

$$\mathbf{a}_j''^{(k)} = -\mathbf{T}^{-1}\Delta_{j,j}\mathbf{a}_j''^{(k-1)} - \mathbf{T}^{-1}\sum_{\substack{i=0\\ i\neq j}}^{P_1-1}\mathbf{R}_{j,i}\mathbf{a}_i''^{(k-1)} \tag{12}$$

where $\mathbf{a}_0''^{(0)} = \mathbf{T}^{-1}\mathcal{E}^{(0)}$, $\mathbf{a}_j''^{(0)} = \mathbf{T}^{-1}\mathbf{R}_{j,0}\mathbf{a}_0''^{(0)}$ and $j = 1, 2, \ldots, P_1-1$. The forward iteration will provide parameters for the first quadrant spectral estimate.

The second quadrant AR parameters $b(i,j)$ may be estimated from the normal equations using the *backward* iterative Toeplitz approximation [8].

5 Data-Adaptive Algorithm

The data-adaptive Toeplitz approximation algorithm is based on the same formulation found in the fixed data case. The algorithm becomes adaptive in nature when the autocorrelation matrix $\mathbf{R}_n$ is updated for each shift n of the filter mask. This permits near real-time computation of AR parameters since estimation begins with the first iteration [8]. As in the fixed data case only the approximated Toeplitz matrix is inverted. Computation time is shortened further by the fact that $\mathbf{R}_n$ is computed only from data covered by the filter mask which amounts to taking the outer product of a $P_1P_2 \times 1$ vector. The data-adaptive form of the iterative equations are given as

$$\mathbf{a}_j^{(n)} = \mathbf{a}_0^{(0)}\delta(j) - \mathbf{T}^{-1}\Delta_{j,j}\mathbf{a}_j^{(n-1)} - \mathbf{T}^{-1}\sum_{\substack{i=0\\ i\neq j}}^{P_1-1}\mathbf{R}_{j,i}\mathbf{a}_i^{(n-1)} \tag{13}$$

where $\mathbf{a}_0^{(0)} = \mathbf{T_n}^{-1}\mathbf{s}$, for $j = 0$, and $\mathbf{a_O}_j^{(0)} = \mathbf{T_O}^{-1}\mathbf{R_O}_{j,0}\mathbf{a}_0^{(0)}$ for $j = 1, 2, \ldots, P_1 - 1$. The initial parameter estimates $\mathbf{a_O}$ are determined for the initial point being estimated and all subsequent parameters $\mathbf{a}_j$ are a function of the parameters computed at the previous mask positions. In the above iterative equations we update the autocorrelation matrix in accordance with the recursive relationship

$$\mathbf{R}_n = \lambda\mathbf{R}_{n-1} + \left(\frac{n-1}{n}\right)\mathbf{x}_n\mathbf{x}_n^T \tag{14}$$

where the $P_1P_2 \times 1$ vector $\mathbf{x}_n$ is obtained by arranging the 2-D data covered by the filter mask at each shift $n = (n_1, n_2)$ and $0 < \lambda < 1$. The purpose of λ is to nullify the effects of statistical variations in the observed data [8] by lessening the effect of $\mathbf{R}$ computed in the distant past on the current update. The diagonal blocks of $\mathbf{R}_n$ are then used to obtain $\mathbf{T}_n$ in the same manner used for the fixed data case.

6 Experimental Results

The performance of the data-adaptive Toeplitz approximation algorithm is evaluated by examining the estimated spectral densities computed from the first, second and combined quadrant parameter estimates. The algorithm has been used to estimate the parameters of the sinusoidal input array generated by

$$\begin{aligned} x(n_1,n_2) &= A_1\cos(2\pi f_{11}n_1 + 2\pi f_{12}n_2) \\ &\quad + A_2\cos(2\pi f_{21}n_1 + 2\pi f_{22}n_2) \\ &\quad + w(n_1,n_2) \end{aligned} \tag{15}$$

where amplitudes $A_1 = A_2 = \sqrt{2}$ and f_{ij} are normalized frequencies in the range $(0 \le f_{ij} \le 0.5)$. The frequencies chosen are $f_{11} = f_{12} = 0.167$ and $f_{21} = f_{22} = 0.333$. Additionally, the data-adaptive algorithm is tested with off-set frequencies to evaluate its peformance for that case. The noise term $w(n_1,n_2)$ is zero mean and gaussian with the variance σ_w^2 scaled give a desired signal-to-noise ratio (SNR). The SNR (in dB) is defined as [4]

$$\mathrm{SNR} = 10\log_{10}\sum_{i=1}^{N}\frac{A_i^2}{2\sigma_w^2}, \tag{16}$$

where N is the number of sinusoids present and A is the amplitude term. The SNR is varied by holding A_i constant and adjusting the value of σ_w^2. The algorithm's performance using a 3×3 filter mask is evaluated for various sizes of input arrays and the common regions of support. The covariance method was used to produce the results provided. The data-adaptive algorithm is tested with SNR's of 10 dB and 0 dB. It should be noted that the surface plots in all figures have been rotated 90 degrees to show the separation of the spectral peaks more clearly. Contour plots are provided to better judge the accuracy of the estimates. The axes of the contour plots range from 0-30 which is taken from a 32×32 frequency domain array representing a quadrant of a 64×64 point 2-D FFT output. The 32×32 array covers the frequency range from 0 to π. In the following figures the quadrant estimate giving the best resolution is provided.

The first example demonstrates a unique characteristic of the data-adaptive iterative Toeplitz approximation algorithm. In this example the spectral estimate is computed using a 3×3 filter mask with a 4×4 input array. This means that only *four* elements of the input data are available to the filter to determine the parameter estimates. This result is of particular interest because the correlation matrix is actually rank deficient, and the problem cannot be solved by direct inversion. Nevertheless, the iterative Toeplitz approximation method does provide a means to compute the estimate. The results are given in Figure 1.

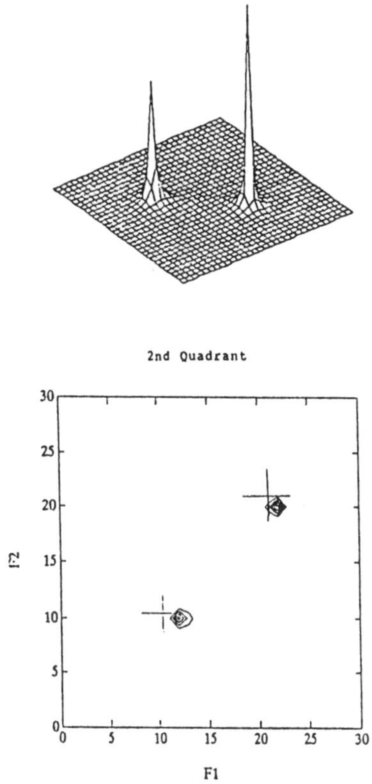

Figure 1: 4 × 4 input array (Covariance Method), SNR=10 dB.

As a further test of the algorithm, a signal with offset frequencies $f_{11} = 0.250$, $f_{12} = 0.125$, $f_{21} = 0.300$, and $f_{22} = 0.400$ is generated and processed. The spectral estimates resulting from the computed model parameters are plotted in Figure 2. For this case, the best spectral estimate was obtained using first quadrant suppport.

The spectral estimates for a signal-to-noise ratio of 0 dB is ploted in Figure 3. The accuracy of the estimates is noticeably degraded in this case. However, a reasonable indication of the signal frequency content can be seen when the combined-quadrant method is used.

The final case using NSHP support is included for completeness. The filter mask had dimensions $P_1 = P_2 = 4$ and $L_2 = -2$. The parameters were estimated for a 16×16 input array. The NSHP spectral estimate is plotted in Figure 4. The resulting spectral estimate shown is quite accurate, but this comes at the cost

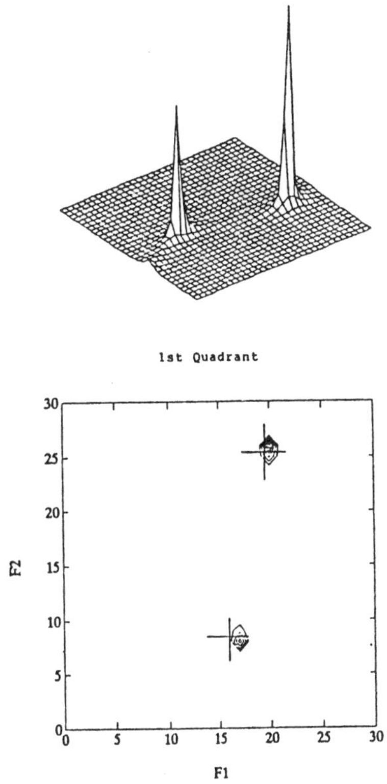

Figure 2: 8 × 8 input array, offset frequencies, (Covariance Method), SNR=10 dB.

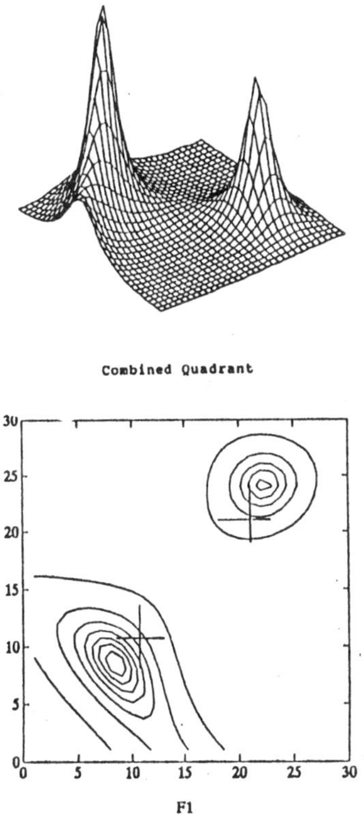

Figure 3: 8 × 8 input array (Covariance Method), SNR=0dB.

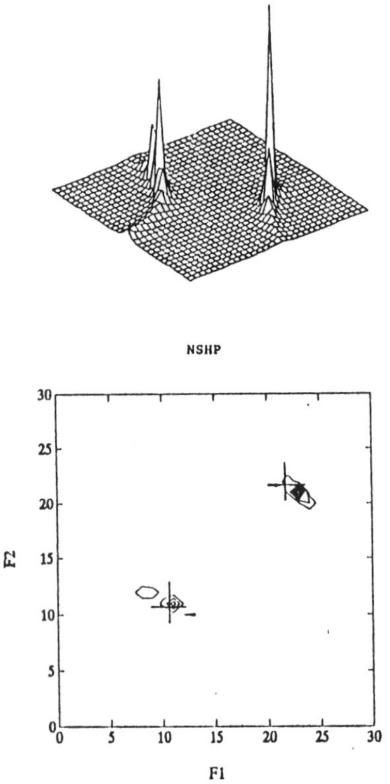

Figure 4: 16 x 16 input array (Covariance Method), SNR=10 dB; nonsymmetric half-plane support.

of greater complexity for the filter. NSHP support may be used as an alternative to combined-quadrant support in some cases.

7 CONCLUSIONS

The results indicate that high-resolution 2–D spectral estimates can be obtained using the data-adaptive iterative method. The algorithm's performance in a noisy environment is encouraging. This has been accomplished using $\mathcal{O}(P_1^2P_2^2K + P_1P_2^2)$ multiplications, which represents a signficant savings over RLS which would require approximately $\mathcal{O}((1.5P_1P_2)^2K + 4.5(P_1P_2)K)$ multiplications to compute the same parameters. The fact that the algorithm could estimate parameters when the direct inversion method could not be used was noteworthy. In general, the combined-quadrant estimates proved to be the most accurate but this level of complexity is not always required to achieve accurate spectral estimates. Additionally, the iterative methods are guaranteed to converge when the block matrix is symmetric and positive definite. Although other methods may exhibit faster convergence to the true parameters, they do so with greater complexity.

References

[1] Dan E. Dudgeon and Russell M. Mersereau, *Multidimensional Digital Signal Processing*, Prentice Hall Inc., Englewood Cliffs, New Jersey, 1984.

[2] Murali Tummala, "New algorithm for solving block matrix equations with applications in 2-D spectral estimation," *IEEE Trans. on Signal Processing*, Vol. 39, No. 3, pp. 759-764, March 1991.

[3] Roderick C. Wester, *Multidimensional Spectral Estimation Using Iterative Methods.* Master's thesis, Naval Postgraduate School, Monterey, California, June 1990.

[4] Jae S. Lim, *Two-Dimensional Signal and Image Processing*, Prentice Hall Inc., Englewood Cliffs, New Jersey, 1990.

[5] Charles W. Therrien, *Discrete Random Signals and Statistical Signal Processing*, Prentice Hall Inc., Englewood Cliffs, New Jersey, (to be published 1992).

[6] M. Wax and T.Kailath, "Efficient inversion of Toeplitz-block-Toeplitz matrices," *IEEE Trans. Acoust., Speech, Signal Processing*, vol. ASSP-31, no. 5, pp. 1218-1221, Oct. 1983.

[7] H. Akaike, "Block Toeplitz matrix inversion," *SIAM J. Appl. Math.*, vol. 28, no. 128, pp.234-241, Mar. 1973.

[8] John C. Eremic, *Iterative Methods for Estimation of 2-D AR Paramters using a Data-Adaptive Toeplitz Approximation Algorithm.* Master's thesis, Naval Postgraduate School, Monterey, California, September 1991.

A Soft Constrained Generalized Sidelobe Canceller*

Hao Ye **Ronald D. DeGroat**

The University of Texas at Dallas
Erik Jonsson School of Engineering and Computer Science
Richardson, TX 75083–0688
(214) 690–2894 (214) 690–2813 FAX degroat@utdallas.edu

An efficient new beamforming structure similar to a generalized sidelobe canceller is derived from the soft constrained minimum variance (SCMV) method of Van Veen. The new structure greatly reduces the computational load of the SCMV method and provides insight into the nature of SCMV beamforming.

Introduction

A beamformer can be used to estimate a propagating signal that arrives at an array of sensors from a given direction in the presence of sensor noise and interference that propagates from other directions. The beamformer output is generated from a weighted linear combination of the sensor outputs. The general goal is to choose weights which preserve the signal and minimize the effects of interference and noise. A common criterion for choosing the weights is minimization of the output power subject to a constraint on the beamformer response in the signal direction. Different types of constraints result in different beamformers.

The linearly constrained minimum variance (LCMV) beamformer employs a "hard" constraint on the response of the beamformer so that signals from the direction of interest are passed with specified gain and phase. The weights are chosen to minimize output variance subject to the response constraint. The LCMV beamformer was implemented for real time processing by Frost [1] who devised an adaptive algorithm in which the beamformer coefficients adjust to new values as each new sample vector is received.

The generalized sidelobe canceller (GSC) of Griffiths [2] represents an alternative formulation of the adaptive LCMV problem. It consists of a fixed beamformer, and a signal blocking matrix in front of an unconstrained adaptive beamformer. By using a signal blocking matrix, the constrained minimization problem is changed into an unconstrained form. The GSC structure simplifies the LCMV beamforming implementation. It also provides insight into the nature of LCMV beamforming.

* This research was supported in part by the Texas Advanced Technology Program Grant 009741–038, the Texas Advanced Research Program Grant 009741–039 and Texas Instruments.

Due to the rigidity of the hard constraint on the beamformer response, the signal-to-noise ratio (SNR) improvement by LCMV is generally not satisfactory. Recent work with small arrays operating at broad bandwidths has found that relaxation of the constraint can result in improved interference cancellation at the expense of some signal distortion [3, 4]. A soft constrained minimum variance (SCMV) beamformer is proposed by Van Veen [4] in which an inequality or "soft" constraint on the mean squared error between a desired response and the actual response in the signal direction is used to constrain the beamformer weights. The constraint is purposely chosen to permit distortion of the signal with the goal of achieving improved interference cancellation. Under certain assumptions, Van Veen showed that the SNR is a nondecreasing function of the mean square distortion.

Although the SCMV beamformer achieves better interference cancellation than the LCMV method, it requires more computation. With L sensors and M taps in each sensor channel, the computational complexity of the SCMV beamformer is on the order of M^2L^2, compared to ML for the GSC or the Frost beamformer.

The purpose of this paper is to present an efficient new beamforming structure for the SCMV method of Van Veen [4]. The new structure is closely related to the GSC and is a direct (although not necessarily obvious) modification and simplification of Van Veen's general algorithm. The main advantage of the new algorithm is that the computational complexity is reduced from an order of M^2L^2 to $ML+M^2$. Another advantage is that the new structure reveals a simple relationship between SCMV and LCMV beamforming.

It should also be noted that a computationally efficient GSC-like structure is given in [4], but it is based on the assumption that the desired signal is a white noise spatial point source. Our new structure achieves computationally efficiency for any broadband point source.

In this paper we first review Van Veen's SCMV beamformer, its theory and adaptive implementation. We then show that the SCMV method can be modified and simplified to yield an efficient, new beamforming structure which can be described as a soft constrained – generalized sidelobe canceller (SC-GSC). The relationship between the SC-GSC, the

GSC, and the SCMV beamformer is discussed. Simulations confirm that the SC-GSC achieves the same performance as an SCMV beamformer with much less computation.

Concepts of SCMV Beamforming

Assume that the beamformer has L sensors and M taps. If $x(k) = [x_1(k), \cdots, x_L(k)]^T$ represents the sensor outputs sampled at time k, then the total collection of data in the beamformer is given by $X(k) = [x^T(k),\, x^T(k-1), \cdots, x^T(k-M+1)]^T$, and the beamformer output at time k can be expressed as

$$y(k) = W^H X(k) \tag{1}$$

where $X(k)$ and the weight vector W are $N = LM$ dimensional vectors. Superscript H denotes complex conjugate transpose. Assuming zero mean, wide sense stationary data, the output power (or variance) is given by

$$P_{out} = E\left\{|y|^2\right\} = W^H R_X W \tag{2}$$

where $R_X = E\{XX^H\}$ is the N by N spatial data correlation matrix.

Let $\mathrm{d}(\theta,\omega)$ be defined as the array response vector, i.e., $\mathrm{d}(\theta,\omega)$ is a vector that describes the amplitude and phase relationships between a specified reference point and each element of X when the signal arriving at the array is of frequency ω and from direction θ. The array response vector is also known as the steering or direction vector [5].

The response of the beamformer at frequency ω and direction θ is now simply expressed as

$$r(\theta,\omega) = W^H \mathrm{d}(\theta,\omega). \tag{3}$$

Consider a fixed set of beamformer weights W_d which implement a desired response $r_d(\theta,\omega)$ to a signal located within a band of frequencies Ω and range of directions Θ,

$$r_d(\theta,\omega) = W_d^H \mathrm{d}(\theta,\omega); \quad \theta \in \Theta,\ \omega \in \Omega. \tag{4}$$

The weighted mean square error between the desired and actual response over Θ and Ω is expressed as

$$e = \int\limits_{\theta\in\Theta}\int\limits_{\omega\in\Omega} \rho(\theta,\omega)\left|r_d(\theta,\omega) - W^H \mathrm{d}(\theta,\omega)\right|^2 d\omega d\theta \tag{5}$$

where $\rho(\theta,\omega)$ is a nonnegative weighting function. If W is decomposed as $W = W_d - W_a$, it is straightforward to show using (4) that

$$e = W_a^H Q W_a \qquad where$$
$$Q = \int\limits_{\theta\in\Theta}\int\limits_{\omega\in\Omega} \rho(\theta,\omega)\mathrm{d}(\theta,\omega)\mathrm{d}^H(\theta,\omega) d\omega d\theta. \tag{6}$$

Thus, the weights which minimize the output variance subject to an inequality constraint on the mean square response error satisfy

$$\min_{W_a} (W_d - W_a)^H R_X (W_d - W_a) \quad with \quad W_a^H Q W_a \le e_0 \tag{7}$$

where e_0 represents the maximum tolerable mean square response error.

From the method of Lagrange multipliers we obtain

$$W_a = (R_X + \lambda Q)^{-1} R_X W_d \tag{8}$$

where the Lagrange multiplier λ is chosen as large as possible subject to $W_a^H Q W_a \le e_0$.

The Q matrix given in (6) corresponds to the signal correlation matrix resulting from a source with power spectral density $\rho(\theta,\omega)$. If Θ corresponds to a single direction, then Q represents a broadband spatial point source. Van Veen [4] proved that SNR is an nondecreasing function of e_0 if the error weighting $\rho(\theta,\omega)$ is proportional to the signal power spectral density. In other words, as e_0 increases, i.e. as greater levels of distortion are tolerated, the SNR improves (or remains constant). This represents a trade of bias (distortion) for reduced variance (noise power).

In general, the potential SNR improvement increases as the signal bandwidth increases because the hard constraint uses more and more degrees of freedom as the constraint bandwidth increases. Relaxation of the constraints frees up a proportionally larger number of degrees of freedom for interference cancellation. This characteristic of the SCMV beamformer makes it especially suitable for broadband signal processing applications, such as speech processing.

Adaptive SCMV Algorithm due to Van Veen

In addition to the theory of SCMV beamforming, Van Veen developed an adaptive algorithm [4] which is summarized as follows. Define the eigendecomposition of Q in Equation (6) as FDF^H. Transform W_a and X so that $W_a = FW_f$ and $X_f = F^H X$. Equation (7) is now written

$$\min_{W_a} (W_d - FW_f)^H R_X (W_d - FW_f) \quad subject\,to \quad W_f^H D W_f \le e_0. \tag{9}$$

If Q is rank p, then only p diagonal elements of D are nonzero and the constraint in (9) only affects p of the elements of W_f. The remaining $N-p$ are unconstrained. Define W_c and W_u as the p constrained and $N-p$ unconstrained components of W_f respectively, and D_c as the p by p diagonal matrix constructed from the p nonzero elements of D. In other words, transform W_a into constrained and unconstrained partitions

$$W_f = F^H W_a = \begin{bmatrix} F_c^H W_a \\ F_u^H W_a \end{bmatrix} = \begin{bmatrix} W_c \\ W_u \end{bmatrix} \tag{10}$$

where the eigendecomposition of $Q = FDF^H = [F_c, F_u]\begin{bmatrix} D_c & 0 \\ 0 & 0 \end{bmatrix}[F_c, F_u]^H = F_c D_c F_c^H$ defines the partitions of $F = [F_c \ F_u]$. Similarly, let

$$X_f = F^H X = \begin{bmatrix} F_c^H X \\ F_u^H X \end{bmatrix} = \begin{bmatrix} X_c \\ X_u \end{bmatrix}. \tag{11}$$

The constrained and unconstrained partitions of $F = [F_c \ F_u]$ are $ML \times p$ and $ML \times (ML - p)$ matrices respectively. Using these definitions, the constraint in (9) can be rewritten as $W_c^H D_c W_c \leq e_0$. And, Van Veen's gradient based adaptive algorithm (with step size μ) is given by

$$\begin{aligned}
&X_f(k) = \begin{bmatrix} X_c(k) \\ X_u(k) \end{bmatrix} = F^H X(k) \\
&y(k) = W_d^H(k)X(k) - W_c^H(k)X_c(k) - W_u^H(k)X_u(k) \\
&W_u(k+1) = W_u(k) + \mu X_u(k)y(k) \\
&\tilde{W}_c(k+1) = W_c(k) + \mu X_c(k)y(k) \\
&W_c(k+1) = \begin{cases} \tilde{W}_c(k+1) & for \ \ e(k) \leq e_0 \\ \frac{\sqrt{e_0}\tilde{W}_c(k+1)}{\sqrt{e(k)}} & for \ \ e(k) > e_0 \end{cases} \\
&where \quad e(k) = \tilde{W}_c^H(k+1)D_c\tilde{W}_c(k+1).
\end{aligned} \tag{12}$$

Soft Constrained Generalized Sidelobe Canceller

The SCMV beamformer achieves better interference cancellation than LCMV at the expense of much more computation. The bulk of the computation is $O(M^2L^2)$ due to the matrix to vector multiplication in (12a). Our goal then is to avoid this computation. By exploiting the structure of the signal correlation matrix Q, we can modify and simplify the computation of (12a) to produce an efficient new algorithm that is $O(ML + M^2)$.

We assume that the array is already steered towards a broadband point source signal. Under this assumption, the desired signal is aligned at each sensor output so that the signal correlation matrix

$$Q = q \otimes \Pi \tag{13}$$

where q is an $M \times M$ temporal (signal) correlation matrix, $\otimes$ denotes Kronecker product [6], and Π is an $L \times L$ matrix of ones. The rank of Q is given by $p = rank(Q) = rank(q) \leq M$. With our broadband assumption, $p = M$.

If the eigendecomposition of q is fdf^H and $\mathbf{1}$ is an L dimensional vector of ones, it can be verified that

$$F_c = \frac{1}{\sqrt{L}} f \otimes \mathbf{1} \quad and \quad D_c = Ld. \tag{14}$$

To simplify the SCMV method, it is helpful to define a projection matrix based on the unconstrained subspace

$$\begin{aligned}
P = F_u F_u^H &= I_N - F_c F_c^H \\
&= I_N - \frac{1}{L}(f \otimes \mathbf{1})(f^H \otimes \mathbf{1}^H) \\
&= I_N - \frac{1}{L} I_M \otimes \Pi.
\end{aligned} \tag{15}$$

By taking advantage of the simplicity of P and F_c, a new algorithm with much less computation can be developed. Consider the output of the beamformer

$$\begin{aligned}
y = W^H X &= (W_d - W_a)^H X \\
&= W_d^H X - W_a^H F F^H X \\
&= W_d^H X - W_a^H (F_c F_c^H + F_u F_u^H) X \\
&= W_d^H X - W_c^H X_c - W_a^H P X \\
&= W_d^H X - W_c^H X_c - W_a^H P^H P X \\
&= W_d^H X - W_c^H X_c - W_p^H X_p
\end{aligned} \tag{16}$$

where $W_p = PW_a$ and $X_p = PX$. It is important to realize that $W_u^H X_u = W_p^H X_p$ (Proof: $W_u^H X_u = W_a^H F_u F_u^H X_a = W_a^H P X_a = W_a^H P^H P X_a = W_p^H X_p$). Thus, even though the unconstrained part is computed differently, the output of (16) is identical to (12b). Also, (16) shows that the output of the beamformer consists of a fixed distortionless part, a constrained part and an unconstrained part. In the next section, we show that this structure is essentially a soft constrained – generalized sidelobe canceller (SC-GSC).

Except for explicitly demonstrating how the computation of (12a) is avoided and simplified, the only thing needed to complete the new algorithm is to find a way to adaptively estimate the optimum weight vector W_p. We begin with the estimate of W_a. In gradient-descent optimization, W_a is iteratively estimated as

$$W_a(k+1) = W_a(k) + \mu X(k)y(k) \tag{17}$$

Pre-multiplying equation (17) by P and substituting $W_p = PW_a$ and $X_p = PX_a$, we obtain

$$W_p(k+1) = W_p(k) + \mu X_p(k)y(k). \tag{18}$$

The new algorithm is thus summarized as follows

$$\begin{aligned}
&X_c(k) = F_c^H X(k) \quad X_p(k) = PX(k) \\
&y(k) = W_d^H(k)X(k) - W_c^H(k)X_c(k) - W_p^H(k)X_p(k) \\
&W_p(k+1) = W_p(k) + \mu X_p(k)y(k) \\
&\tilde{W}_c(k+1) = W_c(k) + \mu X_c(k)y(k) \\
&W_c(k+1) = \begin{cases} \tilde{W}_c(k+1) & for \ \ e(k) \leq e_0 \\ \frac{\sqrt{e_0}\tilde{W}_c(k+1)}{\sqrt{e(k)}} & for \ \ e(k) > e_0 \end{cases} \\
&where \quad e(k) = \tilde{W}_c^H(k+1)D_c\tilde{W}_c(k+1).
\end{aligned} \tag{19}$$

This algorithm is very similar to the original SCMV algorithm in (12), except that $W_p^H X_p = W_u^H X_u$ is used in place of $W_u^H X_u$, W_p is adaptively tracked instead of W_u and $X_p = PX$ is computed in place of $X_u = F_u^H X$. The computationally expensive multiplication of F^H and $X(k)$ in the old algorithm is avoided in the new algorithm by computing the unconstrained part of the problem with $P = F_u F_u^H$ instead of F_u and exploiting the simple structure of P and F_c as we shall now demonstrate.

Define $\bar{x}(k)$ to be the sum of sensor outputs, $\bar{x}(k) = \sum_{i=1}^{L} x_i(k)$, and $\bar{X}(k) = [\bar{x}(k), \bar{x}(k-1), \cdots, \bar{x}(k-M+1)]^T$. With these definitions and (14) we have

$$\begin{aligned} X_c(k) &= F_c^H X(k) \\ &= \frac{1}{\sqrt{L}}(f^H \otimes 1^H)X(k) \\ &= \frac{1}{\sqrt{L}} f^H \bar{X}(k), \end{aligned} \tag{20}$$

which is $O(M^2)$. Similarly, (15) leads to an $O(ML)$ computation,

$$\begin{aligned} X_p(k) &= PX(k) \\ &= \left(I_N - \frac{1}{L} I_M \otimes \Pi\right) X(k) \\ &= X(k) - \frac{1}{L}\bar{X}(k) \otimes \mathbf{1}. \end{aligned} \tag{21}$$

With the modified algorithm of (19) and the simplifications in (20) and (21), the computation time of the new algorithm is much less than the original SCMV algorithm. At each iteration, the number of multiplications is on the order of $M^2 + ML$ instead of M^2L^2. An accurate comparison between the computational complexity of the SC-GSC, the GSC and the SCMV beamformer can be found in Table 1.

The SC-GSC, the GSC and the SCMV Beamformer

Equation (21) indicates that $X_p(k)$ is the difference between each sensor output and the average of sensor outputs. The sum of each row of the P matrix is zero. Thus the matrix P acts as a blocking matrix which prevents the signals in the look direction from going through the unconstrained path. Also note that $X_c(k)$ is only a function of the sum of the sensor outputs. Therefore, the constrained path is actually an FIR filter rather than a beamformer. Figure 1 shows the overall structure of the SC-GSC beamformer.

This system consists of three sub-systems, a fixed distortionless beamformer, a signal blocking matrix in front of an unconstrained adaptive beamformer, and a constrained FIR filter. The first two parts in fact constitute a generalized sidelobe canceller. This is where the name soft constrained – generalized sidelobe canceller comes from. To satisfy the soft constraints, the constrained FIR filter operates on the sum of the array outputs instead of the individual sensor outputs. This configuration provides some insights into the nature of the SC-GSC, as well as its relationship to the conventional beamformer and the GSC. The core of the beamformer is a conventional fixed beamformer. With a blocking matrix and an unconstrained optimization algorithm, it becomes a GSC which is a form of the LCMV beamformer. The SC-GSC which implements SCMV beamforming is essentially a GSC with a constrained FIR filter operating on the sum of the sensor outputs. Thus, the SC-GSC can be viewed either as a fast implementation of the SCMV beamformer, or as a soft constrained extension of the GSC.

Applications

The SCMV beamformer is most effective with small arrays and broadband signals [4]. The SC-GSC achieves the same performance as the SCMV beamformer with a greatly reduced computational load. This reduction in computation makes it possible to apply soft constrained beamforming to real-time broadband array processing, such as multi-microphone speech acquisition. The SC-GSC is particularly appropriate for speech recognition because the speech recognizer can be trained to tolerate the distortion introduced by the SC-GSC [7]. A multi-sensor speech recognition simulation study is described below.

The multi-sensor speech recognition system feeds the beamformer output into a hidden Markov model (HMM) based speaker independent isolated digit recognizer†. The beamformer serves as a performance enhancing front end to the speech recognizer. Thus, the recognition rate is used to evaluate various beamforming algorithms. In this study, the GSC, the SC-GSC, and the SCMV beamformers are examined.

In Figure 2, we plot the recognition rates versus signal to interference ratios (SIR) for the three beamformers. In the simulations, a four-sensor-four-tap equi-spaced linear array is used. Sensor spacing is chosen to be half of the wave length of the highest frequency. The desired signal comes from the broadside of the array. Signal to sensor noise ratio (SNR) is 30 db. A fan noise coming from 54 degrees away from the broadside of the array is used as interference. A generic signal correlation matrix, q, is estimated from the speech of a large number of different speakers. The recognition system is trained with a beamformer front end, but without an interference signal. A database of ten speakers speaking ten digits each is used in the training process. The performance of the recognition system with different

† We would like to thank Dr. Joe Picone at the Texas Instruments Speech and Image Understanding Laboratory for helping us to implement the HMM based speech recognizer.

beamformers is evaluated on another database which contains a different set of six speakers speaking 10 digits each. It can be seen from the plot that the performance of the SC-GSC and the SCMV beamformer are identical. The main advantage of the SC-GSC over the GSC is that the threshold breakdown in recognition rate occurs at a lower SIR. Above threshold breakdown, the performance of SC-GSC is generally comparable to, or better than that of the GSC. At $SIR = 0db$, the SC-GSC beamformer produces a recognition rate of 87% in contrast to 21% for the (hard constrained) GSC. Further improvement of the SC-GSC recognition rate performance can be expected, if an accurate, time-varying estimate of q is used in place of a fixed generic one. We also note that the performance of the SC-GSC is not significantly degraded if a simpler (suboptimal) blocking matrix is used in place of the projection matrix defined in (15).

Conclusion

A soft constrained generalized sidelobe cancelling structure is developed from Van Veen's soft constrained minimum variance method [4]. The new algorithm achieves the same interference cancellation as the SCMV beamformer with much less computation: $O(ML + M^2)$ vs. $O(M^2L^2)$ for L sensors and M taps. This reduction in computation makes it possible to apply the soft constrained beamformer to real-time broadband array processing. The SC-GSC structure also reveals the inherent connection between LCMV and SCMV beamforming. The SC-GSC is appropriate for broadband signal processing, and is especially appropriate for speech recognition applications.

[1] Otis Lamont Frost III. An algorithm for linearly constrained adaptive array processing. *Proceedings of the IEEE, vol. 60, no. 8*, August 1972. pp. 926-935.

[2] Lloyd J. Griffiths and Charles W. Jim. An alternative approach to linearly constrained adaptive beamforming. *IEEE Trans. Antennas and Propagation, vol. AP-30, no. 1*, January 1981. pp. 27-34.

[3] Y. Kaneda and J. Ohga. Adaptive microphone-array system for noise reduction. *IEEE Trans. ASSP, vol. ASSP-34, no. 6*, December 1986. pp. 1391-1400.

[4] Barry Van Veen. Minimum variance beamforming with soft response constraints. *IEEE Trans. Signal Processing, vol. SP-39, no. 9*, September 1991.

[5] Barry D. Van Veen and Kevin M. Buckley. Beamforming: A versatile approach to spatial filtering. *IEEE ASSP Magazine*, April 1988.

[6] J. W. Brewer. Kronecker products and matrix calculus in system theory. *IEEE Trans. Circuits and System. vol. CAS-25*, September 1978. pp. 772-781.

[7] Hao Ye. A multi-sensor speech recognition system. *Master Thesis, University of Texas at Dallas*, May 1991.

Beam-former	Number of Mults. Per Update	Number of Adds Per Update
GSC	ML - M	2ML - 2M
SC-GSC	$3ML + M^2 + 5M$	$3ML + M^2 + 2M + L$
SCMV	$M^2L^2 + 3ML + 3M$	$M^2L^2 + 2ML + M$

Table 1. Computational complexity comparison.

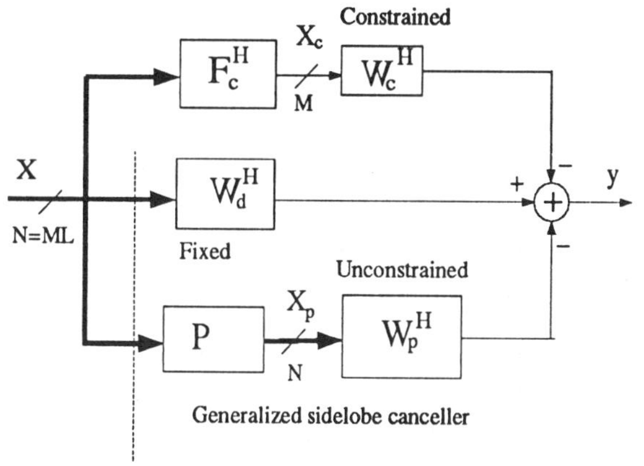

Figure 1. Block diagram of the soft constrained generalized sidelobe canceller. This system consists of three sub-systems, a fixed distortionless beamformer, a blocking matrix in front of an unconstrained adaptive beamformer, and a constrained FIR filter. The first two parts constitute a generalized sidelobe canceller (GSC). The computation is reduced from $O(M^2L^2)$ to $O(ML + M^2)$ by exploiting the structure of F_c and P.

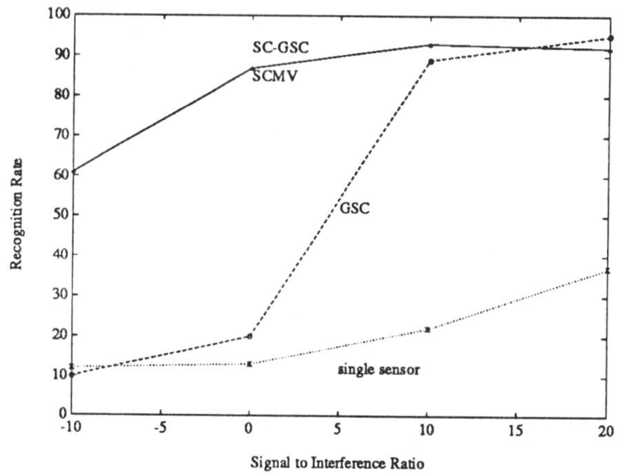

Figure 2. Recognition rates versus signal to interference ratios (SIR) for the GSC, the SC-GSC, and the SCMV beamformer. A four-sensor-four-tap equi-spaced linear array is used. Sensor spacing is half the wave length of the highest frequency. The desired signal comes from broadside. Signal to sensor noise ratio is 30 db. Fan noise coming from 54 degrees away from broadside is used as interference.

Statistical Analysis of SVD-Based Prony Techniques *

William M. Steedly **Ching-Hui J. Ying**
Randolph L. Moses
Department of Electrical Engineering
The Ohio State University
Columbus, Ohio 43210

Abstract

In this paper we present an analysis of parameter variance statistics for the SVD-Prony method applied to damped exponential signals. We derive the covariance matrix of the estimated parameters for this method. The parameters include the magnitudes and angles of the poles, and the magnitudes and angles of the amplitude coefficients. We verify the theoretical results using Monte-Carlo simulations studies. We also compare the variance results to the corresponding Cramér-Rao bounds for several cases.

1 Introduction

The problem of estimating model parameters of noisy exponential signals is an active area of research. These models have a single set of complex poles and one or more set of amplitude coefficients (snapshots). The performance of these parameter estimation methods is often measured by the accuracy of the estimated poles, since these pole locations contain such information as formant frequencies or directions of arrival of signal components.

One of the popular estimation schemes is the SVD-Prony estimation [1]. The multi-snapshot generalization is considered in this paper. This paper provides a complete statistical derivation for the multi-snapshot case where signals consist of arbitrary exponential terms in noise. We have derived its asymptotic statistics, including the covariance matrix of the estimated parameters. The parameters include the magnitudes and angles of the poles, and the magnitudes and angles of the amplitude coefficients.

Using these expressions, it is shown that the angle and magnitude parameters are uncorrelated. It is also shown that if relative magnitude of the pole or amplitude coefficient estimate is considered (*i.e.* $\frac{\hat{\alpha}}{\alpha}$, where α is the true magnitude), then the corresponding angle and relative magnitude variances are equal.

This paper also examines pole estimation accuracy as functions of pole magnitude, data length, and pole separation using the variance expressions. We compare these variance results to the corresponding Cramér-Rao bounds and verify the theoretical results using Monte-Carlo simulations. In addition, we show that the transition from the $1/m^3$ (where m is the number of data points) variance bounds decrease for poles on the unit circle to the variance decrease as poles move off the unit circle is also detailed.

The effects on poles inside and outside the unit circle using backward or forward linear prediction in the SVD-Prony estimation scheme is also detailed.

2 Estimation Procedure Review

2.1 Data Model

Assume we have N "snapshots" of data vectors $y(t)$, each of length m:

$$y(t) = \begin{bmatrix} y_0(t) & y_1(t) & \cdots & y_{m-1}(t) \end{bmatrix}^T, \; t = 1, \ldots, N. \qquad (1)$$

Each data vector is modeled as a noisy exponential sequence

$$y_q(t) = \sum_{i=1}^{n} x_i(t) p_i^q + e_q(t) \qquad q = 0, 1, \ldots, m-1. \quad (2)$$

There are n distinct exponential modes in the data; the n poles $\{p_i\}_{i=1}^{n}$ do not vary from snapshot to snapshot, but the amplitudes $x_i(t)$ may vary. Here, it is assumed that $\{e_q(t)\}$ are uncorrelated zero mean complex white Gaussian noise sequences with variance σ.

*This research was supported in part by the Air Force Office of Scientific Research, the Avionics Division, Wright Laboratories, and the Surveillance Division, Rome Laboratories.

1058-6393/91 $01.00 © 1991 IEEE

Equation 2 may be compactly written as

$$y(t) = Ax(t) + e(t), \tag{3}$$

where $e(t) = \begin{bmatrix} e_0(t) & e_1(t) & \cdots & e_{m-1}(t) \end{bmatrix}^T$, $x(t) = \begin{bmatrix} x_0(t) & x_1(t) & \cdots & x_{n-1}(t) \end{bmatrix}^T$, and A is the $m \times n$ Vandermonde matrix derived from n signal poles

$$A = \begin{bmatrix} 1 & 1 & \cdots & 1 \\ p_1 & p_2 & \cdots & p_n \\ \vdots & \vdots & & \vdots \\ p_1^{m-1} & p_2^{m-1} & \cdots & p_n^{m-1} \end{bmatrix}. \tag{4}$$

2.2 Parameter Estimation

The multi-snapshot backward linear prediction equations are given by:

$$Yb \approx -y, \tag{5}$$

where

$$b = \begin{bmatrix} b_1 & b_2 & \cdots & b_L \end{bmatrix}^T \tag{6}$$

and where

$$\begin{bmatrix} y & Y \end{bmatrix} = \left[\begin{array}{cccc} y_0(1) & y_1(1) & \cdots & y_L(1) \\ y_1(1) & y_2(1) & \cdots & y_{L+1}(1) \\ \vdots & \vdots & & \vdots \\ y_{m-(L+1)}(1) & y_{m-L}(1) & \cdots & y_{m-1}(1) \\ \hline & \vdots & & \\ \hline y_0(N) & y_1(N) & \cdots & y_L(N) \\ y_1(N) & y_2(N) & \cdots & y_{L+1}(N) \\ \vdots & \vdots & & \vdots \\ y_{m-(L+1)}(N) & y_{m-L}(N) & \cdots & y_{m-1}(N) \end{array}\right]. \tag{7}$$

Here L is the order of prediction; b is the coefficient vector of the polynomial $B(z)$ given by

$$B(z) = 1 + b_1 z^{-1} + b_2 z^{-2} + \cdots + b_L z^{-L}. \tag{8}$$

For the noiseless case, L can be any integer greater than or equal to the model order n; in practice, choosing $L > n$ results in more accurate parameter estimates. Note that all of the N snapshots are used simultaneously to estimate a single set of prediction coefficients (and therefore, a single set of poles).

The solution of Equation 5 involves obtaining a singular value decomposition of the matrix $\begin{bmatrix} y & Y \end{bmatrix}$ and truncating all but the first n singular values to arrive at a noise cleaned estimate $\begin{bmatrix} \widehat{y} & \widehat{Y} \end{bmatrix}$ [1]. This leads to the modified linear prediction equation

$$\widehat{Y}\widehat{b} = \widehat{y} \tag{9}$$

from which the linear prediction coefficient vector estimate $\widehat{b}$ is found as

$$\widehat{b} = -\widehat{Y}^{+}\widehat{y}, \tag{10}$$

where $^{+}$ denotes the Moore-Penrose pseudoinverse. Finally, the estimates for the poles are found by

$$\widehat{p}_j = \frac{1}{\text{zero}_j\left(\widehat{B}(z)\right)}, \qquad j = 1, 2, \ldots, L. \tag{11}$$

Once the poles have been determined, the equation for the amplitude coefficients can be formed. Equation 3 leads to the following equation for the amplitude coefficients,

$$\begin{bmatrix} 1 & \cdots & 1 \\ \widehat{p}_1 & \cdots & \widehat{p}_L \\ \vdots & & \vdots \\ \widehat{p}_1^{m-1} & \cdots & \widehat{p}_L^{m-1} \end{bmatrix} \begin{bmatrix} \widehat{x}_1(1) & \cdots & \widehat{x}_1(N) \\ \vdots & & \vdots \\ \widehat{x}_L(1) & \cdots & \widehat{x}_L(N) \end{bmatrix} = \begin{bmatrix} y(1) & \cdots & y(N) \end{bmatrix} \tag{12}$$

or

$$\widehat{A}\widehat{X} = Y_a. \tag{13}$$

The amplitude coefficients can be found from a least squares solution to Equation 13,

$$\widehat{X} = \left(\widehat{A}^{*}\widehat{A}\right)^{-1}\widehat{A}^{*}Y_a = \widehat{A}^{+}Y_a, \tag{14}$$

where * denotes complex conjugate transpose. Because only n singular values of $\widehat{Y}$ are nonzero, there are at most n pole estimates which can correspond to true data modes. Therefore, only the n poles which have the largest energy are retained. Thus estimates $\widehat{A}$ and $\widehat{X}$ become matrices of sizes $m \times n$ and $n \times N$, where the $L - n$ columns and rows, respectively, corresponding to the lowest energy modes are deleted.

3 Statistical Analysis

To analyze parameter statistics we now derive their covariance matrix. This is given in the following theorem.

Theorem: Assume data is given by Equation 2. We define ω_i and α_i, to be the angle and magnitude, respectively, of each pole p_i, thus $p_i = \alpha_i e^{j\omega_i}$. Similarly we define $\gamma(t)$ and $\beta(t)$ to be the angle and magnitude vectors, respectively, of each vector of amplitude coefficients $x(t)$. We now define the following parameter vectors:

$$\begin{aligned} \theta_x &= \begin{bmatrix} \gamma^T(1) & \beta^T(1) & \cdots & \gamma^T(N) & \beta^T(N) \end{bmatrix}^T \\ \theta_p &= \begin{bmatrix} \omega_1 & \cdots & \omega_n & \alpha_1 & \cdots & \alpha_n \end{bmatrix}^T \\ \theta &= \begin{bmatrix} \theta_x^T & \theta_p^T \end{bmatrix}^T, \end{aligned} \tag{15}$$

and let $\widehat{\theta}$ denote the SVD-Prony estimate of θ, which are given by the n highest energy mode estimates found in Equations 11 and 14.

Then the asymptotic (high SNR, where SNR is defined as total signal power divided by total noise power) pdf of $\widehat{\theta}$ is given by

$$\widehat{\theta} \sim N\left(\theta, \Sigma_\theta\right), \tag{16}$$

where

$$\Sigma_\theta = \begin{bmatrix} \overline{U}(1,1) & \widetilde{U}(1,1)T_\beta^{*-1}(1) & \cdots & \overline{U}(1,N) & \widetilde{U}(1,N)T_\beta^{*-1}(N) & \overline{V}(1) & \widetilde{V}(1)T_\alpha^{*-1} \\ -T_\beta^{-1}(1)\widetilde{U}(1,1) & T_\beta^{-1}(1)\overline{U}(1,1)T_\beta^{*-1}(1) & \cdots & -T_\beta^{-1}(1)\widetilde{U}(1,N) & T_\beta^{-1}(1)\overline{U}(1,N)T_\beta^{*-1}(N) & -T_\beta^{-1}(1)\widetilde{V}(1) & T_\beta^{-1}(1)\overline{V}(1)T_\alpha^{*-1} \\ \vdots & \vdots & \ddots & \vdots & \vdots & \vdots & \vdots \\ \overline{U}(N,1) & \widetilde{U}(N,1)T_\beta^{*-1}(1) & \cdots & \overline{U}(N,N) & \widetilde{U}(N,N)T_\beta^{*-1}(N) & \overline{V}(N) & \widetilde{V}(N)T_\alpha^{*-1} \\ -T_\beta^{-1}(N)\widetilde{U}(N,1) & T_\beta^{-1}(N)\overline{U}(N,1)T_\beta^{*-1}(1) & \cdots & -T_\beta^{-1}(N)\widetilde{U}(N,N) & T_\beta^{-1}(N)\overline{U}(N,N)T_\beta^{*-1}(N) & -T_\beta^{-1}(N)\widetilde{V}(N) & T_\beta^{-1}(N)\overline{V}(N)T_\alpha^{*-1} \\ \overline{V^*}(1) & \widetilde{V^*}(1)T_\beta^{*-1}(1) & \cdots & \overline{V^*}(N) & \widetilde{V^*}(N)T_\beta^{*-1}(N) & \overline{Z} & \widetilde{Z}T_\alpha^{*-1} \\ -T_\alpha^{-1}\widetilde{V^*}(1) & T_\alpha^{-1}\overline{V^*}(1)T_\beta^{*-1}(1) & \cdots & -T_\alpha^{-1}\widetilde{V^*}(N) & T_\alpha^{-1}\overline{V^*}(N)T_\beta^{*-1}(N) & -T_\alpha^{-1}\widetilde{Z} & T_\alpha^{-1}\overline{Z}T_\alpha^{*-1} \end{bmatrix} \tag{17}$$

where $\bar{\cdot}$ and $\tilde{\cdot}$ in Equation 17 are real and imaginary part operators, respectively, and where

$$\begin{aligned} U(t,r) &= R(t)ZR^*(r) - R(t)Q(r) - Q^*(t)R^*(r) \\ &\quad + \frac{\sigma}{2}T_x(t)A^+A^{+*}T_x^*(r)\delta_{t,r} \\ V(t) &= -R(t)Z + Q^*(t) \\ Z &= \frac{\sigma}{2}T_pFGS^+DD^*S^{*+}G^*F^*T_p^* \\ Q(t) &= \frac{\sigma}{2}T_pFGS^+D(t)A^{+*}T_x^*(t) \\ R(t) &= T_x(t)A^+CAT_x^{-1}(t) \end{aligned} \tag{18}$$

Here, $T_x(t)$, $T_\beta(t)$, T_p, T_α, C, and F are diagonal matrices given by

$$\begin{aligned} T_x(t) &= \operatorname{diag}\left(\frac{1}{x_1(t)}, \frac{1}{x_2(t)}, \ldots, \frac{1}{x_n(t)}\right) \\ T_\beta(t) &= \operatorname{diag}\left(\frac{1}{\beta_1(t)}, \frac{1}{\beta_2(t)}, \ldots, \frac{1}{\beta_n(t)}\right) \\ T_p &= \operatorname{diag}\left(\frac{1}{p_1}, \frac{1}{p_2}, \ldots, \frac{1}{p_n}\right) \\ T_\alpha &= \operatorname{diag}\left(\frac{1}{\alpha_1}, \frac{1}{\alpha_2}, \ldots, \frac{1}{\alpha_n}\right) \\ C &= \operatorname{diag}(0, 1, \ldots, m-1) \\ F &= \operatorname{diag}\left(\frac{1}{\eta_1}, \frac{1}{\eta_2}, \ldots, \frac{1}{\eta_n}\right) \end{aligned} \tag{19}$$

and η_i and G are given by

$$\eta_i = \begin{bmatrix} b_1 & b_2 & \cdots & b_L \end{bmatrix} \begin{bmatrix} 1 \\ 2p_i \\ \vdots \\ Lp_i^{L-1} \end{bmatrix} \tag{20}$$

$$G = \begin{bmatrix} p_1 & p_1^2 & \cdots & p_1^L \\ p_2 & p_2^2 & \cdots & p_2^L \\ \vdots & \vdots & & \vdots \\ p_n & p_n^2 & \cdots & p_n^L \end{bmatrix}. \tag{21}$$

The matrix D is a $(m-L)N \times mN$ block diagonal matrix given by

$$D = \operatorname{diag}(B, B, \ldots, B), \tag{22}$$

where B is $(m-L) \times m$ and given by

$$B = \begin{bmatrix} 1 & b_1 & b_2 & \cdots & b_L & 0 & 0 & \cdots & 0 \\ 0 & 1 & b_1 & \cdots & b_{L-1} & b_L & 0 & \cdots & 0 \\ \vdots & \ddots & \ddots & \ddots & & \ddots & \ddots & \ddots & \vdots \\ 0 & \cdots & 0 & 1 & b_1 & \cdots & b_{L-1} & b_L & 0 \\ 0 & \cdots & 0 & 0 & 1 & \cdots & b_{L-2} & b_{L-1} & b_L \end{bmatrix}. \tag{23}$$

The matrices $D(t)$ are each given by the tth column block of D and S is the noise free version of Y. □

Proof: See [2].

Corollary: The elements of $\gamma(t)$ are uncorrelated with the corresponding elements of $\beta(t)$ and the elements ω are uncorrelated with corresponding elements of α. □

Proof: Note that the corresponding covariances in Σ_θ are the diagonal elements of the product of the imaginary part of a Hermitian matrix and a diagonal matrix and are thus identically zero. □

We can also note that the angle variances are equal to the magnitude variances except for the transformation matrices $T_x(t)$ and T_p. Without these transformation matrices, Σ_θ would correspond to a relative magnitude parameterization, where magnitude relative to the true magnitude is estimated, rather than absolute magnitude. When real and imaginary parts of the amplitude coefficients and poles are considered as parameters, the resulting variances can be derived using a simple Jacobian coordinate transformation and are equal.

Another property which can be pointed out is absolute phase invariance. We can see that the terms in

Equation 18 remain unchanged as $T_x(t) \rightarrow e^{j\phi(t)}T_x(t)$ and $p_i \rightarrow e^{j\phi}p_i$ due to cancellations. Thus the covariance matrix depends only on relative phase between the poles and between each set of amplitude coefficients for each snapshot.

4 Simulation Studies

Simulations have been performed using the statistical analyses above. Recently a CRB formulation has been developed for multi-snapshot damped exponentials in noise [3]. These CRB results are compared with the variances of the estimated poles using the SVD-Prony method to examine its performance.

4.1 Variance Results for Single Exponential Case

In this simulation, a model with a single pole and with one snapshot of data was chosen. The experiment entailed moving the pole along the positive real axis (the results are independent of the pole angle, so an angle of zero was chosen) from 0.1 to 10 and calculating the variances using Equation 17 for data sets of lengths 2, 5, 10, 20, 50, and 100. For comparative purposes, the amplitude coefficient associated with the pole was chosen to be a positive real number such that the mode energy ($x^2 \sum_{l=0}^{m-1} p^{2l}$) was unity for each pole location and data length. The noise power was also kept constant at $\sigma = 1$. The model order was assumed to be one third of the data length, which has been shown to be optimal [4, 5].

The variances for the pole angle and magnitude appear as the dashed lines in Figures 1 and 2, respectively; the corresponding CRBs appear as the solid lines in these figures. From Figure 1 we see that the pole angle variances are just above the CRBs inside and near the unit circle. Outside the unit circle the variances become much higher (except for the $m = 2$ case) than the CRBs. This is due to the fact that backward linear prediction is used in this SVD-Prony method; note that with backward linear prediction extraneous poles are placed outside the unit circle, thus making estimation of poles outside the unit circle more difficult. Using forward linear prediction will give the opposite results for poles inside and outside the unit circle. Similar observations apply to the pole magnitude variances (*cf.* Figure 2)

From these two figures we see that inside the unit circle the variances for pole angle are higher than the variances for pole magnitude and vice-versa outside the unit circle. This is due to the fact that angular

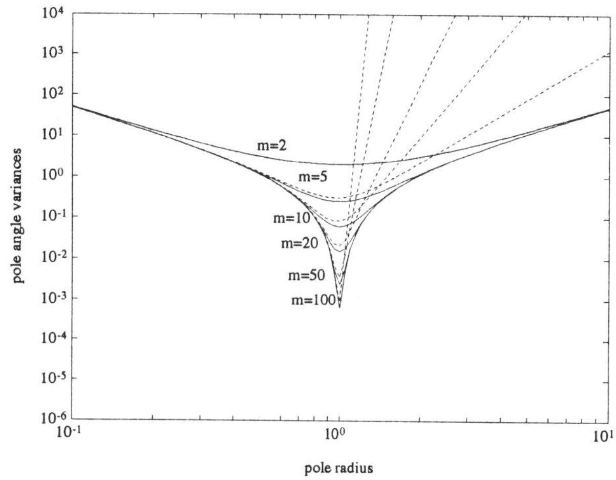

Figure 1: Pole angle variances for single pole data ($n = 1$).

uncertainty becomes greater as a pole moves closer to the origin. Note that the pole angle variance approaches infinity as the pole approaches the origin, which is what one would expect since its angle is undefined at the origin.

We see that the variances for both pole angle and magnitude are asymptotically (as $m \rightarrow \infty$) lowest when the pole is on the unit circle, and that on the unit circle the variances are decreasing by $1/m^2$ (m is the data length). This is consistent with the well-known $1/m^3$ variance decrease, since the amplitude coefficient was adjusted in this experiment to keep the mode energy constant (if the amplitude coefficient is left unchanged, the variance decrease is $1/m^3$). Note, however, that as the number of data points is reduced, the pole magnitude which gives the minimum variance is less than unity.

When the pole is not on the unit circle, the variances do not decrease to zero as $m \rightarrow \infty$. Because of the decay or growth of the exponential mode, and because the mode energy is kept constant, increasing m results in adding data points with lower and lower amplitude. Thus, the variances do not continue to decrease.

The variances for the amplitude coefficient angle and magnitude are also available and appear in [2].

4.2 Some General Observations

Using the covariance matrix, one can determine a confidence region corresponding to an estimate of each pole. This region can be plotted as, for example, a two standard deviation confidence "ellipse" about

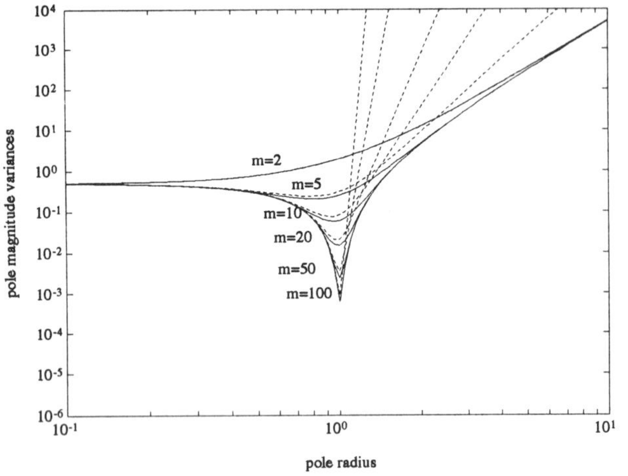

Figure 2: Pole magnitude variances for single pole data ($n = 1$).

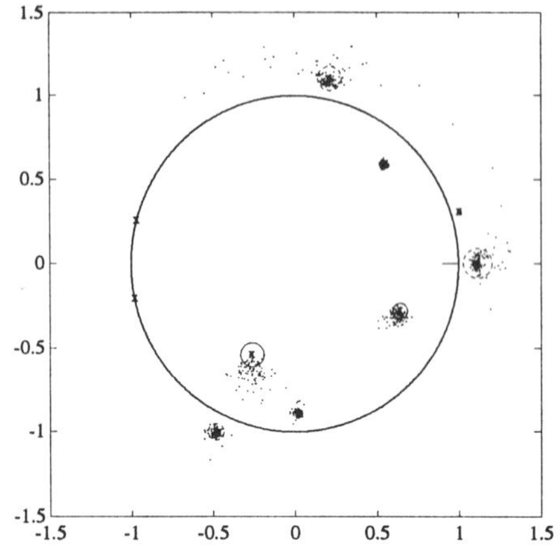

Figure 3: Two standard deviation bounding circles for each pole, using a tenth order model, with $m = 100$ data points, $\sigma = 0.001$, and each mode energy set to unity.

each pole. Figure 3 is an example of such a plot, where the pole locations are indicated by "x" symbols. In this experiment there are ten poles, each with an energy of one. Here, $m = 100$ data points and the noise power is $\sigma = 0.001$. This gives an SNR of 10dB per mode. The dashed ellipses (they are actually circles due to the corollary and the equivalence of the real and imaginary part variances as noted in Section 3) shown in Figure 3 represent the two standard deviation boundaries around the ten pole locations. 87% of poles found in a Monte-Carlo simulation would be expected to fall within these circles. Note that the boundary circles are significantly smaller for the poles which are located close to the unit circle. The solid circles represent the corresponding CRB bounds. One hundred Monte-Carlo simulations were performed and the estimated poles appear as dots in Figure 3. From these estimates, we can see that the statistical analysis is in general confirmed; the exception being some bias in the estimates of the pole closest to the origin.

5 Conclusions

We presented a statistical analysis for estimated poles of the SVD-Prony algorithm. We have provided complete expressions for the covariance matrix of the parameters of an exponential model with one set of poles and multiple sets of amplitude coefficients. The poles of this model may lie anywhere in the complex plane. Using these expressions several useful properties of the covariance matrix were established.

References

[1] M. A. Rahman and K.-B. Yu, "Total least squares approach for frequency estimation using linear prediction," *IEEE Transactions on Acoustics, Speech, and Signal Processing*, vol. ASSP-35, pp. 1440–1454, Oct. 1987.

[2] W. M. Steedly, C. J. Ying, and R. L. Moses, "Statistical analysis of SVD-based Prony techniques," Submitted for publication.

[3] W. M. Steedly and R. L. Moses, "The Cramér-Rao bound for pole and amplitude estimates of damped exponential signals in noise," in *Proceedings of the International Conference on Acoustics, Speech, and Signal Processing*, (Toronto, Ontario), pp. 3569–3572, May 14–17, 1991.

[4] Y. Hua and T. K. Sarkar, "Matrix pencil method for estimating parameters of exponentially damped/undamped sinusoids in noise," *IEEE Transactions on Acoustics, Speech, and Signal Processing*, vol. ASSP-38, pp. 814–824, May 1990.

[5] W. M. Steedly, C. J. Ying, and R. L. Moses, "A modified SVD-Prony method using data decimation for computationally efficient estimation of damped exponentials in noise," *IEEE Transactions on Signal Processing*, Submitted for publication.

Subspace Methods for Identifying Neural Activity from Electromagnetic Measurements of the Brain*

John C. Mosher
Richard Leahy

Signal & Image Processing Institute
Dept. Electrical Engineering-Systems
University of Southern California
Los Angeles, CA 90089-2564

Paul S. Lewis

Los Alamos National Laboratory
Mechanical & Electronic Engineering
Group MEE-3, MS J580
Los Alamos, NM 87545

Abstract

Neural activity in the brain produces weak electromagnetic fields that can be measured outside of the head. Use of subspace-based signal processing techniques allows the underlying neural sources to be localized from measurements of their external fields. In this paper we review the localization problem and construct a mathematical model whose form is analogous to the diversely polarized source form of the direction-of-arrival problem. We present a new source localization technique for this application, based on Schmidt's MUSIC algorithm, and contrast it to existing nonlinear least squares approaches.

1 Introduction

Neural activity within the brain causes minute electrical currents to flow. Measurement of the external electromagnetic fields generated by these currents is the basis of both electroencephalography (EEG) and magnetoencephalography (MEG). In EEG scalp potentials are measured with electrodes, while in MEG the external magnetic field is measured using Superconducting QUantum Interference Device (SQUID) magnetometers. Because of the electrochemical origin of the "source" currents, the frequencies of the external fields are low enough so that propagating wave effects may be ignored.

To identify the underlying neural sources of MEG or EEG measurements, it is necessary to solve an inverse problem—the determination of the three-dimensional current density from the observed external fields. Unfortunately, this general inverse problem is ill-posed because different neural current distributions may produce the same external fields. To overcome this deficiency, *a priori* assumptions must be made about the form of the source currents—source models must be employed. (For an overview see [1].)

In the "localization" of the neural activity evoked in response to a sensory stimulus, such as an auditory sequence or visual scene, a common source model is the current dipole. Each dipole represents the primary current associated with the combined activation of a large number of neurons located in a small volume of the brain. Each localized primary current will of course induce a global return current. To determine the effects of these return currents, the head geometry and internal conductivity must also be modeled. Head models can range from simple conductive spheres to complex, anatomically accurate finite element meshes.

The number of simultaneous sources that can be localized from an instantaneous "snapshot" measurement of the external field is limited. To localize larger numbers of sources, it is necessary to incorporate temporal assumptions into the model, thereby permitting sources to be identified from multiple successive snapshots. Any temporal model must have an underlying physical and anatomical basis. Currently, the only widely accepted temporal constraint is that the source locations are constant throughout an evoked response.

Based on this spatio-temporal model of *statically located* dipoles, neural source localization may be decomposed into the estimation of time invariant parameters using nonlinear least-squares minimization, followed by linear estimation of the associated time varying parameters. Alternatively, a subspace formulation can be used to derive a suboptimal scanning method. The resulting algorithm is a generalization of the well-known MUltiple SIgnal Classification (MUSIC) method, in which the solution (multiple source locations) is found by scanning potential locations using a single source model.

*This work was supported in parts by the TRW Doctoral Fellowship Program, Los Alamos National Laboratory, operated by the University of California for the United States Department of Energy under contract W-7405-ENG-36, and by the Z. A. Kaprielian Innovative Research Fund at the University of Southern California.

1058-6393/91 $01.00 © 1991 IEEE

In this spatio-temporal form, the MEG/EEG inverse problem is similar to the three-dimensional "direction-of-arrival" (DOA) problem for diversely polarized (polarizations may be fixed or rotating) electromagnetic waves impinging on an antenna array. In particular, the locations of the neural sources are analogous to the angles of arrival, the moments of the neural sources are analogous to the polarization, and the magnitude of the neural sources are analogous to the signal strength. Unlike the DOA problem, important normalization issues must be addressed because the fields generated by neural current sources fall off rapidly as the inverse of the distance squared.

2 Spatio-Temporal Model Formulation

In this section we briefly derive the spatio-temporal model formulation for radial magnetic measurements coupled with current dipole source models. (More detailed derivations can be found in [3] and [4].) The head is modeled as a spherically symmetric conductor. Although the example is specific, the same general approach can be applied to a wide variety of source/head models in both MEG and EEG.

Using the Biot-Savart law and a head centered coordinate system, the radial component of the external magnetic field generated by a current dipole source can be expressed as [2]

$$B_i = \frac{\mu_o}{4\pi}\frac{(\vec{R}_i - \vec{L}_j)\times\vec{R}_i}{|\vec{R}_i - \vec{L}_j|^3|\vec{R}_i|}\cdot\widehat{M}_j S_j \quad (1)$$

$$= \vec{g}_i(\vec{L}_j)\cdot\widehat{M}_j S_j \ . \quad (2)$$

B_i is the radial field strength at location $\vec{R}_i$ outside of the head. S_j is the strength and $\widehat{M}_j$ the unit vector moment (or orientation) of the current dipole source located at $\vec{L}_j$ inside the head. The geometry of this model is such that the radial component of the current dipole produces no external field. As a consequence, the moment $\widehat{M}$ has only two tangential components.

The defined $\vec{g}_i(\vec{L}_j)$ is a gain vector, relating the moment intensity of the dipole to the external field. Representing $\vec{g}_i(\vec{L}_j)$ as a 1 x 2 row vector, we define both an individual source gain matrix and an overall magnetic field vector for m measurement locations.

$$\mathbf{G}_j \equiv \begin{bmatrix} \vec{g}_1(\vec{L}_j) \\ \vdots \\ \vec{g}_m(\vec{L}_j) \end{bmatrix} \qquad \underline{B} \equiv \begin{bmatrix} B_1 \\ \vdots \\ B_m \end{bmatrix} \quad (3)$$

Representing the dipole moment $\widehat{M}$ as a 2 x 1 column vector, and assuming that noise and measurement error is additive, the measurements taken of a field generated by p sources may be expressed as

$$\underline{F} \equiv \underline{B} + \underline{N} = \begin{bmatrix} \mathbf{G}_1 \cdots \mathbf{G}_p \end{bmatrix} \begin{bmatrix} \widehat{M}_1 S_1 \\ \vdots \\ \widehat{M}_p S_p \end{bmatrix} + \underline{N} \quad (4)$$

where $\underline{N}$ denotes additive noise or measurement error.

While there is general agreement that source locations should remain constant over the course of a response, there is less consensus on whether the dipole moments ($\widehat{M}$) should remain "fixed" or be allowed to "rotate." In general both cases must be handled, since two fixed dipoles may be so closely located that they appear as a single rotating source, while the moment of a rotating dipole may vary so little in orientation as to appear fixed.

To formulate our general model, we define a *fixed dipole* as one with a time invariant moment—its two moment component time series are linearly dependent, and we define a *rotating dipole* to be one which *must* rotate—its two moment component time series are linearly independent. In this formulation we denote by p the overall number of dipole sources, p_r of which rotate and p_f of which are fixed.

To separate the time invariant and time varying parameters, we first define the hybrid gain matrix

$$\mathbf{H}(\underline{L}, \underline{M}) \equiv \begin{bmatrix} \mathbf{G}_1 \cdots \mathbf{G}_{p_r} & \mathbf{G}_{p_r+1}\widehat{M}_{p_r+1} \cdots \mathbf{G}_p\widehat{M}_p \end{bmatrix} \quad (5)$$

which is a function of both the p source locations, denoted by $\underline{L}$ and the p_f fixed dipole moments, denoted by $\underline{M}$. The dimensions of $\mathbf{H}$ are m x r, where $r \equiv 2p_r + p_f$. Next we define the time varying r x 1 strength vector as

$$\underline{S}(t) = \begin{bmatrix} \widehat{M}_1(t)S_1(t) \\ \vdots \\ \widehat{M}_{p_r}(t)S_{p_r}(t) \\ S_{p_r+1}(t) \\ \vdots \\ S_p(t) \end{bmatrix} \ . \quad (6)$$

Finally, we assume an unambiguous measurement geometry such that rank($\mathbf{H}$) $= r \leq m$. With these definitions, the overall model may be compactly expressed as

$$\underline{F}(t) = \mathbf{H}(\underline{L}, \underline{M})\ \underline{S}(t) + \underline{N}(t) \ . \quad (7)$$

3 Least Squares Source Localization

Using the spatio-temporal model, neural source localization can be set up as a least-squares problem. For n measurement snapshots acquired at times

$t_1 \cdots t_n$, define the $m \times n$ field data matrix $\mathbf{F} \equiv [\underline{F}(t_1) \cdots \underline{F}(t_n)]$, and the $r \times n$ strength matrix $\mathbf{S} \equiv [\underline{S}(t_1) \cdots \underline{S}(t_n)]$. This yields

$$\mathbf{F} = \mathbf{H}(\underline{L}, \underline{M})\, \mathbf{S} + \mathbf{N} \ . \tag{8}$$

Least squares estimation of the nonlinear parameters $(\underline{L}, \underline{M})$ may be separated from estimation of the linear parameters $(\mathbf{S})$ by first solving

$$\min_{\underline{L},\underline{M}} ||\mathbf{F} - \mathbf{H}\mathbf{H}^{+}\mathbf{F}||_F^2 \tag{9}$$

where $\mathbf{H}^{+}$ denotes the pseudo inverse, and then computing [5]

$$\mathbf{S} = \mathbf{H}^{+}\mathbf{F} \ . \tag{10}$$

Any nonlinear minimization or nonlinear least squares algorithm may be used to minimize (9). The Nelder-Meade simplex approach has been applied to EEG measurements [6] and MEG simulations [3, 4]. Cramer-Rao small error bounds have been developed for MEG in [7].

The drawback of this approach lies in the difficulty of reliably minimizing (9) is the presence of local minima. This problem becomes particularly acute as the number of sources increases. Overall, there are $3p_r + 4p_f$ parameters in the minimization. Furthermore, both the number of sources and the split between fixed and rotating dipoles must be known.

4 A Subspace Approach to Source Localization

In this section we introduce a new MEG/EEG inverse algorithm. Although it is suboptimal in a least-squares sense, this algorithm has the strong advantage of "scanning" quickly with a *one* dipole search, rather than the p-dipole search necessary in a complete fit. The algorithm is an extension of the MUSIC direction-finding algorithm for polarized sources [8].

Key to the MUSIC approach are the concepts of *measurement space* and *signal subspace*. Measurement space is m dimensional and consists of all possible noisy measurements generated by (7). From (7) we know that any "perfect" field measurement, uncorrupted by noise, would lie in the column space or signal subspace of the gain matrix $\mathbf{H}$. While the signal subspace is not known *a priori*, it can be estimated from the singular value decomposition (SVD) of $\mathbf{F}$ or from the eigendecomposition of $\mathbf{F}\mathbf{F}^T$ [9].

Furthermore, in the absence of "array ambiguities," the only single sources whose fields can lie in the signal subspace are the actual underlying sources. If this is not the case, then the measurements are ambiguous because a linear combination of two or more sources can generate the same field as another, spatially distinct source.

Once the signal subspace is known, the multiple source localization problem reduces to the geometric problem of determining which single source fields lie within the signal subspace. Since only a single source is involved, this can be accomplished by an exhaustive search or scan over all parameters of the source model.

4.1 Rotating Dipole Model

If all p dipole sources are rotating, then the dimension of the signal subspace (rank of $\mathbf{H}$) will be $r = 2p_r = 2p$. To search all possible source locations, a three-dimensional scan must be performed over the volume of the head. At each scan point in the head, the candidate dipole source "fills up" two dimensions of measurement space. If the candidate corresponds to an actual source, then both of these dimensions will lie within the signal subspace. To determine this, we compute at each location $\vec{L}_i$ a *normalized* metric

$$J_r(i) = \frac{||\mathbf{P}^{\perp}\mathbf{G}_i||_F^2}{||\mathbf{G}_i||_F^2} \tag{11}$$

where $\mathbf{P}^{\perp}$ is the projector onto the *orthogonal complement* of the estimated signal subspace. A value of zero indicates that the source space lies completely within the signal subspace, while a value of one indicates that it is orthogonal. For the rotating p-dipole model, we expect to find p locations where $J_r(i) \approx 0$.

4.2 Fixed Dipole Model

If all p dipole sources are fixed, then the dimension of the signal subspace (rank of $\mathbf{H}$) will be $r = p_f = p$. Each candidate source will now only occupy one dimension of measurement space. In this case, the normalized scan metric is

$$J_f(i) = \frac{||\mathbf{P}^{\perp}\mathbf{G}_i\widehat{M}_i||_F^2}{||\mathbf{G}_i\widehat{M}_i||_F^2} = \frac{\widehat{M}_i^T\mathbf{G}_i^T\mathbf{P}^{\perp}\mathbf{G}_i\widehat{M}_i}{\widehat{M}_i^T\mathbf{G}_i^T\mathbf{G}_i\widehat{M}_i} \tag{12}$$

A straightforward application of this metric would require a four-dimensional scan, over both location and moment orientation. However, scanning over moment orientation $(\widehat{M})$ can be avoided by noting that the minimum $J_f(i)$ for a given $\mathbf{G}_i$ is equal to the smallest *generalized eigenvalue* of [10]

$$\mathbf{G}_i^T\mathbf{P}^{\perp}\mathbf{G}_i\widehat{M}_i = \lambda\mathbf{G}_i^T\mathbf{G}_i\widehat{M}_i \tag{13}$$

Using the SVD $\mathbf{G}_i = \mathbf{U}_{G_i}\mathbf{\Sigma}_{G_i}\mathbf{V}_{G_i}^T$ and letting $\mathbf{U}_{G_i}$ contain only the left singular vectors associated with the non-zero singular values, we can equivalently compute the metric from

$$J_f(i) = \lambda_{min}\{\mathbf{U}_{G_i}^T \mathbf{P}^\perp \mathbf{U}_{G_i}\} \tag{14}$$

where $\lambda_{min}\{\cdot\}$ denotes the minimum eigenvalue of the bracketed term. Note that calculating this cost function requires finding the smallest eigenvalue of only a 2 x 2 matrix.

Thus we need not explicitly scan for, nor calculate, the moment orientation that minimizes $J_f(i)$, but rather calculate just the eigenvalue associated with this moment. Consequently, the fixed dipole scanning procedure is the same as for the rotating model, except that at each location in the 3-D scanning grid we calculate the smallest eigenvalue of the bracketed term. For the fixed p-dipole model we expect to find p locations where $J_f(i) \approx 0$. The optimal moment orientation at each minimum of $J_f(i)$ is found as the eigenvector associated with λ_{min}.

4.3 Fixed and Rotating Dipoles

In the mixed case, we have both rotating and fixed dipoles, along with two related scanning functions $J_r(i)$ and $J_f(i)$. Using the SVD of $\mathbf{G}_i$ defined earlier, the rotating dipole metric can be expressed as

$$J_r(i) = \frac{s_1^2 ||\mathbf{P}^\perp \mathbf{U}_{G_i} \underline{e}_1||_2^2 + s_2^2 ||\mathbf{P}^\perp \mathbf{U}_{G_i} \underline{e}_2||_2^2}{s_1^2 + s_2^2} \tag{15}$$

where s_1 and s_2 are the two non-zero singular values associated with $\mathbf{G}_i$. The vectors $\underline{e}_1 = [1,\ 0]^T$ and $\underline{e}_2 = [0,\ 1]^T$ form an orthogonal unit moment orientation pair that spans the row space of $\mathbf{G}_i$. By comparison, $J_f(i)$ can be expressed as

$$J_f(i) = ||\mathbf{P}^\perp \mathbf{U}_{G_i} \widehat{M}_i^*||_2^2 \tag{16}$$

where $\widehat{M}_i^*$ is the optimal $\widehat{M}_i$ that minimizes $J_f(i)$ over all unit moment orientations at that location. Comparing $\widehat{M}_i^*$ with $\underline{e}_1$ and $\underline{e}_2$ above, we can describe the rotating dipole scanning function as a weighted average measure of how *all* dimensions of $\mathbf{G}_i$ project onto the orthogonal complement of the signal subspace, while the fixed dipole function is a measure of how *one* optimal dimension of $\mathbf{G}_i$ projects.

Thus when $J_r(i) \approx 0$, then $J_f(i) \approx 0$.

Given the above, we proceed as follows. At each location, we calculate the fixed metric $J_f(i)$. If at a location $\vec{L}_i$ we have a *fixed* dipole, then $\mathbf{G}_i\widehat{M}_i$ represents a linear combination of the vectors in $\mathbf{G}_i$ that lies within the signal subspace and we obtain $J_f(i) \approx 0$. Correspondingly, if at a location $\vec{L}_i$ we have a *rotating* dipole, then $\mathbf{G}_i$ itself is already within the signal subspace. So regardless of the fixed orientation $\widehat{M}_i$ we attempt to assign, we still obtain $J_f(i) \approx 0$.

Thus we can use the fixed dipole metric for both rotating and fixed dipoles.

4.4 MEG/EEG MUSIC Algorithm

1. From the SVD of $\mathbf{F}$ estimate the rank of $\mathbf{H}$ and compute $\mathbf{P}^\perp$.

2. Over a fine grid of three-dimensional locations, compute the fixed dipole metric $J_f(i)$ from (14). Form two-dimensional slices through the three-dimensional space, e.g., (x, y) planes for constant z, and plot the function $1/J_f(i)$ as contours, images, or oblique mesh plots.

3. At each sharp "spike" (minimum of J_f), check to see if the entire subspace of $\mathbf{G}_i$ is within the signal subspace by checking the second "larger" eigenvalue of (14). Locations were both eigenvalues are approximately zero indicate rotating dipoles. Alternatively, compute the rotating dipole metric $J_r(i)$. Locations where $J_r(i) \approx J_f(i) \approx 0$ indicate rotating dipoles. If the dipole is fixed, estimate its orientation by calculating the eigenvector associated with λ_{min}.

   Repeat this analysis until p_r rotating dipoles and p_f fixed dipoles are found such that $r = 2p_r + p_f$. Estimates can then be refined by either using a finer grid in source areas, or by using these the estimates as the initialization for a p-dipole least-squares search.

4. Form the hybrid gain matrix $\mathbf{H}(\underline{L}, \mathbf{M})$ and solve for the time series, $\mathbf{S} = \mathbf{H}^+\mathbf{F}$.

5 Simulation Example

We conclude this paper by presenting an example in which the MUSIC algorithm was applied to simulated MEG data. (More extensive applications of the algorithm to simulated, phantom, and human MEG data are described in [4, 11].) Fields were simulated for three dipoles, two with fixed moment orientations and the third with a rotating orientation.

Dipole	L_x	L_y	L_z	Moment
1	-2.9	-1.6	8.3	fixed
2	0.0	3.3	8.4	fixed
3	2.8	-1.7	8.3	rotating

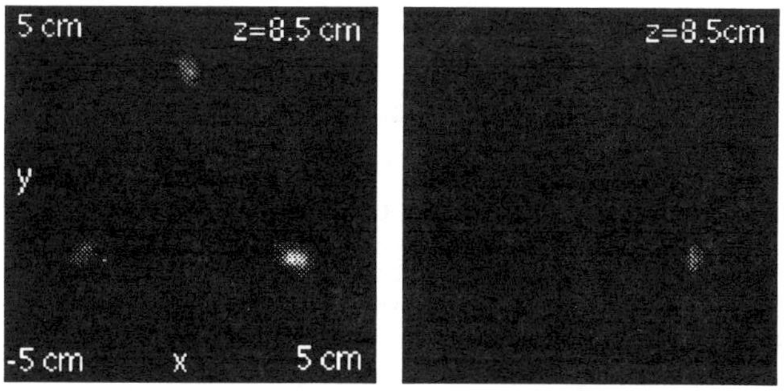

(a) (b)

Figure 1: MUSIC scan results for simulated data. The $z = 8.5$ cm slice is shown from the three dimensional scans of (a) the inverse of the smallest eigenvalue (J_f^{-1}) and (b) the inverse of the remaining eigenvalue.

The data were computed for an array of 37 closely spaced sensors radially oriented and positioned on the surface of an imaginary sphere of radius 12cm. A total of 100 time samples were generated and corrupted by additive white Gaussian noise with an SNR of 10dB.

Our data model had a rank of four for the three dipoles. This was confirmed by the abrupt drop between 4 and 5 in the singular value spectrum of the noisy data matrix $\mathbf{F}$. $\mathbf{P}^{\perp}$ was formed from the singular vectors associated with the singular values 5 to 37. We then scanned the volume $x = -5$ to 5 cm, $y = -5$ to 5 cm, and $z = 6$ to 9.5 cm, at 0.5 cm intervals. The minimum eigenvalue at each position using (14) was found, then the inverses of these values were formed into two-dimensional images for fixed z.

Results for the $z = 8.5$ cm slice are shown in Figure 1a. The cost function shows three distinct peaks which agree well with the true positions. Figure 1b shows the inverse of the second eigenvalue found in the evaluation of (14). This figure shows a distinct peak only for the single rotating dipole. We could refine our estimates by either scanning more finely around the $z = 8.5$ plane or by using these estimates as an initialization point for a full three-dipole least-squares fit.

References

[1] S. J. Williamson, G.-L. Romani, L. Kaufman, and I. Modena, eds., *Biomagnetism: An Interdisciplinary Approach.* New York: Plenum Press, 1983.

[2] J. Sarvas, "Basic mathematical and electromagnetic concepts of the biomagnetic inverse problem," *Physics Medical Biology*, vol. 32, no. 1, pp. 11–22, 1987.

[3] J. C. Mosher, P. S. Lewis, R. Leahy, and M. Singh, "Multiple dipole modeling of spatio-temporal MEG data," in *Digital Image Synthesis and Inverse Optics* (A. F. Gmitro, P. S. Idell, and I. J. LaHaie, eds.), pp. 364–375, Proc. SPIE 1351, San Diego, CA. July 1990.

[4] J. C. Mosher, P. S. Lewis, and R. Leahy, "Multiple Dipole Modeling and Localization from Spatio-temporal MEG Data," *IEEE Transactions on Biomedical Engineering* (in press) 1991.

[5] G. H. Golub and V. Pereyra, "The differentiation of pseudo-inverses and nonlinear least squares problems whose variables separate," *SIAM Journal Numerical Analysis*, vol. 10, pp. 413–432, April 1973.

[6] M. Scherg, "Fundamentals of dipole source potential analysis," in *Auditory Evoked Magnetic Fields and Potentials, Vol. 6* (M. Hoke, F. Grandori, and G. L. Romani, eds.), Adv Audiol. Basel. Karger, 1989.

[7] J. C. Mosher, P. S. Lewis, and R. Leahy, "Spatial localization of neural sources using the magnetoencephalogram," *Fifth ASSP Workshop on Spectrum Estimation and Modeling*, pp. 289–293, Rochester, NY. Oct. 1990.

[8] R. O. Schmidt, "Multiple emitter location and signal parameter estimation," *IEEE Transactions on Antennas and Propagation*, vol. AP-34, pp. 276–280, March 1986. Reprint of the original 1979 paper from the RADC Spectrum Estimation Workshop.

[9] J. A. Cadzow, "Multiple source location – The signal subspace approach," *IEEE Transactions on Acoustics, Speech and Signal Processing*, vol. 38, pp. 1110–1125, July 1990.

[10] G. H. Golub and C. F. V. Loan, *Matrix Computations.* Baltimore, MD: The John Hopkins University Press, 1983.

[11] J. C. Mosher, P. S. Lewis, J. George, R. Leahy, and M. Singh, "Electromagnetic Imagimg of Dynamic Brain Activity," *Proc. 1991 IEEE Medical Imaging Conference*, (in press). Santa Fe, NM. Nov. 1991.

Multiply-Constrained MVDR Matched Field Processing with A-Posteriori Constraints for Enhanced Robustness to Mismatch†

Michael D. Zoltowski*, Gregory M. Kautz*, and S. I. Chou**

*School of Electrical Engineering, Purdue University, West Lafayette, IN 47907
**Naval Ocean Systems Center, San Diego, CA 92152-5000

Abstract

It has been proposed that source localization in passive sonar be accomplished via some form of Matched Field Processing (MFP). Full-coherent, Minimum Variance Distortionless Response (MVDR) MFP assumes complete and perfect characterization of the underwater multipath propagation channel and is known to be extremely sensitive to mismatch between model parameters and actual environmental parameters. We present a minimax, multi-rank signal variation of MVDR MFP referred to as Semi-coherent MVDR MFP. Simulation results are presented demonstrating Semi-coherent MVDR MFP to be both relatively robust to mismatch, with respect to relative amplitudes and phases amongst multipath arrivals, and comparable in performance to Full-coherent MVDR MFP under no mismatch conditions.

1. Introduction

Consider a candidate source location (i. e., a point on a search grid) designated by the position vector $\vec{r} = (R, \theta, z)$ with respect to a cylindrical coordinate system centered at some reference point in the array. Through acoustic ray tracing, we determine the respective arrival angle of each ray path between the candidate source location and some reference point in the array. Let L denote the number of dominant ray paths based on ray tracing prediction. In the case of a linear array, we denote the Lx1 vector composed of the L conical arrival angles as $\underline{\theta}$, i. e., $\underline{\theta} = (\theta_1, \theta_2, \cdots, \theta_L)$, where θ_i is the conical arrival angle associated with the i-th ray path, i=1,...,L. With further modeling, we could determine the relative amplitudes and phases amongst the L multipath arrivals. However, the relative phases of the multipath arrivals can change dramatically with small changes in the ocean parameters, a change in the ocean depth, for example [1]. As a step towards developing a robust procedure, we will not assume knowledge of the relative amplitudes and phases of the multipath arrivals for any source location. Rather, we will only assume knowledge of θ_i and the corresponding steering vector, denoted $\mathbf{a}_i$, i=1,...,L, for each of the L dominant ray paths.

†This work was supported by the Naval Ocean Systems Center (NOSC) under contract no. N66001-87-D-0136 with Office of Naval Technology (ONT) funding.

2. Semi–coherent MVDR MFP

Semi-coherent MVDR MFP is developed as a minimax approach to the source localization problem. Let N denote the number of sensors comprising the array. An Nx1 weight vector is constructed to *minimize* the average power of the corresponding beamformer output under a constraint on the gain and phase response in each of the L multipath arrival directions. The gain and phase response pairs, one for each multipath arrival direction, are jointly determined so as to *maximize* the SNR of the beamformer output. The minimization stage may be mathematically posed as the following constrained optimization problem:

$$\underset{\mathbf{w}(\vec{r})}{\text{Minimize}} \; \mathbf{w}^H(\vec{r})\mathbf{R}_{xx}\mathbf{w}(\vec{r}) \qquad (2.1)$$

$$\text{subject to:} \quad \mathbf{A}^H(\underline{\theta})\mathbf{w}(\vec{r}) = \boldsymbol{\delta}$$

where $\boldsymbol{\delta} = [\delta_1, \delta_2, \cdots, \delta_L]^T$, an Lx1 vector, and $\mathbf{A}(\underline{\theta}) = [\mathbf{a}_1, \mathbf{a}_2, \cdots, \mathbf{a}_L]$, an NxL matrix. The magnitude and phase of δ_i represent the gain and phase response, respectively, in the i-th multipath arrival direction, i=1,...,L. Assuming L < N, $\mathbf{A}^H(\underline{\theta})\mathbf{w}(\vec{r}) = \boldsymbol{\delta}$ represents an underdetermined system of equations. The solution to (2.1) is simply the minimum norm solution to $\mathbf{A}^H(\underline{\theta})\mathbf{w}(\vec{r}) = \boldsymbol{\delta}$ in a Hilbert space with inner product $<\mathbf{x}, \mathbf{y}> = \mathbf{y}^H\mathbf{R}_{xx}\mathbf{x}$:

$$\mathbf{w}(\vec{r}) = \mathbf{R}_{xx}^{-1}\mathbf{A}(\underline{\theta})\left[\mathbf{A}^H(\underline{\theta})\mathbf{R}_{xx}^{-1}\mathbf{A}(\underline{\theta})\right]^{-1}\boldsymbol{\delta} \qquad (2.2)$$

Motivated by the work of Krolik, Lynch, and Swingler in developing "Incoherent" MVDR MFP [1][1], we define the

[1] James Wilson of Neptune Sciences, Inc. is acknowledged for referring the third author to ref. [1].

1058-6393/91 $01.00 © 1991 IEEE

value of the Semi-coherent MVDR MFP ambiguity surface at a point $\vec{r}$ for a given beam response constraint vector $\boldsymbol{\delta}$, denoted $S_{Semi}(\vec{r}\,;\boldsymbol{\delta})$, to be the SNR of the beamformer output obtained with $\mathbf{w}(\vec{r})$ in (2.2). Actually, since we don't have access to signal-only data, we will define it to be the signal plus noise to noise ratio.

For the sake of simplicity, we will assume the noise to be spatially white with the power of the noise at each sensor element equal to σ_n^2. In this case, the output signal plus noise to noise ratio, (S+N)/N, is proportional to $\mathbf{w}^H(\vec{r})\mathbf{R}_{xx}\mathbf{w}(\vec{r})\,/\,\mathbf{w}^H(\vec{r})\mathbf{w}(\vec{r})$. With $\mathbf{w}(\vec{r})$ given by (2.2), $S_{Semi}(\vec{r}\,;\boldsymbol{\delta})$ may be expressed, after some manipulation, as

$$S_{Semi}(\vec{r}\,;\boldsymbol{\delta}) = \frac{\boldsymbol{\delta}^H\mathbf{M}_1^{-1}\mathbf{M}_1\mathbf{M}_1^{-1}\boldsymbol{\delta}}{\boldsymbol{\delta}^H\mathbf{M}_1^{-1}\mathbf{M}_2\mathbf{M}_1^{-1}\boldsymbol{\delta}} \qquad (2.3)$$

where $\mathbf{M}_1$ and $\mathbf{M}_2$ are defined as

$$\mathbf{M}_1 = \mathbf{A}^H(\underline{\theta})\mathbf{R}_{xx}^{-1}\mathbf{A}(\underline{\theta}) \qquad (2.4a)$$

$$\mathbf{M}_2 = \mathbf{A}^H(\underline{\theta})\mathbf{R}_{xx}^{-2}\mathbf{A}(\underline{\theta}) \qquad (2.4b)$$

Note that in accordance with the minimax principle underlying Semi-coherent MVDR MFP espoused previously, $\boldsymbol{\delta}$ is taken to be that value which maximizes $S_{Semi}(\vec{r}\,;\boldsymbol{\delta})$. Since $S_{Semi}(\vec{r};\boldsymbol{\delta})/\sigma_n^2 =$ (S+N)/N˜=˜S/N + 1, this is equivalent to maximizing the beamformer output SNR. Defining $\boldsymbol{\beta} = \mathbf{M}_1^{-1}\boldsymbol{\delta}$, we may alternatively define $S_{Semi}(\vec{r})$ as

$$S_{Semi}(\vec{r}) = \underset{\boldsymbol{\beta}}{\text{Maximum}}\ \frac{\boldsymbol{\beta}^H\mathbf{A}^H(\underline{\theta})\mathbf{R}_{xx}^{-1}\mathbf{A}(\underline{\theta})\boldsymbol{\beta}}{\boldsymbol{\beta}^H\mathbf{A}^H(\underline{\theta})\mathbf{R}_{xx}^{-2}\mathbf{A}(\underline{\theta})\boldsymbol{\beta}} \qquad (2.5)$$

The maximizing $\boldsymbol{\beta}$ is the generalized eigenvector associated with the largest generalized eigenvalue of the LxL matrix pencil $\{\mathbf{A}^H(\underline{\theta})\mathbf{R}_{xx}^{-1}\mathbf{A}(\underline{\theta})\,,\,\mathbf{A}^H(\underline{\theta})\mathbf{R}_{xx}^{-2}\mathbf{A}(\underline{\theta})\}$. In turn, $S_{Semi}(\vec{r})$ is the largest generalized eigenvalue of $\{\mathbf{A}^H(\underline{\theta})\mathbf{R}_{xx}^{-1}\mathbf{A}(\underline{\theta})\,,\,\mathbf{A}^H(\underline{\theta})\mathbf{R}_{xx}^{-2}\mathbf{A}(\underline{\theta})\}$.

3. Comparison of Semi–coherent MVDR MFP With Full–coherent MVDR MFP

Let $\mathbf{c}(\vec{r})$ denote the Lx1 vector composed of the relative complex amplitudes of the L multipath arrivals for a candidate source location $\vec{r}$; $\mathbf{c}(\vec{r})$ is normalized such that $\mathbf{c}^H(\vec{r})\mathbf{c}(\vec{r})=1$. The Full-coherent MVDR MFP beamformer for a candidate search location $\vec{r}$ is the solution to the single constraint MVDR problem

$$\underset{\mathbf{w}(\vec{r})}{\text{Minimize}}\ E\{|\mathbf{w}^H(\vec{r})\mathbf{x}(n)|^2\} = \mathbf{w}^H(\vec{r})\mathbf{R}_{xx}\mathbf{w}(\vec{r})$$

$$\text{subject to:}\quad \mathbf{w}^H(\vec{r})\,\{\mathbf{A}(\underline{\theta})\mathbf{c}(\vec{r})\} = 1 \qquad (3.1)$$

Thus, in addition to knowledge of the arrival angle and corresponding steering vector for each ray path, Full-coherent MVDR MFP also assumes knowledge of the relative amplitudes and phases amongst the L ray paths. In addition, Full-coherent MVDR MFP assumes the multipath arrivals to be 100% correlated. In contrast, proper operation of Semi-coherent MVDR MFP is not premised on such -- the multipath arrivals corresponding to the same source may be partially correlated or 100% correlated.

Similar to the solution to (2.1), the solution to (3.1) is the minimum norm solution to the constraint equation with the norm induced by the inner product $<\mathbf{x}\,,\,\mathbf{y}> = \mathbf{y}^H\mathbf{R}_{xx}\mathbf{x}$.

$$\mathbf{w}(\vec{r}) = \frac{1}{\mathbf{c}^H(\vec{r})\mathbf{A}^H(\underline{\theta})\mathbf{R}_{xx}^{-1}\mathbf{A}(\underline{\theta})\mathbf{c}(\vec{r})}\,\mathbf{R}_{xx}^{-1}\mathbf{A}(\underline{\theta})\mathbf{c}(\vec{r}) \qquad (3.2)$$

Similar to Semi-coherent MVDR MFP, we define the value of the Full-coherent MVDR MFP ambiguity surface at a point $\vec{r}$, denoted $S_{Full}(\vec{r})$, as $\mathbf{w}^H(\vec{r})\mathbf{R}_{xx}\mathbf{w}(\vec{r})\,/\,\mathbf{w}^H(\vec{r})\mathbf{w}(\vec{r})$ with $\mathbf{w}(\vec{r})$ given by (3.2). Again, this is proportional to the output (S+N)/N with the set of weights described by (3.2). Substituting (3.2), we obtain, after some algebraic manipulation,

$$S_{Full}(\vec{r}) = \frac{\mathbf{c}^H(\vec{r})\mathbf{A}^H(\underline{\theta})\mathbf{R}_{xx}^{-1}\mathbf{A}(\underline{\theta})\mathbf{c}(\vec{r})}{\mathbf{c}^H(\vec{r})\mathbf{A}^H(\underline{\theta})\mathbf{R}_{xx}^{-2}\mathbf{A}(\underline{\theta})\mathbf{c}(\vec{r})} \qquad (3.3)$$

Note that the expression for $S_{Full}(\vec{r})$ is very similar to the expression for the objective function in the defining expression for $S_{Semi}(\vec{r})$ in (2.5).

We now examine the performance of Semi-coherent MVDR MFP in the case of a single source at $\vec{r}_S$ and no interferers. We will assume the L multipath arrivals to be 100% correlated. In this case, the covariance matrix has the asymptotic form

$$\mathbf{R}_{xx} = \sigma_S^2\mathbf{A}(\underline{\theta}_S)\mathbf{c}(\vec{r}_S)\mathbf{c}^H(\vec{r}_S)\mathbf{A}^H(\underline{\theta}_S) + \sigma_n^2\mathbf{I} \qquad (3.4)$$

where σ_S^2 is the sum of the square-amplitude of each of the L multipath arrivals. We want to examine the value of $S_{Semi}(\vec{r})$ at $\vec{r}=\vec{r}_S$ when $\mathbf{R}_{xx}$ has the asymptotic form in (3.4). In our analysis we will need expressions for $\mathbf{R}_{xx}^{-1}$ and $\mathbf{R}_{xx}^{-2}$. It is easy to verify that

$$\mathbf{R}_{xx}^{-1} = \frac{1}{\lambda_1}\frac{1}{\alpha_S}\mathbf{A}(\underline{\theta}_S)\mathbf{c}(\vec{r}_S)\mathbf{c}^H(\vec{r}_S)\mathbf{A}^H(\underline{\theta}_S) + \frac{1}{\sigma_n^2}\mathbf{P}^{\perp} \qquad (3.5)$$

where

$$\alpha_s = \mathbf{c}^H(\vec{r}_s)\mathbf{A}^H(\underline{\theta}_s)\mathbf{A}(\underline{\theta}_s)\mathbf{c}(\vec{r}_s) = \|\mathbf{A}(\underline{\theta}_s)\mathbf{c}(\vec{r}_s)\|^2 \quad (3.6a)$$

$$\lambda_1 = \sigma_s^2\alpha_s + \sigma_n^2 \quad (3.6b)$$

and $\mathbf{P}^\perp$ is the projection operator onto the orthogonal complement of the 1-D space spanned by $\mathbf{A}(\underline{\theta}_s)\mathbf{c}(\vec{r}_s)$ such that

$$\mathbf{P}^\perp\mathbf{A}(\underline{\theta}_s)\mathbf{c}(\vec{r}_s) = \mathbf{0} \quad (3.7a)$$

$$\mathbf{c}^H(\vec{r}_s)\mathbf{A}^H(\underline{\theta}_s)\mathbf{P}^\perp\mathbf{A}(\underline{\theta}_s)\mathbf{c}(\vec{r}_s) = 0 \quad (3.7b)$$

Squaring the expression in (3.5), we obtain, after some algebraic manipulation, a similar expression for $\mathbf{R}_{xx}^{-2}$:

$$\mathbf{R}_{xx}^{-2} = \frac{1}{\lambda_1^2}\frac{1}{\alpha_s}\mathbf{A}(\underline{\theta}_s)\mathbf{c}(\vec{r}_s)\mathbf{c}^H(\vec{r}_s)\mathbf{A}^H(\underline{\theta}_s) + \frac{1}{\sigma_n^4}\mathbf{P}^\perp \quad (3.8)$$

$S_{Semi}(\vec{r}_s)$ is the largest generalized eigenvalue of $\{\mathbf{A}^H(\underline{\theta}_s)\mathbf{R}_{xx}^{-1}\mathbf{A}(\underline{\theta}_s)\ ,\ \mathbf{A}^H(\underline{\theta}_s)\mathbf{R}_{xx}^{-2}\mathbf{A}(\underline{\theta}_s)\}$. To evaluate this, we first observe (3.9)

$$\{\mathbf{A}^H(\underline{\theta}_s)\mathbf{R}_{xx}^{-1}\mathbf{A}(\underline{\theta}_s)\}\ \mathbf{c}(\vec{r}_s) = \lambda_1\ \{\mathbf{A}^H(\underline{\theta}_s)\mathbf{R}_{xx}^{-2}\mathbf{A}(\underline{\theta}_s)\}\ \mathbf{c}(\vec{r}_s)$$

where we have employed the asymptotic forms of $\mathbf{R}_{xx}^{-1}$ and $\mathbf{R}_{xx}^{-2}$ in (3.5) and (3.8), respectively, and invoked the property in (3.7). (3.9) dictates that $\mathbf{c}(\vec{r}_s)$ is a generalized eigenvector of $\{\mathbf{A}^H(\underline{\theta}_s)\mathbf{R}_{xx}^{-1}\mathbf{A}(\underline{\theta}_s)\ ,\ \mathbf{A}^H(\underline{\theta}_s)\mathbf{R}_{xx}^{-2}\mathbf{A}(\underline{\theta}_s)\}$ associated with the generalized eigenvalue λ_1, where λ_1 is defined in (3.6b). We now show that the remaining L-1 generalized eigenvalues are equal to the noise power σ_n^2.

Let $\mathbf{e}_i$, i=1,...,L, denote the generalized eigenvectors of $\{\mathbf{A}^H(\underline{\theta}_s)\mathbf{R}_{xx}^{-1}\mathbf{A}(\underline{\theta}_s)\ ,\ \mathbf{A}^H(\underline{\theta}_s)\mathbf{R}_{xx}^{-2}\mathbf{A}(\underline{\theta}_s)\}$. Since $\mathbf{A}^H(\underline{\theta}_s)\mathbf{R}_{xx}^{-1}\mathbf{A}(\underline{\theta}_s)$ and $\mathbf{A}^H(\underline{\theta}_s)\mathbf{R}_{xx}^{-2}\mathbf{A}(\underline{\theta}_s)$ are both Hermitian, the generalized eigenvectors are orthogonal in the metric $\mathbf{A}^H(\underline{\theta}_s)\mathbf{R}_{xx}^{-2}\mathbf{A}(\underline{\theta}_s)$, i. e., $\mathbf{e}_i^H\mathbf{A}^H(\underline{\theta}_s)\mathbf{R}_{xx}^{-2}\mathbf{A}(\underline{\theta}_s)\mathbf{e}_j = 0$ for $i \neq j$. Since $\mathbf{c}(\vec{r}_s)$ is a generalized eigenvector, it follows that

$$\mathbf{c}^H(\vec{r}_s)\mathbf{A}^H(\underline{\theta}_s)\mathbf{R}_{xx}^{-2}\mathbf{A}(\underline{\theta}_s)\ \mathbf{e}_j = 0 \quad j=2,...,L \quad (3.10)$$

Substituting the expression for $\mathbf{R}_{xx}^{-2}$ in (3.8) above and exploiting the property in (3.7) yields the following simplification of (3.10):

$$\mathbf{c}^H(\vec{r}_s)\mathbf{A}^H(\underline{\theta}_s)\mathbf{A}(\underline{\theta}_s)\ \mathbf{e}_j = 0 \quad j=2,...,L \quad (3.11)$$

Invoking the asymptotic forms of $\mathbf{R}_{xx}^{-1}$ and $\mathbf{R}_{xx}^{-2}$ in (3.5) and (3.8), respectively, and the properties in (3.6) and (3.11), we find, after some manipulation, that for j=2,...,L,

$$\{\mathbf{A}^H(\underline{\theta}_s)\mathbf{R}_{xx}^{-1}\mathbf{A}(\underline{\theta}_s)\}\ \mathbf{e}_j = \sigma_n^2\{\mathbf{A}^H(\underline{\theta}_s)\mathbf{R}_{xx}^{-2}\mathbf{A}(\underline{\theta}_s)\}\ \mathbf{e}_j \quad (3.12)$$

Thus, since $\sigma_s^2\ \alpha_s + \sigma_n^2 > \sigma_n^2$, the largest generalized eigenvalue of $\{\mathbf{A}^H(\underline{\theta}_s)\mathbf{R}_{xx}^{-1}\mathbf{A}(\underline{\theta}_s)\ ,\ \mathbf{A}^H(\underline{\theta}_s)\mathbf{R}_{xx}^{-2}\mathbf{A}(\underline{\theta}_s)\}$, and hence the value of $S_{Semi}(\vec{r}_s)$, is $\lambda_1 = \sigma_s^2\ \alpha_s + \sigma_n^2$, where α_s is defined in (3.6a). In addition, the corresponding generalized eigenvector, and hence the value of $\boldsymbol{\beta}$ which maximizes the objective function in (2.5), is $\mathbf{c}(\vec{r}_s)$.

It follows from previous observations that the signal plus noise to noise ratio, (S+N)/N, achieved with both Semi-coherent MVDR MFP and Full-coherent MVDR MFP at the point $\vec{r}_s$ is $S_{Semi}(\vec{r}_s)/\ \sigma_n^2$ and $S_{Full}(\vec{r}_s)/\ \sigma_n^2$, respectively, and that these are equal if $\boldsymbol{\beta} = \mathbf{c}(\vec{r}_s)$ in (2.5). The previous development demonstrated that in the asymptotic case $\boldsymbol{\beta} = \mathbf{c}(\vec{r}_s)$ in the case of a single source, assuming "adequate" SNR, and that $S_{Semi}(\vec{r}_s) = \sigma_s^2\ \alpha_s + \sigma_n^2$. Since (S+N)/N = S/N + 1, it follows that the asymptotic SNR achieved with both Semi-coherent MVDR MFP and Full-coherent MVDR MFP at the point $\vec{r}_s$ is $(\sigma_s^2\alpha_s + \sigma_n^2)/\sigma_n^2 - 1$ such that

$$SNR_{Full}(\vec{r}_s) = SNR_{Semi}(\vec{r}_s) = \frac{\sigma_s^2}{\sigma_n^2}\|\mathbf{A}(\underline{\theta}_s)\mathbf{c}(\vec{r}_s)\|^2 \quad (3.13)$$

where $\|\mathbf{x}\|$ denotes the 2-norm of the vector $\mathbf{x}$. Again, the fact that Semi-coherent MVDR MFP yielded the same output SNR at the true source location as that achieved with Full-coherent MVDR MFP hinged on $\boldsymbol{\beta}$ being equal to the vector of normalized complex amplitudes $\mathbf{c}(\vec{r}_s)$. We only proved this to be true in the asymptotic case with a single source. In general, $\boldsymbol{\beta}$ will only be approximately equal to $\mathbf{c}(\vec{r}_s)$. This is the reason we refer to the method as "Semi-coherent" in contrast to Full-coherent.

4. Incoherent MVDR MFP

Krolik, Lynch, and Swingler [1] also take a multirank signal approach to the source localization problem. For each point on the search grid, they computing a multiply-constrained MVDR beamforming weight vector according to (2.2) with $\boldsymbol{\delta}$ equal to $\mathbf{1}$, a vector composed of all ones. The value of the "Incoherent" MVDR MFP ambiguity surface at $\vec{r}$ may be computed according to (2.3) with $\boldsymbol{\delta} = \mathbf{1}$, i. e., $S_{Incoh}(\vec{r}) = S_{Semi}(\vec{r}\ ;\ \mathbf{1})$.

5. Simulations

A very simple ocean model was employed in the simulations to illustrate fundamental aspects of each of the three versions of MVDR MFP without getting lost in modeling issues. In this model, an iso-velocity ocean is assumed with a sound speed of 1500 m/s and a nominal depth of 4500 m. The receiving array is vertically oriented and composed of 30 sensors equi-spaced by a half-wavelength (7.5 m) at the source frequency, 100 Hz; the array center is situated at a depth of 100 m. A single source is located at a range of 20 Km and a depth of 1200 m; it is assumed that there are L=11 dominant ray paths between source and array. The 11 multipath signals were pairwise 95% correlated and each was attenuated in accordance with three factors: bottom losses, attenuation proportional to path length, and cylindrical spreading factor [2]. The amplitude of each multipath arrival was selected so that the sum of the square-amplitude of each multipath arrival is equal to what it would be if the multipath signals arrived with equal strength having an SNR of 0 dB per element. The wavefront associated with each multipath arrival is modeled as being planar across the face of the array. Finally, the noise was modeled as being independent from sensor to sensor and of equal power. This simulation scenario is similar to that employed by Krolik, Lynch, and Swingler in [1] except that in their case the array was horizontally oriented and the multipath arrivals were of equal strength with 0 dB SNR per element.

Figure 1 compares the performance of the Full-coherent, "Incoherent", and Semi-coherent versions of MVDR MFP in the simulated ocean, source, and array environment described above. (Thus, any simplification or deficiency in the simulation scenario was common to the data supplied to all three algorithms.) Note that peak heights are in units of dB and are relative to the lowest point on the surface being 0 dB. For both the no mismatch and mismatch cases, the same sample covariance matrix formed from 250 snapshots was used for each of the three versions of MVDR MFP. In the mismatch case each version of MVDR MFP operated under the assumption that the ocean depth was 4500 m when it was in fact 4498 m.

Three main conclusions may be drawn from the simulation results: First, Full-coherent MVDR MFP is extremely sensitive to mismatch. A 2 m error in the assumed ocean depth caused roughly a 15 dB drop in the peak of the Full-coherent MVDR MFP ambiguity surface. Second, the performance of "Incoherent" MVDR MFP is substantially degraded relative to that of Full-coherent MVDR MFP in the no mismatch case. The peak of the "Incoherent" MVDR MFP ambiguity surface with no mismatch is roughly 15 dB less than that of the corresponding Full-coherent MVDR MFP ambiguity surface. Third, the performance of Semi-coherent MVDR MFP is both relatively robust to mismatch, with respect to error in the assumed ocean depth, and comparable in performance to Full-coherent MVDR MFP under no mismatch conditions.

These observations may be explained by examining the extent of changes in the signal arrival parameters when the ocean floor is raised by 2 m. Note that this causes a negligible change in the arrival angles of and the relative attenuations amongst the 11 multipath signals. In contrast, since 2 m is a significant fraction of the wavelength, 15 m, the relative phases amongst the 11 multipath arrivals change dramatically. This extent to which the phase of a given path changes depends on the number of bottom (B) bounces (and/or top (T) bounces) incurred en route from source to array. For example, the 2 m change in ocean depth causes a 34.5° change in the phase of the single bottom bounce path B while causing a 227.2° change in the phase of the triple bottom bounce path BTBTBT.

Full-coherent MVDR MFP assumes complete and perfect knowledge of the emitter steering vector, which is a linear combination of the path steering vectors. The path steering vectors are relatively unchanged when the ocean floor is raised by 2 m. However, the path combining coefficients, which depend on the relative phases amongst the multipath arrivals, change dramatically causing a catastrophic loss in the performance of Full-coherent MVDR MFP.

Interestingly, comparing Figures 1(c) and 1(d), it is observed that "Incoherent" MVDR MFP yielded enhanced performance in the mismatched case when the ocean floor was raised by 2 m. Although this seems counter-intuitive at first, note that in the single source case the signal-only (noise-free) output of the beam formed with the "Incoherent" MVDR weight vector is the direct sum of each multipath signal as measured at the array center. Fortuitously for "Incoherent" MVDR MFP, it turns out that for these parameters, this vector sum was larger when the ocean floor was at 4498 m then at 4500 m. In a sense, with the ocean floor at 4498 m the multipath signals were more "in-phase" than they are with the ocean floor at 4500 m. This demonstrates the strong dependence of "Incoherent" MVDR MFP on the relative phases amongst the multipath arrivals.

References

[1] J. Krolik, J. Lynch, and D. Swingler, "A Robust Incoherent Matched Field Processor for Source Localization in Uncertain Multipath Fields," *Proc. of 1989 IEEE ICASSP, May 1989, pp. 2637-2640.*

[2] W. S. Burdic, **Underwater Acoustic System Analysis,** Prentice-Hall, Englewood Cliffs, NJ, 1984.

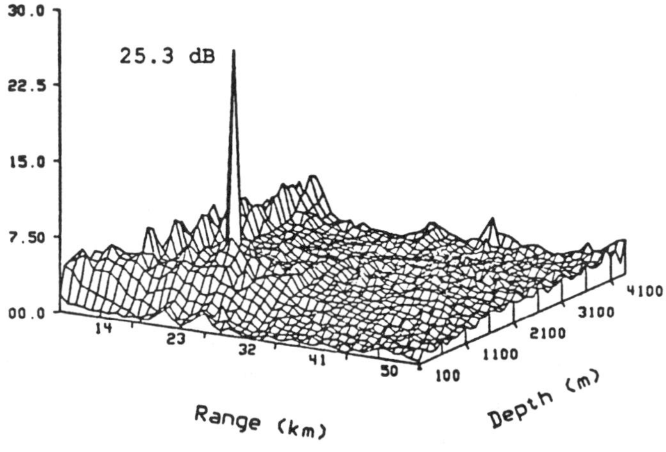

(a) Full-coherent MVDR MFP: no mismatch

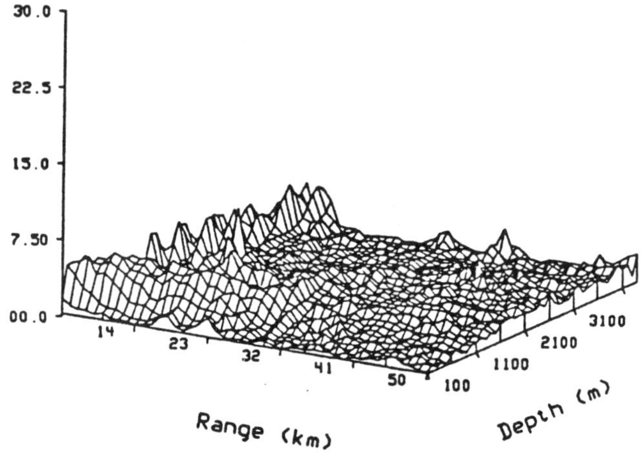

(b) Full-coherent MVDR MFP: mismatch

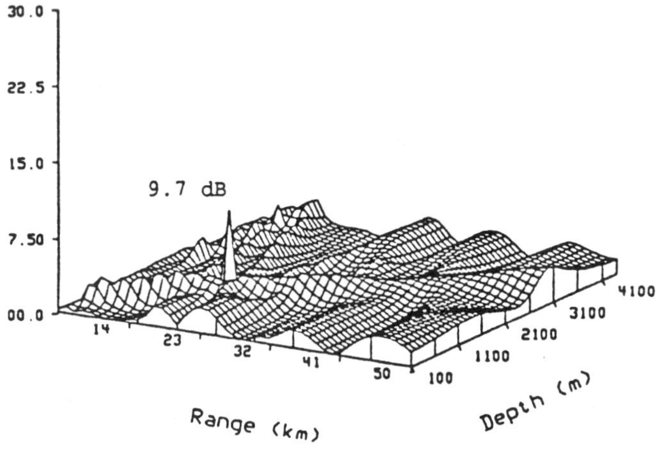

(c) "Incoherent" MVDR MFP: no mismatch

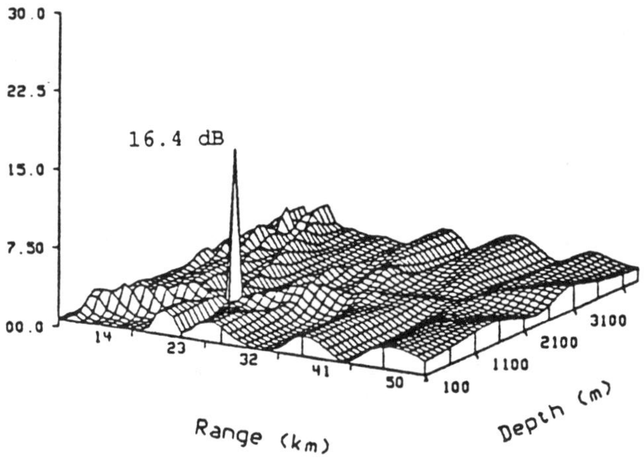

(d) "Incoherent" MVDR MFP: mismatch

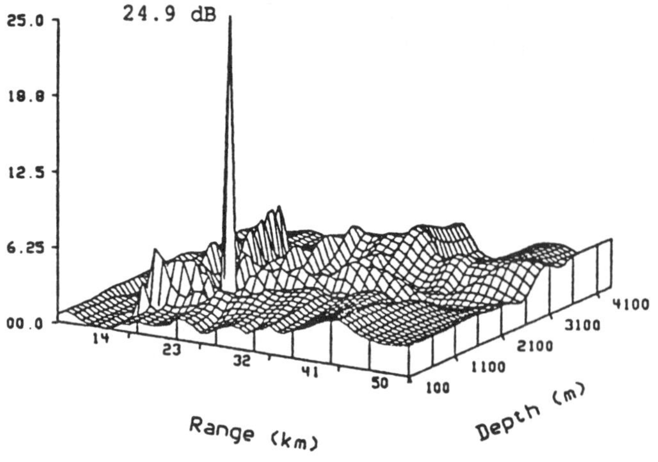

(e) Semi-coherent MVDR MFP: no mismatch

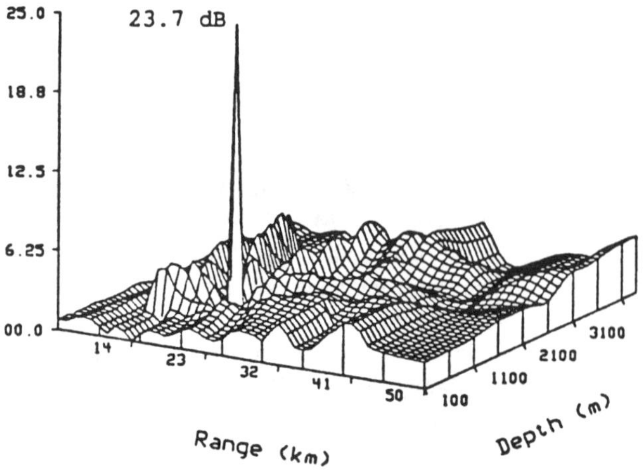

(f) Semi-coherent MVDR MFP: mismatch

Figure 1. Ambiguity surfaces generated via the Full-Coherent, "Incoherent" and Semi-coherent versions of MVDR MFP

Application of CTLS for Estimating the Direction of Arrival of Cyclostationary Signals

Theagenis J. Abatzoglou

Lockheed Research Laboratories
3251 Hanover St, Palo Alto, Ca. 94304

Bart F. Rice

Lockheed Missiles and Space Co., Inc.
1111 Lockheed Way, Sunnyvale, Ca. 94089

Abstract

Cyclic Constrained Total Least Squares (CCTLS) is developed as an alternative to Gardner's Cyclic Music and to the Kailath-Xu technique for estimating the direction of arrival of cyclostationary signals impinging on the elements of an antenna array. CCTLS exploits the fact that a narrowband cyclostationary feature (a symbol rate, for example) of a signal has the same delay at the various array elements as the signal itself. By extracting the narrowband feature at each antenna element, traditional CTLS can be applied. Whereas most direction finding techniques rely on "the narrowband assumption," CCTLS is most appropriate for wideband signals in which estimates of the carrier frequency and symbol rate are available or can be derived. CCTLS is shown to be more interference-tolerant than Cyclic Music and able to separate signals from the same direction with the same symbol rates but with different carrier frequencies.

1. Introduction

Gardner [1] and Kailath and Xu [2] have developed techniques for exploiting cyclostationarity in signals to estimate their directions of arrival. Virtually all "bauded" signals and many others exhibit quadratic stationarity, which may be detected at the output of a delay-and-multiply device. Gardner has adapted the well known MUSIC algorithm to "Cyclic MUSIC," which consists of applying MUSIC to the "cyclic covariance matrix." Kailath and Xu construct a somewhat different matrix, to which MUSIC or other techniques may be applied. In either case, the entries of the matrix are values of the cyclic correlation function

$$R^{\alpha}_{xx}(\tau) = \langle x^*(t-\tau/2)x(t+\tau/2)e^{-2j\pi\alpha t}\rangle$$

where "$\langle\cdot\rangle$" denotes "expected value," x denotes a "snapshot" (an n-vector of signal samples captured a given instant at the n array elements), and α is the "cycle frequency." The quantities $x^*(t-\tau/2)x(t+\tau/2)$ are the outputs of a delay-and-multiply process, which produces a reduction in signal-to-noise ratio (SNR) which must be overcome by processing gain. Assuming that the only interference is additive white Gaussian noise (AWGN), it is easy to show that if the input SNR is $1/\sigma^2$, then (for non-zero delay t) the output SNR is $1/(\sigma^4+2\sigma^2)$. Thus, for example, if the input SNR is -10 dB (σ^2 = 10), the output SNR is about -21 dB. Taking the expected value of $x^*(t-\tau/2)x(t+\tau/2)e^{-2j\pi\alpha t}$ is equivalent to filtering the output of the delay-and-multiply process $x^*(t-\tau/2)x(t+\tau/2)$ with a finite impulse response "resonator" at frequency α. If the resonator bandwidth is sufficiently small, or, equivalently, the averaging period is sufficiently long, the SNR loss inherent in the delay-and-multiply process is overcome. The gain achieved by such filtering is proportional to the filter length, so that, for low input SNR, the amount of averaging which must be done to achieve a desired output SNR is proportional to the square of the input SNR.

It is also worth noting that the entries of the matrices containing $R^{\alpha}_{xx}(\tau)$ are sensitive to pulse shaping, filtering, distortion, and, in general, effects which change the relative strength of the cycle energy at different delays. For example, it is well known that if the pulse shape in a phase-shift-keyed (PSK) or Quadrature-Amplitude-Modulated (QAM) signal is rectangular, the delay for which the symbol-rate frequency is strongest is one-half of the duration of a symbol. This is the delay we used in the implementations of Cyclic MUSIC discussed in section 2. For signals with pulse-shaping or severe bandlimiting (but which still have some excess bandwidth), the optimal delay is $\tau = 0$. In general, the optimal delay for Cyclic Music is a function of these difficult-to-determine parameters. For the Kailath-Xu method, $R^{\alpha}_{xx}(\tau)$ is computed for multiple values of τ. These delays must be spaced closely enough to capture the characteristics of the cyclostationary feature at cycle frequency α regardless of the optimal delay τ, which may take on any value between 0 and one-half of the symbol rate.

In our version of CCTLS, the optimal delay is zero, regardless of bandlimiting or the pulse shaping function. This is due to the filtering which precedes the multiplication. A block diagram of the process is shown in Figure 1. Knowledge of the carrier frequency f_c and the symbol rate α is exploited by filtering the signal using bandpass filters centered at $f_c + \alpha/2$ and $f_c - \alpha/2$. For a given pulse shape, the optimal bandwidth of these filters depends on the SNR, but it is not overly sensitive to this parameter. For PSK, ASK, or QAM signals, the energy in these spectral regions "reinforces," and it is this fact that enables the extraction of symbol rate energy by what has come to be known as "spectral correlation." The outputs of the two filters are mixed (conjugate-multiplied), and the resultant signal is filtered with a "resonator" centered at the symbol rate. The narrowband signals so obtained - one for each element of the array - are then treated as if they are the received signals whose angle of arrival must be determined.

It should be noted that normally the "cycle frequency" α is quite small compared to the usual carrier frequencies or intermediate frequencies used for direction finding. For example, a 2 MHz bandwidth signal may be broadband as far as the "narrow band criterion" is concerned, but the 2 MHz cycle frequency α is low in that, regardless of the angle of arrival, the phase difference between 2 MHz sinusoids at different antenna elements may be too small to measure accurately unless the antenna elements are well separated. In the present example (2 MHz), if the angle of arrival (from broadside) is θ radians and the antenna separation is d meters, then the phase difference between the sinusoids is $\phi = 2\pi(2x10^6)(d \sin\theta)/(3x10^8) = .0133\, d \sin\theta$ radians. If $\sin\theta = .5$ and $d = 4.8$ meters, then $\phi = .032$ radians , or 5.76 degrees.

The prefiltering technique described above has the effect of decreasing noise, interference, and signal components which do not contribute to the energy at the cyclostationary frequency. The SNR is therefore enhanced prior to mixing, so that the degradation in SNR described above is less severe than when a delay-and-multiply is performed. In the simulation described in section 2, for example, the noise bandwidth is reduced by about a factor of 20, or 13 dB, by the prefiltering. Rather than a degradation in SNR from -10 dB to -21 dB, the loss is from about 0 dB to -5 dB.

In addition, CTLS can separate multiple signals from the same direction and with the same symbol rate but with carrier frequencies which differ by at least the baud rate. Such signals may appear spectrally to be a single signal, but if the symbol rate and the carriers are known, CCTLS may be applied multiple times - once for each signal - and the directions of arrival obtained. In the Gardner or Kailath-Xu techniques, conjugate multiplication eliminates carrier information except for a constant phase term, and therefore they are unable to separate the signals. A two-signal case is considered in section 2.

Communications engineers will recognize the filtering and mixing "pre-processing" technique described above as nothing more than traditional "band-edge filtering," which has been used for many years in modems and other receivers for extracting a timing signal from the received signal itself. The zero-crossings of the sinusoids at the symbol rate obtained by this method are synchronized with the centers and edges of the keying instants of the bauded signal from which they are derived (which makes them ideal for use as a timing clock in a receiver), and therefore they experience the same delays at the various array elements as the signals themselves. At this point, any narrowband direction-finding technique could be applied. CTLS has been shown in [4] to provide excellent performance on narrowband signals. The results in section 2 show that it may be applied successfully to wideband cyclostationary signals as well. In addition, we show that CTLS has advantages over Cyclic MUSIC in a multiple signal environment. In general, this advantage should extend to any technique which relies on the cyclic correlation function $R^{\alpha}_{xx}(\tau)$.

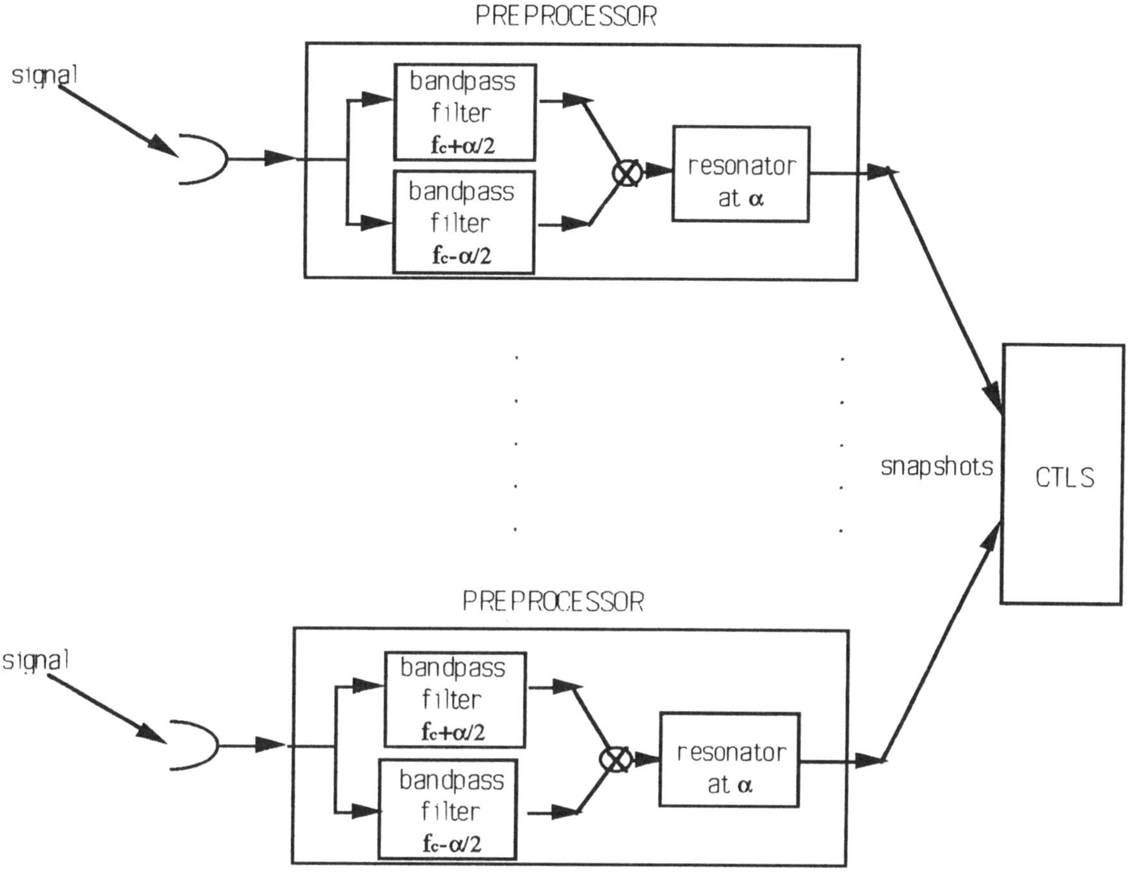

Figure 1. Cyclic CTLS. A narrowband cyclostationary feature- a sinusoid at the the cycle frequency - is extracted from each array element.

2. Superresolution Direction Finding

CTLS can be used to estimate the angular position of farfield sources/emitters when the receiving array is uniform linear or planar. Given an array antenna each receiver is processed according to the functional diagram of Fig. 1. This nearly optimizes reception of cyclostationary signals at known carrier f_c and baud rate α, and filters interfering signals at other baud rates and carrier frequencies. At the output of the resonator of each receiver one can observe a narrowband signal at frequency α whose relative phase is $2\pi\alpha x_n \sin(\theta)/c$. θ is the angle at which the signal wavefront impinges onto the array. To apply CTLS to estimate the angle of arrival of the signals that passed the filtering process of Fig. 1, one must have a uniform linear or planar array. This is not a significant limitation, since an arbitrary geometry array can be interpolated to an equivalent linear array [5]. CTLS can estimate the angle of arrival of wavefronts with optimal resolution and accuracy even in the case of multipath or correlated signals.

When the array is linear uniform and the receiver elements are narrowband relative to the array diameter, the snapshot samples can be modeled as a superposition of sinusoids at frequencies $2\pi d \sin\theta_l/\lambda$, where d is the element spacing θ_l is azimuth angle of the lth source and λ is the wavelength. CTLS can be applied to the forward-backward linear prediction equations (cf. [4]). We concatenate these equations into the matrix form

$$C\begin{pmatrix} x \\ -1 \end{pmatrix} = 0$$

and maximization of

$$\begin{pmatrix} x \\ -1 \end{pmatrix}^* \sum_{m=1}^{M} C_m^* \left(\hat{H}_x \hat{H}_x^*\right)^{-1} C_m \begin{pmatrix} x \\ -1 \end{pmatrix}$$

will yield the CTLS estimate of the vector of linear prediction coefficients x. In the above, C_m is the forward-backward data matrix of the m-th snapshot $\left[y_1^{(m)}\ y_2^{(m)} \cdots y_N^{(m)}\right]^T$.

$$C_m = \begin{bmatrix} y_1^{(m)} & y_2^{(m)} & \cdots & y_{L+1}^{(m)} \\ \vdots & \vdots & \cdots & \vdots \\ y_{N-L}^{(m)} & y_{N-L+1}^{(m)} & \cdots & y_N^{(m)} \end{bmatrix}$$

where $\widehat{H}_x$ is the coefficient noise structure matrix having the Toeplitz-Hankel structure given by

$$H_x = \begin{bmatrix} x_1 & x_2 & \cdots & x_L & -1 & 0 \cdots 0 \\ \vdots & \ddots & \ddots & \ddots & \ddots & \ddots & \vdots \\ 0 \ldots & 0 & x_1 & x_2 & \cdots & x_L & -1 \end{bmatrix}$$

or $$\widehat{H}_x = \begin{bmatrix} H_x & 0 \\ 0 & JH_xJ \end{bmatrix}$$

The directions of arrival are next retrieved from the roots of the linear prediction coefficient polynomial. Figure 2 depicts simulation results showing the MSE of the CTLS, ROOT-MUSIC and Modified Forward-Backward Linear Prediction (MFBLP) [6], methods for estimating the angular position of two closely spaced emitters, when the array is 8-element linear uniform.

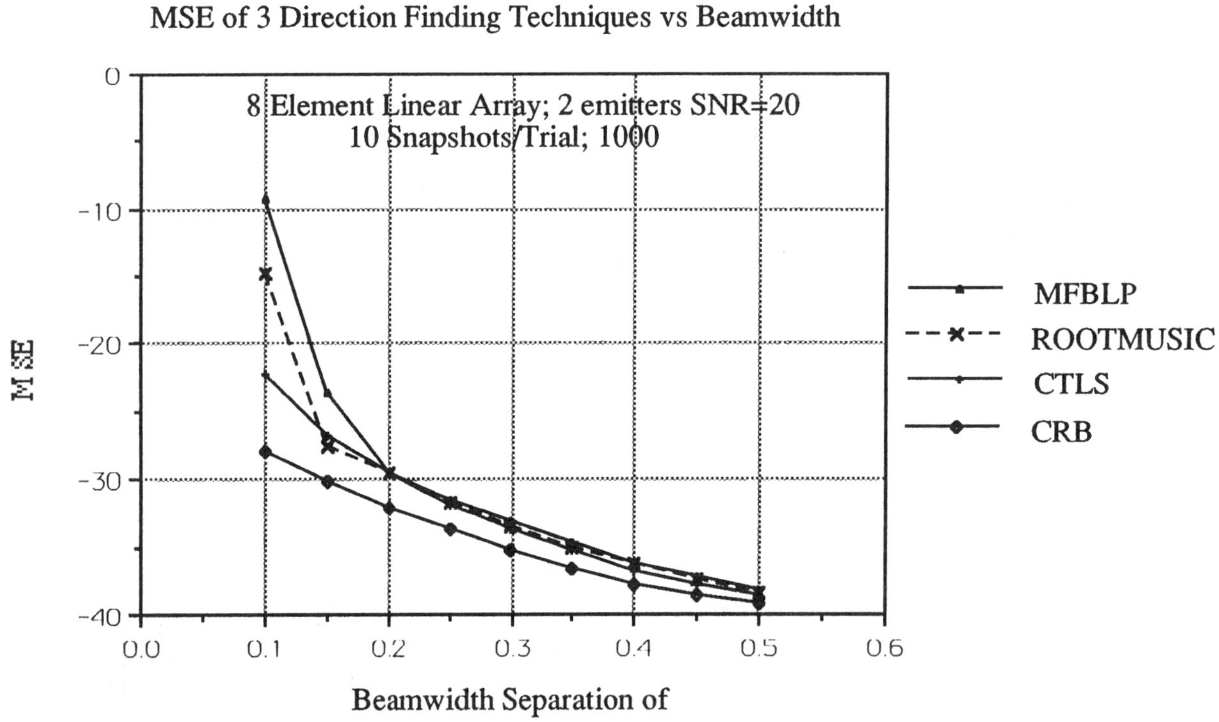

Figure 2. MSE(db) of CTLS, MFBLP and ROOTMUSIC for varying source separation

3. Simulation Results

A. One BPSK signal at a baud rate of 5 MHz. and carrier frequency of 80 MHz impinges on a uniform linear array of 6 receivers at an angle of -53 degrees. The bandpass filters used in the CCTLS technique are centered at 77.5 and 82.5 MHz.. Their bandwidth is about 4 MHz.. The resonator filter is centered at 5 MHz. and has a bandwidth of about 10KHz. 10000 samples of the BPSK signal in the presence of additive white Gaussian noise were processed according to the Cyclic MUSIC method and the CCTLS technique. In the CCTLS approach, 36 snapshots of data were obtained at the output of the resonator filter. The MUSIC method showed good performance down to -10db SNR although a bias of about 2 degrees was present in the angle estimates. The CCTLS technique showed no bias. However, because the effective aperture of the array was cut to 5/80=1/16 of the original one, the angle estimates are a bit unstable at lower SNR (see Table 1).

RMSE vs. SNR	-5db	0db	10db
CYCL. MUSIC	2.2	2.0	1.9
CCTLS	outliers	3.3	1.4

Table 1. RMSE vs. SNR comparison for CYCLIC MUSIC and CCTLS.

B. In this simulation, two BPSK signals at baud rate of 5 MHz. and carrier frequencies of 80 and 86 MHz impinging at the six-element array were generated at angles of -53 and -59 degrees respectively. Gaussian noise was added to the received signal of varying SNRs. The signals were sampled at 200MHz, and 10000 samples/trial were generated for the simulation runs. The Cyclic MUSIC technique was applied to the resulting snapshots and the direction of arrival of the two signals was estimated accordingly. The results of this simulation are shown in Figures 3 and 4. These figures show that Cyclic MUSIC has difficulty resolving the two signals at low SNR and provides a "broadened" direction of arrival estimate.

The CCTLS technique was applied also to the data using two sets of filters. One set had the same characteristics as the one described in simulation A, while the second set consisted two filters centered at 83.5 and 88.5 MHz, respectively. The bandwidth of each filter was 4 MHz. The performance of CCTLS is summarized in Table 2, where the RMSE is given for the estimate of each angle at various SNRs. In this scenario, CCTLS provides better resolution and accuracy than Cyclic MUSIC.

RMSE vs. SNR	-5 dB	0 dB	10 dB
CYCL. MUSIC	4.15	2.35	0.773
CCTLS	8.53	4.86	1.60

Table 2. Performance of the CCTLS technique in the presence of two BPSK signals.

4. Conclusions

The CCTLS technique has been developed here for estimating the angle of arrival of cyclostationary signals impinging on an antenna array. The prefiltering inherent in the CCTLS method results in improved SNR at the input of the direction finding process when compared to cyclic autocorrelation approaches as in Cyclic MUSIC. In a simulation where two BPSK signals, closely spaced in angle and at different carrier frequencies, the CCTLS outperformed the Cyclic MUSIC method at various SNRs.

Figure 3. Cyclic MUSIC, two signals, SNR = 0 dB

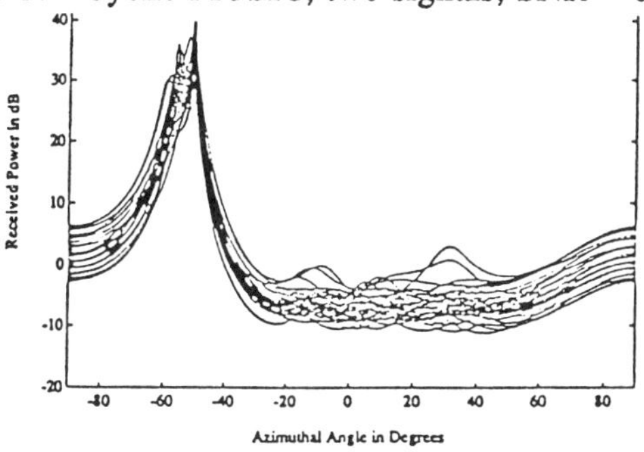

5. References

[1] W. A. Gardner, "Simplification of MUSIC and ESPRIT by Exploitation of Cyclostationarity," Proceedings of the IEEE, 76(7): 845-847, July 1988.

[2] G. Xu and T. Kailath, "Array Signal Processing via Exploitation of Spectral Correlation - A Combination of Temporal and Spatial Processing," Proceedings of the 1989 Asilomar Conference on Signals, Systems, and Computers, pp. 945-949.

[3] W. A. Gardner, "Exploitation of Spectral Redundancy in Cyclostationary Signals", IEEE ASSP Magazine, Vol. 8, pp. 14-37, April 1991.

[4] T. J. Abatzoglou, J. M. Mendel and G. A. Harada, "The Constrained Total Least Squares and its Applications to Harmonic Superresolution", IEEE Transactions in Signal Processing, Vol. SP-39, pp. 1070-1087, May 1991.

[5] B. Friedlander, "Direction Finding with an Interpolated Array", Proceed. 1990 ICASSP, pp. 2951-2954, Albuquerque, New Mexico, April 1990.

[6] D. W. Tufts and R. Kumaresan,"Estimation of Frequencies of Multiple Sinusoids: Making Linear Prediction perform like Maximum Likelihood", Proceedings IEEE,Vol.70, pp.975-989 (1982).

Figure 4. Cyclic MUSIC, two signals, SNR=-5 dB

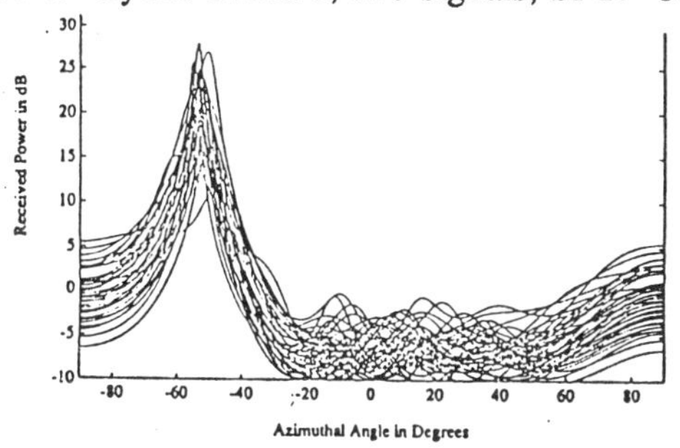

Searching for Convergence Points of the Continuous Time Extended Kalman Filter Used as a Parameter Estimator

L. A. Campbell

The Aerospace Corporation
M1-102, PO Box 92957
Los Angeles, CA 90009-2957

D. M. Wiberg

Electrical Engineering Dept.
University of California
Los Angeles, CA 90024-1594

Abstract

This paper deals with estimation of two stable pole parameters for a two-dimensional continuous-time linear stochastic system with known process noise covariance, using the Extended Kalman Filter. Averaging theory permits algebraic computation of a vector field whose stable stationary points are the estimator's only possible convergence points. Specialized partitioned matrix computations allowed the numerical computation of the vector field and graphical computer search for "spurious" convergence points (not corresponding to the true parameter values), with negative results. This supports the conjecture that none exist, a result known from theory in the one-dimensional case.

1 Continuous Time Averaging Theory

Averaging theory for discrete time parameter estimators has been developed by Ljung [1] and applied to the extended Kalman filter used as a parameter estimator (EKFPE) by Ljung and Söderström [2]. Recently there has been a rigorous derivation of averaging theory for continuous time parameter estimators by DeWolf and Wiberg [3].

To introduce the basic idea of averaging as applied to a continuous time process, let $w(t)$ be a zero mean rapidly fluctuating noise signal, and $x(t)$ an n-vector satisfying

$$dx/dt = t^{-1}f(x,w) \tag{1}$$

Assume that $w(t)$ is stationary with probability density $p(w)$ which does not depend on the state x. The t^{-1} factor will force $x(t)$ to tend toward a constant value when t is large. Therefore $x(t)$ is called the slow variable, and w the fast variable. A simple exponential change of the time scale

$$x_a(t) = x(e^t) \tag{2}$$

converts (1) to

$$dx_a/dt = f(x_a(t), w(e^t)) \tag{3}$$

The rate of fluctuation (i.e. bandwidth) of $w(e^t)$ increases with time. In the limit the noise w averages the function f and asymptotically $x_a(\cdot)$ follows the trajectories of $\bar{x}(\cdot)$ given by

$$d\bar{x}/dt = \bar{f}(\bar{x}) \tag{4}$$

where $\bar{f}$ is defined by

$$\bar{f}(\bar{x}) = E_w f(x,w) = \int f(x,w)\,p(dw) \tag{5}$$

Thus the asymptotic properties of the solution x of the stochastic equation (1) are characterized by the *ordinary* differential equation (4).

Most on-line parameter estimators are not in the form of equation (1), but are slightly more complex. Let θ be the parameter estimate and other visibly slow variables, let z_1 be the states, state estimates, and other variables satisfying linear stochastic differential equations, let z_2 be the variables of the Riccati and other nonlinear ordinary differential equations, and let u be a deterministic input. Then θ will be the slow variable, $z = (z_1^T, z_2^T)^T$ will be the fast variable, and the on-line parameter estimator can (usually) be put in the form

$$\begin{aligned} d\theta &= t^{-1}[F_1(\theta,z)dt + F_2(\theta,z)dw] \\ &+ t^{-2}[F_3(\theta,z,t)dt + F_4(\theta,z,t)dw] \qquad (6)\\ dz_1 &= A_1(\theta,z_2)z_1dt + B_1(\theta,z_2)udt + S(\theta,z_2)dw \\ &+ t^{-1}[G_1(\theta,z,t)dt + G_2(\theta,z,t)dw] \qquad (7)\\ dz_2 &= A_2(\theta,z_2)dt \\ &+ t^{-1}[H_1(\theta,z,t)dt + H_2(\theta,z,t)dw] \qquad (8)\end{aligned}$$

in which the A's, B_1, the F's, G's, H's, and S are locally bounded and have bounded second derivatives, and furthermore, $S, F_3, F_4, G_1, G_2, H_1$, and H_2 are bounded, F_1, F_2, A_1, A_2, and B_1 are limited in growth, and A_1 is asymptotically stable as described in [3].

Then, if equation (8) is well behaved, as stated in [3], and equation (6) has no finite escape, the ordinary differential equation corresponding to (4) for the system (6-8) is

$$d\bar{\theta}/dt = \bar{F}(\bar{\theta}) \tag{9}$$

where $\bar{F}$ is defined by

$$\bar{F}(\theta) = \int F_1(\theta, z)\, P^{\theta}(dz) \tag{10}$$

with $P^{\theta}(\cdot)$ the unique invariant measure of equations (7) and (8) for fixed θ and G_1, G_2, H_1, H_2 set to zero.

The solutions $\theta(t)$ of (6) asymptotically approach the averaged trajectories $\bar{\theta}(t)$ that obey (9), with probability one. In particular, the convergence points of the algorithm (6-8) are the stable stationary points of the ODE (9), given by those solutions θ_∞ of

$$\bar{F}(\theta_\infty) = 0 \tag{11}$$

for which all the eigenvalues of $\partial\bar{F}/\partial\theta(\theta_\infty)$ are in the left half plane.

2 Application to the EKFPE

The above analysis has been applied to the EKFPE used to estimate system parameters including the state noise covariance. In [3] it is concluded that incorrect parameter estimates will result almost always when the state noise covariance is unknown. When the state noise covariance is known, a first order system must have a convergence point at only the true parameter value for the EKFPE [4].

Here algebraic equations are developed for finding the convergence points of an order n EKFPE. For simplicity, this development is for a zero-input EKFPE in canonical form, but it can easily be extended to the general case. Consider the system

$$dx = A(\theta)x dt + \Sigma dv \tag{12}$$

$$dy = e_1^T x dt + \rho dw \qquad (\rho > 0) \tag{13}$$

with $v(\cdot)$ and $w(\cdot)$ uncorrelated standard Wiener processes and

$$A(\theta) = A_0 + e_n\theta^T \tag{14}$$

where A_0 is the zero matrix except for ones on the super diagonal. For the system (12), (13), the EKFPE to estimate θ is

$$d\hat{x} = [A(\hat{\theta})\hat{x} + e_n \mathrm{Tr}(P_{\theta x})]dt + \rho^{-2}P_x e_1(dy - e^T\hat{x}dt) \tag{15}$$

$$d\hat{\theta} = \rho^{-2}P_{\theta x}e_1(dy - e_1^T\hat{x}dt) \tag{16}$$

$$\mathcal{A}(\hat{\theta}) = A(\hat{\theta}) - \rho^{-2}P_x e_1 e_1^T \tag{17}$$

$$\begin{aligned} dP_x/dt &= \mathcal{A}(\hat{\theta})P_x + P_x\mathcal{A}^T(\hat{\theta}) + \Sigma\Sigma^T \\ &+ \rho^{-2}P_x e_1 e_1^T P_x + P_{x\theta}\hat{x}e_n^T + e_n\hat{x}^T P_{\theta x} \end{aligned} \tag{18}$$

$$dP_{x\theta}/dt = \mathcal{A}(\hat{\theta})P_{x\theta} + e_n\hat{x}^T P_\theta \tag{19}$$

$$dP_\theta/dt = -\rho^{-2}P_{\theta x}e_n e_n^T P_{x\theta} \tag{20}$$

These EKFPE equations (15-20) can be put in the form of (6-8) by the change of variables

$$V = t^{-1}P_\theta^{-1} \tag{21}$$

$$W = P_{x\theta}P_\theta^{-1} = tP_{x\theta}V \tag{22}$$

With $\tilde{x} = x - \hat{x}$ and $d_0^1 = e_1 e_1^T$, (6) becomes

$$d\begin{pmatrix} \hat{\theta} \\ V \end{pmatrix} = \frac{1}{t\rho^2}\begin{pmatrix} V^{-1}W^T e_1(e_1^T\tilde{x}dt + \rho dw) \\ (W^T d_0^1 W - \rho^2 V)dt \end{pmatrix} \tag{23}$$

The fast variables z_1 and z_2 can be identified as

$$z_1 = \begin{pmatrix} x \\ \tilde{x} \\ We_1 \\ \cdot \\ \cdot \\ We_n \end{pmatrix} \quad \text{and} \quad z_2 = P_x$$

Let θ_0 be the true value of θ. For any $n \times n$ matrix M, define $\Phi(\theta, M)$ as

$$\begin{pmatrix} A(\theta_0) & 0 & 0 & 0 & \cdot & 0 \\ e_n(\theta_0 - \theta)^T & M & 0 & 0 & \cdot & 0 \\ e_n e_1^T & -e_n e_1^T & M & 0 & \cdot & 0 \\ e_n e_2^T & -e_n e_2^T & 0 & M & \cdot & 0 \\ \cdot & \cdot & \cdot & \cdot & \cdot & \cdot \\ e_n e_n^T & -e_n e_n^T & 0 & 0 & \cdot & M \end{pmatrix}$$

Then (7) becomes

$$dz_1 = \Phi(\hat{\theta}, \mathcal{A}(\hat{\theta}))z_1 + \begin{pmatrix} \Sigma & 0 \\ \Sigma & -\rho^{-1}P_x e_1 \\ 0 & 0 \\ \cdot & \cdot \\ \cdot & \cdot \\ 0 & 0 \end{pmatrix} \begin{pmatrix} dv \\ dw \end{pmatrix} \quad (24)$$

and (8) becomes the matrix differential equation

$$\begin{aligned} dP_x/dt &= \mathcal{A}(\hat{\theta})P_x + P_x\mathcal{A}^T(\hat{\theta}) + \Sigma\Sigma^T + \rho^{-2}P_x d_0^1 P_x \\ &+ t^{-1}[V^{-1}W\hat{x}e_n^T + e_n\hat{x}^T W^T V^{-T}] \end{aligned} \quad (25)$$

Restriction of the estimates $\hat{\theta}$ to a region of θ-space for which the EKFPE is asymptotically stable permits the application of averaging theory as described in the previous section [3]. The ODE (9) then becomes

$$d/dt \begin{pmatrix} \bar{\theta} \\ \bar{V} \end{pmatrix} = \rho^{-2} \begin{pmatrix} \bar{V}^{-1}\mathrm{E}\{W^T d_0^1 \tilde{x}\} \\ \mathrm{E}\{W^T d_0^1 W\} - \rho^2 \bar{V} \end{pmatrix} \quad (26)$$

The expectations E in the above ODE are computed with respect to the steady state variance equations of the fast system (24) and (25). The stationary points of $\bar{\theta}$ are given by $d\bar{\theta}/dt = 0$, and hence, from the first row of (26), they satisfy

$$0 = \mathrm{E}\{W^T d_0^1 \tilde{x}\} = \begin{pmatrix} \mathrm{E}\{W_{11}\tilde{x}_1\} \\ \cdot \\ \cdot \\ \cdot \\ \mathrm{E}\{W_{n1}\tilde{x}_1\} \end{pmatrix} \quad (27)$$

To compute the $\mathrm{E}\{W_{i1}\tilde{x}_1\}$ for $i = 1, \ldots, n$, define the normalized steady state solution $P_\infty(\theta)$ of (25) as

$$P_\infty(\theta) = \rho^{-2} \lim_{t\to\infty} P_x(t;\theta) \quad (28)$$

and define Q as

$$Q = \rho^{-2}\Sigma\Sigma^T \quad (29)$$

Then from (17) and the steady state of (25), $P_\infty = P_\infty(\theta)$ is determined by the algebraic Riccati equation

$$0 = A(\theta)P_\infty + P_\infty A^T(\theta) + Q - P_\infty d_0^1 P_\infty \quad (30)$$

Put

$$\mathcal{A}_\infty(\theta) = A(\theta) - P_\infty(\theta)d_0^1 \quad (31)$$

$$\Lambda(\theta) = \Phi(\theta, \mathcal{A}_\infty(\theta)) \quad (32)$$

and

$$\Gamma(\theta) = \rho^2 \begin{pmatrix} Q & Q & 0 & \cdot & 0 \\ Q & Q + P_\infty d_0^1 P_\infty & 0 & \cdot & 0 \\ 0 & 0 & 0 & \cdot & 0 \\ \cdot & \cdot & \cdot & \cdot & \cdot \\ 0 & 0 & 0 & \cdot & 0 \end{pmatrix} \quad (33)$$

Let $\Omega(\theta)$ be the covariance of z_1

$$\Omega(\theta) = E\{(z_1 - E\{z_1\})(z_1 - E\{z_1\})^T\} \quad (34)$$

Then the steady state variance equation corresponding to (24) is

$$0 = \Omega(\theta)\Lambda^T(\theta) + \Lambda(\theta)\Omega(\theta) + \Gamma(\theta) \quad (35)$$

Since the stationary point condition (27) can be written as

$$\rho^{-2}(0\ 0\ I\ 0 \cdots 0)\Omega(\theta) \begin{pmatrix} 0 \\ e_1 \\ 0 \\ \cdot \\ 0 \end{pmatrix} = 0_n \quad (36)$$

the above two equations determine the convergence points of the EKFPE (15-20) and depend only on Σ and θ_0 and not on ρ.

3 Computing EKFPE Convergence Points

Direct computation of the convergence points θ_∞ involves solution of the algebraic Riccati equation (30) for $P_\infty(\theta)$. This is analytically intractable for $n \geq 2$. Consequently a numerical attack on the problem is required.

In outline, the numerical method chosen is to evaluate (35) for values of θ_i in a grid, and then determine the extent to which the stationary point condition (36) is satisfied. In practice, only a portion of Ω needs to be computed in order to determine the n-vector L_i which is the left side of equation (36). The flow of computation can be listed as:

For grid points $i = 1, 2, \ldots, N$,

1. Select Σ, θ_0, and θ_i.
2. Compute $A(\theta_i)$ and $A(\theta_0)$ from (14).
3. Solve for $P_\infty(\theta_i)$ from (30)
4. Compute $\mathcal{A}_\infty(\theta_i)$ from (31)
5. Compute $\Lambda(\theta_i)$ and $\Gamma(\theta_i)$ from (32) and (33)
6. Solve for Ω at θ_i from (35)
7. Compute the n-vector L_i, defined as the left hand side of (36) (without the factor ρ^{-2})
8. Store (θ_i, L_i)

9. Return to Step 1

10. Display $||L_i||$, for $i = 1, 2, \ldots, N$ for different norms and graphically determine $\min(||L_i||)$

11. Refine the grid size about a minimizing θ_i

4 Numerical Difficulties

In step 3 of the previous section, $P_\infty(\theta_i)$ is computed as the solution of an $n \times n$ algebraic Riccati equation. For low values of n, the solutions found by a call to the MATLAB routine *lqe* (with arguments A, Σ, e_1) were found to be quite accurate.

In step 6 of the previous section, Ω is computed as the solution of a Liapunov equation of order $n(n+2)$. Even for the case $n = 2$, large numerical errors were encountered using the MATLAB routine. Consequently, a block-wise (that is, partitioned) Liapunov equation solver was devised to take advantage of the sparseness of the particular form encountered. A chain of $2(n+2)$ matrix equations results, each of which is only of order $n \times n$.

Partition the $n(n+2) \times n(n+2)$ symmetric matrix Ω as the $n \times n$ matrices V_{ij} for $i \leq j = 1, 2, \ldots, n+2$ and $V_{ji} = V_{ij}^T$ for $i > j$. (A matrix V is defined in (21) – no relation to the blocks V_{ij} defined here is intended.) Then the block-wise solution to (35) can be obtained by equating each of the following expressions to zero:

$$A(\theta_0)V_{1,1} + V_{1,1}A^T(\theta_0) + \Sigma\Sigma^T \quad (37)$$

$$A(\theta_0)V_{1,2} + V_{1,2}\mathcal{A}^T(\theta_i) + V_{1,1}(\theta_0 - \theta_i)e_n^T + \Sigma\Sigma^T \quad (38)$$

$$A(\theta_0)V_{1,j+2} + V_{1,j+2}\mathcal{A}^T(\theta_i) + (V_{1,1} - V_{1,2})e_j e_n^T \quad (39)$$

$$\begin{aligned} &\mathcal{A}(\theta_i)V_{2,2} + V_{2,2}\mathcal{A}^T(\theta_i) + P_\infty(\theta_i)e_1 e_1^T P_\infty(\theta_i) \\ &+ e_n(\theta_0 - \theta_i)^T V_{1,2} + \Sigma\Sigma^T + V_{1,2}^T(\theta_0 - \theta_i)e_n^T \end{aligned} \quad (40)$$

$$\begin{aligned} &\mathcal{A}(\theta_i)V_{2,j+2} + V_{2,j+2}\mathcal{A}^T(\theta_i) \\ &+ (V_{1,2}^T - V_{2,2})e_j e_n^T + e_n(\theta_0 - \theta_i)^T V_{1,j+2} \end{aligned} \quad (41)$$

where $j = 1, 2, \ldots, n$ in expressions (39) and (41).

In order to compute the vector L_i which is the solution of the stationary point condition (36), computation can stop at this point, because the jth component of L_i is $e_1^T V_{2,j+2} e_1$. Thus, for the purpose of computing and storing L_i (as in steps 7 and 8 of the previous section), only $2(n+2)$ matrix equations of order $n \times n$ need be solved (and step 6 need not be carried out completely). These equations are not Liapunov equations, except for the $V_{j,j}$ equations, because the solutions are in general non-symmetric.

For purposes of checking the solution of (35), Ω can be computed in its entirety by using the general formula for $k \leq j = 3, 4, \ldots, n+2$

$$\begin{aligned} 0 &= \mathcal{A}(\theta_i)V_{k,j} + V_{k,j}\mathcal{A}^T(\theta_i) \\ &+ e_n e_{k-2}^T(V_{1,j} - V_{2,j}) + (V_{1,k} - V_{2,k})e_{j-2}e_n^T \end{aligned} \quad (42)$$

Solving for Ω in this manner gave much better solutions to (35) than the MATLAB Liapunov equation solution routine.

5 Results and Conclusions

Although the analysis of the previous sections applies to the estimation of pole parameters of linear systems of arbitrary dimension n, experiments were restricted to the two-dimensional case. As mentioned, even in the case $n = 2$ the partitioned solution of the Liapunov equation was required to accurately solve for the 8×8 matrix computed at each grid point. In the case $n = 2$, there are two pole parameters to estimate. The (true) system is in canonical form, that is, in the form

$$d\begin{pmatrix} x_1 \\ x_2 \end{pmatrix} = \begin{pmatrix} 0 & 1 \\ \theta_1^0 & \theta_2^0 \end{pmatrix} x\,dt + \begin{pmatrix} \sigma_1 \\ \sigma_2 \end{pmatrix} dv \quad (43)$$

$$dy = (1\ 0)\begin{pmatrix} x_1 \\ x_2 \end{pmatrix} + \rho\, dw \quad (44)$$

where θ_1^0 and θ_2^0 are the true pole parameters.

As stated at the end of section 2, the convergence points are independent of ρ. Consequently, we chose to fix ρ at 1, and to vary the remaining parameters $(\theta_1^0, \theta_2^0, \sigma_1, \sigma_2)$ in our experiments. Gridpoints were of the form $(\mu^i\theta_1^0, \mu^j\theta_2^0)$, where μ was a positive multiplicative scaling factor. In the illustrative plots below $\theta_0 = (\theta_1^0, \theta_2^0) = (-1, -2)$, $\sigma = (1, 1)$, $\mu = 2$ and $i, j = -7, \ldots, +7$. The semi-log placement of gridpoints in the spatial dimensions seemed natural in view of the desired stability relations $\theta_1^0 < 0, \theta_2^0 < 0$ and allowed the effective investigation of substantial portions of the plane.

Since $n = 2$ for our experimental cases, the possible convergence points are the zeroes of the two-vector field $L = (L_1, L_2)$. Figure 1 show a plot of the second component of L. Obviously, not a great deal of information can be gleaned from such plots alone, but inspection of the values (by printing out values, rather than visually) does confirm that both components vanish at the center of the grid.

The next figure (Fig. 2) shows the negative of the Euclidean norm of L, -Norm(L) = $-||L||$ = $-\sqrt{L1^2 + L2^2}$. This makes the minimum of $||L||$ at the true value (grid center) visible as a maximum.

Fig. 3 shows the negative of the log likelihood function. This is the asymptotic limit of the output error covariance $E\{C\tilde{x}\tilde{x}^T C^T\} = e_1^T V_{2,2} e_1$ (where $C = e_1^T$ is the output matrix). This function ($J(\theta)$ in [3]) can serve as a criterion function to determine the maximum likelihood values for θ. The fact that it has a maximum at the grid center corresponds to the fact that the true system is necessarily a maximum likelihood solution.

In none of our experiments did we find any indication of points of convergence other than the point corresponding to the true system parameters. Although the possible existence of such additional points cannot be eliminated, since the search was not exhaustive, it appears very unlikely, given the nature of the cost surfaces obtained. We conjecture therefore, that there are no other points of convergence in the two-dimensional case. Note that the analogous result for dimension one has been obtained by a rigorous derivation ([4]).

The above research was supported in part by the Aerospace Sponsored Research program of The Aerospace Corporation.

References

[1] L. Ljung. Analysis of recursive, stochastic algorithms. *IEEE Transactions on Automatic Control*, AC-22(4):551–575, 1977.

[2] L. Ljung and T. Söderström. *Theory and Practice of Recursive Identification.* MIT Press, Cambridge, MA, 1983, 3rd edition 1986.

[3] D. G. DeWolf and D. M. Wiberg. An ordinary differential equation technique for continuous time parameter estimation. In *Proc. 1991 American Conf. on Automatic Control*, volume 2, pages 1390–1397, 1991.

[4] Donald M. Wiberg. Towards a globally convergent approximation of optimal on-line parameter estimation. In *Proc. 8th IFAC Symp. on Identification and System Parameter Estimation*, pages 1165–1170, Aug 1988.

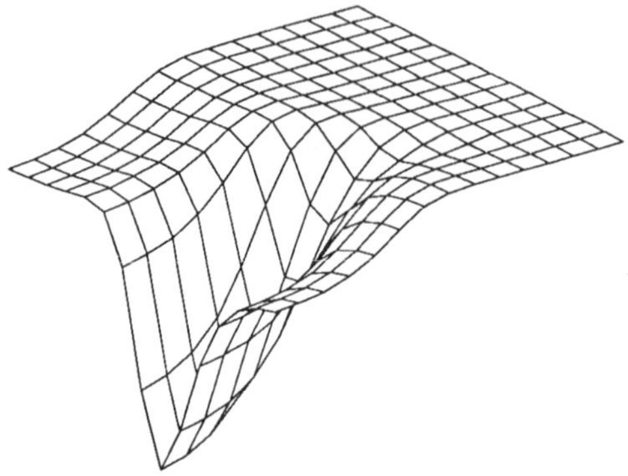

Figure 1: L2 – second component of L = (L1,L2)

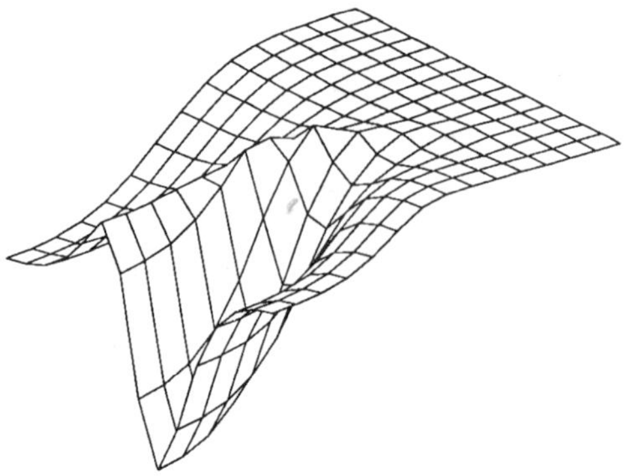

Figure 2: Negative of Norm(L)

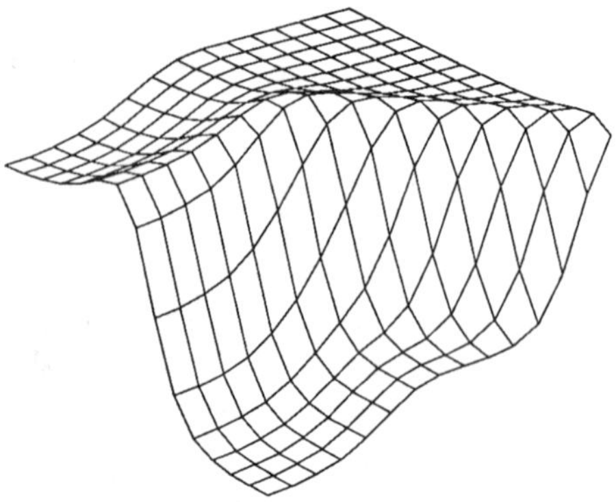

Figure 3: Negative of log likelihood

A Maximum Likelihood Estimation Technique for Spatial-Temporal Modal Analysis*

Michael P. Clark and Louis L. Scharf
University of Colorado at Boulder
Campus Box 425
Boulder, CO 80309

Abstract

We use the data collected by a linear, multisensor array to identify the wavenumber, damping factor, frequency and amplitude of individual point sources in the far-field. The algorithm we propose is based on the principle of maximum likelihood. It allows us not only to identify the spatial and temporal parameters, but also to "pair" the parameters in each domain. The method works by identifying a matrix which characterizes the orthogonal subspace. This matrix has two parts. The first of these is based on a temporal, prediction polynomial. The second uses an interpolating polynomial which predicts spatial modes from temporal modes. The method works by solving for the parameters which characterize the orthogonal subspace, making it a subspace method. The parameters can be identified by any of several methods, including least squares, total least squares, and iterative quadratic maximum likelihood (IQML) [1] [2] [3]. Once the parameters of the orthogonal subspace have been estimated, the temporal polynomial is rooted to obtain the temporal modes. These modes are then plugged into an interpolating polynomial to generate the corresponding spatial modes.

1 The Maximum Likelihood Problem

Consider a linear, equispaced array consisting of N sensors. The data vector of sensor n at times mT for $m = 0,\ 1,\ \cdots,\ M-1$ is

$$\mathbf{y}_n = [y_n(0)\ y_n(1)\ \cdots\ y_n(M-1)]^T. \qquad (1)$$

*This work is supported by the Office of Naval Research, Statistics and Probability Branch under contract N00014-89-J-1070 and by Bonneville Power Administration under contract DEB17990BP07346.

Organize these vectors into the data matrix $[\mathbf{y}_0\ \mathbf{y}_1\ \cdots\ \mathbf{y}_{N-1}]$ and then stack the columns to form

$$\mathbf{y} = \mathrm{vec}([\mathbf{y}_0\ \mathbf{y}_1\ \cdots\ \mathbf{y}_{N-1}]) = [\mathbf{y}_0^T\ \mathbf{y}_1^T\ \cdots\ \mathbf{y}_{N-1}^T]^T. \qquad (2)$$

Let the environment consist of p far-field point sources. Each of these sources is assumed to generate a damped, sinusoidal signal, but the temporal modes of different sources need not be the same. Let each source have amplitude a_i, frequency ω_i, damping factor ρ_i, and direction of arrival ψ_i (as measured from broadside). The propagating waves due to the sources in this environment are represented as

$$s_i(n,m) = a_i \rho_i^m e^{j(\omega_i mT - n\frac{d\omega_i}{c}\sin(\psi_i))}; \qquad 0 \le i \le p-1. \qquad (3)$$

Here n is the sensor number, d is the distance between sensors and mT is the sample time. The vector of array data due solely to source i is

$$\mathbf{s}_i = a_i(\mathbf{h}_i \otimes \mathbf{t}_i) \qquad (4)$$

where $\otimes$ denotes the Kronecker product,

$$\mathbf{h}_i = \Psi_N\left(e^{-j\frac{d\omega_i}{c}\sin(\psi_i)}\right), \qquad (5)$$

$$\mathbf{t}_i = \Psi_M\left(\rho_i e^{j\omega_i T}\right), \qquad (6)$$

and

$$\Psi_n(z) = [1\ z\ z^2\ \cdots\ z^{n-1}]^T. \qquad (7)$$

The vector $\mathbf{h}_i \otimes \mathbf{t}_i$ is termed the *array response vector* for source i.

Since the observed data is a linear superposition of the array response vectors due to each of the sources along with Gaussian noise, the data vector is modeled as

$$\mathbf{y} = F\mathbf{a} + \mathbf{n}, \qquad (8)$$

where

$$F = [\mathbf{h}_0 \otimes \mathbf{t}_0,\ \mathbf{h}_1 \otimes \mathbf{t}_1,\ \cdots,\ \mathbf{h}_{p-1} \otimes \mathbf{t}_{p-1}] \in \mathcal{C}^{NM \times p}, \qquad (9)$$

1058-6393/91 $01.00 © 1991 IEEE

$$\mathbf{a} = [a_0\ a_1\ \cdots\ a_{p-1}]^T \in \mathcal{C}^p, \tag{10}$$

and $\mathbf{n}$ is an additive noise vector which is assumed to be normally distributed with mean $\mathbf{0}$ and spatial-temporal covariance $\sigma^2 R_{nn}$. Note that the columns of the matrix F span the *signal subspace*. F can be shown to have full column rank if $M > p$, $N > p$, and if $\psi_k = \psi_n$ implies $\rho_k e^{j\omega_k} \neq \rho_n e^{j\omega_n}$.

With R_{nn} known the probability density function of the data $\mathbf{y}$ can be written as

$$f_{(F,\mathbf{a},\sigma^2)}(\mathbf{y}) = \frac{(2\pi)^{-\frac{NM}{2}}}{|\sigma^2 R_{nn}|^{\frac{1}{2}}} e^{-\frac{1}{2\sigma^2}(\mathbf{y}-F\mathbf{a})^* R_{nn}^{-1}(\mathbf{y}-F\mathbf{a})}. \tag{11}$$

We can easily show that the ML solutions for σ^2 and $\mathbf{a}$ are given by

$$\hat{\sigma}^2 = \frac{1}{NM}(\mathbf{y} - F\mathbf{a})^* R_{nn}^{-1}(\mathbf{y} - F\mathbf{a}), \tag{12}$$

and

$$\hat{\mathbf{a}} = (F^* R_{nn}^{-1} F)^{-1} F^* R_{nn}^{-1}\mathbf{y}. \tag{13}$$

If we compress the likelihood function we find that the ML problem is equivalent to the maximization problem

$$\max_F \mathbf{y}^* R_{nn}^{-1} F (F^* R_{nn}^{-1} F)^{-1} F^* R_{nn}^{-1}\mathbf{y}. \tag{14}$$

The remaining problem is to estimate the spatial-temporal modes of the structured matrix F.

In many cases we will have some apriori information about the data we observe. For instance, the data might be real, undamped, or both. In these cases we will not get the ML solutions by solving equation 14. Instead we need to constrain the solution to lie in some class of permissible solutions. For the sake of simplicity we will ignore the issue of constraints for the time being. However, we will show in a later section how to deal with them in a straightforward manner.

2 A ML Estimation Technique

Each source has associated with it two modes: its spatial(/temporal) mode $e^{-j\frac{d\theta_i \sin(\psi_i)}{cT}}$, and temporal mode $\rho_i e^{j\theta_i}$. Here T is the sampling interval, d is the distance between sensors and $\theta_i = \omega_i T$. Let $B(z)$ be the degree p polynomial whose roots are the temporal modes of the sources:

$$B(z) = \prod_{i=1}^{p-1}(z - \rho_i e^{j\theta_i}) = b_0 z^p + b_1 z^{p-1} + \cdots + b_p. \tag{15}$$

We shall impose a *nontriviality constraint*, $b_0 = 1$. $B(z)$ can also be written as

$$B(z) = \mathbf{b}^T \Psi_{p+1}(z) \tag{16}$$

where

$$\mathbf{b} = [b_p\ b_{p-1}\ \cdots\ b_0]^T. \tag{17}$$

Define the $(M-p) \times M$-dimensional Toeplitz matrix

$$B = \begin{bmatrix} b_p & \cdots & b_1 & b_0 & & 0 \\ & \ddots & & \ddots & \ddots & \\ 0 & & b_p & \cdots & b_1 & b_0 \end{bmatrix}. \tag{18}$$

As we force $b_0 = 1$, this matrix has full row rank. Further, since the polynomial $B(z)$ was chosen with its zeros corresponding to the temporal modes of the propagating waves

$$B\mathbf{t}_i = \mathbf{0}; \qquad 0 \le i \le p-1. \tag{19}$$

Our geometry is that of a linear, equispaced array and the sources are in the far-field of the array. Thus the angle of arrival of each source is determined by its frequency and its phase delay between neighboring elements. For source i this phase delay is given by

$$\phi_i = -\frac{d\theta_i \sin(\psi_i)}{cT}. \tag{20}$$

Each ϕ_i is actually a normalized wavenumber. Since our method will identify the parameters ϕ_i, the inverse of this function is of primary interest. However, note that some wavenumbers have no real pre-image ψ. Such wavenumbers are said to lie outside the *acoustic cone*. The acoustic cone corresponds to only those wavenumbers which can be associated with acoustic signals propagating in a dispersionless medium. This is important in that our method allows arbitrary spatial modes, some of which may not have wavenumbers in the acoustic cone.

Now, assume that $\rho_i e^{j\theta_i} \neq \rho_j e^{j\theta_j}$ if $i \neq j$ for $0 \le i, j \le p-1$. Then we may let $C(z)$ be the degree $p-1$ Lagrange interpolating polynomial which maps $\rho_i e^{j\theta_i} \mapsto e^{j\phi_i}$ for each $0 \le i \le p-1$. Such a polynomial is useful because if we can identify the polynomials $B(z)$ and $C(z)$, we can plug the temporal modes given by the roots of $B(z)$ into the interpolating polynomial $C(z)$ to obtain the spatial modes. Then, assuming the spatial modes correspond to acoustic signals, the DOA of each source can be easily computed as the scheme has paired temporal and spatial mode estimates.

Write the polynomial $C(z)$ as

$$C(z) = c_0 z^{p-1} + c_1 z^{p-2} + \cdots + c_{p-1} = \mathbf{c}^T \Psi_p(z), \tag{21}$$

where

$$\mathbf{c} = [c_{p-1}\ c_{p-2}\ \cdots\ c_0]^T. \tag{22}$$

Next define the $(M-p+1)\times M$-dimensional Toeplitz matrix

$$C=\begin{bmatrix} c_{p-1} & \cdots & c_1 & c_0 & & 0 \\ & \ddots & & \ddots & \ddots & \\ 0 & & c_{p-1} & \cdots & c_1 & c_0 \end{bmatrix}. \quad (23)$$

Since the polynomial $C(z)$ maps temporal modes into spatial modes we have

$$C\mathbf{t}_i = e^{j\phi_i}\hat{I}\mathbf{t}_i; \qquad 0\le i\le p-1 \quad (24)$$

where

$$\hat{I}=[I_{M-p+1}|\overbrace{0\cdots0}^{p-1}]. \quad (25)$$

We are now at the point where, using B and C as building blocks, we can construct a matrix whose rows span the null space (or the orthogonal subspace) of F. This matrix is

$$V=\left[\frac{I_N\otimes B}{[I_{N-1}|0]\otimes C-[0|I_{N-1}]\otimes\hat{I}}\right]. \quad (26)$$

Note that the top half of this matrix is $N(M-p)\times NM$ while the lower half is $(N-1)(M-p+1)\times NM$. Therefore V is $((2N-1)(M-p)+N-1)\times NM$. Further, $VF=0$. The following theorem gives the row rank of V.

Theorem 1 *V has row rank $NM-p$ if $p\le\frac{M+1}{2}$ and F has full column rank.*

The proof of this theorem is given in [4].

If we satisfy the conditions of the theorem, then V has full row rank. Therefore we can find a $(NM-p)\times((2N-1)(M-p)+N-1)$-dimensional matrix P such that PV has full row rank. For such a P define

$$W=PV. \quad (27)$$

This matrix has all of the properties desired for the orthogonal subspace. That is, its $NM-p$ rows span the orthogonal subspace, are parameterized by the coefficients of $B(z)$ and $C(z)$, and $WF=0$. Therefore, since F has p columns, the projections onto the column space of F and the row space of W form a resolution of the identity. Therefore we can rewrite the maximization problem on the signal subspace 14 as a minimization problem on the orthogonal subspace, to wit:

$$\min_W \mathbf{y}^*W^*(WR_{nn}W^*)^{-1}W\mathbf{y}. \quad (28)$$

We want to solve this optimization problem for the coefficients of $B(z)$ and $C(z)$ contained in the matrix W. It is not at all obvious how to solve for these parameters with the minimization problem written as in equation 28. However, with one simple observation we can change the problem to one for which we have a wealth of intuition. This observation is that the quantity $W\mathbf{y}$ can be written as $Y\mathbf{f}$ where Y contains only data from $\mathbf{y}$ and $\mathbf{f}$ contains only the coefficients of $B(z)$ and $C(z)$. Proof of this fact involves a lot of bookkeeping but is quite straightforward. First remember that $\mathbf{y}$ is partitioned as

$$\mathbf{y}=[\mathbf{y}_0^T\ \mathbf{y}_1^T\ \cdots\ \mathbf{y}_{N-1}^T]^T. \quad (29)$$

Therefore $W\mathbf{y}$ can be written as

$$W\mathbf{y}=PV\mathbf{y}=Y\mathbf{f}, \quad (30)$$

where

$$Y=P\left[\begin{array}{c|c|c} Q_{p+1} & 0 & 0 \\ \hline 0 & Q_p & \mathbf{q}\end{array}\right] \quad (31)$$

and

$$\mathbf{f}=\begin{bmatrix}\mathbf{b}\\ \hline \mathbf{c}\\ \hline 1\end{bmatrix}. \quad (32)$$

Here Q_{p+1} is the data matrix

$$Q_{p+1}=\begin{bmatrix} G_{p+1}(0)\\ G_{p+1}(1)\\ \vdots\\ G_{p+1}(N-1)\end{bmatrix} \quad (33)$$

and each $G_{p+1}(i)$ is a $(M-p)\times(p+1)$-dimensional Hankel matrix

$$G_{p+1}(i)=\begin{bmatrix} y_i(0) & \cdots & y_i(p)\\ \vdots & \iddots & \vdots\\ y_i(M-p-1) & \cdots & y_i(M-1)\end{bmatrix}. \quad (34)$$

Further, $\mathbf{q}$ is the vector

$$\mathbf{q}^T=[\mathbf{q}_1^T\ \mathbf{q}_2^T\ \cdots\ \mathbf{q}_{N-1}^T], \quad (35)$$

where each $\mathbf{q}_i$ is a $(M-p+1)$-dimensional vector

$$\mathbf{q}_i=-[y_{i+1}(0)\ y_{i+1}(1)\ \cdots\ y_{i+1}(M-p)]^T. \quad (36)$$

Substituting 30 into 28 the minimization problem becomes

$$\min_{\mathbf{f}}\mathbf{f}^*Y^*(WR_{nn}W^*)^{-1}Y\mathbf{f}. \quad (37)$$

This optimization problem is exactly what we desire. Its solutions are those for the case where we know the temporal modes to be distinct. The dual formulation of the problem corresponds to the case where we know the spatial modes to be distinct (but don't necessarily know a thing of the temporal modes). Obviously, a dual to the method above is quite easily constructed and this method will solve the dual problem. Finally, note that to this point we have assumed nothing of the structure of the parameters. Of course, if the parameters *do* have some known structure, and we choose to neglect it, we will not be solving the ML problem. Thus we address the issue of constraints next.

2.1 Constraining the Optimization Problem

To account for the structure of the prediction polynomials, we must restate the ML optimization problem 37 as

$$\min_{\substack{\mathbf{b}\in\Theta \\ \mathbf{c}\in\Omega}} \mathbf{f}^* Y^* (W R_{nn} W^*)^{-1} Y \mathbf{f}. \tag{38}$$

Here Θ represents the set of coefficient vectors $\mathbf{b}$ for which $B(z)$ has the desired structure. Further, Ω represents the set of coefficient vectors $\mathbf{c}$ for which $C(z)$ has the desired structure. The sets Θ and Ω may be chosen to reflect many different scenarios. Bresler and Macovski [2] give many possible constraint sets Θ for single dimensional estimation problems. We have formulated the sets Θ and Ω for both the primary and dual methods. The sets include both the real and complex cases and handle both damped and undamped modes [4]. However, for the sake of brevity we do not present them here.

For a given scenario, we can formulate a constrained optimization problem of the type given by equation 38. In order to solve this constrained problem, we rewrite it as an unconstrained problem. The idea which allows this transformation was presented by Starer and Nehorai [5]. Their idea is to use an affine model for the parameter vector $\mathbf{f}$. That is model

$$\mathbf{f} = T\hat{\mathbf{f}} + \mathbf{c} \tag{39}$$

where T is a matrix with full column rank, $\hat{\mathbf{f}}$ is a minimal vector of real parameters which represent the degrees of freedom in $\mathbf{f}$ and $\mathbf{c}$ is a constant vector. It is a trivial matter to generate T and $\mathbf{c}$ for most practical constraint sets.

With this representation for $\mathbf{f}$ we can now rewrite the constrained optimization problem 38 as the unconstrained problem

$$\min_{\hat{\mathbf{f}}} (T\hat{\mathbf{f}} + \mathbf{c})^* Y^* (W R_{nn} W^*)^{-1} Y (T\hat{\mathbf{f}} + \mathbf{c}). \tag{40}$$

3 The Algorithm

We now have all of the pieces we need to write down an algorithm for solving the ML problem. Our algorithm consists of the following steps.

1. Using apriori information construct the constraint matrix T and the constant vector $\mathbf{c}$.
2. Form the data matrix Y from the data vector $\mathbf{y}$.
3. Solve (or find an approximate solution to) the optimization problem,
$$\min_{\hat{\mathbf{f}}} (T\hat{\mathbf{f}} + \mathbf{c})^* Y^* (W R_{nn} W^*)^{-1} Y (T\hat{\mathbf{f}} + \mathbf{c}). \tag{41}$$
4. Form the coefficient vector $\mathbf{f} = T\hat{\mathbf{f}} + \mathbf{c}$.
5. Extract the coefficients of $\mathbf{b}$ and $\mathbf{c}$ from $\mathbf{f}$ using
$$\mathbf{f} = \left[\begin{array}{c} \mathbf{b} \\ \hline \mathbf{c} \\ \hline 1 \end{array}\right]. \tag{42}$$
6. Root the polynomial $B(z)$ to obtain the estimated temporal modes.
7. Plug the roots of $B(z)$ into $C(z)$ to obtain the estimated spatial modes.
8. Form F and compute the estimated source amplitudes using $\hat{\mathbf{a}} = (F^* R_{nn}^{-1} F)^{-1} F^* R_{nn}^{-1} \mathbf{y}$.

The only step in this algorithm which may cause some concern involves solving the optimization problem. However, this problem is of the type soluble by the iterative algorithm of Evans and Fischl [3], the KiSS algorithm of Kumaresan, Scharf and Shaw [1], or the IQML method of Bresler and Macovski [2]. As shown by McClellan and Lee [6], the iteration employed by all three of the above algorithms is equivalent to that of the classic Steiglitz-McBride algorithm, so it matters not which we pick.

In order to reduce computation, we might also reformulate the optimization problem. Notice the optimization problem 40 involves minimizing the weighted 2-norm of the vector $Y(T\hat{\mathbf{f}}+\mathbf{c})$. This vector represents the *equation errors*, $\mathbf{e}$, and can also be expressed as

$$\mathbf{e} = [YT|Y\mathbf{c}]\begin{bmatrix} \hat{\mathbf{f}} \\ 1 \end{bmatrix}. \tag{43}$$

If we work with this system of equations we will not, in general, be solving the exact ML problem. However, the solution will come at considerable computational savings and may be close to the exact solution. There are many well known methods for solving this system of equations. These include least squares Prony [7], min-norm [8] and total least squares [9] methods.

4 An Example

In order to illustrate this new technique we constructed an experimental case. In the far-field of a linear equispaced array are 2 point sources: the first with parameters $a_1 = 0.9e^{-j0.1\pi}$, $\rho_1 = 0.95$, $\theta_1 = -0.25\pi$, $\phi_1 = 0.20\pi$, and the second with $a_2 = 1.0e^{j0.2\pi}$, $\rho_2 = 1.0$, $\theta_2 = 0.5\pi$, $\phi_1 = 0.30\pi$. To circumvent the problem of order selection we let $p = 2$. We ran 100 trials at each of 8 different input signal to noise ratios and at each trial used 5 iterations of IQML to solve the optimization problem. Using the results of these trials we computed the sample MSE's and compared these to the Cramer-Rao (CR) bounds for the unbiased ML problem. Below are shown the CR bounds and resultant MSE's for the parameters θ_1, θ_2, ϕ_1, and ϕ_2.

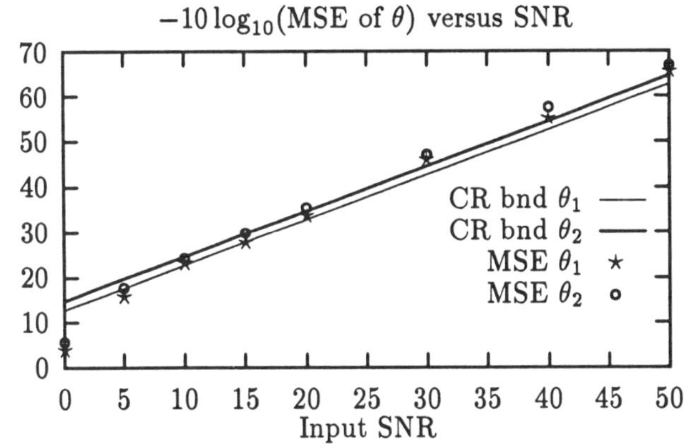

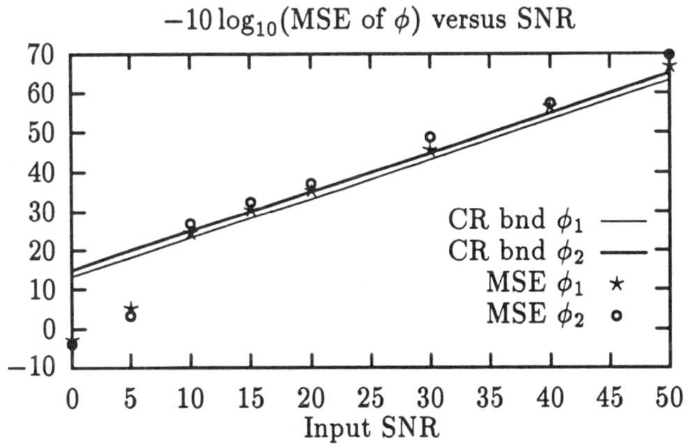

Figure 1: Comparisons between the MSE's of the estimates of θ_1, θ_2, ϕ_1, ϕ_2 and the respective CR bounds.

These figures show that the method does quite well in predicting the source parameters at high to moderate input SNR's. The simulations also show the bias of our estimator to be quite small. Therefore we are justified in comparing the MSE of our estimates to the CR bounds for unbiased estimates.

5 Conclusion

We have identified an optimization problem, based on the principle of maximum likelihood. The solution to this problem yields the maximum likelihood estimates of the spatial-temporal parameters for each source in a far-field environment. The optimization problem was obtained by *characterizing the orthogonal subspace for the space-time data collected by an array of sensors.* This orthogonal subspace was constructed in such a way that we were able to pair the temporal and spatial mode estimates.

References

[1] R. Kumaresan, L. L. Scharf, and A. K. Shaw, "An algorithm for pole-zero modeling and spectral analysis," *IEEE Trans. ASSP*, Vol. ASSP-34, pp.637–640, June 1986.

[2] Y. Bresler and A. Macovski, "Exact maximum likelihood parameter estimation of superimposed signals in noise," *IEEE Trans. ASSP*, Vol. ASSP-34, pp. 1081–1089, Oct. 1986.

[3] A. G. Evans and R. Fischl, "Optimal least squares time-domain synthesis of recuresive digital fiters," *IEEE Trans. Audio Electroacoust.*, Vol. AU-21, pp.61–65, Feb. 1973.

[4] M. P. Clark and L.L. Scharf, "A maximum likelihood method for spatial-temporal modal analysis", to be submitted.

[5] D. Staher and A. Nehorai, "Maximum likelihood estimation of exponential signals in noise using a newton algorithm," *IEEE ASSP Workshop on Spectral Estimation and Modeling*, Minneapolis, MN, pp. 240–245, Aug. 1988.

[6] J. H. McClellan and D. Lee, "Exact Equivalence of the Steiglitz-McBride Iteration and IQML," *IEEE Trans. ASSP*, Vol. ASSP-39, pp. 509–512, Feb. 1991.

[7] F. B. Hildebrand, *Introduction to Numerical Analysis*, p. 379, McGraw-Hill, New York, 1956.

[8] D. Tufts and R. Kumaresan, "Frequency estimation of multiple sinusoids: making linear prediction like maximum likelihood," *Proc. IEEE*, Vol. 70, pp. 975–90, Mar. 1983.

[9] G. H. Golub and C. F. Van Loan, "An analysis of the total least squares problem," *SIAM J. Numer. Anal.*, Vol. 17, No. 6, pp. 883–93, 1980.

Subinterval Cyclic MUSIC - Robust DF with Error in Cycle Frequency Knowledge *

Thomas E. Biedka
E-Systems, Inc. Greenville Division
Greenville, TX 75403

Brian G. Agee
Radix Technologies, Inc.
San Jose, CA 95117

Abstract

Cyclic MUSIC has been proposed as a direction finding technique that reduces the effects of noise and interference by exploiting the cyclostationarity of the signal of interest. Most communication signals are more accurately modeled as cyclostationary rather than stationary. Equivalently they exhibit spectral correlation at cycle frequencies such as baud rates, double carrier frequencies, etc. If these cycle frequencies are not known to the required degree of accuracy, the performance of Cyclic MUSIC can be severely degraded. This paper introduces two techniques for reducing the sensitivity of Cyclic MUSIC to error in cycle frequency knowledge. Both techniques are shown to perform significantly better than conventional Cyclic MUSIC when the cycle frequencies are not known precisely.

1 Introduction

The need to estimate the directions of arrival of signals of interest in crowded environments is becoming increasingly important. Conventional direction finding (DF) techniques such as MUSIC [1] typically perform poorly in high rank environments. This is because these algorithms must estimate the direction of arrival of all signals impinging on the array, regardless of how many of these signals are desired signals. In addition these algorithms require knowledge of the noise and interference covariance matrix, which may not be available. Yet another drawback is that if the direction of arrival of an undesired signal is nearly the same as a desired signal, very long collect times can be required to resolve the signals.

Many signals of interest, including communication signals, are more accurately modeled as cyclostationary rather than stationary. Equivalently they exhibit spectral correlation at cycle frequencies such as baud rates, double carrier frequencies, and sums and differences of these. These cyclostationary properties can be exploited to discriminate against noise and interference. The recently proposed Cyclic MUSIC algorithm [2] [3] is an example of a DF algorithm that exploits cyclostationarity. One of the benefits of this algorithm over MUSIC is that the background noise and interference environment need not be known or estimated. In addition, if an undesired signal is spaced closely to a desired signal, Cyclic MUSIC can require a much shorter collect time to accurately determine the direction of arrival of the desired signal. Because of its ability to discriminate against undesired signals, Cyclic MUSIC is described as a *signal-selective* algorithm.

A well known limitation of algorithms that exploit cyclostationarity is the need to accurately know the cycle frequencies of the signal of interest. This is true of any technique that exploits cyclostationarity and is not limited to DF algorithms. The required accuracy of this knowledge is directly dependent on the collect time. Therefore if highly accurate estimates of the directions of arrival are required, the cycle frequency of the desired signal must be known very precisely. In practice this can be an unrealistic requirement. This issue has been previously addressed by Schell and Gardner in [4]. However, the techniques proposed in [4] can present problems in practical implementations, including large memory requirements. The alternative techniques presented here address some of the limitations of the previously proposed methods.

2 Cyclic MUSIC

The Cyclic MUSIC algorithm assumes that the signals of interest can be accurately described by the narrowband model

$$\mathbf{x}(n) = \sum_{l=1}^{L} \mathbf{a}(\theta_l)s_l(n) + \mathbf{i}(n), \tag{1}$$

where $\mathbf{x}(n)$ is the $M \times 1$ observed signal vector, $\mathbf{a}(\theta_l)$ is the array response vector for a signal incident from direction θ_l, $s_l(n)$ is the sampled complex envelope of the source waveform, and $\mathbf{i}(n)$ is noise and interference that is uncorrelated with the incident signals. The total number of signals in the environment is denoted by L. In addition, the desired signals are assumed to exhibit second-order cyclostationarity. That is, there exists some non-zero frequency α and time lag τ for which the (asymmetric) cyclic autocorrelation function defined by

$$\begin{aligned} \mathrm{R}_{ss}^{\alpha}(\tau) &= \langle s(n)s^*(n-\tau)e^{-j2\pi\alpha n} \rangle_\infty \\ &\neq 0 \end{aligned} \tag{2}$$

where $\langle\cdot\rangle_\infty$ denotes infinite time average. If the signal $s(n)$ is second order cyclostationary, the lag-product

*This work was supported by E-Systems, Inc. Greenville Division.

1058-6393/91 $01.00 © 1991 IEEE

waveform $s(n)s^*(n-\tau)$ will contain a finite strength sine wave component, or *spectral line*. The magnitude and phase of the spectral line at cycle frequency α for lag τ is given by $R_{ss}^{\alpha}(\tau)$. Note that a cycle frequency of zero in the cyclic correlation function gives the conventional correlation function. The signal may also exhibit non-zero *conjugate* cyclic correlation, which may defined by (2) with the conjugation operator removed. Note that a signal can exhibit non-trivial conjugate cyclic correlation at cycle frequency equal to zero. An example of such a signal would be a BPSK signal with carrier frequency equal to zero. A more complete discussion of cyclostationarity may be found in [5],[6],[7],[8].

The Cyclic MUSIC algorithm exploits cyclostationarity to estimate the directions of arrival of only the desired signals by using a finite-time estimate of the cyclic correlation matrix defined by

$$\mathbf{R}_{\mathbf{xx}}^{\alpha}(\tau) = \langle \mathbf{x}(n)\mathbf{x}^H(n-\tau)e^{-j2\pi\alpha n}\rangle_N \quad (3)$$

where $(\cdot)^H$ denotes conjugate transpose and $\langle\cdot\rangle_N$ denotes time average over N samples. In the limit as $N \to \infty$, noise and interference not exhibiting the desired cyclostationary properties will not contribute to the cyclic correlation. Since the cyclic correlation matrix is estimated using a finite collect time, noise and interference will in fact have a non-zero contribution. However, for reasonably long collect times, this contribution is significantly reduced. Once the cyclic correlation matrix has been estimated, an SVD of the matrix is performed, defined by

$$\mathbf{R}_{\mathbf{xx}}^{\alpha}(\tau) = \mathbf{U}\Sigma\mathbf{V}^H. \quad (4)$$

Here $\mathbf{U}$ and $\mathbf{V}$ are orthogonal matrices whose columns are the right and left singular vectors, respectively, and Σ is the diagonal matrix of singular values. Assuming that the number of signals L_α with cycle frequency α is known, their directions of arrival may be found by locating the minima in

$$\left\| [\mathbf{u}_{M-L_\alpha} \ldots \mathbf{u}_M]^H \mathbf{a}(\theta) \right\|^2. \quad (5)$$

The problem addressed in this paper is that where the regenerated spectral line in the lag-product waveform $s(n)s^*(n-\tau)$ is not at α but instead at α_{true}. When this is the case, the effect of the rectangular averaging window implicit in (3) must be considered. The frequency response of a rectangular averaging window of length N is given by

$$\mathrm{H}(f) = e^{-j\pi f(N-1)}\frac{\sin(\pi f N)}{\sin(\pi f)} \quad (6)$$

When the regenerated spectral line falls outside the passband of this filter, the desired cyclic feature will be rejected. The first null of this filter is at $\alpha = 1/N$. Therefore if the cycle frequency error $\Delta\alpha = \alpha - \alpha_{true}$ is larger than the cutoff frequency, the regenerated spectral line will be rejected. In practice performance begins to degrade at $\Delta\alpha = 1/2N$.

2.1 Multi-Cyclic and Adaptive-α Cyclic MUSIC

The problem of how to handle error in the assumed cycle frequency of a desired signal has been previously addressed by Schell and Gardner in [4], where two techniques are considered. One technique, Multi-Cyclic MUSIC, uses the multi-cyclic correlation matrix defined by

$$\mathbf{R}_{\mathbf{xx}}^{A}(\tau) = \left\langle \mathbf{x}(n)\mathbf{x}^H(n-\tau)\left(\sum_{k=1}^{K} e^{-j2\pi\alpha_k n}\right)\right\rangle_N \quad (7)$$

$$= \sum_{k=1}^{K} \mathbf{R}_{\mathbf{xx}}^{\alpha_k}(\tau). \quad (8)$$

This algorithm seeks to cover a desired range of cycle frequencies while avoiding the computationally burdensome approach of calculating a separate cyclic correlation matrix for each cycle frequency. This approach, although attractive, has been observed to perform poorly when applied to the problem at hand. The extra terms in the multi-cyclic correlation matrix increase the noise and interference contributions. In addition, if the true cycle frequency is not exactly equal to one of the cycle frequencies α_k, adding terms together will result in incoherent averaging. This may lead to severe degradation in performance. Schell and Gardner noted degraded performance even for $K = 2$.

The other approach presented in [4] is the Adaptive-α Cyclic MUSIC algorithm. This algorithm calls for estimating the cycle frequencies from the lag product waveform $x_1(n)x_1^*(n-\tau)$ using a simple FFT frequency estimator and then using this estimate of α in the Cyclic MUSIC algorithm. This technique was shown to work well for sufficiently long collect times. However, as was noted in [4], this requires either storing all data to be used to estimate the directions of arrival, or estimating the cycle frequencies based on one data set and the directions of arrival on the next received data set. Consequently, both of these approaches can present problems in practical implementations. Storing the raw data received by all of the sensors can result in large memory requirements if the collect time is long. Using a previous data set to estimate the true cycle frequency will reduce the speed of the DF system and may yield inaccurate estimates if the cycle frequency drifts from collect to collect.

3 Subinterval Techniques

The subinterval techniques for estimating the cyclic correlation matrix are motivated by noting that the frequency-shifted lag-product may be expressed as a sine wave at $\Delta\alpha$ with coefficient $R_{ss}^{\alpha}(\tau)$ plus other terms denoted by $z(n)$

$$s(n)s(n-\tau)e^{-j2\pi\alpha n} = R_{ss}^{\alpha}(\tau)e^{j2\pi\Delta\alpha n} + z(n). \quad (9)$$

If the original estimate α is close to α_{true}, the cycle frequency error $\Delta\alpha$ will be small, and the regenerated spectral line will be at a very low frequency. As a consequence the signal in (9) may be low-pass filtered and decimated without loss of information. The cutoff frequency of the filter should be chosen so that a regenerated spectral line with the largest anticipated error will be captured. An approach for performing this filtering and decimation is discussed below.

The first step in the subinterval algorithms is to calculate a time-variant cyclic correlation matrix for each subinterval of the total collect interval. This matrix is defined for the mth subinterval by

$$\widehat{\mathbf{R}}_{\mathbf{xx}}^{\alpha_0}(m,\tau) = \langle\, w(n)\mathbf{x}(n+mL) \times \mathbf{x}^H(n+mL-\tau)e^{-j2\pi\alpha_0(n+mL)}\,\rangle_{N'} \quad (10)$$

where m is the decimated time index, $w(n)$ is the lag-product window, L is the decimation factor, and the time averaging is over $[mL, mL+N']$. The lag-product window $w(n)$ and subinterval length N' will determine the cycle frequency resolution of each time-variant cyclic correlation matrix. Assuming that the analysis parameters are chosen correctly, the time variant correlation matrices will capture the correct cycle frequency. They may then be combined in some way to recover the available time-bandwidth product and estimate the direction of arrival of the desired signal.

The general approach proposed here is to weight and sum the time-variant cyclic correlation matrices to form a single cyclic correlation matrix. Then this matrix is used to estimate the directions of arrival of the desired signals. Two different methods for choosing the weights are presented. These methods, referred to as Zoom Subinterval Cyclic MUSIC (ZSCM) and Adaptive Subinterval Cyclic MUSIC (ASCM), are described below.

3.1 Zoom Subinterval Cyclic MUSIC

The ZSCM algorithm captures the correct cyclic feature by taking element-by-element DFT of the time-variant cyclic correlation matrices. This is equivalent to trying move the spectral line at $\Delta\alpha$ to DC using different discrete frequency shifts. This element-by-element DFT is defined by

$$\widehat{\mathbf{R}}_{\mathbf{xx}}^{\alpha_0+q/N}(\tau) = \sum_{m=0}^{P-1} \widehat{\mathbf{R}}_{\mathbf{xx}}^{\alpha_0}(m,\tau)e^{-j2\pi mq/P} \quad (11)$$

where q is the fine-α index and $P = N/L$ is the total number of time-variant cyclic correlation matrices. The resulting matrices $\mathbf{R}_{\mathbf{xx}}^{\alpha_0+q/N}(\tau)$ are referred to here as fine-α matrices because they have the same cycle frequency resolution as would be achieved by the estimator of (3). One of the fine-α matrices will capture the correct cycle frequency and may be used, by itself, in the DF algorithm.

Choosing the matrix to use may be based on a simple measure such as magnitude of an arbitrary element. A more sophisticated (although more computationally intensive) approach would be to compare the singular values of each fine-α matrix and use the one with the largest singular value. The latter technique might be expected to exploit the spatial coherence of the desired signal as well as the spectral coherence.

The ZSCM technique may be looked at as the Adaptive-α technique with a data reduction step included to avoid excessive memory requirements. It should also be noted that this technique is very similar in spirit to the FFT Accumulation Method (FAM) used in cyclic spectral analysis [6], which addresses the problem of cycle frequency error in the estimation of cyclic spectra. The FAM is robust to imprecise knowledge of the correct cycle frequency while being less computationally intensive than other approaches.

3.2 Adaptive Subinterval Cyclic MUSIC

The motivation for the ASCM algorithm is based on a desire to determine weights g_m for the time-variant cyclic correlation matrices which are optimal in some sense. An obvious approach is to choose the weights such that the largest singular value in the resulting matrix is maximized. This optimization problem is easily solved for Hermitian matrices, using an alternating maximization approach. However, the resultant techniques do not extend to non-Hermitian matrices (such as the matrices of interest here) in a straightforward manner. Nevertheless, a number of ad hoc techniques have been devised that can perform this maximization in practice. The simplest of these techniques is given (after suitable initialization) by the following iterative algorithm:

1. Form the new cyclic correlation matrix via

$$\widetilde{\mathbf{R}}_{\mathbf{xx}}^{\alpha}(\tau) = \sum_{m=0}^{P-1} g_m^* \widehat{\mathbf{R}}_{\mathbf{xx}}^{\alpha_0}(m,\tau) \quad (12)$$

2. Compute the largest singular value σ_1 and its corresponding right and left singular vectors $\widetilde{\mathbf{R}}_{\mathbf{xx}}^{\alpha}(\tau)\mathbf{v}_1 = \sigma_1\mathbf{u}_1$.

3. Compute the new weights $\mathbf{g} = \{g_m\}$ via

$$g_m = \mathbf{v}_1^H \widehat{\mathbf{R}}_{\mathbf{xx}}^{\alpha_0}(m,\tau)\mathbf{u}_1 \quad (13)$$

4. Normalize the weights $\mathbf{g}$ to have unit 2-norm.

5. Repeat steps (1)-(4) until $|\,\sigma_{old} - \sigma_{new}\,| < \epsilon$.

An analytical treatment of this algorithm shall be performed at a later date: for now simulation results are relied upon to demonstrate this result.

4 Simulation Results

In this section simulation results are presented comparing the performance of conventional Cyclic MUSIC, ASCM, and ZSCM. All simulations are based on a four element circular array of omni-directional sensors with one-half wavelength diameter. The signals obey the narrowband model of (1). The desired signal is a rectangular pulse shaped BPSK signal with a baud rate of 0.25 relative to the sample rate of unity. In all simulations, the Cyclic MUSIC and subinterval Cyclic MUSIC algorithms exploit the *conjugate* cyclic correlation of the BPSK signal at double the carrier frequency. An FM interfering signal with nominally

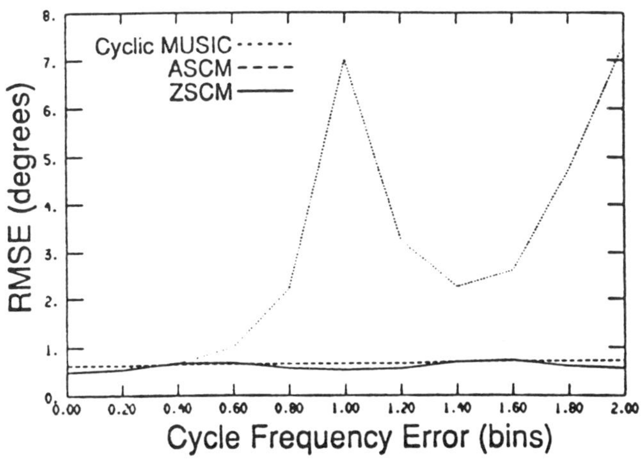

Figure 1: RMSE of Cyclic MUSIC, ZSCM, and ASCM as a Function of Cycle Frequency Error.

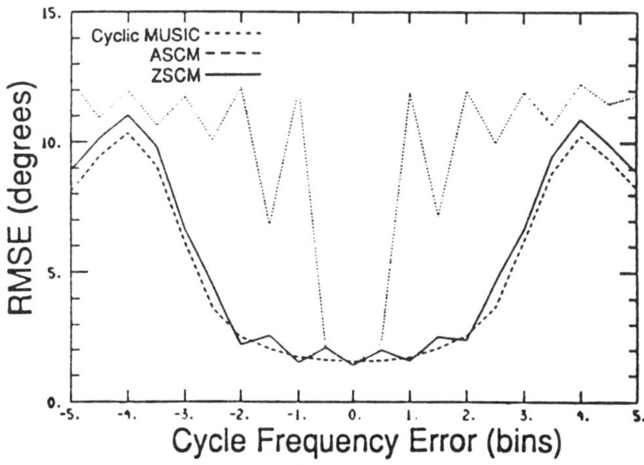

Figure 2: Range of Cycle Frequencies Covered by ZSCM and ASCM with Subinterval Length of $N/4$.

the same bandwidth as the desired signal is also used in these simulations. The noise in all cases is temporally and spatially white, complex, and Gaussian. All algorithms are given knowledge of the rank of the environment. All simulations are conducted with 1000 independent trials.

The results of a comparison of conventional Cyclic MUSIC, ZSCM, and ASCM for varying degree of cycle frequency error are presented in Figure 1. A single desired signal is incident from 50.0 degrees with 10 dB SWNR. An FM interferer is also incident from 40.0 degrees with the same power as the desired signal. In this figure the carrier offset of the BPSK signal is varied from 0 to $1/N$, where the total number of samples N is 1024. Recall that since the doubled carrier frequency is being exploited, the cycle frequency error ranges from 0 to $2/N$. ZSCM and ASCM use a subinterval length N' equal to 128 samples, or 1/8 of the total collect. A rectangular lag-product window with zero overlap is used. ZSCM is configured to select the fine-α matrix with the largest magnitude in an arbitrary cross term. The ASCM algorithm is initialized with $g_m = c$ where c is a real constant.

As Figure 1 shows, the RMSE of Cyclic MUSIC is clearly dependent on the frequency response of the rectangular averaging window. In addition, the performance of Cyclic MUSIC begins to degrade at approximately $1/2N$. Both ZSCM and ASCM appear to perform well in this environment for the indicated range of cycle frequency error. The ASCM algorithm converges in 4-5 iterations in this environment. It should be noted that because ZSCM and Adaptive-α Cyclic MUSIC use essentially the same approach to this problem, the performance of Adaptive-α Cyclic MUSIC would be similar to that of ZSCM. The main benefit of ZSCM is that only the subinterval correlation matrices, and not the raw array data, need be stored in memory.

The range of cycle frequencies covered by ASCM and ZSCM is demonstrated in Figure 2. Here the performance of Cyclic MUSIC, ASCM, and ZSCM is compared over a range of cycle frequency error from $-5/N$ to $5/N$. The environment is similar to that in Figure 1 except that an additional FM signal is incident from 60 degrees. The total collect time is again 1024 samples. The ASCM and ZSCM algorithms use a 256 sample rectangular lag-product window with zero overlap. With this combination of data window and overlap, a cycle frequency error greater than 2 bins results in cyclic aliasing. However, the aliasing is not the main reason for performance degradation beyond a cycle frequency error of 2 bins. Just as conventional Cyclic MUSIC begins to degrade when the rectangular averaging window of length N begins to attenuate the regenerated spectral line, ZSCM and ASCM begin to degrade when the averaging window of length N' begins to attenuate the regenerated spectral line. Analogous to conventional Cyclic MUSIC, the performance degradation begins at $1/2N'$, here equal to 2 bins ($2/N$).

In this more crowded signal environment the differences between ZSCM and ASCM become more apparent. The ASCM algorithm is better able to handle cycle frequency errors that are not bin-centered (i.e., $\Delta\alpha \neq i/N$, where i is an integer), since the frequency shifts which are available to ZSCM are limited to discrete frequencies. ASCM has no such restrictions. It might be thought that adding two fine-α matrices would yield better performance when the error is not bin centered. However, this would result in performance degradation similar to that observed with Multi-Cyclic MUSIC. In contrast, when the error is bin-centered ZSCM performs better than ASCM. As is demonstrated in the results presented below, ZSCM outperforms ASCM when the error is bin-centered because ASCM exhibits a bias towards other signals in the environment.

In this set of simulations, a comparison of Cyclic MUSIC, ASCM, and ZSCM is presented as a function of interference power. The environment is identical to that in Figure 1. The total collect time is again 1024 samples. The subinterval algorithms use a rectangular lag-product window of length 256 samples with zero

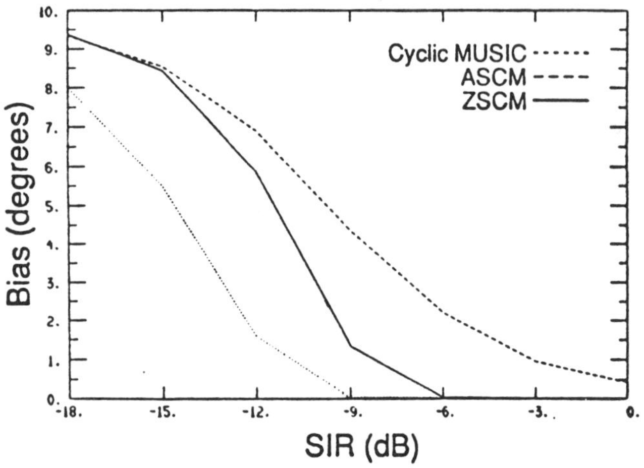

Figure 3: Bias of Cyclic MUSIC, ZSCM, and ASCM as a Function of Signal to Interference Ratio.

overlap. There is now no error in the cycle frequency but ASCM and ZSCM are not given this information. The results in Figure 3 show that the bias of ASCM is significantly higher than that of ZSCM. This may be explained by noting that the performance criterion chosen for ASCM is maximization of the largest singular value of the matrix. This can be achieved by including contributions from other, undesired signals in the signal subspace. Therefore ASCM might be expected to exhibit such a bias. The ZSCM algorithm begins to perform worse than Cyclic MUSIC when it begins to make mistakes in selecting the correct fine-α matrix. This begins to happen at about -9 dB SIR. ZSCM selects the fine-α matrix with the largest magnitude in an arbitrary cross element. It has previously been conjectured that using a detection statistic based on the largest singular value σ_{max} of the fine-α matrices might work better. However, simulation results do not support this conjecture. No significant difference in performance was noted when a detection statistic based on σ_{max} was used, at least for this scenario.

As a final note, a natural question to ask is whether there are any benefits to using a non-rectangular window to estimate the time-variant cyclic correlation matrices. It might be suspected that using windows which have better out-of-band rejection characteristics might improve performance, even when there is no error in cycle frequency. A variety of simulations have been performed, varying the type of lag-product window (e.g., Hamming and Blackman-Harris) and overlap. In no case did the simulations show any significant improvement over using rectangular windows.

5 Conclusions

This paper has presented two techniques for reducing the sensitivity of signal-selective DF algorithms to error in cycle frequency knowledge. Both the Zoom Subinterval Cyclic MUSIC (ZSCM) and Adaptive Subinterval Cyclic MUSIC (ASCM) algorithms appear to perform well. Each has its own benefits and drawbacks. The ZSCM algorithm appears to obtain unbiased estimates of the desired signal when the collect time is sufficiently long for the cycle frequency to be correctly estimated. In this respect it is very similar in performance to the Adaptive-α Cyclic MUSIC algorithm. The main benefit of using ZSCM compared to Adaptive-α Cyclic MUSIC is that only the time-variant correlation matrices, and not the raw data, need be stored. This greatly reduces memory requirements. The main drawback of using ZSCM compared to the Adaptive-α method is the cycle frequency coverage is more limited. The ASCM algorithm appears to perform comparably to ZSCM. However, the iterative technique for determining the weights g_m is computationally much more complex than that used by ZSCM. In addition, ASCM exhibits a bias towards undesired signals in the environment. This bias seems to be a result of the optimization criterion chosen. If so, additional investigation may yield ASCM techniques that can remove this bias, and hopefully improve the performance of this technique.

Acknowledgements

The authors wish to thank E-Systems, Inc. Greenville Division for supporting this work and its publication. Thanks are also due to Mr. Mark F. Kahn and Dr. Stephan V. Schell for many helpful discussions regarding cyclostationarity and its exploitation for signal-selective direction finding.

References

[1] R.O. Schmidt, *A Signal Subspace Approach to Multiple Source Location and Spectral Estimation*, Ph.D. dissertation, Stanford Univ., Stanford, CA, 1981.

[2] S.V. Schell, *Exploitation of Spectral Correlation for Signal-Selective Direction Finding* , Ph.D. dissertation, Univ. of California, Davis, CA, 1989.

[3] S.V. Schell, R.A. Calabretta, W.A. Gardner, and B.G. Agee, "Cyclic MUSIC Algorithms for Signal-Selective Direction Estimation," *Proc. IEEE Int. Conf. Acoust., Speech, Signal Processing,* pp. 2278-2281, Glasgow, Scotland, May 1989.

[4] S.V. Schell and W.A. Gardner, "Progress on Signal-Selective Direction Finding," *Proc. 5th ASSP Workshop on Spectrum Estimation and Modeling,* Rochester, NY, Oct. 1990.

[5] W.A. Gardner, *Statistical Spectral Analysis - A Nonprobabilistic Theory*, Prentice-Hall, Englewood Cliffs, NJ, 1987.

[6] W.A. Brown, *On the Theory of Cyclostationary Signals*, Ph.D. dissertation, Univ. of California, Davis, CA, 1987.

[7] W.A. Gardner, "Spectral Correlation of Modulated Signals: Part I - Analog Modulation ", *IEEE Trans. Comm.*, pp. 584-594, June 1987.

[8] W.A. Gardner, W.A. Brown, III, and C.-K. Chen, "Spectral Correlation of Modulated Signals: Part II - Digital Modulation", *IEEE Trans. Comm.*, pp. 595-601, June 1987.

Strong Consistency of FSD Detection Schemes[†]

Guanghan Xu[†] Richard H. Roy[††] Thomas Kailath[†]

[†]*Information Systems Laboratory*
Stanford University
Stanford, CA 94305
Tel: (415)725-5698

[††]*Systems Research Associates, Inc.*
10030 Phar Lap Drive
Cupertino, CA 95014

Abstract—**Two new detection schemes used in conjunction with the recently developed FSD algorithm are presented. One is based on a series of statistical hypothesis tests; the other employs an information theoretic criterion (MDL). Unlike conventional AIC and MDL detection schemes, the new FSD detection schemes do not require knowledge of all the eigenvalues, and therefore can be carried out at each intermediate step of the FSD algorithm. More importantly, combining numerical analysis and multivariate statistical theory, both detection schemes are shown to be *strongly consistent*.**

I. Introduction

IN MANY RECENTLY DEVELOPED signal processing algorithms, extraction of a *signal subspace* from sensor array measurements is required. This subspace is commonly estimated by performing an eigendecomposition (ED) of the data covariance matrix. Since for large arrays the ED can be computationally complex, efficient techniques for estimating the signal subspace for large arrays are of importance in real-time applications.

Recently, a *fast*, *parallelizable* signal subspace decomposition technique (**FSD**) which achieves a significant reduction in computational complexity for large arrays with few sources was developed. Similar to the ED technique, **FSD** does not require knowledge of the signal subspace dimension *a priori*. The advantage of **FSD** is that the dimension can be estimated in the process of finding the signal subspace. Since previous detection schemes such as AIC and MDL use all the eigenvalues of the covariance matrix, they must be carried out after a full ED. Employing **FSD** and statistically equivalent detection statistics, the ED requirement is mitigated.

Herein the detection schemes are given and their asymptotic properties derived. New statistics formed by so-called Rayleigh-Ritz values at each intermediate step are shown to be asymptotically Chi-square distributed. This result enables the design of statistical hypothesis tests for estimating signal subspace dimension at each step. In addition, a new detection scheme, *i.e.*, **FSD-MDL**, is proposed, which relies on objective information theoretic criteria instead of subjective confidence levels as in the first detection scheme. Using numerical analysis and multivariate statistics, **FSD**-based detection schemes are shown to be *strongly consistent*, *i.e.*, estimation of the signal subspace dimension will be correct with probability one as the number of data samples tends to infinity.

[†]This work was supported in part by the Joint Services Program at Stanford University (US Army, US Navy, US Air Force) under Contract DAAL03-88-C-0011, and by grants from the General Electric Company and Boeing/ARGOSystems Inc.

II. Problem Statement

Consider the problem of estimation of source directions-of-arrival (DOAs) based on the data received from an array of sensors. If there are d noncoherent narrowband plane waves present, the data model for the array output can be written in an obvious notation as follows:

$$\boldsymbol{x}(t) = \mathbf{A}(\boldsymbol{\theta})\boldsymbol{s}(t) + \boldsymbol{n}(t), \tag{1}$$

where $\boldsymbol{\theta}$ is the d-vector of unknown DOAs, and the columns of the $M \times d$ matrix $\mathbf{A}(\boldsymbol{\theta})$ are the *steering or array manifold vectors.* The d-vector $\boldsymbol{s}(t)$ contains the the emitter signal waveforms, and $\boldsymbol{n}(t)$ denotes additive noise. The array output $\boldsymbol{x}(t)$ is observed at N discrete time instances, $t = 1, \ldots, N$, and the objective is to estimate the number of signals d, the DOAs $\boldsymbol{\theta}$, and/or signal waveforms $\boldsymbol{s}(t)$.

The signal component of the data vectors $\boldsymbol{x}(t) = \mathbf{A}(\boldsymbol{\theta})\boldsymbol{s}(t)$ lies in a d-dimensional subspace $\mathcal{R}\{\mathbf{A}(\boldsymbol{\theta})\}$ called the *signal subspace.* Assuming the noise vectors $\boldsymbol{n}(\cdot)$ are spatially white and uncorrelated with the signal, the ideal covariance matrix of $\boldsymbol{x}(t)$, $\mathcal{E}\{\boldsymbol{x}(t)\boldsymbol{x}^*(t)\}$, can be written as follows:

$$\mathbf{R}_{XX} = \mathbf{A}(\boldsymbol{\theta})\mathbf{R}_{SS}\mathbf{A}^*(\boldsymbol{\theta}) + \sigma_n^2 \boldsymbol{I} = \sum_{k=1}^{M} \lambda_k \mathbf{e}_k \mathbf{e}_k^*. \tag{2}$$

where $\mathbf{R}_{SS} \in \mathcal{C}^{d \times d}$ and $\sigma_n^2 \boldsymbol{I}$ are the signal and noise covariance matrices respectively, and standard notation for the ED of $\mathbf{R}_{XX}$ is used. Since $\mathbf{A}(\boldsymbol{\theta})$ is a rank-d matrix, $\lambda_1 \geq \cdots \geq \lambda_d \geq \lambda_{d+1} = \cdots = \lambda_M$, from which d is easily determined. The signal subspace is the column span of the d eigenvectors corresponding to the eigenvalues $\lambda_1, \ldots, \lambda_d$.

In practice, only a sample covariance $\widehat{\mathbf{R}}_{XX}$ estimated from a *finite* number (N) of snapshots is available, *i.e.*, $\widehat{\mathbf{R}}_{XX} = \frac{1}{N}\sum_{t=1}^{N} \boldsymbol{x}(t)\boldsymbol{x}^*(t)$. From this information alone, only an *estimate* of the signal subspace can be obtained. A common estimate is the first d eigenvectors

1058-6393/91 $01.00 © 1991 IEEE

$\{\hat{e}_1, \ldots, \hat{e}_d\}$ of $\widehat{\mathbf{R}}_{XX}$, an estimate easily shown to be the *unstructured* ML estimate of the true signal subspace. Naturally such an estimate is conditioned upon knowledge of d.

Often the dimension of the signal subspace must be estimated as well. Unlike the eigenvalues of the ideal covariance matrix, the $M-d$ smaller eigenvalues are not *exactly* equal. They are *clustered* in a small region of size $O(N^{-1/2})$ however, a property which is exploited for detection purposes. Detection schemes such as Rissanen's Minimum Description Length (MDL) criterion, Akaike's An Information Criterion (AIC), and the standard likelihood ratio tests (LR) are all based on ascertaining whether the $M-d$ smallest eigenvalues $\hat{\lambda}_{\hat{d}+1}, \ldots, \hat{\lambda}_M$ are sufficiently close to one another by testing arithmetic and geometric means for equality. Anderson [2] showed that

$$L(\hat{d}) = -N(M-\hat{d}) \log \left\{ \frac{\left(\prod_{k=\hat{d}+1}^{M} \hat{\lambda}_k\right)^{1/(M-\hat{d})}}{\frac{1}{M-\hat{d}} \sum_{k=\hat{d}+1}^{M} \hat{\lambda}_k} \right\}, \quad (3)$$

is a likelihood ratio satisfying for complex $\widehat{\mathbf{R}}_{XX}$

$$L(\hat{d}) = \begin{cases} O(N) & \hat{d} < d \\ O(1) \overset{N \to \infty}{\sim} \mathcal{X}^2((M-\hat{d})^2 - 1) & \hat{d} = d \end{cases} \quad (4)$$

where $\mathcal{X}^2(p)$ denotes the Chi-square distribution with p degrees of freedom. In fact, the significant change in $L(\hat{d})$ when $\hat{d} = d$ is the basis of these detection schemes. Knowing the limiting distribution of $L(\hat{d})$ when $\hat{d} = d$, a threshold can be set and compared with $L(\hat{d})$ to estimate d which results in the sequential likelihood ratio test. The more objective AIC and MDL information theoretic criteria are given by:

$$\text{AIC}(\hat{d}) = L(\hat{d}) + \hat{d}(2M - \hat{d}), \quad (5)$$

$$\text{MDL}(\hat{d}) = L(\hat{d}) + \frac{1}{2}\hat{d}(2M - \hat{d}) \log N. \quad (6)$$

The minimizers of the AIC and MDL criteria are chosen as estimates of d. MDL has been shown [3] to be strongly consistent, *i.e.*, the probability of correct detection approaches one as $N \to \infty$, while AIC is not.

III. FSD Detection

In [4], a new $O(M^2 d)$ fast subspace decomposition (**FSD**) approach for covariance matrices was developed using the Lanczos algorithm. Partial tridiagonalization of the matrix is performed using the Lanczos algorithm, followed by an ED of the resulting partial tridiagonal matrix $\boldsymbol{T}_j$ at the jth Lanczos step. The following theorem (proved in [5]) involving the eigenvalues of $\boldsymbol{T}_j$, *i.e.*, $\{\theta_k^{(j)}\}$, called Rayleigh-Ritz (RR) values, is the basis of the **FSD** technique.

Theorem 1 *Let $\hat{\lambda}_1 > \hat{\lambda}_2 > \cdots > \hat{\lambda}_M$ be the eigenvalues of the sample covariance matrix $\widehat{\mathbf{R}}_{XX}$ estimated based upon N data vectors from $N(0, \mathbf{R}_{XX})$ where $\mathbf{R}_{XX}$ is the covariance matrix defined in (2) with eigenvalues $\lambda_1 > \cdots > \lambda_d > \lambda_{d+1} = \cdots = \lambda_M = \sigma_n^2$ and eigenvectors $\mathbf{e}_1, \ldots, \mathbf{e}_M$. Denote the RR values by $\theta_1^{(j)} \geq \cdots \geq \theta_j^{(j)}$. Then for $k = 1, \ldots, d$ and $j \geq d+1$, the RR values $\theta_k^{(j)}$ approximate the corresponding eigenvalues $\hat{\lambda}_k$ to order $O(N^{-(j-d)})$, i.e., $\theta_k^{(j)} = \hat{\lambda}_k + O(N^{-(j-d)})$.*

Theorem 1 implies for $j \geq d+2$,

$$\theta_k^{(j)} - \hat{\lambda}_k = O(N^{-2}), \quad k = 1, \ldots, d, \quad (7)$$

$$\lim_{N \to \infty} N(\theta_k^{(j)} - \hat{\lambda}_k) = \lim_{N \to \infty} O(N^{-1}) = 0. \quad (8)$$

Thus, replacing $\hat{\lambda}_k$ by $\theta_k^{(j)}$ in (3) does not change the asymptotic properties of $L(\hat{d})$ (4). The likelihood ratio involves the geometric and arithmetic means of those $M - \hat{d}$ eigenvalues however. Since at the jth Lanczos step, at most j RR values are available, the following matrix identities

$$\text{Tr}(\widehat{\mathbf{R}}_{XX}) = \sum_{k=1}^{M} \hat{\lambda}_k, \quad \text{and} \quad det(\widehat{\mathbf{R}}_{XX}) = \prod_{k=1}^{M} \hat{\lambda}_k, \quad (9)$$

are employed to show

$$\sum_{k=d+1}^{M} \hat{\lambda}_k = \text{Tr}(\widehat{\mathbf{R}}_{XX}) - \sum_{k=1}^{d} \theta_k^{(j)} + O(N^{-(j-d)}) \quad (10)$$

$$\prod_{k=d+1}^{M} \hat{\lambda}_k = = det(\widehat{\mathbf{R}}_{XX}) / \prod_{k=1}^{d} \theta_k^{(j)} + O(N^{-(j-d)}). (11)$$

Thus, $L(\hat{d})$ (3) can be estimated using RR values.

Since the computation of $det(\cdot)$ generally requires $O(M^3)$ flops, it is significantly more expensive than the $O(M^2 d)$ Lanczos procedure. To mitigate this computational problem, a statistically equivalent statistic which can be evaluated in $O(M^2)$ flops is used. The geometric mean of the $M-d$ smallest eigenvalues is replaced by their quadratic mean $\left(\frac{1}{M-d} \sum_{k=d+1}^{M} \hat{\lambda}_k^2\right)^{1/2}$. Knowing that $\|\widehat{\mathbf{R}}_{XX}\|_F^2 = \sum_{k=1}^{M} \hat{\lambda}_k^2$,

$$\sum_{k=d+1}^{M} \hat{\lambda}_k^2 = \|\widehat{\mathbf{R}}_{XX}\|^2 - \sum_{k=1}^{d} {\theta_k^{(j)}}^2 + O(N^{-(j-d)}) \quad (12)$$

Now statistics $\varphi_{\hat{d}}$ for estimating d can be found as follows:

$$\varphi_{\hat{d}} = N(M-\hat{d})\log \frac{\sqrt{\frac{1}{M-\hat{d}}\left(\|\widehat{\mathbf{R}}_{XX}\|^2 - \sum_{k=1}^{\hat{d}} \theta_k^{(j)^2}\right)}}{\frac{1}{M-\hat{d}}\left(\mathrm{Tr}(\widehat{\mathbf{R}}_{XX}) - \sum_{k=1}^{\hat{d}} \theta_k^{(j)}\right)} \quad (13)$$

The following theorem gives the limiting distribution of $\varphi_{\hat{d}}$ under the null hypothesis $H_0 : \hat{d} = d$.

Theorem 2 *For $M > d+1$, the statistic $\varphi_{\hat{d}}$ defined in (13) is asymptotically Chi-squared ($\mathcal{X}^2$) distributed with $\frac{1}{2}(M-d)(M-d+1)-1$ degrees of freedom under $H_0 : \hat{d} = d$, if $\widehat{\mathbf{R}}_{XX}$ is real. If $\widehat{\mathbf{R}}_{XX}$ is complex, $2\varphi_{\hat{d}}$ is asymptotically $\mathcal{X}^2$ with $(M-d)^2-1$ degrees of freedom.*

Proof: The limiting distribution of a related statistic $\tilde{\varphi}_{\hat{d}}$ under H_0 given by

$$\tilde{\varphi}_d \triangleq \log\left(\sqrt{\frac{1}{M-d}\sum_{k=d+1}^{M} \hat{\lambda}_k^2} \Big/ \frac{1}{M-d}\sum_{k=d+1}^{M} \hat{\lambda}_k\right) \quad (14)$$

is derived first. After some algebraic manipulation,

$$\tilde{\varphi}_d = \frac{1}{2}\log\left(1 + \frac{\sum_{k=d+1}^{M}(\hat{\lambda}_k^2 - \sigma^2)}{(M-d)\sigma^2}\right) - \log\left(1 + \frac{\sum_{k=d+1}^{M}(\hat{\lambda}_k - \sigma)}{(M-d)\sigma}\right). \quad (15)$$

Using $\log(1+x) = x - x^2/2 + x^3/3\ldots$, $\hat{\lambda}_k - \sigma = O(N^{-1/2})$ and $\hat{\lambda}_k^2 - \sigma^2 = O(N^{-1/2})$ for $d+1 \le k \le M$, and neglecting $O(N^{-3/2})$ terms, (15) becomes

$$\lim_{N\to\infty}(M-d)N\tilde{\varphi}_d = \lim_{N\to\infty}\frac{N}{2\sigma^2}\left\{\sum_{k=d+1}^{M}(\hat{\lambda}_k-\sigma)^2 - \frac{1}{M-d}\left(\sum_{k=d+1}^{M}(\hat{\lambda}_k-\sigma)\right)^2\right\} \quad (16)$$

The RHS of (16) is asymptotically $\mathcal{X}^2(\frac{1}{2}(M-d)(M-d+1)-1)$ in the real case, and $2(M-d)N\tilde{\varphi}_d$ is $\mathcal{X}^2((M-d)^2-1)$ in the complex case [2, 6]. ∎

The previous theorem leads directly to the proposed detection scheme using the statistic $\varphi_{\hat{d}}$. The choice of the predetermined sequence of numbers $c(N)$ is discussed in the next section. The algorithm is initialized with $j = 1$.

The FSD Detection Scheme

1. Carry out the jth Lanczos step.
2. Set $\hat{d} = 1$.
3. Set null hypothesis $H_0 : d = \hat{d}$.
4. Choose a threshold $\gamma_{\hat{d}} c(N)$ based on the tail area of the $\mathcal{X}^2$ distribution with DOF $\frac{1}{2}(M-\hat{d})(M-\hat{d}+1)-1$ (real $\widehat{\mathbf{R}}_{XX}$) or $(M-\hat{d})^2-1$ (complex $\widehat{\mathbf{R}}_{XX}$).
5. Calculate $\varphi_{\hat{d}}$; if $\varphi_{\hat{d}} \le \gamma_{\hat{d}} c(N)$ accept H_0 and stop.
6. If $\varphi_{\hat{d}} > \gamma_{\hat{d}} c(N)$, reject H_0; if $\hat{d} < j-2$, $\hat{d} := \hat{d}+1$, return to 3. Otherwise, set $j := j+1$ and go to 1.

IV. Strong Consistency

As aforementioned, it is possible to select $c(N)$ so that the detection scheme yields a *strongly consistent* estimate of the signal subspace dimension d. The consistency proof requires the following preliminary results.

Lemma 1 *For the sample covariance matrix $\widehat{\mathbf{R}}_{XX}$ and its eigenvalues $\hat{\lambda}_k$, the following relations hold w.p.1,*

$$\widehat{\mathbf{R}}_{XX} - \mathbf{R}_{XX} = O(\sqrt{\log\log N/N}) \quad (17)$$

$$\hat{\lambda}_k - \lambda_k = O(\sqrt{\log\log N/N}) \quad k = 1,\ldots,M \quad (18)$$

The proof of (17) can be found in [7], while (18) is proved in [8]. From (18), $\hat{\lambda}_k - \sigma = O(\sqrt{\log\log N/N})$. Hence, w.p.1.

$$\epsilon = \hat{\lambda}_{d+1} - \hat{\lambda}_M = (\hat{\lambda}_{d+1}-\sigma) - (\hat{\lambda}_M - \sigma) = O(\sqrt{\log\log N/N}), \quad (19)$$

and strong consistency of the detection scheme is proved as follows.

Theorem 3 *For the statistic $\varphi_{\hat{d}}$ defined in (13), the detection scheme of the previous section is strongly consistent if $c(N)$ satisfies the following conditions:*

$$\lim_{N\to\infty}\frac{c(N)}{N} = 0, \quad \textit{and} \quad \lim_{N\to\infty}\frac{c(N)}{\log\log N} = \infty. \quad (20)$$

Proof: Proving strong consistency is equivalent to demonstrating:

(i) $\lim_{N\to\infty} \varphi_{\hat{d}}/c(N) > \gamma_{\hat{d}}$, *i.e.*, H_0: $\hat{d} = 1, 2, \ldots, d-1$ are rejected w.p.1 as $N \to \infty$.

(ii) $\lim_{N\to\infty} \varphi_{\hat{d}}/c(N) \le \gamma_{\hat{d}}$, *i.e.*, H_0: $\hat{d} = d$ is accepted w.p.1 as $N \to \infty$.

To show (i), assume $\hat{d} < d$ and $j \ge d+1$. From Theorem 1 and Lemma 1, $\theta_k^{(j)} = \lambda_k + O(\epsilon) = \lambda_k + O(\sqrt{\log\log N/N})$ w.p.1. Hence,

$$\tilde{\varphi}_{\hat{d}} \overset{N\to\infty}{\longrightarrow} \log\left(\frac{\sqrt{\frac{1}{M-\hat{d}}\sum_{k=\hat{d}+1}^{M}\lambda_k^2}}{\frac{1}{M-\hat{d}}\sum_{k=\hat{d}+1}^{M}\lambda_k}\right) \quad \text{w.p.1} \quad (21)$$

where $\tilde{\varphi}_{\hat{d}}$ is defined in (14). Since in this case $\hat{d} < d$, $\lambda_{\hat{d}+1}, \ldots, \lambda_M$ are *not* all identical, and the quadratic mean is always *strictly* larger than the arithmetic mean. Hence, $\tilde{\varphi}_{\hat{d}} > 0$ as $N \to \infty$ w.p.1 and by (20),

$$\lim_{N\to\infty} \frac{\varphi_{\hat{d}}}{c(N)} = \lim_{N\to\infty} \frac{N}{c(N)} \cdot (M - \hat{d})\tilde{\varphi}_{\hat{d}} = \infty \qquad \text{w.p.1}$$

For $j \leq d$, the RR values $\theta_k^{(j)}$, $1 \leq k \leq j$ are well-defined functions of the initial Lanczos vector $\boldsymbol{f}$ and $\mathbf{R}_{XX}$ as $N \to \infty$. If $\boldsymbol{f}$ is selected at random, the probability that $\boldsymbol{f}$ makes

$$\log \left(\frac{\sqrt{\frac{1}{M-\hat{d}} \left(\|\mathbf{R}_{XX}\|^2 - \sum_{k=1}^{\hat{d}} \theta_k^{(j)^2} \right)}}{\frac{1}{M-\hat{d}} \left(\operatorname{Tr}(\mathbf{R}_{XX}) - \sum_{k=1}^{\hat{d}} \theta_k^{(j)} \right)} \right) \tag{22}$$

exactly zero is zero. Therefore, (22) is non-zero w.p.1 and H_0 is rejected with probability one.

To show (ii), assume $\hat{d} = d$ and $j \geq d+1$. From (16) and the fact that $\theta_k^{(j)} - \lambda_k = O(\epsilon^{-2(j-d)}) = O(\log\log N/N)$ and $\hat{\lambda}_k - \lambda_k = O(\sqrt{\log\log N/N})$ w.p.1, for $1 \leq k \leq d$ $\varphi_d = O(\log\log N)$

$$\lim_{N\to\infty} \frac{\varphi_d}{c(N)} = \lim_{N\to\infty} \frac{O(\log\log N)}{c(N)} = 0, \quad \text{w.p.1} \tag{23}$$

which means for any positive number γ_d, $\varphi_d/c(N) < \gamma_d$ almost surely as $N \to \infty$. ■

Sequences $c(N)$ which satisfy the conditions in (20) are not difficult to find; $\log N$ is a good candidate. The unique advantage of the **FSD** detection scheme is that it does not require knowledge of all the eigenvalues, rather the RR values of the $(\hat{d}+2)^{th}$ Lanczos step. Also *only* RR values (not RR vectors) need to be computed at each intermediate Lanczos step. Since the eigenvalues of a jxj tridiagonal matrix can be found in $9j^2$ flops, the detection scheme requires $O(j^2)$ flops at the jth Lanczos step leading to a total computational complexity of $O(d^3)$ plus $O(M^2 d)$ plus $M^2/2$ flops for computing $\|\widehat{\mathbf{R}}_{XX}\|^2$.

V. FSD-MDL Detection Scheme

The **FSD** detection scheme of the previous section, though shown to be strongly consistent, is still a subjective criterion; many choices for $c(N)$ and the confidence level are possible. In the spirit of the MDL criterion, and based on the fact that φ_d and $L(d)$ have the same limiting distribution, *i.e.*, $\mathcal{X}^2((M-d)^2-1)$, a new **FSD**-MDL detection criterion is created by replacing $L(\hat{d})$ with $\varphi_{\hat{d}}$ in (6). The detailed **FSD**-MDL procedure is:

The FSD-MDL Detection Scheme

1. Set $\hat{d} = 0$.
2. Set null hypothesis $H_0 : d = \hat{d}$.
3. Choose the threshold $\gamma_{\hat{d}}$ based on the tail area of the $\mathcal{X}^2((M-\hat{d})^2 - 1)$ and evaluate $\varphi_{\hat{d}}$.
4. If $\varphi_{\hat{d}} > \gamma_{\hat{d}}$, reject H_0; if $\hat{d} < m-2$, $\hat{d} := \hat{d}+1$, return to 2. Otherwise, return to the Lanczos algorithm.
5. If $\varphi_{\hat{d}} \leq \gamma_{\hat{d}}$, evaluate the partial MDL sequence, *i.e.*, $MDL(k) = \varphi_k + k(M-k/2)\log(N)$, $k = 0, 1, \ldots, \hat{d}$. Estimate d as the minimizer of $MDL(k)$.

Comparing the two detection schemes, they differ only in **FSD**-MDL does not require $c(N)$, and once $\varphi_{\hat{d}} \leq \gamma_{\hat{d}}$, the MDL criterion is evaluated to determine d. Naturally this is not the original MDL criterion, but it is nonetheless *strongly consistent*.

Theorem 4 *The* **FSD**-*MDL detection scheme is strongly consistent.*

Proof: From the proof of Theorem 3, it is not difficult to see that the proof of (i) still holds since now $c(N) = 1$ and $N/c(N) = N \to \infty$. Thus, $\varphi_{\hat{d}} > \gamma_{\hat{d}}$ w.p.1 as $N \to \infty$ for $\hat{d} < d$ and d is never underestimated.

Since $c(N)/\log\log N \to \infty$ does not hold, the proof of (ii) must be modified. Suppose that $\gamma_{\hat{d}}$ is chosen based on the confidence level α. Then the probability of over-estimating d is asymptotically $1-\alpha$ (usually a small, but always positive number). Suppose $\overline{d}(\geq d)$ is the first one such that $\varphi_{\overline{d}} \leq \gamma_{\overline{d}}$. Since this implies $j \geq \hat{d}+2 \geq d+2$, the d largest RR values are asymptotically equivalent to the corresponding eigenvalues of $\widehat{\mathbf{R}}_{XX}$. Therefore, w.p.1

$$\varphi_{\hat{d}} = N(M-\hat{d})\log \frac{\sqrt{\frac{1}{M-\hat{d}} \sum_{k=\hat{d}+1}^{M} \hat{\lambda}_k^2}}{\frac{1}{M-\hat{d}} \sum_{k=\hat{d}+1}^{M} \hat{\lambda}_k} = \begin{cases} O(N), & \hat{d} < d \\ O(\log\log N) & \hat{d} = d \end{cases}$$

From the proof of Theorem 3, for $\hat{d} = d+1, \ldots, \overline{d}$, w.p.1

$$\varphi_{\hat{d}} = N(M-\hat{d})\log \frac{\sqrt{\frac{1}{M-\hat{d}} \sum_{k=\hat{d}+1}^{M} \hat{\lambda}_k^2 + \sum_{k=d+1}^{\hat{d}} (\theta_k^{(j)^2} - \hat{\lambda}_k^2)}}{\frac{1}{M-\hat{d}} \sum_{k=\hat{d}+1}^{M} \hat{\lambda}_k + \sum_{k=d+1}^{\hat{d}} (\theta_k^{(j)} - \hat{\lambda}_k)}$$

Using the fact that $\theta_k^{(j)} - \hat{\lambda}_k = O(N^{-1/2})$ for $k \geq d+1$ (*cf.* [5]), $\varphi_{\hat{d}} = O(N^{-1/2})$. Thus,

$$\text{MDL}(\hat{d}) = \varphi_{\hat{d}} + \hat{d}(M - \hat{d}/2)\log N = \begin{cases} O(N), & \hat{d} < d \\ O(\log N), & \hat{d} = d \\ O(N^{1/2}), & d < \hat{d} \leq \bar{d}. \end{cases}$$

Therefore, *asymptotically* ($N \to \infty$) the minimizer of $\text{MDL}(\hat{d})$ is d w.p.1. ■

VI. Simulation Results

To indicate the validity of the strong consistency and the robustness of the detection schemes to array errors, extensive simulations were conducted, a few of the results of which are presented here.

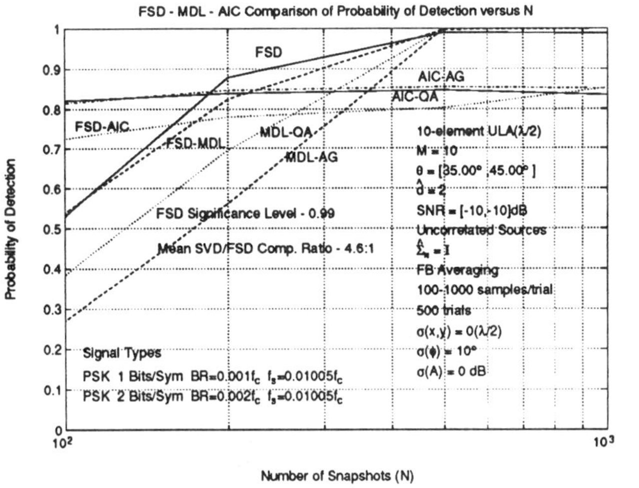

Figure 1: Comparison of Probability of Detection Using Various Techniques in the Presence of Sensor Phase Errors ($\sigma(\phi) = 10°$)

The scenario used for these investigations included two LPI PSK signals at 35° and 45° with a common SNR of $-10\ dB$. The signal at 35° was a BPSK source with $BR = 0.001f_c$ and that at 45° was a QPSK source with $BR = 0.002f_c$. A 10-element $\lambda/2$ ULA was used. Sensor phase errors introduced were assumed to be angle-dependent Gaussian random variables with a standard deviation of 10° which governed the expected variation in phase among Monte Carlo trials. That is, sensor phases were fixed for each trial, but were changed according to a Gaussian distribution from trial to trial. This is important to keep in mind when interpreting the results presented herein. A distinctly different set of results and conclusions arise from setting the gains and phases once (from a given distribution) for all trials.

As indicated in Figure 1, **FSD** detection asymptotically achieves 99% detection, while **FSD**-MDL along with MDL achieves 100%. Perhaps somewhat surprising is the relative insensitivity to sensor phase variations as indicated in Figure 2 where DOA estimation errors are shown along with various theoretical bounds assuming no phase errors.

VII. Conclusions and Remarks

Herein, strong consistency of two **FSD**-based detection schemes has been shown. In addition to having the same asymptotic strong consistency as MDL, the low information-to-noise ratio performance of **FSD**-MDL

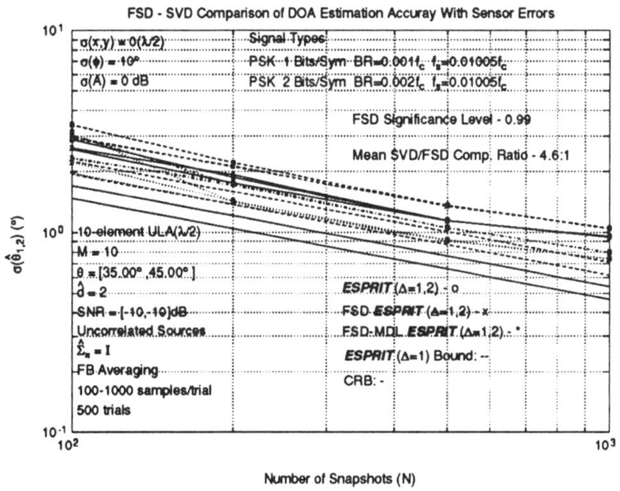

Figure 2: Comparison of DOA Estimation Error Using Various Techniques in the Presence of Sensor Phase Errors ($\sigma(\phi) = 10°$)

seems to be superior to MDL as evidenced by numerous simulation results only one of which was presented here. For scenarios where $M \gg d$, the computational savings of **FSD**-based techniques is significant. Based on the numerous simulations conducted to date, the **FSD** detection capability and **FSD**-based ***ESPRIT*** have proven to be remarkably insensitive to array imperfections.

References

[1] G. Bienvenu and L. Kopp, "Optimality of High Resolution Array Processing", *IEEE Trans. ASSP*, **ASSP-31**(5):953–964, October 1983.

[2] T. W. Anderson, Asymptotic theory for principal component analysis, *Ann. Math. Statist.*, **34**:122–148, 1963.

[3] M. Wax and T. Kailath, "Detection of Signals by Information Theoretic Criteria", *IEEE Trans. ASSP*, **ASSP-33**(2):387–392, April 1985.

[4] G. Xu, R. Roy, and T. Kailath, Fast ***ESPRIT*** – application of fast signal subspace decomposition in array processing, In *24th Asilomar Conference on Signals, Systems, and Computers*, Pacific Grove, CA, November 1990, accepted.

[5] G. Xu and T. Kailath, Fast signal subspace decomposition – part ii: Sample covariance matrices, *submitted to IEEE Trans. ASSP December 1990.*

[6] R.P. Gupta, "Asymptotic Theory for Principal Component Analysis in the Complex Case", *Journal of the Indian Stat. Assoc.*, **3**:97–106, 1965.

[7] K.V. Mardia, J.T. Kent, and J.M. Bibby, *Multivariate Analysis*, Academic Press, London, UK, 1979.

[8] L.C. Zhao, P.R. Krishnaiah, and Z.D. Bai, "On Detection of the Number of Signals in Presence of White Noise", *Journal of Multivariate Analysis*, **20**(1):1–25, October 1986.

On Target Manoeuvres in Bearings Only Tracking

J. Holst

Department of Mathematical Statistics
Lund Institute of Technology
Box 118, S-221 00 Lund
Sweden

A. Holtsberg

Department of Mathematical Statistics
Lund Institute of Technology
Box 118, S-221 00 Lund
Sweden

Abstract

Manoeuvre detection is an essential part of target motion analysis but represents a deflection from the basic assumptions on which standard TMA algorithms are built. In this paper the target is assumed at a random instant to switch between two linear motions by changing its course. The proposed manoeuvre handling algorithm detects the change, estimates the time of change and recomputes the target motion estimates up to detection time. The algorithm performance is illustrated in some representative TMA scenarios.

1 Introduction

Bearing Only Tracking algorithms try to estimate the position, velocity and course of a target using bearing measurements to the target and knowledge of own position, but in absence of measurements of the distance to the target. Tracking algorithms that aim at solving the estimation problem are based on the assumption that the target has constant course and velocity for a period of time. Furthermore in order to render the problem observable, the movement of the platform itself has to deviate from linear motion with constant velocity. In reality targets often make manoeuvres, and trackers therefore must be equipped with a facility that allow them to detect and correct for a target manoeuvre.

In a passive long range sonar system a target manoeuvre can be considered to take place almost at a single time point, compared to the time scale of the problem. In such a system a straightforward way to solve the problem is to run the tracker until a target manoeuvre is signaled by a detector, then estimate the manoeuvre time and character and finally restart the tracker at the estimated manoeuvre time using the extra information that a manoeuvre took place. It is then of great importance that the detector gives fast and accurate detection of the target manoeuvre.

In section 2 we discuss the prediction errors obtained when following a manoeuvring target. A target manoeuvre detection algorithm is proposed in section 3 together with a manoeuvre time estimator. In section 4 some suggestions are made about how to restart a Kalman filter type tracker after manoeuvre detection and in section 5 simulations show the performance of such a tracker.

2 The prediction errors

The detector and the manoeuvre time estimator are based on a sequence of prediction errors, i.e. differences between the observed measurements and the one step ahead predictors of the bearings obtained e.g. from a Kalman tracker. It could be noted that residuals are less valuable in this context.

In the passive sonar situation a target manoeuvre where the target changes course but keeps the same velocity is most common. Furthermore there may be other means for detecting a change of velocity than only using bearings measurements. Therefore we assume that the target moves with constant velocity.

To illustrate the different type of behaviour of the prediction error sequence that in principle may occur just after a manoeuvre we consider the scenarios in figure 1. There are three targets in the geographical overview shown in the left part of the picture, each makes a 45 degree manoeuvre at time $t = 50$. The platform and the targets move with constant speed.

1058-6393/91 $01.00 © 1991 IEEE

The right pictures show the expected prediction errors if the prediction before the manoeuvre is exactly correct and the prediction after the manoeuvre is computed using the erroneous assumption that the targets did not make the manoeuvres.

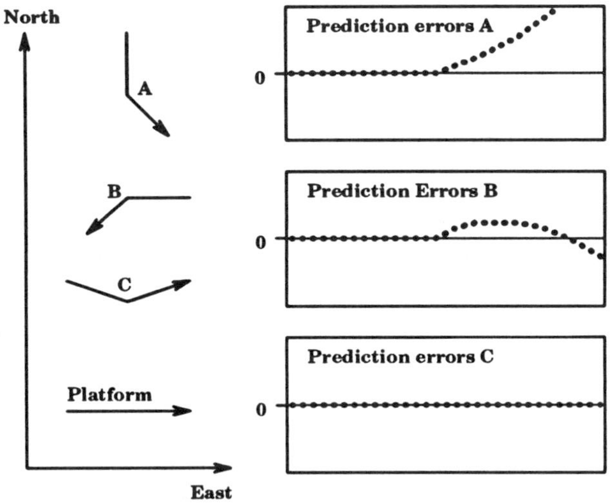

Figure 1: The different types of behaviour that the prediction error sequence may show after a target manoeuvre are given in the right part of the picture. The left part gives a geographical overview.

Target A is a typical example, where the mean of the prediction errors drifts away from zero approximately linearly. For target B the mean of the bearing prediction error sequence might be described as growing with a second order tendency. Furthermore, it changes sign after the manoeuvre, and it can be shown that the bearing prediction error sequence has at most one sign shift when observed from a nonaccelerating platform. The prediction error sequence obtained for certain symmetric target manoeuvres, as in the target C case, may not render the manoeuvre visible at all from the platform unless the platform makes a manoeuvre itself after the target manoeuvre. This shows that there might be situations where the systematic changes in the prediction error sequence easily can be masked by noise. In all cases this description of the bearing prediction error is limited to the time period just after the manoeuvre.

3 An algoritm to detect a manoeuvre

The algorithm for manoeuvre detection that will be discussed is based upon the sequential probability ratio test originally proposed by Wald (1947), [6] and mechanized in a CUSUM (CUmulative SUM) detector. The aim of the test is to detect a shift of the mean value in sequence of random varibles. CUSUM detectors are studied e.g. Dobben de Bruyn (1968) [2], Lucas (1976) [4], and Ryan (1989) [5].

Hence, assume that we have a sequence of random variables $\{e(t)\}$, which are assumed to be independent, Gaussian with known variance, σ_e. We also assume that the mean value of the sequence is zero under basic conditions, i.e. before any manoeuvre.

When the mean value of the variables changes, also the drift of the cumulative sum changes. The basic idea is to analyze the cumulative sum to see whether its performance is in accordance with the null hypothesis or not. In order to account for the size of the mean value shift that the detector should be able to handle with a reasonable response time, a negative correction is introduced by subtracting a value k from every $e(t)$. In general k is chosen as

$$k = K \cdot |\text{alternative mean value}|/\sigma_e$$

and a standard value is $K = 0.5$, i.e. k is chosen as half the smallest mean value change for which we want the detector to work decently fast, the variability taken into account. A detection level h for the random walk is used to signal the mean value shift, and we believe that the mean value has changed, if the cumulative sum becomes greater than h. The value of h reflects roughly the amount of false alarms that may be accepted. Smaller h gives faster detection, at the expense of higher risk for false alarms. Systematic ways of choosing h and k are discussed e.g. in [2] and [5].

If the cumulative sum becomes smaller than 0, the procedure is restarted by resetting it to 0.

This algorithm can be expressed as follows. Use a detector signal $S(t)$ and start with $S(0) = 0$. For every new measurement, update the detector signal as

$$S(t) = max(0, S(t-1) + e(t)/\sigma_e - k)$$
$$\text{alarm if } S(t) > h.$$

If $S(t)$ is greater than h, we believe that the mean value of $e(t)$ has become greater than 0 which signals that a manoeuvre probably has occured. To detect for negative means, an identical filter is run on $\{-e(t)\}$.

However, the target manoeuvre detection problem is slightly different from the standard CUSUM applications. Firstly, after a target manoeuvre the mean value does not shift to a new constant value, but grows with time. This implies a more careful tuning of the parameters in the detector. Secondly, the bearing measurements might have systematic errors that depend on the bearing, which can be difficult to distinguish from a prediction bias due to target manoeuvres. The detector must therefore be robust for small mean deviations. This can be achieved by choosing both k and h large enough.

The performance of a CUSUM detector used for manoeuvre detection in the scenarios A and B in figure 1 is shown in figure 2. The measurement noise has standard deviation 0.8 degrees. The error sequences were rescaled to variance 1 before entered into the CUSUM-detector. The parameter k was chosen to 0.8 in the simulations.

Target A

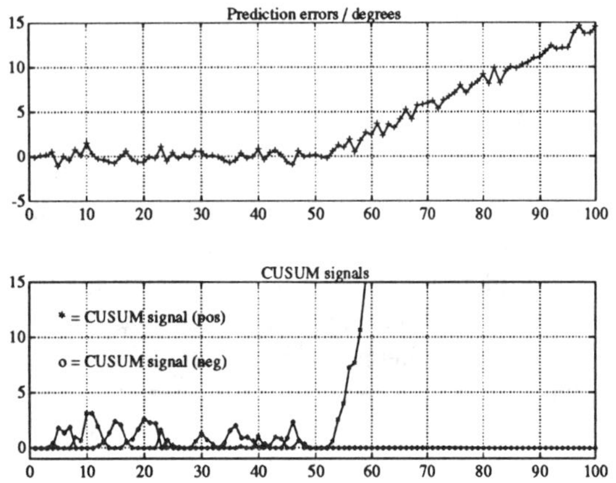

Target B

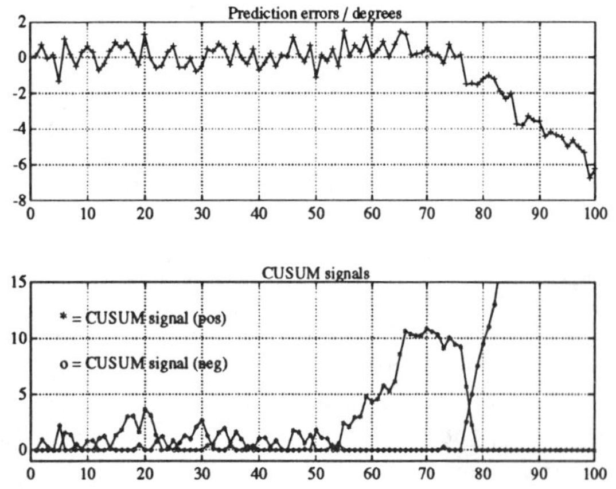

Figure 2: A CUSUM detector used for detecting of the manoeuvres in the scenarios A and B in figure 1. The measurement errors had standard deviation 0.8°. The solid line corresponds to the positive and the dashed line to the negative CUSUM detector signals, respectively.

4 An estimator of the manoeuvre time

When a manoeuvre has been detected, the manoeuvre time t_m has to be estimated. A straightforward way to do this is to use the starting time for that CUSUM sequence with constant sign that eventually alarmed, as the manoeuvre time estimate, i.e. simply to use a counter. However, due to the possible sign shift in the prediction error sequence this technique might give manouevre time estimates that are too late.

This section will present an alternative technique built upon the observation that the mean of the prediction error sequence will drift away from 0 possibly with at sign shift after a target manoeuvre, cf section 2. Assume that the mean value of the prediction error sequence can be described by a second order polynomial during a reasonably long period after the manoeuvre, i.e.

$$E(e(t)) = \alpha_1[t - t_m] + \alpha_2[t - t_m]^2 \qquad t > t_m.$$

The maximum likelihood estimation of α_1,α_2 and t_m is in this situation identical to least squares estimation since the sequence $\{e(t)\}$ is assumed to be Gaussian white noise. Assume that the manoeuvre occurs during a single sampling interval. That gives a discrete space to search for t_m. Furthermore, assume that t_m is greater than a lower limit t_l. These assumptions give a reasonably small number of time points to search.

Let the sample points be integers for ease of notation. Denote the detection time t_d. The least squares estimators for t_m, α_1 and α_2, called $\hat{t}_m$, $\hat{\alpha}_1$ and $\hat{\alpha}_2$, minimize the expression

$$r(t_m, \alpha_1, \alpha_2) = \sum_{i=t_l}^{t_m} e(i)^2 + \sum_{i=t_m+1}^{t_d} (e(i) - \alpha_1[i - t_m] - \alpha_2[i - t_m]^2)^2.$$

Define $r_c(t_m)$, as the minimum residual sum of squares for a given t_m, i.e. fix t_m and minimize $r(t_m, \alpha_1, \alpha_2)$ as a function of the two other parameters. Then to find the parameter values that minimize $r_c(t_m)$ for a given t_m is easy. The amount of computations needed when computing $r_c(t_m)$ for all values of t_m could however be reduced by organizing the computations recursively, i.e. to compute $r_c(t_m - 1)$ given $r_c(t_m)$ was computed before. This algorithm is discussed in Holtsberg and Holst (1989), [3].

The discrimination between a first or a second order polynomial expansion of the prediction error mean value function is accomplished via standard model validation methods like AIC or F-test.

This method for manoeuvre time estimation is used in the simulations shown below. It circumvents the problem of too late manoeuvre time estimates.

5 Restarting the tracker

Since the estimated target motion for times between the manoeuvre time t_m and the detection time t_d is incorrect, the state estimation must be restarted at time t_m and the target motion reestimated up to t_d. This means that the tracking algorithm must be equipped with a bank of previous bearing measurements, state estimates and covariance matrices from the Kalman filter, as well as platform positions. The state may be represented in a cartesian coordinate system, or e.g. by Modified Polar Coordinates, see Aidala and Hammel (1983), [1]. The amount of stored data is a design parameter for the algorithm.

It is often reasonable to believe that the target has the same velocity after the manoeuvre as before, i.e. only the course has been changed. The unknown new course leads to an increased uncertainty about the components of the velocity vector after the manoeuvre. This implies that there is a need of increasing those parts of the covariance matrix at t_m that are connected to the velocity. In case the estimates of t_m or at t_m are very imprecise, it might be necessary also to increase the position error parts of the estimation error covariance matrix.

If, on the other hand, all estimates at t_m and of t_m are precise, it might be worthwhile also to consider that there in fact are just two distinct new target course alternatives after the detection, taking the bearing at the detection time t_d into account and assuming unchanged velocity, cf. [3]. Each of the new target course alternatives demands a tracking filter of its own, and the incorrect alternative is eventually ruled out, e.g. by CUSUM detection. The increase of the error covariance matrix at detection time could be tuned to suit the new course alternatives.

6 Simulation and discussion

The performance of the manoeuvre handling algorithm is illustrate by the simulations in figure 3. Target starts with the course 230 degrees. At time $t = 70$ it makes a manoeuvre and its course is decreased by 40 degrees. The true movements of the platform and the target are seen in the upper left picture, showing that the target and the platform moves approximately towards each other.

The tracker is a linearized Kalman filter in Modified Polar Coordinates, and we use the sampling interval as the time unit. The measurement noise standard deviation is 0.3^o. The initial estimates of the two velocity coordinates are zero.

When the target makes a manoeuvre at time $t = 70$ it can be seen that the CUSUM-signal immediately respond. In other scenarios the mean value of the prediction error sequence might change sign before the manoeuvre was detected, and the CUSUM-signal that then finally alarmed for a course change was not the one that at first made a longer departure from zero.

The error sequence that was entered into the detector was scaled to variance 1. The k-value in the CUSUM-detector was chosen to 0.5 and the detection limit 12. The manoeuvre time was estimated by both of the algorithms discussed above. In this case they gave the same manoeuvre time estimate.

The new course estimate at manoeuvre time was taken as the value before the manoeuvre, and the velocity part of the error covariance matrix was reset to its initial value. Simulations where the algorithm that takes the two possible alternative new target courses into consideration usually results in a better precision of the resulting estimates. The procedure used here is however more robust against erroneous estimates.

Acknowledgements

This research has been done in cooperations with the Swedish Defence Material Administration and Kockums AB. The cooperation and support is greatfully acknowledged.

References

[1] Aidala, V.J. and S. E. Hammel (1987): "Utilization of Modified Polar Coordinates for Bearings-Only Tracking". *IEEE Trans. Automat. Contr.*, Vol. AC-28, pp. 283-294.

[2] van Dobben de Bruyn, C.S. (1968): *Cumulative Sum Tests, Theory and Practice.* Griffin's Statistical Monographs and Courses, No 24.

[3] Holtsberg, A. and J. Holst (1989): *On target manoeuvres in bearings only tracking.* TFMS-3057,

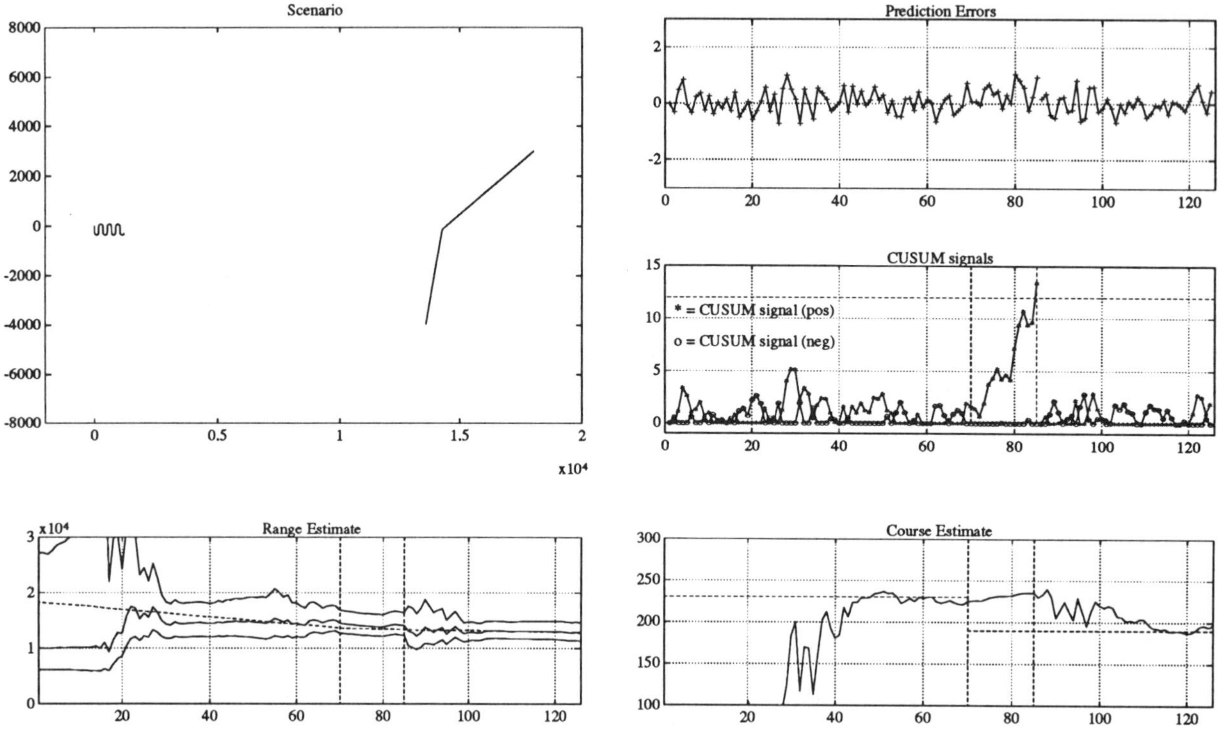

Figure 3: The upper left picture gives an overview of the scenario. The target changes its course from 230 to 190 degrees at time $t = 70$. The CUSUM detector alarms very quickly for a mean value shift just after the correct value of the manoeuvre time. The vertical dotted lines mark the correct manoeuvre time and the detection time respectively.

Dept of Mathematical Statistics, Lund Institute of Technology, Lund, Sweden.

[4] Lucas, J. (1976): "The Design and Use of V-Mask Control Schemes". *Journal of Quality Technology,* no 8, pp. 1-12.

[5] Ryan, T.P. (1989): *Statistical Methods for Quality Improvement.* Wiley, New York.

[6] Wald, A. (1947): *Sequential Analysis.* Wiley, New York.

Updating MUSIC and Root–MUSIC with The Rank-Revealing URV Decomposition

M. F. Griffin

Advanced Digital Systems
United Technologies Research Center
E. Hartford, CT 06108

G. W. Stewart

Department of Computer Science
University of Maryland
College Park, MD 20742

Abstract

In many practical direction-of-arrival (DOA) problems the number of sources and their directions from an antenna array do not remain stationary over many snapshots. Hence a practical DOA algorithm must be able to track changes with a minimal number of snapshots. In this paper we describe DOA algorithms that possess the virtues of subspace-based DOA algorithms but are not expensive to compute or difficult to update. The new DOA algorithms are based on a new decomposition, called the URV decomposition, which can be updated in $O(N^2)$ and serves as an intermediary between the QR decomposition and the singular value decomposition.

1 Introduction

Let X_m be a complex data matrix consisting of m data snapshot vectors x_l from an N-element antenna array. An estimate of the autocovariance matrix at the mth snapshot can be written $C_m = X_m X_m^H$ or alternatively

$$C_m = C_{m-1} + x_m x_m^H = (X_{m-1}, x_m) \begin{pmatrix} X_{m-1}^H \\ x_m^H \end{pmatrix},$$

The signal subspace approach to high resolution direction-of-arrival (DOA) estimation usually requires a spectral decomposition of C_m or a singular value decomposition (SVD) of X_m, from which the orthogonal (noise) subspace can be estimated and used in a frequency estimation algorithm such as MUSIC or root-MUSIC.

Since the *ab initio* calculation of any decomposition is expensive, it is desirable to calculate the orthogonal subspace of X_m from X_{m-1}, a process called updating. Unfortunately, the SVD is difficult to update and all known SVD updating schemes require $O(N^3)$ operations for a $m \times N$ matrix [1]. Because of this difficulty, rank revealing QR decompositions [2] have received renewed interest. However, the QR decomposition does not provide an explicit basis for the orthogonal subspace. Recently, a new rank revealing decomposition, the URV decomposition (URVD) [3], has been developed and applied [4]. It can be updated in $O(N^2)$ time and provides a basis for the orthogonal subspace. The purpose of this paper is to extend our original algorithm to include root-MUSIC, essentially for statistical performance evaluation and not as a fast algorithm. Note that root-MUSIC as a computational algorithm involves finding the roots of a polynomial, thus destroying the advantage of the URVD.

Since the URVD has applications beyond DOA estimation, we will adopt a neutral notation and work with an $n \times p$ matrix A. Owing to space limitations, we can give only a sketch of the decomposition and the updating algorithm. The reader is referred to [3] for details.[1] Throughout this summary $\|\cdot\|$ will denote the Frobenius norm and the smallest singular value of a matrix will be written as $\inf(A)$.

2 The URV Decomposition

Suppose that A has rank k. Then there are orthogonal matrices U and V such that

$$A = U \begin{pmatrix} R & 0 \\ 0 & 0 \end{pmatrix} V^H,$$

where R is an upper triangular matrix of order k. This decomposition is called the URV decomposition. Unlike the SVD, the URVD is not unique; in fact, the SVD is itself a URVD. However, we will be concerned with the case where R is fully triangular.

[1] This report may be obtained by anonymous ftp from thales.cs.umd.edu. Get the file `pub/reports/README` for further details.

1058-6393/91 $01.00 © 1991 IEEE

Now suppose A is nearly of rank k in the sense that its singular values satisfy $\sigma_1 \geq \cdots \geq \sigma_k > \sigma_{k+1} \geq \cdots \geq \sigma_p$, where σ_k is large compared to σ_{k+1}. It can be shown that there is a rank revealing URVD of A of the form

$$A = U \begin{pmatrix} R & F \\ 0 & G \end{pmatrix} V^H,$$

where

1. R and G are upper triangular,
2. $\inf(R) \cong \sigma_k$,
3. $\sqrt{||F||^2 + ||G||^2} \cong \sqrt{\sigma_{k+1}^2 + \cdots \sigma_p^2}$.

3 Deflation and Refinement

For a dynamic DOA problem it is expected that the number of sources, i.e., rank of the data matrix, or A, will change over time. An increase in rank usually makes itself felt in the usual way. On the other hand, a decrease in rank can hide itself in the matrix R of the URVD. Thus any updating algorithm must be able to detect rank degeneracy in R and act accordingly. An algorithm to compute a rank revealing URVD of a $k \times k$ upper triangular matrix R is given next.

The first step is to determine if R is defective in rank. This problem can be solved by means of a condition estimator [5], which produce in $O(k^2)$ time a vector w of norm one such that

$$||Rw|| \cong \inf(R).$$

The next step uses plane rotations [6] to reduce all but the last component of w to zero. Specifically, we determine a sequence $V_1^H, V_2^H, \ldots, V_{K-1}^H$, of rotations that eliminate the first $K-1$ component of w, so that w is zero except for one in its last component. Let $Q^H = V_{K-1}^H V_{K-2}^H \cdots V_1^H$ denote the product of the rotations.

Next determine an orthogonal matrix P such that $P^H RQ$ is upper triangular. This may be done by applying $V_1, V_2, \ldots, V_{K-1}$ from the right to R. The result of applying a rotation V_i to two columns of R is to place a nonzero element below the diagonal of R. A left rotation then eliminates this element and restores triangularity. The matrix P^H is the product of the left rotations. Of course, to get a complete update of the decomposition the left and right rotations must be multiplied into U and V, respectively. However, for the DOA problem updating U is not needed, and therefore can be omitted.

It can be shown that at this point the norm of the last column of R is bounded by the norm of w, so that R is in URV form. However, an additional refinement step will improve the estimate of the orthogonal subspace. The step consists of using rotations to reduce the first $p-1$ elements of the last column to zero and then to reduce R back to triangular form. Again the rotations must be multiplied into U and V.

4 Updating The URVD

In this section an algorithm is sketched to update a rank revealing URVD of A when a row z^H is appended to A; i.e., when A is replaced by

$$\begin{pmatrix} A \\ z^H \end{pmatrix}. \tag{1}$$

The updating procedure determines if the rank has either increased, decreased, or remained the same. To decide what is small, assume a user supplied tolerance, tol, and that

$$\nu \stackrel{\text{def}}{=} \sqrt{||F||^2 + ||G||^2} \leq \text{tol}. \tag{2}$$

The first step is to compute $(x^H y^H) = z^H V$, where x is of dimension k. The problem then becomes one of updating

$$\hat{A} = \begin{pmatrix} R & F \\ 0 & G \\ x^H & y^H \end{pmatrix}.$$

There are two cases to consider. The first, and simplest occurs when

$$\nu_{\text{new}} = \sqrt{\nu^2 + ||y||^2} \leq \text{tol}. \tag{3}$$

In this case $\hat{A}$ is reduced to triangular form by a sequence of rotations applied from the left. Since ν_{new} is less than or equal to the prescribed tolerance, the new URVD satisfies (2), and the approximate rank does not increase. However, it is possible for the rank to decrease. Hence the rank is checked. If it has decreased, R is deflated. This step is $O(p^2)$.

If (3) is not satisfied, there is a possibility that the rank has increased. Since the increase in rank can be at most one, the problem is to transform the matrix to upper triangular form without destroying all the small values in F and G.

The first step is to reduce y^H so that it has only one nonzero component and G remains upper triangular. This is done by a sequence of rotations applied alternately from the right and the left — the right rotations to zero an element of y^H and the left rotations to maintain the matrix in upper triangular form. Finally, the entire matrix is reduced to triangular form in the usual way. Then k is increased by one, and the new R is checked for degeneracy and if necessary deflated as described before. The result is the updated URVD.

5 Updating The DOA

Let us return to the problem of obtaining a high resolution DOA estimate, i.e., MUSIC and root-MUSIC, by updating the covariance matrix C_m on a per snapshot basis.

The signals consist of the standard M narrowband sources impinging on a uniform linear array composed of N ($M < N$) identical, equally spaced, sensors. The narrowband signals with known angular frequency ω impinge on the array from directions, $\theta_1, \theta_2, \ldots, \theta_M$. In the interest of space the reader is referred to [7] for additional details on the signal model and signal covariance matrix.

As mentioned, we may obtain the estimate of the orthogonal subspace required for DOA estimation from the covariance matrix or from the data matrix. In the later case, the problem becomes one of computing a DOA estimate from the updated data matrix

$$\hat{X}_m^H = \begin{pmatrix} \beta X_{m-1}^H \\ x_m^H \end{pmatrix}, \qquad (4)$$

where $\beta < 1$ is "forgetting factor." Comparing (4) with (1), we see that the problem of updating the orthogonal subspace is one of updating a URVD. Note that because $\beta < 1$, the triangular factor of the URVD remains bounded as long as the signals remain bounded. Thus a tolerance can be chosen based on the size of the noise (for one such choice see [3]).

The first step is to update $\hat{X}_m^H$ using the procedure described in the last section. Since the matrix U is not needed for MUSIC or root-MUSIC algorithms, it is not necessary to update U in the first step. After V and k are updated, we compute MUSIC by

$$S_m(\theta) = a^H(\theta) V_o(m) V_o^H(m) a(\theta),$$

where

$$a^H(\theta) = (1 \;\; e^{-j\omega} \;\; e^{-j2\omega} \;\; \ldots \;\; e^{-j(p-1)\omega}),$$

$\omega = 2\pi d sin\theta/\lambda$, and $V_o(m)$ contain the singular vectors of the orthogonal subspace ($V = [V_s \; V_o]$) for the mth snapshot. The spacing between array sensors is d, and λ is the wavelength of the propagating wave.

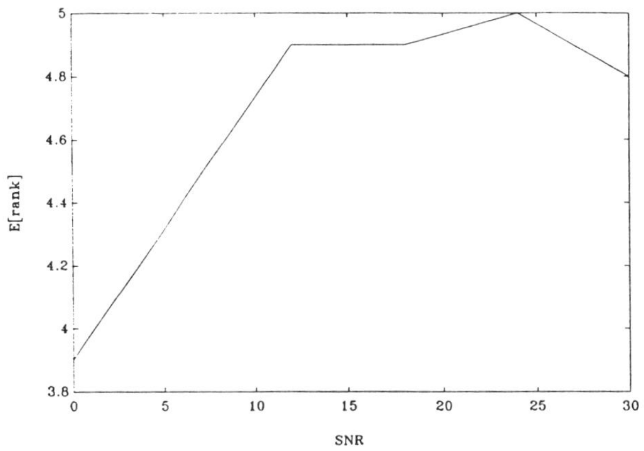

Figure 1. Rank Estimate of 4 Sources.

The M angles θ_i at which the function $S_m(\theta)$ goes to zero correspond to the DOA estimates. Root-MUSIC on the other hand searches for roots of the polynomial

$$P_m(z) = a^T(z^{-1}) V_o(m) V_o^H(m) a(z),$$

where $z = e^{j\omega}$. The direction estimates are those M largest amplitudes of the roots inside the unit circle. These steps are repeated for each snapshot. Of course, β and the tolerance must be chosen beforehand.

6 Some Simulation Results

The data for the simulation consist of four uncorrelated sources impinging on a 10-element array from -15°, 0°, 10°, and 20°. Unless otherwise noted the forgetting factor β was chosen to be 0.79, representing a 10 snapshot effective window. The tolerance was equal to 2.5. The narrowband source frequency was set to 0.2 and $\lambda/d = 2$. Each of the four sources assumed an uniformly distributed random phase term from $-\pi$ to π. Moreover, white Gaussian noise of varying variance and uncorrelated with the sources was added to the data snapshots. Unless otherwise noted the number of updates was 25, i.e., 25 total snapshots. Several experiments were conducted for various signal strengths, i.e., signal-to-noise ratio (SNR), but we shall only show some of our results.

First lets take a look at the ensemble estimate of rank. Figure 1, is the mean rank for 6 values of SNRs, computed by taking the rank at the 25th iteration, 'converged rank estimate', and averaging over 500 records. The rank estimate remained within ± 1

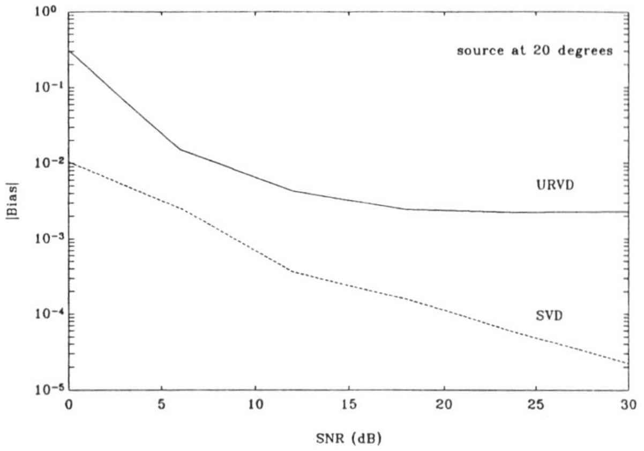

Figure 2. SVD and URVD Bias.

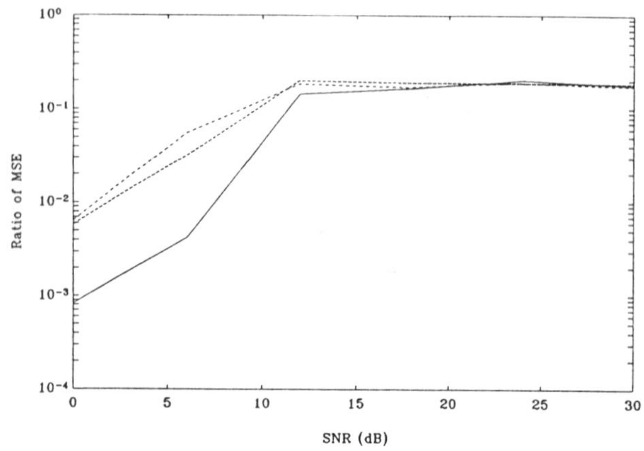

Figure 3. Ratio of SVD and URVD MSE.

of the true rank. Changing tolerance and β will cause the standard deviation of rank to increase, suggesting an optimum value of β and tolerance for a given SNR. Next we compare the bias of the source at 20 degrees for both URVD and SVD based algorithms. What we mean by SVD algorithm is that the orthogonal subspace used in root-MUSIC was computed from the SVD–with known rank of four–of the estimated covariance matrix, computed from all 25 snapshots of the data set. Figure 2 shows the absolute value of the bias for 6 different SNRs. As expected the bias from the updating algorithm is more since we have an effective window over the data set. Again the bias computation was based on the converged estimate of the source angle. Along with bias is the mean square error (MSE) shown as a ratio of the MSE for SVD over MSE for URVD methods in Figure 3. Here the ratio of the converged MSE for each source is computed and plotted together. Though the URVD MSE tends towards the SVD algorithm MSE, beyond 12 dB the ratio remains fixed around 0.19.

In addition to showing the behavior of the MSE versus SNR, Figure 4 illustrates six MSE curves, six different SNRs, as a function of update for the source at 0°. As the SNR increases it becomes clear that the number of updates to achieve convergence is 5-6. The corresponding bias estimates are shown in Figure 5. The bias decreases for high SNR but there is much more statistical variation; here we need to increase the number of records in the ensemble.

We now show how the MSE varies with changing values of β and tolerance. To simplify the experiment we fix the SNR, here at 10 dB, and fix one of two

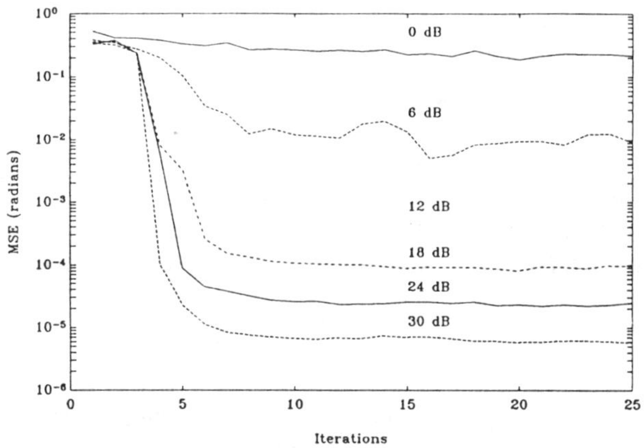

Figure 4. URVD MSE versus SNR and Updates.

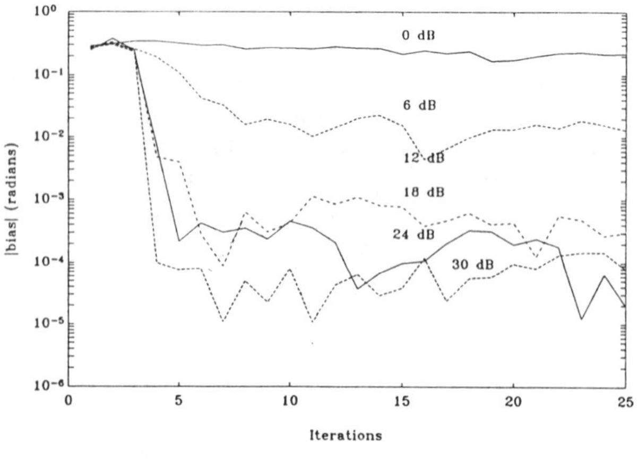

Figure 5. URVD Bias versus SNR and Updates.

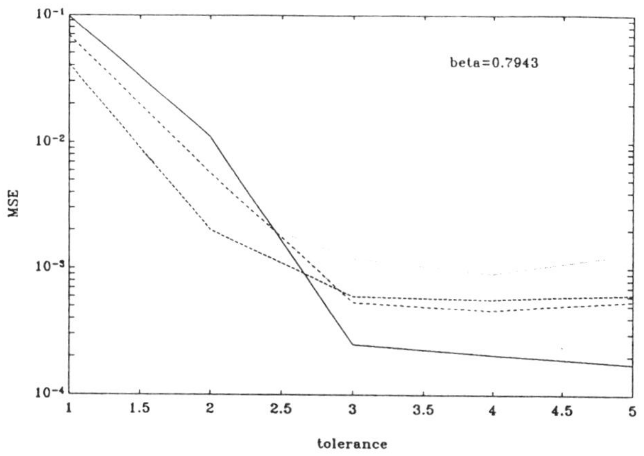

Figure 6. URVD MSE versus Tolerance.

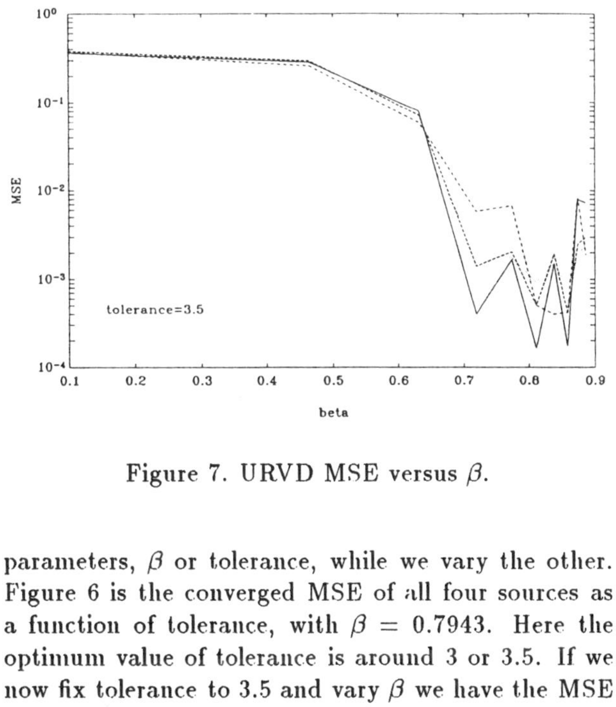

Figure 7. URVD MSE versus β.

parameters, β or tolerance, while we vary the other. Figure 6 is the converged MSE of all four sources as a function of tolerance, with $\beta = 0.7943$. Here the optimum value of tolerance is around 3 or 3.5. If we now fix tolerance to 3.5 and vary β we have the MSE plotted in Figure 7. β, for four sources, between 0.8 and 0.85 is close to optimal. Of course to correctly find the best β and tolerance for a given SNR we need to perform a multidimensional search.

7 Summary and Conclusions

In this paper we have outlined a new matrix decomposition, the URVD, for updating subspace-based DOA algorithms, specifically MUSIC. We have shown, briefly, how to update the URVD and how it may be used for MUSIC. In addition to MUSIC we describe how to update it's variant root-MUSIC. We noted, however, that root-MUSIC was only used here as a means for statistical performance evaluation. We then presented several simulation results in terms of MSE and bias of the direction estimates.

Owing to space limitations the results in this paper represent only the static behavior of the updating algorithm and not its dynamic behavior, which will be presented elsewhere. As clearly seen from the results, there exists a degree of error in using the URVD updating technique over updating a full SVD or spectral decomposition. This error is offset by the reduction in computations offered by the URVD.

References

[1] J. R. Bunch and C. P. Nielson, "Updating The Singular Value Decomposition," *Numerishce Mathematik*, 31, pp. 111–129, 1978.

[2] T. F. Chan, "Rank Revealing QR Factorizations," *Linear Algebra and Its Applications*, 88/89, pp. 67–82, 1987.

[3] G. W. Stewart, "An Updating Algorithm For Subspace Tracking," *University of Maryland Computer Science Technical Report CS-TR 2494*, 1990.

[4] G. Adams, M. F. Griffin, and G. W. Stewart, "Direction-of-Arrival Estimation Using The Rank-Revealing URV Decomposition," *Proc. ICASSP-91*, Toronto, Canada, 1991.

[5] N. J. Higham, "A Survey of Condition Number Estimation for Triangular Matrices," *Siam Review*, 29, pp. 575–596, 1987.

[6] G. H. Golub and C. F. Van Loan, *Matrix Computations*, Johns Hopkins University Press, Baltimore Maryland, 1989.

[7] S. Haykin, ed., *Array Signal Processing*, Prentice-Hall, Englewood Cliffs, New Jersey, 1984.

Signal Subspace Sphericity and Inter-Signal Coherence — *MUSIC* Induces Inter-Signal Coherence†

Darel A. Linebarger and Ronald D. DeGroat
Center for Engineering Mathematics
The University of Texas at Dallas, MP33
Richardson, Texas 75083–0688 USA
(214) 690–2991 dl@utdallas.edu

Abstract

New insights on the role and importance of signal subspace sphericity are presented in this paper. First, using signal subspace arguments with the true correlation matrix, we show that MUSIC effectively induces inter-signal coherence due to its requirement for subspace sphericity. We also show in the finite data case that some commonly used direction finding methods yield results identical to those of MUSIC if subspace sphericity is induced. Lastly, we demonstrate that sphericity induced coherence can be desirable. This is accomplished with an infinitely averaged example using spatial smoothing to reduce the coherence. However, we find that reducing the sphericity induced coherence has a detrimental effect on resolution.

Introduction

In this paper, we will be studying a relationship between signal subspace sphericity and inter-signal coherence. Before introducing our results, we briefly describe some necessary background information. We make only the standard narrowband assumptions concerning the array and sensor outputs; we assume the sensor outputs are modeled by

$$x_m(k) = n_m(k) + \sum_{i=1}^{P} s_{im}(k) \tag{1}$$

where x_m is the output of the mth sensor, s_{im} is the ith signal at the mth sensor, P is the number of signals, and n_m is a noise process at sensor m. We represent the array output at time k with the vector $\mathbf{x}(k)$. We assume there are M sensors, so $\mathbf{x}$ is $M \times 1$. The spatial correlation matrix, a fundamental quantity used by most direction finding techniques, is defined by

$$R = E\left[\mathbf{x}(k)\mathbf{x}(k)^H\right] \tag{2}$$

where E is the expectation operator. The ij^{th} element of R is the correlation between the outputs of the ith and j^{th} sensors. The signals are assumed to be uncorrelated from the noise and the noise is assumed to be uncorrelated from sensor to sensor, which implies that R is a sum of two terms:

$$R = \sigma_n^2 I + SCS^H. \tag{3}$$

The ij^{th} element of the $P \times P$ matrix C represents the inter-signal coherence between the ith and j^{th} signals,

$$C_{ij} = \frac{E\left[\sigma_i \sigma_j^*\right]}{\overline{\sigma}_i \overline{\sigma}_j} \tag{4}$$

where σ_i is a random complex amplitude associated with the ith signal. (The signals, although modeled as narrowband, are allowed a random complex amplitude.) The quantity $\overline{\sigma}_i$ is defined by $\overline{\sigma}_i^2 = E\left[|\sigma_i|^2\right]$. If all the signals are completely incoherent, $C = I$. Throughout most of this paper, we assume the signals are incoherent. Our assumptions of incoherent signals and sensor noise imply that R has the following form:

$$R = \sigma_n^2 I + SS^H. \tag{5}$$

If the correlation matrix is described by eq. (5), then its eigenvectors can be partitioned into those that span a "signal+noise" subspace and those that span a "noise only" subspace:

$$\begin{aligned} R &= V \Lambda V^H \\ &= [V_s V_n] \begin{bmatrix} \Lambda_s & 0 \\ 0 & \sigma_n^2 I_{M-P} \end{bmatrix} [V_s V_n]^H \\ &= V_s \Lambda_s V_s^H + \sigma_n^2 V_n V_n^H \end{aligned} \tag{6}$$

† This work was supported by Texas Advanced Technology Grant 009741-038, Texas Instruments, and Cray Research, Inc.

1058-6393/91 $01.00 © 1991 IEEE

where the diagonal matrix Λ contains the eigenvalues of R ordered from largest to smallest, $\lambda_1 \geq \lambda_2 \geq \cdots \geq \lambda_P > \lambda_{P+1} = \lambda_{P+2} = \cdots = \lambda_M = \sigma_n^2 \geq 0$, and the columns of the orthogonal matrix V are the eigenvectors of R. I_{M-P} is an $(M-P) \times (M-P)$ identity matrix. The $M \times P$ matrix V_s contains the eigenvectors that span the same space as the columns of S (the "signal+noise" space) and the $M \times (M-P)$ matrix V_n contains vectors spanning the orthogonal complement of the span of S (the "noise only" space).

Our notion of subspace sphericity can be described in terms of the eigenstructure of the correlation matrix. If we consider R as described in eq. (6), we note that the eigenvalues of the noise subspace are equal. Thus, we say that noise subspace is spherical. We define $R_{\odot}$ as an infinitely averaged correlation matrix with an induced spherical signal subspace. Thus, if R is defined by eq. (6), we define $R_{\odot}$ as

$$R_{\odot} = (\sigma_s^2 + \sigma_n^2) V_s V_s^H + \sigma_n^2 V_n V_n^H \tag{7}$$

where $\sigma_s^2 > 0$ is an arbitrary positive constant that can be thought of as representing the combined or average signal power. (The actual value of σ_s^2 does not affect the following development.)

MUSIC [1] is usually formulated in terms of the noise subspace. Using the "spectral form" of *MUSIC*, the direction estimates are obtained by solving the following optimization problem:

$$\widehat{\theta}_{MUSIC} = \min_{\theta} \quad \mathbf{e}_\theta^H V_n V_n^H \mathbf{e}_\theta^H \tag{8}$$

where $\mathbf{e}_\theta$ is a steering vector for the direction specified by θ. It has been noted [2] that there is a signal subspace formulation of the *MUSIC* algorithm. We also focus on the signal subspace implementation of *MUSIC*. As $V_n V_n^H = I - V_s V_s^H$, *MUSIC* can be formulated in terms of the signal subspace:

$$\widehat{\theta}_{MUSIC} = \max_{\theta} \quad \mathbf{e}_\theta^H V_s V_s^H \mathbf{e}_\theta^H \tag{9}$$

In this form, we seek the steering vector $\mathbf{e}_\theta$ that has maximum projection into the signal subspace; $V_s V_s^H$ is a projection matrix. By contrast, the noise subspace version of *MUSIC* finds the steering vector that has minimum projection onto the noise subspace. It is interesting to note that both of these implementations of *MUSIC* make use of spherical representations of the signal and noise subspaces, but only the noise subspace is originally spherical. The sphericity in the signal subspace version is induced.

In the following, we first demonstrate that *MUSIC* effectively induces coherence due to its implicit requirement for sphericity of the signal subspace. Then we show that if conventional direction finding methods are used after inducing this same coherence, they will have exactly the same performance as *MUSIC*. Lastly, we show that if a spatial averaging technique is used to reduce the sphericity induced coherence, the performance of the direction finding technique is worsened. In fact, signals that are resolved using the spherical signal subspace may not be resolved if the induced coherence is reduced too much. This result is perhaps counter-intuitive as much work has been devoted to reducing inter-signal coherence under the assumption that it is resolution inhibiting. However, the coherence we will be discussing is definitely resolution enhancing.

Subspace Sphericity Induces Coherence

In the following, we demonstrate that the sphericity imposed by the spectral version of *MUSIC* can be thought of as inducing a particular variety of inter-signal coherence. If we consider the signal subspace version of *MUSIC* (as described in eq. (9)), we note that $V_s V_s^H$ is a projection matrix that can also be written in terms of S, the matrix of signal direction vectors:

$$V_s V_s^H = S(S^H S)^{-1} S^H. \tag{10}$$

If we regarded $(S^H S)^{-1}$ as an inter-signal coherence matrix, $S(S^H S)^{-1} S^H$ could be regarded as a no-noise version of the original correlation matrix, except with an induced coherence. This coherence was induced by forcing signal subspace sphericity. The only modification required to consider $(S^H S)^{-1}$ an inter-signal coherence matrix is to normalize its diagonal elements. So, if we write

$$V_s V_s^H = SDD^{-1}(S^H S)^{-1} D^{-1} DS^H. \tag{11}$$

where the ith element of the diagonal matrix D is just the square root of the ith element on the main diagonal of $(S^H S)^{-1}$, we have

$$V_s V_s^H = S_{\odot} C_{\odot} S_{\odot}^H \tag{12}$$

where $C_{\odot} = D^{-1}(S^H S)^{-1} D^{-1}$ is a valid inter-signal coherence matrix; it is $P \times P$ and positive definite. Also, the only difference between S and $S_{\odot} = SD$ is that the amplitudes of the signals have been modified.

<u>*Example*</u>: If we consider two equi-powered signals ($\mathbf{s}_1^H \mathbf{s}_1 = \mathbf{s}_2^H \mathbf{s}_2 = E_s$) and explicitly calculate

$$C_{\odot} = D^{-1}\left([\mathbf{s}_1\ \mathbf{s}_2]^H [\mathbf{s}_1\ \mathbf{s}_2]\right)^{-1} D^{-1} = \begin{bmatrix} 1 & c \\ c^* & 1 \end{bmatrix}, \tag{13}$$

we see that the induced inter-signal coherence is given by

$$c = -\frac{\mathbf{s}_1^H \mathbf{s}_2}{E_s}. \qquad (14)$$

Thus, more closely spaced signals are more highly coherent after the signal subspace is made spherical.

Sphericity Improves Resolution

We saw in the previous section that if we induce signal subspace sphericity, it is as if we have introduced inter-signal coherence. In this section, we consider correlation matrices corresponding to the same conditions as described in the introduction, but obtained from finite data. In this case, neither the noise or the signal subspaces are spherical in general. With finite data, we assume that a correlation matrix estimate $\widehat{R}$ is available and that $\widehat{R}_{\odot}$ is defined by

$$\widehat{R}_{\odot} = (\sigma_s^2 + \sigma_n^2)\widehat{V}_s\widehat{V}_s^H + \sigma_n^2\widehat{V}_n\widehat{V}_n^H \qquad (15)$$

where $\widehat{V}_s$ and $\widehat{V}_n$ represent the estimated signal and noise subspaces, respectively, obtained from the eigendecomposition of $\widehat{R}$. Similarly, σ_s^2 and σ_n^2 are arbitrary positive constants that are related to the signal and noise powers. Using these definitions, we state the following theorem:

Theorem: Define conventional direction finding by

$$\widehat{\theta}_{CONV} = \max_{\theta} \quad \mathbf{e}_\theta^H \widehat{R} \mathbf{e}_\theta \qquad (16)$$

and the minimum variance distortionless response (*MVDR*, Capon's method) direction estimator by

$$\widehat{\theta}_{MVDR} = \min_{\theta} \quad \mathbf{e}_\theta^H \widehat{R}^{-1} \mathbf{e}_\theta. \qquad (17)$$

Then, if $\widehat{R}_{\odot}$ is used in place of $\widehat{R}$ with either conventional direction finding or *MVDR*, bearing estimates identical to those obtained with *MUSIC* result.

Proof: First, we address conventional direction finding. In the folllowing proof, we use the following relationship: $V_s V_s^H + V_n V_n^H = I$.

$$\begin{aligned}
\widehat{\theta}_{\odot,CONV} &= \max_{\theta} \quad \mathbf{e}_\theta^H \widehat{R}_{\odot} \mathbf{e}_\theta \\
&= \max_{\theta} \quad \mathbf{e}_\theta^H \left((\sigma_s^2 + \sigma_n^2)\widehat{V}_s\widehat{V}_s^H + \sigma_n^2\widehat{V}_n\widehat{V}_n^H \right) \mathbf{e}_\theta \\
&= \max_{\theta} \quad \mathbf{e}_\theta^H \left((\sigma_s^2 + \sigma_n^2)\widehat{V}_s\widehat{V}_s^H + \sigma_n^2\widehat{V}_n\widehat{V}_n^H - \sigma_n^2 I \right) \mathbf{e}_\theta \\
&= \max_{\theta} \ \mathbf{e}_\theta^H \left(\sigma_s^2 \widehat{V}_s\widehat{V}_s^H \right) \mathbf{e}_\theta \\
&= \widehat{\theta}_{MUSIC}
\end{aligned} \qquad (18)$$

Similarly for *MVDR*,

$$\begin{aligned}
\widehat{\theta}_{\odot,MVDR} &= \min_{\theta} \ \mathbf{e}_\theta^H \widehat{R}_{\odot}^{-1} \mathbf{e}_\theta \\
&= \min_{\theta} \ \mathbf{e}_\theta^H \left(\frac{1}{\sigma_n^2}\widehat{V}_n\widehat{V}_n^H + \frac{1}{\sigma_s^2 + \sigma_n^2}\widehat{V}_s\widehat{V}_s^H \right) \mathbf{e}_\theta \\
&= \min_{\theta} \ \mathbf{e}_\theta^H \left(\left(\frac{1}{\sigma_n^2} - \frac{1}{\sigma_s^2 + \sigma_n^2} \right) \widehat{V}_n\widehat{V}_n^H + \frac{1}{\sigma_s^2 + \sigma_n^2} I \right) \mathbf{e}_\theta \\
&= \min_{\theta} \ \mathbf{e}_\theta^H \left(\left(\frac{1}{\sigma_n^2} - \frac{1}{\sigma_s^2 + \sigma_n^2} \right) \widehat{V}_n\widehat{V}_n^H \right) \mathbf{e}_\theta \\
&= \widehat{\theta}_{MUSIC}
\end{aligned} \qquad (19)$$

■

The equivalence between these implementations when $\widehat{R}_{\odot}$ is substituted for $\widehat{R}$ is illustrated with an example in Figure 1. The correlation matrix was estimated in the usual fashion ($\widehat{R} = \frac{1}{K}\sum_{k=1}^{K} \mathbf{x}_k \mathbf{x}_k^H$) from 100 ($K = 100$) snapshots. For the noise subspace implementation of *MUSIC*, we have plotted the reciprocal of the expression in eq. (8). Clearly, the sphericalized *MVDR* and conventional methods produce the same direction estimates as *MUSIC*. They must, because when $\widehat{R}_{\odot}$ is substituted for $\widehat{R}$, they are both *mathematically equivalent* to *MUSIC* from a direction finding point of view. (Note: we are not saying that the "spectral" plots of these three methods are the same, only that the appropriate maxima or minima occur at the same places for a given set of data.)

Sphericity and Spatial Smoothing

We have observed that forcing spherical subspaces allows lower performance methods (conventional and *MVDR*) to perform identically to the higher performance *MUSIC* method. Yet, we have also shown that forcing sphericity is equivalent to inducing inter-signal coherence. That inter-signal coherence might lead to increased resolution is contrary to the conclusions of previous work. In work such as [3], coherence was assumed to be undesirable and techniques such as subaperture averaging were discussed due to their ability to reduce inter-signal coherence. We are not arguing that coherence is good in general. However, we do want to emphasize the special relationship that exists between sphericity and coherence and the desirability of sphericity.

Inter-signal coherence is known to cause difficulties for adaptive methods like *MVDR* [4] or eigenstructure

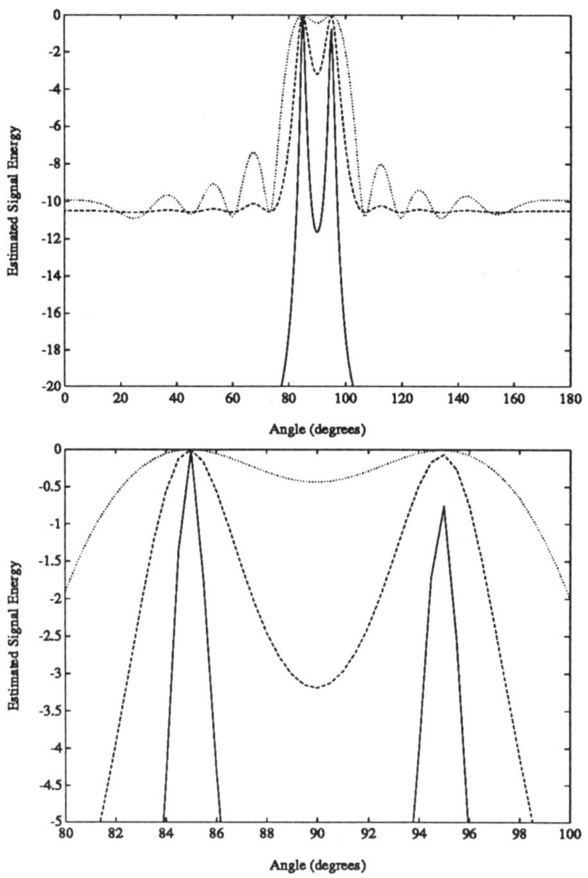

Figures 1a and 1b: In these figures, we illustrate the equivalence between conventional direction finding, MVDR, and MUSIC, for direction finding when $\widehat{R}_{\odot}$ is used in place of $\widehat{R}$. In this example, the correlation matrix is estimated from 100 snapshots taken with a 10 element linear equally spaced array with half-wavelength spacing. The signals are located at plus or minus 5 degrees from broadside. Broadside is represented as 90 degrees. The signal to noise ratio is 0 dB for each source. The solid line corresponds to MUSIC, the dashed line corresponds to MVDR with $\widehat{R}_{\odot}$, and the dotted line corresponds to conventional direction finding with $\widehat{R}_{\odot}$. In panel a), it is hard to see that the signals have been accurately located using the spherical version of the correlation matrix. Thus, in panel b), we zoom in on the peaks, making it apparent that the peak locations are the same for all three methods.

based methods such as *MUSIC* [1] and *EV* [5]. In the extreme case of completely coherent signals, the rank of the matrix C is diminished. If there are two signals, and they are completely coherent, C has rank one and the signal subspace is collapsed. If the SCS^H term in the spatial correlation matrix does not have rank P, partitioning the eigenvectors into those representing signal and noise subspaces is impossible. This is a well known problem, and in the case of linear equally spaced arrays, can be dealt with using subaperture averaging [3]. Subaperture averaging forms a new smaller spatial correlation matrix by averaging overlapping submatrices along the diagonal of the original matrix R. This operation is equivalent to averaging the correlation matrices obtained from overlapping subarrays of the original array. It has been proven that this operation will restore the rank of the signal correlation matrix [6], and thus the signal and noise subspaces can be correctly partitioned. It has also been proven that though the inter-signal coherence is always reduced, the signals that were originally coherent remain coherent after subaperture averaging [7]. As subaperture averaging always reduces inter-signal coherence, and we have proven that the inter-signal coherence induced by signal subspace sphericity is necessary for improved (asymptotically unbiased) resolution, we claim that subaperture averaging is most beneficial when it causes the signal subspace to be more nearly spherical. An obvious example of this is the completely coherent case where the rank of the signal subspace is restored and therefore made more spherical.

We demonstrate the desirability of a spherical signal subspace over reduced inter-signal incoherence in the following example. We assume that the true correlation matrix is known and has no inter-signal coherence until sphericity is induced. The technique used for demonstration purposes is *MVDR*. Due to the close spacing, the signals are unresolved using *MVDR* with the original correlation matrix. (See Fig. 2.) However, after forcing a spherical signal subspace, i.e. replacing R with $R_{\odot}$, the signals are resolved and located accurately. However, we have introduced inter-signal coherence. In order to show that this induced coherence is desirable, we apply subaperture averaging to $R_{\odot}$ in order to reduce the induced coherence. The array used in this example is linear equally spaced with 10 sensors and half-wavelength spacing. When we subaperture average with two 9 element subarrays, the signals remain resolved, but the signal direction estimates are now inaccurate. Also, the signal subspace is no longer spherical. If we subaperture average with three 8 element subarrays, the signals are no longer resolved and the signal subspace is even less spherical. More subaperture averaging implies less inter-signal coherence. Clearly, this example shows that signal subspace sphericity and the inter-signal coherence associated with it can be desirable. It also supports our claim that signal subspace sphericity is a more fundamental characteristic than coherence.

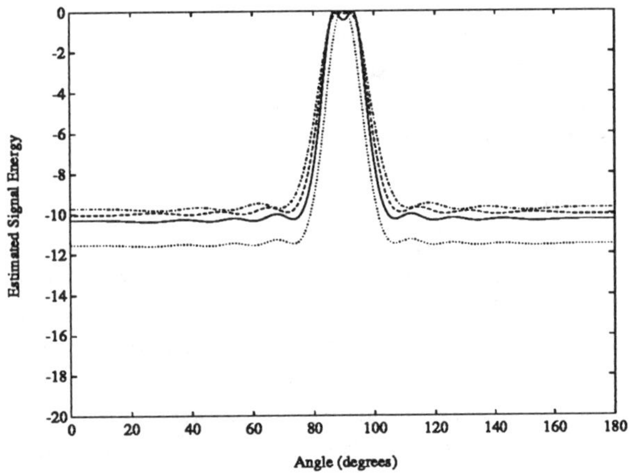

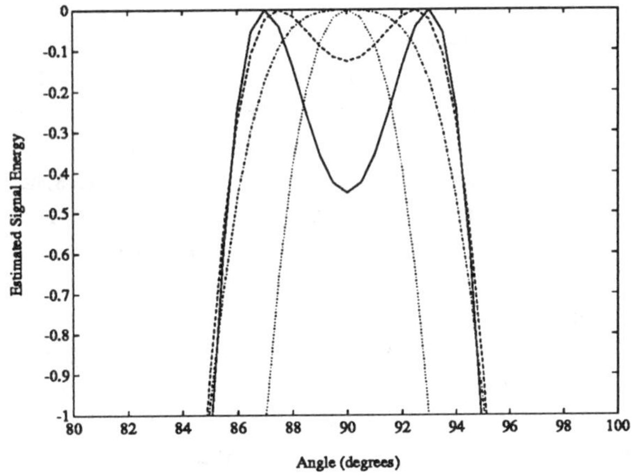

Figures 2a and 2b: The ideal correlation matrices R and $R_{\odot}$ used to generate these plots result from an equally spaced 10 element linear array with half-wavelength spacing. The signals are located at plus or minus 3 degrees from broadside. Broadside is represented as 90 degrees. The signal to noise ratio is 0 dB for each source. In this example, we use MVDR and spatial smoothing. Panel a) contains the entire beampattern while panel b) contains a zoom of the portion of the beampattern containing the signals. There are four cases illustrated on the plots:
1) Using MVDR on R (dotted line), the signals are not resolved. In this case, the signal subspace is not spherical; the ratio of the two signal subspace eigenvalues is given by $\lambda_1/\lambda_2 = 3.48$.
2) Using MVDR on $R_{\odot}$ (solid line), the signals are resolved and accurately located at ± 3 *degrees relative to broadside. In this case,* $\lambda_1/\lambda_2 = 1.00$.
3) Using spatial smoothing on $R_{\odot}$ with 2 subarrays (9 elements in each subarray, see the dashed line), the signals are still resolved, but inaccurately located. In this case, $\lambda_1/\lambda_2 = 1.25$.
4) Using spatial smoothing on $R_{\odot}$ with 3 subarrays (8 elements per subarray, see the dot-dashed line), the signals are not resolved. In this case, $\lambda_1/\lambda_2 = 1.61$.

Conclusions

Our results show that *MUSIC* effectively induces inter-signal coherence by its implicit assumption of signal subspace sphericity. Our results also show that low performance methods (like *MVDR* and conventional) can attain *MUSIC*-like performance if the signal and noise subspaces are individually spherical. Since standard (non-spherical) *MVDR* and conventional direction finding do not perform as well as their spherical counterparts (which are equivalent to *MUSIC*), we must conclude that subspace sphericity is desirable. We would also like to note that subspace sphericity can lead to significant computational advantages [8]. In general, signal subspace sphericity is reduced by high levels of coherence. However, our results indicate that inter-signal coherence is beneficial when it causes the signal subspace to be more nearly spherical.

References

[1] R. O. Schmidt, "Multiple Emitter Location and Signal Parameter Estimation," *IEEE Transactions on Antennas and Propagation*, vol. AP-34, no. 3, pp. 276–280, 1986.

[2] S. Kay and C. Demeure, "The High-Resolution Spectrum Estimator - A Subjective Entity," *Proc. IEEE*, vol. 72, Dec. 1984.

[3] T. Shan and T. Kailath, "Adaptive Beamforming for Coherent Signals and Interference," *IEEE Trans. ASSP*, vol. ASSP-33, pp. 527–536, June 1985.

[4] N. L. Owsley, "Sonar Array Processing," in *Array Signal Processing* (S. Haykin, ed.), pp. 115–193, Prentice-Hall, 1985.

[5] D. Johnson and S. DeGraaf, "Improving the resolution of bearing in passive sonar arrays by eigenvalue analysis," *IEEE Trans. ASSP*, vol. ASSP-30, pp. 638–647, April 1982.

[6] T. Shan, M. Wax, and T. Kailath, "On Spatial Smoothing for Direction-of-Arrival Estimation of Coherent Signals," *IEEE Trans. ASSP*, vol. ASSP-33, pp. 806–811, August 1985.

[7] D. A. Linebarger and D. H. Johnson, "The Effect of Spatial Averaging on Spatial Correlation Matrices in the Presence of Coherent Signals," *IEEE Trans. ASSP*, vol. 38, pp. 880–884, May 1990.

[8] R. DeGroat, "Non-Iterative Subspace Tracking," *to appear in IEEE Trans. Sig. Proc.*, March 1992.

DIRECTION FINDING FOR DIVERSELY POLARIZED SIGNALS USING POLYNOMIAL ROOTING

Anthony J. Weiss

Electrical Engineering Dept.
Tel-Aviv University
Tel-Aviv, 69978, Israel

Benjamin Friedlander

Signal Processing Technology, Ltd.
703 Coastland Drive
Palo Alto, CA 94303, USA

Abstract

We present a direction finding algorithm for diversely polarized arrays, which is based on polynomial rooting. Using polynomial rooting instead of search reduces significantly the computational requirements of the algorithm and enhances resolution. The basic method is limited to linear uniformly-spaced arrays. However, using an interpolation technique, the algorithm is extended to a larger class of arrays. The performance of the proposed algorithm is analyzed. Computer simulations are used to demonstrate the performance of the new technique and to verify the performance analysis.

1. Introduction

Previous works on direction finding using diversely polarized arrays, such as [2], presented estimation algorithms involving a search procedure. In this paper we derive a direction finding algorithm which does not require a search; instead it requires finding the roots of a polynomial. The algorithm is similar to the Root-MUSIC algorithm [5] which replaces the search procedure in the MUSIC algorithm with polynomial rooting, in the case of uniformly polarized arrays.

As in the case of Root-MUSIC, the basic algorithm is restricted to linear uniformly-spaced arrays. However, by using the idea of array interpolation [3], we were able to extend the polynomial rooting algorithm to arbitrary array geometries, provided the array consists of only two types of polarized elements (*e.g.*, vertically polarized and horizontally polarized). This limitation is not very restrictive since it is possible to handle arbitrarily polarized signals using only two types of elements.

In order to evaluate the performance of the algorithm, we derive asymptotic formulas for the variance of the directions-of-arrival and signal polarizations. These variances are then compared to the Cramer-Rao Bound (CRB) to assess the efficiency of the proposed estimator. We found that the variance of the directions-of-arrival is, in general, close to the CRB, except when the signals are correlated.

This work was supported by the Army Research Office under contract no. DAAL03-91-C-0022, sponsored by U.S. Army Communications Electronics Command, Center for Signals Warfare.

2. Problem Formulation

We begin by considering the polarization of an incoming signal. Suppose we have an antenna at the origin of a spherical coordinate system. Assume that a signal is arriving from direction θ, ϕ, where θ is the elevation angle and ϕ is the azimuth angle. Let this signal be a transverse electromagnetic (TEM) wave, and consider the polarization ellipse produced by the electric field in a fixed transverse plane. The unit vectors $\hat{\phi}, \hat{\theta}$ and $-\hat{r}$, in that order, form a right-handed coordinate system. ($-\hat{r}$ is a unit vector pointing in the signal propagation direction; $\hat{\phi}$ is a unit vector parallel to the $X - Y$ plane and perpendicular to $\hat{r}$; $\hat{\theta}$ is a unit vector perpendicular to $\hat{r}$ and lies in the $Z - \hat{r}$ plane). Suppose the electric field is given by

$$\mathcal{E} = E_\phi \hat{\phi} + E_\theta \hat{\theta} \,, \tag{1}$$

where E_ϕ and E_θ are the horizontal and vertical components of the field. For a given state of polarization we have

$$\begin{aligned} E_\phi &= a(t) \cos\gamma \; e^{j[\omega_0 t + \psi(t)]}, \\ E_\theta &= a(t) \sin\gamma \; e^{j[\omega_0 t + \psi(t) + \eta]}, \end{aligned} \tag{2}$$

where $a(t)$ is the signal amplitude, $\psi(t)$ the signal phase, ω_0 the signal frequency, and γ and η define the polarization.

We shall characterize the antenna in terms of its response to linearly polarized signals in the x, y and z directions. Let V_x be the complex voltage induced at the antenna output terminals by an incoming electromagnetic signal with a unit electric field polarized entirely in the x-direction. Similarly, let V_y and V_z be the output voltages induced by signals with unit electric fields polarized in the y and z directions, respectively. In general, V_x, V_y and V_z are complex numbers with different magnitudes and phases. The total

1058-6393/91 $01.00 © 1991 IEEE

output voltage from this antenna in response to the electromagnetic signal in eq (1) is then

$$
\begin{aligned}
s(t) = {} & a(t)\ e^{j[\omega_0 t+\psi(t)]} \\
& \Big[V_x(\sin\gamma\cos\theta\cos\phi\ e^{j\eta} - \cos\gamma\sin\phi) \\
& + V_y(\sin\gamma\cos\theta\sin\phi\ e^{j\eta} + \cos\gamma\cos\phi) \\
& - V_z(\sin\gamma\sin\theta\ e^{j\eta})\Big].
\end{aligned} \tag{3}
$$

To simplify the exposition we assume that the antennas and sources are coplanar, in which case $\theta = 90°$ and (3) reduces to

$$
\begin{aligned}
s(t) = a(t)\ e^{j[\omega_0 t+\psi(t)]} \Big[& -V_x\cos\gamma\sin\phi + V_y\cos\gamma\cos\phi \\
& - V_z\sin\gamma\ e^{j\eta}\Big].
\end{aligned} \tag{4}
$$

The antenna elements are usually designed to be insensitive to the Direction of Arrival (DOA) of the signal and hence we may assume that (4) can be rewritten as

$$s(t) = a(t)\ e^{j[\omega_0 t+\psi(t)]}\left[V_h\cos\gamma + V_v\sin\gamma\ e^{j\eta}\right], \tag{5}$$

where V_h is the complex voltage induced at the antenna output by a signal with a unit electric field polarized horizontally and $V_v = -V_z$.

Consider an array of M diversely polarized antennas. The array observes N polarized sources. The signal at the output of the mth sensor is given by

$$
\begin{aligned}
x_m(t) = {} & \sum_{n=1}^{N} s_n(t)\Big[V_h^{(m)}\cos\gamma_n \\
& + V_v^{(m)}\sin\gamma_n e^{j\eta_n}\Big]\ e^{-j\omega_0\tau_{mn}} + n_m(t)
\end{aligned} \tag{6}
$$

where $s_n(t) = a_n(t)\ e^{j[\omega_0 t+\psi_n(t)]}$ is the n-th signal, γ_n, η_n are its polarization parameters, $V_h^{(m)}$, $V_v^{(m)}$ are the sensitivities of the m-th antenna to horizontally and vertically polarized signals, respectively; $n_m(t)$ is the noise at the m-th antenna output, and τ_{mn} is the delay associated with the signal propagation time from the n-th source to the m-th antenna.

The signals from the sensor outputs are passed through a receiver where they are amplified, shifted to baseband, low-pass filtered, sampled, and digitized. We denote the samples of the receiver outputs by $X_i(k)$, where i is the sensor number and k is the index of different (independent) samples. The vector of sampled outputs, which we denote by $\mathbf{X}(k)$, is given by

$$\mathbf{X}(k) = A\,\mathbf{S}(k) + \mathbf{N}(k), \qquad k = 1,2,\ldots,K, \tag{7}$$

where

$$
\begin{aligned}
\mathbf{X}(k) &= [X_1(k), X_2(k), \ldots, X_M(k)]^T \\
\mathbf{S}(k) &= [S_1(k), S_2(k), \ldots, S_N(k)]^T \\
\mathbf{N}(k) &= [N_1(k), N_2(k), \ldots, N_M(k)]^T \\
A_{mn} &= \left[V_h^{(m)}\cos\gamma_n + V_v^{(m)}\sin\gamma_n e^{j\eta_n}\right] e^{-j\omega_0\tau_{mn}}
\end{aligned} \tag{8}
$$

To further simplify the exposition we assume that the sources are far enough from the observing array so that the signal wave fronts are effectively planar over the array. It is easy to verify that the delays τ_{mn} are given in this case by

$$
\begin{aligned}
\tau_{mn} &= -\frac{d_{mn}}{c}, \\
d_{mn} &= \tilde{x}_m\sin\phi_n + \tilde{y}_m\cos\phi_n,
\end{aligned} \tag{9}
$$

where c is the propagation velocity, d_{mn} is the distance from sensor m to sensor number one (reference sensor) in the direction of the n-th source, $(\tilde{x}_m, \tilde{y}_m)$ are the coordinates of the m-th sensor, ϕ_n is the DOA of the n-th source relative to the $\tilde{y}$ axis, and the origin of the Cartesian coordinate system coincides with sensor number one.

The problem that we are addressing here is how to estimate the DOAs (*i.e.* $\phi_1, \phi_2, \cdots, \phi_N$) and the signal polarizations (*i.e.* $\gamma_1, \eta_1, \gamma_2, \eta_2, \cdots, \gamma_N, \eta_N$) efficiently.

3. DOA Estimation by Polynomial Rooting

In this section we describe a computationally efficient procedure for DOA estimation. We concentrate on *linear equispaced arrays.* We start by describing the array manifold as a product of a matrix and a vector. The vector is defined by the *signal polarization* and is given by

$$\mathbf{q}(\gamma,\eta) \triangleq [\cos\gamma \quad e^{j\eta}\sin\gamma]^T \tag{10}$$

and the matrix is defined by the sensor locations the *sensor polarizations* and the signal DOA as follows

$$W(\phi) \triangleq \mathrm{diag}\{\mathbf{a}(\phi)\}[\mathbf{V}_h \quad \mathbf{V}_v] \tag{11}$$

where

$$\mathbf{V}_h \triangleq \left[V_h^{(1)}, V_h^{(2)}, \cdots, V_h^{(M)}\right]^T \tag{12}$$

$$\mathbf{V}_v \triangleq \left[V_v^{(1)}, V_v^{(2)}, \cdots, V_v^{(M)}\right]^T \tag{13}$$

$$\mathbf{a}(\phi) \triangleq \left[e^{jf_1(\phi)}, e^{jf_2(\phi)}, \cdots, e^{jf_M(\phi)}\right]^T \tag{14}$$

and

$$f_m(\phi) \triangleq \frac{\omega_0}{c}\left[\tilde{x}_m\sin\phi_n + \tilde{y}_m\cos\phi_n\right]. \tag{15}$$

Since we concentrate here on uniform linear arrays, we can write

$$f_m(\phi) = (m-1)\pi d \sin\phi\,, \quad m = 1, 2, \cdots, M; \quad (16)$$

where d is the sensor spacing in half wavelength units, and the array is placed along the x axis.

Using $\mathbf{q}(\gamma,\eta)$ and $W(\phi)$ the array manifold is given by

$$\mathcal{A}(\phi,\gamma,\eta) = W(\phi)\mathbf{q}(\gamma,\eta) \quad (17)$$

As is well known from the MUSIC algorithm, estimates of the DOAs are obtained by minimizing the cost function

$$D(\phi,\gamma,\eta) = \mathcal{A}^H G G^H \mathcal{A} \quad (18)$$

where G is a matrix whose columns are the eigenvectors of the data covariance matrix, corresponding to the N smallest eigenvalues. The minima of $D(\phi,\gamma,\eta)$ can be found by a multidimensional search over ϕ,γ and η. However, more efficient minimization methods are available, as we show next.

Let us rewrite the cost function as

$$D(\phi,\gamma,\eta) = \mathbf{q}^H W^H G G^H W \mathbf{q} \quad (19)$$

Assuming for the moment that W is known, the minimum over $\mathbf{q}$ is obtained when $\mathbf{q}$ is equal to the eigenvector corresponding to the smallest eigenvalue of the matrix

$$C \triangleq W^H G G^H W\,. \quad (20)$$

Thus, for every given ϕ we can immediately compute $\mathbf{q}$. The quadratic form in (19) is non-negative. Thus, the smallest eigenvalue is always non-negative, and is equal to zero when ϕ is equal to one of the observed DOAs. Hence, it suffice to find the values of ϕ for which the smallest eigenvalue of C is minimal [2]. An alternative approach is to define the complex variable z,

$$z \triangleq e^{j\pi d \sin(\phi)} \quad (21)$$

so that $\mathbf{a}(\phi)$ becomes

$$\mathbf{a}(z) = \left[1, z, z^2, \cdots, z^{M-1}\right]^T \quad (22)$$

Now the 2×2 matrix C is a function of z. Recalling that the smallest eigenvalue will equal zero if z corresponds to a true DOA, we can estimate the DOAs by finding the values of z for which

$$\det\{C\} = 0 \quad (23)$$

Note that the left side of (23) is a polynomial of z and therefore the DOA estimation problem is transformed into polynomial rooting problem which can be solved using computationally efficient root-solving algorithms.

As mentioned earlier we replaced the cost function of Ferrara and Parks [2], (*i.e.* the smallest eigenvalue of the matrix C) with a different cost function defined as the determinant of C. In general, each cost function yields different estimates. However, in [1] we show that the first and second moments of the estimates associated with these different cost functions are asymptotically the same.

Once the problem is posed as a polynomial rooting problem the roots are obtained by one of the excellent polynomial rooting algorithms available. The DOAs are obtained using (21)

$$\hat{\phi}_n = \arcsin\left(\frac{1}{\pi d}\arg(z_n)\right)\,, \quad (24)$$

where z_n represent one of the roots that is on or near the unit circle. Since we assume that the number of sources N is known, we select the N roots that are nearest to the unit circle.

In some situations it is of interest to estimate the polarization parameters of each signal. This can be done by evaluating the matrix C for each of the estimated DOAs and then estimating the polarization vector, $\mathbf{q}$, as the eigenvector associated with the smallest eigenvalue of $C(\phi_n)$. Once $\hat{\mathbf{q}}$, the estimate of $\mathbf{q}$, is given then the polarization parameters are evaluated by

$$\hat{\gamma} = \mathrm{arctg}\left(|\hat{q}_2|/|\hat{q}_1|\right) \quad (25)$$

$$j\hat{\eta} = \ln\left(\frac{\hat{q}_2}{\hat{q}_1}\cdot\frac{|\hat{q}_1|}{|\hat{q}_2|}\right) \quad (26)$$

Numerical simulation and performance analysis indicates that the estimation accuracy of the polarization parameters is very close to the ultimate performance as predicted by the Cramer Rao lower bound (see below.)

4. Interpolated Arrays

In this section we outline an approach which makes it possible to extend the polynomial rooting algorithm to arbitrary array geometries, using the idea of array interpolation. This approach is not completely general since we introduce the assumption that the array consists of elements with only two types of polarization. This assumption holds in most practical situations.

The interpolated array approach is based on the idea that the array manifold of a virtual array, whose element locations are chosen by the user, can be obtained by linear interpolation of the array manifold of the real array within a limited sector of directions. In other words, we assume that there exists a constant matrix B such that

$$B\mathcal{A}(\phi,\gamma,\eta) \approx \bar{\mathcal{A}}(\phi,\gamma,\eta) \tag{27}$$

for angles ϕ within the sector, where $\mathcal{A}(\phi,\gamma,\eta)$ is the manifold of the real array and $\bar{\mathcal{A}}(\phi,\gamma,\eta)$ is the manifold of the virtual array. The interpolation matrix B is computed by performing a least squares fit in equation (27); see [3] for a much more detailed discussion of the virtual array design and the interpolation accuracy. In practice, the field of view will be divided into several sectors, and a different interpolation matrix will be calculated for each sector. To simplify the discussion and the notation we will consider in this paper only a single sector. Reference [3] describes how to combine estimates produced for the different sectors.

Consider an array composed of two subarrays, one with M_1 elements and the other with M_2 elements. The antennas in each subarray are the same. In this case, the vectors $\mathbf{V}_h$ and $\mathbf{V}_v$ (see (12), (13)) are given by:

$$Q^T \triangleq [\mathbf{V}_h \mathbf{V}_v]^T = \begin{bmatrix} V_h^{(1)} & V_h^{(1)} & \cdots & V_h^{(1)} & V_h^{(2)} & V_h^{(2)} & \cdots & V_h^{(2)} \\ V_v^{(1)} & V_v^{(1)} & \cdots & V_v^{(1)} & V_v^{(2)} & V_v^{(2)} & \cdots & V_v^{(2)} \end{bmatrix} \tag{28}$$

Here $V_h^{(j)}, V_v^{(j)}$ describe the sensitivity to polarization of each of the antennas in the j–th subarray ($j = 1, 2$.)

Now

$$\mathcal{A}(\phi,\gamma,\eta) = \text{diag}\{\mathbf{a}(\phi)\}\, Q\mathbf{q} = \text{diag}\{\mathbf{a}(\phi)\}\, [\tilde{q}_1,\cdots,\tilde{q}_1,\tilde{q}_2,\cdots,\tilde{q}_2]^T, \tag{29}$$

where

$$\tilde{q}_1 \triangleq V_h^{(1)} q_1 + V_v^{(1)} q_2, \tag{30}$$

and

$$\tilde{q}_2 \triangleq V_h^{(2)} q_1 + V_v^{(2)} q_2. \tag{31}$$

Thus, the array manifold can be partitioned as follows

$$\mathcal{A}(\phi,\gamma,\eta) = [\tilde{q}_1 \mathbf{a}_1^T(\phi), \tilde{q}_2 \mathbf{a}_2^T(\phi)]^T \tag{32}$$

where $\mathbf{a}_1(\phi)$ is a column vector associated with the first M_1 elements of $\mathbf{a}(\phi)$ and $\mathbf{a}_2(\phi)$ is a column vector associated with the last M_2 elements of $\mathbf{a}(\phi)$.

The special form of $\mathcal{A}(\phi,\gamma,\eta)$ allows us to apply the linear interpolation procedure discussed earlier. In this case the interpolation matrix B is block diagonal with two blocks, the upper left block of size $M_1 \times M_1$ and the lower right block of size $M_2 \times M_2$.

The covariance matrix of the actual array is given by

$$R_x = AR_sA^H + \sigma^2 I \tag{33}$$

where R_s is the signal covariance and σ^2 is the noise variance. Using the interpolation matrix B we obtain the covariance matrix associated with the virtual array

$$\bar{R}_x = BR_xB^H = \bar{A}R_s\bar{A}^H + \sigma^2 R_n \tag{34}$$

where

$$\bar{A} = BA \tag{35}$$

and

$$R_n \triangleq BB^H \tag{36}$$

Denote by $\tilde{G}$ the matrix whose columns are the eigenvectors of $R_n^{-1/2}\bar{R}_x R_n^{-1/2}$ corresponding to the N smallest eigenvalues. The DOA estimates are obtained by minimizing the cost function

$$\tilde{D}(\phi,\gamma,\eta) \triangleq \bar{\mathcal{A}}^H R_n^{-1/2}\tilde{G}\tilde{G}^H R_n^{-1/2}\bar{\mathcal{A}} \tag{37}$$

If the virtual array is chosen to be a linear uniformly spaced array, we can use the polynomial rooting algorithm described in the previous section to estimate the DOAs.

5. Numerical Examples

In this section we present some numerical examples to evaluate the performance of the polynomial rooting algorithm and to validate the analytical results presented in [1].

Consider a uniform linear array consisting of 6 pairs of circularly polarized antennas. One of the antennas in each pair is Left Hand Circular (LHC) and the other is Right Hand Circular (RHC). The spacing between the antenna pairs is one half wavelength.

There are two uncorrelated narrowband sources: one located 10 degrees off broadside and the other located at variable angles. The Signal to Noise Ratio (SNR) of both signals is 10 dB. The signals are linearly polarized, with $\gamma_1 = 20°, \eta_1 = 0°$ and $\gamma_2 = 70°, \eta_2 = 0°$. We assumed that the array collected 200 snapshots for processing.

Figures 1 to 3 depict the Cramer Rao Lower bound (solid line) calculated according to the formulas derived in [4], the predicted performance (dashed line) according to the performance analysis given in [1] and the results of 1000 Monte Carlo experiments (small circles).

We observe that the performance analysis curve coincides with the Cramer Rao bound. Moreover, the simulation results verify the performance analysis and indicate that the performance of the algorithm coincides in this case with the performance predicted by the Cramer Rao Bound. In [1] we present examples involving non-uniform arrays and correlated signals.

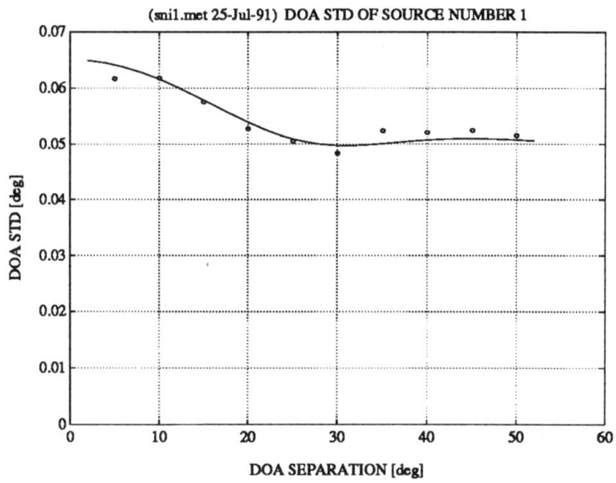

Figure 1: Standard deviation of ϕ_1, the DOA of source 1, in the presence of a second source, versus the angular separation. Linear uniformly spaced array.

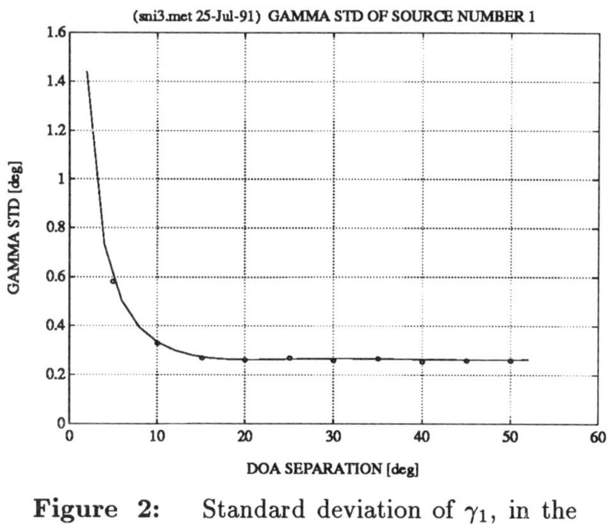

Figure 2: Standard deviation of γ_1, in the presence of a second source, versus the angular separation. Linear uniformly spaced array.

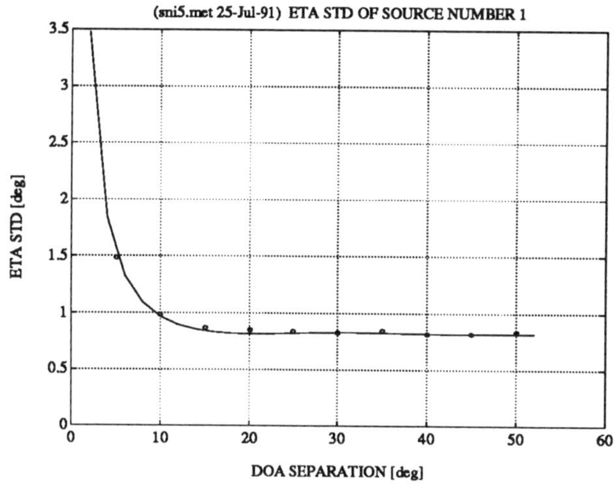

Figure 3: Standard deviation of η_1, in the presence of a second source, versus the angular separation. Linear uniformly spaced array.

References

[1] A.J. Weiss, and B. Friedlander, "Direction Finding for Diversely Polarized Signals Using Polynomial Rooting," *IEEE Transactions on Signal-Processing,* Submitted for publication.

[2] E.R. Ferrara, T.M. Parks, "Direction Finding with an Array of Antennas Having Diverse Polarizations," *IEEE Trans. on Antennas and Propagation,* vol. AP-31, no. 2, pp. 231–236, March 1983.

[3] B. Friedlander, "The Interpolated Root-MUSIC Algorithm for Direction Finding," submitted for publication. See also B. Friedlander, "Direction Finding with an Interpolated Array," Intl. Conference on Acoustics, Speech and Signal Processing, Albuquerque, New Mexico, April 3–6, 1990.

[4] B. Friedlander, and A. J. Weiss, "Performance of Diversely Polarized Antenna Arrays for Correlated Signals," Submitted for publication.

[5] A. J. Barabell, "Improving the Resolution Performance of Eigenstructure Based Direction Finding Algorithms," Int'l Conf. on Acoustics Speech and Signal Processing, Boston, MA, pp. 336–339, 1983.

Tracking Moving Sources Using The Rank-Revealing QR Factorization

Monique P. Fargues

Department of Electrical and Computer Engineering
Naval Postgraduate School
Monterey, CA 93943

Abstract

Subspace decomposition methods are a powerful tool used in different areas of signal processing in which the signal information is usually obtained via eigen-based methods. These techniques are numerically very stable but expensive to update. The rank revealing QR (RRQR) factorization described earlier by Chan provides an attractive alternative to accomplish subspace selection. We propose to add an updating capability to the RRQR factorization, and to apply it to the Direction Of Arrival problem. This technique allows for tracking of moving sources by taking advantage of the simplicity of the regular QR updating scheme and the rank-revealing property of the RRQR factorization.

1. Introduction

Subspace techniques are a powerful tool used in signal processing in which the signal information is obtained via eigen-based methods. Eigen-based updating schemes, which would allow for tracking of time-varying information, have also been proposed [1,2,3,4]. Eigendecomposition techniques are numerically stable but are computationally intensive, which is a drawback for real-time applications. Several researchers have proposed various alternatives in an attempt to decrease the computational load associated with the signal information estimation [5,6,7,8,9]. The rank-revealing QR algorithm described by Chan [9] provides an attractive alternative to accomplish subspace selection, and was applied to the Direction of Arrival (DOA) estimation problem recently [8]. In this paper, we investigate an updating scheme for the rank-revealing QR (RRQR) algorithm, and apply it to the DOA problem. Sections 2 and 3 briefly review subspace methods and the RRQR technique. In Section 4 we show how the RRQR algorithm can be used to update signal and noise subspaces. Section 5 presents experimental results, and comparisons with eigen-based signal and noise subspaces. Finally, conclusions are presented in Section 6.

The work of M.P. Fargues was funded by the Naval Postgraduate School. The United States Government assumes no responsibility for the information provided.

2. Subspace Methods

Assume a linear equi-spaced array of N sensors receiving p narrowband signals. Under the assumption of non-dispersive propagation, sensors without distortion, and envelope variations that are slow relative to the carrier frequencies of the narrowband signals, the signal received by the passive array is:

$$x(t) = M(\theta)s(t) + n(t)$$

where $s(t)$ represents the narrowband signal, the mode matrix $M(\theta)$ is defined as: $M(\theta) = [m(\theta_1), \dots, m(\theta_p)]$. The mode vector $m(\theta_i)$ represents the delay with which the signal impinges on each of the sensors, and $n(t)$ is additive noise. The mode vector is generally a nonlinear function of the signal arrival angle θ, center frequency Ω, array element spacing d, and sensor response. For the above case, $m(\theta)$ is given by the following vector expression:

$$m(\theta) = \left[\, 1, e^{-2\pi jz}, \dots, e^{-2\pi j(N-1)z} \right]^t$$

where the normalized angle z is defined as $z = d\sin(\theta)/\Omega$. Under the assumption that the zero mean noise $n(t)$ is white and uncorrelated with the signals, the theoretical spatial correlation for the received signal is given by:

$$R = E\left[x(t)x(t)^H\right] = MSM^H + \sigma^2 I$$

The sources are assumed to be uncorrelated, thus the signal correlation matrix is given by $S = diag(S_1, \dots, S_p)$, where S_i represents the power associated with each signal. The estimated correlation matrix obtained using n samples of the sensor output $x(t)$ is given by:

$$R = \frac{1}{n}\sum_{i=1}^{n} x(i)x^H(i) = \frac{1}{n}X^H X$$

where $X = [x(1), \ldots, x(n)]^H$. Numerous high resolution techniques [15], based on the EVD or SVD have been proposed to identify the DOA information (number and location of sources) from the signal or noise subspace obtained from R. The minimum-norm method [11] is used in this work to do this.

Reilly, et al [5] proposed to apply a QR factorization with a pre-selected pivoting scheme to the noise-free (signal) correlation matrix $R_s = R - \sigma^2 I = MSM^H$ in an attempt to decrease the computational load associated with obtaining the DOA information. Thus, their work can be viewed as a QR factorization with "partial" pivoting. This specific "partial" pivoting could explain the loss of performance they found when comparing their technique to the MUSIC algorithm. Note that R_s is theoretically rank deficient when $p < N$. Applying the QR algorithm with column pivoting to R_s[14] leads to

$$R_s\Pi = QU = Q\begin{bmatrix} U_{11} & U_{12} \\ 0 & U_{22} \end{bmatrix}$$

where Q is orthonormal, Π is a permutation matrix, U_{11} is $p \times p$, U_{12} is $p \times (N-p)$, and U_{12} is $(N-p) \times (N-p)$. When R_s has rank p, $U_{22} \equiv 0$, and $\text{Range}(R_s) = \text{Range}(q_1, \ldots, q_p)$. Thus, the QR algorithm with column pivoting can theoretically be used as a tool to determine the number of sources and their locations. However, when R is estimated, R_s becomes nearly rank deficient. Recall that the QR method with column pivoting is not completely reliable in such a case, as it may not produce a matrix U_{22} with small norm. The Rank-Revealing QR (RRQR) factorization introduced by Chan [9] does not have this problem; it is designed to produce a block U_{22} with small norm when the matrix to be factorized is nearly rank deficient.

3. The Rank Revealing QR Factorization

The RRQR algorithm is derived in detail in [9]. Basically, the idea is to compute a QR factorization of A in the following form:

$$A\Pi = QU = Q\begin{bmatrix} U_{11} & U_{12} \\ 0 & U_{22} \end{bmatrix}$$

with $\|U_{22}\|$ small, so that any rank deficiency of A will be revealed by the smaller lower right block in U. The algorithm derived by Chan computes a permutation matrix Π which guarantees U_{22} to be small in norm when A is nearly rank deficient. Furthermore, Chan's derivation gives upper and lower bounds for the singular values of A, and an approximate null space for A. The RRQR algorithm applied to a N dimensional square matrix A is summarized below:

1. Compute a QR factorization of A: $A\Pi = QU$
2. Initialize W (of dimension $N \times p$) to zero
3. for $i = N, \ldots, p+1$
   a. Partition $U = \begin{bmatrix} U_{11} & U_{12} \\ 0 & U_{22} \end{bmatrix}$
   b. Compute an approximation to the minimum singular value and the right singular vector v corresponding to U_{11}
   c. Compute a permutation P (of dimension $i \times i$) such that $|(P^t v)_i| = \|P^t v\|_\infty$
   d. Assign $v \equiv [v, 0]^t$ (of dimension $N \times 1$) to the ith column of W.
   e. Compute $W \leftarrow \tilde{P}^t W$, where $$\tilde{P} \equiv \begin{bmatrix} P^t & 0 \\ 0 & I \end{bmatrix}$$
   f. Compute the QR factorization: $U_{11}P = Q_1\tilde{U}_{11}$
   g. $\Pi \leftarrow \Pi\tilde{P}$
   h. $Q \leftarrow Q\begin{bmatrix} Q_1 & 0 \\ 0 & I \end{bmatrix}$
   i. $U \leftarrow \begin{bmatrix} \tilde{U}_{11} & Q_1^t U_{12} \\ 0 & U_{22} \end{bmatrix}$
4. end of the for loop

Note that:

1. The QR factorization may be derived cheaply in step (3.f) using Givens rotations.
2. The matrix W is used to store the singular vectors corresponding to the smallest singular values of the successive matrices U_{11}. Chan and Hansen [18] showed that the matrix $W_n = \Pi W$ can be viewed as an approximation to the null space of A. They further showed that W_n can be used as a starting matrix for inverse iteration to compute an accurate estimate of Null(A). Such a null space can be used obtain estimates of the source directions.
3. Prasad and Chandna [8] noted that the bounds to the singular values proposed by Chan [9] seem to be quite tight. Thus, these bounds could potentially be used to determine the number of sources with the MDL [16] or AIC [17] tests in lieu of the eigenvalues.

4. The matrix $Q_s = [q_1, \ldots, q_p]$ is an approximation of the signal subspace. It can be used to obtain estimates of the source directions. The null space span(W_n) and signal subspace span(Q_s) are used in this work to estimate the source locations and their performances are compared.

4. Updating using the RRQR Factorization

Tracking moving sources requires the correlation matrix to be updated and the signal and noise subspaces to be continuously recomputed. The time varying estimated correlation matrix, obtained by adding a $(k+1)^{th}$ snapshot $x(k+1)$ and deleting the old snapshot $x(1)$, is given by:

$$R(k+1) = \frac{1}{n}\left[X^H X + x(k+1)x^H(k+1) - x(1)x^H(1)\right]$$

where $X = [x(1), \ldots, x(n)]^H$.

Several EVD and SVD based algorithms designed to track moving sources have been proposed [1,2,3]. Adam, et al [6] recently presented a new rank-revealing orthogonal transformation used to extract the noise subspace information needed for the DOA parameters. Bischof and Shroff [7] applied the RRQR factorization to the data matrix and used hyperbolic Householder transformations to update the noise subspace information. Our approach is different from these as we apply the RRQR factorization to the correlation matrix directly. The rank-1 modification of the QR factorization (without column pivoting) can be done in $O(n^2)$ operations [14]. However, recall that the noise free correlation matrix R_s, obtained using the estimated received signal correlation matrix R, is nearly rank deficient. Thus, pivoting is needed to insure a correct estimation of the signal and/or noise subspaces.

A basic flowchart of the updating scheme is presented in Figure 1. At each new iteration, we first apply the basic QR updating scheme to R_s. The RRQR restricted pivoting scheme [9] is applied next to reveal the numerical rank of R_s, and to obtain signal and noise subspace estimates. Note that, in our algorithm the initial rank-1 QR update computed at each time iteration before applying Chan's pivoting scheme does not involve pivoting, while Chan's method does use an initial QR decomposition with pivoting. Our simulations indicate no significant difference in the resulting estimated noise and signal subspaces. However, the original QR factorization is used to estimate minimum singular value and vectors needed in Chan's pivoting scheme. The relationships between the initial QR pivoting, the accuracy of the estimated singular vectors used in the RRQR pivoting section, and approximated noise and signal subspaces are presently under study.

5. Simulation Results

We first study the performance of the RRQR factorization by comparing the signal and noise subspaces obtained by updating the noise free correlation matrix R_s with those obtained using the eigendecomposition (EVD) of R. The largest principle angle between estimated subspaces [14] is used for comparison. Next, the minimum norm algorithm [11] is used to track the estimated source locations obtained using the RRQR factorization and the EVD algorithm.

A ten element array is used in the simulations to track two sources. The noise is assumed to be zero-mean gaussian and uncorrelated from sensor to sensor. We consider the case of two sources impinging on the array. The first source is assumed to be fixed at a normalized angle $\theta_1 = 40°$, the second source location θ_2 is linear time-varying. The two sources are assumed to have a SNR equal to 20dB, 10dB, and 6dB successively. 150 points are used to estimate the correlation matrix. Table 1 presents the average distance (for 200 updates) between the signal subspaces generated using the RRQR updating technique, the QR with and without column pivoting (all QR-based techniques are obtained using R_s) and the signal subspace obtained from the eigendecomposition of R. Approximations to the QR-based noise subspace span(W_n), where $W_n = \Pi W$, and span(W_{n2}), where W_{n2} is obtained after applying two inverse iteration steps to W_n, are considered also. Table 1 presents the distance between the estimated QR-based noise subspaces and the noise subspace obtained from the eigendecomposition of R. Table 1 indicates that the span(W_n) can be considered as an approximation to the EVD-based noise subspace. Note that applying 2 inverse iteration steps to W_n leads to a noise estimation identical as that obtained with the eigendecomposition of R. Table 1 also shows that the RRQR updating scheme leads to a better approximation of the signal subspace than the QR without column pivoting even when the SNR is low.

The minimum norm algorithm is applied to compute the estimated source locations next. Figure 2 presents the tracking of the source with linear time varying position θ_2 obtained for the various signal and noise subspaces presented above. Figure 2 shows that the estimated θ_2 computed using span(W_n) has large variance for medium to small SNR, even though the average distance was found to be small in Table 1. Better tracking is obtained using W_{n2}, as expected from the results given in Table 1. Furthermore, Figure 2 shows that a reasonably good approximation of the signal subspace can be obtained using the RRQR updating technique. Note however, that the source estimated using the RRQR technique has a higher variance than that obtained using eigen-based techniques.

6. Conclusions

We have presented a simple technique for updating signal and noise subspaces from the noise-free correlation matrix. This method takes advantage of the simplicity of the regular QR updating technique and the rank-revealing QR factorization proposed recently. Simulations show that the RRQR updating algorithm approximates signal and noise subspaces which can be used to track moving sources.

Initialization

$R_s = X^H X - n\sigma^2 I$

RRQR factorization

$R_s \Pi = QU$

Start updating

$R_s(k+1) = R_s(k) + x(k+1)x^H(k+1) - x(1)x^H(1)$

- Update the above QR factorization
- Apply 2 successive rank-one modifications to $R_s(k)$
  $\rightarrow R_s(k+1)\Pi = Q_1 U_1$

- Apply Chan's pivoting scheme to Π, Q_1, U_1 (steps 2,3 section 3)
  $\rightarrow$ form: Π_2, Q_2, U_2

- Set $\Pi = \Pi_2$
  $Q = Q_2$
  $U = U_2$

- Identify signal subspace Q_s
- Identify approximated noise subspace $W_n = \Pi W$
- option to improve W_n by applying inverse iteration [18]

k=k+1

Figure 1. RRQR updating algorithm

References

[1]: "Efficient, Numerically Stabilized Rank-One Eigenstructure Updating," R.D. DeGroat & R.A. Roberts, IEEE Trans. Acoust., Speech, Signal Processing, Vol. ASSP-38, No. 2, Feb. 1990, pp. 301-316.

[2]: "Tracking a Few Extreme Singular Values and Vectors in Signal processing," P. Comon & G.H. Golub, Proc. IEEE, Vol. 78, No. 8, Aug. 1990, pp. 1327-1343.

[3]: "Adaptive SVD Algorithm with Application to Narrowband Signal Tracking," W. Ferzali & G. Proakis, SVD and Signal Processing, II, R.J. Vaccaro, Ed., Elsevier, 1991

[4]: "An Algorithm for Subspace Computation with Applications in Signal Processing," D.R. Fuhrmann, SIAM J. Matrix Anal. Appl., Vol. 9, No. 2, Ap. 1988, pp. 213-220.

[5]: "Fast QR-based Array-Processing Algorithm," J.P. Reilly, W.G. Chen, & K.M. Wong, Advanced Algorithms and Architectures for Signal Processing III, SPIE, Vol. 975, 1988, pp. 36-47.

[6]: "Direction-of-Arrival Estimation Using the Rank-Revealing URV Decomposition," G. Adams, M.F. Griffin, & G.W. Stewart, Proc. ICASSP-91, pp. 1385-1388.

[7]: "On Updating Signal Subspace," C.H. Bischof & G.M. Shroff, Advanced Signal Processing Algorithms, Architectures, and Implementations, SPIE, 1990, pp.338-348.

[8]: "Direction-of-Arrival Estimation Using Rank Revealing QR Factorization," S. Prasad & B. Chandna, IEEE Trans. Signal Processing, Vol. SP-38, No. 5, May 1991, pp. 1224-1229.

[9]: "Rank Revealing QR Factorization," T.F. Chan, Linear Alg. and its Appl., Vol. 88, No. 89, 1987, pp. 67-82.

[10]: "Computing Truncated Singular Value Decomposition Least Squares Solutions By Rank Revealing QR Factorizations," SIAM J. Sci. Stat. Comput., Vol. 11, No. 3, May 1990, pp. 519-530.

[11]: "Estimating the Angles of Arrival of Multiple Plane Waves," R. Kumaresan & D.W. Tufts, IEEE Trans. on Aerosp. and Elect. Syst., Vol. AES-19, No. 1, 1983, pp. 134-139.

[12]: A Signal Subspace Approach to Multiple Emitter Location and Spectral Estimation, Ph.D. Dissertation, Stanford University, Stanford, CA, Nov. 1981.

[13]: "Applications of Truncated QR Methods to Sinusoidal Frequency Estimation," Proc. ICASSP-90, pp. 2571-2574.

[14]: Matrix Computations, G.H. Golub & C.F Van Loan, Johns Hopkins University Press, 2nd Ed., 1989.

[15]: Modern Spectral Estimation, S.M. Kay, Prentice Hall, 1988.

[16]: "Modeling by Shortest Data Description," J. Rissanen, Automatica, Vol. 14, 1978, pp. 465-471.

[17]: "Determining the Number of Signals by Information Theoretic Criteria," ASSP Workshop on Spectral Estimation II, Nov. 1983.

[18]: "Computing Truncated Singular Value Decomposition Least Squares Solutions by Rank Revealing QR-Factorizations," SIAM J. Sci. Stat. Comput., Vol. 11, No. 3, May 19990, pp. 519-530.

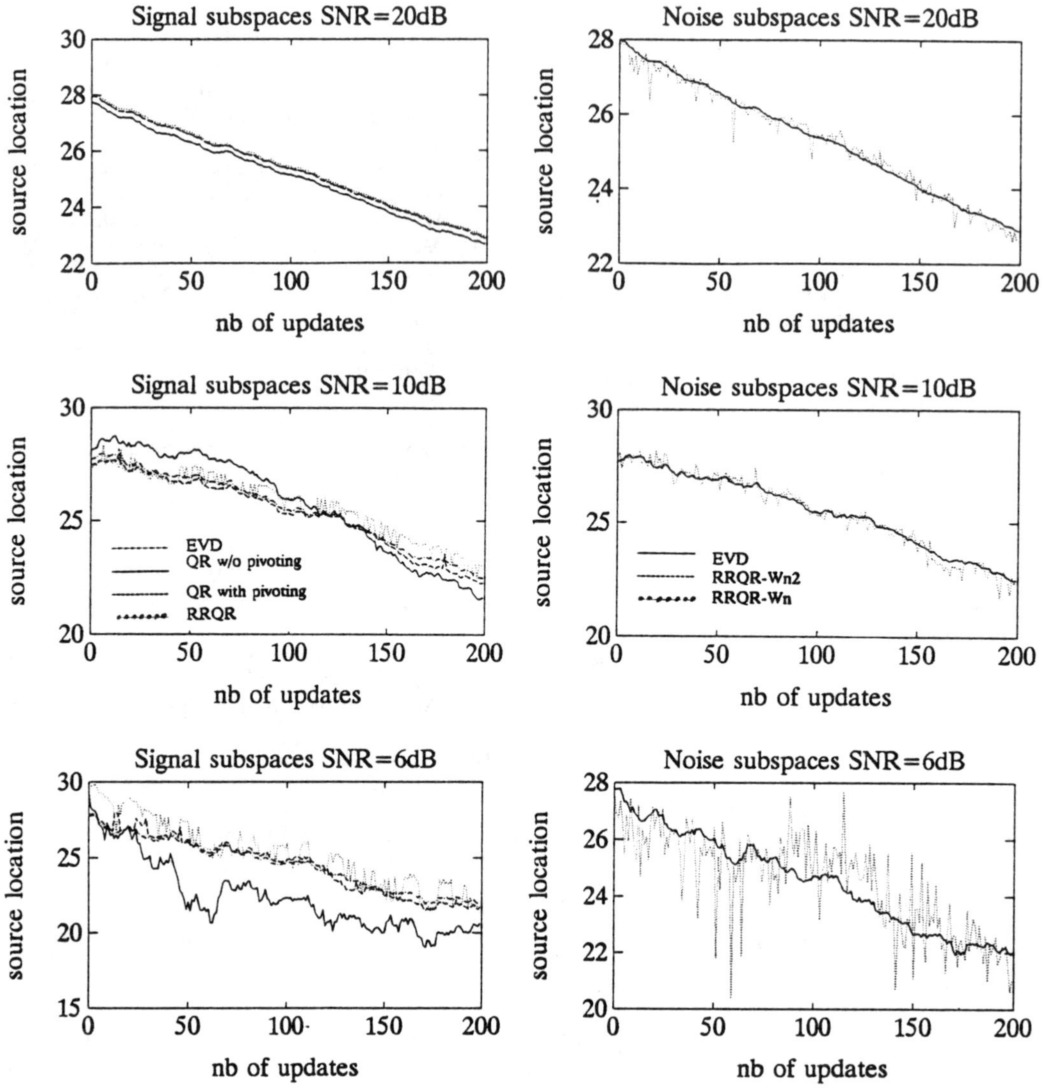

Figure 2. Linear time-varying frequency tracking for QR-based and eigen-based signal and noise subspaces: SNR = 20dB, 10dB, 6dB

Table 1. Distance between signal (noise) QR-based and eigen-based subspaces - 200 updates ($\theta_1 = 40°$, θ_2 linear time-varying)

SNR	20dB	10dB	6dB
Signal subspace	mean, standard deviation (in degrees)		
RRQR	0.164, 0.054	1.721, 0.536	4.2684, 1.193
QR with pivoting	0.3770, 0.094	2.716, 0.448	5.5736, 1.673
QR w/o pivoting	1.4502, 0.352	12.064, 2.216	26.6544, 3.644
Noise subspace	mean, standard deviation (in degree)		
$W_n = \Pi W$	0.0023, 0.0022	0.2859, 0.4912	1.585, 1.975
W_{n2} (2 inv. iteration steps applied to W_n)	0, 0	0, 0	0.0001, 2.10^{-4}

Covariance Structures for Multidimensional Data

Timothy A. Barton

Department of Electrical Engineering
Washington University
St. Louis, MO 63130

Daniel R. Fuhrmann

Department of Electrical Engineering
Washington University
St. Louis, MO 63130

Abstract

A notation for describing covariance structures for a wide class of signal processing problems is established. These structures are used as constraint sets for covariance matrices of naturally multidimensional data organized into column-vector form for signal processing applications. The structures involve a hierarchy of sub-blocks within the matrix, and include block-circulant and block-Toeplitz matrices and their respective generalizations.

1 Introduction

Multidimensional data and structured covariances arise in many signal processing problems. *A priori* knowledge of structure in the true underlying covariance of a random process may be exploited in algorithms in which the covariance must be estimated or manipulated in some way. For example, in wideband array processing, array data is naturally multidimensional in time and space. Depending on the sensor array and organization of the data vectors, complex structures may arise in the data covariance matrix.

In particular, the block-circulant and block-Toeplitz structures of covariances that arise naturally from physical considerations in signal processing problems are of interest. It is known that Toeplitz structure arises in covariances used in time series analysis for discrete wide-sense-stationary processes; if the process is periodic, the covariances may have circulant structure. Exploitation of Toeplitz covariance structure for narrowband uniform linear array data is demonstrated in the work of Gray *et al.* [6] and Fuhrmann [3]. Estimation of Toeplitz constrained covariances was initially investigated by Burg, Luenburger, and Wenger [2], and subsequently by Miller and Snyder [9] and Tourtier and Scharf [10]. Under certain assumptions and processing strategies, the covariance for wideband data taken from an array with arbitrary geometry may have the block-Toeplitz structure. The problem of estimating block-Toeplitz and Toeplitz-block-Toeplitz covariance matrices that arise in wideband array processing has been investigated by Fuhrmann and Barton [4].

Our goal here is the development of an organized notation and a general framework for the discussion of problems that involve multidimensional data and structured covariances under a complex Gaussian model.

The following assumptions and definitions will be used throughout this paper. A set of *multidimensional data* is defined as a set $\mathbf{X}$ of complex samples $x(i_1, \ldots, i_N)$ such that each sample is indexed by a vector of integers. The finite set of such vectors is known as the index set. The *vec* operator is defined as a one-to-one map that organizes the data $x(i_1, \ldots, i_N)$ into a column vector $\mathbf{x}$ with elements $\mathbf{x}(j)$ such that

$$j = \sum_{k=1}^{N} \left\{ \left[\prod_{l=1}^{k-1} n_l \right] i_k \right\} . \qquad (1)$$

Our interest is in the class of problems in which $\mathbf{x}$ can be modeled as a 0-mean complex Gaussian random vector of the Goodman class [5].

2 Covariance Structures and Notation

2.1 Induced Hierarchical Block Structure

The multidimensional data $\mathbf{X}$, consisting of samples $x(i_1, \ldots, i_N)$, will induce a natural block structure on the covariance matrix $\mathbf{R}$. From the definition of the *vec* operator, it is known that the dimension of $\mathbf{R}$ will be $(n_1 \cdots n_N) \times (n_1 \cdots n_N)$. Using the notation for the indices of the multidimensional data $\mathbf{X}$, the covariance can be described as a *block* matrix with *block-size* $(n_1 \cdots n_{N-1}) \times (n_1 \cdots n_{N-1})$ and

1058-6393/91 $01.00 © 1991 IEEE

block-order n_N. Block-order is defined as the number of blocks in a block row of a block matrix. In turn, each of the $(n_1 \cdots n_{N-1}) \times (n_1 \cdots n_{N-1})$ blocks can be described as block matrices with block-size $(n_1 \cdots n_{N-2}) \times (n_1 \cdots n_{N-2})$ and block-order n_{N-1}. This hierarchical block characterization of $\mathbf{R}$ continues to the lowest level in the matrix where the block-size is $n_1 \times n_1$ and the block-order is n_2. It is possible that each structure level in this natural block hierarchical description of $\mathbf{R}$ conforms to a possibly different type of structure.

2.2 Mixed Block Structure

In some signal processing problems, physical considerations impose certain structures on the covariance at different levels in the block hierarchy. These blocks at certain structure levels relate in some way to the other blocks at that structure level. For example, a block-Toeplitz matrix has equal blocks along any block diagonal. However, within each block the structure may be unconstrained or conform to a possibly different block structure. For instance, say that $\mathbf{R}$ is a $Q \times Q$ covariance matrix where $Q = LMNP$, and that it is made up of L^2, $MNP \times MNP$ sub-blocks that conform to a certain block structure. In turn, each of these individual $MNP \times MNP$ blocks may be made up of $NP \times NP$ sub-blocks that may conform to a possibly different block structure at that level. Each of these $NP \times NP$ blocks may be made up of $P \times P$ sub-blocks that may conform to a possibly different block structure at that level, and so on.

2.3 Notation

We now seek to establish a notation for constraint sets of covariance matrices with hierarchical block structure. Circulant and Toeplitz structure plays an important role in many signal processing problems, and the discussion is begun by focusing our attention on these two structures.

First, a notation for the individual types of structure possible at a particular level in the covariance block hierarchy is developed. The types of structure possible are denoted by the symbols in the set $\{\mathcal{C}, \mathcal{T}, \mathcal{U}\}$, where $\mathcal{C}$ represents circulant structure, $\mathcal{T}$ represents Toeplitz structure, and $\mathcal{U}$ represents no structure. An integer subscript on this structure symbol denotes the block-order at that particular structure level within the covariance. In order to specify the entire covariance structure, these structure symbols, along with their block-order subscripts, are concatenated from right to left to form a series of structure symbols. This series of symbols indicates the type of structure and block-order at each structure level, where the rightmost symbol describes the lowest structure level, and the leftmost symbol describes the highest structure level. This notation completely describes the covariance constraint set in that the dimension of the covariances belonging to the set is found by simply multiplying each of the symbol subscripts, and the number of structure levels is denoted by the number of structure symbols.

As an example of the notation just described, consider the set $\mathcal{C}_M \mathcal{T}_N \mathcal{U}_P$. A matrix $\mathbf{R}$ in this set is, at the highest block level, block-circulant such that the matrix has block-size $NP \times NP$ and block-order M. At the intermediate block level, the $NP \times NP$ blocks, of block-order N, are block-Toeplitz made up of blocks of block-size $P \times P$. At the lowest block level, the $P \times P$ blocks are unstructured.

2.4 Block-circulant and Block-Toeplitz Structure Descriptions

2.4.1 Block-circulant Structure Descriptions

An important property of circulant matrices most often exploited in signal processing applications is given by the following theorem.

Theorem 1 *[8],[7] Every $N \times N$ circulant matrix $\mathbf{C}_N$ is unitarily similar to a diagonal matrix $\mathbf{D}_N$ such that*

$$\mathbf{C}_N = \mathbf{W}_N \mathbf{D}_N \mathbf{W}_N^\dagger \tag{2}$$

where $\mathbf{W}_N$ is the $N \times N$ normalized discrete Fourier transform (DFT) matrix. The matrix $\mathbf{D}_N$ has as diagonal entries the elements of the DFT of the first column of $\mathbf{C}_N$.

This theorem says that all circulant matrices have known eigenvectors, and that these eigenvectors are the columns of the appropriate size normalized DFT matrix. Of particular importance is the fact that a circulant matrix has a spectral decomposition with fixed unitary matrices.

The concept of diagonalization of a circulant matrix shown in Theorem 1 can be extended to include block-diagonalization of block-circulant matrices. The matrix Kronecker product and its relevant properties are used to define similarity transformations that take block matrices with circulant structure at different levels in the matrix into block-diagonal form.

Theorem 2 *[4] Let $\mathbf{C}$ be a block-circulant matrix of block-order M and block-size N. Define the block matrix $\mathbf{V} = \mathbf{W}_M \otimes \mathbf{I}_N$ where $\mathbf{W}_M$ is the normalized*

$M \times M$ DFT matrix and $\mathbf{I}_N$ is the $N \times N$ identity matrix. The matrix $\mathbf{D} = \mathbf{V}\mathbf{C}\mathbf{V}^\dagger$ is then a block-diagonal matrix with block-order M and block-size N.

As Theorem 1 establishes an isomorphism between the set of circulant matrices and the set of diagonal $N \times N$ matrices, Theorem 2 establishes an isomorphism between the set of block-circulant matrices of block-order M and block-size N and the set of block-diagonal matrices of block-order M and block-size N.

Theorems 1 and 2 will now be used to establish an isomorphism between the set of circulant-block-circulant matrices of block-order M and block-size N and the set of $MN \times MN$ diagonal matrices.

Theorem 3 *[4] Let $\mathbf{C}$ be a circulant-block-circulant matrix of block-order M and block-size N. Define a block-matrix $\mathbf{V} = \mathbf{W}_M \otimes \mathbf{W}_N$ where $\mathbf{W}_M$ and $\mathbf{W}_N$ are the normalized $M \times M$ and $N \times N$ DFT matrices respectively. The matrix $\mathbf{E} = \mathbf{V}\mathbf{C}\mathbf{V}^\dagger$ is then an $MN \times MN$ diagonal matrix.*

Note that the Kronecker product $\mathbf{W}_M \otimes \mathbf{W}_N$ defines a matrix whose columns are the eigenvectors of the circulant-block-circulant matrix $\mathbf{C}$.

As an extension of the ideas presented in Theorems 1, 2, and 3, it can be easily shown that a covariance matrix with the block-circulant property at any level in its block hierarchy can be block-diagonalized by the proper Kronecker products of the appropriate size normalized DFT and identity matrices. For example, a covariance in the constraint set $\mathcal{C}_M\mathcal{C}_N\mathcal{U}_P$ can be block-diagonalized by the similarity transformation defined by $\mathbf{W}_M \otimes \mathbf{W}_N \otimes \mathbf{I}_P$, and a covariance in the constraint set $\mathcal{C}_M\mathcal{C}_N\mathcal{C}_P$ can be diagonalized by the similarity transformation defined by $\mathbf{W}_M \otimes \mathbf{W}_N \otimes \mathbf{W}_P$.

2.4.2 Block-Toeplitz Structure Descriptions

An important property of Toeplitz matrices relevant for array signal processing is that any non-negative definite Hermitian Toeplitz matrix denoting an autocorrelation (autocovariance matrix in the zero mean case) matrix may be written

$$\mathbf{T}_N = \frac{1}{2\pi}\int_{-\pi}^{\pi} S(\theta)\mathbf{d}(\theta)\mathbf{d}^\dagger(\theta)d\theta \qquad (3)$$

where $S(\theta)$ is a positive spectrum, and $\mathbf{d}(\theta)$ is an N-element Vandermonde vector of the form $\mathbf{d}(\theta) = [1, e^{-j\theta}, e^{-j2\theta}, \ldots, e^{-j(N-1)\theta}]^T$. This matrix form may be easily derived by noting that the first column of the autocorrelation matrix is the inverse Fourier transform of $S(\theta)$. Note that $S(\theta)$ is not necessarily uniquely determined for a given Toeplitz matrix $\mathbf{T}_N$.

In the Toeplitz constrained covariance estimation problem considered by Miller and Snyder [9], the class of non-negative definite Toeplitz matrices was restricted to those having non-negative definite circulant extensions. Matrices in this class can be written

$$\mathbf{T}_N = \sum_{i=0}^{P-1} \sigma_i \mathbf{d}(\theta_i)\mathbf{d}^\dagger(\theta_i). \qquad (4)$$

for some value of $P > N$. For sufficiently large P, this expression is a good approximation to the integral in Eqn. (3). If $\mathbf{T}_P$ denotes the non-negative definite P-length circularly extended matrix with the non-negative definite Toeplitz matrix $\mathbf{T}_N$ as its upper left corner such that $P \geq N$, using Theorem 1 we may write

$$\mathbf{T}_P = \mathbf{W}_P\mathbf{\Sigma}_P\mathbf{W}_P^\dagger. \qquad (5)$$

$\mathbf{W}_P$ is the normalized $P \times P$ DFT matrix and $\mathbf{\Sigma}_P$ is a positive diagonal matrix such that $\mathbf{\Sigma}_P = diag(\sigma_0, \ldots, \sigma_{P-1})$. We may therefore write

$$\mathbf{T}_N = \tilde{\mathbf{W}}_N\mathbf{\Sigma}_P\tilde{\mathbf{W}}_N^\dagger, \qquad (6)$$

where $\tilde{\mathbf{W}}_N = [\mathbf{I}_N \ \mathbf{0}_{N\times(P-N)}]\mathbf{W}_P$. Eqn. (6) is simply the matrix equivalent of Eqn. (4), where the columns of $\tilde{\mathbf{W}}_N$, denoted by $\mathbf{d}(\theta_i)$, are Vandermonde vectors of the form $\mathbf{d}(\theta) = [1, e^{-j\theta}, e^{-j2\theta}, \ldots, e^{-j(N-1)\theta}]^T$ on a fine grid in θ. The only difference in this description for a Toeplitz covariance matrix and the description given in Theorem 1 for a circulant covariance is that $\mathbf{W}_N$ in the circulant covariance description is a square unitary matrix, and $\tilde{\mathbf{W}}_N$ in the Toeplitz covariance description is a rectangular matrix with more columns than rows.

Eqns. (3) and (4) have a special interpretation in array signal processing. Under certain reasonable assumptions these equations can represent the covariance matrix of a uniform linear array under a narrowband signal model, where $S(\theta)$ represents a positive spatial (in terms of electrical angle) power spectral density and $\mathbf{d}(\theta)$ represents an array direction vector from the array manifold. The quantities are only a function of the directions of arrival of the signals in the array environment and that the integral and sum are only taken over the appropriate regions of spatial support (electrical angle). It is important to note that the array manifold for a uniform linear array consists of the set of all Vandermonde vectors of the form $\mathbf{d}(\theta) = [1, e^{-j\theta}, e^{-j2\theta}, \ldots, e^{-j(N-1)\theta}]^T$, where θ ranges over all the possible electrical angles for a given sensor spacing. The $\mathbf{d}(\theta_i)$ used in Eqn. (4) is the direction vector for the ith particular value of electrical angle used in the approximation to Eqn. (3).

Analogous to Eqn. (6), descriptions of covariance matrices that have block-Toeplitz structure at any level in the block hierarchy may be developed. This will be useful in describing the structure of a covariance matrix for wideband data in array signal processing problems using a uniform linear array, or in narrowband problems using a rectangular array.

As in the development of Eqn. (6), we are limited to describing those covariances that have a non-negative block-circulant extension of the appropriate length at each block-Toeplitz structure level. For example, in [4] it has been shown that a covariance matrix $\mathbf{R}$ in the constraint set $\mathcal{T}_M\mathcal{U}_N$ can be written

$$\mathbf{R} = \tilde{\mathbf{V}}\mathbf{\Sigma}\tilde{\mathbf{V}}^{\dagger}, \tag{7}$$

where $\tilde{\mathbf{V}} = [\mathbf{I}_M \ \ \mathbf{0}_{M\times(P-M)}]\mathbf{W}_P \otimes \mathbf{I}_N$ and $\mathbf{\Sigma}$ is a block-diagonal matrix of block-order P and block-size N. Also shown in [4] is the fact that a covariance matrix $\mathbf{R}$ in the constraint set $\mathcal{T}_M\mathcal{T}_N$ can be written in the form of Eqn. (7), where

$$\tilde{\mathbf{V}} = [\mathbf{I}_M \ \ \mathbf{0}_{M\times(P-M)}]\mathbf{W}_P \otimes [\mathbf{I}_N \ \ \mathbf{0}_{N\times(Q-N)}]\mathbf{W}_Q \tag{8}$$

and $\mathbf{\Sigma}$ is a real positive $PQ \times PQ$ diagonal matrix. As with the block-circulant structure, these ideas can be extended to develop descriptions of covariances that have block-Toeplitz structure at any level in its block hierarchy, given that they have a non-negative block-circulant extension at each block-Toeplitz structure level. It is simply a matter of choosing the correct form of the matrices involved in the Kronecker products.

2.4.3 Mixed Block Structure Descriptions

Descriptions of covariance matrices with a mixture of block-circulant and block-Toeplitz structure at any of the different block structure levels may also be developed. This is accomplished by combining the theory involved in the block-diagonalization of a block-circulant matrix, and the block matrix description of a block-Toeplitz matrix given by Eqn. (7). However, as before, we are limited to describing those covariances with mixtures of block-circulant and block-Toeplitz structure that have non-negative definite block-circulant extensions of the appropriate block length at the block-Toeplitz structure levels. As an example of a description of a covariance with this mixture of block structure, consider the following example that was introduced in the previous section. Let $\mathbf{R}$ be a covariance in the constraint set $\mathcal{C}_M\mathcal{T}_N\mathcal{U}_P$, with a non-negative definite Q-length $NP \times NP$ block-circulant extension at the block-Toeplitz structure level. $\mathbf{R}$ may be described by the following relation

$$\mathbf{R} = \tilde{\mathbf{V}}\mathbf{\Sigma}\tilde{\mathbf{V}}^{\dagger} \tag{9}$$

where

$$\tilde{\mathbf{V}} = \mathbf{W}_M \otimes ([\mathbf{I}_N \ \ \mathbf{0}_{N\times(Q-N)}]\mathbf{W}_Q) \otimes \mathbf{I}_P \tag{10}$$

and $\mathbf{\Sigma}$ is a non-negative definite block-diagonal matrix of block-order MQ and block-size P.

3 Generalizations of Block-circulant and Block-Toeplitz Structure

As an extension of the previous covariance matrix descriptions, we may develop descriptions of the covariance for one snapshot of data for an arbitrary array that is made up of a number of identical subarrays with reference sensors located according to some translation array. The subarrays may also be decomposed in a hierarchical manner. If the array manifold for each translation array and subarray is known at each level, the covariance description for one snapshot of data for these types of arrays is developed by simply choosing the appropriate Kronecker products, and has the form

$$\mathbf{R} = (\mathbf{A}_N \otimes \cdots \otimes \mathbf{A}_1)\mathbf{\Sigma}(\mathbf{A}_N \otimes \cdots \otimes \mathbf{A}_1)^{\dagger}. \tag{11}$$

The matrices $\mathbf{A}_i$ involved in the Kronecker products are matrices whose columns are the direction vectors for the translation arrays and subarrays at the different structure levels in the overall array geometry. The reader is referred to [1] for more details in the development of this covariance description.

With all of the above covariance matrix generalizations, it is now possible to develop (with certain restrictions) descriptions of the form

$$\mathbf{R} = \mathbf{A}\mathbf{\Sigma}\mathbf{A}^{\dagger} \tag{12}$$

for covariances in the constraint sets described in Section 2.3, covariances of one snapshot of data for arbitrary array geometries with arbitrary subarrays, and covariances of data vectors consisting of concatenated snapshots from these types of array geometries. Here $\mathbf{A}$ is a known matrix of Kronecker products of the form described above, and $\mathbf{\Sigma}$ is a non-negative definite block-diagonal matrix.

4 Summary and Conclusions

It was shown that multidimensional data and structured covariances arise in many signal processing problems, particularly those in the area of array signal processing. Under a complex Gaussian model, several well known covariance structures were shown to arise naturally from physical considerations involved in the formulation of basic array processing problems.

Multidimensional data was shown to induce a natural hierarchical block structure on the covariance for data organized into vector form by the multidimensional *vec* operator. This then led naturally to descriptions of covariance constraint sets that, due to characteristics of the multidimensional data in certain dimensions, had possibly different block structures at different levels within their block hierarchy. A unifying notation was then developed in order to completely characterize covariances belonging to these constraint sets.

Block-circulant and block-Toeplitz structure were shown to be of particular importance in that they gave rise to several generalizations of covariance structure, and provided a foundation on which to build a generalized notation for covariance structure. Most importantly, these covariance structure generalizations all have the same basic form, in that they are each described by multiplications of matrices made up of Kronecker products of known matrices and a block diagonal matrix (possibly diagonal).

These generalizations made possible the description of covariance constraint sets that characterize the data from an arbitrary array geometry consisting of identical subarrays. This notation was then further generalized to include descriptions of array data covariances that have a hierarchical subarray structure, and covariances for multiple snapshots of data from these types of arrays.

It is desired that the notation and covariance descriptions developed here will lead to straightforward estimation procedures for the covariance matrices in the described constraint sets.

Acknowledgments

This work was supported by the Office of Naval Research under Grant N00014-90-J-1297.

References

[1] T. A. Barton and D. R. Fuhrmann. Covariance Structures for Multidimensional Data, 1991. Research Monograph ESSRL-91-38, Washington University, St. Louis MO.

[2] J. P. Burg, D. G. Luenberger, and D. L. Wenger. Estimation of Structured Covariance Matrices. *Proceedings of the IEEE*, 70(9), September 1982.

[3] D. R. Fuhrmann. Application of Toeplitz Covariance Estimation to Adaptive Beamforming and Detection. *IEEE Transactions on Signal Processing*, 39(10), October 1991.

[4] D. R. Fuhrmann and T. A. Barton. Estimation of Block-Toeplitz Covariance Matrices, 1990. Research Monograph ESSRL-91-31, Washington University, St. Louis MO.

[5] N. R. Goodman. Statistical Analysis Based On A Certain Multivariate Complex Gaussian Distribution (An Introduction). *Annals of Mathematical Statistics*, March 1963.

[6] D. Gray, B. Anderson, and P. Sim. Estimation of Structured Covariances with Application to Array Beamforming. *Circuits, Systems, and Signal Processing*, 6(4):421–447, 1987.

[7] R. M. Gray. Toeplitz and Circulant Matrices: II. Technical report, Information Systems Laboratory, Stanford Electronics Laboratories, Stanford University, Stanford, California, April 1977.

[8] U. Grenander and G. Szego. *Toeplitz Forms and their Applications*. University of California Press, first edition, 1958.

[9] M. I. Miller and D. L. Snyder. The Role of Likelihood and Entropy in Incomplete-Data Problems: Applications to Estimating Point-Process Intensities and Toeplitz Constrained Covariances. *Proceedings of the IEEE*, 75(7), July 1987.

[10] P. J. Tourtier and L. L. Scharf. Maximum Likelihood Identification of Correlation Matrices for Estimation of Power Spectra at Arbitrary Resolutions. *The Proceedings of ICASSP*, pages 2066–2069, 1987.

Fast Updating of Maximum Likelihood Direction of Arrival Estimates

A. LEE SWINDLEHURST
Dept. of Electrical & Computer Engineering
Brigham Young University
Provo, Utah 84602

Abstract

The recently introduced weighted subspace fitting (WSF) and MODE algorithms have been shown to provide direction of arrival (DOA) estimates whose variance is asymptotically equivalent to that achievable by the stochastic (unconditional) maximum likelihood method. Updating the WSF and MODE DOA estimates when new data arrives requires recomputation of an eigendecomposition (EVD), a numerical operation that is quite complex and does not easily lend itself to parallelization. To overcome this problem, a DOA update algorithm is presented in this paper which has asymptotic properties identical to WSF and MODE, but which requires no EVD. The algorithm achieves computationally efficient estimate updates, and unlike the EVD may be easily parallelized. In addition to describing the algorithm and its asymptotic properties, the paper also includes a simulation example to validate the algorithm's performance.

1. Introduction

MAXIMUM LIKELIHOOD methods are a standard approach to solving parameter estimation problems such as those encountered in narrowband direction of arrival (DOA) estimation. In the maximum likelihood (ML) approach, the probability distribution of the observations is expressed as a function of the parameters to be estimated, and the so-called *(log)-likelihood* is maximized with respect to these parameters. Under certain regularity conditions, estimates obtained from the ML approach are both asymptotically unbiased and *efficient*, meaning that they achieve the Cramér-Rao lower bound (CRB) on estimate variance.

Despite its advantages, ML-based methods have not enjoyed much practical application in DOA estimation since they usually lead to non-linear, multidimensional optimization problems that are not easily solved. Even under the "simplest" assumptions of Gaussian signals and noise, maximizing the log-likelihood is quite complicated and does not easily lend itself to numerical solution [1, 2]. As a result, for many years research has focused on more easily implemented one-dimensional techniques such as beamforming, MUSIC [3], and the min-norm algorithm [4]. While these methods achieve ML or near-ML performance in many cases, difficulties arise when the signals have nearly co-incident DOAs or are highly correlated. These difficulties are a direct result of using a one-dimensional optimization procedure to solve what is inherently a multidimensional problem.

Recently, two new multidimensional algorithms have been proposed whose asymptotic second order performance is equivalent to that of ML; *i.e.*, they produce DOA estimates that asymptotically achieve the CRB for Gaussian signals and noise. These algorithms are the *weighted subspace fitting* (WSF) method of Viberg and Ottersten [5, 6], and the *method of direction of arrival estimation* (MODE) algorithm of Stoica and Sharman [7, 8]. One of the principal advantages of these techniques over ML is that the form of their respective criterion functions is more amenable to numerical extremization than is the corresponding log-likelihood.

Both WSF and MODE do, however, require that one compute the d principal eigenvectors and eigenvalues of the data covariance matrix, where d is the number of signals present. Initially, this is not a significant issue since all of the eigenvalues of the covariance are needed in determining the value of d. Once the number of signals is known, however, and an initial DOA estimate is obtained, updating the WSF and MODE estimates in response to additional data from the array requires that the principal eigenspace be completely recomputed. Algorithms have been proposed for performing the eigenspace update efficiently [9, 10, 11, 12, 13], but such techniques are still quite complex and often suffer from a linear buildup of round-off error.

The goal of this paper is to demonstrate that asymptotically efficient DOA estimates can be obtained without resorting to an eigendecomposition. In particular, an algorithm similar in form to WSF and possessing identical asymptotic properties will be de-

[1] This work was supported by the IR&D Program at ESL, Inc.

1058-6393/91 $01.00 © 1991 IEEE

veloped that requires computation of only the sample covariance and its inverse. While computing the inverse of an $m \times m$ matrix takes $O(m^3)$ operations in general and hence would not seem to provide much of a computational advantage, updating the inverse of a matrix after a rank-one modification is a trivial $O(m^2)$ calculation. Hence, the new algorithm is well-suited to the problem of updating the DOA estimates when a new sample, or snapshot, is taken from the array.

After introducing notation and describing the WSF approach in the next section, the new algorithm is presented in Section 3 and its asymptotic equivalence to WSF is established. Some important features of the algorithm are outlined in Section 4, and a simulation example is included in Section 5.

2. Weighted Subspace Fitting

Under the assumption of narrowband signals, the output of an m-element array $\mathbf{x}(t) \in \mathbb{C}^m$ due to d sources can be described by

$$\mathbf{x}(t) = \mathbf{A}(\boldsymbol{\theta}_0)\mathbf{s}(t) + \mathbf{n}(t) , \tag{1}$$

where $\mathbf{s}(t) \in \mathbb{C}^d$ represents the signal amplitude and phase at time t, $\mathbf{n}(t) \in \mathbb{C}^m$ is additive noise, and $\mathbf{A}(\boldsymbol{\theta}_0) \in \mathbb{C}^{m \times d}$ is the matrix of array response vectors for each source parameterized by the DOAs $\boldsymbol{\theta}_0 = [\theta_1, \cdots, \theta_d]^T$:

$$\mathbf{A}(\boldsymbol{\theta}_0) = [\mathbf{a}(\theta_1) \ \cdots \ \mathbf{a}(\theta_d)] \ . \tag{2}$$

In this analysis, it is assumed that the array parameterization $\mathbf{a}(\theta)$ is completely known (*i.e.*, the array is fully calibrated), and is unambiguous (*i.e.*, every matrix of the form (2) is full rank d for distinct θ_i). It will also be assumed without loss of generality that the noise is spatially white, or in other words that $\mathcal{E}\{\mathbf{n}(t)\mathbf{n}^*(t)\} = \sigma^2\mathbf{I}$. If the noise is further assumed to be uncorrelated with the signals, then the covariance of the array data is given by

$$\mathbf{R} \stackrel{\text{def}}{=} \mathcal{E}\{\mathbf{x}(t)\mathbf{x}^*(t)\} = \mathbf{A}(\boldsymbol{\theta}_0)\mathbf{S}\mathbf{A}^*(\boldsymbol{\theta}_0) + \sigma^2\mathbf{I} , \tag{3}$$

where $\mathbf{S} \stackrel{\text{def}}{=} \mathcal{E}\{\mathbf{s}(t)\mathbf{s}^*(t)\}$ is the covariance of the emitter signals. The goal of the narrowband DOA estimation problem is simply to use several observations from the array and exploit the model of (3) to estimate $\boldsymbol{\theta}$.

To implement an ML solution to this problem, the likelihood function of the data must first be determined and then maximized over all relevant model parameters. If the signal and noise waveforms $\mathbf{s}(t)$ and $\mathbf{n}(t)$ are assumed to be stationary, temporally white, zero-mean complex Gaussian random processes, then maximizing the log-likelihood over $\boldsymbol{\theta}$ can be shown to be equivalent to[2] [1, 2]

$$\min_{\boldsymbol{\theta}} \log\left(\det\left[\mathbf{P}_A\hat{\mathbf{R}}\mathbf{P}_A - \frac{\mathrm{Tr}(\mathbf{P}_A^\perp\hat{\mathbf{R}})}{m-d}\mathbf{P}_A^\perp\right]\right) , \tag{4}$$

where $\hat{\mathbf{R}}$ is a sample estimate of (3), and the projection operator $\mathbf{P}_A = \mathbf{I} - \mathbf{P}_A^\perp$ is defined as

$$\mathbf{P}_A = \mathbf{A}(\boldsymbol{\theta}) \left[\mathbf{A}^*(\boldsymbol{\theta})\mathbf{A}(\boldsymbol{\theta})\right]^{-1} \mathbf{A}^*(\boldsymbol{\theta}) \ .$$

This minimization in (4) is sometimes referred to as the *stochastic* or *unconditional* ML approach. While the estimates obtained from (4) will have very desirable properties, the determinant operation makes implementing this minimization quite complicated.

In the WSF approach of Viberg and Ottersten [5], the DOA estimates are obtained from the following much simpler minimization problem:

$$\begin{aligned}\hat{\boldsymbol{\theta}}_{WSF} &= \arg\min_{\boldsymbol{\theta},T} \hat{V}_{WSF}(\boldsymbol{\theta}) = \|\hat{\mathbf{E}}_s\mathbf{W}^{\frac{1}{2}} - \mathbf{A}(\boldsymbol{\theta})\mathbf{T}\|_F^2 \\ &= \arg\min_{\boldsymbol{\theta}} \mathrm{Tr}(\mathbf{P}_A^\perp\hat{\mathbf{E}}_s\mathbf{W}\hat{\mathbf{E}}_s^*) ,\end{aligned} \tag{5}$$

where $\hat{\boldsymbol{\theta}}_{WSF}$ is the estimated angle vector, $\hat{\mathbf{E}}_s$ is a set of vectors obtained from the data which approximately span the "signal" subspace $span\{\mathbf{A}(\boldsymbol{\theta}_0)\mathbf{S}\}$, and $\mathbf{W} = \mathbf{W}^*$ is a positive definite weighting matrix. Ottersten and Viberg showed that for a particular choice of $\mathbf{W}$, the variance of the estimation error $\hat{\boldsymbol{\theta}}_{WSF} - \boldsymbol{\theta}_0$ of (5) is minimized and in fact asymptotically (for large N) achieves the Cramér-Rao lower bound under the assumption of Gaussian signals and noise. WSF is thus asymptotically equivalent to (4), but it is much simpler to work with since it involves a trace operation instead of a determinant. Note that both algorithms are multidimensional in the sense that the elements of the vector $\hat{\boldsymbol{\theta}}$ are searched for "simultaneously." In addition, an estimate of reasonable quality is also required in either case to initialize the search for the DOAs.

The vectors $\hat{\mathbf{E}}_s$ used in (5) can be obtained from an eigendecomposition of the sample covariance $\hat{\mathbf{R}}$ of the array data:

$$\begin{aligned}\hat{\mathbf{R}} &= \frac{1}{N}\sum_{k=1}^{N}\mathbf{x}(t_k)\mathbf{x}(t_k)^* \\ &= \sum_{i=1}^{m}\hat{\lambda}_i\hat{\mathbf{e}}_i\hat{\mathbf{e}}_i^*\end{aligned}$$

[2]Note that equation (4) is expressed in *concentrated* form, where the separable parameters σ^2 and the non-redundant elements of $\mathbf{S}$ have been solved for.

$$= \hat{\mathbf{E}}\hat{\mathbf{\Lambda}}\hat{\mathbf{E}}^*$$

$$= \hat{\mathbf{E}}_s\hat{\mathbf{\Lambda}}_s\hat{\mathbf{E}}_s^* + \hat{\mathbf{E}}_n\hat{\mathbf{\Lambda}}_n\hat{\mathbf{E}}_n^* ,$$

where N is the number of observations, or "snapshots," taken from the array, and where $\hat{\mathbf{\Lambda}} = diag\{\hat{\lambda}_1 \geq \hat{\lambda}_2 \geq \cdots \geq \hat{\lambda}_m\}$ and $\mathbf{E} = [\hat{\mathbf{e}}_1, \cdots, \hat{\mathbf{e}}_m]$ are respectively the eigenvalues and eigenvectors of $\hat{\mathbf{R}}$. If there are no perfectly correlated signals, then

$$\hat{\mathbf{E}}_s = [\hat{\mathbf{e}}_1 \ \cdots \ \hat{\mathbf{e}}_d]$$

$$\hat{\mathbf{E}}_n = [\hat{\mathbf{e}}_{d+1} \ \cdots \ \hat{\mathbf{e}}_m]$$

$$\hat{\mathbf{\Lambda}}_s = \text{diag}\{\hat{\lambda}_1, \cdots, \hat{\lambda}_d\}$$

$$\hat{\mathbf{\Lambda}}_n = \text{diag}\{\hat{\lambda}_{d+1}, \cdots, \hat{\lambda}_m\} .$$

With these definitions, the optimal choice for $\mathbf{W}$ in (5) was shown in [5] to be

$$\mathbf{W} = \hat{\mathbf{W}}_o = (\hat{\mathbf{\Lambda}}_s - \hat{\sigma}^2\mathbf{I})^2\hat{\mathbf{\Lambda}}_s^{-1} , \quad (6)$$

where $\hat{\sigma}$ is a consistent estimate of the noise power. When perfectly coherent signals are present, *i.e.*, when $rank(\mathbf{S}) = d' < d$, then the low rank signal subspace is only of dimension d', and $\mathbf{E}_s$ and $\hat{\mathbf{\Lambda}}$ will contain only the first d' eigenvectors and eigenvalues of $\hat{\mathbf{R}}$. The weighting $\hat{\mathbf{W}}_o$ is $d' \times d'$ in such cases, and provided d and d' are known and $m > 2d - d'$ [14, 15], WSF is still able to obtain unique, minimum variance DOA estimates.

When additional data is obtained from the array, and an updated DOA estimate is desired, a *new* set of eigenvectors $\hat{\mathbf{E}}_s$ and a new weighting matrix $\hat{\mathbf{W}}_o$ must be computed in order to implement the WSF minimization of (5). Since only the d'-dimensional principal eigenspace is needed, this computation requires $O(m^2d')$ operations in general. However, the computation is iterative and relatively complex, and hence is not well-suited to a direct parallel implementation. Thus, in situations where the emitters are rapidly moving and the DOA estimates must be frequently updated, or simply when it is desired to account for an additional observation from the array, recomputing the WSF solution may be prohibitively costly.

3. An Efficient Update Algorithm

The principal objective of this paper is to demonstrate that DOA estimates obtained from the minimization problem

$$\hat{\boldsymbol{\theta}} = \arg\min_{\boldsymbol{\theta}} \hat{V}(\boldsymbol{\theta}) = \arg\min_{\boldsymbol{\theta}} \text{Tr}\left[\mathbf{P}_A^{\perp}(\hat{\mathbf{R}} + \hat{\sigma}^4\hat{\mathbf{R}}^{-1})\right] \quad (7)$$

have asymptotic properties identical to those obtained from WSF and the stochastic ML algorithm, and hence are asymptotically unbiased and statistically efficient. If

$$\mathbf{C}_{WSF} = \mathcal{E}\{(\hat{\boldsymbol{\theta}}_{WSF} - \boldsymbol{\theta}_0)(\hat{\boldsymbol{\theta}}_{WSF} - \boldsymbol{\theta}_0)^T\}$$

$$\mathbf{C} = \mathcal{E}\{(\hat{\boldsymbol{\theta}} - \boldsymbol{\theta}_0)(\hat{\boldsymbol{\theta}} - \boldsymbol{\theta}_0)^T\}$$

are the estimation error covariances for (5) and (7), then this amounts to showing that

$$\mathbf{C} = \mathbf{C}_{WSF} + o(N^{-1}) , \quad (8)$$

which in turn amounts to showing that [5]

$$\hat{V}'(\boldsymbol{\theta}_0) \stackrel{\text{def}}{=} \frac{\partial \hat{V}(\boldsymbol{\theta})}{\partial \boldsymbol{\theta}}\bigg|_{\boldsymbol{\theta}=\boldsymbol{\theta}_0} = \hat{V}'_{WSF}(\boldsymbol{\theta}_0) + o(N^{-1/2}) \quad (9)$$

$$\hat{V}''(\boldsymbol{\theta}_0) \stackrel{\text{def}}{=} \frac{\partial \hat{V}(\boldsymbol{\theta})}{\partial \boldsymbol{\theta}\partial \boldsymbol{\theta}^T}\bigg|_{\boldsymbol{\theta}=\boldsymbol{\theta}_0} = \hat{V}''_{WSF}(\boldsymbol{\theta}_0) + o(1) . \quad (10)$$

These facts are established in the course of proving the following theorem:

Theorem 3.1 *If $\hat{\sigma}^2$ is a consistent estimator of the noise power σ^2, then the DOA estimates obtained from (7) are consistent and have the same asymptotic variance as those obtained from the WSF minimization of (5).*

Proof: To begin with, note that the estimates obtained from (7) are unchanged if the constant $-2\hat{\sigma}^2(m-d)$ is added to $\hat{V}(\boldsymbol{\theta})$:

$$\min_{\boldsymbol{\theta}} \hat{V}(\boldsymbol{\theta}) = \min_{\boldsymbol{\theta}} \text{Tr}\left[\mathbf{P}_A^{\perp}(\hat{\mathbf{R}} + \hat{\sigma}^4\hat{\mathbf{R}}^{-1})\right] - 2\hat{\sigma}^2(m-d)$$

$$= \min_{\boldsymbol{\theta}} \text{Tr}\left[\mathbf{P}_A^{\perp}(\hat{\mathbf{R}} + \hat{\sigma}^4\hat{\mathbf{R}}^{-1} - 2\hat{\sigma}^2\mathbf{I})\right] . \quad (11)$$

Since

$$\hat{\mathbf{R}}^{-1} = \hat{\mathbf{E}}_s\hat{\mathbf{\Lambda}}_s^{-1}\hat{\mathbf{E}}_s^* + \hat{\mathbf{E}}_n\hat{\mathbf{\Lambda}}_n^{-1}\hat{\mathbf{E}}_n^* ,$$

$\hat{V}(\boldsymbol{\theta})$ can be further rewritten as

$$\hat{V}(\boldsymbol{\theta}) = \text{Tr}\left[\mathbf{P}_A^{\perp}\hat{\mathbf{E}}(\hat{\mathbf{\Lambda}} - \hat{\sigma}^2\mathbf{I})^2\hat{\mathbf{\Lambda}}^{-1}\hat{\mathbf{E}}^*\right] \quad (12)$$

$$= \hat{V}_{WSF}(\boldsymbol{\theta}) + \text{Tr}\left[\mathbf{P}_A^{\perp}\hat{\mathbf{E}}_n\hat{\mathbf{W}}_n\hat{\mathbf{E}}_n^*\right] \quad (13)$$

where $\hat{\mathbf{W}}_n = (\hat{\mathbf{\Lambda}}_n - \hat{\sigma}^2\mathbf{I})^2\hat{\mathbf{\Lambda}}_n^{-1}$, and thus the cost function is still guaranteed to be positive. Since $\hat{\mathbf{W}}_n = O(N^{-1})$ when $\hat{\sigma}^2$ is consistent [16], then

$$\lim_{N\to\infty} \hat{V}(\boldsymbol{\theta}) = \hat{V}_{WSF}(\boldsymbol{\theta}) .$$

The consistency of (7) is established by the fact that WSF is consistent, and by the continuity of both

$\hat{V}(\boldsymbol{\theta})$ and $\hat{V}_{WSF}(\boldsymbol{\theta})$ [5]. Along with (13), the fact that $\hat{\mathbf{W}}_n = O(N^{-1})$ also guarantees that the error terms in equations (9) and (10) are at least $o(N^{-1/2})$. Equation (8) thus holds, and the variance of the estimates obtained from (7) is thus asymptotically equivalent to that of WSF. ∎

4. Discussion

The most salient feature of (7) is that it requires no eigendecomposition. This may at first appear to be a moot point since evaluating $\hat{V}(\boldsymbol{\theta})$ does require that $\hat{\mathbf{R}}^{-1}$ be computed. However, updating $\hat{\mathbf{R}}^{-1}$ in response to additional data is considerably less complicated than updating the corresponding dominant eigenspace. For example, suppose $\hat{\mathbf{R}}_N$ represents the value of the array covariance after N snapshots, and suppose an additional observation $\mathbf{z} = \mathbf{x}(N+1)$ is received from the array. If the following consistent estimator of σ^2 is used[3]

$$\hat{\sigma}_N^2 = \frac{1}{m-d}\mathrm{Tr}(\mathbf{P}_A^\perp \hat{\mathbf{R}}_N) ,$$

then the data-dependent quantities in (7) may be updated as follows:

$$\hat{\mathbf{R}}_{N+1} = \frac{N}{N+1}\left[\hat{\mathbf{R}}_N + \frac{1}{N}\mathbf{z}\mathbf{z}^*\right]$$

$$\hat{\mathbf{R}}_{N+1}^{-1} = \frac{N+1}{N}\left[\hat{\mathbf{R}}_N^{-1} - \frac{\hat{\mathbf{R}}_N^{-1}\mathbf{z}\mathbf{z}^*\hat{\mathbf{R}}_N^{-1}}{N+\mathbf{z}^*\hat{\mathbf{R}}_N^{-1}\mathbf{z}}\right]$$

$$\hat{\sigma}_{N+1}^2 = \frac{N}{N+1}\hat{\sigma}_N^2 + \frac{\mathbf{z}^*\mathbf{P}_A^\perp\mathbf{z}}{(m-d)(N+1)} .$$

In situations where the emitters may be rapidly moving and it is desirable to exponentially decrease the contribution of previous snapshots, a constant $0 < \mu < 1$ may be defined and the data-dependent updates obtained as follows:

$$\hat{\mathbf{R}}_{N+1} = \mu\hat{\mathbf{R}}_N + (1-\mu)\mathbf{z}\mathbf{z}^*$$

$$\hat{\mathbf{R}}_{N+1}^{-1} = \mu^{-1}\hat{\mathbf{R}}_N^{-1} - \frac{(1-\mu)\hat{\mathbf{R}}_N^{-1}\mathbf{z}\mathbf{z}^*\hat{\mathbf{R}}_N^{-1}}{\mu\left[\mu+(1-\mu)\mathbf{z}^*\hat{\mathbf{R}}_N^{-1}\mathbf{z}\right]}$$

$$\hat{\sigma}_{N+1}^2 = \mu\hat{\sigma}_N^2 + \frac{(1-\mu)\mathbf{z}^*\mathbf{P}_A^\perp\mathbf{z}}{m-d} .$$

In either case, updating the data-dependent term in the cost function involves a simple, non-iterative $O(m^2)$ calculation that can be easily parallelized.

[3] Since the implementation of multidimensional minimizations like WSF and (7) requires reasonably good initial estimates, there is no loss of generality in assuming $\mathbf{P}_A^\perp$ is available in computing $\hat{\sigma}_N^2$.

In the proof of Theorem 3.1 it was shown that $\hat{V}(\boldsymbol{\theta})$ is essentially equivalent to

$$\mathrm{Tr}(\mathbf{P}_A^\perp\hat{\mathbf{E}}(\hat{\mathbf{\Lambda}} - \hat{\sigma}^2\mathbf{I})^2\hat{\mathbf{\Lambda}}^{-1}\hat{\mathbf{E}}^*) . \quad (14)$$

Comparison of this expression with (5) and (6) shows that (7) can be thought of as implementing a WSF minimization using *all* of the eigenvectors and not just the first d'. The asymptotic equivalence of (14) and $\hat{V}_{WSF}(\boldsymbol{\theta})$ is not surprising since the weights multiplying $\mathbf{e}_{d'+1}, \cdots, \mathbf{e}_m$ in (14) are near zero for large N. Note also that the minimization of (7) requires only that d and not d' be known. Although it is much more difficult to determine d than d' when $d' < d$, not having to continually estimate d' at each update is advantageous in situations where emitters may be fading in and out of coherence.

While adding the constant $-2\hat{\sigma}^2(m-d)$ to $\hat{V}(\boldsymbol{\theta})$ in (11) does not change the resulting DOA estimates, it makes the minimization more amenable to a Gauss-Newton (GN) search. This is because the constant forces $\hat{V}(\boldsymbol{\theta}) \simeq 0$ in the vicinity of the true solution, which is an assumption that the GN technique relies on (*e.g.*, see [6, 17]). For each GN iteration, computing the gradient and approximate Hessian of the cost function in (11) requires $O(m^2d)$ operations, compared with $O(md^2)$ operations for WSF. On the other hand, updating the corresponding data-dependent quantities takes $O(m^2)$ for (11) and $O(m^2d)$ for WSF. Thus, strictly speaking, both algorithms require roughly the same level of computational effort in obtaining an updated DOA estimate via GN. The advantage of (11) lies in the simplicity of its implementation; all of its $O(m^2d)$ and $O(m^2)$ computations are simple matrix/vector products, and are easily parallelized.

If in (11) $\hat{\sigma}^2$ is treated as a parameter to be estimated rather than as a consistent estimate of σ^2, then a somewhat different consistent estimator of σ^2 is obtained:

$$\hat{\sigma}^2 = \frac{m-d}{\mathrm{Tr}(\mathbf{P}_A^\perp\hat{\mathbf{R}}^{-1})} .$$

If this estimate is substituted back into (11), the following concentrated version of the cost function results:

$$\hat{V}(\boldsymbol{\theta}) = \mathrm{Tr}(\mathbf{P}_A^\perp\hat{\mathbf{R}}) - \frac{(m-d)^2}{\mathrm{Tr}(\mathbf{P}_A^\perp\hat{\mathbf{R}}^{-1})} . \quad (15)$$

Equation (15) thus provides another "eigendecomposition-free" alternative to WSF. It is interesting to note the similarity of the cost functions in (7) and (15) with that of the so-called *deterministic* or *conditional* ML algorithm (CML), which implements

$$\hat{\boldsymbol{\theta}} = \arg\min_{\boldsymbol{\theta}} \hat{V}_{CML}(\boldsymbol{\theta}) = \arg\min_{\boldsymbol{\theta}} \mathrm{Tr}(\mathbf{P}_A^\perp\hat{\mathbf{R}}) . \quad (16)$$

The CML approach differs from unconditional ML in that it treats samples of the signal waveforms as deterministic parameters instead of random variables. Ottersten and Viberg have shown that WSF asymptotically outperforms CML independent of whether the signals are deterministic or stochastic [18]. The analysis above demonstrates that the estimates obtained from (7) will also have asymptotically lower variance than CML.

5. A Simulation Example

This purpose of this example is simply to demonstrate that the algorithms described in (5) and (11) have essentially the same performance, and that they both significantly outperform CML. A 12-element uniform circular array with unity gain sensors and a 1-λ radius was assumed in this example, and two emitters were simulated at angles of 0° and 10° with respect to a line through one of the sensors. The emitter signals were generated as constant amplitude planewaves with random phase, uniformly distributed on $[0, 2\pi]$. The 0° source had an SNR of 0 dB, while that of the 10° source was 20 dB.

In each trial, an estimate based on 500 snapshots was obtained for WSF, CML, and the minimization of (11). A total of 1000 trials were conducted for each of the correlation coefficient values 0.8, 0.9, 0.95, and 0.99, and the standard deviation of the estimates was computed. The results are shown in Figure 1, where the connected lines indicate theoretical predictions of the estimate standard deviation (see [5]), and the symbols '*', 'x', and 'o' indicate sample estimates computed from the simulations. Note that the acronym MUD stands for Maximum likelihood UpDate, and refers to the minimization of (11).

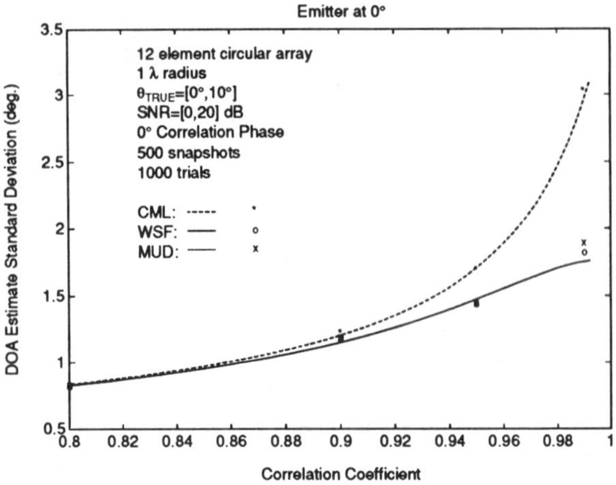

Figure 1: Algorithm Performance vs. Correlation Coefficient

References

[1] J. F. Böhme, "Estimation of Spectral Parameters of Correlated Signals in Wavefields", *Signal Processing*, **10**:329–337, 1986.

[2] A. G. Jaffer, "Maximum Likelihood Direction Finding of Stochastic Sources: A Separable Solution", In *Proc. ICASSP '88*, pages 2893–2896, New York, N.Y., 1988.

[3] R. Schmidt, *A Signal Subspace Approach to Multiple Emitter Location and Spectral Estimation*, PhD thesis, Stanford University, 1981.

[4] R. Kumaresan and D. Tufts, "Estimating the Angle of Arrival of Multiple Plane Waves", *IEEE Trans. on AES*, **19**:134–139, 1983.

[5] M. Viberg and B. Ottersten, "Sensor Array Processing Based on Subspace Fitting", *IEEE Trans. on Sig. Proc.*, **39**(5):1110–1121, May 1991.

[6] M. Viberg, B. Ottersten, and T. Kailath, "Detection and Estimation in Sensor Arrays Using Weighted Subspace Fitting", *IEEE Trans. on Sig. Proc.*, **39**(11), Nov. 1991.

[7] P. Stoica and K. Sharman, "Maximum Likelihood Methods for Direction-of-Arrival Estimation", *IEEE Trans. on ASSP*, **38**(7):1132–1143, July 1990.

[8] P. Stoica and A. Nehorai, "Performance Study of Conditional and Unconditional Direction-of-Arrival Estimation", *IEEE Trans. on ASSP*, **38**(10):1783–1795, October 1990.

[9] J. R. Bunch, C. P. Nielsen, and D. C. Sorensen, "Rank-One Modification of the Symmetric Eigenproblem", *Numerische Mathematik*, **31**:31–48, 1978.

[10] I. Karasalo, "Estimating the Covariance Matrix by Signal Subspace Averaging", *IEEE Trans. on ASSP*, **34**(1):8–12, January 1986.

[11] R. Schreiber, "Implementation of Adaptive Array Algorithms", *IEEE Trans. on ASSP*, **34**(5):1038–1045, October 1986.

[12] R. D. DeGroat and R. A. Roberts, "Efficient, Numerically Stabilized Rank-One Eigenstructure Updating", *IEEE Trans. on ASSP*, **38**(2):301–316, February 1990.

[13] K. Yu, "Recursive Updating the Eigenvalue Decomposition of a Covariance Matrix", *IEEE Trans. on ASSP*, **39**(5):1136–1145, May 1991.

[14] Y. Bresler and A. Macovski, "On the Number of Signals Resolvable by a Uniform Linear Array", *IEEE Trans. on ASSP*, **34**(6):1361–1375, December 1986.

[15] M. Wax and I. Ziskind, "On Unique Localization of Multiple Sources by Passive Sensor Arrays", *IEEE Trans. ASSP*, **ASSP-37**(7):996–1000, July 1989.

[16] P. Stoica and A. Nehorai, "MUSIC, Maximum Likelihood, and Cramér-Rao Bound", *IEEE Trans. on ASSP*, **37**(5):720–741, May 1989.

[17] A. Swindlehurst, B. Ottersten, R. Roy, and T. Kailath, "Multiple Invariance ESPRIT", *to appear in IEEE Trans. on Sig. Proc.*, **40**, April 1992.

[18] B. Ottersten and M. Viberg, "Asymptotic Robustness of Sensor Array Processing Methods", In *Proc. IEEE ICASSP*, volume 5, pages 2635–2638, Albuquerque, NM, 1990.

An Epistemic Utility Based Set-Valued Multiple-Target Tracker

W.C. Stirling and J.B. Thompson
Elect. & Comp. Eng'g Dept.
Brigham Young University
Provo, UT 84602

D.R. Morrell
Elect. & Comp. Eng'g Dept.
Arizona State University
Tempe, AZ 85287-5706

Abstract

The performance of a MTT (multiple target tracking) method should degrade gracefully as the conditions of the collection become less favorable to optimal operation. By stressing the avoidance, rather than the explicit minimization, of error, we obtain a decision rule for trajectory-data association that does not require the resolution of all conflicting hypotheses when the database does not contain sufficient information to do so reliably.

1 Introduction

We develop a multiple target trajectory-data association decision rule that is expressed in terms of two probabilities: one governing the informational value of the association hypothesis, and one governing the subjective belief, or credal, probability of the association hypothesis [1]. We also define a criterion of *serious possibility*, and all trajectory-data associations that are seriously possible are retained; it is not necessary to resolve all conflicting hypotheses before processing more data.

We employ an *set*-valued estimator, rather than the conventional point-valued estimator, to describe the evolution of the vehicle trajectory. The set-valued estimator is based upon the set-valued Kalman filter [2], which computes a convex set of trajectories, all with equal claims to validity, given the observations.

2 Epistemic Utility Theory

For an inquiry under investigation, suppose there are finitely many hypotheses that may be considered. Let U denote this set of possible answers and assume that exactly one element of U is correct, and that all elements of U are consistent. U is said to be an *ultimate partition*, and a *potential answer* is the collection of hypotheses remaining after we have rejected all members of a subset of U. Each element of a potential answer is said to be a *serious possibility.*

Epistemic utility is composed of a convex combination of two probability functions, one measuring the importance of acquiring new information, the other measuring the importance of avoiding error. We define the utility of accepting $g \subset U$ in the interest of avoiding error as $\mathcal{T}(g;\ell) = 1$ if $\ell = true$, and $\mathcal{T}(g;\ell) = 0$ if $\ell = false$. In addition to the cost of error, we also apportion a unit of informational value to each hypothesis $h_i \in U$ by assigning to elements of U non-negative real values such that their sum is unity. If we enumerate $U = \{h_1, h_2, \ldots, h_n\}$, and let $\mathcal{M}(h_j) \geq 0$ denote the value assigned to h_j, then $\sum_{j=1}^{n} \mathcal{M}(h_j) = 1$, and for any set $g \subset U$, we define $m(g) = \sum_{h_j \in g} m(h_j)$ as the informational value of rejecting g.

We may address the conflict that exists between the goals of avoiding error and acquiring information by defining an *epistemic utility function* for acquiring error-free knowledge (that is, making a decision) as the convex combination $u(g;\ell) = \alpha\tau(g;\ell) + (1-\alpha)(1-m(g))$. The quantity α represents the relative importance attached to avoiding error versus acquiring new information. We must restrict $\frac{1}{2} \leq \alpha \leq 1$ to ensure that no erroneous answer is preferred to any correct answer. Since all utility functions that are related by a positive linear transformation are equivalent, we may simplify this utility function by defining $u^a(g;\ell) = \frac{1}{\alpha}u(g;\ell) - \frac{1-\alpha}{\alpha}$. The resulting utility function for accepting g in the interest of both avoiding error and acquiring new knowledge is

$$u^a(g;\ell) = \begin{cases} 1 - bm(g) & \text{if } \ell = t \\ -bm(g) & \text{if } \ell = f \end{cases}, \tag{1}$$

where $b = \frac{1-\alpha}{\alpha}$ is the *coefficient of boldness* and lies in the unit interval. The closer b is to unity, the less caution is exercised that error will be introduced (increased boldness in accepting hypotheses); the closer b is to zero, the lower the risk of error (decreased boldness).

We must also establish a probabilistic measure of belief for the hypotheses that are available. *Credal* probability is probability formed on the basis of subjective judgment, represents the likelihood that an option is true, and is independent of any informational value or demand that might be associated with the option.

For a given ultimate partition U, let $q(g)$ denote the credal probability assignment to any element $g \subset U$. For $g \subset U$, the expected utility is

$$\begin{aligned} E_q u_a(g;\ell) &= [1-bm(g)]q(g) - bm(g)[1-q(g)] \\ &= q(g) - bm(g), \end{aligned}$$

where $E_q(\cdot)$ is mathematical expectation. This expected utility represents a tradeoff between the desire to acquire new knowledge and the desire to avoid error. The choice of b establishes a threshold at which the demand for knowledge renders the risk of error worthwhile.

We may adopt any set of hypotheses in the Boolean algebra generated by the elements of the ultimate partition, U. This expands our possibilities; we are not constrained to select only the elementary hypotheses, h_i, but may choose any subset of them. This decision philosophy may be summarized as follows:

Levi's Rule of Epistemic Utility [3, page 53]: *Given a finite ultimate partition U, an information-determining probability function m defined over the Boolean algebra of elements of U, an expectation-determining probability function q defined over the same algebra, and an index of boldness b, the agent should reject all and only those elements of $h_i \in U$ satisfying $q(h_i) < bm(h_i)$.*

3 Multiple Target Tracking

3.1 Track Initialization

We assume that each track is characterized by a linear stochastic dynamics model of the form

$$\mathbf{x}_{jt} = F_{jt-1}\mathbf{x}_{jt-1} + G_{jt-1}\mathbf{u}_{jt-1}, \qquad (2)$$

where $j = 1, 2, \cdots, \mathcal{T}_t$ and $\mathbf{u}_{jt-1} \sim \mathcal{N}(0, Q_{jt-1})$ (i.e., $\mathbf{u}_{it-1}$ is Gaussian with mean 0 and covariance matrix Q_{jt-1}); there are $\mathcal{T}_t$ active tracks at time t ($\mathcal{T}_t$ is unknown). The number of tracks is allowed to vary since tracks may be initiated or terminated at any time. We assume that all tracks lie in the same state space.

In the interest of brevity, we restrict attention to the outputs of a single sensor, and assume that this output may be characterized by a linear stochastic model of the form

$$\mathbf{z}_{it} = H_{it}\mathbf{x}_{jt} + \mathbf{v}_{it}, \quad i = 1, \cdots s_t, \qquad (3)$$

where s_t is the number of observations vectors at time t and $\mathbf{z}_{it}$ is an r_{it}-dimensional random vector, and $\mathbf{v}_{it} \sim \mathcal{N}(0, R_{it})$. Each observation vector, therefore, lies in an r_{it} dimensional space corresponding to the column space of H_{it}. We assume that each such space is a subset of the state space ($r_{it} \leq n$). We do not, however, require all observations to lie in the same dimensional subspace of the state space, and we permit the dimensionality of these column spaces to be time-varying.

Before data are collected, we characterize the target environment with one set-valued track defined by an initial credal matrix K_0, an initial centroid state $\underline{c}_0$, and a prior covariance matrix Π_0. The initial state-vector set is

$$\mathbf{X}_{0|0} = \left\{\mathbf{x} \sim \mathcal{N}(\underline{x}, P_{0|0}) \ : \ \underline{x} \in \underline{X}_{0|0}\right\},$$

where

$$\underline{X}_{0|0} = \left\{\underline{x} \in \Re^n : (\underline{x} - \underline{c}_{0|0})^T \left[S_{0|0}\right]^{-1} (\underline{x} - \underline{c}_{0|0}) \leq 1\right\},$$

$P_{0|0} = \Pi_0$ is a positive-definite matrix, and $S_{0|0} = K_0 K_0^T$. For sample times $t = 1, 2, \cdots$, we observe s_t observations $\underline{Z}_t = \{\underline{z}_{1t}, \cdots, \underline{z}_{s_t t}\}$. Let us suppose, at time t, that we have $\mathcal{T}_{t-1}$ sets of predicted (from time $t-1$) random variables of the form

$$\mathbf{X}^j_{t|t-1} = \left\{\mathbf{x} \sim \mathcal{N}(\underline{x}, P^j_{t|t-1}) \ : \ \underline{x} \in \underline{X}^j_{t|t-1}\right\},$$

where

$$\underline{X}^j_{t|t-1} = \left\{\underline{x} \in \Re^n : (\underline{x} - \underline{c}^j_{t|t-1})^T \left[S^j_{t|t-1}\right]^{-1} (\underline{x} - \underline{c}^j_{t|t-1}) \leq 1\right\}, \qquad (4)$$

where $S^j_{t|t-1} = \left[K^j_{t|t-1}\right]\left[K^j_{t|t-1}\right]^T$. with

$$\begin{aligned} \underline{c}^j_{t|t-1} &= F_{t-1}\underline{c}^j_{t-1|t-1} & (5) \\ P^j_{t|t-1} &= F_{t-1}P^j_{t-1|t-1}F^T_{t-1} + G_{t-1}Q_{t-1}G^T_{t-1} & (6) \\ K^j_{t|t-1} &= F_{t-1}K^j_{t-1|t-1}, & (7) \end{aligned}$$

for $j = 1, \cdots, \mathcal{T}_{t-1}$. We shall use the same F_t, G_t, and Q_t matrices for each track, thus it will not be necessary to index these matrices with their track identifications.

3.2 The Ultimate Partition

At each time, t, there exist s_t observation sample vectors, $\{\underline{z}_{it}\}_{i=1}^{s_t}$ of the random variables $\{\mathbf{z}_{it}\}_{i=1}^{s_t}$, and $\mathcal{T}_{t-1}$ predicted track sets, $\{\underline{X}^j_{t|t-1}\}_{j=1}^{\mathcal{T}_{t-1}}$. We wish to make decisions regarding the association of each $\underline{z}_{it}$ with each track set $\underline{X}^j_{t|t-1}$.

We must define an ultimate partition for each sample vector, resulting in a set of s_t ultimate partitions of the form $U_{it} = \{h_{it1}, \cdots, h_{it\mathcal{T}_{t-1}}, h_{it\mathcal{T}_t}\}$, $i = 1, \cdots, s_t$, where

$$h_{itj} = \begin{cases} \exists \underline{x} \in \underline{X}^j_{t|t-1} : \underline{z}_{it} = H_{it}\underline{x} & j = 1, \cdots, \mathcal{T}_{t-1} \\ \emptyset & j = \mathcal{T}_t \end{cases},$$

where $\emptyset$ signifies the hypothesis that none of the track sets associate with the observation $\underline{z}_{it}$. We shall say that track set $\underline{X}^j_{t|t-1}$ is *associated* with observation sample vector $\underline{z}_{it}$ if we fail to reject the hypothesis h_{itj}. Each ultimate partition, U_{it}, has the property that exactly one element is true, although each is logically possible.

3.3 Calculation of the m-Function

The m-function measures the information-value of rejecting an association. One such measure of rejecting the association of a predicted track set $\underline{X}^j_{t|t-1}$ with an observation $\underline{z}_{it}$ is the distance between sample values of the observation and and the track set; if the distance is small there is little value in rejecting the association (in other words, there is great value in accepting it). For $\underline{x} \in \underline{X}^j_{t|t-1} \subset \Re^n$ and $\underline{z}_{it} \in \Re^{r_{it}}$, we define the *generalized distance* between them as

$$d(\underline{x}, \underline{z}_{it}) = \|H_{it}\underline{x} - \underline{z}_{it}\| \stackrel{\text{def}}{=} \left[(H_{it}\underline{x} - \underline{z})^T (H_{it} - \underline{z}_{it})\right]^{\frac{1}{2}}.$$

We restrict attention only to a *seriously possible region* for $\underline{z}_{it}$, given by

$$\Sigma_{it} = \{\underline{\varsigma} \in \Re^{r_{it}} : |\varsigma_\ell - z_{it\ell}| \leq \theta \rho_{i\ell}, \ell = 1, r_{it}\}, \quad (8)$$

where $R_{it} = \text{diag}\{\rho^2_{i1}, \cdots, \rho^2_{ir_{it}}\}$. and θ a given constant. Define

$$m_{it}(\underline{x}) = \frac{d(\underline{x}, \underline{z}_{it})}{\int_{\Sigma_{it}} \|\underline{\varsigma} - \underline{z}_{it}\| d\underline{\varsigma}} = \frac{\|H_{it}\underline{x} - \underline{z}_{it}\|}{\int_{\Sigma_{it}} \|\underline{\varsigma} - \underline{z}_{it}\| d\underline{\varsigma}} \quad (9)$$

for all $\underline{x}$ whose projection onto the space spanned by the columns of H_{it} lies in Σ_{it}. We shall denote this space by $\Xi_{it} = \{\underline{x} \in \Re^n : \mathcal{P}_{it}\underline{x} \in \Sigma_{it}\}$, where $\mathcal{P}_{it}$ is the projection operator onto the space spanned by the columns of H_{it}.

The function $m_{it}(\underline{x})$ is a normalized distance between $\underline{z}_{it}$ and $\underline{x} \in \underline{X}^{ij}_{t|t-1}$, and is a measure of the information gained by rejecting the association of $\underline{x} \in \underline{X}^j_{t|t-1}$ and $\underline{z}_{it}$. Let $\underline{B}_{\underline{x}}$ be a ball with center at $\underline{x} \in \underline{X}^j_{t|t-1}$. The information value of rejecting the association of $\underline{B}_{\underline{x}}$ with $\underline{z}_{it}$ is

$$m_{it}(\underline{B}_{\underline{x}}) = \int_{\mathcal{P}_{it}\underline{B}_{\underline{x}} \cap \Xi_{it}} m_{it}(\underline{\xi}) d\mathcal{P}_{it}\underline{\xi}, \quad , i = 1, \cdots, s_t, \quad (10)$$

where $\mathcal{P}_{it}\underline{B}_{\underline{x}}$ is the projection of the ball $\underline{B}_{\underline{x}}$ onto the space spanned by the columns of H_{it}. The vector $\mathcal{P}_{it}\underline{x}$ is the projection of $\underline{x}$ onto the same column space.

For $\underline{x} \in \underline{X}^j_{t|t-1}$ and $\underline{x}' \in \underline{X}^j_{t|t-1}$ with $\underline{x} \neq \underline{x}'$, let the diameter of the balls $\underline{B}_{\underline{x}}$ and $\underline{B}_{\underline{x}'}$ become arbitrarily small. Then the condition $m_{it}(\underline{B}_{\underline{x}}) < m_{it}(\underline{B}_{\underline{x}'})$ indicates that the information value of accepting the association of $\underline{x}$ with $\underline{z}_{it}$ is greater than the value of accepting the association of $\underline{x}'$ and $\underline{z}_{it}$.

3.4 Calculation of the q-Functions

The credal probability, or q-function, is the probability that a given track-data association is correct. This probability is characterized by the conditional distribution of the observation given the track. For each $j = 1, \cdots, \mathcal{T}_{t-1}$, the set of predicted random vectors is given by

$$\mathbf{X}^j_{t|t-1} = \left\{\mathbf{x} \sim \mathcal{N}(\underline{x}, P^j_{t|t-1}) : \underline{x} \in \underline{X}^j_{t|t-1}\right\}.$$

For each observation random vector $\mathbf{z}_{it}$ we represent the corresponding set of filtered random vectors by

$$\mathbf{X}^{ij}_{t|t} = \left\{\hat{\mathbf{x}} : \hat{\mathbf{x}} = \mathbf{x} + W_{it}[\mathbf{z}_{it} - H_{it}\mathbf{x}], \quad \mathbf{x} \in \mathbf{X}^j_{t|t-1}\right\},$$

where W_{it} is the Kalman gain matrix given via the Kalman filter.

The probability distributions of $\{\hat{\mathbf{x}} \in \mathbf{X}^{ij}_{t|t}\}$ are characterized by the family of posterior distributions $\left\{p^{ij}_{\hat{\mathbf{x}}_{t|t}}(\underline{\xi}; \underline{x}); \underline{x} \in \underline{X}^j_{t|t-1}\right\}$ obtained via Bayes rule. We shall restrict consideration to the case where $\mathbf{x} \in \mathbf{X}^j_{t|t-1}$ and $\mathbf{z}_{it}$ are jointly normal, so the mean and variance associated with this conditional distribution are given by the Kalman filter. For each observation $\mathbf{z}_{it} = \underline{z}_{it}$, we calculate the filtered set-valued estimate according to

$$\underline{X}^{ij}_{t|t} = \left\{\underline{x} \in \Re^n : (\underline{x} - \underline{c}^{ij}_{t|t})^T \left[S^{ij}_{t|t}\right]^{-1} (\underline{x} - \underline{c}^{ij}_{t|t}) \leq 1\right\}, \quad (11)$$

for $i = 1, \cdots, s_t$, where $S^{ij}_{t|t} = K^{ij}_{t|t}\left[K^{ij}_{t|t}\right]^T$ and

$$\underline{c}^{ij}_{t|t} = \underline{c}^j_{t|t-1} + W_{ijt}[\underline{z}_{it} - H_{it}\underline{c}^j_{t|t-1}] \quad (12)$$

$$P^{ij}_{t|t} = [I - W_{ijt}H_{it}]P^j_{t|t-1} \quad (13)$$

$$K^{ij}_{t|t} = [I - W_{ijt}H_{it}]K^j_{t|t-1} \quad (14)$$

for $j = 1, \cdots, \mathcal{T}_{t-1}$, with W_{ijt} the Kalman gain defined by $W_{ijt} = P^j_{t|t-1}H^T_{it}\left[H_{it}P^j_{t|t-1}H^T_{it} + R_{it}\right]^{-1}$. The posterior densities assume the form (for $\underline{x} \in \underline{X}^j_{t|t-1}$)

$$p^{ij}_{\hat{\mathbf{x}}_{t|t}}(\underline{\xi}; \underline{x}) = \mathcal{N}\left(\underline{x} + W_{ijt}(\underline{z}_{it} - H_{it}\underline{x}), [I - W_{ijt}H_{it}]P^j_{t|t-1}\right), \quad (15)$$

But $\mathcal{N}\left(\underline{x} + W_{ijt}(\underline{z}_{it} - H_{it}\underline{x}), [I - W_{ijt}H_{it}]P^j_{t|t-1}\right) = \mathcal{N}\left(\hat{\underline{x}}, P^{ij}_{t|t}\right)$, where we define

$$\hat{\underline{x}} = \underline{x} + W_{ijt}(\underline{z}_{it} - H_{it}\underline{x}). \quad (16)$$

The posterior density is

$$p^{ij}_{\hat{\mathbf{x}}_{t|t}}(\underline{\xi}; \underline{x}) = \frac{1}{(2\pi)^{\frac{n}{2}} \left|P^{ij}_{t|t}\right|^{\frac{1}{2}}} e^{-\frac{1}{2}(\underline{\xi} - \hat{\underline{x}})^T [P^{ij}_{t|t}]^{-1} (\underline{\xi} - \hat{\underline{x}})}, \quad (17)$$

with $\hat{x}$ a function of x given by (16).

The family of densities $\{p^{ij}_{\hat{\mathbf{x}}_{t|t}}(\underline{\xi}; \underline{x}), \quad \underline{x} \in \underline{X}^j_{t|t-1}\}$ provides a measure of the truth-value of the association of each $\underline{x} \in \underline{X}^j_{t|t-1}$ with the observation $\underline{z}_{it}$.

Let $\underline{B}_{\underline{x}}$ be a ball with center at $\underline{x}$. The credal probability that the true state lies in $\underline{B}_{\underline{x}}$ conditioned on the observational value $\underline{z}_{it}$ is

$$q_{ijt}(\underline{B}_{\underline{x}};\underline{x}) = \int_{\underline{B}_{\underline{x}}} p^{ij}_{\hat{x}_{t|t}}(\underline{\xi};\underline{x})d\underline{\xi}. \tag{18}$$

For $\underline{x} \in \underline{X}^j_{t|t-1}$ and $\underline{x}' \in \underline{X}^j_{t|t-1}$ with $\underline{x} \neq \underline{x}'$, the condition $q_{ijt}(\underline{B}_{\underline{x}};\underline{x}) > q_{ijt}(\underline{B}_{\underline{x}'};\underline{x}')$ indicates that the association of $\underline{x}$ with $\underline{z}_{it}$ is more credible, or believable, than the association of $\underline{x}'$ and $\underline{z}_{it}$.

3.5 The Association Likelihood Ratio Test

The decision rule may be formulated in terms of the information-determining probability density, $m_{it}(\underline{x})$ (where $\underline{x} \in \underline{X}^j_{t|t-1}$), and the family of subjective belief, or credal, probability density functions functions, $\{p^{ij}_{\hat{x}_{t|t}}(\underline{\xi};\underline{x}), \underline{x} \in \underline{X}^j_{t|t-1}\}$. We desire to apply Levi's rule of expected utility to this problem.

For $\underline{x} \in \underline{X}_{t|t-1}$, let $\underline{B}_{\underline{x}}$ be a ball with center at $\underline{x}$. Using (10) and (18), Levi's rule of expected utility indicates we may not reject the association of $\underline{B}_{\underline{x}}$ and $\underline{z}_{it}$ if

$$q_{ijt}(\underline{B}_{\underline{x}};\underline{x}) \geq bm_{it}(\underline{B}_{\underline{x}}). \tag{19}$$

Now let the radius of $\underline{B}_{\underline{x}}$ go to zero, and define the function

$$q_{ijt}(\underline{x}) \stackrel{\text{def}}{=} p^{ij}_{\hat{x}_{t|t}}(\underline{x};\underline{x}). \tag{20}$$

Since the densities are continuous at $\underline{x}$, a necessary condition for (19) to hold for all balls $\underline{B}_{\underline{x}}$ is that

$$q_{ijt}(\underline{x}) \geq bm_{it}(\underline{x}). \tag{21}$$

If (21) holds for any $\underline{x} \in \underline{X}^j_{t|t-1}$, we may not reject the association of $\underline{x}$ with $\underline{z}_{it}$; consequently, we may not reject the association of the track set $\underline{X}^j_{t|t-1}$ with $\underline{z}_{it}$.

Since

$$p_{\hat{x}_{t|t}}(\underline{\xi};\underline{x}) = \mathcal{N}\{\underbrace{\underline{x} + W_{ijt}(\underline{z}_{it} - H_{it}\underline{x})}_{\hat{\underline{x}}}, \underbrace{[I - W_{ijt}H_{it}]P^j_{t|t-1}}_{P_{t|t}}\},$$

we have

$$q_{ijt}(\underline{x}) = \left.\frac{1}{(2\pi)^{\frac{n}{2}}|P^{ij}_{t|t}|^{\frac{1}{2}}} e^{-\frac{1}{2}(\underline{\xi} - \hat{\underline{x}})^T[P^{ij}_{t|t}]^{-1}(\underline{\xi} - \hat{\underline{x}})}\right|_{\underline{\xi}=\underline{x}}$$

$$= \frac{e^{-\frac{1}{2}(H_{it}\underline{x} - \underline{z}_{it})^T W^T_{ijt}[P^{ij}_{t|t}]^{-1}W_{ijt}(H_{it}\underline{x} - \underline{z}_{it})}}{(2\pi)^{\frac{n}{2}}|P^{ij}_{t|t}|^{\frac{1}{2}}} \tag{22}$$

Given the information-value determining probability density (9) and the credal probability density (22), we may form the Association Likelihood Ratio Test (ALRT):

Let $\underline{X}^j_{t|t-1}$ be a predicted track set, and let $\underline{z}_{it}$ be a sample value of the observation vector $\mathbf{z}_{it}$. We shall say that $\underline{X}^j_{t|t-1}$ and $\underline{z}_{it}$ are associated with boldness *b if $q_{ijt}(\underline{x}) \geq bm_{it}(\underline{x})$ for some $\underline{x} \in \Xi_{it} \cap \underline{X}^j_{t|t-1}$. That is, there exists $\underline{x} \in \Xi_{it} \cap \underline{X}^j_{t|t-1}$ such that*

$$\frac{1}{(2\pi)^2|P^{ij}_{t|t}|^{\frac{1}{2}}} e^{-\frac{1}{2}(H_{it}\underline{x} - \underline{z}_{it})^T W^T_{ijt}[P^{ij}_{t|t}]^{-1}W_{ijt}(H_{it}\underline{x} - \underline{z}_{it})}$$

$$\geq b\frac{||H_{it}\underline{x} - \underline{z}_{it}||}{\int_{\Sigma_{it}} ||H_{it}\underline{\xi} - \underline{z}_{it}||d\underline{\xi}},$$

in which case, $\underline{X}^{ij}_{t|t}$ obtained via (11) is a filtered track set. If $\underline{X}^j_{t|t-1} \cap \Xi_{it} = \emptyset$, then we deem $\underline{z}_{it}$ and $\underline{X}_{t|t-1}$ to be dissociated, *and the set $\underline{X}^{ij}_{t|t}$ has no meaning and is discarded.*

If $\underline{X}^j_{t|t-1}$ survives the ALRT for $\underline{z}_{it}$, then the likelihood that $\underline{X}^j_{t|t-1}$ is associated with $\underline{z}_{it}$ is greater than the information value gained by rejecting the association.

4 Example

To illustrate the concept of the ALRT methodology, consider the case of tracking targets constrained to planar motion. Let $\underline{x} = [x, y, \dot{x}, \dot{y}]^T$ denote the kinematic state of a target in some convenient coordinate system ($n = 4$). The dynamics equation is

$$\mathbf{x}_t = \begin{bmatrix} 1 & 0 & \Delta_t & 0 \\ 0 & 1 & 0 & \Delta_t \\ 0 & 0 & 1 & 0 \\ 0 & 0 & 0 & 1 \end{bmatrix} \mathbf{x}_{t-1} + \begin{bmatrix} u_{x_t} \\ u_{y_t} \\ u_{\dot{x}_t} \\ u_{\dot{y}_t} \end{bmatrix},$$

where Δ_t is the sample interval, and we have set $G \equiv I$. Possible target maneuvers are assumed to be characterized by the process noise, $\mathbf{u}_t$, whose covariance is Q_t. We assume that angle-of-arrival data are available; further, we assume that the sensor is sufficiently far from the target that the linearized model is adequate. For convenience we also assume that the coordinate system is resolved along the azimuth and elevation angles $H_{it} = \begin{bmatrix} 1 & 0 & 0 & 0 \\ 0 & 1 & 0 & 0 \end{bmatrix}$; that is, $r_{it} \equiv 2$. Let $\underline{x} = [x_1, x_2, x_3, x_4]^T \in \underline{X}^j_{t|t-1}$, $\underline{z}_{it} = [z_{it1}, \ z_{it2}]^T$, and suppose $R_{it} = \text{diag}\{\rho_1^2, \rho_2^2\}$. Then the seriously possible region is

$$\Sigma_{it} = \left\{ \begin{bmatrix} \zeta_1 \\ \zeta_2 \end{bmatrix} : |\zeta_1 - z_{it1}| \leq \theta\rho_1, \quad |\zeta_2 - z_{it2}| \leq \theta\rho_2 \right\},$$

and the information-value determining probability density is

$$m_{it}(\underline{x}) = \frac{\sqrt{(x_1 - z_{it1})^2 + (x_2 - z_{it2})^2}}{\int_{-\theta\rho_2}^{\theta\rho_2}\int_{-\theta\rho_1}^{\theta\rho_1}\sqrt{\zeta_1^2+\zeta_2^2}\,d\zeta_1 d\zeta_2}. \qquad (23)$$

Each $\underline{x} \in \underline{X}^j_{t|t-1}$ represents the mean value of a predicted conditional distribution. For each such $\underline{x}$ there exists a filtered conditional density of the form

$$p^{ij}_{\hat{\mathbf{x}}_{t|t}}(\underline{\xi};\underline{x}) = \mathcal{N}\{\underline{x}+W_{ijt}(\underline{z}_{it}-H_{it}\underline{x}), [I-W_{it}H_{it}]P_{t|t-1}\}, \qquad (24)$$

$\underline{x} \in \underline{X}^j_{t|t-1}$, and evaluating this expression at $\underline{\xi} = \underline{x}$,

$$q_{ijt}(\underline{x}) = \frac{(2\pi)^2|P^{ij}_{t|t}|^{\frac{1}{2}}}{e^{-\frac{1}{2}[x_1 - z_{it1},\; x_2 - z_{it2}]W^T_{ijt}[P^{ij}_{t|t}]^{-1}W_{ijt}\begin{bmatrix} z_{it1} - x_1 \\ z_{it2} - x_2\end{bmatrix}}}.$$

The ALRT is: Associate $\underline{X}^j_{t|t-1}$ and $\underline{z}_{it} = [z_{it1}, z_{it2}]^T$ if, for any $\underline{x} = [x_1, x_2, x_3, x_4]^T \in \underline{X}_{t|t-1}$, $q_{ijt}(\underline{x}) \geq b m_{it}(\underline{x})$.

Figure 1(a) illustrates a family of three crossing trajectories. The tracks move generally from left to right at time increases; the lines correspond to the x and y position components and the + symbols correspond to noise-corrupted observations. Figure 1(b) displays the filtered track sets corresponding to this simulation, with the ellipses representing the projections of the track sets onto position space. The initial predicted track set $\mathcal{X}_{0|-1}$ includes the entire field of view, and is not shown. The three large elliptical regions correspond to the the filtered track sets after the first set of observations have been processed. As time increases, the size of these track sets decreases rapidly; for the first few observations corresponding to Tracks 1 and 3, however, there are multiple associations, since the tracks are fairly close and the track sets are still fairly large. As more confidence is obtained in the associations, these tracks become uniquely associated, and the elliptical regions decrease rapidly in size, and will asymptotically become point tracks. At sample ten, Tracks 1 and 2 nearly intersect, and both tracks associate with the observations. These multiple associations persist for a few samples, but as the tracks diverge, the associations again become unique. Track 2 is unambiguously associated and estimated, as evidenced by the radius of the track set converging to zero, and the set-valued estimates asymptotically become point-valued.

References

[1] W. C. Stirling and D. R. Morrell, Convex Bayes Decision Theory, *IEEE Transactions on Systems, Man, and Cybernetics*, **21**(1):173–183, January/February 1991.

[2] D. R. Morrell and W. C. Stirling, Set-Valued Filtering and Smoothing, *IEEE Transactions on Systems, Man, and Cybernetics*, **21**(1):184–193, January/February 1991.

[3] I. Levi, *The Enterprise of Knowledge*, MIT Press, Cambridge, Massachusetts, 1980.

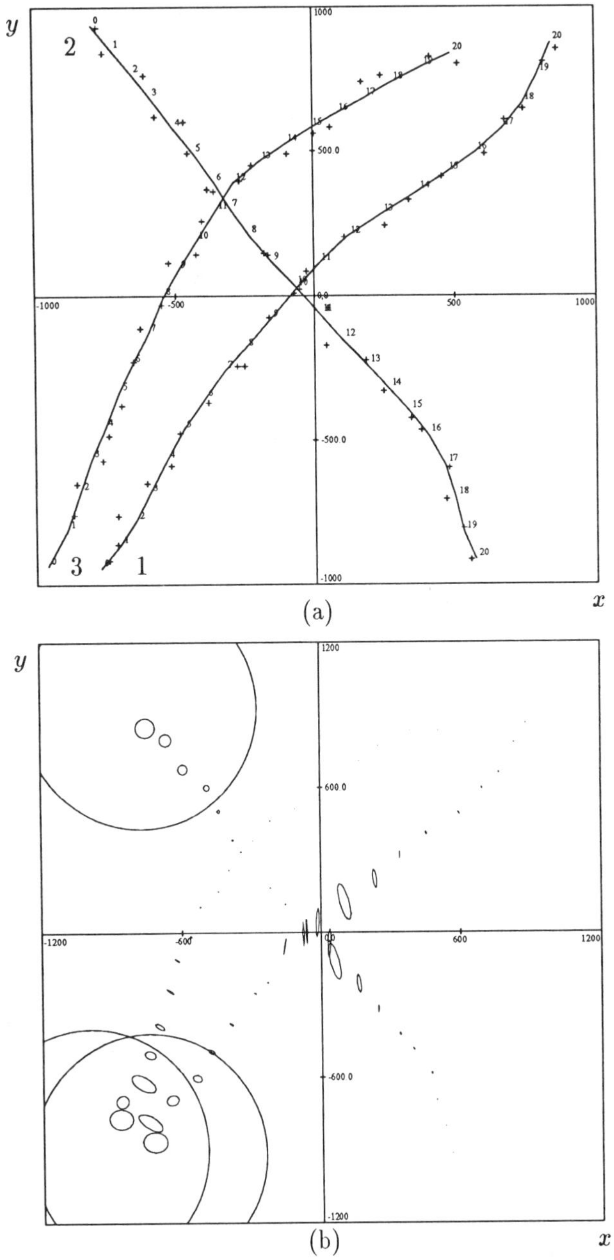

Figure 1: Three Crossing Tracks: (a) simulated trajectories and observations; (b) filtered track sets.

Identifiability Issues for Rotationally Invariant Arrays

A. LEE SWINDLEHURST
Dept. of Electrical & Computer Engineering
Brigham Young University
Provo, Utah 84602

Abstract

The popular ESPRIT algorithm provides a computationally efficient approach for direction of arrival (DOA) estimation in situations where the sensor array is composed of two identical translated subarrays. One of the key advantages of ESPRIT is that the subarrays need not be calibrated in order to obtain the DOA estimates. In this paper, the problem of DOA estimation using sensor arrays composed of two identical, uncalibrated, and rotated subarrays is considered. It is shown that, unlike ESPRIT, such rotationally invariant arrays do not provide an identifiable parameterization of the problem; i.e., unique DOA estimates are not possible without additional calibration information when more than one signal is present.

1. Introduction

THE MAJORITY of algorithms developed for the narrowband direction of arrival (DOA) estimation problem require that the array response be completely known for each DOA. This is usually accomplished by either direct calibration in the field, or by analytical means using information about the position and response of each individual sensor (such as is done with a *uniform linear array*, or ULA, for example). A notable exception is the ESPRIT algorithm [1, 2], which assumes instead that the sensor array is composed of two identical, translated subarrays. The key advantage of this approach is that the DOA estimates can be computed very efficiently, *without* knowledge of the response of the individual subarrays. Hence, an explicit calibration of the array is not required.

Several authors have proposed extensions of ESPRIT to more generally configured arrays. In [3], arrays composed of more than two identical, translated subarrays in parallel are considered. Azimuth and elevation angle estimation using arrays with identical, translated subarrays in more than one direction have been studied in [4, 5, 6, 7, 8, 9]. In this paper, arrays composed of two identical, *rotated* subarrays are investigated. Circular arrays are a special case of this type of array. It will be shown that, although such rotationally invariant arrays give rise to a data model similar to the translationally invariant case, they do not in general lead to an identifiable parameterization of the DOAs. In particular, if the subarrays are uncalibrated, unique DOA estimates are not possible whenever two or more sources are present. This contrasts with ESPRIT, which always gives unique estimates provided the subarrays are unambiguous and separated by no more than a half wavelength.

In the next section, the narrowband data model and the ESPRIT algorithm are briefly described, primarily to introduce notation. Section 3 then presents the type of rotationally invariant array considered in this analysis. Note that the definition of rotationally invariant used here is not the most general definition possible. Finally, the loss of identifiability for this type of array is established in Section 4.

2. The ESPRIT Data Model

Under the assumption of narrowband signals, the output of an M-element array $\mathbf{x}(t) \in \mathbb{C}^M$ due to d emitters can be described by

$$\mathbf{x}(t) = \mathbf{A}(\boldsymbol{\theta})\mathbf{s}(t) + \mathbf{n}(t) , \tag{1}$$

where $\mathbf{s}(t) \in \mathbb{C}^d$ represents the signal amplitude and phase at time t, $\mathbf{n}(t) \in \mathbb{C}^M$ is additive noise, and $\mathbf{A}(\boldsymbol{\theta}) \in \mathbb{C}^{M\times d}$ is the matrix of array response vectors for each source parameterized by the DOAs $\boldsymbol{\theta} = [\theta_1, \cdots, \theta_d]^T$:

$$\mathbf{A}(\boldsymbol{\theta}) = [\mathbf{a}(\theta_1) \ \ \cdots \ \ \mathbf{a}(\theta_d)] \ . \tag{2}$$

It can be assumed without loss of generality that the noise is spatially white, or in other words that $\mathcal{E}\{\mathbf{n}(t)\mathbf{n}^*(t)\} = \sigma^2\mathbf{I}$. There are a number of important applications for which the model of (1) is appropriate, the goal being to use several observations from the array to estimate the DOAs $\theta_1, \cdots, \theta_d$.

As mentioned above, most algorithms developed for this problem require that $\mathbf{a}(\theta)$ be known for all possible θ in order to estimate the DOAs. This, however, is

1058-6393/91 $01.00 © 1991 IEEE

not true of the ESPRIT algorithm. ESPRIT assumes that the M-element array is composed of two identical, translated, m-element subarrays, where $m < M \leq 2m$ in general. For subarrays that do not *overlap* (*i.e.*, share elements), $M = 2m$, though in general, $M \leq 2m$ since overlapping subarrays are allowed. An example of a non-overlapping array that satisfies the ESPRIT requirement is depicted in Figure 1. The arrow superimposed on each array element simply indicates the absolute orientation of the element.

Figure 1: A Translationally Invariant Array

The assumption of translationally invariant subarrays results in a very special structure for $\mathbf{A}(\boldsymbol{\theta})$. To see this, let $\mathbf{J}_0$ and $\mathbf{J}_1$ represent the $m \times M$ selection matrices that assign the elements of $\mathbf{x}(t)$ to the two subarrays. Then

$$\mathbf{J}\mathbf{A}(\boldsymbol{\theta}) = \begin{bmatrix} \mathbf{J}_0 \\ \mathbf{J}_1 \end{bmatrix} \mathbf{A}(\boldsymbol{\theta}) = \begin{bmatrix} \mathbf{A}_e(\boldsymbol{\theta}) \\ \mathbf{A}_e(\boldsymbol{\theta})\boldsymbol{\Phi} \end{bmatrix}, \quad (3)$$

where $\boldsymbol{\Phi}$ is a unitary diagonal matrix with diagonal elements ϕ_i given by

$$\phi_i = \exp\{-j2\pi\Delta \sin\theta_i/\lambda\}, \qquad i = 1, \cdots, d, \quad (4)$$

and where λ is the wavelength of the narrowband signal, Δ is the distance between the subarrays, and $\mathbf{A}_e(\boldsymbol{\theta})$ represents the response of subarray 0 (assumed to be full rank $m > d$ for all $\boldsymbol{\theta}$).

The ESPRIT algorithm exploits the structure in (3) in the following way. If $\mathbf{E}_s = [\mathbf{e}_1 \cdots \mathbf{e}_d] \in \mathbb{C}^{M\times d}$ represents the eigenvectors corresponding to the d largest eigenvalues of the covariance $\mathbf{R} = \mathcal{E}\{\mathbf{x}(t)\mathbf{x}^*(t)\}$, and if no two signals are fully correlated, then it is easily shown that

$$\mathbf{J}\mathbf{E}_s \stackrel{\text{def}}{=} \begin{bmatrix} \mathbf{E}_0 \\ \mathbf{E}_1 \end{bmatrix} = \begin{bmatrix} \mathbf{A}_e(\boldsymbol{\theta}) \\ \mathbf{A}_e(\boldsymbol{\theta})\boldsymbol{\Phi} \end{bmatrix} \mathbf{T}, \quad (5)$$

for some full rank $d \times d$ matrix $\mathbf{T}$. Solving for $\mathbf{A}_e(\boldsymbol{\theta})$ and substituting into the lower block of this equation leads to

$$\mathbf{E}_1 = \mathbf{E}_0\mathbf{T}^{-1}\boldsymbol{\Phi}\mathbf{T} = \mathbf{E}_0\boldsymbol{\Psi}, \quad (6)$$

where the matrix $\boldsymbol{\Psi} = \mathbf{T}^{-1}\boldsymbol{\Phi}\mathbf{T}$ has been defined. Thus, if $m \geq d$ and $\Delta < \lambda/2$, the DOAs may be uniquely determined from the eigenvalues of the operator $\boldsymbol{\Psi}$ that maps $\mathbf{E}_0$ onto $\mathbf{E}_1$. Note that this result is independent of the value of $\mathbf{A}_e(\boldsymbol{\theta})$ (as long as it is full rank), so the array need not be calibrated in order to estimate the DOAs.

Since a perfect measurement of $\mathbf{R}$ cannot be obtained, the sample covariance $\hat{\mathbf{R}}$ defined by

$$\hat{\mathbf{R}} = \frac{1}{N}\sum_{k=1}^{N} \mathbf{x}(t_k)\mathbf{x}^*(t_k) \quad (7)$$

is used as an estimate, and its eigenvectors $\hat{\mathbf{E}}_0$ and $\hat{\mathbf{E}}_1$ will not exactly satisfy the relationship of equation (5). Hence, there will be no operator that exactly maps the columns of $\hat{\mathbf{E}}_0$ onto those of $\hat{\mathbf{E}}_1$. Though a least-squares estimate of $\boldsymbol{\Psi}$ may be easily obtained [10, 11], since both $\hat{\mathbf{E}}_0$ and $\hat{\mathbf{E}}_1$ have errors, a *total*-least-squares (TLS) [12] estimate of $\boldsymbol{\Psi}$ is more appropriate. This is the approach taken by the (TLS)-ESPRIT algorithm. As formulated in [2], the TLS estimate of $\boldsymbol{\Psi}$ is calculated by performing the singular value decomposition (SVD)

$$\begin{bmatrix} \hat{\mathbf{E}}_0^* \\ \hat{\mathbf{E}}_1^* \end{bmatrix} \begin{bmatrix} \hat{\mathbf{E}}_0 & \hat{\mathbf{E}}_1 \end{bmatrix} = \begin{bmatrix} \mathbf{U}_{11} & \mathbf{U}_{12} \\ \mathbf{U}_{21} & \mathbf{U}_{22} \end{bmatrix} \boldsymbol{\Sigma} \begin{bmatrix} \mathbf{U}_{11} & \mathbf{U}_{12} \\ \mathbf{U}_{21} & \mathbf{U}_{22} \end{bmatrix}^* \quad (8)$$

and setting

$$\hat{\boldsymbol{\Psi}}_{TLS} = -\mathbf{U}_{12}\mathbf{U}_{22}^{-1}. \quad (9)$$

The ESPRIT DOA estimates are then calculated using the eigenvalues of $\hat{\boldsymbol{\Psi}}_{TLS}$ and equation (4). The modular nature of the algorithm (it requires successive eigendecompositions or SVDs of order $M, 2d$, and d) coupled with the fact that it requires no explicit gradient search makes it very attractive from a computational standpoint.

3. Rotationally Invariant Arrays

While the class of arrays for which ESPRIT is applicable is an important one, there are many other common array structures that do not possess translationally invariant subarrays. Circular arrays are a prime example. It would be desirable to develop fast algorithms like ESPRIT that exploit the regular structure present in such arrays.

This paper considers the possibility of developing an ESPRIT-like algorithm for 180° rotationally invariant arrays, *i.e.*, arrays composed of two identical rotated subarrays. To illustrate the type of array geometry to be studied, consider the sensor configurations of Figures 2 and 3. The two subarrays consist of the upper six and lower six elements, respectively, and, as before, the arrow superimposed on each sensor indicates its absolute orientation. If Array 1 is rotated 180°, and each of the sensors is forced to maintain its absolute orientation, then another array identical to Array 1 is obtained. The same is also true for Array 2 if the orientation of the individual sensors is allowed to change as the array is rotated. Strictly speaking, the analysis of this paper will only apply to a rotational invariance of the type exhibited by Array 1. However, in many cases the individual antenna elements have response patterns that are themselves 180° rotationally invariant (*i.e.*, they are symmetric about some axis of rotation), in which case Arrays 1 and 2 are equivalent. Thus, the ideas presented below will often apply to arrays that possess the kind of rotational invariance exhibited by Array 2 as well.

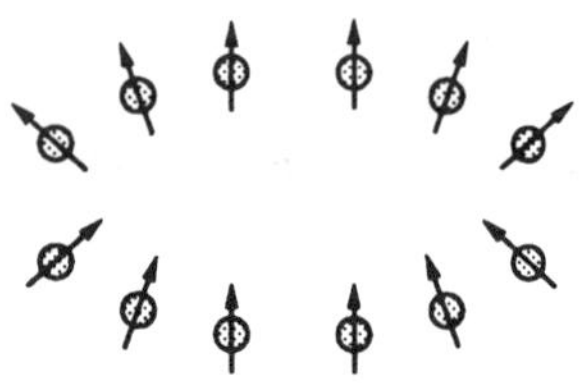

Figure 2: Rotationally Invariant Array 1

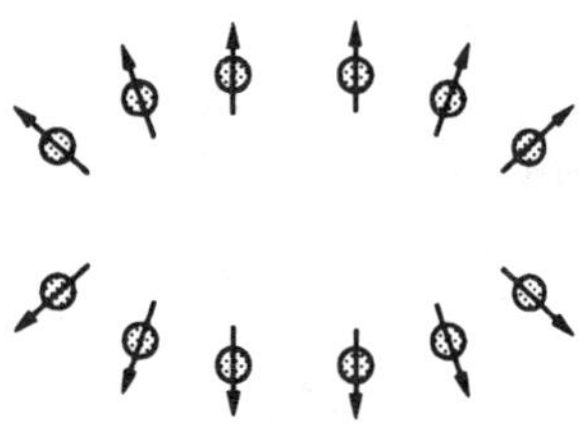

Figure 3: Rotationally Invariant Array 2

Both circular and linear arrays belong to this class. As in the case of linear arrays and translationally invariant subarrays, a circular array possesses a number of possible subarray structures that satisfy the required rotational invariance, some of which lead to overlapping subarrays. An example of a maximally overlapping subarray structure for a circular array is depicted in Figure 4.

As in the case of ESPRIT, rotationally invariant arrays like those depicted in Figures 2 and 4 lead to a very special shift structure for the array response. Instead of equation (3), vectors from the array manifold

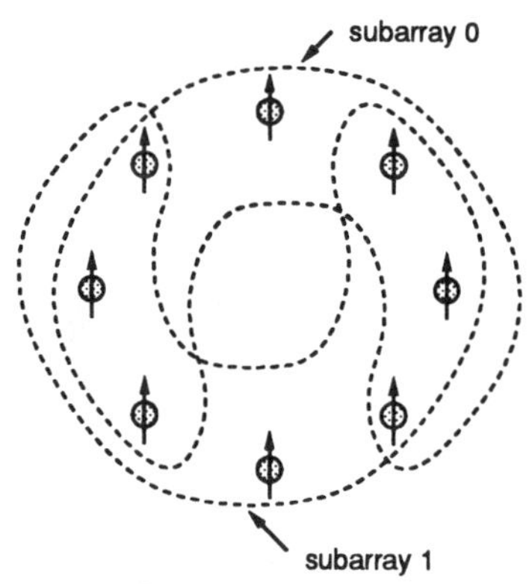

Figure 4: A Possible Subarray Configuration for a Circular Array

are described by the equation

$$\mathbf{JA}(\boldsymbol{\theta}) = \begin{bmatrix} \mathbf{A}_e(\boldsymbol{\theta}) \\ \overline{\mathbf{A}}_e(\boldsymbol{\theta})\boldsymbol{\Phi} \end{bmatrix}, \tag{10}$$

where $\overline{\mathbf{A}}_e(\boldsymbol{\theta})$ denotes the *complex conjugate* of the subarray manifold $\mathbf{A}_e(\boldsymbol{\theta})$, $\boldsymbol{\Phi}$ is defined exactly as in (4), except that Δ is the distance between a given arbitrary sensor (the reference) and its identical twin in the companion subarray, and θ is measured with respect to the normal of the line joining these two sensors. The conjugation of $\mathbf{A}_e$ accounts for the fact that the propagation of a wavefront across one of the subarrays can be considered to be equivalent to the same wavefront propagating *backwards* across the rotated companion subarray. The diagonal matrix $\boldsymbol{\Phi}$ takes care of the bulk phase shift due to the delay in propagation between the subarrays.

This particular definition forces the row of $\mathbf{A}_e(\boldsymbol{\theta})$ corresponding to the reference sensor to be purely real, but this is not a problem since any non-zero phase response for the reference sensor in direction θ_i can be lumped together with the phase of the ith signal waveform. For simplicity, in this paper it will be assumed (without loss of generality) that the reference sensor corresponds to the first row of $\mathbf{A}_e(\boldsymbol{\theta})$, and that this row is composed of all 1's.

For an array satisfying (10), the eigenvectors corresponding to the d largest eigenvalues of the covariance $\mathbf{R}$ will satisfy (compare with equation (5))

$$\mathbf{JE}_s \stackrel{\text{def}}{=} \begin{bmatrix} \mathbf{E}_0 \\ \mathbf{E}_1 \end{bmatrix} = \begin{bmatrix} \mathbf{A}_e(\boldsymbol{\theta}) \\ \overline{\mathbf{A}}_e(\boldsymbol{\theta})\boldsymbol{\Phi} \end{bmatrix} \mathbf{T} \tag{11}$$

for some full rank $d \times d$ matrix $\mathbf{T}$, and the corresponding relationship between $\mathbf{E}_0$ and $\mathbf{E}_1$ is

$$\mathbf{E}_1 = \overline{\mathbf{E}}_0 \overline{\mathbf{T}}^{-1} \boldsymbol{\Phi} \mathbf{T} \tag{12}$$

$$= \overline{\mathbf{E}}_0 \mathbf{\Psi}_c \; .$$

The factorization $\mathbf{\Psi}_c = \overline{\mathbf{T}}^{-1} \mathbf{\Phi} \mathbf{T}$ is referred to as a *coneigendecomposition*, and the DOAs are thus a function of the *coneigenvalues* of $\mathbf{\Psi}_c$ (*e.g.*, see [13]). It is interesting to note that, while the eigenvalues of the ESPRIT operator $\mathbf{\Psi}$ are uniquely defined, the coneigenvalues of a given matrix are *not*. This is fundamentally the reason why unique DOA estimates are not possible when $d > 1$, as demonstrated below.

4. Identifiability

If there is only one source present, $d = 1$ and $\mathbf{E}_s$ is $M \times 1$. The lone element of $\mathbf{\Phi}$ (described by (4)) in this case can be obtained by first normalizing $\mathbf{JE}_s$ so that its first element is one, and then solving, for example,

$$\phi = \frac{\overline{\mathbf{E}}_1^* \mathbf{E}_0}{\mathbf{E}_0^* \mathbf{E}_0} \; .$$

If $\Delta < \lambda/2$, the DOA can be uniquely determined using (4).

To demonstrate that the model of (11) is not identifiable for $d > 1$, it will be shown that there exists a full rank $d \times d$ matrix $\mathbf{T}_1$, a full rank $m \times d$ matrix $\mathbf{A}_1$ with first row all 1's, and a diagonal matrix $\mathbf{\Omega} = diag\{e^{j\omega_1}, \cdots, e^{j\omega_d}\}$ such that

$$\left[\frac{\mathbf{A}_e}{\mathbf{A}_e \mathbf{\Phi}} \right] = \left[\frac{\mathbf{A}_1}{\mathbf{A}_1 \mathbf{\Omega}} \right] \mathbf{T}_1 \qquad (13)$$

for *arbitrary* $\omega_1, \cdots, \omega_d$ satisfying

$$\omega_i \neq \omega_k, \; i \neq k, \quad \text{and} \quad |\omega_i| < \pi \, , \; i, k = 1, \cdots, d \, . \qquad (14)$$

In other words, there are an infinite number of matrices of the form (10) that satisfy the subspace relationship of (11), and hence there are an infinite number of possible DOA estimates given only $\mathbf{JE}_s$.

For a given $\mathbf{\Omega}$ satisfying (14), let $\mathbf{A}_1 = \mathbf{A}_e \mathbf{T}_1^{-1}$ and $\mathbf{T}_1 = \mathbf{\Omega}^{-\frac{1}{2}} \mathbf{X} \mathbf{\Phi}^{\frac{1}{2}}$ for some real-valued $d \times d$ matrix $\mathbf{X}$. It is easily verified that these choices satisfy equation (13). It now remains to choose $\mathbf{X}$ such that the first row of $\mathbf{A}_1$ is all 1's, *i.e.*,

$$[1 \;\; \cdots \;\; 1] = [1 \;\; \cdots \;\; 1] \mathbf{\Omega}^{-\frac{1}{2}} \mathbf{X} \mathbf{\Phi}^{\frac{1}{2}} \; .$$

This is equivalent to choosing $\mathbf{X}$ such that

$$\mathbf{X}^T [\boldsymbol{\omega}_r \;\; \boldsymbol{\omega}_i] = [\boldsymbol{\phi}_r \;\; \boldsymbol{\phi}_i] \, , \qquad (15)$$

where

$$\boldsymbol{\omega}_r = \mathrm{Re}(\mathrm{diag}\{\mathbf{\Omega}^{-\frac{1}{2}}\})$$

$$\boldsymbol{\omega}_i = \mathrm{Im}(\mathrm{diag}\{\mathbf{\Omega}^{-\frac{1}{2}}\})$$

$$\boldsymbol{\phi}_r = \mathrm{Re}(\mathrm{diag}\{\mathbf{\Phi}^{-\frac{1}{2}}\})$$

$$\boldsymbol{\phi}_i = \mathrm{Im}(\mathrm{diag}\{\mathbf{\Phi}^{-\frac{1}{2}}\}) \, ,$$

$Re\{\cdot\}$ and $Im\{\cdot\}$ denote real and imaginary parts respectively, and $diag\{\cdot\}$ here denotes the column vector formed from the diagonal elements of the argument. Provided $d \geq 2$ and the elements of $\mathbf{\Omega}$ satisfy (14), such an $\mathbf{X}$ is always possible since both $[\boldsymbol{\omega}_r \;\; \boldsymbol{\omega}_i]$ and $[\boldsymbol{\phi}_r \;\; \boldsymbol{\phi}_i]$ are rank[1] 2. Consequently, the angles $\omega_1, \cdots, \omega_d$ are equally as valid as $\phi_1, \cdots, \phi_d$ in satisfying the subspace model of (11).

5. Conclusion

Although the shift structure of the steering vectors for an array composed of identical rotated subarrays bears a striking resemblance to that for ESPRIT, unique DOA estimates are not possible when more than one signal is present. This loss of identifiability could only be overcome by somehow incorporating information about the individual subarray responses. However, doing so would probably mean that a computationally efficient solution such as that for ESPRIT would not be possible.

References

[1] A. Paulraj, R. Roy, and T. Kailath, "A Subspace Rotation Approach to Signal Parameter Estimation", *Proceedings of the IEEE*, pages 1044–1045, July 1986.

[2] R. Roy and T. Kailath, "ESPRIT – Estimation of Signal Parameters via Rotational Invariance Techniques", *IEEE Trans. on ASSP*, **37**(7):984–995, July 1989.

[3] A. Swindlehurst, B. Ottersten, R. Roy, and T. Kailath, "Multiple Invariance ESPRIT", *to appear in IEEE Trans. on Sig. Proc.*, **40**, April 1992.

[4] D. V. B. Rao and S. Y. Kung, "A State Space Approach for the 2-D Harmonic Retrieval Problem", In *Proc. IEEE ICASSP*, pages 4.10.1–4.10.4, 1984.

[1] The matrix $[\boldsymbol{\phi}_r \;\; \boldsymbol{\phi}_i]$ is rank 2 since the elements of $\mathbf{\Phi}$ can be assumed to satisfy (14). If this were not true, unique DOA estimates would not be possible even if equation (13) were not considered.

[5] M. Zoltowski and D. Stavrinides, "Sensor Array Signal Processing via a Procrustes Rotations Based Eigenanalysis of the ESPRIT Data Pencil", *IEEE Trans. on ASSP*, **37**(6):832–861, June 1989.

[6] A. Swindlehurst and T. Kailath, "2-D Parameter Estimation Using Arrays with Multidimensional Invariance Structure", In *Proc.* 23rd *Asilomar Conference of Signals, Systems, and Computers*, pages 950–954, Asilomar, CA., November 1989.

[7] R. Johnson and G. Miner, "An Operational System Implementation of the ESPRIT DF Algorithm", *IEEE Trans. on AES*, **27**(1):159–166, January 1991.

[8] A. Swindlehurst and T. Kailath, "Algorithms for Azimuth/Elevation Direction Finding Using Regular Array Geometries", *to appear in IEEE Trans. on AES*, 1992.

[9] P. Ober, E. Deprettere, and A. van der Veen, "Efficient Methods to Compute Azimuth and Elevation in High Resolution DOA Estimation", In *Proc. ICASSP '91*, pages 3349–3352, 1991.

[10] R. Roy, A. Paulraj, and T. Kailath, "ESPRIT – A Subspace Rotation Approach to Estimation of Parameters of Cisoids in Noise", *IEEE Trans. on ASSP*, **34**(4):1340–1342, October 1986.

[11] S.Y. Kung, C.K. Lo, and R. Foka, "A Toeplitz Approximation Approach to Coherent Source Direction Finding", In *Proc. ICASSP 86*, Tokyo, Japan, 1986.

[12] G. H. Golub and C. F. Van Loan, *Matrix Comptutations*, Johns Hopkins University Press, Baltimore, MD., 1984.

[13] R. A. Horn and C. A. Johnson, *Matrix Analysis*, Cambridge University Press, Cambridge, England, 1985.

Detecting Abrupt Changes in ARMA Signals

Kevin J. Delaney Charles W. Therrien*

Abstract

Autoregressive Moving Average (ARMA) models are useful for the short-time description of time series in such diverse fields as speech processing, seismology, and economics. Detecting phonemes, earthquakes or recessions can be viewed as detecting changes in the parameters of the ARMA model. This paper describes a new method for optimal sequential detection of parameter changes in ARMA models. Examples are given of its application.

1 Introduction

This paper describes a new method for detecting changes in time series represented by ARMA models, based upon a method by Nikiforov (Nikiforov, 1986), (Nikiforov and Tikhonov, 1986), (Kushnir, Nikiforov, and Savin, 1983). We first review previous work by Nikiforov, describing a derivation of the sequential change detection method. Following this, we describe the application of Nikiforov's method to AR models and its extension to ARMA models. Finally, we give examples of the algorithm's performance.

2 Sequential change detection method

Hypothesis testing problem This method is based on a Generalized Likelihood Ratio Test (GLRT) which is applied to include cases where the post-change model is unknown. We begin with a hypothesis-testing problem. Under H_0, the hypothesis that there is no change, the parameter vector $\underline{\theta}$ that describes the model remains equal to the parameter vector $\underline{\theta}_0$ of a known model. Under H_1, the hypothesis that a model change has occurred, the model changes at some unknown time τ from $\underline{\theta}_0$ to $\underline{\theta}_1 \neq \underline{\theta}_0$:

$$H_0: \ \underline{\theta} = \quad \underline{\theta}_0 \qquad n = 1, \ldots, t$$

$$H_1: \ \underline{\theta} = \begin{cases} \underline{\theta}_0 & n = 1, \ldots, \tau \\ \underline{\theta}_1 & n = \tau + 1, \ldots, t. \end{cases}$$

In order to make a Maximum Likelihood (ML) decision, we form the log-likelihood ratio based upon $\underline{y}_1^t$, the entire received signal up to time t.

$$\begin{aligned} \ln \Lambda(t) &\triangleq \ln \frac{p(\underline{y}_1^t \mid H_1)}{p(\underline{y}_1^t \mid H_0)} \\ &= \ln S(t) - \ln S(\tau) \end{aligned}$$

where we have defined:

$$S(t) \triangleq \frac{p_1(\underline{y}_1^t)}{p_0(\underline{y}_1^t)}. \tag{1}$$

The ML estimate of the change time τ is that value of $\hat{\tau}$ that minimizes $\ln S(\hat{\tau})$. Substituting $\hat{\tau}_{\mathrm{ML}}$ in place of the unknown τ leads to the GLRT:

$$\ln S(t) - \min_n \ln S(n) \underset{H_0}{\overset{H_1}{\gtrless}} h. \tag{2}$$

The likelihood ratio of (1) assumes that we know the value of the parameter vector $\underline{\theta}$ both before and after the change. However, while we know $\underline{\theta}_0$, the model *before* the change, we do not know $\underline{\theta}_1$, the model *after* the change. Using results from Wald (Wald, 1947, pp. 73-77, 199-200) and Roussas (Roussas, 1972, p. 53), Nikiforov derived a sufficient statistic χ_t which does not require knowledge of the post-change model (Nikiforov, 1986, pp. 227-228):

$$\chi_t \triangleq \frac{1}{t} \Delta_t^T(\underline{\theta}_0) F_1^{-1}(\underline{\theta}_0) \Delta_t(\underline{\theta}_0) \tag{3}$$

*Both authors are with the U.S. Naval Postgraduate School, Code EC/Ti, Monterey CA, 93943. For information, contact Prof. C.W. Therrien at (408) 646-3347 or fax (408) 646-2760.

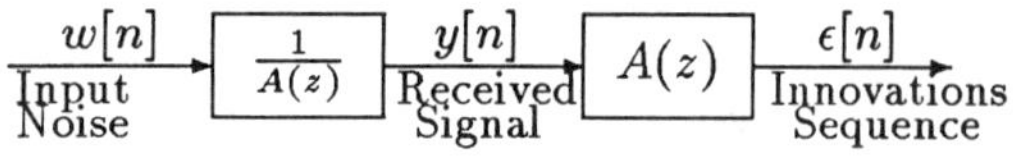

Figure 1: Processing noise-free AR signals

where:

$$\Delta_t(\underline{\theta}) \triangleq \begin{bmatrix} \frac{d\ln p(\underline{y}_t|\theta)}{d\theta_0} \\ \frac{d\ln p(\underline{y}_t|\theta)}{d\theta_1} \\ \vdots \\ \frac{d\ln p(\underline{y}_t|\theta)}{d\theta_R} \end{bmatrix}$$

$$F_1(\underline{\theta}) \triangleq E\left[\Delta_1(\underline{\theta})\Delta_1^T(\underline{\theta}) \mid \underline{\theta}\right].$$

Application to AR models Nikiforov applied the algorithm to detection of AR model changes by deriving Δ_t, F_1 and χ_t for the AR case. By processing the received noise-free signal $\underline{y}$ through the inverse of the filter which created $\underline{y}$, as shown in Figure 1, we produce the white-noise innovations process $\underline{\epsilon}$. Nikiforov derived an expression for χ_t based upon this inverse-filter innovations process. To summarize Nikiforov's derivation:

$$\Delta_t(\underline{\theta}) = \begin{bmatrix} \frac{1}{\sigma^2}\sum_{l=1}^{t} \underline{Y}_{l-1}\epsilon_l \\ \frac{1}{\sigma}\sum_{l=1}^{t}\left(\frac{\epsilon_l^2}{\sigma^2}-1\right) \end{bmatrix}$$

$$F_1(\underline{\theta}_0) = \begin{bmatrix} \frac{R_{\underline{y}}}{\sigma^2} & \underline{0} \\ \underline{0}^T & \frac{2}{\sigma^2} \end{bmatrix}$$

where we have defined the received signal vector $\underline{Y}_{l-1}$ as:

$$\underline{Y}_{l-1} \triangleq \begin{bmatrix} y[l-1] \\ y[l-2] \\ \vdots \\ y[l-P] \end{bmatrix}.$$

3 Extension to noisy ARMA signals

3.1 Derivation

State space approach We have reformulated the change-detection method by allowing the creating filter to be of the more general ARMA type, and for $\underline{y}$

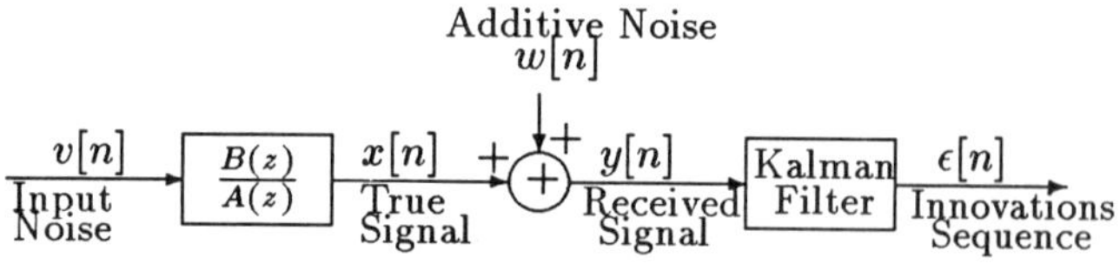

Figure 2: Processing noisy ARMA signals

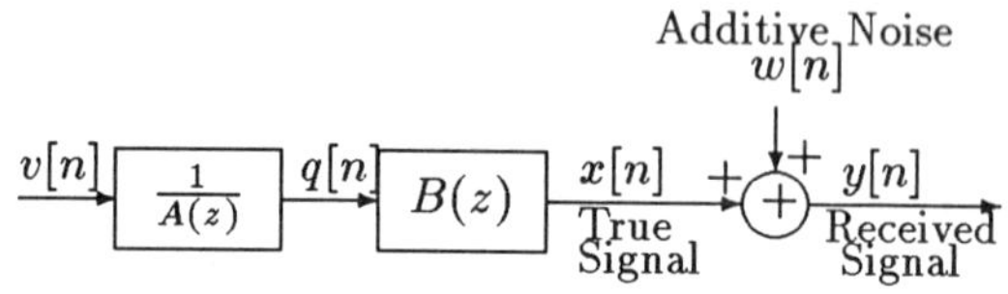

Figure 3: Kalman model

to be contaminated with additive noise. Rather than process $\underline{y}$ through an inverse filter, as in Nikiforov's noise-free case, we employ a Kalman filter to produce the residual process $\underline{\epsilon}$, as shown in Figure 2. From the structure of Figure 3, we can directly write the Kalman model in state-space form as:

$$\begin{aligned} \underline{q}[n+1] &= \Phi\underline{q}[n] + \underline{v}[n] \\ y[n] &= \underline{c}^T\underline{q}[n] + w[n] \end{aligned}$$

where $\underline{v}[n]$ is the driving noise, $\underline{w}[n]$ is the additive noise and:

$$\underline{q}[n] \triangleq \begin{bmatrix} q[n] \\ q[n-1] \\ \vdots \\ q[n-P+1] \end{bmatrix}$$

$$\Phi \triangleq \begin{bmatrix} -a_1 & -a_2 & \cdots & -a_P \\ 1 & 0 & \cdots & 0 \\ \vdots & \ddots & & \vdots \\ 0 & \cdots & 1 & 0 \end{bmatrix}$$

$$\underline{c} \triangleq \begin{bmatrix} b_0 & b_1 & \cdots & b_Q & 0 & \cdots & 0 \end{bmatrix}^T$$

$$\underline{v}[n] \triangleq \begin{bmatrix} v[n] \\ 0 \\ \vdots \\ 0 \end{bmatrix}$$

$$R_v \triangleq E\left[\underline{v}[n]\underline{v}^T[n]\right] = \begin{bmatrix} \sigma_v^2 & 0 & \cdots & 0 \\ 0 & 0 & \cdots & 0 \\ \vdots & & \ddots & \vdots \\ 0 & \cdots & & 0 \end{bmatrix}$$

$$\sigma_\omega^2 \triangleq E\left[\omega^2\right].$$

The above expressions are valid for the case in which $Q+1 \leq P$. If $Q+1 > P$, we rearrange the model to retain more of the history of the state $\underline{q}$ than is required for the AR portion. In this case, $\underline{q}[n]$, Φ, and $\underline{c}$ have alternate definitions; we do not show them here as they do not change the subsequent derivation.

We complete the transformation of the model into Kalman form by computing an estimate of the state $\underline{q}$ through the recursive equations (Haykin, 1989, pp. 349-362):

$$\epsilon[n] = y[n] - \underline{c}^T \hat{\underline{q}}[n] \quad (4)$$

$$\hat{\underline{q}}[n+1] = \Phi \hat{\underline{q}}[n] + \underline{g}_n \epsilon[n] \quad (5)$$

$$\sigma^2_{\epsilon[n]} = \underline{c}^T K_n \underline{c} + \sigma^2_\omega \quad (6)$$

$$\underline{g}_n = \frac{\Phi K_n \underline{c}}{\sigma^2_{\epsilon[n]}} \quad (7)$$

$$K_{n+1} = \left[\Phi - \underline{g}_n \underline{c}^T\right] K_n \left[\Phi - \underline{g}_n \underline{c}^T\right]^T + R_v + \sigma^2_\omega \underline{g}_n \underline{g}_n^T. \quad (8)$$

Computing Δ_t Since χ_t is based upon the gradient of the log-likelihood function with regard to the model parameters, we begin by computing the log-likelihood function of the signal.

Because the innovations process $\underline{\epsilon}$ is linearly derived from the signal $\underline{y}$, we may use the log-likelihood function of $\underline{\epsilon}$ to compute χ_t. Since $\epsilon[l]$ is a white-noise process, we write:

$$p(\underline{\epsilon}) = \frac{1}{2\pi^{t/2}\left(\prod_{l=1}^t \sigma^2_{\epsilon[l]}\right)^{\frac{1}{2}}} e^{-\frac{1}{2}\sum_{l=1}^{t} \frac{\epsilon^2[l]}{\sigma^2_{\epsilon[l]}}}$$

$$\ln p(\underline{\epsilon}) = -\frac{t}{2}\ln 2\pi - \frac{1}{2}\sum_{l=1}^{t} \ln \sigma^2_{\epsilon[l]} - \frac{1}{2}\sum_{l=1}^{t}\frac{\epsilon^2[l]}{\sigma^2_{\epsilon[l]}}$$

$$\frac{d \ln p(\underline{\epsilon})}{d\theta_i} = \sum_{l=1}^{t} \frac{1}{\sigma^2_{\epsilon[l]}}\left(\frac{1}{2}\left(\frac{\epsilon^2[l]}{\sigma^2_{\epsilon[l]}} - 1\right)\frac{d\sigma^2_{\epsilon[l]}}{d\theta_i} - \epsilon[l]\frac{d\epsilon[l]}{d\theta_i}\right) \quad (9)$$

where θ_i is any one of the $P+Q+1$ elements of the model parameter vector $\underline{\theta}$, which we here define as:

$$\underline{\theta} \triangleq \begin{bmatrix} -a_1 & -a_2 & \cdots & -a_P & b_0 & b_1 & \cdots & b_Q \end{bmatrix}^T.$$

In order to use (9) to compute $\frac{d \ln p(\underline{\epsilon})}{d\theta_i}$ and therefore $\Delta_t(\underline{\theta}_0)$, we need expressions for the derivatives $\frac{d\sigma^2_{\epsilon[l]}}{d\theta_i}$ and $\frac{d\epsilon[l]}{d\theta_i}$. We can compute these derivatives recursively along with the recursive computations of (4) - (8). The result is the following expression for the ith element of $\Delta_t(\underline{\theta}_0)$:

$$\sum_{l=1}^{t} \frac{1}{\sigma^2_{\epsilon[l]}}\left(\frac{1}{2}\left(\frac{\epsilon^2[l]}{\sigma^2_{\epsilon[l]}} - 1\right)\frac{d\sigma^2_{\epsilon[l]}}{d\theta_i} + \epsilon[l]\left(\frac{d\underline{c}^T}{d\theta_i}\hat{\underline{q}}[l] + \underline{c}^T\frac{d\hat{\underline{q}}[l]}{d\theta_i}\right)\right) \quad (10)$$

for each value of i: $i = 1, 2, \ldots, R$, where $R \triangleq P + Q + 1$.

Computing F_1 Since $F_1(\underline{\theta}_0) \triangleq E_0\left[\Delta_1(\underline{\theta}_0)\Delta_1^T(\underline{\theta}_0)\right]$ then, defining $F_1(i,j)$ as the $(i,j)^{\text{th}}$ element of $F_1(\underline{\theta}_0)$, and using (10) at $t=1$:

$$F_1(i,j) = \frac{1}{2\sigma^4_{\epsilon[1]}}\frac{d\sigma^2_{\epsilon[1]}}{d\theta_i}\frac{d\sigma^2_{\epsilon[1]}}{d\theta_j} + \frac{1}{\sigma^2_{\epsilon[1]}}\left(\frac{d\underline{c}^T}{d\theta_i}E_0\left[\hat{\underline{q}}[1]\hat{\underline{q}}^T[1]\right]\frac{d\underline{c}}{d\theta_j} + \underline{c}^T E_0\left[\frac{d\hat{\underline{q}}[1]}{d\theta_i}\hat{\underline{q}}^T[1]\right]\frac{d\underline{c}}{d\theta_j} + \frac{d\underline{c}^T}{d\theta_i}E_0\left[\hat{\underline{q}}[1]\frac{d\hat{\underline{q}}[1]}{d\theta_j}\right]\underline{c} + \underline{c}^T E_0\left[\frac{d\hat{\underline{q}}[1]}{d\theta_i}\frac{d\hat{\underline{q}}^T[1]}{d\theta_j}\right]\underline{c}\right). \quad (11)$$

We can compute the indicated expected values in a recursive manner.

Decision rule In analogy with (2), the decision rule based upon χ_t is of the form:

$$L(\chi_t) - \min_n L(\chi_n) \underset{H_0}{\overset{H_1}{\gtrless}} h. \quad (12)$$

To use this rule to accept or reject the change hypothesis H_1, we must compute the log-likelihood ratio based upon the non-central χ^2 statistic χ_t. This log-likelihood ratio $L(\chi_t)$ is (Jackson, 1960, p. 351), (Patnaik, 1949, p. 204):

$$L(\chi_t) = -t\frac{\lambda_1^2}{2} + \ln\left({}_0F_1\left(\frac{r}{2}; \frac{t\lambda_1^2\chi_t}{4}\right)\right) \quad (13)$$

where ${}_0F_1(c;x)$ is the hypergeometric function (Jackson and Bradley, 1961, p. 519):

$${}_0F_1(c;x) \triangleq 1 + \frac{x}{c} + \cdots + \frac{x^n}{c(c+1)\cdots(c+n-1)n!} + \cdots. \quad (14)$$

As an infinite series, the hypergeometric function can take thousands of term to converge; rather than compute it directly we use a polynomial approximation we have derived.

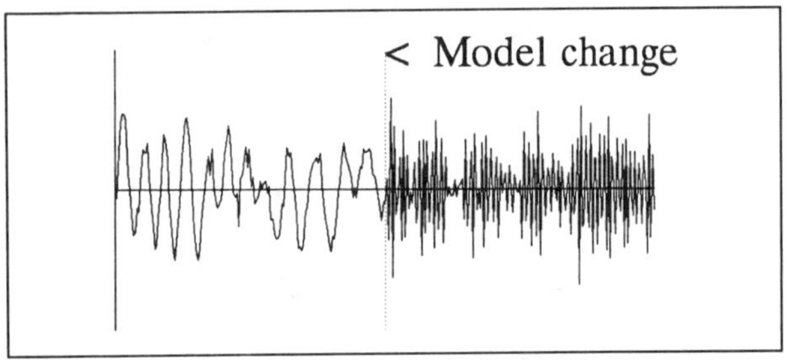

Figure 4: Stochastic ARMA signal undergoes model change.

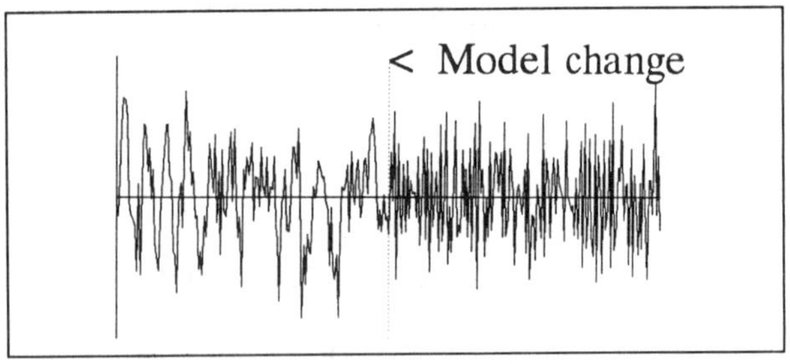

Figure 5: Signal in 3 dB noise.

4 Example

In Figure 4, we see a stochastic signal which undergoes a model change at the time indicated; Figure 5 shows the same signal in 3 dB additive white noise. After computing the statistic χ_t from (3) and its log-likelihood ratio from (13), we detect the model change as shown in Figures 6 and 7.

References

[1] Haykin, S. S., *Modern Filters.* Macmillan, New York, 1989.

[2] Jackson, J. E., "A Short Table of the Hypergeometric Function ${}_0F_1(c;x)$", *Sankhyā*, Vol. 22, Parts 3 & 4, June 1960, pp. 351-356.

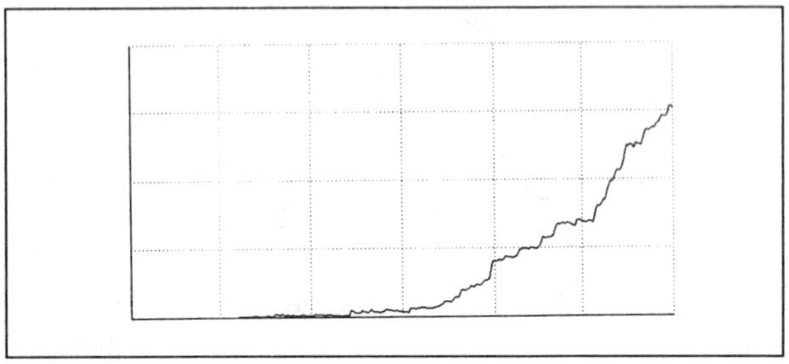

Figure 6: χ_t statistic responds to model change.

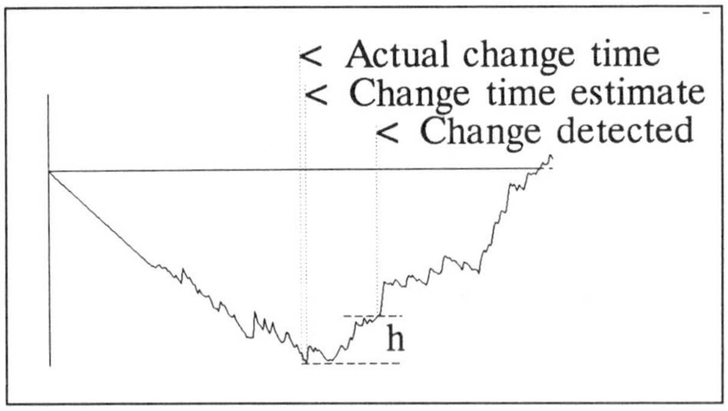

Figure 7: Change detected by Generalized Likelihood Ratio Test.

[3] Jackson, J. E., and Bradley, R. A., "Sequential χ^2- and T^2-Tests and Their Application to an Acceptance Sampling Problem," *Technometrics.* Vol. 3, No. 4, November 1961, pp. 519-534.

[4] Kushnir, A. F., Nikiforov, I. V., and Savin, I. V., "Statistical Adaptive Algorithms for Automatic Detection of Seismic Signals, I. One-Dimensional Case," *Computational Seimsmology*, Vol. 15, 1983, pp. 145-154.

[5] Nikiforov, I. V.,, "Sequential Detection of Changes in Stochastic Systems," *Detection of Abrupt Changes in Signals and Dynamical Systems.* ed. M. Basseville and A. Beneveniste, Springer-Verlag, Berlin, 1986, pp. 218-258.

[6] Nikiforov, I. V., and Tikhonov, I. N., "Application of Change Detection Theory to Seismic Signal Processing," *Detection of Abrupt Changes in Signals and Dynamical Systems.* ed. M. Basseville and A. Beneveniste, Springer-Verlag, Berlin, 1986, pp. 356-373.

[7] Patnaik, P. B., "The Non-Central χ^2 and F-Distributions and Their Applications," *Biometrika*, Vol. 36, 1949, pp. 202-232.

[8] Roussas, G. C., *Contiguity of Probability Measures: Some Applications in Statistics.* Cambridge University Press, London, 1972.

[9] Wald, A., *Sequential Analysis.* John Wiley & Sons, Inc., New York, 1947.

Detection of Fatal Transient Impairments Using an Embedded Secondary Channel

Yong-Hwan Lee
Motorola Codex Corp.
20 Cabot Blvd.
Mansfield, MA 02048

Abstract

Unlike voice communication, voiceband data communication can be seriously affected by transient channel impairments. In some cases, synchronization between the transmitter and the receiver may not be recovered from these kinds of impairments without a reinitialization process. Therefore, it is highly desirable for data communication equipment to have an efficient transient channel impairments detector for fast recovery. In this paper, one such detector is proposed for high speed modems using a secondary channel synchroneously embedded in the main channel. A known code word being multiplexed with the secondary port data is sent through the embedded secondary channel. The received code word is tested for detection of fatal transient impairments. For reliable detection, the confirmation is sequentially performed based on a modified up/down counter scheme. The performance of the proposed detector is analytically evaluated in closed form.

1 Introduction

Although transient disturbances such as impulse noise, gain hits, drop outs and phase hits are regulated, they account for a substantial portion of performance degradation when signals are transmitted through a network system. These types of transient impairments can cause audible clicks or pops and thus may result in unobnoxious effect to listners on the telephone lines. However, it can seriously affect data communication since it can cause loss of system synchronization.

In particular, when a signal is transmitted through a digital carrier system such as T1, it can be corrupted by slips. A slip is an intended insertion or deletion of a single DS1 (T1) data frame in the digital switching operation to avoid loss of network synchronization due to either the lack of frequency synchronization of clocks at various network nodes, or phase wander and jitter on the digital bit stream [1]. Like the above transient noises, slips cause occasional audible clicks and thus do not cause serious degradation in voice communication [2]. However, a single slip can seriously affect data communication. For example, a slip affects Group 3 facsimile transmission, where a maximum of eight horizontal scan lines can be missed [3]. Experimental results on voiceband modems indicate that a single slip causes a long burst of errors over a time period which can be up to several seconds depending on the modem type and data rate [4]. In some cases, the modem never recovers from a slip and requires a forced manual reset. Therefore, it is highly desirable for a modem to have a detector for rapid recovery from a fatal loss of synchronization due to such transient impairments.

High speed modems usually have a secondary channel to support other functions, such as the digital network management. The secondary channel can be implemented using a separate FSK modulation module. However, to maximize the use of the channel bandwidth and to reduce the implementation cost, the secondary channel is frequently embedded into the main channel. Using a time division multiplexing method, for example, the secondary channel data is synchronously multiplexed into the main channel data before being modulated.

The characteristics of a time division multiplexed signal can be used for detection of fatal transient impairments which cause a loss of system synchronization. After being multiplexed with the secondary port data, a predetermined sync code word is repeatedly transmitted through the embedded secondary channel. Only if the receiver is properly operating (*i.e.*, is synchronized with the transmitter), can the sync code word be correctly received (after being demultiplexed from the main channel data). Note that the additional bandwidth of the secondary channel for transmitting this sync code word can be obtained with only a fraction of a dB increase of signal to noise ratio (SNR).

The proposed detector is composed of two processes: the detection and confirmation processes. If the Hamming distance between the received sync code word and the reference exceeds a threshold, it can be declared that loss of synchronization due to a sudden transient impairment has occurred. Since a single decision is not reliable, an additional confirmation process is required. Considering a possible modem operation under high bit error rate (BER) conditions, we adopt a sequential confirmation process based on a up/down counter scheme. Once a loss of synchronization is confirmed, the modem needs to initiate retraining.

Section 2 briefly explains how the sync code word can be efficiently sent through an embedded secondary channel. The proposed sequential detection scheme is

1058-6393/91 $01.00 © 1991 IEEE

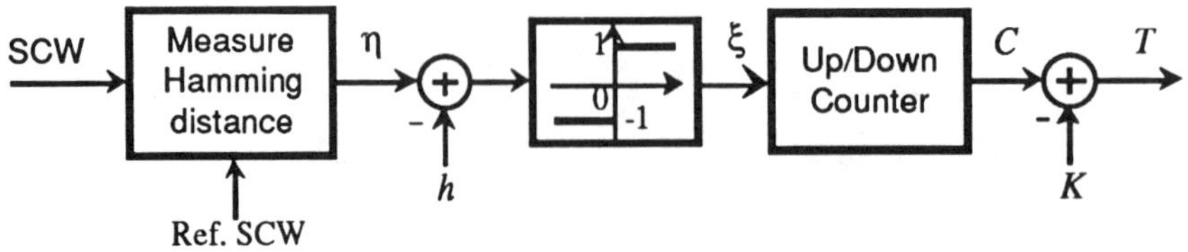

Fig. 1. A block diagram of the proposed detector

described in Section 3. The performance of the detector is analytically evaluated in Section 4. Section 4 also presents some numerical results.

2 The Embedded Secondary Channel

To make embedding the secondary channel independent of the data rate, we consider a baud rate based embedding. Each bit of the secondary channel data is synchronously inserted into the main channel data stream at every L bauds of the main channel data, which costs approximately an additional $3/L$ dB of SNR on the average. With proper choice of the multiplexing index number L, the embedded secondary channel can have a data bandwidth larger than one required for the secondary port data, which is usually less than 150 bps when the secondary port is used for the digital network management controller. This marginal bandwidth can be used for sending the sync code word (SCW).

Assume that the embedded secondary channel bandwidth is $W(= B/L)$ bps, where B is the baud (symbol) rate of the main channel. Assume also that the maximum data rate of the secondary port is $V(< W)$ bps. Then, the rest of $(W - V)$ bits can be used for sending the SCW. Each SCW is multiplexed with the secondary port data before being embedded into the main channel data. For ease of implementation, an M-bit SCW is sent after an 8-bit secondary port data word is transmitted, where M is an integer less than or equal to $8(W - V)/V$. It takes $T(= L(M + 8)/B)$ seconds for a single secondary channel data frame to be sent. For example, when $B = 2400$, $L = 10$, and $V = 110$, the bandwidth of the embedded secondary channel is 240 bps and the bit size M of a sync code word can be up to 9. Note that only an additional 0.3 dB of SNR is required for the main channel to have this embedded secondary channel without performance degradation.

3 The Detector Scheme

Since each bit of embedded secondary channel data is transmitted every L-baud time interval, a single secondary data frame spans a time interval of $(M + 8)L$ bauds. Under the normal modem operating condition, the transmitted SCW can be correctly received only when no transient impairment occurs. The only exception can be when there occurs a bit loss event whose duration is exactly equal to a multiple of the time span of a secondary channel data frame. However, this has very low probability of occurance. Although detection can be performed by bit by bit basis, for ease of implementation, we consider a detection process based on a single SCW. A simplified block diagram of the proposed detector is shown in Fig. 1.

Whenever an M-bit SCW is received, it is compared with the transmitted(reference) SCW by measuring the Hamming distance η, which can be easily calculated using a look-up table. If η is larger than a threshold h, it is assumed that a transient performance occurred during this data frame interval. Here, the threshold level h should be chosen so as to accommodate the channel BER under the worst permissible operating condition.

Since this single test cannot satisfy the required detection performance, an additional confirmation process is required. A sequential test based on the *c-successive* counter or *modified up/down* counter scheme is widely used for confirmation of PCM frame synchronization [5,6]. The c-successive counter scheme declares confirmation if bad frames (*i.e.*, $\eta > h$) are recieved c times in a row. The up/down counter declares confirmation if its content exceeds a threshold, where the content of the counter increases by ℓ, if $\eta > h$, and decreases by n if not. For efficient confirmation even under possible high BER conditions, and for simplicity of implementation, we use a simple up/down counter scheme, where ℓ and n are set to the same value of 1. In the case of the normal operation condition, η is usually less than h, *i.e.*, the content of the counter C is less than zero. Since the confirmation test should detect a true loss of synchronization as fast as possible, the content of the counter is bounded so that C has non-negative value. The updating algorithm can be summarized as follows. Set $C(0) = 0$. Then, at time $t = kT$, $k = 1, 2, \cdots$,

$$\begin{aligned} C(k) &= C(k-1) + \xi(k); \\ C(k) &= 0, \text{ if } C(k) < 0, \end{aligned} \tag{1}$$

where $\xi(k)$ is the quantized output of the tester,

$$\xi(k) = \begin{cases} 1, & \text{if } \eta(k) > h, \\ -1, & \text{otherwise.} \end{cases} \tag{2}$$

If $C(k)$ becomes larger than a threshold K, a loss of synchronization is verified and the modem needs to retrain. The threshold level K should be determined so that the confirmation process guarantees the desired detection criteria such as the true and false detection time. The design of these parameters, M, h and K, will be discussed in the next section.

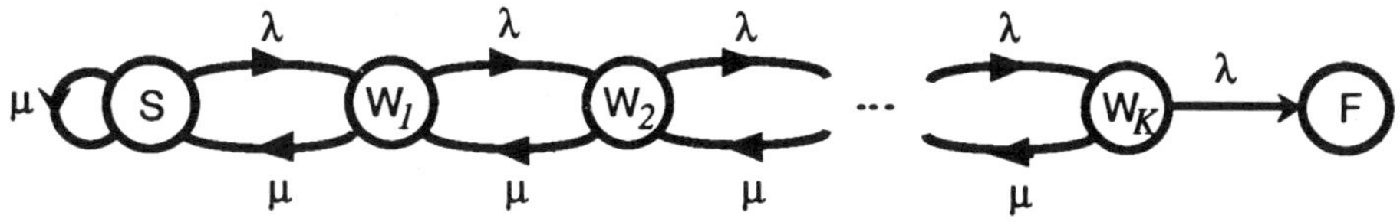

Fig. 2. An equivalent state diagram

Fig. 3. An equivalent state transfer function

4 Performance Analysis

The up/down counter scheme is easy to describe, but it is difficult to analyze because of the boundary conditions at the normal (sync) state S and the final (sync lost) state F. Although statistical analysis of this counter scheme can be fairly well formulated in a matrix form by using a probability generating function (PGF)[5], to the best of the author's knowledge, no formal solution has been obtained in closed form. For analytical evaluation of the detection performance, the detection problem can be modeled as a simple hypothesis test against another simple alternative. Let H_0 be the null hypothesis corresponding to the normal condition and H_1 be the alternative corresponding to the sync lost condition due to a fatal transient impairment.

Since each bit of an SCW is transmitted exactly L-baud time intervals apart after being scrambled, the test statistic ξ of (2) can be treated as an independent binomial random variable. Let q_j be the BER of the received data under the hypothesis H_j, and $p_j = 1-q_j$, for $j = 0,1$. For given threshold h and SCW bit size M, the probability λ_j that a bad frame is detected, *i.e.*, $\xi = 1$, by each test is given by

$$\begin{aligned} \lambda_j &= \text{Prob}\{\xi = 1 | H_j\} \\ &= \sum_{i=h+1}^{M} \binom{M}{i} p_j^{M-i} q_j^i, \end{aligned} \quad (3)$$

under the hypothesis H_j, $j = 0,1$.

For a given threshold K, the confirmation process based on the modified up/down counter by (1) can be represented by an equivalent finite state machine as depicted in Fig. 2, where λ denotes λ_0 and λ_1 under H_0 and H_1, respectively, and $\mu = 1 - \lambda$. When the channel has no fatal transient impairments, the state machine usually resides in the normal state S. When the machine reaches the final state F, loss of synchronization is declared. The states $\{W_i\}, i = 1, \cdots, K$, are intermediate transition states. The average true (under H_1) or false (under H_0) detection time T_D is equivalent to the mean time of first transition from state S to F, which can be estimated using a PGF.

The timing transition of each state for a given value of K is represented by its neighbor states as

$$\begin{aligned} S(z) &= 1 + \mu z^{-1}[S(z) + W_1(z)] \\ W_1(z) &= \lambda z^{-1} S(z) + \mu z^{-1} W_2(z) \\ W_2(z) &= \lambda z^{-1} W_1(z) + \mu z^{-1} W_3(z) \\ &\vdots \\ W_{k-1}(z) &= \lambda z^{-1} W_{K-2}(z) + \mu z^{-1} W_K(z) \\ W_K(z) &= \lambda z^{-1} W_{K-1}(z) \\ F(z) &= \lambda z^{-1} W_K(z), \end{aligned} \quad (4)$$

where z^{-1} denotes the time delay corresponding to the test interval T. For an intermediate state W_i, $i = 1, 2, \cdots, K$, define $G_i(z)$ by

$$G_i(z) = \frac{W_{K-i+1}(z)}{W_{K-i}(z)}. \quad (5)$$

Then, the state flow between the states can be represented by an equivalent transfer function shown in Fig. 3. From Eq. (4), $G_i(z)$ is expressed by

$$G_i(z) = \frac{\lambda z^{-1}}{1 - \mu z^{-1} G_{i-1}(z)}, \text{ for } i = 2, 3, \cdots, K, \quad (6)$$

where $G_1(z) = \lambda z^{-1}$. Let $G_i^N(z)$ and $G_i^D(z)$ be the numerator and denominator of the intermediate state transfer function $G_i(z)$, respectively. Then, it can be shown that the denominator $G_i^D(z)$ can be expressed in terms of a sum of power series as

$$G_i^D(z) = \sum_{j=0}^{K} \alpha_{i,j} (-\mu\lambda)^j z^{-2j}, \quad (7)$$

and the numerator $G_i^N(z)$ is simply given by

$$G_i^N(z) = \lambda z^{-1} G_{i-1}^D(z). \quad (8)$$

Here, the coefficients $\{\alpha_{i,j}\}$ can be recursively calculated using

$$\alpha_{i,j} = \begin{cases} \alpha_{i-2,j-1} + \alpha_{i-1,j}, & j = 1, 2, \cdots, \kappa; \\ 0, & \text{otherwise}, \end{cases} \tag{9}$$

where $\alpha_{i,0} = 1$, for all i, and κ is the largest integer $\leq \frac{i}{2}$. Note that, in Fig. 3, the normal state $S(z)$ is replaced with an equivalent state $W_0(z)$ given by

$$W_0(z) = \frac{1}{1 - \mu z^{-1} - \mu z^{-1} G_K(z)}. \tag{10}$$

Thus, for a given K, the PGF $F(z)$ of timing transition from state S to F is

$$F(z) = W_0(z)\lambda z^{-1} \prod_{i=1}^{K} G_i(z). \tag{11}$$

Using Eqs. (8) and (10), it can be simplified to

$$F(z) = \frac{(\lambda z^{-1})^{K+1}}{(1 - \mu z^{-1})G_K^D(z) - \mu z^{-1} G_K^N(z)}. \tag{12}$$

From Eqs. (7) to (9), $F^D(z)$, the denominator of $F(z)$, is given by

$$\begin{aligned} F^D(z) &= (1 - \mu z^{-1}) \sum_{j=0}^{\kappa_1} \alpha_{K,j}(-\mu\lambda)^j z^{-2j} \\ &\quad - \mu\lambda z^{-2} \sum_{j=0}^{\kappa_2} \alpha_{K-1,j}(-\mu\lambda)^j z^{-2j}, \end{aligned} \tag{13}$$

where κ_1 and κ_2 are the largest integers $\leq K/2$ and $\leq (K-1)/2$, respectively. The average detection time T_D is obtained by taking the derivative of $F(z)$ with respect to z and then evaluating it at $z = 1$, *i.e.*,

$$T_D = -\frac{\partial F(z)}{\partial z}\Big|_{z=1} \cdot T. \tag{14}$$

The detector performance associated with design parameters is evaluated in terms of the false and true detection time. Because of the random nature of the data, the BER q_1 under hypothesis H_1 can be assumed to be 0.5. Assume also BER $q_0 = 10^{-3}$ is the worst permissible operating condition. Fig. 4 plots the detection times versus the threshold level K for various bit sizes M of the sync code word with the threshold $h = 0$. For a given K, it takes at least $(K+1)T$ seconds for the detector to confirm a slip. To examine how many tests are required in addition to (K+1) tests under H_1, Fig. 4.(a) plots the additional average detection time δT normalized by the minimum time $(K+1)T$, *i.e.*, the average true detection time under H_1 will be

$$T_D = (1 + \delta T)(K + 1)T. \tag{15}$$

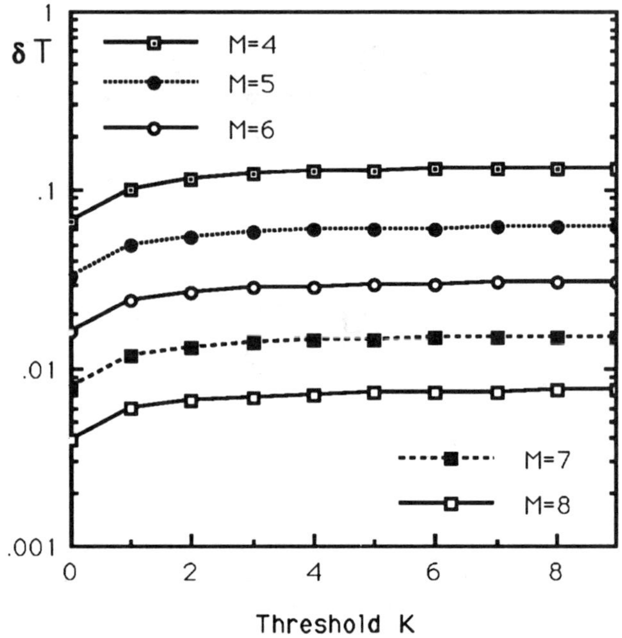

(a) Additional average detection time δT under H_1

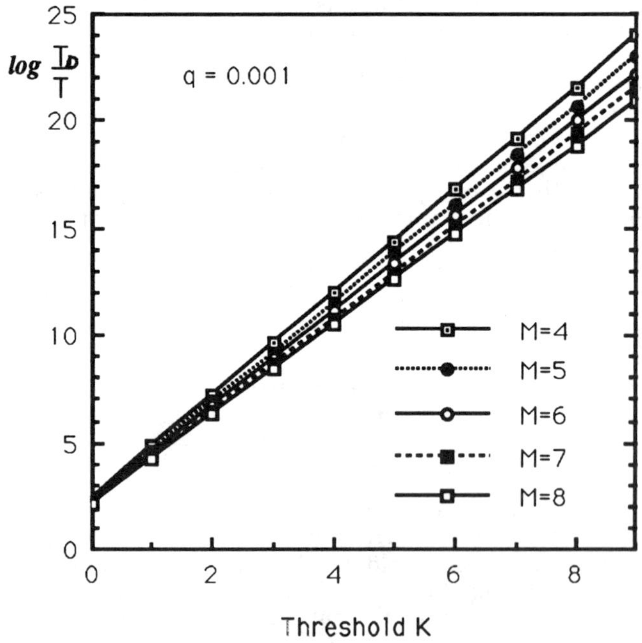

(b) Average false detection time when $q_0 = 10^{-3}$

Fig. 4. Average detection time of the detector

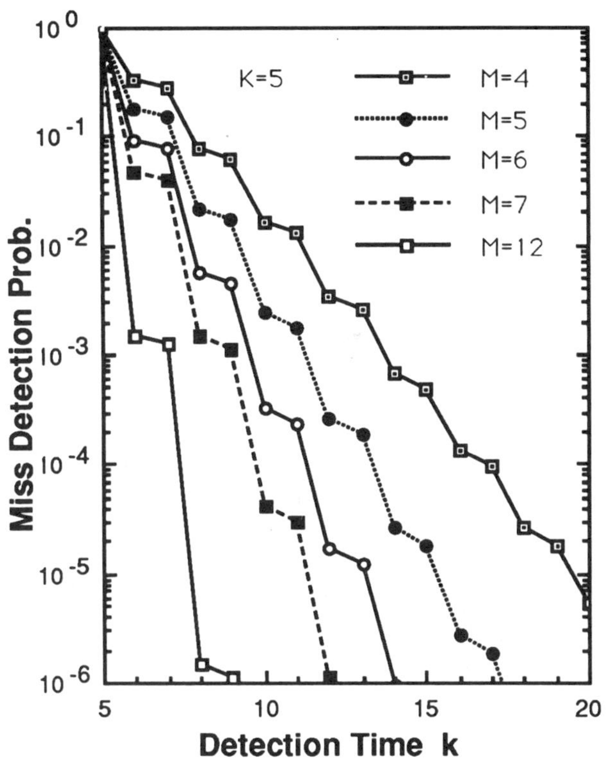

Fig. 5. Miss detection probability when $K = 5$

It can be seen that the amount of the additional test time δT is almost constant for a given SCW size M, *i.e.*, the average detection time is likely to be linearly proportional to the threshold level K. This implies that, for a given K, the smaller the SCW size M, the larger the true detection time T_D. However, since the test interval T also increases as the SCW size M increases, the use of a small M can result in a shorter true detection time T_D than that of a large M. Fig. 4.(b) shows the average false detection time under H_0, which is plotted in the common logarithmic scale. It can be seen that the average false detection time is logarithmically proportional to the threshold K for a given M. In addition, a smaller size of the SCW results in larger false detection time. However, the variance of the detection time using an SCW with small bit size significantly increases as the threshold K increases. It may be desirable not to use an SCW of too small a bit size. The maximum and minimum values of K are thus determined by the desired true and false detection time, respectively.

As another performance indicator of the detector, the miss detection probability associated with various bit sizes M is plotted in Fig. 5, when the threshold levels h and K are set to values of 0 and 5, respectively. The result is obtained by considering all kinds of possible events at each detection time k. It can be seen that the use of smaller bit sizes require more tests to achieve the same detection performance. Moreover, a smaller bit size results in larger variance in detection time. Note that the staircase-like shapes in detection probability curves are due to the up and down counting nature. Note also that a larger bit size is required for a larger threshold K to obtain the same detection performance. Without using an excessively large bit size of the code word, however, the proposed detector can have a resonable performance. For example, when $B = 2400$, $L = 10$, $M = 7$ and $K = 5$, (*i.e.*, $T = 62.5(msec)$), a fatal transient impairment can be detected at $(K+1)$-th test (the minimum trial) with a detection probability larger than 95%. With these values, it takes about $380.4(msec)$ on the average for the detector to confirm a true loss of synchronization. But the average false detection time is larger than several years even when $q_0 = 10^{-3}$. In practice, it may be desirable to set the threshold K to be large enough that the average detection time be longer than the possible recovery time by the modem itself without a retrain.

5 Conclusion

In this paper, using the marginal bandwidth of the embedded secondary channel, an efficient scheme has been proposed for detecting fatal transient impairments which cause loss of synchronization. It can be easily realized without placing a big burden on the main channel operation. The performance of the detector based on a modified up/down counter scheme has been analytically evaluated in terms of the average detection time. The analytical results can be applied to other synchronization problems such as the PCM frame synchronization in a digital carrier system.

References

[1] R. L. Freeman, *Telecommunication System Engineering,* John Wiley & Sons, 2nd Ed.,1989.

[2] M. Decina and U. de Julio, "Performance of integrated digital networks: International standards," *Proc. Int. Conf. on Commun.*, pp. 2D.1.1-2D.1.6, 1982.

[3] J. E. Abate and H. Drucker, "The effect of slips on facsimile transmission," *Proc. Int. Conf. on Commun.*, pp. 32.3.1-32.3.4, 1988.

[4] H. Drucker and A. C. Morton, "The effect of slips on data modems," *Proc. Int. Conf. on Commun.*, pp. 12.4.1-12.4.5, 1987.

[5] P. Bylanski and D. G. W. Ingram, *Digital Transmission systems,* Peter Peregrinus Ltd., 2nd Ed., Chapter 5, 1980.

[6] D. Choi, "Frame alignment in a digital carrier system-A tutorial," *IEEE Commun. Mag.*, pp. 47-54, Feb. 1990.

A Comparison of DFT, PLP and Cochleagram for Alphabet Recognition

Mark Fanty and Ronald Cole

Oregon Graduate Institute
19600 N.W. Von Neumann Dr.
Beaverton, OR 97006

Malcolm Slaney

Apple Computer Inc.
20525 Mariani Ave
Cupertino, CA 95014

Abstract

The English alphabet is a small but difficult vocabulary for speech recognition, with many fine phonetic distinctions, such as M/N and B/V. We use speaker-independent classification of isolated English letters to evaluate the relative performance of the DFT, Perceptual Linear Predictive analysis, and the cochleagram auditory model. Feedforward neural network classifiers were trained using all three representations on 60 speakers and tested on 60 new speakers. Training and testing data were independently modified by adding two levels of Gaussian noise and babble (20 random letter utterances, attenuated and given random offsets). PLP gave the best results, especially when trained or tested on Gaussian noise.

1 Introduction

The English alphabet is a small but difficult vocabulary for speech recognition, with many fine phonetic distinctions, such as M/N, B/V, B/D and T/G. Because the task requires fine phonetic distinctions, it is ideal for comparing signal representations for computer speech recognition. The authors have previously achieved speaker-independent classification rates of 96% on the English alphabet [1] using DFT and a variety of other features. Classification begins with a broad-category (fricative, closure, stop, sonorant) segmentation of the signal. Spectral and other features are extracted from the various segments of the letter and used by a neural network classifier.

Previous studies have shown the effectiveness of auditory models for speech recognition with degraded and undegraded speech [6, 4, 7]. This paper presents comparative results for three representations when used for whole-word classification with neural networks. The Discrete Fourier Transform (DFT) is included as a baseline. Two auditory models, Perceptual Linear Predictive (PLP) analysis and the cochleagram, are used for the same task.

2 The Representations

2.1 DFT

A 128 point FFT is computed on a 10 msec window (extra points folded back in). The signal was Hanning windowed. The result of the FFT is preemphasized to increase the magnitude of the higher frequencies, then converted to decibels. This yields 64 numbers per frame. Based on previous work, we used all 32 coefficients up to 4 kHz, and compressed the coefficients in the range from 4 kHz to 8 kHz down to 8, yielding 40 numbers per frame.

2.2 Perceptual Linear Predictive Analysis

The PLP speech analysis technique [3] estimates an all-pole autoregressive model of the auditory-like short-term speech spectrum. PLP has been shown to be efficient in suppressing speaker-dependent components in the speech signal, and uses fewer coefficients than the DFT. Compared to the conventional linear predictive (LP) analysis of speech (which estimates an all-pole model of the short-term power spectrum), the number of coefficients needed to obtain comparable recognition performance is typically lower for the PLP.

The auditory-like spectrum is obtained by integrating the short-term power spectrum of speech over simulated critical-band auditory masking curves, resampling the integrated spectrum in approximately 1 Bark intervals, modifying the spectral amplitude by

1058-6393/91 $01.00 © 1991 IEEE

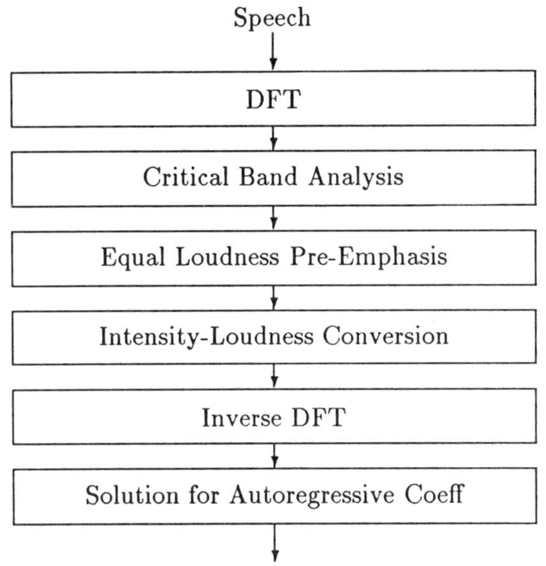

Figure 1: Stages in the PLP analysis.

a simulated fixed equal-loudness curve and compressing it through the cubic root nonlinearity to simulate the intensity-loudness power law of hearing (see Figure 1).

This autoregressive modeling efficiently approximates the spectral peaks in the auditory-like spectrum. The cepstral coefficients of the PLP all pole model are recursively computed, and weighted with an exponential window so all coefficients have a similar range for input to a neural network. Eight cepstral coefficients, including log power, of a seventh order PLP model are produced for each frame.

2.3 Cochleagram

The cochlear model designed by Lyon [5] and described by Slaney [8] converts a sound waveform into a multidimensional vector that represents the information sent from the ear to the brain. This system is illustrated in Figure 1. It is important to remember that the cochlear model used here does not try to accurately model the internal structure of the ear but only to approximate the information contained in the auditory nerve.

The cochlear model in this study uses a relatively simple filter to simulate the response of the outer and middle ears. A cascade of second order filters is used to model the propagation of sound along the basilar membrane. At each point along the cochlea, the basilar membrane responds best to a broad range of frequencies and it is this movement that is sensed by the inner hair cells. The "best" frequency of the cochlea

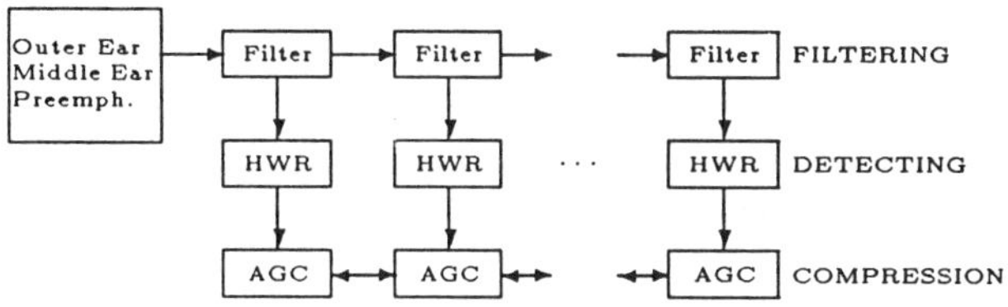

Figure 2: Lyon's Cochlear Model.

varies smoothly from high frequencies at the base to low frequencies at the apex. Inner hair cells only respond to movement of the basilar membrane in one direction. This is simulated in Lyon's model with an array of Half Wave Rectifiers (HWRs) that detect the output of each second order filter. The HWR nonlinearity serves to convert the motion of the basilar membrane at each point along the cochlea into a signal that represents the energy of the acoustic input and retains the fine time structure.

Finally, four stages of Automatic Gain Control (AGC) allow the cochlear model to compress the dynamic range of the input to a level that can be carried on the auditory nerve. The AGC used here also serves to simulate the ear's adaptation to loud sounds.

For these experiments, the 85 outputs per frame are compressed to 42 for efficiency. (Experiments showed little difference in classification accuracy after compression.) The temporal delay of wave propagation in the cochlea is part of the model. The result is that low frequency output for a speech event is delayed with respect to high frequency output. Feature extraction based on boundary locations took the delays into account.

3 Experiments

3.1 Data

A subset of the ISOLET database [2] was used for these experiments. The training set consisted of 60 speakers saying each letter of the English alphabet, in isolation, twice. The test set consisted of 60 different speakers saying each letter twice. Speech was digitized at 16 kHz using 16 bits per sample.

3.2 The Input Features

A rule-based segmenter was used to provide the broad-category segmentation used in all the experiments. This segmentation was fixed at the beginning and used in all experiments. All letters except W have a single sonorant. The feature extraction routine first finds the (longest) sonorant. This is divided into seven equal parts and the spectral representation is averaged over each of those parts. If there is a stop or fricative preceding the sonorant, the spectrum is averaged from three equal parts; if no preceding consonant is present, the spectrum is averaged over the previous 198 msec.

The 240 msec following the sonorant is also divided into three equal parts and the average spectrum is extracted from each interval. A single non-spectral feature is used: the duration of the consonant before the sonorant.

3.3 Network Configurations

Three layer networks, totally connected between layers, were trained and tested using back propagation with conjugate gradient descent. The networks were run through successive sets of 40 or 80 epochs, then tested. Training was halted when performance on the test set leveled off. The PLP networks had 105 inputs; the DFT networks had 521 inputs; the cochleagram networks had 547 inputs. All networks had 48 hidden units and 26 outputs.

3.4 Noise

Three different kinds of noise were added to the signals: Gaussian noise level 1, Gaussian noise level 2, and babble. In each condition, the noise was generated using a different random seed for each utterance (the same seeds were used for each representation) and added to the original signal. Sample values for Gaussian noise were chosen randomly according to a Gaussian probability distribution with mean 0 and standard deviation 500. For noise level 1, the gain was 1.0; for noise level 2, the gain was 2.0. The perceptual effect was to add a fairly strong background hiss.

Babble was the addition of 20 different utterances picked randomly from the training set and added together with random offsets and a gain of 0.1. The perceptual effect was several people talking in the background and one person nearby saying a single letter (the target). Table 1 shows the signal to noise ratio, with the signal being the average power in the undegraded sonorant of the letter and the noise being the average power in the added noise signal.

Table 1: Signal to noise ratio for the three conditions.

noise 1	noise 2	babble
12.5 dB	6.5 dB	14.9 dB

Table 2: Performance of the three representations.

	DFT	PLP	COCH
clean	93.6	93.8	90.9
noise 1	83.0	84.7	79.7
noise 2	74.0	76.4	70.9
babble	79.7	80.0	70.9

3.5 Results

Table 2 compares each representation when trained and tested under identical conditions (e.g. trained with babble added and tested with babble added).

Tables 3 through 5 show the performance, for each representation, under cross-testing. Nets trained under each noise condition were tested on data generated from the other noise conditions. Three patterns emerge: the best performance is usually obtained by training and testing in the same condition—the exception is that that networks trained in the noise 2 condition tested better with noise 1 data; training on noisy speech and testing on undegraded speech produces better results than training on undegraded speech and testing on noisy speech; for the best overall performance across conditions, training on babble is superior (see Table 6).

4 Summary and Conclusions

It is not possible for us to determine which representation is "best," but we conclude, for the task investigated here, PLP is the representation of choice. It has the highest overall accuracy and produces the

Table 3: Using DFT.

train/test	clean	noise 1	noise 2	babble
clean	93.6	33.0	20.0	57.3
noise 1	58.8	83.0	74.4	38.9
noise 2	46.0	78.6	74.0	32.7
babble	87.2	59.7	41.9	79.7

Table 4: Using PLP.

train/test	clean	noise 1	noise 2	babble
clean	93.8	47.4	27.8	55.4
noise 1	68.7	84.7	75.5	46.2
noise 2	53.0	81.6	76.4	36.8
babble	85.6	71.8	58.8	80.0

Table 5: Using Cochleagram.

train/test	clean	noise 1	noise 2	babble
clean	90.9	30.6	19.1	47.2
noise 1	56.9	79.7	69.7	41.2
noise 2	40.5	75.3	70.9	36.5
babble	76.6	61.0	47.3	70.9

least number of features. Its performance is about the same as the DFT when trained and tested on the same conditions, and better when trained on Gaussian noise and tested on clean speech or vice versa.

It is not easy to compare such different representations. For example, the representations have different numbers of coefficients which may require different network configurations to produce optimal results; it is not possible to try all combinations. We ran some additional experiments using the cochleagram representation without improving performance. We ran one of the cochleagram networks with twice as many hidden units. There was little change in the performance. Compressing the cochleagram channels from 85 to 42 was motivated by a desire for smaller networks and by a history of successfully compressing DFT coefficients. A single network was run without compression with little change in performance.

We cannot conclude from this study that the cochleagram is an inappropriate representation for speech recognition. It is designed to mimic the behavior of the first stage of human audition. It may prove very effective in other systems—in particular, those which attempt to more directly mimic higher-level human audition.

Acknowledgements

Research supported by APPLE Computer and ONR grant no. N-00014-91-J1482. We would like to thank Hynek Hermansky and US WEST for making the PLP algorithm available.

Table 6: Average test results across train conditions.

train	DFT	PLP	COCH
clean	51.0	56.1	47.0
noise 1	63.8	68.8	61.9
noise 2	57.8	61.9	55.8
babble	67.1	74.5	64.0

References

[1] R. Cole, Mark Fanty, Yeshwant Muthusamy, and Murali Gopalakrishnan. Speaker-independent recognition of spoken english letters. In *International Joint Conference on Neural Networks*, 1990.

[2] Ronald Cole, Yeshwant Muthusamy, and Mark Fanty. The ISOLET spoken letter database. Technical Report CS/E 90-004, Oregon Graduate Institution, 1990.

[3] H. Hermansky. Perceptual Linear Predictive (plp) analysis of speech. *J. Acoust. Soc. Am.*, 87(4):1738–1752, 1990.

[4] M. J. Hunt and C. Lefebvre. Speaker dependent and independent speech recognition experiments with an auditory model. In *Proceedings IEEE International Conference on Acoustics, Speech, and Signal Processing*, 1989.

[5] R. F. Lyon. A computational model of filtering, detection, and compression in the cochlea. In *Proceedings IEEE International Conference on Acoustics, Speech, and Signal Processing*, 1982.

[6] H. M. Meng and V. W. Zue. A comparative study of acoustic representations of speech for vowel classification using multi-layer perceptrons. In *International Conference on Spoken Language Processing*, 1990.

[7] Y. K. Muthusamy, R. A. Cole, and M. Slaney. Speaker-independent vowel recognition: Spectrograms versus cochleagrams. In *Proceedings IEEE International Conference on Acoustics, Speech, and Signal Processing*, 1990.

[8] Malcolm Slaney. Lyon's cochlear model. Technical Report Apple Technical Report #3, Apple Computer Inc., 1988.

A Comparative Study of Five Spectral Representations for Speaker-Independent Phonetic Recognition

Joseph W. Creekmore
The MITRE Corporation
7525 Colshire Drive
McLean, VA 22102

Mark Fanty
Dept. Comp. Sci. and Eng.
Oregon Graduate Institute
Beaverton, OR 97006

Ronald A. Cole
Dept. Comp. Sci. and Eng.
Oregon Graduate Institute
Beaverton, OR 97006

Abstract

This paper describes a comparative study of five spectral representations for speaker-independent phonetic recognition using the TIMIT data base. A feedforward network is trained to classify 20-msec frames of speech as one of 39 phonetic classes derived from the TIMIT data base. The five representations investigated include the discrete Fourier transform, three representations based on conventional linear predictive coding (LPC), and the cepstral coefficients derived from Perceptual Linear Predictive (PLP) analysis. The PLP cepstral coefficients outperformed the other representations on the task of assigning the correct phonetic label to individual time frames. We show how phonetic context can be exploited by providing spectral information before and after the frame to be classified. The effect of the training set size and distribution is also examined.

1 Introduction

This paper compares five spectral representations of speech for the task of recognizing phonetic labels derived from the TIMIT data base [1]. Features derived from each spectral representation are used to train a neural network to assign phonetic labels to individual frames of speech. In order to compare with other techniques, the 64 TIMIT symbols were grouped into 39 phonetic classes according to a mapping taken from Lee and Hon [2].

The present experiments extend previous research comparing signal representations [3, 4, 5, 6] to new feature sets that examine the ability to capture information derived from phonetic context. The objective is not to attain the highest recognition rate, but rather to evaluate commonly used signal representations. The results have important implications for the choice of signal representations in speech recognition systems.

2 Spectral representations

Five spectral representations of speech were investigated. These include the discrete Fourier transform, three representations based on linear predictive coding (LPC), and the cepstral coefficients derived from Perceptual Linear Predictive (PLP) analysis.

Of the three LPC spectral representations, cepstral coefficients have been the most widely used in speech recognition. We investigate two other LPC spectral representations; LPC spectral envelope samples, and line spectrum frequencies (LSF).

LSF parameters are derived from LPC prediction coefficients through decomposition of the impulse response of the LPC analysis filter into even and odd functions [7]. LSF trajectories follow speech resonant frequencies, and exhibit similarities with the speech spectrogram.

The PLP speech analysis technique computes an all-pole model of the auditory-based speech spectrum [8]. The auditory-based spectrum is obtained by integrating the short-time power spectrum over simulated critical-band auditory masking curves, re-sampling the integrated spectrum at approximately 1 Bark intervals, modifying the spectral amplitude by a simulated fixed equal-loudness curve, and compressing it through the cubic root non-linearity to simulate the intensity-loudness power law of hearing. The PLP spectral representation has been shown to be efficient in suppressing speaker-dependent components in the speech signal by using a low-order model.

2.1 Signal processing

The speech signal was sampled at 16 kHz, and a 20-msec Hamming window was applied every 3 msec. To obtain the discrete Fourier transform representation, a 512-point DFT was computed from the Hamming windowed waveform, and the 256 magnitude coefficients were compressed as follows: 32 coefficients were extracted from the 0-4 kHz region, and 16 coefficients were extracted from the 4-8 kHz region.

For the three LPC representations, a 12th-order LPC analysis was performed using the autocorrelation method. Twelve cepstral coefficients and 12 line spectrum frequencies were derived from the LPC coefficients, and 32 equally-spaced samples of the LPC spectral envelope were also derived from the LPC coefficients.

In addition to the DFT and the three LPC spectral representations, the cepstral coefficients of the 7th-order PLP analysis were extracted.

1058-6393/91 $01.00 © 1991 IEEE

Each of the five spectral representations was augmented with one additional feature, the log power of the speech frame. Note that the dimensionality of the five feature sets varies, and consequently the results presented below also provide information regarding computational complexity and the amount of network training required.

3 Experiments

Phonetically labeled speech from a portion of the "sx" sentences of the TIMIT data base was assigned to training and test sets, with no speakers in common. The training set consisted of 1155 utterances spoken by 230 speakers (160 male and 70 female), while the test set consisted of 580 utterances spoken by 110 speakers (72 male and 28 female). Table 1 enumerates the list of 39 phonetic labels, along with examples. Glottal stops were omitted and the following within-class confusions were not counted: {ih, ix}, {ah, ax, ax-h}, {uw, ux}, {aa, ao}, {l, el}, {er, axr}, {m, em}, {n, en, nx}, {ng, eng}, {sh, zh}, {hh, hv}, {cl, pcl, tcl, kcl, qcl, bcl, dcl, gcl, epi, h#, sil, pau}.

Index	Label	Example	Index	Label	Example	Index	Label	Example
1	iy	beat	14	ow	boat	27	d	dad
2	ih	bit	15	l	led	28	g	gag
3	eh	bet	16	w	wet	29	dx	butter
4	ae	bat	17	r	red	30	p	pop
5	ah	butt	18	y	yet	31	t	tot
6	er	bird	19	m	mom	32	k	kick
7	uw	boot	20	n	non	33	ch	church
8	uh	book	21	ng	sing	34	s	sis
9	aa	cot	22	z	zoo	35	sh	shoe
10	ey	bait	23	v	very	36	f	fief
11	ay	bite	24	dh	they	37	th	thief
12	oy	boy	25	jh	judge	38	hh	hay
13	aw	about	26	b	bob	39	cl	(clos)

Table 1 List of Phonetic Labels Used in the Recognition Task

The recognition task was to assign the correct phonetic label to selected time frames of the speech signal. For each utterance in the training set and test set, 20-msec frames were excised from labeled phonetic segments. The identical training set and test set utterances were used in the evaluation of all five spectral representations. In order to capture the dynamic characteristics of each phonetic class, frames were extracted at random locations within each phonetic segment. To obtain approximately equal numbers of time frames for each label, the speech frames were sampled inversely proportional to the frequency of occurrence of the label. This resulted in approximately 200 time frames for each of the 39 phonetic labels.

The feedforward networks were trained using back-propagation with conjugate gradient optimization [9]. The first set of experiments described below used networks with 24 hidden units. Each network was trained using 80 presentations of the training vectors, and then evaluated on both the test set and training set. This was repeated until test set scores converged, i.e., the experimental results presented below reflect the performance of the network with the highest performance on the test set.

3.1 Influence of context

As in [3], phonetic context was exploited by providing spectral parameters both before and after the frame to be classified. In order to evaluate the effect as well as the type of context, the first set of experiments were conducted using three types of context: adjacent context, near context, and far context, depicted below in Figure 1.

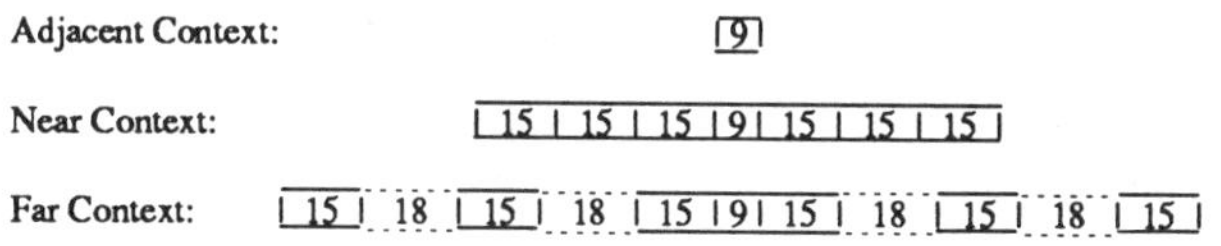

Figure 1 Phonetic Context Intervals (msec)

Solid boxes in the figure illustrate intervals in milliseconds over which each spectral representation is averaged, while dashed lines between boxes illustrate gaps between the sampled intervals. For all three types of context, spectral vectors associated with the frame to be classified were averaged over three adjacent frames (9 msec). Moreover, both near and far context append spectral vectors derived from six additional blocks, each averaged over five adjacent frames (15 msec). Note that the dimension of the feature vector for near and far context is the same, and only the location of the context frames differ.

Distinct training set and test set pairs were constructed for each of the five spectral representations and for each phonetic context scheme. The table below gives the percent of frames correctly recognized for each condition.

	LPC Cep	LPC Env	LSF	PLP	DFT
Adjacent Context	24.4%	25.3%	24.8%	25.1%	22.4%
Near Context	37.7%	40.1%	40.8%	44.1%	40.6%
Far Context	39.1%	41.3%	40.4%	44.8%	41.0%

Table 2 Frame-Based Phonetic Recognition Results

It can be seen that performance increases substantially for all five spectral representations as additional spectra are added. Results for near and far context were comparable, with slight improvements for all but the LSF spectral representation using far context. Note that, with the exception of adjacent context, the PLP spectral representation outperformed the other spectral representations. Moreover, significantly fewer features were required.

Analysis of the confusion matrix obtained for each experiment revealed how frame-based errors were distributed across the 39 classes for the five spectral representations. Figure 2 below shows how the recognition rate varied across phonetic class for the PLP and LPC cepstral representations with far context, using the label indexing given in Table 1. (Plots for the other three representations exhibited the same general shape.) Note that the recognition rate across phonetic class is highly correlated for the two representations.

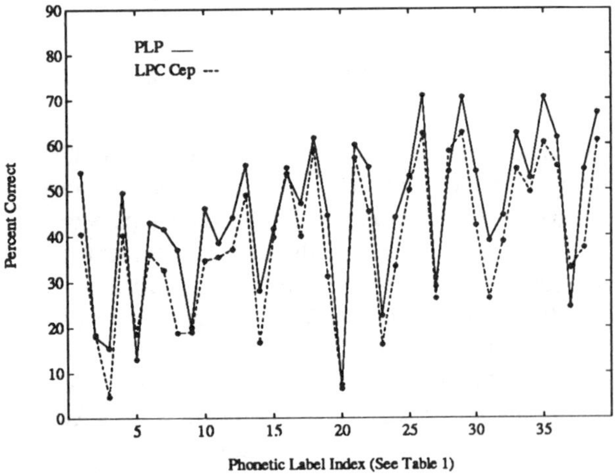

Figure 2 Recognition Rate Across Phonetic Class

3.2 Training set size and distribution

The above results were obtained with training sets composed of equal numbers of tokens for each phonetic class. However, the distribution of phonetic class in the training and test sentences, as well as most speech databases, is far from uniform. Moreover, performance generally improves as additional training tokens are utilized. Consequently, several experiments were performed to determine the impact of the training set size and distribution on performance for the PLP spectral representation with far context.

For these experiments, training and test sets were constructed by sampling speech frames proportional to the frequency of occurrence of labeled frames from each phonetic class in the training sentences. Four separate experiments were conducted using 4,000, 8,000, 16,000, and 32,000 excised training tokens. Each network included 24 hidden units. Recognition performance was obtained for the training set as well as the test set. Figure 3 shows recognition rate as a function of the number of training tokens. For a training set with 32,000 tokens, the difference between training set and test set performance is approximately 3%, implying good generalization.

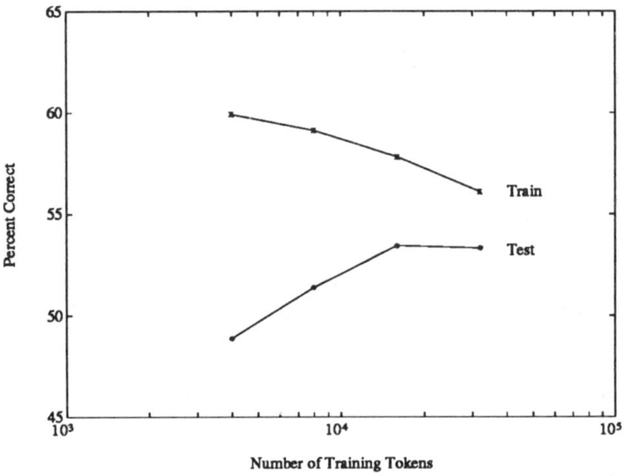

Figure 3 Recognition Rate as a Function of Training Set Size

A comparison of the above test result for 8,000 training tokens with that given in Table 2 for the PLP spectral representation (far context using 7800 training tokens uniformly distributed) reveals a significant increase in performance (44.8% vs. 51.8%). This is at the expense, however, of poor performance on infrequently occurring phonetic classes. Figure 4 shows how the recognition rate varied across phonetic class for the test set after training on 8,000 tokens extracted by frequency of occurrence. A comparison with Figure 2 reveals a markedly different performance on most of the phonetic classes, as well as several phonetic classes whose recognition rate is less than 10%.

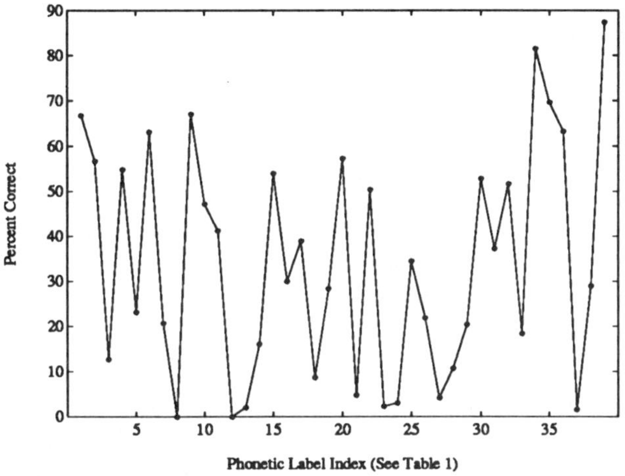

Figure 4 Recognition Rate Across Phonetic Class

Many speech recognition systems employ a phoneme-to-word recognizer whose performance depends not only on the overall phoneme recognition rate, but on the lower recognition rates for individual phonemes as well. The above result suggests that although a higher overall frame recognition rate can be obtained by training the network according to the natural frequency of occurrence of phonetic classes, the decrease in performance due to the lower recognition rates may outweigh the gain in overall recognition rate.

To investigate this, training sets were constructed as follows. In addition to the training set consisting of 8000 tokens extracted by the frequency of occurrence of phonetic class, four additional training sets were constructed. The number of training tokens allocated to each phonetic class was varied in linear steps until the distribution was uniform, so that each training set consisted of approximately 8000 total tokens. After training networks with these sets, performance was measured on the test set constructed by the frequency of occurrence of phonetic class.

Table 3 below summarizes the results for these five experiments, showing both the overall recognition rate as well as the phonetic classes which exhibited the three lowest individual scores. As the table indicates, a significant improvement in the minimum recognition rate can be achieved at the cost of a slight reduction in overall recognition rate.

Training Set Distribution	Overall Recognition Rate	Lowest Recognition Rates
Freq of Occur	51.8%	uh (0.0%), oy (0.0%), th (1.6%)
Step 1	51.6%	uh (0.0%), oy (0.0%), th (3.2%)
Step 2	50.7%	th (1.6%), uh (3.6%), v (5.7%)
Step 3	48.2%	eh (8.6%), ah (10.9%), uh (14.3%)
Uniform	42.6%	n (5.4%), ah (11.9%), eh (14.1%)

Table 3 Effect of Training Set Distribution on Recognition Rate

3.3 Number of hidden units

In addition to the training set size and distribution, the network performance is also affected by the number of hidden units. Subtle distinctions between classes may not be captured by the network if there are too few hidden units; consequently, performance typically improves as more hidden units are included. However, for a fixed training set size, performance often degrades when too many hidden units are included.

Several experiments were performed to determine the impact of the number of hidden units on network performance for the PLP spectral representation with far context. Three of the training set and test set pairs described in the previous section, consisting of 4,000, 8,000, and 16,000 tokens extracted by frequency of occurrence, were used measure recognition performance for networks with 24, 48, 96, and 192 hidden units.

Figure 5 shows the recognition rate as a function of both the number of hidden units and the size of the training set. Note that performance initially improves for all three training sets, but that performance degrades for the training sets with 8,000 and 16,000 tokens as additional hidden units are included. The best performance obtained was 54.8% using a network which included 48 hidden units and was trained with the PLP spectral representation on 16,000 tokens using far context.[1]

[1] In subsequent experiments with the goal of improving the recognition accuracy, we achieved greater than 60% performance.

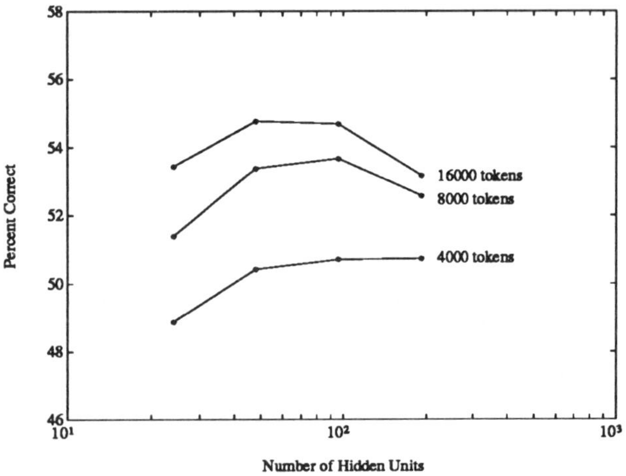

Figure 5 Recognition Rate as a Function of Number of Hidden Units and Training Set Size

4 Summary

We may summarize the above experimental results as follows:

(1) For all five spectral representations, performance improves dramatically when additional spectra from nearby frames are added.

(2) Performance for near context and far context was comparable, with slight improvements for all but the LSF spectral representation using far context.

(3) The PLP outperformed the other spectral representations for both near and far context; moreover, less computation and training time are required.

(4) Phonetic class recognition errors are highly correlated among the five representations.

(5) Performance increases significantly (approximately 7%) when the network is trained by the frequency of occurrence of phonetic classes. This is at the expense of poor performance on infrequently occurring phonetic classes.

(6) The minimum recognition rate on individual phonetic classes can be significantly improved by adjusting the training set distribution with a slight reduction in overall recognition rate.

(7) The performance of the network typically improves as additional hidden units are included; for a fixed training set size, however, performance can degrade when too many hidden units are included.

Acknowledgment

This work was supported in part by Office of Naval Research grant number N00014-91-J1482.

References

[1] L. F. Lamel, R. H. Kassel, and S. Seneff, "Speech Database Development: Design and Analysis of the Acoustic-Phonetic Corpus," Proceedings of Speech Recognition Workshop (DARPA), 1986.

[2] K. F. Lee and H. W. Hon, "Speaker-Independent Phone Recognition using Hidden Markov Models," IEEE Transactions on Acoustics, Speech, and Signal Processing, 37(11), pp. 1641-1648, November, 1989.

[3] R.D.T. Janseen, M. Fanty, and R. A. Cole, "Speaker-Independent Phonetic Classification in Continuous English Letters," Proceedings of the International Joint Conference on Neural Networks, June, 1991, Seattle, Washington.

[4] H. M. Meng and V. W. Zue, "A Comparative Study of Acoustic Representations of Speech for Vowel Classification using Multi-Layer Perceptrons," International Conference on Spoken Language Processing, November 18-22, Kobe, Japan, 1990.

[5] Y. K. Muthusamy and R. A. Cole, "Speaker-Independent Vowel Recognition: Spectrograms Versus Cochleagrams," Proceedings of the IEEE International Conference on Acoustics, Speech, and Signal Processing, Toronto, 1990.

[6] P. Mermelstein and S. Davis, "Comparison of Parametric Representations for Monsyllabic Word Recognition in Continuously Spoken Sentences," IEEE Transactions on Acoustics, Speech, and Signal Processing, vol. ASSP-28, No.4, August, 1980.

[7] G. Kang and L. Fransen, "Low-Bit Rate Speech Encoders Based on Line-Spectrum Frequencies," Naval Research Laboratory Report 8857, January, 1985.

[8] H. Hermansky, "Perceptual Linear Predictive (PLP) Analysis of Speech," J. Acoust. Soc. Am., 87 (4), pp. 1738–1752.

[9] E. Barnhard and R. A. Cole, "A Neural-Net Training Program Based On Conjugate Gradient Optimization," Technical Report CSE 89–014, Oregon Graduate Institute of Science and Technology, Beaverton, OR, July, 1989.

[10] F. Fallside, H. Lucke, T.P. Marsland, P.J. O'Shea, M. St. J. Owen, R.W. Prager, A.J. Robinson, N.H. Russell, "Continuous Speech Recognition for the TIMIT Data Base using Neural Networks," in Proceedings of the IEEE International Conference on Acoustics, Speech, and Signal Processing, Toronto, 1990.

A Walsh Transform Based Endpoint Detection of Isolated Utterances

John Huang and Ben-Dau Tseng
Department of Electrical Engineering
California State University, Chico
Chico, California 95929-0930

Abstract

In order to reduce the processing time of a speech signal, the beginning and end of an utterance must be determined. In this paper, the Walsh spectra of speech data will be used to provide energy distribution of an isolated utterance. The distribution along with an empirical threshold setting is the basis used to detect the endpoints. To demonstrate the efficiency of the proposed method, experimental results along with comparisons to the commonly used endpoint detection algorithm [1] will be included.

1 Introduction

It is important to know the beginning and end of an utterance for the purpose of analyzing speech data efficiently. The commonly used algorithm for determining the endpoints of an isolated utterance utilizes the measures of energy and zero-crossing rate from the digitized time-domain data [1]. According to Parseval's theorem, the equivalent energy distribution in the time-domain may be extracted in the frequency domain.

The Fast Fourier Transform (FFT) has been applied to many time-domain data to obtain their frequency representations. However, the computation time will be too long even for an isolated utterance. The Walsh transform offers an alternative way to obtain frequency (sequence) information about time-domain data, but requires no multiplications. With a minor modificaton of the algorithm, a fast Walsh transform algorithm (FWT) has been developed to further speed up its computation.

In this paper the Walsh transform will be applied to a sequentially collected speech data in 128-point segments. From the Walsh spectra, the energy distribution can be generated and analyzed. The distribution along with an empirical threshold setting will be used to detect the endpoints.

2 Walsh Concept

Given a physically realizable signal, a linear combination of orthogonal periodic functions may be used to approximate the given signal. The conventional, or commonly used trigonmetric Fourier series representation is an application of sinusoidal functions. The Walsh function series is being introduced as an alternative, with the exception that it's signal approximations are based on non-sinusoidal functions [2].

The primary interest is focused on the binary nature of the functions, i.e. the funtions form a set of rectangular waveforms that take on two amplitude values, ±1, defined over a limited time interval. This clearly simplifies computations.

The concept of frequency is occasionally overlooked. Frequency is associated with periodicity, and a more general definition of frequency is sequency specifically applied to sinusoidal waveforms. A more precise

1058-6393/91 $01.00 © 1991 IEEE

definition of sequency is described as 'one half the average number of zero crossings per unit time interval' [3].

Two arguments are required for complete definition of the Walsh function; a normalized time period t as t/T, and an ordering number, n, representing sequency. The function may be expressed as

$$WAL(n,t).$$

Fig. 1 reveals the first eight Walsh functions in order of increasing number of zero crossings. Recall the trigonometric form of the Fourier series [4] as

$$f(t) = a_{\emptyset} + \sum_{n=1}^{\infty} (a_n \cos nw_{\emptyset}t + b_n \sin nw_{\emptyset}t)$$

Decompositon of a time series function f(t) into a Walsh series is represented in a similar manner by

$$f(t) = a_{\emptyset}WAL(\emptyset,t) + \sum_{i=1}^{N/2-1} \sum_{j=1}^{N/2-1} [a_i SAL(i,t) + b_j CAL(j,t)]$$

Walsh functions may be categorized by even and odd functions known as "CAL" and "SAL" [3], respectively

$$CAL(k,t) = WAL(2k,t)$$
$$SAL(k,t) = WAL(2k-1,t)$$

which are similar to the cosine and sine series.

The expected Fast Walsh transform pair [5] can be described for a time series f_t comprised of N samples as

$$F_n = 1/N \sum_{t=\emptyset}^{N-1} f_t Wal(n,t)$$
$$n=1,1,2,...,N-1$$

$$f_t = \sum_{n=\emptyset}^{N-1} F_n WAL(n,t)$$
$$t=\emptyset,1,2,...,N-1.$$

Compare the FWT to the Discrete Fourier transform

$$F_n = 1/N \sum_{i=\emptyset}^{N-1} f_t \exp(-j2\pi if/N)$$
$$n=\emptyset,1,2,...,N-1.$$

It is recognized that $\exp(-j2\pi if/N)$ is complex and can assume N differnt values, and support of the FFT requires $N/2\log_2 N$ complex multiplications. WAL(n,t) is real and limited to binary values, thus, allowing the Walsh transform to be easily implemented by digital methods and requiring only real additions and subtractions. Evaluating the FWT for N values requires $2N\log_2 N$ mathematical operations, where N is a radix 2.

A closer view of the Walsh function set yields a strong correlation with the orthogonal Hadamard matrix of the same size [2] [5]. This case reveals that the matrix serves as a mathematical transform, matrix multiplication in particular, to convert binary information (±1), and expressing it in an alternate form to generate the spectral domain. Figure 1 is a complete function-set for N=8 and may be derived from the relation

$$W_8 = [H]_8 \cdot [f(t)]_8$$

3 Walsh Endpoint Detector

The foundations of the conventionally used endpoint detector is applied here using the energy distribution generated by the Walsh transform. For completeness, the empirical relations [1] governing the energy thresholds are expressed as

$$ITL = MIN(I1,I2)$$
$$ITU = 5 \cdot ITL$$

$$I1 = 0.03 \cdot (IMX-IMN)+IMN$$
$$I2 = 4 \cdot IMN.$$

I1 is depended upon three percent of the peak energy, i.e. difference between the maximum energy, IMX, and the average unvoiced energy, IMN, in addition to the average unvoiced energy. I2 is set to be four times the unvoiced energy. ITL is the minimum of I1 and I2 while ITU is five times ITL.

With reference to Fig. 2, the algorithm is described by assuming that voiced speech lies inside the interval defined by preliminary endpoint estimates N1 and N2. The final endpoints, N1HAT and N2HAT, are determined by searching a 250-msec interval prior to N1 and following N2 for a "dip-and-rise" in energy levels.

It was originally intended to extract zero-crossing information from the Walsh spectrum to assist in detecting weak fricatives outside N1 and N2. The lack of a physically interpretable spectrum is mostly responsible for this inconvenience, and is directly linked to the aperiodic nature of the Walsh funtions.

4 Experimental Results

The speech data was sampled at 10-kHz. A frame size of 128 points corresponding to 12.8-milliseconds was used to calculate one energy record. The test data involved the uttering of digits zero to nine. The results in Table 1 were recorded on-line with the use of a dynamic microphone in a closed room. VISUAL endpoints were hand selected and used as a bench mark for METH 1 [1] and the PROPOSED METHOD. Significant noise levels were introduced to test the performance of weak-fricative handling capabilities of the proposed method - results in Table 2. The first and second columns under each field corresponds to the frame number identifying the begin and end points, respectively.

Tables 1 and 2 reveal the proposed method consistantly identifying beginnig endpoints earlier than visual verification. Refering to Table 1, the endpoint approximation varied for the majority, to within 3 or 4 frames of the hand-selected endpoint. Table 2 shows what appears to be a possible improvement for both methods in approximating the endpoint. Keep in mind that a speaker has a tendancy to talk louder with increasing background noise, as may be the case. Missed endpoints for /six/ and /eight/ was due to energy levels below ITU while searching for weak ending fricatives.

Results were collected for 10 other speakers uttered by a combination of male and females. The proposed endpoint detector appears to "straddle" the visually-located endpoints for some speakers.

5 Conclusion

A definite characteristic of the proposed endpoint detector is consistant early identification of the start of voiced speech. Using the Walsh transformation for speech signal approximation along with digital computer implementation, and based on energy thresholds perse, yields a reasonable endpoint locator for isolated digit utterances.

TABLE 1 – Accuracy of both endpoint detectors.

	METH 1		PROP'D METH		VISUAL	
ZERO	32	64	28	66	31	65
ONE	26	55	23	63	27	60
TWO	32	58	27	57	32	60
THREE	25	49	22	51	23	53
FOUR	28	57	19	61	22	59
FIVE	25	53	20	53	24	60
SIX	28	47*	25	66	20	62
SEVEN	28	58	24	62	26	59
EIGHT	31	45*	28	48*	32	60
NINE	30	59	28	62	31	60

TABLE 2 – Accuracy of both endpoint detectors with heavy background noise.

	METH 1		PROP'D METH		VISUAL	
ZERO	29	81	27	68	29	68
ONE	42	70	40	70	42	74
TWO	51	77	49	78	52	77
THREE	22	47	20	49	23	48
FOUR	28	56	26	57	28	56
FIVE	61	86	59	87	61	87
SIX	32	57*	28	59*	31	74
SEVEN	32	67	29	68	30	67
EIGHT	31	52*	34	53*	36	67
NINE	31	75	27	85	29	74

* Missed endpoint.

FIG. 1 – Walsh Functions for N=8.

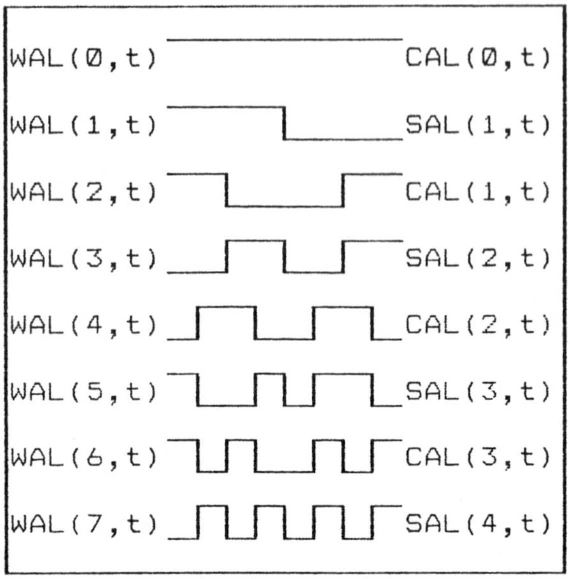

FIG. 2 – Example of Walsh energy data for a digit utterance.

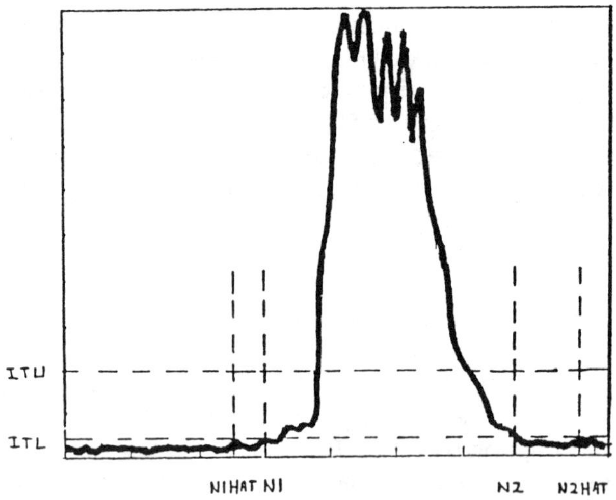

References

[1] L.R. Rabiner and M.R. Sambur, "An algorithm for Determining the End-points of Isolated Utterances", pp. 297-315, B.S.T.J., Vol. 54, No.2, Feb. 1975.

[2] Taylor, Fred J., "Digital Filter Design Handbook", pp. 68-71, Marcel Dekker, Inc., 1983.

[3] Harmuth, H.F., "Transmission of Information by Orthogonal Functions Springer Verlag, 1972.

[4] William H. Hayt, Jr. and Jack E. Kemmerly, "Engineering Circuit Analysis", p. 648, McGraw Hill, Inc., 1978.

[5] K.G. Beauchamp, "Transform for Engineers", p. 39, Oxford University Press, 1987.

A Customized Beamformer System for Acquisition of Speech Signals†

Tod Switzer
Bell Northern Research
1130 E. Arapaho Rd.
Richardson, Texas 75081 USA
(214) 907-7747

D. Linebarger, E. Dowling, Y. Tong, M. Munoz
Center for Engineering Mathematics
The University of Texas at Dallas, MP33
Richardson, Texas 75083–0688 USA
(214) 690–2991 dl@utdallas.edu

Abstract

This paper describes algorithms and hardware used to implement an acoustic beamformer for the purpose of speech acquisition. Beamforming algorithms studied include conventional delay and sum, Frost, and the generalized sidelobe canceller. In each case, parameters related to the algorithm such as the number of sensors, the number of taps, and the rate of adaptation were studied extensively in order to optimize performance with speech signals yet maintain feasibility of real-time implementation on a single DSP chip. Results with simulated and real data will be presented in the paper as well as with an audio tape at the presentation. The hardware implementation will also be presented and discussed. Eight channels of microphone data are simultaneously digitized using TLC32044's, multiplexed in a T1 data frame format, and sent to a TMS320C30 serial port for subsequent processing.

Introduction

The problem of acquiring signals in the presence of noise is common to many areas of communications. We can all relate to the situation in which we are trying to listen to a single speaker in a noisy environment. The noises may be created by machines or by other speakers. The human brain and auditory system together have a remarkable ability to separate the desired sound from the undesired ones. For a machine however, this is a very complex task. It is precisely this problem that we address here. The approach taken is to employ an array of microphones to sample the propagating field of acoustic waves both temporally and spatially. Array processing techniques are then used to combine the output signals of the microphones in such a way as to reinforce the desired signal while canceling out the undesired ones. Applications for this work might include hands free conference phones, hands free cellular phones [1], or voice recognition systems for man-computer interfaces.

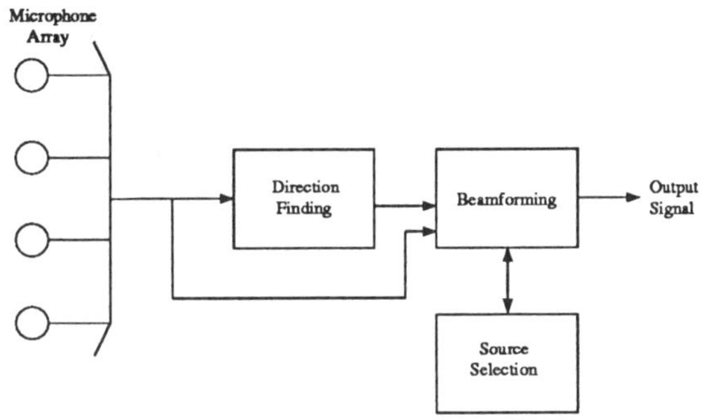

Figure 1: Speech acquisition system using array processing techniques

Figure 1 depicts a speech acquisition system which uses array processing techniques. Data from the array of microphones is first processed by a direction finding algorithm. The purpose of this block is to determine the number of sources present in the field and their directions. This information, along with the original array data, is fed into a beamforming algorithm which forms an estimate of the signal arriving from each direction. These signals are then analyzed by a selection algorithm to determine which of the signals is speech. If more than one speaker is detected, a choice must be made between them. The system output is the estimate of the selected signal. Direction finding and signal selection might be repeated periodically.

A speech acquisition system such as the one described above is currently being developed at the University of Texas at Dallas. The goal of the project is to design a complete system which operates in real time on a single DSP chip. In order to meet this goal, research is underway on the various blocks which will make up the system. The direction finding problem was analyzed in [2]. The hardware platform was designed in [3]. In this paper, we give an overview of the hardware platform

† This work was supported by Texas Advanced Technology Grant 009741-038 and Texas Instruments.

1058-6393/91 $01.00 © 1991 IEEE

and then concentrate on the selection of a beamforming algorithm to be used in the UTD speech acquisition system.

Hardware Overview

The microphone array hardware configuration is depicted in Figure 2. At the front end is a set of eight omni-directional microphones whose outputs are conditioned and amplified by analog circuits. The conditioned signals are fed into an array of eight TLC32044 A/D subsystems, each of which contains an antialiasing filter, a sample and hold circuit, and a data encoding stage. A state machine residing on a programmable logic device ensures that all channels sample the input simultaneously, and buffer their serial digital outputs in their respective shift registers. The state machine next generates framing signals and sequences the data out of the shift registers onto a T1 PCM highway which is terminated by the TMS320C30 serial port. An interrupt service routine on the TMS320C30 moves the data from the serial port into local SRAM of the Spirit30 DSP board that is housed in an IBM-AT card slot.

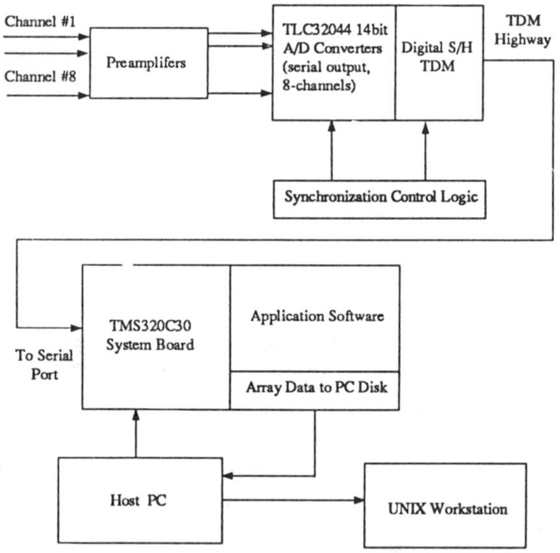

Figure 2: Acoustical Array Processing System Block Diagram

The array hardware was designed to accommodate both real time and off-line processing tasks. The real time capability provides a development platform where signal processing algorithms can be tested and evaluated on live data on the same type hardware that would be used in an application system. The real time tasks are loaded from the IBM-AT and run entirely on the Spirit30 card. On the other hand, off-line tasks only use the array and Spirit30 card to collect data. Here the array is used to feed the Spirit30 as described above, but now the Spirit30 sets up a pair of data buffers in its local SRAM which are alternately transferred over an IBM-AT DMA channel to a ram disk in the PC's main memory. Once collected, the data is transferred to a Unix workstation and converted into Matlab readable form for analysis and algorithm development.

Selecting a Beamforming Algorithm

Beamforming is an array processing technique which is used to focus an array on signals arriving from a particular direction known as the look direction. In our case, we assume that the array has already been focused on the look direction by applying delays to the sensor outputs which cancel the propagation delays experienced by a signal arriving at the array from the desired direction. After the steering delays are applied, the look direction components of the sensor outputs are temporally aligned. Components due to signals arriving from other directions, in contrast, are generally not aligned and cancel through incoherent signal addition.

Beamforming algorithms investigated include the conventional delay and sum as well as the adaptive Frost [4] and generalized sidelobe canceller [5] beamforming algorithms. These algorithms were analyzed in the context of speech acquisition with the above described system [6]. The two adaptive techniques are similar in that the weights are chosen by first constraining the look direction frequency response to be some desired response and then adapting the beamformer weights in order to minimize the energy in the array output. They differ in the way they implement the look direction frequency response constraint. Selection of an algorithm involved testing the three algorithms using array output simulated with a recorded speech signal. For each algorithm, appropriate implementation parameters had to be determined. The parameters with which we were concerned were the number of sensors M, the number of taps per sensor N, and the adaptation rate μ. In the cases of the number of sensors and taps, it was expected that more would be better, except that in our case, we intended to implement the system in real time on a single DSP chip. Therefore we were interested in optimizing these parameters to obtain maximum performance within reasonable limits on computational complexity.

The three algorithms were implemented in Matlab for evaluation purposes. The conventional beamformer used weights equal to $\frac{1}{M}$ where M is the number of sensors in the array. For the two adaptive algorithms, the adaptation step size was normalized as follows

$$\mu_k = \frac{\alpha}{P(k)} \tag{1}$$

where $0 < \alpha < 1$, and $P(k)$ is the instantaneous power in the array snapshot vector $\vec{x}_k$.

$$P(k) = \vec{x}_k^T \vec{x}_k \tag{2}$$

The look direction frequency response was constrained to be flat over the entire spectrum of interest. This constraint is specified with an FIR filter. In our case the filter was the N tap all-pass filter given below.

$$\vec{f} = [1 \quad 0 \quad \dots \quad 0]^T \tag{3}$$

In addition, the GSC implementation used an equal weight conventional beamformer in its mainbeam path and an orthogonal blocking matrix in its sidelobe cancelling path. With the two adaptive algorithms configured as described here, their outputs were identical when used to process the same set of array data. In the following two sections we discuss the results of the simulations tests and the computational complexity of the three algorithms.

Evaluation with Simulated Data

The simulated array data was used to evaluate the algorithms under ideal conditions. The desired signal was a recording of a human speaker arriving at the array from broadside. Two sets of tests were run, each with a different interfering signal arriving at the array from 30 degrees away from broadside. The interfering signal was delayed to reflect the relative propagation delays a real signal would experience in traveling from the source to each of the sensors assuming plane wave propagation. The array output was then formed by adding the delayed interferer to the desired signal for each sensor. This data was then processed using the three beamforming algorithms and the results were compared on the basis of signal to noise and interference ratio (SNIR). Here, we define SNIR as

$$SNIR = 10\log_{10}\left(\frac{\vec{s}^{\,T}\vec{s}}{(\vec{s}-\vec{y})^T(\vec{s}-\vec{y})}\right) \tag{4}$$

where $\vec{s}$ is the desired signal and $\vec{y}$ is the beamformer output. For the adaptive algorithms, exhaustive searches were performed over the three parameters of interest.

The first test used a $1200Hz$ sine wave as the interfering signal. The results are plotted in Figure 3. The solid line at the bottom of the plot shows that the SNIR of a single omni-directional sensor in this environment is $-7.7dB$. The three dashed lines show the performance of conventional beamformers with equal weights and $M = 4$, 8, and 16. The corresponding SNIRs for the conventional beamformers are $-5.9dB$, $.9dB$, and $5.6dB$ respectively. Thus, increasing the number of sensors does improve the beamformer performance. Note, however, that increasing from $M = 4$ to $M = 8$ produces a $6.8dB$ increase in SNIR, while increasing from $M = 8$ to $M = 16$ produces only a $4.6dB$ increase. Listening tests confirm the improvement of the conventional beamformer outputs over the single sensor output, though the interferer is audible even in the 16 sensor case.

The results for the adaptive beamformers with the narrowband interferer, shown with the three dotted curves in Figure 3, reveal that they are very effective for cancelling narrowband stationary interferers. Only one set of curves is given to represent both the Frost adaptive beamformer and the GSC, since the results were identical. The performance for the adaptive beamformers is significantly (more than $10dB$) better than that of the conventional beamformer with the same number of sensors, even when the adaptive beamformers have only one tap per sensor. It will be shown in the next example that with a wideband interferer, it is only the addition of multiple taps per sensor which allows the adaptive algorithms to out perform the conventional beamformer. It can be seen that for each value of M, the SNIR increases with increasing number of taps per sensor N. It is important to note that the SNIR for the 4 sensor 20 tap adaptive beamformer is $11.4dB$ better than the performance of the 16 sensor conventional beamformer. In fact, the interferer cannot be heard in any of the adaptive beamformer outputs with more than 8 taps per sensor.

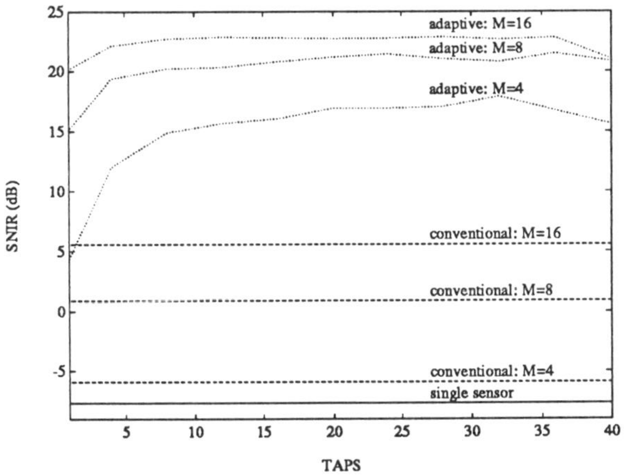

Figure 3: Beamformer performance comparison for the narrowband interferer scenario.

The second test used a recording of the same speaker reading a different sentence as the interfering signal. From the dashed lines in Figure 4, it can be seen that the conventional beamformers were not very good at cancelling the interfering speech signal. With an SNIR of $-1.1dB$ it is difficult to distinguish the four sensor conventional beamformer output from the single sensor output. The same is true of the eight sensor conventional beamformer with an SNIR of $.7dB$. The sixteen sensor conventional beamformer, however, is noticeably better than the other two. Its SNIR of $5.2dB$ makes it possible for a careful listener to understand the desired signal. What remains of the interfering signal is mainly its low frequency components.

Performance of the Frost adaptive beamformer and the GSC is quite good for the competing speaker test. Shown with the dotted curves in Figure 4, these results clearly indicate the advantage of the adaptive algorithms with multiple taps per sensor over the conventional beamformers. It can be seen that the adaptive algorithms with a single tap per sensor ($N = 1$) perform only slightly better than the conventional beamformers with the same number of sensors. As N is increased, beamformer output SNIR improves on each curve up to a maximum on all three curves near $N = 20$. The peak on the 4 sensor curve is $9.9dB$ with $N = 28$ and $\alpha = .8$. Notice that this is $11dB$ better than the four sensor and $4.7dB$ better than the sixteen sensor conventional beamformers. On the eight sensor curve, $12.3dB$ SNIR is the maximum and is achieved with $N = 28$ and $\alpha = .3$. The best SNIR of all is $13.9dB$ produced with the sixteen sensor array with $N = 24$ and $\alpha = .09$. It should be noted that all three curves begin to flatten at $N = 20$ and very little performance improvement is achieved by adding more taps per sensor beyond this point.

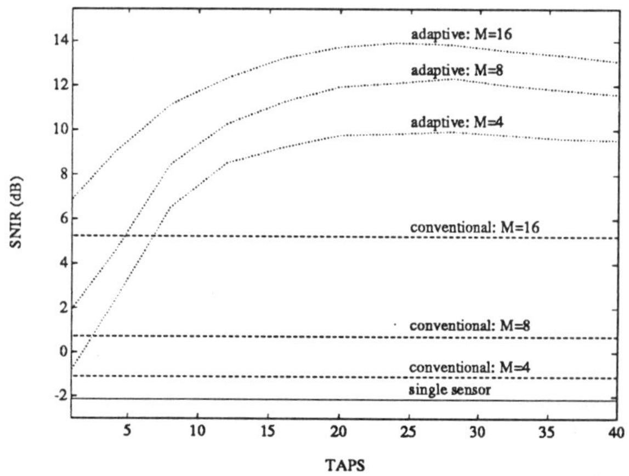

Figure 4: Beamformer performance comparison for the competing speaker scenario.

Beamformer Complexity

We have shown that for the simulated array data the adaptive beamforming algorithms perform better in terms of SNIR than the conventional delay and sum beamformer. Furthermore, the two adaptive algorithms had identical performance and eight sensors with twenty taps per sensor worked well in each test. Next, we examine the computational complexity of the algorithms, as our goal was to select an algorithm which could be accommodated on a fraction of a TMS320C30. In the analysis which follows, we consider the TMS320C30 floating point multiply with accumulate as a single floating point operation (FLOP).

Each algorithm was analyzed to determine how many TMS320C30 FLOPS would be required for a real time implementation. Significant effort went towards finding the most efficient implementation possible. Expressions were derived which give FLOPS as a function of sampling rate (f_s), number of sensors M, and number of taps per sensor N. The expressions for the conventional, Frost, and GSC algorithms are given below.

$$FLOPS_c = Mf_s \tag{5}$$

$$FLOPS_f = f_s[1 + M + N(5M + 4)] \tag{6}$$

$$FLOPS_g = f_s[3 + M + M^2 + N(2M - 1)] \tag{7}$$

These equations were evaluated for an $8kHz$ sampling frequency, eight sensors ($M = 8$), and twenty taps per sensor ($N = 20$) and the results are given in Table I. It can be seen that the conventional beamformer requires far less computations than the adaptive algorithms, the cost being poorer SNIR performance. Either of the two adaptive algorithms is expected to require less than half of the TMS320C30. The GSC is the better choice, however, as it requires half as many computations as the Frost algorithm while yielding identical results.

Algorithm	**FLOPS**	**Percentage of TMS320C30**
Conventional	64 k	< 1 %
Frost	6.8 M	42 %
GSC	3.0 M	18 %

Table I: Complexity of beamforming algorithms

Preliminary Results with Real Array Data

After constructing the array system, we first verified its operation by performing direction finding on data collected with the array. Recordings of a single narrowband signal arriving at the array from various directions were made and the data was processed using conventional direction finding techniques on the Unix workstation. We found that we were able to determine the direction of the interferer in each test to within 3 degrees. This confirmed that the hardware and data collection software were functioning properly.

Our next step was to perform conventional beamforming on data collected with the array. Recordings were made with a configuration similar to that used in the simulations: recorded speech arriving at the array from broadside with a single interferer arriving from another direction. The interferers we used in included narrowband signals as well as male and female speakers. We found that the level of the interferer in the beamformer output was lower than it was in the output of a single sensor, but was still audible.

We have now begun to study the operation of the adaptive algorithms with real array data. Our initial tests were compromised by gain mismatch among the sensor channels and/or reverberation, resulting in severe signal cancellation. After recording data in a low reverberation environment, we obtained reasonable results, though some signal cancellation was still evident. We are currently pursuing sensor gain equalization to improve performance in low reverberation environments and also we are investigating the use of spatial averaging methods to cope with the reverberation found in many typical office environments.

Conclusions

We have discussed the development of an acoustical array for speech acquisition with emphasis on the beamforming algorithms we plan to use in the system. Our simulations showed that the Frost and GSC adaptive beamformers performed better than conventional delay and sum beamforming in terms of SNIR. Because we plan to implement the beamformer on a fraction of a TMS320C30 digital signal processor, we have also studied the complexity of the three beamforming algorithms. We have determined that the GSC algorithm with an orthogonal blocking matrix and twenty taps per sensor can be accommodated on less than one fourth of the processor, and is the best compromise between complexity and performance for our system. Initial testing with conventional direction finding and beamforming algorithms has allowed us to verify the array's functionality. In the future we plan to examine equalization of sensor channels and reverberation induced signal cancellation. We will also move to a custom printed circuit board and investigate better microphones and analog to digital converters.

References

[1] I. Claesson, S. Nordholm, B. Bengtsson, and P. Eriksson, "A multi-dsp implementation of a broad-band adaptive beamformer for use in a hands-free mobile radio telephone," *IEEE Trans. Vehic. Tech.*, vol. 40, pp. 194–202, feb. 1991.

[2] Y.-C. Tseng, "Bearing Estimation for Wideband Signals," Master's thesis, University of Texas at Dallas, 1990.

[3] Y. Tong, "An Acoustical Array Processing System," Master's thesis, University of Texas at Dallas, 1991.

[4] O. L. F. III, "An Algorithm for Linearly Constrained Adaptive Array Processing," in *Proceedings of the IEEE*, pp. 926–935, 1972.

[5] L. J. Griffiths and C. W. Jim, "An Alternative Approach to Linearly Constrained Adaptive Beamforming," *IEEE Transactions on Antennas and Propagation*, vol. AP-30, pp. 27–34, January 1982.

[6] T. Switzer, "Beamforming Techniques for Acquisition of Speech Signals," Master's thesis, University of Texas at Dallas, 1991.

Desired-speech signal cancellation by microphone arrays in reverberant rooms

M.W. Hoffman, M.J. Link, and K.M. Buckley
Department of Electrical Engineering
University of Minnesota

Abstract

This paper addresses the problem of cancellation of a desired-signal in a head worn microphone array processor. The effect of reverberation-induced multipath on a constrained minimum variance beamformer is considered. Constrained power minimization algorithms are particularly susceptible to this desired-signal cancellation. Specifically, this paper evaluates the conditions under which the cancellation of a desired-speech signal occurs for a head worn array of microphones in a reverberant room. The effects of acoustic headshadow are incorporated in a simulated array processor. In addition, room reverberation effects are also modelled. The acoustic parameters investigated include room size, average wall absorption, and speaker-to-listener distance. Distortion-based speech intelligibility measures are employed to assess the desired-speech cancellation.

I INTRODUCTION

The work described herein is part of a larger effort investigating the potential of an array of microphones as a preprocessor for a hearing aid. Currently available single microphone hearing aids suffer from a number of problems. A common complaint of poor performance in reverberant and noisy environments suggests that a spatial filter may enhance listening for the impaired in these difficult circumstances. The advantage of binaural listening over monaural listening in complex sound fields has often been cited as motivation for this application. Past work, [Allen et al. 1977, Peterson 1989, Hoffman and Buckley 1990], has demonstrated both the advantages and the disadvantages of the multi-microphone approach. This paper attempts to establish when, and to what extent, cancellation of the desired-speech signal occurs when a power minimizing array processor is applied in a reverberant environment.

Specifically, linearly constrained minimum variance (LCMV) beamforming [Frost 1972] is considered as a potential candidate for this application. Since desired-signal cancellation can be a problem with this approach, this issue must be addressed. Other approaches to array processing which are not as susceptible to desired signal cancellation include fixed beamforming. While lacking this drawback of LCMV beamforming, fixed beamforming also lacks the most useful asset of LCMV. Namely, that LCMV is adaptive and in many diverse acoustic scenarios it can provide an enhanced desired signal which contains less noise and reverberated energy than that produced by a fixed beamformer. In considering LCMV over other data adaptive beamforming approaches it should be noted that LCMV provides a *relatively* simple, well understood, effective array processing approach. LCMV also has a number of simple, robust extensions.

Recently [Greenberg and Zurek 1991] reported that cancellation of the desired-signal can be a problem with a two channel, modified version of an LCMV based processor under certain reverberant conditions. The modification they made to the Griffiths-Jim beamformer inhibited array adaptation, hence desired-signal cancellation, in the presence of a high power desired signal relative to the background noise. With this modification the desired-signal cancellation reported was dependent on the level of the reverberated (correlated) desired-speech in the LCMV processor.

The parameters effecting desired signal cancellation which are investigated in this paper are the tap length in the LCMV processor (see figure 1) and the direct-to-reverberant energy ratio of the desired speech signal. Many individual parameters affect the direct-to-reverberant energy ratio, such as wall absorption, room size, and source-to-listener distance. The performance measures used to quantitatively describe the

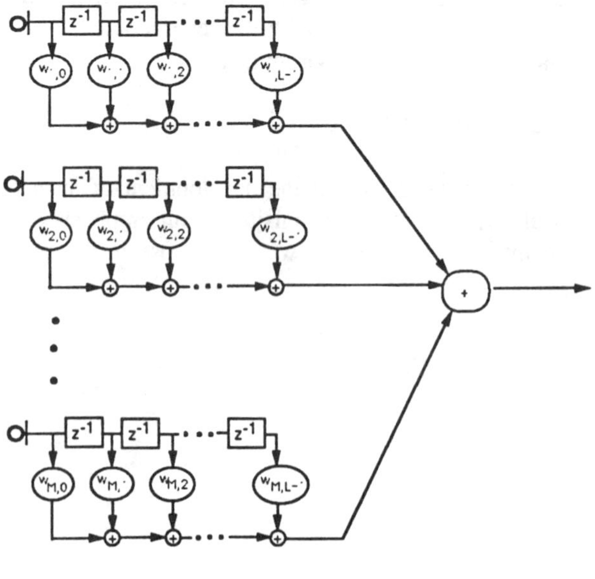

Figure (1) Broadband spatial filter structure.

This work was supported by NSF under contract MIP 9057071.

1058-6393/91 $01.00 © 1991 IEEE

cancellation are based on the Articulation Index [Kryter 1962] and other intelligibility indices. The remainder of the paper is subdivided as follows. In section II LCMV processing is reviewed in the context of the microphone array problem. Also in II, the acoustic model is described. The performance measures mentioned above will be described in more detail in section III. Section IV presents the results of the computer simulations performed. Section V summarizes the main results.

II Array Observation Model

(A) LCMV array processor

Figure 1 illustrates the structure of an M sensor, L tap, spatial filter. Define

$$\mathbf{x}(n) = [\chi(n)^T, \chi(n-1)^T, \cdots, \chi(n-L+1)^T]^T$$

$$\text{and} \quad \chi(n) = [x_1(n), x_2(n), \cdots x_M(n)]^T$$

where ^{T} denotes transpose and $\chi(n)$, is the vector of output signals from each of the M sensors. $\mathbf{x}(n)$ is called the "stacked snapshot" vector, since it consists of the vector of the received signals at each microphone at a point in time, $\chi(n)$, stacked into a single vector. This vector, $\mathbf{x}(n)$, represents all of the received signal present in the beamformer at time instant n. For a stationary input into the array, the data covariance matrix is given by:

$$\mathbf{R} = E[\mathbf{x}(n)\mathbf{x}(n)^T] \tag{1}$$

The dimensions of this matrix are $ML \times ML$. The response of the array to a narrowband source from direction θ at frequency f is denoted by $\mathbf{a}(\theta,f)$. $\mathbf{a}(\theta,f)$, the array response vector, is an $ML\times 1$ vector consisting of the magnitude gains and phases for each of the samples of the "stacked snapshot" vector relative to the array phase reference at a frequency f and from a direction θ. For a stationary signal, $s(n)$, with power spectral density, $S(f)$, impinging on the array from direction θ, the data covariance can be obtained from the Wiener-Hopf relation as:

$$\mathbf{R} = \int_0^B S(f)[\mathbf{a}_c(\theta,f)\mathbf{a}_c(\theta,f)^T + \mathbf{a}_s(\theta,f)\mathbf{a}_s(\theta,f)^T]df \quad . \tag{2}$$

Here B is the bandwidth of interest and the subscripts c and s denote the real (cosine) and imaginary (sine) values used in processing real data.

Consider the standard problem of a linearly constrained minimum variance beamformer. Given an observed array data covariance matrix, $\mathbf{R}$, and a set of linear constraints on the array weights, $\mathbf{w}$, the problem statement is simply:

$$\min_{\mathbf{w}} \quad \mathbf{w}^T \mathbf{R} \mathbf{w} \tag{3}$$

$$\textit{subject.to} \quad \mathbf{C}^T\mathbf{w} = \mathbf{g}$$

where $\mathbf{C} = [\mathbf{c}_1, \mathbf{c}_2, \cdots \mathbf{c}_J]$ is $ML\times J$ and contains as columns the set of J linear constraint vectors; the elements of $\mathbf{g}$ are the specified responses to each of the J constraints; $\mathbf{w}$ is $ML\times 1$; and $\mathbf{R}$ is $ML\times ML$, positive definite and symmetric. The well-known solution to (3) is given by:

$$\mathbf{w}_{opt} = \mathbf{R}^{-1}\mathbf{C}(\mathbf{C}^T\mathbf{R}^{-1}\mathbf{C})^{-1}\mathbf{g} \quad . \tag{4}$$

When the input data is spectrally white noise, the quiescent weight vector, $\mathbf{w}_q$, is given by:

$$\mathbf{w}_q = \mathbf{C}(\mathbf{C}^T\mathbf{C})^{-1}\mathbf{g} \quad . \tag{5}$$

The simplest form of linear constraints for power minimization beamforming would be a set of look direction constraints. These constraints fix the beamformer response in the direction of the desired signal. Subject to this fixed response, output power from the beamformer is minimized. [Cox et al. 1987] proposed a robustness constraint for practical array processing which limits the norm (or length) of the weight vector to prevent desired-signal cancellation due to array imperfections and misteering. This approach can be implemented by the structure shown in figure 2 and known as the Generalized Sidelobe Canceller (GSC). The weight vector $\mathbf{w}_q$ from figure 2 is fixed, while the weight vector $\mathbf{v}$ is subject to a norm constraint. See [Hoffman et al. 1990] for details of how this algorithm can be applied to the microphone array problem to inhibit signal cancellation due to misteering of the array. For the present study, no quadratic constraint has been incorporated.

(B) Acoustic model

Two primary components comprise the acoustic model used for the computer simulations of the hearing aid array. The first component is the room reverberation model which is based upon the method of images [Allen and Berkley,1979]. The second is the model of the effects of acoustic headshadow on the observations of the array of microphones. The simulations generate the composite pulse response from the speaker to each microphone mounted on the head of the listener.

[Allen and Berkley,1979] outline a basic procedure for using the method of images to model a reverberant room transfer function. The computed pulse response, $h_{tot}(n)$, from this method can be written as follows:

$$h_{tot}(n) = \sum_{i=0}^{N} \frac{\beta_i * \delta(n-[n_i]_{rnd})}{4\pi|\mathbf{R}_i|} \tag{6}$$

where, $\delta(n) = 1$ if $n = 0$ and $\delta(n) = 0$ otherwise; β_i is the aggregate frequency independent wall reflection attenuation for the ith reflection; $\mathbf{R}_i$ is the distance from the ith image source to the listener; and $[\ ]_{rnd}$ indicates integer rounding. While (6) is sufficient for estimating the pulse response of a room from a speaker to a single microphone, more accurate accounting of the relative phase between microphones in an array is needed. [Peterson] incorporates this via a time domain interpolation for each reflection in (6). In the acoustic model employed in the following simulations proper accounting for phase is incorporated with the headshadow effects.

In [Hoffman and Buckley 1990], a model of acoustic headshadow was described. Data taken from psycho-acoustical literature for the interaural pressure level and time delay differences was used to form array response vectors incorporating acoustic headshadow effects. These frequency domain data are used here to incorporate the headshadow effects into the room reverberation model. Figure 3 shows a block diagram of the general process. The frequency dependent time delays and pressure levels are stored in a two dimensional table with frequency and azimuthal angle of arrival as the dependent variables. For a specific azimuth, the values of pressure magnitude and time delay are interpolated linearly at each frequency. In the domains of pressure magnitude and time delay the headshadow data are quite smooth. An elevation correction is applied to these data for sources arriving from outside the azimuthal plane. For a specific angle of arrival, the time delay data and pressure magnitude data are coupled with the noninteger portion of the reverberation delay. The amplitude attenuation factors in (6) are incorporated. An IFFT of the data provides a relatively short time domain response which incorporates the frequency dependent effects of the acoustic headshadow for each reflection. The effects of the fractional time delay are included within the IFFT. While this represents a circular convolution in the time domain, the effects are negligible compared to other non-ideal aspects of the model. The overall room/head pulse response is the sum of the appropriately delayed individual contributions from each image source.

III Performance Measures

(A) Distortion and MSE

Distortion can be defined as the mean-squared difference between a desired signal and a received distorted desired signal. For the general case of a received distorted signal plus corrupting noise, this measure is simply the mean squared error (MSE) between the received signal and the original signal transmitted. In the case where no noise is present, the term distortion is sometimes used to characterize the effects that the communications channel has on the desired signal only. In evaluating the effects of reverberation on a desired speech signal without any interfering signals, this study employs this type of measure. Given a composite transfer function from source to listener, $H_C(f)$, the distortion that a channel introduces to a signal with power spectral density $S(f)$ is given by:

$$D = \int_0^B |H_C(f)-H_d(f)|^2 S(f)df \qquad (7)$$

where $H_d(f)$ is the desired channel response consisting of unity magnitude and a constant group delay over B.

For an array with a limited spatial aperture, better beamforming capability exists at higher frequencies. In addition, high level, low frequency sounds have a tendency to mask higher frequency sounds which can contain other information important to speech intelligibility. For these and other reasons, microphone arrays exhibit better performance when using pre-whitened speech versus cases without whitening. For the purposes of measuring distortion, this results in a frequency weighted MSE cost function. Distortion after pre-whitening can be defined as:

$$D_W = \int_0^B |H_C(f)-H_d(f)|^2 S_W(f)df \qquad (8)$$

here $S_W(f)$ is the whitened input spectrum. Rewriting (8) gives:

$$D_W = \int_0^B |H_C(f)-H_d(f)|^2 |H_W(f)|^2 S(f)df$$

$H_W(f)$ defines the preferential weighting of the distortion

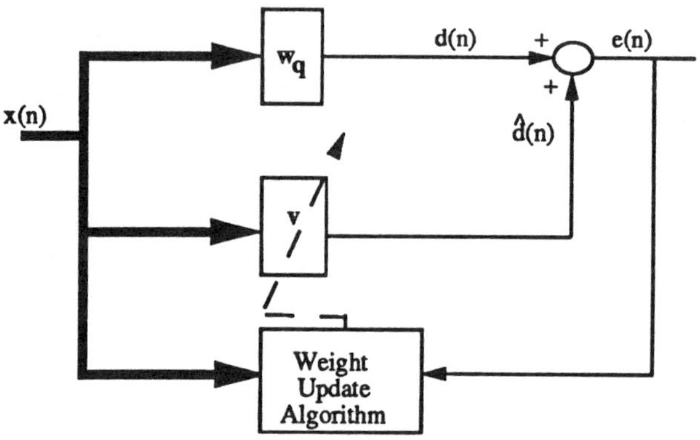

Figure (2) Generalized Sidelobe Canceller.

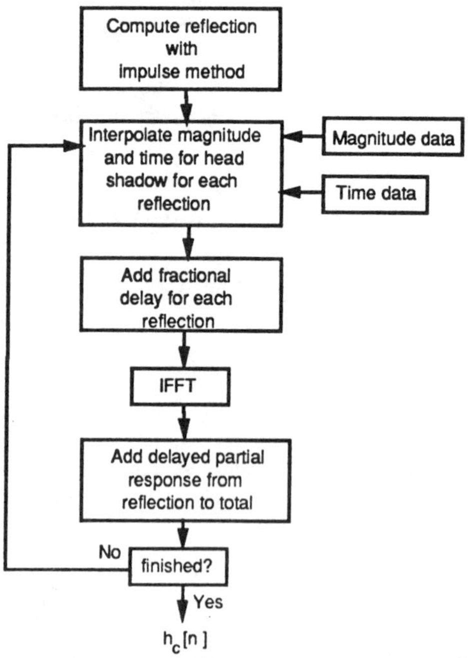

Figure (3) Room reverberation and headshadow model block diagram.

measure. In the present context it is the inverse of the long-time speech spectrum.

Another distortion measure can be defined which treats the first 50 msec or so of reverberated energy as helpful to speech intelligibility. [Houtgast et al. 1980] show a close relationship between this good/bad reverberation model and the Speech Transmission Index (STI). All signal energy arriving within 50 msec of the direct path defines the desired signal, while all other reverberation paths contribute to distortion. The effectiveness of STI in predicting intelligibility in reverberant situations and it's relationship to this definition of distortion motivate the following definition:

$$D_{50} = \int_0^B |H_C(f) - H_{50}(f)|^2 S_W(f) df \qquad (9)$$

$H_{50}(f)$ represents the reverberation transfer function for the first 50 msec of the pulse response. The input signal is again assumed to be pre-whitened.

(B) Articulation Index measures

The Articulation Index (AI) [Kryter 1962] was developed to provide an objective measure of speech intelligibility for a given physical situation. Many empirically based adjustments are made to provide accurate predictions of speech intelligibility under certain situations (e.g., level dependent masking). The use of a simplified version of AI fails to incorporate several important features which are, in fact, integral parts of the Articulation Index. The above caveat having been made, AI is used in the present investigation to indicate the effect the desired signal cancellation will have on the intelligibility of the received/cancelled speech.

[Peterson 1989] developed a measure G_{AI}, called intelligibility weighted gain, to quantify the effects that interfering speech has on the intelligibility of desired speech processed with a microphone array. The incorporation of some features of the Articulation Index in G_{AI} allows the prediction of intelligibility increases when a microphone array replaces a single omni-directional microphone. The application of an AI-based performance measure in the present study is similarly motivated. A simplified version of AI can be defined as follows:

$$\mathrm{AI} = \sum_{i=1}^{15} w_{AI,i} 10 * log_{10}([SNR_i]_{1/3})_{[0,30]} \qquad (10)$$

where

$$[SNR_i]_{1/3} = \frac{\int_{1/3oct_i} |S(f)| df}{\int_{1/3oct_i} |N(f)| df}$$

is the signal to noise ratio of the third octave averaged speech and noise (or distortion) spectra, $S(f)$ and $N(f)$, respectively. The $w_{AI,i}$ in (10) are the empirically derived [Kryter 1962] third octave weights applied to 15 third octave bands of speech. Note that the weighted sum of signal to noise ratios is applied to the dB value of the smoothed SNR clipped to fall between 0 and 30 dB. AI is a value between 0 and 1. AI = 0 corresponds to no intelligibility, AI = 1, to very high intelligibility.

Two distortion measures, D_W and D_{50}, will be used to generate two AI-based measures. AI_{MS} is the AI-measure computed when D_W distortion, i.e., the weighted MSE distortion, is considered the noise in (10). Specifically,

$$[SNR_{MS,i}]_{1/3} = \frac{\int_{1/3oct_i} S_W(f) df}{\int_{1/3oct_i} |H_C(f) - H_d(f)|^2 S_W(f) df} \qquad (11)$$

AI_{50} is the AI-measure computed when D_{50} distortion is considered noise and $|H_{50}(f)|^2 S_W(f)$ is considered the signal:

$$[SNR_{50,i}]_{1/3} = \frac{\int_{1/3oct_i} |H_{50}(f)|^2 S_W(f) df}{\int_{1/3oct_i} |H_C(f) - H_{50}(f)|^2 S_W(f) df} \qquad (12)$$

AI_{MS} and AI_{50} are both computed as per AI in (10). Note that since AI is a function of SNR, pre-whitening has no direct effect on it. However, pre-whitening does effect the LCMV processor, hence the processor output spectra, hence AI.

IV Results

Computer simulations were run on a five microphone, head worn array with the microphones evenly spaced from the left ear to the right ear on an arc around the head. Two reverberant rooms were modelled: both the living room and conference room described in [Peterson 1989]. Average wall absorptions of 0.3, 0.6, and 0.9, along with speaker-to-listener distances ranging from 0.9m to 4.6m, provided simulated reverberant environments with Direct-to-Reverberant energy ratios ranging from -14 to +12 dB. The modelled acoustic headshadow described above was also incorporated. Exact covariances of stationary, pre-whitened signals of interest produced optimum weights for an LCMV beamformer. The number of taps in the processor were varied from 8 to 60, with a sample rate of 10 kHz. The number of linear look direction constraints varied in proportion with the number of taps. Twelve constraints were employed in cases of 8 taps, while 66 were used in cases of 60 taps. While no quadratic constraint was employed, a white noise at a level of ~50 dB below the maximum data covariance eigenvalue was added to ensure numerical stability.

Figure 4 presents computed values of AI_{50} and AI_{MS} for a 30 tap processor over a range of 7 different direct-to_reverberant energy ratios. AI_{50} and AI_{MS} were each computed for two weight vectors, the optimum processor and the quiescent (see Eqn (5)). On the high end of the Dir/Rev axis, the AI_{50} measures predict excellent intelligibility (AI=1), in a situation where it is expected, a fixed weight vector, $\mathbf{w}_q$, in a room with a short reverberation time of <100 msec. The AI_{MS} measure, however, predicts rather poor intelligibility (AI=0.4). For the optimum LCMV weight vector, the performance predicted is very similar to that for the quiescent in both cases. For all combinations of tap length and Dir/Rev, the AI_{MS} provided what seem to be overly pessimistic predictions of intelligibility. With neither measure, however, is there any evidence of a significantly cancelled signal of interest as evidenced by decrease in performance with optimum over quiescent processor. Again this finding was consistent over all trials.

In addition, for the range of tap lengths considered, there was no relationship evidenced between an increase in tap length and a subsequent increase in signal cancellation as depicted in figure 5. While these results seem at odds with [Greenberg and Zurek 1991], it is not inconsistent considering that they adaptively processed speech data, used long tap lengths (100 and 400), and applied a different performance measure (G_{AI}). While longer tap lengths increase the nulling capability of the array, that nulling is achieved at the cost of greater computational complexity and a greater susceptibility to cancellation of the desired signal.

As a final comparison, the AI_{50}'s from the 30 tap processor (shown in figure 4) are replotted in figure 6 against the predicted AI_{50} for the single microphone positioned at the right ear. AI_{50} for the optimum and quiescent are higher throughout the transition range of Dir/Rev values. The consistency of AI_{50} in predicting good intelligibility with a single microphone in the high Dir/Rev cases suggests a useful test case in which to process speech.

V Summary

While high amounts of weighted mean squared error suggest significant amounts of reverberation-related distortion, simulation results did not suggest any significant signal of interest cancellation for an optimum LCMV processor using up to 60 taps. An interesting AI-based performance measure was applied to the simulated room reverberation and headshadow models. Application of AI_{50} to processed speech signals of interest will provide a better idea of its utility.

REFERENCES

[1977] J.B Allen, D.A. Berkley,and J. Blauert, "Multi-microphone signal processing technique to remove room reverberation from speech signals," *Journ. of the Acoust. Soc. of Am.*, vol. 62, No. 4, pp. 912-915, Oct. 1977.

[1989] P.M. Peterson, "Adaptive Array Processing for Multiple Microphone Hearing Aids," *MIT Ph.D. Dissertation, RLE Technical Report No. 541*, Feb., 1989.

[1990] M.W. Hoffman and K.M. Buckley, "Constrained optimum filtering for multi-microphone digital hearing aids," *Proc. Twenty-Fourth Asilomar Conf. on Signals, Systems and Computers*, Nov. 1990.

[1991] J.E. Greenberg and P.M. Zurek, "Adaptive Beamformer Performance in Reverberation," *Proc. 1991 IEEE ASSP Workshop on Appl. of Sig. Proc. to Audio and Acoustics*, Oct., 1991.

[1987] H.Cox, R.M. Zeskind, and M.M. Owen, "Robust adaptive beamforming," *IEEE Trans. on Acoust., Speech and Signal Processing*, vol. ASSP-35, pp. 1365-76, Oct., 1987.

[1979] J.B Allen and D.A. Berkley, "Image method for efficiently simulating small room acoustics," *Journ. of the Acoust. Soc. of Am.*, vol. 65, No. 4, pp. 943-950, Apr. 1979.

[1962] K.D. Kryter, "Methods for the calculation and use of the articulation index," *Journ. of the Acoust. Soc. of Am.*, vol. 34, No. 11, pp. 1689-1697, Nov. 1962.

[1980] T. Houtgast, H.J.M Steeneken and R. Plomp, "Predicting Speech Intelligibility in Rooms from the Modulation Transfer Function. I. General Room Acoustics," *Acustica*, vol. 46, 1980, pp. 60-72.

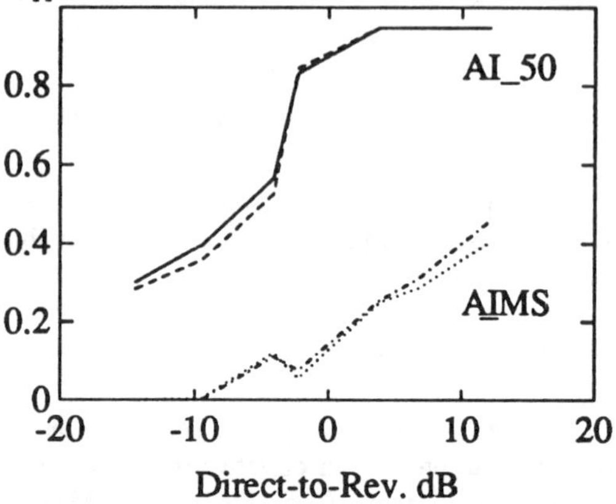

Figure (4) Two AI measures for $\mathbf{w}_q$ and w_{opt} weight vectors. 30 tap per element, 5 element processor.

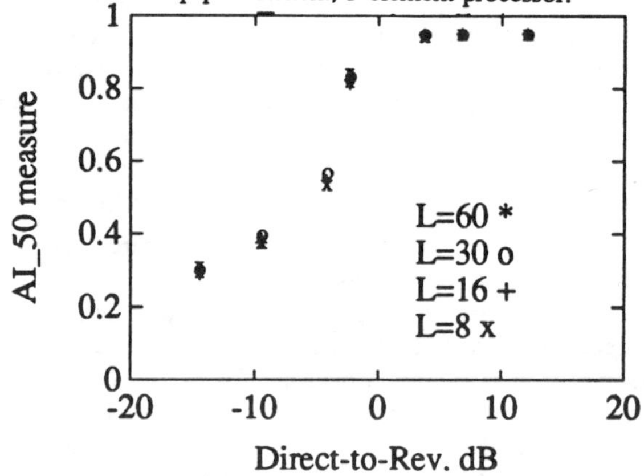

Figure (5) AI_{50} for 8, 16, 30 and 60 taps per element.

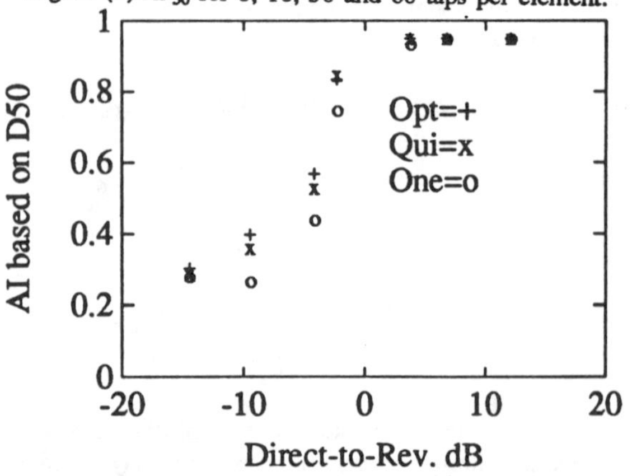

Figure (6) AI_{50} for 30 tap optimum (+) and quiescent (x) processors and single microphone (o).

Use of Hidden Markov Chain Models for Evaluating and Selecting Codes for Fading Channels and Erasure Prediction Decoding Co-operated with Interleaver and HMC

Dongku Kim and Lloyd R. Welch
Communication Sciences Institute, Dept. of Electrical Engineering
University of Southern California, Los Angeles, CA 90089-2565

Abstract

We developed a decoder co-operated with interleaver and hidden Markov chain modeler(DIHMC) for the decoding of linear block codes on the frequency flat correlated Rayleigh(Rician) fading channel. DIHMC acquires the structure of symbol errors at the demodulator output and exploits it in decoding process. DIHMC has two decoder. The first decoder corrects only errors. The second decoder re-decodes only the error words out of the first decoder. By letting the first decoder go ahead of the second decoder, the first decoder can provide each symbols in the error word out of the first decoder with preceding and following error record. DIHMC computes the reliabilities of symbols using the available error record and error structure, symbols having small reliability are erased. The second decoder corrects errors and erasures. Computer simulation shows that the second decoder in DIHMC improves nearly 2dB coding gain compared to the first decoder.

1 Introduction

Since a communication channel with memory has larger capacity than an independent channel, how can memory be effectively exploited to realize the additional capacity? This paper presents a methodology to answer the question. We consider frequency flat correlated Rayleigh and Rician fading communication channels. A channel is represented as shown Figure 4. A frequency flat Rayleigh(Rician) fading channel imposes multiplicative time-varying complex gaussian noise $g(t)$ on the transmitted signal $u(t)$, *i.e.*, received signal $r(t)$ can be expressed as follows,

$$r(t) = g(t)u(t) + n(t)$$

where $n(t)$ is additive white gaussian noise. The envelope of $g(t)$ is Rayleigh(Rician) process if $g(t)$ has zeromean(non-zeromean). We assume that $g(t)$ is 1st order gaussian Markov process in which $\mathrm{Re}\{g(t)\}$ and $\mathrm{Im}\{g(t)\}$ has following second order properties,

$$x(t) = \mathrm{Re}\{g(t)\} \quad y(t) = \mathrm{Im}\{g(t)\}$$
$$E(x(t_i)x(t_j)) = \rho^{|t_i - t_j|}\sigma^2 = E(y(t_i)y(t_j))$$
$$E(x(t_i)y(t_j)) = 0 \;\; \forall t_i, t_j.$$

Errors occur not only due to additive noise but also due to multiplicative disturbance $g(t)$. Since $g(t)$ is correlated, so are errors. A typical technique in correlated channel is interleaving. Interleaving does not require for accurate knowledge of the channel burst statistics(other than the maximum burst length) since interleaving is the technique which does not use error structure but converts long bursts to short burst that can be corrected by decoder. Most error correcting code performance analysis on fading channels has assumed infinite depth interleaving so that deinterleaved symbol errors become purely random and code performance is the same as in the independent channels. However infinite depth interleaving loses channel information and practically impossible because of infinite delay and actual deinterleaved symbols are still correlated after finite depth interleaving. The performance of many block codes are degraded by the remaining correlation in deinterleaved symbols. The effect of the correlation to the performance of linear block codes is analyzed [1]. To control the errors effectively through some coding technique, the statistical structure of errors are required. One of ways of exploiting error structure in erasure decoding process is error forecasting[2]. It uses a helical in/deinterleaver to give the reliability of the present symbol based on only one previous symbol so that decoder can correct erasures as well as errors. Helical interleaving is best described by the following example. We consider interleaving a code with a block length of seven symbols. To each codeword we attach another synchronization symbol. Figure (1.a) shows several codewords with the attached synchronization symbol **S**. The numbering reflects the time ordering of the symbols leaving the encoder and arriving at the interleaver. (Note that the synchronization character represents the symbol numbers divisible by eight.) The codewords are written down the columns of a helical array as shown in Figure (1.b). When complete, the rows of the array are read out and transmitted. Figure (1.c) gives part of the transmitting sequence. Note that consecutive symbols of a codeword are separated by precisely 7 (code length) symbols. At the receiver the process is reversed, *i.e.*, the rows are filled in with the received data and complete columns ending with S are read out. Note that 6 of the 7 symbols in the present code word immediately follow symbols of the previous codeword. Thus decoding of the previous codeword gives information about where errors may occur in the present codeword. We proposed a system which has the spirit of error forecasting on computing reliability information. However reliability of a symbol is computed by exploiting preceding and following error record.

This paper is composed as follows. In section 2, hidden Markov process is described as a way of approximating

1058-6393/91 $01.00

Figure 1: Example of helical interleaving

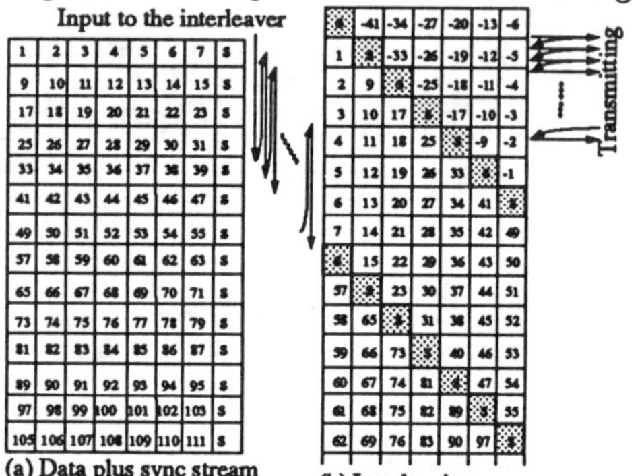

(a) Data plus sync stream (b) Interleaving array

......., 7, 14, 21, 28, 35, 42, 49, S, 15, 22, 29, 36, 43, 50, 57, S, 23, 30, 37, 44, 51, 58, 65, S, 31, 38, 45, 52, 59, 66, 73, S, 39, 46, 53, 60, 67, 74, 81, S, 47, 54, 61, 68,

(c) Transmitted stream(partial)

statistical structure of errors. In section 3, we develop a decoding system(DIHMC) that can acquire the characteristics of error process on the training mode, generate reliability values of symbols and exploits them on decoding process. In section 4, DIHMC is simulated on frequency flat correlated Rayleigh and Rician fading channel.

2 Hidden Markov Chain

Finite state hidden Markov chain has been used to characterize errors on telephone line [3] and HF channel [4]. Hidden Markov chain is a stochastic process generated by two interrelated mechanisms, an underlying Markov chain having a finite number of states, and a set of random functions, one of which is associated with each state. At the discrete instants of time, the process is assumed to be in some state and an observation is generated by the random function corresponding to the current state. The underlying Markov chain then changes states according to its transition probability matrix. The observer sees only the output of the random functions associated with each state and cannot directly observe the states of the underlying Markov chain; hence the term hidden Markov model. Multiplicative disturbance $g(t)$ is assumed to be reasonably constant for at least one symbol duration T. Then error probability of nth symbol is a function of $g(nT)$ and denoted as $P_E(g(nT))$. If we allow complex value of 1st order gaussian process $g(nT)$ to be the state at time nT, error source of actual channel is a continuous state hidden Markov process, *i.e.*, underlying process is in state $g(nT)$ at nth symbol duration and generate either 1 or 0 with symbol error probability $P_E(g(nT))$ or correct symbol probability $1 - P_E(g(nT))$. Actual error structure(continuous state hidden Markov process) is approximated by a finite number of states hidden Markov chain by using Baum-Welch algorithm [5]. A hidden Markov chain in our particular application is identified by following set of parameters,

- $\underline{E}_n = (E_1, E_2, \cdots, E_n)$: sequence of observation from time T to nT, $E_n \in \{1, 0\}$
- N : number of states of underlying Markov chain
- $\{q_0, q_1, \cdots, q_{N-1}\}$: states of underlying Markov chain

Figure 2: forward-backward algorithm

FORWARD ALGORITHM

forward probability, $\alpha_t(i) \triangleq Pr(E_1, \cdots, E_t, s_t = q_i | M)$

$$\alpha_t(i) = \begin{bmatrix} \pi_{q_0} & \cdots & \pi_{q_{N-1}} \end{bmatrix} \prod_{i=1}^{t} AB_i \begin{bmatrix} \delta_{s_t - q_0} \\ \delta_{s_t - q_1} \\ \vdots \\ \delta_{s_t - q_{N-1}} \end{bmatrix}$$

$$\underline{\alpha}_t \triangleq \begin{bmatrix} \alpha_t(0) & \cdots & \alpha_t(n-1) \end{bmatrix}, \underline{\alpha}_{t+1} = \underline{\alpha}_t AB_{t+1}$$

BACKWARD ALGORITHM

backward probability, $\beta_t(i) \triangleq Pr(E_{t+1}, \cdots, E_n | s_t = q_i, M)$

$$\beta_n(j) = 1 \ \forall j$$

$$\beta_t(i) = \begin{bmatrix} \delta_{s_t - q_0} & \cdots & \delta_{s_t - q_{N-1}} \end{bmatrix} \prod_{i=t+1}^{n} AB_i$$

$$\alpha_{t+1}(j) \triangleq \left[\sum_{i=0}^{n-1} \alpha_t(i) a_{ij}\right] b_j(E_{t+1}) \quad 1 \le t \le n-1$$

$$\beta_{t-1}(j) = \sum_i a_{ji} b_i(E_t) \beta_t(i)$$

$$\underline{\beta}_t = \begin{bmatrix} \beta_t(0) & : & \beta_t(n-1) \end{bmatrix}^t, \underline{\beta}_{t-1} = AB_t \underline{\beta}_t$$

$$\underline{\alpha}_t \cdot \underline{\beta}_t = Pr(E_1, E_2, \cdots, E_n | M) \triangleq P$$

- s_t is the state in time t.
- $A = [a_{ij}]$: $N \times N$ state transition matrix where $a_{ij} = Pr(s_{t+1} = q_j | s_t = q_i)$
- $b_j(E_t = k) \triangleq b_{jk}$: probability of observing $E_t = k$ given current state q_j
- $B = [b_{ij}]$:$N \times 2$ event matrix.
- $B_t = \begin{bmatrix} b_0(E_t) & 0 & 0 & \cdots & 0 \\ 0 & b_1(E_t) & 0 & \cdots & 0 \\ \vdots & \vdots & \vdots & \cdots & 0 \\ 0 & 0 & 0 & \cdots & b_{N-1}(E_t) \end{bmatrix}$,
- $\underline{\pi} = [\pi_{q_0}, \pi_{q_1}, \cdots, \pi_{q_{N-1}}]$: initial state probability vector.

The main idea of Baum-Welch algorithm is as follows. First of all, there is a forward algorithm of computing $\alpha_t(i)$ that is a probability of observation upto time t and terminal state i, and also there is a backward algorithm of computing $\beta_t(j)$ that is a conditional probability of observation from time $t+1$ to n given that we just started with a state j at time t. Putting these two together, we can compute state probability, state transition probabilities and probability of event at any state at individual time. Time averages of each of them over all time are directly usuable as the parameters of new models. Theorem says that likelihood increases in value as we keep using new model, *i.e.*, model M_k at kth iteration satifies

$$Pr(\underline{E}|M_0) < \cdots < Pr(\underline{E}|M_k).$$

Table 2 summarize the forward-backward algorithm. Baum-Welch algorithm computes the following.

- Expected number of transitions, γ_{ij}, from state q_i to q_j given $\underline{E}$ and a model:

$$\gamma_{ij} = \frac{1}{P} \sum_{t=1}^{T-1} \alpha_t(i) a_{ij} b_j(E_{t+1}) \beta_{t+1}(j).$$

- Expected number of transitions, γ_i out of q_i:

$$\gamma_i = \frac{1}{P} \sum_{t=1}^{T-1} \alpha_t(i) \beta_t(i)$$

- Parameters of new model are

$$\overline{a}_{ij} = \gamma_{ij}/\gamma_i, \quad \overline{\pi}_i = \frac{1}{P} \alpha_t(1) \beta_t(1)$$

$$\overline{b}_{jk} = \left[\sum_{t \ni E_t = v_k} \alpha_t(j) \beta_t(j)\right] / \left[\sum_{t=1}^{T} \alpha_t(j) \beta_t(j)\right]$$

Figure 3: RS(8,4) 8FSK, 2 state HMC model, correlation=0.98

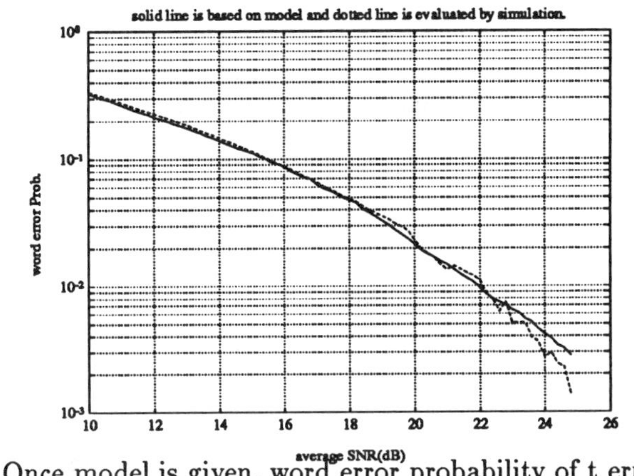

Once model is given, word error probability of t error correcting code of length CN is evaluated by simple recursion formula. Let's define

$$F_t(q_i,k) \triangleq Pr(s_t = q_i, \sum_{m=1}^{t} E_m = k).$$

Then

$$F_t(q_i,k) = \sum_{j=0}^{N-1} F_{t-1}(q_j,k)a_{ji}b_i(0) + \sum_{j=0}^{N-1} F_{t-1}(q_j,k-1)a_{ji}b_i(1).$$

Word error probability is

$$P_w = 1 - \sum_{k=0}^{t}\left[\sum_{i=0}^{N-1} F_{CN}(q_i,k)\right].$$

Figure 3 compares word error probabilities of RS(8,4) based on 2 state HMC model(solid line) and evaluated by simulation(dotted line) on frequency flat Rayleigh channel of $\rho = 0.98$.

3 Error/Erasure Decoder Co-operated with Helical Interleaver and Hidden Markov Chain Modeler

Figure 4 shows an error/erasure decoder co-operated with helical interleaver and hidden Markov chain modeler(DIHMC) developed for decoding linear block codes on correlated error channel. DIHMC is constructed of 5 main blocks which are training error sequence generator, HMC modeler, helical deinterleaver, symbol reliability computer, and block code decoder which consists of the first and second decoders. DIHMC using BCH $(q^m - 1, K)$ code with q-ary symbol in which q is a prime or power of a prime, is described. A training sequence is helically interleaved and transmitted to a receiver. Then code is helically interleaved and transmitted by q-ary modulation technique. Let's denote $\tilde{I}_n$ as a receiver(demodulator) output at time n. $\tilde{I}_n$ is a q-ary symbol. The training sequence error generator compares receiver output and an interleaved training sequence and provides sequence $\{E_n^0\}$ such that

$$E_n^0 = \begin{cases} 0 & \text{if } \tilde{I}_n \text{ is correct,} \\ 1 & \text{if } \tilde{I}_n \text{ is in error.} \end{cases}$$

Figure 4: error/erasure decoder co-operated with HMC and helical in/deinterleaver

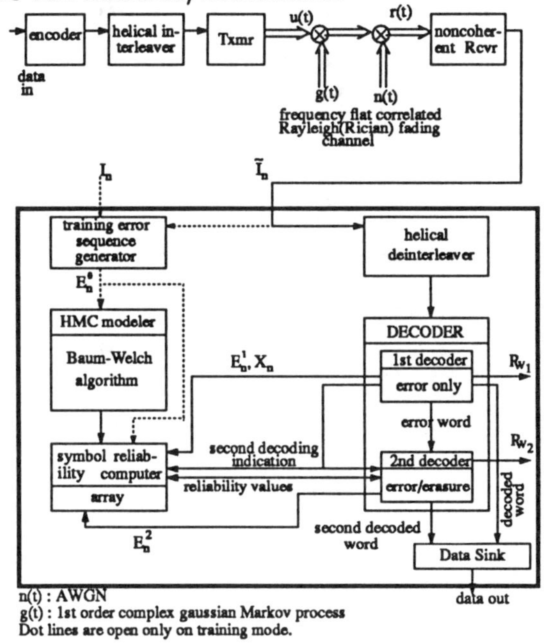

We call $\{E_n^0\}$ training error sequence. E_n^0 is stored in the array of symbol reliability computer. Suppose that the error sequence $\{E_n^0\}$ obtained until the first information symbol is received, is

$$(\cdots, \mathbf{E}_0, E_1^0, E_2^0, E_3^0, E_4^0, E_5^0, E_6^0).$$

Baum-Welch algorithm estimates the parameters of finite state HMC and computes a forward error probabilities $\alpha_6(q_i) = Pr(\{E_n^0 : n \le 6\}, s_6 = q_i | M)$. Therefore HMC modeler provides the finite state hidden Markov model that characterizes a sequence of errors $\{E_n^0 : n \le 6\}$. Deinterleaved words go to decoder. Decoder consists of 2 decoders. The first decoder begins decoding error only. If the number of errors t holds

$$t \le \lfloor \frac{d_{min} - 1}{2} \rfloor,$$

in which d_{min} denotes minimum distance of code. the first decoder decodes successfully and provides error sequence E_n^1 to the array in symbol reliability computer such that

$$E_n^1 = \begin{cases} 0 & \text{if } \tilde{I}_n \text{ is a correct symbol in a decodable word,} \\ 1 & \text{if } \tilde{I}_n \text{ is an error symbol in a decodable word.} \end{cases}$$

If decoding failure occurs, all symbol positions of that word are marked as X_n in the array of symbol reliability computer. Reliability computation and decoding procedure in the the second decoder are described by an example of RS(7,3) code in which $q = 2^3$ and $m = 1$. Figure 5 shows how E_n^1 and X_n are saved in the array of symbol reliability computer. N implies that it is not decoded yet in the first decoder and bold $\mathbf{E}_n$ is an error corresponding to an synchronization symbol. After the first decoder decodes $(q^m - 1)$ code words, the second decoder starts redecoding received words. The second decoder works only on error words of the first decoder and corrects errors and erasures based on reliability values of each symbols. Note that the first

E^0_{-7}	E^0_{-6}	E^0_{-5}	E^0_{-4}	E^0_{-3}	E^0_{-2}	E^0_{-1}
$\mathbf{E}_0$	E^0_1	E^0_2	E^0_3	E^0_4	E^0_5	E^0_6
E^1_7	$\mathbf{E}_8$	E^0_9	E^0_{10}	E^0_{11}	E^0_{12}	E^0_{13}
E^1_{14}	E^1_{15}	$\mathbf{E}_{16}$	E^0_{17}	E^0_{18}	E^0_{19}	E^0_{20}
E^1_{21}	E^1_{22}	E^1_{23}	$\mathbf{E}_{24}$	E^0_{25}	E^0_{26}	E^0_{27}
E^1_{28}	E^1_{29}	E^1_{30}	X_{31}	$\mathbf{E}_{32}$	E^0_{33}	E^0_{34}
E^1_{35}	E^1_{36}	E^1_{37}	X_{38}	E^1_{39}	$\mathbf{E}_{40}$	E^0_{41}
E^1_{42}	E^1_{43}	E^1_{44}	X_{45}	E^1_{46}	E^1_{47}	$\mathbf{E}_{48}$
E^1_{49}	E^1_{50}	E^1_{51}	X_{52}	E^1_{53}	E^1_{54}	E^1_{55}
$\mathbf{E}_{56}$	E^1_{57}	E^1_{58}	X_{59}	E^1_{60}	E^1_{61}	E^1_{62}
E^1_{63}	$\mathbf{E}_{64}$	E^1_{65}	X_{66}	E^1_{67}	E^1_{68}	E^1_{69}
E^1_{70}	E^1_{71}	$\mathbf{E}_{72}$	X_{73}	E^1_{74}	E^1_{75}	E^1_{76}
E^1_{77}	E^1_{78}	E^1_{79}	$\mathbf{E}_{80}$	E^1_{81}	E^1_{82}	E^1_{83}
E^1_{84}	E^1_{85}	E^1_{86}	E^1_{87}	$\mathbf{E}_{88}$	E^1_{89}	E^1_{90}
E^1_{91}	E^1_{92}	E^1_{93}	E^1_{94}	N	$\mathbf{E}_{96}$	E^1_{97}
E^1_{98}	E^1_{99}	E^1_{100}	E^1_{101}	N	N	$\mathbf{E}_{104}$
E^1_{105}	E^1_{106}	E^1_{107}	E^1_{108}	N	N	N
$\mathbf{E}_{112}$	E^1_{113}	E^1_{114}	E^1_{115}	N	N	N
N	$\mathbf{E}_{120}$	E^1_{121}	E^1_{122}	N	N	N
N	N	$\mathbf{E}_{128}$	E^1_{129}	N	N	N
N	N	N	$\mathbf{E}_{135}$	N	N	N

Figure 5: some of errors stored in helical array of symbol reliability computer

decoder always goes ahead of the second decoder by $(q^m - 1)$ code words. Suppose the first decoder finishes code word $(\tilde{I}_{87}, \tilde{I}_{94}, \tilde{I}_{101}, \tilde{I}_{108}, \tilde{I}_{115}, \tilde{I}_{122}, \tilde{I}_{129})$ and $(\tilde{I}_{31}, \tilde{I}_{38}, \tilde{I}_{45}, \tilde{I}_{52}, \tilde{I}_{59}, \tilde{I}_{66}, \tilde{I}_{73})$ is the error word in first decoder. Now each of $(X_{31}, X_{38}, X_{45}, X_{52}, X_{59}, X_{66}, X_{73})$ has its preceding error record as well as following error record. For example, preceding record of X_{31} is all the error sequence obtained until time 30, $\{E_n : n \leq 30\}$, and following record is $(\mathbf{E}_{32}, E^0_{33}, E^0_{34}, E^1_{35}, E^1_{36}, E^1_{37})$. Of course, available following record is longer than this but we use only $q^m - 1$ following error data. Likewise, other symbols in error word have proceeding and following record. Reliability and temporary decision of X_n are defined as follows. Reliability of X_n is

$$L_n = Pr(X_n = 0 | \underline{\hat{E}}_{n-1}, {}_{n+1}\underline{E}_{n+6}, M)$$

and temporary decision is

$$\hat{E}_n = \begin{cases} 1 & \text{if } Ln < 0.5 \\ 0 & \text{otherwise,} \end{cases}$$

where $\underline{\hat{E}}_n$ is the observation up to time n in which positions marked X are replaced with their temporary decision if available. For example, reliability of X_{31} is conditional probability that X_{31} is 0, *i.e.*, a received symbol at time 31 is correct given $(\underline{E}_{30}, \mathbf{E}_{32}, E^0_{33}, E^0_{34}, E^1_{35}, E^1_{36}, E^1_{37})$ and HMC model. Let's denote ${}_{n_1}\underline{E}_{n_2}$ error sequence obtained at time between n_1 and n_2, *e.g.*, ${}_{32}\underline{E}_{37} = (\mathbf{E}_{32}, E^0_{33}, E^0_{34}, E^1_{35}, E^1_{36}, E^1_{37})$. Then reliability value of X_{31} is computed as

$$L_{31} = Pr(X_{31} = 0 | \underline{E}_{30}, {}_{32}\underline{E}_{37}, M)$$

and temporary decision of X_{31} is

$$\hat{E}_{31} = \begin{cases} 1 & \text{if } L_{31} < 0.5 \\ 0 & \text{otherwise.} \end{cases}$$

The reliability of X_{38} is

$$L_{38} = Pr(X_{38} = 0 | \underline{\hat{E}}_{37}, {}_{39}\underline{E}_{44}, M)$$

in which $\underline{\hat{E}}_{37} = (\underline{E}_{30}, \underline{\hat{E}}_{31}, {}_{32}\underline{E}_{37})$, and temporary decision is

$$\hat{E}_{38} = \begin{cases} 1 & \text{if } L_{38} < 0.5 \\ 0 & \text{otherwise.} \end{cases}$$

By the Bay's theorem, L_n can be represented as

$$L_n = \frac{Pr(X_n = 0, \underline{\hat{E}}_{n-1}, {}_{n+1}\underline{E}_{n+6} | M)}{Pr(\underline{\hat{E}}_{n-1}, {}_{n+1}\underline{E}_{n+6} | M)}.$$

where

$$Pr(\underline{\hat{E}}_{n-1}, {}_{n+1}\underline{E}_{n+6} | M) = \sum_{j=0}^{N-1} \alpha_{n+6}(q_j).$$

Since X_n is unknown, $\alpha_n(q_j)$ can not be computed. However, $\alpha_{n+1}(q_j)$ can be computed from $\alpha_{n-1}(q_i)$ by 2-step transition probabilities, *i.e.*,

$$\alpha_{n+1}(q_j) = \left[\sum_{i=0}^{N-1} \alpha_{n-1}(q_i) a^{(2)}_{q_i q_j} \right] b_{q_j}(E^1_{n+1}),$$

where

$a^{(2)}_{q_i q_j}$ denotes 2-step transition probabilities from state q_i to state q_j. So $\alpha_{n+6}(q_i)$ is computed from $\alpha_{n+1}(q_i)$ recursively. When there are more erased symbols in ${}_{n+1}\underline{E}_{n+6}$, the erasured symbols are excluded on computing $Pr(\underline{\hat{E}}_{n-1}, {}_{n+1}\underline{E}_{n+6} | M)$ by using multiple transition probability. Based upon reliability values of each symbol, $(L_{31}, L_{38}, L_{45}, L_{52}, L_{59}, L_{66}, L_{73})$, some of symbols corresponding to small reliability values are erased in the second decoder. In our example, the second decoder can correct a received code word if

$$2t + e \leq 4$$

holds, where t is the number of errors and e is the number of erasures. The second decoder starts decoding with 2 erasures and 1 more error correctability. If decoding failure still occurs, we re-decode with 4 erasures. By increasing the number of erasures, linear block code increases its error correcting capability if more than 50% of erased symbols are indeed errors. Second decoding in our example RS(7,3), is finished here since 4 is the maximum number of erasure we can use in the RS(7,3) code. If the second decoder decodes successfully, it sends error locations information E^2_n to the array in symbol reliability computer in which

$$E^2_n = \begin{cases} 0 & \text{if } \tilde{I}_n \text{ is a correct symbol in a decodable word,} \\ 1 & \text{if } \tilde{I}_n \text{ is an error symbol in a decodable word.} \end{cases}$$

If decoding failure occurs in the second decoding, all the symbols of the error word are remained marked X. We summarize decoding scheme of BCH$(q^m - 1, K)$ code.

1. Training sequence errors obtained until the first information symbol is received, are used in modeling finite state HMC which approximates error sequence.
2. The first decoder decodes deinterleaved received words. If it decodes successfully, it sends error location information E^1_n to the array in the symbol reliability computer. If not, symbols in that word are marked as X_n.
3. After the first decoder finishes decoding $(q^m - 1)$th received word, the second decoder start re-decoding error words out of the first decoder with symbol reliability values. The first decoder always goes ahead of the second decoder by code-length.

4. Reliability values of symbols are defined as

$$Ł_n = Pr(X_n = 0 | \underline{\hat{E}}_{n-1}, {}_{n+1}\underline{E}_{q^m-2} M)$$

and temporary decision of X_n is

$$\hat{E}_n = \begin{cases} 1 & \text{if } Ł_n < 0.5 \\ 0 & \text{otherwise.} \end{cases}$$

5. Symbols having small reliability are erased in the second decoder. Start decoding with 2 erasures. If decoding failure occurs, we increase the number of erasures by 2 until the number of erasures reaches to $d_{min} - 1$. If the second decoder decodes successfully, it sends symbol error location E_n^2 to the array in symbol reliability computer. If not, the erasured word by the first decoder is remained marked X.
6. At the next error word out of the first decoder, 4 and 5 are repeated.

4 Simulation Result on Frequency Flat Correlated Releigh and Rician Fading Channel

Source data is encoded by BCH(63,47) with 8-ary symbol. d_{min} is 11. Therefore decoder can correct a received word if

$$2t + e \leq 10.$$

Each 8-ary symbol is transmitted by 8FSK and noncoherently detected. Helical interleaving depth is 63. Baum-Welch algorithm acquires 28225 long error sequence to estimate 4 state HMC parameters. Note that training sequence must be long enough to cover the phenomenon of fading. If the length is two short, a hidden Markov model might approximate either strong fade or out of fade only. The performance of DIHMC is shown as a function of ρ and received average signal to noise ratio $\frac{E_s}{N_0} E(|g(t)|^2)$ where E_s is signal energy, N_0 is AWGN power density level. Let's denote N_1 as the number of word errors out of the first decoder and N_2 as the number of word errors out of the second decoder. Define error correction ratio, ECR $= (N_1 - N_2)/N_1$, *i.e.*, fraction of the number of errors corrected by the 2nd decoder. N_w denotes the number of simulated words. P_{w_1} and P_{w_2} are word error probabilities of first and second decoder. Table 1 shows simulation results on independent Rayleigh fading channel. Note that DIHMC uses available error structure in decoding process. Since independent errors do not have structure, DIHMC shows no coding gain at all on independent channel. Let's define ρ_E as least square estimate of correlation coefficient of consecutive errors and α as rice factor which is $\alpha = E^2(|g(t)|^2)/2\sigma^2$. Table 2 and 3 show P_{w_1}, P_{w_2}, ECR and ρ_E for Rayleigh and Rician fading channel when correlation coefficient of consecutive samples of $g(t)$, ρ, is 0.99. In Table 2, ECR has a maximum value, 0.83, at R_{av}=20dB, *i.e.*, 83% of word errors out of the first decoder are corrected by the second decoder. Correlation of consecutive errors increases over the low R_{av}. However, many received words are undecodable in the first decoder so that the first decoder could not provide enough error record and it results in inaccurate reliabilities. Over the high R_{av} region, correlation between consecutive errors decreases, *i.e.*, errors are losing their structure and ECR is getting smaller. Therefore DIHMC performs best over the signal-to-ratio on which errors are quite correlated and first decoder can provide enough error record with a few symbols marked X. Table 3 shows that DIHMC performs with high ECR on low R_{av} of Rician channel with 5dB rice factor. 99% of word errors out of the first deocder is corrected by the second decoder at 9dB (only one error occurs on 70,000 words simulation). At 11dB, all errors are corrected by the second decoder. By comparing P_{w_1} and P_{w_2}, it is proven that nearly 2dB coding gain is obtained by DIHMC when $\rho = 0.99$.

Table 1: Independent Rayleigh fading channel, $\rho = 0$

R_{av}	18dB	20dB
P_{w_1}	0.03769	0.00455
P_{w_2}	0.03769	0.00455
ECR	0	0
N_w	30,000	30,000

Table 2: Rayleigh fading channel, $\rho = 0.99$

R_{av}	18dB	19dB	20dB	21dB	22dB
P_{w_1}	3.716e-2	1.563e-2	4.251e-3	1.331e-3	5.92e-4
P_{w_2}	1.289e-2	5.556e-3	7.70e-4	4.15e-4	1.91e-4
N_1	1100	464	126	93	65
N_2	375	163	22	29	21
ECR	6.591e-1	6.487e-1	8.254e-1	6.882e-1	6.769e-1
N_w	30,000	30,000	30,000	70,000	100,000
ρ_E	3.571e-1	3.301e-1	3.12e-1	2.856e-1	2.645e-1

Table 3: Rician fading channel, $\rho = 0.99$ and rice factor=5dB

R_{av}	9dB	10dB	11dB
P_{w_1}	5.54e-3	9.30e-4	2.20e-4
P_{w_2}	7.55e-4	1.40e-5	0
N_1	110	65	22
N_2	15	1	0
ECR	8.64e-1	9.85e-1	1
N_w	20,000	70,000	100,000
ρ_E	4.41e-1	4.44e-1	4.64e-1

References

[1] Dongku Kim and Robert E. Peile. Error statistics and block code performance on correlated frequency flat Rayleigh/Rician fading channel. CSI Report 91-08-02, University of Southern California, 1991.

[2] R. Peile W.W. Wu, D. Haccoun and Y. Hirata. Coding for satellite communication. *IEEE Journal on Selected Areas in Communications*, Vol 5, No. 4:724–748, May 1987.

[3] E.N. Gilbert. Capacity of a Burst-Noise Channel. *Bell System Technical Journal*, pages 1253–1265, Sept 1960.

[4] Bruce D. Fritchman. A binary channel characterization using partitioned Markov chains. *IEEE Transaction on Information Theory*, pages 221–227, April 1967.

[5] Levinson, S.E., Rabiner, L.R. and Sondhi, M.M. An introduction to the application of the theory of probabilistic functions of a Markov porcess to automatic speech recognition. *Bell System Technical Journal*, 62, No. 4, Part 1:1035–1074, April 1983.

Time-Frequency Localization via the Weyl Correspondence

Jayakumar Ramanathan and Pankaj Topiwala

The MITRE Corporation
Burlington Road
Bedford, MA 01730

Abstract

A technique for producing signals whose energy is concentrated in a given region of the time-frequency plane is examined. The degree to which a particular signal is concentrated is measured by integrating its time-frequency distribution over the given region. This technique, using the Wigner distribution, has recently been used as a framework for time-varying filtering. In this paper, the associated subspace projection operator is studied. Estimates for the eigenvalue decay and the smoothness and decay of the eigenfunctions are presented.

1 Introduction

It is well known that the time-frequency characteristics of a square integrable signal cannot be arbitrary. For example, no such signal can be both time and band limited. Heisenberg uncertainty is another restriction. Nevertheless, in many applications it is important to use signals whose time-frequency characteristics are highly localized. Landau, Pollack, and Slepian [7,8] investigated band-limited signals that are as concentrated as possible within a prescribed time span. More recently, there has been interest in finding signals that are localized in general regions of phase space via methods that keep time and frequency on an equal footing [5,6].

In this paper, we ask: Given a joint time-frequency distribution and a bounded region in the phase plane, which signals have their energy most concentrated in the given region [11]? We formulate this question for Cohen's class of time-frequency distributions [10], and show how the work of Daubechies [5] can be set in that context. We study in particular the case of the Wigner distribution as the measure of energy concentration. This leads to a self-adjoint *localization operator* that is easy to study in terms of the Weyl correspondence. In the Weyl correspondence, a function of two variables—called the *symbol*—is associated with an operator on functions of one variable.

The symbol of the localization operator we study is simply the indicator function of a given region in the phase plane. The eigensignals with eigenvalues above a fixed threshold then define a suitable subspace onto which a given signal can be orthogonally projected for localization. This procedure has received recent attention in the engineering literature, especially with regard to time-varying filtering (see for example the papers of Hlawatsch, Kozek, and Krattenthaler [2,3]). This paper reports on new results [4] on the localization operator described above. The asymptotic properties of the eigenvalues are studied and it is established that they are in $O(k^{-3/4})$. In addition, the eigenfunctions with nonzero eigenvalues are shown to have faster than exponential decay in both the time and frequency domains. This, of course, leads to a statement regarding the smoothness of these eigenfunctions. In particular, we show that they are analytic. In the last section, some numerical examples are provided.

2 Localization Operators

2.1 Weyl Correspondence

The basic properties of the Weyl correspondence that we will need are collected in this section; for a thorough treatment, the reader is referred to the book by Folland [1].

Let $f, g \in L^2(\mathbb{R})$. A general time-frequency shift of f is

$$\rho(\tau,\sigma)f(t) = e^{\pi i\tau\sigma}e^{2\pi i\sigma t}f(t+\tau).$$

For example, the time-frequency shifts of the Gaussian $\phi = 2^{1/4}e^{-\pi x^2}$ are known as the "coherent states". The

1058-6393/91 $01.00

cross-ambiguity function of f and g is given by

$$A(f,g)(\tau,\sigma) = \langle \rho(\tau,\sigma)f, g\rangle$$
$$= \int e^{\pi i\tau\sigma} e^{2\pi i\sigma t} f(t+\tau)\overline{g(t)}\, dt$$
$$= \int e^{2\pi i\sigma s} f(s+\tau/2)\overline{g(s-\tau/2)}\, ds.$$

The value of $A(f,g)$ at a particular point is the time-frequency crosscorrelation between a shift of f and g. The Wigner distribution is the two-dimensional Fourier transform of the cross-ambiguity function, thus giving it the interpretation of a time-varying spectrum. The Wigner distribution can be written as

$$W(f,g)(\xi,t) = \int e^{-2\pi i\tau\xi} f(t+\tau/2)\overline{g(t-\tau/2)}\, d\tau.$$

We will denote $W(f,f)$ by W_f. Several useful properties of the Wigner distribution are catalogued below.

THEOREM 1. *Let $f, g \in L^2(\mathbb{R})$. Then*

1) $W(g,f) = \overline{W(f,g)}$,
2) $\langle f_1, f_2\rangle \langle g_1, g_2\rangle = \langle W(f_1,g_2), W(f_2,g_1)\rangle$,
3) $\|W(f,g)\|_\infty \leq \|f\|\|g\|$,
4) $W(\hat{f},\hat{g})(\xi,t) = W(f,g)(t,-\xi)$,
5) $W(\rho(a,b)f, \rho(a,b)g)(\xi,t)$
$= W(f,g)(\xi - b, t + a)$.

The Weyl correspondence uses the Wigner distribution to define a correspondence between functions $S(\xi,t)$ of two variables and operators L_S on $L^2(\mathbb{R})$. It is defined, via duality, by

$$\langle L_S f, g\rangle = \iint S(\xi,t) W(f,g)(\xi,t)\, d\xi dt,$$

where $f, g \in L^2(\mathbb{R})$. $S(\xi,t)$ is called the "symbol" of the operator L_S. Regularity and decay properties of S determine the class of the operator L_S [1]. An important special case is given in the following theorem of Pool [9].

THEOREM 2. *A symbol $S(\xi,t) \in L^2(\mathbb{R}^2)$ gives rise to an operator L_S that is Hilbert-Schmidt on $L^2(\mathbb{R})$. Moreover, the mapping $S \mapsto L_S$ is a unitary operator from $L^2(\mathbb{R}^2)$ to the Hilbert-Schmidt operators on $L^2(\mathbb{R})$.*

LEMMA 3. *If S is real-valued, L_S is a self-adjoint operator with real eigenvalues. L_S has real eigenfunctions if $S(-\xi,t) = S(\xi,t)$.*

An important special case where details of the eigenstructure of the operator L_S are available is the following [4].

PROPOSITION 4. *For a radially symmetric symbol S, the eigenfunctions of the operator L_S are the Hermite functions*

$$h_j(t) = \frac{2^{1/4}}{\sqrt{j!}}\left(\frac{-1}{2\sqrt{\pi}}\right)^j e^{\pi t^2} \frac{d^j}{dt^j} e^{-2\pi t^2};$$

the eigenvalues are

$$\lambda_j = \int_0^\infty (-1)^j 4\pi S(r) e^{-2\pi r^2} L_j^{(0)}(4\pi r^2)\, r dr.$$

Here $L_j^{(0)}$ is the jth Laguerre function.

2.2 Cohen's Class Operators

Let $P(f,g)(\xi,t)$ be a shift-invariant bilinear distribution, called a Cohen's class joint distribution. It can be represented [10] as a smoothed version of the Wigner distribution:

$$P(f,g)(\xi,t) = \iint W(f,g)(a,b)\mu(\xi - a, t - b) da db$$

where μ is an arbitrary function, up to the requirement that

$$\iint P_f(\xi,t) d\xi dt = \|f\|^2.$$

For a choice of a joint distribution $P(\xi,t)$ and a symbol function $S(\xi,t)$, we define a localization operator L_S^P by

$$\langle L_S^P f, g\rangle = \iint P(f,g)(\xi,t) S(\xi,t) d\xi dt.$$

This operator can be interpreted in the Weyl correspondence as $L_{S'}$, for

$$S'(a,b) = \iint \mu(\xi - a, t - b) S(\xi,t) d\xi dt$$
$$= \iint \mu_-(a - \xi, b - t) S(\xi,t) d\xi dt$$

where $\mu_-(a,b) = \mu(-a,-b)$.

2.3 Daubechies Operator

In [5], I. Daubechies introduced a special Cohen's class localization operator, whose underlying distribution is essentially the spectrogram. For a symbol function $S(\xi,t)$, it is defined by

$$\langle D_S f, g\rangle = \iint S(\xi,t) D(f,g)(\xi,t) d\xi dt,$$

where

$$D(f,g)(\xi,t) = \langle f, \rho(a,b)\phi\rangle\langle\rho(a,b)\phi, g\rangle.$$

In particular, we have that

$$D_f(a,b) = |\langle f, \rho(a,b)\phi\rangle|^2,$$

the spectrogram of f with a Gaussian window. By Theorem 1,

$$\begin{aligned} D(f,g)(\xi,t) &= \langle W_f, W_{\rho(a,b)\phi}\rangle \\ &= \iint W(f,g)(\xi,t) 2e^{-2\pi[(a-\xi)^2+(b-t)^2]} d\xi dt. \end{aligned}$$

Thus, $D(f,g)(\xi,t)$ is a Gaussian smoothed Wigner distribution.

3 Localization via a Cutoff

We have studied [4] the eigenstructure of the Weyl localization operator L_{χ_Ω}, where χ_Ω is the characteristic function of some bounded domain in the $t-\xi$ plane. This corresponds to studying the localization of the Wigner energy of a function in a domain. If one choses a different joint distribution, localization in the domain Ω corresponds to a Weyl operator for a smoothed characteristic function. We will assume that $\Omega \subset [-B,B]\times[-T,T]$. Note that

$$\langle L_{\chi_\Omega} f, g\rangle = \iint_\Omega W(f,g).$$

L_{χ_Ω} is self-adjoint, since χ_Ω is real valued. Pool's theorem implies that L_{χ_Ω} is Hilbert-Schmidt. Hence there is an orthonormal basis $\psi_1, \psi_2, \ldots$ of $L^2(\mathbb{R})$ and real numbers $\lambda_1, \lambda_2, \ldots$ such that $L_{\chi_\Omega}\psi_k = \lambda_k\psi_k$. The Hilbert-Schmidt norm of L_{χ_Ω} is $\sum|\lambda_k|^2 = |\Omega|$, the measure of Ω. We will assume that the eigenvalues are arranged in order of decreasing absolute value. It is easy to check that the largest positive eigenvalue corresponds to the maximum energy an L^2 function can have within the domain Ω. The corresponding eigenfunction would then be a time function with energy as concentrated as possible within Ω. Our aim in this paper is to report results [4] on the decay of the eigenvalues of L_{χ_Ω} and the smoothness properties of the eigenfunctions.

THEOREM 5. *If Ω is a bounded domain with piecewise C^1 boundary, then λ_k is in $\mathcal{O}(\frac{1}{k^{3/4}})$, but not in $\mathcal{O}(\frac{1}{k^{1+\epsilon}})$ for any $\epsilon > 0$. The series λ_k is not absolutely convergent.*

THEOREM 6. *Suppose Ω is an open set contained in the interval $[-B,B]\times[-T,T]$ with the property that every crosssection in either the ξ or t direction consists of at most N intervals. Let ψ be an eigenfunction belonging to a nonzero eigenvalue of L_{χ_Ω}. Then:*

(i) *for any $b > 0$, there are constants C_b and $\hat{C}_b$ such that*

$$|\psi(s)| \le C_b e^{-b|s|} \ \forall s \text{ and}$$
$$|\hat{\psi}(\xi)| \le \hat{C}_b e^{-b|\xi|} \ \forall\xi,$$

and

(ii) *ψ and $\hat{\psi}$ are analytic and have all their derivatives in L^2.*

It is instructive to consider the case when Ω is a ball centered at the origin. It is well known that, in this case, the ψ_k are Hermite functions (see Proposition 4). The Hermite functions $h_j(t)$ will satisfy estimates of the form $h_j(t) \le Ce^{-(\pi-\epsilon)t^2}$ for any $\epsilon > 0$. It is directly evident that they will then satisfy the weaker inequalities in Theorem 6. This theorem states that this weaker decay statement holds for general regions in the time-frequency plane. We do not know whether these estimates can be improved.

4 Numerical Examples

For illustration purposes, we provide two numerical examples. We consider localization onto a disc and a rectangle. For a disc of area 100, we compare the Weyl localization operator with the Daubechies operator [5]. Both have the Hermite functions as the eigenfunctions (see Proposition 4). A plot of the eigenvalues is shown in figure 1. The Weyl eigenvalues have

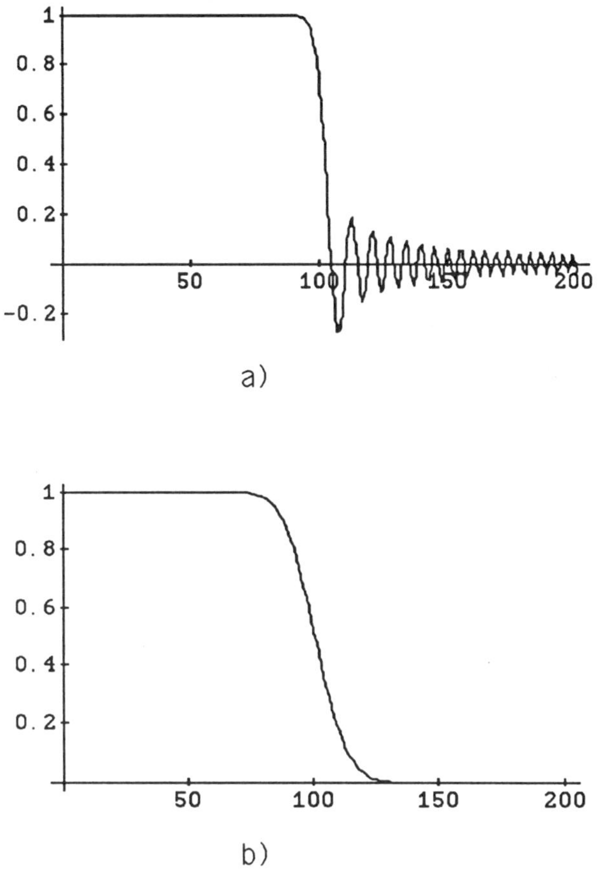

Figure 1. Plot of Eigenvalues for Localization onto a Disc of Area 100, for the (a) Weyl Operator, (b) Daubechies Operator

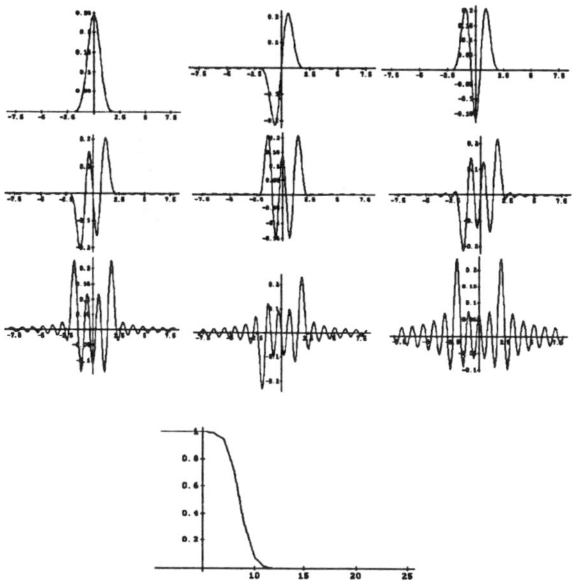

Figure 2. First Nine Prolate Spheroidal Wave-Functions for Localization onto a 4×2 Rectangle, with Eigenvalue Plot

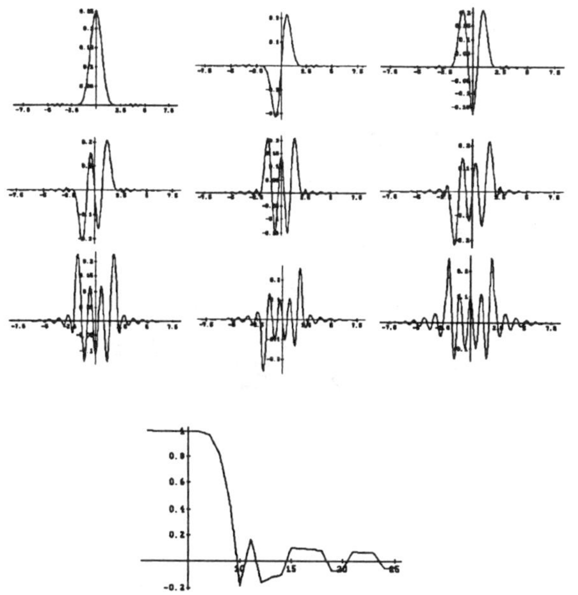

Figure 3. First Nine Weyl Solutions for Localization onto a 4×2 Rectangle, with Eigenvalues in Decreasing Magnitude

a numerically sharper plunge, although they display a Gibbs-like overshoot and oscillation.

Next, we consider localization onto the rectangle $[-2,2] \times [-1,1]$ in the time-frequency plane using the Weyl correspondence. This allows comparison to the prolate spheroidal wavefunctions. The first nine solutions for the prolate spheroidal and Weyl operator cases are depicted in figures 2 and 3, respectively. The time domain $[-2,2]$ corresponds to the middle fourth of each eigenfunction plot. Note that the prolate spheroidal wavefunctions have $\frac{1}{x}$ decay, while the Weyl eigenfunctions have exponential decay. This difference is visible when one compares the last three eigenfunctions in each case. A plot of the eigenvalues is also included.

Finally, we provide a simple illustration of these ideas in the context of filtering Gaussian noise from a corrupted linear FM (chirp) signal. Figure 4a shows the Wigner distribution of a linear FM chirp. Figure 4b shows this signal after 0 dB white Gaussian noise has been introduced. Using the subspace projection technique, the image in figure 4c was obtained.

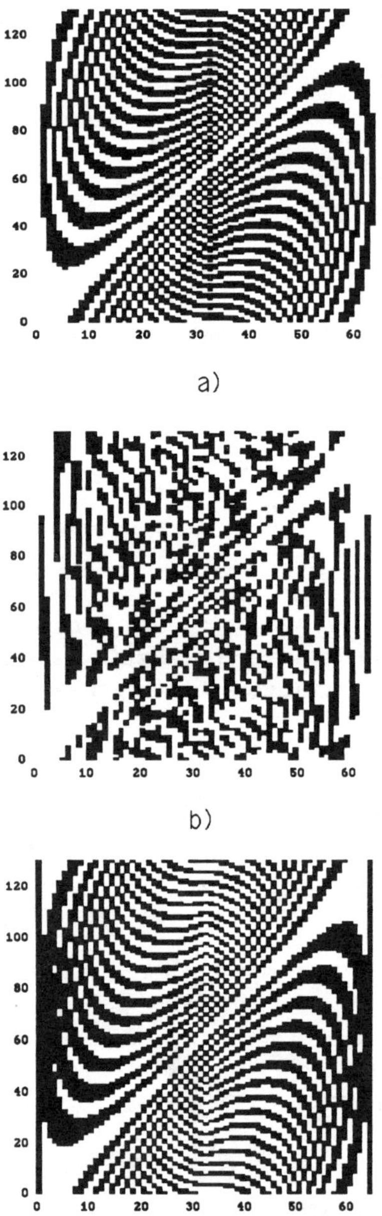

Figure 4. Wigner Distribution of (a) a Chirp Signal, (b) Chirp in 0 dB Noise, and (c) Weyl Filtered Noisy Signal

References

1. Folland, G. B., 1989, *Harmonic Analysis in Phase Space*, Princeton University Press, Princeton, N.J.

2. Hlawatsch, F., Kozek, W. and W. Krattenthaler, 1990, Time Frequency Subspaces and their Application to Time-Varying Filtering, *Proc. ICASSP*, pages 1609–1610

3. Hlawatsch, F., and W. Kozek, 1991, Time-Frequency Analysis of Linear Signal Subspaces, *Proc. ICASSP*, pages 2045–2048

4. Ramanathan, J., and P. Topiwala, Time Frequency Localization via the Weyl Correspondence (submitted for publication)

5. Daubechies, I., 1988, Time-Frequency Localization Operators—A Geometric Phase Space Approach,I, *IEEE Trans. Inf. Thry.*, Vol. 34, pages 605–612

6. Daubechies, I. and T. Paul, 1988, Time-Frequency Localization Operator—A Geometric Phase Space Approach,II, *Inverse Problems*, vol 4, pages 661–680

7. Slepian, D., and H. O. Pollack, 1961, Prolate Spheroidal Wavefunctions, Fourier Analysis and Uncertainty: I, *Bell Syst. Tech. J.*, Vol. 40, pages 43–64

8. Landau, H. J., and H. O. Pollack, 1961, 1962, Prolate Spheroidal Wavefunctions, Fourier Analysis and Uncertainty: II, III, *Bell Syst. Tech. J.*, Vol. 40,41, pages 43–64, 1295–1336

9. Pool, J., 1966, Mathematical Aspects of the Weyl Correspondence, *J. Math. Phys.*, Vol. 7, pages 66–76

10. Cohen, L., 1989, Time-Frequency Distributions—A Review, *Proc. IEEE*, vol 77, pages 941–981

11. Flandrin, P., Maximal Signal Energy Concentration in a Time-Frequency Domain, *Proc. ICASSP*, pages 2176–2179

Statistical Trade-Offs in Modern Time-Frequency Kernel Design

Steven Hearon and Moeness Amin

Department of Electrical Engineering
Villanova University
Villanova, PA 19085

Abstract

When dealing with random processes, reduced spectrum estimate variance becomes an important property which augments the list of desired time-frequency properties of modern distributions. In this paper, the degrees of freedom left in the two-dimensional kernel after satisfying the support, the marginal, and the instantaneous frequency requirements are used to yield a kernel of minimum variance. The average variance over the Nyquist interval of the spectrum estimate of a white noise process is used as the measure to be minimized. We prove that the Born-Jordan kernel [1] has the lowest possible average variance. It is also shown that the cone-shape of the modern t-f kernels is primarily responsible for their high variance. A comparison of the statistical performance of different shapes of kernels is provided.

1. Introduction

Wigner, Pseudo Wigner, and Smoothed Wigner distributions have been shown to be powerful tools in representing the signal in both time and frequency. Modern t-f techniques such as Choi-Williams and the cone shape kernels [2,3] reduce the undesired extraneous components in these time-frequency representations and as such have opened several potential applications to other members of Cohen's general class of distributions[1].

Most of the constraints on kernel selectivity are compatible when considering deterministic signals in noise. In dealing with random processes, a new constraint, namely reduced variance, must be exercised. In this paper, we introduce kernels with minimum variance and show the clear incompatibility between reduced variance and other t-f desired properties, primarily the time support property.

2. The Variance of Time-Frequency Estimates

The time-frequency distribution can be written as

$$\rho(n,\omega) = 2 \sum_{\tau=-\infty}^{\infty} \hat{r}(n,\tau)\, e^{-j2\tau\omega} \tag{1}$$

where n is the time index, ω is the radian frequency, and τ is the lag number. In this expression, the estimated autocorrelation function $\hat{r}(n,\tau)$, is given by

$$\hat{r}(n,\tau) = \sum_{t=-\infty}^{\infty} \phi(t,\tau)x(n+t-\tau)x^{*}(n+t+\tau) \tag{2}$$

where x(t) is the signal at time t, (*) denotes complex conjugation, and $\phi(t,\tau)$ is the time-frequency kernel in the 2-D time domain. The variance of a time-frequency estimate can be determined from Eqn. (1) to be [4]

$$\mathrm{var}[\rho(n,\omega)] = 4 \sum_{\tau_1} \sum_{\tau_2} \mathrm{cov}[\hat{r}(n,\tau_1),\ \hat{r}(n,\tau_2)]e^{-2j\omega(\tau_1-\tau_2)} \tag{3}$$

1058-6393/91 $01.00 © 1991 IEEE

where $\text{cov}[\hat{r}(n_1,\tau),\hat{r}(n_2,\tau)]$ is the covariance between the estimates of the autocorrelation function at τ_1 and τ_2. Using Eqn. (2), we find that

$$E\left[\hat{r}(2\tau_1)\hat{r}(2\tau_2)\right] = \sum_{t_1}\sum_{t_2}\phi(t_1,\tau_1)\phi^*(t_2,\tau_2)\cdot \tag{4}$$

$$E\left[x(t_1-\tau_1)x^*(t_1+\tau_1)x^*(t_2-\tau_2)x(t_2+\tau_2)\right]$$

where $E[\cdot]$ denotes the ensemble average. (We assume stationarity and so have dropped the parameter n from $\hat{r}(n,\tau)$.) For a Gaussian process, the quantity $E\{\cdot\}$ on the r.h.s. of Eqn. (4) can be expressed solely in terms of $r(\tau)$. In this case,

$$\text{cov}\left[\hat{r}(\tau_1),\hat{r}(\tau_2)\right] = \sum_{t_1}\left[\sum_{t_2}\phi(t_1+t_2,\tau_1)\cdot\right.$$

$$\left.\phi(t_2,\tau_2)\right]\left\{r\left[t_1-(\tau_1+\tau_2)\right]r\left[t_1+(\tau_1+\tau_2)\right]\right.$$

$$\left.+\, r\left[t_1-(\tau_1-\tau_2)\right]r\left[t_1+(\tau_1-\tau_2)\right]\right\}. \tag{5}$$

The variance of the spectral estimate for a normal process is obtained by inserting Eqn. (5) into Eqn. (3):

$$\text{var}\left[\rho(\omega)\right] = 4\sum_{\tau_1}\sum_{\tau_2}\sum_{t_1}\left[\sum_{t_2}\phi(t_1+t_2,\tau_1)\cdot\right.$$

$$\left.\phi(\tau,t_2)\right]\left\{r\left[t_1-(\tau_1+\tau_2)\right]r\left[t_1+(\tau_1+\tau_2)\right]\right.$$

$$+\, r[t_1-(t_1-t_2)]\,.$$

$$\left.r[k+(\tau_1-\tau_2)]\right\}e^{-2j\omega(\tau_1-\tau_2)}\,. \tag{6}$$

For a normal white noise process, the autocorrelation function equals

$$r(\tau) = \sigma^2\,\delta(\tau) \tag{7}$$

where σ^2 is the noise power and $\delta(\tau)$ is the Kronecker delta function. From (6,7)

$$Q^{-1}(\omega) = \sum_{\tau}\left[\sum_{t_1}\phi^2(t_1,\tau)\right] + \sum_{\tau}\left[\sum_{t_1}\phi^2(t_1,\tau)\right]e^{-j4\tau\omega} \tag{8}$$

where $Q^{-1}(\omega)$ is the normalized variance given by

$$Q^{-1}(\omega) = \frac{\text{var}\left[\rho(\omega)\right]}{E^2\left[\rho(\omega)\right]}\,.$$

Equation (8) consists of two terms: the first is constant in frequency and is the sum of the squared kernel values in the 2-D time domain, whereas the second is the Fourier transform of the summed-squared kernel values at each lag. Thus, the variance of the flat spectrum is frequency-dependent.

Figure 1 illustrates the theoretical normalized variance of the PSD estimate in dB for a normal white noise process using several popular time-frequency estimators. These curves were generated using Equation (8). The Pseudo-Wigner, Born-Jordan, and Choi-Williams (σ=1) distribution variances assume 65 non-zero lag values (L_{max} = 32) and a rectangular lag window. The periodogram variance assumes an equivalent number of lags and a rectangular data window. Note that the variance for the Born-Jordan and Choi-Williams distributions is intermediate between that of the periodogram and the Pseudo-Wigner distribution.

3. Minimum-Variance Time-Frequency Distributions

It is of interest to determine which kernels give PSD estimates of lowest variance. The quantity we seek to minimize is the average variance $\overline{\text{var}}[\rho(\omega)]$ [4,5], which for white noise processes is obtained from Eqn. (8) as

$$\overline{Q}^{-1} = \sum_{\tau}\left(1+\delta(\tau)\right)\sum_{t}\phi^2(t,\tau)\,. \tag{9}$$

Proper kernel normalization is required so that the estimate of the total power of the process is unbiased. This requirement is met by imposing the condition

$$\sum_t \phi(t,\tau{=}0) = 1. \qquad (10)$$

Other desirable distribution properties have been suggested. Claasen and Mecklenbrauker [6] list several properties: nonnegativity, realness, time and frequency shift invariance, correct time and frequency marginals, time and frequency support properties, and correct instantaneous frequency and group delay; and their associated kernel constraints. One expects that the minimum variance will certainly not decrease but may increase as additional constraints are imposed on the kernel.

The frequency marginal is satisfied if the kernel obeys the condition

$$\sum_t \phi(t,\tau) = 1 \quad \text{for all } \tau . \qquad (11)$$

This implies infinite frequency resolution. In practice only a finite number of lags $-L_{max} \le \tau \le L_{max}$ can be used in the spectrum estimate; Eqn. (11) will hold for only these lags. The effect on the frequency spectrum is to convolve the ideal spectrum with a sinc function. The sinc function has the best resolution for a given number of lag terms.

The variance [Eqn. (9)] subject to the frequency marginal constraint can be made arbitrarily small by incorporating many terms at each lag. However, distributions incorporating large amounts of temporal smoothing, such as the smoothed periodogram, are quite poor at tracking time-varying behavior.

One criterion which limits the amount of smoothing is that the time support property be satisfied. Since this property implies quick response to signals being suddenly extinguished, one might expect that distributions satisfying time support track time-varying signals well. The kernel constraint corresponding to time support is given by

$$\phi(t,\tau) = 0 \text{ for } |t| > |\tau| . \qquad (12)$$

We desire to minimize the variance for a white-noise process (Eqn. 9), subject to the kernel constraints of the frequency-marginal and time-support properties (Eqn. 11 and 12). Interestingly, the result is the Born-Jordan distribution (See Appendix A) whose kernel in the 2-D discrete-time domain is given by

$$\phi(t,\tau) = \frac{1}{2|\tau|+1} \cdot \text{rect}\left(\frac{t}{2\tau}\right) \qquad (13)$$

where rect(x) is equal to 1 for $|x| \le 1/2$ and is 0 otherwise. Its average variance can be found from Equation (9) to be given by the series

$$\overline{var} = 2 \sum_{\tau=0}^{L_{max}} \frac{1}{2\tau+1} , \qquad (14)$$

which, for reasonable values of L_{max} (e.g., $L_{max} \gtrsim 10$), can be written as

$$\overline{var} = \ln\left(2(2L_{max}+1)\right) + \gamma \qquad (15)$$

where $\gamma = 0.577...$ is Euler's constant. Since only even lags are used in the estimate, the number of lags $(2L_{max}+1)$ is related to the frequency resolution Δf by

$$\frac{\Delta f}{F_s} = \frac{\alpha}{2(2L_{max}+1)}, \qquad (16)$$

where F_s is the sampling frequency, and α is a constant approximately equal to 1. Thus the minimum variance for satisfying time-support and achieving a desired frequency resolution is [using (15) and (16)]:

$$\overline{var} = -\ln(\Delta f/Fs) + \gamma + \ln(\alpha). \qquad (17)$$

The above logarithmic dependence of the variance on resolution is not exhibited by every t-f estimator. For instance, the Pseudo-Wigner whose kernel is given by

$$\phi(t,\tau) = \delta(t)\ \mathrm{rect}\left(\frac{\tau}{2L_{max}}\right) \qquad (18)$$

has white-noise average variance equals

$$\overline{Q}^{-1} = \left(2L_{max} + 1\right) , \qquad (19)$$

which is a linear variance-resolution dependence. The periodogram kernel is

$$\phi(t,\tau) = h(t + \tau)h^{*}(t-\tau) , \qquad (20)$$

where h(t) is the data window. In this case, the white-noise average variance is given by

$$\overline{Q}^{-1} = \sum_{\tau=-L_{max}}^{L_{max}} \left(1+\delta(\tau)\right) \sum_{t=-\infty}^{\infty} h^{2}(t+\tau)h^{2}(t-\tau) \qquad (21)$$

which, in contrast with Pseudo-Wigner, is nearly independent of the number of data samples (and hence resolution). For a rectangular data window the average variance equals one-half.

The minimum variance Born-Jordan kernel not only satisfies the frequency marginal and time support, it satisfies all other desirable properties but non-negativity and reduced interference terms. Furthermore, the minimum average variance subject to these properties exhibits a logarithmic dependence with frequency resolution.

4. Simulations

Several time-frequency distributions (Pseudo-Wigner, Choi-Williams, Born-Jordan, periodogram) were run on simulated white noise. The mean and variance of each t-f distribution were determined based upon 600 trials with L_{max}=32. In all cases, the estimated mean of the power spectra were found to be constant in frequency and equal to twice the noise power of the process.

The normalized variance for the distributions simulated is presented in Figure 2. It is evident that there is a strong agreement between these curves and the theoretical results shown in Figure 1.

5. Conclusions

We have presented analytical expressions for the variance of an arbitrary time-frequency distribution for a Gaussian stochastic process. The Born-Jordan distribution has been found to have the lowest average variance for a white-noise Gaussian process, given that the desirable t-f properties are satisfied. Its variance is logarithmic in the frequency resolution. Other time-frequency kernels which in effect use fewer samples at smaller lags in the 2-D time domain, such as the Choi-Williams distribution, exhibit the same logarithmic variance dependence. It is expected that any distribution which exhibits good resolution in both time and frequency will have a logarithmic dependence of variance with resolution, or worse.

Appendix A

Desired: To minimize Eqn. (9) subject to Eqns. (11) and (12). We note that the contribution to $\overline{Q}^{-1}$ at each τ is given by the non-negative quantity

$$\sum_{t=-\infty}^{\infty} \phi^{2}(t,\tau) . \qquad (A\text{-}1)$$

Furthermore, the kernel constraints (11) and (12) are applicable to each τ separately. It thus suffices to minimize Eqn. (A-1).

Combining Eqns. (A-1) and (12), the quantity to be minimized is

$$\sum_{t=-\tau}^{\tau} \phi^{2}(t,\tau) \quad \text{at each } \tau$$

subject to

$$\sum_{t=-\tau}^{\tau} \phi(t,\tau) = 1 \ .$$

Thus the problem reduces to the minimization of the length of a ($2|\tau|$ + 1)-dimensional vector, subject to a linear constraint. The solution is obtained by the method of Lagrange multipliers and is given by

$$\phi(t,\tau) = \frac{1}{2|\tau|+1} \qquad \text{(A-2)}$$

for $|t| < |\tau|$. Eqns. (A-2) and (12) lead directly to Equation (13).

References

1. L. Cohen, "Time-frequency distributions - a review", *Proc. IEEE*, vol.77, No. 7, pp. 941-981, July 1989.

2. H. I. Choi and W. J. Williams "Improved time-frequency representation of multi-component signals using exponential kernels", *IEEE Trans. on Acoustics, Speech and Signal Processing*, vol. 37, No. 6, pp. 862-871, June, 1989.

3. Y. Zhao, L. E. Atlas and R. J. Marks, II, "The use of cone-shaped kernels for generalised time-frequency representations of non-stationary signals", *IEEE Trans. on Acoustics, Speech and Signal Processing*, pp. 1084-1091, July 1990.

4. M. G. Amin, "Time varying spectrum estimation for a general class of nonstationary processes", *Proc. IEEE*, vol. 74, no. 12, pp. 1800-1802, December, 1986.

5. M. G. Amin, "Time-frequency spectrum analysis and estimation for non-stationary random processes", in *Methods and Applications of Time-Frequency Signal Analysis*, Ed. B. Boashash, Chpt. 9, Longman Cheshire, Melbourne, Australia, (ISBN No: 0 582 86873 4), 1991.

6. T. A. C. M. Claasen and W. F. G. Mecklenbrauker, "The Wigner distribution - a tool for time-frequency signal analysis - Part III: Relations with other time-frequency signal transformations", *Philips Jour. Research.*, vol. 35, pp. 372-389, 1980.

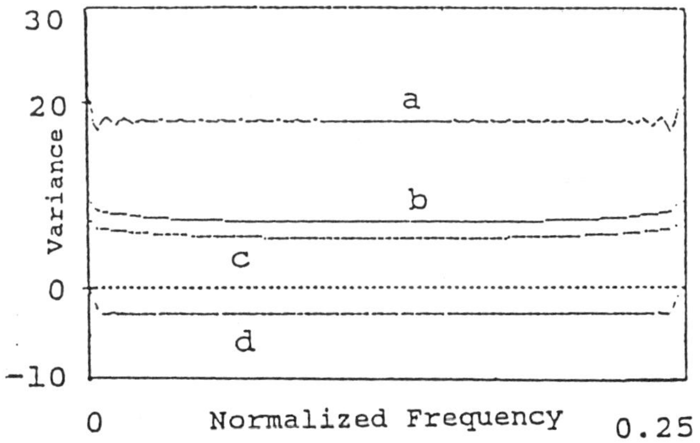

Figure 1 Theoretical normalized variance

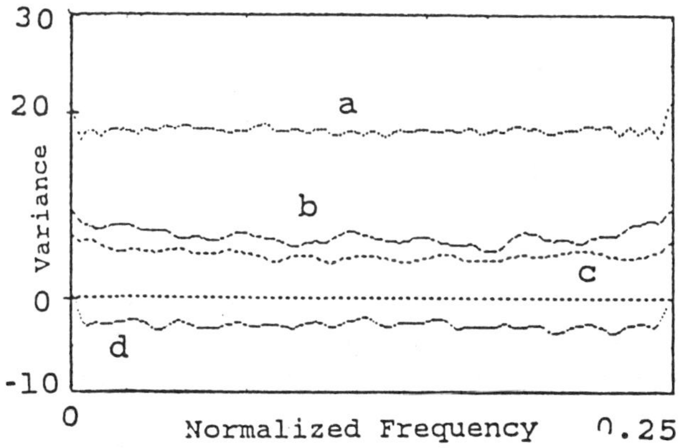

Figure 2 Simulated normalized variance

a) Pseudo-Wigner, b) Born-Jordan, c) Choi-Williams,

d) Periodogram

On a Generalization of the Choi-Williams Time-Frequency Distribution

G. Faye Boudreaux-Bartels and Antonia Papandreou *
Dept. of Electrical Engineering
University of Rhode Island
Kingston, RI, 02881

Abstract

We propose the following generalization of the kernel of the Choi-Williams time-frequency Exponential Distribution (ED): $\phi_{\mathrm{GED}}(\xi,\tau)=e^{-(\frac{\xi}{\xi_1})^{2N}(\frac{\tau}{\tau_1})^{2M}}$. *The order parameters* M *and* N *can be chosen so that the generalized ED (GED) kernel is arbitrarily close to an ideal lowpass filter in the Ambiguity Function (AF) plane. The scaling parameters* τ_1 *and* ξ_1 *are used to vary the kernel's passband and stopband edges. The GED kernel can be used to filter out troublesome oscillatory cross-terms which map away from the origin in the AF plane. We will derive optimum design curves for the order and scaling parameters, prove that the GED kernel satisfies all of the desirable distribution properties of the original ED, and provide examples demonstrating that the GED is better at removing cross-terms while retaining desirable auto-terms than the ED.*

1 Introduction

The analysis of non-stationary signals such as speech, sonar, and bio-acoustical signals can be achieved using time-frequency representations (TFR) that map time signals into 2-D distribution functions of time and frequency. The Wigner Distribution (WD) is one such TFR which possesses many desirable signal analysis properties [1]. However, when multicomponent signals are analyzed, the quadratic WD suffers from interfering cross-terms. Although these interfering cross-terms are essential components of the transformation, they can be detrimental in pattern recognition and analysis applications.

The WD belongs to Cohen's class of time and frequency shift invariant TFR [2], characterized by[1]

$$C_x(t,\omega,\phi)=\frac{1}{2\pi}\int\int e^{j(\xi t-\tau\omega)}\phi(\xi,\tau)A_x(\xi,\tau)d\tau d\xi, \quad (1)$$

with the TFR dependent kernel $\phi_{\mathrm{WD}}(\xi,\tau)=1$. Equation (1) can be interpreted as the inverse Fourier transform of the product of the TFR kernel with the signal's Ambiguity function $A_x(\xi,\tau)$ [1]. If the TFR dependent kernel is designed to be a 2-D lowpass filter in the Ambiguity function (AF) plane, then the oscillatory cross-terms can be removed because they map away from the AF plane's origin whereas the auto-terms map to the origin and can thus be retained. The lowpass filter should also be designed so that the resulting TFR satisfies as many of the desirable properties of the WD as possible.

Other TFR, which have appeared in the literature, can be generated with a different choice of the kernel $\phi(\xi,\tau)$ [2]. Choi and Williams proposed the Exponential Distribution (ED) with the kernel function [3]

$$\phi_{\mathrm{ED}}(\xi,\tau)=e^{-\frac{\xi^2\tau^2}{\sigma}}, \qquad \sigma>0. \quad (2)$$

The exponential kernel decays for increasing values of $|\xi\tau|$ and thus acts as a 2-D lowpass filter in the AF plane with cross-term reduction capabilities. Simultaneously, the ED shares the following TFR properties with the WD: it is real-valued, it preserves time and frequency shifts, and it preserves the marginals, the group delay and the instantaneous frequency of the signal. The ED, with a special modification that enables it to preserve the signal's time and frequency support, has been successfully used to analyze non-stationary signals such as EMG and ERP signals, TMJ sounds, and sounds from muscles and marine mammals [4]. However, the ED lowpass kernel does not have a flat passband and it decays relatively slowly. Thus,

*This work was supported in part by ONR grant N00014-89-J-1812.

[1] Unless otherwise indicated, the limits of integration are assumed to extend from $-\infty$ to $+\infty$.

1058-6393/91 $01.00 © 1991 IEEE

although the ED reduces cross-terms that occur very far from the origin of the AF plane, it fails to reduce low frequency cross-terms that arise when signal components are closely spaced in either time or frequency. Further, the AF of the signal's auto-terms is sometimes distorted because the ED kernel's passband is not very flat. The flatness of the kernel depends on σ, which Choi and Williams choose in an ad hoc manner to be σ=1 for cross-term reduction based on their experimental results; they note that σ=100 approaches the allpass filter of the WD.

2 Generalized ED Properties

Since the cross-term location relative to that of the auto-terms varies, it is reasonable to introduce a lowpass filter with a variably flat passband and a variably narrow transition region in comparison to the fixed (up to a scaling constant) ED lowpass filter. Thus we propose the Generalized Exponential Distribution (GED), which belongs to Cohen's class with the kernel

$$\phi_{\mathrm{GED}}(\xi,\tau) = e^{-(\frac{\xi}{\xi_1})^{2N}(\frac{\tau}{\tau_1})^{2M}} \tag{3}$$

where N, M are positive order parameters, and ξ_1, τ_1 are positive spectral and temporal scaling constants, respectively, chosen such that $\phi_{\mathrm{GED}}(\xi_1,\tau_1)=e^{-1}$ [5, 6]. Note that the GED generalizes the ED since for N=M=1, and for $\xi_1^2\tau_1^2=\sigma$, the GED kernel in (3) simplifies to the Choi-Williams ED kernel in (2).

The GED satisfies the desirable TFR properties that the ED satisfies. Specifically, for a signal $x(t)$, the GED

1. preserves time shifts on $x(t)$ since $\phi_{\mathrm{GED}}(\xi,\tau)$ does not depend on the output parameter t: $y(t) = x(t-\mu) \Rightarrow \mathrm{GED}_y(t,\omega) = \mathrm{GED}_x(t-\mu,\omega)$ where $\mathrm{GED}_x(t,\omega) = \frac{1}{2\pi}\int\int e^{j(\xi t-\tau\omega)}\phi_{\mathrm{GED}}(\xi,\tau)\cdot A_x(\xi,\tau)\,d\tau\,d\xi$
2. preserves frequency shifts on $X(\omega)$ since $\phi_{\mathrm{GED}}(\xi,\tau)$ does not depend on the output parameter ω: $Y(\omega) = X(\omega+v) \Rightarrow \mathrm{GED}_y(t,\omega) = \mathrm{GED}_x(t,\omega+v)$
3. is real-valued since the kernel satisfies the relationship: $\phi^*_{\mathrm{GED}}(\xi,\tau) = \phi_{\mathrm{GED}}(-\xi,-\tau)$
4. preserves the instantaneous signal energy, $|x(t)|^2$, and the energy density spectrum, $|X(\omega)|^2$, since the kernel satisfies the relationships: $\phi_{\mathrm{GED}}(\xi,0)=1,\ \forall\xi$ and $\phi_{\mathrm{GED}}(0,\tau)=1,\ \forall\tau$
5. preserves the instantaneous frequency and the group delay of the signal since for $N, M > 0.5$ the kernel satisfies the relationships:
$\phi_{\mathrm{GED}}(0,\tau) = 1, \forall\tau$, and $\frac{d}{d\xi}\phi_{\mathrm{GED}}(\xi,\tau)|_{\xi=0} = 0, \forall\tau$
$\phi_{\mathrm{GED}}(\xi,0) = 1, \forall\xi$, and $\frac{d}{d\tau}\phi_{\mathrm{GED}}(\xi,\tau)|_{\tau=0} = 0, \forall\xi$

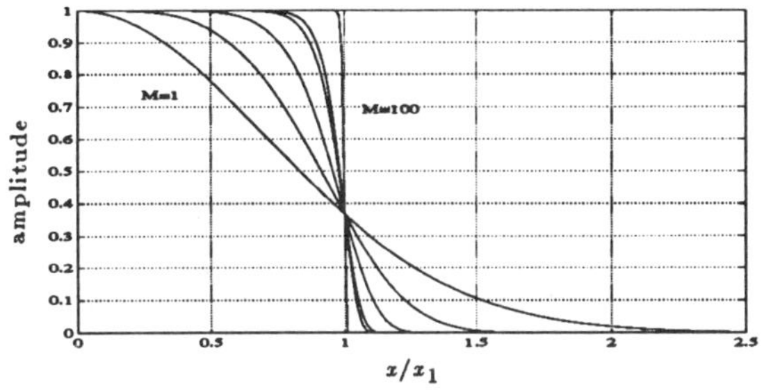

Figure 1: Plot of $\theta_M(x)$ for increasing M.

2.1 GED Lowpass Filter Nature

The kernel of the GED has a variable flatter passband and a narrower transition region than that of the ED which can result in less auto-term distortion and more low frequency cross-term reduction. The improved lowpass filter nature of the GED kernel over the ED kernel can be demonstrated with the 1-D lowpass function $\theta_M(x)=e^{-(\frac{x}{x_1})^{2M}}$ plotted in fig. 1 for $M \geq 1$. Note that although all the curves intersect at the point x=x_1, the passband becomes flatter and the transition region narrower for increasing values of M, approaching an ideal lowpass filter with cutoff x=x_1 for $M \geq 100$. When M=1, the lowpass filter is Gaussian with a wide transition region and a non-flat passband. When $x_1=\tau_1\xi_1^{N/M}$, $\phi_{\mathrm{GED}}(\xi,\tau)=\theta_M(\tau\xi^{N/M})$, implying that the 1-D function $\theta_M(x)$ plotted in fig. 1 is a slice of the 2-D GED kernel; likewise, for M=1 and $x_1^2=\sigma$, $\theta_1(\tau\xi)$ is a slice of the ED kernel. Consequently, for $M > 1$, fig. 1 reveals that the GED kernel has a variable, flatter passband and a variable, narrower transition region than the ED.

Considering equal amplitude contours of the 2-D GED kernel provides further insight into its lowpass filter nature. Specifically, if we take the natural logarithm and the logarithm of $\phi_{\mathrm{GED}}(\xi,\tau)=k$ in (3) we obtain the iso-contour relationship

$$\log\left(\frac{\tau}{\tau_1}\right) = \frac{1}{2M}\log(-\ln k) - \frac{N}{M}\log\left(\frac{\xi}{\xi_1}\right). \tag{4}$$

In the normalized $(\log\frac{\xi}{\xi_1}, \log\frac{\tau}{\tau_1})$ plane, (4) is simply the equation of a straight line with slope $-N/M$ and intercept dependent upon the amplitude level k and the order M. The linear iso-contours are plotted in fig. 2a-b for increasing N (fixed M=1) and for increasing M (fixed N=1), respectively. The curves for the ED correspond to N=M=1, assuming that $\sigma=\xi_1^2\tau_1^2$. A time versus frequency trade-off exists in fig. 2a, since as N increases the temporal passband increases for $\xi < \xi_1$

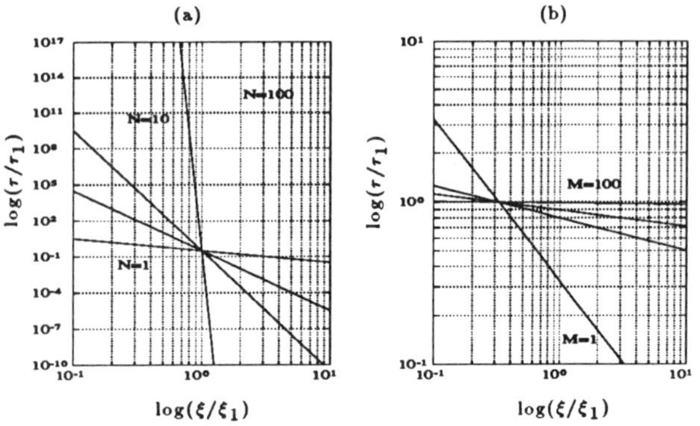

Figure 2: TF linear iso-contours for ϕ_{GED}=0.9. (a) M=1, N=1, 5, 10, 100, (b)M=1,5,10,100, N=1.

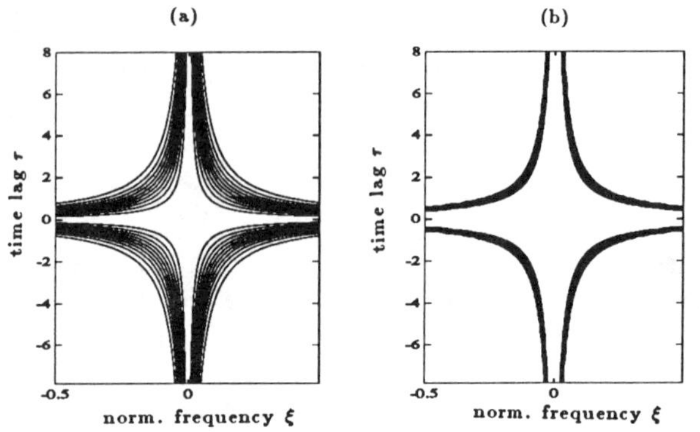

Figure 3: Contour plots of (a) ϕ_{ED}, N=M=1, $\sigma=\xi_1^2\tau_1^2=25^2$, (b) ϕ_{GED}, N=M=5, τ_1=2, ξ_1=12.5.

where as it decreases for $\xi > \xi_1$. Fig. 2b shows a similar trade-off for the spectral passband; the temporal bandwidth approaches τ_1 for all frequency for large M.

Contour plots of the GED (N=M=5, τ_1=2, ξ_1=12.5) and the ED (N=M=1,$\sigma=\xi_1^2\tau_1^2=(12.5)^2(2)^2$) kernels are shown in fig. 3. Although the two kernels have the same ratio N/M and scaling parameters τ_1 and ξ_1, and hence the same slope in (4) and fig. 2, the GED kernel has a flatter passband and a narrower transition region than the ED. Consequently, the flatness of the passband and the width of the transition region depend on the actual values of N and M, as the next section demonstrates.

2.2 GED Parameter Design

The GED kernel is completely determined by (1) the order parameters N and M which provide information on its spectral and temporal passband and stopband edges and on its transition width, and (2) the scaling factors τ_1 and ξ_1 that simply scale the axes of the iso-contours in fig. 2. Thus, using the linear iso-contours, design curves and equations can be obtained for the optimal parameters of the kernel. Equation (4) in the un-normalized $(\log\xi, \log\tau)$ plane,

$$\log(\tau) = \left(\log(\tau_1\xi_1^{\frac{N}{M}}) + \frac{1}{2M}\log(-\ln k)\right) - \frac{N}{M}\log(\xi)$$
$$= b + a\ \log(\xi), \tag{5}$$

is the equation of a straight line with slope

$$a = -N/M \tag{6}$$

and $\log\tau$-axis intercept $b = \log(-\ln k)^{\frac{1}{2M}} + \log(\tau_1\xi_1^{\frac{N}{M}})$ for a given iso-contour amplitude k. Since decreasing k from the passband amplitude edge k_p to the stopband edge k_s does not change the slope of the iso-contour but simply increases the $\log\tau$-axis intercept, the passband and stopband iso-contours result in two parallel lines in the $(\log\xi, \log\tau)$ plane. Design criteria are then obtained given k_p, k_s, a set of $L_p \geq 2$ passband points $\{(\xi_{p_i}, \tau_{p_i})\}_{i=1}^{L_p}$, and a set of L_s stopband points $\{(\xi_{s_l}, \tau_{s_l})\}_{l=1}^{L_s}$ in the first quadrant of the AF plane. The passband edge is defined by a straight line in the $(\log\xi, \log\tau)$ plane such that all passband points substituted into (5) satisfy

$$\log(\tau_{p_i}) \geq b_p - \frac{N}{M}\log(\xi_{p_i}), \quad i = 1, 2, \ldots, L_p \tag{7}$$

and the stopband edge is similarly defined such that

$$\log(\tau_{s_l}) \leq b_s - \frac{N}{M}\log(\xi_{s_l}), \quad l = 1, 2, \ldots, L_s \tag{8}$$

where b_p and b_s are the passband and stopband intercepts of the parallel, linear specification curves.

$$b_p = \frac{1}{2M}\ \log(-\ln k_p) + \log\left(\tau_1\xi_1^{\frac{N}{M}}\right) \tag{9}$$

$$b_s = \frac{1}{2M}\ \log(-\ln k_s) + \log\left(\tau_1\xi_1^{\frac{N}{M}}\right) \tag{10}$$

The "critical" passband line is found by searching through all possible lines in the $(\log\xi, \log\tau)$ plane connecting any pair of the given passband points such that the resulting slope is negative and all other passband points are either on the line or underneath it [2] as shown in fig. 4. Given two passband points used to form the critical line, $(\xi_{p_{i_c}}, \tau_{p_{i_c}})$ and $(\xi_{p_{j_c}}, \tau_{p_{j_c}})$, then one can derive its slope in the $(\log\xi, \log\tau)$ plane as

$$a = \frac{\log(\tau_{p_{i_c}}) - \log(\tau_{p_{j_c}})}{\log(\xi_{p_{i_c}}) - \log(\xi_{p_{j_c}})} = \frac{\log\left(\frac{\tau_{p_{i_c}}}{\tau_{p_{j_c}}}\right)}{\log\left(\frac{\xi_{p_{i_c}}}{\xi_{p_{j_c}}}\right)}. \tag{11}$$

[2] Convex passband regions, such as circles or ellipses, will give rise to several permissible critical lines. In this case, choose the slope which best fits the stopband criteria [6].

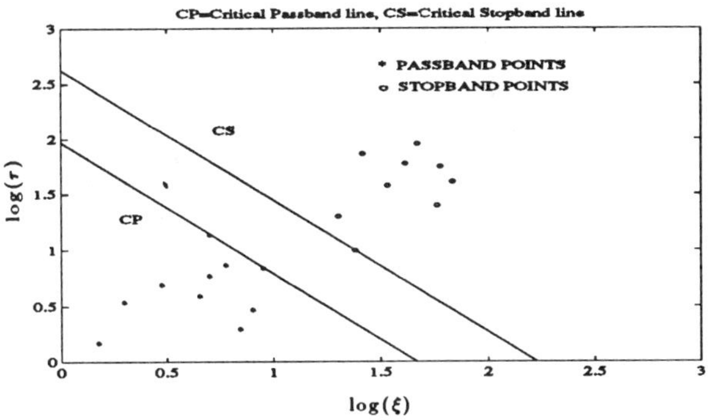

Figure 4: Critical passband and stopband lines.

Combining (6) and (11), the $\log\tau$-axis intercept in (5) is given by

$$b_p = \log(\tau_{p_{i_c}}) - a\log(\xi_{p_{i_c}}) = \log\left(\tau_{p_{i_c}}\xi_{p_{i_c}}^{\frac{N}{M}}\right). \quad (12)$$

Thus, the lower bound of equation (7) is satisfied for at least two equations and exceeded for all others. The critical stopband line must have the same slope as the critical passband line and it must have all stopband points either on it or above it. This results in the line with the minimum stopband intercept

$$b_s = \min_{l=1,2,\ldots,L_s}\left\{\log\left(\tau_{s_l}\xi_{s_l}^{\frac{N}{M}}\right)\right\} = \log\left(\tau_{s_{l_c}}\xi_{s_{l_c}}^{\frac{N}{M}}\right) \quad (13)$$

where the point $(\xi_{s_{l_c}}, \tau_{s_{l_c}})$ is the stopband point that minimizes equation (13)[3]. The difference of the intercepts b_p and b_s in (12-13) corresponds to the vertical distance between the two critical lines. Thus, subtracting the intercepts in equations (9-10) and in (12-13) and then equating the two differences

$$\begin{aligned} b_p - b_s &= \frac{1}{2M}\Big(\log(-\ln k_p) - \log(-\ln k_s)\Big) \quad (14)\\ &= \log\left(\tau_{p_{i_c}}\xi_{p_{i_c}}^{\frac{N}{M}}\right) - \log\left(\tau_{s_{l_c}}\xi_{s_{l_c}}^{\frac{N}{M}}\right)\end{aligned}$$

can be used to solve for the temporal order parameter M and for the spectral order parameter N using (6) and (11).

$$M = \frac{\frac{1}{2}\log\left(\frac{\ln k_p}{\ln k_s}\right)}{\log\left(\frac{\tau_{p_{i_c}}}{\tau_{s_{l_c}}}\frac{\xi_{p_{i_c}}}{\xi_{s_{l_c}}}\right)^{\frac{N}{M}}}, \quad N = -aM = -M\frac{\log\frac{\tau_{p_{i_c}}}{\tau_{p_{j_c}}}}{\log\frac{\xi_{p_{i_c}}}{\xi_{p_{j_c}}}} \quad (15)$$

[3] In the proposed method, the critical stopband slope is fixed, and only one degree of freedom remains to choose its intercept. Thus, the critical stopband line may pass through only one stopband edge point. If the stopband is more important than the passband, then the algorithm can be changed to determine the slope $a = -N/M$ by finding the critical lines that satisfy the upper bound of at least two stopband equations and the lower bound of one passband equation.

Note that N and M in (15) are not necessarily integers. If N and M are rounded to integer values, then one should check that the set of passband and stopband constraints in (7-8) are still satisfied. If non-integer values are used, N and M should be rational numbers with an odd denominator. Due to the nature of the GED kernel, if N is a rational number, then $e^{-(\frac{\xi}{\xi_1})^{2N}}$ will be real for negative ξ provided it is computed as $e^{-[(\frac{\xi}{\xi_1})^2]^N}$. The latter computation, however, will only yield the right results when $2N=2p/q$, p and q are relatively prime integers, and q is odd. This is because the identity $r^{a/b}=(r^a)^{1/b}=(r^{1/b})^a$ holds only for relatively prime a and b. We developed an efficient algorithm that given N and an error tolerance $\epsilon > 0$, can approximate $N \approx p/q$ by finding relatively prime integers $2p$ and q such that $|N - p/q| < \epsilon$ [6]. Note also that the 5th GED property restricts $M > 0.5$ and $N > 0.5$. As a result, for a rational M or N, p cannot be equal to 1 unless q also equals 1.

The scaling constants τ_1 and ξ_1 are dependent on one other due to the time-frequency trade-off. As a result only one of them can be arbitrarily chosen. Equating (9) and (12) results in

$$\log(\tau_1\xi_1^{\frac{N}{M}}) = b_p - \log(-\ln k_p)^{\frac{1}{2M}} = \log\frac{\tau_{p_{i_c}}\xi_{p_{i_c}}^{\frac{N}{M}}}{(-\ln k_p)^{\frac{1}{2M}}}. \quad (16)$$

If τ_1 equals one of the critical passband τ points, then

$$\tau_1 = \tau_{p_{i_c}}, \qquad \xi_1 = \xi_{p_{i_c}}/(-\ln k_p)^{\frac{1}{2N}}. \quad (17)$$

Thus, the GED optimal parameters can be obtained using equations (15) and (17). In order to demonstrate the effectiveness of the design equations described above, consider the GED of $x(t) = e^{-j\frac{1}{4}t^2}\cdot(e^{-\pi\frac{(t-2.2)^2}{2(4)}}e^{-j20.2t} + e^{-\pi\frac{(t+2.2)^2}{2(4)}}e^{j20.2t})$. The contour plot of the signal's WD is shown in fig. 5a and the designed GED kernel superimposed with the signal's AF in 5b. The optimal GED parameters are: N=11.0781 $\approx$ (565/51), M=24.9594 $\approx$ (624/25), τ_1=3.3662, and ξ_1=3.3499. As expected from the design constraints, the GED in 5c removed the cross-terms completely and retained the auto-terms.

2.3 Choi-Williams Parameter Design

Choi and Williams choose the value of σ in an ad hoc manner. However, we can provide design equations for the optimal value of σ. Since N=M=1 in the ED, the slope of the corresponding ED critical lines is fixed at $-N/M$=-1. As a result, there is only one degree of freedom left that can rarely satisfy both passband and stopband constraints. From (2-3), $\sigma=\xi_1^2\tau_1^2$.

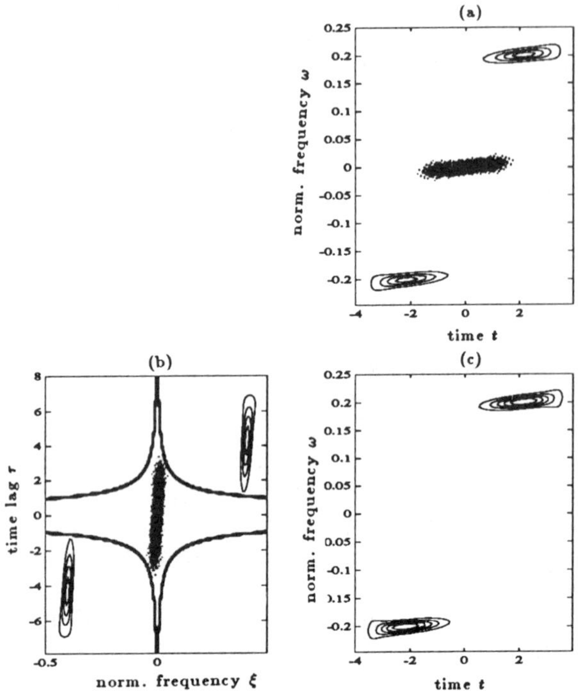

Figure 5: (a) WD of the sum of two Gaussian chirps, (b) designed GED kernel superimposed on the AF of the signal, (c) resulting GED.

Using (17), the optimal[4]

$$\sigma = \xi_1^2 \tau_1^2 = (\tau_{p_{i_{ED}}}^2 \xi_{p_{i_{ED}}}^2)/(-\ln k), \quad (18)$$

where k, $\tau_{p_{i_{ED}}}$, and $\xi_{p_{i_{ED}}}$ are selected from either passband or stopband constraints. For stopband constraints to hold in (18), $k=k_s$ and $(\xi_{i_{ED}}, \tau_{i_{ED}})=(\xi_{s_{l_c}}, \tau_{s_{l_c}})$ in (13). However, for passband constraints, $k=k_p$, and the critical passband line has slope$=-1$ which passes through the outermost passband point $(\xi_{i_{ED}}, \tau_{i_{ED}})=(\xi_{p_{i_c}}, \tau_{p_{i_c}})$ which maximizes the passband intercept

$$b_p = \max_{i=1,\ldots,L_p} \left\{ \log \tau_{p_i} + \log \xi_{p_i} \right\} = \log(\tau_{p_{i_{ED}}} \xi_{p_{i_{ED}}}). \quad (19)$$

In order to demonstrate the trade-off between the passband versus stopband constraints consider the sum of two Gaussian chirps with variance $4/\pi$ (sec.^{2}). Equation (18) based on passband constraints results in the optimal value of $\sigma=(116.6477)^2$. The kernel superimposed with the AF of the signal is shown in fig. 6a. Indeed, the auto-terms are maintained but the cross-terms are not completely removed. Equation (18) based on stopband constraints results in the optimal value of $\sigma=(6.5068)^2$. The kernel superimposed with the AF of the signal, shown in fig. 6b, shows that the auto-terms are distorted in the transition region but the cross-terms are completely removed.

[4]In discretizing (1), Choi and Williams (C-W) inconsistently sampled $x(t+\tau/2)$ every $\tau=2mT$ but $\phi_{ED}(\xi,\tau)$ every $\tau=mT$ (where T is the sampling period). We assumed that $\tau=2mT$ for *both* the signal and the kernel. Thus, for the C-W discrete time implementation in [3], (18) should be divided by 4 to consistently equate GED with ED.

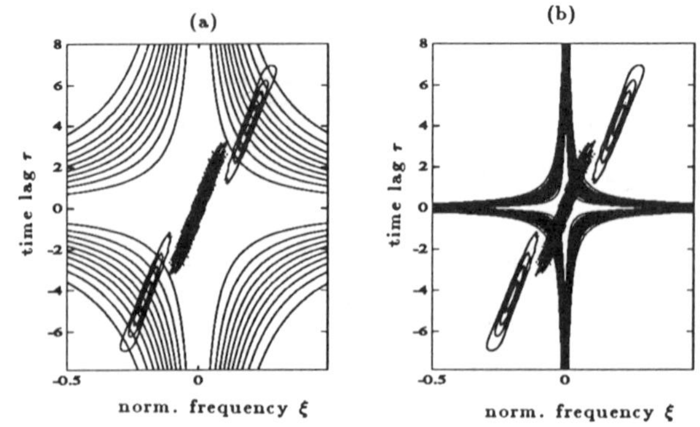

Figure 6: AF of two Gaussian chirps superimposed with (a) passband designed ED kernel, (b) stopband designed ED kernel.

2.4 Simulations

In order to demonstrate the superior nature of the GED over the ED, consider the sum of four Gaussian signals with variance $1/\pi$ (sec^2). Since the signal has four components, the WD (shown in fig. 7c) has four auto-terms and each pair of auto-terms gives rise to one oscillatory cross-term, centered around the mid point of the line joining the centers of the two auto-terms. The AF of the signal superimposed with the GED kernel with $N=10$, $M=1$, $\tau_1=8$ and $\xi_1=6.25$ is shown in 7b and the corresponding one for the ED with $\sigma=(6.25)^2(8)^2$ in 7a. The resulting ED and GED are shown in 7d and 7e, respectively. The GED has succeeded in removing all the cross-terms leaving the auto-terms unaffected whereas the ED has only removed some of the cross-terms. Because the GED and the ED kernels equal 1 for $\tau=0$ or $\xi=0$, both TFR have trouble removing cross-terms that fall along the τ or ξ axes in the AF plane [6].

3 Computational issues

The GED is obtained using (1) which involves computing the AF of a signal by taking the 1-D Fourier transform (FT) of $x(t+\frac{\tau}{2})x^*(t-\frac{\tau}{2})$ with respect to t, then multiplying the AF with the kernel function and taking the 2-D FT of the resulting product. The

ED, although interpreted in the AF domain, is actually computed in the (t, τ) plane using an 1-D convolution and an 1-D FT. This is shown by rewriting (1) for the ED as

$$ED_x(t, \omega) = \int e^{-j\tau\omega} K_x(t, \tau)\, d\tau \qquad (20)$$

$$K_x(t, \tau) = \Phi_{\mathrm{ED}}(t, \tau) *_t x(t + \frac{\tau}{2}) x^*(t - \frac{\tau}{2})$$

where $\Phi_{\mathrm{ED}}(t, \tau)$ is the 1-D inverse FT with respect to ξ of the kernel function in (2) [3], and $*_t$ indicates convolution in the parameter t. Since we have found $\Phi_{\mathrm{GED}}(t, \tau)$ in closed form only for $N{=}1$ [6], we implement the GED in the AF plane. Currently we are working on a Butterworth TFR which has a 2-D lowpass Butterworth kernel that has similar properties as the GED kernel [7]. The FT of its kernel exists in closed form and one could use it as an approximation to the FT of the GED kernel.

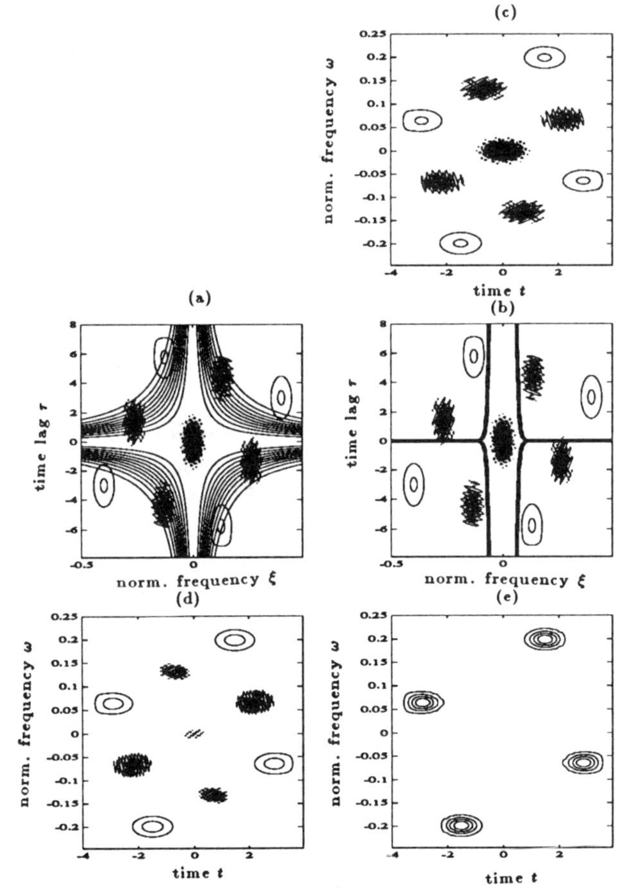

Figure 7: AF of the sum of two Gaussian chirps superimposed with the (a) ED kernel ($\sigma{=}(6.25)^2(8)^2$), (b) GED kernel ($N{=}10$, $M{=}1$, $\tau_1{=}8$, $\xi_1{=}6.25$), (c) WD (d) ED, (e) GED of the signal.

From our examples, we have found that one advantage in doing the computation in the AF plane rather than the (t, τ) plane in (20) is that for large values of σ, fewer aliasing and truncation errors seem to occur. Choi and Williams suggested values of σ between 1 and 100 [3] in order to avoid errors caused by large values of σ that result in the broadening of $\Phi_{\mathrm{ED}}(t, \tau)$ in the (t, τ) plane.

4 Conclusion

The Choi-Williams ED has received a lot of attention in the literature due to its ability to reduce cross-terms while simultaneously maintaining many desirable TFR properties. However, its kernel is a 2-D lowpass filter in the AF plane with a wide transition region and a non-flat passband. Our Generalized ED has a lowpass kernel with a variable, flatter passband and a narrower transition region than the ED. As a result, it reduces more low frequency cross-terms than the ED and provides less auto-term distortion, while simultaneously maintaining the desirable TFR properties of the ED. Further, given a set of passband and stopband constraints, we provide optimal design equations for the GED parameters as well as the ED scaling parameter σ, which previously had been chosen in an ad hoc manner.

References

[1] T. A. C. M. Claasen and W. F. G. Mecklenbräuker, "The Wigner distribution—A tool for time-frequency signal analysis, Part III: Relations with time-frequency signal transformations," *Philips J. Res.*, vol. 35, pp. 372–389, 1980.

[2] L. Cohen, "Generalized phase-space distribution functions," *Math. Phys.*, vol. 7, pp. 781–786, 1966.

[3] H. Choi and W. J. Williams, "Improved time-frequency representation of multicomponent signals using exponential kernels," *IEEE Trans. Acoust., Speech, Sig. Proc.*, vol. ASSP-37, pp. 862–871, June 1989.

[4] C. Zheng, S. E. Wildmalm and W. J. Williams, "New time-frequency analyses of EMG and TMJ sound signals," *IEEE Engineering in Medicine and Biology society 11th annual intern. conf.*, pp. 741–742, June 1989.

[5] G. F. Boudreaux-Bartels and A. Papandreou, "Generalized reduced interference distribution," *Third Biennial ASSP Miniconf.*, Boston, MA, April 1991.

[6] A. Papandreou, "A generalization of Choi-Williams exponential distribution for time-frequency analysis," MS thesis, University of RI, Kingston, RI, May 1991.

[7] A. Papandreou and G. Faye Boudreaux-Bartels, "Generalization of the Choi-Williams Distribution and the Butterworth Distribution for Time-Frequency Analysis," submitted to *IEEE Trans. on Sig. Proc.*.

Estimation of Cyclic Polyspectra*

Chad M. Spooner William A. Gardner
Department of Electrical and Computer Engineering
University of California
Davis, CA 95616

Abstract

The term polyspectra *refers to the spectral cumulants of a stationary random process or time-series, which are equivalent to the multidimensional Fourier transforms of the temporal cumulants. The Fourier coefficients of periodic (and multiply-periodic) cumulants of cyclostationary random process or time-series are called* cyclic cumulants. *The multidimensional Fourier transforms of cyclic cumulants are called* cyclic polyspectra, *and are themselves spectral cumulants. In this paper the general formulas for spectral and temporal cumulants of cyclostationary time-series are presented, some applications are briefly discussed, and methods for measuring cumulants from finite-length data records are described and shown to produce consistent estimators.*

1 Introduction

The power spectral density function is of great significance in the mathematical analysis of signal processing problems that involve wide-sense stationary time-series. Since the power spectral density is the Fourier transform of the autocorrelation function, it provides a complete frequency-domain description of the second-order statistical characteristics of such time-series. Similarly, the collection of cyclic spectral density functions, indexed by the cycle frequency α, completely describes the second-order statistical behavior of cyclostationary time-series in the frequency-domain [1]. The set of polyspectra through order n are sufficient to describe the nth-order statistical characteristics, in the frequency domain, of nth-order stationary time-series [2]. In the case of an nth-order cyclostationary time-series, we require knowledge of the collections of mth-order cyclic polyspectra for all cycle frequencies for orders $m = 1$ to $m = n$ [3] [4].

*This work was supported by the U. S. Army Research Office under contract DAAL03-91-C-0018

The implementation of signal processing algorithms that are based on such idealized functions must naturally entail measurements of the functions that use only finite segments of data. Thus we are led to consider measurement techniques for the power spectrum [1][5], the cyclic spectrum [1][6], the polyspectrum [8][9], and the cyclic polyspectrum. Each of these frequency-domain functions can be characterized as the Fourier transform of a time-domain function. For the PSD it is the autocorrelation, for the cyclic spectrum it is the cyclic autocorrelation, for the polyspectrum it is the temporal cumulant function, and for the cyclic polyspectrum it is the cyclic temporal cumulant function. Thus, measurements of the time-domain parameters can be used to indirectly obtain measurements of the corresponding frequency-domain parameters. It is also sometimes feasible to measure the frequency-domain parameters directly in the frequency domain. Both of these approaches are considered herein.

In this paper we review the definitions of and relations between cyclic cumulants, cyclic moments, and cyclic polyspectra, and we consider nonparametric estimation of both cyclic cumulants and cyclic polyspectra. It is shown that cyclic polyspectra can be estimated consistently by first measuring the cyclic cumulant, multiplying it by a tapering window, and then Fourier transforming it. Measurement of cyclic polyspectra directly in the frequency domain is shown to be relatively difficult due to the fact that infinite-strength spectral functions (containing Dirac delta functions) must be estimated and combined to obtain estimates of finite-strength spectral functions (in which all Dirac deltas cancel each other). Examples are provided to illustrate the theory.

As explained in [10], there is a duality between the statistical theory of parameter (moment or cumulant) estimation formulated within the stochastic-process framework, where moments and cumulants are defined by ensemble averages, and the alternative statistical theory formulated within the nonstochastic framework

1058-6393/91 $01.00 © 1991 IEEE

of time-series, where moments and cumulants are defined in terms of time averages. Consequently, the consistency of the estimates presented here within the nonstochastic framework of time-series applies equally well when these estimators are interpreted as estimators for parameters of stochastic processes, provided that the mathematical models of the processes exhibit appropriate cycloergodic properties. Otherwise, there is no reason to expect that estimators that rely on only a single time-series should be consistent. Moreover, the absence of cycloergodic properties in a mathematical model is often indicative of an inappropriate model.

Cyclic cumulants and cyclic polyspectra can be used to perform signal processing tasks that cannot be accomplished by using techniques that involve only second-order statistics. This is so because the higher-order cyclic cumulants can reflect underlying periodicities that the cyclic spectrum cannot. Some applications include signal-selective time-delay estimation, synchronization, and weak-signal detection. Cyclic cumulants also play a fundamental role in the design of low-probability-of-intercept signals.

2 The Parameters of Higher-Order Cyclostationarity

For the time-series $x(t)$ for $-\infty < t < \infty$, we define the nth-order lag-product time-series by

$$L_{\boldsymbol{x}}(t, \boldsymbol{\tau})_n \triangleq \prod_{j=1}^{n} x(t+\tau_j), \tag{1}$$

where $\boldsymbol{x} \triangleq [x(t+\tau_1)\cdots x(t+\tau_n)]^\dagger$ and $\boldsymbol{\tau} \triangleq [\tau_1 \cdots \tau_n]^\dagger$, and $[\cdot]^\dagger$ denotes matrix transposition. The *cyclic temporal moment function* (CTMF) of order n is defined by the limiting time average

$$\begin{aligned} R_{\boldsymbol{x}}^{\alpha}(\boldsymbol{\tau})_n &\triangleq \lim_{T\to\infty} \frac{1}{T} \int_{v-T/2}^{v+T/2} L_{\boldsymbol{x}}(t, \boldsymbol{\tau})_n \, e^{-i2\pi\alpha t} \, dt \\ &\equiv \, < L_{\boldsymbol{x}}(t, \boldsymbol{\tau})_n e^{-i2\pi\alpha t} >_t, \end{aligned} \tag{2}$$

and is simply the Fourier coefficient associated with the component $e^{i2\pi\alpha t}$ in the time-series $L_{\boldsymbol{x}}(t,\boldsymbol{\tau})_n$. It can be seen that the CTMF is a Fourier coefficient of a moment function because the nth-order fraction-of-time probabilistic moment (the *temporal moment function* [TMF]) associated with the lag product (1) can be expressed as [1][12]

$$R_{\boldsymbol{x}}(t, \boldsymbol{\tau})_n = \sum_{\alpha} R_{\boldsymbol{x}}^{\alpha}(\boldsymbol{\tau})_n \, e^{i2\pi\alpha t}, \tag{3}$$

where the sum is over all real numbers α, called *cycle frequencies*, for which $R_{\boldsymbol{x}}^{\alpha}(\boldsymbol{\tau})_n \not\equiv 0$. The functions (2) and (3) exist and are well-behaved for appropriate models of many time-series including amplitude modulated (AM), pulse-amplitude-modulated (PAM), and phase-shift-keyed modulated (PSK) signals. It is assumed in this paper that the limit (2) exists for all α pointwise in v and in temporal mean-square over v, and is continuous in $\boldsymbol{\tau}$, and that the series (3) converges.

The *temporal cumulant function* (TCF) of order n for the set of time-series translates $\{x(t+\tau_j)\}_{j=1}^{n}$ is defined by

$$C_{\boldsymbol{x}}(t, \boldsymbol{\tau})_n = \sum_{P=\{\nu_k\}_{k=1}^{p}} \left[k(p) \prod_{j=1}^{p} R_{\boldsymbol{x}}(t, \boldsymbol{\tau}_{\nu_j})_{n_j} \right], \tag{4}$$

which is completely analogous to its stochastic-process counterpart in [7]. The sum in (4) is over all distinct partitions P of the set of indices $\{1, 2, \cdots, n\}$, where each partition $\{\nu_k\}_{k=1}^{p}$ has p elements, $1 \le p \le n$, and $k(p) = (-1)^{p-1}(p-1)!$. The vector $\boldsymbol{\tau}_{\nu_j}$ is the vector of n_j lags with indices in the set ν_j. The *cyclic temporal cumulant function* (CTCF) is the Fourier coefficient of the TCF:

$$C_{\boldsymbol{x}}^{\alpha}(\boldsymbol{\tau})_n \triangleq \, < C_{\boldsymbol{x}}(t, \boldsymbol{\tau})_n \, e^{-i2\pi\alpha t} >_t . \tag{5}$$

Combining (3)–(5) reveals that the CTCF is given by the following explicit function of lower-order CTMFs:

$$C_{\boldsymbol{x}}^{\alpha}(\boldsymbol{\tau})_n = \sum_{P} \left[k(p) \sum_{\boldsymbol{\alpha}^\dagger \mathbf{1} = \alpha} \left(\prod_{j=1}^{p} R_{\boldsymbol{x}}^{\alpha_j}(\boldsymbol{\tau}_{\nu_j})_{n_j} \right) \right], \tag{6}$$

where $\mathbf{1} = [1 \cdots 1]^\dagger$, and $\boldsymbol{\alpha} = [\alpha_1 \cdots \alpha_p]^\dagger$. The CTCF was derived in [4] as the solution to the problem of removing from the Fourier coefficient $R_{\boldsymbol{x}}^{\alpha}(\boldsymbol{\tau})_n$ all contributions from Fourier coefficients $R_{\boldsymbol{x}}^{\alpha_j}(\boldsymbol{\tau}_{\nu_j})_{n_j}$ of lower order. This is equivalent to removing from the finite-strength additive sine-wave component of frequency α in the lag product time-series $L_{\boldsymbol{x}}(t, \boldsymbol{\tau})_n$ all contributions from products of sine-wave components in lag products $L_{\boldsymbol{x}}(t, \boldsymbol{\tau}_{\nu_j})_{n_j}$ of lower order that can be obtained by factoring $L_{\boldsymbol{x}}(t, \boldsymbol{\tau})_n$.

The CTMF and the CTCF are not in general integrable due to the presence of sinusoidal components. These components formally result in Dirac deltas in the n-dimensional Fourier transform of the CTCF. However, a reduced-dimension version of the CTCF is absolutely integrable for many time-series of interest[1]

[1] A condition for integrability of the reduced-dimension ver-

and, therefore, it is strictly Fourier transformable [4]. The reduced-dimension CTCF is simply the CTCF associated with the n variables $\{x(t+\tau_j)\}_{j=1}^{n}$ with $\tau_n = 0$. We use the notation

$$\bar{C}_{\boldsymbol{x}}^{\alpha}(\boldsymbol{u})_n \triangleq C_{\boldsymbol{x}}^{\alpha}(\boldsymbol{\tau})_n|_{\tau_i=u_i;\tau_n=0}, \tag{7}$$

where $\boldsymbol{u}$ is an (n-1)-dimensional vector for the reduced-dimension CTCF. The (n-1)-dimensional Fourier transform of (7) is denoted by $\bar{P}_{\boldsymbol{x}}^{\alpha}(\boldsymbol{f}')_n$:

$$\bar{P}_{\boldsymbol{x}}^{\alpha}(\boldsymbol{f}')_n = \int_{-\infty}^{\infty} \bar{C}_{\boldsymbol{x}}^{\alpha}(\boldsymbol{u})_n \, e^{-i2\pi \boldsymbol{u}^{\dagger}\boldsymbol{f}'} \, d\boldsymbol{u}, \tag{8}$$

where $\boldsymbol{f}' = [f_1 \cdots f_{n-1}]^{\dagger}$.

The cyclic polyspectrum is defined as follows. Consider the n complex-demodulate time-series $X_T(t, f_j)$ for $j = 1, \cdots, n$, associated with narrow bandpass filtered versions of $x(t)$, where

$$X_T(t, f) = \int_{t-T/2}^{t+T/2} x(v) e^{-i2\pi f v} \, dv. \tag{9}$$

The limit as $T \to \infty$ of the limiting time-average of the product of these spectral components is called the *spectral moment function* (SMF)

$$S_{\boldsymbol{x}}(\boldsymbol{f})_n \triangleq \lim_{T\to\infty} < \prod_{j=1}^{n} X_T(t, f_j) >_t, \tag{10}$$

and it can be shown that Dirac deltas can be factored out as follows:

$$S_{\boldsymbol{x}}(\boldsymbol{f})_n = \sum_{\alpha} \bar{S}_{\boldsymbol{x}}^{\alpha}(\boldsymbol{f}')_n \delta(\boldsymbol{f}^{\dagger}\boldsymbol{1} - \alpha), \tag{11}$$

where $\delta(\cdot)$ is the Dirac delta function. However, the factor $\bar{S}_{\boldsymbol{x}}^{\alpha}(\boldsymbol{f}')_n$ typically contains additional Dirac deltas. The *spectral cumulant function* (SCF) is given by

$$P_{\boldsymbol{x}}(\boldsymbol{f})_n = \sum_{P=\{\nu_k\}_{k=1}^{p}} \left[k(p) \prod_{j=1}^{n} S_{\boldsymbol{x}}(\boldsymbol{f}_{\nu_j}) \right], \tag{12}$$

where $\boldsymbol{f}_{\nu_j}$ is the vector of frequencies with subscripts in the set ν_j, and it follows from (11) that Dirac deltas can again be factored out:

$$P_{\boldsymbol{x}}(\boldsymbol{f})_n = \sum_{\alpha} \bar{P}_{\boldsymbol{x}}^{\alpha}(\boldsymbol{f}')_n \delta(\boldsymbol{f}^{\dagger}\boldsymbol{1} - \alpha). \tag{13}$$

---

sion of the CTCF is sufficiently rapid decay of the statistical dependence of the time-samples of the time-series, as their separation in time increases. This can be expressed as a condition on the rate of decay of the CTCF as $\boldsymbol{\tau} \to \infty$ in all directions.

Analogous to the definition of the cyclic spectrum (and power spectrum) in [1], the factor $\bar{P}_{\boldsymbol{x}}^{\alpha}(\boldsymbol{f}')_n$ is defined to be the *cyclic polyspectrum* (CP) and is given explicitly by

$$\bar{P}_{\boldsymbol{x}}^{\alpha}(\boldsymbol{f}')_n = \sum_{P=\{\nu_k\}_{k=1}^{p}} k(p) \sum_{\{\alpha_j\}\ni\boldsymbol{\alpha}^{\dagger}\boldsymbol{1}=\alpha} \times \bar{S}_{\boldsymbol{x}}^{\alpha_p}(\boldsymbol{f}'_{\nu_p}) \prod_{j=1}^{p-1} \left\{ \bar{S}_{\boldsymbol{x}}^{\alpha_j}(\boldsymbol{f}'_{\nu_j}) \delta(\boldsymbol{f}_{\nu_j}^{\dagger}\boldsymbol{1} - \alpha_j) \right\} \tag{14}$$

where $\delta_{(.)}$ is the Kronecker delta function. As first shown in [4], the CP is the $(n-1)$-dimensional Fourier transform of the CTCF $C_{\boldsymbol{x}}^{\alpha}(\boldsymbol{\tau})_n$ (cf. (8)). This is a generalization of the Wiener relation between the power spectrum and autocorrelation from second-order stationary time-series (cf. [1]) to nth-order cyclostationary time-series. Within the stochastic-process framework of generally nonstationary processes, it should be called the *cyclic Shiryaev-Kolmogorov* relation [11], which is the generalization of the Wiener-Khinchin relation (cf. [1]). Consequently, we can call the relation (8) for nonstochastic time-series the SKG (Shiryaev-Kolmogorov-Gardner) relation. The SKG relation suggests a time-domain-based method for estimating the CP in addition to the frequency-domain-based method suggested by the definitions (10), (11), and (14). Because the CTCF can be absolutely integrable, the CP—unlike the SMF—contains no Dirac deltas. That is, all Dirac deltas present in the individual terms in (14) cancel.

3 Time-Domain Estimation Methods

To estimate the CP, we first consider estimation of the CTCF $\bar{C}_{\boldsymbol{x}}^{\alpha}(\boldsymbol{u})_n$, which in turn means that we must consider estimates of the CTMFs $R_{\boldsymbol{x}}^{\alpha}(\boldsymbol{\tau}_{\nu_j})_{n_j}$, which we can then combine according to (6). Given the finite segment of data $x(v)$ for $t - T/2 \le v \le t + T/2$, which can be expressed as $x(v)\,\mathrm{rect}[(v-t)/T]$, where

$$\mathrm{rect}(v) \triangleq \begin{cases} 1, & |v| \le 1/2, \\ 0, & \text{otherwise,} \end{cases}$$

we can form the lag product

$$L_{\boldsymbol{x}_T}(v, t, \boldsymbol{\tau})_n \triangleq \prod_{j=1}^{n} x(v+\tau_j)\,\mathrm{rect}\left[\frac{(v+\tau_j-t)}{T}\right],$$

and the CTMF estimate

$$R_{\boldsymbol{x}_T}^{\alpha}(t, \boldsymbol{\tau})_n \triangleq \frac{1}{T}\int_{-\infty}^{\infty} L_{\boldsymbol{x}_T}(v, t, \boldsymbol{\tau})_n \, e^{-i2\pi\alpha v} \, dv$$
$$= \frac{1}{T}\int_{t_l}^{t_u} \prod_{j=1}^{n} x(v+\tau_j)\, e^{-i2\pi\alpha v} \, dv, \tag{15}$$

where $t_l = t - T/2 - \min\{\tau_j\}$ and $t_u = t + T/2 - \max\{\tau_j\}$. Clearly this estimate is strongly consistent (converges pointwise in $\boldsymbol{\tau}$ for every t) since the limit (2) is assumed to exist,

$$\lim_{T\to\infty} R^{\alpha}_{\boldsymbol{x}_T}(t,\boldsymbol{\tau})_n = R^{\alpha}_{\boldsymbol{x}}(\boldsymbol{\tau})_n. \tag{16}$$

The estimator for the cyclic cumulant is (6) with each CTMF replaced by the estimated CTMF (15),

$$C^{\alpha}_{\boldsymbol{x}_T}(t,\boldsymbol{\tau})_n = \sum_{P=\{\nu_k\}_{k=1}^{p}} k(p) \times \left[\sum_{\{\alpha_j\}\ni\boldsymbol{\alpha}^{\dagger}\boldsymbol{1}=\alpha} \left(\prod_{j=1}^{p} R^{\alpha_j}_{\boldsymbol{x}_T}(t,\boldsymbol{\tau}_{\nu_j})_{n_j}\right)\right]. \tag{17}$$

The estimate of the reduced-dimension CTCF, denoted by $\bar{C}^{\alpha}_{\boldsymbol{x}_T}(t,\boldsymbol{u})_n$, is given by (17) with $u_i = \tau_i, i = 1,\cdots,n-1$, and $\tau_n = 0$. Since the limits defining all of the lower-order CTMFs are assumed to exist, the reduced-dimension cyclic cumulant estimator is strongly consistent,

$$\lim_{T\to\infty} \bar{C}^{\alpha}_{\boldsymbol{x}_T}(t,\boldsymbol{u})_n = \bar{C}^{\alpha}_{\boldsymbol{x}}(\boldsymbol{u})_n.$$

Because we know that in the $n = 2$ case the limit of the transform of the estimated reduced-dimension CTCF does not exist unless the CTCF estimate is first tapered [1], we introduce a cumulant tapering window $w_{1/\Delta f}(\boldsymbol{u})$ before transforming in $\boldsymbol{u}$ to obtain an estimate of the CP:

$$\bar{P}^{\alpha}_{\boldsymbol{x}_T}(t,\boldsymbol{f}')_{\Delta f} = \int_{-\infty}^{\infty} w_{1/\Delta f}(\boldsymbol{u})\bar{C}^{\alpha}_{\boldsymbol{x}_T}(t,\boldsymbol{u})_n\, e^{-i2\pi\boldsymbol{u}^{\dagger}\boldsymbol{f}'}\, d\boldsymbol{u}. \tag{18}$$

The tapering window is assumed to be absolutely integrable with approximate width $1/\Delta f$ in each dimension. It is can be taken to be the product of $n-1$ one-dimensional window functions such as

$$w_{1/\Delta f}(\boldsymbol{u}) = \prod_{j=1}^{n-1} \text{rect}(u_j \Delta f).$$

Since (18) can be expressed as the multidimensional convolution

$$\bar{P}^{\alpha}_{\boldsymbol{x}_T}(t,\boldsymbol{f}')_{\Delta f} = W_{\Delta f}(\boldsymbol{f}') \otimes \bar{P}^{\alpha}_{\boldsymbol{x}_T}(t,\boldsymbol{f}')_n, \tag{19}$$

where

$$W_{\Delta f}(\boldsymbol{f}') \triangleq \mathcal{F}\{w_{1/\Delta f}(\boldsymbol{u})\},$$

$$\bar{P}^{\alpha}_{\boldsymbol{x}_T}(t,\boldsymbol{f}')_n \triangleq \mathcal{F}\{\bar{C}^{\alpha}_{\boldsymbol{x}_T}(t,\boldsymbol{u})_n\}, \tag{20}$$

we can propose the alternative time-domain estimation technique that first computes the cyclic cumulant estimate (17), then Fourier transforms to obtain (20), and finally frequency smooths as in (19). This is the generalization from the spectrum to the cyclic polyspectrum of the Wiener-Daniell method (cf. [1]). The estimator (18) is the generalization from the spectrum to the cyclic polyspectrum of the Blackman-Tukey method (cf. [1]).

As long as the tapering window $w_{1/\Delta f}(\cdot)$ has finite support, we can approximate the Riemann integral in (18) to within any desired tolerance with a finite sum and then interchange the order of the limit as $T \to \infty$ and the sum. The resultant limit then exists pointwise in $\boldsymbol{u}$ for every t and equals the reduced-dimension CTCF. Since the Fourier transform of the reduced-dimension CTCF is assumed to exist, then the Fourier transform of its product with the tapering window exists and equals the convolution of the CP with the transformed window. Finally, the effect of this convolution on the CP vanishes in the limit as the width parameter $1/\Delta f$ approaches infinity. Hence the CP estimate (18) is strongly consistent:

$$\bar{P}^{\alpha}_{\boldsymbol{x}}(\boldsymbol{f}')_n = \lim_{\Delta f\to 0}\lim_{T\to\infty} \bar{P}^{\alpha}_{\boldsymbol{x}_T}(t,\boldsymbol{f}')_{\Delta f}. \tag{21}$$

The order of the limits in (21) reflects the fact that $T \gg 1/\Delta f$ is required to obtain an estimate with low temporal variance (cf. [1] for the $n = 2$ case).

4 Frequency-Domain Estimation Methods

A frequency-domain method is one in which the first step is to transform the data as in (9). The motivation for doing this is to utilize the power and speed of the FFT. This approach leads to the well-known computationally efficient methods of Bartlett and Welch (the time-averaged periodogram method) and Wiener and Daniell (the frequency-smoothed periodogram method) in the $n = 2$ case for $\alpha = 0$ (stationary time-series) (cf. [1]). For $\alpha \neq 0$ (for cyclostationary time-series) in the $n = 2$ case, these methods also can be computationally attractive [1][13]. However, for the $n > 2$ case, there is a serious computational impediment. The time-averaging approach follows from (10) and (11) and uses in (14) the estimates

$$\bar{S}^{\alpha}_{\boldsymbol{x}_T}(t,\boldsymbol{f}')_{\Delta t} \triangleq g_{\Delta t}(t) \otimes \bar{S}^{\alpha}_{\boldsymbol{x}_T}(t,\boldsymbol{f}'), \tag{22}$$

where $g(\cdot)$ is a one-dimensional function with approximate width Δt, which should be much greater than the reciprocal $T = 1/\Delta f$ of the spectral resolution width Δf (to obtain low temporal variance, cf. [1] for the $n = 2$ case), and

$$\bar{S}^{\alpha}_{\boldsymbol{x}_T}(t, \boldsymbol{f}') \triangleq \frac{1}{T} X_T(t, \alpha - \mathbf{1}^{\dagger}\boldsymbol{f}') \prod_{k=1}^{n-1} X_T(t, f_k).$$

The frequency-smoothing approach replaces the time-averaged estimates of SMFs used in (14) with the following frequency-smoothed estimates:

$$\begin{aligned} \bar{S}^{\alpha}_{\boldsymbol{x}_T}(t, \boldsymbol{f}')_{\Delta f} &= \mathcal{F}\{\bar{R}^{\alpha}_{\boldsymbol{x}_T}(t, \boldsymbol{u})_n w_{1/\Delta f}(\boldsymbol{u})\} \\ &= W_{\Delta f}(\boldsymbol{f}') \otimes \bar{S}^{\alpha}_{\boldsymbol{x}_T}(t, \boldsymbol{f}'). \end{aligned} \quad (23)$$

The computational impediment in these two methods is most easily illustrated for the $n = 2$ case (for which it can be removed). For $n = 2$, (14) reduces to

$$\bar{P}^{\alpha}_{\boldsymbol{x}}(f_1)_2 = \bar{S}^{\alpha}_{\boldsymbol{x}}(f_1)_2 - \sum_{\alpha_1+\alpha_2=\alpha} \bar{S}^{\alpha_1}_{x} \bar{S}^{\alpha_2}_{x} \delta(\alpha - f_1) \quad (24)$$

where

$$\bar{S}^{\alpha_k}_{x} = \langle x(t) e^{-i2\pi\alpha_k t} \rangle_t, \quad k = 1, 2.$$

For $\alpha = 0$, $\bar{S}^{0}_{\boldsymbol{x}}(f_1)_2$ is the usual power spectrum. The spectral lines in $\bar{S}^{0}_{\boldsymbol{x}}(f_1)_2$ correspond precisely to the terms $|\bar{S}^{\alpha_k}_{x}|^2 \delta(f_1)$ that are subtracted off (because of the assumed temporal mean-square convergence in (2), cf. [1] for this $n = 2$ case). Even though the CP in (24) contains no Dirac deltas, the second-order SMF $\bar{S}^{\alpha}_{\boldsymbol{x}}(f_1)_2$ to be estimated to form the CP does in general contain Dirac deltas. However, the linear combination of the products of SMFs that defines the CP results in all Dirac deltas cancelling. Nevertheless, when estimates of these SMFs must be made, a huge dynamic range for computation is required. This can be avoided in the $n = 2$ case by simply detecting the spectral lines and then estimating the corresponding sine-wave components in the data and subtracting them from the data. The second order SMFs then contain no Dirac deltas. However, for $n > 2$, this approach is not adequate. For example, for third-order SMFs, the sine-wave components present in second-order lag products but absent in the data itself cannot be removed by preprocessing the data.

In general, all frequency-domain-based methods of estimating the CP suffer from the problem of requiring excessive dynamic range for computation because all such methods entail the estimation of SMFs, which in general contain Dirac deltas that are supposed to cancel each other when combined to form the CP estimate. However, if the time-series does not exhibit lower-order cyclostationarity, then the following equivalence holds

$$\bar{S}^{\alpha}_{\boldsymbol{x}}(\boldsymbol{f}')_n \equiv \bar{P}^{\alpha}_{\boldsymbol{x}}(\boldsymbol{f}')_n, \quad \alpha \neq 0. \quad (25)$$

In this case, the CP can be estimated in the frequency domain using either the time-averaging method (22) or the frequency-smoothing method (23).

Estimation of the polyspectrum for stationary stochastic processes is treated in [9], where the frequency-domain technique (23) with $\alpha = 0$ is proposed. The authors of [9] state that this can be done only if the set of frequencies $\boldsymbol{f}$ in (11) is such that no proper subset sums to zero. Since the set itself must sum to $\alpha = 0$ (zero is the only cycle frequency for stationary signals), this condition is equivalent to stating that there can be no partition of the frequencies for which each subset sums to zero. Thus, none of the terms in (12) that contain more than one factor are nonzero. Or, in other words, (25) holds with $\alpha = 0$. In the case of a zero-mean stationary time-series with no additive sine-wave components, the polyspectra of orders 1, 2, and 3 can all be estimated for any $\boldsymbol{f}$ using the frequency-smoothed higher-order periodogram method (23) of Brillinger and Rosenblatt because no lower-order SMF products are nonzero. This explains the common appearance of frequency-domain techniques for estimation of the bispectrum in the literature. However, for estimation of cyclic polyspectra, restriction to specific sets of frequencies $\boldsymbol{f}$ for which equality in (25) holds can be excessively limiting in practice.

5 Examples

In this section we provide examples of the estimation of the CTCF $\bar{C}^{\alpha}_{\boldsymbol{x}}(\boldsymbol{u})_n$ and CP $\bar{P}^{\alpha}_{\boldsymbol{x}}(\boldsymbol{f}')_n$ for a PAM time-series in two corruptive environments for the case of $n = 4$. The two environments are additive white Gaussian noise and cyclostationary interference. The PAM time-series is modeled by

$$x(t) = \sum_{m=\infty}^{\infty} a_m p(t + mT_0), \quad (26)$$

where $\{a_m\}$ is an iid symbol sequence, and $p(t)$ is a pulse function. It can be shown that the CTCF for $x(t)$ is given by

$$\bar{C}^{\alpha}_{\boldsymbol{x}}(\boldsymbol{u})_n = \frac{C_{a,n}}{T_0} \int_{-\infty}^{\infty} p(v) \prod_{j=1}^{n-1} p(v + \tau_j) \, e^{-i2\pi\alpha v} \, dv,$$

for $\alpha = k/T_0$ for all integers k, where $C_{a,n}$ is the nth-order cumulant for the symbol variables $\{a_m\}$.

Estimation for a discrete-time version of (26) was carried out on a digital computer. The PAM signal that was generated is considered to be a sampled version of the continuous model (26), with sampling increment T_s, which we take to be equal to unity for convenience. The symbol interval T_0 is equal to 8, the symbols are binary-valued (± 1), and occur with equal probability. The pulse function is a rectangle with unity height and width equal to T_0.

Estimates (15) of the CTMF are produced by applying an FFT to the lag product time-series, and choosing the bin that corresponds to the desired value of α. Such estimates are combined as in (17) with $n = 4$ (with $t = 0$). The sets of α values $\{\alpha_j\}$ associated with $n_j = 2$ used in (17) is $\{k/T_0 : |k| \leq 3\}$. Partitions with an element ν_j such that $n_j = 1$ are not used to form the CTCF estimate, since the signal is known to have no additive sine-wave components (i.e., $R_{\boldsymbol{x}}^{\alpha_j}(\tau_{\nu_j})_{n_j} \equiv 0$ for $n_j = 1$). Thus, for a given $\boldsymbol{u} = [u_1\ u_2\ u_3]^\dagger$, the value of the reduced-dimension CTCF is computed by (17) with P restricted to the four partitions $\{\{1,2,3,4\}\}, \{\{1,2\},\{3,4\}\}, \{\{1,3\},\{2,4\}\}$, and $\{\{1,4\},\{2,3\}\}$, with $\tau_i = u_i, i = 1,2,3$, and $\tau_4 = 0$.

Several cases are considered. In case (a), 4096 samples of the PAM time-series are used to estimate the CTCF for $\alpha = 1/T_0 = 0.125$ and $|u_i| \leq 2T_0$, for $i = 1,2,3$; in case (b), WGN with spectral height $N_0 = 1$ is added to the signal in case (a) (SNR = 0dB), and the CTCF is again measured; in case (c), case (b) is repeated for a data record length of 16384 samples. Figure 1 shows one-dimensional slices of the resulting CTCF estimates ($|\bar{C}_{\boldsymbol{x}_T}^{\alpha}([0\ 0\ \tau])_4|$) for cases (a)–(c), along with the theoretical CTCF for reference.

Because the CTCF is calculated for a limited range of $\boldsymbol{u}$ (because of the large computational time required for multidimensional data sets), it has already been effectively windowed and a three-dimensional FFT routine can be applied to yield an estimate of the CP. However, to increase the spectral resolution, the CTCF estimates are first zero-padded by a factor of 2, which increases the size of the data set by a factor of 8, and then transformed. One-dimensional slices of the CP estimates ($|\bar{P}_{\boldsymbol{x}_T}^{\alpha}([0\ 0\ f])_{\Delta f}|$) are shown in Figure 2 for cases (a)–(c), along with the theoretical CP for reference. These figures show that the estimators do indeed produce the proper functions, and that the estimates are asymptotically unaffected by the presence of WGN. Next we show that the estimates of the CTCF and the CP are similarly unaffected by the presence of nonGaussian cyclostationary interference, that is, by the presence of another PAM signal.

The signal of interest is the same PAM signal as in the previous example ($T_0 = 8$), and the interferer is the same kind of PAM signal, but with $T_0 = 5$. Both signals have unity power. Thus, the interferer completely spectrally overlaps the signal of interest. The CP was estimated as in Figure 2 for 8192 samples of the PAM + PAM time-series. One-dimensional slices of the theoretical CP for the signal of interest and the estimated CP are shown in Figure 3.

6 Conclusions

We have considered the problem of estimating cyclic polyspectra, which are higher-order statistics of a particular class of nonstationary time-series, namely cyclostationary time-series. The cyclic polyspectrum can be viewed as the Fourier transform of a higher-order cyclic temporal cumulant, or as a cumulant of spectral components of the time-series. In the latter case, the cyclic polyspectrum is seen to be a linear combination of products of spectral moment functions which include Dirac deltas, all of which cancel out in the cyclic polyspectrum. Because of the impracticality of this cancellation using finite precision arithmetic, it is concluded that the cyclic polyspectrum is best measured by first measuring the cyclic temporal cumulant, tapering it, and then transforming it. The noise tolerance of the cyclic polyspectrum, which results from the tolerance to Gaussian noise exhibited by any cumulant (of order greater than 2) combined with the tolerance to any stationary noise exhibited by any cyclic statistic is illustrated qualitatively by considering estimation for a noisy pulse-amplitude-modulated time-series. Also, the tolerance to cyclostationary interference with cycle frequencies unequal to those of the signal of interest is illustrated by considering estimation for a time-series consisting of the sum of two PAM signals.

References

[1] W. A. Gardner. *Statistical Spectral Analysis: A Nonprobabilistic Theory.* Englewood Cliffs, New Jersey: Prentice-Hall 1987.

[2] D. R. Brillinger. "An Introduction to Polyspectra," *Annals of Mathematical Statistics,* Vol. 36, pp. 1351–1374, 1965.

[3] W. A. Gardner. "Spectral Characterization of N-th Order Cyclostationarity," in *Proceedings of the IEEE/ASSP Workshop on Spectrum Estimation,* Rochester, NY, pp. 251–255, 1990.

[4] W. A. Gardner and C. M. Spooner. "Higher-Order Cyclostationarity, Cyclic Cumulants, and Cyclic Polyspectra," in *Proceedings of the International Symposium on Information Theory and its Applications (ISITA),* Honolulu, Hawaii, pp. 355–358, November 1990.

[5] G. M. Jenkins and D. G. Watts. *Spectral Analysis and its Applications,* San Francisco: Holden-Day, 1968.

[6] W. A. Gardner. "Measurement of Spectral Correlation," *IEEE Transactions on Acoustics, Speech, and Signal Processing,* Vol. 34, No. 5, pp. 1111–1123, Oct. 1986.

[7] V. P. Leonov and A. N. Shiryaev. "On a Method of Calculation of Semi-Invariants," *Theory of Probability and Its Applications,* Vol. 4, No. 3, pp. 319–328, 1959.

[8] D. R. Brillinger and M. Rosenblatt. "Computation and Interpretation of *k*–Th Order Spectra," in *Spectral Analysis of Time Series,* B. Harris, ed., New York: Wiley 1967.

[9] D. R. Brillinger and M. Rosenblatt. "Asymptotic Theory of Estimates of *k*–Th Order Spectra," in *Spectral Analysis of Time Series,* B. Harris, ed., New York: Wiley 1967.

[10] W. A. Gardner. "Two Alternative Philosophies for Estimation of the Parameters of Time-Series," *IEEE Transactions on Information Theory,* Vol. 37, No. 1, pp. 216–218, Jan. 1991.

[11] A. N. Shiryaev. "Some Problems in the Spectral Theory of Higher-Order Moments – I," *Theory of Probability and Its Applications,* Vol. 5, No. 3, pp. 265–284, 1960.

[12] W. A. Gardner and W. A. Brown. "Fraction-of-Time Probability for Time-Series that Exhibit Cyclostationarity," *Signal Processing,* Vol. 23, No. 3, pp. 273–292, June 1991.

[13] R. S. Roberts, W. A. Brown, and H. H. Loomis. "Computationally Efficient Algorithms for Cyclic Spectral Analysis," *IEEE Signal Processing Magazine,* Vol. 8, No. 2, pp. 38–49, April 1991.

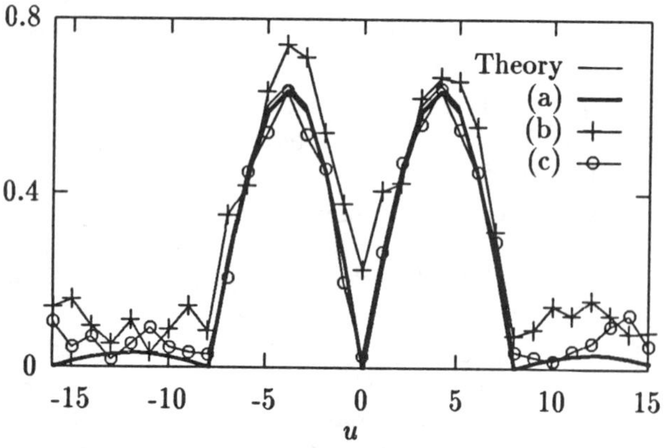

Figure 1: CTCF estimates for binary PAM in WGN (see text).

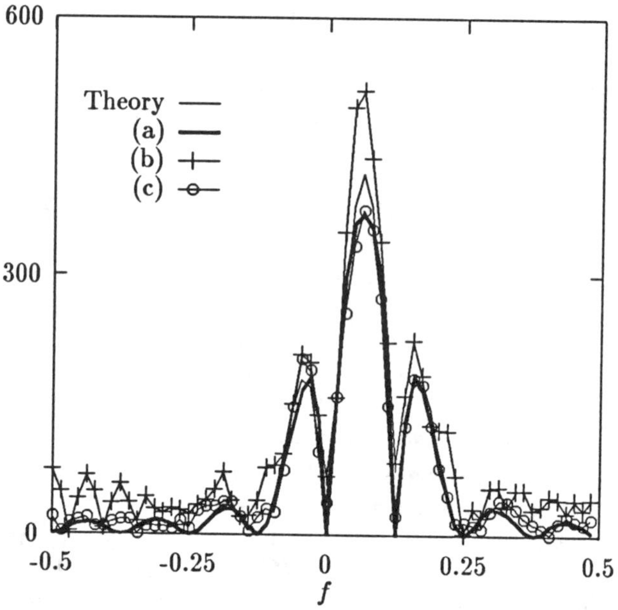

Figure 2: CP estimates for binary PAM in WGN (see text).

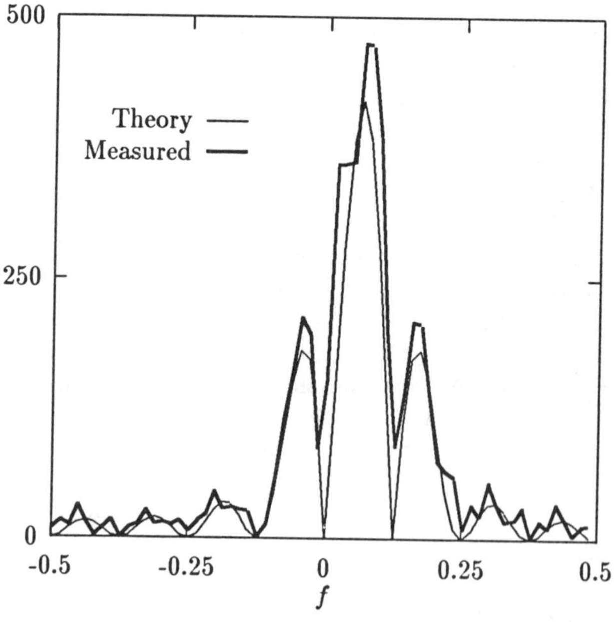

Figure 3: The CP for PAM + PAM (see text).

Polyspectral Analysis of (Almost) Cyclostationary Signals : LPTV System Identification and Related Applications

Georgios B. Giannakis and Amod V. Dandawate
Department of Electrical Engineering
University of Virginia
Charlottesville, VA, 22903-2442

Abstract

Polyspectral estimators are proposed for (almost) cyclostationary signals and are shown to be consistent and asymptotically normal. These estimators are employed for identification of linear (almost) periodically time-varying systems. Both non-parametric and parametric approaches are described for input-output and output only identification. Statistical analysis of nonstationary signals with missing observations is treated and tests are developed for checking the presence of cycle frequencies. Finally, frequency estimation and detection of coupling are addressed in cyclic domain without resorting to phase randomization. All the methods are proven to be insensitive to stationary noise and employ consistent single record estimators.

1 Introduction – Problem statement

Most signals encountered in real life are time-varying and require tools which are appropriate for nonstationary signal processing. For example, consider transmission through an AGN channel. The received signal is given by

$$y(t) = x(t)g(t) + v(t), \tag{1}$$

where $x(t)$ is a 0-mean nonGaussian stationary message signal, $g(t)$ is a *deterministic* carrier and $v(t)$ is zero mean additive Gaussian channel noise (perhaps *nonstationary*) with *unknown* spectrum. This model is appropriate for several AM signal transmission schemes including BPSK, PAM and DSBSC; further it is useful for modeling missing observations (see e.g., [19]). In (1) $y(t)$ is *nonstationary* even if $v(t)$ is stationary due to the presence of the deterministic $g(t)$. It is of interest to separate and estimate the statistics of $x(t)$ from those of $v(t)$ and $g(t)$, given a finite amount of data from a *single record* of $y(t)$. One is faced with a similar proposition when estimating the sinusoidal frequencies and detecting presence of coupling from a *single* record of the observed signal

$$y(t) = \sum_{k=0}^{P-1} \rho_k e^{j\omega_k t} + v(t), \tag{2}$$

where ρ_k and ω_k are the amplitudes and frequencies of the sinusoids respectively. The sinusoids are deterministic, which makes $y(t)$ nonstationary.

Another problem of interest is identification of linear (almost) periodically time-varying systems (LPTV); see [1] for theory on almost periodic functions. Let $h(t;\tau)$ be the impulse response of an LPTV system with nonGaussian input $x(t)$ and observed output

$$y(t) = \sum_{\tau} h(t;\tau)x(t-\tau) + v(t). \tag{3}$$

We wish to estimate $h(t,\tau)$ using a *single record* of its input and or output data. Notice that (3) reduces to (1) when $h(t;\tau) = g(t)\delta(\tau)$.

From the description of these problems it seems desirable to employ a tool that separates stationary from nonstationary signals, Gaussian from nonGaussian signals, and is capable of handling deterministic, stationary and nonstationary signals on a *common framework*. Further, for this tool to be useful in practice it is important that the relevant *nonstationary* statistics be *consistently* estimated using a *single record* of the available *noisy* data. Higher-order statistics (HOS) are capable of separating nonGaussian signals from Gaussian and have received considerable interest in *stationary* signal processing lately [15]. The theory of *2nd-order* (almost) periodic statistics has also been developing[11], [19], [17], [18], [7], [6], [16] and it has been established that they are capable of separating *2nd-order* statistics of stationary and nonstationary signals. In this paper we *combine* the two theories and develop polyspectra based noise insensitive algorithms for (almost) periodic signals and treat deterministic, stationary and nonstationary signals on a common framework. We develop consistent and asymptotically normal polyspectral estimators using *single* record. Time-domain estimation of higher-order statistics was addressed in [3], [10], [4], [5] (see also [14] for 2nd-order results on consistent time- and frequency-domain estimation). The proposed estimators are employed for HOS-based identification, spectral analysis with missing observations, simultaneous frequency estimation and detection of coupling, and finally for developing tests for cyclostationarity.

The problem of missing observations and amplitude modulation (AM) was studied by Parzen in [19] for estimating *2nd-order* statistics of the message signal, under *noise-free* conditions. Consistent single record observations were developed under the assumption that the message signal was Gaussian. We also treat the stochastic carrier case in (1) which was not addressed in [19].

HOS-based algorithms developed for dealing with the sinusoids in noise problem require availability of *multiple* records for statistical estimation [22], [20] as opposed to the single record estimation of this paper. Also, several approaches introduce phase randomization to force the sinusoids to be wide-sense stationary, contrary to the approach adopted here (see also [24]).

A 2nd-order, input-output LPTV system identification approach was reported in [7]. These approaches although insensitive to any stationary noise are susceptible to (almost) periodic Gaussian noise of the same cycle as that of the signal. On the other hand, the polyspectra based algorithms of this paper are insensitive *any* stationary noise as well as (almost) periodic Gaussian noise.

2 NS-polyspectral estimation

Let $x(t)$ be a nonstationary (NS) signal and let $m_{kx}(t;\underline{\tau})$ be its kth-order moment defined as:

$$m_{kx}(t;\underline{\tau}) = E\{x(t)x(t+\tau_1)\cdots x(t+\tau_{k-1})\} \quad (4)$$

where $\underline{\tau} \stackrel{\triangle}{=} [\tau_1, ..., \tau_{k-1}]$. We next define a NS-cumulant of $x(t)$. Let us use the following notation

$$m_x(t;\mathrm{l}) \stackrel{\triangle}{=} E\{x(t+\tau_{l_0})\cdots x(t+\tau_{l_{k-1}})\}; \quad \mathrm{l} \stackrel{\triangle}{=} (l_0, \ldots, l_{k-1}). \quad (5)$$

To express the kth-order *cumulant* of a process $x(t)$ consider the set of all partitions $\mathcal{P} = \{\nu_1, \ldots, \nu_P\}$ of $\{0, 1, \ldots, k-1\}$ [2], where the pth-partition $\nu_p \stackrel{\triangle}{=} \{\mathrm{l}_1, \ldots, \mathrm{l}_{N_p}\}$, with the l's consisting of some of $\{0, 1, \ldots k-1\}$ depending upon the partition. We use the convention that $\nu_1 = \{\mathrm{l}_1 = (0,\ldots,k-1)\}$ and $\nu_P = \{\mathrm{l}_1 = 0, \ldots, \mathrm{l}_k = k-1\}$. With $\tau_0 = 0$, the kth-order cumulant is defined as

$$c_{kx}(t;\tau_1,\ldots,\tau_{k-1}) \stackrel{\triangle}{=} \sum_{p=1}^{P} (-1)^{N_p}(N_p-1)! \prod_{m_p=1}^{N_p} m_x(t;\mathrm{l}_{m_p}). \quad (6)$$

We will be interested in a class of NS-process whose statistics accept a Fourier series representation; i.e.,

$$\begin{aligned} c_{kx}(t;\underline{\tau}) &= \sum_{\alpha\in\mathcal{A}} \mathcal{C}_{kx}(\alpha;\underline{\tau})e^{j\alpha t} \\ \mathcal{C}_{kx}(\alpha;\underline{\tau}) &= \lim_{T\to\infty}\frac{1}{T}\sum_{t=0}^{T-1} c_{kx}(t;\underline{\tau})e^{-j\alpha t} \quad (7) \\ \mathcal{A} &\stackrel{\triangle}{=} \{\alpha : \mathcal{C}_{kx}(\alpha;\underline{\tau}) \not\equiv 0\}, \end{aligned}$$

where α is the cyclic frequency. This class of signals may be referred to as a class of *persistently nonstationary* signals [3], [10], [4], [5]. The statistics in (7) may be called *cyclic-cumulants* or cyclic-HOS as an extension to the terminology used by Gardner [6] for 2nd-order statistics.

Let $S_{kx}(t;\underline{\omega})$ denote the *time-varying polyspectrum* of $x(t)$ defined as

$$S_{kx}(t;\underline{\omega}) = \sum_{\underline{\tau}} c_{kx}(t;\underline{\tau})e^{-j\underline{\omega}'\underline{\tau}}. \quad (8)$$

Using (7) and (8) it follows that,

$$\begin{aligned} S_{kx}(t;\underline{\omega}) &= \sum_{\alpha\in\mathcal{A}} \mathcal{S}_{kx}(\alpha;\underline{\omega})e^{j\alpha t} \\ \mathcal{S}_{kx}(\alpha;\underline{\omega}) &= \lim_{T\to\infty}\frac{1}{T}\sum_{t=0}^{T-1} S_{kx}(t;\underline{\omega})e^{-j\alpha t} \quad (9) \end{aligned}$$

We refer to $\mathcal{S}_{kx}(\alpha;\underline{\omega})$ as the Cyclic Polyspectrum (CP).

• Properties [5]:

(i) Stationary signal: If $x(t)$ is stationary then $c_{kx}(t;\underline{\tau}) = c_{kx}(\underline{\tau})$ and hence from (7)

$$\begin{aligned} \mathcal{C}_{kx}(\alpha;\underline{\tau}) = c_{kx}(\underline{\tau})\eta(\alpha) \equiv 0 \quad \alpha \neq 0(mod 2\pi); \\ \mathcal{S}_{kx}(\alpha;\underline{\omega}) = S_{kx}(\underline{\omega})\eta(\alpha) \equiv 0 \quad \alpha \neq 0(mod 2\pi); \quad (10) \end{aligned}$$

where $\eta()$ is the discrete (Kronecker) delta train.

(ii) Gaussian noise insensitivity : We now demonstrate the ability of the NS-HOS to *separate* Gaussian and non-Gaussian signals. Let $y(t) = x(t) + v(t)$ where $x(t)$ is a persistently NS signal and $v(t)$ is additive Gaussian noise of *unknown* covariance perhaps NS (AGN/UC) independent of $x(t)$. We are interested in extracting the statistics of $x(t)$ from those of $y(t)$. Using the insensitivity of cumulants to AGN/UC [5] it follows that

$$\begin{aligned} c_{ky}(t;\underline{\tau}) \equiv c_{kx}(t;\underline{\tau}); \quad \mathcal{C}_{ky}(\alpha;\underline{\tau}) \equiv \mathcal{C}_{kx}(\alpha;\underline{\tau}) \\ S_{ky}(t;\underline{\omega}) \equiv S_{kx}(t;\underline{\omega}); \quad \mathcal{S}_{ky}(\alpha;\underline{\omega}) \equiv \mathcal{S}_{kx}(\alpha;\underline{\omega}). \quad (11) \end{aligned}$$

(iii) Stationary noise insensitivity: Let $y(t) = x(t) + v(t)$ with $x(t)$ a persistently NS-signal and $v(t)$, *any* stationary noise independent of $x(t)$. Using property (i) and $c_{ky}(t;\underline{\tau}) = c_{kx}(t;\underline{\tau}) + c_{kv}(t;\underline{\tau})$, it follows that :

$$\mathcal{C}_{ky}(\alpha;\underline{\tau}) = \mathcal{C}_{kx}(\alpha;\underline{\tau}); \quad \mathcal{S}_{ky}(\alpha;\underline{\omega}) = \mathcal{S}_{kx}(\alpha;\underline{\omega}) \quad \alpha \neq 0 \quad (12)$$

Properties (ii) and (iii) depict the noise-insensitivity of NS-HOS to AGN/UC perhaps NS and to *any* stationary noise. These properties will be exploited in the remaining sections of this paper to derive noise-insensitive algorithms. For implementation of these algorithms in practice it is necessary to define consistent estimators for the NS-HOS. Strongly consistent and asymptotically normal estimators for the *time-domain* NS-HOS were given in [5].

In this paper we present results on NS-polyspectral estimators. This work shows that the pioneering work of Brillinger-Rosenblatt [2] can be extended to (almost) cyclostationary-polyspectral estimation. Our aim is to develop *periodogram*-type estimators for the CP $\mathcal{S}_{kx}$ of the *(almost) periodic* process $x(t)$. Let $X_T(\omega)$ be the finite-Fourier transform of $x(t)$ defined as

$$X_T(\omega) = \sum_{t=0}^{T-1} x(t)e^{-j\omega t}. \quad (13)$$

• Second-order : Let the cyclic 2nd-order periodogram be

$$I_{2x}^{(T)}(\alpha;\omega) \stackrel{\triangle}{=} \frac{1}{T}X_T(\omega)X_T(\alpha-\omega). \quad (14)$$

The asymptotic properties of this periodogram can be summarized in the following theorem,

<u>Theorem A</u>: If $x(t)$ is a zero-mean persistently NS signal then the 2nd-order periodogram given by (14) is asymptotically unbiased with asymptotic covariance given as follows: [1]

$$\begin{aligned}\lim_{T\to\infty}\ & cov\{I_{2x}^{(T)}(\alpha;\lambda), I_{2x}^{(T)}(\beta;\mu)\} \\ &= \mathcal{S}_{2x}(\lambda-\mu;-\mu)\mathcal{S}_{2x}(\alpha-\beta-\lambda+\mu;\mu-\beta) \\ &+ \mathcal{S}_{2x}(\alpha-\lambda-\mu;-\mu)\mathcal{S}_{2x}(\lambda+\mu-\beta;\mu-\beta).\end{aligned} \tag{15}$$

Variance expressions for the inconsistent cyclic periodogram of *periodically* time-varying processes were derived by Hurd in [12] and are apparently in some minor disagreement with the expressions of this paper. Theorem A deals with covariance expressions for *different* cycle frequencies and for *different* frequencies. As in the stationary case the 2nd-order periodogram is an inconsistent estimator of NS-spectra. It is however interesting to note that if λ and μ are such that $\left[\{(\lambda-\mu) or (\alpha-\beta-\lambda+\mu)\} and \{(\lambda-\beta+\mu) or (-\mu+\alpha-\lambda)\}\right] \notin \mathcal{A}$ then the covariance expression of Theorem A tends to zero asymptotically. Loosely speaking this amounts to saying that is λ and μ are not separated by a cycle frequency then the 2nd-order periodograms at these two frequencies are asymptotically independent, however, if λ and μ are indeed separated by a cycle frequency then they are correlated, which quantifies the spectral correlation claims for the APC-signals. Similar results hold for the general kth-order case. Remembering that the periodogram ordinates are unbiased and asymptotically independent for certain frequencies, we *average* such frequencies around the neighborhood of an ordinate of interest and appeal to a law of large numbers type argument to achieve a local ensemble average of frequency in that neighborhood. By assuming that the NS-polyspectra are smooth we may then argue that the local average yields the desired NS-polyspectra ordinate asymptotically.

• <u>*kth-order*</u>: We now construct consistent NS-polyspectra estimators for the kth-order case ($k \geq 2$). Averaging of NS-periodogram ordinates can be performed using a continuous weight function $W(u_0,\ldots,u_{k-1})$ that lives on $u_0+\cdots+u_{k-1}=\alpha$ and is such that

(W1) $\int\cdots\int_{-\infty}^{\infty} W(\underline{u})\delta(\sum_{l=0}^{k-1} u_l)du_0\cdots du_{k-1}=1$.

(W2) $W(u_0,\ldots,u_{k-1}) = W(-u_0,\ldots,-u_{k-1})$

(W3) and bandwidth B_T satisfies $B_T \to 0$ and $B_T^{k-1}T\to\infty$ as $T\to\infty$. Define the *principal manifold* $\mathcal{P}(\underline{u}) \triangleq \{u_0,\ldots,u_{k-1} : \sum_j u_j \in \mathcal{A}\}$ and the *proper submanifold* $\mathcal{P}_s(\underline{u}) \triangleq \{u_0,\ldots,u_{k-1} : \sum_{j\in J} u_j \in \mathcal{A}$ for some nonempty proper subsets J of $[0,\ldots,k-1]\}$. Let $\phi(\underline{u})$ be an indicator function that is non-zero and equal to unity only when $\underline{u} \in \mathcal{P}(\underline{u})\cap\bar{\mathcal{P}}_s(\underline{u})$ where $^-$ denotes a complement. Let $W_T(u_0,\ldots,u_{k-1}) \triangleq B_T^{k-1}W(B_T^{-1}u_0,\ldots,B_T^{-1}u_{k-1})$. Define the *smoothed* NS-periodogram

$$\begin{aligned}\hat{\mathcal{S}}_{kx}^{(T)}&(\alpha;\omega_0,\ldots,\omega_{k-1}) \\ &\triangleq T^{-k+1}\sum_{s_0,\ldots,s_{k-1}} W_T(\omega_0-\frac{2\pi s_0}{T},\ldots\omega_{k-1}-\frac{2\pi s_{k-1}}{T}) \\ &\qquad \phi(\frac{2\pi s_0}{T},\ldots,\frac{2\pi s_{k-1}}{T})I_{kx}^{(T)}(\alpha;\frac{2\pi s_0}{T},\ldots,\frac{2\pi s_{k-1}}{T})\end{aligned} \tag{16}$$

where the kth-order NS-periodogram is defined as $I_{kx}^{(T)}(\alpha;\lambda_0,\ldots,\lambda_{k-1}) = \frac{1}{T}X_T(\lambda_0)\cdots X_T(\lambda_{k-1})$ and lives only on $\mathcal{P}(\underline{\lambda})$.

We now summarize the asymptotic properties of the kth-order smoothed NS-periodogram in the following theorem:
<u>Theorem B</u>: Let $x(t)$ be a persistently nonstationary process so that, **(A1)** $\sum_{\underline{\tau}}|\tau_l c_{kx}(t;\underline{\tau})| < \infty$, $l = 0,\ldots,k-1$ **(A2)** $\sum_{\underline{\tau}}|\tau_l \mathcal{C}_{kx}(\alpha;\underline{\tau})| < \infty$, $l=0,\ldots,k-1$ and let the window function W, along with **(W1)**-**(W3)**, satisfies the following condition : **(W4)** $|W(\underline{u})| \leq A(1+[\sum_{l=0}^{k-1}u_l^2]^{1/2})^{-(k+\epsilon-1)}$ and $|\frac{\partial}{\partial u_l}W(\underline{u})| \leq A(1+[\sum_{l=0}^{k-1}u_l^2]^{1/2})^{-(k+\epsilon-1)} l=0,\ldots,k-1$. Then $\hat{\mathcal{S}}_{kx}^{(T)}$ as defined by (16) is consistent and asymptotically normal with covariance

$$\begin{aligned}\lim_{T\to\infty}\ & B_T^{k-1}T cov\{\hat{\mathcal{S}}_{kx}^{(T)}(\chi;\omega_0,\ldots,\omega_{k-1}), \\ &\hat{\mathcal{S}}_{kx}^{(T)}(\psi;\mu_0,\ldots,\mu_{k-1})\} = \sum_P\sum_{\gamma_0,\ldots,\gamma_{k-1}}\eta_{\sum_i\gamma_i-\chi-\psi} \\ &\eta_{\omega_0+\mu_{P^{-1}(0)}+\gamma_{P^{-1}(0)}}\cdots\eta_{\omega_{k-1}+\mu_{P^{-1}(k-1)}-\gamma_{P^{-1}(k-1)}} \\ &\mathcal{S}_{2x}(\gamma_0;\mu_{P^{-1}(0)}+\gamma_{P^{-1}(0)})\cdots \\ &\mathcal{S}_{2x}(\gamma_{k-1};\mu_{P^{-1}(k-1)}+\gamma_{P^{-1}(k-1)}) \\ &\int\cdots\int_{-\infty}^{\infty}W(\tau_0,\ldots,\tau_{k-1})W(-\tau_{P^{-1}(0)},\ldots,-\tau_{P^{-1}(0)}) \\ &\eta(\sum_i\tau_i-\chi)d\tau_0\cdots d\tau_{k-1}\end{aligned}$$

where the summation over P includes only the terms in the indecomposable partitions taken from the following table

$$\begin{matrix}\omega_0 & \omega_1 & \cdots & \omega_{k-1} \\ \mu_0 & \mu_1 & \cdots & \mu_{k-1}\end{matrix}, \tag{17}$$

Although the results presented here are for auto-polyspectra of 1-d processes, analogous results hold for m-d cross-polyspectra that are useful for multivariate and multidimensional processes.

3 Nonparametric identification

Let $h(t;\tau)$, be an LPTV system with period T_0, input $x(t)$ and output $y(t)$ so that,

$$y(t) = \sum_{\tau=-\infty}^{\infty} h(t;\tau)x(t-\tau), \tag{18}$$

with $h(t+nT_0;\tau) \equiv h(t;\tau)\ \forall n,\tau,t$. The system accepts a Fourier series expansion w.r.t. t; thus,

$$h(t;\tau)=\sum_\beta \mathcal{H}(\beta;\tau)e^{j\beta t} \quad \mathcal{H}(\beta;\tau)\triangleq \lim_{T\to\infty}\frac{1}{T}\sum_{t=0}^{T-1}h(t;\tau)e^{-j\beta t}. \tag{19}$$

Let *noisy* measurements of the I/O data be available,

$$\bar{y}(t) = y(t)+v_y(t)\ ; \quad \bar{x}(t) = x(t)+v_x(t). \tag{20}$$

Following assumptions are made,

[1] For Proofs see Univ. of Virginia Rep. No. 192444B/EE91/126, Oct., 1991.

($\mathcal{A}1$) $\sum_{\tau} \sup_{t} |h(t;\tau)| < \infty$

($\mathcal{A}2$) $x(t)$ is an NLNS process as in [10, Thm. 1]

($\mathcal{A}3$) $v_x(t)$ and $v_y(t)$ are zero-mean stationary or NS AGN/UC, independent of both $x(t)$ and $y(t)$.

- I/O identification of LPTV systems

If $x(t)$ is stationary then it can be shown [4] that

$$\mathcal{H}(\alpha; -\omega_1 - \cdots - \omega_{k-1}) = \mathcal{S}_{\bar{y}\bar{x}}(\alpha; \underline{\omega}) / \mathcal{S}_{k\bar{x}}(0; \underline{\omega}), \quad (21)$$

providing a noise-insensitive polyspectra-based algorithm. I/O identification for LPTV systems using *2nd-order* cyclostationary statistics have been considered in [6]. However estimators for the ideal statistics required by the 2nd-order algorithm have not been considered. Also these approaches are susceptible corruption of the input $x(t)$ by cyclostationary AGN of *unknown* covariance and of the same cycle frequency as that of the input as opposed to the HOS statistics of this paper.

- Output-only identification

If $h(t;\tau)$ models a seasonal time-series described by a sinusoidal periodicity, for example, then assuming that $x(t)$ is i.i.d., considerably simplifies the expression for H and yields a simple frequency domain, *output only* identification scheme; e.g., for $k = 3$, $\mathcal{S}_{3y}(\alpha; \omega_1, \omega_2) = \gamma\ \mathcal{H}(\alpha; \omega_1 - \alpha)\mathcal{H}(\alpha; \omega_2 - \alpha)\mathcal{H}(\alpha; -\omega_1 - \omega_2)$. In this case, the log-bispectrum methods of [21] may be used for recovering $\mathcal{H}(\alpha; \omega)$ from $\mathcal{S}_{3y}(\alpha; \omega_1, \omega_2)$.

Third-order cumulant based algorithms for I/O and output-only identification can be found in [3].

- TV-MA estimation – a linear method

Let $y(t) = \sum_{\tau=0}^{q} h(t;\tau)x(t-\tau)$, represent an MA($q$) system h with i.i.d. input $x(t)$. With $k = 3$, it can be shown that , [4],

$$h(t;\tau) = c_{3y}(t; q, \tau) / c_{3y}(t; q, 0). \quad (22)$$

After estimating c_{3y}, one can use the linear equations of (22) to solve for $h(t;\tau)$. Consistent HOS-estimators of Section 2 and [3] should be employed for the required statistics. Using the asymptotic normality of these HOS-estimators one can do performance evaluation for these algorithms.

4 Parametric identification

Let $y(t)$ be generated by the following time varying ARMA(p_t,q_t) model,

$$\sum_{i=0}^{p_t} a(i)y(t-i) = \sum_{i=0}^{q_t} b(i)\epsilon(t-i), \quad (23)$$

where, a_t's and b_t's represent the PTV AR and MA coefficients, p_t and q_t represent the TV AR and MA orders and finally $\epsilon(t)$ represents i.i.d. noise.

The *noisy* measurement of $y(t)$ is given as,

$$\bar{y}(t) = y(t) + v_y(t) \quad (24)$$

where $v_y(t)$ is zero-mean stationary or NS AGN/UC, independent of $y(t)$.

The following "TV normal equations" can be derived for $\tau_1 \geq q_t, \ldots, \tau_{k-1} \geq q_t$ in the cumulant domain,

$$\sum_{i=0}^{p_t} a_t(i) cum\{\bar{y}(t-i), \bar{y}(t-\tau_1), \ldots, \bar{y}(t-\tau_{k-1})\} \equiv 0. \quad (25)$$

Using the strongly consistent estimators of [3] the required cumulants, $cum\{\bar{y}(t-i), \bar{y}(t-\tau_1), \ldots, \bar{y}(t-\tau_{k-1})\}$, are estimated and (25) is solved for a_t's. Using the estimated a_t's $\bar{y}$ is inverse filtered to obtain an AR compensated process. Using the methods outlined in [3], on the AR compensated process, the MA part can be estimated. Other approaches using cumulant or polyspectral matching criteria are also possible for estimating the MA part.

A special case of the system identification results when a_t, b_t, p_t and q_t are periodic in t [25]. The resulting ARMA system is periodically time varying and has an important application in recasting multichannel systems into simple one dimensional periodic ARMA models, thus considerably reducing the required memory storage for multichannel system identification.

It is well known that HOS preserve phase information and 2nd-order statistics are "phase blind." Therefore all the HOS-algorithms described in this and the previous section have the capability of preserving *time-varying phase* information and hence are capable of identifying *noncausal* and *nonminimum phase* systems. This comes as an additional merit in favor of cyclic HOS over 2nd-order statistics when phase information is important.

5 Missing Observations/AM signals

Let $x(t)$ be a series of interest and suppose that $y(t)$ is available instead of $x(t)$ such that

$$y(t) = g(t)x(t) + v(t), \quad (26)$$

where $g(t)$ is a known function modulated by 0-mean non-Gaussian $x(t)$ and $v(t)$ is additive noise. The goal is to estimate statistics of $x(t)$ from a given stretch of data $\{y(t)\}_{t=0}^{T-1}$ from a *single* record.

The problem of spectral analysis with missing observations and amplitude modulation has been of interest in communication and radar signal processing (see e.g., [19] and references therein). In these situations $g(t)$ is *deterministic* and known; e.g., communication carrier, or a model for systematically missing observations; it is assumed that $x(t)$ is zero-mean stationary and $v(t)$ is also zero-mean independent of $x(t)$. The 2nd-order cumulant of $y(t)$ is given as

$$c_{2y}(t;\tau) = c_{2x}(\tau)g(t)g(t+\tau) + c_{2v}(t;\tau). \quad (27)$$

In [19] $v(t)$ was assumed to be absent. Further, *assuming* that $x(t)$ is Gaussian it was shown under certain regularity conditions that,

$$\lim_{T\to\infty} \frac{1}{T} \sum_{t=0}^{T-1} y(t)y(t+\tau) \quad (28)$$

$$= c_{2x}(\tau) \lim_{T\to\infty} \frac{1}{T} \sum_{t=0}^{T-1} g(t)g(t+\tau),$$

where the equality is to be interpreted in the MSS. From (28) it is clear that if $g(t)$ is known then c_{2x} can be consistently estimated from the sample correlation of $y(t)$. However when $\lim_{T\to\infty} \frac{1}{T} \sum_{t=0}^{T-1} g(t)g(t+\tau) = 0$ this estimator is not useful. We now present an estimator for c_{2x} which imposes looser conditions on $g(t)$ and treats the approach [19] as a special case. Also the estimator proposed in this paper can be shown to be strongly consistent under certain general conditions and are also insensitive to the noise $v(t)$. Let $\hat{\mathcal{C}}_{2y}^{(T)}$ denote an estimator of $\mathcal{C}_{2y}$ defined as

$$\hat{\mathcal{C}}_{2y}^{(T)}(\alpha;\tau) = \frac{1}{T} \sum_{t=0}^{T-1} y(t)y(t+\tau)e^{-j\alpha t}. \tag{29}$$

Notice that the estimator employed in [19] is a particular case of $\hat{\mathcal{C}}_{2y}^{(T)}(\alpha;\tau)$ for $\alpha = 0$. Under fairly general conditions it can be shown [3] that for a large class of *nonlinear nonstationary* processes

$$\lim_{T\to\infty} \hat{\mathcal{C}}_{2y}^{(T)}(\alpha;\tau) \overset{a.s.}{=} \mathcal{C}_{2y}(\alpha;\tau) = c_{2x}(\tau)\mathcal{M}_{2g}(\alpha;\tau) + \mathcal{C}_{2v}(t;\tau). \tag{30}$$

If $v(t)$ is *any* stationary noise then

$$\lim_{T\to\infty} \hat{\mathcal{C}}_{2y}^{(T)}(\alpha;\tau) \overset{a.s.}{=} \mathcal{C}_{2y}(\alpha;\tau) = c_{2x}(\tau)\mathcal{M}_{2g}(\alpha;\tau) \quad \alpha \neq 0.$$

Hence, a consistent estimator of $c_{2x}(\tau)$ is given as

$$\hat{c}_{2x}(\tau) = \hat{\mathcal{C}}_{2y}^{(T)}(\alpha;\tau)/\mathcal{M}_{2g}(\alpha;\tau), \tag{31}$$

provided that $\mathcal{M}_{2g}(\alpha;\tau) \neq 0$ for this value of α. Notice that $\hat{\mathcal{C}}_{2y}^{(T)}$ is insensitive to $v(t)$. Variance of $\hat{c}_{2x}(\tau)$ can be obtained in terms of the variance of $\hat{\mathcal{C}}_{2y}^{(T)}$ and $\mathcal{M}_{2g}$. Variance expressions for estimators of the type $\hat{\mathcal{C}}_{2y}^{(T)}$ can be found in [5]. The following interesting example illustrates the main features of this section.

Example: Let $g(t) = e^{j\omega_0 t}$ therefore $\mathcal{M}_{2g}(\alpha;\tau) = e^{j\omega_0\tau}\delta(\alpha - 2\omega_0)$ therefore $e^{-j\omega_0\tau}\hat{\mathcal{C}}_{2y}^{(T)}(2\omega_0;\tau)$ is a consistent estimator of $c_{2x}(\tau)$. The estimator of [19] cannot be used in this case since $\mathcal{M}_{2g}(0;\tau) \equiv 0$. Also the approach of this paper is insensitive to presence of *any* stationary $v(t)$ since $\mathcal{C}_{2v}(\alpha;\tau) \equiv 0, \ \alpha \neq 0$.

Estimation of cumulants: The aim is to estimate the cumulant statistics of $x(t)$ using finite stretch of data $y(t)$. Cumulants of $x(t)$ can be used for *non-minimum phase* ARMA identification when $x(t)$ is an ARMA process. As pointed out in Section 3, 2nd-order statistics cannot be used to identify non-minimum phase systems because they do not preserve phase information.

It follows from (26) that

$$c_{ky}(t;\underline{\tau}) = g(t)g(t+\tau_1)\cdots g(t+\tau_{k-1})c_{kx}(\underline{\tau}) + c_{kv}(t;\underline{\tau}) \tag{32}$$

$$\mathcal{C}_{ky}(\alpha;\underline{\tau}) = \mathcal{M}_{kg}(t;\underline{\tau})c_{kx}(\underline{\tau}) + \mathcal{C}_{kv}(\alpha;\underline{\tau}) \tag{33}$$

If $v(t)$ is AGN/UC then $\mathcal{C}_{kv} \equiv 0$; also if $v(t)$ is stationary then the following holds true

$$\mathcal{C}_{ky}(\alpha;\underline{\tau}) = \mathcal{M}_{kg}(t;\underline{\tau})c_{kx}(\underline{\tau}), \quad \alpha \neq 0. \tag{34}$$

Analogous to the 2nd-order case one can define an estimator $\hat{c}_{kx}$ of c_{kx} by replacing $\mathcal{C}_{ky}$ with $\hat{\mathcal{C}}_{ky}$; thus

$$\hat{c}_{kx}(\underline{\tau}) = \hat{\mathcal{C}}_{ky}(\alpha;\underline{\tau})/\mathcal{M}_{kg}(t;\underline{\tau}). \tag{35}$$

Variance expressions can be derived for this estimator by using the results of [5]. Identification of ARMA models when $x(t)$ is an ARMA process and the available observations are missing will be considered in future. Also, recovery of signal correlation as a function of SNR with specific type of $g(t)$'s can be considered along the lines of [19].

The case where $g(t)$ is stochastic is considered in [5] (see also [23] for stationary processes).

6 Test for cyclostationarity

In order to efficiently exploit the ability of (almost) periodic HOS to separate a (almost) periodic process from a stationary process one must have knowledge about the cycle frequencies of the (almost) periodic process. Often, in practice, this knowledge is not available and therefore it is important to develop tests for presence of cycle frequencies. Since $\mathcal{S}_{kx}(\alpha;\underline{\omega})$ is non-zero only if α is a cycle frequency, one can a define a kth-order cyclostationarity test with the following detection statistic

$$\mathcal{D}_{kx}^{(T)}(\alpha;\underline{\omega}) = \frac{\left|\hat{\mathcal{S}}_{kx}^{(T)}(\alpha;\underline{\omega})\right|^2}{var\{\hat{\mathcal{S}}_{kx}^{(T)}(\alpha;\underline{\omega})\}}, \tag{36}$$

where the numerator and the denominator of (36) can be *consistently* estimated using the estimators of Section 2. One can plot $\mathcal{D}_{kx}^{(T)}$ as a function of α for fixed $\underline{\omega}$'s and check for the "peaks" to test for presence of cyclic frequency α. Other possibility of averaging the test statistic over different $\underline{\omega}$ and over different α's exist and may have applications in detection and classification of nonGaussian cyclostationary signals in Gaussian noise.

Using the *asymptotic normality* and the *variance* expressions of the cumulant spectral estimators one can do performance evaluation using χ^2 type tests. Second-order tests for presence of cycle frequencies were also reported in [13], however, consistency and other asymptotic properties of the statistic were not derived. Also the variance normalization used in their statistics seems to be inappropriate for deriving statistical tests.

7 Coupling and frequency estimation

Let $x(t)$ be a nonstationary signal given as

$$x(t) = s(t) + v(t) = \sum_{k=0}^{P-1} \rho_k e^{j\omega_k t} + v(t) \tag{37}$$

with $0 < \omega_0, \ldots, \omega_{k-1} < \pi$, $v(t)$ is stochastic. The aim is to be able to estimate ω_k's and detect coupling i.e., determine whether $\omega_k = \omega_l + \cdots + \omega_n \quad k, l, n \in [0, P-1]$

from a finite stretch of data $\{x(t)\}_{t=0}^{T-1}$. We will glance at the cyclic-mean, cyclic-correlation and cyclic third-order moment as tools for frequency estimation and detection of coupling for sinusoids in noise. It should be noted that we *do not* impose any phase randomization assumption on the sinusoids.

Cyclic-mean

Let $v(t)$ have constant mean although it can be nonstationary; e.g., $v(t) = y(t)e^{j\lambda_0 t}$ with $E\{y(t)\} = 0$. Now,

$$m(t) \triangleq E\{x(t)\} = s(t) + m_v \tag{38}$$

where m_v is the mean of $v(t)$.

If $\mathcal{M}(\alpha) \triangleq \lim_{T\to\infty} \frac{1}{T}\sum_{t=0}^{T-1} m(t)e^{-j\alpha t}$ represents the cyclic mean then

$$\mathcal{M}(\alpha) = \lim_{T\to\infty} \frac{1}{T}\sum_{t=0}^{T-1} s(t)e^{-j\alpha t} + m_v \eta(\alpha) \tag{39}$$

Using (37) and (39) it follows that for $\alpha \neq 0$

$$\mathcal{M}(\alpha) = \begin{cases} \rho_k & \alpha = \omega_k \\ 0 & \text{else} \end{cases} . \tag{40}$$

The cyclic mean $\mathcal{M}(\alpha)$ can be consistently estimated using $\frac{1}{T}\sum_{t=0}^{T-1} x(t)e^{-j\alpha t}$. Therefore, regardless of whether $v(t)$ is stationary or nonstationary as long as it has constant mean the sinusoidal frequencies and the ρ's can be consistently estimated. However, this method does not explicitly indicate presence of coupling. Coupling is important when one wishes to detect nonlinearity.

Cyclic-correlation

We now demonstrate that second-order coupling can be detected using cyclic-correlations. We assume that $v(t)$ is stationary. It can be shown that for $\alpha \neq 0$, $\alpha \neq \lambda$

$$\mathcal{S}_{2x}(\alpha;\lambda) = \sum_{k_0,k_1=0}^{P-1} \rho_{k_0}\rho_{k_1}\eta(\omega_{k_0} + \omega_{k_1} - \alpha)\eta(\lambda - \omega_{k_1}) + m_v \sum_{k=0}^{P-1} \rho_k \eta(\alpha - \omega_k - \lambda)\delta(\alpha - \omega_k). \tag{41}$$

The first term in the r.h.s. of (41) peaks for $\alpha = \omega_{k_0} + \omega_{k_1}$ and $\lambda = \omega_{k_1}$. By locating these peaks we can solve for ω_{k_1} from the value of λ and then solve for $\omega_{k_0} = \alpha - \omega_{k_1}$ as long as it is not zero. The second term indicates the presence of coupling. This term peaks for $\alpha = \omega_k$ at $\lambda = 0$. Therefore if and only if there exists second order coupling i.e., for some k, k_0, k_1, $\omega_k = \omega_{k_0} + \omega_{k_1}$ then there will be peaks at $(\alpha = \omega_{k_0} + \omega_{k_1}, \lambda = \omega_{k_1})$ *and* $(\alpha = \omega_{k_0} + \omega_{k_1}, \lambda = 0)$, provided that $m_v \neq 0$. Thus the cyclic-correlation can be used to simultaneously estimate frequencies and detect coupling provided $v(t)$ is non-zero mean. However, if $m_v = 0$ this method can be used only to estimate frequencies and not for detecting coupling. Next we describe a method for simultaneous frequency estimation and detection of third-order coupling that works even when $v(t)$ is zero-mean.

Cyclic-third-order moment

Third-order coupling can be detected using cyclic-third-order moments. It can be shown that for $\alpha \neq 0$, $\alpha \neq \lambda$ $\lambda_1 \neq \alpha$ and $\lambda_2 \neq \alpha$,

$$\mathcal{S}_{3x}(\alpha;\lambda_1,\lambda_2) = \sum_{k_0,k_1,k_2=0}^{P-1} \alpha_{k_0}\alpha_{k_1}\alpha_{k_2}\eta(\lambda_1 - \omega_{k_1}) \eta(\lambda_2 - \omega_{k_1})\eta(\omega_{k_0} + \omega_{k_1} + \omega_{k_2} - \alpha) + \eta(\lambda_1 + \lambda_2)\sum_{k=0}^{P-1} \alpha_k \eta(\alpha - \omega_k) S_{2v}(\lambda_2) \tag{42}$$

The first term in the r.h.s. of (42) peaks at $(\alpha = \omega_{k_0} + \omega_{k_1} + \omega_{k_2}, \lambda_1 = \omega_{k_1}, \lambda_1 = \omega_{k_2})$ and the second term peaks at $(\alpha = \omega_k, \lambda_1 = -\lambda_2)$. Therefore if one avoids $\lambda_1 + \lambda_2 = 0$, with $\alpha \neq 0$, $\lambda_1 \neq \alpha$ and $\lambda_2 \neq \alpha$ and searches for the peaks, these will be due to the first term of (42) and one can solve for ω_{k_1} and ω_{k_2} from the values of λ_1 and λ_2 and then solve for $\omega_{k_0} = \alpha - \omega_{k_1} - \omega_{k_2}$. Further if $\mathcal{S}_{3x}(\alpha;0,0)$ is non-zero then there exists a frequency $\omega_k = \omega_{k_0} + \omega_{k_1} + \omega_{k_2}$ indicating presence of third-order coupling.

References

[1] A.S. Besicovitch, Almost periodic functions, Dover publications, 1954.

[2] D.R. Brillinger and M. Rosenblatt, in Spectral Analysis of Time Series, ed. B. Harris, pp. 153-232, Wiley, 1967.

[3] A.V. Dandawate and G.B. Giannakis, Proceedings of 25th Conf. on Information Sciences and Systems, pp. 976-983, March, 1991; also Univ. of Virginia Rep. No. 192444B/EE91/125, May 1991.

[4] A.V. Dandawate and G.B. Giannakis, Proceedings of Higher-Order Spectral Analysis Workshop, France, July 10-12, 1991.

[5] A.V. Dandawate and G.B. Giannakis, "Ergodicity and asymptotic normality for cumulant estimators of nonstationary signals," Submitted to IEEE Trans.-IT, (See also Univ. of Virginia Rep. No. 192444B/EE91/125, June 1991.

[6] W.A. Gardner Statistical spectral analysis: A nonprobabilistic theory, Prentice Hall, 1988.

[7] W.A. Gardner Introduction to random processes, second ed., McGraw-Hill, N.Y., 1990.

[8] W.A. Gardner, IEEE Signal Processing Magazine, pp. 14-36, April 1991.

[9] W.A. Gardner and C.M. Spooner, International Symposium on Information Theory and Applications, Hawaii, November 1990.

[10] G.B. Giannakis and A.V. Dandawate, Proceedings of Higher-Order Spectral Analysis Workshop, France, July 10-12, 1991.

[11] E.G. Gladyšev, Soviet Mathematics, Vol., 2, pp. 385-388, 1961.

[12] H.L. Hurd, IEEE Trans. IT, Vol., 35, pp. 350-359, March, 1989.

[13] H.L. Hurd and N.L. Gerr, Journal of time series analysis, pp. 337-350, vol. 12, No. 4, 1991.

[14] H.L. Hurd and J. Leskow, Tech. Rep. # 308 and #330, Dept. of statistics, Univ of North Carolina, Chapel Hill, March 1991.

[15] IEEE Trans. on AC, Vol. 35, pp. 4-56, January 1990; and IEEE-ASSP, Vol. 38, pp. 1238-1317, July 1990

[16] V.Z. Marmarelis, Proc. of the IEEE, pp. 841-842, July 1981; see also IEEE Trans. on Information Theory, pp. 937-938, November 1983.

[17] H. Ogura, IEEE Trans. on Information Theory, pp. 143-149, March 1971.

[18] M. Pagano, The Annals of Statistics, pp. 1310-1317, 1978.

[19] E. Parzen, SANKHYĀ : The Indian journal of statistics : Series A, pp. 383-392, 1963.

[20] M.R. Raghuveer and C.L. Nikias, IEEE Trans. on ASSP, pp. 1213-1230, Oct., 1985

[21] M. Rangoussi and G.B. Giannakis, IEEE-CAS, p. 281, 1991.

[22] A. Swami and J.M. Mendel, IEEE Trans. on SP, pp. 1099-1109, May 1991.

[23] A. Swami, 5th ASSP Workshop on Spectral Estimation and Modeling, pp. 212-216, October 1990.

[24] A. Swami, Proceedings of Higher-Order Spectral Analysis Workshop, France, July 10-12, 1991.

[25] A.V. Vecchia, Water Resources Bulletin, p. 721, oct., 1985.

Wideband Source Modeling and Localization: A HOS-based Approach

Sanyogita Shamsunder and Georgios B. Giannakis†
Department of Electrical Engineering
University of Virginia
Charlottesville, VA 22903-2442

Abstract

A polyspectra based solution to the nonparametric multichannel system identification problem is proposed. As an application, the problems of wideband source bearing estimation and identification are considered. Simultaneous detection and parameter estimation of wideband sources arriving at a sensor array is performed using higher order statistics (HOS). Source detection and consistent parameter estimation is approached via a polyspectra-based information theoretic criterion. The proposed methods besides being insensitive to additive Gaussian noise are not restricted to minimum phase equivalents and are asymptotically optimal for non Gaussian signals with unknown distributions.

1 Introduction

In applications such as, passive sonar, seismic and biomedical signal processing, the output of N_x sensors is the set of vector observations $\{\mathbf{x}(t)\}$ which result from the superposition of delayed far-field source signals $s_k(t)$, corrupted by additive noise, and can be written as:

$$\mathbf{x}(t) = P(t;\theta) * \mathbf{s}(t) + \mathbf{v}(t) \quad (1)$$

$$x_i(t) = \sum_{k=1}^{N_s} s_k(t - \tau_i(\theta_k)) + v_i(t), \quad (2)$$

where, $s_k(t)$ is the kth source signal and the delay undergone by it in traveling to the ith sensor is $\tau_i(\theta_k)$; $v_i(t)$ is the Gaussian sensor noise. Given the noisy observations, the task usually is to estimate the $\tau_i(\theta_k)$ which convey information about θ_k. Our *objective* in this paper is twofold: **(i)** to estimate the source bearings and, **(ii)** to model and characterize the source signals themselves.

If the signals are wideband with identical bandwidth BW and T is large, the Fourier Series coefficients of $\mathbf{x}(t)$ in (1) can be approximately represented as (see e.g., [16]),

$$\mathcal{X}(\omega_n) = \mathcal{P}(\omega_n;\theta)\mathcal{S}(\omega_n) + \mathcal{V}(\omega_n), \; \omega_n = 2\pi n/T. \quad (3)$$

[0]† Research supported by LABCOM contract 5-25254

The ith column of the $N_x \times N_s$ matrix $\mathcal{P}(\omega_n;\theta)$ is $\mathbf{p}_i(\omega_n) \triangleq [1, \; \exp(-j\omega_n\tau_2(\theta_i)) \; \cdots \; \exp(-j\omega_n\tau_{N_s}(\theta_i))]'$. Using the Kronecker product $\otimes$ notation, the spectral vector and the spectral matrix are given by:

$$\mathbf{c}_{2x}(\omega) \triangleq E[\mathcal{X}(\omega)\otimes\mathcal{X}^*(\omega)] = \mathcal{P}_2(\omega)\mathbf{c}_{2s}(\omega) + \mathbf{c}_{2v}(\omega) \quad (4)$$

$$\mathcal{C}_{2x}(\omega) \triangleq E[\mathcal{X}(\omega)\mathcal{X}^H(\omega)] = \mathcal{P}(\omega)\mathcal{C}_{2s}(\omega)\mathcal{P}^H(\omega) + \mathcal{C}_{2v}(\omega), \quad (5)$$

where, in (4) $*$ denotes conjugation, $\mathcal{P}_2(\omega) \triangleq \mathcal{P}(\omega;\theta) \otimes\mathcal{P}(\omega;\theta)$. The dependence of $P(t), \mathcal{P}(\omega)$ on θ is implicit and has been dropped for convenience. Since higher than second-order cumulants of Gaussian processes vanish, the 3rd-order cumulant of $\{\mathbf{x}(t)\}$ defined as:

$$\mathbf{c}_{3x}(i_1,i_2) \triangleq E[\mathbf{x}(t)\otimes\mathbf{x}(t+i_1)\otimes\mathbf{x}(t+i_2)] \quad (6)$$

$$= \sum_{\tau_0,\tau_1,\tau_2} P(\tau_0)\otimes P(\tau_1)\otimes P(\tau_2)\mathbf{c}_{3s}(i_1+\tau_0-\tau_1, i_2+\tau_0-\tau_2) \quad (7)$$

"removes" the AGN $\mathbf{v}(t)$. Higher-order cumulants can be similarly defined [11]. With $\mathcal{X}(\omega)$ as in (3), the bispectrum of $\mathbf{x}(t)$ is defined as [12, eq.(9)],

$$\mathbf{c}_{3x}(\omega_l,\omega_m) \triangleq E[\mathcal{X}(\omega_l)\otimes\mathcal{X}(\omega_m)\otimes\mathcal{X}^*(\omega_l+\omega_m)]$$

$$= \mathcal{P}_3(\omega_l,\omega_m)\mathbf{c}_{3s}(\omega_l,\omega_m) \quad (8)$$

$$= \sum_{i_1=-\infty}^{\infty}\sum_{i_2=-\infty}^{\infty} \mathbf{c}_{3x}(i_1,i_2)\exp[-j(i_1\omega_l + i_2\omega_m)]. \quad (9)$$

Here, $\mathcal{P}_3(\omega_l,\omega_m) \triangleq \mathcal{P}(\omega_l)\otimes\mathcal{P}(\omega_m)\otimes\mathcal{P}^*(\omega_l+\omega_m)$ and $\mathbf{c}_{3s}(\omega_l,\omega_m)$ is the input (cross-) bispectrum vector.

In practice, sample averages replace ensemble averages in (4-9). The sample estimates of (8-9) are consistent when smoothed in the frequency domain with a window of support $[-M, M]$ with $M \to \infty$ and $M^2/T \to 0$ as $T \to \infty$ [1].

We approach the source modeling problem in two different ways using the 2nd- and higher-order spectra of $\mathbf{x}(t)$. In the *first* method, we model the source signals $\mathbf{s}(t)$ as outputs of a MIMO linear system $H(t)$ driven by non Gaussian i.i.d.vector input $\mathbf{w}(t)$. The

1058-6393/91 $01.00 © 1991 IEEE

overall model for the sensor outputs hence is:

$$\mathbf{x}(t) = P(t) * \mathbf{s}(t) + \mathbf{v}(t) \quad (10)$$
$$= P(t) * H(t) * \mathbf{w}(t) + \mathbf{v}(t) \quad (11)$$
$$= D(t) * \mathbf{w}(t) + \mathbf{v}(t) = \mathbf{u}(t) + \mathbf{v}(t) \quad (12)$$

where $D(t)$ is the composite response which combines the source model as well as the propagation mechanism. We employ higher-order spectra(polyspectra) to identify the system transfer function $\mathcal{D}(\omega)$ following which, we estimate the DOAs using a HOS-based MUSIC spectrum. Having obtained the DOAs, one can specify $\mathcal{P}(\omega)$ and hence $\mathcal{H}(\omega) = \mathcal{P}^{-1}(\omega)\mathcal{D}(\omega)$. When the time-delays are not integers, one can model $P(t)$ appropriately using sinc functions [10].

In the second method, we propose a HOS-based simultaneous detection and parameter estimation algorithm which is based on an information theoretic criterion. The classical AIC assumes Gaussian observations with Gaussian noise of known covariance. The proposed polyspectra-based AIC(MDL) is applicable to non Gaussian observations observed in noise. The estimator is consistent, efficient and is developed using the asymptotic normality of the sample bispectral estimator.

The advantages of employing HOS is that we can identify the system uniquely (within a permutation matrix [5, 12]) unlike second-order statistics (SOS) based methods which only yield $H(t)$ within a unimodular matrix. This enables us to analyze the source signals and to perform deconvolution. The added advantage here is the immunity of HOS-based methods to additive Gaussian noise. A bicepstrum-based method for modeling was proposed in [2] for the special case of a single-input, three-output system which in fact, can be handled using the existing SISO approaches. The proposed algorithms are linear unlike the rather complex, nonlinear algorithms reported in [14].

The assumptions made with regards to the models (1)-(2) are:
(A1) $\{\mathbf{s}(t)\}$ are non Gaussian, stationary, zero-mean, with finite moments and identical bandwidths.
(A2) $\{\mathbf{v}(t)\}$ is Gaussian, zero-mean with unknown covariance structure.
(A3) $\mathcal{P}(\omega)$ is an $N_x \times N_s$ matrix, $N_x > N_s$.

2 Source modeling and localization

In this section, we address the problem of both source modeling and bearing estimation. In section 2.1, we model the source signals as multivariate linear processes and propose a frequency-domain reconstruction algorithm which yields $\mathcal{D}(\omega)$ as a solution of linear equations. In section 2.2, we assume the sources to be modeled as multichannel ARMA (MC-ARMA) processes. The ARMA parameters are estimated via cumulant based linear equations. Subsequent to the estimation of $\mathcal{D}(\omega)$, the DOAs (or $\mathcal{P}(\omega)$) are obtained using SOS or HOS based methods. In this direction, we propose a frequency-domain trispectral MUSIC-like objective function (see also [7]).

2.1 Nonparametric modeling of $\mathcal{D}(\omega)$

Let $\mathbf{u}(t)$ be the N_x-variate output of a discrete-time LTI, FIR model $\{D(t)\}_{t=0}^{Q}$, driven by a N_s-variate non Gaussian, i.i.d. (space and time), zero-mean process $\mathbf{w}(t)$ ($N_s \leq N_x$). We assume that the components of $\mathbf{w}(t)$ given by $w_i(t)$ have finite and non-zero 2nd-, 3rd- and 4th-order cumulants. The output $\mathbf{u}(t)$ may be corrupted by additive Gaussian noise $\mathbf{v}(t)$, (of unknown covariance, but independent of $\mathbf{w}(t)$) so that $\mathbf{x}(t) = \mathbf{u}(t) + \mathbf{v}(t)$ is observed. The ith output is then given by:

$$x_i(t) = u_i(t) + v_i(t) = \sum_{k=1}^{N_s}\sum_{\tau=0}^{Q} d_{ik}(\tau) w_k(t-\tau) + v_i(t). \quad (13)$$

Using the multi-linearity property of the cumulant spectra [1], and with $\omega_3 = 0$, the cross-trispectrum (4th-order spectrum) of $x_a(t)$ and $x_b(t)$ is,

$$C_{abbb}(\omega_1, \omega_2, 0) = \sum_{i=0}^{N_s} d_{bi}(\omega_1) d_{bi}(\omega_2) d_{bi}^*(\omega_1 + \omega_2) d_{ai}^*(0), \quad (14)$$

where, in (14), we have assumed w.l.o.g. that the 4th-order cumulant of $w_i(t)$, $\gamma_{4wi} \equiv 1 \, \forall \, i$. In vector form, (14) is,

$$C_{abbb}(\omega_1, \omega_2, 0) = [d_{a1}(0) d_{a2}(0) \cdots d_{aN_s}(0)]^* \times \underbrace{\begin{bmatrix} d_{b1}(\omega_1) d_{b1}(\omega_2) d_{b1}^*(\omega_1 + \omega_2) \\ d_{b2}(\omega_1) d_{b2}(\omega_2) d_{b2}^*(\omega_1 + \omega_2) \\ \vdots \\ d_{bp}(\omega_1) d_{bp}(\omega_2) d_{bN_s}^*(\omega_1 + \omega_2) \end{bmatrix}}_{\mathbf{d}_{3b}(\omega_1, \omega_2)}. \quad (15)$$

Collecting equations like (15) for $a = 1, \ldots, N_x$, we have

$$\mathcal{D}^*(0)\mathbf{d}_{3b}(\omega_1, \omega_2) = [C_{1bbb}(\omega_1, \omega_2, 0), \cdots, C_{N_x bbb}(\omega_1, \omega_2, 0)]', \quad (16)$$

where $\mathcal{D}(0) = \sum_{i=0}^{Q} D(i)$ is assumed to be of full rank N_s. Defining the trispectral vector on the r.h.s. of (16) as $\mathcal{C}_{bbb}(\omega_1, \omega_2, 0)$, we can solve for the triple products as:

$$\mathbf{d}_{3b}(\omega_1, \omega_2) = \mathcal{D}^{\dagger}(0)\mathcal{C}_{bbb}(\omega_1, \omega_2, 0) = \mathcal{G}_b(\omega_1, \omega_2). \quad (17)$$

We know from [1] that the nonredundant region of support for the trispectrum is $\mathcal{S}_4 = \{\omega_1, \omega_2, \omega_3 : |\omega_3| \leq \omega_2 \leq \omega_1 \leq \pi, 2\omega_1 + \omega_3 \leq \pi\}$. If $\mathcal{C}_{bbb}$ is sampled at $T = 2^{\nu} \geq Q$ equispaced FFT points $\{2\pi k/T\}_{k=0}^{T-1}$, we have that $\omega = 2\pi k/T$, which means that $\mathcal{S}_4 = \{k_1, k_2, k_3 : |k_3| \leq k_2 \leq k_1 \leq Q, 2k_1 + k_3 = Q\}$. The r.h.s. of (15) is essentially the auto bispectrum of $d_{bi}(t)$. Therefore, using the results of [8], it follows that there are at least $M_\lambda \triangleq [(T/4)+1]^2$ nonredundant lags in $\mathcal{S}_4$ (there will be more, but we consider here only a subset of $\mathcal{S}_4$), of which $[(T/2)+1]$ lie on the $k_2 = 0$ frequency axis. Note that

$$d_{bl}(k_1)d_{bl}(k_2)d^*_{bl}(k_1+k_2) = \mathcal{G}_{bl}(k_1,k_2). \quad (18)$$

where r.h.s is the lth element of $\mathcal{G}_b(k_1,k_2)$. We will make use of this relation in the next sections for amplitude and phase reconstruction.

• *Phase Reconstruction*

Evaluating (18) on the frequency grid $\{2\pi k/T\}_{k=0}^{T-1}$ and equating the phase on both sides we have:

$$\phi_{3bi}(k_1,k_2) = \phi_{bi}(k_1) + \phi_{bi}(k_2) - \phi_{bi}(k_1+k_2), \quad (19)$$

for $0 \le k_1, k_2 \le T-1$, where, $\phi_{3bi}(k_1,k_2) = \arg(\mathcal{G}_{bi}(k_1,k_2))$ and $\phi_{bi}(k) = \arg(d_{bi}(k))$. Concatenating (19) for $M_\phi = M_\lambda - [(T/2)-1]$ nonredundant trispectrum points, yields the following system of linear equations to be solved for $\bar{\phi}_{bi} = [\phi_{bi}(1), \ldots, \phi_{bi}(T/2-1)]'$:

$$\underbrace{\begin{bmatrix} 2 & -1 & 0 & 0 & \cdots & 0 \\ 1 & 1 & -1 & 0 & \cdots & 0 \\ 1 & 0 & 1 & -1 & \cdots & 0 \\ \vdots & \vdots & \vdots & \vdots & \vdots & \vdots \\ 0 & 0 & 0 & 2 & \cdots & -1 \end{bmatrix}}_{\mathcal{A}_\phi} \underbrace{\begin{bmatrix} \phi_{bi}(1) \\ \phi_{bi}(2) \\ \vdots \\ \phi_{bi}(T/2-1) \end{bmatrix}}_{\bar{\phi}_{bi}} = \underbrace{[\phi_{3bi}(1,1)\ \phi_{3bi}(1,2)\ \cdots\ \phi_{3bi}(T/4,T/4)]'}_{\bar{\phi}_{3bi}}. \quad (20)$$

Collecting (20) for $b = 1,\ldots,N_x$; $i = 1,\ldots,N_s$, we obtain the super-matrix of unknowns $\mathbf{\Phi}$ as:

$$\mathbf{\Phi} = [\mathcal{A}'_\phi \mathcal{A}_\phi]^\dagger \mathcal{A}'_\phi \mathbf{\Phi}_3 \quad (21)$$

where, $\mathbf{\Phi} \triangleq [\bar{\phi}_{11}\bar{\phi}_{12}\cdots\ \bar{\phi}_{1N_s}\cdots\bar{\phi}_{N_xN_s}]$ and $\mathbf{\Phi}_3$ is defined compatibly with $\mathbf{\Phi}$ using $\bar{\phi}_{3bi}$ instead of $\bar{\phi}_{bi}$.

In practice, sample trispectral phase esimates $\hat{\phi}_{3bi}$ replace ϕ_{3bi} and hence $\hat{\mathbf{\Phi}}$ is obtained as the LS solution. Note that only the principal argument vector $\bar{\mathbf{\Phi}}_3$ of the true trispectrum phase $\mathbf{\Phi}_3$ can be obtained from the data. Phase-unwrapping algorithms proposed in [8] may be used to get the true phase $\mathbf{\Phi}$ from $\bar{\mathbf{\Phi}}$.

• *Log-magnitude Reconstruction*

Along the same lines, if we consider the log-amplitude on both sides of (18), we obtain:

$$ln|\mathcal{G}_{bi}(k_1,k_2)| = ln|d_{bi}(k_1)| + ln|d_{bi}(k_2)| + ln|d_{bi}(k_1+k_2)|, \quad (22)$$

or $\lambda_{3bi}(k_1,k_2) = \lambda_{bi}(k_1)+\lambda_{bi}(k_2)+\lambda_{bi}(k_1+k_2)$ where $\lambda_{3bi}(k_1,k_2) \triangleq ln|\mathcal{G}_{bi}(k_1,k_2)|$ and the other symbols are similarly defined.

Concatenating (22) for $(k_1,k_2) \in M_\lambda$ results in :

$$\underbrace{\begin{bmatrix} 3 & 0 & 0 & 0 & \cdots & 0 \\ 1 & 2 & 0 & 0 & \cdots & 0 \\ \vdots & \vdots & \vdots & \vdots & \vdots & \vdots \\ 1 & 0 & 0 & 0 & \cdots & 0 \\ 0 & 0 & 0 & 2 & \cdots & 1 \end{bmatrix}}_{\mathcal{A}_\lambda} \underbrace{\begin{bmatrix} \lambda_{bi}(1) \\ \lambda_{bi}(2) \\ \vdots \\ \lambda_{bi}(T/2-1) \end{bmatrix}}_{\bar{\lambda}_{bi}} = \underbrace{[\lambda_{3bi}(1,1)\ \lambda_{3bi}(1,2)\ \cdots\ \lambda_{3bi}(T/4,T/4)]'}_{\bar{\lambda}_{3bi}}. \quad (23)$$

where, $\mathcal{A}_\lambda$ is full rank [8]. Collecting (23) for $b = 1,\ldots,N_x; i = 1,\ldots,N_s$, we again obtain the following LS solution for $\mathbf{\Lambda} \triangleq [\bar{\lambda}_{11}\bar{\lambda}_{12}\cdots\bar{\lambda}_{21}\cdots\bar{\lambda}_{N_xN_s}]$: $\mathbf{\Lambda} = [\mathcal{A}'_\lambda \mathcal{A}_\lambda]^\dagger \mathcal{A}'_\lambda \mathbf{\Lambda}_3$, where $\mathbf{\Lambda}_3$ is again defined like $\mathbf{\Phi}_3$ but with $\bar{\lambda}_{3bi}$ instead of $\bar{\phi}_{3bi}$.

• *Estimating D(0)*

In the above development, we have assumed that $\mathcal{D}(0)$, the mean of the impulse response matrices is known. We now see how it can be estimated using 2nd and 3rd-order auto- and cross-spectra of $\mathbf{x}(t)$. Consider the cross-bispectrum of $x_a(t), x_b(t)$ which we define as:

$$C_{abb}(\omega_1,\omega_2) \triangleq E[\mathcal{X}_a(\omega_1)\mathcal{X}_b(\omega_2)\mathcal{X}^*_b(\omega_1+\omega_2)] = \sum_{i=1}^{p} d_{ai}(\omega_1)d_{bi}(\omega_2)d^*_{bi}(\omega_1+\omega_2)\gamma_{3w}. \quad (24)$$

With $\omega_1 = 0$ and $a = 1,\ldots,N_x$, we have:

$$\underbrace{\begin{bmatrix} C_{1bb}(\omega,0) \\ \vdots \\ C_{N_xbb}(\omega,0) \end{bmatrix}}_{\mathbf{c}_{bb}(\omega,0)} = \mathcal{D}(0) \underbrace{\begin{bmatrix} |d_{b1}(\omega)|^2 \\ \vdots \\ |d_{bp}(\omega)|^2 \end{bmatrix}}_{\mathbf{d}_{2b}(\omega)}. \quad (25)$$

Collecting equations for $b = 1,\ldots,N_s$ results in:

$$\mathcal{C}_{3x}(\omega,0) = \mathcal{D}(0)\mathbf{d}_2(\omega). \quad (26)$$

Post-multiplying (26) by Γ_2, the N_s-element vector of 1's and observing from (4) that $\mathbf{d}_2(\omega)\Gamma_2$ is in fact $\mathbf{c}_{2x}(\omega)$ we get:

$$\mathcal{C}_3(\omega) \triangleq \mathcal{C}_{3x}(\omega,0)\Gamma_2 = \mathcal{D}(0)\mathbf{d}_2(\omega)\Gamma_2 = \mathcal{D}(0)\mathbf{c}_{2x}(\omega). \quad (27)$$

Note that the entries of $\mathcal{C}_3(\omega)$ are a subset of the entries of $\mathbf{c}_{3x}(\omega,0)$ as defined in (9). Concatenating (27) for $\omega = \omega_k, k = 1,\ldots,N_s$ and defining $\bar{\mathcal{C}}_3 = [\mathcal{C}_3(\omega_1)\ \mathcal{C}_3(\omega_2)\ \ldots\ \mathcal{C}_3(\omega_{N_x})]$ and $\bar{\mathcal{C}}_2 = [\mathbf{c}_{2x}(\omega_1)\ \mathbf{c}_{2x}(\omega_2)\ \ldots\ \mathbf{c}_{2x}(\omega_{N_x})]$ we have:

$$\mathcal{D}(0) = \bar{\mathcal{C}}_3\bar{\mathcal{C}}'_2[\bar{\mathcal{C}}'_2\bar{\mathcal{C}}_2]^\dagger, \quad (28)$$

which is the LS solution.

Having estimated the amplitude and phase, the impulse response matrices $D(t)$ may be obtained as the inverse FFT of $\mathcal{D}(\omega)$. Performance evaluation can be done on the lines of [8].

2.2 Parametric modeling of $\mathcal{D}(\omega)$

If $\mathbf{x}(t)$ can be modeled as MC-ARMA(n_a, n_b) process so that $\mathcal{D}(z) = \mathcal{A}^{-1}(z)\mathcal{B}(z)$ [4, 10], the following theorem furnishes a *unique* linear equations based solution to the AR parameters; the MA matrices may then be identified as proposed in [5]. Using correlations, identifiability is possible only within a unimodular matrix; hence we use cumulants. In [10] rational subspaces were used for DOA estimation and SOS-based methods for system-identification. These methods are not only sensitive to AGN, but also yield only the spectra.

Theorem 1: Under the assumption that $B(0)$, $B(n_b), A(n_a)$ are of full rank N_s, $\mathbf{w}(t)$ is a N_s-variate non Gaussian i.i.d. processes and the system is stable, the $N_x \times N_s$ AR matrices $\{A(k)\}_{k=1}^{n_a}$ can be obtained as:

$$\mathcal{A} \triangleq [A(1) \cdots A(n_a)] = -[(\mathcal{C}'\mathcal{C})^{\dagger}\mathcal{C}'\mathbf{c}]' \qquad (29)$$

where, $\mathcal{C} = [\bar{\mathbf{c}}_{kx}(m - n_a, n) \cdots \bar{\mathbf{c}}_{kx}(m-1, n)]$ and $\mathbf{c} = [\bar{\mathbf{c}}_{kx}(m,n)]$ collected for $m = n_b + 1, \ldots n_b + n_a; n = n_b - n_a + 1, \ldots, n_b + (N_x - 1)n_a$. The matrix $\bar{\mathbf{c}}_{kx}(m,n) = \mathbf{unvec}_{N_x \times N_x^{k-1}}[\mathbf{c}_{kx}(0, \ldots, m, n)]$ [12].

Proof: It can be shown using the definition of $\mathbf{c}_{kx}$, the recursion between the ARMA parameters and the IR matrices and after some algebra that:

$$\sum_{j=0}^{n_a} [I_{k-2} \otimes A(j) \otimes I]\mathbf{c}_{kx}(n_b - j, n) \qquad (30)$$
$$= [D_{k-2}(0) \otimes B(n_b) \otimes D(n)]\Gamma_{kw}$$

where I_{k-2} is the identity matrix of size $k-2$ and Γ_{kw} is the kth cumulant of $\mathbf{w}(t)$. After "unvecing", we obtain:

$$\sum_{j=0}^{n_a} \bar{\mathbf{c}}_{kx}(n_b - j, n)[I_{k-2} \otimes A'(j)] = D(n)L, \qquad (31)$$

where $L \triangleq \bar{\Gamma}_{kw}\ [D'_{k-2}(0) \otimes B'(n_b)]$ and $\bar{\Gamma}_{kw} \triangleq \mathbf{unvec}_{N_s \times N_s^{k-1}}[\Gamma_{kw}]$. In the vector form (31) is:

$$\tilde{\mathbf{C}}(m)\tilde{A}' = D(m)L \qquad (32)$$

where $\tilde{\mathbf{C}}(m) \triangleq [\bar{\mathbf{c}}_{kx}(n_b, m)\ \bar{\mathbf{c}}_{kx}(n_b - 1, m), \ldots, \bar{\mathbf{c}}_{kx}(n_b - n_a, m)]'$ and $\tilde{A}' \triangleq I_{k-2} \otimes [A(n_a), \cdots, I]$. Collecting (32) for $m = n_b + 1 - n_a, \ldots n_b + (N_x - 1)n_a$, we get:

$$\mathcal{C}\mathcal{A}' = \mathcal{D}\mathcal{L}. \qquad (33)$$

The matrix $\mathcal{C}$ is the super-matrix of the cumulants, $\mathcal{D}$ is the block Hankel matrix of the system IR, $\mathcal{A}$ and $\mathcal{L}$ are the block diagonal matrices formed using $\tilde{A}$ and L respectively. Using some linear algebra and Sylvester's inequality, it follows that $\mathcal{C}$ has rank $N_x n_a$ if rank($\mathcal{D}$) $= N_x n_a$. Since $A(n_a)$ is full rank, it follows that rank ($\mathcal{D}$) $= N_x n_a$. Therefore, $\mathcal{C}$ has rank $N_x n_a$ iff $\mathcal{D}$ has rank $N_x n_a$, which implies that the $[\mathcal{C}'\mathcal{C}]^{-1}\mathcal{C}'\bar{\mathbf{c}}$ yields a unique set of AR matrices [13]. □

If the system is noncausal, then the above theorem does not hold. Theorem 2 stated below establishes the identifiability of noncausal models via a cumulant-matching approach which generalizes the results of [6].

Theorem 2: Let θ_1 and θ_2 be ARMA parameters corresponding to ARMA processes of order (n_{a1}, n_{b1}) and (n_{a2}, n_{b2}) respectively. Let also $n_a \triangleq max(n_{a1}, n_{a2})$, $n_b \triangleq max(n_{b1}, n_{b2})$. If for $k \geq 3$, it holds that,

$$\mathbf{c}_{kx}(i_1, \ldots, i_{k-1}|\theta_1) \equiv \mathbf{c}_{kx}(i_1, \ldots i_{k-1}|\theta_2) \qquad (34)$$

$\forall\, i \in \mathcal{I}_k(n_b + 3n_a) \triangleq \{i : |i_1|, |i_2| \leq n_b + 3n_a, 0 \leq i_l \leq n_a\ \ l \geq 3\}$ then, $\theta_1 \equiv \theta_2$.

Proof: Follows the steps in [6] with cumulant vectors and matrices instead of scalar cumulants. See [13] for details. □

Hence, cumulant-matching may be used to obtain the AR and MA parameters of multichannel processes.

2.3 Estimation of $\mathcal{P}(\omega)$

We propose in this section, a MUSIC-like objective function based on the fourth-order instead of the second-order spectral matrix. We define the 4th-order spectral matrix as:

$$\mathcal{C}_{4x}(\omega, 0, 0) \triangleq E\{[\mathcal{X}(\omega) \otimes \mathcal{X}^*(0)][\mathcal{X}(0) \otimes \mathcal{X}^*(\omega)]^H\} \qquad (35)$$
$$= [\mathcal{P}(\omega) \otimes \mathcal{P}^*(0)]\mathcal{C}_{4s}(\omega, 0, 0)[\mathcal{P}(0) \otimes \mathcal{P}^*(\omega)]^H \qquad (36)$$

where the second equality follows from (3). Assuming that the source signals are not completely coherent we have that the source trispectral matrix $\mathcal{C}_{4s}$ has rank N_s. Since the direction vectors (columns of $\mathcal{P}(\omega)$) are assumed to be linearly independent, we have that the rank of the $N_x{}^2 \times N_x{}^2$ matrix $\mathcal{C}_{4x}$ is N_s. Performing SVD on $\mathcal{C}_{4x}$ we have:

$$\mathcal{C}_{4x} = [U\ U_{\perp}] \begin{bmatrix} \Lambda & 0 \\ 0 & 0 \end{bmatrix} \begin{bmatrix} V^H \\ V_{\perp}^H \end{bmatrix} \qquad (37)$$

where Λ is the diagonal matrix of the non-zero singular values of $\mathcal{C}_{4x}(\omega, 0, 0)$. Along the lines of [16, 7], we conclude that for almost all ω, the left eigenvectors $U_{\perp}$ (corresponding to the zero eigenvalues) of $\mathcal{C}_{4x}$ are orthogonal to the columns of $\mathcal{P}(\omega) \otimes \mathcal{P}^*(0)$ which are of the form $\mathbf{p}_i(\omega) \otimes \mathbf{p}_j(0)$, $i, j \in [1, N_s]$. Averaging over all $\omega \in BW$, the common bandwidth of the source signals, we propose:

$$J(\theta) = \frac{1}{\sum_{\omega \in BW} [\mathbf{p}(\theta,\omega) \otimes \mathbf{p}^*(\theta,\omega)]^H U_{\perp} U_{\perp}^H [\mathbf{p}(\theta,\omega) \otimes \mathbf{p}^*(\theta,\omega)]}, \qquad (38)$$

the trispectral MUSIC objective function, whose peaks yield the DOAs.

The DOAs specify the array manifold and hence $\mathcal{P}(\omega)$. Finally, $\mathcal{H}(\omega)$ is given by $\mathcal{P}^{-1}(\omega)\mathcal{D}(\omega)$. The transfer function $\mathcal{H}(\omega)$ can be used to completely characterize the source signals unlike in the SOS based approach, where only the spectra are available. These characterizations can be used for instance in source classification, detection of jamming and analyzing multipath.

3 Detection and parameter estimation

If the number of sources is known or if the source signal characterization (beyond spectral), is essential, the approach of Section 2 can be used. In this section we take a look at a HOS-based information theoretic criterion for simultaneous estimation of N_s and θ. The AIC(MDL) criterion of [15] yields only N_s and further assumes spatially independent noise and/or knowledge of the noise covariance. Here, we propose an information theoretic criterion based on the asymptotic normality of the bispectral estimator; the criterion is consistent, insensitive to additive Gaussian noise and yields θ and N_s simultaneously. In [9], a consistent, cumulant-based AIC was proposed for narrowband signals and in [3], a bispectrum based beamforming technique was reported for bearing estimation *only* of narrowband signals.

The ensemble averaged bispectrum $\mathbf{c}_{3x}(\omega_1, \omega_2)$ as defined in Sec. 1 can be consistently estimated using smoothed sample based estimator $\hat{\mathbf{c}}_{3x}(\omega_1, \omega_2)$. The asymptotic distribution of estimation error has been shown to be [1]:

$$\sqrt{T}[\mathbf{c}_{3x}(\omega_1, \omega_2) - \hat{\mathbf{c}}_{3x}(\omega_1, \omega_2)] \underset{T\to\infty}{\longrightarrow} \mathcal{N}(\mathbf{0}, \Sigma_3(\omega_1, \omega_2)). \quad (39)$$

The exact expression for the variance Σ_3 depends on the estimator used (bicorrelogram vs. biperiodogram).
Theorem 3: Under (A1)-(A3), if **(i)** $\mathbf{c}_{3s}$ is known, N_s and θ can be consistently estimated as,

$$[\hat{N}_s, \hat{\theta}] = \arg\min_{N_s} [\arg\min_{\theta} V_T(\theta, N_s)] \quad (40)$$

where,

$$V_T(\theta, N_s) \stackrel{\triangle}{=} \sum_{(\omega_1,\omega_2)\in BW} [\hat{\mathbf{c}}_{3x}(\omega_1, \omega_2) - \mathbf{c}_{3x}(\omega_1, \omega_2; \theta)] \times$$
$$\Sigma_3^{-1}[\hat{\mathbf{c}}_{3x}(\omega_1, \omega_2) - \mathbf{c}_{3x}(\omega_1, \omega_2; \theta)]^H + N_s \frac{log(T)}{T} \quad (41)$$
$$\Sigma_3(\omega_1, \omega_2) \stackrel{\triangle}{=} \lim_{T\to\infty} \frac{T}{M^2} E\{[\hat{\mathbf{c}}_{3x}(\omega_1, \omega_2) - \mathbf{c}_{3x}(\omega_1, \omega_2)] \times$$
$$[\hat{\mathbf{c}}_{3x}(\omega_1, \omega_2) - \mathbf{c}_{3x}(\omega_1, \omega_2)]^H\}.$$

(ii) With $\mathbf{c}_{3s}$ unknown, its LS estimator $\hat{\mathbf{c}}_{3s} = \mathcal{P}_3[\mathcal{P}_3^H \Sigma_3^{-1} \mathcal{P}_3]^{-1} \mathcal{P}_3^H \Sigma_3^{-1} \hat{\mathbf{c}}_{3x}$ is employed in (41). □

Remarks

- The covariance matrix $\Sigma_3(\omega_1, \omega_2)$ for large T was shown to be [1]:

$$\Sigma_3(\omega_1, \omega_2) \approx \frac{M^2}{T} \mathbf{c}_{2x}(\omega_1) \otimes \mathbf{c}_{2x}(\omega_2) \otimes \mathbf{c}_{2x}^*(\omega_1 + \omega_2) \mathcal{E}_{W,M} \quad (42)$$

where, $\mathbf{c}_{2x}$ is the spectral vector defined in (4) and $\mathcal{E}_{W,M}$ is the energy of the window W with support $[-M, M]$ and $(\omega_1, \omega_2) : 0 \le \omega_2 \le \omega_1 \le \pi, |2\omega_1 + \omega_2| \le \pi$. Consistent estimators of Σ_3 can be obtained via the windowed periodogram. For simplicity $\Sigma_3 \equiv I$ may be used.

- For wideband signals with finite bandwidth BW, identifiability may be possible by including fewer bifrequency points in the estimation algorithm as in [6]. It may be possible to estimate the parameters of more sources using fewer sensors similar to the time-domain case reported in [9].

- Arguing as in [6], it can be shown that θ obtained via (41) is asymptotically normal i.e.,

$$\sqrt{T}[\hat{\theta} - \theta] \underset{T\to\infty}{\longrightarrow} \mathcal{N}(\mathbf{0}, \mathbf{Q}) \quad (43)$$

where, $\mathbf{Q} = \lim_{T\to\infty} TE[(\hat{\theta} - \theta)(\hat{\theta} - \theta)']$. Confidence intervals and CR-bounds for the multivariate estimator can be developed along the lines of [6].

4 Conclusions

A multichannel system identification approach was proposed for modeling of multiple wideband sources arriving at a sensor array. By employing higher-order spectra, it was shown that the system impulse response can be uniquely determined. Subsequently, a trispectrum-based MUSIC-like criterion was used to determine the source bearings. Simultaneous source detection and parameter estimation strategy was proposed based on the asymptotic normality of the bispectrum estimator. The proposed algorithms yield the true systems underlying the phenomena rather than just the spectra; as a byproduct insensitivity to Gaussian noise is also obtained. Performance analysis of the proposed methods is currently under investigation.

References

[1] D. R. Brillinger and M. Rosenblatt, Spec. Anal. of Time Ser., B. Harris (Ed.), Wiley-NY 1967, pp. 153-188, pp. 189-232.

[2] D. Brooks and C. L. Nikias, ICASSP'91, pp. 3433-3436.

[3] P. Forster and C.L. Nikias, IEEE T-SP, vol.39, Sept. 1991, pp.1994-2006.

[4] B. Friedlander and B. Porat, IEEE T- IT, vol. 29, Apr. 1983, pp.412-425.

[5] G. B. Giannakis, Y. Inouye, and J.M. Mendel, IEEE T- AC, 1989, p.783.

[6] G. B. Giannakis and M. K. Tsatsanis, IEEE T-IT, Mar. 1992.

[7] B. Porat and B. Friedlander, IEEE T-SP, Sept. 1991, pp.2016-2024.

[8] M. Rangoussi and G. B. Giannakis, IEEE T- CAS., 1991, pp.281-296.

[9] S. Shamsunder and G. B. Giannakis, Intl. Wkshp. on HOSA, France, July 1991, pp.265-268; see also CISS 1991, pp. 600-606.

[10] G. Su and M. Morf, IEEE T-ASSP, Dec. 1983, pp. 1502-1522.

[11] A. Swami, Ph.D. Diss., Univ.of S. Cal., CA Oct 1988.

[12] A. Swami, G.B. Giannakis, J. M. Mendel, ICASSP'89, pp.2182-2185.

[13] A. Swami, G. B. Giannakis and S. Shamsunder, UVA rep. 192444B/EE91/127, Nov. 1991.

[14] L. Tong, Y. Inouye and R. Liu, ICASSP'91, Canada, pp.3445-3448.

[15] M. Wax and T. Kailath, IEEE T-ASSP, April 1985, p.387.

[16] M. Wax, T.J. Shan, and T. Kailath, IEEE T-ASSP, pp. 817-827.

Cramer-Rao Bounds for Deterministic Modal Analysis

L. T. McWhorter L. L. Scharf
Department of Electrical and Computer Engineering
University of Colorado
Boulder, CO 80309

Abstract

In this paper Cramer-Rao bounds are computed for any unbiased estimator of the mode parameters that characterize an ARMA impulse response. The bound can be computed for any of the standard parametric descriptions, including autoregressive moving average parameters, poles and residues, and poles and zeros. It will be demonstrated that the bounds for any characterization can be obtained from any other characterization via simple transformations. Asymptotic, frequency domain versions of the Cramer-Rao bounds are also derived. The bound is computed for a fourth-order system to illustrate the coupling between estimation errors and illuminate the influence of mode locations on the ability to identify mode parameters.

1 Introduction

Deterministic modal analysis techniques are used to identify the mode parmeters in the model

$$\mathbf{y} = \mathbf{x}(\underline{\theta}) + \mathbf{n}, \tag{1}$$

where $\mathbf{y} = [y_0\, y_1 \ldots y_{N-1}]^T$, $\mathbf{x}(\underline{\theta}) = [x_0\, x_1 \ldots x_{N-1}]^T$, and $\mathbf{n} = [n_0\, n_1 \ldots n_{N-1}]^T$ are the measurement data, modal signal, and noise vector respectively. The modal signal $\mathbf{x}(\underline{\theta})$, parametrized by $\underline{\theta}$, can be modeled as a sum of damped sinusoids (modes), or equivalently, as the deterministic impulse response of an ARMA system of the appropriate order. The ARMA system can, in turn, be parameterized by its AR and MA coefficients, poles and residues, poles and zeros, etc. The Cramer-Rao (CR) bound, regardless of the representation utilized, provides an effective measure for evaluating estimators of modal parameters.

The problem of calculating CR bounds for the modal signal model has been addressed in the literature by Tufts, et. al., in [1], and [2], Yau and Bresler [7], Stoica and Nehorai [3], and Scharf [4]. Friedlander and Porat, [5] and [6], have studied CR bounds for *stochastic* ARMA time series problems, but not for *deterministic* ARMA impulse responses. But in spite of this work and the numerous performance analyses that have been performed for specific algorithms, no general study of the CR bound for deterministic modal analyses problems has been published. In this paper we derive finite and asymptotic formulae for the Fisher information matrix when the modal signal is parameterized by its equivalent AR and MA coefficients. We also outline the construction of transformation matrices that allow the Fisher information matrix to be calculated for the corresponding pole-residue and pole-zero characterizations.

In the final part of this paper we apply our results to a fourth-order system in order to analyze the effects of closely spaced modes on our ability to identify mode parameters. We find, for example, that identification errors are correlated so that the high-frequency modulation associated with the average frequency of two closely spaced modes is well approximated at the expense of poor estimation accuracy for the difference (or beat) frequency between the modes.

2 Preliminaries

In this section we will derive the finite and asymptotic forms of the Fisher information matrix for a modal signal parameterized by its AR and MA coefficients. We have assumed in this formulation that the modal signal $\{x_t\}$ is generated by exciting the stable linear system $B(z)/A(z)$ with the unit pulse sequence $\{\delta_t\}$ where

$$\frac{B(z)}{A(z)} = \frac{\sum_{i=0}^{p-1} b_i z^{-i}}{\sum_{i=0}^{p} a_i z^{-i}} \;;\; (a_0 = 1). \tag{2}$$

1058-6393/91 $01.00 © 1991 IEEE

This transfer function model can be written in its analysis form

$$\begin{bmatrix} 1 & & & & \\ a_1 & 1 & 0 & & \\ \vdots & & & & \\ a_p & \cdots & 1 & & \\ \ddots & & & & \\ 0 & a_p & \cdots & 1 \end{bmatrix} \begin{bmatrix} x_0 \\ x_1 \\ \vdots \\ x_p \\ \vdots \\ x_{N-1} \end{bmatrix} = \begin{bmatrix} b_0 \\ b_1 \\ \vdots \\ b_{p-1} \\ 0 \end{bmatrix} \quad (3)$$

$$\mathbf{G}^{-1}\mathbf{x}(\underline{\theta}) = \begin{bmatrix} \mathbf{b} \\ 0 \end{bmatrix},$$

where the parameter vector $\underline{\theta}^T = [\mathbf{b}^T \ \mathbf{a}^T]$ is a composition of the MA parameters $\mathbf{b} = [b_0 \, b_1 \ldots b_{p-1}]^T$ and the AR parameters $\mathbf{a} = [a_1 \, a_2 \ldots a_p]^T$. If we denote the unit pulse response of the AR system $1/A(z)$ as $\{h_t\}$, the synthesis form for $\mathbf{x}(\underline{\theta})$ is given by

$$\mathbf{x}(\underline{\theta}) = \mathbf{G}^{-1}\begin{bmatrix} \mathbf{b} \\ 0 \end{bmatrix} = \mathbf{H}(\mathbf{a})\mathbf{b} \quad (4)$$

$$= \begin{bmatrix} h_0 & & 0 \\ h_1 & \ddots & \\ \vdots & & \\ h_{p-1} & \cdots & h_0 \\ \vdots & & \vdots \\ h_{N-1} & \cdots & h_{N-p} \end{bmatrix} \mathbf{b},$$

where

$$\mathbf{H}(\mathbf{a}) = \mathbf{H} = [\mathbf{h}_0 \, \mathbf{h}_1 \ldots \mathbf{h}_p]. \quad (5)$$

The Fisher information matrix for this linear separable model is (see [4])

$$\mathbf{J}(\underline{\theta}) = \frac{1}{\sigma^2}\left[\frac{\partial \mathbf{x}^T(\underline{\theta})}{\partial \underline{\theta}}\right]\left[\frac{\partial \mathbf{x}^T(\underline{\theta})}{\partial \underline{\theta}}\right]^T \quad (6)$$

$$= \frac{1}{\sigma^2}\begin{bmatrix} \left\{\mathbf{h}_i{}^T\mathbf{h}_j\right\} & \left\{\mathbf{h}_i{}^T\frac{\partial \mathbf{x}(\theta)}{\partial a_j}\right\} \\ \left\{\frac{\partial \mathbf{x}^T(\theta)}{\partial a_i}\mathbf{h}_j\right\} & \left\{\frac{\partial \mathbf{x}^T(\theta)}{\partial a_i}\frac{\partial \mathbf{x}(\theta)}{\partial a_j}\right\} \end{bmatrix},$$

when the noise is gaussian with zero mean and variance σ^2. Equation (6) is easily generalized to account for shaped inputs and colored gaussian noise. The Fisher matrix can be simplified by letting $\mathbf{k}_i = \frac{\partial \mathbf{x}(\theta)}{\partial a_i}$, and using a matrix identity in conjunction with equation (4) to obtain

$$\mathbf{k}_i = \frac{\partial \mathbf{x}(\underline{\theta})}{\partial a_i} = \frac{\partial \mathbf{G}}{\partial a_i}\begin{bmatrix} \mathbf{b} \\ 0 \end{bmatrix} \quad (7)$$

$$= -\mathbf{G}\frac{\partial \mathbf{G}^{-1}}{\partial a_i}\mathbf{G}\begin{bmatrix} \mathbf{b} \\ 0 \end{bmatrix} = -\mathbf{G}\frac{\partial \mathbf{G}^{-1}}{\partial a_i}\mathbf{H}\mathbf{b}.$$

The partial derivative $\frac{\partial \mathbf{G}^{-1}}{\partial a_i}$ is easy to compute, and the resultant matrix is a shift operator of the form,

$$\mathbf{Z}^i = \frac{\partial \mathbf{G}^{-1}}{\partial a_i} = \begin{bmatrix} 0 & \cdots & 0 \\ \vdots & & \\ 0 & 0 & \vdots \\ 1 & & \\ & \ddots & \\ 0 & 1 & 0 \end{bmatrix}. \quad (8)$$

Since the matrix $\mathbf{G}$ is a Toeplitz convolution matrix and the corresponding system $1/A(z)$ is shift invariant, the matrices $\mathbf{G}$ and $\mathbf{Z}^{-i}$ commute and equation (8) reduces to

$$\mathbf{k}_i = -\mathbf{Z}^{-i}\mathbf{G}\mathbf{H}\mathbf{b} = \mathbf{Z}^{-i}\mathbf{k} \quad (9)$$

$$= [0 \ldots 0 \, k_0 \, k_1 \ldots k_{N-i-1}]^T$$

where $\mathbf{k} = [k_0 \, k_1 \ldots k_{N-1}]^T$ is a length N snapshot of the signal $\{k_t\} = -[B(z)/A^2(z)]\{\delta_t\} = -[B(z)/A(z)]\{h_t\}$. The Fisher information matrix can now be written as

$$\mathbf{J}(\underline{\theta}) = \frac{1}{\sigma^2}\begin{bmatrix} \{\mathbf{h}_i{}^T\mathbf{h}_j\} & \{\mathbf{h}_i{}^T\mathbf{k}_j\} \\ \{\mathbf{k}_i{}^T\mathbf{h}_j\} & \{\mathbf{k}_i{}^T\mathbf{k}_j\} \end{bmatrix} \quad (10)$$

$$= \frac{1}{\sigma^2}\begin{bmatrix} \mathbf{H}^T\mathbf{H} & \mathbf{H}^T\mathbf{K} \\ \mathbf{K}^T\mathbf{H} & \mathbf{K}^T\mathbf{K} \end{bmatrix}$$

where $\mathbf{H}$ is defined in equation (4) and

$$\mathbf{K} = \begin{bmatrix} 0 & & \\ k_0 & 0 & \\ k_1 & k_0 & \\ \vdots & \ddots & \\ k_{p-1} & \cdots & k_0 \\ \vdots & & \vdots \\ k_{N-2} & \cdots & k_{N-p-1} \end{bmatrix}. \quad (11)$$

The asymptotic form for the Fisher information matrix is readily obtained by allowing $N \to \infty$ in equation (10) and using the representation for $\mathbf{k}_i$ described in equation (9). The resulting matrix has partitions

$$\mathbf{J}_\infty(\underline{\theta}) = \frac{1}{\sigma^2}\begin{bmatrix} \mathbf{R}_{hh} & \mathbf{R}_{hk} \\ \mathbf{R}_{kh} & \mathbf{R}_{kk} \end{bmatrix}. \quad (12)$$

The matrices $\mathbf{R}_{hh}$, $\mathbf{R}_{hk}$, and $\mathbf{R}_{kk}$ have (i,j) elements $r_{hh}(i-j)$, $r_{hk}(i-j+1)$, and $r_{kk}(i-j)$ respectively. The sequences r_{hh}, r_{kk}, and r_{hk} are the deterministic auto- and cross-correlation sequences for $\{h_t\}$ and $\{k_t\}$. The autocorrelation sequences can be constructed using the algorithms presented in [8] and [9]. Both autocorrelation sequences in the asymptotic Fisher matrix have terms of the form

$$r_m = \frac{1}{2\pi j} \oint \frac{F(z)F(z^{-1})}{G(z)G(z^{-1})} z^m \frac{dz}{z}, \qquad (13)$$

where $F(z) = 1, G(z) = A(z)$ for $r_m = r_{hh}(m)$ and $F(z) = B(z), G(z) = A^2(z)$ for the sequence $r_m = r_{kk}(m)$. From this equation it is apparent that $\{r_m\}$ can be determined from the impulse response of the system $F(z)F(z^{-1})/G(z)G(z^{-1})$. Since we have constrained the roots of $A(z)$ to lie inside the unit circle, the system $F(z)F(z^{-1})/G(z)G(z^{-1})$ can be easily factored into its causal $(Y(z)/G(z))$ and anti-causal $(Y(z^{-1})/G(z))$ parts. The resulting equation

$$\frac{F(z)F(z^{-1})}{G(z)G(z^{-1})} = \frac{Y(z)}{G(z)} + \frac{Y(z^{-1})}{G(z^{-1})} \qquad (14)$$

or

$$F(z)F(z^{-1}) = Y(z)G(z^{-1}) + Y(z^{-1})G(z) \qquad (15)$$

is a symmetric form of Euclid's equation. Define the order of the polynomial $G(z)$ to be q. To generate $\{r_m\}$ we need only to determine $Y(z)$ and compute the elements of the impulse response of the system $Y(z)/G(z)$, since $\{r_m\}$ is symmetric. To determine $Y(z)$, the first step is to compute the product $F(z)F(z^{-1})$ and decompose this square into its causal $(C(z))$ and anti-causal $(C(z^{-1}))$ pieces

$$F(z)F(z^{-1}) = C(z) + C(z^{-1}). \qquad (16)$$

We let the polynomials $F(z)$ and $C(z)$ have the same order as $G(z)$, namely q, and allow some of the coefficients to be zero. From equation (16) we can derive the following expression for the coefficients of $C(z)$

$$c_i = \begin{cases} \frac{1}{2}\sum_{k=0}^{q} f_k{}^2 & i = 0 \\ \sum_{k=0}^{q} f_k f_{k+i} & 1 \le i \le m \end{cases} \qquad (17)$$

where $f_k = \delta_k$ for r_{hh} and $f_k = b_k$ for r_{kk}. Equation (15) can now be written as

$$Y(z)G(z^{-1}) + Y(z^{-1})G(z) = C(z) + C(z^{-1}). \qquad (18)$$

The solution to this equation, $Y(z)$, will exist when $G(z)$ is minimum phase. Equation (18) can be expressed in matrix notation as

$$\mathbf{y}^T\mathbf{M} = \mathbf{c}^T \qquad (19)$$

where M is the Jury matrix with elements

$$\mathbf{M}_{ij} = \begin{cases} g_{i-1} & j = 1 \\ g_{i+j-2} + g_{i-j} & j \neq 1 \end{cases} \qquad (20)$$

and $\mathbf{c}^T = [c_0\ c_1 \ldots c_q]$, $\mathbf{y}^T = [y_0\ y_1 \ldots y_q]$, and $\mathbf{g}^T = [g_0\ g_1\ \ldots g_q]$ are vectors containing the coefficients of $C(z)$, $Y(z)$, and $G(z)$ respectively. The polynomial $Y(z)$ can be obtained by inverting $\mathbf{M}$ in equation (19). If we denote the impulse response sequence of $Y(z)/G(z)$ as $\{\alpha_m\}$, the autocorrelation sequence obeys the relation

$$r_m = \begin{cases} 2\alpha_0 & m = 0 \\ \alpha_m & m \neq 0 \end{cases}. \qquad (21)$$

Computation of the cross-correlation sequences is more dificult because this problem lacks the symmetry of the autocorrelation case. For this reason, and in the interest of brevity, we refer the reader to [9] for these algorithms.

We now have finite and asymptotic forms of the Fisher matrix for the modal sigmal parameterized by $\underline{\theta}^T = [\mathbf{b}^T\ \mathbf{a}^T]$. To obtain the Fisher information matrix for any equivalent parameterization $\underline{\gamma} = \mathbf{g}^{-1}(\underline{\theta})$, we only need to compute the "sensitivity matrix" $\mathbf{S}(\underline{\gamma})$ and let (see [4])

$$\mathbf{J}(\underline{\gamma}) = \mathbf{S}(\underline{\gamma})\mathbf{J}(\underline{\theta} = \mathbf{g}(\underline{\gamma}))\mathbf{S}^T(\underline{\gamma}) \qquad (22)$$

where

$$\mathbf{S}(\underline{\gamma}) = \frac{\partial \underline{\theta}^T}{\partial \underline{\gamma}} = \frac{\partial \mathbf{g}^T(\underline{\gamma})}{\partial \underline{\gamma}}. \qquad (23)$$

The pole-residue parameterization of the modal signal is important because the parameters which characterize each mode (amplitude, phase, damping factor and frequency) give an insightful physical description of the modes.

The sensitivity matrices in the section are derived assuming the $A(z)$ and $B(z)$ have real coefficients, or equivalently, the poles and zeros occur in conjugate pairs. For the pole-residue model, a sample of the signal is

$$(\mathbf{x}(\underline{\gamma}))_k = \sum_{i=1}^{q} A_i \rho_i^k \cos(\theta_i k + \phi_i) \qquad (24)$$

where

$$\begin{aligned} \frac{B(z)}{A(z)} &= \sum_{i=1}^{q} \frac{\beta_i(z)}{\alpha_i(z)} \\ \beta_i(z) &= A_i \cos\phi_i - A_i\rho_i \cos(\theta_i - \phi_i)z^{-1} \\ \alpha_i(z) &= 1 - 2\rho_i \cos(\theta_i)z^{-1} + \rho_i^2 z^{-2}. \end{aligned} \qquad (25)$$

If the order of A(z) (denoted as p) is even we define $q = p/2$, otherwise $q = (p-1)/2$. If we define

$\underline{\gamma}^T = [\underline{A}^T\ \underline{\phi}^T\ \underline{\rho}^T \underline{\theta}^T]$, the sensitivity matrix $\mathbf{S}(\underline{\gamma})$ has partitions

$$\mathbf{S}(\underline{\gamma}) = \begin{bmatrix} \frac{\partial \mathbf{b}^T}{\partial \underline{A}} & \frac{\partial \mathbf{a}^T}{\partial \underline{A}} \\ \frac{\partial \mathbf{b}^T}{\partial \underline{\phi}} & \frac{\partial \mathbf{a}^T}{\partial \underline{\phi}} \\ \frac{\partial \mathbf{b}^T}{\partial \underline{\rho}} & \frac{\partial \mathbf{a}^T}{\partial \underline{\rho}} \\ \frac{\partial \mathbf{b}^T}{\partial \underline{\theta}} & \frac{\partial \mathbf{a}^T}{\partial \underline{\theta}} \end{bmatrix}. \tag{26}$$

where the parameter vectors in $\underline{\gamma}$ all have a form similar to $\underline{A} = [A_0\ A_1 \ldots A_q]^T$. Using equation (25) as a starting point, we can compute the matrix $\mathbf{S}(\underline{\gamma})$ using the techniques of [10].

The sensitivity matrix for the pole-zero parameterization is similar to the sensitivity matrix for the pole-residue parameterization. In fact, the partitions which are functions of the pole parameters are identical. The partitions for the zero parameters have the same structure as the partitions for the pole parameters but with zero magnitude replacing pole magnitude and zero angle replacing pole frequency.

3 Results

In this section a modal signal consisting of two independent modes is analyzed. Since the effects of mode amplitude are easily determined, the modes in this example have unit amplitude. We will be concentrating on the effects of mode frequency spacing, so we have also set the phase offset of each mode to zero. With these constraints in force, the modal signal

$$x_t = \rho_1^t \cos(\theta_1 t) + \rho_2^t \cos(\theta_2 t) \tag{27}$$

can be characterized by the frequencies θ_1 and θ_2 and the damping coefficients ρ_1 and ρ_2. It will be convenient for subsequent discussions to also characterize the signal in terms of the mean frequency $\theta_m = (\theta_1 + \theta_2)/2$ and the difference or beat frequency $\theta_d = |\theta_2 - \theta_1|$:

$$x_t = \rho_1^t \cos((\theta_m + \theta_d)t) + \rho_2^t \cos((\theta_m - \theta_d)t). \tag{28}$$

Figures (1) and (2) were generated for infinite data with the mode parameters $\rho_1 = \rho_2 = 0.9$, $\theta_m = \pi/2$, and θ_d variable. The CR bounds were calculated assuming that the estimator is unbiased and that the noise variance is normalized to 1. It is evident from the figure that, as $\theta_d \to 0$ (i.e. the mode frequencies approach each other), the mode parameters θ_d and θ_m become harder to identify. A similar effect can be observed for ρ_1 and ρ_2. Note that this pole proximity effect is stronger for the beat frequency θ_d than for the

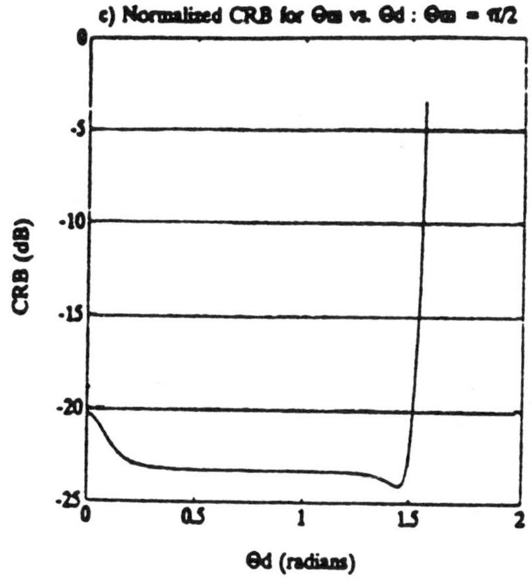

Figure 1: CRB(θ_m) vs. θ_d

mean frequency θ_m. As $\theta_d \to \pi/2$, the conjugate poles for mode 1 approach each other on the real axis at ρ_1. The conjugate poles for mode 2 approach each other on the real axis at $-\rho_2$. Again we notice a similar pole proximity effect.

These same effects can be observed by plotting the concentration ellipses for the Fisher information matrix. The concentration ellipse is defined by

$$\mathbf{e}^T \mathbf{J} \mathbf{e} = 1, \tag{29}$$

where $\mathbf{J}$ is the Fisher information matrix and $\mathbf{e}$ is the parameter error vector. The ellipse constructed from the error covariance matrix of any unbiased estimator will be larger than the ellipse constructed in equation(29) [4]. The error ellipses of Figure (3) were constructed by setting the errors in ρ_1 and ρ_2 equal to zero. The size differences in the ellipses as θ_d is varied corroborate the pole proximity effects depicted in the previous figures. These ellipses also establish that the estimation errors for θ_d and θ_m are uncorrelated for large data sets. We can state this because the axes of the ellipses are approximately $e(\theta_d) = 0$ and $e(\theta_m) = 0$.

Acknowledgements

This work was supported by Bonneville Power Administration under contract #DEBI7990BPO7346.

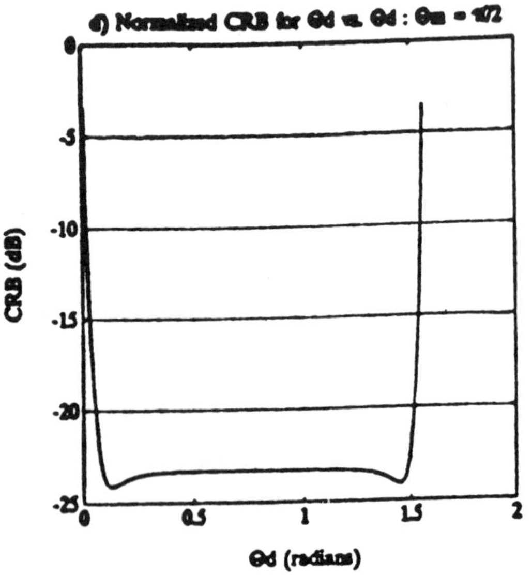

Figure 2: CRB(θ_d) vs. θ_d

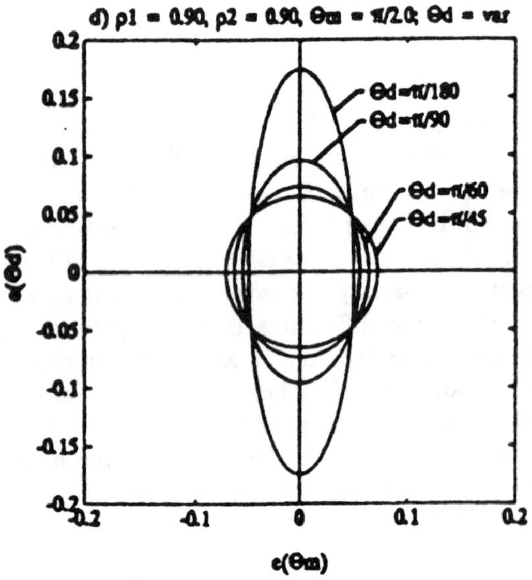

Figure 3: Ellipses for (θ_d, θ_m)

References

[1] Tufts, D. and R. Kumaresan, "Frequency Estimation of Multiple Sinusoids: Making Linear Prediction Work like Maximum Likelihood," *Proc IEEE*, Vol 70, pp. 975-990, March 1983

[2] Tufts, D., F. Giannella, I. Kirsteius, and L. L. Scharf, "Cramer-Rao Bounds on the Accuracy of Autoregressive Parameter Estimators," *Proc IEEE Second Workshop on Spectrum Analysis*, Tampa, FL, November 1983

[3] Stoica, P, and A. Nehorai, "MUSIC, Maximum Likelihood, and Cramer-Rao Bound," *IEEE Trans ASSP*, Vol ASSP-37:5, pp. 720-741, May 1989

[4] Scharf, L. L., *Statistical Signal Processing*, (Reading, MA: Addison-Wesley, 1991)

[5] Friedlander, B., "On the Computation of the Cramer-Rao Bound for ARMA Parameter Estimation," *IEEE Trans ASSP*, Vol ASSP-32:4, pp. 721-727, August 1984

[6] Friedlander, B., and B. Porat, "A General Lower Bound for Parametric Spectrum Estimation," *IEEE Trans ASSP*, Vol ASSP-32:4, pp. 728-733, August 1984

[7] Yau, S. F., and Y. Bresler, "Worst-Case Cramer-Rao Bounds for Parametric Estimation of Superimposed Signals," *Proc 5th Spectrum Estimation Workshop*, Rochester, NY, pp. 187-191, October 1990

[8] Dugre, J.-P., A. A. Beex, and L. L. Scharf, "Generating Covariance Sequences and the Calculation of Quantization and Rounding Error Variances in Digital Filters," *IEEE Trans ASSP*, Vol ASSP-28, pp. 102-104, February 1980

[9] Demeure, C., and C. T. Mullis, "The Euclid Algorithm and the Fast Computation of Cross-Covariance and Autocovariance Sequences," *IEEE Trans ASSP*, Vol ASSP-37:4, pp. 545-552, April 1989

[10] McWhorter, L. T., and L. L. Scharf, "Cramer-Rao Bounds for Modal Analysis" Submitted for Publication, June 1990

Time Varying Higher Order Spectra

Boualem Boashash and Branko Ristic
Centre for Signal Processing Research
Queensland University of Technology
GPO Box 2434, Brisbane, Q 4001, Australia

Abstract

The Wigner-Ville Distribution (WVD) is a second order time-frequency representation in a sense that it is able to give ideal energy concentration for quadratic (2nd order) FM signals, and its expectation value is a second order time-varying spectrum. The WVD may be extended so as to achieve ideal energy concentration for higher order phase laws, and such that its expectation value is a time-varying higher order spectrum.

The paper presents two methods for generalizing the WVD in order to achieve the ideal energy concentration for polynomial frequency (phase) signals. By taking the expected value of the GWVD, time-varying higher-order spectra can be defined.

1 Introduction

The Wigner-Ville Distribution of a linear FM signal shows a continuum of delta functions in the time-frequency plane along the signal's instantaneous frequency (IF) law. It is this ideal energy concentration which makes the WVD useful for the IF estimation [1], [3]. For *non-linear* FM signals this ideal concentration is no longer assured, and smeared spectral representations often result. In this paper we show how to design *Generalised Wigner-Ville Distributions (GWVDs)* which will exhibit a continuum of delta functions along the IF law, for IF laws of arbitrary polynomial form. To explain how this is achieved, one needs to first look closely at the mechanism by which the WVD attains ideal concentration for a linear FM signal.

In [2], it was shown that the WVD kernel is essentially a signal which is constructed from the central finite difference derived IF estimate. Therefore, the reason why the WVD yields good energy concentration for linear FM signals is that the central finite difference estimator is unbiased for such signals [3].

Thus linear FM signals are transformed into sinusoids in the WVD kernel, with the frequency of the sinusoid being equal to the instantaneous frequency at that value of time. Fourier transformation of the kernel leads quite naturally, then, to a row of delta functions along the true IF law of the signal. However, this is not true for non-linear FM signals [2].

One can generalise the WVD in order to be able to analyse non-linear FM signals more effectively by replacing the central finite difference IF estimator inherent in the WVD formulation with an alternative phase difference estimator [2] which is unbiased for polynomial frequency (or phase) laws of arbitrary order. The problem of designing these polynomial phase model based "phase differentiating filters" is discussed in detail in [3], and summarized in section 2.2.

One method to generalize the WVD by incorporating the generalized phase difference estimators [3] is presented in [2] and summarized below:

$$W^g(t,f) = \underset{\tau \to f}{\mathcal{F}} \left[K^g(t,\tau)\right] \tag{1}$$

where $K^g(t,\tau)$ is the *generalised kernel product* , and is given by

$$K^g(t,\tau) = \prod_{k=1}^{q/2} [z(t+c_k\tau)]^{b_k} [z^*(t+c_{-k}\tau)]^{-b_k} \tag{2}$$

where b_k and the c_k may be varied to yield unbiased IF estimates for signals with arbitrary polynomial FM laws. The procedure for determining the b_k and c_k is given in [2].

The conventional WVD may be recovered from the generalised WVD (GWVD) by setting $q = 2$, $b_{-1} = -1$, $b_0 = 0$, $b_1 = 1$, $c_{-1} = -1/2$, $c_0 = 0$, $c_1 = 1/2$. Note that the generalised kernel product (GKP) is a *multi-linear* one, as opposed to the usual bilinear one. While the standard bilinear kernel product is guaranteed to transform linear FM signals into sinusoids, the multi-linear product can be designed to

1058-6393/91 $01.00 © 1991 IEEE

guarantee transformation of higher order FM signals into sinusoids. These sinusoids manifest as delta functions when Fourier transformed.

The GWVD designed for a fourth order polynomial phase law is the Fourier transform of [2]:

$$K^g(t,\tau) = [z(t{+}.675\tau)z^*(t{-}.675\tau)]^2[z^*(t{+}.85\tau)z(t{-}.85\tau)] \quad (3)$$

It was shown in [2] that the GWVD peak based IF estimate variance meets the Cramer-Rao lower variance bounds at high SNR. In addition, by taking the expected value of the GWVD, the time-varying higher-order spectra can be defined [2] (section 3).

2 Another method to generalize the Wigner-Ville distribution

2.1 Background

Let us consider the signal

$$z(t) = A(t)e^{j\phi(t)} \qquad -T/2 \le t \le T/2 \quad (4)$$

The instantaneous frequency of this monocomponent FM signal is defined as

$$f_i(t) = \frac{1}{2\pi}\frac{d\phi(t)}{dt}$$

that is

$$f_i(t) = \frac{1}{2\pi}\lim_{\tau\to 0}\frac{\phi(t+\tau/2)-\phi(t-\tau/2)}{\tau}$$

A central difference estimator of the instantaneous frequency can be defined as:

$$\hat{f}_i(t,\tau) = \frac{1}{2\pi}\frac{\phi(t+\frac{\tau}{2})-\phi(t-\frac{\tau}{2})}{\tau} \quad (5)$$

where t is the time at which the estimate is taken and τ is the time lag. This estimator is unbiased only if the IF is constant or a linear function of time. Consider the case when:

(i) $\phi(t) = a_0 + a_1 t + a_2 t^2$

(ii) $A(t) = 1$,

the IF estimate equals the correct value:

$$\hat{f}_i(t,\tau) = f_i(t) = a_1 + 2a_2 t \quad (6)$$

For an FM signal with the polynomial form of its phase:

$$\phi(t) = \sum_{m=0}^{q} a_m t^m \quad (7)$$

where q is the polynomial order ($q \in Z^+$), the general form of the continuous-time phase difference estimator can be defined as [3]:

$$\hat{f}_i^{<r>}(t,\tau) = \frac{1}{2\pi\tau}\sum_{k=0}^{2r} d_k^{<r>}\,\phi(t+\frac{k-r}{2r}\tau) \qquad r \in Z^+ \quad (8)$$

where

$$r = \lceil \frac{q}{2} \rceil \quad (9)$$

and coefficients $d_k^{<r>}$ are derived in detail in [3] for the discrete-time polynomial phase case. The IF estimator given by (8) gives the exact value for signals with polynomial phase (and frequency). Suppose $A(t) = 1$ in (4) and phase $\phi(t)$ is given by (7) then the IF estimate using (8) is:

$$\hat{f}_i^{<r>}(t,\tau) = f_i(t) = \frac{1}{2\pi}\sum_{m=1}^{q} m\,a_m t^{m-1} \quad (10)$$

2.2 Generalized phase difference coefficients

Here we briefly summarize the procedure given in [3]. The polynomial phase in discrete-time form is:

$$\phi(n) = \sum_{m=0}^{q} a_m n^m \quad (11)$$

The corresponding IF is:

$$f_i(n) = \frac{1}{2\pi}\sum_{m=1}^{q} m a_m n^{m-1} \quad (12)$$

The q-th order phase difference estimator is defined as [3]:

$$\hat{f}_i^{<q>}(n+q/2) = \frac{1}{2\pi}\sum_{k=0}^{q} b_k^{<q>}\phi(n+k) \quad (13)$$

The $b_k^{<q>}$ coefficients are to be found so that $f_i(n+q/2) = \hat{f}_i^{<q>}(n+q/2)$. After several lines of mathematical manipulations the following matrix equation is obtained:

$$\underline{Q}\underline{b} = \underline{\xi} \quad (14)$$

where

$$\underline{Q} = \begin{bmatrix} 1 & 1 & \dots & 1 \\ 0 & 1 & \dots & q \\ \dots & \dots & \dots & \dots \\ 0 & 1 & \dots & q^q \end{bmatrix} \quad (15)$$

$$\underline{b} = [b_0^{<q>}\; b_1^{<q>}\; \dots\; b_q^{<q>}]^T \quad (16)$$

$$\underline{\xi} = [0\ 1\ q\ \dots\ \frac{q^q}{2^{q-1}}]^T \quad (17)$$

The coefficients of the polynomial phase estimators given in Tab.1 are derived by solving the matrix equation (14) for $q = 1, 2, \dots, 6$.

Comparing (8) and (13) one can note the relationship:

$$d_k^{<r>} = q\, b_k^{<\lceil\frac{q}{2}\rceil>} \qquad (k = 0, \dots, q) \quad (18)$$

Note that the estimator given in (8) does not use $b_k^{<q>}$ coefficients for odd values of q. The reason is that there is no advantage in using the estimator with odd values of q, only a disadvantage – a need for a higher degree of interpolation of the input signal. This will be clarified later in the text.

Tab. 2 shows the values of coefficients $d_k^{<r>}$.

2.3 Another Form of the Generalized Kernel Product

Based on the general form of the phase difference estimator given in (8) it is straightforward to define another form of the GKP as:

$$\mathcal{K}^{<r>}(t,\tau) = \prod_{k=0}^{2r} \left[z(t + \frac{k-r}{2r}\tau)\right]^{d_k^{<r>}} \quad (19)$$

where the subscript $< r >$ denotes the order of generalization. The GWVD is defined as the Fourier transform of the GKP $\mathcal{K}^{<r>}(t,\tau)$ with respect to τ (see eq.(1)).

$$\mathcal{G}^{<r>}(t,f) = \underset{\tau \to f}{\mathcal{F}} \left[\mathcal{K}^{<r>}(t,\tau)\right] \quad (20)$$

Example 1: Suppose the signal to be analyzed is a linear FM ($q = 2$). Then, by (9) $r = 1$ and the GKP is given by:

$$\mathcal{K}^{<1>}(t,\tau) = z(t + \tau/2) z^*(t - \tau/2) \quad (21)$$

where * denotes the complex-conjugate operation. Note that the above expression is identical to the kernel of the conventional WVD.

Example 2: Suppose the signal to be analyzed is a quadratic FM ($q = 3$) or cubic FM signal ($q = 4$). Then, by (9) $r = 2$ and the GKP is given by:

$$\mathcal{K}^{<2>}(t,\tau) = \begin{array}{l}\left[z^*(t + \frac{\tau}{2})\right]^{1/3} z^{1/3}(t - \frac{\tau}{2}) \times \\ z^{8/3}(t + \frac{\tau}{4}) \left[z^*(t - \frac{\tau}{4})\right]^{8/3}\end{array} \quad (22)$$

The argument of $\mathcal{K}^{<2>}(t,\tau)$ for a signal given by (4) with assumption that $A(t) = 1$ is

$$arg\{\mathcal{K}^{<2>}(t,\tau)\} = \begin{array}{l} -\frac{1}{3}\phi(t + \tau/2) + \frac{1}{3}\phi(t - \tau/2) + \\ +\frac{8}{3}\phi(t + \tau/4) - \frac{8}{3}\phi(t - \tau/4)\end{array} \quad (23)$$

For cubic FM it becomes:

$$arg\{\mathcal{K}^{<2>}(t,\tau)\} = 2\pi\tau(a_1 + 2a_2 t + 3a_3 t^2 + 4a_4 t^3) \quad (24)$$

If we assume that $T \to \infty$ then

$$\mathcal{G}^{<2>}(t,f) = \delta(f - (a_1 + 2a_2 t + 3a_3 t^2 + 4a_4 t^3)) \quad (25)$$

Compare the GKP in (22) to the corresponding one in (3). We note that there is no need for a high degree of interpolation of the input signal in a discrete-time form for the GKP in (22). However the input signal samples are raised to non-integer powers, which is a serious implementation problem for complex signals. Namely it involves the phase unwrapping procedure which is known to work well only for high SNRs and monocomponent signals. Alternatively, the signal may be interpolated 4 times and a cubed kernel product $\left[\mathcal{K}^{<2>}(t,\tau)\right]^3$ calculated. Then, by frequency scaling of the kernel's Fourier transform, we could find the GWVD. The problem with this approach is that for multicomponent signals the energy of the cross-terms is substantially greater than the energy of the auto-terms. Fig. 1 and Fig. 2 show the GWVD and the WVD of a quadratic FM signal respectively.

Here we emphasize once again that there is no need to introduce a special GKP for odd values of q. Consider $q = 3$ (quadratic FM) and the prospective kernel based on coefficients $b_k^{<3>}$ given in Tab.1. It would be a 4 member kernel product (same as the one in (22)) and it would require a higher degree of interpolation of the input signal compared to the GKP in (22).

3 Time-Varying Higher Order Spectra Based on Generalised WVDs

The vast majority of the research which has been done in the area of higher order spectra (HOS) has assumed that the signals under consideration were stationary. To deal with *non-stationary* signals Gerr [4] has proposed a third order Wigner Distribution, while Thatcher and Amin [5] have devised a third order spectrogram, or "Running Bispectrum". There is a wide scope for other forms of extension which can take advantage of the available expertise in time-varying spectral analysis. One of these, described below, is based on GWVDs. A description of various other extensions is provided in [2].

The generalised Wigner-Ville Distributions are very closely linked to Higher Order Spectra (HOS). Consider the GWVD formulation as given by eq.(1) and (2). The generalized kernel product is:

$$K^g(t,\tau) = \prod_{k=1}^{q/2} z^{b_k}(t + c_k\tau)[z^*(t + c_{-k}\tau)]^{-b_{-k}} \quad (26)$$

For random signal $z_r(t)$, the expectation operator $\xi\{.\}$ must be introduced:

$$K^g(t,\tau) = \xi\{\prod_{k=1}^{q/2} z_r^{b_k}(t+c_k\tau)[z_r^*(t+c_{-k}\tau)]^{-b_{-k}}\} \quad (27)$$

The resulting GWVD designed for the fourth order polynomial FM law is:

$$\xi\{W^g_{z_r}(t,f)\} = \underset{\tau\rightarrow f}{\mathcal{F}} \xi\{z_r^2(t + .675\tau)$$
$$[z_r^*(t - .675\tau)]^2 z_r^*(t + .85\tau) z_r(t - .85\tau)\} \quad (28)$$

This expression is a spectrum of a higher order (6th order) moment, in which there is only one frequency variable. It is in fact a *time-varying higher order spectrum*. It may be rewritten as

$$\xi\{W^g_{z_r}(t,f)\} = \underset{\tau\rightarrow f}{\mathcal{F}} [m_6'(\tau)] \quad (29)$$

where $m_6'(\tau)$ is a "reduced" sixth order moment. It is obtained from the conventional sixth order moment, defined below,

$$m(\tau_1,\tau_2,\tau_3,\tau_4,\tau_5,\tau_6) =$$
$$\xi\{z_r(t+\tau_1)z_r^*(t+\tau_2)z_r(t+\tau_3)$$
$$z_r^*(t+\tau_4)z_r(t+\tau_5)z_r^*(t+\tau_6)\} \quad (30)$$

by making the following substitutions:

$$\tau_1 = 0.675\tau,\ \tau_2 = -0.675\tau,\ \tau_3 = 0.675\tau$$
$$\tau_4 = -0.675\tau, \tau_5 = -0.85\tau,\ \tau_6 = 0.85\tau \quad (31)$$

Because the GWVD spectrum reveals information about the 6th order moment of random processes, it will be able to discriminate between (non-stationary) processes with differing 6th order moments.

4 Generalization of the ambiguity function for non-linear polynomial FM signals

There is a duality between the energy density domain and the correlation domain for bilinear time-frequency representations [1]. For example, the WVD is related to the ambiguity function (AF) by the 2D Fourier transform duality. However, the question is what is a dual function of the GWVD in the *generalized* correlation domain.

The model of signal to be analyzed is given by eq.(4). Assume $A(t) = 1$ and the phase $\phi(t)$ given by (7). Then,

$$\mathcal{K}^{<r>}(t,\tau) = \exp j2\pi\tau f_i(t) \quad (32)$$

where $f_i(t)$ is given by (10) and repeated here:

$$f_i(t) = \frac{1}{2\pi}\sum_{m=1}^{q} m\, a_m t^{m-1} \quad (33)$$

An attempt to apply the conventional (2D Fourier transform) approach to the definition of the dual of the GWVD in the correlation domain does not seem to be appropriate. The reason is that the sinusoidal component of the GKP in (32) is *nonstationary* with respect to t. The natural way to process the GKP being a non-stationary signal would be by means of non-stationary signal analysis: using the GWVD. However, there is a problem in this approach. Namely, the GKP signal is a complex signal with possibly non-monotone phase even for the monocomponent case, which means that its first derivative (that is its IF by definition) could change sign. Note that for a class of practical real signals this is never the case since their corresponding analytic signals have everywhere positive IFs.

5 Conclusion

The paper presents two methods of generalizing the WVD in order to achieve ideal energy concentration for polynomial phase signals. The expected values of such generalized WVDs are the Fourier transforms of some particular higher-order moments and/or cumulants. The ensemble averaged GWVDs therefore have interpretation as time-varying higher-order spectra.

References

[1] B. Boashash. Time-frequency signal analysis. In S. Haykin, editor, *Advances in Spectral Estimation and Array Processing*, volume 1 of 2, chapter 9, pages 418–517. Prentice Hall, Englewood Cliffs, New Jersey, 1990.

[2] B. Boashash and P. O'Shea. Time-varying higher order spectra. In Franklin T. Luk, editor, *Advanced Signal Processing Algorithms, Architectures and Implementations*, San Diego, USA, July 1991. Proceedings of SPIE.

[3] B. Boashash, P. J. O'Shea, and M. J. Arnold. Algorithms for instantaneous frequency estimation: A comparative study. In Franklin T. Luk, editor, *Advanced Signal Processing Algorithms, Architectures and Implementations*, pages 24–46, San Diego, USA, August 1990. Proceedings of SPIE 1348.

[4] N. L. Gerr. Introducing a third order Wigner distribution. *Proceedings of the IEEE*, 76:290–292, March 1988.

[5] J. D. Thatcher and M. G. Amin. The running bispectrum. In *Proceedings of the Workshop on Higher-order Spectral Analysis*, pages 36–40, Vail, Colorado, USA, June 1989.

q	$b_0^{<q>}$	$b_1^{<q>}$	$b_2^{<q>}$	$b_3^{<q>}$	$b_4^{<q>}$	$b_5^{<q>}$	$b_6^{<q>}$
1	-1	1					
2	-1/2	0	1/2				
3	-1/3	-1/2	1	-1/6			
4	1/12	-2/3	0	2/3	-1/12		
5	-1/20	-1/2	-1/3	1	-1/4	-1/30	
6	-1/60	3/20	-3/4	0	3/4	-3/20	1/60

Table 1. Coefficients of the general phase difference

r	$d_0^{<r>}$	$d_1^{<r>}$	$d_2^{<r>}$	$d_3^{<r>}$	$d_4^{<r>}$	$d_5^{<r>}$	$d_6^{<r>}$
1	-1	0	1				
2	1/3	−8/3	0	8/3	−1/3		
3	−1/10	9/10	−9/2	0	9/2	−9/10	1/10

Table 2. - The values of the coefficients $d_k^{<r>}$ where $r = \lceil \frac{q}{2} \rceil$

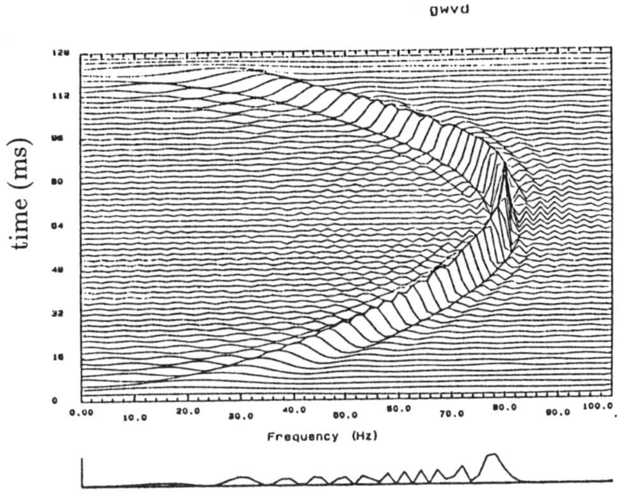

Figure 1: Generalized WVD of a quadratic FM signal.

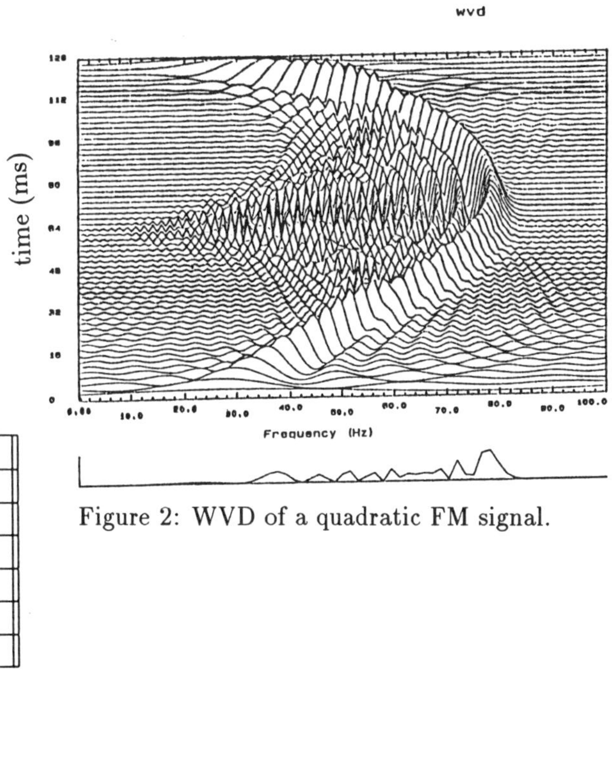

Figure 2: WVD of a quadratic FM signal.

Area–Efficient VLSI Implementation of FIR Digital Filters using Shifted Partial Products

Christopher Young and Douglas L. Jones*
Coordinated Science Laboratory
University of Illinois at Urbana-Champaign
Urbana, IL, 61801

Abstract

Many signal processing algorithms involve multiplications of data that can be organized as scalar–vector products. This paper describes a novel method of computing scalar–vector products based on shifted partial products of the vector elements. The method achieves a several–fold reduction in the computation required to implement a scalar–vector product. Of interest is the application of this idea to the highly parallel, area–efficient VLSI implementation of an FIR digital filter in the transpose form. The reduction in computation translates into an overall decrease in chip area required for a VLSI layout of the filter. As an example, using these ideas in a representative VLSI technology, a pipelined design yields a 30% decrease in chip area per tap and a 20% decrease in overall chip area compared to a modified Booth's multiplier implementation.

1 Introduction

Multiplications dominate the overall computational cost for many signal processing algorithms, whether this cost is measured in terms of computation time or chip area. In a conventional approach, multiplication can be considered a sequence of additions and shifts. The sequential multiplication of two b–bit numbers has a cost of between $b/2$ and b additions, depending on the choice of multiplier implementation. Many different methods to perform a multiplication exist, including the well–known modified Booth's multiplier. Past research has focused on developing efficient ways to reduce the computational cost, and some innovative techniques have been developed [1].

In common signal processing computations, multiplications can often be organized as a single number multiplying many numbers in parallel, or a *scalar–vector product.* This is true of a finite impulse response (FIR) digital filter in the transpose form, shown in Figure 1, where the scalar input multiplies many different coefficients. It is this application that is of particular interest in this work.

A simple yet novel method based on *shifted partial products* (SPP) of the vector elements is described [2–3]. The shifted partial products method provides substantial reductions in the number of additions required to implement a scalar–vector product compared to a simple binary approach. This savings in computation thus reduces the chip area required to implement an FIR filter, showing that the method compares favorably with other currently used techniques for reducing chip area.

The proposed method is coefficient–independent and regular, which allows for a straightforward implementation in VLSI. Section Two describes the method in detail. Section Three presents an efficient FIR filter architecture based on shifted partial products and discusses the difficulties associated with implementing the filter in VLSI using shifted partial products. It will be seen that the reductions in chip area compared to a Booth's multiplier implementation are possible.

2 Scalar Vector Product Computation Using Shifted Partial Products

The number of additions required to calculate a scalar–vector product is far less than the total number of additions required if each multiplication is computed separately. The method described here is referred to as the *shifted partial products method.*

The key concept is to regard multiplication as a summation of appropriately shifted partial products. Consider first the multiplication of two numbers, h and X, where h is a b–bit binary number, and X is a binary number of arbitrary bit–length. In a digit–

*Research supported by NSF grant no. MIP 8908777.

1058-6393/91 $01.00 © 1991 IEEE

coded multiplier, the b-bit binary number h is regarded as a $\lceil b/R \rceil$-digit number with R-bit digits d_k, where $\lceil . \rceil$ is the greatest integer value. Only $\lceil b/R \rceil - 1$ additions are required to compute the product of the digit-coded number with another number X, according to

$$X \cdot h = X \cdot \sum_{k=1}^{\lceil b/R \rceil - 1} d_k 2^{Rk} = \sum_{k=1}^{\lceil b/R \rceil - 1} d_k X \cdot 2^{Rk}. \quad (1)$$

The products $0X$, $1X$, $2X$, $3X$, ... $(2^R - 1)X$ are known as the set of *partial products*.

In general, since the value of h is not known in advance, the entire set of partial products $0X$, $1X$, $2X$, $3X$, ... $(2^R - 1)X$ must be computed. If only the multiplication of X with h is required, then the additional cost of calculating the partial products would not make this method worthwhile. However, for a scalar-vector product, this method becomes attractive. Since the same partial products can be used for all of the multiplications, the cost of calculating the partial products becomes a small fraction of the overall computation.

For the case of the FIR digital filter in the transpose form, all multiplications appear as a scalar-vector product. Thus, shifted partial products can be used in the implementation in VLSI of an FIR digital filter. This method also allows for a regular design, a desirable feature in VLSI implementations.

When implementing an FIR digital filter in VLSI, it is desirable to reduce the number of partial products explicitly calculated because the cost of creating partial products is quite expensive with regard to chip area. In general, for each input X that arrives, the partial products $1X$, $2X$, $3X, \ldots$, $(2^R - 1)X$ are needed.

In order to minimize chip area, not every partial product need be explicitly calculated. Only one quarter of the total partial products must be explicitly computed in this method, if we obtain even products by shifting odd products and use signed digit coding [2].

The number of additions required in the scalar-vector multiplication for a vector with N elements using shifted partial products is

$$C_s(N, b, R) = N\left(\lceil \frac{b+1}{R} \rceil - 1\right) + 2^{R-2} - 1 \quad (2)$$

compared to

$$C_o(N, b) = N(b - 1) \quad (3)$$

using a conventional approach (i.e., $R = 1$). A comparison of equations (3) and (4) shows that significant savings in the number of required additions are possible for large values of N.

3 Implementation Issues in the CAD VLSI Layout of an FIR Digital Filter

It is anticipated that the reduced number of additions using SPP will result in a reduction of chip area, and a quantification of this savings is desired. The best way to implement this filter and the problems in doing so are addressed in this section. A comparison will be made between a SPP VLSI implementation and a Booth's VLSI implementation.

The design issues are relatively independent of process technology. FIR filters in the transpose form were designed using a static logic 2.0 micron CMOS, single-poly, double-metal process supported by MOSIS. The CAD tool used is MAGIC, which supports a Manhattan geometry. Substantial efforts were made to optimize the design in terms of minimizing chip area. Many different layouts were attempted, and although the final design may not be optimal, it is sufficient to accurately compare to Booth's method in terms of chip area.

Two different filter comparisons were made: one used custom-designed cells, and the other used a modification of cells from a standard cell library. The custom-designed cells minimized chip area (using minimum size transistors), whereas the modified standard cells emphasized speed (using larger transistors). Both comparisons demonstrated area reductions using SPP.

3.1 Major Issues When Using SPP for a Chip Layout

For each input X that arrives at the filter, the partial products must first be generated. These generated partial products are then distributed in parallel to all of the locations on the chip where a filter coefficient multiplication occurs. When using SPP, reduction in the number of additions required for a multiplication is offset by the need to calculate and distribute the partial product values. Since there are many partial product lines, the chip area needed to distribute these is not insignificant.

To minimize this increase, it is most efficient to bus these lines over the active circuit elements (i.e., adder and edge-triggered register cells) instead of devoting area exclusively to bussing them. Therefore, metal 2 lines were used to bus these partial product lines, and the active circuit elements were designed using this metal. Note that there can be significant RC time delays for these long metal 2 lines, but this can be reduced by buffering.

It is desirable to have the taps laid out in one direction only because of the number of partial product lines, which will require chip area to "turn" the partial products lines from one row to the next (see Figure 2). This is possible for short length filters, but for longer length filters, taps can not be laid out in only one direction due to die sizes. This is also the case for the Booth's implementation, but fewer lines need to be "turned", so the chip area is less significant.

The second issue to be addressed when using SPP is the latency delay due to the necessity of creating the partial products. Since it will take at least one addition to make a partial product (e.g., 5X = 1X + 4X), and the time to do so will depend on the bit length, the latency is increased. Latency is not a problem for the Booth's implementation since only terms that are obtained without calculation are needed (i.e., 1X, 0X, -1X, -2X, 2X).

Throughput is usually a more important consideration than latency. By accepting latency, pipelining to multiples of $\lceil b/R \rceil$ bit level improves the throughput. Pipelining to this bit level introduces more registers which increases the chip area, but this favors SPP over Booth's since there are fewer terms to pipeline through. We note that $b > R$ for most applications. Pipelining to a smaller bit level would not be advantageous because of the increase in registers to hold partial products, and more metal 2 lines would be needed.

3.2 Layout Overview

Throughout this section, $R = 4$, $b = 16$, and $b/R = 4$-bit level pipelining is assumed. The coefficient and data bitlengths are arbitrary, but the value of R does have a bearing on the choice of pipelining and bit length.

The data is input as a 16–bit two's complement number, but to prevent overflow only values no greater in magnitude than 15–bits are allowed. This is necessary to prevent overflow when recoding the coefficients to use only small odd–valued partial products. The choice of input bit length can be increased or decreased, but it is convenient to choose a multiple of R. The filter coefficients are also 16–bits long but are signed numbers (sign bit included).

Carry–save adders are used as opposed to ripple carry adders. Using these adders, the propagation delay of a tap is in the direction of the data flow, and thus is not dependent on the wordlength [4]. Thus, the SPP method produces an inherently faster design than a Booth's implementation for large wordlengths when pipelining.

Figure 2 shows the floor plan of a filter using SPP. Three rows of taps, with area in one row to calculate the partial products, are shown. The partial product lines are routed parallel to the data flow. Because of Manhattan geometry constraints, chip area is required to redirect the partial product lines and data lines to the next row. A final stage is needed to collapse the carry–save adder outputs. With the larger frame sizes available commercially, a smaller fraction of the chip would be diverted to corner turning of the partial product metal 2 lines and partial product generation, yielding greater overall reduction in chip area. The layout for the Booth's implementation is similar, except that there is very little chip area needed for partial products generation (only -1X is calculated) and the area needed to turn the data lines is smaller.

3.3 Physical Details of the Tap

Figure 3 shows the block diagram of a single 4–bit level pipelined tap using SPP. Overall, four columns of adders (two to do a multiplication, two to do an add with previous register contents) are required to implement a tap. Nine columns of adders (seven to do a multiplication, two to do an add with previous register contents) are required for a Booth's multiplier to perform the same multiplication with the filter coefficient and add with the previous register's contents. The layout is uniform and has simple interconnections. Partial product and data lines using metal 2 run across these adders and registers.

Overflow adders and registers are placed at the top of each tap. The number of adders and registers is unlimited and will depend both on the accuracy desired for the filter and the length of the filter. Figure 3 shows four overflow bits, allowing for the possibility of filter lengths up to $2^4 = 16$.

In addition to the area consumed by the adders and D–latches, area is required to shift the temporary outputs of previous adders to inputs of the next stage. The outputs need to be shifted by b/R bits, since the partial product lines are bussed horizontally. Area must also be devoted to selecting shifted partial products.

3.4 Design of Adders and Registers

In order to bus the partial products and data lines in the most area–efficient manner, "tall, thin" full adders and edge–triggered registers were designed. Tall and thin adders and registers allow many partial product lines to horizontally pass over the active elements with a minimum increase in area. The adder and register cells are designed using only a single metal, allowing metal 2 lines to horizontally traverse the width of the adder with no interference. However, adder and register cells are larger than if metal 2 were used in their design.

To connect a partial product line to an adder, metal 1 adder input lines run vertically the length of the adder; thus, connections to the appropriate partial product line can be accomplished simply by placing a contact at the appropriate intersection. In this manner, a minimal amount of extra chip area is required for bussing the partial product lines. However, since the input vertical metal 1 lines need to run the length of the adder in order to connect to any possible partial product line, the width of the adder increases.

A CAD tool was designed which requires the user to input the coefficients and length of the filter. The filter then chooses the appropriate types of taps and connects the partial product lines at the appropriate points. This allows easy wiring of the taps, which is a time consuming process if done individually.

To simplify the implementation of a CAD tool in our design, all taps were made the same size. This allowed the tap locations to be coefficient-independent, simplifying the programming required to connect partial product lines at the appropriate locations. The difference in sizes for these taps is less than 3 percent, but a designer could choose to optimize each possible tap and thus reduce area slightly.

Many different adder and register cell sizes were designed. The "tallest, thinnest" adder and register we could design that still maintained minimum total area supports seven partial product lines, allowing a value of $R = 4$. A larger value of R may be used if an increase in chip area solely for interconnects can be justified.

Though a large value of R is desirable in regards to reducing the number of additions, it was not the best choice in terms of area efficiency. Thus, for the final choice of adder and register cell sizes, the optimal choice of R was $R = 4$, and only the four partial products $1X$, $3X$, $5X$, and $7X$ were needed. Other horizontal lines were required for shifting intermediate values and bussing previous tap values to the next tap. Area is required at each tap location for shifting the bussed subset of partial products to create the even-valued and higher odd-valued partial products; nonetheless, the overall chip area was decreased.

3.5 Comparison in Chip Area

Table 1 shows a comparison in chip area using custom-designed cells with filters of length 32. Comparing tap sizes for the shifted partial products method and modified Booth's method, a 30% area reduction is possible when pipelining. Overall, a 19% area reduction is achieved. Different pipelining levels also result in reduction of chip area.

For the designed filter, the difference between the 19% overall reduction and the 30% reduction in tap size was due to the chip area required for the creation of the partial products and the increased number of lines which must be redirected to the next row using the shifted partial products method. With larger dies and longer filters, overall reduction will approach 30%.

For a non-pipelined tap, a 20% area reduction is gained. This reduction is not as large as the pipelined case, because more registers are needed for the Booth's method when pipelining, and these registers are removed. A 12% reduction in the overall filter layout is possible compared to a filter designed using Booth multiplication.

The partial product generator requires the area of 2.5 taps, which is a substantial. The area required to "turn" lines is not small for the SPP case, as expected, but is also substantial for the Booth's implementation case. A better way to reduce these areas is not known to the authors.

Table 2 shows the comparison of values for the filters built with the modified cell library. The results are quite similar to those obtained with the custom-designed cells, with a 29% reduction per tap for the pipelined case.

Acknowledgements

Thanks to Wendy Knepp for reviewing this manuscript, and to David Lindley at MS State for help in obtaining the standard cell library.

References

[1] K. Hwang, *Computer Arithmetic: Principles, Architecture, and Design,* John Wiley, 1979.

[2] D.L. Jones, "Efficient Scalar-Vector Product Computation for Signal Processing," *Proc. 23rd Asilomar Conference on Signals, Systems, and Computers*, Pacific Grove, CA, pp.28–32, October 30–November 1, 1989.

[3] C. Young and D.L. Jones, "Area-Efficient VLSI Implementation of Digital Filters via Multiple Product Intercoding," *32nd Midwest Symposium on Circuits and Systems*, Champaign, Illinois, pp.373–376, August 14-16, 1989.

[4] P. Yang, R. Jain, T. Yoshino, W. Gass, A. Shah, "A Functional Silicon Compiler For High Speed FIR Digital Filters," *Proc. 1990 International Conference on Acoustics, Speech, and Signal Processing*, Albuquerque, NM, pp.1329–1332, April 3–6, 1990.

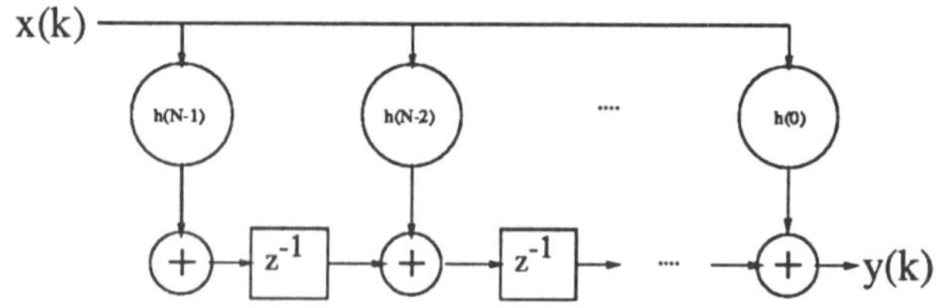

Figure 1: FIR filter in transpose form

Type	Pipeline	λ^2 (SPP)	λ^2 (Booths)	% Area
Overall	none	53,965,440	61,483,008	87.8%
Tap	none	1,439,491	1,796,928	80.1%
Overall	4	70,236,745	87,134,624	80.6%
Tap	4	1,729,406	2,467,920	70.1%
Spacing	none	3,388,224	1,327,104	255%
Spacing	4	4,356,288	2,096,640	207%
Partial	none	1,125,280	0	NA
Partial	4	4,202,900	0	NA
Ender	4	1,980,277	1,871,264	106%
Adder	4	7,228	5,934	121%
Register	4	8,645	8,272	105%

Table 1: Comparison of Area for Custom Design Filters

Type	Pipeline	λ^2 (SPP)	λ^2 (Booths)	% Area
Tap	none	2,098,228	2,809,318	74.7%
Tap	4	2,330,622	3,281,695	71.0%
Adder	none,4	13,860	11,304	123%
Register	4	8,484	7,272	117%

Table 2: Comparison of Area for Modified Cell Library Filters

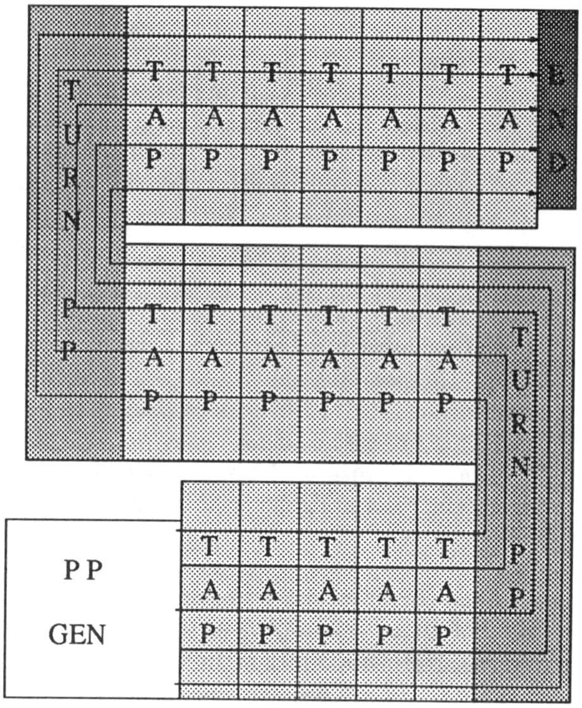

Figure 2: Filter layout

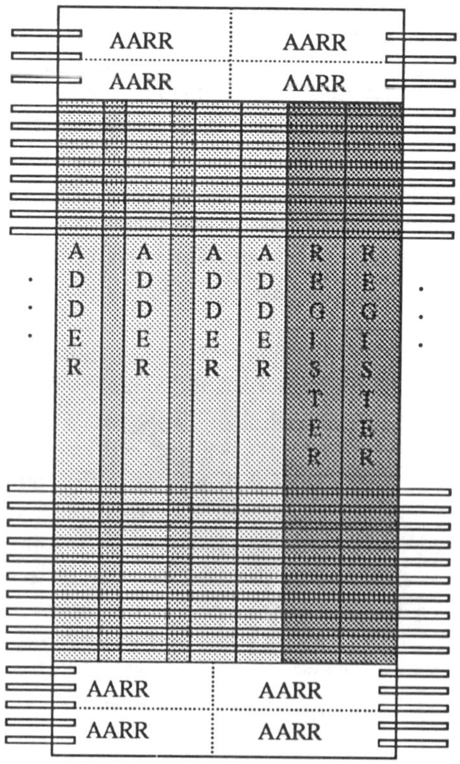

Figure 3: Tap using SPP

An Architecture for Parallel Multipliers

Zhongde Wang, G.A. Jullien, W. C. Miller

VLSI Research Group, Department of Electrical Engineering
University of Windsor, Windsor, Ontario, Canada N9B 3P4

Abstract

An architecture for parallel multipliers, based on two-bit full adders, is proposed. Multipliers built in this way possess the same structure as previously published five-counter multipliers, but with less cells and one less stage. The architecture is particularly suited to n-FET complex block pipelined dynamic logic, built with Switching Trees, and it is shown that the two-bit adder cell requires smaller area than the five-counter cell. Therefore, this new multiplier architecture gains in both speed and area efficiency.

1 Introduction

Multipliers are of vital importance for modern computers and for a variety of signal processors. Although column compression techniques [1,2], which reduce the critical path delay to $O(\log_{1.5}n)$ cell delays, have been proved to provide time optimal techniques for building multipliers, the resulting irregular interconnections between cells produce awkward structures for implementation in silicon. Parallel multipliers, which require a computation time of 2(n-1) cell delays, are often adopted for silicon implementation because of the regular structure and simple interconnection between cells [3,4].

Conventionally, parallel multipliers are built with a binary full adder as the basic cell [3,4]. An nxn bit parallel multiplier built in this way requires 2(n-1) cell delays to complete the computation [3]. Recently, Nakamura proposed several algorithms for parallel multipliers which require only n+1 cell delays to perform the multiplication operation [5]. Under the assumption that the cell delay for a binary full adder and for the five-counter, the cell proposed by Nakamura, are the same, multipliers with architectures proposed by Nakamura will yield about half the latency of conventional parallel multipliers. Maden and Guy [7] proposed another architecture for parallel multipliers which uses (2,2,3) counters as basic cells. Among these different multiplier architectures, the five counter multiplier appears to be the most useful and a VLSI implementation of this type of multiplier has been reported [6].

In this paper, we propose a multiplier architecture which uses a two-bit full adder as the basic building cell. In a dynamic logic complex block implementation, the multiplier possesses almost the same structure as the five-counter multiplier, but with fewer cells and a slightly lower critical path delay.

2 The Component Blocks

Our implementation procedure, at the circuit level, is to use complex dynamic blocks. Although these complex circuits have been proposed, and used, over the past 8 years, we believe that we have developed a new design approach that allows graph theoretic minimization of the blocks directly from a truth table or minterm description. The logic is either pipelined at each complex block evaluation node [10], or placed in a dynamic combinational chain using some form of differential logic (e.g. DCVSL [9]). The pipelined approach uses a true single-phase clocked D-latch [10] in which the complex block is embedded as shown in Fig. 1.

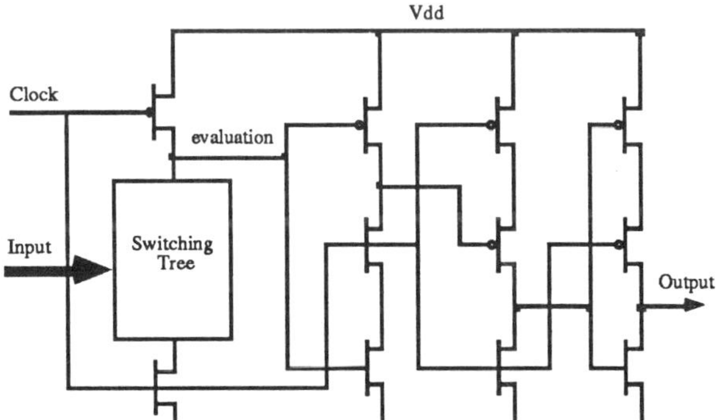

Fig. 1 Embedding the Tree in a TSPC D-Latch

The complex block starts as a binary tree and retains much of the original tree characteristics, even when fully

1058-6393/91 $01.00 © 1991 IEEE

minimized. The important characteristics are: minimized drain/source capacitance at each node (3 in the full tree) and fixed path structure (a path of height n with the equivalent of one extra transistor connection at each node). Techniques based on logic gate minimization and mapping do not necessarily yield these structures, particularly for complex, merged blocks.

The numbers of inputs and outputs of the two-bit full adder are identical to those of the five-counter of ref. [6]. Our particular graph theoretic design techniques will yield a tree height and complexity strongly dependent on the number of inputs for these type of adder/counter circuits [8]. For basic comparison purposes, therefore, it will be reasonable to assume that the complexity of both cells is the same.

We will now discuss each of the components used in the original multiplier [5] and in our new architecture.

2.2 Five-Counter

The five-counter is shown, in block form, in Fig. 2.

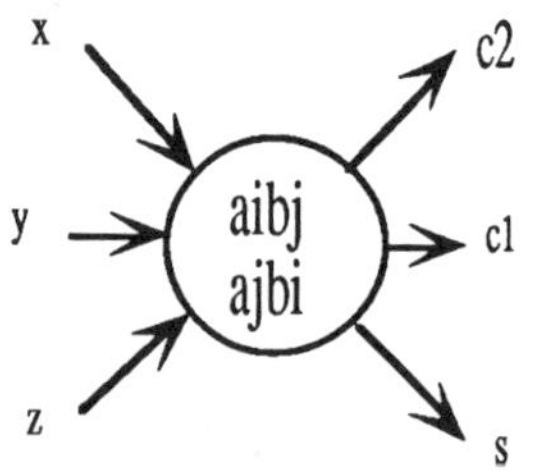

Fig. 2 Five-Counter

We have retained the diagrammatic form used in [5], where two of the inputs are the bit level multiplication (logical AND) of appropriate input bits to the multiplier represented by the logical expressions in the centre of the circle. The other 3 1-bit inputs are shown in conventional form from the left. The 5 single-bit weight inputs are summed (or counted) to produce the 3-bit weighted output on the right of the diagram. This is sometimes referred to as a 5-3 compressor, and is conventionally used in binary arithmetic circuits.

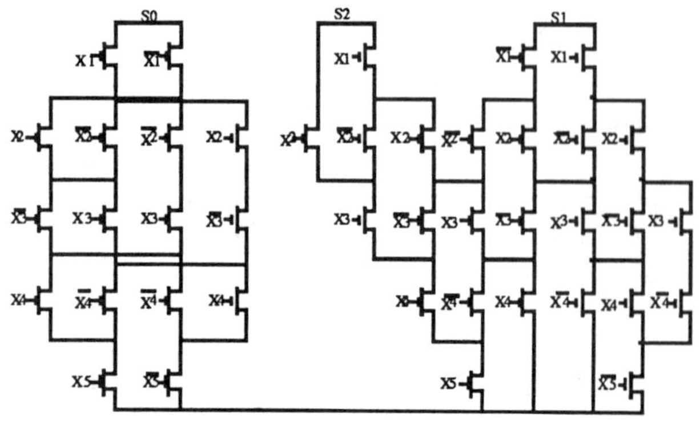

Fig. 3 n-FET Complex block for the five-counter

The relationship between inputs and outputs is given by eqn. (1).

$$x+y+z+a_i b_j+a_j b_i = 4c_2+2c_1+s \qquad (1)$$

The merged n-FET circuit is shown in Fig. 3. This circuit is inserted into the *Switching Tree* block in Fig. 1. Because the merged trees use common inputs and ground switch, these 'costs' are effectively amortized over the large number of transistors making up the block. For differential dynamic logic, such as DCVSL [9], a set of complement outputs can be provided, with full merging, to allow the differential signal generation.

We have established, through the fabrication of more complex test cells, that the designs resulting from the switching tree graph theoretic minimization procedure have acceptable charge sharing problems up to 6 inputs for these types of adder/counter circuits. We have also demonstrated that our mask extracted analog simulations are accurate in detecting such charge sharing problems. All blocks used in this architecture have been simulated using mask extracted data, and perform with appropriate noise margins associated with the evaluation nodes driving the D-latch output stage. We have techniques, that can be applied at layout time, to eliminate charge sharing problems, for particularly heavily merged designs, if this is required.

2.3 Two-Bit Full Adder

The two-bit full adder, shown in Fig. 4, produces a three-bit output number $c_0 s_1 s_0$ from two two-bit input numbers $m_1 m_0$, $n_1 n_0$ and a single-bit carry-in c_i. The relationship between inputs and outputs is given by eqn. (2).

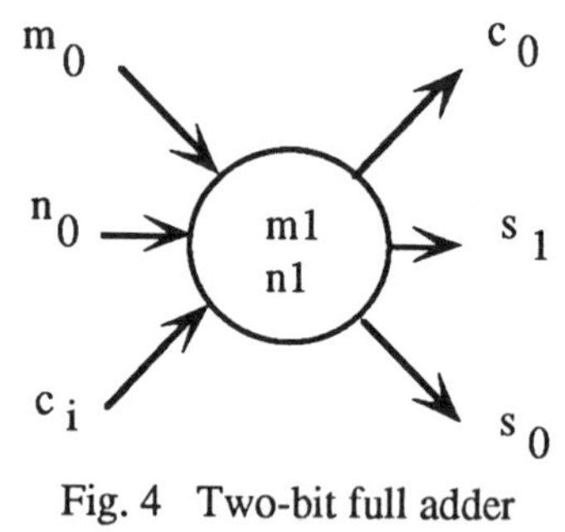

Fig. 4 Two-bit full adder

$$4c_0+2s_1+s_0 = 2(m_1+n_1)+m_0+n_0+c_i \qquad (2)$$

The logic expressions for s_0, s_1, c_0 are given by:

$$s_0 = m_0(n_0 c_i + \bar{n}_0 \bar{c}_i) + \bar{m}_0(n_0 \bar{c}_i + \bar{n}_0 c_i) \qquad (3)$$

$$s_1 = (m_1 n_1 + \bar{m}_1 \bar{n}_1)[m_0 n_0 + (m_0 + n_0)c_i]$$

$$+ (m_1 \bar{n}_1 + \bar{m}_1 n_1)[\bar{m}_0 \bar{n}_0 + (\bar{m}_0 + \bar{n}_0)\bar{c}_i] \qquad (4)$$

$$c_o = m_1 n_1 + (m_1 + n_1)[m_0 n_0 + (m_0 + n_0)c_i] \quad (5)$$

The merged two-bit adder block is shown in Fig. 5.

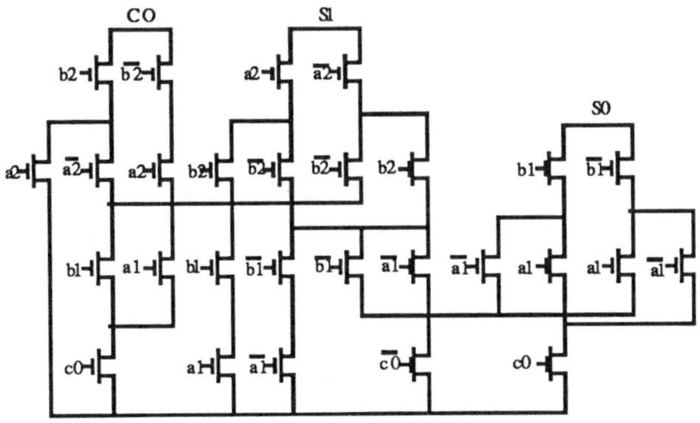

Fig. 5 n-FET Complex Block for 2-bit Adder

3 The Multiplier Architectures

Since this new architecture represents an improvement on an existing architecture, we will start by discussing the existing multiplier structure [5].

3.1 The Five-Counter Multiplier

Fig. 6 shows the architecture of a 6*6 multiplier built with five-counter cells. In addition to the network of five-counter cells, an input AND plane (implied in Fig. 6) is used to generate the partial product elements and standard full adders and XOR gates are used to form final sums of the partial products.

There is a redundancy associated with the fact that although the three outputs of the basic cell can accommodate a count of seven equally weighted inputs, the maximum count is only five. This redundancy reflects in an inefficiency in the overall architecture. There is also considerable redundancy associated with the five-counters on the first stage, i.e. cells on the most left column in Fig. 6, since these have only two active inputs.

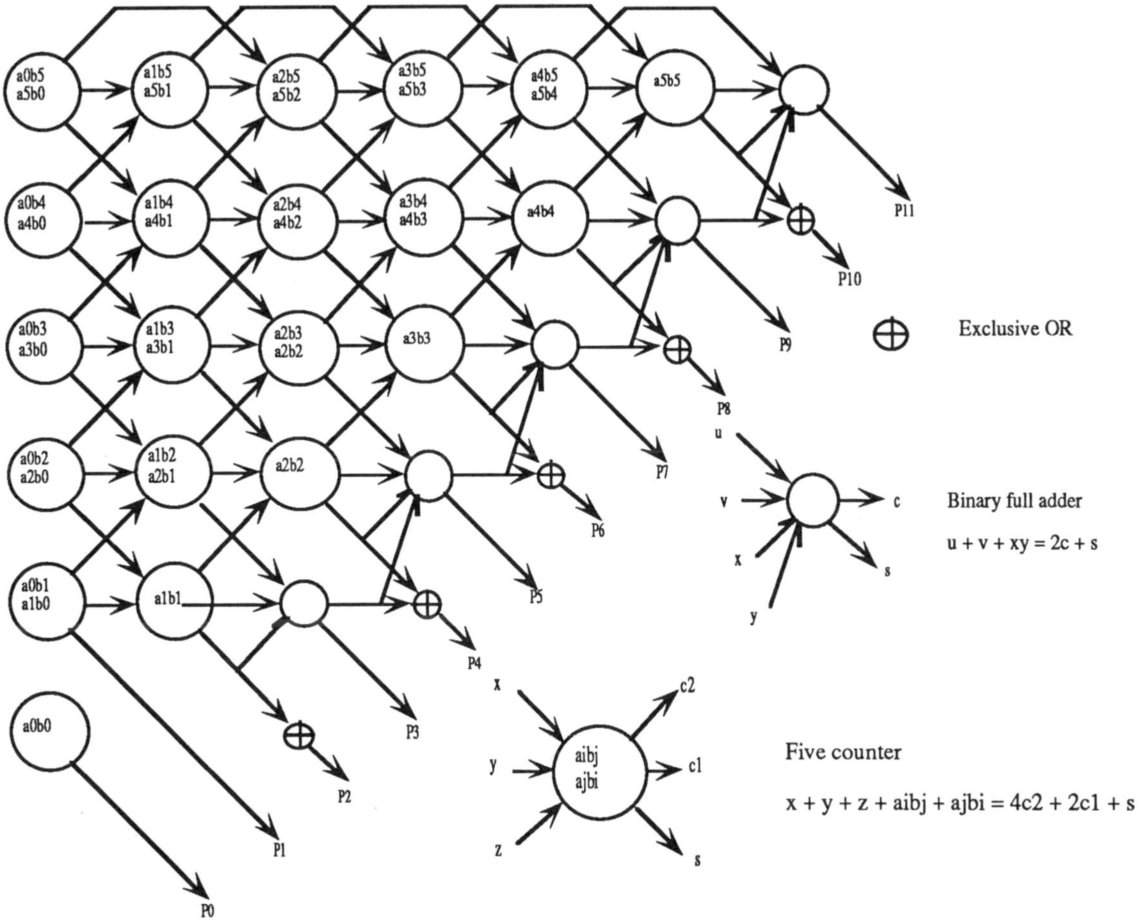

Fig. 6 Five-Counter based 6 x 6 multiplier

The use of the 2-bit adder (the output of which completely spans the 3 bits), in our new architecture, removes the first redundancy, and also helps eliminate the second.

3.2 The New Two-Bit Full Adder Multiplier

Fig. 7 shows the architecture of a 6x6 two-bit full adder multiplier. The two-bit full adder block is used to add five independent one-bit numbers. Three of them are generated by cells of the previous stage, and they possess the same weight. The other two numbers, each with a weight twice the weight of the previous stage inputs, are elements of the partial products of the two n bit input numbers:

$$b_i \cdot (a_{n-1} a_{n-2} \ldots\ldots a_0); \qquad i = 0 \ldots n-1$$

The structure is similar to that of inner part of the five-counter multiplier (starting one column in); the internal product elements are identical. Each cell has symmetric partial product elements for the most significant inputs: $a_i b_j$ and $a_j b_i$ for off diagonal cells; $a_i b_i$ for cells on the diagonal. The first column has lower significant inputs generated from the a_0 and b0 AND planes at the input (implied in both Fig. 6 and Fig. 7). Here we see the elimination of the underutilized first column of the five-counter multiplier. The two-bit full adder cell is symbolized as C_{ij}. The weight of $a_i b_j$ and $a_j b_i$ is 2^{i+j}. Consequently, the weight of the other three inputs from the previous stage should be 2^{i+j-1}. The weight of the output of this cell, referring to Fig 7, is 2^{i+j+1}, 2^{i+j}, and 2^{i+j-1} for c_o, s_1 and s_0, respectively.

Theorem 1 proves that the correct weight is received from the prior stages, proving that every cell in Fig. 7 is correctly arranged.

Theorem 1: The weight of all the inputs of C_{ij} from the previous stages are 2^{i+j-1}.

Proof: The three inputs of C_{ij} from the previous stage are:

s_o from $C_{i-1,j+1}$
s_1 from $C_{i-1,j}$
c_o from $C_{i-1,j-1}$

Their weights are:

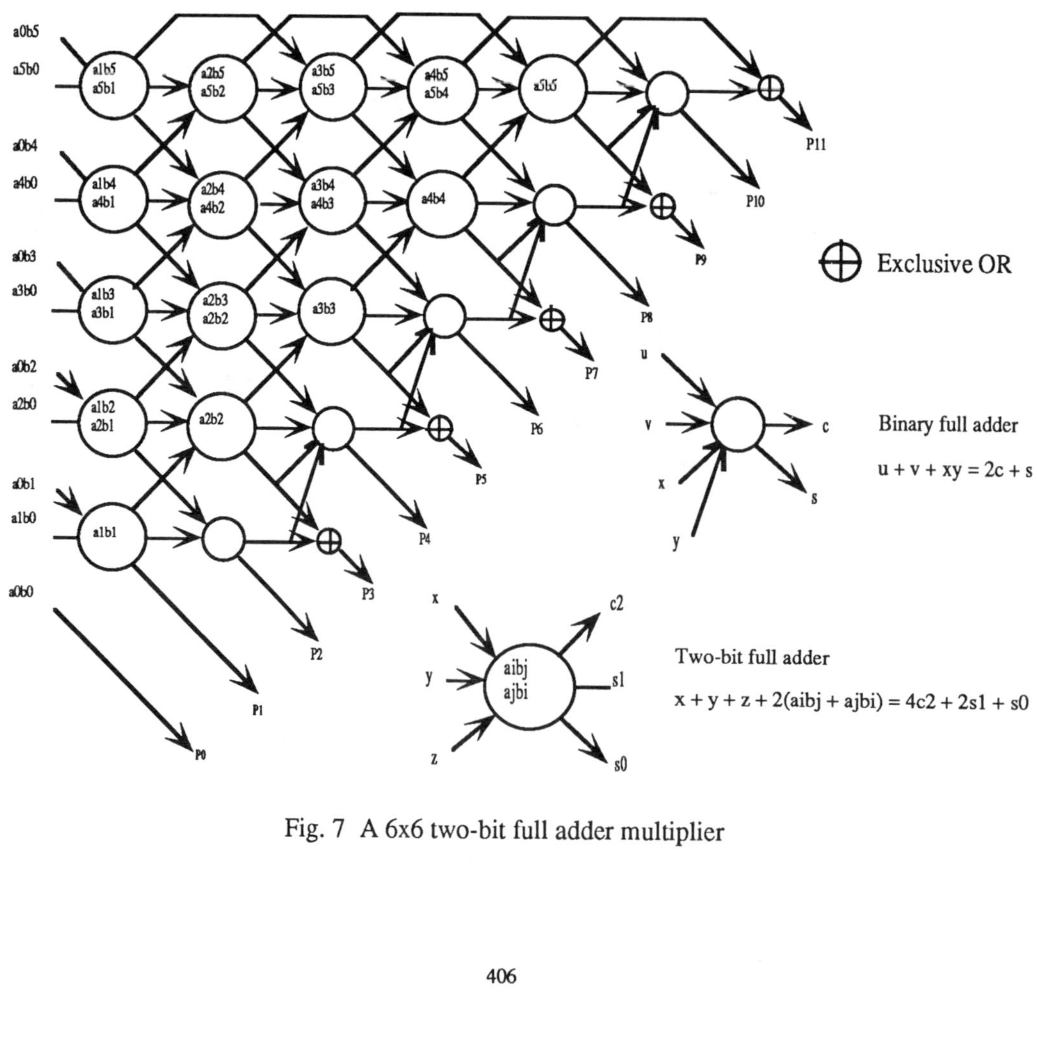

Fig. 7 A 6x6 two-bit full adder multiplier

$$2^{(i-1)+(j+1)-1} = 2^{i+j-1}$$
$$2^{(i-1)+j} = 2^{i+j-1}$$
$$2^{(i-1)+(j-1)+1} = 2^{i+j-1}$$

thus they all have the same weight of 2^{i+j-1}. ❑

4 Comparison of the architectures

There several advantage of the 2-bit adder cell over the five-counter cell.

The first advantage is the fact that the two-bit adder logic block requires fewer transistors than the five-counter counter block. As can be shown by Figs 3 and 5, a merged tree for a two-bit adder requires 28 working transistors; while a merged tree for a five-counter requires 41 transistors, a 46% increase.

The second advantage is concerned with evaluation time of the main cells used in the two architectures. The two-bit adder cell has a critical discharge path of only 4 transistors, compared to a path of 5 transistors for the five-counter cell. For reference, the discharge path for the binary adder is 3. We obviously have a better match between the full adder and 2-bit adder cells for the new architecture, and a pipelined version will be able to be clocked at a speed closer to that of the full adder cell.

A third advantage of the two-bit adder over the five-counter is the fact that the two-bit full adder cell can handle inputs with different weights, whereas the five-counter has to have inputs with the same weight. All cells on the first stage of the two-bit full adder multiplier can add two pairs of partial products together. As a result, in comparison with the five-counter multiplier, the two-bit full adder multiplier requires one less stage of cells. i.e., n less cells for an nxn bit multiplier.

For a strictly combinational implementation, this latter advantage will shorten the evaluation time in addition to the decrease offered by the decrease in cell evaluation time. An estimate of a 20% reduction in cell delay together with a 15% decrease in cells in the critical path will result in close to a 30% decrease in evaluation time for the new architecture. As the number of multiplier bits increases this decrease in time will asymptotically approach 20%, as the efficiency of the first stage reduction becomes insignificant.

If we consider the first and third comparison we can quickly determine that, for the 6-bit multiplier, the area reduction is almost 50%. This is a somewhat rough guess, since cell area is not necessarily a linear function of transistor count, but, clearly, the area savings will be substantial.

5 Conclusions

A new architecture for parallel multipliers has been developed based on a previously published architecture that used five-counters as the basic building block. The new architecture removes the redundancy associated with an incompletely spanned output range of the five-counter cell, by changing the building block to a two-bit full adder. This cell has the same number of inputs, but produces an output that completely spans the output dynamic range. We also show that a dynamic logic implementation of the five-counter has a 46% hardware increase over the two-bit full adder cell proposed for this new architecture. The resulting two-bit adder cell architecture shows an almost 50% decrease in silicon area and almost a 30% decrease in evaluation time compared to the five-counter cell architecture.

6 References

[1] C. S. Wallace: A suggestion for a fast multiplier, IEEE Trans. Elec. Comput., vol. EC-13, pp. 14-17, 1964.

[2] L. Dadda: Some schemes for parallel multipliers, Alta Freq., vol. 34, pp. 349-356, 1965.

[3] A. Habibi and P. A. windy: Fast multipliers, IEEE Trans. Computers, vol. C-19, 153-157.

[4] J. Y. Lee et al: A High-speed high-density silicon 8x8-bit parallel multiplier, IEEE J. Solid-State Cir. vol. SC-22, pp.35-40, 1987,

[5] S. Nakamura: Algorithms for iterative array multiplication, IEEE Trans. Comput. vol. C-35, 713-719, 1986.

[6] S. Nakamura and K-,Y. Chu: A single chip parallel multiplier by MOS technology, ibid, C-37, 274-282, 1988.

[7] B. Maden and C. G. Guy: Parallel architectures for high speed multipliers, Proc. ISCAS'89, pp. 142-145, 1989.

[8] D. Zhang, G.A. Jullien, W.C. Miller and E. Swartzlander, 1991, "Arithmetic for Digital Neural Networks." Proc. 10th International Symposium on Computer Arithmetic, pp. 58-63.

[9] M. K. Chu and D. I. Pulfrey. "Design procedures for differential cascode-Voltage switch circuits." *IEEE J. Solid-State Circuits.* vol.SC-21, pp. 1082-1087, 1986.

[10] J. Yuan and C. Svennson. "High-Speed CMOS Circuit Technique." *IEEE. J. Solid-State Circuits.* vol. 24, pp. 62-70, 1989.

A MASSIVELY PARALLEL RNS ARCHITECTURE

Khaled M. Elleithy & Magdy A. Bayoumi*

Computer Engineering Department
King Fahd University of Petroleum and Minerals
Dhahran 31261, Saudi Arabia

Abstract

In this paper parallelism on the algorithmic, architectural, and arithmetic levels is exploited in the design of a Residue Number System (RNS) based architecture. The architecture is based on modulo processors. Each modulo processor is implemented by two dimensional systolic array composed of very simple cells. The decoding stage is implemented using a 2-D array, too. The decoding bottleneck is eliminated. The whole architecture is pipelined which leads to high throughput rate.

1. Introduction

Parallelism can be achieved at different levels; algorithmic, architectural, language, and arithmetic. Parallelism on the arithmetic level has not been exploited at the same degree as on the first three levels. On the arithmetic level, parallelism is achieved at a fine granularity which means a great impact of technology on the followed approach to achieve such parallelism. In the same time, the arithmetic parallelism (AP) pattern and mechanism can influence the architectural parallelism strategy granularity. AP is gaining recognition as technology is approaching its limits and parallel architectures have not achieved the expected performance in solving many problems in artificial intelligence, neural networks, and complex DSP applications.

In this paper, Residue Number System (RNS)[1] has been used to achieve arithmetic parallelism. The algebraic properties of RNS provide both high speed computation and parallel operations[2]. RNS is very efficient for algorithmic specific architectures in which multiplication and addition are employed extensively. Figure 1 shows a general RNS-based architecture. Orthogonal parallelism is achieved through the set of independent modulo processors, they operate in an SIMD fashion. The decoding stage can represent a computational bottleneck.

In our implementation, parallelism is achieved in three directions; each modulo processor is implemented by two dimensional systolic array composed of very simple cells. The decoding stage is implemented using a 2-D array, too. The decoding bottleneck will be eliminated. The whole architecture will be pipelined which leads to high throughput rate.

* Second author is with The Center for Advanced Computer Studies, University of SW Louisiana, USA.

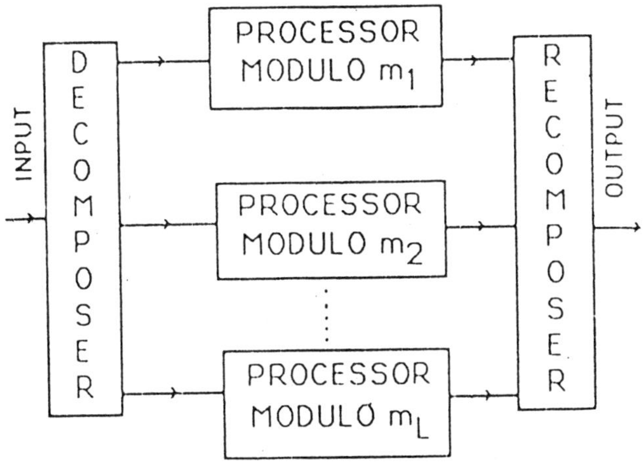

Figure 1. General RNS Based Architectures.

2. Modulo Adder

The modulo adder performs addition in time complexity O(1)[3]. Figure 2 shows the modulo adder architecture. It is composed of fine 1-D arrays independent of the size of the moduli. The modulo addition based on the following algorithm:

Procedure:

```
begin
  Do in parallel
    begin
      Call Sum(temp_1 , A_S , A_C , B_S)
      Call Carry(temp_2 , A_S , A_C , B_S)
    end
  Do in parallel
    begin
      Call Sum(temp_3 , temp_1 , temp_2 , B_C)
      Call Carry(temp_4 , temp_1 , temp_2 , B_C)
    end
  Case ( temp_2[n+1]+temp_4[n+1] ) of
    0: Do in parallel
        begin
          Result_S := temp_3
          Result_C := temp_4
        end
        exit
    1: Do in parallel
        begin
          Call Sum(temp_5 , temp_3 , temp_4 , (2^n - m))
          Call Carry(temp_6 , temp_3 , temp_4 , (2^n - m))
        end
    2: Do in parallel
        begin
          Call Sum(temp_5 , temp_3 , temp_4 , 2*(2^n - m))
          Call Carry(temp_6 , temp_3 , temp_4 , 2*(2^n - m))
        end
  end case
  Case ( temp_6[n+1] ) of
    0: Do in parallel
        begin
          Result_S := temp_5
          Result_C := temp_6
        end
        exit
    1: Do in parallel
        begin
          Call Sum(temp_7 , temp_5 , temp_6 , (2^n - m))
          Call Carry(temp_8 , temp_5 , temp_6 , (2^n - m))
        end
  end case
  Case ( temp_8[n+1] ) of
    0: Do in parallel
        begin
          Result_S := temp_7
          Result_C := temp_8
        end
    1: Do in parallel
        begin
          Call Sum(temp_9 , temp_7 , temp_8 , (2^n - m))
          Call Carry(temp_10 , temp_7 , temp_8 , (2^n - m))
        end
      Do in parallel
        begin
          Result_S := temp_9
          Result_C := temp_10
        end
  end case
end.

Sum (A , B , C , D)
begin
  Do in parallel (1 ≤ i ≤ n)
    A[i] := (B[i]∧C[i]) ∨ (B[i]∧D[i]) ∨ (C[i]∧D[i])
end

Carry (A , B , C , D)
begin
  A[1] := 0
  Do in parallel (1 ≤ i ≤ n)
    A[i+1] := B[i] ⊕ C[i] ⊕ D[i]
end
```

3. Modulo Multiplier

The modulo multiplier is using the modulo adder as a computational kernel. The multiplier consists of two stages as shown in Figure 3. In the first stage an array of AND gates is used to obtain the partial products. The second stage of the adder is a binary tree of modulo

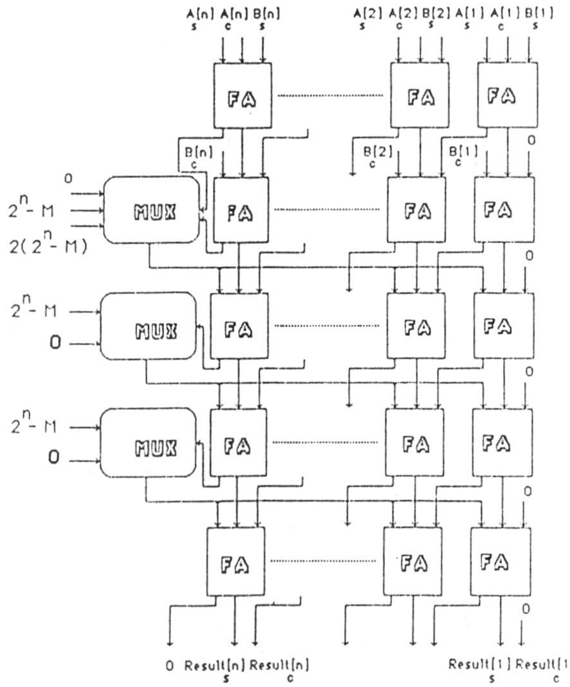

Figure 2. Different Stages of the Modulo Adder.

adders used to perform the addition of the n partial products. The modulo multiplier has a time complexity of O(logn).

4. The decoding stage

The proposed decoding stage is based on the Chinese Remainder Theorem(CRT)[4], which states that:

$$\left|X\right|_M = \left|\sum_{j=1}^{N} \hat{m}_j \left|\frac{r_j}{m_j}\right|_{m_j}\right|_M$$

where

$$M = \prod_{j=1}^{N} m_j, \quad \hat{m}_j = \frac{M}{m_j}$$

and

$$GCD(m_j, m_k) = 1 \qquad \text{for } j \neq k$$

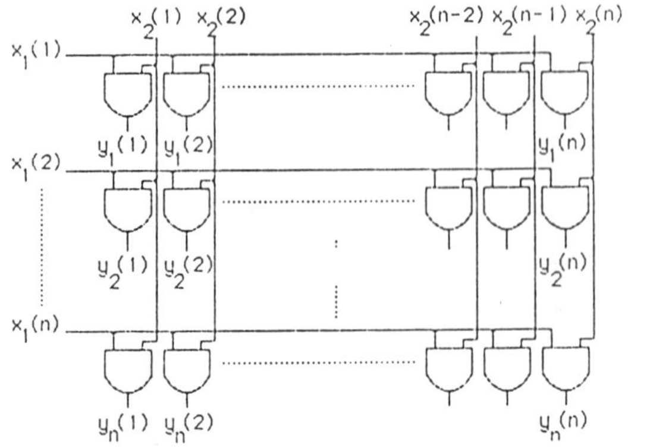

(a) Partial Product Generator.

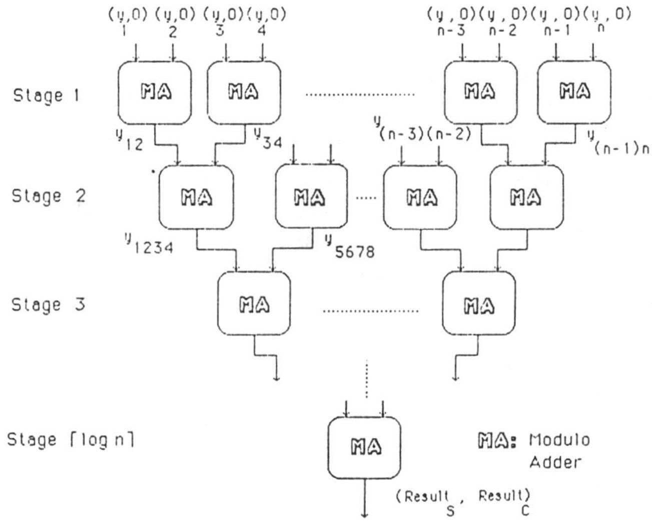

(b) Addition of Partial Products Using Modulo Adders.

Figure 3. Modulo Multiplier Architecture.

Although the CRT provides a direct, fast, and simple conversion formula, the lack of large and fast modulo M adder has held back this approach. Systolizing the CRT will overcome this problem.

The proposed systolic CRT architecture can produce both sign-magnitude and 2's complement data representation. The overall structure is shown in Figure 4. The Inputs to the residue decoder are the residues and a control line, C, which determines the output to be in sign magnitude or 2's complement representation. The partial sum generator computes the partial sums,

$$t_i = \left| \hat{m}_i \left| \frac{r_i}{m_i} \right|_{m_i} \right|_M$$

The partial sum addition is computed using the developed modulo adder. The range determinator is reponsible for the correctness of data. The input data to this stage is decomposed into groups of bits to be processed in parallel. Figures 5,6 show the implementation of the range determinator for both sign magnitude and two's complement cases, respectively. The overall time complexity of the CRT systolic architecture is O(log n), where N is the number of moduli.

5. Conclusions

We have presented a SIMD architecture which exhibits two level parallelism; architectural (systolic array) and arithmetic (RNS). It is a good candidate for specific algorithms. High speed algorithms for modulo addition, modulo multiplication, and RNS decoding have been presented in the paper.

Acknowledgements

The first author wishes to thank King Fahd University of Petroleum and Minerals for support.

References

[1] F. J. Taylor, "Residue Arithmetic: A Tutorial with Examples," IEEE Computer Magazine, pp. 50-62,May 1984.

[2] K. M. Elleithy, M. A. Bayoumi, and K. P. Lee, "O(log n) Architectures for RNS Arithmetic Decoding," Proc. of the 9th Symposium on Computer Arithmetic, pp. 202-209, Sep. 1989.

[3] K. M. Elleithy and M. A. Bayoumi, "A O(1) Algorithm for modulo Addition," IEEE Transactions on Circuits and Systems, vol. 37, no. 5, pp. 628-631, May. 1990.

[4] N. S. Szabo and R. I. Tanaka, Residue Arithmetic and its Applications to Computer Technology, New York: McGraw-Hill, 1967.

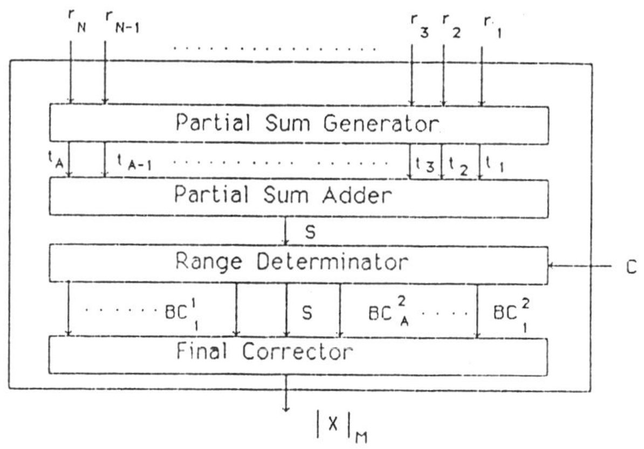

Figure 4. Block Diagram of the Residue Decoder.

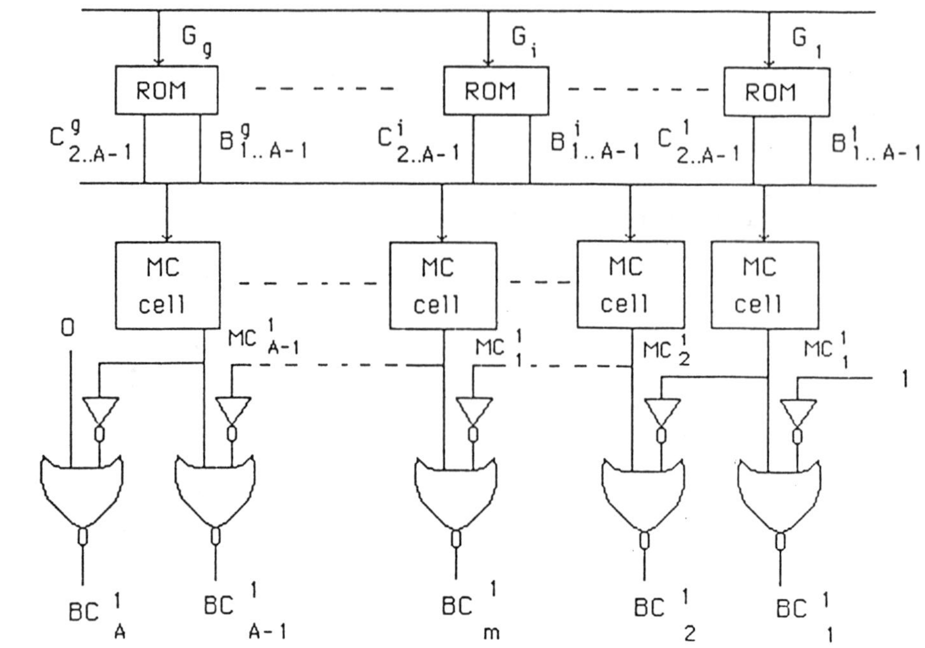

Figure 5. First stage of Range Determinator.

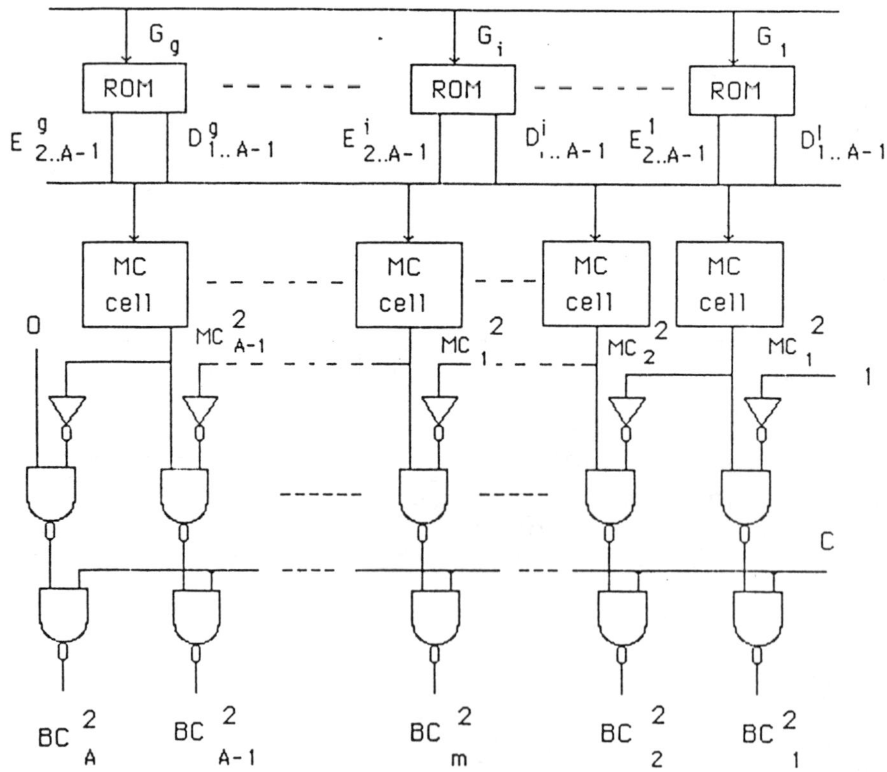

Figure 6. Second stage of Range Determinator.

ROM Table Reduction Techniques For Computing The Squaring Operation Using Modular Arithmetic

Alexander Skavantzos

Louisiana State University, Baton Rouge, Louisiana

Abstract

Modular arithmetic is necessary when using the Residue Number system. Such a system is suitable for high-speed Signal Processing, due to its carry free and parallel nature. In this paper, we present techniques for minimizing the size of the look-up table ROMs used for computing the squaring operation, when arithmetic modulo m is performed. Such squaring operations are necessary when implementing the quarter squared and the one over eight squared algorithms. These squared-law algorithms are useful in computing multiplicative intensive operations, such as convolutions, correlations and complex multiplications. The presented memory compression schemes can result in significant ROM savings, with the only extra cost of a small overhead.

1. Introduction

A number of important signal processing and communication applications use algorithms which are based on complex operations. Such operations are complex multiplications, linear and cyclic convolutions, autocorrelations and cross-correlations. Since speed and cost are the main design factors, Digital Signal Processing (DSP) engineers and scientists have been looking at various speed/cost efficient techniques for implementing their algorithms.

Modular arithmetic has been quite common in DSP applications, since it can be combined with the use of the Residue Number System (RNS); an integer system capable of performing parallel, carry-free, high-speed arithmetic [1]-[2]. Since the RNS is an integer system, it can easily handle only the simple operations of addition, subtraction and multiplication. Divisions do not always produce integer results and as such, they are not preferred operations in such a system, while magnitude comparisons are difficult because of the unweighted nature of the RNS. As a result, the Residue Number System has been receiving significant attention for Digital Signal Processing applications, where the dominant operations are the operations of addition, subtraction and multiplication.

An RNS is defined by a set of relatively prime integers $P = \{m_1,...,m_L\}$, where P is called the set of moduli. In such a system, arithmetic is performed "modulo m_i", while the computations between two different moduli m_i and m_j are independent from each other. Since RNS implementations of DSP tasks rely on additions, subtractions and multiplications, the focus of many RNS researchers has been on efficient implementations of the above arithmetic operations "modulo m". References [3]-[4] represent some of the existing work on addition and subtraction in RNS. On the other hand, multipliers "modulo m" can be implemented using several techniques. Among these techniques, one can refer to the look-up table technique using ROMs [5]-[6], the index calculus technique which can only be used with prime moduli [7] and the quarter squared algorithm technique [8]-[10].

Recently, a new squared law algorithm has been developed for modulo m arithmetic. It is "the one over eight squared algorithm" and can be used in computing convolutions and correlations [11].

The following points should be clarified here.

1. Both the quarter squared and the one over eight squared algorithms can be used with arithmetic modulo m, for any modulus m.
2. These two algorithms require only squaring operations and additions, but not two-operand multiplications.
3. When the two algorithms are used with arithmetic modulo m, ROM look-up tables can be used to perform the squaring operations modulo m.
4. Both the algorithms reduce the size of the ROM tables from 2^{2n} words to 2^n words, where n is the wordlength used.

This paper presents techniques for minimizing

1058-6393/91 $01.00 © 1991 IEEE

the size of the look-up ROMs used for computing the useful squaring operations, when arithmetic modulo m is performed. Designs for the cases of $m=2^n$ and $m=2^n-1$ are presented in the paper. The case of $m=2^n+1$ is a simple modification of the case of $m=2^n-1$. The techniques can be modified to apply to any general modulus $m=2^k-k$, but in this case the extra overhead increases.

2. The Memory Compression Schemes

We first present the memory compression schemes when the modulus used is $m=2^n$. Consider a number A belonging in Z_{2^n}, where $Z_{2^n} = \{0,1,...,2^n-1\}$. Then, A has an n-bit binary representation as in

$$A = a_{n-1}a_{n-2} \cdots a_1 a_0 \qquad (1)$$

Our task is to compute $<A^2>_{2^n}$, where $<x>_m$ denotes the operation x modulo m. If the table look-up approach is used, then a ROM table of size $2^n \times n$ bits needs to be used, before any compression is applied.

Alternatively, consider the decomposition of A into a 1-bit upper byte $A_{H1} = a_{n-1}$ and an (n-1)-bit lower byte $A_{L1} = a_{n-2} \cdots a_0$. Then

$$A = A_{H1}2^{n-1} + A_{L1} \qquad (2)$$

while

$$A^2 = A_{H1}^2\, 2^{2n-2} + A_{H1}A_{L1}\, 2^n + A_{L1}^2 \qquad (3)$$

and for $n>2$ one gets

$$<A^2>_{2^n} = <A_{L1}^2>_{2^n} \qquad (4)$$

According to (4) the desired square can be computed using a ROM table of size $2^{n-1}\times n$ bits, or 50% reduction in the size of the ROM has been achieved without any extra overhead.

Consider now decomposing A into a 2-bit upper byte $A_{H2} = a_{n-1}\, a_{n-2}$ and an $(n-2)$-bit lower byte $A_{L2} = a_{n-3} \cdots a_0$. We then get

$$A = A_{H2}\, 2^{n-2} + A_{L2} \qquad (5)$$

or

$$A^2 = A_{H2}^2\, 2^{2n-4} + A_{H2}A_{L2}\, 2^{n-1} + A_{L2}^2 \qquad (6)$$

while for $n>4$ we get

$$<A^2>_{2^n} = <A_{H2}A_{L2}\, 2^{n-1} + A_{L2}^2>_{2^n} \qquad (7)$$

The amount $<A_{H2}A_{L2}\, 2^{n-1}>_{2^n}$ is given below.

$$<A_{H2}A_{L2}\, 2^{n-1}>_{2^n} = \left\{ \begin{array}{ll} 0 & \text{if } a_{n-2}=0 \\ <A_{L2}\, 2^{n-1}>_{2^n} & \text{if } a_{n-2}=1 \end{array} \right\} \qquad (8)$$

Since
$<A_{L2}\, 2^{n-1}>_{2^n} = <(a_{n-3}a_{n-4}....a_0)2^{n-1}>_{2^n} = <(a_{n-3}a_{n-4} \cdots a_1)2^n + a_o 2^{n-1}>_{2^n} = a_o\, 2^{n-1}$, then (7) and (8) give

$$<A^2>_{2^n} = \left\{ \begin{array}{ll} <A_{L2}^2>_{2^n} & \text{if } a_{n-2}=0 \\ <a_o 2^{n-1} + A_{L2}^2>_{2^n} & \text{if } a_{n-2}=1 \end{array} \right\} \qquad (9)$$

Suppose that $<A_{L2}^2>_{2^n}$ is represented in binary as $<A_{L2}^2>_{2^n} = b_{n-1}b_{n-2} \cdots b_1 b_o$. Then (9) gives

$$<A^2>_{2^n} = \qquad (10)$$

$$\left\{ \begin{array}{ll} <A_{L2}^2>_{2^n} = b_{n-1}b_{n-2} \cdots b_1 b_o & \text{if } a_{n-2}=0 \\ c_{n-1}b_{n-2}b_{n-3} \cdots b_1 b_o & \text{if } a_{n-2}=1 \end{array} \right\}$$

where

$$c_{n-1} = a_o \oplus b_{n-1} \qquad (11)$$

and where $\oplus$ denotes the EXCLUSIVE-OR operation.

Figure 1 shows the computation of $<A^2>_{2^n}$ using a $2^{n-2}\times n$ ROM and the extra overhead of one exclusive-or gate and a 2-to-1 multiplexer. In this case, a 75% reduction in the size of the ROM has been achieved.

The computation of $<A^2>_{2^n}$ based on decomposing A into a 3-bit upper byte $A_{H3} = a_{n-1}a_{n-2}a_{n-3}$ and an $(n-3)$-bit lower byte $A_{L3} = a_{n-4} \cdots a_o$ is shown in Figure 2. The derivations are not shown due to restrictions in the size of the paper. Figure 2 demonstrates that the desired square can be computed using a ROM of size $2^{n-3}\times n$ bits plus the extra overhead of few gates and a 4-to-1 multiplexer. An 87.5% savings in ROM bits has been achieved in this case.

We now present the memory compression schemes for arithmetic modulo 2^n-1. Our number A that needs to be squared is belonging in $Z_{2^n-1} = \{0,1, \ldots, 2^n-2\}$ and thus A assumes an n-bit binary representation as in (1). In our effort to minimize the size of the paper we will not show derivations. Figures 3 and 4 show the obtained result.

Figure 3 shows the computation of $<A^2>_{2^n-1}$ based on decomposing A into a 1-bit upper byte $A_{H1} = a_{n-1}$ and an $(n-1)$-bit lower byte $A_{L1} = a_{n-2} \cdots a_o$. It relies on a ROM of size $2^{n-1} \times n$ bits instead of $2^n \times n$ bits that would have otherwise been required. The extra overhead is a modulo 2^n-1 adder and a 2-to-1 multiplexer. Figure 4 shows the computation of $<A^2>_{2^n-1}$ based on a decomposition of A into a 2-bit upper byte $A_{H2} = a_{n-1}a_{n-2}$ and an $(n-2)$-bit lower byte $A_{L2} = a_{n-3} \cdots a_o$. Here the required ROM is of size $2^{n-2} \times n$ bits and the extra overhead two modulo 2^n-1 adders and a 4-to-1 multiplexer.

3. Conclusion

We have presented memory compression schemes for minimizing the size of the table look-up ROMs used for computing the squaring operation. These techniques resulted in significant savings in ROM bits at an extra cost of a small overhead. The presentation considered modulo 2^n and modulo 2^n-1 arithmetic. Simple modifications of the case of arithmetic modulo 2^n-1 can result in arithmetic modulo 2^n+1. For example, by replacing the ROM and the adder of Figure 3 with a ROM and an adder performing arithmetic modulo 2^n+1 and by negating modulo 2^n+1 the left input to the adder, the scheme of Figure 3 would then be computing $<A^2>_{2^n+1}$ for any $A < 2^n$. The presented techniques can be modified to apply to any general modulus $m = 2^n - k$, but in this case the extra overhead increases. The overall result is that efficient implementations of the quarter squared and the one over eight squared algorithms can be obtained, using reduced address space ROM tables.

Acknowledgment

This work has been supported in part by NSF Grant No. CCR8910187.

References

[1] F.J. Taylor, "Residue arithmetic: A tutorial with examples," *IEEE Computer*, vol. 17, no. 5, pp. 50-62, May 1984.

[2] M.A. Soderstrand, W.K. Jenkins, G.A. Jullien and F.J. Taylor, Eds., Residue Number System Arithmetic: Modern Applications in Digital Signal Processing, New York: IEEE Press, 1986.

[3] M.A. Soderstrand, "A New Hardware Implementation of Modulo Adders for Residue Number Systems," *Proceedings of the 26th Midwest Symposium on Circuits and Systems*, Puebla, Mexico, pp. 412-415, August 1983.

[4] M.A. Bayoumi, G.A. Jullien and W.C. Miller, "A VLSI Implementation of Residue Adders," *IEEE Transactions on Circuits and Systems*, vol. CAS-34, no. 3, pp. 284-288, March 1987.

[5] G.A. Jullien, "Residue Number Scaling and Other Operations Using ROM Arrays," *IEEE Transactions on Computers*, vol. C-27, no. 4, pp. 325-336, April 1978.

[6] C.H. Huang and F.J. Taylor, "A Memory Compression Scheme for Modular Arithmetic," *IEEE Transactions on Acoustics, Speech, and Signal Processing*, vol. ASSP-27, no. 6, pp. 608-611, December 1979.

[7] G.A. Jullien, "Implementation of Multiplication Modulo a Prime Number, with Applications to Number Theoretic Transforms," *IEEE Transactions on Computers*, vol. C-29, no. 10, pp. 899-905, October 1980.

[8] M.A. Soderstrand and E.L. Fields, "Multipliers for Residue Number Arithmetic Digital Filters," *Electronics Letters*, vol. 13, no. 6, pp. 164-166, March 17, 1977.

[9] M.A. Soderstrand and C. Vernia, "A High-Speed Low-Cost Modulo Pi Multiplier with RNS Arithmetic Applications," *Proceedings of the IEEE*, vol. 68, no. 4, pp. 529-532, April 1980.

[10] F.J. Taylor, "Large Moduli Multipliers for Signal Processing," *IEEE Transactions on Circuits and Systems*, vol. CAS-28, no. 7, pp. 731-736, July 1981.

[11] A. Skavantzos, "Novel Approach for Implementing Convolutions with Small Tables," *IEE Proceedings-E*, vol. 138, no. 4, pp. 255-259, July 1991.

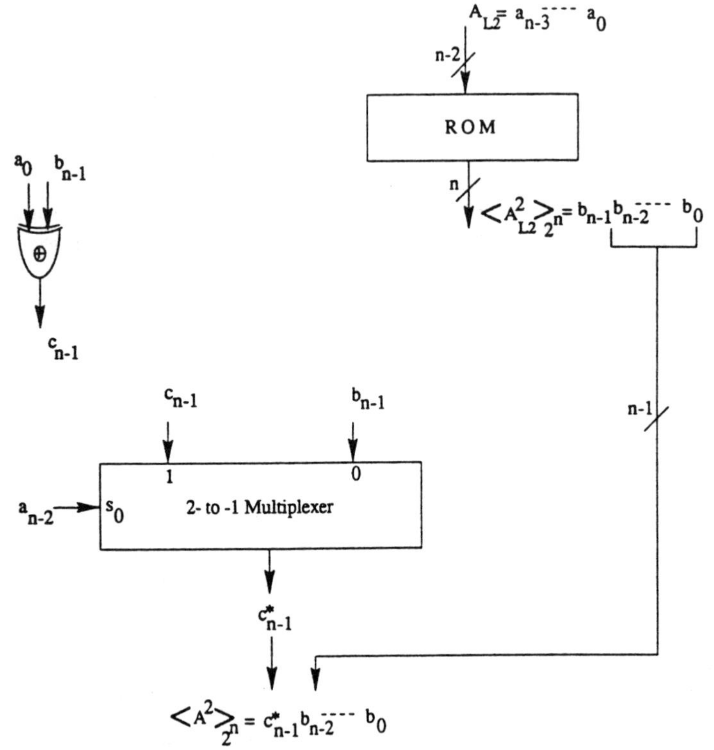

Figure 1: The computation of $<A^2>_{2^n}$ based on a $2^{n-2} \times n$ ROM.

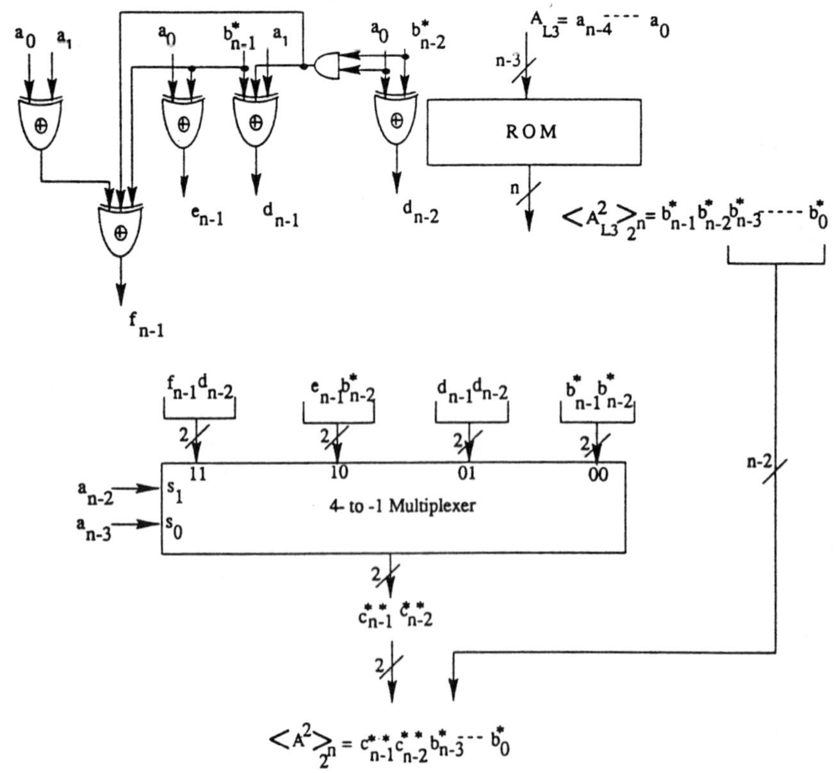

Figure 2: The computation of $<A^2>_{2^n}$ based on a $2^{n-3} \times n$ ROM.

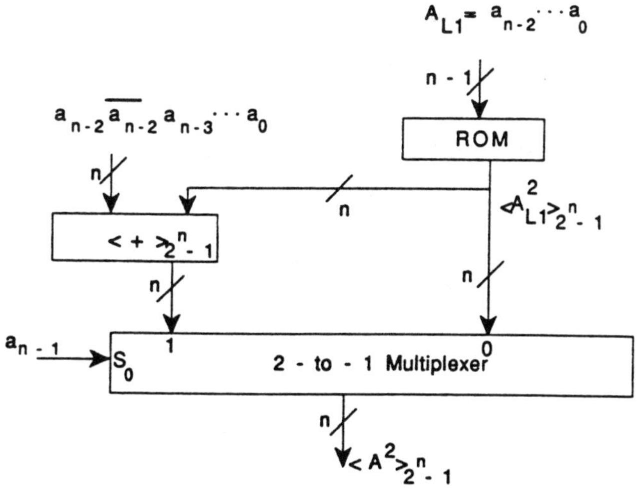

Figure 3: The computation of $<A^2>_{2^n-1}$ based on a $2^{n-1}\times n$ ROM.

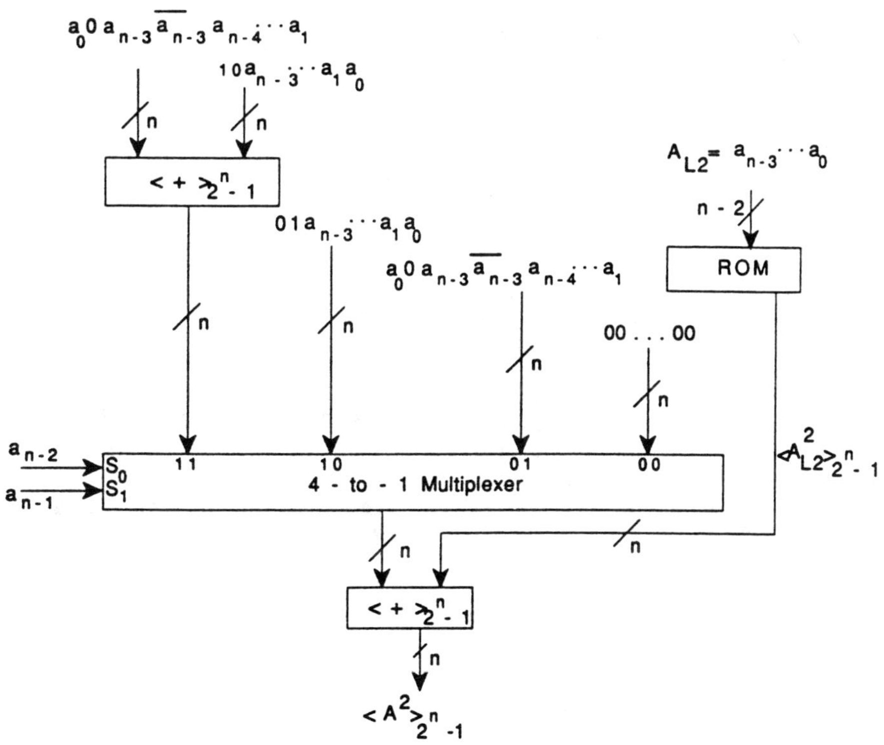

Figure 4: The computation of $<A^2>_{2^n-1}$ based on a $2^{n-2}\times n$ ROM.

Algorithm and Architectures for Radix-4 Division with Over-Redundant Digit Set and Simple Digit Selection Hardware

Paolo Montuschi and *Luigi Ciminiera*
Dipartimento di Automatica e Informatica
Politecnico di Torino
Corso Duca degli Abruzzi 24
10129 Torino (Italy)

Abstract

In this paper we analyze a SRT radix 4 division algorithm which has a trivial selection function. Our algorithm uses an over-redundant digit set together with the same prescaling considered in [7] where, however, it is shown that the digit selection requires the knowledge of 6 assimilated bits of the partial remainder in order to work correctly. Architectures for implementing the proposed algorithm are also presented, which can be considered as alternatives to the existing architectures for radix 4 division. We provide also an architectural solution to the problem of on-the-fly translating the quotient from over-redundant to non-redundant form.

1 Introduction

Among the four basic arithmetic operations $(+, -, *, /)$, division certainly presents the most relevant difficulties of implementation. The methodologies for computing a division can be coarsely classified in recursive and digit-by-digit algorithms. The former directly or indirectly refer to the Newton method of tangents and find their ideal implementation in software. It is in the area of digit-by-digit methods (also known as direct methods) that we will concentrate our attention, since they are more suitable for VLSI implementations. A number of papers can be found in the literature on both conventional [1], [6], [9], [11], and on-line [5] division. The fundamental idea of direct methods is the same as the classic paper and pencil algorithm, that is: the computation of the final result is carried out by performing several iterations where, like in the classical paper and pencil method, a new digit of the result is computed according to a set of selection rules stored in a digit selection table, by inspecting the value of the current partial remainder. At the end of each iteration, the partial remainder is updated accordingly, in order to get ready for the next iteration. The main characteristics of SRT methods [11] is that each selected digit of the result is picked up by a set (referred to as digit set) which is redundant. Key works have been provided for SRT division by Ercegovac and Lang with the introduction of some new concepts such as prescaling [7]. It should be observed that if the operands are prescaled so that the divisor is close to 1, the quotient digit at a given step is the most significant digit of the remainder. This idea was first discussed by Svoboda in [12] and later used by several people, although in these researches the partial remainders were represented in carry assimilated form. Ercegovac described a similar scheme using redundant remainders in his Ph.D. thesis, in [3] and, in collaboration with Lang, in [6]. In this paper we analyze a SRT radix 4 division algorithm and we use the same prescaling considered in a recent paper [7], where it is shown that the digit selection requires the knowledge of 6 assimilated bits of the partial remainder, in order to work correctly. However, our work is based on a different idea, i.e. the use of an over-redundant digit set [5]. In the existing literature a limitation has always been imposed [1], [2], [11], on the choice of a redundant digit set: the maximum value for a digit, s, must be smaller than the radix. Viceversa, in our case, we use an over-redundant the digit set, Δ_s, where $s \geq B$, given the radix B. We have shown that, by carefully selecting an over-redundant digit set, it is possible to design a new radix 4 division unit without a digit selection table. Furthermore, the introduction of an over-redundant digit set influences the design of parts of the division unit other than the digit selection logic, such as the sub-unit used to on-the-fly translate the quotient from redundant to non-redundant form since a new digit may influence the final values of the previous ones in different ways, as $s \geq B$. From a high level, in our unit part of the complexity of the digit selection table has migrated to the prescaling and to the on-the-fly conversion units, thus defining the profile of an alternative architecture for radix 4 division.

The algorithm for computing the division x/d radix $B = 2^b$, relies on the following formulae

$$w_i = B(w_{i-1} - y_i d) = B^{i+1}(x - Y_i d) \qquad (1)$$

and

$$Y_i = Y_{i-1} + B^{-i} y_i \qquad (2)$$

where: i is the iteration step, w_{i-1} is the shifted partial remainder value at step $i-1$ (with $w_0 = Bx$), y_i is the digit of the partially developed result which has been computed at i-th iteration of (1) (with $y_i \in \Delta_s$), Δ_s is the digit set, i.e. the set of possible values for y_i ($\Delta_s = \{-s, -s+1, \cdots, 0, +1, \cdots + s\}$), d is the divisor, and Y_i is the value of the quotient after the i-th iteration, with $Y_0 = 0$. For this algorithm we define the index of redundancy [6] as $\eta = s/(B-1)$ and consider

1058-6393/91 $01.00

x and d normalized to $1/2 \le d, x < 1$. It turns out that the result $Y = x/d$ is normalized to $1/2 < Y < 2$. It is assumed for x and d to be available in their full precision and in carry assimilated form, at the beginning of the iterations, thus excluding from this analysis the on-line algorithms [5]. Moreover, we assume to store the partial remainders w_i in signed digit form [2]. Even when an over-redundant digit set is considered, the digit selection rules for radix 4 division [6] are

$$\text{select } y_i = k \quad \text{if} \quad (-\eta + k)d \le w_{i-1} < (\eta + k)d \quad (3)$$

2 The algorithm

Let us examine in the detail radix 4 division with over redundant digit set $\Delta_4 = \{\pm4, \pm3, \pm2, \pm1, 0\}$ and how the selection rules (3) are implemented by a digit selection table. It is known [6] that, in general, such a digit selection table must receive in input a truncation of the divisor $d_{min} \le d \le d_{MAX}$ and a truncation to the t-th fractional bit of the partial remainder (denoted with $\widehat{w_{i-1}}$). It can be shown [10] the minimum number of bits to be input to the digit selection table is obtained when the value of the ratio d_{MAX}/d_{min} is minimized. In our algorithm we will implicitly consider the prescaling of [7], where the range of the divisor evolves from $[1/2, 1)$ to $[63/64, 9/8)$. From (3), by performing an assimilation and then a truncation of the redundant remainder w_{i-1} to its most significant 4 integer bits, we get

$$\begin{array}{lll} \text{select } y_i = +4 & \text{if} & 4 \le \widehat{w_{i-1}} \le 6 \\ \text{select } y_i = -4 & \text{if} & -6 \le \widehat{w_{i-1}} \le -4 \\ \text{select } y_i = \pm3 & \text{if} & \widehat{w_{i-1}} = \pm3 \\ \text{select } y_i = \pm2 & \text{if} & \widehat{w_{i-1}} = \pm2 \\ \text{select } y_i = \pm1 & \text{if} & \widehat{w_{i-1}} = \pm1 \\ \text{select } y_i = 0 & \text{if} & \widehat{w_{i-1}} = 0 \end{array} \quad (4)$$

From the set of selection rules (4) we observe that: *i)* it is possible to design a division unit with the minimum number of input lines, i.e. 4 with digit set Δ_4, and *ii)* this unit does not require any digit selection table since $\widehat{w_{i-1}}$ reflects exactly the quotient digit to be selected.

3 The architectures

3.1 General model

The high level sketch of the proposed architecture is shown in Fig. 1. The partial remainder is stored in signed digit form inside the block "Remainder". The four most significant bit positions of the partial remainder are assimilated by CLA, i.e. a carry lookahead adder of length 4 and represent the selected digit. Actually, as can be seen from the selection rules (4), 13 are the possible combinations. This is due to the selection rules related to the digits ±4 that cover 3 possible combinations each. It turns out that it should be available one multiplexer 13-to-1. It is possible to operate a "folding" of the extra combinations in the real digits to be selected, i.e. to convert the combinations $-6, -5, -4$ of the partial remainder into the digit -4 of the result, and similarly for the positive combinations. Anyway, it

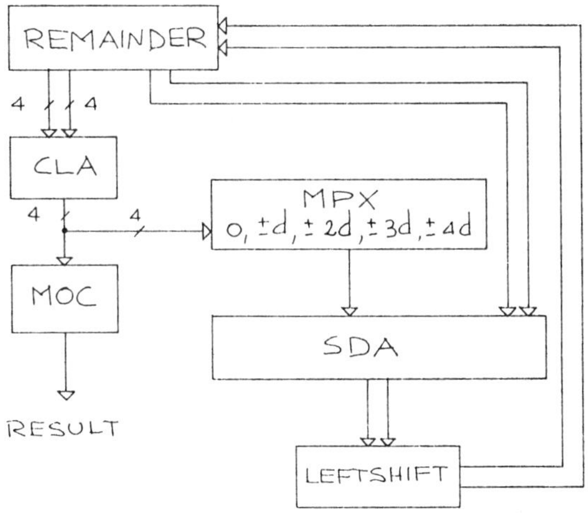

Figure 1: Architecture of the proposed radix-4 division unit

is still necessary one multiplexer 9-to-1, unless a different solution is found. After the correct multiple of the divisor has been selected, the updating of the partial remainder occurs by means of the signed digit adder SDA and by the shift of two positions left of the result. Finally, the selected digit is on-the-fly converted from redundant into non-redundant form by block MOC (i.e. Modified On-the-fly Conversion).

3.2 Implementations

The need for a large multiplexer together with the need of generating the non-trivial multiples $\pm3d$ in the architecture of Fig. 1, may impose some attention in the implementation of the digit selection hardware. A possible solution is to replace the large multiplexer with a tree of cascading smaller multiplexers, (Fig. 2). In such a case, the high level description of the architecture of Fig. 1 still remains valid, while the delay in the selection of the multiple of the divisor for the updating process is twice the delay of a multiplexer 4-to-1. The substitution of the large multiplexer with a tree of smaller multiplexers, does not eliminate the need to generate the multiple $\pm3d$ before the iterations start. Among all the possibilities, a solution following a methodology similar to the technique adopted in [9] is presented here. The digit selection rules of the algorithm are split into two subsets with the transformation

$$y_i = y_i' + y_i'' \quad (5)$$

where $y_i' \in \{\pm2, 0\}$ and $y_i'' \in \Delta_2$. One possibility is

$$\textit{select } y_i' = +2 \quad \textit{if} \quad 2 \le \widehat{w_{i-1}} \le 6$$

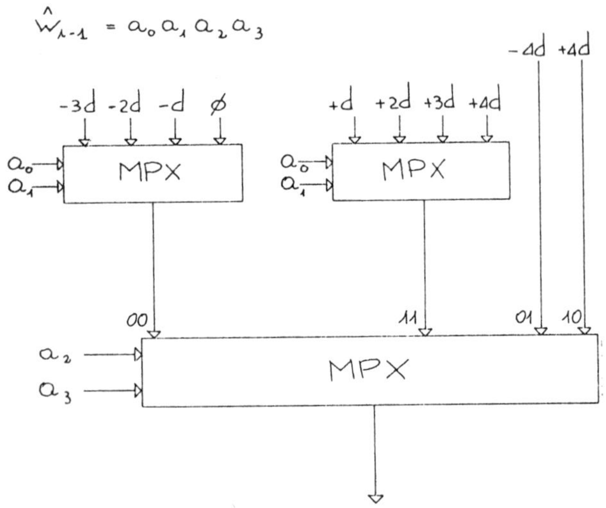

Figure 2: Tree of cascading multiplexers for implementing the 13-to-1 multiplexer

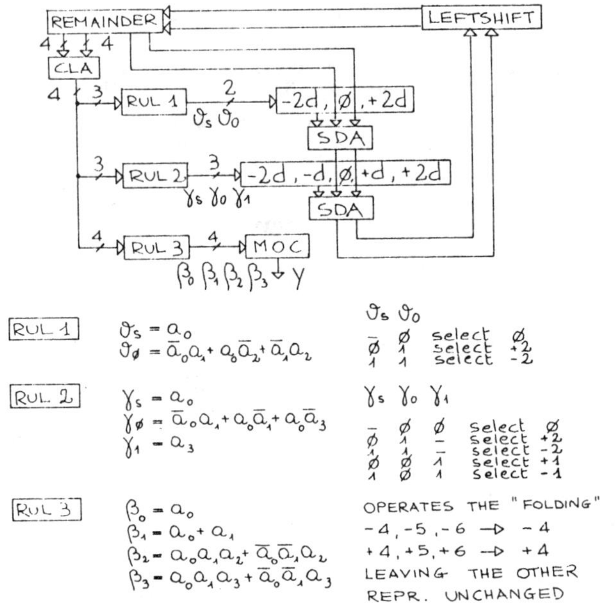

Figure 3: Architecture of the variated radix-4 division unit

$$\begin{array}{llll} select\ y_i' = 0 & if & -2 \le \widehat{w_{i-1}} \le +1 & \\ select\ y_i' = -2 & if & -6 \le \widehat{w_{i-1}} \le -3 & (6) \end{array}$$

together with

$$\begin{array}{lll} select\ y_i'' = +2 & if & 4 \le \widehat{w_{i-1}} \le 6 \\ select\ y_i'' = +1 & if & \widehat{w_{i-1}} = 3 \ or\ \widehat{w_{i-1}} = 1 \\ select\ y_i'' = 0 & if & \widehat{w_{i-1}} = 2 \ or\ \widehat{w_{i-1}} = 0 \\ select\ y_i'' = -1 & if & \widehat{w_{i-1}} = -1 \ or\ \widehat{w_{i-1}} = -3 \quad (7) \\ select\ y_i'' = -2 & if & \widehat{w_{i-1}} = -2 \ or - 6 \le \widehat{w_{i-1}} \le -4 \end{array}$$

By combining (6) with (7) by means of (5) the selection rules (4) are obtained. In this case the selection problem of y_i has been split into the two sub-problems, of determining y_i' and y_i'', which are separate and then can be carried out in parallel. Moreover, each sub-selection procedure makes reference to a simpler and more limited digit set than the one required by the original algorithm. The high level scheme of the radix 4 division unit implementing (6) and (7) is reported in Fig. 3, where the four most significant bits of $\widehat{w_{i-1}}$ have been denoted with $a_0a_1a_2a_3$. In Fig. 3, three new blocks are identified: RUL1 provides the input to the multiplexer $(-2, 0, +2)$, while RUL2 provides the input to the multiplexer $(-2, -1, 0, +1, +2)$, and RUL3 operates the "folding" of the combinations $-6, -5, -4$ of the partial remainder into the digit -4 of the result (and similarly for the positive values). The implementation of RUL1, RUL2 and RUL3 requires a very small number of gates. The maximum advantages are obtained when all block RUL1, multiplexer $(-2, 0, +2)$ and the signed digit adder SDA execute in parallel with the block the RUL2 and multiplexer $(-2, -1, 0, +1, +2)$, since for both architectures it is (reasonably) assumed that the duration of one step of on-the-fly conversion (together with the "folding" provided by RUL3) is smaller than the duration of an iteration. The evaluation of this is highly technology dependent. However, by modifying the selection rules for y_i' and y_i'' it is possible to find the most convenient solution for any given implementation technology.

3.3 On-the-fly conversion block

The role of the on-the-fly conversion block MOC is the same as the unit described in [4], that is, to convert the result from a redundant into the non-redundant representation. To achieve this goal in [4] is used a pair of co-operating registers. However, the architecture presented in [4] cannot be directly used when an over-redundant digit set is considered. We use an on-line recoding of the digits y_i's of the over-redundant digit set into digits $\overline{y_i}$'s of the maximally redundant digit set in a similar way to the methods outlined in [2] and [8]. This requires the introduction of an on-line delay of one digit. Let us consider the digit $y_i \in \Delta_4$ of the result which has been selected during iteration i. The on-line recoding unit (referred to below as block M), has (internally) one register labeled α and, being given the digit $y_i \in \Delta_4$ as input during the iteration i, produces a recoded digit $\overline{y_{i-1}} \in \Delta_3$ as output. At the beginning of the operations, the register α is initialized with the digit 0, and the operating principles of the on-line recoding unit during the generic iteration i are:

- if $y_i \geq 3$ then

$$\begin{aligned} \overline{y_{i-1}} &= \alpha_i + 1 \\ \alpha_{i+1} &= y_i - 4 \end{aligned} \quad (8)$$

- if $-2 \leq y_i \leq 2$ then

$$\begin{aligned} \overline{y_{i-1}} &= \alpha_i \\ \alpha_{i+1} &= y_i \end{aligned} \quad (9)$$

- if $y_i \leq -3$ then

$$\begin{aligned} \overline{y_{i-1}} &= \alpha_i - 1 \\ \alpha_{i+1} &= y_i + 4 \end{aligned} \quad (10)$$

From (8), (9) and (10), it follows $|\alpha| \leq 2$. This implies that $\overline{y_{i-1}} \in \Delta_3$, i.e. the maximally redundant digit set; hence it is possible to use the same unit as in [4]. Let us consider the practical example of recoding (0).(3)(−4)(1), where the positions are weighted by the base 4. Initially we have $\alpha_1 = 0$. Since $y_1 = 3$, we get $\overline{y_0} = 1$, $\alpha_2 = -1$. Then, $y_2 = -4$ implies $\overline{y_1} = -2$, $\alpha_3 = 0$. Finally, from $y_3 = +1$ we get $\overline{y_2} = 0$ and $\overline{y_3} = \alpha_4 = 1$. Actually, the result of the recoding is (1).(−2)(0)(1), which can be on-the-fly converted by the unit in [4]. The recoding unit M can be implemented by an increment/decrement three bit register, and a few gates of logic. However, it has the small penalty represented by the on-line delay of one digit.

3.4 Correctness of the result

In order to obtain a correct result, each division algorithm must ensure that the remainder is bounded. According to the iteration rules (1) and (2), a correct result after n iterations which is expressed on $n \log_2(B)$ bits, can be obtained if the shifted final remainder can be guaranteed to belong to the interval

$$[-Bd, Bd] \quad (11)$$

However, in general, with the division algorithm defined by (1) and (2) it is ensured that the final shifted remainder belongs to the interval $[-\eta Bd, \eta Bd]$ which is contained in the interval (11) only in the case of non-over-redundant digit sets. We show now that for our algorithm the supplementary iteration required by the on-line recoding, also suffices to guarantee that the shifted final remainder belongs to the interval (11). After $n+1$ iterations we have obtained digits up to y_{n+1} and partial remainders up to w_{n+1}. Denoting with $\overline{Y_n}$ the recoded result obtained from the recoded digits $\overline{y_i}$'s so that $\overline{Y_n} = \sum_{i=0}^{n} \overline{y_i} 4^{-i}$. Since the recoding algorithm is based on (8), (9) and (10), the relationship between Y_{n+1} and $\overline{Y_n}$ is

$$Y_{n+1} = \overline{Y_n} + \alpha_{n+2} 4^{-(n+1)} \quad (\textit{with } |\alpha_{n+2}| \leq 2) \quad (12)$$

Moreover, from the iteration rule (1) we have

$$w_{n+1} = 4^{n+2}(x - Y_{n+1}d) \quad (13)$$

By substituting (12) into (13) we get

$$w_{n+1} = 4[4^{n+1}(x - \overline{Y_n}d) - \alpha_{n+2}d] \quad (14)$$

From the viewpoint of the recoded result $\overline{Y_n}$, the shifted final remainder $\overline{w}_n$ which is derived from multiple iterations of (1), can be expressed as $\overline{w}_n = 4^{n+1}(x - \overline{Y_n}d)$, and hence from (14) it follows that the relationship between w_{n+1} and $\overline{w}_n$ is

$$w_{n+1} = 4(\overline{w}_n - \alpha_{n+2}d) \quad (15)$$

Since our algorithm is based on (1) with $B = 4$ and $\eta = 4/3$, after $n+1$ iterations we have

$$-16/3d \leq w_{n+1} \leq 16/3d \quad (16)$$

By substituting (15) into (16) we get

$$-(4/3 - \alpha_{n+2})d \leq \overline{w}_n \leq (4/3 + \alpha_{n+2})d \quad (17)$$

We study now (17) in the worst case conditions, i.e. when $|\alpha_{n+2}| = 2$. In such a case (17) becomes

$$-4d < -10/3d \leq \overline{w}_n \leq 10/3d < 4d$$

In conclusion, we have therefore shown that after $n+1$ iterations of the recoded algorithm we have that the shifted final remainder (in presence of a recoded result) belongs to the interval (11), and hence the result is correct.

4 Evaluation and conclusions

The low level evaluation of the proposed architecture is very technology dependent, and therefore only a high level evaluation will be considered. However, among all the possible existing implementations, a fair comparison can be carried out with the unit presented in [7], since both architectures perform radix 4 division by adopting the same prescaling of the operands. By means of the symbols defined in Table 1 it is possible to write down the expression of the duration of an iteration for the architecture of [7] as $t_{EL} = t_{CLA,6} + t_{PLA} + t_{MPX,5} + t_{CSA} + t_{SH}$. Conversely, the duration of an iteration for the proposed architecture is $t_{PA} = t_{CLA,4} + t_{DSHW} + t_{SDA} + t_{SH}$, where $t_{DSHW} = t_{MPX,13}$ when a multiplexer 13-to-1 is used, and $t_{DSHW} = t_{MPX,9} + t_{folding}$ when the architecture uses a folding and the multiplexer 9-to-1, in both cases either directly or implemented as a tree of multiplexers. Conversely, when the architecture of Fig. 3 is used, $t_{DSHW} = \max(t_{RUL1} + t_{MPX,3} + t_{SDA}, t_{RUL2} + t_{MPX,5})$. For both architectures it is (reasonably) assumed that the duration of one step of on-the-fly conversion (together with the on-line recoding for our architecture) is smaller than the duration of an iteration. A fair comparison between the proposed architecture and the unit of [7], however, cannot be limited only to the single iteration times, but must consider the duration of a whole computation of division. For the architecture of [7], the computation of a result expressed on $2n$ bits takes the time

$$T_{EL} = t_{PR} + nt_{EL} \quad (18)$$

Table 1: Delays of functional blocks

$t_{CLA,x}$	delay of a CLA of x bits
t_{SDA}	delay of a SDA
t_{CSA}	delay of a CSA
$t_{MPX,x}$	delay of a multiplexer with x inputs
t_{RULx}	delay of the block RULx of the variation of proposed architecture
t_{PLA}	delay of the digit selection table of [7]
t_{PR}	delay of the prescaling operation in the architecture of [7]
$t_{\pm 3d}$	delay for obtaining the term $\pm 3d$
t_{SH}	delay of the register loading and of the shift register

The same result is computed by our architecture in

$$T_{PA} = t_{PR} + (n+1)t_{PA} + t_{\pm 3d} \quad (19)$$

where the contribution of the term $t_{\pm 3d}$ must be set equal to 0 in the case that the architectural solution with a two-stage digit selection-updating described in section 3.2 is used. From (18) and (19) it is interesting to observe the main differences existing between the architecture of [7] and the proposed architecture. In the architecture of [7] some of the complexity of the digit selection table has migrated to the hardware implementing the prescaling. In order to achieve a global speedup in the computation time, it is intended that in the implementation technology of the architecture, the advantages of having a faster iteration time more than compensate the delay introduced by the delay step. On the other hand, in the proposed architecture some more complexity of the digit selection table has further migrated to the multiplexer, to the multiple generation hardware (to generate the terms $\pm 3d$) and to the on-the-fly conversion unit. In the implementation described in section 3.2 very small digit selection tables have been used to avoid the large multiplexer and the multiple generation hardware to generate the terms $\pm 3d$. While it is reasonable to assume that the on-the-fly conversion operates in parallel with the digit selection step, it is necessary to measure accurately the delays of the new digit selection hardware and of the new multiple selector, in order to conveniently determine the role of the proposed architecture, given the implementation technology details. From the computations which we have developed, it is clear that the problem of radix 4 division has an intrinsic complexity, which cannot be decreased by all the architectures proposed in the literature, but can be moved from one functional unit to the other. The best advantages in terms of execution time, are obtained when all the independent functional units operate in parallel, for the specific implementation technology chosen. From these considerations and from the above expressions (18) and (19), it follows that the proposed architecture offers a valid alternative to the existing units for radix 4 division. It is left to the designer the task of choosing the best architecture among the possible solutions the, once the implementation technology has been defined.

Acknowledgements

We thank Prof. M. D. Ercegovac and Prof. T. Lang for helpful comments. We thank also C. Chiariglione for having prepared the figures of this paper, and E. Macii.

References

[1] D. E. Atkins, "Higher-Radix Division Using Estimates of the Divisor and Partial Remainders," IEEE Trans. Comput., Vol.C-17, pp.925-934, October 1968.

[2] A. Avižienis, "Signed-Digit Number Representation for Fast Parallel Arithmetic," IRE Trans. Electron. Comput., Vol.EC-10, pp.389-400, September 1961.

[3] M. D. Ercegovac, "A General Hardware-Oriented Method for Evaluation of Functions and Computations in a Digital Computer," IEEE Trans. on Comput., Vol.C-26, pp.667-680, July 1977.

[4] M. D. Ercegovac and T. Lang, "On-the-Fly Conversion of Redundant into Conventional Representations," IEEE Trans. Comput., Vol.C-36, pp.895-897, July 1987.

[5] M. D. Ercegovac and T. Lang, "On-Line Scheme for Computing Rotation Factors," Journal of Parallel and Distributed Computing, Vol.5, pp.209-227, 1988.

[6] M. D. Ercegovac and T. Lang, "Division," Internal Report CS252, Computer Science Department, UCLA, 1988.

[7] M. D. Ercegovac and T. Lang, "Simple Radix-4 Division with Operands Scaling," IEEE Trans. Comput., Vol.C-39, pp.1204-1208, September 1990.

[8] M. D. Ercegovac and T. Lang, "Fast Multiplication Without Carry Propagate Addition," IEEE Trans. Comput., Vol.C-39, pp.1385-1390, November 1990.

[9] P. Montuschi and L. Ciminiera, "Design of a Radix 4 Division Unit with Simple Selection Table," to appear IEEE Transactions on Computers.

[10] P. Montuschi and L. Ciminiera, "Radix-4 Division without Digit Selection Table," I.R. DAI/ARC 1-91, Dipartimento di Automatica e Informatica, Politecnico di Torino, Torino, Italy.

[11] J. E. Robertson, "A New Class of Digital Division Methods," IRE Trans. Electron. Comput., Vol.EC-7, pp.218-222, September 1958.

[12] A. Svoboda, "An Algorithm for Division," Inf. Proc. Mach., Vol.9, pp.25-32, 1963.

Hardware Control of a Real-Time Multichannel Digital Signal Processing System

William A. Weeks
Constance T. Chittenden
Paul M. Heilman

SRI International
Menlo Park, CA 94025

Abstract

SRI International has developed a high-performance parallel signal processing system for use in real-time applications. The system consists of multiple analog input and output channels linked by an array of digital signal processing (DSP) elements. System performance requires that multiple output samples, which depend on the input data, be generated immediately following the acquisition of the input samples. Custom architectures and algorithms were developed to meet system requirements of up to 100-kHz sample rate per channel and over 14,000 MFLOPS. Two innovative features of SRI's system are an ensemble of DSP boards, each of which contains eight AT&T DSP32C processors, and a network of multiplexed data paths and scheduling hardware. The latter feature solves the I/O bottleneck problem common to commercial DSP approaches and results in higher data throughput than is possible with run-time software scheduling. Planned improvements to the design include the addition of logic control elements, which will allow for greater flexibility in configuring the system to perform a wider variety of signal processing tasks.

1 Introduction

Since 1984, SRI has produced an evolving series of multichannel signal processing systems for use in a variety of real-time applications. The basic system is shown in Figure 1. Data acquired from a number of analog input signals are digitally processed to produce a set of analog output signals. System performance requirements for sample rate, sustained data throughput, and group delay could not be met with off-the-shelf components. Thus, SRI has developed the current generation of custom state-of-the-art signal processing systems to meet all the requirements. Key developments include the design of an array of DSP elements to execute finite impulse response (FIR) filters, the design of a data transfer system to route the data between the data converters and the processing elements, and techniques for minimizing the group delay between input and output samples.

An earlier paper [1] gave an overview of the system architecture and discussed details of the DSP array and some of the signal processing issues. This paper details how data transfers and individual processes are controlled during run-

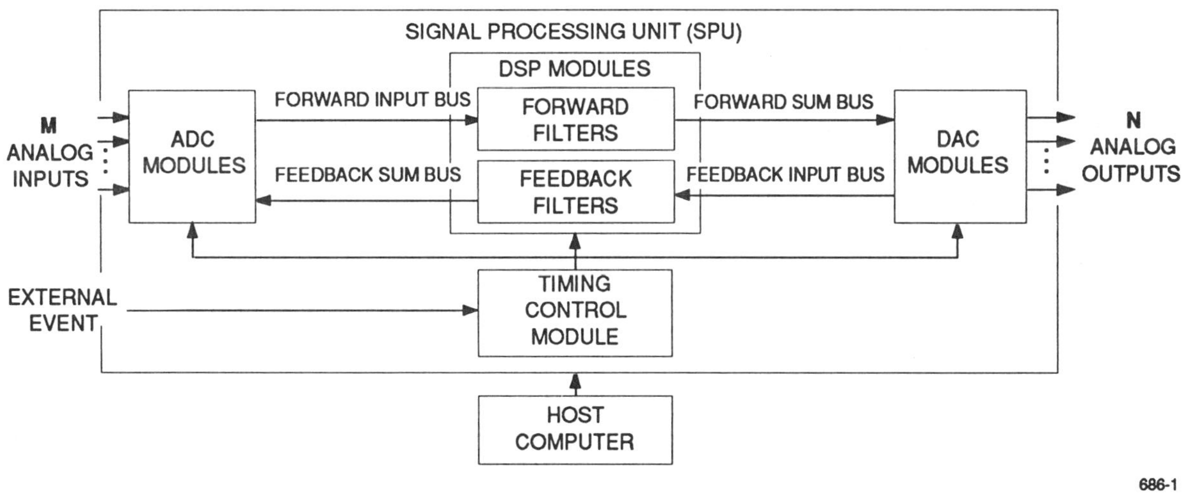

FIGURE 1 SYSTEM BLOCK DIAGRAM

1058-6393/91 $01.00 © 1991 IEEE

time operation of the system, and presents some future enhancements.

2 System Requirements

As shown in Figure 1, the processor has M inputs and N outputs. The digital array connects all the inputs to all the outputs, thereby defining MxN processing channels. In addition, since this system is designed for use in real-world applications, each channel also contains a digital feedback filter to mitigate the coupling effects of the output signals on the inputs. Figure 2 is a block diagram of a single channel. The filters shown in Figure 2 are multi-tap FIR filters. Typical filter lengths are in the 200- to 500-tap range.

Table 1 summarizes the sustained data handling requirements for the successive generations of SRI processors. When design started on the first system in 1984, it was possible to purchase an off-the-shelf array processor which was specified to operate at 200 MFLOPS. However, no

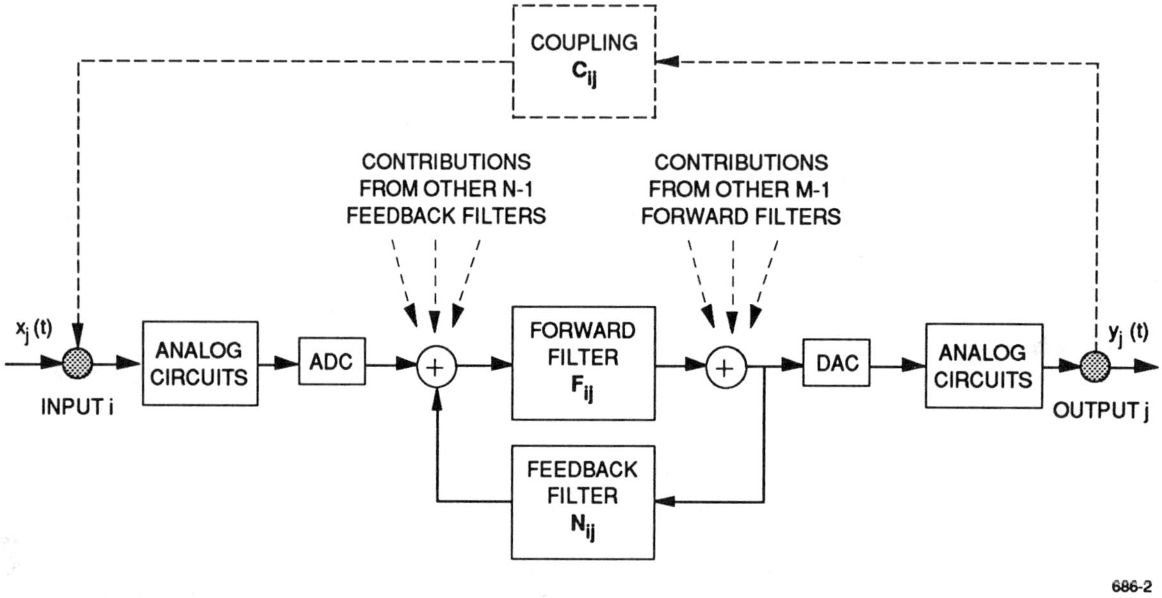

FIGURE 2 SCHEMATIC OF SINGLE CHANNEL

Table 1

REAL-TIME DATA PROCESSING REQUIREMENTS

YEAR	NO. OF INPUTS	NO. OF OUTPUTS	NO. OF PROCESSING CHANNELS	MAX SAMPLE RATE (kHz)	PEAK OPERATIONS (MFLOPS)	AVG. DATA TRANSFERS PER SECOND	PROCESSOR USED	DATA BUS ARCHITECTURE
1985	2	2	4	25	128	200K	FUJITSU MB 8764	POINT-TO-POINT
1986	2	4	8	25	256	300K	FUJITSU MB 8764	POINT-TO-POINT
1989	6	8	48	25	2,400	1,200K	AT&T DSP32	TIME MULTIPLEX
1990	2	24	48	50	4,800	2,400K	AT&T DSP32C	TIME MULTIPLEX
1990	6	8	48	100	4,800	2,800K	AT&T DSP32C	TIME MULTIPLEX
1992 (proposed)	6	24	144	100	14,400	6,000K	AT&T DSP32C	TIME MULTIPLEX
1993 (concept)	40	40	1,600	100	>80,000	>16,000K	?	?

686-3

commercially available hardware could sustain that rate uninterrupted, especially when combined with the 200-kb/s data transfer rate and group delay requirements. With each successive generation, the search for off-the-shelf processors failed to identify any that would meet the ever- expanding requirements.

The following sections detail the design and control of the third and most recent generation of processors based on the AT&T DSP32C 32-bit floating point DSP chip [2].

3 System Architecture

3.1 Overview

The signal processing functions are performed by the signal processing unit (SPU), which is based on a modified version of the CAMAC (IEEE 583-1982) instrumentation standard. A DEC 3600 mini-computer functions as the host computer to configure the SPU's operational parameters and to supervise the system's run-time operation. The host computer communicates with the processing modules over the CAMAC dataway. The real-time processing is performed by three types of custom-designed CAMAC modules: the analog-to-digital converter (ADC), digital signal processing (DSP), and digital-to-analog converter (DAC) modules. A fourth custom module, the timing control module (TCM), generates signals to control the sequencing of individual module processes. Data transfers between the processing modules occur over the four time-division-multiplexed data buses shown in Figure 1. The transfers are controlled by circuitry on the modules.

Figure 3 shows a partial view of the resulting signal processing topology for a 6-input, 8-output system which implements the channel algorithm shown in Figure 2. This configuration has 48 forward filters (F_{ij}) and 48 feedback filters (N_{ij}). Since each filter is composed of two DSP chips, this corresponds to 192 DSP chips operating in parallel.

3.2 Processing Modules

Figure 4 shows block diagrams of the three processing modules. The ADC modules contain dual 400-kHz, 16-bit ADCs and a 16-bit integer adder for summing the feedback filter output with the converted input values. The DAC modules contain RAM-based lookup tables to optionally remap the digital filter outputs before they are converted to analog values to correct for nonlinearitics. They also contain 16-bit DACs with integral high-speed deglitchers. Most of the digital signal processing is performed by the DSP modules, each of which contains eight 25-MHz AT&T DSP32C chips configured in pairs to function as four digital filters. Each DSP32C device also includes 8K of 32-bit local memory.

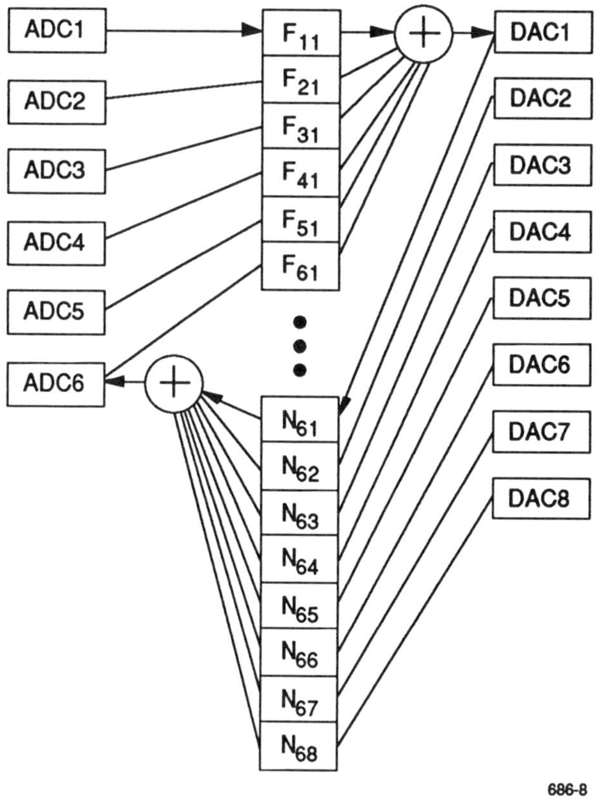

FIGURE 3 SIGNAL PROCESSING TOPOLOGY FOR A 6x8 SYSTEM

3.3 Data Buses

A key element in the architecture is the set of four independent time-division-multiplexed data buses which allow the system to obtain the required data throughput. The selection of active data channels on each bus and the assignment of the bus time slots are reconfigurable by the host computer. This flexibility permits the same hardware to satisfy a variety of signal processing needs.

The forward input bus routes 16-bit integer data from the ADCs to the DSP modules performing the forward filtering operations. Likewise, the feedback input bus sends the 16-bit DAC input values to the feedback DSP modules.

On the outputs of the DSP modules are 20-bit integer adders with pipeline registers. These adders are linked by the daisy-chained sum buses which convey the combined results from groups of DSP modules to effect the extended summations necessary in the cross-channel processing shown in Figure 2. The forward sum bus conveys the outputs of the forward filters to the DAC modules, and the feedback sum bus carries the feedback data to the ADC modules.

In Figure 4(c), the input sum and output sum bus connections depend on the DSP function. On forward-filter DSP modules, the output summers are linked by the daisy-chained forward sum bus. Likewise, the feedback filter DSPs are linked by the feedback sum bus.

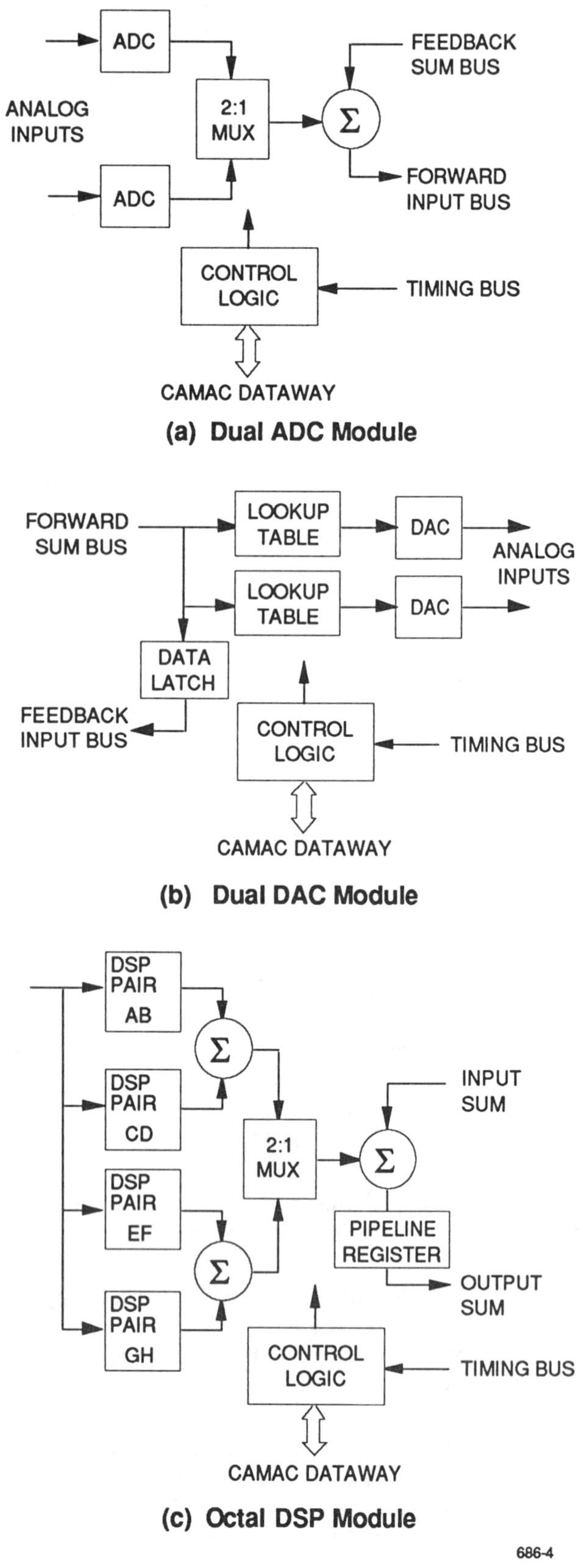

FIGURE 4 PROCESSING MODULES

4 Real-Time Control

4.1 Timing Control Module (TCM)

The heart of the real-time control of the SPU is the timing control module (Figure 5). The functions of the TCM are to: (1) generate a 16-MHz state clock which is common to all the SPU modules; (2) synchronize the SPU processing to an external event; (3) control the sample rate; and (4) generate a sequence of timing pulses. These output pulses, which occur at the data sample rate and are in phase with the state clock, are used by the processing modules as triggers for their internal processes and data transfers.

4.2 Module Timing

Each of the processing modules contains circuit elements to control its run-time behavior. These elements are: (1) a programmable stand-alone microsequencer (SAM) [3] from Altera Corporation; (2) a set of timing control registers (TCRs), which are read by the SAM to control the time-multiplexing operations of the module; and (3) programmable logic devices (PLDs), also from Altera Corp. [3], which control the enabling of the SAM and perform logic operations on some of the SAM outputs.

Figure 6 shows the generic block diagram of a processing module's timing control logic. The SAM outputs control the data flow between the time-multiplexed data buses and the module input and output registers. They also control the timing of processes within the modules such as the start of data conversion or the start of a digital filter. The actual number and purpose of the SAM outputs depends on the

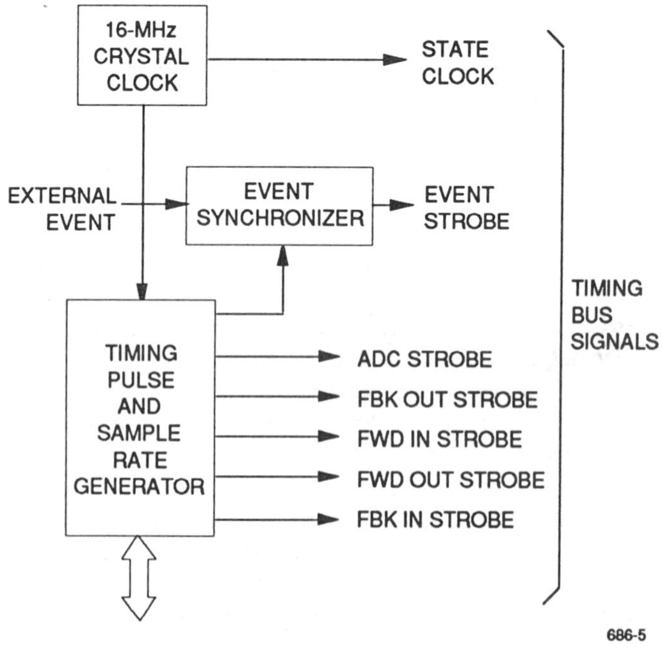

FIGURE 5 TIMING CONTROL MODULE

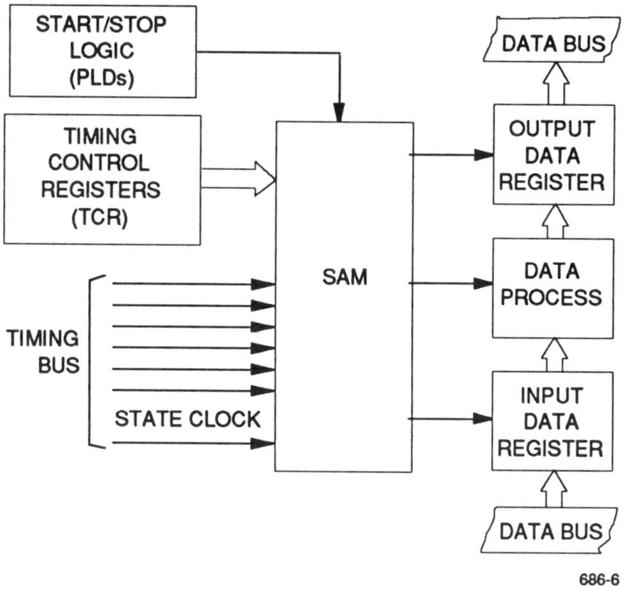

FIGURE 6 GENERIC PROCESSING MODULE TIMING CONTROL LOGIC

module type. In some cases the SAM outputs are gated or modified by the PLD devices based on operational modes that have been set up by the host computer.

4.3 Operation

Prior to executing run-time operations on the SPU, the user specifies various parameters to the host computer such as the SPU operation to be performed, numbers of input and output channels, sample rate, and filter filenames. The host computer then loads the DSP memories and module registers with the appropriate values and enables the TCM to start generating timing pulses.

Once started, the SPU will execute its real-time operations without any interaction from the host computer. Figure 7 shows the timing flow between TCM timing pulses, module processes, and data transfer for a typical filtering configuration. The multiple process strobes and data paths allow the different parts of the hardware to function in parallel, each at their top performance. This makes it possible to have a multichannel system with all the DAC outputs settling to their final values in less than one sample interval following the ADC conversion strobe.

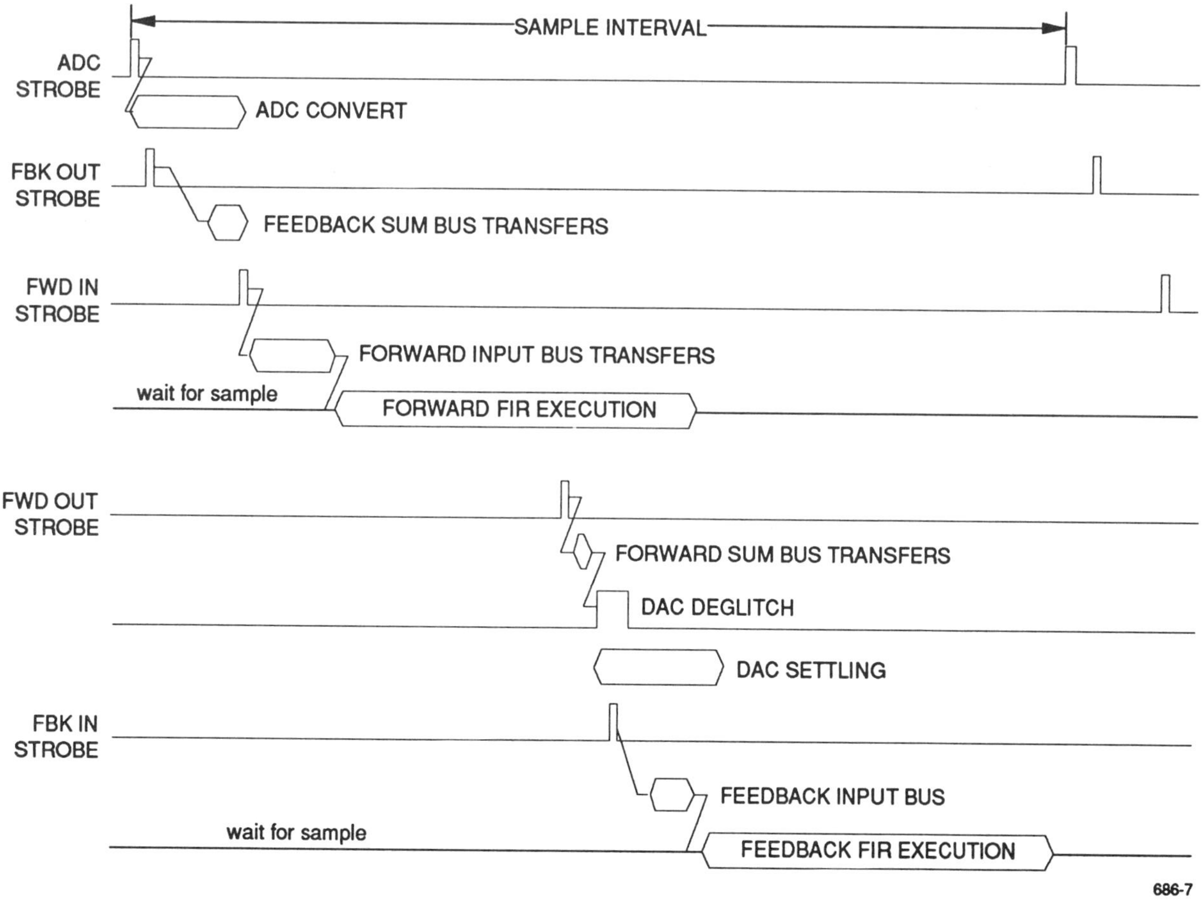

FIGURE 7 PROCESS AND DATA TRANSFER TIMING FOR REAL-TIME FILTERING

5 Future Configurations

The previous section discussed how certain run-time parameters specified by the user are used by the host computer to set up the hardware into various configurations. Using the existing hard-wired architecture, many different real-time processes can be executed on a varying number of input and output channels.

In situations where the present architecture is not adequate for a specific task, it is possible to enhance the basic system design. A number of "hooks" have been designed into the system to allow for such enhancements. Among the parts of the system that can be modified to accommodate such enhancements are: (1) the layout of the data buses; (2) the DSP32C processor firmware routines; (3) the sequencing programs for the processing module SAMs; (4) the programming of the PLDs; and (5) the control functions on the host computer. While implementation of those changes would require some design effort, the process is certainly less costly in time and money than designing a new system from scratch.

The third-generation system described in this paper is designed to execute many FIR filters fast. SRI is currently investigating some major revisions to the architecture presented here. The fourth-generation system will be more flexible than the previous system and will be capable of executing more general classes of real-time signal processing algorithms. Among the changes being considered are: (1) modifying the TCM and module sequencers and the DSP firmware to allow each DSP pair to execute more than one digital filter per sample; (2) replacing the ADC and DAC modules with a general purpose I/O module which includes satellite DSP elements for pre- and post-data processing; (3) adding global memory and crossbar communication switches on the DSP modules; and (4) switching the host interface from CAMAC to VME to take advantage of VME's more current technology. In addition, major changes in the data bus structure are being researched to allow for more flexibility and connectivity between processing elements.

6 Conclusions

SRI has developed a series of real-time signal processors to meet expanding data processing requirements. The latest generation uses high-performance processing modules and a time-division-multiplexed architecture to provide enhanced system response time, data throughput, and flexibility. The architecture incorporates hardware that is configurable by software to control and coordinate parallel processing and data transfer paths and to schedule individual processes. This real-time architecture is a framework upon which SRI plans to build even more sophisticated systems.

The SRI processor can be applied to a variety of problems that require multichannel cross-connected real-time signal processing. It is particularly well-suited to process-control applications requiring forward and feedback processing of multiple channels of information, leading to fast, accurate, multichannel response. The software reconfigurability of the hardware allows it to be tailored to particular processing algorithms. In future generations, SRI plans to expand the capability of mapping general signal processing algorithms into the hardware by enhancing the connectivity between the processing elements.

References

[1] W.A. Weeks and B.L. Curless, "A Real-Time, Multichannel System with Parallel Digital Signal Processors," *Proceedings of the ICASSP 90: International Conference on Acoustics, Speech, and Signal Processing,* Albuquerque, New Mexico, April 1990.

[2] *WE DSP32C Digital Signal Processor Information Manual,* The AT&T Documentation Management Organization, 1988.

[3] *User-Configurable Logic Databook,* Altera Corporation, September 1988.

OVERSAMPLED DIGITAL TO ANALOG CONVERTERS WITH VERY SMALL COMPUTATIONAL LOAD

fred harris
Department of E&CE
San Diego State University
San Diego, CA 92182-0190

Bill McKnight
ASW Signal Processing Branch
Naval Ocean Systems Center
San Diego, CA 92152-5000

1. ABSTRACT
We present a non-traditional method of performing the decimation or interpolation required in an oversampled ADC or DAC. This technique uses a very efficient polyphase all-pass structure to perform the resampling task via a sequence of half-band filters. The arithmetic workload is reduced by a factor of five relative to the standard form of interpolation.

2. INTRODUCTION
The ability of the sigma-delta ($\Sigma\Delta$) converter to obtain high resolution analog-to-digital conversion (ADC) with low cost hardware motivates the application of the same techniques to the reverse task of digital-to-analog conversion (DAC). One approach to this task reverses the process by reversing the signal flow through the individual processing blocks of the ADC system, essentially backing through the ADC. The forward direction (ADC), shown in Fig. 1 realizes 64-to-1 decimation with a 4-th order, 16 tap, sinx/x filter as an initial 16-to-1 decimator followed by a 225 tap FIR filter as a final 4-to-1 decimator stage. The FIR filter also equalizes the mainlobe gain of the preceding sinx/x filter. Note that the input to the sinx/x filter is a one bit sequence at $64 \bullet f_s$, the input to the FIR filter is a 16 bit sequence (with noise levels equivalent to 12 bit resolution) at $4 \bullet f_s$, and the output of the FIR filter is a 16 bit sequence at f_s.

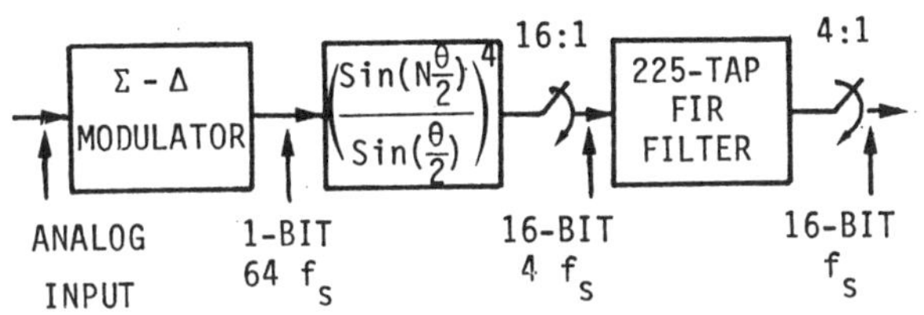

FIGURE 1. DECIMATING CHAIN FOR SIGMA-DELTA ($\Sigma\Delta$) ADC

The sinx/x filter has the transfer function shown in (1).

$$H(Z) = \left\{\frac{1 - Z^{-16}}{1 - Z^{-1}}\right\}^4 \qquad (1)$$

This can be factored into the form shown in (2).

$$H(Z) = \left\{\frac{1}{1 - Z^{-1}}\right\}^4 \{1 - Z^{-16}\}^4 \qquad (2)$$

The first term in (2) represents a cascade of four digital integrators while the second represents a cascade of four 16-point comb filters. The integrators supply the denominator zeros necessary to cancel the numerator zeros at DC (Z=1). The 16-to-1 desampler following the comb filter can be factored to the input of the comb and the Z^{-16} delay term at the input rate becomes a Z^{-1} delay term at the desampled rate. This modified structure is shown in figure 2.

The computational burden for each

1058-6393/91 $01.00 © 1991 IEEE

output data point is obtained by operating the four integrators 64 times, the four differentiators 4 times, and the FIR filter once for a total of 272 adds for the Sinx/x and 225 multiplies and additions for the FIR.

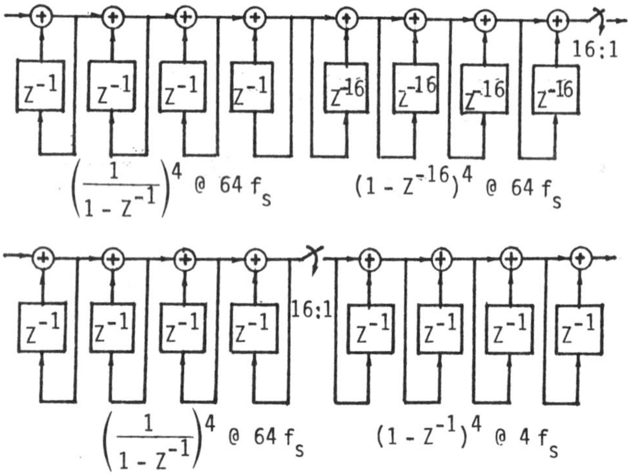

FIGURE 2. DESAMPLED SINX/X DECIMATING CHAINS FOR ΣΔ ADC

One form of the ΣΔ DAC transposes (or reverses the block diagram flow of) the structure which performs the ΣΔ ADC. This structure is shown in Figure 3 where we note one significant difference between the ADC and the DAC. In the ADC, there is bit growth from 1 to 16 bits as we move through the decimator chain. In the DAC, on the other hand, the number of bits increases from 16 to about 28 through the interpolator chain and is converted from 28 to 1 bit at the noise shaping digital sigma-delta block. This means that, even though the workload for the DAC appears to be the same as that for the ADC, the integrators in the DAC are working harder (moving more bits per add) than are those in the ADC.

3. NEW INTERPOLATING STRUCTURE

We now consider an alternate interpolating structure to replace the FIR and SINX/X filter chain. The basic structure of the

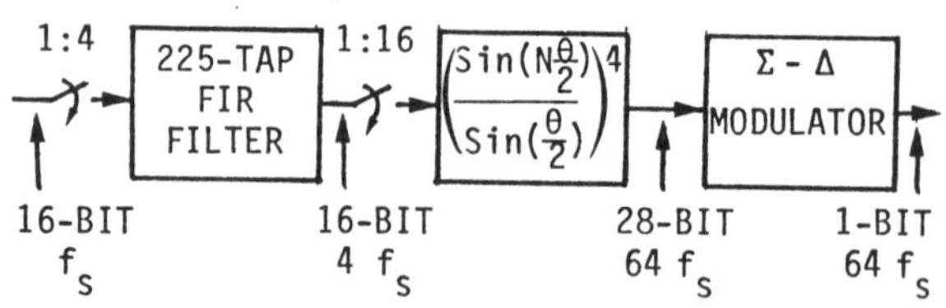

FIGURE 3. RESAMPLED SINX/X INTERPOLATING CHAIN FOR ΣΔ DAC

new interpolator chain is the two-path all-pass network [1] of the form shown in Figure 4. The two paths each contain one or more second order all-pass sub networks of the form shown in Figure 5. Each sub network has the transfer function shown in (3) and the composite network has the transfer function shown in (4).

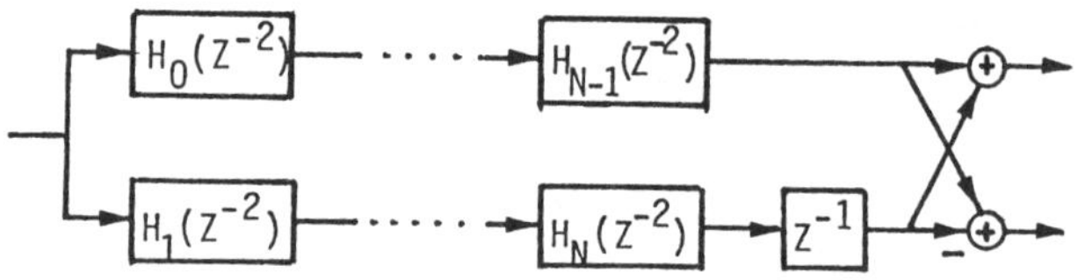

FIGURE 4. TWO-PATH ALL-PASS FILTER STRUCTURE.

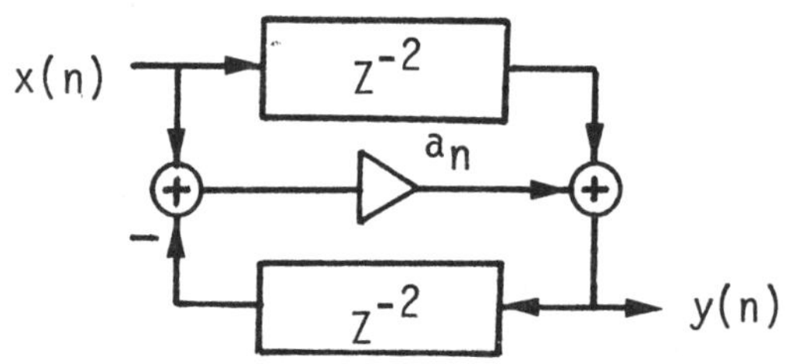

FIGURE 5. SECOND ORDER ALL PASS SUB NETWORK.

$$H_n(Z^{-2}) = \prod^{i} \frac{\alpha_{n,i} + Z^{-2}}{1 + \alpha_{n,i}\, Z^{-2}} \tag{3}$$

$$H(Z) = \sum_{n=0}^{1} H_n(Z^{-2})\, Z^{-n} \tag{4}$$

The single multiplication in each segment forms a pair of conjugate poles and reciprocal zeros. These all-pass sub networks have unity gain at all frequencies but exhibit a frequency dependent phase shift. The phase shift of each path is adjusted (by the pole-zero placement) to have matching phase shifts in the first quadrant and to have matching phase slopes with a 180° phase difference in the second quadrant. When the two paths are summed, signals in the first quadrant add constructively and those in the second quadrant add destructively. This destructive cancellation results in zeros on the unit circle, hence the stopband zeros are obtained without additional multiplication.

The number of segments permitted in the two path filter is either 2N or 2N+1 split between the two paths with path-0 having at most one more segment than path-1. The permitted number of segments, the required arithmetic per output point (when used as a filter or as an interpolator), and the number of poles formed by these filters is shown in TABLE 1.

PATH 0	PATH 1	FILTER MULT/ADD	INTRP/DEC MULT/ADD	NO. POLE/ZERO
1	0	1/3	0.5/1	3
1	1	2/5	1/2	5
2	1	3/7	1.5/3	7
2	2	4/9	2/4	9
3	2	5/11	2.5/5	11
3	3	6/13	3/6	13
4	3	7/15	3.5/7	15
4	4	8/17	4/8	17
.	.	. .	. .	.

TABLE 1. STAGE OPTIONS, WORKLOAD, AND PERFORMANCE COMPLEXITY

Figure 6 presents the pole-zero diagram and frequency response of a 9-pole, 9-zero filter obtained by a filter with two sub segments per path. The technique used to obtain coefficients for this filter is reported in [2].

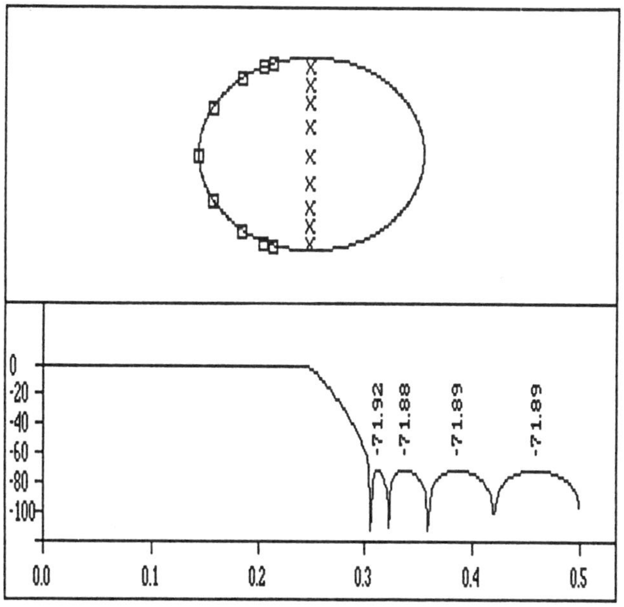

FIGURE 6. POLE-ZERO DIAGRAM AND FREQUENCY RESPONSE OF TWO-PATH ALL-PASS FILTER

Since the two-path filter is a half bandwidth filter it can be used as a 2-to-1 interpolator or decimator. Interpolation is performed by halfband filtering input data multiplexed 2-to-1 with zero valued samples. Since alternate input values are zero they need not be processed. Finite impulse response (FIR) filters are partitioned into polyphase sub filters to take advantage of the known positions of these zero valued samples.

A similar polyphase partition of a recursive filter is generally not possible. Such a partition, however, is feasible for the two-path all-pass structure because the filter is a polynomial in Z^{-2}. The non zero values of the zero packed data are also described by a polynomial in z^{-2}. By interpreting the two time delay units (Z^{-2}) at the output rate as one time delay unit (Z^{-1}) at the input rate we obtain a polyphase partition of this filter. Thus for each input sample, alternate output samples are obtained successively from the two separate filter paths. The output commutation between these two paths replaces the single delay in the lower path of the filter. The interpolator architecture is shown in Figure 7.

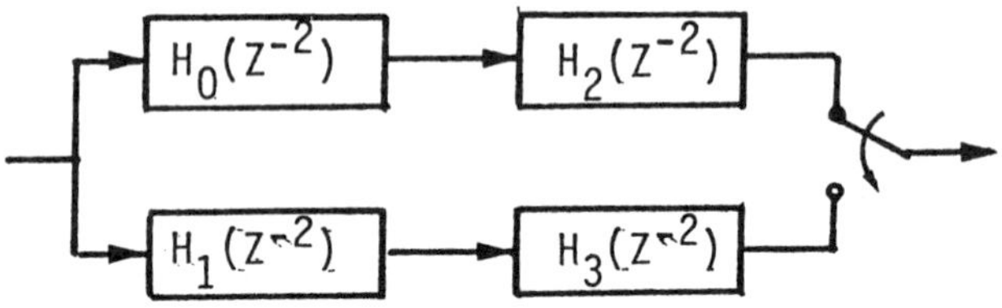

FIGURE 7. RESAMPLING TWO-PATH FILTERS

Note that by using the filter paths alternately the workload per output point is extremely low, such as 3 multiplications and 6 additions per output for a 13-pole, 13-zero filter (see Table 1).

4. DESIGN EXAMPLE

We are now prepared to use the two-path all-pass networks as a high order interpolator for the oversampled DAC. As a specific example, let us use a sampling rate which is 10% above Nyquist. This is a severe requirement, since most systems use 20%, but the 10% figure matches that of a compact disk (CD) recording which operates at a sample rate of 44.1 KHz for a 20.0 KHz bandwidth.

Each filter must supply 96 dB rejection of the replicate spectrum at each filtering stage. The normalized cut-off frequencies for the filter set and the number of stages required to meet that specification are listed in TABLE 2. Also listed is the workload per output point and the output rate for that filter relative to the first stage input rate.

STAGE	F_{stop}	ORDER	MULT/ADD	RATE• f_s
1	0.273	17	4/8	2
2	0.387	7	1.5/3	4
3	0.443	5	1/2	8
4	0.472	5	1/2	16
5	0.486	3	0.5/1	32
6	0.493	3	0.5/1	64

TABLE 2. STAGE CHARACTERISTICS OF HALF BAND INTERPOLATOR CHAIN

The workload per output point for this filter chain is 90 multiplications and 156 additions as opposed to the standard technique of 225 multiplications and 497 additions. The frequency responses and poles & zeros of the first three filter stages are shown in Figures 8.

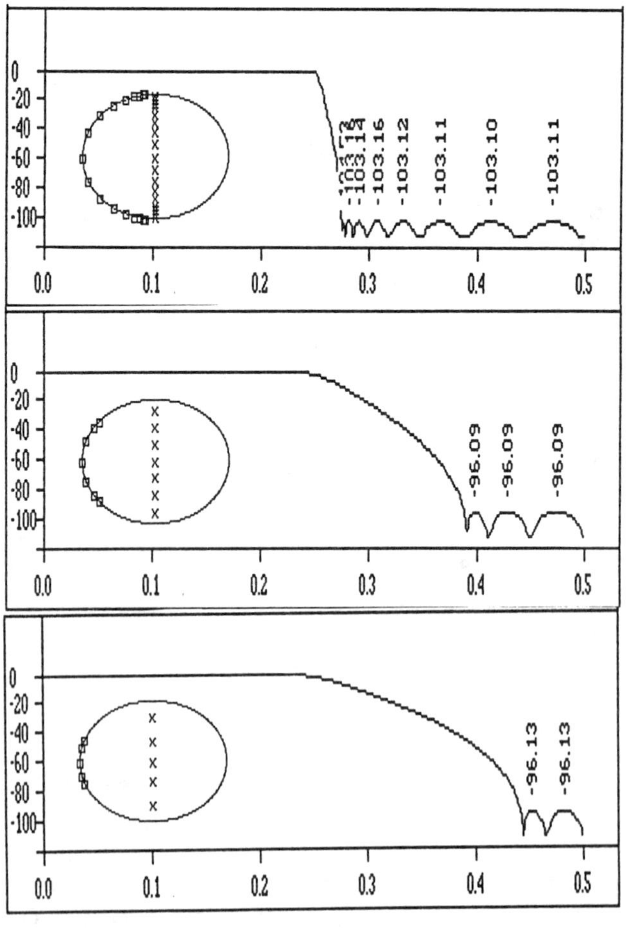

FIGURE 8. FREQUENCY RESPONSE AND POLES & ZEROS OF INTERPOLATOR FIRST 3-STAGES

The workload can be further reduced by replacing the last two stages of the half-band filter with a 4-tap sinx/x filter of the form shown in (5).

$$H(Z) = \left\{ \frac{1 - Z^{-4}}{1 - Z^{-1}} \right\}^4 \qquad (5)$$

This can be factored into the form shown in (6).

$$H(Z) = \left\{ \frac{1}{1 - Z^{-1}} \right\}^4 \{1 - Z^{-4}\}^4 \qquad (6)$$

This filter which places fourth order zeros at the spectral locations (1/4 and 1/2 of $64 \cdot f_s$), requires 8 additions per output point and must be operated 4 times for each input for a total of 20 additions. Replacing the final two stages listed in Table-1 with this filter results in a total workload of 42 multiplications and 96 additions to perform the 1-to-64 interpolation.
These comparisons are tabulated in Table 3.

STRUCTURE	MULT	ADDS
225-PNT FIR & 4 STAGE 16-PNT SINX/X	225	497
CASCADE HALf-BAND POLYPHASE ALL-PASS	90	156
HYBRID, 4-STAGES HALF-BAND & 4 STAGE 4-PNT SINX/X	42	96

TABLE 3. WORKLOAD COMPARISON FOR THREE INTERPOLATING STRUCTURES

4. COMMENTS AND CONCLUSIONS

The multiplications in the all-pass structure can be replaced by a pair of binary shifts and adds. The resultant filters offer less attenuation but can be cascaded to obtain the required rejection. The present form of filter does not exhibit linear phase but can be equalized by a prefilter (operating at the low input rate). Another alternative is to use a modified linear phase form of the filter. Both options are a topic of a following paper.

We have shown it is possible to perform high order, high quality, interpolation (or decimation) with a cascade of polyphase all-pass half-band filters. The filter set exhibits superior sensitivity to coefficient quantization as well as controlled bit growth (1-bit per stage as a filter, 0-bits per stage as an interpolator).

ACKNOWLEDGEMENT
This work was sponsored in part by the ASW Signal Processing Branch, at the Naval Ocean Systems Center (NOSC). The programming for the filter design routine was performed by former SDSU graduate student Maximilien d'oreye de Lantremange.

REFERENCES

1. harris, f.j., d'oreye de Lantremange, M., and Constantinides, A.G., "Digital Signal Processing With Efficient Polyphase All-Pass Filters", Intl' Conf. on Signal Proc., Florence, Italy, 4-6 Sept. 1991.

2. harris, f.j., d'oreye de Lantremange, M., and Constantinides, A.G., "Design and Implementation of Efficient Resampling Filters Using Polyphase All-Pass Filters" 25-th Asilomar Conference on Signals, Systems, and Computers, Pacific Grove, CA, 4-6 Nov., 1991

3. Reagalia, P.A., Mitra, S.K., and Vaidyanathan, P.P., "The Digital All-Pass Filter: A versatile Signal Processing Building Block", Proc. of IEEE, Vol. 76, No. 1, Jan 1988, pp. 19-37.

Training Sequence Size and Vector Quantizer Performance

Pamela C. Cosman Keren O. Perlmutter Sharon M. Perlmutter
Richard A. Olshen* and Robert M. Gray

Information Systems Laboratory
Durand Building
Stanford University
Stanford, CA 94305-4055

*Division of Biostatistics
Health Research and Policy Building
Stanford University
Stanford, CA 94305-5092

Abstract

We examine vector quantizer performance as a function of training sequence size for both tree-structured and full-search vector quantizers. The performance is measured by the mean-squared error between the input image and the quantizer output at a given bit rate. The training sequence size is measured either by the number of training images, or by the number of training vectors. When the training vectors are counted, they are selected randomly from among the training images. For every training sequence size, vector quantizers are developed from several different training sequences and the distortion is calculated for different test sequences in a cross validation procedure. Preliminary results suggest that plots of distortion vs. number of training images (n) *follow an algebraic decay of the form* $An^{-\alpha} + B$ *as expected from analogous results of learning theory.*

1 Introduction

Vector quantizers are generally trained on a sequence of images. Insufficient training data causes the vector quantizer to become "over-trained," in that it becomes very good at coding the training vectors it was supplied with, but is likely to be highly unreliable when coding anything else. Too much training data, on the other hand, should not degrade VQ performance, but can slow down research and development efforts. The amount of training data needed to obtain reliably a preassigned level of performance from the vector quantizer has not been studied and researchers employ "rules of thumb" to decide how much training data to use. The purpose of this work was to address this question, and generally to study the form of the relationship between performance and training sequence size. The study of training sequence size for vector quantization using learning theory is independently being pursued by David Cohn at the University of Washington. Separate training and test images were chosen from a set of 12 mid-sagittal MRI brain scans of different individuals. The two vector quantizers considered were the unbalanced tree-structured quantizer (TSVQ) and the full-search quantizer (FVQ). The quantizers were developed using the Generalized Lloyd Algorithm, and the TSVQs were grown one node at a time in a "greedy" fashion as described in [5, 4, 3] and then pruned back using the generalized Breiman, Friedman, Olshen and Stone (BFOS) algorithm [1]. In all cases the vector dimension was 4 (2×2 pixel block). The performance was measured by the mean-squared error (mse) between the input image and the quantizer output at a given rate.

2 Approach

We used two sampling approaches in this study. First, images were considered as sampling units, which implies that with a total of 12 images available, one would choose n as the actual training images. This is the approach generally used in practice. Second, vectors were considered as sampling units, which implies that we would randomly choose n images' worth of training vectors from all 12 images, rather than a specific set of n images, to create a training sequence of size n. With the first approach, including a highly unusual image as one of the training images could result in high variability. With the second approach, we take only some of the vectors from the unusual image for all training sequence sizes, thus lessening the effect of a very unusual image. On the other hand, if the images all have very similar pixel intensity distributions,

1058-6393/91 $01.00 © 1991 IEEE

then one specific image might be more representative of the distribution of another image than a random sampling of 5 images would be. It is thus not obvious *a priori* which sampling method will yield better or more trustworthy results.

2.1 Images as Sampling Units

Training vectors were obtained from a subset n of the $I = 12$ images, where $n \in \{\frac{1}{8}, \frac{1}{4}, \frac{1}{2}, 1, 2, 3, 4, 5, 6, 7, 8, 9, 10\}$. A training sequence from less than 1 image was obtained by randomly choosing a subset of the training vectors from 1 image. Four of the 12 images were used separately as the test images, and the results for the 4 images at a specific size n were then averaged to obtain each data point on the distortion vs. training sequence size plot.

In order to avoid biased results, images used as a source for training vectors were not used as a source for the corresponding test vectors. In addition, we used a split sample scheme, similar to cross-validation, where we took more than one subset of size n as a source for training vectors whenever possible, and then averaged over the results obtained for the different subsets. The number of different training sequences used to obtain each data point varied slightly due to the limited number of images available, and because we wanted to ensure that particular training vectors never composed more than one training sequence for a specific training sequence size n. Four different training sequences were used separately to obtain results for the training sequence sizes of $\frac{1}{8}$, $\frac{1}{4}$, $\frac{1}{2}$, 1 and 2. Two different training sequences were used separately to obtain results for training sequence sizes 3 and 4. Only one training sequence was used to obtain results for the sizes 5, 6, 7, 8, 9, and 10.

2.2 Vectors as Sampling Units

All training vectors from all 12 images were randomly reassigned to one of 12 new fictitious "images." For an m-fold cross-validation for $n < I$, we pick

$$t = v[\frac{m}{m-1}][\frac{n}{I}] \qquad (1)$$

vectors from each new "image." n denotes the number of training images, v denotes the number of vectors in each "image," and I denotes the total number of images available. In our study, we used m = 10-fold cross-validation, with v = 16384 vectors/image, and I = 12. Nine-tenths of these t vectors are obtained from each new "image," and these are combined together to form the training sequence. The remaining one-tenth of the t vectors from each new "image" are combined to form the test sequence. Ten different training sequences of size n and ten corresponding different test sequences are then produced by choosing different nine-tenth and one-tenth combinations. For example, to investigate the distortion when one training image's worth of vectors is used, i.e., $n = 1$, we picked 16384(10/9)(1/12) or approximately 1520 training vectors from each image (rounding ensured an equal number of training vectors from each image).

2.2.1 Bias-Variance Tradeoff of Nesting

Training sequences of different sizes were nested. For example, half of the training vectors that form a training sequence of size 2 are those vectors that composed the training sequence of size 1.

The decision to nest the increasing sets of training vectors or not is a choice of trading off bias and variance that is familiar in the statistical estimation of functions and also in signal processing. The tradeoff that we describe is a very subtle analogue of the problem that arises when one filters a periodogram to estimate a spectral density. An identity filter which simply passes the signal without change, i.e., a filter with a delta function as its pulse response and a constant transfer function, has little bias but a variance that does not tend to 0 with increasing sample size. On the other hand, a long comb filter for smoothing the input signal, i.e., a filter with a very narrow bandwidth, will have little variability, but will have a persistent bias that does not tend to 0 with increasing sample size. Either situation is unacceptable in practice. For us the bias-variance tradeoff arises in designing the successive training units and their relationships to each other. In the end we implemented only one possibility; that is, increasing sets of training vectors were nested.

We are interested in the mse of the prediction of pixel intensity with respect to its experimental value. If we let the number $\hat{y}$ represent our prediction of the actual intensity value y (a constant for present purposes), then the mse can be expressed as a function of its variance and bias thus:

$$\begin{aligned} E[(\hat{y}-y)^2] &= E[(\hat{y}-E\hat{y}+E\hat{y}-y)^2] \\ &= E[(\hat{y}-E\hat{y})^2]+(E\hat{y}-y)^2 \\ &= \mathrm{var}(\hat{y})+(\mathrm{bias})^2 \end{aligned} \qquad (2)$$

since the cross product terms vanish. When we nest and include a previous sequence in an expanded sequence, as opposed to selecting completely new random sets, we obtain a decrease in variability. For example, if we choose a subset of training vectors that formed two images' worth of training vectors, and these vectors well represented the test sequence, the distortion could be quite low. If we then randomly choose four images' worth of training vectors from all the training vectors (not necessarily including those used for size 2), and these training vectors badly represented the test sequence, the distortion could be quite high. This potential high variability could produce an mse that would not decay as we naturally expect from an increase in the number of training vectors. However, nesting to decrease this potential variability may lead to an increase in bias that may result if one of the original smaller sequences is in some way unusual. We thus forego the opportunity to restart with a new set of training vectors and avoid such a potential bias, so that we can be faithful to the expected decay.

3 Learning Theory

Minimax results and work in progress in the theory of nonparametric function estimation suggest that, at least for numbers of training images or vectors below a threshold, the plots of distortion versus training sequence size (at a fixed bit rate) should be of the form

$$y(n) = An^{-\alpha} + B \qquad (3)$$

where y = distortion and n = training sequence size. We are interested in the norm $\| y - \hat{y} \|$, where $\hat{y}$ is the prediction of pixel intensity y subject to a bit rate constraint. If we consider a particular class of distributions of pixel vector intensities, then the cited heuristic deals with the best case of the worst situation, that is, with

$$\min_{\text{predictors } \hat{y}} \quad \max_{\text{distributions}} \quad \| y - \hat{y} \|. \qquad (4)$$

Previous work of Stone [6] and others deal with (what amount to) pixel intensity distributions smoother than what applies to our imaging problems. In addition, this work treats only cases for which $B = 0$, that is, where, with increasing sample size, distortion tends to 0. That is clearly impossible in our setting since, even with an infinite amount of training data, the quantizer cannot represent the original image in a distortionless way when the rate is less than the entropy of the source. And, too, we stretch to suggest that TSVQ or FVQ are algorithms that even for large samples do as well as can be in the minimax sense of (4). If the

type	rate		B	A	alpha	fit
TSVQ	1 bpp	V	279	16.3	0.573	351
	1 bpp	I	248	36.0	0.245	189
	2 bpp	V	45.3	37.2	0.533	39.9
	2 bpp	I	31.1	43.7	0.366	15.7
FVQ	1 bpp	V	363	23.2	0.240	341
	1 bpp	I	380	0.028	3.38	374
	2 bpp	V	95.4	11.1	0.389	10.2
	2 bpp	I	18.4	85.7	0.035	22.6

Table 1: Fit parameters to $y(n) = B + An^{-\alpha}$

code design were indeed optimal, B could be interpreted as the operational distortion-rate function of the source for the given bit rate, code structure, and blocklength. Notwithstanding, especially in view of recent unpublished work of colleagues David Donoho and Iain Johnstone, we can make the guess that the model (3) will fit and that α will be between 0 and .5; a special argument not reproduced here suggests further that α ought to be about 1/3. The cited research of Donoho and Johnstone (unlike that of Stone) indicates that there will be a threshold above which α=1 applies. The existence of this threshold, which will be smaller at lower bit rates, renders fitting problematical when the threshold is less than the smallest sampling unit. This offers a possible explanation of the result α=3.38 for FVQ, I at 1 bpp in Table 1.

4 Results

The results for TSVQ at 1 bpp and 2 bpp are shown in Figures 1 and 2, respectively, and the results for FVQ at 1 bpp and 2 bpp are shown in Figures 3 and 4, respectively. In all figures, the x-axis has units of training sequence size (in number of images), where each image (256 x 256 pixels) provides 16384 training vectors. The y-axis has units of distortion, as measured by the mse between the input image and the quantizer output. The data points are shown as x's for the vector sampling approach, and as o's for the case of images as sampling units. The solid curve is fitted to the x's, and the dashed curve is fitted to the o's. Table 1 shows the fit parameters for the 8 cases. V and I denote the two sampling methods: vectors and images. The column marked fit shows the residual sum of squares between the fitted and actual values.

In general, the TSVQs performed better than the FVQs. This is expected because we used variable-rate TSVQ, which is designed to minimize average distortion for a given average rate rather than the more re-

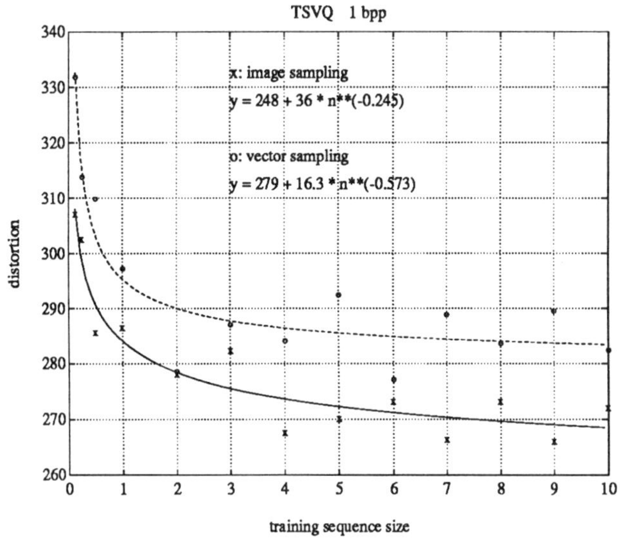

Figure 1: Distortion vs. Training Sequence Size for 1 bpp TSVQ

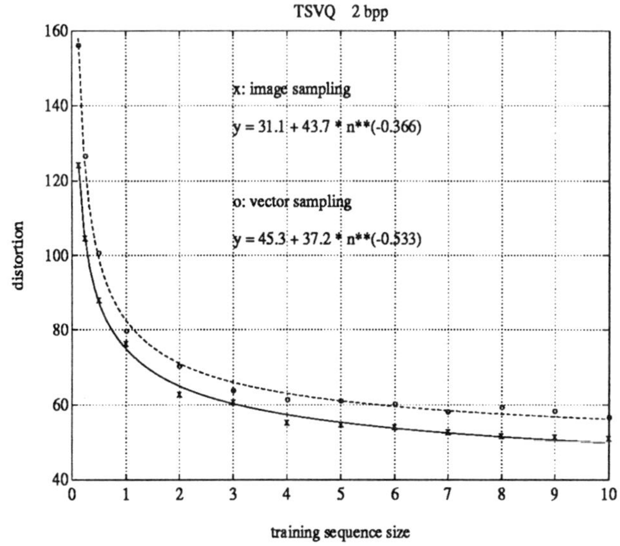

Figure 2: Distortion vs. Training Sequence Size for 2 bpp TSVQ

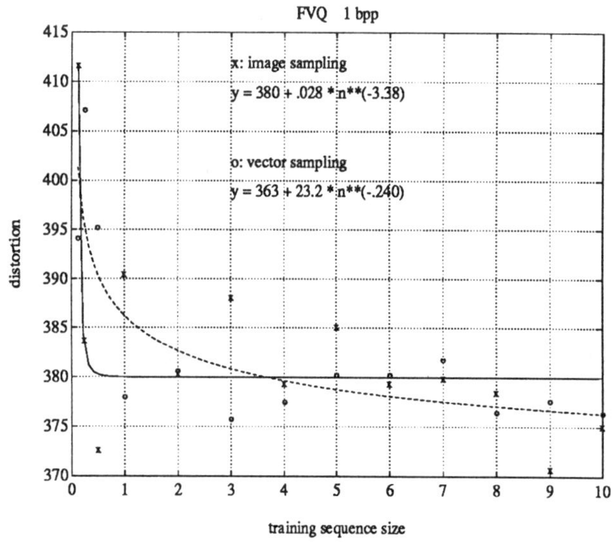

Figure 3: Distortion vs. Training Sequence Size for 1 bpp FVQ

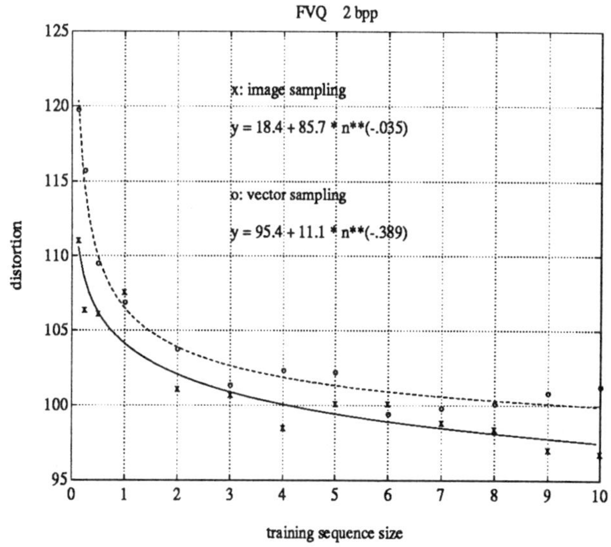

Figure 4: Distortion vs. Training Sequence Size for 2 bpp FVQ

stricted minimization of average distortion for a given fixed rate. Also as expected, the distortions at 2 bpp were considerably lower than the distortions at 1 bpp.

The differences between using images and vectors as sampling units were not substantial, although image sampling generally produced lower distortion.

The results from both TSVQ and FVQ at 1 bpp indicate that any small number of training images is sufficient for the quantizer to perform near the distortion asymptote. For 2 bpp, the results for TSVQ using both approaches and for FVQ using vector sampling fit the model of $y(n) = An^{-\alpha} + B$ very well.

For FVQ with image sampling the decay in distortion at 1 bpp was extremely rapid and, as a result, an α of about 3.4 was obtained. At 2 bpp, the distortion asymptote was much lower than expected, and α, although between 0 and 0.5, was surprisingly small. Thus, both the 1 bpp and 2 bpp results seem to indicate that the model is not applicable to FVQ with image sampling. In contrast, the A, B and α for vector sampling were reasonable values. Similar to the results for the 1 bpp TSVQ, the distortion from a training sequence size of 1 image or greater is quite close to the distortion asymptote.

The curves can be used to determine how much training data is needed to obtain results close to the asymptote for the image type considered. The experimenter first decides how much distortion above the asymptote is tolerable, and then chooses the training sequence size corresponding to that distortion.

5 Future Research and Conclusions

There are a number of issues in this work that can be further explored. Bootstrapping [2] (or some other sample reuse method) is the only plausible way to quantify the variability of estimates of A, B, α, and points on the curves. With the 10-fold cross-validation, the 10 training sequences of any particular size n were not independent. Variances computed for A, B, and α using these ten different values would thus underestimate the true variances.

The study can be extended to investigate results with different vector sizes, image sizes, and other types of images, e.g., the USC database images. In addition, the importance of variability between the images used could be considered.

The differences between the images as sampling units or vectors as sampling units can be further explored. In particular, given more training data than one is able to use, should one choose images or vectors from the set?

It is likely that the results of this preliminary study are highly dependent upon the statistics of the particular images used, yet we find it intriguing that the results fit the algebraic decay model predicted by heuristic arguments from nonparametric function estimation. Ultimately this work may provide researchers with some guidelines for both method and quantity when choosing training data for a VQ.

6 Acknowledgments

This work was supported in part by the National Institutes of Health under Grants CA49697-02 and CA55325-01, by the National Science Foundation under Grant DMS-9101548 and by Graduate Fellowships from the National Science Foundation. We thank Iain Johnstone, Jerry Halpern, and Eve Riskin for their helpful comments.

References

[1] L. Breiman, J. H. Friedman, R. A. Olshen, and C. J. Stone. *Classification and Regression Trees.* The Wadsworth Statistics/Probability Series. Wadsworth, Belmont, California, 1984.

[2] B. Efron. *The Jackknife, the Bootstrap, and Other Resampling Plans.* Society for Industrial and Applied Mathematics, Philadelphia, 1982.

[3] A. Gersho and R. M. Gray. *Vector Quantization and Signal Compression.* Kluwer Academic Publishers, Boston, 1992.

[4] E. A. Riskin and R. M. Gray. A greedy tree growing algorithm for the design of variable rate vector quantizers. Presented at the 1990 Picture Coding Symposium, Cambridge, MA, 1990.

[5] E. A. Riskin and R. M. Gray. A greedy tree growing algorithm for the design of variable rate vector quantizers. *IEEE Transactions on Signal Processing*, 39(11):2500–2507, 1991.

[6] C. J. Stone. Optimal global rates of convergence for nonparametric regression. *Annals of Statistics*, 10:1040–1053, 1982.

Classification Using Vector Quantization

Karen L. Oehler   Pamela C. Cosman   Robert M. Gray   Jack May*

Information Systems Laboratory
Department of Electrical Engineering
Stanford University
Stanford, CA 94305-4055

*ESL Incorporated
MS 408
495 Java Drive
Sunnyvale, CA 94086-3510

Abstract

We describe a simple technique for combining vector quantization and low level classification of images. The goal is to classify automatically certain simple features in an image as part of the compression process in order to enhance their appearance in the reconstructed image. Images in the training sequence are divided into blocks and each block is classified into a particular class by a human observer. This a priori knowledge is then used when designing the codebook so that both small average distortion and accurate implicit classification are achieved. The codebook can also be designed to have different average distortions for the different classes. The technique is a variation on a variable rate tree-structured vector quantizer which is grown by splitting a single terminal node at each iteration. The splitting criterion selection allows tradeoffs among compression rate, distortion and misclassification rate.

1 Introduction

Because digitized images require tremendous memory storage and transmission bandwidth, image compression is often a necessity. Vector quantization (VQ) has for several years been used for image compression [1, 2, 3, 4]. We describe a simple technique for combining vector quantization and low level classification of images in order to enhance their appearance in the reconstructed image. The cluster analysis and nearest neighbor techniques used for vector quantization have also long been used for traditional pattern classification [5]. The two applications differ primarily in the measures of algorithm quality and cost and in the fact that pattern classification often implies an *a priori* set of desired patterns while in quantization the design algorithm can choose the templates to best approximate the possible inputs. Because of the similarity of the two methods, it is natural to consider combinations that both compress and classify. One possible use of this combination is to highlight regions in the reconstructed image belonging to a specific class, in order to draw attention to certain features. Another use is to develop codebooks which have improved representations for those classes that are deemed important.

2 Codebook structure

A tree-structured vector quantizer (TSVQ) is used because it has the advantage of greatly reduced search complexity over full search VQ. Traditionally, a balanced TSVQ is designed one layer at a time and implements a fixed rate code [6]. Makhoul *et al.* [7] introduced an alternative design algorithm that grows the tree one node at a time by splitting the node that contributes most to the overall distortion of the coder. This method results in an unbalanced tree because the node that is split can be at any depth. Because unbalanced trees generally produce lower distortion than balanced trees, we consider unbalanced TSVQ.

Another unbalanced tree design algorithm is presented in [8]. Here, a technique from classification and regression tree design is extended to variable rate coding. In [9], decision trees are grown by splitting one node at a time. An "impurity function" measures the lack of homogeneity of a particular node in a tree T. The "goodness" of a candidate split of node t is defined as the decrease in node impurity that it effects and is given by

$$\Delta i(s,t) = i(t) - p_L i(t_L) - p_R i(t_R). \qquad (1)$$

Here, s is a binary test (a nearest neighbor selection in a TSVQ), $i(t)$ is the impurity measured at node t of tree T (average distortion in a TSVQ), p_L is the proportion of the samples in node t that go to the left child t_L, and p_R is the proportion that go to the right child t_R.

For a TSVQ with a squared error distortion, the

impurity function becomes

$$i(t) = d(t) = \frac{1}{||\mathcal{T}_t||} \sum_{k:x_k \in \mathcal{T}_t} d(\mathbf{x}_k, \hat{\mathbf{x}}_t) \qquad (2)$$

where $\mathcal{T}_t$ is the set of all training vectors $\mathbf{x}_k$ mapping into node t, $||\mathcal{T}_t||$ is the number of vectors in $\mathcal{T}_t$, $\hat{\mathbf{x}}_t$ is the reproduction codevector associated with node t, and $d(\mathbf{x}, \mathbf{y}) = (\mathbf{x} - \mathbf{y})^t(\mathbf{x} - \mathbf{y})$. In growing an unbalanced TSVQ, the terminal node measuring the highest $\Delta d(s,t)$ is always split. Let D and R stand for the distortion and rate, respectively, measured by T, and let D' and R' stand for the distortion and rate of the tree after t is split into t_L and t_R. Let $\Delta D = D' - D$ and $\Delta R = R' - R$ be the change in the distortion and rate, respectively, due to splitting t. The splitting criterion from [8] is

Criterion Split the node that provides the largest ratio of decrease in distortion to increase in rate, i.e. split the node with the largest value of

$$-\frac{\Delta D}{\Delta R} = d(t) - p_L d(t_L) - p_R d(t_R) = \Delta d(s,t). \qquad (3)$$

It was found that such trees outperform balanced trees of equivalent bit rates by up to 3.5 dB at high rates on a magnetic resonance image database [8]. Observe that average squared error is simply one example of a node impurity measure that can be used to grow TSVQs. We will here consider other measures.

3 Hand-labeled classification

The problem chosen here is to highlight areas in images that are deemed to belong to a specific category. In aerial images, image sublocks that are suspected of being "manmade" as opposed to "natural" are highlighted to attract attention of human observers. The classification of the training set of aerial photographs is done by hand labeling those features (man-made and natural regions) that are to be recognized in subsequent images. This *a priori* knowledge is used when designing the TSVQ codebook so that both small average distortion and accurate implicit classification are achieved. Hence, the TSVQ encoder is designed to classify image subblocks while compressing the image. Several codebook design methods are possible with differing compression and classification abilities on the test images. The splitting criterion selection allows tradeoffs among compression rate, distortion, and misclassification rate.

The new approach taken here is that the classification is done simultaneously as the image is encoded. One codebook search is sufficient to both encode a subblock and to classify it into one of the previously determined categories. Stored with each codeword in the codebook is a class label representing the best class prediction for image subblocks that will be represented by that codeword. Thus, once the encoder selects the most appropriate codeword, the preliminary classification of the subblock at hand is simply a matter of memory lookup; no other computations are required. In effect, we are getting this classification knowledge "for free." A simple means of incorporating classification into a TSVQ is to classify each terminal node according to a majority vote of the *a priori* class assignments of the training vectors that were represented by that node after encoding. This makes sense if the costs of misclassifying the image subblocks of the various classes are equal. However, if the error costs are unbalanced, then a more sophisticated weighting function would be appropriate.

It is important to point out similarities and differences between this approach and Classified Vector Quantizers (CVQ) [10]. Both approaches divide the training sequence vectors among different categories or classes. While the CVQ algorithm constructs a VQ codebook for each class, the algorithm described here constructs only one VQ codebook using the class information. In [10], the key goal was to code edges with higher visual quality and to reduce the complexity of the overall encoder by creating many independent subcoders. Here the key goal is to recognize and "highlight" specific classes; complexity is not reduced by the classification information, but the encoder is already low complexity due to its tree structure.

3.1 Alternative splitting criteria

The codebooks are constructed by splitting one node at a time. The splitting criterion, i.e. the methodology for deciding which node to split next, can have a strong effect on the quality of the codebook. Three different criteria are investigated:

Criterion 1 Ignore the classification information and split the node with the largest goodness of split as defined in Equation 3 where the distortion measure is mean-squared error. This design yields an ordinary VQ for comparison [8].

Criterion 2 Split the node that has the greatest percentage of misclassified training vectors, i.e. split the node with the largest value of

$$\lambda_2 = \frac{\text{(Number of misclassified vectors)}}{\text{(Total number of vectors)}}. \quad (4)$$

This corresponds to measuring impurity by the Hamming distance between the node class, c_t, and the hand-labeled class, $c(x)$. Thus, if $d_H(x,y) = 0$ if $x = y$ and $d_H(x,y) = 1$ if $x \neq y$, then the impurity of node t is given by

$$d(t) = d_H(t) \equiv \frac{1}{\|\mathcal{T}_t\|} \sum_{k:x_k \in \mathcal{T}_t} d_H(c(\mathbf{x}_k), c_t) \quad (5)$$

from Equation 2. The node with the highest such impurity is split. Thus the encoder nearest neighbor mapping and the centroid reproduction levels are chosen to minimize squared error, but the tree is grown to reduce classification error.

Criterion 3 Split the node that has the greatest number of misclassified training vectors. This is equivalent to splitting the node with the largest impurity defined by the partial Hamming distortion $d(t) = \|\mathcal{T}_t\| d_H(t)$.

Note that the latter two splitting criteria are not formatted in the traditional "goodness" of split criterion as are criterion 1 and [8], but are more similar to the splitting criterion used in [7]. Because the classification of each node is based on majority rule, splitting criteria based on classification error can lead to a desirable split having a zero "goodness" of split.

3.2 Experimental Results

The training set consisted of 5 images provided by ESL, Inc. The images were 512 × 512 pixels of 8 bit grayscale consisting of aerial photography of the San Francisco Bay area. Each 16 × 16 pixel subblock in the training set was assigned to be either "man-made" or "natural" based on the perceptions of a human observer. While the training vectors were classified in 16 × 16 vectors to simplify the task of the human classifier (even using the 16 × 16 subblocks, over 5000 decisions had to be made), the codebook construction and image encoding were carried out using 4 × 4 pixel vectors. This coarse resolution is partially responsible for the relatively low classification abilities demonstrated with the data. Classification ability is defined as the percentage of vectors classified correctly by the TSVQ, compared to the classification standard created by the human observer.

Sets of codebooks having either the same rate or the same number of nodes were constructed to allow for comparison. Results for codebooks at a rate of 0.5 bpp are shown in Table 1. In general, the first splitting criterion provided the lowest mean squared error in the encoded image at the expense of relatively poor classification ability. The latter two splitting methods provided poorer encoded images (much more blocky in appearance) but better classification ability. Criterion 3 classified more vectors as man-made than criterion 2. For a given number of nodes, criterion 2 produced a higher rate code than 1, and 3 higher than 2. Likewise, for a given rate the tree structure produced by criterion 1 has substantially more nodes than criterion 2, and 2 more nodes than 3. Choosing the splitting criterion involves a tradeoff between compression and classification quality; or similarly, a tradeoff between rate and memory requirements (memory being proportional to the number of nodes.)

Images outside the training sequence were encoded with the resulting codebooks. Because the simultane-

Figure 1: Original aerial image

Criterion:	1	2	3
Training PSNR (dB)	26.7	25.4	25.1
Training classification ability	0.72	0.74	0.70
Test PSNR (dB)	23.4	22.8	22.7
Test classification ability	0.71	0.74	0.75

Table 1: PSNR and classification ability using TSVQ codebooks grown using various splitting criteria for both the training sequence and the test image.

Figure 2: Image compressed at 0.5 bpp using compression/classification encoder designed using Criterion 2.

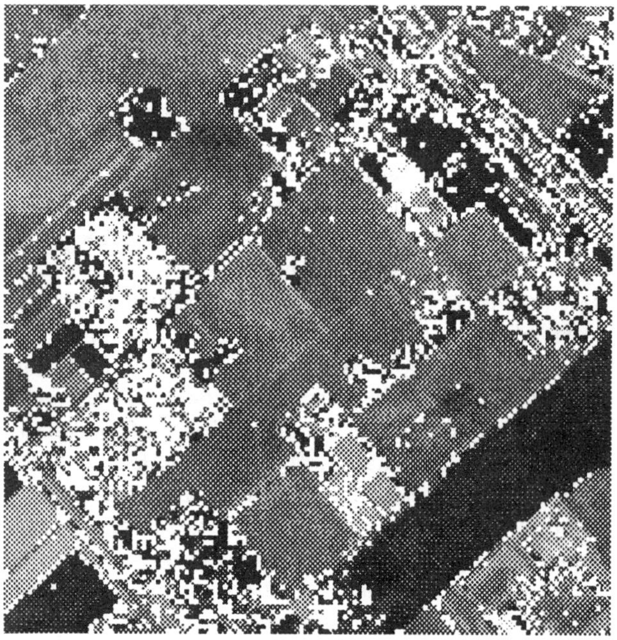

Figure 3: Natural sublocks using encoder designed using Criterion 1. Man-made subblocks are replaced by solid white subblocks.

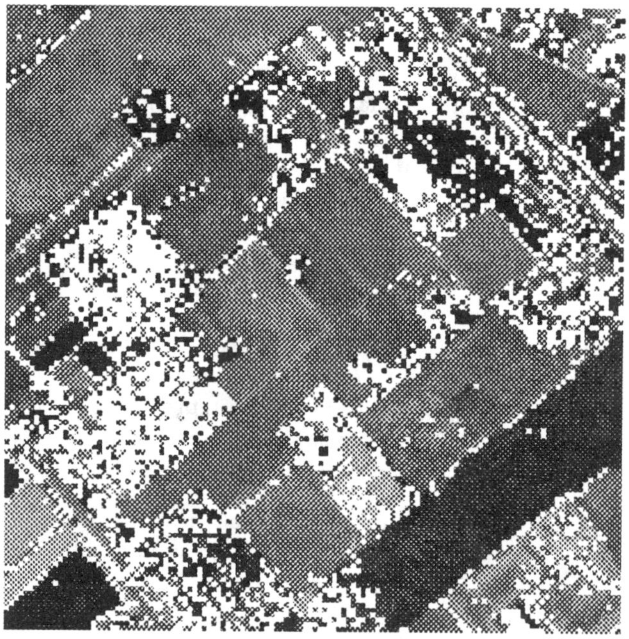

Figure 4: Natural sublocks using encoder designed using Criterion 2. Man-made subblocks are replaced by solid white subblocks.

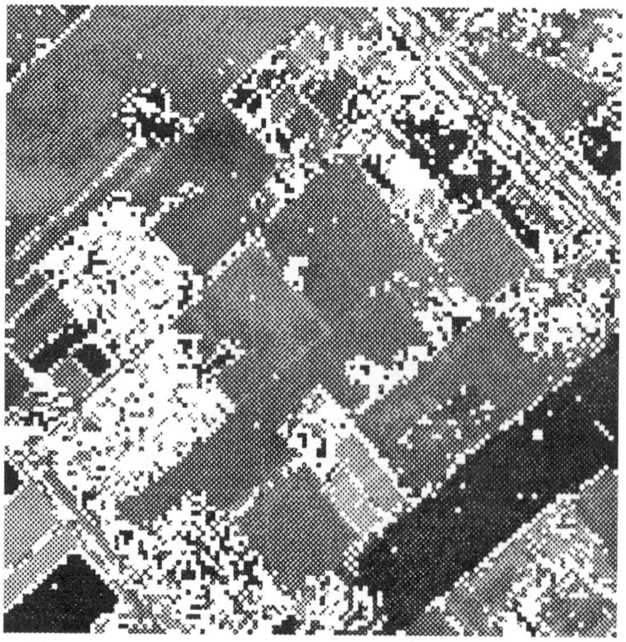

Figure 5: Natural sublocks using encoder designed using Criterion 3. Man-made subblocks are replaced by solid white subblocks.

ous display of the compression and classification results is difficult in a grayscale format, the results of compression and classification are shown separately. Figure 1 shows the original test image (not in the training set) that was used to evaluate the different codebooks. Figure 2 shows the image after compression at 0.5 bpp using the codebook constructed with Criterion 2 (results from the other criteria are similar.) Figures 3, 4 and 5 demonstrate classification from codebooks constructed with Criteria 1, 2, and 3 respectively. The images show the subblocks that the encoder classified as natural, whereas the man-made subblocks are replaced by solid white subblocks. Although this information is not as amenable to grayscale display, the images shown here reflect the accuracy of the classification encoder. The classification ability was modest; at 0.5 bpp the best classification encoder still had 25% misclassification rate on the training sequence, that is, one in four decisions disagreed with the hand-labeled classes. However, the resulting error rate was partly due to the fact that the hand-labeling of the training and test images was performed at one-fourth the resolution that the classifier was trying to achieve. This resolution mismatch not only degrades the training sequence but also produces an imperfect "gold standard" classification for the test image. Since the classification ability is computed with respect to the hand-labeling of the training and test images, it is also affected by the human observer's perception and consistency limitations.

Ideally, the compressed images would be viewed on a color monitor so that classification information can be indicated by color superimposed on the grayscale compressed image. Such a contrast makes the natural and man-made features of the image easier for a human viewer to differentiate.

3.3 Comparison with other classification methods

The classification results were compared with results obtained using the CART™ algorithm, described in [9]. CART uses training data to construct a decision tree for classification. Here, we considered decision trees constructed as a series of binary splits, where each split is conducted along one coordinate in the training data. The training vectors were preprocessed to obtain new vectors consisting of the minimum, maximum, average and standard deviation of pixel values within the vector. CART trees developed with these preprocessed vectors showed better classification ability than trees developed with the original training vectors. This was due to the fact that splitting along a single coordinate in the image vector is not as efficient as splitting along a function of several coordinates. A classification tree was trained using preprocessed vectors from the same 5 images as in Section 3.2, and was applied to the same test image. A compressed version of the test image (compressed to 0.5 bpp using the "standard" TSVQ tree built with Criterion 1) was also classified using the CART tree. The results are shown in Table 2. This preprocessing allowed CART to classify test images very well, considering the limitations on the training sequence. However, there was notable degradation in classification ability on the compressed image.

	CART
Test classification ability	0.81
Comp. Test classification ability	0.72

Table 2: Classification ability using CART algorithm for both the test image and a 0.5 bpp compressed version of the test image.

On the test image, CART performed better than the vector quantizer classifier; however, CART performed worse on the compressed test image. In general, the two classification methods are comparable.

4 Automatic classification based on input-weighted distortion measures

In the previous experiment, we used the classified training vectors to grow the tree in such a way that it attempts to keep vectors of the same class together. Instead of, or in addition to, this classification purpose, classified training vectors can also be used for an enhancement purpose: the tree can be grown so that it does a better job of encoding certain classes. We consider two examples of this type of enhancement. Both examples involve an automatic classification of the training vectors: in one case a classification by brightness, and in the other, by texture. A weight is associated with each class, and the tree is grown with a weighted distortion measure that causes the tree to have "growth spurts" for certain types of inputs that have been declared *a priori* to be important, and to become "stunted" for inputs that are less important. Further information on this approach can be found in [11].

The most common distortion measure for growing and pruning TSVQs is the squared error. We now consider the splitting criterion of Equation 3 with instead

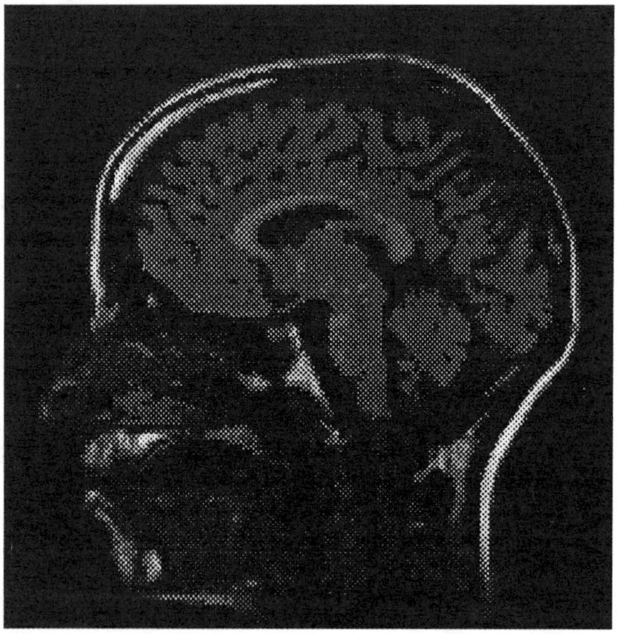

Figure 6: Compressed image at 0.75 bpp: regular distortion measure.

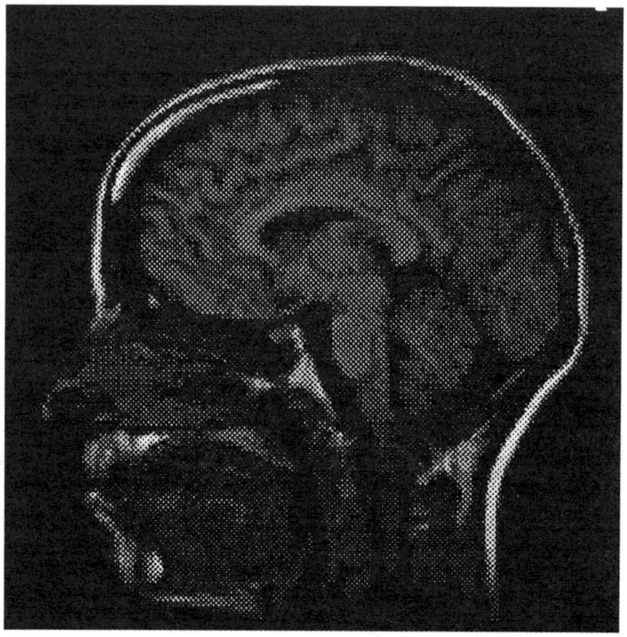

Figure 7: Compressed image at 0.75 bpp: weighted distortion measure.

an input-weighted distortion measure of the form

$$d(\mathbf{x},\mathbf{y}) = (\mathbf{x}-\mathbf{y})^t \mathbf{W}_x (\mathbf{x}-\mathbf{y}), \qquad (6)$$

where $\mathbf{W}_x$ is a strictly positive definite weighting matrix that depends on the input $\mathbf{x}$. In our examples, $\mathbf{W}_x$ will equal $w_x\mathbf{I}$, where $\mathbf{I}$ is the identity matrix, and w_x is a weight assigned to each training vector $\mathbf{x}$ according to its classification. Training vectors which are more important according to some criterion will be assigned higher weights. Thus nodes having high concentrations of training vectors from important classes will have an inflated ΔD and therefore will split earlier and more often than they would in the unweighted case.

4.1 Classification based on intensity

A classification based on intensity, in which bright training vectors are weighted more heavily than dark ones, is appropriate for MR brain scans because the bright parts of the image correspond typically to what is medically most important in the image. A training sequence consisting of 8 MR brain scans was blocked into 2×2 pixel blocks, and each vector was automatically assigned a weight proportional to its energy. A tree was grown to 2 bpp using the weighted distortion measure for splitting, and pruned back to 0.75 bpp. The tree was evaluated on images not in the training sequence. The same training sequence without weights assigned was used to grow an unweighted unbalanced tree according to the original greedy growing algorithm. The compressed images at bit rates of 0.75 bpp produced by the two different trees are shown in Figures 6 and 7. The image made from the weighted tree looks better in the bright regions (e.g., the cortex and cerebellum) which generally correspond to the diagnostically important part of these images.

4.2 Classification based on texture

Due to pattern masking, the human visual system is generally less sensitive to noise in highly active or textured regions of an image, and we used our class-weighted distortion measure to effect a redistribution of quantization noise into regions of more texture. The training sequence consisted of six USC database images, blocked into 4×4 vectors. A weight was assigned to each training vector according to how many of the pairs of adjacent pixels had differences exceeding some threshold. Highly textured vectors having many pairs exceeding the threshold were assigned low weights, and highly homogeneous vectors were assigned large weights. A tree was grown to 2 bpp using the weighted distortion for the splitting criterion, and pruned back to 0.54 bpp. The test image showed clouds, a lake, and trees. A comparison of compressed images encoded from the weighted and unweighted trees showed that the one from the weighted tree had less distortion in the cloud regions, where distortion is most noticeable, and it had more in the areas of trees, where the high texture masks the noise.

5 Conclusions

We have used classification information in conjunction with a greedy growing algorithm to grow unbalanced TSVQ. Hand-labeling of aerial images was used to grow TSVQ which combine moderate classification ability with encoding, useful in highlighting the reconstructed images. The classification schemes used so far have classified each training vector without regard to its context in the training image. It might be possible to develop classifiers which incorporate more contextual information. Classified training sequences were also used to redirect quantization noise to different parts of compressed images, by using a class-weighted distortion measure. A weighting based on brightness for medical images and one based on texture for outdoor scenes both produced weighted trees that significantly improved the perceptual compressed image quality.

Acknowledgements

This work was supported in part by ESL, Incorporated, the National Science Foundation under Grant MIP-9016974, the National Institutes of Health under Grant CA49697-02, and a National Science Foundation Graduate Fellowship.

References

[1] R. M. Gray. Vector quantization. *IEEE ASSP Magazine*, 1,No. 2:4–29, April 1984.

[2] N. M. Nasrabadi and R. A. King. Image coding using vector quantization: A review. *IEEE Transactions on Communications*, COM-36:957–971, August 1988.

[3] H. Abut, editor. *Vector Quantization.* IEEE Reprint Collection. IEEE Press, Piscataway, New Jersey, May 1990.

[4] A. Gersho and R. M. Gray. *Vector Quantization and Signal Compression.* Kluwer Academic Publishers, Boston, 1992.

[5] M. R. Anderberg. *Cluster Analysis for Applications.* Academic Press, San Diego, 1973.

[6] A. Buzo, A. H. Gray, Jr., R. M. Gray, and J. D. Markel. Speech coding based upon vector quantization. *IEEE Trans. Acoust. Speech Signal Process.*, ASSP-28:562–574, October 1980.

[7] J. Makhoul, S. Roucos, and H. Gish. Vector quantization in speech coding. *Proc. IEEE*, 73. No. 11:1551–1587, November 1985.

[8] E. A. Riskin and R. M. Gray. A greedy tree growing algorithm for the design of variable rate vector quantizers. *IEEE Trans. Signal Process.*, November 1991. To appear.

[9] L. Breiman, J. H. Friedman, R. A. Olshen, and C. J. Stone. *Classification and Regression Trees.* Wadsworth, Belmont,California, 1984.

[10] B. Ramamurthi and A. Gersho. Classified vector quantization of images. *IEEE Trans. Comm.*, COM-34:1105–1115, November 1986.

[11] P. C. Cosman, K. L. Oehler, A. A. Heaton, and R. M. Gray. Tree-structured vector quantization with input-weighted distortion measures. In *Proceedings Visual Communications and Image Processing '91*, Boston, Massachusetts, November 1991. SPIE- The International Society for Optical Enginering.

Transform Coding of Monochrome and Color Images Using Trellis Coded Quantization*

Kai-Loong Tong and Michael W. Marcellin
Department of Electrical and Computer Engineering
The University of Arizona
Tucson, AZ 85721

Abstract

Trellis coded quantization (TCQ) is incorporated into a transform coding structure for encoding monochrome and color images at 1 bit/pixel. Both fixed-rate and entropy-constrained designs are considered. For monochrome images, a simple 4-state fixed-rate TCQ-based system provides gains in peak-signal-to-noise ratio of up to 2.4 dB over scalar quantizer based systems, while the entropy-constrained design provides gains of up to 7.0 dB. Gains in PSNR for color images of up to 1.6 dB and 3.9 dB are achieved for the 4-state fixed-rate and entropy-constrained TCQ-based systems, respectively. The subjective quality of the images produced by the TCQ-based systems is significantly better than that of the scalar quantizer-based systems. The high frequency background noise, fuzziness, and block artifacts produced by the scalar quantizer-based systems are virtually eliminated by the TCQ-based systems.

1 Introduction

Trellis coded quantization (TCQ) was recently introduced [1] as an effective scheme for encoding memoryless sources with low to moderate complexity. The motivation for TCQ comes from Ungerboeck's formulation of trellis coded modulation (TCM) [2]. As in TCM, an expanded codebook is partitioned into subsets, and these subsets are used to label the branches of an appropriate trellis. For a given data sequence, the Viterbi algorithm [3] is then used to find the minimum mean squared error (MSE) path through the trellis.

Entropy-constrained TCQ was introduced in [4]. In that work, near optimal performance (in a rate-distortion theory sense) for encoding memoryless sources was achieved at encoding rates greater than about 1.5 bits/sample. [5] introduced a new approach to encoding codeword sequences into bit sequences which enabled entropy-constrained TCQ to achieve near optimal performance at all non-negative encoding rates.

In this work, TCQ is incorporated into a transform coding structure for encoding monochrome and color images at an average encoding rate of 1 bit/pixel. Both fixed-rate and entropy-constrained TCQ-based systems are investigated. 256 × 256 pixel, 8-bit gray level images and 24-bit color images are encoded using both TCQ- and scalar quantizer-based schemes to provide comparisons on the basis of peak-signal-to-noise ratio (PSNR) and subjective image quality. Comparisons are also provided for several image coding schemes in the literature.

In the monochrome case, the two-dimensional discrete cosine transform (DCT) is used to transform all 8 × 8 blocks of an image. Like coefficients from each block are collected and encoded using an optimum rate-allocation method. This rate allocation algorithm requires the transmission of a small amount of side information, consisting of the sample mean of the "DC" coefficient and the sample standard deviation of all coefficients to be encoded. The rates obtained by this method are adjusted to integers in the fixed-rate scheme and to multiples of 0.25 in the entropy-constrained scheme. The side information is quantized uniformly using encoding rates derived from sample statistics computed over a large collection of images. The procedural flow for color images is similar, except for the conversion of the RGB planes into NTSC transmission primaries (Y, I, and Q). The I and Q planes are sub-sampled (2:1 in each direction) before the DCT is applied.

Iterative codebook optimization algorithms are used to minimize MSE in both the fixed-rate and entropy-constrained TCQ systems. A large set of images (different from those used to test the resulting systems) are used as training data for these algorithms. In the fixed-rate case, the codebooks are designed for minimum MSE, while in the entropy-constrained case,

*This work was supported in part by the National Science Foundation under Grant No. NCR-8821764.

1058-6393/91 $01.00 © 1991 IEEE

the goal is to minimize MSE, but subject to a constraint on entropy. In each case, a different codebook is generated for each encoding rate used.

2 Rate Allocation and Side Information

The MSE for many source coders can be reasonably modeled by

$$D(R) \approx \alpha\sigma^2 2^{-\beta R} \tag{1}$$

where D is the MSE (or distortion), R is the encoding rate in bits/sample, σ^2 is the variance of the source, and α and β are parameters that depend on the particular encoder structure and the probability density function of the source. For encoding rates above 0.5 bits/sample, the MSE performance of TCQ can be approximated by selecting α as 1.26 and 1.03 for Gaussian and Laplacian sources, respectively, while β can be chosen as 1.88 for both sources. These sources are of interest in DCT-based image coding since for a large class of images, the DC coefficient is (roughly) Gaussian distributed and the other coefficients are Laplacian distributed [6].

Let $L = M^2$ be the number of coefficients obtained by transforming a $M \times M$ subblock of image data. The average distortion incurred by encoding the ith coefficient with R_i bits/coefficient can be determined with (1). Since the DCT is a unitary transform, the overall average distortion introduced in the image by encoding the first $L' \leq L$ coefficients using rates $R_i > 0$ (with the remaining $L - L'$ rates being 0.0) is given by

$$D = \frac{1}{L}\left[\sum_{i=1}^{L'} \alpha_i\sigma_i^2 2^{-\beta R_i} + \sum_{i=L'+1}^{L} \sigma_i^2\right]. \tag{2}$$

Using the Lagrange multiplier method to minimize (2) subject to the average rate constraint

$$\frac{1}{L}\sum_{i=1}^{L'} R_i = R, \tag{3}$$

yields the optimum rates

$$R_i = \begin{cases} \frac{1}{\beta}\log_2 \frac{\alpha_i\sigma_i^2}{(\prod_{j=1}^{L'} \alpha_j\sigma_j^2)^{\frac{1}{L'}}} + \frac{RL}{L'} & i = 1, 2, \ldots, L' \\ 0 & i = L'+1, \ldots, L. \end{cases} \tag{4}$$

The value of L' is chosen by setting $L' = L$ and computing the rates, R_i, $i = 1, 2, \ldots, L'$ from (4). If any rates are negative, L' is decreased by one and the rates are computed again. This process is repeated until all encoding rates are non-negative.[1] An alternate approach to rate allocation that can handle a more complicated distortion model than that of (1) is given in [7].

To determine a nominal rate-allocation scheme for monochrome images, ten 256 x 256 pixel training images were used. These images were divided into 8 x 8 subblocks and transformed. All like coefficients were collected and used to estimate the variance of each coefficient. These variance estimates were substituted into (4) to obtain a rate-allocation matrix (for an average encoding rate of 0.995 bits/pixel) given by

$$\begin{bmatrix} 6.3 & 4.3 & 3.4 & 2.8 & 2.2 & 1.7 & 1.2 & 0.5 \\ 4.1 & 3.2 & 2.6 & 2.0 & 1.3 & 0.8 & 0.4 & 0.2 \\ 3.1 & 2.5 & 2.1 & 1.5 & 1.0 & 0.6 & 0.2 & 0.0 \\ 2.3 & 1.8 & 1.5 & 1.1 & 0.6 & 0.3 & 0.0 & 0.0 \\ 1.8 & 1.1 & 1.0 & 0.6 & 0.2 & 0.0 & 0.0 & 0.0 \\ 1.3 & 0.7 & 0.4 & 0.1 & 0.0 & 0.0 & 0.0 & 0.0 \\ 0.7 & 0.1 & 0.0 & 0.0 & 0.0 & 0.0 & 0.0 & 0.0 \\ 0.6 & 0.0 & 0.0 & 0.0 & 0.0 & 0.0 & 0.0 & 0.0 \end{bmatrix} \tag{5}$$

The average encoding rate of 0.995 bits/pixel was chosen to allow a small amount of side information to be transmitted while maintaining an overall average encoding rate of 1.0 bit/pixel. This side information consists of the sample mean of the DC coefficient (the means of the other coefficients are assumed to be zero) and the sample standard deviation of each coefficient to be encoded. These quantities were encoded with uniform scalar quantizers using a total of 344 bits/image $\approx$ 0.005 bits/pixel. Since the sample standard deviations are available at both the encoder and decoder, (4) can be used to determine the optimal rate allocation on an image by image basis.

3 Fixed-Rate TCQ of Transform Coefficients

The procedural flow for the TCQ transform coder is described as follows. All 8 x 8 blocks of a 256 x 256 monochrome image are transformed using a two-dimensional DCT and like coefficients are collected into sequences of data to be encoded using TCQ. Since there are 1024 8 x 8 blocks, and 64 DCT coefficients per block, there are 64 sequences of coefficients, each of length 1024. Each sequence is normalized by subtracting its mean and dividing by its standard deviation. The normalized version of the sequences are then encoded using TCQ at rates determined by the optimum rate-allocation scheme discussed in the previous

[1] Note that the variances are sorted so that $\sigma_i^2 \geq \sigma_{i+1}^2$ for all i before applying (4). This insures that the coefficients assigned an encoding rate of zero are those with the smallest variances.

Table 1: Fixed-Rate PSNR for Transform Coding of Monochrome Images at 1 bit/pixel. (Rate $(R_i + 1)$ Lloyd-Max Codebooks)

Image	Trellis Size (States)				Scalar
	4	16	64	256	Quant.
Tree	29.46	29.59	29.65	29.71	28.00
Couple	34.44	34.55	34.59	34.62	32.79
Girl	34.49	34.60	34.67	34.70	33.00
Lena	30.28	30.35	30.39	30.41	28.95

section (all rates are first adjusted to integer values). At the receiving end, the resulting bit sequence and normalization parameters (side information) are used to reconstruct the quantized coefficients. The inverse DCT is performed to obtain the final reconstructed image.

Table 1 lists the objective performance achieved by encoding the ith DCT coefficient at a rate of R_i (bits/coefficient) using TCQ with codebooks from rate $(R_i + 1)$ Lloyd-Max scalar quantizers. Gaussian codebooks were used for the DC coefficient while Laplacian codebooks were used for the others. Objective performance is reported by tabulating peak-signal-to-noise ratios (PSNR) obtained when encoding the four images ("Tree," "Couple," "Girl," and "Lena") from the USC data base.

Also included in Table 1, for the purpose of comparison, is the performance obtained when using Lloyd-Max scalar quantization rather than TCQ. Note that the TCQ-based systems outperform the scalar quantizer-based systems with PSNR gains ranging from 1.46 dB to 1.83 dB for a 256-state trellis.

In [8], the generalized Lloyd algorithm [9] was incorporated into the TCQ structure to design optimal codebooks. Based on the nominal rate-allocation matrix shown in (5), the transform coefficients, X_{ij}, $i, j = 1, \ldots, 8$, from our training data were divided into several sets. For example, the set consisting of all X_{11} coefficients was used as the training data for the design of all codebooks for rates of 5 bits/coefficient and larger. Three and four bit codebooks were designed using a set of data consisting of all the X_{12}, X_{13}, X_{21}, X_{22}, and X_{31} coefficients.

Table 2 lists PSNR values obtained by the TCQ-based systems having codebooks designed using this method. As before, the size of the codebook used to encode each coefficient was twice that required to encode coefficients using a scalar quantizer. The performance obtained using these optimized codebooks is at least as good as that obtained by the Lloyd-Max codebooks in every case. For a 256-state trellis, improvements in PSNR range from 0.21 dB to 0.86 dB.

Table 2: Fixed-Rate PSNR for Transform Coding of Monochrome Images at 1 bit/pixel. (Rate $(R_i + 1)$ Generalized Lloyd Algorithm Codebooks)

Image	Trellis Size (States)			
	4	16	64	256
Tree	29.54	29.75	29.84	29.92
Couple	35.18	35.36	35.43	35.48
Girl	35.16	35.32	35.44	35.49
Lena	30.57	30.68	30.75	30.78

4 Entropy-Constrained TCQ of Transform Coefficients

The procedural flow for the entropy-constrained TCQ system is the same as that of the fixed-rate TCQ. The optimal rate-allocation scheme discussed in previous sections is also used in the entropy-constrained TCQ system. The optimization is carried out using the distortion model of (1) with $\alpha = 0.32$ and $\beta = 2.0$. These values were chosen based on simulation results obtained using training data. Also, for the entropy-constrained system, rates are adjusted to multiples of 0.25 rather than integers (as was done in the fixed-rate system). The side information (consisting of 344 bits/images) used by the entropy-constrained scheme is identical to that of the fixed-rate scheme.

In [4] and [5], it was shown that for encoding memoryless sources with smooth densities, near-optimal performance (in a rate-distortion theory sense) can be obtained by employing the codebook design algorithm and encoding rule from [10]. In [5], it was found that for rates greater than about 2.5 bits/sample, the optimized codebooks do not provide a significant improvement in MSE over uniform codebooks (although there may be some advantage in codebook size). Thus, in our simulations with images, optimized codebooks were used for encoding rates less than or equal to 2.5 bits/sample while uniform codebooks were used for all other rates. To obtain training data for the optimization procedure, we divided the transform coefficients for the training images into several sets (based on a nominal rate-allocation matrix similar to (5)) and applied the optimization process discussed previously.

Table 3 presents PSNR values obtained using a 4-state entropy-constrained TCQ system. The PSNR for the entropy-constrained system is significantly higher

Table 3: Entropy-Constrained TCQ PSNR for Transform Coding of Monochrome Images (4 states).

	Image			
	Tree	Couple	Girl	Lena
Entropy	1.04	0.85	0.79	0.96
PSNR	33.14	39.77	39.05	33.86

than that of the fixed-rate systems for every image. In fact, the 4-state entropy-constrained TCQ system outperforms the 256-state fixed-rate TCQ system of Table 2 by 3.08 dB to 4.29 dB.

5 TCQ of Color Transform Coefficients

The procedure for transform coding of color images differs only slightly from that for monochrome images. The main difference is that the YIQ-transform is applied to the image before the DCT. Since most of the energy is contained in the Y plane, the I and Q planes are subsampled (2:1 in each direction). Thus, the dimensions of the Y plane are $M \times M$ while the I and Q planes are each $\frac{M}{2} \times \frac{M}{2}$. The same rate-allocation algorithm discussed before may be used, but now $L = M^2 + 2(\frac{M}{2})^2$.

Side information (339 bits/image — consisting of the average value for each of the three DC coefficients and the sample standard deviation of all coefficients to be encoded) is transmitted for normalization, rate allocation, and reconstruction of the YIQ coefficients at the receiver as in the monochrome case. Before transforming the reconstructed coefficients back to the RGB domain at the receiver, the I and Q planes are up-sampled by a factor of 1:2 in each direction.

The four 256 × 256 pixel, 24-bit color images ("Tree," "Couple," "Girl," and "Lena") that were encoded in our simulations are from the USC image data base. Tables 4 and 5 list the PSNR values obtained when using fixed-rate TCQ with Lloyd-Max codebooks and codebooks obtained by the generalized Lloyd algorithm, respectively. Also, included in Table 4 is the performance achieved when using scalar quantization rather than TCQ. The 256-state trellis, with generalized Lloyd algorithm codebooks, provides improvements of 0.96 dB to 1.76 dB over the scalar quantizer-based scheme.

For our entropy-constrained color image coding system, fixed-rate TCQ is replaced by entropy-constrained TCQ. As in the monochrome case, optimized codebooks were used for encoding rates of less than or equal to 2.5 bits/sample, while uniform

Table 4: Fixed-Rate PSNR for Transform Coding of Color Images at 1 bit/pixel. (Rate (R_i+1) Lloyd-Max Codebooks)

	Trellis Size (States)				Scalar
Image	4	16	64	256	Quant.
Tree	25.51	25.61	25.65	25.69	24.78
Couple	30.98	31.06	31.10	31.12	29.89
Girl	29.81	29.88	29.92	29.95	28.87
Lena	27.39	27.47	27.52	27.54	26.52

Table 5: Fixed-Rate PSNR for Transform Coding of Color Images at 1 bit/pixel. (Rate (R_i+1) Generalized Lloyd Algorithm Codebooks)

	Trellis Size (States)			
Image	4	16	64	256
Tree	25.51	25.63	25.67	25.74
Couple	31.47	31.58	31.62	31.65
Girl	30.17	30.24	30.31	30.36
Lena	27.60	27.72	27.76	27.84

codebooks were used for higher rates. Table 6 lists the PSNR values obtained using a 4-state entropy-constrained TCQ system. Comparison with Table 5 reveals that the entropy-constrained TCQ system outperforms the 4-state fixed-rate TCQ system by 1.84 dB to 2.35 dB.

6 Conclusion

The application of transform coding using TCQ to encode monochrome and color images has been investigated. Specifically, TCQ was used to encode transform coefficients resulting from applying an 8×8 DCT to 256 × 256 pixel 8-bit gray level, and 24-bit color images. For color images, the red, green, and blue planes were transformed into the NTSC transmission primaries specified as Y, I, and Q before the DCT

Table 6: Entropy-Constrained TCQ PSNR for Transform Coding of Color Images (4 states).

	Image			
	Tree	Couple	Girl	Lena
Entropy	1.04	0.92	0.79	0.95
PSNR	27.70	33.82	32.07	29.44

was applied. Rate-allocation schemes for encoding the transform coefficients and side information were developed and discussed. The generalized Lloyd algorithm [9][8] and the entropy constrained VQ algorithm [10][4][5] were used to design optimal fixed-rate and entropy-constrained codebooks.

In comparisons with other results from the literature, we find that our entropy-constrained TCQ systems are quite competitive. For the monochrome "Lena" image, our PSNR value of 33.86 dB is below that of the adaptive DCT-based scheme of [11] (34.6 dB), but above those of the adaptive subband coding scheme of [12] (32.5 dB) and the pyramid VQ-based DCT scheme of [13] (32.2 dB). Similarly, for the monochrome "Girl" image, we find that the entropy-constrained TCQ system performs better than the lattice VQ-based DCT scheme of [14] (38.8 dB) and the pyramid VQ-based DCT schemes of [15][16] (35.72 dB and 35.92 dB, respectively). We were not able to find results in the literature for the proper images and bit-rates needed for a meaningful comparison with the color image coding results presented here.

Subjectively, images produced by the TCQ-based schemes are significantly superior in perceptual quality as compared to those produced by the scalar quantizer-based schemes. They are much clearer and show a striking improvement in background noise content. Additionally, the block artifacts present in the images encoded using scalar quantization are almost nonexistent in the images encoded using TCQ.

References

[1] M. W. Marcellin and T. R. Fischer, "Trellis coded quantization of memoryless and Gauss-Markov sources," *IEEE Trans. Commun.*, vol. COM-38, pp. 82–93, Jan. 1990.

[2] G. Ungerboeck, "Channel coding with multi-level/phase signals," *IEEE Trans. Inform. Th.*, vol. IT-28, pp. 55–67, Jan. 1982.

[3] G. D. Forney, Jr., "The Viterbi algorithm," *Proc. IEEE* (Invited Paper), vol. 61, pp. 268–278, Mar. 1973.

[4] T. R. Fischer and M. Wang, "Entropy-constrained trellis coded quantization," to appear in *IEEE Trans. Inform. Th.*

[5] M. W. Marcellin, "On entropy-constrained trellis coded quantization," submitted to *IEEE Trans. Commun.*

[6] R. C. Reininger and J. D. Gibson, "Distributions of the two-dimensional DCT coefficients for images," *IEEE Trans. Commun.*, vol. COM-31, pp. 835–839, June 1983.

[7] Y. Shoham and A. Gersho, "Efficient bit allocation for an arbitrary set of quantizers," *IEEE Trans. Acoust., Speech, and Signal Proc.*, vol. ASSP-36, pp. 1445–1453, Sept. 1988.

[8] T. R. Fischer, M. W. Marcellin, and M. Wang, "Trellis coded vector quantization," to appear in *IEEE Trans. Inform. Th.*

[9] Y. Linde, A. Buzo, and R. M. Gray, "An algorithm for vector quantizer design," *IEEE Trans. Commun.*, vol. COM-28, pp. 84–95, Jan. 1980.

[10] P. A. Chou, T. Lookabaugh, and R. M. Gray, "Entropy-constrained vector quantization," *IEEE Trans. Acoust., Speech, and Signal Proc.*, vol. ASSP-37, pp. 31–42, Jan. 1989.

[11] W. A. Pearlman, "Adaptive cosine transform image coding with constant block distortion," *IEEE Trans. Commun.*, vol. COM-38, pp. 698–703, May 1990.

[12] J. W. Woods and S. D. O'Neil, "Subband coding of images," *IEEE Trans. Acoust., Speech, and Signal Proc.*, vol. ASSP-34, pp. 1278–1288, Oct. 1986.

[13] M. E. Blain and T. R. Fischer, "A comparison of vector quantization techniques in transform and subband coding of imagery," *Signal Processing: Image Communication*, vol. 3, pp. 91–105, Feb. 1991.

[14] D. G. Jeong and J. D. Gibson, "Lattice vector quantization for image coding," in *Conf. Proceedings,* 1989 Int. Conf. on Acoust., Speech, and Signal Proc., May 1989.

[15] M. E. Blain, "Transform coding of images using a pyramid vector quantizer," M.S. thesis, Texas A&M University, Aug. 1987.

[16] H.-C. Tseng and T. R. Fischer, "Transform and hybrid transform/DPCM coding of images using pyramid vector quantization," *IEEE Trans. Commun.*, vol. COM-35, pp. 79–86, Jan. 1987.

A HIGH QUALITY DIGITAL HDTV CODEC

Arun Netravali, Eric Petajan, Scott Knauer, Kim Matthews, Robert J. Safranek, and Peter Westerink
AT&T Bell Laboratories
600 Mountain Avenue
Murray Hill, New Jersey 07974

Abstract

A digital video codec has been developed for the Zenith/AT&T HDTV system for terrestrial broadcast over NTSC taboo channels. The codec works on an image progressively scanned with 1575 scan lines every 1/30th of a second and achieves a compression ratio of approximately 50 to 1. The transparent image quality is achieved using motion compensated transform coding coupled with a perceptual criterion to determine the quantization accuracy required for each transform coefficient. The combination of a sophisticated encoded video format and advanced bit error protection techniques results in a highly robust reception and decoding of the compressed video signal.

1. Introduction

The Zenith/AT&T Digital Spectrum Compatible (DSC)-HDTV system uses a video compression algorithm which is optimized for the terrestrial broadcast environment. A very high compression ratio is achieved which enables robust transmission without sacrificing the image quality. The algorithm is designed for practical implementation and results in a decoder which is realizable in a small number of VLSI chips.

The complete DSC-HDTV system is shown in Figure 1. The encoder accepts YUV 4:2:2 with 9 bits per sample. The image is progressively scanned at 59.94 Hz with 720 active lines and 1280 active pels per line resulting in an uncompressed video bit-rate of 994 Mbits per second. After the removal of redundant information from the video signal the encoder provides a 17 Mbit/second coded video bit stream to the transmission system in a robust format. After error correction processing the video decoder receives the bit stream and outputs YUV 4:2:2 for display (after matrixing to RGB).

Three basic types of redundancy are exploited in the video compression process. Motion compensation removes temporal redundancy, spatial frequency transformation removes spatial redundancy, and perceptual weighting removes amplitude redundancy by putting quantization noise in less visible areas.

Temporal processing occurs in two stages. The motion of objects from frame to frame is estimated using hierarchical block matching. Using the motion vectors, a displaced frame difference (DFD) is computed which generally contains a small fraction of the information in the original frame. The DFD is transformed using a two dimensional transform to remove the spatial redundancy. Each new frame of DFD is analyzed prior to coding to determine its rate versus perceptual distortion characteristics and the dynamic range of each coefficient. Quantization of the transform coefficients is performed based on the perceptual importance of each coefficient, the precomputed dynamic range of the coefficients, and the rate versus distortion characteristics. The perceptual criterion uses a model of the human visual system to determine a human observer's sensitivity to color, brightness, spatial frequency and spatial-temporal masking. This information is used to minimize the perception of coding artifacts throughout the picture. Parameters of the coder are optimized to handle the scene changes that occur frequently in entertainment/sports events, and channel changes made by the viewer. The motion vectors, compressed transform coefficients and other coding overhead bits are packed into a format which is highly immune to transmission errors. In case of transmission errors, the decoder uses a recovery technique which masks the errors. If loss of signal is detected or the channel is changed, the decoder switches to a special mode which quickly builds the picture to full quality.

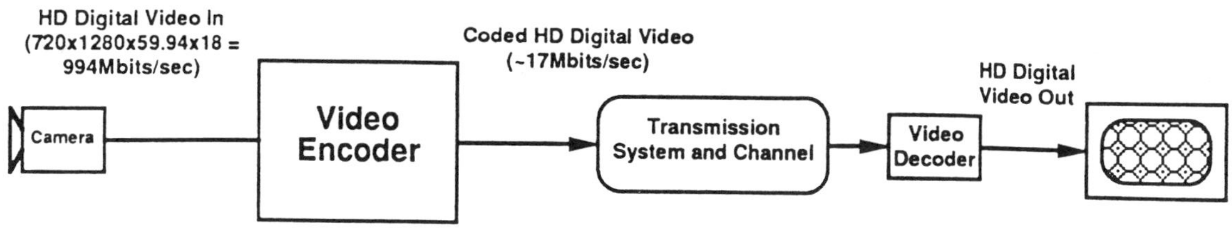

Figure 1. DSC-HDTV system

In choosing the video compression algorithm the need for a modular architecture and low cost was considered. Particular attention has been given to minimizing the decoder circuits that will be part of every HDTV receiver. Some elements of the encoding algorithms that affect picture quality can be altered without requiring modifications to the decoder. This feature provides an opportunity for future improvement without affecting the installed base of equipment. Some coding applications may also require an inexpensive encoder. The DSC-HDTV system has a modular architecture which allows a subset of the encoder to be used at a much lower cost.

2 . Encoder

The encoder is shown in Figure 2. Motion from frame to frame is estimated using a hierarchical block-matching motion estimator. The motion estimator produces motion vectors, which are compressed and sent to the output buffer for transmission. Each frame is analyzed before being processed in the encoder loop. The motion vectors and control parameters resulting from the forward analysis are input to the encoder loop which outputs the compressed prediction error to the channel buffer. The encoder loop control parameters are weighted by the buffer state which is fed back from the channel buffer.

In the predictive encoding loop, the generally sparse differences between the new image data and the motion-compensated predicted image data are encoded using adaptive transform coding. The parameters of the encoding are controlled in part by forward analysis. The data output from the encoder consists of some global parameters of the video frame computed by the forward analyzer and transform coefficients that have been selected and quantized according to a perceptual criterion.

Each frame is composed of a luminance frame and two chrominance difference frames which are half the resolution of the luminance frame horizontally and vertically. The compression algorithm produces a chrominance bit-rate which is generally a small fraction of the total bit-rate, without perceptible chrominance distortion.

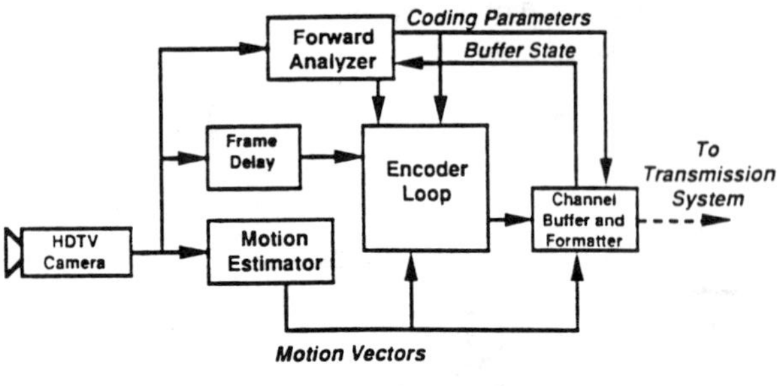

Figure 2. Encoder

The output buffer has an output rate of approximately 17 Mbits/sec and has a varying input rate that depends on the image content. The buffer history is used to control the parameters of the coding algorithm so that the average input rate equals the average output rate. The feedback mechanism involves adjustment of the allowable distortion level, since increasing the distortion level (for a given image or image sequence) causes the encoder to produce a lower output bit rate.

The encoded video is packed into a special format before transmission which maximizes immunity to transmission errors by masking the loss of data in the Decoder. The duration and extent of picture degradation due to any one error or group of errors is limited.

2.1 Motion Estimation

Motion is estimated in stages on a block by block basis. At each stage the best block match is defined to be that which has the least absolute difference between blocks. The results from one stage are used as a starting point for the next stage to minimize the number of block matches per image. The motion estimator is capable of handling large frame-to-frame displacements typical of entertainment and sport scenes. Finally, the resolution of motion estimation is adapted spatially to those places in the picture which could have the most benefit within the limit of the compressed motion vector bit rate.

The motion estimator compares a block of pels in the current frame with a block in the previous frame by forming the sum of the absolute differences between the pels, known as the prediction error. Each block in the current image is compared to displaced blocks at different locations in the previous image and the displacement vector that gives the minimum prediction error is chosen as being the best representation of the motion of that block. This is the motion vector for that block. The end result of motion estimation is to associate a motion vector with every block of pels in the image.

The motion estimator is shown in Figure 3. To reduce the complexity of the search, hierarchical motion estimation is used in which a first stage of coarse estimation is refined by a second, finer estimation. The first stage matching is performed on the images after they have been decimated by a factor of two both vertically and horizontally. This reduces both the block size and the search area by factors of four. The motion vectors that are generated are passed to the second stage which performs a search centered around this coarse estimate. The motion vectors generated from the first stage are used by the finer motion estimator that can estimate the motion of small blocks to within sub-pixel accuracy.

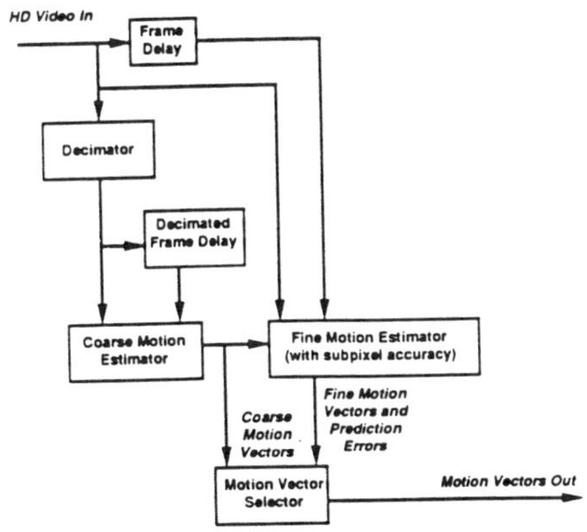

Figure 3. Motion Estimator

The second motion estimator stage generates the prediction errors of the small blocks for each location within the search window. The prediction errors of the coarse blocks are derived from the sums of the appropriate small block prediction errors. The final stage of the motion estimator uses the prediction errors to generate the motion vectors by finding the minimum prediction error for all blocks in every location. The resulting motion vectors are then passed on to the vector selector stage.

Given the motion vectors from the coarse and fine motion estimator stages, the motion vector selector must select the set of motion vectors that will give the best prediction of the next frame while limiting the bit rate of the compressed motion vector data to be in a range. This is achieved by sending two resolutions of motion vectors. The first set represents the motion vectors of the coarse blocks which are unconditionally transmitted, and the second set represents the motion vectors for small blocks. However, not all of the small block motion vectors are transmitted, but only those which can be sent within the bandwidth remaining after the coarse motion vectors have been sent.

2.2 Encoder Loop

The encoder loop is shown in Figure 4. The encoder loop generates a transformed and quantized displaced frame difference (DFD) using the motion vectors, perceptual thresholds and loop control parameters. The motion compensated predictor applies the motion vectors to the predicted frame which is stored in a buffer. The resulting displaced frame (DF) is subtracted from the input frame to generate the DFD. The DFD is then spatially transformed and adaptively quantized by the Quantizer Vector Selector. The quantized transform coefficients and coded selection vectors are passed to the Channel Buffer and Formatter. The DFD coefficient quantization and spatial transform are inverted resulting in an estimate

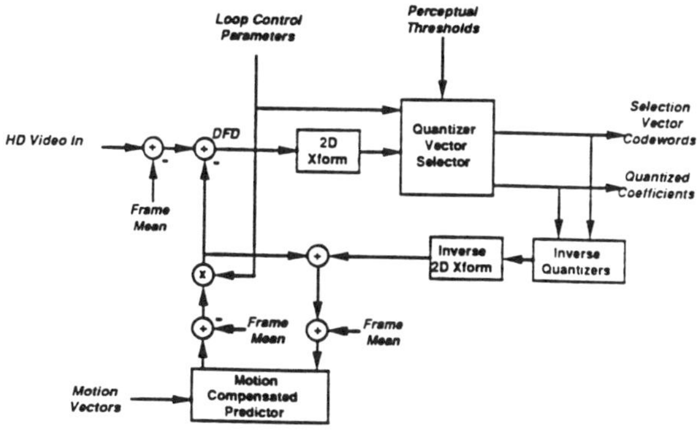

Figure 4. Encoder Loop

of the DFD as received by the decoder. This estimated DFD is added to the DF resulting in the next reconstructed frame which replaces the current reconstructed frame in the buffer.

The mean of each frame is calculated in the Forward Analyzer and subtracted from each frame before differencing with the zero-mean DF. This results in a zero-mean DFD for greater transform efficiency. The mean is added back to the reconstructed frame in order to accurately estimate the quantization effects of mean addition prior to display in the receiver.

2.2.1 Adaptive Postprocessing

Under certain circumstances, coarse quantization of the transformed DFD coefficients can produce random noise from frame to frame resulting in increased visibility. To avoid this potential flickering of quantized information, postprocessing is applied at the receiver. At each pixel a weighted average of temporally successive pixels is stored in the reconstructed frame buffer if it differs from the current pixel by an amount less than some threshold. If the pixel differs by an amount larger than the threshold, the new pixel is used without averaging.

2.2.2 Spatial Transform

Blocks of pixels containing the displaced frame difference are transformed using a spatial transform. The transform is implemented using an algorithm that minimizes the number of multiplications required. Sufficient accuracy is maintained throughout to ensure that the transform is not a source of quantization noise.

2.2.3 Adaptive Quantization of Transform Coefficients

A principal means of reducing the number of bits needed to represent an image is to control the number of bits used to represent the individual transform coefficients. The algorithm described here recognizes that transform coefficients vary in importance based on the limitations of human vision. By adjusting the coarseness of quantization of individual coefficients to local regions of the image, the algorithm can minimize the amount of transmitted information, while retaining the flexibility to apply better coefficient precision where needed.

A set of nonuniform quantizers is used to quantize the coefficients. Each quantizer has a zero level and is adaptively scalable to maximize its efficiency given the amplitudes of coefficients. A coefficient may also be dropped or forced to zero. The quantized coefficients are variable length encoded using a codebook matched to each quantizer.

2.2.4 Vector Quantization of Selection Patterns

A variety of quantizers are available for each coefficient. However, constraints on the selection of quantizers is needed to accommodate the quantizer selection overhead. Within the bit budget, Vector Quantization is used to represent the possible combinations or patterns of quantizers which can be applied to a given block of coefficients. Coding efficiency is achieved by compressing and transmitting the index associated with a given quantizer selection pattern instead of the pattern itself.

The quantizer vector selector is shown in Figure 5. Unquantized transform coefficients are input to the quantizer selector and buffered to accommodate quantizer selector delay. Each entry in the quantizer selection vector codebook is retrieved in sequence and applied to a given vector of coefficients. Each element of a selection vector contains a quantizer code which selects a quantizer for the corresponding coefficient. Each coefficient is quantized and variable length encoded in order to compute the number of bits per coded coefficient. The quantization error is compared to a perceptual error threshold and the result is summed to produce the selection error for the vector. The optimal vector is selected by considering both selection error and the bit-rate.

The selected vector is now applied to the buffered unquantized coefficients via the quantizers as shown in Figure 5. The vector selects which quantizer to apply to each coefficient. The quantized coefficients are sent to the inverse quantizers and inverse transform in the encoding loop, and to the channel buffer and formatter for transmission after variable length encoding.

Additional coding efficiency is achieved by applying the same quantizer selection vector to all of the coefficients in a spatial block. The block size for the chrominance DFD coefficients is larger than the luminance block size, and the

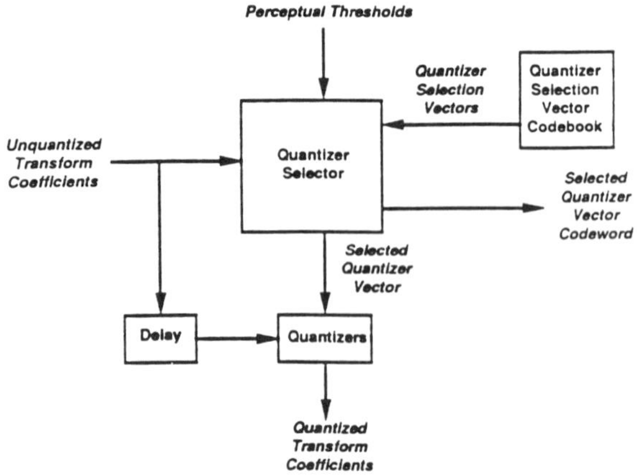

Figure 5. Quantizer Vector Selector

number of chrominance selection vectors can be smaller than the number of luminance vectors with no perceptible chrominance quantization noise.

2.2.5 Selection of Quantizer Patterns.

An important principle of the proposed coding algorithm is that the perceptible distortions should be minimized. Moreover, perceptibility of errors and distortions should be equally visible throughout the image. This is accomplished using adaptive quantization. Quantizer selection patterns are chosen as follows. A target distortion is defined for every coefficient. This target distortion is a combination of local perceptual thresholds that depend on the local scene content, and on a global target distortion that is influenced by the history of channel buffer fullness. A selection error is computed for every coefficient using the target distortion and the actual quantization for that coefficient. For every coefficient, the information rate (number of bits) for a particular quantizer is computed. For every allowable pattern of quantizers, the absolute value of the selection errors are summed and the information rates are summed for all of the coefficients in a block. The pattern of quantizers with minimum selection error is selected. If the minimum selection errors are within a small range, the pattern with minimum information rate is chosen. Variable Length Coding is applied to pattern indices because certain patterns are statistically much more likely than others.

2.3 Forward Analyzer

The forward analyzer is show in Figure 6. The original frame is transformed and analyzed by the perceptual weight calculator which passes weights to the buffer controller. The perceptual thresholds are computed by the buffer controller using the weights and the buffer state.

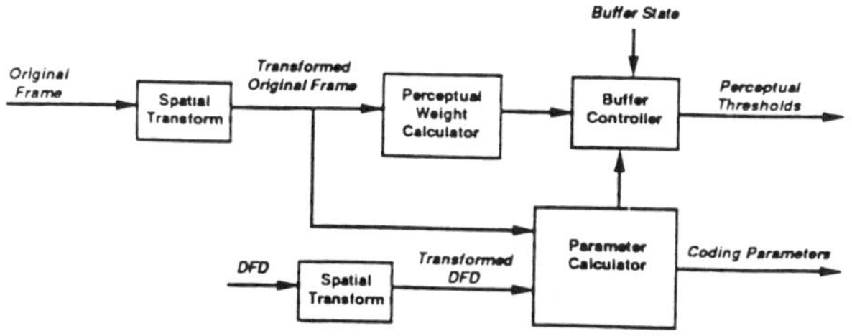

Figure 6. Forward Analyzer

The forward analyzer also approximates the operation of the encoding loop in order to optimize encoding loop control parameters. These parameters depend on the results of analysis of the entire original frame and DFD and would be available one frame too late if computed in the loop. Examples of control parameters are the image mean and the quantizer scaling which increase the efficiency of the quantizers.

3. Decoder

The Decoder is shown in Figure 7. The compressed video data enters the buffer which is complementary to the compressed video buffer at the encoder. The decoding loop uses the motion vectors, transform coefficient data, and other side information to reconstruct the HDTV images. Channel changes and severe transmission errors are detected in the decoder causing a fast picture recovery process to be initiated. Less severe transmission errors are handled gracefully by several algorithms depending on the type of error.

Processing and memory in the decoder are minimized. Processing consists of one inverse spatial transform and a variable length decoder which are realizable in a few VLSI chips. Memory in the decoder consists of one full frame and a few compressed frames (< 1.5 Mbytes).

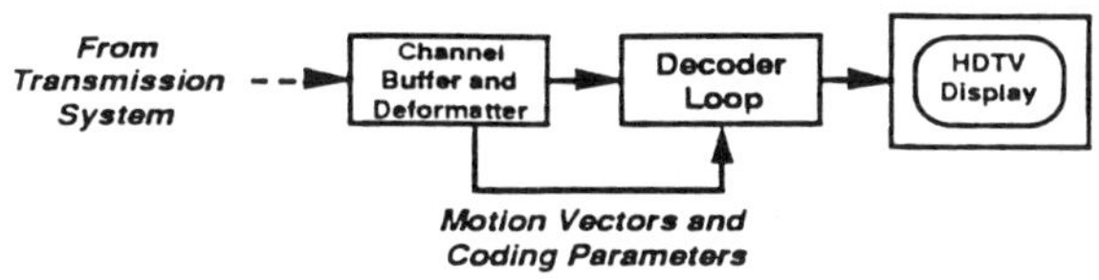

Figure 7. Decoder

4. Summary

The video compression algorithm includes a unique set of features that maximizes picture quality within the available channel bandwidth. The principal features are:

- Hierarchical, subpixel motion estimation within a motion vector budget
- Handling of scene changes with no perceptible distortion
- Quick buildup of picture after channel change by the viewer
- Adaptive post-processing of reconstructed images to improve picture quality
- Perceptual quantization and dropping of transform coefficients
- Vector quantization of quantizer selection patterns
- Smooth control of quantization based on buffer fullness history
- Forward analyzer to determine coding parameters of each new frame
- Randomizing and limiting effects of transmission errors
- Simple receiver

Transparent image quality is achieved using motion compensated transform coding coupled with a perceptual criterion to determine the quantization accuracy required for each transform coefficient. The combination of a sophisticated encoded video format and advanced bit error protection techniques results in a highly robust reception and decoding of the compressed video signal.

The DSC-HDTV Video Coder algorithm is optimized to simplify the decoder. The motion estimator, forward analyzer, vector quantizer, and buffer state control are all functions which exist only in the encoder. An additional attribute of the DSC-HDTV Video Coder algorithm is the ability to improve the encoder in the future without modifying the Decoder. The Decoder is realizable in a small number of VLSI chips. Frame memory and processing in the receiver are minimized while maintaining transparent image quality.

Exploiting Scene Structure in Video Coding *

Michael T. Orchard
Coordinated Science Laboratory and
Department of Electrical and Computer Engineering
University of Illinois at Urbana-Champaign
Urbana, IL 61801

Abstract

Second-generation coding techniques [1, 2] increase coding efficiency by operating on groups of pixels which reflect the structure of the scene (contours, objects, etc.), in contrast to the square blocks used by standard methods. For image coding, this requires transmitting information about scene structure to the receiver.

In video coding, information about scene structure can be estimated from previous frames in the video sequence. A technique has been proposed [3, 4] which exploits such information about scene structure to improve motion compensation accuracy, without transmitting structural information to the decoder. The more accurate motion compensation has been shown to increase overall coding efficiency by 20-35%. This paper reviews this approach to motion compensation, and proposes a new recursive (in time) algorithm for computing the required motion field segmentation. The new algorithm reduces the computational complexity of motion field segmentation while increasing its reliability.

1 Introduction

Standard approaches to coding images and video sequences group together blocks of pixels and operate on each block independently. The manner in which pixels are grouped is independent of the scene content, and thus it is often incorrect to assume that the pixels in a given block represent a common object in the scene. Second-generation coding techniques [1, 2] have achieved higher coding efficiencies by segmenting an image into objects, and grouping pixels according to the object they represent. These techniques, however, have high computational complexity and require transmission of segmentation information as well as the encoded groups of pixels.

Standard video coding techniques perform both motion compensation and residual error coding based on blocks of pixels. The block-based motion compensation techniques perform very poorly at boundaries of moving objects, where the assumption that blocks move with uniform translational motion is violated. The problems encountered at these boundaries result in block artifacts which both reduce image quality and increase coding efficiency.

Recently, a technique has been proposed [3, 4] for improving motion compensation accuracy at object boundaries by incorporating information about the structure of the motion field. The required structural information, represented by a segmentation of the motion field, would require a significant bit-rate if it were computed at the encoder and transmitted to the decoder. However, [4] showed that, in most cases, the required segmentation can be reliably predicted from previous frames of the video sequence, thereby eliminating the need for transmitting side information. This paper reviews this new approach to motion compensation, and proposes a new recursive (in time) algorithm for computing the motion field segmentation. The new algorithm improves the reliability of the motion field segmentation, while reducing the computational requirements.

2 Block-based Motion Compensation

We begin by defining some notation. $S = \{s_i\}$ denotes a lattice of N pixels and $I^n(s)$ denotes the intensity at pixel $s = (x, y)^t$ of frame n. Let $\hat{I}^n(s)$ denote the pixel intensity of the decoded frame corresponding to $I^n(s)$. Motion vectors are two dimensional vectors relating two pixels on the lattice; i.e. $s_1 = s_0 + v$. Let v_a^n represent a motion vector representing object motion from frame $n-1$ to n. The subscript a can represent either a pixel or group of pixels to which the motion vector is assigned. We call a set of motion vectors $\{v_s^n\}_{s \in S}$ for all pixels in the lattice a motion field at frame n. Throughout, we omit the pixel argument when referring to an entire frame, e.g., I^n, v^n, etc., and, when the frame number is clear from context, we omit the frame designator n.

The displaced frame difference $DFD^n(s, v)$ produced by a motion vector v at pixel s of frame n is defined as:

$$DFD^n(s, v) = I^n(s) - \hat{I}^{n-1}(s - v). \tag{1}$$

The $DFD^n(s, v)$ represents the error produced by predicting $I^n(s)$ with a pixel from the previously decoded frame. We use DFD^n to refer to an entire frame of displaced frame differences when the motion vectors used are understood from the context.

Block-based approaches to motion compensation assume that blocks of pixels in a frame move with

*This work has been partially supported the Ricoh Company, Inc.

1058-6393/91 $01.00 © 1991 IEEE

uniform translational motion. The lattice S is partitioned into a rectangular grid of blocks $\{b_{i,j}\}$, each of size $B \times B$. Let R denote the lattice of blocks formed by this partitioning. Each block, $b_{i,j}$, is assigned a single motion vector, $v^n_{b_{i,j}}$, and the following constraint is imposed:

$$\text{if } s \in b_{i,j}, \quad v^n_s = v^n_{b_{i,j}}. \tag{2}$$

This constraint allows the motion field to be transmitted with a reduced set of motion vectors $\{v^n_{b_{i,j}}\}_{b_{i,j} \in R}$, but it restricts the motion field constructed at the decoder to be one which satisfies (2). When the block lattice is understood from context, we use v^n_R to refer to the set of block motion vectors $\{v^n_{b_{i,j}}\}_{b_{i,j} \in R}$. By selecting larger block sizes, the bit rate required for transmitting motion vectors can be reduced to arbitrarily small levels but at the expense of more restricted motion fields.

A common problem with block-based motion compensation methods is that blocks located on boundaries of moving objects are not compensated accurately. Blocks on these boundaries contain objects moving in at least two different directions, and a single motion vector applied to the entire block compensates at least one of the objects incorrectly. In frames with multiple moving objects, artifacts produced by applying the wrong motion vector to portions of boundary blocks account for a significant percentage of the bit rate used to transmit the *DFD*, and overall picture quality is degraded.

3 Predictive Motion Field Segmentation

This section reviews a recently proposed algorithm [3, 4] for improving the accuracy of standard block-based motion compensation along boundaries of moving objects by predicting a segmentation of the motion field of each frame. The new method, which we refer to as the predictive motion field segmentation (PMFS) method, operates with the identical set of block motion vectors, v^n_R, as the standard method, but it relaxes the constraint imposed by 2, allowing more than one motion vector to be assigned to pixels within each block. This less constrained approach to assigning motion vectors raises two questions: where will the extra motion vectors needed for each block come from, and how does the decoder know which motion vector to apply at each pixel. The PMFS method answers these questions in a way which introduces no significant increase in the information which needs to be transmitted to the decoder.

The PMFS method generalizes the constraint of the standard method by allowing pixels in block $b_{i,j}$ to be assigned a motion vector from a neighborhood of $b_{i,j}$. This approach is motivated by the assumption that any portion of $b_{i,j}$ not moving with velocity $v^n_{b_{i,j}}$ is connected to an object which appears in another block. This reflects an underlying assumption that object sizes are large compared with the size of blocks. Defining the neighbors of $b_{i,j}$ as

$$\mathcal{N}_{b_{i,j}} = \{b_{i-1,j}, b_{i,j-1}, b_{i+1,j}, b_{i,j+1}\},$$

the pixels in each block can be assigned one of two motion vectors: $v_{b_{i,j}}$ or an alternate vector $v_{b_{i,j},alt}$ selected from $\{v_b : b \in \mathcal{N}_{b_{i,j}}\}$. We will omit the details of how this alternate vector is chosen from the neighboring vectors, but we note that the choice is based on the block structure predicted from previous frames. This method of sharing neighboring block motion vectors makes it possible to assign multiple motion vectors to each block, while not increasing the total number of motion vectors transmitted to the decoder.

Given two motion vectors for each block, the complete specification of a motion compensation scheme requires information assigning one of the two vectors to each pixel in the block. If no constraints are placed on the assignment of vectors, one bit per pixel would be required to complete the specification. However, the PMFS method assumes that this assignment should reflect the local (since blocks are small) structure of objects in the scene, and that the local object structure can be predicted from previous frames of the sequence. More specifically, the PMFS method constrains the assignment to be one which can be specified by a segmentation of the block into two connected regions of pixels, one region assigned to each motion vector. Instead of sending this segmentation to the decoder, it is predicted from a segmentation of the motion field of the previous frame of the sequence. The remainder of this section outlines the general framework of the predictive segmentation algorithm, omitting many of the details provided in [4].

Let X denote the random block segmentation assigning a binary random variable to each pixel of b. (Since the segmentation is performed identically for every block, we omit the subscript on $b_{i,j}$.) Let X_s denote the binary random variable at pixel s and let $v(X_s)$ denote the motion vector assigned at that pixel. We adopt the convention that $v(0) = v_b$ and $v(1) = v_{b,alt}$

Given frames I^n and $\hat{I}^{n-1}$ and the motion vectors v_b and $v_{b,alt}$, an optimal segmentation of block b would be the X which assigned motion vectors to pixels in b so as to minimize some error norm involving $DFD^n(s, v(x_s))$ for $s \in b$. However, since I^n is not available to the decoder, $DFD^n(s, v(x_s))$ cannot be evaluated at the decoder, and the above optimization procedure can only be computed at the encoder and transmitted over the channel. To avoid this overhead, the PMFS method assumes that the segmentation itself, representing the shape of objects in a scene, does not change significantly from frame to frame. Therefore, the PMFS method computes a motion field segmentation of a block $b^\star$ in frame $\hat{I}^{n-1}$ defined by

$$b^\star = \{s^\star; s^\star = s - v^n_b, s \in b\}, \tag{3}$$

and uses this segmentation for block b. Note that $b^\star$ is simply the block in $\hat{I}^{n-1}$ used by standard motion compensation methods to compensate block b. The segmentation of $b^{\star}$ is based on frames $\hat{I}^{n-1}$ and $\hat{I}^{n-2}$ and the block motion vectors v_R^{n-1} of the previous frame. Two vectors $v_{b^\star}, v_{b^\star,alt}$ are selected from v_R^{n-1} for performing the segmentation. (The details of this selection are omitted.) By computing the segmentation based on motion vectors v_R^{n-1} from the previous frame, the method does not assume constant (or even continuous) motion over time. It only assumes that the shape of objects do not change significantly when moving from frame to frame.

The PMFS method models each pixel as the sum of a correctly compensated (by either $v_{b^\star}$ or $v_{b^\star,alt}$) pixel intensity and an independent zero-mean Gaussian random variable with variance σ^2. That is, given the correct segmentation x,

$$\hat{I}^{n-1}(s) = \hat{I}^{n-2}(s - v(x_s)) + e(s) \quad (4)$$

for each $s \in b^\star$, with $e(s) \sim N(0, \sigma^2)$. Using this model, the segmentation problem is formulated as a maximum *a posteriori* (MAP) estimation of X given $\hat{I}^{n-1}$, $\hat{I}^{n-2}$, $v_{b^\star}$, and $v_{b^\star,alt}$. Namely,

$$\hat{x}_{b^\star} = \underset{x}{\operatorname{argmax}} \{P(X = x | \hat{I}^{n-1}, \hat{I}^{n-2}, v_{b^\star}, v_{b^\star,alt})\}. \quad (5)$$

Applying the model of (4), (5) can be written

$$\hat{x}_{b^\star} = \arg\min_x \{\sum_{s\in b^\star} (DFD^{n-1}(s, v(x_s))^2 - \ln P(X = x)\}. \quad (6)$$

The PMFS method models the *a priori* distribution of the segmentation field, $P(X = x)$, with a Markov Random Field (MRF) model [5]. The MRF model imposes a smoothness constraint on the segmentation boundary, motivated by both both physical and practical considerations. Given a assumed small block size, some degree of contour smoothness can be expected at sub-block resolutions. In addition, coding errors contribute to the roughness of contours, and our studies have shown that object boundaries can be predicted more accurately from frame to frame after filtering out high spatial frequencies corresponding to these errors. Skipping over the details of the MRF model [4], the MAP segmentation becomes

$$\hat{x}_{b^\star} = \arg\min_x \{\sum_{s\in b^\star} (DFD^{n-1}(s, v(x_s))^2 + C(t_1(x) + t_2(x)\sqrt{2})\}, \quad (7)$$

where $t_1(x)$ counts the number of horizontal and vertical neighboring pixel pairs in X with different values, $t_1(x)$ counts the number of diagonal neighboring pixel pairs in X with different values, and C is a parameter adjusting the degree of smoothness imposed by the model.

The cost function defined by (7) is not convex and, in general, has many local minima. Typical approaches to computing $\hat{x}_{b^\star}$ are based on iterative search procedures which have high computational complexity and can get trapped in local minima of (7) far removed from the true global minimum. The PMFS method of [4] applies a multiresolution algorithm to minimizing (7). The multiresolution algorithm performs the minimization in stages, making decisions on the gross appearance of the segmentation in early stages, and further refining the segmentation in each subsequent stage. This algorithm reduces the computational complexity of straightforward search procedures, and helps avoid getting trapped in local minima far removed from the global minimum. However, standard multiresolution approaches still require an iterative search procedure at each stage of the algorithm. Instead, the PMFS method makes some simplifying approximations on the dependencies of segmentation estimates at adjacent resolution stages, leading to a completely noniterative multiresolution algorithm. The resulting algorithm has a very regular structure, suitable for parallel implementation, but, as a consequence of the simplifying assumptions, provides only an approximation of the minimum to (7). However, simulations show that the segmentations produced by the modified multiresolution algorithm have costs nearly as low as those produced by the original algorithm.

4 A Recursive Approach to Predictive Motion Field Segmentation

The PMFS method described in the last section demonstrates the feasibility of using structural information from previous frames in a sequence to improve coding efficiency. However, although its multiresolution minimization algorithm is designed to reduce computational complexity, its decoder complexity remains high enough to call into question its practical feasibility. The complexity of PMFS method is similar to typical motion estimation algorithms, but the PMFS differs from such algorithms by placing a large portion of the computational requirements at the decoder, which is often constrained to be much simpler than the encoder.

This section proposes a new recursive approach to computing the predicted motion field segmentation which reduces the computational complexity of the previous method, while improving the reliability of the resulting segmentations. The PMFS method computes each predicted segmentation based on the two previously decoded frames, ignoring the segmentations which had been computed during the decoding of those frames, and were used for motion compensating those frames. Consequently, the search procedure for minimizing (7) does not benefit from an initial guess of $\hat{x}_{b^\star}$. The new recursive approach reduces computational complexity by using the previously computed segmentation to provide an initial guess to the search procedure. In addition, the *a priori* model for $P(X = x)$ used in the PMFS method is modified to incorporate a dependency between the new segmentation and the previously computed segmentation. This dependency reflects our assumption that object shapes do not change significantly from frame to frame.

The PMFS method predicts the segmentation of block b in frame I^n by computing the segmentation of block $b^\star$ (defined by (3)) in frame $\hat{I}^{n-1}$. During the decoding of $\hat{I}^{n-1}$, the pixels of $b^\star$ were assigned motion vectors according to some segmentation predicted from earlier frames. In selecting he vectors $v_{b^\star}, v_{b^\star,alt}$, this previous segmentation is used to find two vectors of v_B^{n-1} which best fit the pixels of $b^\star$. (Skipping details, this process involves merging a small number of regions from the previous segmentation.) Consequently, no additional computations are required to generate a segmentation, which we will denote $x^\star$, representing the previous segmentation of the block $b^\star$. Note: we do not use $\hat{x}_{b^\star} = x^\star$ because $x^\star$ is our best estimate of $x_{b^\star}$ before frame $\hat{I}^{n-1}$ is available. Thus, $x^\star$ would be a poor estimate of $\hat{x}_{b^\star}$ given $\hat{I}^{n-1}$. However, assuming only small changes in local object structure from frame to frame, $x^\star$ provides a good initial guess for any algorithm searching for $\hat{x}_{b^\star}$.

Incorporating $x^\star$ into our MAP formulation, (5) becomes

$$\hat{x}_{b^\star} = \underset{x}{\operatorname{argmax}}\{P(X = x|\hat{I}^{n-1}, \hat{I}^{n-2}, v_{b^\star}, v_{b^\star,alt}, x^\star)\},$$

$$\hat{x}_{b^\star} = \arg\min_x \{\sum_{s\in b^\star}(DFD^{n-1}(s, v(x_s))^2 - \ln P(x^\star|X = x) - \ln P(X = x)\}.$$

A precise definition of the conditional probability $P(x^\star|X = x)$ requires a model for local object deformation over time in a video sequence. To minimize complexity, we use a very simple MRF for this model, defined by:

$$-\ln P(x^\star|X = x) = C_{b^\star} t_3(x, x^\star), \quad (8)$$

where $t_3(x, x^\star)$ is a function counting the number of pixels at which x and $x^\star$ differ, and $C_{b^\star}$ is a parameter reflecting our confidence in the previous segmentation $x^\star$. As an additional term in the cost expression of (7), (10) encourages $\hat{x}_{b^\star}$ to be close to $x^\star$ when we have high confidence in $x^\star$, and will have no effect when our confidence in $x^\star$ is low.

We define the confidence at each pixel C_s by a first-order Markov (in time) process evolving along the motion trajectories. The process is driven by the difference between the *DFD* produced by the segmented motion compensation and that produced by standard block-based motion compensation (i.e. applying a single vector to the entire block). More specifically, we define the confidence at each pixel as

$$C_s^n = \alpha C_{s-v(x_s)}^{n-1} + (1-\alpha)(|DFD^{n-1}(s, v(x_s))| - |DFD^{n-1}(s, v_b)|), \quad (9)$$

where α is a constant less than 1 (typically: $\alpha = .8$). The block confidence parameter $C_{b^\star}$ is then set proportional to the sum of C_s for all $s \in b^\star$. Loosely speaking, the confidence parameter measures the degree to which the segmentation $x^\star$ and all its ancestors (defined along motion trajectories) have reduced the *DFD* compared with standard motion compensation. The Markov process used to define C_s serves to filter out high-frequency noise due to coding errors and to propagate high confidence segmentations for some time before they are forgotten. This improves the performance of the PMFS method during periods when objects come to a stop before reversing their direction. In this situation, the PMFS method predicts a motion-field segmentation between two moving frames (the frame at which the object begins moving in the new direction) using two frames which reflect no motion, producing very poor segmentations, often unrelated to any structure in the scene. In the new recursive approach, the segmentations computed before the objects stopped are 'remembered', thus producing a more reliable predicted segmentation.

Using the new approach, the MAP segmentation is given by

$$\hat{x}_{b^\star} = \arg\min_x \{\sum_{s\in b^\star}(DFD^{n-1}(s, v(x_s))^2 + C_{b^\star} t_3(x, x^\star) + C(t_1(x) + t_2(x)\sqrt{2})\}. \quad (10)$$

By using $x^\star$ as an initial guess in our search, the multiresolution framework becomes unnecessary. The primary advantage of the multiresolution approach is its ability to identify the course features of the segmentation in early stages of processing. Since we begin our search with a very good guess of the segmentation, these early processing stages are unnecessary. With the new approach, we perform an iterative, pixel-by-pixel, greedy search for $\hat{x}_{b^\star}$. The computational savings are due to the fact that only a few pixels along object boundaries are involved in the search, and a local minimum to the cost function (12) is typically found in very few iterations. Although the greedy search does increase the likelihood of getting trapped in a local minimum of (12), the segmentation associated with this minimum is never too far from the optimum segmentation.

5 Practical Considerations

Typical video codecs allow a variety of operational modes, of which one is selected for decoding each macroblock (a large block) of pixels. The operational mode dictates the set of techniques to be used in encoding and decoding the macroblock, and the mode is selected to optimize coding efficiency. Both the PMFS method and modified PMFS method described in the previous section are intended to be used as one mode of operation in such a codec. The new modes would be used at blocks containing moving object boundaries, and, therefore, the increased computational complexity would be limited to those blocks.

The modified PMFS described in the previous section cannot completely replace the PMFS method as an operational mode in a video codec. A number of situations arise in which no previous segmentation is available to the decoder. As examples, no previous segmentations are available at scene changes, or when objects which had been at rest begin to move for the

Figure 1: Typical test sequence frame from Tennis

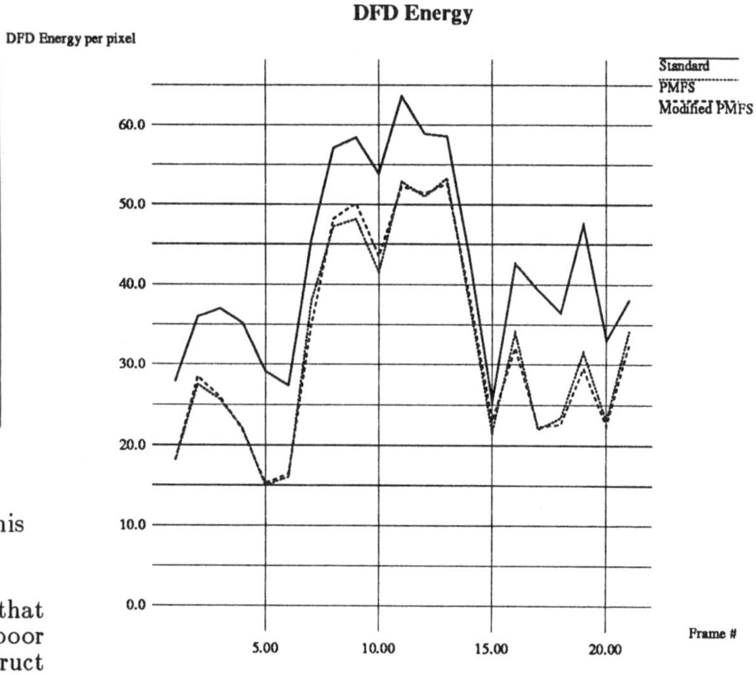

Figure 2: Average Pixel *DFD* Energy Produced by Motion Compensation Methods

first time. In addition, if the encoder anticipates that the modified PMFS method will get trapped in a poor local minimum for a particular block, it can instruct the decoder to use the original PMFS method for that block. Therefore, a practical codec would implement the new approach by adding two new modes to its operational modes. While the modified PMFS method would usually be used, occasional application of the PMFS method would be required.

6 Results

The PMFS method and the modified PMFS method were applied to motion compensate the table tennis sequence (sample frame shown in figure 1). Figure 2 shows the average *DFD* energy of the motion-compensated frame produced by standard block-based motion compensation, the original PMFS method, and the modified PMFS method. The results show the two PMFS methods to give virtually identical reductions in *DFD* energy. These tests were performed assuming infinite bandwidth is available for correcting the motion compensated frame. Therefore, the effects of coding errors are removed, and the increased reliability of the modified PMFS method is not evident.

Next, we compare the computational complexity of the original PMFS method with the modified PMFS method proposed here. Table 1 shows the maximum, minimum, and average number of pixel iterations needed by each of the two approaches. A pixel iteration represents the evaluation of either cost function (7) or (12) in deciding whether to change the assignment of one pixel. For the original PMFS method, iterations at all levels of the multiresolution structure are treated equally. Since the original PMFS method is non-iterative, its maximum, minimum, and average values are identical. It should be noted that, in addition to the larger number of pixel iterations, the original PMFS method requires computations for setting up the multiresolution heirarchy which are not required by the modified PMFS.

Algorithm	Pixel Iterations per Block		
	Maximum	Minimum	Average
Original PMFS	511	511	511
Modified PMFS	217	39	57

Table 1: Computational Complexity of PMFS Methods

References

[1] M. Kunt, A. Ikonomopoulos, M. Kocher, "Second-generation image-coding techniques," *Proceedings of the IEEE,* vol. 73, no. 4, pp. 549-574, April, 1985.

[2] Gilge, M., "Region-Oriented Transform Coding (ROTC) of Images," *Proc. ICASSP, 1990,* vol. 4, pp. 2245-2248 Albuquerque, NM, April, 1990

[3] Orchard, M. T., "Predictive Motion Field Segmentation for Image Sequence Coding," *Proc. ICASSP, 1990,* vol. 4, pp. 1977-1980 Albuquerque, NM, April, 1990

[4] Orchard, M. T., "Predictive Motion Field Segmentation for Image Sequence Coding," (in review) *IEEE Trans. Circuits Syst. for Video Technology,*

[5] Bouman, C. and Liu, B., "Multiple Resolution Segmentation of Textured Images," *IEEE Trans. PAMI,* vol. 13, no. 2, pp. 99-113, Feb. 1991

Multiresolution Transform Matrices

T. G. Marshall, Jr.

Center for Digital Signal Processing Research
Bureau of Engineering Research
Rutgers University
Piscataway, New Jersey 08854
(908)932-3116 Office

Abstract

Matrix descriptions of subband coders are introduced and used to obtain transform matrices of pyramidal and related algorithms. These matrices are generalizations of the Haar matrix, and they are all shown to have a stratified structure related to circulant matrix structure. An implementation of a class of biorthogonal digital filter banks (DFB's) with an efficient predictive structure is introduced. DFB's and pyramids optimal for quantization are also introduced.

Introduction

Multiresolution methods for image processing have been under development since the 1970's, but these methods date back at least to early work of Haar in the 1920's. The prototype multiresolution transform matrix is the Haar matrix [1] shown in (1) with spacing to emphasize its stratification. Its corresponding fast algorithm is now referred to as a pyramidal algorithm (or simply a pyramid) [2]. Recent research in the area of wavelets [3-8] has identified new orthogonal and biorthogonal multiresolution transforms, generalizing the Haar transform.

$$\mathbf{A} = \begin{bmatrix} 1 & 1 & 1 & 1 & 1 & 1 & 1 & 1 \\ \\ 1 & 1 & 1 & 1 & -1 & -1 & -1 & -1 \\ \\ 1 & 1 & -1 & -1 & 0 & 0 & 0 & 0 \\ 0 & 0 & 0 & 0 & 1 & 1 & -1 & -1 \\ \\ 1 & -1 & 0 & 0 & 0 & 0 & 0 & 0 \\ 0 & 0 & 1 & -1 & 0 & 0 & 0 & 0 \\ 0 & 0 & 0 & 0 & 1 & -1 & 0 & 0 \\ 0 & 0 & 0 & 0 & 0 & 0 & 1 & -1 \end{bmatrix} \quad (1)$$

In this paper transform matrices of multiresolution transforms are introduced and illustrated with a new pyramidal algorithm with reduced computational load. This algorithm is based upon optimal linear estimation methods and coset predictors, recently introduced [9].

Matrix Descriptions

Coders Based On Biorthogonal Pairs of Matrices

In this and succeeding sections matrices will be introduced which describe multiresolution transforms of signal vectors. This is illustrated in Figure 1a, in which **x**, **u**, **v**, and **y**, are all assumed to be row vectors of length N of signal components, e.g. $\mathbf{x} = [x_0, x_1, \ldots, x_{N-1}]$. If $\mathbf{v} = \mathbf{u}$, it will be assumed that the output, **y**, equals **x**, and, since they are row vectors, $\mathbf{y} = \mathbf{x}\mathbf{A}^*\mathbf{B}$, and therefore $\mathbf{A}^*\mathbf{B} = \mathbf{I}_N$, the unit matrix of order N. This describes a perfect reconstruction (PR) coder. An adjoint coder, shown in Figure 1b, is obtained by taking the Hermitian transpose giving

---

This work was supported by a grant from the David Sarnoff Research Center, a Division of SRI.

1058-6393/91 $01.00 © 1991 IEEE

$B^*A = I_N$. A pair of matrices, **A** and **B**, satisfying $A^* = B^{-1}$ are said to be bi-orthogonal. In the unitary case, **A** = **B**, and the two coders coincide.

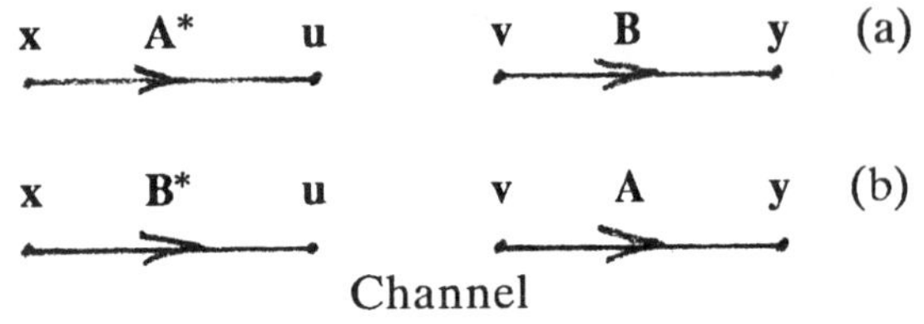

Figure 1. A Coder (a) and its Adjoint (b) with a Pair of Biorthogonal Matrices

Digital Filter Banks

A matrix description of a two-channel analysis/synthesis digital filter bank (DFB) is shown in Figure 2(a). The vectors u_i and v_i are of length M = N/2, and therefore, the H_i and G_i are M × N rate increasing matrices and their Hermitian conjugates, H_i^* and G_i^*, rate decreasing. For this DFB to be perfect reconstruction,

$$[H_0^*, H_1^*] \begin{bmatrix} G_0 \\ G_1 \end{bmatrix} = H^*G = I_N , \qquad (2)$$

which defines **H** and **G**. Equivalently,

$$H_0^*G_0 + H_1^*G_1 = P_{1,0} + P_{0,1} = I_N , \qquad (3)$$

which defines the two matrices, $P_{i,j}$. These necessarily satisfy $P_{i,j}P_{i,j} = P_{i,j}$, i.e., they are idempotent matrices or projectors [10]. Since a matrix necessarily commutes with its inverse

$$H^*G = GH^* = I_N , \qquad (4)$$

giving four conditions equivalent to (3),

$$G_iH_j^* = \delta_{i,j}I_M , \qquad (5)$$

for i,j = 0,1 and $\delta_{i,j}$ the Kronecker delta. From j = i, it is seen that G_i is a left inverse of H_i^*, for i = 1,2. Equation (5) states *necessary and sufficient conditions for* the structure in Figure 2a to provide *perfect reconstruction.*

From the preceding discussion, it is evident that the associated synthesis/analysis DFB, shown in Figure 2b, is also PR. In addition to these two DFB's are their adjoints obtained by interchanging the H_i and G_i. It is evident that all these fit the description of Figure 1 if the definitions $u = [u_0, u_1]$ and $v = [v_0, v_1]$ are adopted in Figure 2a with similar definitions for **x** and **y** in Figure 2b.

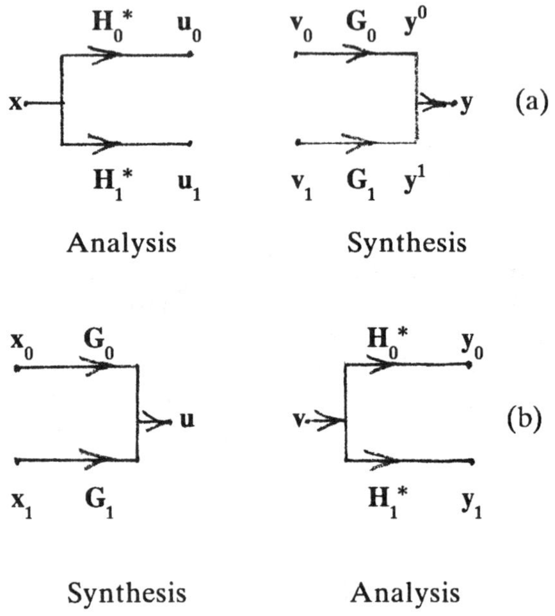

Figure 2. Analysis/Synthesis and Synthesis/Analysis DFB's Resulting from Equation (4).

A general expression for a left inverse will prove useful in identifying the G_i, which are left inverses of H_i^*,

$$G_i = K_i^{-1}H_iJ_i , \qquad (6)$$

in which

$$K_i = H_iJ_iH_i^* , \qquad (7)$$

for any J_i such that K_i is invertible. It is evident that $G_iH_i^* = I_M$, however, the converse is easily shown by assuming $J_i = G_i^*G_i$ if G_i is known, indicating that any left inverse can be expressed as in (6) [11].

It can be seen that given $\mathbf{H}_0$ of full rank that any complementary matrix, $\mathbf{H}_1$, may be chosen such that $\mathbf{H}$ is invertible. Equation 2 then uniquely defines $\mathbf{G}$ and hence $\mathbf{G}_0$ and $\mathbf{G}_1$. There are therefore many matrices satisfying (2) - (5). Further properties of these matrices, appropriate to their roles in DFB's and wavelet transforms, are to be discussed in the next section, and these give guidance in their determination.

Circulant and Rate-Changing Matrices

Permutation Matrices

The matrices and projectors just described have been under investigation in a related problem, that of designing error-cor-recting codes that work over the real number field rather than over finite fields [12-14]. In this application as well as the application at hand, two types of matrices appear, incidence matrices, denoted $\mathbf{M}_i$, and Toeplitz matrices associated with a z-transform, H(z), denoted H(S). In the finite block length case for one-dimensional systems, $\mathbf{S}$ is the circular shift matrix

$$\mathbf{S} = \begin{bmatrix} 0 & 0 & 0 & 0 & 0 & 0 & 0 & 1 \\ 1 & 0 & 0 & 0 & 0 & 0 & 0 & 0 \\ 0 & 1 & 0 & 0 & 0 & 0 & 0 & 0 \\ 0 & 0 & 1 & 0 & 0 & 0 & 0 & 0 \\ & \cdot & & \cdot & & & \cdot & \\ & \cdot & & & \cdot & & \cdot & \\ & \cdot & & & & \cdot & \cdot & \\ 0 & 0 & 0 & 0 & 0 & 0 & 1 & 0 \end{bmatrix} \quad (8)$$

H(S) is then a polynomial in powers of $\mathbf{S}$, a circulant matrix [15]. $\mathbf{S}$ is a permutation matrix, each of the N rows being a distinct length N unit vector, denoted $\mathbf{e}_i$. A second permutation matrix for the one-dimensional case, $\mathbf{M}$, consists of two $M \times N$ incidence matrices, $\mathbf{M}_i$,

$$\mathbf{M} = \begin{bmatrix} \mathbf{M}_0 \\ \mathbf{M}_1 \end{bmatrix}, \quad (9)$$

in which

$$\mathbf{M}_0^T = [\mathbf{e}_0^T, \mathbf{e}_2^T, \ldots, \mathbf{e}_{N-2}^T] \quad (10)$$

and $\mathbf{M}_1 = \mathbf{M}_0\mathbf{S}^{-1}$ consists of the remaining unit vectors. The multidimensional case was discussed in Reference [9], and examples of the $\mathbf{M}_i$ were given for quincunx sampling lattices.

z-Transform Specification of Matrices

The DFB matrices will be determined and shown to yield the z-transforms in the diagram of Figure 3, in which q = 1 - p, p = 0 or 1. A delay and advance in one of the paths is explicitly shown, as in [16]. The

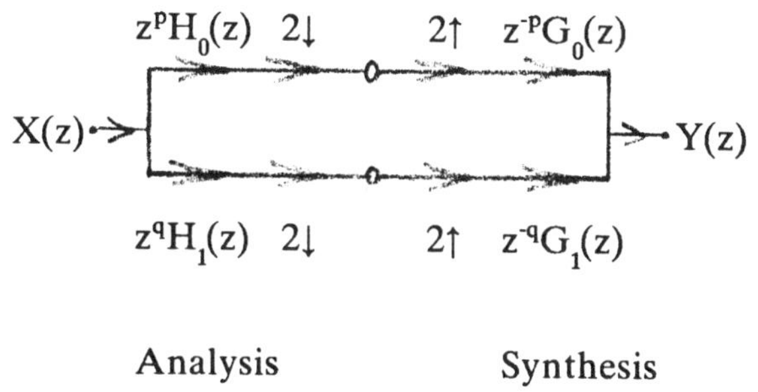

Figure 3. z-transform description of a DFB

arrows indicate down and up sampling by a factor of 2, in the even locations. The integers p and q can actually be chosen to be any pair of positive or negative integers of opposite parity, without affecting the perfect reconstruction property, a flexibility that can be used to obtain causal solutions (if, a constant delay is also accepted instead of the perfect reconstruction property, assumed here).

To begin, assume

$$\mathbf{H}_0^* = H_0(\mathbf{S})\mathbf{M}_p^*, \quad (11)$$

in which the circulant matrix, $H_0(\mathbf{S})$, describes the filtering and the $\mathbf{M}_p^*$ the downsampling with samples in even or odd locations preserved as p is even or odd. The left inverse of $\mathbf{H}_0^*$, $\mathbf{G}_0$, will be determined from (6). In order to avoid recursive filtering, it will be required that

$$\mathbf{K}_0 = \mathbf{M}_p H_0(\mathbf{S}^*)J_0(\mathbf{S})H_0(\mathbf{S})\mathbf{M}_p^* = \mathbf{I}_m \,. \quad (12)$$

Define

$$N_0(z) = H_0(z^{-1})J_0(z)H_0(z) \,, \quad (13)$$

assume $J_0(z) = J_0(z^{-1})$, and observe that (12) implies the half-band condition, well-known in perfect reconstruction DFB design [16],

$$N_0(z) + N_0(-z) = 1 \,. \quad (14)$$

The presence of the factor $J_0(z)$ ensures that the following development applies to general biorthogonal DFB's [6-7]. To summarize, *the condition* $\mathbf{K}_0 = \mathbf{M}_p\mathbf{N}_0\mathbf{M}_p^* = \mathbf{I}_m$ *is equivalent to* $N_0(z)$ is half-band.

In a similar manner, $H_1(z)$ and $G_1(z)$ are related by

$$N_1(z) = H_1(z^{-1})J_1(z)H_1(z) \,, \quad (15)$$

which is also half-band. Comparing (14) with (3) implies that

$$N_1(z) = N_0(-z) \,. \quad (16)$$

Equation (5) with $i \neq j$ indicates that these will be satisfied if

$$G_1(z) = H_0(-z) \,, \quad (17)$$

since $\mathbf{G}_1\mathbf{H}_0^*$ becomes $\mathbf{M}_q G_1(\mathbf{S})H_0(\mathbf{S})\mathbf{M}_p^*$ which is $\mathbf{0}_M$ since $G_1(\mathbf{S})H_0(\mathbf{S})$ consists of only even powers of $\mathbf{S}$, and $\mathbf{M}_q\mathbf{S}^k\mathbf{M}_p^* = \mathbf{0}_M$ for any even k. Lastly,

$$H_1(z) = H_0(-z^{-1})J_0(-z) \quad (18)$$

follows from (15) - (17).

The matrices for a biorthogonal, perfect reconstruction DFB are then of the form:

$$\mathbf{H}_0^* = H_0(\mathbf{S})\mathbf{M}_p^* \quad (19a)$$

$$\mathbf{H}_1^* = H_1(\mathbf{S})\mathbf{M}_q^* = H_0(-\mathbf{S}^{-1})J_0(-\mathbf{S})\mathbf{M}_q^* \quad (19b)$$

$$\mathbf{G}_0 = \mathbf{M}_p G_0(\mathbf{S}) = \mathbf{M}_p H_0(\mathbf{S}^{-1})J_0(\mathbf{S}) \quad (19c)$$

$$\mathbf{G}_1 = \mathbf{M}_q G_1(\mathbf{S}) = \mathbf{M}_q H_0(-\mathbf{S}) \,. \quad (19d)$$

If $J_0(\mathbf{S}) = \mathbf{I}_N$, the DFB is orthogonal, as in [16]. One can specify $\mathbf{H}_1^*$ first, instead of (11), if it is desired that $H_1(z)$ be of lower order than $H_0(z)$, effectively swapping the $J_0(z)$ factor between $\mathbf{H}_0^*$ and $\mathbf{G}_0$ and the $J_0(-z)$ factor between $\mathbf{H}_1^*$ and $\mathbf{G}_1$.

The definitions of the z-transforms in Figure 3 are apparent from (19), and they follow because (19) holds for $\mathbf{S}$ of any size which then specifies the z-transforms arbitrarily densely on the unit circle, and hence over the whole z-plane except points of singularity, the only finite possibility being z = 0. The equivalence of (12) and (14) also depends upon this argument.

Multiresolution Structures and Matrices

Pyramid Structures

Multiresolution DFB's can be formed by repeatedly introducing A/S DFB's that transform both $\mathbf{u}_0$ and $\mathbf{u}_1$ in Figure 2a, giving tree structures, or that transform only $\mathbf{u}_0$ giving octave band structures [17]. The wavelet transform has been shown to be equivalent to the latter case [4], and, for demonstrating convergence to signals with arbitrarily high bandwidth, is considered in the limit for an infinite number of resolution levels. In the applications, only a finite number of levels need be employed, which permits a matrix description to be introduced that generalizes the Haar matrix.

Following the notation of the wavelet literature, the resolution is identified by an integer, m, and 2^{-m} indicates the resolution. In our notation, with the original N, the length of $\mathbf{x}$ and $\mathbf{y}$ in Figure 1, denoted $N^{(0)}$,

$$N^{(m)} = 2^{-m}N^{(0)}. \quad (20)$$

Figure 4 illustrates a system with L resolution levels, indexed from 0 to L - 1, provided by concatenating L - 1 DFB's in a pyramidal construction.

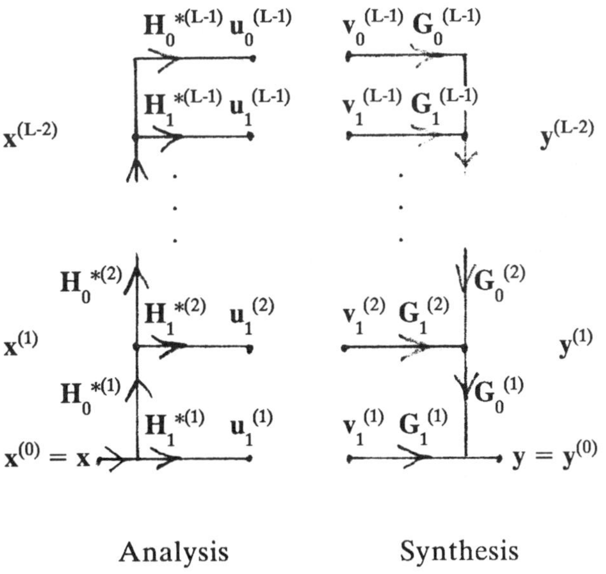

Figure 4. Pyramidal Multiresolution Structure Realizing Figure 1a.

The DFB at level L - 2 is conventional, but at higher resolution levels, only the lower, high-frequency path is present.

Stratified g-circulant Matrices

The analysis output and synthesis input is seen to be of the form

$$\mathbf{v} = [\mathbf{v}_0^{(L-1)}, \mathbf{v}_1^{(L-1)}, \mathbf{v}_1^{(L-2)}, \ldots, \mathbf{v}_1^{(2)}, \mathbf{v}_1^{(1)}] \,. \quad (21)$$

Each input $\mathbf{v}_1^{(m)}$ can be seen to contribute to the output $\mathbf{y}$ a vector $\mathbf{q}^{(m)}$, of length $N^{(0)}$, i.e., $\mathbf{y} = \Sigma_m \mathbf{q}^{(m)}$. Tracing the typical path gives

$$\mathbf{q}^{(m)} = \mathbf{v}_1^{(m)} \mathbf{B}_1^{(m)}$$

$$= \mathbf{v}_1^{(m)} [\mathbf{G}_1^{(m)} \mathbf{G}_0^{(m-1)} \ldots \mathbf{G}_0^{(2)} \mathbf{G}_0^{(1)}] \,, \quad (22)$$

which defines the $N^{(m)} \times N^{(0)}$ matrix $\mathbf{B}_1^{(m)}$. The matrix, $\mathbf{B}$, which maps $\mathbf{u}$ to $\mathbf{y}$ therefore has the following stratified structure (shown transposed), e.g., the Haar matrix in Eq. 1:

$$\mathbf{B}^T = [\mathbf{B}_0^{(L-1)T}, \mathbf{B}_1^{(L-1)T}, \mathbf{B}_1^{(L-2)T}, \ldots \mathbf{B}_1^{(2)T}, \mathbf{B}_1^{(1)T}] \,. \quad (23)$$

$\mathbf{B}$ bears an interesting relationship to a g-circulant matrix which is a matrix in which each row is a cyclic right shift of the row above it by g, $g = 2^m$ in this case [14]. Assuming g divides N, an $N \times N$ g-circulant has N/g identical strata of this form, and $\mathbf{B}$ has only one stratum from each of the L g-circulants. The Haar matrix in Figure 1 also illustrates this. Some properties will be stated.

Properties:

1. The synthesis matrix $\mathbf{B}$ for the pyramid in Figure 4 having L resolution levels is a statified g-circulant matrix with L strata, and $g = 2^m$ for levels $0 < m < L$ and $g = 2^{(L-1)}$ for level L.

2. The matrix $\mathbf{B} = \mathbf{B}_{L-2}\mathbf{B}_{L-1} \ldots \mathbf{B}_1\mathbf{B}_0$, in which $\mathbf{B}_n = \text{diag}(\mathbf{G}^{(n)}, \mathbf{I}_{N(0)-N(n)})$, with $\mathbf{G}^{(m)}$ being defined, as in (2), of dimension $N^{(n)}$. This description, like Figure 2, defines the fast pyramidal algorithm for the synthesis branch.

3. The matrix $\mathbf{B}$ will be orthogonal if each of the DFB's composing it are orthogonal.

Optimal Biorthogonal DFB Implementation with Coset Predictors

The problem of making a minimum-variance estimate of $\mathbf{y}$ in Fig. 2a with $\mathbf{y}^0$ was considered in [9] for the special case of $\mathbf{H}_0 = \mathbf{M}_0$, and an efficient coset predictor structure was introduced for implementation. It is significant that the variance of $\mathbf{u}_1^{(m)}$ is also minimized, as this facilitates the quantization of $\mathbf{u}$. The following properties generalize the former results:

Properties:

4. For a given $\mathbf{H}_0$, $\mathbf{y}^0$ will be a minimum-variance estimate of $\mathbf{y}$ if $\mathbf{J}_0 = \mathbf{R}_{xx}$, the correlation matrix of $\mathbf{x}$ (or $J_0(z) = R_{xx}(z)$).

5. If a biorthogonal DFB is defined by (19), if $N_0(z)$ and $H_0(z)H_0(z^{-1})$ are both half-band, and if $\mathbf{C}_{01} = \mathbf{G}_0\mathbf{G}_1^*$, then the DFB is implemented by the structure in Figure 5.

If $\mathcal{J}_0(z) = R_{xx}(z))$ is a good model for the signal, **x**, then the DFB designed with these matrices in (19) should be optimal in the pyramidal structure. $\mathbf{R}_{xx}$ could be chosen differently at each resolution level. An efficient realization of the DFB's in the pyramid using coset predictors has the structure shown in Figure 5.

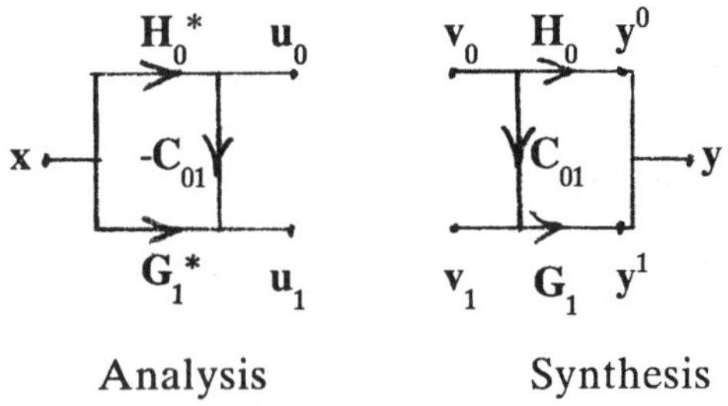

Figure 5. Biorthogonal DFB Implemented with Low-Rate Coset Predictors.

References

[1] A. K. Jain, *Fundamentals of Digital Image Processing.* Englewood Cliffs, NJ: Prentice Hall, 1989.

[2] P. J. Burt and E. H. Adelson, " The Laplacian pryamid as a compact image code," *IEEE Trans. Commun.,* vol. COM-31, April 1983, pp. 532-540.

[3] I. Daubechies, "Orthonormal bases of compactly supported wavelets," *Comm. on Pure and Appl. Math.,* vol. 41, 1988, pp. 909-996.

[4] S. Mallat, "A theory for multiresolution signal decomposition: the wavelet representation," *IEEE Trans. on Pattern Anal. and Mach. Intel.,* vol. 11, Jul. 1989, pp. 674-693.

[5] M. Antonini, M. Barlaud, P. Mathieu, and I. Daubechies, "Image coding using vector quantization in the wavelet transform domain," *Proc. IEEE Int. Conf. Acoust., Speech, Signal Processing,* Albuquerque NM, Apr. 1990, pp. 2297-2300.

[6] M. Vetterli and C. Herley, "Wavelets and filter banks: theory and design," *Center for Telecomm. Res. TR-206,* Columbia University, NYC, 1990.

[7] M. Vetterli and C. Herley, "Wavelets and filter banks: relationships and new results," *Proc. IEEE Int. Conf. Acoust., Speech, Signal Processing,* Toronto Ont., June 1991, pp. 1723-1726.

[8] A. N. Akansu and Y. Liu, "Tree structures for time-frequency signal analysis," *Proc. IEEE Int. Conf. Acoust., Speech, Signal Processing,* Toronto Ont., June 1991, pp. 2037-2040.

[9] T. G. Marshall, Jr., "Rate conversion operators for television signals," *Proc. Asilomar Conf. on Signals, Syst., and Computers,* Pacific Grove, CA, Nov. 1990.

[10] A. Ben-Israel and T. N. E. Greville, *Generalized Inverses: Theory and Applications.* New York: Wiley-Interscience, 1974.

[11] C. R. Rao and S. J. Mitra, *Generalized Inverses of Matrices and Its Applications.* New York, NY: Wiley, 1971.

[12] T. G. Marshall, Jr., and H.-Y. Kim, "Optimum projectors for error-trapping and interpolation in multidimensional signals," *32nd Midwest Symp. on Circuits and Systems,* Urbana, IL, Aug. 1989.

[13] T. G. Marshall, Jr., "Optimal error estimation and interpolation in frequency constrained image sequences," *IEEE Int. Symp. on Circuits and Systems,* New Orleans, May 1990.

[14] T. G. Marshall, Jr., "Image filtering with orthogonal projectors for burst error control," *IEEE Trans. on Circuits and Syst. for Video Systems,* vol. 1, Sep. 1991.

[15] P. J. Davis, *Circulant Matrices.* New York: Wiley-Interscience, 1979.

[16] M. J. T. Smith and T. P. Barnwell III, "A procedure for designing exact reconstruction filter banks for tree structured subband coders," in *Proc. IEEE Int. Conf. Acoust., Speech, Signal Processing,* San Diego, CA, Mar. 1984, pp. 27.1.1-27.1.4.

[17] R. E. Crochier and L. R. Rabiner, *Multirate Digital Signal Processing.* Englewood Cliffs, NJ: Prentice-Hall, 1983.

Fractal Scanning For Image Compression

K.S. Thyagarajan
Dept. of Elec. & Comp. Engg.
San Diego State University
San Diego, Calif. 92182

Shankar Chatterjee
Computer Laboratory
for Analyzing Spatial Signals
Dept. of Elec. & Comp. Engg.
University of California, San Diego
La Jolla, Calif. 92093-0407

Abstract

This paper describes an alternative scanning technique called fractal scanning for images which increases pixel correlation. The increased correlation can be better exploited to achieve higher image compression. The scanning procedure described here uses a basic 4×4 pattern recursively to scan an image. The scanned image is vector quantized to achieve image compression.

1 Introduction

Due to the escalating need for digital image and video data transmission, existing digital channels with bandwidth constraint cannot carry these signals without some form of signal compression. Digital image compression refers to any process by which the original raw data rate is reduced. Current compression or source coding techniques fall into the broad categories of predictive coding, transform coding, vector quantization and a combination of these [1]. These techniques basically try to either remove the correlation among the pixels in an image, as in predictive and transform coding, or exploit the correlation among the pixels, as in a vector quantizer. These are classified as lossy compression schemes for, the original information is not completely recoverable from the compressed image. However, there are schemes, such as Huffman coding and arithmetic coding, which are lossless. But the achievable compression ratio will usually be higher for the lossy schemes than for the lossless schemes.

The achievable compression ratio by the lossy techniques mentioned above depends on the degree of pixel correlation. Therefore, if one can increase the pixel correlation by a known reversible process, then the usual compression techniques mentioned above will perform better. This paper discusses such a technique to increase the existing pixel correlation and the use of a vector quantizer to achieve higher data compression. In section 2, a scanning procedure called fractal scanning, different from the usual raster scan is described which increases the image correlation. Section 3 outlines various possible quantization procedures with an emphasis on the design of vector quantizers for the rescanned images. Experimental results are given in section 4 and conclusions in section 5.

2 Scanning Procedures

Digital video images are obtained and stored using raster scans, whereby images are scanned from left to right and top to bottom according to the natural order of the object images. There is a growing interest in the use of video scrambling techniques that have been shown to decrease transmission bandwidth requirements. For example, Wyner [2], [3] proposed a technique to scramble discrete-time analog sequences using a large family of linear, orthogonal, invertible transformations. The scrambled sequence had negligible bandwidth expansion. However, these transformations are not computationally efficient and do not exploit the two-dimensional nature of the video signals. Matias and Shamir [4] have advanced a video scrambling technique which scans a picture stored in a buffer along a pseudo-random space filling curve or fractal. They have shown that such a scanning procedure actually reduces the bandwidth required for transmission. Walach and Karnin [5] have come up with a scanning procedure based on fractals. The idea they have used is to traverse the entire scan line using a fixed length "yardstick". Compression is then achieved by transmitting only the sign bit and the horizontal distance spanned by the yardstick. This is equivalent to delta modulation but the slope overload problem is somewhat alleviated by the reduction in bandwidth. More recently, Yang, Wu and Mills [6] have devised a method that uses a scanning procedure called Peano scanning [7] along with a fractal-based coding to achieve im-

1058-6393/91 $01.00

age compression. Distortion free compression of two-dimensional data can be obtained by such scanning which is better than any other scanning procedure [8].

This paper describes the scanning procedure used in [6] to increase the signal correlation followed by vector quantization to obtain data compression. In particular, a basic 4 × 4 scan pattern, shown in figure 1, is used. Since the scan line has many bents, such a scanning procedure may be called a fractal scanning. This basic pattern can be used recursively to scan the entire image. Figure 2 shows another possible 4 × 4 basic pattern which can also be used recursively to traverse through all the image points. It should be noted that *all* the pixel points in any quadrant is traversed before scanning an adjacent quadrant in the image. This may be useful in coding if there are not too much variations in pixel intensties in a particular quadrant, in addition to the fact the scan is defined recursively for each quadrant, [8]. Since this procedure requires a full frame buffer, one can use a larger pattern, say a 16 × 16 pattern, generated recursively using the basic 4 × 4 pattern and then use the larger pattern to scan the image. Figure 3, for example, shows a 16 × 16 pattern obtained from 4 basic 4×4 patterns. As can be seen from figure 3, each 4 × 4 pixel block contains the same 4 × 4 basic pattern with possible rotations by 90 and 180 degrees. The advantage is that a buffer of only 16 rows is required as opposed to a full frame buffer [6]. The shortcoming is that true reduction in bandwidth in both dimensions may not be achieved. The complete scanning used in this work based on repeating the 16 × 16 patterns is shown in figure 4 [6]. Figure 5 is the girl's image using normal scanning while figure 6 is the image using fractal scanning as described above. The fractal-scanned image shows horizontal strips of more or less flat areas implying thereby an increase in correlation in the horizontal direction. It is shown [4] that the autocorrelation function for fractal-scanned images is of the form

$$r(n) = A_n \sum_D \left(\frac{D}{n^\nu}\right)^\theta \exp\left\{-\frac{D^\delta}{n^\nu} - \alpha D\right\} \quad (1)$$

where $\nu = 0.75$, $\theta = 1.93 \pm 0.27$, $\delta = 4.6 \pm 0.06$, A_n is a normalizing constant and D is the Euclidean distance between two pixels separated by n. For the usual raster-scan images, the autocorrelation function can be written as, $r(D) = e^{-\alpha D}$. Figure 7 shows the autocorrelation for the normal and fractal-scanned images. From the figure it is clear that the correlation has increased over much longer lags for the scanned image compared to the normal image.

3 Compression of Fractal-scanned Images

In order to compress images obtained via fractal scanning as mentioned in the last section, there are several possible schemes available. One can use simple predictive techniques like the 1-D DPCM or more complex 2-D adaptive DPCM. However, it is not possible to get fractional bit rates with scalar quantizers inherent in the DPCMs. Vector quantization (VQ) is used widely to compress both speech and image signals [9]. Since VQs can achieve very low bit rates especially for correlated signals, it is often an excellent vehicle for image compression. Locally optimal VQs can be designed using the popular LBG iterative algorithm [10]. For images, typical vector dimensions are 16 obtained from 4 × 4 pixel blocks. For details on VQ, the reader is referred to [9]. In the next section, experimental results are described.

4 Experimental results

In order to demonstrate the possible improvement in the visual quality of image compression of fractal-scanned images over the usual raster-scanned images under the same compression scheme, we used VQs on digital images of 512 × 512 pixels at 8 b/pixel gray scale resolution. Two locally optimal tree structured VQs were designed with 256 codevectors for both the normal and the fractal-scanned images, resulting in 0.5 bits/pixel. The training set consisting of four images included human faces and vegetables. The original girl's image of figure 5 was used as the test image. For the VQ of the normal images, 4 × 4 blocks were used, while for the fractal-scanned images, block sizes of 16 × 1 were used since it resulted in the largest peak signal-to-noise ratio (PSNR). PSNR is defined as

$$PSNR = \frac{10 \log_{10} 255^2}{10 \log_{10} MSE}$$

where MSE is the mean squared error between the original image and the reconstructed image. Large PSNR is due to the fact that the fractal-scanned images have increased correlation along the horizontal direction but not along the vertical direction. Figures 8 and 9 show the vector quantized versions of 5 and 6 respectively. The PSNR was about 28 dB for both scan patterns. However for smaller codebook (e.g., 128 codewords), PSNR was about 2 dB more for fractal-scanned image than that for the raster image with a corresponding visual quality improvement.

5 Conclusions

In this paper, the use of a fractal scanning technique for video bandwidth compression has been described. In particular, a basic 4 × 4 scan pattern reported in [6] has been used to recursively scan the whole image. Autocorrelations for the normal and the rescanned images show that the correlation was increased for the rescanned images over a longer lag than for the normal images.

In order to achieve high compression, vector qaunizers were designed for the rescanned images. It was found that the PSNR for the rescanned images were higher than that obtained for the normal images by about 1.5 dB. Moreover, the visual quality was significantly better for the rescanned images as compared to the normal images. It was further found that block sizes of 16×1 were the best for the rescanned images in terms of PSNR and visual quality. Since the scan pattern was restricted to a size of 16 × 16, full bandwidth reduction was not achieved. However,storage requirement was reduced. Further improvements may be obtained by exploring other random scanning procedures that could increase the fractal dimension, thereby increasing the two-dimensional correlation of images. In addition, multiresolution schemes, such as subband coding, can be used to further improve compression and quality of reconstruction.

References

[1] A.K. Jain, *Fundamentals of Digital Image Processing*, Prentice-Hall, 1989.

[2] A.D. Wyner, An analog scrambling scheme which does not expand bandwidth, part 1: Discrete time, *IEEE Trans. Inform. Theory*, vol. IT-25, pp. 261–274, May 1979.

[3] A.D. Wyner, An analog scrambling scheme which does not expand bandwidth, part 2: Continuous time, *IEEE Trans. Inform. Theory*, vol. IT-25, pp. 415–425, July 1979.

[4] Y. Matias and A. Shamir, A video scrambling technique based on space filling curves, Technical report, Weizmann Inst. of Sci., Israel, 1986.

[5] E. Walach and E. Karnin, A fractal based approach to image compression, In *Proc. of ICASSP-86*, pages 529–532, 1986.

[6] K. Yang, L. Wu and M. Mills, Fractal based image coding scheme using Peano scan, In *Proc. of ICASSP-88*, pages 2301–2304, 1988.

[7] M. Gardner, Mathematical games, *Sci. Amer.*, pages 124–133, Dec. 1976.

[8] A. Lempel and J. Ziv, Compression of two-dimensional data, *IEEE Trans. Info. Theory*, vol. IT-32, pp. 2–8, Jan. 1986.

[9] A. Gersho and R.M. Gray, *Vector Quantization and Signal Compression*, Kluwer Academic Publishers, 1990.

[10] Y. Linde, A. Buzo and R.M. Gray, An algorithm for vector quantizer design, *IEEE Trans. Comm.*, vol.28, pp. 84–95, Jan. 1980.

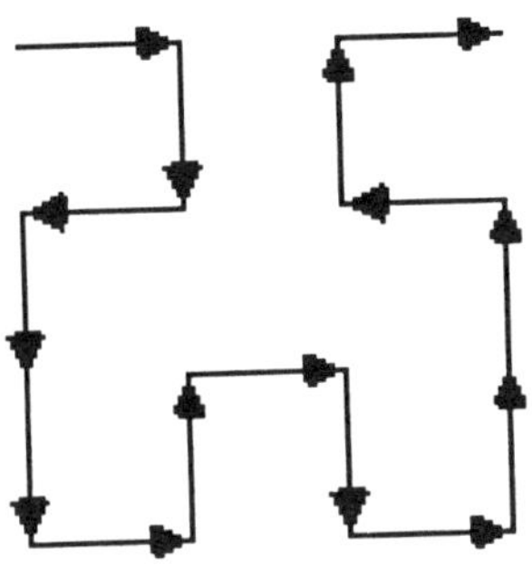

Figure 1: Basic 4 × 4 scan pattern

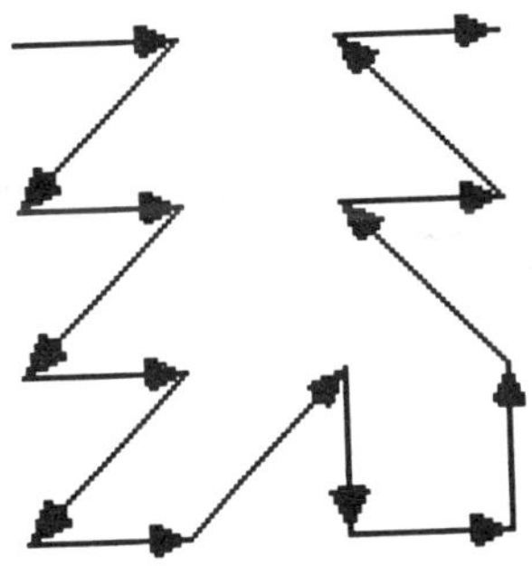

Figure 2: Another basic 4 × 4 scan pattern

Figure 3: A 16 × 16 pattern genrated recursively form the basic pattern in fig. 1.

Figure 5: Original Girl image (512 × 512)

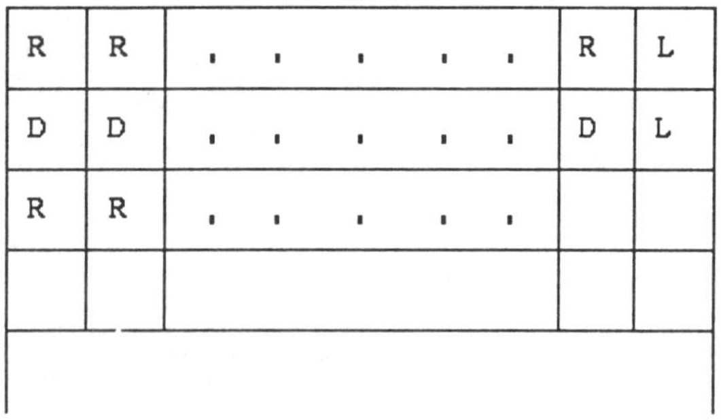

R: scan starts at top left corner, ends at top right corner
L: scan starts at top left corner, ends at bottom letf corner
D: scan starts at bottom right corner, ends at bottom left corner

Figure 4: A scan of the whole image using 16×16 blocks recursively

Figure 6: Fractal scanned version of image in fig. 5

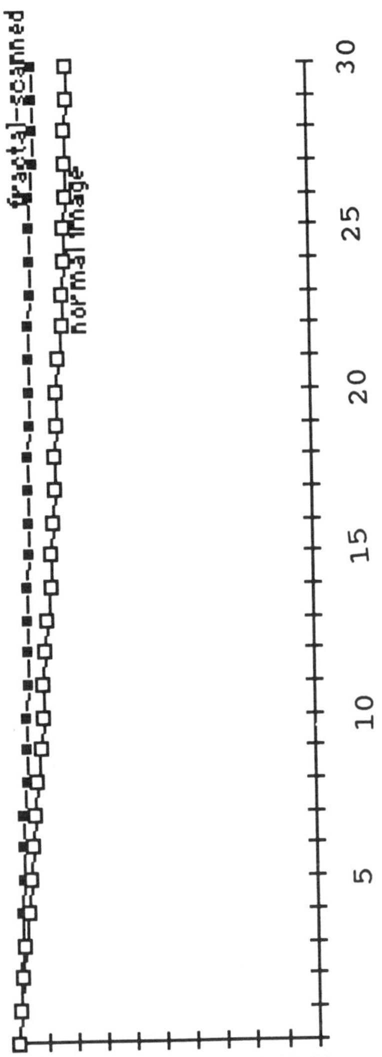

Figure 7: Autocorrelation (horizontal direction) of the images in figs. 5 and 6

Figure 8: Reconstructed image after tree-search VQ of fig. 5, 0.5 b/pixel

Figure 9: Reconstructed image after tree-search VQ of fig. 6, 0.5 b/pixel

The C-Chain Code - A New Method For Coding 3D Curves

Yu-Wei Chen Samuel C. Lee
School of EECS
University of Oklahoma
Norman, Oklahoma 73019

Abstract

In this paper, the C-chain code, a new three-dimensional (3D) chain code for encoding 3D curves, is introduced. One special feature of this chain code is that quantization points are allocated on the surfaces, rather than at the corners of the cube, as were used by Freeman's chain code. By doing so, it is shown that the C-chain code has longer average link lengths and a wider range of variations in link length, when compared with Freeman's chain code using the same cube size and similar number of quantization points, yet the maximum quantization errors of the two chain codes remain the same. It is also shown that through estimating the frequencies of individual quantization points by their spatial configurations, the efficiency of labeling quantization points of the C-chain code can be improved by applying the Huffman coding. Examples are given to illustrate this new 3D chain code as to show its major advantages over the Freeman's chain code. Applications of the C-chain code to the encoding of 3D objects are also included.

I. Introduction

In pass fifteen years, there were considerable research interest in encoding 2-D and 3-D objects using chain codes. Among various chain code schemes in encoding 2-D line drawings, two received most attention. One was introduced by Freeman [1] who latter extended it into a 3-D scheme, known as cubic lattices [2]. A 26-point cubic lattices is shown in Figure 1(a). The other 2-D encoding scheme known as the multiple grid (MG) chain coding method, recently proposed by Minami and Shinohara [3], was shown to have many advantages over the Freeman chain code. One of the special features of the MG chain code is that the quantization points of this chain code are allocated at the perimeter of the square rather than at its corners.

Similar to the MG chain code, the C-chain code proposed in this paper is a 3-D chain code which places its quantization points at the centers of grids of the side surfaces of the cube, not at its corners, to achieve longer and more varieties of the link lengths. It will be shown in this paper that the C-chain code also has many advantages over the Freeman 3-D chain code when applied to encoding of 3-D space curves.

II. The C-Chain Code

For a quantization cube of the C-chain code, each *surface* of the cube is divided into four equal *subsurfaces*, and the quantization points are set at the centers of these subsurfaces. The labeling of quantization points of the six surfaces for encoding space curves in the six different directions is shown in Figure 1(b). Referring to Figure 1(c), Six square cubes C_0, C_1, C_2, C_3, C_4, and C_5, with edge length 4q are used for quantization of space curves. These six quantization cubes have same configurations of labeling. The only differences among these quantization cube are their coordinate systems. During quantization process, a curve crossing two subsurfaces of the cube is approximated by a straight line called a *link*, which is the line connecting the two quantization points on these two subsurfaces. The first quantization point on the subsurface where the curve enters to the cube is labeled as '00.' The new quantization cube for next quantization is determined by the orientation of the link. It can be easily shown that when these quantization orientations are used to encoding the space curve, the link lengths will be longest.

III. An algorithm for encoding 3-D space curves using the C-chain code

To encode a curve, the point '00' of the subsurface '00' of cube C_0 is superimposed with the starting point of the curve to be encoded. The first quantization point is determined by the quantization point of the subsurface that the curve crosses through. The next quantization cube is chosen as the cube C_{5-i}, where i is the surface which the curve passes through. After the quantization point is found, reset the current quantization cube to be cube C_0. Then rotate the new quantization cube to match the relative orientation of the previous quantization cube, and let the point '00' of the new cube coincide with the quantization point just found. The reason to select the next quantization cube this way is that it will make the next link to be the longest. The algorithm is described as follows.

Algorithm

Let C be the next quantization cube, and (X_o, Y_o, Z_o) be the origin of the next quantization cube.

1. *initialize the C and (X_o, Y_o, Z_o).*
   $C = C_0$
   $(X_o, Y_o, Z_o) = (0, 0, 0)$
2. *Compute the crossing point of the curve through cube C.*
3. *Determine the quantization point and let L_1L_0 be the quantization point.*
4. *Select the next quantization cube, C_i, where $i = 5 - L_1$.*
5. *Compute the origin and the orientation of the next quantization cube.*
   case i of
   0: *if* $L_0 = 0$, $(X_o,Y_o,Z_o) = (X_o,Y_o,Z_o)+(0,4q,0)^*$
   if $L_0 = 1$, $(X_o,Y_o,Z_o) = (X_o,Y_o,Z_o)+(0,4q,2q)^*$
   if $L_0 = 2$, $(X_o,Y_o,Z_o) = (X_o,Y_o,Z_o)+(2q,4q,2q)^*$
   if $L_0 = 3$, $(X_o,Y_o,Z_o) = (X_o,Y_o,Z_o)+(2q,4q,0)^*$
   rotate C_i such that X-, Y-, Z-axis of C_i is parallel to X-, Y-, Z-axis of C_0.
   1: *if* $L_0 = 0$, $(X_o,Y_o,Z_o) = (X_o,Y_o,Z_o)+(0,0,0)^*$
   if $L_0 = 1$, $(X_o,Y_o,Z_o) = (X_o,Y_o,Z_o)+(0,2q,0)^*$
   if $L_0 = 2$, $(X_o,Y_o,Z_o) = (X_o,Y_o,Z_o)+(2q,2q,0)^*$
   if $L_0 = 3$, $(X_o,Y_o,Z_o) = (X_o,Y_o,Z_o)+(2q,0,0)^*$
   rotate C_i such that X-, Y-, Z-axis of C_i is parallel to X-, (-) Z-, Y-axis of C_0.
   2: *if* $L_0 = 0$, $(X_o,Y_o,Z_o) = (X_o,Y_o,Z_o)+(0,0,0)^*$
   if $L_0 = 1$, $(X_o,Y_o,Z_o) = (X_o,Y_o,Z_o)+(0,0,2q)^*$
   if $L_0 = 2$, $(X_o,Y_o,Z_o) = (X_o,Y_o,Z_o)+(0,2q,2q)^*$
   if $L_0 = 3$, $(X_o,Y_o,Z_o) = (X_o,Y_o,Z_o)+(0,2q,0)^*$
   rotate C_i such that X-, Y-, Z-axis of C_i is parallel to Y-, (-) X-, Z-axis of C_0.
   3: *if* $L_0 = 0$, $(X_o,Y_o,Z_o) = (X_o,Y_o,Z_o)+(4q,0,0)^*$
   if $L_0 = 1$, $(X_o,Y_o,Z_o) = (X_o,Y_o,Z_o)+(4q,2q,0)^*$
   if $L_0 = 2$, $(X_o,Y_o,Z_o) = (X_o,Y_o,Z_o)+(4q,2q,2q)^*$
   if $L_0 = 3$, $(X_o,Y_o,Z_o) = (X_o,Y_o,Z_o)+(4q,0,2q)^*$
   rotate C_i such that X-, Y-, Z-axis of C_i is parallel to Z-, X-, Y-axis of C_0.
   4: *if* $L_0 = 0$, $(X_o,Y_o,Z_o) = (X_o,Y_o,Z_o)+(0,0,4q)^*$
   if $L_0 = 1$, $(X_o,Y_o,Z_o) = (X_o,Y_o,Z_o)+(0,2q,4q)^*$
   if $L_0 = 2$, $(X_o,Y_o,Z_o) = (X_o,Y_o,Z_o)+(2q,2q,4q)^*$
   if $L_0 = 3$, $(X_o,Y_o,Z_o) = (X_o,Y_o,Z_o)+(2q,0,4q)^*$
   rotate C_i such that X-, Y-, Z-axis of C_i is parallel to Y-, Z-, X-axis of C_0.
   5: *if* $L_0 = 0$, $(X_o,Y_o,Z_o) = (X_o,Y_o,Z_o)+(0,0,0)^*$
   if $L_0 = 1$, $(X_o,Y_o,Z_o) = (X_o,Y_o,Z_o)+(2q,0,0)^*$
   if $L_0 = 2$, $(X_o,Y_o,Z_o) = (X_o,Y_o,Z_o)+(2q,0,2q)^*$
   if $L_0 = 3$, $(X_o,Y_o,Z_o) = (X_o,Y_o,Z_o)+(0,0,2q)^*$
   rotate C_i such that X-, Y-, Z-axis of C_i is parallel to Z-, (-) Y-, X-axis of C_0.

   * *The coordinate is relative to the current cube, C_0, coordinate system.*
6. *If not finished, go to 2.*

With this implementation, the C-chain code has a greater number of links with different lengths than the 26-point Freeman chain code does. In the Freeman chain code, there are only 3 different link lengths, which are $2q$, $2\sqrt{2}q$, and $2\sqrt{3}q$. On the other hand, the C-chain code has 12 different link lengths, which are 0, $\sqrt{2}q$, $2q$, $\sqrt{6}q$, $2\sqrt{2}q$, $\sqrt{10}q$, $\sqrt{14}q$, $4q$, $3\sqrt{2}q$, $2\sqrt{5}q$, $\sqrt{22}q$, and $2\sqrt{6}q$. An advantage of C-chain code over Freeman's chain code is that on the average link length which a bit can represent is longer than that of the Freeman's. This is discussed in [5]. A short discuss is given in Appendix A. The maximum quantization error of the C-chain code is $2\sqrt{2}q$, which is also that of the 26-point Freeman chain code. Some examples are shown as follows.

Example 1.

$x = 1.1 * t$
$L : y = 1.1 * t \quad 0 \le t \le 30$
$z = 1.2 * t \quad$ fineness 2

C-chain code :
12 21 53 22 21 53 22 21 53 22 21 53 22 21 53 22 21 53 22 21 53 22 21 53 22 21
Freeman's code :
11 11 11 11 11 18 02 11 11 11 11 11 11 11 11 11 11 18 12 11 11 11 11 11 11 11 11 11 11 18 12 11 11 11 11

Example 2.

$x = 1.1 * t$
$L : y = 2.1 * t \quad 0 \le t \le 30$
$z = 3.1 * t \quad$ fineness 2

C-chain code :
12 50 53 22 11 22 11 22 11 22 11 22 11 22 11 22 11 22 11 22 12 50 53 22 11 22 11 22 11
Freeman's code :
12 02 12 12 02 12 12 02 12 11 01 12 11 01 12 11 12 01 11 12 01 11 12 02 12 12 02 12 12 02 12 12 02 12 12 02 12 12 02 12 11 01 12 11 01 12

Example 3.

$$L: \begin{cases} x = 1.1 * t & \\ y = 5.1 * t & 0 \le t \le 30 \\ z = 1.1 * t & \text{fineness } 2 \end{cases}$$

C-chain code :
51 53 51 53 51 53 51 51 53 51 53 51 53 51 51 53 51 53 51 53 51 51 53 51 53 51 53 51

Freeman's chain code :
18 18 11 18 18 18 11 18 18 18 18 11 18 18 18 18 11 18 18 18 11 18 18 18 18 11 18 18 18 18 11 18 18 18 11 18 18 18 18 18 11 18 18 18 18 11 18 18 18 18 11 18 18 18 18 11

IV. Conclusion

In this paper, we have shown that the C-chain code is more efficient than the Freeman chain code in that the former has more varieties of link lengths and longer average link length per bit. By calculating the frequencies of occurrence of the quantization points and using the huffman code, we have obtained a set of statistically optimal binary codes for coding the 24 quantization points. The advantages of using the C-chain code to encode 3-D curves are illustrated by examples. It should noted that just like the Freeman chain code can be generalized into multiple cubes forms [4], the C-chain code also can be easily extended to multiple cube forms to have more link length varieties.

Apendix A.

In this apendix, the estimating probabilities of the occurrences of the 24-point C-chain code is presented.

Let A and B be two subsurfaces of two sequent quantization cubes. Assume that lines starting from a point of '00' subsurface first passes through subsurface A then crossing subsurface B. Let S_A and S_B be the two surface areas of unit spheres covering these two subsurfaces A and B, respectively (see Figure 2(a)). The probability of the occurrence of such lines is defined by

$$p = \frac{S_B}{S_A}$$

The area S of the spherical polygon form by S_A and S_B is

$$S = \sum_{i=1}^{n} \alpha_i - (n-2)\pi$$

where n and α_i are the number and the angle of the spherical polygon, respectively. (see Figure 2(b))

Applying the above formula, the conditional probability measure *dp* of straight lines pass subsurfaces A and B can be derived as

$$dp = \frac{S_B}{S_A}\,dxdy = \frac{\sum_{i=1}^{n_B} \alpha_i - (n_B-2)\cdot\pi}{\sum_{i=1}^{n_A} \alpha_i - (n_A-2)\cdot\pi}\,dxdy$$

By integrating *dp* over the subsurface '00', the total probability measure can then be obtained, which is

$$Q = \int_0^{2q}\int_0^{2q} \frac{\sum_{i=1}^{n_B} \alpha_i - (n_B-2)\cdot\pi}{\sum_{i=1}^{n_A} \alpha_i - (n_A-2)\cdot\pi}\,dxdy$$

The calculated state transition matrix is a 24x24 matrix. In the matrix, P_{ij} is the probability of lines from the subsurface '00' passing the subsurfaces (i div 4)(i mod 4) of first quantization cube and subsurface (j div 4)(j mod 4) of second quantization cube. If we delete the first 4 rows and first 4 columns of this matrix, since they are unreachable states, this results in a Markov transition matrix. Denote it by P. If we assume that the current quantization point after n steps does not depend on the initial quantization point, then the stationary distribution can be found as

$$\lim_{n\to\infty} P^{(n)} = \lim_{n\to\infty} P^{(0)}P^{(n)} = P^{(0)} I\, \Pi = \Pi$$

where $\Pi = (\pi, \pi, \bullet\bullet\bullet, \pi)$

In our example, π may be obtained from the matrix of Table II with first 4 rows and first 4 columns deleted, and the result is

$$\pi^t = (0.004, 0.004, 0.178, 0.091, \\ 0.004, 0.091, 0.178, 0.004, \\ 0.001, 0.004, 0.001, 0.002, \\ 0.001, 0.002, 0.001, 0.004, \\ 0.066, 0.012, 0.066, 0.286)$$

Arranging the quantization points in the descending order of their occurrence probabilities, we obtain the follows

Probability	code(link length)
0.286	53 ($2\sqrt{6}$)
0.178	12 ($\sqrt{22}$), 22 ($\sqrt{22}$)
0.091	13 ($3\sqrt{2}$), 21 ($3\sqrt{2}$)
0.066	50 ($2\sqrt{5}$), 52 ($2\sqrt{5}$)
0.012	51 (4)

0.004	10 ($\sqrt{10}$), 11 ($\sqrt{14}$), 20 ($\sqrt{10}$), 23 ($\sqrt{14}$), 31 ($\sqrt{2}$), 43 ($\sqrt{2}$)
0.002	33 ($3\sqrt{2}$), 41 ($3\sqrt{2}$)
0.001	30 ($\sqrt{10}$), 32 ($\sqrt{6}$), 40 ($\sqrt{10}$), 42 ($\sqrt{6}$)
0.000	00 (0), 01 (2), 02 ($2\sqrt{2}$), 03 (2)

From the above result, we can get a clear picture that the links have longer lengths tend to have more frequency of occurrences. The average link length is calculated as $4.58q$ which is longer than the longest link length, $2\sqrt{3}q$, of the Freeman's scheme.

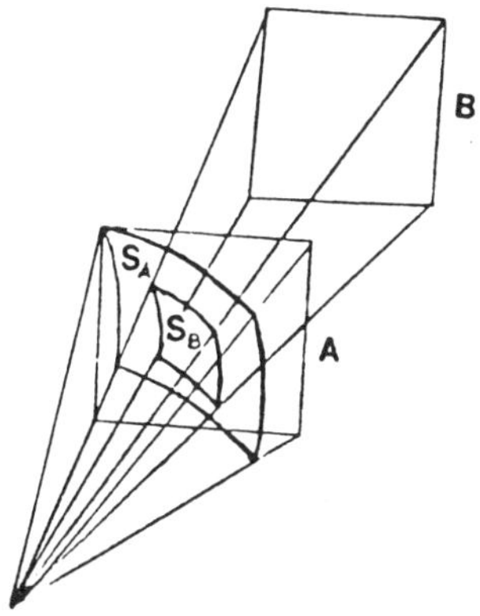

Figure 2(a)

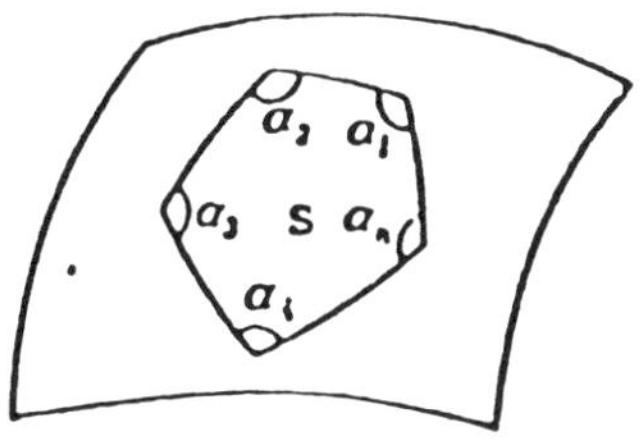

Figure 2(b)

REFERENCES

[1] H. Freeman, "On the encoding of arbitrary geometric configurations," IRE Trans. Electron. Comput., vol. EC-10, June 1961.

[2] H. Freeman, "Computer processing of line-drawing," Computing Surveys, vol. 6, no. 1, pp. 57-96, 1974.

[3] T. Minami and K. Shinohara, "Encoding of line Drawings with a multiple grid chain code," IEEE Trans. Pattern Anal. Machine Intell., vol 8, no. PAMI 2, 1986.

[4] H. Freeman and J.A. Saghri, "Generalized chain codes for planar curves," 4th Int. Joint conf. Pattern Recognition, Kyoto, Japan, Nov. 1978.

[5] Y.-W. Chen and S. C. Lee, "The analysis of C-chain code," 29th Annual Allerton Conference on Communication, Control and Computing.

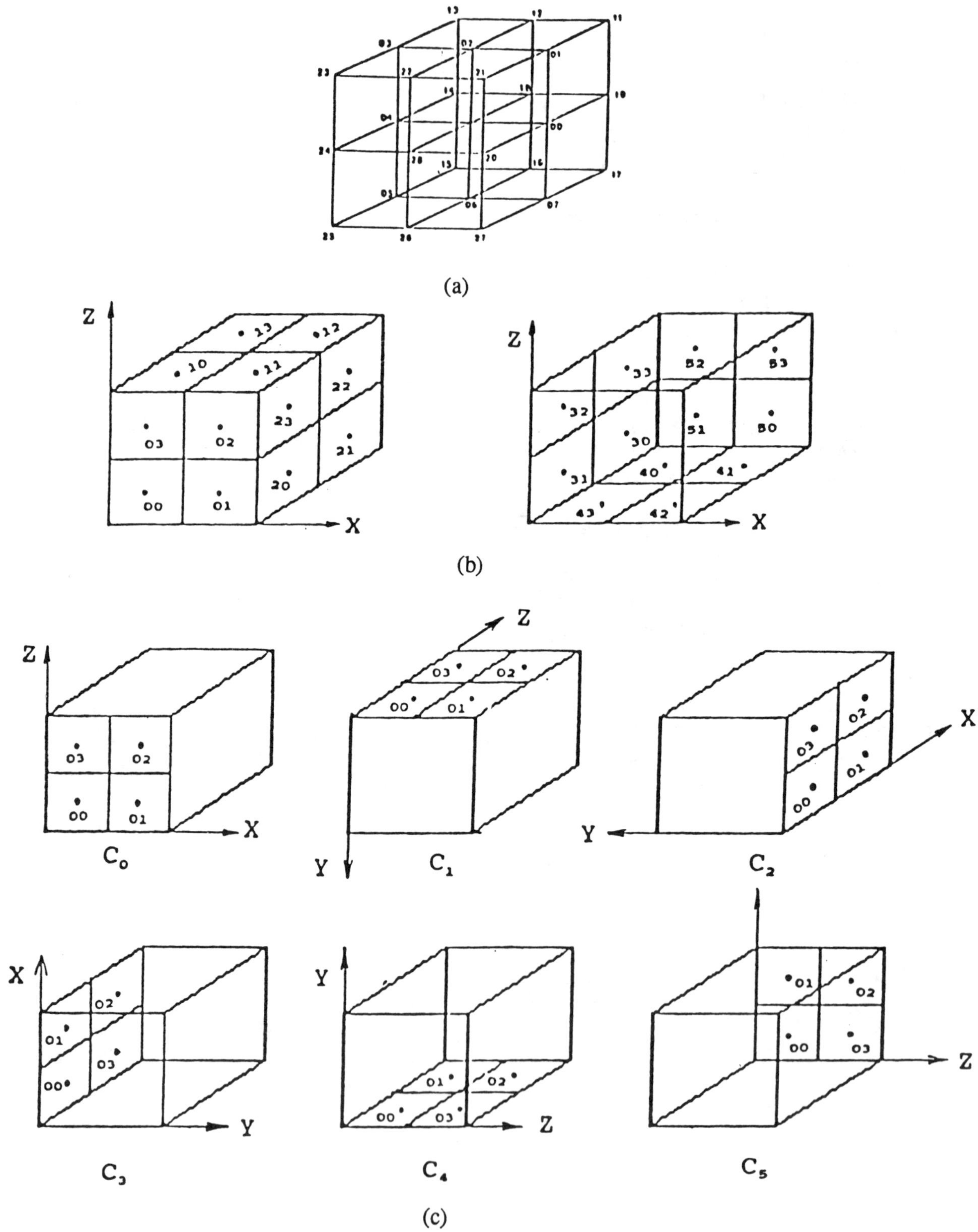

Figure 1. (a) The Freeman Chain Code
(b) The C-Chain Code
(c) The Six Quantization orientation cubes

Vector Quantization And Progressive Image Transmission Using Kohonen Self-organizing Feature Map

Wei Gong, IEEE student member, K. R. Rao and M. T. Manry, IEEE senior member
Department of Electrical Engineering, University of Texas at Arlington
P.O.Box 19016, Arlington, Tx 76019

Abstract

In this paper, vector quantization is implemented using a modified Kohonen self-organizing feature map algorithm, called KVQ. To alleviate edge distortion a classification technique is applied to vector quantization. The classification technique is based on the edge detection since the human visual system is more sensitive to edges. A new simple spatial domain progressive image transmission (PIT) is presented, which uses separating mean KVQ. In simulation results, very good intermediate images can be obtained at reasonable bit rates.

1 Introduction

Shannon's rate-distortion theory [1,2] points out that better performance is achievable theoretically when coding of individual scalars is replaced by coding of vectors, even when the source is uncorrelated. Vector quantizers (VQs) [3,4,5,6], which are designed using clustering algorithms, provide one method for realizing Shannon's theory. Although VQ has a high data compression rate, there are a few problems:(1) VQ can seriously distort edges since the distortion function generally used is the Euclidean distance,(2) the quality of conventional VQ codebooks is affected by the initial codebook and also by the number of training vectors. Large numbers of vectors are required to represent images more accurately,(3) the conventional VQ requires a long training time.

Neural networks have a lot of attractive features, such as learning, inferencing, selection, tracking and parallelism. A well-known unsupervised network is Kohonen's self-organizing feature map [8,9], which can be used to perform clustering or VQ. KVQ is able to start with a poor initial codebook, and yet achieve a good quality final codebook. KVQ is less sensitive to the ratio of the number of training vectors to the number of codewords than the conventional VQ. Also KVQ needs less training time than a conventional VQ [10]. In this paper we modify KVQ algorithm. To alleviate edge distortion, a classification technique is applied. A PIT system in spatial domain is developed using the KVQ.

A short description of the KVQ is given in section 2. Section 3 gives a modified version of the separating mean KVQ (SMKVQ) algorithm. Section 4 details the implmentation of image coding using the SMKVQ, with and without classification. Section 5 describes a PIT in the spatial domain using KVQ. Conclusions are given.

2 Kohonen's Self-Organizing Feature Map

Figure 1 shows the architecture of the KVQ. In KVQ [8,9,11], every input element is connected to all output nodes by connecting weights $w_k(i,j)$. Each output node corresponds to one cluster. Weights at the each output node are the cluster centers. After training, all weights will be organized such that topologically close output nodes are sensitive to inputs which are physically similar.

The KVQ training algorithm is summarized as follows: (1) Given the number of clusters Nc, the number of vectors in the training set Nv, and the number of iterations Nit, initial weights $w_k(i,j)$ are chosen to be small random numbers; (2) Start at iter=1 and set k=0, where iter is the index of iteration and k is the index of the input vectors; (3) For m=1,...,Nv, present new input vector X_m, k=(iter-1)•Nv+m; calculate learning factor Z(k),Euclidean distance neighborhood around the selected output node Dis(k), and the Euclidean distance between the *m*th input and *i*th cluster center d(i),i=1,...,Nc. They are defined as

$$d(i)=\sum_{n=1}^{N}[x_m(n)-w_k(i,n)]^2 \tag{1}$$

$$Dis(k)=a_1+a_2\cdot\exp(-\frac{k}{T_1}) \tag{2}$$

$$Z(k)=a_3\cdot\exp(-\frac{k}{T_2}) \tag{3}$$

where N is the number of elements in a vector, a_1, a_2 and a_3 are positive constants, T_1 and T_2 are time constants; (4) Select the output node j_s with minimum $d(j_s)$, and update its contents and its neighboring Euclidean distance vector patterns by

$$w_k(j,n)\leftarrow w_k(j,n)+z(k)\cdot(x_m(n)-w_k(j,n)),\ n=1,\ldots,N$$
$$\text{where, }\min_{j=1}^{Nc}[d(j)]=d(j_s)\ \text{ for some } j_s, \tag{4}$$
$$[j_s-Dis(k)]\le j\le[j_s+Dis(k)]$$

increment m by 1, go back to step (3) until m > Nv; (5) Increment iter and go back to step (2) until iter=Nit.

1058-6393/91 $01.00 © 1991 IEEE

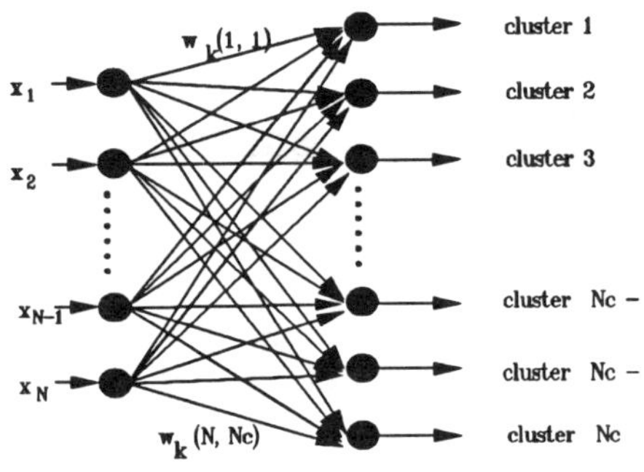

Figure 1. Architecture of KNN

After using this algorithm, adjacent clusters are related to each other. In other words, $w_k(j,i)$ is related to $w_k(j+1,i)$ etc. The ordering process is called self-organization. The weights at each output node correspond to the codebook for this cluster.

3 Modified Separating Mean Kohonen Vector Quantization

As mentioned earlier, KVQ codebook design is not too sensitive to initial weights and the number of training vectors. The significant part of this algorithm is the adjustment of the KVQ's weights, that is the choice of the Z(j) and Dis(j). One of the important characteristics of KVQ is that after training KVQ, the topologically close output nodes are sensitive to inputs that are physically similar [8,11]. According to this characteristic, it is reasonable that the learning factor function is affected by both the training time (represented by iteration number iter), and the distance between the updated cluster and the selected cluster, which is defined as the absolute difference value between the number of updated cluster j and the number of the selected cluster j_s, $|j-j_s|$. Here, the learning factor function is modified as,

$$Z_n(j,iter) = a_3 \cdot \exp(-\frac{iter}{T_2}) \cdot \exp(-|j-j_s| \cdot 0.1) \qquad (5)$$

$$\text{where,} \quad \min_{j=1}^{Nc} [d(j)] = d(j_s) \quad \text{for some } j_s ,$$
$$[j_s - Dis(k)] \leq j \leq [j_s + Dis(k)].$$

4 Image Coding Using SMKVQ

Figure 2 shows the SMKVQ scheme. The SMKVQ procedure is as follows. The first step is training the KVQ. The training images are divided into 4x4 blocks. The vectors for training are obtained by subtracting the block means. The mean saparated vectors are used to train the KVQ, resulting in the codebooks. The means of all blocks are processed into difference means [12]. The difference mean is the difference of the mean values between the current block and the previous block. For the left top block of the image the mean value is chosen as the difference mean. The mean value of the rightmost block at the previous row is chosen as the reference mean value for the leftmost block in the next row. Then the difference means of the image blocks are quantized into 8 bits, and are Huffman coded.

In computer simulations, the training image is 'girl' and the test images are 'girl', 'boat', 'martha', and 'baboon'. Table 1 shows the results for the modified SMKVQ and the traditional SMKVQ, where SNR is defined as,

$$SNR = 10 \cdot \log_{10}(\frac{255^2}{MSE}) \qquad (6)$$

where MSE is the mean-square error of the grey levels between the original image and the reconstructed image. The size of codebook is chosen as 256, i.e. 8 bits. The results in Table 1 show that the modified SMKVQ performs better than the traditional SMKVQ.

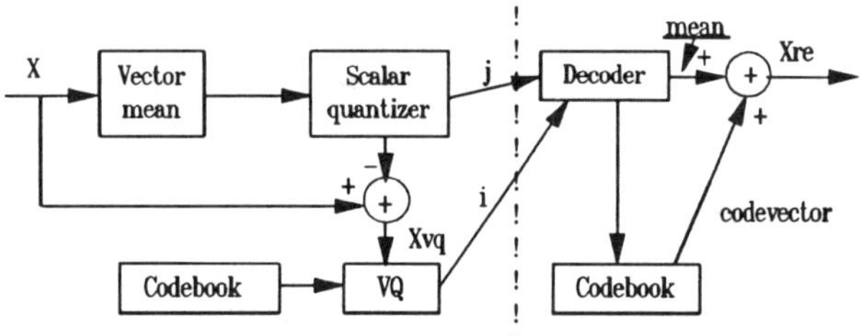

Figure 2. Separating mean VQ

Table 1. Comparison of Two SMKVQ's

Coding scheme	SMKVQ (traditional version)			SMKVQ (modified version)		
Image	MSE	SNR (dB)	Bit rate (BPP)	MSE	SNR (dB)	Bit rate (BPP)
Girl	51.26	31.03	0.824	48.70	31.26	0.824
Boat	44.26	31.63	0.824	37.04	32.44	0.824
Martha	60.17	30.34	0.813	57.63	30.52	0.813
Baboon	365.7	22.49	0.871	348.4	22.71	0.871

As mentioned earlier, edges in an image are very important parts of the perceptual information. If even a few edge blocks are poorly coded, the perceived quality of the image is degraded. Clasified SMKVQ is used to alleviate edge distortion. Also the average computations is significantly lower than that of ordinary coding algorithms since the computations involve a tree-search.

Figure 3 illustrates the CSMKVQ scheme. The image is divided into 4x4 blocks, which are classified into four different classes (two edge classes and two non-edge classes) [12]. In each class, codebooks are designed using SMKVQ. Overhead information includes the classification map. At the receiver, indices of clusters for input vectors of image and overhead information are decoded to get the locations and the corresponding codewords for reconstructing the image.

In computer simulation, the training image 'girl' was divided into 4 x 4 blocks. After all blocks were classified, four separate SMKVQ codebooks were obtained by using the scheme of Figure 2. The codebook size for each class is 128, i.e. 7 bits. Table 2 gives the comparison results for the CSMKVQ and SMKVQ. It shows that the classified version of the modified SMKVQ performs better than the modified SMKVQ.

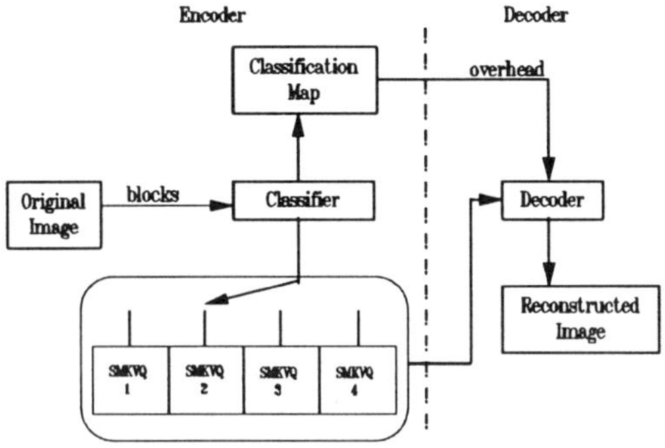

Figure 3. Classified SMKVQ

Table 2. Comparison of CSMKVQ & SMKVQ

Coding scheme	CSMKVQ			SMKVQ (modified version)		
Image	MSE	SNR (dB)	Bit rate (BPP)	MSE	SNR (dB)	Bit rate (BPP)
Girl	44.60	31.64	0.887	48.70	31.26	0.824
Boat	35.03	32.69	0.887	37.04	32.44	0.824
Martha	53.75	30.83	0.875	57.63	30.52	0.813
Baboon	359.3	22.58	0.932	348.4	22.71	0.871

5 PIT System Using SMKVQ

We have described so far the SMKVQ and CSMKVQ. The next thing is to use them to implement PIT. The basic task in a PIT system design is to determine the transmission sequence (TS) of the blocks so that good intermediate images can be obtained.

Transmission Scheme Of PIT In the SMKVQ PIT system, the cluster indices of blocks are transmitted in a special order instead of in raster fashion. The proposed scheme is shown in Figure 4. At the transmitter, the original 512 x 512 pel image is divided into blocks of size NxN, which are classified into four classes. All blocks in each class are quantized by SMKVQ. Four codebooks are obtained. In each of the four codebooks, the codewords have been rearranged to form the ordered codebooks, which are described below. The classification map is transformed into the TS according to the ordered codebook. Overhead information is transmitted first. Then the indices of the cluster numbers for the input image blocks are transmitted progressively according to the TS.

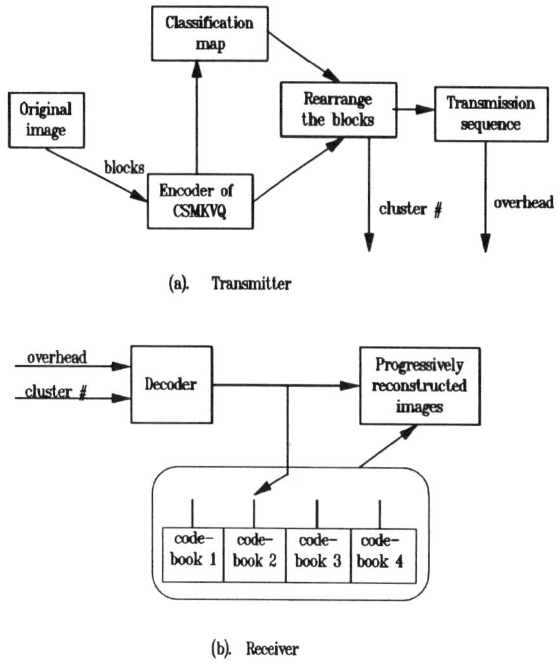

Figure 4. PIT scheme using SMKVQ

The receiver searches the codebook and puts the blocks into the proper locations, and the intermediate images are obtained progressively.

Transmission Sequence In VQ PIT systems, we transmit the order of the block numbers needed to get good intermediate images. The TS is the new classification map which is formed by combining the old classification map with the ordered codebooks. Assume that we transmit the image in 5 stages, with the first stage consisting of the means of all the (4x4) blocks in the image followed by four intermediate images. Figure 5 gives the procedure for ordering the codebooks. First, ordered codebooks, $C_{iv}(j)$, i=1,2,3,4, and j=1,2,...,128, are formed by arranging codewords, $C_i(j)$, i=1,2,3,4, and j=1,2,...,128, (which are obtained using SMKVQ) in order of decreasing variance inside each codebook. The ordered codebooks are stored at the transmitter. To form the reordered codebooks for the receiver, the first one fourth of the codewords in every ordered codebook , $C_{iv}(1)$ through $C_{iv}(32)$, i=1,2,3,4, forms the first new codebook for the second stage. The remaining three codebooks for stages 3-5 are formed as shown in Figure 6. The reordered codebooks are stored at the receiver.

The TS is generated as shown in Figure 7. The classification map of the test image is calculated first. Each input block is vector quantized by a SMKVQ using the ordered code books. The corresponding cluster number $\hat{j}$ and class number $\hat{i}$ are used to form the new class (codebook) number i,

$$i = \lceil 0.5 + \frac{4 \cdot (\hat{j} - 1)}{N_{cs}} \rceil \tag{7}$$

where $\lceil\ \rceil$ denotes rounding of the number to the nearest integer, and N_{cs} is the size of the codebook. The output cluster number j in the *ith* reordered codebook is calculated as

$$j = \hat{j} + (\hat{i} - i) \cdot \frac{N_{cs}}{4} \tag{8}$$

After all the blocks go through above procedure, the TS is obtained from the new class numbers of all image blocks.

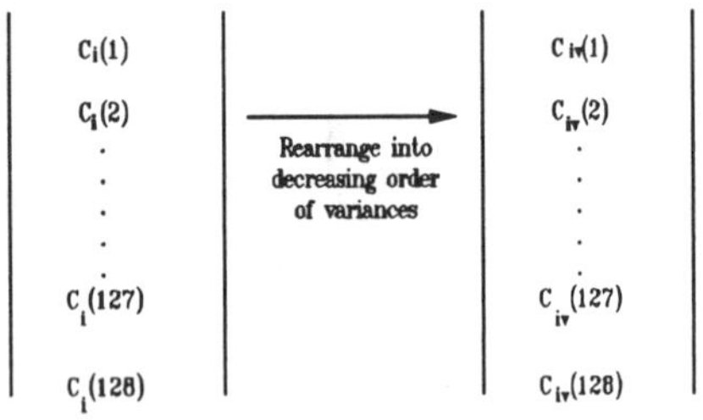

Figure 5. $C_{iv}(j)$, i=1,2,3,4, j=1,...128, are codevectors, after rearranging $C_i(j)$ in decreasing order of variances.

2*nd* stage	3*rd* stage
$C_{1v}(1)$ through $C_{1v}(32)$	$C_{1v}(33)$ through $C_{1v}(64)$
$C_{2v}(1)$ through $C_{2v}(32)$	$C_{2v}(33)$ through $C_{2v}(64)$
$C_{3v}(1)$ through $C_{3v}(32)$	$C_{3v}(33)$ through $C_{3v}(64)$
$C_{4v}(1)$ through $C_{4v}(32)$	$C_{4v}(33)$ through $C_{4v}(64)$
4*th* stage	5*th* stage
$C_{1v}(65)$ through $C_{1v}(96)$	$C_{1v}(97)$ through $C_{1v}(128)$
$C_{2v}(65)$ through $C_{2v}(96)$	$C_{2v}(97)$ through $C_{2v}(128)$
$C_{3v}(65)$ through $C_{3v}(96)$	$C_{3v}(97)$ through $C_{3v}(128)$
$C_{4v}(65)$ through $C_{4v}(96)$	$C_{4v}(97)$ through $C_{4v}(128)$

Fig.6. Codevectors rearranged into 4 stages for the PIT

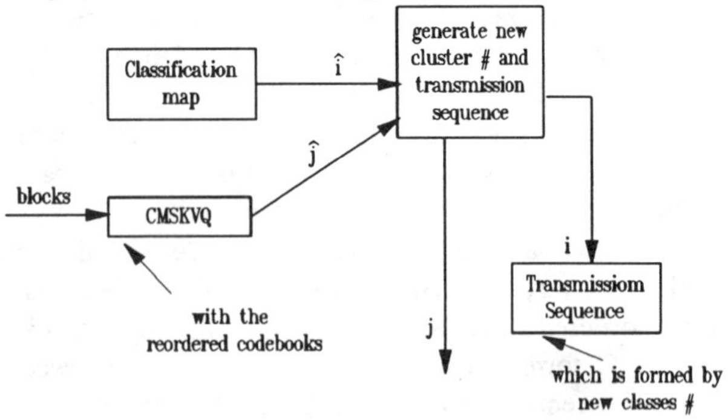

Figure 7. Transmission sequence generator

Quantization And Overhead Information In the PIT system, we only quantize the difference means of the blocks [12]. The difference mean of each block is rouneded to an integer. The values of the difference means are quantized to 8 bits and then Huffman coded. The only overhead information in this PIT system is the transmission sequence. For an Ix x Iy image, the transmission sequence is an Mrb x Mcb array where Mrb=Ix/N and Mcb=Iy/N, and the block size is (NxN). The total number of blocks for an image is Mrb x Mcb. For each block $\log_2(I_{stage}-1)$ bits are required for each element in the transmission sequence array, where I_{stage} is the total number of progressive stages. The total number of overhead bits is

$$B_{total} = Mrb \cdot Mcb \cdot \log_2 (I_{stage}-1) \quad \text{bits.}$$

Decoder Scheme Of PIT Figure 4b shows the decoder scheme. Since the overhead information is simple, the decoder works very efficiently. At the receiver, the overhead information and cluster number are decoded to get the corresponding codeword and the location of the block.

In the first stage, the progressively reconstructed image is obtained by transmitting the difference means of all blocks. Then the overhead information is transmitted. After that, for each received entry the receiver knows the received block's location exactly. At stage 2, the progressively reconstructed image is obtained by transmitting the cluster indices of class 1 in the TS. After getting the cluster number, the receiver puts the corresponding codeword into the location specified by the transmission sequence. New intermediate images are obtained by adding the resulting block codewords to the previous intermediate image. This process is repeated to get intermediate images for stages 3 through 5, except that the class indices correspond to different stages.

Simulation Results Computer simulations were carried out using the proposed PIT system. The training image 'girl' was divided into blocks of size 4x4 which were classified into four classes. For each class the SMKVQ was obtained by training a KVQ, which has 16 inputs and 128 outputs, i.e. codebook size is 128 and the input block is of size 4x4. After training, four SMKVQ codebooks corresponding to the four classes were obtained. Table 3 shows the bit rates for transmitting each intermediate image and the SNR of the intermediate image with respect to the original image. The bit rate of the second stage includes the bit rate for the overhead information. For a 5 stage PIT the bit rate for sending the overhead is 0.125 bit/pixel. The intermediate images for 'girl' are shown in Figures 8 through 10.

6 Conclusions

In this paper VQ is implemented using the SMKVQ. SMKVQ is applied to image coding. Also, to alleviate edge distortion, a classification technique is applied to image coding with SMKVQ. PIT is implemented using SMKVQ. Simulation results show that the modified SMKVQ works better for image coding than the traditional SMKVQ. Classified SMKVQ reduces the edge distortion for coded

Table 3. PIT Results With SMKVQ

Image	Girl			Boat		
number of stage	MSE	SNR (dB)	Bit rate (BPP)	MSE	SNR (dB)	Bit rate (BPP)
1	132.1	26.92	0.324	97.59	28.23	0.324
2	74.09	29.43	0.480	74.72	29.40	0.468
3	57.06	30.56	0.531	49.12	31.22	0.543
4	46.52	31.45	0.658	39.93	32.12	0.662
5	44.61	31.64	0.886	35.03	32.69	0.886

images. For PIT in spatial domain using SMKVQ, very good intermediate images at reasonable bit rates are obtained.

References

[1] R. G. Gallager, **Information Theory and Reliable Communication**, John Wiley, 1968.

[2] K. S. Shanmugam, **Digital and Analog Communication Systems**, John Wiley, 1979.

[3] R. M. Gray, "Vector quantization," IEEE ASSP Magazine, vol.1, pp.4-29, April 1984.

[4] N. M. Nasrabadi and R. A.King, "Image coding using vector quantization: A review," IEEE Trans. Commun., vol.COM-36, pp.957-971, August 1988.

[5] Y. H. Pao, **Adaptive Pattern Recognition and Neural Networks**, Addison-Wesley, 1989.

[6] N. M. Nasrabadi and Y. Feng, "Vector quantization of images based upon the Kohonen self-organizing feature maps," IEEE International Conference On Neural Networks, pp.I-101 to I-107,1988.

[7] B. Ramamurthi and A. Gersho, "Classified vector quantization of images," IEEE Trans. Commun., vol. COM-34., pp.1105-1115, Nov. 1986.

[8] T. Kohonen, "The self-organizing map," Proc. IEEE, special issue on Neural Networks I, pp.1464-1480, September 1990.

[9] T. Kohonen, **Self-Organization and Associative Memory**, Springer-Verlag, 1984.

[10] J. D. Auliffe, L. Altlas and C. Rivera, "A comparison of the LBG algorithm and Kohonen neural network paradigm for image vector quantization," IEEE ICASSP, pp. 2293-2296, April 1990.

[11] R. P. Lippmann, "An introduction to computing with neural models," IEEE ASSP Magazine, pp.4-22, April 1987.

[12] W. Gong, K. R. Rao and M. T. Manry, "Image transform coding using self-supervised backpropagation neural network," under review.

Figure 8. Original 'girl' image

Figure 9. 1st stage reconstructed image with SNR=26.92dB and bit rate=0.324BPP

Figure 10. 5th stage reconstructed image with SNR=31.64dB and bit rate=0.886BPP

Application of unsupervised neural networks to the enhancement of polarization targets in dual-polarized radar images

Andrew Ukrainec Simon Haykin
Communications Research Laboratory
McMaster University
1280 Main Street West
Hamilton, Ontario, Canada L8S 4K1

Abstract

We discuss a new approach to constrast enhancement using an unsupervised neural network with a mutual information learning criterion. The learning algorithm used is presented, which is based on minimizing the mutual information between the network outputs. The aim is to apply this neural network learning paradigm to dual-polarized images, and thus enhance an image of a radar reflector target in radar clutter. The operation of the neural network is demonstrated on radar data collected in the field.

Traditionally, techniques based on eigenvector decompositions, such as principal components analysis (PCA), have been used to enhance dual-polarized radar images to maximize the contrast, and target-to-clutter ratio of polarization targets [1], [2]. The contrast enhancement provides better features for human visualization, as well as for further detection processing. The PCA approach applies a linear transformation to the input data to generate uncorrelated features. This processing is sub-optimum in the case of distributions other than multivariate Gaussian.

For more general data distributions, which are common in a radar environment, the use of nonlinear transformations can provide superior performance. It has been shown previously that an unsupervised neural network is able to learn important features of the input data under a mutual information learning criterion [3],[4]. An unsupervised mutual information based learning algorithm is thus formulated and applied to the enhancement of the radar polarization images. The neural network processed images show highly improved contrast enhancement over those processed using just PCA.

1 Introduction

The motivation of this paper is to design a processor so as to provide contrast enhancement for dual polarized radar images. The example radar scene used to illustrate the processing is shown in Figure 1. An incoherent radar transmits in a horizontally polarized fashion, and receives both horizontal and vertical polar returns. The radar data were collected in field trials at the Centre for Inland Waters (CCIW) in Burlington, Ontario. The target of interest is a cooperative, polarization twisting reflector target that was developed at CRL for use in precise radar-aided navigation (PRAN). Before processing can begin, the raw radar data are range-normalized and scaled to approximately to the interval $[0, 1]$.

Images of the input data samples $x_{\mathrm{HH}}(n)$ and $x_{\mathrm{HV}}(n)$, as well as images of the traditional principal component analysis (PCA) projections of the radar data are included to provide a benchmark against which to compare the neural network-based result. Next, the mutual information-based approach is presented, including a discussion of the assumptions made regarding the cost function. The network-processed scene shows marked improvement over that of the PCA approach. Finally, the advantages and disadvantages of the approach are summarized, and suggestions of some possible future improvements are made.

2 Traditional techniques

Figure 2 is a scatter plot of the horizontal (HH) channel versus the vertically (HV) polarized channel. It graphically illustrates the dependence of the HV polar return. To remove this dependence from the data and enhance the information that is unique to each channel, we first take a look at a traditional linear technique based on eigenvector decomposition.

1058-6393/91 $01.00 © 1991 IEEE

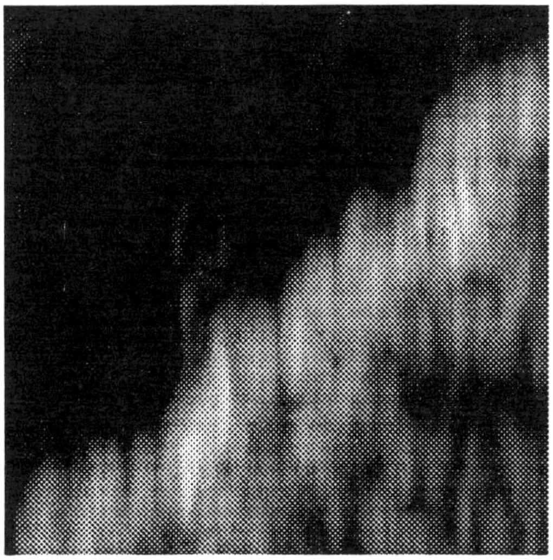

Horizontal polar return.

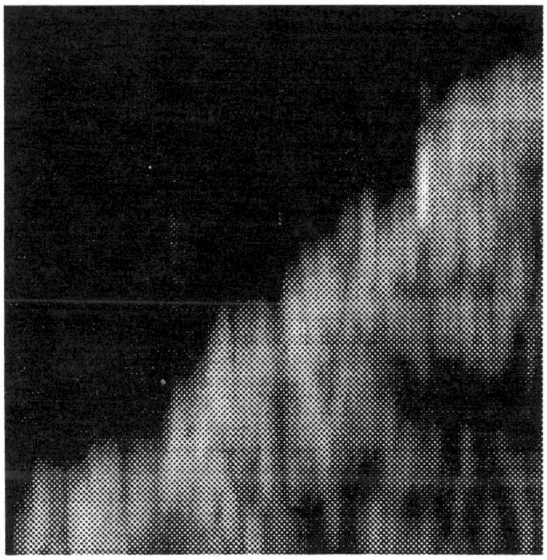

Vertical polar return.

Figure 1: Lassalle park scene, horizontal and vertical polar radar returns. The range coordinate is along the bottom of the image, and azimuth increases down the image.

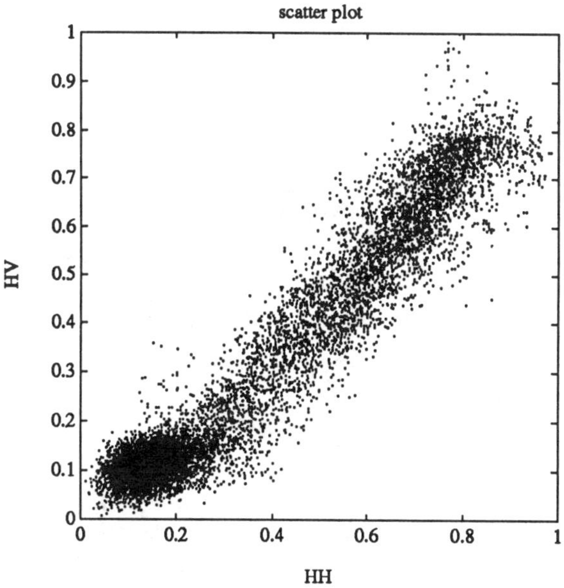

Figure 2: Scatter plot of HH and HV images.

A popular technique used to enhance the contrast of correlated images is the principal component analysis (PCA) [5]. It has been used in the past to enhance the contrast and target-to-clutter ratio of polarization targets [1],[2]. PCA is a data–dependent transformation that applies a linear, orthogonal rotation to the original data, so as to remove the correlation between the dependent variables. Expressed in matrix notation,

$$\mathbf{Y} = \mathbf{U}^T\mathbf{X}, \tag{1}$$

where $\mathbf{X} = [\mathbf{x}_{HH} \mid \mathbf{x}_{HV}]$ is the original data, $\mathbf{U}$ is the matrix of eigenvectors, and $\mathbf{Y} = [\mathbf{y}_{HH} \mid \mathbf{y}_{HV}]$ is the projected data. Applying this analysis to the 2×2 autocorrelation matrix of the dual polarized samples gives the result shown in Figure 3. The images are the absolute value of the projected data, since both positive and negative values are present in the projections.

The PCA approach has been cast into a connectionist framework, using a Hebbian learning rule to achieve optimal (in the linear sense) unsupervised learning [6]. The linear network has the advantage of being guaranteed (under reasonable conditions) to converge to a single, global maximum.

3 Mutual information approach

It is percieved that a nonlinear mapping is needed to account for the more general distributions common

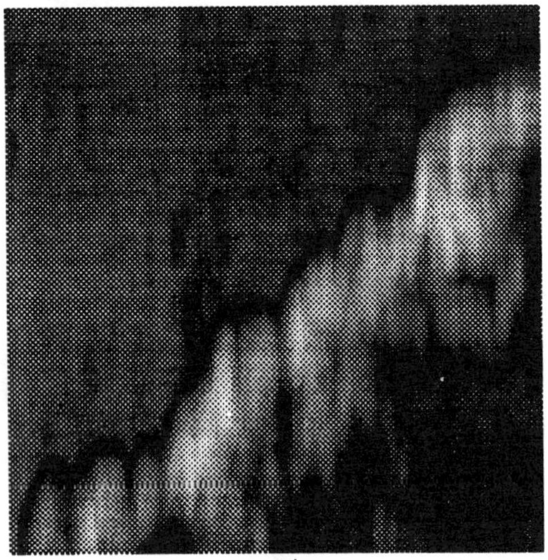

HH′

HV′

Figure 3: PCA projected HH and HV polar returns.

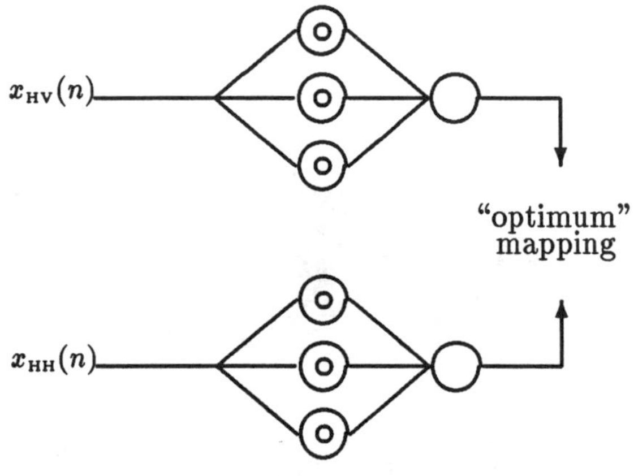

Figure 4: Unsupervised neural net architecture.

to a radar sensing environment. Regions of differing clutter will have their own distributions, and may vary with radar range. In the case of PCA, dependence between channels was removed by removing correlation between them. The solution can also be cast as a variational problem, a minimization of a quadratic functional with constraints [7]. For the neural network, the variational approach is taken, and the task is generalized to that of minimizing the mutual information between the outputs of the neural network. An example of the neural network architecture is shown in Figure 4. Although any nonlinear network could be used (e.g. multi–layer perceptron, polynomial network), a radial basis function (RBF) network was chosen because it has the advantage of providing a set of fixed basis functions, or in other words, a non-adaptive hidden layer.

The mutual information between y_1 and y_2 is given by

$$\mathrm{I}(y_1, y_2) = \iint f(y_1, y_2) \log \frac{f(y_1, y_2)}{g(y_1)h(y_2)} dy_1 dy_2, \quad (2)$$

where $f(y_1, y_2)$ is the joint probability density function of y_1 and y_2, and $g(y_1)$ and $h(y_2)$ are the corresponding individual distributions.

To numerically estimate $\mathrm{I}(y_1, y_2)$ requires either a priori assumed distribution model, or a model–free estimate. Previous researchers [3] have suggested using the Gaussian distribution model assumption when attempting to estimate mutual information. The cost function used in this paper is

$$\mathrm{I}(y_1, y_2) = \log \frac{\mathrm{var}(y_1 + y_2)}{\mathrm{var}(y_1 - y_2)}, \quad (3)$$

which was suggested in [4]. A constraint in the form

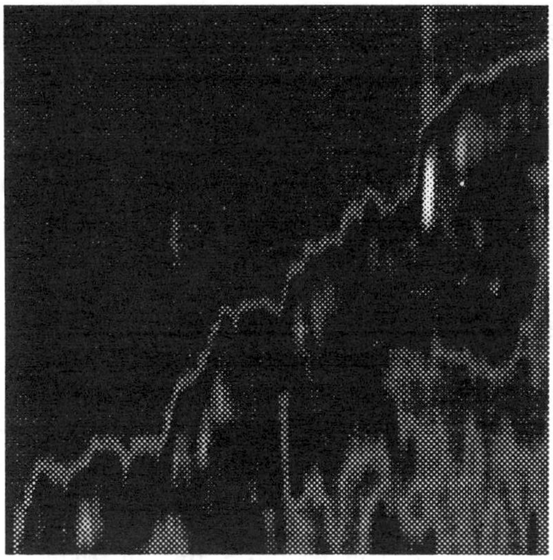

Figure 5: Nonlinear projection.

Figure 6: Nonlinear projection.

of a penalty function is introduced to avoid trivial solutions,

$$P(y_1, y_2) = \rho[\det(\mathbf{Y}^T\mathbf{Y}) - 1], \tag{4}$$

where ρ is the Lagrange multiplier. A stationary point is reached when

$$\nabla\mathbf{I} + \rho\nabla\mathbf{P} = \mathbf{0}. \tag{5}$$

This constraint forces the resultant output mapping to have a constant volume. A conjugate–gradient optimization routine was used to find a solution [8].

The RBF network consists of gaussian, radially symmetric basis functions. The centres were chosen at random from the data set, and the widths were chosen using a heuristic. Training was done on 6.25% of the dataset, and then tested on the entire dataset.

In Figure 5 we see the projection learned by the network configured as shown in Figure 4, with 40 basis functions in the hidden layer. Although there is improvement in the visibility of the target, there is still dependence in the structure of the background. The network has no cross-channel connections, and therefore is a separable mapping, $g(x_1, x_2) = g_1(x_1)g_2(x_2)$.

An improved result was achieved using a fully interconnected network containing 60 basis functions, with cross–connections between inputs and the hidden layer, and the hidden layer and the outputs. The resultant nonlinear projection is shown in Figure 6. Only the channel with enhanced vertical polar response is shown. The corresponding scatter plot is shown in Figure 7. The contour and mesh plots illustrate the range of the learned nonlinear mapping, and are shown in Figure 8 spanning the domain $[0,1] \times [0,1] \Rightarrow y$.

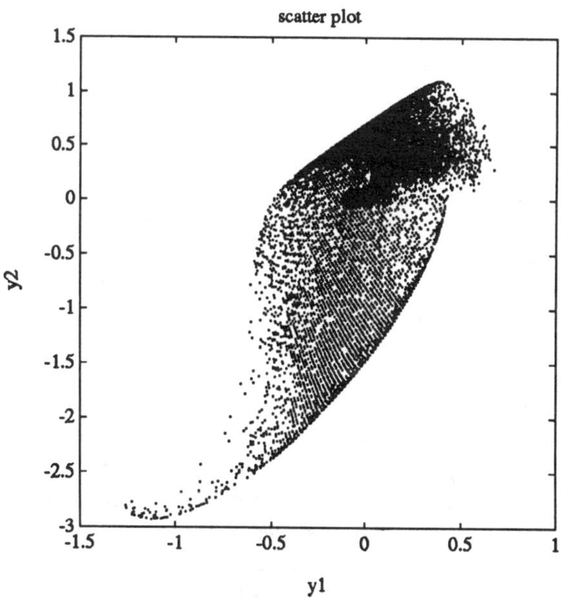

Figure 7: Scatter plot of y_1 and y_2.

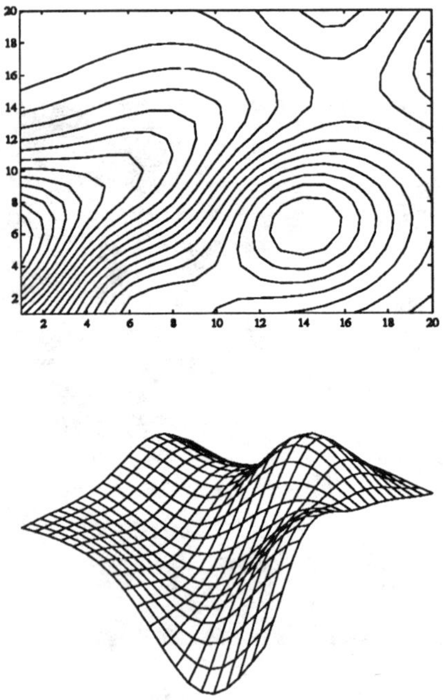

Figure 8: Contour and mesh plots corresponding to the resultant nonlinear mapping of input space.

4 Summary and Conclusions

It is shown that an unsupervised neural network can be trained to provide a contrast enhancement to dual–polarized radar images that exceeds the performance capabilities of the standard PCA method. It does so by minimizing a mutual information–based cost function, thus making use of the nonlinear transformations possible with a neural network.

Some problems remain. In the case of a linear network, one is guaranteed of finding a unique solution to the PCA variational problem. The neural network approach, however, can have many suboptimal local minima. Further work is needed to design better constraints, or penalty functions, into the cost function. Another limitation is the Gaussian assumption for the mutual information measure. Model–free estimates of the probability densities are possible, but the trade–off is greatly increased computations, since the densities would need to be re–estimated for each iteration of the optimization.

Acknowledgments

The researchers would like to thank the National Science and Engineering Research Council (NSERC) and the Telecommunications Research Institute of Ontario (TRIO) for funding this work, and CCIW for providing the radar site facilities.

References

[1] A. A. Swartz, H. Yueh, J. A. Kong, L. Novak, and R. T. Shin, "The optimal polarizations for achieving maximum contrast in radar images," NASA Contractor Report NASA-CR-183349, MIT, 1988.

[2] E. Lewis, B. Currie, and S. Haykin, *Detection and Classification of Ice.* Letchworth, UK: Research Studies Press, 1987.

[3] S. Becker and G. E. Hinton, "Spatial coherence as an internal teacher for a neural network," Tech. Rep. CRG-TR-89-7, Department of Computer Science, University of Toronto, Toronto, Ont, dec 1989.

[4] R. S. Zemel and G. E. Hinton, "Discovering viewpoint-invariant relationships that characterize objects," in *Advances in Neural Information Processing Systems 3* (R. P. Lippmann, J. E. Moody, and D. S. Touretzky, eds.), (San Mateo, CA), Morgan Kaufmann Publishers, 1991.

[5] C. Chatfield and A. J. Collings, eds., *Introduction to Multivariate Analysis.* London: Chapman and Hall, 1980.

[6] T. D. Sanger, "An optimal principle for unsupervised learning," in *Advances in Neural Information Processing Systems 1* (D. S. Touretzky, ed.), (San Mateo, CA), Morgan Kaufmann Publishers, 1989.

[7] L. E. Franks, *Signal Theory.* Stroudsburg, PA: Dowden & Culver, Revised ed., 1981.

[8] P. E. Gill, W. Murray, and M. H. Wright, *Practical Optimization.* London: Academic Press, 1981.

Circuit Complexity for Neural Computation

Kai-Yeung Siu
Dept. of Electrical & Computer Engineering
University of California
Irvine, CA 92717

Vwani Roychowdhury
School of Electrical Engineering
Purdue University
West Lafayete, IN 47907

Thomas Kailath
Dept. of Electrical Engineering
Stanford University
Stanford, CA 94305

Abstract

One common model for ANNs is the threshold circuit - *multilayer feedforward network of linear threshold gates. We provide a rigorous analysis of this model via a circuit complexity theoretic approach by focusing on the basic computational properties of threshold circuits with binary inputs. The model of threshold circuits is shown to be computationally more powerful than the conventional model of AND-OR logic circuits. In particular, we present the best known results on the depth of threshold circuits in implementing common arithmetic functions such as multiplication, division, and sorting. Moreover, we investigate the issues of depth-size tradeoffs by demonstrating that a small increase in the depth can significantly decrease the size required in the threshold circuit for the class of symmetric functions.*

1 Introduction

Recent interest in the application of artificial neural networks [5, 3] has spurred research interest in the theoretical study of such networks. While neural networks have found wide application in many areas, the behavior and the limitation of these networks are far from being understood. One common model of a neural network is a *threshold circuit*. Incidentally, the study of threshold circuits, motivated by some other complexity theoretic issues, has also gained much interest in the area of computer science.

Threshold circuits are Boolean circuits in which each gate computes a linear threshold function, whereas in the conventional model of *unbounded fan-in* Boolean circuits only AND, OR, NOT gates are allowed. A Boolean circuit is usually arranged in layers such that all gates in the same layer are computed concurrently and the circuit is computed layer by layer in some increasing *depth* order. We define the *depth* as the number of layers in the circuit. Thus each layer represents a unit delay and the depth represents the overall delay in the computation of the circuit. In the rest of this paper, we shall use the terms "threshold circuits" and "neural networks" interchangeably.

Theoretical computer scientists have used *unbounded fan-in* Boolean circuits as a model to understand fundamental issues of parallel computation. To be more specific, this computational model should be referred to as *unbounded fan-in* parallelism, since the number of inputs to each gate in the Boolean circuit is not bounded by a constant. The theoretical study of unbounded fan-in parallelism may give us insights into devising faster algorithms for various computational problems than would be possible with bounded fan-in parallelism. In fact, any nondegenerate Boolean function of n variables requires at least $\Omega(\log n)$ depth to compute in a bounded fan-in circuit. On the other hand, in some practical situations, (for example large fan-in circuits such as programmable logic arrays (PLAs) or multiple processors simultaneously accessing a shared bus), unbounded fan-in parallelism seems to be a natural model. For example, a PLA can be considered as a depth-2 circuit of (arbitrary fan-in) AND/OR gates.

In the Boolean circuit model, the amount of resources is usually measured by the number of gates, and is considered to be 'reasonable' as long as it is bounded by a polynomial (as opposed to exponential) in the number of the inputs. For example, a Boolean circuit for computing the sum of two n-bit numbers with $O(n^3)$ gates is 'reasonable', though circuit designers might consider the size of the circuit impractical for moderately large n. One of the most important theoretical issues in parallel computation is the following: *Given that the number of gates in the Boolean circuit is bounded by a polynomial in the size of inputs, what is the minimum depth (i.e. number of layers) that is needed to compute certain functions?*

It was known that [13, 4] for many simple functions, such as the *parity* of n Boolean variables, or the multiplication of two n-bit numbers, any *fixed* depth

1058-6393/91 $01.00 © 1991 IEEE

(*i.e.* independent of n) conventional Boolean circuit of unbounded fan-in AND/OR gates (AND-OR circuit) computing these functions must have exponential (in n) number of gates. Another way of interpreting these results is that circuits of AND/OR gates computing these 'hard' functions which use polynomial amount of chip area must have ***unbounded delay*** (*i.e.* delay that increases with n). In fact, the lower bound results imply that the minimum possible delay for multipliers (with polynomial number of AND/OR gates) is $\Omega(\log n / \log\log n)$. These results also give theoretical justification why it is impossible for circuit designers to implement fast parity circuit or multiplier in small chip area using AND, OR gates as the basic building blocks.

The model of threshold circuits seems so much more powerful than the conventional model of AND-OR circuits. In fact, it was known [2, 1, 7] that many common arithmetic functions such as multiplication, division and sorting can be computed by polynomial-size threshold circuits of 'constant' depth. However, the issue of how small these 'constants' can be was not examined before. For example, it was not clear from previous known results that the 'constant' of a threshold circuit for division can be reduced to below 20. Because of the fundamental importance of these arithmetic operations, for both practical and theoretical reasons, it is interesting to determine the smallest possible depth of the circuits that compute these functions.

One of our main contributions is to demonstrate that small constant depth neural networks for multiplication, sorting, division and related arithmetic functions can be constructed. These results have the following implication on their practical significance: *Suppose we can use analog devices to build threshold gates with a cost (in terms of delay and chip area) that is comparable to that of AND, OR, logic gates, then we can compute many basic functions much faster than using conventional circuits.*

Another important issue in neural computation is the trade-off between the *size* and the *depth* of neural networks. This issue arises naturally in the implementation of neural networks. Sometimes when the time for the computation is not as crucial, one might want to have a smaller network at the expense of a slight increase in the time for computation. It is interesting to investigate if increasing the depth by a small additive constant can significantly reduce the size of the network. In fact, our most recent work [12] has given a positive answer to this question. The best known prior results [6] in computing any symmetric Boolean function of n variables with neural networks require $O(n)$ threshold gates and two layers. It has been an open problem for 30 years to determine if the number of threshold gates can be further reduced for an arbitrary symmetric Boolean function. We are able to show that by increasing the depth by one, i.e. by using a 3-layer neural network, one can reduce the number of threshold gates from $O(n)$ to $O(\sqrt{n})$. For functions with 'periodic structures' such as the parity, we have extended the depth-size tradeoffs results to higher depth: for every integer $d > 0$, one can construct neural networks of depth-$(d+1)$ and size $O(dn^{1/d})$ computing them. Moreover, using classical results from the theory of rational approximations and harmonic analysis, we obtained lower bound results by showing that our networks are almost optimal in the size ($\Omega(dn^{1/d-\epsilon})$ threshold gates are needed for any fixed $\epsilon > 0$).

In the following section, a summary of all these results will be presented. Because of space limitation, we refer the interested readers to our related journal papers [8, 9, 10, 12, 11] for the detailed descriptions and proofs of our results.

2 Summary of Main Results

Definition 1 A *linear threshold function* $f(X) : \{0,1\}^n \rightarrow \{0,1\}$ is a Boolean function such that

$$f(X) = sgn(F(X)) = \begin{cases} 1 & \text{if } F(X) \geq 0 \\ 0 & \text{if } F(X) < 0 \end{cases}$$

where

$$F(X) = \sum_{i=1}^{n} w_i \cdot x_i + w_0$$

$$X = (x_1, \ldots, x_n) \in \{0,1\}^n$$

Our model of neural network is simply a multilayer feed-forward network of linear threshold gates, *i.e.* a threshold circuit.

Note that the above definition does not impose any restriction on the *fan-in* and *fan-out* of each threshold gate in the network, *i.e.* each gate can have arbitrarily many inputs from previous layer and can feed its output to arbitrarily many gates in subsequent layers. Since neural networks are characterized by their massive parallelism, the unbounded fan-in parallelism of threshold circuits seems a natural assumption.

Given two input n-bit (unsigned) integers, X and Y, the product of X and Y (denoted by

$PROD(X,Y)$) will be a $2n$-bit integer, the (truncated) quotient of X divided by Y (denoted by $DIV_k(X,Y)$) will be an $(n+k)$-bit number (where the last k bits constitue the truncated fractional part of the quotient). Given n input n-bit positive integers z_i, $i = 1, ..., n$, their multiple product $\Pi_{i=1}^{n} z_i$ will be an n^2-bit integer denoted by $MPROD(z_1, ..., z_n)$, and the sorted list denoted by $SORTING((z_1, ..., z_n)$ is the same list of integers sorted in nondecreasing order. While the computation of these functions requires any fixed depth AND-OR circuit to have an exponential size, the following theorem states that these functions can all be computed by polynomial-size neural networks of small constant depth, independent of the input size n. These are the best known results on the depth of neural networks in computing these arithmetic functions.

Theorem 1

1. $PROD(X,Y)$ and $SORTING((z_1, ..., z_n)$ can be computed by polynomial-size neural networks of depth-4.
2. $DIV_k(X,Y)$ can be computed by polynomial-size neural networks of depth-5.
3. $MPROD(z_1, ..., z_n)$ can be computed by polynomial-size neural networks of depth-6.

Definition 2 A Boolean function $f : \{1,0\}^n \rightarrow \{1,0\}$ is said to be *symmetric* if

$$f(x_1, \ldots, x_n) = f(x_{(1)}, \ldots, x_{(n)})$$

for any permutation $(x_{(1)}, \ldots, x_{(n)})$ of $(x_1, \ldots, x_n)$, or equivalently, the value of f only depends on the number of 1's in the input variables.

An example of a symmetric function is the PARITY function, which is defined to be 1 if the number of 1's in the inputs is *odd* and is 0 otherwise. Symmetric functions constitute an important class of Boolean functions and come up very often in the design of logic circuits. As mentioned before, the best known prior result is that any symmetric function (of n variables) can be computed by a depth-2 neural network of size $O(n)$. The following results show that a significant reduction in the size can be achieved if the depth is increased by a small constant.

Theorem 2 Any symmetric function of n variables can be computed in a depth-3 neural network with $2\sqrt{n} + O(1)$ threshold gates.

Theorem 3 For any integer $d > 0$, the PARITY function of n variables can be computed in a depth-$(d+1)$ neural network with $O(dn^{1/d})$ threshold gates.

Using novel techniques from harmonic analysis and the theory of rational approximations, we are able to show that the sizes of our neural networks for symmetric functions are *almost optimal*. (For details of the proof, see [11])

Theorem 4 For any fixed $\epsilon > 0$ and any integer $d > 0$, almost all symmetric functions of n variables (including the PARITY function) require any depth-(d+1) neural network at least $\Omega(dn^{1/d-\epsilon})$ threshold gates to compute.

Acknowledgements

This work was done while the the first author was a research student associate at IBM Almaden Research Center, San Jose, CA and supported in part by the Joint Services Program at Stanford University (US Army, US Navy, US Air Force) under Contract DAAL03-88-C-0011, the SDIO/IST, managed by the Army Research Office under Contract DAAL03-87-0033, and the Department of the Navy, NASA Headquarters, Center for Aeronautics and Space Information Sciences under Grant NAGW-419-S6.

References

[1] P. W. Beame, S. A. Cook, and H. J. Hoover. Log depth circuits for division and related problems. *Siam J. Comput.*, 15:994–1003, 1986.

[2] A. K. Chandra, L. Stockmeyer, and U. Vishkin. Constant depth reducibility. *Siam J. Comput.*, 13:423–439, 1984.

[3] J. L. McClelland D. E. Rumelhart and the PDP Research Group. *Parallel Distributed Processing: Explorations in the Microstructure of Cognition, vol. 1.* MIT Press, 1986.

[4] J. Hastad. Almost optimal lower bounds for small depth circuits. *ACM Symp. Theor. Computing*, 18:6–20, 1986.

[5] J. J. Hopfield. Neural Networks and physical systems with emergent collective computational abilities. *Proceedings of the National Academy of Sciences*, 79:2554–2558, 1982.

[6] R. Minnick. Linear-Input Logic. *IEEE Trans. on Electronic Computers*, EC 10, 1961.

[7] J. Reif. On Threshold Circuits and Polynomial Computation. In *Structure in Complexity Theory Symp.*, pages 118–123, 1987.

[8] K. Y. Siu and J. Bruck. Neural Computation of Arithmetic Functions. Special Issue on Neural Networks, *Proc. IEEE*, 78, No.10:1669–1675, October 1990.

[9] K.-Y. Siu and J. Bruck. On the Power of Threshold Circuits with Small Weights. *SIAM J. Discrete Math.*, 4(3):423–435, August 1991.

[10] K.-Y. Siu, J. Bruck, and T. Kailath. Depth-Efficient Neural Networks for Division and Related Problems. *IEEE Trans. Information Theory.* in press.

[11] K. Y. Siu, V. P. Roychowdhury, and T. Kailath. Computing with Almost Optimal Size Threshold Circuits. submitted to JCSS, 1990.

[12] K. Y. Siu, V. P. Roychowdhury, and T. Kailath. Depth-Size Tradeoffs for Neural Computation. *IEEE Transactions on Computers* , Special Issue on Neural Networks, to appear December 1991.

[13] A. Yao. Separating the polynomial-time hierarchy by oracles. *IEEE Symp. Found. Comp. Sci.*, pages 1–10, 1985.

The Learning Parsimony of Projection Pursuit and Back-Propagation Networks

Jenq-Neng Hwang, Hang Li,

Information Processing Laboratory
Dept. of Electrical Engr., FT-10
University of Washington
Seattle, WA 98195

Doug Martin, Jim Schimert

Department of Statistics
Mail Stop: GN-22
University of Washington
Seattle, WA 98195

Abstract

The learning parsimony of two types of feedforward network learning methods for model-free regression problems are studied and compared in this paper: one is the popular back-propagation learning (BPL) which is well known in the artificial neural networks literature; the other is the projection pursuit learning (PPL) which has emerged in recent years in the statistical estimation literature. Algorithmically, both the BPL and the PPL parametrically form projections of the data in directions determined from interconnection weights. However, unlike a BPL which employs a fixed set of (usually sigmoidal) nonlinear nodal functions to perform an explicit parametric estimate of all the weights simultaneously at each iteration, the PPL non-parametrically estimates the unknown nonlinear mapping sequentially (neuron-by-neuron and layer-by-layer) at each iteration. In terms of learning efficiency, both methods have comparable training speed while the PPL is more parsimonious, i.e., in order to achieve comparable accuracy, the PPL requires less number of hidden neurons to approximate the desired true function.

1 Introduction

A *network* is a function represented by the composition of many basic functions. The basic functions (in our discussion, the artificial neural units) are typically nonlinear functions of a few variables. A *network learning* estimates an unknown function from representative observations of the relevant variables. The study of learning parsimony is to study how many parameters in the networks should be required for appropriate approximations. The parsimony properties of two types of network learning methods are studied and compared in this paper: one is the popular *back-propagation learning* (BPL) which is well known in the artificial neural network literatures; the other is the *projection pursuit learning* (PPL) which has emerged in recent years in the statistical estimation literatures [3].

There is a huge statistical literature on the asymptotic theory of *non-parametric* estimation in univariate model-free regression problems. Suffice it to say that the work has little relevance to real problems where the dimension p is larger, say $p > 5$. The reason is that a form of "curse of dimensionality" comes into play. All fully non-parametric methods perform some kind of smoothing, and in a high dimensional setting one faces the following dilemma: if one smooths enough to reduce variability of the functional estimate to an acceptable level, then one suffers an unacceptably large bias, and conversely. This fact has led to a number of more or less "parsimonious" *semi-parametric* methods (e.g., the regression tree, projection pursuit regression [3]).

In this paper we perform a thorough comparative study algorithmically as well as experimentally between BPL's and PPL's, especially on the learning efficiency (parsimony) of both methods.

2 A Two-layer Perceptron

A multilayer perceptron is a feed-forward neural network which has one or more layers of *hidden neurons* between the input and output layers. Several recent results have shown that a two-layer (one hidden layer) perceptron with sigmoidal nodes can in principle represent any Borel-measurable function to any desired accuracy, assuming "enough" hidden neurons are used. This, along with the fact that theoretical results are known for the PPL in the analogous two-layer case, justifies focusing on the two-layer perceptron for our studies here.

2.1 Mathematical Formulation

A two-layer perceptron can be mathematically formulated as follows:

$$
\begin{aligned}
u_k &= \sum_{j=1}^{p} w_{kj} x_j - \theta_k = \mathbf{w}_k^T \mathbf{x} - \theta_k, \quad k = 1, m \\
y_i &= \sum_{k=1}^{m} \beta_{ik} f_k(u_k) = \sum_{k=1}^{m} \beta_{ik} \sigma(u_k), \quad i = 1, q \quad (1)
\end{aligned}
$$

where u_k denotes the weighted sum input of the kth neuron in the hidden layer; θ_k denotes the bias of the kth neuron in the hidden layer; w_{kj} denotes the input-layer weight linked between the kth hidden neuron and the jth neuron of the input layer (or j^{th} element of the input vector $\mathbf{x}$); β_{ik} denotes the output-layer weight linked between the ith output neuron and the kth hidden neuron; f_k is the nonlinear activation function, which is usually assumed to be a fixed monotonically increasing (logistic) sigmoidal function, $\sigma(u) = 1/(1+e^{-u})$.

The above formulation defines quite explicitly the parametric representation of functions which are being used to approximate $\{g_i(\mathbf{x}),\ i = 1, 2, \cdots, q\}$. Specifically,

$$\hat{g}_i(\mathbf{x}) = \sum_{k=1}^{m} \beta_{ik}\sigma(\mathbf{w}_k^T\mathbf{x} - \theta_k) \tag{2}$$

A simple reparametrization allows us to write $\hat{g}_i(\mathbf{x})$ in the form:

$$\hat{g}_i(\mathbf{x}) = \sum_{k=1}^{m} \beta_{ik}\sigma(\frac{\alpha_k^T\mathbf{x} - \mu_k}{s_k}) \tag{3}$$

where α_k is a unit length version of weight vector $\mathbf{w}_k$. This formulation reveals how $\{\hat{g}_i\}$ are built up as a linear combination of sigmoids evaluated at translates (by μ_k) and scaled (by s_k) projection of $\mathbf{x}$ onto the unit length vector α_k. However, we will use Eq. (2) for computational purpose in the later discussion.

2.2 Back-Propagation Learning

Historically, the learning of a multilayer perceptron uses back-propagation learning (BPL), a *simple* iterative gradient descent algorithm designed to minimize the mean squared error, $E = \sum_l \sum_i (t_{li} - \hat{g}_i(\mathbf{x}_l))^2$, between the the desired response (target) vectors $\{\mathbf{t}_l\}$ and the actual resesponse vectors $\{\hat{\mathbf{g}}(\mathbf{x}_l)\}$. There are two common types of BPL: *batch* learning and *sequential* learning. The batch BPL updates the weights after the presentation of the complete set of training data. Hence, a training iteration incorporates one sweep through all the training patterns. On the other hand, the sequential BPL adjusts the network parameters as training patterns are presented, rather than after a complete pass through the training set.

While the two-layer perceptron provides a very powerful nonlinear modeling capability, the BPL training can be slow and inefficient since only the first derivative (or gradient) information about the training error is utilized. To speed up the training process, several *second-order* optimization algorithms, which take advantage of second derivative (or Hessian matrix) information, have been proposed for training perceptrons [4]. For example, the Newton method, the Quasi-Newton method, the conjugate gradient method, as well as the Gauss-Newton method, which is also used in the PPL [3].

The fixed nonlinear nodal (sigmoidal) function is a monotone nondecreasing differentiable function with very simple first derivative form, and possesses nice properties for numerical computation. However, it does not interpolate/extrapolate well in regression applications. Several attempts have been proposed to improve the choice of nonlinear nodal functions; e.g., the Gaussian or bell-shaped function, the locally tuned radial basis functions [2], and semi-parametric nonlinear functions used in PPL's and hidden Markov models.

3 Projection Pursuit Learning

The projection pursuit learning (PPL) is a statistical procedure proposed for multivariate data analysis using a two-layer network given in Eq. (1). This procedure derives its name from the fact that it interprets high dimensional data through well-chosen lower-dimensional projections, which is equivalent to the scalar nonlinear neuron-output of the weighted sum of high dimensional data in perceptron formulation [3].

Similar to a BPL perceptron, a PPL network forms projections of the data in directions determined from the interconnection weights. However, unlike a BPL perceptron, which employs a fixed set of nonlinear (sigmoidal) functions, a PPL non-parametrically estimates the nonlinear nodal functions based on optimization approach with a one-dimensional data-smoother. Therefore, it is important to note that a PPL network is a semi-parametric learning network, which consists of both parametrically and non-parametrically estimated elements. This is in contrast to a BPL perceptron, which is a completely parametric model.

In comparison with a batch BPL, which employs either 1st-order gradient descent or 2nd-order Newton-like methods to estimate the weights of all layers *simultaneously* after all the training patterns are presented, a PPL learns neuron-by-neuron and layer-by-layer *sequentially* after all the training patterns are presented. Specifically, it applies least squares to estimate the output-layer weights, a one-dimensional data smoother to estimate the nonlinear functions of each hidden neuron, and the Gauss-Newton method to estimate the input-layer weights.

The PPL procedure uses the batch learning technique to iteratively minimize the mean squared error, E, over all the training data. All the parameters to be estimated are hierarchically divided into m groups (each associated with one hidden neuron), and each group, say the kth group, is further divided into three subgroups: the output-layer weights, $\{\beta_{ik},\ i = 1, \cdots, q\}$, connected to the kth hidden neuron; the nonlinear function, $f_k(u)$, of the kth hidden neuron; and the input-layer weights, $\{w_{kj},\ j = 1, \cdots, p\}$, connected to the kth hidden neuron. The PPL starts from updating the parameters associated with the first hidden neuron (group) by updating each subgroup, $\{\beta_{i1}\}$, $f_1(u)$, and $\{w_{1j}\}$ consecutively (layer-by-layer) to minimize the mean squared error E; a similar modular training concept was also proposed. It then updates the parameters associated with the second hidden neuron by consecutively updating $\{\beta_{i2}\}$, $f_2(u)$, and $\{w_{2j}\}$. A complete updating pass ends at the updating of the parameters associated with the mth (the last) hidden neuron by consecutively updating $\{\beta_{im}\}$, $f_m(u)$, and $\{w_{mj}\}$. Repeated updating passes are made over all the groups until convergence (i.e., $\frac{|E^{(new)}-E^{(old)}|}{E^{(old)}}$ smaller than a prespecified small constant, $\xi = 0.005$).

4 Learning Parsimony

Having compared the "parametric" BPL and the "semi-parametric" PPL from algorithmic and theoretical viewpoints, we now make a more practical comparison in learning parsimony via a simulation study. For simplicity of comparison, we confine the simulations to the two-dimensional univariate case, i.e., $p = 2,\ q = 1$. This is an important situation in practice, because the models can be visualized graphically as functions $y = g(x_1, x_2)$.

4.1 Protocols of the Simulations

Nonlinear Functions: There are five nonlinear functions $g^{(j)} : [0,1]^2 \to \mathbb{R}$ to be investigated, which are *scaled* such that the standard deviation is 1 (for a large regular grid of 2500 points on $[0,1]^2$), and *translated* to make the range nonnegative.

- *Simple Interaction Function:*

$$g^{(1)}(x_1, x_2) = 10.391((x_1 - .4) \cdot (x_2 - .6) + .36)$$

- *Radial Function:*

$$\begin{aligned} g^{(2)}(x_1, x_2) &= 24.234\left(r^2(.75 - r^2)\right), \text{ where} \\ r^2 &= (x_1 - \frac{1}{2})^2 + (x_2 - \frac{1}{2})^2 \end{aligned}$$

- *Harmonic Function:*

$$g^{(3)}(x_1, x_2) = 42.659\left((2 + x_1)/20 + Re(z^5)\right)$$

where $z = x_1 + ix_2 - \frac{1}{2}(1 + i)$, or equivalently, with $\tilde{x}_1 = x_1 - .5$, $\tilde{x}_2 = x_2 - .5$,

$$\begin{aligned} g^{(3)}(x_1, x_2) &= 42.659(.1 + \tilde{x}_1(.05 + \tilde{x}_1^4 - \\ &\quad 10\tilde{x}_1^2\tilde{x}_2^2 + 5\tilde{x}_2^4)) \end{aligned}$$

- *Additive Function:*

$$\begin{aligned} g^{(4)}(x_1, x_2) &= 1.3356(1.5(1 - x_1) + \\ &\quad e^{2x_1 - 1} sin(3\pi(x_1 - .6)^2) + \\ &\quad e^{3(x_2 - .5)} sin(4\pi(x_2 - .9)^2)) \end{aligned}$$

- *Complicated Interaction Function:*

$$\begin{aligned} g^{(5)}(x_1, x_2) &= 1.9(1.35 + e^{x_1} sin(13(x_1 - .6)^2) \\ &\quad e^{-x_2} sin(7x_2)) \end{aligned}$$

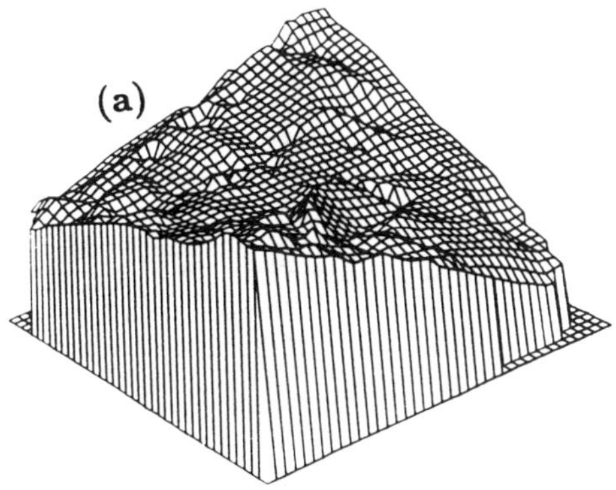

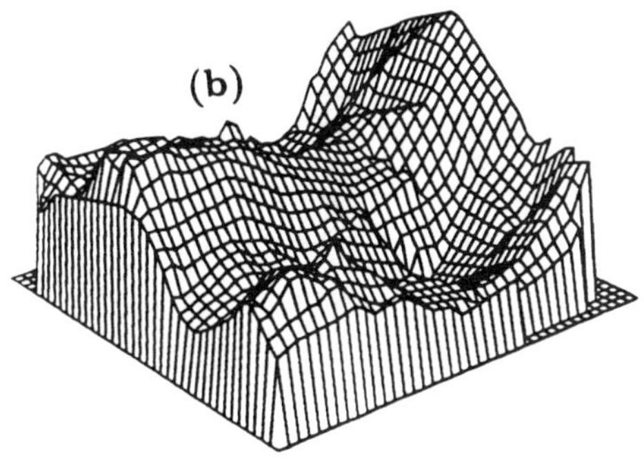

Figure 1: Examples of noisy training data set for (a) $g^{(1)}$ function. (b) $g^{(4)}$ function.

Training and Test Data: Two independent variables (or carriers) are set as $(x_{l1},\ x_{l2}) \sim U([0,1]^2)$, i.e., the abscissa values $\{(x_{l1},\ x_{l2})\}$ are generated as uniform random variates on $[0,1]$ and independent from each other. There are $n = 225$ pairs of abscissa values generated, and we use the same set of $\{(x_{l1},\ x_{l2})\}$ for experiments of all five different functions, thus eliminating an unnecessarily extra random component of the simulation. In addition to one set of noiseless training data, another set of noisy training data is also generated by adding *iid* Gaussian noises to the noiseless set:

$$y_l^{(j)} = g^{(j)}(\mathbf{x}_l) + .25\varepsilon_l, \qquad l = 1, \cdots, n, \qquad j = 1, \cdots, 5. \tag{4}$$

where $\varepsilon_l \sim \mathcal{N}(0,1)$, the zero mean unit variance Gaussian noise. Note that the S/N ratio is roughly $\frac{1}{.25} = 4$ by the scaling of the $g^{(j)}$'s, (approximately only, because of small size of samples, $\mathbf{x}_l$'s and ε_l's). Examples of noisy training data sets for $g^{(1)}$ and $g^{(4)}$ is shown in Figures 1 (a), (b).

Independent Test Data Set: The assessment of performance was done by comparing the fitted models with the "true" function counterparts on a large independent test set. Throughout all the simulations, we used the same set of test data for performance assessment, i.e., $\{(x_{l1}, x_{l2})\}$, of size $N = 10000$, namely a regularly spaced grid on $[0,1]^2$, defined by its marginals

$$x_{li} := (2l-1)/200, \quad \text{for } l = 1, \cdots, 100, \quad i = 1, 2.$$

Implementation Variations: To have a fair comparison, the BPL was implemented using batch Gauss-Newton method on two-layer perceptrons with linear output neurons and nonlinear sigmoidal hidden neurons [4], where 5 and 10 hidden neurons were tried. The PPL simulations were conducted using the *S-Plus* package [1], where 3 and 5 hidden neurons were tried (with 5 and 7 maximum working hidden neurons used separately to avoid the overfitting [1]).

Error Measures: The absolute L_2 error measures were used to compare the performance on independent test data of all functions:

$$L_2 = \frac{\sum_{l=1}^{N}(\hat{g}(\mathbf{x}_l) - g(\mathbf{x}_l))^2}{\sum_{l=1}^{N}(g(\mathbf{x}_l) - \text{mean}_l\{g(\mathbf{x}_j)\})^2}. \quad (5)$$

Note that this error measure is proportional to the usual "mean squared error" (MSE), and a more intuitive value is its *square root*.

4.2 Simulation Results

In comparison with BPL, the PPL is more parsimonious in training all types of nonlinear functions, i.e., in order to achieve comparable accuracy to the BPL's for a two-layer perceptrons, the PPL's require less number of hidden neurons (more parsimonious) to approximate the desired true function. For example, all functions except $g^{(1)}$ cannot be reasonably well approximated by BPL networks with 3 hidden neurons. On the other hand, all functions except $g^{(4)}$ and $g^{(5)}$ can be reasonably well approximated by PPL networks with 3 hidden neurons. Several factors may contribute to this favorable performance. First, the batch Gauss-Newton BPL updates all the weights in the network simultaneously while the PPL updates sequentially (neuron-by-neuron and layer-by-layer), which allows the most recent updating information to be used in the subsequent updating. That is, more important projection directions can be determined first so that the less important projections can have a easier search (the same argument used in favoring the Gauss-Seidel method over the Jacobi method in iterative linear equation solver). Second, the data-smoothing technique creates more pertinent nonlinear nodal functions, so the network adapts more efficiently to the observation data without using too many terms (hidden neurons) of interpolative projections.

The above two factors can be verified by the following observations: according to Figure 1(a), the 3-D surface of $g^{(1)}$ can be "theoretically" constructed using two projections, i.e., $x_1x_2 = \frac{1}{2}[(x_1+x_2)^2 - (x_1-x_2)^2]$. This fact is clearly reflected in the weights trained by the PPL networks with 2, 3, and 4 hidden neurons,

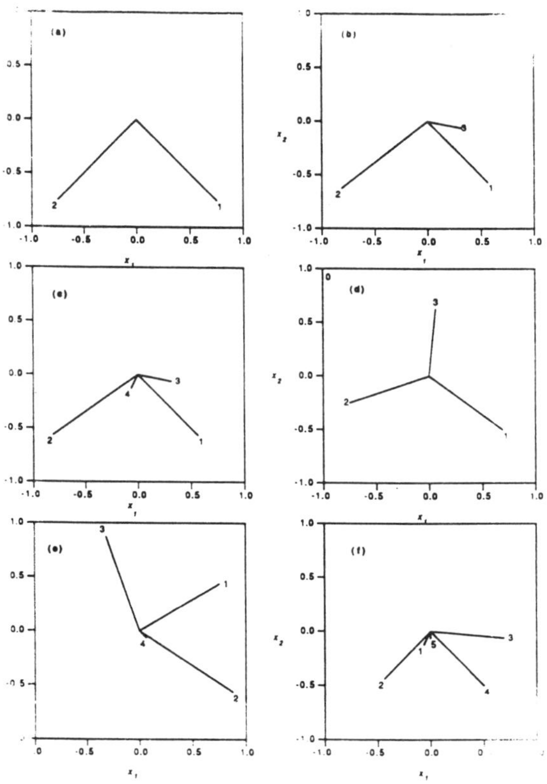

Figure 2: The trained projection directions with normalized magnitudes. (a) Trained by PPL's with 2 hidden neurons. (b) Trained by PPL's with 3 hidden neurons. (c) Trained by PPL's with 4 hidden neurons. (d) Trained by BPL's with 3 hidden neurons. (e) Trained by BPL's with 4 hidden neurons. (f) Trained by BPL's with 5 hidden neurons.

where the two major projection directions (normalized from the weights of the first two neurons in the trained PPL networks) consistently indicates the desired projection directions (see Figures 2(a),(b),(c)). In these figures, each enumerated 2-D vector denotes the projection direction in the $[x_1, x_2]$ 2-D space associated with each enumerated hidden neuron, and the length of the vector denotes the normalized (by the variance of $f_k()$ over all the training patterns) magnitude of β_k corresponding to the same neuron. On the other hand, the networks trained by BPL could only construct the correct 3-D surface of $g^{(1)}$ with at least 3 hidden neurons, and the projection directions are not consistent when the number of hidden neurons increases (due to the implicit equal-importance assumption of each hidden neuron). Figures 2(d),(e),(f) show the projection directions and the corresponding normalized magnitude of the hidden neurons in the trained BPL network.

To further support the parsimony property, we take $g^{(4)}$ as an illustrative example. Figures 3(a),(b) show the approximated functions trained (from noisy data) by PPL's with 3 and 5 hidden neurons, respectively. Note that the PPL network can well approximate $g^{(4)}$

using 5 hidden neurons. On the other hand, the BPL's cannot reasonably well approximate $g^{(4)}$ using 5 hidden neurons and require approximately 10 hidden neurons to approximate this function under the same training convergence criterion (see Figures 3(c),(d)).

5 Concluding Remarks

We have examined the theoretical and algorithmic similarities/differences of BPL's and PPL's. In our simulation study, PPL's and batch Gauss Newton BPL's have quite comparable training speed, and achieve comparable accuracy for independent test data, but PPL's are more parsimonious. Through a better understanding of these factors, we should be able to better control the operational mechanisms involved in existing learning networks, and create new powerful learning networks.

References

[1] R. A. Becker, J. M. Chambers, and A. R. Wilks. The New S Language: A Programming Environment for Data Analysis and Graphics. Wadsworth & Brooks/Cole 1988.

[2] D.L. Donoho and I.M. Johnstone. Projection-based approximation and a duality with kernel methods. The Annals of Statistics, Vol. 17, No. 1, pp. 58–106, 1989.

[3] J.H. Friedman. Classification and multiple regression through projection pursuit. Technical Report No. 12, Department of Statistics, Stanford University, January 1985.

[4] J. N. Hwang and P. S. Lewis. From nonlinear optimization to neural network learning. In Proc. *24th Asilomar Conf. on Signals, Systems, & Computers*, pp. 985-989, Pacific Grove, CA, November 1990.

[5] M. Maechler, D. Martin, J. Schimert, M. Csoppenszky and J. N. Hwang. Projection pursuit learning networks for regression. in Proc. *2nd Int'l Conf. Tools for AI*, pp. 350-358, Washington D.C., November 1990.

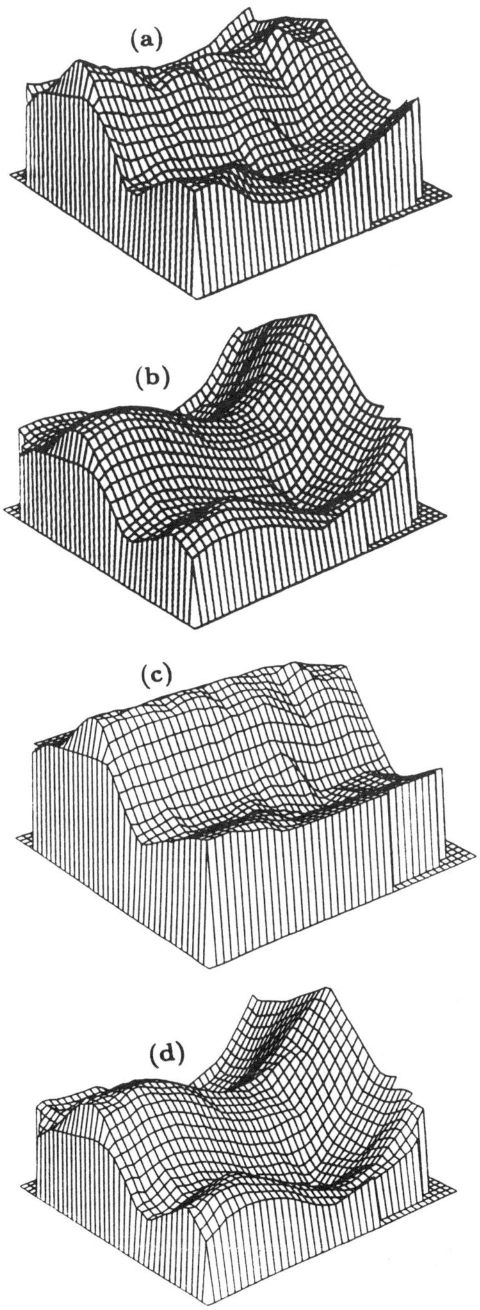

Figure 3: A 3-D perspective plots of the reconstructed $g^{(4)}$ function using 225 noisy training data. (a) The reconstructed surface of a trained PPL network with 3 hidden neurons. (b) The reconstructed surface of a trained PPL network with 5 hidden neurons. (c) The reconstructed surface of a trained BPL network with 5 hidden neurons. (d) The reconstructed surface of a trained BPL network with 10 hidden neurons.

Adaptive Equalization for PAM and QAM Signals with Neural Networks*

Marcia Peng, Chrysostomos L. Nikias† and John G. Proakis

Dept. of Electrical and Computer Engineering
Northeastern University
Boston, MA 02115

† Dept. of Electrical Engineering-Systems
University of Southern California
Los Angeles, CA 90089-2564

Abstract

In this paper, we investigate the application of neural networks to adaptive and blind equalization problems. In particular, the purpose of the paper is twofold: (i) to introduce a new realization structure of a multilayer perceptron with a backpropagation training algorithm and show that it works well for both PAM and QAM signals of any constellation size (e.g. 4–PAM, 8–PAM, 16–QAM and 64–QAM), and (ii) to demonstrate the performance of self-organizing maps as blind equalizers and establish that they are simply not powerful enough for this problem, especially when the intersymbol interference (ISI) is large.

1 Introduction

Equalization techniques for combating intersymbol interference (ISI) due to band-limited channels may be divided into two general categories: linear and nonlinear equalizers. Nonlinear equalizers find use in applications where the channel distortion is too severe for a linear equalizer to handle. In particular, a linear equalizer does not perform well on channels with deep spectral nulls in their amplitude characteristics or with nonlinear distortions. In an attempt to compensate for the channel distortion, the linear equalizer places a large gain in the vicinity of the spectral null and, as a consequence, significantly enhances the additive noise present in the received signal.

Neural Networks(NNWs) are mathematical models of theorized mind and brain activities. The fundamental idea of NNWs is to organize many simple identical processing elements into layers to perform more sophisticated tasks. The properties of NNWs include: massive parallelism; high computation rates; great capability for non-linear problems, continuous adaptation; inherent fault tolerance and ease for VLSI implementation, etc.. All these properties make NNWs attractive for various applications. Recently, Gibson, Siu and Cowan [1]– [4] have effectively utilized multi-layer perceptrons(MLPs) as adaptive equalizers for both finite linear and non-linear channel models. **Their applications, however, have been limited to binary {0,1} or bipolar {-1,1} valued data and real valued channel models, due to the sigmoid nonlinear activation function in a MLP processing element.** There is a great need, however, to develop MLP equalizers that can deal with bigger signal constellations, such as 4–PAM, 8–PAM, as well as with complex channel models and 16–QAM, 64–QAM or 256–QAM signals.

Another application of NNWs to channel equalization was reported by Kohonen et al. [6] [7] based on self-organizing maps(SOMs), which are unsupervised learning, feed forward recall NNW paradigms. From a blind equalization point of view, SOMs have several advantages over MLPs. First, they can learn without supervision, which is the most appealing property for blind equalization. Second, there are no restrictions on the constellations of the concerned signal. That is, they work equally well with 256–QAM as they do with 2–PAM. Finally, SOM-based algorithms are very simple and they have much less computation complexity than any other NNW algorithm. However, after theoretical analysis and numerous simulations, we have found that equalizers based on SOMs work well only when **the distortion is topology-preserving**. In the case of ISI, however, the distortions are far from topology-preserving which implies that the SOMs are simply not powerful enough for practical blind equal-

*This work was supported in part by the National Science Foundation under grant MIP-8904109 and by the Office of Naval Research under contract N-00014-91-J-1124.

1058-6393/91 $01.00 © 1991 IEEE

ization problems.

The paper is organized as follows. In Section 2, the equalization problem in data communication systems is defined. A brief overview of the NNW algorithms for this problem is given in Section 3. A new MLP structure for adaptive equalization of PAM and QAM signals is described in Section 4, and its performance along with the simulation results of SOMs as blind equalizers is demonstrated in Section 5. Section 6 is devoted to conclusions.

2 Equalization problem

Let us consider a (discrete-time) linear channel whose impulse response $\{f(i)\}$ is unknown and possibly time-varying. The input data $\{x(i)\}$ are assumed to be independent and identically distributed (i.i.d) random variables, with non-Gaussian probability density function. From Figure 1, it follows that the model we consider is

$$y(i) = f(i) * x(i), \tag{1}$$

where "$*$" denotes linear convolution and $\{y(i)\}$ is the received sequence. The problem is to reconstruct (or restore) the input sequence $\{x(i)\}$ from the received sequence $\{y(i)\}$ or, equivalently, to identify the inverse filter (equalizer) $\{u(i)\}$ for the channel.

If linear equalizer is used, the output sequence $\{\hat{x}(i)\}$ from the equalizer is given by

$$\hat{x}(i) = u(i) * y(i) = u(i) * f(i) * x(i). \tag{2}$$

So, to achieve

$$\hat{x}(i) = x(i-D)e^{j\theta}, \tag{3}$$

where D is a constant delay and θ is a constant phase shift, it is required that

$$U(\omega) = \frac{1}{F(\omega)} \cdot e^{j(\theta-\omega D)}. \tag{4}$$

We can see from Eq.(4), that if $F(\omega)$ has nulls, $U(\omega)$ will place a large gain around the nulls, and consequently will enhance the additive noise in a real communication system. It is believed that this problem can be solved by using nonlinear equalization technique, which can be described mathematically as a nonlinear multi-dimensional function mapping of the received signal $\{y(i)\}$,

$$\hat{x}(i) = \Psi\left(y(i), y(i-1), \cdots, y(i-m)\right). \tag{5}$$

3 Neural networks

3.1 Multi-layer perceptron

The multi-layer perceptron(MLP), introduced by Rosenblatt [5], is one of the most widely used implementations of NNWs. It comprises a number of nodes which are arranged in layers, as shown in Figure 2. A node receives a number of inputs $x_1, x_2, \cdots, x_n$, which are then multiplied by a set of weights $w_1, w_2, \cdots, w_n$ and the resultant values are summed up. A constant v is added to this weighted sum of inputs, known as the node threshold, and the output of the node is obtained by evaluating a nonlinear(sigmoid) function, $f(\cdot)$, which is called activation function.

The architecture of a perceptron can be described by a sequence of integers $n_0, n_1, \cdots, n_k$ where n_0 is the dimension of the input to the network, and the number of nodes in each successive layer, ordered from input to output, is $n_1, n_2, \cdots, n_k$. In this notation, the MLP produces a nonlinear mapping $g = R^{n_0} \rightarrow R^{n_k}$.

The updating of the connection coefficients of the MLP is done iteratively by using backpropagation(BP) algorithm with the following formula:

$$(w_{i+1}, v_{i+1}) = (w_i, v_i) + \Delta_i \tag{6}$$

and

$$\Delta_i = -(\beta, \eta) \cdot \frac{\mathrm{d}e^2}{\mathrm{d}(w_i, v_i)} + \alpha \cdot \Delta_{i-1}. \tag{7}$$

3.2 Self-organizing feature maps

The topology of self-organizing feature map(SOM), which is introduced by Kohonen in 1981 [8], consists of two layers of nodes, referred to as input layer and output layer, which are fully connected with different connection weights. The inputs to the SOM can be any continuous values, whereas each of the output-layer node represent a pattern class that the input vector may belong to. That means the outputs of SOMs are discrete values, and therefore, the SOM is sometimes also referred to as learning vector quantizer.

The SOM works iteratively as follows. First, find the set of connection coefficients W_g which is the closest to the input vector A_k,

$$\| A_k - W_g \| = \min_{j=1}^{p} \| A_k - W_j \| . \tag{8}$$

Second, perform the following quantization of the output-layer node:

$$b_g = \begin{cases} 1, & \text{if } \|A_k - W_g\| = \min \|A_k - W_j\| \\ 0, & \text{otherwise.} \end{cases} \tag{9}$$

and then move W_g closer to A_k using the equation

$$\Delta W_{ij} = \begin{cases} \alpha(k) \cdot [a_i^k - W_{ig}], & j = g \\ \beta(k) \cdot [a_i^k - W_{ij}], & j \in N_r, j \neq g \\ 0, & j \notin N_r, \end{cases} \qquad (10)$$

where N_r is the topological neighborhood of the winning node b_g which consists b_g itself and its direct neighbors up to the depth, $1, 2, \cdots$, and $\alpha(k)$ and $\beta(k)$ are the learning rates at time k.

4 The MLPs in channel equalization for PAM and QAM signals

As mentioned in Sec.1, the applications of MLP in equalization problems so far, have been limited to binary {0,1} or bipolar {-1,1} valued data and real valued channel models. In this section, we introduce for the first time a **new implementation structure** of MLP which works well with L–PAM ($L > 2$) and N–QAM ($N > 4$) signals.

Looking into a MLP structure, we find out that it is the sigmoid function of the output layer nodes that confines the network outputs to the range $[-1, 1]$. In our equalization problem, the signals are equally spaced and symmetric with respect to either the original point of the coordinate, or to the x and y axes. Thus **we can just scale up the node function of the output layer by a constant factor C which is large enough to cover our maximum signal range**, e.g., $[-15, 15]$ for 16–PAM or 256–QAM signals. So, for the output layer, we have

$$f_M(x) = C \cdot \frac{1 - e^{-\alpha x}}{1 + e^{-\alpha x}}, \qquad (C \geq 1) \qquad (11)$$

as the activation function. For the hidden layers, we still use the sigmoid function

$$f_i(x) = \frac{1 - e^{-\alpha x}}{1 + e^{-\alpha x}}. \qquad (12)$$

The idea of adding another constant α comes from the thought that a smaller α, equivalently, a lower slope in Figure 3, would avoid high vibration, and in turn, decrease the chance of divergence in the course of weight adjustment.

For complex channel models and QAM signals, we use complex connection coefficients to get the weighted sum to which a complex threshold is added. Then the sigmoid functions of the real and the imaginary parts of the threshold added weighted sum are evaluated separately. Again, for the output layer nodes, the outputs are multiplied by a constant C. Using the steepest descent formula (Eq.6,7), we get the adaptation algorithm of our new MLP equalizer which is described in Table 1.

5 Simulations

5.1 MLP equalizer

The communication system employed in our simulations is shown in Figure 4. The equalizer is implemented by the new MLP structure with only one output node. The input data to the system x_i are assumed to be independent of each other. The delayed input sequence x_{i-d}, where d is channel dependent, is used as the training sequence. The performance of MLP equalizers is evaluated by calculating the mean square error (MSE) $E[(x - \hat{x})^2]$ and the average symbol error rate (SER) of the quantizer output. The eye pattern of equalizer outputs around certain number of iterations is shown in Figure 5.

Figure 6 illustrates the performance comparison between MLP and LMS-based linear transversal equalizer with the same number of inputs. The structure (the number of nodes in the hidden layer) of the MLP has been fine-tuned through experiment. The step size μ of the LMS-based equalizer is also optimized (the biggest value without causing divergence). From Fig.6, it appears that the new structure of MLP works no much better, as a channel equalizer, than the simple linear adaptive equalizer. As matter of fact, both methods end up giving similar results.

5.2 SOM equalizer

The SOM employed in our problem is illustrated in Figure 7. In the simulations, we chose a SOM with a linear-array topology for L–PAM. For QAM input signals, we equalize the real and imaginary parts separately. The output MSE performance of SOM equalizer for 16–PAM and 256–QAM constellations is shown in Figure 8 and 9, respectively. From these results, we see that the SOM equalizer seems to work better for QAM than for PAM signals. The reason for this, we believe, is that, the introduction of one more dimension of the QAM signals makes the topology relations more stable than PAM's, i.e., as long as the real and imaginary topology orders are not destroyed simultaneously, it is possible for the corresponding cell to remain in the correct order in a mesh topology. But we must say, even for QAM constellations, the results are not satisfactory. In Figure 10, a 4–PAM signal after channel distortion is illustrated. The distortion has totally destroyed the linear neighbor order of the input sequence. This breaks the topology-preserving assumption, and may weaken the SOM's ability to trace the input correctly, and thus could be the reason for those unsatisfactory results.

6 Conclusions

From the results shown in Figure 5 and 6, we conclude that the new structures of MLP, introduced in Section IV, are capable of dealing with signals beyond the range $[-1, 1]$, as well as with complex data, while maintaining almost the same computational complexity as the original MLP. **Our work provides the first application evidence that MLP is not only a specialized method for pattern classification, but also potentially capable for a broader class of problems**, such as arbitrary functional mapping. However, the performance of new-structured MLP equalizer is not better than that of the LMS-based linear equalizer.

It is desired to apply NNWs in blind equalization. All the existing MLPs, including our new implementation structure, however, do not work as blind equalizers because of their need for a supervised training. The SOMs do work as blind equalizers but only for very small ISI. As shown in Fig. 8 and 9, in the presence of severe ISI, SOMs do not open the eyes of the equalizer output signals.

Acknowledgements

The authors would like to thank Mr. E. Zervas for his generous help in programming of the simulations.

References

[1] S. Chen, G. J. Gibson, C. F. N. Cowan and P. M. Grant, "*Adaptive Equalization of Finite Nonlinear Channels Using Multilayer Perceptrons*", Signal Processing 20, 1990, *pp.*107–109.

[2] G. J. Gibson, S. Siu and C. F. N. Cowan, "*Application of Multilayer Perceptrons as Adaptive Channel Equalizers*". Proc. IEEE ICASSP, Glasgow, Scotland, May 23–26, 1989, *pp.*1183–1186.

[3] G. J. Gibson, S. Siu and C. F. N. Cowan, "*The Application of Nonlinear Structures to the Reconstruction of Binary Signals*". IEEE Trans. Signal Processing, Vol. 39, No. 8, August 1991.

[4] S. Siu, G. J. Gibson and C. F. N. Cowan, "*Decision Feedback Equalization Using Neural Network Structures*". IEE Int. Conf. Neural Networks, London, 1989.

[5] F. Rosenblatt, **Principles of Neurodynamics: Perceptrons and the Theory of Brain Mechanisms**, Washington DC: Spartan Books, 1962.

[6] T. Kohonen, K. Raivio, O. Simula, O. Ventä and J. Henriksson, "*An Adaptive Discrete-Signal Detector based on Self-Organizing Maps*", Proc. of IJCNN, Washington, D.C., Jan. 15–19, 1990, Vol.1.

[7] T. Kohonen, K. Raivio and O. Simula, "*Combining Linear Equalization and Self-Organizing Adaptation in Dynamic Discrete-Signal Detection*", IJCNN-90-San Diego, San Diego, 1990, Vol.1.

[8] T. Kohonen, "*Automatic Formation of Topological Maps in a Self-Organizing System*", In E. Oja and O. Simula (Ed.), Proc. 2nd Scandinavian Conf. on Image Analysis, *pp.* 214–220, 1981.

Table 1. Complex MLP adaptation algorithm

1). Assign small random complex numbers to all the connections and thresholds.

2). Forward propagate inputs through the network:

$$\tilde{a}_{i+1,j} = \sum_{l=1}^{n_i} a_{i,l} \cdot w^*_{i,l,j} + v_{i,j} = \tilde{a}^I_{i+1,j} + j \cdot \tilde{a}^Q_{i+1,j},$$

$$a_{i+1,j} = f\left(\tilde{a}^I_{i+1,j}\right) + j \cdot f\left(\tilde{a}^Q_{i+1,j}\right),$$

where $i = 1, \cdots, M$ (M is the number of layers), $f(\cdot)$ is the sigmoid function, and get the output,

$$\hat{x} = C \cdot a_{M1}.$$

3). Present the training signal to find the output error,

$$\overline{e}_M = e^I_M \left[1 - \left(\hat{x}^I/C\right)^2\right]/C + je^Q_M \left[1 - \left(\hat{x}^Q/C\right)^2\right]/C$$

where $e_M = x_{i-d} - \hat{x}$.

4). Find the backpropagation error,

$$\overline{e}_{ij} = \underline{e}^I_{ij} \cdot \left[1 - (a^I_{ij})^2\right] + j \cdot \underline{e}^Q_{ij} \cdot \left[1 - (a^Q_{ij})^2\right],$$

where

$$\underline{e}_{i,j} = \sum_{l=1}^{n_{i+1}} w_{i.j.l} \cdot \overline{e}_{i+1,l}.$$

5). Adjust connections and thresholds:

$$w_{i,j,k}(n+1) = w_{i,j,k}(n) + \eta \cdot \overline{e}^*_{i+1,j}(n) \cdot a_{ij}(n),$$

$$v_{ij}(n+1) = v_{ij}(n) + \beta \cdot \overline{e}_{ij}(n).$$

where "*" denotes conjugate operator. The momentum term can also be added.

6). Back to step 2.

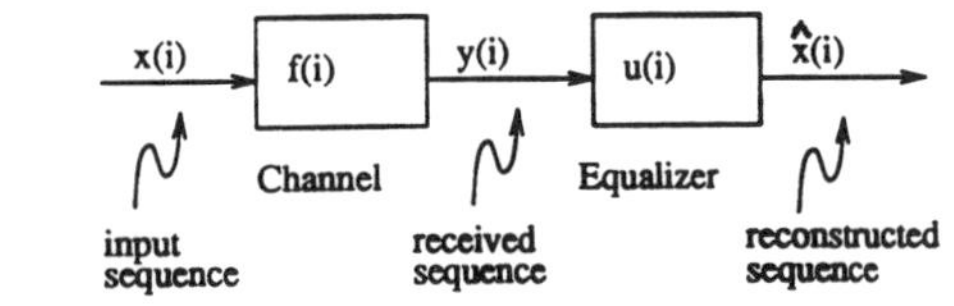

Figure 1. Simplized communication system.

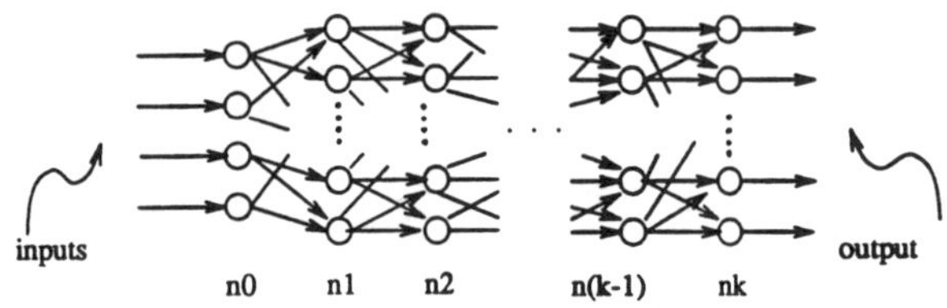

Figure 2. Multi-layer perceptron architecture.

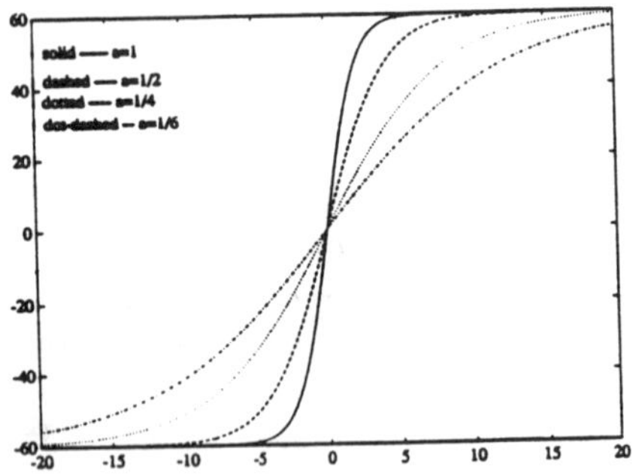

Figure 3. Sigmoid function with different a.

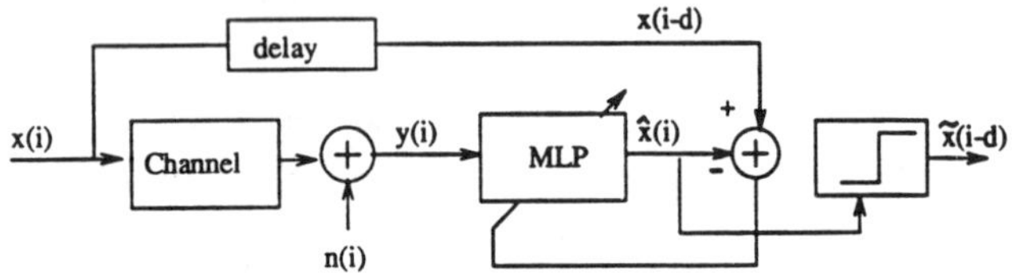

Figure 4. Communication system for simulation.

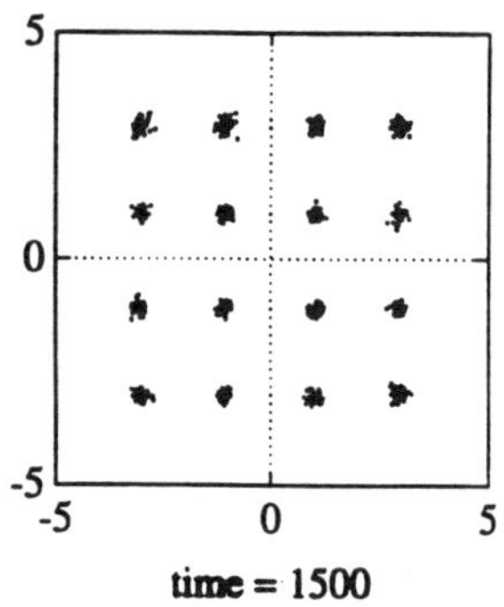

Figure 5. Eye pattern of 1000 equalizer outputs.

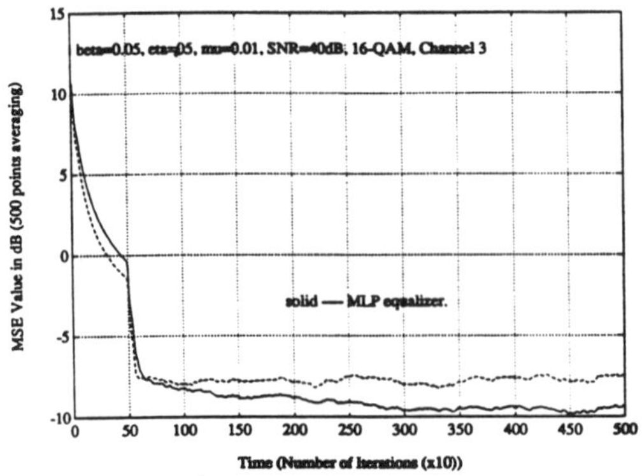

Figure 6. MSE comparison of MLP and LMS equalizers.

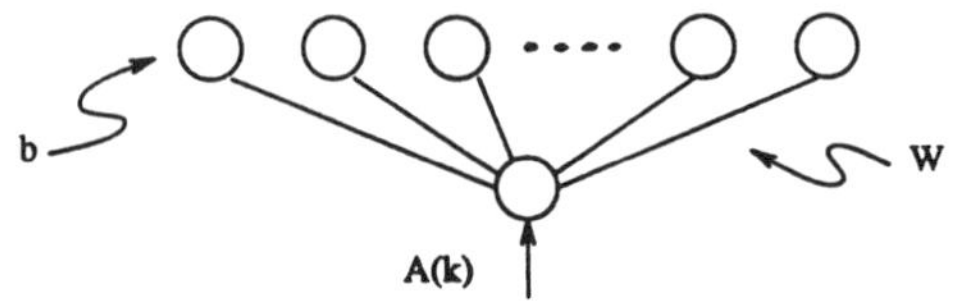

Figure 7. SOM architecture.

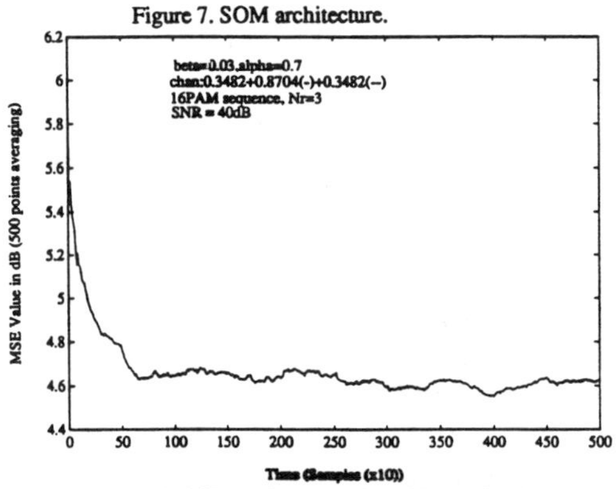

Figure 8. SOM performance for 16-PAM signal.

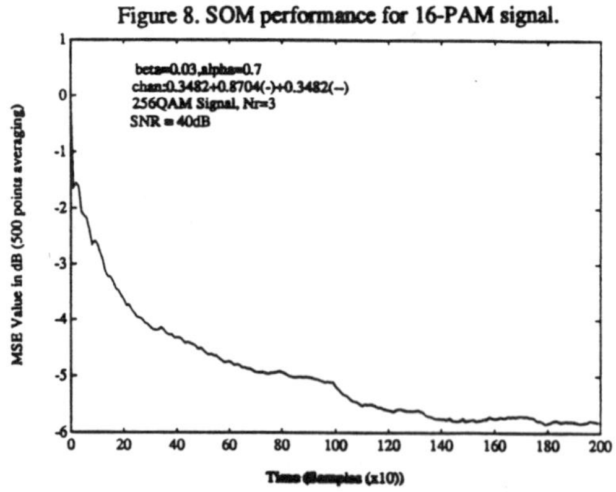

Figure 9. SOM performance for 256-QAM signal.

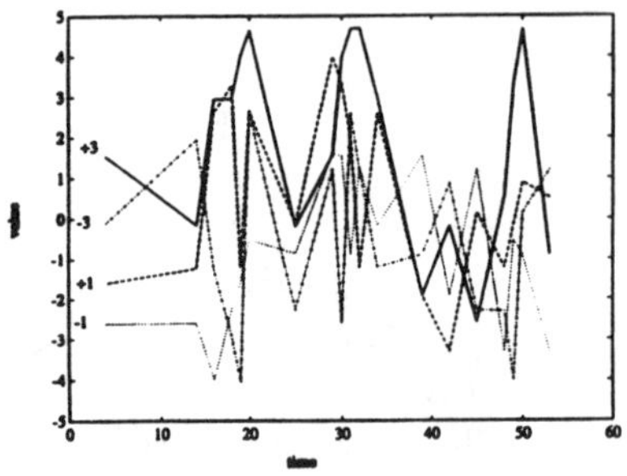

Figure 10. Channel distortion of 4-PAM signal.

Integrated Models of Signals and Background for an HMM/Neural Net Ocean Acoustic Events Classifier

William Y. Huang
Code 535
Marine Sciences and Technology
Naval Ocean Systems Center
San Diego, CA 92152–5000

Richard C. Rose
Speech Systems Technology
Lincoln Laboratory, M.I.T.
Lexington, MA 02173

Abstract

This paper investigates the use of Hidden Markov models (HMM's) for the classification and detection of ocean acoustic events in a nonstationary ocean background. A statistical formalism is described for integrating models for dynamic acoustic events and ocean background into a unified statistical framework. In this framework, both signal processes and background processes are modeled as HMM's, and signal classification is performed by obtaining the likelihood of a corrupted observation sequence through a combined state space of signal and background. Techniques are presented for estimating the acoustic event model parameters from training exemplars that are observed in these difficult background conditions. Finally, a novel neural network technique is proposed for the automatic learning of the nonlinear mechanism through which signal and background observations interact. Experimental results are presented.

1 Introduction

The ocean acoustic events that are of interest in this work can generally by characterized as short duration non-stationary events whose spectral energy evolve according to some characteristic temporal structure. The detection and classification of these acoustic events in an ocean environment is complicated by the presence of background signals that are not well modeled as traditional wideband or impulsive noise processes. In fact, the ocean background may itself contain acoustic events which are similar in nature to those events that we are trying to detect. Existing techniques that have been developed for ocean signal classification do not explicitly account for the desired signal having been observed in this difficult ocean background environment [1, 2, 3, 4, 5]. Failure to do so, however, can result in severe performance degradation, especially when a significant mismatch in the background characteristics exists between the training and testing of the classifier.

The principal motivation for applying HMM techniques to classification and detection of acoustic events is that they provide a means for temporal integration of short–time frame based spectral measurements. When temporal information is an important part of the signal representation, as is the case in ocean acoustic events [6], frame based static classifiers can provide poor event classification performance. This point was illustrated by a study comparing the performance of selected static pattern classifiers in classifying vowels sounds as spoken by a large population of speakers[7]. It was often the case in this study that a classifier that achieved a very low classification error rate when classifying independent speech frames, achieved a very high error rate when classifying the entire utterance. Hence, if the signal model suffers from an impoverished representation of temporal information in a signal, the performance of the resulting classifier is also likely to suffer. This issue has been addressed in previous work by applying heuristic rules [8] or neural networks with time delayed inputs [9, 2]. Recently, hidden Markov models have been applied with some success to the problem of ocean acoustic event classification [10, 11].

The principal contribution of this paper is to extend the definition of the HMM to incorporate the effects of ocean background. By modeling component sources, the system developed here provides a robust way to train and classify signals under differing background conditions. In Section 2, the form of the model is introduced. In Section 3 a maximum likelihood technique for estimating the acoustic event model parameters is introduced, along with a proposed hybrid neural network approach for estimating the process of signal corruption by background. Finally, in Section 4 a set of experiments is performed to evaluate the effectiveness of the approach in detecting an ocean acoustic event in the presence of actual ocean background.

2 Modeling Assumptions

We define an ocean acoustic event $a^i \in \{A\}$, $i = 1, \ldots, M$, taken from a set of possible events $\{A\}$. It is assumed that an acoustic event is produced with prior probability $P(a)$ and that there is an acoustic channel which produces D dimensional signal vectors, $X = (\vec{x}_1, \vec{x}_2, \ldots, \vec{x}_T)$ with probability $P(\vec{x} \mid a)$. We depart from the traditional HMM model development by assuming that the signal vectors are observed in the presence of of an ocean background process b which gives rise to D dimensional background observation vectors Z. The output sequence $Y = (\vec{y}_1, \vec{y}_2, \ldots, \vec{y}_T)$ is then observed as a component–wise function of signal

1058-6393/91 $01.00 © 1991 IEEE

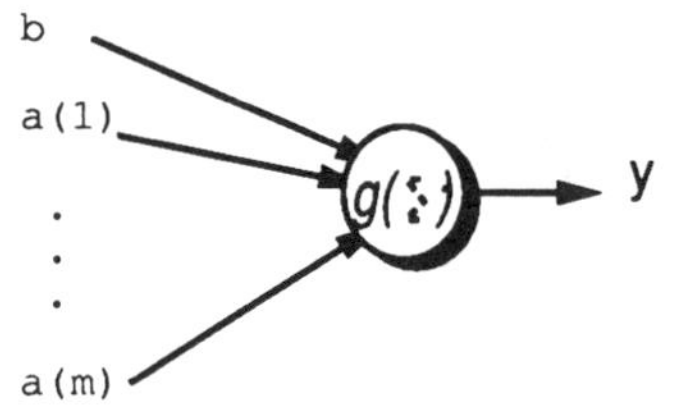

Figure 1: The observed signal, y, is a composite of signal from a background, b, and M signal generators, a^1, a^2, ..., a^M

and background, $y_t = g(z_t, x_t)$. We consider those functions $g(\cdot)$ for which the equation $y = g(z, x)$ define a one dimensional contour in the $x - z$ plane.

Both the signal and the background processes are represented by HMM models. The choice of the topology of the HMM models that are used for signal and background was made experimentally. The background HMM is a 4 state fully connected model, and the signal event model is a N state left–to–right model containing $N - 1$ non–null states and a "null state" that always emits a zero, to account for the "no signal" condition. The definition of this null signal state allows the background model parameters to be estimated from those observations that have been classified as representing background and not signal. The observation probabilities for all states, both signal and background, consist of single Gaussian densities.

The goal in acoustic event classification is to choose that event $\hat{a}$ by maximizing $P(a|Y)$, which from Bayes rule,

$$P(a \mid Y) = \frac{P(Y \mid a)P(a)}{P(Y)}, \tag{1}$$

is equivalent to maximizing $P(Y \mid a)P(a)$. Estimating $P(Y \mid a)$ is accomplished using a probabilistic HMM to represent the acoustic event a. The prior probability $P(a)$ is estimated from higher level non–acoustic source of knowledge. These higher level hierarchical sources of knowledge have been shown in [6] to be critical in acoustic event classification by humans. The success of HMM's in continuous speech recognition is partly attributable to the ability of HMM's to combine these hierarchical sources of knowledge. It is expected that HMM's will provide similar benefits in the area of ocean acoustic event classification.

3 Estimating Model Parameters

3.1 Maximum Likelihood Formulation

The ML parameter estimation employed here is based on Rose et al. [12] and is similar to approaches taken in [13] and [14]. The noise corrupted observations Y arise from underlying state sequences $I = (i_1, i_2, \ldots, i_t)$, of the signal HMM, and $J = (j_1, j_2, \ldots, j_T)$ of the background HMM. The likelihood of the output sequence given signal model λ, which consists of the mean $\vec{\mu}_k$ and standard deviation $\vec{\sigma}_k$ of the Gaussian HMM observation probabilities $p_a(\vec{x}_t \mid k)$ and transition probabilities $p_a(k \mid l)$ is given as

$$P(Y \mid \lambda) = \sum_I \sum_J \oint_C P(X, Z, I, J \mid \lambda) dX dY \tag{2}$$

where the summation is over all possible state sequences in the signal–background state space, and the notation $\oint_C$ refers to the integral along the contour $\mathcal{C}_t$ in the signal background observation space defined by $y_t = g(x_t, z_t)$. Expanding the observation probability in terms of the joint probability of the hidden state sequences and hidden data sequences allows us to isolate pertinent terms relating to the signal model parameters. The complete data likelihood in Eq. 2 is given as

$$P(X, Z, I, J \mid \lambda) = \prod_{t=1}^{T} p_a(i_{t+1} \mid i_t) p_a(\vec{x}_{t+1} \mid i_t) p_b(j_{t+1} \mid j_t) p_b(\vec{z}_{t+1} \mid j_t) \tag{3}$$

Given an initial estimate of the acoustic signal model parameters λ, and following the method of Baum et al. [15] it is possible to find a new set of model parameters $\hat{\lambda}$ such that $P(Y \mid \hat{\lambda}) \geq P(Y \mid \lambda)$. This is done by maximizing the auxiliary Q function

$$Q(\lambda, \hat{\lambda}) = \sum_I \sum_J \oint_C P(X, Y, I, J|\lambda) \log P(X, Y, I, J|\hat{\lambda}) dX dY. \tag{4}$$

In our simulations the probabilities obtained in the forward backward algorithm were replaced by a state sequence produced by the process of Viterbi training. Viterbi decoding in this context involves selecting the single path through the signal–background state space illustrated by the diagram in Figure 2 that maximizes $P(Y \mid \lambda)$. In this case, the summation over all possible state sequences in Eq. 2 is replaced by a max over all I and J.

Space does not permit a detailed description of the steps leading to the expressions for the ML parameter estimates. Taking the partial derivative of Equation 4 with respect to the signal component mean yields the estimate

$$\hat{\mu}_{a_i, ML} = \frac{\sum_t \sum_j P(i_t, j_t|Y, \lambda) E(\vec{x}_i|\vec{y}_t, i_t, j_t, \lambda)}{\sum_t \sum_j P(i_t, j_t|Y, \lambda)}, \tag{5}$$

where

$$E(\vec{x}_t|\vec{y}_t, i_t, j_t, \lambda) = \frac{\oint_C \vec{x}_t p_a(\vec{x}_t \mid i_t) p_b(\vec{z}_t \mid j_t)\, d\vec{x}_t d\vec{y}_t}{\oint_C p_a(\vec{x}_t \mid i_t) p_b(\vec{z}_t \mid j_t)\, d\vec{x}_t d\vec{y}_t}, \tag{6}$$

and $P(i_t = i, j_t = j \mid Y) = 1$ if the optimum Viterbi path passes through states i and j at time t, and equals 0 otherwise.

The contour integral depends on the definition of the noise corruption function $g()$. While the choice this function can be very general, not all choices of $g()$ will lead to a contour integral that has a closed form solution. For instance if $g()$ represents an additive function of signal and background

$$g(x, z) = x + z, \tag{7}$$

then, using the notation x to refer to a single component of the vector $\vec{x}$, it can be shown that $E(x \mid y)$ of Eq. 5 is

$$E(x|y) = \frac{\sigma_a^2}{\sigma_b^2 + \sigma_a^2}\left[y + \left(\frac{\sigma_b^2}{\sigma_a^2}\mu_a - \mu_b\right)\right]. \tag{8}$$

3.2 A Neural Net Mapping Function

For spectral parameters, the additive assumption of Eq. 7 is inaccurate. If the observations are averages of N magnitude squared spectral frames, then the noise corrupted observation probability is non–central χ^2 distributed with $2N$ degrees of freedom. and parameters $\sigma^2 = \sigma_b^2 + \sigma_a^2$ and $\delta = \mu\sqrt{2N}$, where $\mu = \mu_b + \mu_a$ [16]. No closed form solution for $E(x|y)$ has been found for this case; nor have closed form solutions been found for other important noise corruption functions. Numerical evaluation of the integrals required for $E(x|y)$ can involve costly series expansions. Reasonably accurate approximations can be obtained using the trapezoidal rule and approximating the non–central χ^2 density with 50 summands [16]. However, it may be simpler and faster to approximate $E(x|y)$ by training the parameters of a general mapping function using actual observations. Neural nets can be trained using a minimum mean squared error criteria to approximate a mapping from clean to noise corrupted observations that converges with infinite data to $E(x|y)$ [17]. This is true for a broad class of networks including the Multi–Layered Perceptron (MLP) trained using back–propagation, and the Radial Basis Function (RBF). This suggests the use of an HMM–neural net hybrid system where the HMM provides the temporal decoding and the neural network performs the mapping from the space of uncorrupted signal vectors to the space of observable noise corrupted observation vectors.

3.3 Algorithm Summary

The Viterbi lattice that results from the composition of signal and background HMM's is multi–dimensional (Fig. 2). Viterbi decoding on a multi–dimensional lattice is a straight–forward extension of the standard 2–D case. The complete algorithm, summarized in Fig. 3, has one Expectation Maximization (EM) estimator embedded in another. Convergence is guaranteed since $P(Y)$ cannot decrease at any step in Fig. 3, by virtue of the properties of Viterbi decoding and ML estimation.

However, if neural nets are used to estimate $E(x|y)$, convergence in the inner EM estimator is not guaranteed. The system illustrated in Fig. 4, simulates the ocean environment internally during training. The internal simulation is necessary to generate training data for the neural net mapping function. Neural net training for estimating $E(x|y)$ has the problem shown

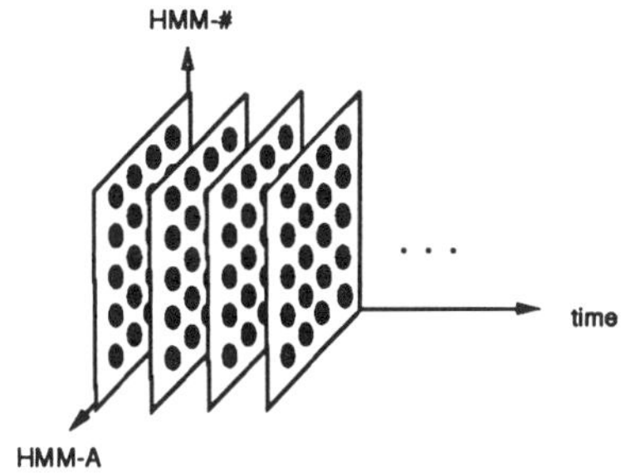

Figure 2: Multi-dimensional Viterbi lattice. In this example, an HMM for background (labeled b), containing 4 nodes, and an HMM for source a, with four nodes, formed a Cartesian product with 16 possible states (or state combinations) per time frame. Viterbi decoding on this lattice is equivalent to decoding on a standard 2–D lattice with 16 nodes.

1. Estimate $P(i_t, j_t \mid Y)$ via Viterbi decoding
2. Estimate the probability density functions for signal and background
   (i) Compute $\hat{\mu}_{b,ML}$, $\hat{\sigma}_{b,ML}$ for decoded "non–signal" observations
   (ii) Compute $\hat{\mu}_{a,ML}$, $\hat{\sigma}_{a^{(1)},ML}$
3. Obtain new mapping function based on $g()$ and new source densities
4. Go to 2 until convergence in source densities
5. Go to 1 until overall convergence is achieved

Figure 3: Training algorithm summary.

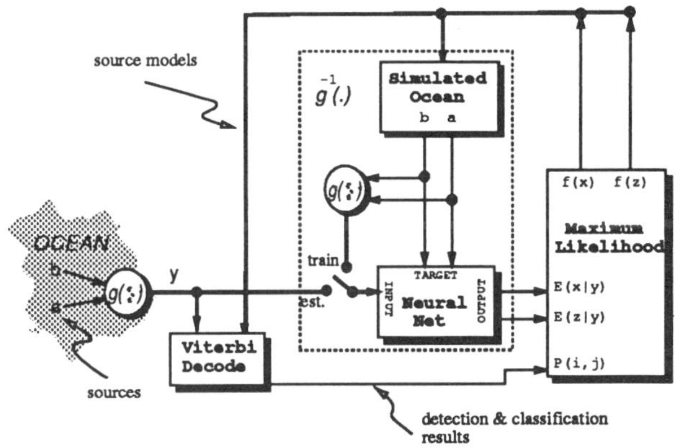

Figure 4: Algorithm of the HMM–neural net hybrid system. The input to the neural net switches to simulated ocean during neural net training, and to observed ocean when estimating source model parameters.

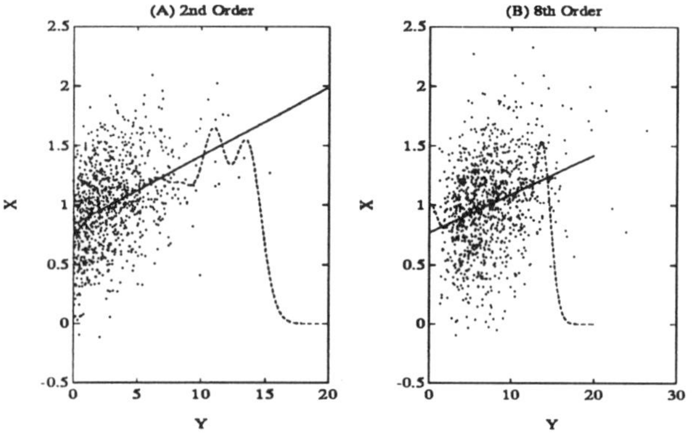

Figure 5: Scatterplot for 1000 samples of signal (a) vs. its (A) 2nd order non–central χ^2–distribution (a single frame of magnitude–squared spectra), and (B) 8th the order case (4 spectral frame averages). The corresponding estimate of $E(x|y)$ are plotted for the LMS (solid line) and RBF (dashed line) techniques. $\mu_a = 1$, $\sigma_a^2 = .14$, $\mu_b = .3$, $\sigma_b^2 = .5$

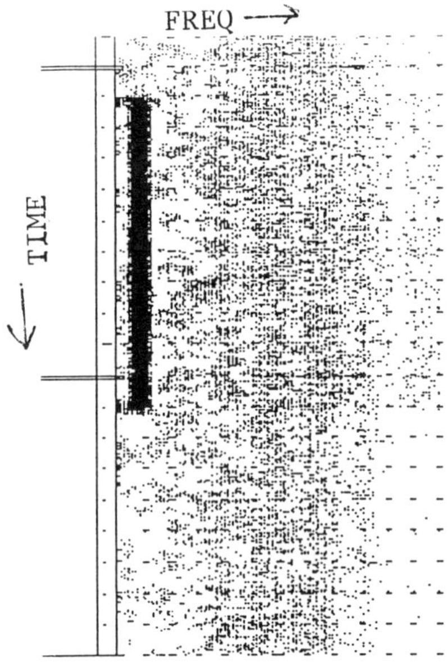

Figure 6: Time vs. frequency (via wavelet decomposition) for a sample in the test set. Horizontal ticks mark .1 sec intervals. Solid horizontal long–dashes on the left side are the Viterbi decoder's estimates of the starting and stopping boundaries for signal "A" (a 10 msec broadband event).

Table 1: Summary of experiments performed on additive normal components. Results labeled with "LMS" were obtained using neural net.

TRAINING SNR	*GAIN*	*FALSE ALARM RATE* (sec^{-1})	*MISS* (*% ERROR*)
high	0	0	0%
low	0	.041	0%
low (LMS)	0	0	0%
high	-6dB	.123	5%
low	-6dB	.165	20%
low (LMS)	-6dB	.082	20%

in Fig. 5. The neural net does not perform well in regions with limited training data. These regions from a mismatch between the real the simulated ocean. This is illustrated by the noise in the Radial Basis Function (RBF) output for large y's in Fig. 5. For large values of y, $E(x|y)$ is nearly linear. A linear mapping, implemented as using the LMS algorithm, does a good job in this case, and in the case of approximating Eq. 8, which is linear to begin with. However, the linear estimator is inaccurate for small values of y, especially in the second order case in Fig. 5, where the RBF correctly estimated a slight downward nonlinearity. The linear LMS estimator is used for the work presented here.

4 Experimental Results

The purpose of the preliminary synthetic data experiment described in this section is to validate the algorithms presented above. The DARPA Standard Phase I database is designed to test conventional classifiers. It does not have the adverse background conditions that this algorithm is designed to handle. Therefore, an adverse condition testing set was constructed by mixing different parts of the Phase I dataset. The Phase I signals used were signal "A", a 10msec wideband pulse, "E", a 1 second low frequency tonal, and the "quiet ocean" background. The testing set consists of 24.3 seconds of data, containing 10 samples of E, and 20 samples of A. The E samples overlapped A samples 9 times, to simulate an adverse, highly nonstationary background. The signals were mixed into the background with a gain of either 0 or -6dB. Fig. 6

shows a sample from this testing set. The solid horizontal long–dashes at the left are the Viterbi decoder estimates of the starting and ending times for signal A events that it detected. The Viterbi decoder used models that were trained either offline, in a high SNR and no interference condition, or in adverse conditions that are similar to the test set.

To evaluate the robustness of the neural net approximation of $E(x|y)$, relative to the closed form solution, this first set of experiments involve mixing signals in background *after* spectral decomposition. This was necessary since Eq. 8 works only for additive normal components, and it is found that spectral components are not well approximated by Eq. 7.

Tab. 1 summarizes the results on this database. Models were trained either under similar adverse conditions, or they were trained with no noise and with no other signal (high "training SNR" in Tab. 1). The results in Tab. 1 show that it is possible to (1) train models of signal A under adverse background conditions, and to (2) train models of A under high SNR conditions, and then detect and classify A under adverse background conditions.[1]

5 Summary

The techniques presented here uses Hidden Markov Models and the Maximum Likelihood (ML) formalism to address the issues of temporal structures and nonstationary backgrounds. Temporal structures provide an important cue for human acoustic events classifiers [6]; but is not well exploited by static, frame-based classifiers. HMM's can provide a succinct model for temporal structures. The ML technique integrates HMM models of component processes to provide a robust way to handle highly nonstationary background. In the experiments presented here, signal models were trained offline under high SNR conditions, and then used to detect and classify signal events under an adverse, highly nonstationary background. The experiments also demonstrates an ability to train for signal parameters when a training set of isolated events is not available.

References

[1] S. Beck, L. Deuser, R. Still, and J. Whiteley, "A hybrid neural network classifier of short duration acoustic signals," in *Internation Joint Conference on Neural Networks*, pp. I-119–124, IEEE, July 8–12 1991.

[2] L. I. Perlovsky, "Model based classification of transient signals using the MLANS neural network," in *IEEE Conference on Neural Networks for Ocean Engineering*, pp. 239–246, 1991.

[1]Tab. 1 shows that the use of neural net (LMS) to estimate $E(x|y)$ lead to better results than when the closed form solution is used. This is probably a procedural artifact: The closed form system was trained to five iterations, yielding the results reported in Tab. 1. Neural net training was initialized with the final parameters from the closed form training. $P(Y)$ increased during each of 5 neural net iterations. Therefore, the results reported for neural net training is slightly better. Tests using Radial Basis Function (RBF) and Multi-Layered Perceptron (MLP) had very high false alarm rates, for reasons discussed above.

[3] J. C. Solinsky and E. A. Nash, "Neural-network performance assessment in sonar applications," in *Proceedings of the IEEE Conference on Neural Networks for Ocean Engineering*, pp. 1–12, IEEE, August 1991.

[4] D. Montana and K. Theriault, "Neural-network-based classification of acoustic transients," in *Proceedings of the IEEE Conference on Neural Networks for Ocean Engineering*, pp. 247–254, IEEE, August 1991.

[5] F. L. Casselman, D. F. Freeman, D. A. Kerrigan, S. E. Lane, N. H. Millstrom, and W. G. Nichols, "A neural network-based passive sonar detection and classification design with a low false alarm rate," in *Proceedings of the IEEE Conference on Neural Networks for Ocean Engineering*, pp. 49–55, IEEE, August 1991.

[6] J. H. Howard and J. J. O'Hare, "Human classification of complex sounds," *Naval Research Review*, vol. xxxvi, pp. 26–31, 1984.

[7] W. Huang and R. Lippmann, "HMM speech recognition with neural net discrimination," in *Advances in Neural Information Processing Systems 2* (D. S. Touretzky, ed.), pp. 194–202, San Mateo, CA: Morgan Kaufmann, 1990.

[8] D. W. Cottle and D. J. Hamilton, "All neural network sonar discrimination system," in *Proceedings of the IEEE Conference on Neural Networks for Ocean Engineering*, pp. 13–19, IEEE, August 1991.

[9] Y.-H. Pao, T. L. Hemminger, D. J. Adams, and S. Clary, "An episodal neural-net computing approach to the detection and interpretation of underwater acoustic transients," in *Proceedings of the IEEE Conference on Neural Networks for Ocean Engineering*, pp. 21–28, IEEE, August 1991.

[10] M. K. Shields and C. W. Therrien, "A hidden Markov model approach to the classification of acoustic transients," in *Procedings of the International Conference on Acoustics, Speech and Signal Processing*, pp. 2731–2734, April 1990.

[11] J. P. Woodard, "Modeling and classification of acoustic transients by speech recognition techniques," *Journal of Underwater Acoustics*, Oct 1990.

[12] R. Rose, J. Fitzmaurice, E. Hofstetter, and D. Reynolds, "Robust speaker identification in noisy environments using noise adaptive speaker models," in *Proceedings of the International Conference on Acoustics, Speech and Signal Processing*, May 1991.

[13] A. Nádas, D. Nahamoo, and M. A. Picheny, "Speech recognition using noise-adaptive prototypes," *IEEE Transactions in ASSP*, vol. 37, pp. 1495–1503, 1989.

[14] A. P. Varga and R. E. Moore, "Hidden Markov model decomposition of speech and noise," in *International Conference on Acoustics, Speech, and Signal Processing*, IEEE, 1990.

[15] L. E. Baum, T. Petrie, G. Soules, and N. Weiss, "A maximization technique occurring in the statistical analysis of probabilistic functions of Markov chains," *Annals of Mathematical Statistics*, vol. 41, no. 1, pp. 164–171, 1972.

A VLSI Array Architecture for the on-line training of Recurrent Neural Networks.

George Kechriotis and Elias S. Manolakos
Electrical and Computer Engineering Dept.
Northeastern University, Boston, MA 02115

Abstract

For Recurrent Neural Networks that need continuous training in order to adapt to a rapidly changing environment, the Real Time Recurrent Learning (RTRL) algorithm [1], has shown a lot of potential in many application domains. By careful reformulation we have been able to localize the global dependencies and systematically match both phases (retrieving and learning) of the algorithm to the universal systolic ring array architecture recently proposed by Kung, Hwang and Vlontzos [2]. The control complexity and memory requirements of the Processing Elements (PEs) remain reasonably low.

1 Introduction - Motivation.

Recurrent Neural Networks (RNNs), the most general case of neural networks, in which every unit can be connected to any other unit, have shown the ability to implement dynamical systems of arbitrary complexity.

RNNs can be trained to solve benchmark problems like the XOR, to learn state space trajectories and follow limit cycles of dynamical systems. Other examples of applications of RNNs found in the literature include speech processing, encoding sequential structures for learning simple grammar rules and the emulation of a Turing machine for parenthesis balancing.

For the training of the RNNs several algorithms have been proposed [3]. The most powerful algorithm seems to be the *Real Time Recurrent Learning Algorithm* (RTRL) proposed by Williams and Zisper [1], [3], which is numerically stable and can be used to update the weights of the RNN in real-time. However, the computations and storage requirements of the RTRL are considerably large. For a network of n units the computations required belong to the class of $O(n^4)$. In this paper we propose a VLSI *ring array* for the efficient implementation of both the learning and retrieving phases of RTRL, that reduces the computational requirements to $O(n^3)$.

2 The Real-Time Recurrent Learning Algorithm.

In this section we briefly describe the RTRL algorithm. For a more detailed discussion on the RTRL see [1, 3]. Assume that the network consists of n fully connected semilinear units and m input units. Let $w_{ij}[t]$ be the weight (i.e the synapse strength) of the connection from the jth to the ith unit at time t. The net input to the kth unit $s_k[t+1]$, at time $t+1$ is given by the weighted average of the activations $y_l[t]$ of all the other units in the network and the external inputs $x_l^{net}[t]$ at time t. In matrix form :

$$\vec{s}[t+1] = \mathbf{W}[t]\vec{\mathbf{x}}[t] \tag{1}$$

where $\mathbf{W}[t] \in \mathbf{R}^{n\times(n+m)}$ is the *weight* matrix, with elements $w_{ij}[t]$, and $\vec{\mathbf{x}}[t] \in \mathbf{R}^{n+m}$ is a vector whose n first elements are the activations of the n units, and its m last elements are the external inputs to the network. The output of the kth unit is given by:

$$y_k[t+1] = f(s_k[t+1]) \tag{2}$$

where $f(\cdot)$, is a nonlinear function. Assume w.l.o.g that for every unit in the network there exists a specified desired value $d_k[t]$ at time t. Then, the *overall network error* at time t is defined as:

$$J[t] = -\frac{1}{2}\sum_{k=1}^{n}(y_k[t]-d_k[t])^2 = -\frac{1}{2}\sum_{k=1}^{n}e_k^2[t] \tag{3}$$

The error is minimized by moving along a constant positive multiple of the performance measure gradient :

$$w_{ij}[t+1] = w_{ij}[t] + \Delta w_{ij}[t+1] = w_{ij}[t] + \alpha\frac{\partial J(t+1)}{\partial w_{ij}[t]} \tag{4}$$

where, $i = 1,2,\ldots,n$ and $j = 1,2,\ldots,n+m$. Defining, as in Williams [3], the terms : $p_{ij}^k[t+1] = \frac{\partial y_k[t+1]}{\partial w_{ij}[t]}$, $k = 1,2,\ldots,n$, and differentiating (3) one can get that:

$$\frac{\partial J[t+1]}{\partial w_{ij}[t]} = \sum_{k=1}^{n} e_k[t+1]p_{ij}^k[t+1] \tag{5}$$

1058-6393/91 $01.00 © 1991 IEEE

Step	Multiplications	Additions
1	$n(n+m)$	$n(n+m-1)$
2	–	n
3	$n^2(n+m)(n+2)$	$n^3(n+m)+n$
4	$n^2(n+m)$	$n(n+m)(n-1)$
5	$n(n+m)$	$n(n+m)$

Table 1: Number of Multiplications and Additions per step of the RTRL algorithm.

Differentiating equations (1) and (2) for the network dynamics yields the following system of recursive equations :

$$p_{ij}^k[t+1] = f'(s_k[t+1])(\sum_{l=1}^{n} w_{kl}[t]p_{ij}^l[t] + \delta_{ik}x_j[t]) \quad (6)$$

with the initial condition : $p_{ij}^k[t_1] = 0$. Here δ_{ik} is the Kroneker delta and $x_j[t]$ is the jth element of the vector $\vec{\mathbf{x}}[t]$. If the activation function $f(\cdot)$ is the *sigmoid* nonlinearity, then it holds that:

$$f'(s_k[t+1]) = y_k[t+1]\cdot(1-y_k[t+1]) \quad (7)$$

The RTRL Algorithm is summarized below:

1 : Forward Step: (Matrix–Vector multiplication) Compute $y_k[t+1]$, from the weights $w_{ij}[t]$, the previous outputs $y_k[t]$ and the net inputs $x_k^{net}[t]$, using equations (1) and (2).

2 : Error vector computation: Find $e_k[t+1]$, subtracting $y_k[t+1]$ from $d_k[t+1]$.

3 : Sensitivities computation: Find recursively $p_{ij}^k[t+1]$ from $w_{ij}[t]$, $p_{ij}^k[t]$, $y_k[t]$, $x_k^{net}[t]$, and $y_k[t+1]$, using equations (6) and (6).

4 : Gradients computation: Find $\frac{\partial J[t+1]}{\partial w_{ij}[t]}$, from $e_k[t+1]$ and $p_{ij}^k[t+1]$, using equation (5).

5 : Weights Update: Compute $w_{ij}[t+1]$, from $w_{ij}[t]$ and $\frac{\partial J[t+1]}{\partial w_{ij}[t]}$, using equation (4).

If we assume that the network consists of n fully connected neuron units, with m external inputs, and w.l.o.g the target outputs are specified for all the n units, then the number of additions and multiplications needed for each of the steps described above is shown in Table 1. As we can see during each time update we have to perform $O(n^4)$ additions and $O(n^4)$ multiplications.

2.1 Matrix formulation of the RTRL Algorithm.

In order to express equation (6) in a matrix format we can rearrange the terms $\{p_{ij}^k\}$ into a vector $\vec{\mathbf{p}}^k$ and the elements $\{w_{ij}\}$ of the weight matrix $\mathbf{W}$ into another vector $\vec{\mathbf{w}}$ as follows:

$$\vec{\mathbf{p}}^k = [\, p_{11}^k p_{12}^k \cdots p_{1,n+m}^k \mid p_{21}^k p_{22}^k \cdots p_{2,n+m}^k \mid \cdots \cdots \mid p_{n,1}^k p_{n,2}^k \cdots p_{n,n+m}^k \,]^T$$

$$\vec{\mathbf{w}} = [\, w_{11}w_{12}\cdots w_{1,n+m} \mid w_{21}w_{22}\cdots w_{2,n+m} \mid \cdots \cdots \mid w_{n,1}w_{n,2}\cdots w_{n,m+n} \,]^T$$

where $\vec{\mathbf{p}}^k$ and $\vec{\mathbf{w}}$ belong to $\mathbf{R}^{(n^2+nm)\times 1}$. Furthermore let $\mathbf{P}$ be the $[(n^2+nm)\times n]$ matrix with columns $\vec{\mathbf{p}}^k$, I.e : $\mathbf{P} = [\vec{\mathbf{p}}^1\vec{\mathbf{p}}^2\ldots\vec{\mathbf{p}}^n]$ and let $\mathbf{X}$ be the block diagonal matrix : $\mathbf{X} = diag(\vec{\mathbf{x}},\vec{\mathbf{x}},\ldots,\vec{\mathbf{x}}) \in \mathbf{R}^{(n^2+nm)\times n}$. Furthermore, let $\mathbf{F} = diag(f_1',\ldots,f_n') = diag(y_1(1-y_1),\ldots y_n(1-y_n))$. Then, the recursive equation (6) that is in the heart of the learning phase becomes:

$$\mathbf{P}[t+1] = (\mathbf{P}[t]\cdot\mathbf{W}_1^T[t] + \mathbf{X}[t])\cdot\mathbf{F}[t+1] \quad (8)$$

where $\mathbf{W}_1$ is the $n\times n$ principal submatrix of $\mathbf{W}$, and the initial condition becomes : $\mathbf{P}[t_1] = \mathbf{0}$. Moreover, the weights update equation becomes :

$$\vec{\mathbf{w}}[t+1] = \vec{\mathbf{w}}[t] + \alpha\cdot\mathbf{P}[t+1]\cdot\vec{\mathbf{e}}[t+1] \quad (9)$$

Therefore the learning phase computations can be expressed in a consecutive matrix–matrix multiplication form, that will be exploited in section 4.

3 Mapping the RTRL Algorithm to an Array Architecture

To systematically derive a VLSI array architecture for the efficient implementation of both the alternating phases of the RTRL we are going to use the *Canonical Mapping Methodology* (CMM) introduced by S.Y. Kung [4]. To transform an algorithm to a Systolic or Wavefront array we have first to express it in Single Assignment Form, i.e in a form where no variable is assigned more than once. This form can be localized by eliminating token *broadcasting* to more than one computational sites. Then we end-up with a locally recursive algorithm that can be represented by a *Dependencies Graph* (DG). A DG vertex models a primitive operation and a directed arc depicts a data/results dependency (precedence relationship). By choosing a *projection direction vector* and a *schedule vector* we can derive *Signal Flow Graphs* (SFG) that provide behavioral level specification for the array Processing Elements (PEs). As we will see shortly, the CMM provides useful geometrical intuition that facilitates the design of systolic/wavefront arrays with desired optimality characteristics.

3.1 The retrieving phase.

In order to perform efficiently the computations associated with consecutive retrieving phases in an one dimensional array, we would like to cascade many instances of the DG for a single matrix vector multiplication. Therefore we have to align the directions of the input and the output data of each DG. For the case of a square weight matrix and a vector consisting only of external inputs (feed forward neural net), Kung, Hwang and Vlontzos [2] have shown to how derive a *ring systolic* array. In our case, the weight matrix $\mathbf{W}$ is an $n \times (n+m)$ matrix and the vector to multiply with consists of the $\vec{\mathbf{y}}[t]$ vector of unit activations concatenated with the external inputs $\vec{\mathbf{x}}^{net}[t]$. This leads to a slightly different DG but following a similar procedure we obtain a modified ring systolic array for the retrieving phase of the RTRL.

The DG when $n = 3$, $m = 2$ (3 units, 2 input lines) is shown in Figure 1a. The oval nodes (Figure 1b) model a simple multiply and accumulate operation and the square ones the computation of the nonlinear activation function. As in [2] we rearrange the ordering of the $\{w_{ij}\}$ elements so that the direction of the inputs $y_i[t]$, $i = 1, 2 \ldots n$ in the DG becomes aligned with that of the outputs $y_i[t+1]$. In this DG, the ith row of the $\mathbf{W}$ matrix is placed on the ith column of the $[w_{ij}]$ data array. Furthermore the ith column of the $[w_{ij}]$ array is circularly shifted up by $i-1$ positions. As we can see the left most node in each row j of the DG, is connected to the right most node of the row $j+m+1$. Since the inputs (from the top) and the outputs (from the bottom) are now aligned, many copies of that DG can be cascaded top-down, with the inputs and outputs perfectly interfaced.

The DG is projected along the vertical direction ($\vec{d} = [i, j]^T = [0, 1]^T$) and the *default schedule* [4] was used. ($\vec{s} = \vec{d} = [0, 1]^T$). The projection direction and schedule hyperplanes are shown at Figure 1a. Projecting along the direction $\vec{d} = \vec{s}$ yields the SFG of Figure 1c. The final *Ring Systolic Array* implementation is shown at Figure 2, with the appropriate number of extra buffers in the spiral communication link. The operation of the ring systolic array can be better understood by following an example, where the operations performed at each schedule time period are described :

t = 1 : The computations $y_1 w_{11}$, $y_2 w_{22}$, $y_3 w_{33}$, are performed by the three PEs and the results are added to update the previous values of the internal accumulators. Concurently every PE transmits its corresponding y input data to the neighboring PE in a *left circular* fashion; PE_3 sends y_3 to PE_2, PE_1 sends y_1 to the first buffer on the spiral link.

t = 2 : All PEs receive their input values from the right neighbor. Especially PE_3 (PE_n in general) receives the external input x_1^{net} from the upper horizontal link. The computations $y_2 w_{12}$, $y_3 w_{23}$, $x_1^{net} w_{34}$ are performed by the three PEs and the results are again accumulated. Every PE transmits it's input data to the left neighbor in a left circular fashion.

t = 3 : PE_3 receives x_2^{net} from outside. (In general PE_n receives inputs from outside during schedule time periods $2, \ldots m+1$). All PEs perform a multiply-accumulate, while they transmit the input data to the left.

...

t = 5 : At the end of this period, the accumulator of PE_i contains the output $s_i[t+1]$, $i = 1, 2, \ldots n$.

t = 6 : PE_i performs the nonlinear operation $y_i[t+1] = f(s_i[t+1])$, $i = 1, 2, \ldots n$.

The next two time periods are optional and they are performed only in the case where a learning phase is to follow the retrieving phase.

t = 7 : Every PE_i, $i = 1, 2, \ldots, n$ is supplied with the target value $d_i[t+1]$ and computes the error $e_i[t+1] = d_i[t+1] - y_i[t+1]$.

t = 8 : Every PE_i, $i = 1, 2, \ldots, n$ computes the derivative $f_i'[t+1] = y_i[t+1] \cdot (1 - y_i[t+1])$

4 The Learning Phase.

While the derivation of the ring systolic array for the retrieving phase was a quite straightforward extension of the ideas presented in [2] and [5], the matching of the learning phase steps in the same ring array is a more challenging and important problem, since most of the computational work in the RTRL algorithm is spent in that phase. As discussed in section 2 the main computational effort in the $\{p_{ij}^k\}$ terms updating can be expressed in matrix-matrix multiplication form (8).

4.1 DG for the consecutive matrix-matrix multiplication.

The sequence of consecutive matrix-matrix multiplications that we would like to compute is of the form $\mathbf{C}[t+1] = \mathbf{C}[t] \cdot \mathbf{B}$ ($\mathbf{C} = \mathbf{A} \cdot \mathbf{B}$ in general). Therefore it is essential to allign the direction of arrival of the input matrix $\mathbf{A}$ (here $\mathbf{C}[t]$), with that of the output matrix $\mathbf{C}$ (here $\mathbf{C}[t+1]$), so that multiple copies of the DG can be cascaded and the overall structure then collapsed to the same one dimensional ring array architecture. It is for that purpose why we rearrange the ordering of the $\{b_{ij}\}$ elements (in our case the weights). Assuming that all three matrices belong to $\mathbf{R}^{3\times3}$, the modified DG is shown at Figure 3. In this DG, the jth column of the $[b_{ij}]$ data array is circularly shifted-up by $j-1$ positions. Although the resulting DG is not totally localized, due to the presence of the spiral communication arcs, the direction of the $\mathbf{C}[t]$ inputs (from the bottom) and the $\mathbf{C}[t+1]$ outputs (from the top) are now parallel.

4.2 Ring systolic array for the learning phase.

Similarly the DG for the calculation of the product $\mathbf{P}[t+1] = \mathbf{P}[t] \cdot \mathbf{W}_1^T$ is shown at Figure 4. Note here that $\mathbf{P}[t+1]$ corresponds only to the intermediate result and not the final value of the updated $\mathbf{P}$ matrix. Choosing the projection direction $\vec{\mathbf{d}} = [i,j,k]^T = [1,0,0]^T$ and the default schedule $\vec{\mathbf{s}} = \vec{\mathbf{d}} = [1,0,0]^T$, we obtain the two-dimensioanl SFG of Figure 5. To obtain an one-dimensional systolic array, we project again along the direction $\vec{\mathbf{d}} = [j,k]^T = [1,0]^T$ and we choose again the default schedule $\vec{\mathbf{s}} = \vec{\mathbf{d}} = [1,0]^T$. The resulting ring array is shown at Figure 6. The operation of the array can be better explained by following an example on Figure 8. So let's assume that $n = 3$, $m = 2$ (hence $\mathbf{W}_1^T \in \mathbf{R}^{3\times3}$). Also let p_{ij} denote the ij^{th} element of matrix $\mathbf{P}[t]$, where $i = 1,2,\ldots 15$ (in general $i = 1,2,\ldots n(n+m)$) and $j = 1,2,3$ (in general $j = 1,2,\ldots n$). The operations in the array for every schedule time period are described bellow:

t = 1 : The products $p_{11}w_{11}$; $p_{12}w_{22}$; $p_{13}w_{33}$; are computed in PE_1, PE_2, PE_3 respectively and the results are stored in an accumulator of each PE. At the same time, p_{11} is transmitted to PE_3, p_{12} to PE_1, and p_{13} to PE_2 in a left circular fashion.

t = 2 : The products $p_{12}w_{12}$; $p_{13}w_{23}$; $p_{11}w_{31}$; are computed, and added to the previous values of the accumulators. p_{12}, p_{13} and p_{11} are transmitted once more in a left circular fashion.

t = 3 : At the end of the scheduling period $t_c = n$ in general the values $p_{11}[t]$, $p_{12}[t]$, $p_{11}[t]$, have traveled once around the ring and the new partial results of $p_{11}[t+1]$,$p_{12}[t+1]$ and $p_{13}[t+1]$ have been computed in the corresponding PEs , PE_1, PE_2 and PE_3. Also notice that inside every PE the weights involved in the computations during the first 3 periods are fetched in a circular manner.

t = 4 : ($t = n+1$ in general) $x_1[t] = y_1[t]$ is added to the current value of $p_{11}[t+1]$ in PE_1 (see (8)). Note that all $\{x_i\}$s are available in every PE since they have travelled once around the whole ring array during the retrieving phase.

t = 5 : The products $p_{1i}[t+1] \cdot f_i'$ are computed in parallel in all PEs , $\{PE_i,\ i = 1,2,3\}$

t = 7 : The products $\{p_{1i}[t+1] \cdot e_i[t+1]\}$ are computed in parallel in all PEs , $\{PE_i,\ i = 1,2,3\}$

t = 7,8 : During these steps (at times $n+4$, $n+5$, $\ldots$, $2n+2$ in general) the sum $\sum_{i=1}^{3} p_{1i}[t+1] \cdot e_i[t+1]$ is computed in PE_1. This is accomplished by adding and transmitting the partial sums leftwards from one PE to the other, till we reach PE_1. This accumulation produces Δw_{11} which is now stored in PE_1.

t = 9 : During that period, (period 2n+3 in general) the next row of the $\mathbf{P}$ matrix p_{21}, p_{22}, p_{23} can enter the ring. As before, after $2n+2$ more steps, the update term Δw_{12} becomes available in PE_2, and the quantities $p_{21}[t+1]$, $p_{22}[t+1]$ and $p_{23}[t+1]$ are calculated and stored in the the corresponding PEs. Note that this scheme requires $(n^2+nm)\cdot(2n+2)$ time steps for the calculation of all the Δw_{ij} terms. Moreover, after we have computed all the Δw_{ij} terms we need $n+m$ more steps to update in parallel all the w_{ij}'s, if desired. Thus, the total time for updating the $\mathbf{W}$ matrix once, is $(n^2+nm)\cdot(2n+2)+(m+n) = (2n^2+2n+1)\cdot(n+m)$ schedule time periods.

5 Conclusions

In this paper we have shown how to derive systematically a VLSI systolic array for the RTRL algorithm. Our main goal was to match the RTRL to the ring architecture proposed by Kung, Hwang and Vlontzos that was shown to be generally applicable to many neural network models. Although this task was a straight forward extension for the retrieving phase, skillful reformulation of the learning phase was necessary.

One of the main advantages of our systematic derivation is that if the ring array is proven to be inadequate for real-time performance in certain nonstationary environments, it can be easily scaled up to a mesh where the number of the required computations will be further reduced to $O(n^2)$.

References

[1] D.Zipser R.J. Williams. Experimental analysis of the Real-Time Recurrent Learning algorithm. *Connection Science*, 1(1), 1989.

[2] J.N. Hwang S.Y. Kung, J. Vlontzos. VLSI array processors for neural network simulation. *Journal of Neural Network Computing*, 5–20, Spring 1990.

[3] D.Zipser R.J. Williams. Gradient-based learning algorithms for recurrent networks. In Y. Chauvin and D.E. Rumelhart, editors, *Back-propagation : Theory, Architectures and Applications*, Hillsdale, NJ : Erlbaum Publ., 1991.

[4] S. Y. Kung. *VLSI Array Processors*. Prentice Hall Inc., N.J., 1988.

[5] S.Y. Kung and J.N. Hwang. A unified systolic architecture for artificial neural networks. *Journal of Parallel and Distributed Computing*, 358–387, April 1989.

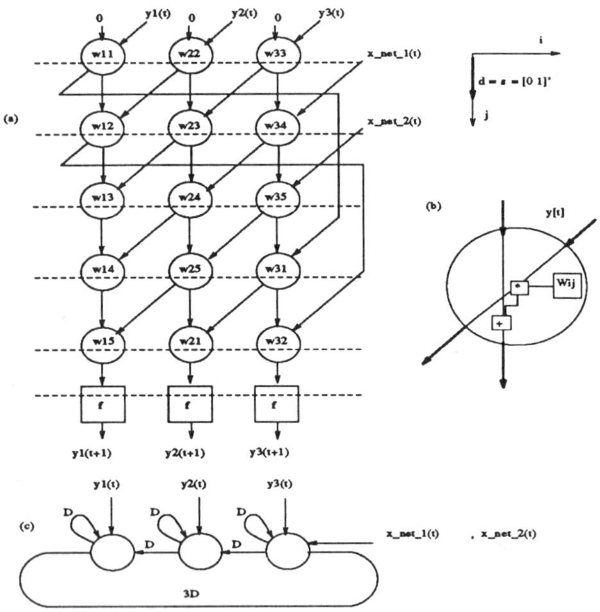

Figure 1: (a) The DG for the consecutive matrix - vector multiplications. (b) The structure of the DG node. (c) The systolized SFG.

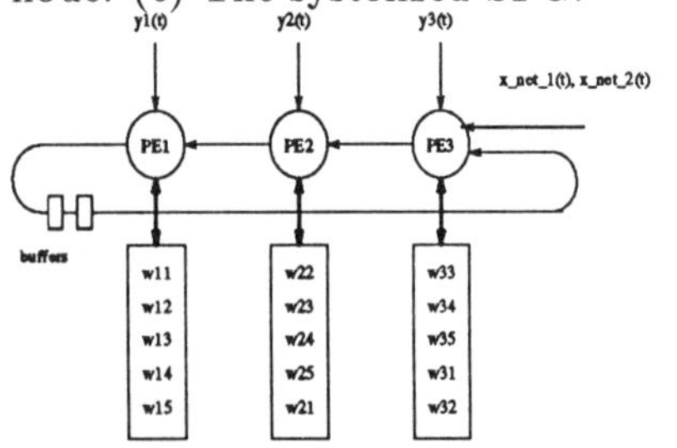

Figure 2: Systolic array for the retrieving phase.

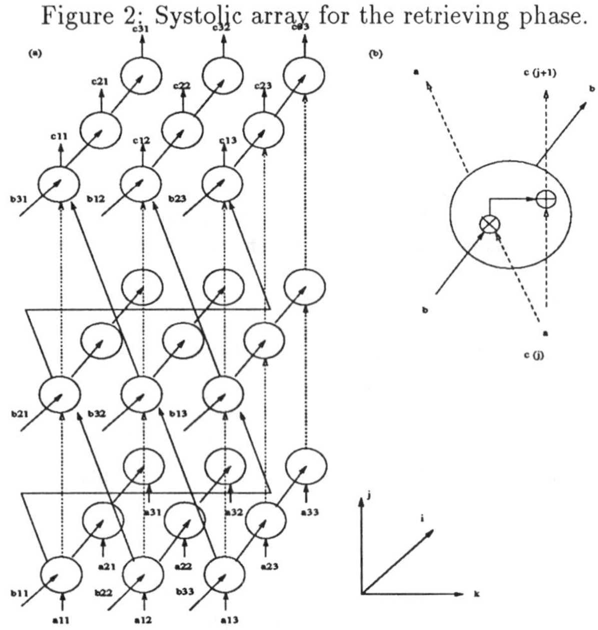

Figure 3: (a) The DG planes for consecutive matrix - matrix multiplication. (b) The DG node.

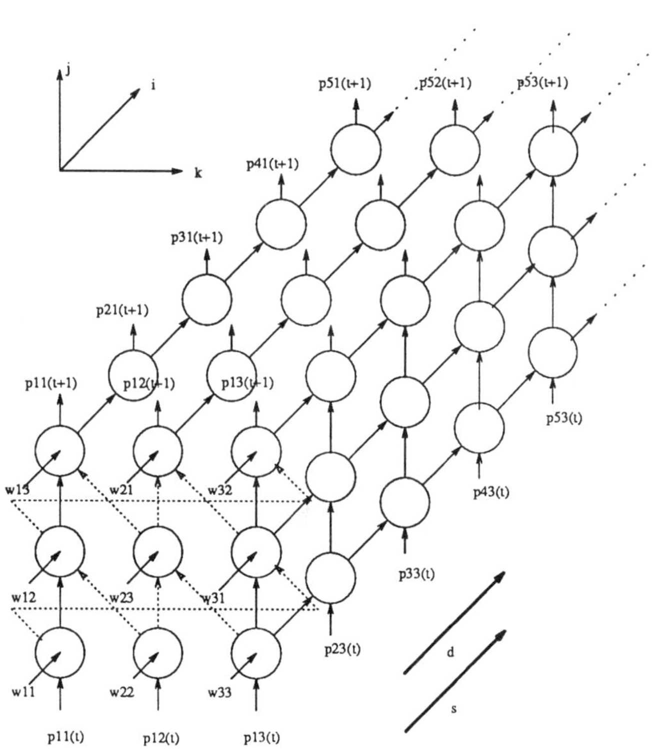

Figure 4: DG for the updating of the $\{p_{ij}\}$s; $n = 3, m = 2$.

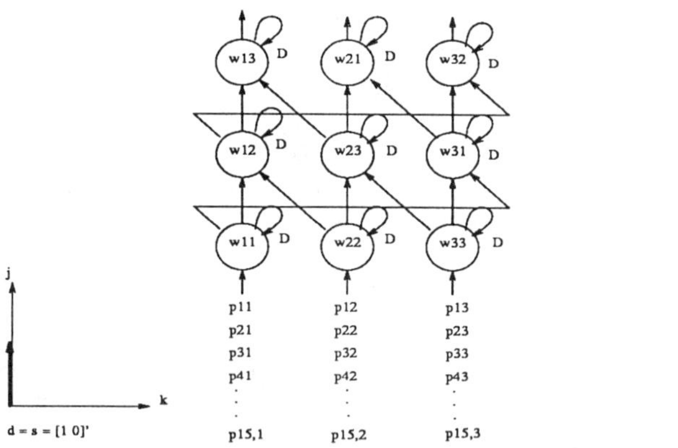

Figure 5: 2-D SFG for the updating of the $\{p_{ij}\}$s.

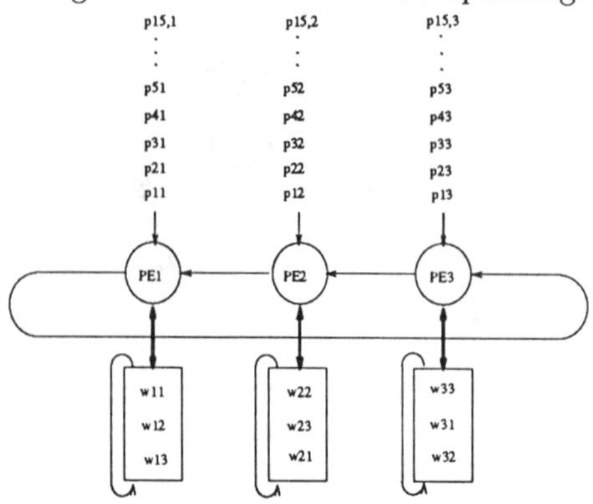

Figure 6: Ring Systolic array for the learning phase.

Concept Formation and Statistical Learning in Nonhomogeneous Neural Nets

Richard L. Tutwiler Leon H. Sibul

Pennsylvania State University
Applied Research Laboratory
Post Office Box 30
State College, PA 16804 USA

Abstract

This paper presents an analysis of complex nonhomogeneous neural nets, an adaptive statistical learning algorithm, and the potential use of these types of systems to perform a general sensor fusion problem. The three main contributions of this paper are the following. First, an extension to the theory of ***Statistical Neurodynamics*** *is introduced to include the analysis of complex nonhomogeneous neuron pools consisting of three subnets. Second, a statistical learning algorithm is developed based on the differential geometric theory of statistical inference for the adaptive updating of the synaptic interconnection weights. The statistical learning algorithm is merged with the subnets of nonhomogeneous nets and it is shown how these ensembles of nets can be applied to solve a general sensor fusion problem.*

Statistical Neurodynamics

The theory of *Statistical Neurodynamics* developed by Amari [1, 2] treats an ensemble of neural nets as a probability space whose sample space is a function of the dimensionality of the synaptic interconnection matrix and the threshold vector. Therefore, the weight matrix and the threshold vector are treated as random variables. The probability measure is defined on the Borel sets of the space. In our case we assume that they are both normally distributed and i.i.d. A complex net is composed of a series of simple nets which are called subnets. Amari's theory treats these networks from a macroscopic point of view. Our analysis derives the state equations describing the dynamic behavior of the complex net consisting of three subnets. Figure 1 illustrates a general block diagram of the complex net consisting of three subnets.

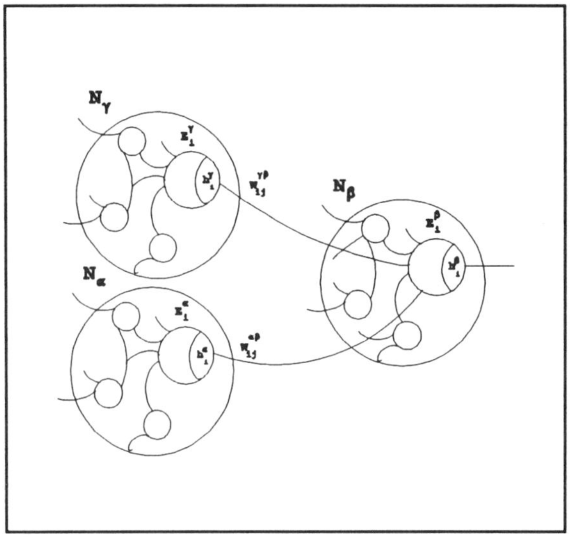

Figure 1. *Complex Net (Three Subnets)*

The *microstate transition law* for the nets is given by

$$u_i \equiv \sum_{j=1}^{n} w_{ij} x_j - h_i \qquad (1)$$

$$i = 1,\dots,n$$

$$\tilde{X}_i = sgn(u_i) \qquad (2)$$

The *state transition operator* is defined by

$$\tilde{X}_i = T_\omega X_i = sgn(u_i) \qquad (3)$$

T_ω is a nonlinear operator depending on ω that maps a state to its next state. If S = {set of all states}, then $T_\omega : S \to S$. The *activity level* of the net yields the *macrostate transition equation*

$$Z = a(X) = \frac{1}{n}\sum_{i=1}^{n} \tilde{X}_i = \frac{1}{n}\sum_{i=1}^{n} sgn(u_i) \;, -1 \le Z \le 1 \qquad (4)$$

The explicit form of the macroscopic transition

1058-6393/91 $01.00 © 1991 IEEE

function is expressed through the ensemble mean $\varphi(Z)$, which for n large (n = number of neurons) maps to

$$\varphi(Z) = \Phi(WZ-H)$$

$$W = \frac{nE\{w_{ij}\}}{[n\sigma_w^2 + \sigma_h^2]^{1/2}}, \quad H = \frac{E\{h_i\}}{[n\sigma_w^2 + \sigma_h^2]^{1/2}} \tag{5}$$

with

$$\Phi(u) = 2\int_o^u \frac{1}{\sqrt{2\pi}} e^{-v^2/2}\, dv \tag{6}$$

Referring to Figure 1 (and letting $\alpha = 1$, $\gamma = 2$, $\beta = 3$) the *equilibrium* macrostate transition equations for the three subnet case are given by

$$Z_1 = \Phi(W_{11}Z_1 + W_{12}Z_2 + W_{13}Z_3 - H_1) = \Phi(\eta) = 0$$

$$Z_2 = \Phi(W_{21}Z_1 + W_{22}Z_2 + W_{23}Z_3 - H_2) = \Phi(\phi) = 0$$

$$Z_3 = \Phi(W_{31}Z_1 + W_{32}Z_2 + W_{33}Z_3 - H_3) = \Phi(\psi) = 0 \tag{7}$$

The stability of the equilibrium states is conducted by evaluating the generalized Jacobian matrix J at the solutions of the simultaneous equilibrium equations given above. The generalized Jacobian is given by

$$J = \sqrt{\frac{2}{\pi}} \begin{bmatrix} W_{11}\, e^{-\eta} & W_{21}\, e^{-\eta} & W_{31}\, e^{-\eta} \\ W_{21}\, e^{-\phi} & W_{22}\, e^{-\phi} & W_{32}\, e^{-\phi} \\ W_{31}\, e^{-\psi} & W_{23}\, e^{-\psi} & W_{33}\, e^{-\psi} \end{bmatrix} \tag{8}$$

Decomposing J by standard eigensystem analysis yields

$$J = U\Lambda U^T$$

$$|\lambda_i| > 1 \rightarrow Z_i \; is \; UNSTABLE \tag{9}$$

$$|\lambda_i| \leq 1 \rightarrow Z_i \; is \; STABLE$$

Statistical Learning

The statistical learning algorithm [3, 4, 5, 6] is essentially a nonlinear recursive density estimation procedure. Differential geometry of statistical inference treats the probability space as a statistical manifold. The objective of the estimation procedure is to project the true density orthogonally onto a prespecified approximation family. The differential geometric analogy is the alignment of parallel tangent spaces. The approximating density is chosen such that the projection of the tangent vectors coincides. When this is accomplished the divergence also known as the Kullback-Leibler distance is minimized. The main goal is to continuously adapt the interconnection matrix to train the network to identify specific conditional statistical estimates.

The assumption taken is that the n-dimensional manifold includes all the possible posterior densities together. We consider the family of normal distributions $N(\mu,\sigma^2)$, the family of densities given by

$$p(x;\mu,\sigma) = \frac{1}{\sqrt{2\pi\sigma^2}}\, e^{-\{\frac{(x-\mu)^2}{2\sigma^2}\}} \tag{10}$$

with respect to the Lebesque measure on $\mathbb{R}$. The *Univariate Gaussian Manifold* is defined as

$$\mathcal{F} = \{p(\theta;w) \,|\, W \in \mathbb{R}^n\} \tag{11}$$

The submanifolds of approximating distributions are given by

$$\mathbf{A} = \{p(\theta;s) \,|\, s \in \mathbb{R}^n\} \subset \mathcal{F} \tag{12}$$

If $W=(w_1,\ldots,w_N)$ is a coordinate system in $\mathcal{F}$ and $s=(s_1,\ldots,s_M)$ is a coordinate system in $\mathbf{A}$, a relationship between coordinate systems is defined by the mapping $W=w(s)$. If the mapping is *differentiable* and regular, then $\mathbf{A}$ is *embedded* in $\mathcal{F}$. Therefore, the submanifolds are *foliations* of $\mathcal{F}$ as shown in Figure 2a. where

$$\mathcal{F} = \bigcup_{u \in A} A(u) \tag{13}$$

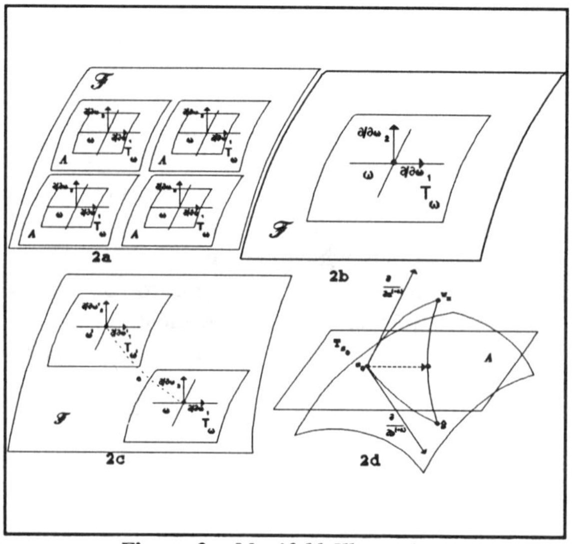

Figure 2. ***Manifold Illustrations***

A *foliation* is a partition of $\mathcal{F}$ into submanifolds $\mathbf{A}(u)$ of dimension N-M. The tangent space of $\mathcal{F}$ at point

p is a vector space obtained by local linearization around p. This space is composed of tangent vectors of all smooth curves passing through p. The *natural basis* of the tangent space T_w, is formed by $\frac{\partial}{\partial W_i}$ $(i = 1,N)$, as illustrated in Figure 2b. The concepts of *affine connection* and *geodesic*, shown in Figure 2c, specifies the way of transferring the tangent space over the manifold $\mathcal{F}$. The *affine connection* describes the relationship between adjacent tangent spaces. In addition, if the tangent vector is displaced parallelly along the curve c (shown in Figure 2c), the curve forms a *geodesic* with respect to the *affine connection*. A detailed explanation of these concepts is beyond the scope of this paper and the interested reader is referred to references [3]-[6]. The principle of approximation defines geodesics connecting; (1) the prior density $p(\theta) \in \boldsymbol{A}$ and the true posterior density $p(\theta;w_x) \in \mathcal{F}$, and (2), the prior density$p(\theta) \in \boldsymbol{A}$ and the approximating density $p(\theta;\hat{s}) \in \boldsymbol{A}$.. The approximating density $p(\theta;\hat{s})$ is chosen such that the *projection* of the tangent vectors of the *geodesics* defined by (1) and (2) onto the tangent space of $\boldsymbol{A}$ at point$p(\theta)$ coincide as illustrated in Fig. 2d.

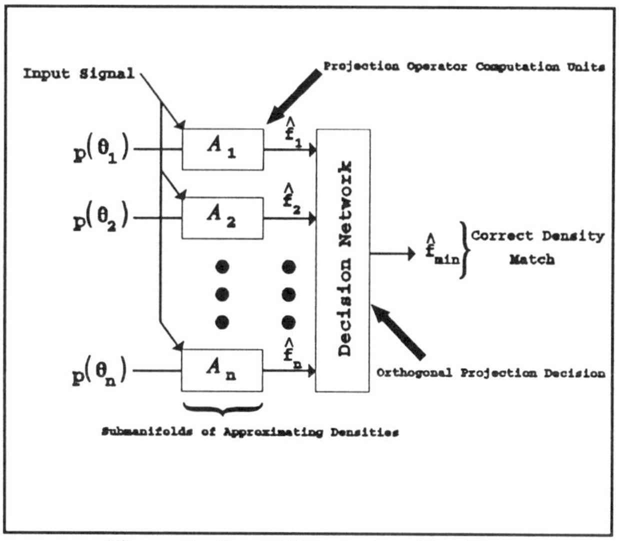

Figure 3. *Parallel Projection Algorithm*

Parallel Projection Algorithm (PPA), shown in Figure 3, places a grid
over both $\mathcal{F}$ and all *foliations* of $\mathcal{F}$. The projection algorithm is computed in parallel and the decision network searches for the minimum projection function which is the true *orthogonal projection* of $\boldsymbol{A}$ onto $\mathcal{F}$. The calculations performed in parallel at each processing stage are the following

Divergence Measure

$$D = \int p(\theta;s)\ell n\left(\frac{p(\theta;\hat{s})}{p(\theta;w_x)}\right)d\lambda(\theta) \tag{14}$$

Inner Product Calculation of Tangent Vectors

$$\left\langle \frac{\partial}{\partial b^{(+1)}}, \frac{\partial}{s_i} \right\rangle = E\left\{ \ell n\left(\frac{p(\theta;\hat{s})}{p(\theta)}\right)\left(\frac{\partial}{\partial s_i}\ell n(p(\theta;s))\right)\right\} \tag{15}$$

$$\left\langle \frac{\partial}{\partial a^{(+1)}}, \frac{\partial}{s_i} \right\rangle = E\left\{ \left(\ell n\left(\frac{p(\theta;w_x)}{p(\theta)}\right)\right)\left(\frac{\partial}{\partial s_i}\ell n(p(\theta;s))\right)\right\}$$

Angle between the Tangent Vectors

$$\cos(\alpha) = \frac{\frac{\partial}{\partial b^{(+1)}} \cdot \frac{\partial}{\partial a^{(+1)}}}{\left|\frac{\partial}{\partial b^{(+1)}}\right| \left|\frac{\partial}{\partial a^{(+1)}}\right|} \tag{16}$$

Therefore, at convergence the following conditions would hold i.) D = 0 in equation (14), ii) The inner product calculations in equation (15) would be equal, and (iii) $\cos(\alpha) = 1$ in equation (16).

Classification Procedure

The objective behind the recognition scheme is to train each subnet to recognize specific segments of the distribution function as shown in Figure 4.

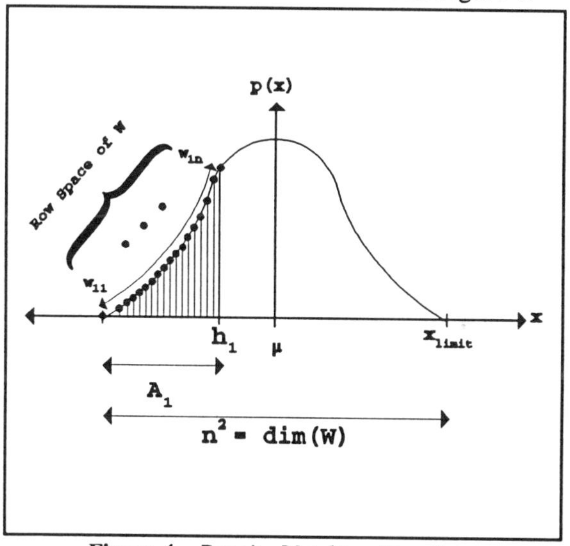

Figure 4. *Density Matching Functions*

Assuming a density estimate has been calculated according to the *PPA* described in the previous section, the synaptic weights are given by the quantized levels of the density function. The number of quantization levels is proportioned to the dimension of the synaptic interconnection matrix.

Given that $dim(\underline{\underline{W}}) \sim n^2$, choose x_{limit} such that $p(x) \cong 0$. To accomplish this set equation (10) equal to ϵ and solve for x_{limit}. The value of ϵ would be equal to the machine precision of the hardware used. The stepsize Δ for the weight increment is given by $\Delta = \frac{|2\,x_{limit}|}{n^2}$. Knowing these values $\underline{\underline{W}}$ can be calculated and mapped to the neuron plane. This implies that the row space of $\underline{\underline{W}}$ corresponds to specific segments of $p(x)$ needed to be classified. Since $\sum_{j=1}^{n} w_{ij} x_j$ approximates the area under the curve of $p(x)$, the thresholds h_i's can be precomputed based on standard probability theory. From Figure 4

$$h_1 = P_r(X \le x_1) = A_1 = \int_{-\infty}^{x_1} f(t)\,dt = \int_{-\infty}^{x_1} \frac{1}{\sqrt{2\pi\sigma^2}} e^{-\left\{\frac{(x-\mu)^2}{2\sigma^2}\right\}} dx$$

where the value x_1 corresponds to the value of x used to calculate w_{1n}. Therefore, knowing n, x_{limit}, Δ, $\underline{\underline{w}}$, one can precompute the h_i's and index them via a lookup table for each subnet. Correct classification implies that the macrostate activity level given by equation (4) will be equal to one.

Application to General Sensor Fusion

The objective of fusion in estimation theory is to combine various estimates to obtain a good estimate. The basic concept related in [7] is that there exists a Borel measurable function that performs the mapping $g: \mathbb{R}^n \to \mathbb{R}$. This mapping allows conditional estimates from n type sensors to be fused into a total conditional estimate. If the random variables are Gaussian as ours are by assumption, then further simplifications may be made as mentioned in [7].

Let Z be the *activity level* of the net with $Prob(\mathbf{Z} = 1) = \frac{1}{2}$ and $Prob(\mathbf{Z} = -1) = \frac{1}{2}$. Then

$$E\{X|Z = z\} = \tanh(Z/\sigma_i^2) \tag{17}$$

This implies that

$$E\{X|Z_1,\cdots,Z_n\} = \tanh\left(\sum_{i=1}^{n} \tanh^{-1}(E\{X|Z_i\})\right) \tag{18}$$

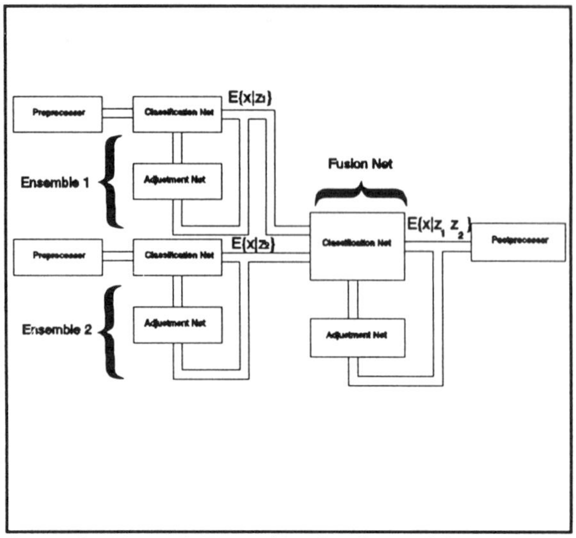

Figure 5. *Network Ensembles for General Sensor Fusion*

Figure 5 illustrates the concepts behind the nonhomogeneous network ensembles for the general sensor fusion problem. The *classification net* performs the functions mentioned in the previous section for correct density matching and the adjustment net performs the density estimation via the *parallel projection algorithm*. The inputs to the net are the conditional estimates $E\{X|Z_1\}$ and $E\{X|Z_2\}$ from these respective ensembles. The conditional estimates are given by equation (17). However, the output mapping function for the macrostate activity level is given by equation (6). The main concept of *Statistical Neurodynamics* treats both the $\underline{\underline{W}}$ and $\underline{h}$ as normally distributed random variables that are i.i.d. This assumption implies that

$$P_{\underline{\underline{W}}\underline{h}}(x, y) = P_{\underline{\underline{W}}}(x) P_{\underline{h}}(y) \tag{19}$$

and

$$\mu_{\underline{\underline{W}}} = \mu_{\underline{h}} \qquad \sigma_W^2 = \sigma_h^2 \tag{20}$$

Therefore, $WZ-H$ reduces to the following

$$WZ - H = \frac{\bar{w}(nZ-1)}{\sqrt{(n+1)\sigma_w^2}} \tag{21}$$

Furthermore, assuming n >> 1 (n = # of neurons) equation (21) reduces to

$$WZ - H = \frac{\sqrt{n}\,\bar{w}Z}{\sigma_w} \tag{22}$$

Since n, $\bar{w}$, and σ_w^2 are known we can compute a scale factor (S.F.) equal to $\frac{1}{\sqrt{n w}\sigma_w}$, which applied to equation (22) yields

$$WZ-H = S.F. \frac{\sqrt{nw}Z}{\sigma_w} = \frac{Z}{\sigma_w^2} \qquad (23)$$

which is the same argument as in equation (17). Equations (6) and (17) are both S shaped curves. Transcendental solutions for $erf(x) - \tanh(x) = 0$ exist only for $x = 0$. However, since we know the dimension of the system ahead of time, we can compute a polynomial fit of $\tanh(x)$ to $erf(x)$ using a *Linear Mean Square Error* estimation technique. This implies that if $\hat{y} = erf(x)$ and $y = \tanh(x)$, then $\hat{y} = \underline{c}\, y$ where $\underline{c}$ is the vector of polynomial coefficients. This illustrates the following point. If $\tilde{T}_\omega: S \to S$ is the *macrostate transition operator* defined by equation (6), and $g: \mathbb{R}^n \to \mathbb{R}$ is the Borel measurable fusion function defined by equation (17), then $\tilde{T}_\omega$ is *isomorphic* to g to within linear scaling constant. The isomorphism allows the *fusion* function to be *imbedded* in the *nonlinear output mapping* function of the neural network.

The fusion network takes the conditional estimates from each subnet and fuzes the estimates. The fusion net equations are given by

$$u_i = \sum_{j=1}^{n} w_{ij} x_j - h_i \qquad i = 1, m$$

with $\underline{\underline{W}} = I_{nxn}$ and $\underline{h} = 0 \; \forall \, i$. In our particular case, $n = 2$ and $m = 1$. Therefore, given u_i

$$\tilde{X}_1 = \Pi(u_1)$$

$$\Pi(u_i) = \{1 \cdot u_i \; \forall \, i\}$$

$$Z = \frac{1}{m} \sum_{i=1}^{m} \tilde{X}_i = \tilde{X}_1 \qquad m = 1$$

The output mapping is given by equation (6) with $\Phi(WZ - H) = \Phi(WZ)$ since $\underline{h} = 0 \; \forall i$. Letting $W = \frac{\sqrt{nw}Z}{\sigma_w}$, and defining the scale factor to be $S.F. = \frac{\sigma_w}{\sqrt{nw}}$ $\Phi(WZ)$ reduces to $\Phi(Z)$ and equation (6) is used to calculate the fuzed estimate $E\{X|Z_1 Z_2\}$.

Conclusions

An analysis of a complex nonhomogeneous neural net structure applicable to statistical inference procedures has been described. The three main points elucidated in this paper are the extension of the theory of *Statistical Neurodynamics* to include the analysis of complex nonhomogeneous neuron pools consisting of three subnets, the development of a statistical learning algorithm based on the orthogonal alignment of tangent spaces with respect to the submanifold of approximating probability density to the statistical manifold containing the true probability density, and the application of these structures to a general sensor fusion problem.

References

[1] S. Amari," A Mathematical Theory of Nerve Nets", *Advan. in Biophys.*, Vol. 6, pp. 75-120, 1974.

[2] S. Amari, K. Yoshida, and K. Kanatani," A Mathematical Foundation for Statistical Neurodynamics", *SIAM J. Appl. Math.*, Vol. 33, No. 1, pp. 95-126, July 1977.

[3] S. Amari, *Differential-Geometrical Methods in Statistics*, Springer-Verlag, 1990.

[4] S. Amari, O.E. Barndorff-Nielsen, R.E. Kass, S.L. Lauritzen, and C.R. Rao, *Differential Geometry and Statistical Inference*, Institute of Mathematical Statistics Lecture Notes - Monograph Series, S.S. Gupta (Series Editor), 1987.

[5] R. Kulhavy," Recursive Nonlinear Estimation: A Geometric Approach", *Automatica*, Vol. 26, No. 3, pp. 545-555, 1990.

[6] R. Kulhavy, "Differential Geometry of Recursive Estimation," *11th IFAC World Congress*, Tallinn, 1990.

[7] E.B. Hall, A. E. Wessel, and G.L. Wise," Some Aspects of Fusion in Estimation Theory", *IEEE Trans. Information Theory*, Vol. 37, No.2, March 1983.

State Assignment Selection Tests for Pass Transistor Asynchronous Sequential Circuits

Suresh K. Gopalakrishnan

California Design Center
Hewlett Packard Co.
Santa Clara, CA 95051

Gary K. Maki

Nasa Space Engineering Research Center
University of Idaho
Moscow, Idaho 83843

Abstract

Asynchronous sequential circuits avoid many problems facing high speed VLSI synchronous sequential circuits and therefore are becoming a viable alternative to clocked synchronous circuits. State assignment generation is a more complex process in the asynchronous case and produces many valid assignments. This paper presents easy to use procedures for selecting the best assignments using the number of transistors and the speed of circuit operation as metrics.

1 Introduction

As the integration density of VLSI circuits increase, synchronous circuits face significant clock and power distribution problems. An asynchronous circuit on the other hand does not use a clock and hence the clock distribution problems are completely eliminated. Further, the modules within an asynchronous circuit switch over a period of time instead of at a clock edge thereby reducing instantaneous current requirement and the complexity of the power distribution problem. These advantages project asynchronous controllers as a viable alternative. The major factor preventing the widespread use of asynchronous sequential circuits is that a complex design procedure is required which must account for problems like critical races and hazards.

It has been shown that pass transistor logic has definite advantages over conventional CMOS gate logic structures in factors like DC power requirement and compactness [1]. These advantages have been exploited in various integrated circuits. Recently, a new design for asynchronous sequential circuits using pass transistors was presented [2].

The standard methods for generating a critical race free internal state assignment for asynchronous sequential circuits can produce a large number of state assignments [3]. The logic designer is then faced with the problem of selecting the 'best' one of these state assignments for implementation. A strong set of tests for state assignment selection was presented by Maki, Sawin and Jeng [3] for implementing asynchronous sequential circuits using NAND gates and SR flip-flops. No tests have been attempted for VLSI sequential circuits. The original tests were based on partition algebra and can be easily extended to pass transistor asynchronous sequential circuits designed using the procedure outlined in [2].

The metric used in the previous work for selecting a 'good' assignment is the number of gate inputs needed to implement the state assignment. Similar metrics could be chosen for a pass transistor implementation also, but the number of transistors is a better metric which would directly correspond to the area required to implement the circuit. Another valuable metric is the speed of operation, which would give the designer a means of estimating the delay through the circuit before it is implemented. With these two metrics the designer can choose the circuit that would have the least delay or fewest number of transistors or a combination of both from a set of state assignments without realizing each assignment first.

2 Asynchronous Sequential Circuit Basics

Two block partitions $\tau_1, \tau_2, \ldots \tau_n$, induced by the internal state variables $y_1, y_2, \ldots y_n$, respectively, are called the τ partitions of that assignment. The first

This research was supported in part by the NASA Space Engineering Research Center Program under grant NAGW-1406.

1058-6393/91 $01.00 © 1991 IEEE

y_1	y_2	y_3		I_p
0	0	0	A	A/1
0	1	0	B	A
1	1	0	C	C/0
1	0	0	D	D/1
0	1	1	E	F
1	0	1	F	F/0

Table 1: Partial flow table

block of τ partitions is coded '1' and the second block is coded '0' by the corresponding state variables.

A next state partition η_i^p partitions the states where $Y_i = 1$ from those states where $Y_i = 0$ under input I_p. For the state assignment given in Table 1, the τ and η partitions are:

$$\begin{array}{llll} \tau_1 = & \{CDF; ABE\} & \eta_1 = \{CDEF; AB\} \\ \tau_2 = & \{BCE; ADF\} & \eta_2 = \{C; ABDEF\} \\ \tau_3 = & \{EF; ABCD\} & \eta_3 = \{EF; ABCD\} \end{array} \quad (1)$$

The next state equations are assumed to be in the form shown in Equation 2 [2,3].

$$Y_i = I_1(f_i^1) + I_2(f_i^2) + \cdots + I_m(f_i^m) \quad (2)$$

The f_i^p expressions corresponding to the next state partitions of the partial flow table shown in Table 1 are given below.

$$\begin{array}{lll} f_1^p & = & y_1' y_3' \\ f_2^p & = & y_1 y_2 \\ f_3^p & = & y_3 \end{array} \quad (3)$$

Theorems that relate the next state partitions and their corresponding sum of products expression were presented by Maki, Sawin and Jeng [3]. These theorems are stated for completeness.

Theorem 1 *If all the states of η_i^p appear in only one block, then f_i^p = 1 or 0.*

Theorem 2 *If $\eta_i^p \leq \tau_j$, then $f_i^p = y_j$ or y_j'.*

Theorem 3 *If $\eta_i^p = \eta_j^p$, then $f_i^p = f_j^p$ or $f_i^p = f_j^{p'}$.*

3 State Assignment Selection Procedures

The next state equations are shown in Equation 2. The general form of the pass transistor network needed to implement this set of design equations is shown in Figure 1. The f_i^p terms can be implemented using Binary Tree Structured (BTS) networks [4]. This would lead to a very area efficient implementation. The basic structure shown in Figure 1 depicts that each state variable and output has one transistor controlled by each input and have transistors to implement the buffer and f_i^p or g_i^p terms. The total number of transistors required for implementing the buffer is six, four transistors for two inverters, a weak PMOS transistor for restoring the degraded '1' due to the NMOS network used for implementing f_i^p and g_i^p terms and an NMOS transistor for initializing the circuit to a known state.

Consider a circuit with 'ℓ' input states and 'm' outputs. Let the number of state variables needed to encode a state machine be 'n'. Let (f_i^p) and (g_i^p) denote the number of transistors needed to implement f_i^p and g_i^p respectively. The total transistor count is given by:

$$Total = (\ell+6)(m+n) + \sum_{i=1}^{n}\sum_{p=1}^{\ell}(f_i^p) + \sum_{k=1}^{m}\sum_{p=1}^{l}(g_k^p) \quad (4)$$

The problem reduces to finding the transistor count for generating the f_i^p and g_k^p terms. There are four different cases to be considered; these are summarized in Table 2. For Case 1 and 2 the transistor count needed to implement $f_i^p(g_k^p)$ is zero as all these signals are available at the power supply rails or the buffer outputs. For Case 3(a) if $f_j^p(g_m^p)$ is already generated then this signal can be passed by the transistor controlled by I_p without using any more transistors. For Case 3(b) since $f_j^p(g_m^p)$is already generated the number of additional transistors need to generate $f_i^p(g_k^p)$ is the transistor count for an inverter, namely 2. In Case 4 there is no signal that can be tapped for implementing $f_i^p(g_k^p)$ and hence a new transistor network is needed.

Procedure 1 *Procedure for selecting the minimum transistor state assignment.*

Step 1: Form the α and τ partitions.

Step 2: Form the η and γ partitions.

Step 3: Use Table 2 and estimate the transistor count for the f_i^p and g_k^p terms.

Step 4: Use Equation 8 to estimate the total transistor count for each assignment.

Case No.	Condition	Transistor Count
Case 1	(a) $f_i^p(g_k^p)=0$	0
	(b) $f_i^p(g_k^p)=1$	0
Case 2	(a) $f_i^p(g_k^p) = y_j$	0
	(b) $f_i^p(g_k^p) = y_j'$	0
Case 3	(a) $f_i^p(g_k^p) = f_j^p$	Additional Transistors = 0
	(b) $f_i^p(g_k^p) = f_j^{p'}$	Additional Transistors = 2
Case 4	None of the above	$(f_i^p)((g_k^p))$

Table 2: Transistor count for $f_i^p(g_k^p)$ terms

Assign 1			Assign 2							
y_1	y_2	y_3	y_1	y_2	y_3	y_4		I_1	I_2	I_3
1	1	1	1	0	0	1	A	B	C	A
1	0	0	1	0	1	0	B	**B**	D	C
0	1	0	0	0	0	0	C	E	**C**	C
0	0	0	0	0	1	0	D	E	**D**	C
0	0	1	0	1	0	1	E	**E**	**E**	A

Table 3: Flow table with two state assignments

	I_1	I_2	I_3
η_1	AB;CDE	-;ABCDE	AE;BCD
η_2	-;ABCDE	AC;BDE	ABCDE;-
η_3	CDE;AB	E;ABCD	AE;BCD

Table 4: Next state partitions for the first state assignment

Another important metric is to select the assignment which will produce the fastest circuit. This test is based on the delay through the pass transistor chain of the circuit. It is assumed that the f_i^p terms are generated using a BTS network for the sake of layout regularity. This gives a method of estimating the delay through the pass transistor network used for implementing the f_i^p terms directly from the number of transistors needed to realize each f_i^p.

From the delay characteristics of the pass transistor chain, it can be seen that the delay through the chain increases as the number of transistors in series increases [5]. The speed then can be inferred by estimating the maximum transistor count for generating an f_i^p term. The f_i^p with the maximum number of transistors will be the critical path for the circuit under consideration. Thus the critical paths of the different circuits under consideration can be compared and the fastest circuit selected. However, in some cases the critical paths may also have the same number of transistors. In this case the circuit that does not share any of its f_i^p terms will be the fastest since sharing terms leads to a larger capacitive load at common nodes and hence increases delay.

Procedure 2 *Procedure for selecting the state assignment for the best speed.*

Step 1: Form α and τ partitions.

Step 2: Form η and γ partitions.

Step 3: Estimate the transistor count for implementing the f_i^p and g_k^p terms using Table 2. Find the maximum transistor count for an f_i^p or g_k^p term.

Step 4: Compare the maximum transistor counts for the different state assignments. Choose the assignment with the minimum maximum count for an f_i^p ro g_k^p term. If a tie exists then choose the circuit that has the minimum sharing of the maximum count f_i^p or g_k^p terms.

It is possible that there may be assignments that are equal in both metrics in which case any one of these assignments can be selected for implementation. Procedures 1 and 2 are illustrated by the following example which compare two state assignments for an asynchronous sequential machine for the constraints of transistor count and speed. The comparison is done for two state assignments with a different number of variables to illustrate that the procedures can be employed even when the assignments are different.

To illustrate the above procedures, consider the flow table shown in Table 3. The τ partitions for the first state assignment given in Equation 5.

$$\begin{aligned} \tau_1 &= \{AB; CDE\} \\ \tau_2 &= \{AC; BDE\} \\ \tau_3 &= \{AE; BCD\} \end{aligned} \qquad (5)$$

The next state partitions for the first state assignment are given in Table 4. The transistor counts corresponding to the next state partitions in Table 4 are given in Table 5. The total transistor count then becomes equal to **29**, for the first state assignment (using Equation 8 with $\ell = 3$, m = 0, n = 3 and $(f_3^2) = 2$).

The τ partitions for the second state assignment

	I_1	I_2	I_3
f_1	0	0	0
f_2	0	0	0
f_3	0	(f_3^2)	0

Table 5: Transistor count for the product terms for the first state assignment

	I_1	I_2	I_3
η_1	AB;CDE	-;ABCDE	AE;BCD
η_2	CDE;AB	E;ABCD	-;ABCDE
η_3	AB;CDE	BD;ACE	-;ABCDE
η_4	CDE;AB	E;ABCD	AE;BCD

Table 6: Next state partitions for the second state assignment

are given in Equation 6.

$$\begin{aligned} \tau_1 &= \{AB; CDE\} \\ \tau_2 &= \{E; ABCD\} \\ \tau_3 &= \{BD; ACE\} \\ \tau_4 &= \{AE; BCD\} \end{aligned} \tag{6}$$

The next state partitions for the first state assignment are given in Table 6. The transistor counts corresponding to the next state partitions in Table 6 are given in Table 7. The total transistor count then becomes equal to **36**, for the second state assignment.

Thus the first state assignment is better than the second state assignment when transistor counts are compared.

From Tables 5 and 7, it can be seen that the maximum transistor count for generating an f_i^p term for the first and second state assignments are two and zero respectively. Therefore, the second state assignment would lead to a faster circuit though it uses more transistors than the first state assignment.

	I_1	I_2	I_3
f_1	0	0	0
f_2	0	0	0
f_3	0	0	0
f_4	0	0	0

Table 7: Transistor count for the product terms for the second state assignment

4 Summary

State assignment selection procedures using minimum transistor count or maximum speed of operation as selection metrics for pass transistor asynchronous sequential circuits were presented in this paper. These procedures help the designer choose a state assignment from the set of valid state assignments produced by the standard procedures for generating critical race free state assignments.

References

[1] D. Radhakrishnan, S. Whitaker and G. Maki, "Formal Design Procedures for Pass Transistor Switching Circuits", IEEE JSSC, Vol. SC-20, pp. 531-536, April, 1985.

[2] S. Gopalakrishnan and G. Maki, "VLSI Asynchronous Sequential Circuit Design", Proc. ICCD-90 Sept. 1990, pp. 238-242.

[3] G. Maki, D. Sawin and B. Jeng, "Improved State Assignment Selection Tests", IEEE Transactions on Computers, Dec. 1972, pp. 1443-1449.

[4] G. Peterson and G. Maki, "Binary Tree Structured Logic Circuits: Design and Fault Detection", Proceedings of IEEE International Conference on Computer Design: VLSI in Computers, Port Chester, NY, Oct., 1984, pp. 671-676.

[5] N. Weste and K. Eshraghian, *Principles of CMOS VLSI Design*, Reading, Mass., Addison-Wesley, 1985.

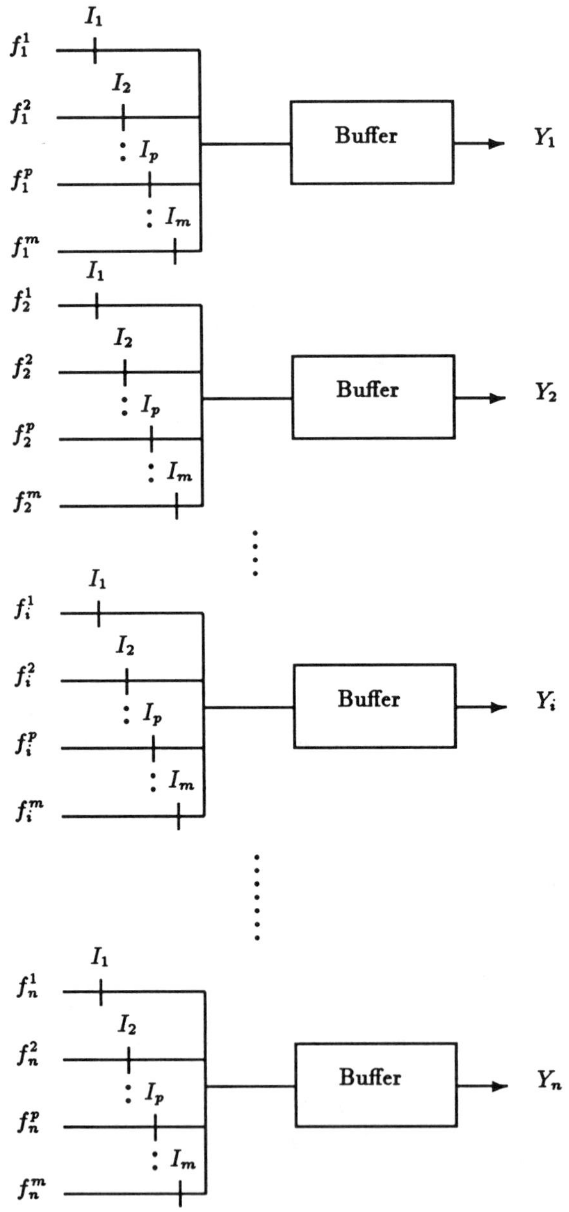

Figure 1: The general pass transistor network

A Single-chip QR Decomposition Processor for Extended Square-root Kalman Filters

Sung H. Chun

Advanced Micro Devices, Inc.
901 Thompson Place, P.O.Box 3453, MS 59
Sunnyvale, CA 94088-3453

Abstract

A single-chip architecture of a *dedicated-hardware* QR decomposition processor (QRDP) for the extended square-root Kalman filter is presented. As a slave peripheral to a 32-bit host processor, this chip acts as a numeric coprocessor which exclusively performs the QR decompositions for measurement and time updates by using the sequences of *fast Givens rotations* and *elementary matrices*. QRDP can be programmed to implement the square-root Kalman filter with various dimensions. The maximum allowable numbers of *measurements*, *process states*, and *noises* are 6, 12, and 12, respectively.

The application of QRDP considered in this paper is the target-tracking Kalman filters. For the 9-state case, simulation study shows that QRDP can achieve the sub-millisecond performance for both QR decompositions for measurement and time updates. Based on this run-time performance, QRDP can be a *cost-effective* alternative solution to the parallel implementation of the Kalman filter without degrading the speed performance.

1. Introduction

Recently, a number of parallel algorithms have been developed to handle the time-complexity of the Kalman filter which requires $O(N^3)$ floating-point operations at each iteration. In general, these parallel implementations on array processors are still expensive. Furthermore, it is difficult to construct the efficient parallelism due to the complexity of the interconnection topology and the *inherent sequential characteristics* of the Kalman filtering algorithm.

In the square-root implementation of the Kalman filter, QR decomposition procedures for measurement and time updates are major computational bottlenecks. In this paper, these procedures are isolated from the remaining computational stream, and implemented on a single-chip, *dedicated*, hardware processor, which will be called the QR decomposition processor (QRDP). QRDP acts as a slave peripheral on the 32-bit system bus. However, QRDP can be treated as a *black box* hardware processor. Once it is activated by the host processor, no assistance from the host processor and no external memory support are required during the entire course of the QR decomposition. QRDP includes a 32-bit floating-point datapath and 2976 bytes of on-chip memory. The architecture of QRDP is fully optimized for the high-speed execution of the QR decomposition procedures. Chief among these optimizations are :

- Hierarchically organized control state machines and an efficiently optimized datapath which maximize the computational throughput of the QR decomposition procedures.
- Efficient memory storage scheme in which the address of a matrix element is directly given by the combination of the corresponding row and column indexes.
- Fast memory addressing scheme for the generation of repetitive addressing sequences of a sophisticated nature.
- Flexible architecture in which the order of the Kalman filter can be changed by a single CPU command.

The QR decomposition procedures are controlled by a state machine which is *fully optimized* for the repetitive computations of *fast Givens rotations* and *elementary matrices*. At each machine cycle, this state machine generates *multi-tasking* control signals for address construction, memory access, register transfer, and floating-point operation.

2. Square-root Kalman Filter

Let the dynamic behavior of a process and the related measurement model satisfy the following relationships :

$$\mathbf{x}_{t+1} = F_t \mathbf{x}_t + G_t \mathbf{u}_t , \qquad \mathbf{y}_t = h(\mathbf{x},t) + \mathbf{v}_t \qquad t \geq 0, \tag{1}$$

where

$\mathbf{x}_t$: process state vector (N×1)
$\mathbf{u}_t$: process noise vector (M×1)
$\mathbf{y}_t$: sensor measurement output vector (P×1)
$\mathbf{v}_t$: measurement noise vector (P×1)
F_t : state transition matrix (N×N)
G_t : process noise matrix (N×M)
$h(\cdot)$: vector of nonlinear measurement equations (P×1).

P, N, and M denote the numbers of measurements, process states, and noises, respectively. The vectors $\mathbf{u}_t$ and $\mathbf{v}_t$ are zero-mean, independent white noises with known statistics :

$E[\mathbf{u}_t \mathbf{u}_s^T] = Q_t \delta_{ts}$: process noise covariance matrix (M×M)
$E[\mathbf{v}_t \mathbf{v}_s^T] = R_t \delta_{ts}$: measurement error covariance matrix (P×P).

The matrices F_t, G_t, $h(.)$, Q_t, R_t are assumed to be known *a priori*. The estimation problem is to determine the one-step predicted state estimate $\hat{\mathbf{x}}_{t+1|t}$ at time t which is the "*best*" in the sense that $E[(\mathbf{x}_{t+1} - \hat{\mathbf{x}}_{t+1|t})^T(\mathbf{x}_{t+1} - \hat{\mathbf{x}}_{t+1|t})]$ is minimum. The Kalman filter recursion for the least square predicted estimate of $\mathbf{x}_{t+1}$ is given as follows [6], [11] :

Measurement update :

$$R_{e,t} = R_t + H_t P_{t|t-1} H_t^T \tag{2a}$$

$$K_t = P_{t|t-1} H_t^T R_{e,t}^{-1} \tag{2b}$$

$$P_{t|t} = P_{t|t-1} - K_t H_t P_{t|t-1} \tag{2c}$$

$$\mathbf{e}_t = \mathbf{y}_t - H_t \hat{\mathbf{x}}_{t|t-1} \tag{2d}$$

$$\hat{\mathbf{x}}_{t|t} = \hat{\mathbf{x}}_{t|t-1} + K_t \mathbf{e}_t , \tag{2e}$$

Time update :

$$P_{t+1|t} = F_t P_{t|t} F_t^T + G_t Q_t G_t^T \tag{2f}$$

$$\hat{\mathbf{x}}_{t+1|t} = F_t \hat{\mathbf{x}}_{t|t} + \text{forcing function}, \tag{2g}$$

Jacobian update :

$$H_{t+1} = \left.\frac{\partial H(\mathbf{x})}{\partial \mathbf{x}}\right|_{\mathbf{x}=\hat{\mathbf{x}}_{t+1|t}}, \tag{2h}$$

where

$P_{t|t}$: filtered state error covariance matrix (NxN)
$P_{t+1|t}$: predicted state error covariance matrix (NxN)
$R_{e,t}$: innovation covariance matrix (PxP)
K_t : Kalman gain (optimal) matrix (NxP)
$\mathbf{e}_t$: vector of innovation (Px1)
H_t : measurement matrix (PxN).

In the above recursion, $P_{t|t}$ and $P_{t+1|t}$ may lose the *positive-definiteness* under the finite precision arithmetics, and this lack of *positive-definiteness* may cause the recursion to diverge. A numerically stable way to carry out the above recursion is to use the so-called *square-root Kalman filter* [3], [4], [12], [14], [15]. The idea is to maintain the state estimate error covariances in the factorized forms, and obtain the state estimates and error covariances updates from the triangularizations of particular block matrices. This will guarantee, at least, the *positive-semidefiniteness* of error covariances. The algorithm is based on the following algebra :

Block LDU and UDL factorizations :

$$\begin{bmatrix} P & Q \\ R & S \end{bmatrix} = \begin{bmatrix} I & 0 \\ RP^{-1} & I \end{bmatrix}\begin{bmatrix} P & 0 \\ 0 & S-RP^{-1}Q \end{bmatrix}\begin{bmatrix} I & -P^{-1}Q \\ 0 & I \end{bmatrix}$$

$$= \begin{bmatrix} I & QS^{-1} \\ 0 & I \end{bmatrix}\begin{bmatrix} P-QS^{-1}R & 0 \\ 0 & S \end{bmatrix}\begin{bmatrix} I & 0 \\ -S^{-1}R & I \end{bmatrix}.$$

Lemma : If $AD_1A^T = BD_2B^T$, then there exists $\mathbf{\Theta}$ such that $A\mathbf{\Theta} = B$ and $\mathbf{\Theta}D_2\mathbf{\Theta}^T = D_1$.

Above Lemma can be easily proved by using the *singular value decomposition*, $A = U\Sigma V^T$ and $\mathbf{\Theta} = V\Sigma^+U^TB$. Here, $\mathbf{\Theta}$ is highly nonunique and is often implemented as a sequence of *Householder reflections* or *Givens rotations.* QRDP *implicitly* constructs $\mathbf{\Theta}$ by using a sequence of *fast Givens rotations* which are square-root free.

2.1 Square-root algorithm for measurement update

Consider the following block matrix identities obtained from the measurement update recursions (Eqs. 2a, 2c),

$$\begin{bmatrix} R_{e,t} & H_tP_{t|t-1} \\ P_{t|t-1}H_t^T & P_{t|t-1} \end{bmatrix} = \begin{bmatrix} I_p & H_t \\ 0 & I_n \end{bmatrix}\begin{bmatrix} R_t & 0 \\ 0 & P_{t|t-1} \end{bmatrix}\begin{bmatrix} I_p & 0 \\ H_t^T & I_n \end{bmatrix}$$

$$= \begin{bmatrix} I_p & 0 \\ P_{t|t-1}H_t^TR_{e,t}^{-1} & I_n \end{bmatrix}\begin{bmatrix} R_{e,t} & 0 \\ 0 & P_{t|t} \end{bmatrix}\begin{bmatrix} I_p & R_{e,t}^{-1}H_tP_{t|t-1} \\ 0 & I_n \end{bmatrix}. \tag{3a}$$

Let $R_t = L_{R_t}L_{R_t}{}^T = (L_{R_t}D_{R_t}^{1/2})(L_{R_t}D_{R_t}^{1/2})^T = L_{R_t}D_{R_t}L_{R_t}{}^T$. Similarly, let $R_{e,t} = L_{R_{e,t}}D_{R_{e,t}}L_{R_{e,t}}{}^T$, $P_{t|t} = L_{P_{t|t}}D_{P_{t|t}}L_{P_{t|t}}{}^T$, and $P_{t|t-1} = L_{P_{t|t-1}}D_{P_{t|t-1}}L_{P_{t|t-1}}{}^T$. Here, L_{R_t}, $L_{R_{e,t}}$, $L_{P_{t|t}}$, $L_{P_{t|t-1}}$ are *Cholesky factors* of R_t, $R_{e,t}$, $P_{t|t}$, and $P_{t|t-1}$, respectively, whose diagonals are all 1s'.

Then, Eq. 3a can be written as

$$\begin{bmatrix} L_{R_t} & H_tL_{P_{t|t-1}} \\ 0 & L_{P_{t|t-1}} \end{bmatrix}\begin{bmatrix} D_{R_t} & 0 \\ 0 & D_{P_{t|t-1}} \end{bmatrix}\begin{bmatrix} L_{R_t} & H_tL_{P_{t|t-1}} \\ 0 & L_{P_{t|t-1}} \end{bmatrix}^T$$

$$= \begin{bmatrix} L_{R_{e,t}} & 0 \\ \bar{K}_t & L_{P_{t|t}} \end{bmatrix}\begin{bmatrix} D_{R_{e,t}} & 0 \\ 0 & D_{P_{t|t}} \end{bmatrix}\begin{bmatrix} L_{R_{e,t}} & 0 \\ \bar{K}_t & L_{P_{t|t}} \end{bmatrix}^T.$$

Hence, given $L_{R_t}, D_{R_t}, L_{P_{t|t-1}}, D_{P_{t|t-1}}$, and H_t, the above Lemma indicates that we can find $\mathbf{\Theta}$ such that

$$\begin{bmatrix} L_{R_t} & H_tL_{P_{t|t-1}} \\ 0 & L_{P_{t|t-1}} \end{bmatrix}\mathbf{\Theta} = \begin{bmatrix} L_{R_{e,t}} & 0 \\ \bar{K}_t & L_{P_{t|t}} \end{bmatrix} \quad \text{and}$$

$$\mathbf{\Theta}\begin{bmatrix} D_{R_{e,t}} & 0 \\ 0 & D_{P_{t|t}} \end{bmatrix}\mathbf{\Theta}^T = \begin{bmatrix} D_{R_t} & 0 \\ 0 & D_{P_{t|t-1}} \end{bmatrix}. \tag{3b}$$

2.2 Combined state and measurement updates

Consider the following (2x2) block matrix

$$\begin{bmatrix} L_{R_{e,t}} & -\mathbf{e}_t \\ P_{t|t-1}H_t^TL_{R_{e,t}}{}^{-T} & \hat{\mathbf{x}}_{t|t-1} \end{bmatrix}. \tag{3c}$$

Block matrices in the first column result from the QR decomposition for measurement update. Assuming $L_{Re,t}$ is nonsingular, the *Schur complement* [8] of $L_{Re,t}$ in the above block matrix is given by

$$S = \hat{\mathbf{x}}_{t|t-1} - P_{t|t-1}H_t{}^TL_{R_{e,t}}{}^{-T}L_{R_{e,t}}{}^{-1}(-\mathbf{e}_t). \tag{3d}$$

Note that S is equivalent to the filtered state estimate $\hat{\mathbf{x}}_{t|t}$. Above *Schur complement* can be found by using the Gaussian elimination [8]. After P steps of Gaussian elimination to zero-out $-\mathbf{e}_t$ and the strictly upper triangular portion of $L_{Re,t}$, $A_{2,2}$ entry of the above block matrix will become $\hat{\mathbf{x}}_{t|t}$. Hence, we can avoid the explicit matrix inversion to obtain $\hat{\mathbf{x}}_{t|t}$. Here, $L_{Re,t}$ is already lower triangular and $-\mathbf{e}_t$ has only one column. Hence, it is more efficient to apply one step of column-wise Gaussian elimination to zero-out P elements of $-\mathbf{e}_t$, using the diagonal elements of $L_{Re,t}$ as pivot elements. This column-wise Gaussian elimination can be performed by post-multipling the above block matrix (Eq. 3c) with *elementary matrices* [12].

• Implementation of QR decomp. for measurement update†

QRDP implements the QR decomposition for the combined state and measurement updates by using the sequences of *fast Givens rotations* and *elementary matrices* :

1) Form the block matrix A_t and the corresponding diagonal matrix D_{A_t} as follows :

$$A_t \equiv \begin{bmatrix} L_{R_t} & -\mathbf{e}_t & H_tL_{P_{t|t-1}} \\ 0 & \hat{\mathbf{x}}_{t|t-1} & L_{P_{t|t-1}} \end{bmatrix} \in R^{(P+N)\times(P+N+1)}. \tag{4a}$$

† The exact expression is "QR decomposition and Gaussian elimination for the combined state and measurement updates". However, we will call it simply "QR decomposition for measurement update".

$$D_{A_t} \equiv \begin{bmatrix} D_{R_t} & & 0 \\ & 1 & \\ 0 & & D_{P_{t|t-1}} \end{bmatrix} \in R^{(P+N+1)\times(P+N+1)} . \quad (4b)$$

2) Post-multiply A_t and D_{A_t} with a sequence of *fast Givens rotations* M_{A_t} so as to zero-out $H_t L_{P_{t|t-1}}$:

$$\begin{aligned} A_t M_{A_t} = A_t &\times M_{1,P+N+1} \times M_{1,P+N} \times \cdots \times M_{1,P+2} \\ &\times M_{2,P+N+1} \times M_{2,P+N} \times \cdots \times M_{2,P+2} \\ &\vdots \\ &\times M_{P,P+N+1} \times M_{P,P+N} \times \cdots \times M_{P,P+2} . \end{aligned} \quad (4c)$$

Same as for $D_{A_t} M_{A_t}$. (4d)

3) Post-multiply $A_t M_{A_t}$ with a sequence of *elementary matrices* E_{A_t} so as to zero-out $-\mathbf{e}_t$:

$$A_t M_{A_t} E_{A_t} = A_t M_{A_t} \times E_{1,P+1} \times E_{2,P+1} \times \cdots \times E_{P,P+1} . \quad (4e)$$

Then, entries of the resulting block matrices $\hat{A}_t$ and $\hat{D}_{A_t}$ become

$$\hat{A}_t = \begin{bmatrix} L_{R_{e,t}} & 0 & 0 \\ \bar{K}_t' & \hat{\mathbf{x}}_{t|t} & L_{P_{t|t}} \end{bmatrix} \text{ and } \hat{D}_{A_t} = \begin{bmatrix} D_{R_{e,t}} & & 0 \\ & 1 & \\ 0 & & D_{P_{t|t}} \end{bmatrix} . \quad (4f)$$

Note that, in the above procedure, we update diagonal elements d_i instead of $d_i^{1/2}$ (square-root free).

2.3 Square-root algorithm for time update

Let $Q_t = L_{Q_t} D_{Q_t} L_{Q_t}^T$, $P_{t|t} = L_{P_{t|t}} D_{P_{t|t}} L_{P_{t|t}}^T$, and $P_{t+1|t} = L_{P_{t+1|t}} D_{P_{t+1|t}} L_{P_{t+1|t}}^T$, where $L_{Q_t}, L_{P_{t|t}}, L_{P_{t+1|t}}$ are *Cholesky factors* of Q_t, $P_{t|t}$, and $P_{t+1|t}$, respectively.

Then, the time update equation (Eq. 2f) can be written as

$$\begin{bmatrix} F_t L_{P_{t|t}} & G_t L_{Q_t} \end{bmatrix} \begin{bmatrix} D_{P_{t|t}} & 0 \\ 0 & D_{Q_t} \end{bmatrix} \begin{bmatrix} F_t L_{P_{t|t}} & G_t L_{Q_t} \end{bmatrix}^T$$

$$= \begin{bmatrix} L_{P_{t+1|t}} & 0 \end{bmatrix} \begin{bmatrix} D_{P_{t+1|t}} & 0 \\ 0 & 0 \end{bmatrix} \begin{bmatrix} L_{P_{t+1|t}} & 0 \end{bmatrix}^T .$$

Hence, given $F_t L_{P_{t|t}}$, $D_{P_{t|t}}$, $G_t L_{Q_t}$, and D_{Q_t}, the above Lemma indicates that we can find Θ such that

$$\begin{bmatrix} F_t L_{P_{t|t}} & G_t L_{Q_t} \end{bmatrix} \Theta = \begin{bmatrix} L_{P_{t+1|t}} & 0 \end{bmatrix} \text{ and}$$

$$\Theta \begin{bmatrix} D_{P_{t+1|t}} & 0 \\ 0 & 0 \end{bmatrix} \Theta^T = \begin{bmatrix} D_{P_{t|t}} & 0 \\ 0 & D_{Q_t} \end{bmatrix} . \quad (5)$$

• Implementation of QR decomp. for time update

QRDP implements the QR decomposition for time update by using a sequence of *fast Givens rotations* :

1) Form the block matrix B_t and the corresponding diagonal matrix D_{B_t} as follows :

$$B_t \equiv \begin{bmatrix} F_t L_{P_{t|t}} & G_t L_{Q_t} \end{bmatrix} \in R^{N\times(N+M)} . \quad (6a)$$

$$D_{B_t} \equiv \begin{bmatrix} D_{P_{t|t}} & 0 \\ 0 & D_{Q_t} \end{bmatrix} \in R^{(N+M)\times(N+M)} . \quad (6b)$$

2) Post-multiply B_t and D_{B_t} with a sequence of *fast Givens rotations* M_{B_t} so as to zero-out $G_t L_{Q_t}$ and the strictly upper triangular portion of $F_t L_{P_{t|t}}$:

$$\begin{aligned} B_t M_{B_t} = B_t &\times M_{1,N+M} \times M_{1,N+M-1} \times \cdots \times M_{1,2} \\ &\times M_{2,N+M} \times M_{2,N+M-1} \times \cdots \times M_{2,3} \\ &\vdots \\ &\times M_{N,N+M} \times M_{N,N+M-1} \cdots \times M_{N,N+1} . \end{aligned} \quad (6c)$$

Same as for $D_{B_t} M_{B_t}$. (6d)

Then, entries of the resulting block matrices $\hat{B}_t$ and $\hat{D}_{B_t}$ become

$$\hat{B}_t = \begin{bmatrix} L_{P_{t+1|t}} & 0 \end{bmatrix} \text{ and } \hat{D}_{B_t} = \begin{bmatrix} D_{P_{t+1|t}} & 0 \\ 0 & 0 \end{bmatrix} . \quad (6e)$$

2.4 Fast Givens rotation and elementary matrix

Among the several versions of *fast Givens rotations*, QRDP implements the *two-multiplication form* [8]. Suppose that $\hat{\mathbf{a}}$ and $\hat{\mathbf{b}}$ are two column vectors under the Givens rotation $[\hat{\mathbf{a}}\ \hat{\mathbf{b}}]\,\tilde{G} = [\hat{\mathbf{a}}'\ \hat{\mathbf{b}}']$, and define the vectors $\mathbf{a}$ and $\mathbf{b}$ such that

$$[\hat{\mathbf{a}}\ \hat{\mathbf{b}}] \equiv [\mathbf{a}\ \mathbf{b}] \begin{bmatrix} \sqrt{d_i} & 0 \\ 0 & \sqrt{d_j} \end{bmatrix} \text{ and } [\hat{\mathbf{a}}'\ \hat{\mathbf{b}}'] \equiv [\mathbf{a}'\ \mathbf{b}'] \begin{bmatrix} \sqrt{d_i}' & 0 \\ 0 & \sqrt{d_j}' \end{bmatrix} .$$

Furthermore, let b_i be the element to be annihilated in $\mathbf{b}$ and a_i be the corresponding pivot element in $\mathbf{a}$. Then, the matrix $M_{i,j}$ for the *two-multiplication form* of *fast Givens rotation* $[\mathbf{a}\ \mathbf{b}]\,\tilde{M} = [\mathbf{a}'\ \mathbf{b}']$ and the resulting updated values a_i', d_i', and d_j' are given by

$$M_{i,j} = \begin{bmatrix} \mathbf{I} & & & & \mathbf{O} \\ & 1 & & M_a & \\ & & \mathbf{I} & & \\ & M_b & & 1 & \\ \mathbf{O} & & & & \mathbf{I} \end{bmatrix} , \quad (7)$$

where $M_a = -\dfrac{b_i}{a_i}$, $M_b = \dfrac{b_i d_j}{a_i d_i}$.

$$a_i' = a_i + \frac{b_i^2 d_j}{a_i d_i}, \quad d_i' = \frac{a_i^2 d_i^2}{a_i^2 d_i + b_i^2 d_j}, \quad d_j' = \frac{a_i^2 d_i d_j}{a_i^2 d_i + b_i^2 d_j} . \quad (8)$$

The *elementary matrix* $E_{i,j}$ for the Gaussian elimination is given by

$$E_{i,j} = \begin{bmatrix} \mathbf{I} & & & & \mathbf{O} \\ & 1 & & \alpha & \\ & & \mathbf{I} & & \\ & 0 & & 1 & \\ \mathbf{O} & & & & \mathbf{I} \end{bmatrix} , \quad \text{where } \alpha = -\frac{b_i}{a_i} . \quad (9)$$

Note that diagonal elements d_i and d_j are not affected during the Gaussian elimination by *elementary matrices*.

2.5 Extended Kalman filter for target-tracking

The extended Kalman filter recursively estimates the state variables of a time-varying dynamic system from noisy measurements, and outputs the state estimates and corresponding estimate error covariances. In target-tracking applications, depending upon the target model, the number of state variables to be estimated are 3 (constant location), 6 (constant velocity), or 9 (constant acceleration) [2]. The deviation of a maneuvering target from the straight line trajectory is termed as a target acceleration or maneuver variable. The target equation of motion can be expressed in terms of the white noise which drives the autoregressive process of the target acceleration [17].

Assuming the large acceleration time constant and fast sampling time, the state transition matrix F_t can be simplified to a constant matrix [1], [17]. The maneuver excitation covariance Q_t and measurement noise covariance R_t can be used for compensating the target dynamic uncertainty and measurement model uncertainty, respectively. Increasing Q_t effectively increases the bandwidth of the Kalman filter for the noise rejection thus making the filter more responsive for the future target dynamics at the expense of larger estimate error covariances. In multi-target tracking systems, depending upon the prioritization criterion and correlation of the incoming track data, the order of the Kalman filter can be changed at each sampling time.

2.6 Algorithm implementation

The following procedure summarizes the computational sequence of the extended square-root Kalman filter :

At time t :

1) Update R_t and compute $R_t = L_{Rt} D_{Rt} L_{Rt}^T$.
2) Compute $\mathbf{e}_t = \mathbf{y}_t - H_t \hat{\mathbf{x}}_{t|t-1}$.
3) Compute $H_t L_{Pt|t-1}$.
4) Compute the **QR decomposition for measurement update** (Eqs. 4c, 4d, 4e) and obtain $\hat{\mathbf{x}}_{t|t}, L_{Pt|t}, D_{Pt|t}$.
5) Update F_t, G_t, Q_t and compute $Q_t = L_{Qt} D_{Qt} L_{Qt}^T$. Compute $G_t L_{Qt}$.
6) Compute $F_t L_{Pt|t}$.
7) Compute the **QR decomposition for the time update** (Eqs. 6c, 6d) and obtain $L_{Pt+1|t}, D_{Pt+1|t}$.
8) Update $\hat{\mathbf{x}}_{t+1|t} = F_t \hat{\mathbf{x}}_{t|t}$ + forcing function.
9) Compute $H_{t+1} = \left.\dfrac{\partial h(\mathbf{x},t)}{\partial \mathbf{x}}\right|_{\mathbf{x}=\hat{\mathbf{x}}_{t+1|t}}$.

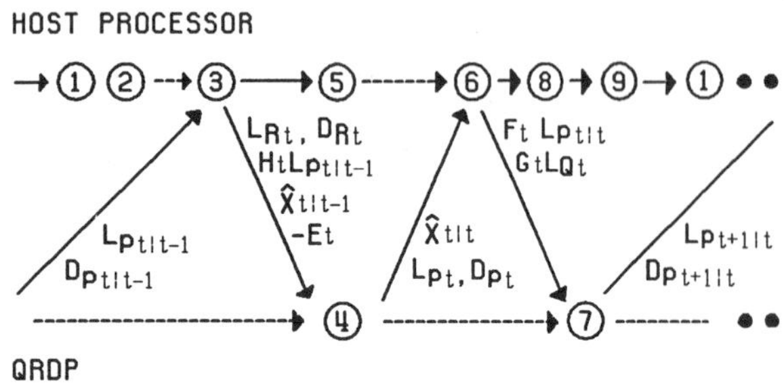

Fig. 2-1 Job partitioning between QRDP and host processor

The QR decomposition procedures require extensive computational efforts. QRDP is a *dedicated* hardware processor for the QR decompositions. At each sampling time, QRDP executes the QR decomposition for measurement update (Eqs. 4c, 4d, 4e, 7, 8, 9) and time update (Eqs. 6c, 6d, 7, 8). By using QRDP and host processor (CPU), the above computational sequence can be partitioned into two sequential procedures which can be executed *partially in parallel*. Fig. 2-1 illustrates the local parallelism between QRDP and host processor.

3. QRDP Architecture

QRDP is composed of five major blocks :

- Address generation unit (AGU).
- Memory unit (MU).
- Floating-point computational unit (FPU).
- Control unit (CU).
- Interface unit (IU).

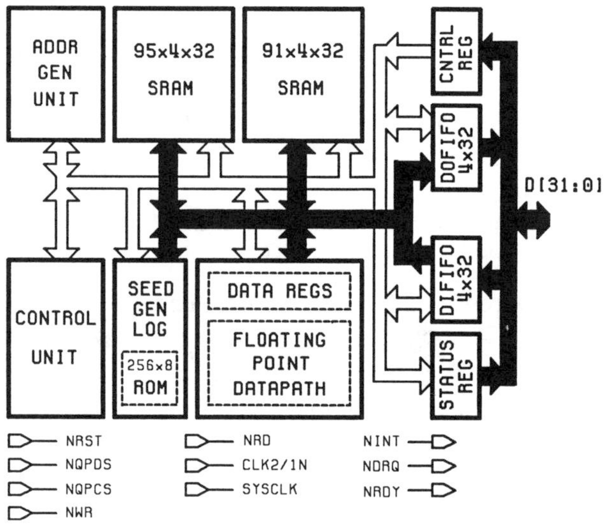

Fig. 3-1 QRDP block diagram

At each sampling time, QRDP performs :

- Receive $L_{Rt}, D_{Rt}, -\mathbf{e}_t, \hat{\mathbf{x}}_{t|t-1}, H_t L_{Pt|t-1}$ from system memory and store into Mmem.
- Execute the **QR decomposition for measurement update.**
- Assert CPU interrupt (QR decomposition for measurement update has been completed).
- Send $\hat{\mathbf{x}}_{t|t}, L_{Pt|t}, D_{Pt|t}$ from Mmem to system memory.
- Assert CPU interrupt (data transfer has been completed).
- Receive $F_t L_{Pt|t}, D_{Pt|t}, G_t L_{Qt}, D_{Qt}$ from system memory and store into Tmem.
- Execute the **QR decomposition for time update.**
- Assert CPU interrupt (QR decomposition for time update has been completed).
- Send $L_{Pt+1|t}, D_{Pt+1|t}$ from Mmem to system memory.
- Assert CPU interrupt (data transfer has been completed).

3.1 Address generation and memory units

The memory addressing and storage schemes of QRDP are efficiently organized for indexing the matrix elements in the sophisticated ways of row, column, and diagonal fashions.

Fig. 3-2 shows the measurement and time update matrices† with their corresponding row and column indexes. Note that the row index 0 is assigned to the diagonal matrices. The measurement and time update matrices are stored in the 2976 bytes of on-chip SRAM memory organized as two modules of 95×4×32-bit words (Mmem) and 91×4×32-bit words (Tmem).

The 10-bit memory address (addr[9:0]) for each matrix element is directly given by the combination of the corresponding row index (mapped into addr[9:5]) and column index (mapped into addr[4:0]). For example, [0,5] element of the measurement update matrix, which is $D_{P_{t|t-1}}(1)$, is stored at @0000000101 (005_H) in Mmem. [9,18] element of time update matrix, which is $G_t L_{Q_t}(9,9)$, is stored at @0100110010 (132_H) in Tmem.

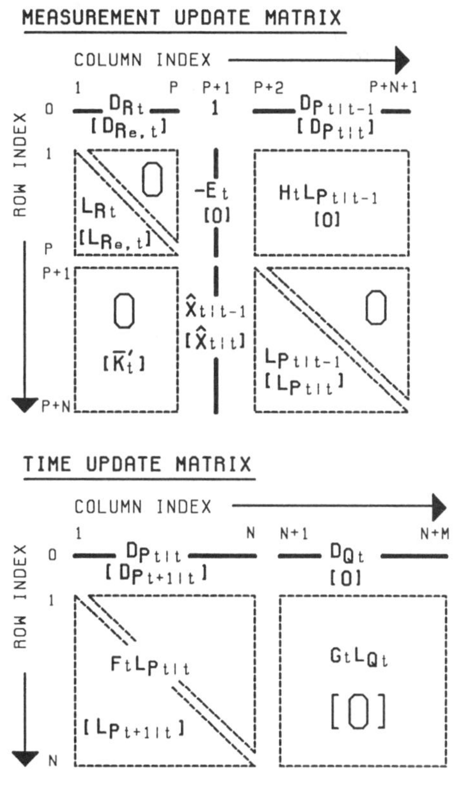

Fig. 3-2 Measurement and time update matrices

If we organize Mmem and Tmem such that row address = addr[9:2] and column address = addr[1:0] (four columns each in Mmem and Tmem) and exclude the unused memory space by limiting the dimension of the Kalman filter as N_{max} = 12, M_{max} = 12, and P_{max} = 6, the above addressing scheme yields the ***discontinuous*** memory address spaces which are the design specifications for the row address decoders of Mmem and Tmem. In this case, the corresponding memory sizes are given as follows. The redundant columns result from the four fixed columns :

Mmem : $(0 \sim P+N) \times [(1 \sim P+N+1) + 1 \text{ (redundant)}] \times 32$
$= 19 \times 20 \times 32 = 95 \times 4 \times 32$.

† For convenience, the initial set-up matrices (Eqs. 4a,4b) for measurement update will be called the measurement update matrix and the initial set-up matrices (Eqs. 6a,6b) for time update will be called the time update matrix.

Tmem : $(0 \sim N) \times [(1 \sim N+M) + 4 \text{ (redundants)}] \times 32$
$= 13 \times 28 \times 32 = 91 \times 4 \times 32$.

When the order of the Kalman filter is changed, the above addressing scheme provides the automatic scaling for the memory address construction, thus avoids adjusting the values of index registers which is typically required by the conventional memory addressing scheme.

AGU contains five 5-bit presettable counters which act as pointers for row, column, and diagonal indexings of measurement and time update matrices. The contents of any two counters are combined together to form the 10-bit memory address. AGU also includes four comparing registers in which the final indexing values of row, column, and diagonal elements are stored. The following example sequentially accesses D_{Rt}, L_{Rt}, and $\hat{x}_{t|t-1}$ in Mmem in the order of :

$D_{Rt}(1,1), D_{Rt}(2,2), D_{Rt}(3,3), \cdots\cdots\cdots D_{Rt}(P,P)$,
$L_{Rt}(1,1), L_{Rt}(2,1), L_{Rt}(2,2), \cdots\cdots\cdots L_{Rt}(P,P-1), L_{Rt}(P,P)$,
$x_{t|t-1}(1), x_{t|t-1}(2), \cdots\cdots x_{t|t-1}(N)$.

```
      set CCNT ← 1, CCMPCNT ← P
LP0   read Mmem@0C and store at · · · ·
      if (CCMPEQ), then set RCNT ← 1, RCMP ← P,
         CCNT ← 1, TCNT ← 1 and goto LP1
      else CCNT++ and goto LP0
LP1   read Mmem@RC and store at · · · ·
      if (CTEQ & !RCMPEQ), then RCNT++, TCNT++,
         CCNT ← 1 and goto LP1
      else if (CTEQ & RCMPEQ), then set RCNT ← P+1,
         RCMP ← P+N, CCNT ← P+1 and goto LP2
      else CCNT++ and goto LP1
LP2   read Mmem@RC and store at · · · ·
      if (RCMPEQ) then exit
      else RCNT++ and goto LP2
```

Here, @0C means addr[9:5]←0 and addr[4:0]←[CCNT]. @RC means addr[9:5]←[RCNT] and addr[4:0]←[CCNT]. CCMPEQ, RCMPEQ, and TCEQ are true when [CCNT]=[CCMPCNT], [RCNT]=[RCMP], and [TCNT]=[CCNT], respectively.

3.2 Floating-point computational unit

The floating-point computational unit consists of a fully combinatorial, 32-bit floating-point datapath (arithmetic unit and multiplier) which follows the IEEE 754 floating-point standard, a seed value generation logic for the floating-point inversion, and eight data registers for storing intermediate computational results.

The floating-point datapath receives the operand either from the memory or feedback registers and performs the operations $(\pm A) + B$, $(\pm A) - B$, $(\pm A) \times B$, $2 - A$ in one machine cycle. Internally, it appends eight guard bits and one sticky bit. The resulting operand is rounded to the closest even number and protected by the saturated overflow and sudden underflow methods. The seed value generation logic provides an initial value for the Newton-Raphson iterations. It consists of a seed exponent generation logic and a seed fraction look-up ROM (256x8-bit) whose output is accurate to eight bits of precision. Since Newton-Raphson iterations converge quadratically, the number of bits of precision doubles at each iteration. Two iterations should provide the full precision for the 23-bit of mantissa†. The output of a seed exponent logic and content of a seed fraction look-up ROM are based on the following for-

† Quantization errors introduced by rounding, however, can prevent the LSB from being accurate (provides 22 bits of accuracy).

mulae :

seed exponent = 253_{10} − input biased exponent .
seed fraction = [2/(1 + input fraction)] − 1 .

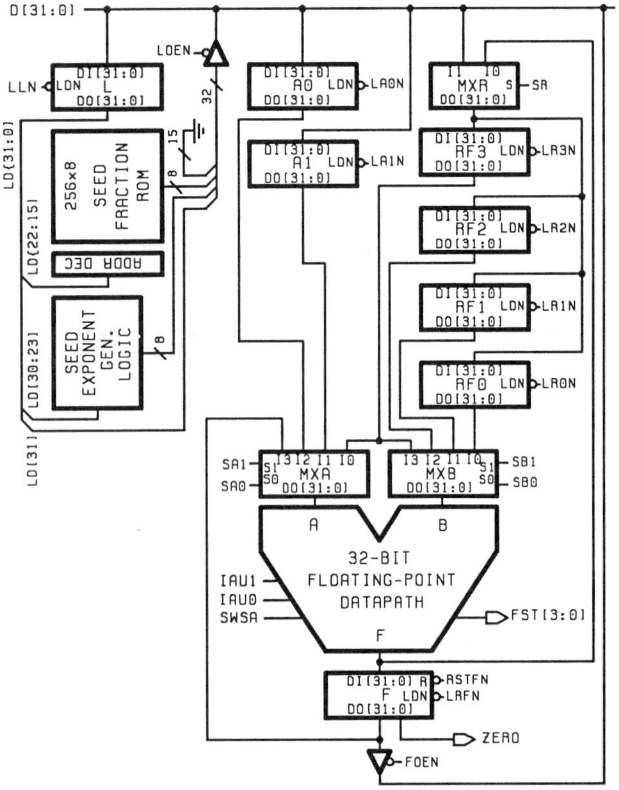

Fig. 3-3 Floating-point computational unit

3.3 Control unit

Fig. 3-4 summarizes the hierarchical structure of QRDP control scheme. The control unit contains five, hierarchically organized, state machines which perform :

- Top-level control (TOPCTRL).
- QR decompositions control (QRCTRL).
- Floating-point inversion control (IVCTRL).
- Data-in FIFO control (DICTRL).
- Data-out FIFO control (DOCTRL).

TOPCTRL directly receives the CPU commands through the control register and activates the appropriate down-level state machines QRCTRL, DICTRL, or DOCTRL. Initially, TOPCTRL transfers the configuration parameters (P, N, M, P+N, N+M) provided by the CPU command RFMGO, into the registers in AGU. TOPCTRL also asserts the CPU interrupt by setting the appropriate condition bit in the status register.

QRCTRL is activated by TOPCTRL and performs the QR decompositions. During the QR decomposition for measurement update, QRCTRL implements Eqs. 4c, 4d by using a sequence of *fast Givens rotations* and implements Eq. 4e by using a sequence of *elementary matrices.* The resulting matrices $\hat{\mathbf{x}}_{t|t}$, $L_{Pt|t}$, and $D_{Pt|t}$ are overwritten in Mmem. During the QR decomposition for time update, QRCTRL implements Eqs. 6c, 6d by using a sequence of *fast Givens rotations.* The resulting matrices $L_{Pt+1|t}$ and $D_{Pt+1|t}$ are stored in Mmem instead of Tmem for the next measurement update. QRCTRL activates IVCTRL whenever the floating-point inversion is required.

The following procedure is a detailed, register-transfer level description of QRCTRL's control flow. ICNT, TCNT, CCNT, RCNT, SCNT, ICMP, RCMP, and CCMPCNT are 5-bit counters and registers in AGU. ICMPEQ, RCMPEQ, and CCMPEQ are true when [ICNT]=[ICMP], [RCNT]=[RCMP], and [CCNT]=[CCMPCNT], respectively. OMFLG is the output of the control flag MFLG. The inversion seed value is generated by placing the operand to be inverted in the register L which drives the seed value generation logic. QRCTRL activates IVCTRL by asserting INVGO. ITRDN and INVDN are asserted by IVCTRL after one and two Newton-Raphson cycles, respectively. IVCTRL returns the inverted operand in the FPU register RF3. During measurement update, the control signal UPC is held low and Umem denotes Mmem. During time update, UPC is held high. In this case, Umem denotes Tmem and !Umem denotes Mmem. Each state corresponds to one machine cycle :

ST00 if (!QRGO) then wait (QRGO)
  else if (UPC) then ICNT ← 1, ICMP ← N,
    TCNT ← P+2, CCMPCNT ← 2
  else ICNT ← 1, ICMP ← P //set diag. indexing
ST01 if (UPC) then CCNT ← N+M
  else CCNT ← P+N+1, CCMPCNT ← P+2 //set clm. indexing
ST02 A0, RF2, L ← Umem@II (a_i) //addr[9:0] ← {[ICNT],[ICNT]}
ST03 RF2 ← A0 × RF2 (a_i^2)
ST04 RF3 ← SEED VALUE for $1/a_i$
ST05 assert INVGO
ST06 if (!ITRDN) then assert INVGO and goto ST06
  else if (!INVDN) then goto ST06
  else A0, RF1 ← Umem@IC (b_i) //addr[9:0] ← {[ICNT],[CCNT]}
ST07 RF1 ← A0 × RF1 (b_i^2)
ST08 RF0 ← − A0 × RF3 ($-b_i/a_i = M_a$)
ST09 A0, L ← Umem@0I (d_i) //addr[9:0] ← {00000,[ICNT]}
ST10 RF2 ← A0 × RF2 ($a_i^2 d_i$)
ST11 RF3 ← SEED VALUE for $1/d_i$
ST12 assert INVGO
ST13 if (!ITRDN) then assert INVGO and goto ST13
  else if (!INVDN) then goto ST13
  else RF3 ← − RF0 × RF3 ($b_i/a_i d_i$), A1 ← Umem@0C (d_j)
ST14 F ← A1 × RF1 ($b_i^2 d_j$)
ST15 F ← F + RF2 ($a_i^2 d_i + b_i^2 d_j$)
ST16 RF1 ← A1 × RF3 ($b_i d_j / a_i d_i = M_b$), A0, L ← F
ST17 RF3 ← SEED VALUE for $1/(a_i^2 d_i + b_i^2 d_j)$
ST18 assert INVGO
ST19 if (!ITRDN) then assert INVGO and goto ST19
  else if (!INVDN) then goto ST19
  else A0 ← Umem@0I (d_i)
    and RF3 ← RF2 × RF3 [$(a_i^2 d_i/(a_i^2 d_i + b_i^2 d_j)$)]
ST20 F ← A0 × RF3 [$a_i^2 d_i^2/(a_i^2 d_i + b_i^2 d_j) = d_i'$]
ST21 if (UPC & CCMPEQ) then !Umem@0T ← F (d_i')
  else Umem@0I ← F (d_i')
ST22 if (ZERO) then assert SSNGRN and goto ST00 //zero := ![F]
    //singular error : $d_i' = 0$
  else F ← A1 × RF3 [$a_i^2 d_i d_j/(a_i^2 d_i + b_i^2 d_j) = d_j'$]
ST23 Umem@0C ← F (d_j')
ST24 if (ZERO) then assert SSNGRN and goto ST00
    //singular error : $d_j' = 0$
  else if (UPC) then RCNT ← I, RCMP ← N, SCNT ← P+I
    and set MFLG
  else RCNT ← I, RCMP ← P and reset MFLG
    //set row indexing
ST25 A0 ← Umem@RC (b_i)
ST26 RF2 ← A0 × RF1 ($M_b b_i$), A1 ← Umem@RI (a_i)

ST27 F ← A1 + RF2 ($a_i + M_b b_i = a_i'$) //fast Givens rotation
ST28 if (UPC & CCMPEQ) then !Umem@ST ← F (a_i'), SCNT++
else Umem@RI ← F (a_i')
ST29 RF2 ← A1 × RF0 ($M_a a_i$)
ST30 F ← A0 + RF2 ($M_a a_i + b_i = b_i'$) //fast Givens rotation
ST31 Umem@RC ← F (b_i')
ST32 if (!RCMPEQ) then RCNT++ and goto ST25
else if (!OMFLG) then RCNT ← C–1, RCMP ← P+N,
set MFLG and goto ST25
else if (!CCMPEQ) then CCNT – – and goto ST02
else if (!UPC) then CCNT – – , A0, L ← Umem@II (a_i)
and goto ST33
else if (!ICMPEQ) then ICNT++, TCNT++, CCMPCNT++
and goto ST01
else assert QRDN and goto ST00 //end of QR (time update)
ST33 A1 ← Umem@IC (b_i)
ST34 RF3 ← SEED VALUE for $1/a_i$
ST35 assert INVGO
ST36 if (!ITRDN) then assert INVGO and goto ST36
else if (!INVDN) then goto ST36
else RF0 ← – A1 × RF3 ($-b_i/a_i = \alpha$), RCNT ← I,
RCMP ←P and reset MFLG //set row indexing
ST37 A0 ← Umem@RI (a_i)
ST38 RF1 ← A0 × RF0 (αa_i), A1 ← Umem@RC (b_i)
ST39 F ← A1 + RF1 ($\alpha a_i + b_i = b_i'$) //Gaussian elimination
ST40 Umem@RC ← F (b_i')
ST41 if (!OMFLG & RCMPEQ) then RCNT++, RCMP ← P+N,
set MFLG and goto ST37
else if (!OMFLG & !RCMPEQ) then RCNT++ and goto ST37
else Umem@RI ← "0" //overwrite $\bar{K}'_t$ by 0s'
ST42 if (!RCMPEQ) then RCNT++ and goto ST37
else if (!ICMPEQ) then ICNT++ and goto ST01
else assert QRDN and goto ST00 //end of QR (msrmnt update)

In the above procedure, QRCTRL does not perform the dummy computations for the upper triangular zero elements. At the end of the QR decomposition for measurement update, the $\bar{K}'_t$ portion is overwritten by 0s' for the next iteration. All the necessary preset paths for each counter and register in AGU are individually implemented and are connected to the set and reset nodes of flip-flops (instead of D input) through the multiplexer. This ensures that the multi-preset operation can be achieved in one machine cycle, and preset values are stable at the beginning of the next machine cycle.

IVCTRL is activated by QRCTRL and executes the two cycles of Newton-Raphson iteration for the floating-point inversion by implementing $x_{k+1} = x_k(2 - Ax_k)$, where x_0 is a seed value for 1/A. Before activating IVCTRL, QRCTRL places the operand to be inverted (A) and its seed value (x_0) in the FPU registers A0 and RF3, respectively. Register-transfer level description of IVCTRL's control flow is given as follows :

ST0 if (!INVGO) then wait (INVGO)
else F ← A0 × RF3 //Ax_k
ST1 F ← 2 – F //2 – Ax_k
ST2 RF3 ← F × RF3 //x_k(2 – Ax_k)
ST3 if (!ITRDN) then set IVFLG and goto ST0
else assert INVDN, reset IVFLG and goto ST0

Here, ITRDN is the output of the control flag IVFLG.

DICTRL and **DOCTRL** are activated by TOPCTRL and control the data-receive and data-transfer operations, respectively. As long as Data-in FIFO (DIFIFO) is not empty, DICTRL successively moves the 32-bit data from DIFIFO to either Mmem or Tmem at every other machine cycle. As long as Data-out FIFO (DOFIFO) is not full, DOCTRL successively moves the 32-bit data from Mmem to DOFIFO at every other machine cycle. The host processor should transmit and receive the elements of measurement and time update matrices in the specific orders defined by DICTRL and DOCTRL.

Outputs of state machines pass through the buffer logic which combines the control signals through the pipeline delays and generates the multi-tasking control signals for register updates in AGU, memory accesses, data transfers between registers, and floating-point operations.

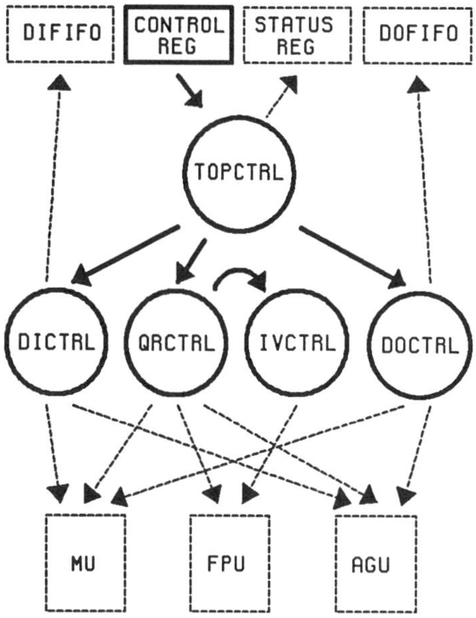

Fig. 3-4 QRDP control scheme

3.4 Interface unit

The interface unit includes the control register, status register, DIFIFO, and DOFIFO. QRDP supports the register access by the host processor and the DMA-controlled data transfer. The external 32-bit databus is shared by the CPU access and the DMA-controlled data transfer.

Control register receives the CPU commands which include five bits of specific GO control commands and two control bits for enabling/masking the CPU interrupt on the overflow and underflow conditions. There are five CPU commands: 1) RFMGO always includes the configuration parameters (P, N, M, P+N, N+M) for specifying the Kalman filter dimension and orders QRDP to receive the entire elements of initial measurement update matrix (D_{Ro}, 1, D_{Po}, L_{Ro}, $-\mathbf{e}_o$, $H_o L_{Po}$, 0s', $\hat{\mathbf{x}}_o$, L_{Po}, 0s'). 2) RMGO orders QRDP to receive the part of measurement update matrix (L_{Rt}, D_{Rt}, $-\mathbf{e}_t$, $\hat{\mathbf{x}}_{t|t-1}$, $H_t L_{Pt|t-1}$). 3) RTGO orders QRDP to receive the entire elements of time update matrix ($F_t L_{Pt|t}$, $D_{Pt|t}$, $G_t L_{Qt}$, D_{Qt}). 4,5) SMGO and STGO order QRDP to send the resulting matrices after the QR decomposition for measurement and time updates, respectively.

Upon receiving RFMGO or RMGO, external DMA controller and internal state machine DICTRL simultaneously transfer the 32-bit data from system memory to DIFIFO in QRDP and from DIFIFO to Mmem, respectively. After receiving all data, QRDP immediately starts the QR decomposition for measurement update. In case of receiving RTGO, DICTRL transfers data from DIFIFO to Tmem and, after receiving all data, QRDP immediately starts the QR decomposition for time update. Upon receiving SMGO or STGO, DOCTRL and DMA controller transfer the 32-bit data from Mmem to DOFIFO and from DOFIFO to system memory, respectively.

In summary, the host processor controls QRDP by sequentially issuing the commands RFMGO → SMGO → RTGO → STGO → RMGO → SMGO → STGO → RMGO · · · · . When the order of Kalman filter is changed, the host processor should provide the entire elements of the new, initial measurement update matrix in the system memory and assert RFMGO to QRDP which includes the new configuration parameters.

Status register indicates the status of QRDP. There are eight condition bits : 1,2) MRDY and TRDY inform that QR decomposition for measurement update and time update have been completed, respectively, and resulting matrices are available. 3,4) MODN and TODN inform that the data transfer initiated by the CPU command SMGO and STGO are completed, respectively. 5) SNGR informs that the state estimate error covariance $P_{t|t}$ or $P_{t+1|t}$ becomes singular. In this condition, QRDP immediately resets the internal logic. Hence, the host processor should provide the new, initial measurement update matrix and re-initialize QRDP by asserting RFMGO command. 6,7,8) OVF, UDF, and NAN inform that overflow, underflow, and NAN operand have arisen, respectively. In these conditions, QRDP still continues the computation by using a saturated overflow, sudden underflow, or quiet NAN operand.

DIFIFO and **DOFIFO** are 32-bit wide, four-deep input and output FIFOs and support the DMA-controlled data transfer in the burst mode, in which the entire data transfer is completed without an interruption (one-cycle transfer with no cycle-stealing and no handshaking). To achieve the maximum data transfer rate, the data transfer rate of the external DMA controller should be synchronized to that of QRDP. The DMA controller can suspend the data transfer at any time, for example, at the page boundary of system memory or during the DRAM refresh cycle. During this time, QRDP keeps the Data-Request/Ready signal active and stays in the waiting status until DIFIFO is not empty or DOFIFO is not full.

3.5 Design considerations

There are two critical paths which limit the operating frequency of QRDP. One is the fully combinatorial floating-point datapath which performs the floating-point operation including mantissa alignment, addition/multiplication, renormalization, rounding, and exception handling in one machine cycle. The following table summarizes the physical dimensions of state machines in the two-level logic implementation case. Here, # of I/Os include the state feedbacks and # of minterms are counted for the maximum ones. The state feedback paths of QRCTRL are also major critical paths. To speed up, QRCTRL should be implemented as a dynamic PLA. Assuming the implementation by 0.8 μm CMOS technology, the target operating frequency can be 20 MHz.

	Ins	Outs	P. terms	Min. terms
TOPCTRL	15	28	19	11
QRCTRL	15	60	65	37
IVCTRL	4	11	5	3
DICTRL	12	29	26	14
DOCTRL	10	27	13	7

The operating frequency of QRDP can be improved by providing a pipeline stage inside the floating-point datapath. However, this will increase the complexity of the control state machine QRCTRL which is also a critical path. In this case, QRCTRL should be split into two separate state machines, one for implementing a sequence of *fast Givens rotations* and another for implementing a sequence of *elementary matrices.*

3.6 Performance analysis

The proposed QRDP architecture has been implemented in the behavioral and register-transfer levels by using the Verilog-XL† HDL. The following table shows the number of machine cycles required for performing each of the procedures. Procedures RFMGO, RMGO, RTGO, SMGO, and STGO are directly initiated by the host processor and are for the DMA-controlled data transfers between QRDP and system memory. Procedure QRMGO (QR decomp. for measurement update / QRGO with UPC : low) is immediately executed after completing RFMGO or RMGO. Procedure QRTGO (QR decomp. for time update / QRGO with UPC : high) is immediately executed after completing RTGO. Simulations have been conducted for N = M = 3,6,9,12 with fixed P = 3. For the data-transfer procedures, full synchronization of the data transfer rates between QRDP and external DMA controller is assumed.

	N=M=3	N=M=6	N=M=9	N=M=12
RFMGO	99	201	339	513
RMGO	49	73	97	121
QRMGO	**835**	**1816**	**3013**	**4426**
RTGO	49	169	361	625
QRTGO	**776**	**3972**	**10669**	**21947**
SMGO	25	67	127	205
STGO	19	55	109	181

P = 3 Unit : machine cycle

Run-times for the data-transfer procedures are small compared to the run-times for the QR decompositions. As the dimension of the Kalman filter is increased, run-time for the QR decomposition for time update is drastically increased. If we assume 50 nsec of machine cycle, for the 9-state case (P = 3, N = M = 9), procedure QRMGO takes 0.151 msec and procedure QRTGO takes 0.533 msec.

The performance of QRDP is summarized in the following table. MIPS and MFLOPS ratings are measured as additional metrics for the performance evaluation, although they are dependent on the instruction set and type of the task.

	QRMGO	QRTGO
Total run-time (machine cycle)	3013	10669
Total run-time (millisecond)	**0.151** msec	**0.533** msec
# of instructions	4326	14190
MIPS rating	**28.72**	**26.60**
# of floating-point additions	606	2109
# of floating-point multiplications	1020	3864
Ratio of total # of floating-point oprns. to total run-time (m. cycle)	53.9 %	55.9 %
MFLOPS (normalized) rating	**10.79**	**11.20**

P = 3, N = M = 9, Fsysclk = 20 MHz

A floating-point addition or multiplication is performed in one machine cycle. A floating-point inversion requires two floating-point additions and four floating-point multiplications

† Verilog is a mixed-level hardware description language and interactive simulator. Verilog is a trademark of Cadence design systems, Inc.

(two Newton-Raphson iterations). In measuring MIPS, each of the following major operations is counted individually :

- Memory read and register load.
- Memory write.
- Register-to-register transfer.
- Floating-point operation and register load.
- Register load.
- Conditional branch.

The MFLOPS rating of QRDP is comparable to that of an array processor (single processing element) which is typically capable of performing 10 MFLOPS at the 20 MHz system clock. High MIPS rating of QRDP is due to the nature of the *dedicated* architecture.

4. Summary and Conclusions

QRDP executes the QR decomposition procedures by using a state machine which is fully optimized for the repetitive computations of *fast Givens rotations* and *elementary matrices.* This hardware design approach for the control scheme, which is supported by the dedicated memory and datapath organizations, provides very fast multi-tasking operations for the matrix indexings and floating-point operations, although this prohibits QRDP from having the flexible programming environment.

In target-tracking applications, the sampling interval is typically on the order of 10^{-1} sec ~ 10^{-2} sec. For the 9-state case, performance evaluation shows that QRDP can achieve the submillisecond performances for both measurement and time updates at the 20MHz system clock. The speed of the proposed architecture can be further improved by providing a pipeline stage, which implements the *MAC* function, inside the floating-point datapath. The 2976 bytes of on-chip memory directly dictates the maximum order of the square-root Kalman filter as $N_{max} = M_{max} = 12$, and $P_{max} = 6$. This internal memory size can be increased, perhaps as much as twice. In this case, the maximum allowable order of the Kalman filter can reach up to 20.

The system level evaluation incorporating a 32-bit host processor, a 32-bit DMA controller, and a system memory, remains to be performed. This three-chip set (QRDP, host processor, and DMA controller) for the implementation of the extended square-root Kalman filter can be a *cost-effective*, alternative solution to the parallel implementation on array processors without considerably degrading the speed performance. The performance comparison with parallel algorithms implementing on array processors such as Intel's Hypercube or iWarp† machine would be an interesting task.

The development of a *dedicated* hardware chip such as QRDP requires a long design cycle for both product definition and implementation. Proposed future enhancements of the QRDP architecture include : 1) A pipeline stage will be added in the floating-point datapath to increase the operating frequency and computational throughput. 2) QRDP will be designed as a *scalable* processor so that a number of mesh-connected QRDP chips could be used for the QR decomposition of an arbitrarily large matrix. 3) QRDP will be equipped with the *flexible* programming environment. 4) The architecture of QRDP will be modified such that QRDP can be configured as an "on-the-fly" chip although this will considerably increase the off-chip data communication complexity.

† iWarp is a mesh-connected systolic array processor. iWarp is a trade mark of Intel Corporation.

Acknowledgement

I would like to thank Advanced Micro Devices, Inc. for the usage of the CAD tools : Verilog-XL and SUN 4/490 for the algorithm and logic simulations, Mentor/Apollo workstation for the schematic captures and logic syntheses.

References

[1] R. Baheti, Efficient approximation of Kalman filter for target tracking, *IEEE Trans. Aerospace and Electronic Systems*, Vol.AES-22, pp.3-14, Jan. 1986.

[2] R. Baheti, D. O'Hallaron and H. Itzkowitz, Mapping extended Kalman filters onto linear arrays, *IEEE Trans. Automatic Control*, Vol.AC-35, pp.1310-1319, Dec. 1990.

[3] G. Bierman, Measurement updating using the *U-D* factorization, *Proc. IEEE Conf. Decision and Control*, Houston, Texas, pp.337-346, Dec. 1975.

[4] G. Bierman, *Factorization methods for discrete estimation*, Academic Press, New York, 1977.

[5] M. Birman *et. el.*, Design of a high-speed arithmetic datapath, *Proc. ICCD Conf. VLSI in Computers and Processors*, pp.214-216, 1988.

[6] A. Gelb *et. el.*, *Applied optimal estimation*, MIT Press, Massachusetts, 1974.

[7] W. Gentleman, Least squares computations by Givens transformations without square-roots, *J. Inst. Math. Applics.*, Vol.12, pp.329-336, 1973.

[8] G. Golub and C. Van Loan, *Matrix computations*, Johns Hopkins Univ. Press, Maryland, 1983.

[9] J. Hennesy and D. Patterson, *Computer architecture : a quantitative approach*, Morgan Kaufmann Publishing, California, 1990.

[10] K. Hwang, *Computer arithmetics : principles, architecture, and design*, John Wiley and Sons, 1978.

[11] A. Jazwinski, *Stochastic processes and filtering theory*, Academic Press, New York, 1970.

[12] J. Jover and T. Kailath, A parallel architecture for Kalman filter measurement update and parameter estimation, *Automatica*, Vol.22, pp.43-57, Jan. 1986.

[13] C. Mead and L. Conway, *Intro. to VLSI system*, Addison-Wesley Publishing, Massachusetts, 1980.

[14] M. Morf and T. Kailath, Square-root algorithms for linear least-squares estimation and control, *IEEE Trans. Automatic Control*, Vol.AC-20, pp.487-497, Aug. 1975.

[15] M. Morf, G. Siduh, and T. Kailath, Some new algorithms for recursive estimation in constant, linear, discrete-time systems, *IEEE Trans. Automatic Control*, Vol.AC-19, pp.315-323, Aug. 1974.

[16] S. Samudrala, M. Gavrielov, and V. Peng, On the implementation of shifters, multipliers, and dividers in VLSI floating-point units, *Proc. 8th IEEE Symp. Computer Arithmetic*, pp.95-102, 1987.

[17] R. Singer, Estimating optimal tracking filter performance for manned maneuvering targets, *IEEE Trans. Aerospace and Electronic Systems*, Vol.AES-6, pp.473-483, Jul. 1970.

[18] G. Stewart, *Intro. to matrix computations*, Academic Press, New York, 1973.

[19] S. Waser and M. Flynn, *Intro. to arithmetic for digital systems designers*, CBS College Publishing, New York, 1982.

Partitioning Considerations in systolic array design *

Prashanth Kuchibhotla
Dept. of AMES
University of California, San Diego
La Jolla, CA 92037-0411

Bhaskar D. Rao
Dept. of ECE
University of California, San Diego
La Jolla, CA 92037-0407

Abstract

Various methods for mapping algorithms into systolic arrays have been developed in the past few years. In this paper, efficient scheduling techniques are developed for methods in partitioning problems that do not match the array size exactly. In particular, scheduling for the Locally Parallel-Globally Sequential (LPGS) technique and the Locally Sequential-Globally Parallel (LSGP) technique are developed. The scheduling procedure developed exploits the fact that after LPGS and LSGP partitioning, the locality constraints get modified allowing one to use certain inter-connections that were not available before. The scheduling method allows us to develop a flexible scheduling order for LPGS that is useful in evaluating a trade-off between execution time and size of partitioning buffers. The scheduling techniques are illustrated with the help of matrix multiplication and QR decomposition examples.

1 Introduction

Recent advances in VLSI technology and design has made possible the realization of low cost Application Specific Array Processors (ASAPs) for computationally intensive image and signal processing algorithms. A well known approach to VLSI implementation of a signal processing algorithm is to design a *systolic* array for that algorithm. A systolic array is an ASAP made up of regularly-connected processing elements, all of which are connected locally and exploit the inherently regular nature of signal processing algorithms in order to achieve the high throughput demands of the tasks.

Considerable effort has been expended by researchers to find procedures to map algorithms onto a systolic array [1,2,3]. These methods aim to systematically map a given algorithm into a high-level hardware processor array form. Very often the array that is obtained by these mapping procedures is suited only for a particular size of the problem. An array designed for that size only could be very inefficient or even unusable if the target problem size were bigger. It is critical that the design system should incorporate a technique for *partitioning* a larger problem as well as provide extra hardware devices that can handle this partitioning.

In section 2, we review the main issues involved in partitioning systolic algorithms, along with a brief description of the two main categories of partitioning methods. In section 3, we discuss a scheduling method based on modification of the locality constraints of partitioned Dependence Graphs, as well as a scheduling order for LPGS scheme when there is a constraint on the size of the buffers that are needed for partitioning. The benefits of the proposed procedures are illustrated with the help of matrix multiplication and QR decomposition algorithms in Section 4.

2 Mapping & Partitioning issues

General partitioning methods are often derived from the mapping procedures for full-size arrays. Most mapping procedures are only applicable to a class of algorithms that can be represented by nested loops with uniform dependencies. The algorithms are written in the form of for-loops and in Single-Assignment form; i.e. each variable is assigned a value only one time during the algorithm, no variable is overwritten. The algorithm is then represented by an indexed dependence graph (DG) in an Index space. The DG contains nodes at various points in the Index space representing all the computations that are indexed by that index point. The edges represent the data dependencies between computations at various index points. This DG form is useful in removing unwanted data broadcasts by allowing dependencies between adjacent nodes only. This spatial locality preserves the local nature of inter-processor connections characteristic of a systolic array.

*This work was supported in part by the UCSD/NSF Integrated Circuits And Systems Research Center and by the ARMY Research Office under Grant No. DAAL-03-90-G-0095.

1058-6393/91 $01.00 © 1991 IEEE

Example: In a regular format, the algorithm for the multiplication of two matrices A & B would be;

```
for i = 1, 2, ..., n
    for j = 1, 2, ..., n
        for k = 1, 2, ..., n
            C(i, j) = C(i, j) + A(i, k) * B(k, j)
        end
    end
end                                              ...(1)
```

Variables A & B have to be broadcast simultaneously to various indexed nodes. The same algorithm in localized Single Assignment form is shown below;

```
for i = 1, 2, ..., n
    for j = 1, 2, ..., n
        for k = 1, 2, ..., n
            A(k, i, j) = A(k, i, j - 1)
            B(k, i, j) = B(k, i - 1, j)
            C(k, i, j) = C(k - 1, i, j) + A(k, i, j) * B(k, i, j)
        end
    end
end                                              ...(2)
```

The mapping techniques then allocate sets of nodes to be computed on the processors as well as map dependencies onto inter-processor connections in the systolic array. The design process also schedules the computational nodes such that the dependency ordering is preserved while ensuring that there are no processor resource conflicts. For typical algorithms, the mapping and scheduling is a linear operation.

2.1 Partitioning issues

Usually, full-size systolic arrays do not have any or have very little local memory. The critical idea is to access a required data element from main memory only once . The systolic array then performs all the computations that have to done using that data element. However, general partitioning methods break up the problem DG into partitions of the size that can be handled by the target array of smaller size. Data items that were used by one partition might be needed by other partitions in the DG. Hence that item might have to be accessed more than once.

In order to avoid reaccessing the main memory, extra storage buffers (either local or external to the processors) are needed to transfer data in between the partitions. The amount of data transfer, and hence the size of these buffers depends on the partitioning scheme and the scheduling algorithm that is used. Therefore, for problems of varying sizes and a fixed systolic array, the size of the buffers required for partitioning is not fixed. There is an upper bound on the problem size that the array can handle that is dependent on the size of the buffers built into the array architecture.

The standard partitioning methods which schedule and allocate processors are [2];

(1) Locally Parallel-Globally Sequential (LPGS): In this technique, the size of the DG partitions match the size of the available array. The array then executes (in parallel) each of these partitions one after the other (globally sequential).

(2) Locally Sequential-Globally Parallel (LSGP): In LSGP, each of the DG partitions is executed by a specific processor in the available array. The tasks performed by all the nodes in a partition are done sequentially in the assigned processor (locally sequential). Since the partitions are handled separately by different processors, they can run in parallel (globally parallel).

The storage requirements of the two methods are different in that the LPGS scheme needs external buffers while the LSGP scheme needs buffers internal to the processors [4]. The issue of scheduling algorithms for these methods is addressed in the next section.

3 Scheduling & Resource allocation

Scheduling the partitions of partitioned DGs using either LPGS or LSGP scheme is a non-linear operation. Some attempts have been made to formalize scheduling of a partitioned system [1]. Although research work has gone into optimizing dedicated full-size array design, limited attention has been paid to the problem of optimal partitioning of algorithms. The question of what scheduling algorithm & task allocation exactly constitutes the *best* partitioning in terms of execution speed, processor efficiency and hardware overhead (buffers, etc.) has yet to be fully addressed.

Approaching the problem from a partitioning viewpoint, we try to optimize the execution time under the constraint of a fixed buffer size. Another approach we take is to optimize the buffer size with a constraint on the maximum acceptable execution time.

3.1 Relocalizing Partitioned DGs

In standard systolic array design, it is important that the need for data broadcasts be eliminated. The algorithm is therefore localized such that any broadcasts that may occur are reduced to a sequence of local broadcasts. The example in Section 2 shows the localization of the MM algorithm. Such an algorithm allows the designer to construct a DG whose edges are spatially local.

The DG is then mapped onto a systolic array with local inter-connections. A linear schedule can be found for scheduling the localized DG on the systolic array [1,2]. The cost that is incurred for requiring elimination of broadcasts is a reduction in the speedup that can be achieved by the processor set using data broadcasts. The absence of broadcasts now forces the processors to wait for the data to arrive via local inter-connections.

When a DG is partitioned, the allocation of computational nodes to processors is non-linear. The nodes assigned to a processor could be adjacent (as in LSGP) or disjoint (as in LPGS). An important question in the context of partitioning is "Is it possible to obtain the best possible computational scheme using the completely localized DG for scheduling the partitions of nodes". The answer, as we shall see, is no. *The complete localization that is necessary for full-size systolic array design is an overkill for the partitioning problem.*

In typical systolic algorithms, there are data variables that are propagated and used throughout the array but their value is not modified anywhere. For these *transmittent* variables, the localization is artificial. The local dependencies they create are not strictly necessary. We can keep track of such dependencies and they can be exploited in the partitioned DG where the definition of locality gets modified.

For example, consider the simple DG shown in Fig 1a. The transmittent variable A is broadcast to the nodes numbered 1, 2, 3 and 4. The systolization procedure localizes the broadcasts by propagating the variable as shown in Fig 1b.

In LSGP partitioning, each of the demarcated sets of nodes are assigned to a single processor (Fig 2a). Such a partitioning now makes nodes 1 and 3 *adjacent.* Now we can allow a 'non-local' connection between 1 and 3 as shown in the redrawn DG in Fig 2b. Similarly, we observe that when the DG is partitioned by the LPGS technique (Fig 2c), nodes 1 and 3 are assigned to the *same* processor. Therefore they can be considered 'local' to each other.

A similar analysis can be applied when partitioning other DGs where transmittent data streams that could have been broadcast are localized by the systolization procedure. The above approach *relocalizes* the DG so that some of the speedup that has been lost is regained. It is important to note that this approach cannot work if the data values are modified at each node they pass through. However, there is a large class of algorithms where unchanged, transmittent variables are used, and the relocalizing approach is applicable for this class. A scheduling order which exploits the *new* locality is described in the next sub-section.

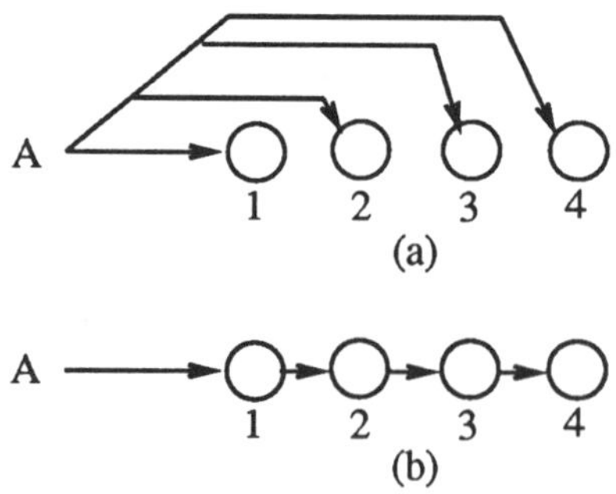

Figure 1: Non-local & Localized DGs

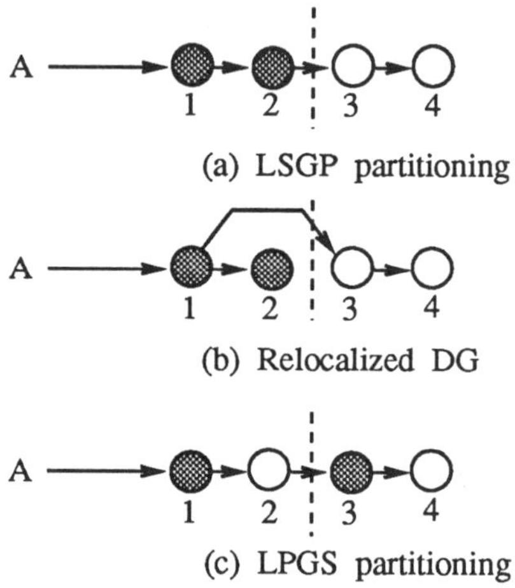

Figure 2: Partitioned and relocalized DGs

3.2 A Scheduling order for LPGS

In standard LPGS scheduling, each of the partitions for the DG are executed on the target array in strict sequential order. *The relocalized DG allows us to relax the rigid sequential restriction and overlap the computations in different partitions, while maintaining the dependency ordering in the DG.*

We can schedule *a part of a partition* (called a *block*) first, then execute a block from another partition, and so on. Then we *return to the first partition*, execute another block of this partition and resume the ordering again. The size of the block that is selected can depend on the amount of FIFO word storage that is available for partitioning. This will be discussed later.

For example, consider the index-space of the DG for the MM problem (Fig 3). The DG has a node at each point in the cube and the dependence edges $d_1 = [1\ 0\ 0]^T$, $d_2 = [0\ 1\ 0]^T$ & $d_3 = [0\ 0\ 1]^T$ are uniform. The two partitioning planes corresponding to $i = n/2$ & $j = n/2$ are shown. Standard LPGS scheduling tells us

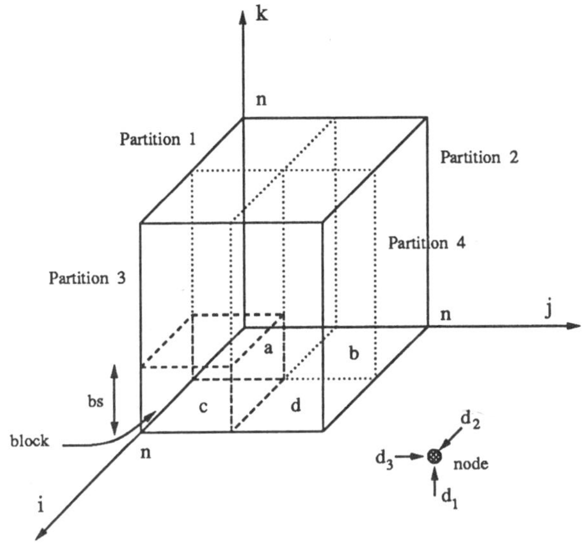

Figure 3: Index space for Matrix Multiplication (MM)

to schedule partitions 1, 2, 3 and 4 in that order. Consider *a layer* of the index space corresponding to $k = 1$. The part (marked a) of partition 1 in this layer, can be scheduled first, the part in partition 2 (marked b) next, then the part c in partition 3, and part d in partition 4. The execution order then starts again with the layer in partition 1 corresponding to $k = 2$, etc.

3.3 Buffer Structures for LPGS

The main motivation behind the modification to the standard LPGS ordering is to reduce the buffer sizes that are required by the partitioning, hopefully without compromising speed, while being able to maximize the size of the problem the array can handle. Following the analysis used in [4] for computing buffer size complexity, we find that if we execute smaller blocks of the partitioned DG, the amount of data that has to be transferred between partitions is smaller. Therefore, the size of the required FIFO buffer structure is smaller than what we would need if the entire partition were executed before going to the next partition.

For example, assume that an $n \times n$ MM algorithm is implemented on a smaller $m \times m$ systolic array. For standard LPGS ordering, we will need $2m$ FIFOs, each with about kn words of storage ($k = n/m$). If we order the computations a layer at a time, m FIFOs, each with only k words of storage, are required.

The cost of this storage reduction is an increase in execution time of the array. However, when there is a hard limit on the storage that is available to the designer and there is not enough storage available to handle a large problem, the proposed ordering scheme is a solution to the problem. The size of the available buffer can be used to select the size of the block.

Ideally, the desirable size of the block (bs), in Fig 3, is its largest possible value (bs = size of partition = n). This gives the best possible speedup for the partitioned problem. If the storage limitations do not allow this, then one will have to look for a trade-off to find the largest block size for which the scheduling will allow the limited storage capacity to be used efficiently.

The proposed scheduling ordering, when combined with the relocalization procedure, results in an improved performance over standard LPGS scheduling [2,1] in terms of a speedup vs. storage trade-off.

The buffer size requirement when a single layer is executed at a time (i.e., $bs = 1$) is the limiting factor in the maximum size of the problem the array can handle. Beyond that size, some of the input data has to reaccessed from the main memory, and consequently the slow I/O speed between the main memory and the array could severely degrade the array's performance.

3.4 Nonlocality inside LSGP partitioning

The LSGP partitioning technique creates sets of adjacent computational nodes, each set being executed by single processor. *Since all the nodes inside a set are executed in a serial fashion on the same processor, the dependency structure between these nodes need not be uniform or local.* Data broadcasts are allowed inside this set. This leads to an interesting observation regarding some algorithms in modern signal processing (like Householder Transformations) that are difficult to map onto systolic arrays because of their non-local dependency structure. Although a full-size systolic array with local interconnections may be difficult to design for such problems, a *partitioned* algorithm using the LSGP method could be designed and mapped onto a smaller systolic array. This is possible if the non-uniform or non-local dependencies are clustered into the same LSGP set (or sets).

4 Results

A schedule based on Relocalized DGs and modified ordering was computed for two common matrix-based tasks in signal processing; matrix multiplication (MM) and QR decomposition (QRD). The full-size array used for $n \times n$ MM is a square array of $n \times n$ processors connected via a simple mesh. The available smaller array is assumed to be of size $m \times m$.

Table 1 : LPGS Partitioning, $(n = 100, m = 10)$

block size	MM		QRD	
(bs)	T_{exec}	buffer size	T_{exec}	buffer size
1	10018	100	18118	90
2	10018	200	10116	180
4	10018	400	7062	360
5	10018	500	6510	450
10	10018	1000	5500	900
20	10018	2100	5480	1900
25	10018	2650	5470	2400
50	10018	5400	5420	4900
100	10018	10900	5320	9900

Table 2 : LSGP Partitioning of MM, $(m = 10)$

n	T_{exec}		buffer size	
	Std. DG	Reloc. DG	Std. DG	Reloc. DG
10	28	28	100	100
20	116	98	200	200
30	324	288	300	300
40	712	658	400	400
50	1340	1268	500	500
60	2268	2178	600	600
70	3556	3448	700	700
80	5264	5138	800	800
90	7452	7308	900	900
100	10180	10018	1000	1000

The full-size array required by the QRD to triangularize a $p \times n$ matrix is the classic QR-triarray of side n processors as described in [5]. Let the given array have m processors on it's side.

LPGS partitioning: Table 1 shows the performance of the scheduling algorithm when LPGS partitioning is applied to the MM and QRD problems. For the MM, both input variables A and B are both of the transmittent type; and the relocalizing procedure is applicable to all edges propagating A and B. The block size $bs = 100$ corresponds to the standard way of scheduling LPGS partitioning; whereas $bs = 1$ corresponds to executing the partitions a layer at a time. As anticipated, the buffer word storage requirement is minimized for $bs = 1$. Since both A and B are transmittent, there is no loss in speed when the size of the block is reduced.

For the QRD, the variable A (corresponding to the input matrix A) is *not* of the transmittent type and cannot be relocalized. But the angles that are generated in the diagonal processors (see [5]) are transmittent and therefore can be relocalized. As the size of the block increases, T_{exec} decreases, but one pays a higher price in terms of storage requirements. However, it is interesting to notice that for $bs \geq m$, the decrease in T_{exec} is very small, while the storage requirement still rises steeply. A trade-off point between T_{exec} and storage requirements would be to have a block size $bs = m$.

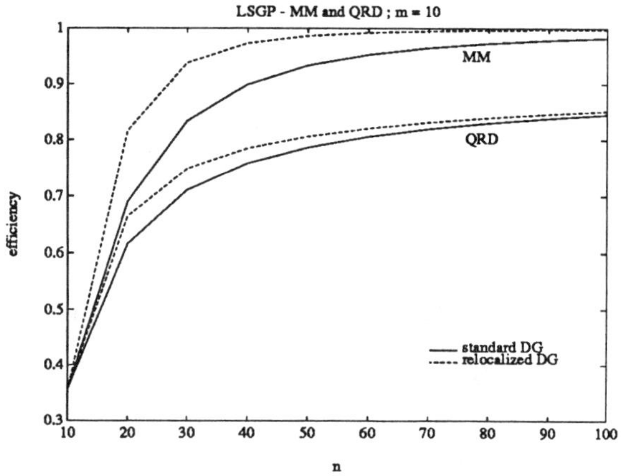

Figure 4: LSGP performance

LSGP partitioning: Table 2 shows the performance of the proposed LSGP scheduling algorithm using relocalization for MM. As the problem size n increases the T_{exec} using the proposed procedure is lower when compared to standard LSGP scheduling. Fig 4 shows a comparison of the efficiencies, defined as,

$$efficiency = T_{serial}/(\sharp processors \times T_{parallel}).$$

The relocalized scheduling is a marked improvement in terms of the efficiency, especially for low values of n/m (Fig 4). A similar trend can be seen in the performance results for QRD. The improvement in efficiency is not as marked as it is for the MM because the non-symmetric nature of the DG for QRD leads to unbalanced allocation of tasks to processors. This is an inherent limitation in the LSGP technique when applied to the QRD.

References

[1] D. I. Moldovan and J. A. B. Fortes. "Partitioning and mapping algorithms into fixed size systolic arrays". *IEEE Trans. on Computers*, C-35, Jan. 1986.

[2] S. Y. Kung. *VLSI Array Processors.* Prentice Hall, Englewood Cliffs, NJ, 1988.

[3] S. K. Rao. *Regular Iterative Algorithms and their Implementations on Processor Arrays.* PhD thesis, Stanford University, Stanford, CA, 1985.

[4] W. P. Burleson. "The partitioning problem on VLSI arrays: I/O and and local memory complexity". In *Proc. IEEE ICASSP*, pages 1217–1220, May 1991.

[5] W. M. Gentleman and H. T. Kung. "Matrix Triangularization by Systolic arrays". In *Proc. SPIE Real Time Signal Processing IV*, pages 19–26, 1981.

MSS: Message-based Systolic Structures For Real-Time Signal Processing Applications[1]

*A. O. Barbir * , J. L. Aravena** and H. Barada****

* Computer Science, Western Carolina University, Cullowhee, NC 28723
** Electrical & Comp. Engr., Louisiana State University, Baton Rouge, LA 70803
*** Computer Science and Elect. Engr., Lehigh University, Bethlehem, PA 18015

Abstract

The paper presents initial results of a research aimed at developing a class of VLSI structures better suited for the parallel implementation of dense and structured sparse matrix-matrix and triple matrix product algorithms. The paper proposes a structure for the efficient evaluation of such algorithms that consists of systolic nodes of fixed dimension connected via a global system network. The nodes have their own local memory and can access a common memory. Communication among the nodes is performed by message passing. This class of arrays will be termed message-based systolic structures (MSS). In addition, the paper develops a nonconventional method that analyzes the evolution of results in the MSS array. The methodolgy enables the designer to use the compute time as one more design parameter. This results in the ability of generating a family of low complexity high concurrence algorithms in the structure. This class of algorithms offer an attractive alternative for increasing the response time of the dedicated architecture.

I. Introduction

Recent advances in VLSI technology have made it possible to design and fabricate specialized parallel architectures for numerous applications. Parallel structures such as systolic arrays [1-5] offer attractive improvement on the computational speed as opposed to the general purpose sequential computer. However, the systolic approach for improving the computation time suffers from several problems. First, the architecture is size and problem dependent. Second, the arrays are designed for the generic algorithms and hence can not take advantage of special realizations that can be evaluated more efficiently. In addition, VLSI limitations exists that bound the number of processing elements (PEs) and I/O bandwidth in a wafer.

In 2-D large systolic arrays clock skew is unbounded, while it is bounded in linear (1-D) arrays regardless of the size [3]. As a result, multi-linear arrays [3], were recently proposed by the authors as a mean to offset the clock skew problem in arbitrary large 2-D arrays. Multi-linear arrays consists mainly of a collection of linear arrays that are designed to solve a fixed size problem. They are a good design choice since they offer : a) bounded clock skew, b) higher throughput than linear arrays, c) adaptability to changes in technology.

With the above limitations it is wise to assume that if systolic arrays were ever manufactured on a single wafer they will be of fixed size and dimension. In this case, the individual systolic chips may be connected in a cascaded fashion to simulate a larger array, or alternatively, one might consider building an architecture that combines the best features of systolic arrays such as parallelism and pipelining, and the rich connectivity offered by other topologies such as the mesh and *n*-cube. The MSS array takes advantage of the parallelism offered within an array and the high connectivity of other topologies [2]. The proposed MSS structure is a generalization of the multilinear arrays approach. It is an efficient parallel structure that is better suited for the implementation of dense and structured sparse matrix-matrix and triple matrix product.

The second part of the paper develops a nonconventional method that analyzes the evolution of results in the MSS. The work improves on previous results by showing that the dedicated array can be used to generate other classes, of potentially useful transformations. These realizations have restricted structure, but can be evaluated faster. The approach proposed considers compute time as a primary constraint and the MSS architecture as a device capable of implementing families of transformations parameterized by the compute cycle number. The methodology is applicable to any computing structure and it leads to a

[1] Work was funded in part by NSF Grant USE-9152384.

1058-6393/91 $01.00 © 1991 IEEE

class of algorithms with reduced complexity that is determined by the alloted compute time and the computing structure.

The paper is organized as follows: Section 2 introduces the new MSS architecture to implement matix-matrix product, and develops the necessary analysis tools for determining the class of algorithms generated in the structure as a function of the compute cycle. Section 3 modifies the structure to include triple matrix product and extends the procedures to cover the time constrained analysis of the triple matrix product algorithm. Section 4 concludes with some closing remarks and directions for future work.

2. MSS for Matrix Multiplication

The MSS array for computing the product, **C** = **AB**, is a specialized form of a class of MSIMD (multiple single instruction multiple data). The general architecture of the array is given in Figure 1. In the figure, the structure consists of r nodes (SN) each consisting of a systolic array of fixed dimension, and a global interconnection network termed the system interconnection network (SIN). The topology of SIN is algorithm dependent and may vary from one architecture to another. Communication algorithms among the SN are message-based and are similar to the protocols used in the n-cube [4]. For an initial analysis the communication time, t_m, will be modeled by $t_m = h + a_k$, where h is a constant overhead time, a is a constant and k is the size of the data package being transmitted.

It is convenient to introduce the specific MSS array with an example. Here, **C** = **AB**, where, **A**, **B** and **C** ε $\mathbf{R}^{nxn}$ and $n = 4$. The matrices are partitioned into sub-blocks such that

$$\begin{bmatrix} C_{11} & C_{12} \\ C_{21} & C_{22} \end{bmatrix} = \begin{bmatrix} A_{11} & A_{12} \\ A_{21} & A_{22} \end{bmatrix} \begin{bmatrix} B_{11} & B_{12} \\ B_{21} & B_{22} \end{bmatrix}.$$

The matrix product can be re-written using sub-block notations as

$$C_{ij} = \sum_{k=1}^{2} A_{ik} B_{kj} \quad ,$$

where, each sub-block is treated as an $[a_{ij}]$ or $[b_{ij}]$ entry.

The MSS array for evaluating the resultant matrix **C** ε $\mathbf{R}^{4x4}$ is depicted in Figure 2. For this particular example, the SNs are homogeneous, each consisting of $2x2$ mesh [5] connected systolic arrays, with processing elements (PEs) of the multiply-add type. Such an array will multiply two kxk matrices in $3k\text{-}1$ cycles. Here, it is assumed that the compute cycle is the time needed to perform a one multiply-add instruction. The SIN is similar to a fully connected mesh [4], with bi-directional links communicating via messages [4].

We will use the algorithm given in [4] to multiplying the sub-block matrices. The initial loading of the matrices is depicted in Figure 3a. The computation of the partial results progresses in two phases. Figure 3b depicts the partial results after phase I, while Figure 3(c,d) depicts the final results after phase II. Each phase is preceded by a specific SN node broadcasting the proper sub-blocks to selected SN nodes in the array. The movement of the data elements in the systolic nodes for phase I and II are given in Figure 4(a,b). It is noted that the operations of all systolic nodes in the figure during each phase are performed concurrently.

For simplicity, we assume that each kxk sub-block broadcast will take a fixed increment of time denoted by t_k. This includes message initiation, actual transfer time and other factors. For the general case, with a mesh connection of size p, and $n = pk$, the execution time of the pre-loaded MSS array is $T = p\ ((3k\text{-}1) + t_k)$. Since $t_k = h + a_k$ the execution time is $T = p\ ((3k+h\text{-}1) + a_k)$. This expression allows a simple analysis of partition strategies. Since the overhead time is at least one compute cycle, one can see that the total execution time increases linearly with the number of nodes. But in this case, the complexity and cost of each node becomes smaller as p increases. Thus we have the usual tradeoff of speed vs cost and the application of the conventional dictum of minimizing the amount of interprocessor communication.

Before designing an algorithm specific MSS, one must specify the following: i) decide on the SIN, ii) decide on the connectivity of PEs inside the SN, iii) decide on the types of PEs and their functionality for each SN. Here, we can use one or few different types of systolic arrays with different types of processing elements and functionality. The designing of an algorithm specific MSS consists of the following steps:

1: Divide the problem into several subproblems of smaller dimension not necessary of equal size.
2: Design systolic nodes for each subproblem.
3: Draw a graph showing the dependencies, timing and the communication between the subproblems.
4: Map he flow graph into an MSS array.

3. MSS for Triple Matrix Product

Consider the linear filtering algorithm of the form

$$\mathbf{c} = \mathbf{A}\ \mathbf{b}, \quad \mathbf{c}\ \varepsilon\ \mathbf{E}^m \quad , \mathbf{b}\ \varepsilon\ \mathbf{E}^n$$

The mxn matrix **A**, completely defines the transformation. For the case under consideration, assume that $m = n$ and, let $n = kl$. For a given n vector, **b**, define the kxl matrix **B** by using the elements in the vector as entries in the matrix. Given a kxk matrix, **L**, and an lxl matrix, **R**, compute the product **C** = **LxR**. By rearranging the entries in **C** as an n vector, **c**, it can be shown [5] that this procedure defines a linear transformation on the original vector. It can be given a representation as an nxn matrix acting on **b**.

Using tensor product notations, the $klxkl$ matrix

$\mathbf{L} \otimes \mathbf{R}$ has the block partition form $\mathbf{L} \otimes \mathbf{R} = [\, r_{ij} \mathbf{L} \,]$. Hence, the linear transformation $\mathbf{c} = [\mathbf{L} \otimes \mathbf{R}]\mathbf{b}$ can be evaluated as $C = \mathbf{L x R}^t$ [5], provided that the vectors $\mathbf{c}$, $\mathbf{b}$, are column by column ordering of the corresponding matrices. A simple operation count will show that the computation using triple matrix structure uses less operations than the conventional row-column mode for the matrix-vector form. The use of dedicated architectures increases the efficiency of the triple matrix approach [5].

Consider now the triple matrix structure $\mathbf{C} = \mathbf{ABW}$, with $\mathbf{A}$, $\mathbf{B}$, and $\mathbf{W}$ in $\mathbf{R}^{nxn}$ implemented with the following sequence

$$\mathbf{Z} = \mathbf{AB}, \text{ and } \mathbf{C} = \mathbf{ZW} .$$

The same MSS array given in Figure 2 can be used to implement the sequential triple matrix product. Here, it is assumed that the multiplicand matrices $\mathbf{A}$, $\mathbf{B}$, and $\mathbf{W}$ are decomposed into sub-blocks and pre-loaded into the memories of the SNs in the same manner as in Figure 3.

Since the sub-blocks of $\mathbf{Z} = \mathbf{AB}$ are stored in the SNs in the same manner as the sub-blocks of the matrix $\mathbf{A}$, the array can implement the triple matrix product by performing twice the algorithm of Figure 3, with the sub-blocks of the matrix $\mathbf{A}$, replaced by the proper sub-blocks of the matrix $\mathbf{Z}$. In this case, the MSS array will perform the triple matrix product $\mathbf{C} = \mathbf{ABW}$ in four phases, and the overall execution time of the structure is $T = 4\,(\,(3k-1\,) + t_k\,)$.

The working assumption that motivates this study is that for some real-time applications, even the response of a specialized architecture such as the MSS array, may not be sufficient. In this regard, the paper considers the option of stopping the array operation before its normal execution time. The research presented hare views the partial computations as potentially useful transformations which can be evaluated at a faster rate.

The above basic ideas will be illustrated by a simple 2x2 matrix product. Let $\mathbf{C} = \mathbf{AB}$ with $\mathbf{A}$, $\mathbf{B}$ and $\mathbf{C}$ ε $\mathbf{R}^{2x2}$. Using vector notation, we have

$\underline{\mathbf{A}} = col\ (a_{11}, a_{21}, a_{12}, a_{22}\,)$,

$\underline{\mathbf{B}} = col\ (b_{11}, b_{21}, bl_{2}, b_{22}\,)$,

$\underline{\mathbf{C}} = col\ (c_{11}, c_{21}, c_{12}, c_{22}\,)$,

$\underline{\mathbf{A}} \otimes \underline{\mathbf{B}} = col\ (a_{11}\, b_{11}, a_{21}\, b_{11}, a_{12}\, b_{11}, a_{22}\, b_{11}, a_{11}b_{21}, a_{21}\, b_{21}, a_{12}\, b_{21}, a_{22}\, b_{21}, a_{11}\, b_{12}, a_{21}b_{12}, a_{12}b_{12}, a_{22}\, b_{12}, a_{11}\, b_{22}, a_{21}b_{22}, a_{12}b_{22}, a_{22}\, b_{22}\,)$.

For a moment assume that the operation is implemented in the systolic array given in Figure 5. The function of each processing element is of the multiply-add type. The array produces the final results in 4 ($2n$ in general) cycles. At cycle $n = 4$, the operation $\mathbf{C} = \mathbf{AB}$ can be represented as $\underline{\mathbf{C}} = \mathbf{A}_m\, [\, \underline{\mathbf{A}} \otimes \underline{\mathbf{B}}]$, where

$$\mathbf{A}_m = \begin{bmatrix} 1000001000000000 \\ 0100000100000000 \\ 0000000010000010 \\ 0000000001000001 \end{bmatrix}.$$

However, after only two compute cycles, the multiplication is not yet completed. The partial products can be represented as $\underline{\mathbf{C}} = \mathbf{A}_2\, [\, \underline{\mathbf{A}} \otimes \underline{\mathbf{B}}]$, with

$$\mathbf{A}_2 = \begin{bmatrix} 1000001000000000 \\ 0100000000000000 \\ 0000000010000000 \\ 0000000001000001 \end{bmatrix}.$$

For the general nxn case, the operation $\mathbf{C}=\mathbf{AB}$ defines a bilinear map on the factors, given by $\mathbf{C} = \mathbf{M}(\mathbf{A},\mathbf{B})$. The map $\mathbf{M}(.,.)$ is independent of the architecture performing the product. Since the transformation is bilinear, it can be used to define a transformation $\mathbf{L}_n\,(\mathbf{A})$ acting on the matrix $\mathbf{B}$. At cycle $n = 4$ we have

$$\mathbf{L}_4\,(\mathbf{A}) = \begin{bmatrix} \mathbf{A} & 0 \\ 0 & \mathbf{A} \end{bmatrix},$$

and $\mathbf{A}_4\,(\mathbf{A} \otimes \mathbf{B}) = \mathbf{L}_4(\mathbf{A})\underline{\mathbf{B}}$. Alternatively, the operation can be viewed as a transformation $\mathbf{R}_n\,(\mathbf{B})$ acting on the matrix $\mathbf{A}$. Therefore, at compute cycle $n = 4$ one has

$$\mathbf{R}_4\,(\mathbf{B}) = \begin{bmatrix} b_{11} & 0 & b_{21} & 0 \\ 0 & b_{11} & 0 & b_{21} \\ b_{12} & 0 & b_{22} & 0 \\ 0 & b_{12} & 0 & b_{22} \end{bmatrix}.$$

The SNs in the MSS array evaluating the matrix product $\mathbf{C} = \mathbf{AB}$ will execute a number of multiplication plus addition computation at every compute cycle. After, $\tau < T$ compute cycles, it will have evaluated a subset of the total operations. Each entry of the matrix product will be partially evaluated. Therefore, if the MSS array operations are stopped after $\tau = 7 + 2t_k$, the matrix product is not completed at this stage. However, the array performs the partial transformation $\mathbf{L}_\tau\,(\mathbf{A})$ given in Figure 6. This partial transforma tion is not the complete one, but it may represents a close enough approximation for some specific applications. The above analysis gives the designer the option of stopping the array before the normal execution time to examine the usefulness of the partial results. This option is not available in regular array design, where the structure either produces results on time or fails to operate properly.

In the MSS array depicted in Figure 2, if the product $\mathbf{Z} = \mathbf{AB}$ is interrupted after n_1 cycles, where $0 \leq n_1 < 2((3k-1)+t_k\,)$, then the partial results will correspond to the linear transformation $\underline{\mathbf{Z}}\,(n_1) = \mathbf{L}\, n_1\, (\mathbf{A})\, \underline{\mathbf{B}}$. In particular, $\underline{\mathbf{Z}}\,(n_1) = [\mathbf{L} \otimes \mathbf{I}]\, \underline{\mathbf{B}}$, when $n_1 = 2((3k-1)+t_k\,)$. These partial results will feed the computations of the second stage of the product, namely $\mathbf{C} = \mathbf{ZW}$. If the product $\mathbf{C} = \mathbf{ZW}$ is constrained to n_2 cycles, where, $2((3k-1)+t_k \leq n_2 < 4((3k-1)+t_k)$, the final result will correspond to the linear transformation given by

$$\mathbf{C}(n_1, n_2) = \mathbf{R}_{n2}(\mathbf{W})\ \underline{\mathbf{Z}}(n_1)$$

$$\mathbf{C}(n_1, n_2) = \mathbf{R}_{n2}(\mathbf{W})\mathbf{L}_{n1}(\mathbf{A})\ \underline{\mathbf{B}},$$

$$\mathbf{C}(n_1, n_2) = \mathbf{T}(n_1, n_2)\ \underline{\mathbf{B}}.$$

The transformation $\mathbf{T}(n_1, n_2)$ characterizes the constrained implementa tion of the triple matrix product. By noting that $\mathbf{L}_{n1}(.)$ and $\mathbf{R}_{n2}(.)$ are linear operators, then the transformation $\mathbf{T}(n_1, n_2)$ is clearly a bilinear function of the matrices **A** and **W**. The next theorem relates $\mathbf{T}(.,.)$ to the tensor product $\mathbf{A} \otimes \mathbf{W}$.

Theorem 1: If the triple matrix product $\mathbf{C} = \mathbf{ABW}$ implemented on the MSS array of Figure 3, as

$$\mathbf{Z} = \mathbf{AB} \text{ and } \mathbf{C} = \mathbf{ZW},$$

is constrained by allocating n_1 cycles to the product $\mathbf{Z} = \mathbf{AB}$, and n_2 cycles to the product $\mathbf{C} = \mathbf{ZW}$, then the transformation on the data vector **w** is represented by

$$\mathbf{T}(n_1, n_2) = \mathbf{L}_{n1}(\mathbf{Z})\ \mathbf{R}_{n2}(\mathbf{W}).$$

Furthermore, there exist a linear transformation $\mathbf{M}(n_1, n_2)$ such that

$$\underline{\mathbf{T}}(n_1, n_2) = \mathbf{M}(n_1, n_2)\ [\underline{\mathbf{A}} \otimes \underline{\mathbf{W}}]$$

The matrix $\mathbf{M}(n_1, n_2)$ will depend only on the values of n_1 and n_2, and the specific MSS array evaluating the triple product. For notational convenience, to every kxl matrix, **A**, we associate the kl-vector $\underline{\mathbf{A}}$, obtained by stacking the column of the matrix. The following results follows by inspection.

Lemma 1: Let $\underline{\mathbf{A}}$, $\underline{\mathbf{W}}$, $\underline{[\mathbf{A} \otimes \mathbf{W}]}$ be the vector representation of the corresponding matrices. Furthermore, let $\underline{\mathbf{A}} \otimes \underline{\mathbf{W}}$ be the vector tensor product. Then there exists a unique permutation matrix, **P**, such that

$$\underline{[\mathbf{A} \otimes \mathbf{W}]} = \mathbf{P}\ [\underline{\mathbf{A}} \otimes \underline{\mathbf{W}}].$$

From the analysis, it is clear that the transformation $\mathbf{T}(n_1, n_2)$ and the matrix $[\mathbf{A} \otimes \mathbf{W}]$ are closely related. This is because the nonzero entries in $\mathbf{T}(.,.)$ match exactly the selected entries in $[\mathbf{A} \otimes \mathbf{W}]$. These entries correspond to the terms already computed in the allowed compute time. Hence, one has

$$\underline{\mathbf{T}}(n_1, n_2) = \mathbf{D}(n_1, n_2)\ \underline{[\mathbf{A} \otimes \mathbf{W}]}$$

The matrix $\mathbf{D}(n_1, n_2)$ is diagonal. An entry in the diagonal will be nonzero only if the corresponding entry of $[\mathbf{A} \otimes \mathbf{W}]$ appears in $\mathbf{T}(.,.)$. The next result follows immediately,

Lemma 2: The matrix $\mathbf{M}(n_1, n_2)$ has the structure

$$\mathbf{M}(n_1, n_2) = \mathbf{D}(n_1, n_2)\ \mathbf{P}.$$

The matrix **P** is a fixed permutation.

To illustrate the use of the above analysis tools, assume that the triple matrix product is constrained such that $\tau_1 = 2 + t_k$ cycles were available to compute the product **Z**, and $\tau_2 = 1 + t_k$ cycles were available to evaluate the resultant matrix **C**. The resulting $\mathbf{T}(\tau_1, \tau_2)$ is given in Figure 7.

The reduction in computational complexity offered by the above approximation produces a fast algorithm for evaluating the product $\mathbf{C} = \mathbf{ABW}$. In spite of the fact that the constrained approximations will introduce errors if the exact product is required. However, in [5], it was shown that any linear transformation can be expressed exactly as a sum of triple matrix structures. In this case, if one triple matrix structure does not produce acceptable results, the designer has the option of using more than one array concurrently. This approximation and related questions have been considered in [5].

4. Conclusion

The paper has introduced a new structure for evaluating the matrix-matrix and triple matrix product that is better suited for real-time linear filtering applications. The new array combines the best features of systolic arrays and the rich connectivity networks of other topologies, such as the mesh and the n-cube. In addition, the paper developed a nonconventional approach that analyzes the evolution of partial results in the MSS array. The number of compute cycle was used as a design parameter to determine families of transformations implementable in the structure. The approach indicated that a single architecture is capable of evaluating a large number of potentially useful algorithms, that can be evaluated at a faster rate. The approach is applicable to a wide class of VLSI arrays implementing similar problems.

5. References

1- H.T. Kung, " *Special-purpose devices for signal and image processing : an opportunity in VLSI* ", Proc. SPIE, vol. 241, July 1980, pp. 76-84.

2- J. L. Aravena, " *Triple Matrix Product Algorithms and Architectures for Fast Signal Processing*", IEEE Trans.Circ. Syst., Vol. CAS-31, Jan 1988.

3- H. R. Barada, " *A comprehensive methodology for algorithm characterization, regularization and mapping into optimal VLSI arrays* ", Ph.D. dissertation, Louisiana State Univ., Aug. 1989.

4- A. O. Barbir, " *Systolic Array Implementations with Reduced Compute Time* ", Ph.D. Dissertation, Louisiana State University, December 1989.

5- J. L. Aravena and A. O. Barbir, " *A Class of low Complexity High Concurrence Algorithms* ", IEEE Trans. Parallel & Dis. Systems, Oct., 1991.

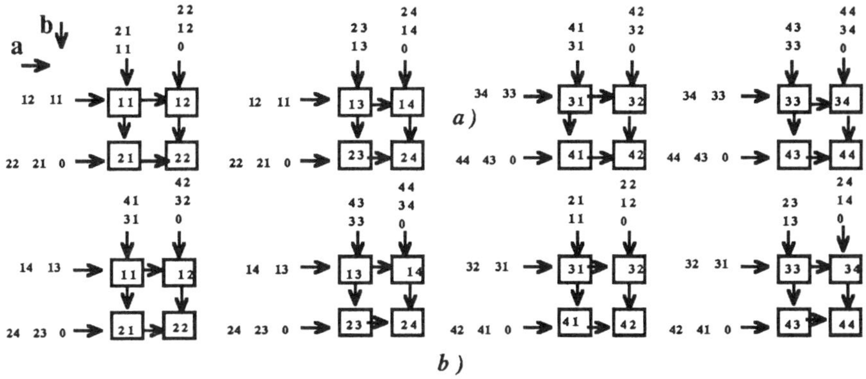

Figure 4. a) Data Movement in MSS for phase I.
b) Data Movement in MSS for Phase II.

$$T(\tau_1, \tau_2) = L_{\tau_1}(A)\, R_{\tau_2}(W) =$$

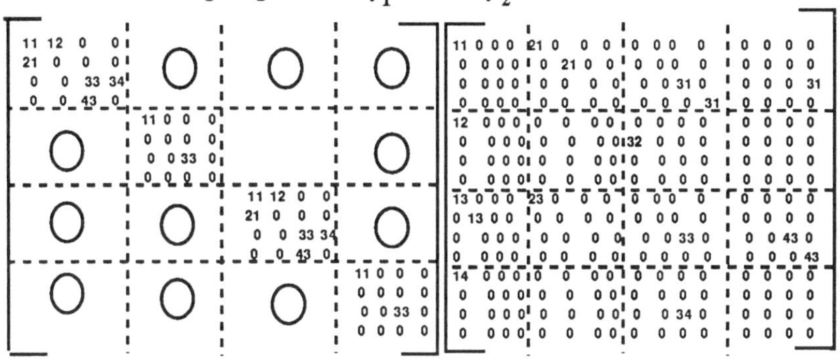

Figure 7. $T(\tau_1, \tau_2)$ for C=ABW.

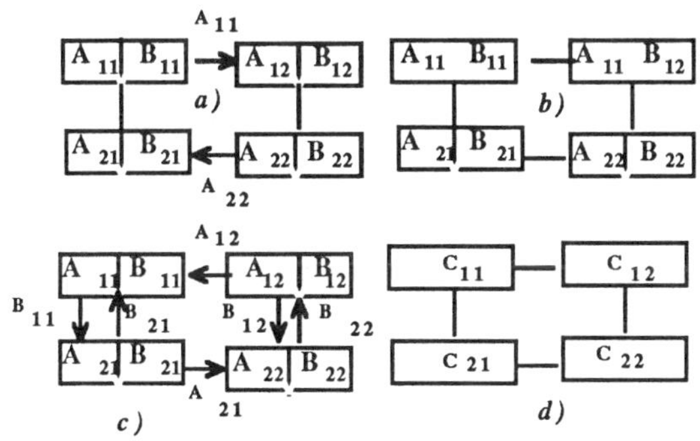

Figure 3. C=AB on MSS.

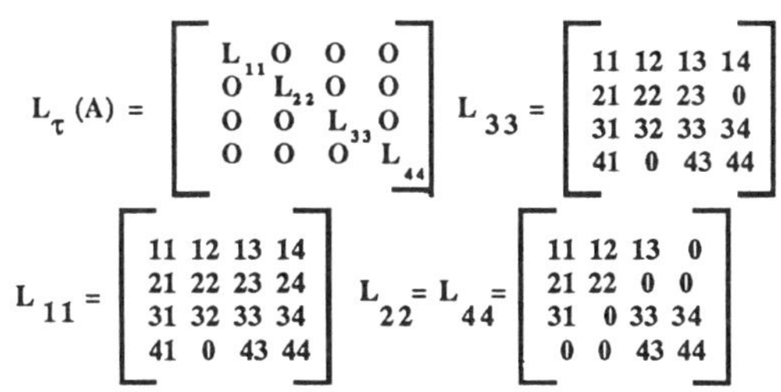

Figure 6. Partial Transformation on A.

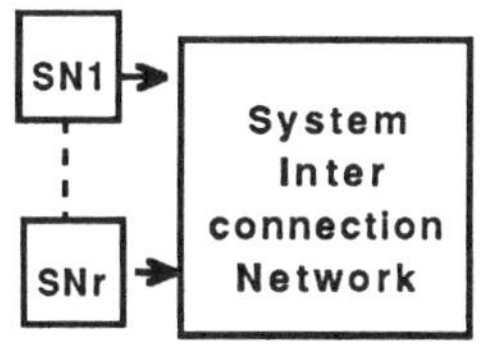

Figure 1. Architecture of MSS Array.

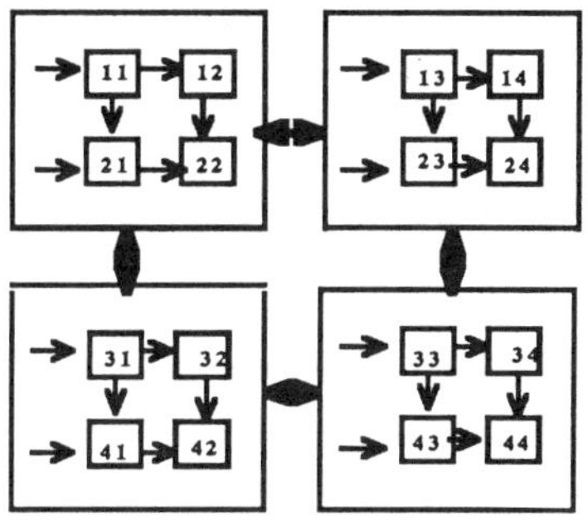

Figure 2. MSS for C=AB.

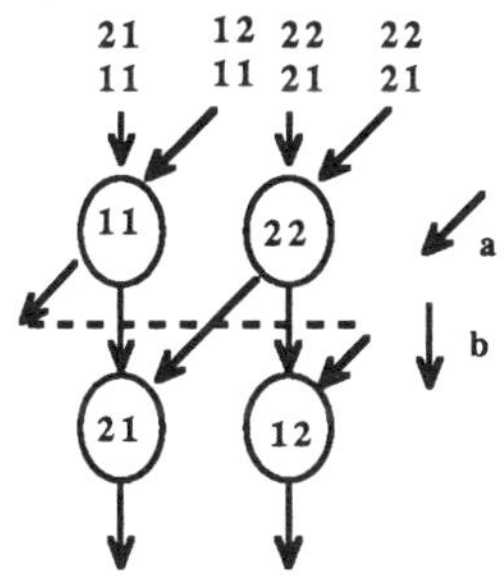

Figure 5. Cylindrical Array for C=AB.

Meta-Evolutionary Programming

D. B. Fogel
ORINCON Corporation
9363 Towne Centre Dr.
San Diego, CA 92121

L.J. Fogel
ORINCON Corporation
9363 Towne Centre Dr.
San Diego, CA 92121

J.W. Atmar
AICS Research, Inc.
P. O. Box 4691
University Park, NM 88003

Abstract

Evolutionary Programming is a stochastic optimization technique that is useful for discovering the extrema of a nonlinear function. To implement such a search, several "high-level" parameters must be chosen such as the the amount of mutational noise, the severity of the mutation noise, and so forth. This paper addresses incorporating a meta-level evolutionary programming that can simultaneously evolve optimal settings for these parameters while a search for the appropriate extrema is being conducted. Experimental results from function optimization problems are presented.

1 Introduction

Natural evolution can be viewed as an optimization process. Recently, there has been an increased interest in simulating natural evolution on a computer in order to address various optimization problems. These efforts have been directed at system modeling [1,2], artificial neural network training [3], automatic control [4], path planning [5], fire control strategies [6], and other combinatorial optimization problems.

Classic methods for discovering a "best" solution to such problems often rely on the use of a gradient-based search (e.g. recursive least squares estimation). Unfortunately, the error response surface to be examined is often a general nonlinear function possessing multiple local optima. Gradient methods are guaranteed to converge to locally optimal solutions if the step size of the search tends to zero. But such solutions are often far removed from the global optimum and may not provide for adequate system performance.

Natural evolution contends with many difficult optimization problems. Real world environments are nonlinear, time-varying, stochastic, chaotic and are a function of the extant population. Simulating this process can lead to algorithms that are capable of discovering globally optimum solutions despite pathologies in the given response surface.

Previous research in simulated evolution has largely centered around two paradigms: evolutionary programming [7] and genetic algorithms [8]. D. Fogel [1] indicated through a series of function optimization experiments that evolutionary programming typically provides for a more efficient and effective search for the extrema of a given topography than does the genetic algorithm. L. Fogel [9] speculated that by incorporating a hierarchical evolutionary programming, the search could be made more efficient and robust.

In this paper, a brief review of efforts in simulated evolution is given. A new technique for simultaneously evolving candidate solutions and the appropriate mutation noise is offered. This method is evaluated in a series of Monte Carlo experiments over a range of response surfaces. Conclusions regarding this "meta-evolutionary programming" approach are offered.

2 Simulated Evolution

There have been many attempts to simulate evolution on a computer. One of the initial efforts was made by Fogel [10], who conceived of using simulated evolution to develop artificial intelligence [11-14].

Such "evolutionary programming" is generally conducted as follows: a population of possible solutions to the task at hand are chosen at random. The representation for these solutions generally follows in a top-down fashion from the task. For example, in a traveling salesman problem, the solution may be coded simply as an ordered list of cities to be visited, while in a neural network training problem the solution may be coded as a real-valued vector of weight and bias terms). Each parent is scored with respect to a chosen but arbitrary well-defined payoff function. Offspring are created from each parent through random mutation. The specific mutation operations follow naturally from the coding. For example, if the solution is coded as a real-valued vector, a multivariate Gaussian perturbation may be applied. The mutation operation is only restricted in that suitable mutations should yield offspring which are probabilistically highly correlated with their parents in terms of their behavior.

1058-6393/91 $01.00 © 1991 IEEE

Competition for survival is executed in a probabilistic manner. This can be accomplished in a variety of ways. One convenient mechanism is to require each contending solution (both parents and offspring) to compete against a randomly selected subset of the remaining population in individual encounters. Those solutions having greater fitness, or less error, are awarded "wins." After the competition, those solutions with the greatest number of wins are retained to become parents of the next generation. This sequence of steps is repeated until either a sufficiently worthwhile solution is discovered or the available computer time is exhausted.

Despite numerous successful demonstrations of the evolutionary programming technique by Fogel (as already cited) and others (e.g. [15-18]), many reviews of the method were not favorable (e.g. [19-21]). In the early 1970s, the prevailing view of artificial intelligence was one of modeling human behavior through the use of expert systems. A catalyst of this perspective was the work of Minsky and Papert [22]. Dogmatically repeated criticisms and misstatements regarding evolutionary programming led to a belief that "all the attempts [of researchers from 1958-1970] to get programs to 'evolve' failed miserably...." [23].

An alternative method of simulating evolution, the "genetic algorithm," was proposed by Holland [8]. There are many similarities between the genetic algorithm and evolutionary programming. Both methods operate on a population of contending solutions, alter the coding of the parents at each generation to create offspring and compare the contending solutions according to their fitness. Holland [8] proposed three elements to simulated evolution which differ from the evolutionary programming [7] formulation:

1. The coding for each solution should be a string of bits {0,1},
2. The number of offspring created from each parent should be exponentially related to the parent's relative fitness, and
3. Offspring should be created by using "genetic operators" such as crossover and inversion, and bit mutations.

Holland [8] suggested that the alphabet of coding strings be over the set {0,1} in order to maximize the number of substrings ("building blocks") that could be searched. He speculated [8, p. 71] that "many detectors deciding among few attributes are preferable to few detectors with a range of many attributes." This restriction has been challenged by Antonisse [24] and much of the research in genetic algorithms no longer uses bit strings (e.g. [25]). It is unreasonable to require a single specific representation to address all optimization problems.

Reproducing the best parents in proportion to their relative fitness was in fact first suggested by Fogel [7,11], but is not required for successful simulated evolution. Simply reproducing (i.e. copying) the observed best parents to generate one offspring per parent will result in an exponential increase in the frequency of the best parents over time. Even with relatively large population sizes of several thousand, only 10 to 12 iterations are required for a single parent to dominate its competitors.

The crucial difference between evolutionary programming and genetic algorithms then must be focused on the use of specific mutation operations: crossover and inversion. Crossover is defined by selecting two parents, choosing a random position (or positions) and creating two offspring by splicing together the selected information from each parent. Inversion reverses the positional ordering of the components of a single parent, but is rarely used in current research. Davis [25] views the use of crossover operations as the "distinguishing feature" of genetic algorithms.

Holland [8] described evolution based on mutation in the absence of other genetic operators as being an "enumerative search," in large measure equivalent to blindly selecting solutions at random. Clearly, such a claim is incorrect. There is now considerable recent evidence that the use of macromutations such as crossover are not required for successful adaptation [1-6,26] and further, that the genetic algorithm relying on crossover and bit flipping does not perform as well as evolutionary programming at generalized function optimization (e.g. [1,27-29]). This follows previous research by Bremermann [30,31] and Reed et al. [32] that indicated crossover provided no benefit for searching spaces in which the components of the coding for contending solutions were interactive.

A more detailed review of research in simulated evolution from 1958-1990 may be found in [1] and [33].

3 Meta-Evolutionary Programming

For each evolutionary programming application, there are many parameters that must be chosen *a priori*. Experience with the algorithm can facilitate an appropriate choice for these parameters, but it is desirable to minimize the involvement of a human operator while maintaining an efficient search procedure. Fogel [9] speculated that the evolutionary parameters, such as the severity of mutation, could be simultaneously evolved along with the contending solution.

Specifically, the "standard" evolutionary programming approach to a multi-dimensional real-valued optimization problem [1] is to code contending solutions as real-valued vectors. The typical mutation operation is to perturb each component of the parent vector by adding a Gaussian random variable with zero mean and a variance that is an affine function of the parent's error score. Useful settings for the constant of proportionality and the offset value are determined through experimentation.

While the corresponding decrease in mutation severity accompanies a decrease in "behavioral error," simply setting the variance of the mutation noise to be a monotonic function of the height of the response surface can lead to problems in convergence. Any surface may be raised an arbitrary distance in its surrounding hyperspace. In the limit, this may result in the evolutionary algorithm approaching a completely random search. A meta-level evolution to optimize the appropriate mutation severity should in principal be capable of discovering the appropriate degree of perturbation for the given response surface regardless of its location in its surrounding hyperspace.

4 Experiments

To provide a basis for further study, a set of simple function optimization tasks in $\mathbf{R}^2$ were chosen. An evolutionary program was written to evolve vectors in $\mathbf{R}^2$. This algorithm proceeded as follows:

1) Distribute 200 solutions in accordance with a uniform distribution in (x,y) ranging over [-50,50],
2) Evaluate each "parent" solution with respect to $F(x,y)$,
3) Generate a single "offspring" from each parent by adding a Gaussian random variable with zero mean and variance equal to $F(x,y)$ to each component,
4) Evaluate each offspring solution with respect to $F(x,y)$,
5) Compete each solution against 10 randomly chosen solutions from the extant population. For each competition in which the solution being conditioned on has a lower error score than its opponent, assign a "win,"
6) Select those solutions with the greatest number of wins to be parents for the next generation, and
7) Go to step 3.

To incorporate the meta-level evolution of the mutation variance, this was made an evolving parameter according to the following change to steps 1 and 3:

1) In addition to distributing 200 solutions in accordance with a uniform distribution in (x,y) ranging over [-50,50], attach a perturbation variable to each (x,y), namely (p_x,p_y) chosen with respect to a uniform distribution ranging over [0,25], and
3) Generate a single "offspring" from each parent by adding a Gaussian random variable with zero mean and variance equal to the parent's corresponding perturbation term associated with each component. Generate new perturbation terms for each offspring by adding a Gaussian random variable with zero mean and standard deviation equal to the corresponding parent's perturbation term divided by six (scaling to assure mostly positive terms). Should a perturbation term become negative, set it to a small positive value, in this case 0.001.

For comparison, the standard genetic algorithm using a one-point crossover (probability = 0.8) and bit mutation (probability = 0.01/bit) was also developed. The coding for each component of the solution vector was taken to be a string of 14 bits (six decimal places of accuracy over [-50,50]). The population size was set at 200 parents to provide for comparisons with the evolutionary programming methods outlined above.

Each procedure was evaluated over 1,000 Monte Carlo trials on two test functions. The first function was taken from [34]. It is:

$$F(x,y) = x^2 + 2y^2 - 0.3\cos(3\pi x) - 0.4\cos(4\pi y) + 0.7.$$

It has numerous local minima and a global minimum of zero at the origin. The function is illustrated in Figure 1 with x,y ranging over [-1,1]. Outside this range, the quadratic terms dominate the trigonometric terms and the overall shape of the response surface is quadratic. Figures 2-4 indicate the optimization of the best score in the population at each generation, averaged over all 1,000 trials. Evolution was halted after 100 generations.

While all three methods indicate rapid initial optimization, the genetic algorithm stalls prematurely and cannot discover solutions with $F(x,y) < 0.01$ in under 20,000 evaluating offspring. Little further improvement may be expected after the 90th generation with the genetic algorithm. The standard evolutionary programming technique clearly outperforms the meta-evolutionary programming, but this is to be expected as $F(x,y)$ resides close to the (x,y) plane and the distance to the global minimum generally varies with the height of the response surface. The meta-evolutionary programming does appear to be converging to the global optimum and this performance does not depend upon the height of $F(x,y)$.

The second function studied was the Rosenbrock function:

$$F(x,y) = 100(x^2 - y)^2 + (1 - x)^2.$$

It has a parabolic valley along the curve $y = x^2$, with a unique minimum of zero at the point (1,1). There are no local optima associated with the function. The function is illustrated in Figure 5 with x, y ranging over [-4,4]. Figures 6-8 indicate the optimization of the best score in the population at each generation, averaged over all 1,000 trials. Evolution was halted after 100 generations.

Again, all three methods indicate fairly rapid initial optimization, although the standard evolutionary programming method is outperformed in the first 20 generations by both the meta-evolutionary technique and the genetic algorithm. Apparently, the rate at which $F(x,y)$ increases is sufficiently large so as to make it difficult for the standard evolutionary programming method to fall into the parabolic valley. But as indicated in Figure 7, once the standard method does discover the valley, optimization proceeds more rapidly, and while the scale on Figure 8 prohibits explanation, the standard evolutionary programming outperform meta-evolutionary programming before the 100th generation. Across these experiments, the meta-evolutionary programming is able to rapidly approach the global optimum without utilizing the height of $F(x,y)$ to guide the mutation noise. The genetic algorithm, on the other, becomes stalled at points on the response surface that cannot be considered "close" to the optimum. It should be noted that these are not points of local optima (none exist).

5 Conclusions

While past efforts in evolutionary programming have been broadly successful (e.g. [1-3,26,28,29]), considerable effort has been directed at finding suitable parameter values in each specific application. As the representation and objective function generally vary by application, no standard heuristics have been developed for optimally setting evolutionary parameters *a priori*. This suggests the use of a meta-level evolutionary programming that can evolve these parameters while concurrently searching for the optimum solution.

The preliminary experiments reported here indicate the suitability of such a procedure. Meta-evolutionary programming was able to converge to points on each of two response surfaces that were close to the global optimum. The standard evolutionary algorithm did discover lower points on both functions but it is not robust to varying the height of $F(x,y)$. The meta-approach would generate the same evolution regardless of any offset to $F(x,y)$.

The poor performance of the genetic algorithm on these functions was not unexpected. Similar behavior has been described in [1,27,28] and others. The meta-evolutionary programming discovered points for both functions that were superior to those attained with the genetic algorithm. These results are statistically significant ($t = 35.4$, $P \ll 10^{-6}$ for the Bohachevsky function; $t = 3.77$, $P < 0.0001$ for the Rosenbrock function).

Hierarchic evolution has relatively little additional accompanying computational cost. Moreover, the algorithm is inherently parallelizable. It is expected that a significant increase in efficiency could be obtained if the proposed algorithm were implemented on a distributed processing architecture. Analysis in [3] indicates that parallel implementations of evolutionary programming and a gradient search technique (e.g. backpropagation) require similar amounts of computation. Evolutionary programming is more robust, however.

References

[1] D.B. Fogel, *System Identification through Simulated Evolution: A Machine Learning Approach to Modeling,* Needham: Ginn Press, 1991.

[2] D.B. Fogel, "Evolutionary Modeling of Underwater Acoustics," *Proc. of OCEANS'91*, J. R. Vadus and P. C. Yuen (chairs), Honolulu, HI, pp. 453-457, 1991.

[3] D.B. Fogel, L.J. Fogel and V.W. Porto, "Evolving Neural Networks," *Biological Cybernetics*, Vol. 63, pp. 487-493, 1990.

[4] A.V. Sebald and D.B. Fogel, "Design of SLAYR Neural Networks Using Evolutionary Programming," *Proc. of 24th Asilomar Conf. on Signals, Systems & Computers*, R.R. Chen (Ed.), Pacific Grove, CA, pp. 1020-1024, 1990.

[5] J.R. McDonnell and W.C. Page, "Mobile Robot Path Planning Using Evolutionary Programming," *Proc. of 24th Asilomar Conf. on Signals, Systems & Computers*, R.R. Chen (Ed.), Pacific Grove, CA, pp. 1025-1029, 1990.

[6] L.J. Fogel, D.B. Fogel and J.W. Atmar, "Robotics and Artificial Intelligence: Evolutionary Programming for ASAT Battle Management," Contract No. DASG60-90-C-0071, U.S. Army Strategic Defense Command, Huntsville, AL, 1990.

[7] L.J. Fogel, A.J. Owens and M.J. Walsh, *Artificial Intelligence Through Simulated Evolution*, New York: John Wiley & Sons, 1966.

[8] J.H. Holland, *Adaptation in Natural and Artificial Systems*, Ann Arbor: Univ. of Michigan Press, 1975.

[9] L.J. Fogel, "The Future of Evolutionary Programming," *Proc. of 24th Asilomar Conf. on Signals, Systems & Computers*, R. R. Chen (Ed.), Pacific Grove, CA, pp. 1036-1038, 1990.

[10] L.J. Fogel, "Autonomous Automata," *Industrial Research*, Vol. 4, pp. 14-19, 1962.

[11] L.J. Fogel, "On the Organization of Intellect," Doctoral Dissertation, UCLA, 1964.

[12] L.J. Fogel, "Extending Communication and Control Through Simulated Evolution," *Bioengineering — an Engineering View, Proc. Symp. Eng. Significance of the Biol. Sciences,* G. Bugliarello (Ed.), San Francisco, San Francisco Press, pp. 286-304, 1968.

[13] L.J. Fogel and G.H. Burgin, "Competitive Goal-Seeking Through Evolutionary Programming," Final Report under Contract No. AF 19(628)-5927, Air Force Cambridge Research Labs., 1969.

[14] M.J. Walsh, G.H. Burgin and L.J. Fogel, "Prediction and Control Through the Use of Automata and Their Evolution," Final Report under Contract No. N00014-66-C-0284, 1970.

[15] B.E. Lutter and R.C. Huntsinger, "Engineering Applications of Finite Automata," *Simulation*, Vol 13, pp. 5-11, 1969.

[16] G.H. Burgin, "System Identification through Quasilinearization and Evolutionary Programming," *Journal of Cybernetics*, Vol 2:3, pp. 4-23, 1974.

[17] D.W. Dearholt, "Some Experiments on Generalization Using Evolving Automata," *9th Int. Conf. on System Science*, Honolulu, HI, pp. 131-133, 1976.

[18] J.W. Atmar, "Speculation on the Evolution of Intelligence and Its Possible Realization in Machine Form," Doctoral Dissertation, New Mexico State Univ., Las Cruces, 1976.

[19] D. Michie, "Future for Integrated Cognitive Systems," *Nature*, Vol. 228, pp. 717-722, 1970.

[20] B. Chandrasekaran and L.H. Recher, "Artificial Intelligence — a Case for Agnosticism," *IEEE SMC*, Vol. SMC-4:1, pp. 88-94, 1974.

[21] R.J. Solomonoff, "Some Recent Work in Artificial Intelligence," *Proc. of the IEEE*, Vol. 54:12, pp. 1687-1697.

[22] M.L. Minsky and S. Papert, *Perceptrons: An Introduction to Computational Geometry*, MIT Press, Cambridge, 1969.

[23] D.B. Lenat, "The Role of Heuristics in Learning by Discovery: Three Case Studies," *Machine Learning*, R.S. Michalski, J.G. Carbonell and T.M. Mitchell (Eds.), Tioga Publishing Co., Palo Alto, pp. 243-306, 1983.

[24] J. Antonisse, "A New Interpretation of Schema Notation that Overturns the Binary Encoding Constraint," *Proc of Third Intern. Conf. on Genetic Algorithms*, J.D. Schaffer (Chair), George Mason Univ., pp. 86-91, 1989.

[25] L. Davis, *Handbook of Genetic Algorithms*, Van Nostrand Reinhold, New York, 1991.

[26] V.W. Porto, "Evolutionary Methods for Training Neural Netwokrs for Underwater Pattern Classification," *Proc. of the 24th Asilomar Conf. on Signals, Systems and Computers*, R.R. Chen (Ed.), Pacific Grove, CA, pp. 1015-1019, 1990.

[27] D.B. Fogel and J.W. Atmar, "Comparing Genetic Operators with Gaussian Mutations in Simulated Evolutionary Processes Using Linear Systems," *Biological Cybernetics*, Vol. 63, pp. 111-114, 1990.

[28] D.B. Fogel and A.V. Sebald, "Training Neural Networks through Evolutionary Adaptation," *Proc. of Intern. AMSE Conf. Neural Networks Methodologies and Applications*, R. Swiniarski (Chair), San Diego, 1991.

[29] D.B. Fogel, L.J. Fogel and V.W. Porto, "Evolutionary Methods for Training Neural Networks," *Proc. of Conf. on Neural Networks and Ocean Engineering*, P. Simpson (Ed.), Washington DC, pp. 317-327, 1991.

[30] H.J. Bremermann, "Optimization through Evolution and Recombination," *Self-Organizing Systems*, M.C. Yovits, G.T. Jacobi, G.D. Goldstine (Eds.), Washington DC, Spartan Books, pp. 93-106, 1962.

[31] H.J. Bremermann, "Quantitative Analysis of Goal-Seeking Self-Organizing Systems," *Progress in Theoretical Biology*, Vol. 1, Academic Press, New York, pp. 59-77, 1967.

[32] J. Reed, R. Toombs and N.A. Barricelli, "Simulation of Biological Evolution and Machine Learning," *Journ. of Theoretical Biology*, Vol. 17, pp. 319-342, 1967.

[33] D.B. Fogel, "Simulated Evolution: A 30-Year Perspective," *Proc. of the 24th Asilomar Conf. on Signals, Systems and Computers*, R.R. Chen (Ed.), Pacific Grove, CA, pp. 1009-1014, 1990.

[34] I.O. Bohachevsky, M.E. Johnson and M.L. Stein, "Generalized Simulated Annealing for Function Optimization," *Technometrics*, Vol. 28, pp. 209-218, 1986.

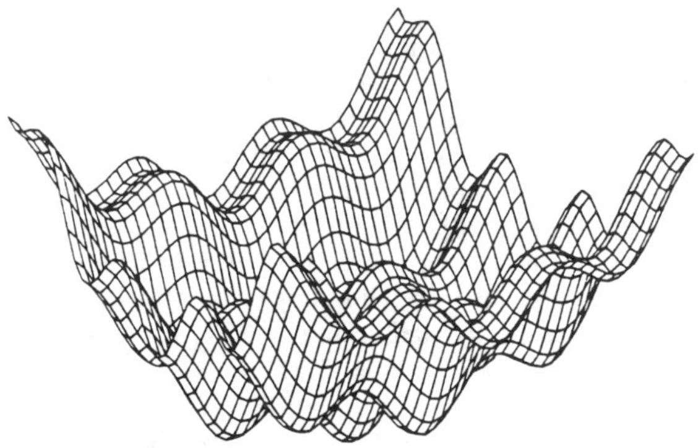

Figure 1. Bohachevsky Function. $(x,y) \varepsilon [-1,1]$.

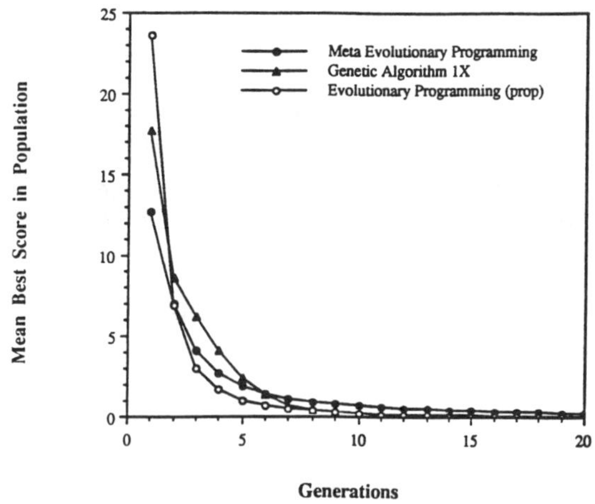

Figure 2. Bohacheveky Function.
Optimization of Best Score in Population.
1,00 Monte Carlo Trials
Generations 0-20.
"1X" Refers to One-point Crossover.
"prop" Refers to Proportional Variance.

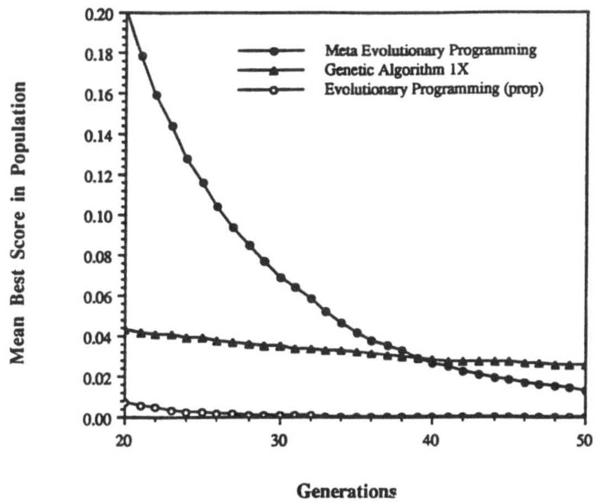

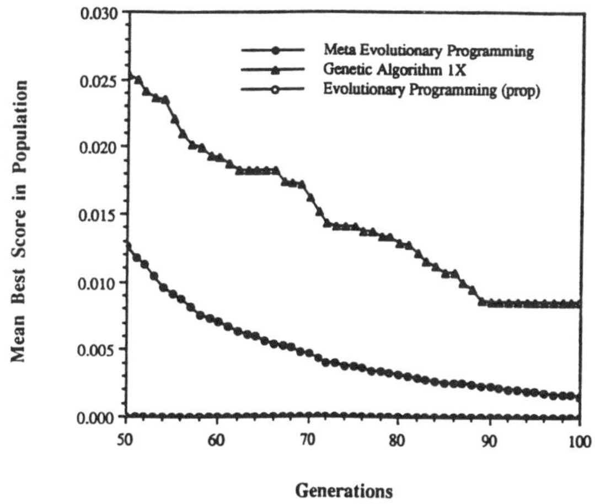

Figures 3 and 4. Optimization of Best Score in Population. 1,000 Monte Carlo Trials.

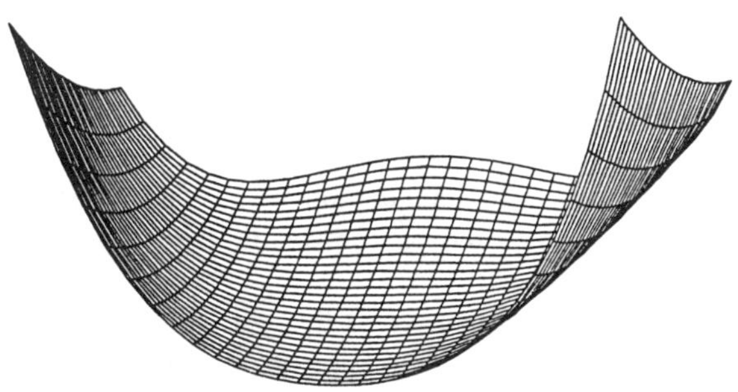

Figure 5. Rosenbrock Function. $(x,y) \varepsilon [-4,4]$.

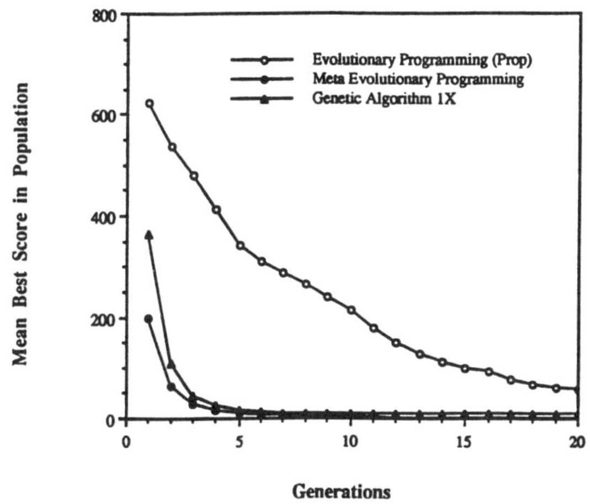

Figure 6. Optimization on Rosenbrock Function. 1,000 Monte Carlo Trials. Best Scores.

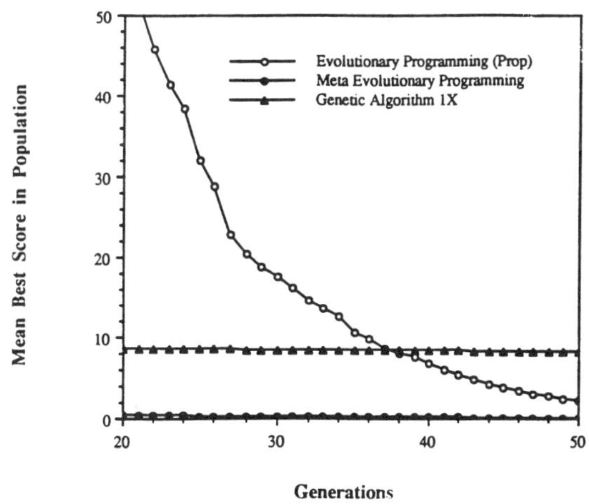

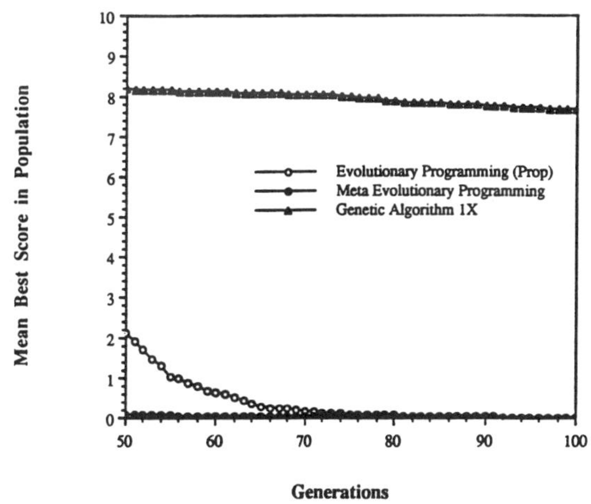

Figures 7 and 8. Optimization on Rosenbrock Function. 1,000 Monte Carlo Trials. Best Scores.

Multi-Output System Identification using Evolutionary Programming

B.L. Andersen, W.C. Page, and J.R. McDonnell
Naval Ocean Systems Center
San Diego, CA

Abstract

Evolutionary programming (EP) has been demonstrated to be an effective method of system identification of single-input-single-output (SISO) systems. This paper investigates the use of EP in system identification of single-input-multi-output (SIMO) systems. EP is used to identify parameters of a linear, time-invariant system. Specifically, this paper examines the identification of SIMO systems whose measurements can contain different amounts of noise. A cost function is proposed to take into account disparate noisy observations.

I. Introduction

System identification is the process of developing an accurate model of a dynamic system. This involves selecting the appropriate model structure and choosing the model parameters which optimize an objective function. The usual approach taken in system identification is to assume a model structure based on the process physics and then to determine the model parameters. Oftentimes, nonlinear models are linearized so that an appropriate state space formulation may be used.

Typically, methods of system identification involve minimizing the objective function using gradient descent techniques. Use of the gradient may result in the identified parameters being only locally optimal. Continuity constraints are another limitation associated with pure gradient search techniques. A Kalman filter approach is given in [1] along with cautions on its application to system identification. Many algorithms have been proposed which attempt to alleviate the inherent problems associated with gradient methods. Recent approaches have included simulated annealing, genetic algorithms, neural networks and evolutionary programming. The latter method has proven extremely successful for function optimization.

Evolutionary programming (EP) has been successfully applied to the problem of system identification. Fogel [2] has proposed the use of EP in system identification for evolving ARMAX process coefficients. His investigations extend to simultaneous parameter and model structure estimation. This paper extends the work on SISO systems to SIMO systems. The extension to multi-input, multi-output (MIMO) systems is obvious.

Identification of a SIMO system could be divided into identifying a set of SISO system transfer functions between the input and each of the outputs. However, each of these transfer functions would be identified using only the measurements of its output and not the full set of observations which contains more information. Also, it is not clear how one might combine these individual transfer functions with parameter estimates of varying degrees of uncertainty into one combined system model that would be required to control all the states of the system.

The remainder of this paper is organized as follows. Section II contains an overview of the EP paradigm. Section III formulates EP for SIMO system identification and describes the model structure. Section IV presents a linear, time-invariant system that is used for computer experiments. Section V discusses the results and Section VI provides a summary of the results and conclusions.

II. Evolutionary Programming

EP is a parallel stochastic search technique that maps natural evolution to an effective multi-agent search strategy. This search technique is not based on gradient methods and is therefore not susceptible to entrapment in local minima. Search by simulated evolution was first described by Fogel [3]. This investigation follows the search strategy outlined in Fogel [2] and described by the sequence below

1. Create the initial population P consisting of N parents.
2. Assess the fitness of each parent p_i in the population.
3. Mutate each element in the population by a $N(0,\sigma)$ random variable, where σ is

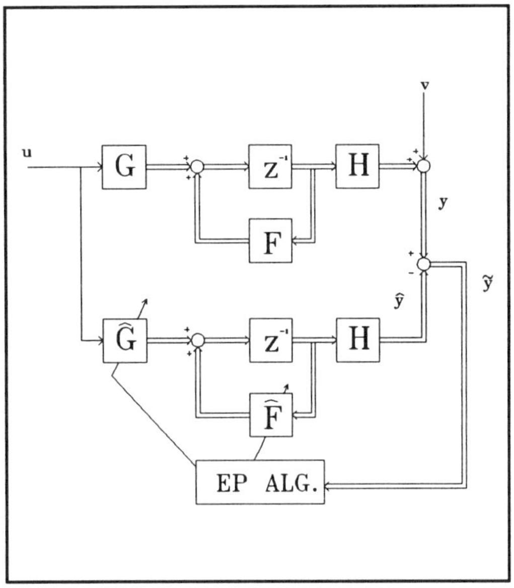

Figure 1. Evolutionary programming parameter estimation block diagram.

proportional to the fitness.

4. Assign a new fitness score to each element of the population.
5. Hold a stochastic competition between the members of the population.
6. Rank the members of the population according to the results (number of "wins") of step 5.
7. Repeat step 3 with the highest ranked *N* members from the population.

This algorithm is applied to system identification in the following section.

III. EP for System Identification

The linear time-invariant SIMO structure is assumed to be of the form

$$x(k+1) = Fx(k)+Gu(k)$$

$$y(k) = Hx(k)+v(k)$$

where $x(k)$ is the state vector of length n, $u(k)$ is the input, $y(k)$ is the vector of measurements of length l, $v(k)$ is the vector of $N(0,\sigma_v)$ measurement noise. F is the system matrix, G is the input weighting matrix and H is the measurement scaling matrix. The block diagram for estimating parameters using EP is shown in Figure 1.

The vectors to be evolved will consist of the elements of the F and G matrices. The H (measurement) matrix will not be evolved because finding all three of the matrices is an underdefined problem since an infinite number of combinations of F, G and H matrices apply to the same system dynamics. For this reason, the measurements will be assigned to the first l states of the system. This results in an H matrix of the form

$$H = \left[I_{l\,x\,l} \mid 0_{l\,x\,(n-l)}\right]$$

Given this formulation, the method of applying EP to the identification problem is straightforward. The remaining issues to be resolved are the mutation strategy and the objective function.

Two mutation strategies were investigated. The first incorporates a standard deviation of the mutation for each vector that is linearly proportional to the fitness score as suggested by Fogel [2]. This results in the size of the mutations decreasing as the population approaches the optimum value. In addition to ensuring convergence, this strategy also assures an effective search by allowing larger mutations for elements that have poor fitness scores. Atmar [4] suggests a variety of mechanisms to accelerate the search process.

The second mutation strategy investigated is based on the number of iterations. The mutation of any element in the vector can be described by the perturbation

$$\delta = \mu e^{-\alpha\tau k}$$

where μ is a $N(0,\sigma)$ random variable (r.v.), τ is a $U[0,1]$ r.v., α is a constant and k is the generation number. This mutation is a function of the number of iterations only. A random walk is maintained over a high number of generations while still preserving the capability of infinitesimal perturbations to achieve optimum values as k increases. Reasonable results were obtained using this mutation strategy for optimizing mobile manipulator configurations [5].

The objective function must be carefully chosen for this search. Not only will the measurements generally be of different units, but the amount of confidence in each measurement may differ due to sensor characteristics. A cost function has been formulated to take into account disparate sensor noise and the difference between state estimates and observations. The error vector between observation and estimate is given by

$$\tilde{y} = y - \hat{y}$$

where y is the observation and $\hat{y}$ is the estimate. The cost function incorporates sensor uncertainty (noise) using the diagonal standard deviation matrix

$$\Omega = \begin{bmatrix} \frac{1}{\sigma_1} & 0 & 0 & . & . & . \\ 0 & \frac{1}{\sigma_2} & 0 & . & . & . \\ . & . & . & . & . & . \\ 0 & 0 & . & . & . & \frac{1}{\sigma_n} \end{bmatrix}$$

where σ_i corresponds to the noise characteristics of the ith sensor. The cost associated with each parameter estimate can be described as

$$\Gamma = \frac{1}{n}\tilde{y}^T R \tilde{y}$$

where the variances are selected according to sensor specifications, n is the length of the error vector and $R = \Omega^T\Omega$ or the identity matrix I depending on the experiment. Using $R = \Omega^T\Omega$ accomplishes two things. First, it allows the cost to be a dimensionless quantity. Second, it scales each measurement by the amount of confidence in its accuracy. If the measurement is noisy, its contribution to the cost function is reduced since the prediction errors could be due primarily to measurement noise. If an observation is relatively noise-free, its contribution to the cost function will be large since the error is due primarily to estimation errors, assuming the model structure is correct. For these trials, it is assumed that the variances are constant but not necessarily equal. If the variances are equal, then the cost function is simply scaled by this magnitude. Using $R = I$ and scaling the cost function will yield the same results.

IV. Simulated System

A simple second-order system based on spring-mass-damper dynamics was used for computer experiments. Parameters were chosen so that the system was critically damped. The continuous state equations can be written as

$$\dot{x} = \begin{bmatrix} 0 & 1 \\ -1 & -2 \end{bmatrix} x + \begin{bmatrix} 0 \\ 1 \end{bmatrix} u$$

$$y = \begin{bmatrix} 1 & 0 \\ 0 & 1 \end{bmatrix} x + \nu$$

The equivalent discretized state-space formulation, assuming a zero-order-hold and a sample rate of T=0.05 sec, is given by

$$x(k+1) = \begin{bmatrix} 0.9988 & 0.0476 \\ -0.0476 & 0.9037 \end{bmatrix} x(k) + \begin{bmatrix} 0.0012 \\ 0.0476 \end{bmatrix} u(k)$$

$$y(k) = \begin{bmatrix} 1 & 0 \\ 0 & 1 \end{bmatrix} x(k) + \nu(k)$$

The forcing function is ensured to be persistently exciting by defining $u(k)$ as an r.v. with a uniform distribution. The system was propagated forward 100-200 time steps for each simulation from an initial state randomly chosen from a U[-1,1] distribution.

V. Results

Simulations were conducted using both mutation strategies. In general, the second mutation strategy outperformed the first. The reason for this is still under investigation. The state variables had to be constrained to prevent numeric overflow. This is not unreasonable given that the time integration of an unstable system is not bounded. However, it does have implications in system identification of unstable plants.

The first experiment for the second mutation strategy used numerically equal measurement noise terms with N(0,0.05) statistics and α=0.01. The resulting parameters are given in Table I under experiment 1. The average percentage error for all parameters is approximately 8%.

The magnitude of the forcing function was increased tenfold in an attempt to obtain better system identification. The parameters resulting from this experiment are given in Table I in the experiment 2 column. As expected, these parameters are have a lower overall average error than those obtained in experiment 1. For this case, the average overall parameter error is roughly 3.14%.

Figure 2 shows the mean and best population scores using the second mutation strategy. The cost function is simply the mean sum squared error (MSE) of both observations. At roughly 300 iterations the identification process converged. This run had a signal-to-noise ratio of approximately 20 and equivalent noise statistics on each measurement. Figure 3 illustrates the convergence of the first parameter (f_{11}) in the highest ranked F matrix. The other parameters converge in a similar fashion.

An experiment was conducted to compare the effect of the two cost functions using the second mutation strategy. The standard deviation of the position measurement noise was 0.01 and the standard deviation of the velocity measurement noise was 0.05. In both cases α was set at 0.05. The best values of the parameters are shown in Table I. The parameter values under the experiment 3 heading are for the cost function that included the variance term ($R = \Omega^T\Omega$). The parameter values given under the experiment 4 heading are for R equal to the identity matrix. The experiment that incorporated a cost function without the variance term unexpectedly had better results than the experiment which incorporated a confidence weighting term. The reasons for this result are still being investigated.

VI. Conclusions

EP is a powerful search technique that can be applied to system identification. This research has demonstrated that EP can be extended to identify linear SIMO systems of known order. The extension to identification of MIMO systems is obvious. It is also clear that the scaling of the cost function must be chosen carefully to account for disparate sensor data and measurement noise. The results given in this work have yet to be statistically verified and therefore should not be generalized. Investigations on the algorithmic properties of EP are being continued for validation in further experiments.

The formulation described here is a general one for the identification of state-space models, but there are still additional considerations to be addressed. First, the model ignored process noise, the terms that account for differences from the actual dynamic equations. Second, this work assumed that the order of the system is known. To identify the order of the model as well as the parameters, a multi-dimensional equivalent of Akaike's information criterion (AIC) or minimum defining length (MDL) principle similar to that used by Fogel [6] for neural networks must be developed.

The application of EP for adaptive control is readily apparent. Issues to be addressed in using EP for adaptive control pertain to the bandwidth of the process being controlled. That is, of course, assuming a traditional adaptive control architecture. EP could be implemented in an arrangement such that system identification would not have to be explicit.

References

[1] B.D.O. Andersen and J.B. Moore, *Optimal Filtering*, Prentice-Hall, 1979.

[2] D.B. Fogel, *System Identification through Simulated Evolution: A Machine Learning Approach to Modeling*, Ginn Press, Needham, MA, 1991.

[3] L.J. Fogel, A.J. Owens, and M.J. Walsh, *Artificial Intelligence through Simulated Evolution*, John Wiley and Sons, New York, 1966.

[4] W. Atmar, "Natural Processes which Accelerate the Evolutionary Search", IEEE Proc. of Twenty-Fourth Asilomar Conf. on Signals, Systems, and Computers, 1990.

[5] J.R. McDonnell, W.C. Page, and B.L. Andersen, "Configuration Optimization for Platform-Manipulator Systems using Evolutionary Programming", Proc. AMSE Conf. on Neural Networks, San Diego, 1991.

[6] D. B. Fogel, "An Information Criterion for Optimal Neural Net Selection", *IEEE Trans. on Neural Networks*, Vol. 2, No. 5, 1991.

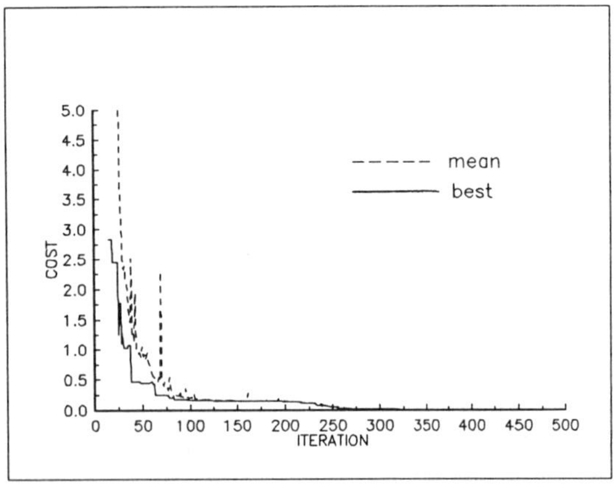

Figure 2. The mean and minimum (best) population costs.

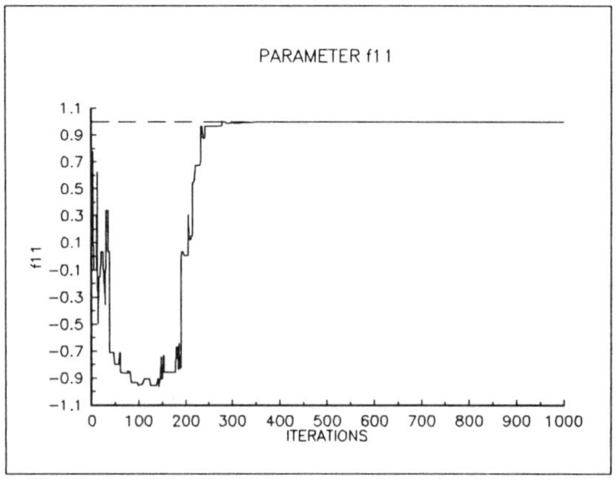

Figure 3. Evolution of parameter f_{11}.

Para-meters	Actual values	Final Best Value Statistics							
		1st Experiment		2nd Experiment		3rd Experiment		4th Experiment	
		value	%error	value	%error	value	%error	value	%error
f_{11}	0.9988	0.9965	0.23	0.9986	0.02	1.0193	2.05	0.9973	0.15
f_{12}	0.0476	0.0527	10.7	0.0480	0.84	0.1015	113.2	0.0422	11.3
f_{21}	-0.0476	-0.0449	5.67	-0.0481	1.05	-0.0745	56.51	-0.504	5.88
f_{22}	0.9037	0.90163	0.23	0.9033	0.04	0.8315	7.99	0.9062	0.28
g_1	0.0012	0.0009	25.0	0.0010	16.7	-0.0152	1367	0.0009	25.0
g_2	0.0476	0.0507	6.51	0.0477	0.21	0.0480	0.84	0.0440	7.56

Table I. Comparison of final value results of experiments and actual parameters.

Minimax Design of CMAC Encoded Neural Network Controllers Using Evolutionary Programming

A V Sebald J Schlenzig and D B Fogel

University of CA, San Diego
La Jolla CA 92093
(sebald@ucsd.edu)

Abstract

This paper describes the use of evolutionary programming for computer-aided design and testing of Cerebellar Model Arithmetic Computer (CMAC) encoded neural network regulators. The design and testing problem is viewed as a game in that the controller parameters are to be chosen with a minimax criterion, i.e. to minimize the loss associated with their use on the worst possible plant parameters. The technique permits analysis of neural strategies against a set of plants. This gives both the best choice of control parameters and identification of the plant configuration which is most difficult for the best controller to handle.

I. Introduction

CMAC encoders[1] have been used in regulators proposed for robotics and other problems[2]. They are members of a class of Single **LAY**er **R**egulators (**SLAYR**) which have recently been proposed [3-5] as a potential solution to the adaptive control problem in highly complex and uncertain environments such as those encountered in closed loop control of drug infusion. These algorithms are generalizations of the work by Albus, Miller and Barto et. al. [1,2,5]. As discussed in [4], these algorithms have two important advantages. First, they have no hidden nodes and can learn in real time on the actual plant being controlled. Second, their architecture lends itself very well to exploitation of human experience and, permits the design engineer and the neural net to cooperate nicely. The network uses the human's experience but learns those things the human does not know.

Since the combination of neural controllers and human plants does not easily lend itself to mathematical design or testing techniques, the utility of the neural net is heavily dependent on one's ability to design and test it in detail. Sebald has proposed[6,7] that the design problem be viewed as a game and that numerical minimax techniques be used as a computer-aided design and testing tool.

Finally, Fogel and Fogel [8-15] have developed an interesting numerical optimization tool termed evolutionary programming, which has solved very complex problems like the high dimensional travelling salesman without converging to poor locally optimal solutions. In [15], this technique has been compared with a variety of numerical optimization approaches applied to travelling salesman, adaptive control and system identification problems with excellent results.

This paper integrates the above by giving results of a minimax design and testing analysis of CMAC SLAYR algorithm using evolutionary programming.

Advantages of the method

SLAYR neural controllers are very convenient. First, they do not have hidden nodes and therefore can learn without difficult algorithms such as back propagation which are needed in more standard three layer network architectures. This means that they have a much easier time of learning on-line in real time using the system being controlled as a kind of teacher. This makes them very agile and lessens the danger associated with a network trained off-line suddenly encountering unforeseen circumstances.

CMAC provides a reasonably sophisticated categorization scheme which converts real measurements of the plant's state vector (e.g. value and rate of change of velocity) into pointers to a set of weights. Albus's very astute idea was that if suitably done, the CMAC encoding scheme could provide generalization in that small movements in state space would correspond with small movements in pointer space.

1058-6393/91 $01.00

This potentially could be of great help. SLAYR scemes all suffer from the fact that learning (ie weight adjustment) only takes place for encoded positions which have been visited before. Since the state space is typically quite large, there tend to be substantial numbers of encoded positions which have not been previously visited. This shows up as jitter on the control signal. The generalization in Albus' CMAC tends to alleviate the problem, since nearby encoded positions differ only slightly in corresponding control value and therefore, it is sufficient to have visited a nearby encoded location.

Why is design of SLAYR networks non trivial? All SLAYR algorithms including those using CMAC encoders have a large number of parameters to be optimized. If one simply picks these parameters with common sense, one is never certain whether performance problems are due to the inherent limitations of CMAC encoders etc, or whether more exhaustive optimization of the controller parameters. This is complicated by the fact that, for plants like the ones encountered in our work, theoretical optimizations based on linear time invariant plant assumptions simply are of no use.

Minimax is a good way to view the design problem in its true level of complexity. The control engineer determines the control architecture and parameters, while nature controls which plant will present itself to be controlled. The control designer's task is to pick a fixed (and perhaps intelligent) controller which must perform adequately no matter which plant shows up to be controlled. The minimax solution is also useful since it yields the best controller and the toughest plant. The latter information is invaluable for improving on the underlying control architecture.

Finally, the process requires a very efficient numerical optimization algorithm capable of efficiently searching extremely high dimensional optimizations. Both simulated annealing and evolutionary programming are being considered for this task. This paper deals with the experiment in evolutionary programming.

II. Problem Statement

Given a plant, $X_{k+1} = f(X_k, \theta, u_k)$, a control strategy, $u_k = g(X_k, \gamma)$, and a performance index $J(\theta,\gamma)$, where $q \in \Theta$, and $\gamma \in \Gamma$,

find $\theta^* \in \Theta \ni$

$$\max_{\gamma \in \Gamma} J(\theta^*, \gamma) \leq \max_{\gamma \in \Gamma} J(\theta, \gamma), \quad \forall \theta \in \Theta .$$

III. Method

The candidate plant set was:

$$X_{k+1} = \alpha X_k + b u_{k-d} + D0 + D1*k + D2 * k^2 + D3 * k^3, \quad (1)$$

where d is the transport delay for the infusion, and u is the input. This plant is motivated by the closed loop drug infusion problem (see [16]).

The Plant Design Parameters were:
α, b, D0, d, D1, D2 and D3.

The first three parameters (α, b and D0) in the plant were constrained (in order of their appearance in (1)), to the vectors:

Upper = [0.9672, -0.25, 5.740]
Lower = [0.9355, -5.00, 2.565].

Furthermore, the steady state solution, $D0/(1-\alpha)$, was constrained to the interval [80,120] (again for biomedical reasons). Finally, the remaining D parameters were constrained such that the entire solution trajectory remain in the interval [80, 120]. For this experiment, the initial value of X was 100. The sign of these terms is consistent with the fact that increasing the drug decreases the system output (blood pressure). Note that setting these constraints is neither unimportant nor simple to achieve.

The delay, d, was constrained to be in the interval [15,60].

The control:
has three components. First, the state vector:

$$(X_k, \Delta X_k \triangleq X_k - X_{k-1}) \quad (2)$$

of the system (1) is determined. Second, a CMAC encoder converts each point in state space to a coded word. Third, this coded word is hashed to provide a set of weights whose sum is the control (u_k in (1).) See Figure 1.

The Controller Design Parameters were:

1) $N \triangleq$ the number of weights selected (see Fig 1), (This determines the number of weights summed to determine the control.)

2) $M \triangleq$ the memory size for the hash coder output. (This essentially determines the probability of collisions due to hash coding of the CMAC encoder output).

3) $\hat{W} \triangleq$ a parameter in the weight learning law (Eqn (3)),

4) $b_0 \triangleq$ the initial learning rate, and

5) $D_a \triangleq$ the delay assumed by the controller.

6) $\Gamma \triangleq$ The number of timesteps between learning rate changes.

These six parameters were constrained to lie in the following vector limits:

upper = [75, 2400, 0.050, 0.100, 60, 50] and
lower = [30, 1000, 0.001, 0.006, 15, 10].

The weight learning scheme:
All weights, W_i, were initialized to zero. Each time the weight is visited, W_i is changed by:

$$\Delta W_i \triangleq \begin{cases} W_i * [\ z_k (\beta / N)] & \text{if } W_i >= 1.0e\text{-}4 \\ \hat{W} & \text{if not} \end{cases} \quad (3)$$

where:

W_i is the value of the ith weight,
$\hat{W}$ is a parameter to be optimized,
$\zeta_k \triangleq$ Target - X_k,
N is the number of weights selected by the CMAC encoder design and
β is the learning rate which satisfies (4).

$\beta = \beta_0$ at start, and once every Γ time steps:

$$\beta = \beta + 2 * |\Delta X_k| / N \text{ if: } \quad (4)$$
$$\Delta X_k > 0 \text{ and } X_k > \text{Target or}$$
$$\Delta X_k < 0 \text{ and } X_k < \text{Target}$$

where $\Delta X_k \triangleq X_k - X_{k-1}$ and,
β is never permitted to exceed 1.0.

Note that the learning scheme (3) only changes the weights if the current error, ζ_k, is non zero. However, the index, i, of the weights to be changed is determined by the delayed state vector:

$$(X_{k-D_a}, \Delta X_{k-D_a} \triangleq X_{k-D_a} - X_{k-D_a-1}) \quad (5)$$

The delayed state is used to account for the fact that the pure delay requires blame for error to be allocated to the weights pointed to d steps ago. Here, Da is used in place of d since the controller only "knows" D_a.

The performance index was the sum of the squares of the deviation of X from the target (70). It proved extremely useful to add a substantial cost if the controlled X was deemed grossly unacceptable. We defined unacceptable as X<50 at any time or X>120 after 200 time steps.

IV Results

After 10 generations in each inner loop and 10 generations in the outer loop, the following control solution was found:
control: [75, 1929, 0.004390, 0.02587, 44, 50].

The best patient for this controller was

[0.93558, -1.81309, 5.15344, 0.001385, -3.021e-06, 1.5046e-09, 15].

The overall cost for this pair was 11935.

The inside (i.e. max) loop was then run for 5000 generations holding the control parameters at the "optimum" values given above. The histogram of cost values for each organism encountered is given in Fig 3. Of these, 22 or 0.44% of the patients "died".

We use quotes when referring to optimal solutions since the results can be claimed to be minimax only after more extensive runs are conducted.

We looked at over 150 of the controlled trajectories emanating from the "optimal" controller. Very interesting insights emerge regarding choices of penalty function and improvements in the controller architecture.

Computer time: This analysis required 10 hours of background cpu time on a SUN 4/260 machine. Each of the generations in this analysis involved a population of 50 organisms.

V Conclusions

1)Evolutionary programming is a very useful tool for performing minimax computer aided design and testing of neural controllers.

2) On the plants tested, CMAC encoders generated useful SLAYR controllers, when designed with a minimax design criterion . These plants are suggestive of the types of systems encountered in closed loop control of drug infusion, namely human patients, and therefore represent a reasonably challenging test bed.

3) Minimax testing provides an analysis over a very broad range of plants and therefore provides much insight into the robustness of a control strategy and into the kinds of plants with which it has difficulties.

4) Minimax design also forces the designer to think about and carefully specify the class of plants to be encountered, as well as the true performance measures needed to penalize bad behavior and ignore unimportant changes in the trajectory. Both of these

areas are difficult and deserve more attention than they typically get in the design process.

References

[1] Albus, JS, "Mechanisms of planning and problem solving in the brain", **Math. Biosci.**, vol 45, Aug. 1979, pp. 247-293.

[2] Miller, W. T. III, "Sensor based control of robotic manipulators using a general learning algorithm", **IEEE Jrnl of Robotics and Automation**, vol RA3 , no 2, April, 1987, pp. 157-165.

[3] Sebald AV, Ph.D., J. Schlenzig, "Use of SLAYR neural strategies for closed loop control of drug infusion", **Proceedings, 12th Annual International Conference of the IEEE Engineering in Medicine and Biology Society**, November 1990, Philadelphia.

[4] Sebald, AV, C A Sebald and J Schlenzig, "Use of Neural Net Control Strategies in Difficult Adaptive Control Problems: Closed Loop Control of Drug Infusion" **Proceedings, 1989 Asilomar Conference on Circuits and Systems**, November, 1989.

[5] Barto A, R Sutton and Anderson, " Neuronlike elements that can solve difficult learning control problems", **IEEE Transactions on Systems Man and Cybernetics,** 13, 1983, pp 835- 846.

[6] AV Sebald, M Quinn, N Ty Smith, A Karimi and G Schnurer, "Engineering Implications of Closed Loop Control during Cardiac Surgery: An Engineering Perspective", **Journal of Clinical Monitoring**, 1990, Also delivered to the **1988 Fifth Annual Symposium on Computers in Anesthesia**, San Diego CA, May 1988.

[7] Karimi A. and A.V. Sebald, "Computer Aided Design of Closed-Loop Controllers for Biomedical Applications", **Proceedings, 10th Annual International Conference of the IEEE Engineering in Medicine and Biology Society**, pg 1404-1405, November 1988.

[8] L. J. Fogel, "Autonomous Automata," **Industrial Research**, Vol. 4, pp. 14-19, 1962.

[9] L. J. Fogel, "On the Organization of Intellect," Doctoral dissertation, UCLA, 1964.

[10] L. J. Fogel, A. J. Owens, and M. J. Walsh, **Artificial Intelligence through Simulated Evolution**, John Wiley & Sons, New York, 1966.

[11] D. B. Fogel, "An Evolutionary Approach to the Traveling Salesman Problem," **Biological Cybernetics**, Vol. 60, No. 2, pp. 139-144, 1988.

[12] D. B. Fogel, "A Parallel Processing Approach to a Multiple Traveling Salesman Problem using Evolutionary Programming," **Fourth Annual Symposium on Parallel Processing**, Fullerton, CA, pp. 318-326, 1990.

[13] D. B. Fogel and L. J. Fogel, "Evolutionary Programming for Voice Feature Analysis," **23rd Asilomar Conf. on Signals, Systems & Computers**, Pacific Grove, CA, pp. 381-383, 1989.

[14] D. B. Fogel, L. J. Fogel, and V. W. Porto, "Evolving Neural Networks," **Biological Cybernetics**, to appear, 1990.

[15] D. B. Fogel, "System Identification through Simulated Evolution," Master's thesis, UCSD, 1990. Also avilable as **System Identification through Simulated Evolution: A Machine Learning Approach to Modelling**, Ginn Press, 1991.

[16] Sebald, AV , G. Schnurer, M. Parti , NT Smith, ML Quinn,"A Dynamic Empirical Model Of The Human Response To Sodium Nitroprusside During Cardiac Surgery", **Proceedings, 10th Annual International Conference of the IEEE Engineering in Medicine and Biology Society**, pg 517-518, November 1988.

[17] J. H. Holland, **Adaptation in Natural and Artificial Systems**, University of Michigan Press, Ann Arbor, 1975.

[16] D. B. Fogel and J. W. Atmar, "Comparing Genetic Operators with Gaussian Mutations in Simulated Evolutionary Processes Using Linear Systems," **Biological Cybernetics**, Vol. 63, No. 2, pp. 111-114, 1990.

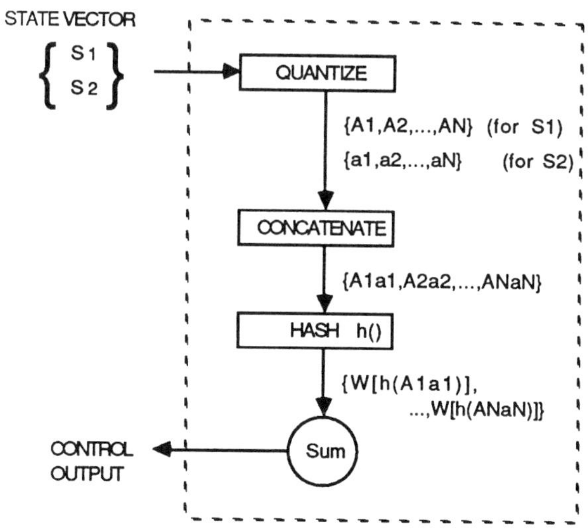

Figure 1: The CMAC encoder / hash scheme used in this study.

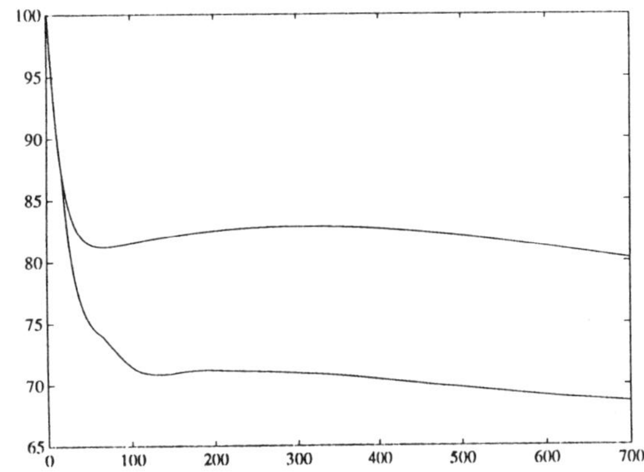

Figure 2: The controlled (lower) and uncontrolled (upper) pressure trajectories for the best controller used on the best plant.

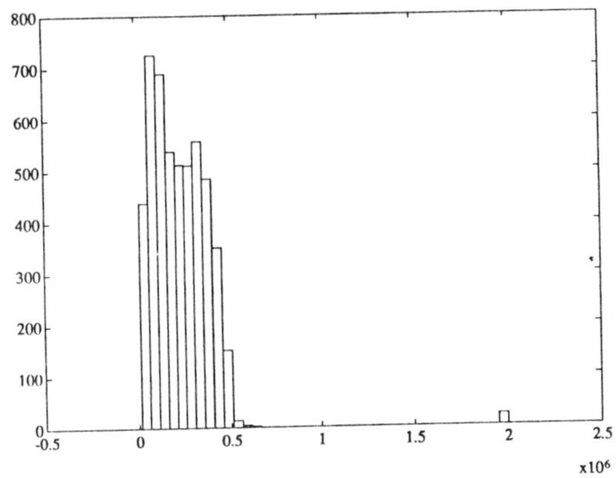

Figure 3: The histogram of costs encountered when the inner (i.e. max) optimization was performed over 5000 generations while holding the controller parameters fixed at the "optimal" values. Note that the minimax cost estimate from the smaller run was 487997 while the highest cost found in the 5000 generation case was "death".

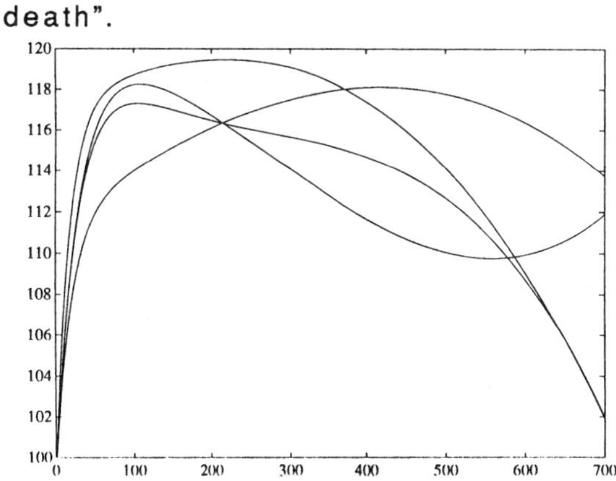

Figure 4: A sampling of the uncontrolled pressures for patients who "died" even with the best controller.

Beamspace Root-MUSIC for Rectangular Arrays, Circular Arrays, and Nonredundant Linear Arrays

Michael D. Zoltowski and Cherian P. Mathews

School of Electrical Engineering, Purdue University, West Lafayette, IN 47907

Abstract

Beamspace Root-MUSIC is a computationally efficient, beamspace implementation of Root-MUSIC developed recently for use in conjunction with a uniformly-spaced linear array (ULA). Employing B real-valued beams having all but B-1 nulls in common, the major computations in Beamspace Root-MUSIC are a BxB real-valued eigenvalue decomposition (EVD) and the rooting of a polynomial of order either 2B-2 or B-1. Beamspace Root-MUSIC is here extended for other common array geometries including the rectangular array, the circular array, and the minimum redundancy linear array.

1. Introduction

Root-MUSIC offers improved performance over Spectral MUSIC in the case of a uniformly-spaced linear array (ULA) of N sensors. At the same time, operation in beamspace offers a number of advantages over operation in element space. These observations motivated the recent development of Beamspace Root-MUSIC [1]. Employing B real-valued beams having N-B (out-of-band) nulls in common, the major computations in Beamspace Root-MUSIC are a BxB real-valued eigenvalue decomposition (EVD) and the rooting of a polynomial of order on the order of the number of beams, B. Simulations results and experiments with real data show the performance of Beamspace Root-MUSIC, in terms of arrival angle estimator accuracy, to be comparable to that of element space Root-MUSIC despite the dramatic reduction in computational complexity [1].

Beamspace Root-MUSIC offers advantages over Beamspace Spectral MUSIC as well. Source angles are determined via the roots of small-order polynomials rather than through spectral searching. Aside from the computational advantages of such, this leads to improved performance since only the polar angles of the signal roots are used in estimating the respective arrival angles [1]. Up to the present point in time, though, Beamspace Root-MUSIC was limited to use in conjunction with a ULA. The main contribution of this work is the development of means by which Beamspace Root-MUSIC may be used in conjunction with Uniform Rectangular Arrays (URA's), Uniform Circular Arrays (UCA's) [2,3], and Minimum Redundancy Linear Arrays (MRLA's) [4,5].

2. Overview of Beamspace Root–MUSIC

Beamspace Root-MUSIC employs a set of B real-valued beams exhibiting a large number of common nulls. Such a set of beams is produced by the NxB matrix beamformer

$$\mathbf{W} = \frac{1}{\sqrt{N}}\left[\, \mathbf{a}(u_1) \,\vdots\, \mathbf{a}(u_2) \,\vdots\, \ldots \,\vdots\, \mathbf{a}(u_B) \right] \tag{1}$$

where each of the columns is of the form

$$\mathbf{a}(u) = \exp[-j((N-1)/2)u]\ \mathbf{v}(u) \tag{2}$$

where, in turn, $\mathbf{v}(u)$ has the Vandermonde structure

$$\mathbf{v}(u) = [1\ ,\ \exp[\,ju]\ ,\ \exp[\,j2u]\ ,\ \cdots\ ,\ \exp[\,j(N-1)u]]^T \tag{3}$$

The electrical angle u is related to the arrival angle with respect to the normal to the ULA, denoted θ, as

$$u = \frac{2\pi}{\lambda} d \sin\theta \tag{4}$$

where λ is the wavelength and d is the interelement spacing. The pointing angles of the B beams produced by the columns of $\mathbf{W}$ are

$$u_i = \frac{2\pi}{\lambda} d \sin\theta_o + (i-1)\frac{2\pi}{N} \quad i=1,\ldots,B \tag{5}$$

where θ_o is the pointing angle of the first beam. It is easy to verify that the B columns of $\mathbf{W}$ are mutually orthonormal and produce B real-valued beams having N-B nulls in common [1]. The real-valued nature of the beams is due to the fact that each of the columns of $\mathbf{W}$ is conjugate centro-symmetric.

Each of the beams produced by the columns of $\mathbf{W}$ has a peak sidelobe of -13.5 dB. Beams with reduced sidelobe levels may be synthesized by weighting and summing adjacent beams. For example, tapering at the element level according to the cosine window may be alternatively achieved by adding each adjacent pair of

1058-6393/91 $01.00 © 1991 IEEE

beams out of the B beams produced by **W**. This yields B-1 beams with each beam having a peak sidelobe of -26 dB; the B-1 beams have the same N-B nulls in common as the original B beams. In accordance with the concept of discrete prolate spheroidal sequences (DPSS) *applied in beamspace*, the optimal J Bx1 weight vectors for combining the B beams produced by **W** in (1) to yield J < B beams having N-B nulls in common may be computed as the J "largest" eigenvectors of the BxB real-valued, symmetric matrix

$$\mathbf{B} = \mathbf{W}^H \left\{ \int_{u_o-\pi/N}^{u_B+\pi/N} \mathbf{a}(u)\mathbf{a}^H(u)du \right\} \mathbf{W} \qquad (6)$$

Letting **T** denote a BxJ whose columns are the J "largest" eigenvectors of **B**, this yields a new beamforming matrix $\mathbf{W}_T = \mathbf{W}_V\mathbf{T}$. It is easy to verify that the J columns of $\mathbf{W}_T$ are mutually orthonormal, conjugate-centrosymmetric and produce J real-valued beams having N-B nulls in common.

An algorithmic summary of Beamspace Root-MUSIC is provided below. In this summary, x(n) denotes the Nx1 element space snapshot vector; M denotes the number of snapshots. **T** is BxJ composed of J real-valued beamspace DPSS's determined as the J "largest" eigenvectors of **B** in (6). Finally, **Q** is a BxB matrix whose columns are related on a one-to-one basis with the columns of **W** in (1): the roots of the polynomial formed with the k-th column of **Q** are related on a one-to-one basis via the transformation $z = \exp[\,ju]$ with the B-1 nulls of the beam generated by the k-th column of **W** that are not in the set of N-B common nulls. A simple procedure for constructing **Q** is described in [1].

Algorithmic Summary of Beamspace Root-MUSIC

1. form BxB beamspace covariance matrix as

$$\mathbf{y}(n) = \mathbf{W}^H\mathbf{x}(n) \quad n=1,\ldots,M, \quad \hat{\mathbf{R}}_{yy} = \frac{1}{M}\sum_{n=1}^{M} \mathbf{y}(n)\mathbf{y}^H(n)$$

2. compute EVD of $\mathbf{T}^H\mathrm{Re}\{\hat{\mathbf{R}}_{yy}\}\mathbf{T}$, estimate number of sources, K, and place J-$\hat{K}$ "smallest" eigenvectors as the columns of $\mathbf{E}_N$

3. with $\mathbf{P} = \mathbf{Q}\mathbf{T}\,\mathbf{E}_N\mathbf{E}_N^H\mathbf{T}^T\,\mathbf{Q}^H$, construct

$$p(z) = p_0 + p_1 z + \cdots + p_{B-1}z^{B-1} + \cdots + p_1^* z^{2B-3} + p_0^* z^{2B-2}$$

with the coefficients computed according to

$$p_j = \sum_{i=0}^{j} \mathbf{P}_{B-j+i,\, i+1} \qquad j=0,1,\ldots,B-1$$

4. root p(z) and select $\hat{K}$ signal roots: $\hat{z}_k$, k=1,2,...,$\hat{K}$.

$$\hat{\theta}_k = \sin^{-1}\left[\frac{\lambda}{2\pi d}\mathrm{angle}\{\hat{z}_k\}\right] \qquad k=1,2,\ldots,\hat{K}$$

Step 3 involves the rooting of a polynomial of order 2B-2. By suitably defining FINE Beamspace Root-MUSIC or Min-Norm Beamspace Root-MUSIC, we may work with a polynomial of order B-1. In this case, we root the polynomial $p(z) = p_1 + p_2 z + \cdots + p_B z^{B-1}$, where p_i, i=1,...,B, are the components of the Bx1 vector $\mathbf{p} = \mathbf{E}_N\boldsymbol{\alpha}$. For FINE Beamspace Root-MUSIC, $\boldsymbol{\alpha}$ is computed as the "largest" eigenvector of **B** in (6). For Min-Norm Beamspace Root-MUSIC, $\boldsymbol{\alpha} = \mathbf{E}_N\boldsymbol{\delta}_1/\boldsymbol{\delta}_1^T\mathbf{E}_N\mathbf{E}_N^H\boldsymbol{\delta}_1$, where $\boldsymbol{\delta}_1$ is a Bx1 vector the first component of which is one and remaining components of which are zero.

Note that in the case of correlated sources, taking the real part of $\hat{\mathbf{R}}_{yy}$ has the same decorrelating effect as performing a single forward-backward average in element space. For the case of uncorrelated sources, it has been shown that with complex Gaussian noise taking the real part of $\hat{\mathbf{R}}_{yy}$ has the effect of artificially doubling the number of snapshots [1].

3. Beamspace Root–MUSIC: Rectangular Arrays

A straightforward approach to applying Beamspace Root-MUSIC in the case of a Uniform Rectangular Array (URA) is based on the observation that a URA may be viewed as an array of identical ULA's in parallel. Applying the **same** beamforming weight vector to each ULA in parallel, forming a beam pointed to the same elevation angle, for example, yields a ULA of "super-element" outputs to which Beamspace ROOT-MUSIC may be applied to determine the azimuth angles of sources at that elevation angle for each of a number of azimuthal sectors. B beams spanning an azimuthal sector at that elevation angle are formed by applying B weight vectors to the ULA of "super-element" outputs. As the procedure is repeated each time the elevation angle is incremented in the process of scanning, the computational benefits of Beamspace Root-MUSIC are reaped many times.

4. Beamspace Root–MUSIC for Circular Arrays

Consider a UCA of N elements with radius R. The UCA manifold is denoted $\mathbf{c}(\theta, \phi)$, where θ and ϕ are the azimuthal and elevation angular variables, respectively. The i-th component of $\mathbf{c}(\theta, \phi)$, denoted $c_i(\theta, \phi)$, i=1,...,N, is given by

$$c_i(\theta, \phi) = \exp\left[j\frac{2\pi}{\lambda} R \sin\phi \cos(\theta - \frac{2\pi i}{N}) \right] \quad (7)$$

Motivated by the work of Weiss and Friedlander [2], we consider the interpolated array concept as a means of facilitating the use of Beamspace Root-MUSIC in conjunction with the UCA. For a particular azimuthal sector at a particular elevation angle, ϕ_o, an interpolation matrix $\mathbf{M}$ is constructed which approximately maps $\mathbf{c}(\theta, \phi)$ for any azimuth angle within the sector into the ULA manifold for that angle, i. e.,

$$\mathbf{M}\,\mathbf{c}(\theta, \phi_o) \approx \mathbf{a}(\theta) \quad \text{for } \theta \in \text{azimuthal sector} \quad (8)$$

where $\mathbf{a}(\theta)$ is given by (2) with u replaced by θ. That is, $\mathbf{a}(\theta) = \exp[-j((N-1)/2)\theta]\mathbf{v}(\theta)$, where $\mathbf{v}(\theta) = [1, \exp[j\theta], \exp[j2\theta], \cdots, \exp[j(N-1)\theta]]^T$. Note that there is a one-to-one mapping between θ and $\mathbf{a}(\theta)$ for $0 < \theta < 360°$. The efficacy of mapping $\mathbf{c}(\theta, \phi_o)$ into $\mathbf{a}(\theta)$ for a subsector of the entire 360° azimuthal sector is based on the work of Hong and Tewfik [3]. They showed that if N is much greater than $2\pi R/\lambda$, $\mathbf{c}(\theta, \phi_o)$ may be mapped into $\mathbf{a}(\theta)$ over the entire 360° azimuthal sector. However, in their approach the maximum number of sources that may be resolved is N/2 -1, roughly one-half the number of elements. In addition, their method requires that N be odd. The present approach has neither of these limitations.

$\mathbf{M}$ is computed as the least square error solution to the overdetermined system of equations $\mathbf{M}\,[\mathbf{c}(\theta_1, \phi_o) \vdots \cdots \vdots \mathbf{c}(\theta_I, \phi_o)] = [\mathbf{a}(\theta_1) \vdots \cdots \vdots \mathbf{a}(\theta_I)]$, where θ_i, i=1,...,I, is a dense set of angles encompassing the azimuthal sector of interest. $\mathbf{M}$ is then applied to each snapshot and beams encompassing the azimuthal sector for which $\mathbf{M}$ is designed are formed on the interpolated array data. Note that since the interpolation matrix is only designed to work well for a specific sector, it makes sense to work in beamspace in this manner.

Given $\mathbf{M}$ as constructed above, the only modifications to the Beamspace Root-MUSIC algorithm outlined previously is to solve for the noise eigenvectors as the "smallest" generalized eigenvectors of the matrix pencil $\{\mathbf{T}^H \mathrm{Re}\{\mathbf{W}^H \mathbf{M} \hat{\mathbf{R}}_{xx} \mathbf{M}^H \mathbf{W}\}\mathbf{T},\ \mathbf{T}^H \mathbf{W}^H \mathbf{M}\mathbf{M}^H \mathbf{W}\mathbf{T}\}$, where $\hat{\mathbf{R}}_{xx}$ is element space sample correlation matrix, and the signal roots are of the form $z = \exp[j\theta]$, i. e., the azimuth angle estimates are equal to the phase angles of the signal roots. Since both $\mathbf{W}$ and $\mathbf{T}$ are composed of orthonormal columns, a generalized EVD would not be necessary if $\mathbf{M}$ was unitary. We are currently assessing the efficacy of imposing a unitary constraint on $\mathbf{M}$.

5. Beamspace Root–MUSIC for Minimum Redundancy Linear Arrays

The MRLA is a nonuniformly spaced linear array structure in which the number of sensor pairs measuring the same spatial correlation lag is minimized for a given number of elements, N [4,5]. The MRLA is attractive in that it is able to detect a number of sources greater than N, assuming the sources to be uncorrelated [4]. In addition, compared to a ULA having the same number of elements, the elements in an MRLA may be spaced over a larger aperture yielding enhanced resolution capability without ambiguities due to grating lobes. Due to the nonuniform spacing of the elements, Beamspace Root-MUSIC may not be applied in the usual manner with beams formed directly on the elemental data. Rather, beams must be formed on the augmented autocorrelation sequence. A procedure for using Beamspace Root-MUSIC in conjunction with an MRLA is described below.

Let L denote the total number of possible spatial correlation lag values measurable with an N element MRLA and let d denote the spacing between the two closest elements. Based on the work of Pillai, Bar-Ness, and Haber [4], the first step is to construct an LxL "augmented" Toeplitz-Hermitian covariance matrix, denoted $\hat{R}_{xx}$, with the i-th component of the first column equal to $r_{xx}[(i-1)d]$, i=1,....,L, where $r_{xx}[(i-1)d]$ is the spatial correlation between two elements separated by the distance (i–1)d. Again, L, and hence the dimension of $\hat{R}_{xx}$, may be much greater than N. The BxB beamspace covariance matrix, $\hat{\mathbf{R}}_{yy}$, is then formed as

$$\hat{\mathbf{R}}_{yy} = \mathbf{W}^H \hat{\mathbf{R}}_{xx} \mathbf{W} \quad (9)$$

where $\mathbf{W}$ is an LxB beamforming defined in accordance with equations (1) through (5) with N replaced by L. That is, beams are formed on the augmented autocorrelation sequence effectively assuming an L element ULA with interelement spacing d. With $\hat{\mathbf{R}}_{yy}$ constructed according to (9), the Beamspace Root-MUSIC algorithm outlined previously may be applied starting with step 2. Note that since $\hat{\mathbf{R}}_{xx}$ is Toeplitz-Hermitian and the columns of $\mathbf{W}$ are conjugate centro-symmetric, it is easy to show that $\hat{\mathbf{R}}_{yy} = \mathbf{W}^H \hat{\mathbf{R}}_{xx} \mathbf{W}$ is automatically real-valued so it is not necessary to take the real part in step 2.

Recall that for the MRLA, $\hat{\mathbf{R}}_{xx}$ is LxL and $\mathbf{W}$ is LxB. Formation of $\hat{\mathbf{R}}_{yy}$ according to (9) involves BL(L+B) complex multiplications. Exploiting the Toeplitz-Hermitian structure of $\hat{\mathbf{R}}_{xx}$ and the fact that each column of $\mathbf{W}$ is a scalar multiple of a Vandermonde vector of the form in (3), Seth Silverstein[1] has recently derived formulas by which the elements of the BxB matrix $\hat{\mathbf{R}}_{yy} = \mathbf{W}^H\hat{\mathbf{R}}_{xx}\mathbf{W}$ may be computed from the components of the Bx1 vectors $\mathbf{W}^H\mathbf{r}$ and $\mathbf{W}^H\{\mathbf{n} \circ \mathbf{r}\}$, where $\mathbf{r}$ is the first column of $\hat{\mathbf{R}}_{xx}$, i. e., the L pt. spatial autocorrelation sequence, $\mathbf{n} = [0, 1, \cdots, L-1]^T$, and $\circ$ denotes the Schur or pointwise matrix product. In this way the computational complexity is reduced to 2BL complex multiplications.

6. Simluations: Circular Array

Simulations were conducted employing an N=13 element UCA with a radius of $R = \lambda$. Three equi-powered, uncorrelated sources were located at an elevation angle $\phi_o = 90^\circ$, corresponding to ground waves, with respective azimuth angles of $\theta_1 = 70^\circ$, $\theta_2 = 60^\circ$, and $\theta_3 = 20^\circ$. For $\phi_o = 90^\circ$, the mainlobe of the azimuthal beam pattern pointed to any θ has a 3 dB width of 20° such that the first two sources were separated by a half-beamwidth. An interpolation matrix was designed for the azimuthal sector $0^\circ < \theta < 180^\circ$, half of the entire 360° azimuth sector, at the elevation angle $\phi_o = 90^\circ$. B=7 beams were formed encompassing this sector with respective pointing angles, $\theta_i = 2\pi i/N$, i=0,1,...,6. Applying the Beamspace Root-MUSIC algorithm outlined in Section 2, we obtain 2B-2 = 12 roots per run.

The results of 250 independent runs for the respective cases of SNR = 8 dB per source per element and SNR = 15 dB per source per element are superimposed in Figures 1(a) and 1(b). The sample bias and sample standard deviation for each source angle for each of these two cases is listed in the corresponding figure as well. For each case, the two sources separated by a half-beamwidth are clearly resolved in all 250 runs. As would be expected, estimator accuracy increases with increasing SNR. Interestingly, though, the third source has a lower sample std. dev. but a higher sample bias than the two sources separated by a half-beamwidth. It is conjectured that the higher bias is due to the fact that the the half-beamwidth sources are at the center of the sector while the third source is towards the edge of the sector. Similar phenomena have been observed in the case of beamspace processing with a ULA.

7. Simluations: MRLA

Simulations were conducted employing an N=8 element MRLA with elements located at 0, 2d, 5d, 7d, 13d, 19d, 22d, and 23d, where $d = \lambda/2$ [5]. This array is able to measure L=24 equi-spaced spatial correlation lag values. Loosely speaking, in the case of uncorrelated sources this array has the resolving power and detection capability of an L=24 element ULA with $d = \lambda/2$. Two equi-powered, uncorrelated sources were located at bearing angles of $\theta_1 = -3.74^\circ$ and $\theta_2 = -1.25^\circ$. The broadside 3 dB beamwidth for a 24 element ULA with $d = \lambda/2$ is roughly 5.0° such that the two sources were separated by a half-beamwidth. B=4 beams were formed with respective pointing angles roughly equal to −10.0°, −5.0°, 0.0°, and 5.0°. Applying the Beamspace Root-MUSIC algorithm outlined in Section 2, with $\hat{R}_{yy}$ computed according to (9), we obtain 2B-2 = 6 roots per run.

The results of 250 independent runs for the respective cases of SNR = 3 dB per source per element and SNR = 9 dB per source per element are superimposed in Figures 2(a) and 2(b). The sample bias and sample standard deviation for each source angle for each of these two cases is listed in the corresponding figure as well. For each case, the two sources are clearly resolved and accurately localized in all 250 runs. The main point of these simulations is to demonstrate that Beamspace Root-MUSIC, although developed for a ULA, may be used in conjunction with an MRLA which is an nonuniformly-spaced linear array structure.

References

[1] M. D. Zoltowski, G. M. Kautz, and S. D. Silverstein, "Beamspace Root-MUSIC," accepted for publication in *IEEE Trans. on Signal Processing.*

[2] A. Weiss and B. Friedlander, "Performance Analysis of Spatial Smoothing with Interpolated Arrays," *Proc. of ICASSP '91*, pp. 1377-1380.

[3] W. Hong and A. H. Tewfik, "On the Equivalence of Uniform Circular Arrays and Uniform Linear Arrays," *Proc. of Fifth ASSP Workshop on Spect. Est. and Modeling*, Oct. 1990, pp. 139-143.

[4] S. U. Pillai, Y. Bar-Ness, and F. Haber, "A New Approach to Array Geometry for Improved Spatial Spectrum Estimation," *IEEE Proc.*, Oct. 1985, pp. 1522-1524.

[5] D. Pearson, S. U. Pillai, and Y. Lee, "An Algorithm for Near Optimal Placement of Sensor Elements," *IEEE Trans. on Information Theory*, Nov. 1990, pp. 1280-1284.

[1] private correspondence with Seth Silverstein, Senior Scientist, Corporate Research & Development Center, General Electric Co.

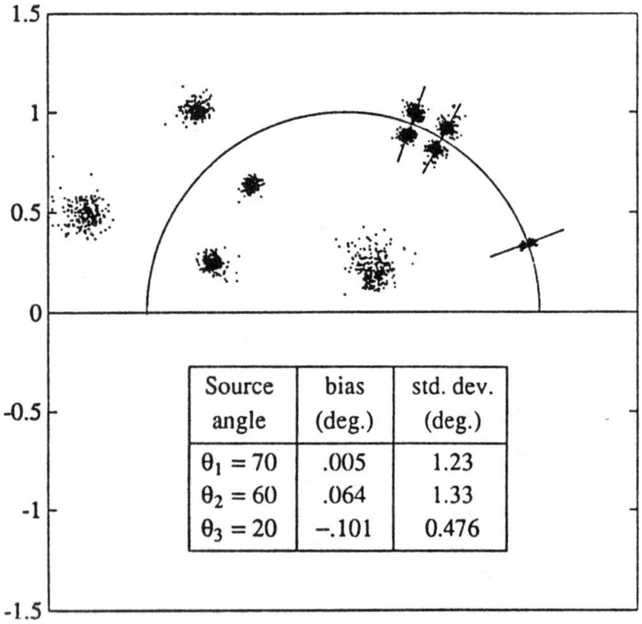

Source angle	bias (deg.)	std. dev. (deg.)
$\theta_1 = 70$	.005	1.23
$\theta_2 = 60$	.064	1.33
$\theta_3 = 20$	−.101	0.476

(a) UCA: SNR = 8 dB

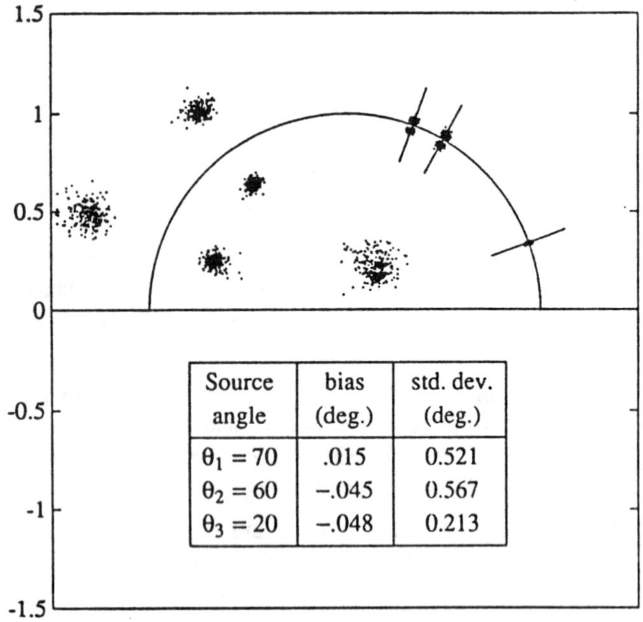

Source angle	bias (deg.)	std. dev. (deg.)
$\theta_1 = 70$	.015	0.521
$\theta_2 = 60$	−.045	0.567
$\theta_3 = 20$	−.048	0.213

(b) UCA: SNR = 15 dB

Figure 1. Simulations for N=13 element UCA with radius R = λ.

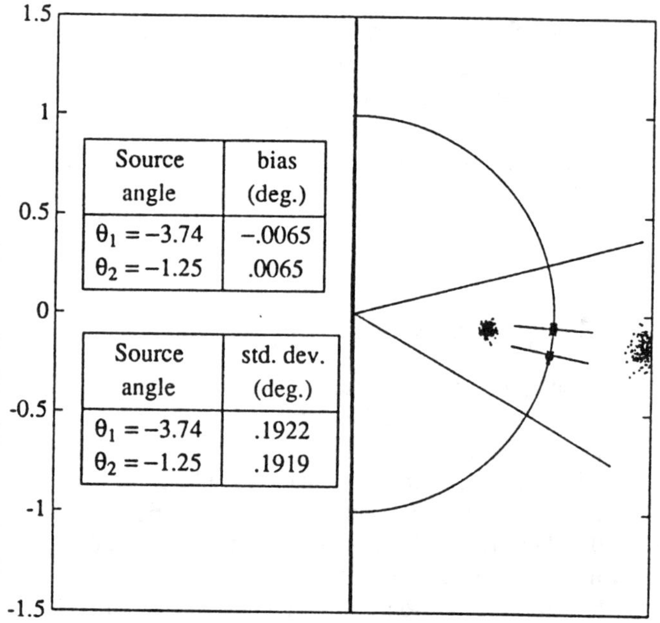

Source angle	bias (deg.)
$\theta_1 = -3.74$	−.0065
$\theta_2 = -1.25$	.0065

Source angle	std. dev. (deg.)
$\theta_1 = -3.74$	.1922
$\theta_2 = -1.25$	.1919

(a) MRLA: SNR = 3 dB

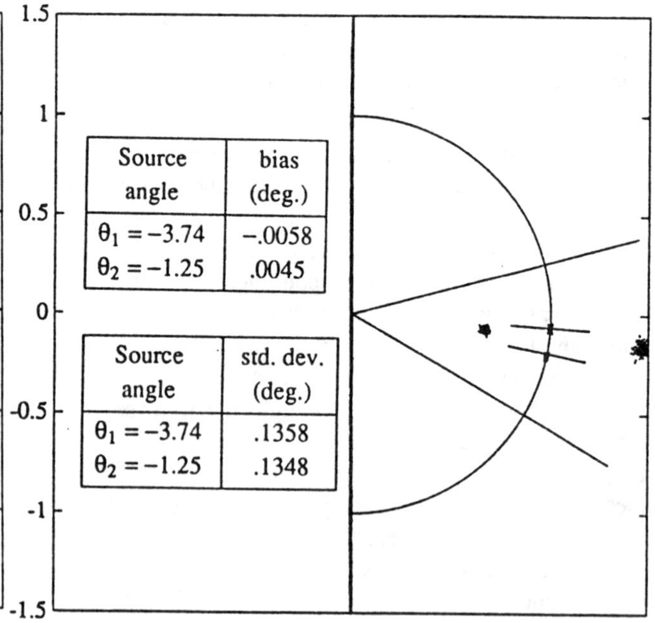

Source angle	bias (deg.)
$\theta_1 = -3.74$	−.0058
$\theta_2 = -1.25$	.0045

Source angle	std. dev. (deg.)
$\theta_1 = -3.74$	.1358
$\theta_2 = -1.25$	.1348

(b) MRLA: SNR = 9 dB

Figure 2. Simulations for N=8 element MRLA "equivalent" to L=24 element ULA.

Comparison of Resolution Enhanced Eigenspace Based Spectral, Rooting and ESPRIT Direction-of-Arrival Estimators

J. Marks, R. Hamza and K. Buckley
Department of Electrical Engineering
University of Minnesota

Abstract

In this paper we compare spectral, rooting and ESPRIT approaches to eigenspace based Direction od Arrival (DOA) estimation with a Uniform Linear Array (ULA). We consider three performance measures: DOA estimate variance; bias; and resolvability. Since resolvability is a principal interest, we consider resolution enhances MUSIC, Root-MUSIC and ESPRIT algorithms which incorporate DOA sector focusing. Comparison is based on Monte Carlo simulation. A statistical comparison will appear elsewhere.

1. Introduction

Our objective is to provide a useful comparison of several popular eigenspace source localization approaches. Specifically, we compare spectral, rooting and ESPRIT methods. For this comparison, then, we naturally consider DOA estimation with a ULA array. In comparing approaches to source localization, numerous criteria are relevant, such as location estimate accuracy, ability to resolve closely spaced sources, robustness to signal and noise observation inaccuracies, computational requirements, and realm of applicability. To keep our task manageable, we limit our consideration to resolution, and DOA estimate variance and bias.

Much has been written on statistical analysis of asymptotic DOA estimate variance for spectral, rooting and ESPRIT eigenspace methods, and on enhancements of these methods to reduce asymptotic variance (e.g., [3-5]). Thus, in terms of asymptotic variance, the relative performance of spectral, rooting and ESPRIT approaches is well understood. On the other hand, investigators have considered issues such as resolvability, resolution enhancement, and DOA estimate accuracy in the resolution threshold region for these approaches (e.g. [1,2,8,10,11]). However, a comprehensive investigation and resulting understanding of relative performance of resolution enhanced algorithms for these methods, operating in SNR around the resolution threshold, is not available in the literature. Our aim is to begin to fill this void.

Numerous spectral, rooting and ESPRIT algorithms have been proposed. Algorithms we consider here are based on an eigenspace resolution enhancement approach termed CLOSEST [1], which is based on focusing for a DOA sector of interest. For spectral based DOA estimation, the FINE algorithm, based on the CLOSEST approach, was shown to enhance resolution (compared to MUSIC and MIN-NORM) by effectively reducing location-parameter estimate bias, without increasing estimate variance (compare to MUSIC which is the estimate variance standard for these eigenspace classes). In this paper we introduce rooting and ESPRIT algorithms, termed Root-FINE and Resolution Enhanced (RE) ESPRIT, which incorporate sector focusing. We note that there has been considerable interest recently in beam-space preprocessing for eigenspace location estimators (e.g. [1,6-8]), and resolution enhancement via this approach has been demonstrated (e.g. [1,7]). For single source-cluster situations, comparable resolution enhancement and DOA variance can be realized with beam-space preprocessing and sector focusing. However, performance differs significantly in terms of computational requirements and resolvability and variance sensitivity in multiple source-cluster situations [8]. In this paper we consider only sector focusing applied in element space.

This paper is organized as follows. In Section 2 background is provided. In Section 3 the algorithms are described. Section 4 addresses empirical resolution, and resolution enhancement. In section 5 FINE, RT-FINE and RE-ESPRIT are compared. Conclusions are then drawn.

2. Array Observation and DOA Sector Representation

Consider a ULA of K sensors. Let θ represent the DOA of an impinging source. Let

$$\mathbf{a}(\theta) = [\,1,\ e^{-j\omega},\ e^{-j2\omega},\ \ldots,\ e^{-j(K-1)\omega}]^T, \qquad (1)$$

$\omega = \pi \sin(\theta)$, denote a K dimensional array response vector. Consider $D < K$ noncoherent sources impinging from DOA's θ_i ; $i = 1,2, \ldots D$. Let $\mathbf{g}(t)$ denote an array observation vector,

$$\mathbf{g}(t) = \mathbf{A}\,\mathbf{s}(t) + \mathbf{n}(t), \qquad (2)$$

where $\mathbf{A}$ is the K by D matrix whose linearly independent columns are the array response vectors $\mathbf{a}(\theta_i)$; $i=1,2, \ldots D$. $\mathbf{s}(t)$ is the D vector of source time-series as observed at the array phase center, and $\mathbf{n}(t)$ is additive noise uncorrelated with the sources. The covariance matrix of $\mathbf{g}(t)$ is,

$$\mathbf{R_g} = \mathbf{A}\,\mathbf{P_s}\,\mathbf{A}^H + \sigma_n^2\,\mathbf{I}_K, \qquad (3)$$

where $\mathbf{I}_K$ is the K dimensional identity, and the D by D source covariance matrix $\mathbf{P_s}$ is assumed full rank.

This work was supported by ONR under contract N00014-90-J-1049.

1058-6393/91 $01.00 © 1991 IEEE

Let $\hat{R}_g$ denote the sample covariance matrix estimate of R_g obtained by averaging the outer products of N observation vectors ($\hat{x}$ denotes an estimate of the true value x). Let $\{\hat{\lambda}_i;\ i=1, 2, \ldots, K;\ \lambda_i \geq \hat{\lambda}_{i+1}\}$ and $\{\hat{e}_i;\ i=1, 2, \ldots, K\}$ be the ordered eigenvalues and corresponding normalized eigenvectors of $\hat{R}_g$. The estimated signal subspace and noise-only subspace are, respectively, the ranges of the matrices

$$\hat{E}_s = [\,\hat{e}_1, \hat{e}_2, \ldots, \hat{e}_D], \text{ and } \hat{E}_n = [\hat{e}_{D+1}, \ldots, \hat{e}_K\,]. \quad (4)$$

Suppose that Θ is a DOA sector-of-interest. A low rank subspace representation of this sector can be obtained as follows. Form a spatially extended source covariance matrix as,

$$R_\Theta = \int_\Theta a(\theta)a^H(\theta)d\,\theta \quad (5)$$

Let $\{\eta_i;\ i= 1,2,\ldots,K;\eta_i \geq \eta_{i+1}\}$ and $\{u_i;\ i= 1, 2, \ldots, K\}$ be the ordered eigenvalues and corresponding eigenvectors of R_Θ. For a FOV that fills the observation space, for a reasonable size Θ, R_Θ will typically be low rank. Let D_Θ be chosen based on some percentage of the trace of R_Θ. Form the K by D_Θ matrix,

$$U_\Theta = [u_1, u_2, \cdots, u_{D_\Theta}]. \quad (6)$$

The range of this matrix, Ω_Θ, is termed the source representation subspace for Θ.

3. Resolution Enhanced Algorithms

A) Spatial-Spectrum Estimation

With a CLOSEST spatial-spectrum estimator, a spatial-spectrum estimate is formed by projecting onto a vector or vector set in the estimated noise subspace which is in some sense closest to the portion of the array manifold corresponding to a sector-of-interest. *The FINE algorithm* is so named because it employs the FIrstpriNcipal vEctor between the estimated noise subspace and a subspace that represents the sector-of-interest. The FINE vector $\hat{e}_F$ is the vector in the estimated noise subspace which has minimum angle to Ω_Θ. $\hat{e}_F$ is computed as follows.

1) Generate the noise subspace eigenvector matrix $\hat{E}_n$.
2) Form the Singular Value Decomposition (SVD) $U_\Theta^H \hat{E}_n = Y\,X\,Z^H$. Let the singular values along the diagonal of X be arranged in nonincreasing order. $\hat{e}_F$ is the first I column of $E_n Z$. The FINE spatial-spectrum estimator is

$$\hat{P}_{FINES}(\theta) = \frac{a^H(\theta)\,a(\theta)}{\left|a^H(\theta)\,\hat{e}_F\right|^2} \quad (7)$$

For $D_\Theta = 1$, $\hat{e}_F$ is just the projection of u_1 on the estimated noise subspace. $D_\Theta = 1$ is effective as long as Θ is not much larger than two 3 dB beamwidths.

B) Rooting Methods

For linear equispaced arrays, one can formulate the array response vector in terms of powers of the complex variable z, and thus formulate the null-spectrum used in spectral methods as a polynomial in z, with roots in the complex z-plane. Consider the case where D, the number of sources, has been estimated correctly. If the SNR is not high enough, there is no guarantee that there will be D (or more) peaks in the spectrum. However, there will always be $(K-1)$ unique roots in the z-plane. The D roots with radii nearest to the unit circle are used as the DOA estimates. The DOA estimate is a function of only the angle of the root. However, the radii of the roots plays a role in the resolution of these methods. For algorithms that use a single vector in the Noise Subspace, one uses that vector as the coefficients of the polynomial to be rooted.

Corresponding rooting versions exist for spectral algorithms such as MUSIC, MIN-NORM, and FINE. Root-FINE can form estimates for the entire field of view (FOV) by dividing the FOV into sectors and finding the FINE vector for each sector. For each sector, roots are determined from the FINE vector and disgarded unless they have angles corresponding to the sector under consideration. One repeats the process for each sector and concatenates the resulting roots. Finally, one selects the D roots nearest the unit circle as candidate estimates. It is possible that one will have less than D roots because out of sector roots are discarded, however, this occurs only at very low SNR, i.e. when root-FINE can not resolve the sources.

C) ESPRIT

The basic idea behind ESPRIT is to exploit the array displacement invariance structure between two identical subarrays. Since we are only concerned with ULA, a natural choice is the overlapping structure. Let $x(t)$,the first subarray output, consist of all elements of $g(t)$ excluding the last h sensors outputs. Let $y(t)$, the second subarray output, consist of all elements of $g(t)$ excluding the first h components.The two subarrays will not be overlapped if $h \geq K/2$. The shift invariance property on the subarray manifolds implies that

$$A_y = [\,O_{m,h}\ \ I_m\,]\,A = [\,I_m,\ O_{m,h}]\,A\Phi = A_x\,. \quad (8)$$

The DOA's can then be estimated from $\Phi = \text{diag}\,\{e^{j2\pi\,(h\,d/\lambda)\sin\theta_i};\ i=1,\ldots,D\}$. The dimension of each subarray is specified by $m = K-h$.

In practice, Φ is obtained as the eigenvalues of Ψ, where Ψ is determined from

$$\hat{E}_x\,\Psi = \hat{E}_y\,. \quad (9)$$

$\hat{E}_x$ contains the first h rows of $\hat{E}_s$ (associated with the first subarray) and $\hat{E}_y$ contains the last h rows of $\hat{E}_s$ (associated with the second subarray).

A least squares estimate of Ψ can be obtained from (9) as

$$\hat{\Psi}_{ls} = \left(\hat{E}_x^H\,\hat{E}_x\right)^{-1}\hat{E}_x^H\,\hat{E}_y, \quad (10.a)$$

$$\text{or} \quad \hat{\Psi}_{ls} = \left(\left(\hat{E}_y^H\,\hat{E}_y\right)^{-1}\hat{E}_y^H\,\hat{E}_x\right)^{-1}. \quad (10.b)$$

This implicitly assumes that there are no measurement errors in either $\hat{E}_x$ in (10.a) or $\hat{E}_y$ in (10.b) [3]. With noise appearing in both signal subspaces, both equations of (10) might result in two different estimates.

A symmetric treatment of the uncertainties in both submatrices is then proposed which uses Total Least Squares concepts[9]. Since $\hat{E}_x$ and $\hat{E}_y$ share a common column space, the rank of the matrix $\hat{E}_{xy} = [\hat{E}_x \ \hat{E}_y]$ must be equal to D. Let $F = [\Psi_x^T \ \ \Psi_y^T]^T$ spans the null space of E_{xy}.

$$\widehat{\Psi}_{tls} = -\widehat{\Psi}_x\left(\widehat{\Psi}_y\right)^{-1}. \tag{11}$$

Resolution Enhanced (RE) ESPRIT

Viberg and Ottersten [4] proposed Weighted Subspace Fitting as means to generalize the deterministic Maximum likelihood and the subspace methods into a common class. Their study leads to a natural choice of a weighted matrix that would result in an optimal solution (i.e $\tilde{E}_z = E_z W$). Tilde is used through out the paper to indicate a weighted matrix. This type of weighting has been shown [5] to be ineffective with ESPRIT. Left weighted ESPRIT (i.e. $\tilde{E}_z = W E_z$) has been considered by Stoica and Nehorai [5] and by Ottersten *et al.* [3] as a mean to reduce the estimation error variance.

In this paper we propose to incorporate sector weighting into ESPRIT, via left weighting, to improve resolution. Since we are dealing with a ULA, we will also employ the Viberg *et al.* weighting as follows. The solution to *ESPRIT* is still given by (10) or (11), but now with $\hat{E}_x$ and $\hat{E}_y$ are replaced by

$$\tilde{E}_x = U_{\Theta J}^H J \hat{E}_x \quad \text{and} \quad \tilde{E}_y = U_{\Theta J}^H J \hat{E}_y, \tag{12}$$

where J is chosen to be the same as $\tilde{J}$ in equation (24) of [3] . The columns of $U_{\Theta J}$ are the first principal eigenvectors associated with the largest eigenvalues of

$$R_{\Theta J} = \int_{\Theta} J\, a(\theta)\, a^H(\theta)\, J^T \, d\theta . \tag{13}$$

Notice that the redundancy of the rows is also included in $R_{\Theta J}$ so as to match the dimension and structure of $J\hat{E}_x$ (and $J\hat{E}_y$) . Computation can be reduced to the following form

$$\begin{bmatrix} \tilde{E}_x \\ \tilde{E}_y \end{bmatrix} = \begin{bmatrix} U_{\Theta\Sigma}^H \Sigma^{1/2} \hat{E}_x \\ U_{\Theta\Sigma}^H \Sigma^{1/2} \hat{E}_y \end{bmatrix} = \begin{bmatrix} W_{D_\theta,m} & O_{m,h} \\ O_{m,h} & W_{D_\theta,m} \end{bmatrix} \hat{E}_s, \tag{14}$$

where $U_{\Theta\Sigma}$ is obtained from (13), and where J is replaced by the $m \times m$ matrix $\Sigma^{1/2}$ ($\Sigma = J^T J = \text{diag}\{1,2,3,4, \ldots, \omega, \omega, \omega, \ldots, 3,2,1\}$; $\omega = min(ms, m-ms+1)$ [3]). The $D_\theta x m$ matrix $W = U_{\Theta\Sigma}^T \Sigma^{1/2}$ was also introduced in (14). Recall that m is the size of the original overlapping subarrays (i.e. $m = K - h$). Since the dimension of the $\tilde{E}_x$ (and $\tilde{E}_y$) is reduced to $D_\theta x D$, RE-ESPRIT accounts more fully for the noise structure of the eigen problem.

One advantage of RE-ESPRIT is that if D_θ is selected to be equal to D, there is no need for the Total Least Squares criteria. That is, the weighted TLS version is equivalent to weighted LS-ESPRIT. The prove is outlined in the following. Since $\tilde{E}_{xy} = [\tilde{E}_x \ \tilde{E}_y]$ has dimension equal to $D \times 2D$, its rank is exactly determined to be D and hence the singular vectors $[\tilde{\Psi}_x^T \ \tilde{\Psi}_x^T]^T$, associated with its D zero singular values, will perfectly span its null space (i.e. the following equation holds exactly),

$$[\ \tilde{E}_x \quad \tilde{E}_y\ ] \begin{bmatrix} \tilde{\Psi}_x \\ \tilde{\Psi}_y \end{bmatrix} = \tilde{E}_x \tilde{\Psi}_x + \tilde{E}_y \tilde{\Psi}_y = 0 ,$$

$$\text{or} \quad \tilde{E}_x \tilde{\Psi}_x \left(\tilde{\Psi}_y\right)^{-1} + \tilde{E}_y = 0 . \tag{15}$$

Moreover, $\tilde{E}_x = W \hat{E}_x$ is a full rank square matrix. Similar to (10), the enhanced left weighted TLS -ESPRIT estimates are determined from

$$\tilde{\Psi}_{tls} = -\tilde{\Psi}_x\left(\tilde{\Psi}_y\right)^{-1} = \left(\tilde{E}_x\right)^{-1} \tilde{E}_y = \tilde{\Psi}_{ls} . \tag{16}$$

Equation (16) states that both weighted TLS-ESPRIT and weighted LS -ESPRIT reduce to the same estimate. We term (16) the RE-ESPRIT estimate. Notice that RE-ESPRIT eliminates the additional SVD required by TLS-ESPRIT, while improving its resolution and preserving its accuracy (as we shall illustrate).

4. Resolution

To proceed in comparing the three approaches considered, according to the criteria of interest here, definitions of resolution must first be identified. Consider a single trial were N snapshots are used to generate a single spectrum (or a set of roots or a set of singular values). Qualitatively, D sources should be considered resolved if and only if exactly D sources are indicated, and if the DOA estimates obtained are accurate.

The sector focused algorithms compared in this paper are not prone to give false peaks. By this we mean, for *spectral* (or Root) FINE, if D sources are impinging on the array with SNR above resolution thresholds, then if we construct a *spatial-spectrum* (or root plot) for the entire FOV by dividing the FOV into contiguous sectors, computing and concatenating *spatial-spectrum for* (or only roots within) each sector, we resolve exactly D spectral peaks (or roots). With ESPRIT, the issue of resolution is more subtle since if you ask for a certain number of DOA estimates you automatically get exactly that number of estimates. As with spectral and rooting algorithms, we assume that the total number of sources is known. For multiple source-cluster cases, we have developed RE-ESPRIT algorithms both for situations where the number of sources in each sector is known, and when it is not. Below we consider only single cluster cases.

Quantitatively, in this paper we will consider two closely spaced sources to be resolved only if exactly two sources are observed within a tight angular evaluation region around the true source angles. This tight region is defined in terms of the 3dB beamwidth of the array (at broadside). Specifically, given DOA's θ_1 and θ_2 ($\theta_1 < \theta_2$), and 3dB beamwidth BW, the evaluation region we use is $\theta_1-(BW/2) < \theta < \theta_2+(BW/2)$. If exactly two sources are detected in the evaluation region, then the sources are considered resolved only if the two source power estimates derived from the DOA estimates are within 10dB of the true source powers. (This power test, first suggested in [10], has the effect of reducing empirical probability of resolution at very low SNR.)

To illustrate the effectiveness of these definitions of resolution, and to show the improved resolution achieved with sector focusing, we compared RT-MUSIC with RT-FINE, and TLS-ESPRIT with RE-ESPRIT. We considered the following example, which we refer to as CASE 1. Let $K = 16$, $D = 2$, $\theta_1 = 14^\circ$, $\theta_2 = 16^\circ$, and $N = 100$. The two sources were equi-powered and uncorrelated. For this situation, the evaluation sector was $10.7^\circ < \theta < 19.3^\circ$. For TLS ESPRIT, Viberg and Ottersten's weighting for uniform linear arrays was used. For both sector focused algorithms the source representation subspace was constructed for $7.7^\circ < \theta < 22.6^\circ$ (i.e. the 1-st zero-crossing extent of a beam pointed at 15°). For rooting and ESPRIT algorithms respectively, Figures 1 & 2 show results of randomly selected trials. These figures compare established and new sector-focused algorithms for SNR's of: a) -10 dB (-6dB for rooting), ; and b) 10 dB. For 10 dB SNR, using the criteria stated above, all trials resolved. For the lower SNR, only the sector enhanced algorithms resolved the sources (in both trials). Clearly, upon inspection of the resulting roots and singular values, the criteria generated reasonable results. Figure 3 shows empirical probability of resolution vs. SNR, based on 800 trials/SNR. This figure suggests that, in terms of resolution, ESPRIT and rooting algorithms perform similarly, and sector focusing substantially improves the resolution of both rooting and ESPRIT algorithms.

5. Simulation Comparison

In this section we compare spectral FINE, Root-FINE and RE-ESPRIT. We used the 16 element array described in section 4, and two equi-powered sources with 14° and 16° DOA's. The evaluation sector was again $10.7^\circ < \theta < 19.3^\circ$. The source representation subspace used for all three algorithms was as described in section 4. N=100 snapshots/trail were simulated.

First, uncorrelated sources were considered (that is, CASE I was used). Figure 4 shows probability of resolution, and standard deviation (STD) and mean of the 14° source estimate, vs. SNR. Results are based on 800 trials/SNR. Root-FINE and RE-ESPRIT provide better resolution than spectral FINE. Spectral FINE and Root-FINE result in lower variance than RE-ESPRIT in and above the spectral FINE resolution threshold region. (Note that spectral FINE provides estimate variance comparable to that of spectral MUSIC [1,11].) Estimate bias for all algorithms is small relative to STD.

For CASE II, CASE I was altered only in that two correlated sources were simulated (correlation was .95). The curves shown in Figure 5 are based on 500 trials/SNR. Relative performance is as in CASE I.

6. Summary and Conclusions

In this paper we compared resolution enhanced versions of MUSIC, Root-MUSIC and ESPRIT using resolution and estimator accuracy as criteria. Our objective was to show relative performance under idealistic and simplified situations, so as to further our understanding of the merits of these algorithms. This is of course useful when deciding which algorithm to employ in application, possibly as part of a source detection/estimation/copy scheme. Considering these criteria (only) we conclude that root-FINE provides the best combination of resolution and estimation accuracy.

References

[1] Buckley, K. M. and X. L. Xu, "Spatial-spectrum estimation in a location sector," *Trans. on ASSP*, ASSP-38, No. 11, pp. 1842-1852, Nov. 1990.

[2] Xu, X. L. and K. M. Buckley, "Bias analysis of the MUSIC location estimator," *Trans. on SP*, to appear Nov. 1992.

[3] Ottersten, B., M. Viberg and T. Kailath, "Performance analysis of the total least squares ESPRIT algorithm," *IEEE Trans. on SP*, Vol. 39, No. 5, pp. 1122-1134, May 1991.

[4] Viberg, M. and B. Ottersten, "Sensor array processing based on subspace fitting," *IEEE Trans. on SP*, Vol. 39, No. 5, pp. 1110-1121, May 1991.

[5] Stoica, P. and A. Nehorai, "Performance comparison of subspace rotation and MUSIC methods of direction estimation," *IEEE Trans. on SP*, Vol. 39, No. 2, pp. 446-453, Feb. 1991.

[6] Xu, G., S. Silverstein, R. Roy and T. Kailath, "Parallel implementation and performance analysis of beamspace ESPRIT," *Proc. ICASSP-91*, pp. 1497-1500, May 1991.

[7] Hamza, R. "A comparison of ESPRIT algorithms and solutions for the related ambiguity problem," Masters Degree Thesis, University of Minnesota, Aug. 1989.

[8] Xu, X. L. and K. M. Buckley, "Statistical performance comparison of MUSIC in element-space and beam-space," *Proc. ICASSP'89*, pp. 2124-2127, May 1989.

[9] Roy, R. and T. Kailath, "ESPRIT- estimation of signal parameters via rotational invariance techniques," *IEEE Trans. on ASSP*, Vol. 37, No. 7, pp. 984-995, July 1989.

[10] Kaveh, M. and A. J. Barabell [1986], "The statistical performance of the MUSIC and the minimum-norm algorithms in resolving plane waves in noise," *IEEE Trans. on ASSP*, Vol. ASSP-34, No. 2, pp.331-341, April 1986.

[11] X. L. Xu and K. M. Buckley, "Statistical analysis of eigenspace-based source location estimates," *Proc. ICASSP-91*, pp. 3041-3044, May 1991.

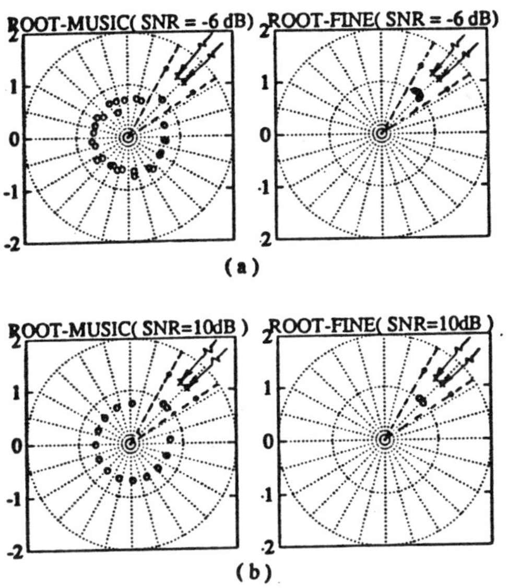

Figure 1. For RT-MUSIC and RT-FINE, roots estimates for two randomly selected trials.
a) SNR = - 6 dB; b) SNR = 10 dB.

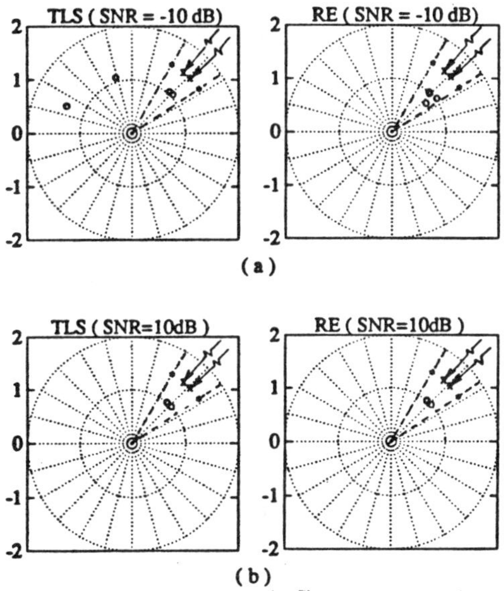

Figure 2. For TLS-ESPRIT and RE-ESPRIT, singular values estimates for two randomly selected trials.
a) SNR = - 10 dB; b) SNR = 10 dB.

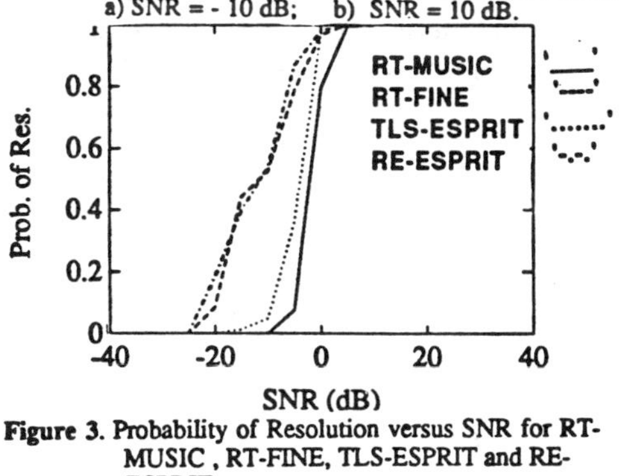

Figure 3. Probability of Resolution versus SNR for RT-MUSIC , RT-FINE, TLS-ESPRIT and RE-ESPRIT.

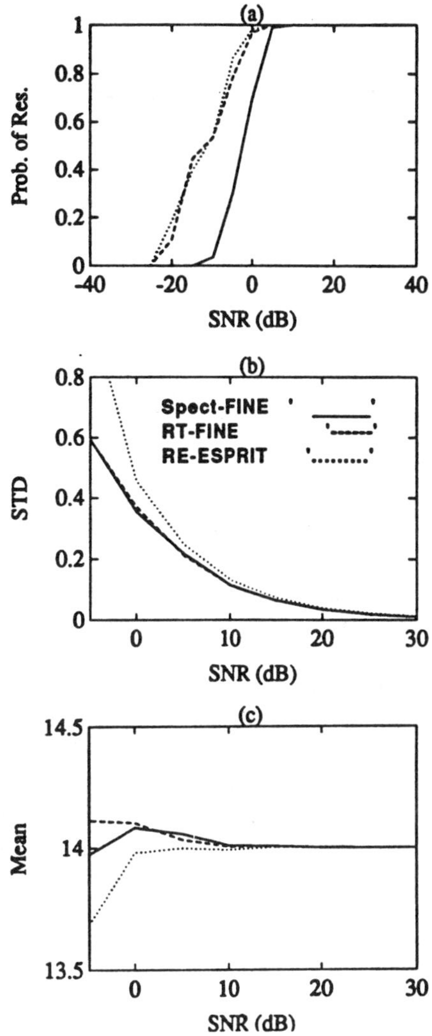

Figure 4. Case I comparison based on empirical resolution and estimate Standard Deviations (STD) and mean of Spectral-FINE, Root-FINE and RE-ESPRIT.
a) Probability o Resolution; b) STD; c) Mean.

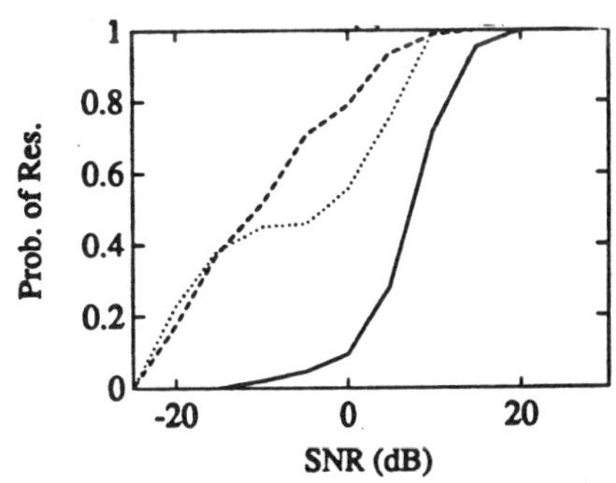

Figure 5. Case II comparison based on empirical resolution and estimate Standard Deviations (STD) and mean of Spectral-FINE, Root-FINE and RE-ESPRIT.

Vector Sensor Processing for Electromagnetic Source Localization*

Arye Nehorai and Eytan Paldi
Department of Electrical Engineering
Yale University
New Haven, CT 06520

Abstract

A new approach is presented for localization of electromagnetic sources using sensors whose output is a vector consisting of the complete six electric and magnetic field components measured at the sensors. Two types of source transmissions are considered: i) Single signal transmission (SST), and ii) dual signal transmission (DST). The model is given in terms of several parameters including the wave direction of arrival (DOA) and state of polarization. A compact expression is derived for the Cramér-Rao bound (CRB) on the estimation errors of these parameters for the multi-source multi- vector sensor model. The advantage of using vector sensors is highlighted by explicit evaluation of the CRB and the associated mean-square angular error (MSAE) for source localization with a single vector sensor. A simple algorithm is proposed for estimating the source DOA with this sensor.

1 Introduction

The localization of source signals using sensor data processing has attracted significant attention lately. Most existing methods employ sensor arrays in which the output data of each sensor is a scalar corresponding, for example, to magnitude of pressure in the acoustic case, or to magnitude of electric field in the electromagnetic case. This paper considers new methods for multiple electromagnetic source localization using sensors whose data output is a *vector* corresponding to the complete electric and magnetic fields measured at the sensor. These sensors, which will be called vector sensors, can consist for example of two lumped orthogonal triads of scalar sensors that measure the electric and magnetic field components. The main advantage of these sensors over the traditional scalar sensors is that they offer greater observability for direction of arrival (DOA) estimation resulting in a smaller requirement on the array aperture. They permit estimation of some of the source parameters that have not been accessible before.

Section 2 derives the measurement model. The electromagnetic sources considered can originate from two types of transmissions: i) Single signal transmission (SST), in which only a single signal message is transmitted, and ii) dual signal transmission (DST), in which two separate signal messages are transmitted simultaneously [1], [2]. The great interest in the latter case is due to the fact that it makes full use of the two spatial degrees of freedom present in transversal electromagnetic plane wave transmission.

Section 3 analyzes the minimum attainable variance of unbiased DOA estimators for a general vector sensor array model and multi-electromagnetic sources that are assumed to be stochastic and stationary. A compact expression for the corresponding Cramér-Rao bound (CRB) on the DOA estimation error is presented, which extends previous results for the case of scalar sensor measurements that appear in [3] (see also [4], [5].)

A significant property of the vector sensors is that they enable DOA (azimuth and elevation) estimation of an electromagnetic source with a *single* (lumped) vector sensor and a single snapshot. This result is explicitly shown in detail by investigating the CRB expression for this problem in Section 4. A bound on the associated mean-square angular error (MSAE, to be defined later) which is invariant to the reference coordinate system is used for an in-depth performance study. A simple algorithm is proposed for estimating the source DOA with this sensor. Section 5 discusses some performance results for DOA estimation of two sources using a single vector sensor.

*This work was supported by the Air Force Office of Scientific Research under Grant no. AFOSR-90-0164, the Office of Naval Research under Grant no. N00014-91-J-1298, and the HTI fellowship.

1058-6393/91 $01.00 © 1991 IEEE

The main difference between the present paper and previous papers is in our use of vector sensors with complete electric and magnetic data. Most previous papers used scalar sensors. Attempts to extend scalar sensor array processing were made in [6]–[9], but these references used only part of the electric data and did not consider the use of magnetic data. This simplifying model reduced the sensors' observability of source DOAs. This results in that all the information on source location is coded in the time difference of arrivals of the source waves between spatially distributed sensors.

The use of the complete electromagnetic vector sensor data enables, for example, source parameter estimation with a single lumped sensor (even with a single time sample) where time delays are not used at all. In fact, this is shown to be possible for at least two sources. As a result, the derived CRB expressions for this problem are applicable to wide-band sources. The source DOA parameters considered include azimuth and elevation, while generally previous papers usually considered only azimuth. This paper also considers estimation of wave ellipticity and orientation (that will be defined later) and estimation of direction to DST sources.

2 The Measurement Model

2.1 Single-Source Single-Sensor Model

Throughout the paper it will be assumed that the medium is nonconductive, homogeneous and isotropic. In addition, the following will also be assumed:

A1: Plane wave at the sensor.

A2: Band-limited spectrum: The signal has a spectrum including only frequencies ω satisfying $0 < \omega_{\min} \leq |\omega| \leq \omega_{\max} < \infty$.

Let $\mathcal{E}(t)$ and $\mathcal{H}(t)$ be the vector phasor representations (or complex envelopes) of the electric and magnetic fields at the sensor. Also, let $\boldsymbol{u}$ be the unit vector from the sensor to source, i.e.

$$\boldsymbol{u} = \begin{bmatrix} \cos\theta_1 \cos\theta_2 \\ \sin\theta_1 \cos\theta_2 \\ \sin\theta_2 \end{bmatrix} \tag{2.1}$$

where θ_1 and θ_2 denote, respectively, the azimuth and elevation angles of $\boldsymbol{u}$. Then, the Maxwell equations for band-limited plane waves are equivalent to:

$$\boldsymbol{u} \times \mathcal{E}(t) = -\eta\mathcal{H}(t) \tag{2.2a}$$

$$\boldsymbol{u} \cdot \mathcal{E}(t) = 0 \tag{2.2b}$$

where "$\times$" and "$\cdot$" are the cross and inner product between two vectors and η is the intrinsic impedance of the medium.

Let $\boldsymbol{y}_E(t)$ be the measured electric phasor field vector at the sensor at time t and $\boldsymbol{e}_E(t)$ its noise component. Then the electric part of the measurement will be

$$\boldsymbol{y}_E(t) = \mathcal{E}(t) + \boldsymbol{e}_E(t) \tag{2.3}$$

Similarly, from (2.2a), after appropriate scaling, the magnetic part of the measurement will be taken as

$$\boldsymbol{y}_H(t) = \boldsymbol{u} \times \mathcal{E}(t) + \boldsymbol{e}_H(t) \tag{2.4}$$

Define the matrix cross product operator that maps a vector $\boldsymbol{v} \in \mathbb{R}^{3\times 1}$ to $\boldsymbol{u} \times \boldsymbol{v}$ by

$$(\boldsymbol{u}\times) \triangleq \begin{bmatrix} 0 & -u_z & u_y \\ u_z & 0 & -u_x \\ -u_y & u_x & 0 \end{bmatrix} \tag{2.5}$$

where u_x, u_y, u_z are the x, y, z components of $\boldsymbol{u}$. With this definition (2.3), (2.4), can be combined to

$$\begin{bmatrix} \boldsymbol{y}_E(t) \\ \boldsymbol{y}_H(t) \end{bmatrix} = \begin{bmatrix} I_3 \\ (\boldsymbol{u}\times) \end{bmatrix} \mathcal{E}(t) + \begin{bmatrix} \boldsymbol{e}_E(t) \\ \boldsymbol{e}_H(t) \end{bmatrix} \tag{2.6}$$

where I_3 denotes the 3×3 identity matrix. For notational convenience the dimension subscript of the identity matrix will be omitted when its value is clear from the context.

From the constraint (2.2b), $\mathcal{E}(t)$ belongs to the orthogonal complement of $\boldsymbol{u}$, hence

$$\mathcal{E}(t) = V\boldsymbol{\xi}(t) \tag{2.7}$$

where V is a 3×2 matrix whose columns span the orthogonal complement of $\boldsymbol{u}$ and $\boldsymbol{\xi}(t) \in \mathbb{C}^{2\times 1}$. It is easy to check that

$$V = \begin{bmatrix} -\sin\theta_1 & -\cos\theta_1 \sin\theta_2 \\ \cos\theta_1 & -\sin\theta_1 \sin\theta_2 \\ 0 & \cos\theta_2 \end{bmatrix} \tag{2.8}$$

satisfies this requirement. Combining (2.6) and (2.7) we now have

$$\begin{bmatrix} \boldsymbol{y}_E(t) \\ \boldsymbol{y}_H(t) \end{bmatrix} = \begin{bmatrix} I \\ (\boldsymbol{u}\times) \end{bmatrix} V\boldsymbol{\xi}(t) + \begin{bmatrix} \boldsymbol{e}_E(t) \\ \boldsymbol{e}_H(t) \end{bmatrix} \tag{2.9}$$

The measured signals in the sensor reference frame can be further related to the original source signal using the following lemma.

Lemma 2.1: Every vector $\boldsymbol{\xi} = [\xi_1, \xi_2]^T \in \mathbb{C}^2$, where the superscript T denotes transpose, has the representation

$$\boldsymbol{\xi} = \|\boldsymbol{\xi}\| e^{i\varphi} Q \boldsymbol{w} \tag{2.10}$$

where

$$Q = \begin{bmatrix} \cos\theta_3 & \sin\theta_3 \\ -\sin\theta_3 & \cos\theta_3 \end{bmatrix} \tag{2.11a}$$

$$\boldsymbol{w} = \begin{bmatrix} \cos\theta_4 \\ i\sin\theta_4 \end{bmatrix} \tag{2.11b}$$

and where $\varphi \in (-\pi, \pi]$, $\theta_3 \in (-\pi/2, \pi/2]$, $\theta_4 \in [-\pi/4, \pi/4]$. Moreover, $\|\boldsymbol{\xi}\|, \varphi, \theta_3, \theta_4$ in (2.10) are uniquely determined if and only if $\xi_1^2 + \xi_2^2 \neq 0$.

Proof: See [10].

Let ω_c be the reference frequency of the signal phasor representation. In the narrow-band SST case the incoming electric wave signal $\text{Re}\{e^{i\omega_c t}\|\boldsymbol{\xi}(t)\|e^{i\varphi(t)}Q\boldsymbol{w}\}$, where $\text{Re}\{\cdot\}$ denotes the real part operator, moves on a quasistationary ellipse whose major and minor semi-axes' lengths are proportional, respectively, to $\cos\theta_4$ and $\sin\theta_4$. The ellipse's spin direction (right or left handed with respect to $\boldsymbol{u}$) and eccentricity are determined, respectively, by the sign and magnitude of θ_4. The variable θ_3 is the rotation angle between the ellipse axes and the $(\boldsymbol{v}_1, \boldsymbol{v}_2)$ coordinates.

Single Signal Transmission (SST) Model

Suppose that a single modulated signal is transmitted. Then, using (2.10), this is a special case of (2.9) with $\boldsymbol{\xi}(t) = Q\boldsymbol{w}s(t)$ where $s(t)$ denotes the complex envelope of the (scalar) transmitted signal. Thus, the measurement model here is

$$\begin{bmatrix} \boldsymbol{y}_E(t) \\ \boldsymbol{y}_H(t) \end{bmatrix} = \begin{bmatrix} I \\ (\boldsymbol{u}\times) \end{bmatrix} VQ\boldsymbol{w}s(t) + \begin{bmatrix} \boldsymbol{e}_E(t) \\ \boldsymbol{e}_H(t) \end{bmatrix}. \tag{2.12}$$

Dual Signal Transmission (DST) Models

Methods of transmissions in which two separate signals are transmitted simultaneously will be called dual signal transmissions. Various DST forms exist, see, for example [10], and all of them can be modeled by (2.9) with $\boldsymbol{\xi}(t)$ being a linear transformation of the two dimensional source signal.

2.2 Multi-Source Multi-Sensor Model

Suppose waves from n distant electromagnetic sources are impinging on an array of m vector sensors. Introduce the following additional assumptions:

A3: Plane wave across the array.

A4: Narrow-band signal assumption: The minimum modulating wave length is much larger than the array size.

Let

$$\boldsymbol{y}_{EH}(t) \triangleq \Big[(\boldsymbol{y}_E^{(1)}(t))^T, (\boldsymbol{y}_H^{(1)}(t))^T \ldots (\boldsymbol{y}_E^{(m)}(t))^T, (\boldsymbol{y}_H^{(m)}(t))^T\Big]^T \tag{2.13a}$$

where $\boldsymbol{y}_E^{(j)}(t)$ and $\boldsymbol{y}_H^{(j)}(t)$ are, respectively, the phasor measured electric and magnetic vector fields at the jth sensor. Using a similar notation for the noise vector $\boldsymbol{e}_{EH}(t)$, then, with A4 and from (2.9) we have

$$\boldsymbol{y}_{EH}(t) = \sum_{k=1}^{n} [\boldsymbol{e}_k \otimes I_6] \begin{bmatrix} I_3 \\ (\boldsymbol{u}_k\times) \end{bmatrix} V_k \boldsymbol{\xi}_k(t) + \boldsymbol{e}_{EH}(t) \tag{2.14}$$

where $\boldsymbol{e}_k$ denotes the kth column of the matrix $E \in \mathbb{C}^{m\times n}$ whose (j,k) entry is $E_{jk} = e^{-i\omega_c\tau_{jk}}$ and τ_{jk} is the differential delay of the kth signal between the jth sensor and the coordinate reference; $\otimes$ is the Kronecker product. The rest of the notation in (2.14) is similar to the single source case. The vector $\boldsymbol{\xi}_k(t)$ can have either the SST or the DST form described above.

3 Cramér-Rao Bound for a Vector Sensor Array Model

Consider the problem of finding the parameter vector $\boldsymbol{\theta}$ in the following discrete-time vector sensor array model associated with n vector sources and m vector sensors:

$$\boldsymbol{y}(t) = A(\boldsymbol{\theta})\boldsymbol{x}(t) + \boldsymbol{e}(t) \qquad t = 1, 2, \ldots \tag{3.1}$$

where $\{\boldsymbol{y}(t)\} \in \mathbb{C}^{\overline{\mu}\times 1}$ are the vectors of sensor outputs (or snapshots), $\{\boldsymbol{x}(t)\} \in \mathbb{C}^{\overline{\nu}\times 1}$ are the unknown source signals, $\boldsymbol{e}(t) \in \mathbb{C}^{\overline{\mu}\times 1}$ is an additive noise, and the transfer matrix $A(\theta) \in \mathbb{C}^{\overline{\mu}\times\overline{\nu}}$ and vector $\boldsymbol{\theta} \in \mathbb{R}^{\overline{q}\times 1}$ are given by

$$A(\boldsymbol{\theta}) = \Big[A_1(\boldsymbol{\theta}^{(1)}) \cdots A_n(\boldsymbol{\theta}^{(n)})\Big] \tag{3.2a}$$

$$\boldsymbol{\theta} = \Big[(\boldsymbol{\theta}^{(1)})^T, \cdots, (\boldsymbol{\theta}^{(n)})^T\Big]^T \tag{3.2b}$$

where $A_k(\boldsymbol{\theta}^{(k)}) \in \mathbb{C}^{\overline{\mu}\times\nu_k}$, thus $\overline{\nu} = \sum_{k=1}^{n} \nu_k$, and the parameter vector of the kth source $\boldsymbol{\theta}^{(k)} \in \mathbb{R}^{q_k\times 1}$, thus $\overline{q} = \sum_{k=1}^{n} q_k$. The following notation will also be used:

$$\boldsymbol{y}(t) = \left[(\boldsymbol{y}^{(1)}(t))^T, \cdots, (\boldsymbol{y}^{(m)}(t))^T\right]^T \quad (3.3a)$$

$$\boldsymbol{x}(t) = \left[(\boldsymbol{x}^{(1)}(t))^T, \cdots, (\boldsymbol{x}^{(n)}(t))^T\right]^T \quad (3.3b)$$

where $\boldsymbol{y}^{(j)}(t) \in \mathbb{C}^{\mu_j\times 1}$ is the vector measurement of the jth sensor, thus $\overline{\mu} = \sum_{j=1}^{m} \mu_j$, and $\boldsymbol{x}^{(k)}(t) \in \mathbb{C}^{\nu_k\times 1}$ is the vector signal of the kth source.

It will be shown that the single- vector source single- vector sensor models are special cases of (3.1) with appropriate choices of matrices.

The matrix A in (3.1) is assumed to have full rank $\overline{\nu}$ (this implies $\overline{\mu} \geq \overline{\nu}$) and continuous full rank Jacobian $\partial A/\partial\theta$. Other assumptions on the model (3.1) are:

A5: The source signal sequence $\{\boldsymbol{x}(1), \boldsymbol{x}(2), \ldots\}$ is a sample from a temporally uncorrelated stationary (complex) Gaussian process with zero mean and with

$$\mathrm{E}\,\boldsymbol{x}(t)\boldsymbol{x}^*(s) = P\delta_{t,s}$$

$$\mathrm{E}\,\boldsymbol{x}(t)\boldsymbol{x}^T(s) = 0 \qquad \text{(for all } t \text{ and } s).$$

where E is the expectation operator, the superscript * denotes the conjugate transpose and $\delta_{t,s}$ is the Kronecker delta.

A6: The noise $\boldsymbol{e}(t)$ is Gaussian with zero mean and the following covariances

$$\mathrm{E}\,\boldsymbol{e}(t)\boldsymbol{e}^*(s) = \sigma^2 I\delta_{t,s}$$

$$\mathrm{E}\,\boldsymbol{e}(t)\boldsymbol{e}^T(s) = 0 \qquad \text{(for all } t \text{ and } s).$$

It is also assumed that the signals $\boldsymbol{x}(t)$ and the noise $\boldsymbol{e}(s)$ are uncorrelated for all t and s.

The unknown parameters in the model (3.1) include the vector $\boldsymbol{\theta}$, the covariance matrix P and the noise variance σ^2. The problem of estimating $\boldsymbol{\theta}$ in (3.1) from N snapshots $\boldsymbol{y}(1), \ldots, \boldsymbol{y}(N)$ and the statistical performance of the corresponding estimation methods are the main concerns of this paper.

Theorem 3.1: The Cramér-Rao lower bound on the covariance matrix of any (locally) unbiased estimator of the vector $\boldsymbol{\theta}$ in the model (3.1) under assumptions A5 and A6 with P and σ^2 unknown and $\nu_k = \nu$ for all k is

$$\mathrm{CRB}(\boldsymbol{\theta}) = \frac{\sigma^2}{2N}\left\{\mathrm{Re}\left[\mathrm{btr}\left((1 \boxtimes U) \boxdot (D^*\Pi_c D)^{bT}\right)\right]\right\}^{-1} \quad (3.4)$$

where

$$U = P\left(A^*AP + \sigma^2 I\right)^{-1} A^*AP \quad (3.5a)$$

$$\Pi_c = I - A(A^*A)^{-1}A^* \quad (3.5b)$$

$$D = \left[D_1^{(1)} \cdots D_{q_1}^{(1)} \cdots D_1^{(n)} \cdots D_{q_n}^{(n)}\right] \quad (3.5c)$$

$$D_\ell^{(k)} = \frac{\partial A_k}{\partial \theta_\ell^{(k)}} \quad (3.5d)$$

and where 1 denotes a $\overline{q}\times\overline{q}$ matrix with all entries equal to one, and the block trace operator $\mathrm{btr}(\cdot)$, the block Kronecker product $\boxtimes$, the block Schur-Hadamard product $\boxdot$ and the block transpose operator bT are as defined in Appendix A with blocks of dimension $\nu\times\nu$, except for the matrix 1 which has blocks of dimensions $q_i \times q_j$. Furthermore, the CRB (3.4) remains the same, independently of whether σ^2 is known or unknown.

Proof: See [10].

4 Single-Source Single-Sensor Analysis

This section analyzes the performance of parameter estimation of an electromagnetic source whose covariance is unknown using a single (lumped) vector sensor. It should be emphasized that assumptions A3 and A4 are not needed in the following analyses of this section since they are done for a single lumped vector sensor. In other words, all the results of this section are valid for *wide-band* sources.

4.1 Definition of the Mean-Square Angular Error (MSAE)

The angular error, say δ, corresponding to a direction error $\Delta\boldsymbol{u}$ in $\boldsymbol{u}$, can be shown to be $\delta = 2\arcsin(\|\Delta\boldsymbol{u}\|/2)$. It follows that under the assumption that $\mathrm{E}\left[|\Delta\theta_1|^3 + |\Delta\theta_2|^3\right] = o(1/N)$, the normalized asymptotic mean-square angular error associated with the CRB is (see [10])

$$\mathrm{MSAE}_{CR} \triangleq N[\cos^2\theta_2 \cdot \mathrm{CRB}(\theta_1) + \mathrm{CRB}(\theta_2)] \quad (4.1)$$

where $\mathrm{CRB}(\theta_i)$ is the (i,i) entry of $\mathrm{CRB}(\boldsymbol{\theta})$. The MSAE is a quality measure useful for gaining physical insight and for performance comparisons. It has the desirable property of being invariant to the choice of reference coordinate frame. Note that it is also not a function of N.

4.2 DST Source Analysis

Suppose the noise vector of (2.9) is Gaussian with zero mean and the following covariances:

$$\mathrm{E}\begin{bmatrix} e_E(t) \\ e_H(t) \end{bmatrix} [e_E^*(s),\ \ e_H^*(s)] = \begin{bmatrix} \sigma_E^2 I_3 & 0 \\ 0 & \sigma_H^2 I_3 \end{bmatrix} \delta_{t,s}$$

$$\mathrm{E}\begin{bmatrix} e_E(t) \\ e_H(t) \end{bmatrix} [e_E^T(s),\ \ e_H^T(s)] = 0 \qquad \text{(for all } t \text{ and } s).$$

To relate the model (2.9) to (3.1), define a scaled measurement $y(t) \triangleq [ry_E^T(t),\ y_H^T(t)]^T$ where $r \triangleq \sigma_H/\sigma_E$. The resulting scaled noise vector $e(t) \triangleq [re_E^T(t),\ e_H^T(t)]^T$ then satisfies assumption A6 with $\sigma = \sigma_H$. Assume further that the signal $\xi(t)$ satisfies assumption A5 with $x(t) = \xi(t)$. Then, under these assumptions, the scaled version of the DST source (2.9) can be viewed as a special case of (3.1) with $m = n = 1$ and

$$\begin{aligned} A &= \begin{bmatrix} rV \\ (u\times)V \end{bmatrix}, \qquad x(t) = \xi(t), \qquad \sigma^2 = \sigma_H^2 \\ \theta &= [\theta_1\ \ \theta_2\ \ r]^T \end{aligned} \tag{4.2}$$

The unknown parameters here are θ, P, σ^2.

From the discussion above the CRB expression (3.4) is applicable to the present problem with the special choice of the variables in (4.2). This application was done in [10], and substitution of the (1,1) and (2,2) entries of the resulting CRB matrix in (4.1) gives

$$\mathrm{MSAE}_{CR}^{D} = \frac{\left(\sigma_E^2 + \sigma_H^2\right) \sigma_E^2 \sigma_H^2 \,\mathrm{tr}\, U}{2\left[\sigma_E^2\sigma_H^2 \left(\mathrm{tr}\, U\right)^2 + \left(\sigma_E^2 - \sigma_H^2\right)^2 \det\left(\mathrm{Re}\, U\right)\right]} \tag{4.3}$$

where $U = P\left[P + \sigma_{\parallel}^2 I_2\right]^{-1} P$ and $\sigma_{\parallel}^2 \triangleq \sigma_E^2 \cdot \sigma_H^2/(\sigma_E^2 + \sigma_H^2)$. The result (4.3) is independent of whether σ_E^2, σ_H^2 are known or unknown.

4.3 SST Source (DST Model) Analysis

Consider the MSAE for a single signal transmission source when the estimation is done under the assumption that the source is of a dual signal transmission type. In this case, the model (2.9) has to be used but with a signal in the SST form. The signal covariance is then $P = \sigma_s^2 Qw(Qw)^*$ where $\sigma_s^2 = \mathrm{E}\, s^2(t)$. Using these expressions we find that

$$\mathrm{MSAE}_{CR}^{S} = \frac{(1+\varrho)(\sigma_E^2 + \sigma_H^2)^2}{2\varrho^2\left[\sigma_E^2\sigma_H^2 + (\sigma_E^2 - \sigma_H^2)^2 \sin^2\theta_4 \cos^2\theta_4\right]} \tag{4.4}$$

where $\varrho \triangleq \sigma_s^2/\sigma_{\parallel}^2$ is an effective SNR. The result (4.4) is independent of whether σ_E^2, σ_H^2 are known or unknown.

Table 1 summarizes several special cases of (4.4). The elliptical polarization column corresponds to an arbitrary polarization angle $\theta_4 \in [-\pi/4, \pi/4]$. The circular and linear polarization columns are obtained, respectively, as special cases of (4.4) with $|\theta_4| = \pi/4$ and $\theta_4 = 0$. The row of precise (noise-free) electric measurement (with noisy magnetic measurements) is obtained by substituting $\sigma_E^2 = 0$ in (4.4). The row of electric measurement only is obtained by deriving the corresponding CRB and MSAE. Alternatively, the MSAE can be found for this case by taking the limit of (4.4) as $\sigma_H^2 \to \infty$. The results show that when $\sigma_H^2 \neq \sigma_E^2$ the MSAE is minimized for circular polarization and maximized for linear polarization sources. When only the electric field is measured, the MSAE increases as the polarization changes from circular to linear. As the source becomes closer to being linearly polarized, it becomes more important to measure both the electric and magnetic fields to get good direction estimates using a single vector sensor.

4.4 SST Source (SST Model) Analysis

To get a statistical model for the SST measurement model (2.12) as a special case of (3.1), we will make the same assumptions on the noise and use a similar data scaling as was done in the DST source case. Assume also that the signal envelope $s(t)$ satisfies assumption A5 with $x(t) = s(t)$ in (2.12). Then the resulting statistical model becomes a special case of (3.1) with

$$\begin{aligned} A &= \begin{bmatrix} rV \\ (u\times)V \end{bmatrix} Qw, \qquad x(t) = s(t), \qquad \sigma^2 = \sigma_H^2 \\ \theta &= [\theta_1\ \ \theta_2\ \ \theta_3\ \ \theta_4\ \ r]^T \end{aligned} \tag{4.5}$$

The unknown parameters are θ, P, σ^2.

The MSAE for an SST source under the SST model was calculated in [10]. The result was shown to coincide with (4.4). That is, *the MSAE for an SST source is the same under both the SST and the DST models.*

4.5 A Cross Product Based DOA Estimator

We propose here a simple algorithm for estimating the DOA of a single source using the measurements of a single sensor. The motivation for this algorithm

	Elliptical	Circular	Linear
General MSAE^S_{CR}	(4.4)	$\frac{2(1+\varrho)}{\varrho^2}$	$\frac{(1+\varrho)(\sigma_E^2+\sigma_H^2)}{2\varrho\sigma_s^2}$
Precise Electric Measurement	0	0	$\frac{\sigma_H^2}{2\sigma_s^2}$
Electric Measurement Only	$\frac{2(1+\varrho)}{\varrho^2 \sin^2(2\theta_4)}$	$\frac{2(1+\varrho)}{\varrho^2}$	∞

Table 1: Normalized Asymptotic CRB Mean-Square Angular Error for an SST Source, Single Sensor.

stems from the expression for the average Poynting vector. This algorithm computes

$$\widehat{\boldsymbol{s}} = \frac{1}{N}\sum_{t=1}^{N} \text{Re}\,\{\boldsymbol{y}_E(t) \times \overline{\boldsymbol{y}}_H(t)\} \tag{4.6}$$

and then $\widehat{\boldsymbol{u}} = \widehat{\boldsymbol{s}}/\|\widehat{\boldsymbol{s}}\|$. The performance of the estimator $\widehat{\boldsymbol{u}}$ is analyzed in [10] under the previous assumptions on $\boldsymbol{\xi}(t), \boldsymbol{e}_E(t), \boldsymbol{e}_H(t)$, except that the Gaussian assumption is omitted. The results are as follows:

a) If $\|\boldsymbol{\xi}(t)\|^2, \|\boldsymbol{e}_E(t)\|, \|\boldsymbol{e}_H(t)\|$ have finite first order moments, then $\widehat{\boldsymbol{u}} \longrightarrow \boldsymbol{u}$ almost surely.

b) If $\|\boldsymbol{\xi}(t)\|^2, \|\boldsymbol{e}_E(t)\|, \|\boldsymbol{e}_H(t)\|$ have finite second order moments, then $\sqrt{N}(\widehat{\boldsymbol{u}} - \boldsymbol{u})$ is asymptotically normal and its MSAE for both DST and SST is

$$\text{MSAE} = \frac{1}{2}\varrho^{-1}(1 + 4\varrho^{-1})(r + r^{-1})^2 \tag{4.7}$$

where $\varrho \triangleq \text{tr}\,(P)/\sigma_{\|}^2$. For the SST case the ratio of the MSAE of this estimator to the CRB MSAE is

$$\frac{\text{MSAE}}{\text{MSAE}^S_{CR}} = \frac{\varrho+4}{\varrho+1}\left[1 + (r - r^{-1})^2 \sin^2\theta_4 \cos^2\theta_4\right] \tag{4.8}$$

Hence this estimator is nearly efficient if $\varrho \gg 1$ and $r \simeq 1$, or $\theta_4 \simeq 0$. This estimator can be improved by using weighted averages of cross products between all possible pairs of real and imaginary parts of $y_E(t)$ and $y_H(s)$ taken at different times t and s. A similar algorithm suitable for real time applications can also be developed in the time domain without preprocessing needed for phasor representation.

5 Results for Two Sources, Single Sensor

In [10] we have analyzed the CRB for two sources and a single vector sensor model. This was done by scaling the measurement model as in the single source case and making the appropriate statistical assumptions matching the general assumptions of (3.4). As in Section 4, the results are applicable to wide-band sources since a single sensor is used and hence A3 and A4 are not needed.

The following are common to all the two source transmission cases:

1. Identification of two sources' DOAs, is possible with *only one* vector sensor.
2. When only the electric field is measured, the DOAs are nonidentifiable.
3. When the electric measurement is precise, the CRB variances are generally nonzero.
4. In general r is not decoupled from the remaining parameters. Knowledge of r can sometimes be vital for identifiability.
5. The CRB is very sensitive to the model. For example, if an SST signal with linear polarization is transmitted, the two source DST model does not allow identification while the two source SST model is identifiable.
6. The MSAE^S_{CR} increases without bound with decreasing source separation for sources with equal

spin direction, but it remains bounded for sources with opposite spin direction.

Properties 2 to 5 are in general different from the single source case. Property 2 shows that it is necessary to include the magnetic mesaurement to be able to estimate the direction of more than one source using a single sensor.

6 Summary

A new approach has been presented for the localization of electromagnetic sources using vector sensors. It was shown that the DOA of one or two sources can be estimated using a single lumped vector sensor. A simple algorithm and its performance were presented for the DOA estimation of a single source. Several interesting phenomena were observed for the case of two sources. Further research will include development of estimation algorithms, extended models, performance analysis, communication applications, practical issues and implementations.

Appendix A. Definitions of Some Block Matrix Operators

The following notation will be used for a blockwise partitioned matrix A:

$$A = \begin{bmatrix} A_{<11>} & \cdots & A_{<1n>} \\ \vdots & & \vdots \\ A_{<m1>} & \cdots & A_{<mn>} \end{bmatrix} \triangleq [A_{<ij>}] \quad \text{(A.1)}$$

with the block entries $A_{<ij>}$ of dimensions $\mu_i \times \nu_j$. The matrices in the following definitions are assumed to be of compatible dimensions. These dimensions are not specified here due to lack of space.

Definition: *Block transopse.*

$$\left(A^{bT}\right)_{<ij>} = A_{<ji>} \quad \text{(A.2)}$$

Definition: *Block Kronecker product.*

$$(A \boxtimes B)_{<ij>} = A_{<ij>} \otimes B_{<ij>} \quad \text{(A.3)}$$

where $\otimes$ denotes the usual Kronecker product.

Definition: *Block Schur-Hadamard product.*

$$(A \boxdot B)_{<ij>} = A_{<ij>} B_{<ij>} \quad \text{(A.4)}$$

Definition: *Block trace operator.*

$$(\mathrm{btr}\,[A])_{ij} = \mathrm{tr}\ A_{<ij>} \quad \text{(A.5)}$$

References

[1] M. Schwartz, W. R. Bennett, and S. Stein, *Communication Systems and Techniques,* McGraw-Hill, 1966.

[2] B.E. Keiser, *Broadband Coding, Modulation, and Transmission Engineering,* Englewood Cliffs, New Jersey: Prentice Hall, 1989.

[3] P. Stoica and A. Nehorai, "Performance Study of Conditional and Unconditional Direction-of-Arrival Estimation," *IEEE Trans. Acoust., Speech, Signal Processing,* Vol. ASSP-38, pp. 1783-1795, Oct. 1990.

[4] B. E. Ottersten, "Parametric Subspace Fitting Methods for Array Signal Processing," Ph.D. Dissertation, Stanford Univ., Stanford, CA, Dec. 1989.

[5] M. Viberg, "Subspace Fitting Concepts in Sensor Array Processing," Ph.D. Dissertation, Linköping University, Linköping, Sweden, Oct. 1989.

[6] R. O. Schmidt, "A Signal Subspace Approach to Multiple Emitter Location and Spectral Estimation," Ph.D. Dissertation, Stanford Univ., Stanford, CA, Nov. 1981.

[7] E. R. Ferrara, Jr. and T. M. Parks, "Direction Finding with an Array of Antennas Having Diverse Polarization," *IEEE Trans. on Antennas Propagat.*, Vol. AP-31, pp. 231-236, March 1983.

[8] I. Ziskind and M. Wax, "Maximum Likelihood Localization of Diversely Polarized Sources by Simulated Anealing," *IEEE Trans. on Antennas Propagat.*, Vol. AP-38, pp. 1111-1114, July 1990.

[9] A.J. Weiss and B. Friedlander, "Performance Analysis of Diversely Polarized Antenna Arrays," *IEEE Trans. on Signal Processing,* Vol. SP-39, pp. 1589-1603, July 1991.

[10] A. Nehorai and E. Paldi, "Vector Sensor Processing for Electromagnetic Source Localization," full version in preparation.

Direction Finding Using Uniform Arrays and The Constrained Total Least Squares Method

T. J. Abatzoglou L. K. Lam

Lockheed Research and Development Division, (O/91-50,B/251)
Lockheed Missiles and Space Co
Palo Alto, California, 94304

Abstract

Direction finding using snapshots collected from an uniform array is equivalent to estimating the 2-dimensional frequencies of sinusoids from a set of double indexed samples. The frequencies here correspond to the directional cosines of the impinging wavefronts on the array. A minimal set of multilinear prediction equations is derived based on the annihilator method. The snapshot samples are used in the above equations as well as information on the number of sources/emitters present. A generalized multilinear version of the Constrained Total least Squares (CTLS) technique is used to solve the above equations with optimal accuracy. The directional cosines are extracted as the solutions of the above multilinear set of equations. It can be seen from a set of examples that the derived root mean square error (RMSE) of our estimates are very close to the corresponding Cramer-Rao bound (CRB). We derive the corresponding maximum likelihood estimator for this problem in closed form and also the corresponding CRB analytically.

1. Introduction

The problem of estimating 2 dimensional frequencies of sinusoids arises naturally in radar range-doppler estimation of point scatterers, [2], estimation of directional cosines of wavefronts impinging on a planar uniform array, etc.. As long as the separation of the frequencies of the sinusoids is more than the Fourier Transform frequency bin-width in each dimension, the 2-dimensional discrete Fourier Transform can be used to adequately resolve and estimate the various frequencies present in the data. However many cases of interest require the estimation of frequencies which are separated by a small fraction of the bin-width. Therefore for sometime now there has been an effort to develop statistically efficient and computationally realizable algorithms to estimate these frequencies, [3], [4], [5]. In [4], an algorithm is developed which produces alternating estimates of the unknown frequencies along each frequency dimension. This algorithm has good convergence and statistical properties but in the example provided in [4], its performance is about 5-db lower than the corresponding CRB. In [5] the matrix enhancement and matrix pencil (MEMP) method is used to estimate these 2-dimensional frequencies. This technique does use the coupling between the 2-dimensional frequencies in the construction of the matrix pencil, however it does estimate the frequencies one dimension at a time. Its performance is rather good and in an example involving 3 closely spaced sinusoids which have pairs of common frequencies in each dimension the MEMP technique comes within 5db of the CRB.

In this paper we present a new technique for estimating the 2-dimensional frequencies which is based on a generalization of the CTLS technique, [6], to nonlinear equations whose coefficients have been perturbed by additive noise. Multilinear prediction on the data is used to arrive at these equations and the frequencies are extracted directly from the solution of these equations. Simulation results compare the resolution and accuracy of this technique with the corresponding CRB for a variety of signal scenarios.

2. Data Samples

The data samples can be modeled as a superposition of 2-dimensional sinusoids as follows:

$$X = \sum_{l=1}^{L} z_l \vec{e}_1(\omega_{1l})\, \vec{e}_2(\omega_{2l})^T + V$$

L = is the number of sinusoids assumed known

z_l = complex amplitude of lth sinusoid

$(\omega_{1l}, \omega_{2l})$ = is the 2-d frequency of lth sinusoid

$\vec{e}_1$ and $\vec{e}_2$ are 1-d frequency steering vectors defined by:

$\vec{e}_1(\omega) = [\exp(j\omega) \cdots \exp(Mj\omega)]^T$;

$\vec{e}_2(\omega) = [\exp(j\omega) \cdots \exp(Nj\omega)]^T$

V is a M×N matrix of white noise gaussian samples

1058-6393/91 $01.00 © 1991 IEEE

X is called the M×N data matrix; note that the index for the first component of the 2-d frequency runs from 1 to M while the second one goes from 1 to N.
The problem here is to estimate the unknown frequencies $(\omega_{11}, \omega_{21})$ with optimal accuracy and the least effort.

3. The Maximum Likelihood Estimator (MLE) and CRB

The data matrix X can be represented as

$$X = S(\vec{z},\vec{\omega}) + V \qquad \text{where}$$

$$S(\vec{z},\vec{\omega}) = \left[\vec{e}_1(\omega_{11})\,\vec{e}_2(\omega_{21})^T \,\vdots\, \cdots \,\vdots\, \vec{e}_1(\omega_{1L})\,\vec{e}_2(\omega_{2L})^T\right]\begin{bmatrix} z_1 I \\ \vdots \\ z_L I \end{bmatrix}$$

The MLE can be obtained by minimizing the log-likelihood function

$$LL(\vec{z},\vec{\omega}) = tr\left((X - S(\vec{z},\vec{\omega}))^*(X - S(\vec{z},\vec{\omega}))\right)$$

Here * stands for complex conjugate transpose. As is it done in the 1-dimensional case, the redundant variables $\vec{z}$ are removed from the minimization process by minimizing $LL(\vec{z},\vec{\omega})$ first with respect to $\vec{z}$ alone. The resulting minimum value of $\vec{z}$ is then a function of $\vec{\omega}$ and this is substituted into the LL function. Finally it suffices to maximize the following function of $\vec{\omega}$ to obtain the MLE.

$$\underset{\vec{\omega}_1,\vec{\omega}_2}{MAX}\left\{\vec{b}(X,\vec{\omega}_1,\vec{\omega}_2)^*\tilde{A}(\vec{\omega}_1,\vec{\omega}_2)^{-1}\vec{b}(X,\vec{\omega}_1,\vec{\omega}_2)\right\}$$

Here

$$\tilde{A}(\vec{\omega}_1,\vec{\omega}_2) = A_2(\vec{\omega}_2)^*A_2(\vec{\omega}_2) \odot A_1(\vec{\omega}_1)^*A_1(\vec{\omega}_1)$$

where A_1 and A_2 are the 1-dimensional frequency steering matrices,

$$A_1(\vec{\omega}_1) = [\vec{e}_1(\omega_{11}) \cdots \vec{e}_1(\omega_{1L})]\ ;$$

$$A_2(\vec{\omega}_2) = [\vec{e}_2(\omega_{21}) \cdots \vec{e}_2(\omega_{2L})]\ ,$$

$\odot$ denotes the Hadamard product between 2 matrices.

$$\vec{b}(X,\vec{\omega}_1,\vec{\omega}_2) = diag\left(\overline{A_2(\vec{\omega}_2)}A_1(\vec{\omega}_1)^*X\right)$$

The maximization above can be a formidable task unless L = 1. The main contribution of superresolution signal processing is to mimic the performance of maximum likelihood estimators while using fast and robust algorithms. Given the observation X, the CRB for $\vec{\omega},\vec{z}$ are obtained from inverting the Fisher Information matrix J which is computed as:

$$J = \frac{1}{\sigma^2}E\left\{\frac{\partial^2 LL(\vec{z},\vec{\omega})}{\partial\alpha_i\partial\alpha_j}\right\} = \frac{2}{\sigma^2}Re\left\{tr\left\{\frac{\partial S(\vec{z},\vec{\omega})^*}{\partial\alpha_i}\frac{\partial S(\vec{z},\vec{\omega})}{\partial\alpha_j}\right\}\right\}$$

where

$$\vec{\alpha} = [\alpha_1 \cdots \alpha_{4L}] = \left[\vec{\omega}_1^T\ \vec{\omega}_2^T\ \vec{z}_R^T\ \vec{z}_I^T\right].$$

$\vec{z}_R$ and $\vec{z}_I$ are the real and imaginary components of the amplitude vector $\vec{z}$. Following techniques developed in [8] we isolate the CRB for $\vec{\omega}_1$ and $\vec{\omega}_2$ from that of $\vec{z}$ and express it in terms of a 2×2 block matrix where each block is L×L.

$$CRB^{-1}(\vec{\omega}_1,\vec{\omega}_2) = \frac{2}{\sigma^2}Re\begin{Bmatrix} B_{11} & B_{12} \\ B_{21} & B_{22}\end{Bmatrix}$$

with

$$B_{11} = (Z^*A_1'^*A_1')\odot(A_2^*A_2Z) - \left[(Z^*A_2^*A_2)\odot A_1'^*A_1\right]\tilde{A}^{-1}\left[(A_1^*A_1')\odot(A_2^*A_2Z)\right]$$

$$B_{12} = (Z^*A_1'^*A_1)\odot(A_2^*A_2'Z) - \left[(Z^*A_2^*A_2)\odot A_1'^*A_1\right]\tilde{A}^{-1}\left[(A_1^*A_1)\odot(A_2^*A_2'Z)\right]$$

$$B_{21} = B_{12}^*$$

$$B_{22} = (Z^*A_1^*A_1)\odot(A_2'^*A_2'Z) - \left[(Z^*A_2'^*A_2)\odot A_1^*A_1\right]\tilde{A}^{-1}\left[(A_1^*A_1)\odot(A_2^*A_2'Z)\right]$$

In the above expressions $Z = diag(z_1,\cdots,z_L)$ and A_1' is the derivative of the frequency steering matrix A_1 as defined by

$$A_1' = \left[\frac{d\vec{e}_1(\omega_{11})}{d\omega_{11}} \cdots \frac{d\vec{e}_1(\omega_{1L})}{d\omega_{1L}}\right]$$

Similar definition holds for A_2'.

Special Case : L=1

The CRB for ω_1 and ω_2 are :

$$\sigma^2_{\omega_1} = \frac{\sigma^2}{2|z|^2}\frac{1}{2N}\frac{1}{S_2(M)}\ , \quad \sigma^2_{\omega_2} = \frac{\sigma^2}{2|z|^2}\frac{1}{2M}\frac{1}{S_2(N)}$$

where $S_2(K) = \sum_{-(K-1)/2}^{(K-1)/2} k^2$.

Here z is the complex envelope of the single sinusoid and σ is the standard deviation of the noise samples.

4. CTLS and 2-Dimensional Frequency Estimation

a. Introduction

Many system analysis problems can be solved by obtaining a solution of a linear system of equations whose coefficients are noisy and satisfy arbitrary linear relations among themselves. Say the equation is $C\begin{pmatrix} x \\ -1 \end{pmatrix} \approx 0$.

Abatzoglou and Mendel in [5], introduced the CTLS method which generalizes the Total Least Squares (TLS) technique, [7], to the case that the entries of ΔC can have arbitrary covariance and satisfy arbitrary linear equations (constraints). The formulation of the CTLS problem is:

Let $C\begin{pmatrix} x \\ -1 \end{pmatrix} \approx 0$ be an overdetermined inconsistent linear system of equations. Let the noise columns in ΔC be parametrically represented by $\Delta C_i = G_i\vec{u}$, $i=1,\cdots,L+1$. The CTLS solution is obtained from $\underset{\vec{u},x}{\mathrm{Min}} \|\vec{u}\|^2$ subject to

$$\left(C + \left[G_1\vec{u} \vdots \cdots \vdots G_{L+1}\vec{u}\right]\right)\begin{pmatrix} x \\ -1 \end{pmatrix} = 0,$$

where $\vec{u}$ is a white noise vector of minimal dimension. Gi are the noise structure matrices which define the linear combinations of noise components in C. This formulation handles arbitrary linear relations among the entries of ΔC; also nonwhite noise vectors can always prewhitened before applying the CTLS formulation.

Assume M (# of rows of C) $\leq$ K (dimension of $\vec{u}$), then the CTLS solution x_{CTLS} is obtained from the minimum of $F(x) = \begin{pmatrix} x \\ -1 \end{pmatrix}^* C^* H_x^{\dagger *} H_x^{\dagger} C \begin{pmatrix} x \\ -1 \end{pmatrix}$.

Here

$$H_x = \sum_{i=1}^{L} x_i G_i - G_{L+1}$$

and $H_x^{\dagger}$ is the pseudoinverse of H_x. When K < M the CTLS solution is extracted from the minimum of F(x) over all x satisfying $\left(I - H_x H_x^{\dagger}\right) C \begin{pmatrix} x \\ -1 \end{pmatrix} = 0$. One of the most powerful and useful properties of the CTLS solution is that it is equivalent to a MLE [6].

b. Generalization of CTLS to nonlinear unknowns

If an equation of the form $C\begin{pmatrix} Q(x) \\ -1 \end{pmatrix} = 0$ must be solved for x where Q is a nonlinear function of x and C is a matrix which contains noise perturbations as described above then the CTLS formulation can be applied to this problem and the resulting solution is obtained from the minimum of

$$F_Q(x) = \begin{pmatrix} Q(x) \\ -1 \end{pmatrix}^* C^* H_{Q(x)}^{\dagger *} H_{Q(x)}^{\dagger} C \begin{pmatrix} Q(x) \\ -1 \end{pmatrix}$$

where

$$H_{Q(x)} = Q_1(x)G_1 + \cdots Q_{L'}(x)G_{L'} - G_{L'+1}$$

and $L'+1$ is the number of columns of C

A version of the IQML algorithm can be used to minimize the function $F_Q(x)$, i.e., using the IQML method, [10], with the single coordinate descent method, [9].

c. Use of CTLS and the Annihilator technique to estimate 2-d frequencies of sinusoids

The samples for 2-dimensional frequency estimation of sinusoids problem are given as:

$$y_{m,n} = \sum_{l=1}^{L} z_l \exp(j(m\omega_{1,l} + n\omega_{2,l})) + v_{m,n},$$

$$m = 1, ..., M; \quad n = 1, ..., N.$$

$(\omega_{1,l}, \omega_{2,l})$ is the 2-dimensional frequency of the lth sinusoid; $v_{m,n}$ is additive noise. CTLS can be applied to a set of 2-dimensional linear difference equations where the samples satisfy in the absence of additive noise. The set can be solved for the linear prediction coefficients with maximal accuracy, however there is no unique way of extracting the unknown frequencies from the linear prediction coefficients. To circumvent this difficulty we propose to model the linear prediction equations directly in terms of the exponential frequencies $a_{i,l} = \exp(j\omega_{i,l})$; i = 1, 2; l = 1, ..., L. and then solve for the $a_{i,l}$ directly using nonlinear CTLS. We proceed with a description of the Annihilator method for L = 2:

Let D_1 and D_2 be the 1 unit delay operators with respect the first and second frequency component respectively; then assuming that L = 2, which is the # of sinusoids, we can write

$$(1-a_{11}D_1)(1-a_{12}D_1)y_{m,n}=0 \Leftrightarrow y_{m,n}-(a_{11}+a_{12})y_{m-1,n}+a_{11}a_{12}y_{m-2,n}=0$$

$$(1-a_{21}D_2)(1-a_{22}D_2)y_{m,n}=0 \Leftrightarrow y_{m,n}-(a_{21}+a_{22})y_{m,n-1}+a_{21}a_{22}y_{m,n-2}=0$$

$$(1-a_{11}D_1)(1-a_{22}D_2)y_{m,n}=0 \Leftrightarrow y_{m,n}-a_{11}y_{m-1,n}-a_{22}y_{m,n-1}+a_{11}a_{22}y_{m-1,n-1}=0$$

$$(1-a_{12}D_1)(1-a_{21}D_2)y_{m,n}=0 \Leftrightarrow y_{m,n}-a_{12}y_{m-1,n}-a_{21}y_{m,n-1}+a_{21}a_{12}y_{m-1,n-1}=0$$

The above are the forward equations which can be used alone if the 2-dimensional frequencies contain decaying factors. If these frequencies are pure then the corresponding backward equations can be added to the ones above. In any case the above equations can be concatenated to the form.

$$Y\begin{bmatrix} Q(\vec{a}) \\ -1 \end{bmatrix}=0; \text{ where } \vec{a}=\begin{bmatrix} a_{11} \\ a_{12} \\ a_{21} \\ a_{22} \end{bmatrix}, \text{ and}$$

$$Q(\vec{a})=[-(a_{11}+a_{12}),\ a_{11}a_{12},\ -(a_{21}+a_{22}),\ a_{21}a_{22},$$
$$-a_{11},\ -a_{22},\ a_{11}a_{22},\ -a_{12},\ -a_{21},\ a_{12}a_{21}]^T$$

It is not hard to show that the CTLS algorithm can be reduced to a minimization over $\vec{a}$ of the function

$$F_Q(\vec{a})=\begin{pmatrix} Q(\vec{a}) \\ -1 \end{pmatrix}^* Y^* H_Q^{\dagger *} H_Q^{\dagger} Y \begin{pmatrix} Q(\vec{a}) \\ -1 \end{pmatrix}$$

where $H_Q=Q_1(\vec{a})G_1+\ldots+Q_{L'}(\vec{a})G_{L'}-G_{L'+1}$

$Q(\vec{a})$ is a multilinear function of $\vec{a}$ and minimization of F_Q can be accomplished for example by the coordinate descent method where the optimization of F_Q is performed with respect to a single coordinate of $\vec{a}$ at a time and also incorporating some of the features of the IQML algorithm. Also starting points to apply a Newton-type descent algorithm can be provided by suboptimal estimators of frequency.

d. Implementation of the CTLS Method

If the linear prediction equations are used for all possible pairs of indices then the resulting data matrix Y has dimension $[8(N-1)(M-1)-4]\times 11$. Since the number of equations $[8(N-1)(M-1)-4]$ exceeds 2NM, the number of independent random variables in the data, there is a redundancy in the equations. The CTLS technique can still be applied at a price of increased instability. For the simulations below only the mixed indices forward-backward prediction equations were used to estimate the $a_{i,l}$; $i = 1, 2$; $l = 1, \ldots, L$. In this case Y has the size: $2(M-1)(N-1)\times 4$ and the CTLS algorithm can be applied in straightforward fashion.

Also $Q(\vec{a})=[a_1\ a_2\ -a_1a_2]^T$ where now the pair (a_1,a_2) cna be either (a_{11},a_{22}) or (a_{12},a_{21}).

The minimization of $F_Q(\vec{a})$ is accomplished by the IQML algorithm where $F_Q(\vec{a})$ is minimized at each step with respect to $\vec{a}$ while holding H_Q fixed. The new value of $\vec{a}$ is used to update H_Q. Convergence occurs when successive values of a change very little. The minimization step is implemented by the coordinate descent method where the objective function is minimized one variable at a time and cycled through the variables until convergence occurs.

e. Simulation Results

The performance of the CTLS method is investigated based on simulating three test cases. For each of the test cases and for each signal to noise ratio (SNR) parameter, the root mean squared errors (RMSE) of the frequency estimates are computed based on 100 trials. The form of the simulated data is

$$y_{m,n}=z_1\exp(j2\pi(mf_{1,1}+nf_{2,1}))$$
$$+z_2\exp(j2\pi(mf_{1,2}+nf_{2,2})+j\phi)+v_{m,n}$$

where m = 1, 2.. 8, and n = 1, 2, ..5 and ϕ is the phase of the second sinusoid relative to the first. SNR in units of dB defines the amplitudes of the sinusoids, which are $z_1=z_2=10^{SNR/20}$. The noise term $v_{m,n}$ is generated by a complex random number of variance one. The Fourier transform bin-widths are denoted by Δf_1 and Δf_2, and respectively they are equal to 0.125 and 0.2 for our cases.

Case 1: The parameter used are
$f_{1,1}=0.30$ $f_{1,2}=0.70$
$f_{2,1}=0.40$ $f_{2,2}=0.50$ $\phi=\pi/4$.

In terms of Fourier transform bin-widths, the two sinusoids are separated by 3.2 Δf_1 and 0.5 Δf_2. The CTLS frequency estimates obtained are close to the Cramer-Rao bound for SNR from 30 dB down to 9 dB.

In figureS 1, 2, and 3 the RMSE of the estimated frequencies are plotted against the SNR for the respective cases.

Case 2 : The parameter used are

$$f_{1,1} = 0.60 \quad f_{1,2} = 0.63$$
$$f_{2,1} = 0.40 \quad f_{2,2} = 0.50 \quad \phi = \pi/4$$

The sinusoids are separated by 0.23 Δf_1 and 0.5 Δf_2, thus they are not well separated. The results obtained are shown in figure 2. The estimates are close to the Cramer-Rao bound for SNR from 30 dB down to 15 dB.

Case 3 : The parameter used are

$$f_{1,1} = 0.60 \quad f_{1,2} = 0.63$$
$$f_{2,1} = 0.40 \quad f_{2,2} = 0.45 \quad \phi = \pi/2$$

The sinusoids are separated by 0.23 Δf_1 and 0.25 Δf_2. This is the worst case separation of the three cases. The estimates obtained are close to the Cramer-Rao bound at high SNR, but begin to break away at SNR = 24 dB as shown in figure 3.

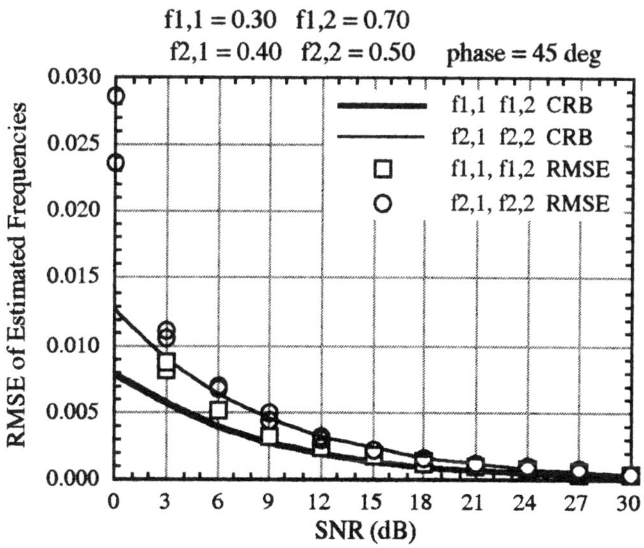

Figure 1. Simulation results for Case 1.

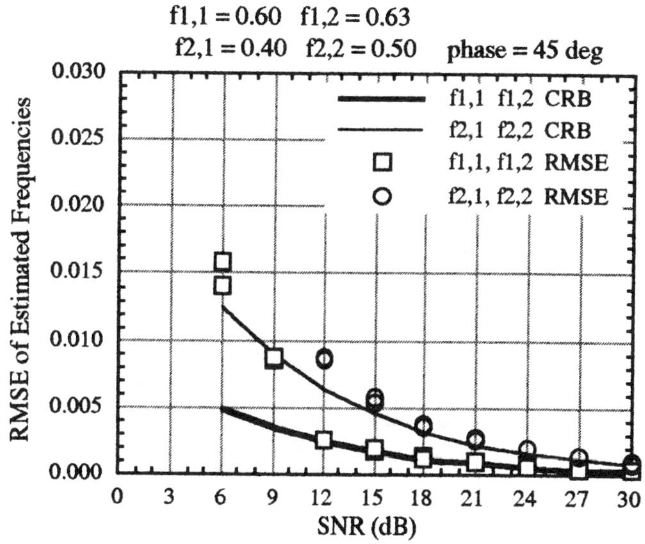

Figure 2. Simulation results for Case 2.

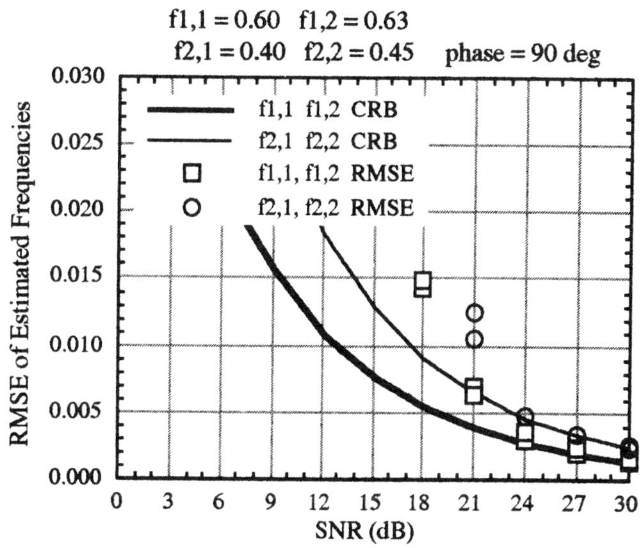

Figure 3. Simulation results for Case 3.

Conclusions

We have derived characteristic properties of the problem of estimating the 2-dimensional frequencies of superimposed sinusoids in gaussian noise. We derived the MLE in closed form and the corresponding CRB was derived analytically and separated from the CRB of the complex amplitudes. A nonlinear generalization of the CTLS method was applied to the data to estimate the frequencies based in linear prediction. Simulation results with 2 sinusoids which use only the mixed indices linear prediction equations show great promise and practically attain the CRB for a number of signal scenarios. Investigations continue for determining an optimal set of prediction equations to use for this problem.

REFERENCES

[1] (Deleted)

[2] W. F. Gabriel, "Superresolution Techniques and ISAR Imaging", IEEE National Radar Conference Digest, pp. 48-55, March 1989.

[3] R. Kumaresan and D. W. Tufts, "A Two-Dimensional Technique for Frequency-

Wavenumber Estimation", Proc. IEEE, Vol. 69, pp. 1515-1517, Nov. 1981.

[4] A. K. Shaw and R. Kumaresan, "Frequency-wavenumber Estimation by Structured Matrix Approximation", Proceedings of Third ASSP Workshop on Spectrum Estimation and Modeling, pp. 81-84, Nov. ,1986.

[5] Y. B. Hua, "Estimating Two-Dimensional Frequencies by Matrix Enhancement and Matrix Pencil", 1991 ICASSP Proceedings, pp. 3073-3076, Toronto, Can., May 1991.

[5] T. J. Abatzoglou and J. M. Mendel, "Constrained Total Least Squares" ,Proceed. 1987 IEEE ICASSP,Dallas,Texas,pp.1485-1488 ,April 1987.

[6] T. J. Abatzoglou, J. M. Mendel and G. A. Harada," The Constrained Total Least Squares Technique and its Applications to Harmonic Superresolution", IEEE Trans. Sign. Proc., Vol. SP-39, pp.1070-1087, May 1991.

[7] G. H. Golub and C. F. Van Loan," An Analysis of the Total Least Squares Problem", SIAM J. Num. Anal.,Vol.17, pp.883-893,1980

[8] P. Stoica and A. Nehorai, "MUSIC,Maximum Likelihood,and Cramer-Rao Bound" ,IEEE Trans. Acoust.,Speech Sign. Proc.,Vol. ASSP-37,pp.720-741,June 1989.

[9] B. O'Donnel and T. J. Abatzoglou, "The Coordinate Descent Method", Journal of Optimization Theory and Applications, vol. 36, pp. 163_174, 1982.

[10] Y. Bresler and A. Macovski," Exact Maximum Likelihood Parameter Estimation of Superimposed Exponential Signals in Noise," IEEE Trans. Acoust. Speech Sign. Proc., Vol. ASSP-34, pp.1081-1089, Oct. 1986.

Constrained Iterative Deconvolution Applied to Bearing and Amplitude Estimation of Coherent Sources

D. J. Becker and D. A. Linebarger
Erik Jonsson School of Electrical Engineering and Computer Science
University of Texas at Dallas
Richardson, TX 75083-0688

Abstract

Until now, Constrained Iterative Deconvolution was applied to the amplitude and bearing of only incoherent sources. We introduce a CID technique for coherent sources which applies two constraints at each iteration of the deconvolution: the estimated source energy must be non-negative, and coherence may exist only within a predetermined separation in the spatial frequency ("far-field") domain.

1 Introduction

The magnitude and bearing estimation of unknown far-field sources is a classic problem of array processing [2]. The sensors of the array, located in the spatial domain, spatially sample the propagating signals from sources located in the spatial frequency (far-field) domain.

Since the array is finite, the sensor outputs are spatial samples of a window of the infinite spatial distribution of the propagating signals. For example, the spatial distribution of sensors $h(z)$ for a uniformly weighted, linear, equally-spaced (LES) array is:

$$h(z) = \sum_{m=0}^{M-1} \delta(z - md) \tag{1}$$

where d is the sensor spacing and M is the number of sensors. Let $x(z)$ be the spatial distribution of the propagating signals along the z-axis. The sensor outputs $y(z)$ are then:

$$y(z) = x(z)h(z) \tag{2}$$

In the spatial frequency domain Eq. (2) becomes a convolution:

$$\begin{aligned} Y(\theta) &= \int_{-\infty}^{\infty} y(z)e^{-jz\theta}\,dz && (3) \\ &= X(\theta) * H(\theta) && (4) \end{aligned}$$

where θ is the spatial frequency variable, $X(\theta)$ is the spatial frequency distribution of sources, and $H(\theta)$ is the spatial frequency distribution of sensor weights.

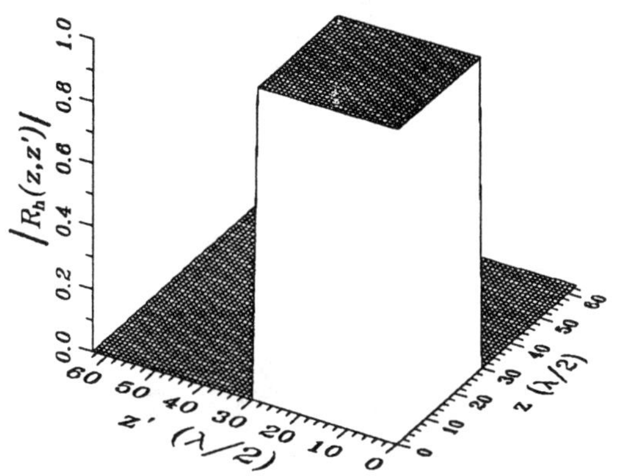

Figure 1: Magnitude of the spatial autocorrelation function (SAF) of sensors for an LES array of 30 sensors with $\lambda/2$ spacing.

We now derive the spatial autocorrelation function (SAF) of the sensor outputs $\mathbf{R}_y(z, z')$:

$$\begin{aligned} \mathbf{R}_y(z, z') &= E\{y(z)y^*(z')\} && (5) \\ &= E\{x(z)h(z) \cdot x^*(z')h^*(z')\} && (6) \\ &= E\{x(z)x^*(z')\}h(z)h^*(z') && (7) \\ &= \mathbf{R}_x(z, z') \cdot \mathbf{R}_h(z, z') && (8) \end{aligned}$$

where $\mathbf{R}_x(z, z')$ is the SAF of the sources and $\mathbf{R}_h(z, z')$ is the SAF of the sensor weights. Figure 1 shows the magnitude of the SAF for an LES array of 30 sensors along the z-axis. Figure 2 shows the magnitude of part of the SAF for two coherent monochromatic sources (of wavelength λ) in the far-field with 20 dB SNR isotropic noise. Figure 3 shows the magnitude of the SAF of the sensor outputs after infinite averaging. Therefore, the SAF of the sensor outputs is a windowed version of the SAF of the sources.

The spatial frequency autocorrelation function (SFAF) is related to the SAF as follows:

$$\mathbf{R}_Y(\theta, \theta') = E\{Y(\theta)Y^*(\theta')\} \tag{9}$$

1058-6393/91 $01.00 © 1991 IEEE

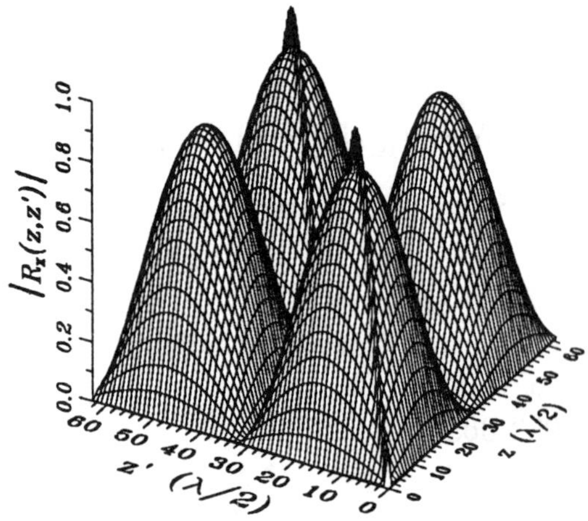

Figure 2: Magnitude of part of the spatial autocorrelation function (SAF) of two coherent sources in the presence of isotropic noise (SNR = 20dB).

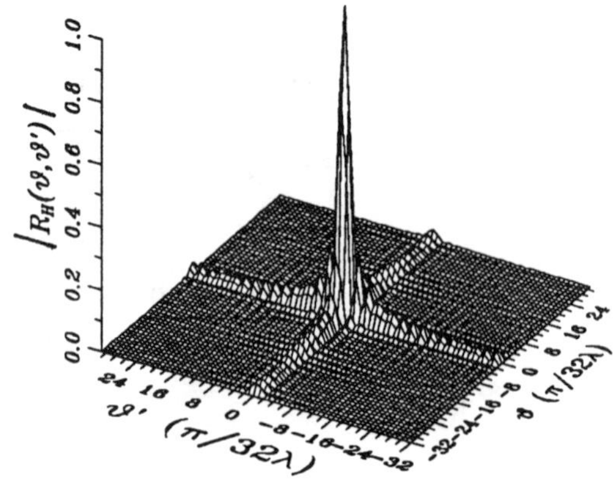

Figure 4: Magnitude of the spatial frequency autocorrelation function (SFAF) of sensors.

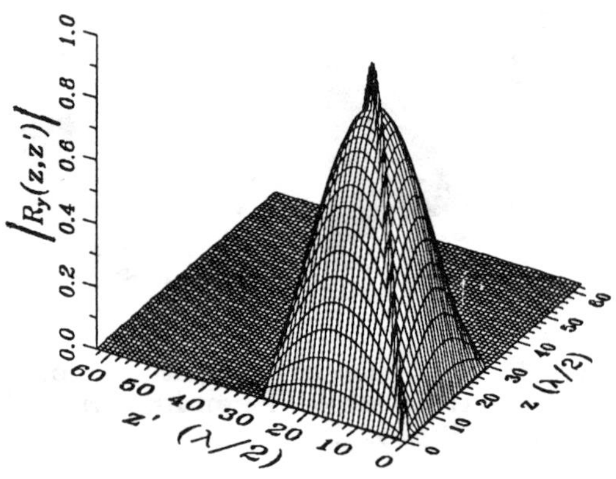

Figure 3: Magnitude of the spatial autocorrelation function (SAF) for the sensor outputs. Note that this is a windowed version of Fig. 2, with the window given by Fig. 1.

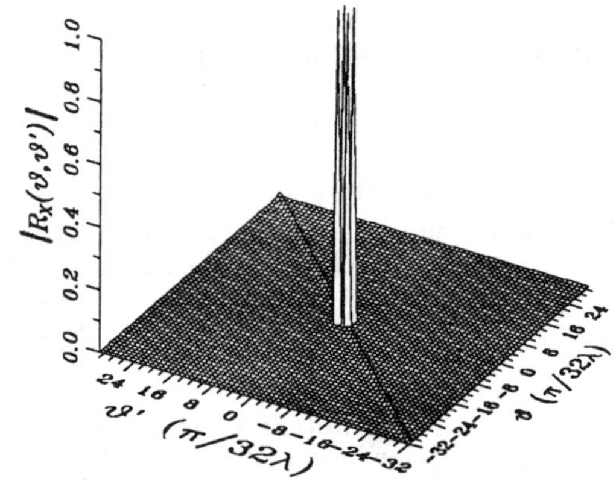

Figure 5: Magnitude of the spatial frequency autocorrelation function (SFAF) of sources.

$$= \frac{1}{2\pi}\iint_{-\infty}^{\infty} \mathbf{R}_y(z,z')e^{j(z'\theta'-z\theta)}dz dz' \quad (10)$$

and from Eqs. (8) and (10), it may be shown that

$$\mathbf{R}_Y(\theta,\theta') = \mathbf{R}_X(\theta,\theta') * \mathbf{R}_H(\theta,\theta') \quad (11)$$

where

$$\mathbf{R}_X(\theta,\theta') = E\{X(\theta)X^*(\theta')\} \quad (12)$$

$$\mathbf{R}_H(\theta,\theta') = H(\theta)H^*(\theta') \quad (13)$$

Figure 4 shows the SFAF for the LES array of Fig. 1. Figure 5 shows the SFAF of the sources of Fig. 2, and Fig. 6 shows the SFAF of the sensor outputs of Fig. 3. Therefore, the SFAF of the sensor outputs is a convolved (blurred) version of the SFAF of the sources.

(For clarity, we will now drop the arguments from the SAF and SFAF notation.)

2 Constrained Iterative Deconvolution

To increase the resolution, we try to 'un-window' what has been spatially windowed, in estimating $\mathbf{R}_x$. In the spatial frequency domain, this process is called deconvolution.

Constrained Iterative Deconvolution (CID) is a deconvolution technique which iteratively estimates $\mathbf{R}_x$, while applying non-linear constraints to the estimate to recover missing values of $\mathbf{R}_x$ [3].

Each CID iteration has four steps [1]93-134:

$$\bar{\mathbf{R}}_x^k = \hat{\mathbf{R}}_x^{k-1} + \mathbf{R}_y - \hat{\mathbf{R}}_x^{k-1} \cdot \mathbf{R}_h \quad (14)$$

$$\bar{\mathbf{R}}_X^k = \frac{1}{2\pi}\iint_{-\infty}^{\infty} \bar{\mathbf{R}}_x^k e^{j(z'\theta'-z\theta)}dz\, dz' \quad (15)$$

$$\hat{\mathbf{R}}_X^k = C\{\bar{\mathbf{R}}_X^k\} \quad (16)$$

$$\hat{\mathbf{R}}_x^k = \frac{1}{2\pi}\iint_{-\infty}^{\infty} \hat{\mathbf{R}}_X^k e^{j(z\theta-z'\theta')}d\theta\, d\theta' \quad (17)$$

where k is the iteration number, C is the constraint

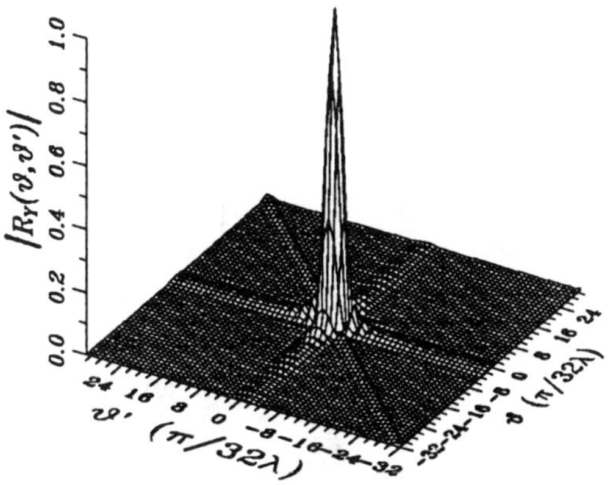

Figure 6: Magnitude of the spatial frequency autocorrelation function (SFAF) of sensor outputs.

operator, and the initial estimate $\hat{\mathbf{R}}_x^0 = \mathbf{R}_y$.

In Eq. (14), the estimate $\hat{\mathbf{R}}_x$ is updated by subtracting off the error term $\hat{\mathbf{R}}_x^{k-1} \cdot \mathbf{R}_h - \mathbf{R}_y$, resulting in the intermediate estimate $\bar{\mathbf{R}}_x^k$. Equation (15) transforms $\bar{\mathbf{R}}_x^k$ to the SFAF $\bar{\mathbf{R}}_X^k$.

The constraints are applied in Eq. (16). The first constraint operates on the energy of $\bar{\mathbf{R}}_X$ (for $\theta = \theta'$), setting all values below the noise floor equal to the noise floor. It may be expressed as

$$C\{\cdot\} = \begin{cases} \sigma_N^2, & \bar{\mathbf{R}}_X^k < \sigma_N^2 \text{ and } \theta = \theta', \\ \bar{\mathbf{R}}_X^k, & \text{otherwise.} \end{cases} \qquad (18)$$

where σ_N^2 is the noise floor.

The second constraint operates when $|\theta - \theta'| > \Delta$, where Δ is a predetermined coherence limit (i.e., the sources are known to be incoherent beyond the spatial frequency separation Δ). This constraint is reasonable since sources widely separated in the spatial frequency domain tend to be incoherent.

Therefore, all $\bar{\mathbf{R}}_X$ for $|\theta - \theta'| > \Delta$ are set to zero. This constraint may be expressed as

$$C\{\cdot\} = \begin{cases} 0, & |\theta - \theta'| > \Delta, \\ \bar{\mathbf{R}}_X^k, & \text{otherwise.} \end{cases} \qquad (19)$$

Equation (17) transforms $\hat{\mathbf{R}}_X^k$ back to the SAF $\hat{\mathbf{R}}_x^k$ and the process repeats.

3 Convergence

We now investigate the convergence properties of CID, following the analysis of Silverman [4]. The iterations are said to converge when:

1. $\mathbf{R}_y = \mathbf{R}_h \cdot \hat{\mathbf{R}}_x^k$ (i.e., the estimate is consistent with the measured data).

2. $\hat{\mathbf{R}}_X^k = \bar{\mathbf{R}}_X^k$ (i.e., the constraints are satisfied).

Convergence to a unique solution is assured if each iteration is a contraction. From the Contraction Mapping Theorem [5]243-275, each iteration is a contraction if

$$||\hat{\mathbf{R}}_x^{k+1} - \hat{\mathbf{R}}_x^k|| \le r||\hat{\mathbf{R}}_x^k - \hat{\mathbf{R}}_x^{k-1}|| \qquad (20)$$

with $0 \le r < 1$. Let us separate Eq. (20) into two parts

$$||\bar{\mathbf{R}}_x^{k+1} - \bar{\mathbf{R}}_x^k|| \le r_1||\hat{\mathbf{R}}_x^k - \hat{\mathbf{R}}_x^{k-1}|| \qquad (21)$$

$$||\hat{\mathbf{R}}_X^{k+1} - \hat{\mathbf{R}}_X^k|| \le r_2||\bar{\mathbf{R}}_X^{k+1} - \bar{\mathbf{R}}_X^k|| \qquad (22)$$

where $r_1 r_2 = r$.

Eq. (21) may be rewritten as

$$||(1 - \mathbf{R}_h) \cdot (\hat{\mathbf{R}}_x^k - \hat{\mathbf{R}}_x^{k-1})|| \le r_1||\hat{\mathbf{R}}_x^k - \hat{\mathbf{R}}_x^{k-1}|| \qquad (23)$$

and it is clear that $r_1 < 1$ except when $\hat{\mathbf{R}}_x^k = \hat{\mathbf{R}}_x^{k+1}$ (when the constraint has no effect on the estimate). In that case, $r_1 = 1$.

Likewise, Eq. (22) becomes

$$||C\{\bar{\mathbf{R}}_X^{k+1}\} - C\{\bar{\mathbf{R}}_X^k\}|| \le r_2||\bar{\mathbf{R}}_X^{k+1} - \bar{\mathbf{R}}_X^k|| \qquad (24)$$

We see from Eq. (24) that if the constraint operator affects the estimate (i.e., $\bar{\mathbf{R}}_X^k < 0$ for any $\theta = \theta'$ or $\bar{\mathbf{R}}_X^k \neq 0$ for $|\theta - \theta'| > \Delta$), then $r_2 < 1$.

If the constraint has no effect, then $r_2 = 1$. Therefore, as long as the constraint has an effect, the iterations are contractions. When the constraints no longer have an effect, the iterations have converged.

Consequently, the solution is not unique, since the iterations become non-expansive ($r_1 r_2 = 1$) when the constraint fails to have an effect on the estimate.

4 Example

Consider the case of two coherent, monochromatic sources in the spatial frequency domain, with equal amplitudes and 180° out of phase. They are located at ±1.8° off broadside of an LES array of 30 sensors. The SNR is 20 dB.

Figures 7 and 8 show the magnitudes of $\hat{\mathbf{R}}_x$ and $\hat{\mathbf{R}}_X$ after 500 iterations (with Δ corresponding to a separation of 3.6° at broadside). Figure 9 shows the diagonals of $\hat{\mathbf{R}}_X$ and $\mathbf{R}_Y$. It is clear that the two sources have been resolved without a great enhancement of the noise, whereas conventional beamforming failed to resolve them.

5 Summary and Conclusions

This paper described a modified Constrained Iterative Deconvolution algorithm which can be applied to bearing and amplitude estimation of coherent sources. Our technique has shown, via the Contraction Mapping Theorem, to converge to a non-unique solution, but one consistent with both the measured data and *a priori* knowledge of the sources.

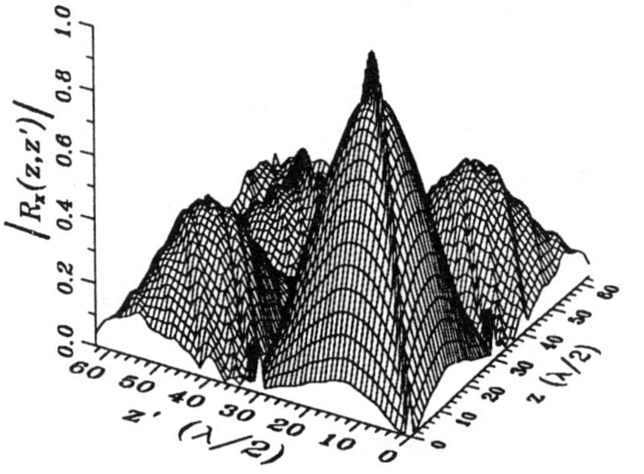

Figure 7: Magnitude of the estimated spatial autocorrelation function (SAF) of the sources.

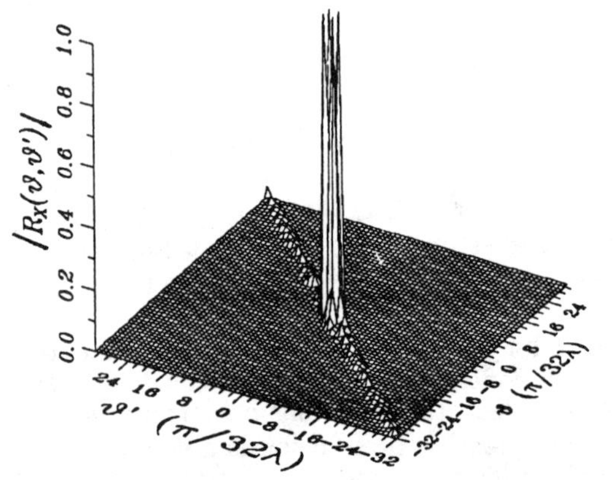

Figure 8: Magnitude of the estimated spatial frequency autocorrelation function (SFAF) of the sources.

References

[1] Peter A. Jansson. *Deconvolution: With Applications in Spectroscopy*, pages 93–134. Academic Press, 1984.

[2] Don H. Johnson. The application of spectral estimation methods to bearing estimation problems. *IEEE Proceedings*, 70(9):1018–1028, September 1982.

[3] M.A. Richards, C.E. Morris, and Monson H. Hayes. Iterative enhancement of noncoherent radar data. *ICASSP Proceedings*, pages 1929–1931, 1986.

[4] H.F. Silverman and A.E. Pearson. On deconvolution using the discrete fourier transform. *IEEE Trans. Audio Electroacoust.*, AU-21(2):112–118, April 1973.

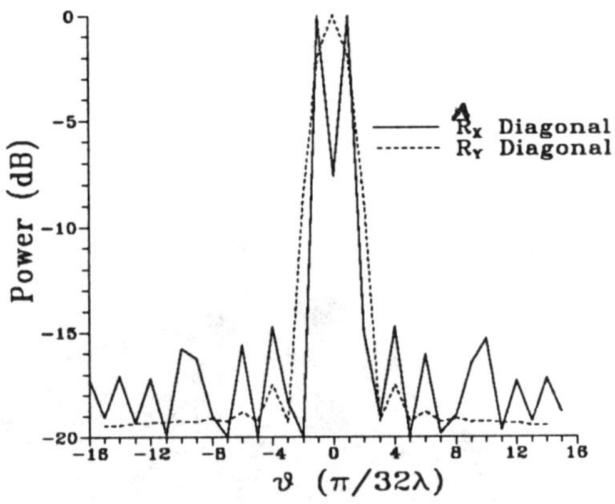

Figure 9: Estimated energy of the estimate (solid line) and the result given by traditional beamforming (dotted line).

[5] I. Stakgold. *Green's Functions and Boundary Value Problems*, pages 243–275. Wiley - Interscience, 1979.

DIRECTION FINDING FOR WIDEBAND SIGNALS USING AN INTERPOLATED ARRAY

Benjamin Friedlander

Signal Processing Technology, Ltd.
703 Coastland Drive
Palo Alto, CA 94303, USA

Anthony J. Weiss

Dept. of Electrical Engineering
Tel Aviv University
Tel Aviv, 69978, Israel

Abstract

We derive a new direction finding algorithm for multiple wideband signals received by an *arbitrary* array, and analyze its performance. Using an interpolation technique, a set of *virtual* arrays is generated, each for a different frequency band, having the same array manifold. The covariance matrices of these arrays are added up to produce a composite covariance matrix. Direction of arrival estimates are obtained by eigendecomposition of this composite covariance matrix using the narrowband MUSIC algorithm or its variants. By choosing the virtual arrays to be linear and equally spaced we are able to apply the root-MUSIC algorithm to this problem. Closed form expressions for the asymptotic covariance matrix of the DOA estimation errors are derived using a perturbation analysis.

1. Introduction

In this paper we consider a coherent processing technique for wideband direction finding, based on the following idea. Consider a planar array with sensors located at coordinates $\{x_i, y_i\}$. The response of this array to a narrowband signal at frequency f_0 is the same as the response of an array with sensors located at coordinates $\{\alpha x_i, \alpha y_i\}$, to a narrowband signal at frequency f_0/α. In other words, "stretching" an array by a factor of α and reducing the frequency by the same factor, leaves the array response unchanged. Next consider a wide frequency band, which is divided into multiple narrow bands with center frequencies f_k. Assume that each band is received by a *different* array, where all the arrays are "stretched" or "compressed" versions of a nominal array, and the amount of stretching or compression is selected so as to make all the arrays have the same response. Because of the fact that all arrays have the same response (at their respective operating frequencies), it is possible to combine the covariance matrices for the different frequencies in a straightforward manner, by simple addition. The conventional (*i.e.* narrowband) MUSIC algorithm [5] or similar techniques can then be applied to the composite covariance matrix. This idea is described in more detail below.

This work was supported by the Army Research Office under contract no. DAAL03-91-C-0022, sponsored by U.S. Army Communications Electronics Command, Center for Signals Warfare.

The difficulty with the approach described above is, of course, that we have only a single array, rather than a collection of arrays "tuned" to different frequencies. It is not necessary, however, to have multiple *physical* arrays. It is possible instead to use *virtual* arrays whose outputs are estimated from the outputs of the *real* array. These outputs are obtained by a straightforward linear interpolation technique, with interpolator coefficients selected so as to minimize the interpolation error for a signal arriving from a given sector (*i.e.* a range of bearing angles), at a particular frequency. Different sets of interpolator coefficients will be used to provide good estimates for different sectors and different frequencies. The size of the sector will be chosen to give sufficiently good estimates of the virtual array outputs. The design of the interpolator needs to be performed only once, off-line, and thus does not impose a computational burden on the algorithm.

Given measurements of the real array outputs, the interpolator can be used in principle to compute the outputs of the virtual array. Actually, it is not necessary to compute the outputs of the virtual array, since it is possible to compute directly the sample covariance matrix corresponding to these outputs. The techniques considered here require only the sample covariance matrix, not the "raw" data.

To summarize: using interpolation we synthesize a sequence of covariance matrices corresponding to a sequence of stretched and compressed versions of a nominal array. These covariance matrices are added up to produce a composite covariance matrix which is used by a signal subspace based direction finding algorithm. The virtual array is chosen to be linear and uniformly spaced, to make it possible to use the root-MUSIC algorithm.

The interpolated array concept has been explored recently in [3] for narrowband direction finding appli-

1058-6393/91 $01.00 © 1991 IEEE

cations. Krolik and Swingler [2] presented an interpolation based algorithm for wideband direction finding. However, their approach uses a different interpolation procedure tailored specifically for linear uniformly spaced arrays, and does not apply to arbitrary array geometries as the technique presented here.

2. Problem Formulation and Proposed Solution

Consider N radiating sources observed by an arbitrary array of M sensors. The signal at the output of the m-th sensor can be described by

$$x_m(t) = \sum_{n=1}^{N} s_n(t-\tau_{mn}) + v_m(t) \tag{1}$$
$$-T/2 \leq t \leq T/2 \qquad m = 1,2,\cdots,M$$

where $\{s_n(t)\}_{n=1}^N$ are the radiated signals, $\{v_m(t)\}_{m=1}^M$ are sample waveform from additive noise processes and T is the observation interval. The parameters $\{\tau_{mn}\}$ are delays associated with the signal propagation time from the n-th source to the m-th sensor. These parameters are of interest since they contain information about the source location relative to the array.

A convenient separation of the parameters to be estimated is obtained by representing the signals using fourier coefficients defined by

$$X_m(\omega_\ell) = \frac{1}{\sqrt{T}} \int_{-T/2}^{T/2} x_m(t) e^{-jw_\ell t}\, dt \tag{2}$$

where $w_\ell = \frac{2\pi}{T}(\ell_1 + \ell)$, $\ell = 1,2,\cdots,L$; and ℓ_1 is a constant. In principle, the number of coefficients required to capture all the signal information is infinite. However, if we consider signals with energy concentrated in a finite spectral band, we can use only $L < \infty$ coefficients.

Taking the Fourier coefficients of (1) we obtain

$$X_m(\omega_\ell) = \sum_{n=1}^{N} e^{-jw_\ell \tau_{mn}} S_n(\omega_\ell) + V_m(\omega_\ell) \tag{3}$$

where $S_n(\omega_\ell)$ and $V_m(\omega_\ell)$ are the Fourier coefficients of $s_n(t)$ and $v_m(t)$, respectively. Equation (3) may be expressed using vector notation as follows,

$$\mathbf{X}(\omega_\ell) = A(\omega_\ell)\mathbf{S}(\omega_\ell) + \mathbf{V}(\omega_\ell) \tag{4}$$

where

$$\begin{aligned}
\mathbf{X}(\omega_\ell) &\triangleq [X_1(\omega_\ell), X_2(\omega_\ell), \cdots, X_M(\omega_\ell)]^T \\
\mathbf{S}(\omega_\ell) &\triangleq [S_1(\omega_\ell), S_2(\omega_\ell), \cdots, S_N(\omega_\ell)]^T \\
\mathbf{V}(\omega_\ell) &\triangleq [V_1(\omega_\ell), V_2(\omega_\ell), \cdots, V_M(\omega_\ell)]^T \\
[A(\omega_\ell)]_{mn} &\triangleq e^{-j\omega_\ell \tau_{mn}} \\
& m = 1,2,\cdots,M; \quad n = 1,2,\cdots,N.
\end{aligned}$$

In order to simplify the exposition we assume that the sensors and sources are coplanar and the sources are far enough from the observing array so that the signal wavefronts are effectively planar over the array. In this case we have

$$\begin{aligned}
\tau_{mn} &= -\, d_{mn}/c \\
d_{mn} &= \tilde{x}_m \sin\theta_n + \tilde{y}_m \cos\theta_n
\end{aligned} \tag{5}$$

where c is the propagation velocity, d_{mn} is the distance from sensor m to a reference sensor (selected here to be sensor number 1) in the direction of the n-th source, $(\tilde{x}_m, \tilde{y}_m)$ are the coordinates of the m-th sensor, θ_n is the DOA of the n-th source relative to the $\tilde{y}$-axis, and the origin of the Cartesian coordinate system coincides with the reference sensor.

The main problem that we are addressing here is how to estimate $\{\theta_n\}_{n=1}^N$ from the data vectors $\{\mathbf{X}(\omega_\ell)\}_{\ell=1}^L$. In practice, we have N_s snapshots of $\mathbf{X}(\omega_\ell)$ which we denote by $\{\mathbf{X}(\omega_\ell, j)\}_{j=1}^{N_s}$.

Before we turn to describe our estimation approach we define the following covariance matrices,

$$R_s(\omega_\ell) \triangleq E\left\{\mathbf{S}(\omega_\ell)\mathbf{S}^H(\omega_\ell)\right\} \tag{6}$$

$$R_n(\omega_\ell) \triangleq E\left\{\mathbf{V}(\omega_\ell)\mathbf{V}^H(\omega_\ell)\right\} \tag{7}$$

$$\begin{aligned}
R_\ell &\triangleq E\left\{\mathbf{X}(\omega_\ell)\mathbf{X}^H(\omega_\ell)\right\} \\
&= A(\omega_\ell)R_s(\omega_\ell)A^H(\omega_\ell) + R_n(\omega_\ell)
\end{aligned} \tag{8}$$

We shall also make extensive use of the sample covariance matrix

$$\hat{R}_\ell \triangleq \frac{1}{N_s}\sum_{j=1}^{N_s} \mathbf{X}(\omega_\ell, j)\mathbf{X}^H(\omega_\ell, j) \tag{9}$$

as an estimate of R_ℓ.

The Composite Covariance matrix Assume, for a moment, that we use different arrays for collecting data at different frequencies. The data covariance matrix of the ℓ-th array is given by

$$\bar{R}_\ell = \bar{A}(\omega_\ell)R_s(\omega_\ell)\bar{A}^H(\omega_\ell) + R_v(\omega_\ell)\,, \tag{10}$$

where $\bar{A}(\omega_\ell)$ is the array manifold of the ℓ-th array at frequency ω_ℓ, and $R_v(\omega_\ell)$ is the noise covariance of this array. By design, we can make all these array manifolds equal, *i.e.*

$$\bar{A}(\omega_\ell) = \bar{A}\,. \tag{11}$$

The composite covariance matrix $\bar{R}$ is simply the sum of the covariance matrices of the different arrays,

$$\bar{R} \triangleq \sum_{\ell=1}^{L} \bar{R}_\ell = \bar{A}\bar{R}_s\bar{A}^H + \bar{R}_n\,, \qquad (12)$$

where

$$\bar{R}_s \triangleq \sum_{\ell=1}^{L} R_s(\omega_\ell)\,, \qquad (13)$$

$$\bar{R}_n \triangleq \sum_{\ell=1}^{L} R_v(\omega_\ell)\,. \qquad (14)$$

The composite covariance matrix (12) has the same structure as the covariance matrix arising in the case of narrowband signals. Therefore, the directions of arrival can be estimated by eigendecomposition of $\bar{R}$ exactly as in the narrowband case.

Since in practice we do not have multiple physical arrays we propose to use virtual arrays whose output can be computed by *linear* interpolation, of the single physical array output as described next.

The Interpolated Array

The interpolated array approach is based on the idea that the array manifold of a virtual array, whose element locations are chosen by the user, can be obtained by linear interpolation of the array manifold of the real array within a limited sector. In other words, we assume that there exists a constant matrix B such that

$$B\mathbf{a}(\theta) \approx \bar{\mathbf{a}}(\theta) \qquad (15)$$

for angles θ within the sector, where $\mathbf{a}(\theta)$ is the real array manifold and $\bar{\mathbf{a}}(\theta)$ is the virtual array manifold. The interpolation matrix B is computed by performing a least squares fit in equation (15); see [3] for a much more detailed discussion of the virtual array design and the interpolation accuracy. In practice, the field of view will be divided into several sectors, and a different interpolation matrix will be calculated for each sector. To simplify the discussion and the notation we will consider in this paper only a single sector. Reference [3] describes how to combine estimates produced for the different sectors.

In the context of the wideband problem, $\bar{\mathbf{a}}(\theta)$ is the virtual array manifold (recall that it is the same for all virtual arrays), and $\mathbf{a}(\theta)$ is the real array manifold at a particular frequency. If we denote the real array manifold at ω_ℓ by $\mathbf{a}_\ell(\theta)$, equation (15) becomes

$$B_\ell\mathbf{a}_\ell(\theta) \approx \bar{\mathbf{a}}(\theta) \qquad (16)$$

In other words, for each frequency we need to compute a single interpolation matrix which map the real array manifold to a fixed array manifold, for a given sector.

Given the interpolation matrix B_ℓ we can compute the covariance (or sample covariance) matrix $\bar{R}_\ell$ of the virtual array from the covariance (or sample covariance) matrix R of the real array:

$$\bar{R}_\ell = B_\ell R B_\ell^H \qquad (17)$$

The composite covariance matrix used by the DOA estimation algorithm is the sum of the matrices $\bar{R}_\ell$.

The Wideband Direction Finding Algorithm

We are now ready to describe the wideband direction finding algorithm proposed and analyzed in this paper. As was mentioned before, we choose the virtual arrays to be uniformly spaced linear arrays which permit the use of the Root-MUSIC algorithm. The physical (real) array is arbitrary.

Consider a linear uniformly spaced array with omnidirectional sensors located on the x-axis. The array manifold of this array is given by

$$\bar{\mathbf{a}}(\theta) = \left[1, e^{j2\pi(d/\lambda)\sin\theta}, \ldots, e^{j2\pi(d/\lambda)(M-1)\sin\theta}\right]^T \qquad (18)$$

where λ is the wavelength of the received signals, and d is the element spacing. Since

$$d/\lambda = d\frac{\omega}{2\pi c} \qquad (19)$$

it is clear that if we select the element spacing according to the rule

$$d\omega = \text{constant}. \qquad (20)$$

the array manifold will be the same for every frequency. We therefore select our interpolated arrays as follows:

$$d_\ell = \frac{d_o\omega_o}{\omega_\ell} \qquad (21)$$

where d_ℓ is the element spacing for the interpolated array at frequency ω_ℓ, and d_o is the element spacing at the center frequency ω_o.

Since the original array manifold is different for different frequencies we will need L interpolation matrices, one for each frequency. Thus,

$$\bar{A} = B_\ell A(\omega_\ell) \quad \forall\ell \qquad (22)$$

Since all the interpolated arrays have the same manifold and therefore the same direction matrix $A(\omega_o) =$

$\bar{A}$ we can process all the data coherently by forming a single "covariance" matrix as follows

$$\begin{aligned}\bar{R} &= \sum_{\ell=1}^{L} \bar{R}_\ell \\ &= \sum_{\ell=1}^{L} B_\ell R_\ell B_\ell^H \\ &= \sum_{\ell=1}^{L} B_\ell A(\omega_\ell) R_s(\omega_\ell) A^H(\omega_\ell) B_\ell^H + B_\ell R_n(\omega_\ell) B_\ell^H \\ &= \bar{A}\left\{\sum_{\ell=1}^{L} R_s(\omega_\ell)\right\}\bar{A}^H + \sum_{\ell=1}^{L} B_\ell R_n(\omega_\ell) B_\ell^H \\ &= \bar{A}\bar{R}_s\bar{A}^H + \bar{R}_n \end{aligned} \tag{23}$$

where

$$\bar{R}_n \triangleq \sum_{\ell=1}^{L} B_\ell R_n(\omega_\ell) B_\ell^H \tag{24}$$

It is now clear that one can apply the Root-MUSIC method (or any other MUSIC Variant) to obtain a good estimate of the DOAs using $\bar{R}$. The algorithm may be summarized as follows:

a) Obtain an estimate of R_ℓ using the sample covariance in each frequency bin.

b) Compute $\bar{R}_n$ using (24), and estimate $\bar{R}$ using

$$\hat{\bar{R}} = \sum_{\ell=1}^{L} B_\ell \hat{R}_\ell B_\ell^H \tag{25}$$

c) Prewhiten the noise by modifying $\hat{\bar{R}}$ as follows

$$\hat{\tilde{R}} = \bar{R}_n^{-1/2}\,\hat{\bar{R}}\,\bar{R}_n^{-1/2} \tag{26}$$

Note that the expected value of $\hat{\tilde{R}}$ is

$$\tilde{R} = \bar{R}_n^{-1/2}\bar{A}\bar{R}_s\bar{A}^H\bar{R}_n^{-1/2} + I \tag{27}$$

d) Perform SVD on $\hat{\tilde{R}}$ and obtain the noise subspace estimate $\hat{G}$ (G is a matrix whose columns are the eigenvectors of $\tilde{R}$ associated with the M-N smallest eigenvalues).

The next steps are specific to Root-MUSIC.

e) Find the roots of the polynomial

$$D(z) = \bar{\mathbf{a}}^T(1/z)\bar{R}_n^{-1/2}\hat{G}\hat{G}^H\bar{R}_n^{-1/2}\bar{\mathbf{a}}(z) \tag{28}$$

where

$$\bar{\mathbf{a}}(z) = [1, z, z^2, \cdots, z^{M-1}]^T \tag{29}$$

f) Estimate the DOAs using the relation

$$\omega_i = 2\pi(d_o/\lambda_o)\sin\theta_i \tag{30}$$

where ω_i is the angle of the root corresponding to the i-th signal.

A detailed explanation of why Root-MUSIC work can be found in many texts including [3, 4].

3. Performance Analysis

In [1] we analyze the performance of the proposed algorithm by developing formulas for the asymptotic variance of the DOA estimates. The performance analysis is based on evaluating the perturbations in the estimates of $\{\theta_n\}_{n=1}^{N}$, due to perturbations in the matrix

$$G \triangleq [\mathbf{u}_{N+1}, \mathbf{u}_{N+2} \cdots \mathbf{u}_M] \tag{31}$$

where $\{\mathbf{u}_m\}_{m=1}^{M}$ are the eigenvectors of $\tilde{R}$. The perturbations in G are due to perturbations in $\tilde{R}$.

The main result in [1] is

$$E\{\delta\theta_i\delta\theta_j\} = \frac{\mathbf{n}}{2\left[\dot{\mathbf{v}}^H(\theta_i)GG^H\dot{\mathbf{v}}(\theta_i)\right]\left[\dot{\mathbf{v}}^H(\theta_j)GG^H\dot{\mathbf{v}}(\theta_j)\right]} \tag{32}$$

where

$$\mathbf{n} = \sum_{k,t=1}^{N} \mathrm{Re}\left\{\mathbf{u}_k^H V(\theta_i) E\left\{\delta\mathbf{u}_k\delta\mathbf{u}_t^T\right\} V^T(\theta_j)\mathbf{u}_t^* + \mathbf{u}_k^H V(\theta_i) E\left\{\delta\mathbf{u}_k\delta\mathbf{u}_t^H\right\} V(\theta_j)\mathbf{u}_t\right\}$$

$$\mathbf{v}(\theta) = \bar{R}_n^{-1/2}\bar{\mathbf{a}}(\theta)$$

$$\dot{\mathbf{v}}(\theta) = \frac{\partial\mathbf{v}(\theta)}{\partial\theta}$$

$$V(\theta) = \dot{\mathbf{v}}(\theta)\mathbf{v}^H(\theta) + \mathbf{v}(\theta)\dot{\mathbf{v}}^H(\theta)$$

Equation (32) can be used to evaluate the covariance matrix of the DOA estimation errors for any given scenario.

4. The Cramer Rao Bound

In this section we present the CRB for estimating the directions of arrival of N wideband sources observed by an array of M sensors. The signals may be correlated or even coherent.

We assume that the noise variance is the same in all the frequency bins (*i.e.* $R_n(\omega_\ell) = \sigma^2 \cdot I \;\; \forall\ell$), and for notational convenience we use $\eta = \sigma^2$. We further assume that the data consists of N_s independent realizations of $\mathbf{X}(\omega_\ell)$ for each frequency bin, and that the observation time T of each realization is much longer than the correlation times of the signals and noise and therefore

$$E\{\mathbf{X}(\omega_\ell)\} = 0 \tag{33}$$

$$E\{(\mathbf{X}(\omega_\ell)\mathbf{X}^H(\omega_k)\} = \delta_{k\ell}\left[A(\omega_\ell)R_s(\omega_\ell)A^H(\omega_\ell) + \eta \cdot I\right] \triangleq \delta_{k\ell}R_x(\omega_\ell) \tag{34}$$

The parameter vector to be estimated is defined by

$$\mathbf{p} = [\theta, \mu_1, \cdots, \mu_L, \eta] \tag{35}$$

where θ is a vector of the N DOAs, η is the unknown noise variance and μ_ℓ is a vector describing all the components of $R_s(\omega_\ell)$.

Under the above model the FIM is given by

$$F = N_s \sum_{\ell=1}^{L} F(\omega_\ell) \tag{36}$$

where

$$[F(\omega_\ell)]_{mn} = \mathrm{tr}\left\{R_x^{-1}(\omega_\ell)\frac{\partial R_x(\omega_\ell)}{\partial p_m}R_x^{-1}(\omega_\ell)\frac{\partial R_x(\omega_\ell)}{\partial p_n}\right\} \tag{37}$$

In [1] we show that the matrices $F(\omega_\ell)$ can be partitioned as follows

$$\begin{aligned}F_{\theta\theta}(\omega_\ell) = 2\mathrm{Re}\Big\{&\left(R_s(\omega_\ell)A^H(\omega_\ell)R_x^{-1}(\omega_\ell)A(\omega_\ell)R_s(\omega_\ell)\right)\\ &\times\left(\dot{A}_\theta^H(\omega_\ell)R_x^{-1}(\omega_\ell)\dot{A}_\theta(\omega_\ell)\right)^T\\ &+\left(R_s(\omega_\ell)A^H(\omega_\ell)R_x^{-1}(\omega_\ell)\dot{A}_\theta(\omega_\ell)\right)\\ &\times\left(R_s(\omega_\ell)A^H(\omega_\ell)R_x^{-1}(\omega_\ell)\dot{A}_\theta(\omega_\ell)\right)^T\Big\}\end{aligned} \tag{38}$$

$$\begin{aligned}F_{\theta\mu_k}(\omega_\ell) = \delta_{k\ell}Q_4\Big\{&\left(A^H(\omega_\ell)R_x^{-1}(\omega_\ell)A(\omega_\ell)R_s(\omega_\ell)\right)^T\\ &\otimes\left(\dot{A}_\theta^H(\omega_\ell)R_x^{-1}(\omega_\ell)A(\omega_\ell)\right)\\ &+\left(A^H(\omega_\ell)R_x^{-1}(\omega_\ell)\dot{A}_\theta(\omega_\ell)\right)^T\\ &\otimes\left(R_s(\omega_\ell)A^H(\omega_\ell)R_x^{-1}(\omega_\ell)A(\omega_\ell)\right)\Big\}Q_t^H\end{aligned} \tag{39}$$

$$F_{\theta\eta}(\omega_\ell) = 2\mathrm{Re}\left\{\mathrm{diag}\left(R_s(\omega_\ell)A^H(\omega_\ell)R_x^{-2}(\omega_\ell)\dot{A}_\theta(\omega_\ell)\right)\right\} \tag{40}$$

$$\begin{aligned}F_{\mu_k\mu_p}(\omega_\ell) = \delta_{k\ell}\delta_{p\ell}Q_t&\Big[\left(A^H(\omega_\ell)R_x^{-1}(\omega_\ell)A(\omega_\ell)\right)^*\\ &\otimes\left(A^H(\omega_\ell)R_x^{-1}(\omega_\ell)A(\omega_\ell)\right)\Big]Q_t^H\end{aligned} \tag{41}$$

$$\begin{aligned}F_{\mu_k\eta}(\omega_\ell) = \delta_{k\ell}Q_t&\Big[\left(R_x^{-1}(\omega_\ell)A(\omega_\ell)\right)^T\\ &\otimes\left(R_x^{-1}(\omega_\ell)A(\omega_\ell)\right)^H\Big]I(:)\end{aligned} \tag{42}$$

$$F_{\eta\eta}(\omega_\ell) = \mathrm{tr}\left\{R_x^{-2}(\omega_\ell)\right\} \tag{43}$$

where $\times$ denote the Hadamard product, diag(X) is a column vector equal to the diagonal of X, and Q_t, Q_4 are constant matrices defined in [1]. Finally

$$\dot{A}_\theta(\omega_\ell) \triangleq \sum_{n=1}^{N}\frac{\partial A(\omega_\ell)}{\partial\theta_n}$$

With these compact formulas the CRB can be easily evaluated numerically for any given problem. It should be emphasized that these formulas provide a bound on estimating the DOAs, the signal covariance at all frequencies, and the noise variance, for Gaussian signals and Gaussian noise. Since the proposed estimation method does not make use of any knowledge of the signal covariance it should be compared to the above bound. However, if the signal covariance is known, the above bound can be easily modified to accommodate this case by eliminating the rows and columns of F associated with $\{R_s(\omega_\ell)\}_{\ell=1}^{L}$. It may be useful for algorithms that assume known signal covariance and make use of this knowledge.

References

[1] B. Friedlander, and A. J. Weiss, "Direction Finding for Wideband Signals Using Interpolated Arrays," Submitted for publication.

[2] J. Krolik, and D.N. Swingler, "Focussed Wideband Array Processing via Spatial Resampling," *IEEE Trans. Acoust., Speech, Signal Processing*, Vol. ASSP-38, No. 2 February 1990.

[3] B. Friedlander, "The Interpolated Root-MUSIC Algorithm for Direction Finding," submitted for publication. See also B. Friedlander, "Direction Finding with an Interpolated Array," Intl. Conference on Acoustics, Speech and Signal Processing, Albuquerque, New Mexico, April 3–6, 1990.

[4] B.D. Rao and K.V.S. Hari, "Peformance Analysis of Root-MUSIC," *IEEE Trans.Acoust., Speech, Signal Processing*, Vol. ASSP-37, No. 12, pp. 1939–1949, December 1989.

[5] R.O. Schmidt, "Multiple Emmitter Location and Signal Parameter Estimation," in *Proc. RADC Spectrum Estimation Workshop,"* Griffiths AFB, New York, 1979, pp. 243–258. Reprinted in *IEEE Trans. Antennas and Propag.*, vol. AP-34, no. 3, pp. 276–380, March 1986.

Estimation of Array Element Phases in Adverse Interference Conditions

Eric K.L. Hung

Radar Division, Defence Research Establishment Ottawa
Ottawa, Ontario, Canada K1A 0K2

Abstract

A method to estimate the element phases of a sensor array is presented. It uses a movable calibration source in the farfield zone of the array. Snapshots of the array element outputs are taken and individually transformed to make the signal appear to come from a fixed direction. The array output power from this direction is then calculated with the Capon method and maximized with respect to the estimates of element phases.

This method is tolerant to many types of interference, including specular multipath interference. Computer simulations indicate that the accuracy of phase estimates is high when reasonably accurate estimates of the source direction can be obtained from the array snapshots.

1 Introduction

The outputs of a sensor array require accurate compensations in gains and phases if they are to be analyzed by high-resolution direction-finding methods such as the MUSIC method [1]. Otherwise, the full potential of these methods in resolving target directions cannot be realized.

In the absence of mutual coupling among the array elements, the compensations become corrections for the unequal element gains and phases introduced by the elements and the electronics. Two methods to estimate these gains and phases have been developed, one by Paulraj and Kailath [2] and the other by Friedlander and Weiss [3]. Both assume that the number of signal sources is known, this number is less than the number of array elements, and specular multipath interference is absent.

This paper contains a method to estimate array element phases when the above assumptions are not satisfied. It ignores the estimation of element gains, because highly accurate estimates of these gains can be obtained by monitoring the array element output powers. The algorithm is presented in Section 2. Its performance is studied with computer simulations in Section 3. Some concluding remarks are given in Section 4.

2 Algorithm

The algorithm is developed for the case where the correlation matrix of array element outputs can be written as a sum of two terms: one for the direct signal from a calibration source and one for noise-plus-interference. A technique to make the correlation matrix possess this property, even when specular multipath interference is present, is then described.

2.A Signal Model

The array is narrowband and has N elements. The element gains are known and are equal to unity. Mutual coupling is absent. An array snapshot is given by

$$\underline{x} = s_c \underline{b}(\phi,\theta_c) + \underline{v}\,, \tag{1}$$

where s_c is the complex amplitude of a calibration signal at the array elements, $\underline{b}(\phi,\theta_c)$ is the array manifold vector for the signal direction, $\underline{\phi}$ represents the array element phases $\{\phi_1,\phi_2,\cdots,\phi_N\}$, θ_c is the signal direction, and $\underline{y}$ denotes the contributions of noise and interference sources. The expression for an array manifold vector is

$$\underline{b}(\underline{\phi},\theta) = \underline{\Gamma}(\underline{\phi})\ \underline{a}(\theta)\ . \tag{2}$$

Here,

$$\underline{\Gamma}(\underline{\phi}) = \mathrm{diag}\{\exp[j\phi_1],\exp[j\phi_2],\cdots,\exp[j\phi_N]\} \tag{3}$$

is a diagonal matrix associated with the element phases, and $\underline{a}(\theta)$ is the array steering vector for direction θ. The uniqueness condition for the element phases is defined as

$$\phi_1 + \phi_2 + \cdots + \phi_N = 0\ . \tag{4}$$

The correlation matrix of array element outputs is the expected value

$$\underline{R}_x = E\{\underline{x}\,\underline{x}^H\}\ . \tag{5}$$

Superscripts H denotes the conjugate-transpose. It has the decomposition

$$\underline{R}_x = |s_c|^2\, \underline{b}(\underline{\phi},\theta_c)\ \underline{b}^H(\underline{\phi},\theta_c) + \underline{R}_y\ , \tag{6}$$

where the first term denotes the contribution from the calibration source and $\underline{R}_y = E\{\underline{y}\underline{y}^H\}$ is the correlation matrix for noise-plus-interference.

2.B Theory

The Capon method [4,5] is used to calculate the array output power for direction θ_c. Given that $\hat{\underline{\phi}}$ is the estimate of $\underline{\phi}$ in this calculation, the power is

$$\mathcal{P}(\hat{\underline{\phi}},\underline{R}_x) = \frac{1}{F(\hat{\underline{\phi}},\underline{R}_x,\hat{\underline{\phi}})}\ , \tag{7}$$

where

$$F(\underline{\alpha},\underline{R},\underline{\beta}) = \underline{b}^H(\underline{\alpha},\theta_c)\ \underline{R}^{-1}\ \underline{b}(\underline{\beta},\theta_c)\ . \tag{8}$$

Matrix $\underline{R}_x$ in (7) is given by (6), application of the matrix inversion lemma leads to

$$\mathcal{P}(\hat{\underline{\phi}},\underline{R}_x) = \frac{1 + |s_c|^2\, F(\underline{\phi},\underline{R}_y,\underline{\phi})}{F(\hat{\underline{\phi}},\underline{R}_y,\hat{\underline{\phi}}) + |s_c|^2\, G(\underline{\phi},\underline{R}_y,\hat{\underline{\phi}})}\ , \tag{9}$$

with

$$G(\underline{\phi},\underline{R}_y,\hat{\underline{\phi}}) = F(\underline{\phi},\underline{R}_y,\underline{\phi})\ F(\hat{\underline{\phi}},\underline{R}_y,\hat{\underline{\phi}}) - |F(\underline{\phi},\underline{R}_y,\hat{\underline{\phi}})|^2$$

$$\geq 0\ . \tag{10}$$

The last line is obtained with the Schwarz's inequality. The equality sign holds if and only if $\hat{\underline{\phi}}=\underline{\phi}$.

It can be shown that $\mathcal{P}(\hat{\underline{\phi}},\underline{R}_x)$, as a function of $\hat{\underline{\phi}}$, has a local maximum close to $\underline{\phi}$ if the rate at which $F(\hat{\underline{\phi}},\underline{R}_y,\hat{\underline{\phi}})$ changes in this neighbourhood is small compared with that of $|s_c|^2 G(\underline{\phi},\underline{R}_y,\hat{\underline{\phi}})$. Increasing $|s_c|$ would make the deviation of $\hat{\underline{\phi}}$ from $\underline{\phi}$ smaller. Making $F(\hat{\underline{\phi}},\underline{R}_y,\hat{\underline{\phi}})$ independent of $\hat{\underline{\phi}}$ in the neighbourhood of $\underline{\phi}$ would make $\hat{\underline{\phi}}=\underline{\phi}$. Hence, an estimate of $\underline{\phi}$ can be obtained by maximizing $\mathcal{P}(\hat{\underline{\phi}},\underline{R}_x)$ with respect to $\hat{\underline{\phi}}$.

2.C Estimation Procedure

The procedure is based on the use of a calibration source in the farfield zone of the array. The source is moved over a range of directions centred at θ_c. Snapshots of the array outputs are taken and transformed individually to make the source direction appear fixed at θ_c. Matrix $\underline{R}_x$ is calculated with the transformed snapshots. Estimate $\hat{\underline{\phi}}$ is obtained by maximizing $\mathcal{P}(\hat{\underline{\phi}},\underline{R}_x)$ with respect to $\hat{\underline{\phi}}$.

A transformed array snapshot $\underline{x}_m$, say, is obtained from a measured snapshot $\underline{x}'_m$ as

$$\underline{x}_m = \underline{A}(\theta_c)\ \underline{A}^{-1}(\theta_m)\ \underline{x}'_m\ , \tag{11}$$

where θ_m is the source direction in $\underline{x}'_m$ and $\underline{A}(\theta)$ is a diagonal matrix such that $A_{nn}(\theta)=a_n(\theta)$ for n=1 to N.

PROCEDURE:

Step 1.

Take a set of M array snapshots $\{\underline{x}'_1,\underline{x}'_2,\cdots,\underline{x}'_M\}$ with the calibration source moved over a range of directions.

Step 2.

Measure or estimate the corresponding true source directions. Denote them by $\{\theta_1,\theta_2,\cdots,\theta_M\}$, respectively. Identify θ_c as the average of these direction.

Step 3.

Shift the source direction in each snapshot to θ_c, using (11). Denote the transformed snapshots by $\{\underline{x}_1,\underline{x}_2,\cdots,\underline{x}_M\}$.

Step 4.

Calculate an estimate of the correlation matrix $\underline{R}_x$ as

$$\hat{\underline{R}}_x = \frac{1}{M} \sum_{m=1}^{M} \underline{x}_m \underline{x}_m^H . \tag{12}$$

Step 5.

Find $\underline{\hat{\phi}}$ which maximizes the $\mathcal{P}(\underline{\phi},\hat{\underline{R}}_x)$ given by (2), (3), (7), (8), and (12).

A few notes are in order. Firstly, a calibration source is used, because a stronger signal, equivalent to a larger $|s_c|$, makes the product $|s_c|^2G(\underline{\phi},\underline{R}_y,\underline{\hat{\phi}})$ in (9) change more rapidly with $\underline{\hat{\phi}}$. Secondly, the transformation in Step 3 incorporates an attempt to make $\underline{R}_y$ behave like a white noise correlation matrix by making each interference source appear to move over a range of directions. Thirdly, using a moving source reduces the correlation between the calibration signal and the specular multipath interference, if present, in $\underline{R}_x$. The reduction would be more effective if the source moves over a larger range of directions. Fourthly, the angular separation between the calibration signal and specular multipath interference should be as large as practicably possible, so that reasonably accurate estimates of the source directions can be obtained in Step 2.

3 Performance Evaluations

The approach to locate the global maximum of $\mathcal{P}(\underline{\hat{\phi}},\hat{\underline{R}}_x)$, as a function of $\underline{\hat{\phi}}$, is described. Two computer simulations of element phase estimation are then given.

3.A Location of Global Maximum

The results of a preliminary study indicate that $\mathcal{P}(\underline{\hat{\phi}},\underline{R}_x)$ has a large number of local maxima in the neighbourhood of the true $\underline{\phi}$. The global maximum is usually not the one closest to $\underline{\phi}$, but its distance from $\underline{\phi}$ generally decreases when signal to noise ratio (SNR) or signal to noise-plus-interference ratio (SNIR) increases. It also appears that many of the local maxima are false peaks introduced by the finite precision in computer calculations.

The technique to locate the global maximum here is hill-climbing with random searches near each local maximum for a position with a larger value. There is no guarantee that this would result in the location of the global maximum. The problem is addressed here by calculating a number of estimates, using different initial values of $\underline{\hat{\phi}}$. If this number is large, it is reasonable to identify the estimate with the largest power as the position of the global maximum.

Let $\{\underline{\hat{\phi}}_k,\mathcal{P}(\underline{\hat{\phi}}_k,\hat{\underline{R}}_x);\ k=1,2,\cdots,K\}$ be K sets of element phase estimates and corresponding maximum values obtained by using the $\hat{\underline{R}}_x$ given by (12) and K different initial values of $\underline{\hat{\phi}}_k$. Let kbig be the index of the pair with the largest maximum. The maximum power at $\underline{\hat{\phi}}_k$, relative to the largest maximum, is

$$\rho_k = \frac{\mathcal{P}(\underline{\hat{\phi}}_k,\hat{\underline{R}}_x)}{\mathcal{P}(\underline{\hat{\phi}}_{kbig},\hat{\underline{R}}_x)} , \tag{13}$$

the deviation of $\underline{\hat{\phi}}_k$ from the true value is

$$\underline{\delta}_k = \underline{\hat{\phi}}_k - \underline{\phi} , \tag{14}$$

and the root mean square (RMS) deviation, or RMS error, for $\underline{\hat{\phi}}_k$ is

$$\epsilon_k = \frac{|\underline{\delta}_k|}{\sqrt{N}} . \tag{15}$$

When (ϵ_1,ρ_1) to (ϵ_K,ρ_K) are plotted in a scatter diagram, the point where $\rho=1$ represents the pair $(\epsilon_{kbig},\rho_{kbig})$, and the number of other points close to this point is a useful parameter to check if K is large enough.

3.B Examples

Two computer simulations of element phase estimation are presented. In each example, the array is uniform, linear, and has eight (N=8) isotropic elements. The N true phases comprising the components of the true $\underline{\phi}$ are generated as samples of a Gaussian variate with zero mean and eight degrees standard deviation. Their values are the same in both examples and are given in Column 2 of Table 1. Four hundred (M=400) array snapshots are generated while the calibration signal direction changes from θ_1 to θ_M. The signal directions are estimated from the individual array snapshots, using the root modified covariance method [6] and a third order filter. No smoothing to improve their accuracies is made. The matrix $\hat{\underline{R}}_x$, given by (12), is used to calculate five hundred (K=500) pairs of $\{\underline{\hat{\phi}}_k, \mathcal{P}(\underline{\hat{\phi}}_k, \underline{R}_x)\}$. In each pair, the components of $\underline{\hat{\phi}}_k$ are initiated as samples of a random variate uniformly distributed in the range [-180,180) degrees.

Example 1.

In this simulation the number of interfering signals is larger than the number of array elements. Specular multipath interference is absent. The array is shown in Fig. 1. Noise is white. The signal scenario is shown in Fig. 2. The calibration signal has a 30 dB SNR at the array elements. It moves in direction from -0.5 to 0.5 beamwidth during array snapshot measurement and is represented by a horizontal line with arrow heads at the ends. Ten 10 dB interfering signals are presented. They are uncorrelated with each other and with the calibration signal.

Direction in beamwidth is defined as

$$\theta_{BW} = N\,d\sin\theta\ . \tag{16}$$

where d is the element spacing in wavelengths, and θ is corresponding direction in degrees.

In the results, the ten largest maxima are separated from each other by a Euclidean distance of less than 0.30 degree. The dependence of RMS phase error ϵ on relative power ρ is plotted in Fig. 3. The selected $\underline{\hat{\phi}}$, i.e. $\underline{\hat{\phi}}_{kbig}$, is shown in Column 3 of Table 1. Its deviation from the true $\underline{\phi}$ in Column 2 is the $\underline{\delta}$ in Column 4. This deviation is decomposed into a linear component $\underline{\delta}_L$ in Column 5 and a random component $\underline{\delta}'$ in Column 6, using the least squares linear curve fitting method. The linear component decreases in steps of 0.3224 degrees down the table, indicating that the element phase estimates would produce a rotational bias of 0.10 degree in signal direction estimates. If a reference signal is used to remove this bias, the new estimate is

$$\underline{\hat{\phi}}' = \underline{\hat{\phi}} - \underline{\delta}_L \tag{17}$$

and the error in this new estimate is the random component in Column 6, i.e.

$$\underline{\delta}' = \underline{\hat{\phi}}' - \underline{\phi}\ . \tag{18}$$

The reference signal reduces the RMS phase error from $\epsilon = |\underline{\delta}|/\sqrt{N} = 0.76$ degree to $\epsilon' = |\underline{\delta}'|/\sqrt{N} = 0.17$ degree. In Fig. 4 are the new results after the rotational bias in each $\underline{\hat{\phi}}_k$ has been removed. Here, the RMS phase error ϵ' is between 0.16 and 0.25 degrees; the average is 0.19 degree; and the value at the largest maximum is 0.17 degree.

The rotational bias is mainly caused by the errors in estimating the source directions in Step 2 of the element phase estimation procedure.

Example 2.

In this simulation, a strong specular multipath interference is present. Other interference are absent. The array is a radar antenna operating at a frequency of 10 GHz. It has eight isotropic elements spaced at a vertical distance of 12.5 cm, or 4.17 wavelengths. The experimental setup is shown in Fig. 5. The array is centred at a height of 15 m. The calibration source is mounted on a tower 500 m in front of the array. It moves up in height in equal steps from 10 to 25 m while the four hundred array snapshots are taken. A flat earth model is used. The coefficient of reflection at the surface is -1, corresponding to 0 dB attenuation and 180 degrees phase change. Noise is white, and SNR is 30 dB.

With rotational biases removed, the ten largest maxima are separated from each other by a Euclidean distance of 0.18 degree or less. The dependence of RMS phase error ϵ' on relative power ρ is plotted in Fig. 6. Apparently, $\mathcal{P}(\underline{\hat{\phi}}, \hat{\underline{R}}_x)$ has two ridges converging at the global maximum. Most of the local maxima found by the hill-climbing method are on these ridges. The RMS error ϵ' at the largest maximum is 0.10 degree.

In this example, the specular multipath interference is just as strong as the direct signal at the array elements; SNIR is close to 0 dB; θ_1=-0.38 BW; and θ_M=0.76 BW. As the calibration source moves up in height from 10 to 25 m, the angular separation between the direct and multipath signals increases from 1.52 to 3.81 BW and the phase difference between these signals at the array centre increases from 7020 to 17820 degrees for a total of 10800 degrees.

4 Conclusions

A method which uses a movable source to estimate the element phases of a sensor array has been described. Computer simulation using a uniform linear array with eight isotropic elements indicate that RMS phase errors less than 0.2 degree are achievable.

The method assumes that mutual coupling among the array elements is absent. Future development work should include the removal of this assumption.

References

1. R.O. Schmidt, "Multiple Emitter Location and Signal Parameter Estimation", IEEE Trans. Antennas and Progag., vol. AP-34, pp. 276-280, March 1986.
2. A. Paulraj and T. Kailath, "Direction of Arrival Estimation by Eigenstructure Methods with Unknown Sensor Gain and Phase", ICASSP 1985, Tampa, U.S.A., pp. 640-643, 1985.
3. A.J. Weiss and B. Frielander, "Eigenstructure Methods for Direction Finding with Sensor Gain and Phase Uncertainties", Circuits and Systems and Signal Processing, Vol. 9, No. 3, pp. 271-300, 1990.
4. J. Capon, "High-Resolution Frequency-Wavenumber Spectrum Analysis", Proceedings of the IEEE, vol. 57, pp. 1408-1418, August 1969.
5. R.T. Lacoss, "Data Adaptive Spectral Analysis Methods", Geophysics, Vol. 36, pp. 661-675, August 1971.
6. S.L. Marple, Jr., Digital Spectral Analysis With Applications, Prentice-Hall, Englewood Cliffs, New Jersey 07632, U.S.A., 1987.

	ELEMENT PHASE (DEG)				
N	TRUE ϕ	ESTIMATE $\hat{\phi}$	ERROR δ	LINEAR δ_L	RANDOM δ'
1	−6.81	−5.50	1.31	1.13	0.18
2	−1.94	−1.33	0.61	0.81	−0.20
3	7.47	7.96	0.49	0.48	0.01
4	2.74	2.95	0.21	0.16	0.05
5	−7.09	−7.17	−0.08	−0.16	0.08
6	7.21	6.50	−0.71	−0.48	−0.22
7	3.30	2.34	−0.96	−0.81	−0.16
8	−4.87	−5.75	−0.88	−1.13	0.25
			ε=0.76		ε'=0.17

TABLE 1: The true $\underline{\phi}$, estimate $\hat{\underline{\phi}}$, and deviation $\underline{\delta}$ of $\hat{\underline{\phi}}$ from $\underline{\phi}$ in Example 1. The errors in estimates are decomposed into a sum of a linear component $\underline{\delta}_L$ and a random component $\underline{\delta}'$. The RMS errors derived from $\underline{\delta}$ and $\underline{\delta}'$ are ϵ and ϵ', respectively.

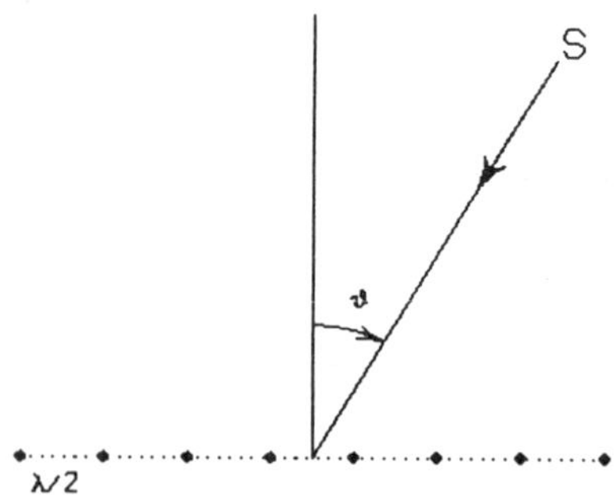

Fig.1 Array configuration in Example 1.

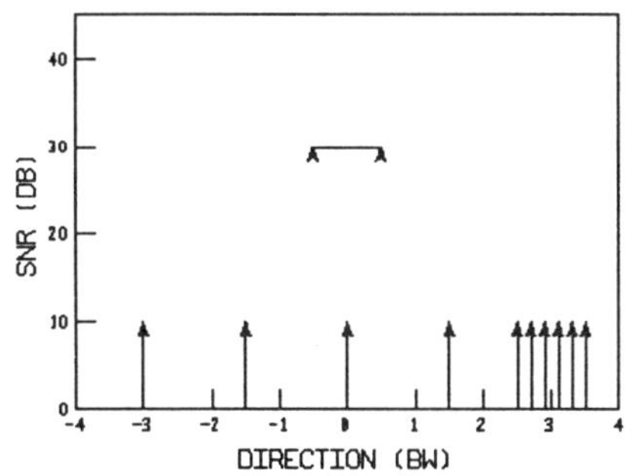

Fig.2 Pilot signal and interferences in Example 1.

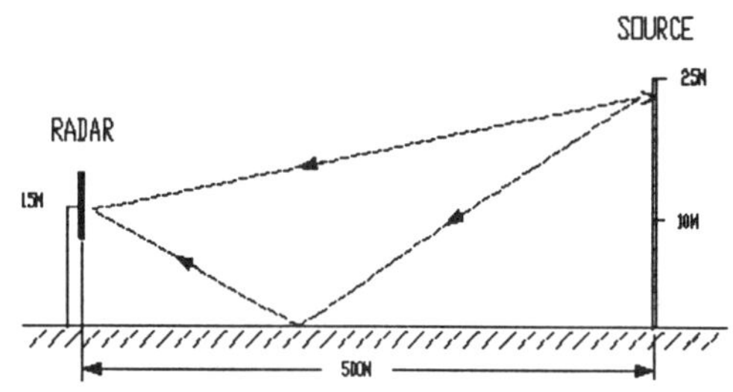

Fig.5 Experimental setup for Example 2.

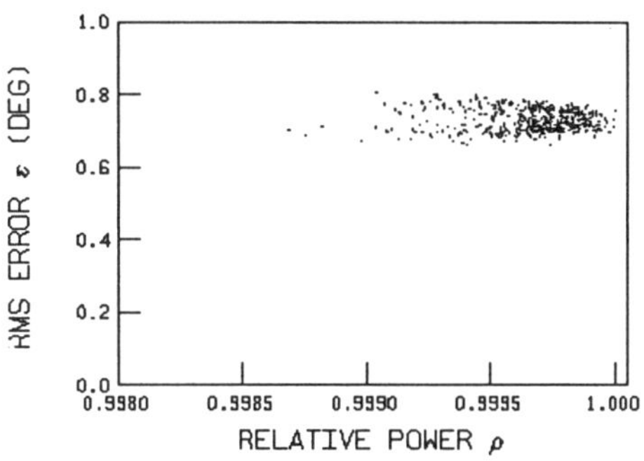

Fig.3 Dependence of RMS phase error ε on relative maximum power ρ in Example 1.

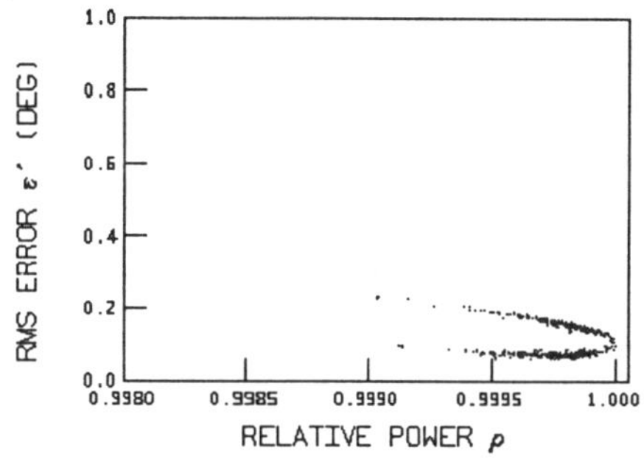

Fig.6 Dependence of RMS phase error ε' on relative maximum power ρ in Example 2. Rotational bias is removed.

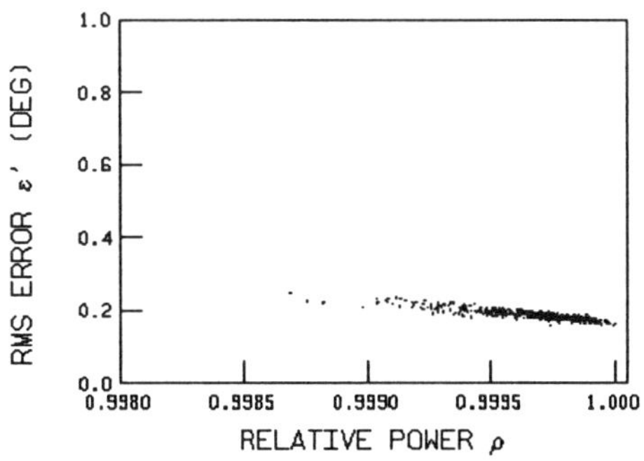

Fig.4 Dependence of RMS phase error ε' on relative maximum power ρ in Example 1. Rotational bias is removed

On Determining the Radar Threshold from Experimental Data

Prakash R. Chakravarthi
Donald D. Weiner
Deptt. of Electrical and Computer Engg.
Syracuse University
Syracuse, NY 13244-1240

Aydin Ozturk
Deptt. of Computer Engg.
Ege University
Izmir, Turkey

Abstract

In order to set the radar threshold for small false alarm probabilities, it is necessary to know the tail of the probability density function for the test statistic under the "no target" assumption. In this paper, it is shown that the Generalized Pareto Distribution (GPD) can be used to approximate the extreme tail of the density function. As a result, fixing the threshold is equivalent to estimating the two parameters of the GPD. We show for a variety of probability density functions (pdf's) that accurate results can be obtained with orders of magnitude fewer samples than are required by Monte Carlo simulation. [1]

1 Introduction

It is well known [6] that the optimum radar receiver performs a likelihood ratio test (LRT) when deciding between the two hypotheses

$$H_1 : r_i = s_i + c_i + n_i \quad (1)$$
$$H_0 : r_i = c_i + n_i \quad i = 1, 2...N \quad (2)$$

where N pulses are being processed and r_i,s_i,c_i and n_i respectively denote the total received signal, the target return, the clutter and background noise for the ith received pulse. Let L denote the sufficient statistic of a Likelihood Ratio Test. This paper develops a new approach for experimentally determining the extreme tail of the pdf of L under the *no signal* hypothesis, $f_L(l|H_0)$ where the number of samples is several orders of magnitude smaller than required by the Monte Carlo simulation. Once the tail of $f_L(l|H_0)$ has been estimated, the threshold η is determined by use of $P_F = \int_\eta^\infty f_L(l|H_0)dl$, where P_F is the false alarm probability. Note that the threshold can be related to a percentile of the distribution on L. Let $F_L(l|H_0)$ denote the conditional probability distribution of L given H_0. It follows that

$$F_L(\eta|H_0) = 1 - P_F. \quad (3)$$

By definition the $100 \times p^{th}$ percentile of the random variable L under H_0 is the number $\eta_{100\times p}$ such that

$$F_L(\eta_{100\times p}|H_0) = p. \quad (4)$$

With reference to equation (3), it can be seen that the $100 \times (1 - P_F)^{th}$ percentile of the random variable L is the threshold η. In the following work, it is convenient to express results in terms of the appropriate percentiles. Since it is desired to determine the thresholds under H_0, estimation of the tail of $f_L(l|H_0)$ enables us to estimate large percentiles of L under H_0 where 1-P_F is close to unity.

Based on a concept first suggested by Dumouchel [2], a new estimation procedure is proposed in this paper to estimate the extreme tail. On the basis of numerical results, it is shown that the proposed procedure can effectively be used to estimate the large percentiles in both the upper and lower tails.

2 Estimation of percentiles

Estimation of percentiles has been studied extensively. Many of the proposed percentile estimates assume a known distribution from which the sample is drawn, while others are based on non-parametric methods. Also, a number of procedures have been suggested for the estimation of large percentiles, using the extreme value theory (EVT). Extreme value distributions are obtained as limiting distributions of largest (or smallest) values of sample order statistics. If $X_1 \leq X_2 \leq ... \leq X_n$ are order statistics from a common distribution function F(x), then the cumulative distribution function of the largest order statistic is given by

$$G(x) = [F(x)]^n. \quad (5)$$

It is clear, as $n \to \infty$, that the limiting value of G(x) approaches zero if F(x) is less than 1 and unity if F(x) is equal to 1 for a fixed value of X. A standardized limiting distribution of X_n may be obtained by introducing the linear transformation, $a_n X_n + b_n$, where a_n and b_n are constants depending on the sample size n.

Suppose there exist such sequences a_n and b_n for all n and a distribution function Λ such that

$$lim_{n\to\infty} P[(X_n - b_n)/a_n \leq x] = lim_{n\to\infty} [F(a_n x + b_n)]^n = \Lambda(x) \quad (6)$$

[1] This work was supported by the Rome Laboratory, U.S. Air Force, through a subcontract with Kaman Science Corporation under contract number SCEE-DACS-P35498-2. Dr. Weiner participated on this effort as a member of Rome Laboratory through the AFOSR University Resident Research Program.

1058-6393/91 $01.00 © 1991 IEEE

where $\Lambda(x)$ is a continuous function of x. It is said that $\Lambda(x)$ is an extreme value distribution.

The asymptotic behavior of the largest (or smallest) sample order statistics has been studied extensively [1]. Gnendenko [3] established an important correspondence between the parent distribution F(x) and the type of the limiting distribution $\Lambda(x)$. Specifically, if F(x) is such that a limiting distribution exists, then this limiting distribution must be one of the following types:

$$Type\ 1: \ \Lambda(x) = exp(-e^{-x}) \quad -\infty < x < \infty \tag{7}$$

$$Type\ 2: \ \Lambda(x) = exp(-x^{-\alpha}) \ x > 0, \alpha > 0 \tag{8}$$

$$Type\ 3: \ \Lambda(x) = exp(-(-x^{-\alpha})) \ x < 0, \alpha > . \tag{9}$$

The above results are based on the largest order statistic. The corresponding distributions for the smallest order statistic can be obtained by replacing -x for x in the above equations.

3 Pareto approximation to the tails

In this section we use the Generalised Pareto Distribution (GPD), which has two parameters, to approximate the tails of the probability density function. The two parameters are estimated using an ordered sample least squares (OSLS) approach.

3.1 The generalized Pareto distribution

The GPD is defined for $x > 0$ by the distribution function

$$H(x) = 1 - (1 + \gamma x/\sigma)^{-1/\gamma}, \\ -\infty < \gamma < \infty, \sigma > 0, \gamma x > -\sigma. \tag{10}$$

This distribution has a simple closed form and includes a range of distributions depending upon the choice of γ and σ. For example the exponential distribution results for $\gamma = 0$ and the uniform distribution is obtained when $\gamma = 1$. Pickands [5] showed that this distribution can be obtained as a limiting distribution if and only if the parent distribution is in the domain of attraction of one of the distributions in equations (7-9).

3.2 Estimation methods

Suppose that the sample ordered statistics $X_1 \leq X_2 \leq ... \leq X_n$ are drawn from the distribution function F(x). To estimate the right tail of this distribution it is necessary to determine a value (say x_0) and then use those sample observations which are greater than x_0 to obtain the quantity $z = x - x_0$. If the left tail is to be estimated then we choose x_0 in the left tail and use those observations which are less than x_0 to obtain $z = x_0 - x$. Once the tail observations are chosen, then the GPD can be fitted to these observations by using the standard methods of parameter estimation. A difficulty of this procedure which makes use of only a small portion of the observations in a complete set of samples, is that the choice of x_0 is subjective. One approach to this problem is to make a histogram of the data set and choose x_0 to be near the point of inflection of the histogram on the side of interest. DuMouchel proposed to choose x_0 to be the 90-th percentile of the sample if the right tail is considered or the 10th percentile of the sample if the left tail is considered. Such an approach is less subjective and appears to be satisfactory for many applications. However, it is noted by DuMouchel that " using an even smaller fraction of observations would restrict profitable use of the statistic to much larger sizes. On the other hand, to use more than the upper one tenth of a sample would seem to allow too much dependence on the central part of the distribution."

Let x_0 be chosen as the $100 \times (1-\alpha)^{th}$ percentile of F(x). Now suppose that the right tail of the distribution is to be estimated on the basis of the sample order statistics $X_{n-m+1} \leq X_{n-m+2} \leq ... \leq X_n$ where $m = [\alpha n]$(the square brackets denote the integer part of αn) and $\alpha = 1 - F(x_0)$. The approximated distribution function to be used in estimating the tail can be written as

$$\tilde{F}(x) = (1-\alpha) + \alpha H(x - x_0) \\ = 1 - \alpha(1 + \gamma(x - x_0)/\sigma)^{-1/\gamma} \ x > 0 \tag{11}$$

where H(x) is given in equation (10). When the tail of a given distribution can be approximated by equation (11), then the estimation problem of the distribution in the tail region is reduced to estimation of the parameters of the GPD.

The estimation methods for the GPD have been reviewed by Hosking and Wallis [4]. They considered the method of moments, probability weighted moments and the maximum likelihood method for estimating the parameters and the percentiles. Based on computer simulation experiments, they showed that the probability weighted moment method is more reliable than the method of maximum likelihood for sample sizes less than 500.

3.3 The Ordered sample Least squares method - A new approach

In this study, based on working with many examples, we have found that a least squares method which minimizes the squared distance between the ordered sample and the expected value of the ordered sample is more reliable than the probability weighted moments method.

For the GPD the mean and the variance of the rth order statistics are given by

$$E(X_r) = \mu_r = \frac{\sigma}{\gamma}\{Q_r(\gamma) - 1\} \tag{12}$$

$$Var(X_r) = \sigma_r^2 = \frac{\sigma^2}{\gamma^2}\{Q_r(2\gamma) - (Q_r(\gamma))^2\} \tag{13}$$

where $Q_r(\gamma) = \Pi_{i=1}^{r}(m-i+1)/(m-i+1-\gamma)$.

To find the least squares estimates of the parameters we write the following non-linear model for the rth sample order statistic

$$X_r = E(X_r) + e_r, \quad r = 1, 2, ..., m \tag{14}$$

where the error term e_r has a distribution with mean 0 and variance σ_r^2. Since the order statistics are not independent, the errors also are not independent. Because of the non-linear structure of the model in equation (14) and correlated errors, least squares estimation does not offer a straightforward solution to the estimation problem. Even so, in this study we proceed to use the ordered sample least squares (OSLS) procedure to estimate the parameters.

In equation (12), we note that the scale parameter σ appears linearly whereas the shape parameter γ does not. The least squares estimates are obtained by minimizing the quantity

$$S \doteq \sum_{r=1}^{m} e_r^2 = \sum_{r=1}^{m} (X_r - \sigma(Q_r(\gamma) - 1)/\gamma)^2 \qquad (15)$$

Minimization can be achieved by first minimizing S with respect to σ, then substituting the corresponding value of $\hat{\sigma}$ into equation (15) and minimizing the resulting function with respect to γ. Recall that the GPD is being used to approximate the tail of the underlying distribution. Hence, the ordered statistics X_r, $r = 1, 2, ..., m$, from the GPD actually correspond to the ordered statistics $X_{n-m+1}, X_{n-m+2}...X_n$ from the underlying distribution.

The least squares procedure results in a computationally convenient algorithm. It is noted that the minimization of S is carried out only with respect to the single parameter γ. Furthermore, the underlying criterion is based on minimizing the distance between the empirical and theoretical percentiles. Thus an improvement over the maximum likelihood estimates is expected in terms of the bias. Some numerical comparisons are given in the following section.

3.4 Estimation of percentiles

It should be noted that the GPD distribution estimated from the data is used to approximate the tail of the parent distribution. With reference to equation (11), let x be the $100 \times p^{th}$ percentile of the distribution being estimated. Solution for x results in (denoting x by $\tilde{\eta}_{100\times p}$)

$$\tilde{\eta}_{100\times p} = x_0 + \sigma(q^{-\gamma} - 1)/\gamma \qquad (16)$$

where $\alpha = 1 - F(x_0)$, $q = (1-p)/\alpha$ and $x_0 = F^{-1}(1-\alpha)$. For many applications DuMouchel suggests that $\alpha = 0.1$ be used.

Since the distribution function F(x) is not known, x_0 cannot be determined for a given value of α. Therefore, for a specified value of α, the sample order statistic X_{n-m} where $m = [\alpha n]$ is used as an estimate of x_0. Constructing approximate confidence intervals for the percentiles based on the sample estimators are discussed by Hosking and Wallis [4]. The same approach can be applied in our case to obtain approximate confidence intervals for the percentiles.

4 Numerical results

4.1 Characterization of tail shape

Given a known distribution, DuMouchel proposed a method utilizing maximum likelihood estimation, to estimate the parameters of the GPD. Choosing x_0 to be the 90th percentile of the sample, he then defined the percentiles x_i i=1,2,...1000 associated with $100 \times p_i$ by

$$100 \times p_i = 90.005 + 0.01(i - 1). \qquad (17)$$

The values of x_i were evaluated analytically from the known distribution. The maximum likelihood estimation, OSLS and the probability weighted moments method were then used on the $x_i s$ to estimate the parameters σ and γ. The results are given in Table 1. Since σ is a scale parameter, the shape parameter γ best describes the tail shape. For the Exponential and the Uniform distributions the value of γ can be obtained theoretically. $\gamma = 0$ for the Exponential distribution and is -1 for the Uniform distribution. Since the size of the tail decreases with decreasing γ, the relationship between the tail behavior and the corresponding values of the shape parameter γ can be clearly observed in this table.

Distribution	OSLS	ML	PWM
Gaussian	-0.144	-0.151	-0.174
Weibull(3)	-0.163	-0.168	-0.194
Weibull(.67)	0.108	0.129	0.137
Weibull(.5)	0.201	0.265	0.263
Student-t(3)	0.290	0.260	0.261
Student-t(5)	0.132	0.099	0.090
Student-t(8)	0.031	0.006	-0.010
Lognormal(1)	0.232	0.259	0.258
Chi-square(1)	0.030	0.034	0.044
Chi-square(4)	-0.024	-0.033	-0.034
Chi-square(8)	-0.047	-0.058	-0.064

Table 1: Tail parameter γ describing the upper ten percent of various distributions.

4.2 Empirical properties of the estimators

Seven distributions with widely differing tail behaviors were chosen in order to investigate the adequacy of the approximation of extreme tails by the GPD and to compare the three estimation procedures. The Gamma distribution and Weibull distribution with shape parameter of value 3, have tails lighter than those of the exponential pdf. The tails of the Chi-square and t distributions with degrees of freedom 4 and 8, respectively, are approximately the same as those of the exponential pdf. Finally, the t distribution with 4 degrees of freedom and the Lognormal distribution with shape parameter of value 1 have tails heavier than those of the exponential pdf.

n=25

P_F		10^{-3}	10^{-4}	10^{-5}	10^{-6}
Normal	OSLS	0.0043	-0.0040	-0.0276	-0.0571
Normal	ML	0.0187	-0.0328	0.0358	0.0281
Normal	PWM	-0.0208	-0.0560	-0.1015	-0.1464
Weibull(3)	OSLS	0.0013	-0.0041	-0.0202	-0.0418
Weibull(3)	ML	0.0481	0.0938	0.1374	0.1776
Weibull(3)	PWM	-0.0106	-0.0333	-0.0635	-0.0919
t(4)	OSLS	-0.0792	-0.1658	-0.2727	-0.3872
t(4)	ML	-0.1115	-0.2526	-0.4045	-0.5416
t(4)	PWM	-0.0929	-0.2160	-0.3498	-0.4761
t(8)	OSLS	-0.0186	-0.0572	-0.1164	-0.1975
t(8)	ML	-0.0468	-0.1169	-0.2077	-0.3055
t(8)	PWM	-0.0452	-0.1095	-0.2039	-0.3063
Chi-sq(4)	OSLS	-0.0039	0.0241	0.0333	-0.0088
Chi-sq(4)	ML	0.0943	0.2518	0.4571	0.6185
Chi-sq(4)	PWM	-0.0205	-0.0334	-0.0576	-0.1254
Lognormal	OSLS	-0.0982	-0.0634	0.0016	0.1007
Lognormal	ML	0.1836	0.5932	1.2736	2.4832
Lognormal	PWM	-0.0878	-0.0931	-0.0728	-0.0228
Pareto(-0.25)	OSLS	0.0208	0.0423	0.0631	0.0780
Pareto(-0.25)	ML	0.0523	0.1190	0.1868	0.2479
Pareto(-0.25)	PWM	0.0052	0.0121	0.0199	0.0237

n=50

P_F		10^{-3}	10^{-4}	10^{-5}	10^{-6}
Normal	OSLS	0.0073	-0.0068	-0.0354	-0.0676
Normal	ML	0.0323	0.0497	0.0578	0.0528
Normal	PWM	-0.0118	-0.0459	-0.0861	-0.1318
Weibull(3)	OSLS	-0.0007	-0.0133	-0.0337	-0.0571
Weibull(3)	ML	0.0500	0.0991	0.1436	0.1847
Weibull(3)	PWM	-0.0105	-0.0342	-0.0629	-0.0937
t(4)	OSLS	-0.0646	-0.1800	-0.3209	-0.4501
t(4)	ML	-0.0867	-0.2264	-0.3736	-0.5120
t(4)	PWM	-0.0622	-0.1662	-0.2973	-0.4233
t(8)	OSLS	-0.0222	-0.0841	-0.1723	-0.2694
t(8)	ML	-0.0502	-0.1352	-0.2385	-0.3460
t(8)	PWM	-0.0336	-0.1064	-0.2041	-0.3051
Chi-sq(4)	OSLS	-0.0004	0.0051	0.0060	-0.0498
Chi-sq(4)	ML	0.1134	0.2755	0.4775	0.6368
Chi-sq(4)	PWM	-0.0087	-0.0191	-0.0407	-0.1123
Lognormal	OSLS	-0.0594	-0.0272	0.0458	0.1573
Lognormal	ML	0.2177	0.6336	1.3811	2.6197
Lognormal	PWM	-0.0391	-0.0185	0.0413	0.1480
Pareto(-0.25)	OSLS	0.0100	0.0214	0.0326	0.0404
Pareto(-0.25)	ML	0.0463	0.1011	0.1560	0.2003
Pareto(-0.25)	PWM	-0.0018	-0.0012	-0.0019	-0.0023

n=100

P_F		10^{-3}	10^{-4}	10^{-5}	10^{-6}
Normal	OSLS	-0.0016	-0.0253	-0.0637	-0.1040
Normal	ML	0.0263	0.0306	0.0229	0.0063
Normal	PWM	-0.0181	-0.0549	-0.1022	-0.1524
Weibull(3)	OSLS	-0.0017	-0.0164	-0.0376	-0.0624
Weibull(3)	ML	0.0270	0.0564	0.0840	0.1003
Weibull(3)	PWM	-0.0095	-0.0320	-0.0607	-0.0918
t(4)	OSLS	-0.0441	-0.1421	-0.2680	-0.3922
t(4)	ML	-0.0564	-0.1650	-0.2907	-0.4174
t(4)	PWM	-0.0478	-0.1403	-0.2636	-0.3809
t(8)	OSLS	-0.0134	-0.0751	-0.1606	-0.2578
t(8)	ML	-0.0342	-0.1145	-0.2123	-0.3157
t(8)	PWM	-0.0271	-0.0955	-0.1888	-0.2892
Chi-sq(4)	OSLS	-0.0028	-0.0077	-0.0198	-0.0841
Chi-sq(4)	ML	0.1189	0.2655	0.4581	0.5917
Chi-sq(4)	PWM	-0.0089	-0.0238	-0.0448	-0.1143
Lognormal	OSLS	-0.0542	-0.0876	-0.1089	-0.0940
Lognormal	ML	-0.0251	-0.0068	0.0536	0.1499
Lognormal	PWM	-0.0210	0.0141	0.0924	0.2315
Pareto(-0.25)	OSLS	0.0109	0.0255	0.0350	0.0419
Pareto(-0.25)	ML	0.0544	0.1170	0.1739	0.2215
Pareto(-0.25)	PWM	0.0004	0.0052	0.0084	0.0112

n=1000

P_F		10^{-3}	10^{-4}	10^{-5}	10^{-6}
Normal	OSLS	-0.0013	-0.0244	-0.0613	-0.1010
Normal	ML	0.0017	-0.0259	-0.0626	-0.1075
Normal	PWM	-0.0192	-0.0586	-0.1064	-0.1560
Weibull(3)	OSLS	-0.0023	-0.0175	-0.0381	-0.0627
Weibull(3)	ML	-0.0018	-0.0159	-0.0386	-0.0641
Weibull(3)	PWM	-0.0092	-0.0297	-0.0578	-0.0880
t(4)	OSLS	-0.0044	-0.0605	-0.1574	-0.2690
t(4)	ML	-0.0137	-0.1018	-0.2167	-0.3326
t(4)	PWM	-0.0176	-0.1104	-0.2277	-0.3479
t(8)	OSLS	-0.0021	-0.0452	-0.1167	-0.2001
t(8)	ML	-0.0117	-0.0664	-0.1464	-0.2404
t(8)	PWM	-0.0219	-0.0896	-0.1825	-0.2857
Chi-sq(4)	OSLS	0.0012	-0.0057	-0.0167	-0.0826
Chi-sq(4)	ML	-0.0021	-0.0152	-0.0354	-0.1026
Chi-sq(4)	PWM	-0.0011	-0.0080	-0.0263	-0.0934
Lognormal	OSLS	-0.0221	-0.0259	0.0055	0.0646
Lognormal	ML	0.0063	0.0616	0.1767	0.3495
Lognormal	PWM	-0.0004	0.0567	0.1683	0.3400
Pareto(-0.25)	OSLS	0.0002	0.0012	0.0007	0.0003
Pareto(-0.25)	ML	-0.0010	-0.0044	-0.0061	-0.0081
Pareto(-0.25)	PWM	-0.0011	-0.0007	-0.0006	-0.0035

Table 2: Median of the normalized bias values for different false alarm probabilities.

A Monte Carlo experiment was performed to investigate the bias of the proposed percentile estimates. Four sample sizes $m = 25, 50, 100$ and 1000 were considered. Each set of samples was obtained by generating n observations and taking the largest $m = 0.1n$ observations. For example, a set of samples of size 25 was obtained by selecting the largest 25 observations from a collection of size 250 samples. Let η and $\hat{\eta}$ denote the true and estimated percentiles respectively (i.e. the thresholds). For all four different values of m, r=200,000/m trials were performed for each of the seven distributions. The median of the normalised bias values $(\hat{\eta} - \eta)/\eta$ were computed for each distribution and estimation procedure. The results for $P_F = 10^{-k}$, k=3,4,5,6 are given in Table 2. Extremely poor estimates for η were obtained in some of the trials. These poor estimates could severely influence an arithmetic mean of the estimates. To avoid this problem, median values were used in place of the arithmetic means.

The empirical results in Table 2 indicate that the newly proposed ordered sample least squares estimator generally has a smaller normalised bias than the other estimators for small or moderate sample sizes. Overall the second smallest normalised bias is achieved by the probability weighted moments method. The maximum likelihood estimator has the largest normalised bias when $P_F \geq 10^{-5}$, especially for the long tailed distributions. The normalised bias of all three estimators decrease as the sample size increases. When the parent distribution is GPD,all three estimators perform very well.

In terms of the normalised bias, the ordered samples least squares estimator appears to perform better over the other estimators in estimating the large percentiles when $P_F \leq 10^{-6}$. Empirical results were also obtained for the normalised mean square errors. However, there is no clear winner among the three methods with respect to this criterion. In any event, it is seen that the EVT can be used successfully to determine threshold values, when the false alarm probability is very small.

Two practical advantages of estimation based on EVT are: 1) When there is a constraint on the number of samples, the thresholds obtained from EVT are theoretically expected to be closer to the true threholds than those obtained by conventional Monte Carlo techniques. However, in both techniques an increase in sample size offers greater accuracy in estimating percentiles. 2) Because the estimate of the tail of the underlying distribution is in closed form, estimation can be made for thresholds corresponding to extremely small false alarm probabilities independent of the sample size. In experiments with fixed amounts of data, this is an important advantage.

5 Examples

To evaluate the accuracy of the estimates for threshold values, 10000 samples were generated from the Gaussian and Lognormal distributions and the upper tail of these two distributions were modeled as generalized Pareto. Random variables from the two distributions were generated. Choosing α=0.01, the theoretical percentiles (thresholds) of the Gaussian distribution for $P_F = 10^{-k}$ $k = 2, 3...7$ are 2.326, 3.090, 3.719, 4.265, 4.753 and 5.199, respectively. The percentiles estimated are 2.315, 3.223, 3.847, 4.370, 4.855 and 5.292 . For the Lognormal distribution the theoretical percentiles (thresholds) corresponding to $P_F = 10^{-k}$ $k = 2, 3...7$ were 10.240, 21.982, 41.224, 71.157, 115.981 and 181.152. Once again, using α=0.01, the percentiles estimated are 10.449, 22.862, 42.473, 69.216, 112.229 and 183.495. Note that the estimated results are very close to the true thresholds. We must also note here that these results are obtained on the basis of one set of observations from the two known distributions, corresponding to a particular seed value. For a different set of samples the estimates will be different depending on the tail behavior of that set of samples. But, unless the samples are really not a true representative of the distribution from which they are drawn, we expect that the estimates based on different samples should give threshold values that yield false alarm probabilities of the same order.

6 Conclusion

In this paper the ordered sample least squares estimator is proposed for estimation of the generalized Pareto distribution which is used to approximate the extreme tails of a probability density function. Application of the technique to known distributions reveals that excellent results can be had with sample sizes that are orders of magnitude smaller than those required for conventional Monte Carlo techniques. In particular, the thresholds required for very low false alarm probabilities were obtained with a good deal of accuracy, as is shown in the tables and examples.

References

[1] David, H.A. (1970) "Order Statistics", pp 258 John Wiley and Sons, New York .

[2] DuMouchel, W.H. (1983) "Estimating the stable index α in order to measure tail thickness: A critique", *The Annals of Statistics ,vol 11, pp 1019-1031*

[3] Gnedenko, P.B. (1943) "Sur la Distribution Limited du Terme Maximum d'une Serie Aleatoire" *Annals of Mathematics, vol 44, pp 423-453*

[4] Hosking, J.R.M., Wallis, J.R. (1987) "Parameter and Quantile estimation of the Generalized Pareto Distribution" *Technometrics, Vol. 29, No. 3 pp 339-349*

[5] Pickands, J. (1975) "Statistical inference using extreme order statistics" *The Annals of Statistics, vol 3, pp 119-131*

[6] Vantrees, Harry "Detection, Estimation and Modulation Theory" Part I, John Wiley & Sons.

[7] Weissman, I. (1975) "Multivariate extremal process generated by independent non-identically distributed random variables." *Journal of Applied Probability, vol 12, pp 477-487*

AN UNDERWATER ACOUSTIC MODEM FOR THE VERTICAL SOUND CHANNEL UTILIZING A ZERO-CROSSING RECEIVER

John P. Fumo, Bretton L. Douglas
Department of Electrical and Computer Engineering
University of California, Santa Barbara
Sonatech Inc.
Santa Barbara, CA 93106

ABSTRACT

A zero-crossing based time-domain receiver in an underwater acoustic modem is presented for application in the vertical sound channel. Zero-crossing techniques can provide cost effective solutions to high data rate telemetry in the ocean environment. A comparison of the matched-filter receiver to a zero-crossing receiver is given for an ideal and a noisy environment. Results from computer simulations, laboratory breadboard tests, and an at-sea experiment are presented.

INTRODUCTION

Underwater acoustic telemetry is used in a variety of applications. Some of these include communication with bottom mounted instrumentation and remotely operated vehicles for use in surveys, oil exploration, minesweeping, and deep-sea recovery. In the past, a great deal of work has gone into optimizing a telemetry system by minimizing its bit error rate due to noise and doppler [1][2]. For many undersea applications, other factors such as the cost, maintenance, and power consumption are more important. In these situations the optimal system will be a low cost, easily installed system with a long lifetime and an acceptable bit error rate.

The typical receiver for a frequency shift keying (FSK) modem, is a matched filter followed by an envelope detector followed by a comparator. This receiver is optimum in that it minimizes the bit error probability for the given signal and noise, but matched filter receivers have some undesirable properties. The mixing and filtering circuits in analog matched filter receivers require tuning, which makes them difficult to manufacture, and they use a significant amount of power, which gives them a short life-span on the sea bottom. The digital matched filters require analog to digital converters and digital signal processors which both have a high power consumption.

An alternative receiver with a slightly higher bit error rate is a zero crossing based receiver. In this receiver, all the processing is done digitally using low-power MOS devices. The circuit requires very little tuning, so it is easy to build, easy to install, easy to maintain, and has a long bottom life.

This work addresses communication in the vertical sound channel where a direct path from source to receiver is available, rather than the highly reverberant shallow water channel. In the vertical sound channel the multipath problem is relatively minor. This is due to the large difference between the signal strength of the direct path and the signal strength of the multipath signals. The most serious problems due to multipath are surface reflections for a surface modem and bottom reflections for a near-bottom modem.

The purpose of the system is to allow transmission of data at 2400 baud between two modems that may be moving relative to one another. This relative motion presents several problems. First, if the two modems are moving towards each other, the frequency seen at the receiver will be higher due to doppler, and the bit stream will be received at a higher rate than it was transmitted. A good receiver must be able to withstand moderate frequency shifts and time-skewing. Second, each modem must be able to receive the other modem's signal as they move relative to each other, so the transducers must have a wide beam pattern.

ZERO-CROSSING BASED SYSTEM

The physical realization of the system is shown in Figure 1. The transducers were designed to provide a 45 degree beam pattern from center in order to allow relative motion between the two modems. The front-to-back ratio of the transducers was approximately 40 dB. This helped eliminate the most serious multipath problem, a single surface or bottom reflection.

The two carrier frequencies were moderately high, 51.546 kHz and 48.544 kHz. They were separated sufficiently in frequency to allow 2400 baud transmission. Several advantages were obtained by transmitting data in this frequency range. The attenuation of sound at these frequencies is relatively high, so the signal from multiple surface-to-bottom reflections will be very small compared to the direct path signal. This is also a very quiet frequency range in terms of ambient sea noise [3][4][5].

The FSK transmitter, shown in Figure 2, is comprised of a master oscillator, a data stream synchronizer and a power amplifier. The master oscillator is divided into two discrete frequencies which can be selected from the output of the universal asynchronous receiver transmitter (UART) in the transmitter. The power amplifier passes the fundamental frequency from the square wave output of the digital portion of the transmitter and amplifies the signal to appropriate levels for the projector.

The receiver, shown in Figure 3, is comprised of a hydrophone, a passive filter network, a hard-limiter, a master oscillator, a digital state machine, a period counter, an EPROM, and a UART. The transfer function of the hydrophone provides some bandpass filtering by virtue of its frequency response. The passive filter network provides additional filtering which keeps only those frequencies centered around the two carrier frequencies. The limiter converts the filtered analog signal into a digital PCM data stream, which is used to reset the counter at each negative to positive transition, or rising edge. This process enables the determination of the time interval between rising edges of the data stream. The counts from each time interval are latched at the address lines of an EPROM which has been programmed to make a decision about which bit was sent based on the number of counts in the last interval between rising edges. By properly selecting groups of addresses, A robust decision rule can be obtained which can provide clock-jitter rejection and doppler suppression.

The decisions are sent from the EPROM to the UART as a stream of 1's and 0's. The UART will use the data stream and reference clock to decide which frequency was sent for each bit. The messages are encoded with a software protocol to assist in message reconstruction. This protocol provides a synchronization packet to allow the software to automatically synchronize to the message stream. This type of message synchronization is similar to the auto-baud selection schemes used in telecommunication hardware.

ZERO-CROSSING ANALYSIS

The discrete-time received signal for a FSK system can be expressed by the following equation.

$$s_i(t_k) = \sin(\omega_i t_k + \phi) + n(t_k) \tag{1}$$

where t_k is a time sample, ω_i is the angular frequency associated with bit i, ϕ is a random phase term, and $n(t_k)$ is a noise term. The goal of an FSK receiver is to use the signal $s_i(t_k)$ to determine whether $\sin(\omega_0 t_k + \phi)$ or $\sin(\omega_1 t_k + \phi)$ was sent. Zero-crossing techniques were introduced in an attempt to simplify the hardware involved in determining which frequency was sent [6]. Most of the previous work in the analysis of zero-crossings was concerned with the number of zero-crossings in a given time interval for the received signal and filtered versions of the signal [7][8][9]. In this case we will look at the distribution of the intervals between negative to positive transitions rather than just counting the number of zero crossings in a given time frame. Figure 4 shows the difference in the two techniques.

The interval between negative to positive transitions can be treated as a random variable, L, whose probability density function depends on the frequency of the signal, the signal to noise ratio, and the spectral distribution of the noise. The relationship of the probability density function to the signal to noise ratio is shown in Figure 5 where the signal is a single sinusoid in white Gaussian noise. In this simulation, a histogram of interval lengths is used to approximate the probability density function of the time interval. When the SNR is high, most time intervals are approximately equal to $2\pi/\omega$ seconds, or $L = 2\pi/\omega t_s$ clock periods, where t_s is the sampling frequency. When the SNR is low, the time intervals will be determined by the noise. For a white-noise only received signal, the probability distribution for L can be expressed by the following formula

$$p(L) = \begin{cases} (L-1)\left(\frac{1}{2}\right)^L, & L \geq 2 \\ 0, & otherwise \end{cases} \tag{2}$$

For SNR's in between, peaks in the pdf of L occur near $L = 2\pi/\omega t_s$ and $L = \pi/\omega t_s$. These correspond to the time intervals associated with a full cycle and a half cycle of the sinusoid. In Figure 5, this type of pdf occurs at SNR's around 35 dB.

The pdf of the time interval can be changed by a passive filter such as the frequency response of a transducer or an LRC circuit. Figure 6 shows the results of bandpass filtering the single frequency sinusoid in white noise when the sinusoid is at one of the 3 dB points. The filtering has two effects. First it improves the signal-to noise ratio because it eliminates the out of band noise. It also "colors" the noise so that the pdf of the time intervals at low SNR's is very similar to the pdf at high SNR's. The only effect that the noise causes is a spreading of the pdf around the peak at $L = 2\pi/\omega t_s$.

DECISION RULE

In the time period that a single bit is being sent, several different time intervals between transitions will be available to make a decision about which frequency was sent. The hardware limits the complexity of the decision rule, because at any given time, only the last interval between transitions is available to make a decision. As a result, the best decision rule is a two step process. At each time instant, the EPROM must decide which signal has been sent based on the last interval length L. A lookup table can be used to make a maximum likelihood decision about which signal was sent using a single statistic L

First, for each frequency ω_i and SNR s, a pdf $p_i(L|s)$ can be estimated through experiments or simulations. Next the pdf of the signal to noise ratios $p(s)$ can be used to estimate $p_0(L)$ and $p_1(L)$. When $p_0(L)$ and $p_1(L)$ are known, a maximum likelihood decision rule can be determined and stored as a lookup table in an EPROM.

The results of the first decision will be a series of 1's and 0's coming from the EPROM. The UART must then decide whether a 1 or a 0 was sent. The UART is clocked at 64 times the bit rate, and it has a built-in voting mechanism which will choose a 0 if its input reads 0 for 32 or more of the 64 cycles, otherwise it will choose a 1. So the decision rule consists of two parts. First the EPROM makes a decision at each time instant based on the last interval between transitions. Second the UART counts the number of 0's it receives over the time interval that the bit was sent, then decides which signal was sent.

COMPUTER SIMULATIONS

The purpose of these simulations was to compare the bit error rate of the zero-crossing based system to that of the optimum matched filter FSK receiver. First $p_0(L|s)$ was estimated for a 48.544 kHz signal at 50 different signal to noise ratios ranging from –30 dB to +20 dB. Next an estimate of $p_0(L)$ was obtained by giving each SNR probability, $p(s)$, an equal weight. The process was repeated for a 51.546 kHz signal to estimate $p_1(L)$. A maximum likelihood estimator was obtained from $p_0(L)$ and $p_1(L)$. Finally an input stream of 0's and 1's was simulated at 50 different signal to noise ratios and the bit error rate was estimated based on 10 000 bits sent at each SNR. The results are shown in Figure 7. Comparing the simulated bit error rate to the theoretical bit error rate of an optimum matched filter shows that using this implementation is equivalent to about a 5 dB loss in SNR.

BREADBOARD EXPERIMENTS

The transceiver was fabricated using standard construction techniques. Along with the two signal frequencies, white and colored noise were introduced into the receiver circuit. A computer was connected to the receiver's data interface for data recording. The performance was tested over several long time intervals to enable the calculation of a bit error rate. The interface between the transmitter and receiver was established electrically rather than acoustically. The transmitter was then connected to a projector, and the receiver connected to a hydrophone so the acoustic transmission of data could be tested. The acoustic components were placed in the 5 meter acoustic test facility at Sonatech, Inc. This provided an extremely hostile environment due to severe multipath and reverberation within the tank. The system performed well in this environment, however it was possible to position the transceiver pairs to selectively cancel the telemetry messages, as expected.

SEA TRIALS

The telemetry system was prepared for a sea-trial in early February 1991. This experiment was carried out in the Santa Barbara channel approximately 2 miles from shore in water depths of approximately 200 feet. Figure 1 depicts the sea-trial setup. The ocean environment appeared to be moderate, Sea State 2, during these trial experiments. The modems where instrumented to provide three standard data rates, 1200, 2400, and 4800 baud, and two transmit power levels, low and high, via magnetic switches on the pressure housings. The modem performed extremely well in all configurations; 512 byte messages were received error free. The measured SNR was +19 dB at the synchronizer input. The results of the computer simulations and breadboard tests show that the performance of the system was roughly equivalent to that of an optimal matched filter operating at +14 dB. This indicates an expected bit error rate of 3.51 x 10^{-6}.

CONCLUSIONS

The zero-crossing based binary FSK receiver performed well in computer simulations, breadboard experiments, and sea trials. A modified zero-crossing receiver could provide a multi-symbol transmission scheme by virtue of its ability to determine the time associated with the interval between negative to positive transitions of the filtered received signal. This has a significant impact over the analog receiver approach where each additional symbol requires an additional filter network and comparator. Symbols could be encoded as different time intervals and the associated frequency bands encoded in the EPROM lookup table. The performance of a zero-crossing receiver is roughly equal to that of an optimum matched filter operating at a 5 dB lower SNR. In applications with sufficient SNR's, zero-crossing receivers are a good alternative to matched filtering receivers due to their low cost, low maintenance, and low power consumption.

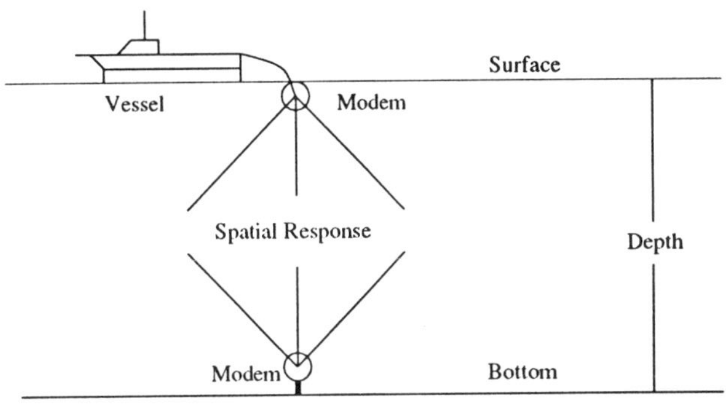

Figure 1 Sea test

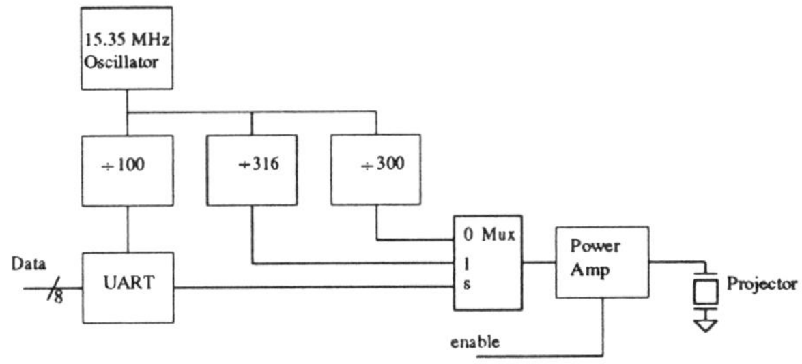

Figure 2 FSK transmitter

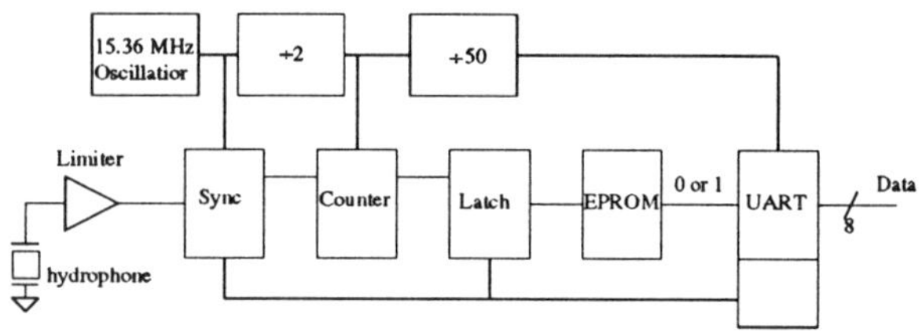

Figure 3 Zero-crossing FSK receiver

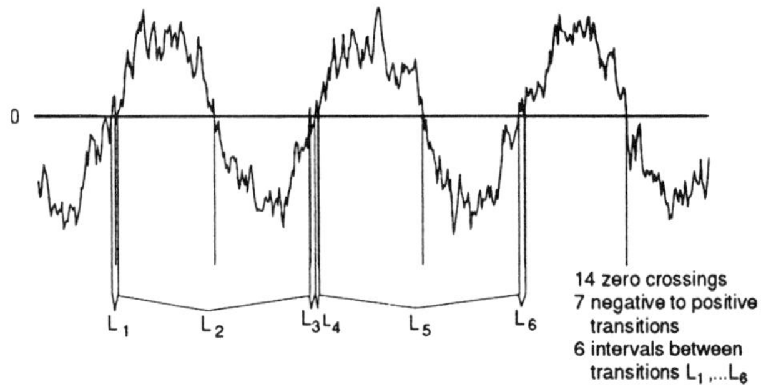

Figure 4 Zero-crossing techniques

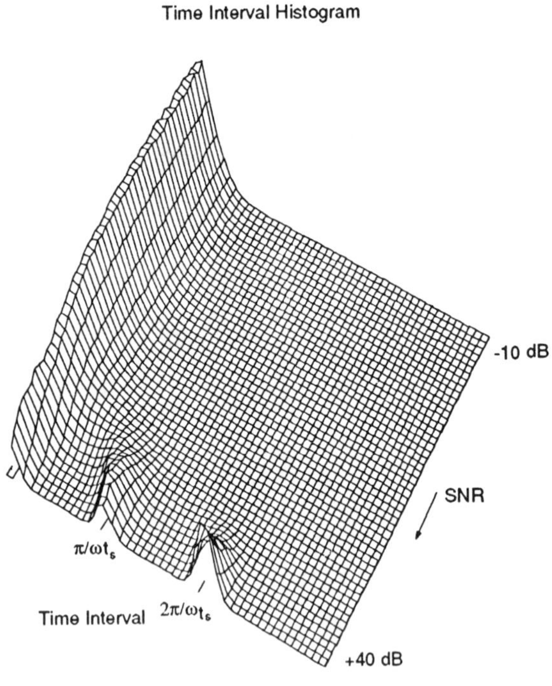

Figure 5 Sinusoid in white noise

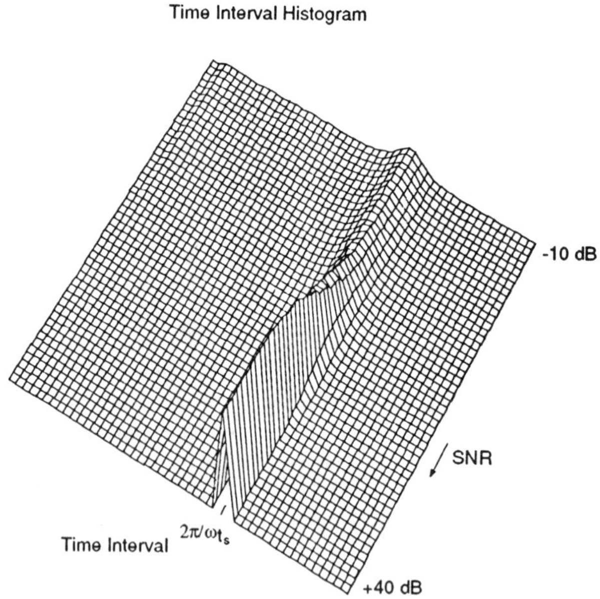

Figure 6 Filtered sinusoid in white noise

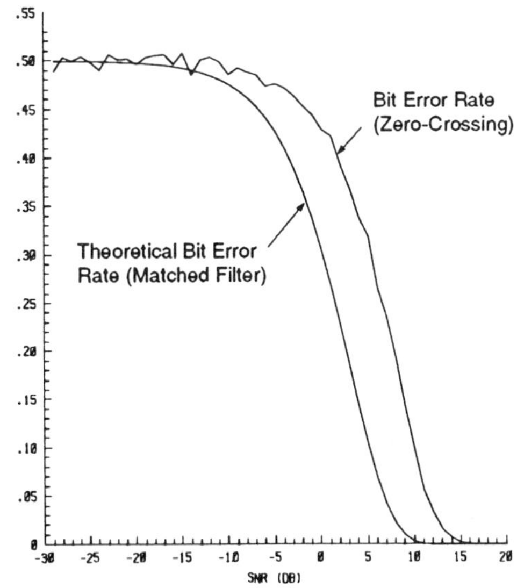

Figure 7 Bit error rate vs. SNR for zero-crossing and matched filter receivers

REFERENCES

[1] G. Mackelburg, S. J. Watson, and A. Gordon. Benthic 4800 bits/second acoustic telemetry. In *Oceans '81*, 1981.

[2] J. Catipovic, A. B. Baggeroer, K. von der Heydt, and D. Keolsch. Design and performance analysis of a digital acoustic telemetry system for the short range underwater channel. *IEEE Journal of Oceanic Engineering*, vol. OE-9(no. 4):pp. 242–252, Oct. 1984.

[3] R. J. Urick. *Principles of Underwater Sound*. McGraw-Hill, New York, 1975.

[4] W. S. Burdic. *Underwater Acoustic System Analysis*. Prentice-Hall, Inc. New Jersey, 1984.

[5] R. F. W. Coates. *Underwater Acoustic Systems*. Halsted Press, New York, 1989.

[6] S. L. Freeny, J. F. Kaiser, and H. S. McDonald. *Applications of Digital Signal Processing*, chapter Some Applications of Digital Signal Processing in Telecommunications, pages 1–26. Prentice-Hall, New Jersey, 1978.

[7] B. Kedem. Spectral analysis and discrimination by zero-crossings. *IEEE Proceedings*, vol. 74(no. 11), November 1986.

[8] M. Kac. On the average number of real roots of a random algebraic equation. *Bull. Amer. Math. Soc.*, vol. 49:pp. 314–320, Apr. 1943.

[9] S. O. Rice. Mathematical analysis of random noise. *Bell Systems Technical Journal*, vol. 24:pp. 46–56, Jan. 1945.

Target Tracking and Identification using an Extended-Kalman-Filter-Based Associative Memory

Dan S. Cheng and Allen R. Stubberud
Department of Electrical and Computer Engineering
University of California, Irvine
Irvine, CA 92717

Abstract

The joint problem of tracking and identification of a target from an airplane is considered. A new powerful combined technique using the Extended Kalman Filter and an Associative Memory(EKFAM) is employed for solving the problem. As we track a target we are able to identify it, as one of a finite set of targets, using its feature pattern. Targets can also be identified when we have only a partial feature vector. We demonstrate these techniques through an example and discuss future research studies.

1 Development of the System Model

In this study, we have developed a method to track and identify a finite set of ground based targets, using the extended Kalman filter combined with an associative memory(EKFAM), in real time from an airplane. For tracking, we use the extended Kalman filter technique to estimate relative angles, angular rates, ranges, and range rates between the airplane and the target as measured by IR, Visual, or Radar Devices. We assume that feature vectors for each target for several sets of aspect angles have been stored in an extended-Kalman-filter based associative memory. From the estimated relative motion of the airplane and target, we use the associative memory for identification. Suppose the angular motion of the ground based target is governed by Newton's Second law of the angular motion which can be written as:

$$\tau = I \cdot \ddot{\theta} \quad ....(1)$$

where τ is the applied torque producing the rotation and I is the moment of inertia of the target about its z-axis in the three dimensional real space. We assume each target type is assigned a unique integer, then the set of all target types is the set of the integers, namely, $T = (T_1, T_2, T_3, \ldots, T_N)$, where T_i is an integer, $T_i \neq T_j$, for each $i \neq j$. Let t_k be the discrete time variable, so that

$$\Delta_k = t_{k+1} - t_k$$

and discretize equation (1) as follow.
Let $x_1 = \theta$, $x_2 = \dot{x}_1$. Then,

$$\begin{pmatrix} \dot{x}_1 \\ \dot{x}_2 \end{pmatrix} = \begin{pmatrix} 0 & 1 \\ 0 & 0 \end{pmatrix} \begin{pmatrix} x_1 \\ x_2 \end{pmatrix} + \begin{pmatrix} 0 \\ \frac{\tau}{I} \end{pmatrix}$$

Thus we discretize this system as

$$\begin{pmatrix} x_1(k+1) \\ x_2(k+1) \end{pmatrix} = \begin{pmatrix} 1 & \Delta_k \\ 0 & 1 \end{pmatrix} \begin{pmatrix} x_1(k) \\ x_2(k) \end{pmatrix} + \begin{pmatrix} \frac{\tau}{I}\frac{\Delta_k^2}{2} \\ \frac{\tau}{I}\Delta_k \end{pmatrix}$$

where we assume that τ is unknown, but target dependent. Since the target is denoted by a fixed integer, we can define the dynamics of the target type by the difference equation

$$T(k+1) = T(k)$$

The combined dynamics of the identification system are

$$\begin{pmatrix} T(k+1) \\ x_1(k+1) \\ x_2(k+1) \end{pmatrix} = \begin{pmatrix} 1 & 0 & 0 \\ 0 & 1 & \Delta_k \\ 0 & 0 & 1 \end{pmatrix} \begin{pmatrix} T(k) \\ x_1(k) \\ x_2(k) \end{pmatrix} + \begin{pmatrix} 0 \\ w_1(k) \\ w_2(k) \end{pmatrix}_T$$

1058-6393/91 $01.00 © 1991 IEEE

where $w_1(k)$, $w_2(k)$ are the unknown, but target dependent inputs. Therefore, in general, this system is nonlinear. Let

$$A_k = \begin{pmatrix} 1 & 0 & 0 \\ 0 & 1 & \Delta_k \\ 0 & 0 & 1 \end{pmatrix},$$

$$X(k+1) = \begin{pmatrix} T(k+1) \\ x_1(k+1) \\ x_2(k+1) \end{pmatrix},$$

$$W_T(k) = \begin{pmatrix} 0 \\ w_1(k) \\ w_2(k) \end{pmatrix}_T$$

Then,

$$X(k+1) = A_k \cdot X(k) + W_T(k) \quad \text{............(2)}$$

Now, let F(k+1) be the measurement of the target feature vectors, then

$$F(k+1) = h(X(k+1)) + V(k) \quad \text{...............(3)}$$

where $h(\bullet)$ defines the relationship between the state variables and the feature vector and V(k) is white noise with covariance R(k). This function is approximated by a random access memory (RAM) which stores a finite set of feature vectors and an appropriate interpolation scheme. Since $W_T(k)$ is unknown, we do not know its probabilistic properties, thus we assume it is a white noise with covariance matrices Q(k). Note that since this "noise" is target dependent, it could also be used to identify the target. This is not done in this paper. Let H(k+1) be the Jacobian matrix of h(X(k+1)), then, the well-known literature of the extended Kalman filter algorithm for equations (2) and (3) gives the filter equations

$$P(k+1|k) = A_k \cdot P(k|k) + P(k|k) \cdot A_k^T + Q(k)$$

$$K(k+1) = P(k+1|k) \cdot H^T(k+1) \cdot$$

$$[H(k+1) \cdot P(k+1|k) \cdot H^T(k+1) + R(k+1)]^{-1}$$

$$P(k+1|k+1) = [I - K(k+1) \cdot H(k+1)] \cdot P(k+1|k)$$

$$\cdot [I - K(k+1) \cdot H(k+1)]^T$$

$$+ K(k+1) \cdot R(k+1) \cdot K^T(k+1)$$

Note then that this identification filter can be combined with the tracking filter to generate a single tracking/identification filter. The technique of EKFAM as depicted in the figure 1 and used for the target identification has given promising results. It is illustrated by an example in the next section.

2 Example

In this section, we provide an example to illustrate the technique described in the above. Three types of moving targets with feature vectors consisting of the 9 signals from a square array of 3 x 3 pixels are considered. A random number generator was used to generate the elements of the feature vectors. The features of target one consist of elements from the interval $[0,1]$. Target two consist of elements from the interval $[0, 0.5]$. Target three consist of elements from the interval $[0.5, 1]$. These feature vectors were stored in the RAM as a function of range, $r(t)$, azimuth angle, $\theta(t)$, and elevation angle, $\phi(t)$. The feature vectors were stored at aspect angles, at 10^0 intervals and for a fixed range. Each target type was assigned an integer, that is, $T = (1,2,3)$. $V(k)$ and $W(k)$ as used in this example are zero mean white gaussian noises with the covariance matrices R(k) and Q(k). $R(k)$, $P(0|0)$, and $Q(k)$ are diagonal matrices with 1.01, 0.16, 0.01 in diagonals respectively. The table 1 is used for the EKFAM training for the tracking.

The figure 2 are the absolute-errors versus the number of the iterations for each target described in the above table. The dotted line is for target 1; the solid line is for target 2; the * line is for target 3.

The feature vectors of targets $(1,2,3)$ obtained from tracking were:

$$\hat{f}_1 = (.73\ \ .33\ \ .65\ \ .25\ \ .17\ \ .07\ \ .32\ \ .78\ \ .29)$$

$$\hat{f}_2 = (.13\ \ .34\ \ .17\ \ .37\ \ .46\ \ .48\ \ .34\ \ .10\ \ .35)$$

$$\hat{f}_3 = (.79\ \ .82\ \ .77\ \ .92\ \ .78\ \ .86\ \ .68\ \ .73\ \ .91)$$

The true feature vectors for the targets $(1,2,3)$ are:

$$f_1 = (.75\ \ .27\ \ .67\ \ .26\ \ .09\ \ .03\ \ .32\ \ .79\ \ .30)$$

$$f_2 = (.12\ \ .36\ \ .16\ \ .37\ \ .46\ \ .48\ \ .34\ \ .10\ \ .35)$$

$$f_3 = (.80\ \ .82\ \ .77\ \ .92\ \ .76\ \ .87\ \ .69\ \ .73\ \ .92)$$

3 Conclusion and Future Studies

The feasibility of this technique has been demonstrated in the above example. For real time applications, the inversion of $H(k+1) \cdot P(k+1|k) \cdot H^T(k+1) + R(k+1)$ to form the Kalman gain may produce an excessive computational burden,

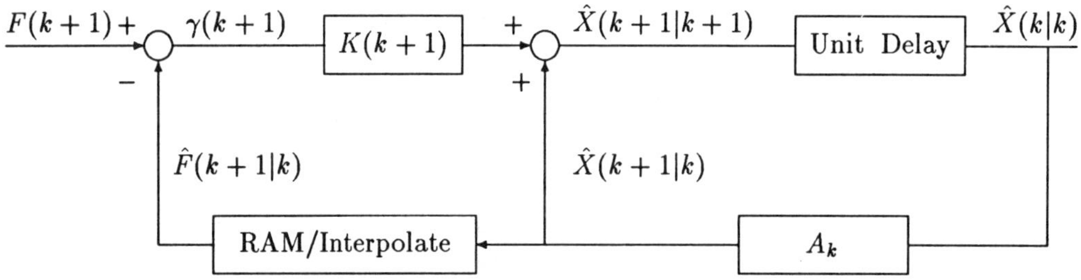

Figure 1: This diagram is for target tracking and Identification

T	$\hat{T}$	θ	ϕ	$x_1(0)$	$x_2(0)$	$\hat{x}_1(0)$	$\hat{x}_2(0)$
1	.8	10^0	-10^0	.5	.1	.5	.3
2	1.5	10^0	-10^0	.5	.1	.5	.3
3	1.5	70^0	-10^0	.5	.1	.5	.3

Table 1: EKFAM Training for the Tracking

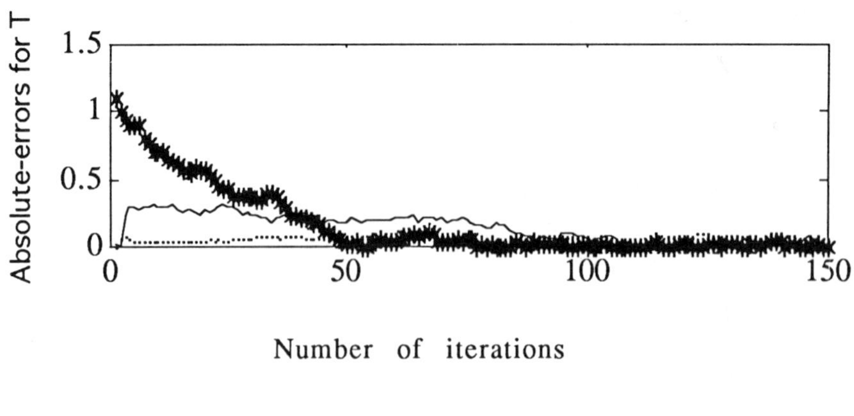

Figure 2: EKFAM Training for the Target T

because the size of this matrix is $n \times n$ with n the number of elements in the array (feature vector). Using certain characteristics of the model, we are able to reduce this burden so that target tracking and identification can be still performed using the filter depicted in the figure 1. We can extend these techniques to solve the detection problem as well. Our efforts for future studies will be toward performing the target detection, identification, and tracking problem using a Neural-Network-based associative memory.

Acknowledgements

This project is partially supported by Ford Aerospace and Communication Corporation and University of California MICRO under the research project, "Neural Network Applications".

References

[1] A. Gelb, et al, Applied Optimal Estimation. The MIT press, 1989.

[2] T. Kohonen, *Self-Organization and Associative Memory*, Berlin: Springer-Verlag, 1989.

[3] W. C. Martin and A. R. Stubberud, "The Innovations Process with Applications to Identifications." *In Control and Dynamics Systems (Vol 12).* Edited by C. T. Leondes, 1976.

[4] A. R. Stubberud and R. J. Thomas, "Associative Recall Using a Constraction Operator." *IEEE Transactions on Circuits and Systems*, Vol 36, PP. 682-686, May 1989.

[5] A. R. Stubberud and G. H. Xia, "A Fixed Complexity Nonlinear Estimation Technique." *Proceedings of the 1986 IEEE Conference on Decision and Control, Athens, Greece. PP1032-1034*, 1986.

[6] A. R. Stubberud, "Implementation of an Associative Memory with a Digital Filter." *Twenty-First Annual Asilomar Conference on Systems and Computers.*, Nov 2-4, 1987, Pacific Grove, CA.

[7] H. W. Sorenson and A. R. Stubberud, "Linear Estimation Theory", *Chapter 1, Theory and Applications of Kalman Filter, NATO-AGARD, Edited by C. T. Leondes*, London, 1970.

Noisy Input/Output System Identification Using Cumulants and the Steiglitz-McBride Algorithm

John M. M. Anderson and Georgios B. Giannakis

Dept. of Electrical Engineering
University of Virginia
Charlottesville, VA 22903-2442

ABSTRACT

Transforming the input/output system identification problem into the high *SNR* cumulant domain, we extend the Steiglitz-McBride algorithm yielding an auto-cumulant and cross-cumulant based approach for *ARMA* modeling. The auto-cumulant approach requires that the *ARMA* parameters be estimated by first estimating the cumulants of the *ARMA* parameters. On the other hand, the cross-cumulant formulation permits the *ARMA* parameters to be estimated directly. Possible convergence points and convergence issues are investigated. Finally, simulations are presented to illustrate the performance of these algorithms.

I. Introduction

For years, investigators have considered the problem of identifying a linear time-invariant (LTI) system from the measurements of its input/output (I/O). Using auto-correlations and cross-correlations, frequency-domain [1], instrumental variables [16] and prediction-error [13] methods have been applied to typical I/O applications such as econometrics [4], noise canceller design [10] and time-delay estimation [15].

Recently, several cumulant based methods have been proposed for I/O system identification. The majority of these methods lack an optimality criterion [3,7,10,21] or require a nonlinear minimization [11,12,22]; although, in [5] a novel cost function, which is proportional to the MSE based on noiseless data, yields batch, least-mean-squares and recursive least-squares algorithms. Due to the fact that cumulants of zero-mean, Gaussian random processes are zero [2], I/O system identification methods based on cumulants suppress the effect of additive Gaussian noise.

In this paper, we focus our attention on an *ARMA* method proposed by Steiglitz and McBride [19], which has been successfully applied to both linear [6,18,19] and non-linear systems [8,14]. Although the method converges locally [20], it has a major limitation - it assumes that the noise is white and only contaminates the output. This method has been extended to handle correlated noise [17]; however, the noise statistics must be known and the noise is assumed to contaminate only the output or input. Typically, however the noise is colored and a priori knowledge of the noise correlation matrix is unavailable; in addition, noise may be present at both the input and output.

We obviate the limitations mentioned above by transforming the problem from the data-domain to the high SNR cumulant-domain. In Section II, we discuss two cumulant-domain extensions of the Steiglitz-McBride algorithm. The first method uses auto- cumulants to estimate the cumulants of the *ARMA* parameters, which are in turn used to estimate the *ARMA* parameters. On the other hand, the second method utilizes cross-cumulants and estimates the *ARMA* parameters directly. Section III, contains analysis of the possible convergence points while in Section IV it is shown that the algorithms converge in one step. Simulations presented in Section V, demonstrate some of the potential advantages of the proposed algorithms.

With this background, let us now formulate the *ARMA* I/O system identification problem.

II. Problem Statement

The following model is assumed

$$w(n) = \frac{B(q)}{A(q)} u(n) , \tag{1}$$

where $u(n)$ and $w(n)$ are the input and output, respectively. And, $A(.)$ and $B(.)$ are relatively prime polynomials in q^{-1}, where q is the unit advance operator (i.e. $q\,s(t) = s(t+1)$),

$$A(q) = 1 + a(1)q^{-1} + \cdots + a(n_a)q^{-n_a}$$
$$B(q) = 1 + b(1)q^{-1} + \cdots + b(n_b)q^{-n_b} \tag{2}$$

The input $u(n)$ is assumed to be a linear process

$$u(n) = G(q)e(n) , \tag{3}$$

where $e(n)$ is independent and identically distributed (i.i.d.) driving noise. Measurements of $u(n)$ and $w(n)$ are contaminated by the input and output noises $v_u(n)$ and v_w, respectively yielding the observed I/O data records

$$x(n) = u(n) + v_u(n) \qquad y(n) = w(n) + v_w(n) . \tag{4}$$

In addition to those stated above, we also make the following assumptions:

1058-6393/91 $01.00 © 1991 IEEE

- $A(q)$ and $G(q)$ are stable LTI systems
- $e(n)$ is i.i.d., zero-mean, non-Gaussian with finite moments, and $\gamma_{3e} \triangleq E[e^3(n)] \neq 0$
- $v_u(n)$, $v_w(n)$ are zero-mean, stationary (correlated) processes which are either Gaussian of *unknown* covariance or linear processes generated by symmetrically distributed noise
- Driving noise $e(n)$ is independent of $v_u(n)$ and $v_w(n)$
- Input is persistently exciting of a sufficient order

Given finite I/O data records corrupted by noise, the objective is to estimate the coefficients of $A(.)$ and $B(.)$.

III.Cumulant Based Steiglitz-McBride Algorithms

A. Auto-cumulant method

Consider the 3rd-order cumulant of the noisy output data

$$c_{3y}(\tau_1,\tau_2) \triangleq E[y(n)y(n+\tau_1)y(n+\tau_2)]\,. \qquad (5)$$

Since the signals and noise sources are independent, and the cumulants of Gaussian processes are zero [2]

$$c_{3y}(\tau_1,\tau_2) = \frac{B_3(q_1,q_2)}{A_3(q_1,q_2)}\, c_{3i}(\tau_1,\tau_2)\,, \qquad (6)$$

where $A_3(q_1,q_2)$ and $B_3(q_1,q_2)$ are the 3rd-order bispectra operator of the *AR* and *MA* parameters, respectively. And, for the same reasons $c_{3x}(\tau_1,\tau_2) = c_{3i}(\tau_1,\tau_2)$; consequently,

$$c_{3y}(\tau_1,\tau_2) = \frac{B_3(q_1,q_2)}{A_3(q_1,q_2)}\, c_{3x}(\tau_1,\tau_2)\,. \qquad (7)$$

This equation forms the basis of our method and for observed I/O data $x(n)$, $y(n)$ we define the error signal

$$e(\tau_1,\tau_2;\theta) \triangleq \hat{A}_3(q_1,q_2)c_{3y}(\tau_1,\tau_2) - \hat{B}_3(q_1,q_2)c_{3x}(\tau_1,\tau_2)\,. \qquad (8)$$

Observe that the error depends on θ, a vector containing the 3rd-order cumulants of the *ARMA* parameters

$$\theta \triangleq [\,\{\hat{a}_3(i,j)\}\ \ \{\hat{b}_3(k,l)\}\,]^T\,, \qquad (9)$$

where the $\hat{a}_3$'s and $\hat{b}_3$'s have extent $\hat{N}_{a_3}$ and $\hat{N}_{b_3}$, respectively. With the aim of solving for θ in a least-squares manner, we express eq. (8) as

$$e(\tau_1,\tau_2;\theta) = \hat{a}_3(0,0)c_{3y}(\tau_1,\tau_2) - \phi_3(\tau_1,\tau_2)\,\theta\,, \qquad (10)$$

where

$$\phi_3(\tau_1,\tau_2) \triangleq [\{-c_{3y}(\tau_1-i,\tau_2-j)\}\ \{c_{3x}(\tau_1-k,\tau_2-l)\}]^T \qquad (11)$$

for $(i,j) \in \hat{N}_{a_3} - (0,0)$, $(k,l) \in \hat{N}_{b_3}$. For convenience, we will assume that $\hat{a}_3(0,0) = 1$. Note, that the ordering of $\phi_3(\tau_1,\tau_2)$ must coincide with the ordering of θ. We choose as our cost function,

$$J(\theta) \triangleq \frac{1}{M^2}\sum_{\tau_1=0}^{M-1}\sum_{\tau_2=0}^{M-1}\frac{1}{2}[e(\tau_1,\tau_2;\theta)]^2\,, \qquad (12)$$

where M is chosen so that only "reliable" lags are included. Minimizing this criterion, we obtain the desired least-squares solution

$$\hat{\theta}_M^{\,ls} = \left[\frac{1}{M^2}\sum_{\tau_1=0}^{M-1}\sum_{\tau_2=0}^{M-1}\phi_3(\tau_1,\tau_2)\phi_3(\tau_1,\tau_2)^T\right]^{-1}\cdot \left[\frac{1}{M^2}\sum_{\tau_1=0}^{M-1}\sum_{\tau_2=0}^{M-1}\phi_3(\tau_1,\tau_2)c_{3y}(\tau_1,\tau_2)\right]. \qquad (13)$$

Given this initial estimate an outline of the proposed algorithm follows:

Step 1: Initialization

Replace the true cumulants with sample cumulants and compute an initial estimate $\hat{\theta}_M^{\,0} = \hat{\theta}_M^{\,ls}$

Step 2: Cumulant Filtering

Given the estimates of the cumulant of the *ARMA* parameters at the ith iteration (i.e. $\hat{A}_3^{\,i}(q_1,q_2)$, $\hat{B}_3^{\,i}(q_1,q_2)$), we obtain new data by filtering the I/O cumulants

$$c^i_{3y}(\tau_1,\tau_2) = \frac{1}{\hat{A}_3^{\,i}(q_1,q_2)}\, c_{3y}(\tau_1,\tau_2) \qquad (14a)$$

$$c^i_{3x}(\tau_1,\tau_2) = \frac{1}{\hat{A}_3^{\,i}(q_1,q_2)}\, c_{3x}(\tau_1,\tau_2) \qquad (14b)$$

Step 3: Update Least-Squares Solution

Compute a new least-squares solution based on the filtered I/O cumulant data

$$\hat{\theta}_M^{\,i+1} = \left[\frac{1}{M^2}\sum_{\tau_1=0}^{M-1}\sum_{\tau_2=0}^{M-1}\phi_3^{\,i}(\tau_1,\tau_2)\phi_3^{\,i}(\tau_1,\tau_2)^T\right]^{-1}\cdot \left[\frac{1}{M^2}\sum_{\tau_1=0}^{M-1}\sum_{\tau_2=0}^{M-1}\phi_3^{\,i}(\tau_1,\tau_2)c^i_{3y}(\tau_1,\tau_2)\right], \qquad (15)$$

where

$$\phi_3^{\,i}(\tau_1,\tau_2) \triangleq [\{\frac{-c_{3y}(\tau_1-i,\tau_2-j)}{\hat{A}_3^{\,i}(q_1,q_2)}\}\ \{\frac{c_{3x}(\tau_1-k,\tau_2-l)}{\hat{A}_3^{\,i}(q_1,q_2)}\}]^T \qquad (16)$$

for $(i,j) \in \hat{N}_{a_3} - (0,0)$, $(k,l) \in \hat{N}_{b_3}$.

Step 4: Iterate and compute *ARMA* parameters

Iterate back and forth between the filtering and least-squares step until the algorithm converges. Using a cumulant based *MA* parameter estimation scheme, compute an estimate for the *ARMA* parameters from the converged cumulant estimates.

At this point, it may be unclear as to the motivation for the filtering step which we will now explain. The error residual we would like to minimize is

$$e_d(\tau_1,\tau_2;\theta) \triangleq c_{3y}(\tau_1,\tau_2) - \frac{\hat{B}_3(q_1,q_2)}{\hat{A}_3(q_1,q_2)}\, c_{3x}(\tau_1,\tau_2)\,. \qquad (17)$$

Unfortunately, the criterion which results is highly non-linear. We instead choose a meaningful error residual that leads to a linear solution, which approximates the desired non-linear solution.

B. Cross-cumulant method

The discussion to follow is nearly identical to the previous presentation. Consider the 3rd-order cross-cumulant between the noisy input and output data

$$c_{xyx}(\tau_1,\tau_2) \triangleq E[x(n)y(n+\tau_1)x(n+\tau_2)] . \quad (18)$$

Since the signals and noise sources are independent, and the cumulants of Gaussian processes are zero [2]

$$c_{xyx}(\tau_1,\tau_2) = \frac{B(q)}{A(q)} c_{3i}(\tau_1,\tau_2) . \quad (19)$$

And, for the same reasons $c_{3x}(\tau_1,\tau_2) = c_{3i}(\tau_1,\tau_2)$; consequently,

$$c_{xyx}(\tau_1,\tau_2) = \frac{B(q)}{A(q)} c_{3x}(\tau_1,\tau_2) . \quad (20)$$

This equation forms the basis of our method and for observed I/O data $x(n)$, $y(n)$ we define the error signal

$$e(\tau_1,\tau_2;\theta) \triangleq \hat{A}(q) c_{xyx}(\tau_1,\tau_2) - \hat{B}(q) c_{3x}(\tau_1,\tau_2) . \quad (21)$$

Observe that the error depends on θ, a vector that contains the *ARMA* parameters,

$$\theta \triangleq [\hat{a}(1) \dots \hat{a}(\hat{n}_a)\, \hat{b}(0) \dots \hat{b}(\hat{n}_b)]^T , \quad (22)$$

where $\hat{n}_a$ and $\hat{n}_b$ are the *AR* and *MA* model orders, respectively. With the aim of solving for θ in a least-squares manner, we express eq. (21) as

$$e(\tau_1,\tau_2;\theta) = c_{xyx}(\tau_1,\tau_2) - \phi_3(\tau_1,\tau_2)\theta , \quad (23)$$

where

$$\phi_3(\tau_1,\tau_2) \triangleq [-c_{xyx}(\tau_1-1,\tau_2) \dots -c_{xyx}(\tau_1-\hat{n}_a,\tau_2) \quad c_{3x}(\tau_1,\tau_2) \dots c_{3x}(\tau_1-\hat{n}_b,\tau_2)]^T \quad (24)$$

We choose as our cost function

$$J(\theta) \triangleq \frac{1}{M} \sum_{\tau_1=0}^{M-1} \frac{1}{2} [e(\tau_1,\tau_2;\theta)]^2 , \quad (25)$$

where M is chosen so that only "reliable" lags are included. Minimizing this criterion, we obtain the desired least-squares solution

$$\hat{\theta}_M^{ls} = \left[\frac{1}{M} \sum_{\tau_1=0}^{M-1} \phi_3(\tau_1,\tau_2)\phi_3(\tau_1,\tau_2)^T \right]^{-1} \cdot \left[\frac{1}{M} \sum_{\tau_1=0}^{M-1} \phi_3(\tau_1,\tau_2) c_{xyx}(\tau_1,\tau_2) \right] . \quad (26)$$

Given this initial estimate an outline of the proposed algorithm follows:

Step 1: Initialization

Replace the true cumulants with sample cumulants, specify a value for τ_2 (say $\tau_2 = 0$) and compute an initial estimate $\hat{\theta}_M^0 = \hat{\theta}_M^{ls}$

Step 2: Cumulant Filtering

Given the estimates of the *ARMA* parameters at the ith iteration (i.e. $\hat{A}^i(q)$, $\hat{B}^i(q)$), we obtain new data by filtering the I/O cumulants

$$c^i_{xyx}(\tau_1,\tau_2) = \frac{1}{\hat{A}^i(q)} c_{xyx}(\tau_1,\tau_2) \quad (27a)$$

$$c^i_{3x}(\tau_1,\tau_2) = \frac{1}{\hat{A}^i(q)} c_{3x}(\tau_1,\tau_2) \quad (27b)$$

Step 3: Update Least-Squares Solution

Compute a new least-squares solution based on the filtered I/O cumulant data

$$\hat{\theta}_M^{i+1} = \left[\frac{1}{M} \sum_{\tau_1=0}^{M-1} \phi_3^i(\tau_1,\tau_2)\phi_3^i(\tau_1,\tau_2)^T \right]^{-1} \cdot \left[\frac{1}{M} \sum_{\tau_1=0}^{M-1} \phi_3^i(\tau_1,\tau_2) c^i_{xyx}(\tau_1,\tau_2) \right] , \quad (28)$$

where

$$\phi_3^i(\tau_1,\tau_2) \triangleq \left[\frac{-c_{xyx}(\tau_1-1,\tau_2)}{\hat{A}^i(q)} \dots \frac{-c_{xyx}(\tau_1-\hat{n}_a,\tau_2)}{\hat{A}^i(q)} \quad \frac{c_{3x}(\tau_1-1,\tau_2)}{\hat{A}^i(q)} \dots \frac{c_{3x}(\tau_1-\hat{n}_b,\tau_2)}{\hat{A}^i(q)} \right]^T . \quad (29)$$

Step 4: Iterate

Iterate back and forth between the filtering and least-squares step until the algorithm converges.

As in the auto-cumulant method, the chosen error residual and filtering step leads to a linear solution which approximates the non-linear solution based on

$$e_d(\tau_1,\tau_2;\theta) \triangleq c_{xyx}(\tau_1,\tau_2) - \frac{\hat{B}(q)}{\hat{A}(q)} c_{3x}(\tau_1,\tau_2) . \quad (30)$$

IV. Possible Convergence Points

In this section, we examine the possible fixed points of the proposed algorithms. Our analysis is based on the 2nd-order analysis of [20].

A. Auto-cumulant method

Starting from eq. (15), it can be shown that convergence points must satisfy

$$\frac{1}{M^2} \sum_{\tau_1=0}^{M-1} \sum_{\tau_2=0}^{M-1} \phi_3^i(\tau_1,\tau_2)\varepsilon^i(\tau_1,\tau_2) = 0 . \quad (31)$$

where

$$\varepsilon^i(\tau_1,\tau_2) \triangleq c_{3y}(\tau_1,\tau_2) - \frac{\hat{B}_3^{\,i}(q_1,q_2)}{\hat{A}_3^{\,i}(q_1,q_2)} c_{3x}(\tau_1,\tau_2) \,. \tag{32}$$

And, assuming that the data in fact obeys an *ARMA* model then under certain conditions

$$\hat{A}_3^{\,c}(z_1,z_2) B_3(z_1,z_2) = A_3(z_1,z_2) \hat{B}_3^{\,c}(z_1,z_2) \,, \tag{33}$$

where $\hat{A}_3^{\,c}(z_1,z_2)$ and $\hat{B}_3^{\,c}(z_1,z_2)$ are the 2-D *z-transforms* of the converged *AR* and *MA* parameter cumulant estimates, respectively.

B. Cross-cumulant method

Following the analysis above, convergence points for this method must satisfy

$$\frac{1}{M} \sum_{\tau_1=0}^{M-1} \phi_3{}^i(\tau_1,\tau_2) \varepsilon^i(\tau_1,\tau_2) = 0 \,, \tag{34}$$

where

$$\varepsilon^i(\tau_1,\tau_2) \triangleq c_{xyx}(\tau_1,\tau_2) - \frac{\hat{B}^i(q)}{\hat{A}^i(q)} c_{3x}(\tau_1,\tau_2) \,. \tag{35}$$

From this result, it can be shown that when certain conditions are satisfied

$$\hat{A}^c(z) B(z) = A(z) \hat{B}^c(z) \,, \tag{36}$$

where $\hat{A}^c(z)$ and $\hat{B}^c(z)$ are the *z-transforms* of the converged *AR* and *MA* parameter estimates, respectively. Hence, $\hat{A}(z) = A(z) L(z)$ and $\hat{B}(z) = B(z) L(z)$ where $L(z) = 1 + l_1 z^{-1} + \cdots + l_{p^*} z^{-p^*}$ is an arbitrary polynomial with stable roots and $p^* \triangleq min(\hat{n}_a - n_a, \hat{n}_b - n_b) \geq 0$ [20].

V. Convergence Analysis

We now proceed to analyze the convergence of the proposed algorithms. For convenience, we introduce the following notation

$$G_a^i = \left[\frac{1}{M^2} \sum_{\tau_1=0}^{M-1} \sum_{\tau_2=0}^{M-1} \phi_3{}^i(\tau_1,\tau_2) \phi_3{}^i(\tau_1,\tau_2)^T \right]^{-1} \tag{37a}$$

$$G_c^i = \left[\frac{1}{M} \sum_{\tau_1=0}^{M-1} \phi_3{}^i(\tau_1,\tau_2) \phi_3{}^i(\tau_1,\tau_2)^T \right]^{-1} \tag{37b}$$

In the analysis to follow, we will assume that G_a^i and G_c^i are invertible, and the data obeys an *ARMA* model with known order.

A. Auto-cumulant method

To explore the convergence properties of this method, we consider the difference between the ith estimate and the true parameters θ^*

$$\hat{\theta}_M{}^i - \theta^* = G_a^i \left[\frac{1}{M^2} \sum_{\tau_1=0}^{M-1} \sum_{\tau_2=0}^{M-1} \phi_3{}^i(\tau_1,\tau_2) \right] \cdot$$

$$\left[\frac{c_{3y}(\tau_1,\tau_2)}{\hat{A}_3^{\,i}(q_1,q_2)} - \phi_3{}^i(\tau_1,\tau_2)^T \theta^* \right] = 0 \,. \tag{38}$$

Hence, if indeed the data satisfies an *ARMA* model and the true order is known, then this algorithm converges in one step!

B. Cross-cumulant method

As before, we examine the difference between the ith estimate and the true parameters θ^*

$$\hat{\theta}_M{}^i - \theta^* = G_c^i \left[\frac{1}{M^2} \sum_{\tau_1=0}^{M-1} \sum_{\tau_2=0}^{M-1} \phi_3{}^i(\tau_1,\tau_2) \right] \cdot$$

$$\left[\frac{c_{xyx}(\tau_1,\tau_2)}{\hat{A}^i(q)} - \phi_3{}^i(\tau_1,\tau_2)^T \theta^* \right] = 0 \,, \tag{39}$$

which implies that this algorithm also converges in one step!

VI. Simulations

In this section, we compare the performance of the proposed methods with the auto-correlation based Steiglitz-McBride algorithm and a cumulant based method of [5].

In our simulations, we use a test scenario of [5] which we discuss below. In the test cases to be examined, the input $u(t)$ is generated by driving the filter

$$G(z) = 1 - 0.2z^{-1} + 0.3z^{-2} \,,$$

with zero-mean exponentially distributed i.i.d. noise with parameters $\lambda = 1$ and $\gamma_{3e} = 2$. Both the input and the output records are corrupted by mutually correlated, colored Gaussian noise of $SNR = 5\,dB$. The input noise $v_i(n)$ is created by driving a $MA(5)$ model with white Gaussian noise, where the model coefficients are [1, –2.33, 0.75, 0.5, 0.3, –1.4]. In turn, the output noise $v_o(n)$ is created by convolving $v_i(n)$ with a $MA(3)$ model which has coefficients [1, 0.2, –0.3, 0.4].

Note, for the auto-cumulant method we use the algorithm of [9]; hence, the estimate for the $b(0)$ coefficient is 1.0 by default.

Case A.

For our first experiment, the system function

$$H(z) = \frac{1 + 0.5z^{-1}}{1 + 0.5z^{-1} - 0.125z^{-2}}$$

is used to generate the output data, where record lengths of $N = 4000$ are used. The number of lags used in the cost function of (14) and (29) is $M = 20$. Using 100 Monte Carlo runs, we compute the mean and standard deviation of the estimates for the Steiglitz-McBride algorithm and the proposed methods. The results are listed in Table I. We also include in this table the results of [5], which we denote as Delopoulos.

Case B.

We repeat the previous experiment, but change the system function to

$$H(z) = \frac{1 + 0.5z^{-1}}{1 - 1.5z^{-1} + 0.7z^{-2}} .$$

Table II contains the same statistics as Table I for this scenario.

Observing the tables, it is apparent that the bias of the cross-cumulant method is comparable to the method of [5]. But note that in most of the estimates, the method of [5] exhibits higher variance. Comparing the proposed cumulant-based methods against the autocorrelation-based of [19], it is seen that the autocorrelation-based method has a much lower variance, as expected. But, for the cases investigated the proposed methods have a significantly lower bias.

REFERENCES

[1] D. R. Brillinger, *Time Series, data analysis and theory*, Holden Day, 1981.

[2] D. R. Brillinger and M. Rosenblatt, *Spec. Anal. of Time Ser.*, B. Harris, Ed., pp. 189-232, NY, Wiley, 1967.

[3] H. H. Chiang and C. L. Nikias, *IEEE Trans. on Auto. Control*, vol. 35, pp. 36-47, 1990.

[4] M. Deistler, *Time Series and Linear Systems, Lecture notes in Control and Information Science*, pp. 37-86, Springer Verlag, 1986.

[5] A. Delopoulos and G. B. Giannakis, "Consistent identification of stochastic linear systems with noisy input-output data," submitted to *Automatica*, 1991.

[6] W. J. Done and C. K. Rushforth, *Proc. Conf. Acoust., Speech, and Sig. Proc.*, pp. 415-418, 1979.

[7] B. Friedlander and B. Porat, *IEEE Trans. on ASSP*, vol. 37 pp. 485-495, April 1989.

[8] P. G. Gallman, *IEEE Trans. Automat. Contr., vol. AC-20*, pp. 771-775, Dec. 1975.

[9] G. B. Giannakis, *Proceedings of the IEEE*, vol. 75, pp. 13333-1334, 1987.

[10] G. B Giannakis and A. Dandawate, *Circuits, Systems, and Sig. Proc.*, 1991 (to appear).

[11] G. B. Giannakis and M. K. Tsatsanis, *IEEE Trans. on Info. Theory*, 1992 (to appear).

[12] K. S. Lii and M. Rosenblatt, *Annals of Stat.*, pp. 1195-1208, 1982.

[13] L. Ljung, "System identification: theory for the user," Chap. 2 Prentice Hall, 1987.

[14] K. S. Narendra and P. G. Gallman, *IEEE Trans. Automat. Contr., vol AC-11*, pp. 771-775, Dec. 1966.

[15] C. Nikias, and R. Pan, *IEEE Trans. on ASSP*, vol. 36, pp. 1706-1714, November 1988.

[16] T. Soderstrom and P. Stoica, *Instrumental Variable Methods for System Identification*, Lecture notes in Control and Info. Sciences, Ed. A. V. Balakrishnan and M. Thoma, vol. 57, Springer-Verlag, Berlin, 1983.

[17] A. Stefanski and C. N. Weygandt, *IEEE Trans. Automat. Contr., vol. AC-16*, pp. 503-504, Oct. 1971.

[18] K. Steiglitz, *IEEE Trans. Acoust., Speech, and Sig. Proc., vol. ASSP-25*, pp. 229-234, June 1977.

[19] K. Steiglitz and L. E. McBride, *IEEE Trans. Automat. Contr., vol. AC-10*, pp. 461-464, Oct. 1965.

[20] P. Stoica and Torsten Soderstrom, *IEEE Trans. Automat. Contr., vol AC-26*, pp. 712-717, June 1981.

[21] A. Swami, "System identification using cumulants," *Ph.D. Thesis*, Univ. of Southern Calif., October 1988.

[22] J. K. Tugnait, *Proc. 29th IEEE Conf. on Decision and Control*, pp. 33-38, Honolulu, Hawaii, December 1990.

Table I

ARMA parameter identification (Case A)				
$N = 4000$, $M = 20$, 100 MC runs, Colored AGN/UC, $SNR = 5\,dB$				
Coefficient	$b(0) = 1.0000$	$b(1) = 0.5000$	$a(1) = 0.5000$	$a(2) = -0.1250$
Conventional	1.0294±0.0029	0.8330±0.0065	0.7932±0.0056	−0.0116±0.0024
Auto-cumulant	1.0000±0.0000	0.5575±0.2053	0.3576±0.1696	−0.1033±0.1143
Cross-cumulant	1.0038±0.0316	0.4770±0.0911	0.4702±0.1075	−0.1325±0.0443
Delopoulos	1.0000±0.0240	0.4730±0.1280	0.5290±0.1300	−0.1450±0.0990

Table II

ARMA parameter identification (Case B)				
$N = 4000$, $M = 20$, 100 MC runs, Colored AGN/UC, $SNR = 5\,dB$				
Coefficient	$b(0) = 1.0000$	$b(1) = 0.5000$	$a(1) = -1.5000$	$a(2) = 0.7000$
Conventional	1.9953±0.0260	0.9488±0.0512	−0.4874±0.0082	−0.0908±0.0079
Auto-cumulant	1.0000±0.0000	0.1690±4.9377	−1.3718±0.2534	0.4801±0.1817
Cross-cumulant	1.0596±0.1278	0.5377±0.1485	−1.4444±0.0472	0.6496±0.0406
Delopoulos	0.9700±0.2450	0.3690±0.9710	−1.6200±1.1680	0.8430±1.4300

High Resolution Optical A/D Conversion Using Oversampling and Interpolative Coding

Barry L. Shoop and Joseph W. Goodman
Department of Electrical Engineering, Information Systems Laboratory
Stanford University, Stanford, California 94305

Abstract

A new approach to optical A/D conversion based on oversampling and interpolative coding techniques is proposed. This new approach incorporates the high resolution capabilities of classical oversampled A/D conversion with the high speed of optical processing technology to extend the resolution and conversion rates beyond that currently possible with other electronic or optical converters. The proposed optical converter employs optical threshold and arithmetic operators and fundamentally trades bandwidth for improved amplitude resolution. A non-interferometric realization using multiple quantum well electroabsorption modulators, photodetectors, and a novel approach for non-interferometric optical subtraction is described that can provide 8 bit resolution at a conversion rate of approximately 1 GHz.

1 Introduction

One of the most popular uniform, scalar quantization techniques used to implement high speed analog-to-digital (A/D) conversion is the flash quantizer. In this type of architecture, 2^b binary comparisons are performed simultaneously to achieve b-bits resolution. With this exponential hardware complexity, flash A/D conversion becomes impractical beyond about 8 bits resolution. In order to achieve resolution in the range of 12 to 16 bits, other methods must be developed. A technique which has become popular in the audio industry is that of oversampled A/D conversion or sigma-delta ($\Sigma\Delta$) modulation, first introduced by Inose and Yasuda [1] and later popularized by Candy [2]. A fully equivalent technique, called error diffusion coding, was later developed by Floyd and Steinberg [3] for digital image halftoning applications. These two techniques have recently been shown to obey a common, underlying theoretical basis [4] which we will use in developing our optical architecture.

Conventional integrated electrooptic A/D converters [5, 6] have typically used guided-wave structures which are fundamentally transit-time limited in performance. Our optical implementation differs from previous optical techniques in that we use a non-interferometric architecture, and optical quantization and subtraction is performed using multiple quantum well modulators. In this paper we describe the underlying theory of oversampled A/D conversion, describe the operation of the non-interferometric optical architecture, and then compare this converter's performance with that of other optical and electronic A/D techniques.

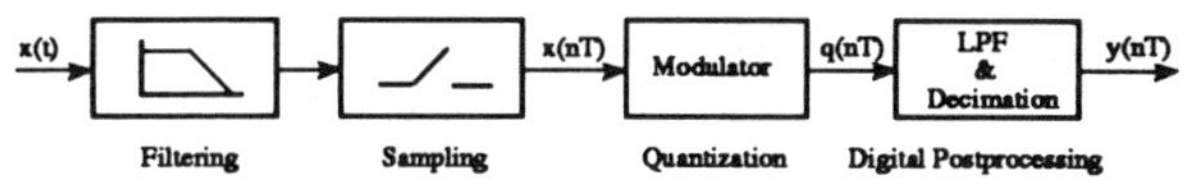

Figure 1: Block diagram of oversampled A/D converter.

2 Oversampled A/D Conversion

Oversampled A/D conversion fundamentally trades sampling speed for improved amplitude resolution by incorporating low rate quantizers and linear filters in a feedback architecture. Fig. 1 shows the block diagram of an oversampled A/D converter. The analog input signal is first bandlimited by an anti-aliasing filter and is then sampled at a frequency much greater than the Nyquist frequency. The sampled signal is then quantized by a modulator which incorporates one-bit quantizers and linear filters in a negative feedback architecture for the purpose of reducing the quantization noise within the signal baseband. This in-band noise reduction is accomplished by spectrally shaping the quantization noise, forcing most of the noise power to frequencies above the signal's cutoff frequency. The output of the modulator, which is a high-rate, single-bit digital word, is subsequently processed by a digital decimation filter which removes the out-of-band noise and generates a high resolution digital approximation to the input analog signal at the signal's Nyquist rate.

There are two general coding techniques associated with oversampling converters [7] which are classified according to the location of the linear filters within the feedback structure and whether or not the signal is spectrally altered by the filters: predictive and interpolative coding. If the feedback is strictly defined between the output and the input of the coder, predictive coding performs the filtering entirely in the feedback path and spectrally shapes both the signal and the quantization noise. Interpolative or noise shaping coding implements the filtering entirely in the feedforward path and spectrally shapes only the quantization noise. $\Sigma\Delta$ modulation, which is an interpolative coding technique, incorporates an infinite impulse re-

1058-6393/91 $01.00 © 1991 IEEE

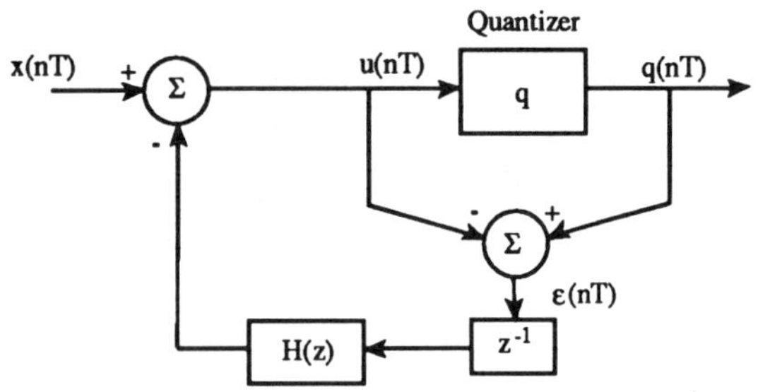

Figure 2: Block diagram of optical oversampled modulator.

sponse (IIR) integrator as the filter in the feedforward path. The error diffusion architecture, which does not meet the strict definition for feedback connectivity but is homomorphic to the $\Sigma\Delta$ architecture, implements a finite impulse response (FIR) transversal filter in the feedback path making it preferable for optical implementation.

2.1 The Modulator

The function of the modulator in an oversampled A/D converter is to quantize the signal and reduce the quantization noise within the signal baseband. Fig. 2 shows the block diagram of a generalized recursive error diffusion modulator. Here, $H(z)$ represents the z-transform of a causal, unity gain filter and z^{-1} represents a unit sample delay.

The input to the modulator, x_n, is generated by oversampling the analog input signal $x(t)$ at $f_S \gg f_N$, where f_S is the sampling frequency of the modulator and f_N is the Nyquist frequency of the analog signal. The input is assumed to be in the range $x_n \in [-b, b)$ where we use the usual definition of the half-open interval $[a, b) = \{x : a \leq x < b\}$. The difference between the modulator input and a delayed version of the quantizer error, ε_{n-1}, is quantized by a binary quantizer, where the quantizer error is defined as the difference between the output and the input to the quantizer, $\varepsilon_n \equiv q(u_n) - u_n$. For $H(z) = 1$, the nonlinear difference equation describing this modulator is

$$u_n = x_n - \varepsilon_{n-1} = x_n + u_{n-1} - q(u_{n-1});\ n = 1, 2, \ldots \quad (1)$$

The memoryless uniform binary quantizer assigns the digital output level according to

$$q(u_n) = \begin{cases} +\frac{\Delta}{2} & \text{if } u_n \geq 0 \\ -\frac{\Delta}{2} & \text{otherwise} \end{cases} \quad (2)$$

where Δ is the quantizer step size and u_n is the quantizer input.

Before continuing, we must establish the relationship between Δ and b so that Δ can be chosen to ensure the quantizer does not overload provided $x_n \in [-b, +b)$; $n = 1, 2, \ldots$. We use mathematical induction to find this relationship. If $\epsilon_0 = \Delta/2$, then to ensure $|u_1| \leq \Delta$ we must have that

$$\left| x_1 - \frac{\Delta}{2} \right| \leq \Delta \quad (3)$$

and therefore

$$b \leq \frac{\Delta}{2}. \quad (4)$$

We now show that this is a sufficient condition to ensure that the quantizer does not overload. We have already proven this for $n = 1$, so we proceed by induction by first assuming that it is true for $n = 1, 2, \ldots, k-1$. By definition of no overload, if $u_{k-1} = x_{k-1} - \varepsilon_{k-2}$ is within the no overload region $[-\Delta/2, +\Delta/2]$ then $|\varepsilon_{k-1}| \leq \Delta/2$. Therefore we need only prove that it is not possible for either $u_{k-1} > \Delta$ or $u_{k-1} < -\Delta$. With the assumption on ε_{k-1} and the range of x_n we have that

$$-b - \frac{\Delta}{2} \leq x_{k-1} - \varepsilon_{k-2} \leq b + \frac{\Delta}{2} \quad (5)$$

and using Eq. (4) we have that

$$|u_k| = |x_k - \varepsilon_{k-1}| \leq b + \frac{\Delta}{2} \leq \Delta \quad (6)$$

proving that the quantizer does not overload for all $n = k$. The upper bound of Eq. (4) is chosen for the quantizer bin width ($\Delta = 2b$), consistent with the most common choice for oversampled quantizer output levels and providing the full extremes of the input as the possible quantizer outputs.

Continuing with the operation of the modulator, since $u_n = x_n - \varepsilon_{n-1}$, Eq. (1) can be written as

$$q(u_n) = \underbrace{x_n}_{signal} + \underbrace{\varepsilon_n - \varepsilon_{n-1}}_{quantization\ error}. \quad (7)$$

In Eq. (7), the quantity ε_n is the quantization error that would have been present at the modulator output *if there were no feedback loop*. However, as a result of the negative feedback, the first-order difference or discrete-time derivative of the error appears at the output instead. The hope is that this difference signal is high frequency and can be removed by the low pass filter in the postprocessor. Eq. (7) also clearly shows that this modulator fits the framework of an intepolative coding modulator. If the quantization noise can be shown to be uncorrelated with the input signal and have statistical moments consistent with a uniform white process, we can justifiably refer to this operation as noise shaping, since only the quantization noise will be effected by the filtering operation. Gray [8] has shown that the quantization noise spectrum of the first-order $\Sigma\Delta$ modulator is in fact correlated with the input and is *not* white but contains discrete spectral components whose amplitudes can have a significant impact on modulator performance. He has also shown that higher-order modulators, such as multi-loop and multi-stage modulators, generate quantization noise which is asymptotically uncorrelated and has white statistics making the noise shaping characterization correct.

Assuming white quantization noise, we can reconsider Fig. 2 in terms of quantization noise shaping

$$E(z) = Q(z) - U(z) = Q(z) - X(z) + z^{-1}E(z)H(z). \quad (8)$$

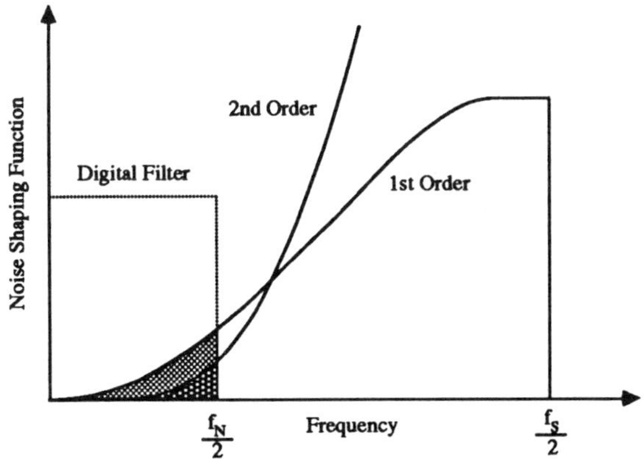

Figure 3: Noise shaping characteristics for first- and second-order filters.

Next, the quantization error of the modulator, $\hat{\varepsilon}_n = q(u_n) - x_n$, can be related to the quantizer error, ε_n, by

$$G(z) = \frac{\hat{E}(z)}{E(z)} = 1 - z^{-1}H(z) \quad (9)$$

where $\hat{E}(z)$ and $E(z)$ are the z-transforms of $\hat{\varepsilon}_n$ and ε_n, respectively. From Eq. (9) it becomes clear that the quantization error is a filtered or spectrally shaped version of the quantizer error. The higher the order N of the filter $H(z)$, the more in-band noise suppression can be expected. Fig. 3 shows the noise shaping characteristics for first-order, $H(z) = 1$, and second-order, $H(z) = 1 - 2z^{-1} - z^{-2}$, noise shaping filters. Here, the filter weights are given by the binomial coefficients

$$w_i = -(-1)^i \begin{pmatrix} N \\ i \end{pmatrix} \qquad \text{for i = 1, 2,..., N.} \quad (10)$$

Since the weights given by Eq. (10) are complex valued, these coefficients require interferometric optical techniques. For modulators with $N \geq 2$ and filter weights described by Eq. (10), stability is also an issue that must be addressed. An alternative method of achieving higher-order noise shaping which fits the framework of a non-interferometric architecture is to use a cascade of first-order modulators. In this case, the quantizer error from the $(N-1)^{st}$-stage is used as the input to the Nth-stage and then the digital output from each stage is combined in such a way that the quantization noise from stages 1 through $(N-1)$ is eliminated. The cascade architecture will be used to achieve higher-order noise shaping in the optical modulator.

2.2 The Postprocessor

As discussed earlier, the function of the postprocessor is to digitally filter and decimate the output of the modulator such that the quantization noise that was shaped by the modulator is removed through low-pass filtering and the output digital signal is decimated to the Nyquist rate of the original sampled signal. An ideal low-pass filter with cutoff frequency $f_x = f_N/2$ is generally used to quantify limiting case performance. Since the ideal low-pass filter can only be approximated in practice, other practical filters have been investigated for this application. A popular decimation filter consists of a sinc^k(f) filter which down-samples the signal to four times the Nyquist frequency, followed by a low-pass baseband filter which further down-samples the signal to the Nyquist frequency. The sinc^k(f) filter is comprised of a cascade of comb filters and can be described by the z-transform

$$D_1(z) = sinc^k(z) \equiv \left[\frac{1}{M_1} \cdot \frac{1 - z^{-M_1}}{1 - z^{-1}}\right]^k \quad (11)$$

which has the frequency response

$$H_1(f) = sinc^k(f) \equiv \underbrace{\left[\frac{sinM_1\pi fT}{M_1 sin\pi fT}\right]^k}_{magnitude} \cdot \underbrace{\frac{e^{-j\pi fTM_1}}{e^{-j\pi fT}}}_{phase} \quad (12)$$

where M_1 is the decimation factor of the filter. Candy[10] has shown that for $k = N + 1$, the sinc^k(f) filter is a practical alternative to the ideal low-pass filter in terms of the trade-offs between the quantization noise power and the oversampling ratio. The use of modulo arithmetic is used within the sinc^k(f) filter to suppress register overflow. A simple linear phase transversal filter is often used for the low-pass baseband filter.

2.3 System Performance

A convenient measure of an A/D converter's performance is the maximum signal-to-quantization noise ratio (SQNR$_{max}$). SQNR$_{max}$ is defined as the ratio of the output power at the frequency of the input sinusoid for a full-scale input to the quantization noise power within the signal baseband. In the case of an oversampled A/D converter, analytic evaluation of this expression requires knowledge of the postprocessor filter transfer function. If we assume an ideal low-pass filter with cutoff frequency f_x, the SQNR$_{max}$ for an Nth-order modulator with a full-scale input range of $\pm\Delta/2$ can be shown to be [9]

$$\text{SQNR}_{max}(M, N) = \frac{3}{2} \cdot \left[\frac{2N+1}{\pi^2 N}\right] \cdot M^{2N+1} \quad (13)$$

where again N represents the order of the modulator and M is the oversampling ratio, defined as

$$M \equiv \frac{f_S}{f_N}. \quad (14)$$

For comparison, the SQNR$_{max}$ of a conventional Nyquist rate uniform PCM quantizer with b-bits resolution can be shown to be

$$\text{SQNR}_{max}(b) = 3 \cdot 2^{2b-1}. \quad (15)$$

Eqs. (13) and (15) provide a complete description of the relationship between the modulator order, oversampling ratio, SQNR, and number of equivalent bits

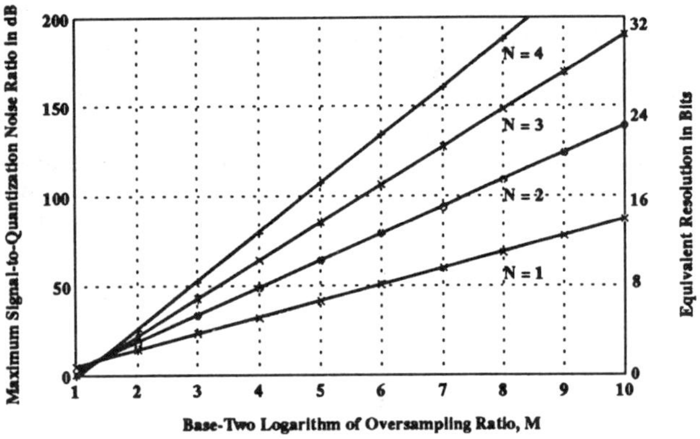

Figure 4: Signal-to-noise ratio for N^{th}-order modulators.

resolution. If we know the modulator sampling rate and the input analog signal bandwidth, the resolution of the converter can be computed directly from these two equations. This is an important result that we will return to later to determine the performance capabilities of our optical oversampled A/D converter. Fig. 4 shows the $SQNR_{max}$(M,N) and equivalent resolution for first- through fourth-order oversampled modulators as a function of oversampling ratio. Results for N=1 are included because, although the additive noise model does not predict the spectral characteristics of the quantization noise, it does yield accurate results for the SQNR and $SQNR_{max}$ [8].

3 The Optical Modulator

Implementation of the optical modulator requires basically two functional elements: a one-bit quantizer and two subtraction nodes. In implementing these functions, speed and low power consumption are important specifications. Quantization, or simple thresholding in this case, can be performed by a whole host of optically bistable devices which are currently available. Temporal, non-interferometric optical subtraction, however, is a much more difficult function to implement. Multiple quantum well (MQW) electroabsorption modulators provide both of these functions and meet the speed and power constraints.

3.1 Optical Devices

The MQW elctroabsorption devices used in this realization rely on the quantum confined Stark effect (QCSE) [11] and the associated shift of the exciton peaks with applied voltage to produce an electroabsorption modulator. The MQWs are placed within the intrinsic region of a reverse biased $p\ i\ n$ diode which facilitates the application of high field strengths while requiring negligible current flow. We focus our attention on the reflective type of device in which the conventional modulator is placed within a Fabry-Perot cavity. These devices provide higher contrast ratios at lower operating voltages than their transmissive counterparts. For optically bistable operation, the modulator requires an unstable operating point which, in turn, requires decreasing responsivity S (in amps/watt) with increasing applied field. Non-interferometric optical subtraction requires a stable operating point, corresponding to $dS/dV > 0$. For a review of MQW electroabsorption modulator operation see [12] and [13].

3.2 Realization

The first-order non-interferometric optical modulator is shown in Fig. 5. Here, x'_n is the non-negative modulator input generated by sampling $x(t)$ and then biasing x_n by $3\Delta/2$ to ensure $x'_n \geq 0$. In this configuration, two reflection electroabsorption modulators (REAMs) are operated at stable operating points to implement non-interferometric optical subtraction. In this mode the optical absorption is directly proportional to the electric current in the external circuit and can be described by

$$P_{a_i} = \frac{h\nu}{\eta e} \cdot I_{C_i}, \quad i = 1,2 \tag{16}$$

where h is Plank's constant, ν is the optical frequency, η is the internal quantum efficiency and e is electron charge. In this case, the external currents I_{ci}, $i = 1,2$ are generated by optical signals ε'_{n-1} and u'_n through photodiodes PD_1 and PD_2 respectively, where PD_1 and PD_2 are assumed to be integrated with their respective modulator to ensure maximum efficieny and speed. The symmetric modulator (S-REAM) provides binary quantization and optical gain. Here, the threshold is set by the supply voltage and the ratio of u'_n and the optical reference. In this case, the quantizer output levels are $q(u'_n) \in \{\Delta, 2\Delta\}$, and as before $\Delta = 2b$. Optical gain is achieved by first writing a low power signal onto the S-REAM and then reading it out using the higher power clock. The quantizer error sequences are related by $\varepsilon'_n = \varepsilon_n + \Delta/2$. The use of highly monochromatic signals throughout the modulator ensures optimum modulator performance. As described earlier, this first-order architecture can also be cascaded to an Nth-order modulator to provide increased performance.

3.3 Postprocessing

One method for accomplishing the postprocessing functions is to convert the optical signal to an electronic signal by means of a photodetector and then perform the decimation filtering electronically. Using high-speed bipolar silicon technology or materials such as GaAs, decimation filters are possible which would permit the full potential of the optical modulator to be realized. Although the simplicity of the decimation filter makes optical implementation possible, the lack of optical digital technology makes it impractical at present. Currently, electronic postprocessing is the prefered method of decimation filtering for our optical oversampled A/D converter. The optical modulator provides high-speed operation for the analog portion of the A/D converter while high-speed electronics easily implements the digital functions of the postprocessor. As optical digital technology becomes more mature, optical decimation filters and therefore, all-optical A/D conversion will be possible.

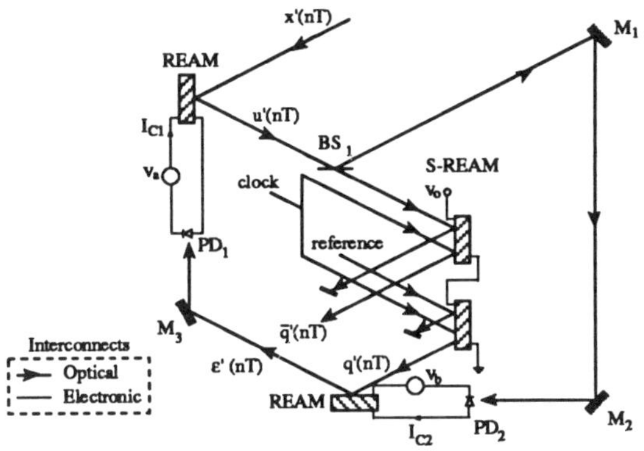

Figure 5: First-order non-interferometric optical modulator.

4 Summary

We have proposed a new approach to optical A/D conversion based on oversampling techniques and have described a non-interferometric realization using MQW devices. MQW modulator switching speeds as fast as 33 psec have been demonstrated [14], and theoretical speeds have been projected to about $1-5$ psec[15]. Based on current device parameters, a two-stage optical oversampled A/D converter that operates at 15 Gbps sampling rate is capable of providing resolution of 16 and 8 bits at a Nyquist conversion rate of 117 MHz and 938 MHz, respectively. This performance compares very favorably with that of current electronic and electrooptic A/D converters. Bell *et al.* recently demonstrated a hybrid electronic - electrooptic A/D converter using demultiplexing techniques that achieves resolution on the order of 6 bits at 4 GHz [16]. Current electronic technology supports approximately 8 bits resolution at 100 MHz for conventional Nyquist rate converters and 12 bits at 1.5 MHz for oversampled converters [17]. With the availability of 2-D arrays of MQW modulators, this architecture could also be extended to include high-speed, high-resolution A/D conversion of images.

References

[1] H. Inose and Y. Yasuda, "A unity bit coding method by negative feedback," *Proceedings of the IEEE*, vol. 51, pp. 1524-1535, 1963.

[2] J. C. Candy, "A use of limit cycle oscillations to obtain robust analog-to-digital converters," *IEEE Transactions on Communications*, vol. COM-22, pp. 298-305, 1974.

[3] R. Floyd and L. Steinberg, "An adaptive algorithm for spatial gray scale," *SID Digest*, vol. 36, 1975.

[4] D. Anastassiou, "Error diffusion coding for A/D conversion," *IEEE Transactions on Circuits and Systems*, vol. CAS-36, no. 9, pp. 1175-1186, 1989.

[5] H. F. Taylor, "An electrooptic analog-to-digital converter," *Proceedings of the IEEE*, vol. 63, pp. 1524-1525, 1975.

[6] R. A. Becker, C. E. Woodward, F. J. Leonberger, and R. W. Williamson, "Wide-band electrooptic guided-wave analog-to-digital converters," *Proceedings of the IEEE*, vol. 72, no. 7, pp. 802-819, 1984.

[7] S. K. Tewksbury and R. W. Hallock, "Oversampled, linear predictive and noise shaping coders of order N > 1," *IEEE Transactions on Circuits and Systems*, vol. CAS-25, no. 7, pp. 436-447, 1978.

[8] R. M. Gray, "Quantization noise spectra," *IEEE Transactions on Information Theory*, vol. 36, no. 9, pp. 1220-1244, 1990.

[9] B. E. Boser, "Design and implementation of oversampled analog-to-digital converters," Ph.D. dissertation, Stanford University, Stanford, CA., 1988.

[10] J. Candy, "Decimation for sigma-delta modulation," *IEEE Transactions on Communications*, vol. COM-34, 72-76, 1986.

[11] D. A. B. Miller, D. S. Chelma, T. C. Damen, A. C. Gossard, W. Wiegmann, T. H. Wood, and C. A. Burrus, "Bandedge electroabsorption in quantum well structures: The quantum confined Stark effect," *Physics Review Letters*, vol. 53, pp. 2173-2177, 1984.

[12] D. A. B Miller, "Quantum-well self-electrooptic effect devices," *Optical and Quantum Electronics*, vol. 22 , pp. S61-S98, 1990.

[13] R. J. Simes, R. H. Yan, C. C. Barron, D. Derrickson, D. G. Lishan, J. Karin, L. A. Coldren, M. Rodwell, S. Elliott, and B. Hughes, "High-frequency electrooptic Fabry-Perot modulators," *IEEE Photonics Technology Letters*, vol. 3, pp. 513-515, 1991.

[14] G. D. Boyd, A. M. Fox, D. A. B. Miller, L. M. F. Chirovsky, L. A. D'Asaro, J. M. Kuo, R. F. Kopf, and A. L. Lentine, "33 ps optical switching of symmetric self-electro-optic effect devices," *Applied Physics Letters*, vol 57, pp. 1843-1845, 1990.

[15] D. A. B. Miller, Personal communications.

[16] J. A. Bell, M. C. Hamilton, D. A. Leep, T. D. Morgan, H. F Taylor, and Y.-H. Lee, "Extention of electronic A/D converters to multi-gigahertz sampling rates using optical sampling and demultiplexing techniques," 23rd Asilomar Conf. Signals, Systems, and Computers, 289-293, 1989.

[17] B. A. Wooley, Personal communications.

A Note on the Derivative Constraints for Minimum Variance Beamforming

Ching-Yih Tseng
Signal and Image Processing Institute
University of Southern California
Los Angeles, CA 90089-2564

Abstract

A new set of first and second order derivative constraints is presented which can be used in minimum variance beamformers to achieve flat main beam responses. In contrast to other derivative constraint methods which constrain both the magnitude (power) and phase responses of the beamformer, the new method only constrains the magnitude response. The constraint on the phase is shown to be the reason that the conventional magnitude-phase derivative constraint methods are sensitive to the location of the spatial reference point used to define the array element positions. This problem, which is a major drawback of the conventional methods, is eliminated in the new approach. A simulation example which clearly illustrates the superiority of the new approach is given.

1 Introduction

The use of derivative constraints in minimum variance beamformers to achieve flat main beams has been studied by several authors including Applebaum and Chapman [1], Er and Cantoni [2], and Buckley and Griffiths [3]. In the first work, Applebaum and Chapman derived a set of linear constraints to control the shape of the main beam in narrow band arrays. These constraints are obtained by directly setting the derivatives of the beamformer response in the look direction to be zero. As a result, both the magnitude and phase of the beamformer response are constrained. The phase constraint is, however, unnecessary since the derivative constraints are used to control only the shape of the main beam.

Derivative constraints defined with respect to only the magnitude response were considered first by Er and Cantoni [2] and later by Buckley and Griffiths [3]. However, probably because of their nonlinear nature, these derivative constraints were not implemented in [2] or [3]. Instead, they were replaced by a set of linear constraints which are sufficient but not necessary to ensure flat main beam patterns — the weights which satisfy the linear constraints also satisfy the original nonlinear constraints but the contrary does not hold. Unfortunately, although the resulting linear constraints are easy to implement, they are in fact equivalent to the derivative constraints defined with respect to both the magnitude and phase responses. Consequently, the phase response is also constrained in these two recent methods.

In this paper, it is shown that the unnecessary phase constraint is a major drawback of the previous derivative constraint methods. The phase constraint causes the beamformer response to be a function of the location of the spatial reference point used to define the array element positions. This phenomenon is inherently undesirable since selecting a spatial reference point for the array should be nothing more than a matter of notational convenience. Buckley and Griffiths [3] observed this reference point problem in simulation examples and cited it as a surprising and inexplicable result. By considering the constraint on the phase, however, this paper clearly demonstrates the underlying reason. A significant objection to derivative constraint methods which are sensitive to the reference point location is that, in general, it is difficult to determine the reference point which results in the optimal beam pattern for an arbitrary array and signal environment. With an inappropriately chosen reference point, it is not uncommon to achieve a beamformer response which has undesirable sidelobes. To alleviate this problem, derivative constraints which are phase independent should be employed.

Although the formulations of phase independent derivative constraints have been given in [2], they appear in nonlinear form which does not provide useful insight for implementation. This paper proposes a process to reformulate the phase independent deriva-

1058-6393/91 $01.00 © 1991 IEEE

tive constraints in linear form, by explicitly expressing them in terms of the real and imaginary parts of the weights, such that efficient implementations can be achieved. For simplicity, the discussion focuses on the case in which the constraints are imposed at a single response point and do not involve derivatives beyond the second order. The response point is specified by three parameters, corresponding to the frequency and the azimuth and elevation angles. Derivatives with respect to these three parameters are considered and thus the approach is applicable to both narrow and broad band arrays with arbitrary geometries. In the first order derivative case, the phase independent derivative constraints are formulated as linear constraints with zero constraint values. In the second order case, the constraints are also linear while the constraint values are variables which need to be determined.

2 Derivative Constraint Formulations

Consider an array of N antenna elements distributed at known spatial locations. The beamformer output is formed by applying the complex conjugate of the complex weights, $\mathbf{w}$, to the element outputs and summing the result. The response of the beamformer to a signal at frequency f coming from direction (θ, ϕ) can be expressed as,

$$b(\theta, \phi, f) = \mathbf{w}^\dagger \mathbf{v}(\theta, \phi, f) \tag{1}$$

where † denotes the complex conjugate transpose. The complex vector $\mathbf{v}(\theta, \phi, f)$ is known as the steering vector [4]. For broadband systems, a Finite Impulse Response (FIR) filter may be applied to each element to improve the frequency resolution. The beamformer response in the broadband case can also be expressed in the form as (1) by properly defining the steering vectors [5].

The conventional way to control the response shape is to directly constrain the derivatives of the beamformer response, $b(\theta, \phi, f)$, at a selected response point to be zero. Since $b(\theta, \phi, f)$ is in general complex-valued, this derivative constraint method constrains both the magnitude and phase responses. An alternative way to control the response shape without constraining the phase response is to constrain the derivatives of the power response,

$$F(\theta, \phi, f) = |b(\theta, \phi, f)|^2 . \tag{2}$$

In the following, we compare the first and second order derivatives of $b(\theta, \phi, f)$ and $F(\theta, \phi, f)$ by explicitly expressing them in terms of the corresponding derivatives of $\mathbf{v}(\theta, \phi, f)$. To facilitate the presentation, we use the notations $\mathbf{v}_\theta$ and $\mathbf{v}_{\theta\theta}$ to denote the first and second derivative of $\mathbf{v}$ with respect to θ, respectively. Similar notations are also applied to designate the derivatives with respect to various combinations of θ, ϕ and f.

2.1 Magnitude and phase derivative constraints

Constraining the derivatives of $b(\theta, \phi, f)$ to be zero at a point (θ_0, ϕ_0, f_0) can be easily formulated by directly taking the derivatives of (1). The resulting first and second order derivative constraints are known as[1],

$$\begin{array}{lll} \mathbf{w}^\dagger \mathbf{v}_\theta = 0, & \mathbf{w}^\dagger \mathbf{v}_\phi = 0, & \mathbf{w}^\dagger \mathbf{v}_f = 0, \\ \mathbf{w}^\dagger \mathbf{v}_{\theta\theta} = 0, & \mathbf{w}^\dagger \mathbf{v}_{\phi\phi} = 0, & \mathbf{w}^\dagger \mathbf{v}_{ff} = 0, \\ \mathbf{w}^\dagger \mathbf{v}_{\theta\phi} = 0, & \mathbf{w}^\dagger \mathbf{v}_{\phi f} = 0, & \mathbf{w}^\dagger \mathbf{v}_{f\theta} = 0. \end{array} \tag{3}$$

Unity gain constraint is conventionally assumed at (θ_0, ϕ_0, f_0):

$$\mathbf{w}^\dagger \mathbf{v} = 1 . \tag{4}$$

Notice that the gain and derivative constraints defined above are not unique since they depend on the choice of the spatial reference point. To explore this dependency in more detail, let $\mathbf{v}$ and $\bar{\mathbf{v}}$ be the steering vectors of an array whose element locations are defined with respect to two different reference points. Then,

$$\bar{\mathbf{v}} = \mathbf{v} e^{j2\pi f \Delta\tau(\theta, \phi)} \tag{5}$$

where $\Delta\tau(\theta, \phi)$ is a time delay which depends upon the relative positions of the two reference points. The gain constraints defined with respect to the two reference points thus differ by a fixed phase shift. It follows that if there is only a single gain constraint, moving the spatial reference point shifts the beamformer phase response by a fixed amount but leaves the magnitude response unchanged. When derivative constraints are applied, however, moving the reference point has a more complicated effect. To see this, differentiate with respect to θ on both sides of (5):

$$\bar{\mathbf{v}}_\theta = \mathbf{v}_\theta e^{j2\pi f \Delta\tau(\theta, \phi)} + \mathbf{v} j2\pi f \frac{\partial \Delta\tau(\theta, \phi)}{\partial \theta} e^{j2\pi f \Delta\tau(\theta, \phi)} . \tag{6}$$

Because of the second term on the right-hand side, $\bar{\mathbf{v}}_\theta$ is not simply a phase shifted version of $\mathbf{v}_\theta$. As a

[1]For simplicity, the arguments of the derivative vectors are omitted whenever it is clear from the context that the derivatives are evaluated at the point (θ_0, ϕ_0, f_0).

result, the choice of the spatial reference point can affect both the magnitude and phase responses of the beamformer. This phenomenon is undesirable since the reference point has no physical significance (changing it has no effect on the observed signals). It can be avoided by forming derivative constraints with respect to the magnitude square of $b(\theta, \phi, f)$.

2.2 Phase independent derivative constraints

The power response $F(\theta, \phi, f)$ and its derivatives are invariant to the spatial reference point since $F(\theta, \phi, f)$ is phase independent. In general, these derivative constraints are not linear and are therefore difficult to implement. In practical applications, however, a n-th order derivative constraint is not used unless all the lower order derivative constraints are also imposed. When this assumption is satisfied, the following shows that the first and second order derivative constraints can be expressed in much simpler forms.

By definition, the power response F is given by,

$$F = \mathbf{w}^\dagger \mathbf{v}\mathbf{v}^\dagger \mathbf{w} \ . \tag{7}$$

Taking the first derivative with respect to θ, one finds,

$$\frac{\partial F}{\partial \theta} = 2Real\{\mathbf{w}^\dagger \mathbf{v}_\theta \mathbf{v}^\dagger \mathbf{w}\} \ , \tag{8}$$

where $Real\{\cdot\}$ denotes the real part of the bracketed quantity. Together with the gain constraint in (4), this yields

$$\frac{\partial F}{\partial \theta} = 2Real\{\mathbf{w}^\dagger \mathbf{v}_\theta\} \ . \tag{9}$$

The first derivatives of $F(\theta, \phi, f)$ with respect to ϕ and f have forms similar to this.

Now consider the second derivative of F with respect to θ and ϕ:

$$\frac{\partial^2 F}{\partial \phi \partial \theta} = 2Real\{\mathbf{w}^\dagger \mathbf{v}_{\theta\phi} \mathbf{v}^\dagger \mathbf{w} + \mathbf{w}^\dagger \mathbf{v}_\theta \mathbf{v}_\phi^\dagger \mathbf{w}\} \tag{10}$$

Using the gain constraint, we have,

$$\frac{\partial^2 F}{\partial \phi \partial \theta} = 2Real\{\mathbf{w}^\dagger \mathbf{v}_{\theta\phi} + \mathbf{w}^\dagger \mathbf{v}_\theta \mathbf{v}_\phi^\dagger \mathbf{w}\} \tag{11}$$

Notice that the second term on the right-hand side can be expressed as,

$$Real\{\mathbf{w}^\dagger \mathbf{v}_\theta \mathbf{v}_\phi^\dagger \mathbf{w}\} = Real\{\mathbf{w}^\dagger \mathbf{v}_\theta\} Real\{\mathbf{w}^\dagger \mathbf{v}_\phi\} + Imag\{\mathbf{w}^\dagger \mathbf{v}_\theta\} Imag\{\mathbf{w}^\dagger \mathbf{v}_\phi\} \tag{12}$$

where $Imag\{\cdot\}$ denotes the imaginary part of the bracketed quantity. By assumption, the first order derivative constraints with respect to θ and ϕ are also imposed, so the first term on the right-hand side of (12) is zero. Thus,

$$\frac{\partial^2 F}{\partial \phi \partial \theta} = 2Real\{\mathbf{w}^\dagger \mathbf{v}_{\theta\phi}\} + 2Imag\{\mathbf{w}^\dagger \mathbf{v}_\theta\} Imag\{\mathbf{w}^\dagger \mathbf{v}_\phi\} \tag{13}$$

One can similarly express the five other second derivatives of $F(\theta, \phi, f)$ with respect to the various combinations of θ, ϕ and f.

The phase independent derivative constraints can be readily obtained by setting the derivatives of $F(\theta, \phi, f)$ equal to zero. Specifically, the first derivative constraints (with an assumed gain constraint) are given by,

$$Real\{\mathbf{w}^\dagger \mathbf{v}_\theta\} = 0, \ Real\{\mathbf{w}^\dagger \mathbf{v}_\phi\} = 0, \ Real\{\mathbf{w}^\dagger \mathbf{v}_f\} = 0. \tag{14}$$

In addition, the second derivative constraints (with the gain and first derivative constraints assumed) can be expressed as,

$$\begin{aligned}
Real\{\mathbf{w}^\dagger \mathbf{v}_{\theta\theta}\} &= -|Imag\{\mathbf{w}^\dagger \mathbf{v}_\theta\}|^2 \\
Real\{\mathbf{w}^\dagger \mathbf{v}_{\phi\phi}\} &= -|Imag\{\mathbf{w}^\dagger \mathbf{v}_\phi\}|^2 \\
Real\{\mathbf{w}^\dagger \mathbf{v}_{ff}\} &= -|Imag\{\mathbf{w}^\dagger \mathbf{v}_f\}|^2 \\
Real\{\mathbf{w}^\dagger \mathbf{v}_{\theta\phi}\} &= -Imag\{\mathbf{w}^\dagger \mathbf{v}_\theta\} Imag\{\mathbf{w}^\dagger \mathbf{v}_\phi\} \\
Real\{\mathbf{w}^\dagger \mathbf{v}_{\phi f}\} &= -Imag\{\mathbf{w}^\dagger \mathbf{v}_\phi\} Imag\{\mathbf{w}^\dagger \mathbf{v}_f\} \\
Real\{\mathbf{w}^\dagger \mathbf{v}_{f\theta}\} &= -Imag\{\mathbf{w}^\dagger \mathbf{v}_f\} Imag\{\mathbf{w}^\dagger \mathbf{v}_\theta\}
\end{aligned} \tag{15}$$

Notice that the phase independent derivative constraints are less restrictive than the dependent ones since any $\mathbf{w}$ which satisfies the constraints in (3) also satisfies (14) and (15) but the contrary does not hold.

3 Implementation of the Phase Independent Derivative Constraints

The output of a beamformer at time n is given by,

$$y(n) = \mathbf{w}^\dagger \mathbf{x}(n) \tag{16}$$

where $\mathbf{x}(n)$ is the data vector. Minimum variance beamforming defines the optimal weights as those which minimize the variance of $y(n)$ subject to a prespecified set of constraints on the weights. In particular, we consider the phase independent derivative constraints derived in the previous section. With $\mathbf{R}_{xx}$ denoting the covariance matrix of the data vector $\mathbf{x}(n)$, the problem can be formulated as:

$$\min_{\mathbf{w}} \mathbf{w}^\dagger \mathbf{R}_{xx} \mathbf{w} \tag{17}$$

subject to (4), (14) and (15). We first derive implementation for beamformers with only the gain and the first order derivative constraints. Extension to the second order derivative constraints are then outlined.

3.1 First Order Case

The first order derivative constraints in (14) are not linear in the complex field since they involve the operation of taking the real part. However, notice that,

$$Real\{\mathbf{w}^\dagger \mathbf{v}_\theta\} = [Real\{\mathbf{w}\}\ Imag\{\mathbf{w}\}] \begin{bmatrix} Real\{\mathbf{v}_\theta\} \\ Imag\{\mathbf{v}_\theta\} \end{bmatrix}. \tag{18}$$

This expression is linear in the *real version* of $\mathbf{w}$ with double dimension defined as:

$$\tilde{\mathbf{w}} = \begin{bmatrix} Real\{\mathbf{w}\} \\ Imag\{\mathbf{w}\} \end{bmatrix}. \tag{19}$$

One can therefore *linearize* the constraints by simply expressing them in the real form with doubled dimension. Accordingly, we introduce the notations

$$\tilde{\mathbf{R}}_{xx} = \begin{bmatrix} Real\{\mathbf{R}_{xx}\} & -Imag\{\mathbf{R}_{xx}\} \\ Imag\{\mathbf{R}_{xx}\} & Real\{\mathbf{R}_{xx}\} \end{bmatrix}, \tag{20}$$

and

$$\tilde{\mathbf{v}} = \begin{bmatrix} Real\{\mathbf{v}\} \\ Imag\{\mathbf{v}\} \end{bmatrix}, \quad \hat{\mathbf{v}} = \begin{bmatrix} -Imag\{\mathbf{v}\} \\ Real\{\mathbf{v}\} \end{bmatrix}. \tag{21}$$

Also, counterparts to the derivative vectors are defined in the same manner as (21), e.g., $\tilde{\mathbf{v}}_\theta$ and $\hat{\mathbf{v}}_\theta$ are generated from $\mathbf{v}_\theta$. With these definitions, minimization of (17) subject to the gain constraint in (4) and the first derivative constraints in (14) can be expressed in real form as:

$$\min_{\tilde{\mathbf{w}}} \tilde{\mathbf{w}}^T \tilde{\mathbf{R}}_{xx} \tilde{\mathbf{w}} \tag{22}$$

subject to the linear constraints

$$\tilde{\mathbf{C}}^T \tilde{\mathbf{w}} = \tilde{\mathbf{f}} \tag{23}$$

where

$$\tilde{\mathbf{C}} = [\tilde{\mathbf{v}}\ \hat{\mathbf{v}}\ \tilde{\mathbf{v}}_\theta\ \tilde{\mathbf{v}}_\phi\ \tilde{\mathbf{v}}_f], \tag{24}$$

and

$$\tilde{\mathbf{f}} = [1\ 0\ 0\ 0\ 0]^T . \tag{25}$$

(The real transpose operation is denoted by ^{T}.) Without loss of generality, we assume that $\tilde{\mathbf{C}}$ is full rank. The closed-form solution for the weights of a minimum variance beamformer with gain and first order derivative constraints is then given by,

$$\tilde{\mathbf{w}}_0 = \tilde{\mathbf{R}}_{xx}^{-1} \tilde{\mathbf{C}} (\tilde{\mathbf{C}}^T \tilde{\mathbf{R}}_{xx}^{-1} \tilde{\mathbf{C}})^{-1} \tilde{\mathbf{f}} . \tag{26}$$

3.2 Second Order Case

To implement the second order derivative constraints, we rewrite (15) as:

$$\begin{aligned} Imag\{\mathbf{w}^\dagger \mathbf{v}_\theta\} &= h_1, & Imag\{\mathbf{w}^\dagger \mathbf{v}_\phi\} &= h_2, \\ Imag\{\mathbf{w}^\dagger \mathbf{v}_f\} &= h_3, & Real\{\mathbf{w}^\dagger \mathbf{v}_{\theta\theta}\} &= -h_1^2, \\ Real\{\mathbf{w}^\dagger \mathbf{v}_{\phi\phi}\} &= -h_2^2, & Real\{\mathbf{w}^\dagger \mathbf{v}_{ff}\} &= -h_3^2, \\ Real\{\mathbf{w}^\dagger \mathbf{v}_{\theta\phi}\} &= -h_1 h_2, & Real\{\mathbf{w}^\dagger \mathbf{v}_{\phi f}\} &= -h_2 h_3, \\ Real\{\mathbf{w}^\dagger \mathbf{v}_{f\theta}\} &= -h_3 h_2 & & \end{aligned} \tag{27}$$

in which h_1, h_2 and h_3 are three unknown variables. With this formulation, the constraints are linear with respect to $\tilde{\mathbf{w}}$ as defined in (19). However, the corresponding constraint values are variables which need to be determined.

To solve for the optimal weights, a two-step optimization procedure is used. In the first step, the optimal weight vector is expressed as a function of the variables h_1, h_2 and h_3. In the second step, the output variance is minimized over all possible h_1, h_2 and h_3. An algorithm to implement this optimization procedure has been derived in [5]. Detailed discussion is not given here due to the limitation of the paper length.

4 An Illustrative Example

We considered a five-element equi-spaced linear array with six taps attached to each element. The elements of the array were assumed isotropic and no inter-element mutual coupling effect was considered. The array elements were assumed to be placed on the x-axis ($\phi = 0°$) and the direction θ covers the angles from $-90°$ to $90°$. Since the array is linear, it is sufficient to consider the spatial/frequency response in (1) as a function of θ and f.

A minimum variance beamformer was designed to receive a broadband desired signal in the broadside direction ($\theta = 0°$). For simplicity, it was assumed that the desired signal had a flat spectrum with a bandwidth B and a center frequency f_c where $B = 0.2 f_c$. At each element, the signal was multiplied by $e^{j2\pi f_c t}$, which down shifted the signal spectrum to the baseband. The signal was then sampled with a sampling rate f_s given by $f_s = 5B$. The element spacing of the array was chosen as a half of λ_c, the wavelength corresponding to the signal center frequency. The frequency is normalized so that the signal spectrum is in the interval $[-0.1, 0.1]$; the element positions are also normalized by λ_c so that they are on the x-axis from

0 to 2 with a spacing of 0.5. The problem can now be considered in this normalization framework.

It was desirable to achieve a flat unity gain response in the broadside direction along the frequency range [−0.1, 0.1]. To accomplish this, we considered imposing phase independent derivative constraints at the response point $(\theta_0, f_0) = (0°, 0)$. The gain and the first order derivative constraints with respect to both θ and f were imposed. However, since flat response along the θ direction was not required, a second order derivative was only imposed with respect to f. To prevent the desired signal from being cancelled, additional constraints are necessary. The underlying reason was described in [6] where a process for adding constraints to prevent signal cancellation was proposed. The detailed discussion of specifying the constraints for this example are given in [5]. Besides the new method, the conventional derivative constraint method was also considered in the simulation.

A signal environment which consisted of a 10 dB broadband desired signal as described earlier and 0 dB white noise background was assumed. With the array center as the spatial reference point, the magnitude responses of the new and conventional methods were obained as shown in figure 1 (a) and (b). The array center was chosen since it represents the best location in the interval [0, 2] for the conventional method. The significantly better sidelobe behavior in Figure 1 (a) clearly indicates that the new method is superior than the conventional method.

Acknowledgment

The author is thankful to Professor Lloyd J. Griffiths for his support and encouragement.

References

[1] S. P. Applebaum and D. J. Chapman. "Adaptive arrays with main beam constraints". *IEEE Trans. on Antennas and Propagation*, AP-24:650–662, Sept. 1976.

[2] M. H. Er and A. Cantoni. "Derivative constraints for broad-band element space antenna array processors,". *IEEE Trans. on Acoustics, Speech, and Signal Processing*, ASSP-31:1378–1393, Dec 1983.

[3] K. M. Buckley and L. J. Griffiths. "An adaptive generalized sidelobe canceller with derivative constraints". *IEEE Trans. on Antennas and Propagation*, AP-34:311–319, Mar. 1986.

[4] R. T. Compton, Jr. *Adaptive Antennas*. Prentice-Hall, 1988.

[5] C. Y. Tseng. "Minimum variance beamforming with phase independent derivative constraints". Submitted to IEEE Trans. on Antenna and Propagation.

[6] C. Y. Tseng and L. J. Griffiths. "A novel approach for designing linear constraints in adaptive arrays". In *Proc. of the Asilomar Conference on Signals, Systems and Computers*, pages 124–128, Pacific Grove, CA, Nov 1990.

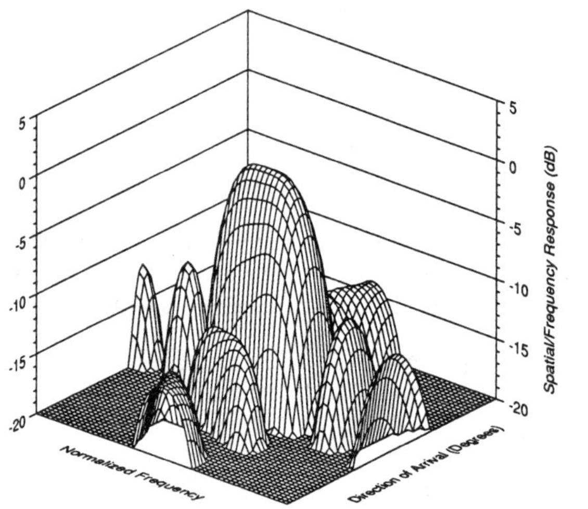

(a)

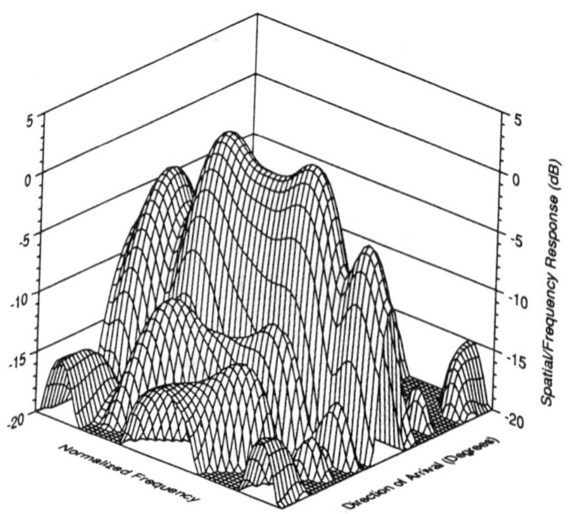

(b)

Figure 1: The magnitude responses for (a) the new method and (b) the conventional method.

Burst Sinusoidal Frequency Estimation with Short Record Length via L_1 Normed Linear Predictive Modeling

Jim Schroeder
Department of Engineering
University of Denver
Denver, CO 80208-0177

ABSTRACT

For certain communications interception scenarios an extremely short duration multi-component sinusoidal burst signal is available for analysis. Transient effects present from uncertain burst signal location within the analysis window may be detrimental to least squares based frequency estimation methods. Such signal transients are manifested as "spikes" within the residual vector produced from a linear predictive model based approach to frequency estimation. For such cases, it is demonstrated that a robust estimator, such as an estimator generated via a L_1 normed error criteria, may produce significantly less biased frequency estimates. Additionally, it is shown that sinusoidal frequency estimates generated via an L_1 normed solution are insensitive to initial signal phase in contrast to least squares based estimates.

I. Introduction

Some difficult communications interception problems require that two (or more) sinusoidal frequencies be accurately estimated from a short length burst signal record. This specific problem for a single frequency sinusoidal burst was considered in [1], but the algorithm is not easily extended to the multi-frequency case. Our intent for this work is to present an L_1 normed linear predictive algorithm that provides relatively unbiased steady-state sinusoidal frequency estimates from short length burst data, and possibly uncertain signal location within the analysis window.

Estimation of sinusoidal frequency location from short record length data necessitates use of modern spectral estimation methods such as linear predictive or autoregressive algorithms, maximum entropy methods, or maximum likelihood estimators [2]. Linear predictive or autoregressive methods are particularly attractive from a computational viewpoint if real-time or near real-time operation is required. Non real-time signal analysis is viable for many scenarios, however, and relatively computationally intensive processing algorithms need not be rejected out of hand. Numerous robust estimation algorithms are available [3,4], but will not be reviewed because of space limitations. Robust estimation algorithms frequently incur excessive computational costs, but may provide less biased frequency estimates wherein the data are contaminated by outliers such as impulsive noise, data dropouts, and signal transients [3,4].

For this work, we assume perfect window alignment so that endpoint detection errors are not present. The only data transient, therefore, is introduced when the sinusoidal signal burst shifts frequency instantaneously from f_1 to f_2 Hz. That is, we are attempting to accurately identify f_1 and f_2 from a burst frequency shift keyed (FSK) waveform. For the pure FSK signal analysis context, wherein the signal repeats for a long time relative to the reciprocal of min $\{f_1, f_2\}$, the problem of estimating f_1 and f_2 is straightforward. Unfortunately, estimating f_1 and f_2 may be difficult for the signal analysis problem that arises when f_1 and f_2 are displacement FSK frequencies relative to a fast frequency hopping (FH) carrier. In such cases, the analysis window typically includes very short record length data; additionally, the FSK waveform may contain intentional phase shift keyed (PSK) components, but we do not consider that possibility here. The hybrid FSK/PSK waveform case, although important, typically occurs over longer dwell times and may be handled via conventional Fourier methods.

1058-6393/91 $01.00 © 1991 IEEE

II. Problem Formulation

The burst frequency shift keyed (FSK) signal model is represented as:

$$s(t)=A\ rect_1((t-t_1)/\tau)\ \cos(2\pi f_1(t-t_1)+\theta_1) \\ +A\ rect_2((t-t_2)/\tau)\ \cos(2\pi f_2(t-t_2)+\theta_2), \quad (1)$$

with amplitude A, frequencies f_1 and f_2, and phases θ_1 and θ_2. $Rect_j((t-t_j)/\tau)$ are rectangular windows both τ seconds wide located on the time axis such that the individual FSK signal tones are concatenated. With exact endpoint detection, the total signal analysis duration is 2τ seconds. With the noise $w(t)$, modeled by zero mean, white Gaussian noise with variance σ^2, the analysis signal is:

$$x(t)=s(t)+w(t). \quad (2)$$

Setting $t=nT$ (T = 1 second for convenience), we retain N samples of $x(t)$, denoted $x(n)$, n = 0,1, ... , N-1. Linear prediction equations may be expressed in matrix form as:

$$X\underline{a}=\underline{y}, \quad (3)$$

where matrix X and vector $\underline{y}$ contain the appropriate data values, and vector $\underline{a}$ contains the unknown prediction coefficients [2]. A wide variety of methods are available for picking X and $\underline{y}$, but to minimize estimated sinusoidal frequency bias and provide robustness against unknown signal phase, we used the forward-backward linear prediction (FBLP) covariance method, detailed in [2]. Numerous other formulations are clearly possible. The least-squares solution is given by:

$$\underline{a}=(X^tX)^{-1}X^t\underline{y}. \quad (4)$$

and estimates of frequencies f_1 and f_2 are derived in the standard manner [2].

To generate an L_p (p=1) normed solution, we used an iteratively reweighted least squares (IRLS) algorithm [4]; that is, we compute the recursion:

$$\underline{a}_{k+1}=(X^tW_kX)^{-1}X^tW_k\underline{y}, \quad (5)$$

where W_k is a diagonal weighting matrix updated at the kth iteration by

$$W_{ii}=abs((X\underline{a}-\underline{y})_i)^{p-2} \quad (6)$$

where the subscript ii denotes the diagonal element of W. Convergence properties of the IRLS algorithm and discussion of asymptotic statistical characteristics of L_p normed minimization are given in [4].

III. Simulation

In this section, via simulation, we demonstrate the robustness of the L_P (p=1) normal linear prediction based sinusoidal frequency estimation algorithm

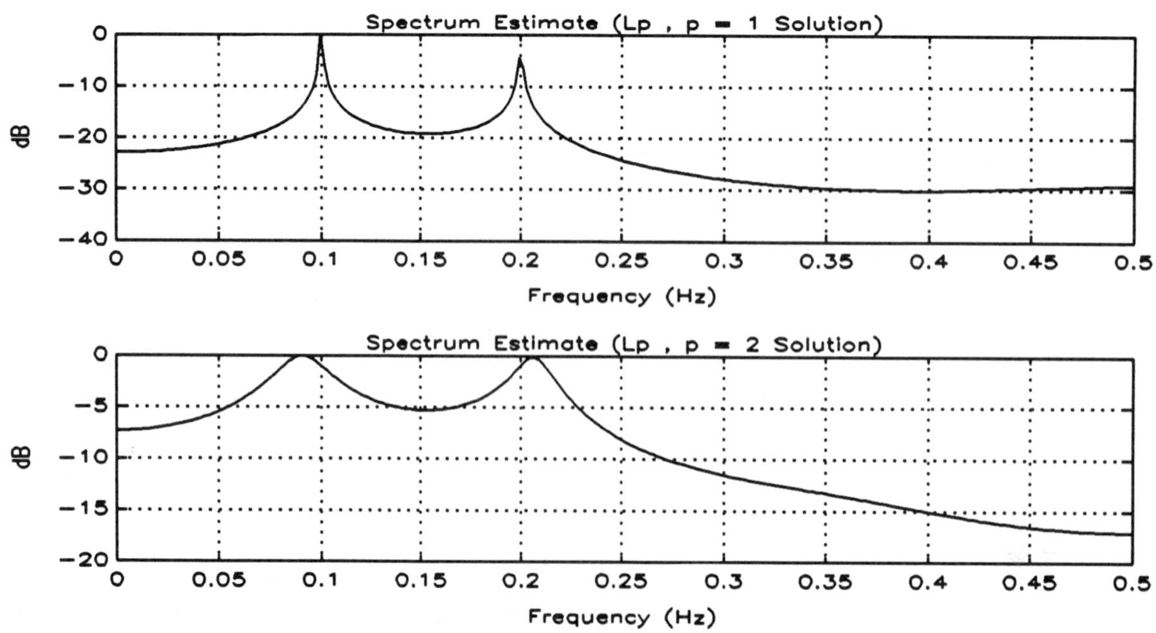

Figure 1. a) (Bottom) Least Squares Sinusoidal Frequency Estimate, N=30, L=7, f_1=.1 Hz, f_2=.2 Hz, No Noise. b) (Top) L_p (p=1) Normed Sinusoidal Frequency Estimate.

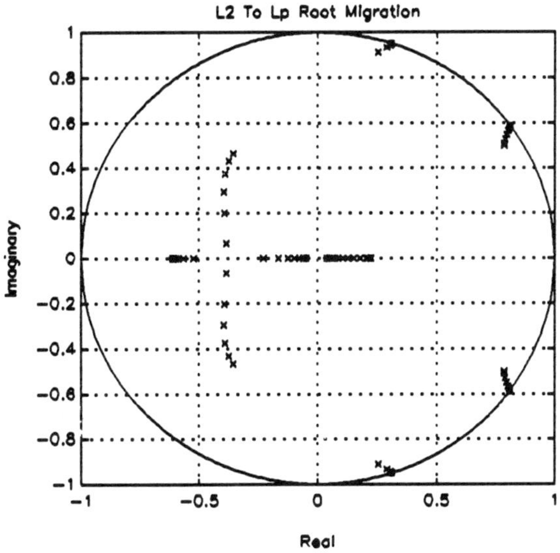

Figure 2. Root Migration From L_p (p=2) Solution to L_p (p=1) Solution; N=30, L=7, f_1=.1 Hz, f_2=.2 Hz, No Noise.

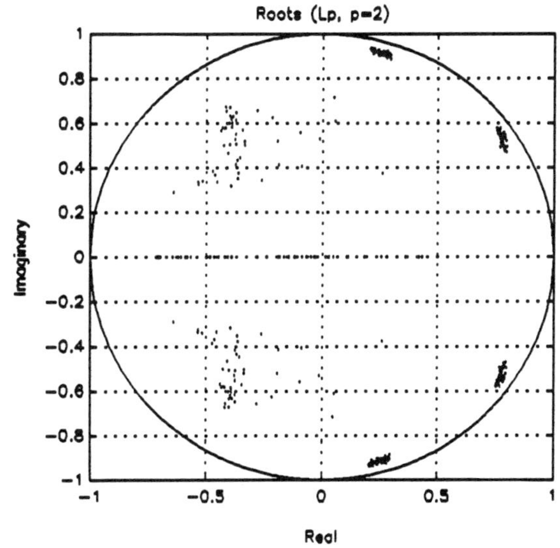

Figure 3. Least Squares Roots, 50 Trials, N=30, L=7, f_1=.1 Hz, f_2=.2 Hz, Phases Uniformly Distributed, No Additive Noise.

proposed in Section II.

The signal model (2) is sampled with period $T=1$ second, $N=30$ samples are retained, f_1 = .1 Hz, f_2 = .2 Hz, and $\Theta_1 = \Theta_2 = 0$ degrees. The first 15 samples of $x(n)$, n = 0,1, ... , 14, correspond to a sinusoid with frequency f_1 = .1 Hz and the last 15 samples of $x(n)$, n = 15, 16, ... , 29, correspond to a sinusoid with frequency f_2 = .2 Hz. The linear predictor is of order seven, no noise has been added, and $A=1$. From Figure 1a. we see that a conventional least squares solution to a set of forward-backward linear prediction (FBLP methods) equations produced biased sinusoidal frequency estimates, while the L_1 based solution to the FBLP equations generated an unbiased frequency estimate. In Figure 2. we plot the trajectories of the prediction error filter roots as the iteratively reweighted least squares (IRLS) algorithm progresses from a least squares solution (initial condition for IRLS algorithm) towards an L_1 solution. Each 'x' on a root trajectory corresponds to a prediction error filter root at the kth iteration. Such a plot displays the transient and steady-state behavior of an iterative algorithm somewhat analogous to the root locus plot used in classical control and system theory. We note that the extraneous roots due to intentional overmodeling appear well behaved.

For results shown in Figure 3. and Figure 4., 50 independent trials were run with the phase of each sinusoid drawn from a uniform distribution; white Gaussian noise has not been added. The least squares based frequency estimates are clearly biased (Figure 3.) with relatively high variance, while the L_1 based estimates (Figure 4.) are essentially unbiased with zero variance (note that the roots are hidden by the unit

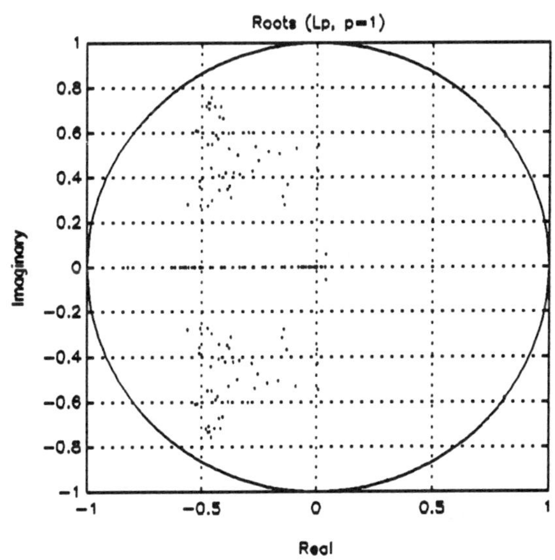

Figure 4. L_p (p=1) Roots, 50 Trials, N=30, L=7, f_1=.1 Hz, f_2=.2 Hz, Phases Uniformly Distributed, No Additive Noise.

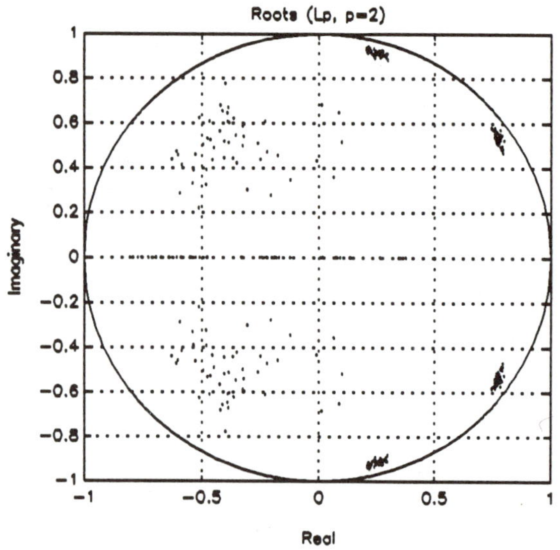

Figure 5. Least Squares Roots, 50 Trials, N=30, L=7, f_1=.1 Hz, f_2=.2Hz, Phases Uniformly Distributed, SNR=30dB.

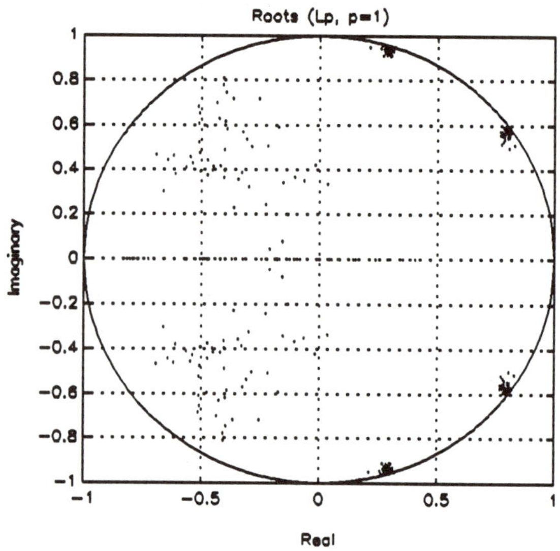

Figure 6. L_p (p=1) Roots, 50 Trials, N=30, L=7, f_1=.1 Hz, f_2=.2 Hz, Phases Uniformly Distributed, SNR=30dB.

circle). It is not necessary for frequency location estimates, and an L_1 linear prediction solution does not guarantee model stability, but for this simulation all roots did fall within the unit circle.

The final simulation results (Figures 5. and 6.) demonstrate the effects of both uniformly distributed random phases and additive white Gaussian noise (30dB), with signal-to-noise ratio (SNR) in dB defined as:

$$SNR=10\ \log(\frac{A^2}{2\sigma^2}), \tag{7}$$

and A=1. The prediction error filter roots for 50 trials produced from a least squares based solution are shown in Figure 5. and the results for the L_1 case are shown in Figure 6. Again the L_1 based solution produces sinusoidal frequency estimates with less bias on the average as compared to a least squares based solution.

IV. Conclusions

We have demonstrated that estimating the two sinusoidal frequencies of a burst sinusoidal frequency shift keyed (FSK) waveform via a least squares based linear predictive algorithm may result in significant frequency bias even in the absence of additional noise. An L_1 solution of the appropriate linear prediction equations by contrast, may be capable of generating frequency estimates with little or no frequency bias. The effects of additive noise, as expected, are detrimental for both L_2 and L_1 based frequency estimation algorithms, however, the L_1 based frequency estimates exhibit less bias than the L_2 biased frequency estimates.

References

[1] Plotkin, E.I., L.M. Roytman, and M.N.S. Swamy, "Estimating the Steady-State Frequency of a Sinewave Burst with Extremely Short Record Length," IEEE Trans. On Acoustics, Speech, and Signal Processing, Vol. ASSP-34, No. 1, pp. 213-15, February, 1986.

[2] Marple, S.L., III, Digital Spectral Analysis, Prentice-Hall, NY, 1989.

[3] Schroeder, J.E., and R. Yarlagadda, "Linear Predictive Spectral Estimation Via the L_1 Norm," Signal Processing, Vol. 17, pp. 19-29, May, 1989.

[4] Schroeder, J.E., R. Yarlagadda, and J.E. Hershey, "L_p Normed Minimization with Applications to Sinusoidal Frequency Estimation," Signal Processing, Vol. 24, Issue 2, pp. 193-216, August, 1991.

Implementation of a Continuous-State Viterbi Algorithm for MAP State Estimation

Darryl R. Morrell and Ross D. Martin
Telecommunication Research Center and
Department of Electrical Engineering
Arizona State University
Tempe, AZ 85287-5706

Abstract

We address the formation of a Viterbi algorithm for MAP state sequence estimation of a discrete-time continuous-valued Markov processes observed in memoryless noise. The usual approach to this problem is to quantize the state space of the Markov process to obtain a standard discrete-state Viterbi algorithm. In this paper, we introduce an alternate approach based on piecewise linear approximation that allows a continuous-state formulation of the Viterbi algorithm. We illustrate metric and survivor calculation using the piecewise linear approximations and provide an example of the algorithm's use.

1 Introduction

The Viterbi algorithm is a well known Maximum *A Posteriori* (MAP) estimator of the state sequence of a discrete-time discrete-state Markov process given a sequence of observations. Recently, it has been applied in several target tracking applications. It is a nonlinear estimator that offers significant robustness when tracking dim or partially observable targets. However, these benefits come at the price of a (usually prohibitive) computational load.

In this paper, we apply piecewise linear approximations to obtain a *continuous-state* implementation of the Viterbi Algorithm. The resulting algorithm provides a computationally tractable method of nonlinear MAP state sequence estimation.

This work was supported by a Faculty Grant in Aid from Arizona State University

2 The Viterbi Algorithm

Let the state of a discrete-time Markov process at time t be denoted x_t and let the evolution of the state be governed by the transition probability density or mass function $p(x_{t+1}|x_t)$. Let z_t be an observation related to the state x_t through the conditional density $p(z_t|x_t)$.

When x_t is defined on a discrete state-space $\mathcal{D}$, the MAP sequence estimate given a sequence of observations is calculated using the Viterbi Algorithm [1]. In the Viterbi Algorithm, there is a stored metric $L_t(x_t)$ for each value x of a discrete state space $\mathcal{D}$ at time t. This metric is proportional to the logarithm of the joint probability of the most probable path to the state x_t from the initial state and the given observations. The metric $L_{t+1}(x_{t+1})$ is calculated recursively from $L_t(x_t)$ by choosing for each state value x_{t+1} the state x_t^o that maximizes the sum $\ln p(z_{t+1}|x_{t+1}) + \ln p(x_{t+1}|x_t^o) + L_t(x_t^o)$:

$$\begin{aligned} L_{t+1}(x_{t+1}) &= \ln p(z_{t+1}|x_{t+1}) + \ln p(x_{t+1}|x_t^o) + L_t(x_t^o) \qquad (1) \\ &= \ln p(z_{t+1}|x_{t+1}) + \max_{x_t}\left[\ln p(x_{t+1}|x_t) + L_t(x_t)\right]. \end{aligned}$$

The recursion is initialized with $L_0(x_0) = \ln p(x_0)$. This recursive update process continues until a decision is required, at which time the state with the largest metric is selected.

In most implementations of the Viterbi algorithm, an estimate of the entire state sequence is required. Thus, the algorithm must store additional information at each step so that the complete estimated sequence can be recovered from the selection of the final state. Typically, this additional information is, for each possible value of x_t, the best state sequence terminating in this value; the best state sequence is called the survivor for x_t. The calculation of the Viterbi Algorithm metric can be thought of as a filtering operation,

1058-6393/91 $01.00 © 1991 IEEE

while the calculation of the survivor sequences can be thought of as a smoothing operation.

The corresponding procedure for a Markov process defined on a continuous state space $\mathcal{C}$ was formulated by Larson and Peschon[2]. Their metric is essentially identical to the Viterbi algorithm metric. However, with a continuous state space, the survivor sequences cannot be computed directly; instead, for each time $t+1$ Larson and Peschon define a functional relationship that provides the best estimate at time t as a function of the estimate at time $t+1$. We adopt this approach; in particular, we define a "survivor" function $S_{t+1}(x_{t+1})$ that consists of the best estimate of the state at time t given that x_{t+1} is the best estimate of the state at $t+1$. Note that $S_{t+1}(x_{t+1})$ is calculated simultaneously with $L_{t+1}(x_{t+1})$ as

$$S_{t+1}(x_{t+1}) = x_t^o, \quad (2)$$

where x_t^o is obtained from (1).

The continuous state Viterbi algorithm holds theoretical interest because many MAP estimation problems are naturally stated on a continuous state space. Approximations to the continuous problem are usually needed to implement an estimation algorithm on a digital computer, and the error incurred by these approximations and the range of possible approximations are best measured with reference to the continuous algorithm. The continuous state Viterbi algorithm also holds interest as a generalization of the Kalman filter, since Larson and Peschon[2] showed that it is equivalent to the Kalman filter for a linear system with Gaussian noise.

For digital implementation, Larson and Peschon converted the continuous problem into a discrete problem by quantizing the continuous state space $\mathcal{C}$. This results in the familiar discrete-state Viterbi algorithm. However, there are other methods of approximating the continuous-state Viterbi algorithm which do not reduce to the discrete-state Viterbi algorithm. In the next section, a method based on piecewise linear approximations of the logarithms of the probability densities in (1) will be discussed.

3 A Piecewise Linear Approximation

We introduce the following piecewise linear approximations to the logarithms of the densities $p(z_{t+1}|x_{t+1})$, $p(x_{t+1}|x_t^o)$, and $p(x_0)$:

$$\ln p(x_{t+1}|x_t) \approx g(cx_{t+1} + d - x_t) \quad (3)$$

$$\ln p(z_{t+1}|x_{t+1}) \approx h_z(x_{t+1}) \quad (4)$$

$$\ln p(x_0) \approx f(x_0) \quad (5)$$

where c is a positive constant, d is a real constant, and $g(\cdot)$, $h_z(\cdot)$, and $f(\cdot)$ are continuous piecewise linear functions of their arguments. ($h_z(\cdot)$ may be a nonlinear function of the observation z.) Note that these piecewise linear functions can be represented by a finite number of parameters; we represent them using a list of the end points of the line segments comprising the functions.

The primary motivation for using piecewise linear approximations is to facilitate the calculation of $L_{t+1}(x_{t+1})$ and $S_{t+1}(x_{t+1})$. In the calculation of $L_{t+1}(x_{t+1})$ from $L_t(x_t)$ according to (1), it is clear that the addition of two piecewise linear functions results in a piecewise linear function. Thus, we must ascertain that the maximum operation in (1) also results in a piecewise linear function of x_{t+1}, and that the maximizing value of x_t can be represented as a piecewise linear function of x_{t+1}.

In order to illustrate that piecewise linear functions result from the maximization operation in (1) and to obtain insight into the result, we investigate the maximization operation for the following two functions:

$$L_t(x_t) = a|x_t - \xi|$$

$$\ln p(x_{t+1}|x_t) = b|cx_{t+1} + d - x_t|, \quad (6)$$

where a and b are negative constants and c is a positive constant. Note that $L_t(x_t)$ has a peak of value zero at $x_t = \xi$ and, for a given value of x_{t+1}, that $\ln p(x_{t+1}|x_t)$ has a peak of value zero at $x_t = cx_{t+1} + d$. Thus, for a given value of x_{t+1}, the sum $L_t(x_t) + \ln p(x_{t+1}|x_t)$ will be a piecewise linear function of x_t and its maximum will occur at either the peak of $L_t(x_t)$ or the peak of $\ln p(x_{t+1}|x_t)$. The point at which the maximum occurs is determined by the relative magnitudes of a and b. If $a > b$, the maximum will occur at $x_t = \xi$, while if $a < b$, the maximum will occur at $x_t = cx_{t+1} + d$. Thus,

$$\max_{x_t} [L_t(x_t) + \ln p(x_{t+1}|x_t)]$$

$$= \begin{cases} b|\xi - cx_{t+1} - d|, & a > b \\ a|cx_{t+1} + d - \xi|, & a < b \end{cases} \quad (7)$$

$$S_{t+1}(x_{t+1}) = \begin{cases} \xi, & a > b \\ cx_{t+1} + d, & a < b \end{cases} \quad (8)$$

Note that (7) and (8) are piecewise linear in x_{t+1}. $L_{t+1}(x_{t+1})$ is obtained by pointwise addition of $\ln p(z_{t+1}|x_{t+1})$ to the right side of (7), clearly resulting in a piecewise linear function of x_{t+1}. These results can be extended in a straight-forward manner

	Time					
	0	1	2	3	4	5
State	0	0.53	1.09	1.74	2.19	2.69
Observation		0.69	2.80	1.55	2.10	2.74

Table 1: State and observation sequences for the example.

to any continuous piecewise linear $L_{t+1}(x_{t+1})$, illustrating that the assumptions (3), (4), and (5) lead to piecewise linear expressions for $L_{t+1}(x_{t+1})$.

4 Simulation Example

A computer program has been implemented that computes (1) and (2) recursively using arbitrary continuous piecewise linear approximations to $\ln p(z_t|x_t)$ and $\ln p(x_0)$; $\ln p(x_{t+1}|x_t)$ is required to have the form given by (6). The sequence of $L_t(x_t)$ and $S_t(x_t)$ computed by the program are processed to obtain both a filtered sequence and a smoothed sequence from an observation sequence.

Two simulations were performed. We used the following approximations for the densities for both simulations:

$$L_0(x_0) = \ln p(x_0) = -2|x_0|$$

$$\ln p(x_{t+1}|x_t) = -8|x_{t+1} + 0.5 - x_t|$$

$$\ln p(z_{t+1}|x_{t+1}) = \begin{cases} -2|x_{t+1} - z_{t+1}|, & |x_{t+1} - z_{t+1}| \leq 0.25 \\ -0.5 - 10|x_{t+1} - z_{t+1}|, & |x_{t+1} - z_{t+1}| > 0.25 \end{cases}$$

For the first example, a state sequence and a corresponding observation sequence were generated by a pseudo-random number generator; these sequences are given in Table 1. The observation sequence consists of the state sequence corrupted by additive white noise. Figure 1 shows $L_t(x_t)$ for times 1 through 5, and Figure 2 shows the corresponding $S_t(x_t)$. Figure 3 compares the actual state trajectory, the estimated state trajectory obtained by choosing the value of x_t that maximizes $L_t(x_t)$ at each t (ie. the filtered state trajectory), and the estimated state trajectory obtained by choosing the value of x_5 that maximizes $L_5(x_5)$ and then working backwards using $S_t(x_t)$ (ie. the smoothed state trajectory). Note the large error in the forward state estimate at time $t = 2$ due to the large deviation between the observation and the state; this error is eliminated in the smoothed state estimate.

A second simulation was run to illustrate longer term behavior of the algorithm. The actual and estimated trajectories from this simulation are shown in Figure 4.

5 Conclusions

In this paper, we apply the Viterbi Algorithm to the problem of MAP state estimation on a continuous state space. We form a MAP estimator using piecewise linear approximations. Preliminary simulations indicate that the computational load of the algorithm developed to compute $L_{t+1}(x_{t+1})$ and $S_{t+1}(x_{t+1})$ is approximately linear in the number of line segments used to represent $L_t(x_t)$, and that the number of line segments remains small. The load is a significant reduction over the computational load required to compute the metric in a discrete state Viterbi algorithm of similar resolution. The memory required by the continuous state algorithm is also smaller than that required by the discrete state algorithm.

We are currently working to extend the results presented here in several ways:

- Develop an algorithm that allows piecewise linear functions of more than two line segments to be used to represent $\ln p(x_{t+1}|x_t)$.
- Modify the linear relationship between x_{t+1} and x_t expressed in (3) to allow non-linear constraints on x_{t+1} and x_t.
- Extend the estimator to higher dimensional state spaces.
- Determine the statistical performance of the estimator.

References

[1] G. D. Fourney, "The Viterbi algorithm," *Proceedings of the IEEE*, Vol. 61, pp. 268–278, March 1973.

[2] R. E. Larson and J. Peschon, "A dynamic programming approach to trajectory estimation," *IEEE Transactions on Automatic Control*, pp. 537–540, July 1966.

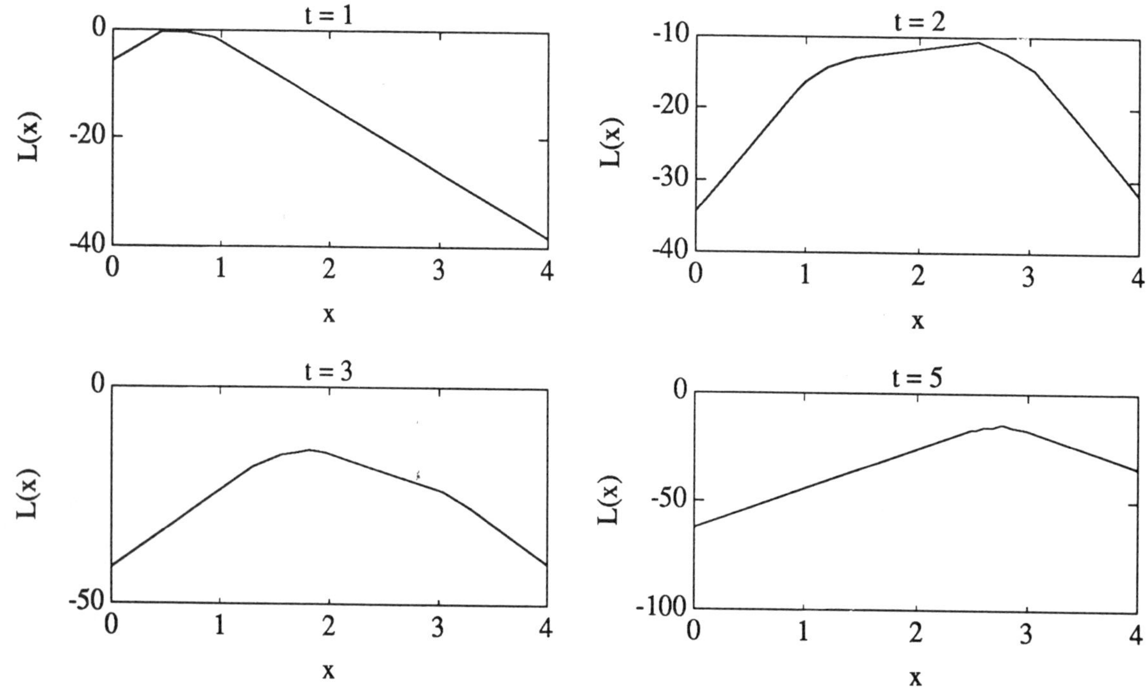

Figure 1: $L_t(x_t)$ for $t = 1$ to $t = 5$ for the first example.

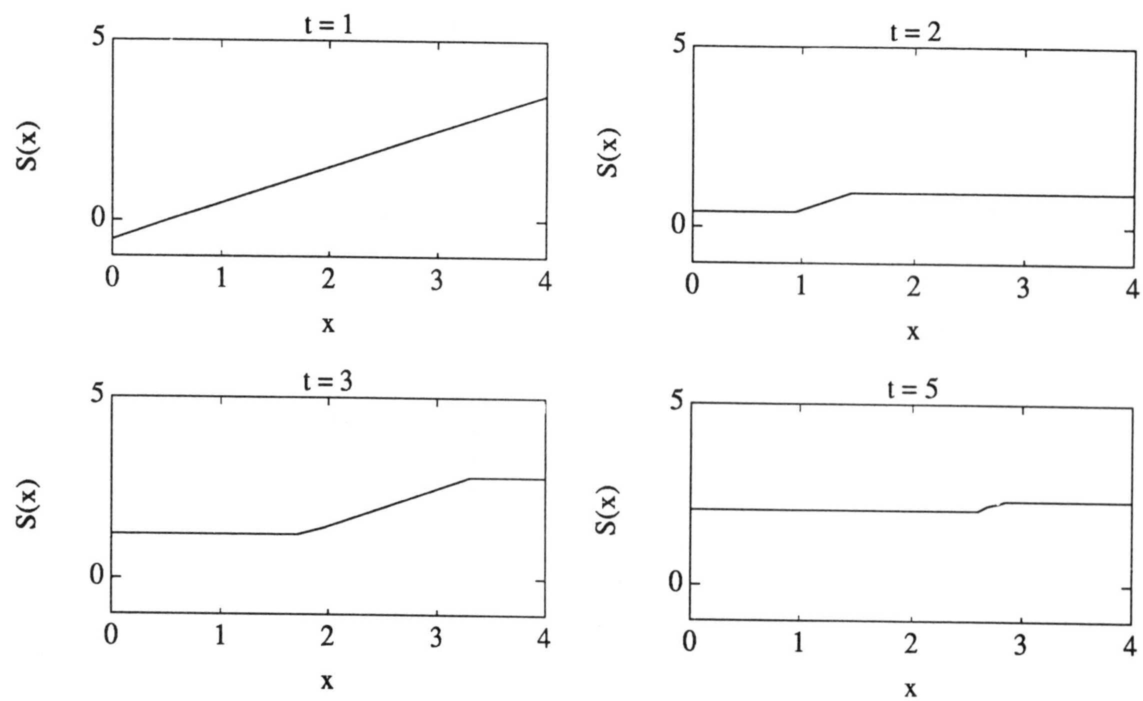

Figure 2: $S_t(x_t)$ for $t = 1$ to $t = 5$ for the first example.

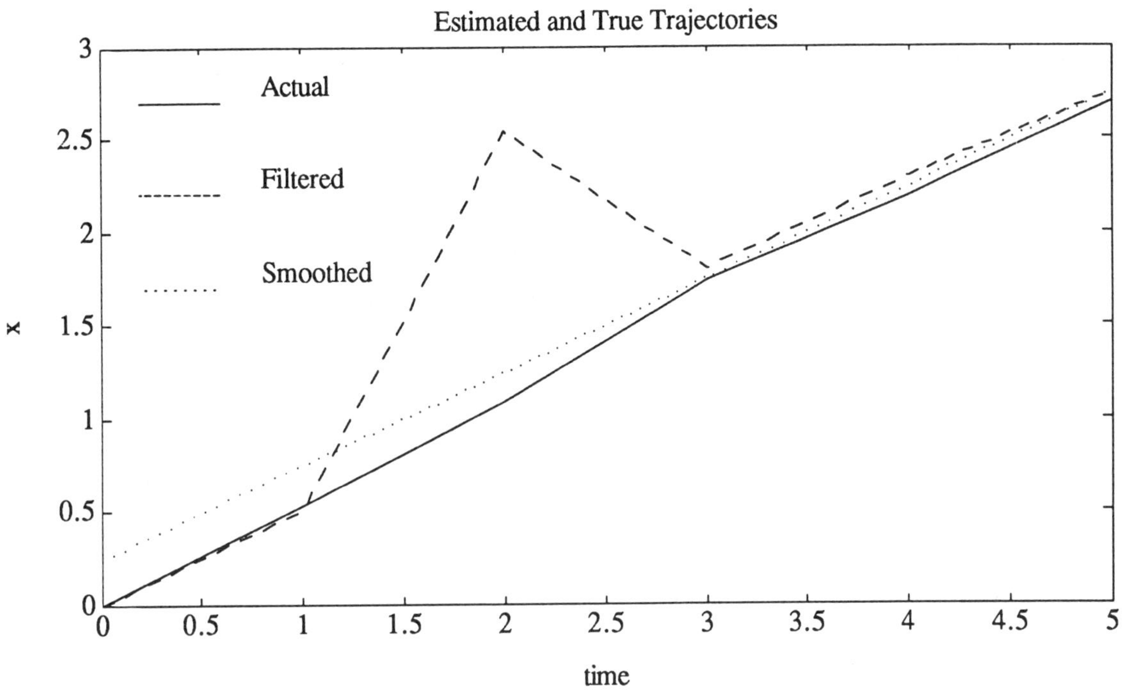

Figure 3: Filtered, smoothed, and actual trajectories for the first example.

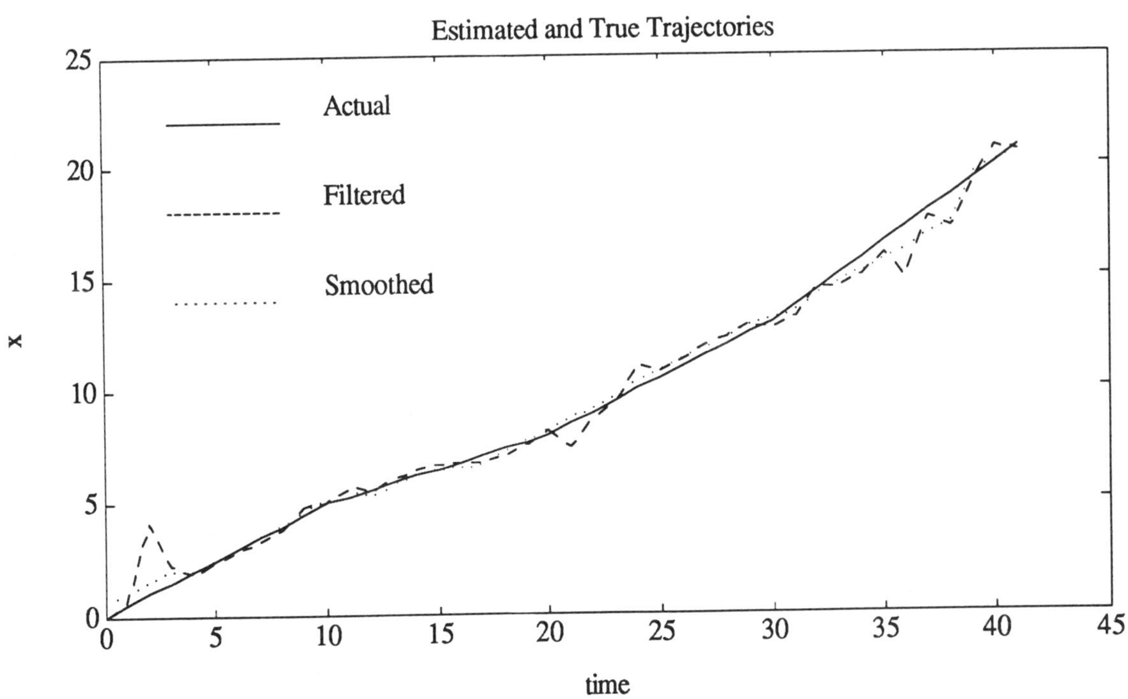

Figure 4: Filtered, smoothed, and actual trajectories for the second example.

Estimating Transfer Functions of Quadratically Nonlinear Systems

Yi-Chyun Tseng and Darel A. Linebarger
Center for Engineering Mathematics
The University of Texas at Dallas, MP 33
Richardson, Texas 75083–0688 USA
(214) 690–2991 dl@utdallas.edu

Abstract

Sinusoidal and white noise input methods have previously been developed for estimating linear or higher order transfer functions of nonlinear systems. Sinusoidal approaches can be more accurate at the cost of more computations. White noise input methods require less computation, but aliasing, which appears in the output due to higher order components in the system, will introduce more errors in the estimation. With a colored noise input method, which is based on partitioning the frequency band of interest into subbands, we can have less aliasing in the output and allow for greater precision in the transfer function estimates.

1. Introduction

Under a very general consideration, a nonlinear time invariant system with the output time function $y(t)$ and input $x(t)$ can be expressed as the sum of a Volterra series:

$$y(t) = \sum_{p=1}^{\infty} \mathbf{H}_p[x(t)] \tag{1}$$

in which

$$\mathbf{H}_p[x(t)] = \int_{-\infty}^{\infty} \cdots \int_{-\infty}^{\infty} h_p(\tau_1, \cdots, \tau_p) x(t-\tau_1) \cdots \cdots x(t-\tau_p) d\tau_1 \cdots d\tau_p \tag{2}$$

where $h_p(\tau_1, \cdots, \tau_p)$ is called the p-th order Volterra kernel [1, 2]. We also assume that the system is causal such that

$$h_p(\tau_1, \cdots, \tau_p) = 0, \quad \text{for any } \tau_i < 0,\ i = 1, 2, \cdots, p. \tag{3}$$

In general, the Volterra kernels are not necessarily symmetric [3], but in this paper, the Volterra kernels are assumed to be symmetric for simplicity. The focus of this paper is on a "weakly nonlinear" system, where the nonlinearity in the system is sufficiently small so that a perturbation expansion may be truncated after the quadratic terms [4].

The estimation of the transfer functions of a discrete time nonlinear system using different inputs has been widely studied [5, 6], but the estimation error, due to the aliasing generated from higher order kernels, has not been discussed. For a system with digital input signals, the frequency range of the input signals must be carefully examined to prevent aliasing. For linear systems, as long as we keep the frequency of the input signal below half of the sampling frequency, the output of the system will be free of aliasing. For a higher order system, the output of the system is composed of the linear response, higher order harmonics, intermodulation frequency components and possibly aliasing due to higher order components. With discrete time nonlinear systems, aliasing can occur even if the input is limited to half of the sampling frequency.

The estimation of the system kernels can be improved by reducing the effect of aliasing. Even though we limit the frequency of the input signal below half of the sampling frequency, the system output will be blurred by the aliasing generated by the higher order kernels. These aliased terms will influence not only the estimation of the higher order kernels, but also the first order kernel. Therefore, if we use the whole signal frequency band from 0 to half of the sampling frequency in our input at one time, the error caused by aliasing will be significant. If we separate the input signal into several frequency bands, and take one or two bands as input to the system at a time and perform the estimation, we will see that either the estimation is free of aliasing, or the aliasing appears only in certain regions. Some of the

1058-6393/91 $01.00 © 1991 IEEE

aliasing effects can be ignored if they fall outside the frequency regions of our interest. Then the estimation will be "aliasing free".

If we want to estimate the transfer function for all the possible frequencies below half of the sampling frequency, it is not possible to have "aliasing free" estimation. However, with the scheme proposed in this paper, we can improve the estimation of both the first and second order Volterra kernels.

2. Estimation of Volterra Kernels

The output of a quadratic system can be expressed as

$$Y(f_m) = H_1(f_m)X(f_m) + \sum_{i=-(\frac{N}{2}-1)}^{\frac{N}{2}} \sum_{j=-(\frac{N}{2}-1)}^{\frac{N}{2}} H_2(f_i, f_j)X(f_i)X(f_j), \; i+j=m \tag{4}$$

where $X(f_m)$ and $Y(f_m)$, respectively, represent the discrete Fourier transform (DFT's) calculated by a length N (assumed even) FFT. Using length N FFT's, $f_{\frac{N}{2}}$ will represent the highest frequency that can be represented without aliasing. $H_1(f_m)$ and $H_2(f_i, f_j)$ represent the linear and quadratic transfer functions. $H_1(f_m)$ and $H_2(f_i, f_j)$ are also the Fourier transforms, respectively, of the first and second order Volterra kernels. Without loss of generality, the quadratic transfer function is assumed symmetric, *i.e.* $H_2(f_i, f_j) = H_2(f_j, f_i)$. To simplify notation, $X(f_m)$ will be denoted as $X(m)$, $H_2(f_i, f_j)$ as $H_2(i,j)$, etc.

As described in [5], the transfer function can be estimated as

$$\mathrm{h}(m) = \left\{E[\mathbf{x}^*(m)\mathbf{x}^T(m)]\right\}^{-1} E[\mathbf{x}^*(m)Y(m)] \tag{5}$$

where h and $\mathbf{x}$ are vectors with the forms

$$\mathrm{h}^T = \begin{cases} [H_1(m), 2H_2(\frac{m+1}{2}, \frac{m-1}{2}), \cdots, \\ \quad \cdots, 2H_2(m,0), \cdots 2H_2(\frac{N}{2}, m-\frac{N}{2})] & \text{for } m \text{ odd} \\ [H_1(m), 2H_2(\frac{m}{2}, \frac{m}{2}), 2H_2(\frac{m}{2}+1, \frac{m}{2}-1), \cdots, \\ \quad \cdots, 2H_2(m,0), \cdots 2H_2(\frac{N}{2}, m-\frac{N}{2})] & \text{for } m \text{ even} \end{cases} \tag{6}$$

and

$$\mathbf{x}^T = \begin{cases} [X(m), X(\frac{m+1}{2})X(\frac{m-1}{2}), \cdots, \\ \quad \cdots, X(m)X(0), \cdots, X(\frac{N}{2})X(m-\frac{N}{2})] & \text{for } m \text{ odd} \\ [X(m), X(\frac{m}{2})X(\frac{m}{2}), X(\frac{m}{2}+1)X(\frac{m}{2}-1), \cdots, \\ \quad \cdots, X(m)X(0), \cdots, X(\frac{N}{2})X(m-\frac{N}{2})] & \text{for } m \text{ even} \end{cases} \tag{7}$$

with the superscripts T and * for transpose and conjugate, respectively. The estimation of the transfer function is given by inputting a random signal, Gaussian or non-Gaussian, and measure the output of the system, then estimate the transfer function by eq. (5). In [5], the input is assumed to be white, but not necessarily Gaussian.

Since the output of the quadratic system contains higher order harmonics and intermodulation components, aliasing might exist in the output if we don't limit the input frequency range. The output of the system can be aliasing free if the input frequency is less than a quarter of the sampling frequency. If the input frequency is higher than a quarter of the sampling frequency, aliasing appears in the output. However, this aliasing effect cannot be detected in the output spectrum if we use a white noise input. Because the scheme for estimating the transfer functions based on eq. (5) depends heavily on the input and output of the system, the estimates of the transfer function will be influenced by this aliasing effect.

By dividing the input into several subbands, we can reduce the aliasing in the output, and hence, have a better estimation. In figure 1, the frequency range of the input signal is equally divided into four subbands. As in [5],

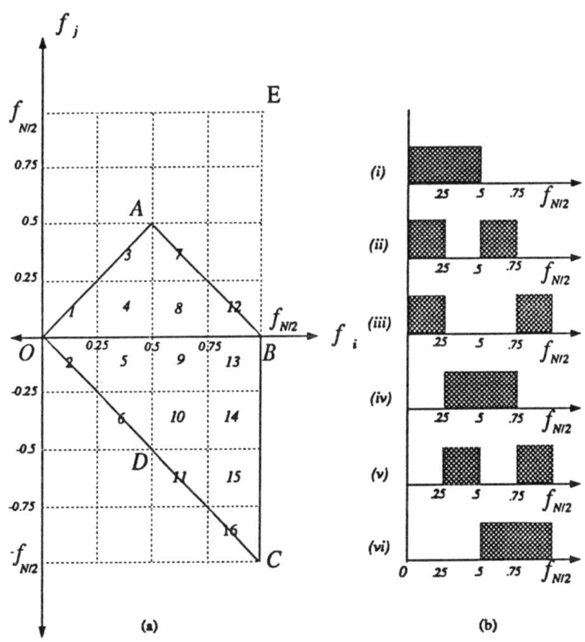

Figure 1: (a) The right hand side of the signal space of quadratic kernels. By use of the symmetric property of the quadratic transfer function, the overall quadratic signal space can be acquired by flipping region OABC along OA axis and OC axis. For colored noise input method, the region OABC is partitioned according to different input frequency bands. The estimation over each subregion of OABC is contributed by different subband inputs shown in (b). Since the input is real, both the positive and negative frequency components of the colored noise are present.

the goal is to estimate the transfer function over the region *OABC*. It is obvious that if we choose the signal frequency subband over $(0...0.5f_{\frac{N}{2}})$, $H_2(f_i, f_j)$ within the region *OAD* can be estimated without aliasing. If we choose subbands which contain any frequency f_o beyond $0.5f_{\frac{N}{2}}$, aliasing will occur. But as long as the aliasing does not fall into the portion of the region *OABC* under consideration, the estimation for both linear and quadratic transfer functions will remain aliasing free.

With a prefiltering stage to limit the input frequency range, we can reduce the influence of the aliasing and have a lower mean squared error in estimation. Instead of taking the whole band as the input, we partition the input signal into several subbands. This is a step towards a sinusoidal input method, by which greater precision in estimating H_1 and H_2 will be obtained at a cost of more computations. Figure 1–(a) shows the right hand side of the signal space of a quadratic system. According to [5], the region *OABC* is the domain over which $H_2(f_i, f_j)$ is to be estimated. However, if we allow the frequency of the input signal to vary from 0 to $f_{\frac{N}{2}}$, we should estimate the quadratic transfer function over region *OEC* instead of over region *OABC*. The estimation of the quadratic transfer function over all frequency space can then be acquired by applying the symmetric property of quadratic transfer functions. For a fair comparison, we limit our discussion into the same region (*OABC*) proposed by [5].

The colored noise input method can be implemented as follows. In our example, we first divide the whole input frequency band into four equal size subbands. Then two of the four noise subbands are combined and used as the input to estimate the transfer functions. Each such input allows estimation of the quadratic transfer function over a subset of the region *OABC*. A list matching the inputs and the corresponding subregions for the output is given in Table 1.

From Table 1, we can see that each estimation based on a subband input provides valid estimates only over some subregions under the region *OABC*. Any output in a subregion other than those listed for a given input in Table 1 can be ignored when using that input since it was presumably induced via aliasing. After repeating this process for all six different input combinations shown in figure 1–(b), we have estimates for all subregions. Take region *1* on figure 1–(a) for example, only the estimations from input *(i)*, *(ii)*, *(iii)* of figure 1–(b) contain valid estimates. The final estimate for subregion *1* is then acquired by averaging the estimates over subregion *1* of three different subband estimates. However, we should note that this partitioning technique is not unique. More complicated partitions can be used, but the computational complexity will be increased accordingly.

Input	Output subregions
i	*1,2,3,4,5,6*
ii	*1,2,8,9,11*
iii	*1,2,12,13,16*
iv	*3,6,7,10,11*
v	*3,6,14,16*
vi	*11,15,16*

Table 1: The estimate over each subregion of the region OABC is contributed by different inputs. For each input combination, only a finite subregions, which are separated with dotted lines in figure 1–(a), under region OABC will have valid estimates as shown in the table. The estimates fall outside the subregions in the list with respect to each input, are ignored since the nonzero estimates over those subregions are introduced only by aliasing generated by the quadratic part of the system.

3. Kernel Estimation of a Known Quadratic System

The known system for our simulation is the same as that in [5]:

$$y(n) = -0.64\,x(n) + x(n-2) + 0.9x^2(n) + x^2(n-1)\,. \tag{8}$$

The theoretical transfer functions are given by

$$H_1(f) = -0.64 + \exp(-j4\pi f) \tag{9}$$

$$H_2(f_i, f_j) = 0.9 + \exp(-j2\pi(f_i + f_j))\,. \tag{10}$$

For a quantitative measure of the quality of the transfer function estimates, the normalized mean squared errors are defined as follows [5]:

$$MSEL = \frac{1}{\left(\frac{N}{2}\right)} \sum_{m=1}^{\frac{N}{2}} \frac{\left|H_1(m) - \hat{H}_1(m)\right|^2}{|H_1(m)|^2} \tag{11}$$

for the linear kernel and

$$MSEQ = \frac{1}{N_q} \sum_{i,j} \frac{\left|H_2(i,j) - \hat{H}_2(i,j)\right|^2}{|H_2(i,j)|^2} \tag{12}$$

for the quadratic kernel. In these error equations, the denotes an estimated quantity, $N/2$ signifies the number of frequencies up to half of the sampling frequency, and N_q is the total number of the frequency points under region *OABC*. The estimation and the errors by different methods are shown in figures 2 to 7.

To estimate the transfer function with white noise input method, the input signal used is white random noise with a Gaussian distribution. Signals with other distributions, such as an exponential distribution or a uniform distribution can be used as well. The input consists of 128,000 sample points generated by PRO-MATLAB random function. The input data is divided into 2,000 segments each with 64 data points each. Each segment is transformed into frequency domain via a 64 point FFT. Then the 2,000 sample spectra are averaged to obtained the spectral estimates. The output spectral estimates are acquired in a similar method. The estimation is acquired by applying eq. (5).

For the colored input method, the input frequency range $\left(0 \cdots f_{\frac{N}{2}}\right)$ was equally divided into four regions. Among those four regions, only two of them were used to do the estimation each time as shown in figure 1–(b). Each of the six estimations contributes only part of the estimation over the region $OABC$, which is divided by the dotted lines in figure 1, according to different input frequency ranges. Some of the regions may be influenced by more than one estimate. The final transfer function estimate is then acquired by taking the average of all related estimates as listed in Table 1. The colored noise is generated by prefiltering the same Gaussian noise for the white noise input method through seventh order Butterworth filters tuned to different frequency band(s).

With the following results, it is obvious that the colored noise input method outperforms the white noise input method in both the estimations of linear and quadratic transfer functions. The true kernels and their estimates with different input methods are shown in figures 2, 3 for linear kernel, and figures 5, 6 and 7 for quadratic kernel. The mean squared errors MSEL and MSEQ are listed in Table 2. The MSEL and MSEQ for colored input method are only 41.67% and 63.38%, respectively, of the white noise input method.

	White Noise Input	Colored Noise Input
MSEL	0.00204	0.00085
MSEQ	0.01436	0.009102

Table 2: The mean squared errors for the linear kernel (MSEL) and for the quadratic kernel (MSEQ) for both white and colored noise inputs are given.

4. Conclusion

In conclusion, at the expense of slightly more computations, our method provides considerably improved transfer function estimates. Also, our method provides flexibility to concentrate on obtaining the best estimate of the transfer functions over certain bands if that is all that is required. Our method is essentially a step towards a sinusoidal input method away from a white noise input method. It can be made to lie anywhere between the single frequency or white noise input methods by controlling the nature of the colored noise.

Reference

[1] M. Schetzen, *The Volterra & Wiener Theories of Nonlinear Systems.* Robert E. Krieger Publishing Company, 1989.

[2] W. J. Rugh, *Nonlinear System Theory.* John Hopkins, 1981.

[3] L. Chua and C. Ng, "Frequency domain analysis of nonlinear systems: general theory," *Electronic Circuits and Systems, vol. 3, No, 4, pp.165-185, Jul.,* 1979.

[4] L. J. Tick, "The Estimation of 'Transfer Functions' of Quadratic Systems," *Technometrics, vol. 3, No. 4, pp. 563-567, Nov.,* 1961.

[5] K. I. Kim and E. J. Powers, "A digital Method of Modeling Quadratically Nonlinear Systems with a general Random Input," *IEEE Trans. ASSP, Vol. 36, No. 11, pp. 1758-1769, Nov.,* 1988.

[6] Y. T. Stephen Boyd and L. O. Chua, "Measuring Volterra Kernels," *IEEE Trans. on Circuits and Systems, vol. CAS-30, No.8, pp. 571-577, Aug.,* 1983.

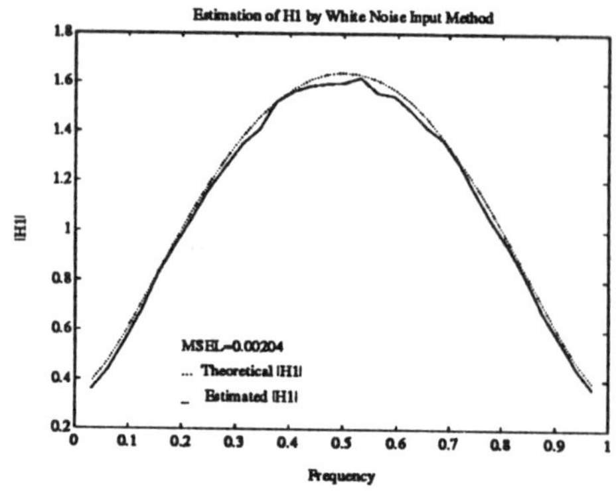

Figure 2: Estimation of first order kernel (H_1) using the white noise input method. The dotted line shows the theoretical values of the first order kernel.

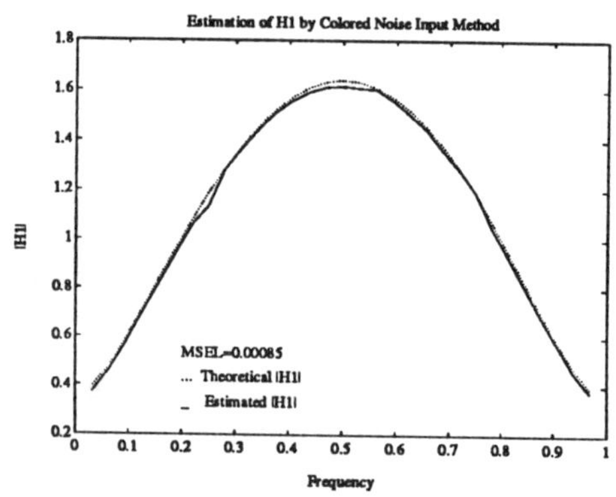

Figure 3: Estimation of first order kernel (H_1) using the colored noise input method. The dotted line shows the theoretical values of the first order kernel.

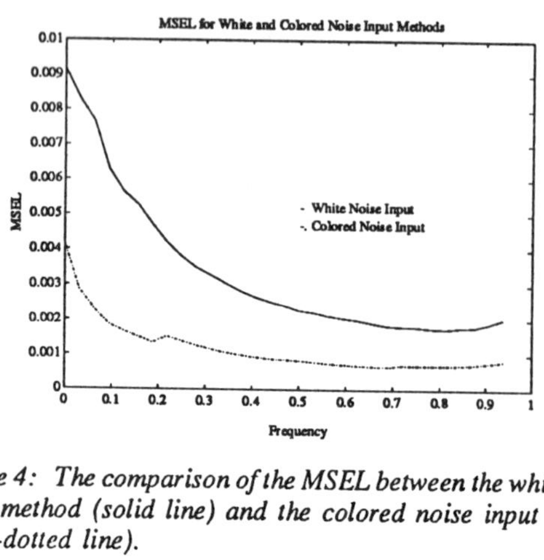

Figure 4: The comparison of the MSEL between the white noise input method (solid line) and the colored noise input method (dash-dotted line).

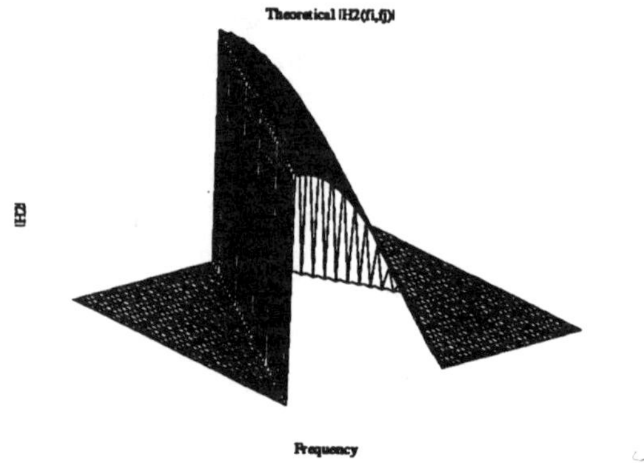

Figure 5: The theoretical values of the second order kernel (H_2) .

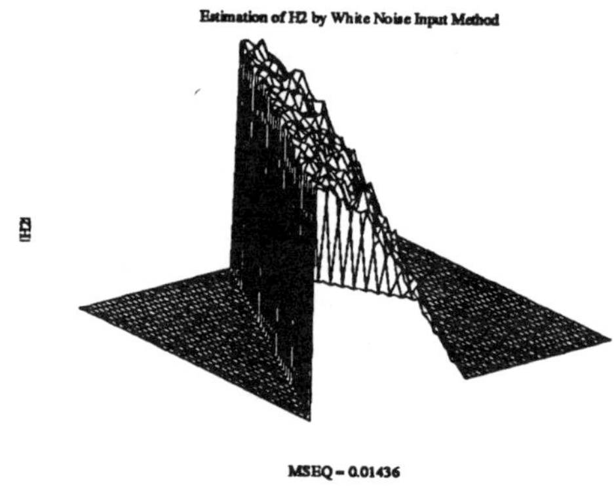

Figure 6: EstimatedH_2 using the white noise input method.

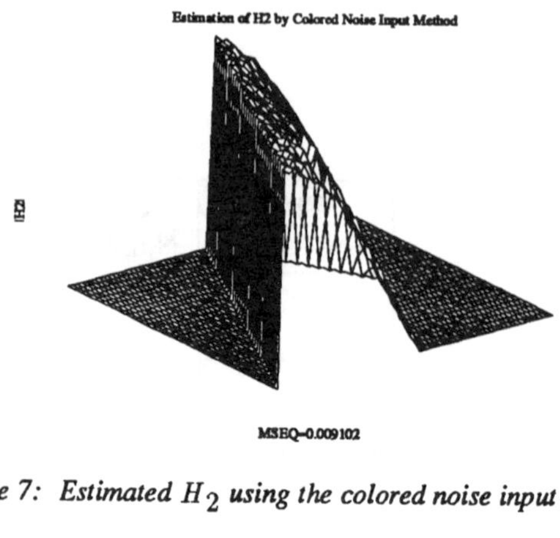

Figure 7: Estimated H_2 using the colored noise input method.

A Unified Formulation for Detection Using Time-Frequency and Time-Scale Methods

Mark L. Fowler and Leon H. Sibul
Applied Research Laboratory
The Pennsylvania State University
P.O. Box 30
State College, PA 16804

Abstract

A unified theory is developed for the detection of signals in nonwhite noise using time-frequency and time-scale signal transforms. This class includes the cross-Wigner distribution and certain other members of Cohen's class, the Gabor transform, and the wavelet transform.

Necessary and sufficient conditions are established such that a signal transform is applicable to the nonwhite noise detection problem. Applying this result to the reproducing kernel Hilbert space (RKHS) associated with a finite-dimensional approximation of the noise covariance leads to weighted correlator structures in the time-frequency and time-scale domains. This extends results previously available only for the white noise case.

1 Introduction

There has been much interest during the past decades in time-frequency representations such as the Wigner distribution and the Gabor transform [1], [2]. More recently, interest has turned to the wavelet transform [3], which is a time-scale representation. In particular, the application of time-frequency and time-scale methods to the problem of detecting a signal in the presence of additive noise has been considered [4] – [7], and advantages of these approaches have been noted. For example, Boashash and O'Shea [5] applied Wigner distribution techniques to the detection and classification of the firings of an underwater diesel engine in the presence of white noise and showed that the resulting detector performed better than did the classical time-domain method. They showed that the improvement is due to time-varying filtering that is inherent to the Wigner-based method. Similarly, Friedlander and Porat [6] demonstrated that the time-frequency localization provided by the Gabor transform yields the capability to detect, in the presence of white noise, partially overlapping transients with unknown decay rates, oscillation rates, and arrival times. Additionally, Tuteur [7] has shown that the wavelet transform is effective in the detection of electrocardiogram signals.

These demonstrations of the effectiveness of the time-frequency methods in overcoming some of the problems suffered by other signal detection techniques are encouraging. However, the existing time-frequency methods for detection have been formulated only for the white noise case, and therefore have limited applicability to practical detection problems. While it is true that a whitening filter could be used prior to the application of the time-frequency methods, a direct formulation for the nonwhite noise case would provide a more versatile and more widely applicable approach. For example, if a whitening filter is used, then the time-frequency correlator uses time-frequency representations of whitened versions of the signal; these representations of the whitened signals are less useful for further processing—such as signaturing and classification—than are the time-frequency representations of the actual signals. Another characteristic of the previously reported techniques is that despite the similar inner product structures of the various time-frequency and time-scale representations, the methods proposed for each transform have been developed using differing formulations. In response to these observations, we have developed a formulation that allows a wide variety of time-frequency and time-scale representations to be applied in a unified way to the detection of signals in nonwhite noise. Furthermore, this unification provides a framework within which the various time-frequency and time-scale detection methods can be compared and contrasted. Therefore, our formulation not only extends the applicability of time-frequency and time-scale methods to the nonwhite noise, but provides insight into the proper choice of a representation for a particular signal/noise environment.

In Section 2 we develop a unified transform approach to detection based on the reproducing kernel Hilbert space (RKHS) formulation of the detection problem. In particular, we give necessary and sufficient conditions for a transform to unitarilly map one RKHS onto another RKHS. In Section 3 we exploit the inherent inner product structure of each of the time-frequency and time-scale representations to show that the results of Section 2 are applicable to each of these signal transforms. The application of the wavelet transform to detection in the presence of a $1/f$ type noise is illustrated by example.

1058-6393/91 $01.00

2 Unified Transform Approach to Detection

The detection problem we consider is as follows. Given a signal r received over the observation interval I, make a decision between

$$\left.\begin{array}{ll} H_1: & r(t) = \gamma(t) + n(t) \\ H_0: & r(t) = n(t) \end{array}\right\} t \in I,$$

where γ is the deterministic signal to be detected, and n is some zero-mean Gaussian noise. The detection problem is typically solved by comparing a sufficient statistic, computed from the received signal r, to some pre-determined threshold. Our results are based on the reproducing kernel Hilbert space (RKHS) formulation of the sufficient statistic. An RKHS $H(K)$ is a special Hilbert space, uniquely associated with the kernel K, that has the property that $K(\cdot, x) \in H(K)$ for all values of x and $< f, K(\cdot, x) >_{H(K)} = f(x)$ for every $f \in H(K)$. Briefly, under the assumptions that I is a bounded interval and that the noise covariance K is continuous on $I \times I$, the sufficient statistic (or likelihood ratio) is given by

$$\Lambda = < r, \gamma >_{H(K)},$$

where $H(K)$ is the RKHS that is uniquely determined by the noise covariance K [8].

The important insight provided by the RKHS formulation is that the sufficient statistic is determined completely by the inner product structure of $H(K)$. Thus, insight into the development of equivalent formulations in alternative domains (e.g., time-frequency and time-scale domains) can be gained by investigating the application of various operators to the space $H(K)$. In particular, we are interested in the question: What types of unitary mappings can be applied to $H(K)$ such that the resulting space is another RKHS? The answer to this question is provided here by developing the following theorem, which is an extension of a theorem given by Chalmers [9] that gives the relationship between the kernels of two RKHSs that are related by a bounded linear operator. Chalmers' result requires *a priori* knowledge that the two spaces are indeed RKHSs; in contrast, our result gives necessary and sufficient conditions for an operator to map an RKHS into another RKHS, and gives the relationship between the kernels.

Notation for the following theorem: $[\mathcal{T} K_1(s,t)]_{s|x}$ means "operate on $K_1(s,t)$ as a function of s and evaluate the resulting function at the point x."

Theorem 1 *Let H_1 be an RKHS with reproducing kernel K_1 defined on some set $I_1 \times I_1$, and let $\mathcal{T}: H_1 \to H_2$ be unitary. The space H_2 is an RKHS of functions defined on some set I_2 if and only if the operator $\mathcal{T}$ can be represented by*

$$(\mathcal{T} f)(x) = < f, T(\cdot, x) >_{H_1}, \qquad (1)$$

for some bivariate function T defined on $I_1 \times I_2$ that has the property that $T(\cdot, x) \in H_1$, $\forall x \in I_2$. Furthermore, the reproducing kernel K_2 for H_2 is given by

$$K_2(y, x) = \left[\mathcal{T}\overline{[\mathcal{T} K_1(s,t)]_{s|x}}\right]_{t|y}, \quad \forall x, y \in I_2. \qquad (2)$$

Outline of proof: Under the assumption that (1) defines a unitary operator, with $T(\cdot, x) \in H_1$, it follows that

$$(\mathcal{T} f)(x) = < \mathcal{T} f, \mathcal{T} T(\cdot, x) >_{H_2}.$$

This establishes that H_2 is an RKHS with reproducing kernel K_2 given by $K_2(y,x) = [\mathcal{T} T(t,x)]_{t|y}$. Conversely, under the assumption that H_2 is an RKHS, the unitary mapping $\mathcal{T}$ can be represented by

$$\begin{aligned} (\mathcal{T} f)(x) &= < \mathcal{T} f, K_2(\cdot, x) >_{H_2} \\ &= < f, \mathcal{T}^{-1} K_2(\cdot, x) >_{H_1} \\ &\triangleq < f, T(\cdot, x) >_{H_1}, \end{aligned}$$

where $T(\cdot, x) \in H_1$. Furthermore, under each of these assumptions, $T(t,x) = \overline{< K_1(\cdot, t), T(\cdot, x) >_{H_1}} = \overline{[\mathcal{T} K_1(s,t)]_{s|x}}$, which establishes (2).
□

This theorem characterizes the transforms that can be applied to the detection problem. Specifically, if $\mathcal{T}$ is a unitary operator that satisfies Theorem 1 with $H_1 = H(K)$, where K is the noise covariance, then the sufficient statistic can be expressed as

$$\Lambda = < \mathcal{T} r, \mathcal{T} \gamma >_{H(\tilde{K})}, \qquad (3)$$

where

$$\tilde{K}(y, x) = \left[\mathcal{T}\overline{[\mathcal{T} K(s,t)]_{s|x}}\right]_{t|y}, \quad \forall x, y \in I_2.$$

Unfortunately, the formulation in (3) can be difficult to apply. In order to use a specific mapping in (3), we must establish that it is a unitary mapping defined on $H(K)$ that satisfies Theorem 1. This can be difficult since the applicability of a specific mapping depends on the interplay between the mapping and the space $H(K)$; therefore, if a specific mapping works for one particular noise, it may not work for some other noise.

By considering a finite-dimensional approximation to the space $H(K)$, it is possible to apply any L^2-unitary operator to a detection problem for which the noise covariance is continuous on the bounded set $I \times I$. Specifically, the covariance K can be approximated by

$$K_N(s,t) = \sum_{n=1}^{N} \lambda_n \, \phi_n(s) \, \overline{\phi_n(t)},$$

where the $\{\lambda_n\}$ and the $\{\phi_n\}$ are the eigenvalues and the eigenfunctions, respectively, of the noise covariance K. Then we can define

$$K_N^{-1}(s,t) = \sum_{n=1}^{N} 1/\lambda_n \, \phi_n(s) \, \overline{\phi_n(t)},$$

which gives rise to an integral operator $\mathcal{K}_N^{-1}$ such that we can form an approximate sufficient statistic by

$$\begin{aligned}\Lambda_N &= <r, \gamma>_{H(K_N)} \\ &= <r, \mathcal{K}_N^{-1}\gamma>_{L^2(I)},\end{aligned} \quad (4)$$

with $\Lambda_N \to \Lambda$ as $N \to \infty$.

By applying Theorem 1 to the finite-dimensional space $H(K_N)$ it is easy to show that it is possible to apply any L^2-unitary operator to (4) to arrive at forms for Λ_N in alternative domains. Namely, if $\mathcal{T}$ maps $L^2(I)$ unitarily onto some Hilbert space H, then

$$\Lambda_N = <\mathcal{T}r, \tilde{\mathcal{K}}_N^{-1}\mathcal{T}\gamma>_H, \quad (5)$$

where $\tilde{\mathcal{K}}_N^{-1}$ is the inverse of the integral operator that corresponds to

$$\tilde{K}_N(y,x) = \left[\mathcal{T}\overline{[\mathcal{T}K_N(s,t)]_{s|x}}\right]_{t|y}. \quad (6)$$

This formulation allows a wide variety of signal transforms to be applied to the detection problem; including the cross-Wigner distribution, the Gabor transform, and the wavelet transform.

3 Application of Time-Frequency and Time-Scale Methods

We are interested here in applying the cross-Wigner distribution, the Gabor transform, and the wavelet transform to the detection problem. It is well known that each of these transforms is, under the right conditions, a unitary operator on $L^2(\mathcal{R})$; therefore, the results of the previous section easily lead to a unified formulation for an approximation to the sufficient statistic in terms of these transforms. In this section we discuss certain aspects of these formulations in detail.

The cross-Wigner distribution (CWD) between signals f and g is defined as

$$W_{f,g}(t,\omega) = \int_{-\infty}^{\infty} e^{-j\omega\tau} f(t+\tau/2)\overline{g(t-\tau/2)}\,d\tau.$$

If g is chosen to be concentrated in time and frequency, then $W_{f,g}$ gives a linear representation of the time-frequency content of f. The CWD is a member of Cohen's class [1], which has a general member given by

$$C_{f,g}^{\Phi}(t,\omega) = 1/2\pi \int_{-\infty}^{\infty}\int_{-\infty}^{\infty}\int_{-\infty}^{\infty} e^{j(\xi t-\tau\omega-\xi u)}\,\Phi(\xi,\tau) \\ f(u+\tau/2)\overline{g(u-\tau/2)}\,du\,d\tau\,d\xi,$$

where the choice of the kernel Φ determines a specific member of the class. Alternatively, a general member of Cohen's class can be expressed in terms of the Wigner distribution by

$$C_{f,g}^{\Phi}(t,\omega) = \\ 1/2\pi \int_{-\infty}^{\infty}\int_{-\infty}^{\infty} \phi(t-\tau,\omega-\xi)\,W_{f,g}(\tau,\xi)\,d\tau\,d\xi,$$

where the bivariate function ϕ is given by

$$\phi(t,\omega) = 1/2\pi \int_{-\infty}^{\infty}\int_{-\infty}^{\infty} e^{j(\xi t-\tau\omega)}\,\Phi(\xi,\tau)\,d\tau\,d\xi. \quad (7)$$

If g is chosen as a fixed analyzing function, then the bivariate function $C_{f,g}^{\Phi}$ can be interpreted as the result of a linear operator acting on f. As seen in the previous section, if this operator is unitary on $L^2(\mathcal{R})$, then the sufficient statistic can be formed in terms of cross-Cohen representations. It is well known that the CWD can be considered as a unitary operator since it satisfies

$$1/2\pi \int_{-\infty}^{\infty}\int_{-\infty}^{\infty} W_{f_1,g_1}(t,\omega)\,\overline{W_{f_2,g_2}(t,\omega)}\,dt\,d\omega \\ = <f_1,f_2>_{L^2}\,\overline{<g_1,g_2>_{L^2}},$$

which is known as Moyal's formula [1]. However, not all members of Cohen's class satisfy Moyal's formula. Previously, Janssen [10] provided a sufficient condition for a member of Cohen's class to satisfy Moyal's formula; namely, if $|\Phi(\xi,\tau)| = 1$ for all ξ and τ, then the corresponding $C_{f,g}^{\Phi}$ satisfies Moyal's formula. (Note: Janssen's condition is necessary and sufficient only for a subclass of Cohen's class.)

The following theorem gives a *necessary and sufficient* condition for a member of Cohen's class to satisfy a form of Moyal's formula and, therefore, be applicable to the detection problem.

Theorem 2 *For a fixed g with $\|g\|_2^2 = 1/2\pi$, $C_{f,g}^{\Phi}$ satisfies*

$$\int_{-\infty}^{\infty}\int_{-\infty}^{\infty} C_{f_1,g}^{\Phi}(t,\omega)\,\overline{C_{f_2,g}^{\Phi}(t,\omega)}\,dt\,d\omega = <f_1,f_2>_{L^2}$$

if and only if the mapping ϕ, defined on a closed subspace of $L^2(\mathcal{R}^2)$ by

$$(\phi F)(t,\omega) = 1/2\pi \int_{-\infty}^{\infty}\int_{-\infty}^{\infty} \phi(t-\tau,\omega-\xi)\,F(\tau,\xi)\,d\tau\,d\xi,$$

where $\phi(t,\omega)$ is defined by (7), is unitary.

Proof: For a fixed g with $\|g\|_2^2 = 1/2\pi$, define the closed subspace $\mathbf{W_g} = \{W_{f,g} | f \in L^2(\mathcal{R})\}$. The operator $W_g : f \mapsto W_{f,g}$ unitarily maps $L^2(\mathcal{R})$ onto $\mathbf{W_g}$. Thus, the mapping of f to $C_{f,g}^{\Phi}$ is then accomplished by applying the composition $\phi \circ W_g$ to the function f. Since the operator W_g is unitary (due to Moyal's formula), it is necessary and sufficient that the mapping ϕ be unitary.
□

The results of Section 2 now imply that any member $C_{f,g}^{\Phi}$ of Cohen's class that satisfies the above theorem may be used to approximate the sufficient statistic via (5) and (6), which reduce to

$$\Lambda_N = \int_{-\infty}^{\infty}\int_{-\infty}^{\infty} C_{r,g}^{\Phi}(t,\omega)$$
$$\int_{-\infty}^{\infty}\int_{-\infty}^{\infty} \tilde{K}_N^{-1}(t,\omega;t',\omega')\,\overline{C_{\gamma,g}^{\Phi}(t',\omega')}\,dt'\,d\omega'\,dt\,d\omega, \tag{8}$$

and

$$\tilde{K}_N^{-1}(t,\omega;t',\omega') = \sum_{n=1}^{N} 1/\lambda_n\, C_{\phi_n,g}^{\Phi}(t,\omega)\,\overline{C_{\phi_n,g}^{\Phi}(t',\omega')}. \tag{9}$$

This formulation results in a weighted correlation between a time-frequency representation of the received signal and a time-frequency representation of a reference signal; the weighting kernel is the inverse kernel of the reproducing kernel in the transformed RKHS, and can be interpreted as the inverse kernel of the covariance of the time-frequency representation of the noise process. This formulation is a direct extension of the unweighted time-frequency domain correlator proposed for the white noise case in [4] and [5].

The Gabor transform is another linear time-frequency representation that is closely related to the cross-Wigner distribution. Let g be an arbitrary finite-energy signal; then, the Gabor transform of a signal f, with respect to g, is given by the bivariate function $G_g f$ defined by

$$(G_g f)(\tau,\omega) = \int_{-\infty}^{\infty} f(t)\,\overline{g(t-\tau)}\,e^{-j\omega t}\,dt.$$

If g is chosen such that $\|g\|_2^2 = 1/2\pi$, then the mapping $G_g : L^2(\mathcal{R}) \to L^2(\mathcal{R}^2)$, defined by $G_g : f \mapsto G_g f$, is a unitary operator. Therefore, it is possible to use the same formulation for solving the detection problem as was used for the cross-Wigner distribution; that is, replace $C_{r,g}^{\Phi}$ and $C_{\gamma,g}^{\Phi}$ in (8) and (9) by $G_g r$ and$G_\gamma r$, respectively. In fact, there really is very little difference between these two time-frequency formulations of the sufficient statistic since

$$W_{f,g}(t,\omega) = 2e^{j\omega t}(G_{\tilde{g}} f)(2t, 2\omega),$$

where $\tilde{g}(x) = g(-x)$. Thus, the CWD is more highly concentrated in the time-frequency plane.

It is interesting to note that the Gabor transform has a discrete form that arises by sampling $(G_g f)(\tau,\omega)$ uniformly in τ and ω to give

$$(G^d f)(m,n) = \int_{-\infty}^{\infty} f(t)\,\overline{g(t-mT)}\,e^{-jn\Omega t}\,dt,$$

where the explicit dependence on g has been suppressed to enhance the readability. By proper choice of the function g and the sampling grid parameters (T,Ω), the discrete form of the Gabor transform can be made to be a unitary operator from $L^2(\mathcal{R})$ onto $\ell^2(\mathcal{Z}^2)$ [3]. The use of the discrete form results in a formulation of the sufficient statistic that is similar to that proposed in [6].

The wavelet transform of a signal f, with respect to a signal g (called the mother wavelet), is given by

$$(W_g f)(\tau,s) = \int_{-\infty}^{\infty} e^{s/2}\,\overline{g(e^s t-\tau)}\,f(t)\,dt,$$

where g is chosen such that

$$c_g \triangleq \int_0^{\infty} \frac{|G(\omega)|^2}{\omega}\,d\omega < \infty.$$

Choosing g such that $c_g = 1$ makes the wavelet transform a unitary operator [11]. Note that the variable s represents "scale" rather than "frequency." The concept of scale arises because as s goes toward positive infinity, $(W_g f)(\tau,s)$ zooms in on the small-scale details of f that are near the time point τ. This zooming property gives the wavelet transform a time resolution that adapts to the frequency range.

Because the wavelet transform is a unitary operator, it can be used for the detection of signals in nonwhite noise by applying the unified formulation developed in Section 2. Like the Gabor transform case, this application results in a formulation identical to (8) and (9), except with $C_{r,g}^{\Phi}$ and $C_{\gamma,g}^{\Phi}$ replaced by $W_g r$ and $W_g \gamma$, respectively. This application of the wavelet transform is illustrated below, where we will use the more practical discrete form of the wavelet transform that arises by sampling $(W_g f)(\tau,s)$ nonuniformly in τ and s to give

$$(W^d f)(m,n) = \int_{-\infty}^{\infty} f(t)\,S^{n/2} g(S^n t - mT)\,dt,$$

which are known as the wavelet coefficients of f. Note that the dependence of the coefficients on g has, for clarity, not been displayed on the left-hand side of this equation; also, when no confusion can arise, $(W^d f)(m,n)$ will be written as W_{mn}. It is possible to choose g and the sampling grid parameters (T,S) such that the set $\{S^{n/2}g(S^n t - mT)\}$ is an orthonormal basis of $L^2(\mathcal{R})$ and the discrete wavelet transform is a unitary operator from $L^2(\mathcal{R})$ onto $\ell^2(\mathcal{Z}^2)$, where $\mathcal{Z}$ is the set of integers [3].

4 Example: Nearly $1/f$ Noise

The discrete form of the wavelet transform was used by Wornell [12] in the analysis of a type of nonstationary process, called nearly $1/f$ noise, having measured spectra satisfying

$$\frac{k_1}{|\omega|^\beta} \le S(\omega) \le \frac{k_2}{|\omega|^\beta},$$

with $0 < k_1 \le k_2 < \infty$, and $\beta \in (0,2)$ a fixed parameter; this includes $1/f$ noise as a special case.

Consider the (real-valued, for convenience) non-stationary noise n defined [12] by the discrete wavelet expansion

$$n(t) = \sum_m \sum_n W_{mn}\, g_{mn}(t), \tag{10}$$

where W_{mn} are the random wavelet coefficients of the discrete wavelet transform, and $\{g_{mn}\}$ is a basis of orthonormal wavelets related to the mother wavelet g according to

$$g_{mn}(t) = 2^{n/2}\, g(2^n t - m). \tag{11}$$

Let the Fourier transform G of g be continuous at $\omega = 0$ and let $|G(\omega)|$ decay at least as fast as $1/\omega$. If the random sequence W_{mn} is such that for arbitrary distinct pairs n' and n, $W_{mn'}$ and W_{mn} are uncorrelated sequences, and if for each fixed n the sequence W_{mn} is white with average power $2^{-\beta n}\sigma^2$, then the noise n is a nearly $1/f$ noise [12].

If the noise n has zero-mean, then

$$E\{W_{mn}\, W_{m'n'}\} = 2^{-\beta n}\, \sigma^2\, \delta_{n,n'}\, \delta_{m,m'},$$

which shows that (10) is a Karhunen–Loève-like expansion of the noise n. Thus, its covariance can be expanded in terms of the ON wavelet basis as follows:

$$\begin{aligned} K(t,s) &= \sum_n \sum_m \sum_{n'} \sum_{m'} \left[2^{-\beta n}\, \sigma^2\, \delta_{n,n'}\, \delta_{m,m'}\right] g_{mn}(t)\, g_{m'n'}(s) \\ &= \sum_n \sum_m 2^{-\beta n}\, \sigma^2\, g_{mn}(t)\, g_{mn}(s). \end{aligned} \tag{12}$$

In practice, the wavelet transform of the received data can be computed only at a finite number of scales n and translations m. Since the effective support width of g_{mn} doubles for each successively coarser scale, the wavelet representation of a finite-duration signal requires half the number of wavelet coefficients at each successively coarser scale. Thus, if at scale N_1, wavelet coefficients are needed for $-M_1 \le m \le M_1$, then at scale $N_1 + P$, the wavelet coefficients are needed for $-2^P M_1 \le m \le 2^P M_1$. Let N_1 be the coarsest scale to be considered and let $N_1 + N$ be the finest scale, where N is some positive integer; then $N_1 \le n \le N_1 + N$. Thus, a finite-term appr the approximate sufficient statistic

$$\begin{aligned} \Lambda_{(N_1,N,M_1)} &= \sum_{n=N_1}^{N_1+N} 2^{\beta n}/\sigma^2 \sum_{m=-2^{n-N_1}M_1}^{2^{n-N_1}M_1} (W^d r)(m,n)(W^d \gamma)(m,n) \\ &\triangleq \sum_{n=N_1}^{N_1+N} (2^{\beta n}/\sigma^2)\, \Lambda_n, \end{aligned} \tag{13}$$

where Λ_n is defined to be

$$\Lambda_n = \sum_{m=-2^{n-N_1}M_1}^{2^{n-N_1}M_1} (W^d r)(m,n)\, (W^d \gamma)(m,n).$$

This has an interesting interpretation. At each scale n, the statistic Λ_n is formed by correlating $(W^d r)(m,n)$ and $(W^d \gamma)(m,n)$—as functions of m—as if making a decision in white noise. The sufficient statistic $\Lambda_{(N_1,N,M_1)}$ is then formed by fusing these "white-noise statistics" together via a linear combination, with the scale n statistic Λ_n weighted by the reciprocal of the average noise power at scale n.

5 Conclusion

We have extended a theorem on the mapping of one RKHS to another, and have used that extension to identify a class of transforms that can be used to formulate the detection problem in alternative domains. In particular, we have shown that a unified formulation for the detection problem exists that allows the use of common time-frequency and time-scale representations such as the cross-Wigner distribution, the Gabor transform, and the wavelet transform. In addition, we have described a subclass of Cohen's class of representations that can be used in this formulation. Although the formulations given here have been for the case of a deterministic signal, the random signal case can be handled by replacing the representation of γ by an appropriate estimate; this results in an estimator-correlator form for the sufficient statistic.

It should be noted that the wavelet transform and the Gabor transform belong to a class of transforms that arise from square-integrable group representations [11]. Formulations of the sufficient statistic in terms of the members of this class can be developed by applying the results of Section 2 [13]. This provides a possible means of choosing a particular alternative domain based on similarities in the group structure of a transform's representation and properties of the additive noise process. For example, the wavelet transform and $1/f$ noise each exhibit a property known as self-similarity. This type of interplay was illustrated here by showing the effectiveness of the wavelet transform for the detection of signals in the presence of $1/f$ type noises.

6 Acknowledgement

This work has been supported by the Office of Naval Research under the Fundamental Research Initiatives Program.

References

[1] L. Cohen, "Time-frequency distributions — a review," *Proc. of IEEE*, vol. 77, pp. 941–980, July 1989.

[2] M. J. Bastiaans, "Gabor's expansion of a signal into Gaussian elementary signals," *Proc. of IEEE*, vol. 68, pp. 538–539, April 1980.

[3] I. Daubechies, "The wavelet transform, time-frequency localization and signal analysis," *IEEE Trans. Inform. Theory*, vol. 36, pp. 961–1005, Sept. 1990.

[4] B. V. K. V. Kumar and C. W. Carroll, "Performance of Wigner distribution function based detection methods," *Opt. Eng.*, vol. 23, pp. 732–737, Nov./Dec. 1984.

[5] B. Boashash and P. O'Shea, "A methodology for detection and classification of some underwater acoustic signals using time-frequency analysis techniques," *IEEE Trans. Acoust., Speech, and Signal Processing*, vol. 38, pp. 1829–1841, Nov. 1990.

[6] B. Friedlander and B. Porat, "Detection of transient signals by the Gabor representation," *IEEE Trans. Acoust., Speech, and Signal Processing*, vol. 37, pp. 169–180, Feb. 1989.

[7] F. B. Tuteur, "Wavelet transformations in signal detection," in *Wavelets: Time-Frequency Methods and Phase Space*. J. M. Combes, A. Grossmann, and Ph. Tchamitchian, Eds. Berlin: Springer-Verlag, 1989, pp. 132–138.

[8] T. Kailath, "RKHS approach to detection and estimation problems—part I: deterministic signals in Gaussian noise," *IEEE Trans. Inform. Theory*, vol. 17, pp. 530–549, Sept. 1971.

[9] B. L. Chalmers, "Images of bounded linear operators on a space with kernel function," *Archiv der Math.*, vol. 26, pp. 313–316, 1975.

[10] A. J. E. M. Janssen, "On the locus and spread of psuedo-density functions in the time-frequency plane," *Philips J. Res.*, vol. 37, pp. 79–110, 1982.

[11] A. Grossmann, J. Morlet, and T. Paul, "Transforms associated to square integrable group representations, I, General results," *J. Math. Phys.*, vol. 26, pp. 2473–2479, Oct. 1985.

[12] G. W. Wornell, "A Karhunen–Loève-like expansion for $1/f$ processes via wavelets," *IEEE Trans. Inform. Theory*, vol. 36, pp. 859–861, July 1990.

[13] M. L. Fowler and L. H. Sibul, "Signal detection using group transforms," in *Proc. of IEEE ICASSP,* 1991.

Author Index

Offices of the IEEE Computer Society

Headquarters Office
1730 Massachusetts Avenue, N.W.
Washington, DC 20036-1903
Phone: (202) 371-0101 — Fax: (202) 728-9614

Publications Office
P.O. Box 3014
10662 Los Vaqueros Circle
Los Alamitos, CA 90720-1264
Membership and General Information: (714) 821-8380
Publication Orders: (800) 272-6657 — Fax: (714) 821-4010

European Office
13, avenue de l'Aquilon
B-1200 Brussels, BELGIUM
Phone: 32-2-770-21-98 — Fax: 32-3-770-85-05

Asian Office
Ooshima Building
2-19-1 Minami-Aoyama, Minato-ku
Tokyo 107, JAPAN
Phone: 81-3-408-3118 — Fax: 81-3-408-3553

IEEE Computer Society

IEEE Computer Society Press Publications

Monographs: A monograph is an authored book consisting of 100-percent original material.
Tutorials: A tutorial is a collection of original materials prepared by the editors, and reprints of the best articles published in a subject area. Tutorials must contain at least five percent of original material (although we recommend 15 to 20 percent of original material).
Reprint collections: A reprint collection contains reprints (divided into sections) with a preface, table of contents, and section introductions discussing the reprints and why they were selected. Collections contain less than five percent of original material.
Technology series: Each technology series is a brief reprint collection — approximately 126-136 pages and containing 12 to 13 papers, each paper focusing on a subset of a specific discipline, such as networks, architecture, software, or robotics.
Submission of proposals: For guidelines on preparing CS Press books, write the Editorial Director, IEEE Computer Society Press, PO Box 3014, 10662 Los Vaqueros Circle, Los Alamitos, CA 90720-1264, or telephone (714) 821-8380.

Purpose

The IEEE Computer Society advances the theory and practice of computer science and engineering, promotes the exchange of technical information among 100,000 members worldwide, and provides a wide range of services to members and nonmembers.

Membership

All members receive the acclaimed monthly magazine *Computer*, discounts, and opportunities to serve (all activities are led by volunteer members). Membership is open to all IEEE members, affiliate society members, and others seriously interested in the computer field.

Publications and Activities

***Computer* magazine:** An authoritative, easy-to-read magazine containing tutorials and in-depth articles on topics across the computer field, plus news, conference reports, book reviews, calendars, calls for papers, interviews, and new products.
Periodicals: The society publishes six magazines and five research transactions. For more details, refer to our membership application or request information as noted above.
Conference proceedings, tutorial texts, and standards documents: The IEEE Computer Society Press publishes more than 100 titles every year.
Standards working groups: Over 100 of these groups produce IEEE standards used throughout the industrial world.
Technical committees: Over 30 TCs publish newsletters, provide interaction with peers in specialty areas, and directly influence standards, conferences, and education.
Conferences/Education: The society holds about 100 conferences each year and sponsors many educational activities, including computing science accreditation.
Chapters: Regular and student chapters worldwide provide the opportunity to interact with colleagues, hear technical experts, and serve the local professional community.

Other IEEE Computer Society Press Titles

MONOGRAPHS

Analyzing Computer Architectures
Written by J.C. Huck and M.J. Flynn
(ISBN 0-8186-8857-2); 206 pages

Branch Strategy Taxonomy and Performance Models
Written by Harvey G. Cragon
(ISBN 0-8186-9111-5); 150 pages

Desktop Publishing for the Writer: Designing, Writing, and Developing
Written by Richard Ziegfeld and John Tarp
(ISBN 0-8186-8840-8); 380 pages

Digital Image Warping
Written by George Wolberg
(ISBN 0-8186-8944-7); 340 pages

Integrating Design and Test—CAE Tools for ATE Programming
Written by K.P. Parker
(ISBN 0-8186-8788-6); 160 pages

JSP and JSD—The Jackson Approach to Software Development (Second Edition)
Written by J.R. Cameron
(ISBN 0-8186-8858-0); 560 pages

National Computer Policies
Written by Ben G. Matley and Thomas A. McDannold
(ISBN 0-8186-8784-3); 192 pages

Physical Level Interfaces and Protocols
Written by Uyless Black
(ISBN 0-8186-8824-2); 240 pages

Protecting Your Proprietary Rights in Computer and High-Technology Industries
Written by Tobey B. Marzouk, Esq.
(ISBN 0-8186-8754-1); 224 pages

X.25 and Related Protocols
Written by Uyless Black
(ISBN 0-8186-8976-5); 304 pages

TUTORIALS

Advanced Computer Architecture
Edited by D.P. Agrawal
(ISBN 0-8186-0667-3); 400 pages

Advances in Distributed System Reliability
Edited by Suresh Rai and Dharma P. Agrawal
(ISBN 0-8186-8907-2); 352 pages

Autonomous Mobile Robots:
Perception, Mapping, and Navigation—Volume 1
Edited by S.S. Iyengar and A. Elfes
(ISBN 0-8186-9018-6); 552 pages

Autonomous Mobile Robots:
Control, Planning, and Architecture—Volume 2
Edited by S.S. Iyengar and A. Elfes
(ISBN 0-8186-9116-6); 536 pages

Broadband Switching: Architectures, Protocols, Design, and Analysis
Edited by C. Dhas, V.K. Konangi, and M. Sreetharan
(ISBN 0-8186-8926-9); 528 pages

Computer and Network Security
Edited by M.D. Abrams and H.J. Podell
(ISBN 0-8186-0756-4); 448 pages

Computer Architecture
Edited by D.D. Gajski, V.M. Milutinovic, H. Siegel, and B.P. Furht
(ISBN 0-8186-0704-1); 602 pages

Computer Arithmetic I
Edited by Earl E. Swartzlander, Jr.
(ISBN 0-8186-8931-5); 398 pages

Computer Arithmetic II
Edited by Earl E. Swartzlander, Jr.
(ISBN 0-8186-8945-5); 412 pages

Computer Communications: Architectures, Protocols and Standards (Second Edition)
Edited by William Stallings
(ISBN 0-8186-0790-4); 448 pages

Computer Graphics Hardware: Image Generation and Display
Edited by H.K. Reghbati and A.Y.C. Lee
(ISBN 0-8186-0753-X); 384 pages

Computer Graphics: Image Synthesis
Edited by Kenneth Joy, Nelson Max, Charles Grant, and Lansing Hatfield
(ISBN 0-8186-8854-8); 380 pages

Computer Vision: Principles
Edited by Rangachar Kasturi and Ramesh Jain
(ISBN 0-8186-9102-6); 700 pages

Computer Vision: Advances and Applications
Edited by Rangachar Kasturi and Ramesh Jain
(ISBN 0-8186-9103-4); 720 pages

Digital Image Processing (Second Edition)
Edited by Rama Chellappa
(ISBN 0-8186-2362-4); 400 pages

Digital Private Branch Exchanges (PBXs)
by Edwin Coover
(ISBN 0-8186-0829-3); 394 pages

Distributed Computing Network Reliability
Edited by Suresh Rai and Dharma Agrawal
(ISBN 0-8186-8908-0); 357 pages

Distributed Software Engineering
Edited by Sol Shatz and Jia-Ping Wang
(ISBN 0-8186-8856-4); 294 pages

Domain Analysis and Software Systems Modeling
Edited by Ruben Prieto-Diaz and Guillermo Arango
(ISBN 0-8186-8996-X); 312 pages

DSP-Based Testing of Analog and Mixed Signal Circuits
Edited by Matthew Mahoney
(ISBN 0-8186-0785-8); 272 pages

Fault Tolerant Computing
Edited by V. Nelson and B. Carroll
(ISBN 0-8186-8677-4); 428 pages

Formal Verification of Hardware Design
Edited by Michael Yoeli
(ISBN 0-8186-9017-8); 340 pages

Hard Real-Time Systems
Edited by J.A. Stankovic and K. Ramamritham
(ISBN 0-8186-0819-6); 624 pages

Integrated Services Digital Networks (ISDN) (Second Edition)
Edited by William Stallings
(ISBN 0-8186-0823-4); 406 pages

Local Network Technology (Third Edition)
Edited by William Stallings
(ISBN 0-8186-0825-0); 512 pages

Microprogramming and Firmware Engineering
Edited by V. Milutinovic
(ISBN 0-8186-0839-0); 416 pages

Modeling and Control of Automated Manufacturing Systems
Edited by A. A. Desrochers
(ISBN 0-8186-8916-1); 384 pages

Nearest Neighbor Pattern Classification Techniques
Edited by Belur V. Dasarathy
(ISBN 0-8186-8930-7); 464 pages

New Paradigms for Software Development
Edited by William Agresti
(ISBN 0-8186-0707-6); 304 pages

Object-Oriented Computing, Volume 1: Concepts
Edited by Gerald E. Petersen
(ISBN 0-8186-0821-8); 214 pages

Object-Oriented Computing, Volume 2: Implementations
Edited by Gerald E. Petersen
(ISBN 0-8186-0822-6); 324 pages

Parallel Architectures for Database Systems
Edited by A.R. Hurson, L.L. Miller, and S.H. Pakzad
(ISBN 0-8186-8838-6); 478 pages

Reduced Instruction Set Computers (RISC) (Second Edition)
Edited by William Stallings
(ISBN 0-8186-8943-9); 448 pages

Software Engineering Project Management
Edited by R. Thayer
(ISBN 0-8186-0751-3); 512 pages

Software Maintenance and Computers
Edited by D.H. Longstreet
(ISBN 0-8186-8898-X); 304 pages

Software Quality Assurance: A Practical Approach
Edited by T.S. Chow
(ISBN 0-8186-0569-3); 506 pages

Software Reusability
Edited by Peter Freeman
(ISBN 0-8186-0750-5); 304 pages

Software Reuse—Emerging Technology
Edited by Will Tracz
(ISBN 0-8186-0846-3); 400 pages

Software Risk Management
Edited by B.W. Boehm
(ISBN 0-8186-8906-4); 508 pages

Standards, Guidelines and Examples on System and Software Requirements Engineering
Edited by Merlin Dorfman and Richard H. Thayer
(ISBN 0-8186-8922-6); 626 pages

System and Software Requirements Engineering
Edited by Richard H. Thayer and Merlin Dorfman
(ISBN 0-8186-8921-8); 740 pages

Test Access Port and Boundary-Scan Architecture
Edited by C. M. Maunder and R. E. Tulloss
(ISBN 0-8186-9070-4); 400 pages

Test Generation for VLSI Chips
Edited by V.D. Agrawal and S.C. Seth
(ISBN 0-8186-8786-X); 416 pages

Visual Programming Environments: Paradigms and Systems
Edited by Ephraim Glinert
(ISBN 0-8186-8973-0); 680 pages

Visual Programming Environments: Applications and Issues
Edited by Ephraim Glinert
(ISBN 0-8186-8974-9); 704 pages

Visualization in Scientific Computing
Edited by G.M. Nielson, B. Shriver, and L. Rosenblum
(ISBN 0-8186-8979-X); 304 pages

VLSI Testing and Validation Techniques
Edited by H. Reghbati
(ISBN 0-8186-0668-1); 616 pages

Volume Visualization
Edited by Arie Kaufman
(ISBN 0-8186-9020-8); 494 pages

REPRINT COLLECTIONS

Distributed Computing Systems: Concepts and Structures
Edited by A. Ananda and B. Srinivasan
(ISBN 0-8186-8975-0); 416 pages

Expert Systems: A Software Methodology for Modern Applications
Edited by Peter Raeth
(ISBN 0-8186-8904-8); 476 pages

Milestones in Software Evolution
Edited by Paul W. Oman and Ted G. Lewis
(ISBN 0-8186-9033-X); 332 pages

Object-Oriented Databases
Edited by Ez Nahouraii and Fred Petry
(ISBN 0-8186-8929-3); 256 pages

Validating and Verifying Knowledge-Based Systems
Edited by Uma G. Gupta
(ISBN 0-8186-8995-1); 400 pages

ARTIFICIAL NEURAL NETWORKS TECHNOLOGY SERIES

Artificial Neural Networks—Concept Learning
Edited by Joachim Diederich
(ISBN 0-8186-2015-3); 160 pages

Artificial Neural Networks—Electronic Implementation
Edited by Nelson Morgan
(ISBN 0-8186-2029-3); 144 pages

Artificial Neural Networks—Theoretical Concepts
Edited by V. Vemuri
(ISBN 0-8186-0855-2); 160 pages

SOFTWARE TECHNOLOGY SERIES

Computer-Aided Software Engineering (CASE)
Edited by E.J. Chikofsky
(ISBN 0-8186-1917-1); 110 pages

Software Reliability Models: Theoretical Development, Evaluation and Applications
Edited by Y.K. Malaiya and P.K. Srimani
(ISBN 0-8186-2110-9); 136 pages

MATHEMATICS TECHNOLOGY SERIES

Computer Algorithms
Edited by Jun-ichi Aoe
(ISBN 0-8186-2123-0); 154 pages

Multiple-Valued Logic in VLSI Design
Edited by Jon T. Butler
(ISBN 0-8186-2127-3); 128 pages

COMMUNICATIONS TECHNOLOGY SERIES

Multicast Communication in Distributed Systems
Edited by Mustaque Ahamad
(ISBN 0-8186-1970-8); 110 pages

ROBOTICS TECHNOLOGY SERIES

Multirobot Systems
Edited by Rajiv Mehrotra and Murali R. Varanasi
(ISBN 0-8186-1977-5); 122 pages

Notes

NOTES

Conference Record of

The Twenty-Fifth Asilomar Conference on Signals, Systems & Computers

Volume 2 of 2

1951-1991
40 YEARS OF SERVICE
IEEE COMPUTER SOCIETY
A member society of the
Institute of Electrical and Electronics Engineers, Inc.

Conference Record

Twenty-Fifth Asilomar Conference on Signals, Systems & Computers

NOVEMBER 4–6, 1991
PACIFIC GROVE, CALIFORNIA

Volume 2 of 2

1951 · 1991
IEEE Computer Society Press

The Institute of Electrical and Electronics Engineers, Inc.

Conference Record of

The Twenty-Fifth Asilomar Conference on Signals, Systems & Computers

Volume 2 of 2

November 4–6, 1991 *Pacific Grove, California*

Edited by
Ray R. Chen
San Jose State University

Sponsored by
Naval Postgraduate School
Monterey, California
and
San Jose State University
San Jose, California

In Cooperation with
IEEE Computer Society

1951–1991
40 YEARS OF SERVICE

IEEE Computer Society Press
Los Alamitos, California

Washington • Brussels • Tokyo

The papers in this book comprise the proceedings of the meeting mentioned on the cover and title page. They reflect the authors' opinions and, in the interests of timely dissemination, are published as presented and without change. Their inclusion in this publication does not necessarily constitute endorsement by the editors, the IEEE Computer Society Press, or the Institute of Electrical and Electronics Engineers, Inc.

1951-1991 IEEE COMPUTER SOCIETY PRESS

Published by the
IEEE Computer Society Press
10662 Los Vaqueros Circle
PO Box 3014
Los Alamitos, CA 90720-1264

IEEE Computer Society Press Order Number 2470
Library of Congress Number 89-640232
ISSN 1058-6393
IEEE Catalog Number 91CH3112-0
ISBN 0-8186-2470-1 (paper)
ISBN 0-8186-2471-X (microfiche)
ISBN 0-8186-2472-8 (case)

Additional copies can be ordered from

IEEE Computer Society Press
Customer Service Center
10662 Los Vaqueros Circle
PO Box 3014
Los Alamitos, CA 90720-1264

IEEE Service Center
445 Hoes Lane
PO Box 1331
Piscataway, NJ 08855-1331

IEEE Computer Society
13, avenue de l'Aquilon
B-1200 Brussels
BELGIUM

IEEE Computer Society
Ooshima Building
2-19-1 Minami-Aoyama
Minato-ku, Tokyo 107
JAPAN

Cover designed by Jack Ballestero

Editorial Production: Edna Straub, Penny Storms, and Lisa O'Conner
Printed in the United States of America by McNaughton & Gunn, Inc.

THE INSTITUTE OF ELECTRICAL AND ELECTRONICS ENGINEERS, INC.

Foreword

The Twenty-Fifth Annual Asilomar Conference on Signals, Systems & Computers was held November 4–6, 1991, at the Asilomar Hotel and Conference Grounds, Pacific Grove, California. The conference was sponsored by the Naval Postgraduate School and San Jose State University, in cooperation with the IEEE Computer Society.

The conference was devoted to the presentation of new ideas and preliminary research results in the areas of theoretical and applied signal processing, computer algorithms, circuits, systems, communications, and image processing.

The papers contained in this record were selected after careful screening from many excellent submissions. The 1991 Conference Committee consists of

General Chair:	Frederic J. Harris *San Diego State University*
Technical Program Chair:	Neil Jablon *AT&T Bell Labs*
Technical Program Committee:	John J. Shynk *University of California, Santa Barbara*
	Hari Krishna *Syracuse University*
	James Ritcey *University of Washington*
	Robert G. Redinbo *University of California, Davis*
	Huseyin Abut *San Diego State University*
	Rabinder N. Madan *Office of Naval Research*
Publications Chair:	Ray R. Chen *San Jose State University*
Publicity Chair:	Avtar Singh *San Jose State University*
NPS Coordinator:	Monique Fargues *Naval Postgraduate School*
Finance Chair:	Ralph Hippenstiel *Naval Postgraduate School*

Gratitude is expressed to the attendees, authors, session chairs, organizers, and reviewers whose interest and efforts contributed much to the success of the conference.

The Twenty-Sixth Annual Asilomar Conference on Signals, Systems & Computers is scheduled for October 26–28, 1992 at the Asilomar Hotel and Conference Grounds, Pacific Grove, California. Further information about this conference may be obtained by contacting the Technical Program Chair, James A. Ritcey, Department of Electrical Engineering FT-10, University of Washington, Seattle, WA 98195.

Ray R. Chen
Editor

Table of Contents

Volume 1

Session MP4: Radar and Sonar Signal Processing
Chair: *J. Ritcey*, University of Washington, Seattle, WA

Session MP5: Knowledge-Based Signal Processing
Chair: *D.D. Sworder*, University of California, San Diego, La Jolla, CA

Session MP6: Adaptive Filtering I
Chair: *N.J. Bershad*, University of California, Irvine, CA

Session MP7: Matrix-Based Signal Processing (Poster Session)
Chair: *M.P. Fargues*, Naval Postgraduate School, Monterey, CA

Session MP8: Speech Processing II
Chair: *H. Abut*, San Diego State University, San Diego, CA

Session TA1: Spectral Estimation
Chair: *R. Hippenstiel*, Naval Postgraduate School, Monterey, CA

Session TA2a: Arithmetic for DSP
Chair: *L. Ciminiera*, Politecnico di Torino, Torino, Italy

Session TA2b: DSP Implementations
Chair: *B.P.M. Tao*, Rockwell International Corporation, Newport Beach, CA

Session TA3: Image Coding
Chair: *K.S. Thyagarajan*, San Diego State University, San Diego, CA

Session TA4: Neural Networks for Signal Processing
Chair: *J.-N. Hwang*, University of Washington, Seattle, WA

Session TA5a: VLSI Implementations
Chair: *M. Tummala*, Naval Postgraduate School, Monterey, CA

Session TA5b: Evolutionary Programming
Chair: *D.B. Fogel*, Orincon Corporation, San Diego, CA

Session TA6: Array Signal Processing
Chairs: *R.N. Madan*, Office of Naval Research, Arlington, VA, and
L.K. Lam, Lockheed Research Laboratories, Palo Alto, CA

Volume 2

Session TP1: Blind Adaptive Processing
Chair: *B.G. Agee*, Radix Technologies, Inc., San Jose, CA

Session TP2: Quadrature-Mirror Filters, Wavelets, and Applications
Chair: *B.C. Levy*, University of California, Davis, CA

Session TP3: Biologically Motivated Signal Processing
Chair: *R.O. Duda*, San Jose State University, San Jose, CA

Session TP4: Military Applications
Chairs: *J.T. Rickard*, Orincon Corporation, San Diego, CA, and
G.M. Dillard, Naval Ocean Systems Center, San Diego, CA

Session TP5: Communication and Estimation
Chair: *J. Burl*, Naval Postgraduate School, Monterey, CA

Session TP6: Signal Subspace Methods
Chair: *A. Nehorai*, Yale University, New Haven, CT

Session TP7: Speech Processing I
Chair: *H. Abut*, San Diego State University, San Diego, CA

Session WA1: Adaptive Beamforming
Chair: *R.P. Gooch*, Applied Signal Technology, Inc., Sunnyvale, CA

Session WA2: Digital Filters
Chair: *F.J. Harris*, San Diego State University, San Diego, CA

Session WA3: Image Processing
Chair: *C.W. Therrien*, Naval Postgraduate School, Monterey, CA

Session WA6: Communication Systems

Chairs: *L.J. Griffiths and M.J. Rude*, University of Southern California, Los Angeles, CA

Session WA7a: Computer Vision (Poster Session)

Chair: *K.S. Arun*, University of Illinois at Urbana-Champaign, Urbana, IL

Session WA7b: Testing and Fault Tolerance (Poster Session)

Chair: *C.-H. Lee*, Naval Postgraduate School, Monterey, CA

Frequency Tracking Performance of Adaptive Lattice Filters

J.R. Zeidler*+, K.C. Chew+, R.A. Axford*, W.H. Ku+

* U.S. Naval Ocean Systems Center, Code 80, San Diego, CA 92152-5000
\+ University of California, San Diego, Dept. of ECE, La Jolla, CA 92093-0407

Abstract

Adaptive linear prediction filters have been proposed for the detection of narrowband signals in nonstationary noise and for the removal of narrowband interference from spread spectrum communications signals. In this paper, analytical results and computer models for the PARCOR coefficients, signal recovery error, and output misadjustment of the stochastic gradient adaptive lattice filter will be obtained for a complex linear chirp FM signal in white, Gaussian noise. From these expressions an optimal adaptation rate is obtained as a function of the filter order and the chirp rate. These results illustrate that short filter lengths are sufficient for tracking chirp signals.

1 Introduction

Adaptive line enhancing is a linear prediction filtering technique for the separation of narrowband and broadband signals with overlapping frequency spectra. The technique can be used to detect and extract narrowband signals imbedded in broadband noise, or, to remove narrowband interference from a desired broadband signal [1]. The adaptive line enhancer (ALE) can be implemented with both transversal (direct-form FIR) and lattice filter structures. Relatively simple computational routines for adjusting the variable parameters of these filters in response to input waveforms with nonstationary statistics are based on gradient search methods. These include the LMS algorithm used with the transversal structure and the stochastic gradient (SG) algorithm for the lattice structure. Because the SG-lattice filter exhibits tracking characteristics that are superior to those of the LMS-transversal filter when the input contains multiple sinusoids with a wide dynamic range of power levels [2], it may be the adaptive filter of choice for many applications including the suppression of narrowband interference in spread spectrum communications receivers.

Spread spectrum (SS) communications techniques have been used extensively in military communications to achieve both low-probability-of-intercept (LPI) and anti-jamming (AJ) performance. Recently, there has been a great deal of interest in the application of direct sequence spread spectrum (DS SS) techniques to commercial multiple access personal communications networks (PCN's) and digital cellular mobile communications systems for the purpose of increasing the efficiency with which the microwave spectrum is utilized [3], [4]. In these applications of code division multiple access (CDMA) techniques, it has been proposed that systems using CDMA be overlaid on top of the spectra of existing narrowband microwave users in urban areas. The resulting issues of mutual interference are of concern to both the CDMA and the narrowband users. Here we will address the issue of rejecting narrowband interference from the broadband DS SS CDMA signals. The use of narrowband interference rejection techniques in CDMA receivers is an important area of research since it will reduce the power requirements of the CDMA transmitters, thus resulting in a lower level of CDMA interference with the existing narrowband systems.

2 Linear Predictive Filtering for Narrowband Interference Rejection in DS SS Systems

Three general techniques have been proposed for the rejection of narrowband interference in DS SS systems [5]. These are linear predictive filtering (i.e. using an ALE), transform domain processing (i.e. notch filtering of the Fourier transform of the received waveform in real-time) and spatial interference rejection (i.e. adaptive antenna arrays with steerable nulls). The linear

1058-6393/91 $01.00 © 1991 IEEE

predictive filtering process is based on the difference between the correlation times of the interfering narrowband signal and the much wider bandwidth DS SS signal. Since the correlation time of the narrowband signal is significantly longer than that of the DS SS signal, its current value can be predicted as a function of a set of samples of its past values. The same is not true of the DS SS signal. Thus, when the *predicted* value of the current sample of the received waveform is subtracted from the *actual* current sample of the received waveform, the relative strength of the interfering narrowband signal in the filter output is reduced. The desired broadband DS SS signal is also affected by the adaptive filter, but to a much lesser degree.

In the application of any signal processing technique for digital communications, the ultimate measures of performance are the effects of the technique on bit error rate (BER) and, particularly in DS SS, on the probabilities of achieving and maintaining correct synchronization. Experimental investigations of the effectiveness of the linear predictive technique for tone-jammer suppression and BER maintenance in DS SS communications systems using both LMS-transversal and SG-lattice ALE's are reported in [6] and [7], respectively. The effects of linear predictive filtering on synchronization acquisition and maintenance in DS SS communications are investigated via analysis in [8].

The adaptive filter can distort the desired DS SS signal in two significant ways. First, since the filter effectively introduces a notch into the frequency response of the receiver, it will unavoidably remove some of the energy of the desired signal. Increasing the order of the filter narrows this notch thus improving the BER performance of the system, all else being equal, as demonstrated in [7] for 3rd and 10th order SG-lattice filters when the interference consists of a *pure tone*. For *finite-bandwidth* narrowband interference, however, optimization of the ALE is more complicated. In this case, the optimum filter order depends not only on the bandwidth of the interferer, but also on the ratio of narrowband interferer power to DS SS signal power as discussed in [9]. One example of a narrowband signal that is a part of the interference environment in which the proposed commercial CDMA systems will operate is that transmitted by a typical Fixed Service (FS) point-to-point microwave link (64 QAM, 45 Mbits/s) with a bandwidth on the order of 10 MHz [10]. In the 1850-1990 MHz band, two 48 MHz wide CDMA PCN channels are proposed. The bandwidth of this FS signal is thus 21% of the proposed CDMA PCN spread bandwidth. This may be approaching the practical limit of fractional bandwidth interference that can be effectively excised from a DS SS signal with an ALE.

Secondly, in order to enable the filter to track faster frequency variations, it is necessary to use a larger convergence factor in the coefficient update equations. As observed by Saulnier et al. in [6] however, using a larger μ in the LMS algorithm can also cause the filter to start tracking the DS PN code to some degree, particularly those sections of it containing long "runs" of +1 or -1. The trade-off between the ability to track faster interference frequency changes and minimizing code distortion is an important issue, particularly since the larger convergence factor actually degrades the BER performance of a DS SS system when the interference frequencies are stable or only slowly varying [6]. In the following we present results that address the issue of determining the optimum value of SG-lattice filter time constant for tracking the frequency variations of a tonal interferer.

3 The Stochastic Gradient Lattice Filter

The lattice filter structure is shown in Figure 1. The lattice order recursion update equations are given by

$$\begin{aligned} e_f(k|n) &= e_f(k|n-1) - K_n^b(k)e_b(k-1|n-1) \\ e_b(k|n) &= e_b(k-1|n-1) - K_n^f(k)e_f(k|n-1) \end{aligned} \qquad (1)$$

where

$e_f(k|n) =$ n^{th} stage forward prediction error,

$e_b(k|n) =$ n^{th} stage backward prediction error,

$K_n^b(k)$, $K_n^f(k) =$ n^{th} stage PARCOR coefficients,

all at iteration k. Note that if $x(k)$ is the input signal, then $e_f(k|0) = e_b(k|0) = x(k)$. The stochastic gradient PARCOR coefficient update formula used here is given by

$$\begin{aligned} K_n(k+1) = [1-\beta|e_b(k-1|n-1)|^2]K_n(k) \\ +\beta e_f(k|n-1)e_b^*(k-1|n-1) \end{aligned} \qquad (2)$$

where β is the adaption rate parameter which is equivalent to μ in the LMS algorithm. Under the assumptions that $E[|e_f(k|n)|^2] = E[|e_b(k|n)|^2]$ and $K_n^f(k) = K_n^{b*}(k)$, the value of the PARCOR coefficient $K_n^f(k)$ that minimizes the mean-squared forward prediction error $E[e_f^2(k|n)]$ is given by [11]

$$K_n^{opt}(k) = \frac{E[e_f(k|n-1)e_b^*(k-1|n-1)]}{E[|e_b(k-1|n-1)|^2]} \quad (3)$$

In the application of the lattice filter for the removal of narrowband interference from a DS SS signal, the forward prediction error signal $e_f(k|n)$ is the desired output. As is apparent from Figure 1 (b), the structure of the lattice filter is such that the forward prediction errors $e_f(k|m)$ for all $m<n$ are also available simultaneously. In this sense then, the nth order lattice filter can be viewed as a multiple order filter. This feature could be exploited for optimum excision of strong narrowband interference of unknown or time-varying bandwidth. In fact, as the bandwidth of the interference is increased, the filter automatically "shuts down" the PARCOR coefficients of stages beyond the optimum filter length.

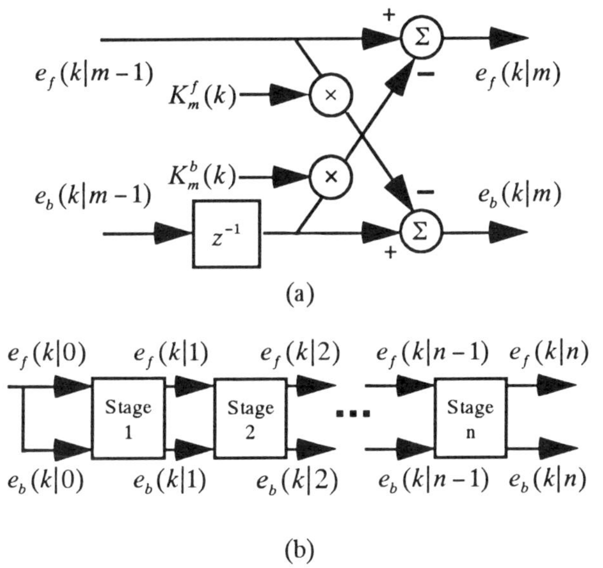

Figure 1 Lattice Filter Structure; (a) single stage; (b) n^{th}-order filter using cascaded stages.

4 Tracking Performance

Analytical and simulation results were presented in [1] for LMS-transversal ALE performance in the detection, parameter estimation, and tracking of stationary and nonstationary multiple tonal and narrowband signals in wideband and bandpass noises. In [2], simulation results were presented demonstrating the comparative performance of the LMS-transversal, SG-lattice, and LS-lattice filters in the tracking of two sinusoidal tones, separated by 10 dB in power, as the frequencies of these tones underwent step, sinusoidal, and linear modulations. In recent analytical work by Bershad and Macchi, the tracking performance of the transversal filter was investigated when the weights were updated using both the RLS [12] and LMS [13] algorithms and the input signal consisted of a complex linear-FM chirp in white Gaussian noise (WGN). The signal models and sampling rate assumptions used in that work are adopted here for evaluation of the tracking performance of the SG-lattice filter in response to a complex linear-FM chirp.

4.1 Input Signal Model

To study the tracking behavior of the SG-lattice filter, the input signal $x(k)$ used here is given by

$$x(k) = s(k) + n(k) \quad (4)$$

where $n(k)$ is complex, zero-mean, WGN with autocorrelation function

$$E[n(i)n^*(j)] = \delta_{ij}P_n \quad (5)$$

and P_n is the noise power. The narrowband signal component $s(k)$ is a complex linear-FM chirp given by

$$s(k) = \sqrt{P_s}\,e^{j(k\omega + k^2\frac{\Psi}{2}+\phi)} \quad (6)$$

where P_s is the signal power, ω is the center frequency, Ψ is the chirp rate and ϕ is a random phase uniformly distributed on $[-\pi, \pi]$.

4.2 Behavior of the PARCOR Coefficients

In order to evaluate (3) explicitly for the input signal defined above, we make the assumption that the statistical expected values of the forward and backward prediction errors of the SG-lattice filter at stage n are equal to those of an nth-order LMS-transversal filter. Then, using the results derived in [12] for the forward prediction error which are extended in [14] for the backward prediction error, the optimal n^{th}-stage PARCOR coefficient for the complex linear-FM chirp in WGN is given by

$$K_n^o(k) = e^{jn\psi k} e^{j(n\omega - \frac{n^2\psi}{2})} \frac{\rho}{1+n\rho} \quad (7)$$

where

$$\rho = \text{SNR} = \frac{P_s}{P_n}$$

In the development leading to (7), the effects of the SG algorithm (2) used to update the PARCOR coefficients were not considered. Note that (7) contains no dependence on the adaption rate parameter β. Using the update formula (2) and the signal model described in Section 4.1, a first-order model for the expected value of the PARCOR coefficients of the nth stage of a multi-stage lattice filter responding to a linear-FM chirp is derived in [14]. The resulting expression is

$$E[K_n(k+1)] = \beta P_s e^{j(n\omega - \frac{n^2\psi}{2})} \frac{e^{jn\psi(k+1)} - Q^k e^{jn\psi}}{e^{jn\psi} - Q} \quad (8)$$

where

$$Q = 1 - \beta E[|e_b(k-1|n-1)|^2], \text{ and}$$

β is selected such that $0<Q<1$. In the expression for Q, the PARCOR coefficient model (8) uses the relation

$$E[|e_b(k-1|n-1)|^2] = \frac{1+n\rho}{1+(n-1)\rho} P_n \quad (9)$$

which is also derived in [14]. The validity of (9) is demonstrated in Figure 2 which shows the quantity

$$[1 - \beta |e_b(k-1|n-1)|^2]^k \quad (10)$$

used here as "ground truth", and

$$\left[1 - \beta \frac{1+n\rho}{\rho(n-1)+1} P_n\right]^k \quad (11)$$

the actual expression used for Q^k in (8), plotted against iteration number k for n=250 stages, noise power P_n from 1.5 to 3.5, and signal power P_s=0.1. Further insight can be gained by noting that for large n, the rhs of (9)→P_n indicating that the filter tends to whiten the backward prediction error as expected.

Figures 3 and 4 show the model-predicted and simulation-result behavior of the real parts of the first 3 PARCOR coefficients as a function of k for chirp rates of $\psi = 1.953125 \times 10^{-5}$ and 1.172×10^{-3}, respectively. In each case the SNR was 10 dB. Additional results concerning the tracking behavior of the PARCOR coefficients are reported in [14] and [15].

4.3 Signal Recovery Error and Output Misadjustment

Although the PARCOR coefficients of the SG-lattice filter can themselves be analyzed in order to detect and estimate the parameters of signals, in the rejection of narrowband interference it is the prediction error signals that contain the signals of interest. The signal recovery error $\eta(k|n)$ at the output of the nth filter stage is defined as

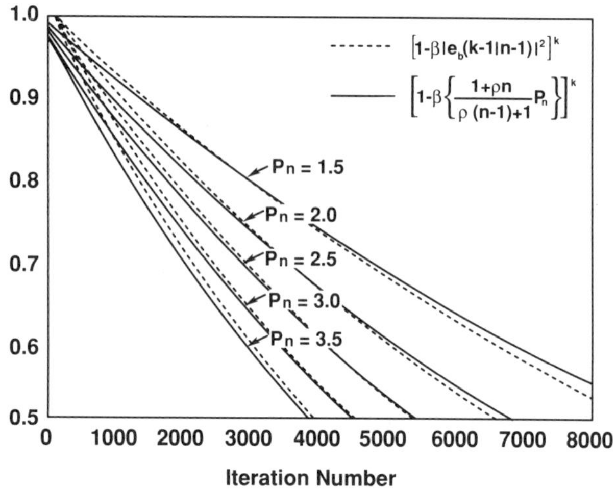

Figure 2 Plots of the expressions in (10) and (11).

$$\eta(k|n) = s(k) - \hat{s}(k|n) \quad (12)$$

where $\hat{s}(k|n)$ is the predicted signal value. In the lattice filter structure used here, the predicted signal value is not directly available. The forward prediction error is given by

$$e_f(k|n) = x(k) - \hat{s}(k|n) = \eta(k|n) + n(k) \quad (13)$$

Hence, in terms of the forward prediction error, the signal recovery error is

$$\eta(k|n) = e_f(k|n) - n(k) \quad (14)$$

Thus the recovery error is a measure of the residual narrowband signal remaining in the forward prediction error signal. Using the first line of (1) in (14) yields a recursive expression for the signal recovery error

$$\eta(k|n) = e_f(k|n-1) - K_n(k)e_b(k-1|n-1) - n(k) \quad (15)$$

The minimum recovery error is given by (see (7))

$$\eta_o(k|n) = e_f(k|n-1) - K_n^o(k)e_b(k-1|n-1) - n(k) \quad (16)$$

The output misadjustment M_n of the nth stage is a measure of the power in that portion of the output $e_f(k|n)$ that is due to the deviation of the PARCOR coefficients from their optimum values. It can be expressed in terms of the signal recovery error as

$$M_n = E[|\eta(k|n) - \eta_o(k|n)|^2] \quad (17)$$

Analytical models of the steady-state (i.e. tracking the chirp) values of the output misadjustments of the first 3 stages of the lattice filter responding to the linear chirp are derived in [14]. In Figure 5, the output misadjustments of the first 3 stages based on these models are compared to simulation results for a signal with chirp rate $\psi = 1.95 \times 10^{-4}$ and SNR=10 dB.

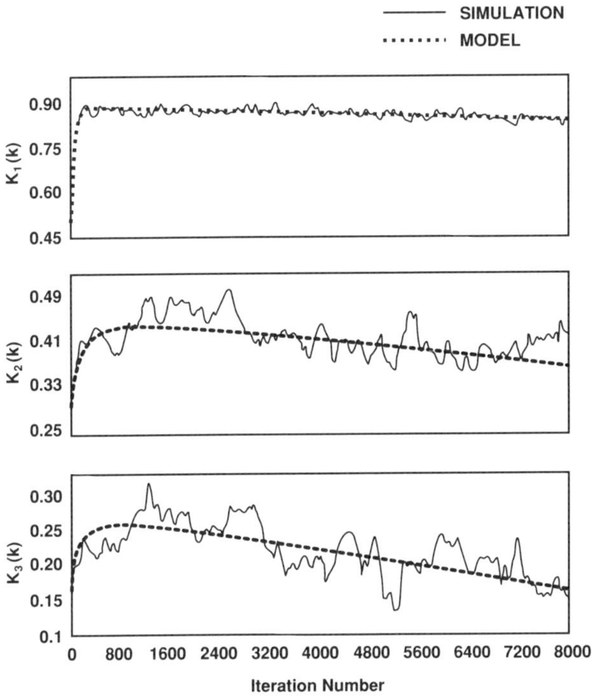

Figure 3 Real parts of the first 3 PARCOR coefficients, $\psi = 1.953125 \times 10^{-5}$.

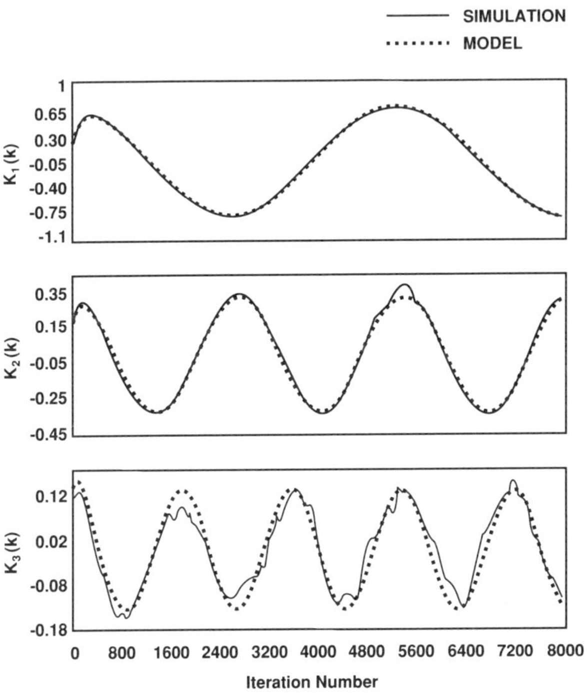

Figure 4 Real parts of the first 3 PARCOR coefficients, $\psi = 1.172 \times 10^{-3}$.

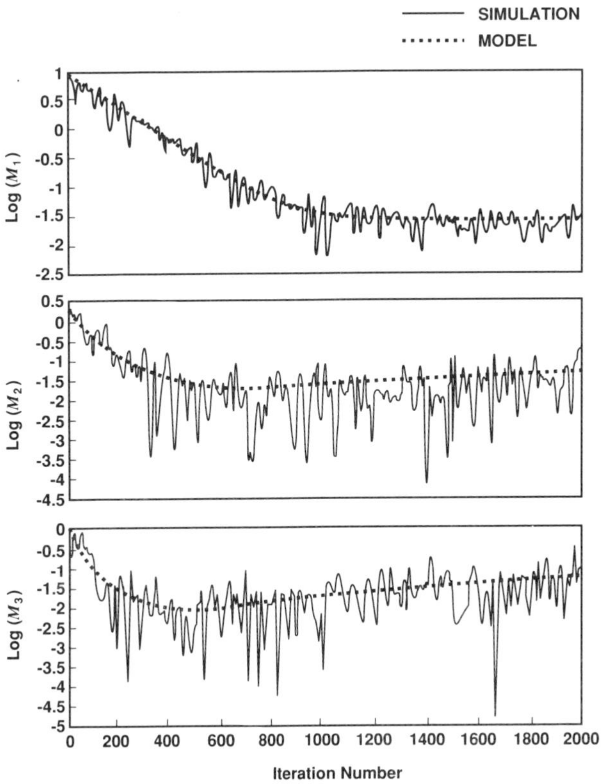

Figure 5 Output misadjustments of the first 3 stages, $\omega = 0.03125$, $P_n = 1.0$, SNR = 10 dB, $\psi = 1.95 \times 10^{-4}$.

In Figure 6, the output misadjustment models of the first 3 stages are plotted vs. β for 5 values of chirp rate equally spaced between 2.5×10^{-5} and 2.5×10^{-4}. The curves exhibit minima indicating optimum values of β, as do similar plots reported for LMS-transversal filters in [16] and [17]. For the high SNR situation presented here, the output misadjustment exhibits converging behavior. Table 1 shows the minimum misadjustment in dB of the first 3 stages for $\psi = 2.5 \times 10^{-5}$. The difference in misadjustment between the first and second stages is 27.6 dB while the difference between the second and third stages is 11 dB, reducing the level to -67 dB at the output of the third stage. Thus relatively short filter lengths appear to be sufficient for the tracking of chirp signals when the SNR as defined in (7) is high. It is also interesting to note that for n ranging from 1 to 3, the dependence of the optimum value of β on ψ decreases with increasing filter order.

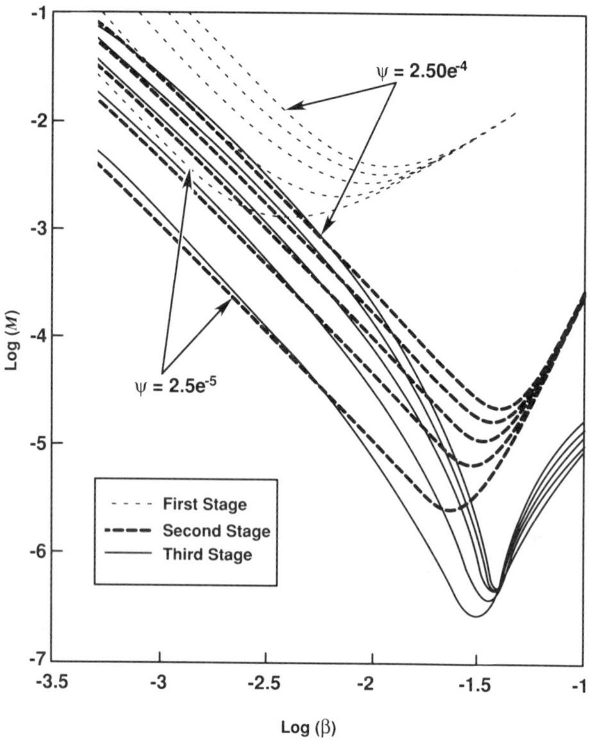

Figure 6 Output misadjustment models of the first three stages.

n(Stage)	M_n^{min} (dB)
1	-28.5
2	-56.1
3	-67.1

Table 1 Minimum misadjustment for first 3 stages.

5 Summary

We have presented new results describing the frequency tracking behavior of the adaptive SG-lattice filter. Results from analysis and simulation are in favorable agreement. Relatively short filter lengths are shown to provide good tracking performance as gauged by the output misadjustments of the first three stages. The application of the SG-lattice filter for the excision of narrowband interference from DS SS signals has been discussed. Further investigation of this application of the SG-lattice filter is planned.

References

[1] J.R. Zeidler, "Performance analysis of LMS adaptive prediction filters," *Proc. IEEE,* Vol. 78, No. 12, pp. 1781-1806, Dec. 1990.

[2] W.S. Hodgkiss and J.A. Presley, "Adaptive tracking of multiple sinusoids whose power levels are widely separated," *IEEE Trans. on Circuits and Systems,* Vol. CAS-28, no. 6, pp. 550-561, Jun. 1981.

[3] D.L. Schilling, L.B. Milstein, R.L. Pickholtz, M. Kullback and F. Miller, "Spread spectrum for commercial communications," *IEEE Commun. Mag.*, pp. 66-79, Apr. 1991.

[4] R.L. Pickholtz, L.B. Milstein and D.L. Schilling, "Spread spectrum for mobile communications," *IEEE Trans. Vehicular Technology,* Vol. 40, No. 2, pp. 313-322, May 1991, (Special issue on digital cellular technologies).

[5] L.B. Milstein, "Interference rejection techniques in spread spectrum communications," *Proc. of the IEEE,*vol. 76, pp. 657-671, Jun. 1988.

[6] G.J. Saulnier, P.K. Das and L.B. Milstein, "An adaptive digital suppression filter for direct-sequence spread-spectrum communications," *IEEE Journ. on Selected Areas in Commun.*, vol. SAC-3, no. 5, pp. 676-686, Sep. 1985.

[7] G.J. Saulnier, K. Yum, and P.K. Das, "The suppression of tone jammers using adaptive lattice filtering," *Proc. IEEE Int. Conf. on Commun.*, pp.869-873, 1987.

[8] L.D. Gottesman and L.B. Milstein, "The performance of a direct sequence spread spectrum acquisition system employing a linear prediction filter," *Proc Globecom,* pp. 45.1.1-45.1.5, 1989.

[9] C.M. Anderson, E.H. Satorius, and J.R. Zeidler, "Adaptive enhancement of finite bandwidth signals in white Gaussian noise," *IEEE Trans. on Acoust., Speech, and Sig. Proc.,* Vol. ASSP-31, No. 1, Feb. 1983.

[10] L.B. Milstein, D.L. Schilling, R.L. Pickholtz, V. Erceg, M. Kullback, E. Kanterakis, D. Fishman, W.H. Biederman, and D. Salerno, "On the feasibility of a CDMA overlay for personal communications networks," to be published in *IEEE Communications Magazine.*

[11] M.L. Honig and D.G. Messerschmitt, *Adaptive Filters: Structures, Algorithms, and Applications,* Kluwer Academic Publishers, Boston, 1984.

[12] O.M. Macchi and N.J. Bershad, "Adaptive recovery of a chirped sinusoid in noise, Part 1: Performance of the RLS algorithm," *IEEE Trans. on Signal Processing,* Vol. 39, No. 3, pp. 583-594, Mar. 1991.

[13] N.J. Bershad and O.M. Macchi, "Adaptive recovery of a chirped sinusoid in noise, Part 2: Performance of the LMS algorithm," *IEEE Trans. on Signal Processing,* Vol. 39, No. 3, pp. 595-602, Mar. 1991.

[14] K.C. Chew, "High performance adaptive and non-adaptive digital signal processors," Ph.D. dissertation, Univ. of California, San Diego, Nov. 1991.

[15] K.C. Chew, J.R. Zeidler, and W.H. Ku, "Tracking model of an adaptive lattice filter for a linear chirp signal in noise," *Proc. IEEE Int. Conf. on Acoustics, Speech and Signal Processing,* pp. 1869-1872, 1991.

[16] O. Macchi, N. Bershad, and M. Mboup, "Steady-state superiority of LMS over LS for time-varying line enhancer in noisy environment," *IEE Proceedings-F,* Vol. 138, No. 4, Aug. 1991.

[17] B. Widrow, J.M. McCool, M.G. Larimore, and R.C. Johnson, "Stationary and nonstationary learning characteristics of the LMS adaptive filter," *Proc. of the IEEE*, Vol. 64, No. 8, Aug. 1976

Two-Dimensional Least-Square ARMA Modeling in the Transform Domain

Wasfy B.Mikhael, Fellow IEEE, and Haoping Yu

Department of Electrical Engineering
P.O.Box 25000
University of Central Florida
Orlando, FL 32816

Abstract:

In this paper, an algorithm for two dimensional(2D) auto-regressive, moving average(ARMA) and moving average(MA) modeling in the Fourier transform domain, which can be implemented in other domains, is presented. The algorithm derived here is based on the equationary model (recursive-like structure) which can match, in the least square(LS) sense, the unknown system. Simulation examples of two dimensional system modeling are given to demonstrate the performance of this algorithm.

I. Introduction:

The technique of modeling and system identification is very important in the field of control systems, communications and signal processing[1]. In speech signal processing, this technique has been used successfully for signal representation[2]. Also, it can be applied in filter design and inverse filtering. Many contributions dealing with signal restoration and noise cancelation for both one dimensional and two dimensional signals have been reported [3][4][5][6]. Most methods and techniques above are developed in the time or spatial domain.

The objective of this contribution is to develop a least square method which can approximate two dimensional transfer functions in the Fourier transform domain. The application of the proposed approach in 2D system modeling for both ARMA and MA structures is given. The algorithm derived here is shown to have excellent accuracy and convergence properties. This method can also be applied in the field of image restoration and noise cancelation. Also, it can be implemented in the Walsh transform domain.

In this paper, the algorithm is introduced first. Then, sample design examples for both ARMA and MA structures are given to illustrate the performance of this method, followed by the conclusions.

II. 2D LS Transform Domain ARMA Modeling Algorithm:

The proposed two-dimensional least square(2D LS) ARMA modeling algorithm is based on the recursive like structure[3], shown in Fig.1. This structure is implemented here in the 2D frequency domain. In Fig.1, $S_d(z_1,z_2)$, which is a 4-dimensional function in general, since both z_1 and z_2 are complex variables, represents the unknown(desired) 2D system transfer function. The algorithm employs the general form of the the ARMA model given by[7]

$$S(z_1,z_2)=\frac{A(z_1,z_2)}{B(z_1,z_2)}=\frac{\sum\limits_{(r1,r2)\in Ra}\sum a(r_1,r_2)\,z_1^{-r1}z_2^{-r2}}{1+\sum\limits_{(l1,l2)\in Rb-(0,0)}\sum b(l_1,l_2)\,z_1^{-l1}z_2^{-l2}} \tag{1}$$

where $S(z_1,z_2)$ is the estimated function, to approximate $S_d(z_1,z_2)$ by choosing the numerator coefficients, $a(r_1,r_2)$, and denominator coefficients, $b(l_1,l_2)$. This is achieved by minimizing

$$E(z_1,z_2)=X(z_1,z_2)[S_d(z_1,z_2)B(z_1,z_2)-A(z_1,z_2)] \tag{2}$$

in the LS sense.

In this paper, Ra represents the region of support of $a(r_1,r_2)$[7], which will be the first quadrant, and Rb-(0,0) has the same definition as Ra while excluding the origin point (0,0). Both $a(r_1,r_2)$ and $b(l_1,l_2)$ are set to be real.

$X(z_1,z_2)$ in (2) has different interpretations depending on the applications. Here, $X(z_1,z_2)$ is assumed as the input of the system and used as a weighting function in this modeling algorithm.

Z_1 and z_2 are evaluated at N_1xN_2 sampling points on the unit surface[7], i.e, $z_1=e^{j\omega_{k1}}$ and $z_2=e^{j\omega_{k2}}$, where $0\le\omega_{k1}<2\pi$ corresponds to $0\le k_1\le N_1-1$. Similarly $0\le\omega_{k2}<2\pi$ for $0\le k_2\le N_2-1$. Thus the discrete frequency domain representation of (2) is obtained as

$$E(\omega_{k_1},\omega_{k_2})=X(\omega_{k_1},\omega_{k_2})[S_d(\omega_{k_1},\omega_{k_2})B(\omega_{k_1},\omega_{k_2})-A(\omega_{k_1},\omega_{k_2})] \tag{3}$$

Substituting $z_1=e^{j\omega_{k1}}$ and $z_2=e^{j\omega_{k2}}$ in $S_d(z_1,z_2)$ yields

$$S_d(\omega_{k_1},\omega_{k_2})=S_d(z_1,z_2)|_{z1=e^{j\omega_{k1}},\ z2=e^{j\omega_{k2}}} \tag{4}$$

which is assumed as an unknown transfer function of a two-dimensional system. Similarly, $X(z_1,z_2)$, $A(z_1,z_2)$ and $B(z_1,z_2)$ are evaluated on the unit surface which results in

$$X(\omega_{k_1},\omega_{k_2})=X(z_1,z_2)|_{z1=e^{j\omega_{k1}},\ z2=e^{j\omega_{k2}}}. \tag{5}$$

$$A(\omega_{k_1},\omega_{k_2})=\sum_{(r1,r2)\in Ra}\sum a(r_1,r_2)\,z_1^{-r1}z_2^{-r2}|_{z1=e^{j\omega_{k1}},\ z2=e^{j\omega_{k2}}}$$

$$=\sum_{(r1,r2)\in Ra}\sum a(r_1,r_2)e^{-j\omega_{k1}r1}e^{-j\omega_{k2}r2} \tag{6}$$

and

$$B(\omega_{k_1},\omega_{k_2})=\sum_{(l1,l2)\in Rb}\sum b(l_1,l_2)\,z_1^{-l1}z_2^{-l2}|_{z1=e^{j\omega_{k1}},\ z2=e^{j\omega_{k2}}}$$

$$=\sum_{(l1,l2)\in Rb}\sum b(l_1,l_2)\,e^{-j\omega_{k1}l1}e^{-j\omega_{k2}l2} \tag{7}$$

Then error measure, ε, becomes

$$\varepsilon=\sum_{k1=0}^{N1-1}\sum_{k2=0}^{N2-1}|X(\omega_{k_1},\omega_{k_2})[S_d(\omega_{k_1},\omega_{k_2})B(\omega_{k_1},\omega_{k_2})-A(\omega_{k_1},\omega_{k_2})]|^2 \tag{8}$$

Substituting from (4),(5),(6) and (7) in (8), where b(0,0)=1, one obtains

$$\varepsilon=\sum_{k1=0}^{N1-1}\sum_{k2=0}^{N2-1}|X(\omega_{k_1},\omega_{k_2})[S_d(\omega_{k_1},\omega_{k_2})\sum_{(l1,l2)\in Rb}\sum b(l_1,l_2)e^{-j\omega_{k1}l1}e^{-j\omega_{k2}l2}$$

$$-\sum_{(r1,r2)\in Ra}\sum a(r_1,r_2)e^{-j\omega_{k1}r1}e^{-j\omega_{k2}r2}]|^2 \tag{9}$$

To minimize ε given by (9), the derivatives of ε with respect to the coefficients are set to be zero, to get the optimal values $a^*(r_1,r_2)$ and $b^*(l_1,l_2)$, as follows:

1058-6393/91 $01.00 © 1991 IEEE

$$\frac{\partial \varepsilon}{\partial a(m,n)} = 0 \quad (10)$$

and

$$\frac{\partial \varepsilon}{\partial b(m'_1,n'_2)} = 0 \quad (11)$$

where $(m,n) \in Ra$ and $(m'_1,n'_2) \in Rb$.

From (9), $\partial\varepsilon / \partial a(m,n)$ is obtained as

$$\frac{\partial \varepsilon}{\partial a(m,n)} = -2 \sum_{k1=0}^{N_1-1} \sum_{k2=0}^{N_2-1} |X(\omega_{k_1},\omega_{k_2})|^2 [\mathrm{Re}\{(W_{k1})^m (W_{k2})^n . S_d(\omega_{k_1},\omega_{k_2}) \sum\sum_{(l1,l2)\in Rb} b(l_1,l_2) \cdot (W_{k1})^{-l1} . (W_{k2})^{-l2} - \sum\sum_{(r1,r2)\in Ra} a(r_1,r_2)(W_{k1})^{-r1}(W_{k2})^{-r2}\}],$$

$$(m,n) \in Ra \quad (12)$$

where $\beta = \angle S_d(\omega_{k_1},\omega_{k_2})$, $W_{k1} = e^{j\omega_{k1}}$, $W_{k2} = e^{j\omega_{k2}}$.

Using (10) and (12) yields

$$\sum_{k1=0}^{N_1-1} \sum_{k2=0}^{N_2-1} |X(\omega_{k_1},\omega_{k_2})|^2 \{|S_d(\omega_{k_1},\omega_{k_2})| \sum\sum_{(l1,l2)\in Rb} b(l_1,l_2) \mathrm{Cos}[\omega_{k_1}(m-l1) + \omega_{k_2}(n-l2)+\beta] - \sum\sum_{(r1,r2)\in Ra} a(r_1,r_2) \mathrm{Cos}[\omega_{k_1}(m-r1) + \omega_{k_2}(n-r2)]\} = 0$$

$$(m,n) \in Ra \quad (13)$$

Similarly, $\partial\varepsilon / \partial b(m'_1,n'_2)$ is derived from (9) as

$$\frac{\partial \varepsilon}{\partial b(m'_1,n'_2)} = -2 \sum_{k1=0}^{N_1-1} \sum_{k2=0}^{N_2-1} |X(\omega_{k_1},\omega_{k_2})|^2 [\mathrm{Re}\{|S_d(\omega_{k_1},\omega_{k_2})|^2 (W_{k1})^{m'} (W_{k2})^{n'} \sum\sum_{(l1,l2)\in Rb} b(l_1,l_2) \cdot (W_{k1})^{-l1} (W_{k2})^{-l2} - S_d(\omega_{k_1},\omega_{k_2})(W_{k1})^{m'}(W_{k2})^{n'} \sum\sum_{(r1,r2)\in Ra} a(r_1,r_2) \cdot (W_{k1})^{-r1}(W_{k2})^{-r2}\}]$$

$$(m',n') = \in Rb-(0,0) \quad (14)$$

Using (11) and (14) results in

$$\sum_{k1=0}^{N_1-1} \sum_{k2=0}^{N_2-1} |X(\omega_{k_1},\omega_{k_2})|^2 \{|S_d(\omega_{k_1},\omega_{k_2})|^2 \sum\sum_{(l1,l2)\in Rb} b(l_1,l_2) \mathrm{Cos}[\omega_{k_1}(m'-l1)+\omega_{k_2}(n'-l2)l1)] - \sum\sum_{(r1,r2)\in Ra} a(r_1,r_2)|S_d(\omega_{k_1},\omega_{k_2})| \mathrm{Cos}[\omega_{k_1}(m'-r1) + \omega_{k_2}(n'-r2)-\beta]\} = 0 \quad (15)$$

To simplify the expressions, the following notations are used for the summations in (13) and (15):

$$T_{11}(l_1,l_2,m,n) = \sum_{k1=0}^{N_1-1} \sum_{k2=0}^{N_2-1} |X(\omega_{k_1},\omega_{k_2})|^2 |S_d(\omega_{k_1},\omega_{k_2})| \mathrm{Cos}[\omega_{k_1}(m-l1)+\omega_{k_2}(n-l2)+\beta]$$

$$(m,n) \in Ra \quad (16)$$

$$T_{12}(r_1,r_2,m,n) = \sum_{k1=0}^{N_1-1} \sum_{k2=0}^{N_2-1} |X(\omega_{k_1},\omega_{k_2})|^2 \mathrm{Cos}[\omega_{k_1}(m-r1) + \omega_{k_2}(n-r2)]$$

$$(m,n) \in Ra \quad (17)$$

$$T_{21}(l_1,l_2,m',n') = \sum_{k1=0}^{N_1-1} \sum_{k2=0}^{N_2-1} |X(\omega_{k_1},\omega_{k_2})|^2 |S_d(\omega_{k_1},\omega_{k_2})|^2 \cdot \mathrm{Cos}[\omega_{k_1}(m'-l1)+\omega_{k_2}(n'-l2)l1)]$$

$$(m',n') \in Rb-(0,0) \quad (18)$$

$$T_{22}(r_1,r_2,m',n') = \sum_{k1=0}^{N_1-1} \sum_{k2=0}^{N_2-1} |X\omega_{k_1},\omega_{k_2})|^2 |S_d(\omega_{k_1},\omega_{k_2})| \cdot \mathrm{Cos}[\omega_{k_1}(m'-r1) + \omega_{k_2}(n'-r2)-\beta]$$

$$(m',n') \in Rb-(0,0) \quad (19)$$

Using (13), (15) and (16-19) gives the following set of linear equations:

$$\sum\sum_{(l1,l2)\in Rb} b(l_1,l_2) T_{11}(l_1,l_2,m,n) - \sum\sum_{(r1,r2)\in Ra} a(r_1,r_2) T_{12}(r_1,r_2,m,n) = -\sum_{k1=0}^{N_1-1} \sum_{k2=0}^{N_2-1} |X(\omega_{k_1},\omega_{k_2})|^2 |S_d(\omega_{k_1},\omega_{k_2})| \mathrm{Cos}[\omega_{k_1} \cdot m + \omega_{k_2} \cdot n+\beta]$$

$$(m,n) \in Ra \quad (20)$$

$$\sum\sum_{(l1,l2)\in Rb-(0,0)} b(l_1,l_2) T_{21}(l_1,l_2,m',n') - \sum\sum_{(r1,r2)\in Ra} a(r_1,r_2) T_{22}(r_1,r_2,m',n') = -\sum_{k1=0}^{N_1-1} \sum_{k2=0}^{N_2-1} |X(\omega_{k_1},\omega_{k_2})|^2 |S_d(\omega_{k_1},\omega_{k_2})|^2 \mathrm{Cos}[\omega_{k_1} \cdot m' + \omega_{k_2} \cdot n']$$

$$(m',n') \in Rb-(0,0) \quad (21)$$

$a^*(r_1,r_2)$ and $b^*(l_1,l_2)$, the optimal value of $a(r_1,r_2)$ and $b(l_1,l_2)$, are obtained from (20) and (21). Here, it is worthwhile to note that the number of sampling points on the unit surface must be greater than the number of the $a(r_1,r_2)$ and $b(l_1,l_2)$ coefficients. Otherwise, (20) and (21) will be illconditionned. This means

$$N_1 \times N_2 \geq N_a \times N_b \quad (22)$$

where N_a and N_b are the number of $a(r_1,r_2)$ and $b(l_1,l_2)$ coefficents respectively.

Sampling the unit surface on a square grid, with equal steps of $2\pi/N$ for both Z_1 and Z_2, yields

$$\omega_{k_1} = 2\pi k_1 / N, \qquad 0 \le k_1 \le N-1 \quad (23)$$

$$\omega_{k_2} = 2\pi k_2 / N, \qquad 0 \le k_2 \le N-1 \quad (24)$$

and

$$Z_1 = e^{j2\pi k_1 / N}, \qquad 0 \le k_1 \le N-1 \quad (25)$$

$$Z_2 = e^{j2\pi k_2 / N}, \qquad 0 \le k_2 \le N-1 \quad (26)$$

Thus, the 2D Fast Fourier Transform(FFT) can be used to implement the above LS algorithm given by (20) and (21). For convenience in the notation, the Fourier transform used here employs 2N X 2N points instead of N X N points. From previous assumptions, $a(r_1,r_2)$ and $b(l_1,l_2)$ are real. Also $S_d(k_1,k_2)$ is an unknown system transfer function whose impulse response is real . As a consequence of the symmetry properties of the discrete Fourier transform, the summation from 0 to N is used instead of from 0 to 2N-1 in evaluating ε. Hence, (8) becomes

$$\varepsilon = \sum_{k1=0}^{N} \sum_{k2=0}^{N} |X(k_1,k_2)[S_d(k_1,k_2)B(k_1,k_2) - A(k_1,k_2)]|^2 \quad (27)$$

Using the FFT in (16) to (19) gives,

$$T_{11}(l_1,l_2,m,n) = \sum_{k1=0}^{N} \sum_{k2=0}^{N} |X(k_1,k_2)|^2 |S_d(k_1,k_2)| \mathrm{Cos}[(k_1m+k_2n-l_1k_1-l_2k_2)\omega+\beta]$$

$$(m,n) \in Ra \quad (28)$$

$$T_{12}(r_1,r_2,m,n) = \sum_{k1=0}^{N} \sum_{k2=0}^{N} |X(k_1,k_2)|^2 \mathrm{Cos}[(k_1m+k_2n-r_1k_1-r_2k_2)\omega]$$

$$(m,n) \in Ra \quad (29)$$

$$T_{21}(l_1,l_2,m',n') = \sum_{k1=0}^{N} \sum_{k2=0}^{N} |X(k_1,k_2)|^2 |S_d(k_1,k_2)|^2 \mathrm{Cos}[(k_1m+k_2n'-l_1k_1-l_2k_2)\omega]$$

$$(m',n') \in Rb-(0,0) \quad (30)$$

$$T_{22}(r_1,r_2,m',n') = \sum_{k1=0}^{N} \sum_{k2=0}^{N} |X(k_1,k_2)|^2 |S_d(k_1,k_2)| \mathrm{Cos}[(k_1m'+k_2n'-r_1k_1-r_2k_2)\omega-\beta]$$

$$(m',n') \in Rb-(0,0) \quad (31)$$

where $\omega = \pi/N$.

Using (23) and (24), and substituting (28-31) in (20) and (21) yield

$$\sum\sum_{(l1,l2)\in Rb} b^*(l_1,l_2) T_{11}(l_1,l_2,m,n) - \sum\sum_{(r1,r2)\in Ra} a^*(r_1,r_2) T_{12}(r_1,r_2,m,n) = -\sum_{k1=0}^{N_1-1} \sum_{k2=0}^{N_2-1} |X(\omega_{k_1},\omega_{k_2})|^2 |S_d(\omega_{k_1},\omega_{k_2})| \mathrm{Cos}[\omega k_1 \cdot m + \omega k_2 \cdot n+\beta]$$

$$(m,n) \in Ra \quad (32)$$

$$\sum_{(l1,l2)\in Rb-(0,0)}^{N_1-1}\sum^{N_2-1} b^*(l_1,l_2)T_{21}(l_1,l_2,m',n') - \sum_{(r1,r2)\in Ra}\sum a^*(r_1,r_2)T_{22}(r_1,r_2,m',n')$$

$$= -\sum_{k1=0}^{N_1-1}\sum_{k2=0}^{N_2-1}|X(\omega_{k_1},\omega_{k_2})|^2|S_d(\omega_{k_1},\omega_{k_2})|^2 Cos[\omega k_1\cdot m' + \omega k_2\cdot n'] \qquad (m',n')\in Rb-(0,0) \qquad (33)$$

Equations (32) and (33) can be solved for the optimal coefficients $a^*(r_1,r_2)$ and $b^*(l_1,l_2)$.

III. Application Examples

In this section, the application of the proposed algorithm for both ARMA and MA structures is presented. To evaluate the accuracy (S/N performance measure) of the modeling, let us define

$$S = \sum_{k1=0}^{N}\sum_{k2=0}^{N} |Sd(k_1,k_2)|^2 \qquad (34)$$

$$N = \sum_{k1=0}^{N}\sum_{k2=0}^{N} |Sd(k1,k2) - S(k1,k2)|^2 \qquad (35)$$

In the following examples, 10log(S/N) (dB) is used as a measure of the performance, and $X(k_1,k_2)$ in Fig.1 is chosen to be unity, which means $X(k_1,k_2)=1$, $0\le k_1<2N$, $1\le k_2<2N$.

III-a Example1. ARMA Modeling:

For illustration, an unknown system transfer function is simulated, which is described by

$$S_d(z_1,z_2)=\frac{1+z_1^{-1}+z_2^{-1}+z_1^{-1}z_2^{-1}}{1+0.5z_1^{-1}+0.25z_2^{-1}+0.125z_1^{-1}z_2^{-1}} \qquad (36)$$

The unknown system given by (36) has a magnitude spectrum as shown in Fig.2. First, both Na and Nb in (22) are set to 2, i.e., $r_1, r_2\in[0,1]$ and $l_1, l_2\in[0,1]$. The proposed algorithm yields the following parameters for $S(z_1,z_2)$:

a(0,0)= 1.00000, a(0,1)= 1.00001
a(1,0)= 1.00001, a(1,1)= 1.00001 (37)
b(0,0)= 1.00000, b(0,1)= 0.250005
b(1,0)= 0.500008, b(1,1)= 0.124999

Using (34), (35), (36) and (37), one gets 10log(S/N)= 114.87 dB. The magnitude spectrum of the error $E(\omega_{k_1},\omega_{k_2})$ in (3) is shown in Fig.3.

Next, a higher order system was used to model the same unknown system(36), i.e., $r_1, r_2\in[0,3]$ and $l_1, l_2\in[0,3]$. For this case, the signal to noise ratio obtained, 10log(S/N), was 116.80 dB corresponding to the following parameters of $S(z_1,z_2)$:

a(0,0)=0.999999, a(0,1)=-.387438 , a(0,2)=-.516985, a(0,4)=.878452
a(1,0)=.126576, a(1,1)=.0009111, a(1,2)=.0084997, a(1,3)=.228688
a(2,0)=.004641, a(2,1)=.189266, a(2,2)=-.009038, a(2,3)=-.179740 (38)
a(3,0)=.919835, a(3,1)=-.189061, a(3,2)=-.638878, a(3,3)=.470023
b(0,0)=1.00000, b(0,1)=-.887437, b(0,2)=.176729, b(0,3)=.435228
b(1,0)=-.623425, b(1,1)=.593189, b(1,2)=.0202903, b(1,3)=-.21608
b(2,0)=.701477, b(2,1)=-.445227, b(2,2)=-.090385, b(2,3)=.153790
b(3,0)=.919835, b(3,1)=-.189061, b(3,2)=-.021109, b(3,3)=.470023

Fig.4 shows the magnitude spectrum of the error. The signal to noise ratio or the magnitude spectrum of the error, is very small. The coefficients in (36) and (38) are quite different due to the common factors between the numerator and denominator of $S(z_1,z_2)$.

III-b Example2. MA Modeling:

For this application, the algorithm derived above is modified by setting the $b(l_1,l_2)$ to zero, where $(l_1,l_2)\in Rb-(0,0)$, and b(0,0)=1. Again, for illustration, an unknown system transfer function is simulated which is described by

$$S_d(z_1,z_2) = 1+0.5z_1^{-1}+0.25z_2^{-1}+0.125z_1^{-1}z_2^{-1} \qquad (39)$$

This unknown system has the magnitude spectrum shown in Fig.5. As in example1, a low order system, for which $r_1, r_2\in[0,1]$, is used first. Then, a higher order system, $r_1, r_2\in[0,3]$, is used. For $r_1, r_2\in[0,1]$, the resulting coefficients of $S(z_1,z_2)$ are given below, which are the same as those of $S_d(z_1,z_2)$ given by (39).

a(0,0)=1.00000 a(0,1)=.500000
a(1,0)=.250000 a(1,1)=.125000 (40)

For $r_1, r_2\in[0,3]$, the resulting coefficients of $S(z_1,z_2)$ are given by

a(0,0)=1.00000, a(0,1)=.500001, a(0,2)=-4.48e-3, a(0,3)=-2.98e-7
a(1,0)=.249999, a(1,1)=.125001, a(1,2)=7.97e-7, a(1,3)=-5.93e-7
a(2,0)=8.70e-7, a(2,1)=-5.96e-7 a(2,2)=1.32e-6, a(2,3)=2.71e-7 (41)
a(3,0)=-2.14e-7, a(3,1)=-9.50e-7, a(3,2)=-3.39e-7, a(3,3)=3.20e-7

Comparing (41) with (40), the coefficients a(i, j), $i, j\in[0,1]$ in (41), are very close to those having the same coordinates in (40), while the others are very small. Thus, the system with the coefficients in (41) has almost the same specifications as the system described by (40). The final S/N is 120.44 dB. Figure 6 shows the performance of this design.

From the above illustrative examples, the accuracy of this method in modeling two dimensional systems in the transform domain is easily seen.

IV.Conclusions:

In this paper, a frequency domain LS technique for the modeling two dimensional ARMA systems is developed. Its application has been successfully demonstrated. Sample examples are given showing the accuracy of this approach for both ARMA and MA structures .

Acknowledgement:

This work is supported by the U.S. Army Program Manager, Training Devices (PM-TRADE) Defence Advanced Research Project Agency (DARPA), Contract#N6133-89-C-0043, the Florida High Technology and Industry Council, Contracts #16-22-717, and 16-22-726, and Harris Semiconductor Digital Signal Processing Prooducts.

Reference:

[1] B.Widrow and S.D.Stearns, *adaptive signal processing*, Englewood cliffs, NJ: Pretice Hall,1985.

[2] Wasfy B. Mikhael and Andreas S. Spanias, " Efficient Modeling of Dominant Transform Components Representing Time-Varying Signals", IEEE Trans. CAS, Vol. CAS-36, No.2, PP.331-334, Feb. 1989.

[3]. W. A. Gardner, " A stable Two-Stage Recursive-Like Adaptive Noise Canceller", Proc. IEEE, Vol 69, PP.487-488, April 1981.

[4] Wasfy B. Mikhael and Shomit M. Ghosh, " Generation of Multidimensional Variable Step Size Sequential Adaptive Gradient Algorithms with Identification and Noise Cancelation Application," accepted for publication, IEEE Trans. CAS.

[5] Hector M. Valenznela " A two-dimensional recursive model for bilinear systems with applications to image reconstruction", IEEE Trans. on CAS Vol. CAS-34, No.7, July 1987.

[6]. Luis F. Chaparro, E.I. Jury "Rational approximation of 2-D linear discrete systems", IEEE Trans. on ASSP, Vol. ASSP-30, No.5, Oct.1982.

[7]. Jae S. Lim, *Two-Dimensionnal Signal and Image Processing*, Engleewood Cliffs, NJ: Prentice-Hall, 1990.

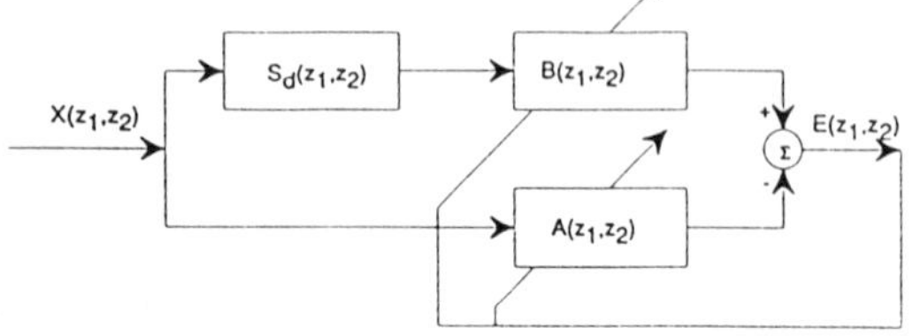

Fig.1. The Block diagram of the 2D equationary model system.

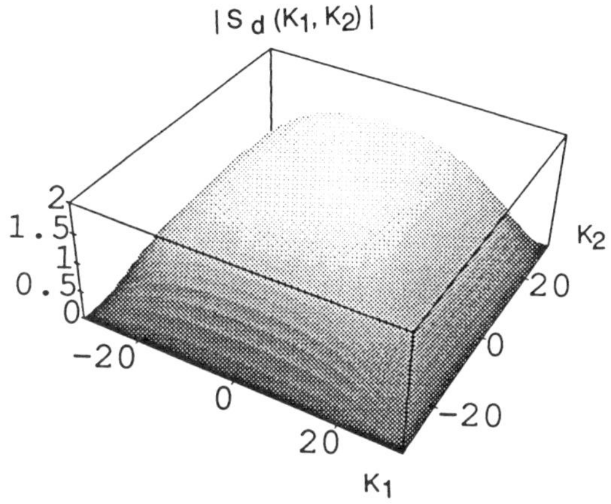

Fig.2. The magnitude of the desired system, Sd(K1, K2), in example 1.

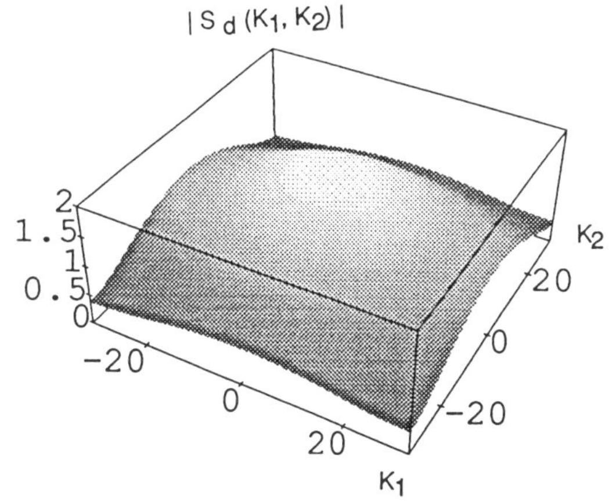

Fig.5. The magnitude spectrum of the desired system, Sd(K1, K2), in example 2.

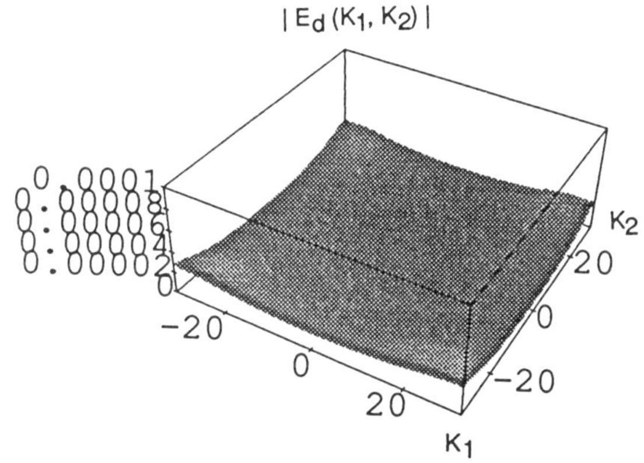

Fig.3. The magnitude spectrum of the error in example 1, where Na=Nb=2.

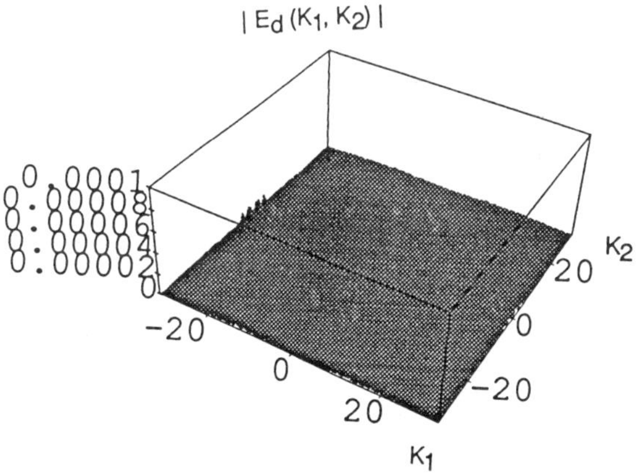

Fig.6. The magnitude spectrum of the error in Example 2, where Na=Nb=4.

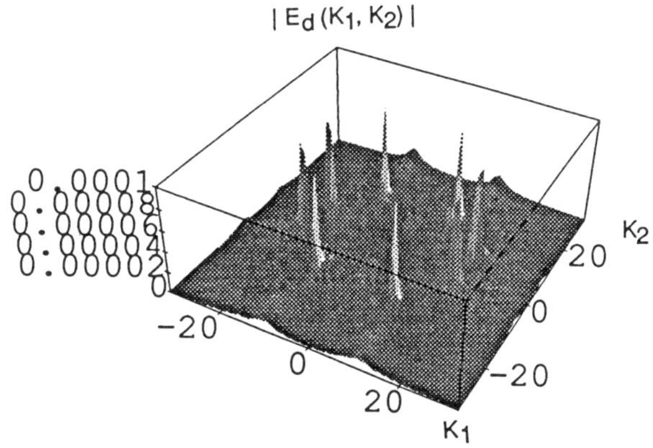

Fig.4. The magnitude spectrum of the error in example 1, where Na=Nb=4.

AN ADAPTIVE TECHNIQUE FOR DESIGNING MINIMUM PHASE MODELS

James L. Rasmussen and Delores M. Etter
Department of Electrical and Computer Engineering
University of Colorado
Boulder, CO 80309

Abstract

We have developed a new filter structure, called a stochastic filter, that consists of a bank of fixed filters with a set of corresponding probabilities. The fixed filters can be viewed as the basis set of filters for the stochastic filter, with the probabilities determining the specific realization represented by the stochastic filter. If the probabilities are allowed to vary with time, the stochastic filter is an implementation of an adaptive filter. We present a configuration, using FIR fixed filters in the stochastic filter, that can be used to adaptively model unknown systems. We will show theoretically that this form of adaptive stochastic filter converges to a minimum phase FIR filter model.

1 INTRODUCTION

A stochastic filter (SF) [1] is composed of a bank of parallel fixed filters. Each fixed filter has a probability associated with it that determines the contribution of that filter to the output. If the probabilities vary with time, then the structure is an adaptive stochastic filter (ASF). In this paper, we use three fixed, second-order filters in the ASF as shown in Figure 1.

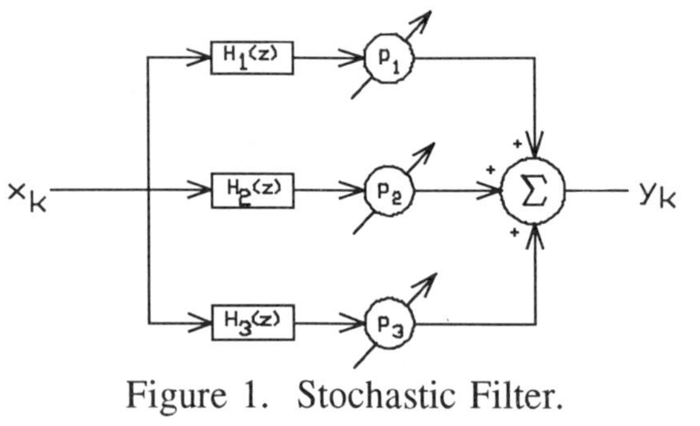

Figure 1. Stochastic Filter.

Each probability functions as a weight or a gain with a value between zero and one. The sum of the probabilities is constrained to equal one. We call this a "stochastic filter" because the probabilities associated with the fixed filters determine the filter's characteristics.

The fixed filters in a SF can be composed of FIR or IIR filters; in this paper we assume they are second-order FIR filters. (Without loss of generality, we assume that the general form of each of the FIR filters is $H_k(z) = 1 + a_{1,k}z^{-1} + a_{2,k}z^{-2}$).

The SF is represented by an FIR filter of the same order as the fixed filters, with coefficients computed as expected values (or weighted averages) of the corresponding coefficients in the SF. Therefore, we represent the SF by the following average filter representation:

$$\begin{aligned} H_\mu(z) &= \sum_{k=1}^{3} H_k(z) p_k \\ &= 1 + (p_1 a_{1,1} + p_2 a_{1,2} + p_3 a_{1,3}) z^{-1} + (p_1 a_{2,1} + p_2 a_{2,2} + p_3 a_{3,2}) z^{-2} \\ &= 1 + a_{1\mu} z^{-1} + a_{2\mu} z^{-2} \\ &\qquad \text{where } a_{k\mu} = \sum_{k=1}^{3} (p_k a_{n,k}) \end{aligned} \tag{1}$$

and p_k represents the probability associated with $H_k(z)$.

We now define the matrix, **H**, whose columns are the coefficients of the fixed filters of the SF.

$$\boldsymbol{H} = \begin{bmatrix} 1 & 1 & 1 \\ a_{1,1} & a_{1,2} & a_{1,3} \\ a_{2,1} & a_{2,2} & a_{2,3} \end{bmatrix} = [\, \boldsymbol{h}_1 \;\; \boldsymbol{h}_2 \;\; \boldsymbol{h}_3 \,].$$

Note that $\mathbf{h}_k$ is also the impulse response of the kth filter. We also represent the probabilities and the impulse response of $H_\mu(z)$ as vectors and then the average filter can

1058-6393/91 $01.00 © 1991 IEEE

be conveniently represented in matrix notation as $\mathbf{h}_\mu = \mathbf{Hp}$.

2 MATHEMATICAL DERIVATION

As shown in Figure 2, any number of the ASF stages can be cascaded to make higher order adaptive filters. Each ASF stage is composed of the three fixed second-order filters as shown in Figure 1. The standard system identification configuration using only one ASF section will be used to show that the adaptive SF converges to a minimum phase filter.

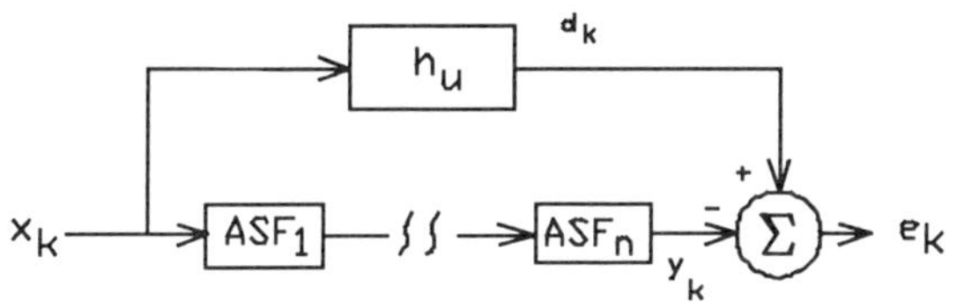

Figure 2. System Identification Configuration.

2.1 Second-Order Derivation

If only one ASF component is used, the error in Figure 2 is given by

$$\begin{aligned} e_k &= d_k - y_k \\ &= d_k - h_\mu^T x_k \\ e_k^2 &= d_k^2 + (h_\mu^T x_k)(x_k^T h_\mu) - 2 d_k x_k^T h_\mu \\ &= d_k^2 + (Hp)^T x_k x_k^T Hp - 2 d_k x_k^T Hp . \end{aligned} \tag{2}$$

The mean-squared error can be shown to be

$$\begin{aligned} \xi &= E(e_k^2) = E(d_k^2) + p^T R' p - 2 q'^T p \\ \text{where} \quad R &= E(\, x_k x_k^T \,) \quad \text{and } R' = H^T R H \\ q^T &= E(\, d_k x_k^T \,) \quad \text{and } q'^T = q^T H . \end{aligned} \tag{3}$$

This mean-squared error is quadratic in **p** and, with the modified **R** matrix and **q** vector, (3) is of the same form as the squared-error expression for the standard linear combiner obtained in [2]. Similarly, the standard LMS algorithm [2] can be implemented using the following equations.

$$\begin{aligned} \nabla &= 2R'p - 2q' \\ p^* &= (R')^{-1} q' \\ p_{k+1} &= p_k + 2\mu e_k H^T x_k . \end{aligned} \tag{4}$$

If we let x_k in Figure 2 be white noise with zero mean and unit variance then (3) can be simplified to the following:

$$\xi = h_u^T h_u + h_\mu^T h_\mu - 2 h_{tu}^T h_\mu , \tag{5}$$

where $\mathbf{h}_{tu}$ is the vector of coefficients of the unknown system impulse response truncated to the first three coefficients.

It can be shown that if the zeros of the three fixed, second-order filters are inside the unit circle, then the zeros of $H_\mu(z)$ will be inside the unit circle for any set of valid probabilities. Therefore, during the convergence process, the zeros of the ASF will never go outside the unit circle.

We have also shown that for any point inside the unit circle there is a unique solution to $\mathbf{Hp} = \mathbf{h}_\mu$. Therefore, if the unknown filter is a second-order FIR filter with its zeros inside the unit circle, then the ASF will converge to the location inside the unit circle such that its zeros are the same as the zeros of the unknown filter. Furthermore, there is a unique set of probabilities that will place the zeros of the ASF at the zeros of the unknown and $\mathbf{h}_\mu$ will exactly match $\mathbf{h}_u$ and e_k will equal 0. The final ASF realization is a minimum phase and minimum error solution.

If the unknown filter is a second-order FIR filter with zeros outside the unit circle, then there is no set of valid probabilities that will produce zeros of the ASF at the zeros of the unknown. The solution achieved will still be minimum phase since the fixed filter's zero locations are in the unit circle. As one would expect, the ASF will converge using the LMS algorithm but it is not obvious that the solution will be a minimum error solution. The optimum solution now lies somewhere higher up on the walls of the quadratic "bowl", not at the bottom of the "bowl" where the true minimum is located. In the following section, the theoretical minimum solution will be derived and simulations will be presented in Section III to illustrate this solution.

2.2 Theoretical Convergence Solution

The theoretical solution to (5) can be obtained by taking the partials with respect to the p_i and by using the constraints that $\sum p_i = 1$ and the $p_i \geq 0$. To demonstrate with a specific examle, let **H** and $\mathbf{h}_u$ be defined as

$$H = \begin{bmatrix} 1 & 1 & 1 \\ 1.8 & -1.8 & 0 \\ .81 & .81 & -.81 \end{bmatrix}; \qquad h_u = [1 \;\; d_1 \;\; d_2]^T . \tag{6}$$

This **H** will allow all locations inside a circle of .9 radius. Using Lagrange multipliers [3], the following system of equations is obtained subject to the above constraints.

$$\begin{bmatrix} a & b & c \\ b & a & c \\ c & c & d \end{bmatrix} \boldsymbol{p} + \lambda \mathbf{1} = \begin{bmatrix} e \\ f \\ g \end{bmatrix} \tag{7}$$

where $a = (1 + 1.8^2 + .81^2)$; $b = (1 - 1.8^2 + .81^2)$; $c = (1 - .81^2)$; $d = (1 + .81^2)$ and $e = (1 + 1.8d_1 + .81d_2)$; $f = (1 - 1.8d_1 + .81d_2)$; $g = (1 - .81d_2)$. Elimination of λ will give an exact, zero error solution if $\mathbf{h}_u$ is second-order and has roots inside the circle of .9 radius. If the zeros lie outside the circle of .9 radius, then it can be shown that p_3 must be zero. The set of equations in (7) must be resolved with p_3 set equal to zero. We then get

$$\begin{bmatrix} a-b & b-a \\ 1 & 1 \end{bmatrix} \boldsymbol{p} = \begin{bmatrix} 3.6 d_1 \\ 1 \end{bmatrix}$$
$$p_1, p_2 = \frac{1}{2} \pm \frac{d_1}{3.6} \tag{8}$$

if the zeros of $\mathbf{h}_u$ lie anywhere in the region bounded by the circle of .9 radius and the lines $x = \pm 1$. If the zeros of $\mathbf{h}_u$ lie outside the lines $x = \pm 1$, then p_3 and either p_1 or p_2 will equal 0. The ideal solution for $\mathbf{h}_\mu$ will then be either $[\ 1\ 0\ 0\ ]^T$ or $[\ 0\ 1\ 0\ ]^T$. These results are illustrated with simulations shown in Section III.

2.3 Fourth-Order Derivation

With two ASF sections cascaded in Figure 1, y_k is

$$y_k = [\boldsymbol{x}_k^T \boldsymbol{h}_{1\mu} \ \ \boldsymbol{x}_{k-1}^T \boldsymbol{h}_{1\mu} \ \ \boldsymbol{x}_{k-2}^T \boldsymbol{h}_{1\mu}]\, \boldsymbol{h}_{2\mu} = \boldsymbol{X}^T \boldsymbol{h}_{1\mu}$$
$$\textit{where } \boldsymbol{X} = [\boldsymbol{x}_k \ \ \boldsymbol{x}_{k-1} \ \ \boldsymbol{x}_{k-2}] \ \textit{and } \boldsymbol{x}_{k-i} = [x_{k-i} \ \ x_{k-i-1} \ \ x_{k-i-2}]^T . \tag{9}$$

The error surface can then be simplified to

$$\xi = E[d_k^2] + \boldsymbol{h}_{2\mu}^T E[\boldsymbol{X}^T \boldsymbol{h}_{1\mu} \boldsymbol{h}_{1\mu}^T \boldsymbol{X}]\boldsymbol{h}_{2\mu} - 2\boldsymbol{h}_{2\mu}^T E[(d_k \boldsymbol{X}_k)^T]\boldsymbol{h}_{1\mu} . \tag{10}$$

If we use white noise with zero mean and unit variance, then (10) can be simplified to

$$\xi = \boldsymbol{h}_u^T \boldsymbol{h}_u + \boldsymbol{h}_{2\mu}^T \boldsymbol{R}_1 \boldsymbol{h}_{2\mu} - 2\boldsymbol{h}_{2\mu}^T \boldsymbol{D} \boldsymbol{h}_{1\mu}$$
$$\textit{where } \boldsymbol{D} = \begin{bmatrix} 1 & d_1 & d_2 \\ d_1 & d_2 & d_3 \\ d_2 & d_3 & d_4 \end{bmatrix} \tag{11}$$

and $\mathbf{R}_1$ is a 3X3 matrix with elements formed with products of the form $\mathbf{h}_{1\mu}{}^T \mathbf{I}_i \mathbf{h}_{1\mu}$ where $\mathbf{I}_i$ is a matrix with ones on the ith diagonal. Thus, it is obvious that the error surface is no longer quadratic since there will now be cross terms of the squared probabilities from each ASF section. If we want to apply the LMS algorithm, a slight modification must be made to make sure that only the inputs to the individual probabilities in each ASF are used in the probability update equations. Taking this into account gives the following

$$\boldsymbol{p}_1(k+1) = \boldsymbol{p}_1(k) + 2\mu e_k \boldsymbol{H}_1^T \boldsymbol{x}_k$$
$$\boldsymbol{p}_2(k+1) = \boldsymbol{p}_2(k) + 2\mu e_k \boldsymbol{H}_2^T \boldsymbol{X} \boldsymbol{h}_{1\mu} . \tag{12}$$

Similarly to the second-order case, if the unknown is a fourth-order system with all zeros inside the unit circle, then the two ASFs will converge to the same zeros as the unknown and will be minimum phase. If the unknown is greater than fourth-order, then the ASFs will converge to a minimum phase filter.

3 SIMULATION RESULTS

Three second-order unknown systems were simulated to illustrate the cases presented previously. These cases are:

Case 1: $\mathbf{h}_u = [\ 1\ 3\ 2.5]^T$ with roots at $-1.5 \pm j.5$.

Case 2: $\mathbf{h}_u = [\ 1\ 1\ 2.5]^T$ with roots at $-.5 \pm j1.5$.

Case 3: $\mathbf{h}_u = [\ 1\ .8\ .25]^T$ with roots at $-.4 \pm j.3$.

The theoretical prediction for convergence for case 1 is $\mathbf{h}_\mu = [\ 1\ 0\ 0\ ]^T$ with roots at -.9. For case 2, the prediction is $\mathbf{h}_\mu = [1\ 1\ .81\ ]^T$ with roots at $-.5 \pm j.7483$. The expected result for case 3 is $\mathbf{h}_u = [1\ .8\ .25\ ]^T$. Figures 3-8 show plots of the root trajectories and the probability trajectories for p1 vs p2 overlayed on the ideal error contours plot. The error contour plot was calculated using (5) with $\mathbf{h}_{\mu\mu}$ equal to the optimum under the imposed constraints.

The circle on each plot represents the ideal convergence point. As predicted, the ASF converges almost exactly to the optimum predicted solution. Finally, Figures 7 and 8 show convergence to the exact minimum error solution if the zeros of the unknown are inside the unit circle.

4 CONCLUSIONS

The equations describing the behavior of a stochastic filter and its error surface were derived for a second-order system. Several simulations were run that illustrated the predicted behavior for unknown filters in several regions of the complex plane. Although this structure has been expanded to higher order systems, the analysis is more complicated than for higher orders of the adaptive linear combiner. However, the algorithm resulting from the imposed constraints by the probabilities is easily implemented. The implementation results in a minimum-phase solution at each iteration of the algorithm and has been proven to converge to a minimum error solution for the second-order case. We are working on the extension and implementation to higher order systems.

REFERENCES

[1] T. J. Canales and D. M. Etter, "Adaptive Stochastic Filters," 32nd Midwest Symposium on Circuits and Systems, 1989

[2] B. Widrow and S.D. Stearns, *Adaptive Signal Processing*, Englewood Cliffs, N.J. Prentice-Hall, 1985.

[3] D.G. Luenberger, *Introduction to Linear and Nonlinear Programming*, Reading, Massachusetts, Addison-Wesley, Chapter 10, 1973.

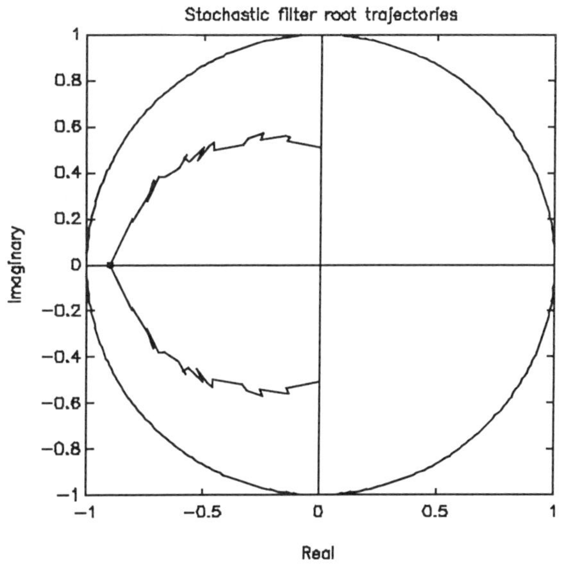

Figure 3. Root trajectories for $\mathbf{h}_u$ = [1 3 2.5]^T.

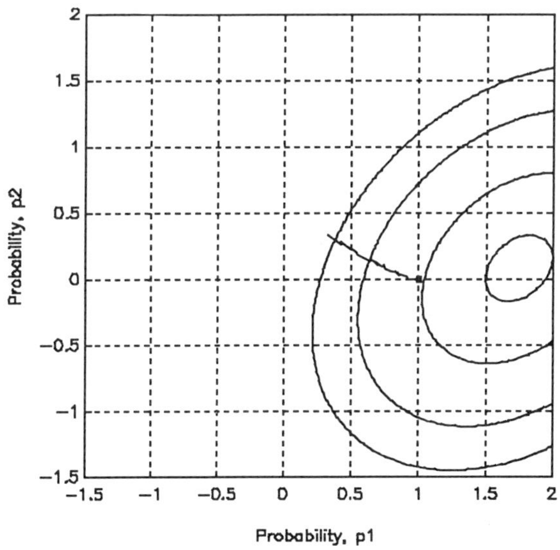

Figure 4. Error Contours for $\mathbf{h}_u$ = [1 3 2.5]^T.

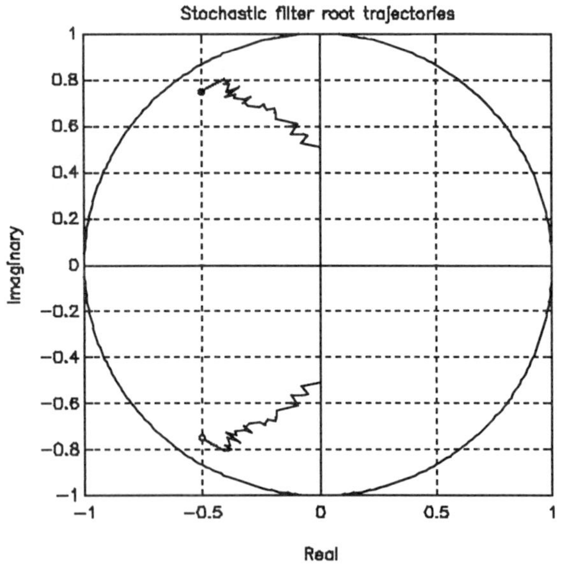

Figure 5. Root trajectories for $\mathbf{h}_u = [\ 1\ 1\ 2.5\ ]^T$.

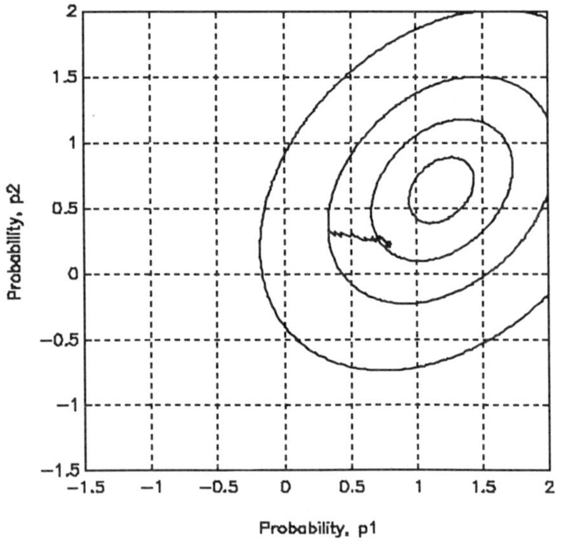

Figure 6. Error Contours for $\mathbf{h}_u = [\ 1\ 1\ 2.5\ ]^T$.

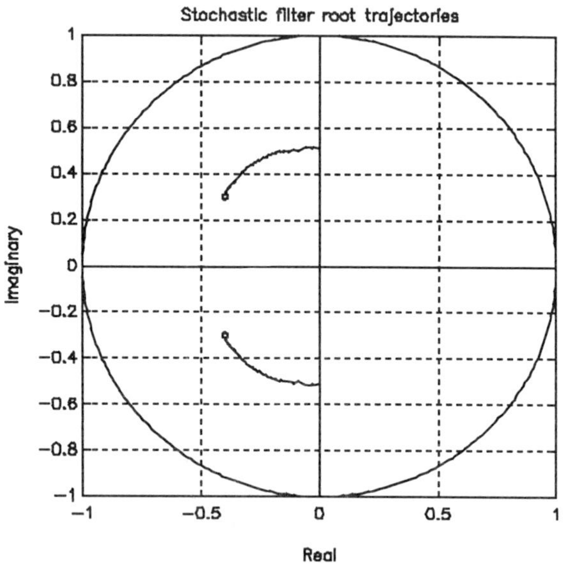

Figure 7. Root trajectories for $\mathbf{h}_u = [\ 1\ .8\ .25\ ]^T$.

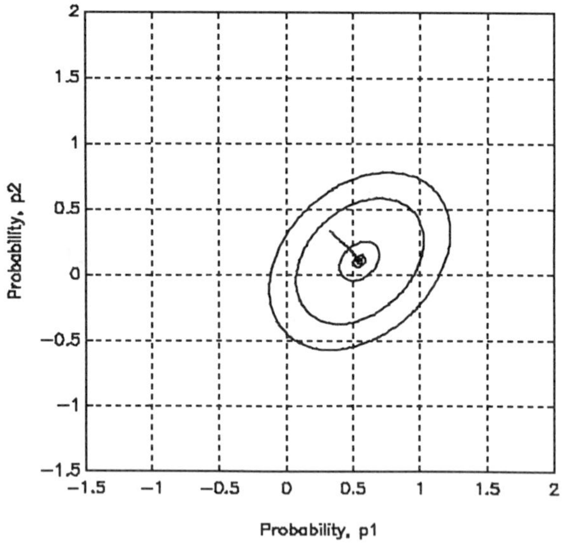

Figure 8. Error Contours for $\mathbf{h}_u = [\ 1\ .8\ .25\ ]^T$.

SEISMIC PHASE DETECTION AND DISCRIMINATION USING ADAPTIVE FILTER COEFFICIENTS

N. Magotra†, *J. Jiang*†, *D. Hush*†, *L. Tan*†

† **Dept. of EECE, University of New Mexico, Albuquerque, NM 87131**

Abstract

This paper presents an approach for the detection and discrimination of seismic phases using multi-channel data. When a seismic event (such as an earthquake or man-made explosion) occurs, the signal propagating from the source has several distinct phases which have varying propagation velocities and frequency content. By detecting and correctly identifying these phases one can estimate the distance between source and receiver. The detection algorithm consists of an adaptive correlation enhancer and a sliding window detector that make the detector insensitive to changes in the background noise level. The discrimination algorithm uses the adaptive correlation enhancer's weights as input discriminants to a neural net.

1. Introduction

The detection and discrimination algorithms presented in this paper represent a continuation of the research effort to detect, locate and discriminate regional seismic events [1, 2]. The term regional implies that the distance between the source and receiver is less than approximately 2000 km. The original motivation for this research was the idea of using seismic information to verify compliance with respect to a nuclear test ban treaty. However, the signal processing algorithms developed as a result of this research can also be applied in the analysis of earthquake data.

The data used in this research consisted of three-channel data obtained from three orthogonally placed seismometers - Vertical (z(m)), North (n(m)) and East (e(m)), provided by Sandia National Laboratories, Albuquerque NM 87185. The original data sampling rate was 200 samples per second (sps). Based on a spectral and polarization analysis [1] of the signals the data were low-pass filtered with a cut-off frequency of 15 Hz using a fourth order Butterworth filter and resampled to 40 sps.

2. Algorithm Description

Detection

To detect the phases of a seismic signal, data from the three channels are combined to provide a measure of polarized energy in the data. The detection parameter used to obtain the results presented in this paper is based on the estimated autocorrelation coefficients ($\hat{R}_{cc}(m,l)$) of each channel (c(m)). Note that c(m) can be set equal to the channel (z(m), e(m), n(m)) whose autocorrelation function needs to be estimated. The autocorrelation coefficients are updated recursively as shown below

$$\hat{R}_{cc}(m,l)=\beta\hat{R}_{cc}(m-1,l)+(1-\beta)c(m)c(m+l) \quad (1)$$

where 'm' is the time index, 'l' is the lag index ($\leq L$), and $0<<\beta<1$ is the smoothing constant.

Equation (1) was derived from the weight update equation of the adaptive correlation enhancer (ACE) described in [3]. The autocorrelation coefficients obtained via equation (1) converge rapidly to their optimal values, a necessary property in order to be able to detect the arrival of different phases in the coda of the seismic signal.

The detection parameter (s(m)) is computed as

$$s(m)=\sum_{\substack{l=-L \\ l\neq 0}}^{L}(\hat{R}_{zz}^{2}(m,l)+\hat{R}_{ee}^{2}(m,l)+(\hat{R}_{nn}^{2}(m,l)) \quad (2)$$

The signal $s(m)$ is then processed through a sliding window detector implemented via the following three equations

1058-6393/91 $01.00 © 1991 IEEE

$$w_1(m)=\beta_1 * w_1(m-1)+(1-\beta_1)*s(m) \quad (3a)$$

$$w_2(m)=\beta_2 * w_2(m-1)+(1-\beta_2)*s(m-\Delta) \quad (3b)$$

$$d(m)=w_1(m)-k_h * w_2(m)>0.0 \quad (3c)$$

In the equations above, β_1 and β_2 are chosen so that $w_1(m)$ represents a short-time average of s(m) and $w_2(m)$ represents a delayed long-time average of s (m). The constant k_h represents a threshold setting parameter. The signal $d(m)$ compares $w_1(m)$ with its past history represented by $w_2(m)$ to look for a sudden increase in the coherent energy level, in the input signal, indicating the presence of an event/phase of interest.

The detection algorithm described above is capable of detecting extremely weak signals of signal-to-noise ratios (SNRs) as low as -10 dB while keeping false alarms down to a minimum. Our research indicates that this detection approach outperforms most of the seismic detectors currently used, which typically monitor the power level of the input signals.

Note in equation (2) that the center autocorrelation coefficient is not included in the summation. With the assumption that the background noise is uncorrelated, most of the noise contribution is in this center coefficient and our experience with exercising the algorithm indicated that deleting this coefficient from the summation made the detector immune to false alarms due to noise bursts and spikes in the data.

Phase Discriminatiom

The task of automatically discriminating between seismic phases is performed by a multilayer perceptron neural network system. The input to the neural net consists of a time averaged value of the estimated crosscorrelation functions of the three data channels.

The three crosscorrelation functions ($R_{EZ}(m,l)$, $R_{ZN}(m,l)$, and $R_{NE}(m,l)$) are obtained recursively via the equations below

$$\hat{R}_{EZ}(m,l)=\beta\hat{R}_{EZ}(m-1,l)+(1-\beta)e(m)z(m+l)$$

$$\hat{R}_{ZN}(m,l)=\beta\hat{R}_{ZN}(m-1,l)+(1-\beta)z(m)n(m+l)$$

$$\hat{R}_{NE}(m,l)=\beta\hat{R}_{NE}(m-1,l)+(1-\beta)n(m)e(m+l)$$

The crosscorrelation functions were then averaged (in time) as shown below.

$$\hat{R}_{avg}(m,l)=\frac{1}{3}\sum_{l=0}^{51}(\hat{R}_{ez}(m,l)+\hat{R}_{zn}(m,l)+(\hat{R}_{ne}(m,l))$$

To generate the training set, we selected a four second window starting at the onset of the phase and output 51 samples (zero lag to a lag value of 50) of $\hat{R}_{avg}(m,l)$ with every input sample, thus producing 160 crosscorrelation functions for each detected phase.

The neural net itself had 2 layers, the input layer had two nodes while the output layer had just one node. The neural net weights were updated (during training) using the back-propagation algorithm [2]. Classification in this system was achieved by presenting each of the 160 autocorrelation signals for an unknown phase to the network and classifying according to a majority vote rule. That is, the phase was assigned to the class that was picked most often by the network. Using multiple patterns to classify a phase increases the robustness of the classifier. The results presented in this paper indicate that a relatively simple neural net can successfully classify seismic phases.

3. Results

The detection and discrimination algorithms outlined above have been successfully tested on field acquired data. On tests using field-acquired data the detector was able to detect the phases to within a second of the true onset time. The 'true' onset times were determined visually. The discriminator was also able to correctly classify most of the phases in the training set and all of the phases in the test data set. Figures 1 - 4 present results for a representative case.

Figure 1 shows the data for an earthquake of magnitude $m_b=3.2$ indicating the onset times of the P_n and L_g phases. Figure 2 shows plots of the vertical data channel as well as the detection parameter (s(m)) and the final detector output for the same event. To illustrate the idea of obtaining time varying crosscorrelation functions, Figure 3 shows a 3D plot of the EZ crosscorrelation functions during the onset of the P_n phase of the earthquake. Figure 4 shows single plots of all three crosscorrelation functions for the same phase.

The performance of the discriminator has been summarized in Tables 1 and 2. To obtain these results the training set consisted of 30 events while the test set consisted of 4 events. Table 1 presents results for the test set using the discriminants described in the preceding section. The discriminator classified all of the test phases correctly. The percentages in Table 1 refer to the percentage of the 160 crosscorrelation functions classified correctly for each phase.

Table 2 presents the corresponding results for thecase where the input discriminants (the crosscorrelation functions) were first transformed via the FFT. All other processing parameters were kept constant. For these results we used a 32 point FFT and then retained 17 FFT coefficients (zero frequency through nyquist) in each discriminant function. As shown in Table 2 the results were similar to those obtained by simply using the correlation functions themselves as the discriminants.

4. Conclusions and Future work

The results presented above indicate that the seismic phase detection/discrimination algorithm works successfully on field-acquired data. We have done some work on extending the discriminator's ability to discriminate between the two phases mentioned in the paper and just noise bursts. The long term goal is to couple the detector and discriminator closely to form a *smart detector*. That is, formulate a detector that will be able to make decisions based on the sequence of phase detections. For example if two P_n phases were detected in sequence the 'smart detector' would be able to make the decision that there were overlapping events.

References

[1] Magotra, N., Ahmed, N., Chael, E., "Single-station seismic event detection and location," IEEE Transactions on Geoscience and Remote Sensing, pp. 15 - 23, January, 1989.

[2] Magotra N., Hush D., Bibbo J., Chael E., "Seismic Signal Discrimination Using Adaptive System Parameters", IEEE Midwest Symposium on Circuits and Systems, Calgary Canada, August 1990.

[3] Magotra, N., Ahmed, N., Chael, E. P., "Single station (3 component) Seismic Data Enhancement," Proc. of 20th IEEE Asilomar Conference on Circuits, Systems and Computers, Nov., 1986.

[5] Magotra N., Bibbo J., "Seismic Event Detection using Single Channel Data," 24 th. Asilomar Conference on Signals, Systems and Computers, Asilomar, CA, November 1990.

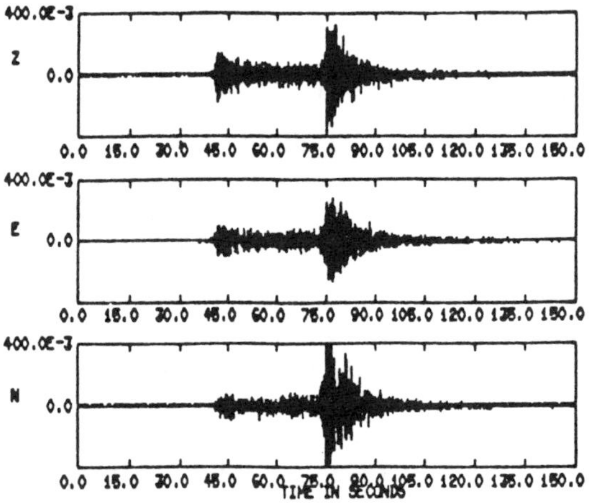

FIGURE 1:Input data(m_b=3.2)

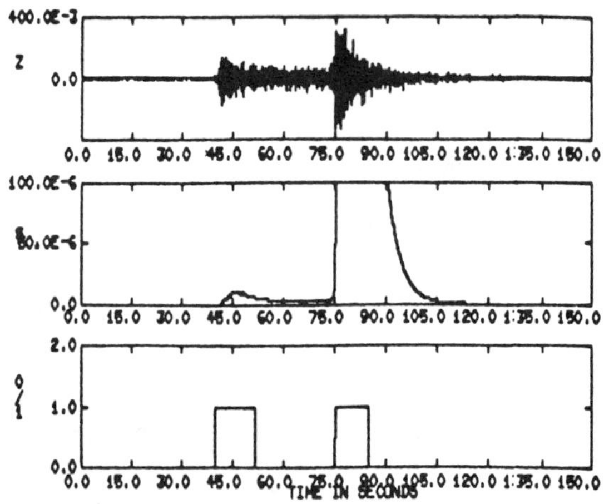

FIGURE 2: Detector outputs

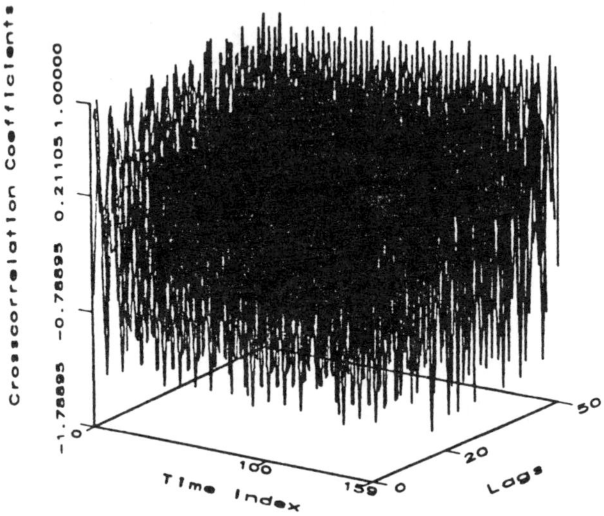

FIGURE 3: EZ Crosscorrelation functions

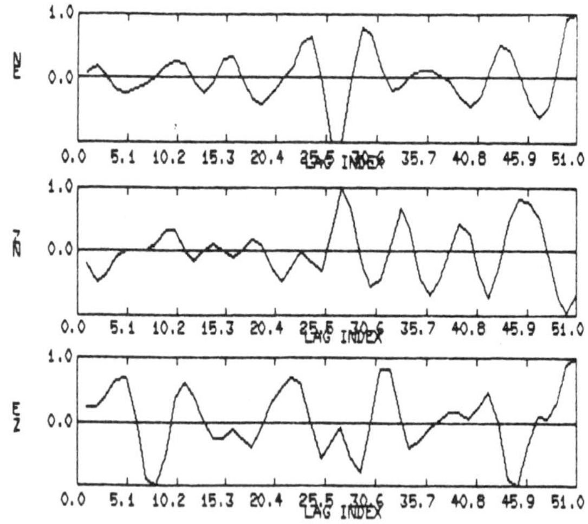

Fig 4: Sample Crosscorrelation Functions

TABLE 1

Results of Using the Autocorrelation Signal After Averaging 3-Channel Data to the Two Layer Perceptron System

% Correct Classification After Run 200 mse=(0.00865)			
Name of Test Signals	Lg	Pn	Lg - Pn (sec.)
e281j0	81.25%	100.0%	40.2
e314j0	100.0%	50.625%	52.5
e324j0	100.0%	95.625%	45.8
e359j1	100.0%	50.0%	41.5

TABLE 2

Results of Using Autocorrelation-FFT Method After Averaging 3-Channel Data to the Two Layer Perceptron System

% Correct Classification After Run 200 mse=(0.024399)			
Name of Test Signals	Lg	Pn	Lg - Pn (sec.)
e281j0	72.5%	100.0%	40.2
e314j0	100.0%	60.625%	52.5
e324j0	100.0%	100.0%	45.8
e359j1	80.0%	100.0%	41.5

Observed Misconvergence in the Constant Modulus Adaptive Algorithm

John R. Treichler — Applied Signal Technology, 160 Sobrante Way, Sunnyvale, CA 94086

Vincent Wolff — Applied Signal Technology, 160 Sobrante Way, Sunnyvale, CA 94086

C. Richard Johnson, Jr. — Cornell University, 305 Phillips Hall, Ithaca, NY 14853

Abstract

The demonstrated ability of the constant modulus algorithm (CMA) to "blindly" adapt digital filters to equalize the effects of multipath-induced dispersion on quadrature amplitude modulated (QAM) signals has encouraged analysis of the algorithm's expected behavior by a number of research groups. Several of these analyses contain proofs of the algorithm's guaranteed convergence to an optimum and satisfactory solution subject to a variety of conditions thought not to be practically taxing. In fact, practical application of CMA has demonstrated a number of circumstances in which CMA fails to converge, or, equally bad from a practical standpoint, converges to a solution which fails to equalize the input signal.

This paper describes several situations in which misconvergence occurs, suggesting that a firmer analytical understanding is needed of the behavior of blind algorithms in the presence of cyclostationary and/or quasi-periodic, non-white inputs. While this analytical understanding is not yet established, the practical experience reported here should be directly useful to those designing new digital communications systems. An excellent example, treated in the paper, is the use of CMA to equalize QAM signals used to broadcast high-definition television (HDTV) signals.

1 Introduction

The constant modulus algorithm (CMA) was developed [1] [3] as a method for adaptively choosing the pulse response of a digital filter in such a way as to remove additive interference from a signal. It is termed a "blind" algorithm since its use does not require a training sequence. Because of this, the application of such algorithms, and CMA in particular, is growing rapidly, especially in situations where a signal with inadequately known parameters must be initially acquired in the presence of interference and dispersion.

The demonstrated ability of CMA to blindly adapt digital filters to equalize the effects of multipath-induced dispersion on quadrature amplitude modulated (QAM) signals has encouraged analysis of the algorithm's expected behavior by a number of research groups [1] [2] [3] [4]. Several of these analyses contain proofs of the algorithm's guaranteed convergence to an optimum and satisfactory solution subject to a variety of conditions thought not to be practically taxing. In fact, practical application of CMA has demonstrated a number of circumstances in which CMA fails to converge, or, equally bad from a practical standpoint, converges to a solution which fails to equalize the input signal.

This paper describes several situations in which misconvergence occurs. These are documented with constellation photographs showing the behavior of an actual demodulator when subjected to a variety of input signals. In some cases it can be shown that conditions stated in the convergence proofs have been violated by the input signal while in others it is not clear that a violation has occured. All cases suggest that a firmer analytical understanding is needed of the behavior of blind algorithms in the presence of cyclostationary and/or quasi-periodic, non-white inputs.

While this analytical understanding is not yet established, the practical experience reported here should be directly useful to those designing new digital communications systems. An excellent example, treated in the paper, is the use of CMA to equalize QAM signals used to broadcast high-definition television (HDTV) signals.

2 Background

Figure 1 shows a simplified model of a wideband PCM communications system. High-speed data is quadrature-amplitude modulated and transmitted at an appropriate radio frequency. When the signal is received it has typically been degraded by the addition of noise and uncorrelated interference and it has been dispersed to some degree by its interactions with the propagation medium and its boundaries. The combination of these effects usually renders the signal undemodulable unless actions are taken to mitigate them. A common approach is the use of an adaptive equalizer, which can, in many cases, equalize the dispersion effects and frequency-selectively remove additive noise and interference.

Proper operation of an adaptive equalizer requires some form of information feedback to it regarding its current degree of success in removing dispersion and interference. Figure 1 shows two points from which this information might come, the demodulator and the equalizer itself. Many communications systems

1058-6393/91 $01.00 © 1991 IEEE

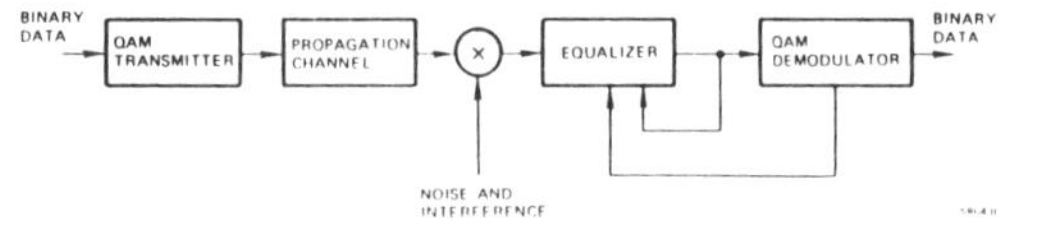

Figure 1: Simplified Model of Signal Flow in a Wideband PCM Communications System

use the former only, a technique termed "decision direction" [6]. This technique works well once the signal has been acquired (and in high SNR environments) because it is equivalent, in essence, to the use of the actual transmitted signal to optimize the equalizer. Unfortunately, the use of this technique alone leads to the need for special procedures, termed "training", during the receiver's acquisition process. This training process is rarely desirable, since it requires the use of time and/or bandwidth which could otherwise be used for data transmission, and is sometimes impossible, as in a one-way-only or broadcast communications system. The constant modulus algorithm (CMA) was developed [1] [3] as a method for adaptively choosing the pulse response of a digital filter in such a way as to remove correlated and uncorrelated additive interference from a constant envelope or QAM signal. It is termed a "blind" algorithm since its use does not require a training sequence. Because of this, the application of such algorithms, and CMA in particular, is growing rapidly, especially in situations where a signal with inadequately known parameters must be initially acquired in the presence of interference and dispersion or where training, either initially or embedded, is impossible or inpractical. Once acquired with CMA or another blind method, it is common to switch over to decision directed algorithms for continued tracking.

The success of CMA in this acquisition mode can be seen by examining Figure 2. Figure 2(a) shows the spectrum of a 33 megabaud, 16-QAM signal from an NEC modem, plus the equalized, carrier-locked constellation. In this case there is no dispersion impressed on the signal other than the pulse shaping imposed at the transmitter. Figure 2(b) shows the spectrum and resulting constellation in the presence of a strong specular multipath. The presence of multipath is indicated in the scalloping seen in the power spectrum of the received signal. Without equalization the signal's "eye" is closed and demodulation is impossible. The CMA-directed equalizer, however, opens the eye and permits successful acquisition of the decision-directed modes. The residual effects of the multipath propagation are still seen after full acquisition, as evidenced by the large constellation "points". The multipath in this case is so strong that an equalizer with hundreds of taps (rather than the 64 used here) would be needed to equalize the signal to the degree obtained with no

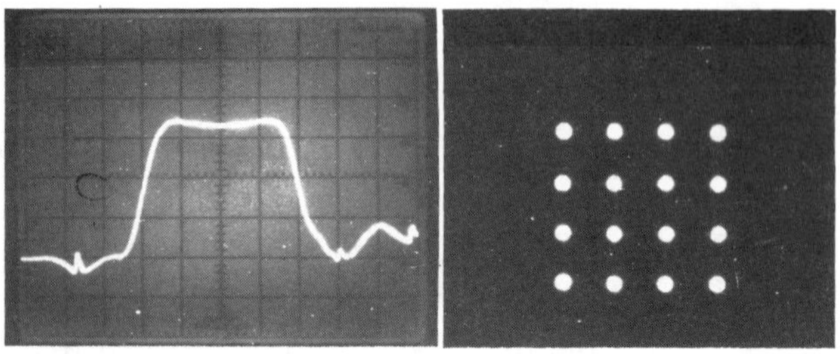

(a) Spectrum and Constellation in the Absence of Multipath

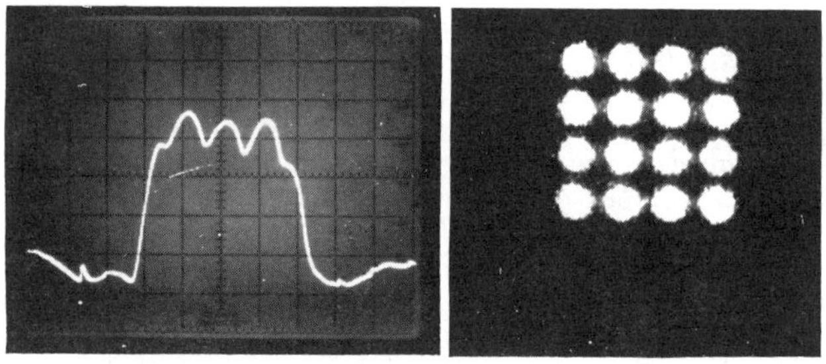

(b) Spectrum and Constellation in the Presence of Heavy Specular Multipath

Figure 2: Successful Operation of a CMA-based Adaptive Equalizer With an NEC 16-QAM Signal, with and without Multipath

multipath in Figure 2(a).

3 Observed Misconvergence and Apparent Cause

3.1 Signal Generation Model

Sixteen-level quadrature-amplitude modulation is commonly used in terrrestrial line-of-sight mircowave transmission of telephone signals. A common design for the transmitter section of a 16-QAM modem is shown in Figure 3. Multiplexing equipment is used to combine up to several thousand PCM telephone channels into a single serial stream, which is then "randomized" in some some fashion to "whiten" the bit stream, making the spectrum of the transmitted signal uniform in amplitude across its assigned band and

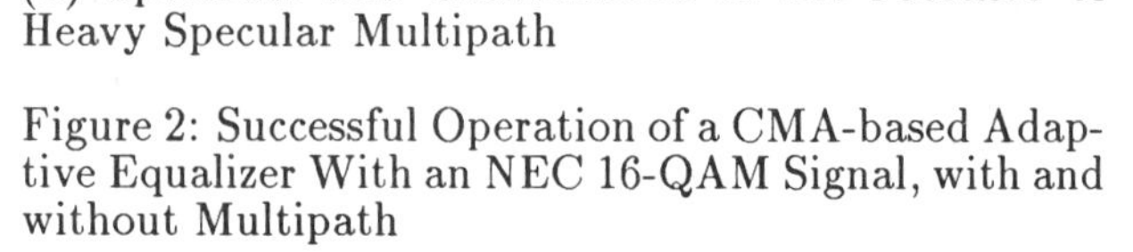

Figure 3: Block Diagram of a Generic 16-QAM Modulator for Digital Radios

thereby reducing any interference to spectrally adjacent signals. A common method of performing this randomization is to cyclically add (modulo-2) a selected sequence of length L, which is typically a section of a pseudorandom (PN) sequence of equal or somewhat greater length. The combined sequence then broken into 4-bit nibbles and each transmitted as one of the 16 possible symbols present in a 16-QAM signal.

An implication of this technique is that when the multiplexed telephone signal is essentially idle, for example, in the middle of the night, the added sequence dominates. As a result the transmitted signal is repetitive with period at least L, and perhaps $\frac{L}{4}$, if L is divisible by 4 as it commonly is.

3.2 Experimental Observations

The implications to blind equalization algorithms can be seen by examining the constellation photographs seen in Figure 4. Figure 4(a) shows the constellation of a properly processed 16-QAM signal operating at 25 megasymbols/sec as demodulated by a Model 250A Equalizer-Demodulator [6]. Achieving this condition requires gain adjustment, symbol timing synchronization, adaptive equalization, and carrier frequency acquistion. Figure 4(b) shows the received signal when only gain adjustment and real-to-quadrature conversion have been done. Clearly the "eye" is closed. After symbol timing synchronization is attained and when L, the additive sequence length, is very long (e.g., greater than 10^6), the constant modulus algorithm (CMA) can be reliably employed to open the eye, even without carrier frequency acquisition. Figure 4(c) shows this constellation. The appearance of the three rings indicates the opening of the eye. These rings correspond to the rotation-induced blurring of the 16-QAM pattern caused by lack of carrier frequency lock.

If the additive sequence length is much lower, then false convergence of CMA can occur. Such a case is shown in Figure 4(d). In this case the length L is 256, four times the length of the adaptive equalizer's regressor vector. (Commercial QAM modems are built which use randomizers as short as 128.) Note that in this situation the algorithm converges to a single "fuzzy" ring, and that the constellation's eye is not open. Thus blind equalization fails and with it the complete acquisition procedure for the demodulator.

Considerable laboratory experimentation has been done using various constellation sizes, equalizer length, PN sequence lengths, and various equalizer parameters. Invariably these experiments indicate that when the sequence length L is long compared to the filter and its principal adaptation modes, the convergence is reliable. When the sequence length is short, however, that is, comensurate in length with the equalizer, then false convergence is common, but even then not guaranteed.

Some specific cases of interest include the following:

- Idle multiplexer tributaries, with tributary-level randomizing: With no randomizing of the composite signal, this often leads to the limiting case of unequal constellation point distribution -

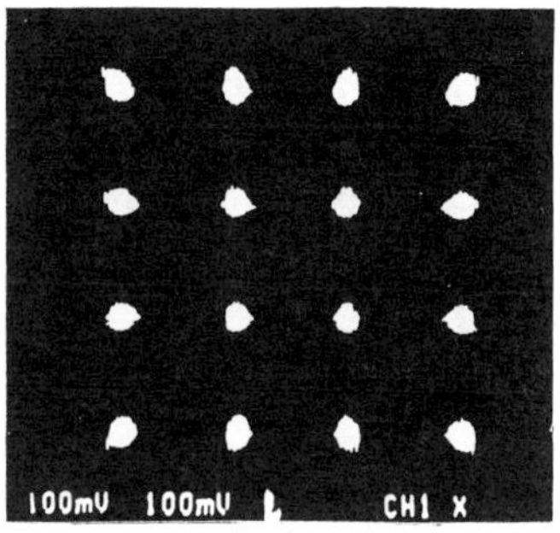

(a) Successful recovery of a 16-QAM, 25 megabaud Signal

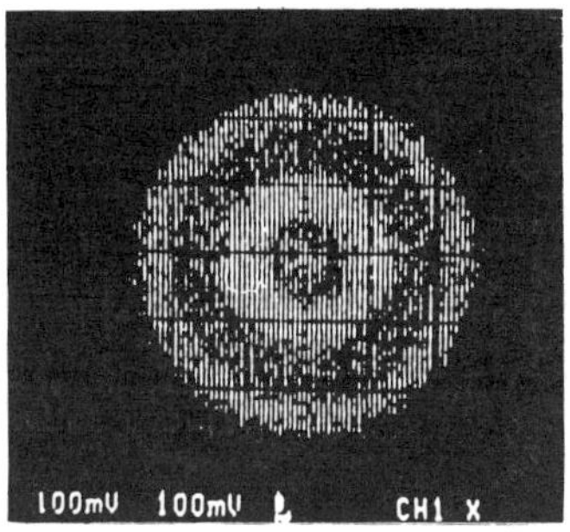

(b) Constellation of Received Signal, with symbol timing achieved but without equalization or carrier recovery

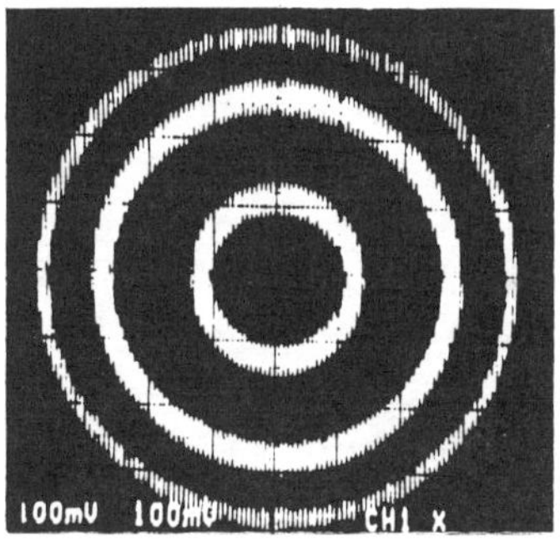

(c) Constellation of the Same Signal After Successful Blind Equalization but Before Carrier Acquistion; Note the open "eye"

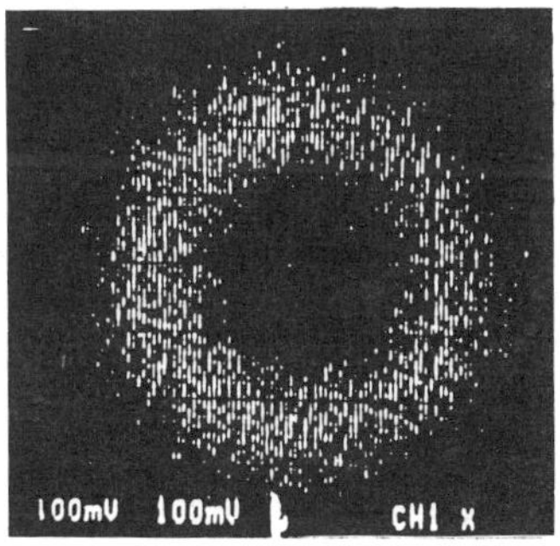

(d) Constellation of the Same Signal After Unsuccessful Blind Equalization; Eye is not open and carrier acquisition is impossible

Figure 4: Particular Instance of CMA Misconvergence Traced to Input Insuffiency

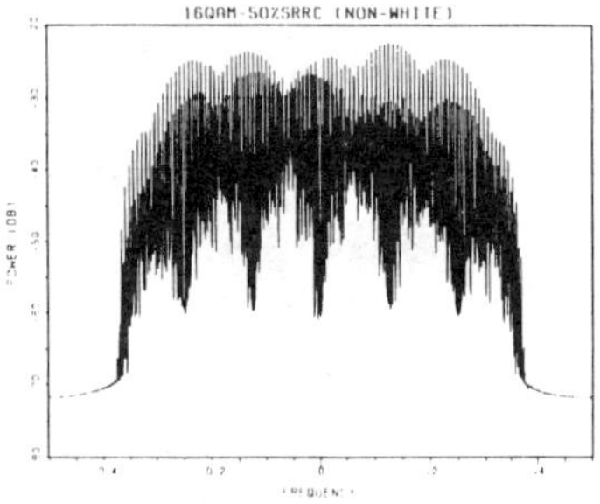

(a) Independent, identically distributed (IID) input

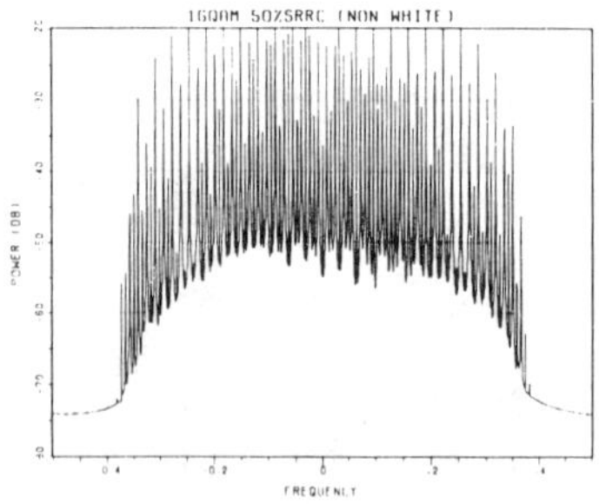

(b) 63-bit PN Periodic Input

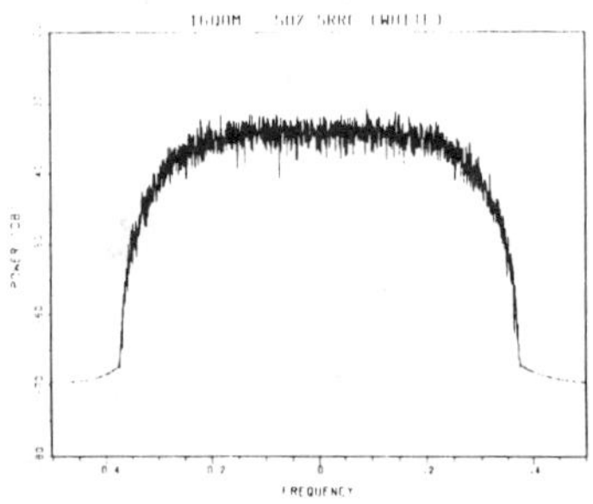

(c) 255-bit PN Periodic Input

Figure 5: The Impact of Periodic PN Inputs on the Spectra of 16-QAM Signals

points which are completely missing. A 16-QAM signal with two missing tributaries can appear as a non-square 8-QAM pattern. The resulting signal is essentially a quadrature partial response (QPR) waveform, a class known to be problematical to equalize.

- Lightly loaded tributaries, PN reset randomizer, sequence length L commensurate with the equalizer's length: In this case the input is essentially periodic, producing a "liney" spectrum. Examples of this phenomenon are shown in Figure 5. Figure 5(a) shows the power spectrum of a 16-QAM modulator driven by a very long $(2^{24} - 1)$ PN sequence. This results in symbols which are independent and identically distributed over the limited period of the spectrum calculation. Figure 5(b) shows the spectrum when driven by a very short sequence, $L = 63$ in this case. The spectrum is no longer relatively smooth but is completely composed of spectral lines. The same effect appears in Figure 5(c) where $L = 255$ and the line spacing is four times closer.

  When confronted with a line-dominated spectrum a CMA-directed adaptive filter can often converge to a solution which isolates a single spectral component. While inappropriate for equalization, the converged solution minimizes the CMA performance criterion even better than the "proper" solution would.

- Lightly loaded tributaries, PN reset randomizer, sequence length L much longer than the equalizer but much shorter than the nominal convergence time: In this case the signal autocorrelation matrix appears to have full rank at all times, but the adaptive algorithm's control loop is driven periodically by the cyclic nature of the input since its frequency is too high for the loop to track.

- And more ...

3.3 Suspected Cause

The demonstrated ability of CMA to blindly adapt digital filters to equalize the effects of multipath-induced dispersion on quadrature amplitude modulated signals has encouraged analysis of the algorithm's expected behavior by a number of research groups [1] [2] [3] [4]. Several of these analyses contain proofs of the algorithm's guaranteed convergence to an optimum and satisfactory solution subject to a variety of conditions thought not to be practically taxing. "Input sufficiency" is immediately assumed. In particular it is always presumed that the transmitted constellation points are identically distributed and appear independently from one another (i.e., the symbols are IID). It is further presumed that the adaptive equalizer has such high order that its length in no way affects the convergence of the adaptive algorithm. Ding [7] has demonstrated the shortcomings in the second assumption but the first was thought to be reasonably representative of practical applications. In fact, it is not and usually will not unless the signal is specifically designed to achieve it. The high-capacity telephony signals described in section 3.1 tend to fail these input sufficiency tests for one or more of the following reasons:

- Input tributaries are sometimes lightly loaded (e.g., in the middle of the night)

- Sometimes tributaries are completely missing (e.g., when a new radio system is installed with excess capacity)

- "Randoming" is not complete or adequate. The degree of randomization needed to avoid adjacent channel interference is much less than that needed to provide statistical stationarity over the convergence interval of an adaptive equalizer. In general a modem designer would choose a short sequence to minimize the time needed for demultiplexer synchronization.

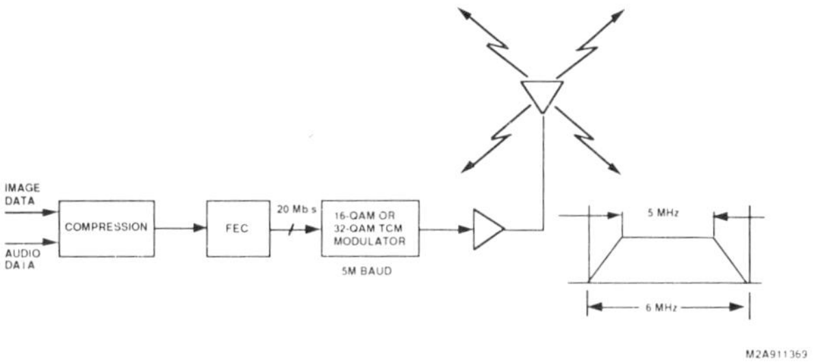

Figure 6: Block Diagram of a Potential HDTV Broadcast Transmitter

- Differential encoding correlates otherwise uncorrelated symbols.

4 Implications to the Design of an HDTV Broadcast System

As a part of the current energetic development of *high-definition television (HDTV)* technology, interest has grown in potential methods for direct broadcast of HDTV signals. To meet the simultaneous constraints of technical feasibility and government regulation, such a broadcast signal must carry at least 20 Mb/s and must be spectrally limited to less than 6 MHz (in the US). These constraints have led to the consideration of using QAM signals. If the symbol were chosen to be about 5 MHz, then a 16-QAM signal could carry 20 Mb/s, as could a 32-QAM signal which employed trellis-coding for improve the received error rate. The simplified block diagram of the transmitter for such a scheme is shown in Figure 6. The audio and video signals are digitized and compressed (i.e., source coded). Some redundacy is added to the compressor output to provide signal synchronization and a basis for forward error correction at the receiver. The resulting data is supplied to a quadrature-amplitude modulator, upconverted, and amplified for broadcast.

Owing to the terrestrial line-of-site nature of the transmitter-to-user propagation path, consideration multipath distortion is expected. By some estimates linear equalizers with hundreds of taps are needed to compensate for the expected range of multipath conditions. The broadcast nature of the HDTV signal (that is, no feedback from users) and the premium on data efficiency (bits used for training can't be used for TV) make blind equalization very attractive, at least for the "acquistion step" of opening the eye so that decision directed carrier recovery and then equalization can follow. The simplicity of CMA makes it then the first candidate to control this acquistion phase.

While all of the theoretical underpinnings are not yet fully understood, the experimental results attained to date and reported in the paper carry lessons to the designer of an HDTV broadcast system. These include, at the least, the following:

- Avoid source coding and FEC designs which result in data sequences which are idle, almost idle, or periodic for long intervals of time
- Avoid the use of "short" randomizers, particularly of the additive (e.g., PN reset) type
- If only one signal constellation is to be employed, consider using a performance function which makes more use of that knowledge, such as the COMA scheme discussed in [8].

References

[1] D.N. Godard, "Self-Recovering Equalization and Carrier Tracking in Two-Dimensional Data Communication Systems", IEEE Trans. on Communications, vol. COM-28, pp. 1867-1875, November, 1980.

[2] G.J. Foschini, "Equalizing Without Altering or Detecting Data", AT&T Technical Journal, vol. 64, pp. 1885-1911, October 1985.

[3] J.R. Treichler and B.G. Agee, "A New Approach to Multipath Correction of Constant Modulus Signals", IEEE Trans. on Acoustics, Speech, and Signal Processing, vol. ASSP-31, pp. 459-472, April 1983.

[4] A.Benveniste and M. Goursat, "Blind Equalizers", IEEE Trans. on Communications, vol. COM-32, pp. 871-883, August 1984.

[5] J.R. Treichler and M.G. Larimore, "The Tone Capture Properties of CMA-based Interference Suppressors", IEEE Trans. on Acoustics, Speech, and Signal Processing, vol. ASSP-33, pp. 946-958, August 1985.

[6] V.G. Wolff, et al, "Specification and Development of an Equalizer-Demodulator for Wideband Digital Microwave Radio Signals", Transactions of the 1988 MILCOM, San Diego, CA.

[7] Z. Ding, "Application Aspects of Blind Adaptive Equalizers in QAM Data Communications", PhD Dissertation, Cornell University, August 1990.

[8] F.L. de Victoria, "An Adaptive Blind Equalization Algorithm for QAM and QPR Modulations: The Concentrically Ordered Modulus Algorithm", Proceedings of the 25th Asilomar Conference on Signals, Systems, and Computers, Pacific Grove, CA, November 4-6, 1991.

A Pipelined LMS Adaptive Filter Architecture[1]

N. R. Shanbhag K. K. Parhi

Dept. of Electrical Engineering
University of Minnesota
Minneapolis, MN 55455

Abstract

A fine-grain pipelined architecture for least mean-square (LMS) filtering is developed by employing a stochastic form of look-ahead. The new architecture offers a trade-off between a variable output latency and the adaptation accuracy. Analytical expressions describing the convergence properties are provided. A comparison with previous work indicates that the new architecture has the least increase in hardware requirements and at the same time has the highest convergence speed in seconds. Simulation results confirm the analytical expressions derived in this paper.

1 Introduction

In the quest for high-speed architectures for recursive signal processing, algorithm transformation techniques based on *look-ahead computation* have been successfully applied to *create* concurrency in non-concurrent signal processing algorithms [1]. These include the techniques of pipelining [2] and parallel processing [3]. While making it feasible to pipeline recursive signal processing algorithms, the *look-ahead computation* [2] requires significant hardware increase. This hardware increase becomes extremely prohibitive when pipelining of adaptive filters is concerned. Existing pipelined adaptive filter architectures [4, 5] maintain exact input-output mapping with the accompanying explosion in hardware complexity. Therefore, a lot of attention has been directed towards block-processing and parallel processing [6, 7, 8] approaches. As suggested before [2], it is more natural to design a new algorithm which is inherently concurrent than creating concurrency in a traditional non-concurrent one. In designing new algorithms, we seek new topologies or architectures, which contain more loop delays, and can operate in functionally correct manner. Thus, the emphasis here is on correct *functionality*, as opposed to preserving the input-output characteristics.

[1]This research was supported by the army research office by contract number DAAL03-90-G-0063.

In this paper, we introduce a new pipelined LMS adaptive filter architecture, referred to as the PIPLMS (to be read as Pipe LMS) filter. The PIPLMS filter is designed on the basis of correct functionality assumption, and is developed using a *stochastic form of look-ahead prediction.* With the stochastic form of look-ahead we look for acceptable convergence behavior rather than invariance with respect to the input-output mapping. It is shown that the use of this approach requires only additional pipeline latches.

In section 2, we derive the PIPLMS architecture. The convergence properties of the PIPLMS are presented in section 3. Comparison of PIPLMS, in terms of hardware requirements and speed, with the corresponding serial architecture (SLMS), the block architecture (BLMS) [6] and the parallel architecture (PARLMS) [8] is carried out in section 4. In section 5, we present simulation results to verify the theoretical analysis done on PIPLMS and compare its performance with the SLMS, BLMS and PARLMS. The conclusions and future research directions are presented in section 6.

2 The Pipelined LMS Architecture

Given a desired speed-up of M, the clock period of PIPLMS should be less than or equal to $1/M$ times that of SLMS.

2.1 Pipelining the Sub-blocks

The block diagram of SLMS in (Fig.1) shows the two major sub-blocks denoted as the input tap filter block (**F**) and the weight update block (**WUD**). We can estimate the required level of pipelining by comparing the computation times of **F** and **WUD** with the desired clock-period. Let M_1 and M_2 represent the level to which the sub-blocks **F** and **WUD** need to be pipelined respectively. Pipelining **F** by M_1 levels is trivial as it is a FIR block.

Instead of employing the look-ahead transformation technique [1] to pipeline the recursive loop in **WUD**, we simply place the desired number (say M_{20}) latches in it. The remaining M_{21} latches, where

$$M_2 = M_{20} + M_{21}, \tag{2.1}$$

1058-6393/91 $01.00 © 1991 IEEE

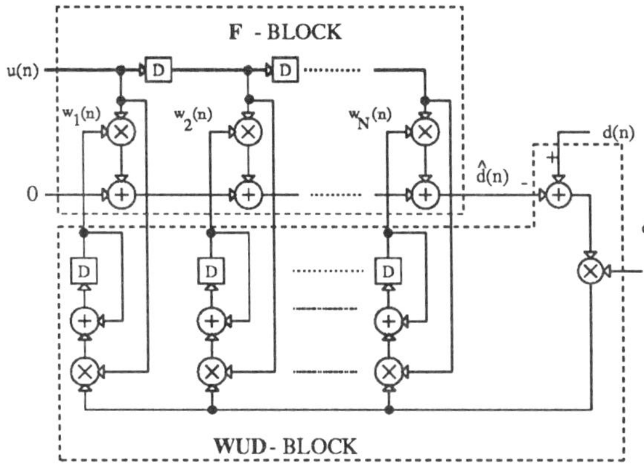

Figure 1: The SLMS architecture.

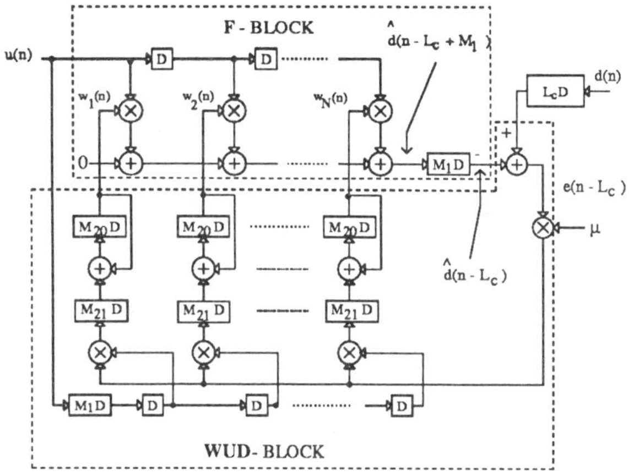

Figure 2: The PIPLMS architecture.

are placed outside the loop (see Fig.2).

Since the PIPLMS structure is not a retimed version of SLMS, a thorough convergence anaylsis needs to be carried out in order to determine the conditions under which the stochastic form of look-ahead prediction, as mentioned earlier, preserves the *overall* convergence of the weight vector and the output. In this way, apart from the pipelining latches, no other hardware overhead is required. This is a remarkable feature considering that an adaptive filter has been pipelined.

2.2 Modified Weight-update Equations

With the pipelining latches introduced in the previous sub-section, it is necessary to derive new weight-update equations (in a fashion similar to that of SLMS). In other words, we start with the expression for the expected value of the output error. We then find its gradient with respect to the weight vector ($\nabla_{\mathbf{W}}(n)$) and substitute its noisy estimate into a modified form of the steepest-descent algorithm shown below,

$$\mathbf{W}(n + M_{20}) = \mathbf{W}(n) + \frac{\mu}{2}(-\nabla_{\mathbf{W}}(n - L_c - M_{21})), \tag{2.2}$$

where μ represents the adaptation constant, L_c is the desired output latency, and we have employed the notation $\mathbf{W}(n)$ to denote the weight vector in block **F** i.e.

$$\mathbf{W}(n) = [w_1(n), w_2(n), \ldots, w_N(n)]^T, \tag{2.3}$$

where T is the transpose operator and N is the filter order.

The following final weight-update equation is obtained

$$\mathbf{W}(n+M_{20}) = \mathbf{W}(n)+\mu\mathbf{U}(n-M_1-M_{21})e^*(n-L_c-M_{21}), \tag{2.4}$$

where $*$ represents the complex conjugation operation.

2.3 Variable Latency Feature

The introduction of the L_c latches in the PIPLMS not only gives rise to a unique variable latency feature but also plays an important role in determining the adaptation accuracy. As the output of the pipelined **F** block is $\hat{d}(n - L_c)$ (Fig.2), it is clear that the output latency of PIPLMS equals L_c, which may be varied from zero to any reasonably large value.

Depending on whether L_c is less than, equal to, or greater than M_1, block **F** behaves as a $(M_1 - L_c)$-step forward, a $(L_c - M_1)$-step backward, or a zero-step predictor, respectively. Hence, only when L_c equals M_1, the adaptation accuracy is guaranteed to be equal to that of SLMS.

3 Convergence Properties of PIPLMS

In order to make the analysis tractable, we consider the following assumption which is general enough to cover all possible situations,

$$M_1 + M_{21} = M_{20}. \tag{3.1}$$

It is necessary to point out that the assumption of (3.1) can cover all possible values of M_1 and M_2. This can be seen to be true if we let $M_{21} = 0$ and $M_{20} = M_2$. Thus, we may move as many as $M_{20} - 1$ latches from the recursive loop in the **WUD** block to pipeline the rest of **WUD** and if necessary the **F** block.

3.1 Bounds on μ for Convergence

The bounds on μ to guarantee the convergence of PIPLMS was found to be tighter by a factor of 8 as compared to those for SLMS. In particular, the bounds on μ for PIPLMS were

$$0 \leq \mu \leq \frac{1}{4\lambda_{max}}, \tag{3.2}$$

TABLE I. Convergence Time-Constants

	$\tau_{k,J}$	$t_{k,J}$
SLMS	$1/(2\mu\lambda_k)$	$T_c/(2\mu\lambda_k)$
BLMS	$L/(2\mu\lambda_k)$	$LT_c/(2\mu\lambda_k)$
PARLMS	$\lceil 1/(2\mu\lambda_k M_B N_1)\rceil M_B N_1$	$\lceil 1/(2\mu\lambda_k M_B N_1)\rceil N_1 T_c$
PIPLMS	$(M_1 + M_{21})/(2\mu\lambda_k)$	$(M_1 + M_{21}) T_c/(2(M_1 + M_2)\mu\lambda_k)$

while those for SLMS were

$$0 \leq \mu \leq \frac{2}{\lambda_{max}}, \tag{3.3}$$

where λ_{max} is the largest eigenvalue of the input correlation matrix $\mathbf{R}$.

3.2 Convergence Speed

The *convergence time-constant* (denoted as $\tau_{k,J}$) and the *convergence time in seconds*, $t_{k,J}$, where index k refers to the kth ($k = 1, 2, \ldots, N$) mode of convergence, for all the architectures under consideration, are compared in Table I. In Table I, L is the block-length in BLMS, and M_B is the number of filters operating in parallel on contiguous blocks of data of length N_1 for PARLMS.

Thus, in spite of having a higher $\tau_{k,J}$ than that of SLMS, PIPLMS has a lower $t_{k,J}$. This is due to its faster clock-speed.

3.3 Adaptation Accuracy

The adaptation accuracy, which is quantified in terms of the misadjustment, is defined as follows

$$\mathcal{M} = \frac{E(J(\infty)) - J_{min}}{J_{min}}, \tag{3.4}$$

where $J(n)$ is the MSE at time instant n, and $E(J(n))$ is its average. The notation J_{min} refers to the minimum mean-squared error, which would be obtained if the filter weight vector $\mathbf{W}(n)$ equalled the optimum Weiner solution $\mathbf{W}_o$.

If $L_c \neq M_1$, then the optimal solution $\mathbf{W}_o$ and hence J_{min}, for PIPLMS, would be different from that of SLMS. Therefore, we need two definitions for the misadjustment of PIPLMS ($\mathcal{M}_{1,PIP}$ and $\mathcal{M}_{2,PIP}$), as shown below

$$\mathcal{M}_{1,PIP} = \frac{E(J(\infty)_{PIP}) - J_{min,S}}{J_{min,S}} \tag{3.5}$$

$$\mathcal{M}_{2,PIP} = \frac{E(J(\infty)_{PIP}) - J_{min,PIP}}{J_{min,PIP}}, \tag{3.6}$$

where $J_{min,S}$ and $J_{min,PIP}$ are the minimum mean-squared errors for SLMS and PIPLMS respectively, while $E(J(\infty)_{PIP})$ is the *average steady-state mean-squared error* of PIPLMS.

TABLE II. Adaptation accuracy.

	Misadjustment
SLMS	M_S
BLMS	M_S/L
PARLMS	M_S
PIPLMS	$M_{1,PIP} = [(M_S + 1)\alpha - 1]$
	$M_{2,PIP} = M_S$

In Table II, we present the misadjustments for all the architectures under consideration, in terms of the misadjustment for SLMS ($\mathcal{M}_S$). The misadjustment $\mathcal{M}_{1,PIP}$ is a linearly increasing function of the factor α, which is the ratio of $J_{min,PIP}$ to $J_{min,S}$. It can be seen that $\mathcal{M}_{1,PIP} = \mathcal{M}_S$, if $\alpha = 1$ and the latter would be true if $L_c = M_1$. The other measure of misadjustment $\mathcal{M}_{2,PIP}$, can be seen to be equal to $\mathcal{M}_S$. Thus, PIPLMS may have a higher misadjustment if we desire an output latency less than M_1.

4 Hardware and Clock-Speed Comparisons

In this section, we present the hardware requirements and the clock speed.

4.1 Hardware Requirements

The hardware requirements for all the four architectures are tabulated in Table III, from which we can see that the extra hardware requirements for PIPLMS is restricted to the $2M_1 + NM_2 + L_c$ latches. This is the only hardware penalty for using PIPLMS to achieve high processing speed. The hardware requirements in BLMS increase linearly with L. The penalty on PARLMS is quite high even for small values of M_B. This is due to the fact that the hardware overhead, which is needed to compute the weight error vector, is proportional to N^2. Therefore, PIPLMS represents an architecture with negligible hardware penalty.

TABLE III. Hardware requirements.

HARDWARE	SLMS	BLMS	PARLMS	PIPLMS
MULTIPLIERS	2N+1	L(2N+1)	$M_B(N^2+3N+4)$	2N+1
ADDERS	2N	2LN	$M_B(N^2+3N+2)$	2N
LATCHES	2N-1	N(L+1)-L	$M_B(2N-1)$	$2N-1+2M_1+NM_2+L_c$

TABLE IV. Speed.

PARAMETERS	SLMS	BLMS	PARLMS	PIPLMS
ITERATION PERIOD	T_c	T_c/L	T_c/M_B	$T_c/(M_1+M_{20})$
CLOCK PERIOD	T_c	T_c	T_c	$T_c/(M_1+M_{20})$

4.2 Speed

For speed comparisons, we employ the iteration period as a measure. The iteration period is simply the ratio of the clock period of the architecture to the number of outputs produced in each clock cycle [3]. The iteration period and the clock period for different architectures are shown in Table IV, where T_c is the time required to compute one serial output.

It can be seen that for each of the high-speed architectures, i.e. BLMS, PARLMS and PIPLMS, the iteration period is a fraction of T_c. However, hardware complexity limits the achievable speed for PARLMS and BLMS. Since the only hardware overhead for PIPLMS are the latches, the increase in hardware complexity would not limit the level of pipelining M.

5 Simulation Results

In this section, we present simulation results to verify the theoretical expressions for adaptation speed and adaptation accuracy for PIPLMS (section 3).

5.1 Experiment A

In this experiment, we attempt to identify a system composed of two delay elements (Fig.3). The noise variances are $\sigma_1^2 = 1$ and $\sigma_2^2 = 0.5$, and the filter order is $N = 4$. The mean squared error expression (eq.(29), [6]), which can be easily derived, is averaged over 32 independent trials and 400 iterations. This experiment is conducted for $\mu = 0.0234$ (experiment A1) and $\mu = 0.0468$ (experiment A2). The results of experiments A1 and A2 are tabulated in Table V. It can be seen (Table V) that the time-constants $\tau_{k,J}$ of BLMS and PIPLMS are nearly the same and that these are greater by a factor of two than the corresponding SLMS time-constants. In this experiment, as was discussed in section 3.2, the convergence time in seconds $t_{k,J}$ for PIPLMS would be one-half and one-fourth of the corresponding values for SLMS and BLMS respectively.

TABLE V. Experiment A.

	Exp.	μ	$\tau_{k,J}$ measured	$\tau_{k,J}$ theoretical	Misadjustment (%) measured	Misadjustment (%) theoretical
SLMS	A1	0.0234	24.0	21.37	4.68	4.68
	A2	0.0468	12.0	10.68	10.45	9.36
BLMS $L=2$	A1	0.0234	43.0	42.74	2.57	2.34
	A2	0.0468	23.0	21.37	4.65	4.68
PARLMS $M_B=2$ $N_1=12$	A1	0.0234	25.0	25.0	4.92	4.68
	A2	0.0468	25.0	25.0	10.45	9.36
PIPLMS $M_1=2$ $M_2=2$	A1	0.0234	44.0	42.74	5.08	4.68
	A2	0.0468	22.0	21.37	10.96	9.36

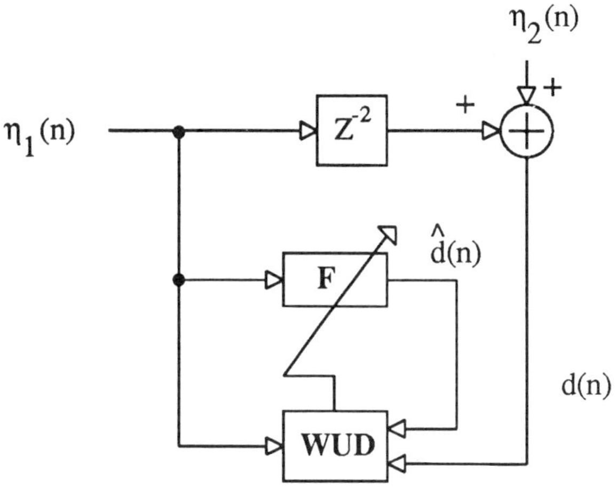

Figure 3: System identification example.

The misadjustment of BLMS is the lowest as expected. It is interesting to note that in this experiment the misadjustment ($\mathcal{M}_{1,PIP}$) for PIPLMS is the same as that of SLMS inspite of having $L_c = 0$. The plots (for experiment A1) depicting the variation of the mean-squared error of PIPLMS and SLMS, with the number of iterations, are shown in Fig.4.

5.2 Experiment B

The linear predictor is used to demonstrate the effect of the variable latency L_c on the misadjustment of the PIPLMS. In this experiment, we perform 1-step forward prediction of a second-order autoregressive (AR) process, whose generating filter has poles at $0.4875 \pm j0.8440$. The variance of the white noise at the input to the generating filter was chosen to be 0.0731 so as to get an output variance of 1. Plots of the mean-squared error for PIPLMS with $L_c = 0$ and $L_c = 2$ along with the corresponding plot for SLMS are shown in Fig.5, where $M_1 = M_{20} = 2$. As $M_1 = 2$, the misadjustment for $L_c = 2$ is the same as $\mathcal{M}_S$. On the other hand for $L_c = 0$, $J_{min,PIP}$ can be calculated to be equal to 0.1427, which results in α being equal to 1.95. When the measured value of 6.29%

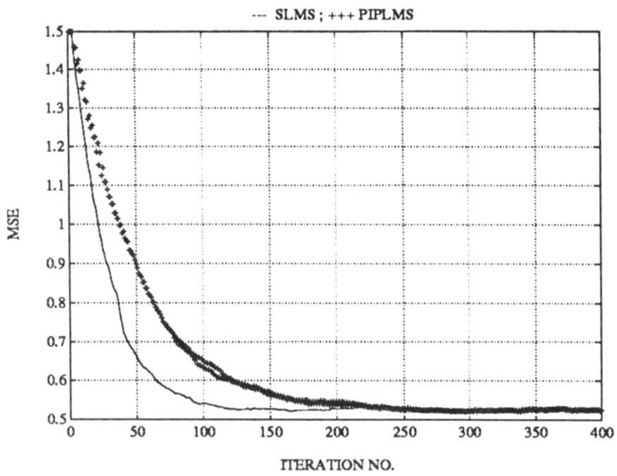

Figure 4: Experiment A : MSE plots for PIPLMS and SLMS.

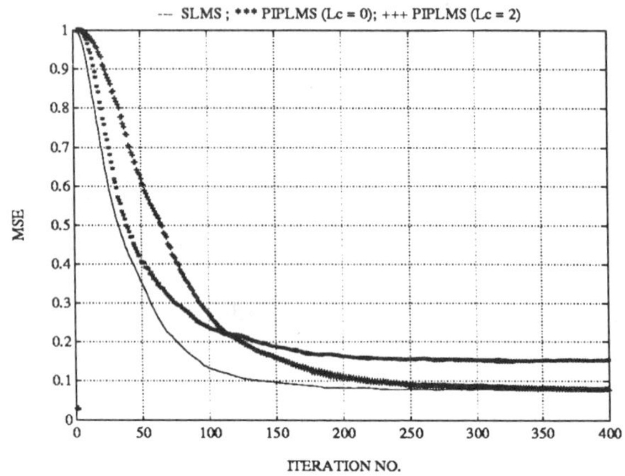

Figure 5: Experiment B : MSE plots for PIPLMS and SLMS.

for $\mathcal{M}_S$ is substituted along with α into the expression for $\mathcal{M}_{1,PIP}$ (Table II), we get a misadjustment of 107.48%. This compares well with the measured value of 110.26% for $\mathcal{M}_{1,PIP}$. Thus, the validity of the analytical expression for $\mathcal{M}_{1,PIP}$ is verified. As expected, the measured value of $\mathcal{M}_{2,PIP}$ was found to be close to $\mathcal{M}_S$. Hence, we may conclude that output latency can be traded-off with the misadjustment in case of PIPLMS.

6 Summary and Conclusions

A novel pipelined LMS architecture (PIPLMS) has been introduced. A stochastic form of look-ahead prediction resulted in enormous hardware savings. In particular, the increase in the hardware requirements are restricted to only pipeline latches. A unique feature of PIPLMS is its ability to trade-off the output latency for the adaptation accuracy.

Current work is being directed towards design of inherently concurrent algorithms for differential pulse code modulation (DPCM), adaptive decision feedback equalizers, and adaptive lattice filters.

References

[1] K.K. Parhi, "Algorithm transformation techniques for concurrent processors", *Proceedings of the IEEE*, vol. 77, pp. 1879-1895, Dec. 1989.

[2] K.K. Parhi and D.G. Messerschmitt, "Pipeline interleaving and parallelism in recursive digital filters - Part I : Pipelining using scattered look-ahead and decomposition", *IEEE Trans. on Acoustics, Speech and Signal Proc.*, vol. 37, pp. 1099-1117, July 1989.

[3] K.K. Parhi and D.G. Messerschmitt, "Pipeline interleaving and parallelism in recursive digital filters - Part II : Pipelined incremental block filtering", *IEEE Trans. on Acoustics, Speech and Signal Proc.*, vol. 37, pp. 1118-1134, July 1989.

[4] K.K. Parhi and D.G. Messerschmitt, "Concurrent cellular VLSI adaptive filter architectures", *IEEE Trans. on Circuits and Systems*, vol. 34, pp. 1141-1151, Oct. 1987.

[5] T. H.-Y. Meng and D.G. Messerschmitt, "Arbitrarily high sampling rate adaptive filters", *IEEE Trans. on Acoustics, Speech and Signal Proc.*, vol. 35, pp. 455-470, April 1987.

[6] G.A. Clark, S.K. Mitra and S.R. Parker, "Block implementation of adaptive digital filters", *IEEE Trans. on Acoustics, Speech and Signal Proc.*, vol. 29, pp. 744-752, June 1981.

[7] J.M. Cioffi, "The block-processing FTF adaptive algorithm", *IEEE Trans. on Acoustics, Speech and Signal Proc.*, vol. 34, pp. 77-90, Feb. 1986.

[8] A. Gatherer and T. H.-Y. Meng, "High sampling rate adaptive decision feedback equalizer", *Proc. ICASSP '90*, pp. 909-912.

A Nonlinear Error Adaptive Notch Filter for Separating Two Sinusoidal Signals

S. C. Douglas and T. H.-Y. Meng
Information Systems Laboratory
Stanford University
Stanford, CA 94305

Abstract

The adaptive notch filter is a simple structure that has found a number of applications in active noise cancellation and vibration control, in which the primary input to the filter is often a sum of sinusoids. For a primary input consisting of two sinusoids, the adaptive notch filter coefficients have sinusoidally-varying components in steady-state which reduce the rejection at the notch frequency due to weight misadjustment. In this paper, we present a simple modification of the adaptive notch filter that removes these sinusoidal variations, improves rejection, and enhances tracking performance. Using a general theory of non-mean-square error stochastic gradient adaptation, we show that the optimum nonlinear error algorithm adapts the system only when the instantaneous error magnitude is greater than the amplitude of the interfering sinusoid. This amplitude can be easily estimated at the output of the notch filter when the system nears convergence. To follow any slow changes in the interfering sinusoid amplitude, we introduce a time-varying *error criterion to keep the algorithm optimal. Using this modified structure with a time-varying nonlinearity, our simulations show a 6 dB reduction of the excess mean-square error, a 10 to 13 dB improvement in rejection at the filter outputs, and an increase in adaptation speed near the solution over the least-mean-square (LMS) adaptive notch filter.*

1 Introduction

The adaptive notch filter [1, 2] is a simple but useful implementation of a time-varying narrowband rejection filter in situations where a sinusoidal reference signal at the desired notch frequency can be easily measured or synthesized. The structure has found use in adaptive sound cancellation [3] and vibration control [4], in which the task of the filter is to separate signals primarily consisting of sinusoidal components. The simplest structure

This research was supported by an NSF Fellowship, an NSF Grant no. 8957058, and an SRC Grant no. 90-MC-500.

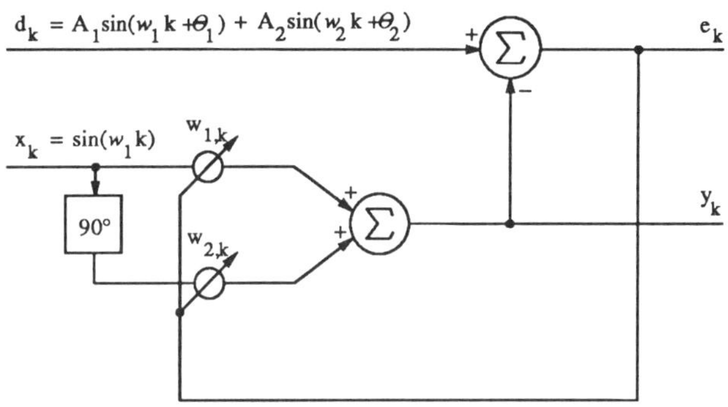

Figure 1: The adaptive notch filter.

for analysis in this situation is the two-weight adaptive notch filter shown in Figure 1. Here, the task is to remove the reference sinusoid of amplitude A_1, frequency ω_1, and arbitrary initial phase θ_1 from the primary input, leaving the interfering sinusoid of amplitude A_2 and frequency ω_2 intact. To cancel the first sinusoid, a reference sinusoid of frequency ω_1, unit amplitude, and zero initial phase is fed into the adaptive structure consisting of two weights w_1 and w_2 as well as a 90^0 phase delay for the second weight. It is not hard to see that if the weights are set at

$$w_1^* = A_1\cos(\theta_1) \quad (1)$$

$$w_2^* = -A_1\sin(\theta_1), \quad (2)$$

then the sinusoid of frequency ω_1 will be completely cancelled at the error output of the adaptive filter, leaving only the interfering sinusoid. The output of the adaptive filter y_k will contain the reference sinusoid of the same amplitude and phase as that in the original signal. Thus, the two sinusoids will be separated at the two outputs.

To determine the filter coefficients on-line to track changes in phases and amplitudes of the sinusoids, an adaptive algorithm such as the popular least-mean-square (LMS) algorithm is employed [2]. At every iteration, the

1058-6393/91 $01.00 © 1991 IEEE

filter coefficients are updated according to the rule

$$\mathbf{W}_{k+1} = \mathbf{W}_k + \mu e_k \mathbf{X}_k \quad (3)$$

$$e_k = d_k - \mathbf{W}_k^t \mathbf{X}_k, \quad (4)$$

in which $\mathbf{W}_k = [w_{1,k} w_{2,k}]^t$ are the time-varying coefficients which adapt to track the possibly time-varying optimum weight solution given by Equations (1)–(2) above. Previous studies have shown that when the adaptive notch filter centered at ω_1 is excited by a sinusoid of frequency ω_2 different from the center frequency, the resulting coefficients oscillate at the sum and the difference frequencies of the two sinusoids [1, 5]. This time-varying steady-state behavior of the filter coefficients reduces the effectiveness of the notch filter at the center frequency. Minimizing this weight fluctuation improves the rejection performance of the system, as minimizing the misadjustment in standard LMS adaptation with stochastic inputs improves the mean-square error performance of the adaptive system.

In this paper, we present a modification of the adaptive notch filter which requires practically no increase in computation and yet yields improved performance of the unmodified adaptive notch filter. In the next section, we describe this modification, in which we allow a memoryless nonlinearity to act upon the error in the stochastic gradient term. Using a theory of non-mean-square error adaptation, we then optimize the form of nonlinearity for the input and noise statistics. We show that the nonlinearity can be implemented as a simple decision to adapt only when the error is greater than the amplitude of the interfering sinusoid, and simulations show that the algorithm achieves both faster adaptation and a lower excess mean-square error (MSE) than the unmodified adaptive notch filter. As the interfering sinusoid's amplitude may be unknown or time-varying, in Section 3 we give a strategy for estimating its amplitude, resulting in an algorithm with a time-varying error criterion which changes according to the signal statistics. We then present simulations showing that the modified algorithm with time-varying nonlinearity achieves better rejection of the interfering sinusoid at the filter output than the standard notch filter. We also indicate the sensitivity of the algorithm's performance to the accuracy of the interfering sinusoid's amplitude estimate.

2 Derivation

We now describe the modified stochastic gradient algorithm which can improve adaptation performance in non-Gaussian noise interference situations. The algorithm modification is shown in Figure 2, in which we insert a memoryless nonlinearity into the error update of the stochastic gradient adaptation algorithm. This nonlinearity is chosen to optimize a non-mean-square error criterion

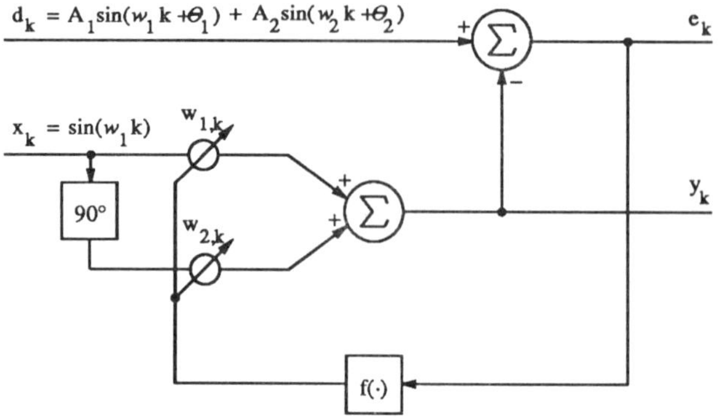

Figure 2: The nonlinear error adaptive notch filter.

matched to the problem at hand. A number of researchers have examined non-mean-square error criteria in stochastic gradient adaptation, either to improve the performance of the algorithm [6, 7, 8] or to simplify the computation of the stochastic gradient estimate [9, 10, 11]. Under slow adaptation conditions near convergence, it has been shown that the optimum error nonlinearity in stochastic gradient adaptation is given by [7, 8]

$$f(e_k) = \gamma \frac{p_n'(e_k)}{p_n(e_k)}, \quad (5)$$

where γ is a constraint parameter, the prime denotes functional differentiation, and $p_n(\cdot)$ is the probability density function of the interfering noise n_k. The stochastic gradient algorithm with this nonlinearity achieves the smallest MSE for a given adaptation rate as the step size μ tends to zero [7, 8]. The reduction in excess MSE that the nonlinear algorithm provides over the LMS algorithm can be expressed as an improvement factor α, defined as

$$\alpha = \frac{\text{final excess MSE of LMS}}{\text{final excess MSE of Modified Algorithm}} \quad (6)$$

$$= \frac{(E[f'(n_k)])^2 E[n_k^2]}{E[f^2(n_k)]}, \quad (7)$$

where the constraint parameter γ is chosen such that the adaptation speeds of the two algorithms are the same. Using the calculus of variations, it can be shown that the nonlinearity in (5) maximizes the performance factor in (7) for all continuous nonlinearities $f(\cdot)$ defined over the region of support of the noise probability density $p_n(\cdot)$.

For the adaptive notch filter with two sinusoids as input, the situation is similar to one discussed in [6], in which the noise $\{n_k\}$ is a sampled sinusoidal signal of amplitude A_2, frequency ω_2, and phase shift θ_2:

$$n_k = A_2 \sin(\omega_2 k + \theta_2). \quad (8)$$

As the phase of this signal is initially unknown, we can consider any one noise sample to be generated from a uniform random variable u_k over the range $[0, 1]$ as follows:

$$n_k \sim A_2 \sin(2\pi u_k). \tag{9}$$

Using a theorem for transforming probability density functions of uniform random variables [12], we may compute the density of n_k as

$$p_n(x) = \frac{1}{\pi\sqrt{A_2^2 - x^2}}, \qquad |x| < A_2. \tag{10}$$

Evaluating the optimum nonlinearity from (5), we find that over the region of support of $p_n(x)$,

$$f(x) = \gamma \frac{x}{A_2^2 - x^2}, \qquad |x| < A_2. \tag{11}$$

Evaluating the quantities $E[f'(x)]$ and $E[f^2(x)]$ for the density given in (10), we find that both integrals are infinite. However, the behavior of α may still be investigated by examining the nonlinearity

$$f_\delta(x) = \begin{cases} \gamma \dfrac{x}{A_2^2 - x^2} & |x| \le (1-\delta)A_2 \\ \gamma \dfrac{x}{A_2^2 \delta(2-\delta)} & (1-\delta)A_2 < |x| < A_2. \end{cases} \tag{12}$$

Clearly, as $\delta \to 0$, the nonlinearity in (12) approaches that in (11). Evaluating $E[f_\delta'(x)]$ and $E[f_\delta^2(x)]$ for this nonlinearity gives, after some calculus,

$$E[f_\delta'(x)] = \frac{2\gamma}{A_2^2 \pi \chi^{\frac{3}{2}}} \left\{ \frac{2}{3}\beta^3 + \beta\chi + \sqrt{\chi}\left(\frac{\pi}{2} - \sin^{-1}(\beta)\right) \right\} \tag{13}$$

$$E[f_\delta^2(x)] = \frac{\gamma^2}{A_2^4 \pi \chi^{\frac{3}{2}}} \left\{ \frac{2}{3}\beta^3 + \beta + \frac{1}{\sqrt{\chi}}\left(\frac{\pi}{2} - \sin^{-1}(\beta)\right) \right\}, \tag{14}$$

where $\beta = 1-\delta$ and $\chi = \delta(2-\delta)$. Plotting α as a function of δ, it can be shown that $\lim_{\delta\to 0} \alpha = \infty$. However, to keep the adaptation speed of both the non-mean-square error and LMS algorithms the same, we require that the adaptation speeds for the two algorithms be the same. Since this adaptation speeds constraint fixes $E[f'(x)]$ to be constant [7, 8], then as $\delta \to 0$, $E[f_\delta'(x)] \to \infty$ means that $\gamma \to 0$. Thus, over the nonzero region $(-A_2, A_2)$ of the probability density function, the optimum nonlinearity is simply given by

$$f(x) = 0, \qquad |x| < A_2. \tag{15}$$

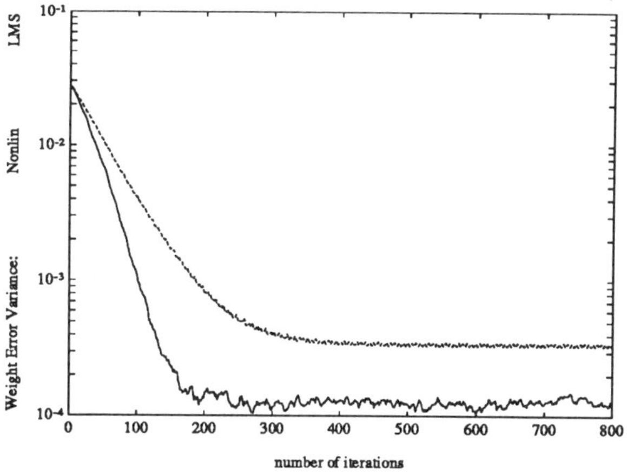

Figure 3: Convergence of weight error variance, LMS and nonlinear adaptive notch filters, $A_1 = A_2 = 1$, $\omega_1 = 1.11$, $\omega_2 = 0.509$, $\mu = 0.02$.

For large errors greater than A_2, the analysis of [7, 8] does not hold. However, we would like desirable properties such as robustness and simplicity in the large error case, and for this reason we choose the nonlinearity to be

$$f(x) = x, \qquad |x| \ge A_2 \tag{16}$$

so that adaptation for the modified algorithm is the same as LMS adaptation when the error is large. Intuitively, when the error is small, the deadzone about the origin effectively turns off the adaptation when the weights are very close to the Weiner solution. However, the large tails of the probability density function $p_n(\cdot)$ near the edges of the deadzone keep the solution near the optimum for small but nonzero step size.

We have simulated the performance of both systems with sinusoidal inputs. Figure 3 plots the convergence of the weight error power for an average of 100 Monte Carlo simulations of the two systems upon the same input data. Initial values of the weights were perturbed from their ideal values by a random variable chosen uniformly in the range $[-0.2, 0.2]$. These simulations indicate that convergence for the nonlinear error algorithm is twice as fast as the unmodified algorithm, and the final weight error power for the nonlinear error algorithm is 4.3 dB less than that of the LMS algorithm. Thus, the modified algorithm performs better both in terms of adaptation speed and final variance in the filter weights.

3 Time-Varying Error Criterion

In practice, the interfering sinusoid amplitude A_2 is generally unknown *a priori*, which makes choosing the width of the deadzone difficult. To overcome this problem, we see that near convergence, the error signal e_k will

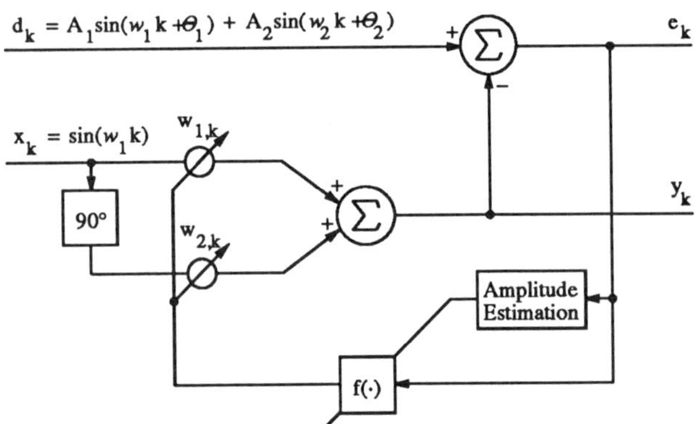

Figure 4: The nonlinear tracking adaptive notch filter.

consist almost wholly of the interfering sinusoid. Thus, we simply estimate the amplitude of the sinusoid by estimating the power at the error output and calculating the amplitude from this measurement. To estimate the error power, we use a simple integrator-type equation,

$$P_{k+1} = (1-\rho)P_k + \rho e_k^2 \qquad (17)$$

where ρ is much less than one, and estimate the instantaneous amplitude $\widehat{A}_{2,k}$ as the square root of $2P_k$. To simplify computation of this estimate, we could iterate upon the estimate of $\sqrt{2P_k}$ directly by using a Taylor series expansion upon the right hand side of (17), which replaces the square root by a divide. Further simplification of the multiply and divide operations could make this calculation even easier. The final adaptive system is shown in Figure 4 with tracking nonlinearity. This system only requires a scaled version of the reference sinusoid and needs no knowledge of the amplitudes or phases of either sinusoid.

Because any on-line estimate of the interfering sinusoid amplitude will not be exact, simulations were run to determine the sensitivity of the nonlinear algorithm's performance with respect to variations in the deadzone width. Figure 5 shows the piecewise-linear sensitivity curve, where $A_2 = 1$ is the correct estimate of the amplitude scaled to unity for clarity. As can be seen, the algorithm has its best performance improvement of 7.4 dB for a deadzone half-width of $A_2 = 1.005$, and adequate performance can be achieved with deadzone half-widths between 0.96 and 1.02. For our simulations, the deadzone half-width actually used in the algorithm was chosen to be nominally 0.99 of the amplitude estimate to fall in the middle of this adequate range.

To determine the tracking performance of the nonlinear error adaptive system, simulations were run with time-varying amplitudes for both reference and interfering sinusoids. Figure 6 shows the nature of the time variation for each sinusoid, chosen to be piecewise linear for simplicity. Figure 7 shows the results of the Monte Carlo simulations, which indicate that the nonlinear error adaptive notch filter has a 6.5 dB average lower weight error power during the time-varying portions of the amplitudes.

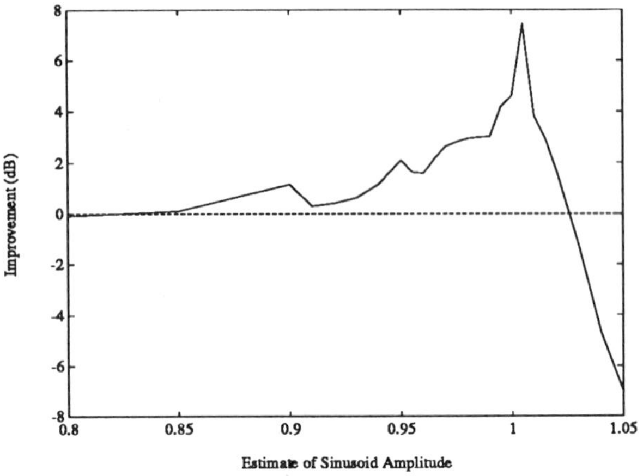

Figure 5: Sensitivity of algorithm performance to amplitude estimate.

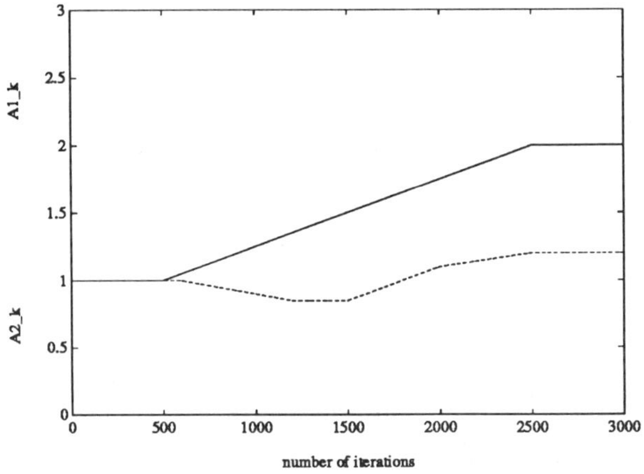

Figure 6: Time variation of sinusoidal amplitudes.

In order to assess the ability of both systems to reject the individual sinusoids at both filter and error output, FFTs were taken of both y_k and e_k signals over the time-varying portion of the amplitudes of the sinusoids. In each case, the signals were normalized by the known instantaneous amplitudes of the sinusoids, and Hamming windows was used in all FFTs to remove end effects. The plots of Figures 8 and 9 show the power spectra of the notch filter and error outputs, respectively, for both the LMS adaptive notch filter and the nonlinear error adaptive notch filter. As can be seen, the nonlinear error adaptive system achieves better isolation of each sinusoid, with a 53.4 dB reduction in the interfering sinusoid as compared to a 39.6 dB reduction for the LMS notch filter y_k output, and an 35.8 dB reduction in the reference sinusoid as

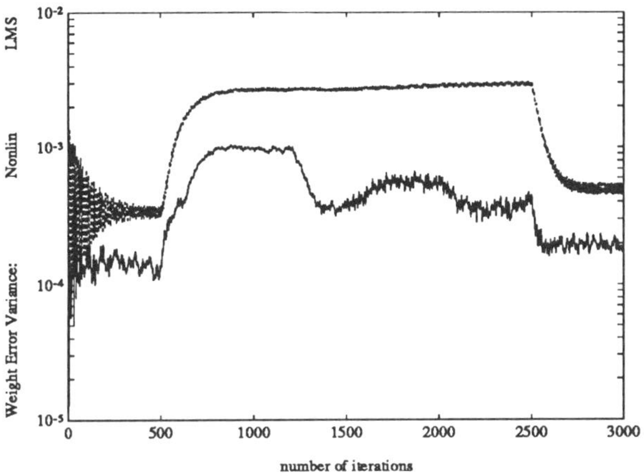

Figure 7: Convergence of weight error variance, LMS and nonlinear tracking adaptive notch filters, $\omega_1 = 1.11$, $\omega_2 = 0.509$, $\mu = 0.02$.

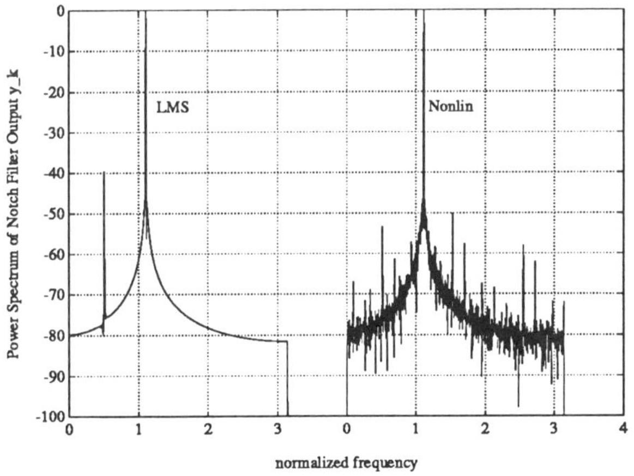

Figure 8: Spectra of y_k output from LMS and nonlinear tracking adaptive notch filters.

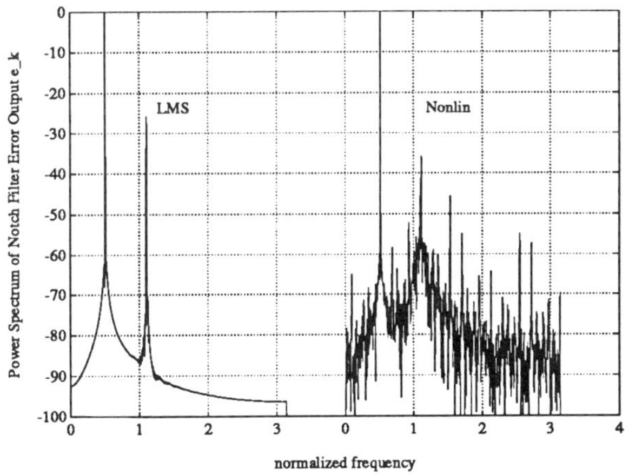

Figure 9: Spectra of e_k output from LMS and nonlinear tracking adaptive notch filters.

compared to a 25.9 dB reduction for the LMS notch filter e_k output. In addition, the nonlinear error notch filter spectra show a number of aperiodic components due to the nonlinear nature of the filter update. Despite the introduction of these components, the final MSE is over 6 dB lower over the period of tracking, as has been confirmed in simulations.

4 Conclusion

In conclusion, we have presented a simple modification of the adaptive notch filter which performs better than the unmodified adaptive notch filter in separating sinusoidal signals. Moreover, we have proposed a structure which adapts both the solution to the filtering problem as well as the algorithm which updates the solution. This two-level adaptation methodology can achieve better performance than one with a fixed adaptation strategy, as can be seen by our simulations.

References

[1] J. R. Glover, Jr., "Adaptive noise cancelling applied to sinusoidal interferences," *IEEE Trans. Acoust., Speech, Signal Processing*, vol. ASSP-25, no. 6, pp. 484-491, December 1977.

[2] B. Widrow *et. al.*, "Adaptive noise cancelling: Principles and applications," *Proc. IEEE*, vol. 63, no. 12, pp. 1692-1716, December 1975.

[3] J.C. Burgess, "Active adaptive sound control in a duct: A computer simulation," *J. Acoust. Soc. Amer.*, vol. 70, no. 3, pp. 715-726, September 1981.

[4] S.J. Elliott, I.M. Stothers, and P.A. Nelson, "A multiple error LMS algorithm and its application to the active control of sound and vibration," *IEEE Trans. Acoust., Speech, Signal Processing*, vol. ASSP-35, no. 10, pp. 1423-1434, October 1987.

[5] S. Conolly and Y. Su, "State-space analysis of the adaptive notch filter," *Proc. IEEE*, vol. 74, no. 1, pp. 219-221, January 1986.

[6] E. Walach and B. Widrow, "The Least Mean Fourth (LMF) adaptive algorithm and its family," *IEEE Trans. Inform. Theory*, vol. IT-30, no. 2, pp. 275-283, March 1984.

[7] S.C. Douglas and T.H.-Y. Meng, "Optimum error nonlinearities for LMS adaptation," *Proc. International Conf. Acoust., Speech, Signal Processing*, Albuquerque, NM, pp. 1421-1424, April 1990.

[8] S.C. Douglas and T.H.-Y. Meng, "Stochastic gradient adaptation under general error criteria," submitted to *IEEE Trans. Signal Processing*, August 1991.

[9] D.L. Duttweiler, "Adaptive filter performance with nonlinearities in the correlation multiplier," *IEEE Trans. Acoust., Speech, Signal Processing*, vol. ASSP-30, no. 4, pp. 578-586, August 1982.

[10] P. Xue and B. Liu, "Adaptive equalizer using finite-bit power-of-two quantizer," *IEEE Trans. Acoust., Speech, Signal Processing*, vol. ASSP-34, no. 6, pp. 1603-1611, December 1986.

[11] S.C. Douglas and T.H.-Y. Meng, "Optimum error quantization for LMS adaptation," *IEEE Pacific Rim Conf. on Communications, Computers, and Signal Processing*, Victoria, B.C., Canada, pp. 704-708, May 1991.

[12] J.W. Goodman, *Statistical Optics* (New York: Wiley, 1985), p. 25.

Comparison of Class 1 and Class 2 Frequency Domain Matrix and Vector Estimation Filters With Adaptive FIR Filters

Erkan Duysal
Electrical & Computer Eng. Dept.
University of Miami
P.O. Box 24-8351
Coral Gables, FL 33124

Claude S. Lindquist
Electrical & Computer Eng. Dept.
University of Miami
P.O. Box 24-8294
Coral Gables, FL 33124

Abstract

Class 1 and 2 frequency domain block matrix and vector estimation filters are compared with adaptive FIR filters based on gradient search algorithms including Newton, steepest descent and LMS. It is showed that algorithm of our Class 1 filter has the same form as Wiener-Hopf equation. It is found that the Class 2 estimation filter has the form of an adaptive FIR filter based on Newton's method. Simulations show that utilizing smoothing techniques can further improve the performance and efficiency of the Class 2 estimation filter.

1 Introduction

In recent years much research has been done in the area of adaptive block matrix filtering systems. Our goal is to compare the formulation of such systems with the adaptive vector filters based on gradient search algorithms. We will focus primarily on the following three algorithms: Newton's, steepest descent and LMS, mainly due to their simplicity, flexibility and robustness.

1.1 Wiener Block Matrix Estimation Systems

The popularity of matrix filtering techniques is increased by implementation of fast transforms and utilization of different implementations in matrix form. Ahmed and Rao [2] utilized matrix filtering in formulating Wiener estimation systems. Lindquist [3] formulated and developed several types of matrix filters including estimation, detection and correlation systems.

Wiener estimation filters are based upon the minimization of MSE error. In frequency domain the optimum gain matrix $[H]$ for a *correlated Wiener estimation filter* is expressed as

$$\begin{aligned}[H_n] \;=\; & \Big[E\{[SS^h]\} + E\{[SN^h]\}\Big] \\ & \Big[E\{[SS^h]\} + E\{[SN^h]\} \\ & +E\{[S^hN]\} + E\{[NN^h]\}\Big]^{-1} \qquad (1)\end{aligned}$$

where $E\{[SS^h]\}$ is the expectation of signal power spectrum, $E\{[NN^h]\}$ is the expectation of noise power spectrum and $E\{[SN^h]\}$, $E\{[S^hN]\}$ are expectation of cross power spectrums. Assuming the signal and noise spectrums are uncorrelated and the noise has zero mean, *uncorrelated Wiener estimation filter* is obtained.

$$[H] = \Big[E\Big\{[SS^h]\Big\}\Big]\Big[E\Big\{[SS^h]\Big\} + E\Big\{[NN^h]\Big\}\Big]^{-1} \qquad (2)$$

If there is a priori information about expected signal and noise spectra and their correlation functions than this type of filter is called as *Class 1* filter. In most cases such apriori information is not available and must be estimated aposteriori from the actual signal and noise processed by the filter. If we have no information about one component of the input, signal or noise this type of filter is called as *Class 2* filter. Thus a Class 2 uncorrelated Wiener estimation filter gain matrix can be expressed as

$$[H]_n = E\Big\{[SS^h]\Big\}\Big\langle R_nR_n\Big\rangle^{-1} \qquad (3)$$

where $\langle R_nR_n\rangle$ is the estimate of the input power spectrum matrix using smoothing algorithm [4].

1.2 Adaptive Vector Filters Using Gradient Search

The performance of an adaptive vector filter improves by adjusting its weight through feedback from output. Widrow [5] showed that the optimum weights for a frequency domain adaptive vector filter is given by the *Wiener-Hopf* equation as

$$[W_{opt}] = \Big([S_{rr}]^t\Big)^{-1}[S_{dr}] \qquad (4)$$

When the power spectrum matrix $[S_{rr}]$ and cross-power spectrum matrix $[S_{dr}]$ are known apriori then optimum tap weight vector can easily be evaluated using Wiener-Hopf equation. When $[S_{rr}]$ and/or $[S_{dr}]$ are not known these terms must be estimated using different search algorithms. The purpose of the search algorithms are to find a weight vector that globally minimizes the performance index. The process begins by choosing a arbitrary initial weight vector $[W_0]$ and iteratively updating it towards the global minimum, $[W_{opt}]$ of the surface. Gradient search and gradient estimate methods are very popular to adjust the weight vector.

The Newton algorithm obtained by substituting the Wiener-Hopf equation into the gradient ∇E_n and then solving for $[W_{n+1}]$ using recursion as

$$[W_{n+1}] = [W_n] - \frac{1}{2}[S_{rr}]^{-1}\nabla E_n \tag{5}$$

where $\nabla E_n = \nabla(E(n)E^*(n))$. A more general recursion equation uses an arbitrary gain u as in Eq. 6

$$[W_{n+1}] = [W_n] - u[S_{rr}]^{-1}\nabla E_n \tag{6}$$

The Newton algorithm given by Eq. 6 is Class 1 since both $[S_{rr}]$ and ∇E_n are known. When $[S_{rr}]$ and/or ∇E_n are not known, they must be estimated by $[\widehat{S}_{rr}]$ or $\widehat{\nabla} E$ which produces Class 2 algorithm.

The steepest descent algorithm moves the weight vector in the opposite direction of the gradient. This algorithm is expressed as

$$[W_{n+1}] = [W_n] - \mu\nabla E_n \tag{7}$$

When ∇E_n is not known apriori it must be estimated by $\widehat{\nabla} E_n$ which is Class 2 algorithm.

$$[W_{n+1}] = [W_n] - \mu\widehat{\nabla} E_n \tag{8}$$

In LMS algorithm the gradient estimate is computed at each iteration using error term at the output and the conjugate of signal spectrum and expressed as

$$\widehat{\nabla} E_n = -2E(n)[R_n]^* \tag{9}$$

Substituting this gradient estimate into weight updating equation Eq. 10 is obtained

$$\begin{aligned} [W_{n+1}] &= [W_n] - \mu\widehat{\nabla} E_n \\ &= [W_n] + 2\mu E(n)[R_n]^* \end{aligned} \tag{10}$$

which is called the *least-mean-square* or *LMS* algorithm. μ is the gain constant that regulates the speed and stability of adaption. As the weight changes at each iteration are based on imperfect gradient estimates, the adaptive process is noisy and the weight vector seeks $[W_{opt}]$ in an oscillatory manner. From practical implementation point LMS algorithm does not require squaring , averaging and differentiation which makes it simple. The excessive noise caused by the lack of averaging is attenuated with time by the adaptive process which acts as a low pass filter from this respect [5],[6].

2 Comparison of Class 1 Filters

A conversion scheme should be introduced to compare block matrix filters with adaptive vector filters. A general block matrix filter can be represented in vector form by selecting a row of the gain matrix at a time. The row is extracted by utilizing a *selector operator* $[U_{0,n}]$ whose all elements except the nth entry are zero. Essentially $[U_{0,n}]$ is a column selector operator but can be used as a row selector by transposing the matrix.

Applying the selector operator to block matrix filter in Eq. 2, correlated Wiener estimation vector filter can be computed as

$$\begin{aligned} [H_n] &= [H]_n^t[U_{0,n}] \\ &= \Big([E\{SS^h\} + E\{SN^h\}][E\{SS^h\} \\ &\quad + E\{SN^h\} + E\{S^hN\} \\ &\quad + E\{NN^h\}]^{-1}\Big)^t[U_{0,n}] \\ &= \Big([E\{SS^h\} + E\{SN^h\} + E\{S^hN\} \\ &\quad + E\{NN^h\}]^{-1}\Big)^t\Big([E\{SS^h\} \\ &\quad + E\{SN^h\}]\Big)^t[U_{0,n}] \\ &= \Big[(E\{SS^h\} + E\{SN^h\} + E\{S^hN\} + \\ &\quad E\{NN^h\})^t\Big]^{-1} \\ &\quad \Big[E\{S(n)S^*\} + E\{S(n)N^*\}\Big] \\ &= \Big[(E\{SS^h\} + E\{SN^h\} + E\{S^hN\} + \\ &\quad E\{NN^h\})^t\Big]^{-1}\Big[E\{S(n)R^*\Big] \end{aligned} \tag{11}$$

If we assume that the signal and noise are uncorrelated and noise has a zero mean, then $E\{SN^h\} = E\{S^hN\} = [0]$. Thus, the Class 1 uncorrelated Wiener estimation filter gain vector can be expressed as

$$[H_n] = \Big[\Big(E\{[SS^h]\} + E\{[NN^h]\}\Big)^t\Big]^{-1} E\{S(n)R^*\} \tag{12}$$

In Wiener estimation filter total input spectrum is $[R] = [S] + [N]$ and the desired response is $D(n) = S(n)$. Therefore $E\{SS^h\} + E\{NN^h\} = [S_{rr}]$ and $E\{S(n)R^*\} = [S_{dr}]$. Thus Eq. 12 can be rewritten as

$$[H_n] = [S_{rr}^t]^{-1}[S_{dr}] \tag{13}$$

which is identical to Eq. . Thus, the Class 1 Wiener estimation gain vector equals the Wiener optimum weight vector or $[H_n] = [W_{opt}]$. Class 1 systems are not dynamically adaptive and they do not require any update since the differential weight vector $[\Delta W_n] = [0]$.

This result leads us to another important observation. Since the Wiener estimation filter gain matrix in Eq. 2 is found by taking the gradient of $E\{E(n)E^*(n)\}$ with respect to gain matrix $[H]_n$, selecting a row of the gradient matrix gives us the gradient of $E\{E(n)E^*(n)\}$ with respect to weight vectors. Thus it is showed that

$$\left[\frac{\partial(E(n)E^*(n))}{\partial[W]_n}\right] = \left[\frac{\partial(E(n)E^*(n))}{\partial[H]}\right]^t [U_{0,n}] \quad (14)$$

3 Comparison of Class 2 Filters

Class 2 uncorrelated Wiener Estimation gain matrix is expressed in Eq. 3. The gain matrix at time $n+1$ can also be expressed as the sum of the previous gain matrix $[H]_n$ and the differential gain matrix $[\Delta H]_n$ as

$$[H]_{n+1} = [H]_n + [\Delta H]_n \quad (15)$$

where $[\Delta H]_n$ is the differential gain matrix and equals

$$\begin{aligned}[\Delta H]_n &= [H]_{n+1} - [H]_n \\ &= E\{SS^h\} \\ &\quad \left[\left\langle R_{n+1}R_{n+1}^h\right\rangle^{-1} - \left\langle R_nR_n^h\right\rangle^{-1}\right] \quad (16)\end{aligned}$$

Therefore the gain matrix in Eq. 15 becomes

$$\begin{aligned}[H]_{n+1} &= [H]_n + E\{SS^h\} \\ &\quad \left[\left\langle R_{n+1}R_{n+1}^h\right\rangle^{-1} - \left\langle R_nR_n^h\right\rangle^{-1}\right] \quad (17)\end{aligned}$$

As shown above, the gain matrix $[H]_n$ for Class 2 algorithm is updated for each iteration n since the input spectrum vector $[R_n]$ is updated continuously. Thus, Class 2 algorithm is considered to be dynamically adaptive.

Applying the selector operator $[U_{0,n}]$ to the recursive equation, Class 2 filter gain vector is obtained as

$$\begin{aligned}[H_{n+1}] &= \left([H]_n + [R_{ss}]\left[\left\langle R_{n+1}R_{n+1}^h\right\rangle^{-1} - \right.\right. \\ &\quad \left.\left.\left\langle R_nR_n^h\right\rangle^{-1}\right]\right)^t [U_{0,n}] \\ &= [H]_n[U_{0,n}] + \left(\left[\left\langle R_{n+1}R_{n+1}^h\right\rangle^{-1} - \right.\right. \\ &\quad \left.\left.\left\langle R_nR_n^h\right\rangle^{-1}\right]\right)^t [S_{ss}]^t[U_{0,n}] \\ &= [H_n] + \left[(\left\langle R_{n+1}R_{n+1}^h\right\rangle^{-1})^t - \right. \\ &\quad \left.\left(\left\langle R_nR_n^h\right\rangle^{-1}\right)^t\right] E\{S(n)S^*\} \quad (18)\end{aligned}$$

Let the weight vectors equal to the gain vectors, $[W_n] \equiv [H_n]$ and $[W_{n+1}] \equiv [H_{n+1}]$. Assuming the input is ergodic, the estimates of the input power spectrum matrices become $[\widehat{S}_{rr}]_{n+1} = \left\langle R_{n+1}R_{n+1}^h\right\rangle$ and $[\widehat{S}_{rr}]_n = \left\langle R_nR_n^h\right\rangle$, and the signal and input cross-power spectrum vector becomes $[S_{dr}] = E\{S(n)S^*\}$. Then Eq. 18 can be written as

$$[W_{n+1}] = [W_n] + \left(\left([\widehat{S}_{rr}]_{n+1}^t\right)^{-1} - \left([\widehat{S}_{rr}]_n^t\right)^{-1}\right)[S_{dr}] \quad (19)$$

Unlike the optimum weight vectors in Class 1 filters, the Class 2 algorithm involves updating the weight vector or the gain vector recursively. Thus, it can be compared with the adaptive gradient search algorithms.

3.1 Comparison with Newton Algorithm

First, let us compare the Class 2 recursive equation with the Newton's algorithm. The weight vector for the Newton's algorithm is written as

$$[W_{n+1}] = [W_n] - \mu[S_{rr}^t]^{-1}\nabla E_n \quad (20)$$

where $[S_{rr}]$ matrix is used to find the direct path, or minimum distance to the optimum weights on error performance surface. Since the gradient is equal to $\nabla E_n = 2([S_{rr}^t][W_n] - [S_{dr}])$, Eq. 20 becomes

$$\begin{aligned}[W_{n+1}] &= [W_n] + 2u[S_{rr}^t]^{-1}\Big([S_{dr}] - [S_{rr}^t][W_n]\Big) \\ &= [W_n] + 2u\Big([S_{rr}^t]^{-1}[S_{dr}] - [W_n]\Big) \quad (21)\end{aligned}$$

where $[W_n]$ is the nth estimate of the weight vector. If we substitute $[W_n] = [\widehat{S}_{rr}^t]_n^{-1}[S_{dr}]$ in Eq. 21 and let the convergence factor u equals to 0.5, then

$$[W_{n+1}] = [W_n] + \Big([S_{rr}^t]^{-1} - [\widehat{S}_{rr}^t]_n^{-1}\Big)[S_{dr}] \quad (22)$$

where $[S_{rr}^t]$ is the exact input power spectrum matrix, and $[\widehat{S}_{rr}^t]_n$ is the estimated input power spectrum matrix.

Comparing Eq. 22 with Eq. 19, it is observed that Newton's algorithm is similar to Class 2 recursive equation. $[\widehat{S}_{rr}]_n$ terms in Eq. 22 and in Eq. 19 are the estimated input power spectrum matrices, but they are computed using different algorithms. Furthermore $[S_{rr}]$ in Eq. 22 is the exact power spectrum matrix and $[\widehat{S}_{rr}]_{n+1}$ in Eq. 19 is the estimated power spectrum matrix for iteration $n+1$. Thus, the Newton's algorithm is equivalent to the Class 2 recursive equation.

3.2 Comparison with Steepest Descent Algorithm

Next, the Class 2 recursive equation is compared with the steepest descent algorithm. The weight vector for the steepest descent algorithm is written as

$$[W_{n+1}] = [W_n] - \mu\nabla E_n \quad (23)$$

The steepest descent algorithm follows a gradient path ∇E_n to find the optimal weight vector which is also the minimum square error point. The steepest descent algorithm is identical to the Newton's algorithm

Type of Algorithm	Recursion Equation
Class 1 Wiener Estimation	$[W_{n+1}] = [W_n] - [0]$
Class 2 Wiener Estimation	$[W_{n+1}] = [W_n] - \left([\widehat{S}^t_{rr}]^{-1}_n - [\widehat{S}^t_{rr}]^{-1}_{n+1}\right)[S_{dr}]$
Newton's	$[W_{n+1}] = [W_n] - \mu[S^t_{rr}]^{-1}\nabla E_n$
Steepest Descent	$[W_{n+1}] = [W_n] - \mu\nabla E_n$
Least Mean Square	$[W_{n+1}] = [W_n] - 2\mu(-E(n))[R_n]^*$

Table 1: General form of recursion equations for adaptive vector filters and gradient search algorithms.

Type of Algorithm	Recursion Equation
Class 1 Wiener Estimation	$[W_{n+1}] = [S^t_{rr}]^{-1}[S_{dr}] - [0]$
Class 2 Wiener Estimation	$[W_{n+1}] = [W_n] - \left([\widehat{S}^t_{rr}]^{-1}_n - [\widehat{S}^t_{rr}]^{-1}_{n+1}\right)[S_{dr}]$
Newton's	$[W_{n+1}] = [W_n] - \left([\widehat{S}^t_{rr}]^{-1}_n - [S^t_{rr}]^{-1}\right)[S_{dr}]$
Steepest Descent	$[W_{n+1}] = [W_n] - \left([S^t_{rr}][\widehat{S}^t_{rr}]^{-1}_n - [I]\right)[S_{dr}]$
Least Mean Square	$[W_{n+1}] = [W_n] - \left([S^t_{rr}][\widehat{S}^t_{rr}]^{-1}_n - [I]\right)[\widehat{S}_{dr}]_n$

Table 2: Comparison of recursion equations for adaptive vector filters and gradient search algorithms.

except for the $[S_{rr}^t]$ term. In Newton's algorithm a weighted gradient, $[S_{rr}]\nabla E_n$ is used to find the direct path to the minimum point on error surface.

Substituting the gradient into Eq. 23, we obtain

$$[W_{n+1}] = [W_n] + 2\mu\Big([S_{dr}] - [S_{rr}^t][W_n]\Big) \quad (24)$$

Let the estimated impulse response $[W_n] = [\widehat{S}_{rr}^t]_n^{-1}[S_{dr}]$ and set the convergence factor u to 0.5, then

$$[W_{n+1}] = [W_n] + \Big([I] - [S_{rr}^t][\widehat{S}_{rr}^t]_n^{-1}\Big)[S_{dr}] \quad (25)$$

where $[S_{rr}^t]$ is the exact input power spectrum matrix and $[\widehat{S}_{rr}^t]_n$ is the estimated input power spectrum matrix.

Comparing Eq. 25 with Eq. 19, both $[\widehat{S}_{rr}]_n$'s in these equations are estimated input power spectrum matrices. They are computed using different algorithms. The $[S_{rr}]$ in Eq. 19 is the exact input power spectrum matrix. Thus, the steepest descent algorithm in Eq. 25 is similar in form to the Class 2 Wiener estimation equation.

3.3 Comparison with LMS Algorithm

For the LMS algorithm, the weight vector is expressed

$$[W_{n+1}] = [W_n] + 2\mu E(n)[R_n]^* \quad (26)$$

where the error $E(n) = D(n) - [W_n]^t[R_n]$. Substituting the error term into Eq. 25, we obtain

$$\begin{aligned}[W_{n+1}] &= [W_n] + 2u\Big(D(n) - [W_n]^t[R_n]\Big)[R_n]^* \\ &= [W_n] + [R_n]^*\Big(D(n) - [R_n]^t[W_n]\Big) \\ &= [W_n] + \Big([S_{dr}] - [S_{rr}^t][W_n]\Big) \\ &= [W_n] + \Big([I] - [S_{rr}^t][\widehat{S}_{rr}^t]_n^{-1}\Big)[S_{dr}] \quad (27)\end{aligned}$$

where the signal and input cross power spectrum vector $[S_{dr}] = S(n)[R_n]$ and the estimated cross power spectrum vector is $[\widehat{S}_{dr}] = [W_n]^t[R_n]$. Thus the error term approaches zero as the product of the cross power spectrum and the inverse of the estimate of cross power spectrum approaches identity matrix.

4 Conclusion

In this paper it is shown that the Class 1 and Class 2 block matrix and vector Wiener estimation filters are either equivalent or very similar to certain adaptive FIR filters with different gradient search algorithms seeking optimum weights.

Utilizing a selector operator block matrix filters are converted to vector filter form. It is observed that Class 1 Wiener estimation filter is equivalent to Wiener-Hopf equation and there is no need to update the weights of the system.

Observation of Class 2 Wiener estimation filter showed that the formulation is equivalent to Newton's search algorithm, which takes the direct path to the optimum weight vector. Using various smoothing techniques in Class 2 estimation filter, advantage of Newton's path can be combined with the computational efficiency and simplicity of the LMS algorithm [1],[3]. Therefore resulting filter's performance is improved while the convergence time and the computational burden are decreased.

Although our algorithms have implemented in the frequency domain they may also be implemented directly in the time domain [7].

References

[1] Duysal, E., "Recursive Implementation of Frequency Domain Estimation Filters," M.S.E.E. Thesis, University of Miami, 102 pp., December 1990.

[2] Ahmed, N. and K.R. Rao, *Orthogonal Transforms for Digital Signal Processing*, Chap 8, Springer-Verlag, 1975.

[3] Lindquist, C.S., *Adaptive and Digital Signal Processing With Signal Processing Applications*, Chaps 4,9,11, Steward and Sons, CA 1989.

[4] Duysal, E. and C.S. Lindquist, "Use of Smoothing in Class 1 Frequency Domain Estimation Matrix Filters," *Proc. 34th Midwest Symposium on Circuits and Systems*, Monterey, CA, May 1991.

[5] Widrow, B., J. McCool, M. Ball,"The Complex LMS Algorithm," Proc. IEEE, Vol. 63, pp. 719-720, April 1975.

[6] Widrow, B. and S.D. Stearns, *Adaptive Signal Processing*, Prentice-Hall, Englewood Cliffs, NJ, 1985.

[7] Ho, J., "Recursive Implementation of Time Domain Estimation Matrix Filters," M.S.E.E. Thesis, University of Miami, 142 pp., December 1990.

Analysis of the Jointly Adapting and Asynchronously Configured V.32 Modem Implementation

Sangil Park, Dion Messer Funderburk and *Garth Hillman*

Motorola Inc.

Digital Signal Processing Operations

Austin, Texas

ABSTRACT

This paper describes an implementation of a V.32 modem using new concepts which make it possible to realize a completely digital V.32 modem. The first new concept incorporated is a jointly adapting echo canceler and equalizer where the error signal is derived from the final data decisions. Another new concept used in this implementation is a sample rate convertor algorithm. The receive modem transmits and receives data at the same rate, determined by the receiving modem. Since the sending modem is typically asynchronous to the receiving modem, a sample rate conversion of the receive data to the rate at which that data was sampled at the sending modem is required in order to recover every data sample transmitted.

Analysis of these concepts is performed and presented as well as a discussion of the importance of these concepts in realizing a completely digital modem. Results of the actual software solutions are shown along with a block diagram of the final implementation.

1.0 Introduction

V.32 (CCITT standard) modems are intended for use on two-wire general switched telephone networks (GSTN) at data rates of 9600 bit per second (bps). The principal characteristics of the modem are defined in [1]; the following are the most critical and computationally intensive blocks: channel separation by echo-cancellation techniques, channel equalization to compensate for intersymbol interferences due to linear distortion on the channel, encoding and decoding of the signal using trellis coded modulation (TCM) for each channel with synchronous line transmission at 2400 baud. The simple and efficient implementation for the TCM block using a 32-state trellis encoding and Viterbi decoding algorithm has been introduced by the authors [1] and is well documented in [2]. The adaptive real-time algorithms for full-duplex voice transmission with electric and acoustic echo-cancellers have also been studied thoroughly by the authors [3,4]. However, it has been shown that the real performance of the V.32 modem is driven by the optimum channel equalization and exceptional echo-cancellation in excess of 60 dB under worst case conditions [5,6]. This paper will discuss the analysis, implementation and simulation of jointly adapting the echo cancellers and channel equalizers to achieve this performance. This paper will also present the problem which occurs when the modem transmitter and receiver operate synchronously on asynchronous data. If data is to be recovered completely, the received data has to be converted to the sample rate used in transmission. This sample rate is not known to the receiver, which makes this a formidable problem. The general sample rate conversion problem is thoroughly discussed and solved by one of the authors in [7]. The problem and solution are also discussed in this paper specifically in reference to the V.32 problem.

The conventional approach performs echo-cancellation and transversal equalization separately [8] as shown in Figure 1. However, this decoupled technique provides only a local minimum of the entire receiver, while any misjudgments in the echo-canceller can deteriorate the equalizer performance. There are at least two recent ideas for improving the overall performance of the system, the first being to jointly optimize the adaptive echo-canceller and transversal equalizer [6]. The overall error output from the different blocks may be used to update both, which makes them a coupled system, as shown in Figure 2. The second idea is presented in [5]; it is to separate the echo canceller into two blocks, the near end echo which is not affected by the cannel distortion and the far end echo which is affected by the same channel distortion which affects the transmitted data itself. By combining these two ideas as shown in Figure 2, the performance improvement of both the echo canceller and equalizer can be realized.

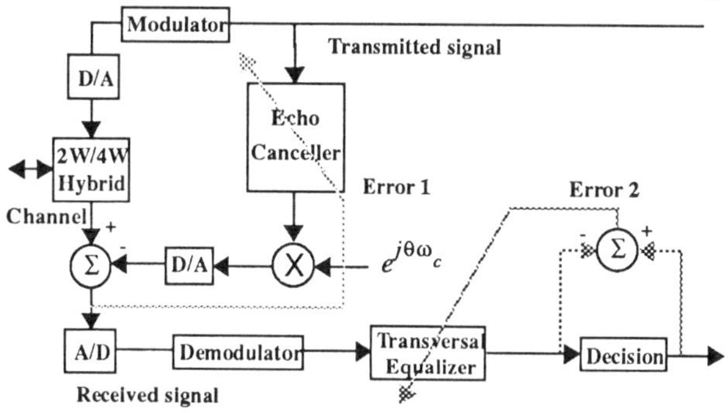

Figure 1. Conventional Decoupled Structure

It has also been shown that coupling the echo-canceller and the channel equalizer in the receiver decreases the residual echo power by a factor equal to the SNR at the equalizer input [6]. Conversely, the same level of residual echo is achieved in the coupled system with an echo canceller

1058-6393/91 $01.00 © 1991 IEEE

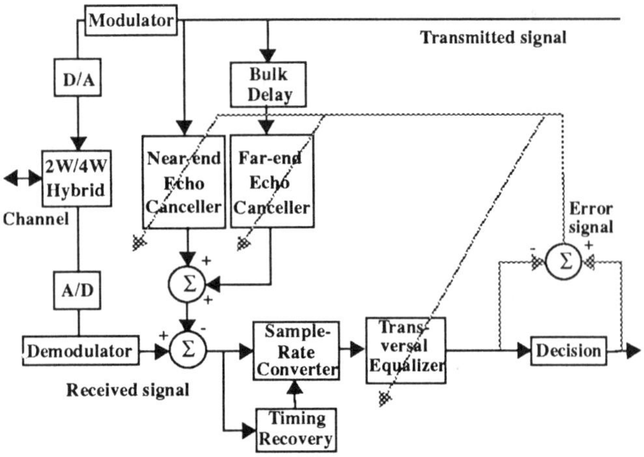

Figure 2. Coupled Structure

convergence parameter much higher than in the decoupled system. But the previous analysis and simulation results in [6] assumed that there were no bounds on the number of coefficients of the echo-canceller and an infinite resolution for coefficients and data word sizes. In a real-time implementation, there are two echoes to attenuate. The near-end echo usually spans up to 20 ms, which is 48 symbol periods at 2400 baud rate. The far-end echo starts typically after 600 ms (1440 symbols) and spans 40 ms, which is 96 symbol periods [5]. If the conventional structure in [8] is used to remove both echoes a complex echo canceller with almost 1500 taps must be implemented.

Another idea is to use synchronous timing for transmission and reception and to subtract the echo digitally from the receive signal rather than in analog, as shown in Figure 2. This also requires that the received data be converted back to it's original symbol rate to account for the asynchronous operation of the sending and receiving modem. Timing recovery of the symbol rate is performed using a nonlinear spectral line method after the signal has been demodulated and the total echo removed [14]. The period will then be compared to the received sample rate period to determine the filter coefficients needed in the sample rate convertor to restore the data to it's original rate. The sample rate convertor is a FIR filter using coefficients that shift the data in time to correspond to the correct sampling rate.

The first objective of this paper is to introduce a new scheme which removes both far-end and near-end echoes in the same echo-canceller structure, while all blocks, including the transversal equalizer create a global error output to be used for updating each adaptive coefficient block, as shown in Figure 2. A second objective of the paper is to recover the original symbol rate of the received data and convert the received data sample rate accordingly.

The new scheme is analyzed for stability characteristics and error performance. The coupled structure with sample rate conversion is implemented in real-time using the Motorola DSP56001, a general purpose DSP, to demonstrate the performance. Since the DSP56001 has 24-bits of data and coefficient precision, it is well suited for high performance communications applications such as V.32 modems [9]. Combined with the trellis encoding and the Viterbi decoding blocks [1,2], this signal conditioning block gives a virtually complete solution to the V.32 data pump implementation.

2.0 Signal Modeling

A quadrature amplitude modulated (QAM) signal is generally represented by [5,11]

$$S(t) = \sum_{n=-\infty}^{\infty} A(n)\, h(t-nT)\, e^{j\omega_c t} \qquad (1)$$

where $h(t - nT)$ is the Nyquist baseband pulse shape at the sample rate of $f_s = 1/T$, $e^{j\omega_c t}$ is the complex signal modulation components at the radian carrier frequency $\omega_c = 2\pi(1800)$, *and* $A(n)$ is the baseband complex signal which is defined by

$$A(n) = a(n) + jb(n), \qquad (2)$$

where $a(n)$ and $b(n)$ represent the real and imaginary values of the 32 QAM constellation.

Since only the real term is transmitted through a phone line, the analytic complex signal (1) has to be processed to obtain a real valued signal as

$$s(t) = Re[S(t)]. \qquad (3)$$

However, the complex signal can be reconstructed by

$$S(t) = s(t) + j\tilde{s}(t) \qquad (4)$$

where $\tilde{s}(t)$ represents the imaginary part of $S(t)$ which can be generated from $s(t)$ using the Hilbert transform [11].

The received signal $x(t)$ shown in Figure 2 is sampled to discrete signal $x(m)$, where m is the sampled signal index, at 7200 Hz which is three times larger than the baseband rate 2400 Hz. Then the signal is demodulated and decimated down to baseband $x(n)$; all remaining processing is performed at the lower baseband rate.

To reconstruct the QAM signal and to decode the original data which has been received, the real valued signal $x(n)$ has to form a complex signal $X(n)$ for further processes, such as echo-cancellation, channel equalization, and Viterbi decoding. Using the discrete-time domain Hilbert transform [11] the complex received signal can be obtained as

$$X(n) = x(n) + j\tilde{x}(n). \qquad (5)$$

By defining $B(n)$ as the incoming baseband signal from the far-end modem, the received signal $X(n)$ can be expressed as

$$X(n) = \sum_i B(n-i)\,F(i) + \sum_i A(n-i)\,G(i) + N(n) \qquad (6)$$

where $F(n)$ and $G(n)$ are the transmission channel impulse response and the echo path impulse response, respectively, while $N(n)$ denotes the additive random (Gaussian and zero-mean) noise. Thus, the adaptive echo-canceller (near-end and far-end) and the adaptive equalizer will try to estimate the actual echo response $G(n)$ and transmission channel response $F(n)$, respectively.

2.1 Adaptive echo cancellation

Adaptive signal processing with an echo cancellation structure has a variety of usages in *telecommunication* applications due to *multi-path,* and *impedance mismatches* in communication channels [10]. There is generally an analog hybrid circuit in the telephone line, which performs 2-wire/4-wire interfaces. The impedance mismatch in the hybrid circuit creates an echo signal which is the replica of transmitted signal attenuated and time-delayed. Since another hybrid circuit is implemented at the far-end connection, there are two echoes which must be cancelled, near-end echo and far-end echo, to achieve full-duplex communications.

It has been shown that the most computationally intensive algorithm in the V.32 modem is the echo cancellation due to its complexity and echo-time span [9]. It is therefore of critical importance to implement it efficiently to achieve real time operation. The Complex Least Means Square (CLMS) algorithm is used to update a Complex Adaptive Filter (CAF), and has been described by the authors in [12]. To attenuate or cancel the echoes, the echo canceller has to estimate the characteristics of the echo path, then generate a replica of the echo, which is subtracted from the received signal. Since the echo path is usually unknown prior to training and time-varying, *adaptive filtering* techniques are the only solution to obtain a good echo replica.

In an ideal situation the echo-canceller can have an infinite number of coefficients, however, to cover the near-end and far-end echoes the time window required is about 600 ms. In other words, 1440-tap echo canceller is required to accommodate up to 600 ms of echoes. This number of coefficients creates such problems as misadjustment error [13], and arithmetic precision error, not to mention the excessive number of instruction cycles it would use. A new structure has been introduced to reduce computational and data storage requirements, and to achieve better numerical performance [5]. Two sets of echo cancellers are used; a near-end echo canceller to span 20 ms or 48 taps and a far-end echo canceller to span 40 ms or 96 taps with a bulk delay which spans up to 600 ms or 1440 symbol periods (taps). The echo response part of the equation in (6) can be approximated and can be divided into two parts as

$$\sum_i A(n-i)G(i) = \sum_{i=0}^{M-1} A(n-i)G(i) + \sum_{i=0}^{N-1} A(n-i-\Delta)G(i-\Delta) \qquad (7)$$

where Δ denotes the bulk delay, and the first part and the second part of (7) represent a near-end echo canceller and far-end echo canceller, respectively, as shown in Figure 2.

2.2 Timing Recovery

Simple timing recovery is performed to recover the period of the transmitted symbols. A nonlinear spectral line method is used. A method of timing recovery which is well suited to this application since the data has been scrambled, and is covered extensively in [14].

Once the period is recovered, the difference between the sampling periods used in the receiving and the sending modems is used to find out the exact time the data should have been sampled. This time value will be used in the sample rate conversion.

2.3 Sample Rate Conversion

2.31 Fractionally Interpolated Data

In this section an interpolation method to create a value between two sampled data points, which can be viewed as fractionally sampled data, is discussed. In order to derive solutions for the *discrete-time domain* interpolation, a mathematical derivation in the *analog domain* is explained followed by *discrete-time domain* equations.

Let $s(t)$ and $s(t-\tau)$ be an analog input signal and its interpolated version, respectively. The Fourier transform (FT) of $s(t)$ is defined as [15]

$$S(\omega) = \int_{-\infty}^{\infty} s(t)e^{-j\omega t}dt \qquad (8)$$

where ω is the radian frequency assumed to satisfy $0 \le |\omega| \le \pi$. The FT of $s(t-\tau)$ can be obtained by

$$FT[s(t-\tau)] \triangleq \int_{-\infty}^{\infty} s(t-\tau)e^{-j\omega t}dt = e^{-j\omega\tau}S(\omega). \qquad (9)$$

Since multiplication in the *frequency domain* is viewed as a convolution operation in the *time domain*, the convolution theorem shows that [15]

$$s(t-\tau) = IFT[e^{-j\omega\tau}] \otimes s(t) \qquad (10)$$

where IFT and $\otimes$ denote the inverse Fourier transform and convolution operator, respectively. Using the IFT definition, $IFT[e^{-j\omega\tau}]$ in (10) can be rewritten as

$$IFT[e^{-j\omega\tau}] \triangleq \frac{1}{2\pi}\int_{-\infty}^{\infty} e^{-j\omega\tau}e^{j\omega t}d\omega = sinc(t-\tau) \qquad (11)$$

where

$$sinc(x) \triangleq \frac{\sin\pi x}{\pi x}. \qquad (12)$$

Thus, the interpolated signal in the *analog-domain* can be found using the convolution operation with two obtainable signals as follows

$$s(t-\tau) = \int_{-\infty}^{\infty} sinc(\upsilon-\tau)s(t-\upsilon)d\upsilon. \qquad (13)$$

Derivation from (8) through (13) can be applied to discrete-time domain signals. When signal $s(t)$ is sampled at the sampling frequency f_s, the discrete signal $s(nT)$ can be obtained and defined as

$$s(t)|_{t=nT} \triangleq s(nT) \qquad (14)$$

where the sampling period $T=1/f_s$. Using the discrete Fourier transform (DFT) representation, (11) can be rewritten as

$$e^{-i\omega\tau} \triangleq DFT[sinc(nT-\tau)] = \sum_{n=-\infty}^{\infty} sinc(nT-\tau)e^{-i\omega nT} \qquad (15)$$

For a simple mathematical derivation purpose, let T=1. When τ is a non-integer (fractional) number, the discrete version of $s(t-\tau)$ in (13) can be obtained by

$$s(n-\tau) = \sum_{m=-\infty}^{\infty} sinc(m-\tau)s(n-m). \qquad (16)$$

Figure 3 shows the discrete (sampled) signal $s(n)$ and a fractionally interpolated value $s(n-\tau)$ at time index n. Using frequency domain interpretation, the delayed signal $s(n-\tau)$ in (15) can be viewed as passing the input signal $s(n)$ through a time-invariant filter whose transfer function is

$$H_d(\omega) = DFT[sinc(m-\tau)]. \qquad (17)$$

Taking DFT of (16) with the definition in (17), we have

$$DFT[s(n-\tau)] = H_d(\omega)S(\omega). \qquad (18)$$

Thus, $s(n-\tau)$ can be exactly realized by the filter which

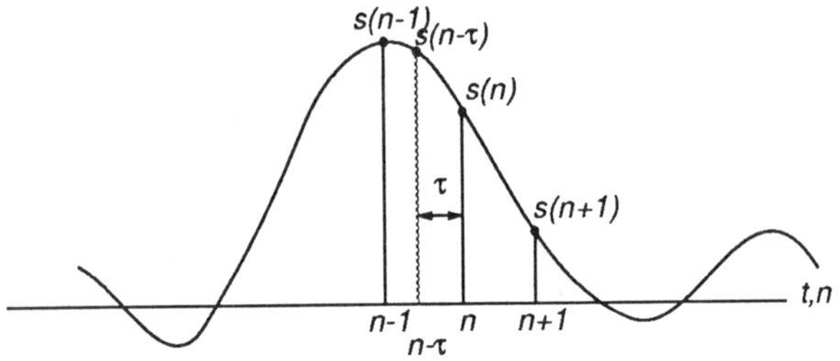

Figure 3. Discrete signal and its delayed value

has an infinite number of FIR filter coefficients whose values are the sinc function.

Clearly, it is not practical to estimate an infinite number of coefficients. However, since the function $sinc\,(m-\tau)$ in (16) approaches zero as $|m|$ increases, and the maximum delay value τ is up to one sample delay, the series summation in (16) can be truncated at a predetermined finite number with predictable performance degradation. Thus, (16) can be written to

$$\hat{s}\,(n-\tau) = \sum_{m=-M}^{M} sinc\,(m-\tau)\,s\,(n-m) \qquad (19)$$

where $\hat{s}\,(n)$ is an approximation through the truncation and M is a design parameter which can be chosen by optimization of the performance criteria [16].

2.32 Time-varying delay function estimation

The above concepts for the constant delay function τ can be extended to the case where the delay τ is a function of the time index n. In other words, the value of τ varies sample-by-sample (time-varying). This is achieved via a bank of time-invariant filters. When the individual set of filter coefficients is formed as a vector, the filter bank then can be formulated as a matrix whose rows correspond to the time varying interpolation value $\tau(n)$. If the time-varying delay function is $\tau(n)$ where n is the index for the new sample rate, then the nth filter has the two-dimensional transfer function $H_d(\omega, n)$ as

$$H_d(\omega, n) \triangleq e^{-j\omega\tau(n)} \quad . \qquad (20)$$

Thus, the individual impulse response function vector can be found by

$$h_d(n, m) = IFT\,[e^{-j\omega\tau(n)}] = sinc\,[m-\tau(n)]\,. \qquad (21)$$

Redefining n to be an output time index and k to be an input time index, the output of the nth filter at time k is given by

$$s_d(k, n) = \sum_{i=-\infty}^{\infty} h_d(n, i)\,s\,(k-i)$$

$$= \sum_{i=-\infty}^{\infty} sinc\,[i-\tau(n)]\,s\,(k-i) = s\,[k-\tau(n)]\,. \qquad (22)$$

The timing relationship between k and n will be discussed in the next section. Since (22) cannot be possible due to the infinite number of coefficients, a similar approach to that described in Section 2.31 can be applied. The converted signal $y(n)$ which is the delayed version of input sample values can be expressed by the approximation (truncation) with a window function $w(m)$ as

$$y\,(n) \triangleq \hat{s}_d\,(k, n) = \hat{s}\,[k-\tau(n)]$$

$$= \sum_{m=-M}^{M} h\,(n, m)\,s\,(k-n) \qquad (23\text{-a})$$

where

$$h\,(n, m) = w\,(m)\,sinc\,[m-\tau(n)] \quad for\ n = 0, 1, \ldots, N\text{-}1 \qquad (23\text{-b})$$

From (23), the time-varying filter coefficients $h(n,m)$ can be interpreted as a weighting function whose maximum value occurs at $m = \tau(n)$. Figure 8 shows the matrix formation of the bank of coefficient vectors where $H(n,m)$ is a N by $2M+1$ matrix and N denotes the total number of different $\tau(n)$ values. Therefore, $y(n)$ in (23) can be viewed as being the output of the time-varying filter which is generating a sample-rate converted (new sample rate) signal.

2.33 Aliasing problems in sampling-rate converters

It is important to note that when the input sample rate is larger than the output sample rate, a large portion of the higher frequency components will be aliased back to lower frequency range at the new sampling rate. This phenomena is due to the decimation (sampling rate reduction). Thus, a band-limiting lowpass filtering must be employed with the conversion process [17,18].

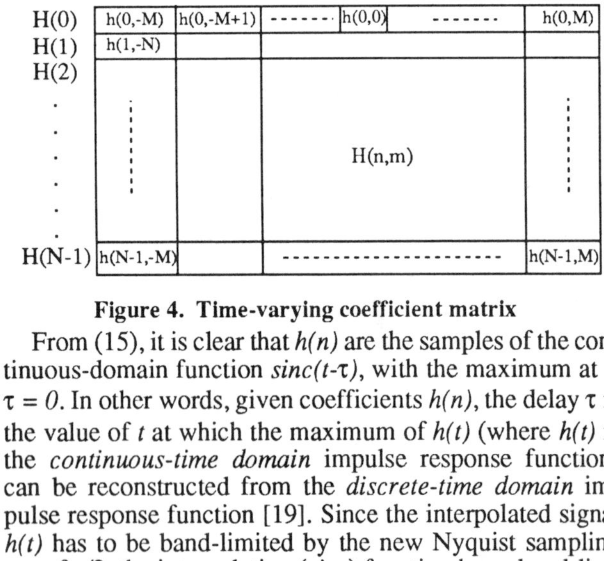

Figure 4. Time-varying coefficient matrix

From (15), it is clear that $h(n)$ are the samples of the continuous-domain function $sinc(t\text{-}\tau)$, with the maximum at $t\text{-}\tau = 0$. In other words, given coefficients $h(n)$, the delay τ is the value of t at which the maximum of $h(t)$ (where $h(t)$ is the *continuous-time domain* impulse response function) can be reconstructed from the *discrete-time domain* impulse response function [19]. Since the interpolated signal $h(t)$ has to be band-limited by the new Nyquist sampling rate, $f_{s2}/2$, the interpolation (*sinc*) function has a band-limiting factor β as [20]

$$h\,(t) = \sum_{m=-\infty}^{\infty} h\,(m)\,\frac{\sin\pi\beta\,(t-m)}{\pi\beta\,(t-m)} \qquad (24)$$

where β is the highest allowed frequency in the new sampling rate, normally less than or equal to $f_{s2}/2$.

Applying this result with discrete-time domain interpretation, we have

$$h\,(n) = w\,(n)\,\frac{\sin\pi\gamma\,(n-\tau)}{\pi\gamma\,(n-\tau)} = w\,(n)\,sinc\,[\gamma\,(n-\tau)] \qquad (25)$$

where $\gamma = f_{s2}/f_{s1}$. However, when $f_{s2} > f_{s1}$, $\gamma = 1$ has to be applied to (25), since there is no need for band-limiting.

2.34 Real-time operation considerations

Consider when the algorithm in (23) is applied in a real-time processor. When the output must be generated at the output sample rate f_{s2}, the value of τ and current data pointer for the input buffer must be determined. Figure 5 shows an example of the input and output timing relationship when the output sampling rate is greater than the input sampling rate. When *y(0)* is ready to output, *s(0)* can be copied to the output buffer because the timing difference between input and output is zero. However, when *s(1)* is ready to output, the value τ_1 has to be estimated, and then the processor has to compute the *2M+1* coefficient FIR filter in (23) whose coefficients correspond to the value of τ_1. Also, note that the same input data

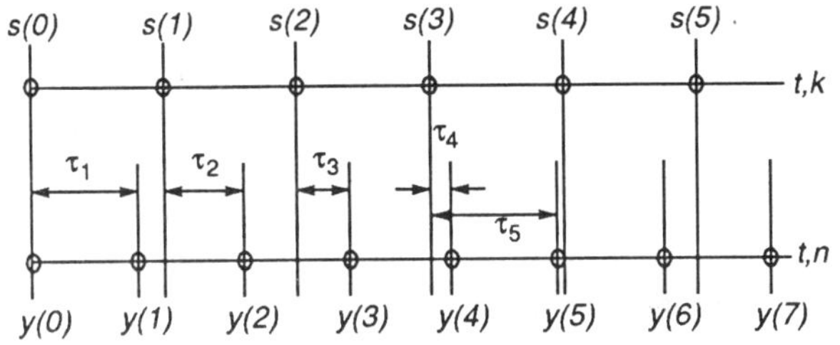

Figure 5. Input and output timing relationship

starting from *s(3)* are used to generate *y(4)* and *y(5)* values with different time delay parameters τ_4 and τ_5 as shown in Figure 5. Thus, the estimation of the time variant delay function and the starting point of the input data vector are very important for this real-time implementation.

Since the current delay value τ_n for the output sample index n is always referenced to the current index k of input data sample, the τ_n can be defined as

$$\tau_n = nT_{s_2} - kT_{s_1} \qquad (26)$$

where $Ts_1 = 1/fs_1$ and $Ts_2 = 1/fs_2$. Thus, τ_n is always a non-negative number with the following range

$$0 \le \tau_n < T_{s_1} = \frac{1}{f_{s_1}}. \qquad (27)$$

2.4 Adaptive Equalizer

In the conventional decoupled system, the Transversal Equalizer (TE) input is the received data symbols after the estimated echo has been removed. The primary purpose of the TE is to eliminate the adjacent symbol interference which is caused by the channel distortion [14,21]. Transmitting a training signal can help to determine this distortion. This is done by transmitting a signal with a known frequency response, *T(f)*. The received signal, *R(f)* must then be the transmitted signal distorted by the channel response (plus some noise which will be ignored for this explanation), *C(f)*:

$$R(f) = T(f) \cdot C(f). \qquad (28)$$

This means that the inverse of the channel response can be estimated during training. Then, if the coefficients of the TE have the frequency response of *1/C(f)*, then the channel distortion can be removed from the input data sequence by filtering with the TE. Since the distortion on the channel can be changing with time, the TE coefficients need to be updated using the same error signal and the method as previously discussed for the echo canceller. The TE uses the same error output for both the coupled and decoupled cases as shown in Figure 1 and Figure 2.

4.0 Simulated Data Results

The error signal output during a training sequence where the NEC is trained and then the FEC is turned on and trained and finally the equalizer is turned on and trained is shown in Figure 6. The real value of the error signal has been plotted as well as magnitude in dB. Rapid convergence for each function is shown as well as 80 dB of echo reduction. Figure 7 shows the sample rate recovery of a tone at 1500 Hz which has been received and sampled at 7200 Hz when the original sample rate was 7216 Hz.

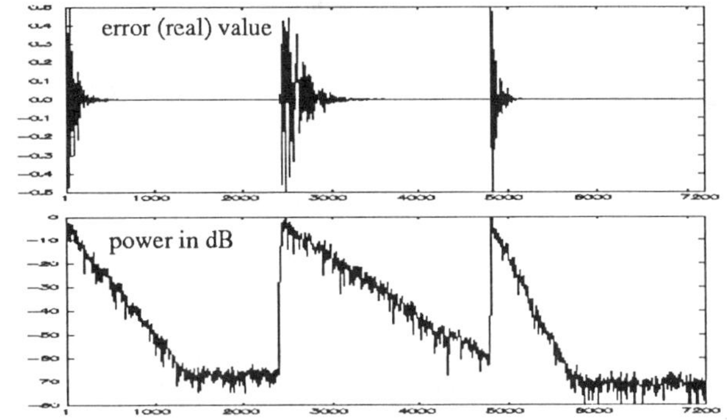

Figure 6. Three-stage training error out-

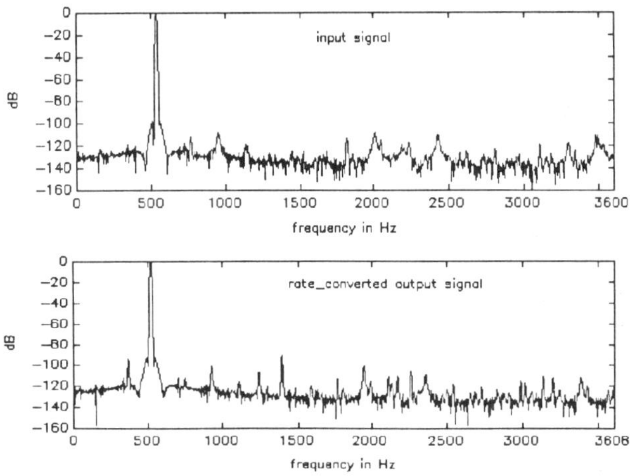

Figure 7. Spectra of the signals when 2M+1=7

5.0 Conclusion

The new structure achieved the performance improvements expected. The echo data symbols are uncorrelated with respect to the received data symbols in a full duplex V.32 system, so using an error term generated after the decision on the symbol has been made, generates more accurate NEC, and FEC coefficients. An important conclusion is that, although the new structure has better performance in normal data transmission, the structure must be decoupled for training. The global error does not converge during training using the coupled system, however, if the structure is decoupled and the error at each block is used to update the coefficients, the error converges rapidly and provides for adequate echo reduction as shown in the results. It has also been shown that it is possible to accurately recover asynchronous data that has been sampled synchronously, which is necessary in a completely digital modem solution.

Without this, it would be impossible to implement a V.32 modem with a single analog to digital and digital to analog convertor. Although this is most important in the V.32 application, it is a breakthrough for all digital receiver systems, and will pave the road for many completely digital receiver designs.

REFERENCES

[1] D. Messer and S. Park, "A simple implementation of the Viterbi algorithm on the Motorola DSP56001," *Proc. International Mobile Satellite Conference*, pp. 205-209, Ottawa, Canada, June, 17-20, 1990.

[2] D. Messer, "Convolutional Encoding and Viterbi Decoding using the DSP56001 with a V.32 Modem Trellis Example," *Motorola DSP Operations Application Report*, ADPR6/D, 1989.

[3] S. Park and G. Hillman, "On acoustic echo-cancellation implementation with multiple cascadable adaptive FIR filter chip," *Proc. Int. Conf. Acoust., Speech, Signal Processing*, pp. 952-955, Glascow, Scotland, May 1989.

[4] S. Park and D. Messer, "Acoustic echo-cancellation for full-duplex voice transmission on fading channels," *Proc. International Mobile Satellite Conference*, pp. 188-193, Ottawa, Canada, June, 17-20, 1990.

[5] J. M. Cioffi, "A fast echo canceller initialization method for the CCITT V.32 modem," *IEEE Trans. on Commun.*, vol. 38, no. 5, pp. 629-638, May 1990.

[6] S. Marcos and O. Macchi, "Joint adaptive echo cancellation and channel equalization for data transmission," *J. of Signal Processing*, vol. 20, no. 1, May 1990.

[7] S. Park, "Low Cost Sample Rate Converters," *Proc. of the National Association of Broadcasters (NAB) Engineering Conference*, pp. 10 - 19, Las Vegas, Nevada, April 14 - 18, 1991.

[8] B. Mulgrew and C. F. N. Cowan, "*Adaptive Filters and Equalizers*," Kluwer Academic Pub., 1988.

[9] A. Fagen, et al., "Single DSP Implementation of a High Speed Echo Cancelling Modem Employing Trellis Coding," Proc. Of the Intnl. ESA Workshop on DSP Techniques Applied to Space Communications, Noordwijk, November 1988.

[10]K.H. Mueller, "New Digital Echo Canceller for Two-wire Full-duplex Data Transmission," *IEEE Trans. Commun.*, Vol. COM-24, pp. 956-962, Sept. 1976.

[11]J-D. Wang and J-J Werner, "Performance analysis of an echo-cancellation arrangement that compensates for frequency offset in the far echo," *IEEE Trans. on Communications*, vol. 36, no. 3, pp. 364-372, March 1988.

[12]S. Park and D. Messer "Analysis and Digital Implementation of Jointly Adapting Echo Canceller, Transversal Equalizer and Decision Feedback Equalizer for V.32 Modems," *Proc. of the International Conference on Signal Processing*, pp 1965-1968, Toronto, Ontario, Canada, May 14 - 17, 1991.

[13]B. Widrow and S.D. Stearns, *Adaptive Signal Processing*, Prentice-Hall, Englewood Cliffs, NJ, 1985.

[14] E.A. Lee and D.G. Messershmitt, *Digital Communication*, Kluwer Academic Publishers, 1988.

[15] A.V. Openheim and R.W. Schafer, *"Digital Signal Processing"*, Prentice-Hall, Englewood Cliffs, NJ, 1975.

[16]D. Youn, N. Ahmed and G. C. Carter, "On using the LMS algorithm for time delay estimation," *IEEE Trans. Acoust., Speech and Signal Proc.*, Vol. ASSP-30, No. 5, pp. 788-801, Oct. 1982.

[17]S. Park, "A real-time implementation of half-band filters to obtain 18 - 20 bit resolution from the DSP56ADC16 16-bit sigma-delta A/D converter," *Proc. of the International Conference on Acoustics, Speech and Signal Processing*, pp. 989-992, Albuquerque, NM, May 23-25, 1990.

[18]S. Park, "Multistage halfband filter design for improving effective resolution from Sigma-Delta analog-to-digital converters," *Preprint of the 89th Audio Engineering Society (AES) Convention*, Preprint #3008, Session #G-5, Los Angeles, CA, Sept. 21-25, 1990.

[19]G.M. Jenkins and D.G. Watts, *Spectral Analysis and Its Applications*, San Francisco, Holden-Day, 1968.

[20]F. A. Reed, *et al.*, "Time Delay Estimation Using the LMS Adaptive Filter - Static Behavior," *IEEE Trans. on Acoust., Speech, Signal Proc.*, vol. ASSP-29, no. 3, pp 561 - 571, Jun1 1981.

[21]S. Qureshi, "Adaptive equalization," *IEEE Communications Magazine*, pp. 9-16, March 1982.

A New Algorithm for Automatic Beamforming

Z. Ding
Department of Electrical Engineering
Auburn University
Auburn, AL 36849

Abstract

A new blind beamforming scheme for a uniform antenna array is presented to automatically capture a narrowband signal with nonzero kurtoses. It simplifies conventional methods that generally estimate the direction of signal arrival first and then design the beampattern accordingly to retrieve the signal of interest. By maximizing the average kurtosis of the output signal, the blind beamformer directly extracts the signal of interest from Gaussian noise and maximizes the resulting SNR. Moreover, minimizing the average kurtosis of the output signal also enables us to identify angles of signal arrival.

1 Introduction

In digital array signal processing, one important problem is to find the direction of arrival of incident plane waves in the presence of additive sensor or background noise. The problem of finding the direction of arrival (DOA) of signals of interest (SOI) has been widely studied. Often, the estimated DOA is used to construct an antenna array such that the SOI can be detected with maximum signal to noise ratio (SNR) in a noisy environment. Many schemes designed to estimate DOA, such as MUSIC [1] and minimum variance [2], require estimating the number of present SOIs and the auto-correlation matrix of the background noise if the noise is not spatially white. The estimated DOA can be very sensitive to differences between the actual auto-correlation matrix and the known auto-correlation matrix.

The idea of blind beamforming is to automatically shape the beampattern to achieve the best reception of SOI in the presence of (colored) Gaussian noise without explicit knowledge of the noise correlation. Based on the observation of the sensor output, a blind beamformer attempts to optimize the gain vector associated with the sensor array. Such schemes are highly desirable in situations where explicit knowledge of the (normal) noise correlation and number of SOIs is difficult to obtain.

Agee [3] proposed the use of the constant modulus algorithm (CMA) in blind adaptive beamforming. It was shown that with the use of CMA, a certain class of non-Gaussian signals with kurtosis less than zero (leptikurtic) can be automatically retrieved by the beamformer through the application of the CMA. However, for signals that have kurtoses greater than zero (platykurtic), CMA causes the beamformer to null the incident signal by capturing the additive noise instead. In short, CMA cannot control the capture of signal or noise.

In practical applications, incoming signals are not necessarily leptikurtic and thus it is not certain, with the use of CMA, whether the adaptive beamformer can receive the signal with maximum SNR or receive noise only. To overcome this uncertainty, this paper presents a new blind beamformer that can automatically receive with maximum SNR not only leptikurtic signals but also platykurtic signals as well. Our method is to maximize the absolute value of the signal kurtosis at the beamformer output. This is conceptually equivalent to minimizing the effect of Gaussian noises in the beamformer output since all Gaussian noises have zero kurtosis. By maximizing the kurtosis of the output signal, the beamformer attempts to focus on the SOI and thereby maximize the SNR at its output. On the other hand, by minimizing the kurtosis of the beamformer output, we can find a parameter vector nearly orthogonal to all signal steering vectors, hence making it possible to estimate the DOA of incoming signals.

The basic setup of blind beamforming systems is presented in §2 and the optimization scheme for iterative update of the beamformer parameter vector is also described and analyzed. It is shown that the optimization strategy is consistent with the maximization of SNR for the incoming SOI. Numerical implementation is outlined in §3. Furthermore, blind beamforming under multiple SOIs is presented in §4 and a new method for estimating DOA is also introduced. Finally, simulations are shown in §5 to confirm the effectiveness of the new blind beamforming system in its ability to automatically capture the incoming signals and to identify the arriving angle of the incoming signals.

1058-6393/91 $01.00 © 1991 IEEE

2 Blind Beamforming Systems

2.1 Basic Setups

Typically in blind adaptive beamforming system as shown in Figure 1, a plane wave, narrowband signal of interest $s(t)$ with wavelength λ arrives at an N-element antenna array in the presence of background interference and noise $\mathbf{n}(t)$. The received signal vector of length N at the array can be written as

$$\mathbf{x}(t) = \mathbf{d}s(t) + \mathbf{n}(t), \qquad (2.1)$$

where the vector

$$\mathbf{d} = [1 \;\; e^{-j\frac{2\pi}{\lambda}r\sin\theta} \;\; e^{-j\frac{2\pi}{\lambda}2r\sin\theta} \;\; \ldots \;\; e^{-j\frac{2\pi}{\lambda}Nr\sin\theta}\;]'$$

is the *steering vector*. The input signal $s(t)$ and the background interference $\mathbf{n}(t)$ are assumed to be independent and of zero mean. The background noise $\mathbf{n}(t)$ is assumed to be stationary, nonremovable, and to have a complex Gaussian distribution with zero mean and autocorrelation matrix

$$R_n = E\{\mathbf{n}(t)\mathbf{n}^H(t)\}.$$

The notation $\mathbf{n}^H$ represents the conjugate transpose of $\mathbf{n}$.

The problem of adaptive beamforming is to design an optimum weight vector

$$\mathbf{w} = [w_1 \;\; w_2 \;\; \ldots \;\; w_N\;]$$

such that the signal to noise ratio in the beamformer output $y(t) = \mathbf{w}^H\mathbf{x}(t)$ is maximized. To avoid getting the trivial all-zero output, the weight vector is required to satisfy certain linear constraint(s) such that $\mathbf{w} \neq 0$. A simple linear constraint would be to fix $w_1 = 1$.

2.2 Optimization Criterion

The adaptation criterion we shall use in optimizing the beamformer weight vector is to maximize the following cost function

$$J(\mathbf{w}) = \frac{|\mathrm{CUM}_4[y(t)]|}{E^2\{|y(t)|^2\}}, \quad y(t) = \mathbf{w}^H\mathbf{x}(t), \qquad (2.2)$$

where

$$\mathrm{CUM}_4[y(t)] = \left|E\{|y(t)|^4\} - 2E^2\{|y(t)|^2\} - |Ey^2(t)|^2\right|$$

is the 4-th order cumulant of the complex signal $y(t)$.

Our next task is to show that maximizing $J(\mathbf{w})$ achieves a unique maximum that corresponds to maximizing the SNR of the beamformer output. First, notice that the beamformer output can be written as

$$y(t) = As(t) + v(t) \qquad (2.3)$$

where $A = \mathbf{w}^H\mathbf{d}$ is the complex variable gain and the interference $v(t) = \mathbf{w}^H\mathbf{n}(t)$ is still Gaussian and independent of the signal $s(t)$. Thus,

$$\begin{aligned} E\{|y(t)|^4\} &= |A|^4E\{|s(t)|^4\} + E\{|v(t)|^4\} \\ &+4|A|^2E\{|s(t)|^2\}E\{|v(t)|^2\} \qquad (2.4) \\ &+A^2E[s^2(t)]E[v^*(t)]^2 + (A^*)^2E[s^*(t)]^2E[v(t)]^2 \end{aligned}$$

and

$$\begin{aligned} E\{|y(t)|^2\} &= |A|^2E\{|s(t)|^2\} + E\{|v(t)|^2\}, \qquad (2.5) \\ E[y^2(t)] &= A^2E[s^2(t)] + E[v^2(t)]. \qquad (2.6) \end{aligned}$$

As a result,

$$\begin{aligned} J(\mathbf{w}) &= \frac{|E\{|y(t)|^4\} - 2E^2\{|y(t)|^2\} - |E[y^2(t)]|^2|}{E\{|y(t)|^2\}|} \\ &= \frac{\left||A|^4\mathrm{CUM}_4[s(t)] + \mathrm{CUM}_4[v(t)]\right|}{\left[|A|^2E\{|s(t)|^2\} + E\{|v(t)|^2\}\right]^2}. \qquad (2.7) \end{aligned}$$

We now simplify the above expression by introducing the following notations. The signal to noise ratio (SNR) can be written as

$$\gamma = \frac{|A|^2E\{|s(t)|^2\}}{E\{|v(t)|^2\}},$$

and the kurtosis of a signal $s(t)$ can be written as [4]

$$K_s = \frac{\mathrm{CUM}_4[s(t)]}{E\{|s(t)|^2\}}.$$

As a result of these two notations, we get

$$J(\mathbf{w}) = \frac{|K_s\gamma^2 + K_v|}{(1+\gamma)^2}, \qquad (2.8)$$

where K_s and K_v are kurtoses of signal $s(t)$ and noise $v(t)$, respectively. Since Gaussian distribution results in zero kurtosis, we consequently have $K_v = 0$. Thus the maximization of $J(\mathbf{w})$ with respect to $\mathbf{w}$ becomes the maximization of

$$J(\mathbf{w}) = \frac{|K_s|\gamma^2}{(1+\gamma)^2} \qquad (2.9)$$

as a function of γ.

To find maxima of the function $J(\mathbf{w})$, take

$$\frac{\partial J(\mathbf{w})}{\partial\gamma} = \frac{2\gamma|K_s|}{(1+\gamma)^3}. \qquad (2.10)$$

The stationary point

$$\gamma = 0$$

is simply a local minimum since a slight change in γ could increase the value of J. It follows then from the positiveness of the derivative (2.10) that unless there are other stationary points in $\mathbf{w}$ space, maximization of $J(\mathbf{w})$ will globally maximize the signal to noise ratio γ. From the chain rule,

$$\frac{\partial J(\mathbf{w})}{\partial \mathbf{w}} = \frac{\partial J(\mathbf{w})}{\partial \gamma} \cdot \frac{\partial \gamma}{\partial \mathbf{w}}. \quad (2.11)$$

Suppose R_n representing the autocorrelation matrix of the input noise $n(t)$ is full rank. Then we have

$$\gamma = \frac{\mathbf{w}^H \mathbf{d}\mathbf{d}^H \mathbf{w} E\{|s(t)|^2\}}{\mathbf{w}^H R_n \mathbf{w}}. \quad (2.12)$$

Thus

$$\frac{\partial J(\mathbf{w})}{\partial \mathbf{w}} = \frac{\partial J(\mathbf{w})}{\partial \gamma} \cdot \frac{\partial \gamma}{\partial \mathbf{w}} = \frac{\partial J(\mathbf{w})}{\partial \gamma} \cdot \frac{2E\{|s(t)|^2\}}{\left(\mathbf{w}^H R_n \mathbf{w}\right)^2}$$
$$\bullet\left[\mathbf{d}(\mathbf{d}^H \mathbf{w})(\mathbf{w}^H R_n \mathbf{w}) - R_n \mathbf{w}(\mathbf{w}^H \mathbf{d})(\mathbf{d}^H \mathbf{w})\right]$$

As a result, the stationary points for $J(\mathbf{w})$ are located at

$$(1). \quad \mathbf{w} = 0, \quad (2.13)$$
$$(2). \quad \mathbf{d}^H \mathbf{w} = 0, \quad (2.14)$$
$$(3). \quad \mathbf{d}(\mathbf{w}^H R_n \mathbf{w}) = (\mathbf{w}^H \mathbf{d}) R_n \mathbf{w}. \quad (2.15)$$

Solution (1) is not reachable due to our parameter constraint. Solution (2) result in $\gamma = 0$, which is clearly not a local maximum. The only other stationary point is at

$$(\mathbf{d}\mathbf{w}^H)(R_n \mathbf{w}) = (\mathbf{w}^H \mathbf{d})(R_n \mathbf{w}). \quad (2.16)$$

This means $(R_n \mathbf{w})$ is an eigenvector of the matrix $\mathbf{d}\mathbf{w}^H$, which is rank 1 with eigenvalue $\mathbf{w}^H \mathbf{d}$. However,

$$(\mathbf{d}\mathbf{w}^H)\mathbf{d} = \mathbf{d}(\mathbf{w}^H \mathbf{d}) = (\mathbf{w}^H \mathbf{d})\mathbf{d}, \quad (2.17)$$

which indicates that the eigenvetor associated with eigenvalue $(\mathbf{w}^H \mathbf{d})$ is in fact $\mathbf{d}$. Thus the solution $\mathbf{w}$ to (3) is

$$R_n \mathbf{w} = \alpha \mathbf{d}, \quad \alpha \in \mathbb{R}, \ \alpha \neq 0 \quad (2.18)$$

or

$$\mathbf{w} = \alpha R_n^{-1} \mathbf{d}, \quad (2.19)$$

at which the SNR becomes,

$$\gamma = \frac{\mathbf{d}^H R_n^{-1} \mathbf{d}\mathbf{d}^H R_n^{-1} \mathbf{d} \alpha^2 E\{|s(t)|^2\}}{\alpha^2 \mathbf{d}^H R_n^{-1} \mathbf{d}} = \mathbf{d}^H R_n^{-1} \mathbf{d} E\{|s(t)|^2\}.$$

This happens to be the global maximum of γ. Hence, we have proved that the sole local maximum of $J(\mathbf{w})$ is at

$$\mathbf{w} = \alpha R_n^{-1} \mathbf{d}, \quad \text{or} \quad \gamma = \gamma_{max} \triangleq \mathbf{d}^H R_n^{-1} \mathbf{d} E\{|s(t)|^2\}.$$

We have hence proved that for single SOI under Gaussian noise, maximization of our cost function achieves the global maximization of SNR for the signal of interest. Our optimization criterion is thus well designed for achieving the most effective beampattern, so long as $K_s \neq 0$.

3 Numerical Method

We have just shown that the maximization of the cost function (2.2) can lead to optimum reception of the signal of interest in the presence of Gaussian noises. Here we describe how the optimization scheme can be implemented in practice. Two prominent methods dominate in solving the maximization or minimization problems, namely, Newton's gradient method and the Newton-Ralphson method. Here we use Newton's gradient method to maximize the cost function (2.2).

In practice, the ensemble statistics in (2.2) need to be replaced by the time averages of actual data. Such approximations are justified if both the signal and noise are ergodic and if the averaging is taken over a reasonably long time. In order to obtain time average statistics, we take snapshots of the incoming signal vector $\mathbf{x}(t)$ at uniform intervals $t_k = kT_0$. A time average approximation of the cost $J(\mathbf{w})$ can be formed over a block of data. We search for the maximum of the cost function by iteratively updating the parameter vector from $\mathbf{w}_i$ to $\mathbf{w}_{i+1}$ using the gradient search,

$$\mathbf{w}_{i+1} = \mathbf{w}_i + \mu \frac{\partial J(\mathbf{w})}{\partial \mathbf{w}},$$

where μ is a small stepsize. The partial differentiation of J with respect to $\mathbf{w}$ is straightforward.

4 Multiple Signal Inputs

Often, the typical environment in adaptive beamforming is the presence of multiple independent signals under Gaussian noise. Let m signals be received by N sensors whose outputs are fed to m separate adaptive beamformers. Each beamformer is attempting to capture, with maximum SNR, a signal of interest.

4.1 Signal Capturing

In adaptive beamforming system of multiple signals, the input signal can be written as a sum of multiple narrowband signals plus Gaussian noise

$$\mathbf{x}(t) = \sum_{i=1}^{m} \mathbf{d}_i s_i(t) + \mathbf{n}(t). \quad (4.1)$$

With the weight vector $\mathbf{w}$, the beamformer output is simply

$$y(t) = \sum_{i=1}^{m} \mathbf{w}^H \mathbf{d}_i s_i(t) + \mathbf{w}^H \mathbf{n}(t)$$
$$= \sum_{i=1}^{m} A_i s_i(t) + v(t), \qquad (4.2)$$

where $v(t)$ is still Gaussian.

For multiple narrowband inputs, our optimization strategy can result in maximum SNR reception of any one of the inputs, provided that the magnitude of its kurtosis is not too much smaller than that of the rest of the signals. When the kurtosis magnitude of a given signal is too small compared with other signals, its behavior is very close to a Gaussian noise which is suppressed by our beamformer. This explains why signals with kurtosis very close to zero can be hard to capture.

Thus, when m input narrowband signals with similar kurtosis are present with Gaussian noise, our optimization of individual beamformer can eventually lead to the reception of any one of the signals with maximum SNR.

If m separate beamformers are built for the same antenna array, then each beamformer following our optimization strategy will converge to capturing one of the incoming signals with maximum achievable SNR. In other words, each beamformer will have m local maxima for the cost function, each corresponding to the maximum SNR reception of a signal of interest. It is therefore advisable to construct m beamformers and design an additional strategy so that no two beamformers converge to identical parameter vector, i.e.,

$$\mathbf{w}_i \neq \mathbf{w}_j, \quad i \neq j.$$

So long as all m beamformers converge to different vector values, their outputs will generate the optimum reception of m different signals of interest present at the sensor array. Clearly, we need to drive parameter vectors of different beamformers to obtain different stationary points in order to separate all the incoming signals. To achieve signal separation, we exploit the fact that all our signals are orthogonal to one another and apply the Gram-Schmidt orthogonalization procedure after each iteration.

4.2 Estimating the Angle of Arrival

If instead of maximizing the cost function $J(\mathbf{w})$, we elect to minimize this cost function, then the SNR will be minimized upon convergence. If $\mathbf{w}$ minimizes the SNR, then $|\mathbf{w}^H \mathbf{d}(\theta)|$ must be the smallest for the angle of signal arrival. Thus the angle of arrival estimation can be accomplished by first finding $\mathbf{w}$ that minimizes the cost function $J(\mathbf{w})$ and then find θ that maximizes the following function

$$\frac{1}{|\mathbf{w}^H \mathbf{d}(\theta)|^2}. \qquad (4.3)$$

This is very much like the MUSIC algorithm except in the simple way that $\mathbf{w}$ can be obtained.

5 Simulations

Three different simulation results are going to be presented. We consider a 16-element antenna array and a narrowband plane-wave input with carrier wavelength λ. Elements of the antenna array are spaced at $r = \lambda/6$ distant apart.

A lowpass complex Gaussian background noise is assumed for $\mathbf{n}(t)$. We further assume that the background noises at different array elements are independent of one another, each with variance $\sigma^2 = 2$. Thus, for each signal, the original SNR at each array element is 0 dB and the maximum obtainable SNR is 12dB. The initial weight vector is set at

$$\mathbf{w} = [1 \ \ 0 \ \ \ldots \ \ 0\,],$$

and the stepsize in updating the weight vector is chosen to be $\mu = 0.02$. In each case, the data sampling frequency is $f_s = 1.6$ KHz and a total of 1024 data samples are taken.

We simulate the situation when three input signals are present with angles of arrival 30°, 60°, and −30° respectively. The signals are sinusoid with frequency 100 Hz, period square wave with frequency 84 Hz, and a signal consisting of two sinusoids of equal magnitude with frequencies $f_{m1} = 145$ Hz and $f_{m2} = 229$ Hz, respectively. Three beamformers are activated and Gram-Schmidt procedure is implemented for signal separation.

Figure 2 illustrates the gradual maximization of the cost function by three beamformers (marked BF_1, BF_2, and BF_2) in 150 iterations. Figure 3 also shows the improvement of SNR for all three beamformers to near their maximum attainable level after merely 150 iterations. Apparently, each beamformer has effectively captured a particular signal and it is clear from Figure 3 that the SNR for each beamformer has approached the maximum attainable value of 12 dB.

Next we find the parameter vector $\mathbf{w}$ that minimizes the cost $J(\mathbf{w})$ when the input SNR at each element is 10dB. The estimated angles of arrival for all three signals after 1,000 iterations are shown in Figure 4. The true angles of arrival are clearly identified as 30°, 60°, and −30°, respectively.

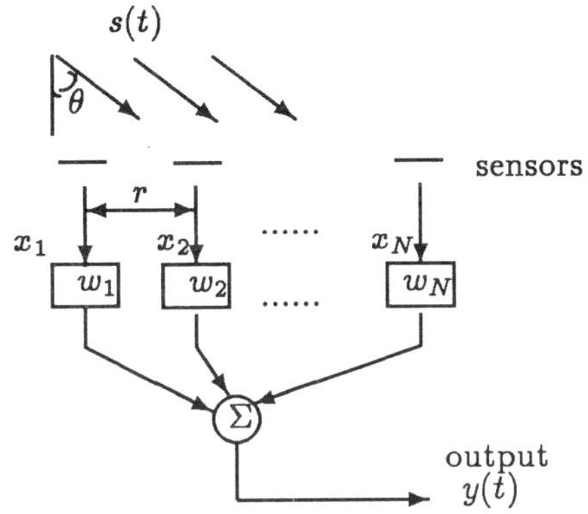

Figure 1: Adaptive beamforming system.

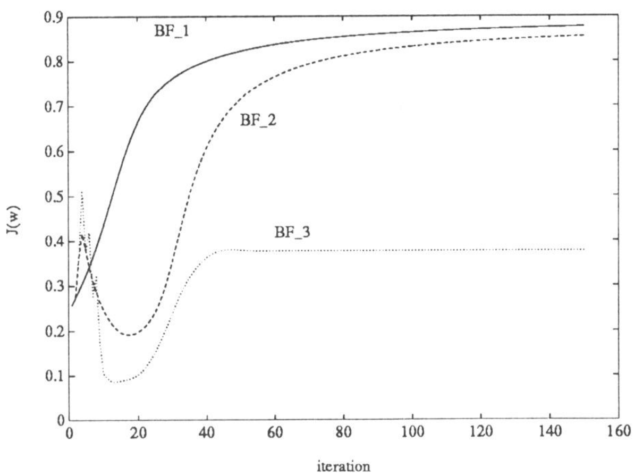

Figure 2: Maximization of the cost function.

6 Conclusions

A blind adaptive beamforming scheme is proposed for the automatic capture of non-mesokurtic signals whose kurtoses are nonzero. Our optimization scheme results in the convergence of beamformer weight vectors to receive signals of interest with maximum SNR. A Gram-Schmidt procedure is performed at the receiver in order to assure that each output port receives a different signal of interest. Furthermore, a new method to estimate the arriving angles of the incoming signals is presented.

References

[1] R. Schmidt and R. Franks. Multiple df signal processing: an experimental system. *IEEE Trans. Antennas and Propagation*, AP-34:281–291, March 1986.

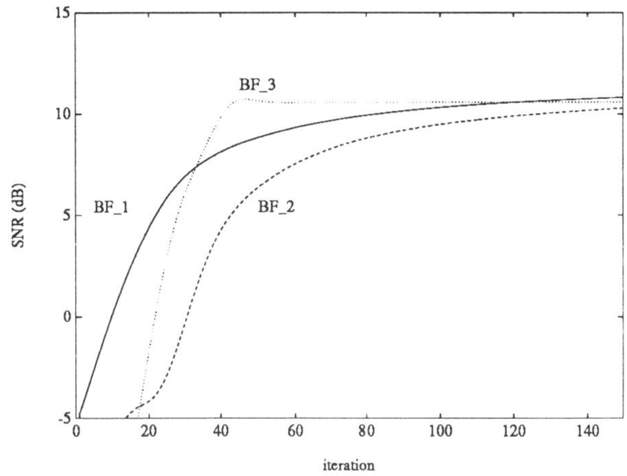

Figure 3: Maximization of the SNR for three signals.

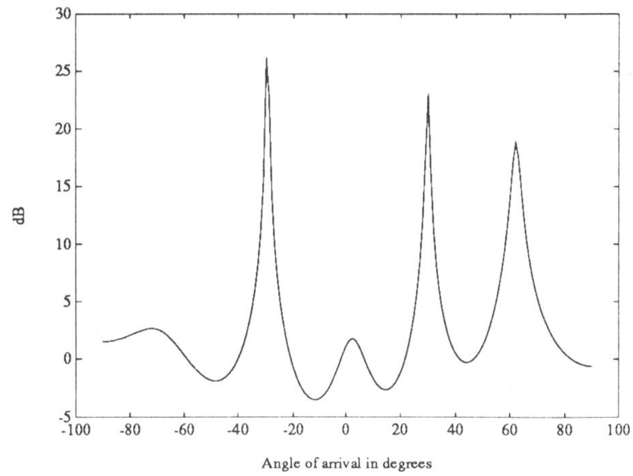

Figure 4: Estimated DOA under 10dB input SNR.

[2] D. W. Tufts and R. Kumaresan. Estimation of frequencies of multiple sinusoids: making linear prediction perform like maximum likelihood. *Proc. IEEE*, 70:975–989, 1982.

[3] B. G. Agee. Convergent behavior of modulus-restoring adaptive arrays in gaussian interference environments. In *Proc. 22nd Asilomar Conference on Signals, Systems and Computers*, Pacific Grove, CA, November 1988.

[4] A. Stuart and J. K. Ord. *Kendall's Advanced Theory of Statistics*, volume V.1: Distribution Theory. Oxford University Press, New York, 5th edition, 1987.

CRIMNO: Criterion with Memory Nonlinearity for Blind Equalization*

Yuanjie Chen†, Chrysostomos L. Nikias†, John G. Proakis‡

†Dept. of EE-Systems
University of Southern California
Los Angeles, CA 90089

‡Dept. of ECE
Northeastern University
Boston, MA 02115

Abstract

A new criterion with memory nonlinearity (CRIMNO) is introduced for blind equalization problems. The basic idea of CRIMNO is to make use of the fact that the transmitted data are statistically independent from each other. It is shown that CRIMNO may not have local minima if its weights are chosen properly, thereby guaranteeing global convergence . An adaptive weight CRIMNO algorithm is also presented and tested with simulation examples of QAM signals. It is shown that the adaptive weight CRIMNO algorithm exhibits faster convergence speed than the Godard algorithm without any significant increase in computational complexity.

1 Introduction

In high speed data communication systems, intersymbol interference caused by channel amplitude and phase distortion may be so severe that channel equalization has to be done in order to make correct decision as to which data symbol is transmitted. Conventionally, equalization is done first through a training mode when a known data sequence is transmitted and is then followed by a decision-directed scheme. However, problems arise in such scenarios as multipoint networks, and multipath fading channels, where the receiver has to perform equalization of the channel without a training mode. When the receiver is "blind" to a training data sequence, the problem is known as blind equalization.

Several algorithms have been proposed for blind equalization. In his original paper[1], Sato proposes the first true blind equalization algorithm for PAM systems, by treating multi-level PAM data as binary ones. In [2], Godard combines blind equalization with carrier recovery and proposes a blind equalization cost function for QAM data based on statistical information about the magnitude of the transmitted data. In [3], Benveniste and Goursat combine Sato's ideas with the decision-directed scheme and obtain a new algorithm which gradually switches to the decision-directed mode when the constellation eye begins to open. In [4], by modifying the decision-directed algorithm in a probabilistic sense, Picchi and Prati obtain the stop-and-go algorithm. In [8], Shalvi and Weinstein generalized Godard's algorithm to a larger class.

Although these algorithms are different, they perform only memoryless nonlinear transformations on the equalizer outputs to generate the desired response. This, in turn, implies that the cost functions they try to minimize with respect to the equalizer coefficients are also memoryless. These algorithms do not employ the fact that *the transmitted data are statistically independent,* which is the essence of the new criterion we introduce in this paper. Since statistical independence of the transmitted data involves more than one data symbol, this results in a memory nonlinear transformation on the equalizer outputs and thus a memory nonlinear cost function.

The paper is organized as follows: In section 2, the new criterion, called CRIMNO, is introduced. In section 3, the corresponding stochastic gradient algorithm are derived. In section 4, the convergence analysis of the CRIMNO algorithms is given. In section 5, an adaptive weight CRIMNO algorithm is proposed for blind equalization. Finally, simulation results and conclusions are given respectively in section 6 and section 7.

*This work was supported in part by the National Science Foundation under grant MIP-8904109 and by the Office of Naval Research under Contract N-00014-91-J-1124.

1058-6393/91 $01.00 © 1991 IEEE

2 Criterion with memory nonlinearity

Consider a synchronous double-sideband quadrature amplitude modulated (QAM) data communication system as shown in Fig. 1. Then the blind equalization problem can be stated as follows:

We have

$$\begin{aligned} y_n &= a_n * h_n = \sum_i a_i h_{n-i} \\ z_n &= y_n * c_n = \sum_i y_i c_{n-i} \end{aligned} \qquad (1)$$

and the following knowledge about the transmitted data

- *The transmitted data $\{a_n\}$ come from a finite discrete set.*
- *The transmitted data are statistically independent from each other.*

The problem is: adjust the equalizer coefficients $\{c_n\}$ to restore the original transmitted data except for a constant delay.

In his original paper[2], Godard proposed a cost function which is independent of the transmitted data, and yet reaches its global minimum at perfect equalization. The Godard's cost function is:

$$D^{(p)} = E\left(|z_n|^p - R_p\right)^2 \qquad (2)$$

where

$$R_p = \frac{E|a_n|^{2p}}{E|a_n|^p} \qquad (3)$$

Note that only the expected value of some function of the *current* equalizer output appears in Godard's cost function. Therefore, Godard's criterion only makes use of the knowledge about the transmitted data constellation and its probability distribution, and it does not directly use the fact that *the transmitted data are statistically independent.*

Assume that perfect equalization is achievable and consider the situation where perfect equalization has indeed been achieved. That is

$$z_n = a_{n-d}$$

where d is a positive integer, which accounts for the constant delay. Since the transmitted data $\{a_n\}$ are statistically independent from each other, so are the equalizer outputs $\{z_n\}$ at perfect equalization. In addition, for most transmitted data constellations, the mean of a_n is zero. Therefore, at perfect equalization, we have

$$E(z_n z_{n-i}^*) = E(a_{n-d} a_{n-i-d}^*) = E(a_{n-d})E(a_{n-i-d}^*) = 0$$

By making use of this property and combining it with Godard's criterion, we obtain a new criterion, called criterion with memory nonlinearity (CRIMNO), which is the minimization of the following cost functions:
CRIMNO Version I: (no update)

$$\begin{aligned} M_1^{(p)} &= w_0 E\left(|z_n|^p - R_p\right)^2 + w_1 \left|E(z_n z_{n-1}^*)\right|^2 \\ &\quad + \cdots + w_M \left|E(z_n z_{n-M}^*)\right|^2 \end{aligned} \qquad (4)$$

CRIMNO Version II: (update)

$$\begin{aligned} M_2^{(p)} &= w_0 E\left(|z_n|^p - R_p\right)^2 + w_1 \left|E(z_n z_{n-1}^{*(n)})\right|^2 \\ &\quad + \cdots + w_M \left|E(z_n z_{n-M}^{*(n)})\right|^2 \end{aligned} \qquad (5)$$

where $z_{n-i}^{(n)}$ is the calculated equalizer output at $t = n - i$ using equalizer coefficients at $t = n$.

The rationale behind the CRIMNO is that since each term reaches its global minimum at perfect equalization, by appropriately combining them, we can increase the convergence speed of the corresponding stochastic gradient algorithms and eliminate the stable local minima inherent in Godard's criterion. The CRIMNO version II is proposed because we hope to increase the convergence speed further by using the most recent equalizer coefficients.
Remarks:

1. The CRIMNO cost functions I and II depend not only on the *current* equalizer output, but also on the *previous* equalizer outputs. As such, they result to the criterion with memory nonlinearity.
2. When $w_0 = 1$, $w_i = 0$ for $i \neq 0$, CRIMNO cost functions reduce to Godard's cost function. Therefore, the CRIMNO algorithms may be seen as generalizations of Godard's algorithm.

3 CRIMNO algorithms

Two versions of CRIMNO were proposed in section 2, whose global minimum corresponds to perfect equalization. Therefore, the blind equalization problem reduces to adjusting equalizer coefficients to seek for the global minimum of the CRIMNO cost functions.

Equalizer coefficients are adjusted using steepest descent method. As is usually done, by substituting the expected value with the current value, and setting p equal to 2, we obtain the following equalizer coefficient adaptation formulas:

CRIMNO version I:

$$C_{n+1} = C_n - \alpha Y_n^* z_n[4w_0|z_n|^2 + 2w_1|z_{n-1}|^2 + \cdots + 2w_M|z_{n-M}|^2 - 4w_0R_2] \qquad (6)$$

CRIMNO version II:

$$\begin{aligned} C_{n+1} &= C_n - \alpha[4w_0Y_n^* z_n(|z_n|^2 - R_2) \\ &+ 2w_1(Y_n^* z_n|z_{n-1}^{(n)}|^2 + Y_{n-1}^* z_{n-1}^{(n)}|z_n|^2) + \cdots \\ &+ 2w_M(Y_n^* z_n|z_{n-M}^{(n)}|^2 + Y_{n-M}^* z_{n-M}^{(n)}|z_n|^2)] \end{aligned} \qquad (7)$$

where
$Y_n = [y_n, \cdots, y_{n+N-1}]^T$; $C_n = [c_{-N_2+1}^{(n)}, \cdots, c_{N_1-1}^{(n)}]^T$; $N = N_1 + N_2 - 1$ is the length of the equalizer; α is the step-size of the algorithm.
Remarks:

1. The first $M+1$ terms in the square bracket of (6) is a weighted sum of $|z_n|^2$, $|z_{n-1}|^2$, $\cdots$, $|z_{n-M}|^2$. As such, it can be regarded as a way of calculating the expected value $E(|z_n|^2)$. Consequently, the CRIMNO algorithm I amounts to substituting the original single point estimate of $E(|z_n|^2)$, $|z_n|^2$, with a more accurate estimate involving a weighted sum of $M+1$ terms. This perhaps explains why the CRIMNO algorithm exhibits faster convergence speed than the original memoryless Godard algorithm.

2. At perfect equalization, $E(|z_n|^2) = E(|z_{n-1}|^2) = \cdots = E(|z_{n-M}|^2)$. Thus the first $M+1$ terms in the square bracket of (6) is an estimate of $\gamma E(|z_n|^2)$, where $\gamma = 4w_0 + 2w_1 + \cdots + 2w_M$. Comparing (6) with the Godard algorithm and noting that the coefficient before R_2 is still $4w_0$, we conclude that the CRIMNO algorithm I will exhibit the problem of constellation eye shrinkage. A solution to this problem is to add an Automatic Gain Control (AGC) unit after the equalizer. The AGC dynamically scales the equalizer output such that it attains the same variance as the transmitted data.

4 Convergence analysis

There are two kinds of equilibria in blind equalization problem. One is with respect to the equalizer coefficients while the other is with respect to the combined impulse response of channel and equalizer. The former includes the latter as a subset. It has been shown in [7] that Godard's criterion has no stable undesired equilibria with respect to the combined impulse response when an infinite length equalizer is used. However, what is required in practice is *no stable local minima with respect to equalizer coefficients for finite length equalizers.* Recently, Ding [5] showed that Godard's criterion does have stable undesired local minima with respect to equalizer coefficients. In this section, we will show that CRIMNO will not have any local minima if the weights are appropriately chosen.

Consider the example used in [5] with an AR(1) channel and 2-PAM data.

$$y_n + 0.6y_{n-1} = a_n, \quad Prob\{a_n = \pm 1\} = 0.5$$

Assume that the equalizer coefficients consist of c_0, c_1 and that the initial states of the received signals y are 0. After some algebra, we obtain the following results:

For the CRIMNO version II (memory size M = 1):

$$M_2^{(2)} = w_0J_0 + w_1J_1 \qquad (8)$$

where

$$\begin{aligned} J_0 &= (c_0^2 - 1)^2 + \kappa_1(c_1 - 0.6c_0)^4 \\ &+\kappa_2(6c_0^2 - 2)(c_1 - 0.6c_0)^2 \\ J_1 &= \kappa_3(c_1 - 0.6c_0)^2(c_0 - 0.6c_1)^2 \end{aligned} \qquad (9)$$

$\kappa_1, \kappa_2, \kappa_3$ are constants,

$$\kappa_1 = \frac{1 + 5 * 0.6^2}{(1 - 0.6^4)(1 - 0.6^2)}$$

$$\kappa_2 = \frac{1}{1 - 0.6^2}$$

$$\kappa_3 = \frac{1}{(1 - 0.6^2)^2}$$

Fig. 2(a) and fig. 2(b) show the 3-D plot and contour map of the CRIMNO II cost function, with memory size $M = 1$ and $w_0 = 1, w_1 = 2$. We see that it has no local minima. Therefore, CRIMNO cost functions may have no local minima if the weights are chosen properly.

5 Adaptive weight CRIMNO algorithm

In this section, an ad hoc way of adjusting the weights on-line is presented

The basic idea is: *we estimate the value of each term in the CRIMNO cost functions and set the weights proportional to the deviations of the corresponding terms from their ideal value at perfect equalization.*

Rewrite the CRIMNO I cost function as

$$M_1^{(p)} = w_0 J_0 + w_1 J_1 + \cdots + w_M J_M \quad (10)$$

and define the deviation of the ith term $D(J_i)$ by

$$D(J_i) \stackrel{\text{def}}{=} \left| J_i - J_i^{(o)} \right| \quad (11)$$

where $J_i^{(o)}$ is the value of J_i at perfect equalization. Then the weights are adjusted using the following formulas:

$$\begin{aligned} w_0 &= \begin{cases} \gamma_0 D(J_0) & \gamma_0 D(J_0) < \lambda \\ \lambda & \gamma_0 D(J_0) \geq \lambda \end{cases} \\ w_i &= \begin{cases} \gamma D(J_i) & \gamma D(J_i) < \lambda \\ \lambda & \gamma D(J_i) \geq \lambda \end{cases} \end{aligned} \quad (12)$$

where γ_0, γ are positive scaling constants; λ is a constraint on the maximum value of the weights for the sake of stability of the algorithm.

The CRIMNO algorithm with weights so adjusted on line, is called adaptive weight CRIMNO algorithm. Remark:

1. When the deviations of all terms vary proportionally, the adaptive weight scheme becomes an adaptive step-size algorithm. Moreover, the adaptation is done automatically. So when the algorithm converges, the weights decrease toward zero. Hence, the adaptive weight CRIMNO algorithm acquires as a byproduct the decreasing step-size, which has been proven to be an optimal strategy for equalization [6].

6 Simulation results

Fig. 3 shows the amplitude and phase characteristics of the channel in our simulation. Fig. 4 shows the performance of the CRIMNO, and the adaptive weight CRIMNO algorithms, compared with that of Godard algorithm. The reason that there is little difference between the performance of the two versions of CRIMNO is the small step-size α used for the stability of the algorithm. Fig. 5 shows the performance of the adaptive weight CRIMNO algorithm for different memory size M. Table 1 summarizes the computational complexity of Godard's, the CRIMNO, and the adaptive weight CRIMNO algorithms.

7 Conclusions

A new criterion with memory nonlinearity and its corresponding blind equalization algorithms have been introduced for blind equalization. The new criterion has been shown to have no local minima if the weights are chosen properly, therefore guaranteeing its global convergence. In addition, the adaptive weight CRIMNO algorithm has been demonstrated to exhibit much faster convergence speed than Godard's algorithm at the expense of little increase in computational complexity.

References

[1] Y. Sato. *A method of self-recovering equalization for multilevel amplitude modulation.* IEEE Trans. Comm., vol. COM-23, pp. 679-682, June 1975.

[2] D.N. Godard. *Self-recovering equalization and carrier tracking in two dimensional data communication systems.* IEEE Trans. Comm., vol. COM-28, pp. 1867-1875, 1980.

[3] A. Benveniste, M. Goursat. *Blind Equalizers.* IEEE Trans. Comm., vol. COM-32, pp. 871-883, 1984.

[4] G. Picchi, G. Prati. *Blind Equalization and Carrier Recovery Using a 'Stop-and-Go' Decision-Directed Algorithm.* IEEE Trans. Comm., Vol. COM-35, pp. 877-887, 1987.

[5] Z. Ding, R.A. Kennedy, B.D.O. Anderson and C.R. Johnson. *Existence and Avoidance of Ill-Convergence of Godard Blind Equalizers in Data Communication Systems.* 23rd Conference on Information Sciences and Systems, Baltimore, MD, March 1989.

[6] G. Ungerboeck. *Theory on the Speed of Convergence in Adaptive Equalizers for Digital Communication.* IBM J. Res. Develop. pp. 546-555, Nov. 1972.

[7] G. J. Foschini. *Equalizing without Altering or Detecting Data.* AT&T Tech. J., Vol. 64, No. 8, pp. 1885-1909, Oct. 1985.

[8] O. Shalvi, and E. Weinstein. *New Criteria for Blind Deconvolution of Nonminimum Phase Systems (Channels).* IEEE Trans. Information Theory, Vol. 36, pp. 312-321, 1990.

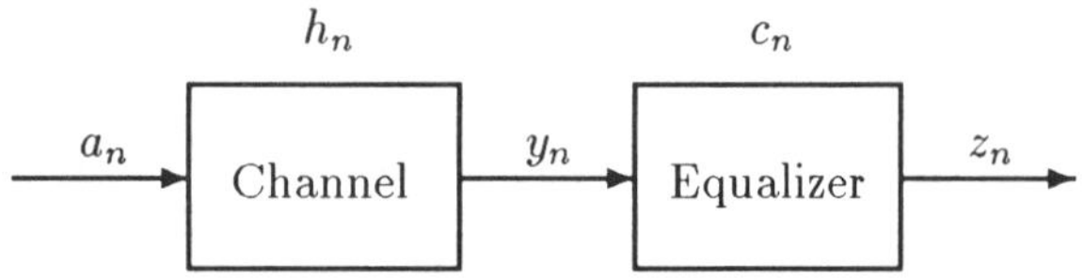

Fig. 1. Diagram of communication systems.

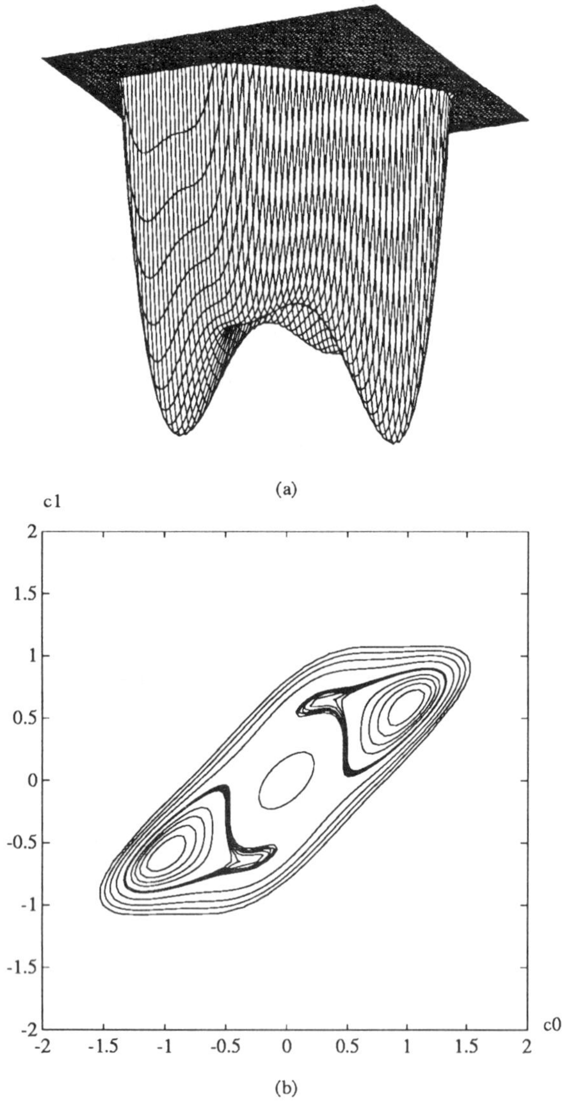

Fig. 2. (a) Plot of CRIMNO cost function;
(b) Contour of CRIMNO cost function.

Table 1. Comparison of Computational Complexity

Algorithm	Real Multiplication
Godard	4N+5
CRIMNO (Version I)	4N+3M+5
CRIMNO (Version II)	MN+8M+4N+5
AW CRIMNO (Version I)	4N+8M+5

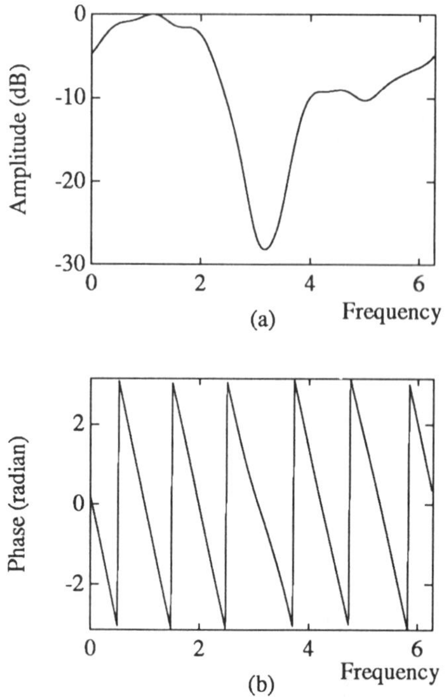

Fig. 3. Channel characteristics.

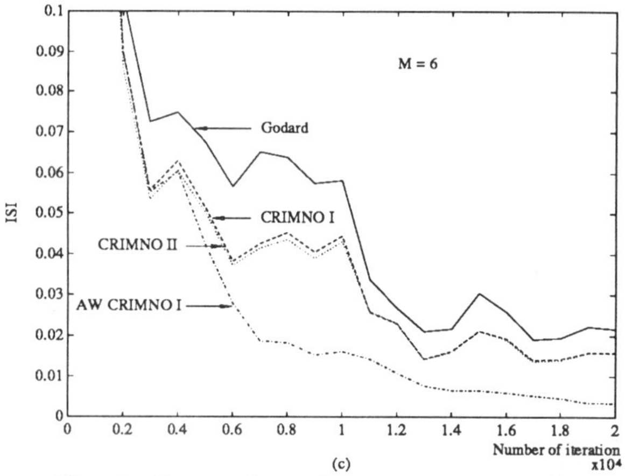

Fig. 4. Comparison of convergence speed.

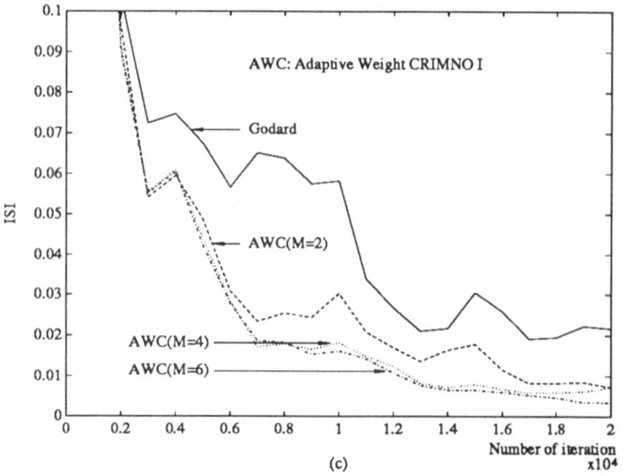

Fig. 5. Convergence speed with memory order.

A New Adaptive Algorithm for Joint Blind Equalization and Carrier Recovery

Z. Ding
Department of Electrical Engineering
Auburn University
Auburn, AL 36849, USA

R. A. Kennedy
Department of Systems Engineering
Australia National University
G.P.O. Box 4, Canberra, ACT 2601, Australia

Abstract

Traditional blind equalization algorithms based on minimizing some specially designed non-MSE cost functions can experience local convergence problems and thereby result in insufficient or no removal of channel distortion. In this paper, we present a new QAM blind equalization scheme that is globally convergent in the equalizer parameter space to a compact set containing the desired ideal equalizer parameter setting. For a generic class of channels, this new algorithm results in the equalizer parameter convergence to a unique global minimum achieving intersymbol interference suppression and carrier phase error removal. Different implementation approaches are assessed and simulation results are shown to confirm the theoretical global convergence of the new algorithm.

1 Introduction

Blind equalization of an unknown, non-minimum phase channel is an important problem in data communication systems. Conventional adaptive equalizers remove the intersymbol interference (ISI) caused by non-ideal channels of limited bandwidth with the aid of a training session which provides the needed reference signal for parameter adaptation. In situations where a training session is impossible or very costly, blind equalizers need to be used to combat the ISI effect.

Many schemes for blind equalization of quadrature amplitude modulation (QAM) system exist in the literature. Typically, they are based on gradient descent minimization of some special non-MSE cost functions that do not require the use of reference signal [1, 2, 3, 4, 5]. They are sometimes known as the "Bussgang" algorithms [6]. However, the lack of unimodality of these cost functions, established in [7, 8, 9], can result in the undesirable (local) convergence of the corresponding algorithms such that the equalizer fails to remove sufficient ISI. Hence, despite the apparent plethora of schemes, there remains a strong practical need for development of new algorithms with more reliable and provable (global) convergence properties.

From recent works [9] it is clear that in order to design blind equalization algorithms that converge to a desirable parameter setting from *arbitrary* initializations (based on gradient descent minimization of some cost function), certain aspects of the performance objective should be relaxed. In designing our new algorithm, we do not attempt to directly identify the exact gain of the channel inverse and focus instead on the elimination of ISI [10]. So long as we can remove the phase error in carrier recovery, ISI elimination without gain identification is a sufficient objective for the ultimate recovery of the channel input in QAM systems.

Blind equalization of pulse amplitude modulation (PAM) systems without gain recovery has been proposed in [11, 12, 13]. However, generalizing these results directly to QAM systems is not straightforward. The difficulty arises since in QAM systems the input signal is constrained to take values in the QAM constellation. Phase also makes tap-fixing constraints unusable since they do not ensure the recovery of the constellation in the equalizer output because of the unknown carrier phase error. In the paper, we achieve this desirable objective of joint blind equalization and carrier phase recovery using a novel non-anchoring tap constraint. Unlike many blind equalization proposals we furnish a theorem that the algorithm we are developing possesses desirable unique parameter convergence for generic channels even when finite impulse response (FIR) equalizers which can only approximate the desired channel inverse are employed.

The major part of this paper is divided into the following sections. In section 2, we describe the fundamental principles of linear blind equalization. Then in section 3, we present the new blind adaptive equalization algorithm by introducing a new cost function with a linear constraint of the complex equalizer parameters. We describe different implementation approaches of the new blind equalization algorithm in section 4.

1058-6393/91 $01.00 © 1991 IEEE

2 Background Information

2.1 Blind Equalization in QAM Systems

In QAM channel equalization system, the original symbol sequence $\{a_k\}$ takes values from a complex signal set $\mathcal{A}$ (known as the QAM constellation) in a manner such that all possible symbol subsequences occur with non-zero probability. This statistical assumption on the input is *weaker* than the typical *i.i.d.* assumption assumed in other works. (Further weak assumptions regarding various symmetries of the constellation will be described later.)

The complex channel input $\{a_k\}$ is transmitted through a nonideal channel, assumed to be linear, causal, and (bounded-input-bounded-output) stable. The channel has transfer function

$$H(z^{-1}) = \sum_{i=0}^{\infty} h_i z^{-i}, \quad h_i \in \mathbb{C}, \tag{2.1}$$

where $\{h_i\}$ represents the channel impulse response sequence.

A *linear* channel equalizer is a linear filter $\Theta(z^{-1})$ that is applied to the channel output x_k in order to eliminate the ISI. Initially we may take this filter as being stable and potentially non-causal (doubly infinite), and of the form:

$$\Theta(z^{-1}) = \sum_{i=-\infty}^{\infty} \theta_i(k) z^{-i}, \quad \theta_i \in \mathbb{C} \tag{2.2}$$

so as to deal with non-minimum phase channels. The time dependence of the equalizer parameters signifies that they are subject to adaptation via an algorithm to be described. The equalizer output can then be written as

$$z_k = \sum_{i=-\infty}^{\infty} \theta_i(k) x_{k-i} = X_k' \theta(k) \tag{2.3}$$

by denoting the adaptive equalizer parameter vector as

$$\theta(k) = [\ \cdots\ \theta_{-2}\ \ \theta_{-1}\ \ \theta_0\ \ \theta_1\ \ \theta_2\ \ \cdots\ ]' \tag{2.4}$$

and the equalizer input signal regressor as

$$X_k = [\ \cdots\ x_{k+2}\ \ x_{k+1}\ \ x_k\ \ x_{k-1}\ \ x_{k-2}\ \ \cdots\ ]'.$$

The ideal objective of any channel equalizer $\Theta(z^{-1})$ is to achieve distortionless reception

$$z_k = \Theta(z^{-1}) x_k = a_{k-\nu}, \quad \nu \in \mathbb{Z}_+, \ \forall k, \tag{2.5}$$

where we can tolerate a finite fixed time delay ν. Such an objective translates into the identification of channel inverse by the equalizer.

In blind equalization, the original sequence is unknown to the receiver except for its probabilistic or statistical properties over the known alphabet $\mathcal{A}$. Usually this signal constellation is $m\pi/2$-rotation-invariant. If the statistics of the input data reflects this same symmetry (which is typical) over $\mathcal{A}$, then a $m\pi/2$ phase rotation does not cause any statistical changes in the channel output. Thus, the data recovered from blind equalization will be intrinsically subjected to a phase ambiguity of $m\pi/2$ and the best possible result would be

$$H(z^{-1})\Theta(z^{-1}) = e^{m\pi/2} z^{-\nu}, \quad m \in \{0, 1, 2, 3\} \tag{2.6}$$

which relaxes the objective (2.5). The remaining phase ambiguity can be resolved through differential encoding of the input data. A further relaxation of the ideal objective (2.5) arises since in practical applications only equalizers with a finite number of adjustable parameters can be implemented rather than those of (2.2). Typically these finite parameters are arranged in the form of a causal FIR transversal filter. Since an FIR equalizer can only approximate the desired impulse response (2.6) a quantizer should be used to recover the original channel input from the equalizer output z_k. This is often regarded as the *practical* objective of blind equalization.

Blind adaptive equalization algorithms are often designed as stochastic gradient descent (SGD) schemes to update the parameter vector by minimizing some special (non-MSE) mean cost functions that do not involve the use of the original input a_k but still reflect the current level of ISI in the equalizer output. Such schemes implicitly involve higher order statistics of the channel and equalizer outputs. These algorithms are sometimes referred to as the "Bussgang" algorithms [6].

Bussgang algorithms have been shown to exhibit local convergence [7, 8, 9]. The lack of unimodality by these algorithms can result in local, undesirable convergence of the equalizer parameters especially for FIR equalizers. Hence this motivates the development of new and better blind equalization algorithms.

2.2 Constrained Blind Equalization

In order to design globally convergent blind equalization algorithms further aspects of the performance objective (2.5) can be relaxed beyond those already indicated. In designing our new adaptation rules, we do not attempt to identify the exact gain of channel inverse and focus instead on the elimination of ISI which is the primary objective of channel equalization [10]. Consequently, it is not essential to recover the exact (complex) gain of the channel inverse because once the ISI is removed such that the equalizer output is

$$z_k = c a_{k-\nu} = |c| e^{j\phi_c} a_{k-\nu}, \quad c \neq 0,$$

then it is straightforward to estimate the unknown gain $|c|$ by power-matching. The constant phase ambiguity ϕ_c can also be readily resolved by the utilization of differential encoding, provided it is a multiple of $\pi/2$.

Blind equalization of pulse amplitude modulation (PAM) systems without gain recovery has been proposed in [11, 12, 13]. Their essential idea is parameter "anchoring". As shown in [11, 13], due to the absence of carrier phase error in PAM systems, "anchoring" [10] the center tap can beneficially constrain the freedom of equalizer parameters such that a convex cost function can be utilized for parameter adaptation.

In the next section, we present a new algorithm for joint blind equalization and carrier phase recovery based on specialized convex cost functions coupled with a constrained equalizer parametrization as a generalization of the algorithm in [12]. For a general class of channels with (independent) real and imaginary distortion, we obtain a unimodal mean cost function with which gradient descent algorithms exhibit parameter convergence that globally minimizes the mean cost. This ideal convergence property, in contrast to the classical "Bussgang" approach, is robust to equalizer length truncation due to the convexity of the cost functions employed. The robustness lies in that: (i) spurious local minima are not created by the truncation; and (ii) the global minimum in the strictly doubly infinite parameter space (2.6) degrades gracefully into a global minimum in the practical finite dimensional parameter space where by selecting the length of the equalizer appropriately one can ensure an arbitrarily small residual ISI term.

3 Globally Convergent Algorithms

3.1 Convexity and Linear Constraint

We assume the QAM constellation is such that the real and imaginary components are independent and identical (*e.g.*, the square type) with

$$M \triangleq \max|Re\{a_k\}| = \max|Im\{a_k\}|$$

(or by rotation can be transformed to such). We can denote the total system (channel and equalizer) as

$$\mathcal{T}(q^{-1}) = H(q^{-1})\Theta(q^{-1}) = \sum_{i=-\infty}^{\infty} t_i q^{-i},$$

where $\{t_i\}$ is the impulse response of the combined system given by

$$t_i = h_i \circledast \theta_i = \sum_{k=0}^{\infty} h_k \theta_{i-k}$$

and $\circledast$ denotes convolution. The equalizer is taken as being doubly infinite and non-causal to permit complete removal of ISI. This assumption will be relaxed (as discussed above) in order to deal the truly implementable finite equalizer parameter space and avoid the potential weaknesses of this simplification as discussed in [14].

The cost function to be minimized is designed as

$$\begin{aligned} J(\theta) &\triangleq \frac{1}{M}\max|Re\{z_k\}| \\ &= \frac{1}{M}\max_{\{a_k\}}\left|\sum Re\{t_i\}Re\{a_{k-i}\} - Im\{t_i\}Im\{a_{k-i}\}\right| \\ &= \sum_{i=-\infty}^{\infty}\Big[|Re\{t_i\}| + |Im\{t_i\}|\Big] \\ &= \frac{1}{M}\max|Im\{z_k\}| \qquad (3.1) \end{aligned}$$

where $z_k = \sum_{i=-\infty}^{\infty} t_i a_{k-i}$.

The convexity of $J(\theta)$ with respect to θ follows from the triangle inequality. Notice that by fixing the real or the imaginary part of the parameter vector θ, the cost function $J(\theta) = J(Re\{\theta\}, Im\{\theta\})$ is also convex in terms of $Re\{\theta\}$ and $Im\{\theta\}$, an observation which will be important later. From

$$J(-\theta) = \sum_i \Big[|Re\{-t_i\}| + |Im\{-t_i\}|\Big] = J(\theta).$$

we have by convexity

$$J(0) \le 0.5J(-\theta) + (1-0.5)J(\theta) = J(\theta), \quad \forall\theta \ne 0.$$

Thus if the range of θ is unrestricted, then the convex cost function (3.1) achieves a trivial global minimum at $\theta = 0$, with equalizer output $z_k = 0$.

In order to make the cost function $J(\theta)$ useful, we constrain the equalizer parameters with the following linear constraint

$$Re\{\theta_0\} + Im\{\theta_0\} = 1 \qquad (3.2)$$

in the equalizer parameter vector (2.4). Due to the linearity of the constraint, the convexity of the cost function (with respect to the real and imaginary parts of the equalizer coefficients) is maintained and global convergence is therefore assured. We now show that under this parametrization, simultaneous removal of ISI and carrier phase error globally minimizes the cost function.

3.2 Global Convergence

Suppose the ideal non-causal channel inverse (as before) is given by

$$\check{h}(z^{-1}) = H^{-1}(z^{-1}) = \sum_{i=-\infty}^{\infty}\check{h}_i z^{-i}, \quad \check{h}_i \in \mathcal{C}.$$

such that $\check{h}(z^{-1})$ exists on the unit circle. Then we have

$$h_i \circledast \check{h}_i = \delta(i),$$

where $\delta(i)$ is the Kronecker Delta function.

Our theoretical demonstration of global convergence proceeds in a number of steps. First we present a lemma which identifies the specific parameter setting from the countable class of ideal equalizers—*i.e.*, ones leading to zero residual ISI—that minimizes the linearly constrained cost (3.1) and (3.2). In what follows it is more convenient to express the constraint (3.2) in polar form:

$$\theta_0^{(\phi)} \triangleq \frac{e^{j\phi}}{\cos\phi + \sin\phi}; \qquad -\frac{\pi}{4} < \phi < \frac{3\pi}{4}.$$

Armed with this candidate the second step (our main theorem) demonstrates that it minimizes globally the given cost function over arbitrary equalizer parameter settings which are subject to the linear constraint (3.2). The proofs are omitted here due to space limitation.

Lemma 3.1 *For ideal equalizer parameter settings of the form*

$$\theta_i(q,\phi) \triangleq \frac{\check{h}_{i+q}}{\check{h}_q} \cdot \theta_0^{(\phi)}; \qquad -\frac{\pi}{4} < \phi < \frac{3\pi}{4}.$$

parametrized by $q \in \mathbb{Z}$, and ϕ, the convex cost (3.1) *subject to the linear constraint* (3.2) *is minimized by the following (q,ϕ)-pair:*

$$m = \arg\max_q \left\{ \left|Re\{\check{h}_q\}\right| + \left|Im\{\check{h}_q\}\right| \right\} \quad (3.3)$$

$$\phi^* = \begin{cases} 0 < \phi_m + k\pi/2 < \pi/2 & \phi_m \neq m\pi/2 \\ \phi \quad \text{for any } \phi \in [0,\pi/2], & \text{otherwise} \end{cases}$$

where $\check{h}_m = \left|\check{h}_m\right| e^{j\phi_m}$.

We defer comments regarding this lemma until after the next result. Now we present the main theorem.

Theorem 3.1 *For a linear channel with inverse impulse response sequence $\{\check{h}_i\}$, a doubly infinite non-causal equalizer with parameter setting*

$$\theta_i(m,\phi^*) = \frac{\check{h}_{i+m} e^{jk\pi/2}}{\left|Re\{\check{h}_m\}\right| + \left|Im\{\check{h}_m\}\right|} \quad (3.4)$$

subject to the constraint

$$Re\{\theta_0\} + Im\{\theta_0\} = 1, \quad (3.5)$$

where

$$m = \arg\max_q \left\{ \left|Re\{\check{h}_q\}\right| + \left|Im\{\check{h}_q\}\right| \right\}$$

and where $k \in \{1,2,3,4\}$ is chosen such that (3.5) *is satisfied, minimizes the blind convex cost*

$$\max_{\{a_k\}} \left|Re\{z_k\}\right| = \sum_{i=-\infty}^{\infty} \left[\left|Re\{t_i\}\right| + \left|Im\{t_i\}\right| \right].$$

3.3 Remarks and Comments

Despite the unimodality of the cost function with respect to the equalizer parameters there is a potential problem of non-uniqueness of the (global) minimum, *i.e.*, the minimum might be achieved by all points in a compact set (actually a convex polytope) rather than at a single point. There are two manifestations of this non-uniqueness: (i) as in Lemma 3.1 there can be a phase ambiguity over a $\pi/2$ range if $\check{h}_m$ is purely real or purely imaginary which is a non-generic property; and (ii) as in (3.3) when the maximum is realized by more than one m which is again a non-generic property. That these are the only ways non-uniqueness arises can be inferred from the proof of Theorem 3.1.

The linear constraint (3.2) can be changed to any of three other possibilities according to various sign possibilities. However, less obviously, more general but not necessarily arbitrary linear constraints on the equalizer parameters (distributed over a number of equalizer parameters. This fact is highlighted in the context that the non-generic channels which suffer from the non-uniqueness problem indicated above can be "cured" (to yield a point minimum) by using these more general linear constraints. We will not further investigate this variation here.

Convexity is an essential property in the above formulation because it guarantees that when we use causal finite dimensional practical equalizers the global convergence tendencies of the algorithm, or unimodality, are preserved. This behavior is a consequence of convexity since truncation (setting many taps zero) is a form of linear constraint which does not destroy convexity. Convexity also ensures that one can approximate arbitrarily closely the performance of the ideal non-implementable doubly infinite non-causal equalizer parameter setting.

4 Implementation

The adaptation of the equalizer parameter based on the minimization of the cost function which is equivalent to

$$J(\theta) = \max\left|Re\{z_k\}\right| = \left\|Re\{z_k\}\right\|_\infty$$

requires some work since this cost function can not be exactly evaluated in practice with finite data length. Here we propose two algorithms: a gradient descent approach and an iterative minimization approach.

4.1 l_p Approximation

One natural approximation to l_∞ is the l_p norm with large p. Given large enough p, the global minimum of the l_p norm will be close to global minima of the l_∞ norm. In other words, the minimization of $\max|Re(z_k)|$ can approximated by minimizing

$$\left\|Re\{z_k\}\right\|_p = \left(E\{|Re\{z_k\}|^p\}\right)^{1/p},$$

or equivalently by minimizing $E\{|Re\{z_k\}|^p\}$ for a fixed p. Thus in practical implementation we minimize

$$E\{|Re\{z_k\}|^p\}$$

for which stochastic gradient descent minimization scheme can be derived. Similarly the minimization of $\max|Im(z_k)|$ can be approximated through the minimization of $E\{|Im(z_k)|^p\}$ for a fixed p. To speed up the convergence rate, we choose to minimize the sum of the two convex cost functions (which is still convex)

$$\widehat{J}(\theta) \triangleq E\{|Re\{z_k\}|^p\} + E\{|Im\{z_k\}|^p\}.$$

In practice we only implement finite-dimensional parameter vector θ for the equalizer. Since the sum of two l_p-norms $\widehat{J}$ is strictly convex, linear constraints such as truncation preserve convexity.

4.2 Block Approximation

As with many other blind equalization algorithms, such as CMA, a gradient descent approach can be very slow in convergence. The main advantages of these algorithms are their potential tracking ability and computational simplicity. The disadvantage is that they do not fully utilize the available data, which results in rather slow convergence rate. Here we present a simple block minimization scheme. The algorithm simply collects a sequence of data $\{x_k\}$ and generates a sequence of outputs $\{z_k\}$ through θ. Then one finds the z_k in the sequence that has the maximum $|Re\{z_k\}|$ and maximum $|Im\{z_k\}|$, respectively. The equalizer parameters are then updated to minimize $\max[Re\{z_k\}]^2$ and $\max[Im\{z_k\}]^2$

5 Conclusion

This paper presents a new adaptive algorithm for the joint blind equalization and carrier recovery of QAM signals. Theoretical analysis shows the convexity of the corresponding cost function under linear constraints, which guarantees the global convergence of the equalizer parameters regardless of the number of equalizer parameters used. The robustness to truncation of equalizer length is in direct contrast with what are known as the "Bussgang" algorithms. Simulation results not shown here have demonstrated the effectiveness of this algorithm in removing ISI and correcting the carrier phase error.

References

[1] Y. Sato. *IEEE Trans. on Communications*, COM-23:679–682, June 1975.

[2] D.N. Godard. *IEEE Trans. on Communications*, COM-28:1867–1875, November 1980.

[3] A. Benveniste, M. Goursat, and G. Ruget. *IEEE Trans. on Automatic Control*, AC-25:385–399, June 1980.

[4] J.R. Treichler and M.G. Agee. *IEEE Trans. on Acoustics, Speech, and Signal Processings*, ASSP-31:349–472, April 1983.

[5] O. Shalvi and E. Weinstein. *IEEE Trans. on Info. Theory*, IT-36:312–321, March 1990.

[6] S. Bellini and F. Rocca. In E. Biglieri and G. Prati, editors, *Digital Communications*, pages 251–263, North-Holland, 1986. Elsevier Science Publishers B.V.

[7] J.E. Mazo. *Bell Syst. Tech. Journal*, 59:1857–1876, December 1980.

[8] Z. Ding, R.A. Kennedy, B.D.O. Anderson, and C.R. Johnson, Jr. *IEEE Trans. on Communications*, September 1991.

[9] Z. Ding, C.R. Johnson, Jr., and R.A. Kennedy. In *Proc. 29th IEEE Conf. on Decision and Control*, pages 225–230, Honolulu, Dec 1990.

[10] S. Verdú, B.D.O. Anderson, and R.A. Kennedy. In *Proc. 25th Conf. on Info. Sci., and Systems*, pages 774–779, Baltimore, MD, March 1991.

[11] S. Vembu, S. Verdú, R.A. Kennedy, and W.A. Sethares. In *Proc. 25th Conf. on Info. Sci., and Systems*, pages 792–797, Baltimore, MD, March 1991.

[12] W.T. Rupprecht. Zeitdiskrete berechnung von verzerrungsmassen als funktionen der filterkoeffizienten adaptiver datenleitungsentzerrer. *ntz Archiv*, pages 251–258, July 1985.

[13] W.A. Sethares, R.A. Kennedy, and Z. Gu. In *Proceedings of IEEE ICASSP-91*, pages 1529–1532, Toronto, Canada, 1991.

[14] R.A. Kennedy, B.D.O. Anderson, Z. Ding, and C.R. Johnson, Jr. In *Proc. 29th IEEE Conference on Decision and Control*, pages 3194–3199, Honolulu, HI, 1990.

THE DEMOD-REMOD TECHNIQUE FOR DEMODULATING CO-CHANNEL FSK SIGNALS

Richard Gooch, Christine Jorgensen, and Michael Ready
Applied Signal Technology
160 Sobrante Way
Sunnyvale, CA. 94086

Abstract

This paper describes the demod-remod method of demodulating two cochannel MEFSK signals. The technique uses a conventional demodulator to obtain an estimate of the primary data sequence. The primary data is then remodulated and subtracted from the received signal to obtain an estimate of the secondary signal. Finally, the secondary signal is demodulated using conventional demodulation techniques. The paper includes a description of phase noise, its effect on demod-remod performance, and methods used to mitigate it. Experimental results showing the range of signal power separations and carrier-to-noise ratios over which the demod-remod technique works are presented.

Introduction

Increased frequency reusage and spectral crowding are causing cochannel interference to become a major limitation on the performance of modern digital communication systems. Figure 1 illustrates a typical cochannel scenario. Two frequency-shift-keyed (FSK) signals, separated in power by 6 dB and possessing the same carrier frequency, illuminate the receiving antenna. The stronger signal is termed the primary signal, and the weaker the secondary. The polar plots and eye diagrams show that the primary signal suffers severe distortion due to the presence of the secondary signal. In cochannel environments such as this, conventional FSK demod techniques generally lock onto the stronger FSK signal and do not recover data from the cochannel signal. It is the goal of the demod-remod technique described in this paper to recover both the primary and secondary FSK data streams. A similar technique was used to demodulate cochannel QAM signals in [1].

The block diagram of the demod-remod technique is shown in Figure 2. The basic premise associated with the technique is that the primary FSK signal is strong enough to permit demodulation using conventional methods. The demodulated bit stream is then adaptively remodulated and subtracted from the aggregate received signal. Provided the remodulated signal accurately mimics the primary signal component of the received signal, the difference should consist of the weaker secondary FSK signal plus noise. This difference signal is then demodulated using a second conventional FSK demodulator. The heart (and soul) of the demod-remod technique lies with the adaptive remodulation process and is the main focus of this paper.

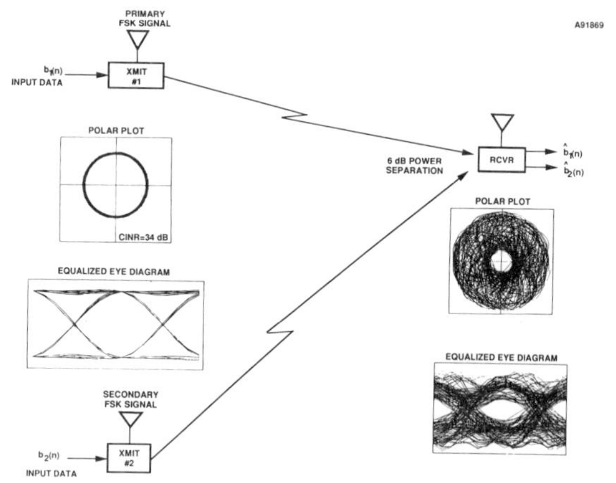

Figure 1: Cochannel FSK signal scenario.

The paper begins by reviewing Manchester-encoded FSK (MEFSK) signal generation and goes on to describe the MEFSK demodulation and remodulation processes in detail. Next, experimental results are presented that demonstrate the performance of the demod-remod algorithm for two different signal power separations. Phase noise is shown to be a major limitation on the performance of the technique. The characteristics of phase noise, as well as methods used to combat it, are also described. Finally, quantitative guidelines defining the scenarios for which the demod-

1058-6393/91 $01.00 © 1991 IEEE

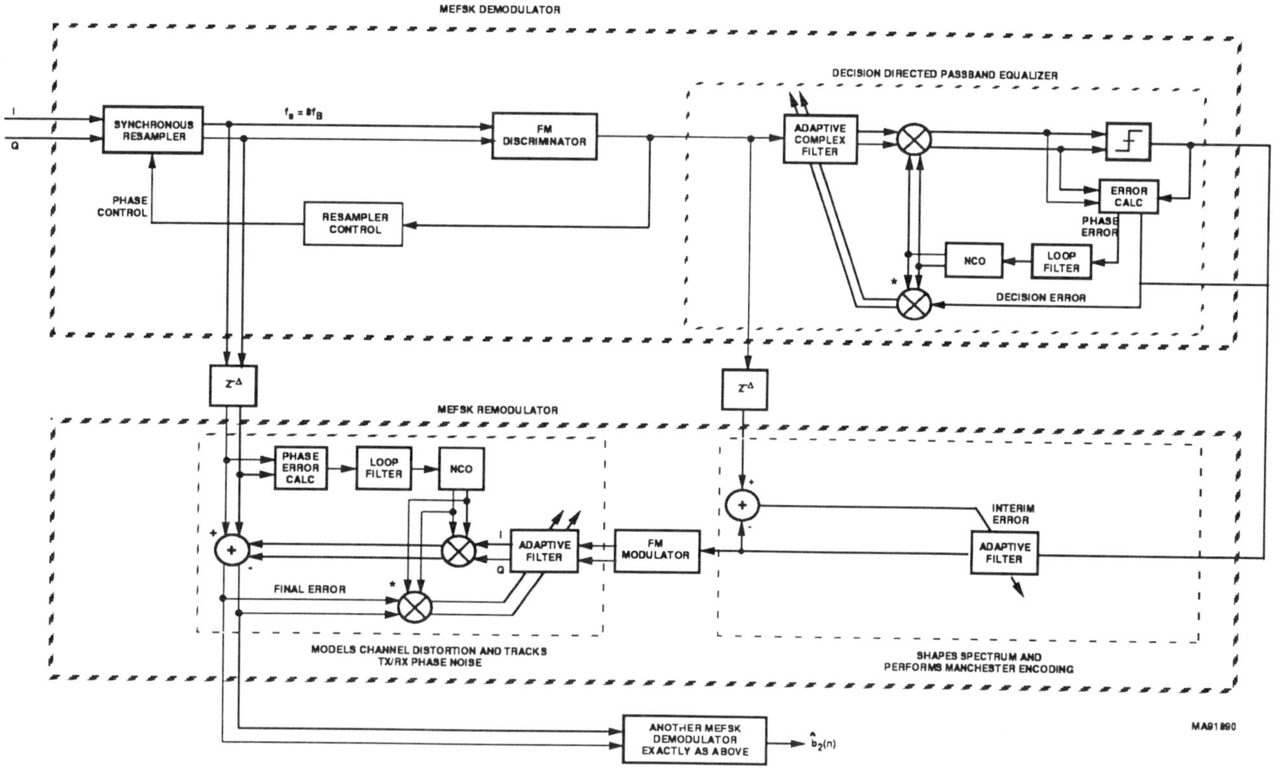

Figure 2: Block diagram of the demod-remod technique.

remod technique can effectively demodulate both the primary and secondary signals are given.

MEFSK Demodulator

This paper is concerned with a specific type of FSK signaling known as Manchester-encoded FSK (MEFSK). To provide a basis for discussion, a MEFSK modulator and demodulator are first described. Manchester encoding exclusive-or's the input bit stream with a 50% duty-cycle bit clock. The resulting signal is lowpass filtered to suppress frequency components greater than twice the bit rate. At this point, the waveform can be interpreted as a BPSK signal with carrier frequency equal to the symbol rate. The signal is then frequency modulated to create the Manchester-encoded FSK signal.

The block diagram of a Manchester-encoded FSK demodulator is shown in the upper half of Figure 2. The received signal is assumed to have been previously digitized and converted to complex I/Q form. The signal is first resampled to eight samples/bit synchronous with the primary signal symbol rate. The resampler phase control signal is generated by extracting the symbol clock and comparing it to a reference. The phase difference is then used to speed-up or slow-down the resampling rate. Care must be taken in choosing the resampler rate change small enough that resampling jitter is not a limiting factor in the remodulated signal accuracy.

The signal is then FM discriminated with a delay-line discriminator in which the complex input signal is delayed, conjugated, and multiplied by itself. The phase of the resulting product comprises the discriminator output. It can be shown that this process is mathematically equivalent to differentiating the (unwrapped) phase of the input signal. The output of the FM discriminator is the BPSK signal.

The output of the FM discriminator is fed to an adaptive fractionally-spaced passband equalizer [2] to generate a complex equalized output signal decimated to one sample per bit. The equalizer is initialized with a complex bandpass filter with a center frequency equal to the symbol rate and bandwidth equal to twice the symbol rate. The converged equalizer performs quadrature splitting, matched filtering, and intersymbol interference (ISI) equalization.

The equalizer output is fed into a complex mixer where BPSK carrier recovery (or equivalently Manchester decoding) is performed. The mixer output is hardlimited to generate a bit decision which corresponds to the transmitted bit. The error signals for both the phase lock loop and the adaptive equalizer are formed by comparing the hard bit decision with the mixer output. This process is known as decision-directed adaptation and carrier recovery [3].

MEFSK Remodulator

The estimated primary data sequence is fed into the adaptive remodulator where it is reshaped into a replica of the primary component of the received waveform. This reshaping process is comprised of a filter followed by an FM modulator followed by a second filter and a phase locked loop (PLL). The first filter generates the Manchester encoding and spectral shaping. The FM modulator converts the shaped data signal into an FSK signal. The second filter models linear distortion imparted on the signal by transmitter or receiver filters as well as propagation-induced distortion. The PLL tracks phase noise imparted on the primary FSK signal by both the transmitter and the receiver.

The objective of the remodulator is to mimic the primary signal component of the received signal. Ideally, the coefficients of both the pre and post FM modulation (pre-m and post-m) filters should be jointly chosen to minimize the variance of the "final error." Unfortunately, optimization of the pre-modulation filter based upon the final error variance is a difficult nonlinear optimization problem. In order to simplify the optimization, an interim error signal is used. As shown in Figure 2, this interim error is the difference between the FM discriminator output and the pre-modulation filter output. The variance of the interim error is a quadratic function of the filter coefficients and can therefore be readily minimized using least mean square (LMS) techniques, e.g. the LMS algorithm [4].

Optimization of the pre-m filter using the interim error is biased in the presence of uncorrelated additive interference since the output of the FM discriminator will consist of the demodulated primary signal plus a distortion component correlated with both the primary and secondary signals. Consequently, minimization of the interim error variance will result in a biased filter solution that attempts to model both the undistorted primary signal component and the distortion components. The bias will become worse as the primary-to-secondary signal power ratio (PSR) decreases. Fortunately, in situations where the PSR is low, the required remodulation accuracy (and subsequent primary signal cancellation) will be less stringent. The experimental results presented in the next section show that interim filter bias is non-problematic.

The pre-m adaptive filter also must account for carrier frequency offset in the receiver tuning as well as frequency drift in both the transmitter and receiver. Carrier offset manifests itself in terms of a dc offset in the FM discriminator output. Frequency drift shows up as a slow variation of that dc offset. The frequency offset can be modeled by injecting a constant value into the adaptive filter output. This value can be adaptively determined by incorporating an additional weight in the adaptive filter with the weight input tied to a constant value. The LMS adaptation will then choose the weight value that results in a best match to the dc offset in the desired response. Frequency drift (or slow dc offset variation) is tracked by a time-varying weight value. The time constant of the dc tracker is determined by the step size and the constant value tied to the weight input.

Once the pre-m adaptive filter converges to its optimal value, the filter output signal should closely mimic the FM discriminator output component corresponding to the primary signal. The FM modulator integrates the input signal and then exponentiates the complex result. This process perfectly inverts the FM discriminator operation modulo a constant phase offset corresponding to the intial state of the integrator. The appropriate phase offset is determined by the complex channel filter following the FM modulator.

For the moment ignore the PLL in the post-m modeling process used to track phase noise. The post-m filter coefficients are determined using the LMS algorithm applied to the final error. This filter time-aligns the steady-state amplitude and phase offset of the FM modulated signal with the primary signal component of the received signal. Small misalignment of either the time, amplitude, or phase can severely degrade cancellation.

When the remodulation process was applied to simulated signals, the final fit was almost perfect. However when applied to snapshots of real FSK signals taken in the laboratory, it was discovered that phase noise in both the transmitter and receiver caused a significant error. The PLL phase noise tracker was added to the remodulator to mitigate this degradation.

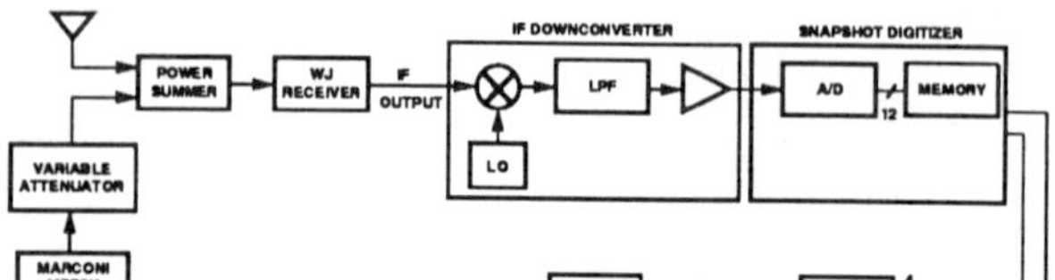

Figure 3: Experimental test setup.

To mitigate phase noise, the number-control oscillator (NCO) output is mixed with the post-m filter output to generate the final remodulator output. The NCO is controlled by the phase error between the remodulated signal and the aggregate received signal. The loop filter realizes a critically-damped second-order response with a user-specified time constant, τ.

The final error signal is fed to a second demodulator identical to the first. As the secondary FSK signal may have a symbol rate different from the primary signal, a second synchronous resampler is employed.

Experimental Results

To provide a realistic test of the demod-remod technique, a cochannel signal scenario was created by adding a MEFSK signal received off-the-air to a second MEFSK signal generated using a test generator as shown in Figure 3. The off-the-air signal was always the secondary signal while the test generator provided the primary signal. The test generator output power was adjusted to control the PSR. The composite signal was fed into a receiver, downconverted to baseband, and snapshot digitized. The noise floor was set by adding artificially-generated white Gaussian noise to the snapshot.

Typical processing results from two snapshots are presented in this section for a low PSR case and a high PSR case. These cases were chosen to illustrate the performance limitations of the demod-remod technique at both ends of the possible range of PSRs. Case #1 consists of a composite signal which has a PSR of 3 dB and a primary signal carrier-to-noise ratio (CNR) of 18 dB. Case #2 consists of a composite signal which has a PSR of 18 dB and a primary signal CNR of 30 dB.

Performance of the primary and secondary demodulators for case #1 can be observed in the eye patterns shown in Figures 4(a) and 4(b), respectively. The wide-open eye shown in Figure 4(b) indicates that BER of the secondary signal will be relatively low. The fuzzy, narrow eye shown in Figure 4(a) indicates the presence of considerable interference and an increased likelihood of bit errors in the primary signal demodulation.

Associated with each eye diagram is an "equalized SNR" that provides a quantitative measure of the eye opening. The equalized SNRs of the primary and secondary signals are 8 dB and 11 dB, respectively. For low PSRs, the equalized SNR of the primary demodulator is limited by cochannel interference, while the secondary signal SNR is limited by the additive noise and remodulation errors (caused by bit errors in the estimated primary data sequence). In this case, the primary demodulator makes few bit errors, and the AWGN is well below the secondary signal power. As the PSR drops below 3 dB, primary signal bit errors will degrade the secondary equalized SNR considerably.

The remodulator performance is quantified through the interim fit and the final fit. The interim fit measures the accuracy with which the first adaptive filter models the Manchester encoding and spectral shaping. It is defined as the ratio of the primary signal power at the FM discriminator output to the interim error signal power (see Figure 2).

For case #1, the interim fit is 3 dB. While this sounds like a poor fit, the interim error power is dominated by the presence of the secondary signal, and, consequently, a 3 dB interim fit does not necessarily imply a poor final fit.

The final fit measures how well the remodulator matches the primary signal. The final fit is defined as the ratio of the received primary signal power to the final error power (see Figure 2). For case #1, the final fit was 5 dB, indicating that the final error was mainly dominated by the secondary signal.

The primary and secondary eye patterns for case #2 are shown in Figures 5(a) and 5(b), respectively. The primary equalized SNR is 21 dB, and the secondary equalized SNR is 8 dB. In this case, the

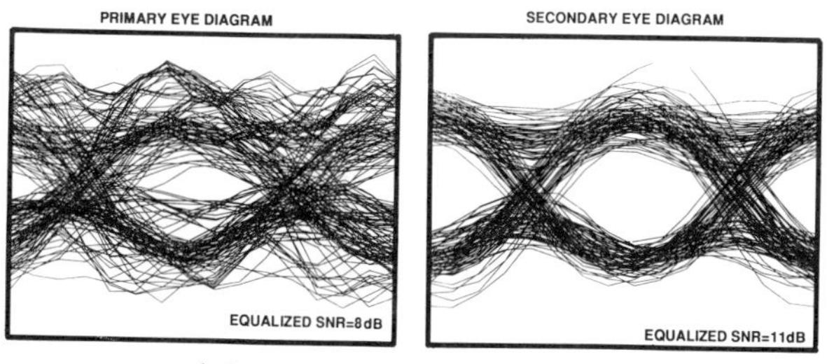

Figure 4: (a) Primary and (b) secondary demodulator eye diagrams for the 3 dB power separation case.

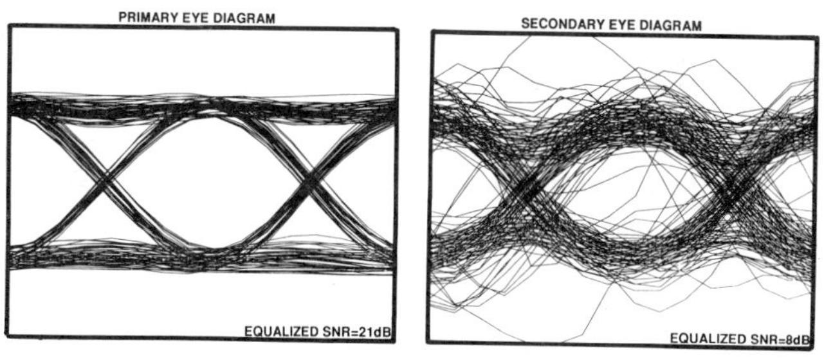

Figure 5: (a) Primary and (b) secondary demodulator eye diagrams for the 18 dB power separation case.

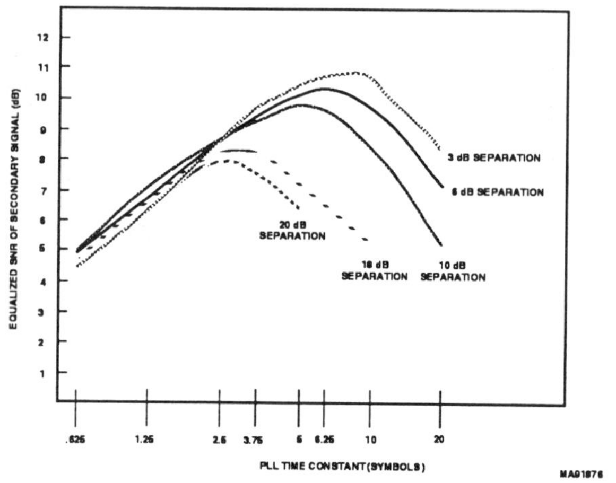

Figure 6: Secondary equalized SNR as a function of PLL time constant.

cochannel interference is low, and the primary demodulator is capable of achieving a high equalized SNR and a wide-open eye. The interim fit is 18 dB indicating that the primary data sequence has been remodulated to provide a near-perfect match. The final fit was also 18 dB indicating that the final error was dominated mainly by the secondary signal. However, as the eye diagram of Figure 5(b) shows, the secondary demodulator is making many bit errors. The source of the demodulation error is not AWGN, but rather untracked phase noise from the primary signal. The performance limitations induced by this phase noise are addressed in the next section.

Performance Limitations

The performance of the demod-remod technique is limited by additive white Gaussian noise, secondary signal interference, and phase noise on the primary signal. Phase noise is caused by a random, time-varying carrier frequency and phase drift. This drift originates in the transmitter, the receiver, or both, and exists on all real signals.

Phase noise fundamentally limits the ability to remodulate and cancel a real FSK signal. The PLL in the post-m filter can, however, partially track this phase noise. Phase noise consists of both high and low frequency components. The PLL time constant (the reciprocal of the loop filter bandwidth) controls how much phase noise is tracked. The smaller the PLL time constant (the wider the loop bandwidth), the more closely the PLL is able to track the phase noise. Although decreasing the PLL time constant causes the PLL to track more phase noise, it also makes the PLL more sensitive to additive noise, creating jitter on the NCO. Since the secondary signal is uncorrelated with the primary signal, it appears as noise to the PLL and contributes to the jitter. Increasing the loop time constant reduces the jitter but also reduces the ability to track phase noise on the primary signal. Untracked phase noise and jitter both degrade the secondary demodulation performance. Figure 6 shows the secondary equalized SNR as a function of the PLL time constant and PSR. The optimal time constant provides the best compromise between the amount of phase noise tracking and jitter reduction. As expected, the optimal time constant decreases as the PSR increases since the jitter caused by the secondary signal decreases.

Figure 7 illustrates the PSR/CNR region of operability for the experimental test setup of Figure 3. The shaded region shows the range of scenarios over which both the primary and secondary signals can be demodulated with an equalized SNR of 7 dB or better.

The upper boundary of the shaded region is set by phase noise limitations. When the secondary signal power is low (high PSR), the distortion caused by untracked phase noise significantly degrades secondary demodulation. In the 18 dB case described earlier, phase noise causes most of the distortion seen in Figure 5(b). As the PSR increases above 20 dB, the secondary signal suffers too much phase noise distortion to be demodulated. Thus, phase noise limits the maximum PSR to 20 dB, regardless of the CNR. Note that the phase noise limitation depends directly on the phase noise characteristics of the transmitter and receiver.

The lower boundary of the shaded region determines the minimum power separation required to demodulate the primary signal. In the previous section, it was shown that the primary demodulator was capable of functioning at a PSR of 3 dB. Below 2 dB, the

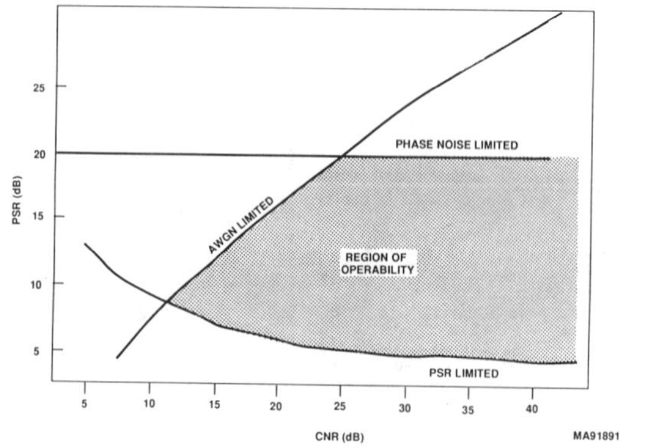

Figure 7: Typical region of operation of the demod-remod technique.

cochannel interference on the primary signal becomes so severe that incorrect bit decisions are frequently made, and neither the primary nor the secondary signals can be demodulated. Thus, secondary interference limits the minimum power separation to 2 dB. As the CNR decreases below 20 dB, the primary demodulator performance is degraded by Gaussian noise as well as cochannel interference, and the PSR must increase in order to maintain the same 7 dB performance. This explains the rise in the lower boundary as the CNR decreases.

Finally, the boundary on the left defines the maximum power separation that allows demodulation of the secondary signal in the presence of additive white Gaussian noise. As the CNR decreases, the additive noise power degrades the secondary demodulation. To maintain the same 7 dB performance, the maximum PSR must decrease as the CNR decreases to boost the secondary signal power relative to the noise. This explains the rise in the left boundary as the CNR increases.

The region enclosed by these boundaries represents the region over which the demod-remod technique can accurately demodulate both cochannel signals for the experimental scenario discussed above. Although this region is specific to this example, the general shape and principles used to establish these boundaries are general to all cochannel situations.

Summary

This paper has presented a method for demodulating two cochannel FSK signals based upon a demod-remod approach. Experimental results indicate that this technique is effective over a wide range of PSRs and CNRs. The technique was evaluated on live signals under realistic scenarios. Phase noise in the received signal causes a fundamental limitation in the technique. This limitation is partially mitigated by a tracking PLL. Further performance evaluation on live signals as a function of the demod- remod parameters is necessary before implementation should be considered.

Bibliography

1. R. P. Gooch and B. J. Sublett, "Demodulation of Cochannel QAM Signals, "*IEEE International Conference on Acoustics, Speech, and Signal Processing*, May 1989

2. K. H. Mueller and J. J. Werner, "A Hardware Efficient Passband Equalizer Structure for Data Transmission, "*IEEE Transactions on Communications*, March 1982

3. D. D. Falconer, "Jointly Adaptive Equalization and Carrier Recovery in Two-Dimensional Data Communication Systems, " *Bell Systems Technical Journal*, March 1976

4. B. Widrow and S. Stearns, *Adaptive Signal Processing*, Prentice Hall, 1985

Recursive Bayesian Algorithms for Blind Equalization

Ronald A. Iltis　　John J. Shynk　　K. Giridhar

Center for Information Processing Research
Dept. of Electrical and Computer Engineering
University of California
Santa Barbara, CA 93106

Abstract

A new blind equalization algorithm is presented based on a suboptimum Bayesian symbol sequence estimator. It is shown that a parallel bank of Kalman filters can be used to update a suboptimum Bayesian formula for the sequence probabilities. Two methods are used to reduce the computational complexity of the algorithm. First, it is shown that the Kalman filters can be replaced by simpler least-mean-square (LMS) adaptive filters. Second, the technique of reduced-state sequence estimation is adopted to reduce the number of symbol subsequences considered in the Bayesian updating, and hence the number of parallel filters required. The performance properties of the resulting algorithms are evaluated through bit-error rate simulations, and these are compared to the bounds of optimum maximum-likelihood sequence estimation. It is shown that the Kalman filter and LMS-based algorithms achieve blind start-up and rapid convergence (within 200 iterations) for both BPSK and QPSK modulation formats.

1 Introduction

Blind equalization algorithms attempt to determine the transmitted symbol sequence in the presence of ISI and without prior knowledge of the channel impulse response. Most effort in the development of blind equalizers has focussed on "property restoral" algorithms, such as CMA (constant modulus algorithm), in which a nonlinear function of a linear equalizer output is forced to a constant magnitude [1]. While CMA-type methods have the advantage of computational simplicity, their convergence cannot be guaranteed. Specifically, it has been shown that the CMA algorithm can converge to undesired local minima [2], and furthermore, convergence can be very slow. With the advent of programmable digital signal processing (DSP) devices, more sophisticated blind equalization algorithms should be considered, especially in relatively narrowband systems such as HF digital radio and voiceband modems.

In this paper, we present a new set of blind equalization algorithms, which are approximations to the optimum MAP sequence estimator for a priori unknown channels. The resulting algorithms can be implemented using a bank of Kalman or LMS adaptive filters, in which each filter maintains a channel estimate conditioned on a given data subsequence. In order to reduce the algorithm complexity, the technique of reduced state sequence estimation (RSSE) [3] is employed to minimize the number of channel estimators. The paper is organized as follows. The optimum MAP sequence estimator is discussed in Section 2, and the suboptimum parallel Kalman estimator is derived in Section 3. The LMS implementation is developed in Section 4. Computer simulations and conclusions follow in Sections 5 and 6, respectively.

2 Development of the Blind Equalization Algorithm

The following discrete-time channel and signal model is assumed, where $r(k)$ denotes the output of a matched filter at time k, $d(k)$ is the transmitted symbol, and the $b_n(k)$ represent the channel coefficients:

$$r(k) = \sum_{n=0}^{N_b} b_n(k)d(k-n) + n(k). \tag{1}$$

The additive noise sequence $n(k)$ is complex white Gaussian with variance σ_n^2. For convenience in the derivation of the suboptimum MAP sequence estimator and blind equalizer, the following sequences

*This work was sponsored in part by the University of California MICRO Program, Sonatech, Inc., and Applied Signal Technology, Inc.

1058-6393/91 $01.00 © 1991 IEEE

are also defined. The cumulative measurements r^k represent the matched filter samples $\{r(k), r(k-1), \ldots, r(0)\}$. A cumulative data sequence is similarly defined by $d_i^k = \{d_i(k), d_i(k-1), \ldots, d_i(0)\}$, for the ith of M^{k+1} possible sequences. A data subsequence comprising the data symbols associated with the channel coefficient vector $\mathbf{b}(k) \in \mathcal{C}^{N_b+1}$ at time k will also be required, and is given by $d_i^{k,N_b} = \{d_i(k), d_i(k-1), \ldots, d_i(k-N_b)\}$. It is further assumed that the vector of coefficients $b_n(k)$, denoted by $\mathbf{b}(k)$, evolves according to the following autoregressive model:

$$\mathbf{b}(k+1) = \mathbf{F}\mathbf{b}(k) + \mathbf{w}(k). \tag{2}$$

In (2), $\mathbf{F} \in \mathcal{C}^{N_b+1 \times N_b+1}$ is the one-step transition matrix, and $\mathbf{w}(k) \in \mathcal{C}^{N_b+1}$ is a circular white Gaussian process with covariance matrix $\mathbf{Q} \in \mathcal{C}^{N_b+1 \times N_b+1}$.

The optimum MAP sequence estimator is now reviewed. The MAP estimator can be written in the following recursive form:

$$p(d_i^k|r^k) = \frac{1}{c} p(r(k)|d_i^k, r^{k-1}) p(d_i^{k-1}|r^{k-1}), \tag{3}$$

for $i = 1, 2, \ldots, M^{k+1}$, where $p(d_i^k|r^k)$ represents the probability of the ith possible data sequence given cumulative measurements r^k. The likelihood $p(r(k)|d_i^k, r^{k-1})$ is given in terms of the Kalman filter (KF) innovations by

$$p(r(k)|d_i^k, r^{k-1}) = \mathcal{N}(r(k); \hat{s}_i(k), \sigma_i^2(k|k-1)) \tag{4}$$

where $\mathcal{N}(x; \bar{x}, \sigma_x^2)$ denotes a univariate circular Gaussian density with mean $\bar{x}$ and covariance σ_x^2. The estimated signal, $\hat{s}_i(k)$, is given by

$$\hat{s}_i(k) = \sum_{n=0}^{N_b} \hat{b}_{i,n}(k|k-1) d_i(k-n) \tag{5}$$

where $\hat{b}_{i,n}(k|k-1)$ is generated by the Kalman filter equations discussed in Section 3. From (3) and (4), it is seen that the optimum MAP sequence estimator requires a bank of M^{k+1} Kalman filter channel estimators. Unfortunately, the number of channel estimators required increases exponentially with time. We thus seek a simplification of the optimum MAP sequence estimator in which the number of metrics evaluated remains fixed.

3 Blind Equalization Using Parallel Kalman Channel Estimators

The suboptimum MAP sequence estimator, or blind equalizer, is now derived. In order to simplify the recursion in (3), the conditional probability density function of the channel coefficient vector, $\mathbf{b}(k)$, is approximated as complex Gaussian.

$$p(\mathbf{b}(k)|d_i^{k,N_b}, r^{k-1}) \approx \mathcal{N}(\mathbf{b}(k); \hat{\mathbf{b}}_i(k|k-1), \mathbf{P}_i(k|k-1)) \tag{6}$$

where $\mathcal{N}(\mathbf{x}, \bar{\mathbf{x}}, \mathbf{P})$ represents a circular multivariate Gaussian density with mean vector $\bar{\mathbf{x}}$ and covariance matrix $\mathbf{P}$. (Note that the true density of $\mathbf{b}(k)$ conditioned on r^{k-1}, d^{k,N_b} is a Gaussian sum.)

The MAP estimate of the subsequence d_i^{k,N_b} is given by

$$p(d_i^{k,N_b}|r^k) = \frac{1}{c} p(r(k)|d_i^{k,N_b}, r^{k-1}) \times \sum_{\{j: d_j^{k-1,N_b} \in d_i^{k,N_b}\}} p(d_j^{k-1,N_b}|r^{k-1}) \tag{7}$$

where c is a normalization constant. (In the above, subsequence $d_j^{k-1,N_b} \in d_i^{k,N_b}$ implies that the first N_b symbols in subsequence d_j^{k-1,N_b} equal the last N_b symbols in d_i^{k,N_b}.) Furthermore, under approximation (6), the likelihood $p(r(k)|d_i^{k,N_b}, r^{k-1})$ is also Gaussian. Thus,

$$p(r(k)|d_i^{k,N_b}, r^{k-1}) = \mathcal{N}(r(k); \hat{s}_i(k), \sigma_i^2(k|k-1)). \tag{8}$$

The mean and innovations covariance, $\hat{s}_i(k)$ and $\sigma_i^2(k|k-1)$, are determined using the KF equations as follows:

$$\begin{aligned} \hat{s}_i(k) &= \sum_{n=0}^{N_b} \hat{b}_{i,n}(k|k-1) d_i(k-n) \qquad (9) \\ \sigma_i^2(k|k-1) &= \mathbf{h}_i(k)\mathbf{P}_i(k|k-1)\mathbf{h}_i^H(k) + \sigma_n^2 \\ \mathbf{h}_i(k) &= [d_i(k), d_i(k-1), \ldots, d_i(k-N_b)]. \end{aligned}$$

At this point, one step in the recursion of (7) can be completed. However, the approximate predicted mean $\hat{\mathbf{b}}_i(k+1|k)$ and covariance $\mathbf{P}_i(k+1|k)$ must be computed for the next iteration. As shown in [4], the predicted mean and covariance are completely determined by the mean vector $\hat{\mathbf{b}}_i(k|k)$ and covariance matrix $\mathbf{P}_i(k|k)$. These quantities are given by the conventional Kalman filter equations. The measurement update is given by

$$\hat{\mathbf{b}}_i(k|k) = \hat{\mathbf{b}}_i(k|k-1) + \frac{1}{\sigma_i^2(k|k-1)} \times \mathbf{P}_i(k|k-1)\mathbf{h}_i^H(k)\left[r(k) - \hat{s}_i(k)\right]. \tag{10}$$

The new predicted estimate, $\hat{\mathbf{b}}_i(k+1|k)$, can be computed in terms of the measurement update as shown in Table 1. The associated error covariances are derived in [4], and are also given in Table 1. The blind equalizer structure is thus seen to consist of a bank of M^{N_b+1} conditional channel estimators. The symbol decisions are made using the MAP probability metrics themselves. The suboptimal decision rule used in the results presented here considers only the metric with the largest value, $\hat{d}_i^{k,N_b} = \arg\max_{d_i^{k,N_b}} p(d_i^{k,N_b}|r^k)$.

One possible drawback to the algorithm in Table 1 is that M^{N_b+1} Kalman measurement updates must be computed at each iteration, where M is the symbol alphabet size, and $N_b + 1$ is the length of the channel impulse response. To alleviate this problem, we consider a simplified algorithm using the concept of reduced state sequence estimation (RSSE), first introduced in [3]. In RSSE, the symbol subsequences d^{k,N_b} are grouped into reduced state sequences, denoted here by $D^{k,N_b} = \{D(k), D(k-1), \ldots, D(k-N_b)\}$. (Although the D^{k,N_b} are not, strictly speaking, states in a VA trellis, each D^{k,N_b} corresponds to an individual Kalman channel estimator, and thus a reduction in the number of symbol subsequences can greatly simplify the algorithm.) Each symbol, $D(k-i)$, is actually a subset of the symbols $d(k-i)$, and has dimensionality $M_i \leq M$. For example, in the case of QPSK with $N_b + 1 = 3$, we could choose $M_0 = 4, M_1 = 2$, and $M_2 = 1$, which results in a dimensionality of 8 for the reduced subsequence D^{k,N_b}, as opposed to 64 for the original subsequence d^{k,N_b}. Thus, eight instead of 64 Kalman channel estimators are required. A recursion for the MAP metrics of the reduced state subsequences is derived in [4], as well as the modified Kalman filter equations.

4 Blind Equalization Using Parallel LMS Adaptive Filters

In order to reduce the complexity even further, we propose using scalar *gradient* algorithms for the Kalman filter measurement updates. The LMS-based algorithm consists of $N = M^{N_b+1}$ single-input, single-output adaptive FIR filters comprised of the coefficient estimates $\hat{\mathbf{b}}_i(k|k-1)$, $i = 1, \ldots, M^{N_b+1}$. The filter inputs are determined by all possible subsequences, $\mathbf{d}_i(k)$, and each filter output, $\hat{s}_i(k)$, is generated according to the inner product in (9). Thus, the output of the ith FIR filter corresponds to an estimate of the current received symbol, $r(k)$, assuming that the ith subsequence was transmitted. The filter outputs are then compared to the received sample, $r(k)$, to generate a set of innovations or prediction error signals, $e_i(k|k-1)$, and these correspond to the last term of the conditional measurement update in Table 1.

The algorithm is simplified by setting all covariance matrices and the system matrix $\mathbf{F}$ equal to the identity matrix. Table 2 outlines the simplified algorithm. Further simplification of the algorithm (for PSK signal formats) can be achieved by setting $\sigma_i(k)$ equal to the constant value σ for all k. The resulting measurement update is equivalent to the LMS algorithm with step size $\mu(k) = \alpha(k)/\sigma^2$. Since the parameters are now fixed, the conditional innovations covariances in Table 2 are no longer computed. Thus, the simplified algorithm requires only (i) the conditional measurement updates with parameter $\mu(k)$, (ii) the weighting probability updates with parameter σ, and, finally, (iii) the one-step predictions based on the results of (i) and (ii).

5 Computer Simulations

The new blind equalization algorithm was simulated for BPSK and QPSK signals using the KF implementation as well as the simpler LMS version. A complex channel was chosen with the transfer function

$$H(z) = 0.444487 + (-0.0488658 - j0.776700)z^{-1} + (-0.440101 + j0.0555976)z^{-2}.$$

This channel has infinite nulls on the unit circle, which is difficult to equalize when the equalizer is an FIR filter.

Figure 1 shows the evolution of the probability metrics of the KF equalizer for SNRs of 20dB and 10dB. Since the channel has three coefficients and $M = 2$, there are 8 possible subsequences, and thus 8 probability metrics. Observe that for the higher SNR, one of the metrics converges to 1 in less than 20 iterations. Figure 2 shows the trajectories for the corresponding average coefficient errors, which were obtained by averaging the squared error between the actual channel coefficients and those of the channel estimator with the *largest* probability metric at each iteration. The bit-error rate (BER) performance curves for BPSK are given in Figure 3. The solid curve, corresponding to zero ISI, is the performance of binary signaling on an additive white Gaussian noise channel (AWGN). The upper limit on the symbol error probability, assuming coherent detection, is given by [5], and depends on the worst-case ISI. The BER for the new algorithm was estimated using Monte-Carlo simulations. Finally, to obtain a tight lower bound on the performance of our algorithm, we simulated the algorithm of Abend and Fritchman [6], which assumes perfect knowledge of the channel and realizes the optimum MAP sequence estimator. This bound is shown for the BPSK case by the dashed line in Fig. 3. Figure 4 shows the corresponding BER performance of the algorithm for QPSK signaling. The reduced state sequence estimator (RSSE) was also simulated in order to decrease the complexity of the filter bank for QPSK signaling, and to observe the effect on performance. The subsets for the 3 feedback taps were defined with $M_0 = 4$, $M_1 = 2$, and $M_2 = 1$. Figure 5 shows the trajectories of the probability metrics of the 8 subsets ($M_0 M_1 M_2 = 8$). For this case, we used a real channel with the transfer function $H(z) = 0.24 + 0.84z^{-1} + 0.48z^{-2}$. The corresponding coefficient estimation error trajectory is also shown in the figure.

6 Conclusions

A new set of blind equalization algorithms has been presented that are approximations to the true MAP sequence estimator for a priori unknown channels. The algorithm consists of M^{N_b+1} conditional Kalman or LMS channel estimators, whose innovations are used to update the MAP sequence probabilities. The Kalman filter based algorithm provided excellent blind start-up performance, with the probability metrics and channel estimates both converging in about twenty iterations. For both the KF and LMS algorithms, the simulated bit-error rate was indistinguishable from that of the optimum MAP sequence estimator. The performance of an RSSE version of the algorithm was found to be poor at intermediate SNRs (10dB). While the RSSE algorithm could compute accurate channel estimates at high SNR (greater than 20dB), its use seems to be more appropriate for ISI-limited channels (rather than thermal noise-limited channels).

To conclude, the Kalman and LMS-based versions of the algorithm, while computationally complex, provide extremely rapid start-up. For relatively narrowband applications, the algorithms developed here are readily implementable in programmable digital signal processing devices, and may provide much better performance than property-restoral type algorithms.

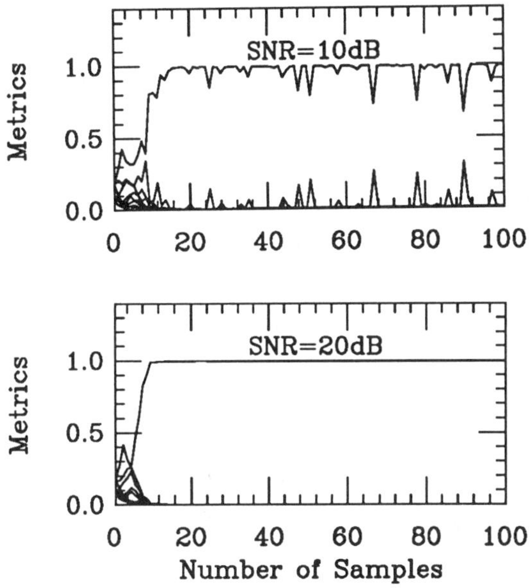

Figure 1: Evolution of the probability metrics (KF, BPSK).

References

[1] J.R. Treichler and B.G. Agee, "A new approach to multipath correction of constant modulus signals," *IEEE Transactions on Acoustics, Speech, and Signal Processing*, vol. ASSP-31, pp. 459–472, Apr. 1983.

[2] Z. Ding, C.R. Johnson, and R.A. Kennedy, "On the admissibility of blind adaptive equalizers," *Proceedings of IEEE Int. Conf. on Acoustics, Speech, and Signal Processing*, pages 1707–1710, Albuquerque, NM, Apr. 1990.

[3] V.M. Eyuboglu and S.U.H. Qureshi, "Reduced-state sequence estimation with set partitioning and decision feedback," *IEEE Transactions on Communications*, vol. 36, pp. 13–20, Jan. 1988.

[4] R.A. Iltis, J.J. Shynk, and K. Giridhar, "Bayesian algorithms for blind equalization using parallel adaptive filtering," *IEEE Transactions on Communications*, submitted, 1991.

[5] J.G. Proakis, *Digital Communications*, McGraw-Hill, New York, Second Ed., 1989.

[6] K. Abend and B.D. Fritchman, "Statistical detection for communication channels with intersymbol interference," *Proceedings of the IEEE*, vol. 58, pp. 779–785, May 1970.

Table 1: Blind Equalization Algorithm

Given: $\hat{\mathbf{b}}_i(k|k-1), \mathbf{P}_i(k|k-1)$

Compute Conditional Measurement Update

$$\hat{\mathbf{b}}_i(k|k) = \hat{\mathbf{b}}_i(k|k-1) + \frac{1}{\sigma_i^2(k|k-1)}\mathbf{P}_i(k|k-1)\mathbf{h}_i^H(k)\left(r(k)-\hat{s}_i(k)\right)$$

Compute Conditional Error Covariances

$$\mathbf{P}_i(k|k) = \left[\mathbf{I} - \frac{1}{\sigma_i^2(k|k-1)} \times \mathbf{P}_i(k|k-1)\mathbf{h}_i^H(k)\mathbf{h}_i(k)\right]\mathbf{P}_i(k|k-1)$$

Update Weighting Probabilities

$$p(d_i^{k,N_b}|r^k) = \frac{1}{c}\mathcal{N}\left(r(k);\hat{s}_i(k),\sigma_i^2(k|k-1)\right) \times \sum_{\{j:d_j^{k-1,N_b}\in d_i^{k,N_b}\}} p(d_j^{k-1,N_b}|r^{k-1})$$

Compute One-Step Predictions

$$\hat{\mathbf{b}}_i(k+1|k) = \sum_{\{j:d_j^{k,N_b}\in d_i^{k+1,N_b}\}} \mathbf{F}\,\hat{\mathbf{b}}_j(k|k)\frac{p(d_j^{k,N_b}|r^k)}{\sum_{\{m:d_m^{k,N_b}\in d_i^{k+1,N_b}\}} p(d_m^{k,N_b}|r^k)}$$

Compute Covariance Outer Product Terms

$$\nu_{j,i}(k) = \left[\hat{\mathbf{b}}_i(k+1|k) - \mathbf{F}\,\hat{\mathbf{b}}_j(k|k)\right]^T$$

Update Error Covariances

$$\mathbf{P}_i(k+1|k) = \sum_{\{j:d_j^{k,N_b}\in d_i^{k+1,N_b}\}} \left(\mathbf{F}\,\mathbf{P}_j(k|k)\mathbf{F}^T + \mathbf{Q} + \nu_{j,i}(k)\nu_{j,i}^H(k)\right) \times \frac{p(d_j^{k,N_b}|r^{k-1})}{\sum_{\{m:d_m^{k,N_b}\in d_i^{k+1,N_b}\}} p(d_m^{k,N_b}|r^k)}$$

Table 2: Simplified Blind Equalization Algorithm

Define Observation Vectors

$$\mathbf{h}_i(k) = [d_i(k),\ldots,d_i(k-N_b)]$$

Compute Signal Estimates

$$\hat{s}_i(k) = \sum_{n=0}^{N_b}\hat{b}_{i,n}(k|k-1)d_i(k-n)$$

Compute Conditional Measurement Update

$$\hat{\mathbf{b}}_i(k|k) = \hat{\mathbf{b}}_i(k|k-1) + \frac{\alpha(k)}{\sigma^2}\mathbf{h}_i^H(k)\left(r(k)-\hat{s}_i(k)\right)$$

Update Weighting Probabilities

$$p(d_i^{k,N_b}|r^k) = \frac{1}{c}\mathcal{N}\left(r(k);\hat{s}_i(k),\sigma^2\right) \times \sum_{\{j:d_j^{k-1,N_b}\in d_i^{k,N_b}\}} p(d_j^{k-1,N_b}|r^{k-1})$$

Compute One-Step Predictions

$$\hat{\mathbf{b}}_i(k+1|k) = \sum_{\{j:d_j^{k,N_b}\in d_i^{k+1,N_b}\}} \hat{\mathbf{b}}_j(k|k)\frac{p(d_j^{k,N_b}|r^k)}{\sum_{\{m:d_m^{k,N_b}\in d_i^{k+1,N_b}\}} p(d_m^{k,N_b}|r^k)}$$

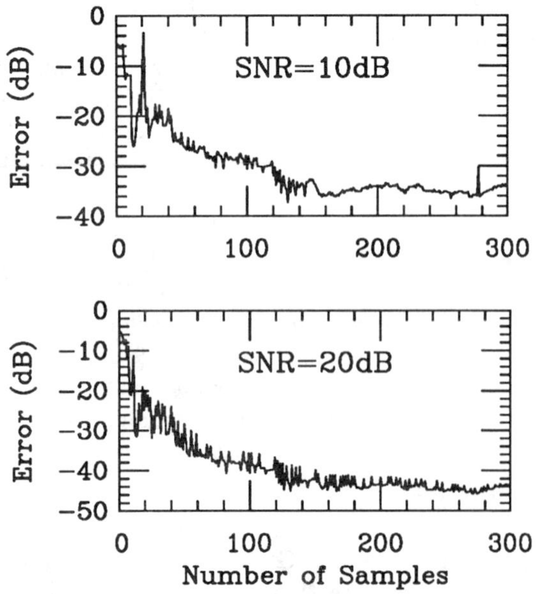

Figure 2: Coefficient error trajectories (KF, BPSK).

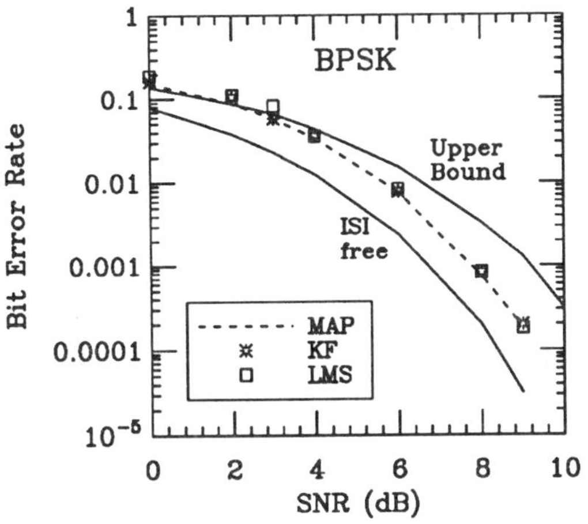

Figure 3: BER performance curves (BPSK).

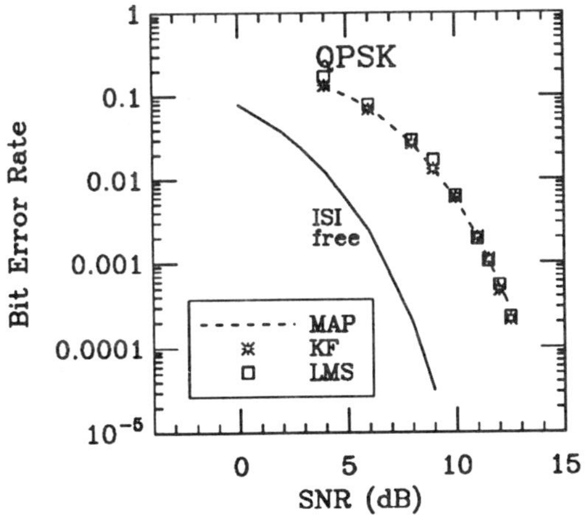

Figure 4: BER performance curves (QPSK).

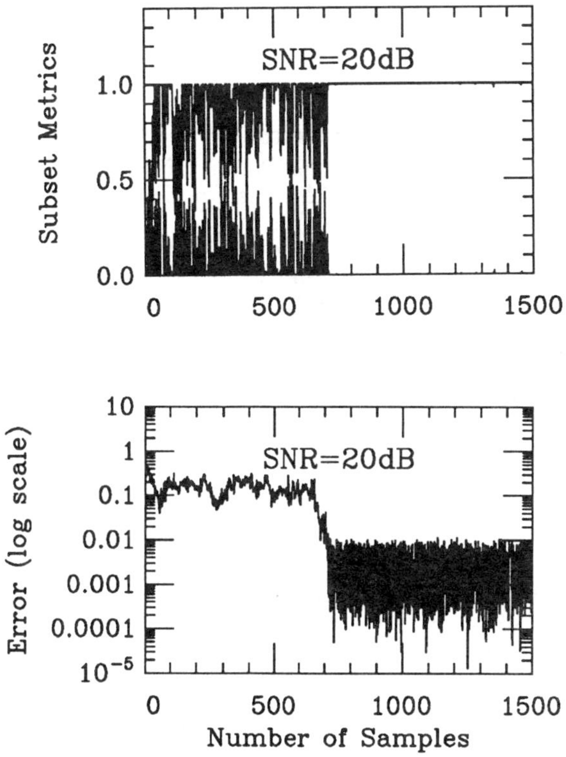

Figure 5: Evolution of the probability metrics and the corresponding coefficient error trajectories (KF, RSSE, QPSK).

Maximum-Likelihood Approaches to Blind Adaptive Signal Extraction Using Narrowband Antenna Arrays

Brian G. Agee
Radix Technologies, Inc.
4020 Moorpark Ave., Suite 212
San Jose, California 95117

Abstract

A unifying approach to design of blind adaptive signal extraction algorithms is presented, based on maximum-likelihood (ML) estimation of unknown signals with known modulation properties. The ML estimator of a signal-of-interest (SOI) waveform is derived, under the assumption that the SOI has known modulation properties, and that the SOI is received by a narrowband antenna array in the presence of temporally-white interference with unknown spatial covariance and complex Gaussian probabilistic fraction-of-time-distribution. It is shown that the ML waveform estimate can be computed from the dominant mode of a generalized eigenequation if the SOI is a burst, narrowband, or spread spectrum waveform, or if the SOI is a perfectly conjugate self-coherent waveform, such as a DSB-AM or BPSK signal. It is also shown that the ML waveform estimate can be computed using an alternating projections formula if the SOI is a constant modulus waveform, such as an FM or CPFSK signal. It is demonstrated that many of the blind array adaptation algorithms developed to date can be interpreted as maximum-likelihood waveform estimators.

1 Introduction

In the past decade, a number of techniques have been advanced for adapting filters, demodulators, and antenna arrays to extract unknown signals of interest (SOIs) from data corrupted by noise and interference signals. These techniques, referred to as *blind adaptation techniques* due to their ability to adapt the receiver processor without knowledge of the actual content of the SOIs, instead operate by exploiting known *properties* of the SOI waveforms or symbol sequences that distinguish them from the background interference, or are destroyed by propagation through the transmission channel. Example techniques include:

- *anticipatory techniques*, such as the Maximin algorithm [1] and the burst-acquisition algorithm reported in [2], which exploit the known temporal or frequency support of the transmitted signal waveforms;

- *coherence exploiting techniques*, such as the self-coherence restoral algorithms [3,4] and the higher-order spectral analysis and cumulant approaches [5,6], which exploit the known spatial, spectral, or higher-order self-coherence properties of the transmitted signal waveforms; and

- *modulus restoral techniques*, such as the constant modulus algorithm [8], known modulus algorithm [9], and multiple modulus algorithm [10], which exploit the known modulus properties of the transmitted signal waveforms or symbol sequences.

The diversity of approaches for blind adaptation has spawned a great deal of discussion on the relative merits of these algorithms. To a large extent, the applicability of one class of approaches over another is self-evident; for example, modulus restoral techniques are the obvious choice in environments containing stationary SOIs with well-known modulus properties, such as audio FM waveforms. However, this determination is much harder to make if the SOI properties do not hold exactly, e.g., if the FM waveforms are harshly bandlimited, or if the SOIs possess multiple exploitable properties, and the "best" algorithm can be very difficult to discern within a given class of techniques. These observations underscore the need to develop methodologies for developing blind adaptation algorithms that links these algorithms to some optimality criterion.

This paper presents such an approach, by developing blind adaptation algorithms as *maximum-likelihood (ML) estimators* of unknown signals with known modulation properties. The general ML estimator of a single signal-of-interest (SOI) with known properties is derived, under the assumption that the SOI is received by a narrowband antenna array in the presence of temporally-white complex-Gaussian interference with unknown spatial covariance. It is shown that the ML estimator admits a simple objective function if the SOI can be modelled as a member of a *property set* that encompasses all signals with the properties possessed by the transmitted SOI waveform or symbol sequence. It is also demonstrated that many of the blind array adaptation algorithms developed to date can be interpreted as maximum-likelihood waveform estimators.

2 Basic Data Models

2.1 Received Data Model

The basic ML estimators are derived in the *narrowband single-SOI multisensor environment* where a single unknown SOI with known properties is received by an M-element narrowband antenna array in the presence of random temporally-white interference with temporal mean zero, unknown spatial covariance, and a circularly-symmetric complex-Gaussian (CSCG) cumulative fraction-of-time distribution (CFOTD). The received data signal is also assumed to be downconverted to complex baseband and critically-sampled at a rate that exceeds the complex bandwidth of the SOI. This model obtains, for example, if $s(m)$ is received in the presence of spatially incoherent background noise and arbitrary numbers of spatially coherent co-channel interferers, as long as each of those interferers has a flat frequency distribution over the bandwidth of the receiver and a CSCG CFOTD.

The received data signal $\{\mathbf{x}(m)\}_{m=1}^{N}$ is then modelled by

$$\mathbf{x}(m) = \mathbf{a}s(m) + \mathbf{u}(m), \qquad (1)$$

where $\mathbf{a}$ is the SOI steering vector, $\{s(m)\}_{m=1}^{N}$ is the transmitted SOI data sequence, and $\{\mathbf{i}(m)\}_{m=1}^{N}$ is a sequence of M-dimensional vectors with joint CFOTD

$$f_{\{\mathbf{i}(m)\}_1^N}(\mathbf{z}_1, \ldots, \mathbf{z}_N) = \prod_{m=1}^{N} \pi^{-M} \left|\mathbf{R_{ii}}\right|^{-1} e^{-\mathbf{z}_m^H \mathbf{R}_{\mathbf{ii}}^{-1} \mathbf{z}_m}, \qquad (2)$$

and where $\mathbf{R_{ii}}$ is the temporal interference autocorrelation matrix and $|\cdot|$ denotes the determinant operation. In order to derive the maximum-likelihood waveform estimator, $\mathbf{R_{ii}}$, $\mathbf{a}$ and $\{s(m)\}_1^N$ are assumed to be unknown. In addition, $\{s(m)\}_1^N$ is assumed to possess known properties such that it belongs to some well-defined *property set* $\mathcal{D}_s$. However, no explicit properties are assumed for $\mathbf{a}$ or $\mathbf{R_{ii}}$, except that $\mathbf{R_{ii}}$ is a positive-definite Hermitian matrix. In particular, $\mathbf{a}$ is *not* assumed to be a member of a known "array manifold," i.e., $\mathbf{a}$ is assumed to range over the M-dimensional complex field $\mathcal{C}^M$.

In order to simplify the derivation of the ML estimators, the following matrix notation is also introduced here:

$$\mathbf{X} \triangleq \begin{bmatrix} \mathbf{x}^H(1) \\ \vdots \\ \mathbf{x}^H(N) \end{bmatrix} \qquad \mathbf{s} \triangleq \begin{bmatrix} s^*(1) \\ \vdots \\ s^*(N) \end{bmatrix} \qquad (3)$$

Using (3), the received data environment and interference CFOTD can be expressed in matrix form as $\mathbf{X} = \mathcal{I} + \mathbf{s}\mathbf{a}^H$, where $\mathcal{I}$ is the matrix representation of the CSCG background interference. The goal is then to derive the joint maximum likelihood estimate of $\mathbf{a}$, $\mathbf{s}$ and $\mathbf{R_{ii}}$, under the assumptions stated above, and given $\mathcal{I}$ modelled by (2).

2.2 Transmitted Signal Models

Following the notation given in (3) the SOI data vector $\mathbf{s}$ is assumed to be *unknown*, with known properties that allow it to be modelled as a member of a well-defined *property set* $\mathcal{D}_\mathbf{S}$. Three property sets are considered here:

- *linear projection sets*, where $\mathbf{s}$ is assumed to be a member of a known *linear subspace*, such that $\mathbf{s} = \mathbf{P_S}\mathbf{s}$ for some known projection matrix $\mathbf{P_S}$;
- *perfect conjugate self-coherence signal sets*, where $\mathbf{s}$ is assumed to possess perfect conjugate self-coherence at some value of frequency shift, such that $\mathbf{s} = \mathbf{s}^*$ after removal of that frequency shift; and
- *modulus property sets*, where the elements of $\mathbf{s}$ are assumed to have moduli (complex amplitude) with known properties.

Examples of signals that can be modelled as members of linear projection sets include: burst and transient signals with known (or estimable) limited temporal support; narrowband signals with known (or estimable) limited frequency support; and direct sequence spread spectrum signals with known codes and known (or estimable) code timings. For example, a burst signal that is known to appear in a data collect after sample N_{ON} satisfies $\mathbf{s} = \mathbf{P_S}\mathbf{s}$ for

$$\mathbf{P_S} = \begin{bmatrix} \mathbf{0}_{N_{\mathrm{ON}}} & \mathbf{0} \\ \mathbf{0} & \mathbf{I}_{N-N_{\mathrm{ON}}} \end{bmatrix}$$

where N is the total number of samples in the collect and $\mathbf{0}_L$ and $\mathbf{I}_L$ is the $L \times L$ zero matrix and identity matrix for a given L, respectively. Examples of signals that possess perfect conjugate self-coherence include DSB-AM and BPSK waveforms. Examples of signals that can be modelled as members of modulus property sets include FM, PSK and FSK signals, which have constant moduli over the reception interval. A number of additional signal types can also be modelled as members of one or more of these property sets [3].

3 ML Waveform Estimator

3.1 General Solution

The general maximum-likelihood estimate of $\mathbf{s}$ is derived for the received data model given in (??)-(??), by jointly maximizing the log of the conditional CFOTD of $\mathbf{X}$ given $\mathbf{a}$, $\mathbf{R_{ii}}$, and $\mathbf{s}$,

$$S_{\mathrm{ML}}(\mathbf{s}, \mathbf{a}, \mathbf{R_{ii}}) = -N \ln \left|\mathbf{R_{ii}}\right| - \mathrm{Tr}\left\{\mathbf{R}_{\mathbf{ii}}^{-1}\left(\hat{\mathcal{I}}^H \hat{\mathcal{I}}\right)\right\} \qquad (4)$$

with respect to $\mathbf{R_{ii}} > 0$, $\mathbf{a} \in \mathcal{C}^M$, and $\mathbf{s} \in \mathcal{D}_\mathbf{S}$, where $\hat{\mathcal{I}} = \mathbf{X} - \mathbf{s}\mathbf{a}^H$. Maximization of (4) with respect to $\mathbf{R_{ii}}$ and $\mathbf{a}$ is straightforward, yielding

$$\left.\hat{\mathbf{R}}_{\mathbf{ii}}\right|_{\mathrm{ML}} = \frac{1}{N}\left(\hat{\mathcal{I}}^H \hat{\mathcal{I}}\right) \qquad (5)$$

$$\hat{\mathbf{a}}_{\mathrm{ML}} = \frac{\mathbf{X}^H \mathbf{s}}{\|\mathbf{s}\|^2}. \qquad (6)$$

Substituting (5)-(6) into (4) and removing terms that do not depend on $\mathbf{a}$, $\mathbf{R_{ii}}$, or s yields

$$\mathrm{S_{ML}}(\mathrm{s}) \rightarrow \frac{\mathrm{s}^H \mathbf{X}\left(\mathbf{X}^H \mathbf{X}\right) \mathbf{X}^H \mathrm{s}}{\|\mathrm{s}\|^2} \quad (7)$$

$$\hat{\mathrm{s}}_{\mathrm{ML}} = \arg \max_{\mathrm{s} \in \mathcal{D}_\mathrm{S}} \mathrm{S_{ML}}(\mathrm{s}) \quad (8)$$

Objective function (7) is referred to here as the *concentrated maximum-likelihood objective function.* Note that the ML estimator is ill-posed if no constraints are imposed on s. This underscores the importance of $\mathcal{D}_\mathrm{S}$ in designing a feasible ML waveform estimator.

3.2 Solution for Linear Subspace Property Sets

If s is a member of a linear subspace, such that $\mathrm{s} = \mathbf{P}_\mathrm{S}\mathrm{s}$ for some projection matrix $\mathbf{P}_\mathrm{S}$, then $\mathbf{X}^H\mathrm{s} = \mathbf{X}_\mathrm{S}^H\mathrm{s}$, where $\mathbf{X}_\mathrm{S} \triangleq \mathbf{P}_\mathrm{S}\mathbf{X}$ is the projection of the received data onto $\mathcal{D}_\mathrm{S}$, and $\mathrm{S_{ML}}$ can be expressed as

$$\mathrm{S_{ML}}(\mathrm{s}) = \frac{\mathrm{s}^H \mathbf{X}_\mathrm{S}\left(\mathbf{X}^H \mathbf{X}\right)^{-1} \mathbf{X}_\mathrm{S}^H \mathrm{s}}{\|\mathrm{s}\|^2}. \quad (9)$$

Substituting general representation

$$\mathrm{s} = \mathbf{X}_\mathrm{S}\mathbf{w} + \mathrm{s}_\perp, \qquad \mathbf{X}_\mathrm{S}^H \mathrm{s}_\perp = 0 \quad (10)$$

into (9) yields

$$\mathrm{S_{ML}}(\mathrm{s}) = \frac{\mathbf{w}^H\left(\mathbf{X}_\mathrm{S}^H\mathbf{X}_\mathrm{S}\right)\left(\mathbf{X}^H\mathbf{X}\right)^{-1}\left(\mathbf{X}_\mathrm{S}^H\mathbf{X}_\mathrm{S}\right)\mathbf{w}}{\mathbf{w}^H\left(\mathbf{X}_\mathrm{S}^H\mathbf{X}_\mathrm{S}\right)\mathbf{w} + \|\mathrm{s}_\perp\|^2}, \quad (11)$$

which is maximized with respect to $\mathbf{w}$ and $\mathrm{s}_\perp$ for $\mathrm{s}_\perp = 0$ and

$$\mathbf{w} = \arg \max_{\mathbf{w} \in \mathcal{C}^M} \hat{\mathrm{S}}_{\mathrm{ML}}(\mathbf{w})$$

$$\hat{\mathrm{S}}_{\mathrm{ML}}(\mathbf{w}) = \frac{\mathbf{w}^H\left(\mathbf{X}_\mathrm{S}^H\mathbf{X}_\mathrm{S}\right)\left(\mathbf{X}^H\mathbf{X}\right)^{-1}\left(\mathbf{X}_\mathrm{S}^H\mathbf{X}_\mathrm{S}\right)\mathbf{w}}{\mathbf{w}^H\left(\mathbf{X}_\mathrm{S}^H\mathbf{X}_\mathrm{S}\right)\mathbf{w}}$$

$\hat{\mathrm{S}}_{\mathrm{ML}}$ is maximized by setting $\mathbf{w}$ equal to the dominant mode of

$$\lambda\left(\mathbf{X}_\mathrm{S}^H\mathbf{X}_\mathrm{S}\right)\mathbf{w} = \left(\mathbf{X}_\mathrm{S}^H\mathbf{X}_\mathrm{S}\right)\left(\mathbf{X}^H\mathbf{X}\right)^{-1}\left(\mathbf{X}_\mathrm{S}^H\mathbf{X}_\mathrm{S}\right)\mathbf{w}$$

$$\Leftrightarrow \quad \gamma\left(\mathbf{X}_\perp^H\mathbf{X}_\perp\right)\mathbf{w} = \left(\mathbf{X}_\mathrm{S}^H\mathbf{X}_\mathrm{S}\right)\mathbf{w}, \quad \gamma = \frac{\lambda}{1-\lambda}, \quad (12)$$

where $\mathbf{X}_\perp = \mathbf{X} - \mathbf{X}_\mathrm{S}$ is the projection of $\mathbf{X}$ onto $\mathcal{D}_\mathrm{S}^\perp$, the linear subspace *orthogonal* to $\mathcal{D}_\mathrm{S}$. The maximum-likelihood estimate of s is therefore given by

$$\hat{\mathrm{s}}_{\mathrm{ML}} = \mathbf{X}_\mathrm{S}\hat{\mathbf{w}}_{\mathrm{ML}} \quad (13)$$

$$\Leftrightarrow \quad \hat{s}_{\mathrm{ML}}(m) = \hat{\mathbf{w}}_{\mathrm{ML}}^H \mathbf{x}_s(m) \quad (14)$$

where $\hat{\mathbf{w}}_{\mathrm{ML}}$ is the dominant mode of (12). This weight vector can also be interpreted as the maximal point of the Rayleigh Quotient

$$\hat{\mathrm{S}}_{\mathrm{ML}}(\mathbf{w}) = \frac{\mathbf{w}^H\left(\mathbf{X}_\mathrm{S}^H\mathbf{X}_\mathrm{S}\right)\mathbf{w}}{\mathbf{w}\left(\mathbf{X}_\perp^H\mathbf{X}_\perp\right)\mathbf{w}}. \quad (15)$$

$$\Leftrightarrow \quad \tilde{\mathrm{S}}_{\mathrm{ML}}(\mathbf{w}) = \frac{\mathbf{w}^H\hat{\mathbf{R}}_{\mathbf{x}_s\mathbf{x}_s}\mathbf{w}}{\mathbf{w}^H\hat{\mathbf{R}}_{\mathbf{x}_\perp\mathbf{x}_\perp}\mathbf{w}} \quad (16)$$

$$\hat{\mathbf{R}}_{\mathbf{x}_s\mathbf{x}_s} = \frac{1}{N_s}\mathbf{X}_\mathrm{S}^H\mathbf{X}_\mathrm{S} \quad (17)$$

$$\hat{\mathbf{R}}_{\mathbf{x}_\perp\mathbf{x}_\perp} = \frac{1}{N-N_s}\mathbf{X}_\perp^H\mathbf{X}_\perp \quad (18)$$

where N_s is the *rank* (trace) of $\mathbf{P}_\mathrm{S}$. The ML weight vector can then be described as the dominant mode of $\hat{\mathbf{T}}_{\mathbf{xx}} = \hat{\mathbf{R}}_{\mathbf{x}_\perp\mathbf{x}_\perp}^{-1}\hat{\mathbf{R}}_{\mathbf{x}_s\mathbf{x}_s}$.

The maximum-likelihood weight vector $\hat{\mathbf{w}}_{\mathrm{ML}}$ can be estimated by a variety of means. Example techniques include: eigenstructure approaches, which perform a singular-value decomposition on either $\hat{\mathbf{T}}_{\mathbf{xx}}$, using standard techniques for nonHermitian matrices, and then set $\mathbf{w}$ to the righthand eigenvector associated with the maximum eigenvalue; power-method approaches, which maximize (16) by successive multiplication of $\hat{\mathbf{w}}$ by $\hat{\mathbf{T}}_{\mathbf{xx}}$ from some initial value; and steepest-ascent approaches, which maximize (16) by the gradient of $\tilde{\mathrm{S}}_{\mathrm{ML}}$.

In fact, many of the blind and partially-blind adaptation algorithms reported in the literature can be interpreted as methods for optimizing $\tilde{\mathrm{S}}_{\mathrm{ML}}$. In particular, the real and complex *Maximin algorithms* reported in [1] for blind reception of frequency-hopped signals can be interpreted as practical implementations of the steepest-ascent technique for maximizing $\tilde{\mathrm{S}}_{\mathrm{ML}}$, while the transient exploiting algorithm reported in [2] for detection and extraction of burst SOIs can be interpreted as an implementation of a power-method algorithm for determining the dominant mode of $\hat{\mathbf{T}}_{\mathbf{xx}}$.

The ML objective function $\tilde{\mathrm{S}}_{\mathrm{ML}}$ can also be used as a detection statistic, if $\mathbf{P}_\mathrm{S}$ is not known exactly. For example, if s is a burst signal with unknown ON and OFF transitions, then these transitions can be estimated by computing $\tilde{\mathrm{S}}_{\mathrm{ML}}$ over a hypothesized set of transitions. This concept forms the basis for the detection component of the burst detection algorithm described in [2]. This approach can be used to estimate other signal parameters as well, for example, the transmission frequency of a narrowband signal, or the timing phase of a DSSS waveform. In particular, if $\mathbf{X}_\mathrm{S}$ has rank 1, then $\tilde{\mathrm{S}}_{\mathrm{ML}}$ has a closed form maximum given by

$$\left.\tilde{\mathrm{S}}_{\mathrm{ML}}\right|_{\max} = \mathbf{x}_s^H(m)\hat{\mathbf{R}}_{\mathbf{x}_\perp\mathbf{x}_\perp}^{-1}\mathbf{x}_s(m), \quad (19)$$

and $\tilde{\mathrm{S}}_{\mathrm{ML}}$ reduces to the generalized likelihood ratio statistic reported in [12], and to the maximum-likelihood code timing estimation algorithm reported in [13].

3.3 Solution for the Perfect Conjugate Self-Coherence Property Set

The ML estimator for perfectly conjugate self-coherent SOIs is derived by considering the case where $s = s^*$. Then

$$\begin{aligned} & s^H X \left(X^H X\right)^{-1} X^H s \\ &= s^H \tilde{X} \Lambda_X^{-1} \tilde{X}^H s \end{aligned}$$

$$\begin{aligned} \tilde{X} &\triangleq \frac{1}{\sqrt{2}} [X \; X^*] \\ \Lambda_X &\triangleq \begin{bmatrix} X^H X & 0 \\ 0 & X^T X^* \end{bmatrix} \end{aligned}$$

holds, and S_{ML} can be expressed as

$$S_{ML}(s) = \frac{s^H \tilde{X} \Lambda_X^{-1} \tilde{X}^H s}{\|s\|^2}$$

Following the derivation employed for the linear subspace sets, the ML estimate of s is then given by

$$\begin{aligned} \hat{s}_{ML} &= \tilde{X}\tilde{w}_{ML}, \qquad \tilde{w}_{ML} = \frac{1}{\sqrt{2}} \begin{bmatrix} \hat{w}_{ML} \\ \hat{w}_{ML}^* \end{bmatrix} && (20) \\ &= \mathrm{Re}\{X\hat{w}_{ML}\} && (21) \end{aligned}$$

where $\tilde{w}_{ML}$ is the dominant mode of the eigenequation

$$\begin{aligned} \lambda \left(\tilde{X}^H \tilde{X}\right) \tilde{w} &= \left(\tilde{X}^H \tilde{X}\right) \Lambda_X^{-1} \left(\tilde{X}^H \tilde{X}\right) \tilde{w} \\ \Leftrightarrow \lambda \Lambda_X \tilde{w} &= \left(\tilde{X}^H \tilde{X}\right) \tilde{w} && (22) \end{aligned}$$

Expanding out the terms in (22), and performing straightforward matrix manipulations yields equivalent ML estimation formula

$$\begin{aligned} \hat{s}_{ML}(m) &= \mathrm{Re}\{\hat{w}_{ML}^H x(m)\} && (23) \\ \nu_{max} \hat{w}_{ML} &= \hat{T}_{XX^*} \hat{T}_{XX^*}^* \hat{w}_{ML} && (24) \\ \hat{T}_{XX^*} &\triangleq \hat{R}_{XX}^{-1} \hat{R}_{XX^*} && (25) \end{aligned}$$

where $\nu_{max} = (2\lambda_{max} - 1)^2$. Equation (24) is identical to the cross-SCORE eigenequation given in [4], configured to exploit conjugate self-coherence features at $\alpha = 0$. Thus, the conjugate cross-SCORE algorithm can be seen to provide the ML estimate of the SOI waveform, in environments where the SOI has perfect conjugate self-coherence.

3.4 Solution for Modulus Property Sets

A closed form solution cannot be derived for SOIs that belong to modulus sets. However, an iterative algorithm can be derived to obtain the ML estimate, by posing $\hat{s}_{ML}$ as

$$\hat{s}_{ML} = \arg \min_{s \mathcal{D}_S} \left\{ \min_{w \in \mathcal{C}^M} \hat{F}_{ML}(s, w) \right\} \quad (26)$$

where $\hat{F}_{ML}$ is the *maximum-likelihood cost function*,

$$\hat{F}_{ML}(s, w) \triangleq \frac{\|s - Xw\|^2}{\|s\|^2}. \quad (27)$$

Estimation of $\hat{s}_{ML}$ can therefore be accomplished by *jointly* minimizing $\hat{F}_{ML}$ with respect to beamformer weight $w \in \mathcal{C}^M$ and $s \in \mathcal{D}_S$.

This minimization can be accomplished using an *alternating projections* procedure, that is, by alternately minimizing $\hat{F}_{ML}$ with respect to $s \in \mathcal{D}_S$ for fixed w, and with respect to $w \in \mathcal{C}^M$ for fixed s. Minimization of $\hat{F}_{ML}$ with respect to w yields

$$\begin{aligned} \hat{w}_{opt}(s) &= \left(X^H X\right)^{-1} X^H s \\ &= \hat{R}_{XX}^{-1} \hat{R}_{Xs} && (28) \end{aligned}$$

In many environments, minimization of $\hat{F}_{ML}$ with respect to s is equally straightforward. For example, if s is a member of the set of *constant modulus waveforms*, then $\hat{F}_{ML}$ is easily minimized with repect to s for

$$\begin{aligned} \hat{s}_{opt}(w) &= \mathrm{CL}\{Xw\} \\ \Leftrightarrow s(m) &= \mathrm{CL}\{w^H x(m)\} && (29) \end{aligned}$$

where $\mathrm{CL}\{\cdot\}$ is the *element-by-element complex limiting* operation, given by $\mathrm{CL}\{z\} \triangleq \frac{z}{|z|}$ for complex scalars z. Combining (28) and (29) leads to the *constant modulus property-mapping recursion*,

$$\begin{aligned} \hat{w} &\leftarrow \left(X^H X\right)^{-1} X^H \hat{s} && (30) \\ \hat{s} &\leftarrow \mathrm{CL}\{X\hat{w}\} && (31) \end{aligned}$$

$$\begin{aligned} \Leftrightarrow \hat{w} &\leftarrow \hat{R}_{XX}^{-1} \hat{R}_{X\hat{s}} && (32) \\ \hat{s}(m) &\leftarrow \mathrm{CL}\{\hat{w}^H x(m)\} && (33) \end{aligned}$$

which is continued until s and w converge to fixed values.

Recursion (32)-(33) is identical to the *least-squares CMA (LSCMA)* first introduced in [11]. As shown in [3], the recursion can also be extended to signals with more complex modulus properties. In particular, the algorithm can be extended to signals with known or periodically repeating moduli.

In addition, a link can be established between the ML waveform estimator described above and several of the steepest-descent algorithms previously reported in the literature. In particular, substitution of (29) into $\hat{F}_{ML}$ yields

$$\hat{F}_{ML}(\hat{s}_{opt}, w) = \left\langle (|y(m)| - 1)^2 \right\rangle_N \quad (34)$$

where $y(m) = w^H x(m)$ is the beamformer output signal and $\langle \cdot \rangle_N$ denotes time-averaging over the reception interval. Equation (34) is the familiar *1-2 constant modulus cost function* defined in [8], formulated here in terms of time averaging over the reception interval rather than statistical averaging over the ensemble of probable signal paths. Consequently, any algorithm designed to optimize (34) can be interpreted as a maximum-likelihood estimator of constant modulus waveforms.

4 Conclusions

A general approach for blind adaptive signal processing, based on maximum-likelihood estimation of signals with known properties received by antenna arrays, is presented. ML waveform estimators are developed for signals with limited temporal and/or spectral support over known time/frequency intervals, for example, burst, narrowband, and frequency agile waveforms; for signals with perfect conjugate self-coherence, for example, DSB-AM and BPSK signals; and for signals with known modulus properties, for example, FM, FSK, PSK, and pulsed radar signals.

The techniques have also been shown to correspond to several technioques already reported in the literature. In particular, the Maximin and burst-acquisition/DMP algorithms can be interpreted as steepest-ascent [1] or power-method [2] approaches for computing the ML estimate of time-limited, bandlimited, time-hopped, or spread spectrum waveforms, and the conjugate cross-SCORE algorithm [4] can be interpreted as an approach for computing the maximum-likelihood estimate of DSB-AM and BPSK signal waveforms with perfect conjugate self-coherence. Similarly, the class of "1-2" modulus restoral algorithms can be interpreted as steepest-descent [8,9] or alternating-projections [3,11] approaches for computing the ML estimate of FM, PSK, or pulsed waveforms.

The estimation approach developed here can also be generalized in a number of different manners. The specific approaches presented here are easily extended to transmitted signals with more complex modulus properties, for example, burst BPSK waveforms. The alternating projections algorithm described in Section 3.4 also bears a strong resemblance to the set-theretical signal reconstruction techniques reported in [14,15] and the composite property mapping techniques reported in [16], and it is likely that these more general techniques can be used to further generalize the approaches developed here.

References

[1] K. Bakhru, D. J. Torrieri, "The Maximin Algorithm for Adaptive Arrays and Frequency-Hopping Communications," *IEEE Trans. Ant. and Prop.*, vol. AP-32, no. 9, pp. 919-928, Sept. 1984

[2] B. G. Agee, "Fast Acquisition of Burst and Transient Signals Using a Predictive Adaptive Beamformer," in *Proc. 1989 IEEE Military Communications Conference*, 1989

[3] B. G. Agee, "The Property Restoral Approach to Blind Adaptive Signal Extraction," Ph.D. Dissertation, Dept. of EECS, Univ. of Calif., Davis, 1989

[4] B. G. Agee, S. V. Schell, W. A. Gardner, "Spectral Self-Coherence Restoral: A New Approach to Blind Adaptive Signal Extraction Using Antenna Arrays," *Proc. IEEE*, vol. 78, no. 4, pp. 753-767, April 1990

[5] C. L. Nikias, M. R. Raghuveer, "Bispectrum Estimation: a Digital Signal Processing Framework," *Proc. IEEE*, vol. 75, no. 7, pp. 869-891, July 1987

[6] G. B. Giannakis, "Cumulants: a Powerful Tool in Signal Processing," *Proc. IEEE*, vol. 75, no. 9, pp. 1333-1334, September 1987

[7] D. N. Godard, "Self-Recovering Equalization and Carrier Tracking in Two-Dimensional Data Communication Systems," *IEEE Trans. Comm.*, vol. COM-28, pp. 1867-1875, November 1980

[8] J. R. Treichler, B. G. Agee, "A New Approach to Multipath Correction of Constant Modulus Signals," *IEEE Trans. ASSP*, vol. ASSP-31, pp. 459-472, April 1983

[9] J. R. Treichler, M. L. Larimore, "New Processing Techniques Based on the Constant Modulus Algorithm," in *IEEE Trans. ASSP*, pp. 420-431, April 1985

[10] W. A. Sethares, G. A. Rey, C. R. Johnson, "Approaches to Blind Equalization of Signals With Multiple Modulus," in *Proc. 1989 Intl. Conf. on ASSP*, pp. 972-975, 1989

[11] B. G. Agee, "The Least-Squares CMA: A New Technique for Rapid Correction of Constant Modulus Signals," in *Proc. 1986 Intl. Conf. on ASSP*, pp. 953-956, 1986

[12] E. Kelly, "An Adaptive Detection Algorithm," *IEEE Trans. Aero. and Elect. Systems*, vol. AES-22, pp. 115-127, Mar. 1987

[13] D. Dlugos, R. Scholtz, "Acquisition of Spread Spectrum Signals by an Adaptive Array," *IEEE Trans. ASSP*, vol. 37, no. 8, pp. 1253- 1270, Aug. 1989

[14] D. Youla, "Generalized Image Restoration by the Method of Alternating Orthogonal Projections," *IEEE Trans. Circuits, Syst.*, vol. CAS-25, pp. 694-702, Sept. 1978

[15] D. Youla, H. Webb, "Image Restoration by the Method of Convex Projections: Part 1 – Theory," *IEEE Trans. Med. Imaging*, vol. MI-1, pp. 81-94, Oct. 1982

[16] J. Cadzow, "Signal Enhancement – a Composite Property Mapping Algorithm," *IEEE Trans. ASSP*, January 1989

A New Blind Equalization Algorithm Using Higher Order Statistics in a Decision Feedback Structure *

Anthony G. Bessios and Chrysostomos L. Nikias

Signal and Image Processing Institute
Department of Electrical Engineering-Systems
University of Southern California
Los Angeles, CA 90089-2564

Abstract

In this paper we present a new adaptive blind equalization algorithm named POTEA the Power cepstrum and Tricoherence Equalization Algorithm using a nonlinear structure. It is based on the combined use of second- and fourth-order statistics of the received sequence. A nonlinear decision-feedback equalization (DFE) structure is proposed for equalizing nonminimum-phase channels with severe distortion, since a linear equalizer does not perform well in cases of channels with deep spectral nulls in their amplitude characteristics. The direct form structure of the DFE consists of a feed-forward T-spaced filter (FFF) with an all-pass mixed phase transfer function, and a feed-backward T-spaced filter (FBF) with a minimum phase one. In this paper the performance of POTEA for linear and decision-feedback equalizers is tested, for the cases of $16-QAM$ *and* $64-QAM$.

1 Introduction

In this paper we present a new adaptive blind equalization algorithm named POTEA the Power cepstrum and Tricoherence Equalization Algorithm [4] with a nonlinear structure. It is based on the combined use of second- and fourth-order statistics of the received sequence. The only information blind equalizers require is related to statistical properties of the transmitted sequence.

A zero-forcing (ZF) constraint is imposed, for a complete cancellation of the intersymbol interference at the output of the equalizer. A nonlinear decision-feedback equalization (DFE) structure is proposed for equalizing nonminimum-phase channels with severe distortion. More specifically, in cases of channels with deep spectral nulls in their amplitude characteristics the linear equalizer does not perform well, because it places a large gain in the vicinity of the spectral null, which enhances the additive noise present in the received signal.

The direct form structure of the DFE consists of a feed-forward T-spaced filter (FFF) and a feed-backward T-spaced filter (FBF). The FFF is driven by the samples of the received signal $\{y(k)\}$ while the FBF is driven by decisions at the output of the detector and its coefficients are adjusted to cancel the ISI on the current symbol that results from past detected symbols. In the DFE structure of POTEA , the FFF part has an all-pass mixed phase transfer function while the FBF has a minimum phase one.

An adaptive gradient-type scheme is employed, in order to estimate the cepstral coefficients of the transfer function of the channel $F(z)$. The sums and the differences of the cepstral coefficients are computed, using the cepstrum of the power spectrum (second-order domain) and the cepstrum of the tricoherence index (fourth- order domain), respectively. The sums of the cepstral coefficients of the transfer function of the channel $F(z)$ contain the amplitude information, whereas their difference contain the phase information. This property makes the POTEA algorithm capable of identifying and equalizing a nonminimum-phase channel from its output only.

We should emphasize that the algorithm performs simultaneous channel identification and equalization, because the channel and its inverse filter (equalizer) have the same cepstral coefficients with opposite signs. Finally, like all the Higher-Order Spectra (HOS) based techniques, POTEA is less sensitive to additive Gaussian noise, because HOS of Gaussian processes of order

*This work was supported by the Office of Naval Research under contract N00014-92-J-1034

greater than two are identical to zero [7].

POTEA (with a linear structure) has been shown [4] to have superior performance in terms of convergence than the existing TEA (the Tricepstrum Equalization Algorithm) originally introduced by Hatzinakos and Nikias [5], and accordingly, to converge faster than the Bussgang-type blind equalization algorithms (Stop-and-Go, Godard or CMA, Benveniste-Goursat, Sato etc.) at the expense of higher computational complexity [5]. Another limitation of the Bussgang-type techniques is that they are not globally convergent, as POTEA is, because they are based on minimization of nonconvex cost functions.

In this paper the performance of POTEA for linear equalizers and DFE is tested, especially with channels that have deep spectral nulls. It is demonstrated that when the zeros of the channel transfer function approach the unit circle, the linear equalizer fails to work, while the DFE performs successful equalization in the cases of $16-QAM$ and $64-QAM$.

2 Problem Definition - Preliminaries

Let us assume that the received sequence, $\{y(n)\}$, is a filtered and noise corrupted version of a transmitted sequence $\{x(n)\}$:

$$y(n) = z(n) + w(n) = \sum_{k=-L_2}^{L_1} f(k)x(n-k) + w(n) \quad (1)$$

where $\{f(n)\}$ is the impulse response of the channel. The input complex sequence $\{x(n)\}$ is non-Gaussian white, i.i.d. with mean $E\{x(n)\} = 0$, second-order moment $Q = E\{x(n)x^*(n)\} \neq 0$, skewness $\beta = E\{x(n)^3\} = 0$, and kurtosis $\gamma_x = E\{|x(n)|^4\} - 2E^2\{|x(n)|^2\} \neq 0$. Finally, $\{w(n)\}$ is zero-mean AGN (additive Gaussian noise) and statistically independent from $\{x(n)\}$.

The transfer function of the system $F(z)$ may be expressed by its poles and zeros as follows:

$$F(z) = Gz^{-r}I(z^{-1})O(z), \quad (2)$$

where G is a constant, r is an integer ,

$$I(z^{-1}) = \prod_{i=1}^{L_1}(1-a_i z^{-1}) \quad \text{(minimum phase component)} \quad (3)$$

$$O(z) = \prod_{i=1}^{L_2}(1-b_i z) \quad \text{(maximum phase component)} \quad (4)$$

because $|a_i| < 1, |b_i| < 1$ for all i.

Let $Y(z)$ be the Z-transform of $\{y(n)\}$.The cepstrum of the Power spectrum is defined as : $c_{P_y}(m) \triangleq Z^{-1}[lnP_y(z)]$, where $Z^{-1}[.]$ denotes inverse Z-transform and $P_y(z)$ is the power spectrum of $\{y(n)\}$.

For symmetric QAM signals the second order moments are defined as follows:

$$r(m) = E\{y(n)y^*(n+m)\} \neq 0, \quad (5)$$

or equivalently $Z[r(m)] = QF(z^{-1})F^*(z^{-1})$, where $Q = E\{x(n)x^*(n)\}$.

The cepstrum of the power spectrum of $\{y(n)\}$ is given by the sum of $A^{(k)}$,$B^{(k)}$ (the minimum and maximum phase cepstral coefficients of F(z)) [8], [4]. These are :

$$A^{(k)} = \sum_{i=1}^{L_1} a_i^k, \qquad B^{(k)} = \sum_{i=1}^{L_2} b_i^k, \quad (6)$$

The tricoherence function is given by:

$$t_y(z_1,z_2,z_3) = [\frac{T_y(z_1,z_2,z_3)}{T_y^*(z_1^{-1},z_2^{-1},z_3^{-1})}]^{1/2} \quad (7)$$

where $T_y(z_1,z_2,z_3)$ is the trispectrum; i.e., the Z-transform of the fourth-order cumulants defined by:

$$\begin{aligned} L(m,n,l) \triangleq\ & E\{y(i)y^*(i+m)y(i+n)y^*(i+l)\} \\ & -E\{y(i)y^*(i+l)\}E\{y^*(i+m)y(i+n)\} \\ & -E\{y(i)y^*(i+m)\}E\{y(i+n)y^*(i+l)\} \end{aligned}$$

The cepstrum of the tricoherence is defined as the inverse Z-transform of the complex logarithm of the tricoherence function (7). That is

$$c_{t_y}(m,n,l) \triangleq Z^{-1}[ln[t_y(z_1,z_2,z_3)]]. \quad (8)$$

It can be shown [4] that:

$$c_{t_y}(m,n,l) = jZ^{-1}[\phi(z_1)+\phi(z_2)+\phi(z_3)-\phi(z_1^{-1}z_2^{-1}z_3^{-1})] \quad (9)$$

From the above equation, it is obvious that the cepstrum of the tricoherence contains the phase information of the channel only.

The complex cepstrum of the tricoherence function can be shown [4] to be equal with the difference of the cepstral coefficients A^k, B^k, along the 3 axes and the main diagonal in a 3-D plot.

The channel identification procedure in POTEA is based on the computation of the cepstral parameters $A^{(I)}$ and $B^{(I)}$, respectively, of the received sequence

$\{y(n)\}$ and it can be shown [6] that the cepstral parameters of the output sequence are identical to those of the channel impulse response $\{f(n)\}$ because the channel input is white:

$$c_y(m,n,l) = c_f(m,n,l) + c\delta(m,n,l) \quad (10)$$

which implies that $c_y(m,n,l) = c_f(m,n,l)$ for $(m,n,l) \neq (0,0,0)$, where $c_y(m,n,l)$ is the tricepstrum of the received sequence $\{y(n)\}$ and $c_f(m,n,l)$ is the tricepstrum of the channel $\{f(n)\}$.

3 POTEA with a DFE structure.

In this section the POTEA algorithm is derived for a DFE structure with a FFF part consisting of N_1 taps and having transfer function $U_1(z)$ and a FBF part with N_2 taps and transfer function $U_2(z)$, see Figure 1. Using a zero forcing (ZF) equalization constraint [10] we obtain:

$$U(z) = \frac{1}{F(z)} = U_1(z)\frac{1}{1+U_2(z)} \quad (11)$$

Combining equations (2),(12) we obtain the following:

$$\begin{aligned} U_1(z) &= \frac{O^*(z^{-1})}{G.O(z)} \\ U_2(z) &= I(z^{-1}).O^*(z^{-1}) - 1 \end{aligned} \quad (12)$$

The above infinite impulse response (IIR) ransfer functions, are approximated by equivalent finite impulse response (FIR) filters.

The steps of the POTEA-DFE algorithm are as follows:

- Let N_1, N_2: lengths of FFF and FBF respectively.
  p : Length of minimum and maximum phase cepstral parameters.
  At iteration i=1,2,...

1 Estimate adaptively the $L_y^{(i)}(m,n,l)$ $-M_1 \leq m,n,l \leq M_1$, and $r_y^{(i)}(m)$ $-M_2 \leq m \leq M_2$ from a finite length window of $\{y(n)\}$ and then generate the following functions:

$$\begin{aligned} \theta_1^{(i)}(m,n,l) &= L_y^{*(i)}(-m,-n,-l) * L_y^{(i)}(m,n,l) \\ \theta_2^{(i)}(m,n,l) &= L_y^{*(i)}(-m,-n,-l) * mL_y^{(i)}(m,n,l) \end{aligned}$$

2 Choose p arbitrarily large such that $A^{(p+1)} \approx 0$, $B^{(p+1)} \approx 0$ and then define :$w = p$, $z \leq \frac{w}{2}$, $h \leq z$.

3 Form the following equations :

$$\sum_{k=1}^{p} S^{*(k)}[-r_y(m-k)] + \sum_{k=1}^{p} S^{(k)}[r_y(m+k)] = mr_y(m) \quad (13)$$

where $S^{(k)} = A^{(k)} + B^{*(k)}$, $k = 1,...,p$, $m = 1,...,2p$.

$$\sum_{k=1}^{p} D^{*(k)}[\theta_1(m-k,n-k,l-k) - \theta_1(m-k,n,l)] +$$

$$\sum_{k=1}^{p} D^{(k)}[\theta_1(m+k,n+k,l+k) - \theta_1(m+k,n,l)] =$$

$$= \theta_2(m,n,l) + \theta_2^*(-m,-n,-l)$$

and the following system of equations

$$\begin{aligned} \hat{P}\hat{a} &= \hat{p} \quad (14) \\ \hat{Q}\hat{b} &= \hat{q} \quad (15) \end{aligned}$$

Solve adaptively the above systems employing gradient-type adaptation [11].

4 Calculate $A^{(k)}$ and $B^{(k)}$ as follows:

$$A^{(k)} = \frac{S^{(k)} + D^{(k)}}{2} \quad B^{(k)} = (\frac{S^{(k)} - D^{(k)}}{2})^* \quad (16)$$

5 Calculate the normalized (G=1) equalizer coefficients as follows:

$$\hat{\imath}(i,k) = -\frac{1}{k}\sum_{n=2}^{k+1} A_{(i)}^{(n-1)}\hat{\imath}(i,k-n+1) \quad k = 1,...,N_1 \quad (17)$$

$$\hat{o}(i,k) = \frac{1}{k}\sum_{n=k+1}^{0} B_{(i)}^{(1-n)}\hat{o}(i,k-n+1) \quad (18)$$

$k = -1,...,-N_2$

with initialization: $\hat{\imath}_{eq}(i,0) = \hat{o}_{eq}(i,0) = 1$ The normalized (G=1) estimate $\hat{U}_{norm}(i,z)$ at iteration (i) is given by equations (11),(12).

6 The reconstructed transmitted sequence at iteration (i) is the inverse Z-transform of $\tilde{X}(z)$:

$$\tilde{X}(i,z) = \hat{U}_{norm}(i,z).Y(z) \quad (19)$$

4 Simulation Results

In this section the effectiveness of POTEA under a nonlinear decision feedback structure in terms of rate of convergence and robustness in signal conditioning is demonstrated with simulation examples. For the simulations a real-life channel was used with a transfer function having only a single zero, of the form $1 + az^{-1}$. The values for the parameters in step 2 are : p=6, w=6, z=1 and h=0. The SNR shown in the figures is the signal-to-noise ratio at the input of the equalizer. The performance of linear and nonlinear POTEA was studied comparatively by approaching the zero closer to the unit circle.In Figure 2 the case of $1 + 0.7z^{-1}$ is shown for 16-QAM where both equalizers obtain equalization. In Figure 3 the case of $1+0.8z^{-1}$ is shown for 16-QAM and 64-QAM and the linear equalizer does not perform equalization while the DFE does. Finally in Figure 4 the successful performance of DFE is shown even for the difficult case of $1 + 0.95z^{-1}$.

References

[1] Y. Sato, '**A Method for Self-Recovering Equalization for Multilevel Amplitude-Modulation Systems**,'*IEEE Trans. on Communications*, vol. COM-23, pp. 679-682, June 1975.

[2] D. N. Godard, '**Self Recovering Equalization and Carrier Tracking in two-Dimensional Data communication Systems**,'*IEEE Trans. on Communications*, vol. COM-28, pp. 1867-1875, Nov. 1980.

[3] S. Bellini and F. Rocca, '**Blind Deconvolution: Polyspectra or Bussgang Techniques** ,'in *Digital Communications*, E. Biglieri and G. Prati Eds. North-Holland, 1986, pp. 251-262.

[4] A. Bessios and C. L. Nikias, **Blind Equalization based on cepstra of Power Spectrum and Tricoherence**,*Proc. SPIE '91*, San Diego, CA, July 1991.

[5] D. Hatzinakos and C. L. Nikias, **Blind Equalization Using a Tricepstrum Based Algorithm**, IEEE Trans. on Communications, accepted for publication, to appear in 1991.

[6] Hsing-Hsing Chiang and C. L. Nikias, '**Deconvolution and Identification of Nonminimum Phase FIR Systems Based on Cumulants** ,'*IEEE Trans. on Automatic Control*, vol.35,No.1, January 1990.

[7] C. L. Nikias and M. R. Raghuveer, 'Bispectrum Estimation: A Digital Signal Processing Framework,'*Proc. IEEE*, Vol. 75, No. 7, pp. 869-891, July 1987.

[8] C.L. Nikias and F. Liu,**Bicepstrum computation based on second and third order statistics with applications**, *Proc. ICASSP '90*, pp.2381-2386, Albuquerque,NM,April 1990.

[9] R. Pan and C. L. Nikias, **The Complex Cepstrum of Higher- Order Cumulants and Nonminimum Phase System Identification**, *IEEE Trans. ASSP*, Vol. ASSP-36, pp. 186-205, Feb. 1988.

[10] J. G. Proakis, **Digital Communications**, N. York: McGraw-Hill, 1983

[11] S. Haykin, **Adaptive Filter Theory**, Prentice Hall Inc., Englewood Cliffs, N.J., 1986.

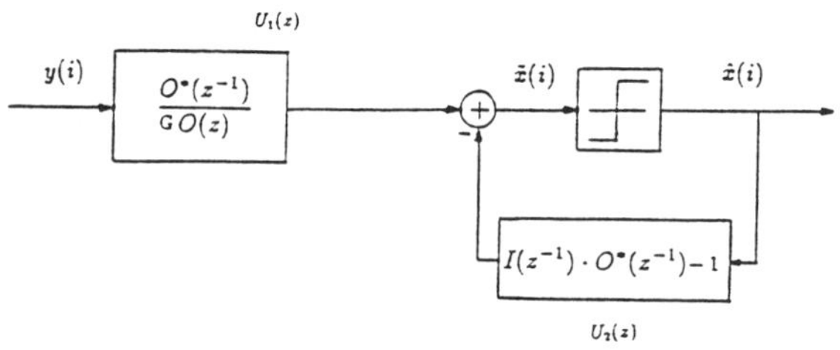

FIGURE 1: DFE structure

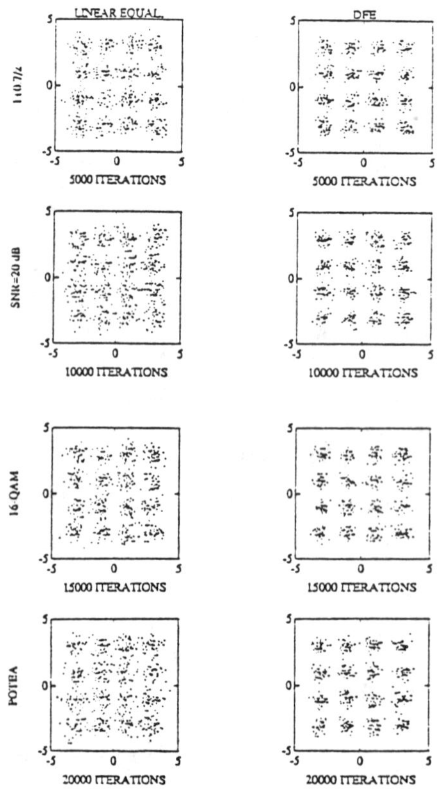

FIGURE 2: Performance of POTEA/DFE

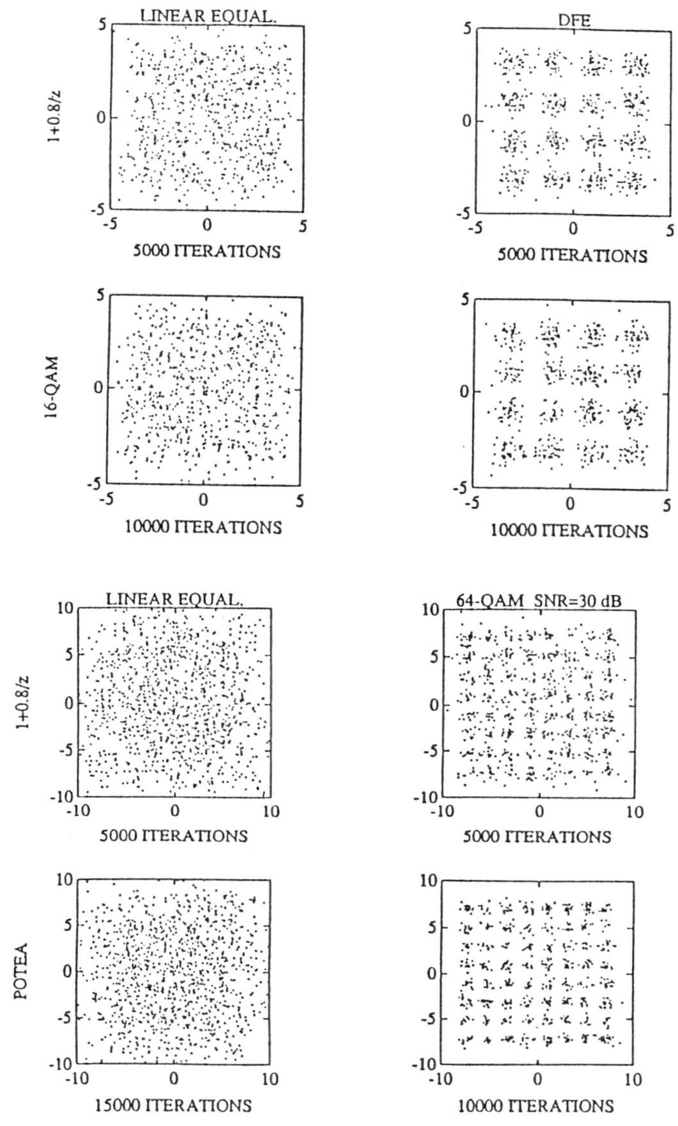

FIGURE 3. Performance of POTEA/DFE

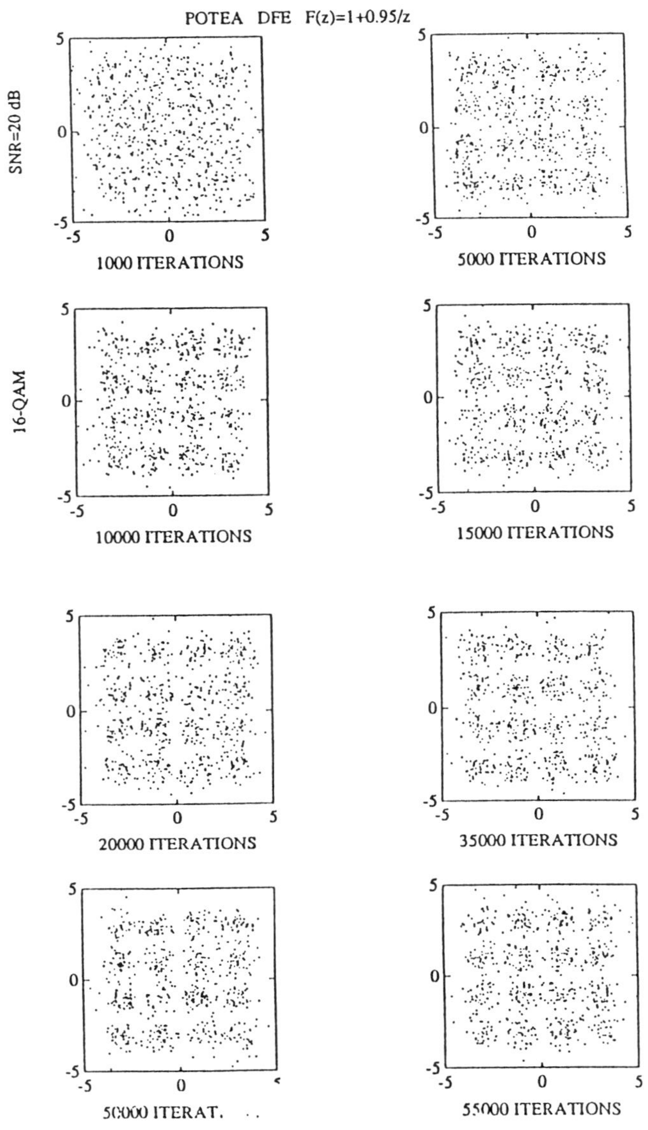

FIGURE 4. Performance of POTEA/DFE

An Adaptive Blind Equalization Algorithm for QAM and QPR Modulations: The Concentric Ordered Modulus Algorithm, with Results for 16QAM and 9QPR

Fernando López de Victoria

Applied Signal Technology, Inc.
Sunnyvale, CA 94086

Abstract

The Concentric Ordered Modulus Algorithm (COMA) represents an approach to the blind equalization of QAM and QPR signals. COMA's error is generated as a weighted minimum difference between a signal sample and a series of radii placed at the constellation points. COMA exhibits desired properties from the Constant Modulus Algorithm (CMA) and Decision Directed Algorithm, such as an invariance to carrier modulation, probability distribution of constellation points and correlation between symbols. COMA's non-linear error generation precludes the use of classical closed-form analysis to insure an optimal weighting function design. Nevertheless, a simplified weighting function design algorithm is presented, which can be extended to high-order modulations.

1 Introduction

The reception of QAM and QPR modulations can represent a problem in urban and suburban areas due to the reflections of signals from structures. Such reception problems can be foreseen with the advent of digital radios and digital format High Definition Television (HDTV). Various equalization methods exist, some of which make use of symbol decisions, to receive QAM and QPR signals. If symbol decisions are used to generate the reference signal of an adaptive equalizer (e.g., Decision Directed error and Decision Feedback Equalizers), then carrier lock must first be performed. As a result, the signal must exhibit a degree of fidelity necessary to allow carrier lock and guarantee convergence prior to signal acquisition. In contrast, the Concentric Ordered Modulus Algorithm (COMA), like the Constant Modulus Algorithm (CMA), can acquire a signal under severe distortion and without carrier lock. These algorithms can be used, for the appropriate constellations, to "clean" the received signal prior to acquisition of carrier and the use of symbol decisions.

CMA, described in [5], has demonstrated excellent performance in the equalization of PSK/FM signals and adequate performance against QAM signals. The premise of the algorithm is based on the transmitted signal envelope being constant; therefore, any deviation from this characteristic can be used as an error in the adaptation process. COMA is an extension of this concept to non-constant, discrete-ordered, modulus signals. It is demonstrated in this paper that COMA provides cluster variance performance similar to Decision Directed equalization, without the need for carrier acquisition. Furthermore, COMA can equalize signals with an initial symbol error rate (SER) larger than that required by Decision Directed equalization.

2 COMA Description

The COMA error is defined as the minimum CMA error of the signal to each of the radii in the constellation, weighted by the function $P(n,y)$. Thus for a complex input sample y,

$$\varepsilon_n = y - r_n(cos(\phi) + i sin(\phi)) \quad (1)$$

and

$$\chi = min[\varepsilon_n] P(n,y) \quad (2)$$

is the COMA error (where $\phi = tan^{-1}(Im\{y\}/Re\{y\})$ and the $min[x]$ is based on the magnitude of the vector x).

The purpose of the weighting function is to guarantee convergence under certain channel conditions and signal characteristics. The design of the weighting function is the critical element in the implementation of COMA.

2.1 COMA's Weighting Function

The weighting function in formula (2) serves to reduce the uncertainty associated with the decision based errors. A simple weighting function is the two valued "gating" function, which takes on the values [0,1], depending on the error magnitude. The design of the weighting function must consider the Radii Decision Boundary (RDB), calculated as the average between nearest neighbor radii, and the minimization of erroneous radius decisions.

The optimal weighting function can only be determined if an appropriate equation can be defined. The suggested optimization equation is derived from the identification of different error regions. Consider the COMA error χ for an arbitrary choice of $P(n,y)$. As a

1058-6393/91 $01.00 © 1991 IEEE

result of a radius crossing other symbol regions, ε_n can be categorized into three different error types:

- EI: ε_n has the correct sign and magnitude,
- EII: ε_n has the correct sign and wrong magnitude,
- EIII: ε_n has the wrong sign and magnitude.

Figure 1 illustrates the occurrence of the three error types in the constellation. Error types EII and EIII are the result of the RDB crossing symbol decision regions. The checkered region in Figure 1 denotes an EIII type error: any input sample y in this region will be erroneously directed to the largest radius shown. The diagonal lines, denoting the EII type errors, will go to the largest radius shown, instead of the radius generated by the corner point. In the figure, the regions of $P(n,y) = 1$ are denoted by the horizontal and vertical lines, with white regions denoting $P(n,y) = 0$ (note: the weighting function is partially shown for discussion purposes; in reality every point has an upper and lower weighting function region). Figure 1 also illustrates the impact of the weighting function on three error types, since the bound on all error types is determined by the RDB for a given radius.

The impact of an EII error type on the adaptation process is equivalent to an EI error type with a slower convergence rate. Although EII's impact on the adaptation process is not appropriate, this error does not create adverse effects. As a result, a candidate equation to obtain the optimal weighting function can be achieved by maximizing,

$$\lambda = \frac{\Omega(EI)}{\Omega(EIII)} \qquad (3).$$

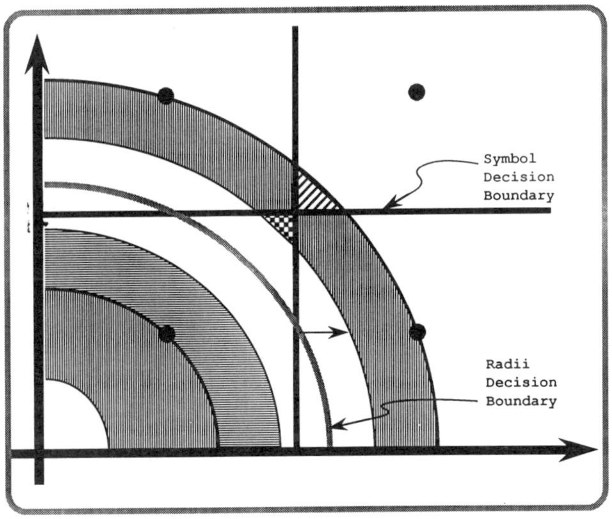

Figure 1 - Weighting Function Effects on Error Types.

2.2 Calculation of λ_{max} and $P(n,y)$

An arbitrary artifice for the calculation of $P(n,y)$ involves the use of Symbol Decision Boundaries (SDB's) in combination with RDB's. This is intuitive, since it requires the upper and lower regions of the weighting function to be confined to the region belonging to the radius generating symbol. This approach is termed the Border Uncertainty Method (BUM). In effect, the upper, or lower, bound on the region where $P(n,y) = 1$ is determined by a radius to a point along the minimum distance line separating two nearest neighbors (symbols). The reduction in the region covered by $P(n, y) = 1$ is also illustrated in Figure 1. The point which dictates the upper, or lower, radius of $W(i, y)$ is calculated based on a value of K_a, or K_b, respectively. Consider the smallest constellation point in Figure 1: the nearest neighbor symbol is given by the edge symbols with equal radius. The BUM dictates that:

$$\rho_a(j) = \sqrt[2]{(p_j+1-K_a)^2 + (q_j+1)^2} \quad \frac{d_{min}}{2} \qquad (4)$$

and similarly,

$$\rho_b(j) = \sqrt[2]{(p_j-1+K_b)^2 + (q_j-1)^2} \quad \frac{d_{min}}{2} \qquad (5)$$

where p_j and q_j are constellation points described by integers and K_a/K_b take on values between [0,1].

For an input sample y, yielding a decision radius n_{min}, the weighting function value is calculated as,

$$\text{if } \rho_a(n_{min}) \geq y \geq \rho_b(n_{min})) \text{ then } P(n_{min}, y) = 1$$
$$\text{else } P(n_{min}, y) = 0 \qquad (6)$$

As seen in equations (4) and (5), the weighting function depends on the values of K_a and K_b,. From Figure 1, the regions of $\rho_a(j)$ and $\rho_b(j)$, due to chosen values of K_a and K_b, reduce the effective regions of all three error types. Therefore, the values of K_a and K_b, have a direct effect on λ.

The effect of K_a and K_b must be weighted by the presence of each error type on a particular radius j. The presence of each error type can be considered as the percentage of the area a particular error covers $\rho_a(j)$ or $\rho_b(j)$. The values of K_a and K_b can be chosen to maximize λ, with the following definitions,

$$\Omega(EI) = \alpha_1 R_a(\rho_a(j)) + \beta_1 R_b(\rho_b(j)) \qquad (7)$$

and

$$\Omega(EIII) = \alpha_3 R_a(\rho_a(j)) + \beta_3 R_b(\rho_b(j)) \qquad (8)$$

where, R_a (R_b) is the ratio of the area covered by $\rho_a(j)$ ($\rho_b(j)$) to the maximum area covered from the radius j to the applicable RDB. α_e (β_e) is the region percentage of $\rho_a(j)$ ($\rho_b(j)$) covered by the error type e. The calculation of the R's can be specified as,

$$R_a(\rho_a(j)) = \frac{(\rho_a(j)^2 - r_j^2)}{(r_u^2 - r_j^2)} \qquad (9)$$

and

$$R_b(\rho_b(j)) = \frac{(r_j^2 - \rho_b(j)^2)}{(r_j^2 - r_l^2)} \qquad (10)$$

where r_u and r_l represent the RDB radius for each region, that is,

$$r_u = \frac{r_j + r_{j+1}}{2} \qquad (11)$$

and

$$r_l = \frac{r_j + r_{j-1}}{2} \quad . \tag{12}$$

The estimation of α_e and β_e is more laborious, since the area is usually bounded by the SDB's. This computation is even more tedious for high order constellations. A simplification on their estimation is based on a rough evaluation of the number of symbols which a given error type covers. The values estimated for QPR and QAM constellations are shown in Table 1 and 2, for the error types EI and EIII respectively. The first ten QPR radii cover the modulations 9, 25 and 49 QPR, with the first nine radii of QAM covering 16 and 64 QAM.

n	QPR		QAM	
	α_1	β_1	α_1	β_1
1	1.00	0.00	1.00	1.00
2	0.50	0.75	0.67	0.80
3	0.50	0.50	0.33	0.33
4	0.33	0.25	0.50	0.50
5	0.50	0.67	0.50	0.50
6	0.20	0.20	0.60	0.60
7	0.25	0.25	0.30	0.40
8	0.34	0.34	0.40	0.17
9	0.33	0.33	0.00	0.10
10	0.00	0.17	-	-

Table 1. α_1 and β_1 values for QPR and QAM (EI)

n	QPR		QAM	
	α_3	β_3	α_3	β_3
1	0.00	0.00	0.00	0.00
2	0.50	0.00	0.33	0.10
3	0.25	0.50	0.67	0.67
4	0.67	0.15	0.35	0.25
5	0.25	0.33	0.10	0.50
6	0.80	0.20	0.40	0.15
7	0.50	0.25	0.10	0.55
8	0.33	0.33	0.60	0.83
9	0.17	0.67	0.00	0.90
10	0.00	0.83	-	-

Table 2. α_3 and β_3 values for QPR and QAM (EIII)

Relaxing the restriction on the values of K_a and K_b, allows the computation of the natural λ (λ_{nat}) for each constellation. λ_{nat} is calculated with the maximum values for ρ_a and ρ_b (i.e., bounded by the RDB instead of the SDB, thus yielding R's $\approx$ 1). The values of K_a and K_b, which yield the desired regions are shown in Table 3, with the values of λ_{nat} for the constellations under consideration listed in Table 4.

The calculation of λ_{max} can also be laborious, specially for high order constellations. In the estimation of this value, a spread sheet was used to calculate all the necessary values (equations 3 through 12) and estimate λ_{max}. This approach has proven successful in the design of the weighting functions for the constellations considered thus far. To fully compute the value of λ_{max}, more accurate estimates of α_e and β_e are necessary.

n	QPR		QAM	
	K_a	K_b	K_a	K_b
1	0.00	0.00	-.05	0.00
2	-.35	0.00	-.16	0.06
3	0.23	0.36	0.42	0.17
4	-.40	0.42	-.20	0.57
5	-.10	0.40	-.07	0.21
6	0.77	0.11	0.63	0.08
7	-.40	0.83	-.09	0.22
8	-.13	0.40	-.04	0.10
9	-.05	0.14	0.00	0.05
10	0.00	0.06	-	-

Table 3. K_a and K_b values for λ_{nat}.

Constellation	λ_{nat}
9 QPR	1.75
25 QPR	1.19
49 QPR	0.95
16 QAM	1.64
64 QAM	1.03

Table 4. Values of λ_{nat} for QPR and QAM.

In the maximization of λ, certain constraints must be imposed to avoid excessive reduction in area covered by $P(n,y)$ = 1 and optimum convergence time These constraints should include, but not limited to,

(a) Ignore ρ's where there is no EIII type errors, such as n = 1 and ρ for the last radius.

(b) Minimum values of K_a and K_b which restrict ρ_a and ρ_b to be contained within the SDB's and RDB's simultaneously.

(c) Applying BUM to K_a and K_b specified in (b). That is, increase the values of K_a and K_b above the minimum values determined in (b). This helps to minimize the occurrence of wrong decisions due to the presence of noise.

(d) Maximum values of K_a and K_b so as not to shut down all regions and maintain a balance of errors in both direction (above and below a radius). Simulations have shown that best results are obtained when the percentage of $W(i,y)$ =1 is $\approx$65%.

The values of K_a and K_b used to calculate λ_b are given in the next section, where simulation results are presented.

3 Results for 16QAM and 9QPR

To assess the performance of COMA under various conditions, a series of simulations were performed. These simulations comprised of 16QAM and 9QPR modulations through a multipath channel and adaptively equalized with a 128 Tap T/2-spaced FIR filter. Results for 25QPR, 49QPR and 64QAM will be shown in [2]. Furthermore, the performance of COMA is compared to CMA for a 16QAM signal.

The pulse shaping used was a 50% excess bandwidth square root raised cosine. The multipath was modeled by a two ray Rummler model [3], described as

$$H(z) = 1 - 0.4762z^{-2}. \quad (13)$$

Additive white gaussian noise was added to each signal, so that the output SNR would provide 10^{-6} BER or better. The equalizer was initialized with a center tap value of 1 and the rest 0. The adaptation of the weights was based on the LMS algorithm [6],

$$W_{n+1} = W_n + 2\mu\chi Y^*_n. \quad (14)$$

A weight update is made once per symbol. The chosen convergence rate was μ=0.001, except for CMA equalization, which was μ=.0001 (since it did not converge for larger μ's). The generation of 9QPR signals was effected as a QPSK signal passed through a 1+D filter [1].

The measure for convergence time was the cluster variance, of 125 symbols at the equalizer output, representative of 10^{-2} SER. This SER value would enable a switch to Decision Directed equalization, after the acquisition of the carrier. Figure 2 and 3 show the cluster variance for CMA and COMA equalization of a 16QAM signal respectively, with a switch to Decision Directed after 25,000 symbols. The COMA performance against the 9QPR signal is demonstrated in Figure 4 (CMA does not converge for QPR signals due to the uneven probability of the constellation points).

Further simulations were performed to provide a comparison in convergence time (measured in symbols) as a function of λ. The results are listed in Table 5. All the values for K_a and K_b satisfy the RDB and SDB constraints, with BUM applied (unless otherwise noted). The values of K_a and K_b which provided the best performance are listed in Table 6.

λ	Time	Comments
1.51	12,250	
1.64	16,375	K_a and K_b for λ_{nat}
1.89	10,125	
2.00	9,625	
2.00	9,500	
2.18	8,000	
n/a	25,000	CMA equalization

Table 5. Convergence Time (in Symbols) for 16QAM

p_i	q_i	K_a	K_b
1	1	0.25	0.25
1	3	0.00	0.40
3	1	0.00	0.40
3	3	0.60	-

Table 6. Values of K_a and K_b for λ= 2.18.

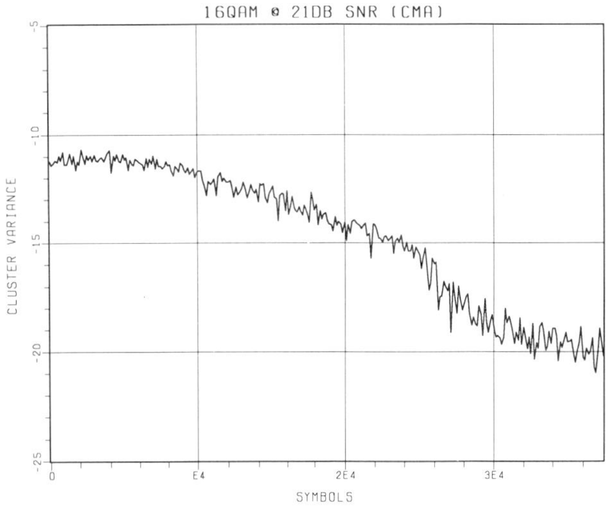

Figure 2 Cluster Variance vs. symbols (CMA -16QAM)

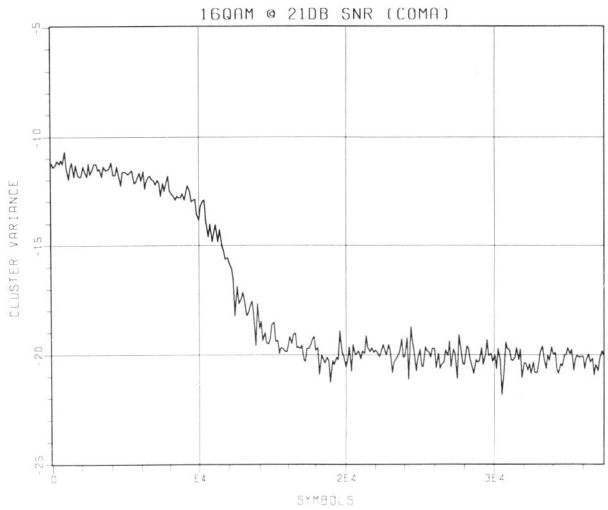

Figure 3 Cluster Variance vs. symbols(COMA-16QAM)

From Table 5, the convergence for the weighting function yielding a value of λ_{nat} , provides an insight into a characteristic of COMA. For the constellations of 25QPR, 49QPR and 64QAM the use of a weighting function based on λ_{nat} does not provide convergence (even for ISI only channels). These facts indicate that

λ_{nat} provides convergence, for a range of initial SER values, if λ_{nat} is higher than a yet to be determined λ^*_{nat}. Indeed, the relationship between λ^*_{nat} and the initial SER can be complex. Therefore, it is recommended to have a weighting function design based on the principles addressed in Section 2, as these principles are in the direction of optimum performance.

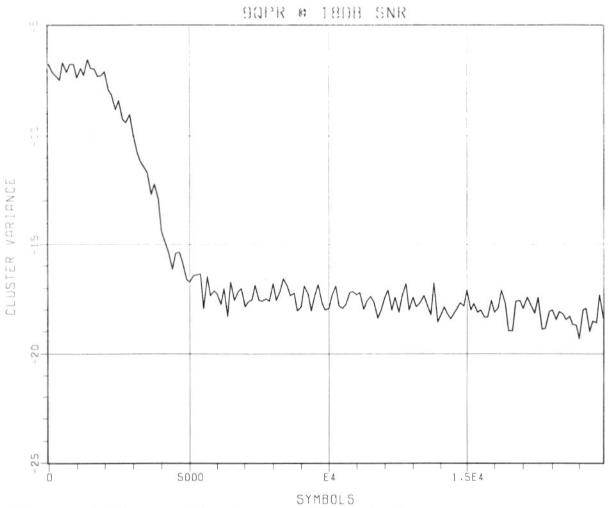

Figure 4 Cluster Variance vs. symbols (COMA-9QPR)

4 General Comments and Conclusion

Although COMA's design is not as simple as CMA, COMA provides benefits not found in CMA and/or Decision Directed, thus creating a role for COMA in the adaptive equalization of certain signals. The design of the weighting function thus far has relied on the desire to equalize a given constellation for known signal characteristics and channels. It is probably through this approach that an optimal weighting function may be found, since the maximization of λ through closed-form techniques is tedious and perhaps not possible. Furthermore, since χ is a decision-based error, there is an upper limit on the initial SER which COMA can provide convergence.

The design of COMA's weighting function is a task to be performed once, since the design for the worst case's initial SER, is the best design for any lesser SER encountered in other channels. Therefore, the closer the calculated value of λ to λ_{max}, a higher initial SER can be processed.

As a result of the COMA performance simulations against many signal characteristics and channels, the following performance manifestations can be made:

(A) The initial SER which the best COMA design has been able to provide convergence is $\approx$0.7, for uncorrelated and identically distributed symbols.

(B) In the presence of precoded symbols, such as QPR modulations, the initial SER lowers to $\approx$0.6. This is due to the correlation between symbols.

(C) If the symbols are non-white (e.g., generated by a Linear Recursive Shift (LRS) register), the initial SER is further lowered from 0.6.

(D) Values to be added to the minimum K_a and K_b, to apply BUM, are not the same for all radii. The values should be chosen based on a reduction of the total area covered by the applicable weighting function.

(E) An Automatic Gain Control (AGC) which normalizes the input signal is adequate to avoid signal collapse. If the signal is not scaled well below the largest radius, then COMA can collapse the signal without converging.

(F) Selection of the nearest neighbor symbol, for the calculation of ρ, should consider the symbol which produces the nearest neighbor, or closest, radius.

An equalizer implementation should include CMA, COMA and Decision Directed errors since each of the algorithms can perform, or improve signal fidelity, in areas where the others fail.

Acknowledgements

Many thanks are due to Dr. Anthony L.D. Chun for listening to ideas and providing reviews of manuscripts (even if it was not convenient). Thanks also to Dr. Jeffrey Harp and Dr. Ernest Tsui for providing possible applications of the algorithm. These applications helped define many of the ideas presented in this paper.

References

[1] P. Kabal, S. Pasupathy, "Partial-Response Signalling", *IEEE Transactions on Communications*, Vol. com-23, No. 9, pp. 921-934, September 1975.

[2] F. López de Victoria, "More on the Concentric Ordered Modulus Algorithm for Blind Equalization of QAM and QPR Modulations, with Results for 64QAM, 25QPR and 49QPR.", to be presented at the *1992 International Conference on Acoustic, Speech, and Signal Processing,* March 23-26, San Francisco

[3] W.D. Rummler, "A New Selective Fade model: Application to Propagation Data," *Bell Syst. Tech. J.*, pp.1037-1071, May-June 1979.

[4] J.J. Shynk, R.P. Gooch, G. Krishnamurthy, C.K. Chan, *"A Comparative Performance Study of Several Blind Equalization Algorithms"*, SPIE, San Diego, July 1991.

[5] J. R. Treichler, B. G. Agee, "A New Approach to Multipath Correction of Constant Modulus Signals", *IEEE Transactions on Acoustic, Speech and Signal Processing*, Vol ASSP-31, pp. 459 - 471, April 1983.

[6] B. Widrow, S. D. Stearns, *Adaptive Signal Processing.* New Jersey: Prentice Hall, Inc.,1985, Chapter 6.

On Parameter Estimation For Noncausal Models Of Non-Gaussian Signals

Jitendra K. Tugnait
Dept. of Electrical Engineering
Auburn University
Auburn, Alabama 36849

Abstract

We consider the problem of estimating the parameters of a stable (stationary), scalar ARMA(p,q) signal model driven by an i.i.d. non-Gaussian sequence. The driving noise sequence is not observed. The signal is allowed to be nonminimum phase and/or noncausal (i.e., poles may lie both inside as well as outside the unit circle). We address the problem of parameter identifiability given the higher order cumulants of the signal on a finite set of lags. The set of lags required to achieve parameter identifiability is the smallest to date.

1 Introduction

The problem of estimating the parameters of noncausal and/or nonminimum phase parameteric models of non-Gaussian signals has been an active area of research for sometime now [1]-[6]. A useful approach has been a fitting error approach based on cumulant matching where the parameters of a hypothesized model are selected to best match (in a least-squares sense) the estimated higher-order cumulants of the data by the theoretical calculated cumulants obtained from the hypothesized model. In this paper we address the issue of parameter identifiability via cumulant matching.

Let $\{x(t)\}$ represent a ARMA(p,q) signal. Denote the k-th order ($k \geq 3$) cumulant of the signal $x(t)$ by

$$c_{kx}(\tau_1, \tau_2, \cdots, \tau_{k-1}) := k - \text{th cumulant of } x(t), x(t+\tau_1), \cdots, x(t+\tau_{k-1}) =: \text{cum}_k(x(t), x(t+\tau_1), \cdots, x(t+\tau_{k-1})). \quad (1)$$

Consider two ARMA(p,q) models parametrized by vectors θ_1 and θ_2, respectively, where θ is the vector of $p+q$ ARMA coefficients. Giannakis [1] recently proved that two ARMA(p,q) models are identical if

$$c_{kx}(\tau_1, \tau_2, \cdots, \tau_{k-1}|\theta_1) = c_{kx}(\tau_1, \tau_2, \cdots, \tau_{k-1}|\theta_2) \quad (2)$$

for

$$\{-q-3p \leq \tau_1 \leq q+3p, \ -q-3p \leq \tau_2 \leq q+3p, \ 0 \leq \tau_l \leq p \text{ for } l = 3, \cdots, k-1\}. \quad (3)$$

[In the opinion of this author, (3) should read as

$$\{-q-3p \leq \tau_1 \leq q+3p, \ -q-3p \leq \tau_2 \leq q+3p, \ 0 \leq \tau_l \leq 2p \text{ for } l = 3, \cdots, k-1\}. \quad (4)$$

But this is a minor point; Giannakis's approach is quite interesting.]

We will show that two ARMA(p,q) models are identical if (2) holds true for

$$\{0 \leq \tau_1 \leq q+3p, \ 0 \leq \tau_l \leq 2p \text{ for } l = 2, 3, \cdots, k-1\} \text{ for any } k \geq 3, \quad (5)$$

and for

$$\{0 \leq \tau_1 \leq q+2p+1, \ 0 \leq \tau_2 \leq q+2p+1, \ 0 \leq \tau_l \leq 2p \text{ for } l = 3, \cdots, k-1\} \text{ if } k \text{ is even.} \quad (6)$$

It is seen that our results are a significant improvement upon that of Giannakis. We do note that we follow the basic approach of Giannakis in proving our results.

2 Model Assumptions

Let $\{x(t)\}$ denote a stationary ARMA(p,q) signal given by

$$\sum_{i=0}^{p} a_i x(t-i) = \sum_{i=0}^{q} b_i w(t-i), \quad a_0 := 1, \ b_0 := 1. \quad (7)$$

The input $\{w(t)\}$ is not observed. The following conditions are assumed to hold.

(H1) $A(z) = \sum_{i=0}^{p} a_i z^{-i} \neq 0$ for $|z| = 1$.

(H2) The random sequence $\{w(t)\}$ is i.i.d., zero-mean and non-Gaussian such that its k-th cumulant γ_{kw} is nonzero.

Let $\theta = (a_1, \cdots, a_p, b_1, \cdots, b_q, \gamma_{kw})$. Let θ_0, p_0 and q_0 denote the true values of θ, p and q, respectively.

This material is based upon work supported by the National Science Foundation under Grant No. MIP-9101457.

1058-6393/91 $01.00 © 1991 IEEE

3 Main Results

An ARMA(p,q) model/process will be called *irreducible* if there are no pole-zero cancellations in its transfer function.

Lemma 1. Given any irreducible ARMA(p,q) process $\{x(t)\}$, the set of cumulants $c_{kx}(\tau_1+q,\tau_2,\cdots,\tau_{k-1})$ cannot be zero for all the lags which satisfy $0\le\tau_i\le p$ $(1\le i\le k-1,\ k\ge 3)$. •

Proof: It follows from (7) that

$$\mathrm{cum}_k\{\sum_{i=0}^{p}a_ix(t+q-i),\sum_{i=0}^{p}a_ix(t-i),\cdots,\sum_{i=0}^{p}a_ix(t-i)\}$$

$$=\mathrm{cum}_k\{\sum_{i=0}^{q}b_iw(t+q-i),\sum_{i=0}^{q}b_iw(t-i),\cdots,\sum_{i=0}^{q}b_iw(t-i)\}. \tag{8}$$

By multi-linearity property of the cumulant operator and (H2), we have

$$\sum_{i_1=0}^{p}\cdots\sum_{i_k=0}^{p}a_{i_1}a_{i_2}\cdots a_{i_k}c_{kx}(i_2-i_1+q,i_2-i_3,\cdots,i_2-i_k)=\gamma_{kw}b_0^{k-1}b_q. \tag{9}$$

It is seen from (9) that the cumulants on the left-side of (9) cannot all be zero since the right-side of (9) is nonzero for a valid ARMA(p,q) model. Now the cumulants $c_{kx}(i_2-i_1+q,i_2-i_3,\cdots,i_2-i_k)$ for $0\le i_j\le p$ $(1\le j\le k)$ can all be expressed in terms of $c_{kx}(\tau_1+q,\tau_2,\cdots,\tau_{k-1})$ for $0\le\tau_i\le p$ by using the various symmetries of the cumulants in their arguments. □

Theorem 1. Consider two ARMA(p,q) models parametrized by θ_1 and θ_2, respectively. Let θ_2 correspond to an irreducible model. Then the two models are identical in the sense of $\theta_1=\theta_2$ if

$$c_{kx}(\tau_1,\tau_2,\cdots,\tau_{k-1}|\theta_1)=c_{kx}(\tau_1,\tau_2,\cdots,\tau_{k-1}|\theta_2)$$

for

$$\{0\le\tau_1\le q+3p,\ 0\le\tau_i\le 2p\ (2\le i\le k-1)\} \tag{10}$$

for any k≥3. •

Proof: Here we will follow the basic approach of Giannakis [1]; we differ in details, however. The proof is by contradiction. Suppose that equality of the appropriate set of cumulants in the hypothesis holds true but that $\theta_1\ne\theta_2$. Let $a_i(\theta)$ denote the AR parameter a_i parametrized by θ. Consider the quantity

$$\sum_{i_1=0}^{p}\sum_{i_2=0}^{p}\cdots\sum_{i_k=0}^{p}a_{i_1}(\theta_1)a_{i_2}(\theta_1)\cdots a_{i_k}(\theta_1)\times$$

$$c_{kx}(i_2-i_1+\tau_1,i_2-i_3+\tau_2,\cdots,i_2-i_k+\tau_{k-1}|\theta_1). \tag{11}$$

By equality of the cumulants in the hypothesis, we have

$$\sum_{i_1=0}^{p}\cdots\sum_{i_k=0}^{p}a_{i_1}(\theta_1)\cdots a_{i_k}(\theta_1)\times$$

$$c_{kx}(i_2-i_1+\tau_1,i_2-i_3+\tau_2,\cdots,i_2-i_k+\tau_{k-1}|\theta_1)$$

$$=\sum_{i_1=0}^{p}\cdots\sum_{i_k=0}^{p}a_{i_1}(\theta_1)\cdots a_{i_k}(\theta_1)\times$$

$$c_{kx}(i_2-i_1+\tau_1,i_2-i_3+\tau_2,\cdots,i_2-i_k+\tau_{k-1}|\theta_2) \tag{12}$$

for all lag (k-1)-tuples that are in the region

$$\{0\le\tau_1\le q+2p,\ 0\le\tau_i\le p\ (2\le i\le k-1)\}. \tag{13}$$

The left-side of (12) is the k-th order cumulant of an MA(q) process with transfer function $B(z|\theta_1)$, at the lag (k-1)-tuple $(\tau_1,\cdots,\tau_{k-1})$. The right-side of (12) is the k-th order cumulant of an ARMA(p,p+q) process with transfer function

$$[A(z|\theta_1)B(z|\theta_2)/A(z|\theta_2)],$$

at the lag (k-1)-tuple $(\tau_1,\cdots,\tau_{k-1})$. This ARMA(p,p+q) model is not necessarily irreducible, i.e., it may have pole-zero cancellations. We will consider two possible cases:

Case 1: Number of irreducible (noncancellable) zeros of the ARMA(p,p+q) process is L such that $q+1\le L\le q+p$. Now the left-side of (12) is zero if $\tau_i>q$ for any $i\in\{1,2,\cdots,k-1\}$. Hence, the right-side of (12) should be zero for these lags. However, by Lemma 1, the right-side of (12) cannot be zero for every lag (k-1)-tuple $(\tau_1,\cdots,\tau_{k-1})$ where $L\le\tau_1\le L+p$ and $0\le\tau_i\le p$ $(2\le i\le k-1)$. This leads to a contradiction since $L\ge q+1$.

Case 2: $L=q$. This implies that ARMA(p,p+q) model is, in fact, an MA(q) model. Two MA(q) models are identical if and only if their k-th order cumulants agree on the set $\{0\le\tau_i\le q\ (1\le i\le k-1)\}$. Hence, the MA coefficients parametrized by θ_1 and θ_2, respectively, must agree. But the AR coefficients for θ_1 and θ_2, respectively, are implicitly identical (else one would not get $L=q$). Again, we have a contradiction. □

Sharper results can be obtained for even k, $(k\ge 4)$.

Lemma 3. Given any irreducible ARMA(p,q) process $\{x(t)\}$, the set of cumulants $c_{kx}(\tau_1+l,\tau_2+l,\tau_3,\cdots,\tau_{k-1})$ cannot be zero for all the lags which satisfy $0\le\tau_i\le p$ $(1\le i\le k-1,\ k\ge 4)$ where l is any fixed interger $0\le l\le q$. •

Proof: Mimicking the proof of Lemma 1, we have

$$\sum_{i_1=0}^{p}\cdots\sum_{i_k=0}^{p}a_{i_1}a_{i_2}\cdots a_{i_k}c_{kx}(i_2-i_1+l,i_2-i_3+l,i_3-i_4,\cdots,i_2-i_k)=\gamma_{kw}\sum_{m=l}^{q}b_m^2\,b_{q-m}^{k-2}. \tag{14}$$

The rest of the proof is as for Lemma 1. □

Theorem 2. Consider two ARMA(p,q) models parametrized by θ_1 and θ_2, respectively. Let θ_2 correspond to an irreducible model. Then the two models are identical in the sense of $\theta_1=\theta_2$ if

$$c_{kx}(\tau_1,\tau_2,\cdots,\tau_{k-1}|\theta_1)=c_{kx}(\tau_1,\tau_2,\cdots,\tau_{k-1}|\theta_2)$$

for

$$\{0 \le \tau_j \le q + 2p + 1,\ 0 \le \tau_i \le 2p$$
$$(j = 1, 2)\ (3 \le i \le k - 1)\} \qquad (15)$$

for any even k≥4. •

Proof: Follow the basic approach of the proof of Theorem 1. In particular, consider (11) and (12) with (13) replaced with

$$\{0 \le \tau_j \le q + p + 1,\ 0 \le \tau_i \le p$$
$$(j = 1, 2;\ 3 \le i \le k - 1)\}. \qquad (16)$$

Case 1: Number of irreducible (noncancellable) zeros of the ARMA(p,p+q) process is L such that $q + 1 \le L \le q + p$. Now the left-side of (12) is zero if $\tau_i > q$ for any $i \in \{1, 2, \cdots, k - 1\}$. Hence, the right-side of (12) should be zero for these lags. However, by Lemma 3, the right-side of (12) cannot be zero for every lag (k-1)-tuple $(\tau_1, \cdots, \tau_{k-1})$ where $l \le \tau_j \le l+p$ $(j = 1, 2)$, $0 \le \tau_i \le p$ $(3 \le i \le k - 1)$ and l is any fixed integer such that $0 \le l \le L$. Pick $l = q + 1$ which is always possible since $L \ge q + 1$. This leads to a contradiction.

Case 2: $L = q$. Mimick Case 2 of the proof of Theorem 1. □

Remark 1. The results of Theorems 1 and 2 show that if θ_2 parametrizes an irreducible ARMA(p,q) model, then so does θ_1 under the respective hypotheses of the Theorems. □

Overparametrization

Let us now compare two models ARMA$(\bar{p}_1, \bar{q}_1)$ and ARMA $(\bar{p}_2\ ,\ \bar{q}_2)$ parametrized by $\bar{\theta}_1$ and $\bar{\theta}_2$, respectively. Suppose that neither is necessarily irreducible, in general. Let us reduce the respective transfer functions to the minimal possible by cancelling out the cancellable pole-zero pairs. Denote the two so reduced models by ARMA$(\tilde{p}_1, \tilde{q}_1)$ and ARMA$(\tilde{p}_2, \tilde{q}_2)$, respectively, and the corresponding parameters by $\tilde{\theta}_1$ and $\tilde{\theta}_2$, respectively.

Define

$$m = \max(\tilde{p}_1, \tilde{p}_2),$$
$$n = \max(\tilde{q}_1, \tilde{q}_2).$$

"Extend" $\tilde{\theta}_1$, $\tilde{\theta}_2$, $\tilde{p}_1$, $\tilde{q}_1$, $\tilde{p}_2$ and $\tilde{q}_2$ to θ_1, θ_2, p_1, q_1, p_2 and q_2, respectively, so that

$$p_2 = p_1 = m,$$

$$q_2 = q_1 = n,$$

and let θ_1 and θ_2 be "extended" versions of $\tilde{\theta}_1$ and $\tilde{\theta}_2$, respectively. As an example, if $m = \tilde{p}_2$ and $n = \tilde{q}_1 > \tilde{q}_2$, then $p_2 = \tilde{p}_2$ and $q_2 = \tilde{q}_2 + (n - \tilde{q}_2)$ so that

$$\sum_{i=1}^{p_2} a_i(\tilde{\theta}_2) z^{-i} = \sum_{i=1}^{p_2} a_i(\theta_2) z^{-i}$$

$$\Longrightarrow a_i(\tilde{\theta}_2) = a_i(\theta_2) \text{ for every } i$$

and

$$\sum_{i=1}^{\tilde{q}_2} b_i(\tilde{\theta}_2) z^{-i} = \sum_{i=1}^{q_2} b_i(\theta_2) z^{-i}$$

$$\Longrightarrow b_i(\tilde{\theta}_2) = b_i(\theta_2) \text{ for } 1 \le i \le \tilde{q}_2$$

and

$$b_i(\theta_2) = 0 \text{ for } \tilde{q}_2 + 1 \le i \le n.$$

Now by Theorems 1 and 2, $\theta_1 = \theta_2$ if the cumulants generated by the two models agree on the sets specified by the respective hypotheses of the two theorems, provided that one takes $p = m$ and $q = n$ in the theorem statements. (This is so because by construction, one of the models parametrized by either θ_1 or θ_2 is irreducible.)

What about the original comparison between the models ARMA$(\bar{p}_1, \bar{q}_1)$ and ARMA $(\bar{p}_2\ ,\ \bar{q}_2)$? In general we have $\theta_1 \ne \theta_2$. It is well known that one can add cancellable pole-zero pairs to an ARMA transfer function without changing its transfer function and without changing the cumulants generated by it. However, it does change θ! So now the equality holds in the following sense:

$$H(z|\theta_1) = H(z|\theta_2)$$

where $H(z|\theta) = A^{-1}(z|\theta)B(z|\theta)$ is the transfer function of the model parametrized by θ.

We summarize the above discussion in Theorems 3 and 4.

Theorem 3. Two models ARMA$(\bar{p}_1, \bar{q}_1)$ and ARMA$(\bar{p}_2\ ,\ \bar{q}_2)$ parametrized by θ_1 and θ_2, respectively, are identical in the sense of $H(z|\theta_1) = H(z|\theta_2)$ if

$$c_{kx}(\tau_1, \tau_2, \cdots, \tau_{k-1}|\theta_1) = c_{kx}(\tau_1, \tau_2, \cdots, \tau_{k-1}|\theta_2)$$

for

$$\{0 \le \tau_1 \le q + 3p,\ 0 \le \tau_i \le 2p \quad (2 \le i \le k - 1)\} \qquad (17)$$

for any k≥3 where

$$p = \max(\bar{p}_1, \bar{p}_2),$$

$$q = \max(\bar{q}_1, \bar{q}_2).\ \bullet$$

Theorem 4. Two models ARMA$(\bar{p}_1, \bar{q}_1)$ and ARMA$(\bar{p}_2\ ,\ \bar{q}_2)$ parametrized by θ_1 and θ_2, respectively, are identical in the sense of $H(z|\theta_1) = H(z|\theta_2)$ if

$$c_{kx}(\tau_1, \tau_2, \cdots, \tau_{k-1}|\theta_1) = c_{kx}(\tau_1, \tau_2, \cdots, \tau_{k-1}|\theta_2)$$

for

$$\{0 \le \tau_j \le q + 2p + 1,\ 0 \le \tau_i \le 2p$$
$$(j = 1, 2)\ (3 \le i \le k - 1)\} \qquad (18)$$

for any even k≥4 where

$$p = \max(\bar{p}_1, \bar{p}_2), \quad q = \max(\bar{q}_1, \bar{q}_2).\ \bullet$$

4 Nonlinear Optimization

The results of Section 3 suggest the following least-squares cumulant matching criterion for consistent parameter estimation (see also [1]-[8]). Choose θ to minimize

$$\sum_{\tau_1}\sum_{\tau_2}\cdots\sum_{\tau_{k-1}}[\hat{c}_{kx}(\tau_1,\tau_2,\cdots,\tau_{k-1}) - c_{kx}(\tau_1,\tau_2,\cdots,\tau_{k-1}|\theta)]^2 \quad (19)$$

where the lags in (19) are selected to include the set of lags indicated in Theorems 1-4 and $\hat{c}_{kx}(\tau_1,\tau_2,\cdots,\tau_{k-1})$ is an estimate of $c_{kx}(\tau_1,\tau_2,\cdots,\tau_{k-1}|\theta)$ obtained by appropriate sample averaging of the moments of the given data record.

Given a particular parameter vector θ, we have[6]

$$c_{kx}(\tau_1,\tau_2,\cdots,\tau_{k-1}|\theta) = \gamma_{kw}\sum_i h(i,\theta)h(i+\tau_1,\theta)\cdots h(i+\tau_{k-1},\theta) \quad (20)$$

where $\{h(i,\theta)\}$ is the impulse response of (7) parametrized by θ. The cumulants (20) can be computed if we know $\{h(i,\theta)\}$. The impulse response can be computed in time-domain by cascading a causal system with an anticausal system (as in [8]) which requires knowledge of the causal and the anticausal poles of the system, i.e., $A(z)$ needs to be parametrized in a factored form. The other option is to carry out the computation in the frequency domain by first calculating $H(e^{j\omega}|\theta)$ ($H(z|\theta)$ is the transfer function of (7)), by the FFT algorithm and then taking its inverse FFT. This method has been used recently in [9] in a slightly different context.

The criterion (19) can be minimized using a Gauss-Newton or a Newton-Raphson type of gradient optimization scheme. The nonlinear optimization requires, in general, availability of a good initial guess since convergence to a local minimum is always a possibility. The approach of [4] may be used to select the initial guess.

5 Simulation Examples

In this section we consider two computer simulation examples to illustrate the proposed approach. The first example consists of a noncausal AR(3) model which has been considered in [8]. The second example involves an allpass noncausal model. Both the examples use noisy measurements where the non-Gaussian noise is such that its third-order cumulants vanish where the signal third-order cumulants do not identically vanish. Note that in the presence of such noise the parameter identifiability results of Section 3 continue to hold.

Example 1. We consider a noncausal AR(3) model given by

$$x(t)-2.05x(t-1)+1.65x(t-2)-0.8125x(t-3) = w(t) \quad (21)$$
$$y(t) = x(t)+v(t)$$

In terms of (7) we have a_0=1.0, a_1=-2.05, a_2=1.65 and a_3=-0.8125 . The sequence {w(t)} consists of zero-mean, i.i.d., exponential random variables. The sequence {v(t)} is an AR(2) process obtained by coloring a zero-mean i.i.d. Laplace (two-sided exponential) sequence $\{\epsilon(t)\}$ as follows:

$$v(t) = 1.4v(t-1) - 0.65v(t-2) + \epsilon(t).$$

The poles of the system (21) are located at

$$0.4 \pm j0.7, \quad 1.25$$

whereas the poles of the noise model are at $0.7 \pm j0.4$.

In order to carry out time-domain computation we need to parametrize the model as a cascade of a second-order causal AR(2) model with coefficients a_{11} and a_{12} with a first-order anticausal AR(1) model with coefficient a_{21}. That is, (21) may be rewritten as

$$x_1(t) = 0.8x_1(t-1) - 0.65x_1(t-2) + w(t) \quad (22)$$
$$x(t) = 0.8x(t+1) + x_1(t) \quad (23)$$

where (22) is causal and (23) is anticausal. In terms of (22) and (23) we have a_{11}=-0.8, a_{12}=0.65 and a_{21}=-0.8.

One hundred independent realizations of the noisy signal were generated with a signal-to-noise ratio (SNR) of 10. Cumulant matching cost criterion with lag pairs in the region $0 \le \tau_2 \le \tau_1 \le L = 12$ was considered for a record length of 2000 for each realization. Both time-domain and frequency-domain optimization approaches of Section 4 with parametrization (22)-(23) and (21), respectively, were used for parameter estimation. The initial guess for frequency domain nonlinear optimization was chosen as ($a_0 := 1.$)

$$\hat{a}_1 = -1.5, \quad \hat{a}_2 = 0., \quad \hat{a}_3 = 0.,$$

whereas that for time-domain optimization was selected as ($a_{i0} := 1.$, i=1,2)

$$\hat{a}_{11} = \hat{a}_{12} = \hat{a}_{21} = 0.$$

Tables 1 and 2 show the results of our simulations for time-domain and frequency-domain optimization approaches, respectively. We show the arithmetic mean and one standard deviation of the parameter estimates averaged over 100 Monte Carlo runs for each approach. It is seen that the results are quite good.

Example 2. We consider an allpass noncausal ARMA(3,3) model given by

$$x(t)-2.0308x(t-1)+2.5231x(t-2)-1.2308x(t-3) = w(t) - 2.05w(t-1) + 1.65w(t-2) - 0.8125w(t-3) \quad (24)$$
$$y(t) = x(t)+v(t)$$

The sequence {w(t)} consists of zero-mean, i.i.d., exponential random variables. The sequence {v(t)} is an AR(2) process as in Example 1. The poles of the system (24) are located at

$$0.6154 \pm j1.0769, \quad 0.8$$

whereas its zeros are at $0.4 \pm j0.7$ and 1.25 .

One hundred independent realizations of the noisy signal were generated with a signal-to-noise ratio (SNR) of 10. Cumulant matching cost criterion with lag pairs in the region $0 \leq \tau_2 \leq \tau_1 \leq L = 12$ was considered for a record length of 2000 for each realization. Both time-domain and frequency-domain optimization approaches were used for parameter estimation. The initial guess for frequency domain nonlinear optimization was chosen as ($a_0 := 1.$)

$$\hat{a}_1 = -1.5,\ \hat{a}_2 = 2.0,\ \hat{a}_3 = 0.,$$

$$\hat{b}_1 = -0.75,\ \hat{b}_2 = 0.5,\ \hat{b}_3 = 0.,$$

whereas that for time-domain optimization was selected as ($a_{i0} := 1.$, i=1,2)

$$\hat{a}_{11} = \hat{a}_{22} = 0.,\ \hat{a}_{21} = -0.7,$$

$$\hat{b}_1 = -0.7,\ \hat{b}_2 = 0. = \hat{b}_3.$$

Tables 3 and 4 show the results of our simulations for time-domain and frequency-domain optimization approaches, respectively. We show the arithmetic mean and one standard deviation of the parameter estimates averaged over 100 Monte Carlo runs for each approach. It is seen that the results are quite good.

References

[1] G.B. Giannakis, "On identifiability, maximum likelihood and novel HOS based criteria," in *Proc. IEEE Fifth ASSP Workshop on Spectral Estimation & Modeling*, Rochester, NY, pp. 217-221, Oct. 1990.

[2] J.K. Tugnait, "Identifiability of AR parameters of noncausal ARMA signals from output cumulants," in *Proc. 27th Allerton Conf. on Communication, Control, and Computing*, pp. 672-681, Univ. of Illinois, Urbana-Champaign, IL Sept. 27-29, 1989.

[3] J.K. Tugnait, "Consistent parameter estimation for noncausal autoregressive models via higher-order statistics," *Automatica*, vol. 26, pp. 51-61, Jan. 1990.

[4] G.B. Giannakis and A. Swami, "On estimating noncausal nonminimum phase ARMA models of non-Gaussian processes," *IEEE Trans. ASSP*, vol. ASSP-38, pp. 478-495, March 1990.

[5] C.L. Nikias and M.R. Raghuveer, "Bispectrum estimation: A digital signal processing framework," *Proc. IEEE*, vol. 75, pp. 869-891, July 1987.

[6] M. Rosenblatt, *Stationary Sequences and Random Fields*. Boston, MA: Birkhäuser, 1985.

[7] J.K. Tugnait, "Identification of linear stochastic systems via second- and fourth- order cumulant matching," *IEEE Trans. Information Theory*, IT-33, pp. 393-407, May 1987.

[8] J.K. Tugnait, "Fitting noncausal autoregressive signal plus noise models to noisy non-Gaussian linear processes," *IEEE Trans. Automatic Control*, AC-32, pp. 547-552, June 1987.

[9] J.K. Tugnait, "Inverse filter criteria for estimation of linear parametric models using higher order statistics," in *Proc. IEEE 1991 Intern. Conf. Acoustics, Speech, Signal Processing*, pp. 3101-3104, Toronto, Canada, May 14-17, 1991.

Table 1. Parameter estimates, Example 1, time-domain optimization, SNR=10., 100 Monte Carlo runs, N = 2000 = data length in each run, σ= one standard deviation.

Parameters	True Values	mean	σ
a_{11}	-0.8	-0.8196	0.1022
a_{12}	0.65	0.6556	0.0657
a_{21}	-0.8	-0.7866	0.0438
γ_{3w}	16.000	16.143	4.733

Table 2. Parameter estimates, Example 1, frequency-domain optimization, SNR=10., 100 Monte Carlo runs, N = 2000 = data length in each run, σ= one standard deviation.

Parameters	True Values	mean	σ
a_1	-2.050	-2.0948	0.1536
a_2	1.650	1.7048	0.2184
a_3	-0.8125	-0.8379	0.1096
γ_{3w}	-35.937	-33.5584	11.0136

Table 3. Parameter estimates, Example 2, time-domain optimization, SNR=10., 100 Monte Carlo runs, N = 2000 = data length in each run, σ= one standard deviation.

Parameters	True Values	mean	σ
a_{11}	-0.8	-0.7971	0.0304
a_{21}	-0.8	-0.8023	0.0316
a_{22}	0.65	0.6443	0.0342
b_1	-2.0500	-2.0402	0.0600
b_2	1.6500	1.6427	0.0716
b_3	-0.8125	-0.8093	0.0466
γ_{3w}	16.000	16.800	3.3754

Table 4. Parameter estimates, Example 2, frequency-domain optimization, SNR=10., 100 Monte Carlo runs, N = 2000 = data length in each run, σ= one standard deviation.

Parameters	True Values	mean	σ
a_1	-2.0308	-2.0452	0.0781
a_2	2.5231	2.5516	0.1409
a_3	-1.2308	-1.2411	0.0856
b_1	-2.0500	-2.0403	0.0601
b_2	1.6500	1.6427	0.0716
b_3	-0.8125	-0.8093	0.0466
γ_{3w}	58.261	64.099	17.779

A Time Domain View of Filter Banks and Wavelets

Kambiz Nayebi, Thomas P. Barnwell III, and Mark J. T. Smith

Georgia Institute of Technology
School of Electrical Engineering
Digital Signal Processing Laboratory
Atlanta, Georgia 30332

Abstract

In this paper, we consider a time domain approach to the reconstruction problem for filter bank and wavelet decompositions. The development of this time domain reconstruction theory has resulted in a unified design approach for uniform, nonuniform, low delay, and efficient analysis-synthesis systems. The theory is based on FIR filters and can also be applied to the design of perfect reconstruction systems based on wavelets for multi-resolution analysis-synthesis systems.

In this paper, the procedure for the design of general decomposition and reconstruction systems, called Generalized Lapped Transforms (GLT), is discussed. GLT's include many classical transforms and the discrete-time wavelet transform (DTWT) as special cases. The new design procedure is used to design wavelets and DTWT systems. Because of the generality of the framework, regularity and phase conditions can easily be imposed on the wavelet. Also, because the design procedure can be used to design nonuniform band systems, systems with higher frequency resolution than dyadic wavelet-based systems can be designed and realized. A number of design examples are included in the paper.

1 Introduction

Signal analysis and reconstruction based on filter banks has been a popular field of research over the last decade. Many different time-frequency representations can be interpreted and implemented as filter bank structures. For example, the discrete short-time Fourier transform (DSTFT), which has long been used to generate spectrograms, can be implemented in a filter bank structure [1]. The discrete wavelet transform is also typically implemented using a tree-structured filter bank system [2]. From a more general point of view, filter bank systems are time-frequency block transforms in which the transformation matrix is $M \times N$ and $M \leq N$ (where N is the length of the longest system filter). From this viewpoint, filter bank systems include classical block transforms (DSTFT, Discrete Cosine Transform, etc.) and Lapped Orthogonal Transforms (LOT's) [3] as special cases.

Recent research on the design of perfect reconstruction analysis-synthesis systems has produced some significant results [3–8]. The invention of Conjugate Quadrature Filters (CQF's), perfect reconstruction filter banks with $N = 2M$, lossless analysis-synthesis systems, perfect reconstructing modulated analysis-synthesis filter banks, and the design of nonuniform filter banks with rational sampling rate changes are among the significant achievements within the last decade.

The time domain formulation introduced in [6,9,10] provides a unified framework for the design of a wide variety of analysis-synthesis systems which includes all known structures based on FIR filter banks. In the past, almost all decomposition-reconstruction systems based on filter banks where *designed by analysis.* This means that the design was accomplished by a thorough and complete analysis of the system of interest. In the new time domain design methodology, a complete analysis of the system is not required in order to accomplish the design. In fact, the only information that is fundamentally necessary in the design process is the desired system structure. Obviously, for the design to be possible, the system structure must be consistent with the desired frequency resolution. This design approach provides an elegant and powerful procedure for designing systems which are not completely understood analytically. For example, it was not known that it was possible to design low and minimum delay systems with perfect reconstruction until they were designed using the time domain design technique [11].

From a signal analysis viewpoint, analysis-synthesis systems are used to project the input signal onto different signal subspaces each of which may have different time and/or frequency characteristics. Enough information is preserved in signal subspaces so that the signal can be reconstructed. It should be clear that the class of analysis-synthesis systems based on filter banks, which is perhaps better called the class of fixed overlapping block transforms, is very large and includes many of the well known transforms as special cases. It is also obvious that there are infinitely many unique time-frequency decompositions that can be achieved using nonuniform band systems. These transforms can be considered to be an invertible map-

1058-6393/91 $01.00 © 1991 IEEE

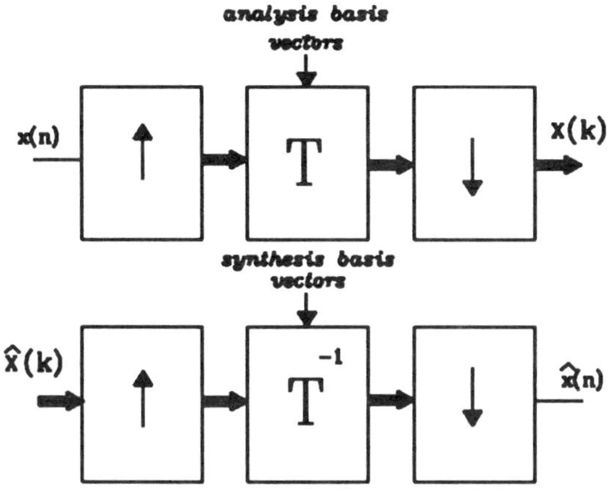

Figure 1: *A simple block diagram of the generalized lapped transform.*

ping of the signal onto the time-frequency plane. The resolution in the time-frequency plane depends of the transformation characteristics.

The ability of the time domain approach to design systems with arbitrary nonuniform decompositions makes it possible to create a very large family of overlapped block transforms with any desired time and frequency resolution (within the constraints of the uncertainty principle). Because of its ability to design fixed lapped transform systems, and because the DTWT is a fixed lapped transform, it can be used to design wavelets with constrained regularity and phase properties. The relationship between compactly supported wavelets and filter banks is described in recent literature including [12,2]. In this next section, a general view of analysis-synthesis systems as lapped transforms is presented. It is shown that the STFT and the DTWT are special cases of this general transform. In the remainder of the paper, a summary of the time domain framework and its applications to the design of wavelets and DTWT systems are presented.

2 Generalized Lapped Transforms

In the DSTFT, the signal is typically viewed through a lowpass analysis window $h(n)$. To analyze the entire signal, the window is shifted in time and the DFT of the windowed signal is computed. In the resulting representation, the signal is mapped onto time-frequency plane in which the time and frequency resolution is fixed. The resolution in the time-frequency plane is determined by the amount of the window shift, the shape of the window function and the length of the DFT. From a filter bank point of view, the DSTFT is the decimated output of a bank of complex modulated filters whose prototype filter impulse response is the analysis window $h(n)$. From a signal decomposition point of view, the DSTFT may be interpreted as the decomposition of the signal into different sub-signals using a set of basis vectors and their time-shifted versions. In the DSTFT, these basis vectors are defined by the analysis window and its frequency modulated versions.

The DTWT offers an alternative time-frequency (known as time-scale) representation in which the resolution in the time-frequency plane is not uniform. As opposed to the STFT in which the signal is viewed from a window of fixed size and shape, the DTWT is based on a series of different windows which are related to a continuous wavelet function $w(t)$.

Each window is of different length and corresponds to a bandpass filter with a center frequency at $\frac{3\pi}{2^{j+2}}$ for $j = 0, 1, \ldots$ In the analysis of the signal, the signal is decomposed into different sub-signals at different resolutions. From a sliding window perspective, the sliding rate of different windows depends on their length. Thus, shorter windows are moved more frequently than longer ones. This results in a time-frequency representation that has high temporal resolution in high frequencies and low temporal resolution (high frequency resolution) in low frequencies. From a filter bank perspective, the DTWT is an octave-band tree-structure in which the low frequency band is further divided into high and low bands and downsampled.

Both the DSTFT and the DTWT can be considered to be special cases of a more general time-frequency transformation which we will refer to as the Generalized Lapped Transform (GLT). The GLT includes DSTFT's, uniform band filter banks, DSTCTs, LOT's, and DTWT's as special cases. These are all examples of fixed lapped transforms, where the basis vectors do not change with time. The GLT also can include systems in which the basis vectors change with time. The time domain design procedure can be used to design any GLT.

In the generalized lapped transform, the characteristics of analysis windows and the resolution in the time-frequency plane are only restricted by the reconstruction requirement of the input signal (equivalently, the invertibility of the transformation). As in the DSTFT and the DTWT, the GLT can be implemented in a filter bank structure whose general form is presented in Figure 1. In such a transform, the input signal is decomposed into a set of sub-signals with different resolutions. To achieve such a multi-resolution time-frequency representation of the signal, the signal is passed through a set of upsamplers before the transformation is applied. The redundancy of the representation is reduced by downsampling the sub-signals to appropriate rates. The basis vectors are designed to have the required time and frequency characteristics.

3 Time Domain Framework

The basic idea behind the time domain formulation of analysis-synthesis structures is to determine the set

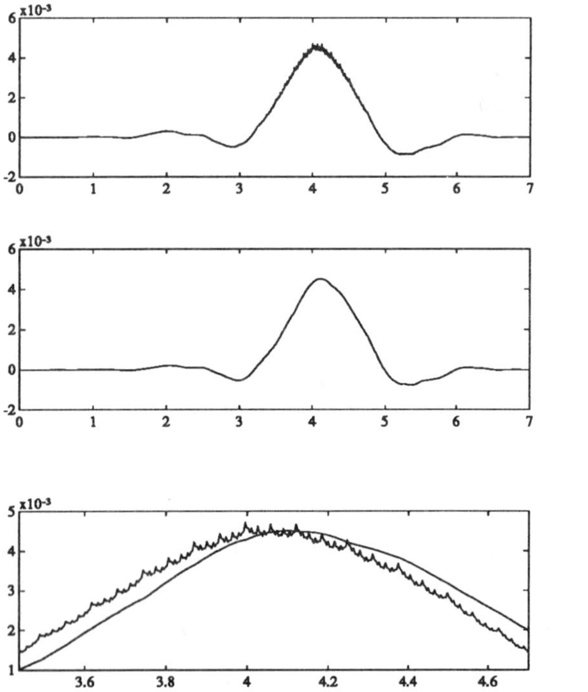

Figure 2: *(a) Scaling function, ϕ_0, generated from CQF-8, (b) Scaling function ϕ_1, (c) A closer view of ϕ_0 and ϕ_1 in the range [3.5,4.7].*

of necessary and sufficient conditions for exact reconstruction of the input at the system output. These conditions are expressed in a proper matrix product form which is used in the design procedure. The conditions can be expressed as

$$\mathbf{A}_i\mathbf{s}_i = \mathbf{b}_i \qquad i = 0, 1, \ldots, T-1, \tag{1}$$

where T is the shift-invariance period of the system. The elements of matrix $\mathbf{A}_i$ are related to the analysis basis vectors and elements of $\mathbf{s}_i$ are related to the synthesis basis vectors [13]. The vector $\mathbf{b}_i$ is constant.

After expressing the reconstruction conditions of the system (i.e. the invertibility conditions of the transform) in a matrix form, a cost function is defined as

$$\epsilon = \sum_{i=0}^{T-1} ||\mathbf{A}_i\mathbf{s}_i - \mathbf{b}_i||. \tag{2}$$

This cost function is minimized (and brought to zero for perfect reconstruction, if possible) using an optimization routine. Other system constraints which can not be directly incorporated into the formulation (such as frequency responses) can also be added to the cost function for minimization. In the next section, the design approach is applied to the design of compactly supported wavelets and DTWT systems.

4 Wavelet Design

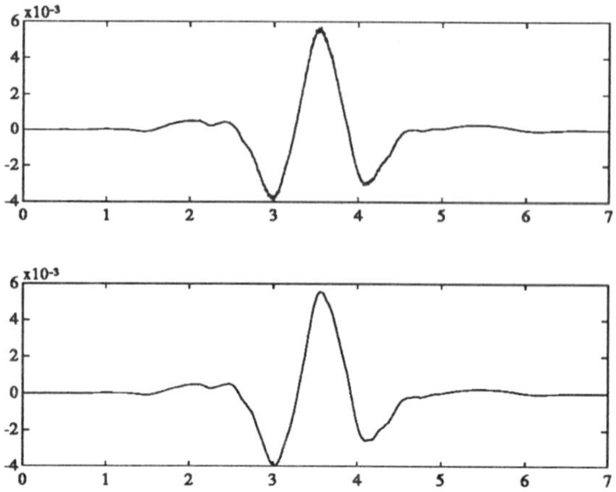

Figure 3: *(a) Wavelet, ψ_0, generated from CQF-8, (b) Wavelet ψ_1.*

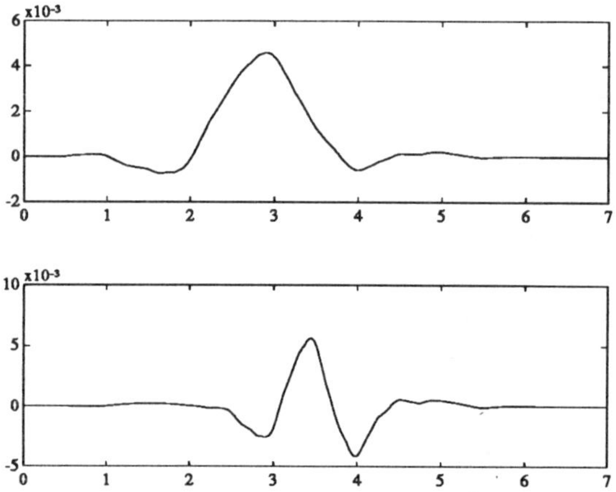

Figure 4: *(a) Synthesis scaling function, (b) Synthesis wavelet, which corresponds to ϕ_0 and ψ_0.*

In this section, the flexibility of the time domain design procedure in designing compactly supported wavelets is illustrated. Examples are designed to exhibit the flexibility of the design approach in imposing regularity condition on the wavelet, low delay or linear phase property on the wavelet, and their effects on the wavelet. Both orthogonal and biorthogonal wavelets are designed and presented. To design orthogonal wavelets, the synthesis filter coefficients are chosen to be the time-reversed version of the analysis filters. In designing biorthogonal wavelets, the synthesis filters are chosen to be

$$\begin{aligned} g_0(n) &= h_1(n)(-1)^n \qquad (3)\\ g_1(n) &= -h_0(n)(-1)^n. \qquad (4)\end{aligned}$$

First, we consider the imposition of regularity constraints on the wavelet. Figure 2a shows the generated scaling function from the 8-tap Conjugate Quadrature Filter (CQF-8). This scaling function is referred to as

ϕ_0 in this paper. The relative irregularity of this function and its corresponding wavelet (ψ_0) shown in Figure 3a is obvious. Scaling functions are generated by iterating the system lowpass filter eight times. To impose regularity on the wavelet, a zero at $z = -1$ is imposed on the lowpass analysis filter, $h_0(n)$. To impose similar regularity on the synthesis side, a zero at $z = 1$ is imposed on the highpass filter, $h_1(n)$. Figure 2b shows the scaling function of this system, referred to as ϕ_1, which is much smoother than ϕ_0. Figure 3b shows the corresponding relatively regular wavelet, ψ_1. Figure 2c shows a closer view of both the scaling functions in the range [3.5,4.7]. Since the synthesis wavelet is different from the analysis wavelet, Figure 4 show the synthesis scaling function and wavelet. To make the wavelet even more regular, two zeros are located at $z = -1$ for $h_0(n)$ and two zeros at $z = 1$ for $h_1(n)$. The resulting scaling function and wavelet are referred to as ϕ_2 and ψ_2, respectively.

To compare the regularity properties of the three wavelets, the spectra of the three scaling functions, $\Phi_0(\omega)$, $\Phi_1(\omega)$, and $\Phi_2(\omega)$, are compared. In Figure 5, Φ_0 and Φ_1 are shown side-by-side for comparison. As seen from this figure, the high frequency components of Φ_1 are significantly smaller than those of Φ_0 which results in a smoother wavelet function. Figure 6 compares Φ_1 and Φ_2. As seen in this figure, since Φ_2 is generated from a lowpass filter with two zeros at $z = -1$, it has a higher decay rate than Φ_1.

To understand the effects of imposing zeros at $z = -1$ to achieve regularity, Figure 7 shows the log magnitude response of lowpass filters used to generate ϕ_0 and ϕ_1. Having the same passband and transition characteristics, a few dB loss in the stopband attenuation is the result of imposing one of the zeros to be at -1.

To exhibit the flexibility of the design approach, three more design examples are shown. Figure 8 shows the analysis and the synthesis wavelets with linear phase. These correspond to a two-band system with 16-tap system filters and two zeros at $z = -1$ for both analysis and synthesis sections. Orthogonal wavelets can also be easily designed. Figure 9a shows the scaling function and Figure 9b shows the corresponding orthogonal wavelet function. The regularity in this case in achieved by imposing two zeros at $z = -1$ for $h_0(n)$. In the last example, the design approach is applied to design low delay wavelet decomposition-reconstruction systems. In these systems, both analysis and synthesis wavelets have low group delay which results in a lower total system delay. Figure 10 shows the analysis and synthesis wavelets of such a system. These wavelets correspond to a two-band system with 16-tap filters and only 10 samples of system delay. This delay is normally 15 samples for CQF and QMF based systems. In this example, two zeros are imposed at $z = -1$ for $h_0(n)$ and at $z = 1$ for $h_1(n)$.

5 Conclusion

In this paper, the generalized lapped transform in its general form is introduced. It is shown that the many well known transforms, including the wavelet transform, are special cases of the GLT. The time domain formulation of the general analysis-synthesis systems based on FIR filters is used to design wavelets with different constraints. Constraints included the regularity, a linear phase, a low delay, and the orthogonality. All these can be imposed using the same design procedure.

References

[1] M. R. Portnoff, "Time-frequency representation of digital signals and systems based on short-time fourier analysis," *IEEE Transactions ASSP*, pp. 55–69, February 1980.

[2] M. Vetterli and C. Herley, "Wavelets and Filter Banks: Relationships and New Results," *Proceedings ICASSP*, pp. 1723 - 1726, April 1990.

[3] H. S. Malvar and D. H. Staelin, "The LOT: Transform Coding without Blocking Effects," *IEEE Transactions on ASSP*, pp. 553–559, April 1989.

[4] M. J. T. Smith and T. P. Barnwell, "Exact reconstrcution techniques for tree-structured subband coders," *IEEE Transaction ASSP*, pp. 434–411, June 1986.

[5] J. P. Princen and A. B. Bradley, "Analysis/Synthesis Filter Bank Design Based on Time Domain Aliasing Cancellation," *IEEE Transactions ASSP*, vol. ASSP-34, October 1986.

[6] K. Nayebi, T. P. Barnwell, and M. J. T. Smith, "The Time Domain Analysis and Design of Exactly Reconstructing FIR Analysis/Synthesis Filter Banks," *Proceedings ICASSP*, pp. 1735–1738, April 1990.

[7] P. P. Vaidyanathan, "Theory and design of M channel maximally decimated QMF with arbitrary M, having perfect reconstruction property," *IEEE Transactions on ASSP*, April 1987.

[8] M. Vetterli and D. L. Gall, "Perfect Reconstruction FIR Filter Banks: Lapped Transforms, Pseudo QMF's and Paraunitary Matrices," *Proceedings ISCAS*, pp. 2249–2253, 1988.

[9] K. Nayebi, T. P. Barnwell, and M. J. T. Smith, "General Time Domain Analysis and Design Framework for Exacly Reconstructing FIR Analysis/Synthesis Filter Banks," *Proceedings ISCAS*, pp. 2022–2025, May 1990.

[10] K. Nayebi, T. P. Barnwell, and M. J. T. Smith, "Time domain filter bank analysis: A new design theory," *To be published Trans. on ASSP.*, June 1992.

[11] K. Nayebi, T. P. Barnwell, and M. J. T. Smith, "Design of low delay FIR analysis-synthesis filter bank systems," *Proc. Conf. on Information Sciences and Systems*, 1991.

[12] I. Duabechies, "Orthonormal bases of compactly supported wavelets," *Comm. in Pure and Applied Math.*, vol. 41, pp. 909–996, 1988.

[13] K. Nayebi, T. P. Barnwell, and M. J. T. Smith, "The design of perfect reconstruction nonuniform band filter banks," *Proceedings ICASSP*, 1991.

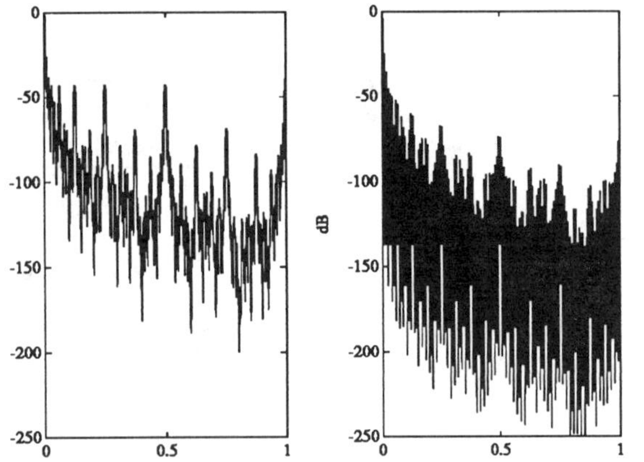

Figure 5: *Magnitude of* $\Phi_0(\omega)$ *(left) and* $\Phi_1(\omega)$ *(right) in dB.*

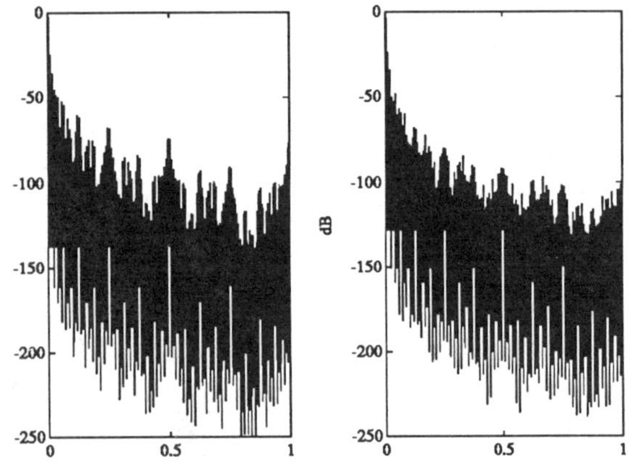

Figure 6: *Magnitude of* $\Phi_1(\omega)$ *(left) and* $\Phi_2(\omega)$ *(right) in dB.*

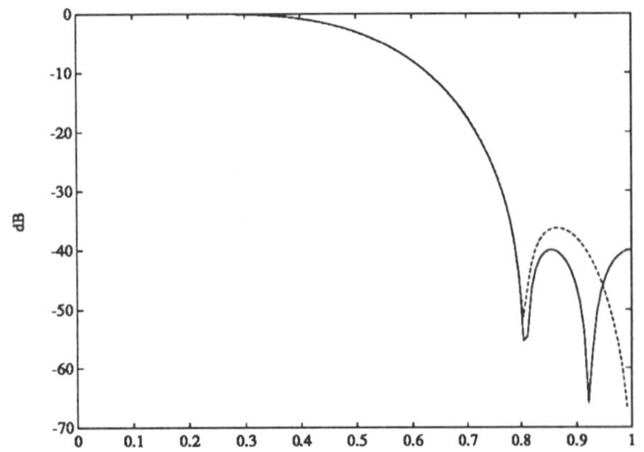

Figure 7: *Frequency response of the lowpass analysis filter with no zeros at* $z=-1$ *(full line) and one zero at* $z=-1$ *(dashed line).*

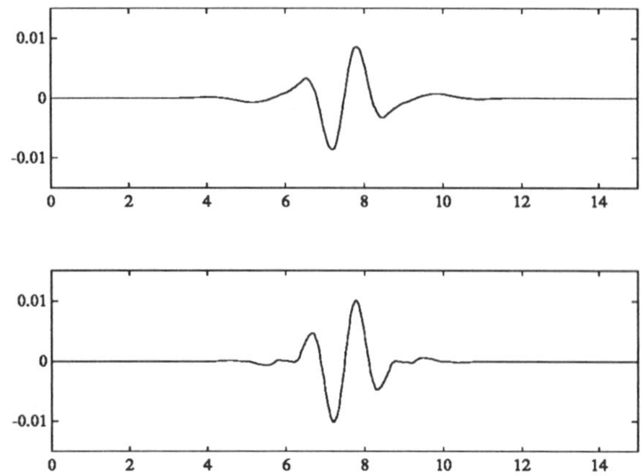

Figure 8: *(a) Analysis wavelet, (b) Synthesis wavelet with linear phase generated from 16-tap filters with two zeros at* $z=-1$.

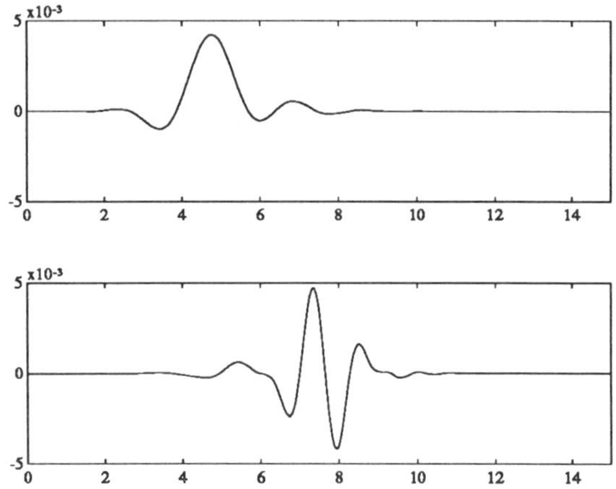

Figure 9: *(a) Scaling function, (b) Orthonormal wavelet generated from 16-tap orthogonal filters with two zeros at* $z=-1$.

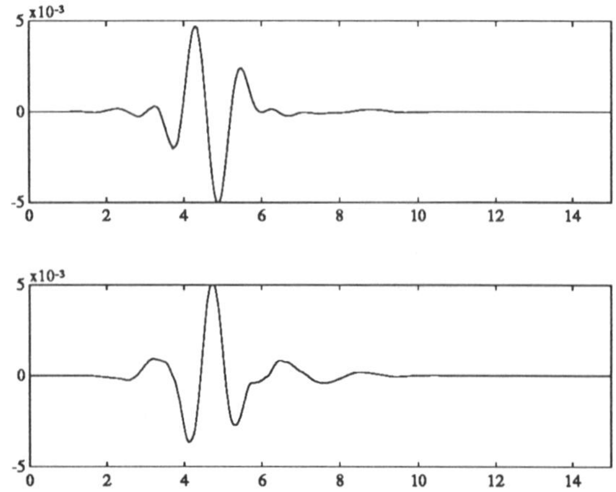

Figure 10: *(a) Analysis low delay wavelet, (b) Synthesis low delay wavelet generated from 16-tap filters of a 10 sample delay system with two zeros at* $z=-1$.

On Bit Allocation Strategies for Paraunitary Subband Coders

A. K. Soman and P. P. Vaidyanathan

Department of Electrical Engineering
California Institute of Technology
Pasadena, CA 91125

Abstract

Quadrature Mirror Filters have been employed in the past to decompose a signal into subbands. The signals in each subband are quantized before transmission. In this paper, we derive conditions on the individual quantizer error variances so as to minimize the overall reconstruction error variance. Our results hold for filter banks having paraunitary polyphase matrices. Under the assumption that the total number of bits over all channels is fixed, it is shown that their distribution which minimizes the reconstruction error variance is the one that makes the individual quantizer variances equal. We extend this analysis to tree-structured filter banks, such as those used for generating orthonormal wavelets. In this case, we show that the quantizer error variance in a particular branch should be proportional to the corresponding decimation factor.

1 Introduction

Consider Fig. 1 which shows the schematic of a well known technique for signal compression called subband coding. Here $x(n)$ is the sequence to be transmitted, and $y(n)$ is the received sequence. Boxes with $\downarrow M$ and $\uparrow M$ stand for decimation and interpolation by a factor of M respectively. This scheme can be redrawn in terms of the polyphase matrices [1] of the analysis and synthesis filters. In the absence of the quantizers the system can be made to have perfect-reconstruction property, i.e., $y(n) = x(n)$. One way to do this is by choosing the polyphase matrix of the analysis filters $\mathbf{E}(z)$ to be a paraunitary matrix (unitary on the unit circle) [1], and then choosing the polyphase matrix of the synthesis filters to be $\widetilde{\mathbf{E}}(z)$. (The matrix $\widetilde{\mathbf{E}}(z)$ is obtained from $\mathbf{E}(z)$ by transposition followed by conjugation of coefficients followed by replacing z by z^{-1}). This is shown in Fig. 2. Since the matrix $\mathbf{E}(z)$ is paraunitary, by definition, it satisfies the equation $\mathbf{E}(z)\widetilde{\mathbf{E}}(z) = \mathbf{I}$. In the presence of quantizers, a natural objective in such a coding scheme is to reduce a suitable measure of the reconstruction error under some constraint on the quantizers. Since we are transmitting M symbols at a time, it is logical to define the error as a *vector*, the difference of the received vector and the transmitted vector, i.e.,

$$\mathbf{r}(n) = \mathbf{y}(n) - \mathbf{x}(n). \qquad (1)$$

Assuming this error to be a zero-mean, Wide Sense Stationary (*WSS*) vector-process, the average reconstruction error variance over a period of M samples is,

$$\sigma_r^2 = (\frac{1}{M})E[\mathbf{r}^T(n)\mathbf{r}(n)]. \qquad (2)$$

This depends on the transform matrix $\mathbf{E}(z)$ and the quantizers Q_i. For a fixed transform matrix $\mathbf{E}(z)$, quantizing the individual signals to different degrees of accuracy will yield different values of σ_r^2. The number of bits R_i allocated to the quantizer Q_i is related to the variance of the error introduced by that quantizer as (assuming uniform probability density function for the error) [3],

$$\sigma_i^2 = \epsilon^2 2^{-2R_i}\sigma_{v_i}^2 \quad , \qquad (3)$$

where ϵ^2 is a constant, and $\sigma_{v_i}^2$ is the variance of the input to the quantizer. Now, suppose the total number of bits available for quantizing all the M signals is fixed, i.e.,

$$R = R_0 + R_1 + R_2 + \ldots + R_{M-1} = constant. \qquad (4)$$

How should these bits be allocated to the individual quantizers so that σ_r^2 is minimized? This problem has been addressed in [2] for the special case where the transform matrix $\mathbf{E}(z)$ is an orthogonal (constant) matrix. It has been shown that the distribution of the R_i that minimizes σ_r^2 is the one that makes all the σ_i^2 equal. In this paper, we address the problem for arbitrary paraunitary transform matrix $\mathbf{E}(z)$.

1.1 Wavelet transforms

Now consider an extension of the subband coding idea described above, namely, tree-structured filter banks. Fig. 3 shows a special case of a 3-level, binary tree-structured filter bank drawn in terms of the polyphase matrices on each level. The input signal $x(n)$ is coded (or 'transformed') by passing it through the analysis bank (Fig. 3(a)). The synthesis bank (Fig. 3(b)) performs the inverse transformation on the quantized versions of $y_k(n)$. One of the ways to ensure perfect reconstruction is to choose the analysis filters such that their polyphase matrix $\mathbf{E}_i(z)$ is paraunitary and then choose the synthesis filters such that

1058-6393/91 $01.00 © 1991 IEEE

$\mathbf{R}_i(z) = \widetilde{\mathbf{E}}_i(z)$. The relation between the above filter bank system and *wavelet transforms* has been known for quite some time [4]-[6]. With reference to Fig. 3(a), the quantities $y_k(n)$ are called the wavelet transform coefficients of the signal $x(n)$. We repeat our question on optimal bit allocation: what is the optimal way to allocate the bits among the quantizers (with their total number being fixed), so that the reconstruction error variance is minimized? We will answer this question in Theorem 2, which shall be proved in section 3.

Our results can be applied to any perfect-reconstruction subband coding system with paraunitary polyphase matrices.

2 Preliminaries

We now state and prove three lemmas which shall be used in the proofs of our main results.

Lemma 1 : Consider a paraunitary multi-input multi-output system $\mathbf{E}(z)$. Let $\mathbf{x}(n)$, the input to this system, be a zero-mean vector WSS process, and let $\mathbf{y}(n)$ be the output vector. Then,

$$E[\mathbf{x}^T(n)\mathbf{x}(n)] = E[\mathbf{y}^T(n)\mathbf{y}(n)]. \quad (5)$$

Proof: Let $\mathbf{S_{xx}}(e^{j\omega})$ and $\mathbf{S_{yy}}(e^{j\omega})$ be the power spectra of the vector sequences $\mathbf{x}(n)$ and $\mathbf{y}(n)$. Then,

$$\mathbf{S_{yy}}(e^{j\omega}) = \mathbf{E}(e^{j\omega})\mathbf{S_{xx}}(e^{j\omega})\mathbf{E}^{\dagger}(e^{j\omega}). \quad (6)$$

Letting $Tr(\mathbf{A})$ denote the trace of the matrix $\mathbf{A}$, we can show

$$Tr(\mathbf{S_{yy}}(e^{j\omega})) = Tr(\mathbf{S_{xx}}(e^{j\omega})). \quad (7)$$

Integrating the diagonal terms on both sides of the above equation gives us the required result.

Lemma 2 : Let $\mathbf{R}(z)$ be a N by N paraunitary matrix. Then its M-fold blocked version [7] $\mathbf{B}(z)$ is also paraunitary.

Proof : Let $\mathbf{s}(n)$ and $\mathbf{y}(n)$ be the input and output of the paraunitary matrix $\mathbf{R}(z)$. Then,

$$\sum_n \mathbf{y}^T(n)\mathbf{y}(n) = \sum_n \mathbf{s}^T(n)\mathbf{s}(n). \quad (8)$$

Consider the system in Fig. 4. $\mathbf{B}(z)$ is the blocked version of $\mathbf{R}(z)$. Now $\mathbf{p}(n)$ is the M-fold blocked version of $\mathbf{s}(n)$, and $\mathbf{t}(n)$ is the M-fold blocked version of the corresponding output $\mathbf{y}(n)$. Hence $\mathbf{p}(n)$ and $\mathbf{t}(n)$ are both vectors of length NM. For an arbitrary vector sequence of length M input to the system $\mathbf{R}(z)$, the total energy at the output equals the total input energy. Thus for the system in Fig. 4, the output energy equals the input energy for any input $\mathbf{p}(n)$, implying $\mathbf{B}(z)$ is paraunitary [1].

Lemma 3 : Consider the system in Fig. 5(a).

(a) Let $\mathbf{s}^T(n) = [s_0(n)\ s_1(n)]$. Then, if $\mathbf{s}(n)$ is a vector Cyclo-Wide Sense Stationary process with period 2^M [9] (abbreviated $(CWSS)_{2^M}$), then $y(n)$ is $(CWSS)_{2^{M+1}}$.

(b) Further, let $\mathbf{s}_0(n)$ and $\mathbf{s}_1(n)$ denote vectors of length 2^M formed by blocking $s_0(n)$ and $s_1(n)$ respectively, and $\mathbf{y}_B(n)$ denote the vector of length 2^{M+1} formed by blocking $y(n)$. Then, if $\mathbf{R}(z)$ is paraunitary,

$$E[\mathbf{y}_B^T(n)\mathbf{y}_B(n)] = E[\mathbf{s}_0{}^T(n)\mathbf{s}_0(n)] + E[\mathbf{s}_1{}^T(n)\mathbf{s}_1(n)] \quad (9)$$

Proof: Note that $\mathbf{y}_B(n)$ can alternatively be defined as the 2^M-fold blocked version of the vector process $\mathbf{y}(n)$ shown in Fig. 5(a). We can redraw the system in Fig. 5(a) as in Fig. 5(b), where $\mathbf{p}(n)$ and $\mathbf{y}_B(n)$ are vectors of length 2^{M+1}. Also, notice that $\mathbf{p}(n)$ is a WSS vector process. Since $\mathbf{y}_B(n)$ is obtained by passing $\mathbf{p}(n)$ through a linear system, $\mathbf{y}_B(n)$ is also WSS. Hence $y(n)$, which is formed by unblocking $\mathbf{y}_B(n)$ is $(CWSS)_{2^{M+1}}$, proving part (a).

Note that since $\mathbf{B}(z)$ is a blocked version of a paraunitary matrix, it is also paraunitary (Lemma 2). Therefore using Lemma 1 , we get

$$E[\mathbf{y}_B^T(n)\mathbf{y}_B(n)] = E[\mathbf{p}^T(n)\mathbf{p}(n)]. \quad (10)$$

But from the definition of these quantites,

$$E[\mathbf{p}^T(n)\mathbf{p}(n)] = E[\mathbf{s}_0^T(n)\mathbf{s}_0(n)] + E[\mathbf{s}_1{}^T(n)\mathbf{s}_1(n)]. \quad (11)$$

Therefore

$$E[\mathbf{y}_B^T(n)\mathbf{y}_B(n)] = E[\mathbf{s}_0^T(n)\mathbf{s}_0(n)] + E[\mathbf{s}_1{}^T(n)\mathbf{s}_1(n)], \quad (12)$$

proving part (b).

3 The Main Results

Consider the system shown in Fig.1. $\mathbf{x}(n)$ and $\mathbf{y}(n)$, both vectors of length M, are the blocked versions of $x(n)$ and $y(n)$ respectively. Let the quantizers be modeled as zero-mean, WSS, mutually uncorrelated noise sources, uncorrelated with the input. Let the total number of bits allocated to all the quantizers be fixed (eq. 4).

Theorem 1: Let $\mathbf{E}(z)$ be a FIR paraunitary matrix. Then the reconstruction error variance of the system in Fig. 1 is minimized when the variances σ_i^2 of the errors introduced by each of the quantizers are equal, i.e., $\sigma_i^2 = \sigma^2$ for all i.

Proof: Let $\mathbf{q}(n) = \mathbf{u}(n) - \mathbf{v}(n)$ and $\mathbf{r}(n) = \mathbf{y}(n) - \mathbf{x}(n)$. Here, $\mathbf{q}(n)$ is the vector whose individual components are the quantizer errors, and $\mathbf{r}(n)$ is the reconstruction error vector. So $\mathbf{r}(n)$ is the output of $\widetilde{\mathbf{E}}(z)$ in response to $\mathbf{q}(n)$, or equivalently, $\mathbf{q}(n)$ is the output of the system $\mathbf{E}(z)$ in response to the input $\mathbf{r}(n)$. Hence if

$$\mathbf{E}(z) = \mathbf{e}(0) + \mathbf{e}(1)z^{-1} + \ldots, \quad (13)$$

we have

$$\mathbf{q}(n) = \sum_m \mathbf{e}(n-m)\mathbf{r}(m). \quad (14)$$

Therefore, applying Lemma 1 we have,

$$E[\mathbf{q}^T(n)\mathbf{q}(n)] = E[\mathbf{r}^T(n)\mathbf{r}(n)]. \quad (15)$$

From the definition of $\mathbf{q}(n)$, we have

$$E[\mathbf{q}^T(n)\mathbf{q}(n)] = \sum_{i=0}^{M-1} \sigma_i^2. \quad (16)$$

so that $E[\mathbf{r}^T(n)\mathbf{r}(n)] = \sum_{i=0}^{M-1}\sigma_i^2$. Our problem is therefore to minimize $\sum_{i=0}^{M-1}\sigma_i^2$. Since the $\sigma_{v_i}^2$ depend only on the input statistics and the paraunitary transformation $\mathbf{E}(z)$, equation (4) is equivalent to the condition $(\prod_{i=0}^{M-1}\sigma_i^2)^{1/M} = constant$. We know [8] that the arithmetic mean of a set of non-negative numbers is always greater than or equal to their geometric mean, i.e., $\sum_{i=0}^{M-1}\sigma_i^2 \geq M(\prod_{i=0}^{M-1}\sigma_i^2)^{1/M} = Mc^{1/M}$, with equality if and only if the σ_i^2 are the same for all i. Thus the above $\sum\sigma_i^2$ in eq. (16) is minimized if and only if all the σ_i^2 are equal, i.e.,

$$\sigma_1^2 = \sigma_2^2 = \sigma_3^2 = \ldots\ldots = \sigma_{M-1}^2. \quad (17)$$

♠

Note that the assumption that the noise sources are mutually uncorrelated is a stricter assumption than we actually need. We only need that $\mathbf{r}(n)$ be a zero-mean WSS vector process.

Comments: In [3], the coding gain of a transform coding system G_{TC} was defined to be the ratio of the error variance in a PCM system to the error variance in the transform coding system. We can extend this idea to the case of paraunitary subband coding systems. Under optimal bit allocation, we have,

$$G_{PU} = \frac{(1/M)\sum_{i=0}^{M-1}\sigma_{v_i}^2}{(\prod_{i=0}^{M-1}\sigma_{v_i}^2)^{1/M}}. \quad (18)$$

By application of Lemma 1, we can see that the numerator in this expression is determined by the input statistics. Hence, in order to maximize the coding gain, the transform matrix $\mathbf{E}(z)$ should be chosen so as to minimize the denominator of the above expression.

For the next result, consider a general L-level tree structured FIR-filter bank such as one used for generating wavelet basis functions (Fig. 3 shows a special case of a three-level binary tree). Let the polyphase matrices $\mathbf{E}_i(z)$ be paraunitary, and let $\mathbf{R}_i(z) = \widetilde{\mathbf{E}}_i(z)$, so that $y(n) = x(n)$ in the absence of quantizers. Let the quantizers be modelled as in Theorem 1 above and let the total number of bits be constant.

Theorem 2: The reconstruction error variance of the tree-structured filter bank is minimized when the variances of the errors introduced by the quantizers are related as

$$2^{L-1}\sigma_0^2 = 2^{L-2}\sigma_1^2 = \ldots = \sigma_{L-1}^2 = \sigma_L^2. \quad (19)$$

Proof: Since the system in the absence of quantizers performs perfect-reconstruction, and since we have assumed that noise and signal are uncorrelated, it is sufficient to consider only the synthesis bank to study the effect of noise on the final reconstruction error. Consider the system in Fig. 6. (The tree can be redrawn in terms of the polyphase matrices on all levels, similar to Fig. 3.) The error-signals $e_i(n)$ are all WSS. Hence in particular, they are $(CWSS)_M$ for any integer M. By repeated application of Lemma 3, it can be shown that the $s_k(n)$ are $(CWSS)_{2^{L-k}}$. Let $\mathbf{e}_k(n)$ and $\mathbf{s}_{k+1}(n)$ denote the (2^{L-k-1})-fold blocked versions of $e_k(n)$ and $s_{k+1}(n)$ respectively for $k = 0, \ldots, L-2$, and let $\mathbf{s}_0(n)$ denote the 2^L-fold blocked version of $s_0(n)$. By applying Lemma 3 to each level of the tree, we get the following set of equalities:

$$E[\mathbf{s}_k^T(n)\mathbf{s}_k(n)] = \left(E[\mathbf{e}_k^T(n)\mathbf{e}_k(n)] + E[\mathbf{s}_{k+1}^T(n)\mathbf{s}_{k+1}(n)]\right),$$

for $k = 0, \ldots, L-2$, and,

$$E[\mathbf{s}_{L-1}^T(n)\mathbf{s}_{L-1}(n)] = E[e_{L-1}^2(n)] + E[e_L^2(n)]. \quad (20)$$

Hence from the above equations, one can write,

$$E[\mathbf{y}^T(n)\mathbf{y}(n)] = E[\mathbf{e}_0^T(n)\mathbf{e}_0(n)] + .. + E[e_L^2(n)] \quad (21)$$

i.e., the variance of the overall reconstruction error is given as

$$\sigma_r^2 = (\frac{1}{2^L})\left(2^{L-1}\sigma_0^2 + 2^{L-2}\sigma_1^2 + \ldots + \sigma_{L-1}^2 + \sigma_L^2\right), \quad (22)$$

where σ_i^2 are the individual quantizer variances. Let

$$2^{L-i-1}\sigma_i^2 = \theta_i^2, \quad for\ i = 0, \ldots, L-1,$$

and

$$\sigma_L^2 = \theta_L^2. \quad (23)$$

Therefore,
our problem again, is minimizing $(\frac{1}{L+1})\sum_{i=0}^{L}\theta_i^2$ under the constraint that $(\prod_{i=0}^{L}\theta_i^2)^{1/(L+1)} = constant$. As in the proof of Theorem 1, we invoke the argument that the arithmetic mean of a given set of numbers is minimized when it is equal to their geometric mean. Hence our objective function is minimized when

$$2^{L-1}\sigma_0^2 = 2^{L-2}\sigma_1^2 = \ldots = \sigma_{L-1}^2 = \sigma_L^2. \qquad (24)$$

♠

Comments: To obtain physical insight into this result, consider the typical appearance of the filter responses in a tree-structured bank. If the polyphase matrices on each level of the tree are paraunitary, these filters have equal energy irrespective of their frequency characteristics. Now consider a block of length 2^L input to each of these filters. On level 'i', there are in this block 2^{L-i-1} non-zero noise samples, and the others are zero. The result of the above theorem states that the sum of the variances of the samples taken over this block of length 2^L should be the same for all the filters. This is indeed appealing intuitively.

Notice that in the proof of the above result, the assumption that the noise sources are mutually uncorrelated is a stronger assumption than we actually need. It would have been enough to assume that the vectors $\mathbf{s}_k(n)$, $k = 0, \ldots, L-1$ are *WSS*.

4 Concluding Remarks

In speech and image coding applications, one might use other criterion for minimizing the reconstruction error, rather than minimizing its variance. For instance, one could attach different (non-negative) weights w_k to each of the sub-bands, and then try to minimize the weighted sum of the variances, i.e., $\sum_k w_k \sigma_k^2$. The fact that the arithmetic mean of a set of non-negative numbers is always greater than or equal to their geometric mean can still be used. It is easy to see that in order to minimize the weighted sum, each of the $w_k\sigma_k^2$ would have to be equal. In other words, the variances σ_k^2 would have to be inversely proportional to the weights w_k, again an intuitively appealing proposition.

Acknowledgements: This work was supported by NSF Grant MIP 8919196 and by Tektronix Inc.

References

[1] P. P. Vaidyanathan, "Multirate Digital Filters, Filter Banks, Polyphase Networks, and Applications: A Tutorial," Proc. of IEEE, Vol.78, No.1, Jan. 1990.

[2] J. J. Huang and P. M. Schultheiss, "Block Quantization of Correlated Gaussian Random Variables," IEEE Trans. on Comm. Syst., pp. 289-296, Sept. 1963.

[3] N. S. Jayant and P. Noll, *Digital Coding of Waveforms- Principles and Applications to Speech and Video,* Prentice-Hall Signal Processing Series, Englewood Cliffs, NJ, 1984.

[4] I. Daubechies, "Orthonormal Bases of Compactly Supported Wavelets," Comm. on Pure and Appl. Math. Vol.4, pp. 909-996, Nov.1988.

[5] M. Vetterli and C. Herley, "Wavelets and Filter Banks: Relationships and New Results," Proc. of ICASSP, pp. 1723-1726, 1990.

[6] P. P. Vaidyanathan, "Lossless systems in Wavelet Transforms," Proc. IEEE Int. Symp. Circuits and Systems, Singapore, 1991.

[7] P.P. Vaidyanathan and S. K. Mitra, "Polyphase Networks, Block Digital Filtering, and Alias-free QMF Banks: A Unified Approach Based on Pseudo-circulants," IEEE Trans. on ASSP, Vol. 36, March 1988.

[8] E. F. Beckenbach and R. E. Bellman, *Inequalities,* Berlin, Springer, 1961.

[9] V. P. Sathe and P. P. Vaidyanathan, "Effects of Multirate Systems on the Statistical Properties of Random Inputs," preprint.

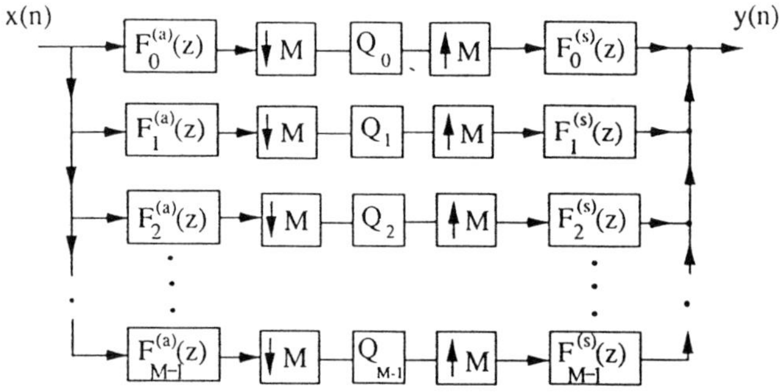

Fig. 1. A typical sub-band coding scheme.

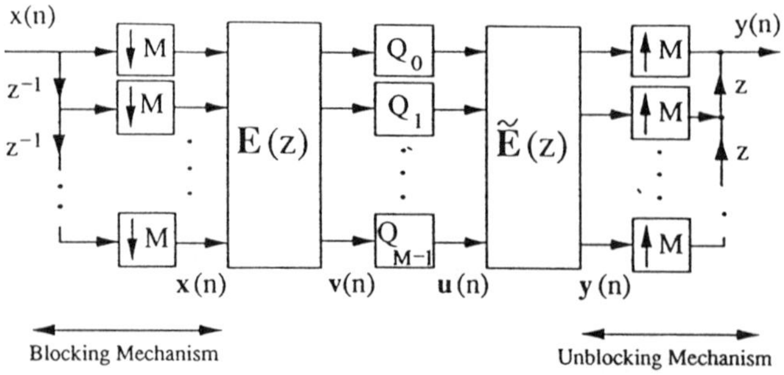

Fig. 2. Sub-band coding scheme showing polyphase matrices

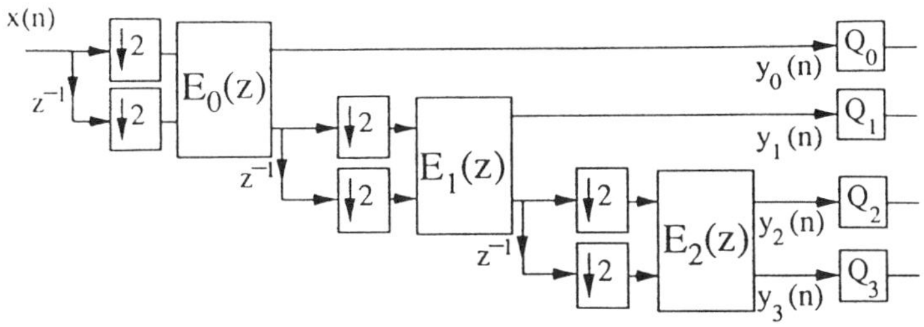

Fig. 3 (a). Tree structured analysis bank

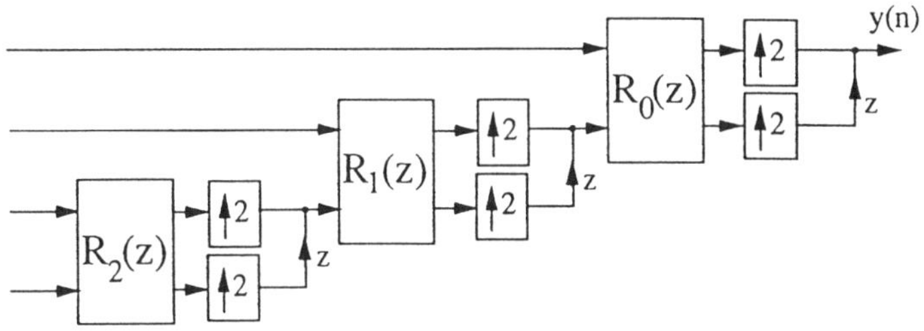

Fig. 3 (b). Tree structured synthesis bank

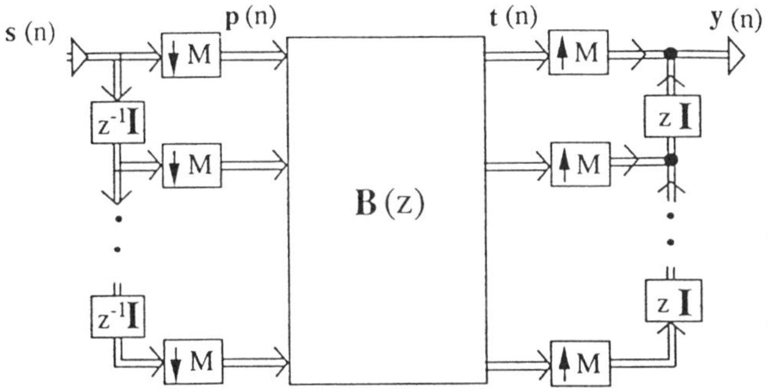

Fig. 4. M-fold blocked version of a MIMO system

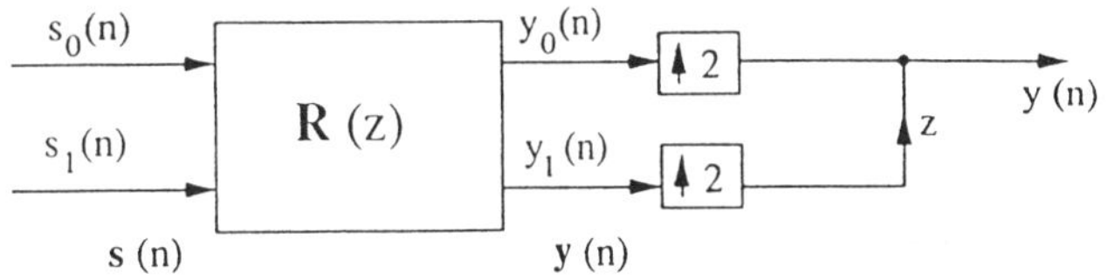

(a)

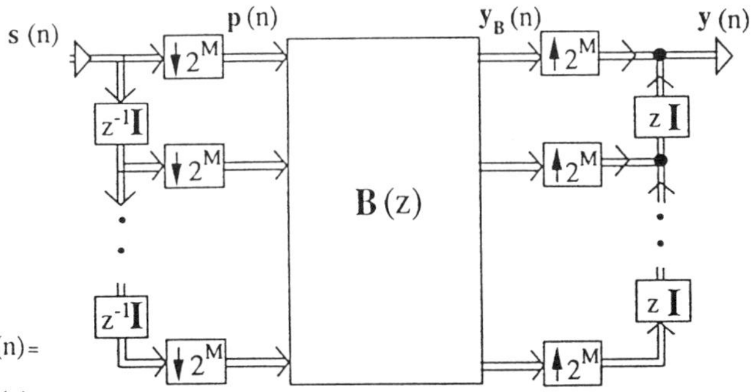

(b)

Fig. 5. (a) A two-input one-output system
(b) Blocked version of the system in (a)

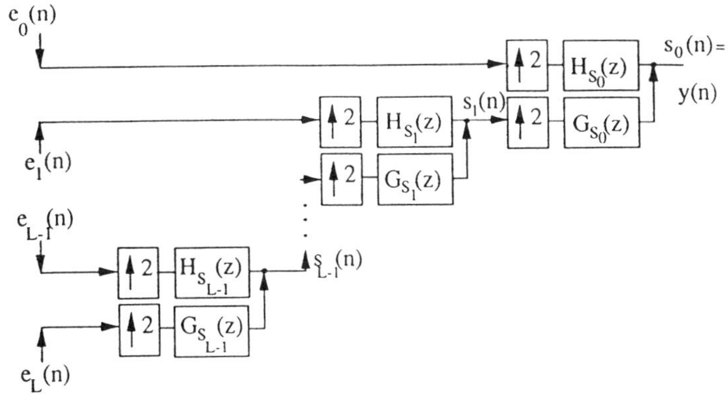

Fig. 6. A binary tree-structured synthesis bank

Perfect Reconstruction Recursive QMF Banks For Image Subband Coding

HRVOJE BABIC†, SANJIT K. MITRA§, CHARLES CREUSERE§, and AMITAV DAS§

†University of Zagreb
Zagreb, Croatia, Yugoslavia

§Department of Electrical & Computer Engineering
University of California, Santa Barbara, CA 93106

Abstract

A scheme for implementing a perfect reconstruction QMF bank is based on an uniform DFT L-band analysis filter bank realized using a polyphase decomposition of a lowpass L-th band filter composed of stable allpass subfilters followed by a L-band synthesis filter bank realized using the transpose of the analysis filter bank with the allpass filters replaced by their inverse filters. In the subband coding of finite-length sequences, the instability problem of the inverse filters can be avoided by using the original stable allpass filter in each branch of the synthesis bank with time-reversed input and output. Conditions for perfect reconstruction are derived and illustrated with examples.

1. Introduction

The basic scheme in the subband coding of images involves the decomposition of the original image into a set of subband images occupying contiguous frequency bands of narrow bandwidth by means of an analysis filter bank. The subband signals are then critically down-sampled and coded separately. The coded signals are next up-sampled and combined by means of a synthesis filter bank, yielding the reconstructed image. To minimize the amplitude and phase distortion introduced by the filtering and the aliasing distortion caused by the sampling rate alteration process, perfect reconstruction (PR) FIR quadrature-mirror filter (QMF) banks have been preferred by most authors [1]-[3]. However, because of the very large computational complexity of FIR filters, the proposed QMF bank based subband coders have so far been limited to 4×4 bands. An elegant approach to the coding of images using IIR filter based QMF banks was proposed recently [4]. The purpose of this paper is to analyze the performance of this new structure with respect to the effect of initial conditions on the reconstructed signal and show how to choose the initial conditions to guarantee perfect reconstruction.

2. The Basic Structure

An L-th band IIR lowpass filter $H_0(z)$ can be realized in a polyphase network form given by

$$H_0(z) = \sum_{k=0}^{L-1} z^{-k} A_k(z^L) \qquad (1)$$

where $A_k(z^L)$ is a stable allpass filter [5]. The above filter can be used as a lowpass prototype to generate a uniform DFT analysis filter bank using the structure indicated in Fig. 1 [6]. Even though a transpose of this structure yields the uniform DFT synthesis bank, a combination of the analysis bank and its transpose with factor-of-L down-samplers and up-samplers does not yield a perfect reconstruction QMF bank. However, perfect reconstruction can be achieved by using instead the structure of Fig. 2 [4]. To verify the perfect reconstruction (PR) property, we can redraw the structure of Fig. 2 making use of the noble identities [6] resulting in the structure of Fig. 3. Since for allpass filters

$$A_k(z)A_k(z^{-1}) = 1 \qquad \text{for } k = 0,1,\ldots,L-1, \qquad (2)$$

Figure 3 reduces to that shown in Fig. 4 establishing the PR property of the QMF bank of Fig.2.

3. Conditions For Perfect Reconstruction

Even though, the QMF structure of Fig. 3 is computationally efficient from a practical point of view, a direct implementation of this structure is not feasible, as the allpass filters in the synthesis bank are unstable. However, in the subband coding of finite-extent sequences, such as digital images, the structure of Fig. 3 can be made realizable using the scheme indicated in Fig. 5. Here, if the input $x_k(m)$ to the allpass filter $A_k(z)$ in the analysis bank is of length M, only the first M samples of the output $v_k(m)$ are computed. These are then time-reversed and fed into the identical allpass section $A_k(z)$ of the synthesis bank and again only first M samples of its output $y_k(m)$ are computed. Finally, these outputs are time-reversed for perfect reconstruction of the input block.

In general, the output of an IIR allpass section will be of infinite length even though its input is finite extent. Moreover, the output will also contain transient terms resulting from the initial conditions present in the allpass section at the beginning of the processing. We now show that by appropriately choosing the initial conditions of the allpass section in the synthesis filter bank, it is always possible to reconstruct exactly the finite extent input sequence of the corresponding allpass section of the analysis filter bank. We first derive the conditions for the case when both allpass sections have identical configurations and then derive the conditions when different structures are used to realize the two allpass sections.

3.1 Identical Allpass Filter Structures

For notational simplicity we drop the subscript "k" from the input and output sequences, and the allpass transfer function in Fig. 5. Let the allpass transfer function in the analysis and synthesis branches be of N-th order and given by

$$A(z) = \frac{a_N + a_{N-1} z^{-1} + \cdots + z^{-N}}{1 + a_1 z^{-1} + \cdots + a_N z^{-N}}, \tag{3}$$

A state-space representation of the above allpass transfer function in the analysis branch is of the form

$$\mathbf{g}(n+1) = \mathbf{A}\ \mathbf{g}(n) + \mathbf{B}\ x(n) \tag{4a}$$
$$v(n) = \mathbf{C}\ \mathbf{g}(n) + D\ x(n) \tag{4b}$$

where $\mathbf{g}(n)$ is the length-N state-variable vector of the analysis stage, and for a direct form II realization, the state space parameters are given by

$$\mathbf{A} = \begin{bmatrix} -a_1 & -a_2 & \cdots & -a_{N-1} & -a_N \\ 1 & 0 & \cdots & 0 & 0 \\ \vdots & \vdots & \ddots & \vdots & \vdots \\ 0 & 0 & \cdots & 0 & 0 \\ 0 & 0 & \cdots & 1 & 0 \end{bmatrix}, \ \mathbf{B} = \begin{bmatrix} 1 \\ 0 \\ \vdots \\ 0 \\ 0 \end{bmatrix}, \tag{5a}$$

$$\mathbf{C} = \left[(a_{N-1} - a_N a_1) \ \cdots \ (1 - a_N{}^2) \right], \ D = a_N. \tag{5b}$$

Likewise, a state-space representation of the allpass transfer function in the synthesis branch is identical in form to that of Eq. (4) and is given by

$$\mathbf{w}(n+1) = \mathbf{A}\ \mathbf{w}(n) + \mathbf{B}\ u(n) \tag{6a}$$
$$y(n) = \mathbf{C}\ \mathbf{w}(n) + D\ u(n) \tag{6b}$$

where $\mathbf{w}(n)$ is the length-N state-variable vector of the synthesis stage. Moreover, for a direct form II realization, the state space parameters are again given by Eq. (5).

The following relations between the state-space matrices of the allpass will be useful later and can be easily verified by direct substitutions of their corresponding expressions from Eqs. (5):

$$\mathbf{A\ P\ A} + \mathbf{B\ C} = \mathbf{P},\ \mathbf{A\ P\ B} + \mathbf{B\ D} = \mathbf{0} \tag{7a}$$
$$\mathbf{C\ P\ A} + \mathbf{D\ C} = \mathbf{0},\ \mathbf{C\ P\ B} + \mathbf{D}^2 = 1 \tag{7b}$$

where $\mathbf{0}$ is a null vector and $\mathbf{P}$ is the exchange matrix

$$\mathbf{P} = \begin{bmatrix} 0 & \cdots & 0 & 1 \\ 0 & \cdots & 1 & 0 \\ \vdots & \ddots & \vdots & \vdots \\ 1 & \cdots & 0 & 0 \end{bmatrix}. \tag{8}$$

By choosing the initial values of the state variables $\mathbf{w}(n)$ in the synthesis branch allpass section as a function of the final values of the state variables $\mathbf{g}(n)$ of the analysis branch allpass section according to

$$\mathbf{w}(0) = \mathbf{P}\ \mathbf{g}(M), \tag{9}$$

where M is the length of the input, we will show that the output y(n) is precisely the time-reversed version of the input sequence x(n), i.e.

$$y(n) = x(M-n-1), \quad \text{for } n = 0, 1, \ldots, M-1, \tag{10}$$

with $u(n) = v(M-n-1)$. We prove the above result in two steps.

<u>Step 1</u>. We first prove by induction that if Eq. (9) holds, then

$$\mathbf{w}(m) = \mathbf{P}\ \mathbf{g}(M-m), \quad \text{for } m = 0, 1, \ldots, M-1. \tag{11}$$

Assume that $\mathbf{w}(\ell) = \mathbf{P}\ \mathbf{g}(M-\ell)$ for $0 \le \ell \le m$. We show that $\mathbf{w}(m+1) = \mathbf{P}\ \mathbf{g}(M-m-1)$. Now from Eq. (6a),

$$\begin{aligned} \mathbf{w}(m+1) &= \mathbf{A}\ \mathbf{w}(m) + \mathbf{B}\ u(m) \\ &= \mathbf{AP}\ \mathbf{g}(M-m) + \mathbf{B}v(M-m-1) \\ &= \mathbf{AP}\{\mathbf{A}\ \mathbf{g}(M-m-1) + \mathbf{B}\ x(M-m-1)\} \\ &\quad + \mathbf{B}\{\mathbf{C}\ \mathbf{g}(M-m-1) + D\ x(M-m-1)\} \\ &= \{\mathbf{APA} + \mathbf{BC}\}\mathbf{g}(M-m-1) \\ &\quad + \{\mathbf{APB} + \mathbf{BD}\}x(M-m-1) = \mathbf{Pg}(M-m-1) \end{aligned} \tag{12}$$

using Eqs. (4), (6), and (7a).

<u>Step 2</u>. We next prove Eq. (10). From Eq. (6b),

$$\begin{aligned} y(m) &= \mathbf{C}\ \mathbf{w}(m) + D\ u(m) \\ &= \mathbf{C}\ \mathbf{w}(m) + D\ v(M-m-1) \\ &= \mathbf{CP}\ \mathbf{g}(M-m) + D\ v(M-m-1) \\ &= \mathbf{CP}\{\mathbf{A}\ \mathbf{g}(M-m-1) + \mathbf{B}\ x(M-m-1)\} \\ &\quad + D\{\mathbf{C}\ \mathbf{g}(M-m-1) + D\ x(M-m-1)\} \\ &= \{\mathbf{CPA} + \mathbf{DC}\}\ \mathbf{g}(M-m-1) \\ &\quad + \{\mathbf{CPB} + D^2\}\ x(M-m-1) = x(M-m-1) \end{aligned} \tag{13}$$

using Eqs. (4), (6), and (7b).

It should be evident from the above discussion that the Condition (9) for perfect reconstruction holds as long as the structures realizing the allpass sections in the analysis and synthesis branches are identical.

3.2 General Case

Now consider the case when different structures are used to realize the allpass sections of the analysis and synthesis branches. If the allpass section of the analysis branch is assumed to be a direct form II realization, its state-space representation is given by Eqs. (4) and (5). If on the other hand, the allpass section in the synthesis branch is realized by a structure other than direct form II, its state space realization is then given by

$$\mathbf{w}'(n+1) = \mathbf{A}'\mathbf{w}'(n) + \mathbf{B}'u(n) \tag{14a}$$
$$y(n) = \mathbf{C}'\mathbf{w}'(n) + D'u(n) \tag{14b}$$

where $\mathbf{w}'(n)$ are the new state-variable vector related to the state variable vector $\mathbf{w}(n)$ of the direct form II realization through a linear transformation $\mathbf{T}$:

$$\mathbf{w}'(n) = \mathbf{T}\,\mathbf{w}(n) \tag{15}$$

Therefore for perfect reconstruction the condition of Eq. (9) should be modified as

$$\mathbf{w}'(0) = \mathbf{T}\,\mathbf{P}\,\mathbf{g}(M) \tag{16}$$

As before, the above modified condition holds for any two different realizations of an allpass transfer function related through a linear transformation $\mathbf{T}$.

4. Illustrative Examples

4.1 Effect of Initial Conditions for a Single Branch

We now present several examples illustrating the effect of initial conditions on the reconstruction of images. We first consider the cascade of Figure 5, and later consider the subband decomposition using the QMF bank of Figure 3.

How an image is effected by using the wrong initial conditions in the allpass filter in the synthesis branch depends strongly on both the image itself and the filter characteristics. For example, consider the effect of setting to zero the initial conditions. In such a case for an image that is dark near its edges, the effect usually is not very noticeable. On the other hand, the effects of improper initial conditions are much more pronounced for an image with very high illuminance near its border, and the difference between the reconstructed images with the proper initial conditions and with zero initial conditions is usually very significant. This difference is illustrated in Figs. 6 (a)-(d) with identical second order allpass filters in the analysis and synthesis stages and with initial conditions set to zero. For comparison, Fig. 6(c) show the image in Fig. 6(b) reconstructed using the proper initial conditions. The allpass filter itself also has a significant effect on the amount of distortion caused in the reconstructed image by incorrect initial conditions. For example, the filter used in the examples of Figs. 6(a)-(c) has poles near the unit circle (at $0.95e^{\pm j\pi/6}$). If the poles are moved to $0.5e^{\pm j\pi/10}$, which is closer to the origin and toward the real axis, the affected areas in the reconstructed image due to incorrect initial conditions is reduced although the actual MSE increased to 471.

As one might expect, the order of the allpass filter also has a significant effect. When an 8th order cascade filter (made up of 4 stages of the 2nd order allpass filter used earlier in Fig. 6(b)) is used to process the image, the distortion becomes much more pronounced and the mean square error increases significantly to 79.

As the final values of the state-variables in the analysis section has to be transmitted to the synthesis section to serve as the initial conditions, in image coding applications these samples need to be included in the overall average bpp calculation. Hence to decrease the average bpp of the coded image one may consider a coarse quantization of the initial conditions. For the filter used to produce Fig. 6(b), we found that an error of 5% was indistinguishable from the case where exact values were used (the MSE was 0.485). With an error of 20% in the initial conditions, the transient effects were noticeable but were still not severe (MSE = 0.9).

4.2 Effect of Initial Conditions on Four-band Subband Decomposition

To observe the effect of initial conditions on an actual subband decomposition, we implemented the basic four band decomposition described in [7] (Fig. 7(c)). Here, each 1D filter has two branches and the transform is a very simple 2x2 DFT. The subband decomposition is then performed separably, first in the horizontal direction and then in the vertical direction. The two allpass filters used were

$$A_0(z) = \frac{0.5+z^{-1}}{1+0.5z^{-1}}, \qquad A_1(z) = 1. \tag{17}$$

Note that one of these filters is of first-order with a single state variable, whereas, the other is a zeroth-order allpass and requires no initial conditions at all. Thus, in this case, one would expect that the effects of neglecting initial conditions would not be very severe. Figure 7(b) illustrates the effect of setting the initial state in the synthesis filter to zero. While the undesired transient response is not too bad, it is still rather significant given the low order of the allpass filters used. For comparison, Fig. 7(d) shows the same image after reconstruction using the correct initial conditions.

4.3 Estimating the Initial Conditions

In most natural images, there is a strong correlation between adjacent lines in large regions of an image. Exploiting this correlation to form an estimate of the next line's initial conditions could be done to eliminate the need for their transmission.

To illustrate the possibility of using such an estimator, we considered how a very simple estimate might reduce the effect of the erroneous initial conditions. In our simulation, we added the simple estimator to the 2nd order analysis/ synthesis system used in the example of Fig. 6. To reconstruct the first line, we set the initial conditions to zero (having no better estimate). For subsequent lines, we set the initial conditions to be equal the internal state of the synthesis filter of the previous line after it had processed a few samples. This internal state will be an estimate the actual initial condition of the previous line. If we assume a strong line to line correlation in this region (i.e. the pixel luminance at the edge does not change rapidly), then the initial conditions also have a strong correlation. Thus, we use the synthesis filter as an estimator of the initial condition of the previous line to estimate that of the current line. As long this strong correlation between lines of the image in the region near edge of the image exists, this estimator will become increasingly accurate as more lines are processed. Figure 7(a) shows the result of using this estimator with no information about initial conditions given. Note that the mean squared error is considerably less than that of Fig. 6(b) (same filter with zero initial conditions for reconstruction) and that the only noticeable distortion is near the bottom, right edge of the picture. This clearly indicates that there is indeed a strong correlation in initial states between lines of the image. Further work is continuing to find a better estimator of initial conditions.

Acknowledgements
This work was supported in part by a University of California MICRO grant with matching support from ESL Inc.

References

[1] J. W. Woods/S. D. O'Neil, "Subband coding of images," *IEEE Trans. Acoustics, Speech, Signal Proc.*, Vol. 34, Oct. 1986, pp. 1278-1288.
[2] H. Gharavi/A. Tatabai, "Subband coding of monochrome and color images," *IEEE Trans. Circuits & Systems*, vol. CAS-35, Feb. 1988, pp. 207-214.
[3] P. H. Westernik et al, "Subband coding of images using vector quantization," *IEEE Trans. Communications*, vol. COM-36, June 1988, pp. 713-719.
[4] J. H. Husøy/T. Ramstad. "Application of an efficient parallel IIR filter bank to image subband coding," *Signal Processing*, vol. 20, Aug. 1990, pp. 279-292.
[5] M. Renfors/T. Saramäki, "Recursive N-th band digital filters - Part I: Design and properties," *IEEE Trans. Circuits & Systems*, vol. 34, January 1987, pp. 24-39.
[6] P. P. Vaidyanathan, "Multirate digital filters, filter banks, polyphase networks, and applications: A tutorial," *Proc. IEEE*, vol. 78, January 1990, pp. 56-93.
[7] T. Ramstad, "IIR filter banks for subband coding of images," *Proc. IEEE Int'l. Symp. on Circuits & Systems*, Helsinki, Finland, June 1988, pp. 827-830.

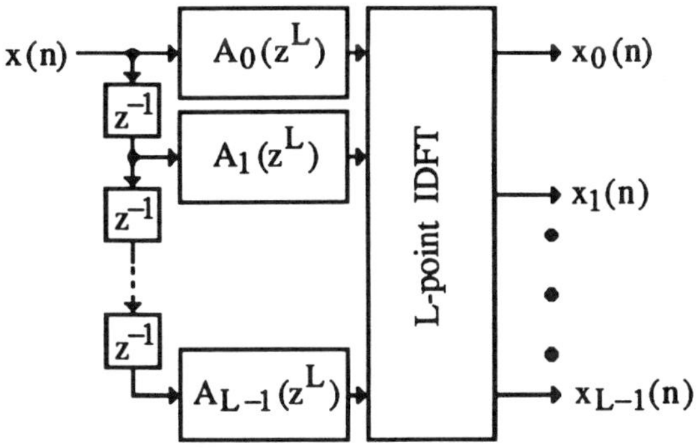

Figure 1 - Uniform DFT analysis filter bank.

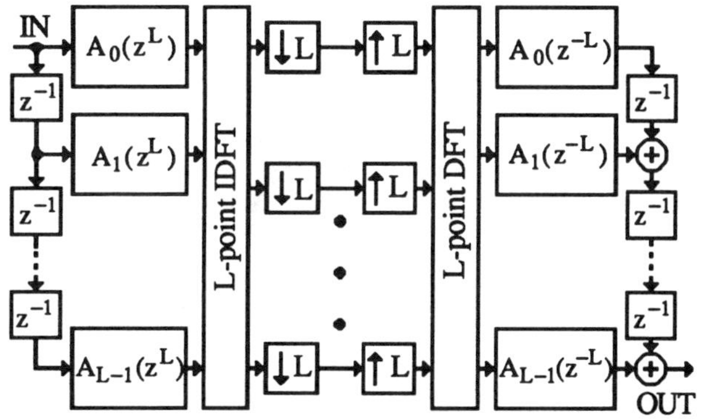

Figure 2 - The basic N-band perfect reconstruction QMF structure.

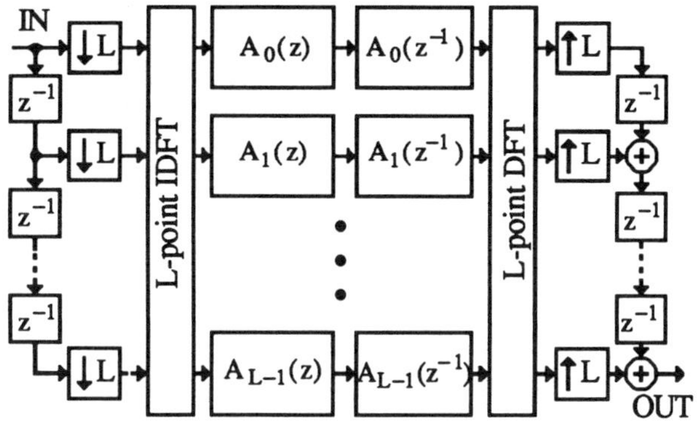

Figure 3 - An equivalent representation of the structure of Figure 2.

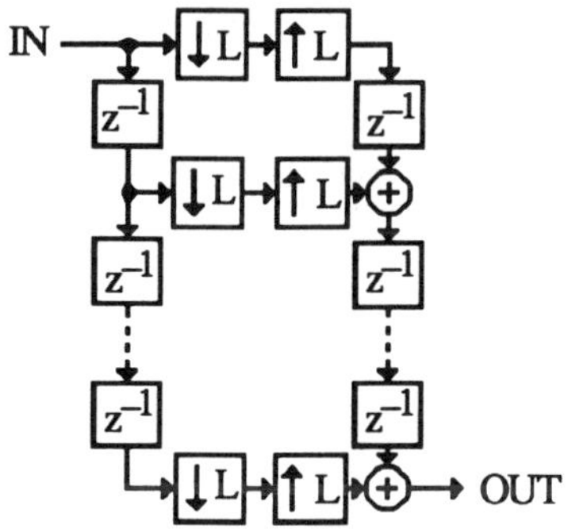

Figure 4 - Further simplification of the structure of Figure 2.

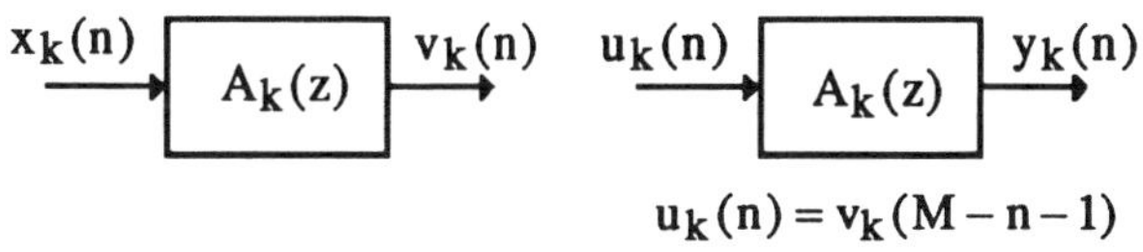

Figure 5 - The k-th subband branch.

Figure 6. (a) Image with dark borders, MSE = 11.7; (b) Image with bright borders, MSE = 20.4; (c) Perfect reconstruction of image of (b), MSE = 0.1; and (d) Zoomed version of part of the image of (b).

Figure 7. (a) Reconstructed image using estimated initial conditions, MSE = 1.48; (b) Reconstructed image using a 4-band subband decomposition with zero initial conditions, MSE = 92; (c) The four subband images; and (d) Perfectly reconstructed image using the correct initial conditions, MSE = 0.2.

State-Space Representations of 2-D FIR Lossless Matrices and their Application to the Design of 2-D Subband Coders*

S. Venkataraman and B. C. Levy
Department of Electrical Engineering and Computer Science
University of California, Davis CA 95616

1 Introduction

Digital multirate filter banks are widely used for subband coding of speech signals [2]. FIR lossless systems form a class of perfect reconstruction filter banks, and 2-D filters of this type have been employed in image coding [11]. The advantages of such filter banks include the fact that the analysis and synthesis filters have the same length, which makes their implementation simple, and the stability of filters is guaranteed [9]. Lately there has been a revival of interest in lossless systems. In addition to subband coding, researchers are finding new applications for such systems. For instance, the design of wavelets for the multiresolution analysis of signals or images [6] is closely tied to the design of lossless filter banks.

The properties of 1-D FIR lossless filter banks have been studied extensively, and many efficient procedures have been developed for their design [3, 9]. The design of lossless filter banks often involves constructing lossless transfer matrices whose rows are formed by the polyphase components of each filter in the filter bank. Perhaps, the most important result in the design of such matrices is the ability to express lossless transfer matrices of a certain degree as a cascade of systems with smaller degrees [3]. However, 2-D FIR lossless systems have not been studied as closely as their 1-D counterparts. This paper provides a general characterization of $p \times p$ lossless 2-D FIR transfer matrices of arbitrary degree in (z_1, z_2). The properties of 2-D state-space realizations of such matrices are examined. It is shown that these matrices are not necessarily decomposable as a product of 1-D matrix transfer functions depending on z_1 and z_2 separately. An approach for the design that does not exclude any lossless filter of given degree is presented. An illustrative example which covers a subset of the ideas discussed is presented. This example consists of a two-band lossless filter bank suitable for hexagonal subsampling.

2 First Level Realization

Given a rational 2-D matrix transfer function, a first-level sate-space realization is a realization with respect to one of its variables, say z_2, involving coefficients that are rational functions of the other variable. An important result of 2-D system theory is that it is always possible to obtain a minimal first-level realization of a 2-D rational transfer matrix, but the same cannot be said of second-level realizations, i.e. state-space realizations with respect to both variables. We consider here 2-D causal FIR lossless transfer matrices. It turns out that the first level state-space realizations of such matrices provide substantial information about their structure, which can then be used to develop a design methodology for their synthesis. Another motivation for using a state-space approach is that alternative representations of lossless transfer matrices, such as cascade lattice structures can be derived easily in the state-space framework.

*This work was supported by the National science Foundation under Grant MIP-9015271.

2.1 Characterization of 2-D FIR Lossless Transfer Matrices

Consider an arbitrary $p \times p$ 2-D FIR quarter plane causal transfer matrix

$$H(z_1, z_2) = \sum_{j=0}^{l} \sum_{i=0}^{k} h_{ij} {z_1}^{-i} {z_2}^{-j} , \tag{1}$$

where k and l denote the highest powers of H in ${z_1}^{-1}$ and ${z_2}^{-1}$, respectively. Let also q and r be the McMillan degrees of H in z_1 and z_2. Then, as shown in [12, 8], it is always possible to construct a minimal first-level state-space realization

$$H(z_1, z_2) = D(z_1) + C(z_1)(z_2 I_r - A(z_1))^{-1} B(z_1) \tag{2}$$

of degree r , where $A(z_1), B(z_1), C(z_1)$ and $D(z_1)$ are causal rational matrices in z_1. In the following, such a realization will be denoted as

$$H(z_1, z_2) \sim M(z_1) = \left[\begin{array}{cc} A(z_1) & B(z_1) \\ C(z_1) & D(z_1) \end{array} \right] . \tag{3}$$

The representation (2) is valid for values of z_1 that do not correspond to poles of $M(z_1)$.

If the transfer matrix (1) is lossless, following a procedure outlined in [12], it can be shown that there exists a minimal realization $M(z_1)$ of size $(r+p) \times (r+p)$ which is paraunitary, i.e.,

$$M(z_1) M^T({z_1}^{-1}) = I_{r+p} . \tag{4}$$

Of course, the converse is also true, i.e. given a paraunitary matrix $M(z_1)$, the corresponding matrix transfer function $H(z_1,z_2)$ is lossless. Furthermore, it can be shown [10] that the causal FIR property of $H(z_1,z_2)$ implies that the above minimal paraunitary realization is causal FIR and of McMillan degree q in z_1.

2.2 Factorizability

The paraunitarity of $M(z_1)$ implies

$$A(z_1)A^T(z_1{}^{-1}) + C(z_1)C^T(z_1{}^{-1}) = I_r \,. \quad (5)$$

It is also shown below that we can always find a rational paraunitary similarity transformation $T(z_1)$, such that after transformation the matrix $A(z_1)$ is triangular. This transformation preserves the paraunitarity of $M(z_1)$, but its entries become rational functions functions of z_1. From identity (5) and the triangularity of $A(z_1)$, the Theorem 2.4 of [1] implies that the 2-D FIR lossless transfer matrix $H(z_1,z_2)$ can be factored minimally as a cascade of r FIR $p\times p$ lossless sections of degree one in z_2, with coefficients which are rational functions of z_1.

In order to express the transfer matrix $H(z_1,z_2)$ as a cascade of degree one transfer matrices in one variable, we must find a similarity transformation $\mathbf{T}(z_1) \triangleq \begin{bmatrix} T(z_1) & 0 \\ 0 & I_p \end{bmatrix}$ such that in the transformed matrix

$$\begin{aligned} \hat{M}(z_1) &= \begin{bmatrix} \hat{A}(z_1) & \hat{B}(z_1) \\ \hat{C}(z_1) & \hat{D}(z_1) \end{bmatrix} \\ &= \mathbf{T}(z_1)M(z_1)\mathbf{T}^{-1}(z_1) \end{aligned} \quad (6)$$

$\hat{A}(z_1)$ is triangular. When $A(z_1)$ is constant, this can be achieved by using the Schur decomposition method. In general, it is not always possible to triangularize a square matrix whose entries are polynomials in $z_1{}^{-1}$ by using a rational transformation $T(z_1)$. Fortunately, here $H(z_1,z_2)$ is FIR, so that $A(z_1)$ is nilpotent. This implies that all the generalized eigenvectors of $A(z_1)$ are polynomial in $z_1{}^{-1}$, so that there exists a rational triangularizing transformation $T(z_1)$.

In addition to triangularizing $A(z_1)$, we seek to preserve the paraunitary property of $M(z_1)$ after applying the similarity transformation. According to [12], $M(z_1)$ will remain paraunitary if and only if $T(z_1)$ is itself a paraunitary matrix of size $r\times r$. The fact that the eigenvectors of $A(z_1)$ are polynomial allows the construction of a rational paraunitary $T(z_1)$ which yields a Schur like decomposition of $A(z_1)$. This implies that in the minimal factorization of $H(z_1,z_2)$, the degree one sections in z_2 will have rational coefficients in z_1.

In summary, given an arbitrary 2-D FIR lossless transfer matrix with minimal paraunitary realization $M(z_1)$, the matrix $A(z_1)$ appearing in such a realization has polynomial eigenvectors. This property implies that $H(z_1,z_2)$ admits a cascade decomposition in degree one sections in z_2 whose coefficients are rational functions of z_1. If the eigenvectors of $A(z_1)$ are constant, the cascade decomposition involves only FIR, i.e. polynomial in $z_1{}^{-1}$, coefficients. Furthermore, if $A(z_1)$ itself is constant, and either $C(z_1)$ or $B(z_1)$ is constant, then $H(z_1,z_2)$ is separable and can be expressed as

$$H(z_1,z_2) = H_1(z_1)H_2(z_2)\,. \quad (7)$$

In the subband coding context, the entries of $H(z_1,z_2)$ represent the polyphase components of the analysis filters. Thus, although the matrix $H(z_1,z_2)$ given by (7) is separable, the polyphase components and analysis filters are not separable, independently of the type of subsampling scheme considered.

3 Synthesis Based on Roesser's Model

The matrix $A(z_1)$ plays a crucial role in determining the structure of the lossless matrix $H(z_1,z_2)$. This role can be described more explicitly by considering the Roesser state-state space model of $H(z_1,z_2)$. Roesser's model provides also a convenient framework for formulating the design of 2-D FIR lossless transfer matrices.

3.1 Roesser's Model for 2-D FIR Lossless Matrices

Roesser's state-space model of a 2-D system [7] is given by

$$\begin{bmatrix} x_1(i+1,j) \\ x_2(i,j+1) \\ y(i,j) \end{bmatrix} = \begin{bmatrix} F_1 & F_2 & G_1 \\ F_3 & F_4 & G_2 \\ H_1 & H_2 & J \end{bmatrix} \begin{bmatrix} x_1(i,j) \\ x_2(i,j) \\ u(i,j) \end{bmatrix}, \quad (8)$$

where $u(i,j)$ and $y(i,j)$ are the input and output vectors, and $x_1(i,j)$ and $x_2(i,j)$ are the horizontal and vertical state vectors. All the matrices are assumed to have compatible sizes. In the following, this model will be denoted as

$$\mathcal{R} \triangleq \begin{bmatrix} F_1 & F_2 & G_1 \\ F_3 & F_4 & G_2 \\ H_1 & H_2 & J \end{bmatrix} \triangleq \begin{bmatrix} F & G \\ H & J \end{bmatrix}. \quad (9)$$

In the case of FIR transfer matrices, the size of F indicates whether the realization is minimal or not. If the transfer matrix $H(z_1,z_2)$ has McMillan degrees q and r in z_1 and z_2, then in a minimal state-space realization, F must have size $(q+r)\times(q+r)$.

The first level realization (2) can be written in terms of Roesser's model (8) as

$$\begin{aligned} M(z_1) &= \begin{bmatrix} A(z_1) & C(z_1) \\ B(z_1) & D(z_1) \end{bmatrix} \\ &= \begin{bmatrix} F_4 & G_2 \\ H_2 & J \end{bmatrix} + \begin{bmatrix} F_3 \\ H_1 \end{bmatrix} \times \\ &\quad (z_1 I_q - F_1)^{-1} \begin{bmatrix} F_2 & G_1 \end{bmatrix}. \end{aligned} \quad (10)$$

As indicated in Section 2.1, we can find a matrix $M(z_1)$ which is causal, paraunitary and of degree q. Therefore, there exists a Roesser realization $\mathcal{R}$ which is orthonormal and of size $(q+r+p)\times(q+r+p)$.

3.2 Design Approach

To obtain a 2-D FIR lossless transfer matrix, one approach would be to construct a $(p+r)\times(p+r)$ causal FIR lossless transfer matrix $M(z_1)$ of degree q by using any of the 1-D design techniques available in the literature. However, to ensure that the 2-D transfer matrix is FIR, the $r \times r$ matrix $A(z_1)$ must be nilpotent.

A design technique similar to the 1-D method proposed in [3] relies on Roesser's state-space model. The quantity that we seek to compute is the $(q+r+p)\times(q+r+p)$ matrix $\mathcal{R}$ given by (10). Losslessness is ensured by making $\mathcal{R}$ orthonormal. Then, if we consider

$$F = \begin{bmatrix} F_1 & F_2 \\ F_3 & F_4 \end{bmatrix}, \tag{11}$$

a set of necessary and sufficient conditions for H to be FIR is that

- F_1 is $q \times q$ and strictly lower (upper) triangular.
- $A(z_1) = F_4 + F_3 (z_1 I_q - F_1)^{-1} F_2$ is $r \times r$ and nilpotent of order r.

$A(z_1)$ is made nilpotent of the required order by setting the traces of A^n for $n = 1, 2, \cdots, r$ equal to zero. These conditions guarantee that in our search for optimum filters, no $p \times p$ 2-D FIR lossless matrix is excluded.

Up to this point, we have considered general 2-D lossless FIR transfer matrices which involved constraints which are not simple to enforce. Let us consider now a subclass of such matrices, associated to a special structure of the matrix F in (11). This subclass corresponds to the case discussed in Section 2.2, where the eigenvectors of the matrix $A(z_1)$ are constant.

Since the matrices F_1 and F_4 in (11) can be assumed strictly lower triangular, the condition that $H(z_1, z_2)$ should be FIR can be expressed as

$$\det\left[(z_2\mathbf{I}_r - F_4) - (F_3(z_1\mathbf{I}_q - F_1)^{-1}F_2)\right] = z_2{}^r. \tag{12}$$

In this expression, the matrices $z_2\mathbf{I}_r - F_4$ and $(z_1\mathbf{I}_q - F_1)^{-1}$ are invertible and lower triangular. Then, a sufficient condition for (12) to be satisfied is that F_2 and F_3 should be both lower triangular, with one of them being strictly lower triangular. In fact, this requirement can be loosened slightly by oberving that one of the two can have several nonzero superdiagonals, if the other one is strictly lower triangular with an equal number of zero subdiagonals. This yields a matrix $A(z_1)$ which is lower triangular. Then, it can be shown [10] that 2-D FIR lossless transfer matrices which satisfy the above conditions for F_1, F_2, F_3 and F_4 can be factored as as a cascade of lossless sections of degree one in z_2, whose coefficients are polynomial in $z_1{}^{-1}$.

3.3 Separable FIR Lossless Transfer Matrices

The class of 2-D lossless transfer matrices considered by Karlsson and Vetterli [4] corresponds to a further subcase of the one we have just considered, where F_1 and F_4 are stricly lower triangular, $F_2 = 0$ and F_3 is arbitrary. From expression (10) for the first-level realization, we see that in this case $A(z_1) = F_4$ is a constant matrix. Consider the factorization

$$\begin{bmatrix} F_3 & G_2 \\ H_1 & J \end{bmatrix} = \begin{bmatrix} L_2 \\ J_2 \end{bmatrix} \begin{bmatrix} L_1 & J_1 \end{bmatrix}, \tag{13}$$

where the matrices appearing on the right hand side of (13)) have dimensions $(q+p)\times t$ and $t \times (p+r)$, respectively. Substituting $F_2 = 0$ and (13) inside the formula for the transfer matrix

$$H(z_1, z_2) = J + \begin{bmatrix} H_1 & H_2 \end{bmatrix} \times \left[\begin{pmatrix} z_1 I_q & 0 \\ 0 & z_2 I_r \end{pmatrix} - F\right]^{-1} \begin{bmatrix} G_1 \\ G_2 \end{bmatrix} \tag{14}$$

yields

$$\begin{aligned} H(z_1, z_2) &= \left[J_2 + H_2 (z_2\mathbf{I}_r - F_4)^{-1} L_2\right] \\ &\quad \times \left[J_1 + L_1 (z_1\mathbf{I}_q - F_1)^{-1} G_1\right] \\ &\triangleq H_2(z_2) H_1(z_1). \end{aligned} \tag{15}$$

Thus the $p \times p$ transfer matrix $H(z_1, z_2)$ can be decomposed as a product of two 1-D transfer matrices $H_1(z_1)$ and $H_2(z_2)$ of sizes $p \times t$ and $t \times p$, respectively, where t is the rank of the matrix appearing on the left hand side of (13).

However, in constructing the factorization (13), we have not used the fact that the 2-D FIR transfer function $H(z_1, z_2)$ is lossless, or equivalently that the Roesser realization $\mathcal{R}$ is orthonormal. When this property is taken into account, it can be shown [10] that $t = p$, so that in (15), the 2-D transfer matrix $H(z_1, z_2)$ has been factored as the cascade of two 1-D $p \times p$ FIR lossless transfer matrices $H_1(z_1)$ and $H_2(z_2)$ of degrees q and r in z_1 and z_2, respectively.

Remark: The number degrees of freedom available for this structure is given by

$$N_s = \frac{1}{2}[(p+q+r)(p+q+r-1)-(q+r)(q+r+1)], \tag{16}$$

All the structures in which unit degree factors in z_1 and z_2 are cascaded have the same degrees of freedom for a given q and r. But the order in which they are cascaded must be construed as degrees of freedom.

4 Design Examples

We design a 2-D FIR lossless filter bank $\mathbf{h}(z_1, z_2) = \begin{bmatrix} \mathbf{h}_0(z_1, z_2) \\ \mathbf{h}_1(z_1, z_2) \end{bmatrix}$ for subband coding of rectangularly sampled 2-D input signals. This filter bank is suitable for hexagonal (quincunx) subsampling by a factor two using the sampling matrix $\mathbf{D}_s = \begin{bmatrix} 1 & 1 \\ -1 & 1 \end{bmatrix}$. The polyphase vectors associated with $\mathbf{D}_s$ are $\{1\}$ and $\{z_1{}^{-1}\}$. $\mathbf{h}(z_1, z_2)$ can be expressed in terms of its polyphase components and polyphase vectors as

$$\begin{bmatrix} \mathbf{h}_0(z_1, z_2) \\ \mathbf{h}_1(z_1, z_2) \end{bmatrix} = H(z_1 z_2{}^{-1}, z_1 z_2) \begin{bmatrix} 1 \\ z_1{}^{-1} \end{bmatrix} \tag{17}$$

where

$$H(z_1, z_2) \triangleq \begin{bmatrix} h_{00}(z_1, z_2) & h_{01}(z_1, z_2) \\ h_{10}(z_1, z_2) & h_{11}(z_1, z_2) \end{bmatrix}. \tag{18}$$

A FIR lossless $H(z_1, z_2)$ was designed, which upon substitution in (17) yields a perfect reconstruction filter bank. 2-D band splitting is accomplished by requiring the passband of $\mathbf{h}_0(z_1, z_2)$ to be restricted to the region enclosed by the square connecting the four points $(-\pi, 0), (0, \pi), (\pi, 0)$ and $(0, -\pi)$ in the 2-D Fourier plane. Three different matrix filters $H(z_1, z_2)$ with $q = r = 2$, $q = r = 4$, and $q = r = 8$ were designed. The low- and high-pass filters $\mathbf{h}_0(z_1, z_2)$ and $\mathbf{h}_1(z_1, z_2)$ corresponding to these three matrix filters have roughly 8, 32 and 64 coefficients, respectively. The state-space methods described in the previous section were employed for the design of a matrix $\mathcal{R}$, where depending on whether $q = r = 2$, $q = r = 4$ or $q = r = 8$, $\mathcal{R}$ has dimension 6×6, 10×10, and 18×18, respectively. The frequency sampling filter design method was employed, and optimization was carried out by using the MINOS nonlinear optimization software package. The 2-D magnitude responses of $\mathbf{h}_0(z_1, z_2)$ for each of the three designs are plotted in Figs. 2, 4, and 6. A cross section of the magnitude response plot along the $\omega_1 = \omega_2$ axis is shown in Figs. 1, 3, and 5, which displays precisely the filter attenuation in the stop band. It can be seen from these figures that the ideal low- and high-pass filter characteristics are approximated increasingly accurately as the filter degrees q and r are increased. The wing-like behavior appearing in the 2-D magnitude response plot is due to the fact that the transition band between the pass and stop bands was kept very narrow.

5 Conclusions

The properties of first-level realizations of 2-D FIR lossless transfer matrices were examined. These properties were used to characterize the structure of 2-D Roesser state-space models, and to develop a general design approach for such transfer matrices. The results described here are directly applicable to the design of 2-D quadrature mirror filters. They can also be employed to generate 2-D nonseparable orthogonal wavelets. For instance, with $\omega = [\ \omega_1 \quad \omega_2\ ]^T$, the inverse Fourier transform of

$$\phi(\omega) = \prod_{k=1}^{\infty} \mathbf{h}_0((\mathbf{D}_s{}^{-T})^k \omega), \tag{19}$$

gives a possible choice [5] for the generating function $\phi(x_1, x_2)$, which along with its translates and dilates, spans the space $L^2(\mathbf{R}^2)$. In the above expression, $\mathbf{h}_0(z_1, z_2)$ is the lowpass filter of a two channel lossless filter bank. Note that regularity constraints need also to be imposed on $\mathbf{h}_0$ to ensure that the function $\phi(x_1, x_2)$ is smooth. The corresponding wavelet $\psi(x_1, x_2)$ generating the wavelet bases can then be expressed in terms of $\phi(x_1, x_2)$.

A number of issues remain to be investigated, such as the relation between 2-D state space models and lattice representations of 2-D lossless transfer matrices. The design methodolgy presented in this paper requires the use of constrained optimization techniques to exploit all the available degrees of freedom. Among the design constraints listed in Section 3.2, the orthonormality property of the matrix $\mathcal{R}$ can easily be imposed structurally by expressing it as the product of planar rotation matrices [9]. On the other hand, it is unclear whether the trace conditions on the matrix $A(z_1)$ can be expressed structurally, thus allowing the use of unconstrained optimization techniques.

References

[1] D. Alpay and I. Gohberg, "On orthogonal matrix polynomials," in *Orthogonal Matrix Valued Polynomials and Applications*, Birkhauser-Verlag, Basel, 1988.

[2] A. Croisier, D. Esteban and C. Galand, "Perfect channel splitting by use of interpolation/decimation/tree decomposition techniques", in *Int. Conf. on Informat. Sci. Syst.*, Patras, Greece, pp. 443–446, 1976.

[3] Z. Doğanata, P. P. Vaidyanathan and T. Q. Nguyen, "General synthesis procedures for FIR lossless transfer matrices, for perfect reconstruction multirate filter bank applications," *IEEE Trans. Acoust., Speech, Signal Proc.*, vol. 36, pp. 1561–1574, Oct. 1988.

[4] G. Karlsson, M. Vetterli, "Theory of two-dimensional multirate filter banks," *IEEE Trans. Acoust., Speech, Signal Proc.*, vol. 38, pp. 925–937, June 1990.

[5] J. Kovacevic and M. Vetterli, "Nonseparable multidimensional perfect reconstruction filter banks and wavelet bases for $\mathbf{R}^n$," technical report no. 221-90-02, Center for Telecommunications Research, Columbia University, NY 10027.

[6] S. Mallat, "A theory of multiresolution signal decomposition: the wavelet representation," *IEEE Trans. Pattern Recog. and Mach. Intell.*, vol. 11, pp. 674–693, July 1989.

[7] R. P. Roesser, "A discrete state-space model for linear image processing," *IEEE Trans. Automat. Contr.*, vol. 20, pp. 1–10, Feb. 1975.

[8] E. D. Sontag, "On first order equations for multidimensional filters," *IEEE Trans. Acoust. Speech, Signal Proc.*, vol. 26, pp. 480–482, Oct. 1978.

[9] P. P. Vaidyanathan, T. Q. Nguyen, Z. Doğanata, and T. Saramäki, "Improved design technique for design of perfect reconstruction FIR QMF banks with lossless polyphase matrices," *IEEE Trans. Acoust. Speech, Signal Proc.*, vol. 37, pp. 1042–1055, July 1989.

[10] S. Venkataraman and B. C. Levy, "State-space representations for 2-D FIR lossless transfer matrices," technical report, Dept. of Elec. and Comp. Eng., UC Davis, October 1991.

[11] J. W. Woods and S. D. O'Neil, "Subband coding of images," *IEEE Trans. Acoust. Speech, Signal Proc.*, vol. 34, pp. 1278-1288, Oct. 1986.

[12] D. C. Youla, "The synthesis of networks containing lumped and distributed elements," in *Symp. on Generalized Networks*, Polytechnic Institute of Brooklyn, April 1966, pp. 289–343.

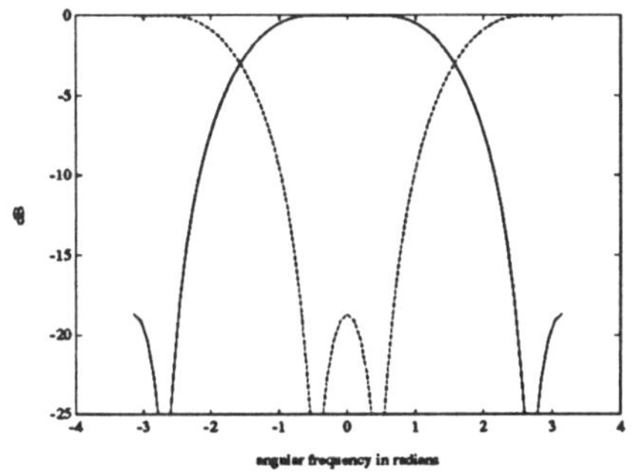

Figure 1: Magnitude response cross section along the $\omega_1 = \omega_2$ axis for the case $q = r = 2$. The middle plot represents the low-pass filter.

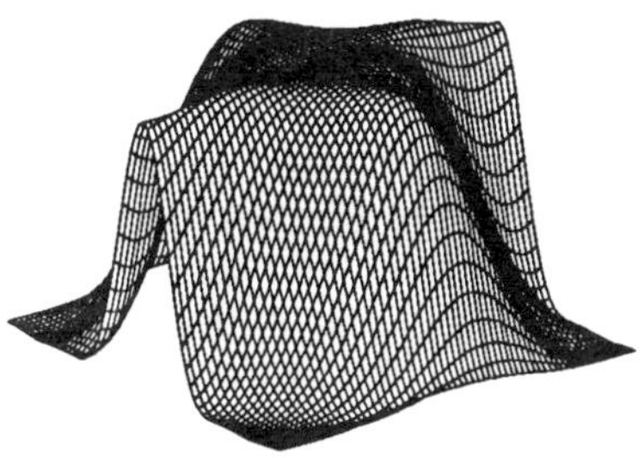

Figure 2: Magnitude response of $\mathbf{h}_0(z_1, z_2)$ for the case $q = r = 2$.

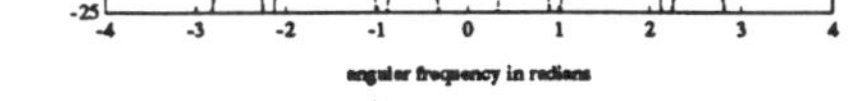

Figure 3: Magnitude response cross section along the $\omega_1 = \omega_2$ axis for the case $q = r = 4$. The middle plot represents the low-pass filter.

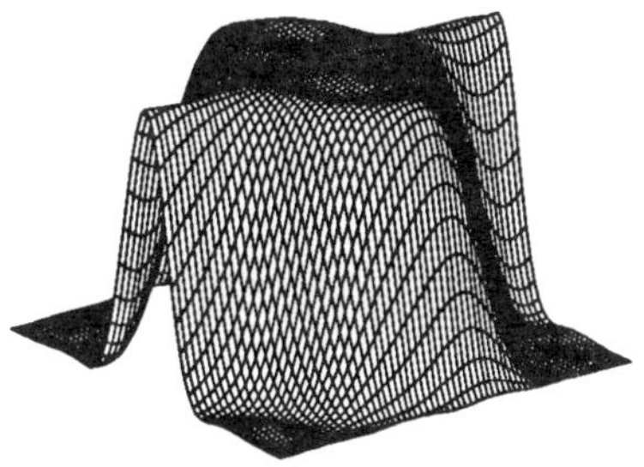

Figure 4: Magnitude response of $\mathbf{h}_0(z_1, z_2)$ for the case $q = r = 4$.

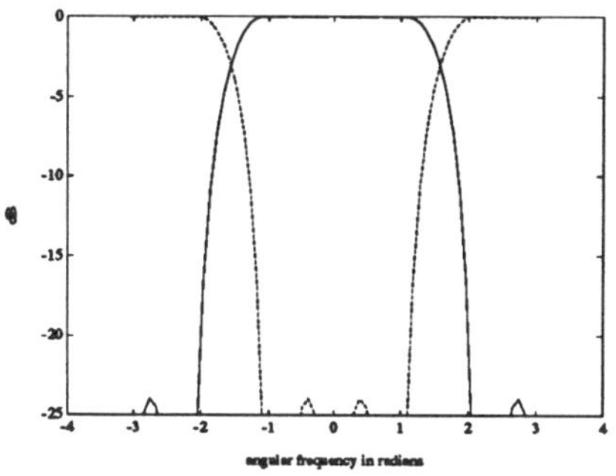

Figure 5: Magnitude response cross section along the $\omega_1 = \omega_2$ axis for the case $q = r = 4$. The middle plot represents the low pass filter.

Figure 6: Magnitude response of $\mathbf{h}_0(z_1, z_2)$ for the case $q = r = 8$.

Adaptive Equalization Using Multirate Filtering Techniques

John J. Shynk† Daniel P. Witmer‡ Michael J. Ready‡
Richard P. Gooch‡ Christina K. Chan†

†Center for Information Processing Research
Dept. of Electrical and Computer Engineering
University of California
Santa Barbara, CA 93106

‡Applied Signal Technology, Inc.
160 Sobrante Way
Sunnyvale, CA 94086

Abstract

This paper presents a computationally efficient blind equalization architecture that is based on multirate and subband adaptive filtering techniques. The received signal is first partitioned by a filter bank into a set of narrowband frequency bins that are approximately nonoverlapping. The adaptive processing is then performed at a lower sampling rate, thus achieving a considerable savings in complexity. Furthermore, the signal power in each frequency bin is normalized to obtain a more uniform convergence rate across the adaptive filter coefficients. The overall convergence rate of the filter is thus dramatically improved compared to that of conventional time- and frequency-domain implementations. Computer simulations of this filter-bank equalizer in a multipath environment are presented to illustrate its transient and steady-state convergence properties.

1 Introduction

Blind equalization algorithms do not require a desired response signal or a training sequence, and they are capable of partially reconstructing the transmitted symbols from only limited knowledge of their *statistical properties* [1]. For example, the constant modulus algorithm (CMA) is a blind strategy that attempts to restore the modulus of the received signal to some predetermined constant [2], [3]. Thus, it is useful for signal constellations such as PSK, and it has even been successfully employed in high-order QAM systems. Once the bit error rate is reduced below a specified threshold, the adaptive algorithm is usually switched to a decision-directed mode whereby the symbol *decisions* are used as the desired response of a gradient-descent algorithm [4]. The three modes of equalizer adaptation are shown in Fig. 1 [5].

The complex LMS (least-mean-square) algorithm has the following weight update [6]:

$$\mathbf{w}(n+1) = \mathbf{w}(n) + 2\mu\mathbf{x}^*(n)e(n) \tag{1}$$

where $\mathbf{w}(n) = [w_0(n), \ldots, w_{N-1}(n)]^T$ is the current weight vector, $\mathbf{x}(n) = [x(n), \ldots, x(n-N+1)]^T$ is the received signal vector (corresponding to the elements of a tapped delay line), and the superscript $*$ denotes complex conjugate. The step size μ is a small positive constant that controls the transient (convergence rate) and steady-state (misadjustment) properties of the algorithm [7]. For normalized versions of this algorithm, μ is often replaced by a time-varying step-size as $\mu(n) = \mu/(\alpha + \mathbf{x}^H(n)\mathbf{x}(n))$ where α is a small positive constant that keeps $\mu(n)$ bounded when the inner product in the denominator is momentarily close to zero. For CMA, the error signal is given by

$$e(n) = -(pq/2)y(n)|y(n)|^{p-2}(|y(n)|^p - r^p)^{q-1} \cdot[\mathrm{sgn}(|y(n)|^p - r^p)]^q \tag{2}$$

where $y(n) = \mathbf{x}^T(n)\mathbf{w}(n)$ is the equalizer output, $\mathrm{sgn}(\cdot)$ is the signum function, and r is a predetermined constant that depends on the constellation of the transmitted symbols. The constants p and q are usually chosen to be either 1 or 2, resulting in four versions of the algorithm. When the algorithm is switched to the decision-directed mode, $e(n) = d(n) - y(n)$ where $d(n)$ is the sequence of symbol decisions at the receiver.

The algorithm in (1) is a *time-domain* implementation that is updated for each new received sample, and thus operates at the transmitted data rate. However, when the length of the equalizer is large (e.g., > 1000 weights), the complexity involved in computing the

*This work was supported by Applied Signal Technology, Inc., and the University of California MICRO Program.

1058-6393/91 $01.00 © 1991 IEEE

equalizer output and updating its weights can be prohibitively high. In this case, it is more cost-effective to implement the equalizer and the adaptive algorithm using *frequency-domain* techniques whereby the data is processed at a much lower sampling rate. Over the last decade, two basic frequency-domain approaches have evolved [8]. One approach is based on efficient FFT implementations of *block* time-domain adaptive algorithms, and they use data sectioning techniques such as the *overlap-save* method [9], [10]. The resulting filters implement either a circular convolution or a linear convolution depending on the block size of the data relative to the size of the FFTs, as well as the type of data sectioning employed.

The second approach is based on *subband* filtering whereby the received signal is processed by a filter bank and the resulting parallel signals are *decimated* before the adaptive processing [11], [12], [13]. The output signals of the parallel adaptive filters are then upsampled to the original data rate, and they are combined by a synthesis filter bank to generate the overall filter output sequence. We refer to an equalizer based on this second approach as a *filter-bank equalizer* (FBE), and that based on the first approach is referred to as a *frequency-domain equalizer* (FDE) [14], [15].

In this paper, we focus on the properties of an FBE architecture for blind equalization and compare its performance to that of an FDE implementation of similar complexity. Section 2 gives a brief overview of an FDE realization of CMA that is based on overlap-save data sectioning. The FBE structure considered here is based on a polyphase filter design, and this is described in Section 3. Computer simulations are then given in Section 4 to illustrate and compare their convergence properties for a multipath channel and QPSK signalling. Conclusions are outlined in Section 5.

2 Frequency-Domain Equalizer

The FDE is comprised of N adaptive weights that are updated according to [9], [10]

$$\mathbf{W}(k+1) = \mathbf{W}(k) + 2\mathbf{G}\mu(k)\mathbf{X}^H(k)\mathbf{E}(k) \quad (3)$$

where $\mathbf{X}(k)$ is the input-signal *matrix* comprised of two blocks of the received signal:

$$\mathbf{X}(k) = \text{diag}\{\mathbf{F}[x(kN/2-N/2),\ldots,x(kN/2-1), x(kN/2),\ldots,x(kN/2+N/2-1)]\} \quad (4)$$

where the operator diag$\{\cdot\}$ forms a diagonal matrix and $N/2$ is the block size. The block time index k is related to the original (nonblock) time index as $n = kN/2$. Thus, the adaptive weights and the internal signals of the FDE are computed only once for each block of $N/2$ input samples. It is convenient to use this matrix definition of $\mathbf{X}(k)$ so that the frequency-domain output vector can be written as

$$\mathbf{Y}(k) = \mathbf{X}(k)\mathbf{W}(k) \quad (5)$$

where $\mathbf{W}(k)$ is the adaptive weight vector defined in terms of the time-domain weights as

$$\mathbf{W}(k) = \mathbf{F}[\mathbf{w}^T(k), 0, ..., 0]^T. \quad (6)$$

$\mathbf{F}$ is the $N \times N$ DFT matrix with components $F_{ml} = e^{-j2\pi ml/N}$. Of course, efficient FFT algorithms would be used to compute $\mathbf{X}(k)$ in (4), but it is convenient to introduce this matrix formulation in order to compactly describe the adaptive algorithm. The output signal vector $\mathbf{y}(k) = [y(kN/2),\ldots,y(kN/2+N/2-1)]^T$ is given by

$$\mathbf{y}(k) = \mathbf{k}\mathbf{F}^{-1}\mathbf{Y}(k) \quad (7)$$

where $\mathbf{k}$ is the following $N/2 \times N$ *constraint* matrix that retains only the last $N/2$ components of $\mathbf{F}^{-1}\mathbf{Y}(k)$:

$$\mathbf{k} = [\ 0 \quad \mathbf{I}\ ]. \quad (8)$$

This constraint is generally necessary so that the filter yields a linear convolution (instead of a circular convolution). Finally, the error vector is defined as

$$\mathbf{E}(k) = \mathbf{F}[0,\ldots,0,\mathbf{e}^T(k)]^T \quad (9)$$

where $\mathbf{e}(k) = [e(kN/2),\ldots,e(kN/2+N/2-1)]^T$ has components given by the CMA error defined in (2). Note that the error terms are computed in the time domain.

Since N frequency-domain weight are required to implement the algorithm and there are only $N/2$ weights in the time domain (see (6)), another constraint must be included in the update to ensure that the *gradient* has only $N/2$ nonzero components when it is inverse transformed. Thus, the operator $\mathbf{G}$ in (3) is given by

$$\mathbf{G} = \mathbf{F}\mathbf{g}\mathbf{F}^{-1} \quad (10)$$

where $\mathbf{g}$ is an $N \times N$ constraint matrix defined as

$$\mathbf{g} = \begin{bmatrix} \mathbf{I} & 0 \\ 0 & 0 \end{bmatrix}. \quad (11)$$

This has the effect of forcing the last $N/2$ components of $\mathbf{F}^{-1}\mu(k)\mathbf{X}^H(k)\mathbf{E}(k)$ to be zero, which is consistent with the definition of the frequency-domain weight vector in (6). Observe that the gradient constraint also premultiplies the time-varying diagonal matrix $\mu(k)$, which is usually full rank with N nonzero diagonal components. If $\mu(k) = \mu\mathbf{I}$ such that each weight is updated with the same step size, then it is clear that μ can be factored to the left of $\mathbf{G}$ in (3). However, one of the main advantages of the FDE is that a different time-varying step size can be used for each weight, thus compensating for the power fluctuations across the frequency bins. In our implementation, we used a normalized version of the algorithm such that the mth diagonal component is $\mu_m(k) = \mu/(\alpha + P_m(k))$ where $P_m(k)$ is the power estimate of the mth frequency bin. It is computed recursively as [16]

$$P_m(k) = \lambda P_m(k-1) + \beta |X_m(k)|^2 \qquad (12)$$

where $\lambda = 1 - \beta$ is a "forgetting" (leakage) factor that smoothes the power estimate, and $X_m(k)$ is the mth diagonal component of $\mathbf{X}(k)$.

3 Filter-Bank Equalizer

A block diagram of the FBE is shown in Fig. 2 [11], [14]. It has a configuration similar to that of the FDE, except the FFTs are replaced by a set of polyphase filter banks. The input signal $x(n)$ is partitioned by an *analysis* filter bank into N parallel signals, defined by the diagonal matrix $\mathbf{X}(k)$, and these are decimated by a factor of $N/2$. Decimation by N leads to a more computationally efficient algorithm, but the resulting aliasing distortion degrades the steady-state performance by "biasing" the converged weight vector. The frequency-domain output vector $\mathbf{Y}(k)$ is computed as in (5), and these signals are then upsampled and recombined by a *synthesis* filter bank to generate the overall filter output at the original data rate.

The adaptive algorithm has the same form as in (3), except that the gradient constraint is removed, i.e., $\mathbf{G} = \mathbf{I}$, yielding

$$\mathbf{W}(k+1) = \mathbf{W}(k) + 2\mu(k)\mathbf{X}^H(k-\Delta)\mathbf{E}(k). \qquad (13)$$

The error is computed in the time-domain according to (2), but there are no constraints on the output signals as in (7). A second analysis filter bank transforms the time-domain error $e(n)$ to $\mathbf{E}(k)$, which is also decimated by a factor of $N/2$. Because of the group delay associated with these filter banks, it is necessary that $\mathbf{X}(k)$ be delayed by Δ before it is cross-correlated with the error vector $\mathbf{E}(k)$ [14]. This type of delay is usually required whenever the error is computed in the time domain. The weights are updated only once for every block of $N/2$ data samples, as is the case for the FDE in the previous section. The step sizes are computed using the power estimates in (12).

The filter banks are based on a polyphase design [17] with a prototype lowpass FIR filter of length QN. Its impulse response coefficients are given by $\{h(n)\}$, and Q is chosen to be an odd integer (typically, $Q =$ 3, 5, 7, or 9). The mth output signal of the analysis filter bank can be written as

$$X_m(k) = \sum_{p=0}^{N-1} e^{-j2\pi mp/N} \times \left[\sum_{q=0}^{Q-1} h(qN+p)x(kN/2 - qN/2 - p)\right]. \qquad (14)$$

The prototype filter is designed such that the $\{X_m(k)\}$ cover the entire bandwidth of the input signal spectrum, approximating the ideal maximally flat magnitude response, but with "minimal" overlap. Thus, the lowpass filter bandwidth is approximately f_s/N where f_s is the sampling rate. Observe that the $\{X_m(k)\}$ are computed as the output of a DFT, so that (14) is a computationally efficient implementation. The input is shifted into a QN-length buffer (tapped-delay line), and for each filter of the filter bank, it is weighted by a subset of the impulse response, $\{h(n)\}$, as determined by the value of m.

The decimated signals are weighted by the adaptive coefficients, $\{\mathbf{W}(k)\}$, to yield the frequency-domain output signals, $\{\mathbf{Y}(k)\}$, and these are upsampled by interleaving them with $N/2-1$ zeros. A bank of interpolating filters with the same prototype lowpass impulse response, $\{h(n)\}$, is then used to reconstruct the overall output as

$$y(n+p) = \sum_{q=0}^{Q-1} h(qN+N-p-1) \times \left[(1/N)\sum_{m=0}^{N-1} Y_m(k-q)e^{j2\pi mp/N}\right]. \qquad (15)$$

Observe that the frequency-domain output signals, $\{Y_m(k)\}$, are processed by an inverse DFT, and these are weighted by the prototype filter coefficients and then summed to yield the overall output.

The impulse response of the prototype filter was obtained using a frequency-sampling design based on the ideal flat response as follows:

$$h(n) = (1/QN) \sum_{m=-Q+1}^{Q-1} H(m)e^{j2\pi mn/QN} \qquad (16)$$

where $\{H(m)\}$ were chosen to satisfy $H(0) = 1$, $H^2(1) + H^2(2) = 1$, and $H(1) - H(2) = 1/2$. Thus, $H(1) = 0.9114$ and $H(2) = 0.4114$. In order that the impulse response be real-valued, it is necessary that $H(m) = H(-m)$. Similar results can be obtained for other values of Q; e.g, for $Q = 5$, $H(0) = H(1) = 1$, $H(2) = 0.9114$, and $H(3) = 0.4114$, and for $Q = 9$, $H(0) = H(1) = H(2) = H(3) = 1$, $H(4) = 0.9114$, and $H(5) = 0.4114$.

By increasing the value of Q, there is less interband energy leakage, and hence less correlation across the frequency bins. This allows the adaptive weights to be adjusted independently of each other since the energy in one bin will have little effect on the adaptive weights of all other bins. As a result, the convergence rate of the adaptive filter improves dramatically when bin normalization is incorporated into the adaptive algorithm. However, when Q is increased, the bulk (transport) delay of the filter banks also increases. Even when Δ is chosen so that $\mathbf{X}(k - \Delta)$ and $\mathbf{E}(k)$ are properly matched in time, the convergence rate of the algorithm can still be adversely affected [18]. Thus, there is a trade-off when choosing the value of Q, and we have found that $Q = 3$ provides satisfactory performance in most cases.

The value of Δ is given by $\Delta = \lfloor (QN-1)/L \rfloor$ where L is the decimation factor and $\lfloor u \rfloor$ retains the greatest integer $\leq u$ [14]. When $L = N/2$ and Q is odd, this expression simplifies to $\Delta = 2Q - 1$. Thus, $\Delta = 5$ for $Q = 3$, and $\Delta = 17$ for $Q = 9$.

4 Computer Simulations

The convergence properties of the FDE and FBE for CMA are now presented for a transmitted signal having a QPSK constellation. The impulse response and frequency response of the channel are shown in Fig 3, and these were obtained from measurements taken in an urban multipath environment using a 22.5 Mbaud digital microwave radio as the probing signal. The transmitted symbols were independent and identically distributed, and since we are interested in $T/2$ fractionally-spaced equalizers, the resulting sequence was interleaved with zeros and passed through a transmit filter with a square-root, raised cosine (SRRC) impulse response. In all simulations, the equalizers had $N = 4096$ adaptive weights, and these were initialized to have the SRRC impulse response. Power normalization based on the update in (12) was used in each case with $\lambda = 0.9$ and $\alpha = 0.01$. For the FBE algorithm, we show results only for a prototype filter with $Q = 3$.

Observe in Fig. 4 that the constellation of the equalized signal has less variance for the FBE compared to the FDE. This indicates that the FBE has improved steady-state convergence properties, as demonstrated by the typical learning curves in Fig. 5. (In these figures, one "record" is equal to 1024 received data samples.)

Based on these simulations and the results of other computer runs, we observed the following:

- For similar complexity (i.e., N adaptive weights), the FBE has N degrees of freedom while the FDE has only $N/2$ degrees of freedom.
- With additive noise only, there was no performance gain observed for the FBE over the FDE.
- For both the FBE and FDE, the equalized SNR improves with increasing N.
- With only a small increase in the computational complexity, the FBE converges dramatically faster than the FDE when power normalization is employed.
- The FBE has an improved steady-state performance compared to the FDE, providing a significant gain in the equalized SNR.
- The convergence rate of the FBE decreases as Q is increased from 3 to 9, and there is no significant change in the steady-state performance.

5 Conclusions

A filter-bank architecture was presented for adaptive equalization applications. When the number of required filter coefficients is large (> 1000), such as in a multipath environment, the computational savings relative to conventional time-domain implementations is considerable. One of the main advantages of the filter bank approach is that the adaptive weights may be updated "independently" because the interbin leakage is negligible. Thus, by incorporating time-varying step sizes that compensate for the power variations across the filter bank bins, the overall convergence rate of the adaptive process is dramatically improved. When designing the filter bank prototype filter, it is desirable to reduce the interbin leakage by having sharp transition regions. However, this generally requires that the prototype filter have a long impulse response, and hence a large transport delay that may adversely affect the gains in the convergence rate provided by the reduced interbin leakage. Thus, there is a trade-off when designing the filter bank, and this depends on the particular application and the signal statistics.

References

[1] A. Benveniste and M. Goursat, "Blind equalizers," *IEEE Trans. Commun.*, vol. COM-32, pp. 871–883, Aug. 1984.

[2] D. N. Godard, "Self-recovering equalization and carrier tracking two-dimensional data communication systems," *IEEE Trans. Commun.*, vol. COM-28, pp. 1867–1875, Nov. 1980.

[3] J. R. Treichler and B. G. Agee, "A new approach to multipath correction of constant modulus signals," *IEEE Trans. Acoust., Speech, Signal Processing*, vol. ASSP-31, pp. 459–472, Apr. 1983.

[4] J. G. Proakis, *Digital Communications*, McGraw-Hill, New York, Second Ed., 1989.

[5] J. J. Shynk, R. P. Gooch, K. Giridhar, and C. K. Chan, "A comparative performance study of several blind equalization algorithms," *SPIE Conf. Adaptive Signal Processing*, San Diego, CA, July 1991.

[6] B. Widrow, J. M. McCool, and M. Ball, "The complex LMS algorithm," *Proc. IEEE*, vol. 63, pp. 719–720, Apr. 1975.

[7] B. Widrow and S. D. Stearns, *Adaptive Signal Processing*, Prentice-Hall, Englewood Cliffs, NJ, 1985.

[8] J. J. Shynk, "Frequency-domain and multirate adaptive filtering," *IEEE Signal Processing Mag.*, vol. 9, Jan. 1992.

[9] E. R. Ferrara, "Fast implementation of LMS adaptive filters," *IEEE Trans. Acoust., Speech, Signal Processing*, vol. ASSP-28, pp. 474–475, Aug. 1980.

[10] G. A. Clark, S. R. Parker, and S. K. Mitra, "A unified approach to time- and frequency-domain realization of FIR adaptive digital filters," *IEEE Trans. Acoust., Speech, Signal Processing*, vol. ASSP-31, pp. 1073–1083, Oct. 1983.

[11] G. C. Copeland, "Transmultiplexers used as adaptive frequency sampling filters," *Proc. IEEE Int. Conf. Acoust., Speech, Signal Processing*, pages 319–322, Paris, France, May 1982.

[12] W. Kellermann, "Kompensation akustischer echos in frequenzteilbandern," *Frequenz*, vol. 39, pp. 209–215, July/Aug. 1985.

[13] A. Gilloire and M. Vetterli, "Adaptive filtering in subbands," *Proc. IEEE Int. Conf. Acoust., Speech, Signal Processing*, pages 1572–1575, New York, NY, Apr. 1988.

[14] C. K. Chan, M. R. Petraglia, and J. J. Shynk, "Frequency-domain implementations of the constant modulus algorithm," *Proc. 23rd Asilomar Conf. Signals, Systems, Computers*, pages 663–669, Pacific Grove, CA, Nov. 1989.

[15] M. J. Ready, S. H. Goldberg, and R. P. Gooch, "Architecture considerations for frequency domain equalizers," *Proc. 23rd Asilomar Conf. Signals, Systems, Computers*, pages 687–691, Pacific Grove, CA, Nov. 1989.

[16] P. C. W. Sommen, P. J. Van Gerwen, H. J. Kotmans, and A. J. E. M. Janssen, "Convergence analysis of a frequency-domain adaptive filter with exponential power averaging and generalized window function," *IEEE Trans. Circuits Systems*, vol. CAS-34, pp. 788–798, July 1987.

[17] R. E. Crochiere and L. R. Rabiner, *Multirate Digital Signal Processing*, Prentice-Hall, Englewood Cliffs, NJ, 1983.

[18] J. R. Treichler, S. L. Wood, and M. G. Larimore, "Convergence rate limitations in certain frequency-domain adaptive filters," *Proc. IEEE Int. Conf. Acoust., Speech, Signal Processing*, pages 960–963, Glasgow, Scotland, May 1989.

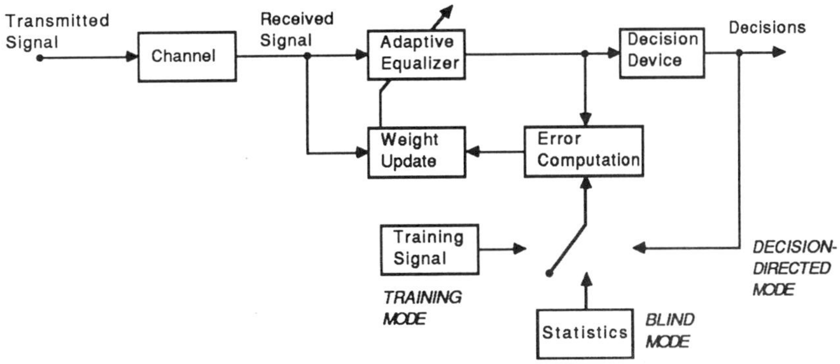

Figure 1: Baseband model of a digital communication system illustrating the three modes of equalizer adaptation.

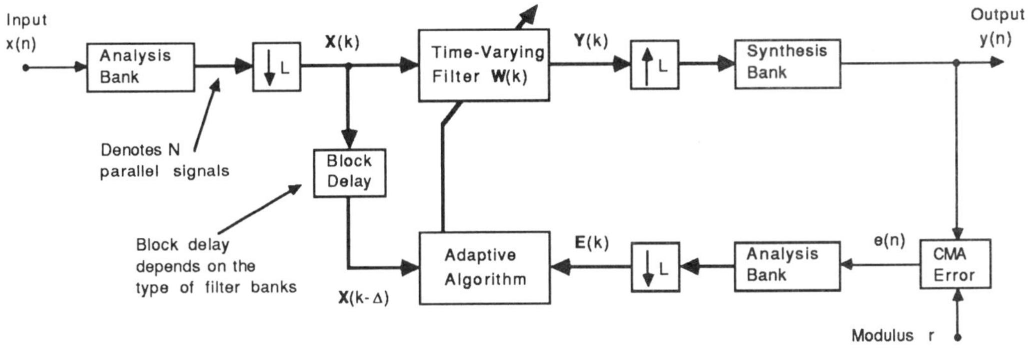

Figure 2: FBE based on polyphase filter banks.

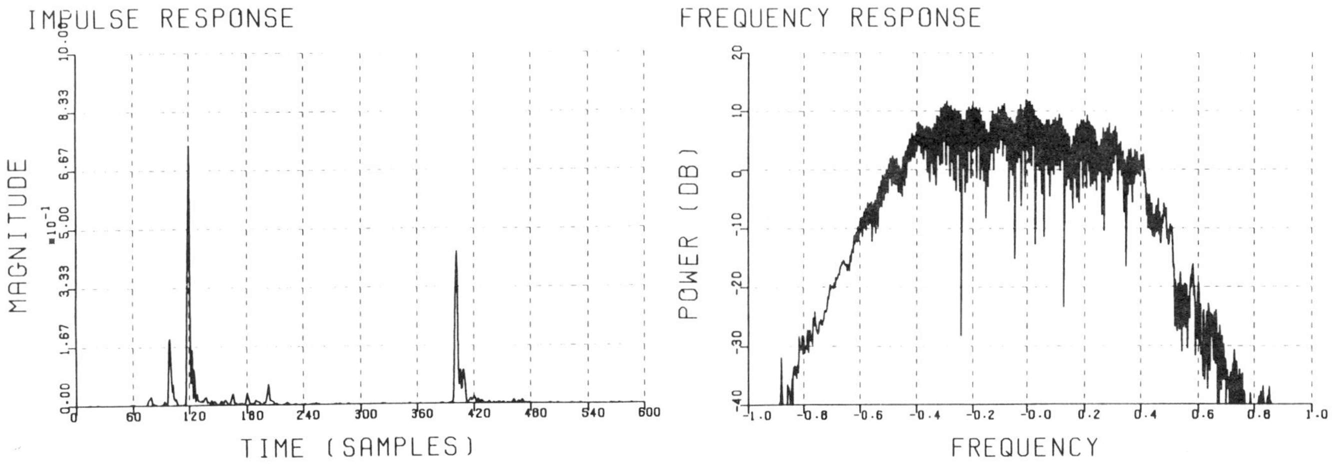

(a) (b)

Figure 3: Channel properties. (a) Impulse response. (b) Frequency response.

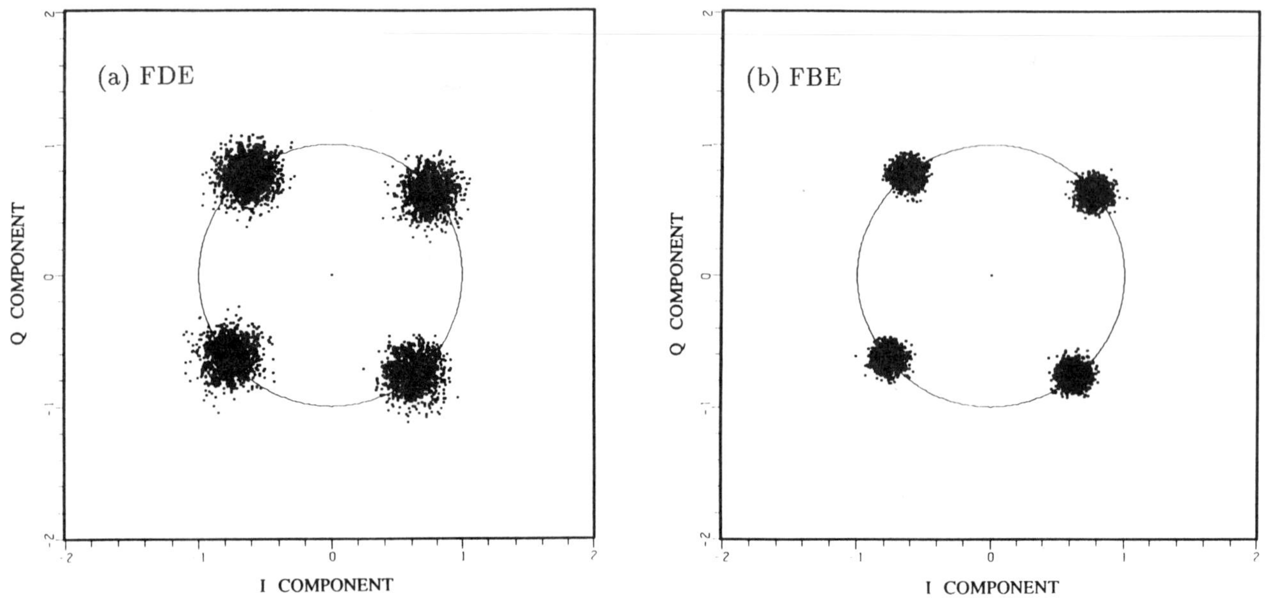

Figure 4: Constellation of the equalized signal. (a) FDE. (b) FBE.

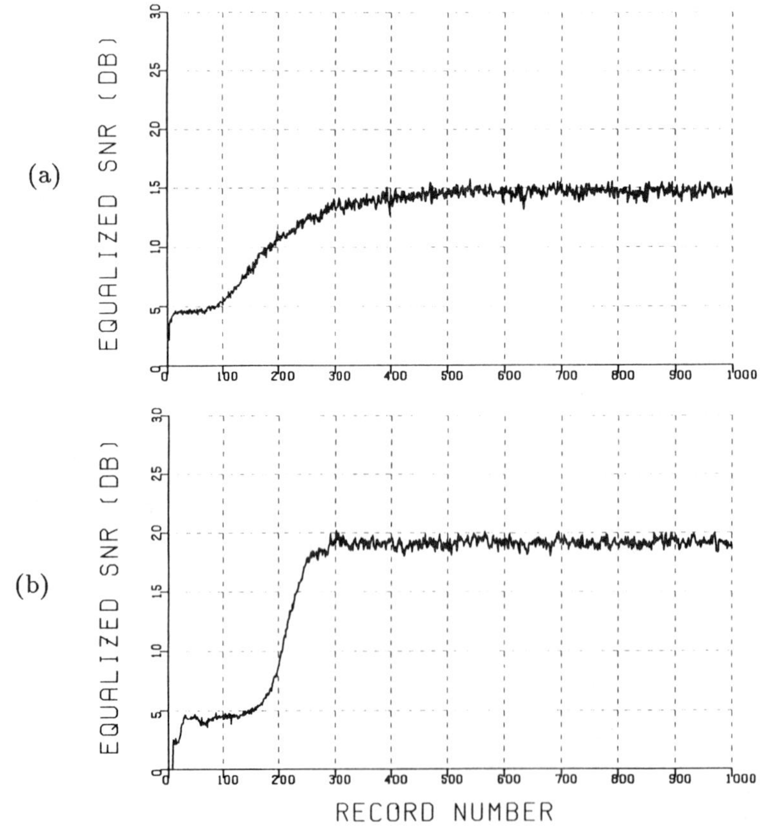

Figure 5: Typical learning curves. (a) FDE. (b) FBE.

PERFORMANCE ANALYSIS OF TRANSIENT DETECTORS BASED ON LINEAR DATA TRANSFORMS

Benjamin Friedlander and Boaz Porat
Signal Processing Technology, Ltd.
703 Coastland Drive
Palo Alto, CA 94303
USA

Abstract

The paper analyzes the performance of transient detection algorithms which are based on non-invertible linear transformations of the data. Examples of such transforms include the short-time Fourier transform and the Wavelet transform. Both linear and nonlinear signal models are considered.

1. Introduction

The need to detect transient signals arises in various applications such as underwater acoustics and seismic surveillance. The processing of these highly non-stationary signals often starts by transforming the time-domain signals using a time-frequency (or time-scale) transform. The Short-Time Fourier Transform (STFT) is by far the most widely used time-frequency transform. More recently, the Gabor and Wavelet transforms were proposed for transient processing applications [1], [2].

The objective of this paper is to study the performance of detection algorithms based on linear time-frequency (or time-scale) transforms, for transient signals characterized by linear and nonlinear parametric signal models, and by a "model mismatch" representing the difference between the model and the actual signal. The transients are assumed to undergo a non-invertible linear transformation prior to the application of the detection algorithm. Closed form expressions are derived for the worst-case detection performance for all possible transients in a given class, and for all possible mismatch signals of a given energy. These expressions make it possible to evaluate and compare the performance of various transient detection algorithms, for both single- and multi-channel data.

2. The Signal Models and Data Transformations

Consider a transient signal received by a single sensor, or by an array of sensors. The family of signals $\{s(t)\}$ is assumed to have finite energy, and to belong to a certain *finite-dimensional* linear space of continuous and continuously differentiable functions. We call this space *the signal space* and denote its dimension by G. Let $\{s_g(t), 1 \leq g \leq G\}$ be a *fixed* basis for the signal space. A specific instance of $s(t)$ can be expressed as $s(t) = \sum_{k=1}^{K} a_k s_{g_k}(t)$, where $K \leq G$ and $\{a_k\}$ are nonzero. Typically, the signal can be assumed to be *sparse*, in the sense that K will be much smaller than G.

This model can be generalized by letting the signal space (hence any signal instance) depend on some parameter vector $\boldsymbol{\theta}$. This vector may contain waveform-shape parameters and/or propagation delays. The signal is then given by $s(t,\theta) = \sum_{k=1}^{K} a_k s_{g_k}(t,\boldsymbol{\theta})$.

Real-life signals will seldom obey the above idealized model, so some allowance should be made for deviations from it. We therefore modify the model to $s(t,\theta) = \sum_{k=1}^{K} a_k s_{g_k}(t,\boldsymbol{\theta}) + e(t,\boldsymbol{\theta})$, where $e(t,\boldsymbol{\theta})$ is the *mismatch signal*. By definition, the mismatch signal is orthogonal to the subspace spanned by $\{s_{g_k}(t,\boldsymbol{\theta}), 1 \leq k \leq K\}$. We will assume that the energy of $e(t,\boldsymbol{\theta})$ is small with respect to the total signal energy. The introduction of $e(t,\boldsymbol{\theta})$ will allow us to study the sensitivity of the detectors to model mismatch.

The received signal is assumed to contain additive white Gaussian noise $w(t)$. The noise components in different sensors are assumed to be independent. After sampling we obtain the discrete-time measurement model

$$\mathbf{y} = \mathbf{S}(\boldsymbol{\theta})\mathbf{a} + \mathbf{e} + \mathbf{w}\,. \tag{1}$$

The dimension of $\mathbf{y}$ will be denoted by N. The matrix $\mathbf{S}(\boldsymbol{\theta})$ contains the sampled basis waveforms as they appear to each sensor (different sensors are assumed to "see" the same signal shape, but with different arrival times; the arrival times appear as parameters in $\boldsymbol{\theta}$). Its dimensions are $N \times K$. $\mathbf{a}$ is the vector of coefficients $\{a_k\}$. The mismatch vector $\mathbf{e}$ is assumed to be orthogonal to the column space of $\mathbf{S}(\boldsymbol{\theta})$ and to have a norm which is small with respect to that of $\mathbf{S}(\boldsymbol{\theta})\mathbf{a}$. The components of the noise vector $\mathbf{w}$ are assumed to have unit variance for convenience.

This work was supported by the Army Research Office under contract no. DAAL03-91-C-0022, sponsored by U.S. Army Communications Electronics Command, Center for Signals Warfare.

1058-6393/91 $01.00 © 1991 IEEE

The received signal vector $\mathbf{y}$ undergoes a linear transformation. Let $\mathbf{R}$ be a full rank $L \times N$ transform matrix and let the transformed measurement $\mathbf{z}$ be

$$\mathbf{z} = \mathbf{RS}(\boldsymbol{\theta})\mathbf{a} + \mathbf{Re} + \mathbf{Rw} . \quad (2)$$

In the case of multiple sensors $\mathbf{R}$ is usually block-diagonal, *i.e.*, the measurements from each sensor are transformed separately. The linear transforms considered in this paper are as follows:

(i) The short-time Fourier transform.
The STFT consists of a sequence of Fourier transforms of windowed segments of the signal. The rows of the transform matrix $\mathbf{R}$ are sampled versions of the windowed Fourier transform operator, for different window positions.

(ii) The wavelet transform.
Two types of wavelets are considered. The first is the Daubechies orthonormal wavelet. The Daubechies wavelet is actually a family of wavelets indexed by an integer parameter (the *order*).

The second wavelet is an oscillating decaying exponential, given by

$$w(t) = t\, e^{-\lambda\omega_0 t} \cos(\omega_0 t - \phi_0)\, u(t) , \quad (3)$$

where λ and ω_0 are user-chosen parameters, and $\phi_0 = 0.5\pi + 2\arctan(1/\lambda)$. This special value of the initial phase is needed to guarantee that $\int_{-\infty}^{+\infty} w(t)dt = 0$, a necessary condition for the admissibility of the wavelet. The rows of the transform matrix $\mathbf{R}$ are the sampled versions of the wavelet transform operator.

The transforms considered here are noninvertible, *i.e.*, we assume that $L < N$. The physical significance of this condition is that the transform is deliberately selective. Stated in another way, transforms are used, among other things, to perform data reduction.

In (2), the transformed noise $\mathbf{Rw}$ is not white, it covariance matrix being $\mathbf{RR}^T$. Only in the special case when the rows of $\mathbf{R}$ are orthonormal will the transformed noise be white. In general, the noise in (2) can by whitened by further multiplying it by $(\mathbf{RR}^T)^{-1/2}$, *i.e.*, by letting

$$\begin{aligned}(\mathbf{RR}^T)^{-1/2}\mathbf{z} =& (\mathbf{RR}^T)^{-1/2}\mathbf{RSa} \\ &+ (\mathbf{RR}^T)^{-1/2}\mathbf{Re} + (\mathbf{RR}^T)^{-1/2}\mathbf{Rw} ,\end{aligned} \quad (4)$$

and redefining $\mathbf{z}$. This operation is invertible, so it causes no loss of information. Also, $(\mathbf{RR}^T)^{-1/2}\mathbf{R}$ is easily shown to have orthonormal rows. Therefore we can usually assume, without loss of generality, that the matrix $\mathbf{R}$ in (2) has orthonormal rows and consequently, that the noise in the transformed model is white.

Note that both the STFT and the Daubechies Wavelet Transform (DWT) give rise to orthonormal $\mathbf{R}$ matrices (this property may be lost, though, after sampling or windowing). The Exponential Wavelet Transform (EWT) is not orthonormal, but can be orthonormalized as described above.

3. Linear Signal Models

In this section we describe the structure of the detector for linear signal models, and analyze its performance. The linear model is a special case of (1), where the parameter $\boldsymbol{\theta}$ is absent. The performance of the detector will depend, among other things, on the signal under consideration, which is determined by a specific choice of $\mathbf{a}$ in the signal model, and a particular mismatch vector $\mathbf{e}$. We will then perform "worst-case" analysis of the detector performance, for all values of $\mathbf{a}$ and $\mathbf{e}$ satisfying certain norm constraints.

As will be shown, the detection statistic for the problem considered here has a noncentral chi-square distribution. Detection performance is, therefore, determined by the number of degrees of freedom of the distribution, and the value of the noncentrality parameter. For a given model, the noncentrality parameter provides a convenient measure of detectability. Changes in the noncentrality parameter reflect a corresponding change in the effective signal to noise ratio.

The linear signal model is

$$\mathbf{y} = \mathbf{Sa} + \mathbf{e} + \mathbf{w} \quad (5)$$

for the raw data vector, and

$$\mathbf{z} = \mathbf{Ry} = \mathbf{RSa} + \mathbf{Re} + \mathbf{Rw} \quad (6)$$

for the transformed data vector. The vector of amplitudes $\mathbf{a}$ is assumed zero under H_0 and unknown under H_1. The matrix $\mathbf{R}$ is assumed to have orthonormal rows. The detector, which does not account for the mismatch, assumes an ideal model of the form

$$\mathbf{z} = \mathbf{RSa} + \bar{\mathbf{w}} , \quad (7)$$

where $\bar{\mathbf{w}}$ is, similarly to $\mathbf{w}$, a vector whose elements are mutually uncorrelated Gaussian random variables with zero-mean and unit-variance.

We consider the Generalized Likelihood Ratio Test (GLRT) [3]. The GLRT statistic is given by

$$\ell = \max_{\mathbf{a}}\{2\log f_1(\mathbf{z})\} - 2\log f_0(\mathbf{z}) , \quad (8)$$

where $f_0(\mathbf{z})$ and $f_1(\mathbf{z})$ are the probability densities under H_0 and H_1 respectively. Using standard techniques it can be shown that, for the model (7),

$$\begin{aligned}\ell &= \mathbf{z}^T\mathbf{R}\mathbf{S}(\mathbf{S}^T\mathbf{R}^T\mathbf{R}\mathbf{S})^{-1}\mathbf{S}^T\mathbf{R}^T\mathbf{z} \\ &= \mathbf{y}^T\mathbf{R}^T\mathbf{R}\mathbf{S}(\mathbf{S}^T\mathbf{R}^T\mathbf{R}\mathbf{S})^{-1}\mathbf{S}^T\mathbf{R}^T\mathbf{R}\mathbf{y}\,.\end{aligned} \tag{9}$$

Since $\mathbf{R}$ has orthonormal rows, the matrix $\mathbf{P}_R = \mathbf{R}^T\mathbf{R}$ is the projection on its row space. Hence columns of the matrix $\mathbf{S}|\mathbf{R} = \mathbf{P}_R\mathbf{S}$ are the projections of the columns of $\mathbf{S}$ on the column space of $\mathbf{R}^T$. Also,

$$\mathbf{P}_{S|R} = \mathbf{P}_R\mathbf{S}(\mathbf{S}^T\mathbf{P}_R\mathbf{S})^{-1}\mathbf{S}^T\mathbf{P}_R \tag{10}$$

is the projection on the column space of $\mathbf{S}|\mathbf{R}$. The detection statistic can thus be expressed as

$$\ell = \mathbf{y}^T\mathbf{P}_{S|R}\mathbf{y}\,. \tag{11}$$

Note that, while the derivation was done under the assumption that $\mathbf{R}$ is orthonormal, the final result holds in general, taking $\mathbf{P}_R = \mathbf{R}^T(\mathbf{R}\mathbf{R}^T)^{-1}\mathbf{R}$.

The distribution of the GLRT

Under H_0 we have $\mathbf{y} = \mathbf{w}$, so

$$\ell = \mathbf{w}^T\mathbf{P}_{S|R}\mathbf{w}\,. \tag{12}$$

This is a chi-square distributed random variable, with a number of degrees of freedom equal to the rank of $\mathbf{P}_{S|R}$. Typically $L > K$, so the rank of this matrix is K or less. In the generic case the rank, hence the number of degrees of freedom, will be K.

Under H_1 we have

$$\ell = (\mathbf{S}\mathbf{a} + \mathbf{e} + \mathbf{w})^T\mathbf{P}_{S|R}(\mathbf{S}\mathbf{a} + \mathbf{e} + \mathbf{w})\,. \tag{13}$$

This has a noncentral chi-square distribution with K degrees of freedom and noncentrality parameter

$$\nu = (\mathbf{S}\mathbf{a} + \mathbf{e})^T\mathbf{P}_{S|R}(\mathbf{S}\mathbf{a} + \mathbf{e})\,. \tag{14}$$

When the mismatch is zero, the noncentrality is

$$\nu = \mathbf{a}^T\mathbf{S}^T\mathbf{P}_{S|R}\mathbf{S}\mathbf{a} = \mathbf{a}^T\mathbf{S}^T\mathbf{P}_R\mathbf{S}\mathbf{a}\,. \tag{15}$$

It is interesting to note that

$$\nu \le \mathbf{a}^T\mathbf{S}^T\mathbf{S}\mathbf{a}\,, \tag{16}$$

where the right-hand side is the signal energy. The noncentrality ν is maximum when the signal happens to be in the column space of $\mathbf{R}^T$, in which case it equals the signal energy. Recall that the analysis assumes that the noise has unit-variance. In general, the maximum value of the noncentrality parameter equals the signal-to-noise ratio (SNR).

For the noncentral chi-square distribution, the detection probability is a monotonically increasing function of the noncentrality (for a fixed number of degrees of freedom). Two interesting special cases in which the noncentrality attains its maximum value are: (a) when $\mathbf{R}^T$ contains $\mathbf{S}$ as a submatrix; (b) when $\mathbf{R}$ is the pseudoinverse of a matrix containing $\mathbf{S}$ as a submatrix.

When the mismatch is nonzero, the noncentrality can be smaller or larger than the value given by (15), depending on the specific value of $\mathbf{e}$. The following question will be of interest here, as well as in the next section. Suppose we fix the energies of the nominal signal and the mismatch to

$$\mathbf{a}^T\mathbf{S}^T\mathbf{S}\mathbf{a} = (1-\alpha)E \quad ; \quad \mathbf{e}^T\mathbf{e} = \alpha E\,, \tag{17}$$

where E is a fixed positive constant and $0 < \alpha < 1$. Note that, since the signal and the mismatch are orthogonal, the energy of their sum is E. In fact, the mismatch is orthogonal not only to the signal, but to the entire signal subspace, *i.e.*,

$$\mathbf{e}^T\mathbf{S} = 0\,. \tag{18}$$

The question is then: what is the minimum noncentrality ν, for all possible choices of nominal signal and mismatch satisfying the constraints (17) and (18)?

The reason for interest in this question is that the noncentrality ν is a measure of the effective SNR, *i.e.*, a change in ν is equivalent to corresponding change in the SNR. The minimum noncentrality provides a measure of robustness of the detector to deviations from the restricted ideal signal model.

To solve this problem, we reformulate it as follows. Let the columns of $\mathbf{V}_1$ be an orthonormal basis for the column space of $\mathbf{S}$, and let the columns of $\mathbf{V}_2$ be an orthonormal basis for the orthogonal complement of $\mathbf{S}$. Let the projection matrix $\mathbf{P}_{S|R}$ be

$$\mathbf{P}_{S|R} = \mathbf{U}\mathbf{U}^T\,, \tag{19}$$

where the columns of $\mathbf{U}$ are an orthonormal basis for the space on which $\mathbf{P}_{S|R}$ projects. Then we can express the signal and the mismatch as

$$\mathbf{S}\mathbf{a} = \mathbf{V}_1\mathbf{x}_1 \quad ; \quad \mathbf{e} = \mathbf{V}_2\mathbf{x}_2 \tag{20}$$

and the noncentrality as

$$\nu = (\mathbf{V}_1\mathbf{x}_1 + \mathbf{V}_2\mathbf{x}_2)^T\mathbf{U}\mathbf{U}^T(\mathbf{V}_1\mathbf{x}_1 + \mathbf{V}_2\mathbf{x}_2)\,, \tag{21}$$

where

$$\mathbf{x}_1^T\mathbf{x}_1 = (1-\alpha)E \quad ; \quad \mathbf{x}_2^T\mathbf{x}_2 = \alpha E \,. \tag{22}$$

The problem is then to find the minimum of (21), subject to the constraints (22).

The constrained minimization problem can be solved by using Lagrange multipliers. The solution, the derivation of which is omitted, is given by

$$\nu_{\min} = \begin{cases} \left[\sqrt{(1-\alpha)\sigma_{\min}^2} - \sqrt{\alpha(1-\sigma_{\min}^2)}\right]^2 E \,, & \alpha < \sigma_{\min}^2 \\ 0\,, & \alpha \geq \sigma_{\min}^2 \end{cases} \tag{23}$$

where $\sigma_{\min}$ is the smallest singular value of $\mathbf{U}^T\mathbf{V}_1$. We remark that $\nu_{\min}$, when plotted against α, has the shape of an elliptical arc whose major axis is the line $\nu_{\min} = (1-\alpha)E$. Larger $\sigma_{\min}$ correspond to larger $\nu_{\min}$ for a fixed α. When $\sigma_{\min} = 1$ (the largest possible value) we get $\nu_{\min} = (1-\alpha)E$. Summing up, the robustness of the detector is determined only by the smallest singular value of $\mathbf{U}^T\mathbf{V}_1$. Any data transform $\mathbf{R}$ for which the column space of $\mathbf{S}$ is included in the column space of $\mathbf{R}^T$ yields a most robust detector.

4. Nonlinear Signal Models

This section discusses the detection problem for the nonlinear signal model

$$\mathbf{y} = \mathbf{S}(\boldsymbol{\theta})\mathbf{a} + \mathbf{e} + \mathbf{w} \,, \tag{24}$$

and the corresponding model for the transformed data. The vector of amplitudes $\mathbf{a}$ is assumed to be zero under H_0 and unknown under H_1. The parameter vector $\boldsymbol{\theta}$ (of dimension M) is meaningless under H_0 (since the signal is absent), and unknown under H_1. The matrix $\mathbf{S}(\boldsymbol{\theta})$ is assumed to have full rank K for all $\boldsymbol{\theta}$. The vector $\mathbf{e}$ represents the mismatch, as in the linear case. The noise $\mathbf{w}$ is a zero-mean Gaussian uncorrelated vector, which is assumed to have unit variance.

We first consider detection based on the raw data, and then proceed to detection from the transformed data. The GLRT detector for the raw data is easily derived to be

$$\ell = \mathbf{y}^T\mathbf{P}_S(\hat{\boldsymbol{\theta}})\mathbf{y} \,, \tag{25}$$

where $\mathbf{P}_S(\hat{\boldsymbol{\theta}})$ is the projection matrix on the column space of $\mathbf{S}(\hat{\boldsymbol{\theta}})$, and $\hat{\boldsymbol{\theta}}$ is the maximum likelihood estimate of $\boldsymbol{\theta}$. Also, the ML estimate can be shown to be the maximizer of (25), so the GLRT is actually given by

$$\ell = \max_{\boldsymbol{\theta}}\{\mathbf{y}^T\mathbf{P}_S(\boldsymbol{\theta})\mathbf{y}\} \,. \tag{26}$$

The asymptotic distribution of nonlinear GLRT detectors is discussed in [4], and is shown to be chi-square under certain assumptions. Unfortunately, those assumptions *are not satisfied* under H_0 for the present problem. Since $\boldsymbol{\theta}$ is meaningless under H_0, the asymptotic distribution of $\hat{\boldsymbol{\theta}}$ is not likely to be normal, hence there is no reason for the distribution of ℓ to be chi-square. Rather, this distribution is likely to be model-dependent and mathematically intractable in most cases. Roughly speaking, the ML estimates $\hat{\boldsymbol{\theta}}$ and $\hat{\mathbf{a}}$ would try, under H_0, to fit the noise $\mathbf{w}$ to the model $\mathbf{S}(\hat{\boldsymbol{\theta}})\hat{\mathbf{a}}$ as close as possible. This may cause ℓ to reach relatively high values and result in unacceptably high probability of false-alarm.

Motivated by the desire to obtain an asymptotically chi-square detector for the nonlinear case, we derive the detector discussed below. Define

$$\mathbf{b}_m = \frac{\partial\mathbf{S}(\boldsymbol{\theta})}{\partial\theta_m}\mathbf{a} \quad ; \quad \mathbf{B}(\boldsymbol{\theta},\mathbf{a}) = [\mathbf{b}_1, \mathbf{b}_2, \ldots, \mathbf{b}_M] \tag{27}$$

$$\mathbf{C}(\boldsymbol{\theta},\mathbf{a}) = [\mathbf{S}(\boldsymbol{\theta}), \mathbf{B}(\boldsymbol{\theta},\mathbf{a})] \quad ; \quad \mathbf{x} = [\Delta\mathbf{a}^T, \Delta\theta_1, ..., \Delta\theta_M]^T \,. \tag{28}$$

Let $\mathbf{y}_1$ be the vector consisting of the measurements at odd time points and $\mathbf{y}_2$ the vector consisting of the measurements at even time points. Then we can write, similarly to (24),

$$\mathbf{y}_1 = \mathbf{S}_1(\boldsymbol{\theta})\mathbf{a} + \mathbf{e}_1 + \mathbf{w}_1 \quad ; \quad \mathbf{y}_2 = \mathbf{S}_2(\boldsymbol{\theta})\mathbf{a} + \mathbf{e}_2 + \mathbf{w}_2 \,. \tag{29}$$

Let $\hat{\boldsymbol{\theta}}$ and $\hat{\mathbf{a}}$ be the ML estimates using the data $\mathbf{y}_2$, and

$$\hat{\mathbf{C}}_1 = [\mathbf{S}_1(\hat{\boldsymbol{\theta}}), \mathbf{B}_1(\hat{\boldsymbol{\theta}},\hat{\mathbf{a}})] \,. \tag{30}$$

The linearized model for $\mathbf{y}_1$ is then

$$\mathbf{y}_1 - \mathbf{S}_1(\hat{\boldsymbol{\theta}})\hat{\mathbf{a}} \approx \mathbf{C}_1\mathbf{x} + \mathbf{w}_1 \,. \tag{31}$$

The GLRT statistic for the linearized model (31) is

$$\ell = \mathbf{y}_1^T\mathbf{P}_{\hat{C}_1}\mathbf{y}_1 \,. \tag{32}$$

Consider now what happens under H_0. Since $\mathbf{P}_{\hat{C}_1}$ depends only on $\mathbf{w}_2$, and since $\mathbf{w}_2$ is independent of $\mathbf{w}_1$, the conditional distribution of ℓ given $\mathbf{w}_2$ is chi-square with $K+M$ degrees of freedom, for *all* values of $\mathbf{w}_2$ (we have used here the fact that the rank of $\mathbf{P}_{\hat{C}_1}$ will be $K+M$ with probability one). Therefore, the unconditional distribution of ℓ will be also chi-square with $K+M$ degrees of freedom.

The asymptotic distribution of ℓ under H_1 can be derived using Taylor series approximations. The result is noncentral chi-square with $K+M$ degrees of freedom and noncentrality equal to the signal energy in the odd measurements (in absence of mismatch). This is about half the total signal energy. In other words, the effective SNR is reduced by 3dB.

The detector for the transformed data is obtained in a similar manner. Partition the matrix $\mathbf{R}$ into odd- and even-numbered columns, compatibly with $\mathbf{y}$, and denote the resulting matrices by $\mathbf{R}_1$ and $\mathbf{R}_2$ respectively. Then the detector for the transformed measurements is given by

$$\ell = \mathbf{y}_1^T \mathbf{P}_{\hat{C}_1|R_1} \mathbf{y}_1 , \tag{33}$$

where $\mathbf{P}_{\hat{C}_1|R_1}$ is the projection on the column space of the projection of $\hat{\mathbf{C}}_1$ on the column space of $\mathbf{R}_1^T$.

The asymptotic distribution of ℓ is as before, except that the noncentrality under H_1 is approximately (including mismatch)

$$\nu = [\mathbf{S}_1(\boldsymbol{\theta})\mathbf{a} + \mathbf{e}_1]^T \mathbf{P}_{C_1|R_1} [\mathbf{S}_1(\boldsymbol{\theta})\mathbf{a} + \mathbf{e}_1] . \tag{34}$$

We now ask the same question as before, namely — what is the minimum noncentrality when $\mathbf{a}$ and $\mathbf{e}_1$ take all possible values subject to the constraints

$$\mathbf{a}^T \mathbf{S}_1^T \mathbf{S}_1 \mathbf{a} = (1-\alpha)E \quad ; \quad \mathbf{e}_1^T \mathbf{e}_1 = \alpha E \quad ; \quad \mathbf{e}_1^T \mathbf{S}_1 = 0 . \tag{35}$$

There is a subtle difference between this problem and the one solved in Section 3, which makes it much harder to solve. Unlike $\mathbf{S}_1$, the matrix $\mathbf{C}_1$ depends on the gain vector $\mathbf{a}$, cf. (27), (28). This makes $\mathbf{P}_{C_1|R_1}$ dependent on $\mathbf{a}$, so the result of Section 3 does not apply here.

There does not appear to exist any simple algorithm for solving the minimization problem posed above. However, if we restrict ourselves to a fixed vector $\mathbf{a}$, we can find the minimum noncentrality over all mismatch vectors $\mathbf{e}_1$ satisfying the above constraints. A solution to this minimization problem is described in [5].

5. Discussion

As the analysis shows, detection performance is a function of the number of degrees of freedom and of the noncentrality parameter, which depends on the signal model, not on the transform being used. The noncentrality parameter ν depends on how well the transform is matched to the signal model. If the space spanned by the rows of $\mathbf{R}$ is "close" to the space spanned by the columns of $\mathbf{S}$, we get ν close to the signal energy (assuming $\alpha = 0$).

Recall that a signal of finite duration T is considered to be narrowband if its bandwidth B is in the order of $1/T$, and broadband if $B >> 1/T$, *i.e.*, if its time-bandwidth product is much larger than 1. Then, the following observations were made from numerical tests reported in [5]:

(i) The detector based on the STFT is generally better than the detectors based on the wavelet transform, for relatively narrowband signals. The DWT and EWT seem generally to be the better detectors for signals with large time-bandwidth product.

(ii) In the case of signals with large time-bandwidth product, there was a very significant difference between the performance of all three detectors and the performance of the optimal detector (*i.e.*, the one where the transform is perfectly matched to the signal space).

(iii) The differences between the three detectors are quite small for large time-bandwidth product. Noticeable differences were observed only in the narrowband case.

In summary, the wavelet transform does not appear to offer clear advantages over the short-time Fourier transform for detection purposes (under our model assumptions). For narrowband signals, it is likely to be inferior to STFT, while for broadband signals it may be slightly superior, but not consistently so.

References

[1] B. Friedlander and B. Porat, "Detection of Transient Signals by the Gabor Representation," *IEEE Trans. Acoustices, Speech and Signal Processing*, vol. ASSP-37, no. 2, pp. 169–180, 1989.

[2] I. Daubechies, "The Wavelet Transform: A Method for Time-Frequency Localization," Ch. 8 in *Advances in Spectrum Analysis and Array Processing, Vol. I*, S. Haykin, Ed., Prentice-Hall, Englewood Cliffs, 1991.

[3] H.L. Van Trees, *Detection, Estimation and Modulation Theory, Part I*, p. 92, John Wiley and Sons, 1968.

[4] M.G. Kendall and A. Stuart, *The Advanced Theory of Statistics, Vol. 2*, pp. 246–247, Charles Griffin & Co. Ltd., 1979.

[5] B. Friedlander and B. Porat, "Performance Analysis of Transient Detectors Based on Linear Data Transforms," *IEEE Transactions on Information Theory*, to appear.

Wavelet Transform Mother Mapper Operator

Randy K. Young

Pennsylvania State University, Applied Research Laboratory
Post Office Box 30
State College, PA 16804 USA
(814) 863-4499 Fax: (814) 863-7841

Abstract

The wavelet transform domain characterization of a signal is highly dependent upon the ***mother wavelet*** *which is chosen as the kernel of the transform. A new operator, the* ***Mother Mapper Operator****, is derived to efficiently evaluate several transforms of the same signal with different mother wavelets. The Mother Mapper Operator* ***directly*** *maps a transform with respect to one mother wavelet to a new transform with respect to a new, different mother wavelet. The Mother Mapper Operator leads to signal representations which involve* ***multiple mother wavelets*** *instead of just one mother wavelet.*

Introduction

Wavelet theory is a functional decomposition analogous to Fourier theory. Unlike a Fourier transform though, the wavelet transform simultaneously localizes in two variables, scale and translation; the Fourier transform localizes only in frequency. The wavelet transform has several advantages over Fourier transforms for wideband and/or nonstationary processes; it may also be applied in either the spatial or temporal domains [1-9, 11, 12]. Each wavelet transform is parameterized by a mother wavelet, $g(x)$. When the mother wavelet is changed the characteristics of the wavelet transform change. The same signal may have very different representations when different mother wavelets are employed.

In Fourier analysis the only signal which can be used for decomposing a signal is a tone; it is the kernel of the Fourier integral transformation. With wavelet transforms the mother wavelet is the kernel of the integral transform. The freedom for choosing this mother wavelet may be used to great advantage in characterizing and compressing signals. A wavelet transform of a signal does not have much meaning unless the mother wavelet is explicitly stated (unlike a Fourier transform whose kernel is known to be a tone).

Suppose a signal, $f(x)$, is represented by a wavelet transform with respect to a particular mother wavelet, $g(x)$, and that this transform domain representation has compact support. Now suppose a second mother wavelet, $g_2(x)$, leads to a transform domain representation with spread out support (noncompact). Obviously, if the information in $f(x)$ is to be saved or transmitted, then the compact representation (provided by the mother wavelet $g(x)$) will lead to efficiencies. Thus many different mother wavelets may be used to create an efficient representation.

For the purpose of efficiently evaluating wavelet transforms with respect to several, different mother wavelets a new operator, the Mother Mapper Operator, is presented. This operator will allow several mother wavelets to be considered without degrading the efficiency of the compression (or other) processor. The operator can also be applied to **wideband cross ambiguity function generation** to enhance the efficiency and structure of the processor [13].

Wavelet Transforms

Before defining the wavelet transform, admissible functions are defined. An L^2 function, g, is an *admissible function* if:

$$c_g \triangleq \int_0^{\infty} \frac{|G(\omega)|^2}{\omega} d\omega < \infty$$

where $G(\omega)$ is the Fourier transform of g. This condition essentially states that the mother wavelets must be "bandpass" processes.

The one-dimensional wavelet transform operator, $\boldsymbol{W}_g$, maps an L^2 (square integrable in its variable) signal which is real or complex as follows: $\boldsymbol{W}_g : L^2(\mathbb{R}) \to L^2(\mathbb{R}\setminus\{0\} \times \mathbb{R})$. The wavelet transform of a function, f, with respect to a given

1058-6393/91 $01.00 © 1991 IEEE

admissible mother wavelet, g, exists and is defined as:

$$(W_g f)(a,b) = |a|^{-\frac{1}{2}} \int f(x)\, g^*\left(\frac{x-b}{a}\right) dx$$
$$= \langle f, g_{a,b} \rangle = \langle f, U(a,b)\, g \rangle$$

where superscript "*" denotes complex conjugation and $\langle \cdot , \cdot \rangle$ denotes an L^2 inner product on the Hilbert space which contains f and g as members. Note that this definition requires that $g(x)$ is an admissible function. The $W_g f(a,b)$'s are the *wavelet transform coefficients.* $g_{a,b}$ is defined by a unitary affine mapping $U(a,b) : g(x) \mapsto \frac{1}{\sqrt{a}}\, g\left(\frac{(x-b)}{a}\right)$.

The $g_{a,b}$ functions are the time scaled and time translated versions of the mother wavelet, $g(x)$, where parameter "a" denotes the time scale and parameter "b" represents a time translation. $g_{a,b}(x)$ is chosen to have a c_g value of unity so that the wavelet transform is isometric ("energy" is the same in each domain) and that any coefficient for a particular (a,b) may be "fairly" compared or combined even across a huge range of scales.

Note that for continuous wavelet transforms the choice of the mother wavelet function is only constrained by the admissibility condition. One is free to choose the mother wavelet for optimal behavior in the particular application of interest.

Mother Mapper Operator

The Mother Mapper Operator is presented and derived, and its properties are detailed. Several interpretations and applications of this structure are also provided.

The Mother Mapper Operator, $MM_{g_2}(\cdot)$, maps $(\mathbf{R}\backslash\{0\} \times \mathbf{R}) \rightarrow (\mathbf{R}\backslash\{0\} \times \mathbf{R})$; i.e., ***it maps a wavelet transform of a function with respect one mother wavelet to a wavelet transform of the same function with respect to a different mother wavelet.*** The operator is defined as:

$$MM_{g_2} : W_{g_1} f(a,b) \mapsto W_{g_2} f(a,b)$$

$$MM_{g_2} : W_g f(a,b) \mapsto$$
$$\frac{1}{c_g} \int \frac{ds}{s^2} \int W_g^* g_2\left(\frac{s}{a}, \frac{(\tau-b)}{a}\right) W_g f(s,\tau)\, d\tau$$

where $W_g f(a,b)$ is the wavelet transform of some function, f, with respect to the mother wavelet, g (g and f are both functions of the independent variable(s) such as time and/or space). The independent variables are not used in this notation to avoid loss of generality. The positive real constant, c_g, is a normalizing term which is a function of the original mother wavelet only. This constant was defined above as the admissibility constant and was discussed there. The kernel of the Mother Mapper Operator is thus the wavelet transform of the new mother wavelet, g_2, with respect to the old mother wavelet, g. In this operator, the kernel is conjugated and then mapped via an affine transformation over all (s,τ). Next, these affine mapped kernels are functionally (or term-by-term in the discrete case) multiplied by the original wavelet transform and then integrated. Since the kernel of this new operator is a wavelet transform of the new mother wavelet with respect to the old mother wavelet, it is possible that it will have approximately compact support (or some type of structured or limited support) which may lead to enhanced efficiencies in implementations.

Mother Mapper Derivation

The *resolution of identity theorem* (which is a special case of the Frobenius-Schur-Godement Theorem in square integrable group theory [7]) for the wavelet transform inner products states that when g is an admissible function then:

$$c_g \langle f, g_2 \rangle = \int_{-\infty}^{\infty} \frac{da}{a^2} \int_{-\infty}^{\infty} \langle f, g_{a,b} \rangle \langle g_{a,b}, g_2 \rangle\, db$$

The admissibility constant, c_g, must be a well defined constant (less than infinity). From the definition of the inner product:

$$\langle g_{a,b}, g_2 \rangle = \langle g_2, g_{a,b} \rangle^*$$

so that:

$$\langle f, g_2 \rangle = \frac{1}{c_g} \int_{-\infty}^{\infty} \frac{da}{a^2} \int_{-\infty}^{\infty} \langle f, g_{a,b} \rangle \langle g_2, g_{a,b} \rangle^*\, db$$

Under the assumption that g_2 is an admissible function, formulate the new wavelet transform (of f with respect to the new mother wavelet g_2) as an inner product:

$$W_{g_2} f(s,\tau) = \langle f, g_{2_{s,\tau}} \rangle$$

where s is the new scale variable and τ is the new translation variable. Now, in the resolution of identity formulation this creates a term $W_g g_{2_{s,\tau}}(a,b)$ which can be expanded as:

$$\langle \sqrt{s}\, g_2(st+\tau), \sqrt{a}\, g(at+b) \rangle$$

$$= \left\langle g_2(t), \sqrt{\frac{a}{s}}\, g\left(\frac{a}{s}t + \frac{(b-\tau)}{s}\right) \right\rangle$$

thus yielding the Mother Mapper Operator:

$$W_{g_2} f(s,\tau) =$$

$$\frac{1}{c_g}\int \frac{da}{a^2}\int (W_g f(a,b))\, W_g^* g_2\left(\frac{a}{s}, \frac{(b-\tau)}{s}\right) db$$

Thus, the Mother Mapper Operator performs the mapping of the wavelet transform with respect to g to a new wavelet transform with respect to mother wavelet, g_2.

Multiresolution Orthogonal Wavelets versus Nonorthogonal Wavelets

The Multiresolution orthogonal wavelet transforms presented by Meyer, Mallat, Daubechies and others [2, 11, 12] yield very specific (and sometimes unique) mother wavelets due to the extensive constraints imposed upon them. Figure 1 identifies some of these constraints and the paths taken to create the Multiresolution orthogonal wavelets. This author believes that these are appropriate for signal/image analysis but may be augmented by the nonorthogonal wavelets to create further efficiencies. By using nonorthogonal wavelet transforms the mother wavelets are simply constrained to be admissible functions (bandpass). The freedom in choosing a mother wavelet may be exploited to better match the signal/image characteristics and to more efficiently concentrate the signal's energy into fewer coefficients.

Mother Mapper Operator Applications

An advantage of using the Mother Mapper Operator to easily and efficiently mapping between transforms with different mother wavelets is that different parameterizations by different mother wavelets may provide efficient representations of the original process. Figures 2a and 1b display wavelet transforms of the same Gaussian windowed tone signal but with respect to different mother wavelets. Figure 2a uses the Morlet, Gaussian weighted tone, and Figure 2b uses a Gaussian weighted linear FM as the mother wavelet, respectively. Figure 2c displays the wavelet transform of a Gaussian weighted linear FM signal with respect to a Gaussian weighted linear FM mother wavelet - thereby creating a

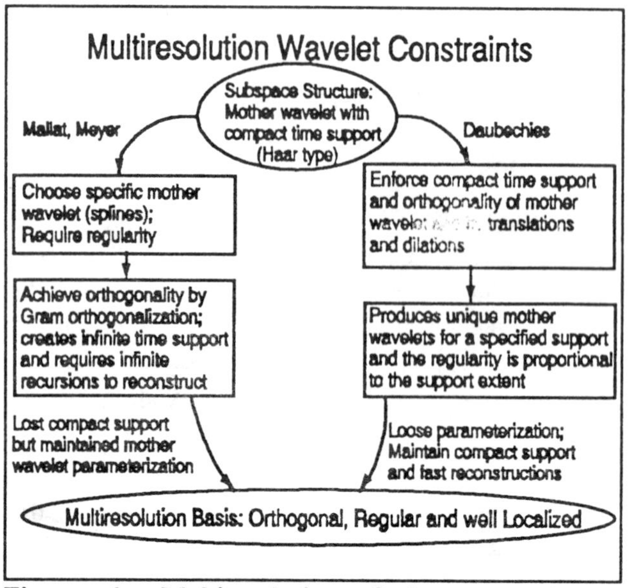

Figure 1: Multiresolution Orthogonal Wavelet Constraints

wideband cross ambiguity function (WBCAF) of the FM signal [13]. These figures emphasize the effects of the mother wavelet choice and how it parameterizes the wavelet transform.

Some processes will be "well represented" by a particular mother wavelet and other processes are poorly represented by that same mother wavelet. The concentration of energy in the transform domain represents the measure of efficiency or performance since a smaller number of coefficients would be required to represent the signal to the same degree of approximation.

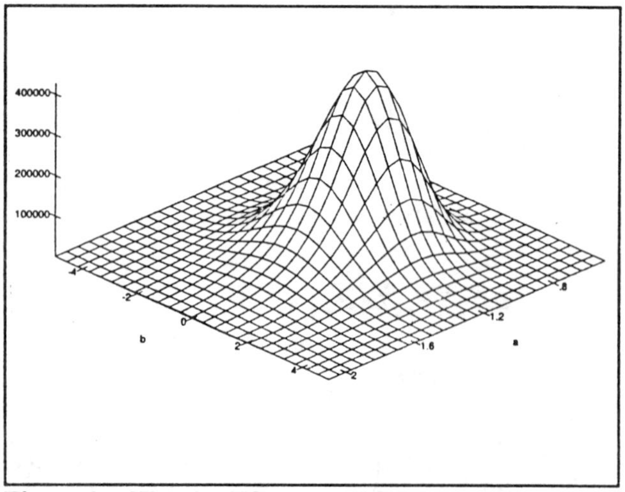

Figure 2a: Wavelet Xform wrt Morlet mother wavelet

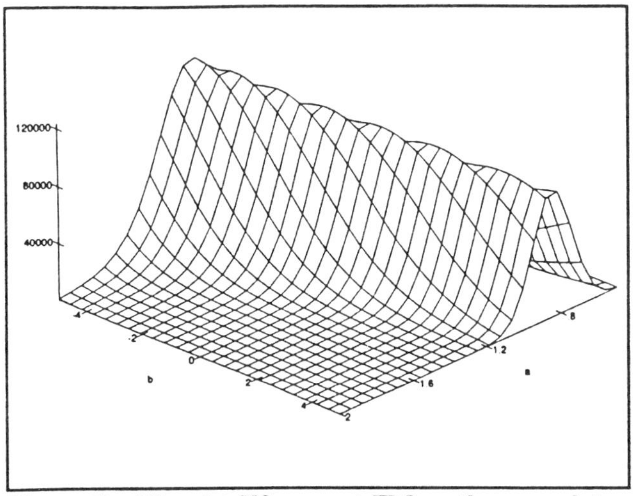

Figure 2b: Wavelet Xform wrt FM mother wavelet

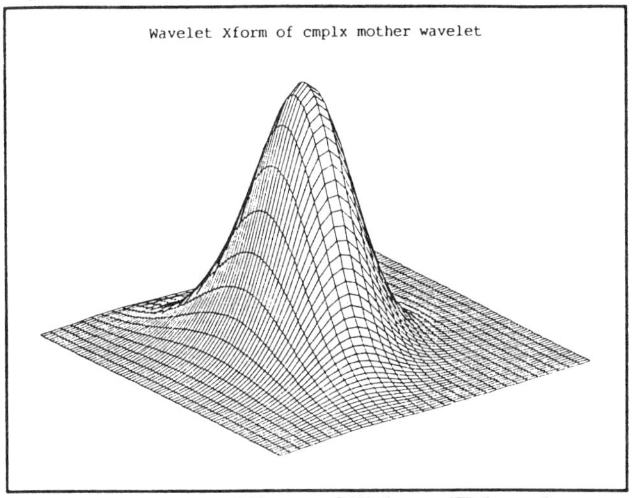

Figure 2c: Wavelet Xform of FM wrt FM mother

The efficiency provided by the Mother Mapper Operator is that it avoids saving the original data or reconstructing that data from a previous wavelet transform. Figure 3 shows the concepts pictorially. Referring to the top row of Figure 3, first take a wavelet transform of the input signal with respect to a initial mother wavelet; if the transform domain representation is "good enough" then stop and use that representation. If the representation is not good enough, then another mother wavelet might be used to decompose the signal (the idea is to proceed to the bottom right corner of Figure 3). At this point, without utilizing the Mother Mapper Operator, either the original data would have had to have been saved or that data would have to be reconstructed from the first wavelet transform. Saving the original data involves large resources as does reconstructing the data from the first transform. Instead, the Mother Mapper Operator provides a more efficient alternative by directly mapping the first wavelet transform to a new wavelet transform with respect to a different mother wavelet kernel. Thus the efficiency of evaluating several transforms with respect to multiple mother wavelets is significantly improved.

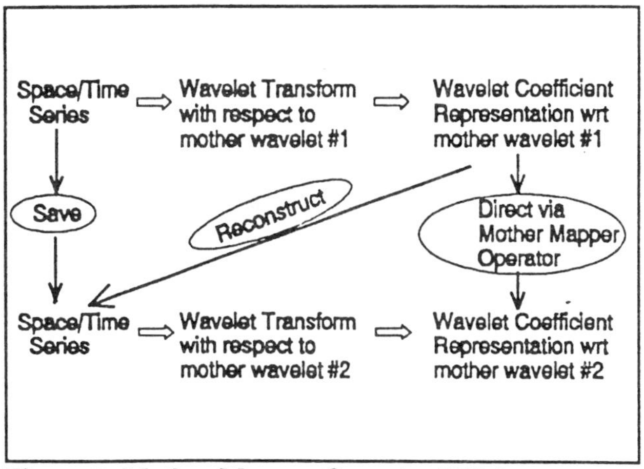

Figure 3: Mother Mapper Operator Efficiency

The "trick" to efficient wavelet representations (a good approximation of the original signal with few wavelet domain coefficients) is to find the "optimum" mother wavelet for the signals being analyzed (the optimality criteria are arbitrary and chosen by the user's applications). Obviously, these mother wavelets themselves may be optimized. Specific optimization criteria or algorithms are not addressed and are deferred to future research; however, the implementation of these algorithms would require efficient evaluations of wavelet transforms with respect to different mother wavelets.

Another advantage of changing mother wavelets is that the mother wavelet parameterizes the transform in a manner similar to parametric spectral estimation. If the signals being transformed are similar to the mother wavelet, then the transform domain representation will tend to be efficient and characterize the salient properties of the signal or system. If one portion of a signal can be very efficiently characterized by a first, arbitrary, mother wavelet but another portion of the signal is inefficiently represented by this first mother wavelet, then the overall representation (the mother wavelet and its associated coefficients) will be inefficient. However, if a second mother wavelet is used which can efficiently characterize the originally poorly represented portion of the signal, then two (or multiple) mother wavelets together may be used to yield an extremely efficient representation. See Figure 4 for a diagram of this multiple mother wavelet decomposition process.

The new representations will involve sets containing the *significant* wavelet coefficients, with one set of coefficients associated to each mother wavelet. Effectively each mother wavelet will extract only that portion of the signal which it can efficiently represent, and it will allow the rest of the signal to be processed by

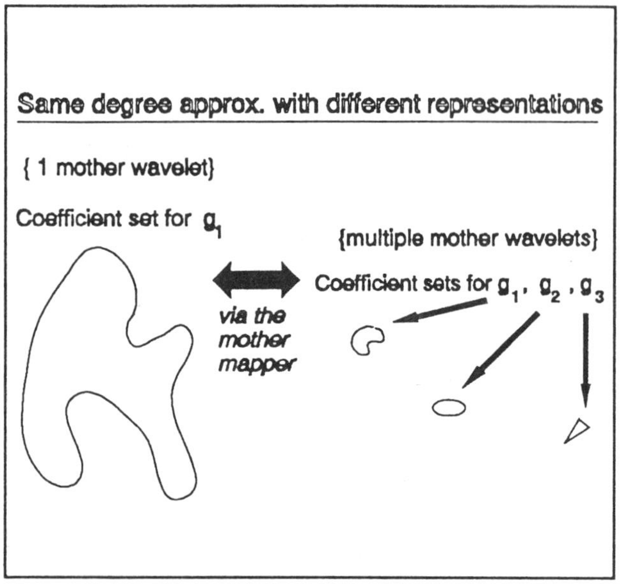

Figure 4: Multiple Mother Wavelet Representation

another different mother wavelet transform. The set of mother wavelets and their associated coefficients (which are hopefully small in number) is the new representation of the signal. Thus a new decomposition, not only in terms of wavelet coefficients but also in terms of mother wavelets, may be defined.

Conclusions

Wavelet theory is extended to include a new operator, the Mother Mapper Operator. This operator allows sets of mother wavelets to be considered rather than just a single mother wavelet. The Mother Mapper Operator can improve the efficiency of implementing several wavelet transforms on the same data set with different mother wavelets (see Figure 4). The Mother Mapper Operator allows a direct mapping from a previously computed wavelet transform to a new transform with respect to a new mother wavelet. The implementation is more efficient since the original space and/or time data does not need to be saved or reconstructed from the original wavelet transform in order to compute the new wavelet transform.

Nonorthogonal wavelet transforms are utilized rather than orthogonal wavelet transforms to provide the freedom in choosing the mother wavelet rather than being highly constrained and yielding a very specific (and possibly unique) mother wavelet.

Acknowledgements

I thank Leon Sibul and Lora Weiss for their ideas and reviews. I also acknowledge Penn State's Applied Research Laboratory for allowing me to complete this research.

References

[1] J. M. Combes, A. Grossman and Ph. Tchamitchian, *Wavelets: Time-Frequency Methods and Phase Space*, Springer-Verlag, New York, NY, 1989, specifically: A. Grossman, R. Kronland-Martinet, J. Morlet's "Reading and Understanding Continuous Wavelet Transforms," pp. 2-20

[2] I. Daubechies, "Orthonormal bases of compactly supported wavelets," *Comm. Pure Appl. Math.*, vol. 41, 1988, pp. 909-996

[3] I. Daubechies, "Time-frequency localization operators: a geometric phase space approach," *IEEE Trans. Inform. Theory*, vol. 34, 1988, pp. 605-612

[4] I. Daubechies, "The wavelet transform, time/frequency localization and signal analysis," *IEEE Trans. Inform. Theory*, vol. 36, Sept. 1990, pp. 961-1005

[5] I. Daubechies, "The Wavelet Transform: A Method for Time-frequency Localization," Chapter 8 of *Advances in Spectrum Analysis and Array Processing: Vol. 1*, (edt. by Simon Haykin) Prentice Hall, Englewood Cliffs, NJ, 1991, pp. 366-417

[6] A. Grossmann and J. Morlet, "Decomposition of Hardy functions into square integrable wavelets of constant shape," *SIAM J. Math. Anal.*, vol. 15, 1984, pp. 723-736

[7] A. Grossmann, J. Morlet, and T. Paul, "Transforms associated to square integrable group representations, I, General results," *J. Math. Physics*, vol. 26, 1985, pp. 2473-2479

[8] A. Grossmann, J. Morlet, and T. Paul, "Transforms Associated to Square Integrable Group Presentations II: Examples," *Am. Inst. Henri Poincare*, vol. 45, no. 3, 1986, pp. 293-309

[9] C. E. Heil and D. F. Walnut, " Continuous and Discrete Wavelet Transforms," *SIAM Review*, vol. 31, no. 4, Dec. 1989, pp. 628-666

[10] J. R. Klauder and E. C. G. Sudarshan, *Fundamentals of Quantum Optics*, W. A. Benjamin, Inc., New York, 1968

[11] S. G. Mallat, "A theory for multiresolution signal decomposition: the wavelet representation," *IEEE Trans. Pattern Anal. Machine Intel.*, vol. 31, 1989, pp. 674-693

[12] S. G. Mallat, "Multiresolution approximation and wavelets," *Trans. Am. Math. Soc.*, vol. 135, 1989, pp. 69-88

[13] R. K. Young, "Wideband Space-Time Processing and Wavelet Theory," Ph.D. Dissertation, Dept. Elect. Engr., Pennsylvania State Univ., University Park, PA, 1991

Fast Multiscale Statistical Signal Processing Algorithms

A. H. Tewfik

Department of Electrical Engineering
University of Minnesota
Minneapolis, MN 55414

M. Kim

Department of Electrical Engineering
University of Minnesota
Minneapolis, MN 55414

Abstract

It is shown that a large set of (not necessarily stationary) correlation matrices may be transformed into a matrix that consists of essentially banded sub-blocks. The transformation is accomplished by pre- and post-multiplication with an orthogonal matrix whose elements are derived from the impulse responses of a suitably designed cascade of alias free multirate analysis filter banks. It is further proved that the Cholesky factor of the transformed matrix also consists of essentially banded sub-blocks. Next, those two observations are combined to show that the linear positive definite systems of equations that arise in statistical signal processing can be solved in $O(\max\{N\log^2(N), N^2\})$ operations while matrix-vector multiplication steps may be implemented in $O(N\log(N))$ operations. An error analysis of the proposed linear positive definite system solver is also provided.

1 Introduction

The past decade has witnessed an increased interest in the multirate processing of signals [1]. In multirate signal processing systems a signal sequence is broken into a number of sub-sequences of various sample spacings. The division into sub-sequences is accomplished by using properly designed perfect reconstruction filter banks to avoid aliasing problems. Recently wavelet analysis has been introduced as a tool for studying signals at multiple resolution scales [2], [3]. Wavelet decompositions have been related to perfect reconstruction filter banks [2]. It has become clear from those studies that it is beneficial to use perfect reconstruction M-band filter banks where $M-1$ of the filters have each at least a large number p of zeros at zero frequency. The remaining filter is also required to have an equal number p of zeros at the frequency $2\pi/M$. The advantage of using 2-band filters with such properties for analyzing and synthesizing fractional Brownian motions was pointed out in [4] and its implication in speeding matrix-matrix and matrix-vector multiplications was indicated in [5], [6].

The main contribution of this paper is a fast $O(\max\{N\log^2(N), N^2\})$ algorithm for solving the positive definite linear systems of equations that arise in statistical signal processing. Motivated by the results of [5], we begin by generalizing the results obtained in [4] to a large class of stochastic processes. In particular we show that if a stochastic process has a correlation matrix with off-diagonal entries whose magnitude does not grow faster than some power of their distance from the main diagonal, then the joint correlation matrix of the outputs of a cascade of M-band perfect reconstruction multirate analysis filter banks having the properties outlined above, will consist of a set of $\log_M^2(N)$ essentially banded sub-blocks. Next we show that the Cholesky decomposition of the joint correlation matrix of the multiscale components also consists of essentially banded sub-blocks. It is proved that the number of non- negligible elements of the Cholesky factor is $O(N\log(N))$ and that this factor can be computed in $O(N\log^2(N))$ operations. This leads to $O(\max\{N\log^2(N), N^2\})$ algorithms for solving the positive definite linear systems of equations.

We also provide an analysis of the numerical accuracy of the proposed technique. This leads to a rule for deciding which elements of the Cholesky decomposition one may neglect (and hence not compute but rather replace by zero) if one wants to achieve any desired level of relative numerical accuracy in the final result.

2 Background

Perfect reconstruction analysis and synthesis filter banks have received a considerable amount of attention in the past decade. The analysis section consists of M filters $H_m(z)$ each followed by decimation by a factor M. The conditions on analysis filters $H_m(z)$ and reconstruction filters $G_m(z)$ for perfect reconstruction and design of such filters are described in [1]. Perfect reconstruction analysis and synthesis filter banks have been shown to be closely related to the discrete wavelet decomposition [2], [7]. It may be shown that computing discrete wavelet decomposition on discrete data is equivalent to successive filtering with 2-band perfect reconstruction analysis filter banks $H_0(z)$ and $H_1(z)$. Motivated by the fact that the regularity property of wavelets leads to efficient representations of signals in the sense that most of the wavelet coefficients will be of negligible magnitude ([3],[5]), we will restrict ourselves here to analysis filter banks that are characterized by filters $H_0(z)$ and

1058-6393/91 $01.00

$H_m(z)$, $m = 1, \cdots, M-1$ that have at least p zeros at $z = e^{j\pi}$ and at $z = 1$ respectively. In particular we consider the decomposition of signal $x(n)$ using a cascade of such M band filter banks. We shall call the output of each filter, $H_m(z)$, $m = 1, \cdots, M-1$, in each filter bank a "multiscale" component of $x(n)$ and we will denote that output by $x_k(n)$ (see Fig. 1). We use the notation $\rho(k)$ to denote the order of the analysis stage in which $x_k(n)$ is computed. We shall then say that $x_k(n)$ is a "higher resolution" component of $x(n)$ than $x_{k'}(n)$ if $\rho(k) < \rho(k')$. Note that each multiscale component can be represented as the output of a filter $W_k(z)$ with

$$\sum_m m^n w_k(m) = 0 \qquad n = 0, 1, \cdots, p-1 \tag{1}$$

where $w_k(n)$ denotes impulse response of the filter $W_k(z)$.

To deal with finite N point segments of data $\{x(n)\}_{n=1}^N$ and finite covariance matrices we will assume that the signal $x(n)$ is periodic with period equal to the length of the data. We can then compute the multiresolution decomposition of a vector $\mathbf{x} = [x(n)]$ which consists of the given samples of $x(\cdot)$, by multiplying $\mathbf{x}$ by an orthogonal circulant matrix $\mathbf{Q}$. In particular, it may be shown that computing $\mathbf{Qx}$ requires $O(N)$ operations [8].

3 Transformation of the Covariance Matrix

Most statistical processing problem involves computing matrix inverses and matrix vector multiplications. Instead of solving this problem directly we shall begin by first decomposing a zero mean stochastic process $x(n)$ with an autocorrelation $R(m,n)$ into its multiscale components. In this section we study asymptotic behavior of the cross correlation between multiscale components of $x(n)$. Denote by $R_w^{j,k}(m,n)$ the cross- correlation between the mth component of a stochastic process $x(\cdot)$ in the jth band and the nth component of $x(\cdot)$ in the kth band. Observe that $R_w^{j,k}(m,n)$ may be written as

$$R_w^{j,k}(m,n) = \sum_{m' \in \mathcal{S}_j} \sum_{n' \in \mathcal{S}_k} R(M^{\rho(j)}m - m', M^{\rho(k)}n - n') w_j(m') w_k(n') \tag{2}$$

where the filters $w_j(m)$ were defined in Section 2 and $\mathcal{S}_k$ is the support of the filter $w_k(n)$. Our objective is to study the asymptotic behavior of $R_w^{j,k}(m,n)$ as $\max(m,n)$ tends to infinity while $\min(m,n)$ is finite and fixed. Without loss of generality we will assume here that $n = \min(m,n)$ and is fixed while $m = \max(m,n)$.

Assume now that all the partial derivatives $\partial^{l+k}R(m,n)/\partial m^k \partial n^l$ of $R(m,n)$ exist for $0 \le l, k \le p$. Following a technique introduced in [3] to study wavelet decompositions of operators and also used in [5], we expand $R(M^{\rho(j)}m - m', M^{\rho(k)}n - n')$ around $(M^{\rho(j)}m, M^{\rho(k)}n)$ and use the fact that the first p moments of $w_j(m)$ and $w_k(n)$ are zero to obtain from (2)

$$|R_w^{j,k}(m,n)| \le C \sup_{(m',n') \in \mathcal{S}_j \times \mathcal{S}_k} \sum_{l=0}^{2p} \left|\frac{\partial^{2p} R(t,s)}{\partial t^l \partial s^{2p-l}}\right|_{t = M^{\rho(j)}m - m', s = M^{\rho(k)}n - n'} \tag{3}$$

where C is a finite constant which depends on the filters $w_j(\cdot)$ and $w_k(\cdot)$. Hence, as long as all terms $|\frac{\partial^{2p} R(m,n)}{\partial m^l \partial n^{2p-l}}|$, $0 \le l \le 2p$, decay faster to zero than $|R(m,n)|$ itself as m tends to infinity, $R_w^{j,k}(m,n)$ will decay faster than $R(m,n)$ as m tends to infinity. In particular, the multiresolution components of $x(n)$ will effectively be partially uncorrelated.

Of special interest to us is the case where $R(m,n)$ behaves asymptotically as

$$R(m,n) = m^{\alpha(n)} e^{-\beta(n)m}(a_0(n) + o(1)) + a_1(n) \quad \text{as } m \to \infty. \tag{4}$$

In the above equation the function $\beta(n)$ may have both a real and an imaginary part. However, its real part is assumed to be a non-negative and non-decreasing function of n and its imaginary part must be zero if its real part is zero. The function $\alpha(n)$ is only assumed to be a non-increasing function of n. The class of processes with covariance functions that behave asymptotically as (4) is quite large. For example it includes all stationary processes that can be obtained by passing white noise through a rational filter. It also includes such nonstationary processes as a sampled version of a fractional Brownian motion, the fractionally differenced white noise process and samples of the underwater background acoustic noise in shallow water taken along any straight line.

As m tends to infinity and since the sums in (2) are finite, we may rewrite (2) as

$$R_w^{j,k}(m,n) = \sum_{n' \in \mathcal{S}_k} w_k(n') \sum_{m' \in \mathcal{S}_j} w_j(m') \Big[(M^{\rho(j)}m - m')^{\alpha(M^{\rho(k)}n - n')} e^{-\beta(M^{\rho(k)}n-n')(M^{\rho(j)}m - m')} (a_0(M^{\rho(k)}n - n') + o(1)) + a_1(M^{\rho(k)}n - n')\Big]$$

where we have used (4). Recalling that $\Re(\beta(\cdot)) > 0$, define the quantities $\alpha_0(M^{\rho(k)}n)$ and $\beta_0(M^{\rho(k)}n)$ as

$$\beta_0(M^{\rho(k)}n) = \beta(M^{\rho(k)}n - n_0) \tag{5}$$

$$n_0 \in \aleph_\beta = \{n' \in \mathcal{S}_k : \Re(\beta(M^{\rho(k)}n - n') = \inf_{n' \in \mathcal{S}_k} \Re(\beta(M^{\rho(k)}n - n')\} \tag{6}$$

$$\alpha_0(M^{\rho(k)}n) = \sup_{n' \in \aleph_\beta} \alpha(M^{\rho(k)}n - n') \tag{7}$$

When $\Re(\beta_0(M^{\rho(k)}n) \neq 0$ it may be readily verified that

$$\lim_{M^{\rho(j)}m\to\infty} \frac{|R_w^{j,k}(m,n)|}{[(M^{\rho(j)}m)^{\alpha_0(M^{\rho(k)}n)}e^{-\Re(\beta_0(M^{\rho(k)}n))(M^{\rho(j)}m))}]} = c_0(M^{\rho(k)}n)$$

where

$$c_0(M^{\rho(k)}n) = |\sum_{n'\in\aleph_\alpha}\sum_{m'\in\mathcal{S}_j} w_k(n')w_j(m') e^{\beta(M^{\rho(k)}n-n')m'}a_0(M^{\rho(k)}n-n')|$$

$$\aleph_\alpha = \{n' \in \aleph_\beta : \alpha(M^{\rho(k)}n-n') = \sup_{n'\in\aleph_\beta} \alpha(M^{\rho(k)}n-n')\}.$$

This of course implies that $|R_w^{j,k}(m,n)|$ decays asymptotically as $(M^{\rho(j)}m)^{\alpha_0(M^{\rho(k)}n)}\ e^{-\Re(\beta_0(M^{\rho(k)}n)M^{\rho(j)}m)}$, i.e. at a rate much faster than $R(m,n)$ as expected since $\beta(\cdot)$ and $\alpha(\cdot)$ are non- decreasing and non-increasing functions of their arguments respectively.

On the other hand, when $\beta(n) = 0$ identically in (4), it may be verified using (1) that

$$\lim_{M^{\rho(j)}m\to\infty} \frac{|R_w^{j,k}(m,n)|}{(M^{\rho(j)}m)^{\alpha_0(M^{\rho(k)}n)-p}} = c_1(M^{\rho(k)}n), \tag{8}$$

where

$$\alpha_0(M^{\rho(k)}n) = \sup_{n'\in\mathcal{S}_k} \alpha(M^{\rho(k)}n-n')$$

$$c_1(M^{\rho(k)}n) = |\sum_{n'\in\mathcal{S}_k}\sum_{m'\in\mathcal{S}_j} w_k(n')w_j(m')m'^p a_0(M^{\rho(k)}n-n')|.$$

Note that in this case and as m tends to infinity, $R(m,n)$ may either decrease hyperbolically (instead of exponentially) fast to a constant which may be a function of n, or it may increase without bound. Stochastic processes with covariance functions that display such an asymptotic behavior include sampled versions of fractional Brownian motions and discrete fractionally differenced white noise processes. However, according to (8) $R_w^{j,k}(m,n)$ decreases to zero asymptotically as $m^{\alpha_0(M^{\rho(k)}n)-p}$ as long as $\alpha_0(\cdot) < p$. In particular, once more $R_w^{j,k}(m,n)$ decreases to zero asymptotically at a rate much faster than $R(m,n)$.

4 Inversion of the Approximate Transformed Covariance Matrix

Most of statistical signal processing algorithms involve solving linear positive definite systems of equations of the form $\mathbf{Rx} = \mathbf{y}$. The result in Section 3 shows that it is clearly beneficial to solve $\mathbf{Rx} = \mathbf{y}$ as $\mathbf{x} = \mathbf{Q}^T(\mathbf{QRQ}^T)^{-1}(\mathbf{Qy})$ especially if all the elements of $\mathbf{QRQ}^T$ that have magnitude less than some pre-specified threshold ϵ are approximated by zero. Specifically, if $\hat{\mathbf{R}}_\mathbf{W}$ denotes the approximation to $\mathbf{QRQ}^T$ obtained by setting all its elements of magnitude less than an ϵ to zero, then each sub-block of $\hat{\mathbf{R}}_\mathbf{W}$ will be banded and $\hat{\mathbf{R}}_\mathbf{W}$ will have $O(\log(N))$ nonzero elements per row. Since $\hat{\mathbf{R}}_\mathbf{W}$ is sparse and has a nice regular structure, one would suspect that it is possible to invert that matrix using less than $O(N^3)$ operations. In this section we show that this is indeed the case and that solving $\hat{\mathbf{R}}_\mathbf{W}\mathbf{x}' = \mathbf{y}'$ requires $O(N\log^2(N))$ operations. It may be shown that computing $\hat{\mathbf{R}}_\mathbf{W}$ itself requires $O(N^2)$ operations [8] and computing an approximate solutions to $\mathbf{Rx} = \mathbf{y}$ will then require $O(\max\{N\log^2(N), N^2\})$ operations.

A popular method of solving positive definite linear systems $\mathbf{Ax} = \mathbf{y}$ is to compute the Cholesky decomposition of the matrix $\mathbf{A}$. Here we show that the Cholesky factor of the approximation to $\mathbf{QRQ}^T$ will also consist of sub-blocks which are essentially banded.

Denote by $\hat{\mathbf{R}}_\mathbf{W}$ the approximation to $\mathbf{QRQ}^T$ and by $\hat{\mathbf{R}}^{(n)}$ its lower $n\times n$ main sub matrix. Now assume that each row and each column in every sub-block of $\hat{\mathbf{R}}_\mathbf{W}$ has at most C_Q entries of magnitude larger than some ϵ_R. Furthermore, assume that the Cholesky factor $\mathbf{G}^{(n)}$ of any $n\times n$ matrix that consists of banded blocks each with at most C_Q entries per row or column of magnitude larger than some ϵ_R, consists of sub-blocks that are banded when entries of $\mathbf{G}^{(n)}$ with a magnitude smaller than a pre-specified threshold ϵ (which of course is related to ϵ_R) are neglected. Note that $\hat{\mathbf{R}}_\mathbf{W}^{(n+1)}$ may be written as

$$\hat{\mathbf{R}}_\mathbf{W}^{(n+1)} = \begin{bmatrix} \alpha & \mathbf{v}^T \\ \mathbf{v} & \hat{\mathbf{R}}_\mathbf{W}^{(n)} \end{bmatrix} = \begin{bmatrix} 1 & 0 \\ \frac{\mathbf{v}}{\alpha} & \mathbf{I}^{(n)} \end{bmatrix} \begin{bmatrix} 1 & 0 \\ 0 & \hat{\mathbf{R}}_\mathbf{W}^{(n)} - \frac{\mathbf{vv}^T}{\alpha} \end{bmatrix} \begin{bmatrix} 1 & \mathbf{v}^T \\ 0 & \mathbf{I}^{(n)} \end{bmatrix}$$

where $\mathbf{I}^{(n)}$ denotes the $n\times n$ identity matrix. Note that the vector $\mathbf{v}$ may be decomposed conformably with $\hat{\mathbf{R}}_\mathbf{W}^{(n+1)}$ and in particular that it consists of groups of at most C_Q elements of magnitude larger than ϵ_R separated by zeros. Hence, $\hat{\mathbf{R}}_\mathbf{W}^{(n)} - \frac{\mathbf{vv}^T}{\alpha}$ also consists of sub- blocks which are banded and such that every row and column in each of its sub-block has at most C_Q entries that are larger than ϵ_R. By assumption $\hat{\mathbf{R}}_\mathbf{W}^{(n)} - \frac{\mathbf{vv}^T}{\alpha}$ has a Cholesky factor $\mathbf{G}_0$ which consists of sub-blocks that are banded when entries of $\mathbf{G}_0$ with a magnitude smaller than a pre-specified threshold ϵ are neglected. Using the induction hypothesis and the sparse structure of $\mathbf{v}$ it then follows that

$$\mathbf{G}^{(n+1)} = \begin{bmatrix} 1 & 0 \\ \frac{\mathbf{v}}{\alpha} & \mathbf{G}_0 \end{bmatrix} \tag{9}$$

is the Cholesky factor of $\hat{\mathbf{R}}_\mathbf{W}^{(n+1)}$ and has the stated sparse structure when its entries of magnitude less than ϵ are neglected.

The above discussion now indicates that the Cholesky factor $\mathbf{G}$ of $\hat{\mathbf{R}}_{\mathbf{W}}$ may be approximated by a matrix $\hat{\mathbf{G}}$ which consists of banded sub-blocks when its entries of magnitude less than ϵ are neglected. In particular, $\hat{\mathbf{G}}$ has $O(N\log(N))$ non- negligible elements. The non-negligible elements $\hat{G}(i,j)$ of this latter matrix may be computed a column at a time via the recursions given in [9]. The computation in each sub-block proceeds until an element of magnitude ϵ or less is computed. At this point all remaining elements in the column being computed within the sub-block are set to zero. Since the matrix $\hat{\mathbf{R}}_{\mathbf{W}}$ itself has at most $O(\log(N))$ element per row or column it is clear that computing all $O(N\log(N))$ non-negligible elements of $\hat{\mathbf{G}}$ will require $O(N\log^2(N))$ operations.

Once $\hat{\mathbf{G}}$ has been computed the solution of the system $\hat{\mathbf{G}}\hat{\mathbf{G}}^T\mathbf{Q}\mathbf{x} = \mathbf{Q}\mathbf{y}$ itself requires $O(N\log(N))$ operations. Hence, instead of solving any system $\mathbf{R}\mathbf{x} = \mathbf{y}$ using $O(N^3)$ operations we may solve the approximate system $\hat{\mathbf{G}}\hat{\mathbf{G}}^T\mathbf{x}' = \mathbf{Q}\mathbf{y}$ and compute an approximation $\hat{\mathbf{x}}$ to $\mathbf{x}$ as $\hat{\mathbf{x}} = \mathbf{Q}^T\mathbf{x}'$ in $O(\max\{N\log^2(N), N^2\})$ operations.

5 Numerical Accuracy of the Approximate Solutions

In this section we study the numerical error introduced by our procedure and derive a criterion for choosing the thresholds ϵ and ϵ_R of Section 4. On a finite precision machine the problem data $(\mathbf{R}, \mathbf{y})$ represents a whole family of problems $(\mathbf{R}+\mathbf{\Delta R}, \mathbf{y}+\mathbf{\Delta y})$ where $\delta\mathbf{R}$ and $\mathbf{\Delta y}$ correspond to the range of problems that are compatible with the finite precision of the data representation. Hence, in practice we will have solved a problem of the form

$$(\mathbf{R}+\delta\mathbf{R})(\mathbf{x}+\delta\mathbf{x}) = (\mathbf{y}+\delta\mathbf{y}). \qquad (10)$$

Then under the assumption that $||\delta\mathbf{R}||||\mathbf{R}^{-1}|| < 1$, we obtain

$$\frac{||\delta\mathbf{x}||}{||\mathbf{x}||} \le \frac{\kappa(\mathbf{R})}{1-\kappa(\mathbf{R})[\frac{||\delta\mathbf{R_a}||}{||\mathbf{R}||}+\frac{||\delta\mathbf{R_n}||}{||\mathbf{R}||}]} \left(\frac{||\delta\mathbf{y}||}{||\mathbf{y}||}+\frac{||\delta\mathbf{R_a}||}{||\mathbf{R}||}+\frac{||\delta\mathbf{R_n}||}{||\mathbf{R}||}\right), \qquad (11)$$

where $\delta\mathbf{R_a}$ and $\delta\mathbf{R_n}$ denote respectively the perturbations in $\mathbf{R}$ which are due to the approximations that we introduced in the procedure and the data uncertainty and finite machine precision. If we can estimate an upper bound on $||\delta\mathbf{R_n}||$ and if we can pick ϵ and ϵ_R so as to guarantee that $||\delta\mathbf{R_a}||/||\mathbf{R}|| < ||\delta\mathbf{R_n}||/||\mathbf{R}||$, then the ultimate accuracy of our results is determined by the machine precision rather than the approximations that we introduced in the procedure.

Now we will derive an upper bound on $||\delta\mathbf{R_a}||$ in terms of ϵ and ϵ_R and we will provide a criterion for choosing the thresholds ϵ and ϵ_R such that accuracy of our solution is determined by the machine precision. If we assume that there are no data uncertainty and other numerical errors in the actual implementation of the Cholesky decomposition approach, then our procedure solves the problem

$$\mathbf{Q^T}\hat{\mathbf{G}}\hat{\mathbf{G}}^\mathbf{T}\mathbf{Q}(\mathbf{x}+\delta\mathbf{x}) = \mathbf{y}. \qquad (12)$$

By subtracting $\mathbf{R}\mathbf{x} = \mathbf{y}$ we find

$$\mathbf{R}\delta\mathbf{x} = (\delta\mathbf{R_a})(\mathbf{x}+\delta\mathbf{x}), \qquad (13)$$

where

$$\delta\mathbf{R_a} = \mathbf{R} - \mathbf{Q^T}\hat{\mathbf{G}}\hat{\mathbf{G}}^\mathbf{T}\mathbf{Q}. \qquad (14)$$

Then relative error bound may be written as

$$\frac{||\delta\mathbf{x}||}{||\mathbf{x}||} \le \frac{\kappa(\mathbf{R})\frac{||\delta\mathbf{R_a}||}{||\mathbf{R}||}}{1-\kappa(\mathbf{R})\frac{||\delta\mathbf{R_a}||}{||\mathbf{R}||}} \qquad (15)$$

provided that $||\delta\mathbf{R_a}||||\mathbf{R}^{-1}|| < 1$. Here $\kappa(\mathbf{R})$ denotes the condition number of the matrix $\mathbf{R}$. It can be verified that

$$||\hat{\mathbf{R}}_\mathbf{W} - \mathbf{Q}\mathbf{R}\mathbf{Q}^T||_p \le N\epsilon_R, \qquad (16)$$
$$||(\hat{\mathbf{G}}\hat{\mathbf{G}}^\mathbf{T} - \hat{\mathbf{R}}_\mathbf{W})||_p \le N\epsilon^2 \qquad (17)$$

where $p = 1$ or $p = \infty$. Since

$$||\mathbf{A}||_2 \le \sqrt{||\mathbf{A}||_1||\mathbf{A}||_\infty}$$
$$||\mathbf{A}||_p \le \sqrt{N}||\mathbf{A}||_2, \qquad p = 1 \text{ or } \infty,$$

for any matrix $\mathbf{A}$, we then get

$$||\delta\mathbf{R_a}||_2 \le N(\epsilon^2+\epsilon_R) \qquad (18)$$
$$||\delta\mathbf{R_a}||_p \le N^{\frac{3}{2}}(\epsilon^2+\epsilon_R) \qquad (19)$$

where $p = 1$ or $p = \infty$. Hence, if we choose the thresholds ϵ and ϵ_R such that

$$\epsilon^2+\epsilon_R \le \epsilon_x\frac{||\mathbf{R}||_2}{N} \qquad (20)$$

or that

$$\epsilon^2+\epsilon_R \le \epsilon_x\frac{||\mathbf{R}||_p}{N^{\frac{3}{2}}} \qquad p = 1 \text{ or } \infty, \qquad (21)$$

we can guarantee that the relative error in the 2 norm or in the 1 or infinity norms respectively, in $\mathbf{x}$ is bounded above by $\epsilon_x\kappa(\mathbf{R})$ for any prescribed ϵ_x.

6 Simulation

Simulations were performed to study the behavior of the linear positive definite system solver proposed in this paper. We considered the discrete stochastic processes with the following covariance function structure:

$$R(m,n) = \begin{cases} m^\alpha e^{-\beta m} \displaystyle\sum_{i=0}^{15} \frac{n^i}{m^i} + n & \text{for } m \geq n \\ n^\alpha e^{-\beta n} \displaystyle\sum_{i=0}^{15} \frac{m^i}{n^i} + m & \text{for } n \geq m. \end{cases} \quad (22)$$

In Fig. 2 we plot magnitude of the entries of the 128×128 correlation matrix corresponding to the stochastic process (22), those of the corresponding transformed correlation matrix, and those of the Cholesky factor. The figure shows banded structure of transformed matrix and its Cholesky factor. The performance of the procedure for the process (22) and for other stochastic processes such as fractional Brownian motion and Gauss-Markov process is given in [8].

References

[1] P. P. Vaidyanathan, "Multirate Digital Filters, Filter Banks, Polyphase Networks and Applications: A Tutorial," *Proc. IEEE*, vol. 78, no. 1, pp. 56-93, Jan. 1990.

[2] I. Daubechies, "Orthogonal Bases of Compactly Supported Wavelets," *Commun. Pure Appl. Math.*, vol. 41, pp. 909-996, 1988.

[3] Y. Meyer, *Ondelettes et Operateurs*, Herman, Paris,France, 1990.

[4] A. H. Tewfik and M. Kim, "Correlation Structure of the Discrete Wavelet Coefficients of Fractional Brownian Motions," to appear in *IEEE Trans. Info. Theory*, January 1992.

[5] G. Beylkin, R. Coifman and V. Rokhlin, "Fast Wavelet Transforms and Numerical Algorithms I," *Comm. on Pure and Appl. Math.*, vol. 44, pp. 141-183, March 1991.

[6] M. V. Wickerhauser, "Nonstandard Matrix Multiplication,"preprint, Dept. of Mathematics, Yale Univ., May 1990.

[7] S. Mallat, "Multifrequency Channel Decomposition of Images and Wavelet Models," *IEEE Trans. Acoust. Speech and Signal Proc.*, vol. ASSP-37, pp. 2091-2110, 1989.

[8] A. H. Tewfik and M. Kim, "Fast Multiscale Statistical Signal Processing Algorithms," submitted to *IEEE Trans. Signal Processing*.

[9] G. H. Golub and C. F. Van Loan, *Matrix Computations*, Johns Hopkins Univ. Press, Baltimore, MD, 1983.

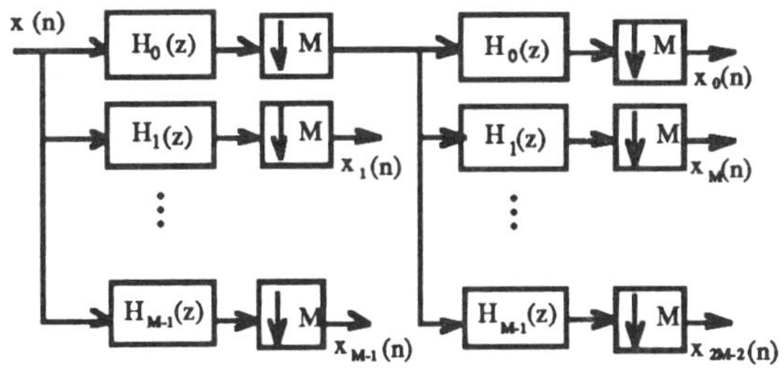

Fig. 1 Multiscale analysis filter bank

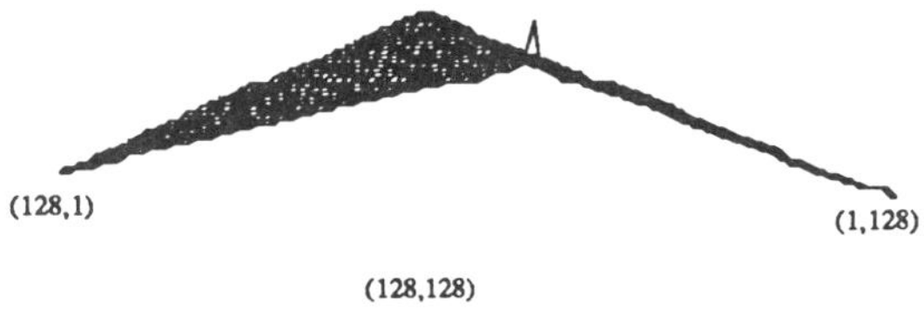

(a) Correlation matrix of the process

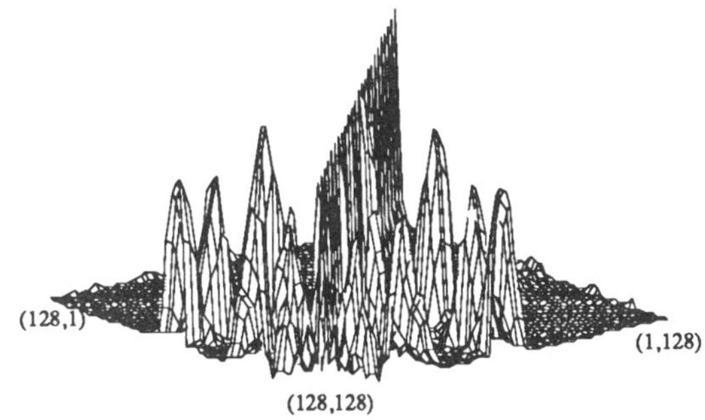

(b) Transformed correlation matrix

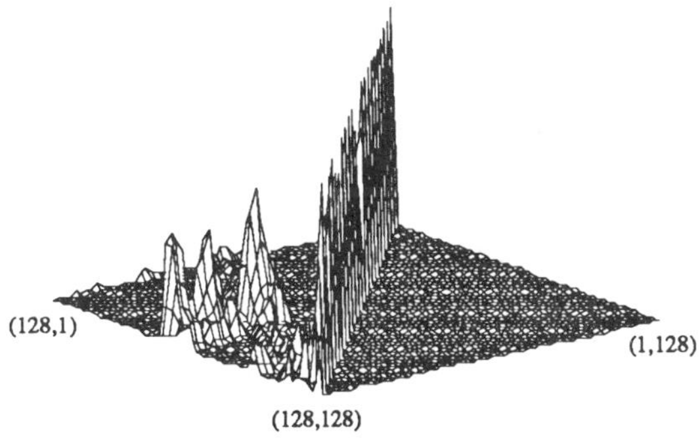

(c) Cholesky factor of the transformed matrix

Fig. 2 Magnitude of correlation matrix, transformed matrix, and Cholesky factor of the transformed matrix

Modeling and Estimation of Multiscale Processes

Kenneth C. Chou

Acoustics and Radar Technology Laboratory
SRI International
Menlo Park, CA 94025

Alan S. Willsky

Lab. for Information and Decision Systems
Massachusetts Institute of Technology
Cambridge, MA 02139

Abstract

In this paper we introduce a class of stochastic processes motivated by the wavelet transform. These processes are represented by Markovian state models in which scale plays the role of a time-like variable. This class of processes is rich enough to model $1/f$-type behavior as well as such standard processes as those belonging to the Gauss-Markov class. This paper presents an efficient smoothing algorithm which allows us to compute estimates based on multiscale data. We give numerical examples to show how our models can be used to smooth noisy data as well as examples of fusing multiscale data.

1 Introduction and Background

In recent years there has been considerable interest and activity in the signal and image processing community in developing multiresolution processing algorithms. One of the more recent areas of investigation in multiscale analysis has been the emerging theory of multiscale representations of signals and wavelet transforms [1, 5, 8].

In this paper we present elements of a framework (more fully developed in [4]) for multiresolution stochastic processes and models aimed at achieving the objectives of describing a rich class of phenomena and of providing the foundation for a theory of optimal multiresolution statistical signal processing. In developing this theoretical framework we have tried to keep in mind the three distinct ways in which multiresolution features can enter into a signal or image analysis problem. First, the phenomenon under investigation may possess features and physically significant effects at multiple scales. For example, $1/f$ -like stochastic processes as in [13, 14] are captured in our framework as are, surprisingly, useful models of many other processes. Second, whether the underlying phenomenon has multiresolution features or not, it may be that the collected *data* are at several different resolutions. Our framework provides a natural way to design algorithms for such multisensor fusion problems in that treating the case of multiple sensors is no more difficult than treating the case of a single sensor.

Finally, whether the phenomenon or data have multiresolution features or not, the signal analysis *algorithm* may have such features. Such an algorithm can be used, for example, to deal with the daunting complexity of many image processing problems. Multigrid relaxation algorithms[2, 9] for solving partial differential equations are of this type, as are a variety of computer vision algorithms. The stochastic models we describe lead to several extremely efficient and highly parallelizable computational structures for signal processing.

While there are several ways in which to introduce and motivate our modeling framework, one that provides a fair amount of insight begins with the *wavelet transform*[5, 8]. However, the key for modeling is not to view the transform as a method for analyzing signals but rather as a mechanism for *synthesizing* or *generating* such signals beginning with coarse representations and adding fine detail one scale at a time.

The multiscale representation of a continuous signal $f(x)$ consists of a sequence of approximations of that signal at finer and finer scales, where the approximations of $f(x)$ at the mth scale consists of a weighted sum of shifted and compressed (or dilated) versions of a basic *scaling function*, $\phi(x)$:

$$f_m(x) = \sum_{n=-\infty}^{+\infty} f(m,n)\phi(2^m x - n) \qquad (1)$$

where $f(m,n)$ are known as "scaling coefficients."

By considering the *incremental detail* added in obtaining the $(m+1)$st scale approximation from the mth, we arrive at the wavelet transform. Such a trans-

1058-6393/91 $01.00

form is based on a single function, $\psi(x)$, that has the property that the full set of its scaled translates $\{2^{m/2}\psi(2^m x - n)\}$ form a complete orthonormal basis for L^2. If $d(m,n)$ denote the wavelet coefficients, then

$$f_{m+1}(x) = f_m(x) + \sum_n d(m,n) 2^{m/2}\psi(2^m x - n) \quad (2)$$

One of the appealing features of the wavelet transforms for the analysis of signals is that they can be computed recursively in scale, from fine to coarse. Specifically, if we have the coefficients $\{f(m+1,\cdot)\}$ of the $(m+1)$st-scale representation we can "peel off" the wavelet coefficients at this scale and at the same time carry the recursion one complete step by calculating the coefficients $\{f(m,\cdot)\}$ at the next somewhat coarser scale:

$$f(m,n) = \sum_k h(2n-k) f(m+1,k) \quad (3)$$

$$d(m,n) = \sum_k g(2n-k) f(m+1,k) \quad (4)$$

where $g(n)$ and $h(n)$ form a *conjugate mirror filter* pair[11]. For convenience we define the operations in eqs.(3,4)) formally in terms of "coarsening" and "differencing" operators, respectively. These operators map l^2 sequences into l^2 sequences and are defined as follows.

$$(H_m f(m+1,\cdot))_n \triangleq \sum_k h(2n-k) f(m+1,k) \quad (5)$$

$$(G_m f(m+1,\cdot))_n \triangleq \sum_k g(2n-k) f(m+1,k) \quad (6)$$

where the operators are indexed with the subscript m to denote that they map sequences at scale $m+1$ to sequences at scale m.[1]

Reversing this process we obtain the synthesis form of the wavelet transform in which we build up finer and finer representations via a coarse-to-fine scale recursion:

$$f(m+1,n) = \sum_k h(2k-n) f(m,k) + \sum_k g(2k-n) d(m,k) \quad (7)$$

[1]Note that for the case of infinite sequences the operators as defined here are precisely equivalent for each scale; i.e. they are not a function of m. However, we adhere to this notation for the reasons that a) we may allow for the QMF filter to *differ* at each scale and b) for the case of finite-length sequences the operators *are* in fact different at every scale because the number of points differ from scale to scale.

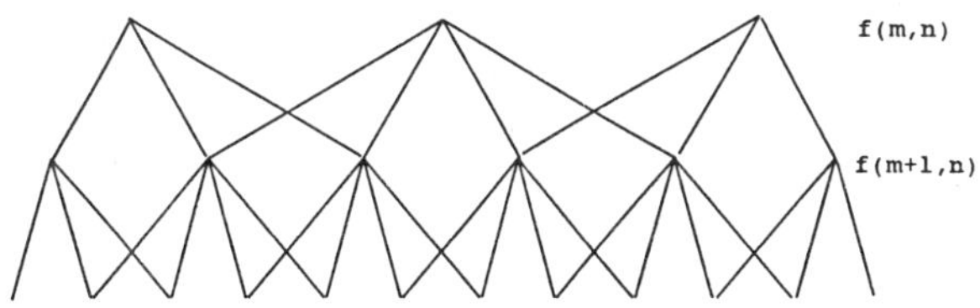

Figure 1: Infinite Lattice Representing Domain of Scaling Coefficients

Expressed in terms of the operators H_m and G_m we have

$$f(m+1,n) = (H_m^* f(m,\cdot))_n + (G_m^* d(m,\cdot))_n \quad (8)$$

where "*" denotes the adjoint of the operator.

Thus, we see that the synthesis form of the wavelet transform defines a *dynamical* relationship between the coefficients $f(m,n)$ at one scale and those at the next. Indeed, this relationship defines an infinite lattice on the points (m,n), where $(m+1,k)$ is connected to (m,n) if $f(m,n)$ influences $f(m+1,k)$. This structure is illustrated in Figure 1 for the case where $h(n)$ is a 4-tap filter, where each level of the lattice represents an approximation of our signal at some scale m. Note that the dynamics in eq.(7) are now with respect to *scale* rather than time.[2]

In developing a class of conveniently parametrized models which describes a rich class of phenomena, we need additional structure to supplement our dynamical multiscale representations. If we view these multiscale representations more abstractly, much as in the notion of a state model, as capturing the features of signals up to a particular scale that are relevant for the "prediction" of finer-scale approximations, we can define rich classes of stochastic processes and models that contain the multiscale wavelet representations of eqs.(3-7) as special (and in a sense degenerate) cases. Carrying this a bit farther, for wavelet transforms to be useful it should be the case that their application simplifies the description or properties of signals. For example, this clearly would be the case for a stochastic process that is whitened by eqs.(3,4), i.e., for which the wavelet coefficients $\{d(m,\cdot)\}$ at a particular scale are white and uncorrelated with the lower resolution

[2]Note also that the fact that scale plays the role of a time-like variable is true whether our representation is of a 1D signal or a 2D signal. This is an important fact which has implications for the development of models and algorithms in 2D in which we use this notion of recursion in scale.

version $\{f(m,\cdot)\}$ of the signal. Results in [6, 12, 13] indicate that one such class of processes, known as *fractional Brownian motion*, is indeed simplified by wavelet analysis and that processes generated by driving the wavelet synthesis equation with uncorrelated coefficients can exhibit fractal behavior.

2 Multiscale Processes on Lattices

In this section we define our class of lattice models, which is directly motivated by the synthesis equation, eq.(7), of the wavelet transform, and analyze its eigenstructure. Included in our class of models is precisely the model used by Wornell[13, 14] to describe $1/f$ processes. The Karhunen-Loeve expansion for the covariance function of our processes at a given scale is achieved exactly by the wavelet transform. We use this fact to derive an efficient smoothing algorithm based on diagonalizing the problem using the wavelet transform.[3]

Consider an infinite lattice corresponding to a wavelet whose scaling filter, $h(n)$, is an FIR filter of length P. Recall that each level of the lattice can be viewed as the domain of an l^2 sequence representing either successively finer approximations of a signal-the scaling coefficients $f(m,n)$, where f gets finer with increasing m, or the corresponding wavelet coefficients, $d(m,n)$. We define models whose state, $x(m) \in l^2$, can in some sense be viewed as the stochastic analog of the scaling coefficients at scale m. Here, $x(m)$ is defined to be an infinite-dimensional random vector whose elements are defined at a particular scale of the lattice. Let H_m and G_m denote the coarsening and differencing operators defined in eqs.(5,6), where the subscript m denotes that the operators map vectors at scale $m+1$ to vectors at scale m.

If we consider eq.(7), driven by uncorrelated wavelet coefficients $d(m,n)$, where the variances are constant along scale but varying from scale to scale, we get the following stochastic-dynamic state model where we define the scale index m from an initial coarse scale, L, to the finest scale, M, and where we assume that the coarsest scaling coefficients $f(L,n)$ are uncorrelated. Thus, with $x(m)$ corresponding to $f(m,\cdot)$ and $w(m)$ to $d(m,\cdot)$ we have[4] for $m = L, L+1, \ldots M-1$

$$\begin{aligned} E[x(L)x(L)^T] &= \tilde{\Lambda}_L \qquad (9)\\ &= \tilde{\lambda}_L I \\ x(m+1) &= H_m^* x(m) + G_m^* w(m) \qquad (10)\\ E[w(i)w(i)^T] &= \Lambda_i \qquad (11)\\ &= \lambda_i I \ , \ i = L, L+1, \ldots M-1 \end{aligned}$$

It is shown in [4] that the eigenstructure of this process at any given scale is given by the wavelet transform; i.e., the wavelet coefficients of $x(m)$ are uncorrelated.

We now consider an important variation of the models we have considered so far. We consider lattice models in which the elements of the state vectors at each scale are no longer simply scalars but rather finite-dimensional *vectors*. In other words, if we let d denote the dimension of these vectors and if $x_i(m)$ denotes the ith entry of $x(m)$, then $x_i(m) \in \mathcal{R}^d$. By abuse of notation we now let H_m, G_m be vector versions of the coarsening and differencing operators.

We specify our model as follows for $m = L, L+1, \ldots M-1$,

$$\begin{aligned} E[x(L)x(L)^T] &= \mathcal{P}_x(L) \qquad (12)\\ x(m+1) &= H_m^* \mathcal{A}(m+1)x(m) \\ &+ \mathcal{B}(m+1)w(m+1) \qquad (13)\\ E[w(i)w(j)^T] &= \mathcal{Q}(i)\delta_{i-j} \ , \ i = L+1, \ldots M \quad (14) \end{aligned}$$

where

$$\mathcal{A}(m) \triangleq diag(\ldots, A(m), \ldots A(m), \ldots) \qquad (15)$$

and where $\mathcal{B}(m), \mathcal{Q}(m), \mathcal{P}_x(L)$ are similarly defined with $B(m), Q(m), P_x(L)$, respectively. The quantities $A(m), B(m), Q(m)$, and $P_x(L)$ are *finite-dimensional* matrices representing the system matrix, the process noise matrix, the process noise covariance matrix, and the initial state covariance matrix, respectively.[5]

It turns out that the block eigenstructure of this process is precisely the same as that of the process defined earlier in eqs.(9-11), modulo the fact that we now must deal with $d \times d$ blocks as the basic elements of our decomposition. In fact, what we will refer to as ***block-eigenvectors*** will have the same structure as in the case of our scalar model, except that H_m, G_m are

[3] We develop our ideas for the case of the infinite lattice, corresponding to the wavelet transform as defined in Section 1. A discussion of the issue of adapting the wavelet transform to the case of finite-length data can be found in [4].

[4] Note that if we let $\lambda_i = \sigma^2 2^{-\gamma i}$ this model is precisely the one considered by Wornell for modeling a $1/f$-type process with spectral parameter γ.

[5] Note that by augmenting our model class to include finite-dimensional state vectors defined on lattices, we allow for the possibility of higher-order models. This allows us to consider a considerably richer class of processes which is parametrized by the matrices corresponding to our state model. Note also that this model bears resemblance to Laplacian Pyramid schemes[3] where the added detail in going from one scale to the next is *not* constrained by the differencing operator G_m.

now the appropriate vector versions and the resulting block-eigenvectors are infinite sequences of $d \times d$ *matrices* rather than an infinite sequence of scalars. The "eigenvalues," however, will differ, as they will in fact now be $d \times d$ *eigenmatrices* which will be functions of the matrices $A(m), B(m), Q(m)$, and $P_x(L)$. The main point is that we can represent the process $x(m)$ using the wavelet transform such that the wavelet coefficients, now vectors, are uncorrelated.

2.1 Efficient Optimal Smoothing

In this section we consider the problem of optimally estimating one of our processes given sensors of varying SNRs and differing resolutions. We formulate this sensor fusion problem as an optimal smoothing problem in which the optimally smoothed estimate is formed by combining noisy measurements of our lattice process at various scales. In other words, each sensor is modeled as a noisy observation of our process at some scale of the lattice. We treat the smoothing problem in a way that is sufficiently general to account for situations in which we have measurements at either one or multiple scales. In the following development we assume the model in eq.(13).

Consider the following multiscale measurements for $m = L, L+1, \ldots M$.

$$y(m) = \mathcal{C}(m)x(m) + v(m) \tag{16}$$

where

$$\begin{aligned} \mathcal{C}(m) &\triangleq diag(\ldots, C(m), \ldots C(m), \ldots) &(17)\\ \mathcal{R}(m) &\triangleq diag(\ldots, R(m), \ldots R(m), \ldots) &(18)\\ E[v(i)v(j)^T] &= \mathcal{R}(i)\delta_{i-j} &(19) \end{aligned}$$

and where $C(m)$ is a $b \times d$ matrix and $R(m)$ is a $b \times b$ matrix. We define the **smoothed** estimate, denoted as $x^s(m)$, to be the expected value of $x(m)$ conditioned on $y(i)$ for $i = L, L+1, \ldots M$; i.e.

$$x^s(m) = E[x(m)|y(L), \ldots y(M)] \tag{20}$$

We define the **coarse-to-fine filtered** estimate, to be the expected value of $x(m)$ conditioned on $y(i)$ for $i = L, L+1, \ldots m$; i.e

$$\hat{x}(m|m) = E[x(m)|y(L), \ldots y(m)] \tag{21}$$

We define the **coarse-to-fine one-step predicted** estimate to be the expected value of $x(m)$ conditioned on $y(i)$ for $i = L, L+1, \ldots m-1$; i.e

$$\hat{x}(m|m-1) = E[x(m)|y(L), \ldots y(m-1)] \tag{22}$$

From standard Kalman filtering theory, we can derive a recursive filter with its associated Riccati equations, where the recursion in the case of our lattice models is in the scale index m in the direction from coarse to fine. We choose to solve the smoothing problem via the Rauch-Tung-Striebel(RTS) algorithm[10]. This gives us a correction sweep, following the Kalman filtering step, that runs recursively from fine to coarse scales with the initial condition of the recursion being given by the final point of the Kalman filter.

Reference [4] shows how the smoothing problem, currently one involving the infinite-dimensional state vector $x(m)$, can be decomposed into a set of standard 1D smoothing problems involving finite-dimensional state vectors. By transforming our state vectors and data, i.e., by representing them in the wavelet basis, we end up with a set of independent 1D RTS smoothing problems which can be computed in parallel. These quantities represent the transformed versions of the predicted, filtered, and smoothed estimates in the RTS algorithm, along with their respective error covariances. The eigenvectors used to transform these quantities involve, as previously defined, $d \times 1$ vector versions of the operators H_i and G_i.

Note that our algorithm is highly efficient in that we have transformed the problem of smoothing infinite dimensional vectors to one of smoothing in parallel a set of *finite* dimensional vectors. Also, the smoothing procedure takes place in *scale* rather than in time, and for finite data of length N this interval is at most of order $logN$. In particular, the number of scale components at each scale increases with finer scales. Since only data at scales at or finer than the scale of a particular component provide information about that component, this leads to smoothers of different length in scale, with their length indexed by scale component. Let us analyze the complexity of our overall algorithm for smoothing our lattice processes. We first transform our data using the wavelet transform which is fast: $O(lN)$ where N is the number of points at the finest scale and l is the length of the QMF filter. We then perform in parallel our 1D smoothers. Note that the different scale components are smoothed using RTS smoothers over different length intervals, where typically the length of the interval is at most $logN$. Even if these smoothers are computed serially, the total computation is $O(lN)$. After performing the parallel 1D smoothers on these transformed variables, an additional inverse transformation is required. This tranformation is performed again using the inverse wavelet transform (the synthesis equations). Thus, the overall procedure is of complexity $O(lN)$.[6]

3 Numerical Examples

We begin by describing the type of Gauss-Markov process which we use in our examples of smoothing processes using lattice models. Consider the following stationary first-order Gauss-Markov process.

$$\dot{x}(t) = -\beta x(t) + w(t) \quad (23)$$
$$E[x^2(t)] = 1 \quad (24)$$

In the numerical examples that follow we use a discretized version of eq.(23) (see [4] for details). We consider the following measurements of $x(t)$.

$$y(t) = x(t) + v(t) \quad (25)$$
$$E[v^2(t)] = R \quad (26)$$

While we would expect our models to yield good approximations for 1/f processes given recent results on this[13, 6], it comes somewhat as a surprise that we are able to estimate so well noisy Gauss-Markov processes using our models. Our ability to use the wavelet transform as a basis to *model* Gauss-Markov processes, however, is supported by recent results by Beylkin et al.[1] in using the wavelet transform to *analyze* classes of kernels that share properties with the covariance functions of Gauss-Markov processes. Also, the work of Golden[7], in which the covariance of Gauss-Markov processes is represented in terms of various wavelet bases, motivates the approximation of the wavelet coefficients of Gauss-Markov processes as uncorrelated. This results in a lattice model precisely as defined in eqs.(9-11). We use this model as an approximation to the Gauss-Markov process to do fixed-interval smoothing.

The results of Beylkin and Golden indicate that this model does very well in approximating Gauss-Markov processes. To determine the variances of the wavelet coefficients, $w(m)$ in eqs.(9-11), we perform a similarity transform on the covariance matrix of the process under investigation using a wavelet transform based on one of the Daubechies FIR filters[5]. In particular, if P_x denotes the true covariance matrix of the process, V the diagonal matrix of wavelet coefficient variances, and W the wavelet transform matrix, then

$$\Lambda = W P_x W^T \quad (27)$$
$$V = W P_{approx} W^T \quad (28)$$

[6] We also point out that our algorithm generalizes Wornell's smoothing algorithm[14] to allow for multiscale measurements and to allow for higher-order models.

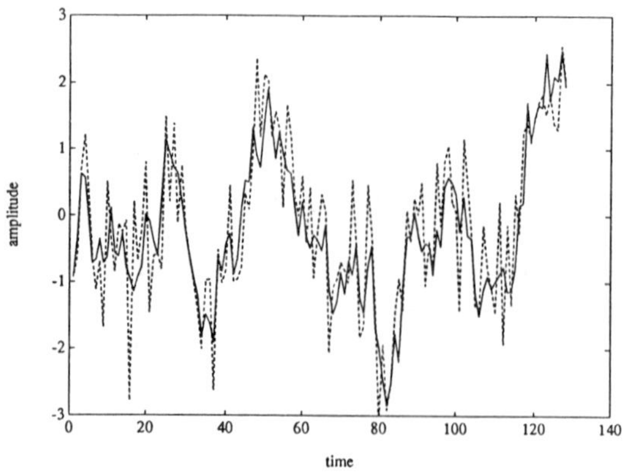

Figure 2: Sample Path of a Stationary Gauss-Markov Process (solid) and its Noisy Version with SNR=1.4142 (dashed)

Thus, this approximate model corresponds to assuming that Λ is diagonal (i.e. to neglecting its off-diagonal elements).

3.1 Smoothing Single-Scale Data

In this section we present examples in which we use our models to smooth processes embedded in noise. These examples not only demonstrate the utility of our framework, but by smoothing processes outside of our class of processes we provide an indication of how rich our models are in approximating other processes. The type of process we consider is the first-order Gauss-Markov process.Since our data is generated from processes other than those described by our model, the use of smoothers based on our multiscale models results in *suboptimal* estimates.

Figures 2 and 3 we compare the performance of the optimal estimator, the one that minimizes the mean-square error, with the performance of our suboptimal estimator based on a lattice model using a 2-tap Daubechies filter. In these examples, the measurement noise variance $R = .5$; i.e., the data is of $SNR = 1.4142$.

3.2 Sensor Fusion

This section provides examples that show how easily our framework handles the problem of fusing multiscale data to form optimal smoothed estimates. In our framework not only is there no added algorithmic complexity to the addition of multiscale measurements, but it is also easy for us to *evaluate the performance* of our smoothers in using multiscale data.

We focus on the problem of using multiscale data to estimate a process in which the data at each of two scales is full in coverage but may vary in SNR from

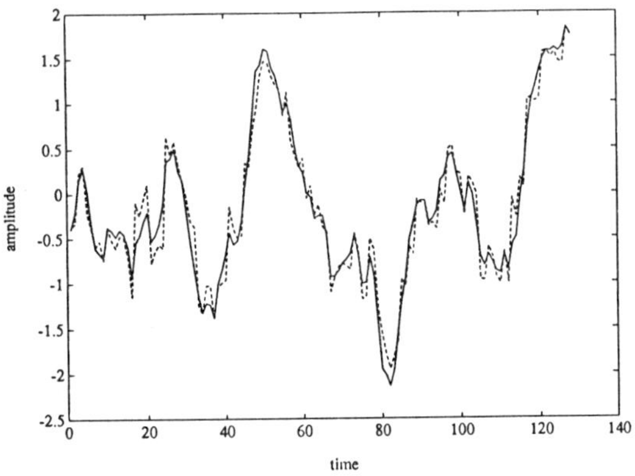

Figure 3: Standard Minimum Mean-Square Error Smoother (solid) versus Multiscale Smoother Using 2-Tap Filter (dashed) (data of SNR=1.4142)

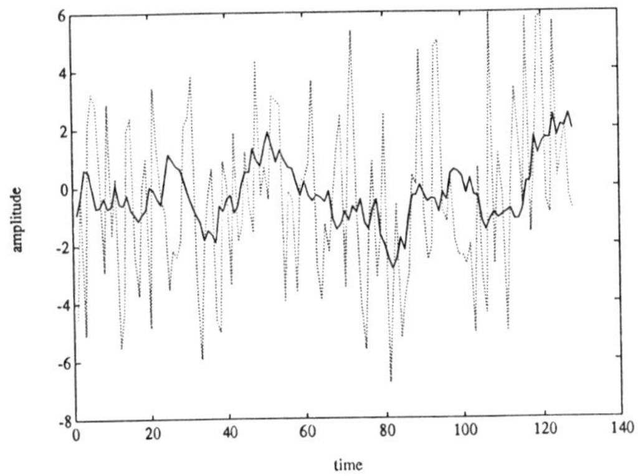

Figure 4: Sample Path of Stationary Gauss-Markov Process (solid), Poor-Quality Data with SNR=.3536 (dotted)

scale to scale. Consider the Gauss-Markov process used in our previous examples as defined in eq.(23). We take our coarse measurements to be the coarsened version of our fine scale signal. In particular, the wavelet-coarsening operator is used to project the fine-scale noiseless signal down to the desired scale, then white noise is added (see [4] for details). The choice of wavelet filter in the definition of the operator, H_i, reflects the way one chooses to model the relationship between the coarse and fine sensors.

Consider the case in which our fine-scale measurements are of extremely poor quality; e.g., our data is of SNR = .3536 (the noise power is eight times the signal power) as illustrated in Figure 4. Figure 5 compares the result of using the standard smoother on this data with the result of using a 4-tap lattice model smoother on the same data. The performance of the two smoothers is comparable, as we would expect from our results in the previous section.

Now we use the same data and consider fusing a higher quality coarse-scale data set to form a

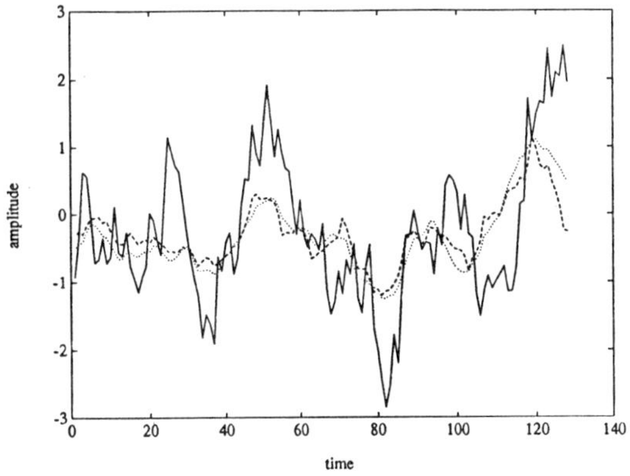

Figure 5: Sample Path of Stationary Gauss-Markov Process (solid), Result of Standard Smoother on Poor Data of SNR=.3536 (dotted), Result of 4-Tap Lattice Smoother on Same Data (dashed)

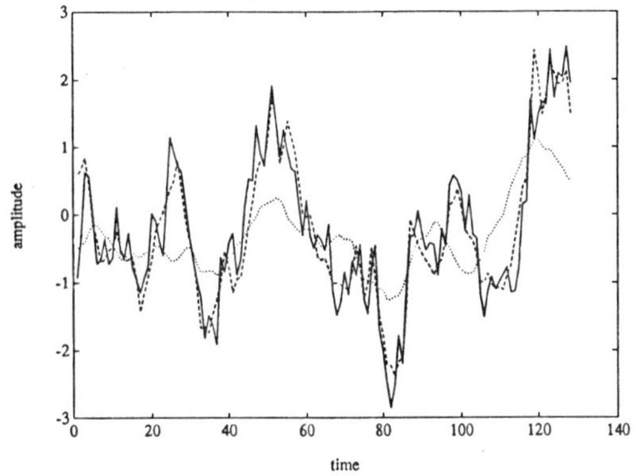

Figure 6: Sample Path of Stationary Gauss-Markov Process (solid), Result of Standard Smoother on Fine Data of SNR = .3536 (dotted), Result of 4-tap Lattice Smoother on Same Data Supplemented with Coarse Data of SNR = 2 (dashed)

smoothed estimate. In particular, we take our coarse data to reside at the scale one level coarser than the original data (scale at which there are 64 points) and the coarsening operator, H_i, corresponds to a 4-Tap filter. The SNR of this coarse data is equal to 2. Figure 6 compares the result of using the standard smoother on the low-quality fine-scale data with the result of using our 4-tap lattice smoother to fuse this low-quality data with high-quality coarse data. Note that the coarse measurement aids dramatically in improving the quality of the estimate over the use of just fine-scale data alone. To quantify this, recall that our smoother computes the smoothing error at each scale. We use these errors as *approximations* to the actual suboptimal errors (note that the computation of the actual error covariance from multiscale data is appreciably more complex than for the case of single scale measurements). The variance reduction[7] in the case of fusing the two measurement sets is 97 percent, versus only 36 percent for the case of using only the poor-quality fine-scale data.

4 Acknowledgments

This work was performed at the Massachusetts Institute of Technology and was supported by NSF Grants MIP-9015281 and INT-9002393, Air Force Office of Scientific Research Grant AFOSR-88-0032, and the Office of Naval Research Grant N00014-91-J-1004.

References

[1] G. Beylkin, R. Coifman, and V. Rokhlin, "Fast Wavelet Transforms and Numerical Algorithms I," to appear in *Comm. Pure and Appl. Math.*

[2] W. Briggs, *A Multigrid Tutorial*, SIAM, Philadelphia, PA, 1987.

[3] P. Burt and E. Adelson, "The Laplacian pyramid as a compact image code," *IEEE Trans. Comm.*, vol. 31, pp. 482–540, 1983.

[4] K. Chou, *A stochastic modeling approach to multiscale signal processing*, PhD thesis, M.I.T., May 1991.

[5] I. Daubechies,"Orthonormal bases of compactly supported wavelets," Comm. on Pure and Applied Math. 91, 1988, pp. 909-996.

[6] P. Flandrin, "On the Spectrum of Fractional Brownian Motions," *IEEE Transactions on Information Theory*, Vol. 35, 1989, pp. 197-199.

[7] S. Golden, *Identifying Multiscale Statistical Models Using the Wavelet Transform*, S.M. Thesis, M.I.T. Dept. of EECS, May 1991.

[8] S.G. Mallat, "A Theory for Multiresolution Signal Decomposition: The Wavelet Representation," *IEEE Transactions on Pattern Anal. and Mach. Intel.*, Vol. PAMI-11, July 1989, pp. 674-693.

[9] S. McCormick, *Multigrid Methods*, Vol. 3 of the SIAM Frontiers Series, SIAM, Philadelphia, 1987.

[10] H. E. Rauch, F. Tung, and C. T. Striebel, "Maximum Likelihood Estimates of Linear Dynamic Systems," AIAA Journal, Vol. 3, No. 8, Aug. 1965, pp. 1445-1450.

[11] M.J. Smith and T.P. Barnwell, "Exact reconstruction techniques for tree-structured subband coders," IEEE Trans. on ASSP 34, 1986, pp. 434-441.

[12] A.H. Tewfik and M. Kim, "Correlation Structure of the Discrete Wavelet Coefficients of Fractional Brownian Motions," submitted to *IEEE Transactions on Information Theory.*

[13] G.W. Wornell, "A Karhunen-Loeve-Like Expansion for 1/f Processes via Wavelets," *IEEE Transactions on Information Theory*, Vol. 36, No. 9, July 1990, pp. 859-861.

[14] G.W. Wornell and A.V. Oppenheim, "Estimation of Fractal Signals from Noisy Measurements Using Wavelets," submitted to *IEEE Transactions on ASSP.*

---

[7]

$$\text{var. red.} \triangleq \frac{p_0 - p_s}{p_0} \quad (29)$$

$$p_0 = \text{average process variance} \quad (30)$$

$$p_s = \text{average smoothing error variance} \quad (31)$$

CCD Correlators for Auditory Models

Richard F. Lyon

Apple Computer, Inc.
Cupertino, CA 95014
and
California Institute of Technology
Pasadena, CA 91125

Abstract

Surface-channel charge-coupled devices (CCDs) provide a mechanism for analog signal delay that can be built using an ordinary double-poly CMOS digital process, such as offered by Orbit through MOSIS. We have applied this technique to implement the correlation processing needed in auditory models of the sort proposed by Licklider for monaural pitch perception and sound separation. The resulting chips take analog audio in and produce analog video out (moving pictures of the correlogram representation of the sound) in real time. These chips present a variety of interesting analog and digital design challenges, which are addressed in this paper. Our first working experimental chips show real-time correlograms with 84 cochlea taps and 70 CCD delay stages per tap. There remains at least one unanticipated and as yet not understood problem with these circuits, resulting in a serious skew in the displayed correlation levels. Several circuit improvements are planned for the next generation of experimental chips.

1 Introduction

Modeling the cochlea has become a popular approach to extending the robustness of speech recognition systems. The cochleagram is a perceptually more relevant version of the commonly used spectrographic or short-time power spectrum family of sound representations. But when smoothed to a time scale comparable with the other representations, the cochleagram does not provide any real new richness of representation, or any particular leverage on the problem of separating out the desired speech from noisy or mixed sound environments.

Before smoothing, the cochlea output has more information, in the form of synchrony to sound. But the information is difficult to use directly, as it is still time-domain Models that use temporal processing of cochlea outputs to arrive at a slowly changing multidimensional representation offer new possibilities for interpreting real-world sound mixtures.

Two such models, for binaural and monaural processing, were proposed long ago, based on theories of how human and animal auditory systems achieve their remarkable levels of performance. The binaural cross-correlation model of Jeffress [1] and the monaural auto-correlation model of Licklider [2] have enjoyed a resurgence of interest in recent years as the high computational requirements of these models have come within reach of experimenters. Patterson's "stabilized auditory image" approach [3] is a more recent example of temporal processing, which may also be described in terms of correlation, and produces results much like Licklider's.

After several years of experimentation with these models using digital computers [4, 5, 6, 7, 8], and after considering the possibilities of custom digital VLSI implementations, we decided instead to pursue analog VLSI as an exciting technology for building low-cost and low-power auditory models. After some work on analog cochleas [9, 10], we spent considerable effort on the problem of analog correlation, eventually focussing on CCDs to implement the required delays [10, 11]. After about a dozen MOSIS test chips, we arrived at the completely working combined cochlea and auto-correlator chip decribed below. The correlation operation dominates the computation and the silicon area (about 80% of the chip), and is the main topic of interest here.

The moving-image output representation of the monaural cochlea and auto-correlation array, called a correlogram, has been described previously [6, 7, 8]. The chips discussed here display the correlogram on a standard TV screen in real time; analog audio input is converted to analog video output using all analog processing, in conjunction with digital timing and addressing.

2 Cochlea Circuits

A progression of variations and improvements to our cochlea circuits has been described before [10], and is continuing. For the chips described here, we selected the third-order "DIF3" stage of Figure 1, in which the first transconductance amplifier is built according to the circuit of Figure 2. The second and third amplifiers of each stage have a second pair of diode-connnected transistors under each differential pair, further expanding their range of input linearity, to make the large-signal behavior of the stage compressive, rather than expansive [10]. Due to the more limited linear voltage range of the first amplifier, the output tends to compress toward about 250 mV peak-to-peak (P-P).

1058-6393/91 $01.00

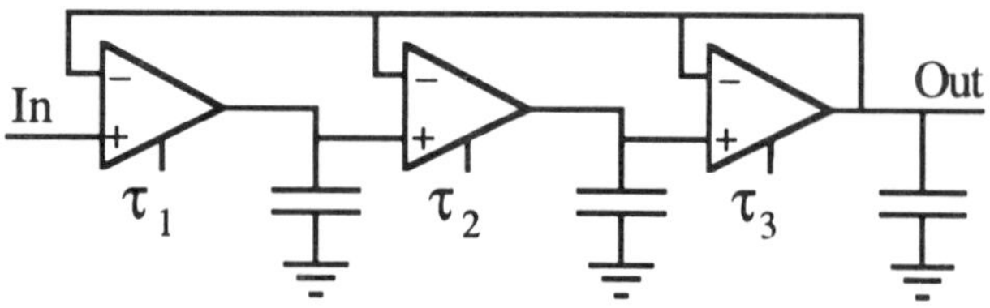

Figure 1: The circuit for the "DIF3" third-order filter, which corresponds well with our model of short-wave hydrodynamics. Three transconductance amplifiers are biased by the time-constant control voltages to operate in the MOS subthreshold (weak-inversion) region. A cascade of these filters constitutes a model of wave propagation in the cochlea.

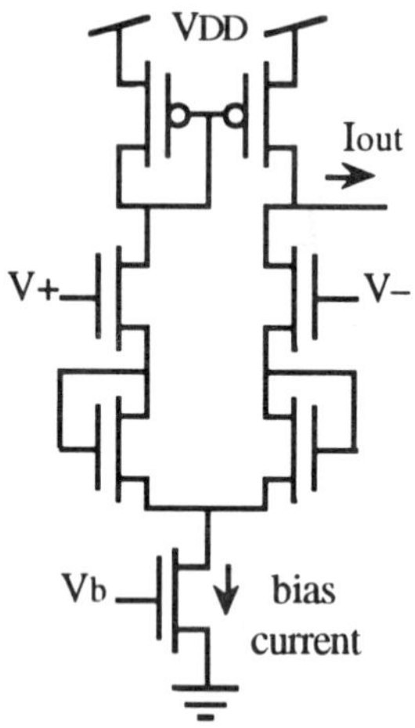

Figure 2: The circuit for the transconductance amplifier using diode-connected transistors to extend the linear input range and reduce the transconductance relative to the bias current.

A cochlea with 168 stages fits along one edge of the correlogram chip. For each pair of stages, a spatial differentiator ("spacediff") circuit is used to provide a bandpass-like signal to each row of the 84-row correlator array. The nonlinear spacediff circuit, shown in Figure 3, is analogous to Mead's hysteretic differentiator ("hysdiff") [12], providing high gain and an output that compresses to about 1V P-P. By using a spatial difference instead of a time difference, the time constant of the differentiator scales naturally with the time constant of the cascaded filter stages (a geometric progression of time constants is controlled by a spatial voltage gradient that sets the subthreshold bias currents in the filter stages).

3 CCD Delay Lines

Jeffress and Licklider discussed the use of neural axonal propagation as a possible delay mechanism for implementing delay-dependent correlation operations [1, 2]. This continuous-time discrete-level mechanism works well when there are enough parallel channels (nerve fibers) to form statistical averages over many fibers, and thereby achieve a nearly continuous-level representation for each frequency region. In VLSI, it is easier to build a continuous-level discrete-time delay mechanism, using clocked CCDs. Compared to continuous-time analog delay mechanisms, the CCD delay line has excellent delay-bandwidth product.

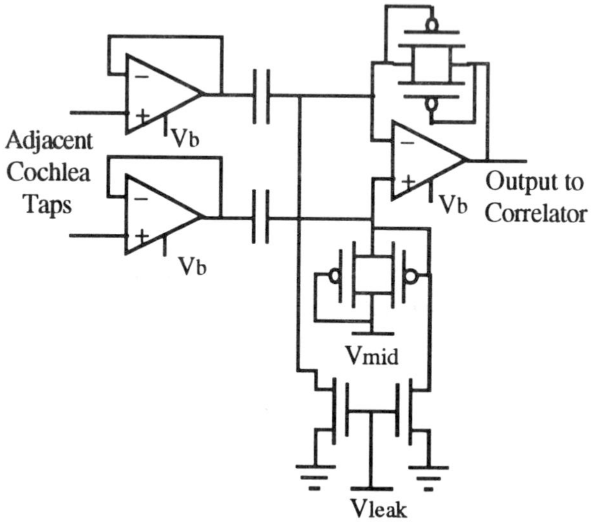

Figure 3: The "spacediff" circuit, a nonlinear spatial differentiator that provides high differential gain to small signals and compresses to about 1V P-P about a reference level V_{mid}. The first two amplifiers are used as followers, so that the cochlea does not see a large nonlinear load.

A CCD is basically just an extended transistor with a compound gate, with different gate segments driven by different clock phases [13]. There are a number of different clocking strategies, including two-phase and three-phase variations, but we chose a simple four-phase scheme with alternating large first-poly storage gates and small second-poly transfer gates, as shown in Figure 4. Charge packets of varying sizes are shifted through the CCD by proper phasing of storage and transfer clocks. For an N-type CCD, the charge carriers being pumped through are negatively charged electrons, so the input (source) current flows outward, as shown. At the far end, the drain terminal is connected to positive V_{DD} as an electron sink.

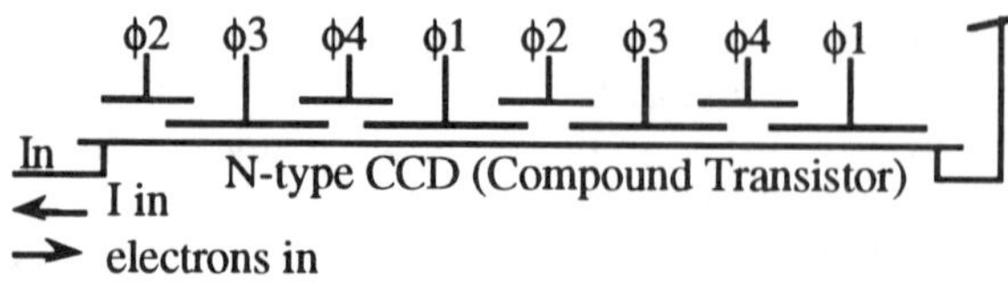

Figure 4: Schematic representation of a four-phase CCD delay line of two stages, showing the relation of poly1 gates for the storage phases (ϕ1 & ϕ3) and poly2 gates for the transfer phases (ϕ2 & ϕ4). The phase numbering we use is arbitrary and historical.

High-quality CCDs are generally built by adding a special implant to a CMOS or NMOS process, so that the charge is held in a "buried channel" below the imperfect $Si-SiO_2$ surface interface. The resulting devices are depletion mode, meaning they need clock voltages that are quite negative relative to the normal enhancement-mode N-type transistors. Special off-chip clock drivers and a special process are neeeded to take advantage of this high-quality approach. We elected instead to employ surface-channel CCDs, which can be built as extended enhancement-mode transistors in any ordinary MOS process that has two levels of poly gates. At the low clock rates needed for audio processing, the charge transfer inefficiency caused by slow trap states at the surface does not seem to be a problem. All the required clock phases are easily generated on-chip, and are driven across the array using limited-speed drivers to prevent charge dispersal.

4 Correlation Detectors

Given a CCD analog delay line, there are still several steps needed to make an auto-correlator. The charge packet must be non-destructively sensed at each delay stage, the signal represented by the delayed charge packet must be multiplied by the undelayed signal, and the instantaneous product must be integrated over a short time interval. All of this must be done with very few transistors, since the core cell is replicated for each pixel of the resulting correlogram output.

Nondestructive charge sensing is achieved using a floating-gate technique. The poly1 $\phi3$ storage gate is used as a floating "sense node"; it is not driven directly, but rather by capacitive coupling from an overlying poly2 electrode (see Figure 5). The voltage to which the poly1 sense node rises when $\phi3$ is driven high on the poly2 electrode depends on how much charge is stored in the channel under the sense node. The voltage on the sense node when the entire charge packet is under it provides one input to the correlation multiplier. To determine the absolute level and timing of this signal, two extra clock phases are used: $\phi5$ resets the floating node to ground during the time that only $\phi1$ is high, and $\phi6$ defines the interval during which we wish to sense the charge, when only $\phi3$ is high. The six clock phases are shown in Figure 6.

Since we view the output of the cochlea as representing something like a rate of neurotransmitter release, or a probability of neuron firing, we do not need to represent negative values; therefore, the multiplication needed is a one-quadrant operation. The simplest one-quadrant multiplier is a single MOS transistor for which the gate and source voltages are arranged to be logarithmically related to the signals to be multiplied. That is, since the MOS transistor in subthreshold and in saturation has a drain current that is exponentially related to both the gate and source voltages, the drain current may be interpreted as the product of the signals whose log voltages are present at gate and source. But the log nonlinearity does not need to be exact to get a reasonable correlation operation. If it is omitted entirely, with proper signal polarities the transistor's drain current will be the product of expontials of the inputs. For inputs swinging a few hundred millivolts, this nonlinear correlation approaches the digital AND operation, or a binary correlation.

In the correlator cell, the floating sense node cannot sink the multiplier's source-terminal current, so it must connect to the (zero-current) gate terminal. The source terminals of all the multiplier transistors in a correlator row are then tied together and driven by the undelayed signal.

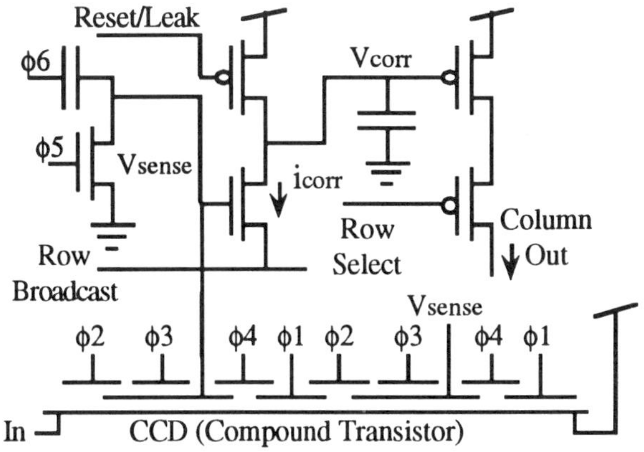

Figure 5: The circuit for two stages of four-phase CCD delay line showing one correlation/integration/readout cell (there is actually one per stage). The $\phi3$ clock couples capacitively to the poly1 floating gate that non-destructively senses the charge in the CCD channel. The correlation current being integrated on the 1 pF storage capacitor is small except during the sense interval when $\phi6$ and $\phi3$ are both high and the other clock phases are low—then it can be as high as 1 nA. If the broadcast and V_{sense} voltages are appropriate logarithms of the input, the current i_{corr} can be an accurate one-quadrant product.

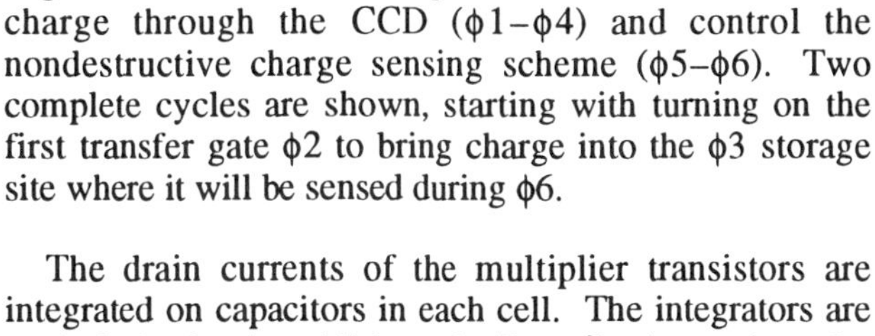

Figure 6: The six clock phase waveforms that shift charge through the CCD ($\phi1$–$\phi4$) and control the nondestructive charge sensing scheme ($\phi5$–$\phi6$). Two complete cycles are shown, starting with turning on the first transfer gate $\phi2$ to bring charge into the $\phi3$ storage site where it will be sensed during $\phi6$.

The drain currents of the multiplier transistors are integrated on capacitors in each cell. The integrators are made leaky by the addition of a Reset/Leak transistor for each cell. These transistors can be driven digitally to reset the integration, or as we use them, can provide a small leakage current that results in a nonlinear lowpass filter.

Two more transistors are used to sense the voltage on the integration capacitor and provide an output current on a column line when the row is selected.

To debug the concepts and tune up the cell design, we built a series of test chips using the cell of figure 5 and variations. For CCD input voltages from near ground (full charge packet) to about 1.5 V (empty charge packet), the sense voltage follows the delayed input signal with a gain of about one-third, which is adequate to provide plenty of gate-voltage variation.

5 Cochlea to CCD Interfaces

Both the charge input terminals of the CCDs and the undelayed "Row Broadcast" signals need to be driven via appropriate interface circuits to represent the approximate logs of the spacediff ouputs from the cochlea. The circuit approaches that we have chosen are based more on getting levels that will work than on getting the nonlinearity right; for loud sounds, the 1 V P-P log-compressed output of the spatial differentiators overdrives the correlator array into a nearly digital mode. Combining this large-signal limiting behavior with a small-signal exponential behavior results in a sigmoidal "squashing" function, which at least potentially resembles the response of an auditory neuron.

Notice that the gate and source voltages at the multiplier need to be of opposite signal polarities; a high correlation current corresponds to a high gate voltage and a low source voltage. Therefore an inverting voltage stage is needed in one interface circuit. To get the operating point of the broadcast source voltage into a reasonable range, we decided to regulate the total correlation current while simply AC coupling the spacediff output onto the broadcast line. The total current is regulated through a series connection of P-type and N-type transistors, so that the low-level (expansive) and high-level (compressive) currents are both adjustable (see Figure 7); the DC operating voltage adjusts itself accordingly.

After experimenting with various inverting and level-shifting voltage input circuits for the CCD charge inputs, we decided to try a current-input technique. Using a current source as the input directly sets the charge packet sizes, settling to whatever voltage works. A source offset voltage (V4) on a current mirror scales the CCD input current relative to the current in an AC-coupled nonlinear voltage-to-current converter, or "current inverter." The complete interface circuits are shown in Figure 7. Note that for the CCD charge input, a full charge packet (low voltage) represents a zero signal, and an empty charge packet represents a maximum signal.

We intend to do further experiments using lower-gain interface circuits: a simple current-output differential pair instead of a spacediff, explicit log compression using diode-connected transistors for at least the broadcast signals, and a lower-gain charge input to get a smoother analog result.

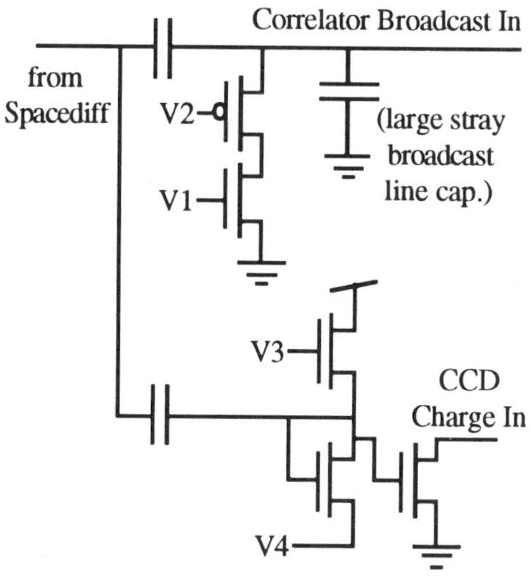

Figure 7: Circuits to interface a cochlea output to a row of the CCD correlator array.

6 Digital and Video Output Circuits

The correlogram chip produces real-time NTSC/RS-170 video output in cooperation with an external digital chip that generates NTSC timing (sync and blank) signals and pixel addresses; the digital chip is an Actel FPGA designed for us by Tanner Research. The correlogram chip has row and column address decoders along two edges to select the pixel of interest at any given time, and a sense amplifier and video output buffer amplifier to drive the pixel's analog correlation signal off chip. External bipolar transistors and a few resistors combine the video with the sync and blank signals to produce the final RS-170 compatible video.

The video timing and pixel rate are controlled by an external crystal oscillator. An independent digital clock is divided by 12 on chip to produce the 6 clock phases needed by the CCD correlators. Due to capacitive coupling in the correlator array, the CCD clocks tend to produce diagonal artifacts in the video; these artifacts are visible but not particularly noticeable, except when the CCD clock frequency is being adjusted, in which case the rapidly changing slopes of the artifacts draw attention to them.

7 Performance Results

We have arrayed these circuits to produce an 84-channel by 70-lag correlator array, which fits with a cochlea in the MOSIS 6.8 by 6.9 mm die size using 2-micron rules. The large number of free parameters (bias voltage "knobs") makes it a challenge to adjust the chip into a reasonable operating range. We have plans to incorporate circuits with fewer degrees of freedom and some self-adjusting biases in the next version to make the chip easier to set up and more robust. In particular, each cell will have a self-adaptive reset/leak current.

Our current chips produce familiar looking correlograms when the input sound is at a reasonable level. Figure 8 shows a still frame captured from a typical speech sound. Having better compression in the cochlea by integrating the AGC ideas that we have talked about [9, 10] would help it to deal better with loud and soft sounds.

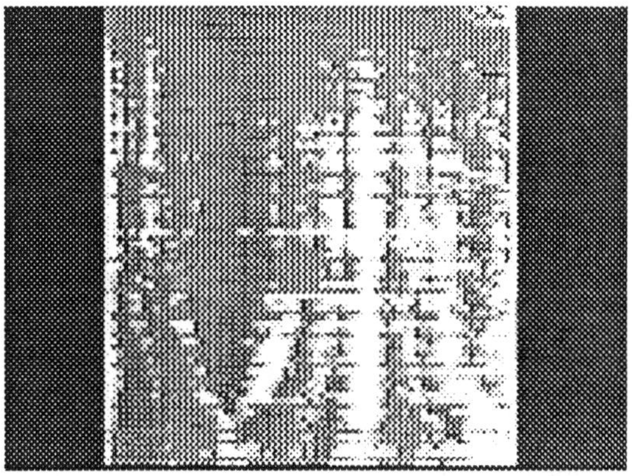

Figure 8: One frame of NTSC video captured from the correlogram chip output.

The current most interesting anomalous behavior is a severe tendency for the correlation currents to be higher toward the long-delay end of the array (toward the right in Figure 8) than they are toward the short-delay end, which makes it hard to get good correlogram images at low sound levels. This skew is consistent with the idea that charges in the CCD are somehow escaping. This idea is the opposite of the more common problem that a CCD accumulates a "dark current" or a "photo current" of additional charges via minority carriers swept up from the bulk of the silicon under the channel.

Our CCD is built using N-type channels in a P-Well CMOS process, because we figured that putting it in an isolated well would give it an environment free of minority carriers injected from the peripheral circuits or from stray light or thermally generated carrier pairs. For a CCD in a well, it may be that some charges that wander off from overfilled storage sites get collected by the well-substrate junction, rather than wandering back as they would in a CCD in substrate. Until we study it more, this is just a speculation. So far we have not been able to avoid the problem by being careful about not using full charge packets or by reducing clock edge rates.

8 Summary and Conclusions

The analog approach to auditory model implementation supports the possibility of putting a cochlea and a correlator array on a single chip, and operating them at a power low enough to consider for a battery-powered portable device. Using CCDs for the correlation delay provides a continuous-level delay line with precise delay timing, excellent delay-bandwidth product, and adequate signal-to-noise ratio. However, it is still not clear that digital custom VLSI approaches are not also competitive for this particular application; the silicon area and power requirements will be higher with a digital approach, but the results can be made cleaner and more predictable. We will continue to pursue the analog path, while comparing both the quality and the implementation to digital alternatives.

Subsequent processing of correlograms to interpret them, to separate and classify sounds, etc., will be explored digitally initially, due to the flexibility of programmable supercomputers. But ultimately, as VLSI densities continue to increase, the in-place local computation approach of Mead's "Silicon Retina" and other "Neural Systems" [12] will probably be the most efficient way to implement the next level or two of processing.

References

[1] Jeffress, Lloyd A., "A Place Theory of Sound Localization," *J. Comp. Physiol. Psychol.*, **41**, pp. 35-39, 1948.

[2] Licklider, J.C.R., "A Duplex Theory of Pitch Perception," *Experientia* **7**: 128–133, 1951.

[3] Patterson, R.D. and Holdsworth, J., "A Functional Model of Neural Activity Patterns and Auditory Images," in Ainsworth, W.A. (ed.), *Advances in Speech, Hearing and Language Processing,* Vol. 3. JAI Press, London, in press.

[4] Lyon, R.F., "A Computational Model of Binaural Localization and Separation," *Proceedings IEEE International Conference on Acoustics, Speech, and Signal Processing*, Boston, April 1983.

[5] Lyon, R.F., "Computational Models of Neural Auditory Processing," *Proceedings IEEE International Conference on Acoustics, Speech, and Signal Processing*, San Diego, March, 1984.

[6] Slaney, M. and Lyon, R.F., "A Perceptual Pitch Detector," *Proceedings IEEE International Conference on Acoustics, Speech, and Signal Processing*, Albuquerque, April, 1990.

[7] Duda, R.O., Lyon, R.F., and Slaney, M., "Correlograms and the Separation of Sound," *24th Asilomar Conference on Signals, Systems and Computers*, IEEE, Maple Press, 1990.

[8] Slaney, M., and Lyon, R.F., "Apple Hearing Demo Reel," Apple Technical Report #25, Apple Computer, Inc. (report and video available from Apple corporate library), Cupertino, 1991.

[9] Lyon, R.F., and Mead, C., "An Analog Electronic Cochlea," *IEEE Trans. ASSP* **36**: 1119–1134, 1988.

[10] Lyon, R.F., "Analog Implementations of Auditory Models," *DARPA Workshop on Speech and Natural Language*, Morgan Kaufmann, San Mateo CA, 1991.

[11] Lyon, R.F., "Analog VLSI Hearing Systems," in Brodersen and Moscovitz (eds.), *VLSI Signal Processing III*, IEEE Press, 1988.

[12] Mead, C., *Analog VLSI and Neural Systems,* Addison-Wesley, Reading MA, 1989.

[13] Séquin, C.H., and Tompsett, M.F., *Charge Transfer Devices,* Academic Press, New York, 1975.

Biologically-based Auditory Signal Processing in Analog VLSI

John Lazzaro
UC Berkeley, EECS
Berkeley, CA, 94720

Abstract

This paper reviews many recent analog silicon implementations of computational models of neural auditory processing. These implementations are based on physiological and psychophysical knowledge about biological auditory systems.

1 Introduction

The questions "How do we hear?" and "How can we improve artificial hearing systems?" are rarely asked by the same researcher. Why is research that holistically addresses hearing systems so rare? In the vision community, a sizable group of researchers use both biological and artificial systems as motivations for their research. Biological motion control and robotic motion control are also considered in a unified framework by many researchers of movement.

The different computational environments of biological auditory systems and artificial hearing systems are a key reason for the separation between the disciplines. Artificial hearing systems are usually specified as sampled digital algorithms, and are usually implemented as programs on a small number of fast serial digital computers. The nature of the computing hardware has affected both the form and the content of algorithms. For example, both serial and shared-memory parallel digital computing architectures insure access to all memory locations in the system, and reward frequent accesses to the same memory location. Efficient algorithms in this architecture exploit these facts.

Likewise, the algorithms that are central to biological auditory processing also exploit the computational architecture of the brain. The cochlea, the sense organ of hearing, uses both electrical and micromechanical techniques to implement the first stage of auditory signal processing. Later processing stages exploit the computational properties of axons, dendrites, and synapses to efficiently perform non-linear time-domain processing of auditory signals. The algorithms are robust to the noisy nature of neural computation, and can be efficiently wired in two-dimensional planes and three-dimensional spaces.

Both artificial and biological auditory algorithms have evolved in harmony with their primary implementation medium. It is not surprising that implementing an algorithm from one system in the computational medium of the other system would be possible but often impractical. For example, a discrete Fourier transform (DFT) could be directly implemented with biological neurons, by forming bistable storage elements and logic gates from small groups of neurons, and then proceeding according to conventional logic design. A clever neural designer could use the three-dimensional wiring space of the brain to offset the slower response time of biological neurons, and leverage nonlinear properties of dendritic trees to perform fast reliable logic. But to our knowledge there are no DFT processors in the brain; the computational properties of the nervous system favor implementing algorithms radically different from those of the digital signal processing tradition.

Likewise, several researchers over the previous decade have implemented computational models of biological auditory processing, with the goal of incorporating these models into a speech recognition system [9] [1]. These projects have shown the promise of the biological approach, sometimes showing clear performance advantages over traditional methods [15]. However, these algorithms are very computationally costly if run on serial digital computers.

Analog integrated circuit technology is a computational medium that shares many of the strengths and weaknesses of neural circuitry [14]. Several research groups have combined analog integrated circuit technology and theories of biological audition, to create silicon microsystems that efficiently implement computational models of biological auditory processing. Some of these systems compute early representations of biological audition [10] [13] [3] [6] [7] [8] [12] [18]; other systems attempt to perform a particular task,

1058-6393/91 $01.00

such as auditory localization [2] [5] or pitch perception [4].

The original publications of many of these systems are in the literature of neural networks and circuit design. In this article, I review these systems from a signal processing perspective.

2 Peripheral Processing

Auditory signal processing begins in the cochlea, the sense organ of hearing. The cochlea acts as a transducer, converting acoustic energy into the first neural representation of audition. The auditory nerve carries this representation to the brainstem; the nerve contains approximately 50,000 fibers. [16] includes a review of the structure and function of the cochlea.

Sound processing begins in the outer and middle ear, where the resonant properties of the ear canal and outer ear gently shape the spectrum of the sound signal. The bones of the middle ear couple the acoustic signal into the cochlea; these bones also implement a gain control mechanism that attenuates high-intensity sounds. This mechanism is under neural control.

Both mechanical and electrical processing occur in the cochlea. Sound energy is coupled into a mechanical traveling-wave structure, the basilar membrane, that converts time-domain information into spatially encoded information by spreading out signals in space according to their time scale (or frequency). Over much of its length, the velocity of propagation along the basilar membrane decreases exponentially with distance. The structure also contains active electromechanical elements; outer hair cells have motile properties, acting to reduce the damping of the passive basilar membrane and thus allowing weaker signals to be heard. Axons from higher brain centers innervate the outer hair cells; these centers may dynamically vary the local damping of the cochlea, providing frequency-specific automatic gain control.

Electrical signal processing in the cochlea occurs during the sensing of the mechanical motion of the basilar membrane. Inner hair cells occur at regular intervals along the basilar membrane, and convert basilar-membrane vibration into a graded electrical signal. Inner hair cells half-wave rectify the mechanical signal, responding to motion in only one direction. Inner hair cells primarily respond to the velocity of basilar-membrane motion, implicitly computing the time derivative of basilar-membrane displacement. Inner hair cells also compress the mechanical signal nonlinearly, reducing a large range of input sound intensities to a manageable excursion of signal level.

Spiral-ganglion neurons connect to each inner hair cell, and produce fixed-width, fixed-height pulses in response to inner-hair-cell electrical activity. The synaptic connection between the inner hair cell and the spiral-ganglion neuron may implement a stage of automatic gain control, exploiting the dynamics of synaptic-transmitter release. Auditory-nerve fibers are axons from spiral-ganglion neurons; these fibers present a neural representation of audition to the brain.

3 Silicon Cochleas

The first integrated circuit model of peripheral processing focused exclusively on the mechanical processing of the cochlea [10] [11]. This circuit is a one-dimensional physical model of the traveling-wave structure formed by the basilar membrane; the circuit also models the effect of the outer haircells on basilar membrane motion. A cascade of second-order sections with exponentially scaled time constants implement this model of basilar membrane motion. An analog, continuous-time circuit implementation of the model computes the pressure at selected discrete points along the basilar membrane in real time.

From a signal processing perspective, the silicon cochlea is a bank of low-pass filters; the response of each filter features a resonant peak and a sharp cutoff. The filter input and outputs are continuous-time voltage signals. The resonant frequencies of the output taps are exponentially spaced over several orders of magnitude. The delay through the silicon cochleas increases exponentially down the structure; the cascade structure enforces a monotonic increase of filter delay.

This cochlea circuit has been used in a complete model of the auditory periphery [3]. In this chip, each silicon cochlea output connects to a circuit that models the signal processing that occurs during inner hair cell transduction. The output from this circuit is a unidirectional current; this output connects to a circuit that models spike generation of the spiral ganglion cells. This circuit produces fixed-width, fixed-height spikes; these spikes are the the final output of the chip.

From a signal processing perspective, the inner hair cell circuit performs several operations. The circuit acts as a time differentiator; this action combines with the cochlea's resonant lowpass response to produce an asymmetric bandpass response. In addition, the circuit acts as a half-wave rectifier; the unidirectional current output produces zero current during half the input waveform cycle. Finally, the circuit nonlinearly compresses the signal.

The spiral ganglion circuit integrates the input current signal until a threshold is reached. At this point a fixed-width, fixed-height spike is produced, and the integration variable is reset. Circuit noise adds a random offset to the integration process.

As a result, the position of each spike statistically encodes the shape of the input waveform. If many spiral ganglion circuits are fed the same input current signal, an average of the spiking outputs would reproduce the shape of the input signal. In addition, the mean rate of spikes over a period of time is a measure of the amplitude of the input current waveform.

Alternative silicon models of peripheral processing have been reported by several research groups [18] [8]. One published effort describes the implementation of a linear model of cochlear mechanics that extends easily to nonlinear modeling [18]. In this novel circuit, traveling waves propagate through a one-dimensional transmission line, implemented with a resistive network constrained by scaled impedances.

Two recent improvements to the inner hair cell and spiral ganglion cell circuits also have been reported. Both improvements aim to model the dynamic adaptation that occurs in the transduction process. One project is a detailed circuit model of current flow in an inner hair cell [8]; the other project is a simple model of dynamic adaptation during spike generation in spiral ganglion neurons [7].

4 Using the Models

Frequency-domain and time-domain issues are both important in systems that use a silicon model of the auditory periphery. Figure 1 shows a frequency-domain characterization of silicon auditory-nerve fiber. This data was taken by presenting sinusoidal signals at different frequencies and amplitudes as input, and noting the average firing rate from a spiking output [3]. Note that the filter bandwidth asymmetrically increases at higher amplitudes; the high-frequency filter cutoff remains fairly stationary, but the low-frequency filter cutoff shifts dramatically downward above 20 dB.

The wide bandwidth of silicon auditory nerve fibers at moderate intensities allows cycle-by-cycle timing information to be accurately preserved in the timing of individual spikes, as shown in Figure 2. This data was taken from a peripheral chip that modeled short-term adaptation [7]; this adaptation enhances changes

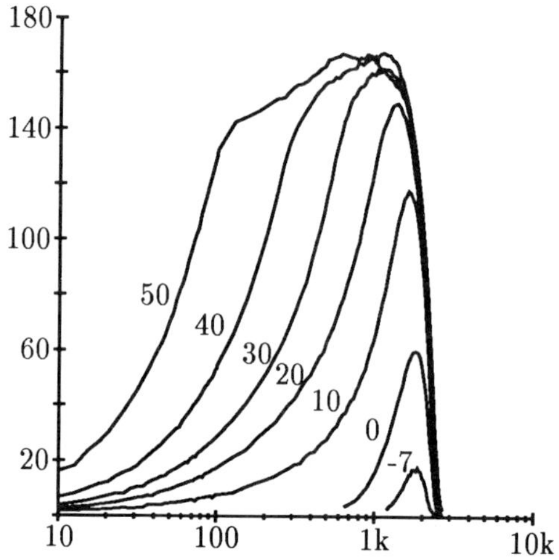

Figure 1: Frequency response of a silicon auditory nerve fiber. Y axis represents fiber activity (spikes/sec), X axis represents input frequency (Hz), labels denote input amplitude (dB).

in the envelope of the sound signal. These temporal properties are robust for high amplitude input signals.

The fundamental properties of silicon auditory nerve fibers, shown in Figures 1 and 2, are also seen in biological auditory nerve fibers. Unfortunately, these circuit models fall short of biological performance in several other functional aspects.

A major shortcoming of the silicon peripheral model is limited dynamic range; this nonlinear aspects of Figure 1 is a direct result of limited dynamic range. Biological auditory nerve fibers, however, show very similar nonlinearity; both silicon and biological fibers encode about 30dB of intensity with mean spike rate, although temporal encoding in both representations continues at high intensity. Why is the limited dynamic range of the silicon model a deficit?

In the biological system, the spiking spiral ganglion neuron is the source of the limited dynamic range; the mechanical motion of the basilar membrane has a much wider dynamic range. As a result, each inner hair cell connects to about ten spiral ganglion neurons. Although most of these ten neurons are sensitive at low sound intensities and saturate at moderate intensities, a few are insensitive at low intensities, and encode sound intensity through mean spike rate at medium sound intensities. In this way, the limited dynamic range of spiral ganglion neurons is overcome.

In the silicon system, however, the silicon cochlea circuit and the inner-hair-cell circuit are the sources of limited dynamic range; the silicon cochlea circuit saturates at moderate sound intensities, and the ca-

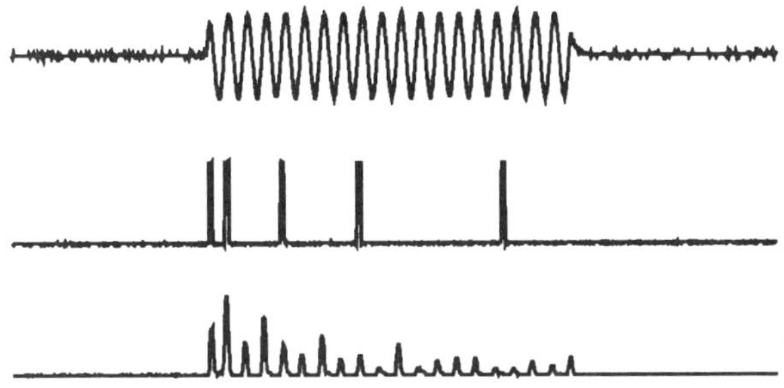

Figure 2: Temporal response of silicon auditory nerve fiber to a 1 kHz, 20 ms tone burst. Top trace shows tone burst input, middle trace shows a sample response from a fiber, bottom trace shows averaged output of 64 responses to tone bursts.

pability of inner-hair-cell circuit to sense small signals is also limited. These two constraints result in the 30dB dynamic range of the silicon periphery model. Expanding the dynamic range of these circuits is an important research issue.

Is the limited dynamic range of silicon cochleas a flaw that prevents their use as a component in larger auditory systems? Although the mean-rate encoding of sound intensity has a 30dB dynamic range, shown by the low-frequency cutoff in Figure 1, the high-frequency cutoff in Figure 1, as well as the temporal coding of Figure 2, are features with a dynamic range of 60dB. Biological models that primarily rely on these features of cochlear representation are well suited to analog VLSI implementation using present silicon cochleas.

Fortunately, many biological models implicitly assume a low dynamic range of mean-rate encoding in the entire auditory nerve. The importance of mean rate encoding in biological audition is a controversial issue in auditory neuroscience. Many theorists believe the temporal encoding shown in Figure 2 is of primary importance in auditory processing, assigning mean-rate encoding a secondary role. These theorists note that the mean firing rate of most auditory fibers are saturated for loud sounds, leaving a small number of fibers to encode the sound. This fact is counterintuitive with practical experience and psychophysical experiments: louder sounds are easier to locate and understand than softer sounds [1].

These theorists have developed models of auditory processing that do not require a large dynamic range of mean-rate encoding. These models have inspired most of the integrated circuits that use silicon cochleas as components. As the dynamic range of silicon cochleas improves, silicon models of auditory theories that require mean-rate encoding with a large dynamic range will also become practical.

5 Correlation Models

Temporal correlation of spike patterns from auditory nerve fibers is a common theme in models of higher function in biological audition. This processing method does not require a large dynamic range of sound encoding, and is realizable with biological neurons; implementation of these algorithms typically requires only time-delay elements and spike-coincidence circuits.

General correlation algorithms often expand the one-dimensional cochlear representation into a two-dimensional representation. In the new representation, one axis represents cochlear place, whereas the other axis represents correlation time delay. This two-dimensional representation directly encodes an aspect of the auditory signal, by extracting information from the cycle-by-cycle time structure of the cochlear representation.

Computing the running autocorrelation function for the cochlear representation, for many linearly-spaced correlation time delays, produces a two-dimensional correlation representation that encodes periodicity pitch information [9]. Computing the running cross-correlation function of the left and right cochlear representations, for many linearly-spaced correlation time delays, produces a two-dimensional correlation representation that encodes auditory localization information [16]. The cross-correlation of two cochlear representations, without any explicit time delays, also produces a two-dimensional representation that encodes auditory localization information [17]; this representation uses the implicit time delay of traveling-wave propagation on the basilar membrane of each cochlea.

These representations easily map to analog VLSI implementation, using coincidence circuits and monostable delay circuits to implement the correlation primitives, and inhibitory circuits to process the final representation. All three representations have been successfully implemented in analog VLSI. A silicon model of pitch perception [4] generates a two-dimensional running autocorrelation function, and estimates pitch from this representation using a simple algorithm. A silicon model of azimuthal auditory localization [2] generates the Jeffress two-dimensional cross-correlation function, and estimates interaural time delay from this representation using a simple algorithm. These single-chip systems each contain over 100,000 transistors.

These two systems estimate properties of auditory signals using simple algorithms on a single chip, and output only the estimate, not the two-dimensional correlation representation. In contrast, a different silicon model of binaural hearing [13] generates a two-dimensional cross-correlation function, using implicit cochlear delays, and outputs this two dimensional representation for display on a video monitor. Another silicon system generates and displays a two-dimensional auto-correlation function, using CCD technology [12].

Other auditory algorithms use temporal correlation to compute new one-dimensional representations of audition. Several algorithms use temporal correlation for spectral enhancement, transforming the nonlinear frequency representation of Figure 1 into an amplitude-independent spectral representation [1]. A spectral enhancement algorithm of this nature has been recently implemented in analog VLSI [6].

Acknowledgements

Thanks to R. Duda for an invitation to present this paper at Asilomar. Thanks to K. Johnson and J. Wawrzynek for financial support during the writing of this document. I also thank the Caltech auditory research community, specifically C. Mead, D. Lyon, M. Konishi, L. Watts, M. Godfrey, and X. Arreguit.

References

[1] S. Greenberg, "The ear as a speech analyzer," *J. Phonetics,* vol 16, pp. 139-149, 1988.

[2] J. P. Lazzaro, C. Mead, "Silicon models of auditory localization," *Neural Computation,* vol. 1, pp. 47–57, 1989.

[3] J. P. Lazzaro, C. Mead, "Circuit models of sensory transduction in the cochlea," in *Analog VLSI Implementations of Neural Networks,* Mead, Ismail, (eds.), Norwell, MA: Kluwer, pp. 85-101, 1989.

[4] J. P. Lazzaro, C. Mead, "Silicon models of pitch perception," *Proc. Natl. Acad. Sci. USA,* vol 86, pp. 9597–9601, 1989.

[5] J. Lazzaro, C. Mead, "Silicon models of auditory localization," in *An Introduction to Neural and Electronic Networks,* Zornetzer, Davis, Lau (eds.), New York: Academic Press, pp. 158–174, 1990.

[6] J. P. Lazzaro, " A silicon model of an auditory neural representation of spectral shape," *IEEE Journal Solid State Circuits,* 26: 772–777, 1991.

[7] J. P. Lazzaro, "Temporal adaptation in a silicon auditory nerve," in *Advances in Neural Information Processing Systems 4,* San Mateo, CA: Morgan Kaufmann Publishers, 1992.

[8] W. Liu, A. Andreou, M. Goldstein, "Analog VLSI implementation of an auditory periphery model," in *Proceedings 25th Annual Conference on Information Sciences and Systems,* Baltimore, MD, 1991.

[9] R. F. Lyon, "Computational models of neural auditory processing," in *IEEE ICASSP,* 1984.

[10] R. F. Lyon, C. Mead, "An analog electronic cochlea," *IEEE Trans. Acoust., Speech, Signal Processing,* vol. 36, pp. 1119–1134, 1988.

[11] R. F. Lyon, "Analog Implementations of Auditory Models," *DARPA Workshop on Speech and Natural Language,* Morgan Kaufmann Publishers, San Mateo CA, 1991.

[12] R. F. Lyon, "CCD correlators for auditory models," in *IEEE Asilomar Conference on Signals, Systems, and Computers,* 1991.

[13] C. A. Mead, X. Arreguit, J. P. Lazzaro, "Analog VLSI models of binaural hearing," *IEEE Journal of Neural Networks,* **2**: 230–236, 1991.

[14] C. Mead, *Analog VLSI and Neural Systems.* Reading, MA: Addison-Wesley, 1989.

[15] Y. Muthusamy, R. A. Cole, M. Slaney, "Speaker-independent vowel recognition:spectrograms versus cochleagrams," in *IEEE ICASSP*, pp. 533-536, 1990.

[16] J. O. Pickles, *An introduction to the physiology of hearing,* San Diego, CA: Academic Press, 1988.

[17] S. A. Shamma, N. Shen, P. Gopalaswarmy, " Stereausis: Binaural processing without neural delays," *J Acoust. Soc. Am,* **86:** 989–1006, 1989.

[18] L. Watts, R. Lyon, C. Mead, "A bidirectional analog VLSI cochlear model," *Advanced Research in VLSI, Proceedings of the 1991 Santa Cruz Conference,* C. Sequin (ed.), Cambridge, MA: MIT Press, pp. 153–163, 1991.

Feature-Map Methods for Extracting Sound Frequency Modulation

David K. Mellinger
Computer Science Department
Stanford University, Stanford, CA 94305
davem@cs.Stanford.EDU

Abstract

The role of frequency modulation (FM) in the auditory system is investigated. Psychoacoustic evidence shows that FM is a cue for non-contextual auditory scene analysis, and neurophysiological evidence shows that it is detected by neurons in some animal auditory systems. Some filters for detecting FM in sound are presented. One method uses correlation of a kernel with a time-frequency representation of cochlear neural impulses of a sound to reveal its FM components. Another correlation method is based on detecting FM in channel-by-channel autocorrelation images, with some data reduction and sharpening applied. The filters are tuned for various spatio-temporal frequencies to extract information about sound with various FM characteristics, producing maps showing time/frequency points at which the features are present. These maps could allow another process to decide what sources are present or to use FM cues for pattern recognition.

1 Introduction

Frequency modulation (FM) is change in frequency of a signal. In an audio signal, it can usually be heard as a change in the pitch or fundamental frequency of a sound. It is reflected spectrally by the motion in frequency of the partials or harmonics that make up a sound signal.

The present study of FM was motivated by the problem of auditory scene analysis, the process by which human and other biological sound-processing systems hear out one source of sound from the mixture of sounds that reach the ears. This separation process is used in the cocktail party effect, in which each of several simultaneous voices may be heard in turn. It is also used in musical listening to hear out individual instruments in an ensemble, as well as in "everyday listening" where one hears such noises as footsteps, cars passing by, and voices despite the interfering presence of many other sounds. The problem of auditory scene analysis is to identify which parts of the sound spectrum over time belong to which sources of sound in the listening environment. Bregman [1] describes many of the characteristics of scene analysis in human hearing. He identifies a number of different cues by which the auditory system is able to separate one sound source from another; one of the more salient of these is frequency modulation.

In addition to its place in scene analysis, FM is used by the auditory system in various other ways. It is used in speech, both in tonal languages to represent words and in other languages to convey other types of meaning such as emphasis. Many animal calls, especially bird calls, include one or more frequency glides used for identification and communication. Animals that use sonar, such as bats, employ FM for discovering the distance to a target.

This paper is an overview of the role of FM in the auditory system. The next section reviews some of the evidence for and characteristics of the auditory system's ability to detect FM. The following sections present some physiologically compatible methods for filtering FM components from a sound mixture.

2 FM in the auditory system

2.1 Psychoacoustic evidence

Chowning [3] noticed that a synthesized singing voice — a vowel sound — became more coherent when identical slight vibrato was added to the synthetic partials. This suggests that there is a mechanism which uses the vibrato as a feature for grouping.

McAdams studied this micromodulation extensively [8]. He found that constant frequency *ratios* between the harmonics of a fundamental frequency are necessary for source formation, while frequency *differences* cause the appearance of several pitched notes instead of fusion into one note.

McAdams also tested the importance of FM for hearing one synthesized voice in a mixture of several

1058-6393/91 $01.00 © 1991 IEEE

[9]. He found that the imposition of identical FM on all harmonics of a voice made the voice stand out significantly more prominently than when the harmonics had no FM or uncorrelated FM.

Rasch [12] played notes with and without 4% vibrato against a background of a masking tone. He found that tones could be decreased in level by 17.5 dB and still be equally audible against the background when vibrato was added. This amount of vibrato, typical of musical instruments, makes notes stand out even when they are much less intense than a masking tone.

Carlyon and Stubbs [2] tested the perception of a mistuned harmonic in a complex tone. Harmonic complexes were more easily heard than inharmonic ones against a noise background, and FM of a partial is also more easily heard in the harmonic case. Their conclusion could be seen as suggesting that detection of FM is solely dependent on harmonicity, but this would be a mistake: They claim only a role for harmonicity, but don't deny a role for FM.

Kay and Matthews studied adaptation to FM tones. They found that playing a modulated tone of one FM rate and depth caused subjects to become less sensitive to FM at the same rate but shallower depth. This suggests that there are adaptive units in the auditory system sensitive to a certain rate of FM. FM tones caused more adaptation than plain and amplitude-modulated tones.

One hypothesis about the salience of frequency-varying sound is that it is due entirely to the passage of a partial into new frequency channels as it moves. In other words, the auditory system has only frequency-spoecific onset detectors, not FM detectors. Gardner and Wilson [4] tested this hypothesis by playing a series of tones which swept downward in frequency until the subject had become adapted to them, followed by an upward-sweeping tone through the same frequency band. The adaptation-inducing down-sweeps did not affect the threshold for up-sweeps. This result suggests that there are mechanisms sensitive to the direction of frequency movement.

2.2 Neurophysiological evidence

Evidence for auditory neurons sensitive to FM comes from several sources. The earliest is probably Whitfield and Evans [16], who recorded units in the auditory cortex. The found many units more responsive to FM tones than to any steady tone, and some that responded only to FM. They also found that FM-responsive units tended to fire at a particular point in the modulation sinusoid — for example, when the tone was rising. Whitfield and Evans also found units responsive to wide frequency sweeps but not to steady-state tones or to narrow FM, and units more responsive to sweeps in one direction than the other. The latter echoes Gardner and Wilson's psychoacoustic study.

Møller extended the above study in a a series of experiments with sweep tones [11], recording responses in the cochlear nucleus. He swept tones at different rates and discovered that for some units, an "optimal rate of frequency change" exists in eliciting a neural response from a unit. These units show a steady increase in firing rate as the sweep rate is increased up to a point, followed by a steady decrease as it is further increased. In other words, units are tuned to detect an optimal FM rate.

Mendelson and Cynader [10] also found units in the cat auditory cortex responsive to the direction and rate of a frequency sweep. They suggest, based on responses of successive units in neural penetrations, that there may be a columnar organization to the auditory cortex — a map of FM sweep rate. They also report the result that, at least in the cat cortex, there are about twice as many units responsive to downward sweeps as to upward ones.

3 Detecting frequency modulation

The methods presented here are intended to be *physiologically compatible*: The computations and data representations needed could be performed in the auditory system, and the evidence above indicates that these methods might be functionally similar to ones used in animal hearing.

3.1 Feature maps

The data representation used in these filters is the *feature map*. A feature map is a rectangular array of up to (currently) three dimensions. The elements of this array are real values (floating-point numbers) that represent intensity of some feature. For example, one feature map for a bit of music is displayed in fig. 1. It is a two-dimensional map with log frequency on the vertical axis and time on the horizontal; darkness indicates intensity of neural firing at each time/frequency point. The sound is a mixture of two notes, a low one of four harmonics with slow vibrato, and a higher one of six harmonics with somewhat faster vibrato. Feature maps for various sound characteristics, including perhaps frequency modulation, have been found in animal auditory systems [10, 15, 13, 5].

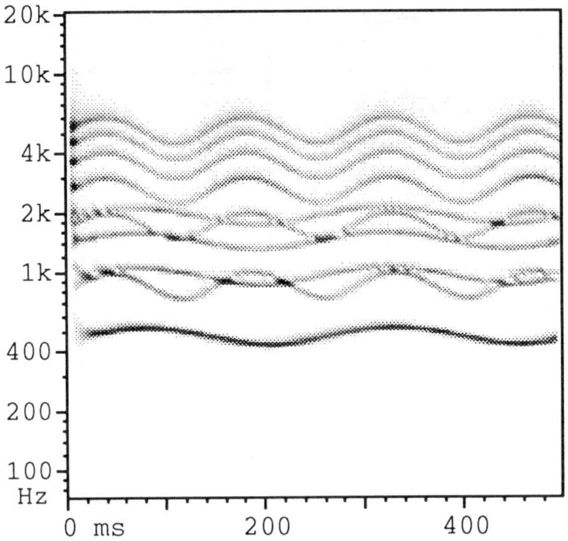

Fig. 1. A feature map for a segment of music.

Fig. 1 was produced by a computational model of the human cochlea [7, 14], which performs the filtering performed by the outer and middle ears, the per-frequency-channel transduction of sound into neural impulses performed by hair cells on the basilar membrane, and the lateral inhibition of early neural processing. Such an image is called a *cochleagram.*

4 FM in the cochleagram

The method described here detects FM in the cochleagram with a two-dimensional correlation operation. The kernel of this correlation operator is chosen to detect FM at a certain rate — for example, at two octaves/s. The kernel in fig. 2, when correlated with the cochleagram of fig. 1, responds most strongly when it is centered on a harmonic that is varying in frequency at the kernel's characteristic rate, as is seen in the result map of fig. 3.

Fig. 2. Correlation kernel. Time is the horizontal axis, height (log f) the vertical; dark is positive, white is negative.

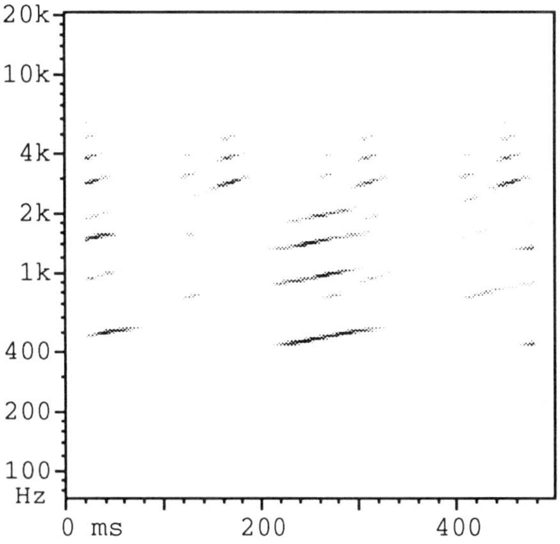

Fig. 3. An FM feature map showing harmonics rising in frequency.

Several factors combine to make this work:

Logarithmic scale. The use of *height*, or logarithm of frequency, in the cochleagram ensures that harmonics that are varying in frequency at a certain rate will have the same slope; this is appropriate because harmonics, when varying in frequency, maintain fixed ratios rather than fixed differences.

Using a log-frequency scale, the correlation is

$$F(h,t) = \int\int_{-\infty}^{\infty} k(h_1,t_1)c(h+h_1,t+t_1)\ dh_1\ dt_1,$$

where h is height (or log f), t is time, $k(h,t)$ is the two-dimensional kernel, $c(h,t)$ is the cochleagram, and $F(h,t)$ is the resulting frequency-modulation feature map.

Kernel shape. The kernel is the Cartesian product of two one-dimensional kernels. In the time (horizontal) dimension, it is Gaussian to localize its effect to a short time period. In the height (vertical) dimension, the kernel is center-surround with the form

$$(\sigma_1^2 - x^2)e^{\frac{-x^2}{2\sigma_2^2}}.$$

This function makes the kernel respond positively to FM only at its characteristic rate.

Zero-sum. The center-surround function f should sum to zero or a negative number: $\int_{-\infty}^{\infty} f(x)\ dx = 0$. If it sums to a positive number, then a crossing harmonic will elicit a positive response, causing smearing in the output image.

Kernel time spread. The spread parameter in the time direction determines how wide the Gaussian function is for the kernel. A short kernel can respond more quickly, while a long kernel is more accurate in detecting FM at only its characteristic rate. Another factor that determines kernel time spread is the response time: The longer a kernel is, the more time it takes to give its response after the sound signal has entered the auditory pathway.

Several lengths seem to work reasonably well; the length used here is 43 ms, which is near to the 33 ms auditory-neuron response time suggested by the work of Whitfield and Evans [16]. The auditory system likely uses several different lengths of time for FM detection.

Frequency Spread. The spread of the kernel's center-surround component in the frequency direction is chosen so that the positive region covers the width of a harmonic in the cochleagram. Sinusoidal components in the cochleagram cover about 4 cochlear channels ($1/e$ size), or 2.2 semitones, so the positive part of the center-surround function covers about 2 semitones.

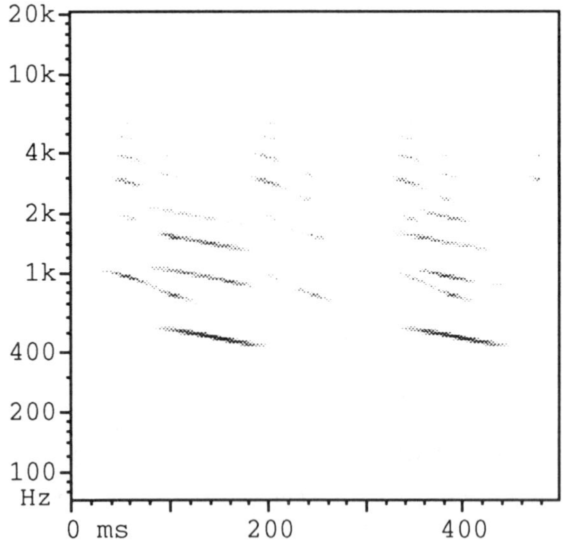

Fig. 4. An FM feature map with harmonics falling in frequency.

Range of Rates. The range of FM rates used here is from -5 to +5 octave/s, corresponding to the highest rates found in music and speech. A set of kernels is used, placed throughout this range at 1.25 octave/s spacing, in order to track FM at different rates. Each kernel filters out those parts of the signal that are varying at its characteristic rate, so that the entire set of kernels covers the range of possible rates.

Figs. 3 and 4 show maps made with these settings.

5 Frequency modulation in the correlogram

A second method for finding FM harmonics relies on the correlogram representation of sound [6]. This is the autocorrelation of each frequency channel of the cochleagram over time. As it introduces the lag, or delay, dimension to the representation, it is a three-dimensional feature map. Fig. 5 shows one instant of a correlogram – a two-dimensional slice.

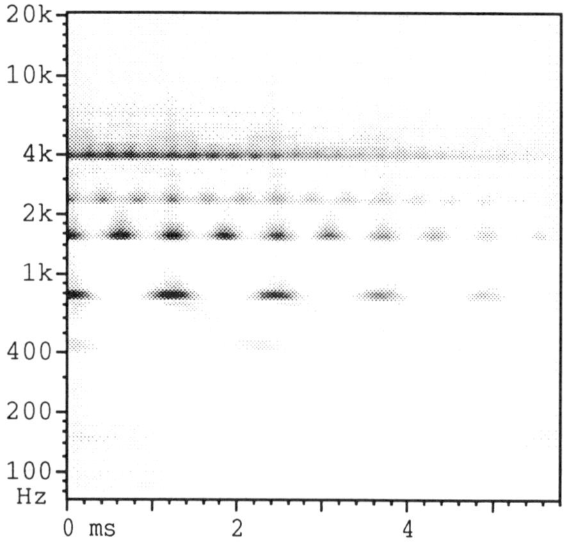

Fig. 5. Correlogram slice: x dimension is lag, y is height.

The technique uses a correlation-like operator on the correlogram as a way to extract frequency-modulated harmonics. FM of a harmonic shows up in the correlogram as motion of the correlation "spots," the regions of high positive response. A two-dimensional dot-product is computed at each position of the feature map to track this motion. This dot-product differs from correlation only in that the kernel used to compute this dot-product varies from place to place within the map. The kernel is multiplied by a two-dimensional slice through the three-dimensional correlogram, with the nature of the slice shown by fig. 6.

The dot-product operation in the correlogram needs many of the same tuning parameters used by the cochleagram. Again, the kernel is Gaussian in the time dimension to localize its effect, and center-surround in the other dimension, lag. (It can be a Gabor function here because correlogram spots repeat at the period of the waveform.) The kernel should be zero-sum and have a certain time-spread to give it a specific precision of detection. Unlike the kernel above, there is no other spread parameter; the width of the kernel in the lag dimension is determined by the spacing of spots in the correlogram. Kernels span the range ±5 octaves/s.

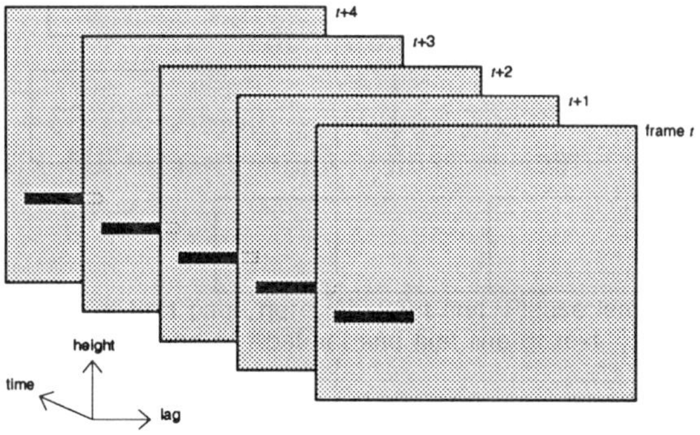

Fig. 6. Black areas show the slice used for one dot-product.

After a dot product is computed at each position of the correlogram, the resulting feature map is reduced to two dimensions (time × height) by summing across the lag dimension. Time-local lateral inhibition within each frequency channel is then performed to sharpen peaks in the image. The result of this processing on a sound with fairly slight vibrato is shown in fig. 7. In this image, the dark areas represent points at which the sound was varying at the characteristic rate of the kernel, 0.5 octave/s.

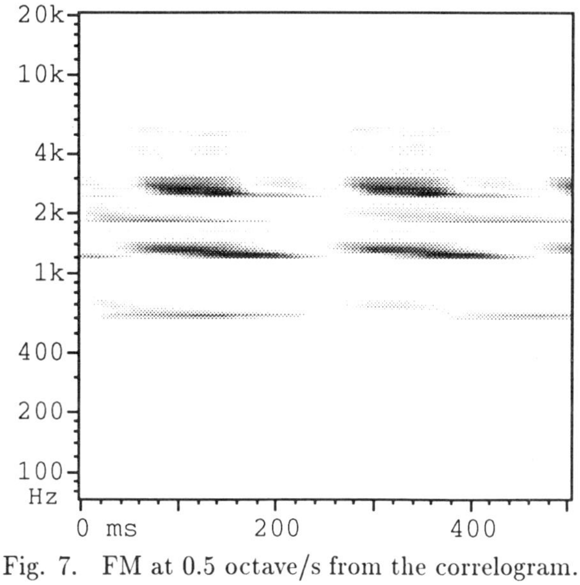

Fig. 7. FM at 0.5 octave/s from the correlogram.

Acknowledgements

I wish to thank the attendees of the CCRMA hearing seminar, for insight and discussion leading to this work, and the National Science Foundation, for partially supporting it with grant IRI-8613574.

References

[1] Albert S. Bregman. *Auditory Scene Analysis.* The MIT Press, Cambridge, MA, 1990.

[2] Robert P. Carlyon and Richard J. Stubbs. Detecting single-cycle frequency modulation imposed on sinusoidal, harmonic, and inharmonic carriers. *J. Acoust. Soc. Am.*, 85(6):2563–2574, 1989.

[3] John M. Chowning. Computer synthesis of the singing voice. In *Sound Generation in Winds, Strings, Computers*, pages 4–13. Royal Swedish Acad. Mus., Stockholm, 1980. Publ. No. 29.

[4] R. B. Gardner and J. P. Wilson. Evidence for direction-specific channels in the processing of frequency modulation. *J. Acoust. Soc. Am.*, 66(3):704–709, 1979.

[5] Eric I. Knudsen. The hearing of the barn owl. *Scientific American*, December 1981.

[6] J. C. R. Licklider. A duplex theory of pitch perception. *Experientia*, 7:128–133, 1951.

[7] Richard F. Lyon. A computational model of filtering, detection, and compression in the cochlea. In *Proc. IEEE Intl. Conf. Acoust., Speech, and Sig. Proc.*, May 1982.

[8] Stephen McAdams. *Spectral Fusion, Spectral Parsing, and the Formation of Auditory Images.* PhD thesis, Stanford University, May 1984.

[9] Stephen McAdams. Segregation of concurrent sounds I: Effects of frequency modulation coherence. *J. Acoust. Soc. Am.*, 86(6):2148–2159, 1989.

[10] J. R. Mendelson and M. S. Cynader. Sensitivity of cat auditory primary cortex (AI) neurons to the direction and rate of frequency modulation. *Brain Research*, 327:331–335, 1985.

[11] Aage R. Møller. Coding of time-varying sounds in the cochlear nucleus. *Audiology*, 17:446–468, 1977.

[12] R. A. Rasch. The perception of simultaneous notes such as in polyphonic music. *Acustica*, 40:21–33, 1978.

[13] Christoph E. Schreiner and Julie R. Mendelson. Functional topography of cat primary auditory cortex: Distribution of integrated excitation. *J. Neurophys.*, 64(5), 1990.

[14] Malcolm Slaney. Lyon's cochlear model. Technical Report 13, Apple Computer, 1988. Avail. from Apple Corp. Library, Cupertino, CA 95014.

[15] Nobuo Suga. Cortical computational maps for auditory imaging. *Neural Networks*, 3:3–21, 1990.

[16] I. C. Whitfield and E. F. Evans. Responses of auditory cortical neurons to stimuli of changing frequency. *J. Neurophys.*, 28:655–672, 1965.

The Challenge of Inverse-E: The RASTA-PLP Method

Hynek Hermansky*, Nelson Morgan**, Aruna Bayya*, Phil Kohn**

* U S WEST Advanced Technologies, 4001 Discovery Drive, Boulder, Colorado 80303
** International Computer Science Institute, 1947 Center Street, Berkeley, California 94704-1105

Abstract

Human perception is generally insensitive to steady-state stimuli, responding largely to relative values of the input stimulus. In spite of that, most speech parameter estimation techniques use absolute spectral values. Consequently, they fail when the spectral values are modified by the frequency response of the communication channel. In response to this challenge, we have developed an analysis technique that is more robust to steady-state factors in speech. The approach appears to work well with realistic channel conditions and is conceptually simple and computationally efficient. The new method is described, and experimental results are reported, showing significant improvements in error rate.

Introduction

In a meeting of researchers working on front ends for speech processing, Spoons [8], the following observation was made: when speech has been filtered by the inverse of the spectral envelope of a steady-state vowel in an utterance, (such as "/iy/"), the speech remains perfectly intelligible in spite of the fact that the short-term spectral envelope of the given vowel has become flat. Most speech parameter estimation techniques (including the Perceptual Linear Predictive (PLP) technique [4] that we have previously used) fail on such a task. Short-term spectral estimates are vulnerable to linear distortions of the signal such as described above. Human perception is generally insensitive to steady-state stimuli, responding largely to relative values of input. We have developed the RelAtive SpecTrAl (RASTA) methodology (first described in [5]) which makes PLP (and possibly also some other short-term spectrum based techniques) more robust to linear spectral distortions. We have also conducted several speech recognition experiments using telephone-quality isolated digits in one case and wideband continuous speech in the other. Experimental results for both cases show significant improvements in error rate.

Approach

To make speech analysis less sensitive to the steady-state factors in speech, we have replaced a conventional short-term absolute spectrum by a spectral estimate in which each frequency channel is band-pass filtered. Since any constant or slowly-varying component in each frequency channel is suppressed by this operation, the new spectral estimate is less sensitive to slow variations in the short-term spectrum. While the approach is simple and efficient, it appears to work well with realistic channel distortions. Furthermore, as described below, the approach provides an intermediate domain that may prove useful in further work.

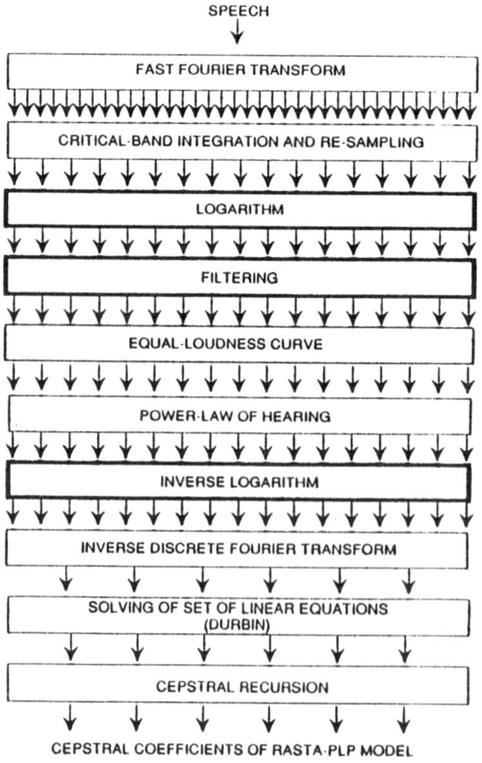

Fig. 1 RASTA-PLP method.
The RASTA processing is indicated by thicker lines.

1058-6393/91 $01.00 © 1991 IEEE

The steps of RASTA-PLP are as follows (see [4] for comparison to the conventional PLP method):

For each analysis frame,

1) Compute the critical-band spectrum (as in the PLP) and take its logarithm.

2) Estimate the temporal derivative of the log critical-band spectrum using regression [3].

3) Nonlinear processing (such as thresholding or median filtering) can be done in this domain. Currently, we do nothing here.

4) Re-integrate the log critical-band temporal derivative using a first order IIR system. The pole position of this system can be adjusted to set the effective window size. Currently, we set this value to 0.98 , providing an exponential integration window with a 3-dB point after 34 frames.

5) In accord with the conventional PLP, add the equal loudness curve and multiply by 0.33 to simulate the power law of hearing.

6) Take the inverse logarithm (exponential function) of this relative log spectrum, yielding a relative auditory spectrum.

6) Compute an all-pole model of this spectrum, following the conventional PLP technique.

The key idea here is to suppress constant factors in each spectral component of the short-term auditory-like spectrum prior to the estimation of the all-pole model.

It can be shown that if the derivative of step (2) is estimated by a simple first difference, and if the the full integration in step (4) is done (pole at $z = 1.0$), then all intermediate terms cancel and the technique is equivalent to subtraction of the log spectrum of the first analysis frame from each new frame. In this special case, the RASTA technique resembles the spectral subtraction or blind deconvolution techniques.

However, in the general case presented here, the whole derivative-reintegration process is equivalent to a bandpass filtering of each frequency channel through an IIR filter with the transfer function

$$H(z)=0.1*\frac{2+z^{-1}-z^{-3}-2z^{-4}}{z^{-4}(1-0.98z^{-1})}. \quad (1)$$

The low cut-off frequency of the filter determines the fastest spectral change of the log spectrum which is ignored in the output, while the high cut-off frequency determines the fastest spectral change which is preserved in the output parameters.

Linear distortions as caused e.g by the telecommunication channel or by using a different microphone should appear as an additive constant in the log spectrum. The high-pass portion of the equivalent band-pass filter is expected to alleviate the effect of the convolutional noise introduced in the channel. The low-pass filtering is expected to help in smoothing out some of fast frame-to-frame spectral changes present in the short-term spectral estimate due to analysis artifacts. In Eq. (1), the low cut-off frequency is 0.26 Hz. The filter slope declines 6dB/oct from 12.8 Hz with sharp zeros at 28.9 Hz and at π (50 Hz). There is no special reason (except historical) for using the particular filter of Eq. (1). Also, the same filter need not be used for all frequency channels. Further, the filtering does not have to be band-pass or even linear.

The result is generally dependent on the starting point of analysis. In our applications we always start analysis well in the silent part which precedes speech.

The whole RASTA-PLP process is illustrated in Fig.1.

The effectiveness of the RASTA processing in PLP analysis is illustrated in Figs. 2 and 3, which show the effect of the simple high-pass filtering (6db/oct preemphasis) of the input speech on the conventional PLP (Fig. 2) and the RASTA-PLP (Fig. 3) analyses. The conventional PLP is significantly affected by the high-pass filtering of speech. This is documented by the substantial cepstral distance between the PLP models estimated from the original and the filtered speech, shown at the bottom of the figure. The RASTA-PLP yields visually almost indistinguishable results for both the original and the filtered speech. As shown again at the bottom of the Fig. 3 the cepstral distance between the respective RASTA-PLP models is much smaller than in the conventional PLP case shown in the Fig. 2.

Isolated Digit Recognition Experiments

A database was derived from connected digits recorded over dialed-up telephone lines. 155 male and female speakers were used for the of the recognizer, data from additional 56 male and female speakers formed the test test. The data were recorded at the Bellcore facility in Morristown, NJ, and represented channel conditions in the

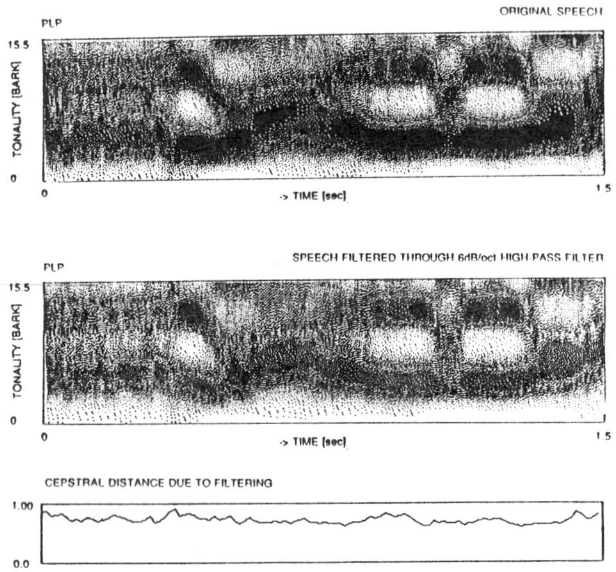

Fig. 2 Spectrograms showing spectra of the 5th order conventional PLP models estimated from the original (upper graph) and the 6db/oct high-pass filtered (lower graph) speech. The cepstral distances between the respective PLP models estimated from the original and the filtered speech shown at the bottom of the figure.

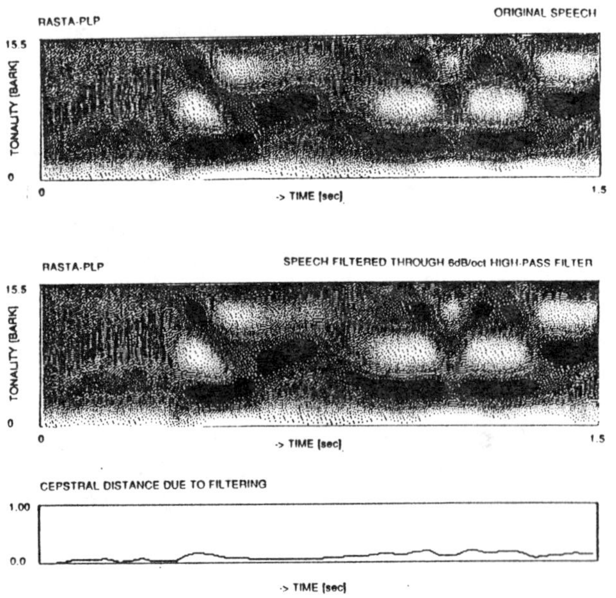

Fig. 3 Spectrograms showing spectra of the 5th order RASTA-PLP models estimated from the original (upper graph) and the 6db/oct high-pass filtered (lower graph) speech. The cepstral distances between the respective RASTA-PLP models estimated from the original and the filtered speech shown at the bottom of the figure.

New Jersey area. An isolated-utterance continuous-density HMM recognizer was used in the experiment. Additional details of the experiment are given in [6].

Three experiments were carried out. In all experiments, the system was trained on the training part of the Bellcore database.

In the first experiment, the test set was a subset of the Bellcore database. Thus, we assume that both the test set and the training set were recorded under similar channel conditions. The first column of Table I shows the percentage error rates on this test data. The RASTA-PLP performs about as well as the standard PLP technique.

In the second experiment, the Bellcore test data set was corrupted by a simulated convolutional noise (pre-emphasis by the first-order differentiation of the signal). The recognizer that had been trained on the uncorrupted data was still used. The recognition tests were run on this data set, using the models obtained from Bellcore data. The results are tabulated in the second column of Table I. The standard PLP technique yielded almost an order of magnitude higher error rate than the error rate on the original Bellcore data. The new approach can be seen to be far more robust to such simulated channel variation.

To extend the result to an experiment with realistic changes in channel conditions, digit strings spoken by four (2 male and 2 female) speakers were recorded over the local telephone lines in the U S WEST speech laboratory. The recognition results on this set are shown in the third column of Table I. As with the previous experiment, the conventional PLP technique yields a very high error rate. A similar test showed that a standard LPC-based system degraded even further, to a 60.7% error rate. The performance of the RASTA-PLP degrades only slightly.

Large Vocabulary Experiments

As a follow-up to the work reported above, we recently applied a simple lowpass filter (a single complex pole pair, with a 3dB point at 2 KHz and a 20 dB loss at 8 KHz) to 300 development test sentences from the October 1989 Resource Management speaker independent continuous speech recognition corpus. This filter was chosen to implement an approximation to the effect of muffled speech that we had observed with a small obstacle between the microphone and the talker's mouth. Both PLP-8 and RASTA-PLP-8 were computed on these data as well as on unfiltered versions of the standard 4000-

Table I : Isolated Digit Error rates

method	same environment	controlled modification	different environment
PLP	4.08%	31.35%	31.10%
RASTA-PLP	3.81%	5.00%	7.64%

sentence test set. The recognizer used was a hybrid recognizer with a neural network trained on the 4000 sentences to predict monophones for each frame, and then used in recognition to estimate likelihoods for a simple context-independent HMM system.

The word error results, shown in Table 2, show that the RASTA extension to PLP had essentially no effect on performance for the clean data, but drastically reduced the error for the filtered case. Note that this result was obtained with absolutely no tuning for the new experiment (other than the choice of 8th order PLP); we believe we could do much better with RASTA parameters adjusted somewhat for the new problem.

Table II : Continuous Speech Word Error rates

method	Unfiltered	Filtered
PLP	17.9 %	64.7%
RASTA-PLP	18.6%	19.2%

Informally we have observed that RASTA-PLP gives a substantial advantage in our live recognition experiments; while the conventional short-term spectrum based front-end is very sensitive to the choice of the microphone or even to the microphone position relative to the mouth, the RASTA-PLP makes the the recognizer much more robust to such factors. Further, even the harmful effect of the constant additive noise background, often present in our live recordings, appears to be reduced.

Discussion

A major current research concern is the significant degradation of high-performance laboratory systems when used in more realistic field environments. We found, for instance, that a simple change of operating environment (Colorado public switched network as opposed to New Jersey) degraded our digit recognizer by an order of magnitude. For digit recognition over the telephone, it appears that our simple technique is quite effective for the drastic reduction of errors due to channels that differ between the training and test sets.

For large-vocabulary speech recognition, researchers have successfully reduced difficulties due to noise and/or spectral deformation using spectral subtraction, estimation, and normalization techniques [1],[2]. These approaches appear to be useful for recognition applications that permit the explicit computation of relative microphone transfer functions (e.g., with blind deconvolution), and typically require an explicit noise or channel spectral estimation phase. The RASTA-PLP computes all estimates on-line, which may prove advantageous for applications such as telecommunications, or in general for unpredictable gross spectral variation for large vocabulary recognition. In these applications, the channel conditions are not known a priori, and it is not generally possible to provide an explicit normalization phase.

Because we have been primarily concerned with convolutional noise in the telephone channel, as well as the effect of changing microhone choice or position for large vocabulary recognition, we conducted our corrections in the log spectral domain. Similar approaches could be used in the magnitude or power spectral domains for additive noise reduction. However, care must be taken to ensure positivity of the enhanced power spectrum, as is also the case for traditional spectral subtraction techniques.

The study reported here made no use of other potential capabilities of the RASTA processing, particularly the ability to apply signal modifiers to the spectral temporal derivative domain. For instance, a threshold imposed on small temporal derivatives could provide a further non-linear smoothing of the spectral estimates, and nonlinear amplitude modifications could enhance or suppress speech transitions. Our current band-pass filter may not be optimal. Further, there is no fundamental reason to use the same filter for all spectral channels. Those issues are topics of our current research.

We also note that a German group of researchers, using a highpass filtering approach, primarily in the power spectral domain, has achieved encouraging results in suppressing the additive noise on a different set of speech recognition problems [7]. Their experience appears to confirm the effectiveness of the RASTA class of techniques.

Summary

A new technique for estimating a robust time-varying spectrum, RASTA-PLP, has been described. A large test was conducted on a speaker-independent telephone digit recognition task using speech that had been corrupted with convolutional noise. Results from this test show an order-of-magnitude improvement in error rate over conventional spectral estimation techniques such as LPC or PLP. Results from similar tests with large vocabulary continuous speech recognition show a halving of error rate with no tuning for the new problem.

Acknowledgements

Thanks to Jordan Cohen for having piqued our interest in the "Inverse-E" problem. Thanks also to Chuck Wooters and Steve Renals of ICSI for assistance with the large vocabulary experiments.

REFERENCES

[1] A. Accero and R. M. Stern : ''Towards Environment-Independent Spoken Language Systems,'' Proc. Speech and Natural Language Workshop, DARPA, June 1990, pp. 157-162

[2] E. Errel and M. Weintraub: ''Recognition of Noisy Speech: Using Minimum-Mean Log-Spectral Distance Estimation,'' Proc. Speech and Natural Language Workshop, DARPA, June 1990, pp. 341-345

[3] S. Furui: ''Speaker-Independent Isolated Word Recognition Based on Emphasized Spectral Dynamics,'' Pros. IEEE Intl. Conf. on Acoustic, Speech & Signal Processing, pp. 1991-1994, Tokyo, Japan 1986

[4] H. Hermansky: ''Perceptual linear predictive (PLP) analysis for speech,'' J. Acoust. Soc. Am., pp. 1738-1752, 1990.

[5] H. Hermansky: ''Auditory model for parametrization of speech in real-life environment based on re-integration of temporal derivative of auditory spectrum,'' U S WEST Advanced Technologies Research Report, File Folder ST 04-01, October 1990.

[6] H. Hermansky, N. Morgan, A. Bayya, P. Kohn: ''Compensation for the effect of the communication channel in auditory-like analysis of speech (RASTA-PLP),'' Proc. of Eurospeech '91, pp.1367-1371, Genova, Italy, 1991.

[7] H. Hirsch, P. Meyer, and H. Ruehl: ''Improved speech recognition using high-pass filtering of subband envelopes,'' Proc. of Eurospeech '91, pp.413-416, Genova, Italy, 1991.

[8] N. Morgan, H. Hermansky and C. Wooters: ''SPOONS '90: SPeech recOgnition frOnt eN workShop,'' International Computer Science Institute Technical report TR-90-045, 1990

Electroencephalogram Pattern Recognition Using Fuzzy Logic

Jung Hu and Benjamin Knapp
Department of Electrical Engineering, San Jose State University
San Jose, CA 95195

Abstract

A fuzzy logic approach to the classification of human sleep using electroencephalogram (EEG) data is presented. In this approach, frequency and amplitude information from an epoch of the EEG signal are extracted into a vector that is then compared to previously taught vectors representing the canonical features of six stages: wakefulness, rapid eye movement (REM) sleep and four nonREM sleep stages. For each stage, membership functions are calculated in each epoch. The stage with the maximum degree of membership is scored and classified. The system is implemented in software using the C programming language. Analysis of about 1101 epoches of the EEG data yielded an overall agreement of 77% between the program and a human scorer.

1 Introduction

Human sleep stages are usually classified according to accepted criteria which are based on the characteristics of the EEG and the electrooculogram (EOG)[1]. For example, a human scorer looks for patterns involving combinations of waveforms, i.e., delta waves, K-complexes, theta, alpha, sigma, and beta spindles in the EEG data. These waveforms have different frequencies and amplitudes.

Numerous attempts have been undertaken to automate the qualification and interpretation of the electroencephalogram (EEG). Those pattern recognition systems were largely based on statistical information that are extracted from EEG segments. In some systems, for example, classification is made by analyzing the number of amplitude, zero crossings, and other features and checking whether a number of those features is above a given threshold. If those thresholds are available, statistical or probabilistic decision making is useful and powerful in recognition of the patterns. However, the situation is often encountered where no statistical information such as a function threshold is available or where insufficient data is available for statistical analysis.

A new approach, fuzzy logic, can be applied to pattern recognition [2]. It plays a basic role in what may well be one of important facets of human thinking. Advantages of fuzzy logic are as the follows:

1) Most of the real world classification problems are fuzzy in nature because there exists no precise boundary between categories of patterns.
2) The conceptual structure of the theory of fuzzy sets may well provide a more natural setting for the formulation and approximate solution of problems in pattern classification than the more traditional approaches based on classical set theory, probability theory, and two-valued logic [3]. This is because the problem of pattern classification is intrinsically unsuited for approaches with precise mathematical formulation from the psycho-physiological aspect.
3) The theory is very flexible and allows easy and simple computation of evidences in complex relations.

Therefore, a fuzzy approach was chosen to implement the recognition of stages in a sleep electroencephalogram (EEG).

2. Fuzzy Pattern Recognition System

Pattern recognition contains two parts, learning and classification. The procedure of learning is to get a good representation of a human teacher for classifying a group of data sets. First, the system make a guess of coming data set. It then compares the classification with the teacher. If refuted, it adjusts its internal parameters. If confirmed, it continues to analyze the next data set. After learned, the system fixes their internal parameters and makes classification. Those two parts are implemented using fuzzy logic here. The concept of fuzzy logic is shown by the following:

A fuzzy set is characterized by a membership function μ_A: [0, 1] which associates with each element y of universe U and represents the grade of membership of y in A[4]. For example, a fuzzy subset *YOUNG* of universe U, which is from age 0 to 100, is expressed as

$$A = 1.0/0+1.0/1+\ldots+0.7/28+0.6/29 +0.5/30+\ldots+0.0/100 \qquad (2.1)$$

where {1.0, 1.0,..., 0.7, 0.6, 0.5,... , 0.0} are the grades of membership functions of ages in U and are assumed to lie in the interval [0,1], with 0 and 1 denoting no membership and full membership, respectively.

A learning system using fuzzy logic is shown in figure 1. The learning mechanism is only active in the estimation, or modification of the partition. The estimated partition is a passive memory, while the classifier is a decision process.

Let us consider a set of data, $x_n(t)$ which are extracted from a training set, that is

$$x(t) = [\, x_1, x_2, \cdots x_n \,] \quad \text{with } 0 < x_i < n \qquad (2.2)$$

1058-6393/91 $01.00

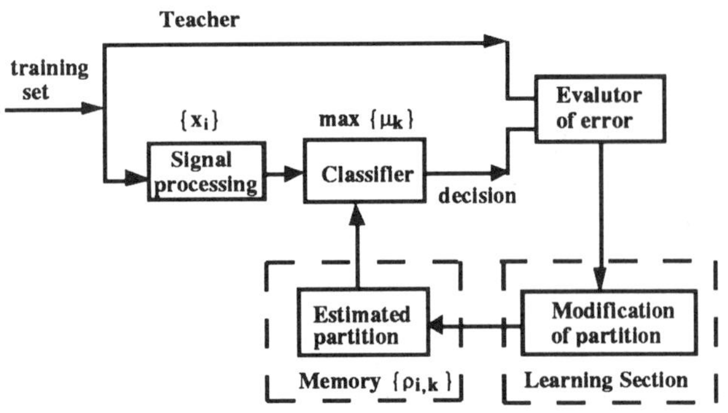

Figure 1. Learning System

where t is the number of times learned by a system. The process of extracting the data set from the training set is completed in the "Signal Processing" step in figure 1. For each class C_k, there corresponds a vector of descriptors characterizing the class.

$$\rho_k = [\rho_{1,k}, \rho_{2,k} \cdots \rho_{n,k}] \text{ with } 0<\rho_{i,k}<1 \qquad (2.3)$$

The estimated parameters set, ρ_k, defines a set of membership functions μ_k. The degree of membership of a set x<U to a class C_k is given by the function:

$$\mu_k = \prod_{i=1}^{n} \mu_{i,k} = \prod_{i=1}^{n} \rho_{i,k}^{x_i} (1 - \rho_{i,k})^{(1-x_i)} \qquad (2.4)$$

The function compares the incoming data to the learned data and shows how similar they are. It has a maximum value when x_i is equal to $\rho_{i,k}$ for all i. In figure 2, for input data $x_i = 0.7$, the grades of membership function $\mu_{i,k}$ are 0,4, 0.3, 0.55, and 0.52 corresponding to different classes with $\rho_{i,k}$ equal to 0.2, 0.3, 0.7, 0.8 respectively. So the maximum value of membership is 0.55 for the class with $\rho_{i,k}$ equal to x_i. Note the maximum value of membership is not necessarily 1. Each data x_i has equal contribution to the membership function μ_k.

The possibility that the element x is in class C_i depends on the degree of membership of x to C_i. The class C_k with the maximum membership is assigned to the element x. That is:

$$C_k = \text{Class } [\mu_k = \max\{\mu_i(x)\}]$$

$$\text{with } 0<i,k\leq \text{ the number of classes} \qquad (2.5)$$

If the decision disagrees with the training set, the parameter $\rho_k(t)$ will be modified into $\rho_k(t+1)$ by the following:

$$\rho_k(t+1) = \rho_k(t) + \frac{1}{N_{k(t)}+1} (x(t) - \rho_k(t)) \qquad (2.6)$$

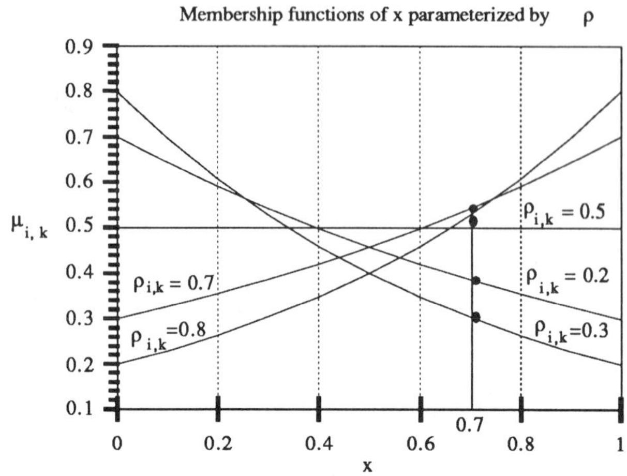

Fig. 2 The membership functions $\mu_{i,k}$ of x_i parametrized by $\rho_{i,k}$

where $N_k(t)$ is the number of training sets assigned to that class at time t. After that, the teaching process will continue until the decision is made correctly according to the given criterion or *teacher*.

3. Results

In our system, each sleep stage is characterized by frequencies and amplitudes, which are extracted using a 4096 point Fast Fourier Transform (FFT).

1101 epoches (30 second each) of the EEG data are tested in the system. An overall agreement of 77% is yielded between the system and the human scorer. The details of this result are shown in table 1. The system matched the human scorer best for the wake stage (92%) and stage 3 (93.48%), but the agreement was poor for stage 1 (23.26%), because stage 1 only has a few epochs in this EEG data. If there are more epochs of stage 1 in the training sets, the percentage of classification of stage 1 would be improved.

4. Conclusion

A fuzzy logic algorithm has been programmed to classify the human sleep stages. Using only frequency characteristics, the system yielded an agreement of 77% with a human scorer on 1101 epoches of data. There are still some confusion between the stages. Note that the system categorizes sleep stages baseed only on the EEG data. Other signatures of sleep stages, like electrooculogram (EOG) and eletromyograph (EMG), are not included in the system.

Future work will involve extracting more features to improve the recognition of the sleep stages. For example, sharp spindle waves help to differentiate stage 1 from stage 2. Another idea is two dimensional pattern recognition, which will use both features in the time and the frequency domains.

Table 1. Computer-human agreement matrix

	Computer						
	Wake	Stage 1	Stage 2	Stage 3	REM	Total	Agree
Human							
Wake	416	9	11	1	15	452	92%
Stage 1	9	10	9	0	14	43	23%
Stage 2	16	27	230	86	22	381	60%
Stage 3	0	0	3	43	0	46	93%
REM	9	15	8	0	147	179	82%
Total	506	95	256	130	118	1101	
Agree	84%	16%	88%	31%	83%		77%

References

[1] Wallance B. Mendelson, "Human Sleep: research and clinical care," Plenum Medical Book Co., New York, 1987.

[2] J. Aguilar Martin and R. Lopez De Mantaras, "The process of classification and learning the meaning of linguistic descriptors of concepts," in *Approximate Reasoning in Decision Analysis*, edited by M. M. Gupta and E. , North-Holland Publishing Company, 1982.

[3] Renato De Mori, "Computer models of speech using fuzzy algorithms," Plenum Press, New York, 1983.

[4] L. A. Zadeh, "Fuzzy sets," *Information and Control*, vol. 8, pp 338-353, June 1965.

A New Approach to Frequency Line Tracking

Harold G. Keane
Senior Principal Engineer

ORINCON Corporation
9363 Towne Centre Drive
San Diego, CA 92121

Abstract

The Dynamic Undersea Detection Extractor (DUDE) is an implemented acoustic feature extractor designed to extract and statistically characterize narrowband signals in noise. DUDE accepts normalized spectral data and produces symbolic representations of narrowband acoustic signals. If the data is beamformed, DUDE also performs multibeam correlation to produce bearing estimates. DUDE distinguishes itself from other feature extractors primarily by its ability to characterize signals that move in frequency over time (i.e., dynamic signals). Among the unique features in DUDE's algorithm is the "line-likeness in a neighborhood" test used to prune spurious signals. Statistical performance analysis results for the DUDE algorithm are also presented.

Background

Acoustic feature extractors accept beamformed and signal processed acoustic data (typically spectral data*). They detect and statistically characterize signals in the noise background. Narrowband feature extractors, such as DUDE and the other extractors mentioned in this paper, attempt to extract "line-like" signals that are narrow in frequency but extend over time.

Many acoustic feature extractors were implemented before DUDE, but they typically use signal models that are more appropriate for 1960's signal processing technology. Many extractors, for example, use a stable (constant in frequency over time) line model with a variety of integration periods to maximize their ability to detect low SNR signals. This technique was appropriate while targeted frequency components were pronounced and largely stable. In fact, stable lines were often the only available sources because processing limits of available computer systems required relatively slow time resolutions that obscured dynamic events. While these extractors are currently useful for exploiting low SNR stable signal components, they can no longer be used exclusively to analyze acoustic data.

As the technology has improved, dynamic signals have gained importance. Modern computing resources have made it possible to reveal details about dynamic signature components, and acoustic experts have begun to gather more information from dynamic signals. To extract these signals, the ADEC[3] algorithm was created. While ADEC can extract some dynamic lines, it uses a signal model that optimizes detection for lines with a constant slope. Many targeted dynamics exhibit lines with a changing slope. While the ADEC algorithm often successfully detects a dynamic signal, it sees that signal as a large set of straight lines with slightly varying constant slopes. As a result, statistical results from ADEC during a dynamic event are suspect.

In recent years, there has been an explosion in computing technology. Improvements in computing resources have made possible advances in beamforming and signal processing technology. Beamforming and signal processing advances have in turn resulted in an orders of magnitude increase in the volume of data these systems produce. Simultaneously, the number of experienced acoustic experts is being reduced because of changing economic imperatives. As a result, the need to automate detection processes has become critical. DUDE meets this requirement by employing a line model that adapts dynamically to the acoustic environment. By dynamically reconfiguring its signal model, DUDE reliably detects and characterizes line-like narrowband signals that are stable, dynamic with a constant slope, and dynamic with a changing slope.

Algorithm

The DUDE algorithm can be divided into two sections. The first section of the DUDE program works on single beam

*The primary exception to this rule is the MAPLE algorithm. MAPLE accepts time series data and extracts stable lines. Information on MAPLE can be found in Joseph J. Wolcin's Journal of the Acoustical Society of America article[1].

1058-6393/91 $01.00 © 1991 IEEE

data. The second section correlates across multiple beams. The initial single-beam step in DUDE's algorithm accepts input data one scan at a time. A scan is a set of normalized magnitude squared (or magnitude) FFT integrands (a.k.a., frequency bin values) from one time period. These frequency bin values (f) are estimates of the signal plus noise value (s+n) in the noise background (n).

$$f = \frac{s+n}{\hat{n}} \text{ ; where } \hat{n} \text{ is an independent noise estimate}$$

The first step in DUDE's algorithm is tracking energy in each new scan within individual beams. The resulting tracks are called Beam Signal Vectors (BSVs). They extend in time within their single beam.

After some predetermined number of scans from all beams, an epoch boundary has been reached. An epoch is a number of scans over which multi-beam correlation takes place. This correlation phase is the second part of DUDE's algorithm. At this point, BSVs are clustered together across beam space within an epoch to form Array Signal Vectors (ASVs). The purpose of this clustering is to group signals from the same energy source together so that bearing and other signal statistics can be computed accurately. ASVs are immediately used to form new and extend existing Array Signal Tracks (ASTs). ASTs represent a history of the energy from a single source. The division between the processing performed on individual scans (BSV) and on epoch boundaries (ASV and AST) is used to separate DUDE into two functions (Figure 1).

Many of the more interesting features of the DUDE algorithm exist at the single beam level. For example, DUDE uses a heuristic algorithm to locate and extend BSV tracks within a beam. The first step in this algorithm is to compare the bin values within a single scan against a threshold. The threshold dynamically adjusts to the changing noise background by multiplying a constant, the *bin factor* (BF), by a periodically updated noise estimate*.

$$f > BF \times \hat{n}$$

Within a single time scan, DUDE allows a potential frequency cluster to drop below that threshold temporarily without declaring two separate clusters if the number of consecutive bins that fall below the threshold is less than the *max ridge width skip* parameter (Figure 2).

Once a set of frequency bins has exceeded the bin factor threshold and been clustered, the cluster must be verified by the *max_path* test. The max_path test is an iterative dynamic programming solution that performs the "line-likeness in a neighborhood" test referred to earlier. This second threshold test examines a region starting at the peak bin of a frequency cluster to determine if a line can be drawn in contiguous bin space through successive scans whose average amplitude exceeds a second user-defined constant, the *path factor* (PF), multiplied by the noise estimate. Max_path reveals the sum of the amplitudes of the largest vertically (time) contiguous bins.

The max_path technique is most clearly explained graphically (Figure 3). Each square in the matrix below represents a frequency bin (f). The rows of the matrix represent scans. The three shaded squares indicate peak bins (i.e., largest amplitudes within their respective frequency clusters). The two leftmost shaded squares have incomplete max_path tree structures attached to show the growth of the max_path function. The rightmost tree is complete for a max_path function whose lag factor (i.e., the number of scans max_path projects forward in time, n_s) is four.

*If no periodically updated noise estimate is provided, a constant noise estimate can be used.

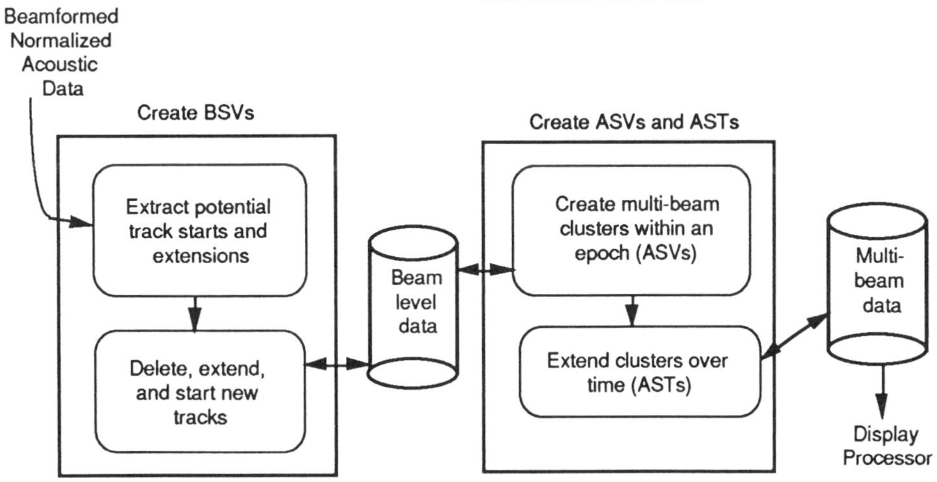

Figure 1: DUDE Data Flow

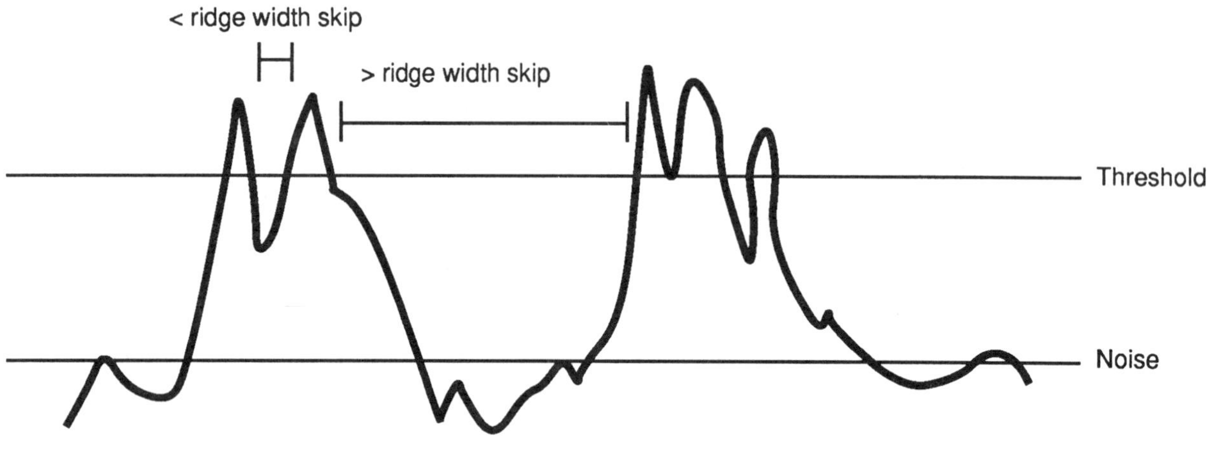

Figure 2: Frequency Amplitudes Within One Scan

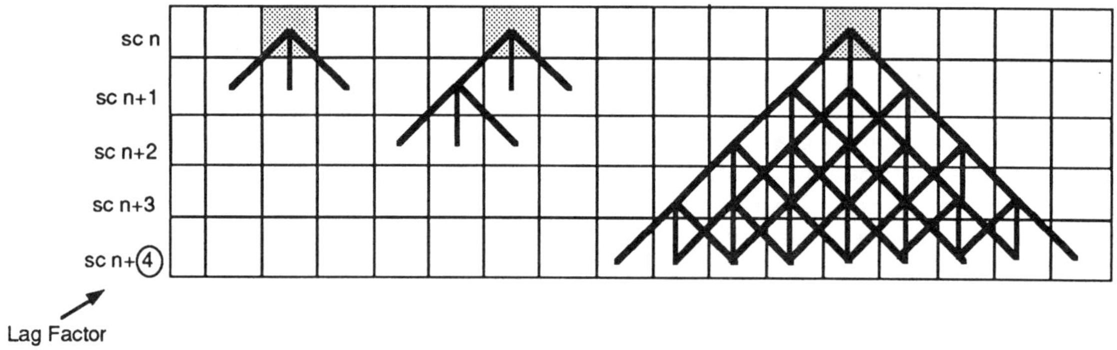

Figure 3: Max_Path Tree Growth

Once the max_path tree is fully formed, it selectively retracts at each level of the hierarchy by pruning the paths that do not contain the maximum amplitude. When that retraction process is complete, the amplitudes along the single remaining path are averaged, creating the max_path score (MPS). This value is compared to the path factor (PF) multiplied by the noise estimate. If the MPS is greater than this product, the associated frequency cluster can be used to extend or start a BSV.

$$MPS = \frac{1}{n_s} \sum_{s=n}^{n+n_s} f_{s,m(s)}$$

where m(s) is the max_path bin for scan s
and $f_{i,j}$ is the frequency component for scan i, bin j

$$MPS > PF \times \hat{n}$$

This test allows DUDE to prune the list of clusters that passed the single scan bin factor test to those clusters that are near an energy pattern that is likely to be line-like, i.e., those clusters most likely to produce a track. As shown above in figure 3, max_path implies that the DUDE algorithm runs behind real-time by at least the number of scans used by max_path (lag factor). Note, however, that this validation step also can be used to examine prior, rather than successive, scans (or both prior and successive scans). Also, the rate of max_path growth (2 bins per scan in figure 3) can also be modified. If max_path growth is set equal to 0, DUDE becomes a near optimal stable line detector.

The max_path path is not used to extend BSVs in time. The max_path test is used only as a threshold test to limit the number of points considered by a linear predictive filter that connects points in time to form a BSV. Figure 4 shows the effectiveness of the max_path test. The number of points that pass the single scan bin factor test in a synthetic gaussian noise file is shown on the single scan test curve. The number of points that pass both the single scan and multiple scan tests is shown on the multiple scan test curve.

The linear predictive filter that associates tracks over time uses an objective function based on statistical estimates of both the frequency clusters that have just passed the single and multiple scan tests and existing tracks. The resulting scores are associated using the Munkres scoring algorithm [3].

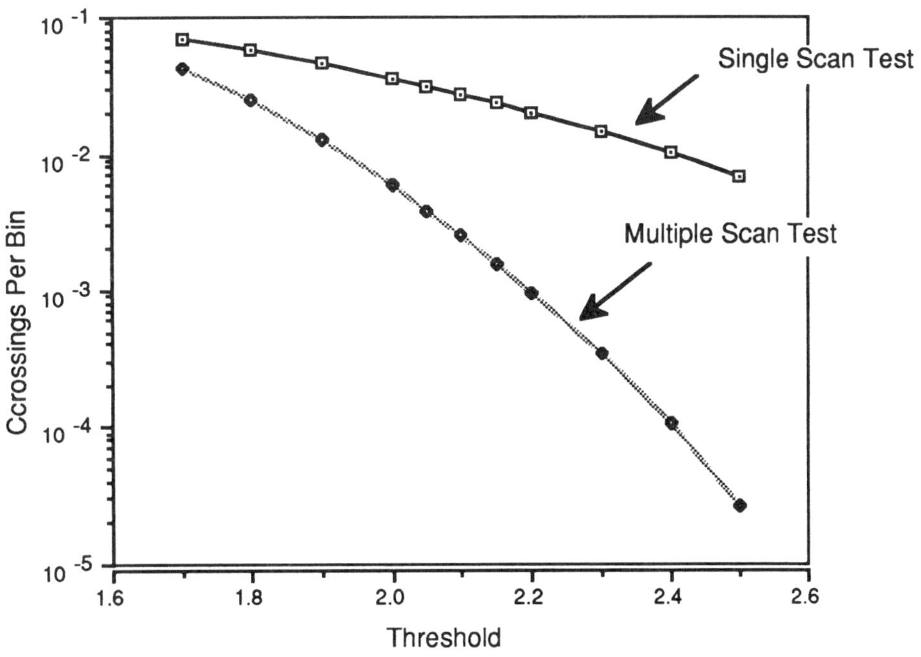

Figure 4: Multiple Scan Test Effectiveness – .7 Sigma Gaussian Data

Statistical Analysis

A preliminary theoretical analysis of the performance of DUDE as a function of the two thresholds, bin factor and path factor, has been carried out[4]. This analysis attempted to characterize the trade-offs in probability of detection and probability of false alarm as the thresholds were changed. As such, this study dealt only with the first order issue of validation of a point or cluster in a particular FFT scan for possible inclusion in a track. DUDE performs a second level of filtering before validating a track. This second level has not been included in the theoretical analysis, and, therefore, further analysis is required to directly apply these analytical results with the overall results of applying DUDE to a real data set.

We have compared DUDE's detection performance to that of an ideal straight line integrator under the assumption of a constant frequency and constant amplitude signal. As might be expected, DUDE is slightly sub-optimum for this case when it is configured to extract dynamic lines. To accommodate dynamics, DUDE considers bins in a trapezoidal region that expands upward and/or downward and is centered on the bin that passed the bin factor test. This provides more opportunities for false alarms than would a straight line integrator. Or conversely, the path_factor threshold must be raised to maintain a constant false alarm rate (FAR) which results in a loss in probability of detection (P_d). On the other hand, DUDE's performance will be only slightly degraded by a line whose frequency is changing with time (slewing) as long as the slew rate is within the maximum rate that is permitted by the max_path region. The straight line integrator's detection performance, however, will degrade very rapidly as a function of slew rate.

The principal question is how much performance against stable lines does DUDE give up to accomplish its goal of tracking highly dynamic lines? The preliminary answer presented here is based on an analytic study that considered a highly simplified, single scan DUDE. The performance baseline was a 2 scan integrator (a 1 scan DUDE looks at 2 scans, one for bin factor and one for path factor). In this case, we found that the P_d vs. P_{fa} receiver operating characteristic (ROC) curve for DUDE at an SNR of 10 dB closely matched that of the integrator for SNR = 9 dB over the range $10^{-7} < P_{fa} < 10^{-3}$. We also found that the best performance occurred when the bin factor is approximately equal to the path factor.

We resorted to a computer simulation for the study of a more realistic version of DUDE in which the max_path test was performed over multiple FFT scans because of the difficulty of analyzing this complex case. The simulation study considered a 4 scan DUDE and compared it to a 5 scan integrator. The DUDE ROC curve for a 5 dB SNR line was calculated with bin factor equal to path factor. It is found to lie between the integrator curves for SNR = 2 dB and SNR = 3 dB over the range $10^{-7} < P_{fa} < 10^{-3}$. Over most of this region, the loss is less than 2.5 dB.

DUDE is a two stage detector that looks for lines of any shape in frequency-time space. We have computed its P_d vs. P_{fa} and SNR relationships for two interesting cases, a single scan DUDE and a 4 scan DUDE. The former calculation was entirely analytic while the latter employed a combination of analysis and simulation. For both the single scan and the 4 scan DUDE, we found that the optimum operating point is at or near bin factor equal to path factor. We also determined that the loss relative to a straight line integrator is about 1 dB for a single scan DUDE while for a 4 scan DUDE, we found a loss of 2 - 2.5 dB.

Clearly, DUDE can provide superior performance in tracking dynamic lines as compared to a straight line integrator. DUDE's performance against stable lines, when it is configured to extract dynamic signals, will always be slightly sub-optimum. The relative importance of the two types of lines in a system application must be used to determine the appropriate algorithm(s), or parameter setting(s), to use.

Conclusion

DUDE represents a novel approach to frequency line tracking. It has been created as a response to improved computing and signal processing technology. Through its ability to detect and characterize dynamic events (Figure 5), DUDE offers the opportunity to automatically extract reliable symbolic and statistical information from any data that contains dynamic narrowband frequency components.

References

[1] Joesph J. Wolcin, "Maximum *a posteriori* estimation of narrow-band signal parameters," J. Acoust. Soc. Am. 68(1), July 1980.

[2] General Electric Corporation, ADEC Computer Performance Specification.

[3] Francois Bourgeois and Jean-Claude Lassalle, "An Extension of the Munkres Algorithm for the Assignment Problem to Rectangular Matrices," Communications of the ACM, Dec. 1971, Volume 14, Number 12.

[4] Joel Young, ORINCON Corporation, "DUDE Performance Analysis," August 1991.

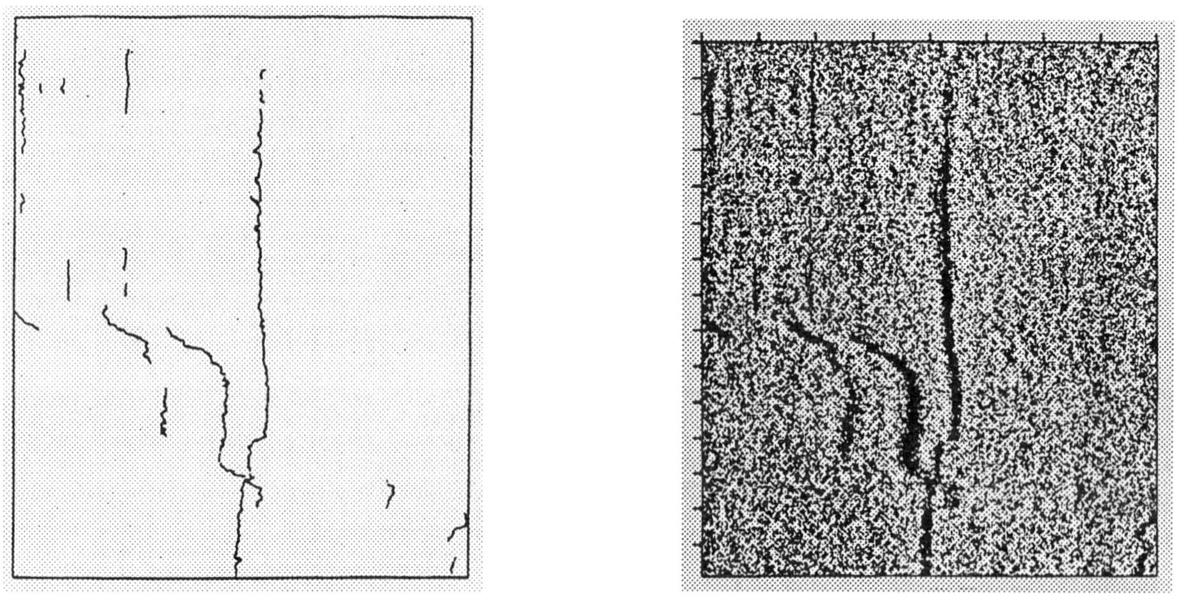

Figure 5: Sample DUDE Performance Results

Adaptive Multidimensional Signal Processing for Ocean Surveillance Using a Surface-Based Radar

Simon Haykin
Communications Research Laboratory
McMaster University
Hamilton, Ontario, Canada, L8S 4K1

Abstract

In this paper we present a philosophical discussion of adaptivity and multidimensionality in the context of ocean surveillance using a surface-based radar. We use this environment to emphasize (1) the basic notions of complexity and efficiency; and (2) the signal-processing roles of chaotic modelling, neural networks, Doppler-polarization (multidimensional) processing, and wavelet-inspired time-frequency analysis. The discussion is supported with new and exciting experimental results (reported in the list of references appended to the paper); these results have been obtained using actual data collected with a world-class quality-instrument facility known as the IPIX radar. The data were collected at a site on the East Coast of Canada.

1 Introduction

"Adaptivity" is a system feature built into the design of a signal processor or filter for the purpose of coping with statistical variations of the environment in which it operates. We may distinguish different classes of adaptive filters: (1) linear and nonlinear; (2) supervised and unsupervised. In a linear adaptive filter the actual response of the filter, representing an estimate of some desired response, is expressed as a linear combination of samples of the filter input. Ordinarily, linear adaptive filters are designed on the basis of second-order statistics. Optimality for such filters is therefore assured only if the underlying statistics are Gaussian. Inability to respond to statistics higher than second-order means a loss of information and therefore a nonoptimal behaviour. Nonlinear adaptive filters overcome this limitation by virtue of some appropriate form of nonlinear processing built into their design.

Supervised forms of adaptive filters require the use of a "teacher" that supplies the desired response. In the course of training the filter, the experience of the teacher is in effect transplanted into the selection of a suitable set of weights or control parameters that characterize the filter. In an unsupervised adaptive filter, on the other hand, learning takes place without a teacher.

Another important system feature that may be built into the design of a signal processor is that of "multidimensionality". The motivation here is to expand the pool of information content about the surrounding environment, and thereby enhance system reliability, by increasing the dimensionality of the observation space. This may be achieved by employing a single system with a multitude of discriminants capable of capturing different features of the environment, or alternatively by using a set of separate sensors that are capable of responding to the environment in different ways.

Inherent in the design of a signal processor are the basic notions of "model complexity" and "system efficiency". We speak of stochastic complexity or deterministic complexity, depending on the viewpoint taken in the mathematical description of the environment. Stochastic complexity refers to the order of a stochastic model of the environment, which hinges on the premise that the model input is a single realization of a stochastic process. Deterministic complexity, on the other hand, arises naturally in the context of chaos theory used to describe the underlying dynamics of the environment. The use of a stochastic or chaotic model is not a matter of choice; rather, it is dictated by physical considerations. Nevertheless, much too often in the past the use of a stochastic model has been invoked without physical justification.

As for the notion of efficiency, it refers to the ability of the system to extract information about the environment from the incoming data. In general, the more efficient an estimator is, the smaller will the variance of the estimation error be; the estimation error is measured with respect to some desired response.

1058-6393/91 $01.00

In this paper we present a philosophical discussion of these ideas in the context of ocean surveillance, supported by experimental results. The discussion is in the spirit that signal processing is the meeting ground between physics and mathematics. Details of the underlying theory and relevant experimental results are however left to references listed at the end of the paper.

The paper is organized as follows. In Section II we outline the operational requirements of a radar system for ocean surveillance, with particular reference to the environmental conditions that prevail on the East Coast of Canada. In Section III we describe an instrument-quality research facility known as the IPIX radar that is used to probe the environment and collect data on it. In Sections IV through VII we present some new and exciting results that illustrate the points made above concerning adaptivity, multidimensionality, complexity, and efficiency. In particular, in Section IV we present experimental evidence supporting the idea that radar backscatter from an ocean surface, commonly referred to as sea clutter, permits chaotic modelling. In that section we also demonstrate that nonlinear filtering is more efficient than linear filtering in suppressing clutter. In Section V we show that the use of Doppler processing, which presumes the availability of both amplitude and phase, can result in a significant improvement in detection performance over a processor that relies on amplitude alone. In Section VI we discuss how a further improvement in detection performance can be attained through the addition of polarization as a radar discriminant. In Section VII we present a new wavelet-inspired idea, referred to as the chirplet transform, for the time-frequency analysis of radar signals. This new method provides a clear visual discrimination between clutter only and clutter plus target signal. The paper concludes with some final remarks in Section VIII.

2 Operational Requirements of Ocean Surveillance

It is generally recognized that radar, in one form or another, is a necessary sensor for the surveillance of an ocean environment. Our concern in this paper is with a surface-based radar that can be ship-borne, located on a drilling platform, or simply shore-based. In any event, the radar is required to provide for local monitoring of the ocean environment, typically within a 10-mile radius of a site of interest. The surface targets of interest include the following:

1. Ships and fishing boats.
2. Icebergs and small fragments of ice (known as growlers).

Ships and icebergs are relatively easy to detect by virtue of their large physical size (and therefore radar cross-section). On the other hand, smaller objects such as a fishing boat or a growler are much more difficult to detect. The visible part of a growler is about the size of a grand piano; however, recognizing that approximately 90 percent of the volume of a growler lies under the ocean surface, we find that a growler can indeed be hazardous to the safety of ships and oil tankers navigating in ice infested waters. There is therefore a special requirement for the reliable detection of such radar targets. Moreover, we would like to be able to discriminate between a growler and a fishing boat. This latter capability is highly desirable in search-and-rescue operations aimed at fishing boats that are in distress on the high seas, and which have no means of communication with home base.

The radar detection of these small surface targets is severely limited by the unavoidable presence of sea clutter. (The effect of receiver noise is usually negligible compared to sea clutter.) Sea clutter refers to unwanted radar backscatter from the ocean surface. Realizing that sea clutter is the result of radar reflections just as a target echo is, it is obvious that we cannot enhance the visibility of a target simply by increasing the transmitted radar power. Rather, we have to use ingenuity in the proper selection of radar discriminants and signal processing tools for the task at hand. In the next section, we describe a unique radar facility for research into this important problem.

3 IPIX Radar

Ordinarily, commercially available marine radars rely solely on amplitude as the radar discriminant, with the result that they are of limited practical use in the detection of small surface targets in the presence of sea clutter. It was the recognition of this fact that prompted the Communications Research Laboratory (CRL) of McMaster University to undertake the design and development of a multidimensional radar for ocean surveillance.

The radar is known as the IPIX (Intelligent PIXel- processing) radar [1]. It is a coherent dual-polarized X-band radar, designed to be of instrumentation-quality for research use. The radar was built as part of the larger research effort with the aim of developing improved techniques for the detection and identification of small targets in an ocean environment by developing algorithms to utilize the available radar features, and quantify the performance improvement afforded by each. The foundation for this research study must be a thorough understanding of the nature of sea clutter and the corresponding behaviour of targets of interest under various sea conditions. To this end, the IPIX radar is being used to build a database of sea clutter and target radar returns to characterize the ocean environment and target behaviour with respect to the available radar parameters, and under varying environmental conditions.

Development of the IPIX radar began in 1984, and a prototype was tested in the summer of 1986. The IPIX radar has the following features:

1. 8-kw pulsed transmitter using a travelling-wave tube (TWT) amplifier operating at 9.39 GHz.
2. Coherence to permit Doppler processing.
3. Dual linear polarization with pulse-to-pulse transmit polarization switching so as to acquire the full polarization scattering matrix.
4. A surface acoustic wave (SAW) pulse compression subsystem, providing improved resolution and range of coverage (a 5-μs transmitted expanded pulse is compressed to an effective pulse width of 32 ns in the receiver, equivalent to a range resolution of 4.8 m and a transmitted peak power of 1.2 MW).
5. Dual-frequency transmission and reception, with one frequency fixed and the other variable over a 500 MHz band; this capability will permit interferometric signal processing in a way not possible with a conventional radar.
6. Interactive computer for highly flexible control of radar parameters and data acquisition (computer involvement includes antenna position control, pulse repetition frequency (PRF) and pulse width selection, down loadable sensitivity time control (STC) curves, windowing of data collection area in a range and azimuth, and transmit polarization selection, among others).
7. Digital data acquisition with in-field processing capability to facilitate algorithm development efforts and provide on-site data validation (four channels of 8-bit digital data are collected; vertical and horizontal polarization, each with in-phase and quadrature signals).
8. Built-in calibration equipment allowing automated in-field system calibration.

The IPIX radar is specifically designed to be a research instrument. Digital control allows experimental parameters to be easily varied, but once selected they remain accurate and stable. Built-in calibration equipment permits quantitative measurements to be made with high confidence. The data acquisition system delivers digital radar data at real-time rates into the address space of a computer, where it may be validated, processed, and stored on removable media for off-line analysis. The computer control system facilitates precise experiment documentation, describing the complete state of the radar system, which is stored with the collected data.

A valuable database collected with the IPIX radar on a field trial conducted in June 1989 on the East Coast of Canada, has been used for extensive signal processing research, the results of which are reported in the rest of the paper.

4 Chaotic Modelling of Sea Clutter

In the development of a mathematical model for sea clutter, it is desirable for the model to satisfy four conditions:

1. Physical plausibility of the model.
2. Consistency of the model with experimental data.
3. Ability of the model to predict useful parameters, given measured data.
4. Minimum number of degrees of freedom needed to describe the model.

In the case of sea clutter, the traditional approach has been to consider a stochastic model largely because of the random appearance of a sea clutter waveform. Indeed, except for Jakeman's K-distribution based on a random walk model [2], all the other stochastic models for sea clutter reported in the literature to-date have been derived in an ad hoc data-fitting fashion. Moreover, not enough attention has been paid to condition 4.

In two recent papers [2,3], we have challenged the traditional stochastic approach for the modelling of sea clutter. In particular, we have used actual data collected with the IPIX radar to show that sea clutter admits a chaotic model that satisfies all the four conditions mentioned above. The chaotic model is not only able to capture the underlying dynamical process responsible for the radar backscattering mechanism, but also it can be implemented using a neural network with a minimum number of degrees of freedom. In effect, the neural network memorizes the essence of the time evolution of the sea clutter process through a set of synaptic weights connecting different layers of neurons in the network.

A comparison has been made between (1) the

chaotic model of sea clutter viewed as a nonlinear predictor, and (2) a linear predictor using the standard recursive-least squares (RLS) algorithm. The following important conclusions can be drawn from this study:

1. The nonlinear predictor is consistently more efficient than the linear predictor in that it always produces a prediction error with smaller variance.
2. The difference between the variances of the linear and nonlinear prediction errors is smallest when the deviation of sea clutter from a Gaussian distribution is the least.

These two points confirm the observations made previously in Section I concerning efficiency and higher order statistics.

Another noteworthy point is that, in accordance with Takens embedding theorem, for the development of a chaotic model for sea clutter, we only need one component (e.g., the amplitude, in-phase, or quadrature component). However, we do not yet know how target detection using an amplitude-based chaotic model compares with one using both in-phase and quadrature components.

5 Doppler Processing

A conventional marine radar is noncoherent; it therefore relies on amplitude processing for target detection. Moreover, cost considerations dictate the use of a simple envelope detector for target detection. The result is that it exhibits an unacceptably poor performance when it is called upon to detect small surface targets such as growlers and fishing boats in the presence of sea clutter.

The use of a coherent radar, and therefore the availability of Doppler information, makes it possible to provide for improved target detection. In the case of a small floating target (e.g., growler), the Doppler information manifests itself in the form of a distinctive "winking" phenomenon [1]. Specifically, the Doppler frequency component of the radar echo from a floating target increases with time, reaches a peak, and then decreases. This phenomenon repeats itself with a frequency that depends on the wave spectrum of the ocean surface and the mass of the target.

In [4], two coherent detectors are investigated, which exploit the time-varying Doppler signature of a floating target. Detector I uses spectral width as the parameter for discriminating between sea clutter alone and a target echo plus sea clutter. For sea clutter, a Gaussian model is fitted to the clutter power spectrum. On the other hand, a two-component Gaussian model (one Gaussian component for the clutter and another Gaussian component for the target echo) is fitted to the target-in-clutter power spectrum. Detector II uses an autoregressive model for the radar returns. If a target echo is present in the data, one of the poles of the transfer function of the model tends to cluster very close to the unit circle in the z-plane, since the corresponding power spectrum is narrow band. In the absence of a target echo, all the poles tend to lie further inside the unit circle. Analysis with actual radar data has shown that the spectral width-based detector is far more robust than the autoregressive model-based detector with respect to changes in signal-to-clutter ratios. Furthermore, both of these coherent detectors are significantly superior in performance to a conventional noncoherent detector.

6 Doppler-Polarization Processing

A further improvement in radar detection performance can be attained through polarization processing as a follow-up to Doppler filtering. Specifically, the time series constituting each element of the polarization matrix is first transformed into the frequency domain by traditional Fourier methods. Then taking the magnitude squared, we get four periodograms pertaining to (in the case of linear polarization) horizontal-horizontal, vertical-vertical, horizontal-vertical, and vertical-horizontal polarizations; the last two elements are the same except for receiver noise and system errors. The periodograms indicate the power of each element of the polarization scattering matrix as a function of frequency. The potential gains of dual-polarization processing lie in the relative powers of the scattering matrix elements because their ratios differ between targets and clutter. Equally important are the phase differences between scattering matrix elements, which also differ between target and clutter. A summary of the observed differences, based on measurements with IPIX radar data, follows:

1. The average (over frequency) ratio of horizontal-horizontal to vertical-vertical periodogram is significantly smaller for ice targets than for sea clutter. At low frequencies, however, the ratio is very small; the vertical-vertical power spectrum, although less powerful than the horizontal-horizontal power spectrum, extends wider into low frequencies. The ratio of horizontal-horizontal

to horizontal-vertical periodogram is greater for ice targets than for sea clutter.

2. The phase difference between the horizontal-horizontal and vertical-vertical polarizations is more random for sea clutter than for targets; but both have the same mean.

For enhanced detection exploiting polarization diversity, the challenge is to combine the outputs from the four polarization filter banks, so that information contained in their relative powers and phase differences can influence the decision-making process in a positive way.

7 Time-Frequency Analysis

For our last topic, we consider the time-frequency analysis of radar signals with particular reference to the target-in-sea clutter problem. In [5] we describe a novel transform that first generalizes Gabor's notion of logons to depict both the Weyl- Heisenberg and the affine (wavelet) transforms, and then further generalizes the notion of a logon to include the expansion of an arbitrary function into a multiscale basis of "chirplets"; a chirplet to a wavelet is like an ordinary chirp to an ordinary wave. The term "chirplet transform" has been coined to refer to this new transform. This new transform is well-suited for the time-frequency portrayal of signals with a chirplet component, which is precisely the case in coherent radar returns from a small floating target over short (less than one second) periods of time. The two-dimensional images obtained by plotting the up-frequency versus the down-frequency exhibit a dramatic difference between clutter alone and target-plus-clutter.

The chirplet transform may well provide the basis of a receptive field design for the visualization of different forms of radar returns from an ocean environment, and therefore their improved classification. In performing the classification here, we see a distinct role for neural networks.

8 Concluding Remarks

In this paper we have touched on the following topics in the context of ocean surveillance using a surface-based radar; chaotic signal processing, neural networks, higher-order statistics, Doppler-polarization processing, and time-frequency analysis using a new wavelet-inspired transform. It is our strong belief that these topics in a broad sense will develop into new corner stones of the ever-expanding field of signal processing.

The experimental results referred to in the paper have all been obtained using the IPIX radar that has evolved into a unique world-class research instrument. The enormous pool of experience gained with this facility is already being put into the building of a new marine radar [6] that will provide the commercial solution to the engineering challenges encountered on the East Coast of Canada.

Acknowledgements

The author of this paper is deeply indebted to his research colleagues (in alphabetical order), Brian Currie, Vytas Kezys, Carl Krasnor, Henry Leung, Steve Mann, Tim Nohara, and Peter Weber for their individual contributions that have made the writing of this paper possible. He is truly grateful to the Natural Sciences and Engineering Research Council of Canada, the Provincial Government of Ontario, Litton Systems (Canada), and the Telecommunications Research Institute of Ontario for financial support.

References

[1] S. Haykin, C. Krasnor, T. Nohara, B. Currie, and D. Hamburger, "A Coherent, Dual-Polarized Radar for Studying the Ocean Environment", Proc. IEEE Transactions on Geoscience and Remote Sensing, Vol. 29, No. 1, pp. 189-191, January 1991.

[2] H. Leung and S. Haykin, "Is there a radar clutter attractor?", Applied Physics Letters, 56(6), pp. 593-595, February 1990.

[3] S. Haykin and H. Leung, "Neural Network Modelling of Radar Backscatter From An Ocean Surface Using Chaos Theory", IEEE Conference on Neural Networks in Ocean Engineering, Washington, DC, August 1991.

[4] T. Nohara, "Detection of Growlers in Sea Clutter Using An X-band Pulse-Doppler Radar", Ph.D. Thesis, McMaster University, 1991.

[5] S. Mann and S. Haykin, "The Chirplet Transform: A Generalization of Gabor's Logon Transform" Vision Interface'91, Calgary.

[6] B. Currie, P. Weber, C. Krasnor, S. Haykin, H. McLaughlin, and R. Worsfold, "An Improved Marine Radar" POAC'91, St. John's, Newfoundland, September 1991.

Advanced Beamforming Concepts: Source Localization Using the Bispectrum, Gabor Transform, Wigner-Ville Distribution, and Nonstationary Signal Representations

Jeffery C. Allen
Naval Ocean Systems Center
San Diego, CA 92152–5000
TP4-3

Abstract

Power spectral estimation provides one approach to direction-finding. This approach readily generalizes to produce a collection of newer direction-finding algorithms. Estimation of the bispectrum yields a bispectral direction finder. Estimates of time-frequency distributions produce Wigner-Ville and Gabor direction-finders. Some types of non-stationary time series admit spectral estimators which can be used to localize a source. Chaotic signals can also be localized using recently developed parameter estimators.

1 Introduction

In 1982, Don Johnson wrote a much-cited paper entitled *The Application of Spectral Estimation Methods to Bearing Estimation Problems* [10]. In the intervening years, considerable development has taken place in time-frequency distributions, higher-order statistics, and estimators for non-stationary random process. This paper extends the theme developed by Johnson to encompass some of these newer results. Following a review of conventional direction-finding (DF), section 3 shows how estimators of the bispectrum can yield DF algorithms. DF using the Gabor Transform and Wigner-Ville Distribution (WVD) are covered in sections 4 and 5. Spectral estimators for time series which are periodically correlated (cyclostationary) also yield a DF algorithm (section 6) as do the estimators for a fractral-like time series (section 7).

Vectors are denoted boldface $\mathbf{x}$. The conjugate transpose is denoted as $\mathbf{x}^H$. The normal or Gaussian distribution with mean μ and variance σ_2 is denoted $\mathcal{N}(\mu, \sigma^2)$. If $\{\mathbf{x}(t)\}$ is a zero-mean, vector-valued random process, the covariance function is

$$R_{\mathbf{x}}(t_1, t_2) = E[\mathbf{x}(t_1)\mathbf{x}(t_2)^H],$$

and the cross spectral density (CSD) matrix is:

$$P_{\mathbf{x}}(f_1, f_2) = \int_{-\infty}^{\infty}\int_{-\infty}^{\infty} e^{-i2\pi(f_1 t_1 - f_2 t_2)} R_{\mathbf{x}}(t_1, t_2)\, dt_1 dt_2.$$

Table 1: Principal Notation

t	time
f	frequency
c	speed of sound
E	expectation operator
$\mathbf{u}$	unit (look) vector
$x_m(t)$	mth sensor output
$\mathbf{x}(t) = [x_m(t)]$	array "snapshots"
$\mathbf{r}_m$	vector to mth sensor
$\mathbf{a}(f, \mathbf{u}) = [\exp(i2\pi f \mathbf{r}_m^H \mathbf{u}/c)]$	steering vector

2 Conventional DF

A quick review of conventional DF is undertaken in this section to provide the nomenclature for the subsequent DF algorithms. Assume a single source is transmitting a signal $s(t)$ in a lossless, plane-wave environment in the direction $-\mathbf{u}_0$. The signal $s(t)$ is received by M omni-directional sensors fixed at positions $\mathbf{r}_m$. Each sensor receives a time-delayed copy of $s(t)$. The time delay function associated with the mth sensor is $\tau_m(\mathbf{u}_0) = \mathbf{r}_m^H \mathbf{u}_0 / c$. If additive noise $(z_m(t))$ is present at the mth sensor, then the noise-corrupted signal at the m-th sensor is determined as [10]:

$$x_m(t) = s(t + \tau_m(\mathbf{u}_0)) + z_m(t). \qquad (1)$$

The classical DF output $x(t, \mathbf{u})$ using look vector $\mathbf{u}$ is determined by weights (w_m) and time delays $(\tau_m(\mathbf{u}))$ as

$$x(t, \mathbf{u}) = \sum_{m=1}^{M} w_m x_m(t - \tau_m(\mathbf{u})). \qquad (2)$$

The fundamental point of Johnson's paper [10] is that the estimates of the power spectrum of $\{x(t, \mathbf{u})\}$ yield bearing estimates. To see this, make the following standard assumptions regarding the stationarity of the signal and the noise process $\mathbf{z}(t) = [z_m(t)]$.

Assumption 2.1 Single source, uncorrelated noise:

1. the plane-wave signal $s(t)$ is a zero-mean, wide-sense stationary (WSS) random process.

2. the noise $\mathbf{z}(t)$ is a zero-mean, wide-sense stationary random process.

3. the signal $s(t)$ and the sensor noise are uncorrelated.

If $s(t)$ WSS, it admits a mean-square Fourier representation

$$s(t) = \int_{-\infty}^{\infty} e^{i2\pi ft} dS(f),$$

where $dS(f)$ is an orthogonally scattered measure [12]. The power spectrum $P_s(f)$ of the signal is determined by $dP_s(f) = E[|dS(f)|^2]$. Likewise, the WSS noise vector $\mathbf{z}(t)$ admits Fourier representation with an orthogonally scattered vector measure $d\mathbf{Z}(f)$ and its CSD matrix is given by $dP_{\mathbf{Z}}(f) = E[d\mathbf{Z}(f)d\mathbf{Z}(f)^H]$. Under the assumptions 2.1, it may be shown that the array output $\{\mathbf{x}(t)\}$ admits an orthogonally scattered vector measure [10]:

$$d\mathbf{X}(f) = \mathbf{a}(f, \mathbf{u}_0)dS(f) + d\mathbf{Z}(f). \quad (3)$$

Likewise, it may also be shown that the CSD matrix of the array output is given as

$$\begin{aligned} dP_{\mathbf{X}}(f) &= E[d\mathbf{X}(f)d\mathbf{X}(f)^H] \\ &= \mathbf{a}(f, \mathbf{u}_0)\mathbf{a}(f, \mathbf{u}_0)^H dP_s(f) + dP_{\mathbf{Z}}(f) \quad (4) \end{aligned}$$

Then the power spectrum $P_x(f, \mathbf{u})$ of the classical delay-sum DF is given by [1]: $P_x(f, \mathbf{u}) = \mathbf{a}(f, \mathbf{u})^H W P_{\mathbf{X}}(f) W \mathbf{a}(f, \mathbf{u})$, where W is the diagonal matrix of the weights. As a special case, suppose the sensor noises are independent and identically distributed so that $P_z(f)$ is a scalar multiple of the identity matrix. In this case, $P_x(f, \mathbf{u}_0)$ is the maximum value and illustrates the idea behind DF: *that the source direction can be determined by examining the maximum $P_x(f, \mathbf{u})$ as a function of the "look vector" $\mathbf{u}$*.

As pointed out by Johnson [10], obtaining the CSD matrix $P_{\mathbf{X}}(f)$ is basic to this classical DF, as well as the MVDR and MUSIC DF. Therefore, *consistent estimation of the CSD matrix is a key point in these DF methods* and is usually justified on explicit WSS assumptions and implicit ergodic assumptions. Before we consider those cases where the array output $\{x(t)\}$ is non-stationary, the next section is devoted to showing how to generalize DF via the power spectrum to DF via the bispectrum.

3 Bispectral DF

Higher-order statistical analysis of time series has been undergoing a tremendous development in the last decade. Excellent review articles of this area are [13], [14], while [19] and [18] are foundational texts. In addition, Forster and Nikias [3] have produced a MVDR-like DF based on the bispectrum while Porat and Friedlander [16] DF using fourth-order cumulants. In this section, we show how to obtain a bispectral MUSIC DF and a mixed bispectral MUSIC DF.

Definition 3.1 [13] If $\{\mathbf{x}(t)\}$ is a zero-mean, third-order stationary random process, then one definition of the third-order cumulant is

$$C_{\mathbf{X}}(\tau_1, \tau_2) = E[\mathbf{x}(t+\tau_1) \otimes \mathbf{x}(t+\tau_2) \otimes \overline{\mathbf{x}(t)}],$$

where $\otimes$ is the Kronecker product. The cross bispectral density tensor (CBT) is:

$$B_{\mathbf{X}}(f_1, f_2) = \int_{-\infty}^{\infty}\int_{-\infty}^{\infty} e^{-i2\pi(f_1\tau_1+f_2\tau_2)} C_{\mathbf{X}}(\tau_1, \tau_2) d\tau_1 d\tau_2.$$

The generalization of equation (3) for the CSD matrix of the array output $\{\mathbf{x}(t)\}$ to the CBT is straightforward [19]:

$$dB_{\mathbf{X}}(f_1, f_2) = E[d\mathbf{X}(f_1) \otimes d\mathbf{X}(f_2) \otimes \overline{d\mathbf{X}(f_1+f_2)}]. \quad (5)$$

For DF using the bispectrum, the classical assumptions 2.1 need to be modified:

Assumption 3.1 Single source, independent Gaussian noise:

1. the plane-wave signal $s(t)$ is a zero-mean, third-order stationary random process.

2. the noise $\mathbf{z}(t)$ is a zero-mean, **Gaussian**, *WSS random process.*

3. the signal $s(t)$ and the sensor noise are independent.

The bispectrum of a Gaussian WSS is zero [19, page 36]. From equation (3), equation (5), and assumption 3.1, it can be shown that

$$dB_{\mathbf{X}}(f_1, f_2) = \mathbf{a}(f_1, \mathbf{u}_0) \otimes \mathbf{a}(f_2, \mathbf{u}_0) \otimes \overline{\mathbf{a}(f_1+f_2, \mathbf{u}_0)} dB_{\mathbf{X}}(f_1, f_2).$$

In this form, the third-order signal subspace is obvious. For computational purposes, we can use the equivalent form for three tensors: $\mathbf{x} \otimes \mathbf{y} \otimes \overline{\mathbf{z}} \sim (\mathbf{x} \otimes \mathbf{y})\mathbf{z}^H$, which permits the application of the SVD to determine the projections for the third-order signal and noise subspaces. Moreover, this equivalence combined with equation (5) indicates that the CBT may be estimated by averaging Kronecker products of the DFT obtained by segmenting the data into time blocks [14].

Bispectral MUSIC performs DF by projecting a steering vector onto the third-order noise subspace and calculating the reciprocal of the norm of the projection. Mixed bispectral MUSIC performs DF by projecting a steering vector onto the second-order noise subspace (standard MUSIC) followed by the projection into the third-order noise subspace and then calculating the reciprocal of the norm.

The following simulation provides an assessment of these DF methods. A 10-sensor linear array is laid along the x-axis. Each sensor is spaced one-half wave length at 11 Hertz. The standard right-hand coordinate system is assumed so that broadside to the array measure $\theta = 90°$. A Gaussian and a bilinear signal were present. The Gaussian signal was generated by AR filtering i.i.d. $\mathcal{N}(0,1)$ noise (AR coefficients: a_0=1, a_1=.9, a_2=.81). This signal was normalized by setting the sample variance to 1. It arrived on the array from 80° azimuth using an FFT to perform the lag. The bilinear signal was generated using the bilinear model of Gabr [5]: $s_t = 1.3 s_{t-3}\epsilon_{t-3} + \epsilon_t$, where $\{\epsilon_t\}$ is i.i.d. $\mathcal{N}(0,1)$ and then normalized by setting the sample variance to 1. It arrived on the array from 100° azimuth using an FFT to perform the lag. Finally, the sensor noise was modeled as i.i.d. $\mathcal{N}(0,1)$. The sample rate was 34.38 Hertz and a total of 16,384 array snapshots were collected.

The classical, MUSIC, bispectral MUSIC and mixed bispectral MUSIC DF were applied to this data. All CSD and CBT estimates were made with 32-sample block averaging assuming f_1=f_2=11.5 Hertz. All MUSIC-type DF assumed two signals present. The results are plotted in figure 1. The classical, MUSIC, and mixed bispectral MUSIC indicate both sources with the primary lobe at the Gaussian source while the bispectral MUSIC appears to discriminate between the Gaussian and the bilinear source.

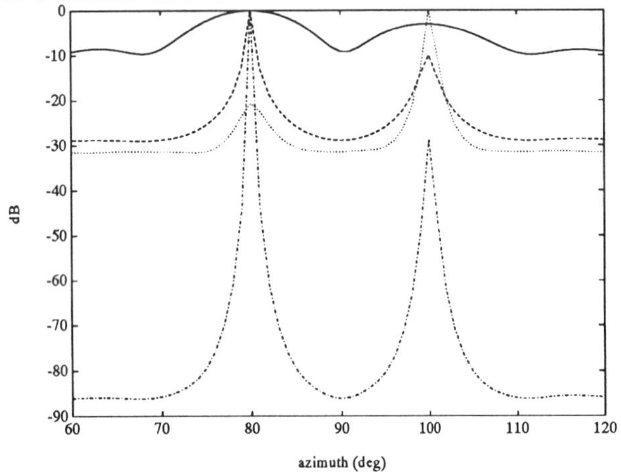

Figure 1: **Bispectral DF on Two Sources** — Gaussian source at 80° azimuth; bilinear source at 100° azimuth; Gaussian sensor noise. The solid line is the classical DF, the dashed line is MUSIC, the dotted line is bispectral MUSIC, and the dash-dot line is mixed bispectral MUSIC.

4 DF via the Gabor Transform

This section shows how the Gabor transform can be used as a pre-filter for removing transient signals corrupting the sensor data. Since a transient can be isolated from the sensor data, it follows that the same processing can apply to transient DF [22, Chapter 4] and admit more advanced transient DF based on minimum-variance or subspace methods. For brevity, we focus on using the Gabor transform as a pre-filter.

The application of the Gabor transform as a means of transient detection was beautifully developed by Friedlander and Porat [4] and we borrow heavily from that paper. Let $y(t)$ be a given continuous-time, real-valued signal. Let $g(t)$ be a fixed, non-negative function of unit area, called the window function. Then the Gabor representation of $y(t)$ using the window function $g(t)$ is

$$y(t) = \sum_{m=-\infty}^{\infty} \sum_{n=-\infty}^{\infty} C_{m,n} g(t - n\alpha) e^{i2\pi m\beta(t-n\alpha)},$$

where α and β are positive, with $\alpha\beta \leq 1$. This representation shows that the signal is to be expanded into *time-shifted* and *frequency-shifted* versions of the window. Therefore, *the application will dictate the choice of the window.* Indeed, this is the basic point [4] for analyzing transient signals. Since they consider a *transient signal* as a signal whose duration is short compared to the observation window, their basic idea is to replace the original Gaussian window of Gabor by the one-sided exponential function: $g(t) = \sqrt{2\lambda}e^{-\lambda t}u(t)$ to better follow transient signals with a sudden "on" time and subsequent exponential decay. Here $u(t)$ is the unit step function and "the parameter λ is used to control the effective width of the window" [4].

The use of the Gabor transform with this window is shown in the following DF simulation. A 10-sensor linear array is laid along the x-axis and each sensor is spaced one-half wave length at 11 Hertz with the first sensor located at the origin. An 11-Hertz plane-wave signal is impinging broadside to the array (azimuth $\theta = 90°$) with an amplitude of 1/10. The signal is corrupted by i.i.d. Gaussian noise $\mathcal{N}(0,1)$ at each sensor. In addition, five one-sided decaying transients were added to the environment (α=1, β=1, λ=1) and relevant parameters are displayed in Table 2: A total of

Table 2: Parameters for Gabor Transients

index	time delay (sec)	azimuth (deg)	frequency (Hertz)
1	13	62	12
2	23	63	12
3	21	110	10
4	5	64	12
5	10	80	13

1,024 array snapshots were collected during the time interval of $0 \leq t \leq 32$ seconds at a sampling rate of 32 Hertz. Figure 2 plots the 11-Hertz signal, the Gaussian noise, and the five transients at the first sensor. Figure 3 is a plot of the discrete Gabor transform of the output of first sensor. The time axis starts at time $t = 0$ in the leftmost corner and moves in one-second increments to the uppermost corner. The frequency axis starts at $f = 0$ in the leftmost corner and moves in one-Hertz increments to the lowest corner. The five transients are clearly displayed.

The effect of the transients on the MVDR DF is shown in Figure 4. MVDR used 11 Hertz in an attempt to find the weak 11-Hertz signal but the transients swamp the output. The transients were removed by taking the Gabor transform of each sensor's output, finding those coefficients exceeding three standard deviations of the Gabor transform, and then subtracting the corresponding Gabor transients (weighted by its Gabor coefficient). Thus, the Gabor transform functioned as a pre-filter and the resulting array output was passed to the DF. With the transients removed, the MVDR DF located the 11-Hertz signal.

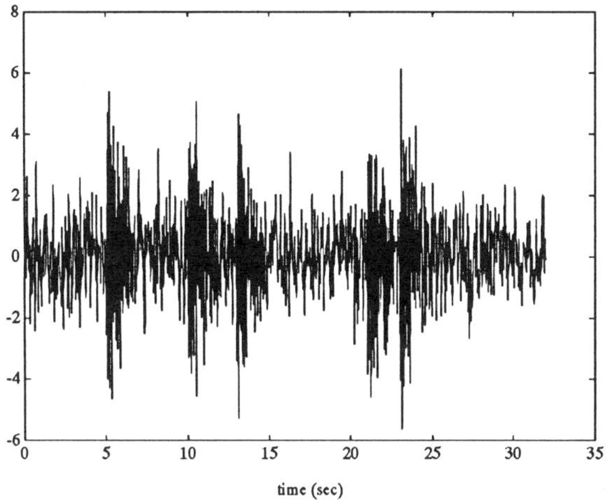

Figure 2: **Time Series of the First Sensor** — The 11-Hertz signal at 1/10 amplitude, additive $\mathcal{N}(0,1)$ noise, and the five Gabor transients.

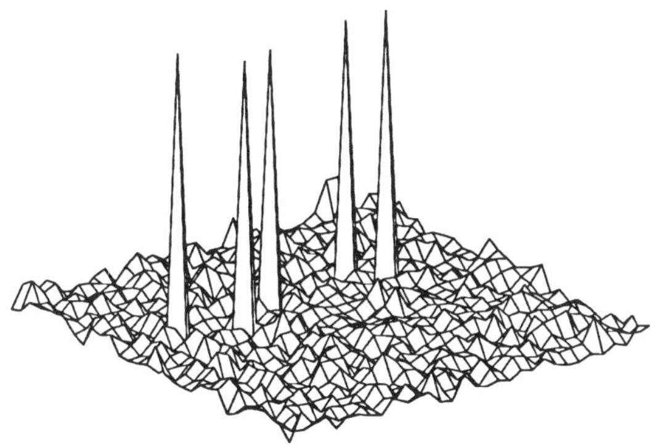

Figure 3: **Gabor Transform of the Time Series of the First Sensor** — The 11-Hertz signal at 1/10 amplitude, additive $\mathcal{N}(0,1)$ noise, and the five Gabor transients.

5 DF via the WVD

This section shows how one *preliminary* estimate of the Wigner-Ville Distribution (WVD) of the array output can be used to DF a source in both time and frequency. This approach is different from that of Swindlehurst and Kailath [21] in which a *spatial* WVD was used to handle the near-field effects and produced both bearing and range estimates. Instead, we show how to use the WVD to DF on a source in the far field, provided the signal admits a WVD sufficiently different from the background noise at a particular time and frequency. As such, this approach could be used to DF on transients.

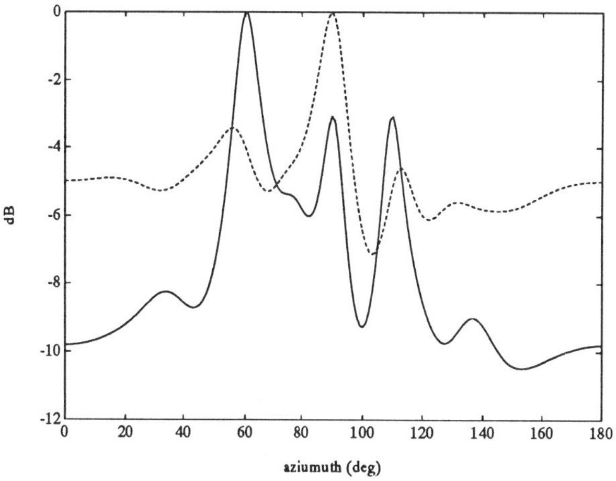

Figure 4: **MVDR using a Gabor Prefilter** — An 11-Hertz signal at 1/10 amplitude from 90° azimuth, additive $\mathcal{N}(0,1)$ noise, and the five Gabor transients. The CSD was estimated at 11 Hertz by averaging 32-sample blocks of the 1,024 array snapshots. The solid line is MVDR without the Gabor prefilter and the dashed is MVDR with the Gabor prefilter.

Given two analytic signals $x(t)$ and $y(t)$, their cross Wigner-Ville Distribution (XWVD) is [2]:

$$W(x,y;t,f) = \int_{-\infty}^{\infty} e^{-i2\pi f} x(t+\tau/2)\bar{y}(t-\tau/2)d\tau.$$

Consider the case of the array output when only one analytic signal is present from the direction $\mathbf{u}_0$. Then $\mathbf{x}(t) = [s(t+\tau_m(\mathbf{u}_0))]$. The XWVD of the mth and m'th sensor outputs is:

$$W(x_m, x_{m'}; t, f) = e^{i2\pi f(\tau_m(\mathbf{u}_0)-\tau_{m'}(\mathbf{u}_0))} \times W(s,s;t+\tau_m(\mathbf{u}_0)/2+\tau_{m'}(\mathbf{u}_0)/2, f).$$

If the XWVD is smoothed in time, then three observations can be made: First, if the time lags are relatively small with respect to the smoothing window, then the phase term is not affected while the time-shifted WVD of the signal may be approximated by the WVD of the signal. In matrix form, we get the estimate of the WVD of the array:

$$\widetilde{W}(\mathbf{x},\mathbf{x};t,f) \approx \mathbf{a}(f,\mathbf{u}_0)\mathbf{a}(f,\mathbf{u}_0)^H W(s,s;t,f),$$

where $\widetilde{W}$ denotes the smoothed WVD of the array output. Second, if noise is present and distributes in a uniform and zero-mean fashion over the time-frequency plane, then the smoothing should have the effect of providing a biased estimate of the WVD of the array but with a lower variance than the "raw" WVD. Third, the form of the estimate permits the application of subspace methods.

The following simulation supports these claims. The 10-sensor linear array spaced at 1/2 wavelength

(f = 11 Hertz) is receiving chirp signals from two sources: one source is broadside to the array (azimuth $\theta = 90°$). and transmits a four-second chirp. The chirp starts at time $t = 0$ at 16 Hertz and sweeps down at 4 Hz/sec. The second source is located at $\theta = 60°$ azimuth and transmits a chirp sweeping up at 4 Hz/sec starting from 0 Hertz. Sensor noise was modeled as i.i.d $\mathcal{N}(0,1)$. A total of N=128 array snapshots were collected at a 32 Hertz sample rate. Figure 5 plots conventional DF at 11 Hertz, the conventional WVD DF, and the MUSIC WVD (one signal). Both WVD DF used time $t = 1.25$ seconds and $f = 11$ Hertz. The conventional DF estimated the CSD matrix using 32-sample time blocks. The XWVD was estimated an FFT size of 64 and rectangular 8-point time-smoothing. The figure shows that both the WVD DF distinguish the two sources while the conventional DF also shows that there are two signals at 11 Hertz at 60° and 90° azimuth.

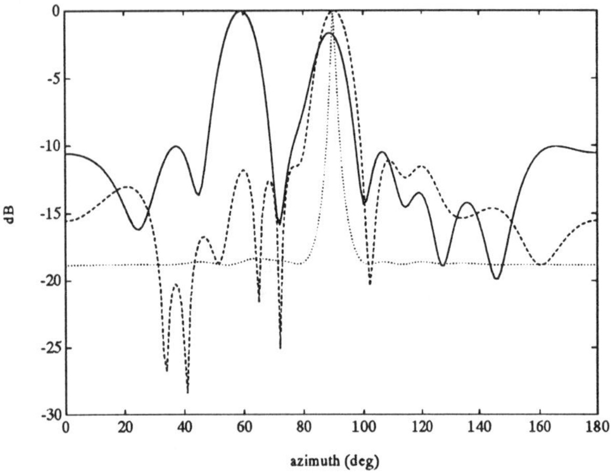

Figure 5: **WVD DF on Two Chirps** — One chirp at 90° azimuth sweeping down from 16 Hertz at -4 Hz/sec. The second chirp at 60° azimuth sweeping up from 0 Hertz at +4 Hz/sec. The solid line is the classical DF, the dashed line is classical WVD DF, and the dotted line is the MUSIC WVD DF (one signal).

6 DF on Cyclostationary Signals

The fundamental stumbling block in treating non-stationary time series is the lack of either a sample ensemble (which would permit a consistent estimation of the CSD matrix), or an ergodic theorem (which can produce a consistent estimate from a sufficiently long realization of the time series). While there is a large literature of attacks on non-stationary time series (see Priestley [17] for an accessible survey), this section will consider harmonizable, cyclostationary time series. These assumptions will permit the estimation of the CSD matrix and readily apply to DF.

Cyclostationary time series have been undergoing an extensive development in the last decade. (the April 1991 issue of the *IEEE Signal Processing Magazine* is devoted to cyclostationary processing). In particular, we single out the excellent Ph.D thesis of Schell [20] which exploits cyclostationarity to perform DF using subspace methods and has performed a Cramer-Rao lower bound analysis. Another approach to DF can be made using spectral estimators obtained by Hurd [7], [8] and is the subject of this section.

Definition 6.1 [15, page 226] A random process $\{x(t)\}$ is called cyclostationary of period $T > 0$ for which the mean and covariance functions are periodic: for all times t, t_1, and t_2, there holds $E[x(t+T)] = E[x(t)]$ and $R_x(t_1,t_2) = R_x(t_1+T, t_2+T)$.

Definition 6.2 [12] A random process $\{x(t)\}$ is called (strongly) harmonizable provided it can be represented in quadratic mean for every time t by the integral

$$x(t) = \int_{-\infty}^{\infty} e^{i2\pi ft} dX(f)$$

where X is a random measure for which the spectral measure defined by $P_x(A \times B) = E[X(A)\overline{X(B)}]$ is of bounded variation.

Hurd [7] points out that when $\{x(t)\}$ is both cyclostationary and harmonizable then the spectrum $P_x(f_1, f_2)$ is confined to diagonal lines in the $f_1 \times f_2$ plane which are parallel to the main diagonal and spaced $1/T$ Hertz apart. In particular, stationary noise will be confined to the main diagonal (f_1=f_2) and DF using $f_1 \neq f_2$ should permit noise suppression. The DF assumptions are as follows:

Assumption 6.1 Single Cyclostationary source, uncorrelated noise:

1. *the plane wave signal $s(t)$ is a strongly harmonizable, cyclostationary random process.*
2. *the noise $\mathbf{z}(t)$ is zero-mean, WSS.*
3. *the signal $s(t)$ and the sensor noise are uncorrelated.*

Since the noise vector $\mathbf{z}(t)$ is WSS and signal $s(t)$ is strongly harmonizable, then the array Then the array output $\mathbf{x}(t)$ is harmonizable with a Fourier measure given by $d\mathbf{X}(f) = \mathbf{a}(f, \mathbf{u}_0)dS(f) + d\mathbf{Z}(f)$, where only $\mathbf{Z}(f)$ is orthogonally scattered. From this representation, the CSD matrix of the array has the form:

$$\begin{aligned} dP_{\mathbf{X}}(f_1, f_2) &= E[d\mathbf{X}(f_1)d\mathbf{X}(f_2)^H] \\ &= dP_s(f_1,f_2)\mathbf{a}(f_1,\mathbf{u}_0)\mathbf{a}(f_2,\mathbf{u}_0)^H + dP_n(f_1,f_2). \end{aligned}$$

Note the following: the noise CSD matrix is supported only on the line of zero frequency difference; the extension to subspace algorithms appears straight-forward; the CSD matrix can be estimated by averaging outer products of FFT vectors of the time segments of the data if the period T is known.

The following simulation examines these claims. A 10-sensor linear array spaced at 1/2 wavelength (f = 11 Hertz) is receiving two signals: the first is a cyclostationary signal generated by amplitude-modulating $\mathcal{N}(0,1)$ noise at 4 Hertz. This AM signal is normalized by setting the sample standard deviation to one and arrives broadside to the array (90°

azimuth). The second signal is an 11-Hertz complex exponential arriving at 75° azimuth. Sensor noise is i.i.d. $\mathcal{N}(0,1)$. A total of 256 array snapshots were collected at a 128 Hertz sample rate. The classical DF estimated the CSD matrix by averaging 32-point blocks at 11 Hertz. The cyclostationary DF worked by picking a look direction and then feeding the output of the conventional delay-sum beamformer into a estimator of the two-dimensional spectrum $P_x(f_1, f_2)$ at the point $(f_1, f_2) = (11, 15)$ Hertz. The spectral estimator performed diagonal smoothing by averaging 11 diagonal bins spaced 0.5 Hertz apart centered about at (11,15). The results are displayed in figure 6. The classical DF localizes the 11-Hertz signal while the cyclostationary DF locates the AM signal.

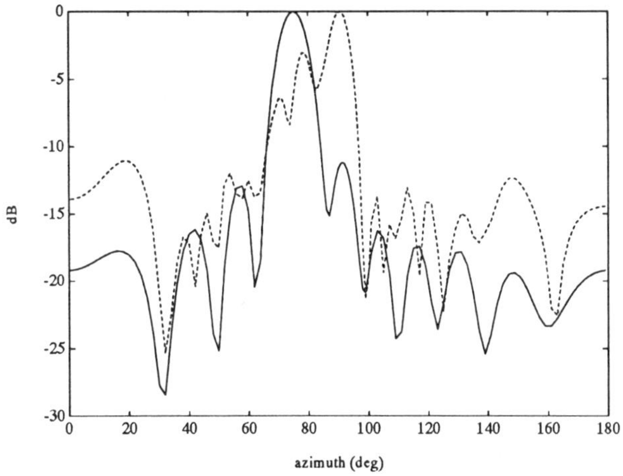

Figure 6: **DF on Cyclostationary Signals** — Gaussian AM signal at 90° azimuth; 11 Hertz signal at 75° azimuth. The solid line is the classical DF; the dashed line is the cyclostationary DF.

7 DF on a Chaotic Signal

In the past decade, there has been a surge of publications in chaotic systems. Of particular interest for sonar applications, we single out the work of K. Karagiannis [11] which observed chaotic behavior in the noise generated by rattling gearboxes. One basic model of chaotic behavior is the conventional logistic map (CLM) [23]:

$$x(t+1) = \mu x(t)(1 - x(t)).$$

One stochastic generalization of the CLM is the Model 2 of [6]: for $t = 1, 2, \ldots$

$$x(t+1) = k(a,b)x(t)^a(1 - x(t))^b u(t+1)^c,$$

where a, b, and c are non-negative, $\{u(t)\}$ are independent and identically distributed uniformly on the interval $[0,1]$, $k(a,b) = (a+b)^{a+b}/(a^a b^b)$, and the initial value is set as $x(0) = u(0)^c$. Estimators of a, b, c were obtained and it was observed that Model 2 provided a good fit to time series generated by the CLM.

This section demonstrates that the Model 2 estimators of a, b, and c can apply to source localization. The basic idea is to assume that a signal $s(t)$ is of the form of the Model 2. A look direction is picked and the time series resulting from the conventional delay-sum beamforming is fed into the estimation scheme. The one-step mean-squared prediction error is calculated and its reciprocal is plotted as a function of the look direction. If the look direction matches the signal direction, then a good fit of the signal is obtained which results in a small prediction error. Otherwise, a poor fit occurs which results in a large prediction error. We compare this Fractal DF with the conventional DF and the MUSIC DF in the following simulations.

We use a 10-sensor linear array with 1/2 wavelength spacing ($f = 11$ Hertz). One signal is placed broadside to the array (azimuth $\theta = 90°$) and is generated from the CLM ($\mu = 4$, $s(0) = 1/3$). An second signal was generated from the CLM and arrives on the array from 95° azimuth. The arrival was simulated using the Fourier method: the FFT of the time series was multiplied by the appropriate phase for each sensor and the inverse FFT determined the signal at each sensor. Sensor noise was modeled as i.i.d zero mean Gaussian with the standard deviation set so the SNR=1. A total of N=128 array snapshots were collected at a 100 Hertz sample rate. The Classical and MUSIC DF were operated at 11 Hertz and the CSD matrix was estimated using 32-point block-averaging. MUSIC assumed two sources. All DFs locate the source at 90° azimuth though the beamwidth of the fractal DF is much narrower than the other DFs. The results are shown in figure 7. The Conventional and MUSIC DFs could not resolve this sources, but the Fractal DF is correctly indicating the signal directions at 90° and 95° azimuth.

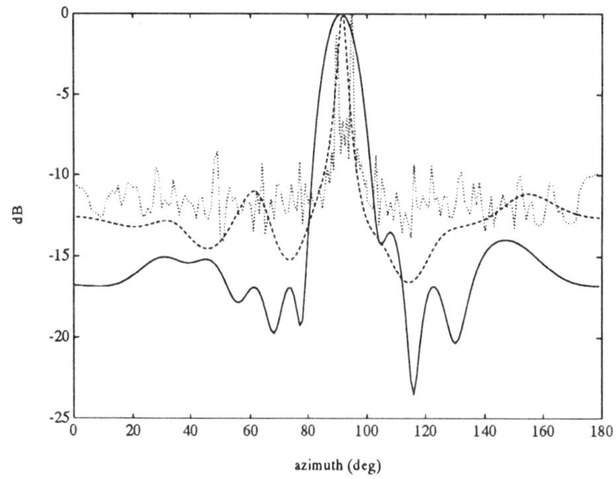

Figure 7: **DF on a Two Chaotic Signals** — Two CLM signals at 90° and 95° azimuth. The solid line is the classical DF, the dashed line is MUSIC, and the dotted line is the Fractal DF.

8 Conclusions

The preceding sections have shown that Johnson's [10] approach to bearing estimation as form of spectral estimation readily generalizes whenever a statistically consistent estimator of a signal attribute can be found.

References

[1] Allen, J. C., *Direction-Finding and Database Techniques*, NOSC Technical Note 1657, Naval Ocean Systems Center, San Diego, CA 92152, 1990

[2] Flandrin, P., "A Time-Frequency Formulation of Optimum Detection", *IEEE Transactions on ASSP*, Volume 36, Number 1, pages 1377–1384, 1988

[3] Forster, P. and C. L. Nikias, "Bearing Estimation in the Bispectrum Domain", *IEEE Transactions on Signal Processing*, Volume 39, Number 9, pages 1994–2006, 1991

[4] Friedlander, B. and B. Porat, "Detection of Transient Signal by the Gabor Representation", *IEEE Transactions on ASSP*, Volume 37, Number 2, pages 169–179, 1989

[5] Gabr, M. M. "On the Third-Order Moment Structure and Bispectral Analysis of Some Bilinear Time Series", *Journal of Time Series Analysis*, Volume 9, Number 1, pages 11–20, 1988

[6] Gerr, N. L. and J. C. Allen, "Multiplicative Models for Bounded Time Series", Submitted to the *Journal of Time Series Analysis*, September, 1991

[7] Hurd, H. L., "Representation of Strongly Harmonizable Periodically Correlated Processes and Their Covariances", *Journal of Multivariate Analysis*, Volume 29, Number 1, pages 53–67, 1989

[8] Hurd, H. L., "Nonparametric Time Series Analysis for Periodically Correlated Processes", *IEEE Transactions on Information Theory*, Volume 35, Number 2, pages 350–359, 1989

[9] Hurd, H. L. and N. L. Gerr, "Graphical Methods for Determining the Presence of Periodic Correlation", *preprint*, 1990

[10] Johnson, D. L., "The Application of Spectral Estimation Methods to Bearing Estimation", *Proceedings of the IEEE*, Volume 70, Number 90, pages 1018–1028, 1982

[11] Karagiannis, K., "Rattling Vibrations in Gearboxes", *Lecture Notes in Physics, Volume 355*, Springer-Verlag, Berlin-New York, pages 175–195, 1990

[12] Loève, M., *Probability Theory*, D. Van Nostrand Company, Inc., New York, 1963

[13] Mendel, J. M., "Tutorial on Higher-Order Statistics (Spectra) in Signal Processing and System Theory: Theoretical Results and Some Applications", *Proceedings of the IEEE*, Volume 79, Number 3, pages 278–304, 1991

[14] Nikias, C. L. and M. R. Raghuveer, "Bispectrum Estimation: A Digital Signal Processing Framework", *Proceedings of the IEEE*, Volume 75, Number 7, pages 869–891, 1987

[15] Papoulis, A., *Probability, Random Variables, and Stochastic Processes*, second edition, McGraw-Hill, New York, 1984

[16] Porat B. and B. Friedlander, "Direction Finding Algorithms Based on High-Order Statistics", *IEEE Transactions on Signal Processing*, Volume 39, Number 9, pages 2016–2024, 1991

[17] Priestley, M. B., *Non-Linear and Non-Stationary Time Series Analysis*, Academic Press, New York, 1988

[18] Rao, T. Subba and M. M. Gabr, *An Introduction to Bispectral Analysis and Bilinear Time Series Models*, Lecture Notes in Statistics, Volume 24, Springer-Verlag, New York, 1984

[19] Rosenblatt, M., *Stationary Sequences and Random Fields*, Birkhäuser, Boston, 1985

[20] Schell, S. V., *Exploitation of Spectral Correlation for Signal-Selective Direction Finding*, Ph.D thesis, Department of Electrical Engineering and Computer Science, University of California, Davis, 1990

[21] Swindlehurst, A. L. and T. Kailath, "Passive Direction-of-Arrival and Range Estimation for Near-Field Sources", *Fourth Annual ASSP Workshop of Spectrum Estimation and Modeling*, Spring Hill Conference Center, Minneapolis, Minnesota, IEEE Inc., pages 123–128, August 3–5, 1988

[22] Squire, B., *A Tutorial Survey of Time-Frequency Analysis and Applications*, Linear Measurements, 1991

[23] Thompson, J. M. T. and H. B. Stewart, *Nonlinear Dynamics and Chaos*, John-Wiley and Sons, New York, 1986

Reduced Variance Distortionless Response (RVDR) Performance with Signal Mismatch

Henry Cox
ORINCON Corporation
1755 Jefferson Davis Hwy., Suite 904
Arlington, VA 22202

Robert M. Zeskind
BBN Systems and Technologies
1300 N 17th Street
Arlington, VA 22209

Abstract

Matched field processing of underwater acoustic fields frequently involves considerable mismatch between the actual received signal and the model of the complicated propagation from various possible source locations to the receiving array. This leads to significant signal suppression in the MVDR processor. The high sidelobes of the conventional processor make it unsatisfactory. In this paper we examine the effectiveness of modifying the diagonal of the spatial correlation matrix. Performance in the face of mismatch is quantified. A simple parametric relationship between MUSIC, MVDR and conventional beamforming is developed.

1 Introduction

The problem of spatial array processing when knowledge of the signal spatial correlation structure is imprecise is one of great practical importance and one of long term interest to us [1-5]. An extensive list of references was provided in [5]. This paper is motivated by the matched field problem of underwater acoustics [6-11] in which processing is based on a model of the complicated propagation between the receiving array and possible source locations rather than the usual plane wave model. Relatively large mismatch between the model and the actual signal of two dB or so is common. We will concentrate on the old approach [1,3,5,11,12] of adding a constant to the diagonal elements of the spatial correlation matrix **R**, prior to the matrix inversion. In particular we will be concerned with how much should be added to protect against significant mismatch.

Let $\tilde{\mathbf{R}} = \mathbf{R} + \varepsilon \mathbf{I}$ be the augmented spatial correlation matrix. A simple and obvious approach is to use $(\mathbf{m}^*(\theta)\tilde{\mathbf{R}}^{-1}\mathbf{m}(\theta))^{-1}$ as the response function where $\mathbf{m}(\theta)$ is the modeled signal directional vector and θ is a signal location parameter. Bold lower case letters are used to denote column vectors. The asterisk means complex conjugate transposition. This approach is useful with small values of ε. However, it is not appropriate as a protection against significant amounts of signal mismatch which would require a large value of ε thereby raising the background response level in directions away from the signal. We will therefore use the augmented matrix in calculation of weights to be applied to the unaugmented matrix **R**.

Specifically we consider the narrow band distortionless response estimator [1,3,5,11,12]

$$\mathbf{P}_{GDR}(\theta,\varepsilon) = \frac{\mathbf{m}^*(\theta)\tilde{\mathbf{R}}^{-1}\mathbf{R}\,\tilde{\mathbf{R}}^{-1}\mathbf{m}(\theta)}{(\mathbf{m}^*(\theta)\tilde{\mathbf{R}}^{-1}\mathbf{m}(\theta))^2} \quad (1)$$

where

$$\tilde{\mathbf{R}} = \mathbf{R} + \varepsilon\mathbf{I} \quad (2)$$

Let
$$\mathbf{w} = \tilde{\mathbf{R}}^{-1}\mathbf{m} / (\mathbf{m}^*\tilde{\mathbf{R}}^{-1}\mathbf{m}) \quad (3)$$

Then
$$\mathbf{P}_{GDR}(\theta,\varepsilon) = \mathbf{w}^*(\theta,\varepsilon)\mathbf{R}\mathbf{w}(\theta,\varepsilon) \quad (4)$$

Thus if **x** is a data vector of element outputs, $\mathbf{P}_{GDR}(\theta,\varepsilon)$ may be interpreted as the average power at the output of a beamformer which uses **w** for weights

$$\mathbf{P}_{GDR}(\theta,\varepsilon) = E[|\mathbf{w}^*(\theta,\varepsilon)\mathbf{x}|2] \quad (5)$$

We call $\mathbf{P}_{GDR}(\theta,\varepsilon)$ the Generalized Distortionless Response (GDR) estimator. When ε is restricted to being positive we call it the Reduced Variance Distortionless Response Estimator (RVDR). When $\varepsilon = 0$, it is the same as Minimum Variance Distortionless Response (MVDR) [13].

2 Single Source

In this section we examine the behavior of (1) when **R** is due to a single source in white noise so that

$$\mathbf{R} = \sigma_n^2\mathbf{I} + \sigma_s^2\mathbf{d}\mathbf{d}^* = \sigma_n^2(\mathbf{I} + r\,\mathbf{d}\mathbf{d}^*) \quad (6)$$

where **d** is the signal direction vector. We use the normalization $\mathbf{d}^*\mathbf{d} = M$ where M is the number of elements in the array. Then r is the element signal-to-noise ratio.

$$\tilde{\mathbf{R}} = \mathbf{R} + \varepsilon\,\mathbf{I} = (\sigma_n^2 / p)(\mathbf{I} + pr\,\mathbf{d}\mathbf{d}^*) \tag{7}$$

where

$$p = \sigma_n^2 / (\sigma_n^2 + \varepsilon) \tag{8}$$

is the fractional change in signal-to-noise due to the augmentation of the diagonal.

When **R** is given by (6),

$$P_{GDR} = \sigma_n^2[\mathbf{w}^*\mathbf{w} + r|\mathbf{w}^*\mathbf{d}|^2] \tag{9}$$

and

$$\tilde{\mathbf{R}}^{-1} = (p/\sigma_n^2)[\mathbf{I} - \mathbf{d}\mathbf{d}^*(pr/1 + Mpr)] \tag{10}$$

We normalize **m** so that $\mathbf{m}^*\mathbf{m} = M$, and using the notation of [2] define

$$\cos^2(\mathbf{m},\mathbf{d}) = |\mathbf{m}^*\mathbf{d}|^2/M = 1 - \sin^2(\mathbf{m},\mathbf{d}) \tag{11}$$

This is the measure of mismatch. For 2 dB of mismatch, $10\log[\cos^2(\mathbf{m},\mathbf{d})] = -2$, $\cos^2(\mathbf{m},\mathbf{d}) = 0.63$, and $\sin^2(\mathbf{m},\mathbf{d}) = 0.37$.

Let

$$g(\theta,\varepsilon) = (1 + Mpr\sin^2(\mathbf{m},\mathbf{d}))^{-1} \tag{12}$$

Then straight forward substitution leads to the following equations:

$$\mathbf{w}^* = [(1 + Mpr)\mathbf{m}^* - pr(\mathbf{m}^*\mathbf{d})\mathbf{d}^*]g\,M^{-1} \tag{13}$$

$$\mathbf{w}^*\mathbf{w} = [1 + (2 + Mpr)Mpr\sin^2(\mathbf{m},\mathbf{d})]M^{-1}g^2 \tag{14}$$

$$\mathbf{w}^*\mathbf{d} = \mathbf{m}^*\mathbf{d}\,M^{-1}g \tag{15}$$

$$|\mathbf{w}^*\mathbf{d}|^2 = \cos^2(\mathbf{m},\mathbf{d})g^2 \tag{16}$$

$$P_{GDR} = (\sigma_n^2/M)\,g^2\{1 + Mr - \sin^2(\mathbf{m},\mathbf{d})\,[(1 + Mr) - (1 + Mpr)^2]\} \tag{17}$$

Under conditions of perfect match $\cos^2(\mathbf{m},\mathbf{d}) = 1$ and $\sin^2(\mathbf{m},\mathbf{d}) = 0$. Then $\mathbf{w}^* = \mathbf{m}^*/M$ the conventional beamformer, and $\mathbf{w}^*\mathbf{w} = 1/M$.

The response is called distortionless because $\mathbf{w}^*\mathbf{d} = 1$ when there is perfect match, that is when $\mathbf{m} = \mathbf{d}$. Then (17) becomes.

$$P_{GDR} = \sigma_n^2(1 + Mr)/M = \sigma_s^2 + \sigma_n^2/M \tag{18}$$

When **m** and **d** are orthogonal there is complete mismatch; $\sin^2(\mathbf{m},\mathbf{d}) = 1$ Then,

$$P_{GDR} = \sigma_n^2/M \tag{19}$$

An important property of the GDR estimator is that, under conditions of perfect match or complete mismatch, it does not depend on ε the amount of diagonal augmentation. The fact that this property is not shared by the estimator $(\mathbf{m}^*\tilde{\mathbf{R}}^{-1}\mathbf{m})^{-1}$ is a principal drawback in its use. The weight vector **w** of course always depends on ε.

When there is no augmentation of the diagonal, $\varepsilon = 0$, $p = 1$ and

$$P_{GDR}(\theta,0) = (\mathbf{m}^*\tilde{\mathbf{R}}^{-1}\mathbf{m})^{-1} = (\sigma_n^2/M)(1 + Mr)g \tag{20}$$

which is the well known result for the minimum variance distortionless response processor [13]. As $\varepsilon \to \infty$ and $p \to 0$, $\mathbf{w}^* \to \mathbf{m}^*/M$ the conventional beamformer. Thus, as ε is varied between 0 and ∞, P_{GDR} provides a continuous parameterization between the MVDR and the conventional array processor [3]. When we apply (1) to data it is useful to also plot

$$\mathbf{w}^*\mathbf{w} = \frac{\mathbf{m}^*\tilde{\mathbf{R}}^{-2}\mathbf{m}}{(\mathbf{m}^*\tilde{\mathbf{R}}^{-1}\mathbf{m})^2} \tag{21}$$

since this is a well known [1,3,5,11,12] sensitivity measure. For the case of a single source in spatially white noise this reduces to (14).

In this case $\mathbf{w}^*\mathbf{w} = 1/M$ for both perfect match and mismatch. It reaches a maximum [11] of $(2 + Mpr)^2/[4M(1 + Mpr)]$

at $\sin^2(\mathbf{m},\mathbf{d}) = (2 + Mpr)^{-1}$.

The question concerning how large ε should be to guard against a specific amount of mismatch can be answered by examining the width of the peak of P_{GDr}. We normalize the peak to unity by dividing (17) by (18).

This yields

$$P_{NORM} = \cos^2(\mathbf{m},\mathbf{d})g^2 + \sin^2(\mathbf{m},\mathbf{d})g^2(1+Mpr)^2/(1+Mr) \quad (22)$$

This quantity is plotted in Fig. 1 for a 10 element half wavelength spaced line array with 10 dB element signal level and zero dB element noise level. The augmentation value ε is varied from 0 to 10000 showing the transition from MVDR to convectional beamformer response.

For large values of ε, $p \approx \sigma_n^2/\varepsilon$. Near the peak when both Mr and ε are large, the first term of (22) dominates over the second. Since the factor $\cos^2(\mathbf{m},\mathbf{d})$ is the conventional response g^2 gives the amount of additional signal suppression over the conventional beamformer. An interesting case is 6 dB of additional suppression so that $g^2 = .25$.. Then the required value of ε is approximately $\varepsilon \approx \sin^2(\mathbf{m},\mathbf{d})\, Mr\sigma_n^2$. For 2 dB of mismatch this becomes $\varepsilon = .37M\sigma_s^2$. This is a large value for ε Since the trace of **R** is $M(\sigma_s^2+\sigma_n^2)$, and $\sigma_s^2 >> \sigma_n^2$ for strong signals, $\varepsilon \approx .37\ \mathrm{tr}(\mathbf{R})$ Thus, in applying (1) to field data we have found it useful to choose ε to be a substantial fraction of the trace of **R**. Figure 2. presents normalized response vs. Mr for several values of Mpr.

3 Multiple Signals

We now consider the situation of K signals in spatially white noise, where K is less than M.

$$\mathbf{R} = \sigma_n^2[\mathbf{I} + \mathbf{U}\mathbf{D}(a_i)\mathbf{U}^*] \quad (23)$$

where **U** is a M by K matrix of K orthonormal column vectors which span the signal subspace. $\mathbf{D}(a_i)$ is a K by K diagonal matrix with a_i as its i-th entry. $\tilde{\mathbf{R}}$ is obtained by adding $\varepsilon\,\mathbf{I}$ to $\mathbf{R}$, thus

$$\tilde{\mathbf{R}} = (\sigma_n^2/p)[\mathbf{I} + \mathbf{U}\mathbf{D}(pa_i)\mathbf{U}^*] \quad (24)$$

Then

$$\tilde{\mathbf{R}}^{-1} = (\mathbf{p}/\sigma_n^2)[\mathbf{I} - \mathbf{U}\mathbf{D}(\frac{pa_i}{1+pa_i})\mathbf{U}^*] \quad (25)$$

Let

$$h(\theta,\varepsilon) = \{1 - M^{-1}\Sigma|\mathbf{m}^*\mathbf{u}_i|^2(pa_i/(1+pa_i))\}^{-1} \quad (26)$$

Then using (29) in (3)

$$\mathbf{w}^* = (\mathbf{m}^*/M)h[\mathbf{I} - \Sigma\mathbf{u}_i\mathbf{u}_i^*(pa_i/(1+pa_i))] \quad (27)$$

This is a beamformer which steers nulls of finite depth in the direction of all signal components. Using (25) and (23), yields

$$P_{GDR} = (\sigma_n^2/M)\,h^2[1 - M^{-1}\Sigma|\mathbf{m}^*\mathbf{u}_i|^2(1-\frac{1+a_i}{(1+pa_i)^2})] \quad (28)$$

and (21) becomes

$$\mathbf{w}^*\mathbf{w} = M^{-1}h^2[1 - M^{-1}\Sigma|\mathbf{m}^*\mathbf{u}_i|^2(1-\frac{1}{(1+pa_i)^2})] \quad (29)$$

Equation (27), (28), and (29) are the general results for K signals in spatially white noise. For $p=1$ they reduce to the MVDR results. For $p=0$ (or $\varepsilon=\infty$) they reduce to conventional beamformer results.

We may generalize by allowing ε to be negative so that the diagonal of **R** is reduced to form $\tilde{\mathbf{R}}$. This is essentially the enhanced minimum variance approach introduced by Owsley [14]. Then as ε approaches $-\sigma_n^2$,p approaches infinity and $\tilde{\mathbf{R}}$. tends toward singularity. In this situation the beamformer nulls approach infinite depth and the response (28) approaches

$$P_{GDR} = (\sigma_n^2/M)\,[1 - M^{-1}\Sigma|\mathbf{m}^*\mathbf{u}_i|^2]^{-1} \quad (30)$$

which is the MUSIC response [14,15]. Thus P_{GDR} provides a continuous parametrization from MUSIC to MVDR to conventional beamforming as ε is varied from $-\sigma_n^2$ to zero to $+\infty$. Unforunately allowing ε to be negative can lead to peak splitting. This is illustrated in Fig 3 which shows the response to a single signal for the same ten element array considered earlier. Now the element level noise is zero dB and the element level signal is -10 dB. Peak splitting is shown

shown for epsilon equal to -0.9. This behavior can be explained by examining (9) and (14). For p sufficiently positive $\mathbf{w}^*\mathbf{w}$ grows faster than the second term in (9) giving rise to peaks at slight amounts of mismatch. The exact location and height of the peaks can be found by differentiating (17).

4 Conclusion

The generalized distortionless response processor provides a continuous tradeoff between resolution and robustness. Achieving robustness to significant signal model errors requires use of a large value for ε, that is a considerable fraction of trace $\mathbf{R}$. Using negative values for ε increases resolution over MVDR and MUSIC is approached as ε approaches $-\sigma_n^2$ and $\tilde{\mathbf{R}}$ approaches singularity. However, negative values of epsilon can lead to peak splitting. An important feature of the GDR approach is that the weights may be used in beamformers to produce beam output time series and are not limited to direction estimation.

References

[1] H. Cox, "*Sensitivity Considerations in Adaptive Beamforming,*" in Signal Processing (Pro, NATO Advanced Study Inst. Signal Processing with particular Reference to Underwater Acoust., Loughborough, U.K., (Aug. 1972.) J.W. R. Griffiths, P.L. Stocklin, and C. Van Schooneveld, Eds. New York and London: Academic, 1973.

[2] H. Cox, "*Resolving Power and Sensitivity to Mismatch of Optimum Array Processors,*" J. Acoust. Soc. Amer., vol. 54 no. 3, pp. 771-785, 1973. Also, in Selected papers in Multidimensional Signal Processing. New York: IEEE Press, 1986

[3] H. Cox, R.M. Zeskind, and T. Kooij, "*Practical Supergain,* " IEEE Trans. Acoust., Speech, Signal Processing, vol. ASSP-34, pp. 393-398 , July 1986.

[4] H. Cox, R.M. Zeskind, and M.M. Owen, "*Effects of Amplitude and Phase Errors on Linear Prediction Array Processor*", IEEE Trans. Acoust., Speech, Signal Processing vol 36 pp. 10-19, Jan 1988.

[5] H. Cox, R.M. Zeskind, and M.M. Owen, "*Robust Adaptive Beamforming*", IEEE Trans. Acoust. Speech, Signal Processing, Vol ASSP 35 pp 1365-1376, Oct. 1987.

[6] H. Cox, R.M. Zeskind, M. Meyers, "*A Subarray Approach to Matched Field Processing*", J. Acoust. Soc. Amer. 87 pp. 168-178, Jan 1990

[7] A. Baggeroer, W.A. Kuperman, and H. Schmidt, "*Matched-field Processing: Source Localization in Correlated Noise as an Optimum Parameter Estimation Problem,*" J. Acoust. Soc. Am 83, 571-587 (1988).

[8] H.P. Bucker, "*Use of Calculated Sound Fields and Matched-field Detection to Locate Sound Sources in Shallow Water,* "J. Acoustic. Soc. Am. 59-368-373 (1976).

[9] H. Schmidt, A.B. Baggeroer, W.A. Kuperman and E.K. Scheer "*Robust Beamforming for Matched Field Processing Under Realistic Environmental Conditions*". in Underwater Acoustic Data Processing Y,T, Chan(ed), Kluwer 1989

[10] J. Krolik, J. Lynch, D. Swingler, "*A Robust Incoherent Matched Field Processor for Source Location in Uncertain Multi-path Fields*". Proc ICASSP-89 pp 2637-2640, IEEE 1989

[11] P.M. Velardo Jr. "*Robust Matched Field Source Localization*" MS Thesis, MIT 1989

[12] E.N Gilbert and S.P. Morgan, "*Optimum Design of Drective Antenna Arrays Subject to Random Variations*""Bell Syst. Tech. J., vol. 34, pp. 637-663, May 1955.

[13] J, Capon, "High-Resolution Frequency-Wave-number Spectrum Analysis, Proc. IEEE, vol 57, pp. 1408-1418, Aug. 1969.

[14] N. Owsley, " Enhanced Minimum Variance Beamforming" in Underwater Acoustic Data Processing Y.T. Chan(ed.) Kluwer 1989.

[15] R.Schmidt, "*Multiple Emitter Location and Signal Parameter Estimation*" Proc. RADC Spectral Estimation Workshop, Rome Air Development Center, Rome NY., 1979.

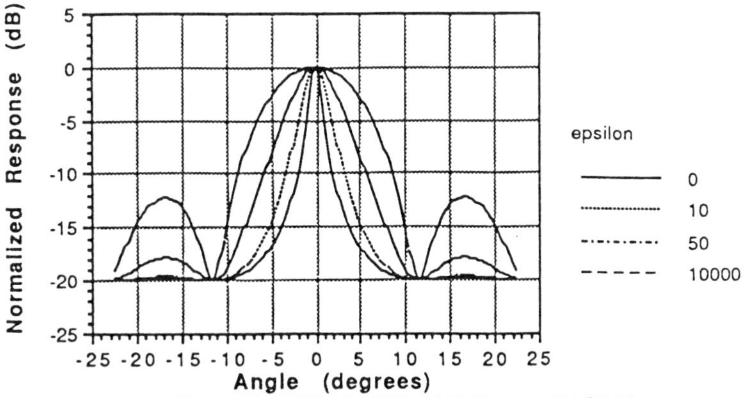

FIGURE 1 NORMALIZED RESPONSE vs ANGLE
Mr = 100

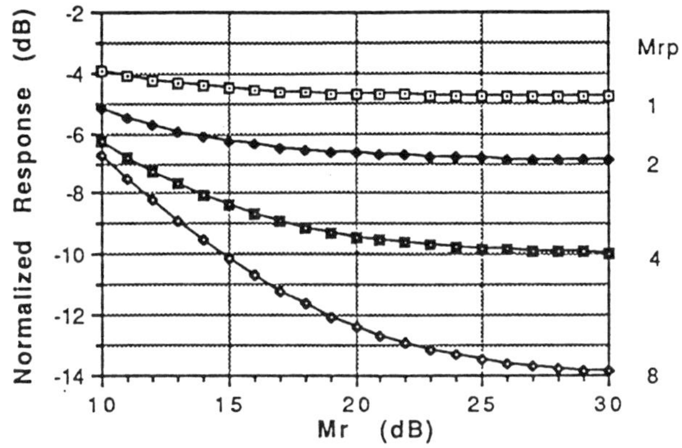

FIGURE 2 NORMALIZED RESPONSE vs
BEAM SNR . 2 dB Mismatch

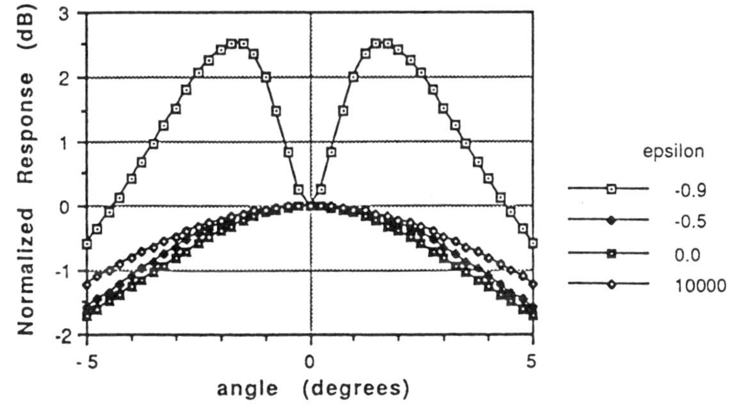

FIGURE 3 NORMALIZED RESPONSE vs ANGLE
SHOWING SPLIT PEAK Mr = 1

Performance Analysis of an Order Statistic-Based Radar Threshold Normalizer in Weibull Noise

R. Rifkin

The MITRE Corporation
Burlington Rd.
Bedford, MA 01730

Abstract

Recent interest has focused on order statistic-based algorithms for calculating detection thresholds. This paper extends previous analyses of these algorithms, to determine closed-form approximations for the signal-to-noise ratio required to achieve a particular probability of detection in environments whose amplitude statistics are modeled by the Weibull distribution. Performance is evaluated in both homogeneous and inhomogenous noise. The analysis shows that the order statistic-based algorithm is quite robust against interference.

1 Introduction

Detection algorithms that operate in nonstationary noise environments generally use an adaptive threshold algorithm to maintain a constant false alarm rate (CFAR). These algorithms estimate the noise energy in the target test cell, and then adjust the detection threshold to reflect changes in this energy at different test cell positions. This maintains a CFAR in all noise backgrounds. The threshold algorithms use detection cells near the target test cell to estimate the background noise level, and then set the threshold to guarantee the desired false alarm probability, P_{fa}.

While simple averaging [1] algorithms perform well in homogeneous, Rayleigh-distributed noise backgrounds, their performance is degraded in either inhomogeneous or non-Rayleigh distributed noise. Non-Rayleigh noise appears in many radar environments, which can often be modeled by the Weibull distribution [2,3]. These environments may include inhomogeneities caused by clutter patches, which can cause the threshold to be seriously overestimated, substantially increasing the signal-to-noise ratio (SNR) required to achieve a particular probability of detection, P_{d}. Averaging algorithms are likely to perform poorly in such environments. Several non-linear estimation techniques have been proposed to reduce the above problems. One such algorithm is based on order statistics (OS) [4,5], which use the kth largest of the $2N$ range cells to determine the threshold.

Analysis of these threshold setting algorithms has generally relied on either Monte Carlo simulation, or closed-form techniques. Monte Carlo simulation has the drawbacks of both requiring substantial computer time, and lacking precision, particularly when the probability density functions used to model the noise have long tails. Closed-form analysis is preferable, but is usually restricted to exponential models for the P_{d} versus SNR relationship for the target, for analytical simplicity. This model precludes examining the performance of the estimation technique in different noise environments. Moreover, closed-form analysis can be difficult to implement when the assumed noise model includes inhomogeneities (i.e., interference in a noise estimation cell).

The approximation technique developed here allows a closed-form analysis to be applied to a variety of noise models. The technique extends previous analyses, yielding closed-form approximations for a class of order statistic-based algorithms under a general noise model. Closed-form approximations are determined for the SNR required to achieve a particular P_{d} for noise environments whose amplitude statistics are modeled by the Weibull distribution. Performance is evaluated when all the estimation cells respond to Weibull noise, and when interference is present in one of the estimation cells. The performance is evaluated as a function of the skewness of the Weibull noise distribution, the number of cells used to estimate the noise energy, the selected order statistic cell, and the required P_{fa}.

Section 2 discusses the theoretical model used

to evaluate the algorithm's performance numerically. The model includes the effects of interference in one of the noise estimation cells. The performance evaluation uses the average detection threshold (ADT) [4,5] as a figure of merit. Section 3 presents results on the algorithm's performance.

2 Mathematical Model

Figure 1 shows the proposed algorithm. The box labeled "test cell" represents the cell that is currently being examined for the presence of a target. To make this determination, the amplitude of the matched filter output for this cell is compared with a threshold. The threshold estimate is based on the $2N$ cell values on both sides of the test cell. Both sets of N cells are sorted according to amplitude, and the amplitudes of the kth largest cells in both sets are forwarded to a "greatest of" (GO) calculator whose output is multiplied by a predetermined constant to obtain the threshold.

The closed-form approach is used here to obtain both the threshold multiplier required to guarantee a particular P_{fa}, and an approximation of the SNR required to guarantee a particular P_{d}. These calculations are performed for various noise environments. The key input to the closed-form approach is the distribution function(s) used to model the noise.

The amplitude distribution of the input noise samples is modeled by the Weibull distribution, whose cumulative distribution function, $Q_{\mathrm{W}}(x)$), is given by:

$$Q_{\mathrm{W}}(x) = \begin{cases} 1 - e^{-x^b/\alpha}, & x \geq 0 \\ 0, & \text{otherwise} \end{cases} \tag{1}$$

where α and b are constants greater than zero and $Q_{\mathrm{W}}(x) = \mathrm{P}(\mathrm{W} \leq x)$. This distribution reduces to the Rayleigh distribution for $b = 2$. Smaller values of b increase the skewness of this distribution, and allow the simulation of "spikey" noise. The Weibull distribution is convenient because it allows the skewness of the distribution to be changed with a single parameter.

Figure 1 identifies the probability functions that describe the random variables along the signal path of the threshold normalizer. The probability density functions $p_{\mathrm{R}}(x)$, $p_{\mathrm{L}}(x)$, $p_{\mathrm{GO}}(x)$, and $p_{\mathrm{T}}(x)$ refer to the density functions of the output of the two order statistic calculators, the GO calculator, and the coefficient multiplier, respectively. The subscripts L and R refer to the leading and lagging estimation cell order statistic calculators, respectively. A similar notation refers to the cumulative distribution functions, given by $Q(x)$.

The following analysis determines the threshold multiplier γ required to guarantee a particular P_{fa}. The P_{fa} is the probability that a particular empty (i.e., non-target) sample of the test cell exceeds the threshold. At a particular threshold level x, this is $1 - Q_{\mathrm{W}}(x)$. Since the threshold is modeled as a random variable, the P_{fa} is given by:

$$P_{\mathrm{fa}} = 1 - \int_0^{\infty} Q_{\mathrm{W}}(x) p_{\mathrm{T}}(x)\, dx. \tag{2}$$

To determine the P_{fa}, the density function of the threshold is determined as follows.

The derivation proceeds backwards from the output of the multiplier. All the $2N$ random variables in the estimation cells are assumed independent, and initially, identically distributed. The cumulative distribution of the random variable at the output of the GO device is then given by [6]:

$$Q_{\mathrm{GO}}(x) = Q_{\mathrm{L}}(x) Q_{\mathrm{R}}(x). \tag{3}$$

The density function $p_{\mathrm{GO}}(x)$ is given by the derivative of equation (3). Since the threshold is obtained by multiplying samples of the random variable out of the GO calculator by γ, the density function at the output of the multiplier is given by:

$$p_{\mathrm{T}}(x) = \frac{d}{dx}\left[Q_{\mathrm{L}}\left(\frac{x}{\gamma}\right) Q_{\mathrm{R}}\left(\frac{x}{\gamma}\right)\right]. \tag{4}$$

The random variables at the input to the order statistic calculator are assumed independent and identically distributed; hence the density function at the output is given by [6]:

$$p_{\mathrm{L,R}}(x) = k\binom{N}{k}[1 - Q_{\mathrm{W}}(x)]^{N-k}[Q_{\mathrm{W}}(x)]^{k-1} p_{\mathrm{W}}(x) \tag{5}$$

where $p_{\mathrm{W}}(x)$ is the density function of the Weibull distribution, obtained from its first derivative. The associated cumulative distribution function is given by:

$$Q_{\mathrm{L,R}}(x) = k\binom{N}{k} \int_0^{Q_{\mathrm{W}}(x)} (1-u)^{N-k} u^{k-1}\, du \tag{6}$$

where $u = Q_{\mathrm{W}}(x)$. The integral is now easily evaluated with a binomial expansion:

$$Q_{\mathrm{L,R}}(x) = k\binom{N}{k} \sum_{i=0}^{N-k} \binom{N-k}{i} (-1)^i \frac{[Q_{\mathrm{W}}(x)]^{i+k}}{i+k}. \tag{7}$$

An equation for the multiplier γ is now obtained by substituting equation (4) into equation (2). With an integration by parts, this becomes:

$$P_{\mathrm{fa}} = \int_0^\infty Q_{\mathrm{L}}\left(\frac{x}{\gamma}\right) Q_{\mathrm{R}}\left(\frac{x}{\gamma}\right) p_{\mathrm{W}}(x)\, dx. \tag{8}$$

With the aid of the first derivative of equation (1) and equation (7), the above equation may be solved for the value of γ that gives a particular P_{fa}. For equivalent distributions in both sets of estimation cells, $Q_{\mathrm{L}}(x) = Q_{\mathrm{R}}(x)$.

3 Figure of Merit

Most analyses that determine the performance of various threshold setting algorithms use the Neyman-Pearson criterion. The threshold is set based on a particular desired P_{fa}. The SNR required for the specified P_{d} is then calculated. The efficacy of a particular threshold algorithm is determined by comparing the required SNR of one algorithm to that of another.

A related figure of merit, the ADT [4,5], is adopted here to evaluate the threshold algorithm. When square-law detection is used, the ADT is defined as:

$$\mathrm{ADT} = \frac{1}{\mu}\overline{\gamma (GO)'} \tag{9}$$

where the overbar is the expectation operator, μ is the average noise energy, and the prime represents the squared value. The ADT represents the expectation of the threshold level divided by the average noise energy. Both $\overline{\gamma(GO)'}$ and μ of the ADT represent energy. To the extent that $\overline{\gamma(GO)'}$ represents the required signal energy, the ADT represents a required SNR. The ADT's main advantage as a figure of merit is that it avoids the complicated evaluations with density functions that are usually required when nonlinear processing is used. For a Gaussian-distributed decision variable, which is a good approximation for Rayleigh-distributed amplitude noise at high SNRs, the ADT represents approximately the required SNR for $P_{\mathrm{d}} = 1/2$. This occurs because under these circumstances the median of the Gaussian distribution represents the signal energy. For Weibull amplitude noise statistics, the decision variable is not Gaussian, but the median approaches even closer to the signal energy. Hence, the ADT approximation improves with more skewed distributions [7].

As mentioned earlier, the order statistic algorithm is assumed to use the magnitude of the matched filter output as an input to the threshold algorithm. Since the ADT figure of merit assumes a square-law device after the integrate and dump filter which follows the matched filter, the effects of a squaring operation on the ADT must be examined. Examination of figure 1 shows that both the order statistic device and the GO device pick the same output, independent of the squaring. When the cumulative probability function satisfies $Q(\gamma GO < 0) = 0$ (true for all amplitude detectors of interest), then

$$Q(\gamma\, GO < x) = Q[(\gamma GO)^2 < x^2] \tag{10}$$

and thus the threshold multiplier γ is squared when the squaring device is included. The ADT is then obtained from

$$\frac{1}{\mu}\gamma^2\overline{(GO)^2}. \tag{11}$$

The second moment of the output of the GO calculator is used to obtain the ADT. Using the first derivative of equation (3) in equation (9), this is given by:

$$\mathrm{ADT} = \frac{\gamma^2}{\mu}\int_0^\infty x^2\left[Q_{\mathrm{L}}(x)p_{\mathrm{R}}(x) + Q_{\mathrm{R}}(x)p_{\mathrm{L}}(x)\right] dx \tag{12}$$

The mean noise energy is given by:

$$\mu = \int_0^\infty \frac{b}{\alpha}x^{b+1}e^{-x^b/\alpha}\, dx \tag{13}$$

The earlier analysis must be modified to evaluate the performance of a threshold algorithm when the amplitudes of one or more of the cells are represented by different density functions (i.e., when interference is present in one of the estimation cells). It is convenient to assume that the mean interference is much greater than that of the noise in the remaining estimation cells. This approximation becomes more accurate as the returned energy from the interference increases relative to the energy in the remaining estimation cells. Then the density function of the order statistic is given approximately by reducing N in the conventional formula (equation (5)) by 1. This result can be understood as follows. Under the above assumptions the interference always occupies the largest cell. The probability of the kth order statistic, $k < N$, lying between x and $x + dx$ is now a function of $N - 1$ variates, since one of the variates will never be less than x, for "moderate" x.

If one of the cells contains interference, the calculation for γ is modified as follows. The cumulative distribution function out of the GO calculator is given by equation (3) with $Q_L(x)$ representing the cumulative distribution function out of the order statistic calculator that contains the interference, obtained from equation (7) with N replaced by $N-1$. The P_{fa} is given by equation (8) with $Q_L(x) = Q_R(x)$. That is, the threshold multiplier is set with no interference present. As in the non-interference case, equation (8) is now solved for γ, given a desired P_{fa}. The ADT is obtained from equation (9), with $Q_L(x)$and$p_L(x)$ a function of $N-1$ estimation cells.

4 Results

This section uses the techniques developed in the previous section to determine both the threshold multiplier γ and the ADT for various noise scenarios. If W is normalized to unity mean, then α in equation (1) can be set to unity, and the Weibull noise distribution is determined solely by the skewness parameter b.

Consider first the threshold multiplier γ. This is determined by selecting P_{fa} and then solving equation (8) for γ. Figure 2 shows a plot of the threshold multiplier vs the order statistic cell k for $2N = 48$ noise estimation cells and various values of the skewness parameter b. Two values for P_{fa} are assumed: $P_{fa} = 10^{-4}$ (solid line) and $P_{fa} = 10^{-6}$ (dashed line). The threshold multiplier is set to guarantee the P_{fa} in a particular Weibull noise environment, as opposed to setting the threshold based on Rayleigh noise statistics and then failing to satisfy the P_{fa} requirement in Weibull noise. Note that since the threshold algorithm uses the amplitude, the ordinate is scaled to amplitude, as opposed to power units. As expected, the threshold multiplier decreases for all cases as a larger value of k is selected. As the skewness of the Weibull distribution increases (i.e., as b decreases), γ increases, raising the threshold to account for noise spikes.

Figure 3 shows the ADT as a function of k for $P_{fa} = 10^{-4}$ and various values of b. The dashed curve in the four pairs that ends at $N = 23$ represents the single interferer case. With interference the best performance (i.e., smallest ADT) occurs near $k = 20$, while with no interference the best performance occurs near $k = 18$. A relatively wide performance plateau exists for k between 14 and 24 for the Rayleigh amplitude ($b = 2$) case, within which the ADT varies less than 0.2 dB. The width of this plateau decreases for more spikey noise. Thus, while smaller values of k can accommodate many interference-dominated noise estimation cells in Rayleigh noise, this option becomes less attractive as the noise becomes more spikey. Essentially, the spikey noise starts to look more like interference. This interference causes a performance loss of only 1 dB or less for $k = 20$. As a comparison, with equal interference and target signal energy, a simple cell averaging detector suffers a loss of tens of dB [3] in required SNR.

Figure 4 considers the selection of k to make P_{fa} insensitive to uncertainty in the noise distribution model. The threshold multiplier γ is calculated to guarantee $P_{fa} = 10^{-4}$ at a particular value of b. Then equation (8) is used to calculate the P_{fa} at different values of b. Figure 4 corresponds to $b = 1.5$. From the figures, $k = 24$ (i.e., the maximum value, N) shows the least sensitivity to changes in b, while $k = 6$ shows the most sensitivity. Similar results are obtained when the threshold is adjusted with $0.5 \leq b \leq 2.0$. Thus, if maintaining a particular P_{fa} in an uncertain noise background is a priority, $k = 24$ (i.e., N) should be selected. Note, however, that this leaves no margin for interference.

The ADT is next calculated as a function of N for the 75th percentile cell. Figure 5 shows this calculation for $P_{fa} = 10^{-4}$ (solid line), $P_{fa} = 10^{-6}$ (dashed line), and four values of b. For $b = 2$ (i.e., Rayleigh amplitude noise statistics) the ADT decreases 2 dB as N is increased from 8 to 24. Greater reductions in ADT are achieved for smaller b. For $b = 0.5$, the ADT decreases 7 dB over the same range of N. Thus, the motivation for increasing N increases with the skewness of the noise distribution.

References

[1] Finn, H. M., and R. S. Johnson, September 1968, "Adaptive Detection Mode with Threshold Control as a Function of Spatially Sampled Noise-Level Estimates," *RCA Review* Vol. 29, pp. 414–464.

[2] Boothe, R. R., June 1969, "The Weibull Distribution Applied to the Ground Noise Backscatter Coefficient," U.S. Army Missile Command, Tech. Report RE-TR-69-15, ASTIA Doc. AD691109.

[3] Schleher, D. C., November 1976 "Radar Detection in Weibull Noise," *IEEE Trans. on Aerosp. Elec. Sys.* Vol. AES-12, pp. 736–743.

[4] Gandhi, P. P., and S. A. Kassam, July 1988, "Analysis of CFAR Processors in Nonhomogeneous Background," *IEEE Trans. Aerosp. Elec. Systems* Vol. 24, pp. 427–445.

[5] Rohling, H., 1983 "Radar CFAR Thresholding in Noise and Multiple Target Situations," *IEEE*

Trans. Aeros. Elec. Sys. Vol. AES-19, pp. 608–621

[6]. Papoulis, A., 1965, *Probability, Random Variables and Stochastic Processes*, McGraw Hill, Inc, New York

[7]. R.Rifkin "Performance Analysis of an Order Statistic-Based Radar Threshold Normalizer in Weibull Clutter," Submitted to *IEEE Trans. Aeros. Elec. Sys.*

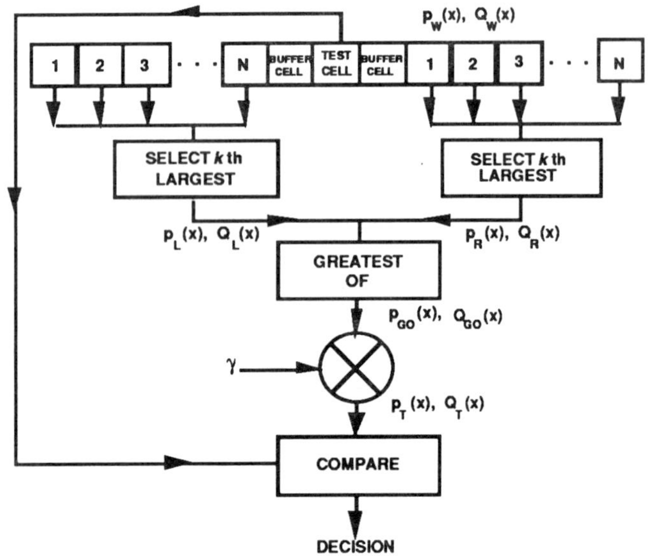

Figure 1. Block diagram of normalizer

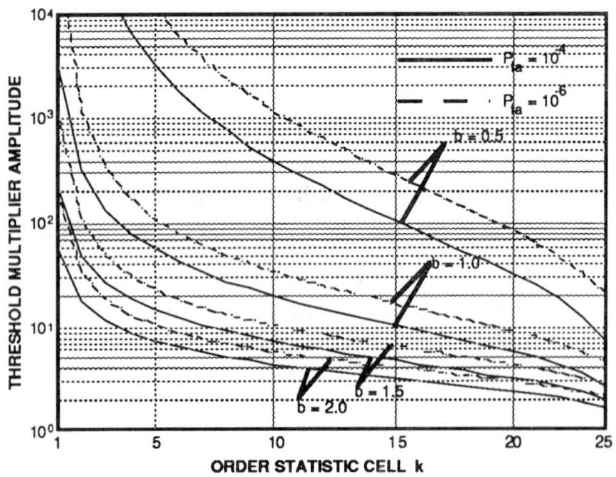

Figure 2. Multiplier for various values of b and P_{fa}

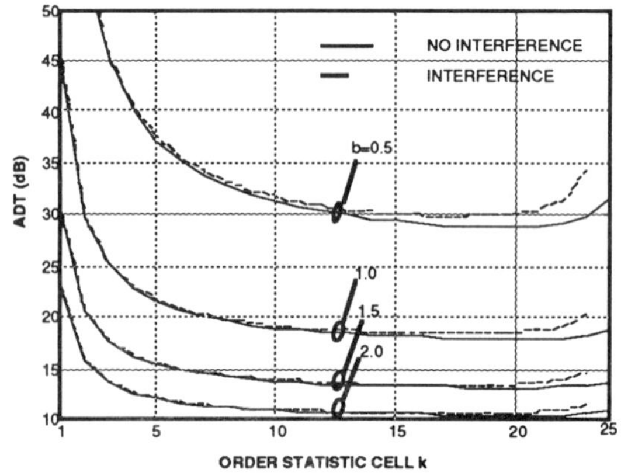

Figure 3. ADT vs order statistic cell for $P_{fa} = 10^{-4}$

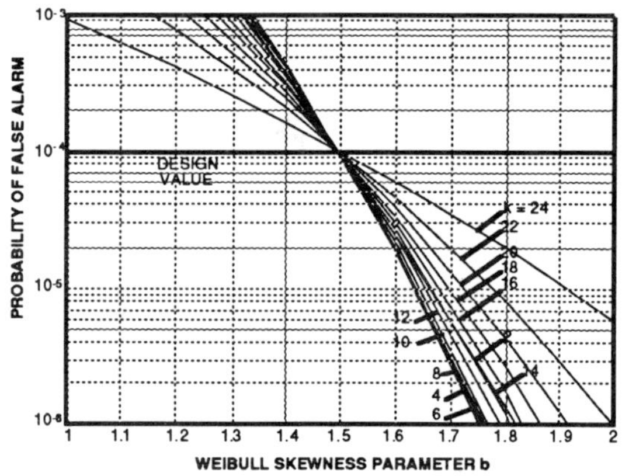

Figure 4. Sensitivity of P_{fa} to b for $P_{fa} = 10^{-4}$, $b = 1.5$

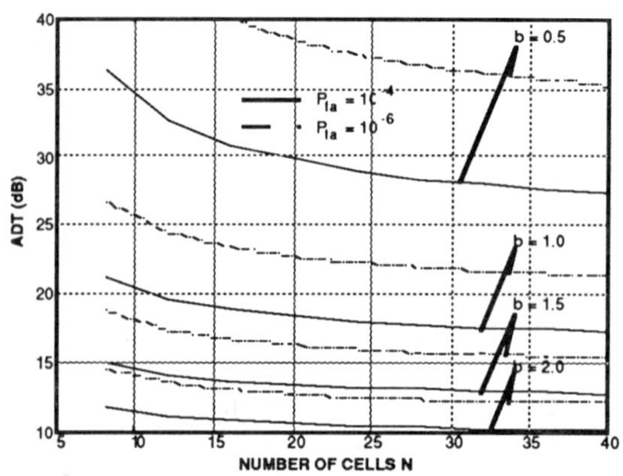

Figure 5. ADT vs N

On the Performance of Order-Statistics CFAR Detectors*

L.M. Novak S.R. Hesse

MIT Lincoln Laboratory
P.O. Box 73
Lexington, Massachusetts 02173-9108

Abstract

The detection performance of order-statistics-based two-parameter CFAR algorithms is compared with that of the classical Goldstein two-parameter CFAR. The performance comparison uses polarimetrically processed, high-resolution (1ft x 1ft) SAR imagery of real targets in clutter. The detectors are shown to provide comparable performance for the case of a single target in clutter. The order-statistics CFAR detectors, however, provide the best performance when the scene contains two closely spaced targets.

1. Introduction

A large database of fully polarimetric, high-resolution (1ft x 1 ft) SAR imagery of clutter and targets in clutter has been gathered by the Lincoln Laboratory MMW synthetic-aperture radar (SAR). This database of real imagery is being used to evaluate the performance of high-resolution, polarimetric target detection algorithms. Figure 1 shows the sensor and lists the important sensor parameters.

Figure 2 shows an example of the quality of the imagery gathered by the Lincoln Laboratory MMW SAR sensor. The image shown in the figure is of a farm area located near Stockbridge, NY. Clearly visible in the image are roads, plowed fields, trees, a river, and a farmhouse. This image was obtained by applying optimal polarimetric speckle-reduction processing to the fully polarimetric 1 ft x 1 ft resolution data, followed by noncoherent spatial averaging of 4 pixel x 4 pixel areas into an effective resolution of approximately 1 m x 1 m. An example of the original 1 ft x 1 ft speckle-reduced imagery of the farmhouse is shown in Figure 3. Our optimal polarimetric speckle-reduction technique, the polarimetric whitening filter (PWF), is described in Reference [1].

We have shown previously that reducing SAR image speckle improves target detection performance: the standard deviation of the clutter background is significantly reduced using PWF processing, increasing the separability of target and clutter [2]. Figure 4 illustrates the effectiveness of polarimetric processing in reducing the standard deviations of various types of clutter. The figure compares single-polarimetric-channel HH data with the corresponding polarimetrically processed data of typical clutter types (grass, trees, shadows, etc.). PWF processing is seen to reduce the clutter standard deviations by approximately 2.5 dB.

The detection performance obtained by running our baseline (Goldstein [3]) two-parameter CFAR detector over single-polarimetric-channel HH-imagery versus polarimetrically processed (PWF) imagery was reported in Reference [2]. It was shown that, at a false alarm density of 1 FA/km^2, only 30% of the targets were detected in single-polarimetric-channel HH-imagery whereas 75% of the targets were detected in polarimetrically processed (PWF) imagery. Therefore, in this paper we have used PWF imagery to compare the various detectors.

This paper describes our most recent target detection studies, the evaluation of order-statistics-based CFAR detectors. We have implemented two-parameter order-statistics CFAR algorithms and have compared the detection probability versus false-alarm probability performance of these CFAR algorithms (P_D versus P_{FA} ROC curves) with that of our baseline two-parameter CFAR. The detection performance curves presented in this paper indicate the following:

(i) The order-statistics-based two-parameter CFAR detectors can achieve performance comparable with that of the classical Goldstein two-parameter CFAR, provided that (1) a sufficient number of samples are used to estimate the clutter statistics (the CFAR losses of both CFAR algorithms are therefore comparable) and (2) there is only a single target in the CFAR detection window.

(ii) It is difficult to detect a target using the classical Goldstein two-parameter CFAR if an adjacent target is in the clutter estimation window. In this case, the adjacent target contaminates the

*This work was sponsored by the Defense Advanced Research Projects Agency.

estimated clutter statistics used in adaptively adjusting the two-parameter CFAR threshold; both the mean and standard deviation estimates become erroneously large. Reference [4] discusses some techniques for alleviating this problem.

The classical two-parameter CFAR calculates the detection statistic:

$$\frac{X_t - \hat{\mu}_c}{\hat{\sigma}_c} > K_{CFAR} \tag{1}$$

where $\hat{\mu}_c$ and $\hat{\sigma}_c$ are the estimated mean and standard deviation of the clutter, respectively. When these estimates become (erroneously) large due to contamination by the adjacent target, the above ratio (i.e., detection statistic) can be significantly reduced and the target may not exceed the detection threshold.

(iii) The order-statistics-based two-parameter CFAR is not as sensitive as the classical two-parameter CFAR is to the presence of two targets in the scene. The estimators of $\hat{\mu}_c$ and $\hat{\sigma}_c$ based on order statistics can be designed to be resistant to outliers falling in the clutter stencil. Therefore, the performance of the order statistics CFAR detector is less seriously degraded than the performance of the classical CFAR detector.

Section 2 of this paper describes the various two-parameter CFAR detectors evaluated. The classical (Goldstein) two-parameter CFAR is used as a baseline against which order-statistics-based two-parameter CFAR detectors are compared.

Section 3 presents the detection performance comparisons of the algorithms described in Section 2. Both hollow-stencil and solid-stencil CFAR algorithms are compared. The order-statistics CFAR detectors are shown to provide better detection performance when the scene contains two closely spaced targets.

Section 4 presents a brief summary and describes possible future studies.

2. Description of CFAR Detectors

This section provides details of the three algorithms compared in this paper.

Algorithm 1: The classical two-parameter CFAR detector is illustrated in Figure 5. This algorithm [2] was used as our baseline detector. As depicted in Figure 5, a hollow stencil consisting of 160 pixels is used to estimate the clutter mean μ_c and standard deviation σ_c. The detector then calculates the ratio

$$\frac{X_t - \hat{\mu}_c}{\hat{\sigma}_c} > K_{CFAR} \tag{2}$$

where X_t is the amplitude of the cell under test, and declares "target present" if the detection statistic (i.e., the ratio) is greater than the CFAR coefficient K_{CFAR}. Estimated statistics of the clutter ($\hat{\mu}_c$ and $\hat{\sigma}_c$) are calculated from the data contained within the stencil. The hollow stencil was designed to keep the target from contaminating the estimated clutter mean and standard deviation.

The detector defined in Equation (2) has been shown theoretically to provide a constant false alarm rate for log-normal and Weibull clutter [3]. Only those cells whose amplitudes stand out from the surrounding cells will be labeled as targets. The higher the constant K_{CFAR} is set, the more a cell under test must stand out from its background to be labeled as a target.

Algorithm 2: An order-statistics-based two-parameter CFAR similar to Algorithm 1 was proposed by Ritcey [5] as a more robust version of Algorithm 1. Robustness against outliers in the clutter stencil is achieved by replacing the clutter mean and standard deviation estimates with order statistics approximations. For this algorithm, the detection statistic is the following ratio:

$$\frac{X_t - X_{50}}{X_{75} - X_{25}} > K_{CFAR} \tag{3}$$

where X_{50} is the median amplitude of the clutter samples contained in the clutter estimation stencil, and X_{75} and X_{25} are the 75th and 25th percentile amplitudes of the data samples contained in the stencil. Thus, the mean is approximated by X_{50}, and the standard deviation is approximated by $X_{75} - X_{25}$. Although the standard deviation could be approximated using other percentiles (e.g., $X_{70} - X_{30}$), we found that the best performance obtained using our data set was achieved using

$$\hat{\sigma}_c \approx X_{75} - X_{25} \tag{4}$$

Algorithm 3: An order statistics CFAR with a solid stencil is shown in Figure 6. Since all the pixels within the solid stencil are used for estimating the clutter statistics, and since the target is much larger than a single 1 ft x 1 ft cell, many target cells will be contained within the solid stencil. Target pixels contained within the stencil will, however, have minimal effect on the performance of the detector if the order statistics approximations to the mean and standard deviation are selected properly. Guard pixels should not be necessary as long as the number of target pixels falling within the clutter stencil is not too great. For example, using the median in place of the mean allows for contamination of up to 50% of the brightest pixels before the estimated mean is affected.

For the solid stencil version of the CFAR detector we used

$$\frac{X_t - X_{50}}{X_{60} - X_{40}} > K_{CFAR} \tag{5}$$

where X_{50} is the median amplitude, and X_{60} and X_{40} are the 60th and 40th percentile amplitudes of the samples contained in the solid stencil

3. Performance Comparison of CFAR Detectors

The preceding section described the two-parameter CFAR detectors that were evaluated in this study. The objectives of the study were (1) to determine whether order statistics CFAR detectors can provide detection performance comparable with that of the baseline CFAR, and (2) to determine whether order statistics CFAR detectors can minimize the detection loss incurred by the baseline CFAR when two closely spaced targets are in the scene.

The detection performance of the algorithms was evaluated using 117 clutter false alarm images obtained from a clutter area near Stockbridge, NY, along with 87 target-in-clutter images. These clutter and target-in-clutter images were obtained by running a prescreener (a two-parameter CFAR) over reduced-resolution (1 m x 1 m) imagery of the type shown in Figure 2. This was done simply for convenience — we needed a set of typical clutter false alarm images along with a set of target-in-clutter images to test our various high-resolution (1 ft x 1 ft) CFAR detectors. By running the prescreener over a 5 Km^2 clutter area, we were able to extract a set of typical difficult clutter objects (each object became a 128 ft x 128 ft "chip" of high-resolution 1 ft x 1 ft data). The 117 clutter chips contained a variety of clutter objects, including manmade non-targets and natural clutter false alarms. For example, three of the clutter objects are (1) the farm house, (2) the vehicle, and (3) the tree that are pointed out in Figure 3. We then ran each detector over each of the 128 ft x 128 ft images and calculated detection probability versus false alarm probability.

3.1 Performance With a Single Target in the Scene

This section summarizes the detection performance obtained using the baseline two-parameter CFAR, the hollow-stencil order-statistics-based two-parameter CFAR, and the solid-stencil order-statistics-based two-parameter CFAR. These algorithms were tested against images containing a single target per chip.

We ran the baseline two-parameter CFAR over the 87 target images and 117 clutter images. Probability of detection versus CFAR coefficient K_{CFAR} was evaluated by calculating the fraction of targets (out of the total of 87) that were detected versus K_{CFAR}. Probability of false alarm was evaluated by calculating the fraction of clutter objects (out of the total of 117) that were detected versus K_{CFAR}. Finally, we combined the results of these two evaluations to obtain the P_D versus P_{FA} ROC curves. Performance for the baseline two-parameter CFAR is shown in Figure 7.

This same procedure was repeated several times using the hollow-stencil order-statistics CFAR detector. In each case, we used the median (X_{50}) as the estimate of the mean (μ_c). We varied the percentiles (X_{75} - X_{25}, X_{70} - X_{30}, etc.) used to estimate the standard deviation (σ_c). The best performance results obtained using the hollow stencil are also shown in Figure 7. As illustrated, with a single target in the scene, the order-statistics two-parameter CFAR provides performance comparable with that of the baseline algorithm. Since 160 samples were used in the clutter stencil to estimate the clutter mean and standard deviation, the CFAR losses of these two algorithms were comparable.

Solid-stencil order-statistics CFAR performance was evaluated using the same target and clutter data. Figure 8 compares the performance obtained using a solid-stencil order-statistics CFAR with the baseline CFAR results. From the results shown, it is clear that target pixels included in the solid stencil resulted in a slight degradation of detection performance.

3.2 Performance With Two Targets in the Scene

Next we evaluated the performance of the three CFAR detectors when there was a second, nearby target in each of the 87 target scenes. This was accomplished by replacing a number of cells within the CFAR stencil with bright target cells, thereby emulating the effect of having a second target in the scene.

We first evaluated the effect of the second target on the detection performance of the baseline two-parameter CFAR. Figure 7 shows the resulting performance: clearly, the baseline algorithm was seriously degraded by the presence of a second, nearby target in the scene. The measured statistics showed that the estimated clutter mean and standard deviation were erroneously large due to contamination by the second target; thus, the detection statistic (ratio) was seriously degraded.

Figure 7 also shows the performance of the hollow-stencil order statistics CFAR for the same clutter and target-in-clutter data. Clearly, this algorithm was much less sensitive to the presence of the second, nearby target in the target scene.

Figure 8 compares the corresponding performance for the solid-stencil order statistics CFAR detector with that of the baseline algorithm. The solid-stencil order statistics CFAR also significantly outperformed the baseline algorithm; its performance was comparable to that of the hollow-stencil order statistics CFAR detector.

4. Summary

The results of our studies, using real clutter and target-in-clutter imagery, indicate that the order-statistics-based two-parameter CFAR detectors proposed by Ritcey [5] have a significant advantage: both the hollow-stencil and solid-stencil versions of the algorithm significantly outperformed the baseline Goldstein CFAR when the scene contained two closely-spaced targets. Also, both versions of the order statistics CFAR provided performance comparable with the baseline CFAR when the scene contained a single target.

The disadvantage of an order statistics CFAR is computational intensiveness: the data contained within the stencil must be ordered at each location in the image, whereas the baseline two-parameter CFAR simply calculates the mean and standard deviation (μ_c and σ_c) from the data within the stencil. However, the significant improvement in the ability to detect targets in scenes containing closely-spaced targets may justify use of the computationally-intensive order-statistics-based two-parameter CFAR.

The baseline two-parameter CFAR was implemented using recursive calculations of the mean and standard deviation as the stencil was moved over a scene. This same concept was applied to the order statistics CFAR: the new data entering the stencil were merged into the already-ordered list and the old data leaving the stencil were deleted from the list. This resulted in significant computational savings.

Much work remains to be done before an order statistics two-parameter CFAR is recommended for use in place of the baseline CFAR. The baseline CFAR provides a true constant false alarm rate in log-normal or Weibull clutter; the order-statistics versions probably will not provide a true constant false alarm rate. Theoretical analyses of false alarm rate for log-normal and Weibull clutter types should be performed to determine the sensitivity of the order statistics algorithms to different clutter types.

Other order-statistics-based CFAR algorithms have recently been proposed (for example, Weber and Haykin [6]). The Haykin algorithm has been shown to theoretically provide a true constant false alarm rate for log-normal, Weibull and gamma-distributed clutter.

We plan to evaluate the Haykin algorithm, as well as the three algorithms discussed in this paper, using all 5 km^2 of Stockbridge clutter data at full (1 ft x 1 ft) resolution. In order to carry out this study, we are developing more computationally efficient, recursive implementations of the order statistics algorithms.

References

[1] L. Novak, M. Burl, R. Chaney, and G. Owirka, "Optimal Processing of Polarimetric Synthetic-Aperture Radar Imagery," Lincoln Laboratory Journal, Vol. 3, 1990.

[2] L. Novak, M. Burl, W. Irving, and G. Owirka, "Optimal Polarimetric Processing for Enhanced Target Detection," IEEE Aerospace & Electronics Systems Society National Telesystems Conference, 1991.

[3] G. Goldstein, "False Alarm Regulation in Log-Normal and Weibull Clutter," IEEE Transactions on Aerospace and Electronic Systems, AES-9, January 1973.

[4] J. Rickard and G. Dillard, "Adaptive Detection Algorithms for Multiple-Target Situations," IEEE Transactions on Aerospace and Electronic Systems, AES-13, July 1977.

[5] J. Ritcey, "An Order-Statistics-Based CFAR for SAR Applications," Electrical Engineering Dept., University of Washington, Seattle, WA., September 1990.

[6] P. Weber and S. Haykin, "Order Statistics CFAR for Two-Parameter Distributions with Variable Skewness," IEEE Transactions on Aerospace and Electronic Systems, AES-21, November 1985

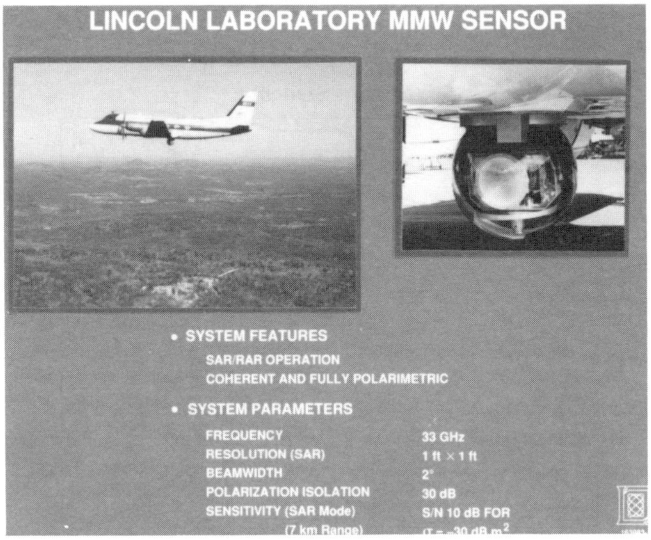

Figure 1. Lincoln Laboratory Millimeter-Wave SAR Sensor. The sensor platform is a Gulfstream G1 aircraft, shown in flight.

Figure 3. SAR image of farmhouse in Stockbridge, NY (1 ft x 1 ft resolution). PWF processing was used to produce this minimum-speckle image.

Figure 2. SAR image of farm area (1 m x 1m resolution). This image was formed by first applying PWF processing to the 1 ft coherent data, then spoiling (4 pixel by 4 pixel noncoherent averaging) to an effective resolution of approximately 1 m.

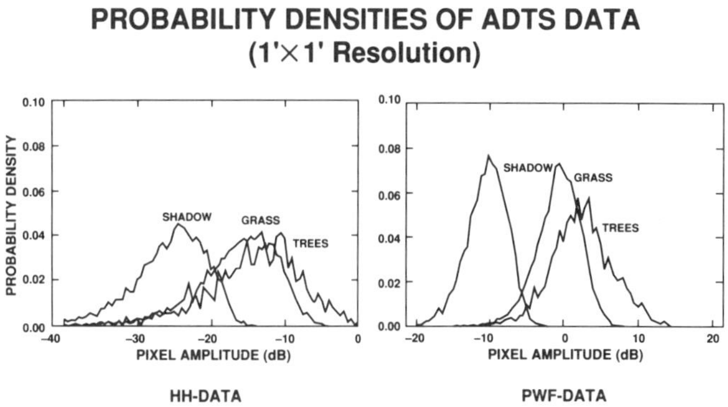

LOG STANDARD DEVIATIONS (dB)

	HH	PWF
TREES	6.7	4.1
GRASS	5.7	3.0
SHADOW	5.5	2.8

Figure 4. Probability density functions of HH data and PWF data (1 ft x 1 ft resolution). The table shows the corresponding standard deviations (dB) for grass, tree, and shadow data.

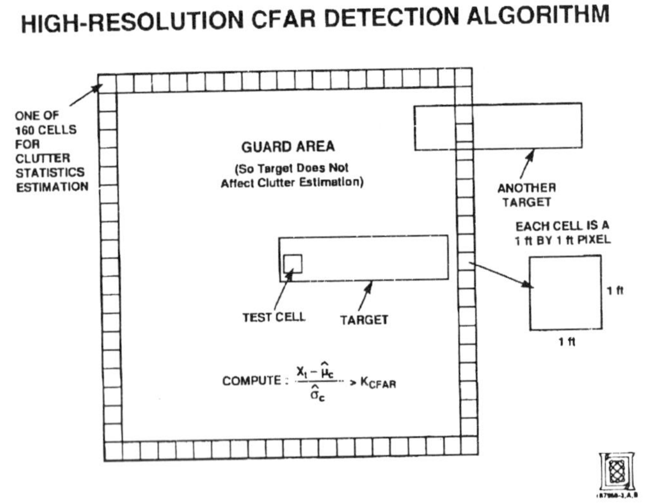

Figure 5. Sketch of hollow-stencil CFAR detector.

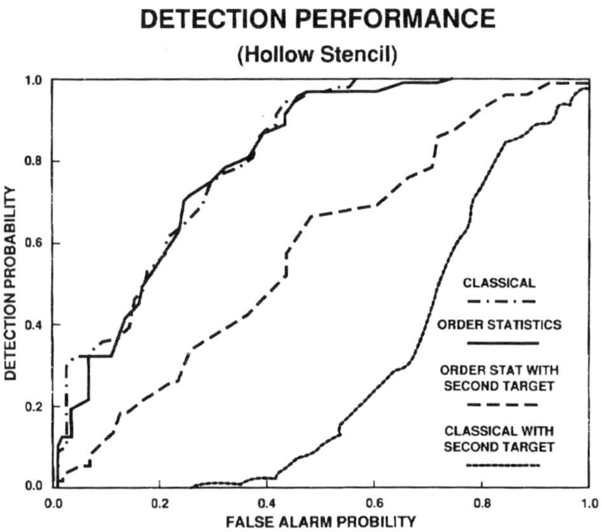

Figure 7. Detection performance comparison of baseline CFAR versus hollow-stencil order-statistics CFAR.

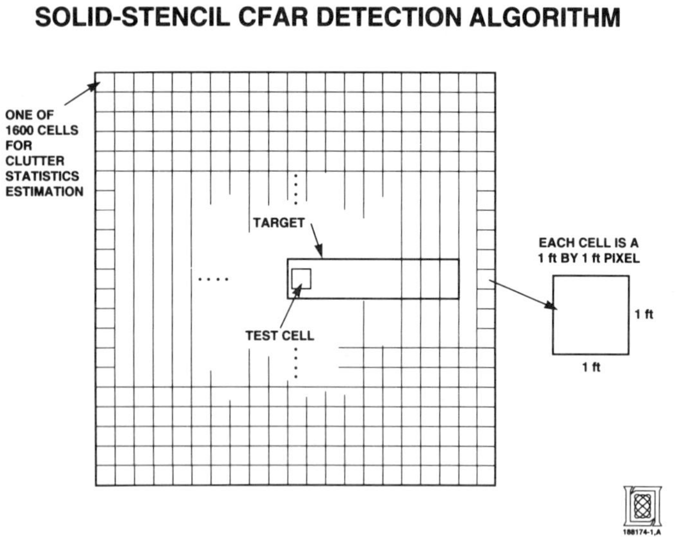

Figure 6. Sketch of solid-stencil CFAR detector.

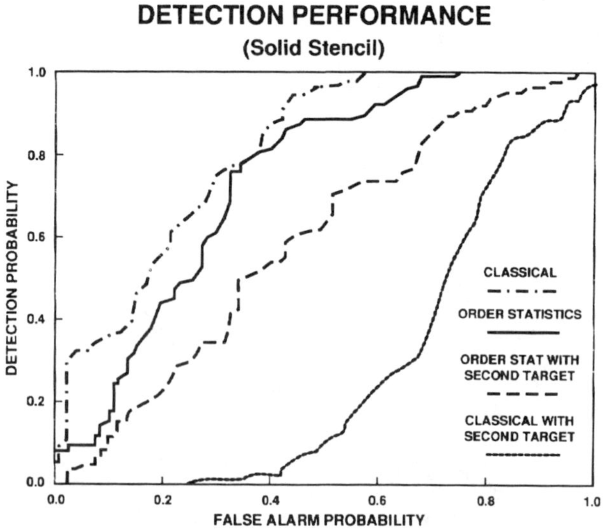

Figure 8. Detection performance comparison of baseline CFAR versus solid-stencil order-statistics CFAR.

Design Considerations for an Adaptive Phased Array Radar

J L Mather

DRA/ED, RSRE, St. Andrews Road, Malvern, Worcs., WR14 3PS, U.K.

Abstract

We examine the system design issues associated with the use of adaptive jammer rejection in the context of hypothetical specifications for a space-based surveillance radar. Two different classes of adaptive processing architecture are considered: the sidelobe canceller and the fully adaptive array. These are shown to achieve different trade-offs between inherent complexity, requirements upon the system specification, flexibility and performance.

1. Introduction

This paper discusses the implications of alternative adaptive nulling architectures for the design of a space-based radar, taking into account typical practical constraints. We briefly describe the space-based radar in section 2, and summarise the beamforming requirements of such a system in section 3. Section 4 defines the sidelobe canceller and fully adaptive sub-arrayed nulling approaches and makes further assumptions about hardware implications. In section 5 we compare the alternative nulling schemes in more detail, taking as a basis the way in which they influence or are influenced by hardware options. Finally, in section 6, we summarise the major advantages and disadvantages of each adaptive scheme for the space-based radar.

2. Space-Based Radar

In order to draw useful conclusions, we must make assumptions about the radar specifications, the jamming threat and the clutter environment. These are made for the purpose of illustration, and are not meant to correspond to any operational design or requirement. Further details are in the Appendix.

The radar will interleave search and tracking functions. Following detection, immediate look-back would confirm and initiate tracking. A phased array will be necessary to achieve the required beam agility. We will assume an active L-band array, aligned to the direction of travel to ameliorate ground clutter returns. The transmission will be a range-ambiguous pulse-Doppler waveform. We assume a few coherent bursts before the beam moves to a new position. Calculations in the Appendix lead us to assume a range of possible target signal-to-noise ratios (SNRs), clutter to noise ratios (CNRs) and jammer to noise ratios (JNRs) at the antenna element.

3. Beamforming Requirements

To integrate adaptive nulling with otherwise fairly "conventional" processing, we need a sum beam and two difference beams for two-axis monopulse, and sidelobe blanking channels. Ground clutter returns of significant Doppler bandwidth will be present in the data, and a clutter cancellation scheme will be necessary. If we consider using the displaced phase centre antenna (DPCA) [1] technique for clutter removal, then all the beamforming hardware (apart from the second difference beam, which would no longer be required) must be duplicated for each of the two (or more) DPCA channels. We must maintain a high gain mainlobe in the sum beams for SNR enhancement, and a notch at boresight in the difference beams for accurate angle estimation. A random sidelobe floor of -15dBi must be achieved, partly for the suppression of sidelobe clutter, and particularly to assist the suppression of sidelobe jamming.

We will assume for subsequent analysis that 32 degrees of freedom will suffice to cope with the likely jamming threat. This choice will influence the settling time of the adaptive process, the peak sidelobe level (which must be minimised) and the predictability of the sidelobe pattern. A smaller number of adaptive channels would be suitable if only limited numbers of jammers are likely to be encountered. This would lead to faster convergence of the adaptive algorithm. Conversely, a larger number of channels would be technologically feasible.

4. Adaptive Cancellation Schemes

4.1. Sidelobe Canceller (SLC)

The SLC uses one or more directional "main beams" and a number of less directional, lower gain "auxiliary beams". The main sum and difference beams would be formed by analogue beamformers, using the majority of elements in the array. The remaining elements, taken singly or in smaller sub-arrays, form auxiliary beams. Each element of the main array will be subject to a fixed analogue beam-tapering weight, designed to lower the quiescent sidelobes [2]. The auxiliary channels need zero gain on boresight for maintenance of the difference beam. Adaptive nulling of signals within the main beam could cause significant skewing of the pattern. This is prevented by the auxiliaries having much lower gain than the main beam.

Associated with each main and auxiliary beam is a channel, containing signal-bandwidth-defining filters and analogue to digital converters (ADCs). We assume the adaptive algorithm to be implemented digitally. DPCA requires an extra, displaced main aperture and hence duplication of each main receiver channel. In practice, the two apertures would overlap considerably. The auxiliaries do not need to be duplicated: they can be adaptively combined with signals from both main apertures. DPCA processing is assumed to be combined with adaptive beamforming. Main beam clutter in particular must be removed from data used by the adaptive weight computer, otherwise severe distortion of the main beam will result.

4.2. Fully Adaptive Array (FAA)

Here, we consider an array which has been divided into a number of equivalent subarrays, each of which contributes a degree of freedom. These are used either for adaptive processing, or to control beamshape via programmable constraints.

The signal from each element would be split and subjected to analogue beam-tapering weights before being fed to two different sub-array outputs, one for a sum beam and the other for a difference beam. Each sub-array would then have its own receiver channel, containing down-converters, signal-bandwidth-defining filters and ADCs. The adaptive algorithm will be implemented digitally, although the weights themselves may be either digital or analogue. Each main beam is formed through directional constraint on the adaptive weight calculation. Normalisation of the weights with respect to the desired signal level at the output fixes the "gain" in the desired direction.

Use of DPCA does not necessarily mean that the number of subarray channels must be doubled. The number was fixed at 32 for adaptivity, not for

This work has been carried out with the support of the Procurement Executive, Ministry of Defence beamforming purposes. It might be possible to redistribute the subarrays so that four sets of eight are each optimised for contributing to either a sum or difference beam at one of two phase centres. As for the SLC, it is necessary to remove clutter from data used by the adaptive weight computer. We assume that DPCA will be combined with the adaptive weight computation, and do not further consider techniques for clutter removal.

5. Comparison of Nulling Schemes

As shown in the Appendix, the desired signal will always be below noise at the input to the adaptive nulling computer. This is important because the presence of significant signal in or around the main beam causes deterioration of the attainable output SNR. The FAA can form nulls in the main beam and would be most sensitive in this respect. For the same reasons, main beam clutter must also be filtered below the noise level if it is not to capture adaptive degrees of freedom. Section 5.2.2 shows that clutter is above noise at the input to the ADC. For comparison of nulling schemes, we assume that it can be suppressed without otherwise affecting the adaptive weights.

5.1. Partitioning of the Phased Array

5.1.1. Degrees of Freedom

For greatest efficiency, in a two-dimensional phased array, sub-arrays should be positioned to give the maximum number of degrees of freedom in all planes normal to the array. For example, if a sixteen phase-centre array is arranged on a regular 4x4 grid, then only four degrees of freedom will be available in the principal planes. Flam [3] concludes that a planar array consisting of 2M carefully placed phase centres should be able to form a minimum of M nulls in a given conical region of interest. Similar design rules will apply to the phase centres of the SLC, in which it is likely that the auxiliaries will be distributed around the periphery of the main array. However, we note that a requirement for randomised phase centres for adaptivity conflicts with the need for regular phase centres for DPCA.

5.1.2. Grating Lobes

Sub-array placement is again important from the point of view of grating lobes. These are images of the main beam, pointing in other directions. They result from having distances between adjacent sub-array phase centres of greater than half of a wavelength (spatial under-sampling). The presence or absence of grating lobes will be governed by the separation of the phase centres as projected onto the plane of interest. Since a grating lobe is an image of the main lobe, its behaviour in a FAA system will be governed by the directional gain constraint applied to form the main beam. If jamming is directed at such a lobe, the system will be unable to adapt against it. This problem will not occur with the SLC, unless individual antenna elements are too widely spaced. However, in either system, creation of a null directed towards a jammer may be accompanied by the formation of grating nulls in other directions. It is possible that these could interact with nulls directed against other jammers, or with the main beam. Randomisation of the phase centre positions will suppress both effects.

5.1.3. Dispersion

Phase dispersion across the aperture, associated with wide-band signals located away from boresight, will increase the number of degrees of freedom required to adapt against a single jammer and will reduce the depth of the resultant null. These degrees of freedom may be spatial (occupying additional phase centres) or temporal (requiring adaptive FIR filters in each adaptive channel). Any increase in the number of degrees of freedom will cause an increase in the amount of hardware required, the digital computation load, and the time taken for the adaptive algorithm to converge and stabilise.

The SLC may be more sensitive to aperture dispersion effects than the FAA because the complex beams of the SLC main and auxiliary arrays will differ and will vary differently with frequency, whereas those of the FAA sub-arrays should match much more closely. This effect therefore may increase the number of degrees of freedom required to null a given number of jammers more significantly with the SLC than with the FAA.

5.1.4. Reliability

Reliability is a complex issue, which we will not consider in detail. However, it is clear that the different architectures will have different reliability characteristics. If the main beam summation node in the SLC fails, then the system will cease to operate. If a sub-array summing node fails in the FAA, then we only lose a single degree of freedom, in principle. Appropriate hardware design should insure against this type of failure. An alternative view might be that the greater hardware complexity of the FAA would lead to a higher probability of some type of failure. Failure of receiver channels during the lifetime of the system will have serious consequences for achievable random sidelobe levels. Although the adaptive control of the sidelobe pattern will work to minimise the effect of this, the dynamic range required of the ADCs may have to be increased to take it into account.

5.2. Receiver Channel and ADC Requirements

5.2.1. Channel Matching

It is important that the receiver channels should be well matched over the operating frequency band, as discussed in connection with dispersion. The match should be maintained as a function of signal amplitude and operating temperature. Taking into account the antenna feeds and RF components, matching may pose a problem in the SLC. The receiver channels for the FAA degrees of freedom are nominally equivalent. Nevertheless, matching the channels of a FAA to the accuracy required for effective suppression of powerful ground-based jamming will still be very difficult. In order to achieve null depths of 50 to 60dB, matching equal to around 0.1° rms in phase and 0.01dB rms in amplitude is likely to be necessary. Such a degree of match is also likely to be necessary for effective DPCA processing.

A major source of receiver channel mismatch is likely to be the signal-band-defining filter. Each channel contains such an analogue filter, and whilst it may be possible to match the filters at band-centre, it will be much more difficult to match the phase characteristics in the band-pass skirts. Although this mismatch can be minimised by careful design, further action may be necessary. One possibility is to calibrate each channel across the band, and then to correct for differences using a digital FIR filter following the ADC. A programmable filter would enable in-service re-calibration. The match would now be limited by the accuracy of the calibration and the resolution of the filter coefficients. Another approach to reducing mismatch problems is to adaptively control analogue weights which precede the filters [4].

The ADCs themselves are another potential source of mismatch. For example, they may have nonlinear operating characteristics, and mismatched aperture times. In the SLC, we shall see that there is apparently potential for using ten-bit converters in the auxiliary channels, whilst using a twelve-bit converter in the main channel. However, apart from the additional cost of manufacturing two different space-qualified ADCs, such a scheme may introduce additional tracking and matching problems associated with the inevitably different characteristics of the two devices. Feedback oversampling ADCs [5] offer improved linearity, and may be available for radar applications in the near future.

5.2.2. Dynamic Range

For a FAA, the dynamic range requirement will be the same for each channel, whereas it will be different for the main and auxiliary channels of the SLC. The SLC, designed to cope with sidelobe jammers, will already have achieved a significant degree of suppression through the tapered sidelobes of the main beam and the correspondingly lower auxiliary channel gains.

Calculations given in the Appendix and [6] derive possible dynamic ranges of signals, as measured at the antenna elements. By following these signals through the different architectures, as illustrated in Figs. 1 and 2, we can see how the dynamic range changes as a result of subsequent processing.

In the main sum-beam channel of the SLC (Fig. 1), we see that the dynamic range, defined by noise at one extreme and the maximum jammer or clutter level at the other, would allow us to use a 12 bit ADC. Although signal levels in the auxiliary channels can be even greater than in the main channel, the ADC range can be lower. Since the only purpose of the auxiliaries is to detect sidelobe jamming, their sensitivity can be reduced, using either an AGC or a fixed attenuator, in which case only 10 bits may be sufficient. The reduced level of airborne main-lobe jamming signals fed to the adaptive processor would not pose a problem: in an SLC we do not attempt to reject main-lobe jamming. In the presence of such jamming, the SLC radar would have to look elsewhere or use alternative countermeasures. However, as mentioned in the previous section, using different ADCs could exacerbate channel matching problems, and an attenuator or AGC may offer a better solution.

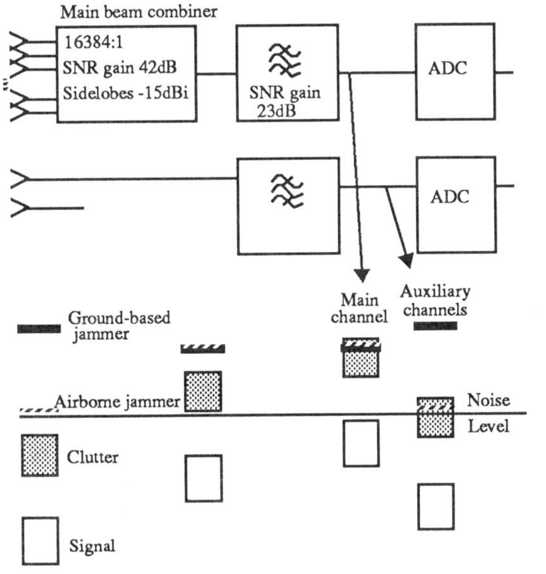

Fig. 1. Signal, clutter and jamming levels with respect to noise, in an adaptive sidelobe canceller. The upper chain corresponds to the main channel, and the lower chain to one of the auxiliaries. The powers shown at element level correspond to those given in the Appendix.

Pohlig [4] has proposed a hybrid analogue and digital nulling scheme in which some cancellation is achieved before digitisation. Analogue weights are controlled at a point preceding the analogue filters. Digital weights are fed forward and improve the convergence characteristics of the technique. Ward et al. [7] have also described a related closed-loop feedback architecture. These methods could reduce the dynamic range needed by the SLC main channel to deal with sidelobe jamming. In both cases, we assume that the (digital) computation of the analogue weights will result in effective nulling, given a starting point which may include signals which have been clipped by ADC overflow.

From Fig. 2, we can see that assumptions made about the sub-arrayed FAA lead to a requirement for a 14 bit ADC in each of the 32 channels. This occurs because we do not have the pre-formed low sidelobe pattern of the SLC with which to gain some immunity from jamming. Since it has not been amplified by a full main-lobe gain, the situation for airborne jamming might appear to be easier than with the SLC. However, it would still be difficult to null such jamming adequately. Attenuating by 34dB, instead of providing 15dB of gain, as the final stage of beamforming should, is equivalent to 49dB of cancellation. At the very least, this would cause severe distortion of the main beam and (more importantly and fundamentally) a significant increase in the noise level following adaptation, due to the increase in the weight norm needed to create such a null. However, the system would still have improved the output over the quiescent un-adapted case.

These estimates of ADC wordlength may have to be further modified to allow for the increase in sidelobe levels resulting from channel failure and degradation over the life of the array. Another influence on the digital wordlength may be the likelihood of higher peak jammer powers caused by constructive interference of multiple jammers or narrowband jamming.

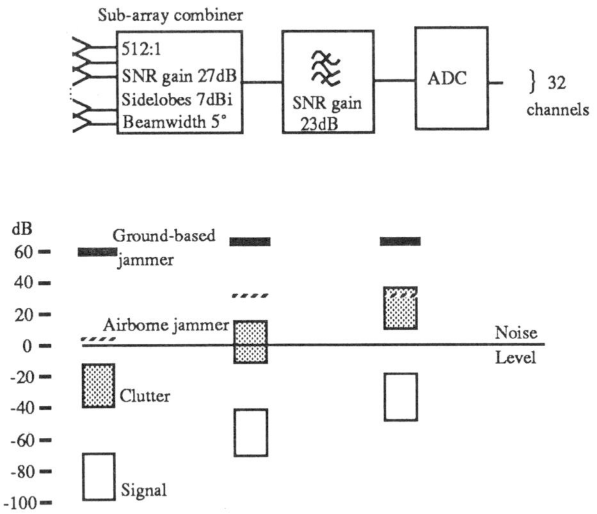

Fig. 2. Signal, clutter and jamming levels with respect to noise, in a subarray channel of a fully adaptive array. The powers shown at element level correspond to those given in the Appendix.

5.2.3. Tapering

The ability to accurately "taper" the aperture, as described in section 4, to achieve low sidelobes implies accurate knowledge of channel calibration and match. In the SLC, aperture tapering partitions sidelobe jammer suppression between a fixed analogue process and an adaptive digital process. This reduces the dynamic range of the receiver channels and the ADCs. This does not happen in an FAA. Gupta and Ksienski [8] have shown how that the output signal to noise plus interference ratio (SNIR) depends on the un-adapted beampattern. Therefore, lower quiescent sidelobes can result in a higher adapted output SNIR. However, aperture tapering to improve sidelobes also increases main beam width. Since the SLC is unable to null main beam jamming, this should be taken into account in deciding the physical aperture. In the FAA, nulling main beam jamming can lead to a lower output SNIR than for sidelobe jammers, as described above.

5.2.4. Bandwidth

To some extent, wide band jamming may be nulled using spatial degrees of freedom, since the time taken for signals to cross the aperture may provide the delay needed for such processing. The number of degrees of freedom required will depend on the dispersion of the jamming signal across the aperture, which in turn depends on the bandwidth and angle of arrival of the jamming signal. Hudson [9] has examined how degrees of freedom are used as a function of bandwidth. If it is known *a priori* that jamming signals are likely to use up multiple spatial degrees of freedom, adding adaptive channels to the system could provide a solution.

Perhaps a better solution to the problem of significant jammer bandwidth would be to simultaneously adaptively adjust the combined spatial and frequency response of the channels. Given a signal requiring more than one degree of freedom to form a null, then there would be a choice of architectures for both the SLC and the FAA. One could use an FFT to break up the signal bandwidth, followed by multiple "narrow band" adaptive cancellers acting on frequency cells containing significant power. Each will have convergence performance similar to the narrow band case. An alternative would be to carry out wide-band nulling using an adaptively weighted FIR filter architecture [10]. Both approaches need a larger number of samples to match the greater number of degrees of freedom.

5.3. Other Processing Issues

5.3.1. Constraints and Robustness

The SLC main beam will remain relatively undistorted following adaptation. Mainlobe nulling will be prevented, either by reducing the gain of the auxiliaries to such jamming, or by constraining the adaptive weight norm (which may restrict the choice of adaptive algorithms in a digital processor). The main beam shape is principally determined by the physical design, electromagnetic properties and the analogue aperture tapering weights.

The FAA main beam is created by the application of adaptive weights. Gain in the desired direction is set by a linear constraint on the calculation of the weight vector. In the absence of jamming, knowledge of the beam-shape will depend on how accurately the antenna and receiver system has been calibrated. Whilst adaption against sidelobe jamming will not greatly perturb this shape, main beam nulling may lead to significant distortion. In order to maintain unbiased monopulse direction-finding, this can be predicted and corrected [11]. The extent to which correction is possible depends on the proximity of the jammer to the constraint direction and on the JNR. The closer the jammer is to the main beam, the higher the weight norm required to null it, and the higher the dynamic range required of the digital part of the system. Although the monopulse direction finding capability may be restored, the increased weight norm will amplify the system noise. Thus output SNIR will be lower than in the presence of a sidelobe jammer of the same power, and monopulse angle estimates will have greater variance.

If the FAA mainbeam constraint is in error, the loss of output SNR will be determined by the rate of change of gain around the nose of the beam (assuming the signal is below thermal noise). In general, this will be much the same for the SLC, although beam skewing in the FAA may lead to higher rates of change. Additional constraints on the FAA, such as derivative or additional gain constraints, help to maintain beam shape, but also reduce output SNIR. However, further "soft" constraints [12] may be useful in order to "tie-up" otherwise un-used degrees of freedom, thus reducing the tendency towards instabilities in the sidelobe pattern (see section 5.3.2).

Sidelobe clutter must be considered, because it will be above the thermal noise level in the SLC main channel and considerably more so in the FAA channels because of their inferior sidelobe suppression. It will have a very wide Doppler spread, and may not be removed efficiently by DPCA because of mis-match. If clutter cannot be rejected, it will capture the degrees of freedom remaining after suppression of the strongest jammers. In this event, its effect would be simply to lower the mean sidelobe level of the adapted beam. Given the limited number of degrees of freedom in a partially-adaptive system, such as the SLC or the FAA being considered here, this would adversely affect the ability to reject weaker jamming signals, and would, in any case, have a limited effect on spatially distributed clutter. Furthermore, a fluctuating sidelobe pattern will result from varying sidelobe clutter.

5.3.2. Convergence Rates

An adaptive system using a direct algorithm to calculate the weight vector converges to within 3dB of the optimum output SNIR within approximately 2N data samples, where N is the number of channels, in the absence of the desired signal [13]. This should be true for both the SLC and FAA systems in the presence of sidelobe jamming. However, the optimum may be different in each case, and, particularly if an auxiliary gain constraint is used in the SLC, the two architectures will produce different beam patterns.

The SLC beam pattern is formed in advance of the adaptive processing. Limitation of auxiliary channel gain results in a pattern which has sidelobes within a few dB of the quiescent pattern. In the FAA the beam pattern is synthesised by the adaptive weights. Although nulls and main-beam are formed quickly, sidelobe levels resulting from different weight vectors may vary (or "jitter") significantly. Achievement of uniformly low and stable sidelobe levels may require many more data snapshots to be used for the weight calculation. Sidelobe "jitter" results from excess degrees of freedom, which model correlation in the sidelobe clutter or random cross-correlations caused by windowing of the noise. Even though this has little effect on output SNR, it can be undesirable. If weights are recalculated at shorter intervals than the coherent dwell, jitter may also affect subsequent signal integration.

5.3.3. Flexibility

FAA beams may be simultaneously optimised to look in a number of different directions, without loss of output SNR or convergence, by using a digital post-processor to apply the constraints. A single adaptive computation is sufficient for the formation of beams covering a range of angles [14]. A FAA is also able to apply simultaneous multiple constraints at the pre-processing stage in order to control main beam gain and shape, or to maintain gain in multiple directions. In the latter case, as the number of constraints increases, there will be a loss of output SNR with respect to the optimum.

The FAA also offers the choice of working directly on the data, or processing in beam space. In principle there is no difference in the resulting adaptive weights. In practice, points to consider would include the effects of grating lobes, impact on dynamic range, the high gain to clutter in each beam, and the flexibility to apply non-interacting constraints in a post-processor. For a system with a large number of degrees of freedom, transforming into beamspace may allow a reduction in the number of degrees of freedom by processing only "significant" beams. This would accelerate convergence and stabilise beam patterns, with only slight loss of output SNR. Alternative approaches to beam stabilisation are given in [15] and [16].

None of these choices are available for the SLC. Therefore, the FAA seems more flexible. The importance of this would depend on the mission requirements and a determination of the practical value of these options.

5.3.4. Choice of Adaptive Algorithm and Architecture

Open-loop computation of the adaptive weight vector typically demands $O(N^3)$ complex numerical operations, where N is the number of adaptive channels. Using combined parallel and pipelined (systolic) processing architectures, these operations can be carried out in $O(N)$ time steps. We have concluded in section 5.2 that the FAA will require longer digital wordlengths than the SLC in all channels. However, as we shall advocate the use of parallel or pipelined processors, possibly working in the data domain, this is unlikely to have a major effect on computation time, although data storage would increase. As the power of DSP chips continues to increase, processing load is no longer thought to be a dominant issue in real-time system design.

For rapid convergence to a stable result, Sample Matrix Inversion (SMI) [13] operating on a covariance estimate, or QR decomposition [17] operating directly on the data, seem to offer the best choices of algorithm. Both compute the adaptive weight vector in open-loop fashion. Both result in an output SNR within 3dB of the optimum when calculated from 2N data snapshots (if signal and mainbeam clutter are excluded from these data). If a covariance domain approach is chosen, QR decomposition provides a numerically secure approach for the necessary matrix triangularisation. Faster adaptive algorithms exist, but result in lower output SNR because of the poorer averaging of the noise used for estimation of the weight vector.

Efficient systolic architectures exist for both narrow-band [17, 18] and wide-band [10] implementations of the QR algorithm. These architectures scale in power to match the increasing number of degrees of freedom. An alternative approach could farm out matrix decompositions onto individual processors [19]. Each would calculate a weight vector appropriate to a different data sample set. In any of these cases, both on-line and off-line architectures are possible [17, 18, 20]. It is likely that the off-line solution with periodic weight update would be used in practice.

A further option with the FAA would be to use a high resolution algorithm in place of the usual adaptive processing. Such an algorithm would provide estimates of jammer directions of arrival, power and cross-correlation, which could be used to form narrow retro-directive receiver beams. This would also improve stability when processing in the presence of a strong desired signal. Alternatively, if the desired signal is below noise, as we have assumed, then the high resolution weight vector may be used directly to null jamming. Techniques such as IMP [21, 22] or PTMF [23] also indicate how to correct the monopulse characteristic in the presence of main beam jamming.

6. Summary

We have examined two different adaptive nulling schemes applied to an imagined space-based phased array radar. We have shown that the design of the "front-end" signal processing (from antenna design through to ADC design and clutter processing) has an impact on the use of an adaptive algorithm. Conversely, adaptivity places constraints and performance criteria on other parts of the system. We can draw conclusions, listed below, regarding the suitability of the different architectures to a space-based radar.

Fully adaptive arrays seem to have a number of broad advantages: 1) Effective main beam nulling capability, but accompanied by distortion of the beamshape and lower output SNIR than for sidelobe jamming. 2) Greater flexibility through the use of programmable constraints, the ability to simultaneously form multiple optimised outputs, and the option of carrying out high resolution processing. 3) Potentially lower sensitivity to aperture dispersion effects. 4) All channels (nominally) the same. The FAA sub-arrays are more likely to track as a function of amplitude, time and temperature.

SLCs appear to have the following advantages: 1) Lower hardware complexity and therefore lower cost. 2) Greater robustness of the main beam shape in the presence of close-in jamming, enabling straightforward monopulse direction-finding (and, conversely, no ability to cope with main-beam jamming). 3) Better control over mean sidelobe levels - the SLC seems to depend less on convergence of the adaptive algorithm to maintain low sidelobes. 4) A lower dynamic range is required of the adaptive circuits if powerful jamming enters only via (attenuated) sidelobes. 5) It is easier to avoid problems with grating lobes.

7. References

[1] G A Shaw, R J McAulay. ASSP Spectrum Estimation Workshop II, Tampa, Florida, USA, 308-312, (1983).

[2] M I Skolnik. *Introduction to radar systems*. McGraw Hill, 1962

[3] R P Flam. IEEE Trans AES-20(2), 199-202, (1984).

[4] S C Pohlig. Proc ICASSP, 1588-1591, (1988).

[5] F Harris. Proc ACSSC, 622-625, (1988).

[6] J L Mather, A C Fairhead, M P Warden. Royal Signals and Radar Establishment Memorandum no 4410, (1991).

[7] C R Ward, P J Hargrave, J G McWhirter, T J Shepherd. IEE Proc 6th Int Conf Ant Prop, Warwick, Apr 1989.

[8] I J Gupta, A A Ksienski. IEEE Trans AP-30(4), 549-553, (1982).

[9] J E Hudson. *Adaptive array principles*. IEE Press, 1981.

[10] I K Proudler, T J Shepherd, J G McWhirter. Proc ICASSP 1990.

[11] U Nickel. *Angle estimation with adaptive phased array radar under mainbeam jamming conditions*. AGARD, October 1990.

[12] J G McWhirter. Royal Signals and Radar Establishment Memorandum no 3939, (1986).

[13] I S Reed, J D Mallet, L E Brennan. IEEE Trans AES-10(6), 853-863, (1974).

[14] J G McWhirter, T J Shepherd. IEE Proc-F 136(2), 75-80, (1989).

[15] D H Brandwood, J D Baker. IEE Proc 6th Int Conf Ant Prop, Warwick, Apr 1989.

[16] W F Gabriel. NRL Report 8920, October 9, 1985.

[17] C R Ward, P J Hargrave. J G McWhirter. IEEE Trans AP-34(3), 338-346, (1986).

[18] J G McWhirter, T J Shepherd. Proc SPIE 696, 80-87, (1986).

[19] M Wells. Contribution no.2, IEE Colloquium on "Adaptive Antennas", Savoy Place, 8th June, (1990), IEE Digest No: 1990/098.

[20] J G McWhirter. Proc SPIE 1152, Aug 1989.

[21] I J Clarke. pp273-277, Underwater Acoustic Data Processing, Y T Chan, Ed., Kluwer Academic (1989).

[22] J L Mather. Proc 24th Asilomar Conf. Signals Systems, and Computers, 368-372, (1990).

[23] U Nickel. IEE Proc-H 134(1), 77-82, (1987).

Appendix

Determination of SNR, CNR And JNR at Antenna Elements

Assume the following radar parameters:

Wavelength λ = 26 cm; Antenna area A = 28 × 9m;

Beamwidths θ_{az} = 1.7°, θ_{el} = 0.5°;

Number of antenna elements 16384 (i.e. 2^{14});

Peak power P = 100kW; Duty ratio = 10%;

Pulse-repetition frequency = 10kHz;

Agile bandwidth, Ba = 200MHz;

Signal bandwidth = 1 MHz;

Target cross-section σ_t = 1 - 10 sq. m.; Noise figure F = 3dB;

Losses L = 10 - 13dB + scanning loss where appropriate

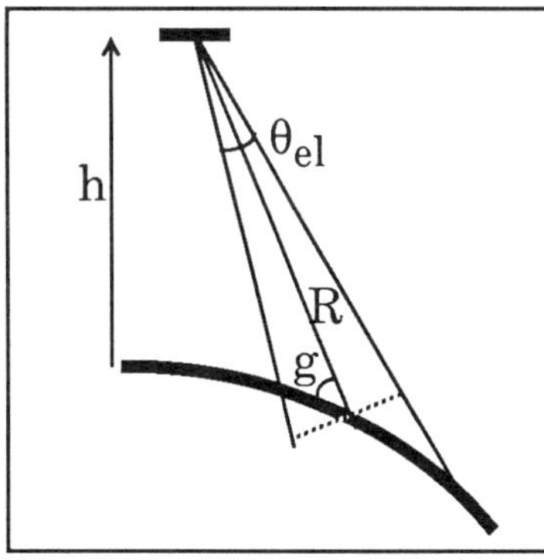

Height, h = 1000 km; Scan angle = 29 - 58°;

Taking the radius of the earth to be 6400km,

Slant range R = 1172 - 2665 km; Grazing angle, g = 56 - 11.3°.

(i) Signal to Noise Ratio (SNR)

To estimate the highest likely SNR, we use the 10m^2 target, slant range of 1172km, and losses of 11dB. These combine with the other parameters to give a figure of -69dB. Using 1m^2 for target cross-section, 2665km for slant range and 16dB for total losses, we estimate a "worst case" SNR of -98dB.

(ii) Clutter to Noise Ratio (CNR)

The effective cross-section σ_c of the ground covered by one range cell can be substituted for σ_t to evaluate mainlobe CNR. The dotted line in the diagram, of length $R\theta_{el}$, approximately represents the length of Earth's surface illuminated by the beam, projected onto a plane perpendicular to the direction of the beam. The length of one range cell and its ambiguities is then $R\theta_{el}$ x duty ratio. (The effect of beam broadening on θ_{el} due to scanning of the beam may be disregarded, because scanning loss will cancel it out.) The area of a range cell is therefore $R^2\theta_{az}\theta_{el}$ x duty ratio. The effective echoing area of each sq. m. of projected area is γ, the normalised backscattering coefficient. We assume γ is constant with grazing angle, and is between 0.1 and 0.01. A beam-weighting factor accounts for reduction in width of ambiguous range cells towards the edge of the beam, and also for reduction in transmit power reaching these cells. The total effective echoing area is

$$\sigma_c = R^2\theta_{az}\theta_{el} \cdot \gamma \cdot \text{duty ratio} \cdot \text{beam-weighting factor}.$$

We deduce values of σ_c which lie between 56 and 66dB for high grazing angles, and between 59 and 69dB at low grazing angles. We thus conclude that mainlobe clutter will be in the range -39dB to -13dB.

(iii) Jammer to Noise Ratio (JNR)

Two jammers are considered. Both emit continuous noise over the 200MHz agile bandwidth of the radar. One is a powerful ground-based system that would normally be outside the look-direction of the radar. The worst case is when it is directly below the radar. We deduce a JNR of 59dB in this case. An airborne stand-off or self-screening jammer can be in the look-direction of the radar. At its closest it might be 1200km from a 1000km-altitude radar.We deduce a JNR of 4dB for this case.

Shift-Keyed Signal Identification Using Time-Frequency Processing

Wayne D. Music[†], Warren L. J. Fox[†], Patrick J. Loughlin[‡], Les E. Atlas[‡], and James W. Pitton[‡]

[†]Applied Physics Laboratory, HN-10
University of Washington
1013 N. E. 40th St.
Seattle, WA 98105

[‡]Department of Electrical Engineering, FT-10
University of Washington
Seattle, WA 98195

Abstract

The use of the cone kernel time-frequency representation (CK-TFR) for shift-keyed signal identification is introduced. In electronic warfare (EW) surveillance systems, the identification of phase shifts in a signal is an important topic due to the prevalent use of phase-shift keyed (PSK) signalling schemes. Examples of TFR patterns for shift-keyed signals are presented. It is observed that, in addition to providing baud rate information about the signal, the CK-TFR clearly distinguishes between phase shifts of 180, +90 and -90 degrees. A "spectral peak splitting" effect is observed in the CK-TFR coincident with phase shifts in a signal. The mathematical foundation for the spectral peak splitting is explored.

1 Background on Time-Frequency Representations

In contrast to the conventional power spectrum S(f) of a signal s(t), time-frequency representations (TFRs) simultaneously convey temporal and spectral information about the signal s(t). These representations are thus useful when s(t) possesses time-varying spectral content. Two common TFRs used in signal analysis are the spectrogram (the magnitude squared of the short time Fourier transform), and the Wigner-Ville distribution (WVD) [1]. Both of these TFRs are members of the general bilinear class of TFRs formulated by Cohen [2].

The spectrogram is the most widely used tool for time-varying spectral analysis. The concept behind it is simple. First, a finite length window centered at some time t is applied to the signal to be analyzed, and the magnitude-squared of the Fourier transform of the windowed segment is computed. The window is then shifted to a new time t and another magnitude-squared Fourier transform is computed. Successive window shifts and Fourier transforms thus generate a time-dependent estimate of the signal power spectrum.

Mathematically, we can express this operation in continuous time as,

$$S(t,f) = \left| \int_{-\infty}^{\infty} s(u)\, h(u-t)\, e^{-j2\pi f u} du \right|^2 \tag{1}$$

where s(t) is the signal and h(t) is a finite-length window. Note that the Fourier transform is a linear operation on the windowed signal (although the spectrogram is quadratic because of the magnitude-squared operation). By this linear Fourier transform relation of equation (1), a larger width window h(t) will improve frequency resolution at the expense of time resolution, and vice versa.

An alternative view of this trade-off can be formulated by expanding the magnitude-squared formulation of equation (1):

$$S(t,f) = \int_{-\infty}^{\infty}\int_{-\infty}^{\infty} s(x)\, s^*(y)\, h(x-t)\, h^*(y-t) \times e^{-j2\pi f(x-y)} dx dy$$

$$= \int_{-\infty}^{\infty}\int_{-\infty}^{\infty} s\left(u+\frac{\tau}{2}\right) s^*\left(u-\frac{\tau}{2}\right) h\left(u-t+\frac{\tau}{2}\right) \times h^*\left(u-t-\frac{\tau}{2}\right) e^{-j2\pi f\tau} du d\tau, \tag{2}$$

where in the last step, the linear mapping [3]

$$x = u+\frac{\tau}{2} \quad ; \quad y = u-\frac{\tau}{2}$$

$$dxdy = |J|\, du d\tau,$$

1058-6393/91 $01.00

$$|J| = \det \begin{bmatrix} \frac{\partial x}{\partial u} & \frac{\partial x}{\partial \tau} \\ \frac{\partial y}{\partial u} & \frac{\partial y}{\partial \tau} \end{bmatrix} = 1,$$

from the (x,y)-plane to the (u,τ)-plane provides a symmetric formulation for the spectrogram.

To now illustrate the time/frequency resolution trade-off of the spectrogram, consider Figure 1, which shows the nonzero support of the spectrogram window function (or "kernel") $h(t+\tau/2)h^*(t-\tau/2)$ of equation (2) in the (t,τ)-plane, where h(t) is zero for $|t| > T/2$. Since this kernel, by equation (2), is correlated in time t with the signal function $s(t+\tau/2)s^*(t-\tau/2)$, the width of the kernel, T, along the t-axis determines the amount of smearing in time. Similarly, the width of the kernel along the τ-axis determines the frequency resolution, since this is the dimension in which the Fourier transform of equation (2) is performed. It is clear from Figure 1 that, for the spectrogram, the width of the kernel in time t and in lag τ are both proportional to T. Thus, it is impossible to improve frequency resolution (which requires increasing T) without inducing a loss of time resolution in the spectrogram.

Equation (2) can be generalized by replacing the 2-D spectrogram kernel $h(t+\tau/2)h^*(t-\tau/2)$ with a generic 2-D kernel, $\phi(t,\tau)$. Doing so yields

$$C_s(t,f;\phi) = \int_{-\infty}^{\infty}\int_{-\infty}^{\infty} \phi(u-t,\tau)\, s\left(u+\frac{\tau}{2}\right) \times s^*\left(u-\frac{\tau}{2}\right) e^{-j2\pi f\tau} du\, d\tau. \qquad (3)$$

This formulation allows for the construction of an infinite number of quadratic time-frequency representations (TFRs), where a particular TFR within this class is uniquely specified by its kernel, $\phi(t,\tau)$. This generalized class of quadratic (bilinear) TFRs was introduced by Cohen [4] in 1966, and has served to unify the discipline of time-varying spectral analysis. Note that unlike the spectrogram, the underlying Fourier transform of the generalized class is quadratic in the signal. (That is, for kernels $\phi(t,\tau)$ other than those in the form of the spectrogram kernel of equation (2), the general form given by equation (3) will not reduce to a magnitude-squared expression for which the Fourier transform is linear in the signal, as in equation (1).)

One choice for the kernel that circumvents the time/frequency resolution trade-off inherent to the spectrogram is illustrated in Figure 2. Note that for this kernel, while the width of the kernel along the τ-axis is again proportional to T (so that frequency resolution is inversely proportional to T), the width of the kernel along the t-axis is independent of T, and indeed narrows to zero at the origin, so that time resolution is independent of frequency resolution. This kernel is $\phi(t,\tau) = g(\tau)\,\mathrm{rect}(t/\tau)$, where $\mathrm{rect}(t/\tau)$ is

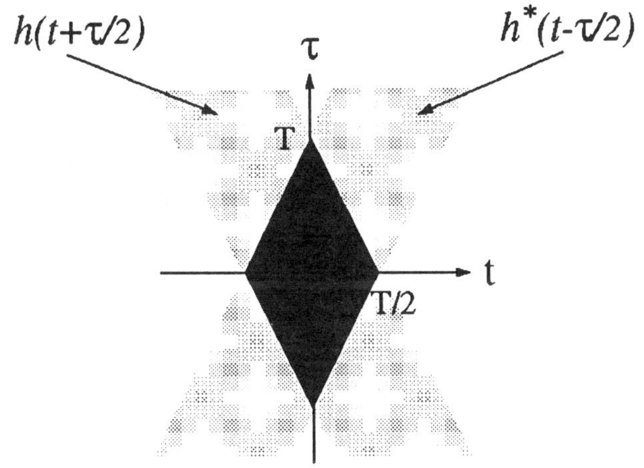

Figure 1: The nonzero support (black region) of the spectrogram kernel, formed by the product $h(t+\tau/2)h^*(t-\tau/2)$ in the (t,τ)-plane. Note that the width of the kernel along the *t*-axis and the width along the τ-axis are both proportional to T. Thus, decreasing T improves time resolution, but also decreases frequency resolution.

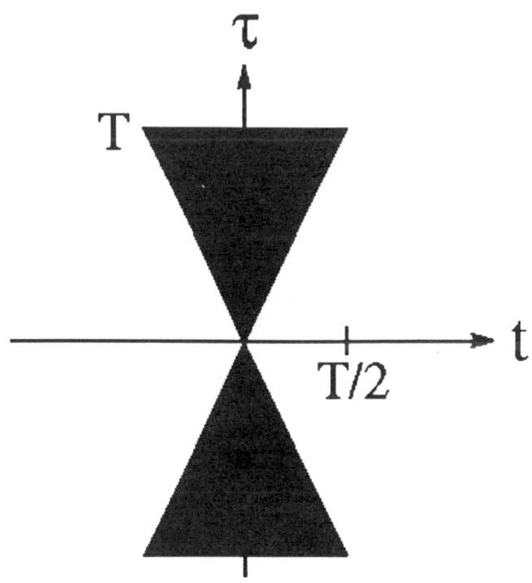

Figure 2: The nonzero support of a kernel that avoids the time/frequency resolution trade-off of the spectrogram. Note that the width of the kernel along the t-axis is independent of T, unlike the spectrogram kernel of figure 1. Thus, here, T can be adjusted to increase frequency resolution without affecting time resolution.

nonzero (and equal to one only for $|t/\tau| \le 1/2$). For bounded $g(\tau)$, this kernel is the "cone kernel," [5,6] which yields time-frequency displays with resolution markedly superior to that of the spectrogram without obscuring its frequency information [7,8,9].

2 Mathematical Analysis of Phase-Shift Effects

Mathematical analysis confirms that Cohen's class can be used to detect and identify phase discontinuities in a signal. For example, the WVD (obtained with the kernel $\phi(t, \tau) = \delta(t)$) of the signal

$$s(t) = e^{j2\pi f_0 t}[u(t_0 - t) + u(t - t_0) e^{jn\pi}], \quad (4)$$

where u(t) is the unit step function and $e^{jn\pi}$ is an arbitrary phase shift occurring at $t = t_0$, is given by:

$$W_s(t, f) = \delta(f - f_0)\cos(n\pi) + 4|t - t_0| \operatorname{sinc}[4(t - t_0)(f - f_0)](1 - \cos(n\pi)) + (\sin(n\pi)\cos[4\pi(t - t_0)(f - f_0)])/[\pi(f - f_0)]. \quad (5)$$

For n = 1, 1/2 and -1/2, corresponding to phase shifts of 180, 90 and -90 degrees, the resultant WVDs are:

$$W_s(t, f)|_{n=1} = 8\psi(t, f) - \delta(f - f_0), \quad (6)$$

$$W_s(t, f)|_{n=1/2} = 4\psi(t, f) + \theta(t, f), \quad (7)$$

and

$$W_s(t, f)|_{n=-1/2} = 4\psi(t, f) - \theta(t, f), \quad (8)$$

where

$$\psi(t, f) = |t - t_0| \operatorname{sinc}[4(t - t_0)(f - f_0)] \quad (9)$$

and

$$\theta(t, f) = \frac{\cos[4\pi(t - t_0)(f - f_0)]}{\pi(f - f_0)} \quad (10)$$

A problem, however, with the WVD is that for multicomponent signals, significant cross terms are generated that can obscure much of the desired information. Smoothing as in equation (3) with a non-impulsive kernel helps to alleviate this problem; however, significant smoothing (such as with the spectrogram) may obliterate important information, such as that pertaining to phase shifts. The CK-TFR smooths such that cross terms are rendered insignificant [10], while preserving much of the information about the phase shifts. It has been observed that CK-TFRs of speech signals have "splitting" of some harmonic peaks at pitch period discontinuities [8]. Thus, for the CK-TFR, 180 degree phase shifts appear as "eyelets" or "splits" in the spectral band at signal frequencies $f = f_0$, while ±90 degree phase shifts appear as either the upper or lower edge of the eyelet, depending upon the sign of the phase shift. The CK-TFR may thus be useful in phase shift keying (SK) applications of parameter estimation and signal identification.

3 Shift-Keyed Signal Applications

In electronic warfare surveillance systems, the identification of phase shifts in a signal is an important topic owing to the prevalent use of SK signalling schemes. Typically, signals-of-interest (SOIs) that are captured are processed for parameter information. These parameters are frequency, phase, amplitude, pulse width, pulse repetition interval, and the shape of the SOI's envelope. These parameters are used to help identify the type of emitter from which the SOI originated, e.g., SAM missile radar or microwave telemetry links.

The CK-TFR offers additional information over that of conventional power spectral density measurements, such as the ability to discriminate between 180, +90 and -90 degree phase shifts. We applied the CK-TFR and the spectrogram to the low probability of intercept DS-BPSK and DS-QPSK (direct sequence binary and quadratic phase shift-keyed) signals found in spread spectrum systems, and to MSK (minimum shift-keyed) signals. These results are shown in the gray-scale plots of Figures 3-7. In Figure 7, note the difference in the time-frequency pattern between a +90 degree and -90 degree phase shift. Note, too, that carrier frequency and baud rate information are clearly conveyed by the spectrogram and CK-TFR displays. Figures 3 and 4 compare the CK-TFR and the spectrogram results for a DS-BPSK signal in additive white Gaussian noise (10 dB SNR, where SNR is defined here as AT/N_0, where $N_0/2$ is the noise variance, A is the signal power, and T is the data bit duration). Observe that the clarity and resolution of the CK-TFR is superior to that of the spectrogram for this case.

When the CK-TFR was applied to an 8-PSK signal, the phase shifts of +45 and -45 degrees were not as apparent in the pattern. However, M-PSK signals, where M is larger than 4, are rarely used. The parameters mentioned earlier would be appended with the additional parameters of the pseudorandom code's period (if the periodicity was short enough for analysis).

4 Other Surveillance Applications

Further usefulness of the CK-TFR can be exploited from the cross-spectrum by obtaining phase features of the

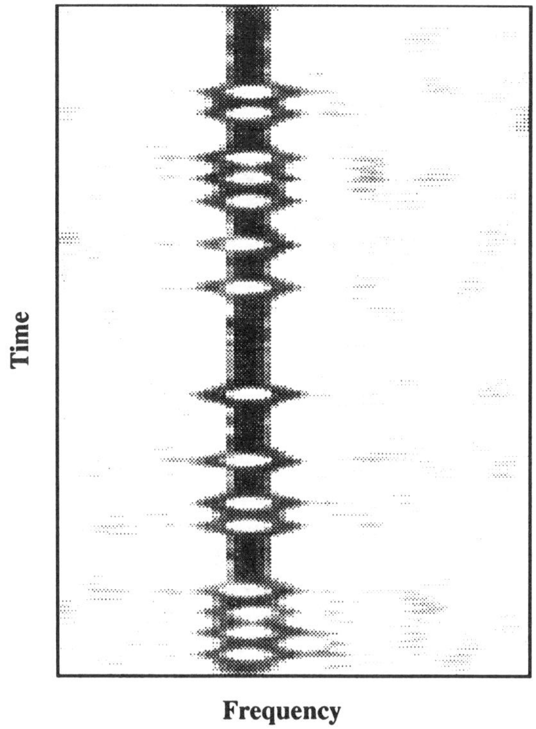

Figure 3: CK-TFR of DS-BPSK signal (SNR = 10 dB).

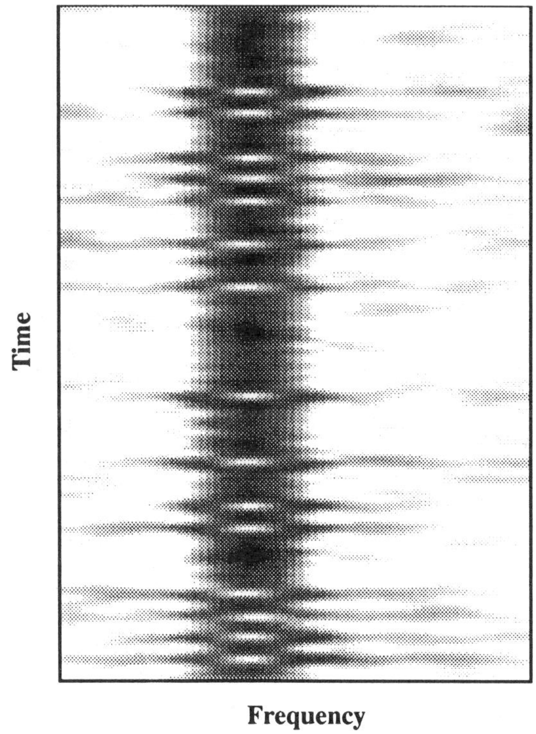

Figure 4: Spectrogram of DS-BPSK signal (SNR = 10 dB).

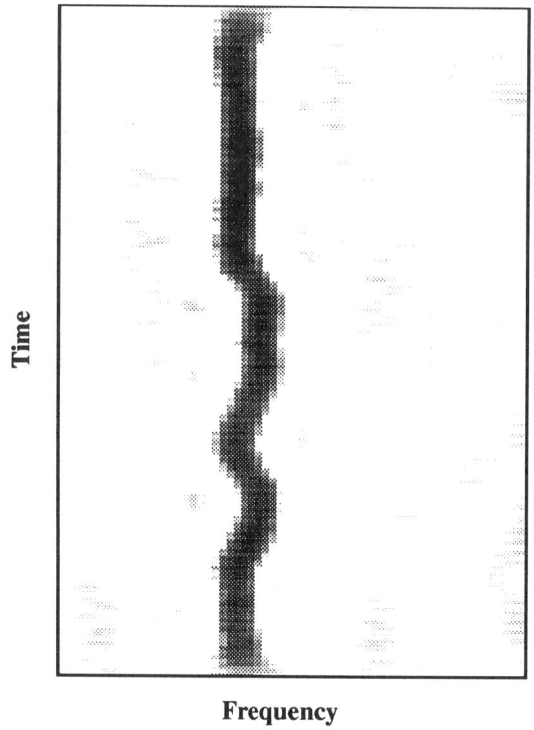

Figure 5: CK-TFR of MSK signal (SNR = 10 dB).

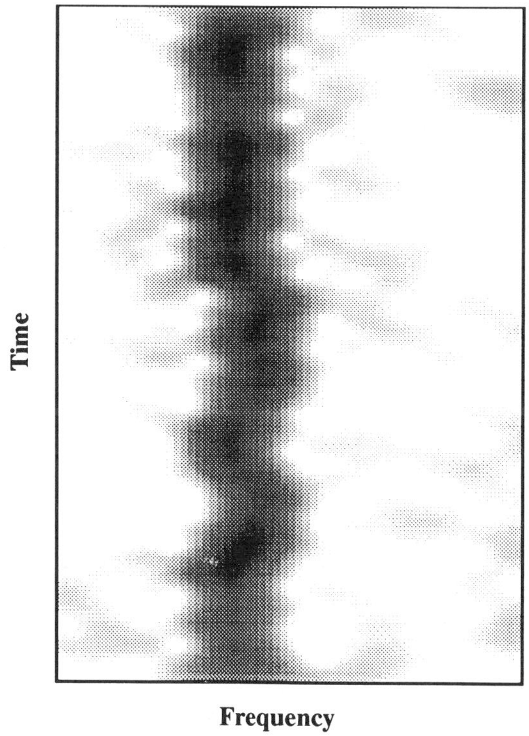

Figure 6: Spectrogram of MSK signal (SNR = 10 dB).

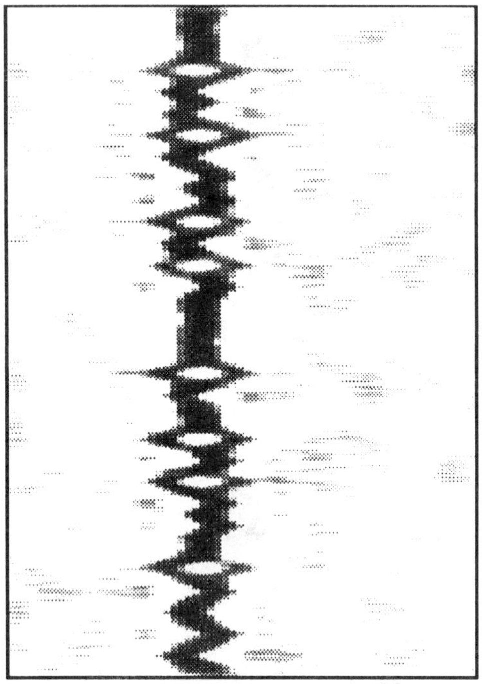

Figure 7: CK-TFR of DS-QPSK signal (SNR = 10 dB).

surface over the time-frequency (t-f) plane. Since the CK-TFR is a high-resolution t-f estimator, it may be an effective tool for representing nonstationary acoustic returns. Thus, spread spectrum signals that use shift-keyed modulation fall in this category, and are considered cyclostationary. In the t-f model, the local behavior of the signal exhibits nonstationarities that are normally masked in the conventional power spectral density. Since a spread spectrum signal is, in general, $r(t) = s(t)d(t)$, then s(t) exhibits high correlation among spectral components with spacing $2/T_0$ (T_0 is the pseudorandom chip rate) even though r(t) is essentially a wideband process. Obtaining this "cyclic" spectrum would help in identifying and classifying the different types of spread spectrum signals.

5 Conclusion

Usage of the CK-TFR for shift-keyed signal identification has been introduced. In particular, spectral "peak splitting" has been observed coincident with direct sequence phase shift-keyed signals, and a mathematical foundation for the splitting has been investigated. Structure of the splitting has been noted to be a possible feature for identification of different unknown a priori phase shift-keyed signals. A comparison of the CK-TFR to that of the spectrogram has shown superior resolution and clarity for the CK-TFR.

References

[1] T. A. C. M. Claasen and W. F. G. Mecklenbrauker, "The Wigner Distribution - A Tool for Time-Frequency Signal Analysis: Part III," Philips J. Res., 35, pp. 372-389, 1980.

[2] L. Cohen, "Time-Frequency Distributions - A Review," Proc. of the IEEE, vol. 77, no. 7, pp. 941-981, 1989.

[3] W. Kaplan, *Advanced Mathematics for Engineers*, sctn. 9-10, Addison-Wesley, Reading, MA, 1981.

[4] L. Cohen, "Generalized Phase-Space Distribution Functions" J. Math. Phys., vol. 7, no. 5, pp.781-786, 1966.

[5] L. Atlas, Y. Zhao and R. Marks II, "Application of the Generalized Time-Frequency Representation to Speech Signal Analysis," Proc. IEEE Pacific Rim Conf. on Comm., Comp. and Sig. Proc., pp. 517-520, 1987.

[6] Y. Zhao, L. Atlas and R. Marks II, "The Use of Cone-Shaped Kernels for Generalized Time-Frequency Representations of Nonstationary Signals," IEEE Trans. ASSP, vol. 38, no. 7, pp. 1084-1091, 1990.

[7] L. Atlas, P. Loughlin and J. Pitton, "Signal Analysis with Cone-Kernel Time-Frequency Representations and their Application to Speech," in *New Methods in Time-Frequency Analysis*, B. Boashash (ed.), Longman Cheshire, Sydney, Australia, 1990 (in press).

[8] L. Atlas, P. Loughlin and J. Pitton, "Truly Nonstationary Techniques for the Analysis and Display of Voiced Speech," Proc. ICASSP'91, pp. 433-436, 1991.

[9] P. Loughlin, J. Pitton and L. Atlas, "New Properties to Alleviate Interference in Time-Frequency Representations," IEEE Proc. ICASSP'91, pp. 433-436, 1991.

[10] P. Loughlin, J. Pitton and L. Atlas, "Bilinear Time-Frequency Representations: New Insights and Properties," submitted to IEEE Trans. Sig. Proc., 1991.

Performance of Fast Frequency-hopped MFSK Receivers with Linear Combining in a Rician Fading Channel with Partial-band Interference

R. Clark Robertson and Kang Yeun Lee
Electrical and Computer Engineering Department, Code EC/RC
Naval Postgraduate School
Monterey, CA 93943

Abstract

An error probability analysis is performed for a conventional M-ary orthogonal frequency-shift keying (MFSK) noncoherent receiver employing fast frequency-hopped (FFH) spread spectrum waveforms transmitted over a Rician fading channel with partial-band interference. The partial-band interference is modeled as a Gaussian process. As might be expected, higher modulation orders and greater diversity improve overall system performance when only wideband interference is encountered. However, the performance of the conventional receiver is found to be significantly degraded by worst case partial-band interference regardless of the modulation order or number of hops per data symbol used for all channel conditions.

1 Introduction

One of the applications of frequency-hopped spread spectrum signals is to reduce receiver performance degradation due to narrowband interference (either intentional or otherwise). This is accomplished by transmitting each symbol at a different carrier frequency according to some apparently random pattern known only by the transmitter and the receiver. The bandwidth of the frequency-hopped signal is generally much larger than the signal bandwidth in the absence of frequency-hopping. Consequently, in order to interfere with the entire spread spectrum band, given a fixed total interference power the noise power spectral density of the narrowband interference is reduced. Rather than interfere with the entire spread-spectrum bandwidth simultaneously with the consequent reduction in noise power spectral density, the narrowband interference may affect only a portion of the total spread spectrum bandwidth at any one time, and the portion of the total spread spectrum bandwidth experiencing narrowband interference may also change in an apparently random manner. This is referred to as partial-band interference. In the case of partial-band interference, only some of the transmitted symbols will be affected by narrowband interference, and the question arises as to whether performance can be improved by implementing diversity in the form of fast frequency-hopping; that is, each information symbol is transmitted multiple times, and each transmission is at a different apparently random carrier frequency within the overall spread spectrum bandwidth.

In this paper the performance of a conventional fast frequency-hopped M-ary orthogonal frequency-shift keyed (FFH/MFSK) noncoherent receiver is investigated. The channel is assumed to experience Rician fading and partial-band interference. By modelling the fading as Rician, in which the signal is considered to consist of a direct signal component and a diffuse signal component, we obtain a general result that is valid in the limit of large direct-to-diffuse signal power ratios for Gaussian channels and in the limit of small direct-to-diffuse signal power ratios for Rayleigh fading channels as well as the general case where the effects of both the direct and diffuse components of the signal must be included in the analysis. The performance of the conventional FFH/MFSK noncoherent receiver in a Rician fading channel with only wideband noise (no partial-band interference) has been previously investigated [1], as has the performance of a conventional noncoherent FFH binary frequency-shift keyed receiver under conditions of worst case partial-band interference for a Gaussian channel [2]. The transmitter is assumed to implement frequency diversity by fast hopping to provide L independent samples of each transmitted symbol. The receiver is assumed to dehop the signal with perfect timing and samples are demodulated with a bandpass filter and a quadratic detector. The L samples at the output of the quadratic detectors are combined to provide the decision statistics for the M channel outputs.

In this paper, we assume that the smallest spacing between frequency hop slots is larger than the coherence bandwidth of the channel [3, 4]. As a result, each hop of a symbol experiences independent fading. In addition, the channel for each hop of a symbol is modeled as a frequency-nonselective, slowly fading Rician process. This implies that the signal bandwidth is much smaller than the coherence bandwidth of the channel and that the hop duration is much smaller than the coherence time of the channel [3, 4]. The latter assumption is equivalent to requiring the hop rate to be large compared to the Doppler spread of the channel. As a result, the signal amplitude can be modeled as a Rician random variable that remains fixed at least for the duration of a single hop.

As mentioned previously, in addition to wideband Gaussian noise, the channel is affected by narrowband noise in the form of partial-band interference. The partial-band interference may be due to either a partial-band jammer or some unintended narrowband noise source. The partial-band interference is modeled as additive Gaussian noise and is assumed to corrupt only a fraction γ, where $1 \geq \gamma \geq 0$, of the entire spread spectrum bandwidth at any one time. In addition, partial-band interference, when present, is assumed to be present in each branch of the MFSK demodulator, and the fraction of the spread spectrum bandwidth experiencing partial-band interference is assumed to be the same for all hops of a symbol. Hence, γ is the probability that partial-band interference is present in all M channels of the receiver, and $1-\gamma$ is the probability that partial-band interference is not present in all M channels of the receiver. If $N_I/2$ is defined to be the average power spectral density of narrowband interference over the entire spread spectrum bandwidth, then $\gamma^{-1}N_I/2$ is the power spectral density of partial-band interference when it is present. Thermal noise and other wideband noise that corrupt the channel are modeled as additive white Gaussian noise, and the power spectral density of this wideband noise is defined as $N_0/2$. Thus, the power spectral density of the total noise is $\gamma^{-1}N_I/2 + N_0/2$ when partial-band interference is present and $N_0/2$ otherwise. If the equivalent noise bandwidth of each bandpass filter in the conventional FFH/MFSK demodulator is B Hz, then for each hop the signal is received with noise of power N_0B with probability $1-\gamma$ when partial-band interference is not present and with noise of power $(\gamma^{-1}N_I + N_0)B$ with probability γ when partial-band interference is present.

The symbol rate is $R_s = R_b/\log_2 M$ where M is the order of the MFSK modulation and R_b is the bit rate. Since the MFSK signal is assumed to perform L hops per symbol, the hop rate is $R_h = LR_s$. The equivalent noise bandwidth of the bandpass filters in each of the M channels of the conventional FFH/MFSK receiver must be at least as wide as the hop rate, and in this paper we use $B = R_h$. The overall system bandwidth is assumed to be very large compared to the hop rate.

2 Conventional FFH/MFSK Receiver Analysis

In this section the probability of bit error for the conventional FFH/MFSK receiver is obtained. The probability of bit error is dependent on the bit energy-to-interference density ratio, the bit energy-to-wideband noise power spectral density ratio, the fraction of the spread bandwidth experiencing partial-band interference, the number of hops L per bit, and the number of symbols M.

2.1 Probability of Bit Error

Due to the symmetry of the conventional FFH/MFSK receiver, the probability of symbol error can be obtained by assuming that the desired signal is present in channel 1 of the receiver. We also assume equally likely M-ary symbols, and we assume that partial-band interference, when present, is uniformly present in all channels. Designating X_i, $i = 1, 2, \ldots, M$ as the random variables that represent the outputs of each of the M channels of the receiver and using the symmetry of the receiver, we can obtain the probability of symbol error conditioned on i out of L hops experiencing partial-band interference as

$$P_s(i) = 1 - \Pr\left[X_1 > X_2 \bigcap X_1 > X_3 \bigcap \ldots \bigcap X_1 > X_M | i\right] \quad (1)$$

Since the output of each channel of the conventional FFH/MFSK is assumed to be independent and since the random variables that represent the output of each channel that do not contain the signal are assumed to be identical, (1) simplifies to

$$P_s(i) = 1 - \int_0^{\infty} f_{X_1}(x_1|i)\left[\int_0^{x_1} f_{X_m}(x_m|i)\,dx_m\right]^{M-1} dx_1 \quad (2)$$

where $m \neq 1$. Given the conditional probability of symbol error, we obtain the probability of symbol error as

$$P_s = \sum_{i=0}^{L} \binom{L}{i} \gamma^i (1-\gamma)^{L-i} P_s(i) \quad (3)$$

where γ is the fraction of the spread bandwidth experiencing partial-band interference. Now the probability of bit error is obtained from (3) as

$$P_b = \frac{M/2}{M-1} P_s \quad (4)$$

where the bit energy-to-noise power spectral density ratio is related to the symbol energy-to-noise power spectral density ratio by

$$\frac{E_s}{N_T} = \left(\frac{E_b}{N_T}\right) \log_2 M \quad (5)$$

In (5), N_T is either N_0 or $\gamma^{-1}N_I + N_0$ depending on whether partial-band interference is absent or present, respectively.

2.2 Decision Variable Probability Density Functions

Since the noise, whether it contains partial-band interference or only wideband interference, is assumed to be Gaussian process, the probability density function at the output of the quadratic detector for one of the M receiver channels when a signal is present in that channel is the probability density function of the square of the envelope of a sine wave with random phase plus a narrowband Gaussian process. The result is a special case of the non-central chi-squared distribution, and assuming the signal is present in channel 1 and for a signal amplitude $\sqrt{2}a_k$, the probability density function for the random variable X_{1k} is given by [5]

$$f_{X_{1k}}(x_{1k}|a_k) =$$

$$\frac{1}{2\sigma_k^2}\exp\left(-\frac{x_{1k}+2a_k^2}{2\sigma_k^2}\right)I_0\left(\frac{a_k\sqrt{2x_{1k}}}{\sigma_k^2}\right)u(x_{1k}) \quad (6)$$

where $u(\bullet)$ is the unit step function, $I_0(\bullet)$ is the modified Bessel function of the first kind and zero order, and σ_k^2 represents the noise power in hop k of a symbol. As previously noted, $\sigma_k^2 = BN_0$ with probability $1-\gamma$ and $\sigma_k^2 = B(N_0 + N_I/\gamma)$ with probability γ. Since a Rician fading channel is assumed, a_k is a Rician random variable representing the fading of hop k of a symbol, and the unconditional probability density function for X_{1k} is found by evaluating

$$f_{X_{1k}}(x_{1k}) = \int_0^\infty f_{X_{1k}}(x_{1k}|a_k)f_{A_k}(a_k)\,da_k \quad (7)$$

The Rician probability density function is [5]

$$f_{A_k}(a_k) = \frac{a_k}{\sigma^2}\exp\left(-\frac{a_k^2+\alpha_k^2}{2\sigma^2}\right)I_0\left(\frac{a_k\alpha_k}{\sigma^2}\right)u(a_k) \quad (8)$$

where α_k^2 is the power of the direct signal component and $2\sigma^2$ is the power of the diffuse signal component. Hence, the total average signal power of hop k of a symbol is $\alpha^2 + 2\sigma^2$ and in this paper is assumed to remain constant from hop to hop. Substituting (8) and (6) into (7) and integrating, we get

$$f_{X_{1k}}(x_{1k}) = \frac{1}{2(\sigma_k^2+2\sigma^2)}$$

$$\times \exp\left[-\frac{1}{2}\left(\frac{x_{1k}+2\alpha_k^2}{\sigma_k^2+2\sigma^2}\right)\right]I_0\left(\frac{\alpha_k\sqrt{2x_{1k}}}{\sigma_k^2+2\sigma^2}\right)u(x_{1k}) \quad (9)$$

The probability density function at the output of the quadratic detector for each of the $M-1$ channels with only Gaussian noise present is a special case of (6) when $a_k \to 0$. Hence,

$$f_{X_{mk}}(x_{mk}) = \frac{1}{2\sigma_k^2}\exp\left[-\frac{x_{mk}}{2\sigma_k^2}\right]u(x_{mk}) \quad (10)$$

The Laplace transform of (10) is

$$F_{X_{mk}}(s) = \frac{1}{2\sigma_k^2 s+1} \quad (11)$$

Designating $\sigma_{k_1}^2$ and $\sigma_{k_2}^2$ as the noise power of hop k of a symbol when partial-band interference is present and when it is not, respectively, and using (11), we obtain the Laplace transform of the probability density function of the decision variable for the channels with no signal when i of L hops experience partial-band interference as

$$F_{X_m|i}(s) = \frac{1}{(\sigma_{k_1}^2 s+1)^i}\frac{1}{(\sigma_{k_2}^2 s+1)^{L-1}} \quad (12)$$

Rather than invert (12) to obtain $f_{X_m}(x_m|i)$, we obtain the Laplace transform of the integral of $f_{X_m}(x_m|i)$ as $F_{X_m|i}(s)/s$. This can be inverted to yield

$$\int_0^{x_1} f_{X_m}(x_m|i)\,dx_m = 1 - \left[K_{21(L-i)} + K_{22(L-i)}x_1\right.$$

$$\left. + \ldots + \frac{K_{2(L-i)(L-i)}x_1^{L-i-1}}{(L-i-1)!}\right]\exp\left[-\frac{x_1}{\sigma_{k_2}^2}\right]u(x_1)$$

$$-\left[K_{11i} + K_{12i}x_1 + \ldots + \frac{K_{1ii}x_1^{i-1}}{(i-1)!}\right]\exp\left[-\frac{x_1}{\sigma_{k_1}^2}\right]u(x_1) \quad (13)$$

where the coefficients K_{1ji}, $j = 1,2,\ldots,i$ and $K_{1k(L-i)}$, $k = 1,2,\ldots,(L-i)$ depend upon L and i.

In order to obtain the probability density function of the decision variable for the channel containing the signal, we first obtain the Laplace transform of (9)

$$F_{X_{1k}}(s) = \frac{\beta_k}{s+\beta_k}\exp\left(-2\sigma_k^2\beta_k\rho_k\right)\exp\left[\frac{2\sigma_k^2\beta_k^2\rho_k}{s+\beta_k}\right] \quad (14)$$

where

$$\beta_k = \frac{1}{2\sigma_k^2(1+\xi_k)} \quad (15)$$

Since the L hops are independent, we obtain the conditional probability density function for the decision variable X_1 given that i hops of a symbol experience partial-band interference as

$$f_{X_1}(x_1|i) = \left[f_{X_{1k}^{(1)}}(x_{1k})\right]^{\otimes i} \otimes \left[f_{X_{1k}^{(2)}}(x_{1k})\right]^{\otimes(L-i)} \quad (16)$$

where $f_{X_{1k}^{(n)}}(x_{1k})$, $n = 1,2$ denotes the probability density function of the random variable X_{1k} for the two possibilities of hop k of a symbol with ($n = 1$) and without ($n = 2$) partial-band interference. Hence, β_k, ρ_k, and ξ_k are now written as $\beta_k^{(n)}$, $\rho_k^{(n)}$, and $\xi_k^{(n)}$, respectively, in order to differentiate between hops with and without partial-band interference. The $c_n$fold convolutions $\left[f_{X_{1k}^{(n)}}(x_{1k}^{(n)})\right]^{\otimes c_n}$ required by (16) are obtained as the inverse Laplace transform of $\left[F_{X_{1k}^{(n)}}(s)\right]^{c_n}$ which is given by

$$\left[f_{X_{1k}^{(n)}}(z_{1k})\right]^{\otimes c_n} = \frac{\beta_k^{(n)} x_{1k}^{(c_n-1)/2}}{\left(2\sigma_{k_n}^2 c_n \rho_k^{(n)}\right)^{(c_n-1)/2}}$$

$$\times \exp\left[-\beta_k^{(n)}\left(x_{1k} + 2\sigma_{k_n}^2 c_n \rho_k^{(n)}\right)\right]$$

$$I_{c_n-1}\left(2\beta_k^{(n)}\sqrt{2\sigma_{k_n}^2 c_n \rho_k^{(n)} x_{1k}}\right) u(x_{1k}) \qquad (17)$$

where $c_1 = i$ and $c_2 = L - i$. The final convolution required by (16) to obtain $f_{X_1}(x_1|i)$ is performed numerically. The conditional probability of symbol error $P_s(i)$ is now found by using this result and (13) in (2). In general, (2) must be evaluated numerically. In the special cases of $i = 0$ (no hops experience partial-band interference) and $i = L$ (all hops experience partial-band interference), $f_{X_1}(x_1|i)$ is obtained analytically, and (2) can then be evaluated analytically [1]. This analytic solution is extremely complicated, however, and numerical integration is preferable even in these cases. This is especially true since the convergence of the numerical integration can be accelerated since

$$\lim_{x_1 \to \infty} \int_0^{x_1} f_{X_m}(x_m|i)\, dx_m = 1 \qquad (18)$$

Hence, by adding and subtracting

$$\int_0^{\infty} f_{X_1}(x_1|i)\, dx_1 = 1 \qquad (19)$$

to the right hand side of (2), we obtain the computationally efficient expression

$$P_s(i) =$$

$$\int_0^{\infty} f_{X_1}(x_1|i)\left\{1 - \left[\int_0^{x_1} f_{X_m}(x_m|i)\, dx_m\right]^{M-1}\right\} dx_1 \qquad (20)$$

since the $\{\bullet\}$ term in (20) approaches zero as $x_1 \to \infty$.

3 Numerical Results

The performance of the conventional FFH/MFSK noncoherent receiver is investigated for various values of E_b/N_0, fraction of spread bandwidth experiencing partial-band interference, diversity order, and modulation order as a function of E_b/N_I. In addition, the effect of a broad range of channel fading, from near the Gaussian limit where $\alpha_k^2/2\sigma^2$ is very large down to near the Rayleigh limit where $\alpha_k^2/2\sigma^2$ is very small, on system performance is examined.

As might be expected, higher modulation orders and greater diversity improve overall system performance when only wideband interference is encountered. However, for all types of channel, from the Gaussian limit to the Rayleigh limit, worst case partial-band interference significantly degrades system performance regardless of the amount of diversity or the modulation order. In fact, except for Rayleigh or near Rayleigh fading, the use of diversity actually degrades receiver performance under conditions of worst case partial-band interference.

As an illustration of the preceding remarks, system performance for several different values of diversity, modulation order, and channel type are shown in Figs. 1–4.

4 Conclusions

The performance of the conventional FFH/MFSK noncoherent receiver is significantly degraded by worst case partial-band interference regardless of the amount of diversity or modulation order and regardless of channel fading conditions.

References

[1] W. C. Lindsey, 'Error probabilities for Rician fading multichannel reception of binary and N-ary signals,' IEEE Trans. on Information Theory, vol. IT-10, pp. 339–350, Oct. 1964.

[2] J. S. Lee, R. H. French, and L. E. Miller, 'Probability of error analyses of a BFSK frequency-hopping system with diversity under partial-band jamming interference-Part I: Performance of square-law linear combining soft decision receiver,' IEEE Trans. Commun., vol. COM-32, pp. 645-653, June 1984.

[3] C. M. Keller and M. B. Pursley, 'Diversity combining for channels with fading and partial-band interference,' IEEE J. Sel. Areas Commun., vol. SAC-5, no. 2, pp. 248-260, Feb. 1987.

[4] J. G. Proakis, Digital Communications, 2nd Ed. New York: McGraw-Hill, 1989.

[5] A. D. Whalen, Detection of Signals in Noise. New York: Academic Press, 1971.

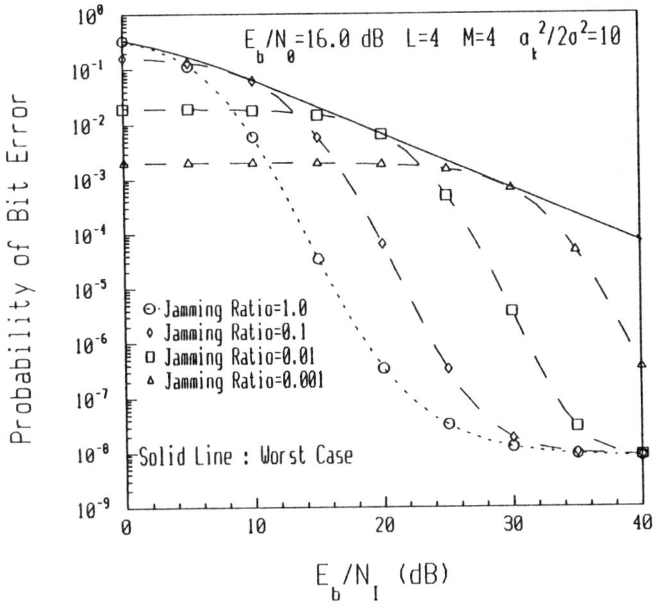

Fig. 1: Performance of the conventional FFH/MFSK noncoherent receiver for $L = 4$, $M = 4$, and moderate Rician fading ($\alpha_k/2\sigma^2 = 10$) with $E_b/N_0 = 16$ dB.

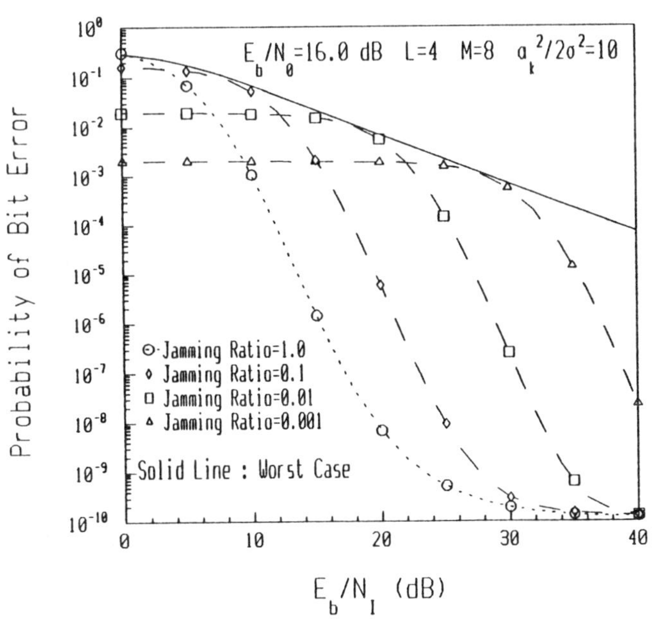

Fig. 2: Performance of the conventional FFH/MFSK noncoherent receiver for $L = 4$, $M = 8$, and moderate Rician fading ($\alpha_k/2\sigma^2 = 10$) with $E_b/N_0 = 16$ dB.

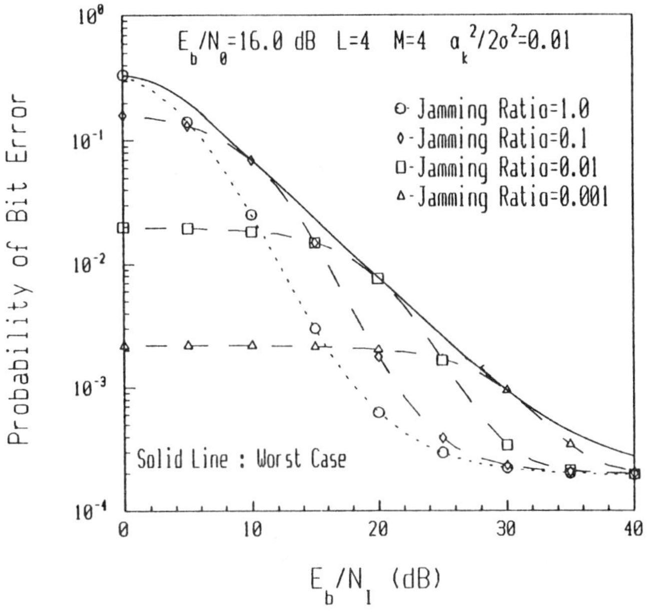

Fig. 3: Performance of the conventional FFH/MFSK noncoherent receiver for $L = 4$, $M = 4$, and nearly Rayleigh fading ($\alpha_k/2\sigma^2 = 0.01$) with $E_b/N_0 = 16$ dB.

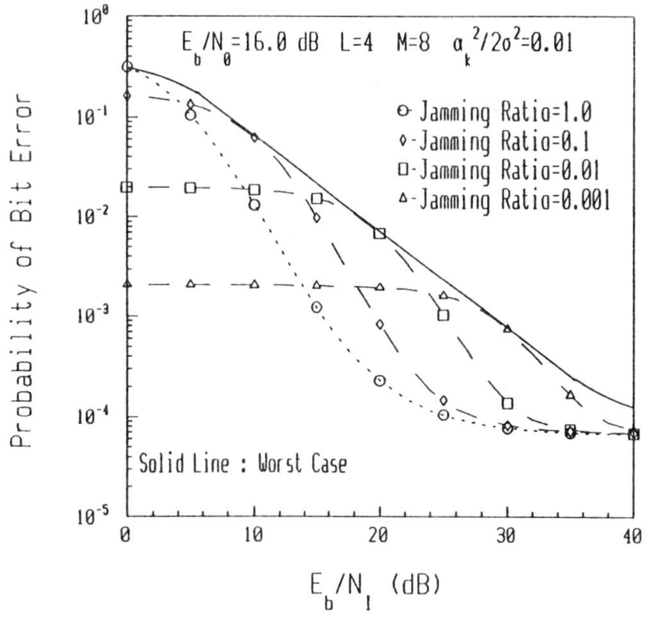

Fig. 4: Performance of the conventional FFH/MFSK noncoherent receiver for $L = 4$, $M = 8$, and nearly Rayleigh fading ($\alpha_k/2\sigma^2 = 0.01$) with $E_b/N_0 = 16$ dB.

A New Approach to Blind Identification and Equalization of Multipath Channels

Lang Tong
Dept of Electrical and Computer Engineering
West Virginia University
Morgantown, WV 26506-6101

Guanghan Xu and Thomas Kailath
Information Systems Laboratory
Stanford University
Stanford, CA 94305

Abstract

A new blind channel identifcation and equalization method is proposed by exploiting the cyclostationarity of communication signals. Identification and equalization of possibly nonminimum phase multipath channels are achieved without using training signals. Unlike most of the adaptive blind equalization methods for which the convergence properties are often problematic, the eigenstructure-based channel estimation algorithm proposed here is asymptotically exact. Based on the second-order properties, the new approach identifies nonminimum phase channels and achieves equalization with less samples than most techniques based on higher-order statistics. Simulations demonstrate promising performance of the proposed algorithm in a blind equalization of a three-ray multipath channel.

1 Introduction

The promise of cellular mobile communication has propelled ever increasing research efforts in high-speed wireless communication in recent years. The implementation of digital communication receivers for universal digital portable communication poses immense engineering challenges, primarily because the environment for wireless communication is much more complicated than the regular telephone channels. In particular, the time varying multipath fading that commonly exists in a wireless communication environment leads to severe intersymbol interference (ISI). Other channel impairments that contribute to ISI include symbol clock residual jitter, carrier phase jitter, etc. In order to achieve high-speed reliable communication, channel identification and equalization are necessary to combat ISI. Traditionally, channel identification and equalization are achieved either by sending training sequences, or by designing the equalizer based on a priori knowledge of the channel. The later approach is clearly not suitable for radio communication environment since little knowledge about such channel can be assumed *a priori*. The standard adaptive approach, though attractive in handling time varying unknown channels, has to waste a fraction of the transmission time slot for a training sequence. Indeed even the so-called the decision feedback equalization (DFE), which does not explicitly use a training sequence, nevertheless requires the sending of known training sequences periodically to avoid catastrophic error propagation [2]. In contrast to the standard adaptive equalization, the blind equalization does not require a training sequence. Instead, the statistical properties of the transmitted signals are exploited. The equalization is performed at the receiver end *without* access to the symbols being transmitted. Instead of choosing the equalizer so that the equalized output sequence $\{\tilde{s}_k\}$ is close to the source symbol sequence $\{s_k\}$ as in the standard equalization formulation, in blind equalization, one chooses the equalizer so that the *statistics* of the equalized output sequence $\{\tilde{s}_k\}$ is close to the *statistics* of the source symbol sequence $\{s_k\}$.

1.1 Existing Blind Channel Identification and Equalization Methods

The innovative idea of self-recovering (blind) adaptive equalization was first proposed by Sato [9], then further developed by Godard [6], Treichler and Agee [10], Benveniste and Goursat [1], Picci and Prati [8], Foschini [4] and more recently Shalvi and Weinstein [11]. Although the blind equalization schemes proposed so far are technically different, they are all derived via some optimization criteria involving certain higher-order (cumulants) statistics of the observation; various gradient-based algorithms are then employed to achieve the optimization. The major problem with the adaptive blind equalization techniques is that their convergence properties, as shown in [7], are often problematic; As pointed out in [7], the global convergence may be jeopardized when the channel has finite impulse response; however, it should be noted that the optimization criteria proposed by Shalvi and Weinstein [11] ensure global optimization whenever there exist ideal equalizers. In addition, the effect of non-Gaussian noise will also affect the convergence and the performance of the cumulant-based approach. Currently, there is no systematic initialization method to insure the convergence.

1.2 A New Blind Channel Identification and Equalization Method

In this paper, we propose a new blind channel identification and equalization method. The main features of our approach can be summarized as follows:

1058-6393/91 $01.00 © 1991 IEEE

1. The algorithm provides *exact* identification of the possibly *non-minimum phase* channel impulse response if the correlation function of the received signal is known exactly; more realistically, when the estimated signal correlation is used, asymptotic exactness can be established.
2. The algorithm can be used to initialize various adaptive schemes and the equalized output can be used to facilitate decision feedback adaptation. The identified channel can be used to implement maximum likelihood sequence estimation [3] to further reduce ISI.
3. The algorithm relies only on the second order statistics of the received signal. Therefore it usually requires smaller sample size than all other schemes suggested to date. In our simulation at a signal-to-noise ratio (SNR) of 30dB, 100 symbols are sufficient to give a good estimate of the channel, which is a much smaller number than for the existing techniques. This fact also implies that the algorithm can be used to estimate channels with relative faster varying fading.
4. There is no restriction imposed on the probability distribution of the source symbols. The random source may be real or complex, continuous or discrete, or even Gaussian, in contrast to the assumptions [11].

It may seem impossible at first glance that a non-minimum phase system can be identified using only the second order statistics of the system output. However, it may be less surprising if one assumes that the system is driven by a *nonstationary* process. This is indeed the case for most communication channels, where the input signals are cyclostationary (or periodically correlated) rather than stationary.

1.3 Problem Statement

When the channel is time invariant, the received complex signal $x(\cdot)$ can be expressed as

$$r(t) = \sum_{k=-\infty}^{\infty} s_k h(t-kT), \qquad (1)$$

$$x(t) = r(t) + n(t), \qquad (2)$$

where

s_k: an information symbol in a signal constellation $\mathcal{S}$

$h(\cdot)$: the "composite" channel impulse response that includes pulse shaping filter, multipath channel, and receiver filters.

T: the symbol interval, $n(\cdot)$: the additive noise.

The objective of *blind* channel identification is to estimate $h(\cdot)$ given only the received signal $x(\cdot)$. Once such identification is achieved, the estimation of the information symbols s_k's becomes more or less a classical problem, and various equalization and sequence estimation techniques can be applied.

We assume the following throughout the development in the sequel:

1. The impulse response $h(t)$ has finite support.
2. $\{s_k\}$ is an i.i.d. sequence drawn from $\mathcal{S}$ with certain distribution not necessarily known.
3. $n(\cdot)$ is a white noise process.

2 A Multichannel Representation

An important observation is that, under the first assumption that the channel impulse response has finite support, the space of signal restricted to any finite time interval I (referred to as an *observation interval*) is a *linear* space spanned by a basis comprised of time-shifted copies of $h(\cdot)$. Specifically, the signal space $\mathcal{R}_{(\mathcal{S},h)}(I)$, or simply $\mathcal{R}(I)$, is defined by

$$\mathcal{R}(I) = \{r(t)|r(t) = \sum_{k=-\infty}^{\infty} s_k h(t-kT), s_k \in \mathcal{S}, t \in I\} \qquad (3)$$

Moreover, when the observation interval I is finite and the channel impulse response has only finite support, $\mathcal{R}(I)$ is *finite* dimensional and its dimension can be computed directly with knowledge of I and the duration of $h(\cdot)$. Consider, for example, the received (noise free) signal $r(\cdot)$ in an observation interval (t_0, t_0+L) for some finite L. The signal space $\mathcal{R}((t_0, t_0+L))$ is spanned by $h(t-K_0T), \ldots, h(t-(K_0+(d-1))T)$ with $h(t-iT)'s$ being restricted to (t_0, t_0+L). The integer K_0 and the dimension d are functions of t_0, L and the duration L_h of the impulse response $h(.)$ as follows

$$K_0 = \lceil \frac{t_0 - L_h}{T} \rceil, \qquad (4)$$

$$d = \lfloor \frac{t_0 - (K_0-1)T + L}{T} \rfloor, \qquad (5)$$

where $\lceil x \rceil$ ($\lfloor x \rfloor$) stands for the smallest (greatest) integer that is greater than (less than) x. If $x(\cdot)$ is sampled in (t_0, t_0+L) with a sampling period Δ, we then have

$$x(t_0+i\Delta) = \sum_{k=0}^{d-1} s_{K_0+k} h(t_0+i\Delta-(K_0+k)T) \qquad (6)$$
$$+n(t_0+i\Delta), i = 1, \ldots, m.$$

A matrix formulation of the above will be convenient. We can write

$$\mathbf{x}(t_0) = \mathbf{H}(t_0)\mathbf{s}(t_0) + \mathbf{n}(t_0), \qquad (7)$$

where

$$\mathbf{x}(t_0) = [x(t_0+\Delta), \ldots, x(t_0+m\Delta)]^T, \qquad (8)$$

$$\mathbf{H}(t_0) = [h(t_0+i\Delta-(K_0+j-1)T)], \qquad (9)$$

$$\mathbf{s}(t_0) = [s_{K_0}, \ldots, s_{K_0+d-1}]^T \qquad (10)$$

$$\mathbf{n}(t_0) = [n(t_0+\Delta), \ldots, n(t_0+m\delta)]^T. \qquad (11)$$

Similarly, if the sampling is performed in any interval of the form $(t_0 + nT, t_0 + nT + L)$, replacing t_0 and K_0 by $t_0 + nT$ and $K_0 + nT$ in (7) yields,

$$\mathbf{x}(t_0+nT) = \mathbf{H}(t_0+nT)\mathbf{s}(t_0+nT)+\mathbf{n}(t_0+nT). \quad (12)$$

Making a closer examination of the $(i,j)^{th}$ entry of $\mathbf{H}(t_0 + nT)$, we see that

$$h(t_0+nT+i\Delta-(K_0+n+j-1)T) = h(t_0+i\Delta-(K_0+j-1)T). \quad (13)$$

Hence for all integer n,

$$\mathbf{H}(t_0) = \mathbf{H}(t_0 + nT). \quad (14)$$

We then have

$$\mathbf{x}(t_0+nT) = \mathbf{H}(t_0)\mathbf{s}(t_0+nT)+\mathbf{n}(t_0+nT), \qquad n = 0, 1, \ldots \quad (15)$$

When no confusion arises, we drop t_0 (or let $t_0 = 0$) for the sake of simplicity, with the understanding that a certain t_0 is fixed throughout; then the multichannel representation of the received signal is

$$\mathbf{x}(iT) = \mathbf{H}\mathbf{s}(iT) + \mathbf{n}(iT), \quad i = 0, 1, \ldots, \quad (16)$$

where $\mathbf{x}(iT)$ and $\mathbf{n}(iT)$ are m-dimensional vectors formed from the m samples of $x(t)$ and $n(t)$ inside the interval $(t_0 + iT, t_0 + iT + L)$ respectively, $\mathbf{H} = \mathbf{H}(t_0)$, independent of i, is given by (9), and $\mathbf{s}(iT)$ is a d-dimensional vector consisting of symbols that have "contribution" to the received signal inside the observation interval $(t_0 + iT, t_0 + iT + L)$ given in (10). In an actual implementation, the vector $\mathbf{x}(iT)$ can be obtained via time-division demultiplexing [5].

3 A New Channel Identification and Equalization Method

3.1 Channel Identifiability

In presenting the ideas of our approach, we ignore the noise for simplicity. The case with noisy data can be treated in a straightforward way and is elaborated in the presentation of the algorithm.

Following the development in the previous section, the *blind* channel identification problem can be phrased as follows:

Consider a vector process $\mathbf{x}(i)$*, i=1, ..., obtained from a linear model*

$$\mathbf{x}(i) = \mathbf{H}\mathbf{s}(i), i = 0, 1, \ldots \quad (17)$$

along with the following constraints:

1. *H is an $m \times d$ complex matrix of full column rank.*
2. *$\mathbf{s}(i)$ is a zero mean stationary process with autocorrelation function*

$$\mathbf{R_s}(k) = E(\mathbf{s}(i)\mathbf{s}^H(i-k)) \quad (18)$$

of the following form:

$$\mathbf{R_s}(k) = \mathbf{J}^k, k \geq 0, \quad (19)$$

$$\mathbf{R_s}(k) = (\mathbf{J}^T)^{|k|}, k < 0, \quad (20)$$

$$\mathbf{J} = \begin{pmatrix} 0 & 0\ldots & 0 & 0 \\ 1 & 0\ldots & 0 & 0 \\ 0 & 1\ldots & 0 & 0 \\ \vdots & \ddots & \vdots & \vdots \\ 0 & 0\ldots & 1 & 0 \end{pmatrix}. \quad (21)$$

The objective of blind channel identification and equalization is to identify $\mathbf{H}$ *(channel identification) and estimate* $\mathbf{s}(i)$ *from* $\mathbf{x}(i)$ *(channel equalization)*

The following theorem settles the issue of channel identifiability.

Theorem 1 *Suppose* $\mathbf{H}$ *and* $\mathbf{s}(i)$ *satisfy the linear model (17) and its constraints. Then* $\mathbf{H}$ *is uniquely determined up to a constant by* $\mathbf{R_x}(0)$ *and* $\mathbf{R_x}(1)$.

The following lemma provides a constructive proof of the above theorem. Its significance is that it offers a computational method to identify $\mathbf{H}$ from $\mathbf{R_x}(0)$ and $\mathbf{R_x}(1)$.

Lemma 1 *Let* $\mathbf{R_x}(0)$ *have the following singular value decomposition,*

$$\mathbf{U}^H\mathbf{R_x}(0)\mathbf{V} = diag(\sigma_1^2, \ldots, \sigma_d^2, 0, \ldots, 0) \quad (22)$$

Let $\mathbf{u}_i$ *denote the* i^{th} *column of* $\mathbf{U}$*, and let*

$$\mathbf{U_s} = [\mathbf{u}_1, \ldots, \mathbf{u}_d], \quad (23)$$

$$\mathbf{\Sigma} = diag(\sigma_1, \ldots, \sigma_d), \quad (24)$$

$$\mathbf{F} = \mathbf{\Sigma}^{-1}\mathbf{U_s}^H \quad (25)$$

Suppose

$$\mathbf{R} = \mathbf{F}\mathbf{R_x}(1)\mathbf{F}^H \quad (26)$$

has the singular value decomposition of the form

$$[\mathbf{y}_1, \ldots, \mathbf{y}_d]^H\mathbf{R}[\mathbf{z}_1, \ldots, \mathbf{z}_d] = diag(\gamma_1^2, \ldots, \gamma_d^2) \quad (27)$$

Then there exists a real phase ϕ *such that*

$$\mathbf{H} = \mathbf{U_s}\mathbf{\Sigma}\mathbf{Q}e^{j\phi} \quad (28)$$

where

$$\mathbf{Q} = [\mathbf{y}_d, \mathbf{R}\mathbf{y}_d, \ldots, \mathbf{R}^{(\mathbf{d-1})}\mathbf{y}_d] \quad (29)$$

or equivalently,

$$\mathbf{Q} = [(\mathbf{R}^\dagger)^{(d-1)}\mathbf{z}_d, (\mathbf{R}^\dagger)^{(d-2)}\mathbf{z}_d, \ldots, \mathbf{z}_d] \quad (30)$$

3.2 Algorithm Implementation

Lemma 1 provides the essential parts of the algorithm. We remark here that the *constant* phase ambiguity is unavoidable and is usually harmless in practice. Nevertheless, such ambiguity can be removed if, for instance, the channel is real. We first outline the algorithm and then address certain technical points of the algorithm.

A Blind Channel Identification and Equalization Algorithm

1. Select a t_0, Δ such that $T = T_s\Delta$, $L = m\Delta$, and form the vectorized observation $\mathbf{x}(i) = [x(t_0+\Delta+iT_s), \ldots, x(t_0+m\Delta+iT_s)]^T, i = 0, 1, \ldots,$

2. Estimate $\hat{\mathbf{R}}_\mathbf{x}(0)$ and $\hat{\mathbf{R}}_\mathbf{x}(1)$ from $\mathbf{x}(i)$ via, for example, time average.

$$\hat{\mathbf{R}}_\mathbf{x}(0) = \frac{1}{N}\sum_{i=1}^{N}\mathbf{x}(i)\mathbf{x}^H(i), \quad (31)$$

$$\hat{\mathbf{R}}_\mathbf{x}(1) = \frac{1}{N}\sum_{i=1}^{N}\mathbf{x}(i)\mathbf{x}^H(i-1). \quad (32)$$

3. From $\hat{\mathbf{R}}_\mathbf{x}(0)$, estimate the noise covariance σ_n and the dimension d of the signal space.

4. Compute the SVD of $\mathbf{R}_0$,

$$\mathbf{R}_0 = \hat{\mathbf{R}}_\mathbf{x}(0) - \sigma_n^2\mathbf{I} \quad (33)$$

and form $\mathbf{U_S}$ which consists of the singular vectors associated with the d largest singular values, $\mathbf{\Sigma}$ which consists of the square-root of the d largest singular values, and then $\mathbf{F} = \mathbf{\Sigma}^{-1}\mathbf{U_S}^H$.

5. Compute the SVD of $\mathbf{R}$,

$$\mathbf{R} \doteq \mathbf{F}(\hat{\mathbf{R}}_\mathbf{x}(1) - \mathbf{R}_\mathbf{n}(1))\mathbf{F}^H, \quad (34)$$

where

$$\mathbf{R}_\mathbf{n}(1) = \sigma^2\mathbf{J}^{T_s}, \quad (35)$$

$$\mathbf{J} = \begin{pmatrix} 0 & 0\ldots & 0 & 0 \\ 1 & 0\ldots & 0 & 0 \\ 0 & 1\ldots & 0 & 0 \\ \vdots & \ddots & \vdots & \vdots \\ 0 & 0\ldots & 1 & 0 \end{pmatrix} \quad (36)$$

Denote $\mathbf{y}_d$ and $\mathbf{z}_d$ as the left and right singular vectors corresponding to the smallest singular value.

6. Form the estimate of $\mathbf{H}$ (and consequently $h(\cdot)$ if necessary) by

$$\hat{\mathbf{H}} = \mathbf{U_S}\mathbf{\Sigma}\mathbf{Q} \quad (37)$$

where

$$\mathbf{Q} = [\mathbf{y}_d, \mathbf{R}\mathbf{y}_d, \ldots, \mathbf{R}^{(d-1)}\mathbf{y}_d] \quad (38)$$

or

$$\mathbf{Q} = [(\mathbf{R}^\dagger)^{(d-1)}\mathbf{z}_d, (\mathbf{R}^\dagger)^{(d-2)}\mathbf{z}_d, \ldots, \mathbf{z}_d] \quad (39)$$

or certain combination of the above.

7. Extract the information symbols

$$\hat{\mathbf{s}}(i) = \hat{\mathbf{H}}^\dagger\mathbf{x}(i) \quad (40)$$
$$= \mathbf{Q}^H\mathbf{F}\mathbf{x}(i), \quad (41)$$

or by implementing various types of equalizer or maximum-likelihood estimation scheme based on the estimated channel.

4 A Simulation Example

In this simulation, a 3-ray multipath channel is constructed from a raise cosine pulses $c(t, \alpha)$ (α is the roll-off factor) and its delays,

$$h(t) = (c(t, 0.11) + 0.8c(t-1, 0.11) - 0.4c(t-3, 0.11))W_{6T}(t), \quad (42)$$

where W_{6T} is a square window of duration 6 symbol intervals, *i.e.* $L_h = 6T$. Among 23 zeros of the system, there are 20 nonminimum phase zeros. The source symbols are drawn from the 16-QAM signal constellation. Figure 1 is the plot of 1000 output symbols of the unequalized channel (obtained by sampling the received signal at $kT, k = 1, 2, \ldots$). The signal to noise ratio is $30dB$. Clearly, the intersymbol interference is severe and a high error rate is expected. 100 symbols are used to estimate $\mathbf{R}_\mathbf{x}(0)$ and $\mathbf{R}_\mathbf{x}(1)$. 1000 symbols are then transmitted and the equalized channel output is shown in Figure 2, which indicates that the channel is equalized rather nicely.

5 Conclusions

A new approach to blind identification and equalization of multipath channels is developed in this paper. By exploiting the nonstationarity of the received signal, we are able to identify the nonminimum-phase channel with only the second-order statistics. This leads to a more accurate estimation with a smaller sample size. The over-sampling of the received signal is similar to that used in fractionally spaced equalization, where it is known to be insensitive to various timing errors. The algorithm achieves channel identification and equalization without interrupt the communication of the information bearing symbols. In addition, it can be easily incorporated into various existing equalization methods.

Acknowledgements

This work was supported by the Advanced Research Projects Agency of the Department of Defense and was monitored by the Air Force Office of Scientific Research under contract F49620-90-C-0014 (the United States Government is authorized to reproduce and distribute reprints for governmental purposes notwithstanding any copyright notation thereon).

References

[1] A. Benveniste and M. Goursat. "Blind Equlizers". *IEEE Trans. Commun.*, COM-32(8):871–883, August 1984.

[2] E. Eleftheriou and D.D. Falconer. "Adaptive Equalization Techniques for HF Channels". *IEEE J. Select. Areas Commun.*, SAC-5(2):238–247, February 1987.

[3] G.D. Forney. "Maximum-Likelihood Sequence Estimation of Digital Sequences in the Presence of Intersymbol Interference". *IEEE Inform. Theory*, IT-18(3):363–378, May 1972.

[4] G.J. Foschini. "Equalizing Without Altering or Detecting Data". *Bell Syst. Tech. J.*, 64:1885–1911, October 1985.

[5] W. Gardner. *Introduction to Random Processes: with Application to Signal and Systems.* McGraw-Hill, Inc., New York, NY, 1990.

[6] D.N. Godard. "Self-Recovering Equalization and Carrier Tracking in Two-Dimensional Data Communication Systems". *IEEE Trans. Commun.*, COM-28(11):1867–1875, November 1980.

[7] C. R. Johnson. "Admissibility in Blind Adaptive Channel Equalization". *IEEE Control Systems Magzine*, pages 3–15, January 1991.

[8] G. Picci and G.Prati. "Blind Equalization and Carrier Recovery Using a Stop-and Go Decision Dirrected Algorithm". *IEEE Trans. Commun.*, COM-35(9):877–887, September 1987.

[9] Y. Sato. "A Method of Self-Recovering Equalization for Multilevel Amplitude-Modulation". *IEEE Trans. Commun.*, COM-23(6):679–682, June 1975.

[10] J.R. Treichler and B.G. Agee. "A New Approach to Multipath Correction of Constant Modulus Signals". *IEEE Trans. ASSP.*, ASSP-31(2):459–472, April 1983.

[11] O. Shalvi E. Weinstein. "New Criteria for Blind Deconvolution of Nonminimum Phase Systems (Channels)". *IEEE Inform. Theory*, IT-36(2):312–320, March 1990.

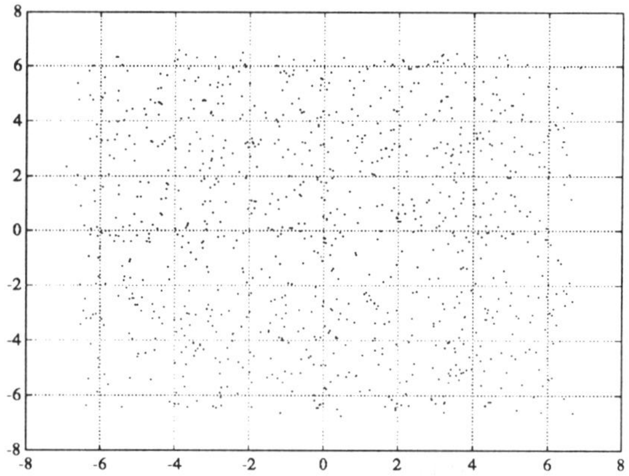

Figure 1: The output sampled at Nyquist instant of unequalized channel. SNR=25dB, 1000 symboles plotted.

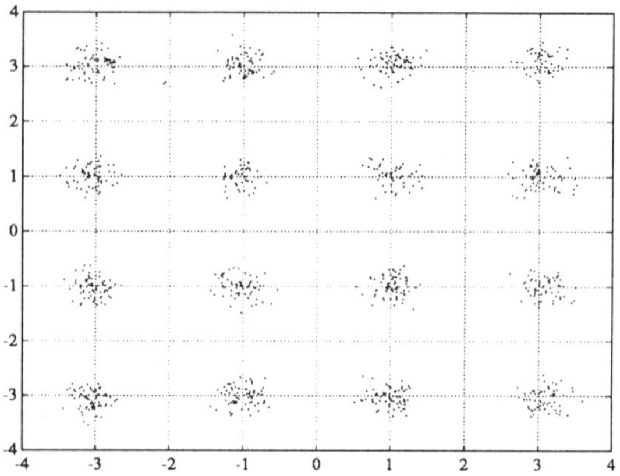

Figure 2: The output of the equalized channel. 1000 symboles are plotted. 100 symboles are used in the estimation. SNR=25dB.

Sigma-Delta Modulation with Powers-of-Two Quantizer

Ping-sheng Lin and **Sawasd Tantaratana**
Department of Electrical and Computer Engineering
University of Massachusetts, Amherst, MA 01003

Abstract

In this paper, we study sigma-delta modulation (SDM) with powers-of-two quantizer. The performances are obtained by simulation. It is observed that the power of the minimum quantization noise stays approximately the same when a powers-of-two quantizer replaces the uniform quantizer, which is traditionally used in SDM. However, the quantization noise of powers-of-two quantizers becomes more stable with respect to the step size, so that a non-optimum choice of the step size has only small effect on the performance. Since the values of the SDM signal represent powers of two, the lowpass filter in the demodulator can be implemented without using multipliers. The application of such SDM for implementing FIR filters is also investigated. The coefficients of FIR filters are encoded using SDM with powers-of-two quantizer. In addition to requiring no multiplier, results show that the ripples of such FIR filters are lower than those based on SDM with uniform quantizer.

1 Introduction

Sigma-delta modulation (SDM) has been receiving considerable attention as an alternative to traditional pulse-code modulation (PCM) high-resolution analog-to-digital (A/D) converter. The 1st-order SDM is depicted in Fig.1. The analog input is first modulated to yield a binary sequence at N ($N \gg 1$) times the Nyquist rate, using a negative feedback loop containing an integrator and a binary (1-bit) quantizer. When the binary signal is demodulated, using a digital decoder followed by a decimator, we obtain a PCM signal at the Nyquist rate.

However, to achieve high-resolution A/D conversion using 1st-order SDM, the system must be designed to operate at a very high oversampling ratio. This ratio typically ranges from 64 to 1024 in practical systems. To reduce the oversampling ratio, higher-order SDM's have been proposed[1][2][3][4]. They can be visualized as either a 1st-order SDM with extra feedback loops or a cascade of 1st-order SDM's with a noise canceling network. The most studied ones are 2nd-order and two-stage SDM's. The output of a 2nd-order SDM is similar to that of a two-stage SDM. Therefore, they have similar performance. In A/D converter we can accomplish the same resolution with a much lower oversampling ratio when higher-order SDM's are used than we can with 1st-order SDM.

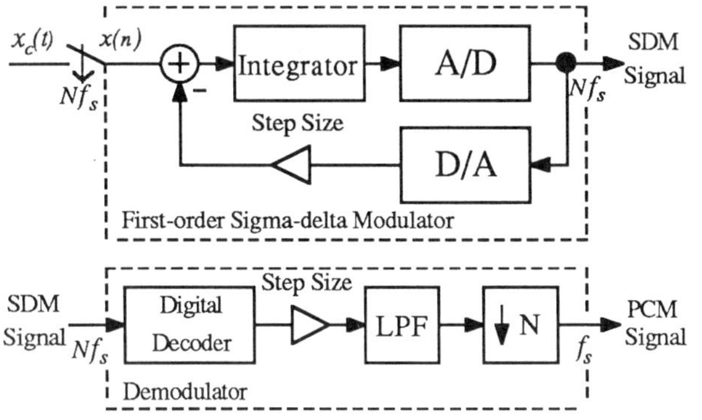

Figure 1: First-order SDM modulator/demodulator

Another approach to reduce the oversampling ratio is to use multi-bit SDM's [5][6]. For example, for small input signals, a ternary (3-level) quantizer improves the SNR by 18 dB over binary quantizer in 1st-order SDM and by 3-4 dB in 2nd-order SDM. The cost in using a ternary quantizer is the additional bit needed to represent the quantizer output. Employing a multi-bit quantizer implies that a digital-to-analog (D/A) converter must be inserted in the feedback loop to provide the analog feedback signal, increasing the complexity of the modulator. It has also been pointed out [7] that the loop delay caused by the conversion time of the D/A converter is an essential parameter in the performance of an SDM.

2 SDM with Powers-of-two Quantizer

In this section we investigate SDM's with powers-of-two quantizer. Such a quantizer has also been termed floating point incremental coder (FLOPTIC) [8]. An n-bit powers-of-two quantizer can have 2^n different output levels which are symmetric about zero. The largest quantizer levels, in magnitude, are usually normalized to -1 and 1 and the remaining 2^{n-1} levels are negative powers of two. The endpoints of the input intervals are chosen as the midpoints between the two corresponding output levels, scaled by the step size Δ. It is illustrated in Fig.2 for $n = 2$, where b is an integer, which is adjustable to minimize the power (or variance) of the quantization noise (Q-noise) $\gamma(n)$, defined as $\gamma(n) = \Delta q(u(n)) - u(n)$, $n = 0, 1, 2, \ldots$, where $u(n)$ and $q(u(n))$ denote the input and output of the quantizer. In the 1st- and

1058-6393/91 $01.00

2nd-order SDM modulators, simulations suggest that the optimum output levels are $\{\pm 1, \pm 1/4\}$, i.e., $b = 2$, for the 4-level (or 2-bit) powers-of-two quantizer, and $\{\pm 1, \pm 1/2, \pm 1/4, \pm 1/16\}$ for the 8-level (or 3-bit) powers-of-two quantizer. As n increases, the power of the Q-noise reduces. For $n > 4$ the reduction is small, so that n is normally chosen as 2, 3, or 4.

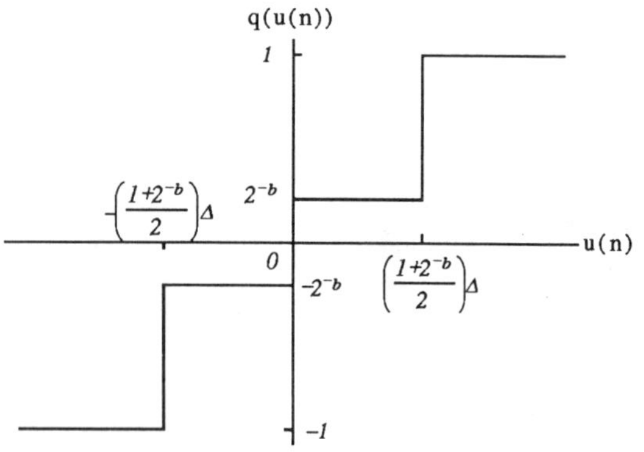

Figure 2: A 4-level powers-of-two quantizer

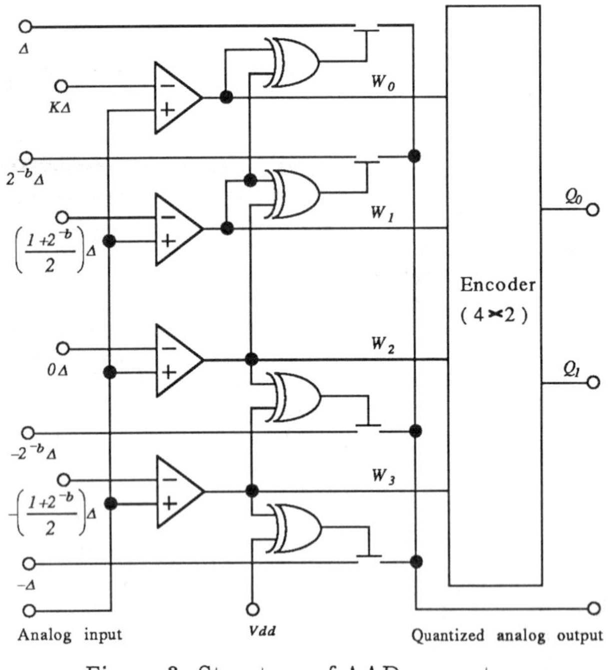

Figure 3: Structure of AAD converter

As mentioned before, it is necessary to place a D/A converter in the feedback loop when a multi-bit quantizer is used in a SDM modulator. To mitigate this problem, we can merge the feedback loop with the A/D converter and directly feedback the analog quantized output instead of the digital coded quantizer output. One possible structure is depicted in Fig.3 by applying an analog-to-(quantized)analog-to-digital (AAD) converter proposed in [9]. This AAD converter generates a quantized analog signal at the outputs of the comparators prior to the digital encoder, removing the requirement of D/A conversion in the feedback loop.

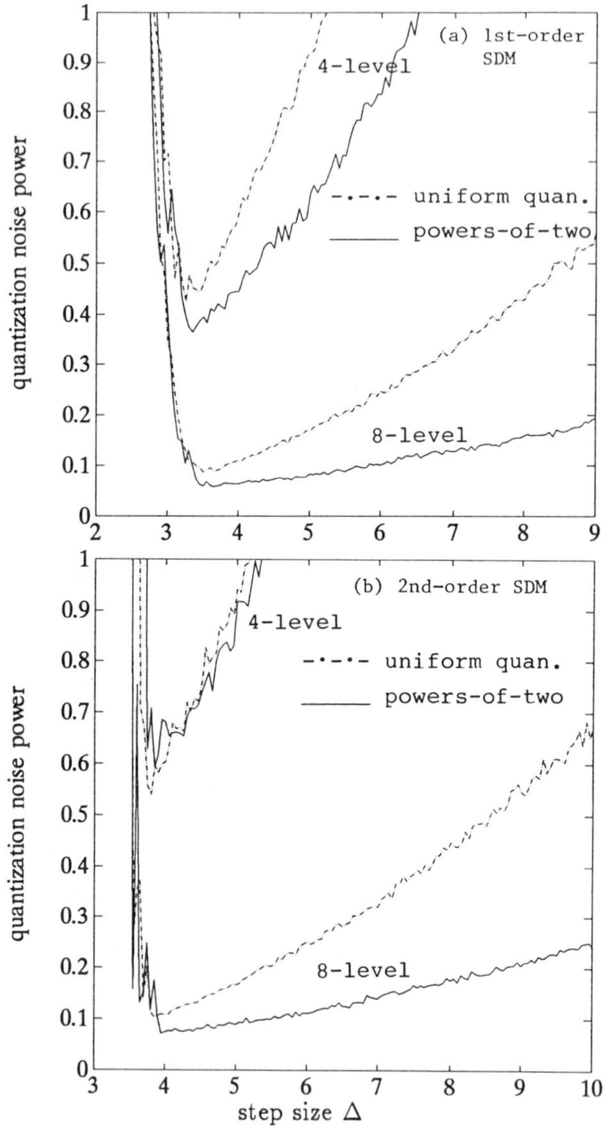

Figure 4: Quantization noise vs. step size

To evaluate the performances of SDM's with powers-of-two quantizer, we drive the SDM modulator by a bandlimited white Gaussian noise (BLWGN) with nominally zero mean and unit variance, using an oversampling ratio N of 16, namely, $x(n)$ is a lowpass signal with a spectrum which is (nominally) flat from (digital) frequency 0 to 1/32 and (nominally) 0 from frequency 1/32 to 0.5. Note that the corresponding Nyquist-rate signal is a white noise. The average power of the Q-noise is obtained for various step size. Results are plotted versus the step size in Fig.4 for the 1st- and 2nd-order SDM modulators. In each case, 4-level and 8-level uniform and powers-of-two quantizers are used in the modulators. Comparing the SDM with uniform quantizer and that with powers-of-two quantizer, we can see that the optimum step sizes, as well as the resulting Q-noises at these optimum step sizes, are approximately the same for the two quantizers. However, in the case of powers-of-two quantizer, the noise increases slowly as the step size increases, especially for 8-level quantizer. Therefore, a step size which is somewhat larger than the optimum one does not serious effect the performance. This phenomenon is useful for choosing a step size slightly larger than the optimum one to reduce the probability of slope overload in practical operating condition.

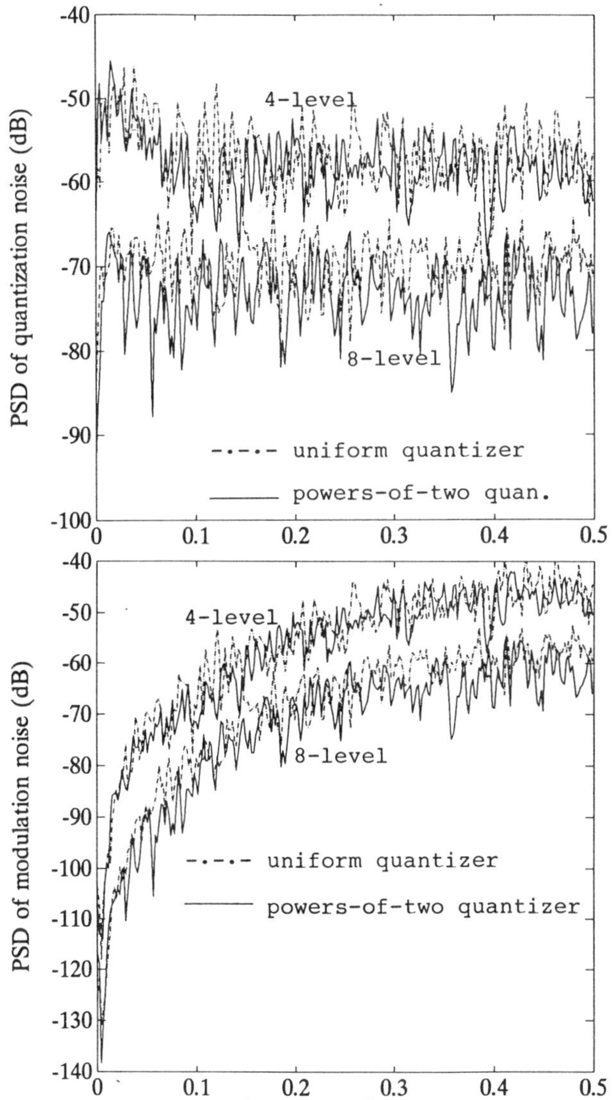

Figure 5: PSD's of quantization and modulation noises for 1st-order SDM

In addition to the Q-noise, we define the modulation noise $\epsilon(n)$ by $\epsilon(n) = \Delta q(u(n)) - x(n)$, $n = 0, 1, \ldots$, where $x(n)$ is the input to the modulator. With the optimal step size, the power spectrum densities (PSD) of the Q-noise and the modulation noise are depicted in Figs. 5-6. Results for the two different quantizers do not differ much since these results correspond to the optimum choices of the step size, which yield approximately the same Q-noise (see Fig.4). Comparing Figs. 5(b) and 6(b), we can see that the 2nd-order SDM modulator pushes more noise towards high frequencies than does the 1st-order SDM modulator, leaving less noise in the low frequency region. This agrees with the z-domain analysis of the noise shaping factor, which is $(1 - z^{-1})^2$ for the 2nd-order SDM modulator and $(1 - z^{-1})$ for the 1st-order SDM modulator. After the lowpass filter in the demodulator filters away the high-frequency noise, the 2nd-order SDM have better SNR than the 1st-order system.

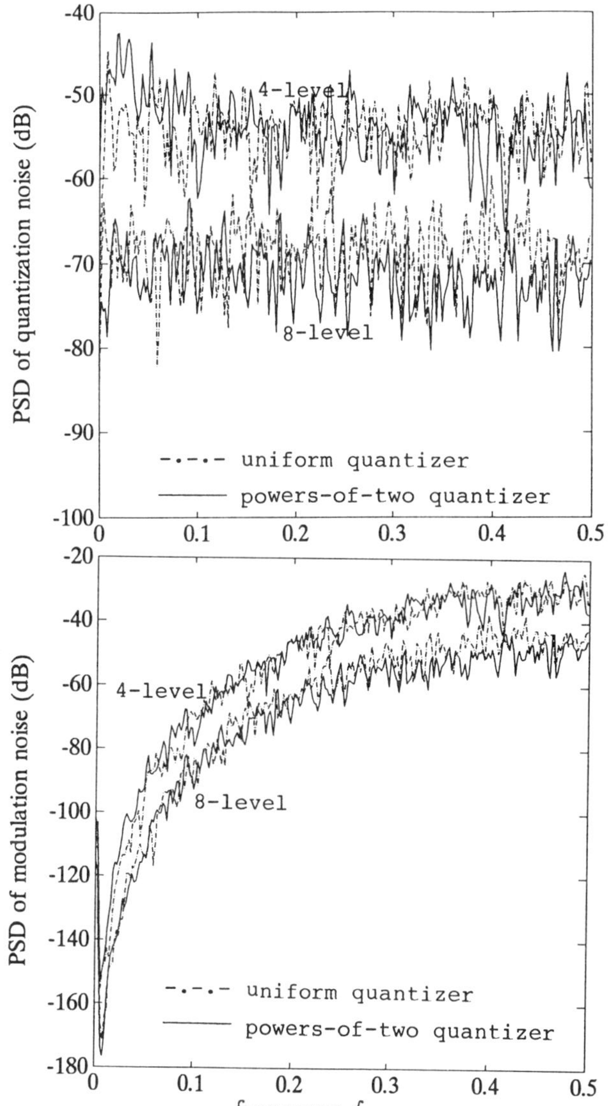

Figure 6: PSD's of quantization and modulation noises for 2nd-order SDM

A traditional realization of the demodulator shown

in Fig.1 uses comb filters followed by decimator (lowpass filter and downsampler), as well as a scaling unit for the step size. Since the output of the SDM modulator is a n-bit code of the quantizer output, a decoder is used prior to the demodulation. To take advantage of the powers-of-two quantizer levels, we can re-arrange the demodulator by moving the step-size scaler to follow to downsampler. Now that the input of the lowpass filter has values which are powers of two, the lowpass filter can be implemented using only shift and add operations, eliminating multipliers. Note that since the Q-noise for powers-of-two quantizer is relatively stable over a wide range of step size (Fig.4). if we choose the step size to be a power of two then no multiplication is needed by the step-size scaler in the demodulator. Therefore, we can totally eliminate multiplication in the demodulation process. This is the main advantage of using powers-of-two quantizer. Note that as the number of levels in the quantizer increases, a system with uniform quantizer becomes more complicated due to the multiplications, while a system with powers-of-two quantizer has only slight increase in complexity since no multiplication is required. As a final note, special hardware such as the digital differential analyzer (DDA) may be used to realize the demodulator [8].

3 Multiplier-free Realization for FIR Filters

Beyond its role in A/D conversion, SDM has been applied to implement FIR filters [10]. The principle behind such a realization is as follows. The input signal $x(n)$ is first interpolated by a factor of N. The resulting signal $\tilde{x}(n)$ is then passed through an FIR filter with the impulse response $h_s(n)$, which is SDM-encoded, of length NK, where K is the length of the target filter. The output of the SDM-based filter is then SDM demodulated to yield a signal equivalent to the output of the target filter. This is depicted in Fig.7(a).

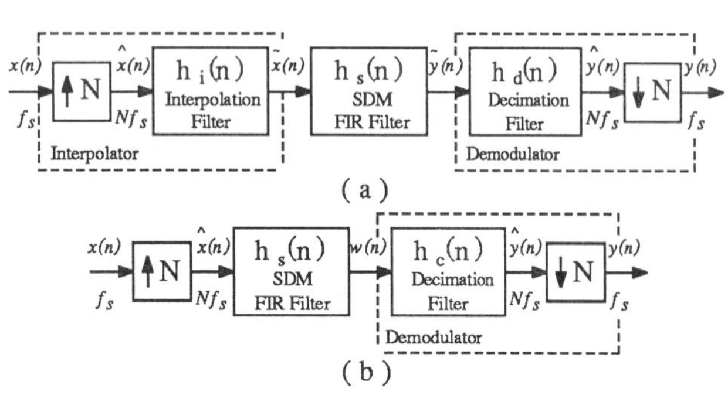

Figure 7: Block diagram of FIR filter based on SDM

Here we study a similar realization with $h_s(n)$ being encoded using the SDM with powers-of-two quantizer. Because the values of $h_s(n)$ are powers of two, multiplication by each coefficient can be implemented by simple shift and add operation. In addition, we can simplify the structure in Fig.7(a) by exchanging the orders of the interpolation filter $h_i(n)$ and the SDM filter $h_s(n)$ and combining $h_i(n)$ and $h_d(n)$ as one filter $h_c(n)$, yielding a structure in Fig.7(b). The filter $h_c(n)$ is a LPF with desired frequency response $H_c(f)$ $= 1$, if $|f| \leq 1/(2N)$, and $= 0$, otherwise. For this filter, we use a structure with coefficients being expressed by one or a sum of powers of two, as presented in [11]. The resulting realization is multiplier-free and can be realized more efficiently than the structure proposed in [10].

We consider two examples of FIR filter implementation. The first target filter is a low-pass FIR filter with impulse response

$$h(n) = w(n)[1.25\mathrm{sinc}^2(\frac{n-32}{4}) - 0.8\mathrm{sinc}^2(\frac{n-32}{5})]$$

where $n = 0, 1, 2, ..., 63$ and $w(n)$ is a Kaiser window function with $\beta = 10.0613$. The second target filter is a double-jump FIR filter with the same length and window function. Its impulse response is

$$h(n) = w(n)[0.168\mathrm{sinc}(\frac{n-32}{1.7857}) + 0.072\mathrm{sinc}(\frac{n-32}{4.1667})$$

$$+0.196\mathrm{sinc}^2(\frac{n-32}{3.5714}) - 0.036\mathrm{sinc}^2(\frac{n-32}{8.333})]$$

For comparison in each case, we obtain the frequency responses when the SDM filter is encoded using 1st- and 2nd-order SDM with uniform and powers-of-two quantizers. The results are displayed in Fig.8. The quantizers used for these results are 4-level. The levels of the powers-of-two quantizers are chosen to be $\{\pm 1, \pm 1/8\}$ for the lowpass filter and $\{\pm 1, \pm 1/16\}$ for the double-jump filter. We can see that the proposed realization of FIR filters using SDM modulation with powers-of-two quantizer has lower ripples in both the passband and the stopband than the filters designed by SDM modulation with uniform quantizers. Note that increasing the number of levels in the powers-of-two quantizers will reduce the ripples, with mild increase in implementation complexity. Equivalently, with the same ripples, a filter can be realized with a smaller oversampling ratio when SDM with powers-of-two quantizer is used for encoding $h_s(n)$.

Note that the length of the SDM filter can be large, since it is N times the length of the target filter, where N is generally large. For the structure in Fig.7(b), we can reduce the length of the filter as follows. Observe that the signal $\hat{x}(n)$ is zero for $N-1$ samples out of N samples. Consequently, each coefficient in the SDM filter $h_s(n)$ is idle for $N-1$ clock cycles out of N clock cycles. Therefore, by lumping N coefficients as one periodically-time-varying (PTV) coefficient, the number of delays can be reduced by a factor of N. Hence, the length of the SDM filter is now K, which is the length of the target filter. The resulting structure is shown in Fig.9, where the coefficients $c_k(m)$ are periodic in m with period N. They are related to the coefficients of the SDM filter by $c_k(m) = h_s(kN+m)$,

$k = 0, 1, ..., K-1$, $m = 0, 1, ..., N-1$. Note that the delays operate at the Nyquist rate (the input rate), while the coefficients $c_k(m)$ operate at N times the Nyquist rate. Although the periodicity of the coefficients slightly increases the implementation complexity, the reduction in the number of delays and coefficients outweighs the cost. Note that $c_k(m)$ are still powers-of-two coefficients, so that only shift and add operations are needed. Another method for efficient realization can be obtained using polyphase filters.

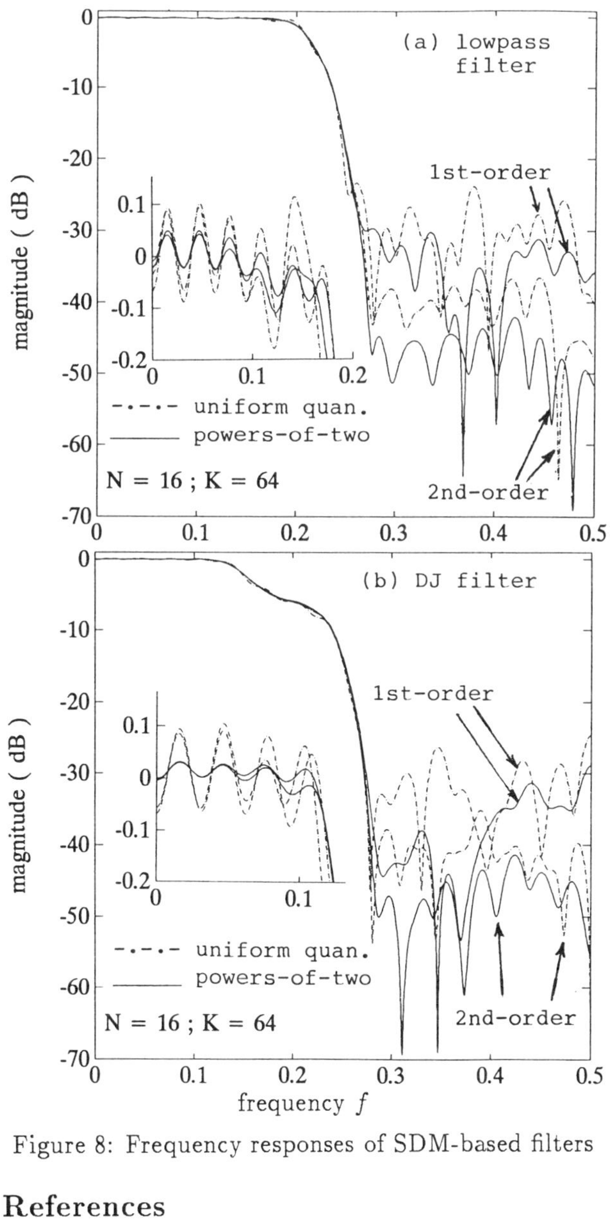

Figure 8: Frequency responses of SDM-based filters

References

[1] J. C. Candy, "A use of double integration in sigma delta modulation," *IEEE Trans. Commun.*, pp. 249-258, Mar. 1985.

[2] K. Uchimura, T. Hayashi, T. Kimura, and A. Iwata, "Oversampling A-to-D and D-to-A converters with multistage noise shaping modulators," *IEEE Trans. on ASSP*, pp. 1899-1905, Dec. 1988.

[3] N. He, A. Buzo, and F. Kuhlmann, "Multiloop sigma-delta quantization: Spectral analysis," *Proc. ICASSP*, New York, pp. 1870-1873, Apr. 1988.

[4] W. Chou, P. W. Wong, and R. M. Gray, "Multistage sigma-delta modulation," *IEEE Trans. Inform. Theory*, pp. 784-796, July 1989.

[5] P. J. A. Naus and E. C. Dijkmans, "Multibit oversampled $\sigma\ \Delta$ A/D converters as front end for CD players," *IEEE J. Solid-State Circuits*, pp. 905-909, July. 1991.

[6] J. J. Paulos, G. T.Brauns, M. B. Steer, and S. H. Ardalan, "Improved signal-to-noise ratio using tri-level delta-sigma modulation," *Proc. ISCAS*, Philadelphia, PA, pp. 463-466, May 1987.

[7] A. Gosslau and A. Gottwald, "Linearization of a sigma-delta modulator by a proper loop delay," *IEEE Proc. ISCAS'90*, pp. 364-367, May 1990.

[8] S. E. Gordon, L. E. Franks, and S. Tantaratana, "Floating point incremental coders in recursive digital filtering," *Proc. Twenty-Third Annual Asilomar Conf. Signals, Syst., and Comp.*, pp. 858-862, Oct. 30 - Nov. 1, 1989.

[9] f.j. harris and B. Mcknight, "New results with oversampled converters," *Proc. Twenty-Third Annual Asilomar Conf. Signals, Syst., and Comp.*, pp. 844-848, Oct. 1989.

[10] P. W. Wong and R. M. Gray, "FIR filters with sigma-delta modulation encoding," *IEEE Trans. on ASSP*, pp. 979-990, June 1990.

[11] T. Saramaki and H. Tenhunen, "Efficient VLSI-realizable decimators for sigma-delta analog-to-digital converter," *Proc. ISCAS*, Espoo, Finland, pp. 1525-1528, June. 1988.

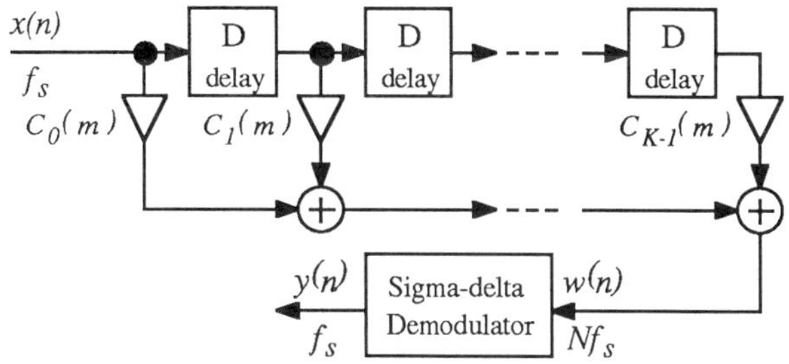

Figure 9: Realization of the SDM filter using PTV coefficients

Reconstruction of Oversampled Band-Limited Signals from $\Sigma\Delta$ Encoded Binary Sequences

Søren Hein and Avideh Zakhor*
Department of Electrical Engineering and Computer Sciences
University of California, Berkeley CA 94720
e-mail: shein@robotics.berkeley.edu and avz.janus.berkeley.edu

Abstract

We consider the application of $\Sigma\Delta$ modulators to analog-to-digital conversion. We have previously shown that for constant input signals, large gains in Signal-to-Noise Ratio (SNR) over linear decoding can be achieved with relatively simple nonlinear decoding. In this paper we consider the more general case of band-limited input signals. The new nonlinear decoding scheme is based on the theory of Projections Onto Convex Sets (POCS), and alternates between a band-limiting operation and a time-domain operation to find a signal invariant under both. For general band-limited inputs, the band-limitation can be based on singular value decomposition of a certain matrix. The time-domain operation results in a quadratic programming problem. We show simulation results for the SNR performance of a POCS-based decoder for the single loop $\Sigma\Delta$ modulator and a specific 4th order interpolative modulator, and present comparisons with linear decoding. Improvements in SNR of up to 20-30 dB can be achieved for the single loop modulator, and up to 10-15 dB for the 4th order modulator.

1 Introduction

$\Sigma\Delta$ modulators are becoming increasingly important as analog-to-digital converters for relatively low-bandwidth applications such as speech and audio. They are ideal for on-chip VLSI implementation because they require fewer and simpler components than Nyquist-rate converters; this is at the expense of using sampling rates many times faster than the Nyquist rate.

The only nonlinear element in a $\Sigma\Delta$ modulator is a 1-bit quantizer. It is well-known that linearizing the quantizer in a modulator can lead to analytical predictions that depart significantly from observed performance [1]. Nonetheless, linearization is routinely used in the analysis and design of $\Sigma\Delta$ modulators because the resulting models are simple, and their shortcomings are perceived to be well-known. In particular, decoding the binary output stream of a modulator has traditionally been done exclusively with linear filtering, and the optimality of this solution has not been questioned. In [2] we suggest that conventional linear decoding does not extract all information from the binary output stream. Specifically, linear decoding only makes use of knowledge of the spectral support of the input, and not of the fact that the input resulted in the particular observed binary output. In [3] we introduce an optimal nonlinear decoding algorithm for the special case of constant inputs. The main result is that significant improvements over linear decoding can be achieved.

In this paper, we consider the more general problem of optimal nonlinear decoding for band-limited inputs [4]. The ideal decoding algorithm should perform significantly better than linear decoding, be simple to implement, be as robust towards circuit imperfections as linear decoding, and have small estimation delay.[1] We take a signal reconstruction point of view, and our main interest will be to derive a conceptually feasible, optimal nonlinear decoder, rather than to minimize computational complexity. Thus our results will serve as upper bounds on attainable performance. Our proposed decoding algorithm is based on the theory of Projections Onto Convex Sets (POCS), and results indicate that our algorithm is up to 20-30 dB superior to linear decoding for the single loop modulator, and about 10-15 dB superior to linear decoding for a specific 4th order $\Sigma\Delta$ modulator [6].

*This work was supported by National Science Foundation grant MIP–9057466 and Analog Devices.

[1] Estimation delay refers to the number of future output samples that a decoder uses to estimate any given input sample. For a linear phase Finite Impulse Response (FIR) filter, the delay is half the filter length.

1058-6393/91 $01.00 © 1991 IEEE

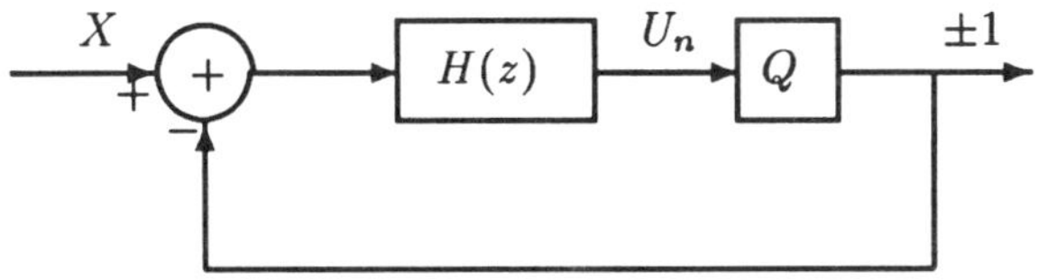

Figure 1: *Discrete-time model of the interpolative $\Sigma\Delta$ modulator.*

The paper is organized as follows. Section 2 presents background information for the decoding problem, and justifies our choice of performance metric. Section 3 describes the reconstruction algorithm mathematically. Section 4 presents numerical results, and section 5 contains conclusions and directions for future research.

2 Background

This section contains a brief introduction to $\Sigma\Delta$ modulation and establishes notation. Figure 1 shows a discrete-time model of the interpolative $\Sigma\Delta$ modulator, consisting of a discrete-time open-loop filter $H(z)$ embedded in a nonlinear negative feedback loop, including also a one-bit quantizer Q. Q is defined by $Q(U) = \text{sign}(U)$, where $\text{sign}(\cdot)$ is the signum function. To avoid race-around, the filter must contain at least one overall delay. The quantizer can be viewed as adding a noise sequence $\{E_n\}$ to its input sequence. The noise sequence is often assumed to be approximately white and uncorrelated with the input sequence $\{X_n\}$, although this assumption is in many respects inadequate. For the special case

$$H(z) = \frac{z^{-1}}{1 - z^{-1}} \tag{1}$$

the interpolative modulator reduces to the single loop modulator. We will also consider the specific 4th order interpolative modulator defined in [6] by

$$H(z) \;=\; z^{-1} \cdot \sum_{i=0}^{4} N_i z^{-i} \Big/ \sum_{i=0}^{4} D_i z^{-i} \tag{2}$$

where $N_0 = 0.8653$, $N_1 = -2.2692$, $N_2 = 2.0064$, $N_3 = -0.59714$, $N_4 = 0.000035$, $D_0 = 1$, $D_1 = -3.99646$, $D_2 = 5.992922$, $D_3 = -3.996460866$, $D_4 = 1.000000433$. $H(z)$ is in general chosen to be a low-pass filter whose pass-band corresponds to the frequency range in which the input signal is concentrated. A digital estimate of the analog input is traditionally formed by low-pass filtering the binary output sequence [6]. The OverSampling Ratio (OSR) is defined as the ratio between the sampling rate and the Nyquist rate of the input.

3 Reconstruction Algorithm

Section 3.1 briefly describes the POCS algorithm, which iteratively enforces the two known constraints on the input signal: First, it resulted in the observed output signal, and second, it is band-limited. Section 3.2 shows that the operation of finding an input sequence that generates the given output sequence can be treated as a Quadratic Programming (QP) problem. Section 3.3 discusses the band-limitation operation. Using the Discrete Fourier Transform (DFT) and its fast implementation, the Fast Fourier Transform (FFT), is only appropriate when all signal frequencies are known to be bin frequencies for the DFT. We take a Singular Value Decomposition (SVD) based approach to accurate band-limitation.

3.1 POCS Introduction

For a given binary output sequence of a known $\Sigma\Delta$ modulator, we have two pieces of information about the input: It resulted in the observed output bit stream, and by assumption it consists of samples of a band-limited infinite-extent sequence. To estimate the input optimally, we must find a signal that satisfies both constraints. The problem formulation and specific characteristics of the constraints make the POCS iterative algorithm a natural choice for achieving our goal [5]. Let us define the set S_1 to be the set of input signals that generate the observed output, and S_2 to be the set of signals that are samples of band-limited infinite-extent sequences with bandwidth equal to the known signal bandwidth. If the sets S_1 and S_2 are convex,[2] and if we denote the orthogonal projections in the sense of some norm onto the sets by P_1 and P_2, respectively, then an element $\hat{\mathbf{x}} \in S_1 \cap S_2$ can be found from any initial guess $\mathbf{x}_0$ by the iteration [5]

$$\mathbf{x}_{n+1} = (P_2 \circ P_1)\mathbf{x}_n,\;\; n \geq 0,\;\; \hat{\mathbf{x}} = P_1\hat{\mathbf{x}} = P_2\hat{\mathbf{x}} = \lim_{n\to\infty} \mathbf{x}_n$$

Therefore the POCS iteration can provide a sequence which is both band-limited and generates the known output sequence. In the following sections we discuss the projections separately.

3.2 Time-Domain Projection

In this section we consider the operation P_1 of projecting a sequence $\mathbf{x} = \{x_0, \cdots, x_{N-1}\}$ onto the space S_1 of sequences that generate a given output sequence; we refer to the operation as time-domain projection.

[2] A set S is said to be convex if for all pairs $\mathbf{a}, \mathbf{b} \in S$, $\alpha\mathbf{a} + (1-\alpha)\mathbf{b} \in S$ for any $0 < \alpha < 1$.

We first derive a time-domain characterization of the modulator, and then show that the projection results in a Quadratic Programming (QP) problem.

Let us denote the given binary output sequence by $\mathbf{y} = \{y_0, \cdots, y_{N-1}\}$, and the impulse response of the open-loop filter $H(z)$ by $\{h_0, h_1, \cdots\}$; as explained in section 2, h_0 must equal zero to avoid delay-free loops. We define the $N \times N$ lower-triangular nonsingular Toeplitz matrix

$$\mathbf{H} = \begin{bmatrix} h_1 & 0 & \cdots & 0 & 0 \\ h_2 & h_1 & \cdots & 0 & 0 \\ \vdots & \vdots & & \vdots & \vdots \\ h_{N-1} & h_{N-1} & \cdots & h_1 & 0 \\ h_N & h_{N-1} & \cdots & h_2 & h_1 \end{bmatrix} \tag{3}$$

Let us further define an $(M+1)$-dimensional state vector $\mathbf{s}$ for an Mth order modulator,[3] and an $N\times(M+1)$ zero-input response matrix $\mathbf{Z}$ whose elements can be obtained from knowledge of the open-loop transfer function $H(z)$. The matrix $\mathbf{Z}$ is defined so that if the initial state of $H(z)$ is $\mathbf{s}$ and the filter is driven open-loop with identically zero input over N samples, the filter output is $\mathbf{Zs}$. The quantizer input vector $\mathbf{u} = \{U_0, U_1, \cdots, U_{N-1}\}$ can now compactly be written

$$\mathbf{u} = \mathbf{H}(\mathbf{x} - \mathbf{y}) + \mathbf{Zs}$$

Our only knowledge of any particular quantizer input U_n is in the form of its sign, the quantizer output $Q(U_n)$. This provides us with a series of bounds on linear combinations of sample values $\mathbf{x}$ and state variables $\mathbf{s}$. Specifically, if we define an $N \times N$ diagonal matrix

$$\mathbf{Q} = -\mathrm{diag}(y_1, y_2, \cdots, y_N) \tag{4}$$

then the bounds imposed by the quantizer outputs can be written as

$$\mathbf{QHx} \leq \mathbf{Q}(\mathbf{Hy} - \mathbf{Zs}) \tag{5}$$

where the inequality sign is to be taken coordinatewise. We will assume the state vector $\mathbf{s}$ to be known. If we consider two sample vectors $\mathbf{x}_1$ and $\mathbf{x}_2$ that both satisfy (5), then for any $0 < \alpha < 1$,

$$\begin{aligned} &\mathbf{Q}(\mathbf{H}[\alpha\mathbf{x}_1 + (1-\alpha)\mathbf{x}_2] \\ \leq\ & \alpha\mathbf{Q}(\mathbf{Hy} - \mathbf{Zs}) + (1-\alpha)\mathbf{Q}(\mathbf{y} - \mathbf{Zs}) = \mathbf{QHy} \end{aligned}$$

[3] The state vector has $M+1$ rather than M elements because of the requirement that the modulator contain no delay-free loops. In [6] this effect is achieved by associating a delay with the quantizer. Equivalently, we have multiplied $H(z)$ in (2) by an overall delay to achieve $h_0 = 0$. This extra dynamic element is also the reason that we need to know the impulse response up to h_N rather than h_{N-1} in (3), and the binary output up to y_N rather than y_{N-1} in (4) below.

so the set of vectors satisfying (5) is convex. If we adopt the 2-norm, the projection P_1 onto the set S_1 is equivalent to the problem of finding the sample vector $\hat{\mathbf{x}}$ that satisfies (5) and minimizes the distance

$$d^2 = \|\hat{\mathbf{x}} - \mathbf{x}\|_2^2 \tag{6}$$

Appendix A justifies our choice of performance metric for $\Sigma\Delta$ modulators as Analog-to-Digital (A/D) converters. The stated problem is a linearly constrained QP problem, and can be solved with a variety of techniques available in the literature [7].

In practice, sample sizes on the order of thousands are desirable to achieve accurate band-limitation projections. Since the computational complexity of the QP problem is $O(N^3)$ and the storage requirement is $O(N^2)$, we derive a simplified approximation to the 2-norm time-domain projection. For simplicity, let us assume the state vector $\mathbf{s}$ to be zero. The idea is to replace the 2-norm in the objective function (6) with the more general $\mathbf{M}$-norm defined by

$$\|\mathbf{x}\|_{\mathbf{M}}^2 \triangleq \|\mathbf{Mx}\|_2^2 = \mathbf{x}^T(\mathbf{M}^T\mathbf{M})\mathbf{x}$$

where $\mathbf{M}$ is any nonsingular $N \times N$ matrix. Setting $\mathbf{M} = \mathbf{H}$ results in the QP problem of minimizing $\|\hat{\mathbf{x}} - \mathbf{x}\|_{\mathbf{H}}^2$ or equivalently $\|\hat{\mathbf{z}} - \mathbf{z}\|_2^2$ subject to $\mathbf{Qz} \leq \mathbf{QHy}$, where we have defined $\mathbf{z} = \mathbf{Hx}$, and the input estimate is found as $\hat{\mathbf{x}} = \mathbf{H}^{-1}\hat{\mathbf{z}}$. Because $\mathbf{Q}$ is diagonal, the QP problem in terms of $\mathbf{z}$ is completely decoupled into N trivial 1-dimensional QP problems. While this approximation to the true 2-norm projection simplifies P_1, we have found numerically that it results in inferior SNR performance. We therefore consider the more general approximation of decoupling the P_1 projection into N/L QP problems with dimension $L > 1$. To see how, we introduce some additional notation. The matrix $\mathbf{H}$ defined in (3) is indexed by its dimension, so $\mathbf{H}_N$ is $N \times N$. Furthermore, $\mathrm{diag}(\mathbf{A}, \mathbf{B}, \mathbf{C})$ denotes a block diagonal matrix with blocks $\mathbf{A}, \mathbf{B}, \mathbf{C}$. If we define the $N \times N$ matrix

$$\mathbf{G} = \mathrm{diag}(\mathbf{H}_L,\ \mathbf{H}_L,\ \cdots,\ \mathbf{H}_L)$$

then the following QP problem decouples into N/L L-dimensional ones: Minimize $\|\hat{\mathbf{z}} - \mathbf{z}\|_2^2$ subject to

$$\mathbf{QGz} \leq \mathbf{QHy}$$

and set $\hat{\mathbf{x}} = \mathbf{H}^{-1}\mathbf{G}\hat{\mathbf{z}}$. Equivalently, the original QP-problem is solved in the $(\mathbf{G}^{-1}\mathbf{H})$-norm. Because $\mathbf{G}$ is block lower triangular, inverting it is straightforward. The choice of QP block size L is a trade-off between the speed with which a solution is obtained, and the closeness of the solution to the true 2-norm solution. We are concerned about closeness because the frequency-domain projection will be done in the 2-norm, and the POCS algorithm only guarantees convergence if the projections are done in strictly the same norm. However, we have found via simulations that a QP block size of 128 is adequate for the specific 4th order $\Sigma\Delta$ modulator considered, and a block size of 64 is adequate for the single loop modulator.

3.3 Frequency-Domain Projection

In this section we consider the operation P_2 of projecting a sequence $\mathbf{x} = \{x_0, \cdots, x_{N-1}\}$ onto the space S_2 of finite sequences that are samples of infinite-extent band-limited sequences; we refer to the operation as frequency-domain projection. This projection raises some general questions about band-limitation that are discussed in more detail in [8], and summarized below.

It is natural to consider the DFT as the frequency-domain projection. However, the DFT is only accurate when the signal frequencies happen to be bin frequencies for the DFT, which in general cannot be assumed. Windowing can be used to suppress the noise, but it has the side-effect of smearing the signal spectrum unacceptably.

Another standard technique for band-limiting finite-extent sequences is linear filtering. In the context of moderate-length block-oriented processing, the disadvantages of linear filtering are large edge effects resulting from padding the block with zeros, and the finite-ripple transfer function over baseband resulting in signal frequency errors.

Our preferred, general technique for band-limitation is described in [8] and is closely related to well-known algorithms for band-limited extrapolation [9]. The technique is based on Singular Value Decomposition (SVD) of an $N \times N$ symmetric Toeplitz matrix $\mathbf{L}$ consisting of samples of the impulse response of an ideal low-pass filter, $\ell_{mn} = \sin[\sigma(m-n)]/[\pi(m-n)]$ where $\sigma = 1/\text{OSR}$. The SVD leads to the truncated Discrete Prolate Spheroidal Sequences (DPSSs) of N-dimensional vectors $\{\mathbf{u}_1, \cdots, \mathbf{u}_N\}$ that have been studied extensively by Slepian [10] and others; they form an orthonormal basis of $\mathcal{R}^N$. Approximately $r = N/\text{OSR}$ of the truncated DPSSs correspond to singular values close to 1, and are analogous to the baseband complex exponentials of the DFT; the remaining truncated DPSSs correspond to singular values close to 0, and are analogous to High Frequency (HF) complex exponentials. Our SVD algorithm projects signals onto the space spanned by the "baseband"

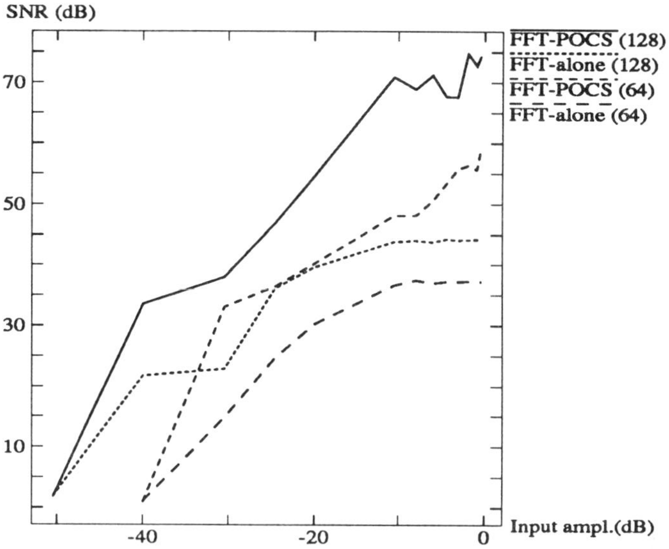

Figure 2: *SNR as a function of input amplitude for the single loop modulator. The amplitude is normalized to the quantizer step size. The OSRs are 64 and 128. The two decoders are an FFT decoder and the FFT-based POCS decoder.*

DPSSs,

$$\hat{\mathbf{x}} = \sum_{n=1}^{r} (\mathbf{x}^T \mathbf{u}_n) \cdot \mathbf{u}_n$$

The projection is in the sense of the 2-norm, and the space is linear and thus convex. The SVD only needs to be done once, and can be implemented efficiently using the Lanczos algorithm [11] and the Toeplitz characteristic of the involved matrix. The projection itself is comparable in speed to an FFT band-limitation for a sample size of 4K. The projection does not require knowledge of the state vector $\mathbf{s}$ for the open-loop filter.

4 Results

In this section we present and discuss numerical results for the single loop modulator and the specific 4th order modulator given by (2) [6]. Throughout, the input signal used is a sinusoid whose amplitude is varied, and whose frequency is $\alpha = 9/32$ times the largest possible signal frequency. The chosen frequency is a bin frequency for the DFT, to facilitate comparison between a DFT technique and the POCS algorithm.

4.1 Single Loop Modulator

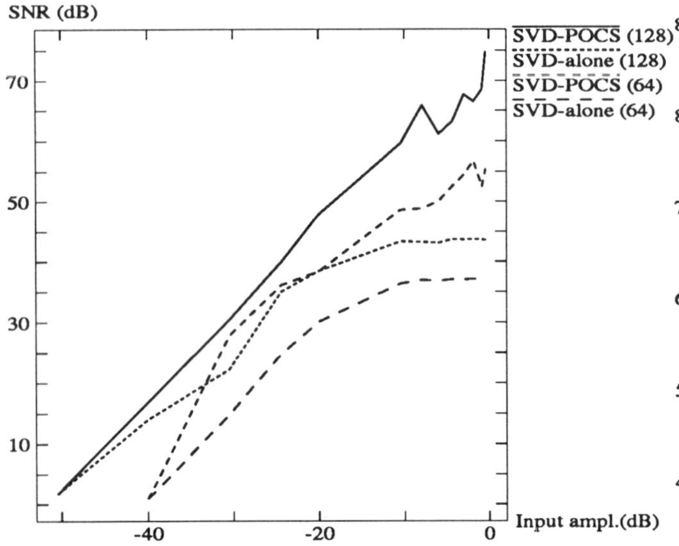

Figure 3: *SNR as a function of input amplitude for the single loop modulator. The two decoders are an SVD-based low-pass decoder and the SVD-based POCS decoder.*

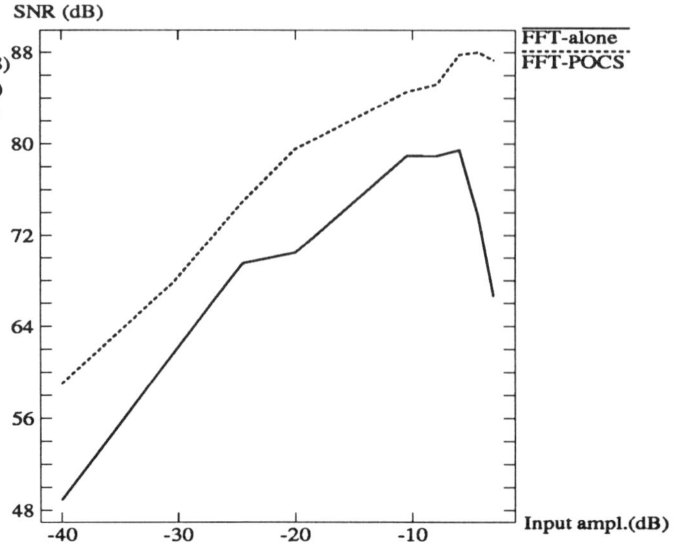

Figure 4: *SNR as a function of input amplitude for a specific 4th order interpolative modulator. The two decoders are an FFT-based low-pass decoder and the FFT-based POCS decoder.*

As a benchmark, we first consider the case where the signal frequency is known to be a bin frequency for the DFT, so that the FFT can be used as the frequency-domain projection. To ease comparison with the SVD-based algorithm below, the SNR is only calculated over the middle half of a 4K block. The QP block size is 64 and the number of POCS iterations is 15. The state vector is set to zero at the beginning of the 4K block. For the two oversampling ratios 64 and 128, Figure 2 shows the SNR performance of the FFT-based POCS algorithm and the FFT method, that is, the low-pass decoder that consists of band-limiting the DFT of the output. The figure shows that particularly for large input amplitudes, the POCS algorithm is clearly superior to linear decoding. For OSR=64, the peak SNR is 37 dB for the linear decoder, and 59 dB for the POCS decoder. For OSR=128, the numbers are 44 dB and 75 dB. The SNR dependence on the OSR cannot reasonably be determined from only these numbers, but it appears that the slope in dB/octave is significantly larger for the POCS decoder than for the linear decoder.

The results in Figure 3 are obtained in exactly the same way as those in Figure 2, except that the frequency-domain projection is done using the SVD-based method. This method is appropriate even when the input frequency is not a bin frequency. The SNR is again only calculated over the middle half of the 4K samples. The number of singular vectors used is 74 for OSR=64 and 39 for OSR=128. We see that the SVD-based POCS decoder is clearly superior to the SVD band-limitation decoder. Comparing Figures 2 and 3, we see that the SVD-based POCS method loses little compared to the FFT-based POCS method; for OSR=64, the loss is generally about 2 dB, and for OSR=128, it is about 4-5 dB.

The observed peak SNRs for the FFT-alone and SVD-alone methods in Figures 2 and 3 can be accounted for almost exclusively by signal frequency errors, rather than HF noise. Adopting the linearized model, it can easily be shown that the error at the signal frequency ω_0 alone would result in an SNR equal to the gain of the open-loop transfer function $H(z)$ at that frequency. Using (1), the open-loop gain is approximately $-20\log_{10}\omega_0$, which equals 37 dB for OSR=64 and 43 dB for OSR = 128, when $\omega_0 = \alpha\pi/\mathrm{OSR} = 9\pi/[32\mathrm{OSR}]$.

4.2 4th Order Modulator

For the 4th order modulator with transfer function (2), we first consider the DFT as the band-limiting operation, assuming the signal frequency is known to be a bin frequency. The sample size is 4K, the QP block size is 256 and the number of POCS iterations is 50. The state vector is set to zero at the beginning of the

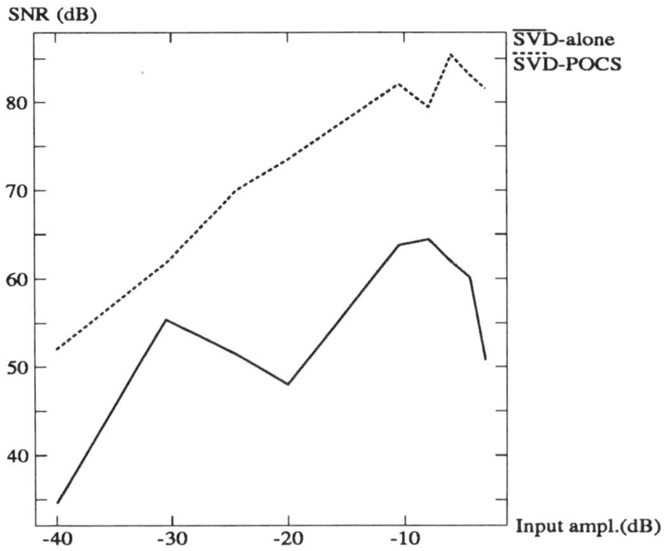

Figure 5: *SNR as a function of input amplitude for a specific 4th order interpolative modulator. The two decoders are an SVD-based low-pass decoder and the SVD-based POCS decoder.*

4K block.[4] Figure 4 shows the SNR performance of the FFT-based POCS algorithm and the FFT band-limiting decoder. The peak SNR is 79 dB for the FFT method, and 88 dB for the POCS algorithm. The gain of the open-loop filter $H(z)$ at the signal frequency is 81 dB, so as in the previous section, signal frequency errors alone can account for most of the observed error on the FFT decoder. The gain in SNR of the POCS method over the FFT-alone method is about 8-10 dB for small and moderate input amplitudes, increasing to above 20 dB for large amplitudes. Simulations indicate that the POCS algorithm converges in a finite number of iterations to input estimates that strictly satisfy both the time-domain and frequency-domain constraints.

The results in Figure 5 are generated using the SVD-based band-limitation. The SNR is calculated over only the middle half of the 4K block. The QP block size is 128, and the number of POCS iterations is 15. The SVD-alone and SVD-POCS curves show simulation results for an SVD-based low-pass decoder and the SVD-based POCS decoder, respectively. We would expect the SVD-alone curve to be inferior to the FFT-alone curve in Figure 4 because the input signal is assumed to have a band-limited DFT. Indeed, the SVD band-limitation is 16-18 dB below the FFT band-limitation in performance. This difference could have been made smaller by a different choice of the number of singular vectors for the SVD band-limitation, but the number was chosen to optimize the SVD-POCS curve rather than the SVD-alone curve in Figure 5. The number of singular vectors in general represents a trade-off between the fidelity with which band-limited signals can be represented, and the amount of HF noise that is accepted as a result [8]. The optimum number for the SVD-POCS curve, 94, is larger than that for the SVD-alone curve, 90 or 91, because the POCS algorithm manages to suppress some additional HF noise that the SVD in itself cannot suppress. The SVD-POCS curve in Figure 5 is seen to mostly be 3-5 dB below the FFT-POCS curve in Figure 4, but still 3-5 dB above the FFT-alone curve in Figure 4 for low and moderate input amplitudes, and up to 15 dB above for large amplitudes. The SVD-based POCS method therefore shows promise for a continuous, overlapping-block technique which is presently being investigated.

[4] The input frequency is chosen as described above, which is close to the frequency used in [6] to generate Figure 12 of that paper. Unlike in [6], we do not apply a DC offset to the input, because we would consider the offset energy to be part of the signal energy, and would count DC error as noise. This would complicate the evaluation of results.

5 Conclusions

We have shown by simulations that for the single loop and a specific 4th order interpolative $\Sigma\Delta$ modulator, significant gains in SNR can be achieved using sophisticated decoding techniques, rather than linear decoding. The presented results are valid for a block-oriented set-up where initial open-loop filter states are periodically reset. We are currently investigating a continuous mode of operation in which filter states are estimated adaptively.

The main differences between the reported work and the approach in [12] are that we consider general interpolative $\Sigma\Delta$ modulators, exemplified with a 4th order one [6], our QP problem is solved in a convenient, scaled 2-norm, and we avoid the assumption that the finite set of samples has a band-limited DFT.

The presented decoding method is computationally quite complex. However, our results can serve as upper bounds on the achievable SNR, and would be useful for future research in simplified, nonlinear decoding algorithms. Such research would also have to address the sensitivity of the algorithm towards circuit imperfections, compared to the sensitivity of linear filtering.

Acknowledgments

We thank Shujaat Nadeem for useful discussions.

A Performance Metric

Denoting the N-sample input sequence by $\mathbf{x} = \{x_0, x_1, \cdots, x_{N-1}\}$ and the decoder estimate by $\hat{\mathbf{x}} = \{\hat{x}_0, \hat{x}_1, \cdots, \hat{x}_{N-1}\}$, we define the SNR as $10\log_{10}(E_{\text{signal}}/E_{\text{noise}})$, where

$$E_{\text{signal}} = \sum_{n=0}^{N-1} x_n^2, \quad E_{\text{noise}} = \sum_{n=0}^{N-1} (x_n - \hat{x}_n)^2$$

This metric counts both high frequency noise and signal frequency errors as noise. The justification is that a decoder should be judged on all the noise that it leaves on its estimates; if not, additional circuitry to improve the accuracy could be lumped together with the decoder.

References

[1] R. M. Gray, "Oversampled Sigma-Delta Modulation", *IEEE Trans. Comm.*, vol. COM–35, pp. 481–489, May 1987.

[2] P. Gray, D. Pederson & A. Zakhor, National Science Foundation proposal, Oct. 1988.

[3] S. Hein & A. Zakhor, "Optimal Decoding for Data Acquisition Applications of Sigma Delta Modulators", *IEEE Trans. Sig. Proc.*, submitted.

[4] S. Hein & A. Zakhor, "Iterative Reconstruction of ΣΔ Encoded Signals", *Proc. Int. Conf. Industr. Appl. Math.*, Washington, D. C., July 1991.

[5] D. C. Youla & H. Webb, "Image Restoration by the Method of Convex Projections: Part 1—Theory", *IEEE Trans. Medical Imaging*, vol. MI–1, no. 2, pp. 81–94, Oct. 1982.

[6] K. C.-H. Chao, S. Nadeem, W. L. Lee & C. G. Sodini, "A Higher Order Topology for Interpolative Modulators for oversampling A/D Converters", *IEEE Trans. Circ. Syst.*, vol. 37, no. 3, pp. 309–318, March 1990.

[7] D. G. Luenberger, *Linear and Nonlinear Programming*, Addison-Wesley, Reading, MA, 1984.

[8] S. Hein & A. Zakhor, "Theoretical and Numerical Aspects of An SVD-Based Method for Band-Limiting Finite Sequences", *IEEE Trans. Sig. Proc.*, submitted.

[9] E. J. Diethorn & D. C. Munson, Jr., "A Linear, Time-Varying System Framework for Noniterative Discrete-Time Band-Limited Extrapolation", *IEEE Trans. Sig. Proc.*, vol. 39 no. 1, pp. 55–68, Jan. 1991.

[10] D. Slepian, "Prolate Spheroidal Wave Functions, Fourier Analysis, and Uncertainty—V: The Discrete Case", *Bell Syst. Tech. J.*, vol. 57 no. 5, pp. 1371–1430, May-June 1978.

[11] B. N. Parlett, *The Symmetric Eigenvalue Problem*, Prentice-Hall, NJ, 1980.

[12] N. T. Thao & M. Vetterli, "Oversampled A/D Conversion Using Alternate Projections", *Proc. 25th Conf. Inf. Sciences and Systems*, Baltimore, MD, March 1991.

Noncoherent Direct-Sequence Code-Tracking in the Presence of Pulsed Jamming*

Haiping Tsou, Raimundo Sampaio-Neto and Robert A. Scholtz

Communication Sciences Institute
Department of Electrical Engineering-Systems
University of Southern California
Los Angeles, CA 90089-2565

Abstract

The performance of a noncoherent direct-sequence code-tracking loop operating in the presence of thermal noise and a pulsed noise jammer is analyzed. In particular, the jammer is modeled as cyclostationary interference resulting from the product of a periodic train of pulses and band-pass white noise. The impact of the jammer's duty factor and repetition frequency on the tracking loop's behavior is examined under both linear and non-linear performance analyses of loop operation.

1 Introduction

This paper addresses the system performance of a reference/difference direct-sequence code tracking loop when operating in the presence of background noise and an on-off wide-band noise jammer, modeled as the product of a periodic pulse train and wide-band bandpass noise. Depending on the ratio of the pulse repetition frequency to the tracker's arm filter bandwidth, the cyclostationary nature of the interfering signal leads to different models of loop feedback noise that are also cyclostationary processes. Under a linear performance analysis, the cyclostationarity of the loop noise translates into a steady-state tracking error whose statistical variance is a periodic function of time.

An approximate nonlinear analysis of the tracking loop behavior that takes into consideration the actual shape of the loop's finite-support S-curve is also performed for a first order loop. Under this analysis, the loop stochastic differential equation of operation is approximated by a diffusion stochastic differential equation with a time-dependent, periodic, diffusion coefficient. Fokker-Planck techniques and computer-generated solutions to the Fokker-Planck equation are used to assess the impact of the different jammer parameters on the statistics of the loop's time to lose lock. Here this is defined as the time it takes for the tracking error process, starting from zero value, to reach one of the boundaries that limit the loop's S-curve support interval.

*This work was funded in part by the Joint Services Electronics Program under contract No. F49620-88-c-0067.

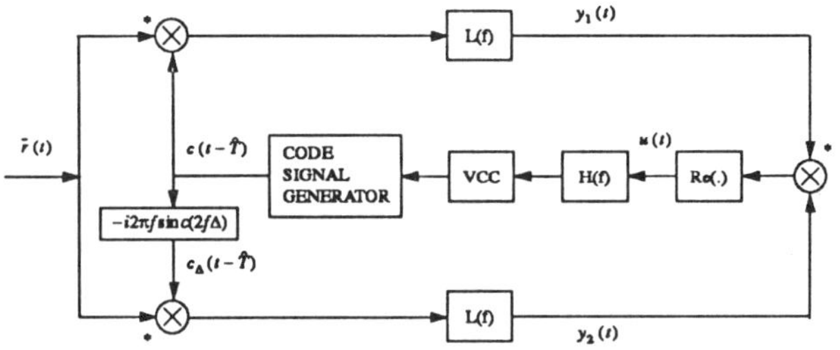

Figure 1: The equivalent baseband block diagram of a non-coherent reference/difference PN code tracking loop.

2 The System Model

The equivalent baseband model for the noncoherent code tracking loop system studied in this paper is shown in Fig. 1 [1]. The received signal $\tilde{r}(t)$ is a complex baseband DS/SS signal corrupted by complex noise $\tilde{n}(t)$ and interference $\tilde{i}(t)$.

$$\tilde{r}(t) = \sqrt{2\mathcal{S}}m(t - T(t))e^{i\theta} + \tilde{n}(t) + \tilde{i}(t) \ . \qquad (1)$$

Here $\mathcal{S}$ is the rms signal power; $T(t)$ is a time-varying delay ; θ is an independent random phase distributed uniformly over a 2π interval. The unit-power modulation signal $m(t) = d(t)c(t)$, where $d(t)$ is the data

1058-6393/91 $01.00 © 1991 IEEE

signal, and $c(t)$ is a PN spreading code waveform which we model as a wide-sense stationary random process. The noise $\tilde{n}(t)$ is the complex envelope of a bandpass white Gaussian noise (WGN) and is assumed to have flat power spectral density $S_{\tilde{n}}(f) = 2N_0$ within a bandwidth much larger than the spread spectrum system bandwidth; the interference $\tilde{i}(t)$ is an independent pulsed noise jammer modeled as

$$\tilde{i}(t) = z(t)\tilde{j}(t) , \tag{2}$$

where $\tilde{j}(t)$ has the same characteristics as $\tilde{n}(t)$ with power spectral density $S_{\tilde{j}}(f) = 2N_j$, and $z(t)$ is a periodic function defined as

$$z(t) = \sum_{n=-\infty}^{\infty} \sqrt{\frac{1}{\rho}} \operatorname{rect}\left(\frac{t-nT_j}{\rho T_j}\right) , \tag{3}$$

where rect(t) is the rectangle function being unity for $|t| < \frac{1}{2}$ and zero elsewhere. $1/T_j$ is the pulse repetition frequency (PRF) of the pulse jammer and ρ is the duty factor of the pulsed jammer with $0 < \rho < 1$. Obviously, the pulsed jammer $\tilde{i}(t)$ is a cyclostationary complex Gaussian random process [2] with period T_j.

The tracking loop must determine an estimate $\hat{T}(t)$ of the delay parameter in the received signal, and despread the DS signal $c(t - T(t))$ to recover a delayed version of the data signal $d(t)$. As shown in Fig. 1, a voltage-controlled clock (VCC) preceded by a baseband loop filter $H(f)$ drives a code signal generator to produce a replica $c(t - \hat{T}(t))$ of the PN spreading signal as the reference for re-estimating $T(t)$ and demodulating the data. Note that the multipliers in Fig. 1 operate on complex signals and have one input arm conjugated in each case, as indicated by the asterisk associated with that arm.

3 The Equivalent Loop Model

Let us assume for the moment that the delays T and $\hat{T}$ are fixed. It is also assumed that the intrinsic noise from PN code signal itself, which is inherently wideband, can be eliminated to a great extent by the lowpass filter $L(f)$. Hence, after being filtered by $L(f)$, the products of the incoming and despreading PN code signals on both arms can be approximated as

$$\rho_1(\epsilon) = E\{c^*(t-T)c(t-\hat{T})\} = R_c(T-\hat{T}) \tag{4}$$

$$\rho_2(\epsilon) = E\{c^*(t-T)c_\Delta(t-\hat{T})\} = R_{c_\Delta c}(T-\hat{T}), \tag{5}$$

where $\epsilon = (T - \hat{T})/T_c$ is the tracking error measured in chip times. The correlator arm filter outputs $y_k(t)$, $k = 1, 2$, are thus given as

$$y_k(t) = \sqrt{2S}\rho_k(\epsilon)(d^*)_L(t-T)e^{-i\theta} + v_k(t) . \tag{6}$$

Here, $(d^*)_L(t)$ is the output of the correlator filter $L(f)$ when driven by the complex conjugated data signal $d^*(t)$. The interferences $v_1(t)$ and $v_2(t)$ are output of the correlator filters when driven by the spread interference $[\tilde{n}(t)+z(t)\tilde{j}(t)]^*c(t-\hat{T})$ and $[\tilde{n}(t)+z(t)\tilde{j}(t)]^*c_\Delta(t - \hat{T})$, respectively. They are uncorrelated, zero mean with autocorrelations given as[1]

$$R_{v_1}(t,\lambda) \triangleq E\{v_1(t+\frac{\lambda}{2})v_1^*(t-\frac{\lambda}{2})\} = R_c(0)w(t,\lambda) \tag{7}$$

$$R_{v_2}(t,\lambda) = R_{c_\Delta}(0)w(t,\lambda) , \tag{8}$$

where $w(t, \lambda)$ is a periodic function of t given as

$$w(t,\lambda) = \int_{-\infty}^{\infty} [2N_0+2N_j z^2(t+\frac{\lambda}{2}-\alpha)]l(\alpha)l^*(\alpha-\lambda)d\alpha . \tag{9}$$

It is assumed that $v_1(t)$ and $v_2(t)$ are jointly Gaussian since the input processes to $L(f)$ have much wider bandwidths than that of $L(f)$ and both $\tilde{n}(t)$ and $\tilde{j}(t)$ are Gaussian. Also note, from eq.(7), that $w(t, \lambda)$ is a legitimate autocorrelation function of the output of the correlator filter when driven by the non-despread cyclostationary noise-and-interference $\tilde{n}(t) + z(t)\tilde{j}(t)$. Hence $w(t, \lambda)$ is Hermitian symmetric with respect to λ and $W(t, f)$, defined as the Fourier transform of $w(t, \lambda)$ with respect to λ, is real.

The feedback control signal $u(t) = Re\{y_1^*(t)y_2(t)\}$, given as

$$\begin{aligned} &2S|(d^*)_L(t-T)|^2 Re\{\rho_1^*(\epsilon)\rho_2(\epsilon)\} \\ &+ \sqrt{2S}Re\{(d^*)_L(t-T)e^{-i\theta}[v_1^*(t)\rho_2(\epsilon) + v_2^*(t)\rho_1(\epsilon)]\} \\ &+ Re\{v_1^*(t)v_2(t)\} , \end{aligned} \tag{10}$$

can be recognized as a sum of three parts from different sources. The first term of $u(t)$, namely,

$$2S|(d^*)_L(t-T)|^2 Re\{\rho_1^*(\epsilon)\rho_2(\epsilon)\} , \tag{11}$$

is the signal $\times$ signal term; the second term, denoted as $n_1(t, \epsilon)$, is the signal $\times$ interference term; and the last term, denoted as $n_2(t)$, is the interference $\times$ interference term.

Note that the data intrinsic noise from data modulation, referred to as the variations of $|(d^*)_L(t)|^2$ about its mean $\alpha = E\{|(d^*)_L(t)|^2\}$, only has secondary effect in this analysis and can be neglected. Hence, the eq.(11) is written as

$$\hat{s}(\epsilon) = 2\alpha S Re\{\rho_1^*(\epsilon)\rho_2(\epsilon)\} , \tag{12}$$

[1] Throughout this paper, the correlation between any cyclostationary processes is defined in this symmetric manner to preserve the Hermitian symmetry of autocorrelation function.

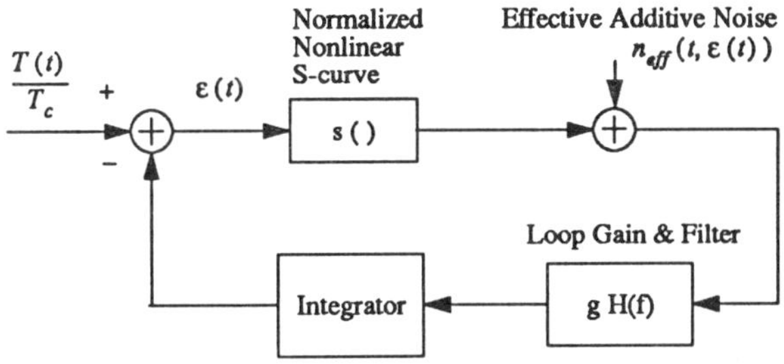

Figure 2: The simplified equivalent loop.

which is an odd function of ϵ providing a stable tracking point at $\epsilon = 0$, and $u(t)$ can be rewritten as

$$u(t) = \hat{s}(\epsilon) + n_1(t,\epsilon) + n_2(t) = \hat{s}(\epsilon) + n_u(t,\epsilon)\ , \quad (13)$$

where $n_u(t,\epsilon)$ is referred to as the loop noise.

We now close the loop, by applying the operating characteristic of the VCC

$$\frac{d}{dt}\hat{T}(t) = g_{vcc}[u(t) * h(t)]\ , \quad (14)$$

and a normalization such that the normalized S-curve of the loop has unit slope at the origin, and get its loop equation as

$$\frac{d}{dt}\epsilon(t) = \frac{d}{dt}\left(\frac{T(t)}{T_c}\right) - g\,[s(\epsilon) + n_{eff}(t,\epsilon)] * h(t)\ , \quad (15)$$

where

$$g = \frac{g_{vcc}}{T_c}(2\alpha DS)$$

$$D = \frac{d}{d\epsilon}Re\{\rho_1^*(\epsilon)\rho_2(\epsilon)\}\Big|_{\epsilon=0}\ ,$$

the normalized S-curve of the loop

$$s(\epsilon) = \frac{\hat{s}(\epsilon)}{2\alpha DS} = \frac{1}{D}Re\{\rho_1^*(\epsilon)\rho_2(\epsilon)\}\ , \quad (16)$$

and the effective additive noise to the loop

$$n_{eff}(t,\epsilon) = \frac{n_u(t,\epsilon)}{2\alpha DS}\ . \quad (17)$$

Fig. 2 shows the block diagram of this equivalent loop.

Note that $n_1(t,\epsilon)$ and $n_2(t)$, are uncorrelated. Also, they are linear and quadratic functions of the cyclostationary complex Gaussian noise processes $v_1(t)$ and $v_2(t)$. Taking into account that the unconjugated correlation functions $E\{v_i(t)v_j(t')\}$ are identically zero for all i, j, t, t', it can be verified that the instantaneous spectral density of $n_{eff}(t,\epsilon)$, defined as the Fourier transform of $R_{n_{eff}}(t,\lambda,\epsilon)$ with respect to λ, can be written as

$$S_{n_{eff}}(t,f,\epsilon) = \frac{C}{(2\alpha DS)^2} \times$$
$$\left\{\frac{S\mathcal{A}(\epsilon)}{C}[S_{d_L}(f) * W(t,f)] + [W(t,f) * W(t,f)]\right\} \quad (18)$$

in which the convolution is with respect to the variable f, and both C and $\mathcal{A}(\epsilon)$ are real-valued given as

$$C = \frac{1}{2}[R_c(0)R_{c_\Delta}(0)]$$
$$\mathcal{A}(\epsilon) = |\rho_1(\epsilon)|^2 R_{c_\Delta}(0) + |\rho_2(\epsilon)|^2 R_c(0)$$

It is interesting to note that $W(t,f)$ is a periodic function of t with period T_j and can be expanded by Fourier series expansion as

$$W(t,f) = (2N_0 + 2N_j)|L(f)|^2 + 2N_j \times$$
$$\sum_{k\neq 0}\text{sinc}(k\rho)L\left(f + \frac{k}{2T_j}\right)L^*\left(f - \frac{k}{2T_j}\right)e^{i2\pi\frac{k}{T_j}t}. \quad (19)$$

Here, the first term of $W(t,f)$ represents the stationary part of the interference and the second term reflects the cyclostationary nature of it. The sinc$(k\rho)$ in the second term suggests that, for a smaller duty factor ρ, a larger portion of the cyclostationary part can pass through the correlator filter $L(f)$. A close inspection also indicates that the time domain transition of $W(t,f)$, and therefore $S_{n_{eff}}(t,f,\epsilon)$, tends to be instantaneous as T_j increases.

4 System Performance Analyses

The Linear Analysis

It is assumed that the tracking loop is operating in the region that the tracking error $\epsilon(t)$ is close to zero. In this situation, we approximate the S-curve $s(\epsilon)$, derived in eq.(16), by $s(\epsilon) = \epsilon$, and the effective loop noise $n_{eff}(t,\epsilon)$ by its value at $\epsilon = 0$.

Under the above approximations and with the additional assumption that the loop is unstressed; i.e., the detuning factor is zero, the loop equation (15) is written as

$$\frac{d}{dt}\epsilon(t) = -g\,[\epsilon(t) + n_{eff}(t,0)] * h(t)\ . \quad (20)$$

It can be shown that it is equivalent to a linear, time-invariant, single filter system whose input is the approximated effective loop noise $n_{eff}(t,0)$ and output is the tracking error $\epsilon(t)$. The system transfer function of this single filter system, denoted as $H_L(f)$, is the closed loop transfer function from the effective noise input to the tracking error output, as seen in Fig. 2 with linearized S-curve $s(\epsilon) = \epsilon$, when only the noise-induced error is considered. It is given as

$$H_L(f) = -\frac{gH(f)}{i2\pi f + gH(f)} .$$

where $H(f)$ is the system transfer function of the loop filter in eq.(20).

The goal of linear analysis is to find the instantaneous tracking jitter $\sigma_\epsilon^2(t)$, defined as

$$\sigma_\epsilon^2(t) = R_\epsilon(t, \lambda = 0) = \int_{-\infty}^{\infty} S_\epsilon(t, f)df , \qquad (21)$$

and to study its relations with different parameters; e.g., the duty factor and the pulse repetition frequency, etc., of the pulse jammer.

It can be shown that $S_\epsilon(t, f)$ equals

$$\sum_k S_{n_{eff}}^{(k)}(f, 0)H_L\left(f + \frac{k}{2T_j}\right) H_L^*\left(f - \frac{k}{2T_j}\right) e^{i2\pi\frac{k}{T_j}t} , \qquad (22)$$

where $S_{n_{eff}}^{(k)}(f, \epsilon)$ is the kth cyclic spectral density [2] of the effective noise, defined as the kth coefficient of the Fourier time series expansion of the periodic spectral density $S_{n_{eff}}(t, f, \epsilon)$.

The Nonlinear Analysis

It is assumed that the effective loop noise $n_{eff}(t, \epsilon)$ has a much wider bandwidth, which is about the bandwidth of the correlator filter $L(f)$ or the data bandwidth, compared to the loop bandwidth. Therefore, we approximate $n_{eff}(t, \epsilon)$ as a cyclostationary process with a flat spectral density of the level at the origin; i.e., $S_{n_{eff}}(t, 0, \epsilon)$ within the desired frequency range. In doing so, we let[2]

$$n_{eff}(t, \epsilon) \approx \sqrt{S_{n_{eff}}(t, 0, \epsilon)}n_w(t) , \qquad (23)$$

where $n_w(t)$ is a white Gaussian noise with $S_{n_w}(f) = 1$.

Now, we will restrict ourselves to the first-order loop, which is again assumed to be unstressed, in order to have a scalar Langevin type stochastic equation from the loop equation eq.(15), although a similar matrix form is always available for a higher order loop. For a first-order loop, its loop gain is

$$g = 4B_L ,$$

where B_L is its loop noise bandwidth. Thus, the loop equation is rewritten, by taking $h(t) = \delta(t)$ and the approximation of eq.(23), as

$$\frac{d}{dt}\epsilon(t) = -4B_L \left\{s(\epsilon) + \sqrt{S_{n_{eff}}(t, 0, \epsilon)}n_w(t)\right\} . \quad (24)$$

By introducing a normalized time variable $\tau = 4B_L t$, the equation of loop operation shown above is further modified as

$$\frac{d}{d\tau}\epsilon(\tau) = -s(\epsilon) - \sqrt{4B_L S_{n_{eff}}(\frac{\tau}{4B_L}, 0, \epsilon)}w(\tau) . \quad (25)$$

where $w(\tau)$ is a white Gaussian process with unit power spectral density.

Eq.(25) is of the type of the Langevin equation and the normalized tracking error $\epsilon(\tau)$ is a Markov process governed by this time-dependent stochastic differential equation [3]. However, the time-dependence of eq.(25) renders that $\epsilon(\tau)$ is not a time-homogeneous Markov process, and, therefore, the renewal process approach [1] [4] can not be applied to get a stationary probability density function for $\epsilon(\tau)$. Instead, we use the mean time for the tracking loop to lose lock since re-acquisition as a measure of nonlinear performance. Assuming that a re-acquisition procedure has occurred at time τ_o with the effect that $\epsilon(\tau_o) = 0$, the mean time it takes for the tracking error process to reach one of the boundaries $\pm\epsilon_B$ that limit the loop's S-curve support interval can be obtained as

$$E\{T_1|\tau_o\} = \frac{1}{4B_L}\int_0^{\infty}\int_{-\epsilon_B}^{\epsilon_B} p(\epsilon, \tau)d\epsilon d\tau , \qquad (26)$$

where $p(\epsilon, \tau)$ is the solution to the corresponding Fokker-Planck equation

$$\frac{\partial p(\epsilon, \tau)}{\partial \tau} = -\frac{\partial}{\partial \epsilon}[k_1(\epsilon, \tau + \tau_o)p(\epsilon, \tau)] + \frac{1}{2}\frac{\partial^2}{\partial \epsilon^2}[k_2(\epsilon, \tau + \tau_o)p(\epsilon, \tau)], \ \tau > 0, \quad (27)$$

with the drift coefficient

$$k_1(\epsilon, \tau) = -s(\epsilon) + B_L\frac{d}{d\epsilon}S_{n_{eff}}(\frac{\tau}{4B_L}, 0, \epsilon) , \qquad (28)$$

the diffusion coefficient

$$k_2(\epsilon, \tau) = 4B_L S_{n_{eff}}(\frac{\tau}{4B_L}, 0, \epsilon) , \qquad (29)$$

and the boundary condition $p(\pm\epsilon_B, \tau) = 0$, for $\tau > 0$; and the initial condition $p(\epsilon, 0) = \delta(\epsilon)$.

[2] It's not always true that the spectral density $S_{n_{eff}}(t, f, \epsilon)$ is non-negative for every t and f. However, it has been verified numerically that, for the case we are interested, it is non-negative for every t at $f = 0$.

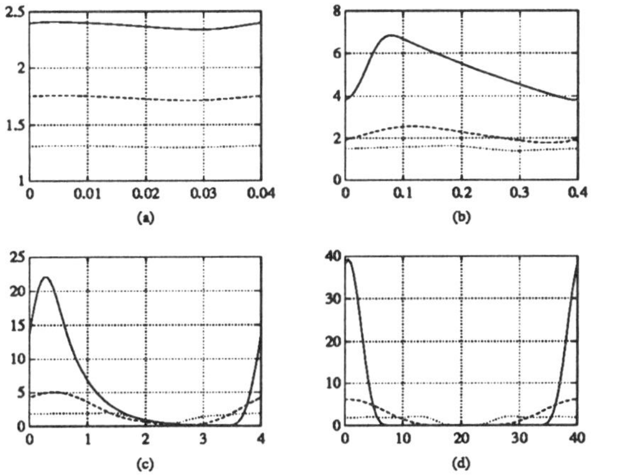

Figure 3: The periodic steady-state variance of the tracking error, for one period, versus the normalized time $\tau = 4B_Lt$. Subplots (a) through (d) are with the cases of PRF/B_L being 100, 10, 1, and 0.1, respectively, with B_{IF}/B_L being fixed at 200. For each subplot, the solid, dashed and dotted curves are associated with ρ=0.2, 0.5, and 0.8, respectively.

PRF/B_L	ρ	max	mean	min	std dev
1000.0	x	1.2369	1.2369	1.2369	0.0000
100.0	0.2	2.4089	2.3737	2.3371	0.0248
	0.5	1.7626	1.7407	1.7174	0.0158
	0.8	1.3154	1.3080	1.2992	0.0057
10.0	0.2	6.8385	5.2092	3.8169	0.9715
	0.5	2.5670	2.1603	1.7767	0.2726
	0.8	1.6446	1.5291	1.4002	0.0756
1.0	0.2	22.0932	5.4097	0.0851	6.9079
	0.5	5.0600	2.1951	0.1838	1.7433
	0.8	1.9701	1.4918	0.4345	0.4954
0.1	0.2	39.2939	5.7130	0.0004	11.6514
	0.5	6.1400	2.2144	0.0025	2.2806
	0.8	2.1406	1.4946	0.0038	0.7337

Table 1: Some of the characteristics of the periodic variances displayed in Fig. 3. Note that, for the case of $PRF > B_{IF}$, the tracking jitter remains a constant irrelevant to jammer's duty factor ρ; hence, the corresponding results also apply to the non-pulsed jammer ($\rho = 1$) case.

5 Results

Numerical results have been obtained for the one-chip early-late Reference/Difference code tracking loop (RDIC) [5]; i.e., the loop shown in Fig. 1 with the time offset $\Delta = T_c/2$, under the assumptions of a square pulse PN code signal; an 8th-order Butterworth low-pass arm filter $L(f)$ with the 3dB cutoff frequency at $B_{IF}/2$, or equivalently, the reciprocal of bit duration T_b; and an all-pass loop filter $H(f)$. Also, the approximation (23) has been used to implement numerical results for both linear and non-linear analyses. In all those the detection signal-to-noise ratio in the absence of interference was set at $E_b/N_0 = 10$ and the interference-to-noise ratio at $N_j/N_0 = 100$.

Fig. 3 and Table 1 illustrate the influence of the jammer's duty factor and pulse repetition frequency on the characteristics of the periodic tracking-error variance for the linearized loop. These results show that, due to the non-linear operation performed by the non-coherent tracker to generate its feedback control signal, the time-average value of the periodic tracking error variance depends strongly on the choice of the aforementioned jammer parameters, increasing significantly with decreasing duty factor. Also, the peak value of the variance can reach levels far above its average value if the pulse repetition frequency is on the order of the loop bandwidth and the duty factor is small. These results indicate that a wide-band pulsed jammer can be much more effective in disrupting loop operation than a non-pulsed jammer ($\rho = 1$) of equal average power.

For the non-linear analysis, it was assumed that re-acquisition is instantaneous and occurs at the center of a jammer's pulse. According to the time origin used in the modeling of $z(t)$ in (3), this corresponds to set $\tau_o = 0 \pmod{T_j}$ in (27). The mean time to lose lock versus the jammer's duty factor and pulse repetition frequency is listed in Table 2. These results show that the jammer's efficiency increases with decreasing duty factor and pulse repetition frequency following the same trend found in the linear analysis.

PRF/B_L	10.0	1.0	0.1
$\rho = 0.2$	0.4259	0.1668	0.1640
$\rho = 0.5$	2.0000	1.0753	1.0751
$\rho = 0.8$	4.7533	3.7035	3.7678
$\rho = 1.0$	5.7855		

Table 2: The mean time to lose lock since re-acquisition measured in the bit duration T_b.

References

[1] R. Sampaio-Neto and R. A. Scholtz. "Precorrelation Filter Design for Spread-Spectrum Code Tracking in Interference,". *IEEE JSAC.*, SAC-3: 662–675, Sep. 1985.

[2] W. A. Gardner. *Introduction to Random Processes.* McGraw-Hill, New York, 2nd edition, 1990.

[3] H. Risken. *The Fokker-Planck Equation.* Springer-Verlag, Berlin; New York, 2nd edition, 1989.

[4] H. Meyr and G. Ascheid. *Synchronization in Digital Communications*, volume 1. John Wiley & Sons, New York, 1990.

[5] R. Sampaio-Neto. *Spread-Spectrum Code Tracking in the Presence of Interference.* PhD dissertation, University of Southern California, Sep. 1983.

Error Probabilities of Coherent Optical Heterodyne FSK-CDMA Signals in a Noncoherent Detection System Degraded by Laser Phase Noise

Kent C. M. Varnum
Naval Ordnance Missile Test Station
White Sands Missile Range
NM 88002

R. Clark Robertson
Electrical and Computer Eng. Dept.
Naval Postgraduate School
Monterey, CA 93943

Abstract

An error probability analysis is performed for noncoherent detection of optical heterodyne frequency-shift keying (FSK) direct sequence spread-spectrum code-division multiple access (CDMA) signals corrupted by laser phase noise, receiver noise, and multiuser noise. Receiver noise is included in the analysis as an additive white Gaussian process. The multiuser noise is also modeled as a Gaussian process.

The application of spread spectrum techniques to an optical channel can increase system simultaneous user capacity significantly with minimal performance degradation for long spreading codes. An FSK-CDMA system with low, moderate, and high user bit rates as compared with the laser linewidth is analyzed over different spreading code lengths for increasing numbers of users. The low bit rate system performs poorly even when the spreading code is long and the number of simultaneous users is very small. On the other hand, a high bit rate system which uses a spreading code with 2^{11} *chips can support up to seventy-five users if the maximum allowed probability of bit error is* 10^{-6}.

1 Introduction

Optical systems are degraded by standard receiver noise, shot noise in the photodetector, and phase noise in the transmitting laser. The impact of receiver and photodetector shot noise is significantly reduced by the application of optical heterodyne techniques which are very similar to standard RF heterodyne techniques. Unlike direct detection systems, optical heterodyne systems mix a locally generated lightwave with the received signal which is then detected by a photodetector. The resulting electrical signal is a replica of the optical signal translated down in frequency, usually to the microwave frequency range. Mixing the incoming optical signal with a local laser provides strong optical input power to the photodetector. The strong local laser condition drastically reduces the effect of the receiver thermal noise and photodetector shot noise. Unfortunately, the addition of a local laser at the receiver increases the effect of the laser phase noise on system performance. Laser phase noise is a noise mechanism inherent to the physical nature of all lasers that impresses random phase and amplitude modulation on the otherwise monochromatic laser output. In optical heterodyne systems, the laser phase noise of the transmit and receive lasers is additive. Current research indicates that in order to attain reasonable bit error performance, the system filter bandwidth must be at least ten times the sum of the laser phase noise bandwidths of the transmitting and local lasers [1]. Current semi-conductor lasers may have a laser phase noise linewidth of up to 50 MHz and require a channel bandwidth of up to 100 MHz. For user bit rates less than or equal to the laser phase noise bandwidth, the channel spacing required in wavelength division multiplexing (WDM) systems to ensure sufficient guardbands results in an extremely inefficient use of available bandwidth.

Future systems will have to accommodate more users with higher bit rates. In this paper we address the high bit rate systems that will be required by future users. As a means of improving the multiuser capacity of high bit rate optical communications systems, we propose the implementation of code-division multiple access (CDMA) techniques in an optical heterodyne FSK system with noncoherent detection. CDMA is a type of spread-spectrum that adds multiuser capability by spreading and despreading each user data signal with a unique digital code. Each system user is assigned a particular code sequence which is used to encode each data bit. In this paper we model the spreading codes with random codes. Random codes are constructed of a sequence of random variables taking values $\{+1, -1\}$ with equal probability, and the sequences assigned to different users are mutually independent [2]. The application of CDMA techniques improves standard optical heterodyne WDM system performance by increasing user capacity on a given WDM channel with minimal impact on system performance.

To illustrate the improvement realized by the application of CDMA techniques this paper analyzes a multiuser optical heterodyne FSK-CDMA system. System performance is measured by the probability of bit error as a function of the combined system laser linewidth, bit time product, and the number of simultaneous users. Both receiver noise and multiuser noise are modeled as additive white Gaussian noise. The ratio of signal power to the noise power from standard receiver noise and shot noise is assumed to be large enough so that the primary degradation in performance is due to laser phase noise and multiuser noise. It is assumed that the user bit stream consists of a mutually independent random series of 'ones' and 'zeros'. The system will only transmit a signal when the user has data to send, otherwise the station will remain idle. Each transmitter FSK modulates a semiconductor laser diode with the user bit stream. It is assumed that there is negligible interference between the two FSK tones. After the user bit stream FSK modulates the transmit laser, the transmitter encodes the FSK bit stream into the spreading sequence through binary phase-shift keying producing a high frequency chip stream consisting of two frequencies, each phase modulated with the user code sequence. The transmitted signal is then optically mixed with other transmitter signals in the common optical fiber channel. The system under consideration is also considered to be a 'power balanced network'. In this type of network every signal, desired or undesired, is transmitted with the same power.

At the receiver, the optical signal is down converted to an intermediate (IF) frequency with a locally generated optical signal and detected with a photodector. The remainder of the receiver consists of two separate receivers, called branches, which demodulate the electrical IF signal at the photodector output. One branch is matched to f_1 and the other is matched to f_0, where f_0 and f_1 are the IF frequencies corresponding to a bit 'zero' and a bit 'one', respectively. Each of the frequency matched branches consists of two sub-branches in quadrature. Each sub-branch of the receiver adds a locally generated IF signal to the incoming composite IF signal. The locally generated IF signal added to the incoming signal is phase modulated with the user code sequence. The resulting electrical signal is then integrated over the bit interval and noncoherently demodulated by a square-law detector. The output of each square-law detector is then added to its quadrature component to form the correlation statistics. The correlation statistics produced by each of the two frequency matched branches are then input to a comparator for bit recovery. As with standard FSK systems, the comparator is a threshold device with the threshold set at zero.

2 Analysis

The performance of the demodulator depends on the statistics of the output waveform $\mu_i(T)$, $i = 0, 1$ of each branch or the receiver corrupted by laser phase noise, receiver shot noise, and multiuser noise. The probability of bit error is

$$P_b = 0.5[P_0 + P_1] \tag{1}$$

where equally likely signalling is assumed and

$$P_0 = \Pr\{\mu_1 > \mu_0 \mid 0\} \tag{2}$$

is the probability of making an error when a 'zero' is sent and

$$P_1 = \Pr\{\mu_0 > \mu_1 \mid 1\} \tag{3}$$

is the probability of making an error when a 'one' is sent.

2.1 Derivation of the Conditional Probability of Bit Error

The IF signal received by an individual user during each T_b interval is expressed

$$s(t) = \mathrm{Re}\,\{r(t)\exp(j\omega_i t)\} \tag{4}$$

where $i = 1, 0$ and

$$r(t) = a(t) S \exp\left[j\theta(t)\right] + n(t) \tag{5}$$

is the complex baseband signal where the power received in the optical pulse is $S^2/2$, $\theta(t)$ is the composite phase noise due to both the transmitting and receiving lasers, and $n(t)$ is the complex noise due to the combination of additive white Gaussian receiver noise and multiuser noise. The user code sequence is represented by a periodic sequence of unit amplitude rectangular pulses $a(t) = \{-1, +1\}$ of duration T_c where $T_c = T_b/N$ and N is the code length.

Due to the symmetry of the receiver, the probability of making an error is the same for both a transmitted 'one' and transmitted 'zero'; that is, $P_1 = P_0$. Because of this symmetry, it is only necessary to examine receiver performance for either a transmitted 'one' or a transmitted 'zero'. The signal is, therefore, assumed to be present in the branch of the receiver corresponding to user bit 'one'. Then for the receiver branch matched to frequency f_1, the output of the ideal baseband integrators, each with an integration time T_b, is

$$\zeta_{1i}(T_b) = \mathrm{Re}[\zeta_1(T_b)] \tag{6}$$

for the in-phase channel and

$$\zeta_{1q}(T_b) = \mathrm{Im}[\zeta_1(T_b)] \tag{7}$$

for the quadrature channel where

$$\zeta_1(T_b) = \frac{S}{T_b}\int_0^{T_b} \exp\left[j\theta(t)\right] dt + \hat{n} \tag{8}$$

and $\hat{n} = \hat{n}_i + j\hat{n}_q$ is a zero mean, additive white Gaussian complex baseband noise sample which includes both receiver noise and multiuser noise. For the receiver branch matched to frequency f_0, the output of the in-phase and quadrature baseband integrators during each T_b interval is $\hat{n}_i$ and $\hat{n}_q$, respectively.

These signals are detected by ideal square-law detectors and the overall decision statistic for each branch is obtained by summing the square-law detector outputs for the in-phase and quadrature channels. For a given of the decision statistic μ_1 for branch 1, the conditional error probability is

$$P_1(\mu_1) = \int_{\mu_1}^{\infty} p\left(\mu_0 \mid 1\right) d\mu_0 \tag{9}$$

where

$$p(\mu_0 \mid 1) = \frac{1}{\sigma^2} \exp\left(-\frac{\mu_0}{\sigma^2}\right) \tag{10}$$

is the probability density function (pdf) for μ_0 when a data bit 'one' is sent and σ^2 is the combined noise power of the in-phase and quadrature square-law detector outputs. The average error probability conditional on the laser phase noise is obtained by averaging over all values of μ_1 to get

$$P_{b|X}(x) = \int_0^\infty P_1(\mu_1) p(\mu_1 \mid 1, x)\, d\mu_1 \tag{11}$$

where X represents the magnitude of the random variable determined by the laser phase noise and

$$p(\mu_1 \mid 1, x) = \frac{1}{\sigma^2} \exp\left(-\frac{\mu_1 + S^2x^2}{\sigma^2}\right) I_0\left(\frac{2Sx\sqrt{\mu_1}}{\sigma^2}\right) \tag{12}$$

is the pdf for μ_1 conditioned by the laser phase noise when a data bit 'one' is sent and $I_0(\bullet)$ represents the modified Bessel function of zero order. The unconditional probability of bit error is thus

$$P_b = \int_0^1 \left[\int_0^\infty p(\mu_1 \mid 1, x)\, P_1(\mu_1)\, d\mu_1\right] p_X(x)\, dx \tag{13}$$

where $p_X(x)$ is the pdf for the random variable X. The expression for the probability of bit error is now be simplified.

Combining (9)–(12), we get

$$P_b = \frac{1}{2} \int_0^1 \exp\left(-\frac{S^2x^2}{2\sigma^2}\right) p_X(x)\, dx \tag{14}$$

which must be evaluated numerically. If the effect of laser phase noise is negligible, then $p_X(x) \approx \delta(x - 1)$ where $\delta(\bullet)$ is the Dirac delta function, and (14) reduces to

$$P_b = \frac{1}{2} \exp\left(-\frac{S^2}{2\sigma^2}\right) \tag{15}$$

which is the probability of bit error for a conventional FSK receiver with noncoherent detection.

2.2 Analytic Expressions for the Noise Terms

The major source of degradation to an optical fiber communication system is laser phase noise. In optical heterodyne systems, both the transmit and local lasers add laser phase noise to the received signal. This causes the introduction of a random frequency deviation to the IF signal related to the sum of the linewidths of both lasers. Due to the filter response of the initial IF filter in each branch of the FSK-CDMA receiver, the pdf of the random variable X determined by the laser phase noise depends strictly upon the laser linewidth β and the bit rate T_b. Direct evaluation of this pdf is computationally intensive since $\theta(t)$ in (8) is a Brownian motion process. The pdf of X can be evaluated through numerical integration and Monte Carlo simulation; however, large amounts of computational time are required [3]. Attempts to simplify the analysis of the probability of bit error in lightwave systems corrupted by laser phase noise through the derivation of a closed form analytical model representing $p_X(x)$ results in a curve fit approximation of the actual pdf. For an integrate and dump filter the pdf of the random variable X determined by the laser phase noise is approximated as [4]

$$p_X(x) = \alpha\left[1 - \exp(-\alpha)\right]^{-1} \exp\left[-\alpha(1 - x)\right] \tag{16}$$

for $0 \le x \le 1$ and zero otherwise where

$$\alpha = \frac{1.6}{\beta T_b}\left(1 + 0.5\sqrt{\beta T_b}\right) \tag{17}$$

The accuracy of this model has been demonstrated over many different system signalling rates [3]. We use this expression to represent the random variable X determined by the laser phase noise.

Spread-spectrum code-division multiple access (CDMA) is an asynchronous multiple access communication scheme in which many users share a common bandwidth. In CDMA each user is assigned a particular code sequence which is used to modulate the carrier depending on the digital data [2]. Under ideal conditions, each particular user code is orthogonal to every other user code, and as a result, invisible to other users. This is not the case in practical systems. A particular user recovers his coded bit stream through a receiver matched to the particular user's code. Other simultaneous user's signals will corrupt the received signal and appear as noise in the particular user's receiver. In many cases of interest, the multiuser noise is represented as a Gaussian random process. The validity of the Gaussian multiuser noise model degrades at lower code lengths and fewer numbers of users, but for small numbers of users the laser phase noise dominates system performance. In addition, system performance is less critical when the number of users is small.

In this paper we assume the use of random signature sequences. Random signature sequences are constructed of a sequence of random variables taking values $\{+1, -1\}$ with equal probability, and all sequences are mutually independent. Analysis using random signature sequences is mathematically simpler.

Much work has been done in recent years to characterize the statistics of direct-sequence spread-spectrum codes. The difficulty in analyzing such systems is the fact that they are asynchronous and proper analysis requires characterizing not only the periodic but also the aperiodic cross-correlation properties. Most current models treat phase shifts, time delays, and data symbols as mutually independent random variables. The Gaussian random variable that describes the multiuser noise for random codes is zero mean with a variance [2, 6]

$$\sigma^2 = \frac{S^2(K-1)}{3N} + \frac{2N_0^s}{T_b} \tag{18}$$

where $N_0^s/2$ is the two-sided spectral density of the additive white Gaussian receiver noise.

3 Numerical Results

The computation of the probability of bit error involves a numerical evaluation of (14) for different lengths of random user signature sequences over the range of simultaneous users the system can support. The ratio of signal power to the noise power due to receiver shot noise (SNR) is taken to be 16 dB. This yields a probability of bit error of 10^{-9} when laser phase noise and multiuser noise are negligible. It is also assumed that the optical signal power of an individual user is normalized to unity and that the transmitter equally balances the active user signals within the composite optical signal. The number of chips in the random user code is varied from 2^8 to 2^{11}. The three cases of a relatively small bit rate, a moderately high bit rate, and a very high bit rate, corresponding to laser linewidth, bit time products of $\beta T_b = 0.3$, 0.1, and 0.02, respectively, are investigated. The resulting system performance is shown in Figs. 1–3.

As can be seen in Figs. 1–3, for an SNR of 16 dB the high bit rate is required in order to obtain a probability of bit error of 10^{-6} or less. Regardless of system bit rate, the probability of bit error increases rapidly with the addition of just a few users for short code lengths. Long code lengths are required in order to support more than just a few simultaneous users without a significant performance degradation. For a bit rate ten times the laser linewidth, a ratio currently a rule of thumb for optical communication systems, the minimum probability of bit error is still significantly larger than that obtained with a high bit rate system. As can be seen by comparing Fig. 2 with Fig. 3, a bit rate fifty times the laser linewidth, five times greater than the customary rule of thumb, is required to eliminate the negative effects of laser phase noise. The results also exhibit the standard CDMA characteristics in that they are fairly steep for low number of users and flat at high usage levels [5]. This behavior is demonstrated for all system bit rates examined (and, thus, for varying effects of laser phase noise on system performance). The steepness of the slope for low numbers of users is more pronounced for the high bit rate system; nevertheless, the high bit rate system is still markedly superior to the system with $\beta T_b = 0.1$. For example, for a user code with 2^{11} chips and a maximum probability of bit error of 10^{-6}, the high bit rate system can support up to seventy-five simultaneous users, while the minimum probability of bit error for the system with $\beta T_b = 0.1$ is greater than 10^{-6} for even a single user due to the effects of laser phase noise. For a user code with 2^9 chips and a maximum probability of bit error of 10^{-5}, the high bit rate system can support up to thirty simultaneous users, while the system with $\beta T_b = 0.1$ can support only about twelve simultaneous users. Due to the limitations on system performance imposed by laser phase noise, the minimum probability of bit error for the low bit rate system is greater than 10^{-3} regardless of the number of simultaneous users and code length.

The effect of laser phase noise on system performance when only one user is transmitting (i.e., there is no multiuser noise) is illustrated in Fig. 4. As can be seen, even a significant increase in signal energy will not substantially improve system performance for systems with a low bit rate. As noted earlier, for a bit rate ten times the laser linewidth, system performance is still significantly poorer than when laser phase noise is absent; although, as already demonstrated, a probability of bit error of less than 10^{-6} is possible with reasonable values of SNR. A bit rate fifty times the laser linewidth is required in order to virtually eliminate the negative effects of laser phase noise if a minimum probability of bit error of 10^{-9} is required; however, virtually identical performance can be obtained at the expense of about 1 dB of signal energy by relaxing the bit rate requirement from fifty to to twenty times the laser linewidth.

4 Conclusions

To improve overall optical communications system performance, we propose the application of CDMA spread spectrum techniques to a standard optical heterodyne FSK system employing Wavelength Division Multiplexing (WDM) to obtain multiple simultaneous user capabilities. This system can greatly improve system user capacity without a substantial increase in the probability of bit error provided the bit rate is high enough and the code is long enough.

References

[1] J. Salz, 'Coherent Lightwave Communications,' AT & T Technical Journal, vol. 64, pp. 2153–2207, Dec. 1985.

[2] M. B. Pursley, 'Performance Evaluation for Phase-Coded Spread-Spectrum Multiple-Access Communication-Part I: System Analysis,' IEEE Trans. on Communications, vol. COM-25, pp. 795–799, Aug. 1977.

[3] G. J. Foschini, L. J. Greenstein, and G. Vannucci, 'Noncoherent Detection of Coherent Lightwave Signals Corrupted by Phase Noise,' IEEE Trans. on Communications, vol. 36, pp. 306–314, Mar. 1988.

[4] G. J. Foschini, G. Vannucci and L. J. Greenstein, 'Envelope Statistics for Filtered Optical Signals Corrupted by Phase Noise,' IEEE Trans. on Communications, vol. 37, pp. 1293–1302, Dec. 1989.

[5] G. L. Turin, 'The Effects of Multipath and Fading on the Performance of Direct-Sequence CDMA Systems,' IEEE J. on Selected Areas in Communications, vol. SAC-2, pp. 597–603, Jul. 1984.

[6] E. A. Geraniotis, 'Performance of Noncoherent Direct-Sequence Spread-Spectrum Multiple-Access Communications,' IEEE J. on Selected Areas In Communications, vol. SAC-3, pp. 687–694, Sept. 1985.

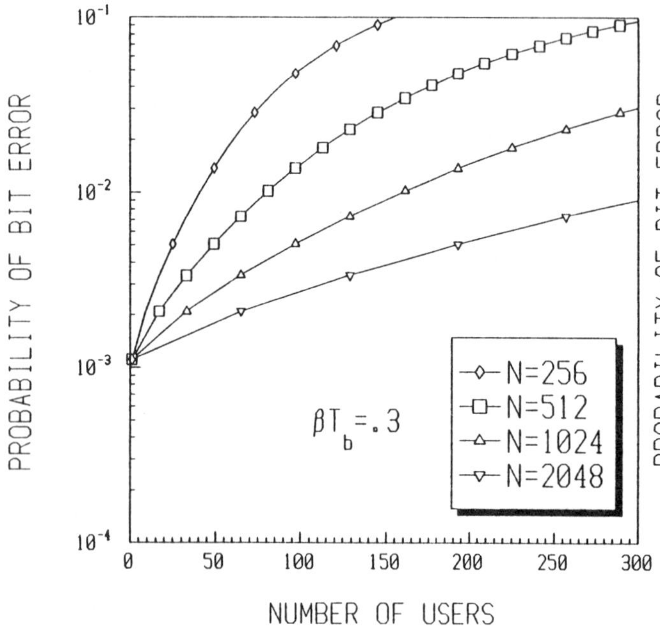

Figure 1: Probability of bit error for high order code lengths and low bit rate ($\beta T_b = 0.3$) with SNR = 16 dB.

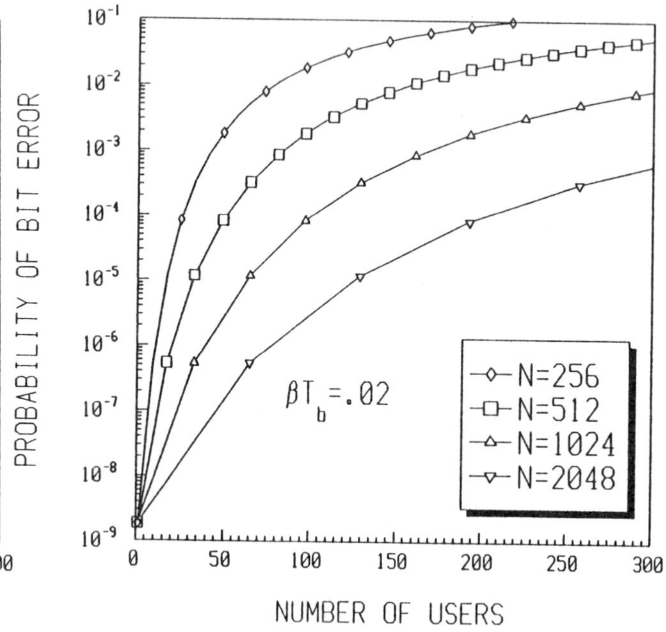

Figure 3: Probability of bit error for high order code lengths and high bit rate ($\beta T_b = 0.02$) with SNR = 16 dB.

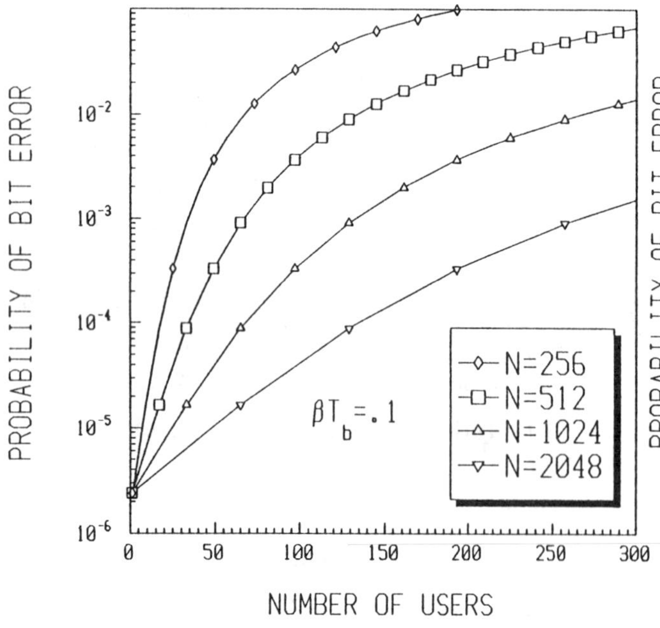

Figure 2: Probability of bit error for high order code lengths and moderate bit rate ($\beta T_b = 0.1$) with SNR = 16 dB.

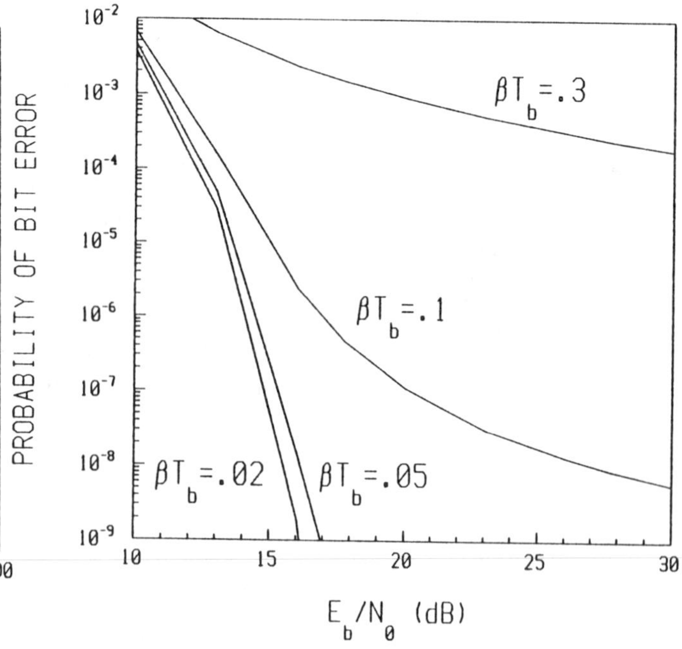

Figure 4: System performance for a single user with βT_b as a parameter.

Estimation and Confidence in Bearings Only Tracking

A. Holtsberg

Department of Mathematical Statistics
Lund Institute of Technology
Box 118, S-221 00 Lund
Sweden

J. Holst

Department of Mathematical Statistics
Lund Institute of Technology
Box 118, S-221 00 Lund
Sweden

Abstract

A new pseudo-linear estimator for bearings only tracking based on Modified Polar Coordinates (MPC) is presented. It is approximately unbiased in a long range scenario contrary to the earlier well known pseudo-linear estimator in Cartesian Coordinates. The description of the new estimator furthermore provides a motivation for expecting the loglikelihood function to be approximately quadratic in MPC. Simulations show the performance of the new estimator and its estimate of the parameter error variance.

1 Introduction

In bearings only tracking the own ship, called the platform, makes a sequence of measurements of the bearing to a target. The sequence of platform positions is assumed to be known. The target moves with constant course and velocity. The notation to be used is the following, where i is the time index.

X	a possible target track
X_c	the true target track
$\hat{X}$	an estimate of the true target track
b_i	bearing measurement
β_i	bearing, function of X
ε_i	residual, function of X
e_i	measurement error
τ	the reference time
p_{xi}, p_{yi}	platform position
r_i	range, function of X
r_{xi}, r_{yi}	relative position, function of X
v_x, v_y	velocity vector, function of X

Note that in order to save space we will not explicitely indicate which quantities that take a target track as argument. Note also that ε_i, the *residual*, depends on a target track argument and is not to be confused with the true *measurement error* e_i ($= \varepsilon_i(X_c)$). The dot will be used consistently to denote derivatives by time with respect to the target motion, not taking into account the platform motion. Specifically the quantity $\dot{\beta}_\tau$ will denote the bearing rate of change we would have had if the platform had been still at the reference time, contrary to what is the standard notation used in e.g [2]. The given sequence of bearing measurements is $\{b_i\}, i = 1...n$. The measurements are assumed to be disturbed by a white Gaussian noise sequence $\{e_i\}, i = 1...n$ with constant but unknown standard deviation σ,

$$b_i = \beta_i(X_c) + e_i.$$

We assume one time unit elapses between every measurement. At time i the residual ε_i associated with the target track X is

$$\varepsilon_i = b_i - \beta_i$$

where

$$\beta_i = \arctan\left(\frac{r_{x\tau} + (i-\tau)v_x - p_{xi}}{r_{y\tau} + (i-\tau)v_y - p_{yi}}\right).$$

There are essentially three types of numerical estimators for the Bearings Only Tracking problem.

- The Maximum Likelihood Estimator.
- Pseudo linear methods.
- Kalman filters.

Here pseudo linear methods will be investigated and the Maximum Likelihood Estimator (MLE) will be used for comparisons.

1058-6393/91 $01.00 © 1991 IEEE

The MLE is the target track X that minimizes the sum of squares of the residuals

$$Q_{MLE} = \sum_{i=1}^{n} \varepsilon_i^2. \quad (1)$$

This estimator is well known to be asymptotically efficient and it is an intuitively very appealing estimator. Unfortunately the MLE is not an explicit estimator, hence one needs a numerical search routine for finding the minimum. This may introduce practical problems, as programming complexity, convergence and computing time.

The pseudo linear methods are chararacterized by modifications of the criterion (1), so that the minimization results in a linear system of equations. The advantage is that no numerical search routine is needed, the solution is simple, explicit and fast. These estimators may however be severely biased.

What we shall call the Cartesian Pseudo Linear Estimator (CPLE) is well known under the name *the* Pseudo Linear Estimator. After defining the CPLE we shall introduce a new estimator, the Modified Polar Pseudo Linear Estimator (MPPLE) which has better properties than the CPLE.

2 The Cartesian PLE

The Cartesian Pseudo Linear Estimator (CPLE) has been discussed and analyzed rather extensively [2-4]. For simplicity of notation we will assume that the own ship position is at the origin at the reference time τ, that is $p_{x\tau} = p_{y\tau} = 0$. The CPLE is then defined as the target track that minimizes

$$\begin{aligned} Q_{CPLE} &= \sum_{i=1}^{n}(r_i \sin \varepsilon_i)^2 \quad (2) \\ &= \sum_{i=1}^{n}(r_i \sin(b_i - \beta_i))^2 \\ &= \sum_{i=1}^{n}(r_i(\sin b_i \cos \beta_i - \cos b_i \sin \beta_i))^2 \\ &= \sum_{i=1}^{n} (\sin b_i \, (r_{y\tau} + (i-\tau)v_y - p_{yi}) \\ &\qquad - \cos b_i \, (r_{x\tau} + (i-\tau)v_x - p_{xi}))^2 . \end{aligned}$$

The minimization of Q_{CPLE} is seen to result in a linear system of equations in the parameters $[v_x, v_y, r_{x\tau}, r_{y\tau}]$. Observe that the quantity $r_i \sin \varepsilon_i$ is the orthogonal distance between the measured line of sight to the target and the target position. This is not a good estimator due to its bias, which is further discussed below.

3 The Modified Polar PLE

The Modified Polar Pseudo Linear Estimator (MPPLE) has not, to our knowledge, been published earlier, but it might be viewed as an extention of a simpler pseudo linear estimator that appears in [5]. We shall define the coordinate system that appears in the MPPLE and then define the estimator itself. Before doing that we shall define the modified polar coordinate system which is used for improving Kalman filtering performance [1]. The reason for this is that we will be able to see the close relationship between this coordinate system and the MPPLE. The modified polar coordiates z are defined as

$$\begin{aligned} z_1 &= \dot{\beta}_\tau = (v_x \cos \beta_\tau - v_y \sin \beta_\tau)/r_\tau \\ z_2 &= \dot{r}_\tau / r_\tau = (v_x \sin \beta_\tau + v_y \cos \beta_\tau)/r_\tau \\ z_3 &= \beta_\tau \\ z_4 &= 1/r_\tau . \end{aligned} \quad (3)$$

Note however that there is a slight difference between our definition and the one in [1] in that v_x and v_y are the components of the target velocity, not of the *relative* target velocity.

The coordinates of the MPPLE will be denoted z'. The MPPLE will need an initial estimate B of the bearing β_τ. A natural choice is the mean of a number of measurements in the neighbourhood of τ. Define r_τ^* as the distance to the target in the B-direction at the reference time τ,

$$r_\tau^* = r_\tau \cos(\beta_\tau - B).$$

The coordinate system z' is defined as

$$\begin{aligned} z'_1 &= (v_x \cos B - v_y \sin B)/r_\tau^* \\ z'_2 &= (v_x \sin B + v_y \cos B)/r_\tau^* \\ z'_3 &= \tan(\beta_\tau - B) \\ z'_4 &= 1/r_\tau^*, \end{aligned} \quad (4)$$

We can see that the coordinate system (4) is very close to a rotated version of the modified polar coordinate system (3) since we choose $B \approx \beta_\tau$.

We shall also need the rotated own ship positions, bearings and bearing measurements for making the notation more compact,

$$p'_{xi} = p_{xi} \cos B - p_{yi} \sin B$$

$$
\begin{aligned}
p'_{yi} &= p_{xi}\sin B + p_{yi}\cos B \\
\beta'_i &= \beta_i - B \\
b'_i &= b_i - B
\end{aligned}
$$

Now we are able to define the criterion that is minimized by the MPPLE,

$$
\begin{aligned}
Q_{MPPLE} &= \sum_{i=1}^{n}\left(\frac{r_i}{r_\tau^*}\sin\varepsilon_i\right)^2 \qquad (5)\\
&= \sum_{i=1}^{n}\left(\frac{r_i}{r_\tau^*}\sin(b'_i-\beta'_i)\right)^2 \\
&= \sum_{i=1}^{n}\left(\frac{r_i}{r_\tau^*}(\sin b'_i\cos\beta'_i - \cos b'_i\sin\beta'_i)\right)^2 \\
&= \sum_{i=1}^{n}(\sin b'_i\,(1+(i-\tau)z'_2 - p'_{yi}z'_4) \\
&\qquad - \cos b'_i(z'_3 + (i-\tau)z'_1 - p'_{xi}z'_4))^2, .
\end{aligned}
$$

It is easily seen that the minimization of this criterion leads to a linear system of equations in the variables $[z'_1, z'_2, z'_3, z'_4]$.

In the next section we will see that using the mid time point as the time reference τ will give smaller bias than other choices of time reference. The estimator with the mid time point as time reference will be called the MPPLE$_m$, the one with the first time point as reference MPPLE$_f$ and the one with the endpoint time reference MPPLE$_e$.

4 The bias of the PLEs

In order to establish an approximate relationship between the CPLE and the MLE we consider all target tracks X that are close to the correct target track in a long range scenario. A long range scenario is a scenario where the individual motion of the target and the platform is much smaller than the distance between them. In such a scenario the range is approximately constant and we can use the approximation

$$
\begin{aligned}
Q_{CPLE} &= \sum_{i=1}^{n} r_i^2\sin^2\varepsilon_i \\
&\approx \bar{r}^2 Q_{MLE},
\end{aligned}
$$

where a bar indicates average over all measurement time points. The approximation $\sin(\varepsilon_i)\approx\varepsilon_i$ for small errors ε_i was also used.

We can now see that if Q_{MLE} is flat in the range direction, that is if there is great uncertainty about the range, then the CPLE underestimates the range since it can decrease the criterion through a lower range at the expence of only slightly higher Q_{MLE}. This is also noted by others through theory and simulations [2-4].

When doing the same calculation for the MPPLE we also note that $r_\tau^* \approx r_\tau$ and that in a long range scenario, by a Taylor expansion to the first degree, $r_i \approx r_\tau + (i-\tau)\dot{r}_\tau - p_{yi}$ (recall that $p_{y\tau}=0$). Then we can write

$$
\begin{aligned}
Q_{MPPLE} &= \sum_{i=1}^{n}(\frac{r_i}{r_\tau^*})^2\sin^2\varepsilon_i \\
&\approx (\frac{\overline{r_i}}{r_\tau})^2\sum_{i=1}^{n}\varepsilon_i^2 \\
&\approx (1-\frac{\overline{p'_{yi}}}{r_\tau} + (\bar{i}-\tau)z'_2)^2 Q_{MLE},
\end{aligned}
$$

where a bar again indicates average over all measurement time points. Unfortunately we can in general ***not*** use the approximation $(\bar{i}-\tau)z'_2 \ll 1$. The reason for this, which can not be deduced from these scetchy calculations, is that Q_{MLE} is very flat in the z'_2-direction in a long range scenario.

We now see that if we use the mid time point as time reference τ, then $\bar{i}=\tau$ and

$$
Q_{MPPLEm} \approx Q_{MLE}, \qquad (6)
$$

where we also used $p'_{yi} \ll r_\tau$. These calculations say that we might hope for the MPPLE$_m$ to perform almost as good as the MLE in a long range scenario, contrary to the CPLE, MPPLE$_f$ or MPPLE$_e$. We might also expect the loglikelihood function to be approximately quadratic in modified polar coordinates with mid time point reference.

In figure 1 we can see the result of a simulation study using the MLE, the CPLE, the MPPLE$_f$, the MPPLE$_m$ and the MPPLE$_e$. The own ship track was a three leg track ending at the origin, velocity 2 m/s. In all pictures the target had a velocity of 6 m/s, ending 20000 meters north of the own ship. The target courses were 90, 150 and 210 degrees respectively. In each picture the results from 20 simulations are shown. The scenarios lasted 20 minutes and measurements were simulated every 10th second and contaminated with white Gaussian noise with a standard deviation of 0.3 degrees. The estimates from four different Pseudo Linear Estimators are plotted against the Maximum Likelihood Estimate for every simulation. The three pictures in figure 1 are typical for a great number of simulations and it can be concluded that the MPPLE$_m$ is the only PLE out of these four that can compete with the MLE.

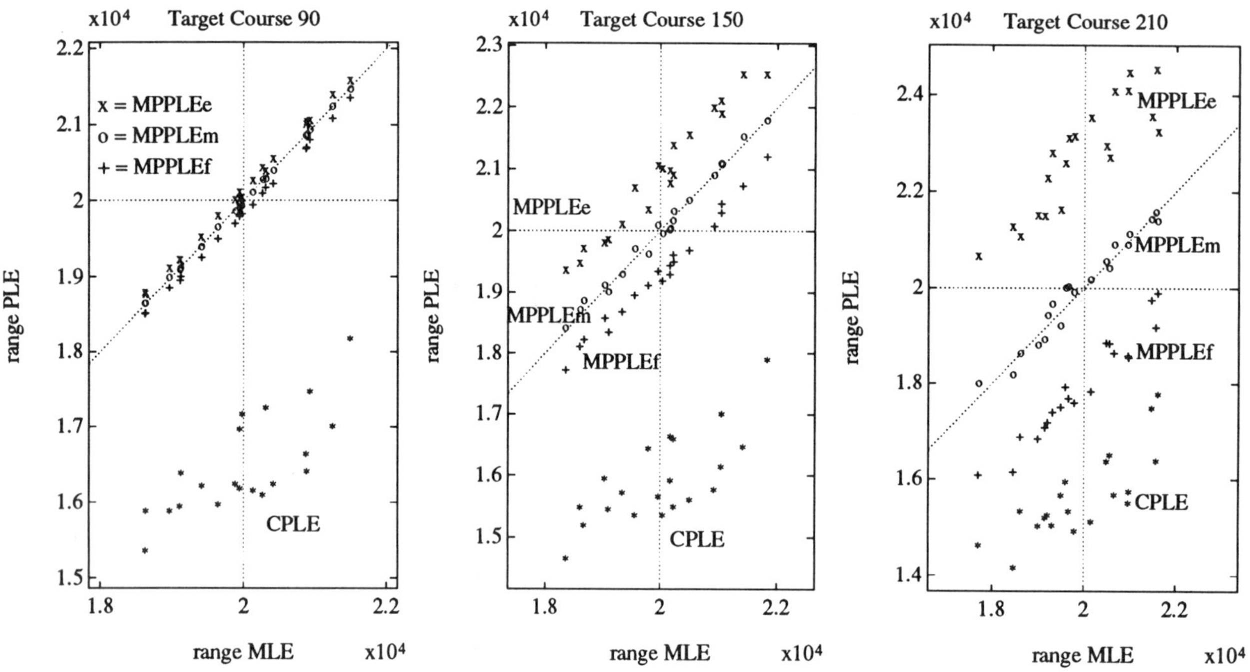

Figure 1: A simulation study that compares the PLEs to the MLE. In each of the three scenarios 20 simulations were carried out. The range at the end time point was estimated by four different Pseudo Linear Estimators. The estimates are plotted against the Maximum Likelihood Estimate for every simulation. The true range was 20000 meters and the own ship made a three leg track. It is seen that the $MPPLE_m$ is by far the best PLE both in the sense that it is closest to the true range and in the sense that it is closest to the MLE.

5 Confidence Intervals

A bearings only tracker must have methods for evaluating the uncertainty of a target parameter estimate. Displaying the standard deviation of a point estimate is common. One often used method for computing the standard deviation of a point estimate is via a Gaussian approximation. Sometimes confidence limits are presented with the underlying assumption that the distribution of estimation error is approximately Gaussian. This often works well for range and velocity but unfortunately the distribution of the error in the target course estimate is seldom close to Gaussian.

Inspired by (6) we will instead use the approximation that the errors in the modified polar coordinates are Gaussian for constructing a confidence interval for the course. The target course c is

$$c = \arctan(\frac{z_1'}{z_2'}) + \arctan(z_3') + B,$$

A conservative approximation will be used. A two dimensional confidence ellipsoid of z_1' and z_2' is covered by a sector as is illustrated in the following picture.

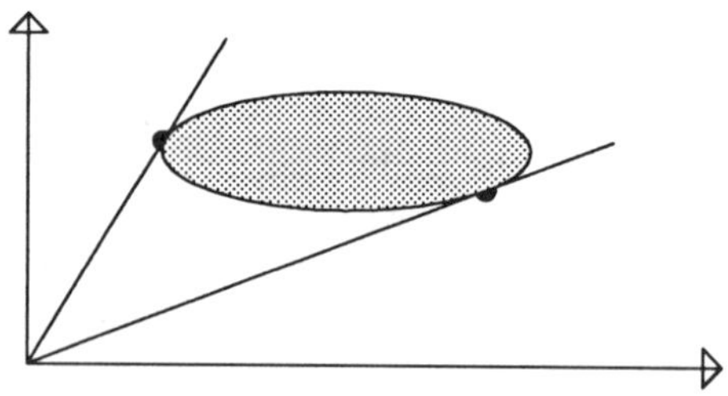

The variance of z_3' is small in comparison with the variance of z_1' and z_2' and is therefore neglected. The details of the outlined calculations may be found in [6].

Simulations were carried out to investigate the performance of this confidence interval. The same scenarios as in figure 1 were used. In each scenario 1000 independent noise sequences were generated and for every such noise sequence the estimate using the $MPPLE_m$

was computed along with its variance matrix based on the normal equations associated with the minimization of (5), see [7]. Then for every estimate the confidence interval for the target course was computed and it was noted if the confidence interval covered the correct course or not. The confidence intervals were aiming at 95 percent confidence. About 25 misses on each side should then be expected. The following table shows the number of times the confidence interval underestimated (upper limit below correct value) and overestimated (vice versa) the correct parameter.

```
------- TARGET COURSE --------
            underest.   overest.
  90 deg:    34 (299)   34 (371)
 150 deg:    27   (0)   41   (0)
 210 deg:    37   (7)   23   (0)
------------------------------
```

The numbers inside the paranthesis show the number of misses for confidence intervals based on direct Gaussian approximation of the course. For the 90 degree scenario the direct Gaussian approximation gives intervals that are far too small, and for the other two scenarios too large.

The result indicates that a confidence interval for the target course based on normality of errors in MPC is better than a direct Gaussian approximation of the target course estimation error.

The same simulations gave the following results for the range estimate at the mid time point.

```
---MID TIME POINT RANGE ------
            underest.   overest.
  90 deg:    22   (9)   21  (29)
 150 deg:    25  (10)   24  (33)
 210 deg:    23   (8)   23  (31)
------------------------------
```

The numbers inside the paranthesis show the number of misses for confidence intervals based on direct Gaussian approximation of the range, with $\pm 2\sigma$-limits. The intervals based on normality of the inverse of the range show perfect symmetry while the direct Gaussian approximation of the range gives scewed intervals. If one instead investigates the range at the end time point the normality property of the inverse of the range is somewhat less perfect.

6 Summary and Conclusions

We have investigated a pseudo linear estimator based on modified polar coordinates. The performance of it is close to that of the maximum likelihood estimator. We have shown that a confidence interval for the target course based on normality of errors in MPC is better than a direct Gaussian approximation of the target course estimation error.

Acknowledgements

This research has been done in cooperation with the Swedish Defence Material Administration and Kockums AB. The cooperation and support is greatfully acknowledged.

References

[1] V. J. Aidala and S. E. Hammel, "Utilization of Modified Polar Coordinates for Bearings-Only Tracking," *IEEE Trans. Automat. Contr.*, Vol. AC-28, pp. 283-294, 1983.

[2] V. J. Aidala and S. C. Nardone, "Biased estimation properties of the pseudolinear tracking filter," *IEEE Trans. Aerosp. Electron. Syst.*, vol AES-18, pp. 432-441, July 1982.

[3] J. Holst, *A note on a least squares method for bearings-only tracking,* TFMS-3047, Dept. of Math. Stat., Lund Inst. of Techn., Lund, Sweden, Jan. 1988.

[4] S. C. Nardone, *Fundamental Properties and Performance of Bearings-Only Target Localization and Motion Analysis,* Ph. D. Thesis, University of Rhode Island, 1982.

[5] O. Carlstedt, *Simulering av taktisk presentation och målfaktorberäkning i ubåt A14* (Simulation of Tactical Presentation and Target Tracking in U-boat A14), Dept. of Information Technology, National Defence Research Establishment, Sweden, Dec. 1970. Classified.

[6] A. Holtsberg, J. Holst, *Confidence Intervals in Bearings Only Tracking,* TFMS-3071, Dept. of Math. Stat., Lund Inst. of Techn., Lund, Sweden, Jan. 1990.

[7] L. Ljung, *System Identification, Theory for the user,* Prentice-Hall, 1987.

On the Steady State Solution of a Two-By-Two Dynamic Jamming Game with Cumulative Power Constraints *

Ranjan K. Mallik, Robert A. Scholtz and George P. Papavassilopoulos
Communication Sciences Institute
Department of Electrical Engineering-Systems
University of Southern California
Los Angeles, CA 90089-2565

Abstract

The process of jamming can be modelled as a two-person zero-sum non-cooperative dynamic game played between a communicator and a jammer over a number of discrete time instants. The simplest case is when, at each instant, the communicator and jammer randomize their strategies between idleness and transmission. The payoff (throughput) matrix is then two-by-two, with one variable parameter. The payoff function is the average throughput summed over time, to be optimized subject to cumulative power constraints. We find an analytical steady-state solution for this game played over an infinite time duration. Results show that when the throughput parameter is lower than a threshold, the optimal strategies are mixed, and the payoff increment constant; otherwise the strategies are pure, with the payoff increment exhibiting oscillatory behavior.

1 The Dynamic Jamming Game Model

The process of communication jamming can be modelled as a two-person zero-sum non-cooperative game played between a *communicator* and a *jammer*. In [Pen86][PS86], the static case, in which the game is played as a one-shot process, was analyzed. However, in most real situations, jamming is a continuous operation performed over time. Therefore in this paper, we consider the game being played over T discrete time instants, which we shall denote as $1,\ldots,T$. The game is dynamic in the sense that *the information and hence the action of each player at every time instant are influenced by the previous actions of both players.* We focus on the particular case in which the communicator's and jammer's *strategies are characterized by the power levels* of their transmitted signals.

Let X_{t_f} be the *communicator's power level at forward time* t_f, and Y_{t_f} the *jammer's power level at time* t_f, for $t_f = 1,2,\ldots,T$. We assume that X_{t_f} is *randomly distributed over two discrete values* $0, P$ ($P > 0$), and Y_{t_f} over the values $0, J$ ($J > 0$). Associated with each pair (X_{t_f}, Y_{t_f}) is a payoff $f(X_{t_f}, Y_{t_f})$ to the communicator, and $-f(X_{t_f}, Y_{t_f})$ to the jammer.

*This research was funded in part by the U.S. Army Research Office under Contract No. DAAL03-88-K-0059.

Hence we have a 2×2 *payoff matrix*

$$G \triangleq \begin{bmatrix} f(0,0) & f(0,J) \\ f(P,0) & f(P,J) \end{bmatrix}.$$

A reasonable measure of the payoff is the normalized instantaneous throughput, which gives rise to the following *payoff matrix* :

$$G = \begin{bmatrix} 0 & 0 \\ 1 & \alpha \end{bmatrix}, \; 0 \le \alpha \le 1. \tag{1}$$

G has only one variable parameter, α, which is the instantaneous throughput as a function of P and J. We shall call α the *throughput parameter.*

The *payoff function*, which we denote as $\mathcal{J}$, is the time average of the mean throughput $\mathbb{E}[f(X_{t_f}, Y_{t_f})]$, i.e.

$$\mathcal{J} \triangleq \frac{1}{T} \sum_{t_f=1}^{T} \mathbb{E}[f(X_{t_f}, Y_{t_f})]. \tag{2}$$

1.1 Constraints

To account for protection from overheating of the communicating (jamming) transmitter, we assume that the communicator's (jammer's) accumulated power at forward time t_f should not exceed a threshold $\hat{P}$ ($\hat{J}$). This gives rise to the following *cumulative power constraints*:

$$\sum_{\tau=1}^{t_f} \delta_1^{t_f-\tau} X_\tau \le \hat{P}, \quad \sum_{\tau=1}^{t_f} \delta_2^{t_f-\tau} Y_\tau \le \hat{J}, \tag{3}$$

$t_f = 1,2,\ldots,T, \quad 0 < \delta_1, \delta_2 < 1.$

Here δ_1, δ_2 are *thermal memory constants,* $\hat{P}$ is the *communicator's peak power accumulation*, and $\hat{J}$ is the *jammer's peak power accumulation* .

The constraints in (3) are effective only when

$$P \le \hat{P} < \frac{P(1-\delta_1^T)}{(1-\delta_1)}, \quad J \le \hat{J} < \frac{J(1-\delta_2^T)}{(1-\delta_2)} \tag{4}$$

are satisfied, and hence these define the interesting ranges of transmitter parameters.

1058-6393/91 $01.00 © 1991 IEEE

Let Z_{t_f} (W_{t_f}) be a random variable representing the *communicator's (jammer's) past power accumulation at forward time* t_f. Thus from (3), $Z_1 \triangleq 0$, $W_1 \triangleq 0$, and

$$Z_{t_f} \triangleq X_{t_f-1} + \delta_1 Z_{t_f-1}, \quad W_{t_f} \triangleq Y_{t_f-1} + \delta_2 W_{t_f-1}, \tag{5}$$

for $t_f = 2, \ldots, T$.

1.2 Conditional Independence Assumption

We assume that (a) each player has knowledge of the opponent's previous actions, (b) X_1 and Y_1 are independent, and (c) for $t_f = 2, \ldots, T$, X_{t_f} and Y_{t_f} are conditionally independent, given $X_1, \ldots, X_{t_f-1}$, $Y_1, \ldots, Y_{t_f-1}$ (and therefore given Z_{t_f}, W_{t_f}).

Thus we have a *dynamic stochastic game model* with a *Markovian evolution* (see [BO82]), in which (Z_{t_f}, W_{t_f}) is the value of the *state of the system at forward time* t_f. The state equation is given by (5).

1.3 Effect of Constraints

For the communicator, at each $t_f = 2, \ldots, T$, we define the *set of past power accumulations* to be

$$\Phi_{t_f} \triangleq \{ z : z = \sum_{\tau=1}^{t_f-1} \delta_1^{t_f-1-\tau} x_\tau, \quad \sum_{\tau=1}^{t'_f} \delta_1^{t'_f-\tau} x_\tau \le \hat{P},\ t'_f = 1, .., t_f-1, \quad x_1, \ldots, x_{t_f-1} \in \{0, P\} \}. \tag{6}$$

Note that $\Phi_1 \triangleq \{0\}$ and $\Phi_2 = \{0, P\}$.

For each $z \in \Phi_1 \cup \cdots \cup \Phi_T$, we define $\mathcal{X}_z$ to be the *communicator's set of allowable power levels at any time instant given that it has accumulated z amount of power in the past.* Thus

$$\mathcal{X}_z \triangleq \{x : x \in \{0, P\}, x + \delta_1 z \le \hat{P}\}. \tag{7}$$

Similarly, for the jammer, constraints (3) determine the analogous sets Ψ_{t_f} and $\mathcal{Y}_w$.

1.4 The Strategies and the Payoff

Let $p_t(x|z,w)$ ($q_t(y|z,w)$) denote the *probability that the communicator (jammer) selects power level x (y) at reverse time t (i.e., forward time $T-t$), given that the communicator and jammer have accumulated z and w amounts of power respectively in the past,* where $x \in \mathcal{X}_z$, $y \in \mathcal{Y}_w$, for each $z \in \Phi_{T-t}$, $w \in \Psi_{T-t}$, $t = 0, \ldots, T-2$. We can then define the *strategy vectors*

$$\mathbf{p}_t(z,w) \triangleq [\cdots p_t(x|z,w) \cdots]', \tag{8a}$$

$$\mathbf{q}_t(z,w) \triangleq [\cdots q_t(y|z,w) \cdots]'. \tag{8b}$$

Let the *initial power selection probabilities* (at reverse time $T-1$) be

$$p_{T-1}(x) \triangleq \text{Prob}(X_1 = x), \quad x \in \Phi_2 = \{0, P\}, \tag{9a}$$

$$q_{T-1}(y) \triangleq \text{Prob}(Y_1 = y), \quad y \in \Psi_2 = \{0, J\}. \tag{9b}$$

The *initial strategy vector* $\mathbf{p}_{T-1}$ for the communicator can then be defined as

$$\mathbf{p}_{T-1} \triangleq [p_{T-1}(0) \ \ p_{T-1}(P)]'. \tag{10}$$

We define the analogous vector $\mathbf{q}_{T-1}$ for the jammer.

The payoff $\mathcal{J}$ in (2) can be expressed in terms of the last term of a sequence $\{S_0, S_1, \ldots, S_{T-1}\}$, defined as follows:

$$\begin{aligned}
S_0 &= \mathbb{E}_{X_T, Y_T | Z_T, W_T}[f(X_T, Y_T)], && \text{(11a)}\\
S_t &= \mathbb{E}_{X_{T-t}, Y_{T-t} | Z_{T-t}, W_{T-t}} [f(X_{T-t}, Y_{T-t}) + S_{t-1}], && \text{(11b)}\\
&\quad \text{for } t = 1, \ldots, (T-2), \\
S_{T-1} &= \mathbb{E}_{X_1, Y_1}[f(X_1, Y_1) + S_{T-2}], \text{ and} && \text{(11c)}\\
\mathcal{J} &= \frac{S_{T-1}}{T}. && \text{(11d)}
\end{aligned}$$

S_t is thus the *accumulated payoff at reverse time t given the past power accumulations Z_{T-t} and W_{T-t}.* Equation (11) can be written in terms of the strategies defined in (8) and (10).

Let $\mathcal{P}_t$ be *the set of all strategy vectors for the communicator at reverse time t, each vector being conditioned upon a past power accumulation $z \in \Phi_{T-t}$ and a $w \in \Psi_{T-t}$, and having a dimension of* $|\mathcal{X}_z|$, for $t = 0, \ldots, T-2$. Thus

$$\mathcal{P}_t \triangleq \{ \mathbf{p}_t(z,w) : z \in \Phi_{T-t}, w \in \Psi_{T-t} \}. \tag{12}$$

The set $\mathcal{P}_t$ contains $|\Phi_{T-t}|.|\Psi_{T-t}|$ probability vectors.

Similarly, for the jammer, we can define the sets $\mathcal{Q}_0, \ldots, \mathcal{Q}_{T-2}$ of strategy vectors.

The *communicator's strategy set* Γ and the *jammer's strategy set* Δ, given by

$$\Gamma \triangleq \{\mathbf{p}_{T-1}\} \cup \mathcal{P}_{T-2} \cup \cdots \cup \mathcal{P}_0,$$

$$\Delta \triangleq \{\mathbf{q}_{T-1}\} \cup \mathcal{Q}_{T-2} \cup \cdots \cup \mathcal{Q}_0,$$

then determine the payoff $\mathcal{J}$.

1.5 Characterization of the Solution

Solving the dynamic game subject to the cumulative power constraints in (3) reduces to the following problem:

Find Γ^*, Δ^* *such that*

$$\mathcal{J}(\Gamma, \Delta^*) \le \mathcal{J}(\Gamma^*, \Delta^*) \le \mathcal{J}(\Gamma^*, \Delta) \quad \forall\ \Gamma, \Delta.$$

Then $\mathcal{J}^* = \mathcal{J}(\Gamma^*, \Delta^*)$ is the *value of the game* and (Γ^*, Δ^*) is the *set of optimal strategies* (see [BG54]).

1.6 The Evolution Equation

When the duration T is finite, dynamic programming (see [Ber76]) can be applied to find $\mathcal{J}^*$ (see (11)). For reverse time $t = 0, 1, 2, \ldots, (T-1)$, we optimize $S_t(z,w)$ with respect to $\mathbf{p}_t(z,w)$ and $\mathbf{q}_t(z,w)$ to obtain $\mathbf{p}_t^*(z,w)$, $\mathbf{q}_t^*(z,w)$ and $S_t^*(z,w)$. Each optimization process involves solving a $|\mathcal{X}_z| \times |\mathcal{Y}_w|$ matrix game, where $z \in \Phi_{T-t}$, $w \in \Psi_{T-t}$.

In terms of the optimum payoffs, equation (11) gives rise to the following evolution equations:

$$S_0^*(z,w) = \max_{\mathbf{p}_0(z,w)} \min_{\mathbf{q}_0(z,w)} \sum_{x\in\mathcal{X}_z} \sum_{y\in\mathcal{Y}_w} p_0(x|z,w) q_0(y|z,w) f(x,y) \quad (13)$$

$z \in \Phi_T$, $w \in \Psi_T$,

$$S_{t+1}^*(z,w) = \max_{\mathbf{p}_{t+1}(z,w)} \min_{\mathbf{q}_{t+1}(z,w)} \sum_{x\in\mathcal{X}_z} \sum_{y\in\mathcal{Y}_w} p_{t+1}(x|z,w) q_{t+1}(y|z,w) [f(x,y) + S_t^*(x+\delta_1 z, y+\delta_2 w)] \quad (14)$$

$z \in \Phi_{T-t-1}$, $w \in \Psi_{T-t-1}$, for $t = 0, 1, \ldots, (T-2)$.

From (7) it is clear that

$$\mathcal{X}_z = \begin{cases} \{0, P\} & \text{if } 0 \le z \le \frac{\hat{P}-P}{\delta_1}, \\ \{0\} & \text{if } \frac{\hat{P}-P}{\delta_1} < z \le \hat{P}. \end{cases} \quad (15)$$

We have an analogous expression for $\mathcal{Y}_w$.

2 Simplification of the Problem

2.1 The Payoff Function

From the structures of $\mathcal{X}_z$ (see (15)) and $\mathcal{Y}_w$, it can be shown that when the conditions

$$\frac{1}{(1-\delta_1^2)} < \frac{\hat{P}}{P} < (1+\delta_1), \quad \frac{1}{(1-\delta_2^2)} < \frac{\hat{J}}{J} < (1+\delta_2) \quad (16)$$

are satisfied, a 2×2 *grid solution* to (14) exists in which $S_t^*(z,w)$ is a *four-valued function* defined as follows:

$$S_t^*(z,w) = \begin{cases} S_{11}^t & \text{if } 0 \le z \le \frac{\hat{P}-P}{\delta_1}, \quad 0 \le w \le \frac{\hat{J}-J}{\delta_2} \\ S_{12}^t & \text{if } 0 \le z \le \frac{\hat{P}-P}{\delta_1}, \quad \frac{\hat{J}-J}{\delta_2} < w \le \hat{J} \\ S_{21}^t & \text{if } \frac{\hat{P}-P}{\delta_1} < z \le \hat{P}, \quad 0 \le w \le \frac{\hat{J}-J}{\delta_2} \\ S_{22}^t & \text{if } \frac{\hat{P}-P}{\delta_1} < z \le \hat{P}, \quad \frac{\hat{J}-J}{\delta_2} < w \le \hat{J}. \end{cases} \quad (17)$$

for $t = 1, \ldots, (T-1)$. From (13) the solution at reverse time $t = 0$ is found to be

$$S_{11}^0 = \alpha, \; S_{12}^0 = 1, \; S_{21}^0 = S_{22}^0 = 0. \quad (18)$$

The inequalities in (16) determine the *operating regions of the players* in the "$\frac{\hat{P}}{P}$ *versus* δ_1" and "$\frac{\hat{J}}{J}$ *versus* δ_2" planes respectively, for which this form of solution exists.

2.2 Optimal Strategies

Let

$$p_t^*(P|z,w) = \begin{cases} p_{11}^t & \text{if } 0 \le z \le \frac{\hat{P}-P}{\delta_1}, \; 0 \le w \le \frac{\hat{J}-J}{\delta_2} \\ p_{12}^t & \text{if } 0 \le z \le \frac{\hat{P}-P}{\delta_1}, \; \frac{\hat{J}-J}{\delta_2} < w \le \hat{J} \\ p_{21}^t & \text{if } \frac{\hat{P}-P}{\delta_1} < z \le \hat{P}, \; 0 \le w \le \frac{\hat{J}-J}{\delta_2} \\ p_{22}^t & \text{if } \frac{\hat{P}-P}{\delta_1} < z \le \hat{P}, \; \frac{\hat{J}-J}{\delta_2} < w \le \hat{J}. \end{cases} \quad (19)$$

Similarly we can define q_{ij}^t's corresponding to $q_t^*(J|z,w)$. The p_{ij}^t's and q_{ij}^t's are the *optimum probabilities of transmission at reverse time t*. Using (8) and (15), we can show that

$$p_{22}^t = p_{21}^t = 0, \quad q_{22}^t = q_{12}^t = 0.$$

2.3 The Simplified Evolution Equation

The quantities z and w are measures of the *communicator's and jammer's power accumulations.* As z increases, the communicator transmits less frequently (i.e., with lower probability), and the throughput decreases. Therefore, for any reverse time index t, $S_t^*(z,w)$ *decreases as z increases.* On the other hand, an increase in w causes the jammer to transmit less frequently, which increases the throughput. Thus $S_t^*(z,w)$ *increases as w increases.*

Substitution of the payoff function of (17) in (14) results in four equations, one for each region of the (z,w) plane. The quantity S_{11}^{t+1} is the value of a 2×2 game whose payoff matrix depends on the S_{ij}^t's. In the other three regions, we need to solve a 2×1, a 1×2 and a 1×1 game, all of which are trivial. When we take into account the fact that $S_t^*(z,w)$ decreases (increases) with increase in z (w), we obtain the conditions

$$S_{12}^t \ge S_{11}^t \ge S_{21}^t = S_{22}^t, \quad S_{11}^t \le (1+S_{21}^t),$$

and (14) simplifies to

$$S_{11}^{t+1} = \text{value}\left(\begin{bmatrix} S_{11}^t & S_{12}^t \\ 1+S_{21}^t & \alpha+S_{22}^t \end{bmatrix}\right) \quad (20a)$$

$$S_{12}^{t+1} = \max(S_{11}^t, 1+S_{21}^t) = 1+S_{21}^t \quad (20b)$$

$$S_{21}^{t+1} = \min(S_{11}^t, S_{12}^t) = S_{11}^t \quad (20c)$$

$$S_{22}^{t+1} = S_{11}^t \quad (20d)$$

Also from (19)

$$\max(S_{11}^t, 1+S_{21}^t) = 1+S_{21}^t \implies p_{12}^t = 1,$$
$$\min(S_{11}^t, S_{12}^t) = S_{11}^t \implies q_{21}^t = 0. \quad (21)$$

3 The Steady State Solution

For a given finite time duration T, the evolution equation (20) can be solved using the initial conditions in (18) to obtain the optimal strategies p_{11}^t, q_{11}^t for $t = 1, \ldots, (T-1)$. The limiting behavior of the optimal strategies when the reverse time index goes to infinity is of particular interest, since it gives us a better analytical feel of the problem. So we attempt to solve the evolution equation for $S_t^*(z,w)$ when $t \to \infty$.

3.1 The Payoff Increments

From (14), it is clear that the *optimum payoff* S_{ij}^t ($i,j \in \{1,2\}$) increases as t increases, but its *increment* (in going from t to $(t+1)$) is bounded and lies in [0,1], since each element of the payoff matrix (see (1)) lies in [0,1].

Let

$$\lambda_{ij}^t \triangleq S_{ij}^{t+1} - S_{ij}^t, \quad i=1,2, \quad j=1,2, \quad t=0,1,2,\ldots \quad (22)$$

The condition $0 \le \lambda_{ij}^t \le 1$ holds.

Substituting $S_{ij}^{t+1} = S_{ij}^t + \lambda_{ij}^t$ in (20) and eliminating S_{11}^t, S_{12}^t and S_{22}^t, the equations simplify to

$$\lambda_{11}^t + \lambda_{22}^t = \text{value}\left(\begin{bmatrix} \lambda_{22}^t & 1 - \lambda_{12}^t \\ 1 & \alpha \end{bmatrix}\right). \quad (23)$$

Note that $S_{21}^t = S_{22}^t$ implies $\lambda_{21}^t = \lambda_{22}^t$.

If we *assume that the optimal strategies exhibit steady state behavior* as $t \to \infty$, i.e.,

$$\lim_{t\to\infty} p_{11}^t = p_{11}\,, \quad \lim_{t\to\infty} q_{11}^t = q_{11}\,,$$

exist, then from (20), we get

$$\lambda_{11}^{t+1} = \text{value}\left(\begin{bmatrix} \lambda_{11}^t & \lambda_{12}^t \\ \lambda_{22}^t & \lambda_{22}^t \end{bmatrix}\right) \quad (24a)$$

$$\lambda_{12}^{t+1} = \lambda_{22}^t \quad (24b)$$

$$\lambda_{22}^{t+1} = \lambda_{11}^t. \quad (24c)$$

3.2 Case 1: Mixed Strategies

Assume that the strategies p_{11} and q_{11} are mixed, i.e. $0 < p_{11}, q_{11} < 1$. We can show from (24) that this implies

$$\lambda_{11}^t = \lambda_{12}^t = \lambda_{21}^t = \lambda_{22}^t \triangleq \lambda \quad \forall t \quad \text{when } t \to \infty. \quad (25)$$

Equation (23) then gives

$$\lambda = \frac{(5-\alpha) - \sqrt{(9-\alpha)(1-\alpha)}}{8}. \quad (26)$$

We see that λ increases as α increases. It can be shown that (25) holds true only when $(1-\lambda) \ge \alpha$.

When $(1-\lambda) = \alpha$, substitution in (26) gives $\alpha = \frac{2}{3}$, $\lambda = \frac{1}{3}$. Therefore, *(25) holds true for* $0 \le \alpha \le \frac{2}{3}$ *only*. In this range, λ increases from $\frac{1}{4}$ at $\alpha = 0$ to $\frac{1}{3}$ at $\alpha = \frac{2}{3}$.

The *steady state mixed strategies* are given by

$$p_{11} = \frac{(1-2\lambda}{2(1-\lambda)-\alpha}, \quad q_{11} = \frac{(1-\lambda}{2(1-\lambda)-\alpha}, \quad (27)$$

where λ is given by (26). Thus p_{11} and q_{11} are functions of α.

3.3 Case 2: Pure Strategies

When we assume that the strategies p_{11} and q_{11} are pure, they take values 0 or 1 only. The range of α for this case is $\frac{2}{3} < \alpha \le 1$. It can be shown that only $p_{11} = q_{11} = 1$ may be possible.

Assuming $p_{11} = q_{11} = 1$, we get the following result:
When $t \to \infty$, if, for any reverse-time index t

$$\lambda_{11}^t = \lambda_{12}^t = \lambda_1\,, \quad \lambda_{22}^t = \lambda_{21}^t = \lambda_2\,,$$

then, at the next reverse-time index $(t+1)$, we have

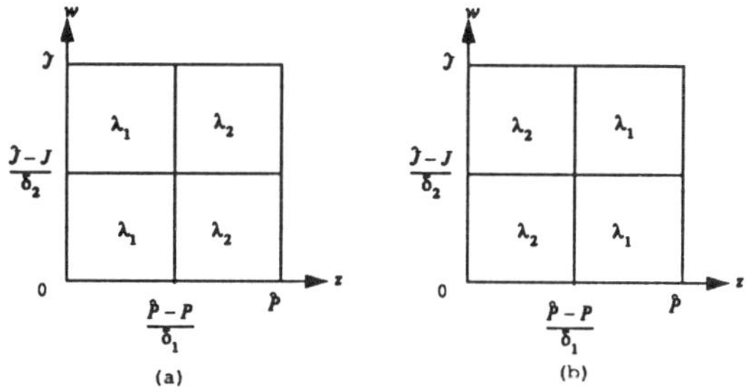

Figure 1: Oscillatory Behavior of Payoff Increment in Steady State

$$\lambda_{11}^{t+1} = \lambda_{12}^{t+1} = \lambda_2\,, \quad \lambda_{22}^{t+1} = \lambda_{21}^{t+1} = \lambda_1\,.$$

The λ_{ij}^t plots on the (z, w) plane are shown in Fig 1. The plot *oscillates* between the pattern in Fig 1a and that in Fig 1b. The *period of oscillation* in terms of the number of reverse time instants is 2.

The next step is to find the steady state payoff increments λ_1 and λ_2 in terms of α. At reverse-time t, let $\lambda_{11}^t = \lambda_{12}^t = \lambda_1$, $\lambda_{22}^t = \lambda_2$. It can be shown that we can have

$$\lambda_1 = (1-\alpha)\,, \quad \lambda_2 = (2\alpha - 1) \quad (28a)$$

$$\text{or} \quad \lambda_1 = (2\alpha - 1)\,, \quad \lambda_2 = (1-\alpha). \quad (28b)$$

Note that *(28) holds true for* $\frac{2}{3} < \alpha \le 1$ *only*. The *steady state pure strategies* are given by

$$p_{11} = 1\,, \quad q_{11} = 1\,. \quad (29)$$

Here the strategies do not vary with variation in α as long as α lies in the range $(\frac{2}{3}, 1]$.

3.4 Results

The optimum steady state transmission probabilities (p_{ij}'s and q_{ij}'s) tell us how the strategies of the players vary with changes in the amounts of power accumulated (z and w) at any time instant as the game progresses. We find that when the communicator's power accumulation (z) exceeds $\frac{\hat{P}-P}{\delta_1}$, it is forced to remain idle (i.e. $p_{21} = p_{22} = 0$), because if it does not, the present power accumulation $x + \delta_1 z$ will exceed $\hat{P}$, violating the power constraint. Since the jammer has nothing to jam in this situation, it remains idle too (i.e. $q_{21} = q_{22} = 0$). When the jammer's power accumulation (w) exceeds $\frac{\hat{J}-J}{\delta_2}$, but the communicator's does not, the jammer remains idle ($q_{12} = 0$) to keep the present power accumulation $y + \delta_2 w$ below $\hat{J}$, while the communicator takes advantage of the situation by transmitting with 100% probability ($p_{12} = 1$) and achieving success.

When z and w satisfy $0 \le z \le \frac{\hat{P}-P}{\delta_1}$ and $0 \le w \le \frac{\hat{J}-J}{\delta_2}$, the communicator and jammer transmit with probabilities p_{11} and q_{11} respectively. These probabilities define the strategies of both players, which depend upon α.

Fig 2 is a plot of p_{11} and q_{11} versus α. *For* $0 \leq \alpha \leq \frac{2}{3}$ *the strategies are mixed, i.e.* $0 < p_{11}, q_{11} < 1$. The probabilities increase as α increases. *For* $\frac{2}{3} < \alpha \leq 1$ *the strategies are pure, i.e.* $p_{11} = q_{11} = 1$. Note that in the mixed strategy zone, $p_{11} < q_{11}$, implying that the communicator has to transmit less frequently compared to the jammer to attain optimum throughput. As a result, there is a jump in the p_{11} curve at $\alpha = \frac{2}{3}$ from $\frac{1}{2}$ to 1. The q_{11} curve is continuous.

The ***steady state payoff increment*** is a single quantity λ (see (25)) when $0 \leq \alpha \leq \frac{2}{3}$, and oscillates between two quantities λ_1 and λ_2 (see (28)) when $\frac{2}{3} < \alpha \leq 1$. A plot of these quantities is shown in Fig 3. As α increases, λ increases from $\frac{1}{4}$ at $\alpha = 0$ to $\frac{1}{3}$ at $\alpha = \frac{2}{3}$, after which it *splits* into two quantities $\lambda_1 = (1 - \alpha)$ and $\lambda_2 = (2\alpha - 1)$. One decreases to 0 at $\alpha = 1$ whereas the other increases to 1.

4 Conclusion

When the throughput parameter α is low, the communicator has a greater need to transmit in order to maximize the payoff. The cumulative power constraint prevents him from transmission at full power with 100% probability at all times, as a result he uses a randomized mixed strategy so that he can transmit most of the time. To counteract the communicator, the jammer uses a mixed strategy too. The payoff increment, which is a measure of the average increase in throughput, also remains constant at all times in steady state because of the possible presence of the same mixed strategy at all times. On the other hand, when the throughput parameter is high (closer to 1), the communicator need not bother so much about transmitting at all times with mixed strategies. So he uses a pure strategy at one time, but this forces him into idleness at the next time instant, to satisfy the power constraint. Again, at the next time instant, he transmits with probability 1, and the phenomenon is repeated. To counteract the communicator, the jammer does the same thing. As a result, the payoff increment oscillates between a high value and a low value. This explains the mathematical results obtained.

It must also be noted that the 2×2 grid structure of the optimum accumulated payoff (see (17)) is not the only possible one that yields a steady state solution in terms of payoff increments and optimal strategies. It covers only small portions in the $\frac{\hat{P}}{P}$ *versus* δ_1 and $\frac{\hat{J}}{J}$ *versus* δ_2 planes (see (16)). In this paper we have discussed the simplest of the infinitely many steady state solutions of this game.

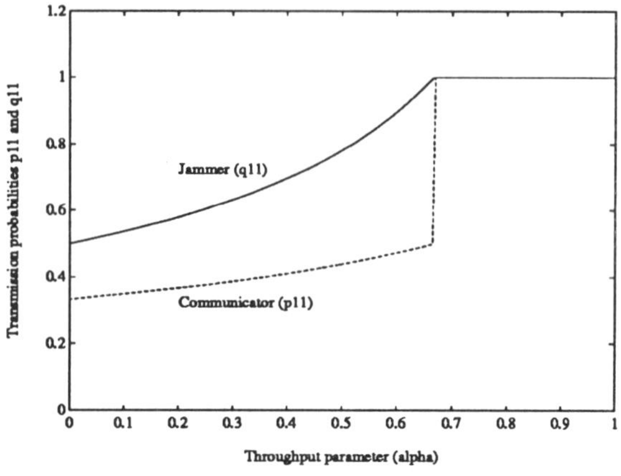

Figure 2: Steady State Transmission Probability versus Throughput Parameter α

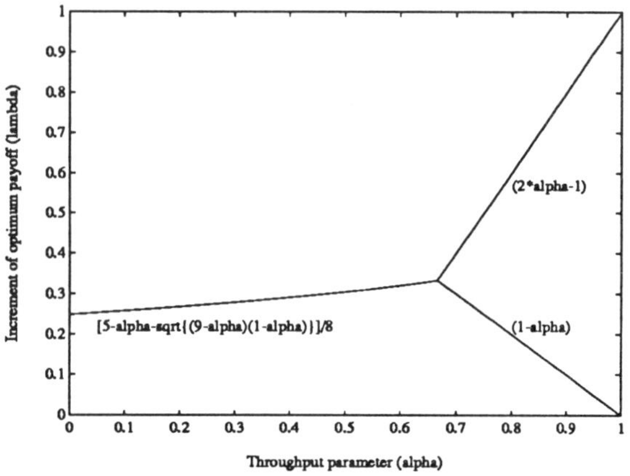

Figure 3: Steady State Payoff Increment versus Throughput Parameter α

References

[Ber76] D. P. Bertsekas. *Dynamic Programming and Stochastic Control.* Academic Press, 1976.

[BG54] D. Blackwell and M. A. Girshick. *Theory of Games and Statistical Decisions.* Dover Pub., 1954.

[BO82] T. Basar and G. J. Olsder. *Dynamic Non-Cooperative Game Theory.* Academic Press, 1982.

[Pen86] Wei-Chung Peng. *Some Communication Jamming Games.* PhD thesis, University of Southern California, 1986.

[PS86] Wei-Chung Peng and Robert A. Scholtz. A Study of Discrete Link Jamming Games. In *MILCOM Record, Vol. 1*, pages 2.6.1–2.6.10, 1986.

Optimum Active Array Shape Calibration

Boon C. Ng

Defense Science Organization

20, Science Park Drive,
Singapore 0511, Republic of Singapore.

Arye Nehorai*

Department of Electrical Engineering
Yale University
New Haven, CT 06520,
USA.

Abstract

This paper examines several array shape calibration algorithms based on maximum likelihood and eigenstructure methods using known calibration source signals. A compact expression for the corresponding Cramér-Rao bound (CRB) on the sensor location parameters is presented. For uncorrelated calibrating sources, a necessary and sufficient condition for the optimal choice of bearings that minimize the CRB is derived. Asymptotic covariances of the estimation errors for the calibration algorithms are also presented and compared with the CRB.

1 Introduction

In the problem of multiple source location, measurements are usually taken from spatially distributed sensors and processed to determine the direction of arrivals of source signals impinging on the array of sensors. In most modern array processing algorithms [1, 2, 8, 12], tha array manifold are assumed to be known analytically or otherwise calibrated and stored in a table. It is clear that this would require the sensor locations to be known precisely. In practice, it is rare that sensor arrays (e.g. hydrophones or antennas) can be deployed with the sensor positions known exactly.

In the presence of sensor position errors, the performances of these array processing algorithms will degrade. There are two approaches to this problem. The first approach uses self-calibrating techniques [3, 4] which do not require calibrating sources. Although attractive, such techniques usually require the knowledge of the location of one sensor and the direction to a second sensor. Furthermore, they suffer from ambiguity problems when the sensors are collinear. The second approach uses calibrating sources with known bearings [6, 7]. This paper adopts the second approach and assumes that the waveforms of the calibrating sources are known. This added assumption is not unreasonable, since in practice there is usually some control over the kind of signals that are used for these calibrating sources. Furthermore, the performance is expected to be better than the case with unknown signal waveforms. We will call this the *active array shape calibration problem.*

In this paper, we will present the Cramér-Rao bound (CRB) on the sensor position errors for the active array shape calibration problem in a compact form. The CRB will be used to reveal information on how to select the number of sources and choose the source bearings so that the best performance can be achieved. We will present Newton-type algorithms for the maximum likelihood estimator as well as eigenstructure-based methods such as MUSIC [1] and MODE [8]. Asymptotic covariances for the ML and eigenstructure-based methods will also be presented.

*The work of A. Nehorai was supported by the US Air Force Office of Scientific Research, under Grant No. AFOSR-90-0164, and the Office of US Naval Research under Grant No. N00014-91-J-1298.

2 The Data Model

Consider an array of m omnidirectional sensing elements in a 2-D plane with n far-field narrowband sources at center frequency f_o. The received signal vector can be written as

$$\mathbf{y}(t) = \mathbf{A}(\omega)\mathbf{s}(t) + \mathbf{e}(t) \tag{1}$$

where

$$\omega = [x_1, x_2, \ldots, x_m, y_1, y_2, \ldots, y_m]^T \tag{2}$$

1058-6393/91 $01.00 © 1991 IEEE

The ith column of $\mathbf{A}(\omega)$ is given by $\mathbf{a}(\omega, \theta_i)$ where

$$\mathbf{a}(\omega, \theta_i) = [e^{j2\pi f_o \tau_1(\theta_i)}, e^{j2\pi f_o \tau_2(\theta_i)}, \dots, e^{j2\pi f_o \tau_m(\theta_i)}]^T \quad (3)$$

and $\tau_k(\theta_i)$ is the propagation delay given by

$$\tau_k(\theta_i) = [\sin\theta_i, \cos\theta_i][x_k, y_k]^T \quad (4)$$

where (x_k, y_k) is the position of the kth sensor and θ_i is the bearing of the ith calibration source.

In the problem of active array shape calibration, the objective is to estimate the position of the sensors with the assumption that the source bearings and waveforms are known exactly, given $\mathbf{y}(t)$, t = 1, 2, ..., N. Since $\mathbf{y}(t)$ has both deterministic and stochastic components, it is convenient to define the generalized expectation operator [15]

$$\overline{E}(\cdot) = \lim_{N\to\infty} \frac{1}{N} \sum_{t=1}^{N} E(\cdot) \quad (5)$$

where E is the usual expectation operator with respect to the stochastic components. Assuming that $\mathbf{s}(t)$ satisfies regularity conditions [5, 9] so that the limit of its sample covariance exists, then

$$\overline{E}\mathbf{s}(t)\mathbf{s}^*(u) = \lim_{N\to\infty} \frac{1}{N} \sum_{t=1}^{N} \mathbf{s}(t)\mathbf{s}^*(u) = \mathbf{P}\delta_{t,u} \quad (6)$$

$$\overline{E}\mathbf{s}(t)\mathbf{s}^T(u) = \lim_{N\to\infty} \frac{1}{N} \sum_{t=1}^{N} \mathbf{s}(t)\mathbf{s}^T(u) = \mathbf{0}\ \forall t, u \quad (7)$$

The noise is assumed to be Gaussian distributed, with zero mean and

$$\overline{E}\mathbf{e}(t)\mathbf{e}^*(u) = \sigma^2 \mathbf{I}\delta_{t,u} \quad (8)$$

$$\overline{E}\mathbf{e}(t)\mathbf{e}^T(u) = \mathbf{0}\ \forall t, u \quad (9)$$

Note that $\mathbf{y}(t) \sim \mathcal{G}(\mathbf{A}(\omega)\mathbf{s}(t), \sigma^2\mathbf{I})$.

3 Cramér-Rao Bound

The Cramér-Rao bound provides a lower bound on the variance of the sensor position errors of any unbiased estimator. It will be useful to investigate how the number of sources and choice of bearings affect this bound. Let α be the deterministic $(2m+1) \times 1$ parameter vector given by

$$\alpha = [\sigma^2\ \omega^T]^T \quad (10)$$

where ω is defined in (2).

Theorem 3.1 *Under the assumption of the signal model in (6)-(9), the CRB covariance matrix for α is given by*

$$\mathbf{C}_{crb}(\alpha) = \begin{bmatrix} \frac{\sigma^4}{mN} & & \mathbf{0}^T \\ & \mathbf{C}_{xx} & \mathbf{C}_{xy} \\ \mathbf{0} & & \\ & \mathbf{C}_{yx} & \mathbf{C}_{yy} \end{bmatrix} \quad (11)$$

where

$$\mathbf{C}_{xx} = \frac{\sigma^2}{2N}\left(\mathbf{A}_x\mathbf{P}_N\mathbf{A}_x^* \odot \mathbf{I} - Re(\mathbf{A}_x\mathbf{P}_N\mathbf{A}_y^* \odot \mathbf{I})\cdot (\mathbf{A}_y\mathbf{P}_N\mathbf{A}_y^* \odot \mathbf{I})^{-1} Re(\mathbf{A}_y\mathbf{P}_N\mathbf{A}_x^* \odot \mathbf{I})\right)^{-1} \quad (12)$$

$$\mathbf{C}_{yy} = \frac{\sigma^2}{2N}\left(\mathbf{A}_y\mathbf{P}_N\mathbf{A}_y^* \odot \mathbf{I} - Re(\mathbf{A}_y\mathbf{P}_N\mathbf{A}_x^* \odot \mathbf{I})\cdot (\mathbf{A}_x\mathbf{P}_N\mathbf{A}_x^* \odot \mathbf{I})^{-1} Re(\mathbf{A}_x\mathbf{P}_N\mathbf{A}_y^* \odot \mathbf{I})\right)^{-1} \quad (13)$$

$$\mathbf{C}_{xy} = \mathbf{C}_{yx}^T = -\left(\mathbf{A}_x\mathbf{P}_N\mathbf{A}_x^* \odot \mathbf{I}\right)^{-1} \left(Re(\mathbf{A}_x\mathbf{P}_N\mathbf{A}_y^*) \odot \mathbf{I}\right) \mathbf{C}_{yy} \quad (14)$$

$$\mathbf{A}_x = \frac{j2\pi}{\lambda_o}\mathbf{A}\mathbf{\Lambda}_{sin} \quad (15)$$

$$\mathbf{A}_y = \frac{j2\pi}{\lambda_o}\mathbf{A}\mathbf{\Lambda}_{cos} \quad (16)$$

$$\mathbf{P}_N = \frac{1}{N}\sum_{t=1}^{N} \mathbf{s}(t)\mathbf{s}^*(t) \quad (17)$$

$$\mathbf{\Lambda}_{sin} = \begin{bmatrix} \sin\theta_1 & & \mathbf{0} \\ & \ddots & \\ \mathbf{0} & & \sin\theta_n \end{bmatrix} \quad (18)$$

$$\mathbf{\Lambda}_{cos} = \begin{bmatrix} \cos\theta_1 & & \mathbf{0} \\ & \ddots & \\ \mathbf{0} & & \cos\theta_n \end{bmatrix} \quad (19)$$

λ_o is the wavelength, and $diag(\mathbf{A})$ =column vector formed from the diagonal of $\mathbf{A}$.

Proof: See [16].

To see how the CRB depends on the characterisitics of the calibrating sources, let

$$\phi_{ij} = \frac{2\pi}{\lambda_o}(x_j \sin\theta_i + y_j \cos\theta_i) \quad (20)$$

and

$$p_{kl} = [\mathbf{P}_N]_{kl} \quad (21)$$

The ii^{th} element of $\mathbf{C}_{xx}$ and $\mathbf{C}_{yy}$ can be shown to be

$$[\mathbf{C}_{xx}]_{ii} = \frac{\sigma^2\lambda_o^2}{8N\pi^2}\left(\sum_{k,l} p_{kl}\sin\theta_k \sin\theta_l e^{j(\phi_{k_1}-\phi_{l_1})} - \frac{Re\left(\sum_{k,l} p_{kl}\sin\theta_k\cos\theta_l e^{j(\phi_{k_1}-\phi_{l_1})}\right)^2}{\sum_{k,l} p_{kl}\cos\theta_k\cos\theta_l e^{j(\phi_{k_1}-\phi_{l_1})}}\right)^{-1} \quad (22)$$

$$[\mathbf{C}_{yy}]_{ii} = \frac{\sigma^2\lambda_o^2}{8N\pi^2}\left(\sum_{k,l} p_{kl}\cos\theta_k \cos\theta_l e^{j(\phi_{k_1}-\phi_{l_1})} - \frac{Re\left(\sum_{k,l} p_{kl}\sin\theta_k\cos\theta_l e^{j(\phi_{k_1}-\phi_{l_1})}\right)^2}{\sum_{k,l} p_{kl}\sin\theta_k\sin\theta_l e^{j(\phi_{k_1}-\phi_{l_1})}}\right)^{-1} \quad (23)$$

For simplicity, suppose the calibrating sources are chosen to be uncorrelated (for finite N) and have equal unity power. The source bearings can be selected to minimize the CRB variance in both the x and y coordinates. Define

$$V(\theta) \triangleq [\mathbf{C}_{xx}]_{ii} + [\mathbf{C}_{yy}]_{ii} \quad (24)$$

Using (22)–(23), (24) can be written as

$$V(\theta) = \frac{\lambda_o^2\sigma^2 n}{8N\pi^2}\left(\sum_{k=1}^{n}\sum_{l=1}^{n}\sin^2\theta_k\cos^2\theta_l - (\sum_{k=1}^{n}\sin\theta_k\cos\theta_k)^2\right)^{-1} \quad (25)$$

Theorem 3.2 *For uncorrelated calibrating sources, the CRBs in both the x and y coordinates are minimized if and only if the bearings of the sources satisfies*

$$\sum_{k=1}^{n} e^{j2\theta_k} = 0 \quad (26)$$

The minimum value of the $V(\theta)$ is given by

$$V_{min}(\theta_{opt}) = \frac{\sigma^2\lambda_o^2}{2Nn\pi^2} \quad (27)$$

Proof: Minimizing $V(\theta)$, is equivalent to maximizing

$$\sum_{k=1}^{n}\sum_{l=1}^{n}\sin^2\theta_k\cos^2\theta_l - (\sum_{k=1}^{n}\sin\theta_k\cos\theta_k)^2$$

$$= \frac{n^2}{4} - \frac{1}{4}\sum_{k=1}^{n}\sum_{l=1}^{n}\cos(2\theta_k - 2\theta_l)$$

$$= \frac{n^2}{4} - \frac{1}{4}Re\sum_{k=1}^{n}\sum_{l=1}^{n} e^{j2(\theta_k-\theta_l)}$$

$$= \frac{n^2}{4} - \frac{1}{4}\left|\sum_{k=1}^{n} e^{j2\theta_k}\right|^2 \quad (28)$$

The results of (26)–(27) follows easily from direct substituition □.

Corollary : A sufficient condition for minimizing the CRB is that the bearings are uniformly distributed.

Remark : Note that the minimum CRB varies uniformly as the reciprocal of the number of sources as well as the number of snapshots. As an example, for $n = 2$, the uncorrelated sources can be selected to be distributed such that they are spatially separated by $\frac{\pi}{2}$; for $n = 3$, the source separation should be $\frac{\pi}{3}$ or $\frac{2\pi}{3}$ and so on.

4 MLE and Eigenstructure-based Calibration Algorithm

Assuming the nominal positions of the sensor to be reasonably close to the true locations, Newton-type algorithms can be devised for calibration algorithms based on maximum likelihood (ML) and eigenstructure-based (ES) estimators.

The ML calibration algorithm for the sensor position is based on the estimates that maximizes the likelihood function of the observed data. That is,

$$\omega_{mle} = \arg\max_{\omega}\left\{(\pi\sigma^2)^{-mN} exp\left(-\frac{1}{\sigma^2}\sum_{t=1}^{N}||\mathbf{y}(t) - \mathbf{A}(\omega)\mathbf{s}(t)||^2\right)\right\}$$

$$= \arg\min_{\omega} Tr(\mathbf{A}(\omega)\mathbf{P}_N\mathbf{A}^*(\omega) - 2\, Re(\mathbf{A}(\omega)\mathbf{P}_{sy})) \quad (29)$$

$$= \arg\min_{\omega} f_{mle}(\omega) \quad (30)$$

where

$$\mathbf{P}_{sy} = \frac{1}{N}\sum_{t=1}^{N}\mathbf{s}(t)\mathbf{y}^*(t) \quad (31)$$

$$f_{mle}(\omega) = Tr(\mathbf{A}(\omega)\mathbf{P}_N\mathbf{A}^*(\omega) - 2\, Re(\mathbf{A}(\omega)\mathbf{P}_{sy})) \quad (32)$$

The approximate gradient and Hessian of the ML cost function in (29) is given by (see [16])

$$f'_{mle}(\omega) = 2Re\ diag\begin{bmatrix}\mathbf{A}_x\Gamma & \mathbf{0}\\ \mathbf{0} & \mathbf{A}_y\Gamma\end{bmatrix} \quad (33)$$

$$f''_{mle}(\omega) = 2Re\left\{\begin{bmatrix} \mathbf{A}_x\mathbf{P}_N\mathbf{A}_x^* & \mathbf{A}_x\mathbf{P}_N\mathbf{A}_y^* \\ \mathbf{A}_y\mathbf{P}_N\mathbf{A}_x^* & \mathbf{A}_y\mathbf{P}_N\mathbf{A}_y^* \end{bmatrix} \odot \begin{bmatrix} \mathbf{I} & \mathbf{I} \\ \mathbf{I} & \mathbf{I} \end{bmatrix}\right\} \quad (34)$$

where $\Gamma = \mathbf{P}_N\mathbf{A}^*(\omega) - \mathbf{P}_{sy}$.

The ES calibration algorithms are based on the class of subspace-based array processing algorithms [8, 9, 12]. The cost function of this class of algorithms have the form

$$f_{es}(\omega) = Tr\left(\mathbf{A}^*(\omega)\hat{\mathbf{G}}\hat{\mathbf{G}}^*\mathbf{A}(\omega)\mathbf{W}\right) \quad (35)$$

where $\mathbf{W}$ is a positive definite weighting matrix and $\hat{\mathbf{G}}$ is a finite sample estimate of $\mathbf{G}$, the matrix of noise eigenvectors. In our case, the approximate gradient and Hessian of the ES cost function in (35) is given by (see [16])

$$f'_{es}(\omega) = 2Re\ diag\begin{bmatrix} \hat{\mathbf{G}}\hat{\mathbf{G}}^*\mathbf{AWA}_x^* & 0 \\ 0 & \hat{\mathbf{G}}\hat{\mathbf{G}}^*\mathbf{AWA}_y^* \end{bmatrix} \quad (36)$$

$$f''_{es}(\omega) = 2Re\left\{\begin{bmatrix} \mathbf{A}_x\mathbf{WA}_x^* & \mathbf{A}_x\mathbf{WA}_y^* \\ \mathbf{A}_y\mathbf{WA}_x^* & \mathbf{A}_y\mathbf{WA}_y^* \end{bmatrix}^T \odot \begin{bmatrix} \hat{\mathbf{G}}\hat{\mathbf{G}}^* & \hat{\mathbf{G}}\hat{\mathbf{G}}^* \\ \hat{\mathbf{G}}\hat{\mathbf{G}}^* & \hat{\mathbf{G}}\hat{\mathbf{G}}^* \end{bmatrix}\right\} \quad (37)$$

Using the above results, iterative solutions to solve for the different estimates can be implemented by the following control equations :

$$\omega_{k+1} = \omega_k + \beta_k[f''(\omega_k)]^{-1}f'(\omega_k) \quad (38)$$

The step length β_k in the search direction is chosen to be the largest number in the sequence $\{1, \frac{1}{2}, \frac{1}{4}, \ldots\}$ that satisfies the inequality $cost(\omega_k) - cost(\omega_{k+1}) > 0$ [11].

Remark : For m sensors, the ML algorithm requires $n \geq 2$ while the ES type of algorithms require $m - 1 \geq n \geq 2$. The dimension of the noise subspace set the upper limit for the latter case.

5 Asymptotic Results of Calibration Algorithms

The asymptotic covariance matrices of the calibration algorithms will be given here. The details can be found in [16].

R1 : The covariances of the ML estimator and the CRB are asymptotically the same, that is

$$\mathbf{C}^{as}_{mle}(\omega) = \mathbf{C}^{as}_{crb}(\omega) \quad (39)$$

where $\mathbf{C}^{as}_{crb}(\omega)$ is the lower block diagonal matrix in (11)-(19) with $\mathbf{P}_N$ replaced by $\mathbf{P}$. This result is not surprising as the MLE is known to be asymptotically efficient when the likelihood function is regular (which is satisfied when the source signal waveforms are known).

R2 : The asymptotic covariance matrix of the ES estimator is given by

$$\mathbf{C}^{as}_{es}(\omega) = \frac{\sigma^2}{2N}\left(Re(\mathbf{D}\odot\mathbf{E})\right)^{-1} Re(\mathbf{D}\odot\mathbf{F})\left(Re(\mathbf{D}\odot\mathbf{E})\right)^{-1} \quad (40)$$

where

$$\mathbf{D} = \begin{bmatrix} \mathbf{GG}^* & \mathbf{GG}^* \\ \mathbf{GG}^* & \mathbf{GG}^* \end{bmatrix} \quad (41)$$

$$\mathbf{E} = \begin{bmatrix} \mathbf{A}_x\mathbf{WA}_x^* & \mathbf{A}_x\mathbf{WA}_y^* \\ \mathbf{A}_y\mathbf{WA}_x^* & \mathbf{A}_y\mathbf{WA}_y^* \end{bmatrix}^T \quad (42)$$

$$\mathbf{F} = \begin{bmatrix} \mathbf{A}_x(\mathbf{AW})^*\mathbf{UAWA}_x^* & \mathbf{A}_x(\mathbf{AW})^*\mathbf{UAWA}_y^* \\ \mathbf{A}_y(\mathbf{AW})^*\mathbf{UAWA}_x^* & \mathbf{A}_y(\mathbf{AW})^*\mathbf{UAWA}_y^* \end{bmatrix}^T \quad (43)$$

$$\mathbf{U} = \mathbf{S\Lambda_S}(\mathbf{\Lambda_S} - \sigma^2\mathbf{I})^{-2}\mathbf{S}^* \quad (44)$$

Here $\mathbf{S}$ is the signal eigenvector matrix and $\mathbf{\Lambda_S}$ is the diagonal matrix of n largest eigenvalues.

MUSIC : The asymptotic covariance of the MUSIC estimator (with $\mathbf{W} = \mathbf{I}$) is given by the expression

$$\mathbf{C}^{as}_{music}(\omega) = \frac{\sigma^2}{2N}\left(Re(\mathbf{D}\odot\mathbf{E}_{music})\right)^{-1} Re(\mathbf{D}\odot\mathbf{F}_{music}) \cdot \left(Re(\mathbf{D}\odot\mathbf{E}_{music})\right)^{-1} \quad (45)$$

where

$$\mathbf{E}_{music} = \begin{bmatrix} \mathbf{A}_x\mathbf{A}_x^* & \mathbf{A}_x\mathbf{A}_y^* \\ \mathbf{A}_y\mathbf{A}_x^* & \mathbf{A}_y\mathbf{A}_y^* \end{bmatrix}^T \quad (46)$$

$$\mathbf{F}_{music} = \begin{bmatrix} \mathbf{A}_x\mathbf{A}^*\mathbf{UAA}_x^* & \mathbf{A}_x\mathbf{A}^*\mathbf{UAA}_y^* \\ \mathbf{A}_y\mathbf{A}^*\mathbf{UAA}_x^* & \mathbf{A}_y\mathbf{A}^*\mathbf{UAA}_y^* \end{bmatrix}^T \quad (47)$$

MODE : The asymptotic covariance of MODE (with $\mathbf{W} = (\mathbf{A}^*\mathbf{UA})^{-1}$) is given by the expression

$$\mathbf{C}^{as}_{mode}(\omega) = \frac{\sigma^2}{2N}\left(Re(\mathbf{D}\odot\mathbf{E}_{mode})\right)^{-1} \quad (48)$$

where

$$\mathbf{E}_{mode} = \begin{bmatrix} \mathbf{A}_x(\mathbf{A}^*\mathbf{UA})^{-1}\mathbf{A}_x^* & \mathbf{A}_x(\mathbf{A}^*\mathbf{UA})^{-1}\mathbf{A}_y^* \\ \mathbf{A}_y(\mathbf{A}^*\mathbf{UA})^{-1}\mathbf{A}_x^* & \mathbf{A}_y(\mathbf{A}^*\mathbf{UA})^{-1}\mathbf{A}_y^* \end{bmatrix}^T \quad (49)$$

R3 : The asymptotic covariance of the class of eigenstructure based estimators satisfy the following relations :

$$\mathbf{C}^{as}_{es}(\omega) \geq \mathbf{C}^{as}_{mle}(\omega) = \mathbf{C}^{as}_{crb}(\omega) \quad (50)$$

The proofs of the assertions R2–R3 can be found in [16]. In general, estimators based on the sample covariance matrix (which includes the class of eigenstructure-based estimators) for the deterministic signal model is *not* asymptotically efficient. This is because the sample covariance matrix is not a sufficient statistic for ω.

6 Summary and Conclusions

Active array shape calibration using the maximum likelihood method and eigenstructure based method were presented. The following results were established :

- The Cramér-Rao bound of the calibration problem.
- A necessary and sufficient condition for the bearings of uncorrelated calibrating sources to achieve minimum variance of the sensor position estimates.
- Newton-type calibration algorithms based on maximum likelihood and eigenstructure of the sample covariance matrix.
- The asymptotically covariances of the estimators.
- The maximum likelihood calibration method is asymptotically efficient.

References

[1] R. O. Schmidt, "A signal subspace approach to multiple source location and spectral estimation," *Ph.D dissertation*, Stanford Univ, 1981.

[2] I. Ziskind and M. Wax, "Maximum likelihood localization of multiple sources by alternating projections," *IEEE Trans. Acoust., Speech, Signal Processing,* Vol. ASSP-36, pp 1553-1560, Oct 1988.

[3] Y. Rockah and P. Schultheiss, "Array shape calibration using sources in unknown locations – Part I : far-field sources," *IEEE Trans. Acoust., Speech, Signal Processing,* Vol. ASSP-35, pp 286-299, Mar 1987.

[4] A. Weiss and B. Friedlander, "Array shape calibration using sources in unknown locations – a maximum likelihood approach," *IEEE Trans. Acoust., Speech, Signal Processing,* Vol. ASSP-37, pp 1958-1966, Dec 1989.

[5] P. Stoica and A. Nehorai, "MUSIC, maximum likelihood and Cramér-Rao bound," *IEEE Trans. Acoust., Speech, Signal Processing,* Vol. ASSP-37, pp 720-741, May 1989.

[6] J. Lo and S. Marple, "Eigenstructure methods for array sensor localization," *IEEE Int. Conf. on Acoust., Speech, Signal Processing,* pp 2260-2263, 1987.

[7] L. Seymour, C. Cowan, and P. Grant, "Bearing estimation in the presence of sensor positioning errors," *IEEE Int. Conf. on Acoust., Speech, Signal Processing,* pp 2264-2267, 1987.

[8] P. Stoica and K. Sharman, "Maximum likelihood methods for direction-of-arrival estimation," *IEEE Trans. Acoust., Speech, Signal Processing,* Vol. ASSP-38, pp 1132-1143, Jul 1990.

[9] P. Stoica and A. Nehorai, "Performance study of conditional and unconditional direction-of-arrival estimation," *IEEE Trans. Acoust., Speech, Signal Processing,* Vol. ASSP-38, pp 1783-1795, Oct 1990.

[10] A. Zeira and A. Nehorai, "Frequency domain Cramér-Rao bounds for Gaussian processes," *IEEE Trans. Acoust., Speech, Signal Processing,* Vol. ASSP-38, pp 1063-1066, June 1990.

[11] A. Björck, *Handbook of numerical analysis vol I : solutions of equations in R^n*, P. Ciarlet and J. Lions edit., North-Holland, 1990.

[12] M. Viberg and B. Ottersten, "Sensor array processing based on subspace fitting," *IEEE Trans. Acoust., Speech, Signal Processing,* Vol. ASSP-39, pp 1110-1121, May 1991.

[13] D. J. Jeffries and D. R. Farrier, "Asymptotic results for eigenvector methods," *IEE Proc,* Vol. 132, Pt F, pp 589-594, Dec 1985.

[14] P. Stoica and A. Nehorai, "MUSIC, maximum likelihood and Cramér-Rao bound : further results and comparisons," *IEEE Trans. Acoust., Speech, Signal Processing,* Vol. ASSP-38, pp 2140-2150, Dec 1990.

[15] L. Ljung, *System identification : Theory for the user*, Prentice-Hall, 1987.

[16] B. C. Ng and A. Nehorai, "Optimum active array shape calibration," *in preparation* 1991.

SENSOR ARRAY PROCESSOR EVALUATION WITH A STANDARDIZED TEST CASE (STC)

Alain C. Barthelemy and Norman L. Owsley

Naval Underwater Systems Center
New London, CT 06320

University of Connecticut
Department of Electrical Engineering
Storrs, CT 06269

Abstract

A standard test case (STC) [15] for the comparative evaluation of the plethora of modern space-time signal processing (STSP) techniques is summarized. The STC is thought to be realistic in the context of superimposed multiple discrete and continuous ambient noise fields for a uniform linear array (ULA) of sensors. Attention is given the MUSIC direction finder (DF), the ESPRIT DF, the linear prediction (LP) based DF and the Enhanced Minimum Variance Distortionless Response (EMVDR) beamformer (BF). The asymptotic estimate of the source subspace dimensionality is used and the performance of several algorithms for the estimation of the dominant source subspace spanning eigenvectors are evaluated with respect to convergence rate and variance. The issue of false alarms induced by overestimating the source subspace dimensionality relative to the actual number of discrete sources is discussed.

1. INTRODUCTION

A requirement in many modern STSP systems is the simultaneous detection and location of multiple angularly extended and/or threshold sources and not simply the angular resolution of two equal strength high SNR sources. Foremost among the modern STSP techniques are those which are based first, on the estimation of the dimensionality of the source subspace and, secondly, on the estimation of an orthonormal (eigenvector) basis for either all or part of the source subspace. Given a mixture of very low SNR and angularly extended sources which are parametrically undefined both dimensionality and basis estimation are formidable problems. The extraction of a "dominant" source subspace which contains sources near a low false alarm detection threshold is extremely difficult.

First, several algorithms for estimating a limited number of eigenvectors are examined. This is motivated by a subsequent consideration of beamforming (BF) techniques which do not require an estimate of the complete source subspace dimensionality and direction finding techniques which do require such an estimate. The EMVDR BF requires only an estimate of a minimal number of source subspace dominant eigenvectors. The DF techniques considered are provided with the information theoretic asymptotic. i.e. correct estimate of the source subspace dimensionality. The BF and DF candidates considered are compared on the basis of temporal detection events as a function of possible source arrival angle. No attempt is made to prewhiten the data because the number and characteristics of the extended sources is assumed to be unknown. Moreover, no attempt is made to resolve correct source detections from false alarms because such techniques have not been proven reliable for sources at the detection threshold.

2. SUBSPACE ESTIMATORS

Dominant subspace techniques such as EMVDR [10], MUSIC [12] and ESPRIT [11] are currently popular as high-resolution methods for direction-of-arrival (DOA) estimation of plane waves by an array of sensors. The computational complexity of the eigendecomposition of the cross spectral density matrix, however, has prohibited real-time implementation.

This section of the paper investigates new classes of dominant subspace algorithms which help alleviate the computational burden. These estimators reduce complexity from $O(N^3)$ to $O(ND^2)$, where D is the dimension of the dominant subspace.

Yang and Kaveh [1] have proposed an adaptive estimator of the complete source or noise subspace based on a gradient search method. Schreiber [2] presented a method which exploits repeated eigenvalues by deflating the source subspace before updating its spectral decomposition. Owsley [3] introduced an adaptive procedure to estimate the source subspace based on orthogonal iteration. Tufts and Melissinos [5] have developed a Prony-Lanczos method to approximate the source subspace. Karasalo [6] developed a method which averages a deflated source subspace.

For notational simplicity, assume that we are estimating source subspaces only, and that their dimension is D.

3. PERFORMANCE METRICS

Four performance metrics were chosen to examine the convergence and accuracy of the proposed estimators. The first metric termed "eigenvector inner product" α is used primarily to examine convergence.

Let $\mathbf{x}(p)$ be the signal vector received from an N sensor ULA at time index p. Let p = 1, 2, ... P, the total number of snapshots. Define $\mathbf{R} = E\{\mathbf{x}(p)\mathbf{x}(p)^H\}$, the cross spectral density matrix (CSDM), where $E\{\bullet\}$ denotes the expectation of $\{\bullet\}$. Define the spectral decomposition of $\mathbf{R}$:

$$\mathbf{R} = \mathbf{M}\Lambda\mathbf{M}^H$$

where $\Lambda = \mathrm{diag}(\lambda_1, \lambda_2, ..., \lambda_N)$ is a diagonal matrix of eigenvalues ordered so that $\lambda_1 \geq \lambda_2 \geq ... \geq \lambda_N$, "H" denotes the Hermitian transpose, and $\mathbf{M} = [\mathbf{m}_1, \mathbf{m}_2, ..., \mathbf{m}_N]$ where $\mathbf{m}_i$ is the i-th normalized eigenvector of $\mathbf{R}$ associated with λ_i. Let $\underline{\mathbf{R}}(p)$ be the estimated CSDM at time index p, obtained recursively from a rank-one update

$$\underline{\mathbf{R}}(p) = \mu\underline{\mathbf{R}}(p-1) + (1-\mu)\mathbf{x}(p)\mathbf{x}(p)^H$$

where $0 \leq \mu \leq 1$. The initial value for $\underline{\mathbf{R}}(p)$

$$\underline{\mathbf{R}}(-1) = \varepsilon\mathbf{I}_N + \frac{1}{\underline{P}}\sum_{p=1}^{\underline{P}}\mathbf{x}(p)\mathbf{x}(p)^H$$

where $\varepsilon << 1$, $\mathbf{I}_N$ is the identity matrix, $\underline{P} << P$ and the summation is the block estimate of the $\underline{P}$-th order CSDM. Let $\underline{\mathbf{M}}(p)$ be a matrix of normalized eigenvectors $\underline{\mathbf{M}}(p) = [\underline{\mathbf{m}}(p)_1, \underline{\mathbf{m}}(p)_2, ..., \underline{\mathbf{m}}(p)_N]$ where $\underline{\mathbf{m}}(p)_i$ is the i-th normalized eigenvector of $\underline{\mathbf{R}}(p)$, obtained from a *dominant subspace estimation algorithm.* The initial value for $\underline{\mathbf{M}}(p)$, $\underline{\mathbf{M}}(-1) \leftarrow \mathrm{SVD}(\underline{\mathbf{R}}(-1))$, was obtained from the MATLAB [8] algorithm.

The inner product

$$\alpha(p)_i = |\underline{\mathbf{m}}(p)_i^H \mathbf{m}_i|, \quad i = 1, 2, ... D$$

where $0 \leq \alpha_i \leq 1$ and D < N is the order of the dominant subspace is used as the *first* metric of convergence.

The second metric β tracks the accuracy of the first D eigenvalues produced by an estimator.

Define the diagonal matrix

$$\Delta(p) = \underline{\mathbf{M}}(p)^H\underline{\mathbf{R}}(p)\underline{\mathbf{M}}(p) = \mathrm{diag}(\underline{\lambda}(p)_1, \underline{\lambda}(p)_2, ..., \underline{\lambda}(p)_D)$$

where $\underline{\lambda}(p)_i$ represents the i-th eigenvalue of $\underline{\mathbf{R}}(p)$. The diagonal elements of $\Delta(p)$

$$\beta(p)_i = \underline{\lambda}(p)_i, \quad i = 1, 2, ..., D$$

were used as the second convergence metric. Graphs associated with this metric are not shown.

The third metric $e_{\alpha,\beta}$ termed "convergence RMS error" compares α and β with their true or steady-state value. Using this metric, it is possible to quantize the performance of an estimator using a single number. Define τ as the number of samples allowed for transient settling. For snapshots $\mathbf{x}(p)$, $p = \tau+1, \tau+2, ..., P$, define two RMS error calculations

$$e(p)_\alpha = [\mathrm{mean}((1 - \alpha(p)_i)^2)]^{1/2}$$

$$e(p)_\beta = [\mathrm{mean}((\beta(p)_i - \lambda_i)^2)]^{1/2}$$

where i = 1, 2, ..., D.

The fourth metric is a bearing-time display which portrays peak information from various spectrum estimators. A block diagram of the technique used is helpful:

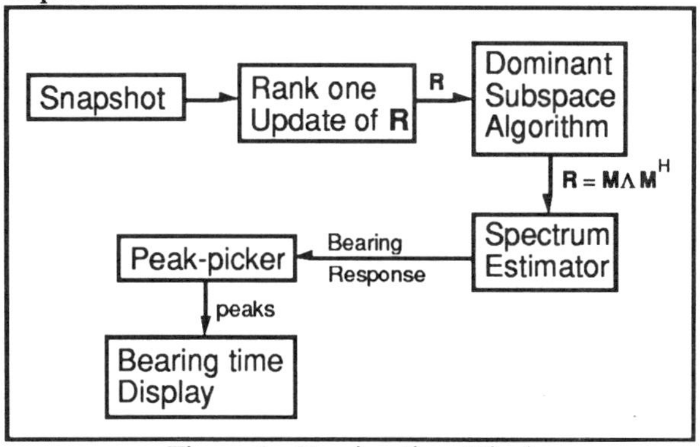

Figure 1 - Bearing-time Display

The spectrum estimators considered are

1) Conventional Beamforming (CBF).
2) MVDR BF [13].
3) LP DF [9].
   - order of predictor = 23.
4) EMVDR BF [10].
   - unit enhancement, number of modes D = 1,3,5.
   - enhancement $e = 10^8$, D = 3.
5) MUSIC DF [12].
   - number of modes D = 13.
6) ESPRIT DF [11].
   - number of modes D = 13.

The peak-picker operates on a three-point rule to calculate local minima. Given three adjacent points **A**, **B** and **C** on a bearing response a peak **B** is detected if **B** > **A** and **B** > **C**. The three point rule is slid along the bearing response curve. Finally, bearing response for the infinitely averaged CSDM is an overlay for the bearing-time display. Since ESPRIT is purely a direction-finding algorithm, the overlay for ESPRIT DOA angles are straight dotted lines.

4. RESULTS

Results for estimator comparison were obtained from the following parameters P = 1000 snapshots, $\underline{P}$ = 3 snapshots, τ = 200 snapshots, μ = 0.99 and the dimension of the dominant subspace D = 3.

The graphs in Figure 2 show the performance of the inner product α metric versus time for the six subspace estimators.

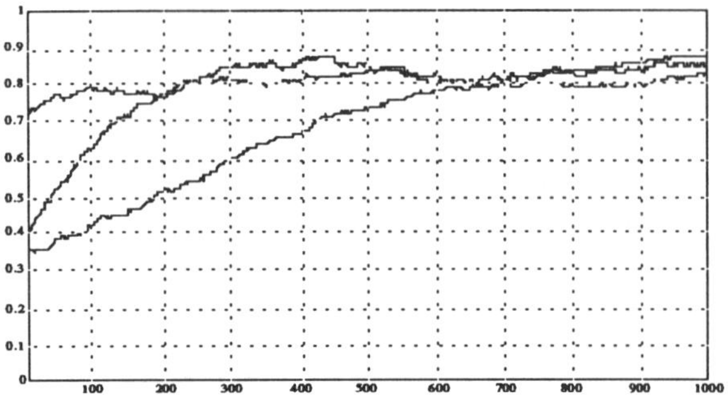

Figure 2a - gradient search [1]

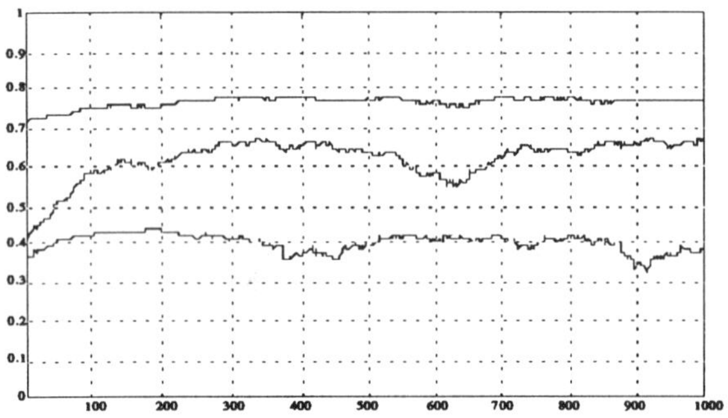

Figure 2b - subspace iteration [2]

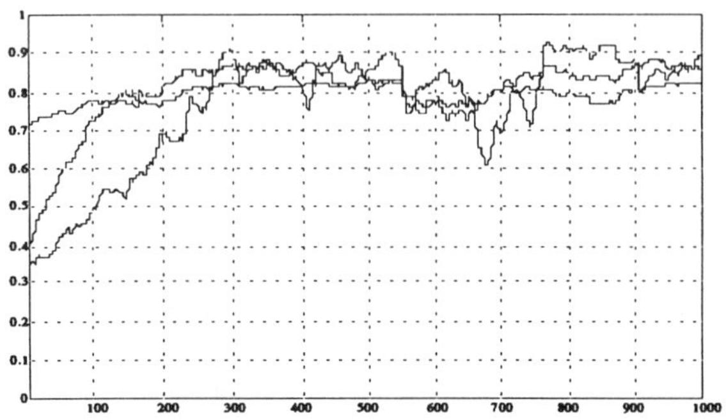

Figure 2c - orthogonal iteration [3,4]

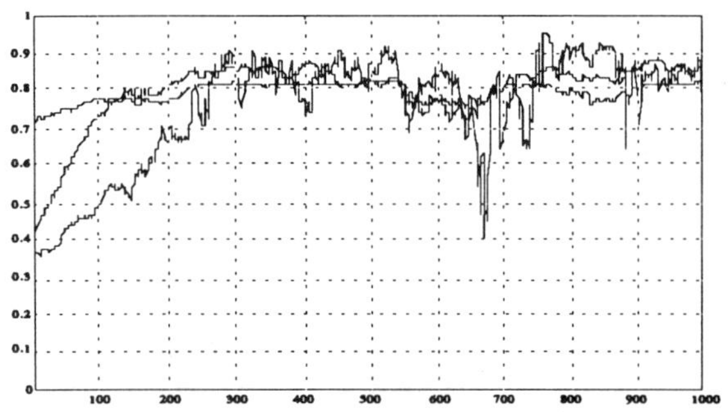

Figure 2d - LINPACK ZSVDC [14]

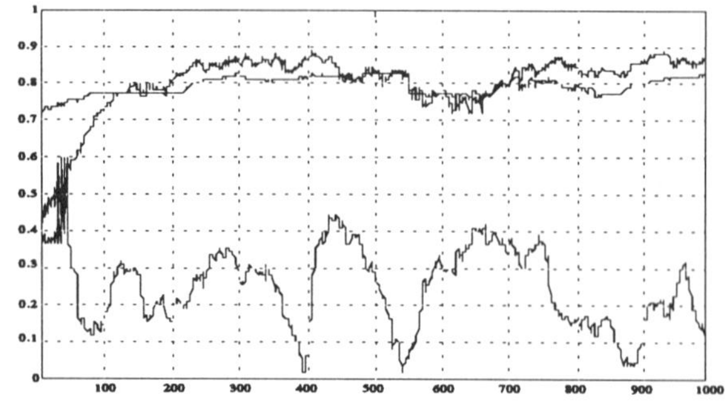

Figure 2e - Prony-Lanczos [5]

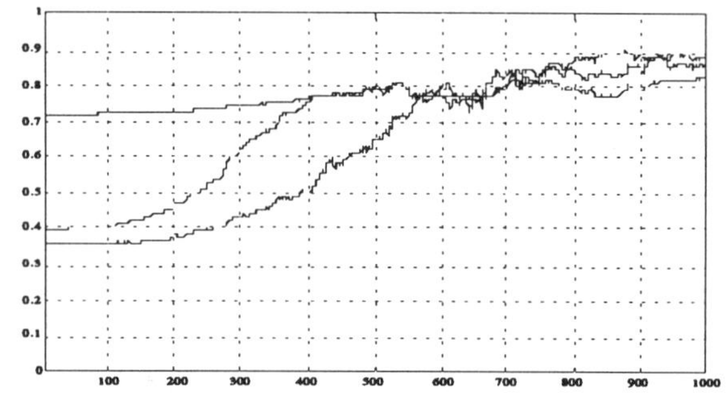

Figure 2f - signal subspace averaging [6]

The following table summarizes performance in terms of RMS error.

	e_{α_i} (dB)			e_{β_i} (dB)		
TYPE	i = 1	i = 2	i = 3	i = 1	i = 2	i = 3
Gradient Search	-7.078	-7.709	-5.661	3.346	1.058	-1.934
Subspace Iteration	-6.378	-4.398	-2.195	5.794	8.072	3.517
Orthogonal Iteration	-7.035	-7.623	-7.375	3.426	0.4441	-1.378
Prony-Lanczos	-7.030	-7.609	-1.181	3.427	0.4528	-0.911
Signal Subspace Averaging	-6.548	-5.849	-4.551	2.338	3.884	0.3675
LINPACK ZSVDC	-7.030	-7.616	-7.033	3.426	0.4416	-1.286

Table 1 - RMS error

The bearing-time displays which depict time versus cos(bearing) were calculated using the gradient search algorithm [1] to supply the eigenstructure estimates to EMVDR, MUSIC and ESPRIT. The parameters P = 600, $\underline{P}$ = 3, μ = 0.99 and 401 bins of angular resolution were used. The source locations and associated variances are

ANGLE	105	90	85	70	45	135	0
COS (ANGLE)	-0.25	0.000	0.087	0.342	0.707	0.707	1.000
σ^2 dB	-20.1	-3.5	-3.5	0.00	-14.1	0.00	-26.9

Table 2 - Source information

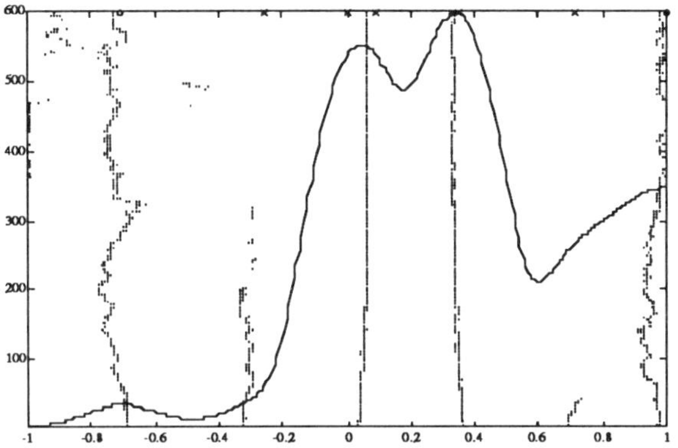

Figure 3a - CBF

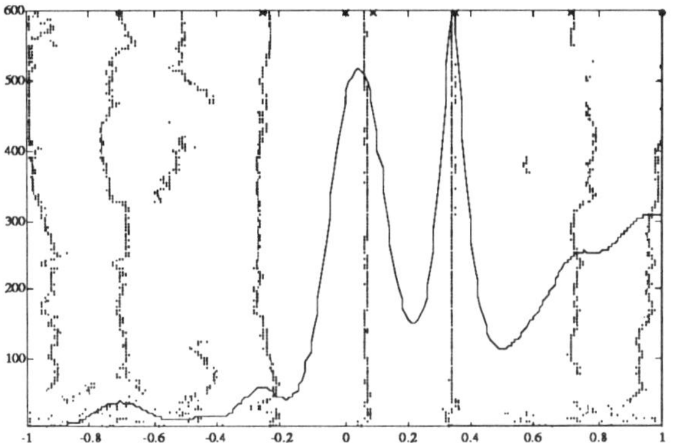

Figure 3b - MVDR

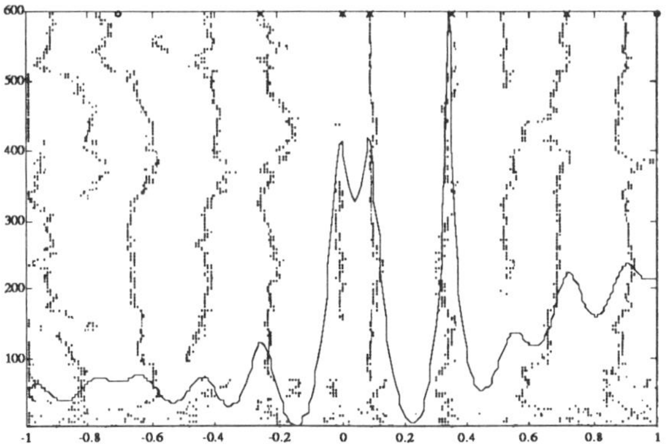

Figure 3c - LP, order = 23

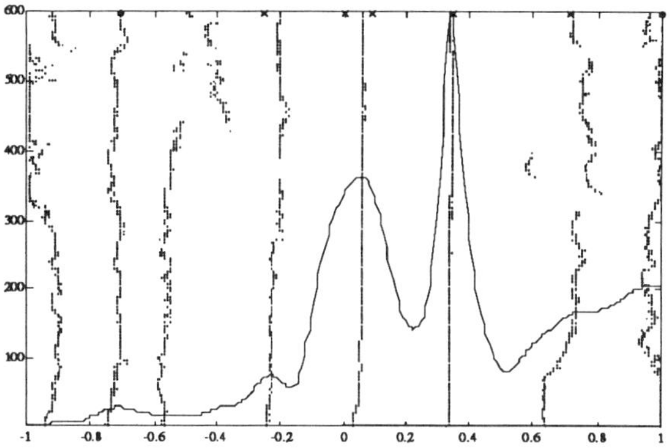

Figure 3d - EMVDR, D = 1

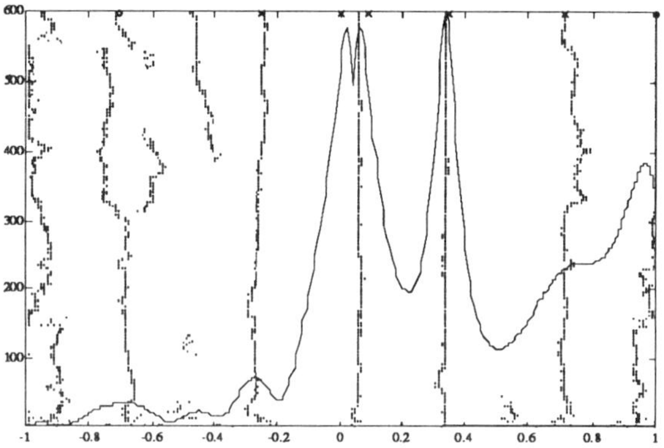

Figure 3e - EMVDR, D = 3, e = 1

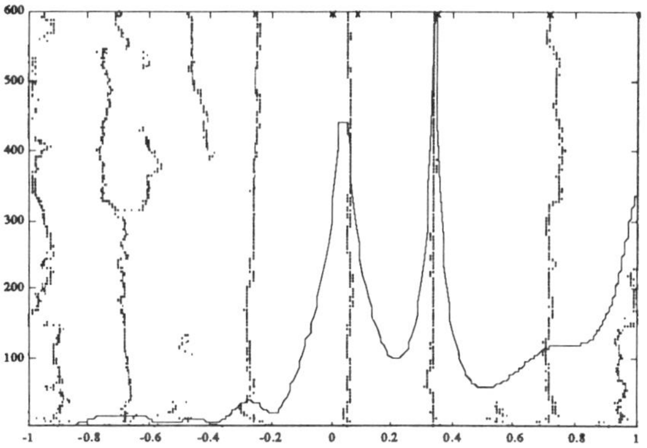

Figure 3f - EMVDR, D = 3, e = 10^8

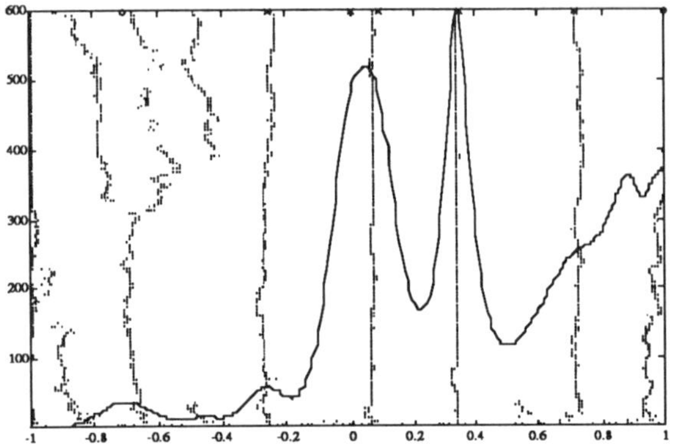

Figure 3g - EMVDR, D = 5

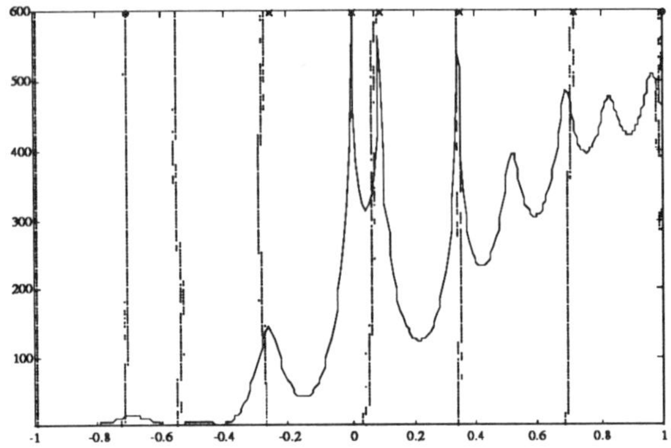

Figure 3h - MUSIC, D = 13

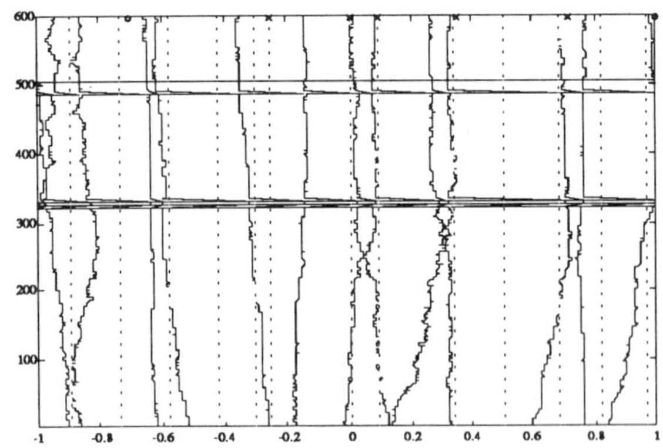

Figure 3i - ESPRIT, D = 13

5. CONCLUSIONS

Modern space-time signal processing techniques have been examined in the context of a standard test case. Six source subspace estimation algorithms have been reviewed and performance metrics established for them. No clear performance advantage can be found between gradient search and orthogonal iteration which are the best of the subspace estimators considered. With regard to detection performance, resolution and false alarm rate appear to be proportional.

Future research in this area will examine other performance metrics, such as angle between subspaces referenced in Golub [4]. Also, response of the eigenstructure estimators will be examined when an impulse is turned on in the acoustic field. Finally, other algorithms will be implemented including "Fast signal-subspace decomposition (FSD)" in [16].

REFERENCES

[1] J. F. Yang and M. Kaveh, "Adaptive eigensubspace algorithms for direction or frequency estimation and tracking," IEEE Trans. ASSP, vol. 36, no. 2, pp. 241-251, Feb. 1988.

[2] R. Schreiber, "Implementation of adaptive array algorithms," IEEE Trans. ASSP, vol. 34, no. 5, pp. 1038-1045, Oct. 1986.

[3] N. L. Owsley, "Adaptive data orthogonalization," in *Proc. Conf. ICASSP*, 1978, pp. 109-112.

[4] G. H. Golub and C. F. Van Loan, *Matrix Computations*, The Johns Hopkins University Press, p.211, 1983.

[5] D. W. Tufts and C. D. Melissinos, "Simple, effective computation of principal eigenvectors and their eigenvalues and application to high-resolution estimation of frequencies," IEEE Trans. ASSP, vol. 34, no. 5, pp. 1046-1053, Oct. 1986.

[6] I. Karasalo, "Estimating the covariance matrix by signal subspace averaging," IEEE Trans. ASSP, vol. 34, no. 1, Feb. 1986.

[7] P. Comon and G. H. Golub, "Tracking a few extreme singular values and vectors in signal processing," IEEE Proc., vol. 78, no. 8, pp. 1327-1343, 1990.

[8] MATLAB, The MathWorks, Inc.

[9] L. Marple, *Digital Spectral Analysis with Applications*, Prentice-Hall Signal Processing Series, 1987.

[10] N. L. Owsley, "Enhanced Minimum Variance Beamforming" in *Underwater Acoustic Data Processing*, Editor Y. Chan, NATO ASI Series, Klewer, 1989, pp. 285-293.

[11] R. Roy and T. Kailath, "ESPRIT - Estimation of Signal Parameters via Rotational Invariance," IEEE Trans. on ASSP, vol. 37, no. 7, pp. 984-995, July 1989.

[12] R. O. Schmidt, "Multiple emitter location and signal parameter estimation," reprinted in IEEE Trans. Antennas Propagat., vol. AP-34, pp. 276-280, Mar. 1986.

[13] N. L. Owsley, *in Array Signal Processing*, Prentice-Hall Signal Processing Series, 1985.

[14] J. J. Dongarra et al., *LINPACK Users' Guide*, SIAM, pp. 11.1-11.23, 1979.

[15] N. L. Owsley, "A Standardized Test Case (STC) for Sensor Array Processor Evaluation," *in Proc. Oceans 1991*.

[16] G. Xu and T. Kailath, "A Fast Algorithm for Signal Subspace Decomposition and Its Performance Analysis," *in Proc. ICASSP '91*, E1.11, pp. 3069-3072, 1991.

Comparison of Spatially Smoothed Weighted Subspace Methods*

Bhaskar D. Rao
Electrical and Computer Engineering Department
Univ. of California, San Diego, R-007
La Jolla, California 92093-0407

Abstract

In this paper, the performance of spatially smoothed weighted MUSIC, and weighted state space methods/ESPRIT are analyzed, and compared for the direction of arrival (DOA) estimation problem. In particular, a key outcome of the analysis is the observation that when a forward-backward spatially smoothed approach is used, optimally weighted MUSIC and optimally weighted state space methods/ESPRIT have identical asymptotic behavior. Also it is shown that the forward-backward covariance matrix is better conditioned than the forward only covariance matrix providing additional justification for the use of the forward-backward approach. The analysis is shown to be extendable to the time series frequency estimation problem, thereby providing a unified framework for dealing with DOA estimation and time series frequency estimation.

1 INTRODUCTION

In this paper, the effect of using a Forward-Backward spatially Smoothed (FBS) covariance matrix on the statistical performance of weighted eigen-based state space based methods or ESPRIT and weighted MUSIC is analyzed. In recent years, a statistical evaluation of subspace based methods has been conducted by a number of researchers with particular attention being paid to the case where the estimate of the covariance matrix was obtained by a straight forward averaging of the outer product of the snapshots. It was shown that MUSIC performs better than eigen-based state space methods with the difference becoming significant as the length of the array increases. However, state space methods are computationally more attractive and there is incentive to seek ways to improve the statistical properties. It is shown that this deficiency can be overcome by using a spatially smoothed forward-backward covariance matrix and proper weighting. A key result of the analysis is that optimally weighted MUSIC and optimally weighted State Space Methods/ESPRIT have identical asymptotic performance. Amongst the covariance estimators, it is shown that combining spatial smoothing with the forward-backward approach is more effective than using forward spatial smoothing alone. Another important attribute of the FBS approach for the problem of narrowband DOA estimation using an ULA is that it is naturally related to the time series frequency estimation problem, i.e. sinusoids in noise problem. This paper capitalizes on this relationship and the analysis framework [1,2] to establish an unified framework for these problems.

*This work was supported by the ARMY Research Office under Grant No. DAAL-03-90-G-0095

2 PROBLEM FORMULATION

The output vector of a ULA of L_1 sensors is given by

$$Y(n) = \left[\sum_{i=1}^{M} p_i(n) \ , \ldots, \sum_{i=1}^{M} p_i(n) \ e^{j(L_1-1)\omega_i} \right]^T + N(n)$$

where $\omega_i = \dfrac{2\pi d}{\lambda} \sin\theta_i$, d being the separation between sensors, λ the wavelength of the incident signal, and θ_i the DOA. Due to the simple mapping between ω_i and θ_i, for simplicity, ω_i will also be often referred to as the DOA.

2.1 FBS Covariance Matrix Estimate

Exploiting the structure in the ULA, the general FBS estimate is given by

$$\hat{\mathbf{R}}_{fbs} = \frac{\hat{\mathbf{R}}_{fs} + \mathbf{J}\hat{\mathbf{R}}_{fs}^{*}\mathbf{J}}{2} \quad where\ \mathbf{J} = \begin{bmatrix} 0 & 0 & 1 \\ \vdots & \vdots & \vdots \\ 0 \ldots & 1 & \ldots 0 \\ \vdots & \vdots & \vdots \\ 1 & 0 & 0 \end{bmatrix}, \quad (1)$$

and $\hat{\mathbf{R}}_{fs}$ is the **Forward Only Smoothing (FS)** covariance estimate given by

$$\hat{\mathbf{R}}_{fs} = \frac{1}{K}\sum_{p=1}^{K}\frac{1}{N}\sum_{n=1}^{N} Y_p(n)\, Y_p^H(n). \qquad (2)$$

$Y_p(n)$ is the output vector of the p th sub-array. An important property of $\mathbf{R}_{fbs}$ and $\hat{\mathbf{R}}_{fbs}$ is that they are centrohermitian, i.e.

$$\mathbf{R}_{fbs} = \mathbf{J}\mathbf{R}^*_{fbs}\mathbf{J} \;\; and \;\; \hat{\mathbf{R}}_{fbs} = \mathbf{J}\hat{\mathbf{R}}^*_{fbs}\mathbf{J}. \qquad (3)$$

Compared to the FS approach, the FBS approach has been found to result in superior estimates. A potential explanation for this observation is provided in the following theorem (proof is presented in Appendix A).

Theorem 1 *For a given number of subarrays K, in the noise free case the FBS covariance matrix is better conditioned than the FS covariance matrix, i.e.*

$$(\lambda_1)_{fbs} \le (\lambda_1)_{fs} \;\; and \;\; (\lambda_M)_{fbs} \ge (\lambda_M)_{fs},$$

resulting in

$$\kappa(\mathbf{R}_{fbs}) \le \kappa(\mathbf{R}_{fs}),$$

where $\kappa(\mathbf{R}) = ||\mathbf{R}||_2 ||\mathbf{R}^+||_2 = \frac{\lambda_1}{\lambda_M}$ denotes the condition number of $\mathbf{R}$. λ_1 is the largest eigenvalue and λ_M the smallest nonzero eigenvalue.

Remark : The result is also true in the presence of white noise. λ_M plays an important role in low SNR and high resolution cases thereby indicating that the FBS approach is preferable to the FS approach.

2.2 Subspace Decomposition

Given any one of the exact covariance matrices, its eigendecomposition is given by

$$\mathbf{R} = \mathbf{R}_s + \sigma_n^2\mathbf{I} = \mathbf{E}\,\mathbf{\Lambda}\,\mathbf{E}^H = \mathbf{E}_s\,\mathbf{\Lambda}_s\,\mathbf{E}_s^H + \sigma_n^2\mathbf{I},$$

where $\mathbf{E} = [\mathbf{E}_s \,,\, \mathbf{E}_n]$, $\mathbf{E}_s = [S_1 \,,\, S_2 \,, \ldots S_M]$, $\mathbf{E}_n = [S_{M+1} \,,\, S_{M+2} \,, \ldots, S_L]$,

$\mathbf{\Lambda} = diag(\lambda_1 \,, \ldots, \lambda_L)$, and $\mathbf{\Lambda}_s = diag(\lambda_1^s \,, \ldots, \lambda_M^s)$.

2.3 Performance Analysis of Subspace Methods

The analysis is conducted under the usual assumptions of circular Gaussianity and independent snapshots [1,2,3,4]. For analysis purposes it is useful to divide the subspace methods into three steps. Step 1 involves the mapping from the data to the covariance matrix estimate, and the statistics of the covariance matrix estimate are necessary for the analysis. Step 2 involves the computing of the eigenvectors from the covariance matrix estimate. Errors in the covariance matrix affect the subspaces, and this affect needs to be quantified in order to carry out the analysis. Step 3 is dependent on the subspace approach used to estimate the DOA, and the weighted subspace methods are the focus of this paper. The three steps can be combined to obtain the overall performance of a method.

3 Eigen-Based State Space Methods

In this approach, a matrix $\mathbf{F}_e$ is estimated using a weighted least squares estimate as shown below [3].

$$\hat{\mathbf{F}}_e = (\hat{\mathbf{E}}_1^H\mathbf{W}\hat{\mathbf{E}}_1)^{-1}\hat{\mathbf{E}}_1^H\mathbf{W}\hat{\mathbf{E}}_2,$$

where $\mathbf{W}$ is a hermitian symmetric, positive semi-definite matrix. The DOA are obtained from the eigenvalues of the matrix $\mathbf{F}_e$. $\mathbf{E}_1$ and $\mathbf{E}_2$ necessary for the computation are obtained by partitionng the signal subspace eigenvectors $\mathbf{E}_s$ as shown below.

$$\mathbf{E}_s = \begin{bmatrix} e_1 \\ \mathbf{E}_2 \end{bmatrix} = \begin{bmatrix} \mathbf{E}_1 \\ e_L \end{bmatrix} \qquad (4)$$

where e_i is the i th row of $\mathbf{E}_s$ and $\mathbf{E}_1, \mathbf{E}_2$ are appropriate $(L-1) \times M$ matrices.

3.1 Mean Squared Error (MSE) Results

First some quantities that appear in the discussion are defined.

$$\mathbf{A} = \begin{bmatrix} a_1 \\ \mathbf{A}_2 \end{bmatrix} = \begin{bmatrix} \mathbf{A}_1 \\ a_L \end{bmatrix} = [V(\omega_1), \ldots, V(\omega_M)] \qquad (5)$$

where $V(\omega) = [1, e^{j\omega}, \ldots, e^{j(L-1)\omega}]^T$. $\beta_i = R_s^+ V(\omega_i)$.

$$\epsilon_i^H = -\, l_i^H \left(\mathbf{A}_1^H\mathbf{W}\mathbf{A}_1\right)^{-1} \mathbf{A}_1^H\mathbf{W}\mathbf{T}, \qquad (6)$$

$$\mathbf{T} = [\mathbf{I}_{L-1\times L-1} \;\; 0_{L-1\times 1}] - z_i^* [0_{L-1\times 1} \;\; \mathbf{I}_{L-1\times L-1}], \qquad (7)$$

l_i is the i th column of $\mathbf{I}_{M\times M}$.

Property 1 *[5] For the FBS approach using a centro-hermitian weighting matrix, i.e. $\mathbf{W} = \mathbf{J}\mathbf{W}^*\mathbf{J}$,*

$$\overline{(\Delta\omega_i)^2_{fbs}} = \overline{|\Delta z_i|^2_{fbs}} = \Gamma_{\epsilon_i\epsilon_i\beta_i\beta_i} \qquad (8)$$

with

$$\Gamma_{\epsilon_i\epsilon_i\beta_i\beta_i} = \epsilon_i^H\, \mathbf{B}_{\beta_i\beta_i}\, \epsilon_i$$

where

$$\mathbf{B}_{\beta_i\beta_i} = \frac{1}{2NK^2}\Big(\sum_{p=1}^{K}\sum_{q=1}^{K}(\beta_i^H \mathbf{R}_{pq}\,\beta_i)\,\mathbf{N}_{qp} + \sum_{p=1}^{K}\sum_{q=1,q>p}^{K} 2\,(\mathbf{N}_{qp}\beta_i\beta_i^H\mathbf{N}_{pq})\Big). \quad (9)$$

and $\mathbf{R}_{pq} = \overline{Y_p(n)Y_q^H(n)}$ *and* $\mathbf{N}_{pq} = \mathbf{P}_n\mathbf{R}_{pq}\mathbf{P}_n = \overline{N_p(n)N_q^H(n)}$ *with* $\mathbf{P}_n = \mathbf{E}_n\mathbf{E}_n^H$.

Theorem 2 *[5] For the FBS approach, the optimum centrohermitian weighting matrix* $\mathbf{W}_{opt}$ *for weighted State Space Method/ESPRIT which minimizes the MSE in a given DOA is given by*

$$\mathbf{W}_{opt} = \left(\mathbf{T}\,\mathbf{B}_{\beta_i\beta_i}\,\mathbf{T}^H\right)^{-1}. \quad (10)$$

The MSE corresponding to $\mathbf{W}_{opt}$ *is then given by*

$$\overline{(\Delta\omega_i)^2_{opt}} = l_i^H\left(\mathbf{A}_1^H(\mathbf{T}\,\mathbf{B}_{\beta_i\beta_i}\,\mathbf{T}^H)^{-1}\mathbf{A}_1\right)^{-1} l_i. \quad (11)$$

4 Comparison with Weighted MUSIC

4.1 Weighted MUSIC

In spectral *Weighted* MUSIC, the DOA are found by locating the peaks of the spatial spectrum $S(\omega) = \frac{1}{D(e^{j\omega})}$, where $D(e^{j\omega})$, the *weighted* null spectrum, is defined as

$$D(e^{j\omega}) = V^H(\omega)\,\mathbf{P}_n\,\mathbf{W}\,\mathbf{P}_n\,V(\omega), \quad (12)$$

$\mathbf{W}$ being an $L \times L$ matrix. $\mathbf{W} = \mathbf{I}$ corresponds to the usual MUSIC procedure.

Theorem 3 *[6] Given the FBS covariance matrix estimate, an optimum weighting matrix* $\mathbf{W}_{opt}$ *for weighted MUSIC is given by*[1]

$$\mathbf{W}_{opt} = (\mathbf{P}_n\mathbf{B}_{\beta_i\beta_i}\mathbf{P}_n)^+.$$

The MSE in DOA corresponding to $\mathbf{W}_{opt}$ *is given by*

$$\overline{(\Delta\omega_i)^2_{opt}} = \frac{1}{V_1^H(\omega_i)\,\mathbf{P}_n\,(\mathbf{P}_n\mathbf{B}_{\beta_i\beta_i}\mathbf{P}_n)^+\,\mathbf{P}_n\,V_1(\omega_i)}, \quad (13)$$

with $V_1(\omega) = \frac{\partial V(\omega)}{\partial\omega}$,

[1] + denotes the Moore-Penrose pseudo inverse

4.2 Comparison

In the Forward Only case (**FO**) approach, it was shown that MUSIC was superior to ESPRIT with the difference becoming significant for large array size [3]. However, we will show that this is not the case in the context of linear arrays. In fact, in the linear array case, using maximal overlap, optimally weighted state space methods/ESPRIT perform as well as optimally weighted MUSIC approach. This is summarized in the next theorem (the proof is presented in Appendix B).

Theorem 4 *The MSE in the DOA obtained using the weighted FBS state space method/ESPRIT approach with an optimal centrohermitian matrix is equal to that obtained using an optimally weighted FBS MUSIC estimate, i.e.*

$$\overline{(\Delta\omega_i)^2_{\substack{music\\opt}}} = \overline{(\Delta\omega_i)^2_{\substack{esprit\\opt}}}.$$

The result is also true for the FO estimates.

5 Non-Gaussian source amplitudes and the Frequency estimation Problem

5.1 Non-Gausian Source Amplitudes

In this section, we shall show that the mean squared error of the subspace estimates is independent of the source amplitude distribution, i.e. the Gaussian assumption is not necessary [4]. For this we need to examine $\Gamma_{\alpha_1\alpha_2\gamma_1\gamma_2}$ and $\Gamma_{\alpha_1\gamma_1\alpha_2\gamma_2}$ where α_i, $i = 1,2$ lie in the noise subspace and γ_i, $i = 1,2$ lie in the signal subspace as these quantities determine the overall performance.

Theorem 5 *Under the assumption that the noise is circularly Gaussian, snapshots are independent and that the source amplitudes belong to a distribution having finite second-order moments,* $\Gamma_{\alpha_1\alpha_2\gamma_1\gamma_2}$ *and* $\Gamma_{\alpha_1\gamma_1\alpha_2\gamma_2}$*, where* α_i*,* $i = 1,2$ *lie in the noise subspace and* γ_i*,* $i = 1,2$ *lie in the signal subspace, are dependent only on the second order moments of the source amplitudes.*

Proof : By definition, we have

$$\Gamma_{\alpha_1\alpha_2\gamma_1\gamma_2} = \overline{(\alpha_1^H(\hat{\mathbf{R}}-\mathbf{R})\gamma_1)(\alpha_2^H(\hat{\mathbf{R}}-\mathbf{R})\gamma_2)^H} \quad (14)$$

For simplicity, only the FS approach is shown, and the same ideas readily extend to the FBS approach. For the FS approach, from eq. 2

$$\hat{\mathbf{R}}_{fs} = \frac{1}{N}\sum_{n=1}^{N}\mathbf{Y}_F(n)\;\mathbf{Y}_F^H(n) \quad (15)$$

where $Y_F(n) = X_F(n) + N_F(n)$ with

$$\mathbf{Y}_F(n) = \frac{1}{\sqrt{K}}[Y_1(n), Y_2(n), \ldots, Y_K(n)]$$

Using the orthogonality of α_1 and γ_1, and that α_1 is a noise eigenvector of $\mathbf{R}_{fs}$,

$$\alpha_1^H(\hat{\mathbf{R}}_{fs} - \mathbf{R}_{fs})\gamma_1 = \alpha_1^H\hat{\mathbf{R}}_{fs}\gamma_1 - \alpha_1^H\mathbf{R}_{fs}\gamma_1 = \alpha_1^H\hat{\mathbf{R}}_{fs}\gamma_1$$

Since α_1 is the noise subspace, $\alpha_1^H\mathbf{X}_F(n) = 0\ \forall n$. Thus from eq. 15

$$\alpha_1^H\hat{\mathbf{R}}_{fs}\gamma_1 = \frac{1}{N}\sum_{n=1}^{N}\alpha_1^H\mathbf{N}_F(n)\left(\mathbf{X}_F(n)^H + \mathbf{N}_F(n)^H\right)\gamma_1$$

Substituting in eq. 14 and noting that the odd moments of the Gaussian noise vectors are zero, we have

$$\Gamma_{\alpha_1\alpha_2\gamma_1\gamma_2} = \frac{1}{N^2}\sum_{n=1}^{N}\sum_{m=1}^{N}$$
$$\overline{\{(\alpha_1^H\mathbf{N}_F(n)\mathbf{X}_F(n)^H\gamma_1)(\alpha_2^H\mathbf{N}_F(m)\mathbf{X}_F(m)^H\gamma_2)^H}$$
$$\overline{+(\alpha_1^H\mathbf{N}_F(n)\mathbf{N}_F(n)^H\gamma_1)(\alpha_2^H\mathbf{N}_F(m)\mathbf{N}_F(m)^H\gamma_2)^H\}}.$$

The results follows since only the first term depends on the signal amplitudes and only on their second order properties. One can prove a similar result for $\Gamma_{\alpha_1\gamma_1\alpha_2\gamma_2}$.
Remark : Note that the Gaussian assumption on the noise can be relaxed in the above theorem to random variables having finite fourth moments. However, to obtain tractable expressions the Gaussian assumption is useful.

5.2 Deterministic Sources and Frequency Estimation

The important special case where the amplitudes are deterministic deserves special attention. Theorem 5 is still valid, and all that needs to be done is to carefully define the covariance matrix $\mathbf{R}$. For the FBS approach we define

$$\mathbf{R}_{fbs} = \frac{1}{N}\sum_{n=1}^{N}(\mathbf{X}_F(n), \mathbf{J}\mathbf{X}_F^*(n))(\mathbf{X}_F(n), \mathbf{J}\mathbf{X}_F^*(n))^H + \sigma_n^2\mathbf{I}$$

and

$$\mathbf{R}_{pq} = \frac{1}{N}\sum_{n=1}^{N}X_p(n)X_q^H(n) + \mathbf{N}_{pq}.$$

Hence, in addition to the direct dependence of the performance on the number of snapshots N, β_i and $\mathbf{R}_{pq}$ also now depend on N. Asymptotically, they can be computed using $\mathbf{R}_{fbs}^{\infty}$ and $\mathbf{R}_{pq}^{\infty}$ where $\mathbf{R}_{fbs}^{\infty} = \lim_{N\to\infty}\mathbf{R}_{fbs}$ and $\mathbf{R}_{pq}^{\infty}$ is defined in a similar manner. However, it is useful to retain the dependence of β_i and $\mathbf{R}_{pq}$ on N as the expression in eq.8 is then a valid first order perturbation expression for all N. This allows for the transition from the asymptotic arguments to the finite number of snapshots case. The special case of $N = 1$ corresponds to the time series case and the results can be used to understand the time series methods.

5.3 Comparison with Previous Results

The results when applied to the frequency estimation problem are no longer asymptotic results and are valid for high SNR situations. However, they are more general then existing results. Normally, the time series analysis results start with data matrices $[\mathbf{Y}_F(1), \mathbf{J}\mathbf{Y}_F^*(1)]$, and treat $[\mathbf{N}_F(1), \mathbf{J}\mathbf{N}_F^*(1)]$ as a perturbation and obtain expression for Δz_i, the error in the estimate of the signal zero, that are linear in $[\mathbf{N}_F(1), \mathbf{J}\mathbf{N}_F^*(1)]$. In contrast, in this approach, we have for the FBS case

$$\begin{aligned}\frac{\Delta z_i}{z_i} &= \frac{1}{2}\epsilon_i^H\left[(\mathbf{N}_F(1), \mathbf{J}\mathbf{N}_F^*(1))(\mathbf{X}_F(1), \mathbf{J}\mathbf{X}_F^*(1))^H\right]\beta_i \\ &+ \frac{1}{2}\epsilon_i^H\left[(\mathbf{N}_F(1), \mathbf{J}\mathbf{N}_F^*(1))(\mathbf{N}_F(1), \mathbf{J}\mathbf{N}_F^*(1))^H\right]\beta_i.\end{aligned}$$

The first term corresponds to the result one obtains using a first order perturbation of the data matrix. The presence of the second term makes the expressions derived in this paper more accurate.

APPENDIX A

Proof of Theorem 1 : Let

$$\mathbf{R}_{fs} = \tilde{\mathbf{E}}_s\tilde{\mathbf{\Lambda}}_s\tilde{\mathbf{E}}_s^H \text{ and } \mathbf{R}_{fbs} = \mathbf{E}_s\mathbf{\Lambda}_s\mathbf{E}_s^H$$

Since the signal eigenvectors of both matrices span the same signal subspace there must exist a unique, nonsingular, unitary matrix φ which satisfies

$$\mathbf{E}_s\varphi = \tilde{\mathbf{E}}_s \tag{16}$$

Therefore one can write using eq. 1 and eq. 16

$$\mathbf{E}_s\mathbf{\Lambda}_s\mathbf{E}_s^H = \frac{1}{2}\left(\mathbf{E}_s\varphi\tilde{\mathbf{\Lambda}}_s\varphi^H\mathbf{E}_s^H + \mathbf{J}\mathbf{E}_s^*\varphi^*\tilde{\mathbf{\Lambda}}_s\varphi^T\mathbf{E}_s^T\mathbf{J}\right) \tag{17}$$

Because of the centro-hermitian nature of $\mathbf{R}_{fbs}$ (c.f. eq. 3), one can choose its eigenvectors to satisfy $\mathbf{J}\mathbf{E}_s^* = \mathbf{E}_s$. Using this in eq. 17

$$\mathbf{\Lambda}_s = \left(\frac{\varphi\tilde{\mathbf{\Lambda}}_s\varphi^H + \varphi^*\tilde{\mathbf{\Lambda}}_s\varphi^T}{2}\right).$$

Then

$$\lambda_i^s = \sum_{k=1}^{M} \tilde{\lambda}_k^s |\varphi_{ik}|^2, \quad i = 1, .., M. \tag{18}$$

The desired result is obtained by combining eq. 18 with $\sum_{k=1}^{M} |\varphi_{ik}|^2 = 1$, a consequence of the fact that φ is an unitary matrix. ∎

APPENDIX B

To prove theorem 4, some preparatory results are needed which are stated below.

Lemma 1 *If* $\mathbf{H}$ *is rectangular and* $\mathbf{S}$ *is symmetric and nonsingular,*

$$(\mathbf{SH})^+ = \mathbf{H}^+\mathbf{S}^{-1}\left[I - (\mathbf{QS}^{-1})^+(\mathbf{QS}^{-1})\right],$$

where $\mathbf{Q} = (\mathbf{I} - \mathbf{HH}^+)$.

Lemma 2 *For* $\mathbf{T}$ *defined in eq. 7*

$$\mathbf{T}^+\mathbf{T} = \mathbf{I} - V(\omega_i)V(\omega_i).$$

Proof : This follows from the fact that the $(L-1)\times L$ matrix $\mathbf{T}$ has rank $L-1$, and that $\mathbf{T}V(\omega_i) = 0$,

Proof of Theorem 4: Let the SVD of $\mathbf{P}_n\mathbf{B}\mathbf{P}_n = \mathbf{U}_b\mathbf{\Sigma}_b\mathbf{U}_b^H$. Then

$$\begin{aligned}
\frac{\overline{(\Delta\omega_i)^2}_{es}}{\overline{(\Delta\omega_i)^2}_{\substack{music \\ opt}}} &= \left(\epsilon_i^H \mathbf{P}_n \mathbf{B}\mathbf{P}_n\epsilon_i\right)\left(V_1^H(\mathbf{P}_n\mathbf{B}\mathbf{P}_n)^+ V_1\right) \\
&= ||\epsilon_i^H\mathbf{U}_b\mathbf{\Sigma}_b^{\frac{1}{2}}||^2\, ||V_1^H\mathbf{P}_n\mathbf{U}_b\mathbf{\Sigma}_b^{-\frac{1}{2}}||^2 \\
&\geq |\epsilon_i^H\mathbf{U}_b\mathbf{U}_b^H\mathbf{P}_nV_1|^2 = |\epsilon_i^HV_1|^2 = 1
\end{aligned} \tag{19}$$

$V_1 = V_1(\omega_i)$, and $\mathbf{B}$ is used instead of $\mathbf{B}_{\beta_i\beta_i}$ for notational simplicity. The last equality was shown in [3]. This shows that optimal MUSIC always has a smaller variance than State Space Methods. Here we will show that Optimal State Space methods have the same mean squared error as optimal MUSIC. From eq. 19, it is equivalent to showing that

$$\epsilon_i^H\mathbf{U}_b\mathbf{\Sigma}_b^{\frac{1}{2}} = c_1^*\, V_1^H\mathbf{P}_n\mathbf{U}_b\mathbf{\Sigma}_b^{-\frac{1}{2}} \quad or \quad \mathbf{P}_n\mathbf{B}\mathbf{P}_n\epsilon_i = c_1\mathbf{P}_nV_1, \tag{20}$$

where c_1 is a scalar quantity whose exact value is not important for the purposes of this proof. In this proof, c_1, c_2, etc will be used to represent scalars. We will show that eq. 20 holds when the optimum choice of $\mathbf{W}$ is used in the state space case, i.e. when $(\mathbf{TBT})^{-1}$ is substituted for $\mathbf{W}$ in the expression for ϵ_i in eq. 6 resulting in

$$\begin{aligned}
(\epsilon_i)_{opt} &= \mathbf{T}^H(\mathbf{TBT}^H)^{-1}\mathbf{A}_1\left(\mathbf{A}_1^H(\mathbf{TBT}^H)^{-1}\mathbf{A}_1\right)^{-1} l_i \\
&= \mathbf{B}_1^{-1}\mathbf{T}_1^+\mathbf{A}_1(\mathbf{A}_1^H(\mathbf{T}_1\mathbf{T}_1^H)^{-1}\mathbf{A}_1)^{-1}l_i,
\end{aligned} \tag{21}$$

where $\mathbf{T}_1 = \mathbf{TB}_1$ with $\mathbf{B}_1$ being a square root of $\mathbf{B}$, i.e. $\mathbf{B} = \mathbf{B}_1\mathbf{B}_1^H$. Since $\mathbf{B}$ is symmetric, $\mathbf{B}_1$ can be chosen to be symmetric, and is assumed to be symmetric for the purpose of simplifying the proof. It can then be shown that

$$\mathbf{A}_1^H(\mathbf{T}_1^+)^H\mathbf{B}_1\epsilon_i = l_i, \quad for \ all \ \mathbf{W}.$$

Note that ϵ_i is contained in the noise subspace and depends on $\mathbf{W}$. By selecting $\mathbf{W}$ properly, $(L-M)$ independent vectors that form a basis for the noise subspace can be generated[2]. Hence, it follows that

$$\mathbf{P}_n\mathbf{B}_1\mathbf{T}_1^+V'(\omega_j) = 0, j \neq i, \ and \ j = 1, 2, ..M, \tag{22}$$

where $\mathbf{A}_1 = [V'(\omega_1), V'(\omega_2), ..., V'(\omega_M)]$ with $V'(\omega) = \frac{1}{\sqrt{L}}\left[1, e^{j\omega}, .., e^{j(L-2)\omega}\right]^T$, which is similar to $V(\omega)$ except that it is a vector of length $(L-1)$. Using eqs. 21 and 22, it can be shown that

$$\mathbf{P}_n\mathbf{B}(\epsilon_i)_{opt} = c_2\mathbf{P}_n\mathbf{B}_1\mathbf{T}_1^+V'(\omega_i). \tag{23}$$

Lemma 1 is used to expand $\mathbf{T}_1^+$ in eq. 23. Then the resulting expression is simplified with the help of lemma 2, and the fact that $\mathbf{T}V_1 = c_2V'(\omega_i)$. This results in

$$\mathbf{P}_n\mathbf{B}(\epsilon_i)_{opt} = c_1\mathbf{P}_nV_1.$$

This completes the proof.

References

[1] H. Clergeot, S. Tressens, and A. Ouamri. *IEEE Trans. ASSP*, ASSP-37(11):1703–1720, Nov. 1989.

[2] B.D. Rao and K.V.S. Hari. *IEE Proceedings F*, 137(6):449–458, December 1990.

[3] P. Stoica and A. Nehorai. *IEEE Trans. ASSP*, ASSP-39(2):446–453, Feb. 1991.

[4] M. Viberg, and B. Ottersten, IEEE Trans. on Signal Processing, May 1991.

[5] B.D. Rao and K.V.S. Hari. In *Proc. ICASSP-91*, pages 3317–3320, Toronto, Canada, May 1991.

[6] B. D. Rao and K.V.S. Hari. In *Proc. of 24th Asilomar Conference* November 1990.

[2]Note that this not true if a maximally overlapping configuration is not used.

Fast Subspace Decomposition of Data Matrices†

Guanghan Xu and Thomas Kailath
Information Systems Laboratory
Stanford University, Stanford, CA 94305
Tel: (415)725-5698 *Email: xu@isl.stanford.edu*

Abstract— **In array signal processing, we often need to estimate a low-dimensional (say d-dimensional) *signal subspace*, by an SVD of a *large-size* $N \times M$ ($N \geq M$) *data* matrix X or by forming a covariance matrix $\hat{R}_x = \frac{1}{N} X^H X$ followed by an eigendecomposition (ED) of $\hat{R}_x$. In either case, the computational complexity is $O(M^2N)+O(M^3)$ multiplications, which are very expensive for large N and M. Moreover, the ED and SVD are difficult to implement in parallel for fast computation. Recently, Xu and Kailath presented a fast $O(M^2d)$ signal subspace decomposition approach based on covariance matrices. In this paper, we shall extend the results in the covariance matrix case and develop a fast $O(MNd)$ and potentially parallelizable (FSD) method to proceed directly from the data matrix. A new detection scheme is also presented here. In most sensor array processing scenarios, $d \ll M \leq N$ and the cost of FSD is almost an order of magnitude less than that of SVD and is even less than the cost of just forming the covariance matrix, *i.e.*, $O(NM^2)$. The performance analysis is also carried over from the covariance case to show that the estimates from FSD are *asymptotically equivalent* to that from the more costly SVD or ED.**

I. Introduction

MOST HIGH-RESOLUTION array processing algorithms require a common step of extracting the signal subspace from either a data or covariance matrix. This step is usually achieved by an eigendecomposition (ED) of a covariance matrix or a singular value decomposition (SVD) of a data matrix. An eigendecomposition requires at least $O(M^3)$ multiplications for an $M \times M$ matrix plus $O(NM^2)$ for forming the covariance. An SVD needs at least $O(NM^2+M^3)$ multiplications for an $N \times M$ ($N \geq M$) matrix; and in many of their more stable instantiations, both eigendecomposition and SVD involve global communication and significant amounts of local storage, properties that make VLSI parallel implementation difficult. Therefore, the step of determining the signal subspace represents a significant barrier that prevents *on-line* realization of a large class of high-resolution array processing algorithms. Recently, a fast $O(M^2d)$ algorithm[1], termed FSD, was presented to estimate the signal subspace from covariance matrices via exploitation of the matrix structure associated with signal subspace algorithms. In this paper, by exploiting the same structure in the data matrix, we present a new *fast* and potentially *parallelizable* algorithm [2] that *directly* proceeds from the data matrix. The new algorithm requires only $O(NMd)$ flops*.

†This work was supported in part by the Joint Services Program at Stanford University (US Army, US Navy, US Air Force) under Contract DAAL03-88-C-0011, and by grants from the General Electric Company and Boeing/ARGOSystems Inc.

II. Problem Statement and Signal Subspace Decomposition

The problem considered here is to estimate source DOAs and/or source waveforms based on the data received from an array of sensors. Suppose there are d noncoherent narrowband plane waves present. As a consequence, the data model for the array output can be written in an obvious notation as follows:

$$\boldsymbol{x}(t) = \mathbf{A}(\boldsymbol{\theta})\boldsymbol{s}(t) + \boldsymbol{n}(t), \tag{1}$$

where the d-vector $\boldsymbol{\theta}$ corresponds to the unknown DOAs and the columns of the $M \times d$ matrix $\mathbf{A}(\boldsymbol{\theta})$ are the *steering vectors* of the sensor array. The d-vector $\boldsymbol{s}(\cdot)$ contains the emitter signal waveforms and $\boldsymbol{n}(t)$ denotes additive noise. The array output $\boldsymbol{x}(t)$ is observed at N discrete time instances, $t = 1, \ldots, N$, and the objective is to estimate the number of signals d, the DOAs $\boldsymbol{\theta}$, and/or signal waveforms $\boldsymbol{s}(\cdot)$.

Clearly, the signal part (or noise-free part) of the data vectors $\boldsymbol{x}(t) = \mathbf{A}(\boldsymbol{\theta})\boldsymbol{s}(t)$ lies in a d-dimensional subspace $span\{\mathbf{A}(\boldsymbol{\theta})\}$, which is called *signal subspace*. Suppose that the noise vectors $\boldsymbol{n}(\cdot)$ are spatially white

*In this paper, additions are not counted for simplicity and a flop is defined to be a real multiplication.

1058-6393/91 $01.00 © 1991 IEEE

and uncorrelated with the signal. Then, the covariance matrix of $\boldsymbol{x}(\cdot)$ can be written as:

$$\boldsymbol{R}_x = E\{\boldsymbol{x}(t)\boldsymbol{x}^H(t)\} = \mathbf{A}(\boldsymbol{\theta})\boldsymbol{R}_s\mathbf{A}^H(\boldsymbol{\theta}) + \sigma_n^2\boldsymbol{I}, \quad (2)$$

where $\boldsymbol{R}_s \in \mathcal{C}^{d\times d}$ and $\sigma_n^2\boldsymbol{I} \in \Re^{M\times M}$ are the signal and noise covariance matrices, respectively. If we perform an eigendecomposition of $\boldsymbol{R}_x$,

$$\boldsymbol{R}_x = \sum_{k=1}^{M} \lambda_k \mathbf{e}_k \mathbf{e}_k^H, \quad (3)$$

it is not difficult to see that all the eigenvalues λ_k have the following distribution: $\lambda_1 \geq \cdots \geq \lambda_d \geq \lambda_{d+1} = \cdots = \lambda_M$, based on which the d can be easily determined. It is also well-known that the signal subspace is the column span of the first d eigenvectors corresponding to the eigenvalues $\lambda_1, \ldots, \lambda_d$.

In practice, however, we usually do not have an ideal covariance matrix $\boldsymbol{R}_x$ but a sample covariance $\hat{\boldsymbol{R}}_x$ estimated from a *finite* number (N) of snapshots.

$$\hat{\boldsymbol{R}}_x = \frac{1}{N}\sum_{t=1}^{N} \boldsymbol{x}(t)\boldsymbol{x}^H(t). \quad (4)$$

In such a circumstance, we cannot *exactly* find the signal subspace but only estimate the signal subspace by the first d eigenvectors $\{\hat{\mathbf{e}}_1, \ldots, \hat{\mathbf{e}}_d\}$ of $\hat{\boldsymbol{R}}_x$. And it has been shown that this estimate is the *unstructured* maximum likelihood estimate of the true signal subspace [3]. Another problem is to estimate the dimension of the signal subspace. Unlike the eigenvalues of an ideal covariance matrix, the $M - d$ smaller eigenvalues of $\hat{\boldsymbol{R}}_x$, *i.e.*, $\hat{\lambda}_{\hat{d}}, \ldots, \hat{\lambda}_M$ are not *exactly* repeated. However, they are clustered in a small region of size $O(N^{-1/2})$ [4] and this property can be utilized for detection purpose. In the following, we list several important detection schemes, *i.e.*, the Minimum Description Length (MDL) criterion, and An Information Criterion (AIC), and the likelihood ratio test (LR). The basic idea of these detection schemes is to find $\hat{d}$ by checking whether $\hat{\lambda}_{\hat{d}+1}, \ldots, \hat{\lambda}_M$ are close to one another or equivalently by checking whether their arithmetic and geometric means are close to each other. According to [4], the following statistic

$$L(\hat{d}) = -N(M-\hat{d})\log\left\{\frac{\left(\prod_{k=\hat{d}+1}^{M}\hat{\lambda}_k\right)^{1/(M-\hat{d})}}{\frac{1}{M-\hat{d}}\sum_{k=\hat{d}+1}^{M}\hat{\lambda}_k}\right\} \quad (5)$$

is a likelihood ratio, which has the following property:

$$L(\hat{d}) = \begin{cases} O(N), & \hat{d} < d \\ O(1) \overset{N\to\infty}{\sim} \mathcal{X}^2((M-\hat{d})^2-1), & \hat{d} = d, \end{cases} \quad (6)$$

where $\mathcal{X}^2(p)$ denotes the chi-square distribution with degrees of freedom p. In fact, the fundamental change of $L(\hat{d})$ when $\hat{d} = d$ is the crucial part of these detection schemes. Knowing the limiting distribution of $L(\hat{d})$ when $\hat{d} = d$ and a desired confidence level (*e.g.*, 99%), we can determine a threshold, which is compared against $L(\hat{d})$ to estimate d. This leads to a sequential likelihood ratio (LR) test. We can also rely on more objective information theoretic criteria, *i.e.*, AIC and MDL criteria:

$$AIC(\hat{d}) = L(\hat{d}) + \hat{d}(2M - \hat{d}) \quad (7)$$

$$MDL(\hat{d}) = L(\hat{d}) + \frac{1}{2}\hat{d}(2M - \hat{d})\log N. \quad (8)$$

The minimizers of the AIC and MDL criteria are chosen as the AIC and MDL estimates of the d. By [5], MDL is strongly consistent, *i.e.*, the probability of correct detection approaches one as $N \to \infty$.

The above results are all based on the ED of the covariance matrix, *i.e.*, $\hat{\boldsymbol{R}}_x = \boldsymbol{X}^H\boldsymbol{X}$, where $\boldsymbol{X} = \frac{1}{N}[\boldsymbol{x}(1), \ldots, \boldsymbol{x}(N)]^H$ is called a data matrix. It is not too difficult to see that the singular values and right singular vectors of $\boldsymbol{X}$ are the eigenvalues of $\hat{\boldsymbol{R}}_x = \boldsymbol{X}^H\boldsymbol{X}$, respectively. Therefore, the signal subspace can be estimated from the data matrix $\boldsymbol{X}$ directly. The latter has additional advantages since squaring the matrix may cause loss of accuracy and squaring of the matrix condition number. At any rate, both approaches require $O(NM^2) + O(M^3)$ flops.

III. FSD for Covariance Matrices

In [1], we presented a new $O(M^2d)$ fast subspace decomposition (FSD) approach for covariance matrices, using the well-known tridiagonalization Lanczos (Tri-Lanczos) algorithm. The basic idea is to partially tridiagonalize a matrix via *e.g.*, the Lanczos algorithm, then to do an ED of the obtained partial tridiagonal matrix $\boldsymbol{T}_j$ at each (say the jth) Lanczos step. In [1], we showed that at the $(d+2)^{th}$ step, the largest d eigenvalues of $\boldsymbol{T}_j$, called Rayleigh-Ritz (RR) values, are *asymptotically equivalent* to the first d eigenvalues of $\hat{\boldsymbol{R}}_x$. Although we do not have all the eigenvalues required by the above detection schemes, with the RR values, the obtainable quadratic sum and the arithmetic sum of all the eigenvalues (*i.e.*, $\|\hat{\boldsymbol{R}}_x\|_F^2$ and $\mathrm{Tr}(\hat{\boldsymbol{R}}_x)$), we developed a detection scheme (Tri-FSD) (at the jth Lanczos step) as follows:

1. Set $\hat{d} = 1$.

2. Set null hypothesis $\mathrm{H}_0 : d = \hat{d}$.

3. Properly choose a threshold $\gamma_{\hat{d}}$ based on the tail area of the $\mathcal{X}^2$ distribution with $(M-\hat{d})^2-1$ degrees of freedom.

4. Evaluate the test statistic $\varphi_{\hat{d}}$.

5. If $2\varphi_{\hat{d}} \leq \gamma_{\hat{d}} c(N)$, accept H_0 and stop.

6. If $\varphi_{\hat{d}} > \gamma_{\hat{d}} c(N)$, reject H_0; if $\hat{d} < j-2$, $\hat{d} := \hat{d}+1$, return to 2. Otherwise, $j := j+1$, continue the $(j+1)^{th}$ Lanczos step.

The statistic $\varphi_{\hat{d}}$ of the above detection scheme is given by

$$\varphi_{\hat{d}} = N(M-\hat{d})\log\left(\frac{\sqrt{\frac{1}{M-\hat{d}}\left(\|\hat{\boldsymbol{R}}_x\|^2 - \sum_{k=1}^{\hat{d}} \theta_k^{(j)2}\right)}}{\frac{1}{M-\hat{d}}\left(\mathrm{Tr}(\hat{\boldsymbol{R}}_x) - \sum_{k=1}^{\hat{d}} \theta_k^{(j)}\right)}\right), \quad (9)$$

where $\{\theta_k^{(j)}\}_{k=1}^{j-2}$ are the RR values at the jth Lanczos step. Once the detection scheme is finished, the whole algorithm terminates and the basis vectors of the signal subspace are estimated by multiplying the eigenvectors of $\boldsymbol{T}_j$ with the so-called Lanczos basis vectors $\boldsymbol{q}_1, \ldots, \boldsymbol{q}_j$. By [6], for $c(N)$ satisfying the following

$$\lim_{N\to\infty}\frac{c(N)}{N} = 0, \quad \lim_{N\to\infty}\frac{c(N)}{\log\log N} = \infty. \quad (10)$$

we showed that the above detection scheme is *strongly consistent.* In [6], we also proved that signal subspace estimate from Tri-FSD is *asymptotically equivalent* to that obtained by the more costly ED.

Though the above FSD algorithm avoids the more costly ED part, it still relies on the covariance matrix $\hat{\boldsymbol{R}}_x$, which is formed at the cost of $O(NM^2)$ flops. Of course, in real-time implementations, $\hat{\boldsymbol{R}}_x$ can be computed in a recursively and/or parallel way. Nevertheless, if this is not possible, then forming the covariance matrix becomes the major part of the whole algorithm. Also, as mentioned before, squaring the data matrix is not a good numerical approach. Motivated by these facts, we shall present another FSD (Bi-FSD) method that *directly* proceeds from the data matrix using the so-called bidiagonalization Lanczos (Bi-Lanczos) method.

IV. FSD for Data Matrices

For Hermitian matrices, Tri-Lanczos gives the Lanczos basis $\boldsymbol{Q}_j$ and the partial tridiagonal matrix $\boldsymbol{T}_j$. The detailed procedure can be summarized as follows:

<u>The Tri-Lanczos Algorithm</u>:[7]

Given $\hat{\boldsymbol{R}}_x$ (Hermitian); $\boldsymbol{r}_0 = \boldsymbol{q}_1$ ($\|\boldsymbol{q}_1\| = 1$); $j = 0$
while $\beta_j^{(t)} \neq 0$
$\quad \boldsymbol{q}_{j+1} = \boldsymbol{r}_j/\beta_j^{(t)}$; $j := j+1$; $\alpha_j^{(t)} = \boldsymbol{q}_j^H \hat{\boldsymbol{R}}_x \boldsymbol{q}_j$
$\quad \boldsymbol{r}_j = \hat{\boldsymbol{R}}_x \boldsymbol{q}_j - \alpha_j^{(t)}\boldsymbol{q}_j - \beta_{j-1}^{(t)}\boldsymbol{q}_{j-1}$; $\beta_j^{(t)} = \|\boldsymbol{r}_j\|$
end

$$\mathbf{T_j} = \begin{bmatrix} \alpha_1^{(t)} & \beta_1^{(t)} & & & \\ \beta_1^{(t)} & \alpha_2^{(t)} & \beta_2^{(t)} & & \\ & \beta_2^{(t)} & \ddots & \ddots & \\ & & \ddots & \alpha_{j-1}^{(t)} & \beta_{j-1}^{(t)} \\ & & & \beta_{j-1}^{(t)} & \alpha_j^{(t)} \end{bmatrix} \quad (11)$$

and $\boldsymbol{Q}_j = [\boldsymbol{q}_1, \ldots, \boldsymbol{q}_j]$. However, the above procedure only partially tridiagonalize $\hat{\boldsymbol{R}}_x$ (*Hermitian*) and is not applicable to a *rectangular* data matrix $\boldsymbol{X}$. However, there is another version of the Lanczos algorithm that bidiagonalizes any rectangular matrix, *i.e.,* the bidiagonalization Lanczos (Bi-Lanczos) algorithm, which is summarized below.

<u>The Bi-Lanczos Algorithm</u>: [7]

Given $\boldsymbol{X}$ ($N \times M$); $\mathbf{p}_0 = \boldsymbol{v}_1$ ($\|\boldsymbol{v}_1\| = 1$); $\beta_0^{(b)} = 1$
$\quad j = 0$; $\boldsymbol{u}_0 = 0$
while $\beta_j^{(b)} \neq 0$
$\quad \boldsymbol{v}_{j+1} = \mathbf{p}_j/\beta_j^{(b)}$; $j := j+1$
$\quad \boldsymbol{r}_j = \boldsymbol{X}\boldsymbol{v}_j - \beta_{j-1}^{(b)}\boldsymbol{u}_{j-1}$; $\alpha_j^{(b)} = \|\boldsymbol{r}_j\|$
$\quad \boldsymbol{u}_j = \boldsymbol{r}_j/\alpha_j^{(b)}$
$\quad \mathbf{p}_j = \boldsymbol{X}^H\boldsymbol{u}_j - \alpha_j^{(b)}\boldsymbol{v}_j$; $\beta_j^{(b)} = \|\mathbf{p}_j\|$
end
Let the partial bidiagonal matrix be

$$\mathbf{B}_j = \begin{bmatrix} \alpha_1^{(b)} & \beta_1^{(b)} & & \cdots & 0 \\ 0 & \alpha_2^{(b)} & \ddots & & \vdots \\ & \ddots & \ddots & \ddots & \\ \vdots & & \ddots & \ddots & \beta_{j-1}^{(b)} \\ 0 & \cdots & & 0 & \alpha_j^{(b)} \end{bmatrix} \quad (12)$$

and $\boldsymbol{V}_j = [\boldsymbol{v}_1, \ldots, \boldsymbol{v}_j]$. Then, the following lemma shows that Bi-Lanczos applied to $\boldsymbol{X}$ yields the same result as Tri-Lanczos applied to $\hat{\boldsymbol{R}}_x$.

Lemma 1 *For any $N \times M$ matrix $\boldsymbol{X}$, apply the Tri-Lanczos algorithm to $\boldsymbol{X}^H\boldsymbol{X}$ and the Bi-Lanczos algorithms to $\boldsymbol{X}$. If both methods use the same initial vector,* i.e. *if $\boldsymbol{q}_1 = \boldsymbol{v}_1$, then*

1. $\boldsymbol{Q}_j = \boldsymbol{V}_j$, $j = 1, 2, \ldots, M$;

2. $\mathbf{T}_j = \mathbf{B}_j^H \mathbf{B}_j$.

The proof of this lemma is in [2]. From Lemma 1, if we perform an SVD of $\mathbf{B}_j$ instead of an ED of $\boldsymbol{T}_j$, then Bi-Lanczos can be used to replace Tri-Lanczos in Tri-FSD. According to the above Bi-Lanczos algorithm, we can also easily see that the major computation of each step is two matrix vector, *i.e.*, $\boldsymbol{X}\boldsymbol{v}_j$ and $\boldsymbol{X}^H\boldsymbol{u}_j$, and the total computational complexity is $O(NMd)$. Now the only problem is in the FSD detection schemes. In Tri-FSD, the detection scheme relies on $\mathrm{Tr}(\boldsymbol{X}^H\boldsymbol{X})$ and $\|\boldsymbol{X}^H\boldsymbol{X}\|_F^2$. Since $\mathrm{Tr}(\boldsymbol{X}^H\boldsymbol{X})=\|\boldsymbol{X}\|_F^2$, it can be found with NM flops. However, $\|\boldsymbol{X}^H\boldsymbol{X}\|_F^2$ cannot be found at the cost lower than $O(NM^2)$. Therefore, such a detection scheme is not useful if we want to find the signal subpace at the cost of $O(NMd)$ flops. Therefore, we shall design a new detection scheme for this algorithm. The new statistic is $\phi_{\hat{d}}$,

$$\phi_{\hat{d}} = \sqrt{N}(M-\hat{d}+1)\log(\hat{\sigma}_{\hat{d}}^{(j)}/\hat{\sigma}_{\hat{d}+1}^{(j)}) \tag{13}$$

where $\hat{\sigma}_{\hat{d}} = \frac{1}{M-\hat{d}}(\|\frac{1}{\sqrt{N}}\boldsymbol{X}\|_F^2 - \sum_{k=1}^{j}\theta_k^{(j)^2})$ and $\{\theta_k^{(j)}\}_{k=1}^{j}$ are the singular values of $\mathbf{B}_j$. The basic idea of this detection scheme is as follows. In [1], $\theta_k^{(j)}$ is *asymptotically equivalent* to $\hat{\lambda}_k$ for $j \geq d+2$ and $1 \leq k \leq d$. Knowing that $\boldsymbol{X}\|_F^2 = \sum_{k=1}^{M}\sigma_k^2 = \sum_{k=1}^{M}\hat{\lambda}_k$, we can see that $\sigma_{\hat{d}}$ is the average of the last $M-\hat{d}$ eigenvalues of $\hat{\boldsymbol{R}}_x$. Since $\hat{\lambda}_{d+1},\ldots,\hat{\lambda}_M$ are clustered in a small region with size $O(N^{-1/2})$, hence, $\hat{\lambda}_{\hat{d}}^{(j)} = \sigma_n^2+O(N^{-1/2})$, for $\hat{d} \geq d$ and $j \geq d+2$ or $\log(\hat{\sigma}_{\hat{d}}^{(j)}/\hat{\sigma}_{\hat{d}+1}^{(j)}) \sim O(N^{-1/2})$. In other cases, $\sigma_k^{(j)}$ contains all the principal eigenvalue component; and it is not difficult to see that $\log(\hat{\sigma}_{\hat{d}}^{(j)}/\hat{\sigma}_{\hat{d}+1}^{(j)}) \sim O(1)$. Therefore,

$$\phi_{\hat{d}} = \sqrt{N}(M-\hat{d}+1)\log(\hat{\sigma}_{\hat{d}}^{(j)}/\hat{\sigma}_{\hat{d}+1}^{(j)}) = \begin{cases} O(\sqrt{N}), & \hat{d}, j \leq d \\ O(1), & j > d \text{ and } \hat{d} \leq d. \end{cases} \tag{14}$$

The fundamental difference between ϕ_{d-1} and ϕ_d can be used for the detection purpose, as was done previously. Nevertheless, the limiting distribution of ϕ_d is no longer $\mathcal{X}^2$. However, it is independent of the signal subspace and has been approximated in [8]. With its approximate distribution, we can determine the threshold $\gamma_{\hat{d}}$ for any desired confidence level (*e.g.*, 99% or 99.9%. The new (Bi-FSD) detection scheme is the same as the one in Section III, except that the $2\varphi_{\hat{d}}$ is replaced by $\phi_{\hat{d}}$. The Bi-FSD detection can also be shown to be *strongly consistent* for $c(N)$ satisfying

$$\lim_{N\to\infty}\frac{c(N)}{\sqrt{N}} = 0, \quad \lim_{N\to\infty}\frac{c(N)}{\sqrt{\log\log N}} = \infty. \tag{15}$$

A detailed proof can be found in [2]. With this detection scheme, the Bi-FSD algorithm is complete and it is listed below.

THE BI-FSD ALGORITHM

1. Properly select $\boldsymbol{v}_1 = \boldsymbol{f}$ ($\|\boldsymbol{f}\| = 1$).
2. Execute the jth Bi-Lanczos step.
3. Compute the singular values of $\mathbf{B}_j$, $\theta_k^{(j)}$, $k \leq j$.
4. Execute the Bi-FSD detection scheme. If H_0 is accepted, go to step 5. Otherwise, $j := j+1$, return to step 2.
5. Pick the d principal right singular vectors $\{\boldsymbol{y}_k^{(j)}\}$ of $\mathbf{B}_j$. The final signal subspace estimate is $span\{\boldsymbol{V}_j\boldsymbol{y}_1^{(j)},\ldots,\boldsymbol{V}_j\boldsymbol{y}_d^{(j)}\}$. Stop.

By Lemma 1, the estimates obtained from Bi-FSD is the same as those from Tri-FSD (without counting the numerical errors); thus the Bi-FSD algorithm is also *asymptotically equivalent* to the most costly SVD of $\boldsymbol{X}$ and ED of $\hat{\boldsymbol{R}}_x$.

V. SIMULATION RESULTS

To verify our theoretic claims, we conducted computer simulations. In the designed simulation scenarios, there are two uncorrelated equal-power sources with DOAs 35° and 50°, impinging on a 20-element uniform linear array (ULA). Bi-FSD and SVD were performed along with many different detection schemes, *i.e.*, the Bi-FSD detection scheme (using the RR values) and the MDL, AIC, LR schemes (using all the eigenvalues). For the Bi-FSD detection schemes, we estimated the limiting distribution of $\phi_{\hat{d}}$ by numerical methods for different choices of $\hat{d}$ and M. To make the Bi-FSD algorithm more robust, we only implement the detection scheme using the reasonably converged RR values (*i.e.*, RR values that are 10% different from counterparts evaluated at the previous step). The 98% confidence level is designed, based on which and the distribution of $\phi_{\hat{d}}$, the threshold $\gamma_{\hat{d}}$ is determined. In this example, we just chose $c(N) = \sqrt{\log N}$. Obviously, such a $c(N)$ satisfies (15). The signal-to-noise-ratios (SNRs) of both sources varied from −16dB to 0dB. Once the signal subspace estimates are found, ESPRIT is carried out to estimate the source DOAs. The root-mean-square-error

(RMSE) of DOA estimates is used as a criterion to evaluate the estimation of the signal subspace.

Figure 1 shows the probability of correct detection by different detection schemes. From this figure, we can see that AIC, LR are not strongly consistent, *i.e.*, the probability of correction detection does not approach to one as the scenario gets close to the asymptotic region (large SNR). On the other hand, the Bi-FSD and MDL schemes always make correct detection when SNR≥ -8dB. It is interesting to note that the FSD algorithm performs slightly better than the MDL criterion in the nonasymptotic region (SNR$= -7$dB). Figure 2 illustrates the results of DOA estimation in cases where the MDL and Bi-FSD detection schemes perform well. From this figure, we can observe that the Bi-FSD algorithm performs almost the same as the more costly SVD for estimating the signal subspace, although the SVD algorithm is slightly better when SNR$= -8$dB. The flops were counted for both methods in the matlab program. A factor of 8.5 computational reduction is achieved by Bi-FSD versus SVD. In other words, the simulation results seem to be consistent with the fact that Bi-FSD achieves the same asymptotic performance as the SVD algorithm with *significantly* less cost.

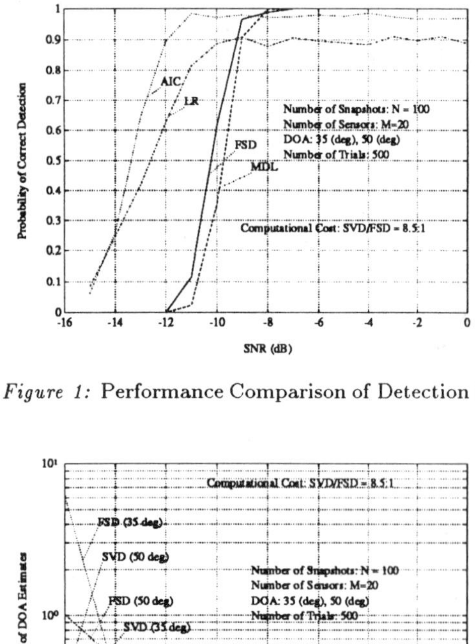

Figure 1: Performance Comparison of Detection Schemes

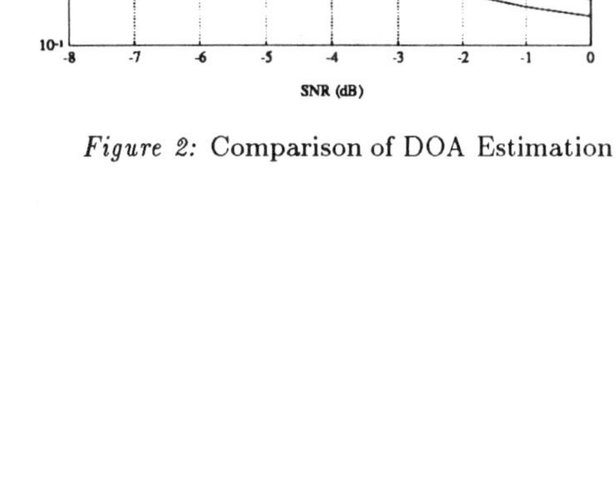

Figure 2: Comparison of DOA Estimation

VI. Concluding Remarks

In this paper, we present a fast subspace decomposition method (Bi-FSD) for (rectangular) data matrices, employing the bi-diagonalization Lanczos algorithm. It only requires $O(NMd)$ flops for a $N \times M$ data matrix and achieves almost an order of magnitude computational reduction over the $O(NM^2+M^3)$ SVD or ED approach. A new detection scheme is also presented that can be implemented at each intermediate step of estimating the signal subspace. Unlike many fast algorithms that trade performance for speed, the rigorous performance analysis shows that Bi-FSD has the *same asymptotic performance* as the more costly SVD and the Bi-FSD detection scheme is *strongly consistent.* Also, the most computationally intensive part (*i.e.*, O(NM) operations) is $O(d)$ matrix-vector products, which can be easily implemented in parallel for even faster computation [9]. All these nice features of the Bi-FSD algorithm make it more possible to implement a class of high-resolution array signal processing algorithms in *real time.*

References

[1] G. Xu, R.H. Roy, and T. Kailath, "Fast ESPRIT – Application of Fast Signal Subspace Decomposition in Array Processing", In *Proc. of the 24th Asilomar Conference on Signals, Systems, and Computers*, pages 961–965, Pacific Grove, CA, Nov. 1990.

[2] G. Xu, *Fast Subspace Decomposition and Its Applications*, PhD thesis, Stanford University, Stanford, CA, September 1991.

[3] G. Bienvenu and L. Kopp, "Optimality of High Resolution Array Processing", *IEEE Trans. ASSP*, **ASSP-31**(5):953–964, October 1983.

[4] T.W. Anderson, "Asymptotic Theory for Principal Component Analysis", Ann. Math. Statist., **34**:122–148, 1963.

[5] M. Wax and T. Kailath, "Detection of Signals by Information Theoretic Criteria", *IEEE Trans. ASSP*, **ASSP-33**(2):387–392, April 1985.

[6] G. Xu and T. Kailath, "Fast Signal Subspace Decomposition – Part II: Sample Covariance Matrices", *Submitted to IEEE Trans. on Signal Processing, December, 1990.*

[7] G.H. Golub and C.F. Van Loan, *Matrix Computations*, The Johns Hopkins University Press, Baltimore, Maryland, 1990.

[8] G. Xu and T. Kailath, "Applications of Fast Signal Subspace Decomposition to Signal Processing and Communications Problems", *Submitted to IEEE Trans. on Signal Processing, Novermber, 1991.*

[9] S.Y. Kung, *VLSI Array Processors*, Prentice Hall, Englewood Cliffs, NJ, 1987.

Beamspace Domain Bearing Estimation for Fast Target Localization Using an Array of Antennas

Ta-Sung Lee
Dept. of Communication Engineering
National Chiao Tung University
Hsinchu, Taiwan, R.O.C.

Michael Zoltowski
School of Electrical Engineering
Purdue University
W. Lafayette, Indiana, U.S.A.

Abstract

We here develop a novel eigen-based bearing estimation scheme for monopulse radar tracking. The eigen-based estimator is chosen in light of its high accuracy. Specifically, we derive the MUSIC estimator of the target angle for the amplitude comparison monopulse radar, when given as data two beams formed by processing the entire set of array signals with two different sets of weights. The conversion to 2-D beamspace domain in this fashion facilitates a simple, closed-form MUSIC procedure when a generalized Butler type beamformer is employed. It is shown that under certain approximations, the MUSIC estimator reduces to the conventional amplitude/sum-and-difference estimator.

1 Introduction

Conventional monopulse radar tracking techniques may be implemented with the use of phased arrays. In this mode of operation, two beams are formed with two sets of weights applied to the array sensors facilitating a direct construction of the sum and difference beams. In spite of the fact that the monopulse scheme was developed on an ad-hoc basis, Davis et. al. [1] show that the amplitude/sum-and-difference technique is, in fact, the Maximum Likelihood (ML) estimator of the target angle given as data the sum and difference beam outputs.

In this paper, we develop a novel eigen-based monopulse bearing estimation technique for accurate, efficient target tracking. We do so by developing a 2-D beamspace domain MUSIC (BDMU) estimation procedure [2]. Beamspace operation represents a means of formulating the array processing problems in terms of a lower dimensionality. As each beam output signal is formed as the weighted sum of a large number of array sensor outputs, it is often adequate to assume Gaussian statistics in beamspace according to the central limit theorem. In order to obtain a closed-form angle estimate, the two beams are chosen as two adjacent generalized Butler beams [3] in the vicinity of the target. A distinctive property of the generalized Butler beamformer is that any set of two beams have all but two nulls in common. Simulation results are presented to demonstrate the performance of the BDMU technique in a one-target environment.

2 BDMU Estimator for One Target

We here consider the single target scenario encountered in monopulse radar tracking. We will here assume that the antenna elements comprising the array are identical and uniformly-spaced by a half-wavelength. We will also assume that the target is in the far field of the array such that the resulting echoes may be modeled as planewaves. Finally, we will invoke the narrowband signal model such that time delay translates into a pure phase shift. Under these conditions, the data snapshot vector $\mathbf{x}$, composed of the complex low-pass signals sensed at each of the M array elements at a given sampling instant may be expressed as

$$\mathbf{x} = \mathbf{a}(u_t)\eta + \mathbf{n} \tag{1}$$

where $u_t=\sin(\theta_t)$ and θ_t is the angle of the target with respect to the broadside of the array. η is the complex envelope of the target echo measured at the center of the array such that the Mx1 array manifold vector is given by

$$\mathbf{a}(u) = \left[e^{-j\pi Lu}, \cdots, e^{-j\pi u}, 1, e^{j\pi u}, \cdots, e^{j\pi Lu}\right]^T \tag{2}$$

with $u=u_t$. Note that we assume M=2L+1 to be odd for convenience. The components of $\mathbf{n}$ represent the additive noise present at the M array element outputs. A 2x1 beamspace data snapshot vector $\mathbf{x}_B$ is formed by operating on $\mathbf{x}$ with an Mx2 matrix beamformer $\mathbf{S} = [\mathbf{s}_1 \mid \mathbf{s}_2]$:

$$\mathbf{x}_B = \mathbf{S}^H\mathbf{x} \tag{3}$$

where $\mathbf{s}_1$ and $\mathbf{s}_2$ are the weight vectors associated with the "left" and "right" beams, and H denotes complex conjugate transpose. $\mathbf{x}_B$ may be expressed in a format similar to that for $\mathbf{x}$:

$$\mathbf{x}_B = \mathbf{b}(u_t)\eta + \mathbf{n}_B \tag{4}$$

where $\mathbf{b}(u_t)=\mathbf{S}^H\mathbf{a}(u_t)$ is the "beamspace" manifold vector evaluated at $u=u_t$ and $\mathbf{n}_B=\mathbf{S}^H\mathbf{n}$ consists of the noise components present at the two beamformer outputs.

We here develop the MUSIC estimator of u_t given as data the beamspace data snapshot vector $\mathbf{x}_B$. The MUSIC algorithm applied in beamspace domain dictates that we determine a 2x1 complex vector $\mathbf{v}$ as the generalized eigenvector (GEV) of the 2x2 matrix pencil $\{\hat{\mathbf{R}}_{BB},\mathbf{Q}_{BB}\}$ associated with the smaller generalized eigenvalue, where $\hat{\mathbf{R}}_{BB}$ is an estimate of the ensemble beamspace domain data correlation matrix $\mathbf{R}_{BB}$ as defined by

$$\mathbf{R}_{BB} = E\{\mathbf{x}_B\mathbf{x}_B^H\} \tag{5}$$

and $\mathbf{Q}_{BB}$ is the ensemble beamspace domain noise correlation matrix given by

$$\mathbf{Q}_{BB} = E\{\mathbf{n}_B\mathbf{n}_B^H\} \tag{6}$$

Note that if the noise is spatially white, then $\mathbf{Q}_{BB}$ is proportional to $\mathbf{S}^H\mathbf{S}$. For brevity, we will refer to $\mathbf{v}$ as the "smaller" GEV of $\{\hat{\mathbf{R}}_{BB},\mathbf{Q}_{BB}\}$. The corresponding value of the target angle estimate $\hat{u}_t$ is then determined as that angle satisfying the orthogonality condition

$$\mathbf{v}^H\mathbf{b}(u) = \mathbf{v}^H\mathbf{S}^H\mathbf{a}(u) = 0 \tag{7}$$

Considering the four components of $\mathbf{R}_{BB}$ as unknown but deterministic, the determination of $\hat{\mathbf{R}}_{BB}$ based on $\mathbf{x}_B$ is a classical parameter estimation problem. This prompts us to develop a Maximum Likelihood (ML) procedure for $\hat{\mathbf{R}}_{BB}$. Since each of the two components of $\mathbf{n}_B$ is a weighted sum of a generally large number of independent random noise, we may invoke the central limit theorem to approximate the noise in beamspace domain as being Gaussian distributed. As a consequence, the ML estimate of $\mathbf{R}_{BB}$, given as data $\mathbf{x}_B$, is the sample beamspace data correlation matrix given by $\hat{\mathbf{R}}_{BB}=\mathbf{x}_B\mathbf{x}_B^H$. In summary, the BDMU procedure for estimating the target angle is described as follows. The ML beamspace data correlation matrix $\hat{\mathbf{R}}_{BB}$ is constructed first as above. A 2x1 vector $\mathbf{v}$ is then determined as the "smaller" GEV of $\{\hat{\mathbf{R}}_{BB},\mathbf{Q}_{BB}\}$. Finally, $\hat{u}_t$ is obtained as the solution to the nonlinear equation $\mathbf{v}^H\mathbf{b}(u)=\mathbf{v}^H\mathbf{S}^H\mathbf{a}(u)=0$ in u. Due to the Vandermonde structure of $\mathbf{a}(u)$, this may be formulated in terms of finding the roots of a polynomial having as coefficients the components of $(\mathbf{Sv})^*$, where * denotes complex conjugate. We expand on this idea by first defining a Vandermonde vector $\mathbf{z}$ as follows:

$$\mathbf{z} = [\ 1\ , z\ , z^2\ , \cdots , z^{M-1}\ ]^T \tag{8}$$

The inner product of any Mx1 vector with the Vandermonde vector $\mathbf{z}$ is then a polynomial of order M–1. The final step above may be formulated as finding $z=e^{j\pi u}$ as a root of the (M–1)-th order polynomial $e(z)=(\mathbf{Sv})^H\mathbf{z}$. Here we refer to e(z) as the signal polynomial.

3 Design of Beamformer

A major drawback of the aforementioned BDMU method is that the final step requires a polynomial rooting of order M–1. As a practical matter, the number of elements M may be of the order of a hundred such that rooting a polynomial of order (M–1) becomes infeasible in real-time monopulse tracking. Moreover, only one out of the M–1 roots are of interest to us and hence directly rooting the signal polynomial is obviously inefficient. A question may arise as to whether one can avoid rooting the large order polynomial and still get the desired root. In the low-angle tracking problem considered in [4], the property of common nulls of the Butler type beamformer was utilized to convert the signal polynomial into a quadratic one whose roots are corresponding to the direct and specular path angles. Note that two adjacent Butler beams have M–2 nulls in common. This is shown in Figure 1. The concept applied here suggests that we form a set of two Butler beams pointed in the vicinity of the target. It follows that the signal polynomial can be reduced to a first order one whose root corresponds to the target angle. However, a major issue regarding the Butler beamforming is its high sidelobe levels due to uniform weighting. This can seriously degrade the accuracy of the monopulse operation when external interferences, such as clutters and jammers exist. As a remedy, we introduce a class of generalized Butler beamforming methods which exhibit practical merits. A popular way of beamforming in practical radars is to employ the Chebyshev weighting. It is suggested [5] that the sum beam be formed first using the Chebyshev weighting. The difference beam is then formed by suitably rearranging the nulls associated with the sum beams. Analysis shows that the target angle estimate thus obtained is less affected by the error in the sum-and-difference ratio compared to the uniform weighting method. An interesting point noteworthy is that the sum and difference beams constructed in the above fashion have the same nulls except for one pair next to the main beam. This is actually a manifestation of the fact that the corresponding left and right beams have M–3 nulls in common. Motivated by the work in [5], we suggest that a set of two beams having M–2 common nulls be formed in such a fashion that they resemble a prescribed set of two Chebyshev beams. It is thus natural that we construct $\mathbf{s}_1$ and $\mathbf{s}_2$ in accordance with

$$\underset{\mathbf{s}_1\,\mathbf{s}_2}{\text{Minimize}}\ \ \|\mathbf{w}_1 - \mathbf{s}_1\|^2 + \|\mathbf{w}_2 - \mathbf{s}_2\|^2 \tag{9}$$

subject to: beam patterns associated with $\mathbf{s}_1$ and $\mathbf{s}_2$ have M–2 common nulls

where $\mathbf{w}_1$ and $\mathbf{w}_2$ are weight vectors associated with two adjacent Chebyshev beams. The solutions for $\mathbf{s}_1$ and $\mathbf{s}_2$ may be obtained via a multi-dimensional search for the common and uncommon null angles. This causes no practical problem since the construction of beamformers can be done prior to radar operation. A more efficient implementation of (9) is currently in progress. As an example, we choose $\mathbf{w}_1$ and $\mathbf{w}_2$ to be weight vectors associated with two Chebyshev beams with –25 dB sidelobes pointed at u=–1/M and u=1/M, respectively.

Note that the pointing angles of the two beams are separated by u=2/M equal to a 3-dB beamwidth of the array. Figure 2 shows the desired beam patterns for the case of M=21. Note also that if strong interferences exist, it may be necessary for each beam to put deep nulls in the interfering directions. In this case, these nulls must belong to the M–2 common nulls associated with the two beams.

4 Derivation of Target Angle Estimate:

The fact that the beam patterns associated with $\mathbf{s}_1$ and $\mathbf{s}_2$ have M–2 nulls in common, or that $\mathbf{s}_1^H\mathbf{z}$ and $\mathbf{s}_2^H\mathbf{z}$ have M–2 roots in common leads to the factorization

$$\mathbf{s}_1^H\mathbf{z} = c(z)q_1(z) \quad ; \quad \mathbf{s}_2^H\mathbf{z} = c(z)q_2(z) \tag{10}$$

where c(z) is a (M–2)-th order polynomial corresponding to the common roots and $q_i(z)$, i=1,2, are first order polynomials corresponding to the uncommon roots associated with the left and right beams, respectively. It follows that $e(z)=(\mathbf{Sv})^H\mathbf{z}$, as a linear combination of $\mathbf{s}_1^H\mathbf{z}$ and $\mathbf{s}_2^H\mathbf{z}$, has the factorization

$$e(z) = (\mathbf{Sv})^H\mathbf{z} = c(z)[v_1^*q_1(z) + v_2^*q_2(z)] \tag{11}$$

where $\mathbf{v}=[v_1,v_2]^T$ is as defined before. In rooting e(z), the factor c(z) is known and thus can be divided out to yield a first order polynomial $d(z)=v_1^*q_1(z)+v_2^*q_2(z)$. Making the practical assumption that the target angle u_t is not equal to any of the common null angles, the root of interest is then that of d(z). Let the two uncommon null angles be u_1 and u_2. The roots of $q_1(z)$ and $q_2(z)$ are simply $e^{j\pi u_1}$ and $e^{j\pi u_2}$, respectively. For convenience, we assume that $\mathbf{s}_1$ and $\mathbf{s}_2$ are properly nomalized so as to exhibit conjugate centro-symmetry (CCS). Note that the CCS property results from that all of the roots of $\mathbf{s}_1^H\mathbf{z}$ and $\mathbf{s}_2^H\mathbf{z}$ are on the unit circle. This guarantees that the coefficients of c(z), $q_1(z)$ and $q_2(z)$ will also exhibit CCS, after properly normalized. We thus have

$$q_1(z) = -je^{-j\pi u_1/2}z + je^{j\pi u_1/2}$$

$$q_2(z) = -je^{-j\pi u_2/2}z + je^{j\pi u_2/2} \tag{12}$$

Substituting (12) into d(z) and solving for the root $\hat{z}_t$ yields

$$\hat{z}_t = \frac{v_1^*e^{j\pi u_1/2} + v_2^*e^{j\pi u_2/2}}{v_1^*e^{-j\pi u_1/2} + v_2^*e^{-j\pi u_2/2}} \tag{13}$$

Note that under noise corruption, v_1 and v_2 are complex such that the desired root $\hat{z}_t$ will not in general lie on the unit circle. The best we can do then is to extract its phase angle and determine the target angle $\hat{u}_t$ according to $\hat{u}_t=\arg\{\hat{z}_t\}/\pi$. Some algebraic manipulations yield the following expression

$$\hat{u}_t = u_o + \frac{1}{\pi}\tan^{-1}\left\{\frac{2\mathrm{Re}\left\{\frac{\Delta_v}{\Sigma_v}\right\}\tan(\frac{\pi\delta_u}{2})}{\left|\frac{\Delta_v}{\Sigma_v}\right|^2 - \tan^2(\frac{\pi\delta_u}{2})}\right\} \tag{14}$$

where $u_o=(u_1+u_2)/2$, $\delta_u=(u_2-u_1)/2$, $\Sigma_v=v_1^*+v_2^*$ and $\Delta_v=v_1^*-v_2^*$. Notice that the expression in (14) involves the ratio of the "sum" to the "difference" of the two components of $\mathbf{v}^*$.

Expression (14) can be readily rewritten as a function of the beamspace data snapshot vector $\mathbf{x}_B$. To begin with, we express the beamspace sample data correlation matrix in the explicit form:

$$\hat{\mathbf{R}}_{BB} = \mathbf{x}_B\mathbf{x}_B^H = \begin{bmatrix} |x_1|^2 & x_1x_2^* \\ x_1^*x_2 & |x_2|^2 \end{bmatrix} \tag{15}$$

where $\mathbf{x}_B=[x_1,x_2]^T$. Due to the fact hat $\hat{\mathbf{R}}_{BB}$ is of rank one whose range space is spanned by $\mathbf{x}_B$, $\mathbf{v}$ is simply a complex vector orthogonal to $\mathbf{x}_B$ and may be chosen to be $\mathbf{v}=[x_2^*,-x_1^*]^T$. Now, substitution of the $\mathbf{v}$ vector thus obtained into (14) yields the following expression for the monopulse BDMU estimate:

$$\hat{u}_t = u_o + \frac{1}{\pi}\tan^{-1}\left\{\frac{2\mathrm{Re}\left\{\frac{\Sigma}{\Delta}\right\}\tan(\frac{\pi\delta_u}{2})}{\left|\frac{\Sigma}{\Delta}\right|^2 - \tan^2(\frac{\pi\delta_u}{2})}\right\} \tag{16}$$

where $\Sigma=x_1+x_2$ and $\Delta=x_2-x_1$ represent the complex-valued sum and difference signals as defined in amplitude-comparison monopulse radars. It is interesting to note that the angle estimate in (16) depends on the data x_1 and x_2 solely through the ratio of the sum signal to the difference signal, leading to an amplitude-comparison/sum-and-difference scheme. Also note that (16) does not depend upon the weight vectors explicitly.

Further simplification of (16) is possible if we take into account the fact that Σ/Δ is purely real under no noise condition, i.e., $\mathbf{n}_B=\mathbf{0}$:

$$\frac{\Sigma}{\Delta} = \frac{x_2 + x_1}{x_2 - x_1} = \frac{(\mathbf{s}_1 + \mathbf{s}_2)^H\mathbf{a}(u_t)}{(\mathbf{s}_1 - \mathbf{s}_2)^H\mathbf{a}(u_t)} \tag{17}$$

The assertion that Σ/Δ is real follows since $\mathbf{s}_1$, $\mathbf{s}_2$ and $\mathbf{a}(u_t)$ all exhibit CCS such that both the numerator and denominator of the far right of (17) are real. Consequently, as long as the SNR is moderately high, Σ/Δ will be approximately real. To incorporate this characteristic property, we replace Σ/Δ in (16) by $\mathrm{Re}\{\Sigma/\Delta\}$ and end up with the simplified expression for $\hat{u}_t$:

$$\hat{u}_t = u_o + \frac{2}{\pi}\tan^{-1}\left\{\mathrm{Re}\left\{\frac{\Delta}{\Sigma}\right\}\tan\left(\frac{\pi\delta_u}{2}\right)\right\} \tag{18}$$

which is similar in form to that derived by Gabriel [6]. It should be noted that the monopulse angle estimate given in [6] was derived on a heuristic basis, i. e., by exploiting the sinusoidal nature of the sum and difference illumination beams.

5 Simulations

Computer simulations were conducted to ascertain the performance of the BDMU estimation scheme developed within in a radar tracking scenario. The array employed was linear, consisting of M=21 identical elements uniformly-spaced by a half-wavelength at the operating frequency. The weight vectors $\mathbf{s}_1$ and $\mathbf{s}_2$ employed were those associated with the patterns shown in Figure 2. Echoes from a single target arrived at $\theta_t=\sin^{-1}(u_t)=5^\circ$ with respect to the broadside of the array and the two beams were steered to the vicinity of the target via progressive phase shifting. The additive noise was modeled as spatially white and uncorrelated with the received echoes. Finally, in all cases, sample means and sample standard deviations were computed from the results of 200 independent trials.

The first set of simulations evaluate the statistical performance of the BDMU method for seven different signal-to-noise ratio (SNR) values. The pointing angle $\theta_o=\sin^{-1}(u_o)$ of the two beams was 5° equal to the target angle. The resulting sample statistics were plotted in Figure 3. As would be expected, the bias and standard deviation drop as the SNR value increases. Note that the theoretical Cramer-Rao Lower Bound (CRLB) for the angle estimates was also included in Figure 3. Impressively, the BDMU estimator produced sample standard deviations nearly attaining the CRLB for SNR above −10 dB, demonstrating the accuracy of the proposed method. As for the situation at −10 dB, it is conceivable that there may exist other angle estimators based on the original data snapshot vector **x** that outperform the BDMU estimator under extremely low SNR. The price is that processing based on **x** requires a much heavier computational load than that of beamspace processing.

The second set of simulations examine the effect of error in the pointing angle of the two beams on the performance of the BDMU estimator considered in the first simulation. For each case, the SNR was fixed at 10 dB. The sample means and sample standard deviations obtained with seven pointing angles are shown in Figure 4. We find that the BDMU estimator performs best when the pointing angle θ_o equals the target angle θ_t as expected. Nevertheless, little difference between the respective sample statistics is observed ascertaining the robustness of the proposed method. As a remark, we find that the BDMU method can in fact handle a target from outside the main beams without ambiguities so long as the SNR is not too low and the target is not in the direction of one of the common nulls. This may again be attributed to the common nulls property of the beamformer.

References

[1] R. C. Davis, L. E. Brennan, and L. S. Reed, "Angle Estimation with Adaptive Arrays in External Noise Fields," *IEEE Trans. Aerospace and Electronic Systems*, vol. AES-12, no. 3, Mar. 1976, pp. 179-186.

[2] K. M. Buckley and X. L. Xu, "Reduced-Dimension Beam-Space Broad-Band Source Localization: Preprocessor Design," *Advanced Algorithms and Architectures for Signal Processing*, vol. 975, 1988, pp. 368-376.

[3] M. I. Skolnik, **Introduction to Radar Systems**, McGraw-Hill, New York, 1980.

[4] M. D. Zoltowski and T. S. Lee, "Maximum Likelihood Based Sensor Array Signal Processing in the Beamspace Domain for Low-angle Radar Tracking," *IEEE Trans. Signal Processing*, vol. 39, Mar. 1991, pp. 656-671.

[5] K. W. Lo and T. B. Vu, "Improving performance of monopulse phased array in direction finding," *IEE Proceedings-F*, vol. 135, No. 6, Dec. 1988, pp. 391-394.

[6] W. F. Gabriel, "A High-Resolution Target-Tracking Concept Using Spectral Techniques," *NRL Report 6109*, May 1984.

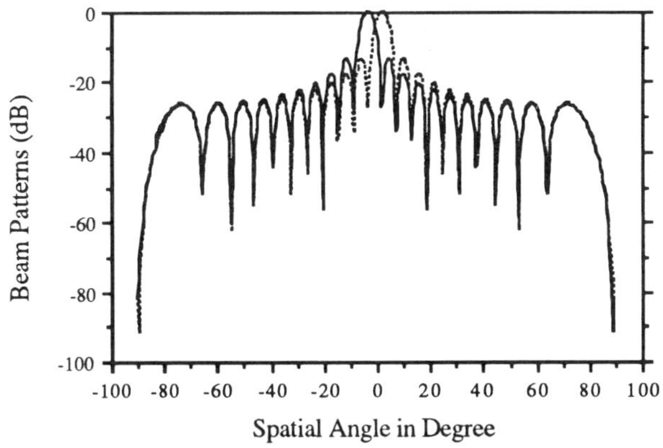

Fig 1 Superposition of two adjacent Butler beam patterns having M–2=19 common nulls

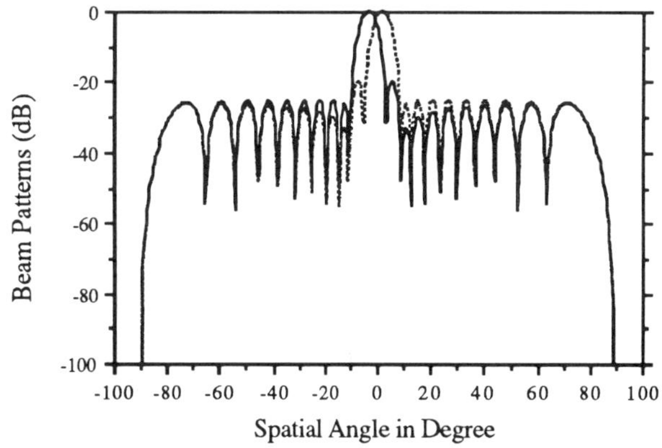

Fig 2 Superposition of two adjacent generalized Butler beam patterns having M–2=19 common nulls

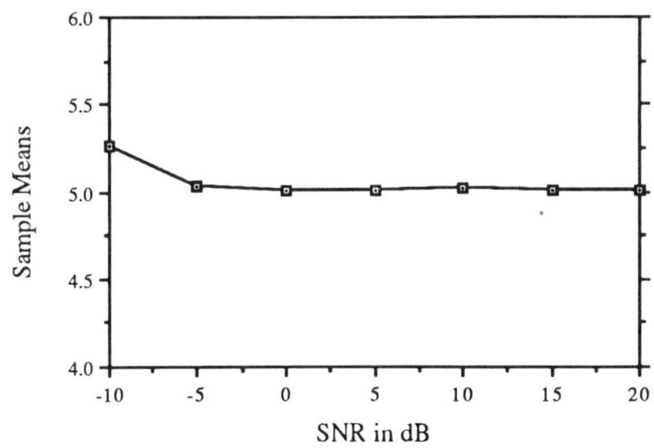

Fig 3 Sample statistics of the BDMU target angle estimates for seven SNR values

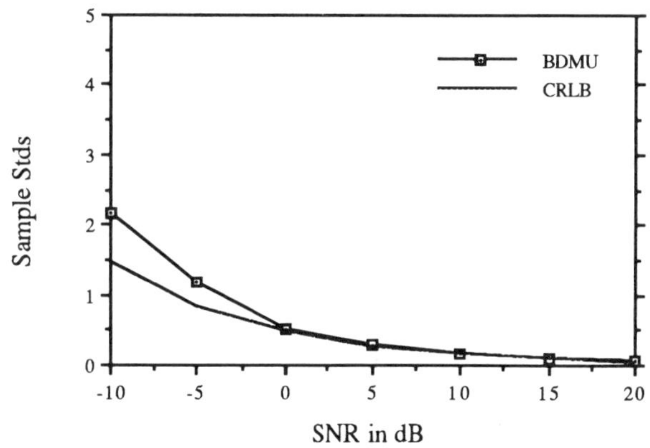

Fig 3 (continued)

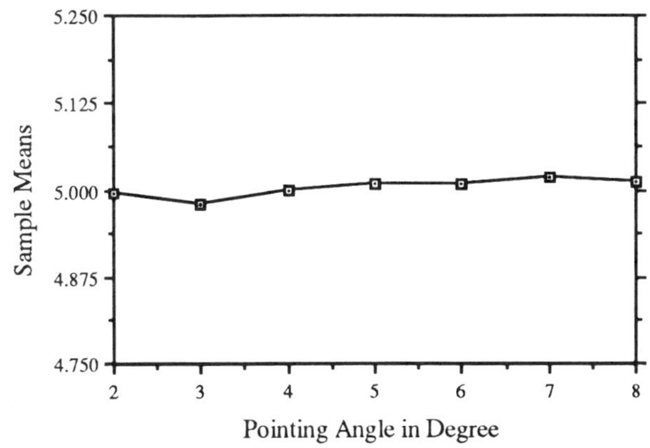

Fig 4 Sample statistics of the BDMU target angle estimates for seven beam pointing angles

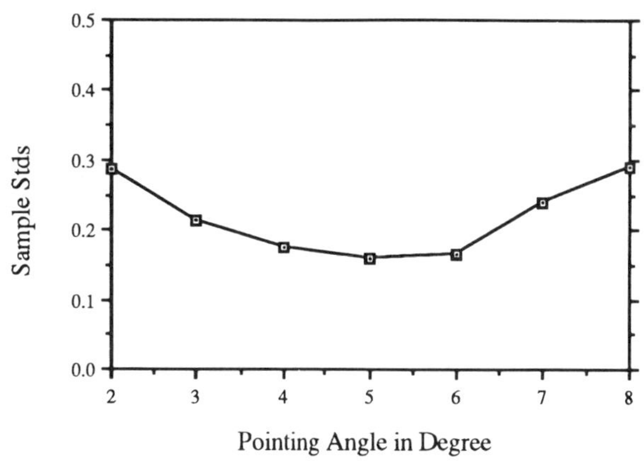

Fig 4 (continued)

Optimal Dimension Reduction for Sensor Array Signal Processing

SÖREN ANDERSON
Department of Electrical Engineering
Linköping University
S-581 83 Linköping, Sweden

Abstract

The computational complexity for direction-of-arrival estimation using sensor arrays increase very rapidly with the number of sensors in the array. One way to lower the amount of computations is to employ some kind of reduction of the data dimension. This is usually accomplished by employing linear transformations for mapping full dimension data into a lower dimensional space. In this paper, a transformation matrix is derived that makes it possible to attain the full-dimension Cramér-Rao bound also in the reduced space. A bound on the dimension of the reduced data set is given, above which it is always possible to obtain the same accuracy for the lower-dimension estimates of the source localizations as that achievable by using the full dimension data. Furthermore, a method is devised for designing the transformation matrix.

1 Introduction

The interest in performance improvement for Direction-Of-Arrival (DOA) estimation algorithms can result in antenna arrays composed of a large number of sensor elements. Since the computational requirements are directly affected by the dimension of the collected data, the burden increases rapidly with the number of sensors. This increase is often proportional to the cube of the dimension, m, of the Element SPace (ESP), *i.e.*, of the dimension of the data. This is the case, *e.g.*, for high-resolution methods based on eigendecomposition of the array covariance matrix, which requires $\mathcal{O}(m^3)$ operations, [1]. Therefore, in order to lower the computational requirements, it is useful to introduce some sort of mapping that reduces the dimension of the data set, from m to r ($r << m$). The space into which the set of full dimensional data is mapped is referred to as the Reduced Dimension BeamSpace (RDBS). One natural approach to reduce the data dimension is to employ a linear (matrix) transformation for mapping the full dimension ESP data into the lower dimensional RDBS, and then apply a signal processing algorithm to this new set of data. The design of the matrix transformation is guided by some subjective criterion as, for instance: selection of spatial sector, [2, 3], maximization of average signal-to-noise ratio within a specified sector, [4]. In [5], the design was made by analytic determination of a transformation matrix through minimization of the RDBS signal-to-noise ratio (SNR) at which two closely spaced emitters can be resolved. The derivation was carried out using the framework of the MUSIC algorithm, [6]. The main purpose of this paper is to analyze under which circumstances it is possible to retain the property of asymptotically achieving the full-dimension CRB, while significantly reducing the dimension of the data set used for the DOA estimation. Somewhat surprisingly, it is possible to, in a very simple way, obtain analytical expressions for the requirements to be met by an optimal transformation matrix. These expressions rely on knowledge of the functional form of the array (calibrated array of arbitrary geometry) and the true locations of the sources present. This last condition is, however, not limiting the possibilities of using the results derived herein to outline a design method.

2 Data Model

Consider an array of arbitrary geometry consisting of m sensor elements. Assume that d wavefronts impinge on the array, due to d far field sources emitting narrowband signals. The output of the sensor array can then be modeled as a weighted superposition of the d wavefronts, corrupted by additive sensor noise, uncorrelated with the emitter signals. That is, the output of the ith sensor is represented by

$$x_i(t) = \sum_{j=1}^{d} a_i(\theta_j)s_j(t) + n_i(t) \ . \qquad (1)$$

Here, $a_i(\theta_j)$ is a complex scalar representing the propagation delay of the jth emitter signal and the gain and phase adjustments by the ith sensor. In matrix notation we obtain

$$\mathbf{x}(t) = [\,\mathbf{a}(\theta_1),\ \ldots,\mathbf{a}(\theta_d)\,]\mathbf{s}(t)+\mathbf{n}(t) = \mathbf{A}(\boldsymbol{\theta}_0)\mathbf{s}(t)+\mathbf{n}(t)\ , \tag{2}$$

where $\boldsymbol{\theta}_0$ is a d-dimensional parameter vector corresponding to the true signal parameters. The *array response* vector $\mathbf{a}(\theta_j) = [\ a_1(\theta_j),\ \ldots, a_m(\theta_j)\ ]^T$ contains the sensor responses to a unit wavefront from the direction θ_j. The additive noise, $\mathbf{n}(t)$, is modeled as a zero-mean stationary, temporally and spatially white Gaussian random process, *i.e.*,

$$\mathrm{E}[\mathbf{n}(t)\mathbf{n}^*(s)] = \sigma^2\mathbf{I}_m\delta_{t,s}\ . \tag{3}$$

The symbol $(\cdot)^*$ denotes the Hermitean transpose and $\delta_{t,s}$ is the Kronecker delta. The noise is uncorrelated with the signal waveforms, modeled as zero-mean stationary, Gaussian random processes with covariance

$$\mathrm{E}[\mathbf{s}(t)\mathbf{s}^*(s)] = \delta_{t,s}\mathbf{S}\ . \tag{4}$$

The covariance matrix of the array output is then

$$\mathbf{R} = \mathrm{E}[\mathbf{x}(t)\mathbf{x}^*(t)] = \mathbf{A}(\boldsymbol{\theta}_0)\mathbf{S}\mathbf{A}^*(\boldsymbol{\theta}_0) + \sigma^2\mathbf{I}_m\ . \tag{5}$$

The ESP data consists of N samples of the array outputs, $\mathbf{x}(1)\ ,\ldots,\ \mathbf{x}(N)$. If we introduce the $m \times r$ matrix $\mathbf{T}$, $d < r \leq m$, and the mapping $\mathbf{x} \mapsto \mathbf{z} = \mathbf{T}^*\mathbf{x}$, from ESP to RDBS, a new set of observations result, consisting of the r-dimensional vectors $\{\mathbf{z}(t)\}_{t=1}^{N}$. Then, the RDBS representation will be

$$\mathbf{z}(t) = \mathbf{T}^*\mathbf{A}(\boldsymbol{\theta}_0)\mathbf{s}(t) + \mathbf{T}^*\mathbf{n}(t)\ . \tag{6}$$

The covariance matrix of the RDBS sensor noise process equals $\mathbf{T}^*\sigma^2\mathbf{I}_m\mathbf{T} = \sigma^2\mathbf{T}^*\mathbf{T}$. We will require $\mathbf{T}^*\mathbf{T} = \mathbf{I}_r$, implying that the RDBS sensor noise is spatially white whenever the ESP noise is. Furthermore, it is assumed that the array manifold has no ambiguites, *i.e.*, any matrix $[\mathbf{a}(\theta_1)\ ,\ldots,\ \mathbf{a}(\theta_m)]$ has full rank, m, for distinct θ_i:s. This ensures unique estimates of the DOAs, [7]. Assume also that $d < r \leq m$, and that the parametrization of array response vector, $\mathbf{a}(\theta)$, is known.

3 RDBS Cramér-Rao Bound

The previously proposed transformation matrix design methods have mainly been focused on obtaining as high as possible RDBS SNR or as low as possible resolution threshold in RDBS. Apart from the results (improved resolution threshold in RDBS for the MUSIC algorithm) in [5], there have not been any real guidelines for how much the ESP dimension actually can be reduced without loss in performance measured by, for example, estimation accuracy. Furthermore, there have not been any discussion if there is a possibility of achieving the ESP CRB, denoted $\mathrm{CRB}_{\mathrm{ESP}}$, which is the best one could hope for, by using the RDBS data set. Now, since there exist estimation methods, *e.g.* the WSF method, [8, 9], that indeed attains the $\mathrm{CRB}_{\mathrm{ESP}}$, the same method can be used in RDBS to reach the $\mathrm{CRB}_{\mathrm{RDBS}}(\mathbf{T})$. Therefore, our aim in this section is to show that there exists a transformation matrix that makes $\mathrm{CRB}_{\mathrm{RDBS}}(\mathbf{T}) = \mathrm{CRB}_{\mathrm{ESP}}$. Of course, the expressions for the CRB's are to be evaluated at $\boldsymbol{\theta}_0$, the true DOAs. However, the result is still useful for deriving practical design methods, as demonstrated in the next section. The emitter waveforms are modeled as zero-mean, stationary Gaussian random processes in Section 2, which means we should consider the expression for the *stochastic*[1] Cramér-Rao bound, [10], namely

$$\mathrm{E}\left[\left(\theta - \theta_0\right)\left(\hat{\theta} - \theta_0\right)^T\right] \ \geq\ \mathrm{CRB}_{\mathrm{ESP}} \tag{7}$$

$$\mathrm{CRB}_{\mathrm{ESP}}^{-1} = \frac{\sigma^2}{2N}\left[\mathrm{Re}\left\{\left(\mathbf{D}^*\mathbf{P}_{\mathbf{A}}^{\perp}\mathbf{D}\right) \odot \left(\mathbf{S}\mathbf{A}^*\mathbf{R}^{-1}\mathbf{A}\mathbf{S}\right)^T\right\}\right]$$

Here, the notation $\mathbf{A} \odot \mathbf{B}$ denotes the Schur product of the matrices $\mathbf{A}$ and $\mathbf{B}$, *i.e.*, $(\mathbf{A} \odot \mathbf{B})_{ij} = \mathbf{A}_{ij}\mathbf{B}_{ij}$, and where

$$\mathbf{P}_{\mathbf{A}}^{\perp} = \mathbf{I} - \mathbf{A}(\mathbf{A}^*\mathbf{A})^{-1}\mathbf{A}^* \tag{8}$$

is the orthogonal projection onto the null space of $\mathbf{A}^*$, and

$$\mathbf{D} \ = \ [\ \mathbf{d}(\theta_1),\ \ldots,\mathbf{d}(\theta_d)\ ] \tag{9}$$

$$\mathbf{d}(\theta_j) \ = \ \frac{\partial}{\partial\theta_j}\mathbf{a}(\theta_j), \quad j = 1,\ldots,d\ . \tag{10}$$

The expression for $\mathrm{CRB}_{\mathrm{RDBS}}(\mathbf{T})$ is obtained by substitution of $\mathbf{A}$, $\mathbf{D}$ and $\mathbf{R}$ by $\mathbf{T}^*\mathbf{A}$, $\mathbf{T}^*\mathbf{D}$ and $\mathbf{R}_{\mathbf{zz}} = \mathbf{T}^*\mathbf{R}\mathbf{T} = \mathbf{T}^*\mathbf{A}\mathbf{R}_{\mathbf{S}}\mathbf{A}^*\mathbf{T} + \sigma^2\mathbf{I}_r$, respectively.

Define $\Re\{\mathbf{M}\}$ to be the range space of the matrix $\mathbf{M}$. Then we have the following result.

Theorem 1 *Assume that* $r \geq 2d$ *and that* $\mathbf{T}$ *is an* $m \times r$ *matrix satisfying* $\mathbf{T}^*\mathbf{T} = \mathbf{I}_r$.
Then, if

$$\Re\left\{[\mathbf{A}(\boldsymbol{\theta}_0)\ \ \mathbf{D}(\boldsymbol{\theta}_0)]\right\} \subseteq \Re\{\mathbf{T}\}\ , \tag{11}$$

the relation $\mathrm{CRB}_{\mathrm{RDBS}}(\mathbf{T}) = \mathrm{CRB}_{\mathrm{ESP}}$ *holds true.* □

[1] Or *unconditional* CRB.

Proof
Introduce first the notation

$$\mathbf{F} = \mathbf{D}^*\mathbf{T}\left(\mathbf{I}_r - \mathbf{T}^*\mathbf{A}(\mathbf{A}^*\mathbf{T}\mathbf{T}^*\mathbf{A})^{-1}\mathbf{A}^*\mathbf{T}\right)\mathbf{T}^*\mathbf{D} \quad (12)$$

and

$$\mathbf{G} = \mathbf{S}\mathbf{A}^*\mathbf{T}\left(\mathbf{T}^*\mathbf{A}\mathbf{S}\mathbf{A}^*\mathbf{T} + \sigma^2\mathbf{I}_r\right)^{-1}\mathbf{T}^*\mathbf{A}\mathbf{S}\,. \quad (13)$$

Here, the dependence on $\boldsymbol{\theta}_0$ for $\mathbf{A}$ and $\mathbf{D}$ has been suppressed for notational convenience. Then the expression for the CRB in RDBS reads

$$\mathrm{CRB}_{\mathrm{RDBS}} = \frac{\sigma^2}{2N}\left[\mathrm{Re}\left(\mathbf{F}\odot\mathbf{G}^T\right)\right]^{-1} \quad (14)$$

Starting with the expression for $\mathbf{F}$ we have

$$\begin{aligned}\mathbf{F} &= \mathbf{D}^*\mathbf{T}\left(\mathbf{I}_r - \mathbf{T}^*\mathbf{A}(\mathbf{A}^*\mathbf{T}\mathbf{T}^*\mathbf{A})^{-1}\mathbf{A}^*\mathbf{T}\right)\mathbf{T}^*\mathbf{D}\\ &= \mathbf{D}^*\mathbf{T}\left(\mathbf{T}^*\mathbf{T} - \mathbf{T}^*\mathbf{A}(\mathbf{A}^*\mathbf{T}\mathbf{T}^*\mathbf{A})^{-1}\mathbf{A}^*\mathbf{T}\right)\mathbf{T}^*\mathbf{D}\\ &= \mathbf{D}^*\mathbf{T}\mathbf{T}^*\left(\mathbf{I}_m - \mathbf{A}(\mathbf{A}^*\mathbf{T}\mathbf{T}^*\mathbf{A})^{-1}\mathbf{A}^*\right)\mathbf{T}\mathbf{T}^*\mathbf{D} \quad (15)\end{aligned}$$

where the requirement that $\mathbf{T}^*\mathbf{T} = \mathbf{I}_r$ is utilized. Now, making use of the assumption (11), there exists a $\mathbf{T}$ such that the relations

$$\begin{aligned}\mathbf{T}\mathbf{T}^*\mathbf{A} &= \mathbf{A}\\ \mathbf{T}\mathbf{T}^*\mathbf{D} &= \mathbf{D} \quad (16)\end{aligned}$$

are satisfied. Clearly, this means that $\mathbf{F} = \mathbf{D}^*\mathbf{P}_{\mathbf{A}}^{\perp}\mathbf{D}$, which is the first part of the CRB expression in ESP. For the second part, we notice first that

$$\begin{aligned}&\left(\mathbf{T}^*\mathbf{A}\mathbf{S}\mathbf{A}^*\mathbf{T} + \sigma^2\mathbf{I}_r\right)^{-1} = \sigma^{-2}\mathbf{I}_r\\ &-\sigma^{-2}\mathbf{T}^*\mathbf{A}\mathbf{S}\left(\sigma^{-2}\mathbf{A}^*\mathbf{T}\mathbf{T}^*\mathbf{A}\mathbf{S} + \mathbf{I}_d\right)^{-1}\mathbf{A}^*\mathbf{T}\sigma^{-2}\end{aligned} \quad (17)$$

by making use of the matrix inversion lemma. Inserting (17) into (13) gives

$$\mathbf{G} = \mathbf{S}\mathbf{A}^*\mathbf{R}^{-1}\mathbf{A}\mathbf{S}\,, \quad (18)$$

by using the relation $\mathbf{T}\mathbf{T}^*\mathbf{A} = \mathbf{A}$ and the matrix inversion lemma. This completes the proof. □

A corollary to the theorem is that also the ESP *deterministic* Cramér-Rao bound, [10], is preserved in RDBS.

A matrix $\mathbf{T}$ of size $m \times 2d$ that makes the relations (11) and (16) satisfied can be obtained by using the Singular-Value-Decomposition (SVD), [1]. Let

$$[\mathbf{A}\ \mathbf{D}] = \mathbf{U}\boldsymbol{\Sigma}\mathbf{V}^* \quad (19)$$

be the SVD of the matrix $[\mathbf{A}\ \mathbf{D}]$, where $\mathbf{U}$ is $m \times 2d$, $\boldsymbol{\Sigma} = \mathrm{diag}[\sigma_1, \ldots, \sigma_{2d}]$, $(\sigma_1 \geq \sigma_2 \geq, \ldots, \geq \sigma_{2d} \geq 0)$, and $\mathbf{V}$ is $2d \times 2d$. Since $\mathbf{U}$ spans[2] the range space of $[\mathbf{A}\ \mathbf{D}]$, one may take $\mathbf{T} = \mathbf{U}$.

[2] The matrix $\mathbf{U}\mathbf{U}^*$ is in fact a projection matrix, projecting onto the range space of $[\mathbf{A}\ \mathbf{D}]$.

4 Transformation Matrix Design

Design Steps.

1. Specify first an interval, $[\theta_a\ ,\ \theta_b]$, within which sources of interest are assumed to be located.
2. Compute the beamwidth, θ_{BW}, of the array, and construct the "design vector" of DOAs as $\boldsymbol{\theta} = \left[\theta_a + \frac{\theta_{BW}}{2}, \theta_a + \theta_{BW}, \ldots, \theta_b - \theta_{BW}, \theta_b - \frac{\theta_{BW}}{2}\right]^T$.
3. Compute the SVD $\mathbf{U}\boldsymbol{\Sigma}\mathbf{V}^* = [\mathbf{A}(\boldsymbol{\theta})\ \mathbf{D}(\boldsymbol{\theta})]$.
4. Take as the transformation matrix $\mathbf{T} = \mathbf{U}$.
5. If possible, reduce the dimension of the transformation matrix (see Eq.(20) and the discussion below).

<u>**Remarks:**</u> Note that the specification of the spatial sector in step 1 above also can include several separated sectors. Moreover, if sectors are specified beforehand, the corresponding $\mathbf{T}$'s can be precomputed and stored (or implemented in analog hardware). The spacing between the design DOAs in step 2 are chosen somewhat arbitrarily. For obtaining enough sector coverage at the end points of the interesting interval it seems necessary to have a slightly smaller spacing there. To examine if a particular spacing is acceptable one can, preferably, evaluate the performance measure (20). It is seen from Theorem 1 that if the matrix $\mathbf{T}\mathbf{T}^*[\mathbf{A}(\boldsymbol{\theta}_0)\ \mathbf{D}(\boldsymbol{\theta}_0)]$ approximates the matrix $[\mathbf{A}(\boldsymbol{\theta}_0)\ \mathbf{D}(\boldsymbol{\theta}_0)]$ well enough, we should be very close to asymptotically attain the ESP CRB also in RDBS. The design procedure guarantees that, due to the $\theta_{BW}/2$-distance between the "design DOAs", the approximation is very good over the whole interval. However, if the spatial sector is large compared to the beamwidth, the RDBS dimension may be unacceptably high. To further reduce the dimension, it is proposed to use only the k "largest" left singular vectors of $[\mathbf{A}(\boldsymbol{\theta}_0)\ \mathbf{D}(\boldsymbol{\theta}_0)]$; the reason being that this choice is the best rank-k approximation, see [1].

Now, if the $(k+1)^{th}$ singular value is small enough compared with the k larger ones, it should be possible to delete $r-k$ columns in the optimal $m \times r$ matrix $\mathbf{T}$, and thus reduce the RDBS dimension, hopefully without too much loss in the RDBS optimal performance.

As a measure of closeness between the ESP and RDBS performance over a spatial interval, we take the following:

$$\eta(k) = \left(\frac{\int_{\theta_a}^{\theta_b}\mathrm{CRB}_{\mathrm{ESP}}(\theta)\,d\theta}{\int_{\theta_a}^{\theta_b}\mathrm{CRB}_{\mathrm{RDBS}}(\theta, \mathbf{T}_k)\,d\theta}\right)^{1/2}. \quad (20)$$

Here, the index k of $\mathbf{T}_k$ indicates the RDBS dimension. Then, $0 \leq \eta(k) \leq 1$, with the upper bound beeing desirable to attain. By depicting the values of $\eta(k)$ obtained for different k's, it is easy to see for which dimensions the performance measure is sufficiently close to 1. To make the measure easy to compute, it can preferably be approximated by taking sums over closely spaced θ's instead of using integrals.

5 Numerical Examples

In the examples to follow, the employed WSF-algorithm is supplied with a detection step, based on the asymptotic χ^2-distribution of the normalized WSF cost function, which estimates the number of emitters present in the actual scenario. The only *a priori* knowledge used for the detection and DOA estimation is the functional form of the array response vector and the spatial sector. In the first example, the noise power will be $\sigma^2 = 1$ and the emitter waveforms are Gaussian distributed. Furthermore, a uniform linear array with half-wavelength element spacing is employed. The estimated root-mean-square (RMS) error curves depicted are based on 100 independent trials for each DOA and $N = 20$ independent snapshots are used in each trial.

Example 5.1
The sector of interest is assumed to be $[-4.3° \ 4.3°]$ and the scenario constitutes an $m = 100$ element ULA as well as an emitter signal for which we have SNR=3 dB. Following the design steps of Section 4, we compute the array beamwidth to approximately 3.6 degrees. Thus, take $\theta_{BW} = 1.8°$.By looking at (not displayed here) the performance measure $\eta(k)$, see (20) it can be seen that it is sufficient to have a dimension of nine for the RDBS. Figure 1 depicts the RMS-error in RDBS and the CRB's versus direction-of-arrival for one single source for an RDBS dimension of nine. Clearly, for DOAs in the interval $[-4.3° \ 4.3°]$, the best possible performance (ESP CRB) can be considered reached. □

Example 5.2
This last example is a direct application of the results derived herein to the problem of estimating the frequencies of complex sinusoidal signals (cisoids) from noisy data. Assume that the collected data points have been generated as, $(t = 0 \ , \ldots, \ m-1)$

$$y(t) = \sum_{k=1}^{d} \alpha_k e^{i(\omega_k t + \varphi_k)} + n(t), \tag{21}$$

where the noise process, $\{n(t)\}$, is assumed to be a temporally white Gaussian process with variance $\sigma^2 =$

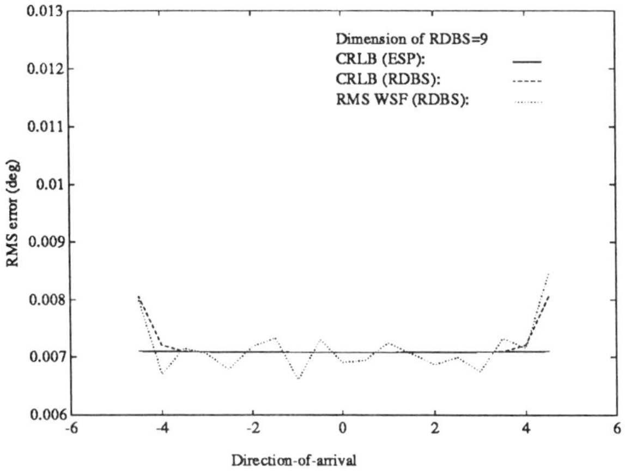

Figure 1: *RMS error of RDBS DOA estimate Scenario: ULA, SNR=3 dB,* $m = 100$, $N = 20$.

1 and the φ_k:s are uniformly distributed in the interval $[0, 2\pi]$ radians. The amplitudes, α_k, are real-valued. The ω_k:s are all assumed to be distinct from each other and satisfy $0 < \omega_k < \pi$. The latter requirement is due to the sampling theorem. We then can write

$$\mathbf{x} = \begin{bmatrix} y(0) \\ \vdots \\ y(m-1) \end{bmatrix} = \mathbf{A}(\boldsymbol{\omega})\mathbf{s} + \mathbf{n}, \tag{22}$$

where

$$\boldsymbol{\omega} = [\omega_1 \ , \ldots, \ \omega_d]^T \tag{23}$$

$$\mathbf{s} = \left[\alpha_1 e^{i\varphi_1} \ , \ldots, \ \alpha_d e^{i\varphi_d}\right]^T \tag{24}$$

$$\mathbf{A}(\boldsymbol{\omega}) = [a(\omega_1) \ , \ldots, \ a(\omega_d)] \tag{25}$$

$$a(\omega) = \left[1 \ , e^{i\omega} \ , \ldots, \ e^{i\omega(m-1)}\right]^T \tag{26}$$

$$\mathbf{n} = [n(0) \ , \ldots, \ n(m-1)]^T . \tag{27}$$

From (26), the array response vector can be seen to have the same form as that of a ULA of dimension m, and from (27), $\mathbf{n}$ satisfies (3). Since the emitted waveforms, s, is a sequence of numbers, the deterministic CRB, [10], is the lower bound on the estimation error to compare the obtained accuracy with. The ML-estimator is as follows.

$$\hat{\omega} = \arg\max_{\omega} \mathrm{Tr}\left[\mathbf{P}_{\mathbf{A}}(\omega)\hat{\mathbf{R}}\right] \tag{28}$$

Here, $\hat{\mathbf{R}} = \mathbf{x}\mathbf{x}^*$. The criterion (28) is only a slight modification of the more general WSF criterion, which means that the method employed earlier can be used also for calculating the deterministic ML-estimates. The number of cisoids will in this example be assumed known . When this is not the case, the number

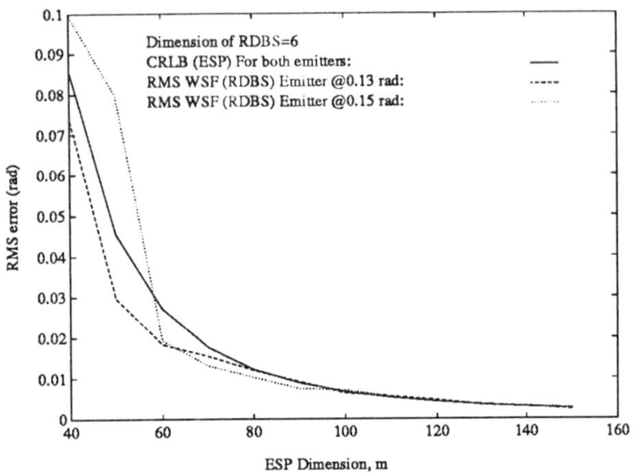

Figure 2: *RMS error of cisoid frequency estimates vs. number of data points. Scenario: SNR = 3 dB, N = 1.*

of signals can be estimated from the set of data, see *e.g.* [11]. In Figure 2, the RMS error curves for the frequency estimates of two cisoids, at 0.13 and 0.15 rad., respectively, are depicted versus the array size, m. The RMS error curves are based on 200 independent trials. The SNR is in this case fixed at 3 dB. The RDBS dimension is 6, the same for all m. Note that this means that the ML-estimation of the cisoid frequencies is based on only six (complex-valued) RDBS observations. However, as expected from the theory, the CRB is attained for the frequencies at hand. □

6 Conclusions

As demonstrated here, it is possible to make a significant reduction of the ESP-dimension without loosing any estimation accuracy at all. There also exists a bound for the RDBS dimension above which it is *always* possible to retain the property of optimally estimating the directions-of-arrival. This bound was shown to be $2d$, *i.e.*, twice the true number of sources present. If this is the *lowest* bound is, however, still an open question. This result also has the consequence that the ESP CRB can be determined computationally less expensive in RDBS, as compared to a direct computation of the ESP expression. Moreover, a practical design strategy for a transformation matrix has been devised that achieve near optimal performance in a given spatial sector, in the sense of ESP CRB. The optimal transformation matrix is easily obtained from a singular value decomposition, assuming knowledge only of the functional form of the array response. This optimal transformation is thus independent of emitter signal correlations as well as of the powers of the sources. Furthermore, *any* efficient estimation method, *i.e.* any method that attains the stochastic (or deterministic) Cramér-Rao bound, can utilize this transformation matrix for obtaining optimal dimension reduction.

References

[1] G.H. Golub and C.F. Van Loan. *Matrix Computations, 2nd edition.* Johns Hopkins University Press, Baltimore, MD., 1989.

[2] P. Forster and G. Vezzosi. "Application of Spheroidal Sequences to Array Processing". *ICASSP 87*, pages 2268–2271, April 1987.

[3] K. Buckley and X.L. Xu. "Spatial-Spectrum Estimation in a Location Sector". *IEEE Trans. ASSP*, pages 1842–1852, Nov. 1988.

[4] B. Van Veen and B. Williams. "Structured Covariance Matrices and Dimensionality Reduction in Array Processing". *Proc. 4:th Workshop Spectr. Est. and Modeling*, Aug. 1988.

[5] H.B. Lee and M.S. Wengrovitz. "Resolution Treshold of Beamspace MUSIC For Two Closely Spaced Emitters.". *IEEE Trans. ASSP*, ASSP-38(9):1545–1559, Sept. 1990.

[6] R. O. Schmidt. *A Signal Subspace Approach to Multiple Emitter Location and Spectral Estimation.* PhD thesis, Stanford Univ., Nov. 1981.

[7] M. Wax and I. Ziskind. "On Unique Localization of Multiple Sources by Passive Sensor Arrays". *IEEE Trans. on ASSP*, ASSP-37, July 1989.

[8] M. Viberg and B. Ottersten. "Sensor Array Processing Based on Subspace Fitting". *IEEE Trans. SP*, SP-39, May 1991.

[9] M. Viberg, B. Ottersten, and T. Kailath. "Detection and Estimation in Sensor Arrays Using Weighted Subspace Fitting". *IEEE Trans. SP*, SP-39, Nov. 1991.

[10] P. Stoica and A. Nehorai. "Performance Study of Conditional and Unconditional Direction-of-Arrival Estimation". *IEEE Trans. ASSP*, ASSP-38:1783–1795, October 1990.

[11] J.-J. Fuchs. "Estimating the Number of Sinusoids in Additive White Noise". In *Proc. ICASSP Dallas 1987, Paper 30.2.*

The Effect of Forward-Backward Data Formulation to Direction-of-Arrival Estimation

Fu Li and Hui Liu
Department of Electrical Engineering
Portland State University
Portland, OR 97207-0751

Abstract

Subspace-based DOA estimation has received significant attention for more than a decade. While most algorithms use standard (forward-only) data formulation, Min-Norm uses forward-backward data formulation. In this paper, we will analyze the performance of DOA estimation using both forward-only and forward-backward data formulations for MUSIC, Min-Norm, State-Space Realization, and ESPRIT in a unified fashion.

1 Introduction

Subspace-based DOA estimation algorithms (MUSIC [1], Min-Norm [2], State-Space Realization [3], ESPRIT [4] for examples) has received tremendous interests over last decade. While most algorithms use a standard sensors versus snapshots (or forward-only) data formulation, Min-Norm algorithm uses a forward-backward data formulation. This forward-backward data formulation was observed experimentally in the past not only reducing the mean-squared DOA estimation error, but also reducing the correlation between the interfered signals. Therefore one question to be answered is that why the other methods do not use forward-backward formulation?

Motivate by their increasing popularity, Subspace-based algorithms have attracted significant effort in their performance analyses (see [5] for a relatively complete references). Among others [6, 7, 8, 9] addressed the performance due to the effect of forward-backward data formulations.

We previously developed a performance analysis of DOA estimation using forward-only data formulation [5, 10]. The unique features of this analysis includes: (1) *unified* for aforementioned algorithms [11, 12]; (2) *unified* for as finite data effect [5], sensor errors [13], and unknown noise fields [14]; (3) based on a *finite* number of measurements. We recently further unified and simplified this analysis [15] in expressing the DOA Estimation error statistics in terms of fundmental parameters such as the array manifold, signal covariance, and number of the snapshots and sensors, rather than data-trial dependent singular values and vectors.

In this paper, we apply our unified analysis to the effects of forward-backward data formulation on the statistical performance of DOA estimation algorithms.

1.1 Notation

The following list is a summary of some of the notation used in this paper

z_k	=	$e^{j\frac{2\pi d}{\lambda}\sin\theta_k}$
$(\cdot)^T$, $(\cdot)^*$, $(\cdot)^H$	=	transpose, complex conjugate, complex-conjugate transpose of matrix $(\cdot)$
$\mathbf{e}_k^H$	=	$[0\cdots1\cdots0]$, 1 in the k-th position
$\mathbf{A}^\dagger$	=	pseudo-inverse of a full-rank matrix $\mathbf{A}$ which has more rows than columns; $\mathbf{A}^\dagger = (\mathbf{A}^H\mathbf{A})^{-1}\mathbf{A}^H$. Note that $\mathbf{A}^\dagger\mathbf{A} = \mathbf{I}$.
$[\cdot](k,k)$	=	the k,k-th element of matrix $[\cdot]$
$\mathbf{M}^\uparrow$	=	the matrix $\mathbf{M}$ with the first row deleted. Note that $(\mathbf{M}_1\mathbf{M}_2)^\uparrow = \mathbf{M}_1^\uparrow\mathbf{M}_2$.
$\mathbf{M}^\downarrow$	=	the matrix $\mathbf{M}$ with the last row deleted. Note that $(\mathbf{M}_1\mathbf{M}_2)^\downarrow = \mathbf{M}_1^\downarrow\mathbf{M}_2$.

2 Problem Formulation

We consider K narrowband plane-waves simultaneously incident on an array of L sensors. For simplicity of notation, we consider a uniform line array in this paper. The signals arriving at all the sensors during M snapshots are

$$\mathbf{Y} = \begin{pmatrix} y_1(1) & \cdots & y_1(M) \\ \vdots & \ddots & \vdots \\ y_L(1) & \cdots & y_L(M) \end{pmatrix} = \mathbf{AS} \qquad (1)$$

where

$$\mathbf{A}(\theta) = \begin{pmatrix} 1 & \ldots & 1 \\ z_1 & \ldots & z_K \\ \vdots & \ddots & \vdots \\ z_1^{L-1} & \ldots & z_K^{L-1} \end{pmatrix} = [\ \mathbf{a}(\theta_1) \quad \cdots \mathbf{a}(\theta_K)\ ], \qquad (2)$$

$$\mathbf{S} = \begin{pmatrix} s_1(1) & \cdots & s_1(M) \\ \vdots & \ddots & \vdots \\ s_K(1) & \cdots & s_K(M) \end{pmatrix} = [\ \mathbf{s}(1) \quad \cdots \mathbf{s}(M)\ ] \qquad (3)$$

$s_k(t)$ is the k-th narrowband signal (with center wavelength λ_c) arriving at angle θ_k, and d is the spacing between adjacent sensors.

The subspace decomposition can be performed on the data matrix $\mathbf{Y}$ by a singular value decomposition (SVD)

$$\mathbf{Y} = \mathbf{U\Sigma V}^H = (\ \mathbf{U}_s \quad \mathbf{U}_o\ )\begin{pmatrix} \mathbf{\Sigma}_s & 0 \\ 0 & 0 \end{pmatrix}\begin{pmatrix} \mathbf{V}_s^H \\ \mathbf{V}_o^H \end{pmatrix} \qquad (4)$$

[1]This research was supported in part by Portland State University through its Faculty Development Program

1058-6393/91 $01.00 © 1991 IEEE

where the vectors in $\mathbf{U}_s$, associated with the K non-zero singular values, span the signal subspace; while the vectors in $\mathbf{U}_o$, associated with the zero singular values, span the orthogonal subspace which is the orthogonal complement of the array manifold

$$\mathbf{a}(\theta_k)^H \mathbf{U}_o = 0, \quad k = 1, \ldots, K. \tag{5}$$

An equivalent approach is to apply eigenvalue decomposition to estimated data covariance matrix

$$\begin{aligned} \mathbf{C} = \mathbf{Y}\mathbf{Y}^H &= \mathbf{A}\mathbf{R}\mathbf{A}^H \\ &= \begin{pmatrix} \mathbf{E}_s & \mathbf{E}_o \end{pmatrix} \begin{pmatrix} \mathbf{\Lambda} & 0 \\ 0 & 0 \end{pmatrix} \begin{pmatrix} \mathbf{E}_s^H \\ \mathbf{E}_o^H \end{pmatrix} \end{aligned} \tag{6}$$

where $\mathbf{R} = \mathbf{S}\mathbf{S}^H$ is estimated signal covariance matrix.

2.1 Forward-Backward Formulation

The forward-backward data formulation utilizes the data as the follows

$$\mathbf{Y}_{FB} = \begin{pmatrix} y_1(1) & \cdots & y_1(M) & | & y_L(1)^* & \cdots & y_L(M)^* \\ \vdots & \ddots & \vdots & | & \vdots & \ddots & \vdots \\ y_L(1) & \cdots & y_L(M) & | & y_1(1)^* & \cdots & y_1(M)^* \end{pmatrix} \tag{7}$$

the subscript FB stands for Forward-Backward, or, in matrix form

$$\mathbf{Y}_{FB} = [\mathbf{Y} \;\; \mathbf{J}\mathbf{Y}^*] = \mathbf{A}[\mathbf{S} \;\; \mathbf{D}^{-L+1}\mathbf{S}^*] \stackrel{\text{def}}{=} \mathbf{A}\mathbf{S}_{FB} \tag{8}$$

where $\mathbf{J}$ is the permutation matrix and D is a diagonal matrix:

$$\mathbf{J} = \begin{pmatrix} 0 & & 1 \\ & \cdot & \\ 1 & & 0 \end{pmatrix} \quad \text{and} \quad \mathbf{D} = \begin{pmatrix} z_1 & & 0 \\ & \ddots & \\ 0 & & z_K \end{pmatrix}.$$

The estimate covariance data matrix is

$$\mathbf{C}_{FB} = \mathbf{Y}_{FB}\mathbf{Y}_{FB}^H = \mathbf{C} + \mathbf{J}\mathbf{C}^*\mathbf{J} = \mathbf{A}\mathbf{R}_{FB}\mathbf{A}^H \tag{9}$$

where estimated signal covariance matrix is

$$\begin{aligned} \mathbf{R}_{FB} &= (\mathbf{S}\mathbf{S}^H + \mathbf{D}^{-L+1}\mathbf{S}^*\mathbf{S}^H\mathbf{D}^{L-1}) \\ &= \mathbf{R} + \mathbf{D}^{-L+1}\mathbf{R}^*\mathbf{D}^{L-1}. \end{aligned} \tag{10}$$

2.2 Previous Performance Analysis

Various data perturbations are always present in practice which results in the perturbation of estimated subspaces. The perturbed data matrix can be written as

$$\tilde{\mathbf{Y}} = \mathbf{Y} + \mathbf{N}$$

where $\mathbf{N}$ is the observation noise. The subspace perturbation approach of this paper can also be used to study the effects of sensor errors [13]. The subspace decomposition of perturbed data by SVD is

$$\tilde{\mathbf{Y}} = \tilde{\mathbf{U}}\tilde{\mathbf{\Sigma}}\tilde{\mathbf{V}}^H = \begin{pmatrix} \tilde{\mathbf{U}}_s & \tilde{\mathbf{U}}_o \end{pmatrix} \begin{pmatrix} \tilde{\mathbf{\Sigma}}_s & 0 \\ 0 & \tilde{\mathbf{\Sigma}}_o \end{pmatrix} \begin{pmatrix} \tilde{\mathbf{V}}_s^H \\ \tilde{\mathbf{V}}_o^H \end{pmatrix}. \tag{11}$$

We can now write

$$\tilde{\mathbf{U}}_o = \mathbf{U}_o + \Delta\mathbf{U}_o \quad \text{and} \quad \tilde{\mathbf{U}}_s = \mathbf{U}_s + \Delta\mathbf{U}_s \tag{12}$$

where $\Delta\mathbf{U}_o$ and $\Delta\mathbf{U}_s$ are the perturbations in the estimated orthogonal- and signal- subspaces. The subspace perturbation expansion gives us, at high SNR (small perturbations), a first-order relationship between subspace perturbation and observation noise (data perturbation)

$$\begin{aligned} \Delta\mathbf{U}_o &= -\mathbf{U}_s\mathbf{\Sigma}_s^{-1}\mathbf{V}_s^H\mathbf{N}^H\mathbf{U}_o \\ \Delta\mathbf{U}_s &= \mathbf{U}_o\mathbf{U}_o^H\mathbf{N}\mathbf{V}_s\mathbf{\Sigma}_s^{-1}. \end{aligned} \tag{13}$$

Further, we can classify the subspace-based DOA estimation algorithms into extrema-searching, polynomial-rooting, and matrix-shifting (subspace-rotation) approaches, then express the DOA estimation error in terms of subspace perturbation. Substituting (13) into these expressions, we can get an unified expression

$$\Delta\theta_k = \frac{Im[\alpha_k^H \mathbf{N}^H \beta_k]}{\gamma_k}. \tag{14}$$

where parameters α_k, β_k, and γ_k for different algorithms are given in Table 1 ($\bar{\mathbf{e}}_1 = \frac{\mathbf{e}_1\mathbf{\Omega}_o}{||\mathbf{e}_1\mathbf{\Omega}_o||}$ and $C_k = \frac{j2\pi d}{\lambda_c}\cos\theta_k$). If the observation noise is assumed to be white circular Gaussian, then the mean-squared estimation error can be derived as

	α_k^H	β_k^H	γ_k
MUSIC	$-j\mathbf{a}^{(1)}(\theta_k)^H\mathbf{\Omega}_o$	$\mathbf{e}_k^H\mathbf{R}^{-1}\mathbf{S}$	$\|\|\mathbf{\Omega}_o\mathbf{a}^{(1)}(\theta_k)\|\|^2$
Min-Norm	$-j\mathbf{a}^{(1)}(\theta_k)^H\frac{\mathbf{e}_1\mathbf{e}_1^H}{\|\|\mathbf{e}_1\mathbf{\Omega}_o\|\|^2}$	$\mathbf{e}_k^H\mathbf{R}^{-1}\mathbf{S}$	$\|\|\alpha_k\|\|^2\|\|\mathbf{e}_1\mathbf{\Omega}_o\|\|^2$
ESPRIT	$C_k\mathbf{e}_k^H\mathbf{A}\downarrow^\dagger(\mathbf{I}\uparrow\lambda_k^{-1} - \mathbf{I}\downarrow)$	$\mathbf{e}_k^H\mathbf{R}^{-1}\mathbf{S}$	1

Table 1: Parameters of DOA Estimation Error

$$E(\Delta\theta_k)^2 = \frac{||\alpha_k||^2||\beta_k||^2\sigma_n^2}{2\gamma_k^2}. \tag{15}$$

We have shown in [15] that

$$||\beta_k||^2 = \frac{\mathbf{R}_s^{-1}(k,k)}{M} \tag{16}$$

and

$$\frac{||\alpha_k||^2}{\gamma_k^2} = \frac{1}{\mathbf{a}^{(1)}(\theta_k)^H\mathbf{\Omega}_o\mathbf{W}\mathbf{\Omega}_o\mathbf{a}^{(1)}(\theta_k)}. \tag{17}$$

Substitute (16) and (17) into (15), we obtain

$$E(\Delta\theta_k)^2 = \frac{\mathbf{R}_s^{-1}(k,k)\sigma_n^2}{2M[\mathbf{a}^{(1)}(\theta_k)^H\mathbf{\Omega}_o\mathbf{W}\mathbf{\Omega}_o\mathbf{a}^{(1)}(\theta_k)]} \tag{18}$$

where the weighting matrix $\mathbf{W}$ can be specified for different algorithms as in Table 2 ($\bar{\alpha}_k = \frac{\alpha_k}{||\alpha_k||}$).

	MUSIC	Min-Norm	SSR & ESPRIT
W	I	$\bar{\mathbf{e}}_1\bar{\mathbf{e}}_1^H$	$\bar{\alpha}_k\bar{\alpha}_k^H$

Table 2: Parameters of MSE Expression

However, when forward-backward formulation is used, $\mathbf{N}_{FB}$ will replace the $\mathbf{N}$ in (14). $\mathbf{N}_{FB}$ is colored even when $\mathbf{N}$ is white. Therefore (18) cannot be applied to the case for forward-backward formulation directly.

3 Statistical Effect on Forward-Backward Formulation

To analyze the effect for forward-backward data formulation, replace the forward-only parameters in (14) by forward-backward parameters.

$$\Delta\theta_k = \frac{Im[\beta_{FBk}^H \mathbf{N}_{FB}^H \alpha_k]}{\gamma_k} \tag{19}$$

where

$$\beta_{FBk}^H = \mathbf{e}_k \mathbf{R}_{FB}^{-1} \mathbf{S}_{FB} (\text{note that } ||\beta_{FBk}||^2 \stackrel{\text{def}}{=} \mathbf{R}_{FB}^{-1}(k,k))$$

and $\mathbf{N}_{FB} = (\mathbf{N} \ \ \mathbf{JN}^*)$ but α_k is the same as in forward-only data formulation. Since

$$\begin{aligned} & [Im(\beta_{FBk}^H \mathbf{N}_{FB}^H \alpha_k)]^2 \\ = & (\frac{\beta_{FBk}^H \mathbf{N}_{FB}^H \alpha_k - \alpha_k^H \mathbf{N}_{FB} \beta_{FBk}}{2j})^2 \\ = & \frac{1}{2} \beta_{FBk}^H \mathbf{N}_{FB}^H \alpha_k \alpha_k^H \mathbf{N}_{FB} \beta_{FBk} \\ - & \frac{1}{2} Re(\alpha_k^H \mathbf{N}_{FB} \beta_{FBk} \alpha_k^H \mathbf{N}_{FB} \beta_{FBk}) \end{aligned} \quad (20)$$

so

$$\begin{aligned} & E[Im(\beta_{FB}^H \mathbf{N}_{FB}^H \alpha_k)]^2 \\ = & \frac{1}{2} \beta_{FBk}^H E(\mathbf{N}_{FB}^H \alpha_k \alpha_k^H \mathbf{N}_{FB}) \beta_{FBk} \\ - & \frac{1}{2} Re[\alpha_k^H E(\mathbf{N}_{FB} \beta_{FBk} \alpha_k^H \mathbf{N}_{FB}) \beta_{FBk}] \end{aligned} \quad (21)$$

where in the first term

$$\begin{aligned} & E(\mathbf{N}_{FB}^H \alpha_k \alpha_k^H \mathbf{N}_{FB}) \\ = & E \begin{pmatrix} \mathbf{N}^H \\ \mathbf{N}^T \mathbf{J} \end{pmatrix} \alpha_k \alpha_k^H (\mathbf{N} \ \mathbf{JN}^*) \\ = & \begin{bmatrix} E(\mathbf{N}^H \alpha_k \alpha_k^H \mathbf{N}) & E(\mathbf{N}^H \alpha_k \alpha_k^H \mathbf{JN}^*) \\ E(\mathbf{N}^T \mathbf{J} \alpha_k \alpha_k^H \mathbf{N}) & E(\mathbf{N}^T \mathbf{J} \alpha_k \alpha_k^H \mathbf{JN}^*) \end{bmatrix} \\ = & \sigma_n^2 ||\alpha_k||^2 \mathbf{I}_{2L \times 2L} \end{aligned} \quad (22)$$

the non-diagonal terms equal to zero because of circular property of $\mathbf{N}$. The first term of (21) becomes

$$\frac{\sigma_n^2 \mathbf{R}_{FB}^{-1}(k,k) ||\alpha_k||^2}{2\gamma_k^2}.$$

Unlike the case of forward-only formulation, the second term in (21) is not zero. To calculate this term we first rewrite β_{FBk} as

$$\beta_{FBk}^H = [\mathbf{e}_k \mathbf{R}_{FB}^{-1} \mathbf{S} \quad \mathbf{e}_k \mathbf{R}_{FB}^{-1} \mathbf{D}^{-L+1} \mathbf{S}^*] = [\beta_{Fk}^H \ \ \beta_{Bk}^H] \quad (23)$$

then

$$\begin{aligned} & E[\mathbf{N}_{FB} \beta_{FBk} \alpha_k^H \mathbf{N}_{FB}] \\ = & E[(\mathbf{N} \ \mathbf{JN}^*) \begin{pmatrix} \beta_{Fk} \\ \beta_{Bk} \end{pmatrix} \alpha_k^H (\mathbf{N} \ \mathbf{JN}^*)] \\ = & E(\mathbf{N} \beta_{Fk} \alpha_k^H \mathbf{N} + \mathbf{JN}^* \beta_{Bk} \alpha_k^H \mathbf{N} \\ & \mathbf{N} \beta_{Fk} \alpha_k^H \mathbf{JN}^* + \mathbf{JN}^* \beta_{Bk} \alpha_k^H \mathbf{JN}^*) \quad (24) \\ = & E(\mathbf{JN}^* \beta_{Bk} \alpha_k^H \mathbf{N} \quad \mathbf{N} \beta_{Fk} \alpha_k^H \mathbf{JN}^*) \quad (25) \end{aligned}$$

from (24) to (25) we used the circular property of $\mathbf{N}$ again. Now we first look at the i,j-th entry of the matrix product $E(\mathbf{JN}^* \beta_{Bk} \alpha_k^H \mathbf{N})$

$$\begin{aligned} & E(\mathbf{JN}^* \beta_{Bk} \alpha_k^H \mathbf{N})_{i,j} \\ = & E[\sum_{m=1}^{M} n_{L-i+1}(m)^* \beta_{Bk}(m)][\sum_{l=1}^{L} \alpha_k(l)^* n_l(j)] \\ = & \sum_{m=1}^{M} \sum_{l=1}^{L} \beta_{Bk}(m) \alpha_k(l)^* E[n_{L-i+1}(m)^* n_l(j)] \\ = & \sum_{m=1}^{M} \sum_{l=1}^{L} \beta_{Bk}(m) \alpha_k(l)^* \sigma^2 \delta(L-i+1-l, j-m) \\ = & \sigma^2 \beta_{Bk}(j) \alpha_k(L-i+1)^* \end{aligned} \quad (26)$$

where the indices in subscript and parenthesis of noise element n in $\mathbf{N}$ are the indices of the column and the row in $\mathbf{N}$. Now we obtain

$$\begin{aligned} E[\mathbf{N}_{FB} \beta_k \alpha_k^H \mathbf{N}_{FB}] = & \sigma_n^2 \begin{bmatrix} \alpha_k(L)^* \\ \vdots \\ \alpha_k(1)^* \end{bmatrix} \times \\ & [\beta_{Bk}(1), \cdots, \beta_{Bk}(M), \beta_{Fk}(1), \cdots, \beta_{Fk}(M)] \\ = & \sigma_n^2 \mathbf{J} \alpha_k^* (\beta_{Bk} \ \beta_{Fk}) \end{aligned} \quad (27)$$

so that

$$\begin{aligned} & \alpha_k^H E[\mathbf{N}_{FB} \beta_k \alpha_k^H \mathbf{N}_{FB}] \beta_k \\ = & \sigma_n^2 (\alpha_k^H \mathbf{J} \alpha_k^*) (\beta_{Bk}^T \ \ \beta_{Fk}^T) \begin{pmatrix} \beta_{Fk} \\ \beta_{Bk} \end{pmatrix} \\ = & 2\sigma_n^2 (\alpha_k^H \mathbf{J} \alpha_k^*) (\beta_{Fk}^T \beta_{Bk}) \end{aligned} \quad (28)$$

notice that from (23)

$$\begin{aligned} \beta_{Fk}^T \beta_{Bk} & = [\mathbf{R}_{FB}^{-1\,*} \mathbf{S}^* \mathbf{S}^T \mathbf{D}^{L-1} \mathbf{R}_{FB}^{-1}](k,k) \\ & = [\mathbf{R}_{FB}^{-1\,*} \mathbf{R}^* \mathbf{D}^{L-1} \mathbf{R}_{FB}^{-1}](k,k). \end{aligned} \quad (29)$$

Because we are only interested in k,k-th element of $[\mathbf{R}_{FB}^{-1\,*} \mathbf{R}^* \mathbf{D}^{L-1} \mathbf{R}_{FB}^{-1}]$, so $\mathbf{D}^{L-1}$ can be substituted by z_k^{L-1}

$$\begin{aligned} \beta_{Fk}^T \beta_{Bk} & = z_k^{L-1} [\mathbf{R}_{FB}^{-1\,*} \mathbf{R}^* \mathbf{R}_{FB}^{-1}](k,k) \\ & \stackrel{\text{def}}{=} z_k^{L-1} [\bar{\mathbf{R}}_{FB}^{-1}](k,k) \end{aligned} \quad (30)$$

Now equation (21) can be expressed as

$$E(\Delta\theta_k)^2 = \frac{\sigma_n^2 ||\alpha_k||^2}{2\gamma_k^2} [\mathbf{R}_{FB}^{-1} + 2Re(\phi_k \bar{\mathbf{R}}_{FB}^{-1})](k,k) \quad (31)$$

or using (17)

$$E(\Delta\theta_k)^2 = \frac{\sigma_n^2 [\mathbf{R}_{FB}^{-1} + 2Re(\phi_k \bar{\mathbf{R}}_{FB}^{-1})](k,k)}{2[\mathbf{a}^{(1)}(\theta_k)^H \mathbf{\Omega}_o \mathbf{W} \mathbf{\Omega}_o \mathbf{a}^{(1)}(\theta_k)]} \quad (32)$$

where we define

$$\phi_k = -z_k^{L-1} \frac{(\alpha_k^H \mathbf{J} \alpha_k^*)}{||\alpha_k||^2}. \quad (33)$$

Using Schwartz Inequality

$$|| - z_k^{L-1} (\alpha_k^H \mathbf{J} \alpha_k^*) || \leq || - z_k^{L-1} || \cdot ||\alpha_k|| \cdot ||\mathbf{J}|| \cdot ||\alpha_k|| = ||\alpha_k||^2$$

it is easy to see $||\phi_k|| \leq 1$. Equality achieved for MUSIC and SSR/ESPRIT as shown below.
For *MUSIC*: Using the properties

$$\mathbf{A} = \mathbf{J} \mathbf{A}^* \mathbf{D}^{L-1} \quad \text{and} \quad \mathbf{A}^\dagger = \mathbf{D}^{1-L} \mathbf{A}^{*\dagger} \mathbf{J} \quad (34)$$

we can get

$$\begin{aligned} & \mathbf{\Omega}_o = \mathbf{I} - \mathbf{A}(\mathbf{A}^H \mathbf{A})^{-1} \mathbf{A}^H \\ = & \mathbf{I} - \mathbf{J} \mathbf{A}^* \mathbf{D}^{L-1} (\mathbf{D}^{1-L} \mathbf{A}^T \mathbf{J} \mathbf{J} \mathbf{A}^* \mathbf{D}^{L-1})^{-1} \mathbf{D}^{1-L} \mathbf{A}^T \mathbf{J} \\ = & \mathbf{I} - \mathbf{J} \mathbf{A}^* (\mathbf{A}^T \mathbf{A}^*)^{-1} \mathbf{A}^T \mathbf{J} = \mathbf{J} \mathbf{\Omega}_o^* \mathbf{J} \end{aligned} \quad (35)$$

or $\mathbf{J}\mathbf{\Omega}_o^* = \mathbf{\Omega}_o\mathbf{J}$ so that we can rewrite

$$\begin{aligned} -\mathbf{J}\alpha_k^* z_k^{L-1} &= j\mathbf{J}\mathbf{\Omega}_o{}^*\mathbf{a}^{(1)}(\theta_k)^* z_k^{L-1} \\ &= j\mathbf{\Omega}_o\mathbf{J}\mathbf{a}^{(1)}(\theta_k)^* z_k^{L-1}. \end{aligned} \tag{36}$$

Since

$$\begin{aligned} &\mathbf{a}^{(1)}(\theta_k) - \mathbf{J}\mathbf{a}^{(1)}(\theta_k)^* z_k^{L-1} \\ &= \begin{pmatrix} 0 \\ C_k z_k \\ \vdots \\ (L-1)C_k z_k \end{pmatrix} - \begin{pmatrix} -(L-1)C_k z_k \\ \vdots \\ -C_k z_k \\ 0 \end{pmatrix} z_k \\ &= (L-1)C_k\mathbf{a}(\theta_k) \end{aligned} \tag{37}$$

pre-multiply (37) by $\mathbf{\Omega}_o$, and notice that $\mathbf{\Omega}_o\mathbf{a}(\theta_k) = 0$, we have

$$j\mathbf{\Omega}_o\mathbf{J}\mathbf{a}^{(1)}(\theta_k)^* z_k^{L-1} = j\mathbf{\Omega}_o\mathbf{a}^{(1)}(\theta_k) = \alpha_k \tag{38}$$

so that

$$-\alpha_k^H\mathbf{J}\alpha_k^* z_k^{L-1} = ||\alpha_k||^2 \tag{39}$$

$\phi_k = 1$ for MUSIC proved.
For *SSR/ESPRIT*:

$$\alpha_k^H = \mathbf{e}_k\mathbf{A}^{\downarrow\dagger}(\mathbf{I}^{\uparrow}z_k^{-1} - \mathbf{I}^{\downarrow})$$

so that

$$\begin{aligned} z_k^{L-1}\alpha_k^H\mathbf{J}\alpha_k^* &= z_k^{L-1}\mathbf{e}_k\mathbf{A}^{\downarrow\dagger}(\mathbf{I}^{\uparrow}z_k^{-1} - \mathbf{I}^{\downarrow})\mathbf{J}\alpha_k^* \\ &= z_k^{L-1}\mathbf{e}_k\mathbf{A}^{\downarrow\dagger}\mathbf{J}(\mathbf{I}^{\downarrow}z_k^{-1} - \mathbf{I}^{\uparrow})\alpha_k^* \end{aligned} \tag{40}$$

where we used $\mathbf{I}^{\downarrow}\mathbf{J} = \mathbf{J}\mathbf{I}^{\uparrow}$ and $\mathbf{I}^{\uparrow}\mathbf{J} = \mathbf{J}\mathbf{I}^{\downarrow}$, further

$$\begin{aligned} -z_k^{L-1}\alpha_k^H\mathbf{J}\alpha_k^* &= \mathbf{e}_k(z_k^{L-2}\mathbf{A}^{\downarrow\dagger}\mathbf{J})(\mathbf{I}^{\uparrow}z_k - \mathbf{I}^{\downarrow})\alpha_k^* \\ &= \mathbf{e}_k(\mathbf{A}^{\downarrow\dagger})^*(\mathbf{I}^{\uparrow}z_k - \mathbf{I}^{\downarrow})\alpha_k^* \\ &= \alpha_k^T\alpha_k^* = ||\alpha_k||^2 \end{aligned} \tag{41}$$

where $z_k^{L-2}\mathbf{A}^{\downarrow\dagger}\mathbf{J} = (\mathbf{A}^{\downarrow\dagger})^*$ is because of (34). $\phi_k = 1$ for SSR/ESPRIT proved.

ϕ_k for different algorithms can be found in Table 3.

MUSIC	Min-Norm	SSR & ESPRIT
$\phi_k = 1$	$\lVert\phi_k\rVert \leq 1$	$\phi_k = 1$

Table 3: ϕ_k for different algorithms

3.1 Uncorrelated Signals

When the signals are uncorrelated, $\mathbf{R}$ is diagonal such that (10) becomes

$$\mathbf{R}_{FB} = \mathbf{R} + \mathbf{D}^{-L+1}\mathbf{R}^*\mathbf{D}^{L-1} = 2\mathbf{R} \tag{42}$$

so that we have

$$\mathbf{R}_{FB}^{-1} = \frac{1}{2}\mathbf{R}^{-1} \tag{43}$$

and

$$\bar{\mathbf{R}}_{FB}^{-1} = \mathbf{R}_{FB}^{-1}{}^*\mathbf{R}^*\mathbf{R}_{FB}^{-1} = \frac{1}{4}\mathbf{R}^{-1} \tag{44}$$

where $\mathbf{R}^* = \mathbf{R}$ for uncorrelated signals, and $\bar{\mathbf{R}}_{FB}^{-1}$ is real and diagonal, so that (32) becomes

$$E(\Delta\theta_k)^2 = \frac{\sigma_n^2[1 + Re(\phi_k)][\mathbf{R}^{-1}](k,k)}{4[\mathbf{a}^{(1)}(\theta_k)^H\mathbf{\Omega}_o\mathbf{W}\mathbf{\Omega}_o\mathbf{a}^{(1)}(\theta_k)]} \tag{45}$$

For MUSIC and SSR/ESPRIT algorithms, $\phi_k = 1$ such that

$$E(\Delta\theta_k)^2 = \frac{\sigma_n^2||\alpha_k||^2}{2\gamma_k^2}\mathbf{R}^{-1}(k,k) \tag{46}$$

which is the same as forward-only data formulations. Forward-backward data formulation brings no improvement for uncorrelated signals when use MUSIC and SSR/ESPRIT algorithms.

For Min-Norm algorithm $Re(\phi_k) \leq ||\phi_k|| \leq 1$

$$\begin{aligned} E(\Delta\theta_k)^2 &= \frac{\sigma_n^2[1 + Re(\phi_k)][\mathbf{R}^{-1}](k,k)}{4[\mathbf{a}^{(1)}(\theta_k)^H\mathbf{\Omega}_o\mathbf{W}\mathbf{\Omega}_o\mathbf{a}^{(1)}(\theta_k)]} \\ &\leq \frac{\sigma_n^2[\mathbf{R}^{-1}](k,k)}{2[\mathbf{a}^{(1)}(\theta_k)^H\mathbf{\Omega}_o\mathbf{W}\mathbf{\Omega}_o\mathbf{a}^{(1)}(\theta_k)]} \end{aligned} \tag{47}$$

forward-backward data formulation does improve performance for uncorrelated signals over forward-only formulation when use Min-Norm algorithm.

3.2 Correlated case

In the case of correlation between sources, as state in [7, 16]. The forward-backward data formulation have the effect to decorrelate the correlation between signal when the number of correlated sources is 2. Similar to [17, 16], we have, in the case of two correlated sources with correlation coefficient ρ. Notice here, for finite data analysis, the ρ is the equivalent correlation coefficient of sample correlation coefficient. With the increase of the snapshot, it will goes to its expected value.

$$\begin{aligned} \mathbf{R}_{FB} &= \mathbf{R}_s + \mathbf{D}^{-L-1}\mathbf{R}_s^*\mathbf{D}^{L-1} \\ &\approx M\begin{pmatrix} \sigma_1^2 & \sigma_1\sigma_2\rho \\ \sigma_1\sigma_2\rho^* & \sigma_2^2 \end{pmatrix} \\ &+ \mathbf{D}^{-L+1}\begin{pmatrix} \sigma_1^2 & \sigma_1\sigma_2\rho^* \\ \sigma_1\sigma_2\rho & \sigma_2^2 \end{pmatrix}\mathbf{D}^{L-1} \\ &= 2M\begin{pmatrix} \sigma_1^2 & \sigma_1\sigma_2\frac{\rho+\rho^* z_1^{1-L}z_2^{L-1}}{2} \\ \sigma_1\sigma_2\frac{\rho^*+\rho z_1^{L-1}z_2^{1-L}}{2} & \sigma_2^2 \end{pmatrix} \\ &\stackrel{\text{def}}{=} 2M\begin{pmatrix} \sigma_1^2 & \sigma_1\sigma_2\overline{\rho} \\ \sigma_1\sigma_2\overline{\rho^*} & \sigma_2^2 \end{pmatrix} \end{aligned} \tag{48}$$

We can see, after forward-backward scheme, the equivalent correlation coefficient goes to $\overline{\rho}$ and

$$|\overline{\rho}|^2 = \frac{|\rho|^2 + Re(\rho^2 z_1^{1-L}z_2^{L-1})}{2} \leq |\rho|^2 \tag{49}$$

so that the sources are decorrelated and the performance will be improved. It is interesting to notice that $\overline{\rho}$ depends the angular separation of the signals. When ρ is real, the maximum decorrelation occurs when two sources are closely spaced.

It has been observed that $\mathbf{R}_{FB}^{-1}(k,k) \leq \frac{1}{2}\mathbf{R}^{-1}(k,k)$, so that it can be shown

$$\begin{aligned} E(\Delta\theta_k)_{FB}^2 &\leq \frac{\sigma_n^2||\alpha_k||^2}{\gamma_k^2}\mathbf{R}_{FB}^{-1}(k,k) \\ &\leq \frac{\sigma_n^2||\alpha_k||^2}{2\gamma_k^2}\mathbf{R}^{-1}(k,k) = E(\Delta\theta_k)_F^2 \end{aligned} \tag{50}$$

where we have used

$$E[Im(\beta^H \mathbf{N}_{FB}^H \alpha_k)]^2 \leq E\ ||\beta^H \mathbf{N}_{FB}^H \alpha_k||^2$$
$$= \sigma_n^2 ||\alpha_k||^2 \mathbf{R}_{FB}^{-1}(k,k) \qquad (51)$$

(50) tells us that the mean-squared error of using forward-backward data formulation will not be larger than it of using forward-only data formulation for all subspace-based DOA estimation algorithms regardless of the coherence of the signals.

References

[1] R. O. Schmidt, *A Signal Subspace Approach to Multiple Emitter Location and Spectral Estimation*. PhD thesis, Stanford University, Stanford, CA, 1981.

[2] R. Kumaresan and D. W. Tufts, "Estimating the angles of arrival of multiple plane waves," *IEEE Trans. Aerosp., Electron. Syst.*, vol. AES-19, pp. 134–139, January 1983.

[3] S. Y. Kung, K. S. Arun, and D. V. Bhaskar Rao, "State-space and singular-value decomposition-based approximation methods for the harmonic retrieval problem," *J. Opt. Soc. Am.*, vol. 73, pp. 1799–1811, December 1983.

[4] A. Paulraj, R. Roy, and T. Kailath, "Estimation of signal parameters via rotational invariance techniques – ESPRIT," in *Proc. 19th Asilomar Conf. on Signals, Systems and Computers*, (Pacific Grove, CA), pp. 83–89, November 1985.

[5] F. Li and R. J. Vaccaro, *SVD and Signal Processing, II*, ch. Analytical Performance Prediction of the Subspace-Based Algorithms for DOA Estimation, pp. 243–260. Elsevier Sicence Publishers, North Holland, 1991.

[6] Y. Hua and T. K. Sarkar, "Perturbation analysis of TK method for harmonic retrieval problems," *IEEE Trans. Acoustics, Speech, and Signal Processing*, vol. ASSP-36, pp. 228–240, February 1988.

[7] S. U. Pillai and B. H. Kwon, "Forward/backward spatial smoothing techniques for coherent signal identification," *IEEE Trans. Acoustics, Speech, and Signal Processing*, vol. ASSP-37, pp. 8–15, January 1989.

[8] D. V. Bhaskar Rao and K. V. S. Hari, "Effect of spatial smoothing on the performance of MUSIC and minimum-norm method," *IEE Proc.*, vol. 137, Pt.F, pp. 449–458, December 1990.

[9] D. V. Bhaskar Rao and K. V. S. Hari, "Effect of Spatial Smoothing on State-Space methods/ESPRIT," in *Proc. 5th IEEE ASSP Workshop on Spectrum Estimation and Modeling*, (Rochester, NY), pp. 377–381, October 1990.

[10] F. Li and R. J. Vaccaro, "Unified analysis for DOA estimation algorithms in array signal processing," to appear in *Signal Processing*, vol. 22, November 1991.

[11] F. Li and R. J. Vaccaro, "Analysis of MUSIC and Min-Norm for arbitrary array geometry," *IEEE Trans. Aerosp., Electron. Syst.*, vol. AES-26, pp. 976–985, November 1990.

[12] F. Li and R. J. Vaccaro, "Performance analysis of the State Space Realization (TAM) and ESPRIT algorithms for DOA estimation," *IEEE Trans. Antenna and Propagation*, vol. AP-39, pp. 418–423, March 1991.

[13] F. Li and R. J. Vaccaro, "Sensitivity analysis of DOA estimation algorithms to sensor errors," to appear in *IEEE Trans. Aerosp., Electron. Syst.*, vol. AES-27, no. July, 1992.

[14] F. Li and R. J. Vaccaro, "Performance degradation of DOA estimators due to unknown noise fields," to appear in *IEEE Trans. Signal Processing*, vol. ASSP-40, March 1992.

[15] F. Li, H. Liu, and R. J. Vaccaro, "Finite data performance analysis for DOA estimation algorithms," in submitted to *ICASSP'92*, 1992.

[16] S. U. Pillai and B. H. Kwon, "Performance analysis of MUSIC-type high resolution estimation for direction finding in correlated and coherent scences," *IEEE Trans. Acoustics, Speech, and Signal Processing*, vol. ASSP-37, pp. 1176–1189, August 1989.

[17] V.U.Reddy, A.Paulraj, and T.Kailath, "Performance analysis of the optimum beamformer in the presence of correlated sources and its behavior under spatial smoothing," *IEEE Trans. Acoustics, Speech, and Signal Processing*, vol. ASSP-35, pp. 927–936, July 1987.

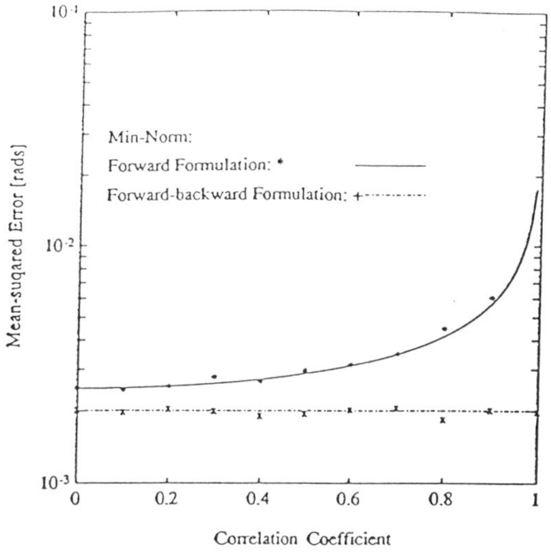

Figure 1: Min-Norm

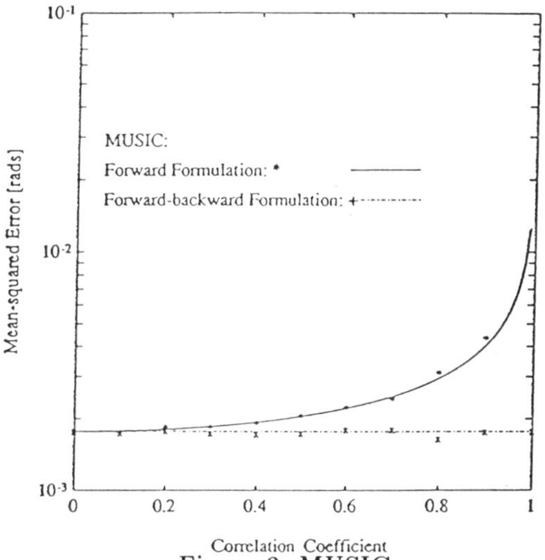

Figure 2: MUSIC

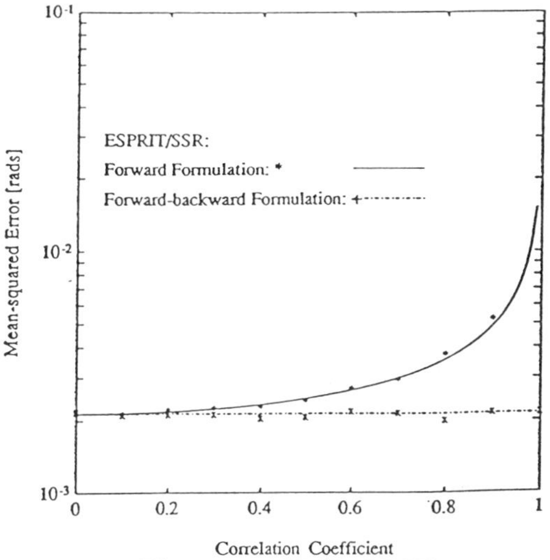

Figure 3: SSR/ESPRIT

Estimation Accuracy of Maximum Likelihood Direction Finding Using Large Arrays*

M. Viberg
Dept of EE
Linköping University
S-581 83 Linköping
Sweden

B. Ottersten
Dept of EE
Royal Institute of Technology
S-100 44 Stockholm
Sweden

A. Nehorai
Dept of EE
Yale University
New Haven, CT 06520
USA

Abstract

This paper analyzes the performance of methods for estimating the parameters of narrowband signals arriving at an array of sensors. The so-called deterministic and stochastic maximum likelihood (ML) methods are considered. A performance analysis is carried out for finite number of snapshots but assuming that the array is composed of a sufficiently large number, m, of sensors. Strong consistency of the parameter estimates is proved and the asymptotic covariance matrix of the estimation error is derived. Unlike the previously studied large (time) sample case, the present analysis shows that the accuracy is the same for the two ML methods. The covariance matrix of the estimation error attains the Cramér-Rao bound (CRB). For many array geometries of practical interest, the array propagation vectors become orthogonal as m is increased. It is shown that also the traditional beamforming method provides consistent (but not necessarily efficient) estimates under this assumption. This is true also in the presence of perfectly correlated emitters.

1 Introduction

Sensor array signal processing has been an active research area for more than three decades. The generic problem is to determine unknown signal parameters from simultaneous measurements of spatially distributed sensors. Several important applications have been reported, for example, interference rejection in radar and radio communication systems and source localization using underwater sonar arrays.

A vast number of methods have been proposed for solving the estimation problem, see e.g. [1, 2, 3]. Asymptotic results for several estimators have recently appeared in the literature, assuming the number of time samples (snapshots), N, to be large, see e.g. [4, 5]. The so-called *stochastic* maximum likelihood (SML) method has been shown to be asymptotically (for large N) efficient, i.e., the estimation error covariance attains the Cramér-Rao bound (CRB). On the other hand, the deterministic maximum likelihood (DML) estimator is not efficient, due to the increasing number of estimated parameters as N increases.

This paper concerns analysis of estimation accuracy for arbitrary N, assuming a large number of sensors, m. Arrays of several hundred antennas are not uncommon in, for example, radar and sonar systems. A condition guaranteeing consistent estimates for $m \to \infty$ is introduced. It is shown that the SML and DML methods are asymptotically identical for large m (unlike the large N case). The asymptotic covariance matrix of the estimation error coincides with the deterministic CRB. Conditions for consistency and statistical efficiency of the traditional beamforming estimates are also discussed.

2 Estimation Problem

This section formulates the DOA estimation problem and briefly describes the methods under consideration.

2.1 Problem Formulation

Consider an array of m sensors, having arbitrary positions and characteristics. Impinging on the array are the waveforms of d ($d < m$) far-field point sources. Under the narrowband assumption, the complex array output, $\mathbf{x}(t)$, is modeled by the following equation

$$\mathbf{x}(t) = \mathbf{A}(\boldsymbol{\theta})\mathbf{s}(t) + \mathbf{n}(t) \ . \qquad (1)$$

The columns of the $m \times d$ matrix $\mathbf{A}$ are the so-called *steering vectors*, $\mathbf{a}(\theta_i)$, $i = 1, \ldots, d$. The steering vectors are functions of the directions-of-arrival (DOAs),

*The work of M. Viberg and B. Ottersten was supported in part by the Swedish Research Council for Engineering Sciences. The work of A. Nehorai was supported by the Air Force Office of Scientific Research under Grant no. AFOSR-90-0164 and the Office of Naval Research under Grant no. N00014-91-J-1298.

1058-6393/91 $01.00 © 1991 IEEE

θ_i and describe the array response to a unit waveform from the direction θ_i. The vector $\mathbf{a}(\theta_i)$ is scaled so that its norm is independent of θ_i. The d-vector $\mathbf{s}(t)$ is composed of the complex emitter waveforms, received at time t, and the m-vector $\mathbf{n}(t)$ accounts for additive measurement noise as well as modeling errors. The array output is sampled at N distinct time instants, and the data matrix is formed

$$\mathbf{X}_N = \mathbf{A}(\boldsymbol{\theta})\mathbf{S}_N + \mathbf{N}_N \ , \tag{2}$$

where $\mathbf{X}_N = [\mathbf{x}(1), \ldots, \mathbf{x}(N)]$, $\mathbf{S}_N = [\mathbf{s}(1), \ldots, \mathbf{s}(N)]$, and $\mathbf{N}_N = [\mathbf{n}(1), \ldots, \mathbf{n}(N)]$. Based on these measurements, the problem of interest is to determine the DOAs of all emitters. The number of signals, d, is assumed to be known.

The main interest herein is to ascertain the accuracy of the ML estimates of $\boldsymbol{\theta}$. Two ML approaches have been studied in the literature. In the so-called *deterministic* model, no restrictions are put on the signal waveforms, i.e., they are modeled as arbitrary sequences, [3, 6]. An alternative is the *stochastic* model, where the signals are assumed to be stationary Gaussian random processes. In both models, the noise term, $\{\mathbf{n}(t)\}$ is assumed to be Gaussian. The noise has zero mean and is spatially and temporally white, i.e.,

$$\mathrm{E}[\mathbf{n}(t)\mathbf{n}^*(s)] = \sigma^2 \mathbf{I}\delta_{t-s}, \tag{3}$$

where δ_t is the Kronecker delta and $(\cdot)^*$ is complex conjugate transpose. Both methods are clearly applicable regardless of the actual distribution of the signal waveforms, although they maximize the likelihood function only under the appropriate assumption. We will here study their performance assuming deterministic signals.

A crucial assumption for most DOA estimation methods is that the functional form of $\mathbf{a}(\cdot)$ is available to the user. For simplicity, it will be assumed that the sensors are omnidirectional and lie in the same plane as the impinging waves. To enable unique identification of the DOAs, it is usually assumed that the array manifold has no ambiguities [1], i.e., that for distinct parameters $\theta_1, \ldots, \theta_m$, the matrix $[\mathbf{a}(\theta_1), \ldots, \mathbf{a}(\theta_m)]$ has full rank. For $m = 2$, the physical interpretation of this assumption is that the sidelobes of the array must be lower than the main lobe. Since the size of the array increases in the present case, the sidelobes may approach the main lobe in height as m increases (but still be lower for any finite m). To prevent this, the following condition is imposed on the array geometry.

Assumption A.1. There exists an $\epsilon > 0$ and an M such that

$$\frac{1}{m}\Big[\mathbf{a}(\theta_1), \ldots, \mathbf{a}(\theta_{2d})\Big]^* \Big[\mathbf{a}(\theta_1), \ldots, \mathbf{a}(\theta_{2d})\Big] \geq \epsilon \mathbf{I} \ , \tag{4}$$

for all $m > M$ and for any set of distinct DOAs $\theta_1, \ldots, \theta_{2d}$. The matrix inequality $\mathbf{X} \geq \mathbf{Y}$ means that $\mathbf{X} - \mathbf{Y}$ is positive semidefinite. □

As will be seen, the ML methods does not require any further assumption on the array geometry to give consistent estimates. However, the conventional beamforming method cannot perfectly separate multiple sources unless the steering vectors become orthogonal when m is increased. This is indeed the case for many array geometries of practical interest, including the linear equi-spaced array. For later reference we thus introduce the following.

Assumption A.2.

$$\lim_{m\to\infty} \frac{1}{m}\mathbf{a}^*(\theta)\mathbf{a}(\eta) = \begin{cases} 1 & , \ \theta = \eta \\ 0 & , \ \theta \neq \eta \end{cases} \tag{5}$$

□

2.2 Estimation Methods

A systematic approach to many parameter estimation problems is the maximum likelihood method. This technique requires a probabilistic setup for the measurements. If the emitter signals are modeled as unknown deterministic quantities, the observation process is distributed as $\mathbf{x}(t) \in \mathrm{N}(\mathbf{A}\mathbf{s}(t), \sigma^2\mathbf{I})$. The unknown parameters in this model are $\boldsymbol{\theta} = \theta_1, \ldots, \theta_d$ as well as $\mathbf{s}(1), \ldots, \mathbf{s}(N)$, σ^2. Introduce the pseudo-inverse of $\mathbf{A}$ and the orthogonal projector onto the nullspace of $\mathbf{A}^*$ by

$$\mathbf{A}^\dagger(\boldsymbol{\theta}) = (\mathbf{A}^*\mathbf{A})^{-1}\mathbf{A}^* \tag{6}$$

$$\mathbf{P}_{\mathbf{A}}^{\perp}(\boldsymbol{\theta}) = \mathbf{I} - \mathbf{A}\mathbf{A}^\dagger = \mathbf{I} - \mathbf{P}_{\mathbf{A}} \ . \tag{7}$$

The so-called DML estimate of $\boldsymbol{\theta}$ is obtained from, [3, 6],

$$\hat{\theta}_{DML} = \arg\min_{\boldsymbol{\theta}} \mathrm{Tr}\{\mathbf{P}_{\mathbf{A}}^{\perp}(\boldsymbol{\theta})\hat{\mathbf{R}}\} \ , \tag{8}$$

where $\mathrm{Tr}\{\cdot\}$ denotes the trace and where $\hat{\mathbf{R}}$ is the sample covariance matrix

$$\hat{\mathbf{R}} = \frac{1}{N}\sum_{t=1}^{N} \mathbf{x}(t)\mathbf{x}^*(t) \ . \tag{9}$$

A different ML estimate is obtained by instead modeling the emitter signals as stationary, temporally white Gaussian random processes with covariance $\mathbf{S}$. The observation process is now distributed as $\mathbf{x}(t) \in \mathrm{N}(\mathbf{0}, \mathbf{A}\mathbf{S}\mathbf{A}^* + \sigma^2\mathbf{I})$.

The unknown parameters in this *stochastic* model are $\boldsymbol{\theta}$, $\mathbf{S}$ and σ^2. For $N \geq d$, the SML estimate has been shown, [2] to be

$$\hat{\theta}_{SML} = \arg\min_{\boldsymbol{\theta}} \log \left|\mathbf{A}\hat{\mathbf{S}}(\boldsymbol{\theta})\mathbf{A}^* + \hat{\sigma}^2(\boldsymbol{\theta})\mathbf{I}\right| \quad (10)$$

$$\hat{\mathbf{S}}(\boldsymbol{\theta}) = \mathbf{A}^{\dagger}(\hat{\mathbf{R}} - \hat{\sigma}^2\mathbf{I})\mathbf{A}^{\dagger *} \quad (11)$$

$$\hat{\sigma}^2(\boldsymbol{\theta}) = \frac{1}{m-d}\mathrm{Tr}\{\mathbf{P}_{\mathbf{A}}^{\perp}\hat{\mathbf{R}}\}, \quad (12)$$

where $|\cdot|$ denotes the determinant.

The exact ML criteria presented above, both suffer from the requirement of a non-linear multidimensional optimization to calculate the estimates. This is particularly prominent when m is large, since the computational cost is at least proportional to m^2. To overcome this difficulty, several suboptimal techniques have been proposed. Perhaps the most natural of these is the traditional "delay-and-sum" beamforming method. This technique can be derived from (8) by setting $\mathbf{A}(\boldsymbol{\theta}) = \mathbf{a}(\theta)$ and searching for d distinct minima in the so-obtained one-dimensional criterion. The beamforming criterion function is thus

$$\mathrm{Tr}\{\mathbf{P}_{\mathbf{a}}^{\perp}(\theta)\hat{\mathbf{R}}\} \propto -\frac{\mathbf{a}^*(\theta)\hat{\mathbf{R}}\mathbf{a}(\theta)}{\mathbf{a}^*(\theta)\mathbf{a}(\theta)} \propto -\mathbf{a}^*(\theta)\hat{\mathbf{R}}\mathbf{a}(\theta) \quad (13)$$

It is easy to show that (13) is in fact also identical to (10) for $d = 1$.

3 Convergence analysis

The analysis of the described methods for large m is different from the large N case in some important respects. A complication is that the steering vectors are often orthogonal for m tending to infinity. This causes a discontinuity in the limiting criterion function. Consider, for example, the beamforming method under Assumption 2. It is easy to show that

$$\lim_{m\to\infty} \frac{1}{m}-\mathbf{a}^*(\theta)\hat{\mathbf{R}}\mathbf{a}(\theta) = \begin{cases} -\hat{\mathbf{S}}_{ii} & , \theta = \theta_i \\ 0 & , \theta \neq \theta_k \text{ for any } k \end{cases}$$

where $\hat{\mathbf{S}}_{ii} = \frac{1}{N}\sum_{t=1}^{N}|s_i(t)|^2$. Since the limiting criterion function is discontinuous, the convergence of the sequence of cost functions cannot be uniform in θ. This makes it difficult to establish consistency of the estimates, since these do not necessarily converge to the minimizing arguments of the limiting criteria. To overcome this, we shall use a convergence result that will be useful for proving consistency despite the lack of uniform convergence.

Lemma 1 *Let $\hat{\boldsymbol{\theta}}_m$ be obtained by*

$$\hat{\boldsymbol{\theta}}_m = \arg\min_{\boldsymbol{\theta}} V_m(\boldsymbol{\theta}) , \quad (14)$$

where the criterion function consists of a "signal" and a "noise" term

$$V_m(\boldsymbol{\theta}) = f_m(\boldsymbol{\theta}) + \varepsilon_m(\boldsymbol{\theta}) . \quad (15)$$

Assume that these terms satisfy the following conditions

i) $\lim_{m\to\infty}\varepsilon_m(\boldsymbol{\theta}) = 0$, *uniformly in* $\boldsymbol{\theta}$.

ii) there exists a point $\bar{\boldsymbol{\theta}}$ such that for any $\delta > 0$ we can find $\epsilon > 0$ and M, satisfying

$$f_m(\boldsymbol{\theta}) \geq f_m(\bar{\boldsymbol{\theta}}) + \epsilon \quad (16)$$

as soon as $|\boldsymbol{\theta} - \bar{\boldsymbol{\theta}}| > \delta$ and $m > M$, see Figure 1.

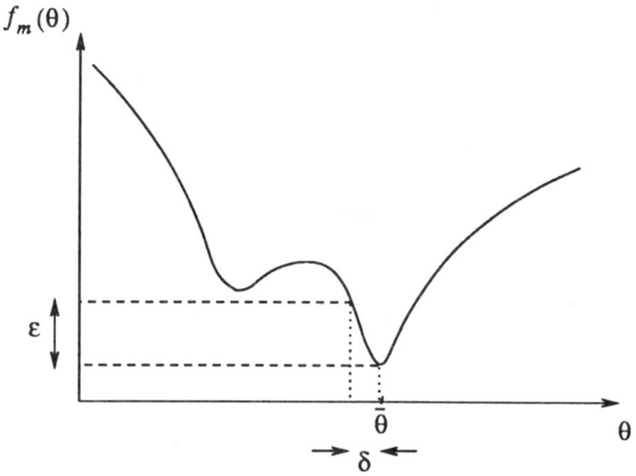

Figure 1: Illustration of condition ii).

Then

$$\lim_{m\to\infty}\hat{\boldsymbol{\theta}}_m = \bar{\boldsymbol{\theta}} . \quad (17)$$

Proof The result is fairly obvious from Figure 1. Without condition *ii)*, the effect of noise could cause a local minimum of $f_m(\theta)$ to become the global minimum of $V_m(\theta)$. To prevent this, the depth of eventual local minima is bounded away from $f_m(\bar{\theta})$ by condition *ii)*. □

Based on the above, let us establish strong consistency of the DML estimate.

Theorem 1 *Under Assumption 1, the DML estimate converges w.p.1 to $\boldsymbol{\theta}_0$ as $m \to \infty$.*

Proof Introduce

$\hat{\mathbf{S}} = N^{-1}\sum_{t=1}^{N}\mathbf{s}(t)\mathbf{s}^*(t) \qquad \hat{\mathbf{Z}} = N^{-1}\sum_{t=1}^{N}\mathbf{s}(t)\mathbf{n}^*(t)$

$\hat{\boldsymbol{\Sigma}} = N^{-1}\sum_{t=1}^{N}\mathbf{n}(t)\mathbf{n}^*(t)$

and subtract the term $\mathrm{Tr}\{\hat{\boldsymbol{\Sigma}}\}/m$ from the DML criterion function (8), to yield

$$V_m(\boldsymbol{\theta}) = \frac{1}{m}\mathrm{Tr}\{\mathbf{P}_{\mathbf{A}}^{\perp}\hat{\mathbf{R}}\} - \frac{1}{m}\mathrm{Tr}\{\hat{\boldsymbol{\Sigma}}\} . \quad (18)$$

By (9) and (3) the sample covariance matrix can be written

$$\hat{\mathbf{R}} = \mathbf{A}(\boldsymbol{\theta}_0)\hat{\mathbf{S}}\mathbf{A}^*(\boldsymbol{\theta}_0) + \mathbf{A}(\boldsymbol{\theta}_0)\hat{\mathbf{Z}} + \hat{\mathbf{Z}}^*\mathbf{A}^*(\boldsymbol{\theta}_0) + \hat{\boldsymbol{\Sigma}} . \quad (19)$$

Hence, we can write, $V_m(\boldsymbol{\theta}) = f_m(\boldsymbol{\theta}) + \varepsilon_m(\boldsymbol{\theta})$, where

$$\begin{aligned} f_m(\boldsymbol{\theta}) &= \frac{1}{m}\mathrm{Tr}\{\mathbf{P}_{\mathbf{A}}^{\perp}\mathbf{A}(\boldsymbol{\theta}_0)\hat{\mathbf{S}}\mathbf{A}^*(\boldsymbol{\theta}_0)\} \\ \varepsilon_m(\boldsymbol{\theta}) &= \frac{2}{m}\mathrm{Re}\left(\mathrm{Tr}\{\mathbf{P}_{\mathbf{A}}^{\perp}\mathbf{A}(\boldsymbol{\theta}_0)\hat{\mathbf{Z}}\}\right) + \frac{1}{m}\mathrm{Tr}\{\mathbf{P}_{\mathbf{A}}\hat{\boldsymbol{\Sigma}}\} \end{aligned}$$

Let us first verify the condition *i)* of Lemma 1. Consider the first term of $\varepsilon_m(\boldsymbol{\theta})$ above. Using the short-hand notation $\mathbf{A}_0 = \mathbf{A}(\boldsymbol{\theta}_0)$ and $\mathbf{A} = \mathbf{A}(\boldsymbol{\theta})$, we have

$$\begin{aligned} &\sup_{\boldsymbol{\theta}} \left|\tfrac{1}{m}\mathrm{Tr}\{\mathbf{P}_{\mathbf{A}}^{\perp}\mathbf{A}_0\hat{\mathbf{Z}}\}\right| \le \left|\mathrm{Tr}\left\{\frac{\hat{\mathbf{Z}}\mathbf{A}_0}{m}\right\}\right| \\ &\quad + \sup_{\boldsymbol{\theta}}\left|\mathrm{Tr}\left\{\frac{\hat{\mathbf{Z}}\mathbf{A}}{m} \times m(\mathbf{A}^*\mathbf{A})^{-1} \times \frac{\mathbf{A}^*\mathbf{A}_0}{m}\right\}\right| \\ &\quad \le C \times \sup_{\boldsymbol{\theta}}\left\|\frac{\hat{\mathbf{Z}}\mathbf{A}}{m}\right\|_F \end{aligned}$$

for some constant C (independent of m). Note that

$$\{\hat{\mathbf{Z}}\mathbf{A}(\boldsymbol{\theta})/m\}_{ij} = \frac{1}{N}\sum_{t=1}^{N} s_i(t) \times \frac{1}{m}\sum_{k=1}^{m} n_k(t)A_{kj}(\boldsymbol{\theta}) .$$

The last sum is readily seen to converge to zero w.p.1 as m tends to infinity, for all values of $\boldsymbol{\theta}$, see e.g. [7]. Thus, the first term of $\varepsilon_m(\boldsymbol{\theta})$ converges almost surely to zero, uniformly in $\boldsymbol{\theta}$. By similar arguments it is straightforward to verify that this is true also for the second term.

To verify condition *ii)*, rewrite the signal term as

$$f_m(\boldsymbol{\theta}) = \frac{1}{m}\mathrm{Tr}\{\mathbf{P}_{\mathbf{A}}^{\perp}\mathbf{A}_0\hat{\mathbf{S}}\mathbf{A}_0^*\} = \frac{1}{m}\|\mathbf{A}_0\mathbf{L} + \mathbf{A}\mathbf{T}\|_F^2 , \quad (20)$$

where $\mathbf{L}$ and $\mathbf{T}$ are defined by the relations

$$\hat{\mathbf{S}} = \mathbf{L}\mathbf{L}^* \qquad \mathbf{T} = -\mathbf{A}^{\dagger}\mathbf{A}_0\mathbf{L} . \quad (21)$$

Let k denote the number of components of $\boldsymbol{\theta}$ that are not present in $\boldsymbol{\theta}_0$. If $|\boldsymbol{\theta} - \boldsymbol{\theta}_0| \ge \delta > 0$, we have $1 \le k \le d$. Assume, without loss of generality, that the unique DOAs in $\boldsymbol{\theta}$ and $\boldsymbol{\theta}_0$ are $\theta_1, \ldots, \theta_k$ and $\theta_{01}, \ldots, \theta_{0k}$ respectively. Let $\mathbf{l}_i^T$ and $\mathbf{t}_i^T$ denote the rows of $\mathbf{L}$ and $\mathbf{T}$, and introduce the matrices

$$\begin{aligned} \mathbf{Q}^T &= \left[\mathbf{l}_1, .., \mathbf{l}_k, \mathbf{t}_1, .., \mathbf{t}_k, \mathbf{l}_{k+1} + \mathbf{t}_{k+1}, .., \mathbf{l}_d + \mathbf{t}_d\right] \\ \mathbf{B} &= [\mathbf{a}(\theta_{01}), \ldots, \mathbf{a}(\theta_{0k}), \mathbf{a}(\theta_1), \ldots, \mathbf{a}(\theta_d)] \end{aligned}$$

We can then rewrite (20) as

$$\begin{aligned} f_m(\boldsymbol{\theta}) &= \frac{1}{m}\|\mathbf{B}\mathbf{Q}\|_F^2 = \mathrm{Tr}\{\mathbf{Q}^*\frac{1}{m}\mathbf{B}^*\mathbf{B}\mathbf{Q}\} \\ &\ge \epsilon\mathrm{Tr}\{\mathbf{Q}^*\mathbf{Q}\} \ge \epsilon\hat{\mathbf{S}}_{11} > 0 = f_m(\boldsymbol{\theta}_0) \quad (22) \end{aligned}$$

In the last expression, $\hat{\mathbf{S}}_{11} = \mathbf{l}_1\mathbf{l}_1^*$ denotes the upper left element of $\hat{\mathbf{S}}$ which is nonzero, and in the first inequality we have used Assumption **A1**. Now, the inequality (22) holds for all m (sufficiently large), which shows that condition ii) of Lemma 1 is satisfied with $\bar{\boldsymbol{\theta}} = \boldsymbol{\theta}_0$. The assertion of Theorem 1 follows. □

Using the above technique, the strong consistency of the SML and beamforming methods can also be verified.

Theorem 2 *Under Assumption 1 the SML estimates converge to the true value w.p.1. If Assumption 2 holds, this is also true for the beamforming estimates.*

Proof The idea of the proof for the SML method is to rewrite the normalized criterion function as

$$\log \frac{1}{m-d}\mathrm{Tr}\{\mathbf{P}_{\mathbf{A}}^{\perp}\hat{\mathbf{R}}\} + \frac{1}{m-d}\log|\mathbf{A}^{\dagger}\hat{\mathbf{R}}\mathbf{A}| \quad (23)$$

The first term obeys condition i) of Lemma 1, by the proof of Theorem 1. It is straightforward to verify that the second term above satisfies condition ii). The proof for the beamforming estimates is similar to the proof for DML. □

It should be noted that although beamforming is a spectral-based method, its consistency for m tending to infinity does not require any assumption of non-coherent sources. This is again an important difference from the large N case. It is well-known that spectral-based techniques loose resolution in the presence of coherent sources. However, the present analysis shows that the resolution is regained by increasing m. In fact, this should not be too surprising; recall the case $N = 1$ and a uniform linear array. This is equivalent to the sinusoids-in-noise problem, for which the DFT (corresponding to beamforming) is known to yield even efficient estimates for large m. When $N = 1$, the "sample signal covariance" matrix is of rank one, which corresponds to perfectly coherent signals.

4 Asymptotic Efficiency

It is well-known that if an ML estimator provides consistent estimates (of all its unknown parameters), it is also asymptotically *efficient*, i.e., the asymptotic covariance of the parameter estimates coincides with the Cramér-Rao bound (CRB). In the deterministic signal model, the unknown parameters are $\boldsymbol{\theta}, \mathbf{S}_N$ and σ^2. We have already shown that $\hat{\boldsymbol{\theta}}_{DML}$ is consistent. For the signal waveforms, we have

$$\hat{\mathbf{S}}_N = \mathbf{A}^{\dagger}(\hat{\boldsymbol{\theta}})\mathbf{X}_N = \mathbf{S}_N + \mathbf{A}^{\dagger}(\hat{\boldsymbol{\theta}})\mathbf{N}_N \to \mathbf{S}_N , \quad (24)$$

since $m(\mathbf{A}^*\mathbf{A})^{-1}$ is bounded and $\mathbf{A}^*\mathbf{N}_N/m$ tends to zero w.p.1. Similarly, it is straightforward to show that also the noise variance estimate is consistent. Hence, the general theory of ML estimation gives the following.

Theorem 3 *Let $\hat{\boldsymbol{\theta}}$ be obtained from (8). Then, asymptotically in m*

$$(\hat{\boldsymbol{\theta}} - \boldsymbol{\theta}_0) \in \text{AsN}(\mathbf{0}, CRB) \ , \tag{25}$$

where CRB is the deterministic CRB [4, 8]

$$CRB = \frac{\sigma^2}{2N} \left[\text{Re}\left\{ (\mathbf{D}^*\mathbf{P}_{\mathbf{A}}^{\perp}\mathbf{D}) \odot \hat{\mathbf{S}}^T \right\} \right]^{-1} \ , \tag{26}$$

and where $\odot$ means elementwise multiplication.

The unknown parameters in the stochastic model include the signal covariance, which obviously cannot be consistently estimated unless $N \to \infty$. Hence, one might be tempted to guess that the DML method should outperform the SML method for large m and small N. However, this is not the case.

Theorem 4 *The DML and SML estimates of $\boldsymbol{\theta}$ are asymptotically (for large m) identical.*

Proof Again, the SML criterion function is rewritten as (23). Using a first order Taylor expansion, it can be verified that the second term does not influence the asymptotic (for large m) behaviour of the estimates.
□

The asymptotic efficiency of the beamforming method is intuitively clear if the CRB is asymptotically diagonal. In view of (26), this is true if the matrices

$$\mathbf{A}^*\mathbf{A} \ , \quad \mathbf{A}^*\mathbf{D} \ , \quad \mathbf{D}^*\mathbf{D}$$

are all diagonal for large m. If this is not the case, the CRB is still diagonal if $N \gg 1$ and the emitter signals are uncorrelated. It is an open problem under what conditions the diagonality of the matrix $\mathbf{A}^*\mathbf{A}$ implies diagonality of the other matrices above.

5 Conclusions

The consistency of the DML and SML methods as the number of sensors tends to infinity is proved. An "identifiability" condition is introduced, which essentially says that the sidelobes of the array must not approach the level of the main lobe for increasing m. The asymptotic efficiency of the DML method follows from its consistency, and it is shown that the DML and SML estimates are asymptotically equivalent.

It is shown that the traditional beamforming method gives consistent estimates if steering vectors corresponding to different DOAs are asymptotically orthogonal. Since this technique coincides with the DML method for one signal, the beamforming estimates are asymptotically efficient if the CRB is asymptotically diagonal.

Interesting questions that deserve further study include a more precise classification of the set of array configurations for which the matrices

$$\mathbf{A}^*\mathbf{A} \ , \quad \mathbf{A}^*\mathbf{D} \ , \quad \mathbf{D}^*\mathbf{D}$$

(with appropriate normalization) are all asymptotically diagonal.

References

[1] R. O. Schmidt. *A Signal Subspace Approach to Multiple Emitter Location and Spectral Estimation.* PhD thesis, Stanford Univ., Stanford, CA, Nov. 1981.

[2] J.F. Böhme. "Estimation of Spectral Parameters of Correlated Signals in Wavefields". *Signal Processing*, 10:329–337, 1986.

[3] M. Wax. *Detection and Estimation of Superimposed Signals.* PhD thesis, Stanford Univ., Stanford, CA, March 1985.

[4] P. Stoica and A. Nehorai. "MUSIC, Maximum Likelihood and Cramér-Rao Bound". *IEEE Trans. ASSP*, ASSP-37:720–741, May 1989.

[5] M. Viberg and B. Ottersten. "Sensor Array Processing Based on Subspace Fitting". *IEEE Trans. SP*, SP-39(5):1110–1121, May 1991.

[6] J.F. Böhme. "Estimation of Source Parameters By Maximum Likelihood and Nonlinear Regression". In *Proc. ICASSP 84*, pages 7.3.1–7.3.4, 1984.

[7] W.F. Stout. *Almost Sure Convergence.* Academic Press, New York, 1974.

[8] H. Clergeot, S. Tressens, and A. Ouamri. "Performance of High Resolution Frequencies Estimation Methods Compared to the Cramér-Rao Bounds". *IEEE Trans. ASSP*, 37(11):1703–1720, November 1989.

New Directions in Low Bit Rate Speech Coding

Bishnu. S. Atal

Speech Research Department
AT&T Bell Laboratories, Murray Hill, N.J. 07974

Abstract

Impressive progress has been made during recent years in coding speech with high quality at low bit rates and at low cost. The new digital cellular system in North America is using coded speech at 8 kb/s, but much lower bit rates are needed to support the increasing demand for cellular telephones. Incremental changes in the present technology are unlikely to produce high quality speech at very low bit rates. Current research at Bell Laboratories is aimed at creating new representations of speech, both in frequency and time, to allow for efficient coding of speech information. This paper reviews the key ideas that support the present speech coding technology and discuss some of the promising new directions.

1. Introduction

Recent success in producing high quality digital speech has sparked a big interest in using the low bit rate speech coding technology for practical applications [1]. New speech transmission standards, with rates ranging from 6.5 kb/s to 13 kb/s, have been adopted to support digital cellular telephone services in North America, Europe and Japan. Rapidly increasing demand for cellular telephones is providing the motivation to decrease the bit rate even further - to 4 kb/s and lower.

The evolution of speech quality at different bit rates is illustrated in Fig. 1. The speech quality on the vertical axis corresponds to an Absolute Category Rating scale, also called the MOS scale [2]. Just ten years ago, little choice, if any, existed for using low bit rate speech coding technology in commercial telephony; the quality of low bit rate speech was below the levels acceptable for such applications. Outstanding progress has been made since then in developing speech coders that provide high quality speech at 8 kb/s or higher. The progress in lowering the bit rate needed for synthesizing high quality speech is expected to continue in future. This paper reviews the current low bit rate speech research - where we are today and what are the new directions that will allow us to code high quality speech at rates all the way down to 2.4 kb/s.

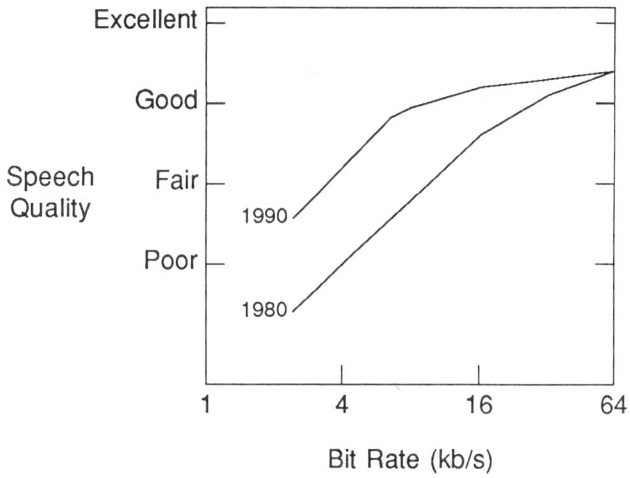

Figure 1. Evolution of speech quality at different bit rates

2. Current Speech Coding Technology

Synthesis of high quality speech at low bit rates using linear predictive coding techniques requires efficient encoding of both the parameters of the LPC filter and its input excitation. Important progress has been made in recent years in reducing the bit rate for encoding the LPC parameters to 1.5 kb/s and lower [3], but the bit rate for encoding the excitation continues to remain high.

The coding of LPC excitation in the present low bit rate coders, such as Code-Excited Linear Prediction (CELP) coders, is handled by selecting the excitation using analysis-by-synthesis matching techniques from an ensemble of waveforms that are stored in a codebook (or

1058-6393/91 $01.00 © 1991 IEEE

several codebooks) [4-6]. The waveform matching is neither necessary nor efficient. Analysis-by-synthesis waveform matching techniques result in wasteful use of bits and are not appropriate for achieving perceptually accurate synthesis of speech.

An important step in encoding the excitation at low bit rates is to create a parametric representation that provides a compact and accurate representation of the excitation for speech synthesis. Lack of an efficient representation of the excitation remains a major obstacle at present in reducing the bit rate.

3. New Research Directions

CELP speech coders use concepts like linear prediction, vector quantization, and noise shaping effectively in producing high quality speech at 8 kb/s and higher. As important as these concepts are, they alone are not sufficient for lowering the bit rate to 4 kb/s or lower. Speech coding invariably introduces quantization noise in the transmitted signal but human perception has limitations and certain kinds of quantizing distortions are not audible. The key to production of high quality speech at low bit rates is in understanding the kinds of distortions that are inaudible to the human ear.

The detailed nature of the LPC excitation waveform is usually very complicated. The important question is - what level of detail in the LPC excitation must be retained to synthesize speech without audible distortion? For periodic (voiced) segments of speech, accurate representation of each period requires knowledge of amplitudes and phases of all the harmonics. But the frequency selectivity of the auditory system decreases at the high frequencies. Is a coarse representation of excitation adequate at high frequencies? Ghitza studied this question in detail using a model of auditory nerve firings to select critical harmonic components [7] and found that approximately ten harmonic components are sufficient. Recent work at Bell Laboratories has also shown that harmonic components with uniform spacing up to 800 Hz and with one-third octave spacing thereafter are sufficient for synthesis of high quality voiced speech [8]. The frequency sampling approach therefore provides an efficient method for representing the details of the excitation waveform within a period.

Successive periods of voiced speech evolve smoothly in time. Kleijn and Granzow have recently found that accurate synthesis of voiced speech is possible from the information about the period specified once every 20 ms [9]. Therefore, the amplitudes and phases of the selected harmonic components need to be transmitted once every 20 ms. For comparison, the excitation parameters are transmitted, typically once every 5 ms, in present CELP (or VSELP) coders.

Preliminary results of studies carried out so far suggest that a small number of parameters transmitted once every 20 ms can provide sufficiently accurate representation of LPC excitation. Proper sampling of excitation in both frequency and time is important for reducing the bit rate without adverse influence on the speech quality.

References

[1] P. Kroon and B. S. Atal, "Predictive coding of speech using analysis-by-synthesis techniques," in *Advances in Speech Signal Processing*, pp. 141-164, S. Furui and M. M. Sondhi, Eds., (Marcel Dekker, New York, N.Y. 1992).

[2] CCITT Series P Recommendations. Telephone transmission quality. Blue Book, vol. V, IXth Plenary Assembly, 1988.

[3] K. Paliwal and B. S. Atal, "Efficient vector quantization of LPC parameters at 24 bits/frame," *IEEE. Trans. Signal Proc.* (to be published, 1992).

[4] M. R. Schroeder and B. S. Atal, "Code-excited linear prediction (CELP): high quality speech at very low bit rates," in *Proc. Int. Conf. on Acoustics, Speech and Signal Proc.*, vol. 1, pp. 937-940, March 1985.

[5] I. M. Trancoso and B. S. Atal, "Efficient search procedures for selecting the optimum innovation in stochastic coders," *IEEE Trans.* ASSP, vol. 38, pp. 385-396, 1990.

[6] I. Gerson and M. Jasuik, "Vector sum excited linear prediction (VSELP)," in *Advances in Speech Coding*, pp. 69-80, B. S. Atal, V. Cuperman and A. Gersho, Eds., (Kluwer Academic Publishers, Norwell, MA, 1990)

[7] O. Ghitza, "Auditory nerve representation criteria for speech analysis/synthesis," *IEEE Trans. ASSP*, vol. ASSP-35, pp. 736-740, June 1987.

[8] S. Gupta and B. S. Atal, "Frequency-domain coding of LPC excitation," *IEEE Workshop on Speech Coding*, Whistler, B.C., Canada, September 1991.

[9] W. B. Kleijn and W. Granzow, "Methods for waveform interpolation in speech coding," *Digital Signal Processing*, vol. 1, pp. 215-230, 1991.

Low Delay Speech Coding

Vladimir Cuperman

Communications Science Laboratory, School of Engineering Science
Simon Fraser University, Burnaby, B.C., Canada V5A 1S6

Abstract

Good speech quality with low-delay coding at 8-16 kb/s has been obtained using backward adaptive analysis-by-synthesis codecs, such as Low-Delay CELP (future 16 kb/s standard), Low-Delay Vector Excitation Coding (LD-VXC), and backward adaptive tree/trellis codecs. This paper presents design and performance trade-offs for the low-delay analysis-by-synthesis codecs at rates of 8-16 kb/s. A number of approaches for improving the speech quality at 8 kb/s are discussed. Backward pitch prediction configuration is compared to a closed-loop forward configuration similar to that used in the conventional forward CELP for the adaptive codebook. Finally, the robustness to transmission errors is discussed and a number of trade-offs for reducing the sensitivity to transmission errors are presented.

1. Introduction

Traditionally, speech coders have been judged based on three major performance criteria: speech quality, rate, and complexity or cost. Recently, communications delay has become an important criterion for speech encoders used in the public switched telephone network (PSTN). In a complex network, the delays of many encoders add together, transforming the delay into a significant impairment of the system. Delay may necessitate the use of echo cancellation and in some applications remains an impairment even after echo cancellation has been performed. For these reasons, the requirements of the new 16 kb/s and 8 kb/s CCITT standards specify low delay as a major requirement.

The delay performance of a speech coder is characterized by the algorithmic delay and the total codec delay. Algorithmic delay is the one-way delay of the encoder and the decoder assuming infinite processing power for the coder implementation. Processing delay is the additional delay due to implementation with a finite processing power. The total codec delay is the sum of algorithmic delay and processing delay. The channel delay caused by transmitting over a finite bandwidth serial channel adds to the total codec delay to give the delay encountered in practical applications. The requirements of the new 8 kb/s CCITT standard specify an algorithmic delay lower than 5 ms (objective 2 ms) and a total codec delay of 10 ms (objective 5 ms). The 16 kb/s standard will have a total delay lower than 2 ms.

The existing speech codecs based on conventional Code Excited Linear Prediction (CELP) [1] or Vector Excitation Coding (VXC) [2] achieve good speech quality at the expense of a substantial delay due to forward adaptation of the short-term predictor, where input speech samples are buffered to compute synthesis filter parameters prior to actual coding of the samples. The input buffering and other processing typically result in a total codec delay of 50 to 60 ms. Hence, the low-delay requirement is not feasible with the established coders that are based on forward adaptive prediction.

Low delay speech coders are based on backward adaptive linear prediction combined with a delayed decision quantization procedure, such as vector quantization or tree/trellis coding. In a backward adaptive configuration, the parameters of the synthesis filter are not derived from the original speech signal, but instead computed by backward adaptation, extracting information only from the reconstructed signal based on the transmitted excitation information. Since both the encoder and decoder have access to the past reconstructed signal, side information is no longer needed for the synthesis filter, and the low-delay requirement can be met with a suitable choice of vector dimension.

Backward adaptation has been widely studied in conjunction with an earlier generation of speech coding systems [3]. The classical ADPCM algorithm based on scalar quantization and backward adaptive linear prediction achieves toll speech quality at 32 kb/s and has an algorithmic delay of only one sample interval, but the

* This work was supported by the National Sciences and Engineering Research Council of Canada and by the Science Council of British Columbia.

1058-6393/91 $1.00 © 1991 IEEE

speech quality degrades quickly for rates under 24 kb/s. Good speech quality was obtained at 16 kb/s by combining block backward adaptive linear prediction with delayed decision quantization based on tree [5, 13] or trellis [12] search. A system based on recursive backward adaptation and analysis-by-synthesis codebook excitation quantization called Vector ADPCM was introduced in [6]. Vector ADPCM, which appears to be the first application of backward adaptation to analysis-by-synthesis excitation coding, evolved into Low-Delay Vector Excitation Coding (LD-VXC) [7]. Lattice LD-VXC (LLD-VXC) achieved toll quality at 16 kb/s with a delay lower than 2 ms [8]. Finally, the emerging 16 kb/s speech coding standard is based on Low-Delay Code Excited Linear Prediction (LD-CELP) which uses codebook based analysis-by-synthesis excitation coding and backward block adaptation of a 50th order short-term predictor; LD-CELP achieves toll quality with a delay lower than 2 ms and good robustness in the presence of transmission errors [10].

Recently, a number of 8 kb/s coders which achieve close to toll speech quality with low delay have been proposed. These include versions of LD-CELP and LD-VXC which use closed-loop quantization of the pitch predictor parameters [11, 9, 17, 18], and a tree encoder based on the fractional rate trees [14].

2. System Overview

The block diagram of an analysis-by-synthesis low-delay speech coder is shown in Fig. 1. At the encoder side, an analysis-by-synthesis technique is used for selecting the optimum excitation sequence. The excitation generator structure may be codebook based as in LD-CELP and LD-VXC, or may consist of a tree (trellis) search module and the corresponding excitation sequence generator. The term codevector will be used for an excitation codebook entry or for an excitation sequence corresponding to a given tree path.

Codebook based analysis-by-synthesis systems use gain-shape or shape-only codebooks. A gain-shape codebook generates the excitation vectors by multiplying vectors in a shape sub-codebook by gain values stored in a gain sub-codebook. In a shape-only codebook the gain sub-codebook has only one entry.

In the diagram shown in Fig. 1, each candidate excitation codevector $\mathbf{c}^{(i)}$ is multiplied by a gain, and the resulting gain scaled vector, **u**, is fed into the synthesis filter. The gain is the product of the predicted gain obtained from the backward adaptive gain predictor and the gain value obtained from the gain codebook. The components of the vector **u** will be denoted by $u(n)$. The output of the synthesis filter, **y**, is then compared to the actual speech signal, **x**, and the best candidate codevector (path) is selected using a perceptually weighted minimum square error criterion based on the weighting filter, $W(z)$.

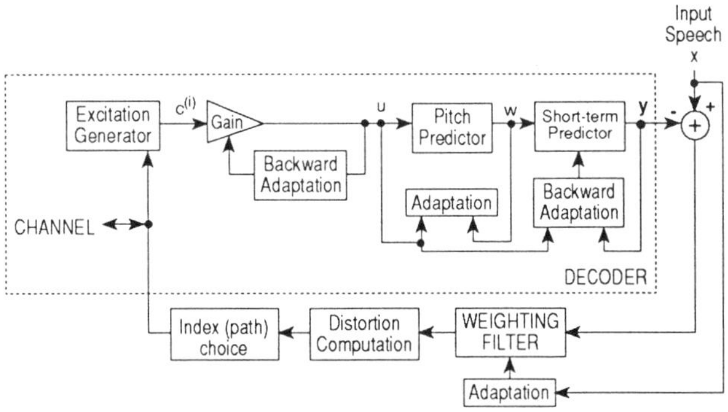

Fig. 1
General Configuration of a Low-Delay Speech Coder

The index of the optimal excitation sequence is then transmitted to the decoder. At the decoder side, the received indices are used to generate the proper excitation sequence. The excitation codevector is then gain scaled using the gain computed in the same way as it is done in the encoder, and fed into the cascade of the pitch and formant synthesis filters. The output of the cascade of synthesis filters is the reconstructed speech which may be postfiltered to reduce the perceived quantization noise.

3. Gain Predictor

Low delay speech coders use backward adaptive linear prediction for excitation gain evaluation. A simple and robust predictor may be obtain by generalizing the Jayant's robust scalar quantizer [10, 14]. Denoting by $\sigma(n)$ the gain at the vector time index n, the corresponding adaptation equation can be written at logarithmic scale

$$\log\sigma(n) = \beta\log\sigma(n-1) + \log M(n-1) \qquad (1)$$

where β is fixed predictor coefficient, $|\beta|<1$, and $M(n-1)$ is a function of the RMS value of the vector $\mathbf{u}(n-1)$.

A further generalization may be obtained by using an adaptive predictor of order M_p [10, 7]

$$\log\sigma(n) = \sum_{i=1}^{M_p} p_i \log|\mathbf{u}(n-i)| \qquad (2)$$

where the predictor coefficients p_i may be computed by applying backward LPC analysis (autocorrelation method) to the sequence of RMS values of the vectors $\mathbf{u}(n-i)$. Alternatively, fixed coefficients p_i optimized on a multi-talker data base may be used [7]. The fixed predictors are more robust in the presence of transmission errors but have slightly lower performance than the adaptive predictors for clean channel conditions.

4. The Short-Term Predictor

The output of the short term predictor, $y(n)$, is computed using the relationship:

$$y(n) = w(n) + \sum_{i=1}^{P} h_i\, y(n-i) + \sum_{i=1}^{Z} g_i\, w(n-i) \quad (3)$$

where $\{g_i\}$ are the coefficients of the all-zero section, and $\{h_i\}$ are the coefficients of the all-pole section.

In block adaptive systems, an all-pole short term predictor is traditionally used (Z = 0). In this case the coefficients h_i can be computed by solving the Wiener-Hopf equations

$$\mathbf{R}_{yy}\mathbf{h} = \mathbf{r}_y \quad (4)$$

where $\mathbf{h} = (h_1, h_2, .., h_P)^T$, $\mathbf{R}_{yy}$ is the autocorrelation matrix of $y(n)$, and $\mathbf{r}_y = (r_{yy}(1), r_{yy}(2), ..., r_{yy}(P))^T$.

The recursive adaptation is based on the application of the *gradient algorithm* for solving the prediction error minimization problem. The corresponding adaptation equations are

$$h_i^{(n+1)} = \lambda_h h_i^{(n)} + \mu_h u(n) y(n-i) \ \text{ for } i = 1, .., P \quad (5)$$

$$g_i^{(n+1)} = \lambda_g g_i^{(n)} + \mu_g u(n) w(n-i) \ \text{ for } i = 1, .., Z \quad (6)$$

where λ_h, λ_g are leakage factors for improving noisy channel performance and μ_h, μ_g are adaptation stepsizes. Simple analytical stability checks exist for coefficients h_i for $P \le 3$. For predictors of order larger than three, the simplest solution is to use the equivalent lattice filter stability check.

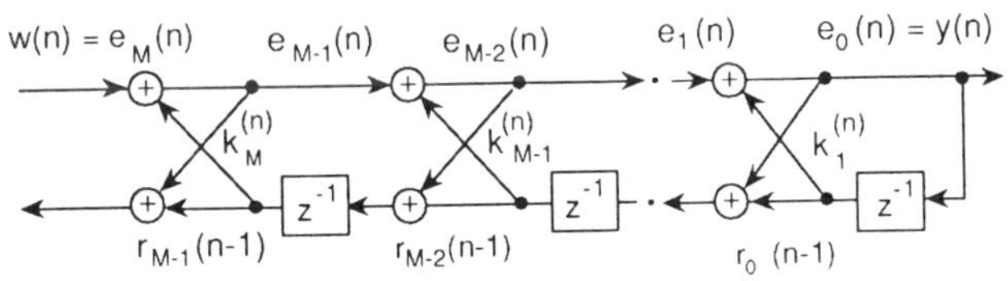

Fig. 2
Lattice Short-Term Predictor

Lattice filters have significant advantages in the implementation of linear predictors. An all-pole lattice filter with recursive adaptation is shown in Fig. 2. The update of the coefficients $k_j^{(n)}$ may be done using the LMS algorithm with leakage factors for improving noisy channel performance. However, a significant reduction in the computational complexity may be obtained by using the so-called sign algorithm [8] which avoids the division operations required by the standard LMS algorithm:

$$k_j^{(n+1)} = (1-\mu)k_j^{(n)} + \lambda sign(e_j(n)) sign(r_j(n-1)) \quad (7)$$

where $1 \le j \le M$. A good choice for the constants is $\lambda = \mu = 2^{-8}$.

In the absence of a pitch predictor, speech quality in low delay coders depends significantly on the short-term predictor order. High order short-term predictors were introduced in LD-CELP where a 50th order block adaptive short-term predictor is used [10]. At 16 kb/s, in the absence of a pitch predictor, the prediction gain saturates for a 20th order predictor for male speakers, but only for a 50th order predictor for female speakers [10]. On the other hand, in the presence of a pitch predictor, the prediction gain saturates at 16 kb/s for a 20th order predictor for male and female speakers [8]. At 8 kb/s there is no performance improvement for predictors of order larger than 10 [11, 9, 18]. The poor performance of high-order predictors at 8 kb/s may be caused by the quantization noise present in the adaptation loop.

The coder robustness in the presence of transmission errors depends significantly on the choice of adaptation signals. Replacing the signal $w(n-i)$ by $u(n-i)$ in (6) (parallel adaptation [7, 15]) results in improved robustness at the expense of a slight performance degradation . The improvement in robustness is due to reduced error propagation. Further reduction of error propagation may be obtained by replacing the signal $y(n-i)$ in (5) by $u(n-i)$, i. e., introducing the residual driven adaptation [3]. However, in this case the performance for clean channels degrades significantly. A good trade-off may be obtained by shaping the residual $u(n)$ by an FIR approximation of the short-term synthesis filter (smoothed error signal adaptation, [14]).

5. The Pitch Predictor

The pitch predictor equation for a three tap predictor is given by

$$w(n) = u(n) + \sum_{i=-1}^{1} a_i\, w(n-k_p-i), \quad (8)$$

where a_i are the filter coefficients and k_p is the current pitch period estimate. Block and/or recursive adaptation may be used for the predictor coefficients a_i.

In block adaptation, the pitch prediction coefficients, a_i, may be calculated from the previous frame of predictor output, $w(n)$, using the autocorrelation [7, 15, 16] or the covariance methods [4, 5, 14].

A robust low-complexity pitch predictor using backward recursive adaptation and a pitch tracker was introduced in [15, 16]. Recursive adaptation allows better tracking of signal characteristics in backward adaptive pitch prediction [15]. The possibility of achieving robustness in the presence of transmission errors with backward recursive pitch prediction was confirmed by the results obtained at 8 kb/s in [14], despite claims that such a predictor may not be robust enough [10].

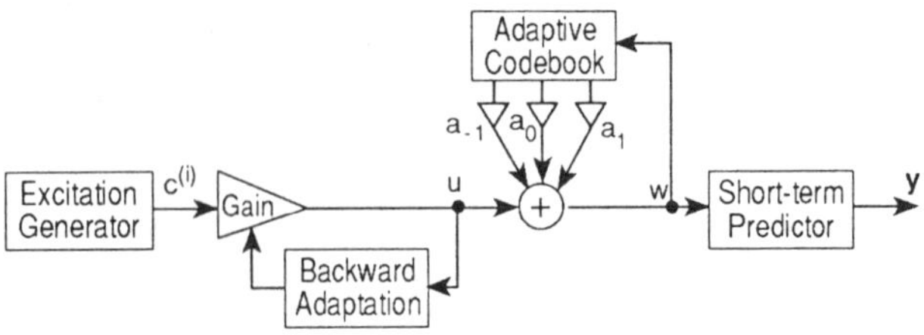

Fig. 3
Closed-Loop Pitch Prediction for Low-Delay Speech Coding

In order to improve speech quality at 8 kb/s, a closed-loop pitch predictor was recently introduced in low-delay speech coders [9, 11, 17, 18]. A "pitch" value is obtained by searching in a codebook of past excitation vectors in a configuration similar to the adaptive codebook used in forward CELP (see Fig. 3). Optimal gain (taps) values a_i are determined during the search procedure for each candidate vector in the excitation codebook. The taps a_i are vector quantized, while the pitch (index) values are quantized using a predictive (delta) coding scheme. The closed-loop procedure requires transmission of the adaptive codebook index ("pitch" value) and quantized taps along with excitation index to the receiver.

6. Excitation Sequence Optimization

The excitation sequence is obtained by minimizing the weighted mean-square error

$$d_w(x,y) = \frac{1}{M} \sum_{j=n_0}^{n_0+M-1} e_w^2(j) \qquad (9)$$

where $e_w(n)$ is generated by passing $e(n) = x(n) - y(n)$ through the weighting filter $W(z)$. For codebook based systems like LD-CELP and LD-VXC, the WMSE is computed for each entry in the excitation codebook and M is the vector dimension. Codebook design algorithms are discussed in [7, 8]. For tree coding, all possible path map sequences to depth M are considered for computing the reconstructed speech $y(n)$ and the path giving the minimum distortion is chosen. The index j refers in this case to "future" instants into the tree, and after selecting the optimum path only the index value at time n_0 is released. A suboptimal search algorithm is employed generally for tree encoding, while codebook encoding uses exhaustive (optimal) search.

7. Weighting Filter

The choice of the excitation sequence is based on a perceptually weighted minimum mean-square error criterion defined by the weighting filter, $W(z)$. Perceptual weighting of the error signal is used to shape the noise spectrum in order to reduce the level of perceived noise.

To derive the weighting filter parameters, the input speech signal is fed into a M_w-order predictor and the coefficients are adapted by the algorithms described in Section 4. The transfer function of the weighting filter is given in terms of the optimal linear predictor for clean speech, $W_1(z)$, as $W(z) = W_1(z/\gamma_1)/W_1(z/\gamma_2)$, where $0<\gamma_2<\gamma_1<1$. Suitable values for γ_1 and γ_2 are 0.9 and 0.4, respectively. The fact that the weighting filter adapts at the same time instants as the short-term predictor and uses an identical adaptation structure was found to have a significant impact on the perceived speech quality. Lower complexity at the expense of a slightly degraded performance may be obtained by using the synthesis filter parameters for designing the weighting filter.

8. Experimental results

In order to study the advantages of closed-loop pitch prediction and the effect of vector dimension upon the performance at 8 kb/s, the following LLD-VXC systems were tested [18]:

System 1 - open-loop pitch predictor, vector dimension 8, 8 bits shape codebook

System 2 - open-loop pitch predictor, vector dimension 10, 8 bits shape and 2 bits gain codebooks

System 3 - closed-loop pitch predictor, vector dimension 18, 8 bits shape codebook, 6 bits adaptive codebook taps, 4 bits pitch for delta pitch encoding

System 4 - closed-loop pitch predictor, vector dimension 20, 8 bits shape codebook, 1 bit gain codebook, 7 bits adaptive codebook taps, 4 bits pitch

All the above systems use a 10th order lattice short-term predictor and a fixed 10th order gain predictor. The signal-to-noise ratio and informal MOS test results for the above systems are given in Table 1.

TABLE I
Performance of 8 kb/s Low Delay Speech Coders

System	MOS	SNR	SEGSNR	
1	3.73	12.1	12.5	
2	3.62	13.4	13.9	
3	3.93	14.1	14.5	
4	3.85	14.2	14.5	

The results in Table 1 indicate that the systems using closed-loop pitch prediction achieve subjective speech quality comparable to the 8 kb/s VSELP standard The systems employing open-loop pitch prediction have a MOS score lower by about 0.2. The results show that the SNR values do not always correlate well with the subjective speech quality estimates. LLD-VXC at 16 kb/s achieves a SEGSNR of about 19 dB and a MOS close to 4.1.

References

[1] B. S. Atal and M. R. Schroeder, "Stochastic Coding of Speech at Very Low Bit Rates", Proc. IEEE Int. Comm. Conf., 1984, pp. 1610-1613.

[2] G. Davidson, M. Yong, and A. Gersho, ''Real-Time Vector Excitation Coding of Speech at 4800 BPS,'' Proc. IEEE Int. Conf. Acoustics, Speech, and Signal Processing, Apr. 1987, pp. 2189-2192.

[3] J. D. Gibson, "Adaptive Prediction in Speech Differential Encoding Systems", Proceedings of the IEEE, vol. 68, no. 4, pp. 488-525, April 1980.

[4] V. Iyengar and P. Kabal, "A Low Delay 16 kbits/sec Speech Coder", Proc. ICASSP, 1988, pp. 243-246.

[5] V. Iyengar, P. Kabal, "A Low Delay 16 kb/s Speech Coder", IEEE Trans. on Signal Processing, vol. 39, May 1991, pp. 1049-1057.

[6] L. Watts and V. Cuperman, "A Vector ADPCM Analysis-by-Synthesis Configuration for 16 kb/s Speech Coding", Proc. IEEE Globecom 1988 Conf., pp. 275-279.

[7] V. Cuperman, A. Gersho, R. Pettigrew, J. J. Shynk and J-H Yao, "Backward Adaptation for Low Delay Vector Excitation Coding of Speech at 16 kb/s", Proc. IEEE Globecom 1989, pp. 34.2.1-34.2.5, see also in "Advances in Speech Coding", B. S. Atal, V. Cuperman, A. Gersho, (editors), Kluwer Academic Publ., 1991.

[8] R. Peng, V. Cuperman, "Variable-Rate Low-Delay Analysis-by-Synthesis Speech Coding at 8-16 Kbit/s", Proc. ICASSP 1991, pp. 29-32.

[9] J-H. Yao, J. J. Shynk, A. Gersho, "Low-Delay Vector Excitation Coding of Speech at 8 Kbit/s", Proceedings of IEEE Globecom'91 Conf., to appear.

[10] J-H. Chen, "A Robust Low-Delay Speech Coder at 16 kb/s", Proc. ICASSP, 1989, pp. 1237-1241, see also in "Advances in Speech Coding", B. S. Atal, V. Cuperman, A. Gersho, (editors), Kluwer Academic Publ., 1991.

[11] J-H. Chen, M. S. Rauchwerk, "An 8 kb/s Low Delay CELP Speech Coder", Proceedings of IEEE Globecom'91 Conf., to appear.

[12] Marcellin M. W., Fischer T. R., "A Trellis-Searched 16 kb/s Speech Coder with Low Delay", in "Advances in Speech Coding", B. S. Atal, V. Cuperman, A. Gersho, (editors), Kluwer Academic Publ., 1991.

[13] J. D. Gibson, Y. C. Cheong, H. C. Woo, W.-W. Chang, "Backward Adaptive Prediction Algorithms in Multi-Tree Speech Coders", in "Advances in Speech Coding", B. S. Atal, V. Cuperman, A. Gersho, (editors), Kluwer Academic Publ., 1991.

[14] H. C. Woo, J. D. Gibson, "Low Delay Tree Coding of Speech a 8 kb/s", Proceedings IEEE Globecom 91 (to appear).

[15] R. Pettigrew and V. Cuperman, "Backward Pitch Prediction for Low-Delay Speech Coding", Proc. IEEE Globecom 1989, pp. 34.3.1-34.3.6, see also in "Advances in Speech Coding", B. S. Atal, V. Cuperman, A. Gersho, (editors), Kluwer Academic Publ., 1991.

[16] V. Cuperman, R. Pettigrew, "Robust low-complexity backward adaptive pitch predictor for low-delay speech coding", IEE Proceedings-I, vol 138, no 4, August 1991, pp. 338-344.

[17] A. Kataoka, T. Moriya, "A Backward Adaptive 8 kb/s Speech Coder Using Conditional Pitch Prediction", Proceedings IEEE Globecom 91 Conf..

[18] A. Husain, V. Cuperman, "Low-Delay Vector Excitation Speech Coding at 8 kb/s", submitted for the 1992 International Workshop on Intelligent Signal Processing and Communications Systems, Taipei.

CELP Coding with Adaptive Excitation Codebooks

U. Kipper, H. Reininger, and D. Wolf

Institut für Angewandte Physik der Universität Frankfurt a. M.
Robert-Mayer-Str. 2-4, 6000 Frankfurt am Main, FRG

Abstract

Code-Excited-Linear-Predictive (CELP) coding has led to speech coding schemes with a good speech quality at 8 kb/s and a fair quality at data rates about 4.8 kb/s. In this paper we propose a new method for improving the speech quality of a CELP codec at low bit rates by adaptation of the excitation codebook (ACELP). The adaptive codebook consists of a set of basic excitation vectors which are adapted to an analysis speech frame by calculating optimum amplitude values. The amplitudes are quantized and encoded at a very low bit rate using a gain shape vector quantizer. The results of simulation experiments show that the mean squared error by CELP with an adaptive codebook is substantially lower than that of CELP with a fixed stochastic codebook. In informal listening tests the quality of the speech processed by the adaptive CELP scheme was superior and less speaker dependent compared to a conventional CELP scheme.

1 Introduction

The coding principle of CELP [1] is analysis by synthesis. A speech signal **s** is encoded by synthesizing a speech signal $\tilde{\mathbf{s}}$ closely resembling **s** and transmitting the parameter for producing $\tilde{\mathbf{s}}$ to a reciever. This concept is realized with a structure shown in Fig. 1. The signal $\tilde{\mathbf{s}}$ is the response of the synthesis filter H(z)=1/P(z)1/A(z) excited by an excitation vector **c** selected from a codebook C. Each excitation vector is amplified with a gain factor g. The difference **d** between the input speech signal **s** and $\tilde{\mathbf{s}}$ is weighted by the filter W(z) and the norm of the weighted difference **w** is calculated. This procedure is repeated with all codebook vectors in order to find that codebook vector which yields the smallest error. The index of this optimum excitation vector and the corresponding gain factor are transmitted to the decoder, together with the encoded filter coefficients of H(z).

CELP schemes achieve at data rates about 8 kb/s a high speech quality. However, at 4.8 kb/s CELP produces only a fair speech quality. Improving the speech quality at 4.8 kb/s is the goal of current research.

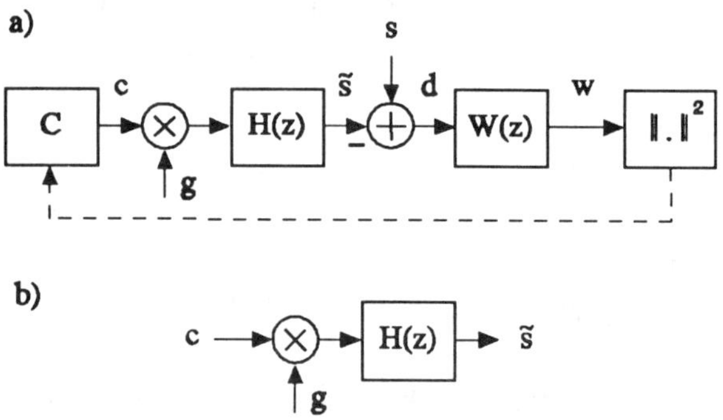

Fig.1: CELP structur: a) coder and b) decoder

Here we focus on the problem of the contents of the excitation codebook. Most CELP schemes use an excitation codebook containing stochastic excitation vectors, i.e. vectors consisting of random numbers. Moreover, the codebook itself is a random but fixed collection of such vectors. This means that no specific information of the speech signal is in such a codebook. The goal of our research was to develop an adaptive excitation sequence [10]. A method for realizing an adaptive excitation sequence arises from Multi-Pulse-Linear-Predictive-Coding (MPLPC) [2,3,7]. The MPLPC-excitation consists of pulses with distances and amplitudes adapted to the speech signal. With a sufficient number of pulses it is possible to achieve a very high speech quality but encoding of the pulse locations and the pulse amplitudes requires a high data rate.

In this paper we present a new type of excitation codebook for CELP which produces a pulse-like excitation and allows the adaptation of the pulse rate and the pulse amplitudes to the speech signal. An efficient encoding of the adaptive excitation is provided by using a regular spacing of the pulses and vector quantization for the pulse amplitudes.

2 CELP with Adaptive Codebook

The adaptive excitation sequence can be regarded as generated by the adaptive excitation codebook depicted in Fig. 2. The adaptive codebook contains excitation vectors which consist of regularly spaced unit pulses, in the following denoted as pulse grids. The excitation grids are grouped into G different grid classes, each comprising all grids with the same number of pulses. Under the assumption that every position of an excitation vector is available as pulse position, the number of the pulse grids within a class is given by the dimension of the excitation vector and the number of pulses specifying a grid class. Therefore, a pulse grid i.e. the pulse positions are uniquely defined by the index of the grid class and the grid number. This allows a very efficient encoding of the pulse positions. Using several pulse grids each with a different number of pulses excitation sequences with variable pulse rates are realized.

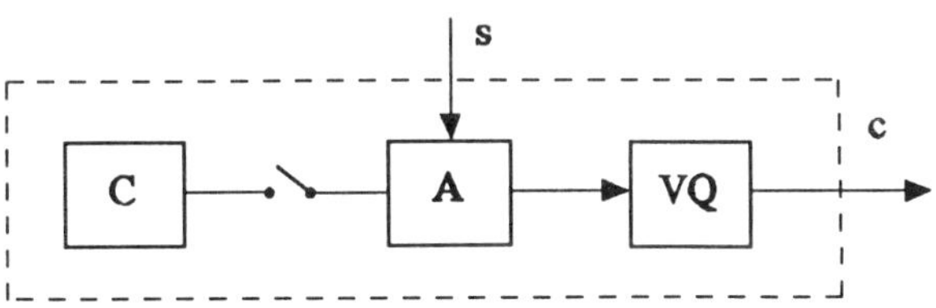

Fig. 2: Adaptive excitation codebook

The optimum pulse amplitudes (A) of all pulse grids are calculated by solving a set of linear equations [3]. For each grid class that pulse grid is determined which minimizes the weighted quadratic error between the original speech signal **s** and the output $\tilde{\mathbf{s}}$ of the synthesis filter H(z). The optimum excitation is found by comparing the errors produced by the best grids of all classes.

In order to realize a very efficient encoding of the pulse amplitudes gain shape vector quantizers are used [9] which allows a high quantization accuracy at very low bit rates as well as non-integer code rates. The gain factor g is quantized by means of an optimum scalar quantizer. In order to obtain a constant data rate for the excitation the quantization accuracy of the pulse amplitudes has to decrease with increasing number of pulses in a grid.

The adaptive excitation sequence can be reconstracted at the decoder by transmitting the index G of the grid class, the grid number NG, the index of the amplitude codebook vector and the quantization level M of the gain factor g.

3 Configuration of the Excitation Codebook

In simulation experiments an ACELP scheme was optimized for a data rate of 4.8 kb/s. For the simulations speech data from English and German male and female speakers were used. All utterances were band-pass limited from 0.3 to 3.4 kHz and sampled at 8 kHz with an amplitude resolution of 16 bit per sample.

For extracting of the parameters of the synthesis filter H(z), which is a cascade of an inverse short-term prediction error filter 1/A(z) an an inverse long-term prediction error filter 1/P(z), each speech segment, consisting of 192 samples with an overlap of 54 samples at both sides, was weighted by a Hanning window [8]. The predictor coefficients of a 10-th order short-term predictor were obtained by the autocorrelation method [4,5] and vector quantized [8]. The coefficient of the first order long-term predictor together with the pitch period, limited to a range of 48 to 176, were adapted every 48 samples according to the closed-loop method [6]. Quantization was performed using optimum scalar quantizers. For transmitting the synthesis filter parameters a data rate of 2.3 kb/s was required. The remaining data rate of about 2.5 kb/s corresponding to a code rate of 15 bits per frame was used for configurating the adaptive excitation codebook. In simulation experiments an excitation codebook containing G=4 different grid classes with 48 dimensional vectors consisting of NP=6, 12, 24, and 48 regularly spaced pulses was found to perform best. For these numbers of pulses the grid class G=1 to G=4 consisted of NG=8, 4, 2, 1, grids, respectively.

The amplitudes of the pulses were encoded by means of a gain shape vector quantizer. Each excitation amplitude A_j was normalized by a gain factor

$$g = \left(\frac{1}{NP} \sum_{j=1}^{NP} A_j^2 \right)^{1/2}$$

The normalized amplitudes were vector quantized on the basis of a quadratic distortion measure using K_G codebook vectors. This encoding requires ld M bits for specifying the gain and ld K_G bits for the codebook index. Table 2 shows also the bit allocation for the encoding of these parameters resulting from simulation experiments. Furthermore, the coderate R is given, which decreases with increasing number NP of amplitudes in a grid.

Table 1: Codebook configuration and bit allocation

G	NP	ld NG	ld M	ld K_G	R
1	6	3	4	6	1.0
2	12	2	4	7	0.58
3	24	1	4	8	0.33
4	48	0	4	9	0.19

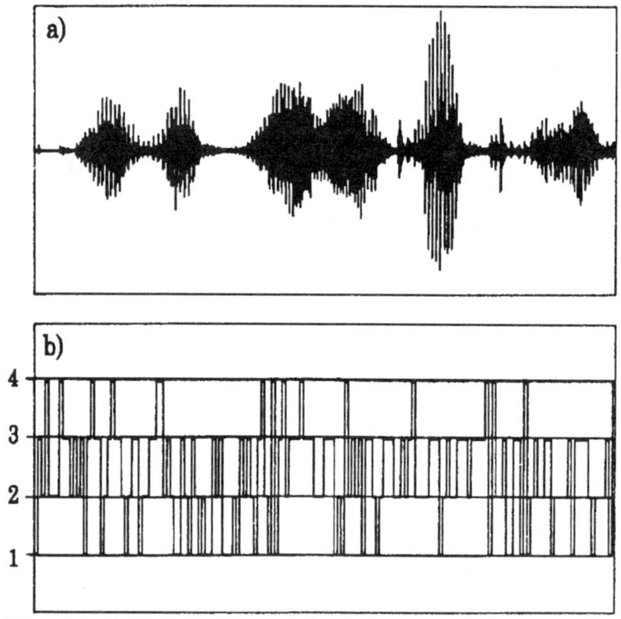

Fig. 3: a) Original speech
b) Optimum grid class per frame

4 Experimental Results

For the encoding of the speech original shown in Fig. 3a an excitation with the grid sequence shown in Fig. 3b results. The strong variation in the number of the best grid is a direct indication of the variation of the pulse rate. Furthermore, it can be seen that no significant preference for a specific gird class excists.

In further experiments we compared ACELP with a conventional CELP scheme using a stochastic excitation codebook. The stochastic codebook contained 1024 48-dimensional vectors, each consisting of gaussian distributed random numbers. The gain factor was quantized with 32 levels leading to a total data rate of 2.5 kb/s for the stochastic excitation. This is equal to the data rate required for quantization of the adaptive excitation sequence. Fig. 4 shows the coding error resulting from the encoding of a speech segment (4a). The coding error of ACELP (4c) is significantly lower than that of CELP with stochastic codebook (4b). In particular, for voiced speech segments the ACELP-excitation is much more appropriate than an stochastic excitation. In terms of SNR an increase of 3 db with the adapted excitation was achieved.

In informal listening tests the quality of speech processed by ACELP was rated higher than that of conventional CELP. In particular the intelligibility of the speech processed by ACELP was found to be superior due to a reduced roughness of voiced parts.

5 Conculsions and Perspectives

An improvement of the speech quality of CELP at low bit rates was achieved by using an adaptive excitation consisting of pulse grids with a variable number of pulses and vector quantized amplitudes. Speech processed with the adaptive excitation codebook sounds remarkably better than that of a conventional CELP scheme and the quality is much less speaker dependent.

Frist experiments directed to the realization of a coding scheme at 2.4 kb/s show that due to the

the excitation is important for obtaining a high speech quality.

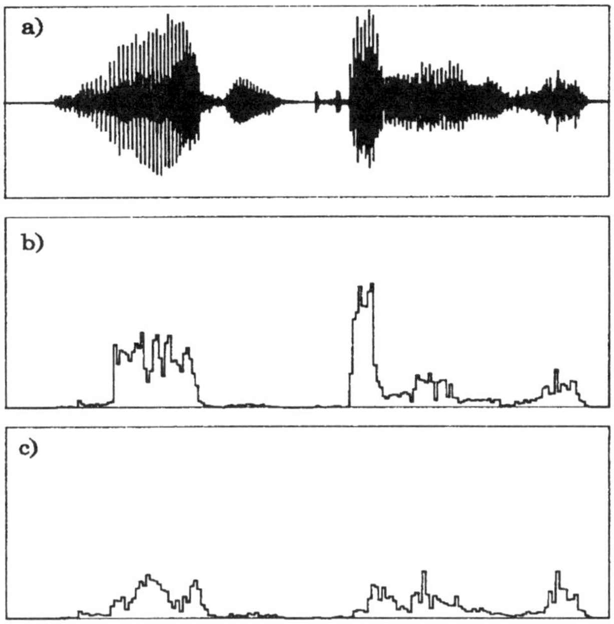

Fig.4:
a) Original speech signal
b) Coding error of CELP with stochastic codebook
c) Coding error of CELP with adaptive codebook

References

[1] Schroeder, M. R., Atal, B. S.: "Code Excited Linear Prediction (CELP): High-Quality Speech at Very Low Bit Rates, Proc. ICASSP, Tampa 1985, pp. 937-940

[2] Atal, B. S., Remde, J. R.: "A New Model of LPC Excitation for Producing Natural-Sounding Speech at Low Bit Rates", Proc. ICASSP, Paris 1982, pp. 614-617

[3] Singhal S., Atal B. S.,: "Improving Performance of Multi-Pulse LPC Coders at Low Bit Rates", Proc. ICASSP, San Diego 1984, pp. 1.3.1-1.3.4

[4] Markel, J. D., Gray Jr., A. H.: "Linear Prediction of Speech", Berlin, Heidelberg: Springer,1976

[5] Jayant, N. S., Noll, P.: "Digital Coding of Waveforms", Englewood Cliffs, N. J.: Prentice Hall, 1984

[6] Siebert, P., Reininger, H., Wolf. D.: "The Efficiency of Long-Term Predictors in CELP Schemes", Proc. Speech'88, Edinburgh 1988, pp. 1033-1038

[7] Kipper, U., Reininger, H., Wolf, D.: "Optimization and Efficient Encoding of Excitation Pulses in Low-Bit-Rate MPLPC Schemes", Proc. URSI-ISSSE, Erlangen 1989, pp. 824-827

[8] Siebert, P., Reininger, H., and Wolf, D.: "Quantization of Predictor Parameters and Gain Factors in a CELP Codec for a Transmission Rate of 4.8 kb/s", Proc. EUSIPCO-88, Grenoble 1988, pp. 1027-1030

[9] Makhoul, J., Roucos, S., and Gish, H.: "Vector Quantization in Speech Coding", Proc. IEEE, Vol. 73, No. 11, Nov. 1985, pp. 1551-1588

[10] Kipper, U., Reininger, H., and Wolf, D.:"Improved CELP Coding Using Adaptive Excitation Codebooks", Proc. ICASSP, Toronto 1991, pp 237-240

FINITE-STATE VQ EXCITATIONS FOR CELP CODING OF SPEECH

Adil Benyassine, Dept. of ECE, NJIT, Newark, NJ 07102

Hüseyin Abut, ECE Department, San Diego State University, San Diego, CA 92182.

ABSTRACT[1]

Today's highly sophisticated digital communication tasks give research engineers impetus to design systems which can reproduce good quality speech at bit rates 4,800 bits per second or lower. Code Excited Linear Prediction (CELP) coding and its derivatives are currently the most frequently used techniques for speech compression at medium-to-low rate range. Until this study, the excitation vectors were always selected from a codebook generated by a Gaussian source or by the ensemble of residual signals collected from the employed speech database. Here we propose an improvement to the excitation model of CELP coders by embedding a labeled-state Finite-State Vector Quantization (FSVQ) in constructing 5 ms long excitation vectors. Our experiments indicate that the quality of the synthesized speech is very similar to those of the full search CELP systems requiring 1000 b/s more.

1. INTRODUCTION

Important advances in algorithmic techniques in speech compression and the availability of efficient low cost signal processors to implement these algorithms resulted in systems which can reproduce reasonably good quality speech at bit rates as low as 4,800 bits per second. The success of these coders has simulated considerable interest both in the scientific research community and in industrial development centers in using the low bit rate speech coding technology for emerging real-life applications. These include the cellular telephone services, the secure telephone, mobile satellite, land mobile communications, and the multimedia applications.

Code Excited Linear Predictive (CELP) coders, Multi-Pulse Linear Predictive Coders (MPLPC), Vector Sum Excited Linear Prediction (VSELP) coders, Vector Excitation Coders (VXC), and their derivatives[1,2] use either an ensemble of pulses or a vector of signal prototypes as the excitation models. The performance of these coders at present degrades rapidly below 4,800 b/s. The rapid growth in demand for cellular communications at half rate (4,000 b/s) has made it necessary to have a careful look at various components of these coders. Important progress has been made in recent years in reducing the bit rate for encoding the LPC parameters [5]. However, the percentage of the bit rate required for encoding the input excitation continues to remain high. Here we propose to improve the excitation model of CELP coders by *embedding* a variation of a Finite-State Vector Quantizer (FSVQ)[3] to structure the excitation process for reducing the bit rate with a "graceful" degradation in synthesized speech quality.

2. FSVQ CELP CODER STRUCTURE

The basic structure of the FSVQ CELP coder is depicted above in Figure 1.The fundamental differences between our system and the basic CELP coding architecture are the inclusion of a *finite-state classifier* in the loop and the way a 40 samples (5.0 ms.) long vector is composed from a number of codebooks. The excitation codebooks could either be populated with *Gaussian codewords* or the *residual prototypes* obtained by filtering the database with the long-term and the short-term filters of the CELP loop. In addition, we have included a *"derailing algorithm"* to guarantee a regular resetting of the finite-state machine at no extra cost and a "state *merging (pruning) technique"* to better populate the codebooks and to reduce the compute time. The remaining blocks in the coder are identical to those of other similar CELP coders.

Analysis

The short-term linear prediction analysis is performed on an open-loop fashion once every 20 ms using a preem-

1. This work has been supported in part by the National Science Foundation Grant ECD-8916669.

1058-6393/91 $1.00

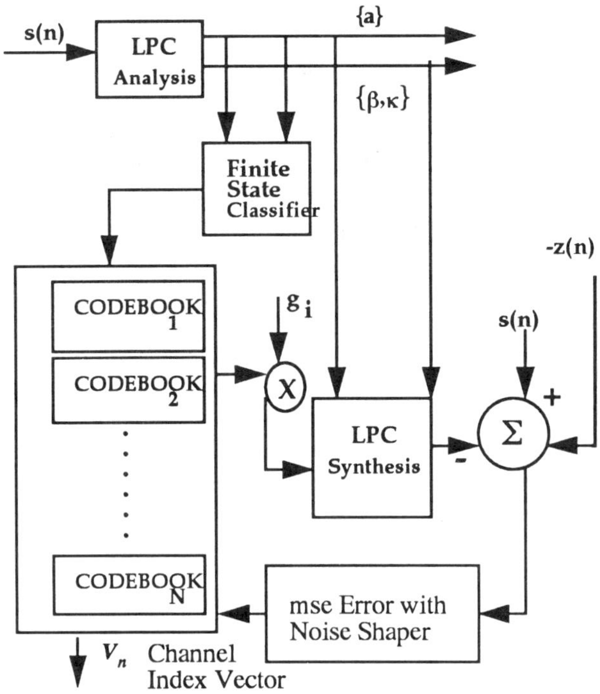

Figure 1. Block Diagram of the FSVQ-CELP Coder.

phasis factor of 0.95, an Hamming window, bandwidth expansion factor of 15 Hz and an autocorrelation analysis of order 16. The analysis window is centered at the end of the last subframe (pitch frame). A memoryless vector quantizer was designed for these coefficients using the basic VQ design algorithm subject to the Itakura-Saito spectral distortion measure [4]. Our long-term analysis was based on an open-loop second-order pitch prediction with quarter sample uniform resolution over 5 ms pitch windows. The quarter sample resolution was achieved via a simple linear interpolation between two tap coefficients. The search was a full search in the pitch lag range of 20 to 147 samples.

FSVQ Classifier

FSVQ can be considered as a VQ version of the backward adaptive quantization technique used in waveform compression. It can be viewed as a finite collection of ordinary vector quantizers. For each incoming vector different codebooks are used, where the codebooks are chosen based on past input vectors. Since the encoder and the decoder both have the same codebook selection procedure, called the next state function, the decoder is able to tract the encoder from the past channel symbols in the absence of channel errors.

A speech coder using finite-state vector excitations in the CELP loop can be described as follows: Suppose that we have a state space S and for each member state S_i in S we have a separate quantizer: an encoder γ_{S_i}, a decoder β_{S_i}, and a codebook C_{S_i}. Given a sequence of input vectors $\{X_n; n = 0, 1, 2, \dots\}$ and an initial state S_0, then the channel index vector, the reproduction vector, and the subsequent state label are defined recursively for n=0,1,2,... as

$$V_n = \gamma_{S_n}(X_n), \quad Y_n = \beta_{S_n}(V_n), \quad S_{n+1} = f(V_n, S_n) \qquad \text{(EQ 1)}$$

The finite-state classifier can either be a Moore machine when its outputs are associated with the states or it can be a Mealy machine when its outputs are associated with the transitions. In FSVQ framework these two structures are referred to as labeled-state and labeled-transition FSVQ's, respectively. However, we will present only a labeled-state FSVQ in a CELP environment. Figure 2 is the FSVQ classifier development block diagram.

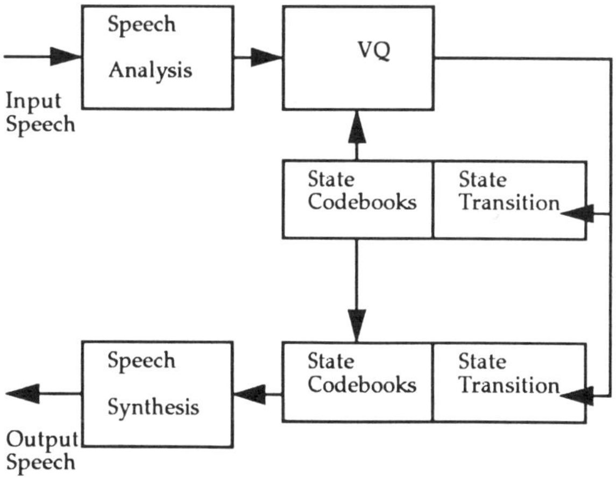

Figure 2. FSVQ Classifier Block Diagram

We have designed the FSVQ classifier of Figure 2 in three steps:

1) Design the state label codebook (VQ).

2) Design each state Codebook

3) Design the next state function.

The design of the state-label codebooks is achieved by employing a memoryless VQ design algorithm to a long sequence training sequence consisting of residual speech signals at the output of the complete linear prediction analysis. We have designed one 40 dimensional codeword for each state. Next, using the VQ just designed we have subdivided the residual signal database into a number of partitions, each partition is labeled with a state label identical to the index of the VQ. As it will be discussed below, we have experimented with various size codebooks for this stage.

The second stage is the design of individual state codebooks. Again using the memoryless VQ design algorithm we have designed one codebook for each state using the appropriate portions of the residual signal database. The ensemble of all these codebooks are called the *supercodebook*. We have limited the size of the supercodebook not to exceed 1024 codewords.

The third stage is the most crucial step in the whole process and there are several techniques to compute the next state function [3,4]. In this study we have chosen the conditional histogram method for its simplicity. Assume that the state codebooks are known, then using these codebooks in the CELP synthesis loop of Figure 1 for every training vector of 40 samples long we compute the relative frequency of each codeword given its predecessor. After determining the conditional histogram, the next state function is easily decided by picking the largest occurrence of each codeword with the same predecessor set. This approach was motivated by the observation that in memoryless VQ speech systems, each state was followed almost always by one of a very small subset of states. Thus, the performance should not be impaired if these subsets are fixed as the individual state codebooks.

State Merging

If the distribution of states is fairly uniform then all the states are equally important. However, if the distribution is highly skewed then it would be difficult to build "good" state codebooks for less probable states. In order to avoid this situation, we have decided to merge less probable states such that the total number of states is fixed and the distribution is fairly smooth. This merging process is completely *ad-hoc* and the specifics of the merging procedure used here are experimentally determined.

In our experiments we have used 4,8, and 16 states and the super codebook was fixed to 1024 codewords of dimension 40, corresponding to the excitation frame size of 5.0 ms. We have merged a 16-state machine into a system having only four states.

Derailing

One particular problem that we have to pay attention closely is when a rare input vector goes into the state machine. Since it does not have any "good" reproduction codeword, the FSVQ cannot track the input closely. Furthermore, channel errors can accumulate and derail the system completely. This problem must be handled by a periodic resetting or by error control or by a combination of the two. We have studied various ways of solving this problem, most of which required additional information to be sent to the decoder. Instead, we have used a simple fixed-time derailing mechanism which did not need any additional bits. After a reasonable number of excitation frames --15 in our final experiments-- we have changed the bit assignment such that the super codebook consisting of all the codebooks is fully searched.

Synthesis Loop

The FSVQ-CELP synthesizer, lower part of Figure 1, is used just like other CELP coders. First zero-input response vector **z(n)** is used to initialize the synthesis filters. Next, given the state codebook, each excitation codeword with its proper gain term is tested in the CELP loop according to the weighted minimum mean square error criterion. The index of the minimum weighted distance codeword is then transmitted through the channel. However, when the finite-state machine is in the resetting mode --the system is derailed-- then the index of the minimum weighted distance codeword in the supercodebook is transmitted. Assuming a noiseless channel, the decoder automatically tracks the coder without any error since it has an exact copy of the encoding procedure.

A noise weighting factor of $\gamma = 0.75$ was used for shaping the error signal. In addition, a post-processing procedure consisting of a deemphasis filter of first order with $\mu = 0.75$ and a single-pole single-zero post-filter were employed for enhancing the quality of the synthesized speech. The transfer function of the post-filter is given by

$$H(z) = (1 + 0.85\, z^{-1}) \div (1 + 0.5\, z^{-1}) \quad \text{(EQ 2)}$$

IMPLEMENTATION AND PERFORMANCE

The complete speech database was 90 seconds of speech consisting of a number of sentences recorded by a variety of speakers (11 male and 11 female). The speech signals were bandlimited to 100-3600 Hz and sampled at 8 kHz[2]. 640,000 samples from this database were used in configuring the finite-state machine and in designing all of the required codebooks. The remaining 80,000 samples from two male and two female speakers were reserved for testing the system. A VQ codebook with 512 codewords (9 bits = 450 bits/s) was designed using the Itakura-Saito spectral distortion measure [4] for the short-term linear prediction coefficient sets $\{a_k; (0 \le k \le 16)\}$ obtained from the autocorrelation analysis. The long-term prediction with quarter sample resolution required 400 bits/s overhead, the pitch lag was coded uniformly into 7 bits $(20 \le k \le 147)$ and the pitch predictor was quantized by a 5-bit 4-dimensional VQ. Since our goal was to exploit any possible structure in the excitation process there was no attempt to use either a non-linear or a differential scheme to reduce the bit rate for the pitch lags. These three components of the long-term prediction process needed 400 bits/s, 1,400 bits/s and 250 bits/s, respectively. However, we are planning to investigate the possible savings there in order to accommodate a better coder for the short-term LPC coefficients [5].

The residual signal vectors were clustered into 16 groups to obtain 16-state label codebook. The histogram of these state labels are shown in Table 1. As it can be seen from this table one state is used more than 61% of the time and a number of states were used less that 1% each. As it is well known in the VQ literature the states with small relative frequencies resulted in empty cells during the state codebooks design stage.

Table 1: Histogram of State Labels S_j vs $P(S_j)$ in Percentage(%)

S_j	1	2	3	4	5	6	7	8
$P(S_j)$	0.7	1.3	7.9	0.9	2.0	2.4	8.6	2.0
S_j	9	10	11	12	13	14	15	16
$P(S_j)$	2.0	1.3	61.5	1.1	1.5	1.3	4.8	0.7

In order to fix the problem of empty cells we have decided to merge the 16-states into a 4-state machine as shown in Table 2. We have then designed one state codebook with up to 256 codewords for each of these four state groups and there were no longer any empty cells in any of the state codebooks. We have labeled these new merged states as *state groups.*

2. This is a copy of the database used by Kroon and Atal [5, page 324]. We would like to acknowledge Peter Kroon for his assistance.

Table 2: Histogram of Merged State Groups SG_j vs $P(SG_j)$ in Percentage(%)

S_j	11	3+7+15	1+2+4+5+8+14	*rest*[3]
SG_j	1	2	3	4
$P(SG_j)$	61.5	21.3	8.2	9.0

As mentioned in the previous section, we have experimented with a number of derailing mechanisms and the derailing times. In Table 3, we tabulate the effects of derailing time on the SNR values of the FSVQ-CELP performance for two different state codebook sizes[4]. To keep in mind, at times when the system is derailed the coder has a different bit assignment strategy. In particular, the supercodebook is used in encoding and decoding to generate the excitation vector. This requires additional 2 bits for 4-state FSVQ and 4 bits for 16-state FSVQ, respectively. These additional bit assignment requirements are met by repeating the short-term spectral coefficients of the previous analysis window. We have also studied interpolation techniques and other mechanisms to compensate the added bits. However, there were no objective or subjective gains with respect to our simple technique.

Table 3. Effects of Derailing Time on SNR

Derail Time in frames	State Codebook Size=256 SNR(dB)	State Codebook Size=128 SNR(dB)
0	8.9	8.1
1	8.7	8.3
5	8.4	7.9
10	8.6	8.4
15	8.9	8.3
20	7.7	8.2
30	8.9	8.1

In order to have a total bit rate in the neighborhood of 4,000 bits/s, we decided to use 128-codeword

3. The term *rest* stands for state labels, 6+9+10+12+13+16
4. We would like to point out that we have used only a subset of the overall database in compiling derailing statistics and hence, the SNR values are different than the overall system performance discussed later.

state codebooks for every 5 ms. long excitation vectors. This corresponds to 1,400 bits/s for the excitation vectors. Finally, we have designed a 4-dimensional 32-level codebook for the gain terms adding another 250 bits/s to the bit rate. Thus, the overall FSVQ-CELP coder bit rate is **4,150 bits/s.**

We have obtained the SNR and segmented SNR values for the proposed system and those for a reference CELP system operating at the same bit rate. The reference system had one 128-level excitation Gaussian codebook with 4-sigma loading and the rest of the system was identical to the FSVQ-CELP. These measurements are tabulated in Table 4. The SNR and segmental SNR values are very similar to each other for both the reference CELP system and the FSVQ-CELP system. So, the proposed system did not yield any improvement in quantitative measurements.

Table 4. Performance of CELP & FSVQ-CELP Systems at 4,150 Bits/s

System	SNR	SegSNR
CELP	7.60	8.44
FSVQ-CELP	7.49	8.41

There might be a number of factors contributing to that. One factor is the assumption on the "goodness" of every state codebook. Another reason could be the way we have grouped the states to form state groups. Yet, another factor could be the next-state function itself. Once a state decision is made then the search is limited to that codebook, which physically eliminates any recourse.

The remedy to that can be to improve the excitation model by employing *a mixture density* approach in forming our 5 ms long excitation vectors. In other words, we can use a percentage of the best match best from every state codebook and the percentages can be computed borrowing the ideas from the *Hidden Markov Modeling (HMM) technique*, which is very frequently used in speech recognition.

The subjective quality of the synthesized speech, however, was uniformly better for the proposed FSVQ-CELP system. In fact the quality of the synthesized speech was very similar to those of significantly more complicated systems operating at 4,800 bits/s or higher. Informal listening tests indicate that this savings of 650 bits/s or more did not cause any noticeable degradation in quality. On the contrary, we believe there was even some improvement. In conclusion, the system proposed here can be a candidate for half-rate speech coders running at 4,000 b/s.

Possible improvement areas are:

1) Better LPC spectrum coding as proposed in [5].
2) Savings on pitch lag bit assignment; such as, the usage of differential coding. Non-linear sub-sample pitch predictors could also result in bit rate reduction without degrading the quality of the reconstructed speech.
3) Mixture density approach for the excitations. We are currently incorporating this idea in our simulations [6].
4) Replacement of residual codebooks with matching Gaussian codebooks. This can be achieved by designing each state codebook from a Gaussian source with statistical and spectral features matching to those of the residual signals falling into that state.

REFERENCES

[1] B.S. Atal, V. Cuperman and A.Gersho, Editors, *Advances in Speech Coding*, Kluwer Academic Publishers, Boston, MA., 1991

[2] L.A. Gerson and M.A. Jasiuk, "Vector Sum Excited Linear Prediction (VSELP) Speech Coding at 8 kbps," Proc. IEEE Int. Conf. on Acoustics, Speech and Signal Processing, pp. 461-464, April 1990.

[3] J. Foster, R.M. Gray, and M. Ostendorf Dunham, "Finite-State Vector Quantization for Waveform Coding," IEEE Trans. on Information Theory, Vol. IT-31, pp. 348-359, May 1985, also in *Vector Quantization*, H. Abut, Ed., IEEE Press, New York, N.Y., 1990.

[4] R.M. Gray, "Vector Quantization," IEEE Acoustics, Speech and Signal Processing Magazine, April 1984, also in *Vector Quantization*, H. Abut, Ed., IEEE Press, New York, N.Y., 1990.

[5] K.K. Paliwal and B.S. Atal, "Efficient Vector Quantization of LPC Parameters at 24 Bits/Frame," to appear in the IEEE Trans. on Signal Processing, 1991.

[6] A. Benyassine and H. Abut, "Mixture Excitations for Half-Rate Finite-State CELP Speech Coders," submitted to IEEE ICASSP-92.

Speech Technology Applications to Telephone Services

Maurizio Copperi

R&D Department
SIP Headquarters
Torino, ITALY 10122

Abstract

This paper presents the performance issues associated with using speech recognition and text-to-speech synthesis technologies in the telephone network, as well as experimental results obtained in preliminary trials. In particular, we discuss both the motivations and the technical approaches for applying these techniques to important automated telephone services, such as the alternate billing service and the reverse directory inquiry.

1 Introduction

The progress of technology in the field of speech processing has laid down the basis for the development of services and facilities that rely upon automated speech input/output systems. In particular, this paper deals with the main applications of speech recognition and Text-To-Speech (TTS) synthesis for the telephone users.

The use of the speech recognition technology in the telephone networks creates specific problems due to both the user reaction and the transmission quality. More precisely, it is almost impossible to get customers to adhere strictly to the procedure developed to elicit their answers, therefore means must be devised to increase the probability of correct customer interactions (man/machine interface). Moreover, the telephone network is typically a hostile environment for speech recognition systems, because of, among other things, bandwidth limitation, channel and background noise, echo, ringing and crosstalk. Even if most service selection automation can be accomplished via caller dual-tone multifrequency inputs, the verification part of a service (e.g. collect call) is often requested from a customer at a dial-pulse telephone. Therefore, to be able to automate the verification to all telephones using one technology, an approach is the use of speaker-independent Automatic Speech Recognition (ASR) techniques.

Research into speaker-independent recognition algorithms, which are robust over telephone lines, has produced promising results, making the provision of the Automated Alternate Billing Service (AABS) feasible [1]. A major element in the growth of AABS and, in general, of automated information services will be the introduction of the Intelligent Network and the related intelligent peripheral processors, equipped with suitable resources including speech recognition, text-to-speech synthesis, speaker verification, etc.

Of course, an important issue to be considered when talking about recognition performance is the measure adopted to evaluate the accuracy in a given task domain. Scoring methodologies should comply with a common framework in order to permit fair comparisons between different systems. This aspect is addressed in [2] for a key wordspotting system used in three applications, including AABS.

As far as speech synthesis is concerned, a number of telephone network applications involves the translation of names and addresses, such as the Automated Customer Name and Address (ACNA) service, also known as the reverse directory service. Conventional TTS synthesizers need specific refinements for achieving good name pronunciation and accentuation, since they are tuned for synthesizing words. However, this technology seems to be promising for conveying to users the information stored in large mass memories, avoiding an operator to speak the various messages required in a service. Current research efforts are focussed on the development of techniques able to offer an acceptable level of quality and intelligibility [3, 4].

This paper is organized as follows. Section 2 describes an experimental collect-call service developed for the Italian telephone network, while section 3 deals with the main issues related to the ACNA service for the Italian language. Both activities have been carried out in a joint effort with CSELT.

2 Collect-call Service

Considering the performance of ASR systems available today, our goal was to test the collect-call service which relies upon isolated word recognition from a small vocabulary. In order to achieve the highest degree of success of the complete service, in addition to the accuracy of the recognizer itself, we have to take into account the human factor aspect of the man/machine dialogue [5]. The requirements of the billing verification procedure are generally characterized by three main items:

- vocabulary to be used in the recognizer;
- prompt announcement and timing to be given to the called party;
- recovery of errors produced by both the recognizer and the user.

To get an insight into the above aspects, a pilot study has been conducted employing 300 naive customers and using three man/machine dialogues with variations in text, timing and quickness [6]. The procedure to obtain a completely automatic billing service consists of the following steps:

- *prologue*, to identify the service, the caller's name, the origin of the call. In this step the called party must quickly realize that he or she is not dealing with a human operator;
- *verification*, to determine billing acceptance or refusal. The possible outcomes are: recognition of Sì (Yes) or No; rejection, which means that the recognition score is poor as very likely the user uttered a word outside the valid vocabulary; time out, that is no recognition due to the large delay in answering; and hang-up, which is obviously considered as a negative response to the query;
- *epilogue*, to notify the result of the verification to the calling party.

Worth noting is that in the case of rejection or time out, the user is required to answer a second question reformulated to elicit a clear-cut response.

2.1 Experimental Set-up

Customers can access the experimental service from the Turin area, dialing a special service number followed by the destination number (local or long distance call within the Italian PSTN) and saying their name for the successive identification to the called party. The hardware system, connected to a special service exchange, is based on off-the-shelf speaker-independent ASR (VOTAN VPC 2000) trained by 16 prototypes for each word Sì, No. These prototypes have been selected via a clustering procedure, based on the Euclidean distance of cepstral features, applied to a training sequence of 1000 repetitions of each key word.

Main functions of the equipment include the management of a single input line, digital recording of the customer's response, control of an external ASR device by a serial port, automatic call-setup monitoring system, complete documentation of incoming and outgoing calls, control of the customer's response via an off-line software procedure, statistics on the service and the recognition results.

2.2 Preliminary Results

In the period January-April 1991, the system managed 3669 accesses with 2106 calls answered. Of these, 1993 came through the billing verification procedure and 1977 were actually billed. The users involved in the completed transactions were 494 male and 1483 female; the relevant traffic was formed by 45 % of local calls and 55 % of long distance calls. It is worth noting that customers were generally unfamiliar with the existing operator assisted collect-call service and unaware of this automatic service.

Table I shows the *customer actions* during the verification step. A large percentage of users (78.2 %) has provided a correct response at the proper time (after beep). The billing re-prompt procedure has been able to recover 59.5 % of previous "time out" (T.O.) caused by no reply and 34.5 % of previous "rejection" (Rj) caused by other speech. It is also evident the attitude of customers to use outliers and/or extraneous speech (indicated as **other** in Table I) only after beep, in contrast with the behavior before beep. In addition, a small but significant number of prompts has not been answered.

Table I - Customer actions (in percentage) vs. the status of the billing verification procedure

Status / Action		Prompt (1993)	Re-prompt from T.O. (173)	Re-prompt from Rj (107)	Re-prompt total (280)
Hang-up		0.8	5.2	7.5	6.0
No reply		5.3	23.1	16.8	20.7
Before beep	**Sì/No**	4.2	0.6	0.9	0.7
Before beep	**Other**	-	-	-	-
After beep	**Sì/No**	78.2	59.5	34.5	50.0
After beep	**Other**	11.5	11.6	40.2	22.5

Table II depicts the *ASR performance* in terms of confusion matrix. Referring to the global number of responses obtained (2478), it results that the word Sì has scored 86.9 % of correct recognition, 0.2 % of substitution error, 5.3 % of time out and 7.5 % of rejection, while the word No has achieved better results. A conspicuous 26.2 % of other speech and 5.5 % of no response have been recognized as either Sì or No, therefore the system must be improved by incorporating much more robust rejection strategies. Typical outliers spoken by the users are "Sì grazie" (Yes thank you) and repetitions of the same key word, such as "Sì Sì", "No No". On the average, a 36 % of **other speech** contains the words Sì, No.

Table II - Confusion matrix of the ASR device

ASR out / User reply	Sì	No	T.O.	Rj
Sì 1769	86.9 % 1538	0.2 % 4	5.3 % 94	7.5 % 133
No 116	0 0	91.4 % 106	0 0	8.6 % 10
Sì (outliers) 85	76.5 % 65	2. 3 % 2	5.9 % 5	15.3 % 13
No (outliers) 9	0 0	55.6 % 5	0 0	44.4 % 4
No response 270	3.7 % 10	1.8 % 5	92.2 % 249	2.2% 6
Other speech 229	10.9 % 25	15.3 % 35	7.0 % 16	66.8 % 153

Table III summarizes the *performance of the billing verification transaction.* Almost 90 % of calls has been correctly automated (77.7 % at the first prompt and 12.2 % at the second one); 9.6 % of calls has been rejected and some 0.5 % incorrectly automated. A breakdown of this last figure shows that a large fraction is due to the interaction with automatic devices, such as fax and answering machines. Moreover, there is an inconsequent behavior of users that, after a negative response to the prompt, still hold on. This fact leads to the conclusion that a simpler verification procedure, permitting only Sì for the positive response and Hang-up for the negative one, may be unreliable.

The results so far collected in the experimental service show that the main shortcomings are caused by:

- customer misunderstanding (e.g. anticipated and missing responses, outliers and extraneous speech, inconsistency of reactions);
- ASR errors (e.g. not triggering, false rejection, interaction with fax and answering machines).

To cope with both types of problems, it is necessary improving the man/machine dialogue as well as the ASR algorithm. This activity is currently underway, and will be reported in a next paper.

Table III - Performance of the billing transaction

Calls / Result	Percentage (100 %)	Number (1993)
Correctly accepted	**80.8**	**1610**
Correctly refused	**9.1**	**181**
Correctly automated at first prompt at 2nd from T.O. at 2nd from Rj	**89.9** 77.7 7.3 4.9	**1791**
Rejected by T.O. - T.O. by Rj - Rj by mixed status	**9.6** 3.4 3.3 2.9	**191**
Incorrectly automated	**0.55**	**11**

3 ACNA Service

In Italy, about 12 % of the total number of calls to the Directory Service refers to the inquiry of the name and address of a user, given the corresponding telephone number. As the answer is unique and the request can be fed into the system using the telephone keypad, this service can be automated. Output messages are generated by means of TTS technology, because prerecorded speech is impractical owing to the huge amount of data to be read by a professional speaker and then stored. Unfortunately, a number of issues must be considered when applying the TTS technique:

- absence of a semantic context to make intelligibility easier;
- expansion and pronunciation of abbreviations, initials, trade marks;
- presence of foreign names with different phonetic transcription;
- lexical stress assignment for surnames.

Fig. 1 shows the block diagram of a text-to-speech synthesizer, in which the arrows depict the information flow from input to output.

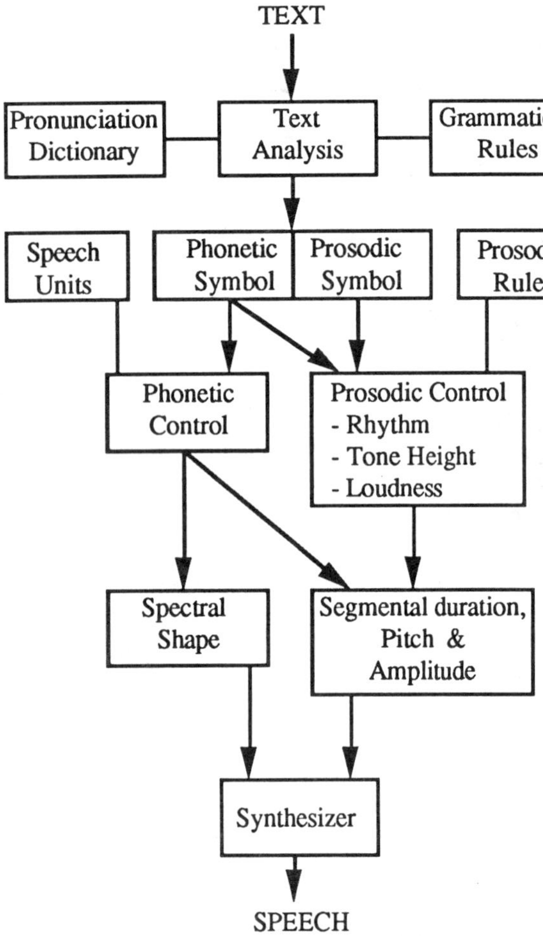

Fig. 1 - Text-to-speech synthesis

In the following, main aspects regarding the particular ACNA application are discussed. A general overview of TTS synthesis is provided in [7].

The grammatical rules exploited by the text analyzer can be simplified because of the rigid nature of the messages to be read. For instance, verbal forms are absent and this makes a syntactic analyzer unnecessary. For what concerns the pronunciation dictionary, special rules should be devised when it comes to assign the stress position which, for names, may occur in different positions with respect to the standard lexical one. The correct pronunciation of Italian names obeys the common text-to-sound rules developed for words, while some difficulty emerges when the system has to manage foreign names that, on the other hand, occur in the telephone directory with a low frequency. A subsystem for predicting the origin of names and defining relevant pronunciation rules could be added to the basic scheme of Fig. 1. Also the prosodic control must be tuned carefully, in order to ensure natural pitch contours for names, addresses and street numbers.

In the experimental system, the speech units are represented by 900 diphones, that basically cover the co-articulation phenomena in Italian and therefore attain a reasonable level of quality. The synthesizer can rely upon a conventional linear predictive model, describing the vocal tract and the excitation source, in which the prosodic control is performed very easily modifying a few parameters of the model. However, a more natural sounding output speech is achievable by exploiting the so-called Time Domain Pitch Synchronous OverLap and Add (TD-PSOLA) method [8]. According to this approach, speech units are windowed and then processed by a least-squares overlap-add synthesis algorithm to produce the output speech. The modification of the pitch in a voiced segment is performed by making the windowed prototype units periodic at the desired synthesis period.

An intelligibility test has been carried out by using a corpus of 180 surnames. Speech stimuli, presented to the listeners through standard telephone sets, were uttered by a "good" male talker and also generated by a TD-PSOLA synthesizer with diphones sampled either at 16 or 8 kHz. Thirty listeners were instructed to write down, in conventional orthography, what they heard.

The results show that natural speech has an intelligibility score of 89 %, while synthetic speech has a score of 75 % (diphones sampled at 16 kHz) or 70 % (diphones sampled at 8 kHz). These values check those obtained in [3] for a different TTS synthesizer and stress the fact that the ACNA task is quite difficult even for natural speech (11 % of errors). There is however room for improving the performance of the experimental system. For instance, by adding new diphones and/or triphones to the inventory and refining the PSOLA algorithm.

4 Conclusions

The application of relatively new technology such as speech recognition in the telephone network gives rise to a number of issues concerning not only the ASR accuracy but also the customer reaction. Even if the continuous effort spent towards improving the robustness of speech recognizers is an important step to be pursued, this may not improve the overall service performance if the human factor component is not fully considered. This fact has been confirmed in preliminary tests and has prompted us to devise new dialogue structures and develop a more effective recognition algorithm, on which we will report at a later date.

In the case of TTS technology, the particular ACNA service requires a high level of performance and represents a challenge for currently available synthesizers. Nevertheless, several studies are in progress to improve the speech intelligibility, providing good pronunciation accuracy and phonetic quality.

Acknowledgements

The author would like to thank F. Canavesio and L. Nebbia of CSELT for many helpful discussions.

References

[1] M.A.J. Murphy et al., "Automation of alternate billed calls using voice recognition", *Proceedings of XIII International Switching Symposium*, pp. 123-127, Stockholm, 1990

[2] J.G.Wilpon, L.G.Miller, P.Modi, "Improvements and applications for key word recognition using Hidden Markov Modeling techniques", *Proceedings of ICASSP*, pp. 309-312, Toronto, 1991

[3] M.F.Spiegel, M.J.Macchi, K.D.Gollhardt, "Synthesis of names by a demisyllable-based speech synthesizer (SPOKESMAN)", *Proceedings of European Conference on Speech Communication and Technology*, pp. 117-120, Paris, 1989

[4] M.Balestri, "A coded dictionary for stress assignment rules in Italian", *Proceedings of European Conference on Speech Communication and Technology*, pp. 1169-1171, Genoa, 1991

[5] L.T.Adams, L.M.Lewin, "The human interaction component of incorporating speech recognition technology into the telephone network", *Proceedings of ICC*, pp. 1-6, Boston, 1989

[6] F.Canavesio, L.Fissore, M.Oreglia, F.Senia, "An experimental automatic collect-call service on the Italian PSTN", *Int. Conf. & Exhibition of Voice Processing Technologies & Applications*, poster, London, 1991

[7] Y.Sagisaka, "Speech synthesis from text", *IEEE Communications Magazine*, pp. 35-41, January 1990

[8] C.Hamon, E.Moulines, F.Charpentier, "A diphone synthesis system based on time-domain prosodic modifications of speech", *Proceedings of ICASSP*, paper 16.S5.7, Glasgow, 1989

Vector-Quantization-Based Speech Recognition and Speaker Recognition Techniques

Sadaoki Furui
NTT Human Interface Laboratories
3-9-11 Midori-cho, Musashino-shi, Tokyo, 180 Japan

Abstract

This paper reviews major methods of applying the vector quantization (VQ) technique to speech and speaker recognition. These include speech recognition based on the combination of VQ and the DTW/HMM technique, VQ-distortion-based recognition, learning VQ algorithms, speaker adaptation by VQ-codebook mapping, and VQ-distortion-based speaker recognition. Not only has it reduced the amount of computation and storage, the VQ technique has also created new ideas of solving various problems in speech/speaker recognition.

1. Introduction

The vector quantization (VQ) technique was first applied to speech coding and image coding. In speech/ speaker recognition, the VQ technique was also first used as an efficient spectral quantization technique, and it has greatly helped to reduce the amount of computation and storage. The VQ technique has also been used as a non-parametric representation method of spectral distribution, and this approach has created various new algorithms for speech/speaker recognition. The following chapters give an overview of these techniques.

2. VQ-Based Speech Recognition

2.1 Combination of VQ and DTW

Speech recognition is essentially based on the comparison between an utterance and a representation (a reference pattern or a reference model) of the vocabulary words obtained by a training phase. Two main distortions are generally observed between them. The first one is a non-linear warping of the time scale, which can be coped with by using the dynamic time warping (DTW) technique. The second difference concerns the pronunciation itself in that a dissimilarity remains in spite of optimal restoration of the time scale.

If the reference patterns or both the input utterance and the reference patterns are represented as sequences of VQ code sequences instead of spectral parameter sequences, the amount of computation and storage for the reference patterns can be greatly reduced [1][2]. Here, the VQ is also implicitly used as a clustering technique to generate good, efficient and reliable templates.

In the SPLIT (strings of phoneme-like templates) system, the first word recognition system based on this technique [1], phoneme-like templates (codewords, prototype vectors) are generated by clustering a set of spectral parameters in training utterances, and each word is represented as a sequence of templates. Since the spectral distance calculation is performed between input utterance frames and phoneme-like templates, the amount of calculation does not depend on the vocabulary size. This method is, therefore, especially effective for speaker-dependent large vocabulary word recognition, as well as speaker-independent word recognition using multiple templates.

2.2 Discrete HMM

Recently, the hidden Markov model (HMM) [3]-[9] has become more popular than the DTW technique. The HMM has the capability of modeling both of the two main distortions described above: the time warping and the pronunciation variations. The HMM which uses VQ-based discrete spectral density is called "discrete HMM" in contrast to "continuous HMM" which uses continuous spectral density functions.

A typical structure of a discrete HMM is based upon a left-to-right Markov chain. The observed spectral sequence of an utterance is assumed to be a stochastic function of the state sequence of the Markov chain. The state sequence itself is unobservable (hidden). The parameters characterizing the HMM are the number of states, the state transition matrix, and the observation probability functions.

Training of the HMM, that is, choosing the parameters to optimally match the observed spectral sequences can be performed with the Baum-Welch algorithm. In the recognition phase for an unknown input, the probability that the observed spectral sequence is generated from an HMM for each vocabulary word is calculated with the forward-backward or the Viterbi algorithm which is similar to the DTW algorithm. The word with the highest probability is selected as the correct recognition.

2.3 Semi-Continuous HMM (SCHMM) and Fuzzy-VQ-Based HMM (FVQHMM)

Although the discrete HMM has various advantages, such that it can model events with any distribution provided enough training data exist, it has a serious problem of causing quantization errors. That is, the VQ operation partitions the spectral space into separate regions according to some distortion measure. Solving this problem by increasing the codebook size creates another problem in that huge amounts of training utterances are necessary and that the amount of computation in the recognition stage becomes very large.

From this point of view, the continuous HMM has been introduced, and is becoming more popular than the discrete HMM. However, the continuous HMM also has a problem in that mixtures of a large number of probability density functions considerably increase not only the computational complexity, but also the number of free parameters that need to be reliably estimated. To cope with

1058-6393/91 $1.00 © 1991 IEEE

these problems, the semi-continuous HMM (SCHMM) [10] and the fuzzy-VQ based HMM (FVQHMM) [11][12], which are somewhere in between the discrete and continuous HMMs, have been proposed.

In the SCHMM, the VQ codebook consists of a mixture of continuous probability density functions (for example, each codeword is represented by a mean vector and a covariance matrix), such that the distributions are overlapped, rather than partitioned. The SCHMM has the modeling ability of large-mixture probability density functions. In addition, the number of free parameters and the computational complexity can be reduced in comparison with the continuous HMM, since all of the probability density functions are tied together in the codebook. The SCHMM thus provides a solution to the conflict between detailed spectral modeling and insufficient training data. The VQ codebook can also be optimized together with the HMM parameters in terms of the maximum likelihood criterion.

The fuzzy VQ is based on a fuzzy K-means method used in pattern recognition. Unlike the standard VQ that generates the index of a single codeword that best matches an input vector, the fuzzy VQ makes a soft decision about which codeword is closest to the input vector. It generates an output vector whose components indicate the relative closeness of each codeword to the input.

It is important that the model, trained by limited data, is robust against variations, such as noise, recording conditions, and speaking styles. Robustness of six types of phoneme-HMMs against speaking-style variations was examined [13]. The six types were discrete HMM, FVQHMM, and single-Gaussian and mixture-Gaussian HMMs with either diagonal or full covariance matrices. Eighteen Japanese-consonant recognition experiments were performed using isolated word utterances, phrase-by-phrase utterances, and sentence utterances. The FVQHMM, the mixture-Gaussian HMM with diagonal covariance matrices and the single Gaussian HMM with full covariance matrices displayed better results than the other three types, when different speaking-style utterances were used in training and testing.

2.4 VQ-Distortion-Based Recognition

In all of the above-mentioned methods, the vector quantizers have been incorporated with a DTW- or HMM-based time alignment mechanism. Recently, simpler recognition algorithms based more directly on VQ have been proposed, in which each word in the vocabulary has its own codebook, designed by clustering the training data for that word. The input test sequence is then encoded by the codebook for each word and the word corresponding to the minimum distortion codebook is chosen as the recognizer output [14][15]. Although this method produced good results with a small amount of computation, especially as a method of preprocessor in large vocabulary recognition [16], the lack of temporal information caused problems in recognizing similar words.

Several variations of the VQ-based recognizer have therefore been proposed to include temporal information. One approach is to divide the words into several sections (or states) and design a sequence of codebooks, one for each section, for each word (multi-section VQ) [17].

Another approach is a conditional histogram approach, which incorporates the relative likelihoods that certain codewords follow others into the distortion measure [18]. The distortion between the input frame and a codeword is defined as the weighted sum of the ordinary spectral distortion and the negative logarithm of the conditional probability of getting this codeword given the predecessor in the same codebook.

Another technique of incorporating temporal information into the codebook has been proposed, in which cepstral and delta-cepstral parameters (short-time regression coefficients) are jointly used as spectral features [19]. Figure 1 shows the block diagram of a word recognizer incorporating a VQ preprocessor and an HMM- or DTW-based postprocessor. A universal codebook for the above-mentioned parameters is constructed based on a multi-speaker, multi-word database, and a separate codebook is designed as a subset of the universal codebook for each word in the vocabulary. These codebooks are used for preprocessing to eliminate word candidates whose distance scores are large. A discrete HMM- or DTW-based recognizer using the universal codebook then resolves the choice among the remaining word candidates. Recently this method was expanded to include hierarchical delta-cepstral parameters extracted over multiple time lengths [20].

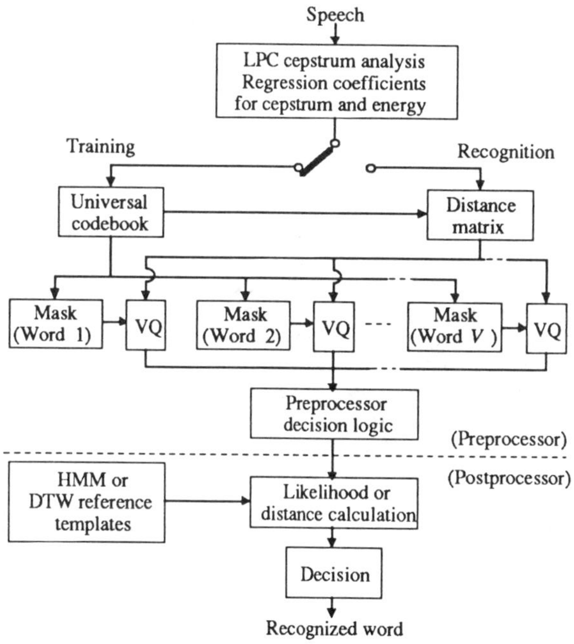

Fig. 1 - Block diagram of a word recognizer incorporating a VQ preprocessor and an HMM-/DTW-based postprocessor.

3. Learning VQ (LVQ)

Learning vector quantization (LVQ) [21][22] is a classifier that is closely linked to Kohonen's work on self-organizing feature maps [23]; the main difference is that LVQ is concerned with finding good category boundaries, while the self-organizing feature maps are designed to find reference vectors that are centroids of the input vectors. The learning in LVQ is, therefore, supervised, while that in self-organizing feature maps is unsupervised.

In LVQ, each category to be learned is assigned a number of reference vectors. Initial configurations of the reference vectors are usually obtained by using the traditional K-means clustering procedure. LVQ training then tries to adjust these positions so that each input vector has a reference vector of the right category as its closest reference vector. Kohonen proposed two versions of LVQ: LVQ1 and LVQ2. The difference between them is the way they select the reference vectors to be adapted.

The adaptation rule for LVQ1 is as follows. If the reference vector closest to the input vector belongs to the same category as the input vector, it is moved closer to the input vector, in proportion to the distance between the two vectors. If the closest reference vector belongs to a category other than that of the input vector, it is moved away, again in proportion to the distance between the two vectors. LVQ2 requires that a number of conditions be met before vector adaptation can occur. These conditions allow the system to pay closer attention to the decision lines of a given categorization problem.

4. Speaker Adaptation by VQ-Codebook Mapping

A number of approaches have been tried in an effort to build speaker-independent recognition systems, typically under HMM-based frameworks. However, since the distributions of feature parameters across speakers are very broad, it is difficult to separate phonemes using speaker-independent methods. Speaker adaptation is a method of automatically adapting reference templates to each new speaker or normalizing interspeaker variations in input utterances.

Since a VQ codebook represents the distribution of given samples in a multi-dimensional spectral space in a non-parametric way, the relationships between spectral distributions of a reference speaker and a new speaker can be represented by correspondences between codewords associated with them. Using mapping rules based on the correspondences, spectra of a new speaker can be adapted to the reference speaker or vice versa. Individual variations on how a word is uttered are modeled by an HMM or multiple sequences of codebook entries in the word dictionary. Speaker-adaptation methods are generally classified into supervised (text-dependent) methods in which training words or sentences are known, and unsupervised (text-independent) methods in which arbitrary utterances can be used.

4.1 Supervised Learning

For supervised adaptation [24], the mapping rules are obtained through DTW or the Viterbi algorithm. First, utterances of a reference speaker are used to create an initial codebook. These utterances are then vector-quantized, that is, converted into sequences of codewords. In the training stage, training utterances of a new speaker are converted into code sequences and time-aligned with the same word or sentence uttered by the reference speaker. The spectral mapping function between the codewords of these two speakers is obtained from alignment functions, that is, the correspondences between the time axes.

Each codeword is included in various words, and each codeword of the reference speaker corresponds to various codewords of the new speaker. Thus, a histogram of correspondences between codewords of the reference speaker and the new speaker, that is, a histogram of co-occurrences of codewords, is calculated using the alignment results of all training words or sentences. The mapping function is weighted by the histogram to find the best correspondence rule. In the recognition stage, input speech is vector-quantized and mapped to the reference speaker's spectrum using the mapping rules at every frame. The similarity between the normalized input speech and each word of the reference speaker is then calculated and used in the recognition decision.

In HMM-based recognition, probabilistic spectral mapping from the reference speaker's spectral space to that of the new speaker has also been investigated [25]. The transformation matrix, which represents the conditional probability of a codeword of the new speaker, given a codeword of the reference speaker, is computed by applying a modified forward-backward algorithm to training utterances.

An approach to speaker adaptation for a large-vocabulary HMM-based speech recognition system has also been tried [26]. The approach is based on the use of a stochastic model, called the "speaker Markov model". The model indicates which codeword of the new speaker is likely to occur and what spectral parameters are generated by the reference speaker if, at the same time, a certain codeword is generated by the new speaker.

4.2 Unsupervised Learning

Figure 2 is a block diagram of an unsupervised codebook adaptation method [27]. The idea of this method is based on an adaptation algorithm for a segment vocoder [28]. First, an initial codebook and a VQ-indexed word dictionary are prepared. The initial codebook is produced by clustering the voices of multiple speakers, and commonly serves as the initial condition for each new speaker.

In the adaptation process, a set of spectra from the

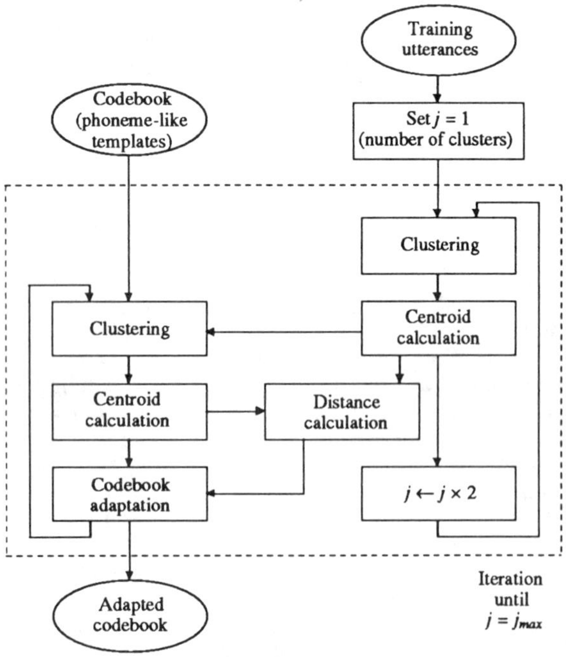

Fig. 2 - Block diagram of an unsupervised codebook adaptation method.

training utterances of a new speaker and the reference codebook elements are clustered hierarchically in an increasing number of clusters. Using the deviation vectors between centroids of the training spectra clusters and the corresponding codebook clusters, either codebook elements or input frame spectra are shifted so that the corresponding centroids coincide. Continuity between adjacent clusters is maintained by determining the shifting vectors to be the weighted-sum of the deviation vectors of adjacent clusters. Adaptation is thus performed hierarchically from global to local individuality. Several modifications to the adaptation method have also been investigated [29].

5. VQ-Distortion-Based Speaker Recognition

VQ-based speaker recognition methods, which are similar to the VQ-distortion-based speech recognition methods, have been investigated. Speaker-specific codebooks are produced in the training stage by clustering the spectral distribution of each reference speaker. In the recognition stage, an input utterance is vector-quantized using the codebook of each reference speaker, and the mean value or the distribution of the quantization error over the entire speech interval is calculated. These values obtained using each reference codebook are examined to make the recognition decision.

Speaker recognition methods can be classified into text-dependent and text-independent methods. The former require the speaker to issue a predetermined utterance, whereas the latter do not rely on a specific text being spoken.

5.1 Text-Dependent Recognition

In the text-dependent speaker recognition, three VQ approaches have been investigated: single section VQ (normal VQ), multi-section VQ, and matrix quantization (MQ) [30]. The multi-section VQ and the MQ approaches are two different ways of incorporating temporal information into the recognition process. Multi-section VQ models a source by dividing it into several independent, time-ordered subsources. On the other hand, MQ models an utterance with a single codebook that contains an unordered set of time-ordered speech spectrum sequences. Although speaker verification performances of the three approaches when using only a single digit per speaker were similar, the multi-section VQ approach did best when 10- or 5-digit sets were used.

5.2 Text-Independent Recognition

A method using two VQ codebooks, containing the cepstral and delta-cepstral parameters, has been examined [31]. The VQ codebooks for each speaker are constructed using the isolated utterances of 10 digits. An unknown speaker then says any one of the 10 digits. Therefore, this method is text-independent but vocabulary-dependent. Distances (distortions) from test vectors to the two VQ codebooks are optimally combined and averaged over the test utterance to make a final recognition decision. The experimental results show that since the cepstral and delta-cepstral parameters are relatively uncorrelated, they provide complementary information for speaker recognition. They also show that the transitional representations and performance are relatively resistant to simple transmission channel variations.

A method using a single codebook for long feature vectors consisting of instantaneous and transitional features representing both cepstral and pitch characteristics has recently been investigated [32]. Figure 3 is a block diagram of the recognition system. Three key techniques were introduced to cope with any temporal and text-dependent spectral variations. First, either an ergodic HMM or a voiced/unvoiced decision was used to classify input speech into broad phonetic classes. Second, a new distance measure, Distortion-Intersection Measure (DIM), was introduced for calculating VQ distortion of input speech using speaker-specific codebooks. DIM is characterized by selective matching using only a stable subset of test speech in the distortion calculation. Third, a new normalization. method, Talker Variability Normalization (TVN), was introduced. TVN normalizes parameter variation taking both inter- and intra-speaker variability into consideration. TVN emphasizes feature parameters that have relatively large inter-speaker variability and small intra-speaker variability. The combination of the three techniques provides highly accurate speaker identification.

A connectionist approach based on the LVQ algorithm has also recently been tried [33].

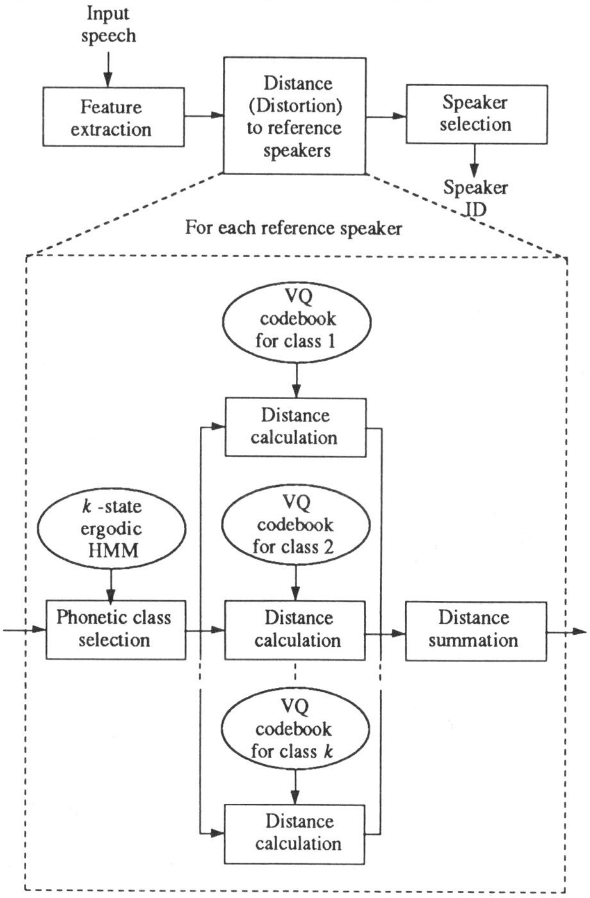

Fig. 3 - Block diagram of text-independent speaker recognition incorporating broad phonetic classification.

6. Summary

This paper reviewed various methods of applying the

VQ technique to speech/speaker recognition. Not only has it reduced the amount of computation and storage, the VQ technique has also created new ideas of solving various problems in speech/speaker recognition, such as speaker adaptation using codebook mapping. Several techniques, such as fuzzy VQ and semi-continuous modeling of HMM, have also been investigated to cope with the quantization distortion problem. The VQ technique has the potential to produce new ideas in speech/speaker recognition in combination with new techniques, such as neural networks and statistical approaches.

References

[1] N. Sugamura, K. Shikano and S. Furui (1983), "Isolated word recognition using phoneme-like templates," Proc. IEEE Int. Conf. Acoust., Speech, Signal Processing, Boston, 16.3.

[2] H. Bourlard, C. J. Wellekens and H. Ney (1984), "Connected digit recognition using vector quantization," Proc. IEEE Int. Conf. Acoust., Speech, Signal Processing, San Diego, 26.10.

[3] L. R. Rabiner, S. E. Levinson and M. M. Sondhi (1982), "On the application of vector quantization and hidden Markov models to speaker-independent, isolated word recognition," Bell System Tech. J., 62, 4, pp.1075-1105.

[4] R. Billi (1982), "Vector quantization and Markov source models applied to speech recognition," Proc. IEEE Int. Conf. Acoust., Speech, Signal Processing, Paris, France, pp. 574-577.

[5] M. A. Bush and G. E. Kopec (1985), "Network-based connected digit recognition using vector quantization," Proc. IEEE Int. Conf. Acoust., Speech, Signal Processing, Tampa, 31.1.

[6] H. Bourlard, Y. Kamp and C. J. Wellekens (1985), "Speaker dependent connected speech recognition via phonemic Markov models," Proc. IEEE Int. Conf. Acoust., Speech, Signal Processing, Tampa, 31.5.

[7] Y. L. Chow, et al. (1987), "BYBLOS: The BBN continuous speech recognition system," Proc. IEEE Int. Conf. Acoust., Speech, Signal Processing, Dallas, 3.7.

[8] K.-F. Lee (1989), "Automatic speech recognition - The development of the SPHINX system," Kluwer Academic Publishers.

[9] L. R. Bahl, et al., (1989), "Large vocabulary natural language continuous speech recognition," Proc. IEEE Int. Conf. Acoust., Speech, Signal Processing, Glasgow, Scotland, S9.8.

[10] X. D. Huang, Y. Ariki and M. A. Jack (1990), "Hidden Markov models for speech recognition," Edinburgh Univ. Press.

[11] H.-P. Tseng, M. J. Sabin and E. A. Lee (1987), "Fuzzy vector quantazation applied to hidden Markov modeling," Proc. IEEE Int. Conf. Acoust., Speech, Signal Processing, Dallas, 15.5.

[12] T. Hanazawa, K. Kita, S. Nakamura, T. Kawabata and K. Shikano (1990), "ATR HMM-LR continuous speech recognition system," Proc. IEEE Int. Conf. Acoust., Speech, Signal Processing, Albuquerque, S2.4.

[13] T. Matsuoka and K. Shikano (1991), "Robust HMM phoneme modeling for different speaking styles," Proc. IEEE Int. Conf. Acoust., Speech, Signal Processing, Toronto, Canada, S5.4.

[14] A. Buzo, H. Martinez and C. Rivera (1982), "Discrete utterance recognition based upon source coding techniques," Proc. IEEE Int. Conf. Acoust., Speech, Signal Processing, Paris, France, pp. 539-542.

[15] J. E. Shore and D. K. Burton (1983), "Discrete utterance speech recognition without time alignment," IEEE Trans. Inform. Theory, 29, 4, pp. 473-491.

[16] K.-C. Pan, F. K. Soong and L. R. Rabiner (1985), "A vector-quantization-based preprocessor for speaker-independent isolated word recognition," IEEE Trans. Acoust., Speech, Signal Processing, 33, 3, pp. 546-560.

[17] D. K. Burton, J. E. Shore and J. T. Buck (1985), "Isolated-word speech recognition using multisection vector quantization codebooks," IEEE Trans. Acoust., Speech, Signal Processing, 33, 4, pp. 837-849.

[18] S.-S. Huang and R. M. Gray (1988), "Spellmode recognition based on vector quantization," Speech Communication, 7, 1, pp. 41-53.

[19] S. Furui (1987), "A VQ-based preprocessor using cepstral dynamic features for large vocabulary word recognition," Proc. IEEE Int. Conf. Acoust., Speech, Signal Processing, Dallas, 27.2.

[20] S. Furui (1990), "On the use of hierarchical spectral dynamics in speech recognition," Proc. IEEE Int. Conf. Acoust., Speech, Signal Processing, Albuquerque, S15a.10.

[21] T. Kohonen (1988), "Self-organization and associative memory (2nd ed.)", Springer Verlag, pp.199-202.

[22] E. McDermott and S. Katagiri (1991), "LVQ-based shift-tolerant phoneme recognition," IEEE Trans. Signal Processing, 39, 6, pp. 1398-1411.

[23] T. Kohonen, G. Barna and R. Chrisley (1988), "Statistical pattern recognition with neural networks: Benchmarking studies," Proc. IEEE Int. Conf. Neural Networks, I, pp.61-68.

[24] K. Shikano, K.-F. Lee and R. Reddy (1986) " Speaker adaptation through vector quantization," Proc. IEEE Int. Conf. Acoust., Speech, Signal Processing, Tokyo, Japan, 49.5.

[25] R. Schwartz, Y.-L.Chow and F. Kubara (1987), "Rapid speaker adaptation using a probabilistic spectral mapping," Proc. IEEE Int. Conf. Acoust., Speech, Signal Processing, Dallas, 15.3.

[26] G. Rigoll (1989), "Speaker adaptation for large vocabulary speech recognition systems using 'speaker Markov models'," Proc. IEEE Int. Conf. Acoust., Speech, Signal Processing, Glasgow, Scotland, S1.2.

[27] S. Furui (1989), "Unsupervised speaker adaptation based on hierarchical spectral clustering," IEEE Trans. Acoust., Speech, Signal Processing, 37, 12, pp. 1923-1930.

[28] Y. Shiraki and M. Honda (1990), "Speaker adaptation algorithms based on piece-wise moving adaptive segment quantization method," Proc. IEEE Int. Conf. Acoust., Speech, Signal Processing, Albuquerque, S12.5.

[29] S. Furui (1989), "Unsupervised speaker adaptation method based on hierarchical spectral clustering," Proc. IEEE Int. Conf. Acoust., Speech, Signal Processing, Glasgow, Scotland, S6.9.

[30] D. K. Burton (1987), "Text-dependent speaker verification using vector quantization source coding," IEEE Trans. Acoust., Speech, Signal Processing, 35, 2, pp.133-143.

[31] F. K. Soong, and A. E. Rosenberg (1988), "On the use of instantaneous and transitional spectral information in speaker recognition," Trans. Acoust., Speech, Signal Processing, 36, 6, pp.871-879.

[32] T. Matsui and S. Furui (1991), "A text-independent speaker recognition method robust against utterance variations," Proc. IEEE Int. Conf. Acoust., Speech, Signal Processing, Toronto, Canada, S6.3.

[33] Y. Bennani, F. Fogelman Soulie and P. Gallinari (1990), "A connectionist approach for automatic speaker identification", Proc. IEEE Int. Conf. Acoust., Speech, Signal Processing, Albuquerque, S5.2.

A Minimum Error Approach to Speech and Pattern Recognition

B. H. Juang *S. Katagiri**

AT&T Bell Laboratories
Murray Hill, New Jersey 07974

*ATR
Kyoto, JAPAN

ABSTRACT. One of the problems in pattern recognition is minimum error classification in which the misclassification probability is to be minimized based on a given set of training samples. Such a problem was, to the best of our knowledge, never directly and sufficiently addressed in the past. In this paper, we propose a new formulation for the minimum error classification problem, together with a fundamental technique for designing a classifier that achieves minimum classification error. Experimental results pertaining to a speech recognition task are provided to show the effectiveness of the new technique.

1. Introduction

Pattern classification, particularly in the area of linear discriminant analysis, is a very well studied topic with most of the original developments completed in the 1960's (see [1]). Consider a set of samples $\mathcal{S} = \{\mathbf{x}_1, \mathbf{x}_2, \mathbf{x}_3, \ldots, \mathbf{x}_N\}$, where each $\mathbf{x}_i$ is a K-dimensional vector and is known to belong to one of M classes C_i, $i = 1, 2, \ldots, M$. A classifier consists of a set of parameters and a decision rule. The task of minimum error classifier design is to find the classifier parameter set, denoted by Λ, and the accompanying decision rule, based on the given set $\mathcal{S}$, such that the probability of misclassifying any $\mathbf{x}$ is minimized.

Bayes decision theory [1] is a fundamental statistical approach to the classification problem and is often the basis of many pattern classification techniques. Suppose we have full knowledge of the *a posteriori* probability $P_\Lambda(C_i|\mathbf{x})$, which is defined by the parameter set Λ. The Bayes decision rule

$$C(\mathbf{x}) = C_i \quad \text{if} \quad P_\Lambda(C_i|\mathbf{x}) = \max_j P_\Lambda(C_j|\mathbf{x}) \qquad (1)$$

where $C(\cdot)$ denotes a classification operation, is known to lead to minimum misclassification probability. The rule is often written in terms of the *a priori* and the conditional probabilities as

$$C(\mathbf{x}) = C_i \quad \text{if} \quad p_\Lambda(\mathbf{x}|C_i)P_\Lambda(C_i) = \max_j p_\Lambda(\mathbf{x}|C_j)P_\Lambda(C_j) \,. \qquad (2)$$

Since the exact probability measure is rarely known in real situations, the problem of optimal classifier design thus becomes that of estimating the *a priori* and the conditional probabilities, defined by the parameter set Λ, using the design samples $\mathcal{S}$. The fundamental assumptions in this approach are that the correct form of the distributions as functions of the parameter set Λ is known and that given a sufficient design sample set there is a good method to estimate the unknown parameters Λ. These assumptions are often unrealistic in real world problems.

An alternative to the Bayes decision approach is to use discriminant functions. This requires a set of discriminant functions, $g_i(\mathbf{x};\Lambda)$, $i = 1,2,\ldots,M$, defined by the parameter set Λ, instead of explicit knowledge of the probability distributions. Unlike the Bayes approach, the problem of "optimal" classifier design becomes that of finding the right parameter set for the discriminant functions to minimize the "sample risk" [1], which is defined as the average cost incurred in classifying the set of design samples $\mathcal{S}$. The cost is usually defined on a pair of class indices (i, j) where i and j are the correct class index and the identified/recognized class index respectively, indicating the penalty in misclassifying a C_i class sample as a C_j class sample.

1058-6393/91 $1.00 © 1991 IEEE

The difficulty associated with the discriminant approach lies in the derivation of a minimum-risk discriminant. A proper discriminant needs to be suitable for incorporation in an objective function for optimization. The sample risk, of which the number of classification errors is one of the simplest cases with a zero-one function as the misclassification cost [1], is obviously a piecewise constant function of the classifier parameter Λ and thus a poor candidate for optimization by a numerical search method. Traditionally, the discriminant based classifier design is formulated as an optimization problem aiming at minimization of some criterion functions that are analytically more tractable than the sample risk. Popular choices of these criterion functions include the perceptron criterion function and sum of squared errors (or minimum squared error MSE), for example. These criterion functions do not generally lead to a minimum error probability classifier, although one can vigorously discuss the convergence properties of the solution as obtained by numerical search algorithms.

In this paper, we propose a new way of deriving the discriminant such that the result of the optimization procedure will be consistent with the minimum sample risk objective. We also propose a descent search algorithm for optimizing the minimum error objective. This combination of a new discriminant, directly related to the minimum error objective, and the descent algorithm would then allow us to circumvent the difficulties encountered in most of the traditional techniques and address the optimal classifier design problem in a straightforward manner.

2. Minimum Classification Error Discriminant

The traditional discriminant formulation can be stated in two steps: definition of the discriminant function and incorporation of the discriminant function in a scalar criterion, such as the perceptron criterion or the mean squared error, suitable for a gradient-type search procedure to find a solution, if the procedure does converge. The inadequacy of this traditional formulation lies in the fact that there is an inconsistency between the chosen scalar criterion function and the desired minimum recognition error probability objective. Here, we propose a new way of deriving the objective criterion.

We use a 3-step procedure [2] to derive the objective criterion. As with the traditional approaches, the form of the discriminant functions $g_i(\mathbf{x};\Lambda)$ are first prescribed. The classifier makes its decision for each input $\mathbf{x}$ by choosing the largest of the discriminants evaluated on $\mathbf{x}$. This decision process is expressed in a functional form such that further optimization can be easily accomplished. We thus in the second step introduce a misclassification measure which allows us to embed the decision process in the error probability formulation. One way of defining the misclassification measure for a k-th class sample is:

$$d_k(\mathbf{x}) = -g_k(\mathbf{x};\Lambda) + \left[\frac{1}{M-1}\sum_{j,j\neq k} g_j(\mathbf{x};\Lambda)^{\eta}\right]^{1/\eta} \quad (3)$$

where η is a positive number. A key concept in the definition of (3) is that the decision rule is being enumerated. By varying the value of η, one can take all the potential classes into consideration, to a various degree, in the search of the classifier parameter Λ. One extreme case is when η approaches ∞, the misclassification measure becomes $d_k(\mathbf{x}) = -g_k(\mathbf{x};\Lambda) + g_i(\mathbf{x};\Lambda)$ where C_i is the class with the largest discriminant value among those classes other than C_k, because $(M-1)^{1/\infty} \cong 1$. Obviously in this case, $d_k(\mathbf{x}) > 0$ implies misclassification and $d_k(\mathbf{x}) \leq 0$ means correct decision. In this way, the decision rule is enumerated as a scalar value.

The misclassification measure is used in the third step where the minimum error objective is formulated. A general form of the cost function can be defined as

$$\ell_k(\mathbf{x};\Lambda) = \ell_k(d_k(\mathbf{x})) \quad (4)$$

which is expressed as a function of the misclassification measure. Note that the cost function ℓ_k and the misclassification measure d_k can be defined individually for each class k for generality. For minimum error classification, the cost function can be any approximate 0-1 functions such as the sigmoid function:

$$\ell_k(d_k) = \frac{1}{1+e^{-\xi(d_k+\alpha)}}, \quad \xi>0. \quad (5)$$

Clearly, when $d_k(\mathbf{x}) < 0$ which implies correct

classification, no cost is incurred. On the other hand, a positive $d_k(\mathbf{x})$ leads to a penalty which becomes essentially a count of classification error if a zero-one cost function or any of the smoothed zero-one functions is used.

This 3-step definition emulates the classification operation as well as the performance evaluation, particularly in terms of classification errors, in a smooth functional form.

3. Descent Methods

The cost function of (4) is defined for each input pattern $\mathbf{x}_i$. This cost function is the basis of the objective that we shall optimize with descent methods. Given a set of labeled training patterns $\mathcal{S} = \{\mathbf{x}_1, \mathbf{x}_2, \ldots, \mathbf{x}_N\}$, there are two ways of defining the performance objective; one is the empirical average cost and the other the expected cost. Optimization of these two different objectives leads to gradient search solutions with different convergence properties.

3.1 Empirical Average Cost and Gradient Descent Algorithm [2]

Given a set of design samples $\mathcal{S}$, we define an empirical average cost as

$$L_0(\Lambda) = \frac{1}{N} \sum_{i=1}^{N} \sum_{k=1}^{M} \ell_k(\mathbf{x}_i;\, \Lambda)\; 1\,(\mathbf{x}_i \in C_k)\,. \quad (6)$$

where $1(\cdot)$ is the indicator function. This well-defined cost function can be conveniently minimized by a gradient descent algorithm, using the following adaptation rule

$$\Lambda_{t+1} = \Lambda_t - \varepsilon\, \nabla\, L_0(\Lambda_t) \quad (7)$$

where Λ_t denotes the parameter set at the tth iteration.

3.2 Expected Cost and Probabilistic Descent Algorithm [2]

The expected cost can be expressed as

$$\begin{aligned} L(\Lambda) &= E\{\ell_k(\mathbf{x};\Lambda)\} \\ &= \sum_k P(C_k) \int \ell_k\,(\mathbf{x};\Lambda) p(\mathbf{x}|\,C_k)\; d\mathbf{x} \quad (8) \end{aligned}$$

where $P(C_k)$ and $p(\mathbf{x}|\,C_k)$ are the class *a priori* and conditional probabilities respectively. Since both the *a priori* and conditional probabilities are unknown, the usual gradient descent algorithm can not be directly used. However, the following theorem provides a basis for minimization of the above expected cost.

Probabilistic Descent Theorem

Given $\mathbf{x} \in C_k$, *if the classifier parameter adjustment* $\delta\Lambda(\mathbf{x}, C_k, \Lambda)$ *is specified as*

$$\delta\Lambda(\mathbf{x}, C_k, \Lambda) = -\varepsilon \mathbf{U}\; \nabla \ell_k(\mathbf{x};\Lambda) \quad (9)$$

where $\mathbf{U}$ *is a positive-definite matrix and* ε *is a small positive real number, then*

$$E[\delta L(\Lambda)] = E[L(\Lambda+\delta\Lambda) - L(\Lambda)] \le 0\,. \quad (10)$$

Furthermore, if an infinite sequence of random samples $\mathbf{x}_i$ *are presented for training and the adjustment rule of* (9) *is utilized with a corresponding step size sequence* ε_i *which satisfies*

$$\text{i)} \sum_{i=1}^{\infty} \varepsilon_i \to \infty; \;\; \textit{and} \;\; \text{ii)} \sum_{i=1}^{\infty} \varepsilon_i^2 < \infty\,, \quad (11)$$

then the parameter sequence Λ_i *according to*

$$\Lambda_{i+1} = \Lambda_i + \delta\Lambda(\mathbf{x}_i, C_k, \Lambda_i) \quad (12)$$

converges with probability one to a Λ^* *which results in a local minimum of* $L(\Lambda)$.

4. Classification Experiment

Several classification experiments were conducted to study the characteristics of the minimum classification error criterion. We report two sets of experimental results here: one involves a set of artificially generated data with mixture distributions and the other pertains to the Fisher's iris data, well known to the pattern recognition community.

The artificially generated data set consists of two classes. The distributions are of Gaussian mixture types with 2 components in each class. This allows careful examination of the loss minimization process while demonstrating a situation more sophisticated than a simple Gaussian distribution. The training tokens (2000 in total) were used to train 3 linear classifiers (LC) with 3 different criterion functions: the perception error criterion (PE), the minimum squared error criterion (MSE) [2] and the minimum classification error (MCE). We further generated an independent set of 2000 tokens according to the same distributions for testing purposes. The classification results (percent correct averaged over 4 independent runs) are: 87.76, 84.75 and 90 for the training set and 88.82, 85.09 and 90.15 for the test set pertaining to PE, MSE and MCE respectively. We have run similar experiments on a number of artificially generated data sets with various distributions. The superior performance of LC+MCE was consistently observed in the experiments.

The other set of experiments involves Fisher's iris data [3]. The iris data consists of four measurements made by E. Anderson on 150 samples of three species of iris. Fifty tokens are available for each of the three species. The task is to classify these measured tokens into the 3 individual species. Fisher used the data in his classic paper on discriminant analysis [3]. Many clustering experiments were studied in the past using this data set. In our current experiment, all the tokens were used for training the classifiers.

To account for the intrinsic characteristics of stochastic training, we performed four runs of classification experiments for each classifier training, using different orders of training data presentation. The average classification error rate is 4% (i.e. 96% correct) when MCE was used as the criterion compared to 14% error for PE and 15.1% error for MSE.

5. Speech Recognition Experiments

The experiments involved recognition of the highly confusible English *E*-set alphabet, namely, b, c, d, e, g, p, t, v, and z. The speech signal was recorded from 100 native Americans, including 50 male and 50 female, through local dial-up telephone lines. The sampling rate was 6.67 kHz and the bandwidth of the anti-aliasing filter was from 100 to 3200 Hz for a digital implementation of analysis processing. Each talker spoke each word twice, producing two sets of databases. One was used as the training set and the other as the test set. (Other parameters include: a 300-sample analysis window, a 200-sample overlap between adjacent analysis windows, an 8-th order linear prediction analysis and a 24 cepstral coefficient representation.)

To normalize the speaking rate variations inherent in the spoken utterances, a conventional speech recognizer with dynamic time warping (DTW) was used as the baseline classifier. The unknown input utterance is first analyzed and then matched to each of the reference templates, resulting in sequences of spectral distortions. Let us denote these distortion sequences by $D_j = \{d_j(i), i=1, 2, \ldots, m_j\}$ where j is the template index and m_j is the length of the jth template. Each vocabulary word may be represented by a multiplicity of reference templates although in this paper we report only the simplest case where we use one reference template for each word. These distortion sequences $\{D_j\}$, $j = 1, 2, \ldots, M$ ($M=9$ in the present case), are thus the input to the classifier to be designed. Note that the total dimension of the input is $M_T = \sum_{j=1}^{M} m_j$.

Traditionally, the classification decision is made based on a simple average distortion

$$y_j = \sum_{i=1}^{m_j} \frac{1}{m_j} d_j(i) . \tag{13}$$

The recognized word k is the one that satisfies $y_k = \min_{1 \le j \le M} y_j$. A more general discriminant function of the distortion sequences is thus

$$y_j = \sum_{i=1}^{m_j} w_{ji}\, d_j(i) + w_{j0} . \tag{14}$$

This can be implemented as a scaled perceptron as well as a fully connected perceptron, taking the entire set of distortion sequences $\{D_j\}$ as the input. The weight functions in these perceptron structures can be trained by the error back-propagation algorithm [4] with the usual sum-squared error criterion.

The recognition rates (% correct) pertaining to these various methods are summarized in the following with reference to the traditional simple uniform weighting method. The traditional method yielded a result of 59.8% accuracy rate which was improved to 62.7% and 63.9% by the scaled perceptron and the fully connected perceptron with the sum-squared error criterion, respectively. The minimum classification error criterion, optimized by the generalized descent algorithm, achieved a recognition rate of 70.0%, the highest among all the competing methods.

6. Summary

In this paper, we present a new formulation of the pattern recognition problem, aiming at achieving a minimum error rate classification. The classical discriminant analysis methodology is blended with the classification rule (traditionally expressed in an operational form) in a new functional form and is used as the design objective criterion to be optimized by numerical search algorithms. The new formulation results in a smooth error function which approximates the empirical error rate for the design sample set arbitrarily closely. We have applied the minimum error formulation to several recognition tasks and demonstrated the advantages of the new method. In a speech recognition experiment involving the English *E*-set vocabulary, it was demonstrated that the new minimum error method achieves the best recognition performance. The proposed learning method and formulation provides a solid analytical ground for the long standing minimum error classifier design problem.

References

1. R. O. Duda and Peter E. Hart, *Pattern Classification and Scene Analysis*, John Wiley & Sons, New York, 1973.

2. B. H. Juang and S. Katagiri, ''Discriminative learning for minimum error classification,'' submitted for publication, IEEE Trans. SP.

3. R. A. Fisher, ''The use of multiple measurements in taxonomic problems,'' *Ann. Eugenics*, 7, Part II, pp. 179-188, 1936.

4. D. Rumelhart, E. Hinton, and J. Williams, ''Learning internal representation by error propagation,'' in *Parallel Distributed Processing*, Rumelhart and McClelland (ed.), vol. 1, pp. 318-364, MIT Press, Cambridge, MA, 1986.

The Stochastic Segment Model for Continuous Speech Recognition

M. Ostendorf V. Digalakis
Electrical, Computer and Systems Engineering
Boston University
Boston, MA, 02215

Abstract

A new direction in speech recognition via statistical methods is to move from frame-based models, such as Hidden Markov Models (HMMs), to segment-based models that provide a better framework for modeling the dynamics of the speech production mechanism. The Stochastic Segment Model (SSM) is a joint model for a sequence of observations, which provides explicit modeling of time correlation as well as a formalism for incorporating segmental features. In this work, the focus is on modeling time correlation within a segment. We consider three Gaussian model variations based on different assumptions about the form of statistical dependency, including a Gauss-Markov model, a dynamical system model and a target state model, all of which can be formulated in terms of the dynamical system model. Evaluation of the different modeling assumptions is in terms of both phoneme classification performance and the predictive power of linear models.

1 Introduction

Most of the existing speaker-independent speech recognition systems today utilize frame-based features in hidden Markov models. Although this approach has been very successful to date, it has some drawbacks. For example, temporal dynamics can only be modeled through the use of additional derivative features, and events that span a longer time scale than 20 msec cannot be represented. Recently, several researchers have investigated segment-based alternatives to hidden Markov models in an attempt to overcome these deficiencies. In a segment-based model, the fundamental score computation (often a probability) is at the level of a phoneme segment, allowing the use of segmental features and more general acoustic models. Of course, the use of segmental models incurs the cost of more complex recognition search algorithms. However, various mechanisms have been developed to overcome this problem [7, 15].

The focus of this paper is on mechanisms for modeling the temporal dynamics of a phoneme, or the statistical dependency of features over time, in the segmental framework. It has been observed experimentally (and shown theoretically [14]) that the joint likelihood of a particular state and observation sequence is dominated by the terms of the output distributions as the dimension of the feature vector increases; hence, the hidden Markov state is not always an effective model of temporal feature dependencies and a better model is of great interest.

The remainder of the paper will begin by providing an overview of different segment-based approaches to speech recognition, focusing on the stochastic segment model to demonstrate the potential of the segmental framework. In the next section, we describe different types of assumptions that can be made in modeling the statistical dependence of features in time, and demonstrate experimentally that a dynamical system model of speech is a robust representation of spectral dynamics for continuous speech recognition. Finally, we conclude with a discussion of the implications of these results for future work in acoustic modeling.

2 Overview of Segment-Based Models

Perhaps the first segment-based model used in speech recognition was proposed by Bush and Kopec [5]. Their approach used vector quantizers designed for specific phone segments, with a recognition search based on minimum quantization error. Although the model was proposed to provide a formalism for incorporating segmental features, the only feature they found to improve recognition performance was segment duration. Segmental features have been more extensively explored by researchers at MIT (e.g., [13]). The segment-based approach is also a natural formalism for neural network classifiers, as shown in [12, 1].

1058-6393/91 $1.00 © 1991 IEEE

Our work is based on the Stochastic Segment Model (SSM) [16, 17], which provides a joint Gaussian model for a sequence of observations. Each segment generates an observation sequence $\{y_t\}_1^L$ of random length L, according to the density $b_{\alpha,L}(y_1, y_2, \ldots, y_L)$. Therefore, the stochastic segment model for a phone α consists of 1) a family of joint density functions (one for every observation length), and 2) a collection of mappings that specify the particular density function for a given observation length. Typically, the model assumes that segments are described by a fixed-length sequence of locally time-invariant regions (or regions of tied distribution parameters).

One version of the segment model used in much of our work is based on the assumption that phonemes have an (unobserved) underlying trajectory in k-dimensional feature space, which is modeled as a sequence of m k-dimensional feature vectors, denoted by the matrix $\mathbf{X}$. An "observed" segment is an L-long vector sequence, $\mathbf{Y}$, obtained by downsampling (linearly in time) the $\mathbf{X}$ trajectory: $\mathbf{Y} = \mathbf{X}T_L$. The matrix T_L is called the time-warping transformation; it warps m-long models linearly into L-long observed segments, $L < m$ and thereby specifies the distribution of $\mathbf{Y}$. In this case, the probability density for an observed segment $\mathbf{Y} = (y_1, \ldots, y_L)$ given phone α is the conditional probability density of all possible fixed length trajectories $\mathbf{X}$ which, sampled by T_L, give $\mathbf{Y}$, i.e., the distribution $p(\mathbf{Y}|\alpha)$ is simply a marginal distribution of $p(\mathbf{X}|\alpha)$ [17]. In other words, we are assuming that segments of different length use marginals of the same basic distribution.

For the case where successive observed frames are assumed independently distributed, the probability of the segment $b_\alpha(\mathbf{Y})$ is the product of the probability of each observation y_i. The **independent-frame** segment model is therefore equivalent to a hidden Markov model with a particular complex topology. If the observation distributions dominate the probability of a phoneme, the difference between the two approaches will be minimal. Indeed, in [6] we were able to show that for equivalent numbers of parameters, an HMM and an independent-frame SSM give the same phone classification performance with context-independent models. Thus, the potential advantages of the segment model which were not utilized in this experiment – the explicit modeling of statistical dependence of observations over time and the use of segmental features – can yield improved recognition performance over HMMs.

Context modeling, or using multiple models for each phone to capture the effects of coarticulation associated with neighboring phones, has been one of the most successful techniques for improving recognition performance. Our approach for robust estimation of context-dependent models is somewhat different from the traditional HMM techniques. Our distribution estimates are based on the assumption that covariance matrices of different distributions are tied across models of similar context [2]. This technique has yielded good results experimentally: a 37% reduction in word recognition error with a combined left and right context model, which is comparable to error reduction observed for discrete HMMs [11]. The context-dependent word recognition results confirm that the independent-frame SSM can achieve performance comparable to HMMs, and therefore we may be able to improve performance by explicitly modeling time correlation.

3 Statistical Dependence in Segment Models

Early SSM experiments in modeling time correlation were not successful, since the dimension of the model becomes large when both time and spectral correlation are used, creating training problems. In addition, we have recently found, as we shall see later, that the assumption of an underlying trajectory specifying a family of marginal distributions is inaccurate. After further investigation into different dependence assumptions, we have developed a new approach based on a dynamical system model of speech that does successfully model statistical dependence of features over time. In this section, we describe this model, together with some of the experiments that led us to a better understanding of the dynamic behavior of speech within a segment.

Linearity Assumptions

The first unsuccessful attempts to model time correlation with a joint Gaussian model suggested the use of a more constrained family of models that can be described with a smaller number of parameters. The initial Gaussian assumption of the SSM, as well as the wide use of Gaussian distributions in continuous-density HMMs, motivated us to restrict the family of models for the segmental density to linear ones. However, it is useful to check the validity of the linearity assumption before adopting it for modeling intrasegmental statistical dependencies.

In order to evaluate this assumption, we compared the performance of a linear regression in explaining the variance of a particular observation within a segment to the performance of a nonlinear regression based on the alternating conditional expectation (ACE) algorithm [3]. The ACE method estimates a set of nonlinear functions for each of the dependent and independent variables so that the percentage of the variance not explained by a linear regression on the transformed variables is minimized. More specifically, we used a frame at a fixed position of all occurences of a particular phonetic segment as the dependent variable, and frames at other fixed positions within the same segment as the independent variables. Our results can be summarized as follows:

- Observations within the same segment are highly correlated, thus the conditional independence assumption given the underlying state sequence often used in HMMs is not valid, and
- Within a segment, the percentage of the variance explained by the linear regression method was in most cases very close to that by the ACE method.

Based on these findings, we concluded that linear models are adequate in modeling intrasegmental dependencies, and therefore the Gaussian assumption is reasonable. Since efficient estimation and recognition algorithms can be derived for these models, we feel that the additional computational burden of using nonlinear models for modeling intrasegmental dependencies is not justified.

Different Correlation Structures

Modeling the statistical dependency between a sequence of vectors in the segment model involves tradeoffs associated with limited amounts of training data, as well as reasonable assumptions about model dynamics. As a result, the obvious approach of using a **full covariance** model of the entire segment is not successful with more than a few features per frame [6].

One alternative is to assume a **Gauss-Markov** structure, that is

$$\mathrm{p}(\mathbf{X}) = \mathrm{p}_1(x_1)\prod_{i=2}^{L}\mathrm{p}_i(x_i|x_{i-1}),$$

where $\mathrm{p}_i(x_i|x_{i-1})$ are conditional Gaussian densities. Under this hypothesis, the number of parameters is less than a factor of three more than the independent-frame model. The Gauss-Markov segment model is analogous to an HMM system which assumes that output distributions are conditioned on the current state and the past observation, as in [4, 9]. As for the HMM systems, our experiments showed that the Gauss-Markov model was useful for cepstral features alone, but performance degraded below that of the independent frame model with the additional use of cepstral derivatives.

A generalization of the Gauss-Markov model is a **dynamical system** (DS) model, in which we assume that observed features are a noisy function of a Gauss-Markov state sequence. For the general case, observations for a phone α are modeled using the stochastic system

$$\begin{aligned} x_{k+1} &= F_k(\alpha)x_k + w_k \\ y_k &= H_k(\alpha)x_k + v_k \end{aligned} \qquad (1)$$

where w_k, v_k are uncorrelated Gaussian vectors with covariances $Q_k(\alpha)$ and $R_k(\alpha)$, respectively. The noise term v_k can be thought of as modeling error or distribution smoothing. Either the driving term w_k or the noise term v_k should have a phone-dependent non-zero mean; our work has included this in the noise term. In order to reduce the number of free parameters in our model, we assume that the dynamical system for a phone segment is locally time-invariant over different regions within the segment, as for the original segment model. The different regions are defined by a fixed time warping; linear time warping has been used thus far.

The dynamical system model was introduced in [8], where training and recognition algorithms are described. Training is equivalent to the maximum likelihood identification of a stochastic dynamical system, and we have developed a simple alternative to the traditional approach which is based on the Expectation-Maximization algorithm. The Expectation step consists of forward and backward Kalman smoothing recursions, and the Maximization step involves a multivariate regression. This procedure is analogous to the Baum-Welch algorithm for HMMs, and the dynamical system can be thought of as a continuous state HMM. Recognition involves computing the probability of the innovations given by Kalman filtering and has complexity similar to a segment model based on a Gauss-Markov assumption.

Many different variations of the dynamical system model are possible, depending on the definition of the time-invariant regions and on simplifying assumptions about the parameters of the model. The **trajectory-invariant** DS model follows our previous work in assuming that the observed features are a down-sampled version of a fixed-length trajectory, so successive observed frames of speech have stronger correlation for longer observations since the underlying trajectory is sampled at shorter intervals (in feature space). For the **correlation-invariant** DS model, the sequence of state transition matrices for a particular observation of a phone depends on the phone length, and we have a complete (albeit noisy) observation of the state sequence. In this case, we assume that it is the correlation between successive frames that is invariant to differences in the segment length. In both of these cases, we have assumed $H_k = I$ and R_k constant for all phones α to reduce the number of parameters in the model.

An alternative form, again a special case of the dynamical system model, assumes a constant **target state** for the phoneme and observations that are transformations of that state. Specifically,

$$y_k = H_k(\alpha)x_0 + v_k,$$

where x_0 is a vector Gaussian with phone-dependent parameters, and the parameters H_k and R_k are also phone-dependent. The correlation between frames in this case is modeled in terms of one state which has a discrete component (index of phone, or more generally of the target distribution) and a hidden continuous component x_0 which represents more global variations of a segment.

Experimental Results

We have implemented a system based on our correlation invariance assumption and performed phoneme classification and recognition experiments on the TIMIT database [10]. The feature vectors included Mel-warped cepstra and their derivatives, computed from a 20 ms window at 100 frames per second. Each segment model had five different distributions (time-invariant regions). We used 61 context-independent phonetic models, but in counting errors we folded homophones together and effectively used the reduced CMU/MIT 39 symbol set.

The classification results for the different correlation assumptions are summarized in Table 1. The results show that explicit modeling of the correlation between successive samples is useful, since the Gauss-Markov and dynamical system models outperform the independent-frame and target-state assumptions. The results also indicate that the smoothing effect of the additive noise term in the DS model is useful, providing a more robust model. Additional experiments were run comparing the trajectory invariance and correlation invariance assumptions. The CI assumption significantly outperformed the TI assumption for short phones (less than 10 frames), and performance was similar for the two models for longer phones. We therefore conclude that the assumption that observed phones correspond to a sampled version of an underlying trajectory does not reflect the correlation structure observed within a segment. This may explain some of the negative results in earlier attempts at SSM time correlation modeling.

Features, Model Assumptions	% correct 61 classes	39 classes
18 Cepstra:		
Independent frame	50.6	63.5
Target state	52.0	63.0
Gauss-Markov	53.9	64.5
Dynamical System (CI)	57.9	68.0
18 Cepstra + 18 Deriv:		
Independent frame	61.5	72.1
Dynamical System (CI)	64.6	73.9

Table 1: Phoneme classification results for context-independent models based on different assumptions about the statistical dependence between features within a segment.

Only the independent frame and dynamical system results are reported for the case using combined cepstra and cepstra derivatives, primarily because earlier results showed that the Gauss-Markov case starts to degrade as the number of features are increased [6]. The gain in performance for the dynamical system is not large when cepstral derivatives are used, but we anticipate greater gains with context-dependent models.

4 Summary

The use of segment-based models for speech recognition is motivated by the potential for using segmental features and modeling the correlation of frame-based features. Our own work with the stochastic segment model has shown that, without these advantages, segment models can yield performance similar to HMMs. Therefore, a robust model of statistical dependence within a segment can push forward the state-of-the-art in speech recognition.

We have described and evaluated several possible modeling assumptions for representing statistical dependency, all of which can be cast in the general dynamical system framework. These are based on an assumption of linearity which, from our analyses, appears to be quite reasonable for cepstral coefficients. Experimental results on the variations investigated here suggest that a correlation invariant dynamical system is a good model of statistical dependence, and in fact is the first approach to successfully model dependence of both cepstra and derivative cepstra for speech recognition.

Acknowledgements

The authors gratefully acknowledge the work of J. Robin Rohlicek who was an important collaborator in much of this work. This work was jointly supported by NSF and DARPA under NSF grant number IRI-8902124.

References

[1] S. Austin, J. Makhoul and R. Schwartz, "Continuous Speech Recognition Using Segmental Neural Nets," *Proc. of the DARPA Workshop on Speech and Natural Language,* pp. 249-252, February 1991.

[2] I. Bechwati, O. Kimball and M. Ostendorf, "Context Modeling with the Stochastic Segment Model," to appear in the *Proc. Int. Conf. Acoust., Speech, Signal Processing,* 1992.

[3] L. Breiman and J. Friedman, "Estimating Optimal Transformations for Multiple Regression and Correlation," *J. Amer. Stat. Assoc.,* Vol. 80, pp. 580-607, September 1985.

[4] P. F. Brown, *The Acoustic Modeling Problem in Automatic Speech Recognition,* Ph.D. Thesis, Computer Science Department, CMU, May 1987.

[5] M.A. Bush and G.E. Kopec. "Network-based connected digit recognition," *IEEE Trans. Acoust., Speech, Signal Processing,* Vol. ASSP-35, No. 10, pp. 1401–1413, October 1987.

[6] V. Digalakis, M. Ostendorf and J. R. Rohlicek, "Improvements in the Stochastic Segment Model for Phoneme Recognition," *Proc. of the DARPA Workshop on Speech and Natural Language,* pp. 332–338, October 1989.

[7] V. Digalakis, M. Ostendorf and J. R. Rohlicek, "Fast Search Algorithms for Connected Phone Recognition Using the Stochastic Segment Model," to appear in *IEEE Trans. on Signal Processing,* December 1992.

[8] V. Digalakis, J. R. Rohlicek, M. Ostendorf, "A Dynamical System Approach to Continuous Speech Recognition," *Proc. Int. Conf. Acoust., Speech, Signal Processing,* May 1991, pp. 289-292.

[9] P. Kenny, M. Lennig and P. Mermelstein, "A Linear Predictive HMM for Vector-Valued Observations with Applications to Speech Recognition," *IEEE Trans. on Acoust., Speech and Signal Proc.,* Vol. ASSP-38(2), pp. 220-225, February 1990.

[10] L.F. Lamel, R. H. Kassel and S. Seneff, "Speech Database Development: Design and Analysis of the Acoustic-Phonetic Corpus," in *Proc. DARPA Speech Recognition Workshop,* Report No. SAIC-86/1546, pp. 100–109, Feb. 1986.

[11] K.-F. Lee, "Context-Dependent Phonetic Hidden Markov Models for Speaker-Independent Continuous Speech Recognition," *IEEE Trans. Acoust., Speech and Signal Processing,* Vol. ASSP-38, No. 11, pp. 599-609, April 1990.

[12] H. C. Leung and V. Zue, "Phonetic Classification Using Multi-Layer Perceptrons," *Proc. Int. Conf. Acoust., Speech, Signal Processing,* April 1990, pp. 525-528.

[13] H. Meng and V. Zue, "Signal Representation Comparison for Phonetic Classification," *Proc. Int. Conf. Acoust., Speech, Signal Processing,* May 1991, pp. 285-288.

[14] N. Merhav and Y. Ephraim, "Hidden Markov Modeling Using the Most Likely State Sequence," *Proc. Int. Conf. Acoust., Speech, Signal Processing,* May 1991, pp. 469-472.

[15] M. Ostendorf, A. Kannan, S. Austin, O. Kimball, R. Schwartz, J. R. Rohlicek, "Integration of Different Recognition Methodologies Through Reevaluation of N-Best Sentence Hypotheses," *Proc. of the DARPA Workshop on Speech and Natural Language,* pp. 83-87, February 1991.

[16] M. Ostendorf and S. Roucos, "A Stochastic Segment Model for Phoneme-based Continuous Speech Recognition," in *IEEE Trans. Acoust., Speech and Signal Processing,* Vol. ASSP-37(12), pp. 1857–1869, December 1989.

[17] S. Roucos, M. Ostendorf, H. Gish, and A. Derr, "Stochastic Segment Modeling Using the Estimate-Maximize Algorithm," *Proc. Int. Conf. Acoust., Speech, Signal Processing,* pp. 127–130, New York, New York, April 1988.

AN EXPERIMENTAL ANGLE OF ARRIVAL ESTIMATION SYSTEM

Robert Lonski and Richard Gooch
Applied Signal Technology
160 Sobrante Way
Sunnyvale, CA. 94086

Abstract

This paper describes an experimental hardware system used to estimate the angles of arrival of multiple cochannel emitters. The system consists of a four-element phased array antenna and a personal computer housing four coherentized digital receivers. The digital receivers are used to capture signal snapshots that are subsequently processed by an adaptive beamforming algorithm implemented on the PC's CPU. Experimental results using the system with the algorithm described in [1] are presented.

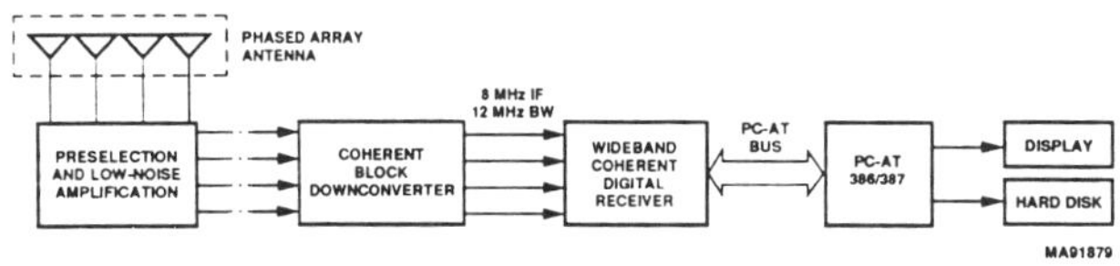

Figure 1: Experimental System Block Diagram

Introduction

As the need for the RF spectrum continues to grow, frequency reusage and spectral crowding are making the implementation of communication systems more complex. Adaptive beamforming techniques have been shown to be useful in such environments [1-4], although they have generally been considered too complex and expensive for integration into practical systems.

With the advent of high-speed Very Large Scale Integrated Circuits (VLSI) and high performance DSP processors, communications systems utilizing adaptive beamformers have now become both technically and economically feasible. In this paper the design of an experimental adaptive beamforming system will be presented. The system utilizes advanced digital signal processing techniques and digital receivers to implement the beamformer "front-end." As shown in this paper, the system provides a valuable platform for the development, test, and verification of angle-of-arrival (AOA) estimation algorithms.

The AOA estimation algorithm implemented in the experimental system is based on the adaptive CM Array first proposed in [2]. The CM Array is ideally suited for real-time implementations due to the computational efficiency of the weight update algorithm.

This paper is organized into five sections. The second section provides a detailed description of the experimental system. The third section provides a brief description of the AOA estimation algorithm presented in [1]. The fourth section presents experimental results obtained in real signal environments. The final section summarizes the results and presents suggestions for future work.

Hardware System

Development of adaptive beamforming systems has been underway at Applied Signal Technology for several years. The experimental system described here was developed for use as a test bed for AOA estimation algorithm development. A programmable architecture was chosen for the system so that various estimation algorithms could be easily implemented and demonstrated.

As shown in Figure 1, the experimental system consists of a four-element phased array antenna assembly, four-channel coherent block downconverter, four coherently synchronized digital receivers, and a 386 personal computer. The outputs of the digital receivers are collected, on a snapshot basis, and processed on the personal computer.

The phased array antenna used in the experimental system is shown in Figure 2. The antenna array consists of four Yagi antennas positioned in a line at half-wavelength spacing. Exact placement of the antennas

1058-6393/91 $1.00 © 1991 IEEE

Figure 2: Phased Array Antenna

is not critical as misalignment is corrected through a calibration procedure. The directional Yagi antennas were chosen to maximize the RF sensitivity of the system. The main beam of the array is approximately 120°, so bearing estimates over a full 360° field of view are not possible. The outputs of the antennas are conditioned using preselect filters to remove unwanted out-of-band energy and low noise amplified to maintain a low noise-figure.

The RF signals received by the antennas are converted to an intermediate frequency (IF) by the downconverter. Since the tuning and channelization functions are performed with digital receivers, the downconverter translates a block of RF energy to an IF suitable for digitization.

The downconverter design is based on a dual-conversion process in which the input signal is mixed down to a first IF centered at 70 MHZ and then to second IF centered at 8 MHz. The bandwidth of the second IF is 12 MHz. IF filtering is performed at 70 MHz to take advantage of the availability of high performance off-the-shelf SAW filters. A diagram of the downconverter is shown in Figure 3.

The 8 MHz IF outputs from the downconverter are each digitized with 8-bit flash converters sampling at 32 million samples per second (MSPS). This process produces four coherent digital replicas of the input signal with up to 15 MHz of bandwidth each. Custom VLSI digital tuners, manufactured by Applied Signal Technology, are used to isolate narrowband signals from the wideband digital input [8]. As shown in Figure 4 the VLSI tuner uses digital multipliers to mix the channel of interest to baseband. Filter and decimate stages follow the mixing operation and produce the narrowband baseband outputs. Since the narrowband filtering is implemented digitally in the VLSI chips, perfect amplitude and phase matching is guaranteed on each channel. Also since the sampling rate of the channelized outputs can be reduced to twice the channel bandwidth, the tuner output data can be fed directly into a DSP processor. Each channel has a TMS320 digital signal processor which collects snapshots of the channelized signal and passes the snapshot via DMA to the 386 CPU for further processing. The digital receiver used in the experimental system is the Applied Signal Technology "B-100" card. The receiver contains a wideband flash A/D converter, the Applied Signal Technology VLSI tuner chip set, and a TMS320C25 DSP processor. The card plugs into a standard PC-ISA backplane.

Phase coherency among the four channels is achieved thru the use of common local oscillators (LO's) for each of the channels in the downconverter and a common clock for all the digital circuitry. Variations in the absolute phase and magnitude among the four channels as a function of frequency is inevitable and must be corrected through a calibration procedure. To minimize the frequency resolution required in the calibration table and to maximize its overall effectiveness, the absolute amplitude and phase variations should be made as small as possible prior to calibration. In Figure 5, measured phase and amplitude tracking is shown across the full 12 MHz of input bandwidth and across a 30 KHz channel bandwidth. The results show that the amplitude variation is less than 2 dB and the phase variation is less than 4° over the 12 MHz of input bandwidth. Over the 30 KHz channel bandwidth, the phase and amplitude variations are negligible.

AOA Estimation Algorithm

The underlying principal of almost all AOA estimation techniques is interferometry. If a wavefront impinging on a two-element array arrives parallel to

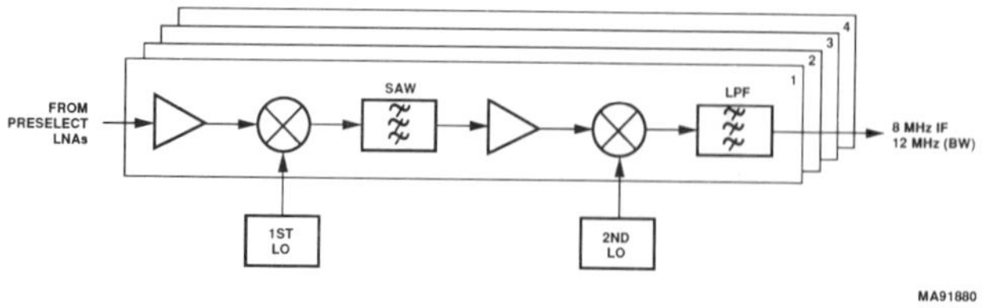

Figure 3: Block Downconverter

the array baseline, the signal received by both sensors will be identical. If the wavefront arrives at an angle Θ, the output from one sensor will be delayed with respect to the other. The delay is proportional to the sine of Θ. For narrow bandwidth signals (relative to the inverse of the propagation delay between sensors), the delay can be measured in terms of a phase shift between the carrier of the outputs of each sensor. By measuring the relative phase shifts between the two element signals, it is possible to calculate the emitter's AOA.

Due to non-ideal antenna characteristics and imprecise antenna placement, it is not realistic to calculate the AOA directly from the phase shifts. In practical DF systems, it is necessary to observe the phase and amplitude values as the frequency and incidence angle of a probe signal are swept across the field of view of the array. The observed phase and amplitude information is stored in a calibration table. An unknown signal's AOA is estimated by searching the calibration table for the nearest amplitude and phase vector and reading off the associated AOA.

The process of estimating an emitter's AOA in a cochannel environment is more complicated. The basic principal of the algorithm used in this system is to isolate one of the cochannel emitters in the environment, then cross correlate the isolated signal with the array signals [1]. The cross-correlation generates a vector of intersensor phase differences, (the direction vector) associated with the isolated signal. The AOA can then be estimated by searching the calibration table for a stored entry that most closely resembles the direction vector.

The process of isolating one of the cochannel emitters is performed using an adaptive CM Array [5,7]. Emitter isolation is accomplished in the CM Array by linearly combining the array signals to steer spatial nulls (zero directional gain) in the directions of the

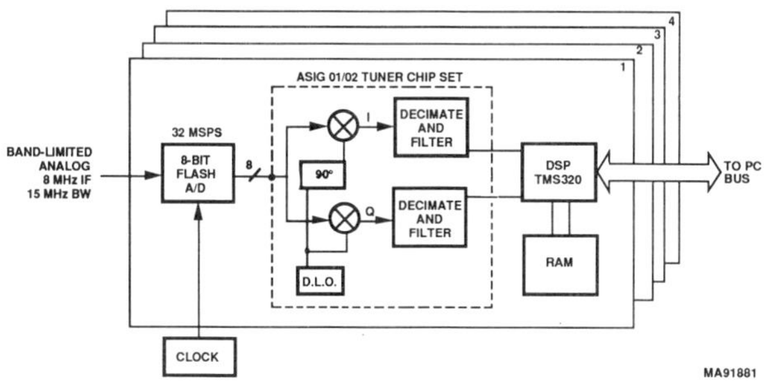

Figure 4: B-100 Digital Receiver

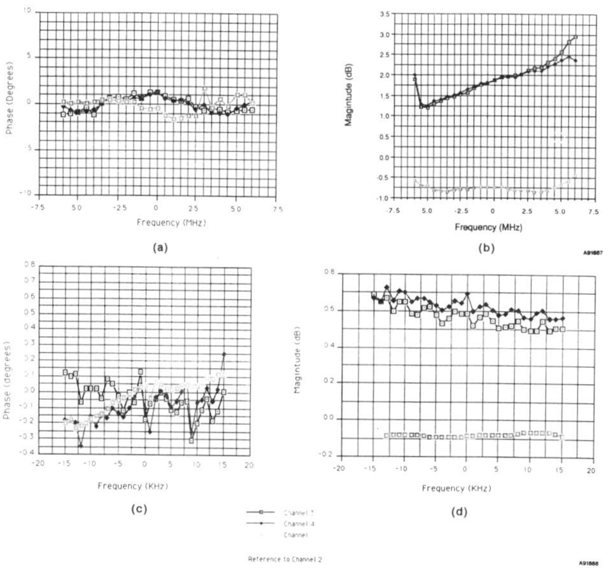

Figure 5: Measured Phase and Amplitude Tracking Performance
(a) wideband phase (b) wideband amplitude
(c) channel phase (b) channel amplitude

interferers and non-zero gain in the direction of the one of the emitters. Adaptation of the CM Array is blind in that it does not require apriori information pertaining to array calibration.

To isolate several cochannel emitters and estimate their AOAs, the cascaded CM Array architecture is used [1,2]. In this implementation each stage captures one of the signals present and produces a multichannel output with the captured signal removed. Cancellation of the captured signal from each of the input channels is accomplished with a single weight LMS adaptive signal canceller. The vector of canceller weights is the cross correlation between the captured signal and the input channels. This cross correlation is the direction vector associated with the isolated signal.

A block diagram of the implementation of the CM Array AOA estimation system is shown in Figure 6. The adaptive CM array and signal cancellers are implemented in "C" on the 386/387 CPU. The algorithm processes the input data on a snapshot basis. Each channel provides the CPU with 1024 data points per snapshot. Each snapshot is used to update the beamformer and canceller weights. This decimated weight updating process allows the system to track the AOA in a dynamic environment. The system is currently capable of providing AOA updates at a once per second rate.

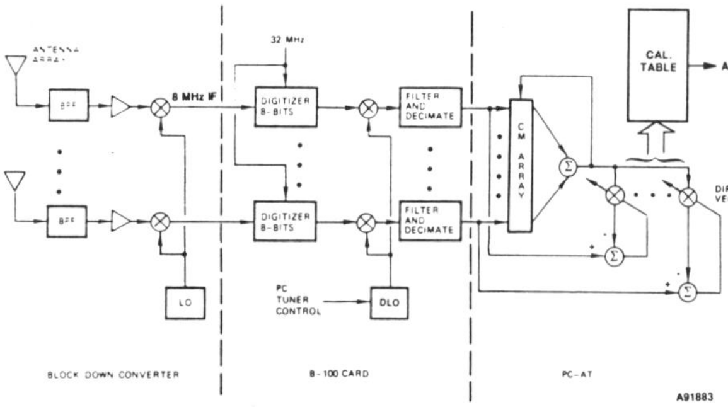

Figure 6: Algorithm Implementation

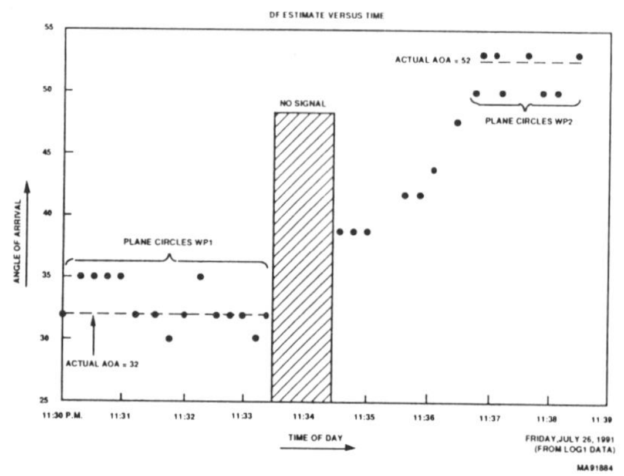

Figure 7: Airborne Emitter Tracking Results

Experimental Results

This section describes experimental results obtained using the previously described hardware system and AOA estimation algorithm. These results were from experiments conducted in *realistic* signal environments. It should be noted that the objective of the tests were to establish the basic functionality of the hardware and algorithms, rather than obtain high-resolution AOA estimates.

Two separate experiments were conducted. The first experiment was designed to test AOA estimation in an noncochannel environment with little multipath. This experiment involved the use of an airplane over flat terrain. The second experiment was designed to stress both the cochannel rejection aspect of the algorithm as well as its immunity to multipath. This experiment was conducted on the company's rooftop in an environment rich in both cochannel interference and multipath reflectors.

Calibration of the system was performed by rotating the phased array antenna in four degree increments, recording the relative phase and amplitudes from each channel in a calibration table. (Four degrees was the best resolution our low-budget Heathkit rotor could supply.) A directional antenna transmitting an un-modulated carrier was used as the source of the calibration signal. Angle-dependent calibration tables were generated every one megahertz across the bandwidth of interest. Signal frequencies falling between calibration frequencies were linearly interpolated using the nearest two calibration tables. There was no attempt to interpolate between angle entries in the calibration tables. The closest four degree entry was selected as the AOA estimate.

For the first experiment, an airplane equipped with a UHF transmitter was used as the signal of interest. The airplane flew a prearranged flight path between two landmarks at known locations. The aircraft circled each of the landmarks before proceeding on to the next one. The results of the experiment are shown in Figure 7. Notice that the DF system has accurately estimated the emitter's angle of arrival at each of the landmarks. The sinusoidal oscillations about the true bearing are indicative of the airplane circling the landmark.

The second experiment employed the geometry shown in Figure 8. The primary emitter was a FM signal located approximately 2.5 miles from the DF antenna. A secondary cochannel FM emitter was located on the roof of the building. To assess DF performance, several AOA estimation measurements were made for each emitter as the position of the secondary emitter was varied. The results shown in Figure 9 demonstrate the system was able to simultaneously estimate the AOAs for both the primary and secondary emitters. The AOA estimation of the secondary emitter was slightly degraded due to the occurrence of correlated multipath caused by scattering off the edges of the roof and the air conditioning unit.

Future Directions

The current experimental system provides the capability to simultaneously estimate the AOAs for mul-

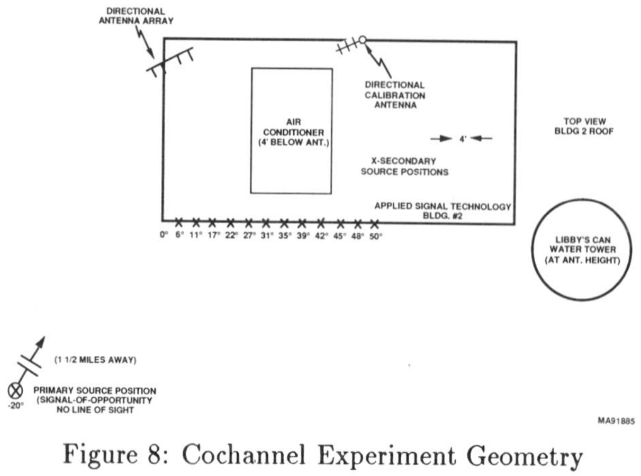

Figure 8: Cochannel Experiment Geometry

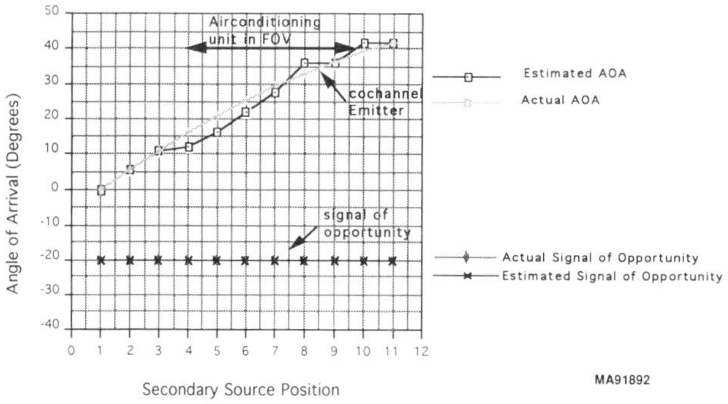

Figure 9: Cochannel AOA Estimation Results

tiple cochannel emitters. Inherent in the AOA estimation process is the isolation of each of the emitters. The logical next enhancement to the system is the inclusion of hardware to provide real-time output of the isolated signals. This enhancement could easily be realized for narrowband signals with bandwidths less than 50 KHz, using another DSP processor. The occurrence of scatter induced correlated multipath in urban signal environments is a common problem. The current CM Adaptive Array algorithm does not address the problem of AOA estimation in that signal environment. Augmentation of the system to provide accurate AOA estimates in correlated multipath environments would be of additional utility to the system.

Conclusions

In this paper an experimental AOA estimation system was presented. The programmable nature of the experimental system provides an excellent test bed for the development of adaptive beamformer and AOA estimation algorithms. The experimental system was based on existing hardware and demonstrates that it is now possible to implement complex algorithms in real time processing systems. The experimental system also demonstrates the functionality of digital receiver technology in implementing adaptive beamformer systems. The experimental results illustrates the utility and practicality of the CM Adaptive Array algorithm for signal COPY and AOA estimation problems. The algorithm is amenable to hardware implementation and was shown to perform well in real signal environments.

Acknowledgements

The authors would like to acknowledge the contributions of Derek Gerlach, Tim Geyer, and Chris Smith who were instrumental in the development and testing of the experimental system.

REFERENCES

1. R. Gooch, B. Sublett and R. Lonski, "Adaptive Beamformers in Communications and Direction Finding Systems," Twenty-fourth Asilomar Conference on Signals, Systems, and Computers, November 1990.
2. B. Sublett and R. Gooch, "Separation and Bearing Estimation of Cochannel Signals," IEEE Military Communications Conference, October 1989.
3. R.T. Compton, "Adaptive Arrays in Packet Radio," Twenty-fifth Asilomar Conference on Signals, Systems, and Computers, November 1991.
4. T. Ohgane, et al., "A Development of GMSK/ TDMA System with CMA Adaptive Array for Land Mobile Communications," Forty-first IEEE Vehicular Technology Conference,July 1991.
5. R. Gooch and J. Lundell, "The CM Array: An Adaptive Beamformer for Constant Modulus Signals," IEEE International Conference on Acoustics, Speech and Signal Processing, May 1986.
6. J.D. Lundell, "The Constant Modulus Adaptive Beamformer," Ph.D. dissertation, Stanford University, December 1988.
7. R. Gooch and B. Sublett, "A Multimode Adaptive Beamformer for Quadrature-Amplitude-Modulated Signals," IEEE International Conference on Acoustics, Speech and Signal Processing, April 1988.
8. J. Gray, J. Treichler, and M. Larimore, "The AST All-Digital Tuner Chip Set," Twenty-second Asilomar Conference on Signals, Systems, and Computers, November 1988.

Mainbeam Nulling with Adaptive Array Interferometry[1]

M. W. Ganz, J. Ward, and B. D. Carlson
M. I. T. Lincoln Laboratory
244 Wood Street
Lexington, MA 02173-0073

Abstract

As interfering signals move close to the steering direction of an adaptive array the main beam typically becomes distorted. Nulls in the main beam often produce patterns that are unacceptable for radar use since the gain may be severely reduced in the desired look direction. In this paper we consider the main beam patterns produced by an adaptive array consisting of two distinct antenna arrays that may be separated by many wavelengths. This type of adaptive array produces patterns that typically have very narrow nulls in the direction of the interference and preserve a main beam shape except for undesired grating nulls within the main beam pattern. These grating nulls can be filled in by averaging the patterns achieved at several operating frequencies. In this paper we discuss the adapted patterns and SINR performance for several types of these interferometric adaptive arrays. The effects of auxiliary element location and signal geometry are considered. Experimental results from an interferometric sidelobe canceller are presented.

1 Introduction

Adaptive array techniques can be used to protect radar systems from sidelobe interference by steering pattern nulls in the direction of the interfering signals. However, when interference arrives from within the main beam of the antenna pattern most adaptive nullling algorithms produce serious degradation of the shape of the main beam. An adaptive null in the main beam typically results in a gain reduction in the desired look direction that is so large as to make the antenna virtually impossible to use for detection of targets in the main beam region.

This paper presents an architecture for an adaptive nulling system that performs well with either sidelobe or main beam interference. For this architecture the antenna aperture is separated into a main aperture and one or more auxiliary apertures whose phase centers are spaced several wavelengths away from the phase center of the main aperture. With sidelobe interference this array produces deep nulls with little or no disruption of the main beam pattern. With main beam interference, the separation between the main and auxiliary arrays causes the adaptive null formed in the main beam to be very narrow. Interferometric grating nulls also appear; however, a significant region of target visibility can be achieved near the grating lobes of this pattern within the main beam. Below we outline this adaptive interferometry technique, its performance limitations, and present some experimental results showing its performance.

[1]This work was sponsored by the Department of the Navy under Air Force Contract F19628-90-C-0002.

2 Problem Formulation

Most adaptive nulling algorithms use feedback algorithms or numeric techniques to set antenna array weights according to

$$\mathbf{w} = \mathbf{\Phi}^{-1}\mathbf{s} \,, \tag{1}$$

where $\mathbf{w}$ is the array weight vector, $\mathbf{\Phi}$ is the covariance matrix of the signals on the array elements, and $\mathbf{s}$ is a steering vector that is typically set equal to the quiescent weights (the desired weight vector in the absence of interference). The quiescent weights are typically chosen to produce a low sidelobe pattern.

Consider first a 25-element half-wavelength uniformly spaced linear adaptive array. We choose the steering vector to be a 55 dB Chebyshev pattern steered to broadside. Figure 1 shows the quiescent pattern, an adapted pattern in response to a strong sidelobe interfering signal at 50°, and an adapted pattern in response to a strong main beam interfering signal at 0.5°. With sidelobe interference, the pattern quality is maintained quite well in both the main beam

1058-6393/91 $1.00 © 1991 IEEE

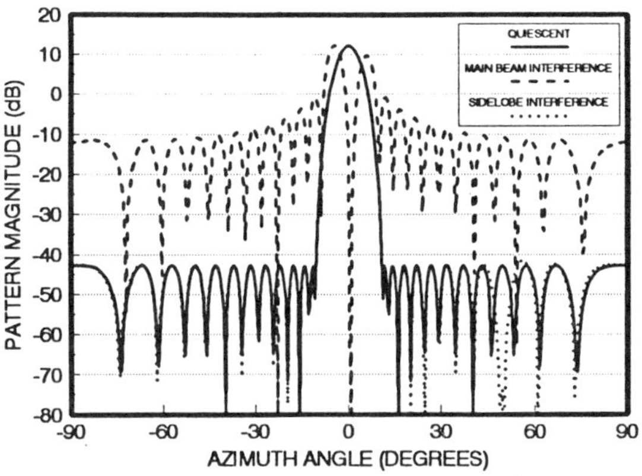

Figure 1: Patterns–uniformly spaced array.

and sidelobe regions. However, when the interference is in the main beam, there is a significant loss of gain for most of the main beam region. In addition, the sidelobes are much higher than the design sidelobes of the quiescent pattern. Our goal is to restore some of the main beam gain and maintain low sidelobes while at the same time suppressing strong main beam interference.

For strong interfering signals, the adapted patterns in Figure 1 are optimum in the sense that they are a least squares fit to the quiescent pattern subject to the constraint that the interference is nulled. An additional implied constraint is the assumed uniform linear array geometry. We shall show that by changing the array geometry, we can improve the performance of the array against main beam interference.

We consider an N-element linear adaptive array whose aperture is split into a main array of N_m elements and an auxiliary array of N_a elements, where the separation between the two arrays is Δ. We call this geometry an adaptive interferometer. For this paper we only consider cases with a single auxiliary array. The case $\Delta = 0.5\lambda$ reduces to the N-element uniform linear adaptive array.

A narrowband plane wave incident on this array produces a signal vector

$$\mathbf{u}(\theta) = \begin{bmatrix} \mathbf{u}_m(\theta) \\ \gamma(\theta)\mathbf{u}_a(\theta) \end{bmatrix},$$

where $\mathbf{u}_m(\theta)$ and $\mathbf{u}_a(\theta)$ are $N_m \times 1$ and $N_a \times 1$ Vandermonde steering vectors for the main and auxiliary arrays, and

$$\gamma(\theta) = e^{j(N_m - 1 + \frac{2\Delta}{\lambda})\pi \sin\theta} .$$

For a fixed number of elements, the total aperture size increases and the array resolution improves as Δ is made larger. As a result, the null on a main beam interference becomes narrower and the array gain close to the interference improves.

For a fully adaptive interferometer, the weight on each element of both the main and auxiliary arrays is computed adaptively using (1). The array output signal-to-interference-plus-noise ratio (SINR) is given by

$$SINR = \frac{\mathbf{w}^H \mathbf{\Phi}_d \mathbf{w}}{\mathbf{w}^H \mathbf{\Phi}_u \mathbf{w}} , \qquad (2)$$

where $\mathbf{\Phi}_d$ and $\mathbf{\Phi}_u$ are the covariance matrices of the desired signal and undesired signals (interference and thermal noise), respectively. When $\mathbf{s} = \mathbf{u}(\theta_d)$, where θ_d is a desired target arrival angle, the fully adaptive weights produce the maximum SINR (for the particular array geometry). However, the quiescent pattern in this case, in addition to having high sidelobes due to the uniform steering vector, also has grating lobes since the quiescent weights on the auxiliary array are non-zero. A more desirable steering vector for a fully adaptive interferometer is $\mathbf{s}^H = [\mathbf{w}_m^H \ \ \mathbf{0}^H]$, where $\mathbf{w}_m$ is an $N_m \times 1$ desired quiescent weight vector and $\mathbf{0}$ is the $N_a \times 1$ zero vector. This steering vector causes only the main array to be used in the absence of interference and produces a desired quiescent pattern.

In a sidelobe canceller interferometer (SLC), the main array has a fixed beamformer $\mathbf{w}_m$ and only the elements of the auxiliary array or the outputs of an auxiliary array beamformer are adaptively weighted. In a system where an adaptive capability is being added to an existing array, this type of architecture is a likely choice. The auxiliary weights for a sidelobe canceller are computed from [1]

$$\mathbf{w}_a = -\mathbf{\Phi}_a^{-1}\mathbf{v} ,$$

where $\mathbf{\Phi}_a$ is the covariance matrix of the auxiliary signals and $\mathbf{v}$ is the cross-correlation vector between the main array output signal and the auxiliary signals. The SINR for a sidelobe canceller can be computed from (2) with $\mathbf{w}^H = [\mathbf{w}_m^H \ \ \mathbf{w}_a^H]$. In the quiescent state the auxiliary weights are zero and the quiescent pattern is just that of the main array.

Although a sidelobe canceller produces a suboptimal SINR, it is a much easier system to build and is less computationally intensive than a fully adaptive array. Moreover, when compared to a fully adaptive array that produces the *same* low sidelobe quiescent pattern, we shall see that the SLC interferometer does *better* in many cases than the fully adaptive interferometer against main beam interference.

3 Performance Comparisons

To gain some understanding of the performance of these adaptive interferometers, we consider a signal scenario consisting of a single desired signal with arrival angle θ_d and a strong interfering signal at θ_i. It can be shown that for this interference scenario, optimum separations between the main and auxiliary arrays exist and are given by

$$\frac{\Delta_o}{\lambda} = \frac{2m+1}{2(\sin\theta_i - \sin\theta_d)} - \frac{N-2}{4}, \tag{3}$$

where m is any integer such that the quantity on the right hand side is positive. The separation corresponding to the smallest such integer is defined to be the single optimum separation since it results in the smallest total aperture size. Note that the Δ_o depends only on N and not on N_m or N_a individually; this means that the optimum occurs at a certain aperture size that depends on the proximity of the interference to the signal. As the interference gets closer to the signal a larger separation is required. Furthermore, it can also be shown that Δ_o is the same for both a fully adaptive and a sidelobe canceller interferometer, and does not depend on whether a uniform or tapered steering vector is used. However, the SINR attained with $\Delta = \Delta_o$ depends both on the allocation of elements between the main and auxiliary arrays and on the type of adaptive interferometer.

Figure 2 shows the SINR as a function of Δ for a scenario where $\theta_d = 0°$, SNR=0 dB, $\theta_i = 0.5°$ and INR=50 dB. For this case we have chosen $N_m = 20$ and $N_a = 5$. As predicted the optimum separation is $\Delta_o = 51.54\lambda$. The SINR of the interferometer at the optimum separation is 10–15 dB better than that attained with a uniformly spaced ($\lambda/2$) array of the same number of elements. The SINR performance of the sidelobe canceller is within 1 dB of the optimum fully adaptive array for most Δ, and is significantly better than that of the fully adaptive interferometer with the same quiescent pattern.

While this result may seem counterintuitive at first, it must be remembered that the fully adaptive array with anything other than the steering vector $\mathbf{s} = \mathbf{u}(\theta_d)$ is *not* optimal in terms of SINR. Figure 3 shows the adapted patterns for a separation of Δ_o. The weights on each configuration are normalized to produce unity noise power at the array output. The sidelobe canceller has higher gain on the desired signal and therefore higher SINR than the fully adaptive array with the low sidelobe steering vector. We see in Figure 3 the grating lobes caused by the separation between

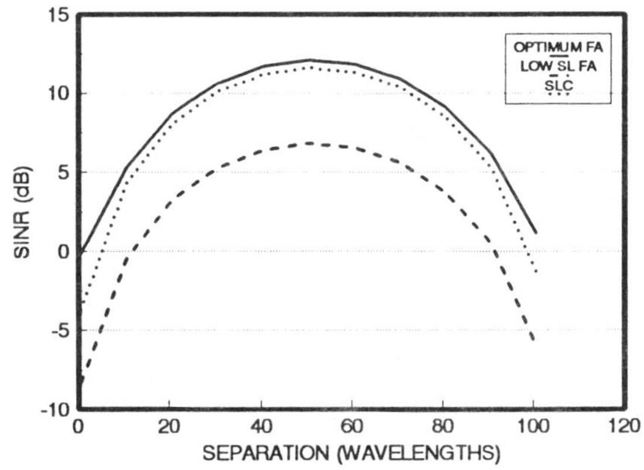

Figure 2: SINR vs. Δ

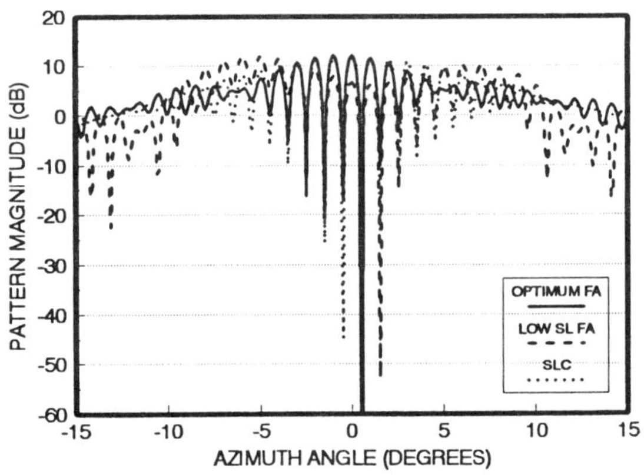

Figure 3: Patterns with main beam interference.

the main and auxiliary arrays. The optimum separation above is that for which a grating lobe peaks at the arrival angle of the desired signal.

Figure 4 shows the adapted patterns for a sidelobe interference at 50°. Both the sidelobe canceller and the low sidelobe fully adaptive cases maintain excellent sidelobes while nulling the interference. With both main beam and sidelobe inteference, the patterns are similar to Figure 3 but with a null on the sidelobe interference as well.

With main beam interference, the adaptive interferometer patterns are dominated in the sidelobe region by the pattern of the auxiliary array. This is easiest to see in the sidelobe canceller case, where large auxiliary array weights are required to match the main array response in the interference direction. The auxiliary array response is thus much larger than the low sidelobe main array response everywhere except in the main beam, and the adapted pattern takes on the auxiliary pattern shape. To simultaneously achieve high SINR mainbeam nulling with a beamshape and sidelobes on the order of those of the main array will require a large auxiliary array of the same size of the main array.

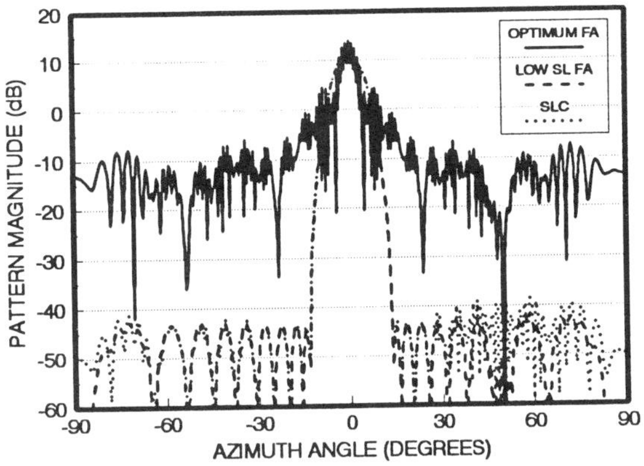

Figure 4: Patterns with sidelobe interference

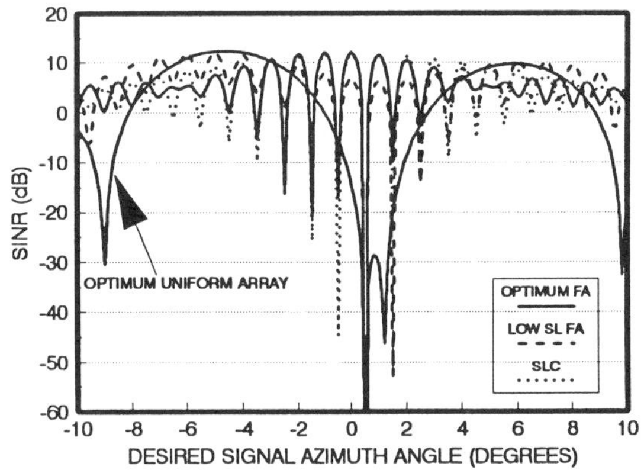

Figure 5: SINR vs. θ_d

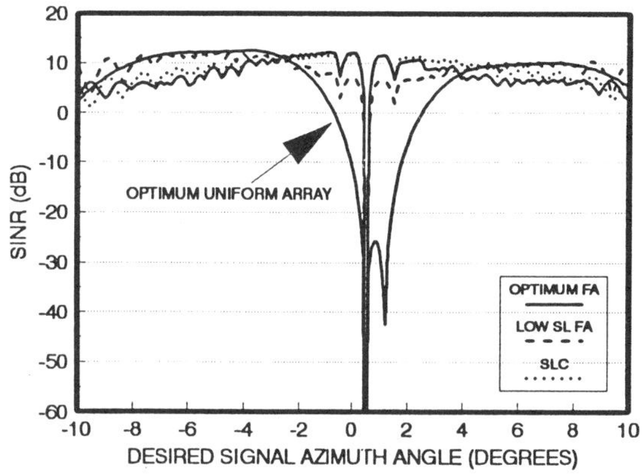

Figure 6: SINR vs. θ_d with frequency diversity.

4 Frequency Diversity

Consideration of a known signal scenario was useful to understand the operation of the adaptive interferometer and the conditions when its performance is best. In practice, neither the desired signal arrival angle nor the interference scenario are known. The purpose of the adaptive array radar is to detect targets of unknown location within the angular region covered by the main beam. One can see from the adapted interferometer patterns above that if the desired signal arrival angle is different from the angle used to steer the beam, it may fall in the nulls between the narrow main beam grating lobes.

This point is illustrated in Figure 5 which shows the SINR as a function of the desired signal arrival angle for the same configurations as those of Figures 3–4. The weights are computed assuming $\theta_d = 0°$. For comparison, the SINR for a 25-element uniformly spaced linear array is also shown. If the desired signal is near the peak of a grating lobe, the interferometer provides much higher SINR than the uniform array. However, if the desired signal falls into a main beam grating null, the benefit of the adaptive interferometer is lost and performance may actually be worse than the uniform linear array.

The degradation due to interferometric grating nulls can be circumvented by the use of frequency diversity in the radar system. Since the spacing between grating lobes will vary as the radar frequency is changed, a desired signal in a main beam grating null at one frequency may not be in a null at a different frequency. By combining adaptive interferometry with multiple looks at different frequencies, one can 'fill in' the grating nulls of Figure 5 and improve the detection in the presence of main beam interference. Figure 6 shows SINR vs. θ_d for the same scenario and configurations as Figure 5, but for adaptive interferometers with frequency diversity. We have assumed that five separate looks are available at frequencies in the range $0.8 \leq f/f_0 \leq 1.2$, where f_0 is the center frequency of the array. This range corresponds to a 40% total operating bandwidth. The SINR shown is the maximum of the SINRs computed for each of the five frequencies chosen. The nulls in SINR within the main beam region are not nearly as deep as with a single frequency. It is clear that the angular region around the interference at which high SINR is achieved is significantly larger than for a uniformly spaced array. The adaptive interferometer with frequency diversity will therefore provide much better detection of targets close to main beam interference.

5 Experimental Results

To test the concepts described above, an experiment was performed at M.I.T. Lincoln Laboratory with a sidelobe canceller interferometer constructed by appending an auxiliary array onto an existing array antenna. This array rests on a rotating pedestal, and is shown in Figure 7. The main array for the

Figure 7: Experimental adaptive interferometer.

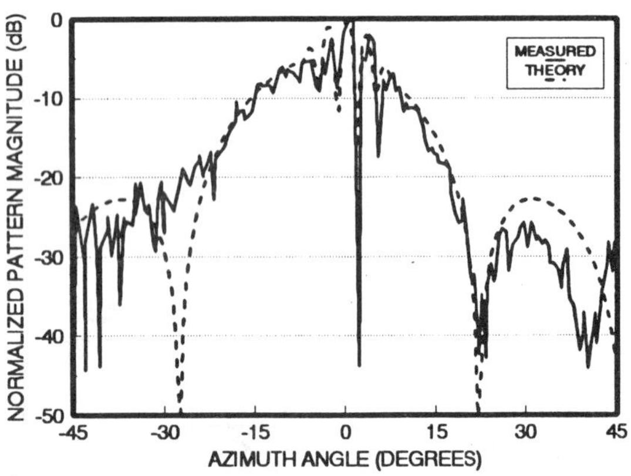

Figure 8: Patterns for the experimental array.

interferometer is taken to be one row of the planar array in Figure 7, and consists of 24 elements with a fixed Chebyshev taper. The auxiliary array is a half-wavelength-spaced linear array of six dipoles. The separation between the main and auxiliary arrays is $15\lambda_o$. The auxiliary element outputs are combined at RF to form a single auxiliary signal that is adaptively weighted and summed with the main array signal to produce the array output.

Adapted patterns for the experimental array were measured as follows. A far-field transmitter served both as a source to measure the patterns and as an "interfering signal". The array was rotated in discrete steps and a block of received data recorded at each step. The data from a particular azimuth corresponding to a main beam interference was used to compute the auxiliary weight. This weight was then applied to each block of snapshots and the average received power was calculated as a function of azimuth. A typical measured pattern is shown in Figure 8. The interference angle is 2.18° and the INR=40 dB per element. We also show the theoretical array pattern with the same auxiliary weight.

The experimental array has put a very narrow null on the interference, and the interferometric grating lobes are clearly visible in the main beam. We also see that the adapted pattern has predominantly the shape of the auxiliary array pattern. There is a good match between the measured and ideal patterns in the main beam region. Outside of this region the match is not as good and the measured pattern quality is poor, primarily because of the quality of our measurement range. Nearby structures and the broad elevation patterns of both the transmitting antenna and the experimental array contribute to a severe mutipath environment. In particular, we believe the missing first null and sidelobe around $-25°$ is due to the presence of reflections from another tower located at this azimuth.

6 Discussion and Conclusions

We have presented the use of an adaptive interferometer array to improve the performance of an adaptive array in the presence of main beam interference. It was shown that for a known signal scenario, an optimum location for the auxiliary array exists. We made the interesting observation that if low sidelobe quiescent patterns are desired, a sidelobe canceller array does better than a fully adaptive array with the same quiescent pattern. In these cases the adapted patterns are dominated by the auxiliary array. The combination of adaptive interferometry and frequency diversity may significantly increase a radar's coverage at angles close to a strong interfering signal.

Although we believe the technique is potentially useful for the difficult mainbeam nulling problem, several issues are yet to be examined. As the separation between main and auxiliary arrays increases, any degradation due to interference bandwidth will get worse. This is more of a problem for a phased array than a rotating array, where a main beam interfering signal is by definition very close to broadside.

Acknowledgements

The efforts of Dr. H. Aumann, R. Torres, F. Willwerth, and J. Charest were instrumental in conducting the experiments and are gratefully acknowledged.

References

[1] R. T. Compton, Jr., *Adaptive Antennas-Concepts and Performance.* Englewood Cliffs, New Jersey: Prentice-Hall, 1988.

Adaptive Arrays in Packet Radio

R. T. Compton, Jr.

Dept. of Elec. Engrg.
Ohio State University
Columbus, OH 43210

J. Ward

Radar Systems Group
MIT Lincoln Laboratory
Lexington, MA 02173

Abstract

This paper describes how a multiple beam adaptive antenna (MBAA) may be used to improve the performance of a random access packet radio communication network. In such networks, packets arriving at a terminal receiver compete for the use of the radio channel. Collisions between uncoordinated packets limit the average throughput and increase the network delay. A multibeam adaptive array used at a packet receiver can eliminate nearly all collisions between packets and hence is a very attractive technique for improving the performance of such networks. The paper describes the procedure used for rapid signal acquisition with the adaptive array and the effect of the array on network performance for a simple slotted ALOHA network.

1 Introduction

ALOHA packet radio systems provide a simple way of multiplexing many users into a single radio channel. However, throughput is limited to 18% in pure ALOHA and to 36% in slotted ALOHA [1]. Carrier Sense Multiple Access (CSMA) [2] permits higher throughputs but requires that all stations in the network be able to hear one another.

In analyzing these systems, it is assumed that when two packets collide, neither is received correctly. In some systems a capture effect [3] may allow one packet to be received correctly even in a collision. However, even with capture, it is usually assumed that a terminal can still receive only one packet at a time.

In this paper we present a method for a slotted system that allows multiple captures in each slot: the use of an adaptive array [4]. We describe an array system where signals from several array elements are combined with more than one set of weights to form several simultaneous receiving patterns. Such an array is called a *multiple-beam adaptive array* (MBAA). Each pattern has its peak response in the direction of one arriving packet and has nulls on the other packets. By using an MBAA, a packet terminal can receive several packets successfully in each slot.

The technique described here may be viewed as an alternative to spread spectrum, which also permits multiple packet receptions in one slot [5]. Spread spectrum allows packets in the same slot to be distinguished either on the basis of their arrival times (*delay* capture [6]) or by means of orthogonal codes (code-division multiple-access (CDMA)). (Recently, Ghez, *et al.* [7, 8] investigated a general multipacket ALOHA model in which multiple packets may be successful in each slot.)

However, spread spectrum systems require much wider bandwidths than conventional ALOHA systems for the same information rate. The MBAA technique described here has the advantage that it can increase the packet rate through a packet receiver without the need for additional bandwidth.

In this paper, we describe how an MBAA can be used in a slotted packet system and determine the performance of a single-hop ALOHA network with an MBAA. We present results for a finite population packet system obtained from a Markov chain model. Our results show that a very significant improvement in both throughput and delay can be achieved with modest MBAA capabilities.

2 Multiple-Beam Adaptive Arrays

An adaptive array is an antenna system that controls its own pattern, by means of feedback, while the antenna operates [4]. In an N_e-element adaptive array, the signal $\tilde{x}_j(t)$ from element j is multiplied by a complex weight w_j and then summed to produce the array output signal $\tilde{s}(t)$. The optimal array weights, which maximize the output SINR, are given by [4]

$$\mathbf{W} = \mathbf{\Phi}^{-1}\mathbf{S}, \tag{1}$$

where $\mathbf{W}$ is the *weight vector*,

$$\mathbf{W} = [w_1,\ w_2,\ \ldots,\ w_{N_e}]^T, \tag{2}$$

$\mathbf{\Phi}$ is the *covariance matrix*,

$$\mathbf{\Phi} = E[\mathbf{X}^*\mathbf{X}^T], \tag{3}$$

and $\mathbf{S}$ is the *steering vector*,

$$\mathbf{S} = E[\mathbf{X}^*\tilde{r}(t)]. \tag{4}$$

In these equations, $\mathbf{X}$ is the *signal vector*,

$$\mathbf{X} = [\tilde{x}_1(t),\ \tilde{x}_2(t),\ \ldots,\ \tilde{x}_{N_e}(t)]^T, \tag{5}$$

$\tilde{r}(t)$ is the reference signal, a signal correlated with the desired signal [4], and T denotes transpose.

By applying multiple weight vectors to the same element signals, an adaptive array can provide *multiple output signals*, with each output resulting from a *different array pattern*, or *beam*. An adaptive array with several simultaneous patterns is called a *multiple-beam adaptive array* (MBAA).

In a packet radio system, the weight vectors used in each slot are chosen so each pattern has a beam on one of the packets and nulls on all other packets. To steer a particular beam toward a given packet, a reference signal $\tilde{r}(t)$ correlated with that signal is used in (4). With a separate beam for each packet, an MBAA permits a terminal to receive multiple packets successfully in each slot.

Two factors limit performance with an adaptive array: the number of degrees of freedom and the angular resolution. An MBAA with N_e elements has $N_e - 1$ degrees of freedom in each pattern [4]. One degree of freedom is needed to point a beam toward a given packet. Each beam can therefore null only up to $N = N_e - 2$ additional packets.

The angular resolution of an array depends primarily on the array aperture size. If the angles of a desired signal and an interfering signal are too close, the output SINR (signal-to-interference-plus-noise ratio) from the array will be too low to receive the desired signal. Let the *resolution width* θ_r of the array be the minimum angular separation between a desired packet and an interfering packet at which the array can maintain a suitable output SINR.

3 A Packet Radio / MBAA System

We consider a simple random access packet radio system in which radio terminals communicate to a central base station. A slotted ALOHA protocol is used [9]. An MBAA is used as the receiving antenna at the base station. We assume free and instantaneous acknowledgements of successful packets.

To permit each MBAA beam to acquire a different packet in each slot, a special acquisition preamble is added to the beginning of each packet. (Every packet has the same preamble.) This preamble consists of three periods of a known pseudonoise (PN) sequence. In addition, the slot width is made slightly longer than the packet width by an *uncertainty interval.* Packet transmission times are randomized over this interval.

The objective is to steer the first beam toward the first packet to arrive in each slot, the second beam toward the second packet in the slot, and so forth. Each beam must also have nulls on any other packets in that slot. To do this, the signal processing in Figure 1 is used. A signal is obtained from the array whose pattern covers all users in the network. (E.g., this output signal could be the signal from one omnidirectional element in the array.) This signal is passed through a filter matched to one period of the preamble PN code. An arriving packet causes a timing spike at the filter output at the end of the first period of its preamble code. The filter output is compared to a threshold. A threshold crossing triggers calculation of a weight vector for that packet. The weight vector is calculated during the second period of the preamble PN code for that packet, as follows. The matrix $\mathbf{\Phi}$ in (3) and the vector $\mathbf{S}$ in (4) are estimated from the samples of the $\tilde{x}_j(t)$ taken during the second period of the preamble PN code. The threshold crossing is used to determine the correct timing. Samples of $\tilde{r}(t)$ for (4) are obtained from a stored replica of the preamble PN code. The weights are then obtained from (1) with the Sample Matrix Inverse algorithm [10]. To prevent two beams from locking on the same packet, the threshold detector for the second packet is not enabled until the first packet is acquired. When the threshold detector for each beam is triggered, a separate weight vector is calculated for that beam.

To analyze the performance of this system, it is necessary to distinguish between *acquired* and *successful* packets. A packet is said to be *acquired* by the MBAA when the threshold has been tripped by a packet and the weight calculations start for that packet. However, packet acquisition alone does not guarantee success. For success, the adaptive array must also be able to null other packets that arrive in the same slot.

Hence, once a packet is acquired, success still depends on the arrival angles of other interfering packets in that slot (because of limited array resolution) and the total number of packets in the slot (because of the finite number of nulls available in the array). Only successful packets contribute to the average throughput.

When multiple packets arrive in the same slot, a given packet is acquired successfully as long as the times of arrival of all other packets differ from that of the given packet by at least one bit. When two packets are separated by more than one bit, the timing spikes at the matched filter output in Fig. 1 can be distinguished and the reference signal $\tilde{r}(t)$ will be correlated with only one of the two packets. If two packets arrive with *less than one bit* separation, the vector **S** will be in error and both packets are lost. The length of the uncertainty interval is chosen just less than one period of the preamble PN sequence, so that all interfering packets in a given slot are present during the second period of each packet preamble, when its weights are calculated. Hence all interfering packets are included in $\mathbf{\Phi}$ and the resulting patterns null all interfering packets [4].

We have analyzed the performance of a single-hop slotted ALOHA system with an MBAA and a finite user population by extending the Markov chain model of [11] to include the effect of the MBAA. This analysis may be found in [12, 13]. For this analysis, we assume that each beam of the K-beam MBAA can form N nulls and has a resolution width of θ_r. In a given slot, packet arrival times are random and uniformly distributed over the uncertainty interval $[0, T_u]$ within each slot. The length of the uncertainty interval T_u is set equal to $(r-1)T_b$, where r is the period of the preamble PN code and T_b is the bit duration. Packet arrival angles are random (independent of arrival times) and uniformly distributed over $[0, 2\pi)$.

At the beginning of every slot, each node is either *unblocked* or *blocked*, depending on whether its last transmitted packet was successful or not. At the end of each slot, all terminals receive immediate feedback as to the fate of their packets. We make the single packet buffer assumption, in which blocked terminals are prevented from generating new packets until their backlogged packet is successfully retransmitted. An unblocked terminal transmits a new packet with probability p_n in each slot. A blocked terminal retransmits its backlogged packet with probability p_r in each slot until successful, at which time it becomes unblocked and again able to generate new packets. We assume p_r is constant and $p_r \geq p_n$.

Under the above assumptions, the state of the network is defined by the number of blocked terminals X_k at the beginning of the slot k. The random process X_k is an irreducible, aperiodic, and ergodic Markov chain. The state space of this Markov chain is $E_X = \{0, \ldots, M\}$, where M is the number of terminals in the network.

The transition probabilities of the Markov chain are found by considering the detailed operation of the adaptive array and enumerating the ways that each possible transition can occur. For an MBAA with K beams, at most K packets can be successful in a slot, so the state can decrease by at most K terminals in a slot. These transition probabilities are derived in [12, 13]. The limiting probability distribution $\boldsymbol{\pi} = [\pi(0), \pi(1), \ldots, \pi(M)]$ is given by the solution to the linear system of equations

$$\boldsymbol{\pi} = \boldsymbol{\pi}\mathbf{P} \tag{6}$$

with $\sum_{i=0}^{M} \pi(i) = 1$, where $\mathbf{P}$ is the transition matrix. The average backlog $\overline{B}$ is the average number of blocked terminals,

$$\overline{B} = \sum_{i=0}^{M} i\pi(i) \ . \tag{7}$$

The average throughput is

$$\overline{S} = \sum_{i=0}^{M} S(i)\pi(i) \ , \tag{8}$$

where $S(i)$ is the conditional throughput, i.e., the average number of successful packets in a slot given that i terminals are blocked at the beginning of the slot. The new packet input rate in state i is

$$S_{\text{in}}(i) = (M-i)p_n \ . \tag{9}$$

In steady state, the average input rate equals the average throughput, so

$$\overline{S}_{\text{in}} = S_{\text{in}}(\overline{B}) = \overline{S} \ . \tag{10}$$

From Little's theorem, the average delay $\overline{D}$ experienced by a new packet is then

$$\overline{D} = \frac{\overline{B}}{\overline{S}_{\text{in}}} = \frac{\overline{B}}{\overline{S}} \ . \tag{11}$$

4 Results

We present typical performance results below for a network of $M = 50$ terminals using an MBAA.

Figure 2 shows the conditional throughput $S(i)$ (defined in (8)) as a function of the number of blocked users for several cases when $p_r = 0.2$ and $p_n = 0.02$ when, on the average, more than one packet is arriving per slot at the central station. $S(j)$ and $S_{\rm in}(j)$ (the load line) are plotted for MBAAs with $K = 1$ to 6 beams. These curves are computed for $N = 8$ nulls and $\theta_r = 5°$, corresponding to a 10-element array. The preamble PN code period is $r = 63$ bits. Figure 2 also shows the $S(j)$ that results for slotted ALOHA without an adaptive array. These curves show how $S(i)$ increases significantly as the adaptive array is added and the number of beams is increased. Conditional throughput levels of nearly 4 packets per slot are possible with a 6-beam MBAA.

Figure 2 also provides insight into the stability of the system. Such a system is stable when there is a single intersection point of the $S(i)$ and $S_{\rm in}(i)$ curves and this intersection point is in a region of high throughput and low delay. It is seen that, without an adaptive array, performance is very poor, with essentially zero throughput and all terminals blocked. This is to be expected, of course, since the total traffic is more than one packet per slot. However, the MBAA has a stabilizing effect in general on the system. All the MBAA curves are stable for this traffic level. As K is increased the average throughput increases from 0.83 packets/slot with $K = 1$ to 0.994 packets/slot for $K > 3$ beams. For this traffic level $S_{\rm in}(i)$, near optimal throughput of Mp_n=1.0 packets per slot is achieved with only 2 beams, and the performance for 3 or more beams is essentially the same as for the 2-beam case. However, for $K \geq 2$, Mp_n is much less than the maximum value of $S(i)$, so the input traffic p_n and M or the retransmission probability p_r could be increased substantially while still maintaining stability.

Fig. 3 shows the average throughtput $\overline{S}$ as a function of the new transmission probability p_n. The retransmission probability p_r is 0.2 for all cases. For low traffic, the average throughput increases linearly with p_n. For higher traffic, more collisions (i.e., non-captures) occur, due to either poor acquisition (a collision in time), insufficient array resolution (a collision in angle), or not enough nulls per beam. Thus, the average throughput peaks and begins to decrease. Increasing p_n further drives the system toward saturation where nearly all terminals are blocked.

Figure 4 shows the average delay $\overline{D}$ vs. p_n. Note that when the input traffic is high, the delay improvement provided by additional beams can be quite large. Fig. 5 shows the average delay $\overline{D}$ vs. the average throughtput $\overline{S}$.

5 Conclusions

We have described the use of a multiple-beam adaptive array in a slotted ALOHA packet radio network. Each beam captures a different packet in a slot by automatically pointing its beam toward that packet and nulling other packets in the slot. Such a system permits average throughputs of 2-4 packets per slot with only modest MBAA capabilities. Moreover, it does this without the need for wide bandwidth as in spread-spectrum packet radio.

References

[1] A. B. Carleial and M. E. Hellman, "Bistable behavior of ALOHA-type systems," *IEEE Transactions on Communications*, vol. COM-23, pp. 401–410, Apr. 1975.

[2] L. Kleinrock and F. A. Tobagi, "Packet switching in radio channels: Part I - carrier sense multiple-access modes and their throughput-delay characteristics," *IEEE Transactions on Communications*, vol. COM-23, p. 1400, Dec. 1975.

[3] D. G. Goodman and A. Saleh, "The near/far effect in local ALOHA radio communications," *IEEE Transactions on Vehicular Technology*, vol. VT-36, pp. 19–27, Feb. 1987.

[4] R. T. Compton, Jr., *Adaptive Antennas - Concepts and Performance*. Englewood Cliffs, NJ: Prentice-Hall, Inc., 1988.

[5] M. B. Pursley, "The role of spread-spectrum in packet radio networks," *Proceedings of the IEEE*, vol. 75, pp. 116–134, Jan. 1987.

[6] D. A. Davis and S. A. Gronemeyer, "Performance of slotted ALOHA random access with delay capture and randomized time of arrival," *IEEE Transactions on Communications*, vol. COM-28, pp. 703–710, May 1980.

[7] S. Ghez, S. Verdu, and S. C. Schwartz, "Stability properties of slotted ALOHA with multipacket reception capability," *IEEE Transactions on Automatic Control*, vol. AC-33, pp. 640–649, July 1988.

[8] S. Ghez, S. Verdu, and S. C. Schwartz, "Optimal decentralized control in the random access multipacket channel," *IEEE Transactions on Automatic Control*, vol. AC-34, pp. 1153–1163, Nov. 1989.

[9] D. Bertsekas and R. Gallager, *Data Networks*. Englewood Cliffs, NJ 07632: Prentice-Hall, Inc., 1987.

[10] I. S. Reed, J. D. Mallett, and L. E. Brennan, "Rapid convergence rate in adaptive arrays," *IEEE Transactions on Aerospace and Electronic Systems*, vol. AES-10, p. 853, Nov. 1974.

[11] C. Namislo, "Analysis of mobile radio slotted ALOHA networks," *IEEE Transactions on Vehicular Technology*, vol. VT-33, pp. 199–204, Aug. 1984.

[12] J. Ward and R. T. Compton, Jr., "High throughput slotted ALOHA packet radio networks with adaptive arrays." To appear in IEEE Transactions on Communications.

[13] J. Ward, *Improving the Performance of Packet Radio Networks with Adaptive Arrays*. PhD thesis, Ohio State University, Columbus, OH, Aug. 1990.

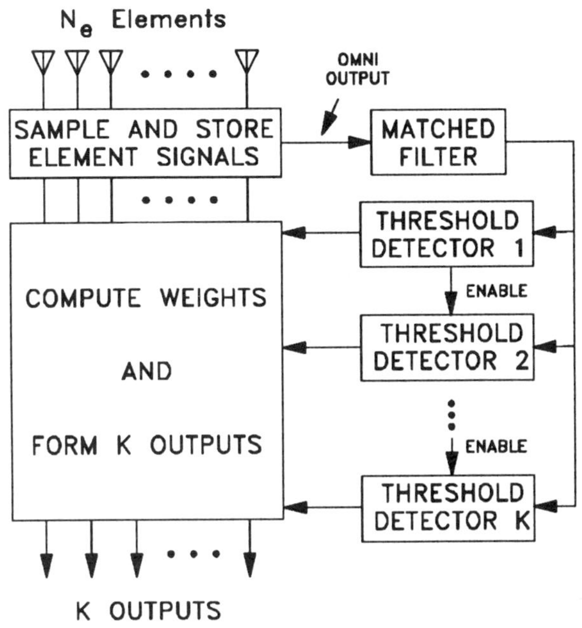

Figure 1: MBAA acquisition circuit

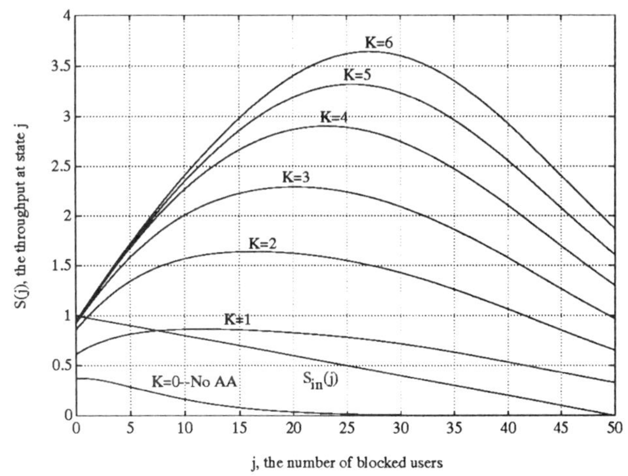

Figure 2: $S(j)$ and $S_{\text{in}}(j)$ for an $M = 50$ terminal system as the number of MBAA beams is increased. $p_n = 0.02,\ p_r = 0.2,\ N = 8,\ \theta_r = 5°,\ r = 63$.

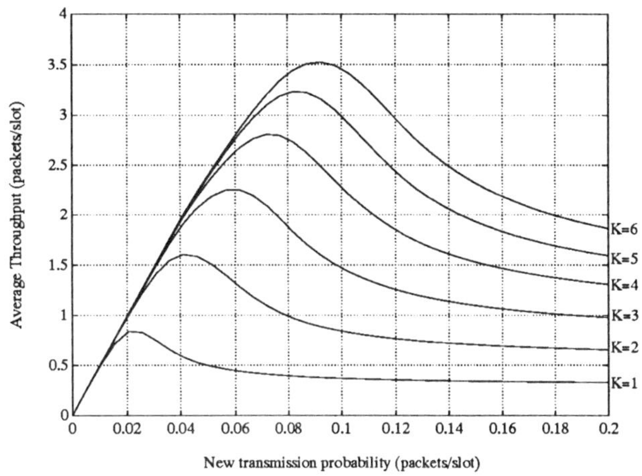

Figure 3: Average throughput $\overline{S}$ vs. p_n. $M = 50,\ p_r = 0.2,\ N = 8,\ \theta_r = 5°,\ r = 63$.

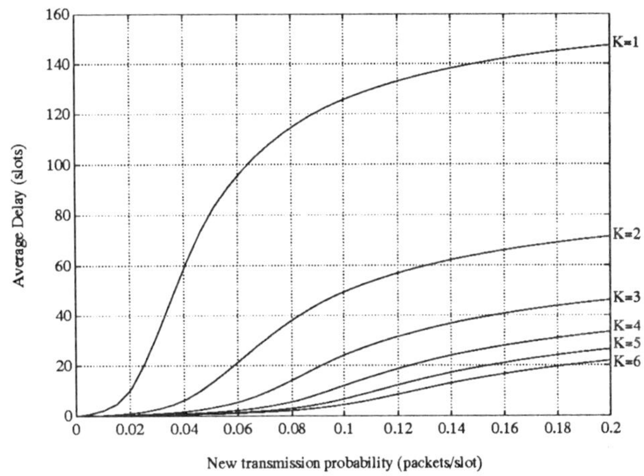

Figure 4: Average delay $\overline{D}$ vs. p_n. $M = 50,\ p_r = 0.2,\ N = 8,\ \theta_r = 5°,\ r = 63$.

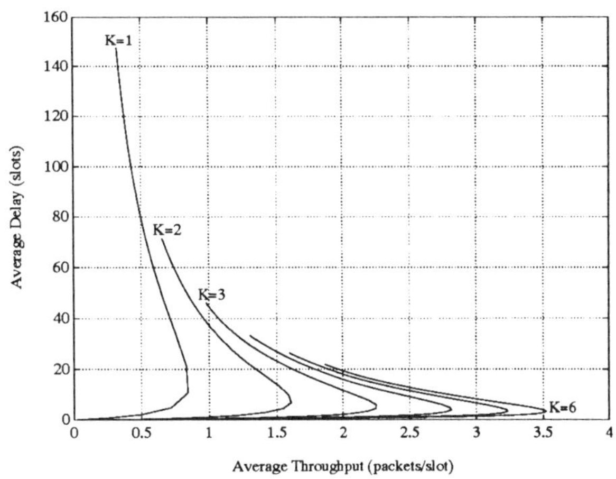

Figure 5: Average delay $\overline{D}$ vs.Average throughput $\overline{S}$. $M = 50,\ p_r = 0.2,\ N = 8,\ \theta_r = 5°,\ r = 63$.

Sensitivity of the Linearly-Constrained Constant-Modulus Cost Function

M. J. Rude

Signal & Image Processing Institute
University of Southern California
Los Angeles, CA 90089-2564

L. J. Griffiths

School of Engineering
University of Southern California
Los Angeles, CA 90089-1450

Abstract

This paper presents results from the application of the Linearly Constrained Constant-Modulus (LCCM) adaptive algorithm to a perturbed array environment. Assuming a narrowband signal model, analytic expressions are derived for the feasible gain values on the signal in the presence of general array perturbations and ambient white noise. The feasible gain values are found to be the roots of a cubic equation that is parameterized by the number of array elements, the SNR, the modulus factor, the kurtosis of the signal of interest, and the angle between the true signal steering vector and the perturbed steering vector. A principal result of the analysis is that LCCM is robust in the presence of array perturbations but is sensitive to the modulus factor value. A simulation is included that compares the sensitivity to perturbations of the techniques of linearly constrained minimum power, unconstrained constant-modulus, and LCCM.

1 Introduction

The Constant-Modulus Algorithm (CMA) is a popular adaptive technique for processing signals in environments where no reference or training signal is available to drive least-squares adaptation [1]. In CMA processing, it is assumed that the complex envelope of the *input* signal is distorted by multipath, interference, and/or channel fluctuations. By minimizing variations in the complex envelope of the *output* signal, CMA can correct for these sources of distortion. This approach may fail, however, in scenarios with multiple co-channel CM signals since each signal is a valid output. The Linearly-Constrained Constant-Modulus (LCCM) algorithm addresses this ambiguity problem by incorporating *a priori* knowledge into the CMA processor through the use of linear constraints on the processor coefficients. It has been demonstrated that this approach has the potential to solve the CMA ambiguity problem for situations where *a priori* knowledge such as the SOI arrival angle or carrier frequency is available. Thus, the CMA ambiguity problem is one motivation for LCCM.

LCCM can also be motivated as an extension to the technique of Linearly-Constrained Power Minimization (LCPM.) One drawback to LCPM is that it can cause signal cancellation when errors exist in the constraint equation or when it is used in environments that contain correlated interference. LCCM addresses this signal cancellation drawback by minimizing the output envelope variations rather than minimizing the output power. The LCCM adaptive recursions are very similar to those that constitute the LCPM algorithm. Only a slight modification in the gradient estimate is required to convert an LCPM algorithm to LCCM.

Previous papers have described the LCCM algorithm and it convergence properties [2, 3]. This paper presents analytical results for the LCCM approach in a perturbed, narrowband array environment. A brief review of the LCCM method follows.

2 The LCCM Approach

The output $y(n)$ of the LCCM adaptive processor is written as

$$y(n) = \mathbf{w}^{\dagger}\mathbf{x}(n) \tag{1}$$

where $\mathbf{x}(n)$ is the vector of complex data, $\mathbf{w}$ is the vector of complex adaptive weights and the superscript $\dagger$ is used to denote a conjugate-transpose operation. For now, the vector $\mathbf{x}(n)$ represents generic data that is either single or multi-channel in origin.

1058-6393/91 $1.00 © 1991 IEEE

The goal of LCCM processing is to minimize fluctuations in the complex envelope of the output $y(n)$ subject to a set of linear constraints on the weight vector $\mathbf{w}$. The first step is to define a set of M linear constraints on the N dimensional weight vector $\mathbf{w}$. Each constraint is expressed in terms of a constraint vector $\mathbf{c}_k$ and a corresponding scalar constraint value f_k. This can be compactly written in matrix notation as

$$\mathbf{C}^\dagger \mathbf{w} = \mathbf{f} \tag{2}$$

where the matrix $\mathbf{C}$ contains the linearly independent $\mathbf{c}_k$ vectors as columns and $\mathbf{f}$ is the vector of constraint values f_k. The general LCCM problem statement is

$$\min_{\mathbf{w}} E\left\{ \left| |y|^2 - \delta \right|^q \right\} \qquad \text{subject to} \qquad \mathbf{C}^\dagger \mathbf{w} = \mathbf{f} \tag{3}$$

where $q = 1$ or 2 and $\delta > 0$.

Clearly the cost functions corresponding to $q = 1$ or 2 do not have equivalent local minima. In practice, however, the local minima are approximately equal. The importance of this is that the $q = 2$ function is more analytically tractable while the $q = 1$ function leads to an adaptive algorithm that has better stability and convergence properties. Therefore, the $q = 2$ function can be used here to study local minima while the $q = 1$ function can be used to implement a well behaved adaptive algorithm that seeks approximately the same local minima.

3 Narrowband Analysis

In this section, the data vector $\mathbf{x}$ is no longer generic but corresponds to a narrowband signal as received by an array of sensors. Let the received narrowband signal be modeled as

$$\mathbf{x}(n) = s(n)\mathbf{d}(\theta_s) + \mathbf{g}(n) \tag{4}$$

where $s(n)$ is the signal of interest with power σ_s^2, $\mathbf{d}(\theta_s)$ is the direction vector, and $\mathbf{g}(n)$ is the vector of complex AWGN with variance of σ_g^2 at each element. The array output is given by

$$y(n) = \mathbf{w}^\dagger \mathbf{x}(n) \tag{5}$$

$$= \gamma s(n) + \hat{g}(n) \tag{6}$$

where $\gamma = \mathbf{w}^\dagger \mathbf{d}(\theta_s)$ and $\hat{g}(n) = \mathbf{w}^\dagger \mathbf{g}(n)$. Since $\hat{g}(n)$ is again a WGN process, let it be redefined as

$$\hat{g}(n) = \beta \tilde{g}(n) \tag{7}$$

where β is such that $\tilde{g}(n)$ has power σ_g^2. Thus, β represents the white noise (amplitude) gain of the weight vector $\mathbf{w}$.

In general, the signal gain γ and the white noise gain β are independent variables. For this analysis, however, it is useful to assume that they are dependent variables. The assumption that accomplishes this dependence is as follows: for a given signal gain, the CM cost function is minimized when the white noise gain is minimized. To be sure, this assumption is not always valid. In particular, with a perfect constraint equation and $\delta >> \sigma_s^2$ the white noise gain β will increase so that the squared output modulus $|y(n)|^2$ will more closely match δ. This effect has been verified experimentally. Clearly, the output noise power can be increased without affecting the signal gain. However, if $\delta < \sigma_s^2$ or if perturbations exist, the assumption is a reasonable approximation.

The dependence between the white noise gain and the signal gain is implemented by modeling β as a function of γ and ν, where ν is a parameter corresponding to generic pertubation. This perturbation coefficient is defined as

$$\nu = \frac{\mathbf{d}^\dagger(\theta_s)\mathbf{d}(\theta_c)}{N} \tag{8}$$

where $\mathbf{d}(\theta_c)$ is the steering vector used in the constraint equation and N is the number of array elements. This definition encompasses a wide range of perturbation sources including steering error, calibration error, and wavefront distortion.

Assuming a signal gain γ, it is straightforward to calculate the minimum noise power using analysis from conventional minimum power beamforming. First, a constraint matrix is formed,

$$\mathbf{C} = \left(\begin{array}{cc} \mathbf{d}(\theta_c) & \mathbf{d}(\theta_s) \end{array} \right) \tag{9}$$

and a vector of constraint values is defined,

$$\mathbf{f} = \left(\begin{array}{c} 1 \\ \gamma^* \end{array} \right). \tag{10}$$

Then, the constraint equation $\mathbf{C}^\dagger \mathbf{w} = \mathbf{f}$ represents both the signal gain γ and the unity steering constraint gain.

The weight vector $\mathbf{w}$ that minimizes output power is given by

$$\mathbf{w}_{min} = \mathbf{R}_{xx}^{-1}\mathbf{C}(\mathbf{C}^\dagger \mathbf{R}_{xx}^{-1}\mathbf{C})^{-1}\mathbf{f}. \tag{11}$$

Since the signal power is constrained to γ, this is also the weight vector that minimizes noise power. Under white noise only input conditions, $\mathbf{R}_{xx} = \sigma_g^2 \mathbf{I}$ and so

$$\mathbf{w}_q = \mathbf{C}(\mathbf{C}^\dagger \mathbf{C})^{-1}\mathbf{f}. \tag{12}$$

This vector, called the quiescent weight vector, is then used to calculate the minimum noise power $\sigma_g^2 \mathbf{w}_q^\dagger \mathbf{I} \mathbf{w}_q$. By comparison with (7), it is clear that $\beta^2 = \mathbf{w}_q^\dagger \mathbf{w}_q$. The inner product $\mathbf{w}_q^\dagger \mathbf{w}_q$ is easily found using the definitions (9) and (10),

$$\beta^2 = \frac{1 - (\nu\gamma^* + \nu^*\gamma) + |\gamma|^2}{N(1 - |\nu|^2)}. \tag{13}$$

It will be assumed that the phase of γ is such that

$$\beta = \frac{1}{\sqrt{N}} \sqrt{\frac{1 - 2|\nu||\gamma| + |\gamma|^2}{1 - |\nu|^2}}, \tag{14}$$

this minimizes β.

Now that β is defined, the output signal

$$y(n) = \gamma s(n) + \beta \tilde{g}(n) \tag{15}$$

can be substituted into the CM cost function

$$\mathcal{J}(\gamma) = E\left\{\left(|y(n)|^2 - \delta\right)^2\right\}. \tag{16}$$

Assuming hat $\tilde{g}(n)$ has uncorrelated and equal power IQ components, the expectation in (16) can be evaluated to yield

$$\begin{aligned}\mathcal{J}(\gamma) &= |\gamma|^4 \overline{|s(n)|^4} + 4\beta^2|\gamma|^2\sigma_s^2\sigma_g^2 + \beta^4\overline{|\tilde{g}(n)|^4} + \delta^2 \\ &\quad - 2\delta(|\gamma|^2\sigma_s^2 + \beta^2\sigma_g^2)\end{aligned} \tag{17}$$

where the overbar $\overline{(\cdot)}$ indicates expectation. Since $\tilde{g}(n)$ is a complex WGN process, $\overline{|\tilde{g}(n)|^4} = 2\sigma_g^4$. Note that the cost function only depends on the magnitude of the signal gain γ and the other complex variables. Therefore, the cost function can be redefined as a real function of real variables with $\phi = |\gamma|$ and $\eta = |\nu|$. This simplifies the analysis significantly.

Differentiating (17) with the complex variables replaced by their magnitudes yields the cubic equation

$$\begin{aligned}\frac{\partial \mathcal{J}(\phi)}{\partial \phi} &= \\ &\left(4\kappa_s\sigma_s^4 + \frac{8\sigma_g^4}{(N(1-\eta^2))^2} + \frac{16\sigma_s^2\sigma_g^2}{N(1-\eta^2)}\right)\phi^3 \\ &- \left(\frac{24\eta\sigma_g^4}{(N(1-\eta^2))^2} + \frac{24\eta\sigma_g^2}{N(1-\eta^2)}\right)\phi^2 \\ &+ \left(\frac{8\sigma_s^2\sigma_g^2}{N(1-\eta^2)} + \frac{8\sigma_g^4(1+2\eta^2)}{(N(1-\eta^2))^2} \right. \\ &\left. - \frac{4\delta\sigma_g^2}{N(1-\eta^2)} - 4\delta\sigma_s^2\right)\phi \\ &+ \left(\frac{4\delta\eta\sigma_g^2}{N(1-\eta^2)} - \frac{8\eta\sigma_g^4}{(N(1-\eta^2))^2}\right),\end{aligned} \tag{18}$$

where

$$\kappa_s = \frac{E\left\{|s(n)|^4\right\}}{\left(\sigma_s^2\right)^2} \tag{19}$$

is the kurtosis of the signal $s(n)$. The roots of this equation are a function of κ_s, σ_s^2, σ_g^2, η, N, and δ. In the next section, the relationship between the signal gain ϕ and these parameters will be investigated numerically.

4 Numerical Results

It is possible to calculate the roots of (18) in closed form since it is a cubic equation but the results are unwieldy and difficult to interpret. It is more useful to calculate the roots numerically for various parameter combinations. In this way, general cost function behavior may be inferred. For some combinations of parameters, two local minima are possible. Scenarios where this is possible are not included here. Every point on the curves to follow corresponds to a unique gain (magnitude) on the signal $s(n)$. Typically, two local minima occur for cases with large perturbation or for signals with high kurtosis values. Two local minima are also possible at very small perturbation values but these can be ignored because the noise model is less accurate for very small pertubation.

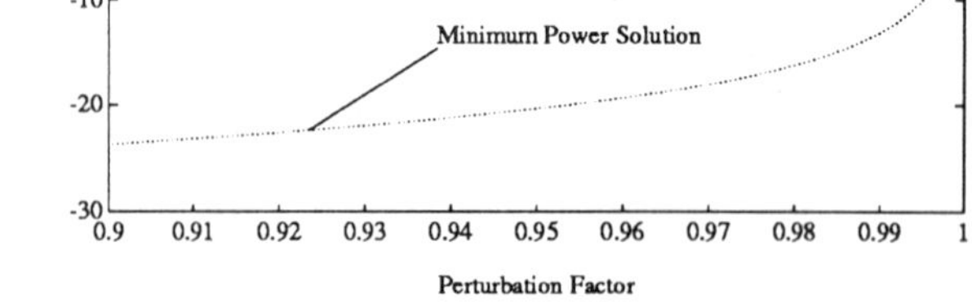

Figure 1: Output SNR vs. the perturbation factor η for three δ values and signal kurtosis = 1.0.

To begin, the relationship between the modulus factor δ and the perturbation factor η will be investigated. Figure 1 plots the output SNR as a function of the perturbation factor. The plots are generated by solving for the roots of (18), finding the root corresponding to the local minimum of the cost function, and then calculating the SNR using this root as the magnitude gain on the signal. The fixed parameters for this plot are $\kappa_s = 1.0$, $\sigma_s^2 = 20$ dB, $\sigma_g^2 = 0$ dB, and $N = 10$. (0 dB $\Leftrightarrow$ 1.) The curve at the bottom indicates the SNR for the conventional LCPM beamformer. Note that output SNR for the conventional beamformer is very sensitive to perturbation at this 20 dB input SNR level, the optimum SNR for this case is 30 dB. In contrast, the SNR curves corresponding to the LCCM analysis are much less sensitive to perturbation. The dashed curve corresponding to $\delta =$ 20 dB is tight to the optimum SNR value of 30 dB. ($\delta = 20$ dB is optimum.) Interestingly, the LCCM SNR curves are not necessarily monotonic for other δ values. For example, the $\delta = 17$ dB curve drops slowly until the perturbation is small (about 0.997) and then rises sharply to the optimum value.

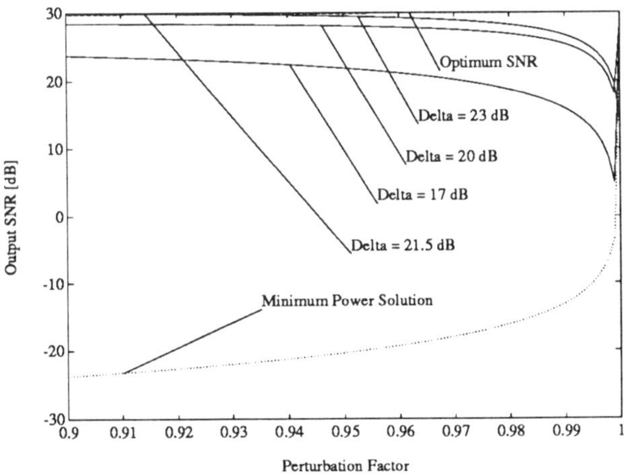

Figure 2: Output SNR vs. the perturbation factor η for three δ values and signal kurtosis = 1.4.

Figure 2 shows output SNR for the same parameters as Figure 1 except that the kurtosis is increased, $k_s = 1.4$. In this case, the optimum value of δ is approximately 21.5 dB (given by the dashed line.) A better indication of the optimum δ as a function of kurtosis is given in Figure 3. The fixed parameters for this plot are $\sigma_s^2 = 20$ dB, $\sigma_g^2 = 0$ dB, $\eta = 0.995$, and $N = 10$. Note how the optimum δ increases as the signal kurtosis increases. For comparison, the optimum SNR for the minimum power beamformer under these same conditions is -10 dB.

These numerical results indicate that LCCM has good and predictable properties in perturbed environments. This is even true for signals that are not con-

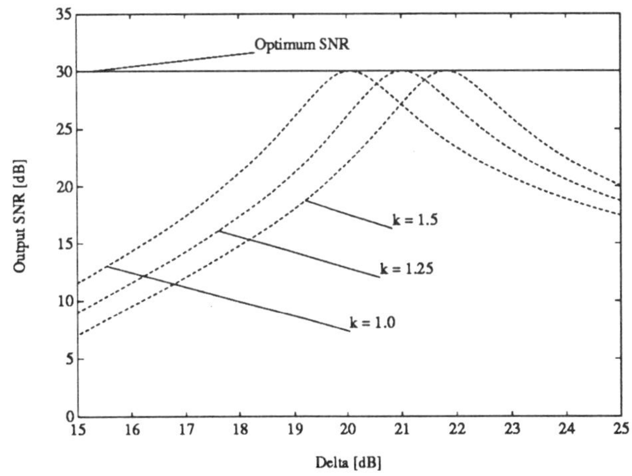

Figure 3: Output SNR as a function of δ for three kurtosis values.

stant modulus, i.e. have kurtosis greater than 1. A major drawback to LCCM in perturbed environments is the selection of the optimal value of δ. However, an adaptive δ version of LCCM was reported in [4]. In that work, the parameter δ was treated as another variable of optimization in the cost function. Preliminary analysis of adaptive δ LCCM is underway but is not included here. Instead, a simulation is presented that demonstrates adaptive δ LCCM in a perturbed scenario.

5 Experimental Results

In this experiment, a four element square array with half-wavelength spacing is used to receive a QPSK signal arriving from a presumably known direction. At each array element, the received signal is quadrature downconverted, matched filtered and synchronously sampled to yield one complex sample per QPSK symbol. These samples are then multiplied by the array weights and added together to form a single channel beam out of the multi-channel input. Finally, the beam is fed into the symbol decision stage of the receiver which completes the detection sequence [5].

It is important to note that the array operates on the matched filter outputs rather than the raw received signal. This is because QPSK signals typically receive pulse shaping before transmission which destroys their constant envelope property. The symbols themselves are still CM however and so LCCM can be effectively applied to the matched filter outputs.

The signal environment consists of a 30 dB QPSK

SOI arriving from 5°, a 30 dB CM interferer arriving from 30° and 0 dB white noise. The presence of the CM interferer causes signal ambiguity for CMA and will often lead to capture. Although the SOI arrives from 5°, the constraint used for LCPM and LCCM assumes that it arrives from 0°, a 5° steering error. This steering error corresponds to a perturbation factor $\eta = 0.995$. All three algorithms used similar operating parameters.

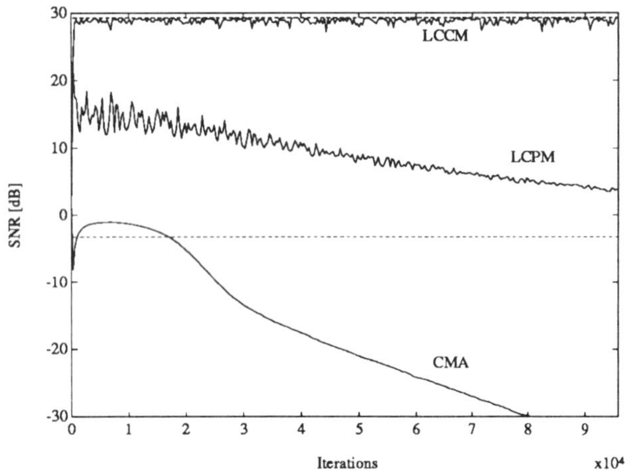

Figure 4: The SNR as a function of iteration for LCCM, LCPM and CMA. The upper dashed line is the maximum SNR possible. The lower dashed line is the converged SNR of LCPM for the 5° steering error in this experiment.

Figure 4 shows SNR as a function of iteration for the three algorithms. The SNR is measured as

$$\mathrm{SNR}(n) = 10 \log \left(\frac{\mathbf{w}^{\dagger}(n)\mathbf{R}_{ss}\mathbf{w}(n)}{\mathbf{w}^{\dagger}(n)\mathbf{R}_{nn}\mathbf{w}(n)} \right) \quad (20)$$

where $\mathbf{R}_{ss}$ is the SOI autocorrelation matrix and $\mathbf{R}_{nn}$ is the matrix of the CM interference plus noise. As can be seen from the plot, the LCCM algorithm converges rapidly to the optimal SNR given by the upper dashed line. The dashed line shows the SNR that is possible when there is no constraint error. LCCM has achieved the maximum SNR even though the constraint equation did not perfectly represent the SOI. The bottom dashed line shows the SNR that corresponds to the minimum power solution with the 5° steering error present, a good example of the problem of signal cancellation. This is the solution that the LCPM algorithm is converging towards. For this experiment, the CM interferer at 30° has captured the CMA algorithm causing a deep null to be placed on the SOI.

6 Conclusions & Further Study

The analysis and experiments presented in this paper suggest that LCCM has good properties in perturbed environments. Signal cancellation is much less of a problem with LCCM than it is with LCPM. Also, an important point to emphasize is that the analysis herein did not assume that the signals were constant modulus. Relatively robust performance of LCCM can be expected even when the signals do not have the constant envelope property.

There are numerous areas of further study for LCCM analysis. One that is underway is to repeat the analysis here for the adaptive δ version of LCCM. Another would be to compare the performance of LCCM with the many existing techniques for dealing with array perturbations.

References

[1] J. R. Treichler and B. G. Agee. "A new approach to multipath correction of constant modulus signals". *IEEE Trans. on Acoustics, Speech, and Signal Processing*, ASSP-31:459–472, Apr. 1983.

[2] M. J. Rude and L. J. Griffiths. "Incorporation of linear constraints into the Constant Modulus Algorithm". In *Proc. of the Int'l. Conference on Acoustics, Speech, and Signal Processing*, pages 968–971, Glasgow, Scotland, UK, May 1989.

[3] M. J. Rude. "A robust algorithm for untrained adaptive signal processing". *Journal of Computers and Electrical Engineering*, 1992. Accepted to 1992 Special Issue on Adaptive Signal Processing.

[4] M. J. Rude and L. J. Griffiths. "A linearly-constrained adaptive algorithm for constant modulus signal processing". In *Signal Processing V: Theories and Applications*, pages 237–240, Barcelona, Spain, 1990. Elsevier Science Publishers B.V.

[5] R.P. Gooch and J. Lundell. "The CM array: An adaptive beamformer for constant modulus signals". In *Proc. of the Int'l. Conference on Acoustics, Speech, and Signal Processing*, pages 2523–2526, Tokyo, Japan, Apr. 1986.

Impact of Mainlobe Interference Angular Extent on Adaptive Beamforming

William P. Ballance and Thomas W. Miller

Hughes Aircraft Company
Fullerton, California

Abstract

Adaptive beamformer performance against mainlobe interference is shown to significantly degrade when the interference has angular extent, even .005 beamwidths. Such angular extent can be caused by antenna rotation, antenna platform pitch or roll, or jammer motion. The degradation is shown to be sensitive to the interference power level. The impact on detection and angle of arrival estimation is addressed. Also, an interesting phenomenon that occurs in adaptive beamforming when using shaded steering constraints is presented.

1 Introduction and Problem Formulation

Modeling interfering sources as point sources, i.e., sources with no angular extent, is common practice in analyzing the behavior of adaptive beamformers [1],[2]. It will be shown, however, that angular extents as small as .005 BW can cause significant performance degradation for mainlobe nulling. An interferer can appear to have angular extent due to antenna rotation, antenna platform pitch or roll as occurs in shipboard and airborne radars, or motion of the interferer/jammer itself. Performance dependence on the power level becomes pronounced, which has implications on performance against pulsed and frequency swept jammers, and on the use of diagonal loading. The following example illustrates how strong interference can be. An S band radar with 1 MHz instantaneous bandwidth , 6 dB noise figure , and 30 dBi gain is jammed by a 10 kW ERP (effective radiated power) jammer located 50 km from the radar. This corresponds roughly to a 70 dB jammer to noise ratio, and serves as an upper limit for the study. In order to readily apply the results of this study to an antenna with any number of elements (N), the following scaling approach is used: target and jammer powers (relative to noise) are specified as that received at the peak of beam when using uniform shading.; and, angular locations and extents are given in terms of beamwidths, BW (defined here as the peak-to-first-null beamwidth, 2/N)). The study uses a 32 element array, sufficiently large to provide results independent of N.

The performance impact on the adaptive beamformer will be described in terms of the impact on detection, quantified by the output signal-to-interference-plus-noise ratio (SINR), and the impact on AOA estimation. The SINR is given by $|\underline{w}^{\dagger}\underline{d}_t|^2/\underline{w}^{\dagger}R\underline{w}$., where $\underline{d}_t$ is the array excitation in the target direction (and therefore includes the effect of gain and phase errors). The weight vector that maximizes the SINR when $\underline{d}_t$ is believed to be $\underline{d}_c$, (where the constraint vector $\underline{d}_c$ does not reflect knowledge of the gain and phase errors, and may be shaded) is [3] $\underline{w}=R^{-1}\underline{d}_c$. Unless otherwise specified, the results presented here are for no gain and phase errors, no shading, and with boresight as the look direction. The SINR performance can be viewed as consisting of two components, a loss of coherent gain in the look direction due to null formation, and residual jammer plus noise output power, which indicates how well the jamming is suppressed. The signal gain is given by $|\underline{w}^{\dagger}d_t|^2/\underline{w}^{\dagger}R_n\underline{w}$., while the output jammer plus noise to noise ratio (JPNR) is given by $\underline{w}^{\dagger}R\underline{w}/\underline{w}^{\dagger}R_n\underline{w}$, (where R_n is the noise covariance matrix). The "relative SINR" is defined to be the SINR relative to the optimum in a

1058-6393/91 $1.00 © 1991 IEEE

benign environment. A similar statement holds for the relative signal gain.

The AOA estimation performance is based on use of the (deterministic) Maximum Likelihood (ML) estimator. The target is present in only one time sample, and is not used in computing the adaptive weight. The AOA estimate $\hat{u}$, given the data vector $\underline{x}$, is that value of u that maximizes the following performance index [4],[5]

$$J(u) = |\underline{d}_c^{\dagger}(u)R^{-1}\underline{x}|^2/\underline{d}_c^{\dagger}(u)R^{-1}\underline{d}_c(u)$$

which can be written as $J(u)=|\underline{w}^{\dagger}(u)\underline{x}|^2/\underline{w}(u)^{\dagger}R\underline{w}(u)$ with $\underline{w}(u) = R^{-1}\underline{d}_c(u)$ (or any scalar times this vector). The latter form for J(u) shows that the estimation procedure can be viewed as beamforming on the data (for multiple angles) and normalizing by the known (estimated) JPNR.

2 SINR Performance

Figure 1 illustrates the effect of adaptation on the antenna pattern. The dashed pattern describes the quiescent pattern, that is, the pattern in the absence of interference, while the solid pattern corresponds to the adapted pattern when a point jammer is present at .5 BW. To suppress the jammer, the adapted pattern places a null at .5 BW, squinting the beam in the process. The gain in the look direction is reduced 2.3 dB (note that if BW were defined as the two-sided -3 dB BW as is often done, this value would be -3 dB). Now consider the effect of nonzero angular extent for the jammer. The solid curve in Figure 2 is the adapted pattern obtained for a 70 dB jammer at .5 BW with .005 BW extent (from .5 BW to .505 BW). The dashed curve is the pattern from Figure 1 for the point jammer. For the point jammer the null width at the -80 dB level is roughly .0001 BW (80 dB nulling puts the jammer 10 dB below the output noise floor, corresponding to .4 dB JPNR). A null of this width would yield a JPNR of 23.6 dB for the jammer with .005 BW extent. This compares to a JPNR of .3 dB for the weight using the wider null in Figure 2. The figure shows, however, that widening the null causes a larger loss in signal gain, -9.3 dB as opposed to -2.3 dB.

Unlike for the case of a point jammer, SINR performance against a jammer with angular extent depends strongly on the jamming power level. For example, if the jammer power in Figure 2 had been 50 dB rather than 70 dB, the relative SINR would have been -4.1 dB rather than -9.6 dB. Whereas for the point jammer, varying the power between 20 and 70 dB causes the relative SINR to differ from -2.3 dB by no more than .1 dB. It should be noted that this strong dependence on jammer angular extent and power level is a mainlobe nulling concern. The impact on sidelobe nulling is negligible in most cases.

Figure 3 plots the relative SINR versus angular extent and jammer power for the jammer at .5 BW. For small extents (e.g., .0001 BW), the performance against a point jammer (-2.3 dB) is approximately obtained. As the angular extent increases, performance degrades. For the 70 dB jammer, the SINR is -2.4 dB for .0001 BW, -6.4 dB for .001 BW, -9.7 dB for .01 BW and -16.7 dB for .1 BW. For lower power levels, more angular extent can be tolerated before suffering the same degradation. For 20 dB, the degradation is almost unnoticeable (for this jammer location). The figure shows that there is a plateau effect with respect to jammer extent. Knowledge of this effect can can be very useful from a system design standpoint. Investigation of the effect for other jammer locations finds that the length of the plateau is shorter when the jammer is closer to the look direction, and longer when it is farther away.

Having studied the relative SINR, it is instructive to view the output JNR. This quantity is of interest because the apparent jammer power in the weight computation can differ significantly from the actual power. This occurs, for example, when the interference is a pulsed jammer or a frequency swept jammer. If the resultant duty factor is 10%, then the the time averaged covariance matrix used to generate the weights will treat the jammer as though it has 10 dB less power. A jammer will also appear to be 10 dB weaker if 10 dB of diagonal loading is used. Diagonal loading is often used to reduce finite data sidelobes, and sometimes to reduce sensitivity to target capture (ie., the undesired nulling of the target due to its presence in the samples used to compute the weight) [2], [3]. For a 70 dB point jammer at .5 BW, the output jammer power is approximately 70 dB below the output noise floor and so a 10 dB misjudgment of the jamming level is inconsequential. As illustrated by Figure 4, however, the output jammer power can be close to, or even above, the output noise floor when the jammer has angular extent. If a 70 dB jammer with .003 BW extent was misjudged to have 60 dB strength, the adaptive processor would believe the output jammer power was 1.6 dB below the noise floor, when it would really be 8.4 dB above it. This results in a SINR of -13.1 dB rather than -6.4 dB (Figure 3 shows a SINR of -9.2 dB would have resulted if no misjudgment were made). Figure 5 shows the SINR versus jammer extent and power that occurs for a 10 dB power misjudgment. A comparison with Figure 3 shows that the performance degrades noticeably, even for lower power jammers.

The previous discussion only addressed performance against a single jammer. For the case of two jammers, the effect of jammer extent is even more pronounced. Consider the stressful case in which the look direction is midway between two jammers that are separated by 1 BW. For two point jammers, the SINR is -7.2 dB. For two 70 dB jammers with .001 BW extent, however, the SINR drops to -27.7 dB, a considerable degradation.

3 AOA Estimation Performance

Now consider the impact of interference angular extent on AOA estimation performance. The plot of the performance index for the cases corresponding to Figure 1 is given in Figure 6 (for a noise-free sample). The dashed curve is for the quiescent case, while the solid curve is for the case of a point jammer at .5 BW. The target strength is 30 dB, as can be seen from the height of the peak for the quiescent case. The peak for the jammer case is 2.3 dB lower, reflecting the relative SINR for this case. Note that what looks like a "null" in the performance index for the jamming case is actually the result of scaling by the denominator in eq 6. When the beamformer looks directly at the jammer, no null is formed (for the case of no gain and phase errors, and no shading), but the beamformer output power is scaled to near unity. The more flattened peak reflects the increased noise sensitivity of the adapted beam due to the deformation of its shape caused by null formation.

To quantify the impact on AOA estimation, we introduce the concept of relative estimation SINR, defined as -20log10 of the ratio of the Cramer-Rao bound on AOA RMS error (unbiased) in a jamming environment to that in a benign environment. A value of -10 dB signifies that the target in the jamming environment must be 10 dB stronger than in a benign environment to achieve the same AOA accuracy. The Cramer-Rao bound is used here because it reflects the asymptotic performance of the AOA estimator. (It is only used here for the case of no gain and phase errors and no shading, since these lead to error biases, as will be shown shortly.) The bound is given by [5]

$$\frac{1}{\sqrt{K}}\frac{1}{\sqrt{2p_t}}\left[\underline{d}'^{\dagger}_t R^{-1}\underline{d}'_t - |\underline{d}'^{\dagger}_t R^{-1}\underline{d}_t|^2/\underline{d}^{\dagger}_t R^{-1}\underline{d}_t\right]^{-1/2}$$

where the prime denotes differentiation with respect to the unknown AOA, p_t is the target power at the element level, and K is the number of target samples (K=1 for the cases considered here).

For the case of the point jammer at .5 BW in Figure 6, the relative estimation SINR is -7.7 dB. Comparing this to the -2.3 dB relative "detection" SINR, one observes 5.4 dB additional effective loss due to the increased noise sensitivity of the adapted beam, i.e., the effect of mainlobe jamming is more pronounced on AOA estimation than on detection performance. Figure 7 shows how the relative estimation SINR (solid curve) varies as a function of jammer angular extent for a 70 dB jammer at .5 BW, and provides the relative detection SINR (dashed curve) for comparison. Note that the relative estimation SINR drops from -7.7 dB for a point jammer to -18.3 dB for a jammer with .01 BW angular extent, a 10.8 dB degradation. Figure 8 shows the relative estimation SINR versus jammer angular extent for other jammer powers. This figure is a companion figure to Figure 3. Clearly, jammer angular extent significantly degrades AOA performance.

The AOA error bias encountered when nulling mainlobe interference in the presence of gain and phase errors is also impacted by jammer angular extent. For example, a bias of -.1 BW occurs for a 70 dB jammer at .5 BW with .001 BW extent, when using a 32 element array with -30 dB errors (i.e., errors corresponding to a -30 dB error sidelobe level). For a point jammer the bias is -.06 BW. Similarly, the use of shading causes a bias that is worsened by jammer angular extent.

4 An Interesting Phenomenon

As an aside, consider the interesting performance that results when using a shaded steering constraint and looking midway between two jammers. Figure 9 shows the relative (detection) SINR as a function of the separation between either jammer and the look direction. The dashed plot is for the unshaded approach, while the solid plot is for a -30 dB Taylor shading. For large separations, e.g., 2 BW, the jammers are in the sidelobes and the patterns are minimally affected. The .7 dB loss at 2 BW for the Taylor case is simply the shading loss. At 1 BW separation, the weight solution for both approaches is the unshaded steering vector, which has natural nulls at $\pm$ 1 BW, the jammer locations (the shaded approach believes it is suffering a .7 dB loss). As the separation decreases, the performance of the two approaches is very close until approximately .5 BW separation. In this vicinity the shaded approach fails (not just for Taylor shading). While the shaded approach believes it is losing approximately 20 dB of signal gain (recall that it believes the signal looks like a Taylor shading) to suppress the two 70 dB jammers, it actually losses well over 60 dB. This shows that shaded constraints must be used with caution when performing mainlobe nulling.

References

[1] F.L.C. Lin and F. Kretschmer Jr., "Angle measurement in the presence of mainbeam interference," *IEEE Mag Aerosp. Electron. Syst.*, pp. 19-25, Nov. 1990.

[2] B.D. Carlson, "Covariance matrix estimation errors and diagonal loading in adaptive arrays, " *IEEE Trans Aerosp. Electron. Syst.*, vol. AES-24, pp. 397-401, July 1988.

[3] J.E. Hudson, *Adaptive Array Principles.* New York: Peter Peregrinus Ltd., 1981.

[4] R.C. Davis, L.E. Brennan and L.S. Reed, "Angle estimation with adaptive arrays in external noise fields," *IEEE Trans Aerosp. Electron. Syst.*, vol. AES-12, pp. 179-186, March 1976.

[5] W.P. Ballance, "Direction finding of low-angle sources in the presence of multipath," Ph.D. dissertation, Univ. of California. Irvine, Irvine, CA, June 1987.

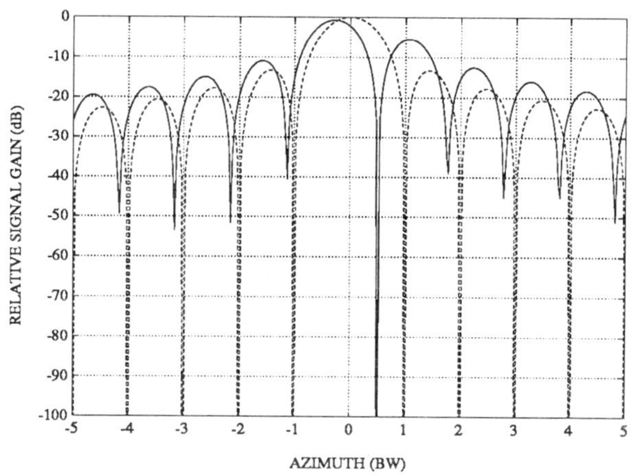

Fig. 1. Adapted pattern for point jammer at .5 BW compared to quiescent pattern.

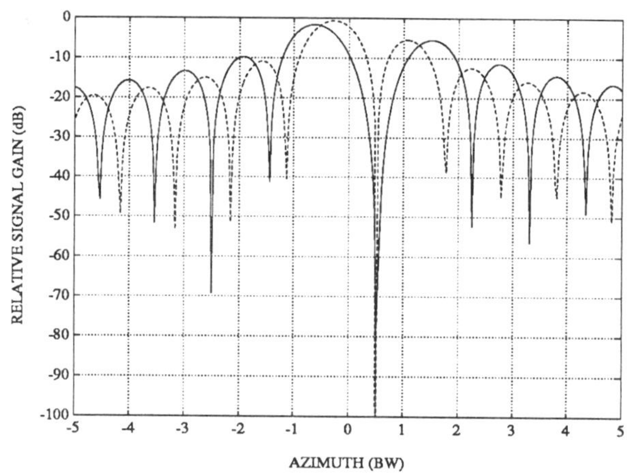

Fig. 2. Adapted pattern for 70 dB jammer at .5 BW with .005 BW angular extent compared to point jammer.

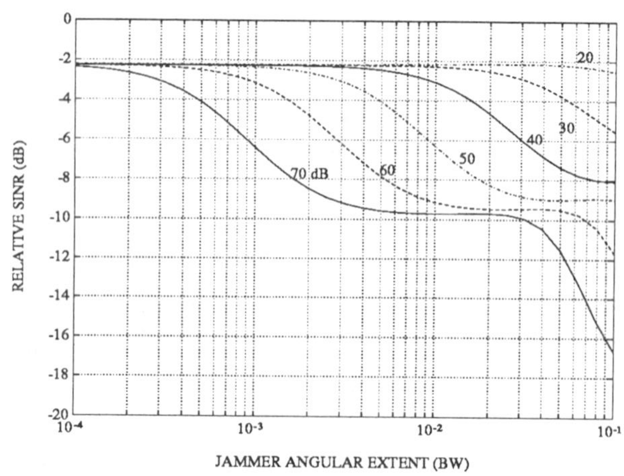

Fig. 3. Relative SINR versus jammer angular extent and power for jammer at .5 BW.

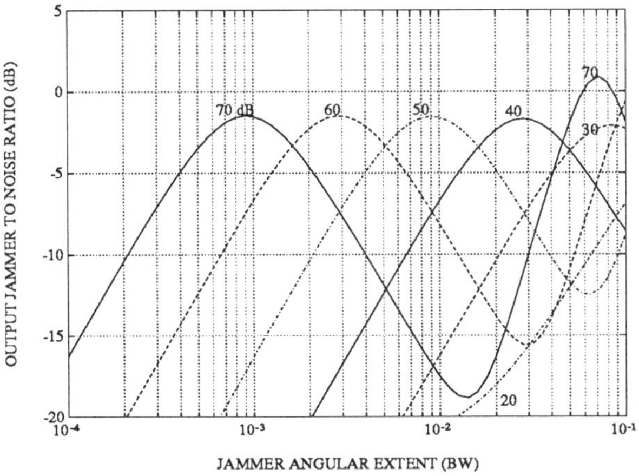

Fig. 4. Output jammer to noise ratio versus jammer angular extent and power for jammer at .5 BW.

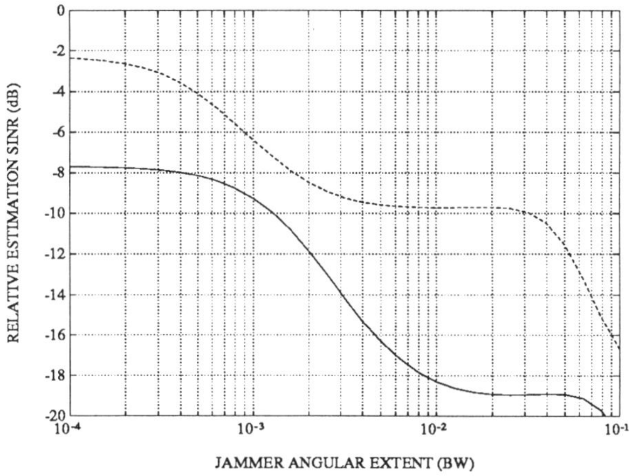

Fig. 7. Relative esimation SINR compared to relative detection SINR for 70 dB jammer at .5 BW.

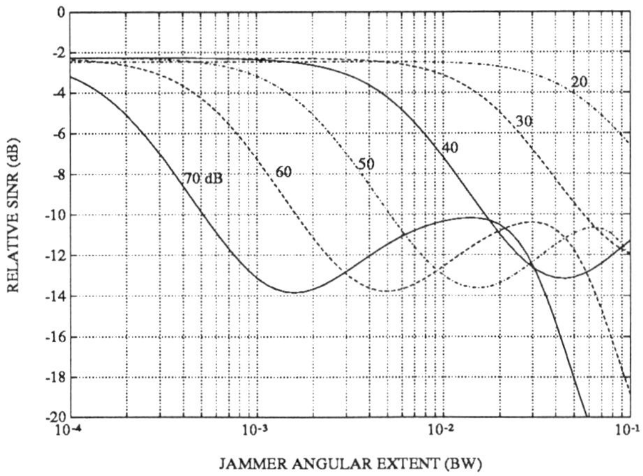

Fig. 5. Relative SINR when jammer has 10% duty factor or 10 dB of diagonal loading is used.

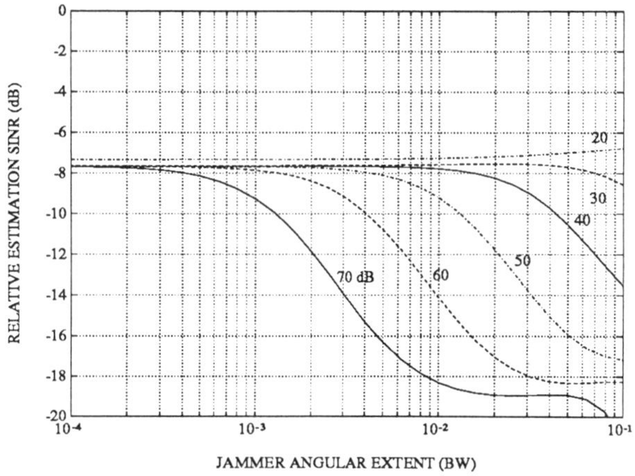

Fig. 8. Relative esimation SINR versus jammer angular extent and power for jammer at .5 BW.

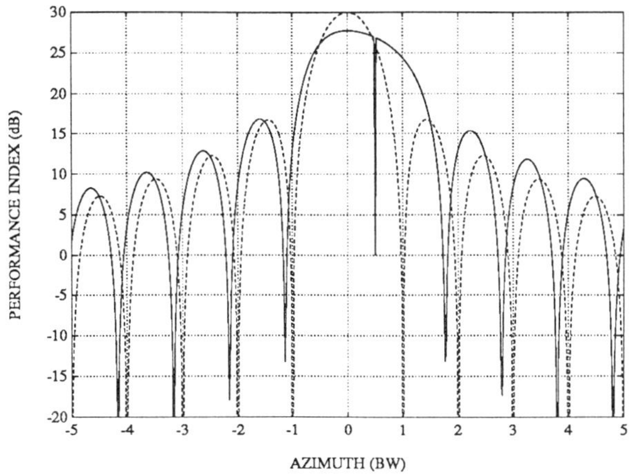

Fig. 6. Performance index for AOA estimatation for point jammer at .5 BW compared to quiescent case.

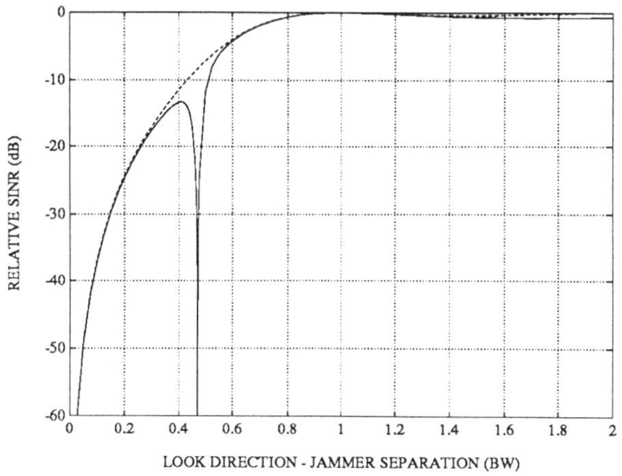

Fig. 9. Relative SINR midway between two point jammers for shaded and unshaded steering constraints.

The Copy/DF Approach to Signal-Specific Emitter Location

Brian G. Agee
Radix Technologies, Inc.
4020 Moorpark Ave., Suite 212
San Jose, California 95117

Abstract

A general approach for performing *signal-specific* superresolution direction finding (DF) of co-channel emitter waveforms is reported. This approach, referred to here as the *copy/DF approach*, uses the *signal copy vectors* provided by a blind or nonblind signal extraction algorithm to construct a spatial spectrum for each signal captured by a narrowband antenna array, where each spectrum has a global maximum in the direction of the signal captured by that copy vector. The approach can be used with any signal extraction algorithm, and is particularly effective in environments containing strong co-channel interference. Two classes of techniques are presented: *ARMA-like* techniques, which form spatial spectra with zeroes in the directions of the received interferers and poles in the directions of the copied signals; and *ML-like* techniques, which approximate the *nonblind maximum-likelihood* estimate of the direction-of-arrival of the copied emitter waveforms.

1 Introduction

Conventional techniques that employ antenna arrays to perform superresolution direction-finding (DF) rely solely on the property that the signals received by the array are *spatially coherent*, i.e., that the signal waveforms arrive at the receiver from discrete directions of arrival. In recent years, however, techniques have been proposed that use additional properties of the signal waveforms to enhance the performance of DF systems in some way or another. Notable among these are the *cyclic MUSIC algorithms* reported in [1,2], which exploit the *spectral self-coherence* property (correlation between frequency-shifted signal replicas at discrete values of frequency shift) possessed by cyclostationary waveforms to perform simultaneous classification and DF of PSK, FSK and PCM QAM signals; and the *CM/MOP* algorithm reported in [3,4], which exploits the *low modulus variation* property possessed by most communication signals to perform simultaneous copy (separation) and DF of the PSK/FSK, PCM QAM and FM signals. More recently, techniques have been proposed for exploiting cumulants of communication waveforms to improve the performance of conventional DF systems [5,6].

These techniques have demonstrated that modulation properties can be exploited to perform *signal-selective DF* of signals with known modulation formats, by using modulation-based classification/copy operations to thin or separate the received signals prior to the DF operation. In addition, in [2] it has been shown that modulation properties can be used to improve the accuracy of DF systems, particularly in dense environments where many signals with different modulation formats (e.g., PCM signals with different keying rates) are received by the array.

However, the signal-selective DF techniques reported to date have also demonstrated some limitations that reduce their applicability in a number of important problem areas. The selectivity of the cyclic MUSIC techniques break down over practical averaging times in the presence of very strong interference, for example, if the receiver is subject to jamming. In addition, the accuracy of these algorithms can be poor if the spectral self-coherence of the signals of interest is weak, for example, if the received signals are PCM with low excess bandwidth. The cumulant techniques suffer from similar problems, especially in environments containing strong co-channel interference, and are also much more computationally complex than the cyclic MUSIC and CM/MOP based DF algorithms. The DF algorithm presented in [3,4] is expected to suffer from similar problems in environments containing strong interferers, especially if the interferers cannot be captured by the copy algorithm, for example, if they are noise jammers or low duty-cycle pulsed radars. Lastly, all of the techniques developed to date have been developed to exploit specific signal properties, and are not easily extended to more general reception scenarios.

This paper presents a general approach for performing signal-selective DF, first reported in [7] and referred to here as the *copy/DF (CDF)* approach, which has the potential to overcome these problems. The approach can be used with any blind or nonblind signal extraction algorithm; in particular, the approach can be used with a SCORE algorithm [8] if the signals of interest possess spectral and/or conjugate self-coherence at known values of frequency separation; with a multitarget CMA [3,9] if the signals of interest have a sufficiently low modulus variation; and with a dominant mode prediction (DMP) or Maximin algorithm [10,11] if the signals of interest are burst or transient waveforms. The approach is also *signal specific*, in the sense that it provides a separate DOA spectrum for each copy vector produced by the signal extraction algorithm, with a global maximum in the direction of the signal captured by that copy vector, regardless of the number of signals captured by the array. Lastly, the approach is particularly effective in environments containing strong co-channel interference.

2 Copy/DF Approach

The basic copy/DF processor is shown in Figure 1. The general copy/DF processor consists of an adaptive copy "front-end" that provides a set of signal copy weights using a blind copy algorithm such as an MT-CMA, DMP, or SCORE algorithm, followed by a signal-specific superresolution DF "back-end" that uses these weights to aid subsequent direction-finding (DF) operations on each signal of interest (SOI) captured by the processor.

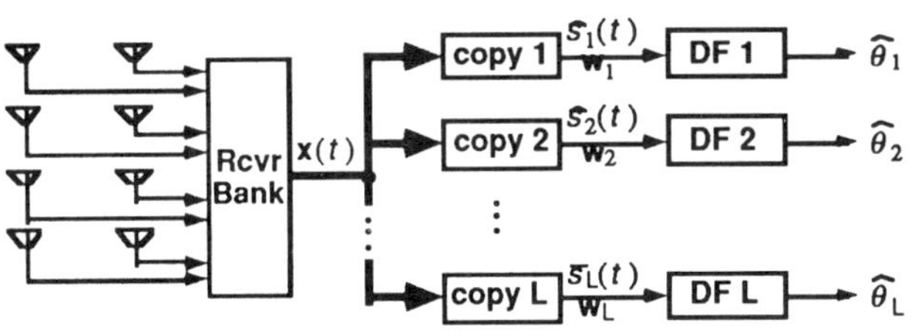

Figure 1: Basic Copy/DF Processor

The algorithms used in the adaptive copy front-end do not rely on the emitter LOBs or the receiver calibration data (array manifold and noise covariance matrix) to construct the signal copy vectors, but instead exploit known properties of the SOI modulation formats, for example, the property that the SOIs have abrupt on/off transitions or low modulus variation (variation of complex envelope magnitude) at the transmission site. In this respect, the copy/DF approach differs fundamentally from conventional "DF/copy" (DF-aided copy) techniques such as MUSIC, which first perform DF using only the spatial self-coherence of the received emitter waveforms, and then compute the subsequent signal copy vectors from these LOB estimates using Weiner filtering techniques.

Two general classes of algorithms are used to operate the superresolution DF back-end:

- *ARMA-like copy/DF algorithms*, which use all of the solutions provided by a multitarget copy algorithm (for example, the signal capture *and* signal rejection solutions of the MT-CMA) to estimate the DOA of each signal captured by the signal copy front-end; and

- *ML-like copy/DF algorithms*, which use only the signal-capture solutions provided by a copy algorithm to estimate the DOA of each signal captured by the signal copy front-end.

Both of these algorithms are motivated by assuming that the received data waveform $\mathbf{x}(m)$ is modelled by

$$\mathbf{x}(m) = \sum_{k=1}^{K} \mathbf{a}(\theta_k) s_k(m) + \mathbf{i}(m) \quad (1)$$

where $\{s_k(m)\}_{k=1}^K$ are the emitter waveforms received by the copy/DF system, $\mathbf{i}(m)$ is the remaining noise and interference received by the array, and $\{\mathbf{a}(\theta)\}$ is the DOA manifold of the array. The copy/DF algorithms then provide a separate DOA spectrum for each copy vector produced by the signal copy front-end, with a global maximum in the direction of the signal captured by that copy vector, regardless of the number of signals captured by the array. These algorithms are described below.

2.1 ARMA-Like Copy/DF Algorithms

The ARMA-like algorithms assume that the blind copy algorithm provides M separate solutions $y_k(m) = \mathbf{w}_k^H \mathbf{x}(m)$, where $\{\mathbf{w}_k\}_{k=1}^M$ are a set of beamformer weight vectors, and where the first K weight vectors $\{\mathbf{w}_k\}_{k=1}^K$ capture each of the emitter waveforms $\{s_k(m)\}_{k=1}^K$ with near-maximum output SINR and the remaining $M-K$ weight vectors $\{\mathbf{w}_k\}_{k=K+1}^M$ reject the K emitter waveforms. The ARMA-like estimate of θ_k is then given by

$$\hat{\theta}_k = \arg\max_{\theta} \mathrm{S}_{k\ell}(\theta)$$

where $\{\mathrm{S}_{k\ell}(\theta)\}_{k=1}^K$ are either of the sets of *ARMA-like DOA spectra*,

$$\mathrm{S}_{k1}(\theta) \triangleq \frac{\left|\hat{\mathbf{w}}_k^H \mathbf{a}(\theta)\right|^2}{\sum_{\ell=K+1}^M \left|\hat{\mathbf{w}}_\ell^H \mathbf{a}(\theta)\right|^2} \quad (2)$$

$$= \frac{\left|\hat{\mathbf{w}}_k^H \mathbf{a}(\theta)\right|^2}{\mathbf{a}^H(\theta) \hat{\mathbf{W}}_\perp \hat{\mathbf{W}}_\perp^H \mathbf{a}(\theta)} \quad (3)$$

$$\hat{\mathbf{W}}_\perp \triangleq [\hat{\mathbf{w}}_{K+1} \cdots \hat{\mathbf{w}}_M] \quad (4)$$

$$\mathrm{S}_{k2}(\theta) \triangleq \frac{\left|\hat{\mathbf{w}}_k^H \mathbf{a}(\theta)\right|^2}{\sum_{\ell \neq k} \left|\hat{\mathbf{w}}_\ell^H \mathbf{a}(\theta)\right|^2} \quad (5)$$

$$= \frac{\tilde{\mathrm{S}}_{k2}(\theta)}{1 - \tilde{\mathrm{S}}_{k2}(\theta)} \quad (6)$$

$$\tilde{\mathrm{S}}_{k2}(\theta) \triangleq \frac{\left|\hat{\mathbf{w}}_k^H \mathbf{a}(\theta)\right|^2}{\mathbf{a}^H(\theta) \hat{\mathbf{W}} \hat{\mathbf{W}}^H \mathbf{a}(\theta)} \quad (7)$$

$$\hat{\mathbf{W}} \triangleq [\hat{\mathbf{w}}_1 \cdots \hat{\mathbf{w}}_M] \quad (8)$$

for each beamformer $\mathbf{w}_k$, and where $\mathbf{a}(\theta)$ is modelled by (1) and the beamformers $\{\hat{\mathbf{w}}_k\}_{k=1}^M$ are normalized to unit output power,

$$\hat{\mathbf{w}}_k = \frac{\mathbf{w}_k}{\sqrt{\mathbf{w}_k^H \hat{\mathbf{R}}_{\mathbf{xx}} \mathbf{w}_k}}. \quad (9)$$

and $\hat{\mathbf{R}}_{\mathbf{xx}} = \left\langle \mathbf{x}(m)\mathbf{x}^H(m) \right\rangle_N$ is the received data autocorrelation matrix computed over the reception interval.

These DOA spectra are referred to as "ARMA-like" because the numerator in (2) and (5) introduces approximate zeroes[1] that offset the zeroes in the denominator of these indicator functions. This can be seen most clearly in (2)-(3). In this case, (3) can be interpreted as the *product* between a conventional MUSIC indicator function and the gain pattern of the antenna array with weight vector $\mathbf{w}_k$. The MUSIC DOA spectrum has peaks in the directions of all of the emitters impinging on the array; however, the gain pattern has nulls in the direction of all of the emitters *except* $s_k(m)$. The product of these functions should then have an enhanced peak in the direction of $s_k(m)$. Consequently, the DOA of $s_k(m)$ should be close to the global maximum of $\hat{S}_{k1}(\theta)$ ($\hat{\theta}_k \approx \theta_k$). A similar result can be obtained for $\hat{S}_{k2}(\theta)$ [7]. For this reason, these DOA spectra are also referred to here as *signal specific*, since each spectrum $S_{k\ell}(\theta)$ has a global maximum in the direction of the specific signal captured by $\mathbf{w}_k$.

2.2 ML-Like Copy/DF Algorithm

The ML-like Copy/DF algorithms are motivated by deriving the *maximum-likelihood* estimate of the DOA of a single *known* signal received with unknown complex gain g and DOA θ_s by an M-element antenna array with array manifold $\{\mathbf{a}(\theta)\}$, in the presence of *temporally*-white complex Gaussian interference with mean zero and unknown spatial covariance $\mathbf{C}$. The received data signal $\mathbf{x}(n)$ is then modelled by

$$\begin{aligned} \mathbf{x}(m) &= g\mathbf{a}(\theta_s)s(m) + \mathbf{i}(m) && (10)\\ \mathbf{i}(m) &\sim \text{i.i.d. } CG(\mathbf{0}, \mathbf{R_{ii}}), && (11) \end{aligned}$$

where $g \in \mathcal{C}$, $\mathbf{a}(\theta_s) \in \{\mathbf{a}(\theta)\}$, and $\mathbf{R_{ii}} \in \mathcal{C}^{M\times M}$. The ML estimate of θ is then computed for an N-sample data collect $\{\mathbf{x}(m), s(m)\}_{n=1}^{N}$ by jointly maximizing the *ML objective function*

$$\begin{aligned} &\mathrm{S_{ML}}\left(\theta, g, \mathbf{R_{ii}} \mid \{\mathbf{x}(m), s(m)\}_{m=1}^{N}\right)\\ &\triangleq \frac{1}{N}\ln\left[\pi^{MN} f_{\{\mathbf{x}(m)\}_{n=1}^{N}}(\mathbf{x}(1), \ldots, \mathbf{x}(N))\right]\\ &= -\mathrm{Tr}\left[\mathbf{R_{ii}}^{-1}\hat{\mathbf{R}}_{\mathbf{ii}}\right] + \ln\left|\mathbf{R_{ii}}\right| \qquad (12) \end{aligned}$$

with respect to θ, g and $\mathbf{R_{ii}}$, where

$$\begin{aligned} \hat{\mathbf{i}}(m) &\triangleq \mathbf{x}(m) - g\mathbf{a}(\theta)s(m)\\ \Rightarrow\ ;\hat{\mathbf{R}}_{\mathbf{ii}} &= \hat{\mathbf{R}}_{\mathbf{ii}} + (g\mathbf{a}(\theta) - \hat{\mathbf{a}}_s)(g\mathbf{a}(\theta) - \hat{\mathbf{a}}_s)^H \hat{R}_{ss}\\ \hat{\mathbf{a}}_s &\triangleq \frac{\hat{\mathbf{R}}_{\mathbf{x}s}}{\hat{R}_{ss}}\\ \hat{\mathbf{R}}_{\mathbf{ii}} &\triangleq \hat{\mathbf{R}}_{\mathbf{xx}} - \hat{\mathbf{a}}_s\hat{\mathbf{a}}_s^H \hat{R}_{ss} \end{aligned}$$

and where $\hat{\mathbf{R}}_{\mathbf{xx}}$, $\hat{\mathbf{R}}_{\mathbf{x}s}$ and $\hat{R}_{ss}$ are the time-average data autocorrelation matrix, data/signal cross-correlation vector, and signal power, respectively, evaluated over $[1, N]$.

[1]The signal copy gain patterns do not necessarily possess true zeroes in general array geometries.

Maximizing (12) with respect to $\mathbf{R_{ii}}$ and g and removing constant terms yields

$$\begin{aligned} \hat{\mathbf{R}}_{\mathbf{ii}}\Big|_{\mathrm{ML}} &= \hat{\mathbf{R}}_{\mathbf{ii}} && (13)\\ \hat{g}_{\mathrm{ML}} &= \frac{\mathbf{a}^H(\theta)\hat{\mathbf{R}}_{\mathbf{xx}}^{-1}\hat{\mathbf{a}}_s}{\hat{\mathbf{a}}_s^H\hat{\mathbf{R}}_{\mathbf{xx}}^{-1}\hat{\mathbf{a}}_s} && (14)\\ \Rightarrow\ \mathrm{S_{ML}} &\to \frac{\left|\hat{\mathbf{a}}_s^H\hat{\mathbf{R}}_{\mathbf{xx}}^{-1}\mathbf{a}(\theta)\right|^2}{\left[\hat{\mathbf{a}}_s^H\hat{\mathbf{R}}_{\mathbf{xx}}^{-1}\hat{\mathbf{a}}_s\right]\left[\mathbf{a}^H(\theta)\hat{\mathbf{R}}_{\mathbf{xx}}^{-1}\mathbf{a}(\theta)\right]}\\ &\quad\times \frac{\hat{\mathbf{a}}_s^H\hat{\mathbf{R}}_{\mathbf{xx}}^{-1}\hat{\mathbf{a}}_s}{\hat{R}_{ss} - \hat{\mathbf{a}}_s^H\hat{\mathbf{R}}_{\mathbf{xx}}^{-1}\hat{\mathbf{a}}_s} && (15)\\ &\to \frac{\left|\hat{\mathbf{w}}_s^H\mathbf{a}(\theta)\right|^2}{\mathbf{a}^H(\theta)\hat{\mathbf{R}}_{\mathbf{xx}}^{-1}\mathbf{a}(\theta)}, && (16) \end{aligned}$$

where $\hat{\mathbf{w}}_s \propto \hat{\mathbf{R}}_{\mathbf{xx}}^{-1}\hat{\mathbf{R}}_{\mathbf{x}s}$ is the *least-squares signal copy vector* that minimizes the time-average mean-square error between $s(n)$ and $\hat{\mathbf{w}}_s^H\mathbf{x}(m)$ subject to a unit-power constraint given by (9).

Motivated by this result, an *ML-like* copy/DF algorithm can be developed for the case where a blind copy algorithm provides K separate solutions $y_k(m) = \mathbf{w}_k^H\mathbf{x}(m)$, where $\{\mathbf{w}_k\}_{k=1}^{K}$ are a set of beamformer weight vectors, each of which captures a separate emitter waveform $s_k(m)$ with near-maximum output SINR. The ML-like estimate of θ_k is then given by

$$\hat{\theta}_k = \arg\max_{\theta} \mathrm{S}_k(\theta)$$

where $\{\mathrm{S}_k(\theta)\}_{k=1}^{K}$ is the set of *ML-like indicator functions*,

$$\mathrm{S}_k(\theta) \triangleq \frac{\left|\hat{\mathbf{w}}_k^H\mathbf{a}(\theta)\right|^2}{\mathbf{a}^H(\theta)\hat{\mathbf{R}}_{\mathbf{xx}}^{-1}\mathbf{a}(\theta)} \qquad (17)$$

for each beamformer $\mathbf{w}_k$, and where $\mathbf{a}(\theta)$ is modelled by (1) and the beamformers $\{\hat{\mathbf{w}}_k\}_{k=1}^{K}$ are normalized to unit output power.

These DOA spectra are referred to as "ML-like" because they are derived from the maximum-likelihood viewpoint described above. However, this algorithm does not provide true ML estimates in practice, because the copy vectors $\{\mathbf{w}_k\}_{k=1}^{K}$ are computed using a blind signal extraction algorithm (i.e., the emitter content is unknown) and the environment actually contains multiple signals with varying statistical distributions (i.e., the background interference is generally non-Gaussian). The ML-like spectrum are more precisely interpreted as *aperture-matching spectra*, by representing (17) as

$$\mathrm{S}_k(\theta) = \frac{\left|\langle\hat{\mathbf{a}}_k, \mathbf{a}(\theta)\rangle\right|^2}{\|\hat{\mathbf{a}}_k\|^2\,\|\mathbf{a}(\theta)\|^2} \qquad (18)$$

where $\hat{\mathbf{a}}_k$ is the unconstrained estimate of $\mathbf{a}(\theta)$, given by

$$\hat{\mathbf{a}}_k \triangleq \hat{\mathbf{R}}_{\mathbf{xx}}\mathbf{w}_k \qquad (19)$$

and where the norm and inner product in (18) are generalized Euclidean norms and inner products with weighting matrix $\hat{\mathbf{R}}_{\mathbf{XX}}^{-1}$, given by

$$\begin{aligned}(\mathbf{u},\mathbf{v}) &\triangleq \mathbf{u}^H \hat{\mathbf{R}}_{\mathbf{XX}}^{-1} \mathbf{v} \\ \|\mathbf{u}\|^2 &\triangleq (\mathbf{u},\mathbf{u}).\end{aligned}$$

If $\hat{\mathbf{a}}_k \propto \mathbf{a}(\theta_k)$, then (18) should have a broad global maximum in the vicinity of θ_k.

The ML-like copy/DF spectra share many of the properties of the ARMA-like spectra, especially the fully-loaded ARMA-like spectrum given in (5). The ML-like spectra are signal-specific, in the sense that they are uniquely defined for each signal copy vector, and have a global maximum in the direction of the signal captured by that copy vector. The global maxima should also be broad, facilitating the integration of these techniques into sequential search strategies. The ML-like spectra also do not require knowledge of the number of received emitters K to operate, and can resolve up to M co-channel emitters. In addition, however, the technique does not require identification of the null-space weight vectors $\{\mathbf{w}_k\}_{k=K+1}^{M}$, giving it some important implementation advantages over the ARMA-like approaches.

3 Technique Demonstration

The ARMA-like and ML-like copy/DF algorithms are demonstrated via computer simulation for a four-element circular antenna array with a half-wavelength diameter, excited by complex-Gaussian background noise, BPSK and QPSK emitters, and one-to-two very strong complex-Gaussian interferers. The PSK emitters are received by the array at 20 dB signal-to-white noise ratios (SWNRs) and at 10° and 30° DOAs, respectively, while the complex-Gaussian interferers are received at 60 dB SWNRs and 60° and (optionally) 90° DOAs. The PSK emitters are captured over a 128 sample (64 baud) reception interval, using a non-blind least-squares algorithm, and using a blind dominant mode prediction algorithm [10], a conjugate and nonconjugate cross-SCORE algorithm [8], and a multitarget CMA [9]. The DF performance of the resultant copy/DF algorithms are then compared to the equivalent performance achieved by the deterministic maximum-likelihood algorithm and the conventional and cyclic MUSIC DF algorithms. The three-signal environment is referred to here as the *underloaded array* environment, because the number of reception paths is less than the number of antenna elements ($K < M$); the four-signal environment is referred to here as the *fully-loaded array* environment, because the number of reception paths is *equal* to the number of antenna sensors ($K = M$). Note that the deterministic ML and conventional MUSIC DF techniques can only function in the underloaded environment[2].

The rms and mean DF error obtained using the ARMA-like and ML-like copy/DF algorithms are listed in Tables 1 and 2, for the BPSK emitter and the single-interferer and two-interferer environments, respectively. The DF errors are computed over 100 independent collects of the environments, and (when applicable) compared to the DF performance obtained by the conventional and cyclic algorithms in the same environment. The rms DF error is expressed in these Tables in terms of the rms DF error relative to the Cramer-Rao lower bound obtained under the single-signal assumption given in (10) (i.e., assuming in either case that the *other* signal is part of the complex-Gaussian interference field),

$$\begin{aligned}\sigma_{\mathrm{CRLB}}^{-2} &= 2N\hat{R}_{ss}\,\|\dot{\mathbf{a}}(\theta_s)\|^2 \\ &\quad\times\left(1-\frac{|(\dot{\mathbf{a}}^H(\theta_s),\mathbf{a}(\theta_s))|^2}{\|\dot{\mathbf{a}}(\theta_s)\|^2\,\|\mathbf{a}(\theta_s)\|^2}\right) \qquad (20)\\ &\rightarrow \begin{cases}0.68^\circ, & \text{underloaded array}\\ 0.74^\circ, & \text{fully-loaded array}\end{cases}\end{aligned}$$

where $\hat{R}_{ss}$ is the emitter power computed over the N-sample collect interval and the norm and inner product given in (20) is given by

$$(\mathbf{u},\mathbf{v}) \triangleq \mathbf{u}^H \mathbf{R}_{\mathbf{XX}}^{-1} \mathbf{v}$$

In both Tables, and "ARMA 1" and "ARMA 2" refer to the ARMA-like DOA spectra given in (3) and (7), respectively, and "C-SCORE" and "P-SCORE" refer to the conjugate cross-SCORE and phase SCORE signal copy algorithms, respectively.

Table 1: SOI 1 Copy/DF Performance, Underloaded Environment

Copy Alg.	DF RMSE Over CRB	DF Bias Rel. to RMSE
Det. ML	90%	13%
MUSIC	88%	5.0%
Conj. Cyclic MUSIC	fails	fails
Nonconj. Cyc. MUSIC	1191%	70%
LS ML-like	0.0%	15%
LS ARMA-like 1	88%	13%
LS ARMA-like 2	80%	88%
DMP ML-like	4.1%	16%
DMP ARMA-like 2	4.1%	16%
C-SCORE ML-like	2.3%	10%
C-SCORE ARMA-like 2	2.3%	0.13
P-SCORE ML-like	104%	55%
P-SCORE ARMA-like 1	496%	0.88
P-SCORE ARMA-like 2	100%	49%
MT-CMA ML-like	5.3%	17%
MT-CMA ARMA-like 1	398%	15%
MT-CMA ARMA-like 2	97%	62%

As these Tables show, the copy/DF algorithms can significantly outperform the conventional DF algorithms in the strong interference environments. The performance of the ML-like copy/DF algorithms is particularly noteworthy here. The ML-like algorithm

[2] In fact, environments where $K = M - 1$ are often referred to as "fully-loaded," since the conventional DF techniques fail for $K \geq M$.

Table 2: Copy/DF Performance, Fully-Loaded Array

Copy Alg.	DF RMSE Over CRB	DF Bias Over CRB
Det. ML	fails	fails
MUSIC	fails	fails
Conj. Cyclic MUSIC	839%	84%
Nonconj. Cyc. MUSIC	fails	fails
LS ML-like	5.3%	2.6%
LS ARMA-like 2	111%	97%
DMP ML-like	12.5%	7.9%
DMP ARMA-like 2	12.5%	7.9%
C-SCORE ML-like	9.8%	3.2%
C-SCORE ARMA-like 2	9.8%	3.2%
P-SCORE ML-like	37.4%	76%
P-SCORE ARMA-like 2	55.1%	57%
MT-CMA ML-like	9.2%	2.0%
MT-CMA ARMA-like 2	123%	61%

came to within 5% of the Cramer-Rao error bound in the underloaded environment, and within 10% of the Cramer Rao bound in the fully-loaded environment, for all of the copy algorithms except the phase SCORE algorithm (which was exploiting a relatively weak signal feature in these examples). In contrast, the conventional ML and MUSIC DF algorithms are only within about 90% of the Cramer Rao bound in the underloaded environment, and completely fail in the fully loaded environment. The copy/DF algorithms performed particularly well in comparison to cyclic MUSIC, which either failed outright or provided a very large DF error in all of the experiments.

The relative performance of the ML-like and ARMA-like copy/DF algorithms is also notable in these examples: the ML-like algorithm significantly outperformed the ARMA-like algorithms in many of these tests, indicating that the ML-like algorithm provides performance as well as implementation advantages over the ARMA-like technique. A possible reason for this result may lie in the quality of the null-space vectors provided by the various copy algorithms. This conjecture is supported by the performance results obtained by the DMP and conjugate cross-SCORE based ARMA-like copy/DF algorithms, which performed identically to the ML-like algorithms.

The signal specific performance of the copy/DF algorithms are illustrated in Figures 2 and 3, showing the fully-loaded ARMA-like DOA spectrum obtaining over a single snapshot in the environment described above when the copy weights are computed using an MT-CMA, and in Figures 4 and 5, showing the equivalent DOA spectra for the ML-like copy/DF algorithm. For comparison with the ARMA-like spectra, the ML-like spectra have been normalized to 0 dB maximum value, and the ML-like DOA spectra have been converted to MUSIC-like spectra via the monotonic transformation $\hat{S}_k(\theta) = \tilde{S}_k(\theta)/[1 - \tilde{S}_k(\theta)]$. In all cases, the copy/DF spectra are clearly maximized in the direction of the emitters captured by each copy vector. This contrasts with the conventional MUSIC DOA spectra, also shown in these Figures, which are dominated by the strong interferers at 90° (in the fully-loaded environment) and 60° (in all of the environments). Moreover, the MUSIC spectrum completely fails to resolve the weak BPSK emitters in the fully-loaded environment (Figures 3 and 5), instead placing a peak at the midpoint between the two emitter DOAs.

4 Conclusions

A new approach to emitter localization, referred to here as the *copy/DF approach*, has been presented. Two classes of copy/DF algorithms have been described: the *ARMA-like* copy/DF algorithms, which form spatial spectra with zeroes in the directions of the received interferers and poles in the directions of the copied signals; and *ML-like* copy/DF algorithms, which approximate the *nonblind maximum-likelihood* estimate of the direction-of-arrival of the copied emitter waveforms. The algorithms have demonstrated both implementation and performance advantages over conventional DF algorithms that exploit only the spatial self-coherence of the received emitter waveforms, as well as the cyclic MUSIC algorithms that additionally exploit spectral and/or conjugate self-coherence features of the received emitter waveforms. The copy/DF algorithms can be used with any blind or nonblind signal copy front-end, do not require knowledge of the background noise covariance matrix or the number of emitters received by the array to operate, and can perform accurate DF in "fully-loaded" environments where the number of received emitters is equal to the number of sensors in the antenna array. The techniques are also *signal specific*, in the sense that they provide a separate DOA spectrum for each emitter captured by the copy front-end, where each spectrum has a global maximum in the direction of the signal captured by the copy front-end. As a consequence, the techniques can be implemented using simple single-dimensional global search techniques, rather than multidimensional or multimodal search algorithms. In addition, the techniques can significantly outperform the conventional and cyclic MUSIC techniques, especially in environments containing strong interference signals. These results indicate that the copy/DF approach can be a powerful technique for emitter localization in environments where the emitters of interest have known modulation properties.

References

[1] W. Gardner, "Simplification of MUSIC and ESPRIT by Exploitation of Cyclostationarity," *Proc. IEEE*, vol. 76, no. 7, pp. 845-847, July 1988

[2] S. Schell, R. Calabretta, W. Gardner, B. Agee, "Cyclic MUSIC Algorithms for Signal-Selective Direction Estimation," in *Proc. 1989 Intl. Conf. on ASSP*, pp. 2278-2281, 1989

[3] B. Sublett, R. Gooch, S. Goldberg, "Separation and Bearing Estimation of Co-Channel Signals," in *Proc. 1989 IEEE Military Communications Conference*, pp. 37.1.1-37.1.6, 1989

[4] R. Gooch "Adaptive Beamformers in Communications and Direction Finding Systems," in *Proc.*

Twenty-Fourth Asilomar Conference on Signals, Systems, and Computers, pp. 11-15, 1990

[5] P. Forster, C. Nikias, "Bearing Estimation in the Bispectrum Domain," *IEEE Trans. Signal Processing*, vol. 39, no. 9, pp. 1994-2006, September 1991

[6] B. Porat, B. Friedlander, "Direction Finding Algorithms Based on High-Order Statistics," *IEEE Trans. Signal Processing*, vol. 39, no. 9, pp. 2016-2024, September 1991

[7] B. Agee, R. Calabretta "ARMA-Like Spatial Spectra for Signal-Selective Direction Finding," in *Proc. Fifth ASSP Workshop on Spectrum Estimation and Modelling*, pp. 134-138, 1990

[8] B. Agee, S. Schell, W. Gardner, "Spectral Self-Coherence Restoral: A New Approach to Blind Adaptive Signal Extraction Using Antenna Arrays," *Proc. IEEE*, vol. 78, no. 4, pp. 753-767, April 1990

[9] B. Agee, "Blind Separation and Capture of Communication Signals Using a Multitarget Constant Modulus Beamformer," in *Proc. 1989 IEEE Military Communications Conference*, pp. 19.2.1-19.2.6, 1989

[10] B. Agee, "Fast Acquisition of Burst and Transient Signals Using a Predictive Adaptive Beamformer," in *Proc. 1989 IEEE Military Communications Conference*, pp. 19.3.1-19.3.5, 1989

[11] K. Bakhru, D. Torrieri, "The Maximin Algorithm for Adaptive Arrays and Frequency-Hopping Communications," *IEEE Trans. Ant. and Prop.*, vol. AP-32, no. 9, pp. 919-928, Sept. 1984

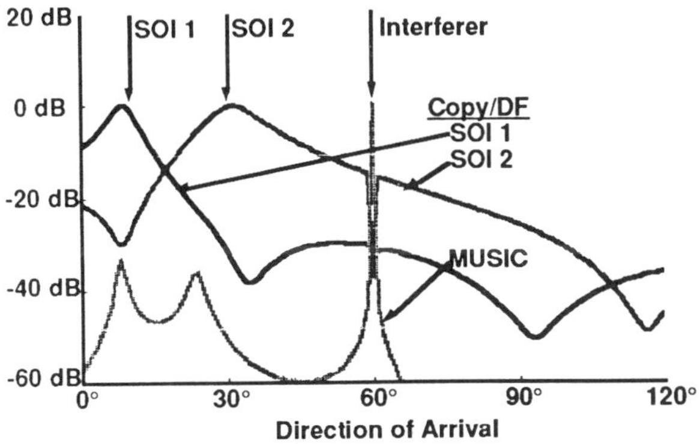

Figure 2: ARMA-Like Spectra, Underloaded Array

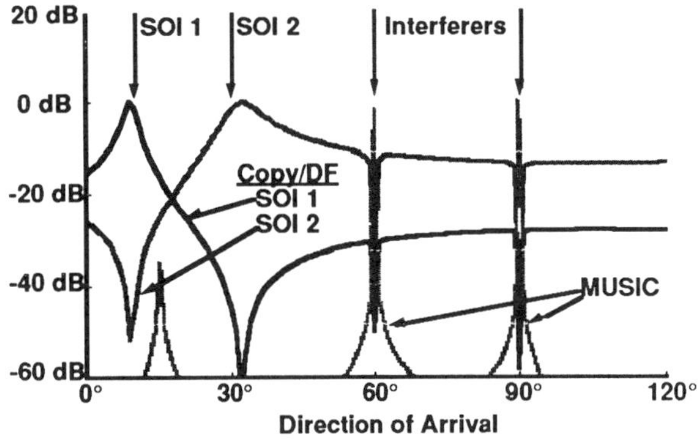

Figure 3: ARMA-Like Spectra, Fully-Loaded Array

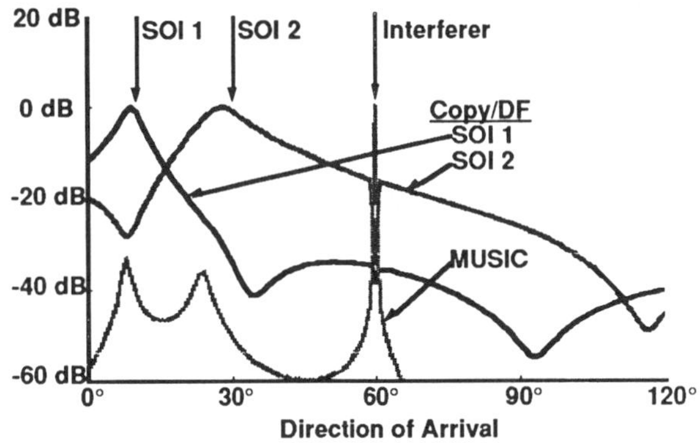

Figure 4: ML-Like Spectra, Underloaded Array

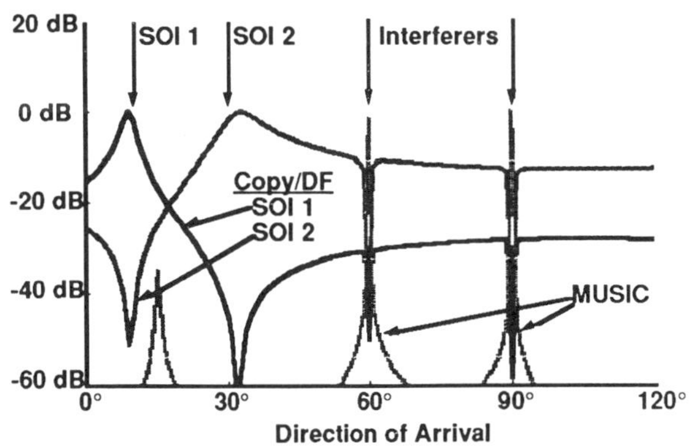

Figure 5: ML-Like Spectra, Fully-Loaded Array

An Iterative Subspace Estimation Method for Adaptive Array Processing

Dean O. Carhoun

The MITRE Corporation*
Bedford, MA 01730

Abstract

We present a new, low-complexity, iterative method of vector subspace estimation applicable to adaptive beamforming and spatial spectral estimation. The new method combines two complementary procedures: the power/deflation method of principal eigenvector estimation, and the Lanczos method of tridiagonalization for computing the orthonormal basis of a Krylov subspace. Side information concerning the spatial coherence of the deflated array data is developed to aid discrimination between the (coherent) signal and (noncoherent) noise subspaces. The effectiveness of the new method is demonstrated by processing a typical example of underwater acoustic receiver data based on Owsley's Standard Test Case (STC) [1]. The algorithm's performance is compared with a baseline method that uses full singular value decomposition of the data.

Introduction and Summary

Adaptive nulling and high-resolution direction-finding are important and related aspects of sensor array processing. Both can be effectively based on eigensystem decompositions of array data matrices or estimated covariance matrices. The eigensystem decomposition is designed to estimate a vector subspace containing the steering vectors, or directions of arrival, of spatially coherent signals while distinguishing the signal subspace from an orthogonal noise subspace attributed to a spatially noncoherent background. In the adaptive nulling application, the spatially filtered output of a coherently formed beam can be obtained by projecting the output of the beam onto the orthogonal complement of the vector subspace containing the spatially coherent interference that is to be rejected [2]. While this can be accomplished in an adaptive sidelobe canceller employing a physical subset of auxiliary elements and using full-rank least squares estimation to determine the optimum auxiliary weight vector, problems of excess degrees of freedom that may distort the desired antenna pattern and increase unnecessarily the background level can be avoided by using reduced-rank principal components estimation based on the Eckart-Young theorem [2,3]. When applied to a fully adaptive array, the procedure is equivalent to adaptive sidelobe cancellation in which the auxiliary sensors are chosen in the eigenspace rather than the element space [4]. For direction-finding purposes, high-resolution algorithms like MUSIC are naturally based on vector subspace decomposition.

Eigensystem estimation, when based on a singular value decomposition of the data, or equivalent processing of a sample covariance matrix, is thought to be computationally costly. This inhibits the use of eigensystem techniques in real-time applications when simpler least-squares techniques are available, such as QR decomposition in the adaptive nulling application. We have shown recently in [4], however, that an iterative procedure based on the power/deflation method [5,6] can be both effective in performance and computationally simple. In a similar vein, Xu, Roy and Kailath [7,8] have advocated the use of an iterative eigensystem estimation procedure based on the computationally simple Lanczos method for obtaining the orthonormal basis of a Krylov subspace [9]. In this paper we present a hybrid algorithm for signal subspace estimation that combines features of both the power/deflation and Lanczos methods.

The Lanczos tridiagonalization method is based on orthogonalizing triplets of adjacent vectors at each step and it is even simpler to implement than modified Gram-Schmidt orthogonalization. By operating either on the array data or the sample covariance matrix, the process results in the formation of a low-order tridiagonal matrix and a corresponding matrix of orthonormal basis vectors defining the transformation to tridiagonal form. The eigenvalues of the tridiagonal matrix are used to approximate the dominant, or signal, eigenvalues of the sample covariance matrix, and the estimates of the signal

* *Work funded by the MITRE Sponsored Research program*

1058-6393/91 $1.00 © 1991 IEEE

eigenvectors are obtained as the product of the Lanczos orthonormal basis with the eigenvectors of the tridiagonal matrix. Xu et al have shown the asymptotic equivalence of these Rayleigh-Ritz values and vectors to the eigenvalues and eigenvectors of the sample covariance matrix.

For algorithm completion, the eigenvalues and eigenvectors of the low order tridiagonal matrix need to be computed. We employ the power/deflation method for this purpose, finding that a small number of iterations may suffice to estimate the signal subspace adequately for the purpose of adaptive beamforming. The Lanczos process is effective for tridiagonalization up to low order, but problems of numerical stability are quickly encountered. These are manifested by growing nonorthogonality of the Lanczos basis vectors and unstable behavior of the tridiagonal entries. Our remedy is to halt the tridiagonalization before these problems arise, compute a small number of signal eigenvectors, use these to deflate the data (or sample covariance) matrix by orthogonal projection, and then repeat the process with the deflated data to estimate the next set of signal eigenvectors. After each cycle of data deflation we derive side information concerning the spatial coherence of the deflated array data. This information could be employed to aid in distinguishing between signal and noise subspaces.

It should be noted that our method is not intended to estimate individual eigenvectors, but rather the subspace spanned by those eigenvectors. For a number of practical examples we have found that a relatively small number of power/deflation iterations leads to an estimated signal subspace that lies quite close to the subspace spanned by the signal eigenvectors, as measured by computing the canonical angles between subspaces [10]. In the following paragraphs we will briefly describe the computational details, and illustrate the performance of the method using Owsley's recommended Standard Test Case benchmark [1]. We will briefly summarize the computational complexity.

Power/Deflation

The power/deflation method for iteratively extracting dominant eigensystem components is a well known technique that repetitively applies the characteristic equation of a system to gradually improving approximations of an eigenvector being estimated. With a random selection of the initial estimate, the process may require a significantly large number of iterations for adequate approximation. The distinguishing feature of the work reported in [4] is the initialization, after each deflation step, with a steering vector derived from a measurement of the deflated sensor array mutual coherence. In practical applications this was seen to provide adequate approximation of a signal subspace after only a few iterations. The use of the complex mutual coherence also provided side information about the spatial coherence of the remaining signal components that could be combined with information-theoretic measures, which are based on the relative size of eigenvalues, to enhance the discrimination between signal and noise subspaces. While useful by itself to estimate signal subspace components, the power/deflation method can also be combined with Fast Subspace Decomposition (FSD) [7,8], essentially a Lanczos method, to avoid problems of gradual nonorthogonality of the computed Lanczos basis vectors while simultaneously being used to complete the algorithm by extracting the eigensystem components of a tridiagonalized matrix.

Fast Subspace Decomposition

The FSD method is described in detail in references [7] and [8]. It requires an initialization vector that is not orthogonal to any of the eigenvectors of the sample covariance matrix of the array data. In the method proposed by Xu et al, the initial vector is chosen at random, satisfying the required condition in a probabilistic sense. In our modified data domain implementation of this algorithm, we choose the initial vector as a steering vector derived from a simple measurement of the complex mutual coherence of the array. Such a vector can be expressed as a linear combination of the eigenvectors, and it has a significant component in the subspace spanned by the principal eigenvector. The Lanczos tridiagonalization procedure is then performed on the selected samples of the array data and intentionally terminated after a few iterations to avoid the deteriorating orthogonality of the basis vectors.

The Rayleigh-Ritz approximation of the first few eigensystem components requires the eigendecomposition of a small, real-valued, tridiagonal matrix formed during the Lanczos step. Because of its simplicity, we employ the power/deflation method for this purpose, again using an initial vector derived from the complex mutual coherence after each deflation. In order to ensure adequate approximation by the subspace spanned by the estimated eigenvectors, the Lanczos step is allowed to run ahead slightly and the corresponding Rayleigh-Ritz approximants are truncated to the desired number of components. The array data matrix is deflated during the power/deflation process; this facilitates repetition of the entire processing cycle to estimate the next subset of eigensystem components.

Standard Test Case Simulation

The Standard Test Case proposed by N. Owsley as a realistic benchmark against which to evaluate processing algorithms is a source simulation for a 24-element, half-wavelength spaced linear array that in its simplest form generates a covariance matrix for five (uncorrelated) spatially coherent sources, two partially coherent (angularly extended) sources modeled by third-order autoregressive processes, and additive white noise. We used this model to observe the performance of the iterative Lanczos algorithm discussed above in comparison with our earlier method of iterated principal components based on the power/deflation method, comparing both with a baseline computation using the LINPACK routine ZSVDC. In our use of this model we generated 1024 snapshots of simulated array data by filtering an array of white gaussian noise with a spatial filter obtained from the Cholesky factorization of the covariance matrix. Figure 1 shows the singular values of the data matrix obtained after applying the LINPACK routine.

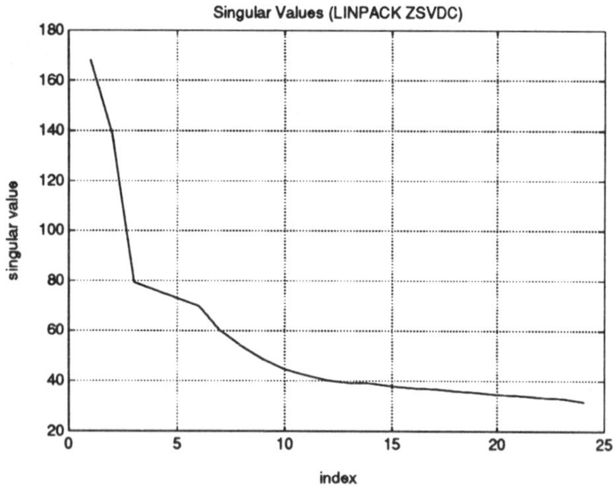

Figure 1. Singular Values Computed Using LINPACK

Figure 2 shows the estimated singular values obtained from the iterated principal components algorithm which applies repetitive power/deflation operations to the data matrix. In this implementation we used four power/deflation steps for each principal component and two iterations of the mutual coherence step to obtain the steering vector for the initial estimate after each deflation.

Figure 3 shows the singular values obtained by running the iterated' Lanczos algorithm repetitively to estimate all 24 singular values. For this computation the Lanczos step ran for 7 iterations and the Rayleigh-Ritz approximation was truncated to 4 components. Six repetitions of the algorithm were performed to estimate all 24 values. Four iterations of power/deflation with two mutual coherence iterations were performed to obtain the eigensystem components of the tridiagonalized Lanczos matrix.

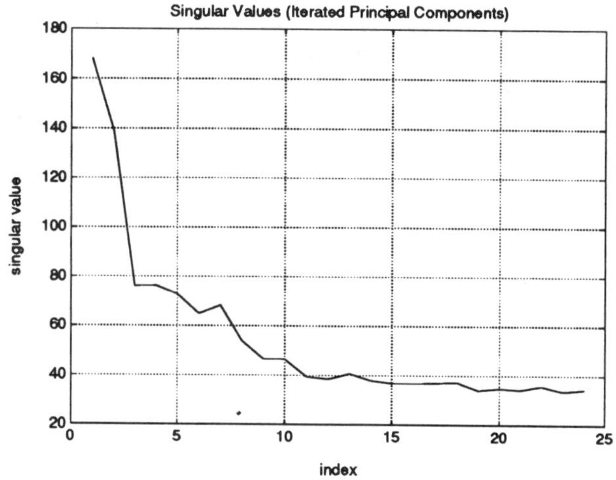

Figure 2. Singular Values Computed Using Iterated Principal Components

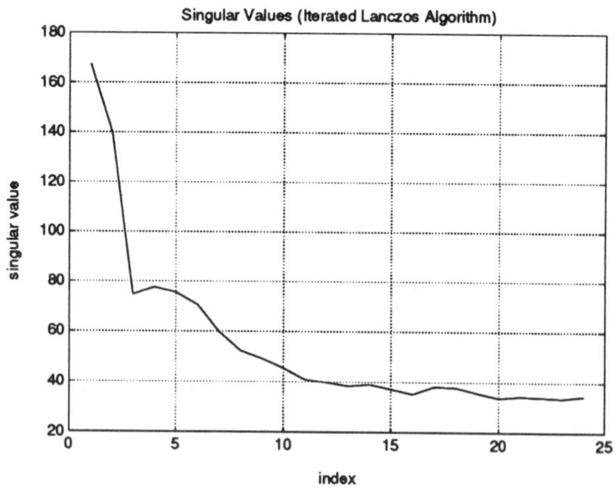

Figure 3. Singular Values Computed Using the Iterated Lanczos Algorithm

Comparison of figures 2 and 3 with figure 1 shows that both iterative algorithms approximate the baseline singular values to within a few percent. More important to the adaptive nulling and direction-finding applications is the accuracy of approximating the subspace spanned by the baseline eigenvectors. While the individual eigenvectors are not specifically approximated, the canonical angles between subspaces can be computed as the inverse cosines of the singular values of the Hermitian product of the matrix of reference eigenvectors with the matrix of approximating eigenvectors [10]. For the STC example, the singular values of this product for the iterated principal components algorithm are all unity to within machine precision. The values for the iterated Lanczos algorithm deviate from unity by less than 0.5 percent. Both methods are accurate.

Enhanced MVDR Response

A practical measure of performance is the ability of the iterative algorithm to be used in a signal processing application to produce an output that is consistent with the output produced by the baseline method. We will compare the enhanced minimimum variance distortionless response (MVDR) array response pattern produced by the different methods. For the STC example scenario in use, the minimum description length (MDL) criterion predicts that the partially coherent signal subspace consists of 12 principal components. For the enhanced MVDR the 12 dominant singular values are amplified relative to the remaining noise singular values before computing the MVDR response [11]. Figure 4 shows the baseline response when the signal eigenvalues are amplified by 20 dB.

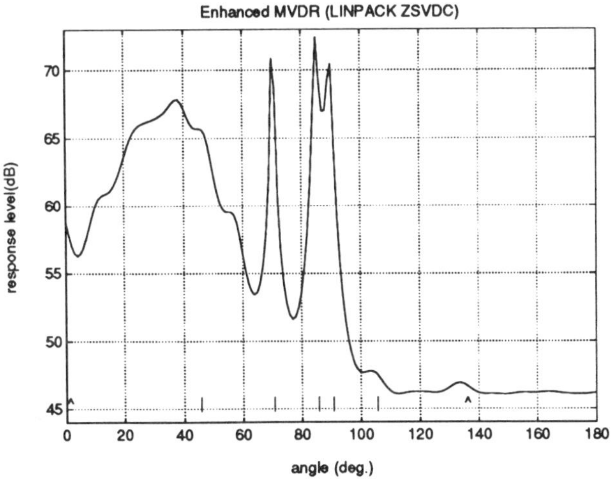

Figure 4. Enhanced MVDR Response (LINPACK ZSVDC)

The registration marks near the bottom of figure 4 indicate the angular locations of the five point sources and the two extended sources. Note the ability of the enhanced MVDR algorithm to resolve the two closely spaced point sources near 90 degrees.

Synthetic Noise Subspace

Although we have shown that the iterated algorithms can be used to extract estimates of all of the eigensystem components, their value lies in the ability to easily extract a few components for reduced-rank applications. In order to apply these reduced-rank estimation techniques to the enhanced MVDR, we must somehow take account of the noise subspace while specifically computing only the signal subspace components. Our approach to this problem is to attach to the signal subspace a synthetic noise subspace that is constructed by specifying constant noise singular values and orthonormalizing a set of arbitrarily selected noise subspace eigenvectors with respect to the signal subspace eigenvector estimates already computed. This synthetic noise subspace procedure allows us to compute the sample covariance matrix inverse directly, by simply augmenting the principal components already estimated by the iterative algorithm. Figure 5 shows the enhanced MVDR response pattern for the STC example, using the iterated principal components algorithm to compute a 12-component signal subspace estimate, attaching a synthetic noise subspace and boosting the signal subspace singular values by 20 dB.

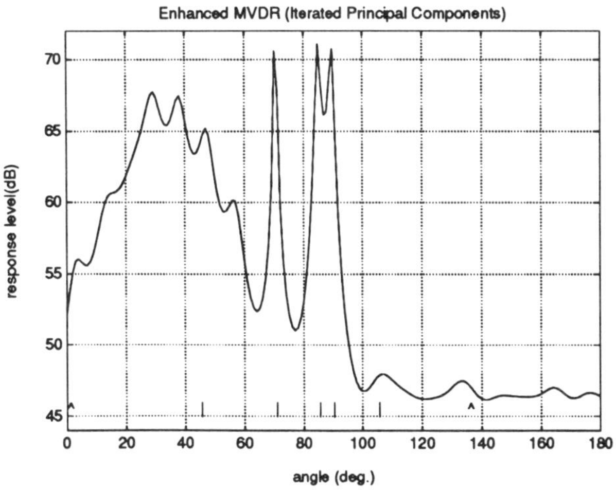

Figure 5. Enhanced MVDR Response (Iterated Principal Components)

Figure 6 shows the enhanced MVDR response pattern for the same example in which the 12 signal subspace components were estimated using the iterated Lanczos algorithm. From figures 4, 5 and 6 it is evident that the iterative algorithms provide about the same enhanced MVDR response as the LINPACK baseline. The iterated principal components example appears to have slightly better resolution.

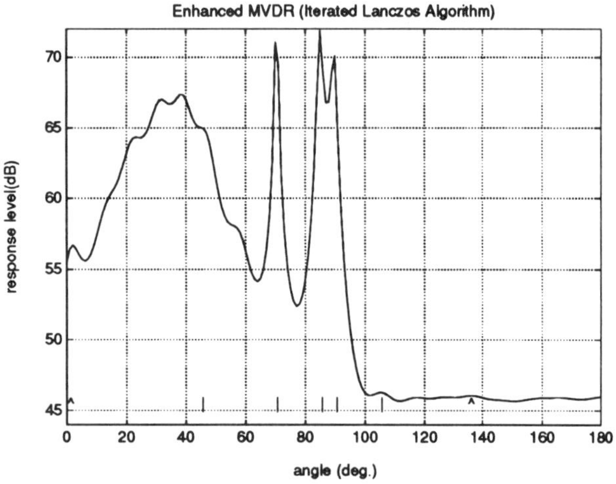

Figure 6. Enhanced MVDR Response (Iterated Lanczos Algorithm)

Complexity

Previously we showed that the power/deflation method required on the order of 15pTN+2pN complex operations, where p is the number of components to be estimated, T is the number of snapshots and N is the number of sensors [4]. From [7] we conclude that the data-domain version of the Lanczos step would require on the order of pTN+5pN complex operations. The combined algorithm, as we employed it in the STC example, used the power/deflation method on a small (7x7) array after each Lanczos step, but required operation on the full array for the mutual coherence computations. The net effect is to multiply the basic Lanczos complexity by a factor of about six. We expect that optimization of the combined algorithm could reduce this factor. In some applications, of course, it might be preferable to reorthogonalize the Lanczos basis vectors and use some other method for the Rayleigh-Ritz approximation.

Conclusions

We have endeavored to show that reduced-rank methods of adaptive beamforming, thought to be costly for real-time implementation because of the need for eigensystem computation, can be made practical by the use of iterative methods of principal eigensystem component estimation that can provide performance equivalent to that achieved by full singular value decomposition, but at lower computational cost. We described a new algorithm based on combining the Lanczos method of tridiagonalization with the power/deflation method of principal component estimation. In comparing the two iterative methods, we concluded that they are roughly equivalent in performance and complexity. Although we did not explore different numbers of samples, we would expect all three methods to decline in performance about equally as the number of snapshots is reduced.

References

1. Owsley, N.L., *A Standard Test Case for Sensor Array Processor Evaluation,* Proceedings of the 25th Annual Asilomar Conference on Signals, Systems and Computers, Pacific Grove, CA, November 1991.

2. Carhoun, D.O., R.A. Games, and R.T. Williams, *A Principal Components Sidelobe Cancellation Algorithm,* Proceedings of the 24th Annual Asilomar Conference on Signals, Systems, and Computers, Pacific Grove, CA, November 1990.

3. Tufts, D.W., R. Kumaresan, and I. Kirsteins, *Data Adaptive Signal Estimation by Singular Value Decomposition of a Data Matrix,* Proc. IEEE, Vol 70, 1982.

4. Carhoun, D.O., *Adaptive Nulling and Spatial Spectral Estimation Using an Iterated Principal Components Decomposition,* Proceedings ICASSP-91, Toronto, Canada, May 1991.

5. Wilkinson, J.H., *The Algebraic Eigenvalue Problem,* Oxford University Press, 1965.

6. Tufts, D.W., and C.P. Melissinos, *Simple, Effective Computation of Principal Eigenvectors and their Eigenvalues and Application to High-Resolution Estimation of Frequencies,* IEEE Trans ASSP, Vol ASSP-34, No 5, October 1986.

7. Guanghan Xu, and T. Kailath, *Fast Signal-subspace Decomposition Without Eigendecomposition,* Proceedings of the 24th Annual Asilomar Conference on Signals, Systems, and Computers, Pacific Grove, CA, November 1990.

8. Guanghan Xu, R.H. Roy and T. Kailath, *Fast **ESPRIT** -- Application of Fast Signal-subspace Decomposition in Array Signal Processing,* Proceedings of the 24th Annual Asilomar Conference on Signals, Systems, and Computers, Pacific Grove, CA, November 1990.

9. Golub, G.H., and C.F. VanLoan, *Matrix Computations,* Johns Hopkins University Press, Baltimore, 1983.

10. Stewart, G.W., *Perturbation Theory for the Singular Value Decomposition,* Second International Workshop on SVD and Signal Processing, U. of Rhode Island, June 1990.

11. Owsley, N.L., *Sonar Array Processing,* Chapter 3 of ***Array Signal Processing,*** S. Haykin, Editor, Prentice Hall, 1985.

Local Modeling and Robust Estimation for High-Resolution Direction Finding[1]

BJÖRN OTTERSTEN
Dept. of Automatic Control
Royal Institute of Technology
S-100 44 Stockholm
Sweden

MATS VIBERG
Dept. of Electrical Engineering
Linköping University
S-581 83 Linköping
Sweden

Abstract

Many signal processing applications are concerned with estimating signal parameters with high accuracy from sensor array measurements. To obtain accurate location estimates, knowledge of the array response is required. Most estimation techniques are sensitive to perturbations in the array model and array calibration is often necessary. This paper discusses the issue of generating the array response parametrization from a finite number of noise contaminated calibration vectors. An estimation technique based on local parametric modeling of the array response is proposed. The potential improvement using the suggested scheme rather than an ideal array model is demonstrated on real data collected from a full-scale hydro-acoustic array.

1 Introduction

High-resolution signal parameter estimation is a problem of significance in many signal processing applications. These include emitter location estimation, jammer suppression, system identification, and time series analysis. Sensor arrays play an important role in areas such as radar, sonar, and electronic surveillance, where the objective is to detect and estimate incoming signals. Signal parameters of interest can include azimuth, elevation, frequency, etc. Though not necessary, we shall herein assume that only the azimuthal angle, termed the *direction-of-arrival* (DOA) is of interest.

Much of the recent work in array processing has focussed on methods for high-resolution DOA estimation, see e.g. [1, 2]. To yield reliable estimates, accurate knowledge of the array parametrization is necessary. Thus, it is assumed that the array response is known for all DOAs of interest. The collection of all array response vectors over the parameter range is often termed the *array manifold*, [1]. See e.g., [3, 4] for an analysis of the MUSIC algorithm in the presence of modeling errors. For certain types of errors, it is possible to estimate a set of array response parameters and the DOAs simultaneously. Such *auto-calibration* techniques are reported in, e.g. [5, 6]. Apart from their limited applicability, these methods suffer from high computational demands and degraded performance, due to the increased number of parameters. Alternative, so-called *robust estimation* techniques are proposed in [7, 8]. However, these require information of the (structure of the) array response uncertainty and will in general also require some sort of array calibration.

In many applications, *array calibration* using sources at known positions appears to be necessary for high-performance DOA estimation. A set of calibration data is then collected in the presence of a single emitter, whos DOA is varied in a known way. This paper addresses some of the practical issues that arise in generating the array response at arbitrary DOAs from the given calibration data set. For high-resolution signal parameter estimation techniques to be successful, the interpolated array manifold has to satisfy certain smoothness conditions. Furthermore, since real-terrain measurements are time consuming and expensive it is clearly desirable to use as few calibration points as possible. To still achieve accurate calibration, a model based approach appears the most practical. The array response model should be flexible enough to accommodate sensor and geometry imperfections without increasing the computational burden too much. Herein, a so-called *local model* is introduced. The array response is modeled as a known nominal response and a multiplicative unknown deviation. The nominal model may be obtained, e.g. from knowledge of the array geometry and the deviations are modeled by a polynomial expansion. For each given DOA, the model is linear in the polynomial coefficients and by appropriate formulation of the optimization criterion, a closed form solution is obtained.

[1]This work was supported in part by the Swedish Research Council for Engineering Sciences

1058-6393/91 $1.00 © 1991 IEEE

2 Problem Formulation

Assume an array of m sensors at which d narrow band plane waves are arriving. The center frequency of the signals is denoted ω. Let the parameter θ_i denote the DOA associated with the ith signal, $s_i(t)$. The time-delay of a signal arriving at sensor k relative to the same signal at sensor 1 (the reference sensor) is given by $\mathbf{r}_k \cdot \mathbf{k}(\theta_i)$, where $\mathbf{r}_k$ is the position vector describing the location of the kth sensor and $\mathbf{k}(\theta_i)$ is the wave-vector of the ith signal. Let the response of the kth receiver at frequency ω and in the direction θ_i be $\rho_k(\theta_i)$. The functional form of $\rho_k(\theta_i)$ as well as the sensor positions, $\mathbf{r}_k$ are in general uncertain. We shall assume that a nominal array response model is available and introduce a parametrized model of the deviation from this model. The response of the kth sensor to a wavefront from θ_i is denoted $a_k(\theta_i, \mathbf{c}) = g_k(\theta_i, \mathbf{c}) f_k(\theta_i)$, where $f_k(\theta_i)$ is known and the unknown array parameters are denoted $\mathbf{c}$. Consider a pth order model of $g_k(\theta_i, \mathbf{c})$. Locally around a point θ, the response of the kth sensor is

$$a_k(\theta_i, \mathbf{c}) = (c_{k1} + c_{k2}\Delta\theta_i + \cdots + c_{kp+1}\Delta\theta_i^p) f_k(\theta_i) \;, \quad (1)$$

where $\Delta\theta_i = \theta - \theta_i$. If the geometry of the array is known, the nominal model could for example be $f_k(\theta_i) = e^{-j\omega \mathbf{r}_k \cdot \mathbf{k}(\theta_i)}$. The individual sensor responses are collected in the m-vector $\mathbf{a}(\theta_i, \mathbf{c})$. Using the first sensor as a reference we can write

$$\mathbf{a}(\theta_i, \mathbf{c}) = (1, \; \mathbf{c}^* \boldsymbol{\phi}_i^*)^* \;, \quad (2)$$

where $(\cdot)^*$ denotes complex conjugate transpose and

$$\begin{aligned} \boldsymbol{\phi}_i &= [\mathbf{F}_i, \Delta\theta_i \mathbf{F}_i, \ldots, \Delta\theta_i^p \mathbf{F}_i] && (3) \\ \mathbf{F}_i &= \mathrm{diag}[f_2(\theta_i), \ldots, f_m(\theta_i)] && (4) \\ \mathbf{c} &= \mathrm{vec}(\mathbf{C}) \;. && (5) \end{aligned}$$

Here $\mathbf{C}$ is the matrix whose elements are c_{kl} for $k = 2, \ldots, m$, $l = 1, \ldots, p+1$ and the $(m-1)(p+1)$-vector $\mathbf{c}$ is obtained by stacking the columns of $\mathbf{C}$.

The output of the the kth sensor is modeled by

$$x_k(t) = \sum_{i=1}^{d} a_k(\theta_i, \mathbf{c}) s_i(t) + n_k(t) \;, \quad (6)$$

where $n_k(t)$ is additive noise. The individual sensor outputs, $x_k(t)$, are collected in the vector $\mathbf{x}(t)$ resulting in

$$\mathbf{x}(t) = \sum_{i=1}^{d} \mathbf{a}(\theta_i, \mathbf{c}) s_i(t) + \mathbf{n}(t) = \mathbf{A}(\boldsymbol{\theta}, \mathbf{c}) \mathbf{s}(t) + \mathbf{n}(t) \quad (7)$$

The problem formulation is now stated as follows: Given an *estimation* data set $\mathbf{x}(t)$ for $t \in \{t_1, \ldots, t_N\}$ the major objective is to estimate the parameter vector $\boldsymbol{\theta} = (\theta_1, \ldots, \theta_d)^T$. Since the array propagation vector also depends on $\mathbf{c}$, which cannot in general be estimated from $\mathbf{x}(t)$ alone, a *calibration* set is used. In the lth calibration set, a single emitter at the known DOA η_l is present. The signal model is thus

$$\mathbf{z}_l(t) = \mathbf{a}(\eta_l, \mathbf{c}) s_l(t) + \mathbf{n}_l(t) \;. \quad (8)$$

Observations of $\mathbf{z}_l(t)$ for $l = 1, \ldots, M$ and $t \in \{t_1, \ldots, t_{N_l}\}$ are assumed available.

3 Array Calibration

In this section the problem of estimating the array parameters and the unknown DOAs is considered. A maximum likelihood approach is used, which is later approximated to yield a closed-form solution.

3.1 Maximum Likelihood Approach

To apply the ML technique, a stochastic framework for the involved signals and noise must be formulated. Following [2], the emitter signals and the noise are modeled as temporally white, independent, zero-mean Gaussian random processes. In addition, the noise is assumed to be spatially white, i.e., the spatial noise covariance is proportional to the identity. Thus we have for the calibration data

$$\begin{aligned} \mathbf{z}_l(t) &\in N(\mathbf{0}, \mathbf{R}_l) && (9) \\ \mathbf{R}_l &= \pi_l \mathbf{a}(\eta_l, \mathbf{c}) \mathbf{a}^*(\eta_l, \mathbf{c}) + \sigma_l^2 \mathbf{I} && (10) \end{aligned}$$

where η_l is known and $\pi_l, \mathbf{c}$ and σ_l^2 are all unknown. The distribution of the estimation data is

$$\begin{aligned} \mathbf{x}(t) &\in N(\mathbf{0}, \mathbf{R}) && (11) \\ \mathbf{R} &= \mathbf{A}(\boldsymbol{\theta}, \mathbf{c}) \mathbf{P} \mathbf{A}^*(\boldsymbol{\theta}, \mathbf{c}) + \sigma^2 \mathbf{I} && (12) \end{aligned}$$

where now $\boldsymbol{\theta}, \mathbf{c}, \mathbf{P}$ and σ^2 are unknown.

An optimal approach requires simultaneous estimation of all unknown parameters. However, it is natural to base the estimate of the array parameters, $\mathbf{c}$, on the calibration data only. This does not significantly change the properties of the estimates if $\sum_{l=1}^{M} N_l \gg N$ and/or if the signal-to-noise ratio (SNR) is much higher in the calibration data. Thus, one can conceptually think of this as two separate steps

1. Estimate $\mathbf{c}$ (and π_l, σ_l^2) from observations of $\mathbf{z}_l(t)$ only.

2. Given $\hat{\mathbf{c}}$, estimate $\boldsymbol{\theta}$ (and $\mathbf{P}, \sigma^2$) using measurements of $\mathbf{x}(t)$.

It is possible to approach this problem in different ways. If the parameter range is small, the manifold model may be a sufficiently accurate *global* model and a fixed set of array parameters can be estimated. Herein, the model is viewed as *locally* valid, i.e., the parameters, $\mathbf{c}$, are functions of θ. This suggests a two step procedure; for every θ estimate $\mathbf{c}$ and then evaluate a cost function (e.g. the MUSIC spatial spectrum) for that θ. Later, we show that with appropriate approximations, $\mathbf{c}$ can be solved for explicitly without loosing accuracy.

The negative log likelihood function (ignoring parameter-independent terms) of the calibration data can be expressed as

$$L(\mathbf{c}, p_l, \sigma_l^2) = \sum_{l=1}^{M} N_l \left(\log|\mathbf{R}_l| + \mathrm{Tr}\{\mathbf{R}_l^{-1}\hat{\mathbf{R}}_l\}\right) \quad (13)$$

where $\hat{\mathbf{R}}_l$ is the sample covariance

$$\hat{\mathbf{R}}_l = \frac{1}{N_l}\sum_{k=1}^{N_l} \mathbf{z}_l(t_k)\mathbf{z}_l^*(t_k) \quad (14)$$

The ML estimates of $\mathbf{c}, \pi_l, \sigma_l^2$, are the minimizing arguments of (13). Using the same technique as in [2], the above can be minimized explicitly with respect to π_l and σ_l^2 resulting in the concentrated criterion

$$\sum_{l=1}^{M} N_l \log\left|\ \hat{\pi}_l(\mathbf{c})\mathbf{a}(\eta_l, \mathbf{c})\mathbf{a}^*(\eta_l, \mathbf{c}) + \hat{\sigma}_l^2(\mathbf{c})\,\mathbf{I}\ \right| \quad (15)$$

where $\hat{\pi}_l(\mathbf{c})$ and $\hat{\sigma}_l^2(\mathbf{c})$ are fairly complicated functions of $\mathbf{c}$ whose precise form is irrelevant.

3.2 Least Squares Approximation

Since for a given θ, $\mathbf{a}(\mathbf{c})$ is linear in $\mathbf{c}$, see (2), one may hope for a closed form solution for $\hat{\mathbf{c}}(\theta)$. However, to achieve this the criterion function (15) must be simplified. Applying the results of [9, 10], shows that each term in the sum (15) can be replaced by the expression

$$N_l\frac{(\hat{\lambda}_l - \hat{\sigma}_l^2)^2}{\hat{\sigma}_l^2\hat{\lambda}_l}|\hat{\mathbf{e}}_l - \mathbf{a}(\eta_l, \mathbf{c})t_l|^2 \quad (16)$$

without affecting the asymptotic (for large N_l) properties of the estimate. In (16), $\hat{\lambda}_l$ is the largest eigenvalue of $\hat{\mathbf{R}}_l$, $\hat{\mathbf{e}}_l$ is the corresponding unit-norm eigenvector, $\hat{\sigma}_l^2$ is the average of the $m-1$ smallest eigenvalues, and t_l is a free "scaling" parameter. Inserting (16) in (15) yields the cost function

$$\sum_{i=1}^{M} N_l\frac{(\hat{\lambda}_l - \hat{\sigma}_l^2)^2}{\hat{\sigma}_l^2\hat{\lambda}_l}|\hat{\mathbf{e}}_l - \mathbf{a}(\eta_l, \mathbf{c})t_l|^2 \quad (17)$$

which is to be minimized with respect to $\mathbf{c}$ and t_l. Making use of the convention $\mathbf{a}_1(\eta_l, \mathbf{c}) = 1$, we see that a consistent (as $N_l \to \infty$) estimate of t_l is simply $\hat{t}_l = \hat{\mathbf{e}}_{l1}$, the first component of $\hat{\mathbf{e}}_l$. Inserting this into (17) and using (2) gives the final criterion

$$\begin{aligned} V(\mathbf{c}) &= \sum_{l=1}^{M} w_l|\hat{\mathbf{e}}_l - \mathbf{a}(\eta_l, \mathbf{c})\hat{\mathbf{e}}_{i1}|^2 \\ &= \sum_{l=1}^{M} w_l|\hat{\mathbf{u}}_l - \hat{\mathbf{e}}_{i1}\boldsymbol{\phi}_l\mathbf{c}|^2 \end{aligned} \quad (18)$$

where

$$w_l = N_l\frac{(\hat{\lambda}_l - \hat{\sigma}_l^2)^2}{\hat{\sigma}_l^2\hat{\lambda}_l} \quad (19)$$

and where $\hat{\mathbf{u}}_l$ contains the $m-1$ last elements of $\hat{\mathbf{e}}_l$. Clearly (18) is quadratic in $\mathbf{c}$ and $\hat{\mathbf{c}}(\theta)$ is obtained as the least squares solution to (18)

$$\hat{\mathbf{c}}(\theta) = \boldsymbol{\Phi}^{\dagger}\mathbf{u} = (\boldsymbol{\Phi}^*\boldsymbol{\Phi})^{-1}\boldsymbol{\Phi}^*\mathbf{u} \quad (20)$$

where we have used the notation

$$\boldsymbol{\Phi} = \begin{bmatrix} \sqrt{w_1}\hat{\mathbf{e}}_{11}\boldsymbol{\phi}_1 \\ \vdots \\ \sqrt{w_M}\hat{\mathbf{e}}_{M1}\boldsymbol{\phi}_M \end{bmatrix}, \quad \mathbf{u} = \begin{bmatrix} \sqrt{w_1}\mathbf{u}_1 \\ \vdots \\ \sqrt{w_M}\mathbf{u}_M \end{bmatrix}$$

Let us again stress that the solution, (20), depends on the DOA of interest (observe that $\boldsymbol{\Phi}$ depends on θ through the $\boldsymbol{\phi}_l$:s). Hence, the principal eigenvector of each calibration data sample covariance matrix must be stored along with the corresponding eigenvalue and the estimated noise variance. To use (15) directly requires storage of the complete sample covariance matrices and in addition a closed form expression for $\hat{\mathbf{c}}$ is unavailable.

3.3 Modeling Errors

The philosophy behind using (2) with θ-dependent coefficients is that a global model (i.e., valid for all DOAs) would be impractically complicated due to imperfect receiver characteristics, mutual coupling effects, etc. However, the present weighting in equation (20) only accounts for finite sample effects. It is useful to apply a further weighting to reduce the bias due to modeling errors. When estimating the array parameters at θ, the calibration points that are close to θ should be given larger weights than those that are further away. Thus the block rows of $\boldsymbol{\Phi}$ and $\mathbf{u}$ are replaced by $\sqrt{q_l w_l}\hat{\mathbf{e}}_{l1}\boldsymbol{\phi}_l$ and $\sqrt{q_l w_l}\mathbf{u}_l$ respectively. A natural choice for q_l is

$$q_l = \mu^{(\Delta\theta_l)^2} = \mu^{(\theta-\eta_l)^2} \quad (21)$$

for some constant μ, $0 < \mu < 1$.

4 DOA Estimation

The focus in this paper is on the calibration problem. However, we shall briefly discuss integration of the calibration procedure with a technique for generating DOA estimates.

Following Section 3.1, the ML estimate of $\boldsymbol{\theta}$ based on measurements of $\mathbf{x}(t)$ is easily formulated. However, to illustrate the joint calibration and DOA estimation we will instead generate "MUSIC spectra", [1],

$$P_{MU}(\theta) = \frac{\mathbf{a}^*(\theta, \hat{\mathbf{c}}(\theta))\mathbf{a}(\theta, \hat{\mathbf{c}}(\theta))}{\mathbf{a}^*(\theta, \hat{\mathbf{c}}(\theta))\hat{\mathbf{E}}_n \hat{\mathbf{E}}_n^* \mathbf{a}(\theta, \hat{\mathbf{c}}(\theta))} \qquad (22)$$

where the matrix $\hat{\mathbf{E}}_n$ contains the $m-d$ "smallest" eigenvectors of the sample covariance. The manifold vectors have the following form

$$\mathbf{a}(\theta, \hat{\mathbf{c}}(\theta)) = (1,\ \mathbf{u}^* \boldsymbol{\Phi}^{\dagger *} \boldsymbol{\phi}^*)^* \ . \qquad (23)$$

When devising a decent algorithm, knowledge of the derivatives of $\mathbf{a}$ are required and can be derived from (23). Further details of such an approach will be published elsewhere. Let us only mention that the main computational requirement in general is forming $\boldsymbol{\Phi}^*\boldsymbol{\Phi}$ since M may be quite large compared to m. The bulk of this operation can however be done off-line. To see this examine the case where $p = 2$

$$\begin{aligned} \boldsymbol{\phi}_l &= [1\ \Delta\theta_l\ \Delta\theta_l^2] \otimes \mathbf{F}_l \\ &= [1\ \theta_l\ \theta_l^2] \begin{bmatrix} 1 & \theta & \theta^2 \\ 0 & -1 & -2\theta \\ 0 & 0 & 1 \end{bmatrix} \otimes \mathbf{F}_l \end{aligned} \qquad (24)$$

where it is important to note that only the 3×3 matrix depends on θ. Using the "Mixed Product Rule" $(A \otimes B)(C \otimes D) = AC \otimes BD$ on $\boldsymbol{\Phi}^*\boldsymbol{\Phi}$ the computational savings become evident.

5 Experimental Results

Below, preliminary results from a project aimed at examining the viability of accurate source localization methods on hydro acoustic sensor array data are presented. The data is collected by a seven-element equispaced linear array with an aperture of approximately one wavelength. The array is mounted horizontally and consists of sensors which are designed to ideally be omni-directional in the horizontal plane. A far field narrow band source is placed, in turn, in 18 different locations around array broadside. The purpose is to estimate the azimuthal angle to the emitter. The noise is assumed spatially white, since no measurements of the noise covariance is available. Cleary, this produces noise modeling errors in addition to the array response errors.

In the experiments to be presented, data from one source location is considered the estimation set,

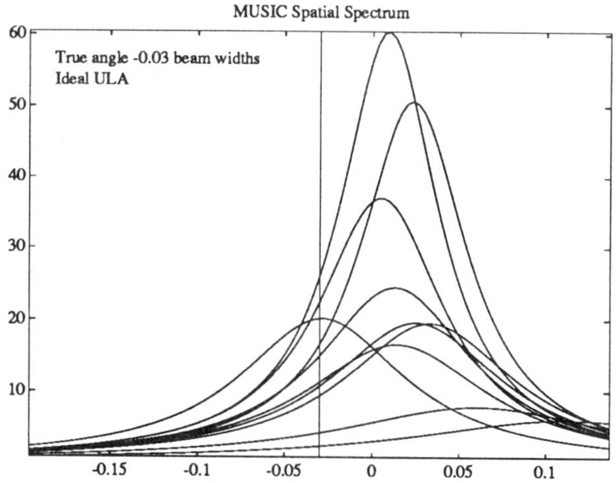

Figure 1: *Ideal uniform linear array manifold, experiment 1. True emitter location indicated by solid line.*

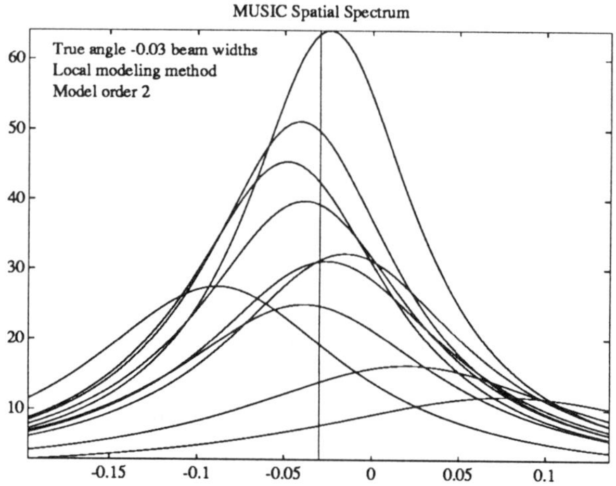

Figure 2: *Locally modeled array manifold, experiment 1. True emitter location indicated by solid line.*

whereas the other data sets are used for calibration. The estimation data is divided into 10 segments of .32 seconds each and the corresponding MUSIC spectra are displayed. The x-axis is normalized in terms of the 3 dB array beam width at broadside, (zero degrees) and the SNR is approximately 6dB. The vertical line indicates the true emitter location. In Figures 1 and 3 the array manifold is assumed to be that of an ideal ULA and in Figures 2 and 4 the local modeling technique is used. No θ-dependent weighting (q_i) is employed.

Note the reduced bias in the peak locations for the local modeling technique. We expect that the performance differences will be even more pronounced when estimating several signal parameters simultaneously by a multi-dimensional technique, e.g. ML. The "variance" seen in the location estimates is due to measure-

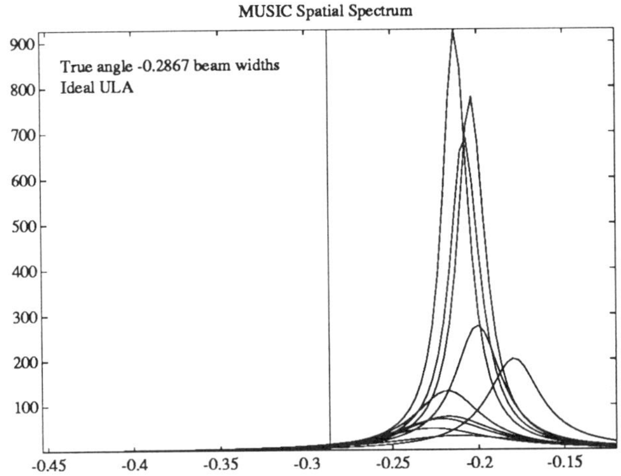

Figure 3: *Ideal uniform linear array manifold, experiment 2. True emitter location indicated by solid line.*

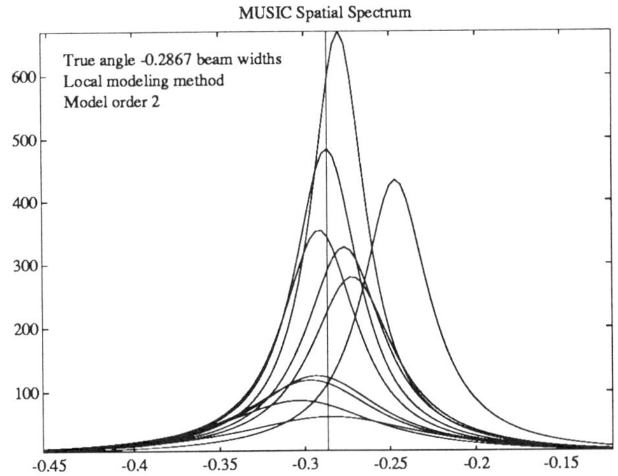

Figure 4: *Locally modeled array manifold, experiment 2. True emitter location indicated by solid line.*

ment noise and time varying conditions.

It should be noted that in the examples above, the estimated DOA is in the center region of the calibration set. When applying the local modeling technique in regions close to the boundary of the calibration set, the performance degrades and is comparable to that of the ideal ULA manifold.

6 Conclusions

High-resolution signal parameter estimation techniques require accurate knowledge of the array characteristics. It is often impossible to obtain an analytical description of the array response and one must resort to measurements, i.e., calibration of the array. This paper addresses the problem of generating the array manifold from a finite number of noisy calibration points in the parameter space. A model based approach motivated by local modeling of the array response is proposed. The ML estimator is formulated and a large sample approximation is described leading to a closed form solution of the parameters in the array model. By applying this technique together with a standard DOA estimation procedure, a significant reduction in estimation bias can be obtained. This is demonstrated on data collected from a large scale hydro acoustic array.

References

[1] R.O. Schmidt. "Multiple Emitter Location and Signal Parameter Estimation". In *Proc. RADC Spectrum Estimation Workshop*, pages 243–258, Rome, NY, 1979.

[2] J.F. Böhme. "Estimation of Spectral Parameters of Correlated Signals in Wavefields". *Signal Processing*, 10:329–337, 1986.

[3] A. Swindlehurst, B. Ottersten, and T. Kailath. "An Analysis of MUSIC and Root-MUSIC in the Presence of Sensor Perturbations". In *Proc. 23rd Asilomar Conf. Sig., Syst., Comput.*, pages 930–934, Nov. 1989.

[4] B. Friedlander. "A Sensitivity Analysis of the MUSIC Algorithm". In *Proc. ICASSP 89*, pages 2811–2814, Glasgow, Scotland, May 1989.

[5] Y. Rockah and P. M. Schultheiss. Array shape calibration using sources in unknown locations – Part I: Far-field sources. *IEEE Trans. on ASSP*, 35:286–299, March 1987.

[6] A. J. Weiss and B. Friedlander. "Array Shape Calibration Using Sources in Unknown Locations - A Maximum Likelihood Approach". *IEEE Trans. on ASSP*, 37(12):1958–1966, Dec. 1989.

[7] A. Swindlehurst. "Robust Algorithms for Direction-Finding in the Presence of Model Errors". In *Proc. 5th ASSP Workshop on Spectral Estimation and Modeling*, pages 362–366, Rochester, NY, October 1990.

[8] B. Wahlberg, B. Ottersten, and M. Viberg. "Robust Signal Parameter Estimation in the Presence of Array Perturbations". In *Proc. ICASSP 91 Conf*, Toronto, Canada, May 1991.

[9] B. Ottersten, B. Wahlberg, M. Viberg, and T. Kailath. "Stochastic Maximum Likelihood Estimation in Sensor Arrays by Weighted Subspace Fitting". In *Proc. 23rd Asilomar Conf. Sig., Syst., Comput.*, pages 599–603, Monterey, CA, Nov. 1989.

[10] P. Stoica and A. Nehorai. "Performance Study of Conditional and Unconditional Direction-of-Arrival Estimation". *IEEE Trans. ASSP*, ASSP-38:1783–1795, October 1990.

Multiplierless FIR Filters With Discrete-Value Sigma-Delta Encoded Coefficients

Scott R. Powell, Paul M. Chau

University of California
Dept. of ECE, R-007
LaJolla, CA 92093-0407

Abstract

This paper presents an oversampled sigma-delta modulation based encoding scheme to reduce an FIR filter's coefficient or input signal samples to values with only a few non-zero bits eliminating the need for multiplications. This new scheme realizes a substantial reduction in sample rate and additions over previous differential digital coding schemes by replacing the direct truncation within a sigma-delta modulator with a more sophisticated quantizing procedure which truncates to the next nearest canonic signed-digit code (number representable as sum or difference of a few powers-of-two). Computer simulations are used to demonstrate and verify the expected performance of this scheme and comparisons are made with previously reported results.

1 Introduction

Over the past two decades, many filter structures have been proposed to reduce the hardware complexity of fixed-point digital FIR filters. Most have concentrated on reducing the number and/or wordlength of coefficient multipliers. Delta modulation [1], sigma-delta modulation [2], and error spectrum shaping [3] techniques use differential encoding schemes to move quantization errors out of the frequency band of interest so that either the coefficient or the input signal wordlength can be truncated to a small number of bits. These filters require no general purpose multipliers.

Another method is to use the full coefficient wordlength but only permit a few of the bits to be non-zero. The resulting sum of signed power-of-two space is also referred to as canonic signed-digit (CSD) code [6]. With this scheme, multiplications can be implemented with a few additions.

This paper combines the above two techniques so that the error introduced by rounding coefficients (or the input signal) to the next nearest CSD code is moved out of the frequency band of interest. Thus, unlike previous differential encoding schemes, the full coefficient wordlength is maintained but restricted to discrete CSD values. Also, unlike the straight CSD truncation schemes, the quantization error is moved out of the frequency band of interest.

2 CSD Sigma-Delta Truncation

Figure 1a repeats the discrete time block diagram of [2] showing a single-loop sigma-delta modulator with a uniform quantizer truncating a 10 bit input sequence into a 2 bit output sequence. Multiplication using this output can be implemented with a single addition. The output sequence can take on only four discrete values. Figure 1b shows the proposed sigma-delta encoding scheme with a non-uniform quantizer truncating a 10 bit input sequence into a 2 bit CSD code. Multiplication using this output requires only one addition, the same as the previous approach of figure 1a.

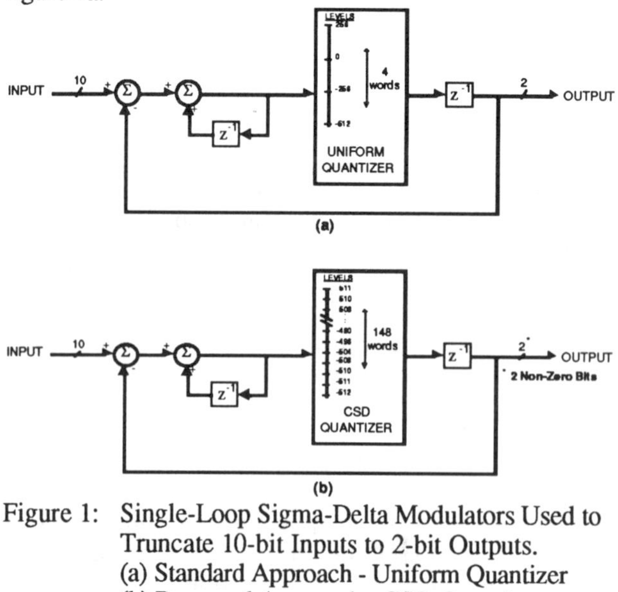

Figure 1: Single-Loop Sigma-Delta Modulators Used to Truncate 10-bit Inputs to 2-bit Outputs.
(a) Standard Approach - Uniform Quantizer
(b) Proposed Approach - CSD Quantizer.

1058-6393/91 $1.00 © 1991 IEEE

Implementing a multiplication operation with samples encoded using uniform quantizer based sigma-delta truncation requires the same number of additions as with CSD based sigma-delta truncation. However, the 2 bit CSD output of figure 1b can take on 148 discrete values, 37 times more than the uniform quantizer case of figure 1a. In general, CSD based sigma-delta truncation has many times more possible discrete-values thereby offering substantially reduced quantization error over previous approaches with no increase in hardware for downstream multiplication operations.

A sinusoid (10 bit) has been applied to the previous 2-bit sigma-delta truncator of figure 1a and the proposed 2-bit CSD sigma-delta truncator of figure 1b. Figure 2 compares the spectral response at the output of each truncator. As expected, the increased number of discrete-values for the CSD case has reduced the overall quantization noise and increased the equivalent number of bits within a given bandwidth. The equivalent number of bits available for a given oversampling ratio (OSR) can be computed by integrating the power spectra of figure 2 to determine rms noise power. Figure 3 compares the equivalent number of bits available as a function of OSR and the number of bits of truncation bits (either CSD or uniform) for the uniform quantization and new CSD quantization sigma-delta encoding methods. As can be seen from figure 3, the CSD based sigma-delta encoding method can be used at a much lower OSR than the uniform quantization method for any given resolution. Thus, when the truncated output is used as FIR filter coefficients, the CSD based sigma-delta scheme requires many times fewer additions per filter output sample than previous approaches.

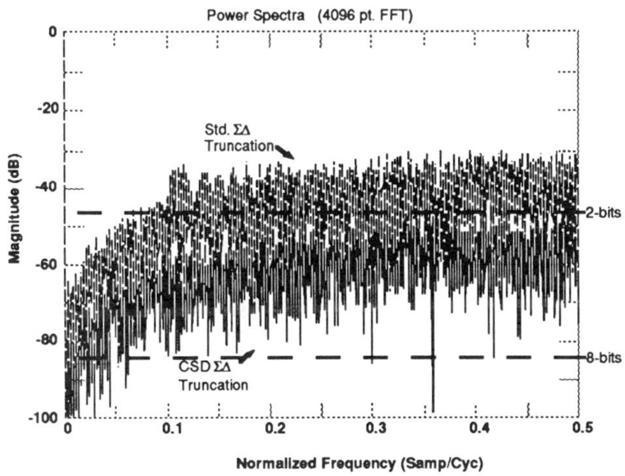

Figure 2: Spectrum of Sinusoidal Response From Standard and CSD Sigma-Delta Truncation

As described in references [1]-[3], differential encoding can be applied either to the input signal or to the filter's impulse response sequence. Although rounding to a CSD value is more difficult than rounding to a binary PCM value, sigma-delta modulation applied to a filter's impulse response is performed off-line where the increased rounding complexity is inconsequential. For applications where sigma-delta modulation is used to encode the input signal, circuits have been recently reported [5] which can perform the CSD rounding in real-time. Since only one sigma-delta modulator is required for this case, the percentage increase in hardware for CSD truncation over using straight truncation is relatively small.

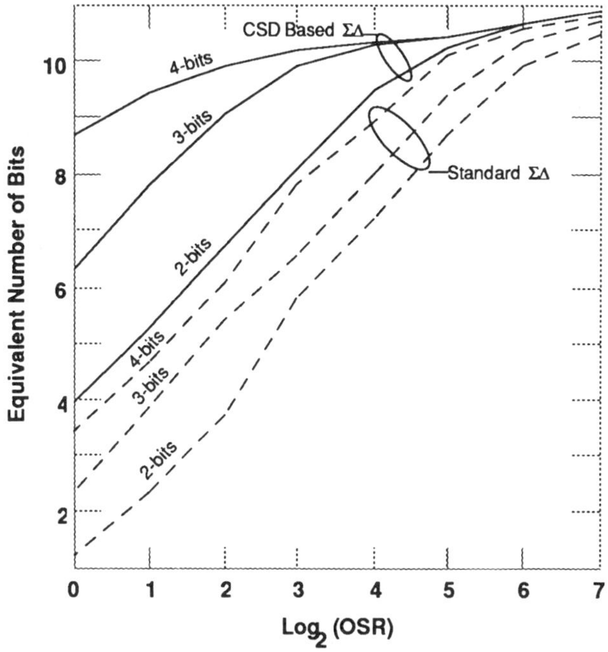

Figure 3: Number of Equivalent Bits v.s. Oversampling Ratio for the Standard and New CSD Based Sigma-Delta Truncation Schemes

3 FIR Filters Based on Sigma-Delta Encoding

A multiplier free FIR filter can be realized by oversampling (interpolating) the desired impulse response by *OSR* times, passing this sequence through the sigma-delta CSD truncator of figure 1b, and using the truncated output as the FIR filter coefficients. Note that this is a multi-rate approach and requires interpolation of the filter's input signal by OSR times as well as decimation of the filter's output signal by OSR times. The interpolation and decimation operations and implementations are described in detail in [2] and are not considered in this paper.

Assuming the CSD truncation within the sigma-delta loop (see figure 1b) is done offline using a look-up table, the cpu time required to perform coefficient truncation increases only linearly with filter length. Although the coefficient truncation is suboptimal, filters with virtually unlimited length can be designed in seconds or minutes. Additionally, the truncation need not be limited to the sum of only 2 powers-of-two. Samueli [6] has shown that permitting some coefficients to have three or more powers-of-two can result in very efficient filters.

To use the CSD sigma-delta truncator of figure 1b to truncate the coefficients of an FIR filter, several implementation details must be addressed. First is input scaling. Since the normalized frequency characteristics of a filter are not affected by multiplying all coefficients by a constant value, we consider the input scaling factor to be a free parameter. For simplicity, we do not attempt to optimize the input scale factor in this paper and simply set it such that the largest filter coefficient value uses the full dynamic range available from the sigma-delta loop. For example, if the sigma-delta CSD truncator of figure 1b were to truncate to the nearest 16 bit binary value which had at most three non-zero bits (a 3 bit CSD code), we would choose to scale the filter coefficients by (assume twos complement):

$$S = (2^{B-1} - 1) / h_{max} \tag{1}$$

Where: S = Input Scale Factor
B = Original Wordlength (16 bits)
h_{max} = Largest Coefficient Value

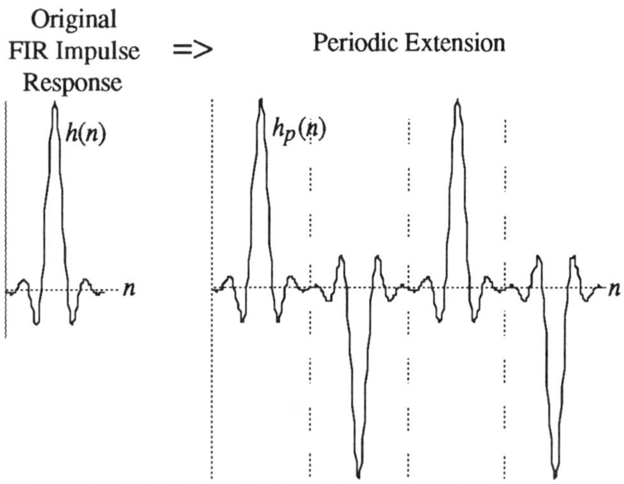

Figure 4: Periodically Extended FIR Impulse Response

Since a sigma-delta loop contains feedback, the output will exhibit a starting and ending transient response when presented with a finite length impulse response as an input signal. To overcome this difficulty, we propose using an input consisting of a periodic extension of the FIR impulse response, as illustrated in figure 4. Periodic inputs to the sigma-delta loop result in periodic outputs. Assuming enough periods are applied to the input (usually one is sufficient), the truncated FIR coefficient values are determined by extracting an appropriate portion of the sigma-delta loop output after steady-state is reached.

Although the output of the sigma-delta CSD truncation loop will be periodic, it will not necessarily be symmetric, even though the input is symmetric. Thus, if a single half-period (positive or negative half) is used as the truncated filter coefficient values, the phase response will not necessarily be linear. We propose the following scheme. First, extract a full period (positive and negative halves) of the sigma-delta truncator output and split the period into four quadrants. Next, consider each quadrant as either the first or last half of a symmetric linear phase FIR impulse response. The negative portions must be multiplied by minus one. Lastly, compute the mean-square error between each quadrant and the infinite precision FIR filter impulse response. Although all quadrants result in linear phase filters with good low frequency response (due to the sigma-delta noise shaping), there is generally one or two quadrants with lower mean-square error values. The quadrant with the lowest mean-square error from the infinite precision case will generally have the best high frequency response characteristics.

To summarize, the proposed design procedure is outlined below:

1) Determine optimal infinite precision filter coefficients (impulse response) $h_i(n)$.
2) Create a periodic extension of $h_i(n)$ (5 periods is sufficient).
3) Scale the periodic extension using (2)
4) Pass the scaled and extended infinite precision input through the CSD sigma-delta truncator of figure 1b and extract the last complete cycle of the output.
5) Divide the extracted cycle into quadrants and use the truncated CSD values to create four candidate linear phase FIR filters.
6) Compute the mean-square error between each of the four filter coefficients and the original (scaled) infinite precision filter coefficients. Select the filter with the lowest mean-square error.
7) Compute the frequency response of the best filter to determine if the desired specifications have been met. If not, increase the number of taps or the number of non-zero CSD bits and iterate back through steps 1-6.

4 Example

As an example, we consider the filter suggested in example 1 of [2] for comparison. In this example, the filter impulse response is encoded with both the previous and the new CSD based sigma-delta schemes. The filter is designed using the Fourier method based on the transfer function:

$$H(f) = \begin{cases} 1.0 & f < 0.20 \\ (0.25 - f)/0.05 & 0.2 \le f \le 0.25 \\ 0 & f > 0.25 \end{cases}$$

Giving an impulse response of

$$h(n) = 1.25\text{sinc}^2(0.25n) - 0.8\text{sinc}^2(0.2n)$$

A 128 tap filter with infinite precision coefficients is determined using the window:

$$w(n) = 0.338946 + 0.481973\cos(\pi n/64) + 0.161054\cos(2\pi n/64) + 0.018127\cos(3\pi n/64)$$
$$n = -64,\ldots,0,\ldots,63$$

The infinite precision impulse response is shown in figure 5a. Figure 5b shows the result of oversampling this impulse response eight times (OSR=8) and passing the resulting sequence through the sigma-delta CSD truncator of figure 1b. Note the similarity between figures 5a and 5b even though the impulse response values in figure 5b are represented with only two non-zero bits. Figure 5c shows the result of oversampling by 8 and using the previously reported sigma-delta uniform truncator of figure 1a. Obviously, the CSD based approach provides a much closer approximation to the ideal impulse response.

As this oversampling rate (OSR) is increased, the truncated filter coefficients represent increasingly better approximations of the ideal frequency response but the filter must be operated at increasingly higher sample rates. Figure 7a shows the frequency response of the sigma-delta CSD truncated filter coefficients at various oversampling rates. As expected, the frequency response improves with increasing OSR. For comparison, figure 7b shows the frequency response of the sigma-delta uniform truncated filter coefficients at the same ORS values. As can be seen from figure 7, the sigma-delta CSD truncated filter has significantly improved filter characteristics (stopband attenuation) for a given OSR. For example, if a 50dB stopband attenuation is desired, the sigma-delta CSD truncated filter need only be oversampled by a factor of 8 whereas the previously reported sigma-delta uniform truncator requires an OSR of at least 512. Although both approaches require only a single addition per tap weight, the sample rate of the sigma-delta CSD truncated filter is reduced by 64 times.

5 Conclusions

A new oversampled sigma-delta differential encoding scheme was described which uses CSD code for the loop truncator. FIR filter coefficient multipliers can be reduced to a few adders. The implementation of the sigma-delta CSD truncation scheme was described in detail. A filter design example was presented and compared to a previously reported sigma-delta truncation scheme. The new scheme offered a 64 times improvement in the sample rate required to implement the filter without increasing the hardware complexity.

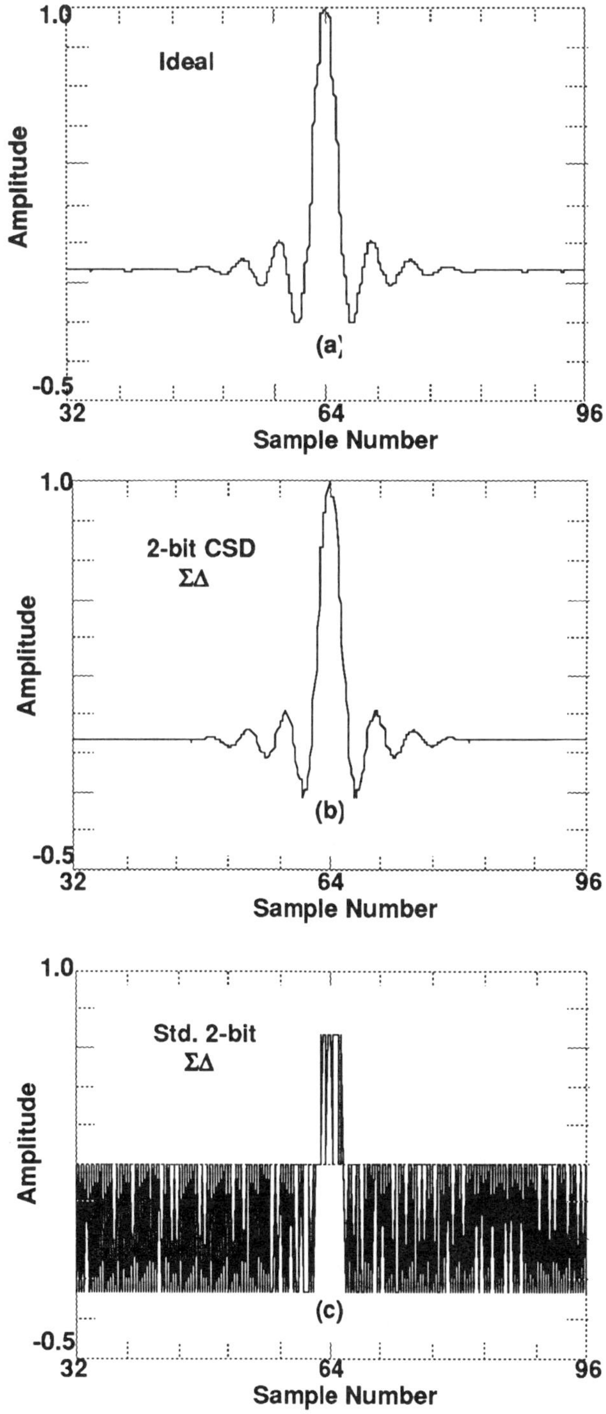

Figure 5: Impulse response for Filter Example.
(a) Ideal (no quantization)
(b) 2-bit CSD Sigma-Delta Truncation
(c) 2-bit Standard Sigma-Delta Truncation

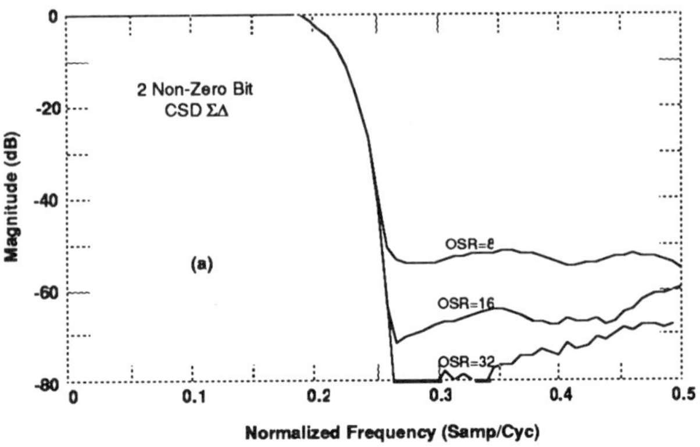

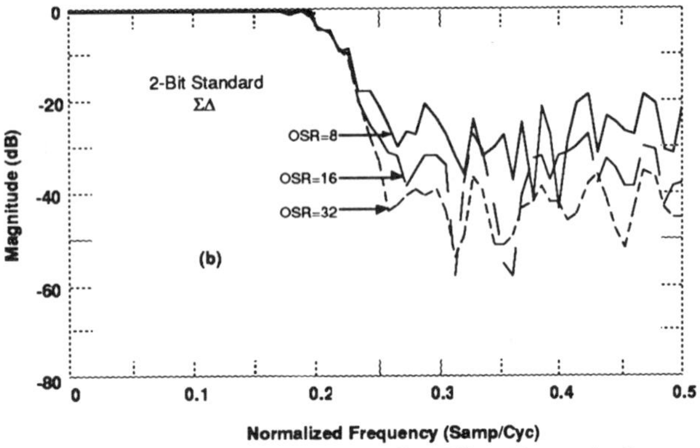

Figure 6: Frequency Response for Filter Example for Various Oversampling Ratios
(a) 2-bit CSD Sigma-Delta Truncation
(b) 2-bit Standard Sigma-Delta Truncation

References

[1] Nevio Benvenuto, L.E. Franks, F.S. Hill, "Realization of FIR Filters Using Coefficients +1, 0, and -1," *IEEE Trans. on Communications*, Vol. COM-33, No. 10, Oct 1985, pp. 1117-1125.

[2] Ping Wah Wong, Robert M. Gray, "FIR Filters with Sigma-Delta Modulation Encoding," *IEEE Trans on Acou. Speech, and Sig. Proc.*, Vol 38, No. 6, June 1990, pp. 979-990.

[3] Jens Jorgen Nielsen, "Design of Linear-Phase Direct-Form FIR Digital Filters with Quantized Coefficients Using Error Spectrum Shaping Techniques," *IEEE Trans. on Acou., Speech, and Sig. Proc.*, Vol 37, No. 7, July 1989, pp. 1020-1026.

[4] Y. C. Lim, "Design of Discrete-Coefficient-Value Linear Phase FIR Filters with Optimum Normalized Peak Ripple Magnitude," *IEEE Trans. on Circuits and Systems*, Vol. 37, No. 12, Dec. 1990, pp. 1480-1486.

[5] Y. C. Lim, J. B. Evans, Bede Liu, "Decomposition of Binary Integers into Signed Power-of-Two Terms," *IEEE Trans. on Circuits and Systems*, Vol. 38, No. 6, June 1991, pp. 667-672.

[6] Henri Samueli, "An Improved Search Algorithm for the Design of Multiplierless FIR Filters with Powers-of-Two Coefficients," *IEEE Trans. on Circuits and Systems*, Vol. 36, No. 7, July 1989, pp. 1044-1047.

Design of Linear Phase FIR Filters with uniformly Flat Band-Response*

S. Unnikrishna Pillai

Department of Electrical Engineering
Polytechnic University
Brooklyn, New York 11201

Rabinder N. Madan

The Office of Naval Research
800 N. Quincy St.
Arlington, Virginia 22105

Abstract

A new procedure that makes use of the maximum entropy method (MEM) for the design of linear phase FIR digital filters is described here. It is shown here that by applying MEM to the inverse of the desired gain function or its square root function, it is possible to generate linear phase FIR filters that match the given gain function to any desired degree of accuracy. Moreover, an iterative algorithm makes the design procedure very efficient, since the higher order filters can be recursively generated from the lower order ones. To minimize the effect of any residual passband distortion, a final averaging scheme on the lower order filters generates a class of passband distortion-free linear phase FIR filters. Simulation results that compare the present procedure with other well known methods are also presented here.

1 Introduction

The Maximum Entropy Method (MEM) is one of the techniques for estimating the power spectrum of a wide sense stationary stochastic process from partial information available about itself. In the ideal case, this information is in the form of knowledge about a portion of its autocorrelation sequence

$$r_k = E[x(nT)\,x^*((n+k)T)]\,, \quad k = 0 \to \infty \qquad (1)$$

where $x(nT)$ represents a discrete-time wide sense stationary stochastic process. Then $r_k = r^*_{-k}$ and the power spectral density $S(\omega)$ associated with this process is given by

$$S(\omega) = \sum_{k=-\infty}^{+\infty} r_k\, e^{-jk\omega} \geq 0\,. \qquad (2)$$

*This research work was supported by the Office of Naval Research under contract N-00014-89-J-1512.

Clearly,

$$r_k = \frac{1}{2\pi}\int_{-\pi}^{\pi} S(\omega)e^{jk\omega}d\omega\,, \quad |k| \geq 0\,. \qquad (3)$$

Given a finite set of autocorrelations $r_0, r_1, \ldots, r_n$ of a process, the spectrum estimation problem refers to evaluating a nonnegative function $S(\omega)$ such that its first $(n+1)$ Fourier coefficients computed using (3) matches with the given autocorrelation sequence (correlation matching property).

It is well known that a necessary and sufficient condition for the existance of a nonnegative function as in (2) is the nonnegativity property of every Hermitian Toeplitz matrix $\mathbf{T}_k$ [1, 2]

$$\mathbf{T}_k = \begin{bmatrix} r_0 & r_1 & \cdots & r_k \\ r_1^* & r_0 & \cdots & r_{k-1} \\ \vdots & \vdots & \ddots & \vdots \\ r_k^* & r_{k-1}^* & \cdots & r_0 \end{bmatrix}, \quad k = 0 \to \infty \qquad (4)$$

generated from the correlations $r_0, r_1 \ldots, r_k$, i.e.,

$$S(\omega) \geq 0 \Longleftrightarrow \mathbf{T}_k \geq 0\,, \quad k = 0 \to \infty\,. \qquad (5)$$

Given a finite set of correlations, the maximum entropy method(MEM) eliminates the remaining ambiguity about the underlying process by choosing that process with maximum entropy [3]. Since maximization of entropy subject to given values of $r_0, r_1, \ldots, r_n$, leads to a Gaussian process [4], the desired MEM-solution can be obtained by maximizing the entropy H of the Gaussian process given by

$$H = \frac{1}{2\pi}\int_{-\pi}^{\pi} \ln S(\omega)d\omega\,. \qquad (6)$$

It is well known that, given $r_0, r_1, \ldots, r_n$, maximization of (6) leads to a stable autoregressive (all-pole) model of order n, i.e.,

$$S(\omega) = \frac{1}{|H_n(e^{j\omega})|^2} \qquad (7)$$

where

$$H_n(z) = a_0^{(n)} + a_1^{(n)} z^{-1} + a_2^{(n)} z^{-2} + \cdots + a_n^{(n)} z^{-n} . \quad (8)$$

The coefficients of the above polynomial satisfy the standard Yule-Walker equations and can be generated through a recursive procedure known as the Levinson recursion algorithm [5]–[6]

$$\sqrt{1-s_n^2} H_n(z) = H_{n-1}(z) - s_n z^{-n} H_{n-1}(1/z) , \quad n = 1 \to \infty . \quad (9)$$

The above update rule (9) starts with

$$H_0(z) = \frac{1}{\sqrt{r_0}} , \quad s_1 = \frac{r_1}{r_0} . \quad (10)$$

Here, s_n, $n = 1 \to \infty$, are known as the reflection coefficients and at stage n it is recursively given by

$$s_n = \left(\sum_{k=0}^{n-1} a_k^{(n-1)} r_{n-k} \right) a_0^{(n-1)} . \quad (11)$$

It can be shown that [5], when (5) is satisfied, these reflection coefficients are bounded by unity, i.e.,

$$|s_n| \le 1 , \quad n = 1 \to \infty , \quad (12)$$

and for every n, the polynomial in (8) has all its roots in $|z| \le 1$. Clearly, the single-step algorithm in (9)–(10) brings in the new information entirely through the reflection coefficients. The unique extension so obtained in (7) has other interesting properties as well. To examine this, notice that for an *arbitrary* stochastic process with integrable power spectal density $S(\omega)$, i.e.,

$$\int_{-\pi}^{\pi} S(\omega) d\omega < \infty , \quad (13)$$

the functional in (6) is referred to as the causality criterion, since the condition

$$H = \frac{1}{2\pi} \int_{-\pi}^{\pi} ln\, S(\omega) d\omega > -\infty , \quad (14)$$

(Paley-Wiener criterion), guarantees the factorization of $S(\omega)$ in terms of a unique function $B(z)$ that is analytic together with its inverse in $|z| > 1$ and has square summable impulse response, i.e.,

$$S(\omega) = |B(e^{j\omega})|^2 \quad (15)$$

where

$$B(z) = \sum_{k=0}^{\infty} b_k z^{-k} , \quad b_0 > 0 , \quad (16)$$

and

$$\sum_{k=0}^{\infty} |b_k|^2 < \infty , \quad (17)$$

Clearly, this minimum phase function $B(z)$, known as the Wiener factor of the process, represents a causal (one-sided) digital filter with square summable impulse response. For Gaussian processes, the Paley-Wiener criterion in (14) refers to its entropy and in general we will refer to H as the entropy functional of that process.

Thus, subject to the additional constraint in[1] (14), the given autocorrelations $r_0 \to r_n$ will generate a positive-definite Toeplitz matrix and in that case the spectrum estimation problem has an infinite number of solutions [5, 6]. To see this, it is best to refer to the well known geometric interpretation for the class of all admissible solutions [6]. Letting Δ_k denote the determinant of $\mathbf{T}_k$, $k = 0 \to \infty$ as in (4), subject to (14) the given covariances satisfy $\Delta_n > 0$ and clearly, any admissible extension r_k, $k = n+1 \to \infty$ must satisfy

$$\Delta_k > 0 , \quad k = n+1 \to \infty . \quad (18)$$

Consequently, at the first step r_{n+1}, must be chosen so that $\Delta_{n+1} > 0$. Using matrix identities, we can expand Δ_{n+1} to give

$$\Delta_{n+1} = \frac{\Delta_n^2 - \Delta_{n-1}^2 |r_{n+1} - \xi_n|^2}{\Delta_{n-1}} , \quad (19)$$

where

$$\xi_n = \mathbf{f}^* \mathbf{T}_{n-1}^{-1} \mathbf{b} , \quad (20)$$

$$\mathbf{f} = [\, r_1 \; r_2 \; \cdots \; r_n \,]^T , \quad \mathbf{b} = [\, r_n \; r_{n-1} \; \cdots \; r_1 \,]^T .$$

From (19)–(20), we obtain the well known relation [6]

$$|r_{n+1} - \xi_n|^2 < R_n^2 \quad (21)$$

where

$$R_n = \frac{\Delta_n}{\Delta_{n-1}} > 0 . \quad (22)$$

Equation (21) represents the interior of a circle with center ξ_n and radius $R_n > 0$ and in turn it implies that every admissible extension must have the unknown correlation r_{n+1} inside the above 'admissible' circle. Repeating (19)–(21) at every stage, we obtain [5, 6]

$$|r_{k+1} - \xi_k|^2 \le R_k \quad (23)$$

where

$$R_k = \frac{\Delta_k}{\Delta_{k-1}} > 0 , \quad k = n \to \infty . \quad (24)$$

Thus, at every stage $k = n \to \infty$, the unknown correlation can be chosen in an infinite number of ways satisfying (23), and since at each stage the present value of the parameters ξ_k and R_k depend on the choice at the previous stages, clearly, there are an infinite number of solutions to the spectrum extention problem.

[1] Moreover, when both (13) and (14) are satisfied, every $\Delta_k > 0$, every $|s_k| < 1$ in (12) and the Levinson polynomials in (9) are strict Hurwitz (all zeros in $|z| < 1$) [5].

To identify the maximum entropy extension described before among the class of all such extensions, it is necessary to connect the entropy H with the extension parameters in (23). Towards this, notice that for any $k \geq n$, equation (19) translates into

$$\frac{\Delta_{k+1}}{\Delta_k} = \frac{\Delta_k}{\Delta_{k-1}}\left[1 - \left(\frac{\Delta_{k-1}|r_{k+1} - \xi_k|}{\Delta_k}\right)^2\right] \quad (25)$$

and using (24), this gives the inequality,

$$R_{k+1} \leq R_k \leq R_n\,, \quad k = n+1 \rightarrow \infty\,, \quad (26)$$

i.e., the radii of the admissible circles in (23) form a monotone nonincreasing positive sequence of numbers bounded by R_n. Letting R_∞denote the limit of this sequence in (26), we have

$$R_\infty \leq R_k \leq R_n = \frac{\Delta_n}{\Delta_{n-1}}\,, \quad k = n+1 \rightarrow \infty\,. \quad (27)$$

Interestingly, this 'final radius' R_∞ is related to the power spectral density function $S(\omega)$ through the relation [1, 5, 6]

$$R_\infty = exp\left[\frac{1}{2\pi}\int_{-\pi}^{\pi} ln\, S(\omega)d\omega\right] = exp[H]\,. \quad (28)$$

From (28), maximization of entropy is equivalent to choosing the extension in (23) such that the radius of the 'final circle' has its maximum possible value R_n. Using (27), clearly R_∞ takes its maximum possible value Δ_n/Δ_{n-1} if and only if *every* $R_k = \Delta_n/\Delta_{n-1}$, $k = n+1 \rightarrow \infty$, i.e.,

$$(R_\infty)_{\max} = \frac{\Delta_n}{\Delta_{n-1}} \text{ iff } R_k = \frac{\Delta_n}{\Delta_{n-1}}\,, \; k = n+1 \rightarrow \infty\,.$$

Then, from (25), we have [6]

$$r_{k+1} = \xi_k\,, \quad k = n \rightarrow \infty\,,$$

and every extension r_{k+1} must be at the centers of the respective admissible circles in the case of maximum entropy method. Since the center of the circle is maximally away from the circumference of the circle, such an extension is also most robust and maximally noncommittal. From previous remarks, maximum entropy extension always leads to an autoregressive solution, and to explore its application in linear phase FIR filter design, we next examine a possible approach.

2 Linear Phase FIR Filter Design

If a given gain function $G(\theta)$ can be written as

$$G(\theta) = |H(e^{j\theta})|^2 \quad (29)$$

where

$$H(z) = h_0 + h_1 z^{-1} + \cdots + h_n z^{-n}\,,$$

then $G(\theta)$ represents a finite impulse response (FIR) filter. If $H(e^{j\theta})$ also has linear phase, i.e.,

$$H(e^{j\theta}) = A(\theta)e^{\psi(\theta)} \text{ and } \psi(\theta) = k\theta\,,$$

then $H(z)$ is a linear phase FIR filter. It is well known that a sufficient condition for a polynomial to have linear phase is that it be reciprocal, i.e.,

$$H(z) = z^{-n}H^*(1/z^*) \triangleq \tilde{H}(z)\,. \quad (30)$$

For any rational function, define

$$H_*(z) = H^*(1/z^*)\,, \quad (31)$$

then (29) can be rewritten as

$$G(\theta) = H(z)H_*(z)|_{z=e^{j\theta}} = H(e^{j\theta})H^*(e^{j\theta})\,.$$

If the given $G(\theta)$ does not represent the gain funtion associated with an FIR filter, nevertheless, it may be possible to obtain an FIR filter $H(z)$, such that (29) is satisfied approximately. More accurately, if the deviation between $G(\theta)$ and $|H(e^{j\theta})|^2$ is negligible for all θ, then $H(z)$ represents a reasonable FIR approximation to the gain funtion $G(\theta)$. In that sense, the quantity

$$\eta \triangleq \sup_{0\leq\theta\leq 2\pi} \left|G(\theta) - |H(e^{j\theta})|^2\right| \quad (32)$$

which represents the supremum error norm can be used as a meaningful measure in selecting $H(z)$.

Since a power spectral density function $S(\theta)$ and a gain function have a one-to-one correspondence, comparing (15) and (29), the filter design problem is conceptually similar to the spectral factorization problem. As a result, given a gain function $G(\theta)$ that satisfies the integrability condition (13) and the causality criterion (14), let

$$G(\theta) = \sum_{k=-\infty}^{+\infty} r_k e^{-jk\theta} \quad (33)$$

represent its Fourier representation. By treating $G(\theta)$ as a spectral density and applying the maximum entropy method to $(n+1)$ of its autocorrelations $r_0, r_1, \ldots, r_n$, one can generate the corresponding Levinson polynomials $H_n(z)$ in (9) and from (7), for any $n \geq 1$, we also have

$$\frac{1}{|H_n(e^{j\theta})|^2} = \sum_{k=-n}^{n} r_k e^{jk\theta} + o(e^{j(n+1)\theta})\,. \quad (34)$$

Clearly, as n increases, the difference

$$G(\theta) - \sum_{k=-n}^{n} r_k e^{jk\theta} \longrightarrow 0 .$$

Since from (33), $r_k \to 0$ as $k \to \infty$, and from (34), $H_n(z)$ represents a reasonable IIR approximation to $G(\theta)$.

Following the above arguments, an FIR approximation is straightforward. Instead of starting with the given gain function $G(\theta)$, define

$$K(\theta) = \frac{1}{G(\theta)} . \qquad (35)$$

Naturally, if $G(\theta)$ satisfies (13) and (14), for $K(\theta)$ to satisfy those equations, it is sufficient that $G(\theta) > 0$ for all θ. Evidently, applying maximum entropy method to $K(\theta)$ allows one to approximate it by $|1/P_n(e^{j\theta})|^2$, where $P_n(z)$ represents the Levinson polynomial associated with $1/G(\theta)$. As n increases, since the sup. error in (32) for $1/G(\theta)$ decreases to zero, we can write

$$K(\theta) = \frac{1}{G(\theta)} \doteq \frac{1}{|P_n(e^{j\theta})|^2} \qquad (36)$$

or

$$G(\theta) \doteq |P_n(e^{j\theta})|^2 . \qquad (37)$$

Thus, $P_n(z)$ is an nth order FIR filter that approximates the given gain function $G(\theta)$. The filter so obtained represents a strict Hurwitz polynomial (see footnote 1). Of course, if $G(\theta)$ in fact represents the gain associated with an nth degree strict Hurwitz polynomial, then at stage n this technique will correctly identify that polynomial uniquely upto sign.

Even though the filter $P_n(z)$ in (37) represents an FIR approximation to $G(\theta)$, its response does not possess linear phase. However, as shown below, the remedy is near by and this can be achieved by beginning with the function

$$K_1(\theta) = \frac{1}{\sqrt{G(\theta)}} . \qquad (38)$$

Clearly when $K_1(\theta)$ satisfies (13) and (14), once again, it is possible to generate its respective Levinson polynomials $H_n(z)$ as in (7) from its Fourier coefficients and for large n, we have

$$K_1(\theta) \doteq \frac{1}{|H_n(e^{j\theta})|^2}$$

and using (38) and (30),

$$\begin{aligned} \sqrt{G(\theta)} &\doteq |H_n(e^{j\theta})|^2 \\ &= \left|H_n(z)\tilde{H}_n(z)\right|_{z=e^{j\theta}} = |A(z)|_{z=e^{j\theta}} \end{aligned}$$

or

$$G(\theta) \doteq \left|A(e^{j\theta})\right|^2 , \qquad (39)$$

where

$$A(z) = H_n(z)\tilde{H}_n(z) = \tilde{A}(z) \qquad (40)$$

represents a symmetric (reciprocal) polynomial of degree $2n$.

Thus, any given power gain function $G(\theta)$ can be approximated by a linear phase FIR filter. The approximation can always be improved by increasing n and to accomplish that one can make use of the recursions in (9)–(11). The sup. error norm in (32) can be utilized to decide upon a stopping criterion.

A further refinement can be employed to minimize any passband distortion in the FIR approximations given by (37) and (40). To illustrate this , refer to (35)–(36) and instead of the approximation in (37), consider a new polynomial approximation of the form

$$G(\theta) \doteq |\sum_{k=0}^{n} \alpha_k P_k(e^{j\theta})|^2 = |Q_n(e^{j\theta})|^2 \qquad (41)$$

such that the error

$$\sqrt{G(\theta)} - \sum_{k=0}^{n} \alpha_k P_k(e^{j\theta})$$

is minimized by selecting the coefficients α_k, $k = 0 \to n$ in some optimal fashion. Here $P_k(z)$ represents the Levinson polynomials of $1/G(\theta)$ as given in (36). Minimization of the square error

$$|\sqrt{G(\theta)} - \sum_{k=0}^{n} \alpha_k P_k(e^{j\theta})|^2$$

that has been weighted with respect to $1/G(\theta)$ gives

$$\frac{\partial}{\partial \alpha_m} \int_{-\pi}^{\pi} |\sqrt{G(\theta)} - \sum_{k=0}^{n} \alpha_k P_k(e^{j\theta})|^2 \frac{1}{G(\theta)} d\theta = 0$$

for $m = 0 \to n$ and hence

$$\sum_{k=0}^{n} \alpha_k \int_{-\pi}^{\pi} \frac{P_k(e^{j\theta})P_m^*(e^{j\theta})}{G(\theta)} d\theta = \int_{-\pi}^{\pi} \frac{P_m^*(e^{j\theta})}{\sqrt{G(\theta)}} d\theta$$

$$\triangleq \beta_m , \quad m = 0 \to n . \qquad (42)$$

Using (8)–(11), it is easy to show that $\tilde{P}_k(z)$ are orthonormal functions with respect to $1/G(\theta)$ and hence it follows that

$$\frac{1}{2\pi} \int_{-\pi}^{\pi} \frac{P_k(e^{j\theta})P_m^*(e^{j\theta})}{G(\theta)} d\theta = \sqrt{\frac{R_k}{R_m}} \le 1 , \ k \ge m , \qquad (43)$$

where R_m is as defined in (24). Substituting (43) into (42), we obtain $n+1$ linear equations in $n+1$

unknowns and these equations can be solved for α_k, $k = 0 \to n$. However, since the off-diagonal elements $\sqrt{R_k/R_m} < 1$ (for $k \neq m$) in (43) and $\beta_m \simeq 1$ in (42), the approximate solution $\alpha_k = 1$, $k = 0 \to n$ turns out to be an excellent choice here in terms of minimizing the ripples in the passband as well as the stopband. In that case, with $P_k(z) = b_0^{(k)} + b_1^{(k)}z^{-1} + \ldots + b_k^{(k)}z^{-k}$, the desired FIR filter of order n is given by

$$Q_n(z) = \sum_{k=0}^{n} P_k(z). \qquad (44)$$

The orthogonal property of the reciprocal polynomials in (43) has effectively decoupled the n linear equations in (42) and hence no matrix operations are involved in the above approximation. These filters, which in a sense combine the maximum entropy as well as the minimization of the mean square error criteria, have been found to possess uniformly low distortion in the passband irrespective of the transition band size as well as the stop band size.

Notice that as before (see (38)– (41)), in this case also, linear phase can be maintained by starting with $\sqrt{G(\theta)}$ rather than $G(\theta)$ in (41). In that case with $H_n(z)$ representing the Levinson polynomials associated with $1/\sqrt{G(\theta)}$, the desired linear phase FIR filter is given by

$$B(z) = Q_n(z)\tilde{Q}_n(z). \qquad (45)$$

and

$$Q(z) = \sum_{k=0}^{n} H_k(z). \qquad (46)$$

Here

$$H_k(z) = a_0^{(k)} + a_1^{(k)}z^{-1} + \cdots + a_k^{(k)}z^{-k}$$

as in (8). Clearly $B(z)$ in (45) possesses linear phase.

Next, numerical results are presented for comparing the performance of the proposed method with other well known design techniques [7]–[9].

3 Simulation and Comparison

The procedure described in (35) or (39) of inverting a given gain function and then applying the maximum entropy method to obtain FIR filters also has other interpretations. To observe this, consider the realization of an ideal low pass gain function $G(\theta)$ (see Fig. 1(a)) using the above technique. Applying (35) to $G(\theta)$ results in Fig. 1(b). Notice that such a $K(\theta)$ has been used in [7], [8] on the basis of low sidelobe characteristics in the stopband region. Implications of Fig. 1(b) has not been made very clear in those cases. However, the application of MEM described here for low-pass FIR filter design suggests $K(\theta)$ in Fig. 1(b) in a natural fashion. Interestingly, the 'don't care' transition band also can be specified in advance in this new scheme and we have used a linear transition band in all our simulations.

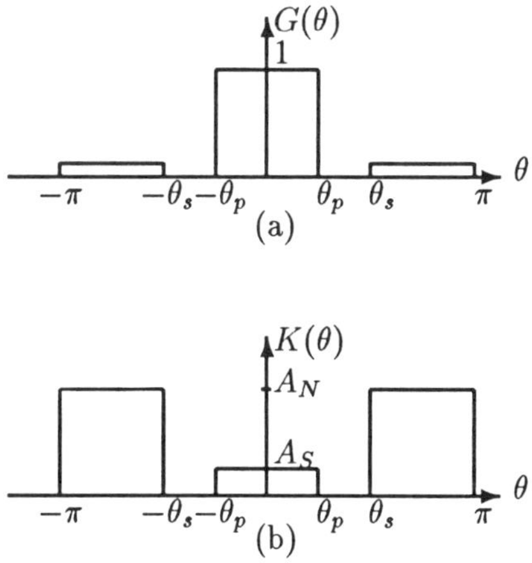

Figure 1: (a) Ideal Low-pass Gain Function $G(\theta)$. (b) $K(\theta) = 1/G(\theta)$.

Fig. 2 shows the magnitude responses $|H(e^{j\theta})|^2$ (in dB) of linear phase low pass FIR filters using the proposed scheme (using (45)) as well as that reported in [7] with details as indicated there. Even though increasing 'the noise spectral height' A_N will help to reduce the stopband sidelobe level, as pointed out in [7], it will nevertheless increase the passband distortion.

Figs. 3(a) and Fig. 3(b) compare the magnitude response $|H(e^{j\theta})|^2$ (in dB) of linear phase FIR filters using the Parks-McClellan technique [9, 10] and the proposed scheme (using (45)). Of these, Fig. 3(a) represents a low pass filter and Fig. 3(b) refers to a bandpass case. Notice that as the transition band becomes smaller and smaller, the passband distortion becomes more and more pronounced in the Parks-McClellan case. As these figures show the pass band distortion is much smaller in the case of the proposed technique and this uniformly superior behavior is exhibited in all cases.

4 Conclusions

This paper describes a new technique to design linear phase FIR digital filters with/without linear phase. The design procedure makes use of the inverse of the desired magnitude gain functions (or its square root) and is able to achieve any desired degree of accuracy by incorporating a higher order model. An iterative procedure makes the implementation of the higher order filters very efficient. To minimize the effect of the residual passband distortion, a final averaging scheme generates a class of distortion free linear phase FIR filters. Since the proposed method is based entirely on 'digital' techniques, it becomes possible to include features such as transition band characterization into the design procedure itself. The new set of filters are seen to be more robust to changes in transition bandwidth as well as stop band level. This is quite valuable from a practical filter design point of view.

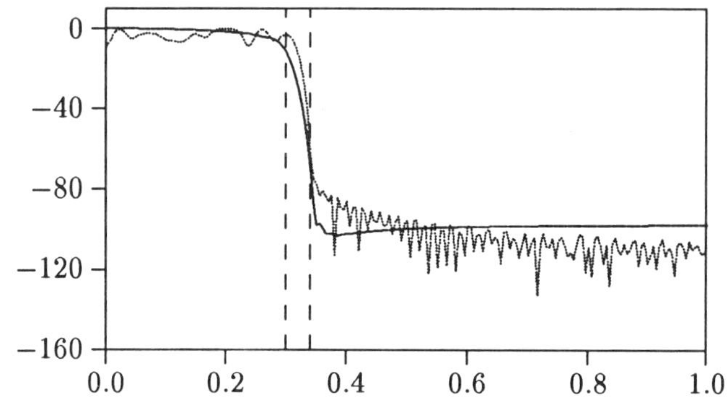

Fig. 2: Linear phase lowpass FIR filter design. Solid curve represents the present technique based on MEM with $A_N = 10^{10}$. Dotted curve represents Farden and Scharf's technique based on MMSE with $A_N = 10^6$. Signal spectral height $A_S = 1$, pass bandwidth BW = 0.3, transition bandwidth TB = 0.04 and filter order N = 128.

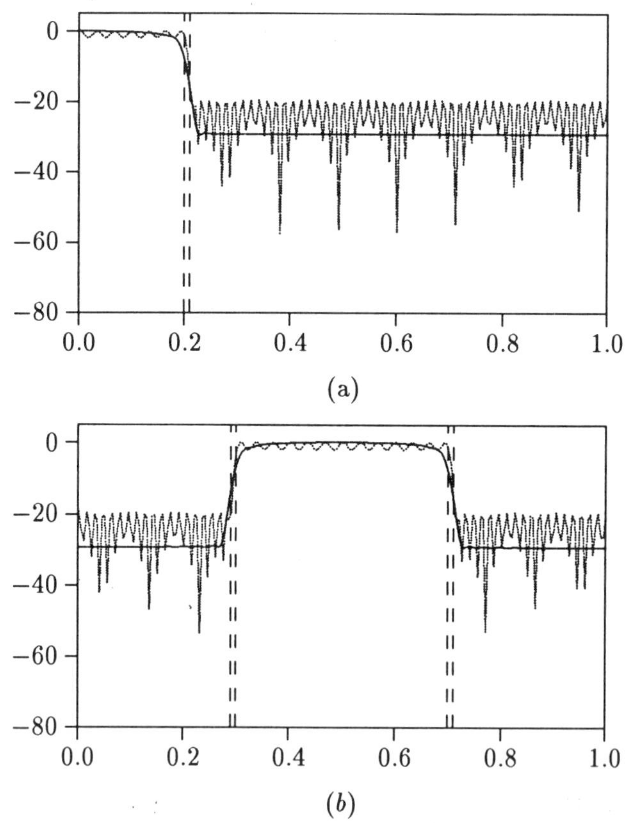

Fig. 3: Linear phase bandpass FIR filer design (a) Lowpass case with pass bandwidth BW = 0.2. (b) Bandpass case with pass bandwidth BW = 0.4. Common transition bandwidth TB = 0.01. Solid curves represent the present technique based on MEM. Dotted curves represent the Parks-McClellan technique. In each case signal spectral height height $A_S = 1$, noise spectral height $A_N = 10^3$ and filter order N = 128.

References

[1] U. Grenander and G. Szegö, *Toeplitz Forms and Their Applications*, New York: Chelsea, 1984.

[2] G. Szegö, *Orthogonal Polynomials*, Amer. Math. Soc., Colloquium Publications, vol. 23, pp. 287–292, 1985.

[3] J. P. Burg, "Maximum entropy analysis," presented at *the 37th Annual Meeting, Soc. Explor. Geophysics*, Oklahoma City, Oklahoma, 1967.

[4] C. R. Rao, *Linear Statistical Inference and Its Applications*, 2nd ed., New York: John Wiley, 1973.

[5] Ya. L. Geronimus, *Polynomials orthogonal on a Circle and their applications*, Amer. Math. Soc., Translation, no. 104, 1954.

[6] D. C. Youla, "The FEE: A new tunable high-resolution spectral estimator, part I," Tech. Report, no. 3, Polytechnic Inst. of New York, Brooklyn, New York, 1980. Also, RADC Report, RADC-TR-81-397, AD A114996, Feb. 1982.

[7] D. C. Farden and L. I. Scharf, "Statistical design of nonrecursive digital filters," *IEEE Trans. Acoust., Speech, Signal Proc.*, vol.ASSP-22, no. 3, pp. 188–196, June 1974.

[8] R. N. Madan, "FIR digital filters in a Linear array processor," *Proc. of the IEEE OCEANS'79 Conference, San Dieago, CA*, pp. 1–5, Sept 17–19, 1979.

[9] J. H. McClellan and T. W. Parks, "A unified approach to the design of optimum FIR linear-phase digital filters," *IEEE Trans. Circuits Systems* , CT-20, pp. 697–701, 1973.

[10] T. W. Parks and C. S. Burrus, *Digital Filter Design*, New York: John Wiley, 1985.

Design of Pipelined Lattice IIR Digital Filters

Jin-Gyun Chung, and Keshab K. Parhi

Department of Electrical Engineering
University of Minnesota
Minneapolis, MN 55455, USA

Abstract

Lattice structures exhibit good finite word length properties. In some real-time applications, high speed is very crucial to the system performance along with finite word length effects. Although lattice filters can be pipelined by cut-set localization procedure, no sample rate increase can be achieved by that method.

In this paper, a pipelining method in lattice digital filters is introduced. This pipelining method is based upon constrained IIR digital filter design method by which pipelined direct-form filters are designed. These direct-form filters are transformed to pipelined lattice digital filters. It is shown that the roundoff error and the number of multiply/add operations of the resulting pipelined lattice filters are smaller than those of the pipelined lattice filters obtained by applying look-ahead on direct-form nonpipelined digital filters.

1. Introduction

Since lattice structures exhibit very good finite word length properties, these structures are preferable for fixed-point finite word-length implementation. Different lattice digital filter structures have been proposed in [1-4].

Achieving high speed in direct-form and and lattice digital filters is difficult because of the feedback loop. Although these filters can be pipelined by cut-set localization procedure proposed in [5], it should be noted that no sample rate increase can be achieved by that method. Cutset transformation can increase the clock speed, but cannot increase the sample rate, since multiple clock cycles are needed to process one sample.

Implementation of high-speed, pipelined, stable, recursive digital filters was considered impossible before. Only recently, scattered look-ahead techniques were proposed to convert a nonpipelined filter to a pipelinable form [6]. The scattered look-ahead technique adds poles at equal angular and radial spacing for each pole of the nonpipelined filter, and these poles are canceled by zeros. This transformation converts the denominator of the transfer function to a pipelined form. This technique, for the first time, made it feasible to implement high speed recursive digital filters using pipelining.

One drawback of the scattered look-ahead technique is the introduction of the canceling zeros. These zeros increase the number of multiplication operations needed to implement the digital filter. Therefore, as suggested in [6], it is of interest to design pipelinable digital filter transfer functions directly from the filter spectrum, as opposed to designing a nonpipelined filter and applying scattered look-ahead on this filter.

To this end, we address the design of pipelined direct-form recursive digital filters from the filter spectrum. By constraining the location of poles at equal angular and radial spacing, we can design a transfer function whose denominator is pipelinable. Fortunately, this is possible by using a variation of the filter design procedure used for decimation filter design [7-10]. The pipelinable filter transfer function can be mapped to many lattice digital filter structures. Pipelining the orthogonal structure in [3] leads to no increase in speed. Therefore, the orthogonal lattice filter structure is not suitable for pipelining. The minimum-noise lattice structure in [4] requires more number of muiltiply-add operations than [1], but leads to less roundoff error. In this paper, we propose the use of pipelining for the basic, normalized, and the minimum-noise lattice structure.

This paper is organized as follows. The Schur algorithm [11] which forms the basis for design of lattice digital filters is reviewed in section 2. Section 3 describes the theory of lattice filter pipelining. Section 4 presents the constrained filter design procedure for design of pipelined direct-form recursive digital filters, which are then reduced to lattice structures. Section 5 presents design examples for several lattice filters.

2. Schur Algorithm

One of the important properties of the Schur algorithm is that *all polynomials expanded from any given polynomial by the Schur algorithm are orthogonal to each other*. Schur polynomial is a polynomial which

This research was spported by the army research office by contract number DAAL03-90-G-0063.

does not have zeros on or outside the unit circle. Therefore, the denominator of a stable IIR filter is a Schur polynomial. Let a polynomial with real coefficients be defined by

$$D(z) = \sum_{i=0}^{N} d_i z^i.$$

Then, initialize Nth-order Schur polynomial $\Phi_N(z)$ as $\Phi_N(z) = D(z)$,

$$\text{where } \Phi_N(z) = \sum_{i=0}^{N} \phi_i z^i.$$

From $\Phi_N(z)$, form the polynomial $\Phi_{N-1}(z)$ as follows:

$$\Phi_{N-1}(z) = \frac{z^{-1}[\phi_n \Phi_N(z) - \phi_0 \Phi_N^*(z)]}{\sqrt{\phi_N^2 - \phi_0^2}},$$

where $\Phi_N^*(z)$ is the reverse polynomial of $\Phi_N(z)$ and is defined by $z^N \Phi_N(z^{-1})$. The degree of $\Phi_{N-1}(z)$ is 1 less than $\Phi_N(z)$.

In order to implement the algorithm, each of the polynomials $\Phi_{N-i}(z)$ is generated from $\Phi_{N-i+1}(z)$, for i=1,2, ... ,N, in the same manner as $\Phi_{N-1}(z)$ is generated from $\Phi_N(z)$. It may be noted that the coefficients of increasing powers of z in $\Phi_{N-1}(z)$ are the N determinants of 2 x 2 submatrices formed by the first column and each succeeding column in the following matrix.

$$\frac{1}{C}\begin{bmatrix} \phi_N & \phi_{N-1} & \phi_{N-2} & \cdots & \phi_1 & \phi_0 \\ \phi_0 & \phi_1 & \phi_2 & \cdots & \phi_{N-1} & \phi_N \end{bmatrix},$$

$$\text{where } C = \sqrt{\phi_N^2 - \phi_0^2}.$$

For each 2 x 2 submatrix, if a column is composed of all zero elements, the determinant of the corresponding submatrix is zero. Using this property of determinant, we can observe an useful fact for pipelining:
If $\Phi_N(z)$ is a Schur polynomial of order N, and has j-consecutive zero coefficients between each two nonzero coefficients of nearest degree, then $\Phi_{N-i}(z)$ of order N-i also has j-consecutive zero coefficients between each two nonzero coefficients of nearest degree for i = 1 to N-j-1 and has only one nonzero coefficient for i = N-j to N.

3. Pipelining of Recursive Digital Lattice Filters

In this section, we study pipelinability of basic, normalized and minimum-noise lattice filters.

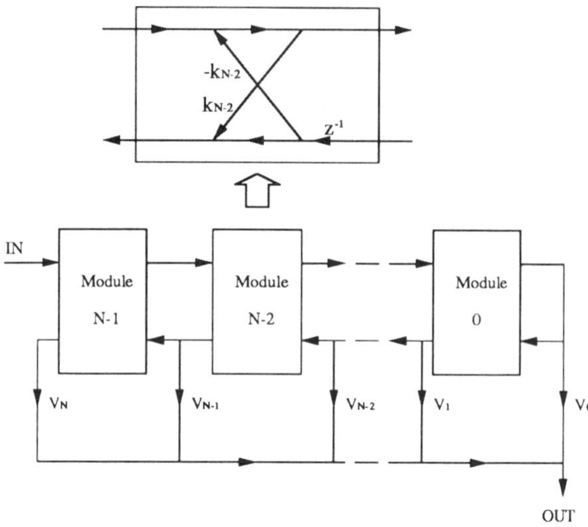

Fig.1 The structure of Basic lattice filter

3.1 Pipelining of Basic Lattice Filter

Fig. 1 shows the structure of the basic lattice filter which is characterized by N k-parameters and $N+1$ tap gain parameters. The multiply-add operations in the feed-forward section can be pipelined at any desired level by placing latches at appropriate feed-forward cutset locations. As can be seen from Fig. 1, the feedback loop in a module is removed if the k-parameter of the module is zero. Therefore, by removing $(M-1)$-consecutive k-parameters, the basic lattice filter can be pipelined by M-stages, leading to M times increase in speed as compared with the original basic lattice filter.

Let the Nth-order denominator of a digital filter transfer function be denoted by

$$D_m(z) = \sum_{i=0}^{m} d_{m,i} z^{-i},$$

where $m = N$, and the leading coefficient, $d_{m,0}$, is assumed to be one. Then, k-parameters are recursively obtained by starting from $D_N(z)$ using:

1) $k_{m-1} = d_{m,m}$

2) $$D_{m-1}(z) = \frac{D_m(z) - k_{m-1} D_m^*(z)}{1 - k_{m-1}^2}$$

for $m = N, N-1, \ldots, 1$.

As can be seen from above steps, $D_m(z)$'s for $m = N, N-1, .., 0$ are Schur polynomials with a different normalization constant. Therefore, if $D_N(z)$ has $(M-1)$-consecutive zero coefficients between each two nonzero coefficients of nearest degree, then $D_m(z)$'s for $m = N, N-1, \ldots, M$, also have $(M-1)$-consecutive zero coefficients between each two nonzero coefficients of nearest degree and $D_m(z)$'s for $m = M-1, M-2, ..., 1$, have only one nonzero coefficients. Then, since k-parameters are the coefficients of z^{-m} of each $D_m(z)$ for $m=N, N-1, ..., 1$, we can obtain $(M - 1)$-consecutive zero k-parameters, which leads to M-stage pipelining. In general, an M-stage pipelined filter contains M-delays in every loop. Since the number of multiply-add operations in every loop remains constant, the M-stage pipelined filter can be clocked at M-times faster rate, as compared with the nonpipelined filter.

3.2 Pipelining of Normalized Lattice Filter

The structure of normalized lattice filter is shown in Fig. 2, where the feedback section is again described by k-parameters only. The k-parameters are calculated by the same procedure as in section 3.1.

As can be seen from Fig. 2, there are four multipliers within a module m: k_m, $-k_m$, and two $\sqrt{1 - k_m^2}$'s. If k_m is zero, then $\sqrt{1 - k_m^2}$ becomes one. Therefore, module m needs no multipliers and the feedback loop of the module is removed, which means the normalized lattice filter can be pipelined.

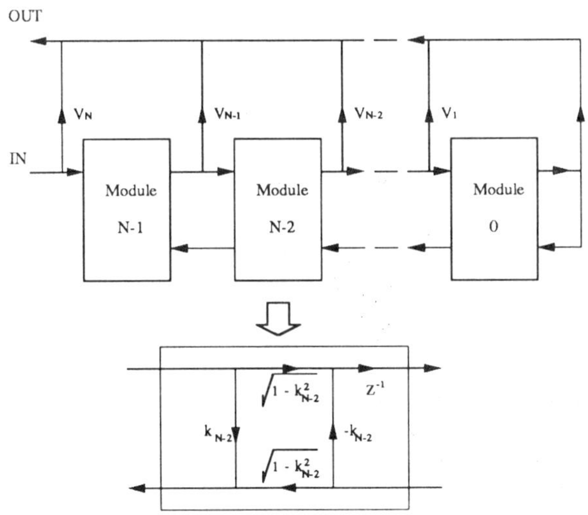

Fig. 2 The structure of Normalized lattice filter

3.3 Pipelining of Minimum-Noise Lattice Filter

The structure of the minimum-noise lattice filter [4] is shown in Fig. 3. Feedback section is described by k-parameters only and the k-parameters are calculated using Schur polynomials as follows:

$$k_m = \frac{\Phi_m(0)}{\Phi_m^*(0)},$$

where $\Phi_m(z)$ is mth-order Schur polynomial obtained from the denominator. If k_m of the module m is zero, then the feedback loop of the module is removed since $\sqrt{1 - k_m^2}$ becomes one. For k_m to be zero, $\Phi_m(0)$, or the constant term of $\Phi_m(z)$, should be zero. As was discussed in section 2, if the denominator has $(M - 1)$-consecutive zero coefficients between each two nonzero coefficients of nearest degree, then the minimum-noise filter can be pipelined.

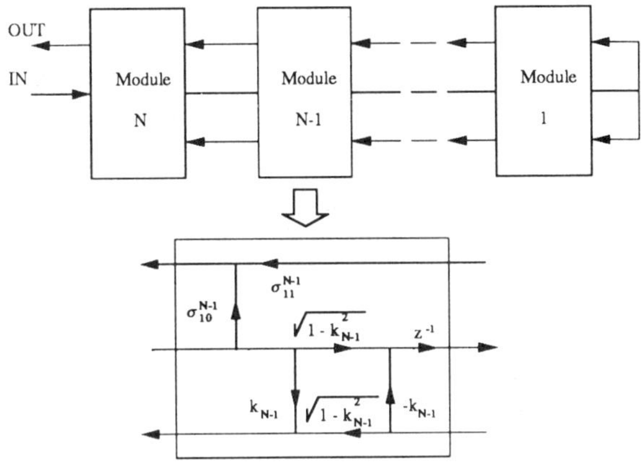

Fig. 3 The structure of Minimum-noise lattice filter

4. Design of Pipelined IIR Lattice Filter Using Constrained Filter Design

It is apparent from previous sections that the basic, normalized, and minimum-noise lattice filters can be pipelined by M-stages if the denominator of a transfer function has $(M-1)$-consecutive zero coefficients between each two nonzero coefficients of nearest degree. The pipelinable transfer functions can be obtained by applying the scattered look-ahead method to the nonpipelined filter transfer functions. However, by using constrained filter design method, we can design inherently pipelinable transfer functions directly from the filter spectrum specifications.

Our procedure is a variation of the procedures in [7] and [9]. The design method in [7] first expresses the magnitude and group delay responses of the filter as functions of the radii and angles of the poles and zeros. Then we obtain formulae for the partial derivatives of the magnitude and group delay with respect to the radius and the angle of a pole and a zero. These derivatives are used in the Fletcher-Powell algorithm to minimize the approximation error [8]. In this algorithm, the minimum value is obtained by only using the first derivative information. To obtain an inherently M-stage pipelinable filter, we recompute the partial derivatives for a denominator in powers of z^M rather than z [9] and substitute these equations into the program in [10].

Then, the resulted denominator is in terms of z^{-M}.

The modified filter design procedure is summarized as follows:

1) For the given filter specifications (pass-band, stop-band, pass-band & stop-band ripples, and M-stage pipelining), we start with 1-complex pole pair. Then, the order of the denominator is $2M$ and the number of zeros are restricted so that the order of the numerator is less than or equal to $2M$. We usually start with M unit-circle zero pairs.

2) If the filter specifications are not satisfied by 1-complex pole pair, we increase the number of poles and zeros and/or adjust initial positions of poles and zeros and weighting factors for the pass-band and stop-band. The filter is redesigned. This procedure is repeated until the filter specifications are satisfied.

3) The pipelined direct-form filter transfer function is then used to obtain the pipelined lattice filters by using the sybthesis procedures of section 3.

5. Design Example of Pipelined Lattice Filters

Consider the lowpass filter with 1)pass-band: 0 - 0.2 π, ripple: 1 dB, and 2)stop-band: 0.3 π - π, ripple: 15 dB. In [12], this example was designed by using the 4th order Chebyshev filter:

$N(z) = 0.0018 + 0.0073z^{-1} + 0.0011z^{-2} + 0.0073z^{-3} + 0.0018z^{-4}$

$D(z) = 1 - 3.0544z^{-1} + 3.829z^{-2} - 2.2925z^{-3} + 0.05z^{-4}$,

where N(z) and D(z) denote numerator and denominator of the transfer function, respectively.

Using the constrained filter design procedure, we obtained the transfer functions for different number of pipelining stages. These filters are listed below.

1) Nonpipelined

$N(z) = 0.0307 + 0.0054z^{-1} + 0.0203z^{-2} + 0.0054z^{-3} + 0.0307z^{-4}$

$D(z) = 1 - 2.5754z^{-1} + 2.868z^{-2} - 1.5335z^{-3} + 0.333z^{-4}$.

2) 2-stage pipelined (M = 2)

$N(z) = 0.0532 + 0.112z^{-1} + 0.1159z^{-2} + 0.0941z^{-3} + 0.1159z^{-4} + 0.112z^{-5} + 0.0532z^{-6}$

$D(z) = 1 - 1.02z^{-2} + 1.065z^{-4} - 0.509z^{-6} + 0.128z^{-8}$.

3) 4-stage pipelined (M = 4)

$N(z) = 0.0754 + 0.132z^{-1} + 0.2018z^{-2} + 0.2411z^{-3} + 0.2815z^{-4} + 0.2411z^{-5} + 0.2018z^{-6} + 0.132z^{-7} + 0.10754z^{-8}$.

$D(z) = 1 + 0.5626z^{-4} + 0.0791z^{-8}$.

4) 8-stage pipelined (M = 8)

$N(z) = 0.0974 + 0.0982z^{-1} + 0.0814z^{-2} + 0.0347z^{-3} + 0.0047z^{-4} - 0.0275z^{-5} + 0.029z^{-6} + 0.0761z^{-7} + 0.065z^{-8} + 0.012z^{-9} - 0.0434z^{-10} - 0.058z^{-11} - 0.055z^{-12} - 0.0034z^{-13} + 0.0082z^{-14} + 0.0098z^{-15} + 0.0098z^{-16}$

$D(z) = 1 - 0.2013z^{-8} + 0.101z^{-16}$.

Fig. 4 and Fig.5 show the magnitude response and the implementation of the 4-stage pipelined normalized lattice filter obtained from the above transfer function.

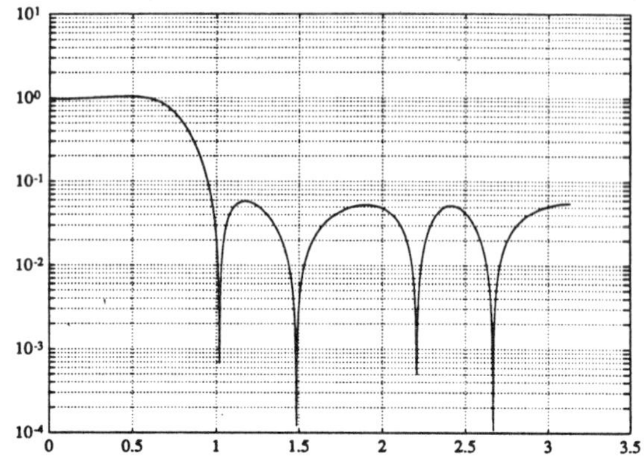

Fig. 4 Magnitude response of 4-stage pipelined filter by Constrained design

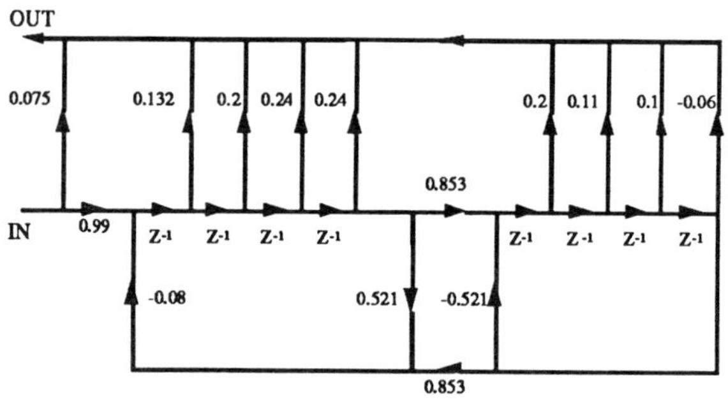

Fig. 5 4-stage pipelined Normalized filter by Constrained design

The feedback loop multiply-add operations can be pipelined by redistributing the delays. Roundoff errors of these pipelined normalized lattice filters and those obtained by using look-ahead method to the nonpipelined Chebyshev filter transfer function are calculated using the method introduced in [4], and are compared in Table 1. Roundoff noise gain factors of the direct-form cascade filters are also given. As can be seen from the table, better roundoff error characteristics are obtained by using constrained filter design. Roundoff noise gain factors of basic and minimum-noise lattice filters are given in Table 2. Note that the roundoff errors of minimum-noise lattice filters decrease as the pipelining stage increases.

Table 1 Roundoff noise gains of Normalized lattice and Direct-form cascade filters

	Normalized lattice		Cascade
	Constrained	Look-ahead	Look-ahead
nonpipe.	5.5246	5.8544	19874.1
2-stage	8.9003	8.1268	20211.9
4-stage	8.3172	16.0518	20227.5
8-stage	16.0489	32.0286	20340.6

Table 2 Roundoff noise gains of Normalized and Minimum-noise lattice filters

	Basic lattice	Minimum-noise
	Constrained	Constrained
nonpipe.	4.5225	2.1524
2-stage	9.0016	2.0216
4-stage	8.5022	1.1348
8-stage	16.1041	0.2815

The number of multipliers and adders needed for the implementation of these filters by normalized lattice filters are compared in Table 3. Note that the 4th order filter by constrained design could be implemented with one complex pole pair, and requires fewer multipliers than all other filters. The table shows that the lattice filters obtained by using the constrained filter design method require less hardware complexity than those obtained using look-ahead on the nonpipelined filter. Table 4 shows the number of multipliers and adders needed for the implementation of basic lattice and minimum-noise lattice filters obtained by using the constrained filter design method.

Table 3 Number of Multipliers and adders of Normalized lattice and Direct-form cascade filters

	Normalized latice				Cascade	
	Constrained		Look-ahead		Look-ahead	
	Multi	Adder	Multi	Adder	Multi	Adder
nonpipe.	19	11	19	11	5	8
2-stage	23	15	23	15	9	12
4-stage	15	11	31	23	13	16
8-stage	23	19	47	39	17	20

Table 4 Number of Multipliers and Adders of Basic and Minimum-noise lattice filters

	Basic lattice (Constrained)		Minimum-noise (Constrained)	
	Multi.	Adder	Multi.	Adder
nonpipe.	13	12	22	11
2-stage	17	16	30	15
4-stage	13	12	22	11
8-stage	21	20	38	19

6. Conclusions

In this paper, it was shown that the lattice filter structures in [1], [2] and [4] are suitable for pipelining. By using constrained filter design and the lattice filter design procedures, we designed pipelined high-speed lattice filters with extremely small roundoff errors. The hardware complexity of these structures is also much less than the pipelined lattice filters obtained by using scattered look-ahead on the nonpipelined filters.

As for further research work, the development of new pipelined lattice filter structures is also of interest. Wave digital filters also possess very good finite word length properties. Design of pipelined wave digital filters needs to be addressed in future. In [13] - [14], most significant bit first redundant arithmetic has been used to pipeline recursive and wave digital filters. Our pipelined filters can be used in combination with the approaches presented in [13] - [14] also.

References

[1] A. H. Gray, Jr. and J. D. Markel, “Digital Lattice and Ladder Filter Synthesis”, *IEEE Trans. on Audio and Electroacoustics*, pp. 491-500, Dec. 1973

[2] A. H. Gray, Jr. and J. D. Markel, “A Normalized Digital Filter”, *IEEE Trans. on Acoustics, speech, and signal processing*, vol. ASSP-33, pp. 258-277, June 1975

[3] S. K. Rao and T. Kailath, “Orthogonal Digital Filters for VLSI Implementation”, *IEEE Trans. on Circuits and systems*, pp. 933-945, Nov. 1984

[4] S. M. Lei and K. Yao, “A Class of Systolizable IIR Digital Filters and Its Design for Proper Scaling and Minimum Output Roundoff Noise”, *IEEE Trans. on Circuits and systems*, pp. 1217-1230, Oct. 1990

[5] S. Y. Kung, "On supercomputing with systolic/ wavefront array processors", *Proc. IEEE*, vol. 72, pp. 867-884, July 1984

[6] K. K. Parhi and D.G. Messerschmitt, "Pipeline Interleaving and Parallelism in Recursive Digital Filters- Part I: Pipelining Using Scattered Look-Ahead and Decomposition", *IEEE Trans. on Acoustics, Speech, and Signal Processing*, pp. 1099-1117, July 1989

[7] A. G. Deczky,"Synthesis of recursive digital filters using the minimum p-error criterion", *IEEE Trans. Audio Electroacoust.*, vol. AU-20, pp. 257-263, Oct. 1972

[8] R. Fletcher and M. J. D. Powell, “A rapidly convergent descent method for minimization", *Comput. J.*, vol. 6, pp. 163-168, 1963

[9] M. A. Richards, "Application of Deczky's program for recursive filter design to the design of recursive decimators", *IEEE Trans. on ASSP*, Oct. 1982

[10] A. G. Deczky, "Program for minimum-p synthesis of recursive digital filters" in *Programs for digital signal processing*, New York: IEEE Press, sec. 6.2 , 1979

[11] J. Schur, “Ueber Potenzreihen die im Innern des Einheitslreises beschrankt sind", *J. Reine Angewandte Mathematik*, vol. 147, pp. 205-232, Berlin 1917

[12] A. V. Oppenheim and R.W. Schafer, *Discrete Time Signal Processing*, Prentice-Hall, 1989

[13] S. C. Knowles, J. G. McWhirther, R. F. Woods, and J. V. McCanny, "Bit-level Systolic Arithmetics for High Performance IIR Filtering", *Journal of VLSI Signal Processing*, Kluwer Academic Press, vol. 1, 1989

[14] R. J. Singh and J. V. McCanny, "A Wave Digital Filter Three-Port Adaptor with Fine-Grained Pipelining", *Proc. of the 1991 Applications-Specific Array Processors*, Sept. 1991, Costa Brava (Spain), IEEE Computer Society Press, pp. 116-128.

The Roundoff Noise of Pipelined Scattered Look-Ahead IIR Digital Filters with Decomposition

William G. Bliss * & Kyung-Hi Chang †
Department of Electrical Engineering
Texas A&M University
College Station, TX, 77843-3128

Abstract

Scattered Look-Ahead (SLA) with power-of-two decomposition has been proposed for high throughput recursive digital filters. We find the roundoff noise of first and second order minimum roundoff noise and direct form SLA structures. We improve on other claimed results by enforcing proper scaling and including non-recursive section noise. For first order filters the Noise Gain (NG) increases by $lg_2 M$ near $z = 0$ and decreases by M near $z = \pm 1$. For second order direct form the NG also increases by $lg_2 M$ near $z = 0$, but near $|z| = 1$ depends on the pole angle; increasing infinitely at $i\pi/M$, and decreasing by M^7 near ± 1. The NG of second order minimum noise filters increases by $\frac{1}{2} lg_2 M$ near $z = 0$ and decreases by M near $|z| = 1$.

1 Introduction

Infinite Impulse Response(IIR) filters are known to be more difficult to implement than FIR filters in real-time high throughput applications due to problematic recursive updates. Many stable look-ahead techniques which break the requirement of Multiply-Add (MA) recursion time less than or equal to the sampling period have been proposed [1], [2], [3]. We consider the pipelined technique called Scattered Look-Ahead with power of two decomposition [4], [5].

Little attention has been given to the numerical properties of SLA filters. Parhi and Messerschmitt originally showed that for the state variable SLA structure, the roundoff noise due to the recursive section of the SLA filter is strictly decreasing with M [5]. This result was based on comparing a non-SLA filter ($M = 1$) with state variable feedback matrix A to a SLA SVD filter with feedback matrix A^M. We note that this analysis is not applicable to the SLA direct form implementations because the implied similarity transformation from A^M to companion form is not considered. In [6] and [7] it was claimed that when considering both the recursive and non-recursive sections, the NG increases with M regardless of pole position, the NG slightly decreases with M for poles near the unit circle independent of Θ, and that the worst case increase in NG occurs near the origin, where the number of extra bits required by SLA is limited to $lg_2 M$. However, the above conclusions were derived using unscaled direct form SLA structures, and are therefore erroneous. It is well known that the effects of scaling must be included in any analysis of finite-length register effects to ensure both that overflows rarely occur and that the dynamic range of registers is efficiently utilized [8].

In this paper we will use the double length accumulation roundoff noise model, i.e., all the double length results of fixed point multiplications are added before they are rounded off. It should be noted that the NG calculated this way can only underestimate the NG with rounding before accumulation by a factor of L, where L is the maximum number of inputs at any accumulation node. Further, only $\frac{1}{2} lg_2 Q$ extra bits in summations are needed at a summing junction with Q inputs to achieve the same order NG as with double length accumulation. We will demonstrate two methods to calculate scaling factors for the SLA structures. The first method we call sectioned K and W, which is a special Factored SVD method [8]. The second method uses a non-canonic SVD which is applicable to any computable SFG. Both Direct Form (DF) and Minimum RoundOff Noise (MRON) structures are examined in SLA implementations. Regions of the z plane where SLA DF filters have lower NG than MRON form are indicated.

*email: bliss@ee.tamu.edu
†This research was supported by a grant from the Texas Advanced Research Program

1058-6393/91 $1.00 © 1991 IEEE

2 Pipelined SLA Filters

SLA allows MA latency M times the sampling period by using M deep pipelining (loop degree M) in the recursive section of the filter. The pipelining of the SLA structure is achieved by inserting $(M-1)$ additional poles. For each original pole in $H(z)$ at radius r and angle θ, the added poles are at the same radius but at angles

$$\theta_i = \theta + i2\pi/M \qquad i = 1, 2, \cdots (M-1).$$

The apparent denominator polynomial of the SLA filter is thus a polynomial in z^{-M}, which allows latency M in the recursion. The sparseness of the denominator polynomial also means that it's computational complexity is low, i.e., only N coefficients for an N'th order filter. To achieve I/O equivalence (in perfect arithmetic) $N(M-1)$ zeroes must be inserted to exactly cancel the added poles. This numerator polynomial isn't sparse, and thus represents a large number of calculations. However, by choosing M a power of two, the numerator was factored as $lg_2 M$ polynomials with N non-trivial coefficients each [5]. The computational complexity of SLA with power of two decomposition is thus $N(1 + lg_2 M)$ for a DF all-pole filter. The non-recursive portion of the filter implementing the canceling zeroes can be pipelined to any required degree to facilitate MA implementation.

3 Noise Gain of First-Order SLA

Define $f_i(k)$ as the impulse response at summing node i and define $g_{ij}(k)$ as the response due to an impulse at node i taking the j'th path to $y(k)$. $K_i = \sum f_i^2(k)$ is the noise power gain from the input to node i. $W_{ij} = \sum g_{ij}^2(k)$ is the noise power gain from node i through the j'th path to the output. Because the SLA structure creates $g_{ij}(k)$ orthogonal sequences over j, then we can also sum powers over j. An l_2 scaled filter has $K_i = \delta^{-2}$ for every summing node i. The NG factor (without $\delta^2 q^2/12$, where q is the quantization level) can then be simply expressed as

$$NG = \sum\nolimits_i \sum\nolimits_{j(i)} K_i W_{ij}$$

Consider the DF filter with single-pole at $z = a$.

$$\begin{aligned} f_1(k) &= \{0 \;\; b \;\; ab \;\; a^2 b \;\; \cdots\} \\ g_{11}(k) &= \{c \;\; ca \;\; ca^2 \;\; ca^3 \;\; \cdots\} \\ K_1 &= b^2 + a^2 b^2 + a^4 b^2 + \cdots \\ W_{11} &= c^2 + c^2 a^2 + c^2 a^4 + \cdots \end{aligned}$$

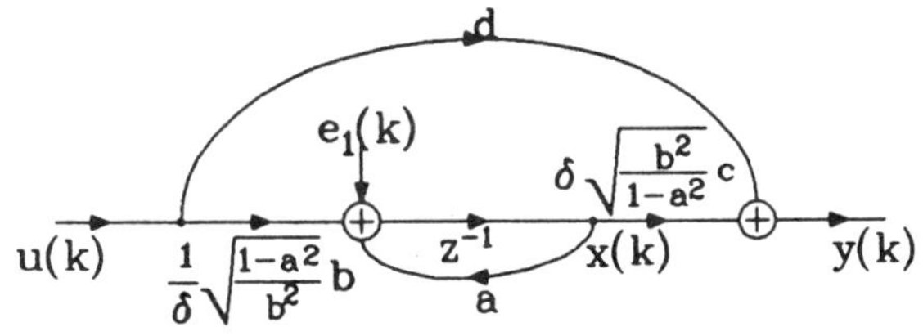

Figure 1: l_2–scaled single pole direct form filter.

The scaled input gain is $b' = b/\delta K_1^{1/2} = (1-a^2)^{1/2}/\delta$ and the scaled output gain is $c' = c\delta K_1^{1/2} = bc\delta(1-a^2)^{1/2}$, while a and d are unchanged, as in Fig.1, where $e_1(k)$ denotes a roundoff equivalent noise source. For the single pole $M = 1$ conventional filter, $NG(M = 1) = b^2 c^2/(1-a^2)^2$.

Fig.2 is the unscaled pipelined single pole filter with $M = 8$ loop pipelining stages. There are three summing nodes (roundoff noise sources), and the pertinent impulse responses are:

$f_1(k) = \{0\ 0\ 0\ 0\ 0\ 0\ 0\ 0\ b\ ab\ a^2b\ a^3b\ \cdots\}$
$f_2(k) = \{b\ ab\ a^2b\ a^3b\ 0\ \cdots\}$
$f_3(k) = \{b\ ab\ 0\ 0\ 0\ \cdots\}$
$g_{11}(k) = \{c\ 0\ 0\ 0\ 0\ 0\ 0\ 0\ ca^8\ \cdots\}$
$g_{21}(k) = \{0\ 0\ 0\ 0\ 0\ 0\ 0\ 0\ 0\ 0\ 0\ 0\ ca^4\ 0\ 0\ 0\ 0\ 0\ 0\ 0\ ca^{12}\ \cdots$
$g_{22}(k) = \{0\ 0\ 0\ 0\ 0\ 0\ 0\ 0\ c\ 0\ 0\ 0\ 0\ 0\ 0\ 0\ ca^8\ \cdots\}$
$g_{31}(k) = \{0\ 0\ 0\ 0\ 0\ 0\ 0\ 0\ 0\ 0\ 0\ 0\ 0\ 0\ ca^6\ 0\ 0\ 0\ 0\ 0\ 0\ 0\ ca^{14}$
$g_{32}(k) = \{0\ 0\ 0\ 0\ 0\ 0\ 0\ 0\ 0\ 0\ ca^2\ 0\ 0\ 0\ 0\ 0\ 0\ 0\ ca^{10}\ \cdots\}$
$g_{33}(k) = \{0\ 0\ 0\ 0\ 0\ 0\ 0\ 0\ 0\ 0\ 0\ 0\ ca^4\ 0\ 0\ 0\ 0\ 0\ 0\ 0\ ca^{12}\ \cdots$
$g_{34}(k) = \{0\ 0\ 0\ 0\ 0\ 0\ 0\ 0\ c\ 0\ 0\ 0\ 0\ 0\ 0\ 0\ ca^8\ \cdots\}$

Therefore, denoting $s = 1/(1-a^{16})$, we find $K_1 = \frac{b^2}{1-a^2}$, $K_2 = b^2(1+a^2+a^4+a^6)$, $K_3 = b^2(1+a^2)$, $W_{11} = c^2 s$, $W_{21} = c^2 a^8 s$, $W_{22} = c^2 s$, $W_{31} = c^2 a^{12} s$, $W_{32} = c^2 a^4 s$, $W_{33} = c^2 a^8 s$, $W_{34} = c^2 s$. Therefore,

$$NG(M = 8) = \frac{b^2 c^2 (3 - 2a^{16})}{(1-a^2)(1-a^{16})} \qquad (1)$$

Following the preceding development, the general form of the NG for $M \geq 2$ is inductively found to be

$$NG(M) = b^2 c^2 \frac{\log_2 M - (\log_2 M - 1)a^{2M}}{(1-a^2)(1-a^{2M})} \qquad (2)$$

The NG goes to infinity as $|a|$ goes to one, but larger M slows the approach. The effect of M can be seen clearly in Fig.3, where the NGs are normalized to the M=1 (no SLA) conventional filter. When the pole goes to the origin, the NG ratio becomes

$$\lim_{a \to 0} \frac{NG(M)}{NG(M=1)} = lg_2 M \qquad (3)$$

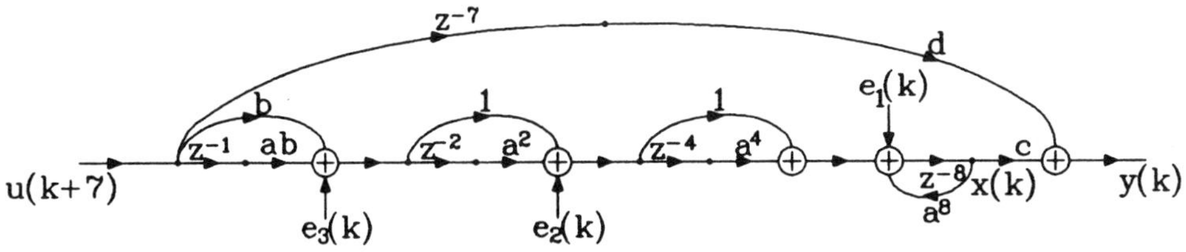

Figure 2: Unscaled first order SLA with M=8 look ahead.

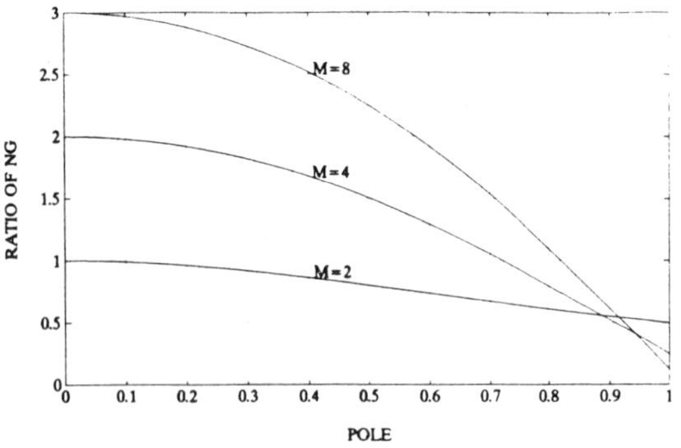

Figure 3: NG ratio with M=1 base.

When the the pole goes to ± 1, the NG ratio becomes:

$$\lim_{a \to \pm 1} \frac{NG(M)}{NG(M=1)} = \frac{1}{M} \tag{4}$$

Thus, at most an extra $\frac{1}{2} lg_2 lg_2 M$ bits are required by first order DF SLA implementations compared to conventional DF. Conversely, at most $\frac{1}{2} lg_2 M$ fewer bits may be sufficient with DF SLA to match conventional performance.

4 Second-Order Minimum Noise SLA

Consider the general SVD form of SLA with power of two decomposition, as in Fig.4 for M=8. Summing nodes now see vector sequences, so extending the previous notation $\mathbf{K}_i = \sum \mathbf{f}_i(k)\mathbf{f}_i^T(k)$ is the covariance matrix of the i'th summing node, and $\mathbf{W}_{ij} = \sum \mathbf{g}_{ij}(k)\mathbf{g}_{ij}^T(k)$ is the matrix of cross power gain from node i through the j'th path to the output. The vector sequences $\mathbf{f}_i$ and $\mathbf{g}_{ij}$ are now the same form as the first order formulations, except $\{a, b, c\}$ are replaced by $\{\mathbf{A}, \mathbf{b}, \mathbf{c}\}$. Because of orthogonality, the NG for an N'th order SVD SLA is

$$NG = \sum_{m=1}^{N} \sum_i \sum_{j(i)} \mathbf{K}_i(m,m)\mathbf{W}_{ij}(m,m)$$

We apply the SVD NG form to the second order sectional optimal MRON structure [10] [11]. Because we consider all-pole filters, we use [12] to find design equations; $a_{11} = a_{22} = r\cos\theta$, $b_1 = c_2 = 0$, $c_1 = 1/b_2 a_{12}$, $a_{21} = -r^2 \sin^2\theta / a_{12}$,

$$a_{12} = [1 + r^2(1 - 3\cos^2\theta) + r^4\cos^2\theta]^{1/2}(1 + r^2)^{-1/2}$$

$$b_2^2 = 1 - \frac{a_{21}^2(1 - a_{12}a_{21} + a_{11}a_{22}) + a_{22}^2(1 + a_{12}a_{21} - a_{11}a_{22})}{1 - a_{12}a_{21} - a_{11}a_{22}}$$

Note we are simply taking the minimum noise $M = 1$ solution and applying the state space SLA algorithm, so there is no claim as to whether or not the new structure minimizes NG. The NG ratio for SLA MRON forms is a smooth function in the unit disk, and it becomes

$$\lim_{r \to 0} \frac{NG(M)}{NG(M=1)} = \frac{1}{2} lg_2 M \tag{5}$$

$$\lim_{r \to 1} \frac{NG(M)}{NG(M=1)} = \frac{1}{M} \tag{6}$$

which closely parallels the first order DF results. Because the number of multiplications in the MRON is $N^2(1 + lg_2 M)$, the performance of the simpler DF SLA filters is examined next.

5 Second-Order Direct Form SLA

We now consider the DF SLA implementation where each section (including non-recursive) has update matrices in companion form, shown in Fig.5 for $M = 8$. The number of multiplications is thus $N(1 + lg_2 M)$. We demonstrate the calculation of NG by explicitly including all summing nodes as state variables. The implementation is thus called a Non-Canonic SVD (NCSVD). For $M = 8$, the equations are: $x_1(k+1) = x_2(k)$

$x_2(k+1) = x_3(k+1)$

...

$$x_{16}(k+1) = -r^{16}x_1(k) + 2r^8\cos 8\theta\ x_9(k) + r^8 x_{17}(k) + 2r^4\cos 4\theta\ x_{21}(k) + x_{25}(k)$$

...

$$x_{25}(k+1) = r^4 x_{26}(k) + 2r^2\cos 2\theta\ x_{28}(k) + x_{30}(k)$$

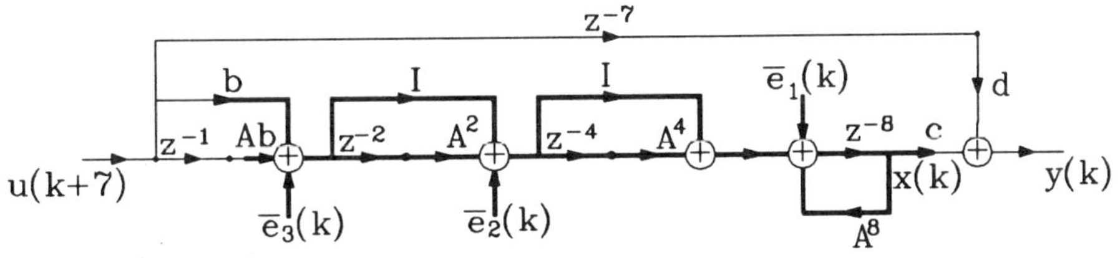

Figure 4: Pipelined SVD SLA filter with M=8 levels of look ahead.

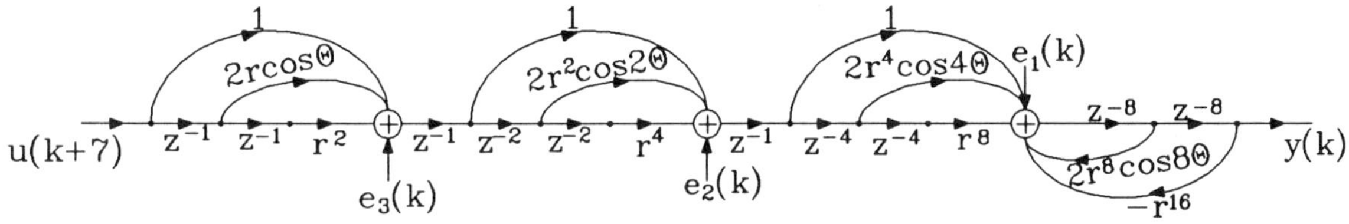

Figure 5: Pipelined Direct Form SLA with M = 8 levels of look ahead.

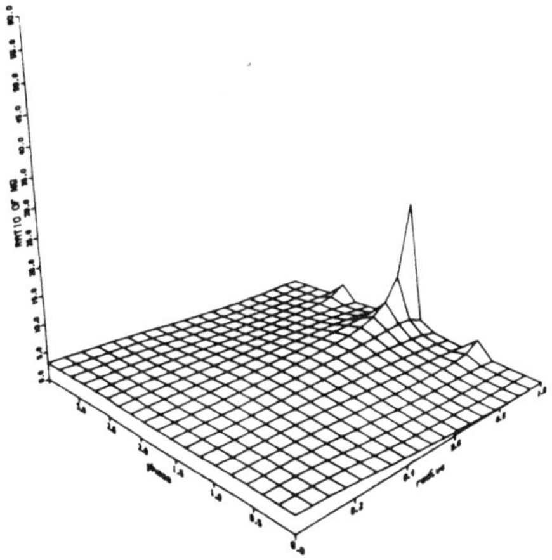

Figure 6: Ratio of NG for M=8 wrt M=1 DF.

...

$x_{30}(k+1) = r^2 x_{31}(k) + 2r\cos\theta\; x_{32}(k) + u(k)$
$x_{31}(k+1) = x_{32}(k)$
$x_{32}(k+1) = u(k)$
$y(k-16) = x_1(k)$

There are only three ($lg_2 M$) summing nodes, the rest are simply pure delays, so there are only three terms in the NG equation with NCSVD

$$NG = \sum_{i=16,25,30} \mathbf{W}_{ii}\mathbf{K}_{ii}$$

The NG ratio for M=8 compared to M=1 non-SLA is shown in Fig.6.

Define $\mathbf{W}^M$ and $\mathbf{K}^M$ as the $\mathbf{W}$ and $\mathbf{K}$ matrices for a given M. Observing that higher M structures have the same recursive power gain to the output, we have $\mathbf{W}^1_{22} = \mathbf{W}^2_{44} = \mathbf{W}^4_{88} = \mathbf{W}^8_{16,16} = \frac{1+r^{2M}}{p(M)}$, where $p(M) = 1 + r^{2M} - 4r^{2M}\cos^2 M\theta - r^{4M} + 4r^{4M}\cos^2 M\theta - r^{6M}$. The NG of second-order pipelined DF SLA structure can then be written as

$$NG(M) = Q_{1,2}(M) + Q_4(M) + \cdots + Q_M(M) \qquad \text{where}$$

$$Q_{1,2}(M) = (1+r^2)(1+r^{2M})/p^2(M)$$
$$Q_4(M) = (1+r^4)(1+4r^2\cos^2\theta + r^4)/p(M)$$
$$Q_8(M) = (1+r^8)(1+4r^2\cos^2\theta + r^4 + 4r^4\cos 2\theta + 4r^4\cos^2 2\theta + 16r^6\cos^2\theta\cos^2 2\theta + r^8 + 4r^8\cos 2\theta + 4r^8\cos^2 2\theta + 4r^{10}\cos^2\theta + r^{12})/p(M)$$

From these results the NG of the second order DF near the origin is the same as the first order SLA (3), the ratio being $lg_2 M$. As the poles approach the unit circle however, unlike the MRON structure, the NG ratio of the DF SLA is seen to be highly dependent on the pole angle

$$\lim_{r\to 1} \frac{NG(M)}{NG(1)} = \frac{(cos^2\theta - 1)^2}{M^3(cos^2 M\theta - 1)^2} \qquad (7)$$

In particular, if $\theta = \pm i\pi/M$ for $i \in \{1, 2, \cdots (M-1)\}$, then the NG ratio becomes infinite. This phenomenon is clearly due to the introduction of extra poles very near to $z = \pm 1$, where the poor numerical performance of DF structures is well known. The number of bits 'saved' by application of DF SLA as the poles near the unit circle is $-\frac{1}{2}log_2(NG\ ratio)$, and is plotted in Fig.7 for $M = 8$.

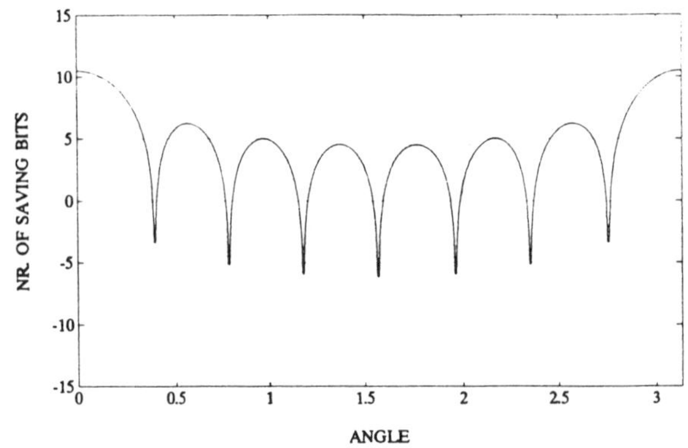

Figure 7: Bits saved by M=8 DF SLA near unit circle.

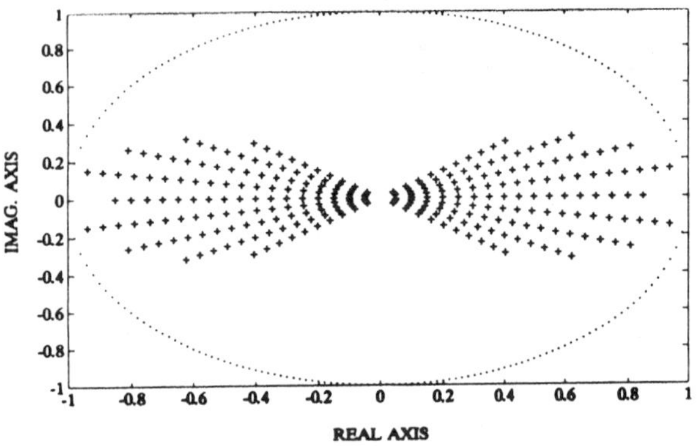

Figure 8: NG(DF)<NG(MRON) in dark areas, M=8.

6 Conclusions

The roundoff noise performance of SLA with power of two decomposition has been investigated for first order and second order all-pole filters. Two methods to calculate scaling parameters and noise gain ratios with respect to the $M=1$ (non-SLA) forms were demonstrated. First order SLA forms increase NG by $lg_2 M$ near zero and decrease NG by M near ± 1. Second order SLA forms using MRON forms have similar NG performance. Second order SLA DF behaves similarly near the origin, but depends strongly on pole angle near the unit circle, where NG ratios vary between infinite increase and decrease by M^7. The overall numerical performance of SLA is quite good, except for small regions of the z plane near the unit circle, where second order DF SLA filters behave poorly. Except for these regions, the need to use MRON instead of DF filters is analogous to the same choice with non-SLA filters. Because DF have fewer summing nodes than MRON structures, DF can have slightly smaller NG. As M increases, this region of the z plane atrophies, as illustrated in Fig.8 for $M = 8$.

References

[1] A.L. Moyer, "An efficient parallel algorithm for digital IIR filters," in *Proc. IEEE Int. Conf. Acoust., Speech, Signal Processing*, Apr. 1976, pp. 525-528.

[2] C.W. Barnes and S. Shinnaka, "Block shift invariance and block implementation of discrete-time filters," *IEEE Trans. Circuits Syst.*, vol. CAS-27, pp. 667-672, Aug. 1980.

[3] R.P. Brent and Z.B. Bing, "A stabilized parallel algorithm for direct-form recursive filters," *IEEE Trans. Computers*, vol. 40, pp. 333-336, Mar. 1991.

[4] K.K. Parhi and D.G. Messerschmitt, "Pipelined VLSI recursive filter architectures using scattered look-ahead and decomposition," in *Proc. IEEE Int. Conf. Acoust., Speech, Signal Processing*, Apr. 1988, pp. 2120-2123.

[5] K.K. Parhi and D.G. Messerschmitt, "Pipeline interleaving and parallelism in recursive digital filters, Part I : Pipelining using scattered look-ahead and decomposition," *IEEE Trans. Acoust., Speech, Signal Processing*, vol. ASSP-37, pp. 1099-1117, Jul. 1989.

[6] K.K. Parhi *et al.*, "Quantization effects in high-speed pipelined recursive filters," in *Proc. IEEE Int. Conf. Acoust., Speech, Signal Processing*, Apr. 1990, pp. 1743-1746.

[7] K.K. Parhi, "Finite word effects in pipelined recursive filters," *IEEE Trans. Signal Processing*, vol. SP-39, pp. 1450-1454, Jun. 1991.

[8] R.A. Roberts and C.T. Mullis, *Digital Signal Processing*. Addison-Wesley, 1987.

[9] A.V. Oppenheim and R.W. Schafer, *Digital Signal Processing*. Prentice-Hall, 1975.

[10] C.T. Mullis and R.A. Roberts, "Synthesis of minimum roundoff noise fixed point digital filters," *IEEE Trans. Circuits Syst.*, vol. CAS-23, pp. 551-562, Sep. 1976.

[11] L.B. Jackson *et al.*, "Optimal synthesis of second-order state-space structures for digital filters," *IEEE Trans. Circuits Syst.*, vol. CAS-26, pp. 149-153, Mar. 1979.

[12] B.W. Bomar, "New second-order state-space structures for realizing low roundoff noise digital filters," *IEEE Trans. Acoust., Speech, Signal Processing*. vol. ASSP-33, pp.106-110, Feb. 1985.

DESIGN AND IMPLEMENTATION OF EFFICIENT RESAMPLING FILTERS USING POLYPHASE RECURSIVE ALL-PASS FILTERS

fred harris
Maximilien d'Oreye de Lantremange

Electrical and Computer Engineering
San Diego State University
San Diego, California 92182-0190

A.G. Constantinides

Signal Processing Section
Electrical Engineering
Imperial College
Exhibition Road
London SW7-2BT, England

1. ABSTRACT
Computationally efficient resampling filters are formed from sets of recursive all-pass sub-filters operating in a polyphase structure. These sub-filters, exhibiting unity magnitude response, have phase shifts which add constructively in the passband(s) and destructively in the stopband(s). The computational burden for these filters is one-fifth to one-tenth of conventional resampling filters.

2. INTRODUCTION
A finite impulse response (FIR) filter can be modeled, as indicated in Fig. 1, as the weighted summation of the contents of a tapped delay line. This summation is shown in (1) and the transfer function of this filter is shown in (2).

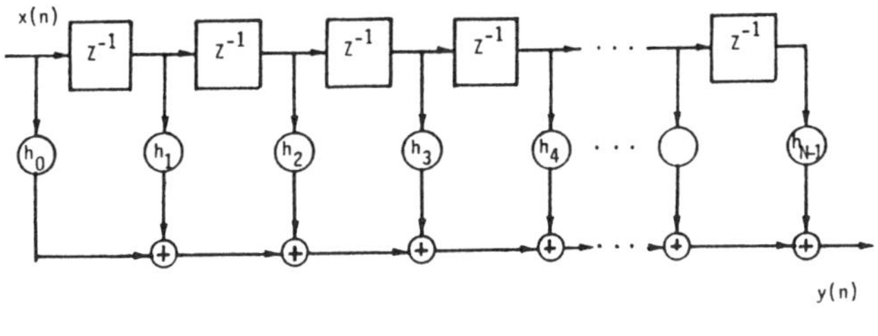

FIGURE 1. TAPPED DELAY LINE FIR FILTER

$$y(n) = \sum_{m=0}^{M-1} h(m-n)\, x(m) \quad (1)$$

$$Y(Z) = X(Z) \sum_{n=0}^{M-1} h(n)\, Z^{-n} \quad (2)$$

The time delays between successive samples in the tapped delay line results in a linearly increasing phase shift proportional to input frequency. The summation of these phase shifted samples is responsible for the frequency dependent gain of the filter. This (weighted) phaser summation is shown in Fig. 2 for an input signal at two different frequencies. Note that it is the phase shift, not the weighting terms, responsible for the frequency dependent gain.

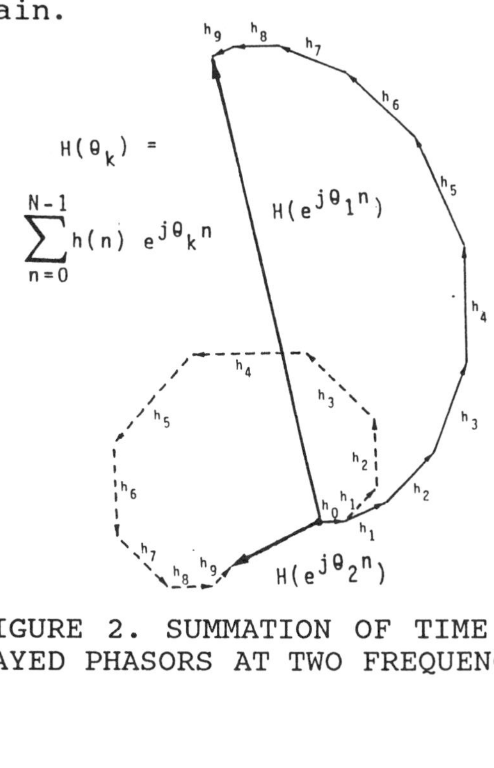

FIGURE 2. SUMMATION OF TIME DELAYED PHASORS AT TWO FREQUENCIES

Bandwidth of a FIR filter can also be controlled by a non uniform phase shift within the weights (as commonly done in spread spectrum communications [1]). The filters described here use (unity gain) all-pass sub-filters to obtain this frequency dependent phase shift. The all-pass sub-filters are, as shown in (3) and Fig. 3, first order polynomials in Z^{-M} where M is the number of taps in the structure. This phase-shift for weights filter form is indicated in Fig. 4a with a modified form reflecting a polyphase structure in 4b. The transfer function of this structure is shown in (4).

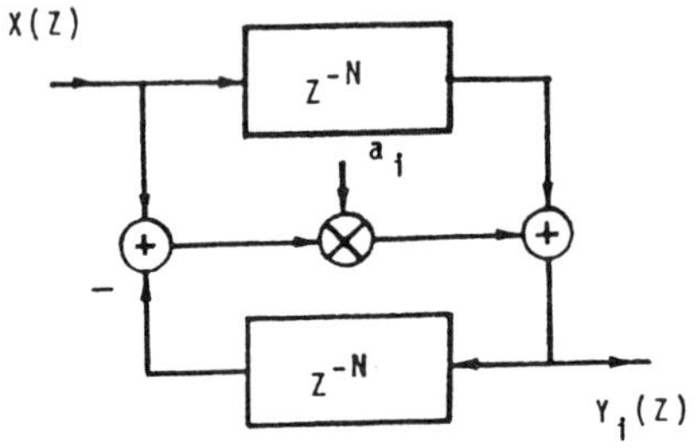

FIGURE 3. ALL-PASS SUBFILTER

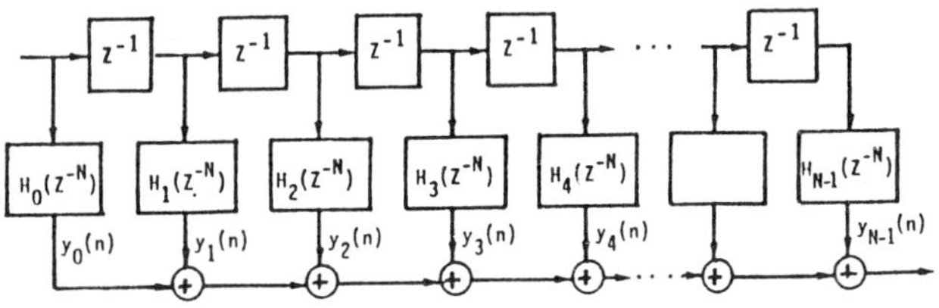

FIGURE 4a. ALL-PASS PHASE SHIFT FOR WEIGHTS FILTER STRUCTURE

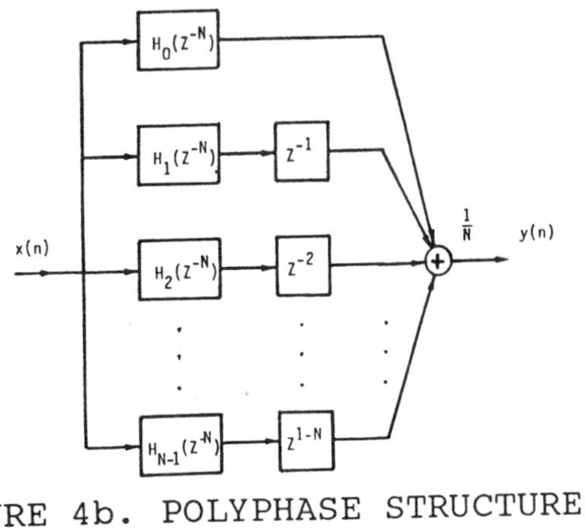

FIGURE 4b. POLYPHASE STRUCTURE

$$H_n(Z^{-M}) = \frac{a_n + Z^{-M}}{1 + a_n Z^{-M}} \qquad (3)$$

$$Y(Z) = X(Z) \sum_{n=0}^{M-1} H_n(Z^{-M}) Z^{-n} \qquad (4)$$

The roots of (3), the M roots of $-a_n$ and (reciprocals), are equally spaced about the origin as shown in Fig. 5 for the indicated degrees M. Note that we realize M poles and M zeros per multiplication in each all-pass stage. Thus we can form two poles **and** two zeros per multiplication in the sub-filters of a two-path filter rather than the one pole (**or** zero) per multiplication of a standard canonic form.

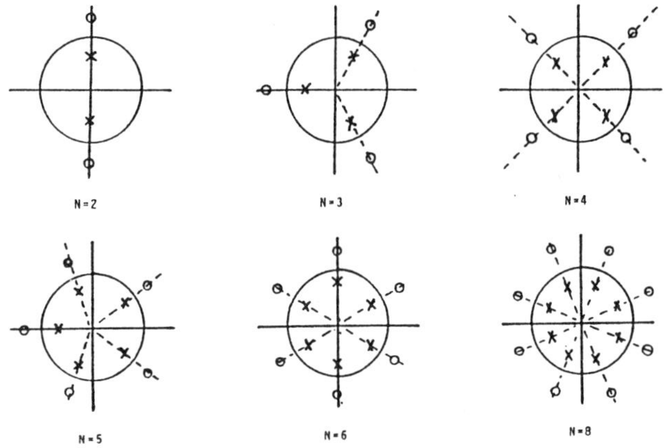

FIGURE 5. POLE-ZERO PLACEMENT FOR ALL-PASS SUBFILTER

As we traverse the unit circle we visualize that the all-pass sub-filters exhibit rapid phase change in the neighborhood of each pole-zero pair. By proper choice of the pole positions, the phase shift of each leg of the M-path filter can be made to match or to differ by multiples of $2\pi/M$ over selected spectral intervals between the pole-zero pairs. This is suggested in Fig. 6 for M=2 and M=4. The coefficients of the all-pass subfilters can be determined by standard algorithms [2] or by a new algorithm developed by harris and d'Oreye and reported in a paper recently submitted for publication (Signal Processing).

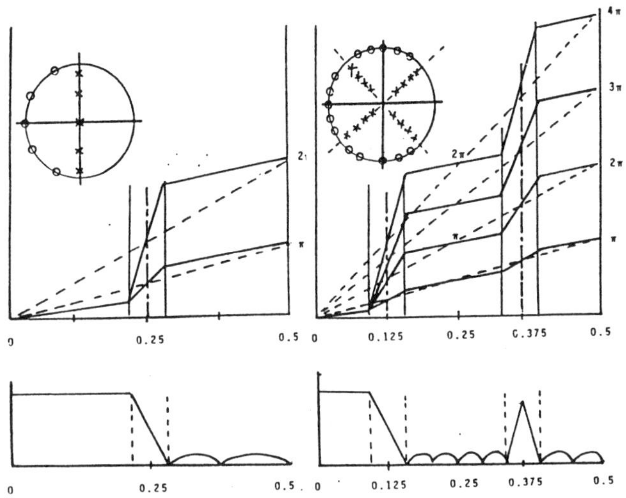

FIGURE 6. TYPICAL PHASE SHIFT VERSUS FREQUENCY FOR ALL-PASS SUB-FILTERS

Figure 7 is an example of an equal ripple design obtained from this new algorithm for a 2-path filter with 2-stages per path and for a 5-path filter with 3-stages per path. Interacting constraints (described in the aforementioned paper) limit the number of stages per path for the 2-path filter to the choices (1,0), (1,1), (2,1), (2,2), (3,2), (3,3), etc. In a similar manner, the 5-path filter choices are (1,1,1,1,0), (2,2,2,1,1), (3,3,2,2,2), etc., so this filter example uses three less stages than allocated to the design.

3. TWO PATH FILTERS

The 2-path version of this filter is simple and is very interesting. This structure is a half bandwidth filter with the 3-dB bandwidth at 0.25 f_s. When the zeros are restricted to the half sampling frequency, the filter is the half bandwidth Butterworth

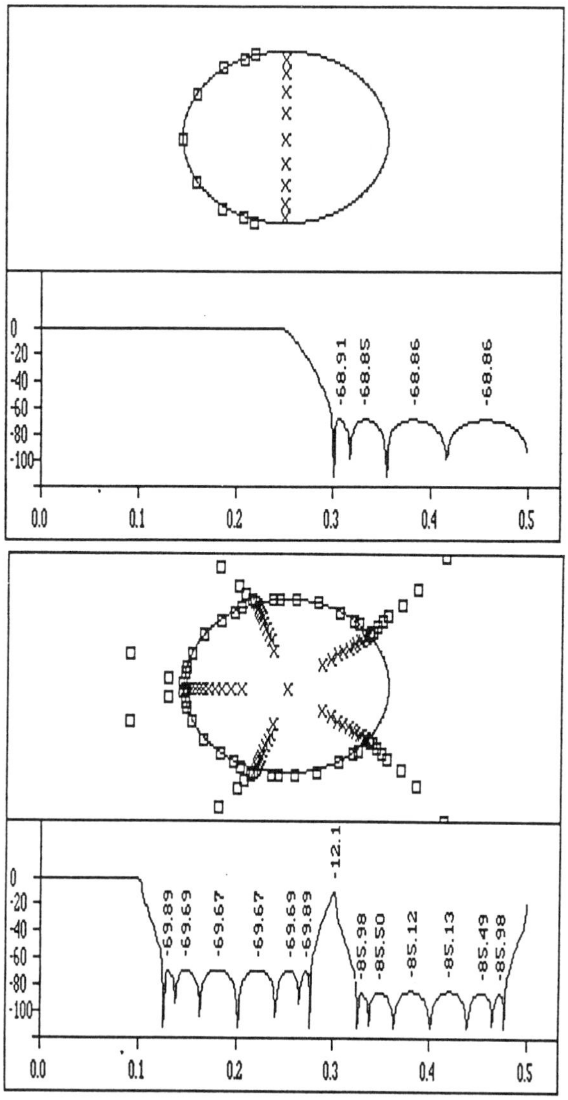

FIGURE 7. FREQUENCY RESPONSE OF 2-PATH AND 5-PATH FILTERS

obtained by the standard bilinear transformation. For this special case, the real parts of the pole locations are zero. If the zeros are distributed around the unit circle (as for equal ripple stop-band behavior) the filter is a constrained Elliptic filter.

The constraint is related to a property of complementary all-pass filters. We define the all-pass sections for the 2-path filter as $A_0(\theta)$ and $A_1(\theta)$ and, as indicated in Fig. 8, the scaled (by 1/2) sum and difference of these paths by $H_A(\theta)$ and $H_B(\theta)$, lowpass and highpass filters, respectively. We know that the all-pass sections satisfy (5) from which we derive the power relationship between the lowpass and highpass filters as in (6).

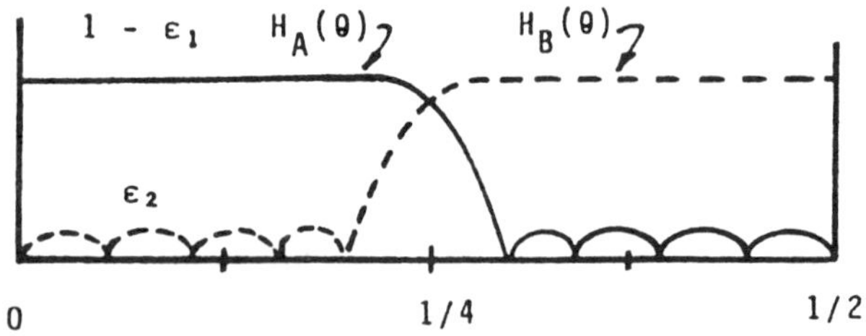

FIGURE 8. COMPLEMENTARY FILTER FREQUENCY RESPONSES

$$|A_0(\theta)|^2 = |A_2(\theta)|^2 = 1 \qquad (5)$$

$$|H_A(\theta)|^2 + |H_B(\theta)|^2 = \qquad (6a)$$

$$|0.5[A_1(\theta) + A_2(\theta)]|^2$$

$$+ |0.5[A_1(\theta) - A_2(\theta)]|^2$$

$$= 0.5 \; [|A_1(\theta)|^2 + |A_1(\theta)|^2] \qquad (6b)$$

$$= 1 \qquad (6c)$$

Now to interpret how this relationship impacts the 2-path filter. For each complementary filter we define the minimum passband gain by $1-\epsilon_1$ and peak stopband gain by ϵ_2. Substituting these gains in (6) we obtain (7).

$$(1-\epsilon_1)^2 + (\epsilon_2)^2 = 1 \qquad (7a)$$

$$(1-\epsilon_1)^2 = 1 - (\epsilon_2)^2 \qquad (7b)$$

$$1-2\epsilon_1 + (\epsilon_1)^2 = 1 - (\epsilon_2)^2 \qquad (7c)$$

For small ϵ_1, we can ignore $(\epsilon_1)^2$ which leads to (8).

$$\epsilon_1 = 0.5 \; (\epsilon_2)^2 \qquad (8)$$

Thus if the stopband attenuation is selected to be 0.001 (-60 dB), the passband ripple is 0.0000005 (-126 dB) or the passband minimum gain is 0.9999995 (-0.00000022 dB). These filters do indeed have flat passbands and cross at their 3-dB points.

There are two significant differences between the 2-path and the standard implementation of the equivalent elliptic filter. The first is a five-to-one savings in multiplications; a fifth order half bandwidth 2-path filter requires only two coefficients as opposed to the direct implementation which requires ten (scaling included). The second is that all-pass structures exhibit unity gain to all internal states hence do not require extended precision registers to store intermediate results as do cascade 2-nd order implementations.

As with any filter design, for a fixed order, the transition bandwidth can be traded for out-of-band attenuation or transition bandwidth can be reduced for a fixed attenuation by increasing filter order. While nomagraphs exist for the elliptic filters which can be implemented by the 2-path structure, a simple approximate relationship for the equal ripple filter is given in (9), where A(dB) is attenuation in dB, Δf is transition bandwidth, and N is the total number of all-pass segments in the filter.

$$A(dB) = (72 \; \Delta f + 10) \; N \qquad (9)$$

4. INTERPOLATION AND DECIMATION

Any half bandwidth filter can be used to perform 2-to-1 resampling (up or down, usually designated interpolation and decimation). Finite impulse response (FIR) filters can be operated with a polyphase partition which avoids processing of inserted zero input points (up-sampling) or the computation of discarded output points (down-sampling). While this option is not generally available for the recursive filter, it can be applied to the recursive all-pass M-path filter. The M-path filter can be partitioned into polyphase segments because the segments are polynomials in Z^{-M}.

How this relationship interacts with the resampling is particularly easy to see when the 2-path filter is used for an interpolator. An input sample applied to the filter makes a contribution to the output at the same time via the upper path but not from the lower path till the next sample due to its extra Z^{-1} delay. The next input sample is zero so at this time, the upper path offers no contribution to the output as there is no Z^{-1} path associated with its Z^{-2} polynomial while the lower path does offer a response to the previous input via its Z^{-1}. Thus the output response of the filter at successive indices are obtained as the response to the same input from alternate paths of the filter.

Up-sampling is performed by delivering non zero packed input data to the 2-path filter and obtaining output samples alternately, via a commutator, from the two sub-filters (in the order 0-1) at an output rate twice the input rate. Down-sampling is accomplished by reversing the flow just described. Alternate input samples are delivered to the two sub-filters via a commutator (in the order 1-0). Each sub-filter, operating at half the input rate, contribute samples to an output adder to form the down-sampled output data. The 2-path resampling structures are shown in figure 9.

The computational efficiency of

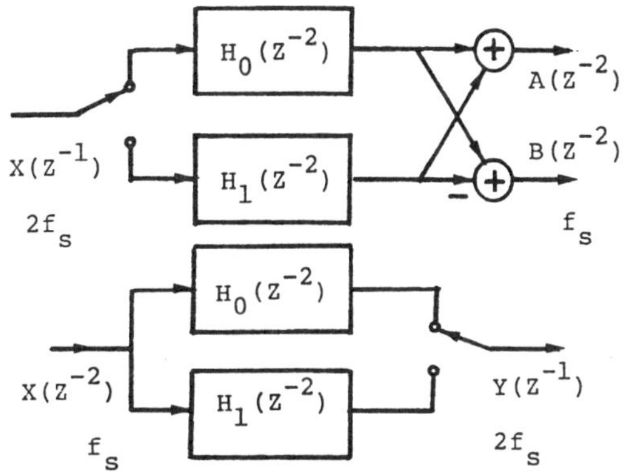

FIGURE 9. UP-SAMPLING AND DOWN-SAMPLING ALL-PASS STRUCTURES

these filters can be seen from the following example. Assume a signal, sampled at 2.56 times the input bandwidth, is to be down-sampled 8-to-1 as part of a zoom transform. (The 2.56 is the standard sampling relationship required for a 400-line spectrum analyzer with a 1024 point FFT.) Specifications for each half-band stage of this filter are seen in Table 1.

Passband Ripple:	0.05 dB
Stopband Attenuation:	80.0 dB
Passband Frequency:	0.1953
Stopband Frequency:	0.3047
Transition Bandwidth:	0.1094

TABLE 1. 2-to-1 RESAMPLING FILTER SPECIFICATIONS

A FIR filter of length 33 (-85 dB) can meet these specifications and when partitioned into a polyphase form, requires 16.5 multiplications per output point. A 2-path all-pass filter which meets these same specifications requires 5-stages, a (3,2) partition, (-89 dB) and thus requires 2.5 multiplications per output point. The frequency response of these filters are presented in figure 10. The 3-stage half band cascade requires seven iterations of the 2-to-1 resampling filter at successively lower sample rates. Translated to the input rate, the computational burden for the FIR filter and all-pass implementations are approximately 173 and 26 multiplications per output point respectively.

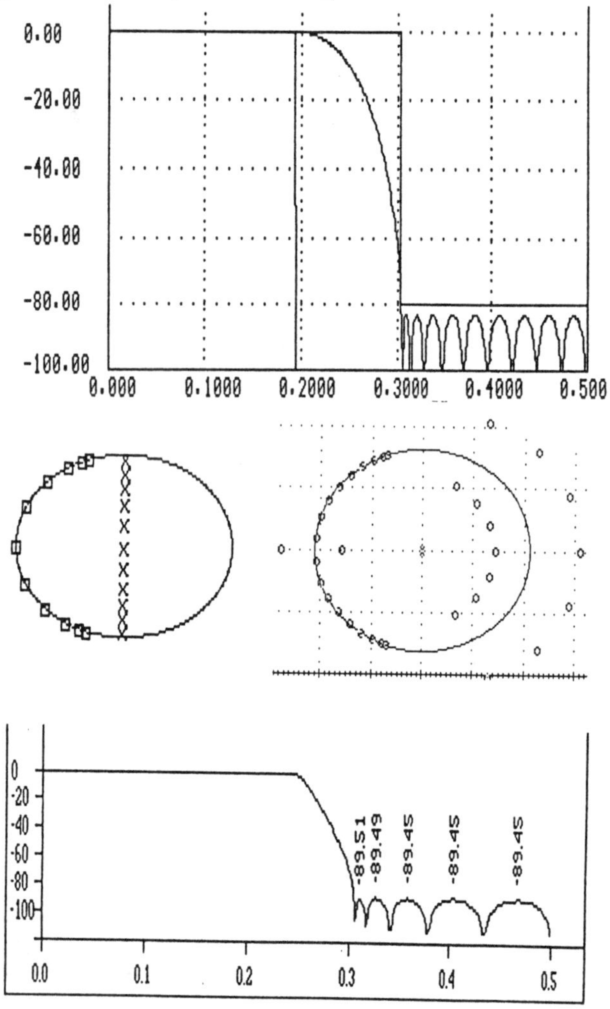

FIGURE 10. RESAMPLING HALF-BAND FILTER RESPONSES (FIR & ALL-PASS)

5. CONCLUSIONS

We have presented the form and usefulness of the polyphase all-pass filter as efficient resampling filters. The primary advantage of these structures is the relatively low workload per output point. A secondary advantage is the low sensitivity to finite arithmetic and the lack of bit growth in internal states. Disadvantages include the non uniform phase response of the filter, as well as newness of idea and difficulty in determining coefficients for the filter. The first objection can be addressed with equalization or with a slightly less efficient linear phase version of the all-pass structure (both of which will be addressed in a following paper). The second is being addressed by excellent survey papers [3,4] and by the increased availability of good design routines.

ACKNOWLEDGEMENT

This work was sponsored in part by the Industry/University Cooperative Research Center (I/UCRC) on Integrated Circuits and Systems (ICAS) at San Diego State University (SDSU) and the University of California, San Diego (UCSD). Programming of the design algorithms was performed by former SDSU graduate student and paper coauthor Maximilien d'orye de Lantremange.

BIBLIOGRAPHY

1. B. Sklar, "Digital Communications, Fundamentals and Applications", Prentice-Hall, 1988, Ch. 10.

2. R.A. Valenzuela and A.G. Constantinides, "Digital Signal Processing Schemes for Efficient Interpolation and Decimation", IEE Proceedings, Vol. 130, Pt. G. No. 6, Dec. 1983, pp. 225-235.

3 P.P. Vaidyanathan and Z. Doganata, "The Role of Lossless Systems in Modern Digital Signal Processing: A Tutorial", IEEE Trans. on Education, Vol. 32, No. 3, Aug. 1989, pp.181-197.

4. f.j. harris, M. d'oreye de Lantremange, and A.G. Constantinides, "Digital Signal Processing with Efficient polyphase Recursive All-Pass Filters", Int. Conf. on Signal Processing, Florence, Italy, 4-6 Sept. 1991.

Multiplier-free FIR Filters with Periodically Time-Varying Ternary Coefficients

Sachin P. Ghanekar, Sawasd Tantaratana, and **Lewis E. Franks**
Department of Electrical and Computer Engineering
University of Massachusetts, Amherst, MA 01003

Abstract

A multiplier-free realization for FIR filters is proposed. The realization uses periodically time-varying (PTV) system, preceded and followed by an upsampler and a downsampler, respectively, to achieve time-invariant operation. The PTV system uses only ternary coefficients. With an upsampling ratio of 3 to 5, we can achieve coefficient precision of 9 to 25 bits, which is sufficient for most applications. The implementation requires only add/subtract operations with a few shift-and-add operations at the input and output of the PTV system.

1 INTRODUCTION

The operating speed of a digital filter is often limited by the speed of the multipliers used in the filter. In addition, multipliers have high hardware complexity and are often expensive to implement using VLSI techniques. A variety of multiplier-free realizations for FIR filters have been proposed in the literature. Some realizations use delta or sigma-delta modulation to encode the target impulse response to avoid coefficient multiplications [1][2], while some others use power-of-two or sum of power-of-two coefficients to replace the full precision coefficients [3]-[5]. The first method results in filters which operate at high oversampling rates and require a large filter length, whereas the second method may require long and variable shift operations for the coefficients to achieve the desired level of ripples.

In this paper, we propose a realization for FIR filters based on a linear system with periodically time-varying (PTV) coefficients taking only ternary values $\{-1, 0, +1\}$. The basic concept is based on distributing the filter operations over space and time. The precision of such realization depends mainly upon the period of the PTV coefficients, which in turn decides the extent of the distribution of computations. The resulting structure consists of an upsampling unit, followed by a PTV system and a downsampling unit. Because of the upsampling and downsampling, there are two signal rates. To avoid notational confusion, we use time index 'm' for signals with higher signal rate and the time index 'n' for signals with normal signal rate (i.e., the rate of the input signal). In all the figures, data lines with faster rate are shown using thin lines and data lines with the normal rate are shown using bold lines.

PTV-FIR Filter

A linear discrete-time PTV system (with period N) can be characterized by a set of N impulse responses. It can be represented by several equivalent filter bank structures [6]-[8]. In our discussion, we will use the equivalent filter bank structure shown in Figure 1(a). The structure consists of a commutator switch followed by a bank of filters. For brevity, we will refer to this structure as the CFB (Commutator and Filter Bank) form. The commutator switch distributes the input data stream $v(m)$ to N different paths. The m-th sample of the data stream in the i-th path is equal to $v(m)$ if $m = jN + i$, $j = ..., 0, 1, ...$, and zero otherwise, i.e., each sample obtained from the input is followed by $N-1$ zeros. The data stream passes through a linear time-invariant (LTI) filter $H_i(z)$, which will be referred to as the i-th component filter. The component filters operate at the same data rate as the rate of $v(m)$. The outputs from the N component filters are added to form the output of the PTV system.

Figure 1(b) depicts the direct form I realization of the PTV-FIR filter of Figure 1(a). A PTV-FIR filter of length L can be represented by a CFB form with each component filter $H_i(z)$ being an FIR filter with length L. The coefficients $c_l(m)$ of the realization are related to the component filter coefficients $h_i(m)$ by

$$c_l(m) = h_{<m-l>_N}(l), \; l \in [0, L-1] \qquad (1)$$

$$h_i(m) = \begin{cases} c_m(m+i), & m \in [0, L-1], i \in [0, N-1] \\ 0, & \text{otherwise} \end{cases} \qquad (2)$$

where $< m >_N$ denotes m modulo N. Note that $c_l(m)$ is periodic with period N.

2 MULTIPLIER-FREE STRUCTURE FOR FIR FILTERS

Consider the structure shown in Figure 2(a). In the figure, the PTV-FIR filter has a period N and it can be represented in the CFB form. We refer to the PTV-FIR filter as the PTV subsystem and the complete structure as the Upsampling-Periodically-Time-Varying-system-Downsampling (UPTVD) structure.

1058-6393/91 $1.00

Note that the Hold unit increases the signal rate by N times and the PTV subsystem operates at this elevated signal rate. The downsampler selects every Nth sample to restore the original signal rate. It can be shown that the UPTVD structure is time-invariant with respect to the input $u(n)$.

Consider a UPTVD structure such that the impulse response of the PTV subsystem has a length of L. The equivalent CFB form for the PTV subsystem contains N FIR filters of length L as the component filters. The impulse response of the UPTVD structure can be obtained as follows. An impulse $\delta(n)$ is injected at the input. The Hold unit converts the input impulse into N different impulses ($\delta(m)$, $\delta(m-1)$,....,$\delta(m-N+1)$) at N times higher sampling rate. Each of these N impulses enters one of the N component filters of the PTV subsystem. The impulse responses from all the component filters are added together and downsampled to form the overall impulse response. Therefore, the impulse response of the UPTVD structure is given by

$$h(n) = \begin{cases} \sum_{i=0}^{N-1} h_i(nN-i), & n \in [0, \mathrm{int}(\frac{L+N-2}{N})] \\ 0, & \text{otherwise} \end{cases} \tag{3}$$

where int(r) is the integer part of r. Note that $h(n) = 0$ for $n \notin [0, \mathrm{int}(\frac{L+N-2}{N})]$ results from the fact that $h_i(m)$ has a finite length of L.

A careful examination of Eq. (3) reveals that only L/N filter coefficients of each component filter ($h_i(m)$) appear in the expression of the overall impulse response of the UPTVD structure. This implies that the remaining coefficients are redundant, as far as the output is concerned. To bring these coefficients to play an active role in determination of the overall response of the UPTVD structure, we add a 'running summer' unit at the output of each component filter. A running summer has a transfer function $1+z^{-1}+\ldots+z^{-(N-1)}$, i.e., the output of the running summer is given by the sum of the N most recent samples. In addition, we introduce scaling units at the input and the output to improve the precision of the impulse response. Combining these ideas, we obtain a UPTVD structure shown in Figure 3(a). Each path is equipped with a scaling unit at the input and another scaling unit with a running summer at the output of the component filter. In actual implementation, the input scaling units in the N paths can be combined into one input PTV scaling unit. We can do similar combinations for the running summer units and the output scaling units. The resulting structure is shown in Figure 3(b). For convenience, we will use Figure 3(a) for analysis, while Figure 3(b) depicts an implementation scheme for the UPTVD structure. Next, we derive an expression for the impulse response of this UPTVD structure.

Impulse Response

The impulse response of the UPTVD structure is obtained by injecting an impulse at the input of the structure, i.e., $u(n) = \delta(n)$. The impulse is converted into a series of N impulses ($\delta(m), \delta(m-1), ..., \delta(m-N+1)$) at N times higher signal rate. Each of these impulses is directed to a different path in the CFB form and processed by an input scaler, a component filter, an output scaler, and a running summer. Consider the i-th path with impulse $\delta(m-i)$. The impulse is scaled by the input scaler, yielding $s_{in}(i)\delta(m-i)$. This scaled impulse is the input to the i-th component filter. Therefore, the output of the i-th component filter is $s_{in}(i)\delta(m-i) * h_i(m) = s_{in}(i)h_i(m-i)$. After scaling by the output scaler and passing through the running summer, the signal at the output of the running summer in the i-th path is given by

$$g_i(m) = \begin{cases} \sum_{j=0}^{N-1} s_{out}(m-j)s_{in}(i)h_i(m-i-j), \\ \qquad m \in [i, i+L+N-2] \\ 0, \qquad \text{otherwise} \end{cases} \tag{4}$$

The impulse response of the UPTVD structure in Figure 3 is the downsampled version of the sum of the outputs of the N running summers, i.e.,

$$\begin{aligned} h(n) &= \sum_{i=0}^{N-1} g_i(nN) \\ &= \begin{cases} \sum_{i=0}^{N-1} \sum_{j=0}^{N-1} s_{out}(N-j)s_{in}(i) \\ \quad \cdot h_i(nN-i-j), \\ \quad n \in [0, \mathrm{int}(\frac{L+2N-3}{N})] \\ 0, \qquad \text{otherwise} \end{cases} \end{aligned} \tag{5}$$

Note that in all the expressions, the component filters $h_i(m)$ are FIR filters of length L, i.e., $h_i(m) = 0$ for $m < 0$ and $m \geq L$. The impulse response (5) is used to approximate the impulse response of a target filter, by appropriate selections of the length L, the scaling factors $s_{in}(m)$ and $s_{out}(m)$, and the coefficients of the PTV subsystem. For multiplier-free implementation, we restrict the coefficients to ternary values and the scaling factors to power-of-two values.

An examination of the impulse response (5) indicates that, in general, each impulse response sample is determined by a distinct set of N^2 component filter coefficients and $2N$ scaling factors (N input and N output scaling factors). However, due to the finite length of the filters, the impulse response samples at $n = 0, 1$ and at $n > \mathrm{int}(\frac{L-1}{N})$ are determined by fewer coefficients. Consequently, full precision for these samples can not be achieved. Therefore, we force these samples to zero by choosing the corresponding coefficients as zero. We call this process 'boundary adjustment', which in effect reduces the length of the overall impulse response of the UPTVD structure to $\mathrm{int}(\frac{L-1}{N}) - 2 + 1$. Therefore, to realize a target filter of length M, the length of the PTV subsystem in the UPTVD structure is selected as $L = (M+1)N + 1$. Since the first two samples of the impulse response are set to zero by the 'boundary adjustment', the impulse response is delayed by two samples. Hence, in the design process, we try to approximate $h_t(n)$ by $h(n+2)$, where $h_t(n)$ is the impulse response of the target filter.

Selection of Scaling Factors and Filter Coefficients

To obtain multiplier-free realization of a target filter, we approximate its impulse response by (5) with the constraints that $h_i(m)$ take only ternary values and the scaling factors are selected from power-of-two values. The closest approximation is obtained when the input and output scaling factors are such that the product term $s_{in}(i)s_{out}(N-j)$ takes the maximum number of different power-of-two values, i.e., $2^{-1}, 2^{-2}, ..., 2^{-N^2}$. One solution is to select the scaling factors as

$$\begin{aligned} s_{in}(i) &= 2^{-i}, \quad i = 0, ..., N-1 \\ &= 1, 2^{-1}, ..., 2^{-N+1} \end{aligned} \quad (6)$$

$$\begin{aligned} s_{out}(N-j) &= 2^{-(jN+1)}, \quad j = 0, ..., N-1 \\ &= 2^{-1}, 2^{-N-1}, ..., 2^{-(N^2-N+1)} \end{aligned} \quad (7)$$

For example, for the case $N = 3$, different combinations of $s_{in}(i) = \{1, 2^{-1}, 2^{-2}\}$ and $s_{out}(N-j) = \{2^{-1}, 2^{-4}, 2^{-7}\}$ generate all the possible power-of-two product terms from 2^{-1} to 2^{-9}. Note that the selection of scaling factors is determined only by the period of the structure (N) and independent of the target filter response.

Since the scaling factors are powers of two, multiplications correspond to shift operations. To avoid truncation, longer registers are needed to store the shifted signals. Therefore, it is desirable that the magnitude of the exponents of the input scaling factors be as small as possible so that scaling by these factors requires a minimum number of additional bits to store the scaled signals. Consequently, the lengths of the registers and adders are minimized.

With the above set of input/output scaling factors, the selection for the component filter coefficients is simple. The expression for the impulse response given by Eq. (5) can now be rewritten as

$$\begin{aligned} h(n) &= \sum_{i=0}^{N-1}\sum_{j=0}^{N-1} 2^{-(jN+i+1)} h_i(nN-j-i) \\ &\equiv \sum_{i=0}^{N-1}\sum_{j=0}^{N-1} s(i,j) h_i(nN-j-i) \end{aligned} \quad (8)$$

where the scaling factor $s(i,j)$ takes all the power-of-two values from 2^{-1} to 2^{-N^2}. Letting $h(n) = h_t(n-2)$, $2 \le n < M+2$, the ternary values for the component filter coefficients can be determined from N^2 bit-long, sign-digit representation of $h(n)$.

The above selection procedure enables us to approximate the target filter impulse response with N^2 bits of coefficient precision. The complete design procedure can be summarized as follows, assuming that the impulse response $h_t(n)$ of the target filter is given.

1. Depending upon the desired precision of the filter, select the period N for the UPTVD structure.
2. Select input and output scaling factors according to Eqs. (6) and (7).
3. Using Eq. (8) and the N^2-bit expansion of the target impulse response, select the ternary component filter coefficients.
4. From the component filter coefficients, obtain the PTV coefficients for the PTV subsystem using Eq. (1).
5. Implement the UPTVD structure as shown in Figure 3(b).

3 DESIGN EXAMPLES

Now, we consider two design examples to examine the precision of the UPTVD structure.

Design Example 1:

In this example, we use a lowpass filter as the target filter. It is designed using Kaiser window (with parameter $\alpha = 5.772$) with normalized cutoff frequencies at 0.3 and 0.35 and allowable ripple of -60 dB. The length of the target filter with these specifications is 75.

For this target filter, multiplier-free UPTVD structures with period $N = 3, 4, 5$ are designed using the process described in the previous section. For example, consider $N = 3$. The length of the PTV subsystem is $(75+1)3+1 = 229$. The periodic (with period 3) scaling factors are $s_{in}(i) = \{1, 2^{-1}, 2^{-2}\}$ and $s_{out}(3-j) = \{2^{-1}, 2^{-4}, 2^{-7}\}$.

Figure 4 shows the frequency responses of the UPTVD structure with period $N = 3, 4, 5$, as well as that of the target filter. For $N = 3$, the UPTVD structure is not a good approximation. However, for $N = 4$ and 5, it has essentially the same frequency response as the target filter.

Design Example 2:

Another target lowpass filter is designed using Parks-McClellan Equiripple design method, with the following design specifications. The passband and stopband cutoff frequencies are at 0.2 and 0.25, respectively, and the allowable ripple level is -80 dB. The length of the target filter is $M = 100$.

For this target filter, multiplier-free UPTVD structures with $N = 3, 4, 5$ are obtained and their frequency responses are plotted in Figure 5, along with the target filter response. The response of the UPTVD structure with $N = 3$ has high ripples. With $N = 4$ the response has slightly large ripples than the ripples of the target filter, while the response with $N = 5$ is indistinguishable from that of the target filter.

Frequency response behavior of the UPTVD structure improves with an increasing value of N. As the ripples get smaller, a larger value of N is needed. When the ripples of the target filter is above -40 dB, it is sufficient to use the UPTVD structure with $N = 3$. Ripples of -70 dB can be achieved with $N = 4$, while a larger N should be used when the ripple specification is below -70 dB.

4 Discussion and Conclusion

Consider a UPTVD structure equivalent to a target FIR filter of length M. The UPTVD structure consists of a PTV subsystem of length $(M + 1)N + 1$ with PTV ternary coefficients. A few accessory units, i.e., Hold, scalers, and downsampler, are also required at the input and output of the subsystem. The PTV subsystem operates at a clock rate which is N times the normal rate. However, the PTV subsystem (PTV-FIR filter) involves only add, subtract or no-operation, which are far less complex as compared to multiply and add operations.

The accessory units can be implemented with minimum hardware. The Hold unit can be implemented using a register which is updated at the end of every Nth clock cycle. Its content is read at every clock cycle to achieve the effect of replicating each sample N times. A direct realization of the output scaler, the running summer, and the downsampler is depicted in Figure 6(a). However, they can be combined and efficiently implemented as a single unit, called shift-accumulate-downsample (S.A.D.) unit. This is possible because of a special relationship between the output scaling factors that $s_{out}(N - j) = 2^{-1}$ for $j = 0$ and $s_{out}(N-j) = 2^{-N}s_{out}(N-j-1)$ for $j \in [1, N-1]$. With this, the S.A.D unit can be implemented as shown in Figure 6(b).

For comparative study, consider a target filter of length M with b-bit coefficient precision. Conventional implementation of such a filter requires M multiplications and $M - 1$ additions per output sample. Since a multiplication consists of shift and add operations, an equivalent of $Mb + M - 1$ additions must be performed. On the other hand, the UPTVD structure with $N = \sqrt{b}$ achieves the same coefficient precision. It is required to perform $L - 1 = (M + 1)\sqrt{b}$ add/subtract operations per output sample. The UPTVD structure reuses the same hardware for N clock cycles with PTV ternary coefficients to compute the final output sample. Due to this sharing of hardware over a period N, the hardware complexity of UPTVD structure is approximately $\frac{1}{\sqrt{b}}$ times that of the target filter with b-bit coefficient precision.

Since the UPTVD structure performs only add/subtract operations, it is much simpler to realize a bit-systolic architecture for such a system. Such an architecture has N times less complexity as compared to the conventional bit-serial architectures which use bit-systolic multipliers and adders. The basic building block for the UPTVD structure consists of a delay register, an add/subtract unit, and a memory ring to store the PTV coefficients. To develop a filter architecture with higher coefficient precision, an architecture with multipliers requires a complete redesign, while the proposed architecture can be easily realized with simple modifications, i.e., by increasing N and the number of basic cells. At the same time, the regularity of the proposed architecture is preserved along the length of the filter as well as across the bit dimension.

Acknowledgements

This work was supported in part by the National Science Foundation under Grant No. MIP-8613299. The authors would like to thank Prof. Wayne P. Burleson for his useful comments.

References

[1] N. Benvenuto, L.E. Franks, and F.S. Hill, Jr., "Realization of Finite Impulse Response Filters Using Coefficients +1,0, and -1," *IEEE Trans. on Commum.*, pp. 1117-1125, Oct. 1985.

[2] P.W. Wong and R.M. Gray, "FIR Filters with Sigma-Delta Modulation Encoding," *IEEE Trans. on ASSP*, pp. 979-990, June 1990.

[3] Y.C. Lim and S.R. Parker, "FIR Filter Design over discrete powers-of-two coefficient space", *IEEE Trans. on ASSP*, vol. ASSP-31, pp. 583-591, June 1983.

[4] Q. Zhao and Y. Tadokoro, "A Simple Design of FIR Filters with powers-of-two coefficients", *IEEE Trans. on Circuits and Systems*, vol. 35, pp. 566-570, May 1988.

[5] H. Samueli, "An improved Search Algorithm for the design of Multiplierless FIR filters with powers-of-two Coefficients", *IEEE Trans. on Circuits and Systems*, vol 36, pp. 1044-1047, July 1989.

[6] Bede Liu and Peter A. Franaszek, "A Class of Time-Varying Digital Filters", *IEEE Trans. on Circuit Theory*, vol. CT-16, no. 4, pp. 467-471, November 1969.

[7] T.A.C.M. Claasen and W.F.G. Mecklenbrauker, "On Stationary Linear Time-Varying Systems", *IEEE Trans. on Circuits and Systems*, vol. CAS-29, pp. 169-184, March 1982.

[8] Rokuya Ishii and Masahiro Kakishita, "A Design Method for a Periodically Time-Varying Digital Filter for Spectrum Scrambling", *IEEE Trans. on Acoustics, Speech and Signal Processing*, vol. 38, no. 7, pp. 1219-1222, July 1990.

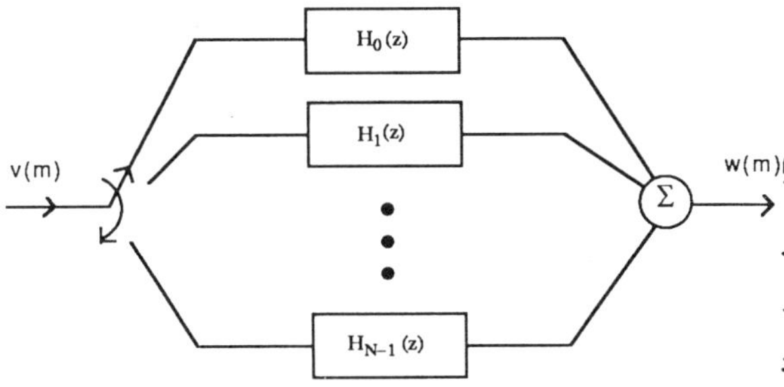

(a) **CFB** form for a PTV-FIR filter

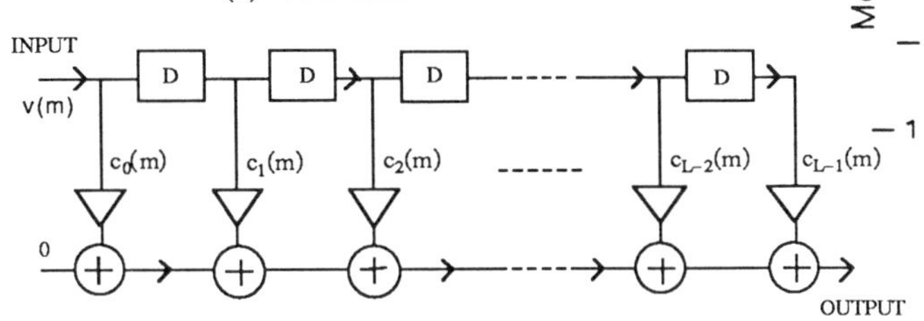

(b) Direct form-I for a PTV-FIR filter

Figure 1 Realizations of Periodically Time-Varying FIR Filter

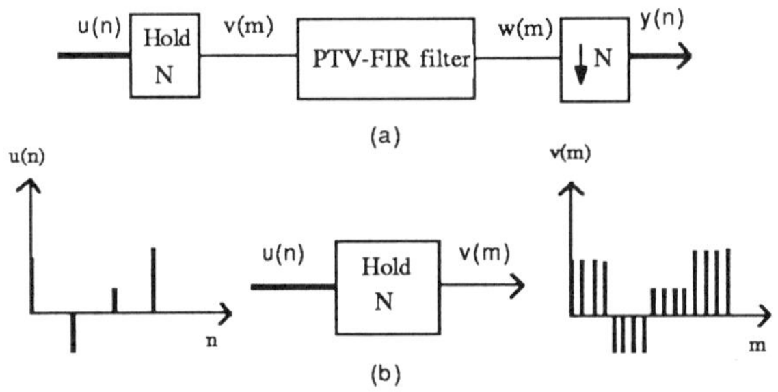

Figure 2: Time invariant structure with PTV filter unit

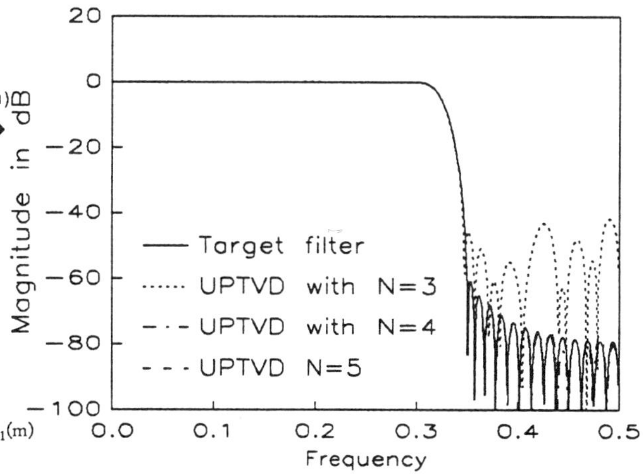

Figure 4 Magnitude Responses for Design Example 1

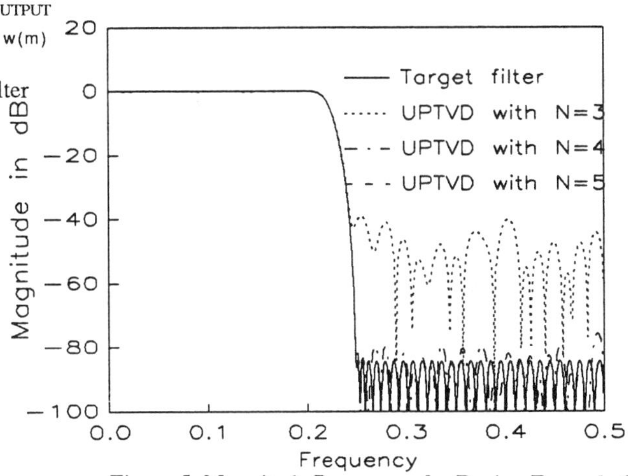

Figure 5 Magnitude Responses for Design Example 2

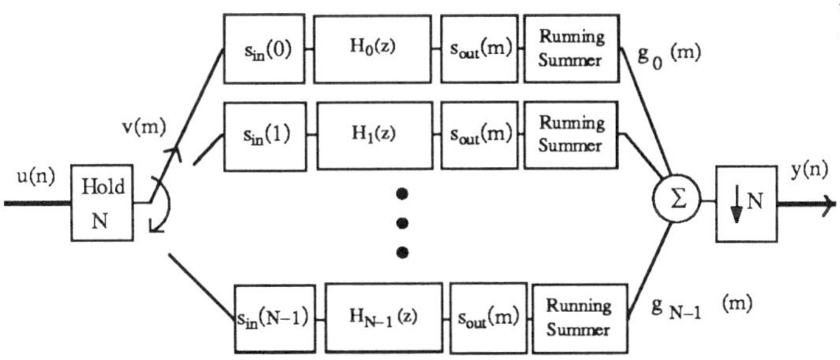

(a) **CFB** form Decomposition

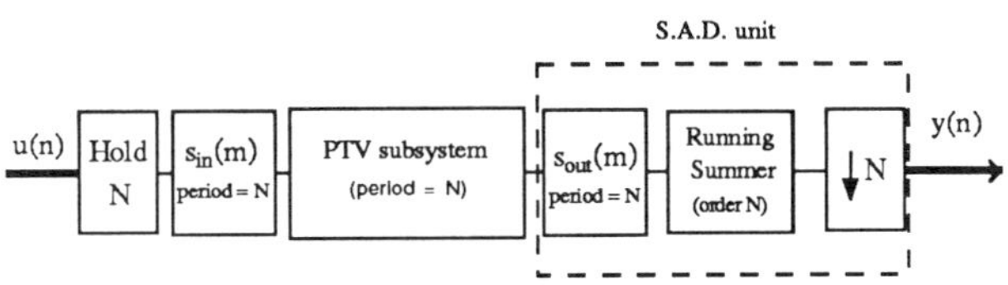

(b) Block diagram for implementation

Figure 3 UPTVD Structure

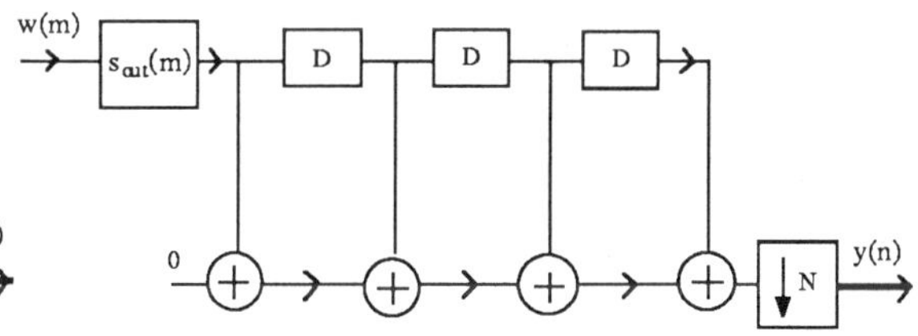

(a) Direct implementation

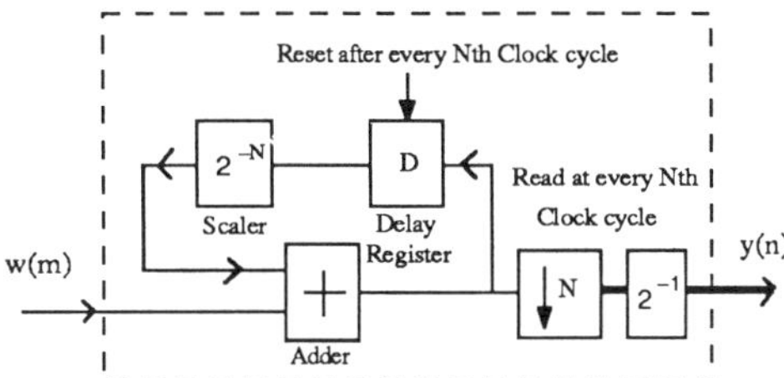

(b) Efficient implementation

Figure 6 Realizations of S.A.D. unit (for N = 4)

A Computational Toolkit for Applied Z-Transforms*

Mark S. Spurbeck Clifford T. Mullis
Department of Electrical and Computer Engineering
University of Colorado
Boulder, CO 80309

Abstract

This paper presents a computational toolkit which operates on rational polynomial structures. This family of structures represents almost all sequences whose z transforms can be written as the ratio of two finite order polynomials in z. The toolkit may be applied to a variety of practical filter design problems. Some of the algorithms developed include spectral factorization, separation into causal and anti-causal subsequences, computation of greatest common denominator, and derivation of equivalent minimum phase polynomials. We briefly develop each algorithm and then present condensed pseudo code versions of each tool.

1 Introduction

The discussion begins by specifying a base FIR polynomial structure. Then we specify a set of primitive computational operators such as addition, multiplication, division etc. which operate on z transform polynomials. With these primitive operators we implement a greatest common denominator algorithm, Euclid's algorithm, and a polynomial factorization algorithm. Building upon this framework we can then define the more general rational polynomial structure which gives us a finite numerical representation of infinite impulse response digital filters. Similarly we develop a set of operators for rational structures.

2 FIR Polynomials

The FIR polynomial structure consists of a finite array of z-transform polynomial coefficients with an order parameter. In this document $\hat{f}$ or $\hat{f}(z)$ will denote a sequence in the z-transform domain while $f(k)$ will represent the kth polynomial coefficient or equivalently the sequence in the time domain.

```
struct FIR {   integer        order;
               double float   coef[ ];   }
```

*This work supported by the office of Navel Research, Mathematics Division, Statistics and Probability Branch, under Contract N00014-89-J1070 and a National Defence Science and Engineering Graduate Fellowship in Electrical Engineering awarded by the U.S. Army Research Office, grant no. DAAL03-91-G-0253

2.1 Primitive FIR Tools

The most basic operations, multiplication, addition, scaling, etc. appear in the following psuedo code.

name: f_sum($\hat{f}, \hat{g}, \hat{h}$) (polynomial addition)
input: $\hat{f}, \hat{g}$ - FIR
output: $\hat{h}$ - FIR satisfying: $\hat{h}(z) = \hat{f}(z) + \hat{g}(z)$

```
f_sum(f̂, ĝ, ĥ) {
   order(ĥ) = max(order(f̂), order(ĝ))
   for(k = 0 to order(ĥ))
      h(k) = 0
      if(k ≤ order(f̂))
         h(k) = h(k) + f(k)
      if(k ≤ order(ĝ))
         h(k) = h(k) + g(k) }
```

name: f_mult($\hat{f}, \hat{g}, \hat{h}$) (sequence convolution)
input: $\hat{f}, \hat{g}$ - FIR
output: $\hat{h}$ - FIR satisfying: $\hat{h}(z) = \hat{f}(z)\hat{g}(z)$

```
f_mult(f̂, ĝ, ĥ) {
   if(order(f̂) = -1 or order(ĝ) = -1)
      order(ĥ) = -1
   else
      N = order(ĥ) = order(f̂) + order(ĝ)
      for(k = 0 to N)
         h(k) = 0
         for(m = 0 to order(f̂))
            n = k - m
            if(n ≤ order(ĝ))
               if(n ≥ 0)
                  h(k) = h(k) + f(m)g(n) }
```

name: f_scale($\hat{f}, \gamma$) (scalar multiply)
input: $\hat{f}$ -FIR γ - real
output: $\hat{g}$ - FIR satisfying: $\hat{g}(z) = \gamma\hat{f}(z)$

name: f_deg($\hat{f}$,d)
input: $\hat{f}$ -FIR
output: d - real satisfying:
d = max(k) such that $|f(k)| >$ machine ϵ

We use f_deg() to find the highest order non-zero coefficient of an FIR polynomial. Another useful primitive tool, f_tilde(), performs sequence time reversal with a right shift by the polynomial order which forces causality (matched filter).

name: f_tilde($\hat{f}, \tilde{f}$)
input: $\hat{f}$ - FIR
output: $\tilde{f}$ - FIR satisfying: $\tilde{f}(z) = z^{-n}\hat{f}(z^{-1})$
where n = order($\hat{f}$)

```
f_tilde(f, f~) {
    N = order(f~) = order(f^)
    for(k = 0 to N)
        f~(k) = f(n - k) }
```

3 Polynomial Equations and Euclid's Algorithm

Next we consider the problem of solving simple polynomial equations. This leads to the development of a polynomial division algorithm, a greatest common denominator algorithm and Euclid's algorithm. Consider the following polynomial equation:

Given: $\hat{a}(z), \hat{b}(z), \hat{c}(z)$
Find: $\hat{x}(z), \hat{y}(z)$
Satisfying: $\hat{a}(z)\hat{y}(z) + \hat{b}(z)\hat{x}(z) = \hat{c}(z)$ (1)

We encounter this equation when performing bulk partial fraction expansion of a polynomial structure:

$$\hat{r}(z) = \frac{\hat{c}(z)}{\hat{a}(z)\hat{b}(z)} = \frac{\hat{x}(z)}{\hat{a}(z)} + \frac{\hat{y}(z)}{\hat{b}(z)}$$

This will allow us to separate a rational z-transform polynomial $\hat{r}(z)$ into its causal and anti-causal components in section 4.

Solving the polynomial equation 1 requires a polynomial division algorithm. Like integers, polynomials do not have multiplicative inverses i. e. for $\hat{a}(z)$ nontrivial $1/\hat{a}(z)$ is not a polynomial [4]. Consequently for any two polynomials $\hat{f}(z)$ and $\hat{g}(z)$ we define a quotient $\hat{q}(z)$ and a remainder $\hat{r}(z)$ satisfying:

$$\hat{f}(z) = \hat{g}(z)\hat{q}(z) + \hat{r}(z) \tag{2}$$

where $\deg(\hat{r}) < \deg(\hat{g})$

name: f_div($\hat{f}, \hat{g}, \hat{q}, \hat{r}$) (polynomial division)
input: $\hat{f}$, $\hat{g}$ - FIR $\hat{g} \neq 0$
output: $\hat{q}$, $\hat{r}$ - FIR satisfying: equation 2 and
order($\hat{r}$) = deg($\hat{r}$) < deg($\hat{g}$)
order($\hat{q}$) = deg($\hat{q}$)

Division Algorithm:

$\hat{r}(z) \longleftarrow \hat{f}(z)$
$\hat{q}(z) \longleftarrow 0$
m $\longleftarrow$ deg($\hat{r}$)
n $\longleftarrow$ deg($\hat{g}$)
while(m $\geq$ n)
$\beta \longleftarrow$ r(m)/g(n)
$\hat{r}(z) \longleftarrow \hat{r}(z) - \beta z^{n-m}\hat{g}(z)$
$\hat{q}(z) \longleftarrow \hat{q}(z) + \beta z^{n-m}$
m $\longleftarrow$ m + 1

3.1 Dual Decomposition and GCD

Next we use the division algorithm to implement a GCD tool. Rewrite equation 1:

$$[\hat{y}(z), \hat{x}(z)] \begin{bmatrix} \hat{a}(z) \\ \hat{b}(z) \end{bmatrix} = \hat{c}(z) \tag{3}$$

Euclid's algorithm progressively simplifies this equation by finding an equivalent equation where the degree of at least some of the polynomials decrease. This simplification is repeated until $\hat{b}(z)$ vanishes. We begin this process by finding the GCD of $\hat{a}(z)$ and $\hat{b}(z)$. First we use the division algorithm to produce:

$$\hat{a}(z) = \hat{q}(z)\hat{b}(z) + \hat{r}(z)$$

Let: $\hat{a}'(z) = \hat{b}(z)$ and $\hat{b}'(z) = \hat{r}(z)$ Then rewrite in matrix form:

$$\begin{bmatrix} \hat{a}'(z) \\ \hat{b}'(z) \end{bmatrix} = \begin{bmatrix} 0 & 1 \\ 1 & -\hat{q}(z) \end{bmatrix} \begin{bmatrix} \hat{a}(z) \\ \hat{b}(z) \end{bmatrix}$$

This process is repeated until there is no remainder. The division algorithm guarantees this result because $\deg(\hat{b}') = \deg(\hat{r}) < \deg(\hat{b})$.

$$\begin{bmatrix} \hat{f}(z) \\ 0 \end{bmatrix} = \begin{bmatrix} \hat{a}^{(m)}(z) \\ 0 \end{bmatrix} = \hat{\mathbf{T}}_m(z) \cdots \hat{\mathbf{T}}_2(z)\hat{\mathbf{T}}_1(z) \begin{bmatrix} \hat{a}(z) \\ \hat{b}(z) \end{bmatrix} \tag{4}$$

where: $\hat{\mathbf{T}}_k(z) = \begin{bmatrix} 0 & 1 \\ 1 & -\hat{q}_k(z) \end{bmatrix}$ and $\det \hat{\mathbf{T}}_k(z) = -1$

Since $\det \hat{\mathbf{T}}_k(z) = -1$, $\hat{\mathbf{T}}_k(z)$ has inverse

$$\hat{\mathbf{T}}_k^{-1}(z) = \begin{bmatrix} \hat{q}_k(z) & 1 \\ 1 & 0 \end{bmatrix}$$

Then we can write

$$\begin{bmatrix} \hat{a}(z) \\ \hat{b}(z) \end{bmatrix} = \hat{\mathbf{T}}^{-1}(z) \begin{bmatrix} \hat{f}(z) \\ 0 \end{bmatrix} \tag{5}$$

$$\text{where: } \hat{\mathbf{T}}(z) = \prod_{k=m}^{1} \begin{bmatrix} 0 & 1 \\ 1 & -\hat{q}_k(z) \end{bmatrix} = \begin{bmatrix} \hat{u}(z) & \hat{v}(z) \\ \hat{s}(z) & \hat{t}(z) \end{bmatrix} \tag{6}$$

$$\hat{\mathbf{T}}^{-1}(z) = \prod_{k=1}^{m} \begin{bmatrix} \hat{q}_k(z) & 1 \\ 1 & 0 \end{bmatrix} = \begin{bmatrix} \hat{t}(z) & -\hat{v}(z) \\ -\hat{s}(z) & \hat{u}(z) \end{bmatrix}$$

Since $\det \hat{\mathbf{T}}(z) = \pm 1$ (unimodular)

$$\Rightarrow \hat{u}(z)\hat{t}(z) - \hat{v}(z)\hat{s}(z) = \pm 1 \tag{7}$$

$\Rightarrow$ $\hat{t}(z)$ and$\hat{s}(z)$ are coprime

Then: $\hat{a}(z) = \hat{t}(z)\hat{f}(z)$, $\hat{b}(z) = -\hat{s}(z)\hat{f}(z)$ and $\dfrac{\hat{b}(z)}{\hat{a}(z)} = -\dfrac{\hat{s}(z)}{\hat{t}(z)}$

Since $\hat{t}(z)$ and $\hat{s}(z)$ are coprime $\hat{f}(z) = \text{GCD}(\hat{a}, \hat{b})$.

name: f_gcd($\hat{a}, \hat{b}, \hat{f}, \hat{u}, \hat{v}, \hat{s}, \hat{t}$)
input: $\hat{a}$, $\hat{b}$ - FIR non-zero
output: $\hat{f}, \hat{u}, \hat{v}, \hat{s}, \hat{t}$ - FIR satisfying: eqn's 4, 6, 7

$\hat{f}(z) \longleftarrow \hat{a}(z)$
$\hat{g}(z) \longleftarrow \hat{b}(z)$
$\hat{u}(z) \longleftarrow \hat{t}(z) \longleftarrow 1$
$\hat{v}(z) \longleftarrow \hat{s}(z) \longleftarrow 0$
while($\hat{g} \neq 0$)
  f_div($\hat{f}, \hat{g}, \hat{q}, \hat{r}$) so that $\hat{f} = \hat{g}\hat{q} + \hat{r}$
  $\hat{f}(z) \longleftarrow \hat{g}(z)$
  $\hat{g}(z) \longleftarrow \hat{r}(z)$
  $\hat{x}(z) \longleftarrow \hat{s}(z)$
  $\hat{y}(z) \longleftarrow \hat{t}(z)$
  $\hat{s}(z) \longleftarrow \hat{u}(z) - \hat{q}(z)\hat{s}(z)$
  $\hat{t}(z) \longleftarrow \hat{u}(z) - \hat{q}(z)\hat{t}(z)$
  $\hat{u}(z) \longleftarrow \hat{x}(z)$
  $\hat{v}(z) \longleftarrow \hat{y}(z)$

3.2 Solving the Polynomial Equation

If $\hat{a}(z)$ and $\hat{b}(z)$ are not coprime then for some λ $\hat{a}(\lambda) = \hat{b}(\lambda) = 0$ and in general $\hat{c}(\lambda) \neq 0$. Then the equation $\hat{a}(z)\hat{y}(z) + \hat{b}(z)\hat{x}(z) = \hat{c}(z)$ has no solution. However if $\deg(\hat{f}) = 0$ then $\hat{a}(z)$ and $\hat{b}(z)$ are coprime[1]. Then using the results of the GCD algorithm, equation 5, we can rewrite equation 3 as follows:

$$[\hat{y}(z), \hat{x}(z)]\hat{\mathbf{T}}^{-1}(z) \begin{bmatrix} f(0) \\ 0 \end{bmatrix} = \hat{c}(z) \tag{8}$$

An immediate solution can be constructed as:

$$[\hat{y}_0(z), \hat{x}_0(z)] = \frac{1}{f(0)}[\hat{c}(z), 0]\hat{\mathbf{T}}(z) \tag{9}$$

Thus $\hat{x}_0(z) = \frac{1}{f(0)}\hat{v}(z)\hat{c}(z)$, $\hat{y}_0(z) = \frac{1}{f(0)}\hat{u}(z)\hat{c}(z)$ is a solution and for any polynomial $\hat{q}(z)$, $\hat{x}(z) = \hat{x}_0(z) - \hat{a}(z)\hat{q}(z)$, $\hat{y}(z) = \hat{y}_0(z)+\hat{b}(z)\hat{q}(z)$ is also a solution pair. In order to specify a unique solution we require $\deg(\hat{x}) \leq \deg(\hat{a})$.

name: f_euclid($\hat{a}, \hat{b}, \hat{c}, \hat{x}, \hat{y}$) (polynomial eqn. solver)
input: $\hat{a}$, $\hat{b}$, $\hat{c}$ - FIR with $\hat{a}$, $\hat{b}$ coprime
output: $\hat{x}, \hat{y}$ - FIR satisfying: $\hat{a}\hat{y} + \hat{b}\hat{x} = \hat{c}$
$\deg(\hat{x}) < \deg(\hat{a})$

f_euclid($\hat{a}, \hat{b}, \hat{c}, \hat{x}, \hat{y}$)
  Execute f_gcd($\hat{a}, \hat{b}, \hat{f}, \hat{u}, \hat{v}, \hat{s}, \hat{t}$) get output $\hat{f}, \hat{u}, \hat{v}, \hat{s}, \hat{t}$
  If $\deg(\hat{f}) > 0$ Then
    Error $\hat{a}$, $\hat{b}$ not coprime
  Else
    $\hat{x}_0(z) \longleftarrow \frac{1}{f(0)}\hat{v}(z)\hat{c}(z)$
    f_div($\hat{x}_0, \hat{a}, \hat{q}, \hat{x}$) so that $\hat{x}_0 = \hat{a}\hat{q} + \hat{x}$
    $\hat{f}(z) \longleftarrow \hat{c}(z) - \hat{b}(z)\hat{x}(z)$
    f_div($\hat{f}, \hat{a}, \hat{y}_0, \hat{r}$) so that $\hat{f} = \hat{a}\hat{y}_0 + \hat{r}$
    (Verify that $\hat{r}(z) = 0$)

[1] $\deg(\hat{f}) = 0 \Rightarrow \hat{f}(z) = f(0)$

4 Factorization

A pratical and robust representation for the general class of LTI digital filters requires a factorization algorithm which separates rational z-transform polynomials into their causal and anti-causal components. Consider the following rational z-transform:

$$\hat{h}(z) = \frac{\hat{b}(z)}{\hat{a}(z)} \tag{10}$$

Our task consists of factoring the denominator polynomial [2] $\hat{a}(z)$ into two polynomials $\hat{a}_1(z)$ and $\tilde{a}_2(z)$ where the roots of $\hat{a}_1(z)$ are inside the unit circle and those of $\tilde{a}_2(z)$ are outside the unit circle.

Given:	$\hat{a}(z)$	
Find:	$\hat{a}_1(z), \hat{a}_2(z)$	
Satisfying:	$\hat{a}(z) = \hat{a}_1(z)\tilde{a}_2(z)$	(11)
Where:	$\hat{a}_1(z)$- monic and min ϕ	
	$\hat{a}_2(z)$- min ϕ	

Monic simply means $a(0) = 1$ and minimum phase (min ϕ) constrains all the roots to be inside the unit circle. Then performing the tilde operation on $\hat{a}_2(z)$ translates all of its roots outside the unit circle without changing the magnitude response. After factoring the denominator we apply Euclids algorithm (f_euclid()) to separate the numerator.

$$\hat{a}_1(z)\hat{x}(z) + \tilde{a}_2(z)\hat{y}(z) = \hat{b}(z)$$

$$\frac{\hat{b}(z)}{\hat{a}_1(z)\tilde{a}_2(z)} = \frac{\hat{y}(z)}{\hat{a}_1(z)} + \frac{\hat{x}(z)}{\tilde{a}_2(z)}$$

This completes the separation into causal and anti-causal components. We address the general factorization problem by first solving the special case of spectral factorization.

4.1 Spectral Factorization

The spectral factorization algorithm plays a key role in Wiener filter construction, l_2 approximation, and causal synthesis of wide-sense stationary sequences. We consider the following problem.

Given:	$\hat{c}(z)$	- FIR order n	
		- $\mathrm{Re}[\hat{c}(z)] > 0$ for all $\lvert z\rvert = 1$	
Find:	$\hat{a}(z)$	- FIR order $\leq n$	
		- minϕ	
Satisfying:	$\hat{a}(z)\hat{a}(z^{-1})$	$= \hat{c}(z) + \hat{c}(z^{-1})$	(12)

Since this equation is quadratic in the coefficients of $\hat{a}(z)$ we cannot solve it with a finite number of primitive operations. However by employing a Newton Raphson algorithm for locating the zero of a quadratic

[2] $\hat{a}(z)$ cannot have any roots on the unit circle $\hat{a}(\lambda) = 0 \Rightarrow |\lambda| \neq 1$, equivalently $\hat{h}(z)$ has no poles on the unit circle

vector valued function we compute a sequence of polynomials which rapidly converge to the solution [3]. The algorithm consists of an iterative procedure which linearizes the function of polynomial coefficients at each iteration and then finds the zero of the linear model. The goal is to find a root of the following quadratic equation:

$$f(a(0)\cdots a(n), c(0)\cdots c(n)) = \hat{a}(z)\hat{a}(z^{-1}) - \hat{c}(z) - \hat{c}(z^{-1})$$

This approach is successful but the details are unenlightening. A polynomial equivalent of Newton's map was recognized by Wilson [1] and goes roughly as follows: Let

$$\hat{e}_0(z) = \hat{a}(z)\hat{a}(z^{-1}) - \hat{c}(z) - \hat{c}(z^{-1})$$

be the current error. Then add a variation $\hat{v}(z)$ to $\hat{a}(z)$ and set the new error equal to the term which is quadratic in the variation (This is equivalent to linearization).

$$\begin{aligned} \hat{e}_1(z) &= [\hat{a}(z) + \hat{v}(z)][\hat{a}(z^{-1}) + \hat{v}(z^{-1})] - \hat{c}(z) - \hat{c}(z^{-1}) \\ &= \hat{v}(z)\hat{v}(z^{-1}) \end{aligned}$$

The Newton iteration then becomes: Solve the following symmetric polynomial equation for $\hat{v}(z)$:

$$\hat{a}(z)\hat{v}(z^{-1}) + \hat{a}(z^{-1})\hat{v}(z) = \hat{c}(z) + \hat{c}(z^{-1}) - \hat{a}(z)\hat{a}(z^{-1})$$

Then replace $\hat{a}(z) \longleftarrow \hat{a}(z) + \hat{v}(z)$

For computational efficiency we avoid computing $\hat{a}(z)\hat{a}(z^{-1})$ by performing the following transformation: Let

$$\hat{v}(z) \longleftarrow \hat{x}(z) - \frac{1}{2}\hat{a}(z)$$

The Newton iteration then becomes: Given $\hat{a}(z)$ and $\hat{c}(z)$ solve the following symmetric polynomial equation for $\hat{v}(z)$:

$$\hat{a}(z)\hat{x}(z^{-1}) + \hat{a}(z^{-1})\hat{x}(z) = \hat{c}(z) + \hat{c}(z^{-1}) \tag{13}$$

Then replace $\hat{a}(z) \longleftarrow \frac{1}{2}\hat{a}(z) + \hat{x}(z)$ The polynomial equation for the Newton iteration is known as the Jury equation (eqn 13). We shall properly employ this equation only when $\hat{a}(z)$ is min ϕ. Using a variation on Euclid's algorithm we solve the Jury equation by successively reducing the problem order until it degenerates to the scalar case. Let

$$\hat{c}(z) = \hat{d}(z) + \gamma\tilde{a}(z) = \hat{d}(z) + \gamma z^{-n}\hat{a}(z^{-1})$$

Where $\gamma = c(n)/a(0) \Rightarrow d(n) = 0$. Substituting back into the Jury equation:

$$\hat{a}(z)[\hat{x}(z^{-1}) - \gamma(z^{n})] + \hat{a}(z^{-1})[\hat{x}(z) - \gamma z^{-n})] = \hat{d}(z) + \hat{d}(z^{-1}) \tag{14}$$

The desired effect is an equivalent problem of reduced order. Set

$$\begin{bmatrix} \hat{b}(z) \\ \hat{b}(z^{-1}) \end{bmatrix} = \begin{bmatrix} 1 & -\beta z^{-n} \\ -\beta z^{n} & 1 \end{bmatrix} \begin{bmatrix} \hat{a}(z) \\ \hat{a}(z^{-1}) \end{bmatrix} \tag{15}$$

Where $\beta = a(n)/a(0) \Rightarrow b(n) = 0 \Rightarrow \deg(\hat{b}) < n$. If $\beta^2 \neq 1$ then:

$$\det \begin{vmatrix} 1 & -\beta z^{-n} \\ -\beta z^{n} & 1 \end{vmatrix} = 1 - \beta^2 \neq 0$$

Since this matrix is unimodular we can invert the system of equations 15:

$$\begin{bmatrix} \hat{a}(z) \\ \hat{a}(z^{-1}) \end{bmatrix} = \frac{1}{1-\beta^2} \begin{bmatrix} 1 & \beta z^{-n} \\ \beta z^{n} & 1 \end{bmatrix} \begin{bmatrix} \hat{b}(z) \\ \hat{b}(z^{-1}) \end{bmatrix} \tag{16}$$

$$\text{equivalently: } \hat{a}(z) = \frac{\hat{b}(z) + \beta z^{-n}\hat{b}(z^{-1})}{1-\beta^2}$$

Then we substitute back into equation 14 and rearrange terms producing:

$$\hat{b}(z)\hat{y}(z^{-1}) + \hat{b}(z^{-1})\hat{y}(z) = \hat{d}(z) + \hat{d}(z^{-1}) \tag{17}$$

where

$$\begin{bmatrix} \hat{y}(z) \\ \hat{y}(z^{-1}) \end{bmatrix} = \frac{1}{1-\beta^2} \begin{bmatrix} 1 & \beta z^{-n} \\ \beta z^{n} & 1 \end{bmatrix} \begin{bmatrix} \hat{x}(z) - \gamma z^{-n} \\ \hat{x}(z^{-1}) - \gamma z^{n} \end{bmatrix} \tag{18}$$

Notice that all of the polynomials in equation 17 have order $< n$ Inverting this system of equations gives:

$$\begin{bmatrix} \hat{x}(z) \\ \hat{x}(z^{-1}) \end{bmatrix} = \gamma \begin{bmatrix} z^{-n} \\ z^{n} \end{bmatrix} + \begin{bmatrix} 1 & -\beta z^{-n} \\ -\beta z^{n} & 1 \end{bmatrix} \begin{bmatrix} \hat{y}(z) \\ \hat{y}(z^{-1}) \end{bmatrix} \tag{19}$$

Given $\hat{y}(z)$ of order $< n$ we could reconstruct $\hat{x}(z)$ from β, γ and $\hat{y}(z)$ using equation 19. Furthermore $\hat{y}(z)$ can be determined recursively by applying an identical process which results in an expression for $\hat{y}(z)$ as a function of a new polynomial $\hat{y}'(z)$ of smaller order. This procedure is repeated until the problem reduces to a scalar equation. Then the solution to equation 13 can be reconstructed by successive application of equation 19.

name: jury_equation($\hat{a}_0, \hat{c}_0, \hat{x}$) (Jury equation solver)
input: $\hat{a}_0$, $\hat{c}_0$ - FIR of same order
output: $\hat{x}$ - FIR satisfying eqn 13

```
a(z) <-- a0(z)
c(z) <-- c0(z)
for(k = order(a0 down to 1)
    γk = c(k)/a(0)
    βk = a(k)/a(0)
    c(z) <-- c(z) - γk z^-k a(z^-1)
    a(z) <-- a(z) - βk z^-k a(z^-1)
x(0) <-- c(0)/a(0)
order(x) <-- 0
for(k = 1 to order(a0)
    x(z) <-- γk z^-k + x(z) - βk z^-k x(z^-1)
```

Throughout this algorithm it is assumed that $a(0) \neq 0$. This is a consequence of $\hat{a}_0(z)$ being minimum phase. It can be shown that successive iterates of the Newton map, $\hat{a}_k(z)$, are guaranteed to be minimum phase and consequently $a_k(0) > 0$ for all k. The Jury equation solver provides the final computational tool required to perform the Newton map for spectral factorization.

name: f_sqrt($\hat{c}, \hat{a}$) (spectral factorization)
input: $\hat{c}$ - FIR $\mathrm{Re}[\hat{c}(e^{j\theta})] > 0$
output: $\hat{a}$ - FIR, min ϕ satisfying eqn 12

$\epsilon \longleftarrow c(0) \times 10^{-15}$
$\alpha \longleftarrow \sqrt{\hat{c}(e^0)}$
$\beta \longleftarrow \sqrt{\hat{c}(e^\pi)}$
$a(0) \longleftarrow (\alpha + \beta)/\sqrt{2}$
$a(1) \longleftarrow (\alpha - \beta)/\sqrt{2}$
$a(k) \longleftarrow 0$ for $2 \leq k \leq \deg(\hat{c})$
$\hat{a} \longleftarrow \sqrt{2\hat{c}(0)}$
count $\longleftarrow$ 0
until(count > 2 and $\alpha < \epsilon + a(0)$) loop:
  $\alpha \longleftarrow a(0)$
  jury_equation($\hat{c}, \hat{a}, \hat{x}$)
  $\hat{a}(z) \longleftarrow \frac{1}{2}\hat{a}(z) + \hat{x}(z)$
  count $\longleftarrow$ count + 1

Next we consider the inverse of the spectral factorization tool f_square(). The f_square() tool computes the product $\hat{b}(z)\hat{b}(z^{-1})$ which is real, symmetric and non-negative. Consequently it has the following representation:

$$\hat{c}(z) + \hat{c}(z^{-1}) = \hat{b}(z)\hat{b}(z^{-1}) \qquad (20)$$

name: f_square($\hat{b}, \hat{c}$) (square magnitude)
input: $\hat{b}$ - FIR
output: $\hat{c}$ - FIR, satisfying eqn 20

f_tilde($\hat{b}, \hat{x}$) so that $\hat{x}(z) = z^{-\mathrm{order}(\hat{b})}\hat{b}(z^{-1})$
f_mult($\hat{b}, \hat{x}, \hat{y}$) so that $\hat{y}(z) = \hat{x}(z)\hat{b}(z)$
order($\hat{y}$) $\longleftarrow$ order($\hat{b}$)
f_tilde($\hat{y}, \hat{c}$) so that $\hat{c}(z) = z^{-\mathrm{order}(\hat{y})}\hat{y}(z^{-1})$
$c(0) \longleftarrow \frac{1}{2}c(0)$

4.2 Minimum Phase Equivalent

The f_sqrt() and f_square() tools are employed to find the minimum phase equivalent of an FIR polynomial.

Given:	$\hat{b}(z)$
Find:	$\hat{a}(z)$ minϕ
Satisfying:	$\hat{a}(z)\hat{a}(z^{-1}) = \hat{b}(z)\hat{b}(z^{-1})$

This is accomplished by first 'squaring' $\hat{b}(z)$ and then computing the 'square root' of the result.

f_square($\hat{b}, \hat{c}$)	:	$\hat{c}(z) + \hat{c}(z^{-1}) = \hat{b}(z)\hat{b}(z^{-1})$
f_sqrt($\hat{c}, \hat{a}$)	:	$\hat{a}(z)\hat{a}(z^{-1}) = \hat{c}(z) + \hat{c}(z^{-1})$ (21)
where:		$\hat{a}(z)$ minϕ

4.3 Causal/Anti-Causal Factorization

Finally we employ the minimum phase extraction technique with the GCD algorithm to perform the general factorization (see eqn 11):

$$\hat{a}(z) = \hat{a}_1(z)\tilde{a}_2(z) \qquad (22)$$

Where: $\hat{a}_1(z)$ - monic and min ϕ
$\hat{a}_2(z)$ - min ϕ

First we find the minimum phase equivalent of $\hat{a}(z)$, call it $\hat{a}_0(z)$, using technique 21. Then:

$$\hat{a}_0(z) = \hat{a}_1(z)\hat{a}_2(z)$$

(Except for a scale factor.) Furthermore:

$$\hat{a}_1(z) = \mathrm{GCD}(\hat{a}(z), \hat{a}_0(z))$$

The GCD algorithm returns polynomials $\hat{f}, \hat{s}, \hat{t}, \hat{u}, \hat{v}$, satisfying:

$$\begin{bmatrix} \hat{a}(z) \\ \hat{a}_0(z) \end{bmatrix} = \begin{bmatrix} \hat{a}_1(z) & \tilde{a}_2(z) \\ \hat{a}_1(z) & \hat{a}_2(z) \end{bmatrix} = \pm \begin{bmatrix} \hat{t}(z) & -\hat{v}(z) \\ -\hat{s}(z) & \hat{u}(z) \end{bmatrix} \begin{bmatrix} \hat{f}(z) \\ 0 \end{bmatrix} \qquad (23)$$

where

$$\begin{cases} \hat{f}(z) = f(0)\hat{a}_1(z), \text{except for order}, \\ \hat{a}_2(z) = \gamma\hat{s}(z), \text{except for scale factor } \gamma, \text{ and order}, \\ \mathrm{order}(\hat{a}_2) = \deg(\hat{a}) - \deg(\hat{f}), \\ \mathrm{order}(\hat{a}_1) = \mathrm{order}(\hat{a}) - \mathrm{order}(\hat{a}_2), \end{cases}$$

This gives us the means to perform polynomial factorization.

name: f_factor($\hat{a}, \hat{a}_1, \hat{a}_2$) (factorization)
input: $\hat{a}$ - FIR $\mathrm{Re}[\hat{a}(z)] > 0$ for all $|z| = 1$
output: $\hat{a}_1, \hat{a}_2$ - FIR, satisfying conditions of eqn 11

f_square($\hat{a}, \hat{c}$)
f_sqrt($\hat{c}, \hat{a}_0$)
f_gcd($\hat{a}, \hat{a}_0, \hat{f}, \hat{u}, \hat{v}, \hat{s}, \hat{t}$)
$\alpha \longleftarrow 1/f(0)$
f_scale($\alpha, \hat{f}, \hat{a}_1$)
f_deg($\hat{a}, d$)
f_deg($\hat{f}, d_1$)
$\gamma \longleftarrow a(d)/[a_1(d_1)s(0)]$
zero pad $\hat{a}_2$ to order $d - d_1$
zero pad $\hat{a}_1$ to order n_1 = order($\hat{a}$) - order($\hat{a}_2$)

5 Summary and Conclusions

This set of algorithms establishes a basic framework for a general purpose z-transform computational toolkit. Rational z-transform structures and their corresponding operators can easily be implemented by straightforward application of the primitives developed here. Furthermore these tools successfully solve a large class of practical filter design problems without resorting to polynomial root finding algorithms.

References

[1] G. T. Wilson, "Factorization of The Covariance Generating Function of a Pure Moving Average Process," *SIAM J. Numer. Anal.*, Vol. 6, pp. 1-7, March 1961.

[2] R. A. Roberts and C. T. Mullis, *Digital Signal Processing*, (Reading, MA: Addison-Wesley, 1987).

[3] C. J. Demeure and C. T. Mullis, "A Newton-Raphson Method for Moving-Average Spectral Factorization using the Euclid Algorithm," *IEEE Trans. Acoust., Speech, Signal Proc.*, Vol. 38, pp. 1697-1709, Oct. 1990.

[4] R. E. Blahut, *Fast Algorithms for Digital Signal Processing*, (Reading, MA: Addison-Wesley, 1985).

Random Access Camera

Neal C. Gallagher
&
Jan P. Allebach
School of Electrical Engineering
Purdue University
West Lafayette, IN 47907-0501, USA

Paul Haugen
&
David Kranz
Cyber Optics Corp.
2331 University Ave, SE
Minneapolis, MN 55414

Abstract

We describe an optical system that allows for the a bit reversed decimation of an image scene. The optical system consists of a successive sequence of lens arrays. The output of one lens array feeds into the input of the next array. The function of each lens array is to divide the input into its four quandrants. The output of each array is the image formed by summing together the four quandrant images.

Our goal is to design an optical front end for an imaging system that allows for the simultaneous imaging of both a wide angle field of view and a zoomed telephoto field of view. In addition we require image acquisition to be optimized for target tracking.

This described optical design utilizes a multi-lens front end. Each lens images one quadrant of the total field of view onto an image plane. All four images are formed at the same position on the image plane, overlapping one another in space; see Fig. 1. The first set of four lenses are at position L_1. These lenses image each of the object plane quadrants onto plane P_1. All four images also overlap on P_1. Behind P_1 is a second set of lenses denoted as L_2. The four lenses at L_2 image each quadrant in plane P_1 onto plane P_2. These lens stages can be pipelined in sequence as shown in Fig. 1.

The imaging properties of this pipelined system may be understood by relating the shaded areas denoted in each image plane. The shaded area in the object plane represents some small region of the original field of view. The image of the shaded area is shown in each image as it is projected at each stage of the optical system plane P_1, P_2, and P_3. A quadrant detector may be used in plane P_3 to detect the image. Alternately another optical stage may be used to image onto a single detector element located in plane P_4. Note that the operation of a single lens imaging systems flips the image left to right and top to bottom.

If the shaded area represents one pixel in the original object plane image, then an image detector array located in plane P_3, needs only four detector elements (a quadrant detector) to image the entire object plane. However, each image plane contains the overlap of the four quadrants from the previous image plane. This means that the shaded area of plane P_3 actually contains the overlapping images of 64 pixels from the original object plane. It may be undesirable to have overlapping images in this fashion. By placing shutters across each lens aperture it is possible to sequence through each of the sixty four images, one at a time. In this way, by sequencing the shutters, each of sixty four pixels, uniformly dispersed across the object plane, may be viewed by a detector element in plane P_3.

Suppose one additional optical stage followed plane P_3, called P_4, as shown in Fig. 1.

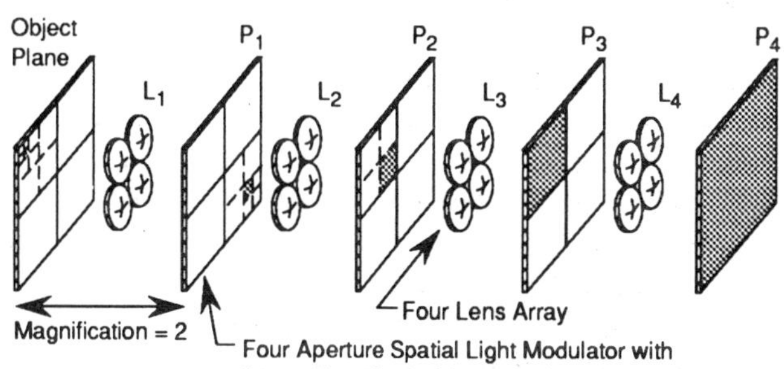

Figure 1. Pipelined parallel optical system for target tracking.

Here an entire 256 (16×16) element image is detected by use of a single time multiplexed point detector. The shutters must be sequenced so that at each optical stage only light from a single lens passes through the system. Each set of open shutter positions represents a single pixel location that is exposed to the detector.

Suppose a (256×256) detector array were used in plane P_4 instead of a single element. Then each "pixel" location in the original image would correspond to a

(256 × 256) array. This means that the original scene would be divided into (4096 × 4096) pixels by the optical system.

The optical system is designed so that each optical stage represents one stage in a two dimensional binary decimation of the image. The most significant digit of the position is controlled by the four lenses L_1 closest to the object plane. Each optical stage represents a progressively less significant digit as we move away from the object plane. For example, suppose we set the shutters so that only one lens is open at each stage, then only one pixel illuminates the detector element in Fig. 1. When each lens in optical stage L_1 is opened, one at a time in turn, we sample the same pixel position in each quadrant of the original image. Then we switch to a different lens in L_2 and sequence through the L_1 lenses again, and so on. The result of this sampling scheme is the so-called bit-reversed sampling scheme [1]. However, it is also possible to utilize the shutter system to perform a random access sample scheme of a field of view to determine target position location.

Random Access Search Methods

In this section, we describe in detail the random access search algorithm. Bit-reversed addressing is a search algorithm that provides optimal mean time to detection of the target. We discuss this scanning algorithm because the random access camera architecture we present in the previous section is a hardware implementation of this algorithm.

In bit-reversed addressing, the object field is segmented into quadrants; an element or group of elements in each quadrant is examined in four successive read sequences. Each of the quadrants are then progressively split into smaller quadrants, with a read from every major quadrant occurring once every four reads. Extending this rule, a quadrant within some square area will be read from every fourth time a pixel is read from that area.

Since the bit-reversed addressing mode uses quadrants, the row and column address indices are determined by reverse mapping of pairs of bits from the binary representation of the pixel read number, ℓ. ℓ is the sequential number associated with a object field element, or pixel. Therefore, if the binary representation of ℓ is given by

$$\ell = \sum_{i=0}^{2k-1} b_i 2^i \qquad (1)$$

where k is an integer. The row index is then given by

$$\alpha_r(\ell) = \sum_{i=0}^{k-1} b_{2k-i-1}\, 2^i \qquad (2)$$

and the column index is given by

$$\alpha_c(\ell) = \sum_{i=0}^{k-1} \mathrm{XOR}(b_{2k-i-1}, b_{2k-i-2}) 2^i \,. \qquad (3)$$

$\mathrm{XOR}(b_i, b_j)$ is the binary exclusive-or function of the bits b_i and b_j. To ensure that the array can be segmented into four equal square areas, the maximum row and column index is given by 2^k. Equations (2) and (3) represent a mapping between reverse ordered pairs of bits from the binary representation of ℓ and the column and row indices. The mapping defined by equations (2) and (3) is listed in Table 1.

Other mappings are possible which result in essentially equivalent scan characteristics. Figure 2 is an example of a bit reverse addressing sequence, showing

b_i	b_{i-1}	$\alpha_r(\ell)^{ith}$	$\alpha_c(\ell)^{ith}$
0	0	0	0
1	0	1	1
0	1	0	1
1	1	1	0

Table 1. Binary mapping of bit pairs in the pixel index, ℓ, to bits in row and column indices.

addressing order over an 8 × 8 object field. Figure 3 illustrates visually how each major quadrant of the 8 × 8 field is examined in every four reads. This figure shows the capability of the bit-reversed algorithm to decrease the average distance between read pixels. Each frame in the figure represents read (black) and unread (white) after four reads from the previous frame.

The random access camera is a direct hardware approach to bit-reversed mapping. The system successively divides the object field into quadrants. Even though the system was designed to perform a bit-reversed mapping, other random access algorithms are possible. Therefore, the camera can theoretically solve many of the problems associated with raster scanned cameras. Conceptually, it is well suited for both searching and tracking of targets. The hardware implementations are broken down into two architectures. We call one architecture, an optical train, and the other, the folded-back camera. The advantages and limitations of the various hardware implementations are also discussed.

0	32	8	40	2	35	11	43
48	16	56	24	51	19	59	27
12	44	4	36	14	47	7	39
60	28	52	23	62	55	31	23
3	34	10	42	1	33	9	41
50	18	58	26	49	17	57	25
14	46	6	38	13	45	5	37
62	30	54	22	61	29	53	21

Figure 2. Actual reversed bit addressing element read sequence on an 8×8 field.

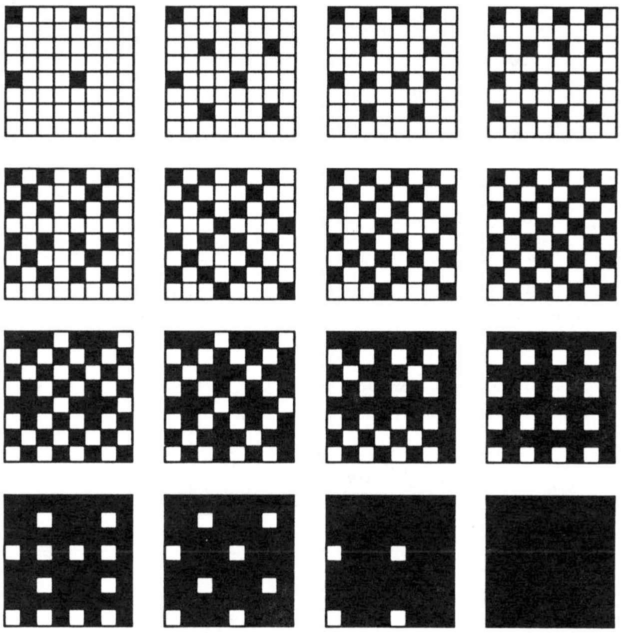

Figure 3. Sequence of elements of an 8×8 object field covered by reverse bit addressing. Shows every 4 reads.

Search Algorithm

A search algorithm attempts to find previously unknown locations for targets with sufficient optical energy above background radiation. Initially, the detector is in a passive search model with all apertures opened. Detection occurs when a signal on the quadrant detector reaches a level indicating a target is within the object field and enables the active search algorithm. The active search mode begins with a scan sequence, which selectively opens individual apertures in the first stage until the threshold radiation energy is again detected. This quadrant is identified with the 2-bit code determining most significant bits of the target address. The appropriate quadrant aperture stays open for the remainder of the search algorithm. At this point, the target quadrant of the first object field is imaged and magnified onto the object field of the second lens stage.

The second stage performs the identical active search sequence described above using the magnified image of the first stage as its object. The quadrants are sequentially opened until the target and the second most significant address bits are determined. The aperture associated with this quadrant will stay open and the sequence is iterated for each successive stage. The final stage provides the second least significant bits of the coordinate address at P_3. Additional resolution generating least significant bit pair is given by the position of the target image of the last lens stage on quadrant detector.

This search mode has the potential to be robust and fast. Initial target detection occurs rapidly since the object field is integrated onto a quadrant detector. After target detection, target location is quickly resolved with only four detector reads required per level.

Random Access Algorithm

Certain cases, such as target tracking or when the signal-to-noise ratio is low, necessitate the use of the camera in the random access mode. In this mode, an individual pixel or pixel groups are tracked in a preselected addressing mode.

If the location of interest is of single pixel size, single quadrants in each stage are selectively gated as determined by the binary coded x-y address pair. With the quadrants selectively gated, the object light propagates through the states, magnified by a factor 2^n, where there are n-1 lens stages. The number stages selectively gated determines how much resolution is used in the addressing. Therefore, groups of $2^p \times 2^p$ pixels can be examined selectively gating the first n-p-1 stages, and gating all quadrants in the last p-1 stages. The quadrant detector provides the final 2×2 area of detection electronically. As an example, if all quadrants in the final stage are gated, with all other stages selectively gated, the resolution is 2^{n-2} and the pixel group is 4×4. Of course, pixel groups can be addressed only on 2^{p-n} boundaries in both the x and y directions.

The most efficient target search sequence examines pixels with a sampling distance equal to the target span. It has been shown that the mean time to detection of a stationary target is

$$\tau = \frac{L}{2}\left[\frac{N}{T} - 1\right] + \frac{N-T+2}{2}$$

where N is the lateral pixel resolution of the camera, and T is the linear target size in pixels.

For targets of unknown size, the bit reverse scanning sequence, as discussed in section 7.1.2, is the most efficient scanning algorithm for a random access camera. A sequential gating of each quadrant of the stage L1 provides an examination of a pixel group in each quadrant. A different gate is activated in stage 2 for the next sweep of the four quadrants. This process is executed for each permutation of gating until all pixels have been examined.

The architecture of this camera lends itself to superpixel scanning. Non-contiguous pixel groups can be scanned by opening more than one aperture in a lens stage. For example, leaving all apertures in stage L1 open, with all other stages selectively gated, a pixel in each quadrant of the object field will be scanned for each combination. By selecting appropriate aperture combinations at each level, any pixel or group of pixels can be addressed making this a true random access architecture.

Limitations and Solutions

The completely free space conception of the random access camera suffers from severe vignetting. Lens array L1 forms an image at P1, but only a small fraction of the light is captured by L2. An actual implementation would require redirection, or amplification of the image to pass it down the train.

A system of relay lenses might be designed to provide the redirection of the image. Whether the relay system can be designed with the proper magnification and still be compatible with the rest of the system is undetermined. The relay lens system still suffers from decreasing signal at each stage and also requires precise alignment for mapping.

Diffusing screens at each plane could be employed for image regeneration. A diffusing screen redirects optical energy towards apertures in the succeeding stage. A diffusing screen approach has the advantage of simple design and construction. However, this implementation is also inherently inefficient, requiring fast lenses.

Addressable image magnifiers at each image field quadrant, such as microchannel spatial light modulators, hold the greatest promise of practical implementation. An image magnifier overcomes decaying signal and is easy to align. Furthermore, image magnifiers can have non-linear transfer functions, thus increasing the signal-to-noise ratio. Image magnifiers are also addressable which eliminates the need for external shutter systems and are generally much faster.

Assuming the optical train is implemented using image magnifiers with linear transfer functions, we can deduce some operating limitations due to background noise. In the passive search mode, a target is captured and amplified by just one quadrant. The amplified background noise in P1 is spread out among the next four quadrants of P2, but so is the noise from the other three quadrants in P1. The result is a wash in the noise after P1. Then the target signal must be greater than the total background energy in the object field quadrant. If the target signal does not satisfy this relation, then stages can be selectively gated; each stage provides a factor of four reduction in signal energy requirements. That is, if the first three stages are selectively gated, the signal must only be $1/4^3$ or 1/64 of the total object field background energy for a S/N=1. The drawback is, of course, that the selectively gated stages must be scanned, increasing the expected time to detection for an existing target. These arguments also ignore image amplifier and electrical noise.

The implementation shown in Figure 1 requires successive repetition of similar arrays of lenses. By feeding back the magnified image through the same lens array, only a single array of lens is required. This implementation of the optical train has the potential of being much smaller and is illustrated in Figure 4. This approach requires an image memory/amplifier to store the

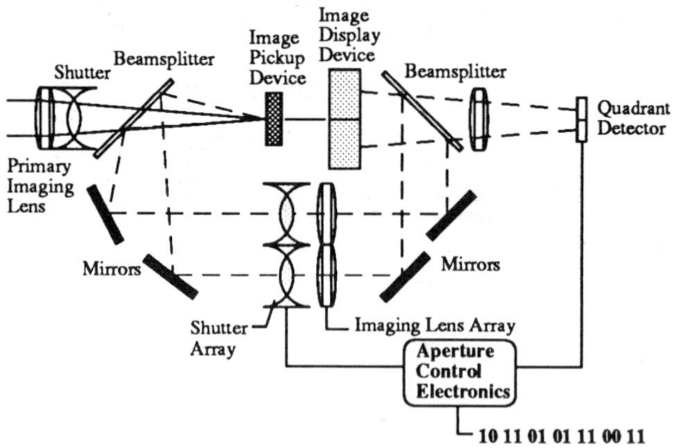

Figure 4. Folded-back implementation of Random Access Camera.

image between each level. Mirrors, beam-splitters and shutters are required to feedback and select the appropriate quadrant for redisplay. The original image is captured in memory and amplified for display. The amplified image is monitored with a quadrant detector through the beamsplitter. Control determines which quadrant the target is in, and selects which shutter to open. This quadrant is magnified by a factor of two and imaged onto the pick-up device. The process is iterated until the target is found to a suitable resolution. The advantage of the feedback implementation is potential compact size. The disadvantage is that this system has image resolution that is limited by the original image acquisition stage.

References

[1] J. P. Allebach, "Effects of Sample Displacement in Time-Sequential Sampling of Spatiotemporal Signals," J. Opt. Soc. Am. Vol. 73, pp. 466-472, April 1983.

Fast Tomographic Reconstruction Using Chirp-z Interpolation*

Kevin J. Moraski and David C. Munson, Jr.
Coordinated Science Laboratory and Department of Electrical and Computer Engineering
University of Illinois at Urbana-Champaign
Urbana, IL, 61801

Abstract

In computerized tomography, samples of the two-dimensional Fourier transform of the object under observation lie on a polar grid. In order to realize the computational advantage of the fast Fourier transform, the data must first be interpolated to a Cartesian grid. In this paper, we present a computationally-efficient direct Fourier tomographic reconstruction algorithm based on chirp-z interpolation. Through computer simulation, we compare the proposed algorithm to the convolution-backprojection algorithm for the parallel-beam geometry case.

1 Introduction

Computerized tomography is an x-ray imaging system that reconstructs a two-dimensional cross-sectional image from its projections. In direct Fourier reconstruction, the transform data data is acquired on a polar grid, so that interpolation from a polar grid to a Cartesian grid must be accomplished in order to achieve the numerical advantage provided by the fast Fourier transform. Various methods of interpolation such as nearest neighbor, two-dimensional linear interpolation, hybrid spline-linear interpolation, circular interpolation, and assorted gridding algorithms have been examined and compared to the convolution-backprojection method [1]-[4]. While these algorithms are arguably more efficient than the convolution-backprojection method, the image quality is not as good as that of convolution-backprojection.

Our proposed algorithm assumes that projection data may be obtained on a set of projection angles with a spacing that is slightly nonuniform. The projection angles are chosen so that the Fourier data will be on a concentric-squares grid [1,5] after performing a chirp-z transform of each projection. Such a grid is shown in Fig. 1. Our approach differs from two related approaches that use the chirp-z transform in conjunction with a concentric-squares grid. In [6], the projection angles are equally spaced and the projections are uniformly sampled. The chirp-z transform is then used to compute samples of the Fourier transform at locations on a concentric-squares grid. Nearest neighbor interpolation follows since the Fourier samples on the perimeters of the squares are nonuniformly spaced. In [7], Lawton's algorithm begins with a concentric-squares grid obtained by choosing projection angles that are equally spaced in $\tan\theta$ and by varying the sample spacing from one projection to the next. Then the same computational trick used in the chirp-z transform is employed for reconstruction of the image. In our approach, we collect projections at the same angles as Lawton, but the sample spacing does not vary between projections.

*This research was supported by the Joint Services Electronics Program (U.S. Army, Navy, Air Force) under contract N00014-90-J1270.

2 Chirp-z Interpolation Algorithm

2.1 Overview of Algorithm

Our proposed algorithm may be divided into the following four steps:

1. Collect uniformly sampled projections at angles that are uniformly spaced in $\tan\theta$.
2. Apply chirp-z transform to each projection to give Fourier data on a concentric squares grid.
3. Perform chirp-z interpolation [8] around perimeters of concentric squares to obtain Fourier data on a Cartesian grid.
4. Apply inverse Fourier transform to obtain spatial domain image.

These steps are described in the following sections.

2.2 Geometry of the Spatial and Fourier Domain Grids

For this paper, we will assume that the original image $f(x,y)$ is $w \times w$ in extent and is nearly bandlimited to a circle of diameter $1/a$. Thus, the sampling period in the spatial domain will be a, and assuming an $N \times N$ reconstructed image with $N \approx w/a$, the spatial domain coordinates will be (ma, na) with

$$a = \frac{w}{N} \quad \text{and} \quad -\frac{N}{2} \leq m,n \leq \frac{N}{2} - 1. \tag{1}$$

The Fourier data grid will be divided into Regions I-IV as shown in Fig. 2. The only requirement on the number of projections and the projection angles θ_i is that the angles θ_i be equally spaced in $\tan\theta$ so that the radial traces in the Fourier domain will intersect the perimeters of the concentric-squares raster at uniform spacings. For this paper, we have imposed two additional constraints. First, there must be projection angles corresponding to $\theta = 0$, $\theta = \pi/4$, $\theta = \pi/2$, and $\theta = 3\pi/4$. Second, the number of projections must be a multiple of four in order to have the same number of radial traces in each of the four regions. If there are L total projections, each region will contain $L/2+1$ radial traces. This is due to counting the projections at $\theta = \pi/4$ and $\theta = 3\pi/4$ in both adjacent regions and the fact that the traces in Regions I and III come from the same projections, and likewise for the traces in Regions II and IV. Since chirp-z interpolation (to be described later) will be performed on the perimeters of the squares in each region separately, it is helpful to incorporate a few additional projections "outside" each region to improve interpolation "inside" each region. This is illustrated in Fig. 3 for Region I. The additional projection angles are chosen so that the spacing in each region continues to be uniform in $\tan\theta$. Essentially data is taken from a slightly larger region in the Fourier domain and interpolated into a smaller region in the Fourier domain. In Region I, the Fourier grid coordinates $(u(i,k), v(i,k))$ are given by

$$u(i,k) = A(k) + iB(k), \qquad v(i,k) = (k+1)C \tag{2}$$

for $0 \leq i \leq \frac{L}{2} + E$, $0 \leq k \leq R-1$, where

$$A(k) = -(k+1+\frac{E}{2})C \tag{3}$$

$$B(k) = \frac{2 \mid A(k) \mid}{\frac{L}{2}+E} \tag{4}$$

$$C = \frac{1}{w\sqrt{2}} \tag{5}$$

$$R = \text{int}(\frac{N}{\sqrt{2}}) \tag{6}$$

and where $\frac{L}{2}+1$ is the number of radial traces within the region and E is the number of radial traces outside the region.

The grids for Regions II-IV are developed analogously.

2.3 Chirp-z Transform, Chirp-z Interpolation, and Fourier Inversion

We will assume that for an $N \times N$ image each projection will have N samples with the samples spaced uniformly apart by a distance a. In addition, we will assume that there are $L + 2E$ projections. The first step of the algorithm uses the chirp-z transform to compute samples of the Fourier transform of each projection at locations on the concentric-squares raster. This is done by

$$P_\theta(m) = \sum_{n=-\frac{N}{2}}^{\frac{N}{2}-1} p_\theta(n)\exp(-j2\pi a\Delta\Omega_\theta mn)$$
$$L_\theta^- \leq m \leq L_\theta^+ \tag{7}$$

where

$$\Delta\Omega_\theta = \begin{cases} \frac{C}{|\sin\theta|} & \text{Region I or III} \\ \frac{C}{|\cos\theta|} & \text{Region II or IV} \end{cases} \tag{8}$$

$$L_\theta^- = -\frac{\frac{1}{2a}}{\Delta\Omega_\theta} \tag{9}$$

$$L_\theta^+ = \frac{\frac{1}{2a}}{\Delta\Omega_\theta}. \tag{10}$$

From $P_\theta(m)$, we obtain the samples of the Fourier transform $F(i,k)$ at the locations on the concentric-squares grid in each region as defined in the previous section.

The second step of the algorithm uses chirp-z interpolation to interpolate along the perimeters of the squares to a Cartesian grid. Since the spacing between samples is different from square to square, the parameters of the chirp-z interpolator vary from square to square. Chirp-z interpolation is an efficient method of interpolating from one uniform grid to another, where the ratio of the grid spacings can be an arbitrary number [8]. In this paper, we will show how chirp-z interpolation can be applied to the perimeter in Fig. 4. The derivation will be done for a perimeter in Region I; the other regions are derived analogously. The continuous-time Fourier transform pairs are

$$f(x,y) = \int_{-\infty}^{\infty}\int_{-\infty}^{\infty} F(u,v)\exp[j2\pi(ux+vy)] \quad (11)$$

$$F(u,v) = \int_{-\infty}^{\infty}\int_{-\infty}^{\infty} f(x,y)\exp[-j2\pi(ux+vy)] \quad (12)$$

An approximate expression for (11) is

$$\begin{aligned} f(x,y) &\approx \Delta u\Delta v \sum_i \sum_k F(i\Delta u, k\Delta v) \\ &\cdot\exp[2\pi(\Delta uxi + \Delta vyk)]. \quad (13)\end{aligned}$$

An approximate expression for (12) is

$$\begin{aligned} F(u,v) &\approx \Delta x\Delta y \sum_m \sum_n f(m\Delta x, n\Delta y) \\ &\cdot\exp[-2\pi(\Delta xum + \Delta yvn)]. \quad (14)\end{aligned}$$

If we combine (13) and (14), we have the a two-dimensional interpolation formula for Cartesian grids in the Fourier domain given by

$$\begin{aligned} F(u,v) &\approx \Delta x\Delta y \sum_m \sum_n \Big[\Delta u\Delta v \sum_i \sum_k \\ &F(i\Delta u, k\Delta v)\cdot\exp[2\pi(\Delta u\Delta xmi + \\ &\Delta v\Delta ynk)]\Big] \cdot \exp[-2\pi(\Delta xmu + \\ &\Delta ynv)]. \quad (15)\end{aligned}$$

For Fig. 4, the v-coordinate is constant. Let $v = pC$ where p is fixed. Then

$$\begin{aligned} F(lC,pC) &\approx \Delta x\Delta y \sum_{m=S^-}^{S^+} \sum_{n=S^-}^{S^+} \Big[C \sum_{i=0}^{\frac{L}{2}+E} \sum_{k=p}^{p} F(i,k) \\ &\cdot B(k)\exp[j2\pi(A(k)+iB(k))\Delta xm] \\ &\cdot\exp[j2\pi C\Delta ynk]\Big] \cdot\exp[-j2\pi\Delta xClm] \\ &\cdot\exp[-j2\pi\Delta ypCn] \\ &= \Delta x\Delta y \sum_{m=S^-}^{S^+} \sum_{n=S^-}^{S^+} \Big[CB(p) \sum_{i=0}^{\frac{L}{2}+E} F(i,p) \\ &\cdot\exp[j2\pi(A(p)+iB(p))\Delta xm] \\ &\cdot\exp[j2\pi C\Delta ypn]\Big] \cdot \exp[-j2\pi\Delta xClm] \\ &\cdot\exp[-j2\pi\Delta ypCn] \\ &= \Delta x\Delta yCB(p) \sum_{m=S^-}^{S^+} \sum_{n=S^-}^{S^+} \sum_{i=0}^{\frac{L}{2}+E} F(i,p) \\ &\cdot\exp[j2\pi(A(p)+iB(p))\Delta xm] \\ &\cdot\exp[-j2\pi\Delta xClm] \\ &= (S^+ - S^- + 1)\Delta x\Delta yCB(p) \sum_{m=S^-}^{S^+} \Big[\\ &\sum_{i=0}^{\frac{L}{2}+E} F(i,p)\cdot\exp[j2\pi(A(p)+ \\ &iB(p))\Delta xm]\Big] \cdot\exp[-j2\pi\Delta xClm]. \quad (16)\end{aligned}$$

To compute (16), we separate the summations and use the same computational trick as used in the chirp-z transform, namely $im = \frac{1}{2}\cdot[i^2 + m^2 - (m-i)^2]$ and $lm = \frac{1}{2}\cdot[l^2 + m^2 - (l-m)^2$. The resulting, efficient algorithm is referred to as *chirp-z interpolation* [8]. Let $S = (S^+ - S^- + 1)$.

$$\begin{aligned} F(lC,pC) &\approx S\Delta x\Delta yCB(p) \sum_{m=S^-}^{S^+} \Big[\sum_{i=0}^{\frac{L}{2}+E} F(i,p) \\ &\cdot\exp[j2\pi B(p)\Delta xim]\Big] \exp[j2\pi A(p)\Delta xm] \\ &\cdot\exp[-j2\pi\Delta xClm] \\ &= S\Delta x\Delta yCB(p)\exp[-j\pi\Delta xCl^2] \\ &\sum_{m=S^-}^{S^+} \Big[\sum_{i=0}^{\frac{L}{2}+E} F(i,p)\cdot\exp[j\pi B(p)\Delta xi^2] \\ &\cdot\exp[-j\pi B(p)\Delta x(i-m)^2]\Big] \\ &\cdot\exp[j2\pi A(p)\Delta xm] \\ &\cdot\exp[j\pi B(p)\Delta xm^2]\cdot\exp[-j\pi\Delta xCm^2] \\ &\cdot\exp[j\pi\Delta xC(l-m)^2]. \quad (17)\end{aligned}$$

Since (17) is a convolution, it may be computed using FFTs. The parameters S^-, S^+, Δx, and Δy were determined by simulation for our test cases.

The final step of the tomographic reconstruction algorithm is to invert the interpolated Fourier data on the Cartesian grid. This is done by

$$\begin{aligned} f(ma,nb) &\approx C^2 \sum_{i=-R}^{R} \sum_{k=-R}^{R} F_{interp}(iC,kC) \\ &\cdot\exp[j2\pi(aCim + aCkn)]. \quad (18)\end{aligned}$$

(18) may be computed using FFTs after using the same trick as used in the chirp-z transform.

3 Simulation Results

The proposed algorithm was used to reconstruct the Shepp and Logan head phantom. The parameters of the image and the algorithm were $w = 2$, $N = 512$, $L = 512$, $E = 10$, $S^{+} = 1536$, $S^{-} = -1536$, $\Delta x = 1/1024$, and $\Delta y = 1/1024$. Fig. 5 shows a comparison between covolution-backprojection and chirp-z interpolation. The convolution-backprojection image was generated using 512 projections with 512 samples per projection. Our simulations of this algorithm demonstrate that high-quality tomographic reconstructions of an $N \times N$ image can be quickly generated using the chirp-z interpolation algorithm with computational complexity $O(N^2 \log_2 N)$. For the above example, the chirp-z algorithm requires roughly one quarter of the computations required by convolution-backprojection. We are currently exploring the possiblity of using different windows in the chirp-z interpolation step in order to improve the quality of the reconstructed image.

References

[1] R. M. Mersereau and A. V. Oppenheim, "Digital reconstruction of mulidimensional signals from their projections," *Proc. IEEE*, Vol. 62, pp. 1319-1338, Oct. 1974.

[2] H. Stark, J. W. Woods, I. Paul, and R. Hingorani, "Direct Fourier Reconstruction in Computer Tomography," *IEEE Trans. Acoust., Speech, Signal Processing*, Vol. ASSP-29, pp. 237-245, Apr. 1981.

[3] J. D. O'Sullivan, "A fast sinc function gridding algorithm for Fourier Inversion in computer tomography," *IEEE Trans. on Medical Imaging*, Vol. MI-4, pp. 200-207, Dec. 1985.

[4] S. Matej and I. Bajla, "A high-speed reconstruction from projections using direct Fourier method with optimized parameters-an experimental analysis," *IEEE Trans. on Medical Imaging*, Vol. 9, pp. 421-429, Dec. 1990.

[5] R. M. Mersereau, "Recovering multidimensional signals from their projections," *Computer Graphics and Image Processing*, Vol. 1, pp. 179-195, Oct. 1974.

[6] F. Natterer, *The Mathematics of Computerized Tomography,* John Wiley & Sons, 1986, pp. 126-127.

[7] W. Lawton, "A new polar Fourier transform for computer-aided tomography and spotlight synthetic aperture radar," *IEEE Trans. Acoust., Speech, Signal Processing,* Vol. 36, pp. 931-933, June 1988.

[8] O. Arikan, *Investigation of Topics in Radar Signal Processing,* Ph.D. Thesis, University of Illinois at Urbana-Champaign, 1990.

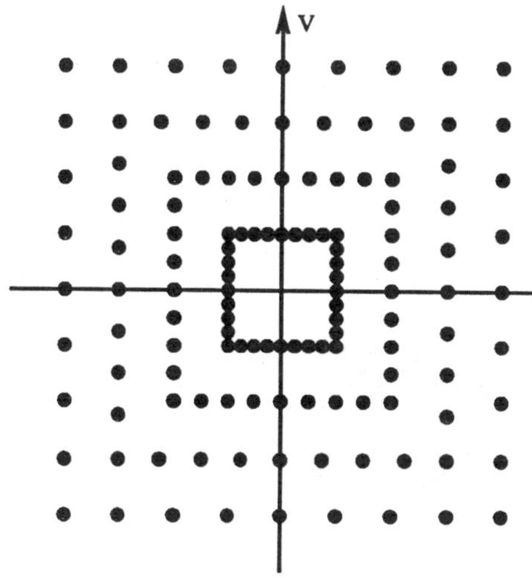

Fig. 1

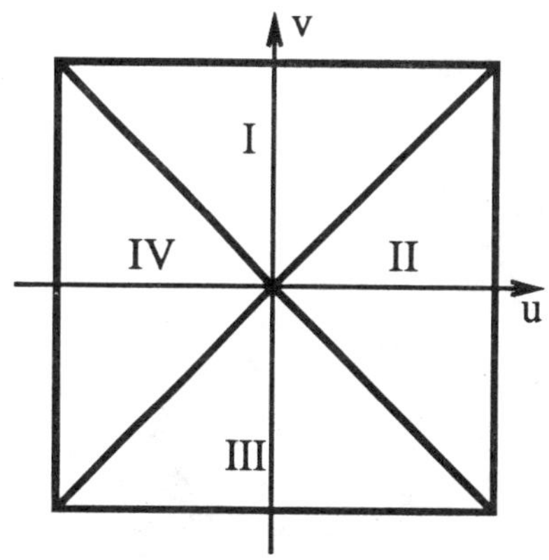

Fig. 2

|extra| region of interpolation |extra|

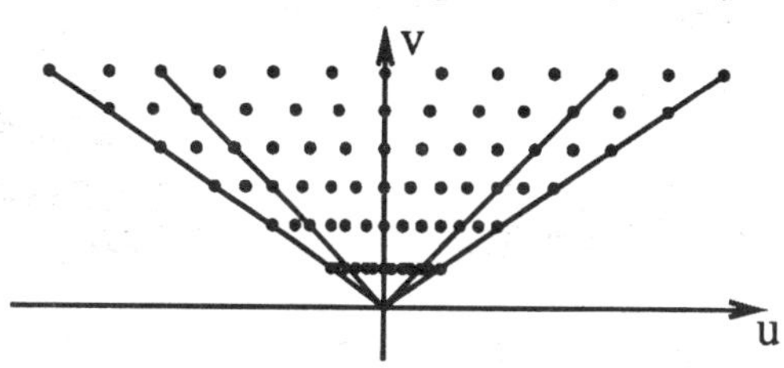

Fig. 3

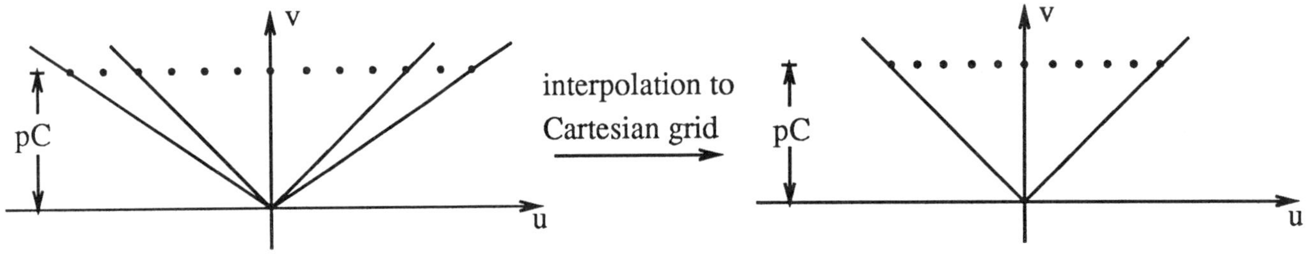

Fig. 4

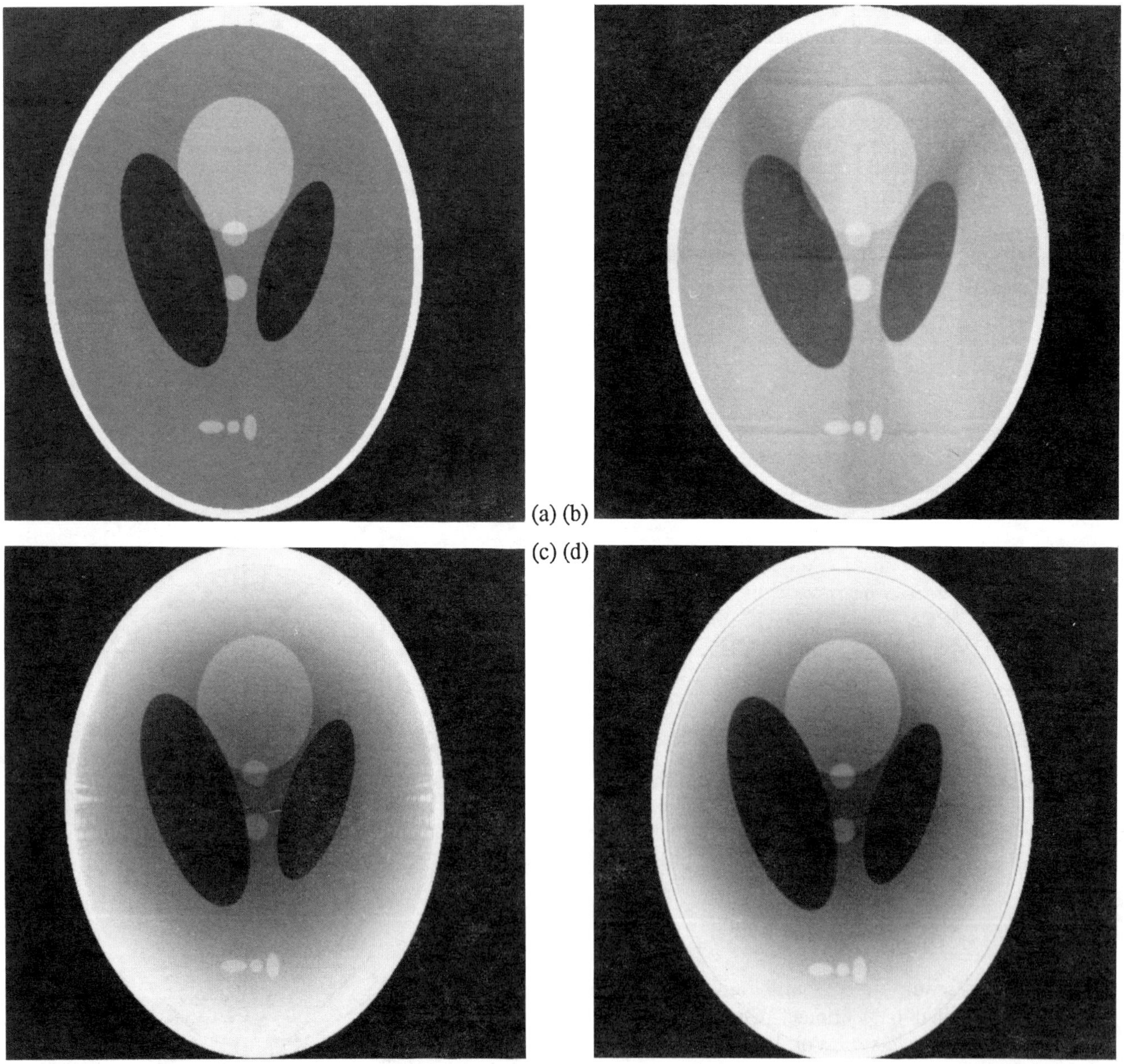

Fig.5 (a) Original image (b) Convolution-backprojection (c) Chirp-z interpolation (d) Chirp-z interpolation with windowing prior to Fourier inversion.

Balloon Motion Estimation Using Two Frames*

Qinfen Zheng
SIPI, EEB-400
Univ. of Southern California
Los Angeles, CA 90089

Rama Chellappa
EE Department
Univ. of Maryland
College Park, MD 20742

B.S. Manjunath
ECE Department
Univ. of California
Santa Barbara, CA93106

Abstract

A computational vision approach is presented for the estimation of 2-D translation, rotation, and scale from two partially overlapping images. An illuminant direction estimation method is first used to obtain an initial estimate of camera rotation. A small number of feature points are then located based on a Gabor wavelet model for detecting local curvature discontinuities. An initial estimate of scale and translation is obtained by pairwise matching of the feature points detected from both frames. Finally, hierarchical feature matching is performed to obtain an accurate estimate of translation, rotation and scale. The approach results in a fast method that produces excellent results even when large rotation has occurred between the two frames and the images are devoid of significant features. Experiments with synthetic and real images show that this algorithm yields accurate results when the scale between the image pair differ by up to 10%, the overlap between the two frames is as small as 35%, and the camera rotation between the two frames is significant.

1 Introduction

Automatic image registration is an important problem in many multi frame based image analysis and applications. Traditional solutions [1, 3, 4, 6, 7, 9] to this problem are unreliable when the rotation of the camera and scale change between the two frames are significant. Registration becomes even more difficult if the images are devoid of significant features and/or the overlap between the two frames is small. In the Mars '94 project we have such a challenging problem. One of the goals of the Mars '94 project is to measure the 3-D wind velocity in Mars. It is proposed to use a downlooking camera attached to a balloon to measure the motion of the balloon and hence determine the wind speed. Figure 1 illustrates the schematic of this project, in which successive image frames are shuttered at times t_1 and t_2. Balloon motion can be determined by measuring the translation, rotation and scale between the image pair. Due to other technical constraints, only two frames will be available for each location, and the rotation and translation between the images could be significant. Also there is a scale change due to the vertical motion of the balloon. A simple and robust registration algorithm is required for on-board motion estimation systems.

In this paper we present a computational vision approach for the estimation of 2-D translation, rotation, and scale from two partially overlapped images. Figure 2 shows the block diagram of our camera motion estimation algorithm. We notice that the illumination on the Mars surface is from the sun and is constant within the time the image pair is taken. By estimating the illuminant direction in each frame, we can estimate the rotation between the two frames and simplify the matching process. Since the common area between the two frames can be much smaller than the image field and in additionally there is scaling between the two frames, methods based on correlation matching become unreliable. In this work we use a feature based matching technique. First we extract a small number of feature points based on a Gabor wavelet model for local curvature analysis. Since no prior knowledge about the translation is available, an initial estimate of scale and translation is obtained by pairwise matching between the neighbors of feature points detected from both the frames. Subsequently, hierarchical correlation matching is performed to obtain an accurate camera motion estimate. Experiments with several real desert balloon images acquired by JPL show that our algorithm works for all the cases tested. The consistency test based on forward and backward motion estimations shows that estimation is quite accurate with the discrepancy in forward and backward estimation of rotation, translation and scale being less than $0.05°$, 0.5 pixel, and 3×10^{-3} respectively [10].

The organization of the paper is as follows: Section 2 formulates the balloon motion estimation as an image registration problem; Section 3 discusses the algorithm. Section 4 presents experimental results on Mojave desert images taken from a flying balloon.

2 Balloon Motion Estimation

Let (x_i, y_i, z_i) be the 3-D coordinates and (X_i, Y_i) be the image frame coordinates, both measured with respect to the position of balloon at time t_i, for $i = 1, 2$. Then the central projection equations are

$$\begin{pmatrix} X_i \\ Y_i \end{pmatrix} = \frac{f}{z_i} \begin{pmatrix} x_i \\ y_i \end{pmatrix}, \qquad \text{for } i = 1, 2 \qquad (1)$$

*The support of the Defense Advanced Research Projects Agency (ARPA Order No. 6989) and the U.S. Army Engineer Topographic Laboratories under Contract DACA 76-89-C-0019 is gratefully acknowledged.

1058-6393/91 $1.00 © 1991 IEEE

where f is the focal length of the camera. The relation between the two image frames can be approximated by

$$\begin{pmatrix} X_2 \\ Y_2 \end{pmatrix} = s \begin{pmatrix} \cos\theta & \sin\theta \\ -\sin\theta & \cos\theta \end{pmatrix} \begin{pmatrix} X_1 \\ Y_1 \end{pmatrix} + \begin{pmatrix} \Delta X_2 \\ \Delta Y_2 \end{pmatrix} \quad (2)$$

where $s = \frac{\sqrt{\delta X_2{}^2 + \delta Y_2{}^2}}{\sqrt{\delta X_1{}^2 + \delta Y_1{}^2}} = \frac{\frac{f}{z_2}\sqrt{\delta x_2{}^2 + \delta y_2{}^2}}{\frac{f}{z_1}\sqrt{\delta x_1{}^2 + \delta y_1{}^2}} = \frac{z_1}{z_2}$ is the scaling factor, θ is the rotation angle between the two image frames, and $(\Delta X_2,\ \Delta Y_2)$ is the image translation measured in the image coordinate of frame t_2. With ΔX_2, ΔY_2, θ, and s determined, the balloon motion $\vec{s} = (\Delta x, \Delta y, \Delta z)$ can be easily determined if the height of the balloon is available using[1]

$$\Delta x = \frac{z_2}{f} \cdot \Delta X_2 \quad (3)$$

$$\Delta y = \frac{z_2}{f} \cdot \Delta Y_2 \quad (4)$$

$$\Delta z = z_2 - z_1 = (1 - s) z_2 \quad (5)$$

3 Algorithm

Step 1: • Estimation of the illuminant azimuth τ_i from frame t_i, $i = 1, 2$;

• Set
$$\begin{cases} s & = 1 \\ \theta & = \tau_1 - \tau_2 \\ \Delta X & = 0 \\ \Delta Y & = 0 \end{cases}$$

Step 2: • Reduce the image size to that of the lowest resolution layer;

• Estimate feature points from each frame.

Step 3: • Apply an affine transformation with parameters $(s,\ \theta,\ \Delta X,\ \Delta Y)$ on the lowest resolution version of frame t_1 and its feature points;

• Do initial matching to obtain estimates $(s'$, θ', $\Delta X'$, $\Delta Y')$;

• Update $(s,\ \theta,\ \Delta X,\ \Delta Y)$.

Step 4: • Reduce the image resolution corresponding to current layer of the matching hierarchy;

• Magnify the coordinates of the feature points corresponding to the resolution of the current layer;

• Apply an affine transform with parameter $(s,\ \theta,\ \Delta X,\ \Delta Y)$ on frame t_1;

• Do matching refinement to obtain $(s'$, θ', $\Delta X'$, $\Delta Y')$;

• Update the estimate of $(s,\ \theta,\ \Delta X,\ \Delta Y)$.

[1] The height of the balloon can be determined by measuring the time lapse between transmitting and receiving of a radio signal from the balloon.

Step 5: • If current level is at the highest resolution then stop;

• Increase image resolution and adjust the translation estimation by

$$\begin{pmatrix} 2\,\Delta X \\ 2\,\Delta Y \end{pmatrix} \Longrightarrow \begin{pmatrix} \Delta X \\ \Delta Y \end{pmatrix};$$

• Go to Step 4.

Some of the key operations in our algorithm are discussed below.

Initial Estimation of Camera Rotation The initial estimation of camera rotation is computed as the difference of illuminant azimuth angles estimated from both of the frames. To estimate the illuminant direction from image we use the local voting illuminant azimuth estimator reported in [11]. To be more specific, for each pixel we first compute

$$\vec{X} = \begin{pmatrix} \tilde{x}_L \\ \tilde{y}_L \end{pmatrix}$$

by solving

$$\vec{X} = (B^t B)^{-1} B^t d\vec{I}$$

where

$$d\vec{I} = \begin{pmatrix} \delta I_1 \\ \delta I_2 \\ \vdots \\ \delta I_N \end{pmatrix}, \quad B = \begin{pmatrix} \delta x_1 & \delta y_1 \\ \delta x_2 & \delta y_2 \\ \vdots & \vdots \\ \delta x_N & \delta y_N \end{pmatrix},$$

δI is the increment in intensity along the direction $\vec{s} = (\delta x, \delta y)$, and N is the number of measured directions for $\vec{s}$.

The estimate of illuminant azimuth is then computed as

$$\tau = \arctan\left(\frac{\mathbf{E}_{x,y}\left\{ \frac{\tilde{y}_L}{\sqrt{\tilde{x}_L^2 + \tilde{y}_L^2}} \right\}}{\mathbf{E}_{x,y}\left\{ \frac{\tilde{x}_L}{\sqrt{\tilde{x}_L^2 + \tilde{y}_L^2}} \right\}} \right)$$

After the illuminant azimuth τ_i, $i = 1, 2$ are computed, the initial estimation of camera rotation is computed as

$$\theta_0 = \tau_1 - \tau_2 \quad (6)$$

Feature Detection We use a method which is based on a biologically motivated model for identifying local curvature discontinuities, and makes use of Gabor wavelet decomposition of the image and local scale interactions between features [8]. The basic wavelet function used in our decomposition is of the form

$$\Phi_\lambda(x, y, \theta) = e^{-(\lambda^2 x'^2 + y'^2) + i\pi x'} \quad (7)$$

$$x' = x\cos\theta + y\sin\theta \qquad y' = -x\sin\theta + y\cos\theta$$

where θ is the preferred spatial orientation and λ is the aspect ratio of the Gaussian. For convenience, we will drop the subscripts in further discussion. In all our experiments, λ is set to 1, and θ is discretized into four orientations.

The corresponding wavelet transformation is obtained by convolving the image data $\mathbf{f}$ with a bank of filters whose responses are simple dilations and translations of the basic wavelet in (7), and are denoted by

$$W_j(x,y,\theta)=\mathbf{f}\bigotimes\Phi(\alpha^j x,\alpha^j y,\theta),\ j=\{0,1,2,..\} \quad (8)$$

Here α denotes the scale parameter. Usually the parameter values used are those corresponding to half-octave ($\alpha=\sqrt{2}$) or octave ($\alpha=2$). Physically, this transformation detects features in the image such as line and step edges. Biologically, this models the processing by simple cells in the visual cortex of mammals. These features by themselves are not good for applications such as obtaining correspondence. The next stage in our feature detection module involves interactions between these simple features (at different spatial frequencies, within each orientation). This step can be identified with the responses of hypercomplex cells in the visual cortex. Hypercomplex cells exhibit end-inhibition. They are sensitive to oriented lines and step edges of short lengths, and their response decreases if the lengths are increased. Using scale interactions to model these cells was first suggested by Hubel and Wiesel [5], who were the first to discover these cells in the visual cortex. Subsequent anatomical studies have also supported this model [2]. We model these interactions as follows:

$$I_{m,n}(x,y)=\max_{\theta} g(||W_m(x,y,\theta)-\gamma W_n(x,y,\theta)||) \quad (9)$$

where g is a non-linear transformation, such as thresholding or a sigmoid non-linearity, γ is a normalizing factor, and $n>m$. In order to identify features at different scales, one has to consider different scale interactions. The final step is to actually localize these features, and this is done by looking at the local maxima of these feature responses.

Scale Estimation Since the Euclidean distance between the feature points only depends on the scale between the two frames, and is invariant to rotation and translation, the scale factor can be estimated prior to the estimation of other parameters. Assume that the matched feature point pairs are $\{(X'_i,Y'_i)\Rightarrow(\hat{X}_i,\hat{Y}_i)\ i=1,\cdots,N'\}$ with N' be the number of matched feature pairs. Then

$$(\hat{d}_1\ \ \hat{d}_2\ \ \cdots\ \ \hat{d}_{N'})=s'(d_1\ \ d_2\ \ \cdots\ \ d_{N'}) \quad (10)$$

where $\hat{d}_i$ is the distance between the feature i and the center of the matched feature points in frame t_1, and d_i is the distance between the feature i and the center of the matched feature points in frame t_2. The scale factor from frame t_1 to frame t_2 is computed as

$$s'=\frac{\sum_{i=1}^{N'} d_i\cdot\hat{d}_i}{\sum_{i=1}^{N'} d_i\cdot d_i}. \quad (11)$$

Rotation and Translation Estimation Assuming that the matched feature point pairs are $\{(X'_i,Y'_i)\Rightarrow(\hat{X}_i,\hat{Y}_i)\ i=1,\cdots,N'\}$, the relation between the matched feature pairs is

$$\begin{pmatrix}\hat{X}_i\\ \hat{Y}_i\end{pmatrix}=s'\begin{pmatrix}\cos\theta' & \sin\theta'\\ -\sin\theta' & \cos\theta'\end{pmatrix}\begin{pmatrix}X'_i\\ Y'_i\end{pmatrix}+\begin{pmatrix}\Delta X'\\ \Delta Y'\end{pmatrix} \quad (12)$$

Note θ', being the residual of initial rotation estimation, is very small. By approximating $\cos\theta'$ and $\sin\theta'$ up to linear terms, (12) can be rewritten as

$$\begin{pmatrix}\hat{X}_i\\ \hat{Y}_i\end{pmatrix}=\begin{pmatrix}1 & \theta'\\ -\theta' & 1\end{pmatrix}\begin{pmatrix}s'X'_i\\ s'Y'_i\end{pmatrix}+\begin{pmatrix}\Delta X'\\ \Delta Y'\end{pmatrix}, \quad (13)$$

or

$$\vec{A}=B\vec{C} \quad (14)$$

where

$$\vec{A}=\begin{pmatrix}\hat{X}_1-s'X'_1\\ \hat{Y}_1-s'Y'_1\\ \vdots\\ \hat{X}_{N'}-s'X'_{N'}\\ \hat{Y}_{N'}-s'Y'_{N'}\end{pmatrix};\quad B=\begin{pmatrix}s'Y'_1 & 1 & 0\\ -s'X'_1 & 0 & 1\\ \vdots & \vdots & \vdots\\ s'Y'_{N'} & 1 & 0\\ -s'X'_{N'} & 0 & 1\end{pmatrix};$$

$$\vec{C}=\begin{pmatrix}\theta'\\ \Delta X'\\ \Delta Y'\end{pmatrix}.$$

The vector $\vec{C}$ can then be computed as

$$\vec{C}=(B^t\,B)^{-1}\,B^t\,\vec{A}. \quad (15)$$

Estimation Refinement In our method, at each level of the matching hierarchy, frame t_1 is first transformed using the estimated camera motion parameters s', θ', $\Delta X'$, and $\Delta Y'$ and then matched to frame t_2. The correction on initial estimation is computed using (11) and (15). After that, the total transformation is obtained by combining the two transformations together [10]:

$$\begin{pmatrix}\theta+\theta'\\ s\,s'\\ s'(\cos\theta'\,\Delta X+\sin\theta'\,\Delta Y)+\Delta X'\\ s'(-\sin\theta'\,\Delta X+\cos\theta'\,\Delta Y)+\Delta Y'\end{pmatrix}\Rightarrow\begin{pmatrix}\theta\\ s\\ \Delta X\\ \Delta Y\end{pmatrix} \quad (16)$$

4 Experiments

We have tested our algorithm on several balloon images of the Mojave desert, whose environment is similar to Mars. The input image size is 512 × 512. The image size for the lowest resolution layer in our implementation is 128 × 128. For feature detection, m and n of (9) are set to 2 and 5 respectively. In the experiments, we compute the local maximum of $I_{m,n}$ and only salient points which are maximum among the radius of $r=\alpha^n$ are selected as feature points. Due to space limitation, only two examples are presented[2]. Figure 3 shows a typical experiment on registration of desert images obtained from a flying balloon. (a)

[2]More examples and accuracy analysis are given in [10]

and (b) are the input images. (c) and (d) show the feature points detected, in which the white crosses indicate the locations of feature points. In our experiments, the number of feature points detected from each frame is between 15 to 30. This can be controlled by automatically adjustive the threshold of the local maximum. The estimated parameters of motion from Ball09 to Ball12 are $s = 0.995574$, $\theta = 1.60708°$, $\Delta X = -250.769$, and $\Delta Y = 146.602$. (e) shows the mosaicking of (a) and (b) using the estimated motion parameters. (f) shows the difference between the transformed (a) and (b) with the zero of difference shifted to 128.

Figures 4 shows another experiment on image pair with significant camera rotation. (a) and (b) are the input images. (c) and (d) show the feature points detected from each frame. The estimated camera motion parameters are $s = 0.991534$, $\theta = 57.1240°$, $\Delta X = 234.276$, and $\Delta Y = -78.1810$. (e) shows the mosaicking of (a) and (b) using the estimated motion parameters. The continuity of features across the frame borders illustrates the correctness of matching. (f) shows the difference between the transformed (a) and (b) with the zero of difference shifted to 128.

Acknowledgment

The balloon images were available through the courtesy of the Jet Propulsion Laboratory, California Institute of Technology, Pasadena, California.

References

[1] P. E. Anuta, "Spatial Registration of Multispectral and Multitemporal Digital Imagery Using FFT Techniques," *IEEE Trans. Geoscience and Electronics*, Vol. GE-8, pp. 353–368, Oct. 1970.

[2] J. Bolz and C. D. Gilbert, "Generation of End-Inhibition in the Visual Cortex via Interlaminar Connections," *Nature*, Vol. 320, pp. 362–365, Mar. 1986.

[3] R. T. Frankot, "'SAR Image Registration by Multi-Resolution Correlation," in *Proc. SPIE*, vol. 220, pp. 195–203, 1983.

[4] T. S. Huang, *Image Sequence Analysis*, Berlin/Heidelberg: Springer-Verlag, 1981.

[5] D. H. Hubel and T. N. Wiesel, "Receptive Fields and Functional Architecture in Two Nonstriate Visual Areas(18 and 19) of the Cat," *J. Neurophysiology*, Vol. 28, pp. 229–289, Mar. 1965.

[6] C. D. Kuglin *et al.*, "Map Matching Techniques for Terminal Guidance Using Fourier Phase Information," in *Proc. SPIE*, vol. 186, 1979.

[7] C. D. Kuglin and D. C. Hines, "The Phase Correlation Image Alignment Method," in *IEEE Int. Conf. on Cybernetics and Society*, pp. 163–165, 1975.

[8] B. S. Manjunath, *"Perceptual Grouping and Segmentation Using Neural Networks"*, Ph.D. dissertation, Department of Electrical Engineering, University of Southern California, Los Angeles, California, Dec. 1991.

[9] Q. Tian and M. N. Huhns, "Algorithms for Subpixel Registration," *Comput. Vision, Graphics, Image Processing*, Vol. 35, pp. 220–233, Aug. 1986.

[10] Q. Zheng and R. Chellappa, "A Computational Vision Approach to Image Registration," Tech. Rep. CAR-TR-583, Center for Automation Research, University of Maryland, College Park, Sept. 1991.

[11] Q. Zheng and R. Chellappa, "Estimation of Illuminant Direction, Albedo and Shape from Shading," *IEEE Trans. Pattern Anal. Machine Intell.*, Vol. PAMI-13, pp. 680–702, July 1991.

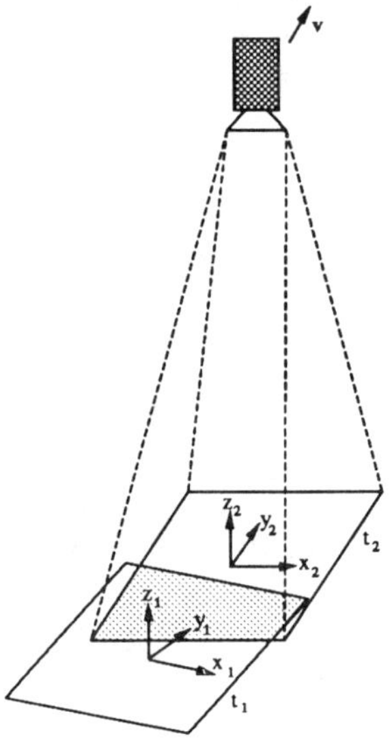

Figure 1: Geometry of Balloon Imagery.

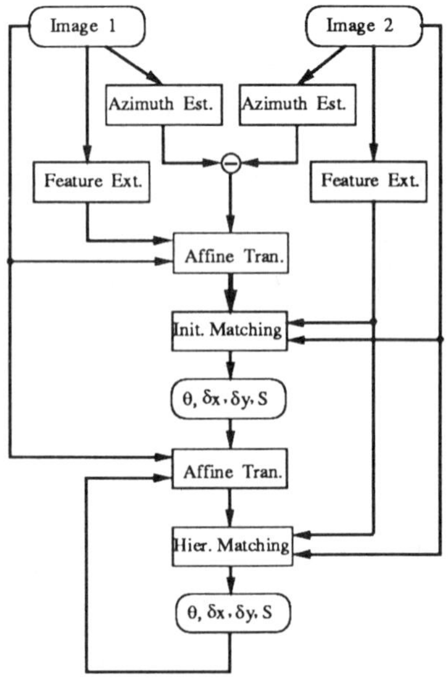

Figure 2: Block diagram of the algorithm

Figure 3: Registration of desert images Bal109 and Bal112.

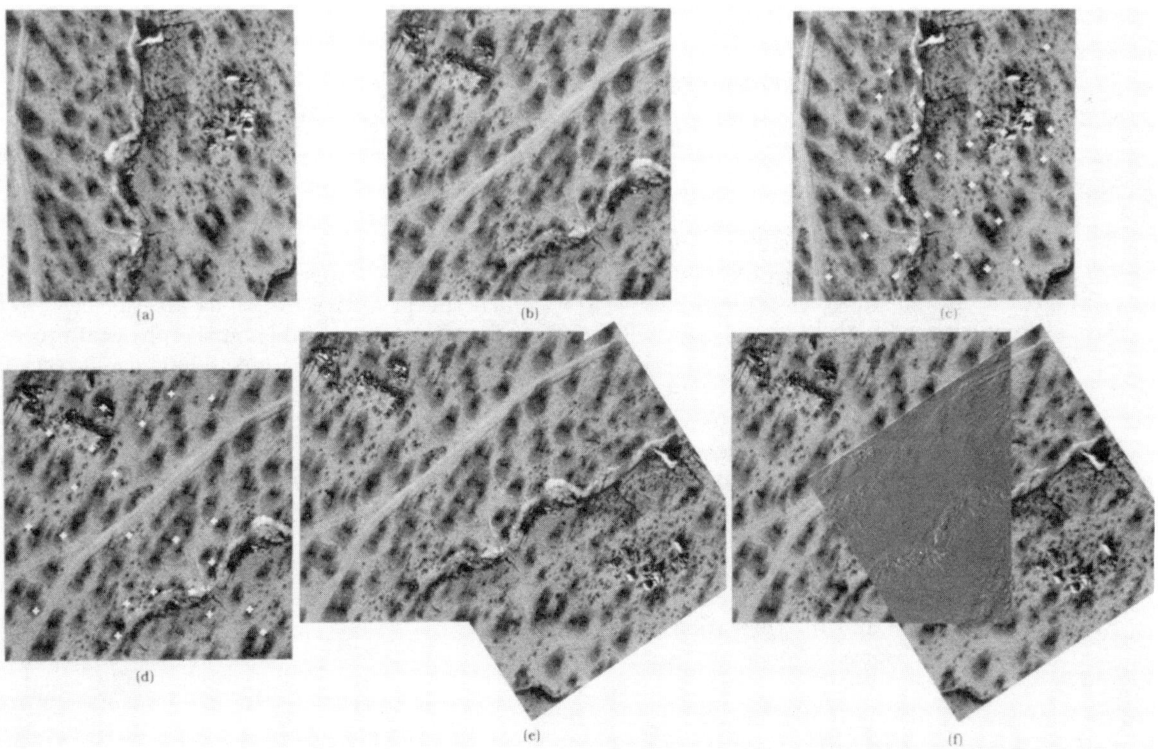

Figure 4: Registration of desert images Bal140 and Bal144.

A Data-Flow Functional Computer for Real-Time Image Processing

Georges M. QUÉNOT, Bertrand ZAVIDOVIQUE

Laboratoire Système de Perception
DGA/Etablissement Technique Central de l'Armement
16 bis Av. Prieur de la Côte d'Or, 94114 Arcueil CEDEX (FRANCE)
Tel: (33 1) 42 31 93 48 Fax: (33 1) 42 31 99 55 Email: quenot@etca.fr

Abstract

We present a data-flow computer, constituted of an array of data-flow processors, dedicated to real-time image processing. A custom processor, able to perform up to 50 millions of operations per second on 25 Mbytes/s data-flows, have been developped for low-level computations and T800 transputers are used for high level ones.

The data-flow computer is programmed in a pure functional style (Backus) programming language with a single formalism for both types of processors.

An experimental system including 256 low-level data-flow processors and 1 transputer has been built and several image processing algorithm have run on it in real time at digital video speed. This system will be extended very soon to 1024 low-level processors and 32 high level processors.

1 Introduction

This paper describes a data-flow functional computer (DFFC) dedicated to real-time image processing. Image processing computations may be roughly divided in two classes: low-level and high level.

By low-level image processing, we mean calculations where the 2D image topology still prevails (a fixed number of operations is involved for each pixel). Such processing is very demanding on computing power and generally involves very simple and very repetitive types of calculations. For example: filters, convolution, histograms, summations, momentums, edge points detection, region growing and labelling.

By contrast, high-level image processing is characterized by the manipulation and interpretation of extracted features such as contours, polygons, sets of points and symbolic trees. Such calculations generally require less raw computing power but also involve much more algorithmic complexity.

For such calculations, Von Neumann style or other classical architectures fail to provide the computing power required by real-time processing. Von Neumann style programming languages also fail to provide an adequate environment for real-time image processing application development.

Many innovative computers have been proposed as alternatives to Von Neumann's. Among them, data-flow architectures [1] [2] appear very well suited to real-time image processing.

Alternative programming languages have also been proposed to the Von Neumann style. Among them the concept of functional programming [3] seems very efficient for the expression of image processing algorithms.

Having exhibited a family of the most frequent operations in image feature extraction, our approach is to integrate the data-flow architecture principle and the functional programming concept together.

This is done through a unification principle which expresses the natural duality between a Control Data-Flow Graph (CDFG) and a functional expression. Both are ways of describing an image processing algorithm. The first representation is relative to a procedure hardware implementation, the second one is its natural software expression.

A functional programming language was defined in which the primitive functions are the basic real-time operators implementable on the physical processors of a data-flow machine. By so doing, a bidirectional and automatic translation between CDFGs and functional expressions becomes possible. The automatic CDFG implementation onto the data-flow machine is feasible using a place and route algorithm (Figure 1)

2 Data-Flow Computer Architecture

We call "data-flow" a structured data set moving serially along a physical link. Our approach is to achieve computations on data-flows, coming directly from digital video cameras, at the pixel acquisition rate and to couple one physical processor with each elementary operation defined in the algorithm.

The core of the data-flow functional computer is a network of processors. Processors may be used simultaneously as data-flow operators or as a cross-bar

1058-6393/91 $1.00 © 1991 IEEE

ALGORITHM:

Y(n) = a.X(n+1) + b.X(n) + c.X(n-1)

INVOLVED FUNCTIONAL OPERATORS:

D : (X(1) X(2) . . . X(n)) = (0 X(1) . . . X(n-1))
id : (X(1) X(2) . . . X(n)) = (X(1) X(2) . . . X(n))
A : (X(1) X(2) . . . X(n)) = (X(2) X(3) . . X(n) 0)

xa : (X(1) X(2) . . . X(n)) = (a.X(1) a.X(2) . . . a.X(n))

+ : ((X(1) X(2) . . . X(n)) (Y(1) Y(2) . . . Y(n))) =
(X(1)+Y(1) X(2)+Y(2) . . . X(n)+Y(n))

FUNCTIONAL EXPRESSION:

DEF F = +o[xa o A, +o[xb o id, xc o D]]

(Y(1) Y(2) . . . Y(n)) = F : (X(1) X(2) . . . X(n))

DATA-FLOW GRAPH:

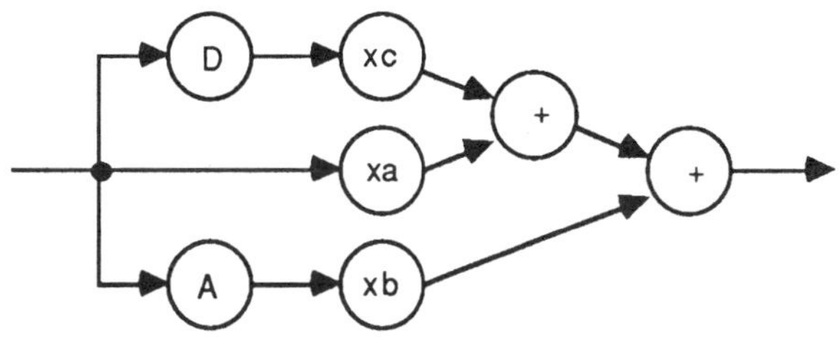

MAPPING ONTO THE DATA-FLOW COMPUTER:

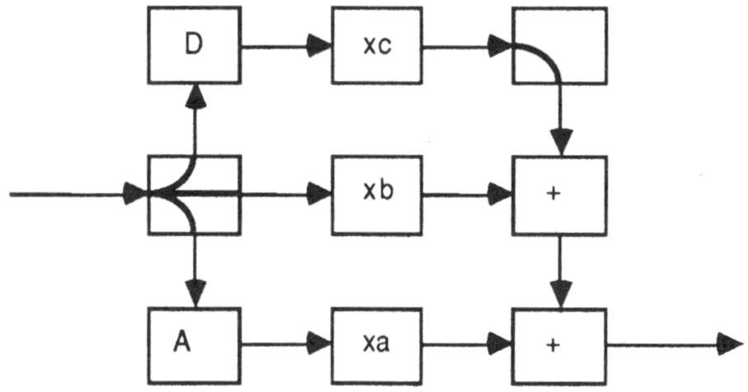

Figure 1: Functional programming and data-flow implementation

routing elements. Two different kinds of processors are integrated in the network. They are respectively dedicated to low-level and high-level image processing. This computer also incorporates digital video input/output subsystems and a global controller coupled with an host workstation (Figure 2).

For the execution of low-level functions, we developed a custom data-flow processor (DFP) [4]. Two coupled DFPs have been included in a single chip. Each DFP has 6 input-output ports (it has been designed to be mesh-connected in 3D networks). Each port is made up of 10 bidirectional lines (9 data and 1 acknowledge). The core of the data-flow processor is interfaced to the outside world through 3 input stacks and 3 output stacks, where each one is a synchronous FIFO with 8 usable 9-bit words, and with a 25 MBytes/second bandwidth. Each I/O port contains a receiving part that can be connected to any input stack and a sending part that can be connected to any output stack (Figure 3).

A datapath is inserted between the input and output stacks. It splits into three stages of pipe-line. The first stage decodes the input data type and generates commands for the following stages. 8-bit operations (input shifts and multiplications) are performed during the second stage, and 16-bit operations (2901-type ALU, absolute value, minimum or maximum, output shift) are performed at the third stage. This datapath is able to perform up to 50 Million of 8- or 16-bit operations per second.

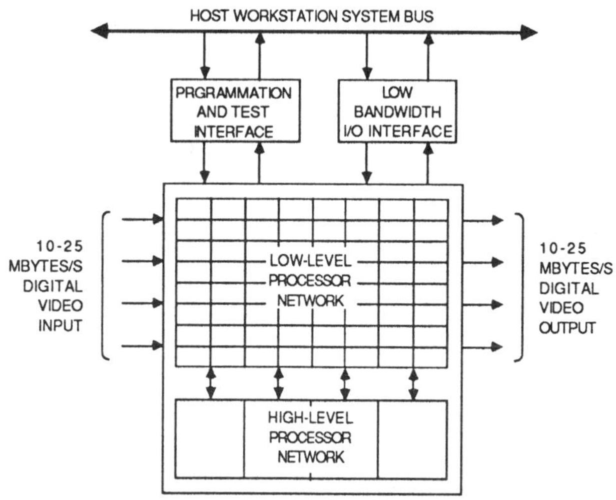

Figure 2: Data-flow computer architecture

A 256 9-bit words data RAM is included in each DFP. It can be used as a dual port RAM operator, as a 256 words FIFO, or as local memory for specific operators such as histogramming. The control part has been designed on the basis of a programable state-machine for which a 64 32-bit word program RAM has been included in the DFP.

A wide variety of data-flow operators can be implemented using the state machine. Among them are additions, substractions, multiplications, line or pixel

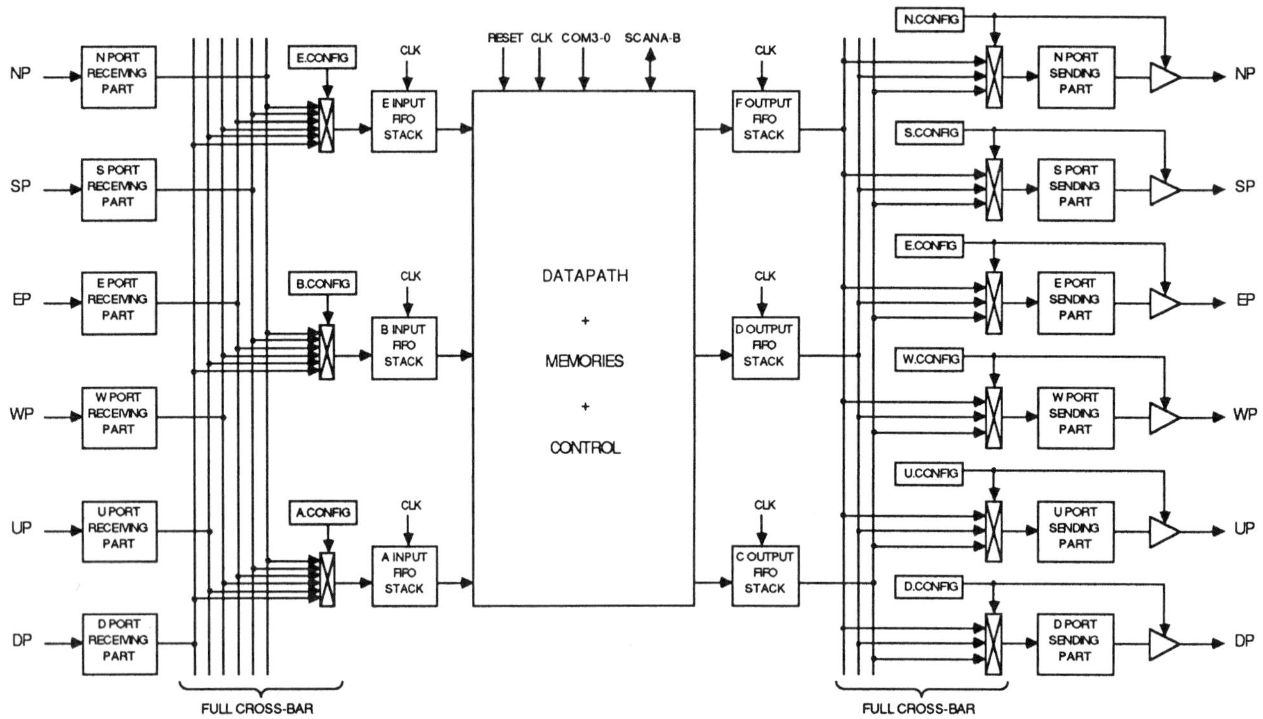

Figure 3: Data-flow processor internal architecture

shifts, derivatives, sommations, scalar products and histograms. More complex operators (convolution for example) can be defined as macro-functions using combinations of basic operators mapped on groups of processors.

For the execution of high-level functions, we choose the T800 transputer because it is a general purpose processor which directly incorporates communication links. It is therefore very well suited for the implementation of complex functions into data-flow operators. Transputer modules including 1 Mbyte of memory are the basic element of the high-level processing network.

Any kind of algorithm may be implemented on a high-level processing element as a C program. There are only two constraints: the first is to follow the data-flow principle (output data-flows are only functions of input data-flows, the control is done locally and the execution is independant of what may occur in other processors) and the second one is to satisfy real-time.

The following table summarizes the characteristics of both processors. Low-level and high-level processors communicate using bidirectional parallel/serial link adaptors.

DFP 8-bit	Low-level Simple operators	High bandwidth 6 × 25 Mo/s, 50 Mops
T800 32-bit	High level Unlimited complexity	Low bandwidth 4 × 2 Mo/s, 3 Mflops

3 Functional Programmimg

The execution of programs on the data-flow functional computer is fully data-driven. There is no global control. Each operator is activated by a completely local firing rule, and operations are performed as soon as all the required data are available (and if, simultaneouley, destination processors for output flows are ready to receive new data). However, control flows may be inserted in data flows.

Every processor is programmed individually. The program of a processor contains two distinct and independant parts. The first one defines the function of the operator assigned to it. The second one defines the routing of data-flows between the physical ports and the logical input-output of the operator. This is the same for low-level or high-level processors.

The first task to perform in order to program the data-flow computer is to define a library of basic operators.

For the low-level processors, the basic operators are defined by the programming of the internal state machine. This is done using a specific assembler.

For the high-level processors, the operators are defined as a C program that uses transputer links as input and output files and that runs as an infinite loop. These operators also appear as data-flow ones.

After the library has been developed, every operator becomes a basic primitive in a proprietary data-flow functional programming language (DFFPL), [5]. Low-level and high level primitives have exactly the same interface relative to this language. The language also incorporates the functional forms necessary to build any legal combinations of the primitives.

An image processing algorithm is described in the DFFPL language as a combination of basic primitives and macrofunctions using functional forms.

A natural duality between Data-Flow Graphs and Functional Expression has been exhibited [6]. Simple transformation rules permits an automatic translation of the whole algorithm in a complete data-flow graph in which the nodes corresponds to operators and macro-functions.

The data-flow graph is then mapped onto the physical network (Figure 1). This can be done automatically by using a place and route algorithm based on simulated annealing. After mapping has been completed, the processors are used simultaneously as operators or as cross-bar routing elements.

The algorithm is executed through a gigantic pipeline where every involved operation is assigned to a physical processor. Its complexity is therefore limited by the size of the available array but it is always executed in real-time.

Normal users generally will not need to develop their own basic operators since the machine will be delivered with a software package that includes a large library of currently used operators and macro-functions. Users will only need to write their image processing algorithms with the functional programming language. The library may be extended as and when required according to the users needs.

Programming the data-flow functional computer is then easy and efficient. Users may describe their algorithms in a text form or in a graphic form. They may place and route their data-flow graph themselves using a graphic editor or let the system do it.

From the theoritical point of view, it may be noticed that we make a pure use of the functional programming concept since the concept of variable never appear in our functional language. Similarily, we make a pure use of the data-flow principle since there is never an address-flow associated to a data-flow.

The common elimination of the variable and address concepts reflects a strong theoritical background in the duality between pure data-flow architecture and pure functional programming.

4 Experimental Results

We have developed a data-flow computer which has 256 DFPs (with 128 biprocessor chips) in a 8 × 8 × 4 tridimensional network and 1 transputer modules. This system is able to run with a 25 MHz clock

Several real-time image processing applications have already run at digital video speed onto this computer including convolutions, histograms, edge detection, region extraction and target tracking.

For example, Figure 4 shows the implementation of a macro-function performing edge point detection on a group of 8 processors and the result of its application in real time of this macro-function on a 25 Hz sequence of 512 × 512 images.

The principle is to get, for each pixel, the absolute value of the normalized derivatives in four directions (1,0), (1,1), (0,1) and (-1,1), to take the maximum of them and to threshold the result.

14 operations per pixel are involved here (4 substractions, 4 absolute value, 2 multiplications, 3 maximums and 1 threshold). DFPs are able to perform on the fly 2 arithmetic operations per pixel.

The data-flow functional computer will be extended soon to 1024 DFPs (in a 8 × 8 × 16 network) and 32 transputer modules. Its theoritical peak power will be of 50 billions of arithmetic operations per second.

5 Conclusion

This paper has described a data-flow functional computer for real-time image processing. Our approach integrates efficiently the data-flow architecture principle and the functional programming concept together.

It leads to the design of a very simple and regular hardware. It also provides a very nice and user-friendly software interface for the development of image processing algorithms. This approach can be maintained even with a system that includes processors of different granularity.

We have developed an experimental data-flow functional computer which has 256 DFPs (with 128 biprocessor chips) in a 8 × 8 × 4 tridimensional network and one T800 transputer.

Convolutions, histograms, edge detection, region extraction and target tracking have run at digital video speed onto this computer. Real-time color image processing algorithms are under development.

The data-flow functional computer is currently being extended to 1024 DFPs in a 8 × 8 × 16 tridimensional network and 32 T800 based transputer modules.

Our data-flow functional computer is mainly dedicated to image processing. However its concept is much more general. The basic processors may be adapted to address many other problems. For example, a 64-bit floating point data-flow processor could be similarily developed and used for the design of a data-flow functional supercomputer dedicated to the resolution of partial derivative equations.

Basic operators:

$\partial 1 : [x_{i,j}] = [\mid x_{i,j} - x_{i+1,j} \mid]$
$\partial 3 : [x_{i,j}] = [\mid x_{i,j} - x_{i,j+1} \mid]$
$\partial 2 : [x_{i,j}] = [\mid x_{i,j} - x_{i+1,j+1} \mid]$
$\partial 4 : [x_{i,j}] = [\mid x_{i,j} - x_{i-1,j+1} \mid]$
$\text{max1} : ([x_{i,j}], [y_{i,j}]) = [\max(x_{i,j}, y_{i,j})]$
$\text{max2} : ([x_{i,j}], [y_{i,j}]) = [\max(x_{i,j}, y_{i,j}/\sqrt{2})]$
$\text{thre} : [x_{i,j}] = [(x_{i,j} > th)]$

Mapping on a 2 × 2 × 2 DFP array:

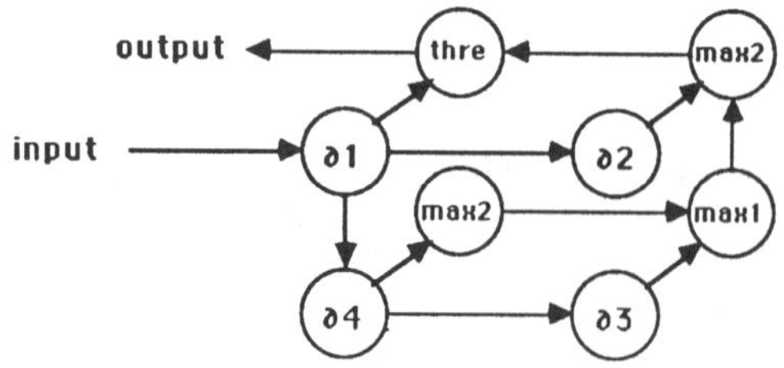

Functional expression:

$\text{edge} = \text{thre}(\text{max2}(\text{max1}(\text{max2}(\partial 1, \partial 2), \partial 3), \partial 4))$

output = edge : **input**

Result of real-time execution (on a view of the data-flow functional computer):

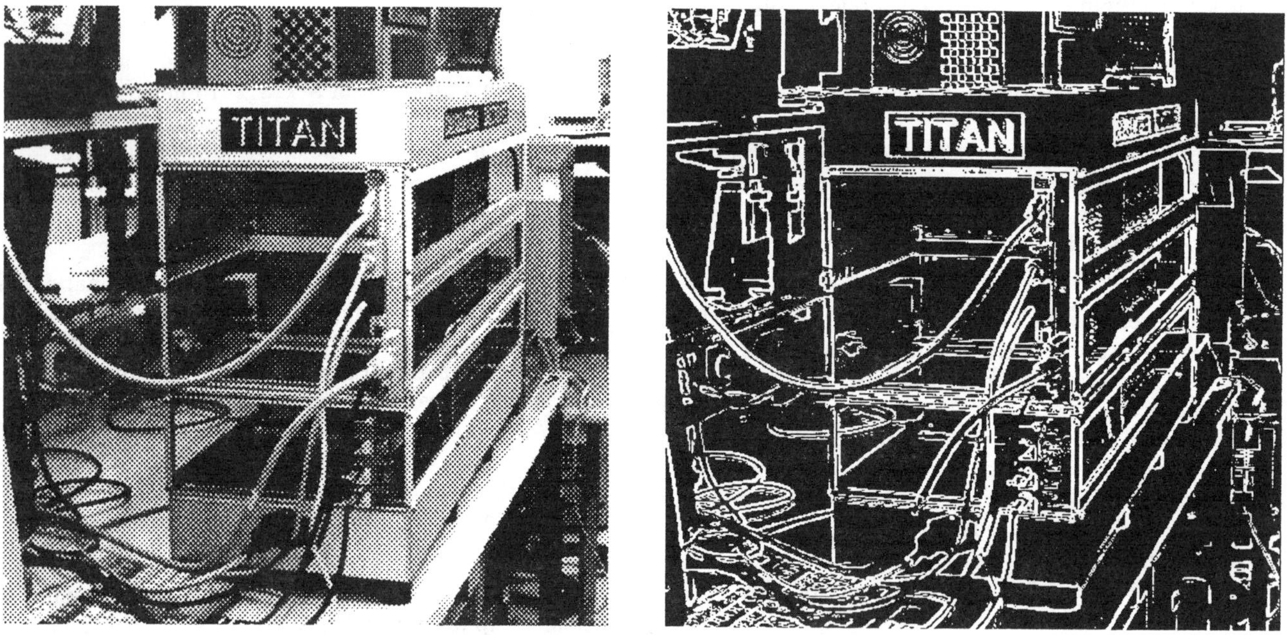

Figure 4: The edge-point detection macro-function

References

[1] J.B. Dennis, "Data Flow Supercomputers," *Computer,* Vol 13, Nov. 1980, pp. 48-56.

[2] K.P. Arvind, D.E. Culler, "Dataflow Architectures," Annual Reviews In Computer Science, Vol. 1, Palo Alto CA, Annual Reviews Inc 1986, pp. 225,253.

[3] J. Backus, "Can programming be liberated from Von Neumann Style? A functional style and its algebra of programs." *Comm. of ACM,* Vol. 21, No 8, August 1978.

[4] G.M. Quénot, B. Zavidovique, "A Data-Flow Processor for Real-Time Low-Level Image Processing," *IEEE Custom Integrated Circuits Conference,* may 1991, San-Diego, CA.

[5] J. Serot, G.M. Quénot, B. Zavidovique, "Real-Time Image Processing using Functional Programming on a Data-Flow Architecture," *CAMP-91,* 15-18 december 1991, Paris, France.

[6] E. Allart, B. Zavidovique, "Image Processing VLSI Design through Functional Match between Algorithm and Architecture," *IEEE Symposium on Circuits and Systems,* 7-9 june 1988, Espoo, Finland.

Two-Dimensional Adaptive Enhancement and Detection of Linear Features in Noisy Images

C. D. Knittle

Communication Systems Division
Sandia National Laboratories
Albuquerque, NM 87185

N. Magotra

Department of ECEE
University of New Mexico
Albuquerque, NM 87131

Abstract

This paper presents results from using a novel two-dimensional adaptive filtering algorithm to enhance linear features embedded in noisy images. The algorithm is called the two-dimensional adaptive correlation enhancer (2DACE). The 2DACE algorithm is a coefficient update scheme that converges to the two-dimensional autocorrelation function of the signal or feature of interest. Since the noise statistics are nonstationary and the duration of a signal or feature can be very short, the algorithm must be capable of rapid convergence to its optimal impulse response, a requirement that 2DACE meets. After the image has been enhanced, a modified Radon transform is employed to explore the improved detectability of the linear features.

1 Introduction

Enhancing and detecting linear features in noisy images presents a significant challenge to current two-dimensional signal processing algorithms. In most cases, the statistics of the signal and background noise are nonstationary or unknown, making the design of fixed filters formidable. To circumvent the problem of unknown or changing image statistics, adaptive filters capable of learning the image statistics are used. Update algorithms subsequently adjust the impulse response according to the learned statistical parameters.

Only recently has interest turned to multi-dimensional adaptive signal processing [1], [2], [3]. The introduction of the two-dimensional least mean square (2DLMS) adaptive filter algorithm [1] is perhaps the most significant contribution thus far. One application of current interest is the extraction of a signal from noise when the signal is present only in a small spatial region or exists for a short period of time. The 2DLMS algorithm has shown potential for enhancing signals of this type [4].

This paper presents a novel two-dimensional adaptive filtering algorithm called the two-dimensional adaptive correlation enhancer (2DACE). It is an extension of a one-dimensional algorithm first reported in [5]. 2DACE is used primarily to enhance features which have alternating bright and dark intensities or appear linear in noisy images. Under short-term stationary conditions, the 2DACE algorithm converges to the two-dimensional autocorrelation function. It is shown for a class of signals the 2DACE algorithm resembles a matched filter. We will show that the 2DACE algorithm significantly improves the signal-to-noise ratio (SNR) in the image. After the image has been enhanced, a modified Radon transform is employed to explore the improved detectability of the linear features. The filter is then applied to synthetic aperture radar (SAR) imagery to enhance surface ship wakes.

2 The 2D Adaptive Correlation Enhancer

The two-dimensional weight update equation for vertical scanning of the image is

$$\mathbf{w}(m,n) = \beta\mathbf{w}(m-1,n)+(1-\beta)\frac{\mathbf{x}(m,n)x(m,n)}{2L^2P(m,n)}. \quad (1)$$

where

$$\mathbf{w}(m,n) = \begin{bmatrix} w_{0,0}(m,n) & \cdots & w_{0,2L}(m,n) \\ w_{1,0}(m,n) & \cdots & w_{1,2L}(m,n) \\ \vdots & \ddots & \vdots \\ w_{2L,0}(m,n) & \cdots & w_{2L,2L}(m,n) \end{bmatrix}$$

is the $(2L+1)$x$(2L+1)$ coefficient matrix,

$$\mathbf{x}(m,n) = \begin{bmatrix} x(m-L,n-L) & \cdots & x(m-L,n+L) \\ x(m-L+1,n-L) & \cdots & x(m-L+1,n+L) \\ \vdots & \ddots & \vdots \\ x(m+L,n-L) & \cdots & x(m+L,n+L) \end{bmatrix}$$

is the input data matrix, β is a smoothing parameter, L is called the two-dimensional maximum lag, and $P(m,n) = \beta P(m-1,n) + (1-\beta)x(m,n)$ is a recursive estimate of the input signal power. For convenience, we let $K = \frac{(1-\beta)}{2L^2 P(m,n)}$. The expected coefficient matrix is

$$\mathrm{E}[\mathbf{w}(m,n)] = \frac{K}{1-\beta}\mathrm{E}[\mathbf{x}(m,n)x(m,n)] \tag{2}$$

where the expectation on the right side of (2) is

$$\mathrm{E}[\mathbf{x}(m,n)x(m,n)] = \begin{bmatrix} R_{xx}(-L,-L) & \cdots & R_{xx}(-L,L) \\ R_{xx}(-L+1,-L) & \cdots & R_{xx}(-L+1,L) \\ \vdots & \ddots & \vdots \\ R_{xx}(L-1,-L) & \cdots & R_{xx}(L-1,L) \\ R_{xx}(L,-L) & \cdots & R_{xx}(L,L) \end{bmatrix} \tag{3}$$

In (3), the function $R_{xx}(i,j)$, $|i,j| \leq L$, represents the two-dimensional autocorrelation function of the input data. Hence, under stationary conditions, the filter coefficient matrix is proportional to the matrix of two-dimensional autocorrelation coefficients of the input signal.

Insight into the stability of the adaptive algorithm can be gained by examining the Z-transform of the weight update equation. Taking the two-dimensional Z-transform of equation (1) and solving for $\mathbf{W}(z_1,z_2)$ yields

$$\mathbf{W}(z_1,z_2) = \frac{K}{1-\beta z_1^{-1}}\mathbf{X}(z_1,z_2) * X(z_1,z_2) \tag{4}$$

where

$$\mathbf{X}(z_1,z_2) = \begin{bmatrix} z_1^{-L}z_2^{-L} & \cdots & z_1^{-L}z_2^{L} \\ z_1^{-L+1}z_2^{-L} & \cdots & z_1^{-L+1}z_2^{L} \\ \vdots & \ddots & \vdots \\ z_1^{L}z_2^{-L} & \cdots & z_1^{L}z_2^{L} \end{bmatrix} X(z_1,z_2).$$

Equation (4) shows that for stability, the restriction

$$\beta < 1 \tag{5}$$

must be adhered to. Notice in equation (4) that the $(1-\beta z_1^{-1})^{-1}$ corresponds to a geometric series with ratio β in the time domain. Hence, an exponential envelope of time constant τ can be fit to the geometric series with ratio β by choosing

$$\beta = e^{\frac{-1}{\tau}}. \tag{6}$$

The time constant is approximately

$$\tau \simeq \frac{1}{1-\beta}. \tag{7}$$

Equation (7) implies that β also controls the rate of adaptation, i.e. larger β implies slower adaptation.

To examine the filter gain, let the input signal be

$$x(m,n) = A\cos(\pi(\gamma_0 m + \alpha_0 n)) + v(m,n) \tag{8}$$

where $v(m,n)$ is a white noise process with variance σ^2, and γ_0 and α_0 are the vertical and horizontal frequencies respectively, normalized with respect to the Nyquist frequency. If $h_{m,n}(i,j)$ denotes the steady state impulse response at location (m,n) then

$$h_{m,n}(i,j) = \frac{K}{1-\beta} \left[\frac{A^2}{2}\cos(\pi\gamma_0(i-L) + \pi\alpha_0(j-L)) + \sigma^2\delta(i-L,j-L)\right] \tag{9}$$

where $i,j = 0,1,2,\ldots,2L$. The magnitude response of the Fourier transform of $h_{m,n}(i,j)$ is

$$|H(e^{j\pi\gamma}, e^{j\pi\alpha})| = \frac{K}{1-\beta}\left[\sigma^2 + \frac{A^2}{4}\left\{\frac{\sin(\frac{\pi(\gamma_0-\gamma)(2L+1)}{2})}{\sin(\frac{\pi(\gamma_0-\gamma)}{2})}\frac{\sin(\frac{\pi(\alpha_0-\alpha)(2L+1)}{2})}{\sin(\frac{\pi(\alpha_0-\alpha)}{2})} + \frac{\sin(\frac{\pi(\gamma_0+\gamma)(2L+1)}{2})}{\sin(\frac{\pi(\gamma_0+\gamma)}{2})}\frac{\sin(\frac{\pi(\alpha_0+\alpha)(2L+1)}{2})}{\sin(\frac{\pi(\alpha_0+\alpha)}{2})}\right\}\right]. \tag{10}$$

Equation (10) represents a two-dimensional bandpass filter with center frequency (γ_0, α_0). The gain at $\gamma = \gamma_0$ and $\alpha = \alpha_0$ is

$$g_{2d} = \frac{K}{1-\beta}\left[\sigma^2 + \frac{A^2}{4}\left\{(2L+1)^2 + \frac{\sin(\pi\gamma_0(2L+1))}{\sin(\pi\gamma_0)}\frac{\sin(\pi\alpha_0(2L+1))}{\sin(\pi\alpha_0)}\right\}\right]. \tag{11}$$

The two-dimensional filter gain, defined as the gain in signal-to-noise ratio, can be shown to be

$$G_{f_{2d}} = \frac{SNR_{out}}{SNR_{in}} \approx 2L^2 \tag{12}$$

under the conditions $\gamma_0, \alpha_0 \gg 0$, $2L \gg 1$, and $\frac{A^2}{2} \gg \sigma^2$.

Note that if the input is given by equation (8), the expected impulse response in equation (9) resembles a matched filter, except for the term containing the impulse. Additionally, if the input consists of only white noise, the impulse response in equation (9) reduces to

$$\begin{aligned} h_{m,n}(i,j) &= \frac{K}{1-\beta}\sigma^2\delta(i-L,j-L) \\ &= \frac{\delta(i-L,j-L)}{2L^2} \end{aligned} \qquad (13)$$

which attenuates all frequencies by $\frac{1}{2L^2}$.

3 The Radon Transform

The Radon transform is a technique which maps a straight line in the image to a single point in the transform space. The magnitude of the point in transform space is directly related to the brightness of the line. The Radon transform for a discrete image is computed using

$$R_g(\rho,\theta) = \sum_{m=0}^{M-1}\sum_{n=0}^{N-1} g(m,n)\delta(\rho - m\cos(\theta) - n\sin(\theta)) \qquad (14)$$

where $g(m,n)$ is the discrete image, M and N are the number of rows and columns in the image, respectively, and $\delta(k)$ is the unit sample function. The Radon transform equation given in equation (14) performs a summation (integration) along the line

$$\rho - m\cos(\theta) - n\sin(\theta) = 0 \qquad (15)$$

where the coordinate system is positioned such that the origin is the center of the image. By simply varying the transform parameters (ρ,θ) an integration is performed along all possible lines in the image. The parameter θ lies in the range $-\frac{\pi}{2} \le \theta \le \frac{\pi}{2}$, with $\theta = 0$ aligning with the positive vertical axis. The shortest distance from the origin to the line, ρ, lies in the range $-\rho_{max} \le \rho \le \rho_{max}$, with positive values of ρ corresponding to lines in the top half plane.

The strength of the Radon transform, with respect to locating linear features in noisy images, lies in the summing process. Summing along a line in the image tends to average the noise samples, thus increasing the signal-to-noise ratio in the transform domain. If the linear feature is, on average, brighter (darker) than the background, the corresponding point in the transform space will be bright (dark). Even if a bright line is not continuous across the image, a bright peak will still exist in the transform space.

Figure 1: Original image corrupted with white noise.

4 Simulation Results

2DACE is best suited for enhancing narrowband signals embedded in broadband noise. Additionally, since it exhibits rapid convergence, it is adept at filtering signals that occupy small spatial regions in the image. As an example, consider the image shown in Fig. 1, where the power of the white noise is increasing linearly with the row number. That is, the noise power increases from $\sigma^2 = 200$ at the top of the image to $\sigma^2 = 450$ at the bottom. The linear features in the image are portions of a two-dimensional sine wave, the top feature having a vertical and horizontal wavelength of 50 pixels/cycle and 250 pixels/cycle, respectively, and the bottom feature exhibiting a vertical wavelength of one-half the top feature. The SNR of the top feature ranges from -6.4 dB to -7.7 dB, while the SNR of the bottom feature is approximately -8.8 dB. Fig. 2 shows the results of processing with the 2DACE filter after local normalization. The filter parameters were $L = 5$ and $\beta = 0.95$. Because the SNR of the bottom feature was extremely poor, the filter was unable to enhance it as well as the top feature. A degradation in filter performance is expected as the SNR decreases.

Figures 3 and 4 display the Radon transform of Figs. 1 and 2, respectively. Detecting the linear features in the original image is transformed to one of isolating local maxima and minima in the transform image. Comparing Figs. 3 and 4, it is evident that the

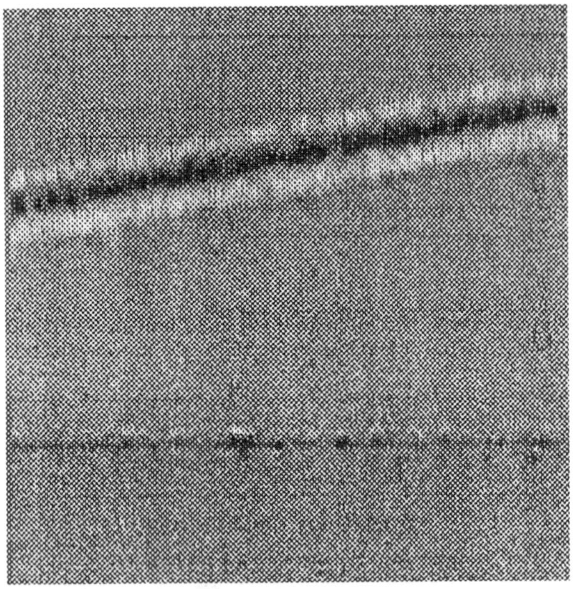

Figure 2: Result of processing with 2DACE.

intensity difference between local minima (maxima) and the background noise is significantly improved in the processed image, leading to higher probability of detection. In fact, if the SNR in the transform domain is defined as

$$SNR \equiv \frac{(signal\ peak - image\ mean)^2}{image\ variance} \tag{16}$$

then 2DACE processing increased the bright-peak SNR by 8.7dB and the dark-peak SNR by 22dB.

5 Application to Surface Ship Wakes in Synthetic Aperture Radar Imagery

Under certain conditions, a SAR is capable of imaging wakes caused by moving surface ships. In high-resolution SAR, segments of the wakes appear as indistinct, bright and dark linear features, as shown in Fig. 5. In this image, the wakes are horizontally oriented and are extremely faint due to radar speckle. The enhanced image shown in Fig. 6 was processed using a combination of horizontal lowpass filtering to reduce the radar speckle and adaptive filtering with 2DACE to enhance the linear ship wakes. The alternating bright and dark wakes appear sinusoidal, allowing 2DACE to form a bandpass filter in the location of the wakes.

Figure 3: Radon transform of original image shown in Fig. 1.

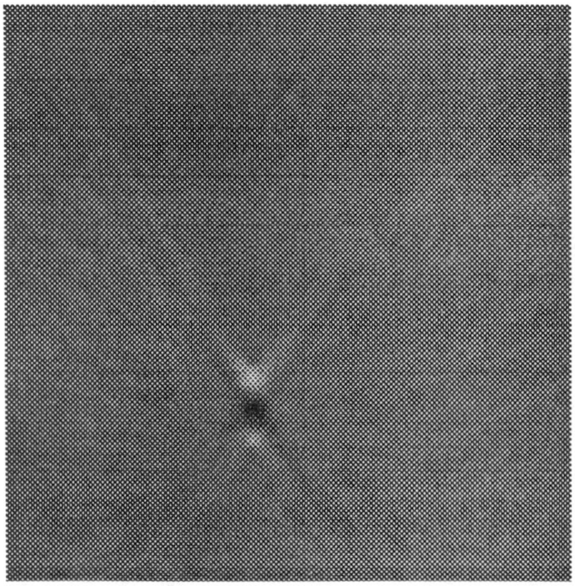

Figure 4: Radon transform of 2DACE-processed image shown in Fig. 2.

Figure 5: SAR image containing surface ship wakes.

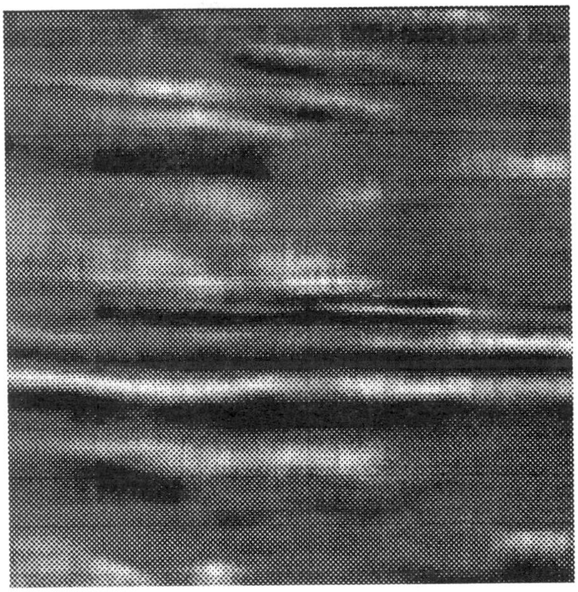

Figure 6: Result of processing Fig. 5 with lowpass filter and 2DACE.

6 Conclusions

This paper has presented a novel two-dimensional adaptive filter called the adaptive correlation enhancer. The filter is best suited for signals with narrowband components in wideband noise. For example, it was shown that if the input signal is sinusoidal, then the filter converges to a matched filter. The Radon transform was used to detect the linear features, and it was shown that preprocessing the image with 2DACE prior to taking the transform yielded a tremendous improvement in signal-to-noise ratio in the transform domain. Finally, it was shown that the adaptive filter significantly enhanced surface ship wakes in SAR images.

Acknowledgements

This work was supported by the U. S. Department of Energy under contract DE-AC0476DP00789 and the Office of the Secretary of Defense.

References

[1] M. M. Hadhoud and D. W. Thomas, "The Two-Dimensional Adaptive LMS (TDLMS) Algorithm", *Trans. on Circuits and Systems*, Vol. 35, pp. 485-494, May, 1988.

[2] J. C. Strait and W. K. Jenkins, "Filter Architectures and Adaptive Algorithms for 2-D Adaptive Digital Signal Processing", *Int. Conf. on ASSP*, pp. 932–935, 1989.

[3] M. Chellali and V. K. Ingle, "Applications of Adaptive Least Squares Filtering to Image Restoration", *SPIE, Visual Communications and Image Processing*, Vol. 1001, pp. 19-24, 1988.

[4] T. Soni, B. D. Rao, J. R. Zeidler and W. H. Ku", "Enhancement of Images Using the 2-D LMS Adaptive Algorithm", *Int. Conf. on ASSP*, pp. 3029–3032, 1991.

[5] N. Magotra, N. Ahmed and E. Chael, "Single Station (Three Component) Seismic Data Enhancement", *20th IEEE Asilomar Conference on Circuits, Systems, and Computers*, pp. 409–413, Nov. 1986.

Improving the Spatial Resolution of Real-Time Radiography Images Using the Maximum Entropy Method

Edward R. Doering and John P. Basart
Center for Nondestructive Evaluation and
Electrical Engineering and Computer Engineering Department
Iowa State University
Ames, Iowa 50011

Abstract

Real-time radiography is a non-destructive evaluation (NDE) technique that permits continuous inspection of an object. This has the advantages of reduced cost and reduced inspection time compared to an equivalent film-based system. However, the real-time technique has reduced spatial resolution and contrast sensitivity due to light scattering in the phosphor conversion screens of the X-ray-to-light converter (image intensifier). We have evaluated the maximum-entropy method (MEM) as a technique to improve the spatial resolution of the real-time image. The point spread function (PSF) of the image intensifier was estimated by differentiating a high resolution scan made by translating the edge of a lead plate in sub-pixel increments relative to single pixel of the CCD video camera. The 1-D PSF was rotated to form the 2-D PSF which was used as the system function for the MEM algorithm. Improvement in the restored image was evaluated quantitatively using a resolution gauge.

1 Introduction

X-ray inspection of industrial components is an important non-destructive evaluation (NDE) technique. Traditionally, X-ray sensitive film is used to record the latent image imparted to the X-ray field by the object under test. Many such images are required to build up a clear idea of what defects may exist in the part. X-ray-to-light conversion devices permit direct observation of the X-ray field, so interaction with the object is much faster. Fluorescent screens were the first X-ray-to-light converters and have been used since the discovery of X-rays nearly 100 years ago. Work with real-time images was limited until progress was made in the ability to intensify the fluoroscopic image and couple digital computers into the system.

Today a real-time inspection system consists of an X-ray source, an sample positioner, a combination X-ray-to-light converter and intensifier (herein referred to as the image intensifier), video camera, frame digitizer, frame processor, and display device. The primary advantages of real-time inspection compared to an equivalent film-based system are high throughput, reduced cost associated with film processing, and interactive inspection. The principal disadvantages of real-time inspection are reduced spatial resolution and reduced contrast sensitivity of the conversion device compared to film. For the case of an image intensifier based on phosphor conversion screens (the most widely used now), the reduced spatial resolution is due to light scattering between the conversion phosphors. This scattering is called phosphor bloom. Typically film achieves 10 to 20 line pairs per millimeter (lp/mm) resolution with 1% contrast sensitivity while real-time systems are limited to 4 lp/mm with 5% to 8% contrast sensitivity [1].

Real-time radiography systems are being improved along two fronts. The first is to improve the hardware itself. This includes development of new detector materials and collimation techniques. These techniques are often more expensive than image intensifier-based methods. The second approach is to apply image processing techniques to improve the image using existing detection schemes. For example, contrast sensitivity is improved by frame averaging that increases the number of photons counted and hence makes the greatest contribution to improving the statistical quality of the image. Background subtraction followed by a contrast stretch further improves detection of low-contrast features. These techniques are easily implemented with dedicated hardware. Improving the resolution of the image requires an image restoration technique, which is more computationally expensive, particularly when trying to minimize noise sensitivity and generation of artifacts.

We have addressed the latter software method by investigating the Maximum Entropy Method (MEM) to restore the real-time image which has been blurred by phosphor bloom. The primary advantages of MEM are positivity of the restored image and minimization of the ringing artifacts and noise sensitivity associated with inverse filters. MEM (or any other image restoration technique) requires an estimate of the point spread function (PSF) responsible for blurring the image. We describe in Section 2 an accurate method for measuring the PSF of the real-time system. We briefly review the MEM restoration technique in Section 3, and discuss quantitative methods for assessing the performance of the MEM with examples in Section 4.

1058-6393/91 $1.00 © 1991 IEEE

2 PSF Measurement

Differentiating the edge trace made by imaging a knife edge is a well-known optical technique to determine the point spread function (PSF) [2]. This technique was applied to the X-ray imaging system by imaging the edge of a lead plate in contact with the input screen of the image intensifier. The edge was aligned with the central axis of the X-ray source to minimize blurring from geometrical unsharpness. The sampling frequency of the system was fixed at 6 pixels/mm and the edge transition region extended over approximately 2 mm, so only 5 to 10 pixels were available to characterize the transition region. To increase the number of samples, a computer-controlled precision stepper motor translated the lead plate by sub-pixel (10 μm) increments. At each step, 64 frames were averaged and the output of a single pixel monitoring the edge transition was saved using 8 bits. Four hundred samples were collected. This method was originally developed to characterize the modulation transfer function (the modulus of the Fourier transform of the PSF) of CCD arrays [3] and its application to characterize real-time radiography systems is more thoroughly described in [4]. The result of this scan was normalized to the range [0,1] and is plotted in Fig. 1. The signal-to-noise ratio (SNR) for this measurement was calculated by taking the signal as 1 and the noise as the standard deviation of the measurements outside the transition region. The samples between +7 and +10 pixels in Fig. 1 contained only digitizer-related noise while the samples between -7 and -10 pixels included photon counting noise. The SNR using the samples between +7 and +10 pixels was typically 50 dB which is consistent with an 8 bit digitizer.

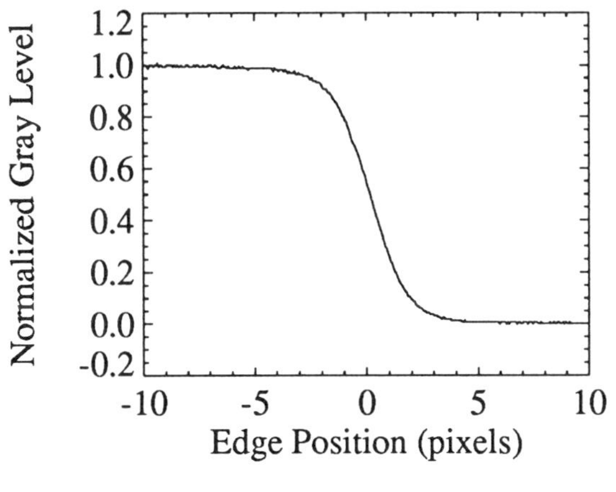

Figure 1: High resolution edge scan measurement.

To find the PSF, the scan in Fig. 1 was numerically differentiated using the first forward difference approximation $y(n) = x(n+1) - x(n)$; this result is plotted in Fig. 2. Since the difference in gray level between adjacent pixels never exceeded 4, the ± 1 gray level quantization noise is prominent. A Gaussian func-

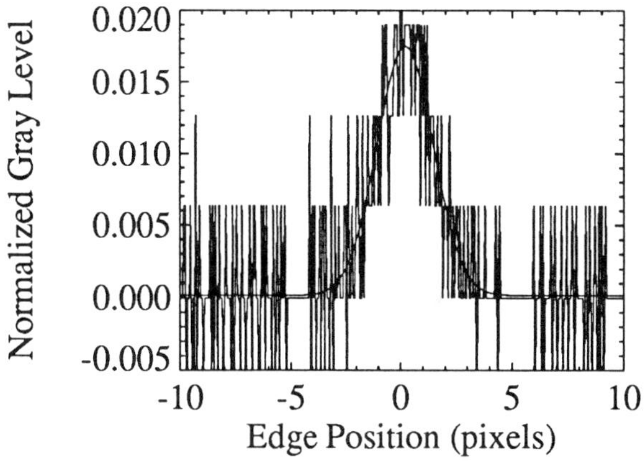

Figure 2: First forward difference of edge scan data

tion was fit to the differentiated scan and yielded an average full-width-at-half-maximum (FWHM) value of 3.2 pixels with 5% variation when the measurement/differentiation process was repeated 9 times. The Gaussian curve is indicated as an overlay in Fig. 2.

A nine pixel uniform-weighted convolution filter was applied to the data prior to differentiation to reduce the effects of quantization noise. This result along with its associated Gaussian fit is shown in Fig. 3. The Gaussian curve follows the measurement

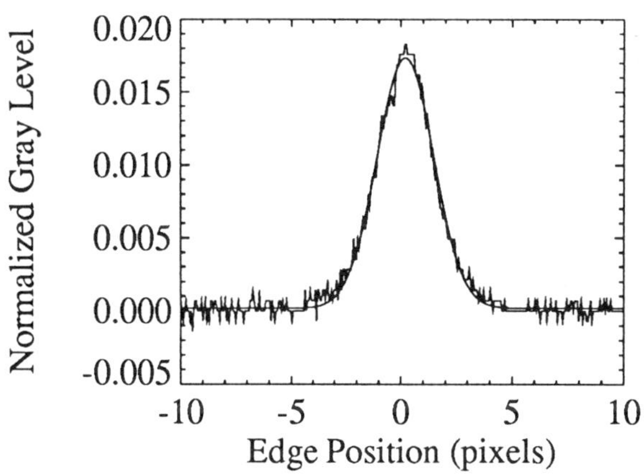

Figure 3: First forward difference of smoothed edge scan data

data very well. The goodness of fit was assessed using the reduced chi-square statistic $\chi^2_\nu = s^2/\sigma^2$ [5], where s^2 is the variance of the fitting error and σ^2 is the variance in the data due to measurement error. The measurement error was estimated by computing the variance of 100 samples outside the edge transition region. χ^2_ν for this data set was 0.95 ($\chi^2_\nu \simeq 1$ indicates a good fit). Data smoothing introduces an error into the

FWHM measurement by widening it, but there was less than 1% change in any of the Gaussian fit parameters with and without smoothing. Thus smoothing is useful to better show the shape of the PSF, but is not required to achieve a good estimate of the FWHM of the spread function.

The measured FWHM value was used to form a 2-D circularly-symmetric Gaussian PSF. The Gaussian model is justified by central limit effects of combined elements of the imaging chain [6]. Also the goodness-of-fit criterion indicates that a Gaussian function adequately models the 1-D PSF. However, it is known that some aspects of the combined PSF are not described by a Gaussian PSF, notably the PSF of the circular aperture of the video camera lens. This PSF has a sinc function-like form in the frequency domain and hence has zeroes which are not modeled with a single Gaussian function. The lens aperture for this study (f/5.8) resulted in the first zero occurring beyond the folding frequency. The 2-D Gaussian is used here as an initial model to be improved upon in the future.

3 MEM Restoration Technique

The Maximum Entropy Method (MEM) has been used successfully for spectral estimation and image restoration [7, 8]. The image restoration method used here closely follows the work of Cornwell and Evans [9] and Zheng and Basart [10]. It has the advantage of requiring only two 2-D FFTs per iteration. The theory is briefly reviewed here.

We start with a model of the image formation process:

$$\mathbf{g} = \mathbf{H}\mathbf{f} + \mathbf{n}, \tag{1}$$

where $\mathbf{f}$ is the original uncorrupted image, $\mathbf{H}$ is a square matrix containing the point spread function, $\mathbf{n}$ is a white noise sequence, and $\mathbf{g}$ is the measured image. The two-dimensional image is lexigraphically ordered to form the column vector $\mathbf{f}$. We seek an estimate of the original image $\hat{\mathbf{f}}$ that incorporates a minimum of additional configuration or image modelling information. Furthermore, the blurred estimate $\mathbf{H}\hat{\mathbf{f}}$ should be consistent with the original measurement in a squared-error sense. The first condition corresponds to an estimated image with maximal entropy and the second condition corresponds to an estimated image that when blurred closely matches the measurement.

The entropy S of the estimated image is calculated as

$$S = -\sum_{i=0}^{N-1} p_i \ln p_i, \tag{2}$$

where $p_i = \hat{f}_i/\Sigma \hat{f}_i$. The chi-squared statistic χ^2 measures the consistency of the blurred estimate with the measurement:

$$\chi^2 = \| \mathbf{g} - \mathbf{H}\hat{\mathbf{f}} \|^2. \tag{3}$$

In general, χ^2 is defined by weighting each squared difference by $1/\sigma_i^2$, where σ_i^2 is the variance associated with each measurement; here the variance of each element is taken as unity. An objective function Q is formed from the entropy and chi-squared functions by:

$$Q(\hat{\mathbf{f}}) = S(\hat{\mathbf{f}}) - \lambda\chi^2(\hat{\mathbf{f}}) - \mu\Sigma f_i \tag{4}$$

where λ and μ are Lagrange multipliers. The additional constraint $\mu\Sigma f_i$ ensures that the energy of the restored image will be the same as the measured image. This function is minimized by finding an image $\hat{\mathbf{f}}$ that satisfies

$$\frac{\partial Q(\hat{\mathbf{f}})}{\partial \hat{\mathbf{f}}} = 0. \tag{5}$$

The Newton-Raphson method is used to iteratively solve this nonlinear optimization problem. A new estimate of the restored image is found using

$$\mathbf{f}^{(n+1)} = \mathbf{f}^{(n)} - \left(\frac{\partial^2 Q^{(n)}}{\partial \hat{f}_i \partial \hat{f}_j}\right)^{-1} \cdot \frac{\partial Q^{(n)}}{\partial \hat{\mathbf{f}}} \tag{6}$$

$\hat{\mathbf{f}}$ is initialized to a constant valued image using the average value of $\mathbf{g}$. This is the maximum entropy image, with a correspondingly large value of χ^2. Thus λ must start at a small value. As the iteration progresses, λ is increased to approach its final value. Fifteen to thirty iterations are typically required to get an acceptably low value of χ^2. At the end of each iteration, any negative values in $\mathbf{f}^{(n)}$ are clipped to zero. This was found to have a major influence on the convergence of the routine.

4 Assessment of Restoration Results

The capability of MEM to improve resolution of the real-time image was assessed by processing the image of a resolution gauge and comparing the input/output image pair based on visual and quantitative means. Quantitative comparison was made by comparing slice profiles across the gauge elements and examining the increase in spectral response of the fundamental frequency.

The resolution gauge is a square-wave pattern of alternating X-ray opaque and X-ray transparent regions increasing continuously in frequency from 1 to 10 lp/mm. The gauge was placed in contact with the input screen of the image intensifier and was digitized using the same experimental setup used to measure the PSF. A 256×256 subimage was extracted for restoration and is shown in Fig. 4(a). The blurring of the image intensifier is evident in the letters, tic marks, and square-wave pattern, which is visible to about 1.5 lp/mm. This image was processed using 20 iterations and the result is indicated in Fig. 4(b). The blurring has been reduced since the contrast of the gauge pattern is higher and the lettering is easier to read.

The improvement in contrast was assessed quantitatively by first examining slice profiles across the square-wave pattern at 1.0 lp/mm (Fig. 5) and at 1.5 lp/mm (Fig. 6). The output slice is offset in these plots to facilitate comparison. At 1.0 lp/mm the contrast increased by a factor of 3 and the regularity of

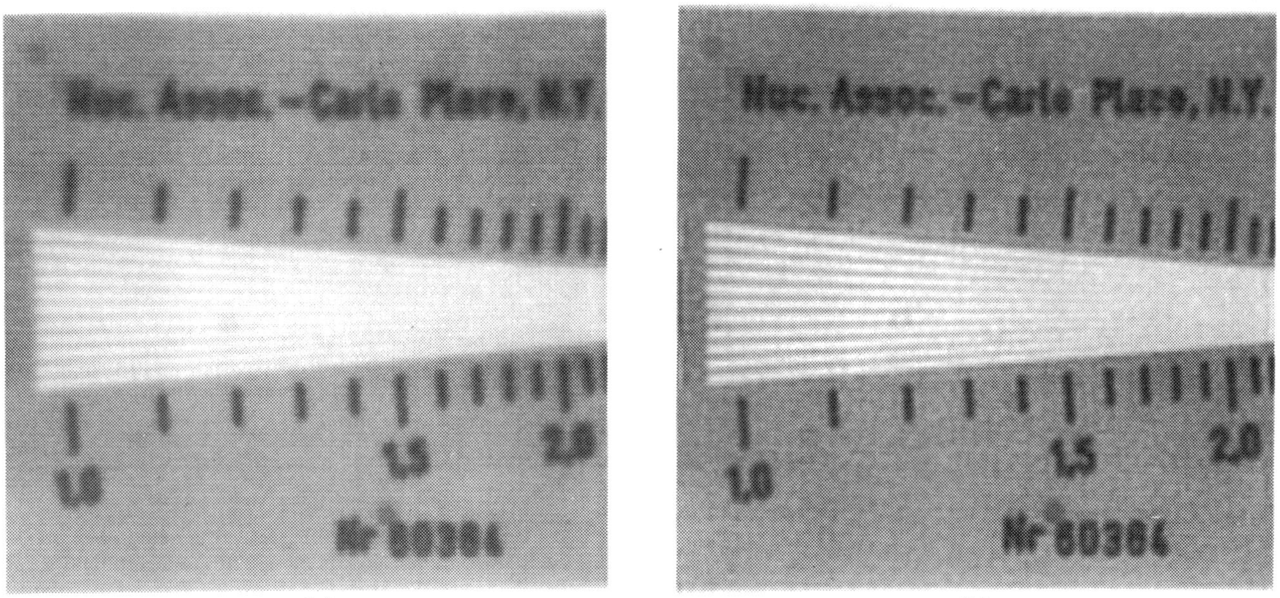

(a) (b)

Figure 4: a) Real-time image of resolution gauge, and b) MEM processed image.

the square-wave signal was improved. At 1.5 lp/mm the contrast improvement was not as significant, but the regularity of the pattern was still improved. MEM smoothed the noise, and did not generate significant artifacts.

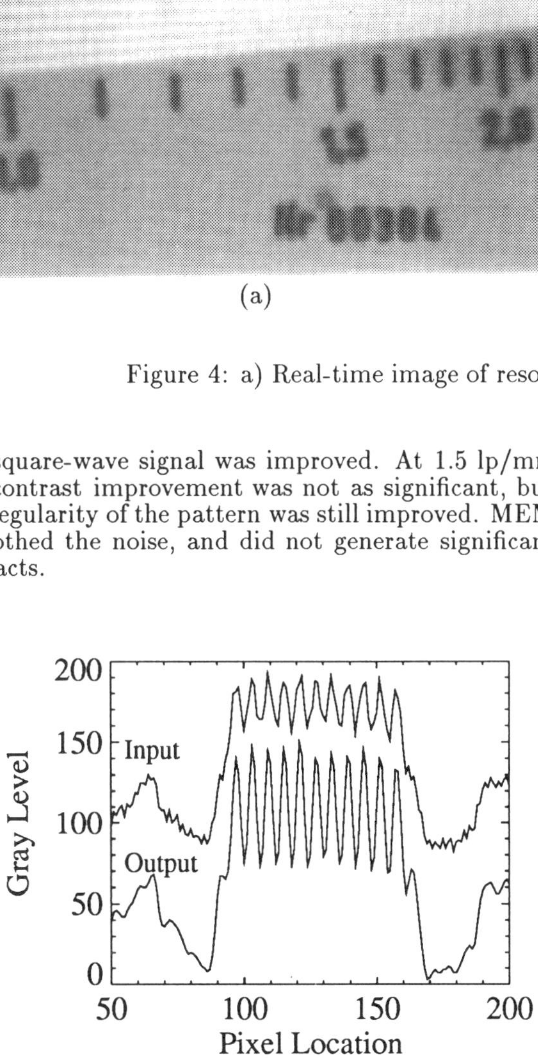

Figure 5: Image profiles at 1.0 lp/mm

Improvement in contrast was also investigated by examining the spectral content of each row slice for the original and processed images. A Hanning window was applied in the spatial domain to isolate the square-wave pattern. This 256 sample record was zero padded to 1024 samples and the 1-D FFT was computed. The amplitude of the fundamental frequency

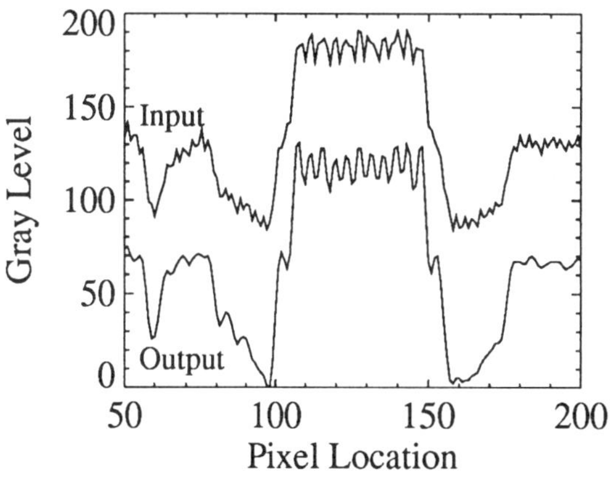

Figure 6: Image profiles at 1.5 lp/mm

was used as a measure of contrast. Fig. 7 shows a plot of amplitude versus frequency for the input and output images. The amplitude was referenced to 0 dB for the input image at 1.0 lp/mm. The MEM processed image had a gain of +8 dB in contrast at 1 lp/mm, a +4 dB gain at 1.5 lp/mm, and broke even at approximately 1.7 lp/mm.

A radiograph was also made of the resolution gauge. The square-wave pattern on the radiograph was visible to 10 lp/mm. This radiograph was digitized using the same sampling frequency and dynamic range as the real-time image of the resolution gauge. Its

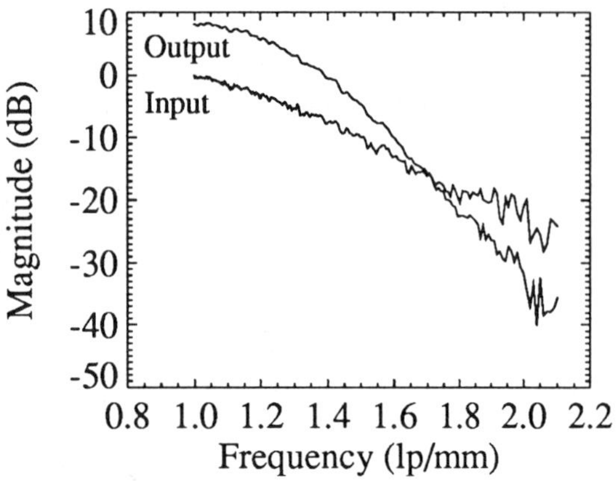

Figure 7: Amplitude of fundamental frequencies for input/output pair

digitized resolution was limited to 1.8 lp/mm. The amplitude of the square-wave pattern for both images at 1.0 lp/mm was 30 gray levels, or 12% of the dynamic range of the 8-bit digitizer. The square-wave amplitude was closer to 80% of the dynamic range at 1.0 lp/mm when the resolution gauge was imaged directly by the video camera. In this image, the square-wave pattern was visible to nearly 10 lp/mm, although aliasing was clearly evident after the folding frequency of 3 lp/mm. This suggests that some of the loss of resolution in the real-time image used for this analysis was due to an insufficient number of gray levels to sample the square-wave pattern. Thus, the success of the MEM restoration at higher frequencies was limited by loss of information in the input image.

5 Summary and Conclusions

The maximum entropy method has been found to improve the resolution of a real-time radiography system by deconvolving the system point spread function. The PSF was measured using a high resolution edge scan followed by numerical differentiation. A Gaussian model was found to fit this data satisfactorily, and was subsequently used as the PSF to be deconvolved by MEM. The MEM image was estimated using a multivariate Newton-Raphson technique. MEM was applied to the image of a resolution gauge, and was found to increase the contrast of the resolution gauge over the visible range of the digitized image by as much as 8 dB. Furthermore, MEM smoothed the image noise, restored the square-wave pattern, and did not generate significant artifacts. This particular MEM implementation is computationally efficient, especially when a hardware accelerator is available to perform the 2-D FFTs.

The relatively low resolution of the digitized resolution gauge radiograph indicated that the video camera and digitizer setup were significant in limiting the potential increase in resolution from MEM restoration. Additional investigation is necessary to determine how increasing the number of quantization levels and sampling frequency would affect the overall performance of this software-enhanced real-time radiography system.

Acknowledgements

This work was sponsored by the NSF Industry/University Cooperative Research Program of the Center for NDE at Iowa State University. The authors also wish to thank Joe Gray for his assistance in collecting data from his X-ray inspection laboratory at the Center for NDE.

References

[1] R. Halmshaw, "An analysis of the performance of X-ray television-fluoroscopic equipment in weld inspection," *Materials Evaluation*, vol. 45, pp. 1298–1302, 1987.

[2] C. S. Williams, *Introduction to the Optical Transfer Function.* New York: Wiley, 1989.

[3] J. Glasser, J. Vaillant, and F. Chazallet, "An accurate method for measuring the spatial resolution of integrated image sensors," in *Image Processing II, SPIE Vol. 1027,* Society of Photo-Optical Instrumentation Engineers, 1988, pp. 40–47.

[4] E. R. Doering, J. Gray, and J. P. Basart, "Point spread function estimation of image intensifier tubes," to be published in *Review of Progress in Quantitative Nondestructive Evaluation,* D. O. Thompson, and D. E. Chementi, Eds. New York: Plenum Press.

[5] P. R. Bevington, *Data Reduction and Error Analysis for the Physical Sciences.* New York: McGraw Hill, 1969.

[6] H. C. Andrews and B. R. Hunt, *Digital Image Restoration.* Englewood Cliffs, NJ: Prentice-Hall, 1977.

[7] S. F. Burch, S. F. Gull and J. Skilling, "Image restoration by a powerful maximum entropy method," *Computer Vision, Graphics, and Image Processing*, vol. 23, pp. 113–128, 1983.

[8] R. A. Gonsalves, J. P. Kennealy, R. M. Korte, and S. D. Price, "FIER: A filtered entropy approach to maximum entropy image reconstruction," in *Maximum Entropy and Bayesian Methods,* Paul F. Fougère, Ed. Boston: Kluwer Academic Publishers, 1990, pp. 369–382.

[9] T. J. Cornwell and K. F. Evans, "A simple maximum entropy deconvolution algorithm," *Astronomy and Astrophysics*, vol. 143, pp. 77–83, 1985.

[10] Y. Zheng and J. P. Basart, "Deblurring NDE X-ray images and infrared images by a maximum entropy deconvolution method," in *Conference Proceedings of the Twenty-first Asilomar Conference on Signals, Systems and Computers,* R. R. Chen, Ed. San Jose, CA: Maple Press, 1988, pp. 72–76.

An Optimal Feature Extraction Method

Jan-Bon Chen Samuel C. Lee

School of EECS
University of Oklahoma
Norman, Oklahoma 73019

Abstract

This paper presents a feature extraction method which can find a minimum set of minimal features for any given set of N binary images (patterns). This method has some similarities to the method of minimizing multiple switching functions. It becomes difficult to apply manually as N becomes large. For large N, a computer implementation of this method is presented. Two examples, one for N = 5 (small N) and the other for N = 10 (large N), are given to illustrate the method. This method can easily be extended and applied to optimal feature extraction of high-dimensional, multi-grey, color images.

1 Introduction

After an extensive literature search, only two articles were found on the subject of the feature extraction for binary patterns [1, 2]. In these articles, two feature extraction algorithms were introduced. One is a sequential algorithm and the other a parallel algorithm. Also was pointed out in [1], the main drawback of these algorithm is that they do not guarantee to yield a solution.

The purpose of this paper is to present a method which will always successfully generate a minimum set of minimal features for a given group of patterns and these given patterns can then be expressed as linear combinations of these extracted features. For example, the four features $F_1, F_2, F_3, and F_4$ of Figure 1(b) constitute a minimum set of features for the five patterns P_1, P_2, P_3, P_4, and P_5 of the same dimension shown in Figure 1(a). Expressed in terms of these four features, the five patterns become $P_1 = [1100]$, $P_2 = [0010]$, $P_3 = [0011]$, $P_4 = [1001]$, $P_5 = [1000]$.

2 The Feature Extraction Method

A binary *pattern* P is defined as a $n \times m$ (binary) cells of 0's and 1's. The cells are numbered by $1, 2, \cdots, n \times m$. The ith cell of a pattern P which is occupied by a 1 is called a feature primitive or simply primitive of P, denoted by p_i as shown in Figure 1(b). By omitting the 0 cells, a pattern can be expressed as $P = \{p_i, p_j, \cdots, p_k\}$ meaning that the ith, jth, $\cdots$, kth cells of P are occupied by 1 and all the remaining cells by 0. Any subset of P, including P itself, is called a feature of P, denoted by F. The features containing a *single* feature primitive $p_1, p_2, \cdots$, or $p_{n\times m}$ are called *primitive features* which are denoted by $f_1, f_2, \cdots$, or $f_{n\times m}$, respectively (see Figure 1(c)). Consider the five 3×3 binary patterns of Figure 1(a). Without loss of generality, the method described below is best illustrated by an example.

Step 1. Construct the primitive table. This table is a tabulation of feature primitives of all the patterns under consideration.

Table 1: The primitive table.

	p_1	p_2	p_3	p_4	p_5	p_6	p_7	p_8	p_9
P_1	1	1	1	1	1		1		
P_2	1	1	1			1			1
P_3	1	1	1			1	1	1	1
P_4	1	1	1	1			1	1	1
P_5	1	1	1	1					

A "1" under column p_i in row P_j means that P_j possesses the feature primitive p_i.

Step 2. Construct the reduced primitive table. From the primitive table, we observe that first three columns are identical. This indicates that the three feature primitives p_1, p_2, and p_3 can be combined into

1058-6393/91 $1.00 © 1991 IEEE

one feature. Denote this composite feature primitive by p_{1-3}. This results in a reduced primitive table.

Table 2: The reduced primitive table.

	p_{1-3}	p_4	p_5	p_6	p_7	p_8	p_9
P_1	1	1	1		1		
P_2	1			1			1
P_3	1			1	1	1	1
P_4	1	1			1	1	1
P_5	1	1					

Step 3. Find the *equivalence* relations among the feature columns of reduced primitive table and determine the Essential and Non-Essential feature columns. The feature column equivalence relation for each feature column is given below.

$$\begin{aligned} col\ p_{1-3} &= col\ p_4 + col\ p_5 + col\ p_6 + col\ p_7 \\ &\quad + col\ p_8 + col\ p_9 \\ col\ p_4 &: None \\ col\ p_5 &: None \\ col\ p_6 &: None \\ col\ p_7 &= col\ p_5 + col\ p_8 \\ col\ p_8 &: None \\ col\ p_9 &= col\ p_6 + col\ p_8 \end{aligned} \quad (1)$$

where *col* stands for column. The feature columns ($col\ p_4, col\ p_5, col\ p_6,$ *and* $col\ p_8$) that cannot be expressed by a "*sum*" of other feature columns are called Essential Feature Columns (EFC) and the others Non-Essential Feature Columns (NEFC). Therefore, we find that the EFC's, and NEFC's of these patterns are as shown in Table 3.

Table 3: The EFC's and NEFC's.

EFC's	NEFC's
$colp_4$	$colp_{1-3}$
$colp_5$	$colp_7$
$colp_6$	$colp_9$
$colp_8$	

Step 4. Remove the NEFC's from the right-hand sides of the *equivalence* relations because they are redundant. The resulting *equivalence* relations for the NEFC's are as follows.

$$\begin{aligned} col\ p_{1-3} &= col\ p_4 + col\ p_5 + col\ p_6 + col\ p_8 \\ col\ p_7 &= col\ p_5 + col\ p_8 \\ col\ p_9 &= col\ p_6 + col\ p_8 \end{aligned} \quad (2)$$

Step 5. Find the "*minimal representations*" of the NEFC's by the EFC's. In order to find a *minimum* set of *minimal* features, the *minimal representations* of the NEFC's in terms of the EFC's are required. Here, by "*minimal* features", we mean that it has the least number of feature primitives. It is easy to see that the solution is:

$$colp_{1-3} = colp_4 + colp_6 \quad (3.1)$$

$$colp_7 = colp_5 + colp_8 \quad (3.2)$$

$$colp_9 = colp_6 + colp_8 \quad (3.3)$$

Step 6. Find the *final primitive table*. The *final primitive table* can be found by removing all NEFC's from *reduced primitive table*. It is shown in Table 4.

Table 4: The final *primitive table* and the coefficient table it implies.

	p_4	p_5	p_6	p_8
P_1	1	1		
P_2			1	
P_3			1	1
P_4	1			1
P_5	1			

$\Downarrow$

	F_1	F_2	F_3	F_4
P_1	1	1	0	0
P_2	0	0	1	0
P_3	0	0	1	1
P_4	1	0	0	1
P_5	1	0	0	0

Step 7. Construct the *minimum* set of *minimal* features. From feature column equivalence relations of (3), we can obtain the following implication relations among the cells of P.

(1) From Eq. (3.1),

a "1" in the 4th cell guarantees a "1" in the 1st, 2nd, and 3rd cells. (4.1)

and

a "1" in the 6th cell guarantees a "1" in the 1st, 2nd, and 3rd cells. (4.2)

(2) From Eq. (3.2),

a "1" in the 5th cell guarantees a "1" in the 7th cell. (4.3)

and

a "1" in the 8th cell guarantees a "1" in the 7th cell. (4.4)

(3) From Eq. (3.3),

a "1" in the 6th cell guarantees a "1" in the 9th cell. (4.5)

and

a "1" in the 8th cell guarantees a "1" in the 9th cell. (4.6)

The number of feature primitives in the final primitive table indicates that the number of minimum features is four. These four features can be obtained from the four primitive features (f_4, f_5, f_6, and f_8) corresponding to the four feature primitives (p_4, p_5, p_6, and p_8) of the final primitive table by using them as four generators and utilizing the above cell implication relations. For example, in constructing the feature F_1 we start with primitive feature f_4 which has a "1" cell at the 4th cell as shown in Figure 2.

Similarly, F_2, F_3, and F_4 can be obtained from f_5, f_6, f_8, respectively. The final set of extracted features are shown in Table 5.

Table 5: The extracted features.

F_1		
1	1	1
1	0	0
0	0	0

F_2		
0	0	0
0	1	0
1	0	0

F_3		
1	1	1
0	0	1
0	0	1

F_4		
0	0	0
0	0	0
1	1	1

3 Computer Implementation

Before the depiction of computer implementation, some notations are necessary to make the job easier. They are listed as follows.

1. The set of patterns which *possess* primitive feature f_i is defined as $C(f_i) = \{P_j | P_j \ possesses \ f_i\}$.
2. A feature f_i is *identical* to the other f_j *iff* $C(f_i) = C(f_j)$.
3. The set $E(f_i)$ of *equivalent* feature columns of a column f_i is defined as $E(f_i) = \{f_j | C(f_i) = \bigcup C(f_j) \ for \ all \ j \neq i\}$.
4. A feature f_i is called a *Essential Feature* (EF) if it has an empty *equivalence*(f_i); otherwise, it is a *Non-Essential Feature* (NEF).

3.1 Algorithm

1. Find $C(f_i)$ for all the *primitive features* f_i according to the given group of N patterns.
2. Combine all the *identical* features to the feature(s) of largest number and remove all the appearances of the features being combined.
3. Find $E(f_i)$ for all the remaining features f_i.
4. Find all the EF's and NEF's; then, delete all the appearances of NEF's from all the $E(NEF)$.
5. For each NEF do the following,
   (a) Find the group of non-empty intersections g_i of $E(NEF)$ with all the patterns; then, find the set of representatives of this group of intersections by the following two-iteration procedure.
      i. Assign $E(NEF)$ to tmp_k. For each EF f_j in the tmp_k, if the intersections of tmp without f_j with g_i for all i are non-empty then delete f_j from tmp_k, where j is in increasing order for the first iteration ($k = 1$) and j in decreasing order for ($k = 2$), if necessary.
      ii. Sum up the number s_k of patterns in $C(EF)$ for all EF's in tmp_k. If s_k equals the number of patterns in $C(NEF)$ tmp_k is the *minimal* representations of NEF; otherwise, do the second iteration.

iii. If the second iteration is done the tmp_k corresponding to the smaller s_k is the *minimal* representations of NEF.

(b) Combine the feature primitives of NEF to each of the EF's in the representations and remove all the appearances of the NEF.

6. Terminate. The resulting EF's are the *minimum* set of *minimal* features for the given set of patterns and the updated patterns contain the coefficients of the linear combinations.

An Example for Large N. An example of ten numerals in 5×5 grid is given in Figure 3(a). By using the above computer implementation of the method, seven features are obtained which are shown in Figure 3(b). Note that these seven features are the *minimum* set of *minimal* features of the given ten patterns.

4 Discussion

Time Complexity. Without loss of generality, let C be the number of cells in a pattern. The time complexities for each step of the algorithm are listed as follows.

Time complexity of Step 1 : $O(CN)$

Time complexity of Step 2 : $O(C^2)$

Time complexity of Step 3 : $O(CN)$

Time complexity of Step 4 : $O(CN)$

Time complexity of Step 5 : $O(C^2N)$

Time complexity of Step 6 : $O(C^2)$

From the list, we see that time complexity of this algorithm is $O(C^2N)$.

N and C. It is possible that the number of final set of features is larger than the number of patterns N if N is smaller than the number of cells C. For example, a group of 3 patterns, P_1 contains p_1 and p_2, P_2 contains p_2 and p_3, and P_3 contains p_3 and p_4, then the number of the resulted features will be 4. Under such circumstances, no feature primitives can be combined; thus, the original patterns have to be used. When $N \leq C$ it is likely to have a smaller number of resulting features depending on how much overlaps of cells exist among the patterns. However, the larger the ratio of N to C is, the more likely the original primitive features will turn out to be the resulting set of features.

5 Conclusion

In this paper an optimal feature extraction method which always yield a minimum set of minimal features for any given set of N binary patterns is presented. Time complexity of the computer algorithm is estimated to be $O(C^2N)$. Because this method begins with C primitive features and tries to combine them the worst case is to end up with the same C primitive features. The larger the ratio of N to C is, the more likely the worst case will occur. Although only two-dimensional were used in illustrating the method, the method is completely general and applicable to high-dimensional, multi-grey, color patterns in a subsequent paper.

References

[1] H. D. Block, N. J. Nilsson, and R. O. Duda, "Determination and Detection of Features in Patterns", in *Computer and Information Sciences – I* (J. T. Tou and R. H. Wilcox, eds.), Spartan Books, Washington, D. C.

[2] J. T. Tou and R. C. Gonzalez, *Pattern Recognition Principles*, Fourth Edition, *Addison Wesley*, 1981, PP. 307.

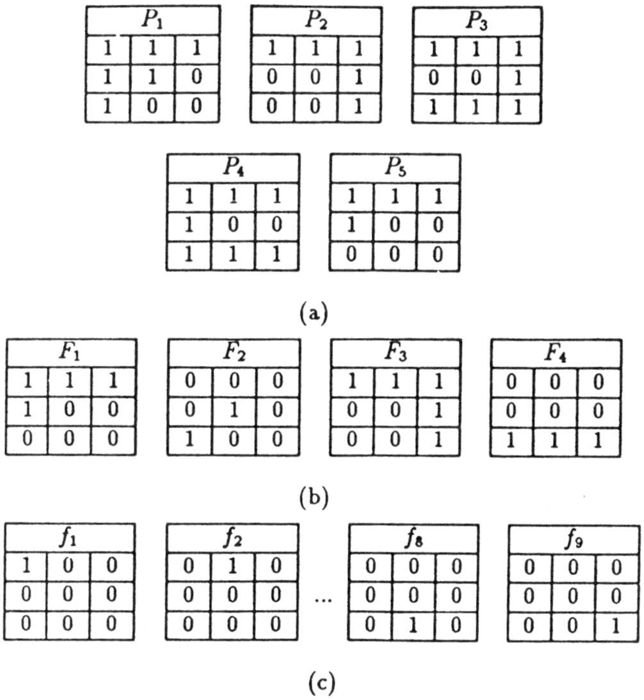

Figure 1: (a) An example of 5 patterns in 3 × 3 grid, (b) Their *minimum* set of *minimal* features, and (c) 9 *primitive features*.

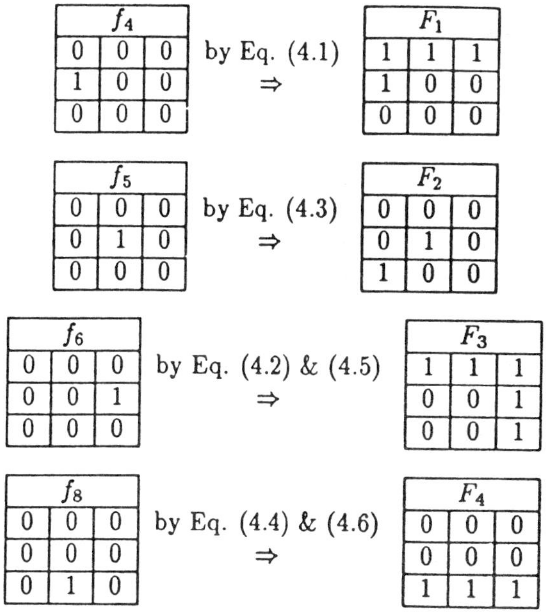

Figure 2: Feature generations by cell implication relations.

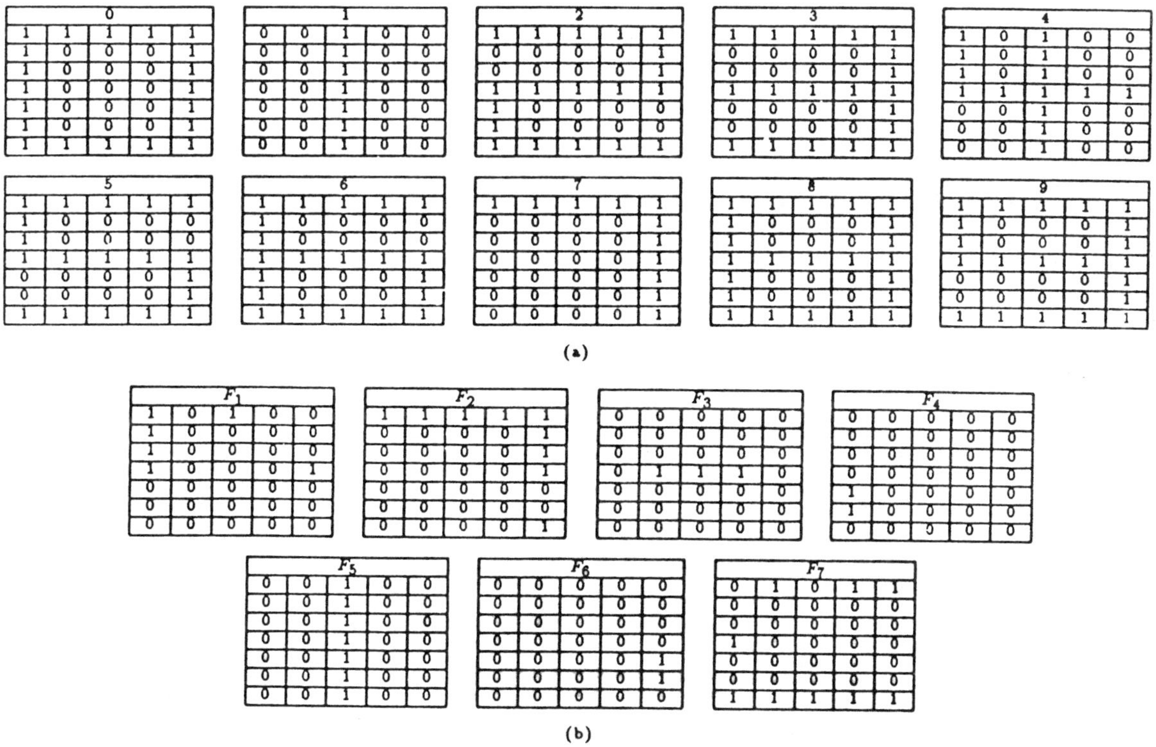

Figure 3: (a) Ten numerals in a 7 × 5 gird (b) Their *minimum* set of seven *minimal* features.

Order Statistic CFAR Detectors for Speckled Area Targets in SAR

James A. Ritcey

Department of Electrical Engineering
University of Washington
Seattle, WA 98195

Huong Du

Department of Electrical Engineering
University of Washington
Seattle, WA 98195

Abstract

Algorithms for the detection of targets in cluttered environments are often required to operate at a constant false alarm rate (CFAR), necessitating an adaptive thresholding approach. We have extended the standard Goldstein's two-parameter detector, the Log T-Test (LTT), and developed a Robust-LTT (RLTT), using order-statistic estimators for the background moments. The LTT uses ML estimates of the local mean and variance, which we replace by the sample median and a reduced range. In both tests, a pixel-under-test is compared to an adaptive threshold composed of these estimators.

The test can be designed to detect both point and area targets. For use with Synthetic Aperature Radar (SAR) images, and because of the interest in speckled area targets, we have developed a block test, which uses the median of a block of pixels-under-test. To evaluate performance we use simulation on a clutter-only SAR image. The method is also evaluated in a quantitative way, using a Receiver Operating Characteric (ROC). An approach to estimate the P_{fa} and P_d using the simulation data is presented. The results show that the RLTT successfully discriminates area targets against the point targets in clutter background.

1 Adaptive Thresholding for Radar CFAR Detection

Modern radar surveillance signal processing for detection is a multi-stage operation in which raw radar data is progressively culled into target tracks with known identities. This progression is necessary to reduce the enormous memory that would be required if all relevant data, in time and space, was made available for automatic decision making. The process in which coherent radar data is culled into threshold crossings or "hits", is crucial as the first data reduction step in any detection and tracking process. This is the function of the CFAR (constant false alarm rate) Detection Processor.

The CFAR detection problems we consider are two-sample situations where reference samples are available, often consisting of resolution cell variates (pixels) $z_1, z_2, \ldots, z_n$ surrounding the pixel-under-test variate, V. The mean-level detector (MLD) is the simplest CFAR technque, but only normalizes the data against a shift in the background location, or mean level. The MLD uses the sample mean $r = \sum_{j=1}^{n} z_j/n$ and implements the adaptive threshold test $V \gtrless R(a) = ar$, rejecting the null hypothesis (H_0) of noise alone when the threshold is exceeded. In this way, a target is cued for further image processing.

The MLD can normalize only a mean shift in the data, although for Gaussian noise and square-law detection, leading to an exponential distribution at the CFAR subsystem, this is adequate. In practice, two-parameter approaches based on that of Goldstein [2] are often required. An important *first step is log-detection* ($X = \log(x)$), converting a scale-shape estimation problem to a location-scale problem, thus allowing standard statistical normalization techniques ($(x-\mu)/\sigma$). Next, the data surrounding the cell-under-test is collected, and *mean and variance estimators determined* ($\hat{\mu}$, $\hat{\sigma}$). Finally, a studentized variate is compared to a threshold ($(x-\hat{\mu})/\hat{\sigma} > a$). This Goldstein scheme maintains CFAR for the Weibull and other two parameter families of clutter distributions, given independent reference samples. Outlier resistance remains a distinct problem.

We refer to the Goldstein approach as the *Log-T Test or LTT*. Given any log-detected pixel-under-test sample V, the detector decides for a target when $T_{LTT} > a$ where the sample mean $\hat{\mu}$ and sample standard deviation $\hat{\sigma}$, computed from the log-detected pixels falling in the reference window, are combined into

1058-6393/91 $1.00

a test statistic through

$$T_{LTT} := \frac{V-\mu}{\sigma}.$$

In this paper, we study the use of robust estimators for μ and σ.

1.1 Order-Statistic CFAR Detectors

One promising approach to outlier resistance is to use estimators based on *order statistics*(*OS*), in place of the sample moments. The order statistics of the reference samples are the sorted data values $z_{(1)} < z_{(2)} < \ldots < z_{(n)}$. In a multiple target environment, the censored mean-level detector (CMLD), which uses the estimator $r = \sum_{j=1}^{k-1} z_{(j)}/k + (n+1-k)z_{(k)}/k$, censoring the largest $(n-k)$ samples, has been advocated in Ritcey [1]. Rohling [3] suggests the OS-CFAR, using a single o.s. $r = z_{(k)}$, which suffers some loss in performance relative to the CMLD, but has advantages in clutter.

Order statistics can be used to form outlier resistant estimators of the variance as well. Here we employ the reduced range, $r_{RRNG} := z_{(k2)} - z_{(k1)}, 1 \leq k1 < k2 \leq n$. Operating sequentially on a one-dimensional data stream, the reduced range estimator acts as an edge detector. The resulting mean and variance estimators are used in the Goldstein CFAR detector structure, which we denote the Robust *LTT*, the *RLTT*.

Under the Weibull clutter model, Weber and Haykin [4] and have proposed order-statistic-based two-parameter CFAR detectors. We have compared our approach against this two-*OS* scheme, and have found that the three *OS* method is superior. For that reason, we do not consider the Weber and Haykin approach.

2 The Robust Log T-Test

2.1 The ADTS Image Data Set

The simulation study has been carried out on imagery collected by *MIT* using the Advanced Detection Technology Sensor (*ADTS*), a fully polarimetric 35 GHz, synthetic aperature radar with 1ft by 1 ft resolution. *ADTS* gathers data coherently at *HH*, *HV*, and *VV* polarizations. For this study we have used only the single channel $|HH|^2$ data, blurred with a 4 : 1 moving average filter and subsampled. Image size is 512 by 512 pixels.

2.2 Basic Definitions and Descriptions

Consider an square image of pixels within which background clutter statistics are gathered from a moving *reference window* of size W by W pixels. At the center of the window, a *test block* of size B by B is tested for the presence of a target. By scanning the window over the entire image, all pixels are eventually tested for target presence. In all cases, the image is preprocessed by computing the log-intensity of each pixel. The block pixels are processed by selecting a single order statistic to measure local target intensity, which is then compared to an adaptive threshold.

The original image contains only background or clutter. To test the algorithms more completely, synthetic targets are placed in the original scene. This synthetic target is composed of a squares whose sides are on the order of $T = 12$ pixels. To model glint, only a percentage p, of the T^2 target pixels are active, or bright. Typically, $p = 0.16$; 16% of the 144 target pixels have been set to the maximum brightness within the scene.

Let W be the dimension of the square solid reference window. *All* pixels within the window are used for background estimation. Note that this includes the pixel (or block) under test! We allow this possibility for two reasons: first, the solid window is more efficient in that it can be much smaller than a hollow window, or stencil, for the same number of reference pixels. Second, when outlier resistant *OS* -based estimators are employed, guard pixels are not be necessary, as long as the number of target pixels falling within the reference does not exceed the breakdown point of the estimator. Ideally, the reference window should contain only background pixels relevant to the pixel-under-test block under clutter-only conditions. This is true even when outlier resistant estimators are used.

The test we use declares for a target at the center of block B when

$$T_B := \frac{\hat{B} - z_{(\tilde{q})}}{z_{(\tilde{t})} - z_{(\tilde{l})}} > a.$$

Nominally, $\hat{B}$ is the median of the test block, $\tilde{l} = .25W^2$, $\tilde{q} = .5W^2$,and $\tilde{t} = .75W^2$. We have experimented with other choices, although this selection gives the best performance. The $z_{(n)}$ are the order stastistics from all data within the reference window.

3 Detection Analysis

The test is of the generic form

$$t = \frac{V - \hat{\mu}}{\hat{\sigma}} \underset{H_0}{\overset{H_1}{\gtrless}} a \tag{1}$$

where V is a function of the block os pixels-under-test, $\hat{\mu}$ and $\hat{\sigma}$ are the mean and standard deviation estimators from gathered from the reference window; a is a threshold parameter. An equivalent test is

$$V \underset{H_0}{\overset{H_1}{\gtrless}} R(a), \quad R(a) := a\hat{\sigma} + \hat{\mu}$$

To evaluate the detection performance of RLTT, we need to find the probability of false alarm (P_{fa}) and probability of detection (P_d). Without assumptions om the distribution of the samples, we find P_{fa}, P_d based on the empirical samplestudies from our image.

We determine P_{fa}, P_d based on kernel estimators of the distributions, and only cite the results here. Let $k(x)$ be the kernel, a pdf, $K(x)$ the cdf, and $\bar{K}(x) := 1 - K(x)$, the edf. We use a Gaussian smoothing kernel, $N(0, \sigma_s^2)$.

The kernel-based density estimator of any iid sample, $x_1, \ldots, x_N$ is given by $\hat{p}(x) := (1/N)\sum_n x_n \delta(x - x_n) \star k(x)$, the convolution of the empirical density function and the smoothing kernel. From the simulated data, we gathered these estimators for both the block medians,$\hat{p}_V(V)$, and the reference, $\hat{f}_(R)$. Note that this is a function of the threshold parameter a. We evaluate detection probabilities,

$$Q(a) = PV > R(a) = \int f(R)\bar{P}_V(R)dR,$$

in terms of the empirical densities. Letting V_n, R_n be the respective samples,

$$P_{fa} = \frac{1}{N}\sum_{n=1}^{N} \bar{K}\left(\frac{R_n - V_n}{\sqrt{2}\sigma_s}\right). \tag{2}$$

Similarly, for P_D, we find

$$P_d = \frac{1}{N}\sum_{n=1}^{N} \bar{K}\left(\frac{R_n - M}{\sqrt{S + \sigma_s^2}}\right). \tag{3}$$

Here the smoothing variance is σ_s^2, the reference samples R_n, and test pixel samples, V_n. For lack of data on true targets, we have used the theoretical model that the log-intensity is $N(M, S)$.

4 Simulation Results

In this section, we apply this methodology of to evaluate the performance of the RLTT for targets with log-normal intensity. All simulations are carried out on the ADTS image. We use the block size $B = 2$, and select the median of the block as the test pixel. From experimentation, the smoothing coefficient is selected as $\sigma_s = 0.04$.

In figure 1, we show the plot of P_{fa} as a function of the threshold for various solid window sizes, $W = 25, 16, 10, 5$. Observe that the range of the threshold is $-4 \leq a \leq 8$. A negative threshold occurs for $P_{fa} \geq 0.5$. This adversely effects the performance at high P_{fa}, but this is of no consequence in practice where $P_{fa} \leq 0.1$.

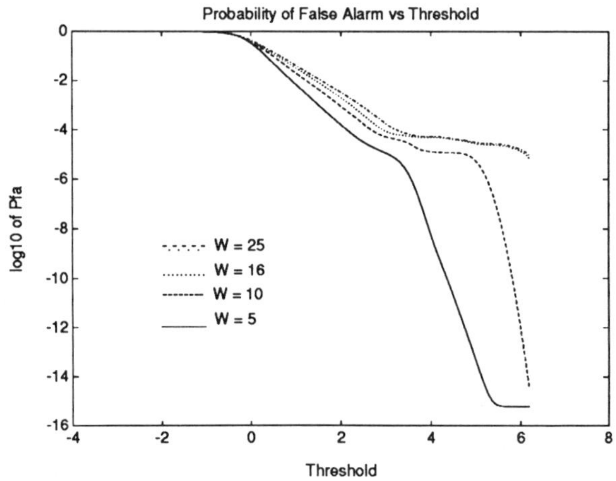

Figure 1: Probability of False Alarm versus Threshold for $W_S = 5, 10, 16, 25$

Uniformily Resolved Targets. To study P_d, we use the target intensity $M = 0.25, 0.75$, corresponding to the signal to clutter ratios, $SCR = 7.64, 9.35$ dB. The SCR is defined as $SCR = 10\log_{10}((M - \mu_C)^2/\sigma_C)$ where μ_C and σ_C denote the mean and standard deviation of the clutter background computed from the ADTS image after preprocessing with the log-detector. In these results, we take the target variance to be zero, $S := 0$. We refer to this as a target model of *uniform resolution* (no glint). Figure 2 shows P_d vs. SCR. The solid, dashed and dotted lines denote each probability of detection corresponding to the probability of false alarm, $P_{fa} = 10^{-4}, 2.5 \times 10^{-5}, 10^{-5}$. For this ADTS image, with $P_{fa} = 10^{-5}$, $P_d = 0.8$, the SCR required is 12 dB.

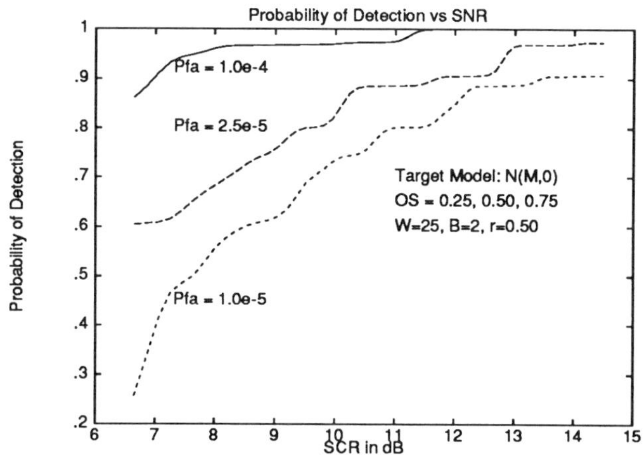

Figure 2: Probability of Detection versus SCR for target with uniform resolution, Gaussian distribution $N(M, 0.0)$

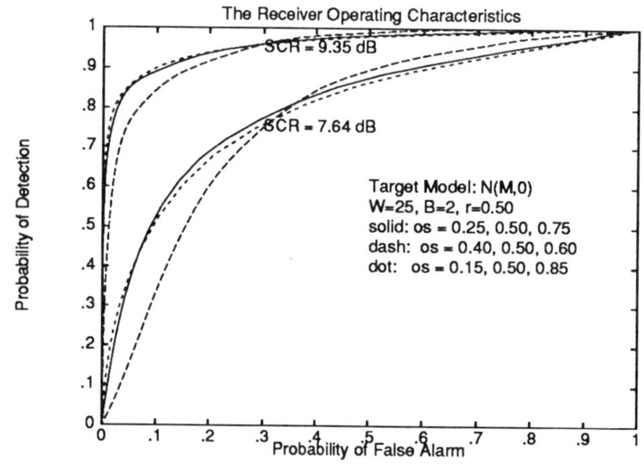

Figure 3: Plot of the Receiver Operating Characteristics for different selections of OS in the reference window, target with uniform resolution, N(M,0)

To study P_d more completely, we construct the ROC curves for $SNR = 7.64, 9.35$ dB. The ROC curve for different values of OS is illustrated in figure 3. The solid line denotes our choice of $OS = \{0.25, 0.50, 0.75\}$, the dashed line is for $OS = \{0.40, 0.50, 0.60\}$, and finally, the dotted line gives the result for $OS = \{0.15, 0.50, 0.85\}$. We observe that both first and second choices give high and comparable values for P_d. $OS = \{0.15, 0.50, 0.85\}$ is not a best selection, it is necessary to increase the window size since to the third percentile used is quite high (0.85). In general, the full range is a more efficient estimator of standard deviation, or spread, tha any reduced range. However, this presumes an iid sample. In out situations, the outlier resistance of the reduced range is more important. The effect on the required threshold parameter is also important.

In figure 4, we show the plot of ROC for different window sizes. The lines from solid to dotted correspond to the ROC for window size $W = 5, 10, 16, 25$. At a given value of P_{fa}, the small window gives high P_d, especially at low SCR. Thus, a small window has better performance than a large window. This is unexpected in the sense that theoretically, the estimators should improve as the sample size increases. Practically, and even despite the use of outlier resistant estimators, it is crucial to maintain a homogeneous reference sample. As the window grows ($W = 25$ is 625 reference pixels), the rate of improvement in efficiency decreases. More important is the fact that we can handle only a 25% level of contamination. This is much more likely with a large window. The design rule is to choose the reference window size large enough to stabilize the variance, but not much larger.

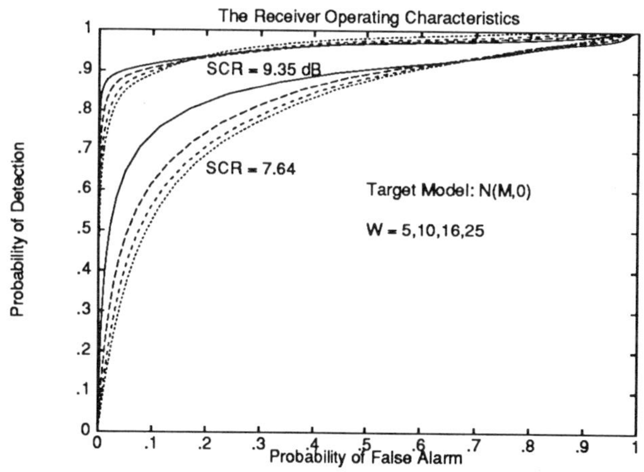

Figure 4: Plot of the Receiver Operating Characteristics for different solid window sizes, target model of N(M,0.0)

Speckled Targets. The same scheme is repeated for targets with the Gaussian log-intensity model, $N(M, 0.50)$. In figure 5, we show the plot of P_d versus SCR. Figure 6 illustrates the ROC curves for various W and choices of the OS's. Again, figure 6 shows that our selection of the order statistic gives high P_d. Comparing with uniform resolution target, N(M,0.0), P_d is lower for the same SCR. This can be simply explained that when $S \neq 0$, there are fewer active pixels. Hence, to achieve high P_d, a target can be treated as a glinting one. Eventually, we need to select the order statistics in the block test with $r > 0.50$. As the result, more bright pixels are deleted, and detection performance is improved.

Notice that in both cases of target distribution, the ROC curves eventually cross. Theoretically, this would indicate a major problem. The curves show abnormal behavior, $P_{fa} > P_d$, especially in the case of N(0.25,0.50). However, these results are invalid, due to the negative threshold at high P_{fa}. In any event, one would never operate with $P_{fa} > P_d$, so, this is of no practical consequence.

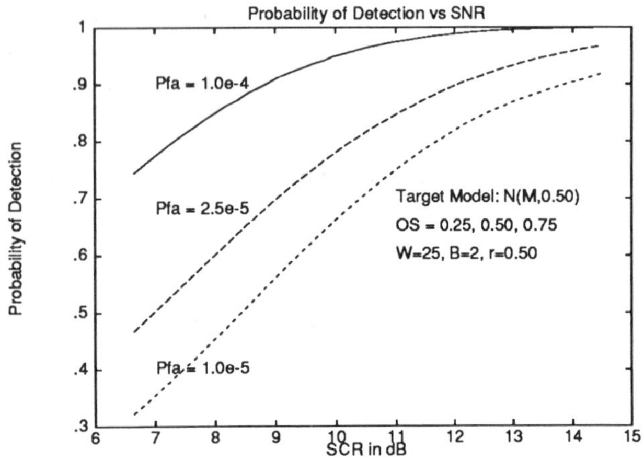

Figure 5: Probability of detection versus SCR for target with Gaussian distribution, N(M,.50)

5 Summary

We have introduced and analyze the performance of the RLTT. The expression for the probability of false alarm and detection are developed, using a Kernel Density Estimate. A combined simulation analytic methodology is used to evaluate the performance of the proposed algorithm. For targets with a Gaussian log-intensity distribution, the detection probability versus SCR, and ROC curves for different choices the algorithm parameters are presented. We study the performance for the RLTT with synthetic targets, especially glinting area targets. The results show that the RLTT successfully discriminate area targets against the point targets in clutter background. The RLTT performs well in the multiple target environments. The methodology and results are useful in

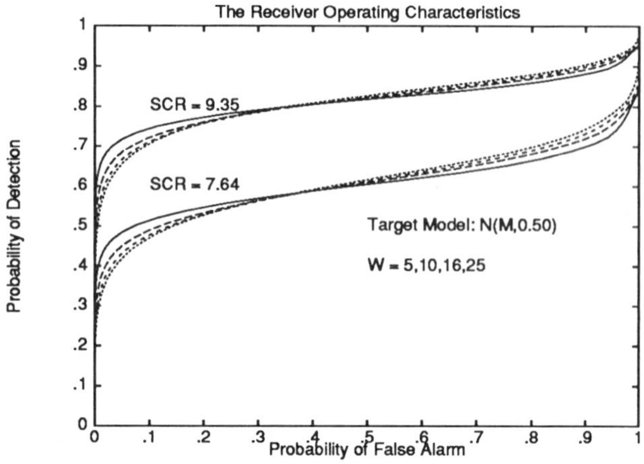

Figure 6: Plot of the Receiver Operating Characteristics for different window sizes, target with Gaussian distribution, N(M,0.50)

modeling target acquisition performance in automatic target recognition systems.

References

[1] J.A. Ritcey, "Performance Analysis of the Censored Mean Level Detector", IEEE Transactions on Aerospace and Electronic Systems, July, 1986.

[2] G.B. Goldstein, "False alarm regulation in lognormal and weibull clutter", IEEE Transactions on Aerospace and Electronic Systems, AES-9, Jan., 1973, pp. 84-92.

[3] H. Rohling, "Radar CFAR thresholding in clutter and multiple target situations", IEEE Transactions on Aerospace and Electronic Systems, AES-19, Sept. 1983, pp. 608-621.

[4] P. Weber and S. Haykin, "Ordered Statistic CFAR Processing for Two-Parameter Distributions with Variable Skewness", IEEE Transactions on Aerospace and Electronic Systems, AES-21, November, 1985, pp. 819-821.

Acknowledgements

The authors would like to acknowledge the support of Dr. Les Novak and MIT Lincoln Lab for their support. This idea was developed in a proposal submiited to Lincoln in October, 1989.

Transient Signal Detection using the Wigner Bispectrum

Javier Rodríguez Fonollosa and Chrysostomos L. Nikias

Signal and Image Processing Institute
University of Southern California
EEB 400
Los Angeles CA 90089-2564

Abstract[1]

Wigner Higher-Order Spectra (WHOS) have been recently defined as extensions of the Wigner Distribution (WD) to higher-order moment spectra domains. A General Class of Time-Frequency Higher-Order Spectra has also been defined in terms of arbitrary higher-order moments of the signal, as generalizations of Cohen's General Class of time-frequency representations. Discrete Time and Frequency Wigner Higher-Order Spectra (DTF-WHOS) expressions have been derived for signal processing applications. Furthermore, the Wigner Bispectrum and Wigner Distribution, particular cases of WHOS for order two and one respectively, have been tested in the context of transient signal detection. The purpose of this paper is to present a new scheme, based on a generalized maximum likelihood ratio theory and the Wigner Bispectrum, for the adaptive detection of transient signals in low SNR.

1 Introduction

Time-frequency distributions are a useful approach to non-parametric analysis of time-varying processes. An excellent review of time-frequency distributions is given in [1] and its applications range from speech analysis and recognition to sonar and radar signal processing. However, Cohen's General Class of time-frequency representations, which include among others the Spectrogram and the Wigner Distribution (WD), are all bilinear with respect to the signal. A definition of time-frequency representations in terms of multilinear expressions, i.e. higher-order moments of the signal, would extend the analysis properties of Cohen's General Class to the time-varying moment spectra domain. This was the main motivation for the formulation of a Wigner Higher-Order Spectra (WHOS). If an appropriate definition is used, the properties of the WD will be preserved or extended to the higher-order domain. Thus, the properties related to the instantaneous power or spectral density function in the WD should be related in WHOS to $(k+1)^{th}$-order moments and $(k+1)^{th}$-order moment spectra. The definition and many properties of WHOS have been reported in [2, 3]. In this paper, we review the definition of WHOS, compare the moment-oriented definition of WHOS with a cumulant-based definition, apply the Wigner Bispectrum (particular case of WHOS for order $k=2$) in the problem of transient signal detection. Finally Reduced Interference Higher-Order Distributions are introduced.

2 Continuous Wigner Higher-Order Spectra (WHOS)

2.1 Definition

Let $x(t)$ be a complex *deterministic* signal, the WHOS of order k of $x(t)$, $W_{kx}(t,f)$, is defined as,

$$W_{kx}(t,f)=\int_{\tau} R_{kx}(t,\tau) exp(-j2\pi f^T\tau)\, d\tau , \quad (1)$$

where $\tau = [\tau_1 ,..., \tau_k]^T, f = [f_1 ,...,f_k]^T, d\tau = d\tau_1 ... d\tau_k$,

$$R_{kx}(t,\tau)= = x^*\left(t-\frac{1}{k+1}\sum_{m=1}^{k}\tau_m\right)\prod_{i=1}^{k} x\left(t+\frac{k}{k+1}\tau_i-\frac{1}{k+1}\sum_{j=1,j\neq i}^{k}\tau_j\right)$$

For k=2, the Wigner Bispectrum is obtained,

$$W_{2x}(t,f_1,f_2)=\int_{\tau_1}\int_{\tau_2} x^*\left(t-\frac{1}{3}\tau_1-\frac{1}{3}\tau_2\right) x\left(t+\frac{2}{3}\tau_1-\frac{1}{3}\tau_2\right)x\left(t+\frac{2}{3}\tau_2-\frac{1}{3}\tau_1\right) exp(-j2\pi f_1\tau_1)\, exp(-j2\pi f_2\tau_2) d\tau_1 d\tau_2 . \quad (2)$$

And the Wigner Distribution follows for k= 1.

2.2 Properties

Most of the properties of WHOS were reported in [2, 3]. Among others, we can mention that integration over frequency (time) gives the instantaneous $(k+1)^{th}$-

[1]This work was supported by the Office of Naval Research under contract N00014-91-J-1034 and the Spanish Ministry of Education and Science.

1058-6393/91 $1.00 © 1991 IEEE

order moment ($(k+1)^{th}$-order spectrum), the instantaneous frequency can be derived, and it satisfies Moyal's formula relating inner products in the time domain with inner products in the time-multifrequency domain.

2.3 Higher-order moment versus cumulant spectra

Most time-frequency representations, including Cohen's general class, are normally defined for non-random signals only. In fact deterministic signals are often the desired signal in many of the applications of time-frequency distributions. For this particular kind of signals, such as energy signals and periodic signals, the higher-order *moment* spectra are defined as the multidimensional Fourier transform of the higher-order *moment* sequence [4], i.e.,

$$r_{kx}(\tau) = \int_t x^*(t) \prod_{i=1}^{k} x(t+\tau_i)\, dt; \qquad (3)$$

$$M_{kx}(f) = \int_\tau r_{kx}(\tau)\, exp\,(-j2\pi f^T \tau)\, d\tau =$$

$$= X^*\left(\sum_{i=1}^{k} f_i\right) \prod_{j=1}^{k} X(f_j), \qquad (4)$$

The definition of WHOS, only for deterministic signals, is consistent with this formulation. Let us consider, however, the stochastic signal case and compare a moment versus a cumulant-based definition of WHOS. For a complex random process $\{x(t)\}$, non-stationary in general, we define,

$$\boldsymbol{x} = [x^*(t-\alpha),\ x(t+\tau_1-\alpha),\dots,x(t+\tau_k-\alpha)]^T.$$

$$\alpha = \frac{1}{k+1}\sum_{i=1}^{k}\tau_i.$$

Higher-order cumulants and moments are denoted by,

$$cum\,\{\boldsymbol{x}\} = cum\,(t,\tau_1,\dots\tau_k) = cum\,(t,\boldsymbol{\tau}),$$

$$m\,\{\boldsymbol{x}\} = m\,(t,\tau_1,\dots\tau_k) = m\,(t,\boldsymbol{\tau}).$$

The cumulant-based WHOS, recently proposed in [5], is defined as,

$$C\_WHOS(t,f) = \int_\tau cum\,(t,\tau)\, exp\,(-j2\pi f^T \tau)\, d\tau, \qquad (5)$$

whereas the moment-based WHOS or expectation of the WHOS defined in [2, 3] as,

$$M\_WHOS\,(t,f) = E\,\{W_{kx}(t,f)\} =$$

$$= \int_\tau m\,(t,\tau)\, exp\,(-j2\pi f^T \tau)\, d\tau.. \qquad (6)$$

Some of the basic properties of $C\_WHOS\,(t,f)$ versus $M\_WHOS\,(t,f)$ are the following:

P1: If $\{x(t)\}$ and $\{y(t)\}$ are independent random processes and $\{z(t)\}$ is defined by $z(t) = x(t) + y(t)$ then,

$$C\_WHOS_z\,(t,f) =$$
$$= C\_WHOS_x\,(t,f) + C\_WHOS_y\,(t,f); \qquad (7)$$

$$M\_WHOS_z\,(t,f) \neq$$
$$\neq M\_WHOS_x\,(t,f) + M\_WHOS_y\,(t,f). \qquad (8)$$

This property follows due to the separation property of cumulants of independent random variables that does not hold for moments.

P2: If $\{x(t)\}$ is complex jointly Gaussian with zero mean and with independent and equally correlated real and imaginary parts (stationary or non-stationary) then ,

$$C\_WHOS_x\,(t,f) = M\_WHOS_x\,(t,f) = 0,$$
$$\text{for } k > 1. \qquad (9)$$

Of course, $C\_WHOS_x\,(t,f)$ is also zero for any Gaussian process and for $k > 1$ since all cumulants are zero.

P3: If $\{x(t)\}$ is deterministic then,

$$C\_WHOS_x\,(t,f) = 0;$$
$$M\_WHOS_x\,(t,f) = W_{kx}\,(t,f). \qquad (10)$$

Deterministic signals can be regarded as Gaussian signals with zero variance and thus are suppressed by the cumulant-based definition. Although additive Gaussian signals are suppressed using the cumulant-based definition, deterministic signals suffer the same cancellation. Using the definition based on moments, one important class of Gaussian noise, stationary or not, is also suppressed in the absence of any other signal. Furthermore, the presence of deterministic signals can be detected using M_WHOS, and it is exactly this property that we exploit in the detection of transients in noise.

3 Discrete Time and Frequency Wigner Higher-Order Spectra

The formulation of the Discrete Time and Frequency WHOS, (DTF-WHOS) was given in [2]. The way in which this definition is related to the WHOS of the signal before sampling, and consequently, the non-aliasing constrains were also discussed. The main result was to show that the DTF-WHOS could be computed by means of N k-dimensional FFTs of order N and that aliasing was avoided sampling at twice Nyquist's frequency *for any value of k.*

4 Detection of transient signals

In [2] we developed a transient detection scheme and applied it with samples of the Wigner Bispectrum and Wigner Distribution. It was shown that the detection statistic for the samples of the Wigner Bispectrum was less sensible to noise than the one based on the Wigner Distribution. However, this scheme was defined in an ad hoc fashion and, thus, no claims of optimality were made at that point. Its main result was to show the possibilities of the Wigner Bispectrum used in transient detection schemes and to motivate the definition of a new scheme to exploit the possibilities of this transformation. The Gabor Representation was employed in [6] for the detection of transient signals in noise. The statistics of the Gabor coefficients were studied and a maximum likelihood ratio was defined. In [3] we proposed a method, in the spirit of [6], where a maximum likelihood ratio was defined over the samples of the Wigner Bispectrum. In this paper we consider the properties of the moment-based definition of WHOS to detect an analytic deterministic signal embedded in complex Gaussian noise using the scheme proposed in [3].

4.1 Detection using the Wigner Bispectrum: Generalized Maximum Likelihood ratio.

A transient signal is a signal whose duration is short compared to the observation period. Usually transients are detected using one filter and comparing the output to one threshold. If the threshold is exceeded then, the presence of one transient is assumed. If the transient waveform is known, a matched filter should be used: If not, different approaches have to be investigated. We propose to define a generalized likelihood test on the distribution of the Wigner Bispectrum coefficients.

Transient detection is a binary detection problem. It has to be determined, every certain period of time, if a transient signal is present (H_1) or not (H_0). Accordingly, the received signal is:

$$H_0 : x(n) = w(n), \tag{11.a}$$

$$H_1 : x(n) = s(n) + w(n), \tag{11.b}$$

where $w(n)$ is a stationary noise process (at least over the observation period) and $s(n)$ is a transient signal.

Our observation space will consist on the coefficients of the Wigner Bispectrum for any particular temporal sample. They are, thus, a non-linear mapping of the received signal $x(n)$ through the definition of the DTF-WHOS for k=2 [2]. Our fundamental assumptions are the following:

- Under H_0 the coefficients of the Wigner Bispectrum at every particular temporal sample are N^2-dimensional Gaussian random vectors, and their components are statistically independent Gaussian random variables with unequal mean and variance.

- Under H_1 the coefficients of the Wigner Bispectrum at every particular temporal sample are N^2-dimensional Gaussian random vectors which statistics only differ in mean from the one under H_0. The mean of each coefficient is assumed to be equal to its mean under H_0 plus the actual value of that coefficient in the absence of noise.

N is the order of the DTF-WHOS. Defining the N^2-dimensional vectors $\boldsymbol{r}(n)$, $\boldsymbol{m}_0$ and $\boldsymbol{m}_1$ as:

$$\boldsymbol{r}(n) = \boldsymbol{W}_{2x}(n, k_1, k_2);\ k_1, k_2 = 0, 1, \dots N\text{-}1; \tag{12}$$

$$\boldsymbol{m}_0 = \mathrm{E}\{\, \boldsymbol{r}(n) \mid H_0\}; \tag{13}$$

$$\boldsymbol{m}_1 = \mathrm{E}\{\, \boldsymbol{r}(n) \mid H_1\}. \tag{14}$$

And the N^2-dimensional diagonal matrix $\boldsymbol{D}$,

$$\boldsymbol{D} = Cov\,(\boldsymbol{r}(n) \mid H_0) = Cov\,(\boldsymbol{r}(n) \mid H_1)\,. \tag{15}$$

The log-probability density function for both hypotheses is:

$$2 \log f_0(\boldsymbol{r}(n)) = -N^2 \log 2\pi - \log\det(\boldsymbol{D}) - (\boldsymbol{r}(n) - \boldsymbol{m}_0)^H \boldsymbol{D}^{-1} (\boldsymbol{r}(n) - \boldsymbol{m}_0); \tag{16.a}$$

$$2 \log f_1(\boldsymbol{r}(n)) = -N^2 \log 2\pi - \log\det(\boldsymbol{D}) - (\boldsymbol{r}(n) - \boldsymbol{m}_1)^H \boldsymbol{D}^{-1} (\boldsymbol{r}(n) - \boldsymbol{m}_1); \tag{16.b}$$

We consider the generalized likelihood ratio:

$$T = max_{\boldsymbol{m}_1}\{\, 2 \log f_1(\boldsymbol{r}(n))\,\} - 2 \log f_0(\boldsymbol{r}(n)) = 2 \log f_1(\boldsymbol{r}(n))\Big|_{\boldsymbol{m}_1 = \boldsymbol{r}(n)} - 2 \log f_0(\boldsymbol{r}(n)) = (\boldsymbol{r}(n) - \boldsymbol{m}_0)^H \boldsymbol{D}^{-1} (\boldsymbol{r}(n) - \boldsymbol{m}_0) \tag{17}$$

For any desired probability of false alarm, the distribution of T under H_0 determines the threshold. The distribution of T under H_1 permits the determination of the detection probability as a function of the coefficients of the Wigner Bispectrum of the transient signal. In our case, T under H_0 is a central chi-square with N^2 degrees of freedom and T under H_1 is a non-central chi-square with non-centrality parameter $\alpha = (\boldsymbol{m}_1 - \boldsymbol{m}_0)^H \boldsymbol{D}^{-1} (\boldsymbol{m}_1 - \boldsymbol{m}_0)$ and N^2 degrees of freedom.

4.2 Estimation of the mean vector m_0 and the covariance matrix D.

A simple adaptive scheme is proposed for the estimation of m_0 and D.

$$m_0(n)= \beta\, m_0(n-1) + (1-\beta)\, r(n), \qquad (18)$$

$$D(n)= \beta\, D(n-1) + (1-\beta)\, Diag(|r(n) - m_0(n)|^2), \qquad (19)$$

where the forgetting factor ß has to be determined as a function of the approximate length of the transient signal and the non-stationarity of the background noise. It has to be big enough such that the effective observation period is much bigger than the transient signal, but small enough for the assumption of stationarity of the noise to hold.

4.3 Simulation results

Simulation results are given for the transient signal $s(n)$ shown in Fig. 1. Complex white Gaussian noise of variance σ_n^2 was added to $s(n)$ to generate $x(n)$. The signal-to-noise ratio, SNR, is defined as:

$$SNR= 10\, log_{10}(max_n\{|s(n)|^2\}/\, \sigma_n^2\,) \qquad (20)$$

The signal $x(n)$, for SNR = -2 dBs is shown in Fig. 2. The likelihood ratio T was calculated every 20 temporal samples and for 10 independent realizations of these signals. The mean plus and minus the standard deviation of the likelihood ratio T is plotted in Fig. 4 and 5 for SNR= 0 dBs and -2 dBs. We observe that the detection is feasible even under severe noise degradation.

5 Reduced Interference Higher-Order Distributions (RIHOD)

The formulation of a General Class of Time-Frequency Higher-Order Spectra was given in [3] as a generalization of Cohen's General Class of time-frequency representations.

$$PHOS_{kx}(t,f)= \int_{\Omega}\int_{\tau}\int_{u} \phi_k(\Omega, \tau)\, R_{kx}(u, \tau)$$

$$exp(j2\pi u\Omega) exp(-j2\pi t\Omega) exp(-j2\pi f^T \tau)\, d\tau\;\; d\Omega\; du. \;(21)$$

The properties of any particular member of the general class can be obtained with the properties of the WHOS and its particular kernel $\phi_k(\Omega, \tau)$. One of the main difficulties in signal analyses using the WD is the presence of interference terms between different time-frequency components of signals. These cross-terms will be even more numerous in the WHOS since the bilinear expression of the WD is substituted by a multilinear form. The Choi-Williams distribution, or in general RID [7] have been conceived to reduce the spurious terms without sacrificing highly desirable properties of time-frequency representations. The equivalent RID in the higher-order moment spectra domain can be obtained using the kernel

$$\phi_{RIHOD}(\Omega, \tau)= exp(-\tau^T \tau\; \Omega^2 /\sigma). \qquad (22)$$

The basic properties of WHOS are preserved if (22) is used since:

$$\phi_{RIHOD}(0, \tau)= 1, \forall\, \tau. \qquad (23.a)$$

$$\frac{\partial}{\partial \Omega}\phi_{RIHOD}(\Omega, \tau)\Big|_{\Omega=0} = 0, \forall\, \tau. \qquad (23.b)$$

$$\phi_{RIHOD}(\Omega, 0)= 1, \forall\, \Omega. \qquad (23.c)$$

$$\frac{\partial}{\partial \tau_i}\phi_{RIHOD}(\Omega, \tau)\Big|_{\tau_i=0} = 0, \forall \Omega \qquad (23.d)$$

Using $\phi_{RIHOD}(\Omega, \tau)$ Reduced Interference Higher-Order Distributions (RIHOD) can be obtained:

$$RIHOD_{kx}(t,f)=$$

$$= \sqrt{\pi\sigma} \int_{\tau} \left(\int_{u} R_{kx}(u, \tau)\, exp(-\frac{(t-u)^2 \sigma}{4\, \tau^T \tau})du. \right)$$

$$exp(-j2\pi f^T \tau)\, d\tau\,. \qquad (24)$$

They can be regarded as the extension of the Choi-Williams or Reduced Interference Distribution to the higher-order moment spectra domain..

5.1 Comparison between the WHOS and the RIHOD

The main goal in the definition of Reduced Interference Distributions is the suppression of spurious terms that appear in the analysis of multicomponent signals. We present here one example where the partial cross-terms cancellation obtained using RIHOD instead of WHOS can be observed. The DTF-WHOS of [2] can be easily modified for a discrete implementation of RIHOD by the inclusion of a temporal convolution prior to the computation of the FFT.

We considered the third order moment spectra domain, i.e. what we call Choi-Williams Bispectrum versus Wigner Bispectrum. Let us define the multicomponent signal $s(n)$ as,

$$s(n) = s_1(n) + s_2(n) \qquad (25.a)$$

$$s_1(n) = exp(\, -\alpha_1 n^2 + j\, \beta_1 n^2 + j\, \omega_1 n); \qquad (25.b)$$

$$s_2(n) = exp(\, -\alpha_2 n^2 + j\, \beta_2 n^2 + j\, \omega_2 n) \qquad (25.c)$$

where, $\alpha_1 = \alpha_2 = 350$; $\beta_1 = \beta_2 = 150$; $\omega_1 = 1000$ and $\omega_2 = 4000$. The discrete Wigner Bispectrum and Choi-Williams Bispectrum of length N= 96 was calculated for $s(n)$ at the temporal origin. Figs. 5, 6 and 7 show the Wigner Bispectrum of $s_1(n)$, $s_2(n)$ and $s(n)$. We can observe the numerous cross-terms that appear when analyzing the multi component signal which might be objectionable for many analysis applications. In Figs. 8, 9 and the Choi-Williams Bispectrum is plotted for the same signals. The number and importance of cross-terms is drastically reduced in this case.

References

[1] L. Cohen, "Time-frequency Distributions. A review", *Proc. of the IEEE*, July 1989.

[2] J.R. Fonollosa and C.L. Nikias, "Wigner Polyspectra: Higher-Order Spectra in time varying signal processing", *Proc. ICASSP 91*.

[3] J.R. Fonollosa and C.L. Nikias, "General Class of time-frequency higher-order Spectra: Definition properties, computation and application to transient signal analysis", *Proc. International Signal Processing Workshop on Higher-Order Statistics*, July 10-12, 1991, Chamrouse, France.

[4] C. L. Nikias, "Higher-Order Spectral Analysis", *Advances in Spectrum Analysis and Array Processing*, S. Haykin, Editor, Prentice-Hall 1991.

[5] G. B. Giannakis and A. V. Dandawate, "Polyspectral analysis of non-stationary signals: bases, consistency and HOS-WV", *Proc. International Signal Processing Workshop on Higher-Order Statistics*, July 10-12, 1991, Chamrouse, France.

[6] B. Friedlander and B. Porat, "Detection of Transient signals by the Gabor Representation", *IEEE Trans. Acoust., Speech, Signal Processing*, February 1989.

[7] W. J. Williams, J. Jeong et al., "Applications of Reduced Interference Distributions", *Proc. of the IEEE 1990 Digital Signal Processing Workshop*. Sep. 16-19,1990, New Paltz, New York.

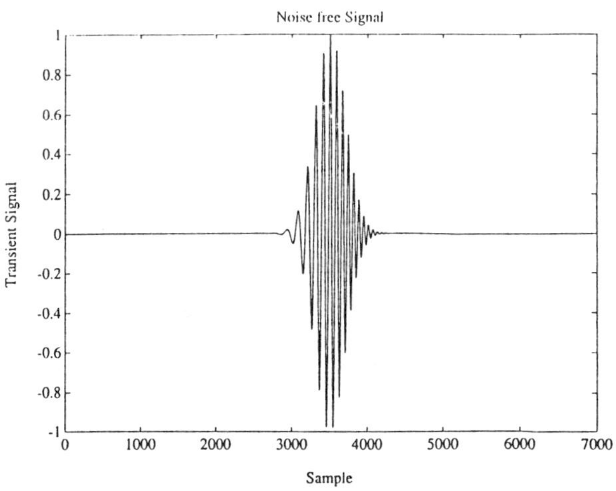

Fig. 1

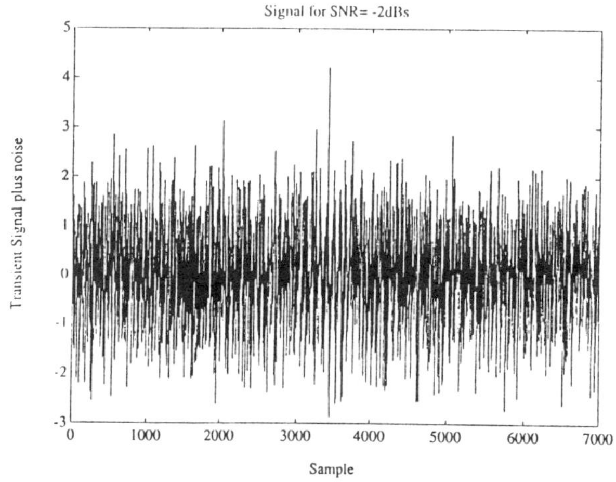

Fig. 2

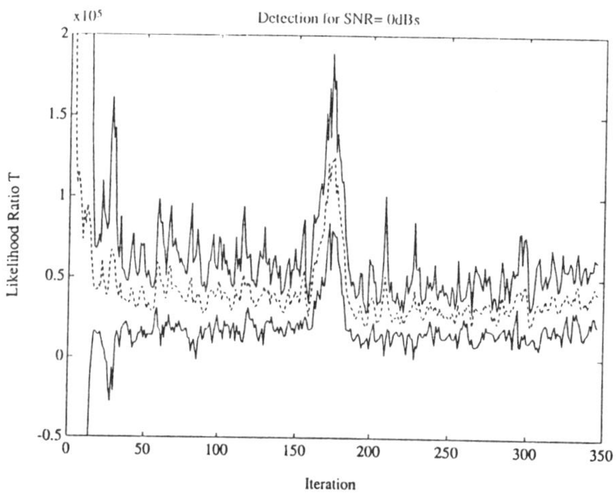

Fig. 3

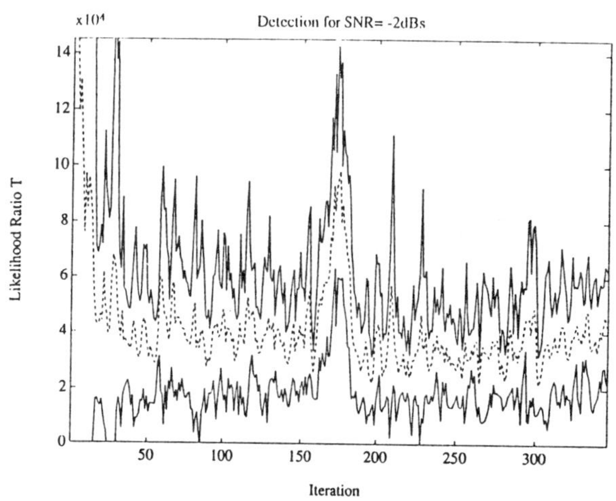

Fig. 4

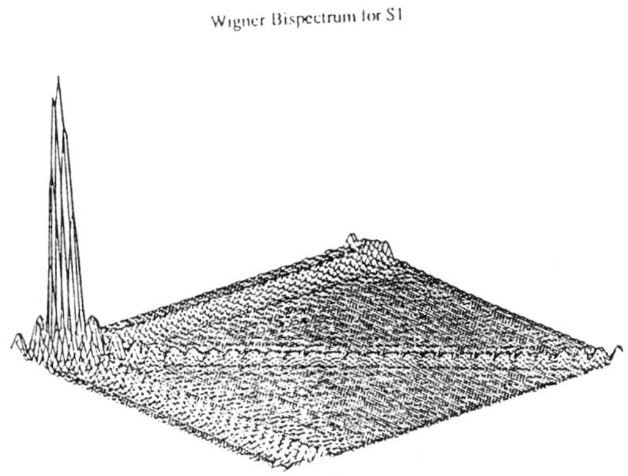

Fig. 5

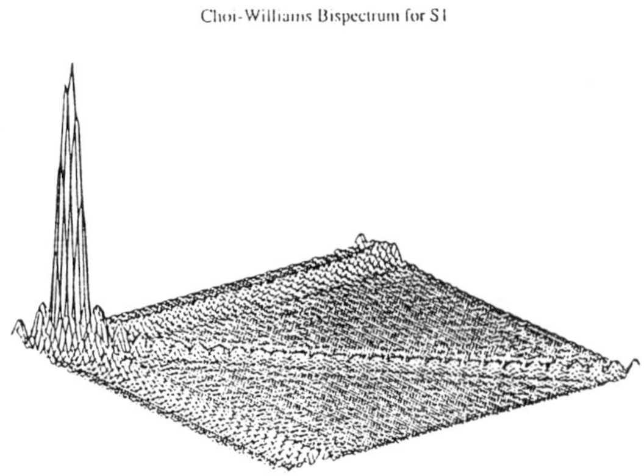

Fig. 8

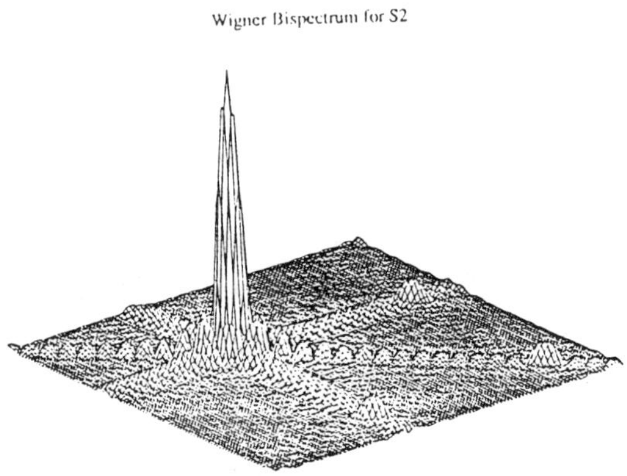

Fig. 6

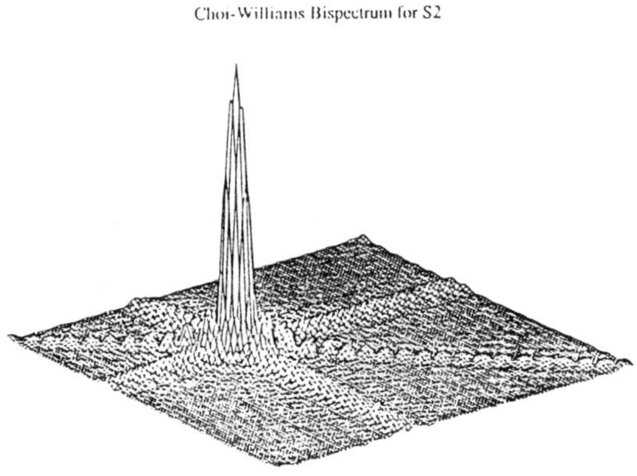

Fig. 9

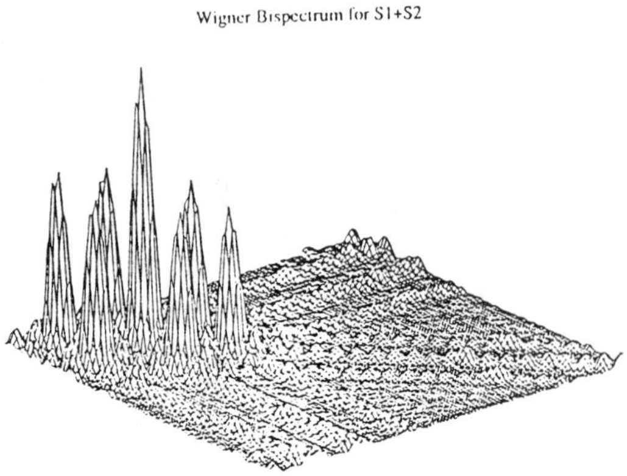

Fig. 7

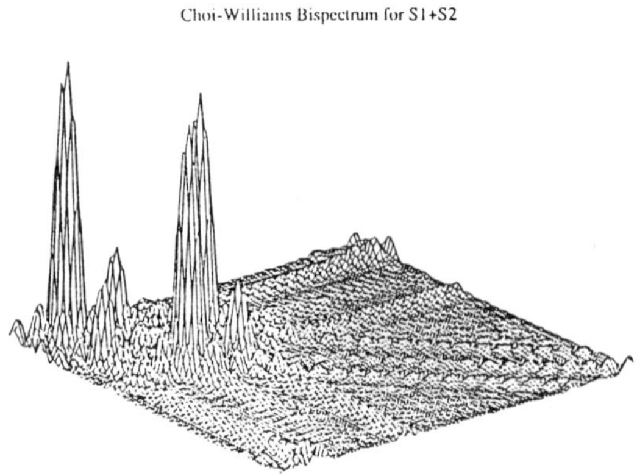

Fig. 10

Detection and Classification of Multicomponent Signals

A.B. Fineberg

CAIP Center
Rutgers University
Piscataway, NJ 08855

R.J. Mammone

CAIP Center
Rutgers University
Piscataway, NJ 08855

Abstract

The multicomponent nature of many naturally occurring signals, such as speech, is exploited to provide a new means of detection and classification. A component of a multicomponent signal is defined in terms of the local bandwidth about the instantaneous frequency in the time-frequency distribution. The components are isolated by an adaptive partitioning algorithm which is constrained to overcome the interference terms often present in such distributions. The redundancy which may be present in the individual components of the signal are discussed along with the means to detect and classify the signal based upon this redundancy. The application of this approach to a vowel classification task using the spectrogram is presented and is shown to provide performance which is favorable compared to conventional methods.

1 Introduction

When studying the time-frequency distributions of many naturally occurring signals, such as speech and music, it is often visually evident that the signal can be decomposed into two or more structural components. These components are readily identified by human inspection as being areas of interest. For instance, in speech the components are typically identified as resonant frequencies of the vocal tract (i.e. formants) which each have an associated bandwidth.

There is much ambiguity in the development of an algorithmic identification procedure because there are an infinite number of ways to represent a signal, $s(t)$, as a sum of components. Therefore some definition must be constructed which, when applied to a component of a multicomponent signal, describes the essence of what is seen when viewing the spectrum of the signal. Additionally, a method of finding a unique representation of each component must also be determined which allows the components to be combined to reconstruct the original signal (with bounded error).

A multicomponent signal representation should have a similar structure for all of its components. This redundancy of structure allows for the elimination of any spectral distortion of one component by calibration against the others. If this structure can be extracted from naturally occurring signals such as speech and music, or generated in man-made signals such as communications, radar, and sonar signals, then detection and classification algorithms can be designed to take advantage of this redundancy and thereby improve performance.

2 Multicomponent Signals

The essence of a multicomponent signal can be found in the signal spectrum. The components which are visualized by human inspection form groups which can be segmented from the rest of the spectrum. These groups are loosely defined as being spectrally compact with respect to the extent of the entire signal spectrum. This local compactness may be witnessed when the local average frequency of each group is separated by more than the sum of the apparent local half-bandwidth of the groups, which corresponds to the condition that

$$\omega_2(t) - \omega_1(t) > \frac{bw_1(t)}{2} + \frac{bw_2(t)}{2} \tag{1}$$

where $\omega_i(t)$ represents the local average frequency of the ith component and $bw_i(t)$ represents the respective local bandwidth of the ith component. Since these are all functions of time, a signal may be considered to be multicomponent at some time and not at others. This definition is similar to that given by Cohen and Lee [1] when the local average frequency is the mean or instantaneous frequency and the respective local bandwidth is given by the variance about the mean.

1058-6393/91 $1.00 © 1991 IEEE

For a signal which can be represented by

$$S(t) = A(t)e^{j\varphi(t)} \tag{2}$$

the instantaneous frequency, $\omega_i(t)$, is the average value of all frequencies which exist in the signal for some short period of time and is given by the derivative of the phase to be

$$\omega_i(t) = \frac{d\varphi(t)}{dt}. \tag{3}$$

The global bandwidth is defined by Cohen as the variance about the mean frequency and is given by [1, 2]

$$\sigma^2 = \int \left(\frac{A'(t)}{A(t)}\right)^2 A^2(t)dt + \int (\varphi'(t) - <\omega>)^2 A^2(t)dt \tag{4}$$

where the prime denotes differentiation, and $<>$ is the expected value operator. It is therefore clear that the global bandwidth is determined by both the amplitude and frequency modulation functions of the signal. The local bandwidth (the bandwidth about the instantaneous frequency) can be found by replacing $<\omega>$ with $\varphi'(t)$. Therefore the local bandwidth is proportional to $\left(\frac{A'(t)}{A(t)}\right)$. This gives a definition which is intuitive and will be shown to be very useful for multicomponent signal analysis.

Analysis of the individual components of a multicomponent signal will require that they be extracted from the composite signal. This is accomplished by adaptively separating each component mode from the composite spectrum. This filtering operation must be performed adaptively in order to accurately track the instantaneous frequency of the component and the instantaneous bandwidth variations. The filters should be constructed with smooth transition regions to minimize the spectral distortion imposed on each component. A new technique for very fast, computationally efficient frequency tracking has been developed [3] which allows the component locations to be tracked sample by sample. The bandwidth adaptation can then be performed by determining the bandwidth bounds which maximize the ratio between the energy captured by the adaptive filter and the total energy. This will help to minimize the reconstruction error.

For analysis of signals from an *ideal* time-frequency distribution this definition of multicomponent signals seems appropriate. It is interesting to note that several naturally occurring signals are of this form such as human speech signals [4], animal calls [5], and music [6]. It is suggested that this form is desirable in nature since it includes inherent redundancy in the signal which can be favorably used for classification in the presence of unknown distortions in the transmission medium.

3 Component Representation

As discussed above, the multicomponent representation of the signal, and therefore of the component, is not unique. If however we decide that a signal meets the criteria established for a multicomponent signal, then a representation of each component is needed which can adequately describe the time-varying nature of the component. Since each component will be spectrally compact, some form of adaptive filtering can be used to extract each component from the composite signal. The component, $s_i(t)$, will then be described by

$$s_i(t) = A_i(t)cos(\omega_i t + \phi_i(t)) \tag{5}$$

where ω_i is the mean frequency of the ith component, $A_i(t)$ is the ith component's envelope or AM modulation function, and $\phi_i(t)$ is the ith component's instantaneous phase or FM modulation function. Therefore in order to completely describe the character of the signal, it is necessary to know not only the mean frequencies of each component, but also the local characteristics which modulate the components.

One method of finding a representation based on the modulation functions $A_i(t)$ and $\phi_i(t)$ can be determined from the analytic representation given by

$$S_i(t) = s_i(t) + j\hat{s}_i(t) \tag{6}$$

where $\hat{s}_i(t)$ is the Hilbert transform of $s_i(t)$ [7], whose frequency components are in phase quadrature with those of $s_i(t)$. The analytic signal $S_i(t)$, for each component i, is given by

$$S_i(t) = A_i(t)e^{j(\omega_i t + \phi_i(t))}. \tag{7}$$

This complex representation yields the modulation functions as given by

$$A_i(t) = \sqrt{(\Re(S_i(t)^2) + \Im(S_i(t)^2))} \tag{8}$$

$$\phi_i(t) = \arctan\left(\frac{\Im(S_i(t))}{\Re(S_i(t))}\right) \tag{9}$$

where $\Re$ represents the real part and $\Im$ represents the imaginary part of the complex signal.

Other representations are possible since the representation given by equation (5) is not unique. Another possibility which may be appropriate would be given by

$$\text{Minimize: } \lambda_1 \int_{-\infty}^{\infty} \omega^2 \left[\int_{-\infty}^{\infty} A_i(t) e^{-j\omega t} dt\right] d\omega$$
$$+\lambda_2 \int_{-\infty}^{\infty} \omega^2 \left[\int_{-\infty}^{\infty} \phi_i(t) e^{-j\omega t} dt\right] d\omega \quad (10)$$

$$\text{Subject to: } s_i(t) = A_i(t)\cos(\omega_i t + \phi_i(t))$$

which attempts to find the representation for which the combined bandwidth of $A_i(t)$ and $\phi_i(t)$ is minimum. This criterion seems appropriate to the multicomponent definition which requires a compactness of each component.

This representation provides one method of solving for the modulation functions and illustrates that the criteria for an optimum decomposition must be carefully chosen. Several optimality constraints and methods for calculating the modulating functions can be developed for specific tasks, however it will be shown that the analytic signal representation is adequate for our purposes.

4 Component Matching

Many distortions experienced by signals traveling through an uncontrollable medium are manifested as modulation of the underlying signal. It is easily shown [8] that phase and frequency modulation are related and indistinguishable from the waveform or from an analytical expression for the waveform. It can also be shown that time modulation produces the same effect on the resulting waveform by the substitution of variables given by

$$t' = t + \frac{1}{\omega}\beta(t) \quad (11)$$

in a representation of a signal, $f(t) = A\cos(\omega t)$. This yields

$$f(t') = A\cos(\omega t') = A\cos(\omega t + \beta(t)). \quad (12)$$

Since these types of distortion, when applied to a multicomponent signal are not component selective, the modulation from the distortion will be imposed on each component. Therefore, if the modulation function corresponding to the first component can be estimated, it would have the same value as the estimates from each other component in the signal. This consistency across the components will be exploited as a means of classifying the underlying signal.

4.1 Relative Modulation Estimation

Typically it is impossible to estimate the absolute modulation functions without any *a priori* knowledge of the undistorted signal. If the problem however is to classify a distorted signal as being a modulated version of one signal from a class of signals, then only the relative modulation functions need to be estimated. This can be seen from the following example which shows the formulation for the analytic signal representations, $S_{ref}(t)$ and $S_{test}(t)$, for some known multicomponent reference and test signals, $s_{ref}(t)$ and $s_{test}(t)$, respectively. The test signal is a distorted version of the reference signal, where t was replaced by $t + cos\Omega t$.

In this case the signals are

$$s_{ref}(t) = \sum_{i=1}^{n} A_i cos(\omega_i t + \phi) \quad (13)$$

and

$$s_{test}(t) = \sum_{i=1}^{n} A_i cos(\omega_i(t + cos\Omega t) + \phi). \quad (14)$$

where n represents the number of components in the signal, ω_i is the average frequency of the component, ϕ is the unknown initial phase offset, and Ω determines the local frequency deviation.

By taking the analytic signal representation of each signal the relative modulation functions for each component can be estimated as

$$S_{ref}(t) = \sum_{i=1}^{n} A_i e^{j(\omega_i t + \phi)} \quad (15)$$

and

$$S_{test}(t) = \sum_{i=1}^{n} A_i e^{j(\omega_i t + \omega_i cos\Omega t + \phi)}. \quad (16)$$

By separating the argument equations (15) and (16) become

$$S_{ref}(t) = \sum_{i=1}^{n} A_i e^{j\omega_i t} e^{j\phi} \quad (17)$$

and

$$S_{test}(t) = \sum_{i=1}^{n} A_i e^{j\omega_i t} e^{j\omega_i cos\Omega t} e^{j\phi}. \quad (18)$$

The relative modulation functions for each component can then be found by dividing the equation for a component of the test signal by the equation for the same component of the reference signal. This is given by

$$\frac{S^i_{test}(t)}{S^i_{ref}(t)} = \frac{A_i e^{j\omega_i t} e^{j\omega_i cos\Omega t} e^{j\phi}}{A_i e^{j\omega_i t} e^{j\phi}} = e^{j\omega_i cos\Omega t}. \quad (19)$$

The modulation function of the ith component relative to the ref reference, $\beta_{i,ref}(t)$, can now be determined from equation (19) by

$$\beta_{i,ref}(t) = \frac{1}{\omega_i} \arctan\left(\frac{\Im(e^{j\omega_i cos\Omega t})}{\Re(e^{j\omega_i cos\Omega t}))}\right). \qquad (20)$$

The relative modulation function can then be calculated for each component of the signal. Since they should all be equal if the test signal was a distorted version of the reference, a consistency classification technique can be developed to verify the similarity of the modulation functions for each component.

4.2 Consistency Classification

If it is known that the test signal will be a distorted version of one of several reference signals, then the relative modulation functions can be determined for each component relative to each reference signal. For example if there are 3 components and 5 reference signals, then there will be a 3×5 array of relative modulation functions. The correct reference can then be determined by finding the distance between the entries in each column. Since the relative modulation functions should be similar along the column corresponding to the correct reference, the column which yields the smallest distance will have the highest probability of corresponding to the correct reference.

Due to the measurement noise and computational precision only an estimate of $\beta_{ij}(t)$ is observable. It is therefore necessary to define a distance measure which quantifies the similarity among the estimates of the relative modulation functions on each component. We have used the L_1 error norm due to its inherent robustness [9]. Therefore the template k, which yields the minimum value of

$$d_j = \sum_{i=1}^{n} \| \beta_{ij}(t) - \frac{1}{n}\sum_{k=1}^{n} \beta_{kj}(t) \| \qquad (21)$$

would be selected. Alternatively, an L_2 or other error norm could be used.

This consistency classification technique will determine the closest reference signal without any *a priori* knowledge of the statistics of the distortion. This is a very powerful tool which can be applied to pattern classification tasks with nonlinearly distorted patterns which correspond to the same class.

5 Vowel Classification Task

A vowel classification task was used to demonstrate the performance of the new method. The data was extracted from one dialect of the TIMIT database [10]. This data consisted of 32 utterances of each of 13 vowels from 12 male and 12 female speakers. Reference utterances were chosen for each vowel from a speaker not in the testing group.

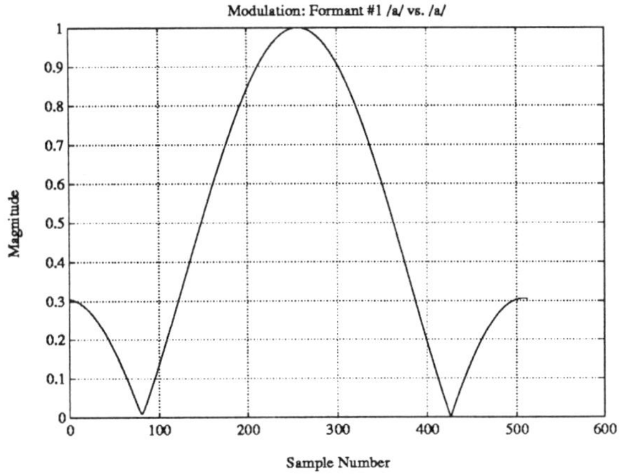

Figure 1: Relative Modulation Function of Formant 1 Test signal /a/ compared to Reference /a/

A spectrogram was generated for each vowel utterance and the first two components were removed from the spectrum by adaptively filtering about the peak of each component. An analytic signal was then generated for each component and the relative modulation function was then calculated for each reference utterance. The relative modulation functions for a matched reference utterance are shown in figures (1) and (2). The relative modulation functions for an unmatched reference utterance are shown in figures (3) and (4). It can be seen that the matched reference utterance provides a similar relative modulation function for component one and component two while the unmatched reference utterance shows little similarity. The consistency classification technique provided 84% correct classification for all test utterances.

6 Summary and Conclusions

A new approach is presented for the analysis of multicomponent signals. This approach has benefits for detection and classification of signals which are, or can be made to be, multicomponent. This technique exploits the spectral redundancy in the components by using the relative modulation of these components to classify the signal.

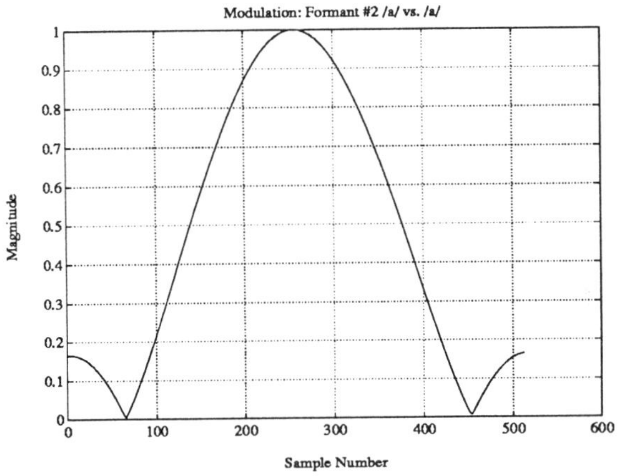

Figure 2: Relative Modulation Function of Formant 2
Test signal /a/ compared to Reference /a/

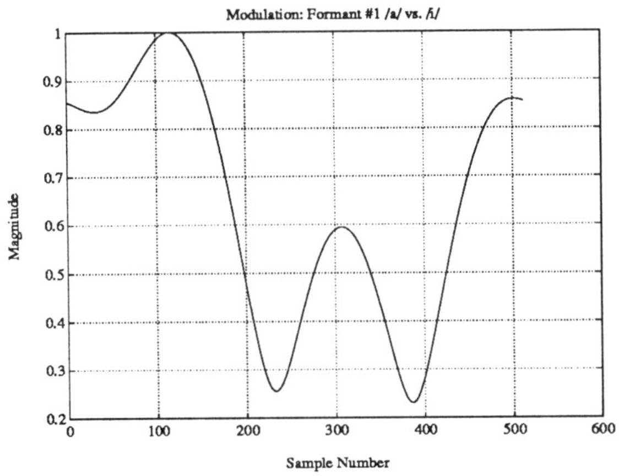

Figure 3: Relative Modulation Function of Formant 1
Test signal /a/ compared to Reference /i/

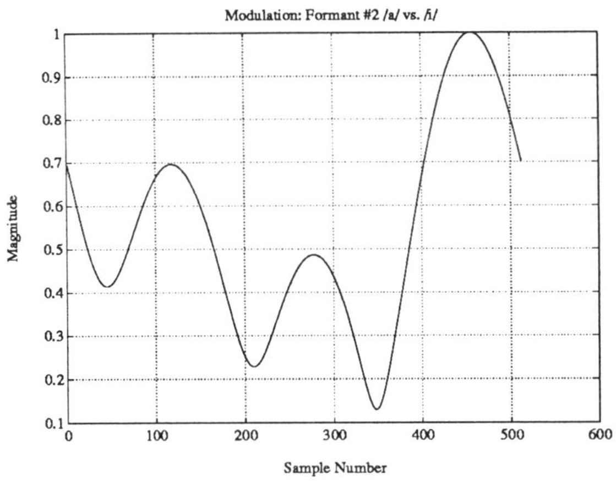

Figure 4: Relative Modulation Function of Formant 2
Test signal /a/ compared to Reference /i/

Acknowledgments

The research reported here was made possible, in part, through the support of the New Jersey Commission on Science and Technology and the Computer Aids for Industrial Productivity (CAIP) Center at Rutgers University.

References

[1] L. Cohen and C. Lee, "Instantaneous frequency, its standard deviation and multicomponent signals," in *Advanced Algorithms and Architectures for Signal Processing III* (F. T. Luk, ed.), pp. 186–208, SPIE, 1988.

[2] L. Mandel, "Interpretation of instantaneous frequency," *Amer. Jour. Phys.*, vol. 42, no. 840, 1974.

[3] A. Fineberg and R. Mammone, "An adaptive technique for high resolution time-varying spectral estimation," in *Proceedings IEEE ICASSP 1991*, (Toronto, Onario), May 1991.

[4] J. Flanagan, *Speech Analysis, Synthesis, and Perception.* New York, NY: Springer-Verlag, 2nd ed., 1972.

[5] J. Lilly, *Man and Dolphin.* New York: Pyramid Books, 1969.

[6] R. Cann, "An analysis/synthesis tutorial," in *Foundations of Computer Music* (C. Roads and J. Strawn, eds.), ch. 9, pp. 114–144, Cambridge, Mass.: MIT Press, 1988.

[7] R. Mammone, R. Rothacker, C. Podilchuk, S. Davidovici, and D. Schilling, "Estimation of carrier frequency, modulation type and bit rate of an unknown modulated signal," in *Proceedings of IEEE ICC-87*, (Seattle, Wash.), June 1987.

[8] H. Taub and D. Schilling, *Principles of Communication Systems.* New York, NY: McGraw Hill, 1986.

[9] R. Mammone, "Spectral extrapolation of constrained signals," *J. Opt. Soc. Am.*, vol. 73, Nov. 1983.

[10] W. Fisher, G. Doddington, and K. Goudie-Marshall, "The DARPA speech recognition research database: Specifications and status," in *DARPA Speech Recognition Workshop*, 1986.

Least-Squares Time-of-Arrival Estimation for Transient Signals in a Multipath Environment

R.J. Vaccaro, T. Manickam, C.S. Ramalingam, R. Kumaresan and D.W. Tufts
Department of Electrical Engineering
The University of Rhode Island
Kingston, RI 02881

Abstract

We consider the problem of estimating the arrival times of overlapping ocean-acoustic signals from a noisy received waveform which consists of scaled and delayed replicas of a deterministic transient signal. We assume that the transmitted signal and the number of paths in the multipath environment are known, and consider algorithms that give least-squares estimates of the amplitude and time delay of each path. The Gauss-Newton algorithm converges quickly to a minimum of the error surface, but if the initial estimates are not sufficiently accurate, only a local minimum will be found. In this paper, we compare several different methods for obtaining initial estimates including coordinate descent and linear prediction. We demonstrate the performance of the complete algorithm (initial estimates followed by a Gauss-Newton search) on experimental data.

1 Introduction

Time delay estimation occurs frequently in the fields of sonar, radar and geophysics. Here, the received waveform consists of delayed and scaled replicas of the transmitted signal, which is the result of multiple reflections and attenuation of the signal in the channel.

The received waveform $r(t)$ can be described as

$$r(t) = \sum_{k=1}^{M} a_k s(t-\tau_k) + w(t) \quad , \quad 0 \leq t \leq T \tag{1}$$

where $s(t)$ is the transmitted signal, a_k the attenuation factor for path k, τ_k the time delay for path k, M the number of different paths, $w(t)$ is the noise component corrupting the received signal, and $s(t)$ is a *deterministic* source signal. The signal waveform itself is used by the algorithms developed in this paper.

In [1] Kirsteins suggests to view the problem in the frequency domain. Since a delay in the time domain is equivalent to multiplication by an exponential in the frequency domain, the corresponding frequency domain problem is one of fitting weighted complex exponentials to the spectrum of the received signal, a problem that has been well studied. However, the algorithm in [1] does not work if the Fourier transform of the source signal has any nulls in the regions of its spectral support.

2 The Least-Squares Estimator

The sampled received waveform (T_s is the sampling period) can be modeled as

$$r(nT_s) = \sum_{k=1}^{M} a_k s(nT_s - \tau_k) + w(nT_s), \quad 0 \leq n \leq N-1. \tag{2}$$

If the noise is white Gaussian, it is well known that the least squares estimator is also the maximum likelihood estimator. The squared error function is given by

$$E(\boldsymbol{\tau}, \tilde{\mathbf{a}}) = \sum_{n=0}^{N-1} \left[r(nT_s) - \sum_{k=1}^{M} a_k s(nT_s - \tau_k) \right]^2 , \tag{3}$$

where $\boldsymbol{\tau}$ and $\tilde{\mathbf{a}}$ are vectors of time-delays and amplitudes, respectively. The objective is to minimize E with respect to $\boldsymbol{\tau}$ and $\tilde{\mathbf{a}}$.

In the above time-domain formulation of the problem, we are restricted to values of τ_k which are integer multiples of the basic sampling interval. If a more accurate estimate is required, then one has to resort to interpolation. This inconvenience can be avoided by transforming the problem to the frequency domain, where τ_k can take on a continuum of values.

[0] This work was supported in part by grants from The Naval Oceanographic and Atmospheric Research Laboratory under contract N0001489K6003 and The Office of Naval Research under contract N0001490J1283.

The data record in (1) exists in the time interval $[0, T]$. If this data record is set to zero outside $[0, T]$, then the summand in (3) will be zero outside the range $0 \leq n \leq N-1$. Hence we can extend the summation range from minus infinity to plus infinity in (3) and invoke Parseval's theorem to get

$$E(\tau, \tilde{\mathbf{a}}) = \frac{T_s}{2\pi} \int_{-\frac{\pi}{T_s}}^{\frac{\pi}{T_s}} \left| R(\omega) - S(\omega) \sum_{k=1}^{M} a_k e^{-j\tau_k \omega} \right|^2 d\omega, \tag{4}$$

where $R(\omega)$ and $S(\omega)$ are the discrete-time Fourier transforms of $r(nT_s)$ and $s(nT_s)$, respectively.

In the above minimization, we need to consider *only those frequency regions which contain significant signal energy.* That is, we can threshold the signal spectrum and consider only those regions which are above this value. This has the advantage of working with regions having good signal-to-noise ratios only. Also, since the received and the modeled signal are real, only half the signal spectrum needs to be considered because of its conjugate symmetry. Therefore the error expression in (4) (after neglecting the scale-factor $T_s/2\pi$) can be approximated by the following:

$$E(\boldsymbol{\lambda}, \ \tilde{\mathbf{a}}) = \|\tilde{\mathbf{r}} - \tilde{\mathbf{P}}(\boldsymbol{\lambda})\tilde{\mathbf{a}}\|^2, \tag{5}$$

where

$$\begin{aligned}
\boldsymbol{\lambda} &= (\lambda_1 \ \lambda_2 \ \cdots \ \lambda_M)^T \quad \text{with} \quad \lambda_k = -\frac{2\pi}{NT_s}\tau_k \\
\tilde{\mathbf{r}} &= (R[q_1] \ R[q_2] \ \cdots \ R[q_L])^T \\
\tilde{\mathbf{a}} &= (a_1 \ a_2 \ \cdots \ a_M)^T \\
\mathbf{S} &= \text{diag}(S[q_1] \ S[q_2] \ \cdots \ S[q_L]) \\
\mathbf{A}(\boldsymbol{\lambda}) &= \begin{pmatrix} e^{j\lambda_1 q_1} & e^{j\lambda_2 q_1} & \cdots & e^{j\lambda_M q_1} \\ e^{j\lambda_1 q_2} & e^{j\lambda_2 q_2} & \cdots & e^{j\lambda_M q_2} \\ \vdots & \vdots & \vdots & \vdots \\ e^{j\lambda_1 q_L} & e^{j\lambda_2 q_L} & \cdots & e^{j\lambda_M q_L} \end{pmatrix} \\
\tilde{\mathbf{P}}(\boldsymbol{\lambda}) &= \mathbf{SA}(\boldsymbol{\lambda})
\end{aligned} \tag{6}$$

$S[q_k]$ are the samples of $S(\omega)$ that lie above the threshold. $R(\omega)$ is also sampled at these points only.

This is the error expression considered in [2]. However, in that work the amplitudes were allowed to be complex. This leads to a biased estimate of the delay parameters if the SNR were not sufficiently high [3]. But an important advantage of allowing complex-valued amplitudes is that the corresponding error surface is reasonably smooth, making it easier to find its global minimum. The key step which helps us to achieve the transition from the biased error surface (due to the complex amplitudes) to the true error surface (due to the real amplitudes) is to rewrite (5) explicitly in terms of real and imaginary parts of each of the terms involved, and add a penalty term consisting of a parameter α times the sum of the squares of the imaginary parts of the amplitudes. The result is,

$$\begin{aligned}
E'_\alpha(\boldsymbol{\lambda}, \ \mathbf{a}) &= \left\| \begin{pmatrix} \mathbf{r}_r \\ \mathbf{r}_i \\ \mathbf{0} \end{pmatrix} - \begin{pmatrix} \mathbf{P}_r & -\mathbf{P}_i \\ \mathbf{P}_i & \mathbf{P}_r \\ \mathbf{0} & \alpha\mathbf{I} \end{pmatrix} \begin{pmatrix} \mathbf{a}_r \\ \mathbf{a}_i \end{pmatrix} \right\|^2 \\
&= \|\mathbf{r} - \mathbf{Pa}\|^2
\end{aligned} \tag{7}$$

where

$$\begin{aligned}
\mathbf{r}_r &= \text{Real}\{\tilde{\mathbf{r}}\} & \mathbf{r}_i &= \text{Imag}\{\tilde{\mathbf{r}}\} \\
\mathbf{P}_r &= \text{Real}\{\tilde{\mathbf{P}}(\boldsymbol{\lambda})\} & \mathbf{P}_i &= \text{Imag}\{\tilde{\mathbf{P}}(\boldsymbol{\lambda})\} \\
\mathbf{a}_r &= \text{Real}\{\tilde{\mathbf{a}}\} & \mathbf{a}_i &= \text{Imag}\{\tilde{\mathbf{a}}\}.
\end{aligned}$$

Since the amplitude vector $\mathbf{a}$ enters linearly in (7), it can be eliminated to yield

$$E_\alpha(\boldsymbol{\lambda}) = \left\| \left[\mathbf{I} - \mathbf{P}(\mathbf{P}^H\mathbf{P})^{-1}\mathbf{P}^H\right] \mathbf{r} \right\|^2. \tag{8}$$

It is easy to see that for $\alpha = 0$, the error function in (7) allows $\tilde{\mathbf{a}}$ to be complex since there is no penalty attached to its imaginary part. As α increases, the penalty on the imaginary part increases, and in the limit as $\alpha \to \infty$, $\tilde{\mathbf{a}}$ is constrained to be purely real. Note that the model for the channel assumes the amplitudes to be real.

With the above modification of the error expression, the procedure to minimize $E_\alpha(\boldsymbol{\lambda})$ is as follows: Initially, α is set equal to zero. $E_0(\boldsymbol{\lambda})$ is then minimized with respect to $\boldsymbol{\lambda}$. The minimum yields a biased estimate of $\boldsymbol{\lambda}$, but one which is reasonably close to the true value. To estimate the true delay values, α is increased gradually and the minimum of $E_\alpha(\boldsymbol{\lambda})$ is found for each α. The minimum of $E_\infty(\boldsymbol{\lambda})$ yields the final time-delay estimates. This method of using a quadratic penalty function to impose a nonlinear constraint is known to converge to a minimum of the true error surface [4].

In [3] we present a Gauss-Newton approach to find the minimum of $E_\alpha(\boldsymbol{\lambda})$. The remainder of this paper considers several approaches to initializing the Gauss-Newton algorithm for $E_0(\boldsymbol{\lambda})$.

3 Initialization Using Coordinate Descent

The first initializing procedure we consider is based on the coordinate descent algorithm employed in nonlinear programming [6]. From (6) it is seen that the k-th column of $\tilde{\mathbf{P}}(\boldsymbol{\lambda})$ is dependent only on λ_k, *i.e.*,

$$\tilde{\mathbf{P}}(\boldsymbol{\lambda}) = [\mathbf{p}_1(\lambda_1) \ \ \mathbf{p}_2(\lambda_2) \ \ \cdots \ \ \mathbf{p}_M(\lambda_M)]. \tag{9}$$

A special case of the coordinate descent algorithm occurs when the columns of $\tilde{\mathbf{P}}(\boldsymbol{\lambda})$ each depend on a single parameter only [5] and is briefly explained below.

The QR decomposition of $\tilde{\mathbf{P}} = \mathbf{Q}(\boldsymbol{\lambda})\mathbf{R}(\boldsymbol{\lambda})$ gives

$$\mathbf{Q}(\boldsymbol{\lambda}) = [\mathbf{q}_1(\lambda_1)\ \ \mathbf{q}_2(\lambda_1,\lambda_2)\ \ \cdots\ \ \mathbf{q}_M(\lambda_1,\ldots,\lambda_M)]\,. \tag{10}$$

The $\mathbf{q}_i$'s are orthogonal and each $\mathbf{q}_i$ is a function of $(\lambda_1,\ldots,\lambda_i)$. Substituting the QR decomposition in (8) with $\alpha = 0$, we get

$$\begin{aligned} E_0(\boldsymbol{\lambda}) &= \| [\mathbf{I} - \mathbf{Q}\mathbf{Q}^H]\ \mathbf{r}\|^2 \\ &= \mathbf{r}^H\mathbf{r} - \mathbf{r}^H\mathbf{Q}\mathbf{Q}^H\mathbf{r} \\ &= \mathbf{r}^H\mathbf{r} - \sum_{i=1}^{M}\mathbf{r}^H\mathbf{q}_i\mathbf{q}_i^H\mathbf{r}. \end{aligned} \tag{11}$$

The coordinate descent algorithm gives the initial estimates of the parameters in a sequential process. First, $\mathbf{r}^H\mathbf{q}_1\mathbf{q}_1^H\mathbf{r}$ is maximized over λ_1. The parameter λ_1^o which maximizes this term is taken as the initial estimate of λ_1. Next, $\mathbf{r}^H\mathbf{q}_2\mathbf{q}_2^H\mathbf{r}$ is maximized with respect to λ_2 while holding the first parameter fixed at λ_1^o. The parameter λ_2^o which maximizes this term is taken as the initial estimate of λ_2. This process is sequentially continued to obtain the M initial estimates λ_i^o, $i = 1,\ldots,M$.

The above process involves a one-dimensional optimization at each step. We employ a coarse grid search followed by a fine search using either a golden-section or a gradient based method. (The fine search can also be done using the Gauss-Newton algorithm.) The spacing of the grid points should be such that there is at least one point in the main lobe of each coordinate descent curve. As a rule of thumb, the coarse grid spacing is chosen as one-third the essential width of the auto-correlation function of the transmitted signal. The coarse grid search does not have to be carried out for all possible delay values. Rather, if the received signal is matched-filtered, only those regions where the matched filtered output is large need to be considered. For an M path signal model, the coordinate descent method requires about $M(M+1)L$ multiplications per grid point.

One could also use the initializing procedure on the original least squares error model, which involves full signal spectra. This is equivalent to applying the coordinate descent algorithm on $E_\infty(\boldsymbol{\lambda})$. However, we have found that this yields poor estimates when the data contain closely spaced paths. The estimates obtained by using coordinate descent algorithm on $E_0(\boldsymbol{\lambda})$ are more robust to noise and the delay spacings.

4 Initialization Using DFT and LP Methods

If the source spectrum does not have any nulls, one can transform the weighted exponential fitting problem into an unweighted one. This can be seen more clearly by rewriting (4) as follows:

$$E(\boldsymbol{\tau},\mathbf{a}) = \frac{T_s}{2\pi}\int_{\mathcal{R}} |S(\omega)|^2 \left| \frac{R(\omega)}{S(\omega)} - \sum_{k=1}^{M} a_k e^{-j\tau_k\omega}\right|^2 d\omega, \tag{12}$$

where $\mathcal{R}$ represents the non-zero regions of $S(\omega)$. The term $|S(\omega)|^2$ can be interpreted as an error weighting. Hence an alternative approach to time-delay estimation is to view it as a weighted least-squares fitting of M undamped complex exponentials. This was first suggested by Kirsteins [1]. DFT Method: If the time-delays are well separated, the location of the peaks of the Fourier transform of $R(\omega)/S(\omega)$ will directly give a very good estimate of $\boldsymbol{\tau}$. Tufts-Kumaresan Method: If the τ_k's are closely spaced one has to resort to high resolution techniques like the Tufts-Kumaresan method [7]. In this method, the so-called forward-backward data matrix is replaced by its rank M approximation using SVD. The prediction error filter co-efficients are obtained using least-squares; subsequent polynomial rooting gives an estimate of the time-delays. FIR Prefiltering Techniques: A related iterative prefiltering technique suggested in [8] and a variation of it were also tried. In the first method, the data is prefiltered by an FIR filter of fixed length L, equal to the prediction error filter order of the TK method. This prefilter is determined iteratively from the data without using any prior information. As a variation of this method, the prefilter order was increased from a low order to L. This approach was taken to prevent the prefilter from mistaking spurious noise peaks as signal.

In [2] the time-delays thus obtained were considered to be *the* ML estimates, wherein the amplitudes were not constrained to be real. But as pointed out in Section 2, these estimates do *not* correspond to the global minimum of (4). In fact, they may not even be at the minimum of the $\alpha = 0$ error surface. That one can improve the estimates obtained from the enhanced linear prediction algorithm by feeding them to a Gauss-Newton procedure was first noted in [9], although in a different context. Hence these estimates are to be viewed merely as very good initial guesses. In using the TK method, it is assumed that the SNR of $R(\omega)/S(\omega)$ is above threshold. The threshold SNR may be high if the source spectrum has regions in $\mathcal{R}$

where it is close to zero.

We tested the performance of the above methods to get good initial time-delay estimates, which were then used as starting values for the Gauss-Newton algorithm. The results are presented in the next section.

5 Example with Experimental Data

The experimental data used here was provided by NUSC. Fig. 1 shows the received signal. It was suspected to contain between two and four multi-path components. The actual transmitted FM pulse was not available and hence was synthesized using its known parameters. Its spectrum is shown in Fig. 2. The region $\mathcal{R}$ contained $N = 307$ points above peak-magnitude/1.5 of the source spectrum. The estimates of the delays (in samples) using the methods described in Section 4 are given in Table I, along with the final ML estimates.

Table I

	τ_1	τ_2	τ_3	τ_4
CD	413.80	414.53	426.37	437.66
TK2	413.55	414.93		
TK3	413.45	414.91		434.67
TK4	413.24	414.37	425.24	436.86
PF1	412.64	414.57	423.38	436.21
PF2	412.64	414.57	423.37	436.21
MLE	**411.66**	**414.86**	**423.09**	**434.24**

The CD plot of $\mathbf{r}^H \mathbf{q}_3 \mathbf{q}_3^H \mathbf{r}$ for the third parameter is shown in Fig. 3. The plots for the other parameters are similar. If the data is matched filtered, the output has significant energy only in the range from sample 390 to sample 440. Hence the grid search needs to be carried only over this interval. Furthermore, the essential width of the signal autocorrelation sequence is about ten samples. A rule-of-thumb for grid spacing is essential-width$\times T_s/3$. The value of $\mathbf{r}^H \mathbf{q}_3 \mathbf{q}_3^H \mathbf{r}$ at these grid points is indicated by crosses in Fig. 3. The grid spacing is seen to be sufficiently small to have at least one grid point near the global maximum.

For the TK algorithm, one has to first divide $R(\omega)$ by $S(\omega)$. By mere inspection of the peaks of the DFT of $R(\omega)/S(\omega)$ it was not possible to resolve all the time-delays. Hence the need to apply high-resolution techniques to estimate them. The results of rank two (TK2), rank three (TK3), and rank four (TK4) approximations using the TK algorithm are shown in Table I. For the fixed-length prefiltering technique L was chosen as $3N/4 = 230$ and for the variation of this method suggested in Section 4 the initial order was chosen to be 50 and its length was increased by 5 at each iteration till it reached 230. The results of both these approaches are also tabulated in Table I, under PF1 and PF2, respectively.

All these estimates were used to initialize a Gauss-Newton routine. The corresponding error residuals are shown in Figs. 4,5, and 6. It can be seen that four paths are able to adequately model the data. The spikes in the error residuals are due to the gradual signal build-up and decay of the received signal, as compared to the synthesized ideal source chirp. From preliminary simulation studies on synthetic data it was observed that if the SNR were sufficiently high, the estimates obtained from linear prediction were quite close to the true values. On the other hand, even though the CD initial estimates were farther away from the true values they were more robust to additive noise. Also, using the TK method a low-rank approximation of order $K < M$ does not always yield the K strongest paths; they depend on the spacing between the τ_k's also.

References

[1] I. P. Kirsteins, "High Resolution Time Delay Estimation," in *IEEE Proc. ICASSP 87*, (Dallas, TX), pp. 451–454, Apr. 1987.

[2] I. P. Kirsteins and A. H. Quazi, "Exact Maximum Likelihood Time Delay Estimation for Deterministic Signals," in *Proc. of EUSIPCO-88*, (Grenoble, France), Sept. 1988.

[3] R.J. Vaccaro, C.S. Ramalingam, D.W. Tufts, and R.L. Field, "Least-squares time-delay estimation for transient signals in a multipath environment," *to appear in JASA*.

[4] P. Gill, W. Murray, and M. Wright, *Practical Optimization*. New York, NY: Academic Press, 1981.

[5] J.A. Cadzow, "Signal processing via least-squares error modeling," *IEEE Sig. Proc. Mag.*, vol. 7, pp. 12–31, Oct. 1990.

[6] D. Luenberger, *Introduction to Linear and Nonlinear Programming*. Reading, MA: Addison-Wesley, 1973.

[7] D. W. Tufts and R. Kumaresan, "Estimation of frequencies of multiple sinusoids: Making linear prediction perform like maximum likelihood," *Proc. IEEE*, vol. 70, pp. 975–989, Sept. 1982.

[8] R. Kumaresan and Y. Feng, "FIR prefiltering improves Prony's methd," *IEEE Trans. Acoust., Speech, and Signal Processing*, vol. ASSP-39, pp. 736–741, Mar. 1991.

[9] S. Parthasarathy and D.W. Tufts, "Maximum-likelihood estimation of parameters of exponentially damped sinusoids," *Proc. of the IEEE*, vol. 73, pp. 1528–1530, 1985.

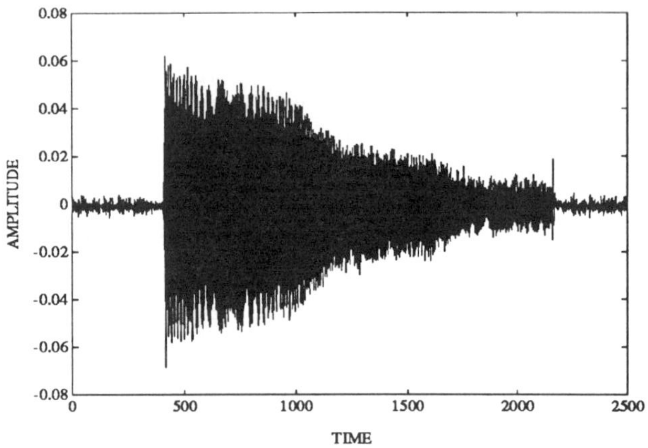

Fig. 1: Received signal containing between two and four multipath components.

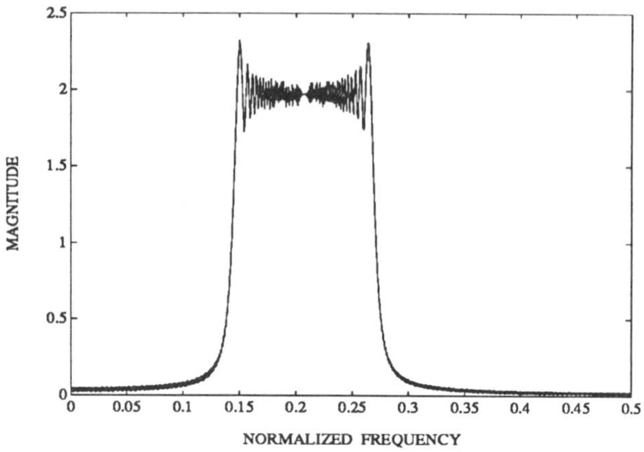

Fig. 2: Magnitude sepctrum of the transmitted source.

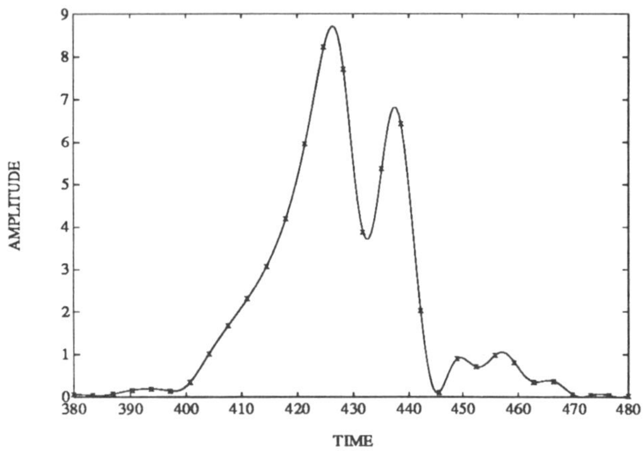

Fig. 3: Coordinate descent curve for τ_3.

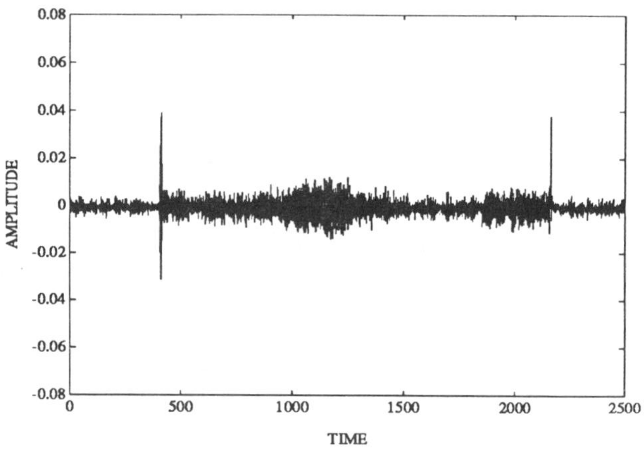

Fig. 4: Residual error for a two-path model.

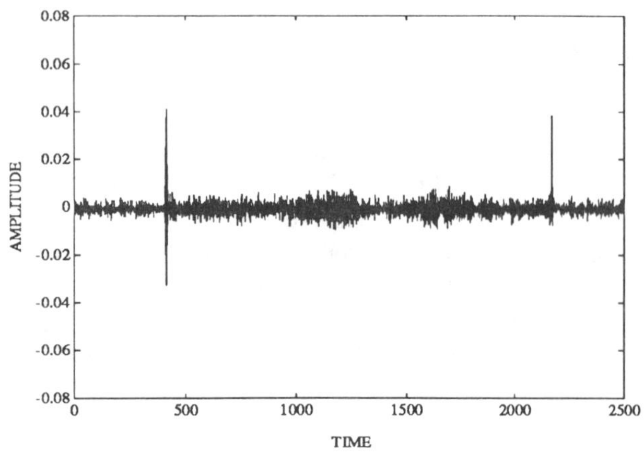

Fig. 5: Residual error for a three-path model.

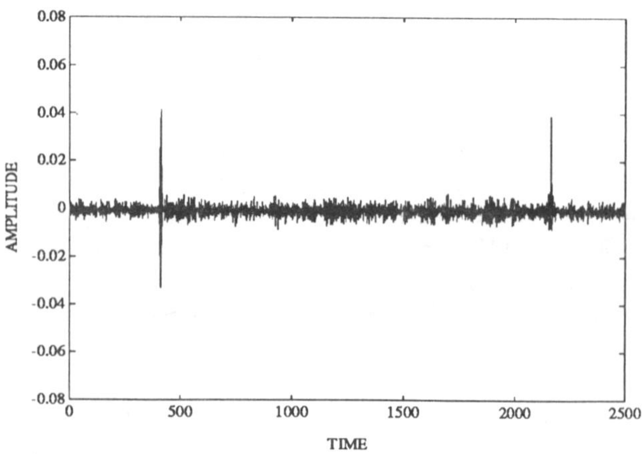

Fig. 6: Residual error for a four-path model.

Detection of Underwater Transient Acoustic Signals using Time-Frequency Distributions and Higher-Order Spectra

Gordon J. Frazer and Boualem Boashash

Centre for Signal Processing Research,
Queensland University of Technology,
Brisbane, QLD. 4001,
AUSTRALIA.

Abstract

In this paper we demonstrate the application of the Wigner-Ville time-frequency distribution, the bispectrum, the time-varying bispectrum and Gerr's third order Wigner distribution, to some underwater acoustic data and demonstrate the merit of including higher-order spectral information when signaturing underwater acoustic sources.

1 Introduction

Underwater signals, both biological and man-made, are often characterised by their acoustic emissions. Analysis in both the time domain and the frequency domain has been a standard approach for signals of this type[1]. More recently, it has been recognised that the time and frequency analysis of signals should be viewed jointly[1] and practical time-frequency analysis tools have been developed and applied to underwater acoustic signals[1][2]. The analysis methods mentioned use either the time series data directly or second moment information such as the power spectrum, the autocorrelation or the time-frequency distribution. These analysis tools do not make use of signaturing features introduced by non-Gaussian signals and nonlinearities inherent in the underwater source or introduced by the propagation medium.

A supplementary analysis technique is to consider higher order cumulant sequences of the data and their corresponding spectra[6]. The most well known higher order spectra are the bispectrum and the trispectrum[10]. With these spectra, non-Gaussianity and non-linearities inherent in the data can be identified and used to augment the signature of the acoustic source derived using the more traditional second order methods. Existing higher-order spectral analysis techniques assume that the signal is stationary and that the ergodic property can be used to produce consistent spectral estimates. Often the signal to be analysed is not stationary and attempts have been made to define a time-varying higher-order spectra [7] [12] [3] [11] [5]. Many analysis methods in current use are optimal for Gaussian and linear models but are suboptimal when the problem is one of analysing non-stationary, non-linear and non-Gaussian signals. The use of time-varying higher-order spectra represents one approach for problems of this type [5].

2 Signal Analysis: A Review of Recent Developments

We now review some of the signal analysis techniques which we applied to underwater acoustic data. We assume that the signal $\{x(t)\}$ is a random process which can be fully described by a probability density function (pdf), $p(x)$.

If we impose the restriction that the pdf of the process, $x(t)$, is not changing with time, then for analysing the signal, one uses the power spectrum of the process defined as;

$$\mathcal{S}(f) = \int_{-\infty}^{\infty} R_x(\tau) e^{-j2\pi f\tau} d\tau \qquad (1)$$

Where $R_x(\tau) = E\{x(t)x(t+\tau)\}$ is the autocorrelation of $x(t)$, and dependent only on the lag variable τ. Power spectral analysis reveals the presence of periodicities in the signal, some of which may not have been obvious in the time domain.

The assumption of stationarity is often not valid. When the signal is nonstationary, i.e. $R_x(t,\tau)$ is time dependent, the power spectrum becomes "smeared" and much of the important information contained in the signal is lost. For this reason, the notion of a time-frequency distribution has arisen[1]. This type of distribution attempts to localise the energy, or energy density, in time and frequency. A class of time-frequency distributions defined below has proven very useful.

$$\rho_z(t,f) = \int_{-\infty}^{\infty}\int_{-\infty}^{\infty}\int_{-\infty}^{\infty} e^{j2\pi v(u-t)}g(v,\tau) \quad (2)$$
$$z(u+\frac{\tau}{2})z^*(u-\frac{\tau}{2})e^{-j2\pi f\tau}dvdud\tau$$

where $z(u)$ is the analytic signal derived from $x(n)$, and the function $g(v,\tau)$ determines the particular time-frequency distribution. A more general class of time-frequency distributions which is optimal for any time-frequency distribution has recently been proposed in [4] and [5]. Two well known members of this class are the Wigner-Ville distribution and the spectrogram. The power spectrum and the time-frequency distribution present second moment properties of the signal. Signals with pdfs other than Gaussian contain information that cannot be analysed using second moment tools, but require higher-order moments.

2.1 Higher-Order Spectral Analysis

For a sequence of real random variables the pdf can be expanded in a power series with the terms in the series, the cumulants, describing the pdf. The rth order cumulants of a sequence $\{x_1,\ldots,x_n\}$ are defined to be;

$$c_{k_1,\ldots,k_n} = (-j)^r \frac{\partial^r \ln\Phi(f_1,f_2,\ldots,f_n)}{\partial f_1^{k_1}\ldots\partial f_n^{k_n}}\Bigg|_{f_1=\ldots=f_n=0} \quad (3)$$

where $r = k_1+\ldots+k_n$ and $\Phi(f_1,\ldots,f_n)$ is the joint first characteristic function given by;

$$\Phi(f_1,f_2,\ldots,f_n) = \mathbf{E}\left\{e^{-j2\pi(f_1x_1+\ldots+f_nx_n)}\right\} \quad (4)$$

A class of spectra can be defined from the higher order cumulant sequences. For zero mean random variables the moment and cumulant sequences are identical for orders one, two and three. For higher orders, however, they are not the same. Two useful higher-order spectra are the bispectrum and the trispectrum which are the Fourier transforms of $c_3(\tau_1,\tau_2)$ and $c_4(\tau_1,\tau_2,\tau_3)$, the third and fourth order cumulant sequences, respectively. Higher-order spectra based on cumulant sequences have the desirable property that for a symmetric pdf, the bispectrum and higher-order spectra are identically zero. In addition, the cumulant of the sum of two random variables is the sum of the cumulants of each random variable. For zero mean signals we use the bispectral definition derived from the moment sequence;

$$\mathcal{B}(f_1,f_2) = \int_{-\infty}^{\infty}\int_{-\infty}^{\infty} E\{x(t)x(t+\tau_1) \quad (5)$$
$$x(t+\tau_2)\}e^{-j2\pi(f_1\tau_1+f_2\tau_2)}d\tau_1 d\tau_2$$

Clearly, the bispectrum and other higher-order spectra present extra information to that contained in the power spectrum, since they use additional terms from the series expansion of the pdf of the process. Naturally the series expansion of the pdf may itself be time-varying. In this case the definitions of higher-order spectra presented so far will smear the spectra and information will be lost.

2.2 Time-Varying Higher-Order Spectral Analysis

There is currently no single unified approach to the analysis of time-varying higher-order spectra. Several extentions to time-varying second moment analysis methods have been proposed [7] [11] [5]. Thatcher and Amin extended the third moment stationary bispectrum, introducing the "running bispectrum" [12] by analogy with the spectrogram. In this paper this is referred to as the time-varying bispectrum. There are no reported applications of these methods to real data and no reports of how the distributions are best interpreted.

The definition for the time-varying bispectrum is a generalisation of the definition of the spectrogram.

$$B(t,f_1,f_2) = \int_{t-\Delta/2}^{t+\Delta/2}\int_{t-\Delta/2}^{t+\Delta/2} x(t) \quad (6)$$
$$x(t+u_1)x(t+u_2)$$
$$e^{-j2\pi f_1u_1-j2\pi f_2u_2}du_1du_2$$

Gerr introduced the third-order Wigner distribution, which he called the Wigner bispectrum. He derived it from the Wigner distribution by retaining the lag centering property of the Wigner-Ville distribution and requiring the following property to be met; if the signal is third order stationary, then the expectation of the new definition is the same as the traditional bispectrum. Gerr's Wigner bispectrum, for a signal $x(t)$, is;

$$\mathrm{W_x(t,f_1,f_2)} = \int_{-\infty}^{\infty}\int_{-\infty}^{\infty} x(t-\frac{2}{3}u_1-\frac{1}{3}u_2) \quad (7)$$

$$x(t+\frac{1}{3}u_1-\frac{1}{3}u_2)x(t+\frac{1}{3}u_1+\frac{2}{3}u_2)$$
$$e^{-j2\pi(f_1+f_2)u_1-j2\pi f_2u_2}du_1du_2$$

Swami generalised the Wigner bispectrum by retaining the same conditions as Gerr but allowing a wider class of centered lags.

Boashash and O'Shea derived a class of higher-order spectra by recognising that the bilinear product in the Wigner-Ville distribution is the first order central finite difference phase estimator. They selected higher order phase estimators and derived the appropriate distributions. Interestingly, they select a subspace of the higher-order distribution and show it to be the optimal time-frequency distribution for signals with polynomial phase law. Their distribution is of the form;

$$\rho^g_{z_s}(t,f) = \int_{-\infty}^{\infty}\int_{-\infty}^{\infty}\int_{-\infty}^{\infty}\phi(\nu,\tau)K^g_{z_s}(\lambda,\tau)\, e^{j2\pi\nu(\lambda-t)}d\nu d\lambda e^{-j2\pi f\tau}d\tau \quad (8)$$

where;

$$K^g_{z_s}(t,\tau) = \prod_{k=1}^{q/2}[z_s(t+c_k\tau)]^{b_k}[z_s^*(t+c_{-k}\tau)]^{-b_{-k}} \quad (9)$$

and $\phi(\nu,\tau)$ is the 2D smoothing function. The conventional WVD may be recovered from the generalised one by setting $q=2$, $b_{-1}=-1$, $b_0=0$, $b_1=1$, $c_{-1}=-1/2$, $c_0=0$, $c_1=1/2$. They define time-varying higher-order spectra by taking the expected value of (9).

3 Analysis of Underwater Data

The data analysed in this paper are digitised recordings of underwater acoustic emissions of whales. The sample plots show a segment of the signal data analysed, the power spectrum of the signal, the spectrogram, the Wigner-Ville distribution, the bispectrum, the mean of the magnitude of the time-varying bispectrum normalised by the signal power in each frame, 1 slice of the time-varying bispectrum and 1 slice of Gerr's Wigner bispectrum.

4 Interpretation

It is difficult to determine much from the time domain plot, (see figure 1) other than that it looks like noise but varies in amplitude over time. The power spectrum (see figure 2) indicates a concentration of energy around the frequency range $\frac{1}{10}\frac{f}{f_s}, \frac{1}{5}\frac{f}{f_s}$. It is not clear from the power spectrum whether there is any relationship between the frequency components visible. Indeed, the power spectrum plot shows little other than the frequency extent of the signal. For example, it cannot indicate whether all frequencies are present all the time, or whether they begin and end at different times.

The Wigner-Ville distribution (WVD) (see figure 4) shows clearly a dominant spectral line starting at about $\frac{1}{3}t_f$ and increasing from approximately $\frac{1}{10}\frac{f}{f_s}$ to approximately $\frac{1}{4}\frac{f}{f_s}$ where it remains before terminating at about $\frac{4}{5}t_f$. Additional lower level signal energy is also visible, some of which will be artifacts introduced by the unsmoothed bilinear term in the definition of the WVD. Nevertheless, this distribution shows considerably more information than either the time signal plot or the power spectrum and clearly indicates that the signal is non-stationary.

The spectrogram (see figure 3) is a smoothed version of the WVD. While it removes the artifacts visible in the WVD, it also smoothes useful signal information and blurs the beginning and end of the signal transients. It shows the dominant time-varying frequencies well. Once again, this distribution is a substantial improvement on the signal plot and the power spectrum.

The bispectrum, as defined, assumes stationarity in the same manner as the power spectrum. Consequently the bispectral plot shown (see figure 5) is the average, or smeared, bispectrum over the complete signal. The critical features visible (all 12 regions of symmetry are shown) are the major peak at approximately $(\frac{1}{10}\frac{f}{f_s}, \frac{1}{5}\frac{f}{f_s})$, indicating phase coupling between these frequencies and the minor peak at approximately $(\frac{1}{5}\frac{f}{f_s}, \frac{1}{5}\frac{f}{f_s})$ which indicates the generation of harmonics. Most definitely the signal should be considered to be generated by some non-linear process, and this knowledge may help detection, analysis, classification and interpretation of signals of this type. Note that this information is not available from any of the previous plots.

The normalised power time-varying bispectrum (see figure 6) shows the mean of the magnitude of the time-varying bispectrum, for successive frames, normalised by the power in the frame of signal for which the bispectrum has been calculated. This gives an indication as to the presence of non-Gaussian and non-linear behaviour throughout the duration of the signal.

The single time slice of the time-varying bispectrum (see figure 7) indicates harmonic behaviour but does not indicate phase coupling which can only be detected when multiple realisations of the process are available.

The single time-slice of Gerr's Wigner bispectrum (see figure 8) is difficult to interpret and no conclusion is drawn from it. To calculate this distribution, the signal must be oversampled or interpolated by a factor of three. This precludes the use of standard FFT based procedures which are suitable for interpolation rates which are powers of two. Mixed radix FFT procedures can be used or alternatively the dft can be implemented directly. As a consequence, the implementation of Gerr's Wigner bispectrum is substantially slower than the time-varying bispectrum.

5 Conclusion

Conventional signal analysis procedures do not utilise all the information available in many practical signal analysis problems. It has been demonstrated that the use of higher-order spectral analysis improves time, frequency and time-frequency analysis methods and provides the analyst with important additional information. Time varying higher-order spectra are not well developed and are difficult to interpret. Phase coupling cannot be detected without some form of ensemble averaging. This important feature of the stationary bispectrum, where for single signal realisation problems, ensemble averages are often replaced by temporal averages, does not carry through to time varying higher-order spectra.

6 Acknowledgements

The authors wish to thank the Defence Science and Technology Organisation, Australia, for financial support and supplying the acoustic data considered in this paper.

References

[1] B. Boashash. Time-frequency signal analysis. In S. Haykin, editor, *Advances in Spectral Estimation and Array Processing*, volume 1 of 2, chapter 9, pages 418–517. Prentice Hall, Englewood Cliffs, New Jersey, 1990.

[2] B. Boashash. *Methods and Applications of Time-Frequency Signal Analysis*. Longman Cheshire, Melbourne, Australia, 1991. ISBN No. 582 86873 4.

[3] B. Boashash and P. J. O'Shea. Time-varying higher order spectra. In *Proceedings of 9th Kobe International Conference on Electronics and Information Sciences*, Kobe, Japan, June 1991.

[4] B. Boashash and P. J. O'Shea. Time-varying higher order spectra. In Constantinides and Cappelini, editors, *Proceedings of Digital Signal Processing Conference*, Florence, Italy, September 1991.

[5] B. Boashash and P. J. O'Shea. Time-varying higher order spectra. In Franklin T. Luk, editor, *Advanced Signal Processing Algorithms, Architectures and Implementations*, San Diego, USA, July 1991. Proceedings of SPIE.

[6] D. R. Brillinger. An introduction to polyspectra. *Ann. Math. Stat.*, 36:1351–1374, 1965.

[7] N. L. Gerr. Introducing a third order Wigner distribution. *Proceedings of the IEEE*, 76:290–292, March 1988.

[8] S. M. Kay. *Modern Spectral Estimation: Theory and Application*. Prentice Hall, Englewood Cliffs, New Jersey, USA, 1987.

[9] J. M. Mendel. Tutorial on higher-order statistics (spectra) in signal processing and system theory: Theoretical results and some applications. *Proceedings of the IEEE*, 79(3):278–305, March 1991.

[10] C. L. Nikias and M. R. Raghuveer. Bispectrum estimation: A digital signal processing framework. *Proceedings of the IEEE*, 75(7):869–891, July 1987.

[11] A. Swami. Third-order Wigner distributions: Definitions and Properties. In *Proceedings of the IEEE International Conference on Acoustics, Speech and Signal Processing*, Toronto, Canada, April 1991.

[12] J. D. Thatcher and M. G. Amin. The running bispectrum. In *Proceedings of the Workshop on Higher-order Spectral Analysis*, pages 36–40, Vail, Colorado, USA, June 1989.

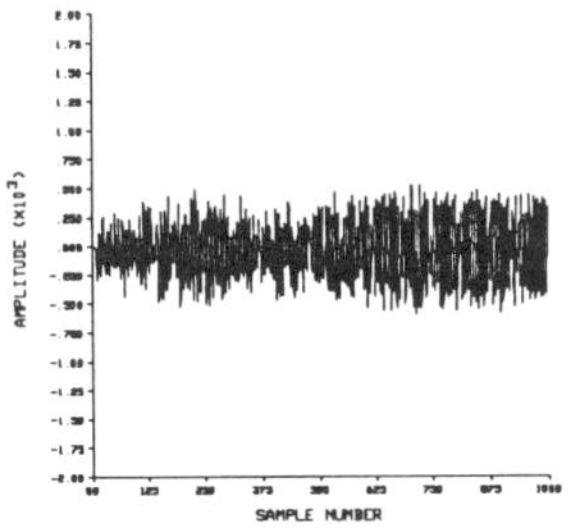

Figure 1: 1024 samples from the underwater acoustic signal data.

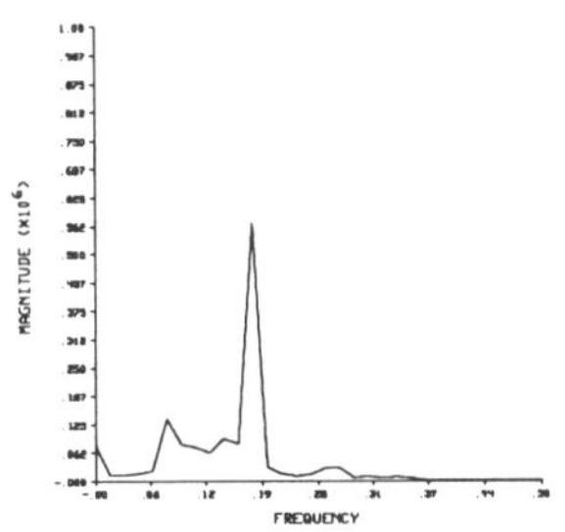

Figure 2: Power spectrum of the underwater acoustic data.

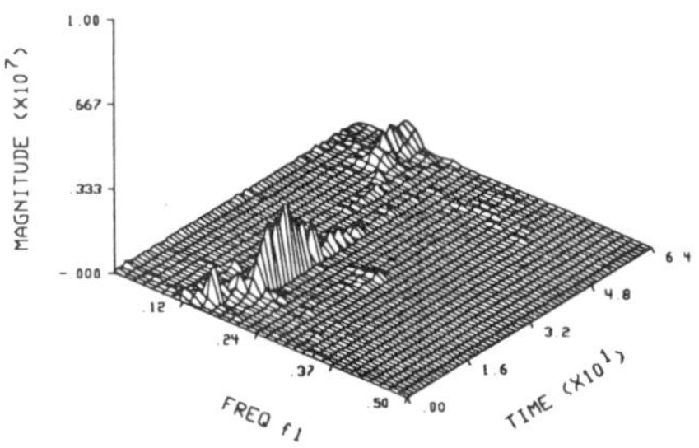

Figure 3: Spectrogram of the underwater acoustic data.

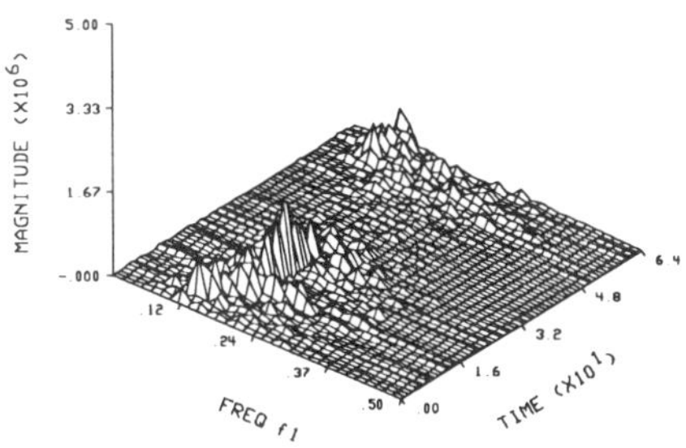

Figure 4: Wigner-Ville distribution of the underwater acoustic data.

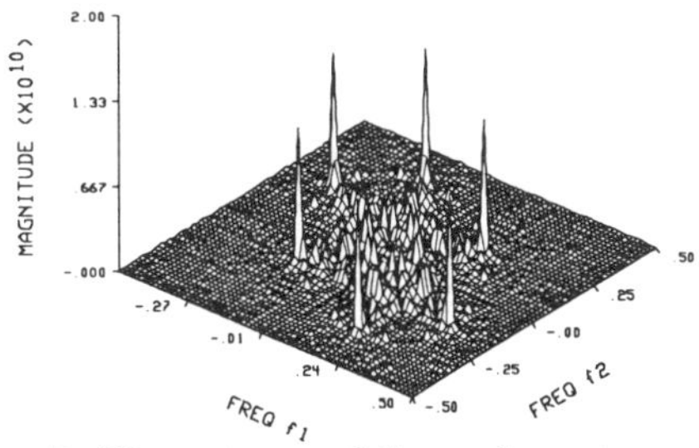

Figure 5: Bispectrum of the underwater acoustic data. (Note, all 12 symmetry regions are shown)

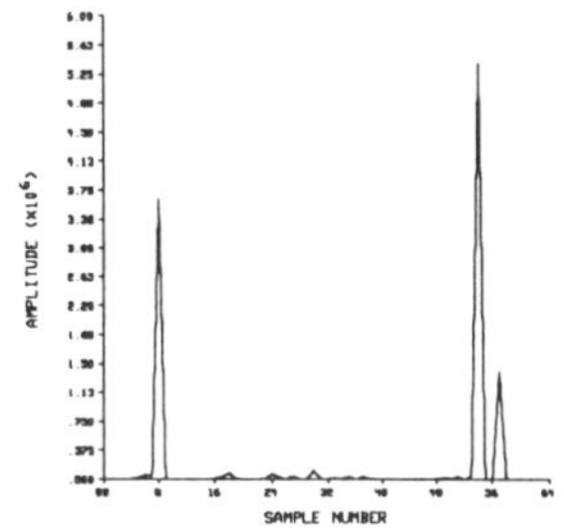

Figure 6: Normalised power time-varying bispectrum mean for the underwater acoustic data.

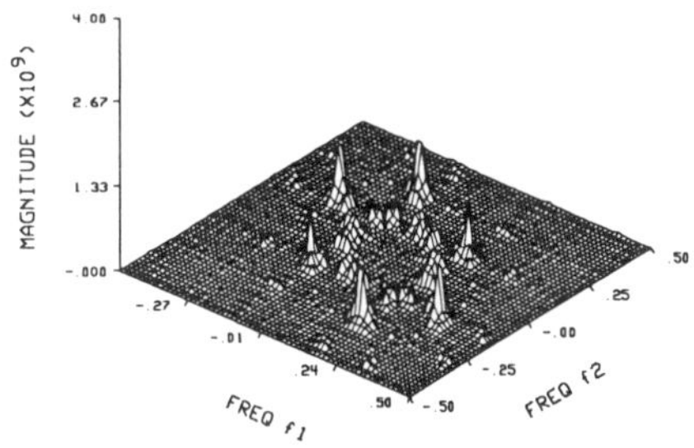

Figure 7: 1 time slice of a time-varying bispectrum of the underwater acoustic data. (Note, all 12 symmetry regions are shown)

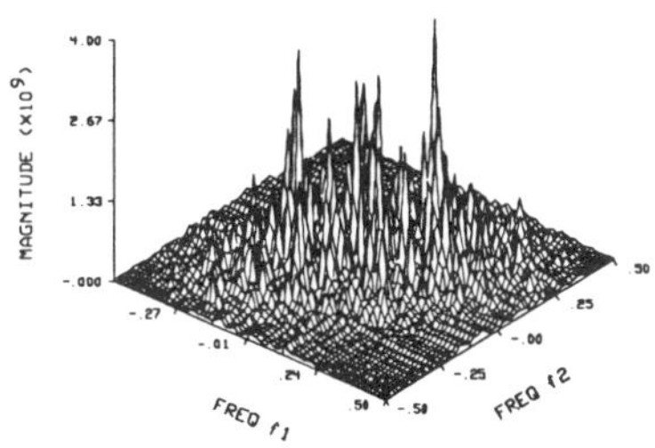

Figure 8: 1 time slice of Gerr's third order Wigner distribution of the underwater acoustic data.

"Singular Value Spectral Characterization of Transient Signals"

N.M. Marinovic

(Paper not received in time for press.)

Applications of Adaptive Time - Frequency Representations to Underwater Acoustic Signal Processing

Richard Baraniuk and Doug Jones
Dept. of Electrical and Computer Engineering
University of Illinois
Urbana, IL 61801

Tom Brotherton and Larry Marple
The ORINCON Corp.
9363 Towne Centre Drive
San Diego, CA 92121

Abstract

There is currently an interest in underwater signal processing to perform detection and classification of transient acoustic events. Traditional FFT based sonar signal processing techniques are not well suited for processing many transient signals of concern, particularly those of short duration and nonstationary nature. Described here is the application of an Adaptive-Optimal-Kernel (AOK) Time-Frequency Representation to the problem. The AOK technique employs a radially-Gaussian signal-dependent smoothing kernel obtained by solving an optimization problem and is ideal for characterizing many acoustic signals of interest. Examples are given which demonstrate the effectiveness of the approach for simulated and real sonar data.

1 Introduction

There is currently an interest in underwater signal processing to perform detection and classification of unknown, transient acoustic events. Traditional sonar signal processing techniques (i.e., FFT based algorithms) are not well suited for processing many transient signals of concern, particularly those of short duration and nonstationary nature [1]. Impulsive events and chirp type signals such as biologic clicks and squeals, underwater explosions, and the operation of mechanical devices, in particular, are not well modeled by standard FFT processing techniques.

Described here is the application of an Adaptive-Optimal-Kernel (AOK) Time-Frequency Representation (TFR), developed by Baraniuk and Jones [2], to the processing of short duration, nonstationary underwater acoustic events. Current bilinear time-frequency representations such as the Wigner [3], Choi-Williams [4], and ZAM-kernel [5] distributions apply a fixed kernel to smooth the cross-components which can obscure the true signal features and complicate interpretation of the representation. Unfortunately, selection of a fixed kernel limits the class of signals that can be analyzed effectively. To overcome this limitation, the AOK uses a radially-Gaussian *signal-dependent* kernel that changes shape to optimally smooth the distribution [2].

The optimal kernel, Φ, for a signal is defined as the solution to the following optimization problem:

$$\max_{\Phi} \int_0^{2\pi} \int_0^{\infty} \left| A(r,\psi)\, \Phi(r,\psi) \right|^2 r\,dr\,d\psi \qquad (1)$$

subject to

$$\Phi(r,\psi) = e^{-\frac{r^2}{2\sigma^2(\psi)}} \qquad (2)$$

$$\frac{1}{2\pi} \int_0^{2\pi} \int_0^{\infty} \left| \Phi(r,\psi) \right|^2 r\,dr\,d\psi \le \alpha, \quad \alpha \ge 0 \qquad (3)$$

A is the ambiguity function (AF) of the signal in polar coordinates. Once the optimal kernel is computed, the AOK Time / Frequency Representation is given by

$$P(t,\omega) = \frac{1}{2\pi} \int_{-\infty}^{\infty} \int_{-\infty}^{\infty} A(\theta,\tau)\, \Phi(\theta,\tau)\, e^{-j\theta t - j\tau\omega}\, d\theta\, d\tau \qquad (4)$$

The AOK is ideal for characterizing many acoustic transient signals of interest. Details of the AOK technique are presented in section 2. Application of the AOK technique to both real and simulated data is presented in section 3. The examples demonstrate the effectiveness of the signal-dependent approach and allow for comparison against other fixed-kernel techniques. Section 4 contains conclusions and a summary of results.

2 The Radially-Gaussian Signal-Dependent Representation

The signal-dependent kernel design procedure is formulated as an optimization problem. In general, the problem formulation involves the signal, and includes a set of constraints that define a class of kernels from which the optimal kernel is chosen, and a performance index

1058-6393/91 $1.00 © 1991 IEEE

that measures the quality of the time-frequency representation for each kernel. The kernel that maximizes the value of the performance measure is selected as the optimal kernel for the signal. Possible constraints could include, for example, constraints that force the kernel to suppress cross-components, and to satisfy the marginal distributions [4]. Clearly, the choice of constraints and performance measure is crucial to the success of the method. However, once a satisfactory set of constraints and measure are found, kernel design for a wide range of signals reduces to solving an optimization problem.

A method for optimal, signal-dependent kernel design is discussed in [6]. In that paper, the constraints force the kernel to be bounded, radially nonincreasing and of finite volume, that is, to be a two-dimensional lowpass filter, while the performance index measures the energy passed from the AF into the TFR. The optimal kernel was shown to yield excellent results for a variety of signals. A fundamental property of this technique is that the optimal kernel is a 1/0 mask, and, hence, can introduce ringing into the TFR. To overcome this problem, a new kernel design procedure using radially-Gaussian functions was introduced in [2]. This class of kernels is inherently tapered, and may potentially yield more desirable results for many signals.

The signal-dependent radially-Gaussian kernel in the (θ,τ)-plane is defined as :

$$\Phi(\theta,\tau) = e^{-\frac{\theta^2+\tau^2}{2\sigma^2(\psi)}} \tag{5}$$

where σ is a parameter which measures spread and varies with radial angle ψ.

The shape of the kernel is completely parameterized by the positive, one-dimensional function σ, so finding the optimal, radially-Gaussian kernel for a signal is equivalent to finding the optimal function, $\sigma(\psi)$, for the signal. Since $\sigma(\psi)$ controls the spread along a radial slice of the kernel at angle ψ, we call it the *spread function*. Clearly, if σ is smooth, then Φ, is also smooth. The kernel is easily expressed in polar coordinates by using $r = \{\theta^2+\tau^2\}^{1/2}$ as the radius variable:

$$\Phi(r,\psi) = e^{-\frac{r^2}{2\sigma^2(\psi)}} \tag{6}$$

In practice, TFR's are computed at discrete time and frequency locations. The polar-sampled, radially-Gaussian kernel is given by

$$\Phi_p(p,q) = e^{-\frac{(p\Delta_r)^2}{2\sigma^2(q\Delta_\psi)}}$$

$$p=0,\ldots,P-1, \text{ and } q=0,\ldots Q-1 \tag{7}$$

where $p=$ and q are the radius and angle indices, and Δ_r and Δ_ψ are the radius and angle step sizes. The discrete kernel is parameterized by a positive spread vector, $\underline{\sigma}$, whose elements correspond to samples of the spread function

$$\sigma_q = \sigma(q\Delta_\psi). \tag{8}$$

We define the optimal discrete kernel as the radially-Gaussian function whose spread vector solves the following optimization problem:

$$\max_{\Phi_p} \Delta_r^2 \Delta_\psi \sum_{q=0}^{Q-1} \sum_{p=1}^{P-1} p|A_p(p,q)\Phi_p(p,q)|^2 \tag{9}$$

subject to

$$\Phi_p(p,q) = e^{-\frac{(p\Delta_r)^2}{2\sigma_q^2}} \tag{10}$$

$$\frac{\Delta_\psi}{2\pi} \sum_{q=0}^{Q-1} \sigma_q^2 \le \alpha, \quad \alpha \ge 0 \tag{11}$$

The constraints and performance index are motivated by a desire to suppress cross-components, and pass auto-components with as little distortion as possible. The first constraint (10) limits the signal-dependent kernel to the class of radially-Gaussian kernels. Thus, the kernel is constrained to be a lowpass filter. Since the AF auto-components are centered at the origin, this encourages the kernel to preferentially pass auto-components. The second constraint (11) limits the volume of the kernel to α, so that cross-components are suppressed. The performance measure (9) expresses a desire to minimize auto-component distortion by passing as much auto-component energy as possible into the TFR for a kernel of fixed volume. An advantage of this formulation is that the constraints are insensitive to both the time-scale and orientation of the signal in time-frequency.

The computation of the discrete, optimal-kernel TFR requires samples of the AF on a polar grid, $A_p(p,q)$. These are computed either directly, or via interpolation from a rectangularly-sampled AF. Solving the optimization problem (9)-(11) for the optimal spread vector, $\underline{\sigma}^*$, determines polar samples of a radially-Gaussian kernel, Φ_p. Interpolation of the kernel to a rectangular grid and multiplication by the rectangularly-sampled AF yields the characteristic function, whose two-dimensional discrete Fourier transform is the optimal-kernel TFR.

The constrained optimization problem (9)-(11) that defines the optimal kernel is nonlinear in the variables σ_q. The solution is computed using a simple iterative solution algorithm that is based on the principle of gradient ascent, with an additional projection step that ensures that the volume constraint (11) remains satisfied.

3 Application to Underwater Acoustic Data

To demonstrate and compare the AOK time / frequency representation with other representations, a simulation was performed. A test data set was generated that was made up of a sequence of signals each of which contains chirps. There are three signals in the data. Each of the simulated signal lasts for 256 samples and is separated from adjacent signals by 256 samples of zeroes. The first signal contains a single chirp. The second signal contains two chirps that have the same chirp rate but are separated in frequency. The third signal contains crossing chirps. Figure 1(e) shows the signal structure in time and frequency.

Simulated white Gaussian noise of varying amplitude levels was than added to the data to generate a received signal at various signal to noise ratios (SNRs). The simulated data was then processed to produce the AOK TFR as well as the STFT spectrum, the Wigner distribution and the ZAM distribution. Figures 1 and 2 shows input signal as well as contour plots for each of the time frequency representations for the same set of signals at two different SNRs. The levels of the contour plots were selected by hand. Parameter settings for each of the techniques were chosen so as to give good results.

Figure 1, the high SNR case, highlights the properties of each of the techniques. The STFT (figure 1(b)) picks up the gross features of the signal, but at a low frequency resolution which is required to track the chirps with the STFT. The STFT does not resolve the two chirps in the second signal. The Wigner distribution (figure 1(c)) has better performance in that frequency estimates are of relatively high resolution. However, the plot contains many artifacts. The ZAM TFR (figure 1(d)) does a good job of tracking the signals and gives relatively high resolution estimates. However the two chirps in the second signal are not resolved. It is clear, however, that the second signal is more complicated then a single chirp. The AOK TFR (figure 1(e)) reproduces the simulated data time / frequency structure exactly.

Figure 2 shows the same set of plots, but now for the low SNR case. The results are dramatic. The signals can still be seen in the STFT and ZAM TFR with some noise coming through. The signals are completely lost in the Wigner TFR. The AOK TFR still gives an almost perfect representation of the signals.

The method was then applied to actual at-sea recorded sonar data shown in figures 3 and 4. The events here are due to man-made operations that result in an impulsive signals shown in figures 3(a) and 4(a). As seen in figure 3, there are no significant differences that can be seen between the STFT, ZAM, and AOK TFRs. The most dramatic results correspond to figure 4. As seen, the AOK TFR (figure 4(e)) indicates that the impulsive signal is made up of three different chirped signal. This signal structure can not be seen in the other TFRs.

For classification of the signal shown in figure 4(a), the AOK TFR gives a unique, unambiguous representation of the signal when compared with the other TFRs. This particularly true if the problem at hand is to use the TFRs as features for classification. It may also offer insight into determining the mechanisms that give rise to signals as shown in figure 4(a).

4 Conclusions

In this paper we have described the application an Adaptive Optimal Kernel Time /Frequency Representation to the processing of underwater acoustic data. The optimal kernel is a signal-dependent radially-Gaussian function. The simulations presented indicate that the technique should work well for a larger set of signal classes then any current fixed-kernel representation. The technique has excellent performance even in the presence of substantial additive noise. This property may be exploited for signal detection. The AOK technique appears to offer unique features that can be used to characterize and automatically classify signals of interest, particularly true when compared to other processing techniques.

References

[1] S.L.Marple and T.W.Brotherton, Detection and Classification of Short Duration Underwater Acoustic Signals by Prony's Method, *IEEE ICASSP '91*, May 1991.

[2] R.G.Baraniuk and D.L.Jones, A Radially Gaussian, Signal-Dependent Time-Frequency Representation, *IEEE ICASSP '91*, May 1991.

[3] P.Flandrin, Some Features of Time-Frequency Representations of Multicomponent Signals, *IEEE ICASSP-1984*, pp. 41. B. 4. 1--41. B. 4. 4, 1984.

[4] H.-I.Choi and W.J.Williams, Improved Time-Frequency Representation of Multicomponent Signals Using Exponential Kernels, *IEEE Transactions on Acoustics, Speech, and Signal Processing* **ASSP-37** (6), pp. 862--871, June 1989.

[5] Y.Zhao, L.E.Atlas, and R.J.Marks, The Use of Cone-Shaped Kernels for Generalized Time-Frequency Representations of Nonstationary Signal, *IEEE Transactions on Acoustics, Speech, and Signal Processing* **ASSP-38** (7), pp. 1084--1091, July 1990.

[6] R.G.Baraniuk and D.L.Jones. Optimal Kernels for Time-Frequency Analysis, *SPIE 1990 Symposium on Optical and Optoelectronic Applied Science and Engineering*, San Diego,California, July 8--13, 1990.

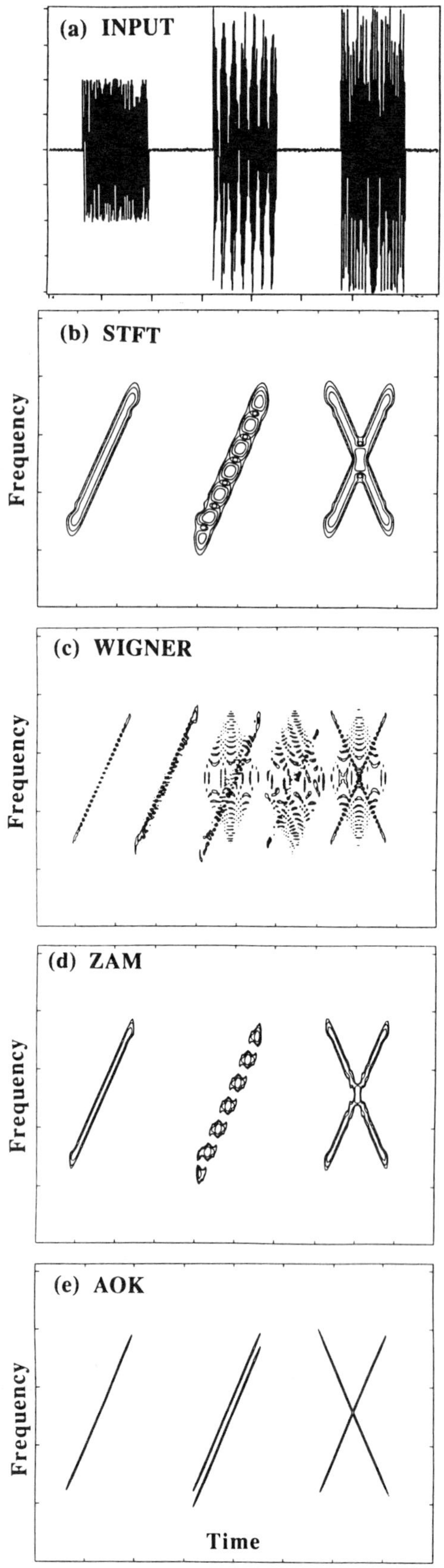

Figure 1 - High SNR Simulation

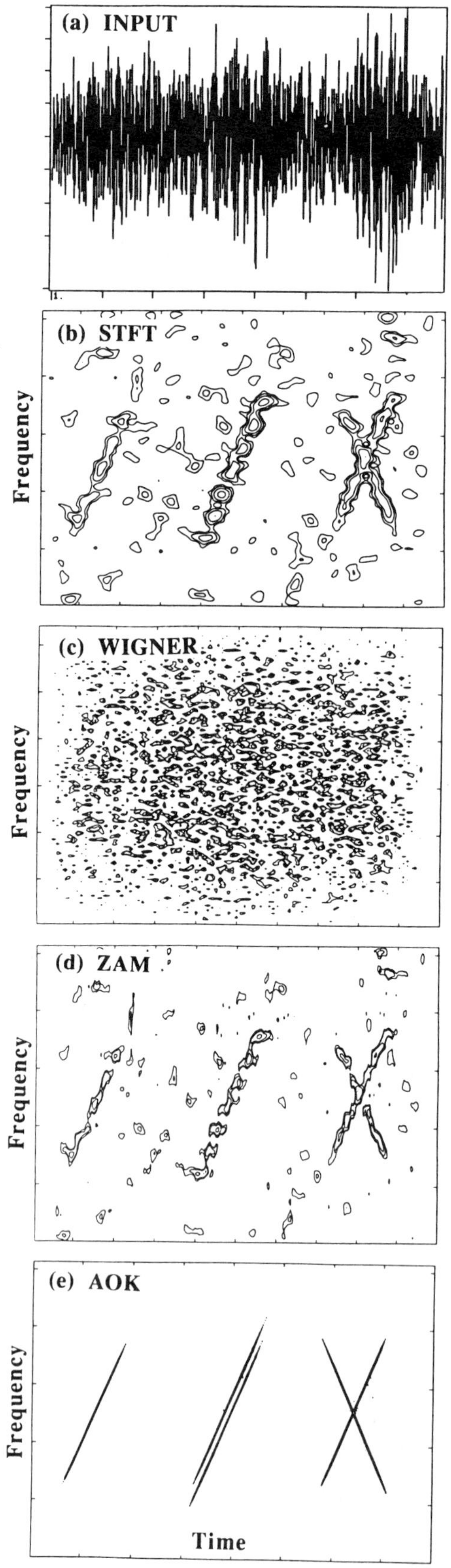

Figure 2 - Low SNR Simulation

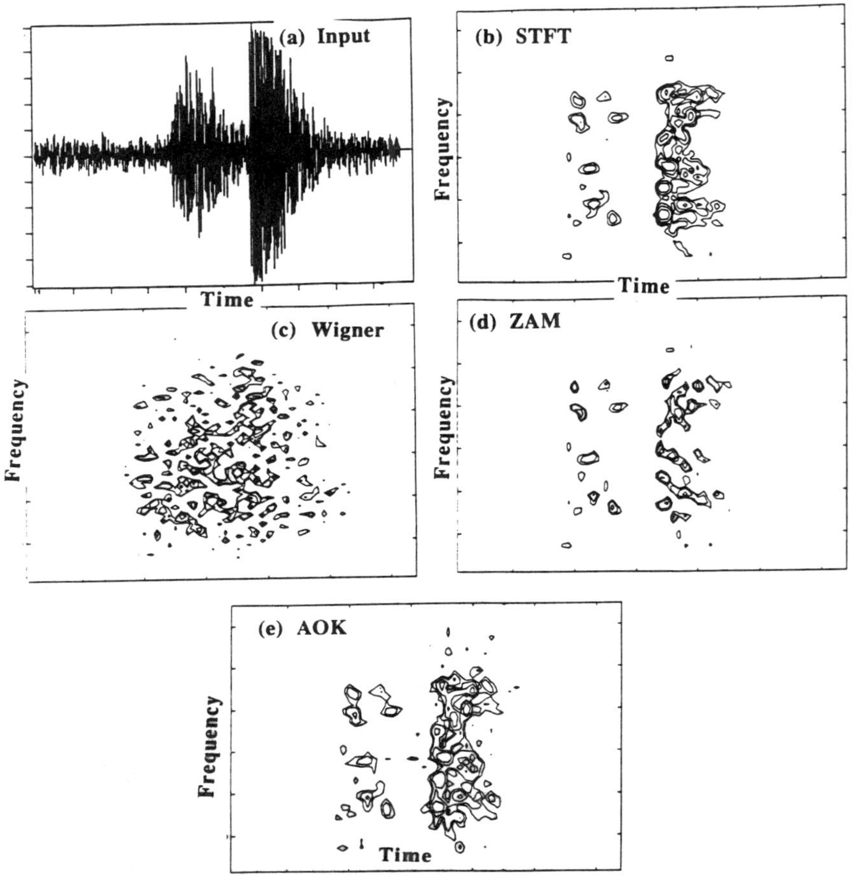

Figure 3 - Acoustic Transient Example 1

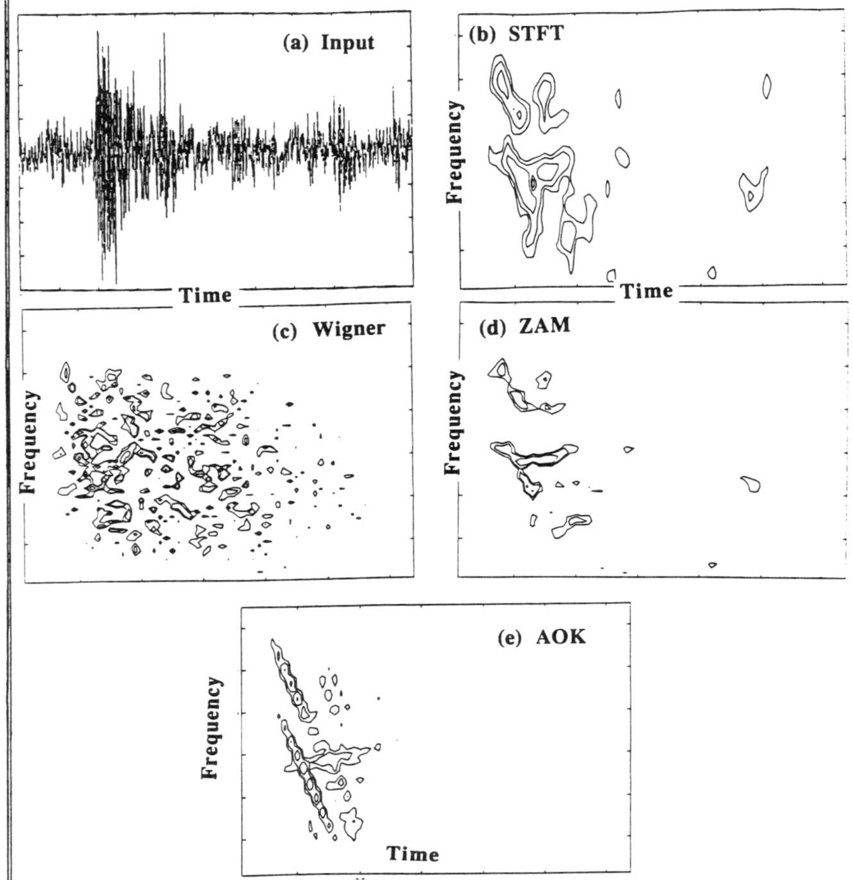

Figure 4 - Acoustic Transient Example 2

Wavelet Preprocessing of Acoustic Signals

William Y. Huang and *Manuel R. Solorzano*
Code 535
Marine Sciences and Technology Dept.
Naval Ocean Systems Center
San Diego, CA 92152–5000

Abstract

This paper describes results using the wavelet transform to preprocess acoustic broadband signals in a system that discriminates between different classes of acoustic bursts. This is motivated by the similarity between the proportional bandwidth filters provided by the wavelet transform and those found in biological hearing systems. The experiment involves comparing statistical pattern classifier effects of wavelet and FFT preprocessed acoustic signals. The data used was from the "DARPA Phase I" database, which consists of artificially generated signals with real ocean background. The results show that the wavelet transform did provide improved performance when classifying in a frame–by–frame basis. The DARPA Phase I database is well matched to proportional bandwidth filtering; i.e., signal classes that contain high frequencies do tend to have shorter duration in this database. It is also noted that the decreasing background levels at high frequencies compensate for the poor match of the wavelet transform for long duration (high frequency) signals.

1 Introduction

An ocean acoustic event classification system includes a pre–processor, a frame–level classifier, and higher level decision logic. For known signal in background, there are a number of ways to optimize at each stage to maximize overall detection/classification probabilities. For short duration ocean acoustic events, however, we look for algorithms that are robust under different training conditions. The goal of this study is to compare wavelet and Fourier preprocessing for a system with the same classifier, in order to identify the characters in the pre–processor that lead to good overall performance.

There are a number of reasons for studying the wavelet transform. Humans are excellent acoustic event classifiers. The wavelet transform provides a proportional bandwidth (bandwidth increases in proportion with center frequency), similar to the filters in the human ear. Also, progress in the wavelet field is at a fast pace. This means that systems that use the wavelet transform is supported by a rich understanding of the different aspects of the wavelet transform. Finally, the wavelet transform projects a signal into a multi–resolution space that are useful for image processing [1], speech coding [2], and sound pattern analysis [3]. For ocean acoustics, this means that wavelet transform may be able to provide a small number of relevant parameters for the classifiers — a property that usually leads to good overall performance.

Nicolas has performed an extensive comparative study between wavelet, FFT, Wigner, and other processors for a database of short duration ocean acoustic events [4]. For the DARPA Phase I dataset, Desai [5] used wavelet transform, with sophisticated feature extraction, to attain 0% error. Beck [6] performed comparative evaluation between wavelet and FFT preprocessing for neural net classification, and found that wavelets lead to better performance for this database. The contribution we make here is to offer some explanations as to why the wavelet transform seem to lead to better performance on the DARPA Phase I database. Section 2 will provide a brief introduction to the wavelet transform, which will leave out much of the mathematical properties but focus on the properties that we feel are important. Section 3 describes our approach to this problem. Section 4 describes the DARPA Phase I database and the experimental performed. Section 4.1 discusses the experimental results.

2 Background

Since the wavelet transform will be compared to the Fourier transform, perhaps the best way to introduce the wavelet transform is to compare and contrast it with the Short Time Fourier Transform (STFT) in the problem of time–frequency localization.

The STFT estimate of a signal, $f(t)$, in the time–frequency domain, for the mth frequency and nth time component, is defined as

$$c_{\mathrm{stft}}[m,n] = \int e^{-j(m\Delta\omega)t} g(t - n\Delta t) f(t) dt \qquad (1)$$

where

$$\begin{aligned} f(t) &= \text{the input signal,} \\ g(t) &= \text{a window,} \\ m\Delta\omega &= \text{frequency shift, and} \\ n\Delta t &= \text{time shift.} \end{aligned}$$

In other words, c_{sftf} is a discrete Fourier component of a windowed input. $g(t)$ is a window, and is required to have finite support in both the time and frequency domain, and centered around $t = 0$ and $\omega = 0$ in the

Table 1: Contrasts between STFT and wavelet transform.

STFT	*WAVELET*
$g(t)$ is centered at $t = 0$, $\omega = 0$ with compact support	$h(t)$ is centered at $t = 0$, with $H(\omega = 0) = 0$, and $H(\omega)$ is centered at k_0
$c_{\text{stft}}[m, n]$ samples $f(t)$ at $\omega = m\Delta\omega$, $t = n\Delta t$	$c_{\text{W}}[m, n]$ samples $f(t)$ at $\omega = k_0/a^m$, $t = bn/a^m$
same bandwidth and time resolution for all m and n, the bandwidth and time resolution of $g(t)$	bandwidth scales as $a^m(BW)$, time window scales as $(Length)/a^m$

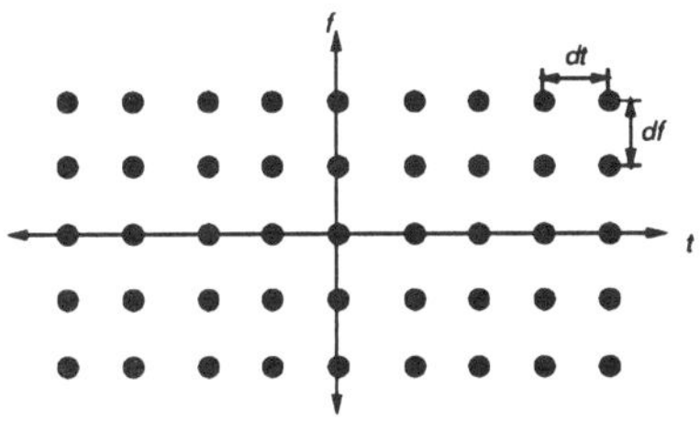

Figure 1: The sampling locations, bandwidth and time window length for STFT in the time–frequency domain.

time and frequency domain (in order to get resolution in time and frequency). The Wavelet transform, on the other hand, is defined as

$$c_{\text{W}}[m, n] = \int \sqrt{a^m} h(a^m t - bn) f(t) dt \qquad (2)$$

where

$$\begin{aligned} h(t) &= \text{wavelet function,} \\ a^m &= \text{a dilation factor, and} \\ nb/a^m &= \text{a time shift.} \end{aligned}$$

Here, $h(t)$, the wavelet function, is required to have finite support in both time and frequency, centered around $t = 0$ in the time domain, and centered around $\omega = k_0 \neq 0$ in the frequency domain. The reason for the requirement of it being not centered at 0 in the frequency domain is that frequency shifting is achieved by scaling to achieve a center frequency of $\omega = k_0/a^m$, and no frequency shifting can be achieved if $k_0 = 0$.

The differences between the wavelet transform and STFT are summarized in Table 1, and illustrated in Figs. 1 and 2. The main point is that wavelet transform provides proportional bandwidths, with wider bandwidth/higher time resolution at high frequencies, and finer bandwidth/wide time windows at lower frequencies.

To see why the wavelet is a biological model, consider the frequency response of the cochlea of a cat,

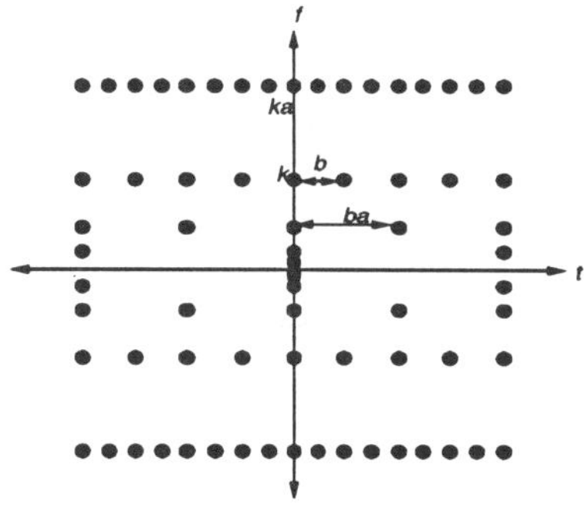

Figure 2: The sampling locations, bandwidth and time window length for wavelet transform in the time–frequency domain.

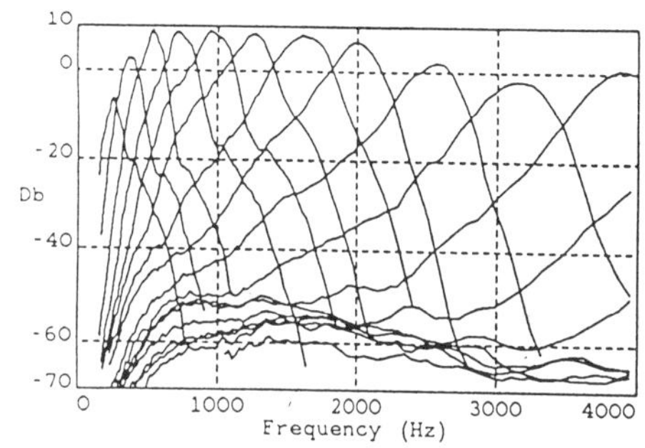

Figure 3: Frequency response of the hair cells at different parts of the cochlea of a cat. After [Bertrand, 1989]

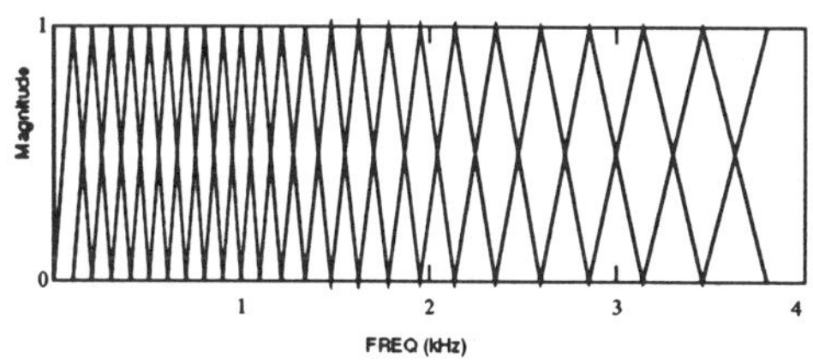

Figure 4: "Mel-scaled" filter bank: Linear spaced triangular filters between 0 and 1kHz, and logarithmic increments in both frequency and bandwidth afterwards.

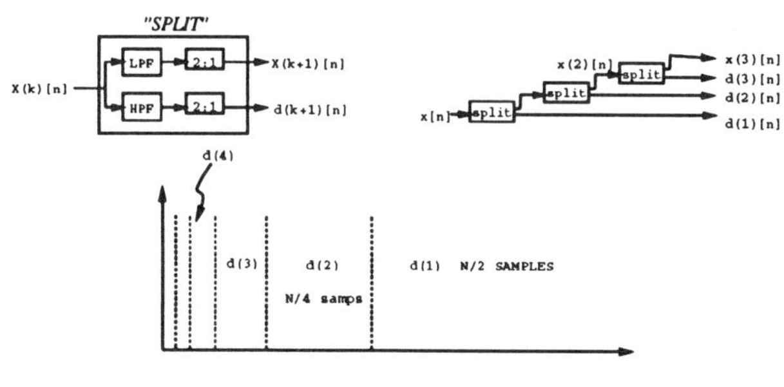

Figure 5: Octave band coder.

shown in Fig. 3. The biological ear, as in the case of wavelet, provide wider bandwidth in higher frequency than in lower frequency. This fact is taken advantage of in many speech processing front–ends. For instance, the "mel-scaled" cepstral coefficients (Fig. 4 leads to improved speech recognition performance [7]. The octave band speech coder, used for speech compression [2], has the same time–frequency resolution and sampling properties as the wavelet transform (Fig. 5).

2.0.1 Implementation

The octave band coder, shown in Fig. 5, represents the way the decimated wavelets are implemented in practice. The wavelet function is designed to pass frequencies from $\omega = \pi/2$ to π. At each stage or "octave", the wavelet function extracts the upper half of the bandwidth. The remaining signal (extracted with a similar low–pass filter) is decimated and then re–applied to the same set of filters. To improve the frequency resolution in the high frequency extraction step, the high frequency filter is divided into several high frequency, non–overlapping narrowband filter that collectively constitute the high frequency filter — and thereby creating several "voices". The "tree wavelet" [8], or "wavelet packets" [9] seeks to improve the resolution in the high frequency filter by splitting not only the low frequency component at the next octave, but also split the high frequency signal.

3 Approach

Performance comparison studies [10, 11] have demonstrated that a number of pattern classifiers will produces results near the Bayes' optimal value. Therefore, for the purpose of this study, evaluation will be restricted to the quadratic Gaussian classifier. Evaluation will be performed based on frame–level classification results. Recent studies have demonstrated that good frame–level recognition rates do not automatically extend to good event–level recognition. In [12], for instance, classifiers that provided best frame–level performance did not lead to good segmental–level performance. However, for the purpose of this study, it was felt that there is a strong correlation between the frame–level scores and segmental level scores so that we can make a valid interpretation from the frame–level scores. Here, the segmentation of data samples (assigning class labels to pre–processor frame outputs) is done by Hidden Markov Models (HMM's) and forced Viterbi decoding, where the HMM's were trained on the training set. Therefore, the segmentation boundaries are those that would lead to the highest recognition rates on both the training and testing set.

4 Experiments

The six signals of the DARPA Phase I dataset was processed by the conventional FFT (summarized in Table 2), and wavelets (Table 3). The frame–by–frame output of the pre–processors being studied are then fed into a quadratic Gaussian classifier.

Table 4 shows the percentage errors resulting from the Gaussian classifier for the various pre–processors.

Table 2: FFT Pre–processing schemes. *NAVG* is the number of FFT frames averaged together, *DATA WINDOW* is the amount of data that each frame of output examines. *FRAME SIZE* is the dimensionality of data supplied to the classifier.

id	*fft size*	*%over-lap*	*navg*	*data window*	*frame size*
ft128	128	50	1	5 msec	65
ft128avg3	128	50	3	10msec	65
ft256	256	75	1	10msec	129
ft64avg4	64	50	4	6.25msec	33

Table 3: Wavelet Pre–processing schemes for the decimated wavelet and the tree wavelet. The Morlet wavelet is a modulated Gaussian pulse. Daubechies wavelet is a 4 point orthogonal wavelet function. ΔT is the time interval between wavelet outputs.

id	*# octaves*	*# voices*	*wavelet*
decwvlt	7	4	Morlet
tree	1	8	Daubechies

id	*window*	*ΔT*	*frame size*
decwvlt	168.4msec	.5msec	28
tree	2.56msec	.04msec	17

Table 4: Final classification percentage errors.

PRE-PROCESSOR	*PERCENTAGE ERROR*
decwvlt	0.14
ft128	7.93
ft128avg3	8.98
ft256	2.58
ft64avg4	6.37
tree	23.40

Table 5: Percentage confusion matrix for DECWVLT.

TRUE CLASS	*ESTIMATED CLASS* #	A	B	C	D	E
#	99.89	0	0	.01	.02	.08
A	0	100	0	0	0	0
B	0	0	100	0	0	0
C	0	0	0	100	0	0
D	0	0	0	0	100	0
E	.19	0	0	0	0	99.81

Table 6: Percentage confusion matrix for FT128.

TRUE CLASS	ESTIMATED CLASS #	A	B	C	D	E
#	91.65	0	0	0	0	8.35
A	0	100	0	0	0	0
B	31.25	0	56.15	0	0	12.5
C	0	0	0	100	0	0
D	0	0	0	0	97.62	2.38
E	6.96	0	0	0	0	93.04

Table 7: Percentage confusion matrix for FT128AVG3.

TRUE CLASS	ESTIMATED CLASS #	A	B	C	D	E
#	88.77	0	0	.15	.31	10.77
A	0	100	0	0	0	0
B	40	0	60	0	0	0
C	0	0	0	100	0	0
D	0	0	0	0	100	0
E	3.70	0	0	0	0	96.30

Table 8: Percentage confusion matrix for FT64AVG4.

TRUE CLASS	ESTIMATED CLASS #	A	B	C	D	E
#	94.87	0	.24	0	0	4.90
A	0	100	0	0	0	0
B	0	0	100	0	0	0
C	0	0	0	100	0	0
D	0	0	0	0	100	0
E	9.15	0	0	0	0	90.85

Table 9: Percentage confusion matrix for FT256.

TRUE CLASS	ESTIMATED CLASS #	A	B	C	D	E
#	98.3	0	.14	0	0	1.56
A	14.29	85.71	0	0	0	0
B	37.5	0	62.5	0	0	0
C	0	0	0	100	0	0
D	4.44	0	0	0	95.56	0
E	5.37	0	0	0	0	94.63

Table 10: Percentage confusion matrix for TREE.

TRUE CLASS	ESTIMATED CLASS #	A	B	C	D	E
#	75.94	0	3.14	0	0	20.92
A	0	96	0	0	0	4
B	1.04	0	98.96	0	0	0
C	0	0	0	100	0	0
D	0	0	0	0	100	0
E	20.97	0	1.33	0	.51	77.19

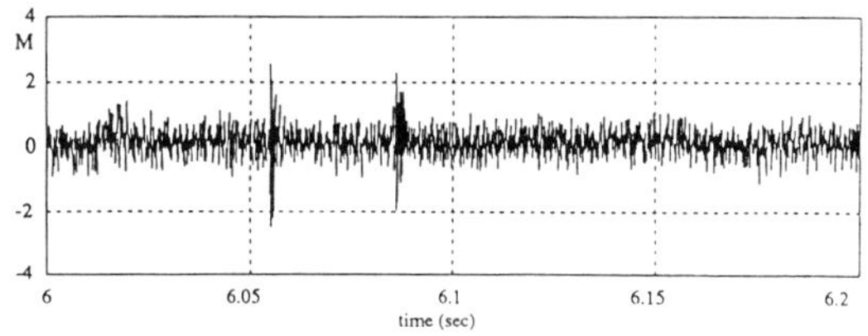

Figure 6: Signal "B".

The confusion matrices are shown in Tables 5 through 8.

4.1 Discussion of Results

The results, according to Table 4, shows that wavelet pre-processing is only marginally superior to FFT. This is consistent with the results of other studies into this problem. In fact, a statistical analysis may reveal that the difference is not significant. However, an examination of the confusion matrices reveal several distinct advantages of the wavelet transform.

First of all, the decimated wavelet performed better than the FFT based methods on signal E. Signal E is a long duration sinusoid. It occurs at a low frequency region that coincides to a region of high levels of ocean noise (high relative to other frequency bands). The time–frequency plots shows that the wavelet transform provided a sharp, distinct features for signal E. This is because at low frequency, the decimated wavelet have a large time window, corresponding to a significant SNR improvement via processing gain. The only other plots that appears to be quite distinct for signal E is that of the 256 point FFT and the 128 point FFT with 3 frame averaging. However, the decimated wavelet resulted in better performance on signal E than the 256 point FFT, because it can integrate over a time window of 168.4 msec's, whereas the 256 point FFT integrates only over 10 msec's.

In the case of signal B, Fig. 6, the wavelet transform was able to demonstrate its multi–resolution advantages. Signal B has important features in both its gross and fine details. Since signal B consists of two pulses that are approximately 30 to 40 msec apart, a pre–processor with a shorter width will likely mistaken the middle part of signal B for noise. Thus, large FFT's, such as the 256 point FFT, fared far better than the 128 point FFT with single frame averaging. The solution to this problem using FFT is to use larger FFT's; however, wavelets allow for classification of long events, such as signal B, without compromising temporal resolution.

Finally, note that the decimated wavelet provided the best overall performance. This may be an indication that it has provided a succinct, parsimonious representation of the essential features of this database. We note that the decimated wavelet output has only 28 points. The FFT can provide a small dimensionality (the 64 point FFT, for instance, provides 33 points), but not without degradation in performance.

The results provided by the tree wavelet are preliminary. It seems that 8 voices are not enough for this problem.

5 Conclusions

The use of the wavelet transform as a front end of an acoustic broadband discriminator is advantageous under at least two conditions. The first condition is for the recognition of a long duration low frequency burst in the ocean, where most of the background noise is also in the low frequency. In this low SNR case, longer wavelet filters in the low frequency provides higher SNR by a processing gain (a stronger signal component resulting from integrating the data over a longer period of time). The second condition is where multi-resolution is required. Pulse trains emitted by dolphins, for instance, have a fine feature for the individual pulses, and a course feature characterized by the frequency and duration of the pulse train. If both features are important characteristics, the wavelet transform provides a good feature extraction.

The experiments here also suggest that wavelet transforms may provide a parsimonious representation of the signal, which can represent all the essential features of a signal with few parameters. However, the these experiments do not imply that wavelet transform is better for all other databases. Using the converse of the arguments made above, one would guess that wavelet might not perform as well as FFT for a long duration high frequency signal when the amount of noise is significant.

Acknowledgement

The authors are grateful to Dr. Mark Shensa for his many helpful discussions.

References

[1] S. G. Mallat, "Multifrequency channel decomposition of images and wavelet models," *IEEE Transactions on Acoustics, Speech, and Signal Processing*, vol. 37, pp. 2091–2110, Dec 1989.

[2] M. J. T. Smith and T. P. Barnwell, "Exact reconstruction techniques for tree-structured sub-band coders," *IEEE Transactions on Acoustics, Speech, and Signal Processing*, vol. 34, pp. 434–440, June 1986.

[3] R. Kronland-Martinet, J. Morlet, and A. Grossman, "Analysis of sound patterns through wavelet transforms," *International Journal of Pattern Recognition and Artificial Intelligence*, vol. 1, no. 2, pp. 274–302.

[4] J.-M. Nicolas, A. Lemer, and J.-C. Delvigne, "Automatic identification of transient biological noises in underwater acoustics using arborescent wavelets and neural networks," in *Colloque "Ondeletts et Premières Applications, Marseille*, June 1989.

[5] M. Desai and D. J. Shazeer, "Acoustic transient analysis using wavelet decomposition," in *Proceedings of the IEEE Conference on Neural Networks for Ocean Engineering*, pp. 29–40, IEEE, August 1991.

[6] S. Beck, L. Deuser, R. Still, and J. Whiteley, "A hybrid neural network classifier of short duration acoustic signals," in *Internation Joint Conference on Neural Networks*, pp. I–119–124, IEEE, July 8–12 1991.

[7] S. B. Davis and P. Mermelstein, "Comparison of parametric representation for monosyllabic word recognition in continuously spoken sentences," *IEEE Transactions on Acoustic, Speech and Signal Processing*, vol. ASSP-28, pp. 357–66, August 1980.

[8] M. J. Shensa, "The discrete wavelet transform," Technical Report 1426, Naval Ocean Systems Center, San Diego, June 1991.

[9] R. R. Coifman, Y. Meyer, S. Quake, and M. V. Wickerhauser, "Signal processing and compression with wavelet packets," tech. rep., Department of Mathematics, Yale University, April 1990.

[10] Y. Lee, "Classifiers: Adaptive modules in pattern recognition systems," Master's thesis, MIT, May 1989.

[11] W. Y. Huang and R. P. Lippmann, "Comparisons between conventional and neural net classifiers," in *1st International Conference on Neural Network*, pp. IV–485, IEEE, June 1987.

[12] W. Huang and R. Lippmann, "HMM speech recognition with neural net discrimination," in *Advances in Neural Information Processing Systems 2* (D. S. Touretzky, ed.), pp. 194–202, San Mateo, CA: Morgan Kaufmann, 1990.

Multi - Feature Fusion Using Neural Networks for Underwater Acoustic Signal Processing

Avi Krieger, Tom Brotherton, and Eric Mears
The ORINCON Corp.
9363 Town Centre Drive
San Diego, CA 92121

Abstract

Currently there is an interest in underwater signal processing to perform detection and classification of transient acoustic events. Traditional FFT based sonar signal processing techniques are not well suited for processing many transient signals of concern due to their short duration and nonstationary nature. To overcome this difficulty, multiple features are used to characterize the events of interest and then fused together to solve the detection and classification problem. Described here is the application of a multi-feature neural net solution to perform the fusion. Two contender neural net architecture approaches are compared and results of processing underwater acoustic data are presented.

1 Introduction and Problem Description

A problem of interest in sonar signal processing is the development of techniques to detect and classify acoustic transient events. The problem is very difficult because every threat source has many classes of associated acoustic events. Moreover, the underwater environment has an even greater number of interference acoustic events, such as various biological signals, seismic activity signals, etc. Experience in processing real data which contain transients has indicated that no single signal processing technique can create sufficient features to enable one to distinguish uniquely among the large classes of transients [1]. Rather, several techniques are required to generate metrics from which transients can be classified. The fusion of these features is difficult because each feature-versus-time plot found from processing real data is a complex pattern to the human eye. Furthermore, the aggregate pattern from all the features is even more complex to perceive. An approach to combine two sets of features to improve classification is described in [2]. However, the neural net in [2] is used to obtain one of the feature sets and not for the overall classification of the two sets.

The results of a study that compares two neural net architectures, each performing the feature fusion for detection and classification of underwater transient signals, is presented here. The candidate architectures considered in this work are described in figures 1 and 2. The impact of the different architectures on the training policies is also considered in this work and the results of various training schemes are presented and an attempt is made to obtain an optimal training policy for each of the competing architectures.

The structure of the paper is as follows. The description of the neural nets and the two architectures used in this study is given in Section 2. A detailed description of the experiment is provided in Section 3. Section 4 contains a detailed description of the results. Summary and conclusions are discussed in Section 5.

2 Architecture

The candidate architectures considered in this work are described in the figures 1 and 2. Figure 1 shows a single N layer perceptron architecture. The input to that network contains all the features, extracted by various techniques, and combine into a single retina. The architecture shown in Figure 2 is a multi-neural net approach. Each of the neural nets is a K layer perceptron architecture. The first layer of neural nets are designed to solve the detection / classification problem to the degree that is possible using a single feature set in each of the networks. The second layer is a single neural net that takes the output from each of the single feature neural nets as it's inputs. Therefore, the first layer of neural nets performs a nonlinear filtering operation on the data in order to improve the overall performance while the second layer uses the 'filtered' information to solve the classification problem. The networks in architecture 2 can be trained by using two different methods. The first method trains each of the two nets in the first layer independently to a certain level. Then the output from these nets are used as inputs to train the next neural net. The second method is to train the three nets together, where only the output of the last net is used to generate the error measure. This approach can be viewed as a modification on the training of the network in architecture 1 when N=2 K. The difference is that there are no links between the nodes corresponding to different features, and so this training method can be treated as a constrained architecture where all the cross links between

1058-6393/91 $1.00 © 1991 IEEE

the two feature sets are forced to be zero. One of the evaluation goals of this study is to compare the two training methods.

To compare the performance of the two architectures, the total number of layers in each of the configurations is selected to be the same which implies that $N = 2K$. In the implementation the basic nets for the second architecture, are three layers perceptron (K=3) so that N is equal to six.

3 Experiment Description

The networks selected in this study are multilayer perceptrons. Training is performed using the back propagation training algorithm. A description of this network and the training algorithm is given in [1]. Two unique feature sets are selected to characterize the different classes. The first feature set is the magnitude squared of the Fourier transform (FT) of the time series. The second feature set is obtained by first computing an estimate of the covariance matrix of the input time series, and then computing the eigenvectors corresponding to the largest and smallest eigenvalues. The input to the neural net classifier is the FT of these two eigenvectors. To synchronize the data stream into the neural nets, the number of data points used to obtain the first feature set via the FFT is selected to be the same as the number of data points used to estimate the covariance matrix. A detailed description of the two algorithms used for the feature extraction in conjunction with the neural net classifier is given in [1].

Three different data sets were used in this study. They were selected to provide a wide spectrum of transient signals that are encountered in the analysis of under water transient acoustic signals. The first data set, denoted as class 1, contains short duration signals characterized as a quickly decaying impulse-like signal. The second set, denoted as class 2, is a longer duration signals that is characterized as long duration, broadband turbulence. The last data set, denoted as class 3, is background ocean noise. In solving detection / classification problems using neural nets the approach we've taken is to always assign a class to each new input retina. The classes assigned to the input signal are class 1, class 2, or noise only (class 3). The three data sets used contain only simulated signals that were obtained by various techniques.

The performance measures used in this study are the probability of correct classification (Pcc) of the three classes via the two architectures. A general discussion on how to evaluate the performance of a neural net classifier with real and synthetic model data is given in [3]. For the second architecture, the effect of the two training schemes on the Pcc of the three classes is also considered.

4 Results

In this study the number of exemplars used for training and testing is as follows. For class 1, there were 7 exemplars each for training and for testing. For class 2, there were 4 exemplars for training and 20 for testing. For class 3, there were 5 exemplars each for training and for testing. Performance results of 5 nets were collected. The five nets are the following.

Two nets, each a three layer perceptron with 256 input nodes, 64 nodes in the hidden layers and 3 output nodes, were used to evaluate performance on the fft and svd feature sets as single feature classifiers. The single feature classifiers form a baseline with which to compare results from the feature fusion classifiers.

Three architectures were used to evaluate the performance of the combined feature set. The first is the unconstrained network which is a 6 layer perceptron that allows cross talk between the features. That is that the first three layers are fully connected with the last three layers in the network. The network has 512 nodes in the input layer and 64, 64, 32, 8, 8, and 3 nodes in each of the hidden and output layers respectively.

The second is denoted as the constrained network. It does not allow cross talk between the first three layers until they are combined in the fourth layer The constrained network has the following topology. There are 256 nodes input and 32, 32, 16, nodes in the first layer neural network for each of the two feature sets. The first neural net layer is combined with three more layers which contain 8, 8, and 3 nodes. Note that the total number of nodes in each of the constrained and unconstrained architectures is the same.

The last architecture, denoted as the fused architecture, is a variation on the training method of the constrained architecture. Here the output from the two nets that were employed for each of the feature sets alone, is combined with a third net with the following topology; 6 input nodes, and 32, 32, and 3 nodes in the hidden and output layers, respectively. Since the data has only three classes, each of the 5 nets has 3 output nodes, one for each class.

Each of the five nets is trained to achieve the same mean squared output error. This is to ensure that all the architectures have the same level of performance on the training data prior to testing.

The test statistic that was used in the experiment is the activation level of the output node for the class of interest. To evaluate the effect of noise on the network performance, the noise level in the test set was varied. A uniformly distributed noise value was added to the feature value at the input to the first layer. Figure 3 summarizes the results obtained in this study. In figure 3 the columns correspond to results obtained for the three classes used, while the rows correspond to the results found by adding different noise values. Noise 0, Noise 1, and Noise 3, refers to no noise, noise with variance 1/12, and noise with variance 1/4, respectively. Noise samples with these different variances was added to the input feature retinas.

For class 1 (column 1 in figure 3), the 5 different methods give consistently good results for low noise case. However, for the large noise case, the performance of the unconstrained and constrained networks deteriorates while that of the fused network remains good. Notice that in this case the svd feature net has the best performance. This is expected due to the impulsive nature of the data in this class [1]. For class 2, the constrained and unconstrained network perform well without noise. However, with additive noise, the performance of the constrained net suffers. The performance of the unconstrained net and the fused net is less affected by the noise. For this class the FFT net performs well when the noise level is low. For class 3, the fused net perform the best for low noise level. The performance of the unconstrained net deteriorates quickly when the noise level is increased.

5 Summary and Conclusions

A study that examines two different neural nets architectures is described in this paper. It is seen that in some cases a neural net, using a single input feature, can perform well. However, the results obtained indicate that the fused network is the most robust with respect to different level of signal with additive noise and across the classes considered. Note that a performance analysis of this type is never completed without the consideration of the probability of false alarm (Pfa). This quantity is needed for the thresholds selection for the various classes. This work describes only the first step in the evaluation process. Larger data sets are needed for each of the classes as well as larger varieties of classes, in addition to the Pfa consideration.

References

[1] T. W. Brotherton, "DARPA Acoustic Transient Signal Processing Survey," Technical Report under Contract No. N66001-87-D0145, Orincon Corporation, San Diego, 1990.

[2] B. P. Yuhas, M. H. Goldstein, Jr., T. J. Sejnowski, and R. E. Jenkins, " Neural Network Models of Sensory Integration for Improved Vowel Recognition," Proceeding of the IEEE, pp. 1658-1668, October 1990.

[3] R. Barton, D. B. Fogel, A. Krieger, " Calibrating the Performance of Neural Networks, " Proc. of the IEEE Conference on Neural Networks for Ocean Engineering, pp. 343-355, August 15-17, 1991.

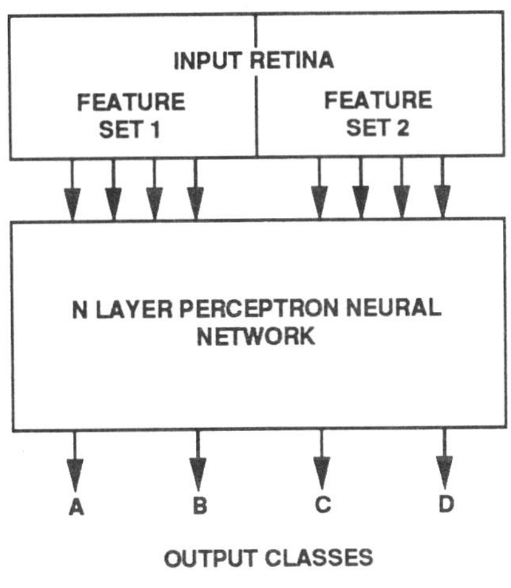

Figure 1 - Single Neural Net Approach

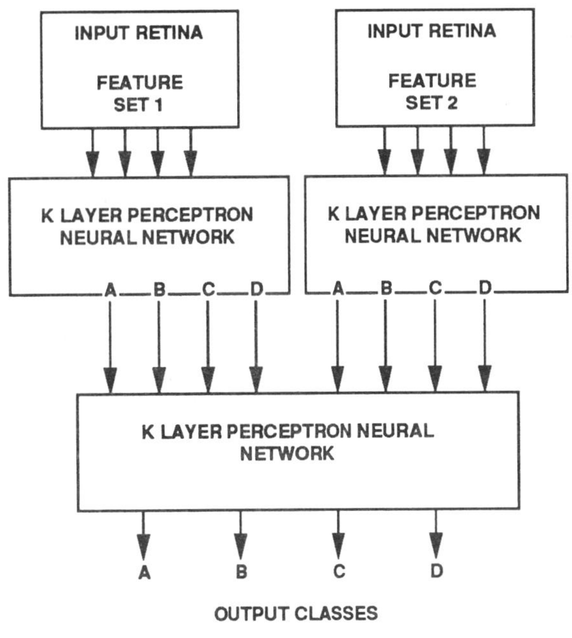

Figure 2 - Multi-Layer Neural Net Approach

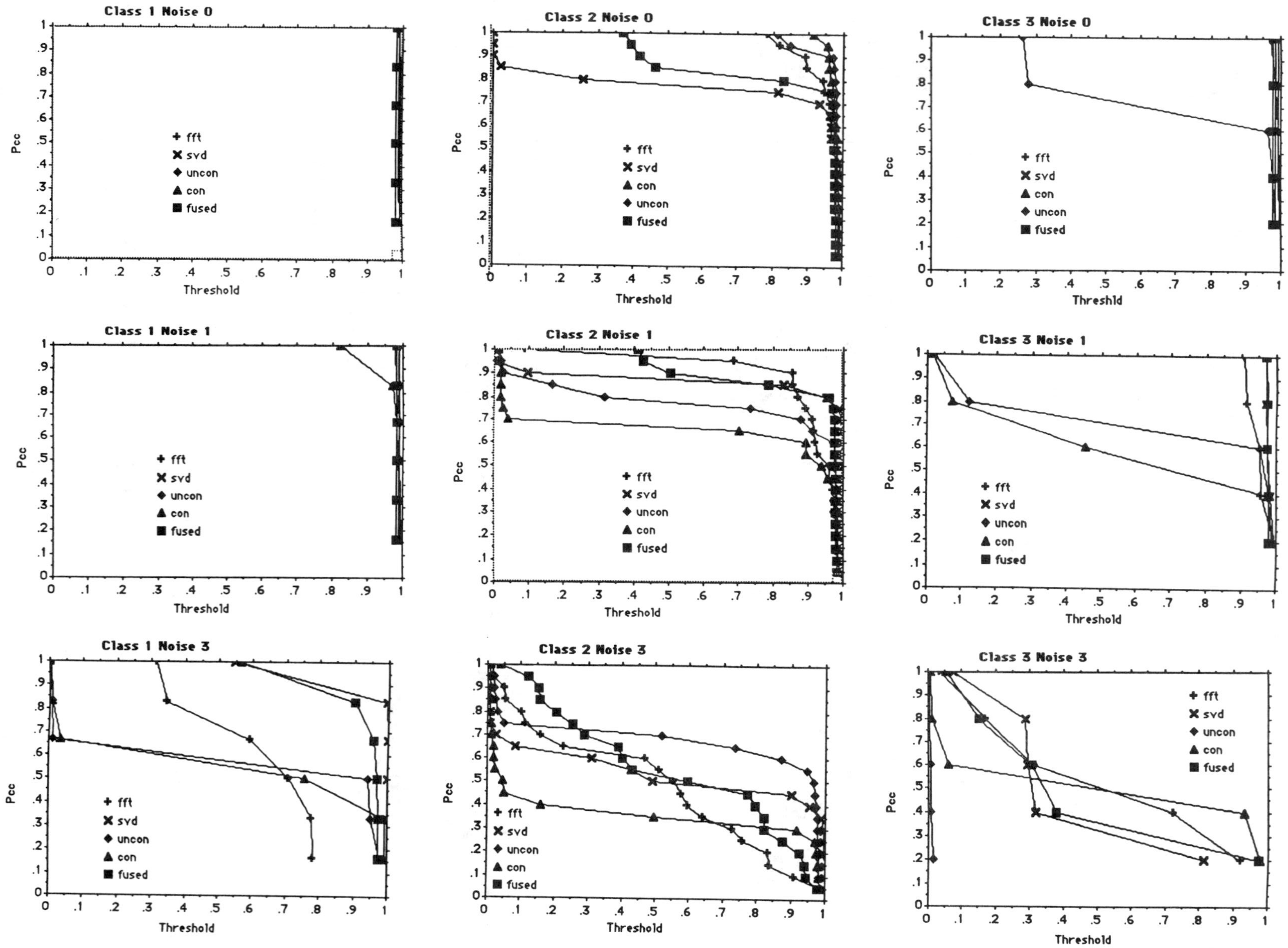

Figure 3 - Neural Net Feature Fusion Pcc Results

NEURAL NETWORK IMPLEMENTATION TO CONTROL SYSTEMS: A SURVEY OF ALGORITHMS AND TECHNIQUES

Philip Tai
Helen A. Ryaciotaki-Boussalis
David Hollaway

California State University, Los Angeles
Los Angeles, CA 90032

ABSTRACT

In the present paper a number of neural network applications which have been proposed in the last twenty years are discussed. Algorithms and techniques developed by different researchers working in the field are also summarized here.

INTRODUCTION

In recent years there is increase in interest shown in the utilization of neural networks for various research fields such as robotics, optimization, linear and non-linear programming, associative memory, image processing, speech and pattern recognition, and computer vision. This is due to the advances in neural training algorithms and also the availability of fast parallel architectures, including very large system integration, electro-optical and dedicated architectures that have made possible extremely fast, relatively low cost microprocessors for implementing these algorithms.

The need to deal with increasingly complex systems, the need to accomplish increasingly demanding design requirements, and the need to control under increased uncertainty have led a number of researchers to a reevaluation of the conventional control methods. The need to control, in a "better way", complex dynamical systems under significant uncertainty made the search of new methods quite active. Neural networks with their massive parallelism and the learning capabilities appear to offer new promising directions toward better understanding and perhaps even solving some of the most difficult control problems.

Interest in neural networks has increased in this decade after a period of inactivity following the shortcomings of early neural networks which were published in the late 1960's. The challenge that the control community faces is to find the best way to fully utilize this powerful new tool in control. The present paper discusses a number of neural network applications which have been proposed in recent years. Designs using neural networks ranging from feed-through systems with no feedback, to fully interconnected systems are briefly described. A summary of algorithms and techniques developed by researchers in the field are also discussed here.

NEURAL NETWORKS AND CONTROL SYSTEMS

In this section we will provide an overview of areas that several researchers had developed algorithms and techniques for control dynamical systems utilizing neural networks. Specifically, the use of neural networks for system identification, adaptive control, modeling of chemical processes, optimization, fault detection, and control of robotic manipulators is described in the sequel.

System identification in the time and frequency domains is examined by Narendra [1] and Chu [19]. In [1] Narendra and Papastathy demonstrate that neural networks can be used for identification and control of nonlinear dynamical systems. Static and dynamic back-propagation methods for the adjustment of parameters are applied. Chu, Shoureshi and Tenirio in [19] present two approaches for utilization of neural networks in identification of dynamical systems. First,

1058-6393/91 $1.00 © 1991 IEEE

a Hopfield network is used to implement a least-squares estimation for time varying and time invariant systems. The second approach utilizes a set of orthogonal basis function and Fourier analysis to construct a dynamic system in terms of its Fourier coefficients. **Design of neural controllers** to control nonlinear dynamical systems is addressed by several researchers. Specifically, Psaltis and Sideris in [2] use a multilayered neural network composed of feedback and feedforward controllers to control a given plant. They use several learning architectures to train the neural controller in order to provide appropriate inputs to the plant so the desired response is obtained. A year later, Anderson in [4] designs a controller that deals with issues of delayed performance evaluation, learning under uncertainty and the learning of nonlinear functions. The following year, Kraft and Campagna in [20] compare a neural network-based controller with two traditional adaptive controllers, a self-tuning regulator and a Lyapunov-based model reference adaptive controller. Their results indicate that the neural network approach functions well in noise, works for linear and nonlinear systems and can be implemented very efficiently for large-scale systems. The same year, Chen [21] shows that neural networks can be combined with self-tuning control algorithms to control a class of unknown discrete-time nonlinear systems. Via simulation he shows that his proposed self-tuning scheme can deal with a large unknown nonlinearity. Further, Nguyen and Widrow in [23] show how a neural network can be used to solve highly nonlinear control problems. Antsaklis, Passino, and Sartoni in [3] use a neural network called the multilayred perceptron to perform numeric-to-symbolic conversion.

Modeling using neural networks is presented by the following articles: Lippman in [5] presents neural network models that can be used for pattern classification. Two years later, Moody and Darken in [13] propose a network architecture which uses a single internal layer of locally-tuned processing units to learn both classification tasks and real-valued function approximations. The proposed networks learn faster than the ones using back-propagation method. The following year, Bhat, Minderman, McAvoy, and Wang [16] use neural networks for modeling non-linear chemical systems. Specifically, they use back-propagation network to model the system.

The utilization of neural networks for **optimization** purposes is addressed by Bavarian and Rauch in 1988 and Segura in 1989. Specifically, Bavarian in [6] uses two simple examples to illustrate control and optimization with neural network architecture. Rouch, and Winarske in [9] introduce the use of neural network computational algorithms to determine optimal traffic routing for communication networks. Further, Segura , and Burl in [12] design a neural network controller with the ability to learn on line the optimal control law for a crew equipment retriever.

Methods to provide **adaptive control** for nonlinear systems are introduced by the following two papers. Guez, Eilbert, and Kam in [8] propose a computing architecture for adaptive control based on computational features of nonlinear neural networks. Two years later, Goldenthal, and Farrell in [14] propose the use of neural networks in adaptive control loops. Specifically, they present an extension of the back-propagation algorithm which adaptively determines the interconnection parameters necessary for the neural network to function as a closed-loop controller and to force the closed-loop system to match a desired reference response.

The article by Naidu, Zafiriou, and McAvoy [18] addresses the use of the back-propagation neural network for sensor **failure detection** in process control systems.

The main emphasis in the next three articles is the **control of robots** using the learning capabilities of neural networks. Kawato, Uno,

Isobe, and Suzuki in [7] based on physiological information and previous madels propose computational theories for the (a) determination of a desired trajectory in the visual coordinates (b) transformation of the trajectory from visual coordinates to body coordiates and and they furtheer introduce a hierarchical neural network model to deal with motor command. The application of this approach to robotics is also outlined. Negata, Sekiguchi, and Asakawa in [22] preaent a mobile robot whose behavior is controlled by a structured hierarchical neural network. Further, a learning algorithm for real-time control is developed , and the "training" of the robot is discussed. The same year, Handelman Lane, and Gelfand in [24] present a methodology for integrating neural networks and knowledge-based systems for the purpose of robotic control, patterning the integration after models of human motor skill acquisition Information about learning algorithms in neuraol networks is given in the next article. Sanner and Akin in [15] develop control algorithms to train neural networks to regulate the pitch attitude of an underwater telerobot.

The **stability** problem of dynamic systems is addressed by Li and Chua. In 1988 Li, Michel, and Porod [10] investigate the dynamic properties of a class of neural networks by studying the qualitative behavior of equiligrium points. Specifically, they obtain results related to the stabillity properties of an equilibrium, asymptotic behavior of solutions. Further, they design a neural network with prespecified equilibrium points which are asymptotically stable. The same year, Chua and Yang in [11] perform an in-depth analysis in the Lyapunov sense, as well as, computeer simulations of the network are performed.

ALGORITHMS

There are at least five algorithms used by researchers to control dynamical systems utilizing neural networks. Specifically, the five algorithms are:

1. Supervised Control /Learning.
2. Inverse Dynamics.
3. Stabilization.
4. Propagation Through Time .
5. Adaptive Critic Systems.

In the sequel we will summarize each of the methods mentioned above.

Supervised Control/Learning

In supervised learning, a neural network is given a set of training inputs X(t) and targets Y(t). To complete the design one needs to determine "H" i.e. design some kind of network to perform supervised learning. Back-propagation is one kind of supervised learning module.

Inverse Dynamics

In inverse dynamics the observed state X(t) of a system is assumed to be a function of current actions u(t) and the prior state if the system:

$$X(t) = F(u(t), X(t-1)) \qquad (1)$$

We do not have to know F but we assume that it is invertible as function of X(t) for any X(t-1). In the training or adaptation phase one is given examples of X(t) which resulted from actual inputs u(t) tried out in some experiment. Supervised learning is used to adapt the network such that

$$u(t) = H(X(t), X(t-1)) \qquad (2)$$

Then in the application phase, given a desired trajectory X*(t) the network H inputs X*(t) and X(t-1) and outputs a control vector u which will lead the system to match the desired trajectory.

Stabilization

Several researchers have worked extensively in the stabilization of dynamic systems utilizing neural networks. There are some promising results in the area but there is space for future work. Standard designs like the self-tuning regulator and model reference adaptive control have been proven to be adequate and stable.

Backpropagation Through Time

Suppose that the model predicts that

$$R(t) = F(R(t-1), u(t))$$

where R(t) is a vector describing the state of reality at time t, u is a vector of controls, and F is implemented by the neural net. Suppose also that there is no random noise in the plant. Further, suppose we try to optimize V(R) which is differentiable over all R(t) for all t.

Using this method one may calculate the derivatives of V with respect to u(t) at all times. By adapting u(t) in response to these derivatives, one obtains the optimal actions or weights.

Adaptive Critic Systems

Adaptive critic systems can be used as in the previous method except that full allowance is made for the presence of noise. This approach requires certain approximations that make it less accurate than backpropagation in time in the case that noise may be neglected.

Since exact use of dynamic programming to optimize the critic in complicated problems is exceedingly complex, the approach is simplified by approximating the functional J(R) as

$$J(R) = \max_{u}[J(F(R,u,w))] + U(R) - Uo$$

where F is represented by a neural network. Several methods may be used to make the above equation approximately true, including, heuristic dynamic programming, dual heuristic programming(DHP) and globalized DHP.

SUMMARY

The use of neural networks in the areas of identification, robotics, detection, adaptive control, modeling, optimization, and robotics has been addressed by several researchers in recent years. In the present paper a survey of articles utilizing neural networks to control linear and nonlinear dynamical systems is provided. A summary of algorithms and techniques used by several researchers working in the field is provided. Further research is needed in all areas mentioned above to fully justify that a powerful tool such as neural networks can be used for analysis and design of complex high order dynamical systems.

REFERENCES

[1] Narendra, K., and K. Parthasarathy, "Identification and Control of Dynamical Systems Using Neural Networks", IEEE Transactions on Neural Networks, Vol. 1, no 1, March 1990.

[2] Psaltis, D., A. Sideris, and A.Yamamura, " A Multilayered Neural Network Controller," IEEE Control Systems Magazine, April 1988.

[3] Passino, K., M. Sartoni, and P. Antsaklis, "Neural Computing for Numeric-to-Symbolic Conversion in Control Systems, " IEEE Control Systems Magazine, April 1989.

[4] Anderson, C., "Learning to Control an Inverted Pendulum Using Neural Networks", IEEE Control Systems Magazine, April 1989.

[5] Lippmann, R., "An Introduction to Computing with Neural Nets, " IEEE ASSP Magazine, April 1987.

[6] Bavarian, B., "Introduction to Neural Networks for Intelligent Control," IEEE Control Systems Magazine, April 1988.

[7] Kawato, M., Y. Uno, M. Isobe, and R. Suzuki,"Hierarchical Neural Network Model for Voluntary Movement with Application to Robotics," IEEE Control

Systems Magazine, April 1988.

[8] Guez, A., J. Eilbert, and M. Kam, "Neural Network Architecture for Control", IEEE Control Systems Magazine, April 1988.

[9] Rauch, H., and T. Winarske, "Neural Networks for Routing Communication Traffic", IEEE Control Systems Magazine, April 1988.

[10] Li, J., A. Michel, and W. Porod, "Qualitative Analysis and Synthesis of a Class of Neural Networks", IEEE Trans. on Circuits and Systems, Vol. 35, no 8, pp. 976-986, August 1988.

[11] Chua, L., and L. Yang, " Cellular Neural Networks: Theory", IEEE Trans. on Circuits and Systems, Vol. 35, no 10, pp. 1257-1272, October 1988.

[12] Segura, C., and J. Burl, " Artificial Neural Network Control for Spacecraft Attitude Stabilization", Proc. 23rd Asilomar Conference on Signals, Systems, and Computers, Oct. 1989.

[13] Moody, J., and C. Darken, " Fast Learning in Networks of Locally-Tuned Processing Units", Neural Computation 1, pp. 281-294, 1989.

[14] Goldenthal, W., and J.Farrell,"Application of Neural Networks to Automatic Control", Proc. of the 1990AIAA Conference, pp. 1108-1112.

[15] Sanner, R., and D. Akin, "Neuromorphic Pitch Attitude Regulation of an UnderwaterTelerobot", IEEE Control Systems Magazine, April 1990.

[16] Bhat, N., P. Minderman, T McAvoy, and N. S. Wang, "Modeling Chemical Process Systems via Neural Computation", IEEE Control Systems Magazine, April 1990.

[17] Huang, S. C., and Y. F. Huang, "Learning Algorithms for Perceptrons Using Back-Propagation with Selective Updates", IEEE Control Systems Magazine, April 1990.

[18] Naidu, S., E. Zafiriou, and T. McAvoy, "Use of Neural Networks for Sensor Failure Detection in a Control System," IEEE Control Systems Magazine, April, 1990.

[19] Chu, S. R., R. Shoureshi, and M. Tenorio, "Neural Networks for System Identification", IEEE Control Systems Magazine, April 1990.

[20] Kraft, L. G., and D. Campagna, " A Comparison Between CMAC Neural Network Control and Two Traditional Adaptive Control Systems", IEEE Control Systems Magazine, April 1990.

[21] Chen F., "Back-Propagation Neural Networks for Nonlinear Self-Tuning Adaptive Control", IEEE Control Systems Magazine, April 1990.

[22] Nagata, S., M. Sekiguchi, and K. Asakawa,"Mobile Robot Control by a Structured Hierarchical Neural Network", IEEE Control Systems Magazine, April 1990.

[23] Nguyen, D., and B. Widrow, "Neural Networks for Self-Learning Control Systems", IEEE Control Systems Magazine, April 1990.

[24] Handelman, D., S. Lane, and J. Gelfand, "Integrating Neural Networks and Knowledge-Based Systems for Intelligent Robotic Control, " IEEE Control Systems Magazine, April 1990.

VOLTERRA CHARACTERIZATION OF NEURAL NETWORKS

N.Z. Hakim*, J.J. Kaufman#, G. Cerf+, H.E. Meadows*

* Department of Electrical Engineering, Columbia University, New York, NY 10027.
# Department of Orthopaedics, The Mount Sinai School of Medicine, New York, NY 10029.
+ NYNEX Science and Technology, 500 Westchester Ave., White Plains, NY 10604.

ABSTRACT

This paper analyzes the application of the Volterra theory of nonlinear systems to neural networks. Expressions to efficiently compute the Volterra kernels of both a single hidden layer feedforward neural network and a recurrent one are presented. We also address the issue of functional representation of the recurrent neural network architecture, and delineate a class of systems that can be approximated by this model. Computer simulations are also presented which indicate that neural networks characterization by their Volterra expansion may be useful in optimal architecture selection and assessment of learning and generalization.

I-INTRODUCTION

Neural networks have become important tools in various research areas such as pattern recognition, adaptive control, and nonlinear modeling and filtering. Neural network models require a learning phase during which they are trained to operate in a desired manner, i.e. to mimic a model ideally suited for the task. Training yields a neural network approximating more or less closely the desired system, according to some error criterion. Together with learning, appropriate choice of the neural network architecture is necessary to provide good performance, not only on the training set, but also on new data. Several neural network structures have been suggested and applied to various tasks. Some are feedforward models, while others provide feedback connections among the elements and constitute truly dynamical systems. The performance of a given model on a particular task is strongly related to its potential to approximate a large class of functions and therefore to its functional representation power. This is a fundamental issue when using an ad hoc structure, such as a neural network, as a black box model. While we know, for example, that feedforward networks can approximate arbitrarily closely continuous nonlinear mappings [1,2,3,4], it is not clear what class of systems they can model, when inserted in a large dynamical structure [5] or, similarly, what class of systems recurrent networks [6,7,8] can approximate.

A coherent framework, in which these different models may be studied and compared, is provided by the Volterra theory of nonlinear systems [9,10,11,12]. Issues that can be addressed within this framework include assessment of learning and generalization, comparison of several architectures with respect to particular tasks, and functional representation power of these models.

The purpose of this paper is to analyze the application of the Volterra theory to neural network modeling of dynamical systems. Section II introduces briefly this theory and section III describes its application to specific neural network architectures. In section IV we present some examples illustrating this approach, and conclude by a short summary and a discussion of the results.

II-THE VOLTERRA THEORY

In contrast to linear systems, which are completely described by their impulse response, nonlinear systems cannot generally be represented by a simple input-output relationship. A large class of nonlinear time invariant systems with continuous input-output relationship, however, admit a functional representation within some convergence domain, called the Volterra expansion [9,10,11,12]. In the single input single output (SISO) stationary discrete-time case, it takes the form

$$y(n)=\sum_{k=0}^{\infty} \sum\sum\ldots\sum_{k_1,k_2,\ldots k_k=1}^{\infty} V_k(k_1,k_2,\ldots,k_k)\, u(n-k_1)u(n-k_2)\ldots u(n-k_k) \qquad (1)$$

where u(.) and y(.) are the input and output of the system, respectively. The term V_k is the k'th order standard Volterra kernel of the system and plays the role of a k-impulse response of the system. The kernels in Eq. 1 are symmetric with respect to their arguments, therefore it is possible to group all those affecting identical products of inputs in a unique kernel with arguments listed in increasing order, resulting in triangular kernel form

$$y(n)= \sum_{k=0}^{\infty} \sum\sum\ldots\sum_{1\le k_1\le k_2\le\ldots\le k_k}^{\infty} h_k(k_1,k_2,\ldots,k_k)\, u(n-k_1)u(n-k_2)\ldots u(n-k_k) \qquad (1a)$$

The symmetric kernels are obtained by multiplying the standard kernels by a combinatorial factor easily derived [13]. Volterra modeling has been of limited practical use, chiefly because convergence is not guaranteed outside a possibly very small domain, and the kernels $h_k(k_1, k_2,\ldots, k_k)$ are not identifiable in general from input-output data. The intimate relation of the Volterra expansion with Taylor series partly explains these properties. It is possible, however, to extend this concept to arbitrary intervals and to continuous functions not admitting a Taylor series expansion, using the Weierstrass theorem of polynomial approximation. This theorem states that any continuous function can be approximated arbitrarily closely over a given interval by a polynomial of sufficiently large order [1,2].

The class of systems admitting a stationary Volterra expansion is fairly large and includes most smooth stationary systems with non-infinite memory of the past [10].

1058-6393/91 $1.00 © 1991 IEEE

III- VOLTERRA CHARACTERIZATION OF NEURAL NETWORKS

In this section, we present expressions to compute the kernels of two neural network structures, discussing existence and uniqueness of the Volterra expansion. We then study the problem of functional representation power of a recurrent neural network model by analyzing the existence of a model that has arbitrary kernel values.

A. KERNEL EXPRESSIONS

We focus here on the problem of characterizing a neural network by its Volterra kernels. We distinguish two network models: A single hidden layer feedforward network with linear output layer, and a fully dynamic recurrent model. The networks are assumed to be stationary, and modeling SISO systems.

1. Feedforward network model

The Stone-Weierstrass theorem can be used to show that feedforward networks are universal function approximators [1,2,3,4]. Using this result, they can also be universal dynamical system approximator, by defining the input as a sufficiently large set of past delayed inputs. They require however a very large number of elements to accurately model long memory systems. Simulations confirmed the power of feedforward networks to approximate a large number of nonlinear mappings [14,15]. These simulations included the prediction of time series generated by the logistic function and the Mackey-Glass equation, two chaotic deterministic iterative mappings. In these studies the input to the network consisted of a window in the input sequence, which provided the static neural network with some finite memory. The feedforward structure we study here consists of an input layer, a single hidden layer of N nonlinear elements (NE) and a linear output layer feedforward network containing one linear summation element. The number of units in the input layer defines the window size w in the input time series, and thus the memory of the neural network. The delay between two adjacent inputs is assumed to be one unit of time. The operation of the network is defined by the input and output weight matrices, M_1 and M_2, respectively. We assume the activation function σ_i of element i, i=1,2,...,N, admits a polynomial approximation in its interval of operation

$$\sigma_i(x_i) = T_{0i} + T_{1i}x_i + T_{2i}x_i^2 + \dots + T_{ni}x_i^n + \dots \tag{2}$$

The standard Volterra kernels are given by

$$\begin{aligned}
V_0 &= M_2\, g_0\\
V_1(k) &= M_2\, g_1(k)\\
V_2(k_1,k_2) &= M_2\, g_2(k_1,k_2)\\
&\dots\\
V_n(k_1,k_2,\dots,k_n) &= M_2\, g_n(k_1,k_2,\dots,k_n)
\end{aligned} \tag{3}$$

where g_0, g_1, ..., g_n are column vectors whose scalar components, indexed by i, i=1,2,...,N, are given by

$$\begin{aligned}
g_{0i} &= T_{0i},\\
g_{1i}(k) &= T_{1i}W_{ik} \quad, 1\le k\le w,\\
g_{2i}(k_1,k_2) &= T_{2i}\, W_{ik1}\, W_{ik2} \quad, 1\le k_1,k_2\le w\\
&\dots\\
g_{ni}(k_1,k_2,\dots,k_n) &= T_{ni}\, W_{ik1}\, W_{ik2} \cdots W_{ikn}\\
&\quad ,1\le k_l\le w,\ l=1,\dots,N
\end{aligned} \tag{4}$$

These expressions are obtained in a straightforward manner by replacing the polynomial expansion of the activation functions into the equations of the feedforward network. The kernels are defined only for arguments less or equal to the size of the window, w. This reflects this structure's finite memory, defined by the number of input elements, making it equivalent to a nonlinear finite impulse response filter.

2. Recurrent neural network model

We derive here the Volterra kernels of a recurrent neural network useful for temporal pattern recognition, dynamical system modeling and identification [6,8]. Consider the following equations for a stationary SISO neural network structure of N NE's with equations

$$\underline{X}(n+1) = \sigma(W\underline{X}(n) + M_1'\, I(n)) \tag{5a}$$
$$y(n+1) = M_2\, \underline{X}(n+1) \tag{5b}$$

where σ is the neurons collective activation function comprised of the individual nonlinearities σ_i, M_1 and M_2 are the input and output connection matrices, and W the recurrent neural network's weight matrix. The sign ' denotes here matrix transpose. The column state vector $\underline{X}$ is composed of the outputs of the NE's and y is the network's output.

Assuming that a Volterra expansion exists for the network in the operating range, we derive equations to recursively compute its kernels. We chose to compute them in triangular form to avoid multiple occurrence of kernels with identical arguments. We also assume that σ admits the polynomial expansion given in Eq. 2. The first three sets of kernels are given by

$$\begin{aligned}
h_0 &= M_2\, g_0\\
h_1(k) &= M_2\, g_1(k) &&, 1\le k\\
h_2(k_1,k_2) &= M_2\, g_2(k_1,k_2) &&, 1\le k_1\le k_2
\end{aligned} \tag{6}$$

where $g_0=[g_{01}\ g_{02} \dots g_{0N}]'$ is an equilibrium point falling within the definition interval of the polynomial expansion of σ. The first order term $g_1(k)=[g_{11}(k)\ g_{12}(k) \dots g_{1N}(k)]'$ is recursively defined by

$$g_{1i}(k+1) = \sigma_i'\Big(\sum_{j=1}^{N} W_{ij}g_{0j}\Big)\Big(\sum_{j=1}^{N} W_{ij}\, g_{1j}(k)\Big) \quad, k\ge 1 \tag{7a}$$

and

$$g_{1i}(1) = \sigma_i'\Big(\sum_{j=1}^{N} W_{ij}g_{0j}\Big) M_{1i} \tag{7b}$$

The second order term $g_2(k_1,k_2)=[g_{21}(k_1,k_2)\ g_{22}(k_1,k_2) \dots g_{2N}(k_1,k_2)]'$ is defined by

$$\begin{aligned}
g_{2i}(k_1+1,k_2+1) &= \sigma_i'\Big(\sum_{j=1}^{N} W_{ij}g_{0j}\Big)\Big[\sum_{j=1}^{N} W_{ij}g_{2j}(k_1,k_2)\\
&+ \sigma_i''\Big(\sum_{j=1}^{N} W_{ij}g_{0j}\Big)\Big[\sum_{j=1}^{N} W_{ij}g_{1j}(k_1)\Big]\Big[\sum_{j=1}^{N} W_{ij}g_{1j}(k_2)\Big]\\
&\qquad 1\le k_1<k_2
\end{aligned} \tag{8a}$$

$$g_{2i}(k+1,k+1) = \sigma_i'(\sum_{j=1}^{N} W_{ij}g_{0j})[\sum_{j=1}^{N} W_{ij}g_{2j}(k,k)$$
$$+ 1/2\,\sigma_i''(\sum_{j=1}^{N} W_{ij}g_{0j})[\sum_{j=1}^{N} W_{ij}g_{1j}(k)][\sum_{j=1}^{N} W_{ij}g_{1j}(k)] \quad (8b)$$

$$g_{2i}(1,k_2+1) = \sigma_i''(\sum_{j=1}^{N} W_{ij}g_{0j})\,[\sum_{j=1}^{N} W_{ij}g_{1j}(k_2)\,]M_{1i} \qquad k_2 \geq 1 \quad (8c)$$

$$g_{2i}(1,1) = 1/2\,\sigma_i''(\sum_{j=1}^{N} W_{ij}g_{0j})\;(M_{1i})^2 \quad (8d)$$

In Eqs. 7 and 8, σ' and σ'' represent the first and second order derivatives of the NE's activation functions, respectively. Eqs. 6,7 and 8 allow to recursively compute the zeroth, first, and second order kernels of the neural network, starting from h_0, $h_1(1)$, $h_1(2)$,...$h_1(k)$,..., followed by $h_2(1,1)$,..., $h_2(1,k)$,..., then $h_2(k_1,k_2)$, for $1 \leq k_1 \leq k_2$. A generalization of these expressions to arbitrary order kernels is presented in [13].

The recursive method presented here offers an efficient way to compute the kernels. These can alternatively be computed in closed form by rewriting the network's equations using tensor products and deriving non-recursive expressions for the kernels [9]. This latter method yields more readily interpretable results but is computationally very expensive, both in memory and processing time.

The derivation of the kernels relies on the existence of a stable equilibrium point g_0; and for every choice of g_0 there is a different expression of the Volterra expansion. This suggests the existence of many such expansions for a recurrent neural network in its operating space, each valid in the neighborhood of a stable equilibrium point. The derivation carried out earlier has therefore to be restricted to the attraction basin associated to a stable point g_0, where there exists a unique Volterra expansion given by Eqs. 6-8. This requires choosing the stable equilibrium point or a point close to it for initial conditions and limiting the amplitude of the input signal. For given values of the weight matrix, it is conceivable that strong nonlinear feedback produces a neural network with no stable equilibrium state at all. In this case it is not possible to characterize it by a Volterra expansion.

B- FUNCTIONAL REPRESENTATION

Section III-A focussed on the problem of deriving a Volterra expansion for neural networks. The converse problem, that of finding a network that has arbitrarily chosen Volterra kernels, is also important and is related to the representation power of the neural network. This section deals with the recurrent model defined by Eq. 5, operating around a stable point used as initial condition, and aims to define a class of models that can be accurately represented by a neural network of this form.

A large class of SISO nonlinear systems can be put in the form

$$\underline{X}(n+1) = \sum_{i=0}^{N} \sum_{j=0}^{N} F_{ij}\,\underline{X}(n)^{(i)}u^j, \quad \text{with } F_{00}=0 \quad (9a)$$

$$y(n+1) = c\,\underline{X}(n+1) \quad (9b)$$

where $\underline{X}^{(i)}$ denotes the Kronecker product of the state vector $\underline{X}$ applied i-1 times [9]. We restrict our scope to structures in this class, and further concentrate on second order systems (N=2). These constitute the subclass of bilinear systems, which is known to represent a fairly large number of nonlinear systems [9].

Theorem: **Given a SISO bilinear system defined by its first and second order coefficients F_{ij}, $i,j \leq 2$, if $\text{rank}(F_{10}) \geq \text{rank}(F_{ij})$, $i,j \leq 2$, then there exists a finite size recurrent neural network of the form of Eq. 5, whose first and second order Volterra kernels are equal to those of the original system.**

The proof of this existence theorem, not provided here, is based on writing a set of constraints that the neural network parameters M_1, M_2, and W should verify, to guarantee first, and second order kernels equal to the desired values. For a sufficiently large number of nonlinear elements in the neural network these constraints are proven to be consistent, which guarantees the existence of a neural network approximating the system up to second order. This network is thus a model of the true system up to an additive constant equal the zeroth order kernel.

This result gives a lower bound on the representation power of the recurrent model. Although the authors only considered bilinear systems, it may be possible to extend these results to higher order systems. As a consequence to this approach, by finding a set of parameters satisfying the constraints appearing in the proof of the theorem, the weight matrix W and the interconnection matrices M_1 and M_2 of the desired network are directly obtained. This is however of restricted practical use since the kernels of the system to learn are in general unknown.

IV. COMPUTER SIMULATIONS

Computer simulations were conducted to illustrate the above results. We present the case where the Volterra kernels were used to assess learning and architecture selection on a time series prediction problem. We also illustrate the functional representation results by finding a neural network with the same first and second order kernel as a given nonlinear system.

A. Logistic Function

In this section we trained both a feedforward network and a recurrent one to predict a time series generated by the logistic mapping, and learning was assessed by means of the Volterra kernels. The logistic iterative mapping is defined as

$$y(n+1) = k\,y(n)\,(1-y(n)) \quad (10)$$

The value of the parameter k was set equal to 4, which yields a chaotic behavior. Let the input u(n) be y(n) for all n. The triangular Volterra kernels of this system are all zero except $h_1(1)=4$, and $h_2(1,1)=-4$.

A feedforward network consisting of one input element, 36 elements in the hidden layer and one linear output element was trained using backpropagation [16] until it achieved a mean square error less than 10^{-5}. The training set contained 300 points. The size of the window being one, the arguments of the kernels take only the value 1. For other argument values, the kernels are null. Computing the kernels, up to second order, was achieved by fitting the sigmoid by a quadratic, and minimizing the mean square error over the operation interval. Using Eqs. 3 and 4, the kernel values we found were

$h_0 = 0.05$
$h_1(1) = 3.71$
$h_2(1,1) = -3.72$

Another experiment was conducted using a similar network, except that the input layer contained 2 elements. The input pattern at time n consisted of the vector [u(n-1) u(n)] and thus, the kernels arguments take now values in {1,2}.

h_0 =	0.01	
h_1 =	3.76	0.26
h_2 =	-3.84	-0.06
	0	-0.19

These results show a much stronger dependence on u(n) than on u(n-1), which is natural, since it is much easier to implement the mapping in Eq. 10, than one involving u(n-1). Furthermore, they suggest that the second input is redundant in the presence of the first one, and indicates that the first neural network is a more appropriate model for the predictor of the time series.

We also trained a recurrent network described by Eq. 5, as a predictor for this time series. The neural network consisted of one input and one output channels with 9 fully connected hidden elements. Learning was performed in batch mode using simulated annealing [17] on 200 data points and achieved a MSE of 4×10^{-4}. For zero initial conditions, the kernels were computed using Eqs. 6, 7, and 8, and their values for arguments in {1,2,3,4,5} were

h_0 =	0.0027				
h_1 =	3.79	0.68	0.12	0.03	0.01
h_2 =	-3.61	-0.69	-0.11	-0.03	0.01
	0	-0.85	0.01	-0.03	0.00
	0	0	-0.10	-0.01	0.00
	0	0	0	-0.02	0.00
	0	0	0	0	0.00

Again, we observe strong dependence of y(n+1) on u(n), but little dependence on delayed inputs. This network gives approximately the same results as the simulations using feedforward networks. Because of their limited dependence on the past of the time series, feedforward nets may be better models for the logistic mapping.

B. Functional Representation Problem

In order to illustrate the results of Section III-B, and to assess the significance of higher order kernels we conducted another computer simulation relating to a simple dynamical system given by

$$x(n+1) = 0.5\, x(n) + u(n) + 0.1\, u(n)^2 \qquad (11a)$$
$$y(n) = x(n) \qquad (11b)$$

This system satisfies the theorem's conditions and therefore it is possible to find a recurrent neural network that has the same first, and second order kernels. We chose a 4 element network obeying Eq. 5, with σ consisting of sigmoids with outputs between -1 and 1. The biases of the 4 sigmoids were arbitrarily fixed at 1, -1, 1, and -1, respectively. Simulated annealing was used to obtain simultaneously M_1, M_2, W, and the operating point of the network, such that the constraints appearing in the proof of the theorem are satisfied.

Table 1 shows the first, and second order kernels of the resulting network and compares them to those of the original system. To assess the performance of the network obtained as described above, we randomly generated a set of 200 points uniformly distributed which served as input sequence to both the dynamical system described by Eq. 11 and the approximating neural network. Disregarding the constant zeroth order kernel, and for inputs ranging between 1 and -1, the output of the neural network was visually indistinguishable from that of the original system (Fig. 1a). For larger input amplitudes, the performance starts to degrade due to the fact that the states start to fall outside the validity interval of the sigmoids polynomial approximation. Fig. 1b, and 1c, show the output of the system (dashed line) and that of the approximating neural network (solid line), when the inputs range between [-2,2], and [-4,4] respectively. In the last case, the higher order kernels cannot be neglected and the network is not a good model for this range of inputs.

V- CONCLUSION

In this paper we focussed on the problem of characterizing neural networks by means of their Volterra kernels. We derived general expressions for these kernels in the case of a feedforward network and a recurrent model. We also used the Volterra kernels to characterize the functional representation power of these networks. Computer simulations were in agreement with the theoretical results.

The derivation of the Volterra kernels has focussed on stationary systems, thus the kernels obtained are independent of time. Both our recursive approach, and the closed form solution can be extended to cover non-stationary systems, in a straightforward matter.

Volterra characterization of neural networks has potential for analyzing both neural network architectures and learning. Deriving the kernels for a given neural network, gives some insight on how this model operates internally. Inspection of the kernels can help assess what the network has learned and estimate its generalization performance. From a theoretical point of view, Volterra modeling constitutes a potentially useful framework to the analysis of the functional representation of a particular architecture. This approach to analyzing neural networks thus represents a method to compare very different architectures with respect to functional representation power and performance.

The combination of both neural networks and Volterra theory, seems to offer great potential in nonlinear system identification and modeling. Learning a neural network predictor, then deriving its kernels, may be an efficient way to obtain mathematical models of time series generated by unknown systems.

ACKNOWLEDGEMENTS

This research was supported by NSF grant #BCS 8914242.

REFERENCES

[1] R. Hecht-Nielsen, "Kolmogorov's Mapping Neural Network Existence Theorem," Proc. 1987 IEEE Intl. Conf. on Neural Networks, IEEE Press, New York, III-11:13, 1987.

[2] N.E. Cotter, "The Stone-Weierstrass Theorem and its Application to neural networks," IEEE Transactions on Neural

Networks, Vol. 1, No. 4, pp. 290-295, December 1990.
[3] G. Cybenko, "Approximations by superposition of a sigmoidal function," Math. Contr., Signals, System., Vol. 2, pp.183-192. 1989.
[4] K. Hornik, M. Stinchcombe, and H. White, "Multilayer feedforward networks are universal approximators," Neural Networks, Vol. 2, pp. 359-366, 1989
[5] K.S. Narendra and K. Parthasarathy, "Identification and Control of Dynamical Systems Using Neural Networks," IEEE Trans. on Neural Networks, Vol. 1, No. 1, pp. 4-27, March 1990.
[6] J.L. Elman, "Finding Structure in Time," Tech. Rep. # 8801, Center for Research in Language, University of California At San Diego, La Jolla, California, 1988.
[7] N.Z. Hakim, J.J. Kaufman, G. Cerf, H.E. Meadows, "A Discrete-Time Recurrent Neural Network Model for System Identification," in preparation.
[8] N.Z. Hakim, J.J. Kaufman, G. Cerf, H.E. Meadows, "A Discrete-Time Neural Network Model for Prediction Error Identification," Proc. IJCNN-90, San Diego, CA, June 1990.
[9] W.J. Rugh, "Nonlinear System Theory: The Volterra/Wiener Approach," Johns Hopkins University press, Baltimore, 1981.
[10] M. Schetzen, "Nonlinear System Modeling Based on the Wiener Theory," Proceedings of the IEEE, Vol. 69, No. 12, pp. 1557-1573, Dec. 1981.
[11] V.Z. Marmarelis (ed.), "Advanced methods of physiological system modeling," Vol. 1, USC Biomedical Simulations Resource, Los Angeles, 1987.
[12] J.L. Casti, "Nonlinear System Theory," Academic Press, Inc., 1985
[13] N.Z. Hakim, J.J. Kaufman, G. Cerf, H.E. Meadows, "Volterra Characterization of Neural Networks," in preparation.
[14] A. Lapedes, R. Farber, "Nonlinear Signal Processing Using Neural Networks: Prediction and System Modeling," Los Alamos National Laboratory, Technical Report LA-UR-87-2662, 1987
[15] Q.Z. Feng, "Neural Networks and Chaotic Time Series Predictions," in "Learning and Recognition: A Modern Approach," World Scientific, Y.C. Lee Ed., 1989.
[16] D.E. Rumelhart and J.L.McClelland, "Parallel Distributed Processing: Exploration in the Microstructure of Cognition, Vol.1, MIT Press, 1986.
[17] S. Kirkpatrick, C.D. Gelatt, M.P. Vecchi, "Optimization by Simulated Annealing," Science, Vol. 220, pp. 671-680, 1983

TABLE 1.
Kernels resulting from solving the set of constraints, as compared to those of the original dynamical system.

FIRST ORDER ($g_1(k)$, k=1,...)

	k=1	2	3	4	5
SYSTEM	1	0.5	0.25	0.125	0.0625
N.N.	1.003	0.5007	0.2499	0.1248	0.0623

SECOND ORDER ($g_2(k_1,k_2)$, k_1,k_2=1,...,5)

		k_1= 1	2	3	4	5
	k_2= 1	0.1	0	0	0	0
	2	0	0.05	0	0	0
SYSTEM	3	0	0	0.025	0	0
	4	0	0	0	0.0125	0
	5	0	0	0	0	0.0063

		k_1= 1	2	3	4	5
	k_2= 1	0.0993	0.0018	0.0009	0.0005	0.0002
	2	0	0.0485	0.0004	0.0002	0.0001
N.N.	3	0	0	0.0244	0.0003	0.0001
	4	0	0	0	0.0123	0.0002
	5	0	0	0	0	0.0061

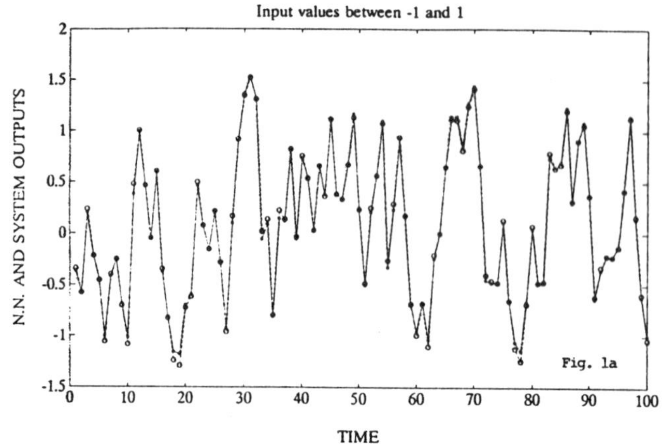

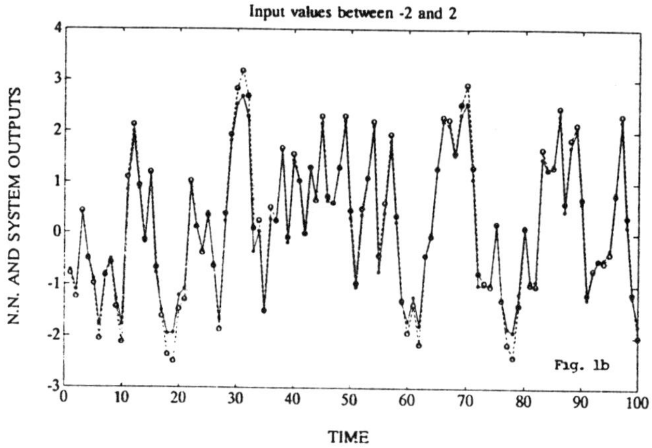

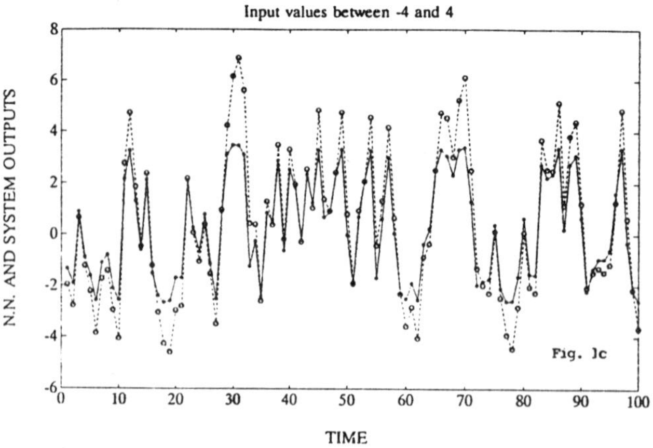

Fig. 1: Comparison of the true system output and that of an approximating neural network for three different input amplitude ranges. The solid lines represent the output of the neural network while the dashed one represents the output of the true system.

Progressive Image Transmission Using Self-supervised Backpropagation Neural Network

Wei Gong, IEEE student member, K. R. Rao and M. T. Manry, IEEE senior member
Department of Electrical Engineering, University of Texas at Arlington
P.O.Box 19016, Arlington, Tx 76019

Abstract

In this paper, a new technique for progressive image transmission (PIT) is presented, which uses a self-supervised backpropagation neural network discrete cosine transform (BPDCT). The transmission sequence is determined using a backpropagation neural network (BPNN) feature importance function. Simulation results show that the PIT system can be successfully implemented using BPNN. Very good intermediate images are obtained at reasonable bit rates.

1 Introduction

Progressive image transmission (PIT) is widely used in many applications, such as teleconferencing, telebrowsing, videotext, electronic mail order shopping, HDTV and medical diagnostics. PIT [1,2,3] involves successively better approximate reconstructions of an image, which are built up gradually until the viewers are satisfied. There are several methods to progressively transmit images. Basically they fall into two categories; those utilizing the spatial and transform domains. In this paper we discuss the transform domain PIT only. Transform-based PIT methods [3,4,5] effectively utilize data compression. The main issues in transform-based PIT schemes are to determine the transmission sequence for the transform coefficients, and to get good intermediate images at reasonable bit rates.

We developed the BPDCT [6]. The BPDCT itself has some attractive features, such as learning and feature importance selection ability since the BPDCT is obtained using the BPNN. This paper uses the BPDCT to implement a PIT system and uses the feature selection ability of the BPNN to determine the transmission sequence (TS).

A short description of the BPDCT is given in section 2. Section 3 develops the algorithm for determining the TS. Section 4 shows the implementation of PIT system using the BPDCT. Conclusions are given in section 6.

2 Modified Discrete Cosine Transform Using Self- Supervised BPNN

The basic idea of the BPDCT is that the DCT is mapped to a BPNN. The performance of the DCT after mapping is improved through BPNN training. First the 2-D forward DCT and 2-D inverse DCT are expressed in lexicographic form. Let x(v) and $x^*(v)$ be the lexicographic form for an input and output of the 2-D inverse DCT of size NxN, $v=1,\ldots,N^2$ respectively, and let T(u) be the lexicographic form for truncated 2-D DCT, $u=1,\ldots,M^2$ (see Figure 1a). The $w_{FD}(u,v)$ and $w_{ID}(u,v)$ are kernel coefficients for the 2-D forward and inverse DCT in lexicographic form respectively. T(u) is written as

$$T(u) = \sum_{v=1}^{N^2} w_{FD}(u,v)x(v) \quad u = 1,2,\ldots,M^2 \tag{1}$$

The lexicographic form for the 2-D inverse DCT is

$$x^*(v) = \sum_{u=1}^{M^2} w_{ID}(v,u)T(u) \quad v=1,\ldots,N^2 \tag{2}$$

Map equations 1 and 2 to a three layer BPNN. The inputs of the first layer are samples of a spatial domain block, x(v). The second layer outputs are outputs of the truncated 2-D forward DCT, T(u). The third layer outputs are samples of the 2-D inverse DCT, $x^*(v)$. The resulting neural network is constructed as shown in Figure 1b and is called the BPDCT. The BPDCT is a three layer self-supervised BPNN with units $N_U(1)=N^2$, $N_U(2)=M^2$, $N_U(3)=N^2$, and NL=3. $N_U(i)$ is the number of units in layer i for i=1,...,NL. NL is the number of layers.

In this BPNN, we choose DCT transform kernel coefficients, $w_{FD}(u,v)$ and $w_{ID}(v,u)$ as initial weights, since the DCT has good compression performance. We also use the input blocks as the desired outputs of the network (this is why it is called self-supervised BP neural network). Here, each unit's activation function is linear, since the DCT is a linear transform. The output of input units is the image block. The outputs of hidden and output units are

$$O(i,j) = net(i,j) = \sum_{k=1}^{N_U(i-1)} w_b(i,i-1,j,k)O(i-1,k) \tag{3}$$

where O(i,j) is the output of unit j at layer i. $w_b(i,Ls,j,Ns)$ is the weight connecting unit Ns at layer Ls to unit j of layer i. net(i,j) is a weighted sum of signals

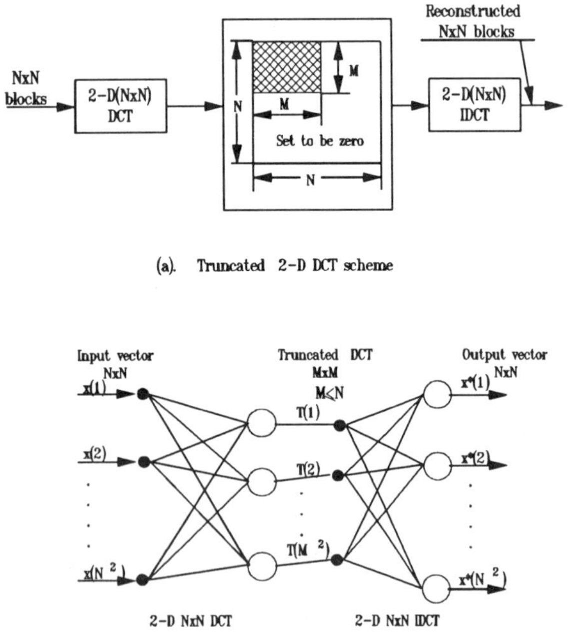

Figure 1. Mapping procedure of BPDCT

received at unit j of layer i. $w_b(2,1,j,k)$ and $w_b(3,2,j,k)$ in BPNN correspond to the DCT kernel coefficients $w_{FD}(u,v)$ and $w_{ID}(v,u)$ respectively.

The output evaluation function is the summation of the mean-square errors over all training blocks,

$$\epsilon = \frac{1}{2}\sum_{p=1}^{Npat} \sum_{j=1}^{N_U(NL)} [\ Tu_p(j)-O_p(NL,j)\ ]^2 \tag{4}$$

where Npat is the total number of training blocks, the subscript p stands for the index of the training blocks, and $Tu_p(j)$ is the desired output at the output layer of BPNN for block p.

The training procedure is performed repeatedly until the mean-square error ϵ is acceptable. After training the BPNN, the input weights from the first layer to second layer, perform the forward 2-D DCT. The output weights, from the second layer to third layer, perform the inverse 2-D DCT. This new transform is the BPDCT.

3 Neural Transmission Sequence

In image transform coding, bit assignment is used to determine which coefficients contribute most heavily to the output. The higher the contribution from a coefficient, the more bits are assigned to it. The bit assignments determines the bit rate and image fidelity. Conventional algorithms assign the bit map according to rate distortion theory [7,8]. It indicates that for a given distortion D, the maximum transmission rate is bounded by source entropy. If the probability density function of the source is $p(x_i)$, where x_i is the *ith* source symbol and $i=1,\ldots,K$, the average bit rate $R_{average}$ is:

$$R_{average} = \sum_{i=1}^{K} p(x_i)\log_2(\frac{1}{p(x_i)}) \tag{5}$$

The TS is determined using the same techniques. More bits are assigned to the high priority coefficients. In practice the probability density function of the source is unknown or not well modelled. Here we propose to use the BPNN feature selection approach [9] to determine the TS. An approach for determining the BPDCT coefficient's contribution [9,10,11] is to calculate the summation of the magnitudes of all weights leaving hidden unit j, $j=1, \ldots, M^2$. The resulting contribution's order function Xm(j) is calculated as

$$Xm(j) = \sum_{i=1}^{N_U(3)} |w_b(3,2,i,j)|\, O_{std}(2,j), \quad j=1,\ \ldots,\ M^2 \tag{6}$$

where $O_{std}(2,j)$ is the standard deviation of O(2,j), measured over all input patterns. Xm(j) tends to be large for high contribution coefficients and small for low contribution coefficients. We have found this approach to be simple but effective. The contribution order function Xm(j) is used for TS assignment as follows. (1) When preprocessing the test image, measure Xm(j) for each BPDCT coefficient. (2) Assign priority to each BPDCT coefficient according to Xm(j) and the required bit rate. The larger the values of Xm(j) in the list, the higher the priority the coefficient is assigned. (3) Transmit the BPDCT coefficients according to the TS.

4 Progressive Image Transmission

The PIT scheme shown in Figure 2 is a modified version of Chen and Smith's adaptive coding system [12]. At the transmitter, the original image of size 512 x 512 pels, each quantized to 8 bits PCM, is divided into NxN blocks which are classified into four classes (two edge classes and two non-edge classes). The classification technique is based on edge detection [6]. All these blocks are transformed using different BPDCT's, according to the classification map. At the same time the TS and standard deviation matrices of each class are generated. The DC coefficients for all blocks are transmitted first. The overhead information is transmitted through the channel. Then some higher order AC coefficients are transmitted progressively according to the TS. At the receiver, the received DC coefficients are used to form the first stage image. The overhead information and certain AC coefficients are decoded, and the inverse quantization and inverse BPDCT (IBPDCT) are applied sequentially to get the progressively reconstructed images.

4.1 Neural Transmission Sequence

We use the method described in section 3 to determine the TS. We want the coefficients with more contribution to the IBPDCT to be transmitted first. We get four contribution orders for the four classes respectively using BPDCT. Then we reorder the four contribution orders into one contribution order, excluding the DC.

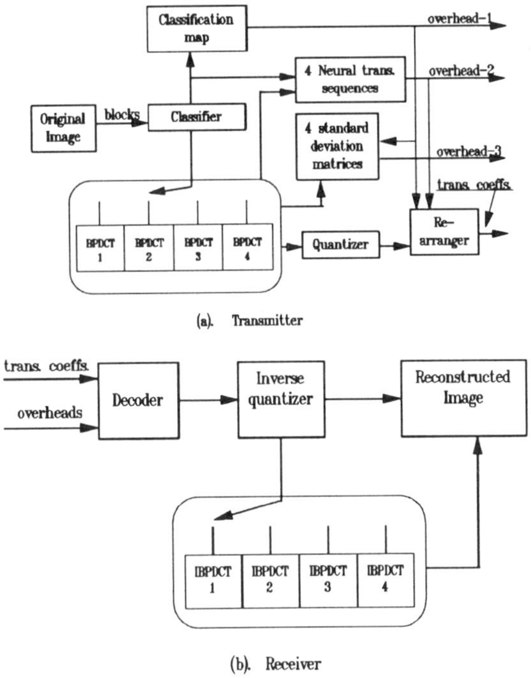

Figure 2. PIT scheme using BPDCT

coefficients. The new contribution order is the TS. The coefficients at specified locations in the image blocks are transmitted from left to right and top to bottom in groups of eight. Figure 3 shows the AC coefficient TS maps of the four classes for the 'girl' images, where N equals 8 and M equals 4

Class 1	Class 2	Class 3	Class 4
0 5 17 40	0 6 23 44	0 9 38 54	0 11 34 53
2 10 20 39	1 12 24 42	4 26 47 56	3 25 41 57
8 16 21 35	7 15 30 37	14 36 50 60	27 46 45 51
22 19 29 31	13 18 28 33	32 49 55 52	43 58 59 48

Figure 3. The AC coefficient TS for "girl" image

4.2 Quantization

The progressively reconstructed intermediate image is heavily affected by the quantization of the transmitted transform coefficients. The optimum quantizer is the Lloyd-Max quantizer [13]. A compandor has overall performance approximately equal to that of the Lloyd-Max quantizer. For optimum quantizer design, the probability density function of the quantized variable is assumed to be known. For DCT coefficients of most image data, it has been shown that the AC coefficients can be assumed to have a Laplacian probability density [14].

For simplicity, the DC coefficients are scaled by N (dimension of image sub-block), and rounded to the nearest integers in the range 0 to 255. This simplifies the receiver since the mean value of each block is obtained by dividing the DC coefficients by N. Also to avoid long codewords and to make DC coefficients follow the Laplacian distribution, we transmit the difference of the mean values between the current block and the previous block. This difference is called the difference mean. The mean value of the left top block of the image is chosen as a reference value, which is transmitted first. The mean value of the rightmost block at the previous row is chosen as the reference mean value for the leftmost block in the next row. Figure 4 shows the histogram of the difference mean for the image 'girl'. Codewords are designed using the Huffman code [7,8].

We quantized AC coefficients using the compandor [14], using 4 bits each (16 levels). To do that, we

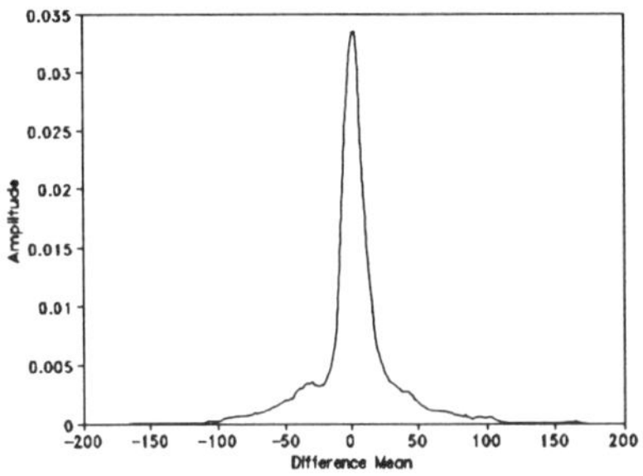

Figure 4. Histogram of difference mean for "girl" image

normalize the coefficients by their standard deviations so that they have unit variance. For four classes we have four standard deviation matrices, one per class, which are transmitted as overhead information, which results in a small amount of side information.

4.3 Overhead Information

In this section, we summarize the overhead information for the proposed PIT system. The overhead information includes the classification map, TS, and standard deviation matrices.

For images with size Ix x Iy the classification map is an Mrb x Mcb array where Mrb equals Ix/N and Mcb equals Iy/N. For each block, $\log_2(I_{clas})$ bits are required for each element in the classification map array where I_{clas} is the number of classes. The number of overhead bits for classification is

$$B_{clas} = Mrb \cdot Mcb \cdot \log_2 (I_{clas}) \quad bits \qquad (7)$$

In BPDCT the NxN input blocks are transformed into MxM blocks. The TS is an array of (M^2-1) AC coefficients (excluding the DC coefficient). $\log_2(I_{clas} \bullet M^2)$ bits are required to identify the class and the transmission order for each individual element in the TS. The number of overhead bits for the TS is

$$B_{trans} = I_{clas} \cdot (M^2-1) \cdot \log_2(I_{clas} \cdot M^2) \qquad bits \qquad (8)$$

The standard deviation matrices are represented by 4 byte floating point numbers. Thus the number of overhead bits for the standard deviation matrices is

$$B_{stand} = 4 \cdot 8 \cdot (M^2-1) \cdot 4 \qquad bits \qquad (9)$$

The total number of overhead bits for Ix x Iy pels is

$$B_{total} = B_{clas} + B_{trans} + B_{stand}. \qquad bits \qquad (10)$$

4.4 Proposed Decoder Scheme

Figure 2b shows the decoder scheme. At the receiver, there are two main operations. The first one is to decode the bit stream which contains the overhead information and coefficients. The other is to apply the inverse operation to the decoded coefficients, and then to obtain the progressively reconstructed image.

The first stage reconstructed image is obtained by transmitting difference mean without the overhead information. Then the overhead information is transmitted. After that, for each received entry, the receiver knows exactly the classification map and location of the coming entry. Starting at stage 2 the intermediate reconstructed images are obtained by decoding some higher order AC coefficients (here the number is eight) in the TS. The decoded AC coefficients are first de-normalized by the standard deviation, then inverse quantized. Then the inverse BPDCT is applied to them. The new intermediate image is obtained by adding the resulting values to the previous intermediate image.

4.5 Simulation Results

Computer simulations are being carried out using the proposed PIT system. The training image is the 512 x 512 pixel 'girl' image, quantized to 8 bit PCM. The training image is divided into 8 x 8 blocks (N=8). They are classified into four classes. For each class the BPDCT was obtained by training the BPNN, which had 16 hidden units, ie. M=4. After training the BPNN, four BPDCT's corresponding to the four classes are obtained. At the first stage, the difference mean of all blocks are transmitted and decoded at the receiver to form the first stage reconstructed image. Starting at the second stage, AC coefficients in groups of eight, according to the TS (see Figure 3), are transmitted to form the progressively reconstructed images. A total of seven stages are used. Table 1 shows an accumulated bit rate at each stage for the PIT and the SNR of the intermediate images with respect to the original image, where SNR is the objective performance measurement,

$$SNR = 10 \cdot \log_{10}\left(\frac{255^2}{MSE}\right) \qquad (11)$$

and MSE is the mean-square error of the grey levels between the original image and coded image. The bit rate at the second stage includes the bit rate for transmitting the overhead information. Figure 5 is the original "girl" image. Figures 6 through 8 show some progressively reconstruncted images of "girl".

Table 1. PIT Results With BPDCT

Image	Girl			Boat		
# of stage	MSE	SNR (dB)	Bit rate (BPP)	MSE	SNR (dB)	Bit rate (BPP)
1	280.0	23.66	0.125	188.9	25.37	0.125
2	105.6	27.90	0.258	85.82	28.79	0.268
3	66.43	29.91	0.362	56.20	30.63	0.393
4	41.82	31.92	0.425	35.34	32.65	0.492
5	32.66	32.99	0.550	28.35	33.61	0.618
6	25.64	34.01	0.657	22.12	34.68	0.766
7	23.13	34.49	0.752	20.19	34.92	0.839

5 Conclusions

In this paper a PIT system is implemented using the BPDCT. The transmission sequence has been successfully determined using BPNN. Good intermediate images at reasonable bit rates are obtained.

References

[1] K. R. Rao and P. Yip, **Discrete Cosine Transform: Algorithms, Advantages, Applications**, Academic Press, San Diego, 1990.

[2] K. H. Tzou, "Progressive image transmission: A review and comparison of techniques," Optical Engineering, vol.26, pp.581- 589, July 1987.

[3] B. Chitprasert and K. R. Rao, "Human visual weighted progressive image transmission," IEEE Trans. Commun., vol.38, pp.1040-1044, July 1990.

[4] J. I. Gimlett, " Use of activity classes in adaptive transform image coding," IEEE Trans. Commun., vol. COM-23, pp.785-786, July 1975.

[5] L. Wang and M. Goldberg, "Progressive image transmission by transform coefficient residual error quantization," IEEE Trans. Commun., vol.COM-36, pp.75-87, January 1988.

[6] W. Gong, K.R. Rao and M. T. Manry, "Image transform coding using self-supervised backpropagation neural network," under review.

[7] R. G. Gallager, **Information Theory and Reliable Communication**, John Wiley, 1968.

[8] K. S. Shanmugam, **Digital and Analog Communication Systems**, John Wiley, 1979.
[9] W. Gong, H. C. Yau and M. T. Manry, "Non-Gaussian feature analysis using neural network," accepted by Progress in Neural Networks, 1991.
[10] W. Gong, "Unconstrained character recognition using neural network," Master's thesis, Department of Electrical Engineering, University Of Texas At Arlington, May 1989.
[11] W. Gong and M. T. Manry, "Analysis of non-Gaussian data using a neural network," Proc. of IJCNN 1989, Washington D. C., June, vol.II, pp.576.
[12] W. H. Chen and C. N. Smith, "Adaptive coding of monochrome and color images," IEEE Trans. Commun., vol.COM-25, pp.1285-1292, June 1977.
[13] A. K. Jain, **Fundamentals of Digital Image Processing**, Prentice-Hall, 1989.
[14]. R. C. Reininger and J. D. Gibson, "Distributions of the two-dimensional DCT coefficients for images," IEEE Trans. Commun., vol. COM-31, pp.835-839, June 1983.

Figure 5. Original 'girl' image

Figure 6. 1st stage reconstructed image with SNR=23.66dB and bit rate=0.125BPP

Figure 7. 2nd stage reconstructed image with SNR=27.90dB and bit rate=0.258BPP

Figure 8. 7th stage reconstructed image with SNR=34.49dB and bit rate=0.752BPP

Detection and Classification of Faults from Helicopter Vibration Data Using Recently Developed Signal Processing and Neural Network Techniques

Manuel R. Solorzano
Dexter K. Ishii
Niel R. Nickolaisen
William Y. Huang
Code 535
Advanced Technology Development Branch
Naval Ocean Systems Center
San Diego, CA 92152–5000

Abstract

The feasibility of using non-traditional methods of helicopter transmission fault classification is studied. Various signal processing and classifier techniques are investigated. All algorithms successfully classified the fault samples from a tail rotor transmission dataset. A hardware neural network system was designed and implemented. The relatively low resolution of the neural network circuitry required extensive pre-processing and scaling of the large dynamic range input signal. Results from the hardware system were similar to those achieved in simulation. A true test of the techniques presented may require a dataset that is statistically richer.

1 Introduction

The goal of this study was to test the feasibility of using neural networks and other signal processing techniques to classify mechanical faults in helicopter transmissions. The final portion of the study would develop a hardware fault detection and classification system.

In order to determine which signal processing and classification methods might provide the best problem solution, several techniques were investigated. These included Fourier, Wavelet, and Wigner- Ville Transforms and Radial Basis Function, Categorization and Regression Trees, Gaussian Mixture, and Neural Network classification paradigms.

The dataset utilized in this study was collected by Mark Hollins of the Naval Air Test Center and consists of data from the tail rotor transmission of a TH-1L helicopter. Several fault types were implanted in the transmission. To allow data for artifact removal, three samples of no-fault (NF) data were collected with an assemble/disassemble cycle between each sample. A total of 5 separate mechanical faults were implanted: Inner Race Fault (IRF), Rolling Element Fault (REF), Outer Race Fault (ORF), Gear Spall (GS), and Gear Fault [1/2 gear tooth missing] (GF). The IRF and ORF conditions have two degrees of severity. Samples of the raw data are shown in Fig. 1.

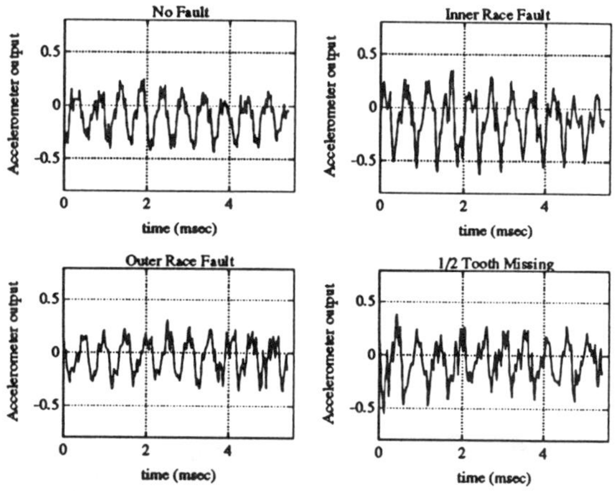

Figure 1: Time series samples of the no–fault, inner race fault, outer race fault, and gear fault from the Hollins database.

2 Algorithms Used

2.1 General Approach

The time series vibrational data was input to various signal processing algorithms. These pre-processors attempted to transform the time series data to a form that would allow classification. The output from the pre-processors was presented to various detection/classification algorithms.

2.2 Pre-processors

The pre-processors studied included the Fourier Transform (FT), Wavelet Transform (WAV), and Wigner-Ville Distribution (WIG).

The FT transforms the data to the frequency domain. A drawback of the FT is that, as frequency

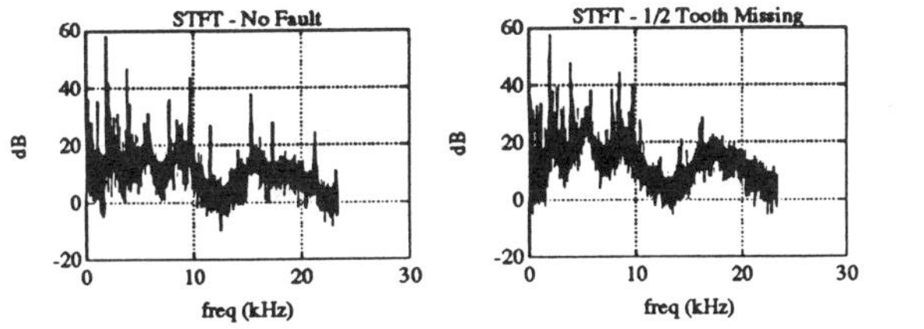

Figure 2: FFT output for the no fault and gear fault data.

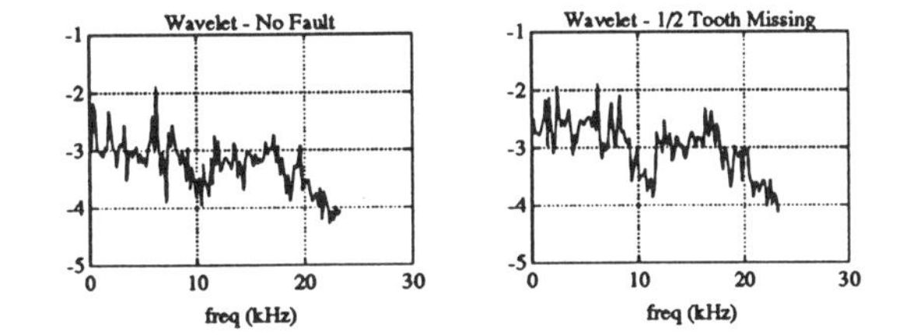

Figure 3: WAV output for the no fault and gear fault data.

resolution is increased, time resolution decreases. The FT is well understood and is easily implemented in hardware. Some FT results, using a 16384 point FFT, are shown in Fig. 2.

The wavelet reflects a cochlear model and employs a constant time/bandwidth product filterbank. WAV is defined as:

$$\mathrm{WAV}(m,n) = \int \sqrt{a^m} h(a^m t - bn) f(t) dt \quad (1)$$

where m and n are indices corresponding to frequency and time, a and b are wavelet parameters, and $h(t)$ is the wavelet function. $h(t)$ must be centered at $t = 0$ with $H(\omega = 0) = 0$ and $H(\omega)$ centered at k_0. The WAV achieves frequency sampling by dilating $h(t)$ by a factor of a^m and time sampling by time shifting $h(t)$ by a factor of nb/a^m. WAV results for a 64 octave wavelet packet using the 10 point Daubechies orthogonal wavelet basis are shown in Fig. 3. Note the reduced dimensionality (as compared to the FFT output) provided by the WAV.

The Wigner-Ville Distribution is one of a class of instantaneous time/frequency representations and is defined by:

$$\mathrm{WIG}(t,\omega) = \int f(t+\frac{\tau}{2}) f^*(t-\frac{\tau}{2}) e^{j\omega\tau} d\tau, \quad (2)$$

The WIG is the Fourier Transform of the instantaneous correlation function. The WIG allows increased frequency resolution without affecting time resolution. This characteristic has made the WIG of interest in studying non-stationary signals. The WIG, however, creates interference patterns that vary with time and are composed of various frequencies. Fig. 4 shows the output of the WIG for several data classes.

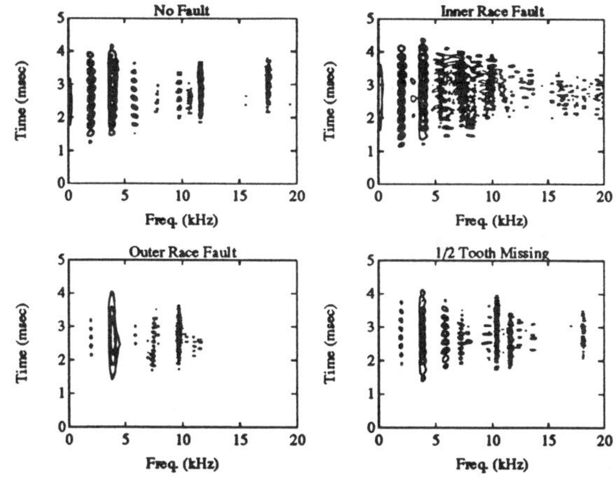

Figure 4: WIG plots (3–D) for the no fault case and 3 fault classes (inner race fault, outer race fault, and gear fault). The interference patterns show dramatic changes under different fault conditions.

Several investigators have proposed methods of reducing the interference patterns created by the WIG. These "improved" Wigner-Ville distributions window the data in time and/or frequency. Reducing the interference patterns increases the mathematical complexity and order of the transform and may limit its ability to be used in a real-time hardware system.

2.3 Classifiers

The classifiers used in this study were: Multi-Layered Perceptron (MLP) trained with Back Propagation (BP), Gaussian Mixture (GM) and Radial Basis Function (RBF), and Categorization and Regression Trees (CART). MLP was chosen because it has excellent classification properties that are well understood. GM and RBF were used because they provide fast training with good results. The CART was explored because it is easily implemented on a serial machine and is somewhat similar to current vibrational analysis procedures.

3 Experiments

3.1 Data

The training and testing data was extracted from the Hollins dataset. For each data class, 15 seconds of data were used for training and 7 seconds were used for testing. Samples were taken at a 20 kHz rate providing 140,000 samples in the 7 second testing set.

3.2 Results

Table 1 shows the results for various pre-processors and classifiers. In general, the neural network based classifiers performed better in differentiating among data classes (six class performance) but all approaches were successful in classifying between fault and no-fault (two class performance) conditions.

4 Hardware Implementation

A system using the FT as a pre-processor and the MLP trained with BP as a classifier was designed and

Table 1: Detection/Classification Results

Classifier	*Pre-Pro.*	*2 Class* (% Correct)	*6 Class* (% Correct)
GM	FFT	100%	92%
GM	WAV	100%	92%
CART	FFT	100%	92%
MLP	FFT	100%	99.6%
MLP	WAV	100%	99.2%

built. The MLP was composed of 62 inputs and 6 outputs. It had one hidden layer of 20 neurons. The system used the available neural network circuitry (the Intel ETANN). Because of the limited dynamic range of the neural network circuit, the data required scaling and compression before it could be presented to the neural network. Because this network device employs analog technology, "chip in the loop" training was required.

After the proper compression and scaling algorithms were determined, the system performed at levels similar to those achieved in simulation (100performance).

5 Conclusion

The primary goal of this study has been satisfied: it appears feasible to use neural network based classifiers to detect and classify mechanical faults in helicopter transmissions. At present, it is difficult to determine which pre-processing and classifying techniques yield the best results since all approaches were successful in solving the problem. It is likely that statistically richer datasets are required before the optimal pre-processor and classifier can be determined. The hardware implementation resulted in performance similar to that achieved in simulation.

References

[1] P. F. Brown. *Maximum Mutual Information Modeling of Speech.* PhD thesis, Carnegie Mellon University, 1989.

[2] Cheeseman et al. Autoclass: A bayesian classification system. *Proc. Fifth Machine Learning Workshop*, pages 54–64, 1988.

[3] R. R. Coifman, Y. Meyer, S. Quake, and M. V. Wickerhauser. Signal processing and compression with wavelet packets. Technical report, Department of Mathematics, Yale University, April 1990.

[4] I. Daubechies. The wavelet transform, time–frequency localization and signal analysis. *IEEE Transactions on Information Theory*, 36(5):961–1005, September 1990.

[5] W. Huang and R. Lippmann. HMM speech recognition with neural net discrimination. In D. S. Touretzky, editor, *Advances in Neural Information Processing Systems 2*, pages 194–202. Morgan Kaufmann, San Mateo, CA, 1990.

[6] Y. Lee and R. Lippmann. Practical characteristics of neural network and conventional pattern classifiers on artificial and speech problems. In *Advances in Neural Information Processing Systems 2*, pages 168–177. Morgan Kaufmann, San Mateo, CA, 1990.

[7] G. J. McLachlan. *Mixture Models.* Marcel Dekker, New York, N. Y., 1988.

A Two-Layer Neural Peak Detector

G. L. Dempsey E. S. McVey
School of Engineering and Applied Science
University of Virginia
Charlottesville, Virginia 22901

Abstract

A neural peak detector based on the Hopfield-Tank optimization network can offer several orders of magnitude reduction in computational time compared to networks based on digital architectures. Applications requiring real-time operation that were previously impractical with digital architectures are possible with the neural approach. Although the circuit implementation of the processing nodes (neurons) is relatively simple, the interconnections present a challenge for VLSI implementation in contrast to the relatively localized connection structure in digital systems. A two-layer network is suggested for the peak detector in order to reduce the neural interconnections and thus allow practical implementation with existing VLSI technology.

1 Introduction

Image processing and waveform analysis problems are applications where a peak detector (PD) that is capable of operating in real-time may be required. The Hough transform [1], a pattern recognition algorithm requiring peak detection, is an example use in an image processing system. Signal processing applications such as sonar, seismic detection, and radar are other possible areas requiring peak detection. Our energy function for the PD was presented in [1] and general circuit implementation considerations in [2].

The neural PD suggested in [1] is a Hopfield-Tank network [3] which consists of a single layer of fully connected neurons. Because of the fully connected structure, which is typical of Hopfield-Tank networks, there is a restriction on the network size due to the external pin limitation of integrated circuits. Although other methods may be available in the future to minimize implementation constraints, such as optical implementation [4], the focus in this paper is implementation using current VLSI technology. A two-layer structure for the PD is suggested which minimizes the number of neuron interconnections and thus makes VLSI implementation feasible. The methods presented here are applicable to other Hopfield-Tank networks which attaches additional importance to the results.

Other designations for PD networks are winner-take-all networks and mutual inhibition networks. Recent work includes Feldman [5], Lazzaro [6], and Majani [7]. Lazzaro *et al.* have shown a CMOS transistor implementation for a winner-take-all network of n neurons with O(n) interconnect complexity. In their network, the winning output is related to the logarithm of its input. Although processing is analog in a neural network, many applications require a digital output format.

Our PD was designed to converge to a digital solution to allow interfacing to digital logic circuitry or a main processing unit (MPU). One neuron is required for each input. The neuron with the peak input converges to logic "1". All other neurons converge to logic "0". A conventional digital logic circuit can be used to encode or condense the PD outputs into a form suitable for a MPU. The PD can also be designed to identify successive peak values, i.e., multiple peak detection [8].

2 Single-Layer Network

Figure 1 shows the circuit implementation for the PD neuron. Each PD neuron is connected to all other PD neurons through equal inhibitory resistances. Note that the input can be represented as a current or voltage. The differential equation for the PD neuron is

$$C_i(dU_i/dt) = -U_i/R_i + I_i + \sum_{\substack{j=1 \\ j\neq i}}^{n} T_{ij} V_j \qquad (1)$$

where U_i is the input voltage of the neuron, I_i is the input current, C_i is the input capacitance, R_i is the input resistance of the neuron

1058-6393/91 $1.00 © 1991 IEEE

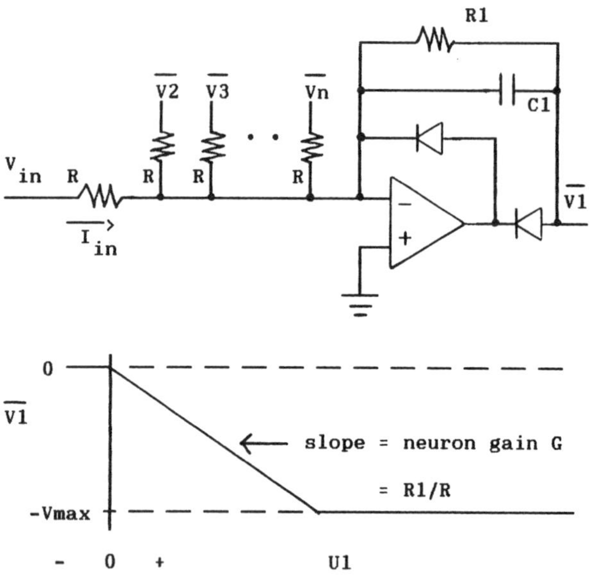

Fig. 1. Peak Detector Neuron.

which is equal to the connection weights 1/T (R in Fig. 1) and n is the number of neurons. The neuron characteristic from U_i to V_i is a linear curve with saturation as shown in Fig. 1. The neuron output voltage is equal to

$$V_i = U_i G_i \tag{2}$$

where G_i is the neuron gain. The neuron is non-linear due to the saturation region.

The half-wave rectifier circuit shown in Fig. 1 provides unidirectional gain and bounds the output at 0 volts and -Vmax (op-amp saturation voltage) as required for the Hopfield-Tank structure. Logic 0 is defined as 0 volts and logic 1 as -Vmax. Because each PD neuron can be described by (1), the network can be simulated by solving n non-linear differential equations using any convenient numerical method.

The energy function for the PD network is [1]

$$E = 0.5\, T \sum_{i=1}^{n} \sum_{\substack{j=1 \\ i \neq j}}^{n} V_i V_j - 0.5 \sum_{i=1}^{n} I_i V_i \tag{3}$$

assuming the neurons have infinite gain and equal to

$$E = 0.5T \sum_{i=1}^{n} \sum_{\substack{j=1 \\ i \neq j}}^{n} V_i V_j - 0.5 \sum_{i=1}^{n} I_i V_i + 0.5 \sum_{i=1}^{n} \frac{V_i^2}{R_i G_i} \tag{4}$$

for finite gain G_i. The energy function is required to determine network structure and component values. The PD network can be designed to guarantee convergence to the correct state if the inputs and network parameters are restricted to the bounds dictated by derived analytical expressions [2]. The equations for neuron gain, input separation, and network component values and restrictions are shown in the next section.

Table I shows the number of internal neuron connections and the number of external pins for an integrated circuit (IC) package for different network sizes. The number of interconnections is equal to n(n-1) and the number of external pins required for the single-layer PD is equal to 2n. The network size is limited to 148 neurons assuming three hundred (a practical value) is the maximum number of external pins for an IC package. Implementing the network with multiple ICs is not possible with the single-layer architecture because of the large number of interconnections.

Table I
Peak Detector Connections

Neurons	Interconnections	External Pins
64	4,032	131
148	21,756	299
256	65,280	515
1024	1,047,552	2051
4096	≈ 17,000,000	8195
16384	≈ 270,000,000	32771

Note: Three power supply pins have been included in the "external pins" column.

3 Two-Layer Peak Detector

In the previous section it was shown that a single-layer network cannot be implemented in an IC if the number of neurons exceeds 148. Biological neural systems offer insight into how nature deals with the large fan-in and fan-out of neurons. The average number of synapses (neuron inputs) in biological neural systems lies in the range from 1000 to 10,000. The maximum number usually does not exceed 100,000 [9]. The number of neuron interconnections (synaptic connections) shown in Table I indicates that the PD is not structured in a manner suggested by nature. Distribution of gain between the synapse, neuron cell, and conductors (dendrites and axon) exists in biological neural systems [10]. Neural layers which may correspond to specific duties are also common. This may be nature's way of minimizing extremely dense connection patterns by computing locally within a layer and then relaying the results to subsequent layers. These theories are used to reduce the spatial density and number of PD neuron interconnections.

The method will be illustrated first for a four PD neuron network as shown in Fig. 2. Each neuron has a feedback connection from the other three neurons. For simplicity, the circuit components have been omitted. A circuit model for the neuron is also shown in the figure where the neuron gain has been set to 3.0. Neurons 1 and 2 have been grouped into section 1 and neurons 3

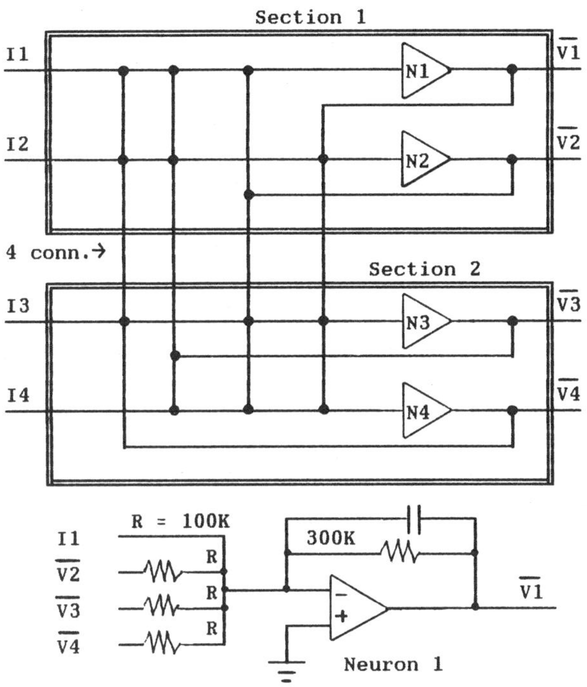

Fig. 2. Peak Detector With Single Layer.

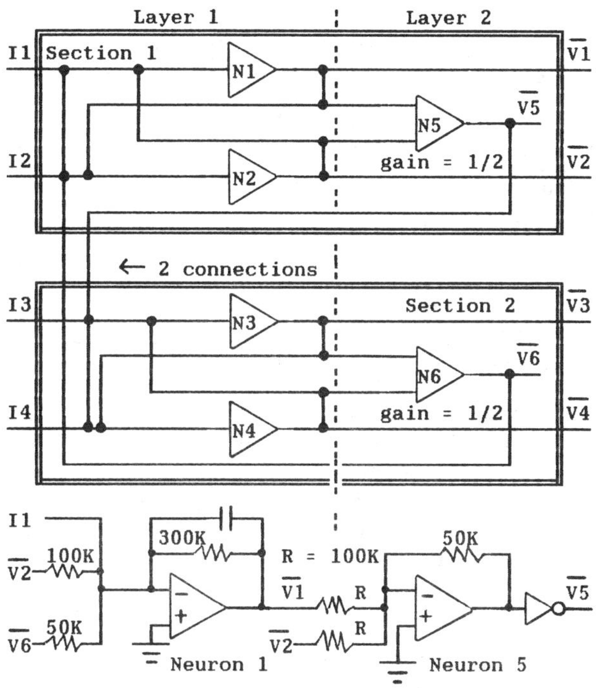

Fig. 3. Peak Detector with Two Layers.

and 4 into section 2. There are four connections between the sections.

The network shown in Fig. 3 performs the same function as the one in Fig. 2 but is arranged in two layers with the original neurons (1-4) in layer 1 and two new neurons in layer 2 which will be designated as summing neurons 5 and 6. Note that the connections between the two sections have been reduced from four to two. A circuit model for the two types of neurons is shown in the figure.

The gain relationship of 3.0 between neurons 1 and 2 is unchanged from the original network (same is true for neurons 3 and 4). For example, neuron 6 sums the outputs from neurons 3 and 4 (a linear operation) but at a gain reduction of 1/2 to maintain linearity (avoids neuron saturation). Neuron 1 sums the output from neuron 6 with its input current and feedback voltage from neuron 2. However, the output from neuron 6 is multiplied by a gain of 6.0 which compensates for the gain reduction in neuron 6. The effect of neurons 3 and 4 on neuron 1 is the same as in the original network in Fig. 2. Distributing the gain in this manner allows a significant reduction in the spatial density and number of neuron interconnections.

Consider a 1024 neuron PD and let each neuron section consist of 26 neurons, 25 PD neurons in the first layer and 1 summing neuron in the second layer. Forty-one sections are required for the network. The neuron in the second layer sums the 25 PD neuron outputs with a gain reduction of 1/25. Each section has 25 normal inputs, 25 normal outputs, 1 output connection from the second layer summing neuron, 40 feedback signals from the other sections' summing neuron, and 3 power supply connections. The total number of external pins required for each section is approximately 100. Each section can now be implemented in an IC. The number of connections to each PD neuron has been decreased from 1023 to 64. The total number of interconnections, including summing neurons, has been reduced from 1,047,552 to 66,561 (≈ 16x reduction). The method can also be used internally in the sections to further reduce the connections between neurons.

Table II is a summary of the connections and components for the IC for 1024 neurons. Although 41 ICs are required for the network, the ICs are relatively simple. It is assumed that the neuron (op-amp) can be implemented with 12 transistors which is the average number used for a general purpose op-amp. The total number of components is approximately 2100.

Because the second layer neuron has a gain reduction of 1/25, the output of this neuron must be increased by 25 in the other IC sections. A total gain of 75 is required for this output, 25 for the loss and 3 for the normal gain. The increased gain is a disadvantage because the network would be more sensitive to input offset

Table II
Peak Detector IC For 1024 Neuron Network

Description	Number
Neuron inputs from source Neuron outputs Misc. inputs/outputs Total pins	25 26 43 94
Neurons Neuron transistors	26 312
Resistors Diodes Capacitors	≈ 1700 50 25
Total components	≈ 2100
Packages to implement design	41

voltages and input bias currents. This issue will be discussed further in the next section.

Table III shows another possible implementation of the network where it has been divided into 8 ICs which each require 267 external pins. Although the number of ICs have been reduced, the complexity of the IC has been increased. Network sizes of 4096 and 16,384 neurons are also shown in the table with examples of partitioning the networks into ICs which consist of 129 neurons. Note that with an external pin limitation of 300, the network size is restricted to 4096 neurons. Recall the single-layer network was restricted to 148 neurons.

4 2-Layer Network With Low Neuron Gain

The disadvantage of the two-layer PD is the requirement of increasing the neuron gain to compensate for the gain reduction due to the second layer summing neuron. Low neuron gain is desired because of practical considerations such as non-ideal op-amp characteristics. For example, input offset voltage of an op-amp is multiplied by the gain. Low neuron gain would also allow a reduction in the complexity of the op-amp.

4.1 Neuron Gain Relationships

From [2], the minimum PD neuron gain is

$$G \geq \frac{Vmax \quad T}{Ipeak\ (min)} \tag{5}$$

where Ipeak(min) is the minimum detectable peak current for the network. The upper bound on input current magnitude is

$$Ipeak(max) = Vmax\ T \tag{6}$$

From (5) and (6), the gain must be greater than or equal to 1.0. Define the following parameters:

$$a = Ipeak - Ic \tag{7}$$

where a is the separation between the peak current and the next closest current Ic and

$$\delta = Uc - Upeak \tag{8}$$

where δ is the separation between the two corresponding neuron input voltages at t = 0. Convergence to a valid solution with random initial

Table III
Implementation Schemes Versus Network Sizes

Neurons	ICs	Neurons/IC	Interconn.	Pins
1024	41	26	66,561	94
1024	8	129	138,240	267
4096	32	129	651,264	291
16384	128	129	4,177,920	387

conditions (non-ideal op-amp characteristics) restricts the neuron gain to

$$G < \frac{a}{\delta\ T} + 1 \tag{9}$$

or the input current separation is required to be

$$a > \delta T\ (G-1) \tag{10}$$

Table IV shows the minimum neuron gain versus minimum detectable peak current for a network with Ipeak(max) equal to 50 ua, T equal to 1/100kΩ, and Vmax equal to 5 volts. From the table, the neuron gain can be minimized by choosing a relatively large value for the minimum peak current. However, the window of detectable peak currents is reduced which is not desirable.

Table IV
Neuron Gain Versus Minimum Peak Current

Minimum Current (ua)	Neuron Gain
0.0	∞
1.0	50.0
10.0	5.0
16.7	3.0
20.0	2.5
30.0	1.7
40.0	1.3
50.0	1.0

It was shown in [8] that the value of peak input current can be approximated after the network converges to a digital solution. A MPU, an encoder to format the PD outputs, and a digital-to-analog converter as shown in Fig. 4 are required. The bias current I_{bias} is connected to each PD neuron as shown. The extra hardware offers the capability to minimize the neuron gain as well as identifying multiple peaks [8].

Consider the following example of minimizing the gain. From Table IV, the minimum gain is equal to 3.0 for a minimum detectable peak current of 16.7 ua. Assume the MPU can control I_{bias} in steps of 1 ua and can approximate the peak input current to within 2 ua. If the sum of the peak input current and I_{bias} is maintained above 40 ua, from Table IV the neuron gain can be set equal to 1.3. With the bias current adjustment, the minimum neuron gain is

$$G \geq \frac{Vmax \quad T}{Ipeak\ (min) + I_{bias}} \tag{11}$$

4.2 Two-Layer Network Modifications

The bias current adjustment can be used in the two-layer network to maintain linearity in the summing operation performed by the second layer summing neurons. The summing neuron gain can be set equal to 1.0, thus the PD neuron's gain would

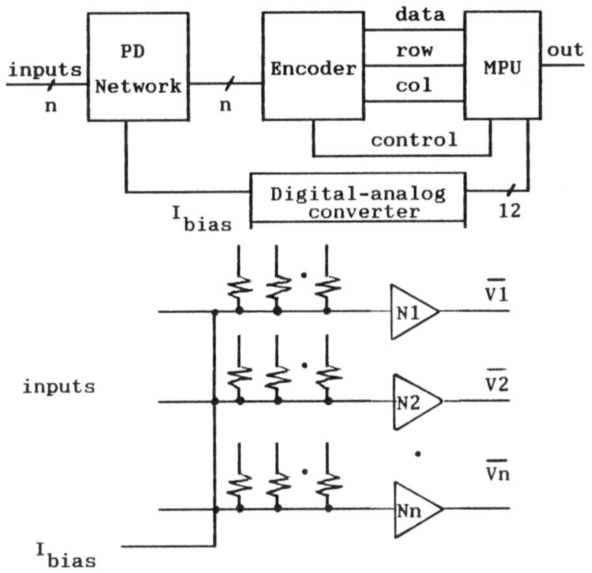

Fig. 4. Neuron Bias Current Adjustment.

be unchanged from the single-layer architecture.

Consider the previous example where Ipeak(min) is 16.7 ua, Ipeak(max) is 50 ua, gain is 3.0, T is 1/100kΩ, and Vmax is 5 volts. The neuron with the peak input current has an upper bound on its output voltage of

$$Vpeak = \frac{(Ipeak + I_{bias})}{T} G \qquad (12)$$

All other neuron output voltages are less than Vpeak because of the mutual inhibition. Zero initial conditions are assumed.

Assume the PD inputs are latched. Select I_{bias} equal to -50 ua (controlled by MPU). This essentially performs a network reset, i.e., all neuron output voltages are forced to logic 0. Now increment the bias current in small steps until it is equal to zero. Between each bias current increment allow enough time for the network to reach steady state (30 us worst case convergence [2]). When the bias current is eventually set equal to zero, the final network convergence is guaranteed to converge to digital values.

The advantage of this method of bias current adjustment is that the neurons not associated with the peak input current have output voltages relatively close to logic 0. After several increments of bias current, normally all neurons will be at logic 0 except for the neuron with the peak current. The summation of the first layer outputs essentially depends only on the neuron with the peak current. Therefore, the summing neuron gain can be made equal to 1.0 and linearity can be maintained. Each PD neuron input has equal gain and the value can be relatively low.

Simulation results show that in cases where the initial conditions (neuron input voltages at t = 0) are equal to zero, the bias current increments are not required. Because of the nature of the PD, mutual neuron inhibition, the neuron outputs are maintained near logic 0 except for the neuron with the peak current. Thus the summing neuron output remains in the linear region of operation. Problems exist for cases where the input voltages at t=0 are non-zero and are relatively large. The summing neurons can saturate and thus the network behavior cannot be described analytically without the bias current adjustment method. Because the MPU can approximate the peak input current, the starting bias current and the number of increments can be optimized.

5 Summary and Conclusions

Investigation of a single-layer PD based on the Hopfield-Tank architecture revealed that the network cannot be implemented in hardware using existing VLSI technology once the network exceeds approximately 150 neurons. We have shown that using a two-layer scheme reduces the number of neuron interconnections and thus makes hardware implementation of large networks feasible.

References

[1] E. S. McVey, G. L. Dempsey, J. Garcia, R. M. Inigo. "Artificial Neural Network Implementation Of the Hough Transform," *23rd Asilomar Conference on Signals, Systems and Computers*, Vol. 1, pp. 128-132, Nov. 1989.

[2] G. L. Dempsey. "Neural Network Implementation of the Hough Transform," Ph.D dissertation, Univ. of Virginia, Charlottesville, VA, January 1991.

[3] J. J. Hopfield. "Neurons with Graded Response have Collective Computational Properties like those of Two-State Neurons," *Proceedings of the National Academy of Sciences*, 81, pp. 3088-3092, May 1984.

[4] N. H. Farhat, D. Psaltis, A. Prata, E. Paek. "Optical Implementation of the Hopfield model," *Applied optics*, Vol. 24, pp. 1469- 475, 1985.

[5] J. A. Feldman and D. H. Ballard. "Connectionist models and their properties," *Cognitive Science*, vol. 6 pp.205-254, 1982.

[6] J. Lazzaro, S. Ryckebusch, M. A. Mahowald, C. A. Mead. "Winner-tale-all networks of O(N) complexity," Advances in Neural Information Processing Systems, D. Touretzky, Ed. San Mateo, CA: Morgan Kaufmann, 1989, pp. 703-711.

[7] E. Majani, R. Erlanson, Y. Abu-Mostafa. "On the k-winners-take-all network," Advances in Neural Information Processing Systems, D. Touretzky, Ed. San Mateo, CA: Morgan Kaufmann, 1989, pp. 634-642.

[8] G. L. Dempsey, E. S. McVey. "Hough Transform System Enhancements Resulting From Neural Network Implementation," *IEEE Southeastcon '91 Conference*, Williamsburg, Virginia, April 1991.

[9] C. R. Noback and R. J. Demarest. The Human Nervous System, Basic Principles of Neurobiology, McGraw-Hill Book Co., 1975.

[10] Neurosciences Research Foundation Inc. Ed. by G. M. Edelman, W. E. Gall, W. M. Cowan. Synaptic Function, John Wiley & Sons, Inc., New York, N. Y., 1987, pp. 647-649.

Image Coding Using Self-Supervised Backpropagation Neural Network

Wei Gong, IEEE student member, K. R. Rao and M. T. Manry,IEEE senior member
Department of Electrical Engineering, University of Texas at Arlington,
P.O.Box 19016, Arlington, Tx 76019

Abstract

In this paper, image data compression is implemented using a self-supervised backpropagation neural network. First, a backpropagation discrete cosine transform (BPDCT) is developed and then used for transform coding. Secondly, to alleviate edge distortion, classification techniques are applied to transform image coding. The classification technique is based on the edge detection since that the human visual system is more sensitive to edges. Simulation results are given.

1 Introduction

Many researchers are interested in image data compression since it is widely used in many applications, such as teleconferencing, telebrowsing, videotext, electronic shopping in mail order, accessing large databases, security services, HDTV, and so on [1]. A fundamental objective of image compression is to represent an image using a minimum number of bits, while maintaining good fidelity. Many approaches [2] are used to reach this goal. One of the methods is transform coding [3,4,5,6]. Compression is obtained by transforming the input image into a domain where a large amount of information is packed into a few coefficients. The advantages of transform coding are that (1) it maps statistically dependent or correlated image data into independent or uncorrelated coefficients and (2) the channel error within each block does not propagate into the other blocks since each block's coefficients are processed independently.

Neural networks have become popular recently because of their parallelism, mapping ability, and learning ability [7,8]. One is the popular backpropagation neural network (BPNN). It transfers input signals from one domain into another domain. The input-output transform is determined by supervised adjustment of the system parameters through training. Conventional transforms can be performed by the BPNN. The performance of transforms for a given task, can be improved by the BPNN's training. The issues addressed above have led us to begin this research, that is to map the discrete cosine transform (DCT) into BPNN.

Section 2 gives a short description of the BPNN. Section 3 develops the BPDCT. Section 4 shows the implementation of image coding using the BPDCT with simulation results. Section 5 gives the classification technique which is applied to image coding using the BPDCT. Conclusions are gives in section 6.

2 Backpropagation Neural Network

Figure 1 shows the architecture of a simple three-layer BP neuarl network (BPNN) [8,12,13]. The general description of the BPNN is given in the following. The output of each unit in the BPNN, is

$$O(i,j) = f(net(i,j)) \tag{1}$$

$$net(i,j) = \sum_{Ls=Lf(i)}^{i-1} \sum_{NS=1}^{N_U(Ls)} w_b(i,Ls,j,Ns)O(Ls,Ns) \tag{2}$$

where $O(i,j)$ is the output of unit j at layer i. $f(\bullet)$ is the activation function. Lf(i) is the number of the first layer which connects to layer i. $N_U(i)$ is the number of units in layer i for i=1,...,NL. NL is the number of layers. $w_b(i,Ls,j,Ns)$ is the weight connecting unit Ns of layer Ls to unit Nd of layer Ld. Ld is between 2 and NL, Ls is between Lo(Ld) and Ld-1, where LO(i) is the last layer which the layer i feeds. net(i,j) is a weighted sum of signals received at unit j of layer i. It is used to calculate the output of unit j at layer i.

In general, the outputs of units in the output layer are different from the target output Tu(j), where $j=1,\ldots,N_U(NL)$. The error between the O(NL,j) and Tu(j) is measured by the mean-square error,

$$\epsilon = \frac{1}{2} \sum_{j=1}^{N_U(NL)} [\ Tu(j) - O(NL,j)\ ]^2 \tag{3}$$

In supervised learning, the goal is to minimize the error ϵ with respect to the weights $w_b(i,Ls,j,Ns)$. Rumelhart and others [8] have developed the backpropagation algorithm. The basic goal is to minimize ϵ with respect to $w_b(Ld,Ls,j,k)$.

3 Modified Discrete Cosine Transform Using Self-Supervised BPNN

The DCT is widely used in image compression since

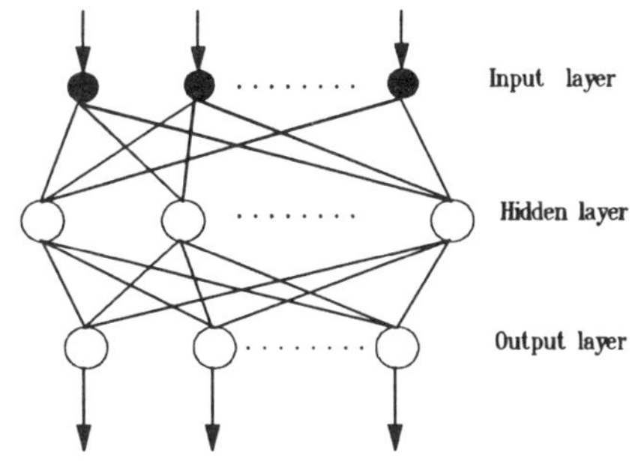

(a). A three-layer BPNN

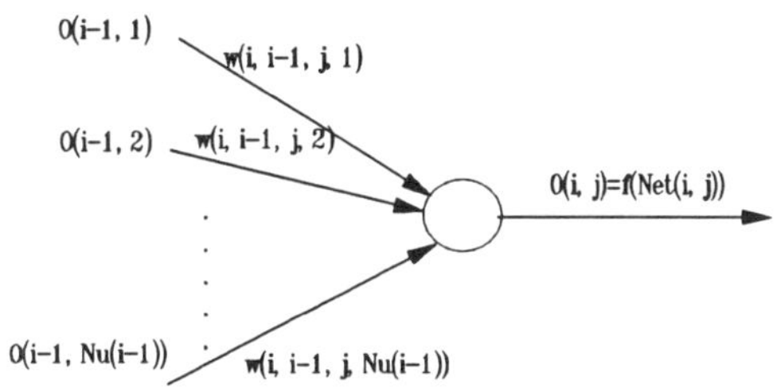

(b). A BPNN unit

Figure 1. Architecture of BPNN

it leads to high data compression rates [1]. In practice there are a few problems: (1) For different input vectors, the distribution of energy among the DCT coefficients changes significantly, (2) Although the DCT concentrates energy well, it is not optimal like the KLT. BPNN can be used in an attempt to alleviate these problems.

Figure 2a shows the truncated 2-D DCT scheme. Input blocks of size NxN are transformed using a 2-D DCT, truncated to size MxM blocks ($M \leq N$), and then inverse 2-D DCT transformed. Here, we map the above processes to the BPNN. The mapping of this 2-D DCT process is described as follows. First the 2-D forward DCT and 2-D inverse DCT are rewritten in lexicographic form. Let x(v) and $x^*(v)$ be the lexicographic form for an input block and output of the 2-D inverse DCT of size of NxN, $v=1,\ldots,N^2$ respectively, and let T(u) be the lexicographic form for the truncated 2-D DCT, $u=1,\ldots,M^2$. The $w_{FD}(u,v)$ and $w_{ID}(u,v)$ are kernel coefficients for the 2-D forward and inverse DCT in lexicographic form respectively. T(u) is written as

$$T(u) = \sum_{v=1}^{N^2} w_{FD}(u,v)x(v), \quad \text{where } u = 1,2,\ldots,M^2 \tag{4}$$

The lexicographic form for the 2-D inverse DCT is

$$x^*(v) = \sum_{u=1}^{M^2} w_{ID}(v,u)T(u), \quad \text{where, } v=1,2,\ldots,N^2 \tag{5}$$

Map equations 4 and 5 to a three layer backpropagation neural network. The inputs of the first layer are samples of a spatial domain block, x(v), $v=1,\ldots,N^2$. The second layer outputs are samples of the 2-D forward DCT, T(u), $u=1,\ldots,M^2$. The third layer outputs are samples of the 2-D inverse DCT, $x^*(v)$, $v=1,\ldots,N^2$. The resulting neural network is constructed as shown as Figure 2b, and is called the BP neural network DCT (BPDCT). The BPDCT is a three layer self-supervised BP network with units $N_U(1)=N^2$, $N_U(2)=M^2$, $N_U(3)=N^2$, and NL=3.

In this BPNN, we choose DCT transform kernel coefficients, $w_{FD}(u,v)$ and $w_{ID}(v,u)$ as initial weights, since the DCT has good compression performance. We also use the input blocks as the desired outputs of the network (this is why it is called self-supervised BP neural network). Here, each unit's activation function is linear, since the DCT is a linear transform. The outputs of input units are the image blocks. The outputs of hidden and output units are

$$O(i,j) = net(i,j) = \sum_{k=1}^{N_U(i-1)} w_b(i,i-1,j,k)O(i-1,k) \tag{6}$$

$w_b(2,1,j,k)$ and $w_b(3,2,j,k)$ in BPNN correspond to the DCT kernel coefficients $w_{FD}(u,v)$ and $w_{ID}(v,u)$ respectively. The output evaluation function is the summation of the mean-square errors over all training blocks,

$$\hat{\epsilon} = \frac{1}{2}\sum_{p=1}^{Npat}\sum_{j=1}^{N_U(NL)} [\, Tu_p(j) - O_p(NL,j)\,]^2 \tag{7}$$

where Npat is the total number of training blocks and the subscript p stands for the index of the training blocks. The mean-square error ϵ is minimized by setting its gradient with respect to the system's unknown weights equal to zero, and then solving for the weights. Since the DCT has a symmetry property, we only take the output weights $w_b(3,2,i,j)$ as unknown variables. The error function is defined as

$$\begin{aligned}\hat{\epsilon} &= \frac{1}{2}\sum_{p=1}^{Npat}\sum_{j=1}^{N_u(3)} [Tu_p(j) - O_p(3,j)]^2 = \sum_{j=1}^{N(3)} \epsilon_j \\ &= \sum_{p=1}^{Npat} \frac{1}{2}[\, Tu_p(j) - \sum_{Ns=1}^{M^2} w_b(3,2,j,Ns)O_p(2,Ns)\,]^2\end{aligned} \tag{8}$$

where ϵ_j is the total error in unit j over all patterns. Setting the gradient of ϵ_j with respect to $w_b(3,2,j,k)$ to zero,

$$\begin{gathered}\frac{\partial \epsilon_j}{\partial w_b(3,2,j,k)} = 0 \\ \sum_{p=1}^{Npat} [\, Tu_p(j) - \sum_{Ns=1}^{M^2} w_b(3,2,j,Ns)O_p(2,Ns)\,] \cdot O_p(2,k) = 0\end{gathered} \tag{9}$$

Solving the M^2 linear equations, we can get the M^2 unknown weights $w_b(3,2,j,k)$ at the output unit j. Since $j=1,\ldots,N^2$, we have N^2 sets of linear equations. After solving all N^2 sets of linear equations, we get improved weights $w_b(3,2,j,k)$. These weights are then copied onto the input layer weights. The training procedure is performed

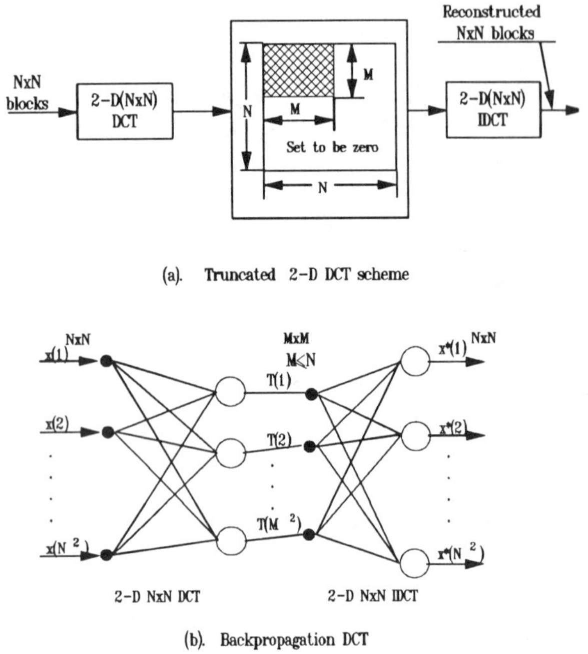

Figure 2. Mapping procedure for BPDCT

repeatedly until the mean-square error ϵ is acceptable.

After training the BPNN, a modified 2-D forward and 2-D inverse DCT are obtained. The input weights from the first layer to second layer, perform the forward 2-D DCTand the output weights, from the second layer to third layer, perform the inverse 2-D DCT. This new transform is the BPDCT.

4 Transform Image Coding Using BPDCT

Figure 3 shows an image transform coding system using the BPDCT. The original image is divided into small blocks of size NxN, and transformed into an MxM set of less correlated coefficients. The BPDCT coefficients of each sub-block are quantized, coded and transmitted. At the receiver the received coefficients are decoded, dequantized, and then the IBPDCT is used to reconstruct the blocks.

The training image for the BPDCT was the 'girl' which is 512 x 512 pels, quantized to 8 bit PCM, where N equals 8 and M equals 4. In these experiments we have not yet used the bit assignment and quantization scheme. We have only compared the truncated DCT and BPDCT. Objective performances are measured by the peak signal to noise ratio (SNR),

$$SNR = 10 \cdot \log_{10}\left(\frac{255^2}{MSE} \right) \qquad (10)$$

where MSE is the mean-square error of the grey levels between the original image and reconstructed image.

In Table 1, the image coding results between the truncated DCT and BPDCT are compared. The simulation results show that the average performance of the image coding using BPDCT is slightly better than using the truncated DCT. Figure 4 is original "girl" image. Figures 5 and 6 are coded "girl" images.

5 Classified Transform Image Coding Using BPDCT

The classification algorithm makes decisions about the activities of the image blocks. Our classification algorithm is based on edge detection since the human visual system is more sensitive to edges. The classifier is a simple rule-based classification algorithm. After the image is preprocessed, a set of representative numbers, called features, are

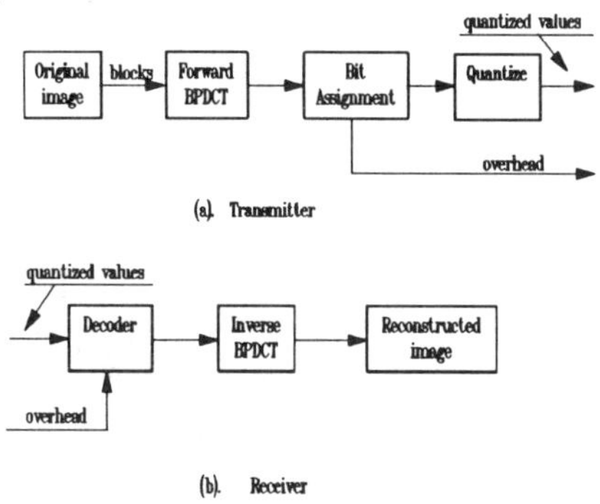

Figure 3. Transform coding scheme using BPDCT

Table 1. Comparison of truncated DCT and BPDCT

Coding scheme	Truncated DCT (8x8 → 4x4)		BPDCT (8x8 → 4x4)	
Image	MSE	SNR(dB)	MSE	SNR(dB)
Girl	22.01	34.70	18.57	35.44
Boat	13.55	36.81	13.48	36.83
Martha	16.81	35.92	15.28	36.29
Baboon	253.90	24.08	263.44	23.92

calculated for each block. Using a set of simple rules, the classifier uses each block's features to decide which class the block belongs to.

In the preprocessing scheme of Figure 7, the Sobel operator is used to obtain the gradient, angle and edge images. The segmentation image is obtained using an edge-guided thresholding algorithm. There are five features. In a given block, feature 1, or Nedge, is the number of edge pixels. Feature 2, or Nedge_1, is the number of edge pixel

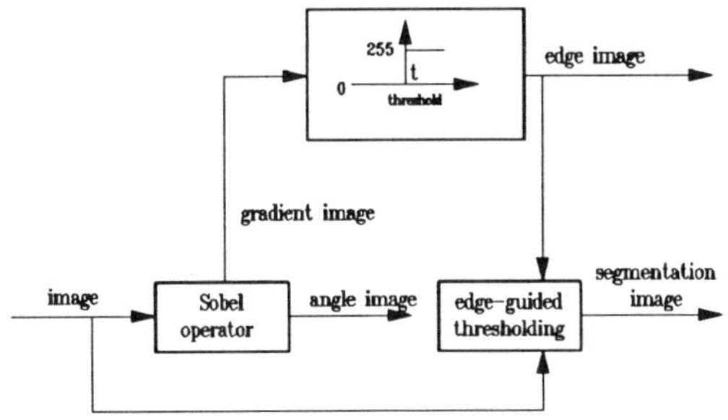

Figure 7. Preprocessing scheme

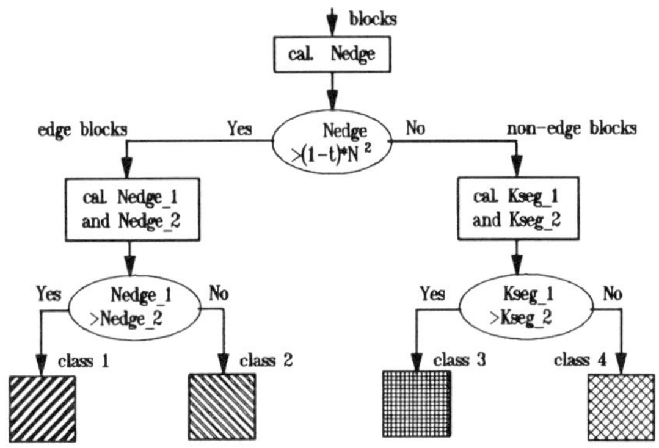

Figure 8. Rule-based classifier

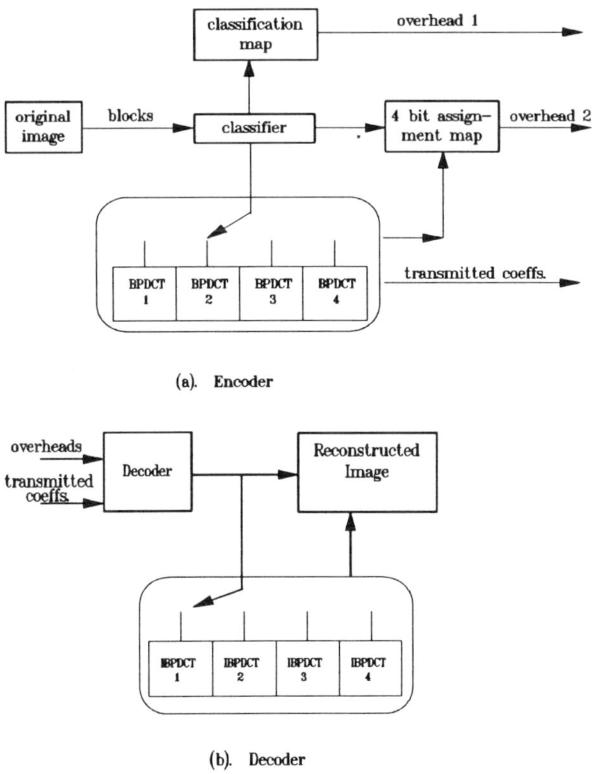

Figure 9. Classified transform coding using BPDCT

Table 2. Comparison Of CBPDCT and BPDCT

Coding scheme	CPBDCT (8x8 → 4x4)		BPDCT (8x8 → 4x4)	
Image	MSE	SNR(dB)	MSE	SNR(dB)
Girl	16.09	36.06	18.57	35.44
Boat	12.21	37.26	13.48	36.83
Martha	15.01	36.36	15.28	36.29
Baboon	251.3	24.13	263.4	23.92

locations where the gradient direction is less than or equal to 90 degrees. Feature 3, or Nedge_2, is the number of edge pixel locations where the gradient direction is greater than 90 degrees. Let Itseg be a threshold such that approximately one half the pixels in the segmentation image are less than or equal to Itesg. Itseg is obtainedusing the cumulative histogram of the segmentation image. In a given block in the segmentation image, feature 4, or Kseg_1, is the number of pixels having gray 2level greater than Itseg. Feature 5, or Kseg_2 equals (N^2 - Kseg_1) where N is the block dimension. The rule-based classifier is shown in Figure 8, where t is the edge threshold.

We experimented with classified transform coding using the BPDCT (CBPDCT) using the system of Figure 9. The blocks of size NxN (here, N is 8), are classified into four classes. For each class, the blocks are transform coded using the BPDCT, so we have four forward and inverse BPDCT's. Overhead information consists of the classification map, which increases the bit rate. At the receiver the quantized BPDCT coefficients and overhead information are decoded to get the location of blocks in the image. Then the corresponding IBPDCT is used to reconstruct the image.

We have compared the performance between image coding using BPDCT and CBPDCT. The simulation results, shown in Table 2, show that the average performance of using the CBPDCT is slightly better than image coding using BPDCT. Figure 10 is the coded "girl" image.

6 Conclusions

In this paper image coding is implemented using a self-supervised BPNN. The resulting BPDCT is sub-optimal for the given training data sets. The simulation results show that the BPDCT works better for image coding than a truncated DCT. Classification techniques improve the performance of the BPDCT.

References

[1] K. R. Rao and P. Yip, **Discrete Cosine Transform: Algorithms, Advantages, Applications**, Academic

Press, San Diego, 1990.

[2] A. K. Jain, **Fundamentals of Digital Image Processing**, Prentice-Hall, 1989.

[3] Y. S. Ho and A. Gersho, "Classified transform coding of images using vector quantization," Proc. ICASSP, pp.1890-1893, April 1989.

[4] W. H. Chen and C. N. Smith, "Adaptive coding of monochrome and color images," IEEE Trans. Commun., vol.COM-25, pp.1285-1292, June 1977.

[5] J. I. Gimlett, " Use of activity classes in adaptive transform image coding," IEEE Trans. Commun., vol. COM-23, pp.785-786, July 1975.

[6] K. N. Ngan, "Adaptive transform coding of video signals," Commun., Radar & Signal Processing, IEE Proc., vol.129, Pt. F, pp.28-40, February 1982.

[7] D. Levine, **Introduction To Neural And Cognitive Modeling**, Lawrence Erlbraum, NJ, 1991.

[8] D. E .Rumelhart and J. L. McClelland, **Parallel Distributed Processing**, vol.I, Cambridge,MA.

[9] P. J. Werbos, "Generalization of backpropagation with application to a recurrent gas market model," Neural networks, vol.1,pp.339-356, April 1988.

[10] P. J. Werbos, "Backpropagation through time: What it does and how to do it," Proc. IEEE, Special issue on neural networks II, pp.1550-1560, October 1990.

[11] P. J. Werbos,"Backpropagation: Past and Future," IJCNN, San Diego, pp.I 343-353, July 1988.

[12] W. Gong, "Unconstrained character recognition using neural network," Master's thesis, Dept. of EE, University Of Texas at Arlington, May 1989.

[13] W. Gong and M. T. Manry, "Analysis of non-Gaussian data using a neural network," Proc. of IJCNN, Wash.D.C., vol.II, pp.576, June 1989.

Figure 4. Original 'girl' image

Figure 5. Coded image using truncated DCT with SNR=34.70dB

Figure 6. Coded image using BPDCT with SNR=35.44dB

Figure 10. Coded image using CBPDCT with SNR=36.06dB

Automatic Speaker Recognition System using the Discrete Hartley Transform and an Artificial Neural Network

Christopher N. Gedo and John C. Eremic

Naval Postgraduate School
Monterey, California 93943

Abstract

A system is presented which will identify speakers that it was trained to recognize with 100% accuracy. This system will also classify inputs that it has not been trained to recognize as unknown. The high degree of confidence offered by this system is attributed to careful processing of the input data sets. Only easily distinguishable segments of the speech samples are passed as inputs to the artificial neural network (ANN). Additionally, sufficient input information is provided with the training inputs to enable the ANN to rapidly train and subsequently classify speakers. This system may be readily implemented in hardware using existing technology. Hardware implementation will render near real-time performance. Applications that would be enhanced by the use of this system include electronic surveillance and computer security.

1 Introduction

The speaker identification system consists of a speech sampler, a signal pre-processor, an ANN, and a network post-processor. The system description begins with a discussion of the techniques used to sample and window data. Pre-processing techniques including definitions of the short-time energy, zero-crossing rate, Discrete Hartley Transform (DHT), and characteristic eigenvector of each speech segment are presented. An example using data from the DARPA TIMIT speech corpus illustrates the pre-processing procedure. Neural network employment including feature vector formulation and post-processing of the network output concludes the system description. Experimental results are provided and demonstrate system effectiveness.

2 Sampling and windowing technique

The speech is sampled at 8 kHz and segmented using a 128 point (16 ms) rectangular window. A short window ensures the stationarity of data within a segment and allows maximum selectivity when choosing segments to process from short-duration speech samples. Each segment overlaps the previous segment by 64 samples. The overlap effectively samples the rectangular window output at its Nyquist rate. The bandwidth of the rectangular window is

$$BW_{rect} = \frac{1}{NT} = \frac{8000}{128} = 62.5 \text{ Hz}. \tag{1}$$

3 Pre-processing technique

The short-time energy and zero-crossing rate (ZCR) are used to differentiate between voiced and unvoiced speech. The discrete convolution

$$E_n = \sum_{m=-\infty}^{\infty} s^2(m)w(n-m) \tag{2}$$

where *s(m)* is the speech signal and *w(m)* is the windowing function defines the short-time energy for the nth segment [1]. It emphasizes the differences in amplitude between voiced speech, unvoiced speech and non-speech portions of the input data set.

The ZCR provides a simple, but accurate, means of spectral measurement [2]. This spectral measure is used to estimate the fundamental vocal tract frequency and characterize the voiced or unvoiced nature of speech segments. The fundamental frequency of a signal as a function of the zero-crossing rate is

$$f_0 = \frac{ZCR}{2} f_s \tag{3}$$

where f_s is the sampling frequency. The ZCR of the nth segment is

$$ZCR_n = \frac{1}{2} \sum_{m=-\infty}^{\infty} \left| \text{sgn}[s(m)] - \text{sgn}[s(m-1)] \right| w(n-m) \tag{4}$$

where the sign function is defined to be

$$\text{sgn}(x) = \begin{Bmatrix} 1, x \geq 0 \\ -1, x < 0 \end{Bmatrix}. \tag{5}$$

As used in the speaker identification system, both the short-time energy and ZCR for each speech segment are compared to thresholds. These thresholds are adjusted until the desired percentage of speech segments pass. This thresholding procedure ensures that the data processed and subsequently presented to the ANN is the most distinctive in nature.

Segments with energy and ZCR above the thresholds are further processed by taking their Discrete Hartley Transform (DHT). The DHT developed by Bracewell [3] is defined to be

$$H[s(m)] = \sum_{m=0}^{N-1} s(m) \cdot \left[\cos\left(\frac{2\pi km}{N}\right) + \sin\left(\frac{2\pi km}{N}\right) \right]. \quad (6)$$

The relationship between the DHT and the DFT is

$$H[s(t)] = M(\omega) \cdot [\cos\phi(\omega) + \sin\phi(\omega)] \quad (7)$$

where $M(\omega)$ is the DFT magnitude and $\phi(\omega)$ is the DFT phase[4]. Equation (7) illustrates the convenient way that the DHT incorporates magnitude and phase information into a single real term.

The DHT coefficients of the selected speech segments are further processed by correlating them. The kth lag of the autocorrelation of a speech segment is defined as

$$r_k = \frac{1}{N} \sum_{m=0}^{N-1} H(m)H(m+k) \quad (8)$$

where $H(m)$ refers to the DHT coefficients for the segment, N is the window length, and $0 \le k \le M-1$. An $M \times M$ dimensional toeplitz matrix of the form

$$\mathbf{R}_n = \begin{bmatrix} r_0 & r_1 & \cdots & r_{M-1} \\ r_1 & r_0 & \cdots & r_{M-2} \\ \vdots & \vdots & \ddots & \vdots \\ r_{M-1} & r_{M-2} & \cdots & r_0 \end{bmatrix} \quad (9)$$

is formed for the nth selected segment. Computation of the eigenvector associated with the largest eigenvalue of $\mathbf{R}_n$ completes the pre-processing of a speech segment. This eigenvector characterizes a processed speech segment and is presented along with the short-time energy and ZCR as the input feature vector to the ANN.

To illustrate the pre-processing procedure, the phoneme *"iy"* is isolated for two male speakers from region one of the DARPA TIMIT speech corpus. A single segment is processed for each speaker. The time domain speech segments and resulting eigenvectors with $M = 16$ are shown in Fig. 1 for the two speakers.

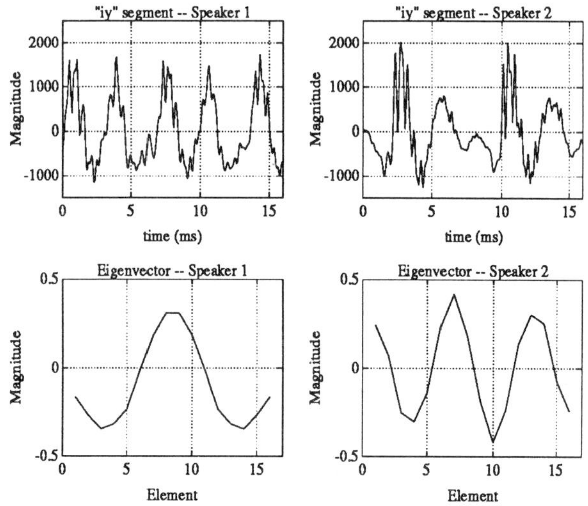

Fig. 1: Time domain speech segments and processed characteristic eigenvectors for two speakers.

To illustrate the influence of the phase on the eigenvectors, the same segments were processed using FFT magnitudes instead of the DHT coefficients. The resulting eigenvectors are shown in Fig. 2.

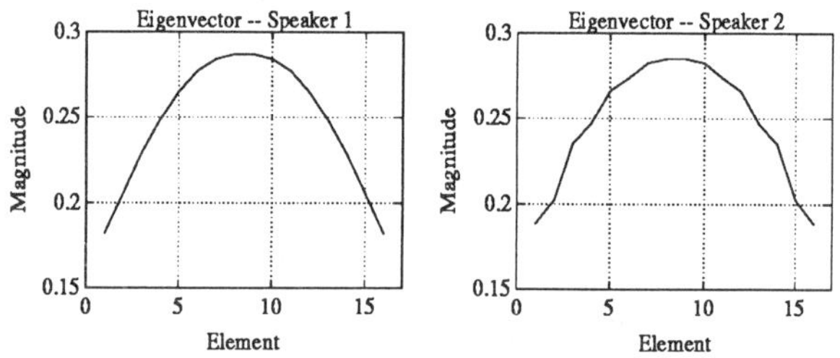

Fig. 2: Eigenvectors derived from FFT magnitudes for the speech segments of Fig. 1.

Comparing the eigenvectors of Fig. 1 to those of Fig. 2, it is clear that including phase information in the pre-processing procedure produces more distinct feature vectors.

4 The artificial neural network

Although the ANN is a major part of this system, it is completely generic in nature and its function is defined completely by the input vectors and the desired output. A standard Back-Propagation (BP) or Learning Vector Quantization (LVQ) network is used for the speaker identification system. Whereas the LVQ network uses a Kohonen layer to compute the Euclidean distance between the input vector and each processing element's (PE's) weight vector, the BP network uses the hyperbolic tangent for its PE transfer function[5].

The number of input PE's is governed by the number of elements in the input vector and the number of desired outputs independent of type of network employed. Each desired output has a corresponding input PE with a linear transfer function. There are normally three or fewer hidden layers. The number of PE's for the first hidden layer is usually the same as the input layer. If additional hidden layers are used, they typically have fewer PE's than the input layer. Figure 3 illustrates a typical BP ANN architecture.

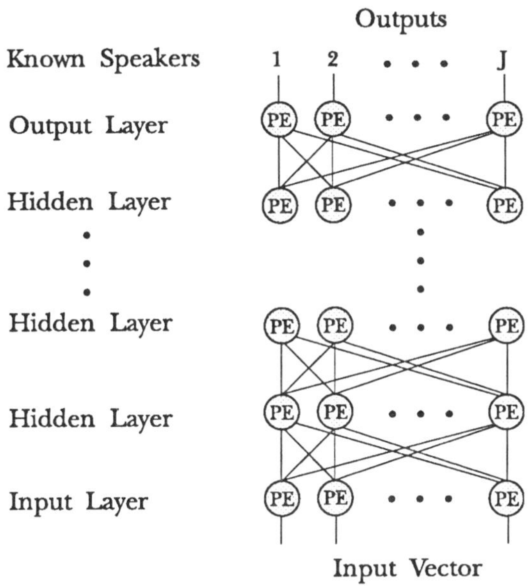

Fig. 3: Typical BP ANN architecture.

The ANN of the speaker identification system must be trained to recognize the known speakers before it can be used. A training file of input vectors consisting of the short-time energy, the ZCR, and the principal eigenvector processed from the correlated DHT coefficients is required. Additionally, each input training vector is appended with a desired output code having the same number of elements as there are outputs. These outputs represent the known speakers. The output code consists of a *one* for the desired output and a *zero* for all other outputs. Figure 4 shows the contents of a training file of input vectors for a two known speaker system ($J = 2$).

The short-time energy and ZCR are used to clarify whether a sample represents the voiced or unvoiced speech of a particular speaker. This accelerates the learning rate of the ANN by reducing the ambiguity caused by having two types of input vectors with very different characteristics representing the same speaker. It is desirable to provide the ANN with as many training inputs as possible to ensure accurate classification.

The ANN is trained until the desired accuracy is obtained. Once trained, the speaker identification system may be used.

Short-Time Energy, ZCR, E(1), E(2), . . . , E(M), 1, 0
Short-Time Energy, ZCR, E(1), E(2), . . . , E(M), 1, 0
•
•
•
Short-Time Energy, ZCR, E(1), E(2), . . . , E(M), 1, 0
Short-Time Energy, ZCR, E(1), E(2), . . . , E(M), 0, 1
Short-Time Energy, ZCR, E(1), E(2), . . . , E(M), 0, 1
•
•
•
Short-Time Energy, ZCR, E(1), E(2), . . . , E(M), 0, 1

Fig. 4: Typical training file of input vectors for a two known speaker system.

A speech sample is collected and processed. A file of input vectors similar to those of Fig. 4 without the training codes is presented to the ANN.

5 ANN post-processing

The outputs of the ANN must be interpreted so that a classification can be assigned to each speech sample. The ANN outputs for the feature vectors of the chosen segments from each speech sample are compared to a threshold, typically > 0.5, and then tallied. This threshold is required for BP networks because the output values normally range from 0 to 1. Conversely, an LVQ network quantizes its outputs at 0 or 1 and does not require a threshold. If a high percentage of the outputs are tallied for one category with a corresponding low percentage for the others, a classification is made. If a sample does not meet this criterion, a classification of unknown is assigned.

6 Experimental results

The system has been tested on a limited case with superior results. Speech samples were obtained from three male speakers who exhibited similar pitch levels in their speech. The ANN was trained to recognize two of the speakers using eight samples of speech per speaker. Each sample was two seconds in length. Samples from the third speaker were reserved to test the system's ability to reject an unknown speaker. With short-time energy and ZCR thresholds set to retain 12% of the speech segments and 16 element eigenvectors, 350 feature vectors per speaker were presented to the ANN for training. A BP ANN was trained for 60,000 cycles and then tested. The results of that test are shown in table 1.

The output value threshold to assign a choice was set at 0.8. The BP ANN correctly identified both speakers with significantly more accuracy than it did the unknown speaker. The same data was presented to an LVQ ANN. In this case, the unknown speaker was rejected with more confidence than with the BP ANN because the outputs were essentially split between the two known speakers. Table 2 shows the results of the test using an LVQ ANN.

Speaker	Choice 1	Choice 2
1	58/72	0/72
2	4/72	55/72
unknown	111/205	44/205

Table 1: Results of test with two known speakers, one unknown speaker, and a BP ANN.

Speaker	Choice 1	Choice 2
1	71/72	1/72
2	11/72	61/72
unknown	102/205	103/205

Table 2: Results of test with two known speakers, one unknown speaker, and an LVQ ANN.

When the output is considered in a statistical sense the ANN demonstrates the ability to classify speakers with no ambiguity. It is interesting to note that in this BP case, the RMS error never converged during the training phase before conducting the test. This indicates that a restrictive RMS error threshold is not necessary to guarantee robust performance of the system.

7 Conclusions

The speaker identification system was successfully tested. Although the test was limited, results indicate that it will perform well in more general cases. It is important to use phase information when pre-processing the data; the DHT conveniently combines phase and magnitude information into a single real term. Quantities such as the short-time energy and ZCR characterize the voiced or unvoiced nature of a speech segment and are necessary to expedite training and testing of the ANN. The thresholding scheme to choose which speech segments to process ensures that only distinctive portions of the speech sample are presented to the ANN for classification.

References

[1] O'Shaughnessy, Douglas, *Speech Communication, Human and Machine*, Reading, Massachusetts: Addison-Wesley, 1990.

[2] Rabiner, L.R. and Schafer, R.W., Digital Processing of Speech signals, Englewood Cliffs, New Jersey, 1978.

[3] "Discrete Hartley Transform", K.N. Bracewell, *J. Opt. Soc. Am.*, Vol 173, No12, pp 1832-1835, Dec 83.

[4] "The Hartley Transform Applied to Speech Coding," E. Chilton, *IEE Colloquium on Speech Coding*, IEE London, Nov 89.

[5] Rumelhart, D.E. and McClelland, J.L., "Parallel Distributed Processing: Explorations in the Micro-Structure of Cognition", Vol I, *Foundations*, MIT Press, 1986.

A Novel Approach to the Integration of Simulation and Implementation to Digital Signal Processing Systems

Sebastian Ritz, Matthias Pankert and Heinrich Meyr

Aachen University of Technology, ERT
Templergraben 55, D-5100 Aachen, Germany

Abstract

This paper presents the integration of simulation and implementation of complex digital signal processing systems based on block diagram oriented simulation and code generation. The basic concept consists of a generic, i.e. reusable block diagram with different underlying semantics. This allows the generation of software adapted to the special needs of both simulation and implementation. In order to achieve application specific implementations, the code generator supports trade-offs between implementation constraints such as code size, throughput or latency. Therefore, a wide spectrum of different optimization strategies have been incorporated in the code generator.

1 Introduction

Computer simulation has become a widespread method for the design and analysis of complex signal processing systems, since analytical tractability is limited to a small class of problems. In the area of communications, measurement and speech processing, block diagram oriented simulation is an accepted standard [1]. By means of evaluation of characteristic performance measures such as bit/symbol error rates, acquisition time or speech quality, block diagram oriented simulation provides efficient support in the selection and optimization of algorithms or systems. Thus, as a result of this first design step, simulation may lead to the specification of the algorithms and systems to be implemented in a second design process.

In contrast to simulation the implementation design process is concerned with efficiency and implementation constraints such as area, time and power consumption (for hardware design) or throughput, code size, latency or precision (for software design). Here appropriate tool support for the synthesis of signal processing systems is required to master the complex design space [2].

In this paper we present a CAD environment which integrates simulation on the one hand and the implementation of digital signal processing systems on the other. The integration is based on the representation of systems and algorithms using generic block diagrams. This means that the block diagram may be used for both purposes by automatically evoking different block diagram oriented simulation or synthesis techniques. The software environment consists of the simulation system COSSAP[1] [3] and the code generator DESCARTES which generates optimized realtime code for programmabel processors, e.g. Digital Signal Processors (DSPs) and transputers. DESCARTES has been motivated by the emerging applications of DSP chips in numerous signal processing areas and their improvement of performance due to their specialized architectures. In the current version DESCARTES generates C code and utilizes the vendors assembler libraries, which comprises most common signal processing algorithms.

In section 2, generic block diagrams are discussed emphasizing different underlying semantics for system simulation and DSP code generation. Section 3 presents the integrated tool architecture. In section 4 our approach to the goal directed optimizing DSP code generation is described, and in section 5 performance examples for the applied optimization strategies are given.

2 Multiple block diagram semantics

A very natural and convenient way of modelling signal processing systems is a description by (possibly

[1] COSSAP is a system level simulation and analysis tool, originally conceived through research started in 1984 at ERT, Aachen University of Technology. Using this reasearch as a base, CADIS GmbH (Aachen) has further devoloped COSSAP to a commercial tool.

graphical) block diagrams. The blocks are behavioral models of the system components (e.g. filters, coders or arithmetic units like adders etc.) and their interconnections define the signal flow in the system. In all state-of-the-art simulation packages [1] the graphical block diagram oriented description may be translated into a simulation program in a one-to-one fashion, where each block is represented by a software module performing the specified signal processing algorithm. The execution of the simulation solely depends on the availability of signal samples to be processed by the blocks. These are scheduled (i.e. activated) in such an order that the required input signal samples of a block are produced by the predecessors of the block. This explicitely distinguishes so called stream-driven scheduling [1,3] from time- or event-driven scheduling techniques [4]. Furthermore, the block diagram oriented representation of signal processing systems is well suited for the exploitation of inherent parallelism either for simulation [5] or implementation [6].

The main concept of integrating block diagram oriented *simulation* and *real-time code generation* is to extend the block diagram oriented representation to a generic specification. According to the actual application, different semantics can be attached to a block diagram. The semantics comprise the scheduling paradigm, the block input/output behavior, the signals and their types, and the implementation of the blocks.

Our first step was to consider two different scheduling paradigms, namely static and dynamic stream-driven scheduling [3,7].

Many algorithms used in communication, measurement or speech processing exhibit a purely static data flow (SDF,[8]) behavior, although, in some cases, they can be modelled more memory-efficiently using dynamic data flow. The SDF scheduling paradigm requires that every block consumes and produces a fixed number of signal samples per activation. This allows for scheduling the block diagramm statically and for allocating static signal buffers, which can lead to an enormous speedup as compared to dynamically scheduled and buffered applications, as shown in section 5.

Dynamic data flow on the other hand always requires runtime scheduling and dynamic buffering. The advantage of increased flexibility in modelling asynchronous systems (e.g. timing synchronization or controlled multiplexing) is paid for by a loss in implementation efficiency. Asynchronous blocks only require the specification of the minimum number of input samples per activation, and the scheduler makes no a priori assumption of how many samples are actually consumed or produced by a block during runtime [3].

For simulation with COSSAP the dynamic scheduling paradigm was chosen to offer the highest flexibility in modelling signal processing systems at the cost of some overhead. For the real-time code generation, however, this overhead will not be always acceptable. Therefore, the primary modelling paradigm is statically scheduled synchronous data flow for the DSP code. Hybrid solutions are applied for heterogenous (i.e. static mixed with dynamic) parts of the signal flow [7]. The different block diagram semantics are summarized in the following table:

	Simulation	Real-Time Code
Scheduling	dynamic	static
Block-I/O	synchr. and asynchr.	synchr.
Block-Impl.	C/FORTRAN	C/Assembler
Signal-Buffer	dynamic	static
Signal-Types	C-types	generic, C-types

While static and dynamic stream-driven scheduling is well suited for the simulation and implementation of asynchronous or synchronous signal flow-oriented applications, a further computational mode has to be introduced in order to deal with control flow-oriented applications. These applications require a scheduling paradigm based on event handling, similar to that in kernels of real-time operating systems. Currently we are working on the extension to the block diagram representation in order to provide control flow combined with signal flow in a hierarchical fashion.

3 Integrated software environment

As shown in Figure 2, the common base is the generic block diagram, which is configured hierarchically by means of a graphical editor. The block diagram editor generates a textual VHDL representation which is parsed into an object-oriented database leaving the hierarchy unflattened. All further software components are organized around this central database. The COSSAP simulation manager translates the block diagram representation in the database into an executable simulation program.

DESCARTES is structured into front end and back end components. The front end comprises the database interface and the target independent transformations and optimizations. Different back ends can be attached to the DESCARTES front end according to the hardware platform and user specified optimization

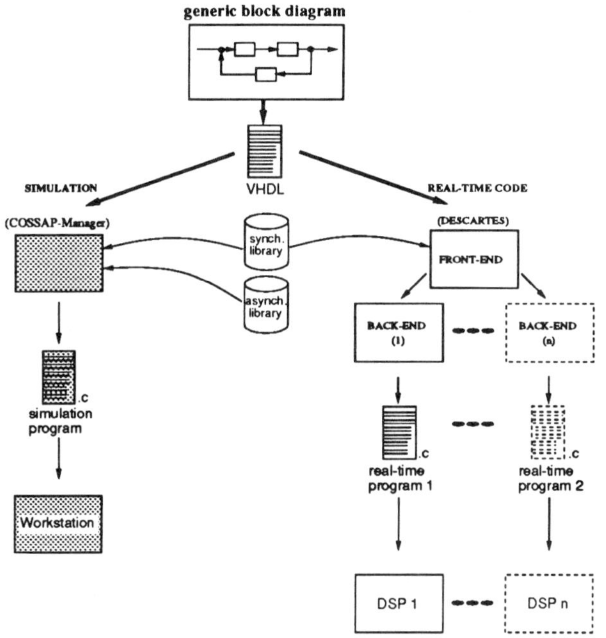

Figure 1: tool architecture

strategies. This structure allows for flexible extendability. Only the back end part has to be extended to support additional hardware or code generation strategies.

The integrated software environment offers different block libraries. While the DESCARTES blocks are required to be synchronous, the COSSAP blocks can be either synchronous or asynchronous. These requirements are reflected by the fact, that DESCARTES relies on an optimized real-time library, which can be used by COSSAP as well. (Note, that a DESCARTES block can be automatically converted into a COSSAP block). The COSSAP library, however, cannot be used with DESCARTES, as its blocks are asynchronous in general.

4 Optimizing code generation

When designing real-time DSP-software, implementation constraints such as ceilings in memory consumption or execution time per sample play an important role. Often such constraints mutually exclude each other. This means that there is no single optimal solution for all purposes. Besides throughput and code size, other constraints may be e.g. latency, arithmetic precision.

Our concept offers a wide variety of optimization strategies which can be combined in order to achieve a customized optimal solution for the user's specific needs. Different classes of optimization techniques are considered:

- optimization at block diagram level
- optimization at block level
- target hardware specific optimization techniques

4.1 Block diagram level optimization

A wide range of admissible schedules is available for processing SDF block diagrams [8]. The scheduling strategy has a direct impact on the signal buffering and therefore on the memory requirements. A schedule can be devized to minimize the buffer size by generating an activation order such that the input samples reach the sinks as fast as possible. This requires that all blocks work with their minimum number of required input samples. In contrast to this scheduling strategy one can improve the execution time per sample by means of block processing. This means that if a block consumes e.g. one input sample at each input port and computes one output sample at each output port these rates can be increased by a blocking factor k. This requires that the implementation of the block has a loop of the length k. For the support of block processing, the scheduling strategies look different, since now every block consumes and produces as many samples per activation as possible. This gain per sample has to be paid for with increased latency and code size due to large buffers (s. section 5). The gain in execution time is mainly due to less overhead considering the number of function calls and initializations per sample and due to a better utilization of the DSP-pipelining.

Another trade-off between code size and execution time per sample is offered through function calls versus in-line code generation. In the first case the implementations of all blocks are defined as C-functions, and the scheduling consists of calls to these functions. For in-line code generation, the blocks' code is expanded in the scheduling order. In general, this leads to faster execution but increased code size.

4.2 Block level optimization

The block processing mentioned above is supported by all blocks of the system library comprising standard DSP algorithms. User-written blocks may have block processing capabilites as well. This is supported by a generic (non-C) loop-construct, which may be mapped by DESCARTES to an ordinary C loop adapted to the optimization strategy of the target C compiler (*do-while(), for()* or *while()*). On the other hand, this loop may be partially or fully unfolded, i.e. the code inside the loop is replicated a specified number of times. Unfolded loops use the internal DSP pipeline more efficiently than ordinary loops. If a block is activated with a blocking factor $k = 1$, then the loop is not generated at all, avoiding any overhead. These generic loops may not only be used for loops for block processing but also for loops where the number of iterations depends on constants or parameters of a block. Another trade-off between code size and computational efficiency is supported by parameter precomputations with inline expansion. Instead of referencing parameters via pointers, their values may be directly expanded in the code. This requires that for every block occuring several times in a block diagramXS the code has to be replicated as often as different parameter sets have been configured for this block.

As mentioned in section 2, multiple implementations can be provided for the same block. All implementations have a common interface, and they all exhibit the same algorithmic behavior but different performance characteristics. The code generator can pick the implementation appropriate to meet its user defined optimization goals.

4.3 Hardware specific optimization

Besides the hardware independent optimization described above, there are many strategies taking special hardware and non standard C compiler features into account. For example assembler instructions can be included at critical parts of the code using in-line assembler macros. Also, some C compiler offer special in-line control directives for controlling DSP-latencies.

Assembler versions or those using vendor supplied, optimized object libraries can be among the multiple implementations of a DESCARTES block, according to the concept presented above. These are, of course, not reusable for different hardware platforms.

5 Performance Analysis

To validate our reasoning about optimizing code generation, we have have simulated and implemented a simple but typical signal processing example on different levels of abstraction. The system consists of a source producing a unit step, a transversal filter system, and a sink writing the samples into a sink. The transversal filter with 40 taps was implemented on three different abstraction levels: (1) fully behavioral (one block), (2) seperate blocks for each filter tap and (3) decomposition into primitives (adder, delay, and multiplier).

In all benchmarks the number of samples processed was so high that at least 90% of the total execution time was spent in the transversal filter system, thus initialization, unit step generation and file output was negligible.

5.1 Comparison of dynamic and static scheduling

For the first benchmark we compare the dynamically scheduled execution generated by COSSAP with the statically scheduled execution of the code generated by DESCARTES. In both cases the generated code ran on a SPARCstation with integer signals. The blocking factor k was choosen to be 1000.

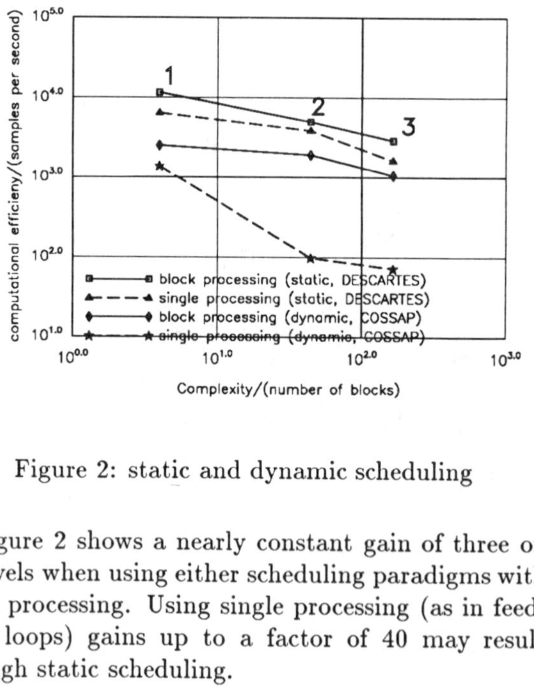

Figure 2: static and dynamic scheduling

Figure 2 shows a nearly constant gain of three on all levels when using either scheduling paradigms with block processing. Using single processing (as in feedback loops) gains up to a factor of 40 may result through static scheduling.

5.2 Efficiency of optimization strategies

Figure 3 illustrates the effect of block processing on the computational effiency. The code generated by DESCARTES runs on a AT&T DSP32C. When the blocking factor is increased to 1000, the computational efficiency is increased by a factor of 3. Loop unfolding further enhances the efficiency but also increases code size as illustrated in the Figure 4.

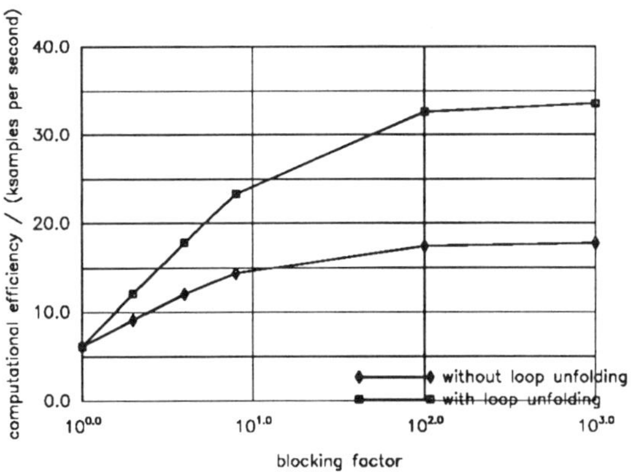

Figure 3: influence of block processing

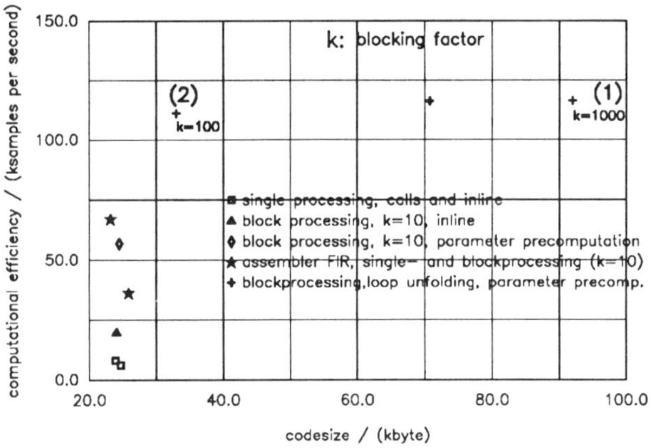

Figure 4: design space on level 1

Point 1 shows generated code, for which block processing ($k = 1000$), loop unfolding of all loops, parameter inline expansion and DSP32C C-compiler specific directives have been applied. Decreasing the blocking factor to 100 with partial loop unfolding leads to code at point 2, descreasing the code size by a factor of more than 2. Even smaller code size may be obtained through single processing, but this also decreases the throughput by a factor of more then 10. Figure 4 shows clearly that through the combination of different optimization strategies code may be generated which meets application dependent constraints.

6 Conclusions

The real-time code generator DESCARTES is a completion of the system level simulation and analysis tool COSSAP. Together it offers an integrated design environment for digital signal processing systems.

All combinations of the described optimization strategies in DESCARTES form a multidimensional design space. With the increasing number of techniques becoming available, the manual handling of this design space gets harder. An automated optimization aid will take the user's design goals as input and searches the design space for solutions that meets these requirements. Thus in DESCARTES the emphasis is now placed on goal directed optimization strategies.

Acknowledgement

The support we were given by the Deutsche Forschungsgemeinschaft (DFG) is gratefully appreciated.

References

[1] Shanmugan, K. S., "An update on software packages for simulation of communication systems (links)," *IEEE Journal on Sel. Areas in Commun.*, vol. 6, pp. 5–12, January 1988.

[2] J. Rabaey, C. Chu, P. Hoang, and M. Potkonjak, "Fast Prototyping of Datapath-Intensive Architectures," *IEEE Desing & Test of Computers*, pp. 40–51, June 1991.

[3] J. Kunkel, "COSSAP: A stream driven simulator," Mar. 1991. IEEE Intl. Workshop on Microelectronics in Communications, Interlaken, Switzerland.

[4] P. Zepter and K. ten Hagen, "Using VHDL with stream driven simulators for digital signal processing applications," in *EURO-VHDL'91 Proceedings*, (Stockholm, Sweden), September 8-11 1991.

[5] H. J. Schlebusch and S. Ritz, "A novel approach to distributed signal-flow driven simulation," *Int. J. Computer Simulation*, 1991. Ablex Publishing Corp., Norwood, NJ, USA, accepted for publication.

[6] Lee, E. A. et al., "Gabriel, a design environment for DSP," *IEEE Micro*, pp. 28–45, October 1990.

[7] J. Buck, S. Ha, E. Lee, and D. Messerschmitt, "Ptolemy: A platform for heterogenous simulation and prototyping," in *Porc. 1991 European Simulation Conf.*, (Copenhagen, Denmark), June 1991.

[8] E. A. Lee and D. G. Messerschmitt, "Static scheduling of synchronous data flow programs for digital signal processing," *IEEE Trans. on Computers*, vol. C-36, No. 1, pp. 24–35, January 1987.

Channel Equalisation using the Non-Canonical LMS

R.W. Stewart, J.J. Soraghan, W.S. Gan, T. S. Durrani
Signal Processing Division
Dept. Electronic and Electrical Engineering
University of Strathclyde
Glasgow
Scotland, U.K.

Abstract

This paper presents the use of an LMS based algorithm called the non-canonical LMS (NCLMS) for channel equalisation. The NCLMS algorithm is not implemented on a standard FIR filter, but uses a structure which we refer to as the non-canonical FIR (NC-FIR). When the NCLMS is implemented the signal flow graph can be shown to be fully pipelinable and the sampling rate is not inhibited by the recursive nature of the algorithm, as in the case for the LMS. Channel equalisation results are given which show that the NCLMS has a reduced excess mean squared error level and an improved performance in an impulsive noise environment over the LMS.

1 Introduction

The realisation of fast adaptive filters is important in many applications. However due to the recursive (feedback) nature of these algorithms an inherent sampling rate limitation is imposed by the feedback of the residual error to the adaption of individual stages. As a consequence of this feedback it is not possible to pipeline and therefore parallelize these algorithms in the conventional way. Therefore it is highly desirable to find a parallel architecture that can be used in very high sampling rate realisations using low cost VLSI technology.

One way of increasing the sampling rate of adaptive filters is to use block computation using parallel hardware. The sampling rate of block adaptive filters is directly proportional to the block size. However as this increases the tracking ability degrades. Therefore there is an *applicational* limit whereby operation at an arbitrarily high speed would have no tracking capability.

Parallel and pipelined computation of the LMS algorithm using a FIR filter is inhibited by the *delayless* loop feedback of the adder line and error feedback (see Figure 3). Therefore all FIR based adaptive filters (LMS, RLS, Kalman) have speeds limited by the hardware, or more precisely proportional to the filter length.

In this paper a non-canonical FIR (NCFIR) structure is derived, presented and analysed for use in parallel adaptive signal processing. The non-canonical FIR essentially replaces the tapped data delay line and ripple through adder line, by a broadcast data line, and a delayed adder line. This circumvents the limiting factor of the adder line latency. Simulation results are given which show a reduced EMSE (excess mean square error) level and an improved performance in an impulsive noise environment for the NCLMS over the LMS algorithm for a channel equalisation example.

2 Canonical and Non-Canonical FIR Filters

Figure 1 shows the canonical FIR filter structure. In order to transform the canonical FIR to an NC-FIR structure, cut-set analysis [4] can be used to modify the temporal functioning of the signal flow graph (SFG) to remove the ripple through adder line, which means that total computation time for one filter iteration is:

$$T_{filter} = T_{mult} + NT_{add} \quad (1)$$

where T_{mult} is the time for a mulitplication, T_{add} is the time for an addition, and N is the filter length. After applying the cut-sets Figure 2 results. where the input signal is supplied in parallel to all N multipliers, and the delay and summation operations are performed in a distributed manner such that the total computation time per iteration is:

$$T_{filter} = T_{mult} + T_{add} \quad (2)$$

The formal transformation ensures that the NCFIR filter of Figure 2 is functionally equivalent to the canonical filter of Figure 1 for fixed weight conditions.

When performing algorithms that update the weight vector (i.e. adaptive) the NCFIR no longer functions identically to the FIR. in terms of adaptation, stability and transient response. Cut set analysis reveals that the class of (weight) adapting algorithms that can be implemented on the NCFIR are completely different from the canonical FIR.

3 The LMS Adaptive FIR Filter

In this section the LMS algorithm [3] will be briefly overviewed before introducing the NCLMS. The LMS algorithm is the most widely used adaptive signal processing algorithm. It uses a special estimate of the gradient that is valid for the FIR filter structure. The LMS's popularity is due to its simplicity and ease of

1058-6393/91 $1.00 © 1991 IEEE

computation, and its avoidance of explicit off-line gradient estimates of the performance surface. The LMS algorithm is widely found in channel equalisation systems.

The output of the N-tap FIR filter of Figure 1 is:

$$y_k = \sum_{i=0}^{N-1} w_i.x_{k-i} = \mathbf{x}_k^T.\mathbf{w} \tag{3}$$

where $\mathbf{w} = [w_0\ w_1\ w_2\ ...\ w_{N-1}]^T$ and $\mathbf{x}_k = [x_k\ x_{k-1}\ x_{k-2}\ ...\ x_{k-N-1}]^T$.

Consider now the canonical FIR LMS adaptive filter of Figure 3. The weight vector $\mathbf{w}_{k+1}^T$ to be used at iteration $k+1$ is calculated at the k-th iteration according to a steepest descent gradient type algorithm:

$$\begin{aligned} \mathbf{w}_{k+1} &= \mathbf{w}_k - \mu\hat{\nabla}_k \\ &= \mathbf{w}_k + 2\mu\epsilon_k.\mathbf{x}_k \end{aligned} \tag{4}$$

where $\hat{\nabla}_k = -2\epsilon_k.\mathbf{x}_k$ is the LMS estimate of the gradient. ϵ_k is calculated simply as, $\epsilon_k = d_k - y_k$, and where d_k is the *desired* response. The step size μ constant regulates the speed and stability of convergence. The output of the LMS FIR filter in Figure 3 is:

$$y_k = \sum_{i=0}^{N-1} w_k^i.x_{k-i} = \mathbf{x}_k^T.\mathbf{w}_k \tag{5}$$

where $\mathbf{w}_k = [w_k^0\ w_k^1\ w_k^2\ ...\ w_k^{N-1}]^T$ and $\mathbf{w}_0 = \mathbf{0}$.

The SFG of Figure 3 has two distinct parts, the FIR structure performing Eq. 5, and the adaptive error feedback calculating the weight updates of Eq. 4 on each iteration.

4 The NCLMS Adaptive Filter:

The FIR structure of Figure 3 can be transformed using cutset analysis, to the non-canonical adaptive FIR structure of Figure 4, noting that the error feedback lines must now also be considered in the transformation. The error feedback line now has $-\Delta$ time delays which in actual fact correspond to a non-causal time advance and are clearly impossible to implement in our causal time world! A consequence of this is that it is impossible to exactly implement any transversal filter based gradient update algorithms such as the LMS algorithm, using the non-canonical structure. Therefore if it is required to perform gradient algorithms on NCFIR structures then the algorithms must be reformulated and reanalysed in terms of convergence performance and stability. Figure 5 shows a derived SFG which implements the algorithm named the NCLMS. Note that this structure contains the NCFIR structure, but the non-causal time advances present in Figure 4 have been removed.

To analyse the NCLMS, first note that the output from the NCFIR structure of Figure 2 can be written as:

$$y_k = \sum_{i=0}^{N-1} h_{k-i}^i.x_{k-i} = \mathbf{x}_k^T.\hat{\mathbf{h}}_k \tag{6}$$

where $\hat{\mathbf{h}}_k = [h_k^0\ h_{k-1}^1\ h_{k-2}^2\ ...\ h_{k-N+1}{}^{N-1}]^T$.

One of the main differences between the NCLMS and LMS algorithms is the updating feedback error signal. In the LMS algorithm, the kth error signal is a function of $\mathbf{w}_k$, whereas in the NCLMS algorithm, the kth error signal is a function of $\hat{\mathbf{h}}_k$ where each element of this coefficient vector is skewed in time. That is each element of the current coefficient vector contributes to the output at a different instant in time. Consider a 5 stage adaptive filter for both the LMS and NCLMS algorithms. A state transition diagram of the coefficient weights for both the LMS and NCLMS algorithms is shown in Figure 7. From Eq. 5, the output of the LMS algorithm is obtained by linearly combining all the weights at the kth time instant with the input tap vector at the kth time instant, while the output of the NCLMS is obtained by forming the sum of the product of the ith coefficient weight at the $(k-i)$th time instant with the $(k-i)$th sample of the input tap vector for $i = 0,\ 1,\ 2,\ 3,\ 4$.

To compare the LMS and NCLMS (temporal) scaling factors [1], can be introduced and Eq. 6 rewritten as:

$$y_k = \sum_{i=0}^{N-1} \alpha_k^i.h_k^i.x_{k-i} \tag{7}$$

$$\alpha_k^i = \frac{w_{k-i}^i}{w_k^i} \tag{8}$$

$$= \frac{i\text{-}th \text{ coefficient weight at } (k\text{-}i)\text{-}th \text{ time instant}}{i\text{-}th \text{ coefficient weight at } k\text{-}th \text{ time instant}} \tag{9}$$

Writing Eq. 7 in matrix-vector form gives:

$$y_k = \mathbf{x}_k^T.\mathbf{A}_k.\mathbf{h}_k \tag{10}$$

Equation 7 now resembles the canonical form in Eq. 3, such that the effect of the current weight values, $\mathbf{h}_k$, on the next output, y_k, does not have a time skewed lag. The matrix of ratio coefficients is a diagonal matrix, $\mathbf{A}_k = diag[\alpha_k^0,\ \alpha_k^1,\ \alpha_k^2,\ ....\ \alpha_k^{N-1}]$.

From Eq. 4 the LMS algorithm can be reformulated for the NCFIR structure. Since this algorithm is different from the standard LMS, the step size is denoted by β. Substituting from Eqs. 4 and 10 gives:

$$\begin{aligned} \mathbf{h}_{k+1} &= \mathbf{h}_k + 2\beta(d_k - \mathbf{x}_k^T.\mathbf{A}_k.\mathbf{h}_k).\mathbf{x}_k \\ &= \mathbf{w}_k + 2\beta(\mathbf{x}_k.d_k - \mathbf{x}_k.\mathbf{x}_k^T.\mathbf{A}_k.\mathbf{h}_k) \end{aligned} \tag{11}$$

By comparing Eqs. 3 and 6, and Eqs. 4 and 11 it can be seen that the output of the NCLMS algorithm is likely to be quite different to the output of the LMS algorithm during adaptation. In the mean the NCLMS algorithm converges to the LMS solution only is $\mathbf{A}_k \rightarrow \mathbf{I}_k$ ($N \times N$ identity matrix). In all simulations where the filter did not go unstable the NCLMS solution was never found to be significantly different from the LMS solution. Note also that the bound on the step size β is not the same as for the LMS algorithm. This was confirmed in both simulations and by analysing Eq. 11.

5 Pipelining the NCLMS

The NCLMS was first proposed in [1]. Recent papers have also considered the NCLMS algorithm. In [6], Harris Semiconductors have described a *fully systolic adaptive filter.* They have in fact implemented the NCLMS which is refered to in the paper as a modified LMS. The motivation for this paper is described as being to overcome the problem of the summation tree, and hence the latency introduced by the FIR and LMS. In actual fact, the implementation by Harris is not truely systolic, as it includes two broadcast lines - one for the error signal and one for the data signal, which would be problematic for large arrays. The NCLMS *cannot* be retimed to give a systolic array due to the recursive nature of the SFG [4]. However after performing *delay scaling* [4] which slows the array down by a factor of two, the NCLMS can be retimed to give a truely systolic implementation.

In Figure 6, after delay scaling by 2, and judicious use of cut sets, a truely systolic version of the NCLMS is obtained. Note that because of the delay scaling this filter now works at only half of the speed of the original filter (after delay scaling input data must be interleaved with blanks) and at any one time only 50% of the hardware is being used. Note however that since the LMS required both the filter update, and the weight update on each iteration, the efficiency of the array can be greatly improved by multiplexing the multiplier between the NCLMS calculation, and the FIR calculation.

6 Channel Equalistion Simulations

The channel equalisation set up shown in Figure 8 was used for the simulation. A random generator which outputs a test signal of ± 1 is input into the linear dispersive channel and the channel noise is represented by an additional Gaussian white noise source at the output of the channel. An impulsive noise generator is also placed at the channel output to simulate the effect of large transient noise. The random input sequence after suitable delay also supplies the desired response to the adaptive equaliser during training mode The channel used in the simulation is a 32 tap simulated telephone channel. The results of 25 runs of ensemble mean square error against number of iterations are presented in Figure 9.

In Figure 9(a), the only additive noise source is the Gaussian white noise generator. For both the LMS and NCLMS algorithms a step size of 0.01 is used. It is observed that the NCLMS algorithm converges to the -3dB point faster than the LMS algorithm. This is a characteristic of using a larger step size and should result in a greater EMSE term in the steady state. However, from Figure 9(a) the EMSE of the NCLMS is in fact less than (approximately 2dB) that of the LMS. It is suggested that the reduced EMSE of the NCLMS algorithm is due to the way in which the NCLMS handles additive noise components. In Figure 9(b), a high gain impulse noise generator is switched on between iteration 500 and 505. It is observed that the LMS loses adaptation while the NCLMS converges after the impulsive period. This is a consequence of using a coefficient weight vector that is skewed in time.

The fast convergence and reduced EMSE have also been observed in extensive simulations with a variety of channels and additive noise variances. Simulations have also shown that the NCLMS algorithm has a different step size bound as compared to the LMS algorithm.

In [5], motivated by the need to provide equalisation and echo cancellation at more than 1.6Mbit/s, a system is presented that used the A100 processor [7] which is a non-canonical FIR, and performs the weight updates on the TMS320C25. The authors also noted that the *modified* LMS algorithm, demonstrated the robustness of the LMS, and deviated only slightly from the true filters convergence rate.

7 Conclusion

One of the advantages of the NCFIR over the FIR is that in a fully parallel VLSI implementation, the throughput of the NCFIR will be almost an order N faster than the FIR for a filter with N taps. This is because the FIR requires N ripple through additions on the adder line to realise the output (see Figure 2) whereas the NCFIR has a delay adder line and the output is available after one addition, at the final adder stage. The NCLMS can also be used to derive a truely parallel and pipelinable (systolic) array, which is impossible with the LMS algorithm. The NCLMS was shown to give faster convergence and a smaller EMSE than the LMS. In addition, the NCLMS appears to be less sensitive to impulsive noise than the LMS.

References

[1] R.W. Stewart. J.J. Soraghan, T.S. Durrani. *Implementation of the LMS Algorithm on the non canonical FIR filter*, Proc. IEEE Colloquim on Digital Signal Processing for VLSI, London 1988.

[2] W.S. Gan, J.J. Soraghan, R.W. Stewart, T.S. Durrani. *Comparitive analysis between non-canonical LMS and LMS adaptive filtering.* Electronics Letters, Vol 27, No.11, pp 947 - 948, May 1991.

[3] B. Widrow, S.D. Stearn. *Adaptive Signal Processing.* Prentice Hall, 1985.

[4] S.Y. Kung. *VLSI Array Processors*, Prentice Hall, 1987.

[5] J.L. Dixon, W.Y. Chen. *The implementation of an adaptive echo canceller for high speed digital data transmission.* Proc. IEEE ICASSP, pp 1597-1600, Toronto, Cananda, May 1991.

[6] D.B. Chester, W.R. Young, M. Petrowski. *A fully systolic adapative filter implementation.* Proc. IEEE ICASSP, pp 2109-2112, Toronto, Cananda, May 1991.

[7] IMS Cascadable Signal Processor Engineering Data Sheet. INMOS Ltd, 1989.

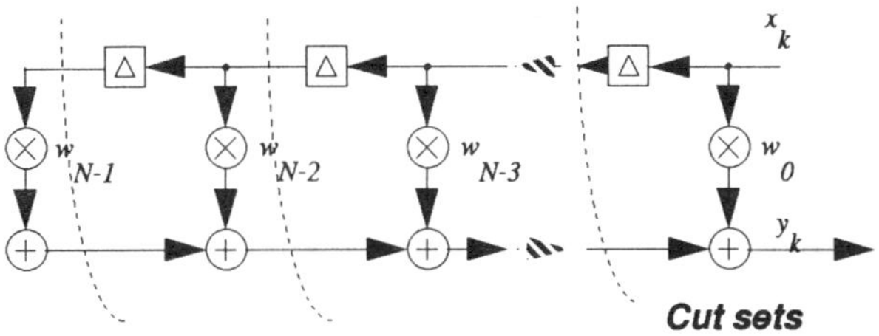

Figure 1: *Canonical FIR filter*

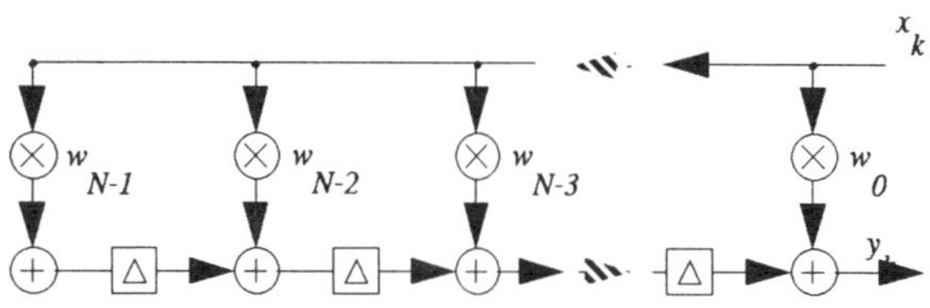

Figure 2: *Non-canonical FIR filter (NCFIR)*

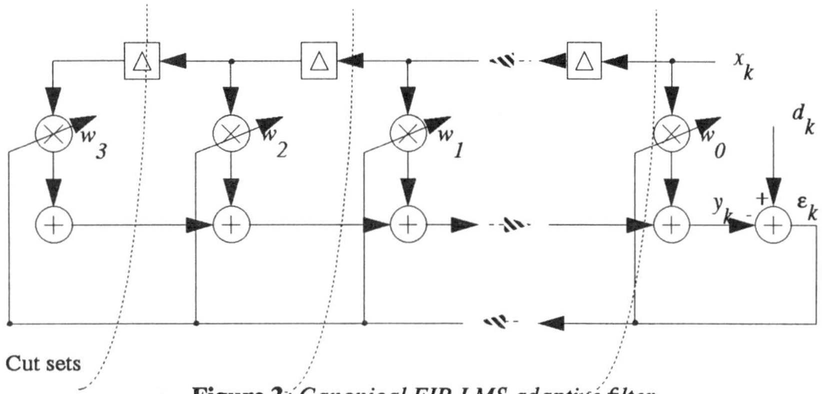

Figure 3: *Canonical FIR LMS adaptive filter*

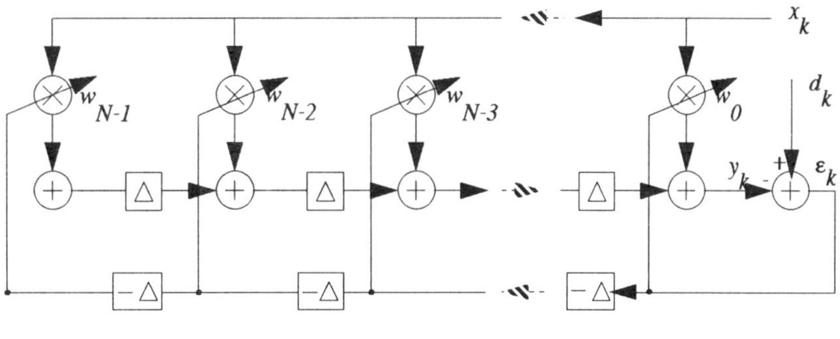

Figure 4: *LMS on NCFIR filter*

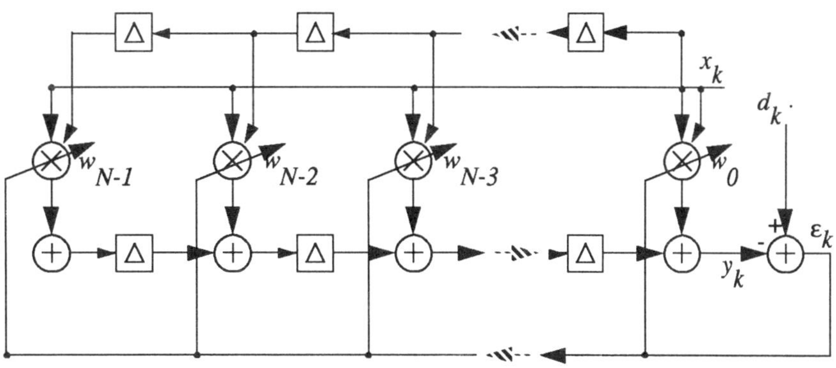

Figure 5: *Non-canonical LMS (NCLMS) adaptive filter*

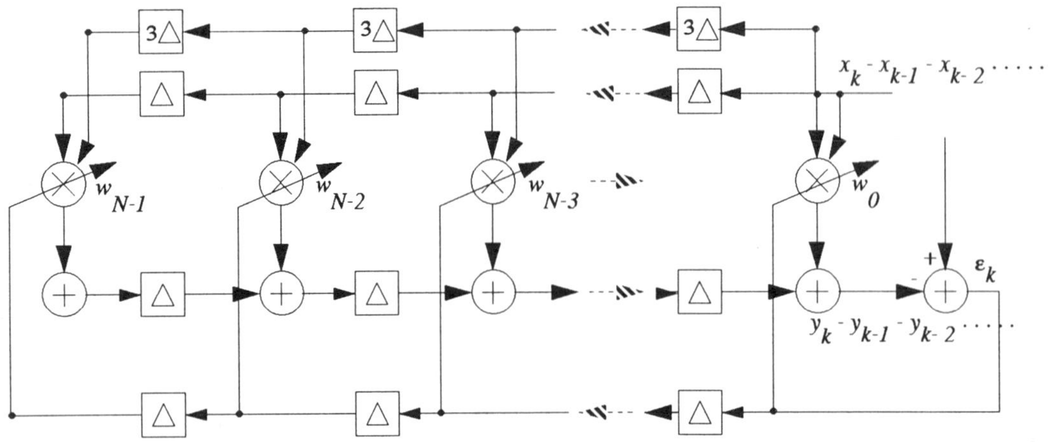

Figure 6: *PIpelined, systolic NCLMS adaptive filter*

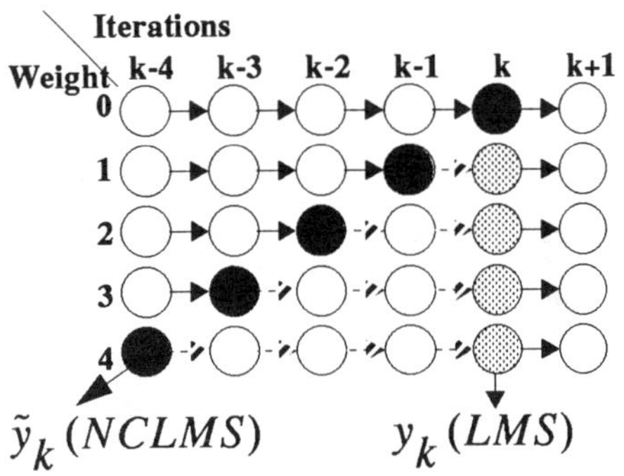

Figure 7: *Computation transition diagram of 5 tap weight vector*

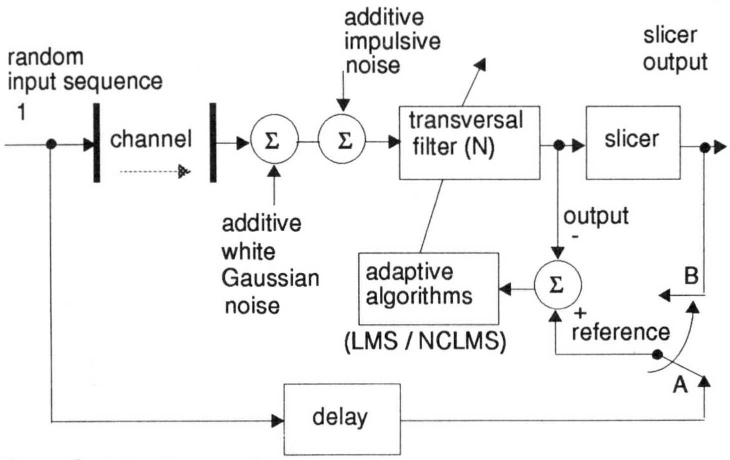

Figure 8: *Adaptive equaliser simulation.*

A : training (learning) period
B : tracking (decision directed) period

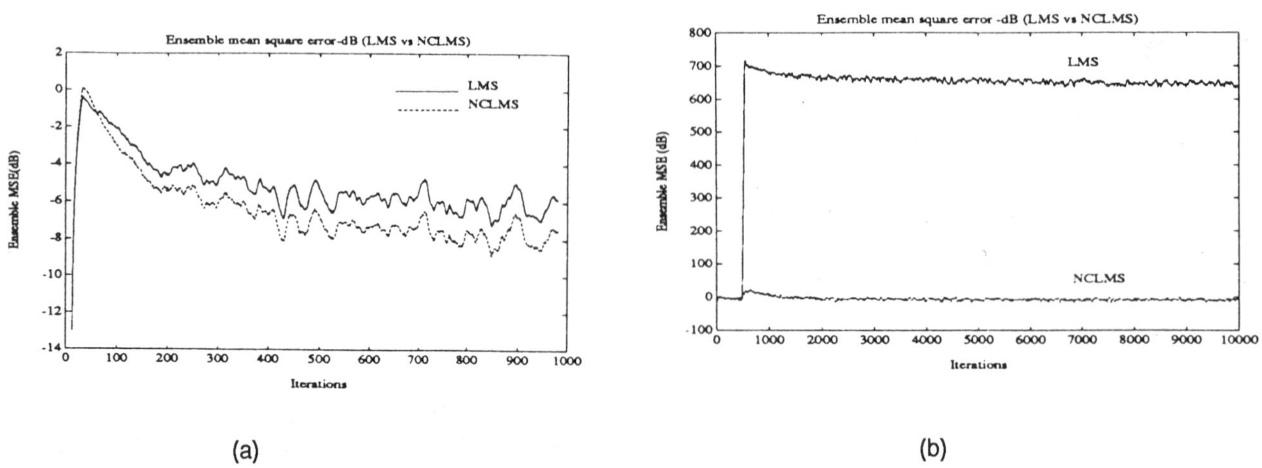

Figure 9: *The ensemble mean square error of 64 stage equaliser for a 32 stage telephone channel under (a) noise variance of 0.01, step size of 0.01 (b) high gain impusive noise from iterations 500 to 505.*

POINTING CORRECTION IN AN ADAPTIVE ARRAY FOR COMMUNICATIONS

Miguel A. Lagunas, J. Fernandez, M. Najar

ETSI de Telecomunicacion, Departamento de TSC
Apdo. 30002, 08080 BARCELONA, SPAIN

ABSTRACT[1]

This work deals with dsp procedures for pointing correction in adaptive arrays which use a spatial reference to form the optimum beam (GSLC). The problem of adequate pointing arises from many effects on the aperture and the receiver front-end of every array channel. Component aging, channel mismatching, sensor location, etc, promote steering errors that may produce the rejection of the desired signal. Currently, high complicated telemetry systems and monopulse radars are used for steering correction. This work describes the existing alternatives as well as further improvements that can be achieved from the use of temporal references combined with the GSLC beamformer.

I. INTRODUCTION

The increasing number of mobile communication systems and ground radar stations promotes the interest of adaptive array beamforming for on-board satellite communications.

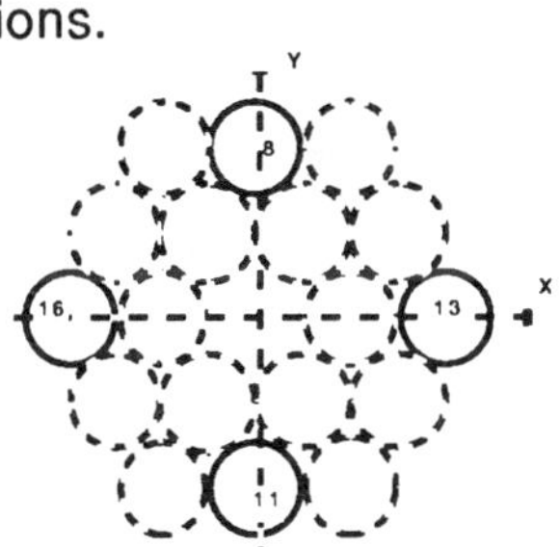

Fig 1. The aperture formed by 18 sensors

[1] This work was supported by the National Research Plan of Spain under project DGT-TIC302, and in part by the CIRIT of Catalunya with grant BV91/1-56

Because the desired direction is known, the adaptive array takes the form of a Generalised Sidelobe Canceller (GSLC). In order to specify the problem, all the experiments supporting this work were done over the so-called MAM antenna. The aperture is formed by 18 sensors located as depicted in fig1, working at a central frequency of 19 Ghz and with a field of view of +/- 10º.

The GSLC was formed by a phased array as quiescent with the corresponding blocking matrix in the lower branch. The adaptive algorithm for the adaptive weights in the lower branch is Kalman type with square-root propagation of the covariance matrix.

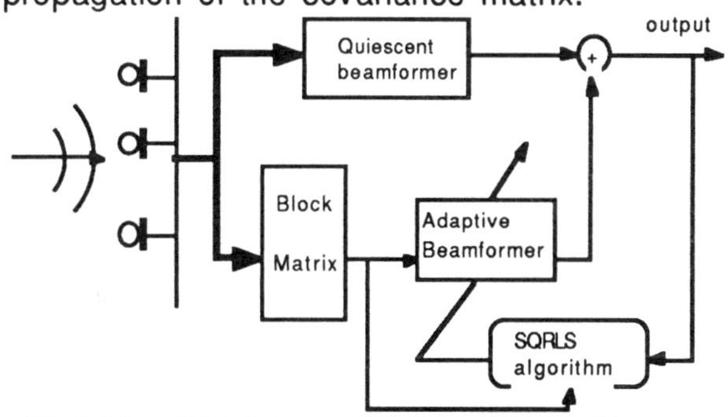

Fig2 The GSLC beamformer with SQ-Kalman adaptive algorithm

The square root version was able to support hard quantization in the overall beamformer, being 6 bits for the snapshots and progressive bit assignment in three layers of Asics from 4 up to 16 bits in the GSLC output. The residual in the diagonal of the square-root matrix recursion was set to 10e-4 /1/ in order to allow tracking of interferers with radial velocity of 0.1º per second.

The first problem, related with pointing, comes from the requirement of automatic steering correction from earth stations. These corrections were usually carried out by

1058-6393/91 $1.00 © 1991 IEEE

corrections were usually carried out by telemetry to a monopulse radar in the satellite. Of course, changing the desired direction in the above structure will motivate the change of both quiescent and blocking matrix. This operation was out of the possibilities of the in-board processor. In order to avoid the mentioned problem, a pre-steering matrix has to be added to the old GSLC structure.

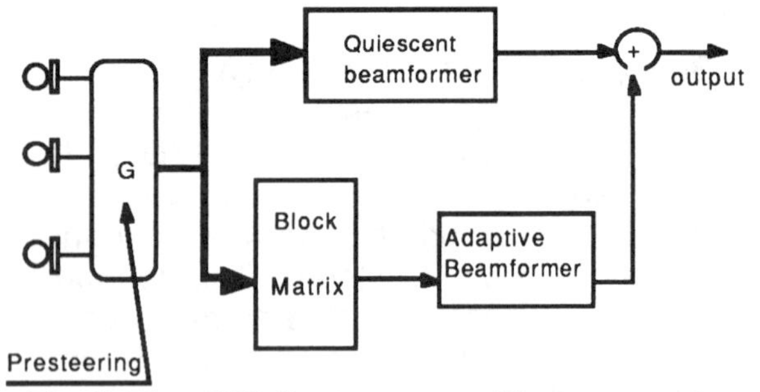

Fig 3 The GSLC structure with pre-steering matrix

The pre-steering stage can be characterised by a diagonal matrix **G** which contains just the phase shifts to convert any direction (Θ_0,Φ_0) to the aperture broadside. Regardless this matrix will play a very important role in the rest of this work, it is important to remark that it is almost mandatory for GSLCs in satellite payloads. The reason is that, even in the case of non-mobile scenarios, missmatching between receivers front-ends, down-conversion and sensor missfunctioning, will produce the same problem demanding for re-steering the array. Just to give an idea, the following table provides the loss in signal to interference ratio (in dBs), for the above aperture, when the desired signal is a BPSK of 3 Mbps (-12.55 dB which corresponds to Spread Spectrum modulation systems) and an unmodulated interference is located at 3º from it with 17.45 dBs. for phase missmatching and 27.44 for phase missmatching.

magnitude miss.	CIR	phase miss.	CIR
0.0	17.32	0º	29.8
1.0	14.91	1º	21.1
2.0	16.75	2º	15.6

Note how phase missmatching almost twice loss (in dB) that magnitude missmatching. The effect is that whenever the desired signal impinges the aperture more than 2º from the steered direction the GSLC starts to look at it as an interference. In fact for phase missmatching close to 2.3º the desired signal is almost nulled out by the algorithm. The values given for mismatch reflect the standard deviation in dBs of a gaussian distribution for the sensors gain around their nominal value; and, in the case of phase, the listed values correspond to the maximum of an uniform phase distribution around the corresponding nominal values.

The aim of this work was to use the presteering matrix elements in order to track the desired signal either for misspointing of the array or beacuse it is moving with radial velocity.

II TIME REFERENCE CONTROL OF A GSLC BEAMFORMER

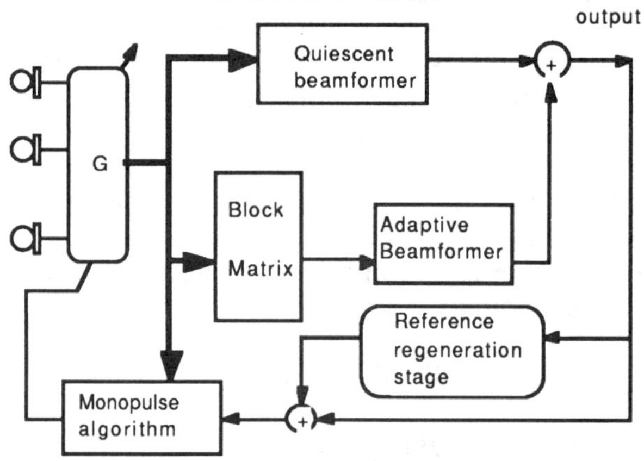

Fig. 4 GSLC with time reference pre-steering

From the above section, a new structure culd be proposed. This structure, in addition to the GSLC loop which works independently of the steering angle, includes a second loop, with lower rate than the GSLC loop, which exploits the time evolution of the desired signal in order to keep steered the overall beamformer response. The structure is shown in figure 4.

The problem we are going to face on this work is the potential misspointing of the pre-steering block with respect the actual doa of the desired source.

The first alternative consist in the use of DOA estimation techniques in order to locate the desired source, assuming it is nearby the alternative is that DOA estimates do not identify the desired source with respect interferences, since they are not sensitive to modulation parameters on the impinging energy; also, the scanning in elevation and azimuth of the DOA estimate could be very time consuming for the satellite payload.

A second approach, within the structure depicted in figure 4, is to set a parallel beamformer with temporal reference which provides the corresponding pointing angles to the GSLC pre-steering.

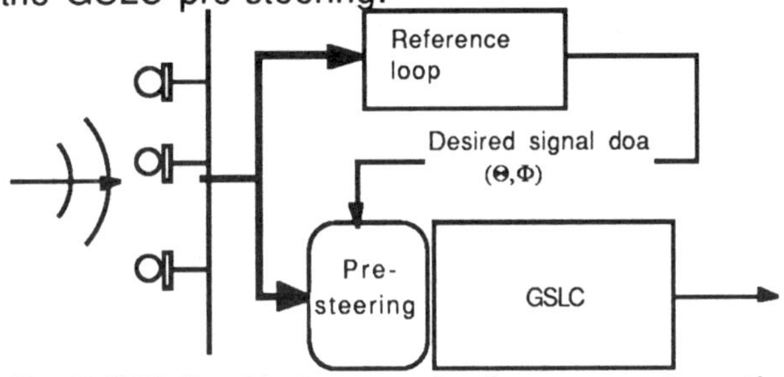

Fig.5 GSLC with temporal reference automatic pointing

Regardless of important issues like reference framing, code access, modulation and protocols, seems to be that if the reference loop is able to regenerate the desired signal the reference loop may provide the adequate steering direction to the GSLC. Of course, in order to keep the communication link the reference loop has to be framed in time (TDM) or in frequency (FDM) with the information channel managed by the GSLC.

The most suggestive approach is to include a coded BPSK signal as temporal reference in the upper zone of the communication bandwidth, as it is indicated in figure 6.

Note that whenever in the upper loop there is not interferences, the temporal reference algorithm will steer the desired signal, and the corresponding weights can be copied directly to the diagonal elements of the pre-steering matrix. The assumption that no significant interferences are present in the temporal reference loop is crucial and it depends very much on the narrowbandwidth assigned to this loop.

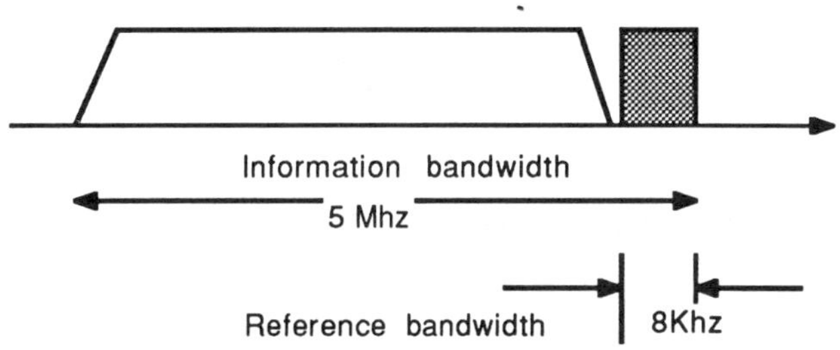

Fig. 6.- 8KHz reference for steering purposes (reference loop) inside the information bandwidth of the communication link (GSLC bandwidth)

The main drawback of this scheme is that the presence of interferences in the steering loop promotes the same problems of pointing we have in the GSLC rendering the problem back to the initial situation. Just to give an idea in figure 7 is depicted the response of a temporal reference loop /2/ where the reference is found at 0º in presence of a severe jammer located in the mainlobe.

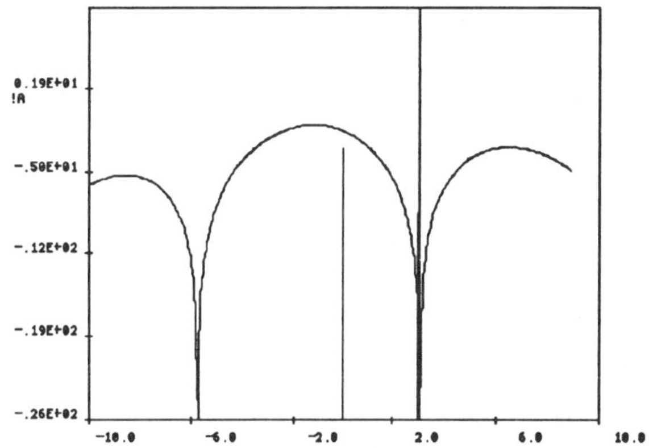

Fig.7 Temporal reference loop response to a desired signal, found at 0º in presence of a severe jammer at 3º

As the reader can see, the presence of interferers in the reference loop introduces bias in the beamformer rendering non valid the overall structure. Furthermore, even in the case of secondary lobe interferences, the pointing is correct but the weights cannot be copied to the diagonal of the pre-steering matrix because it is not sure that the fixed blocking matrix remains valid for the pre-steered snapshot. In summary the temporal reference loop will be adequate whenever the reference is framed in time or frequency in such a way the only it will be present in the temporal reference loop. This situation that may be common in some mobile communications scenarios, fails in the case of collision when two users demand for a dedicated beam simultaneously.

Finally, the authors /3/ and L.J.Griffiths /4/ develop a system which exploits the constant envelope associated to the most used modulations for the desired signal. The basic idea is to control the pre-steering angle in such a way that the envelope or short-time power at the array output is maximised. The algorithm could be a DSD /5/ random search or monopulse. In any case, random search is highly recommended. In the random search a random perturbation is promoted in elevation and azimuth, if the envelope or the short-time power of the output increases the perturbation is retained in the steering angles; if not the steering remains the same. A version of monopulse, handling the output of a short time RMS detector, with symmetric spatial perturbations, was used by the authors with success in data relay satellite scenarios (DRS), where the desired signal impinges the array with SNRs above 3 dB at the sensor level.

In any case, the main problem of these procedures resides in the cases that very low SNRs, at the sensor level, has to be managed. As an example, a BPSK to be detected with a bit error rate of 1e-4 requires an EbNo of 8.7dB at the GSLC output; this means that at the sensor level will be around -3.2dB. With these low signal levels, the envelope is almost independent, due the the output noise, to steering movements of 1º which are enough to promote the rejection of the desired signal by the mispointed GSLC. The mentioned situation becomes more severe when spread spectrum is the original modulation for the received BPSK signal, being the input SNR in the range of -10db -5dBs. None of the previously described techniques may cope this case.

We can say that the reason why the described techniques cannot cover the case of many practical systems in communications applications is due to the fact that working only over the envelope does not provide any modulation or bandwidth compression gain in the reference analysis. Next section will report how the use of good estimates of the desired signal waveform can improve a lot the dynamic range covered by the envelope techniques.

III EKF REFERENCES FOR GSLC POINTING

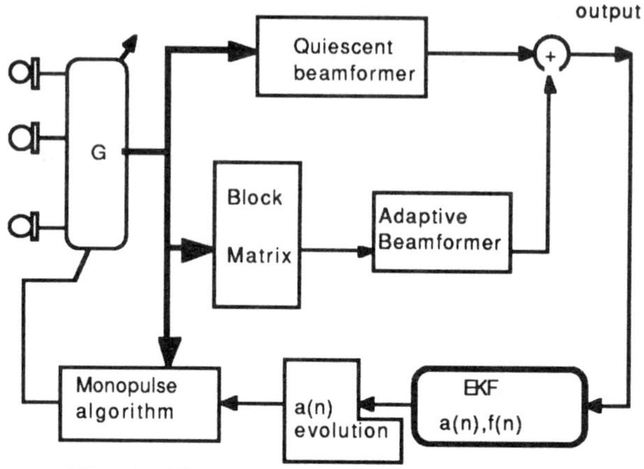

Fig. 8. The EKF loop for pre-steering

The scheme proposed is a slight derivation of the structure shown in figure 4. In the new scheme an Extended Kalman Filter (EKF) has been used to acquire and track the desired signal from the GSLC output. The EKF estimates the magnitude, instantaneous phase and instantaneous frequency of the GSLC output. Tracking the magnitude, almost free of noise, the presteering angle is corrected with the monopulse principle over a random perturbation in elevation and azimuth. At the same time, tracking the instantaneous phase and frequency, the pointing system is able to cope with doppler effects in the received frequency and with the radial velocity of the desired user.

The main advantage of the EKF in following the carrier evolution is that it is able to track 0dB carriers at the output (almost -10 dB level at the sensor level) with 1% error over the actual magnitude value. Doing this, together with the frequency tracking the systems reaches levels of performance unexpected from the previous schemes. Just to provide some insight in the EKF performance, the following table shows the frequency tracking error for different SNR at the array output. The residual mentioned in the table refers to the diagonal value, the same for magnitude phase and frequency, of the covariance matrix propagation equation involved in the EKF /6/.

SNR dB		Residual Q		
		0.001	0.0001	0.00001
-5.5	Convergence	no	no	no
	Tracking error	no	no	no
-1.5	Convergence	10	10	no
	Tracking error	-6.56	-12.48	no
2.5	Convergence	10	10	no
	Tracking error in samples	-6.64	-12.40	no
6.5	Convergence	20	25	no
	Tracking error in samples	-6.64	-12.40	no
10.5	Convergence	20	25	no
	Tracking error in samples	-6.64	-12.32	no
13.5	Convergence	20	25	>1000
	Tracking error in samples	-6.60	-12.26	>3

Note, from this table that the system has convergence rates below 50 snapshots, being the doppler of 1Khz and the doppler rate of 1Khz per second. This doppler rate is far above of the usual 50Hz per second associated with the scenarios used in the application under the scope of this work. The following figures show the EKF performance in frequency tracking.

From this figure is clear that the monopulse explorations in elevation and azimuth can be done every 50 snapshots or more per perturbation without paying attention to the problems derived from time varying position, doppler and other problems. At the same time, the tracking of magnitude fluctuations done at 8 dBs (required for BPSK) at the output is done almost independent of output noise and interference residuals. Experiences carried out by the authors ensures acquisition of adequate steering even for mispointing close to 1º. Regardless that with these errors the aperture may reject in 5 dB or 10 dB the desired signal, the excellent performance of the EKf in regenerating the carrier allows the pointing correction after 1000 snapshots at the sampling level of the EKF bandwidth, which is 8 KHs, meaning about 125 msec. to correct such steering error. With respect to tracking desired sources with radial velocity, satisfactory results have been obtained for sources with radial velocity of 2º per second. The tracking error is lower than 0.05º for this case. Nevertheless, the authors found that the most critical part in this design, before the implementation of the breadbording was the adjustment of the monopulse parameters like size of perturbation, size of the correction and average time to validate the magnitude values to perform the update of the steering.

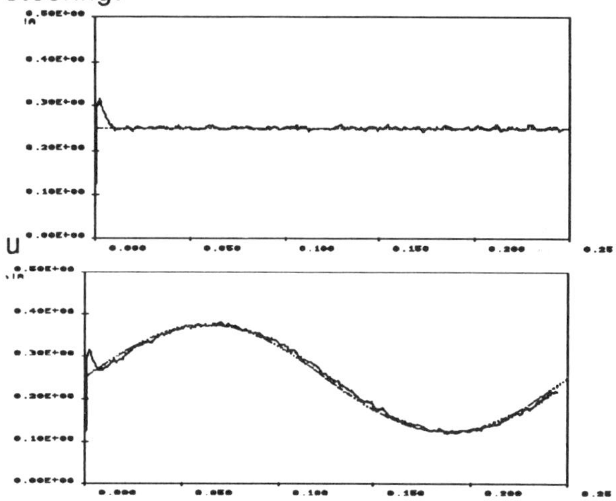

Fig 9. EKF acquisition for 0 db (top)Acquisition and tracking for a doppler rate of 1Khz per second.

VI REFERENCES

|1|J.Fernandez,M.Lagunas,G.Vazquez,"Breadbording of a digital beamforming network" ERA-ESA Technical report, European Spatial Agency, Doc. #1897/01. Contract 8714/90/NL/US(SC), The Netherlands, Dec 1990

|2| M.A. Lagunas. "Direction of Arrival Techniques and Temporal Beamforming" European Spatial Agency, ESA-report,P.O. 112052, October 1991, The Netherlands.

|3| J.Fernandez, "Adaptive beamforming with temporal and spatial references in satellite communications", Signal Proc. V, Eusipco 90, Ed. M.A.Lagunas, ELSEVIER, pp.733-736, Sept 1990.

|4| L.J. Griffiths. " New results in constrained beamforming. Non linear constraint and constant modulus output". Signal Proc. V, Eusipco 90, Ed. M.A.Lagunas, ELSEVIER, pp. 1-8, Sept 1990.

|4| R.A. Monzingo,T.W.Miller. "Introduction to adaptive arrays". N.Y., Wiley, 1980.

Pulse Shape Design Considerations for Self-Synchronizing Systems

Edgar H. Satorius
Jet Propulsion Laboratory
California Institute of Technology
4800 Oak Grove Drive
Pasadena, CA 91109-8099

James J. Mulligan
US Army CECOM Center
for Signals Warfare
Vint Hill Farms Station
Warrenton, VA 22186-5100

Abstract

In this paper, we consider fundamental relationships between signalling pulse shapes used in 1-D (PAM) and 2-D (QAM) bauded transmission systems and their envelope spectra. In particular, we seek to establish necessary and sufficient conditions which guarantee the existence of a signal baud timing component (sinewave) in the mean envelope. As will be shown, excess pulse bandwidth is not sufficient nor is Nyquist's first pulse shaping criterion. On the contrary, additional constraints must be imposed on the pulse to guarantee that timing information can be derived from the envelope function.

I. Introduction

Self-synchronizing communication systems derive symbol timing directly from the information-bearing signal, instead of using a separate subchannel to send synchronization information. One technique that is commonly used to acquire the symbol clock is to filter the envelope of the received data symbols. Here we consider fundamental relationships between signalling pulse shapes used in 1-D (PAM) and 2-D (QAM) bauded transmission systems and their envelope spectra. In particular, we seek to establish necessary and sufficient conditions which guarantee the existence of a signal baud timing component (sinewave) in the mean envelope. The system configuration is depicted in Figure 1. The basic system functions are data recovery and signal characterization. As illustrated in Figure 1, signal characterization can be based on both the envelope as well as the demodulated output. The latter provides specific information regarding the internal structure of the signal (modulation format, etc.) whereas the former, under certain conditions to be discussed here, provides a precise measure of the signal baud rate.

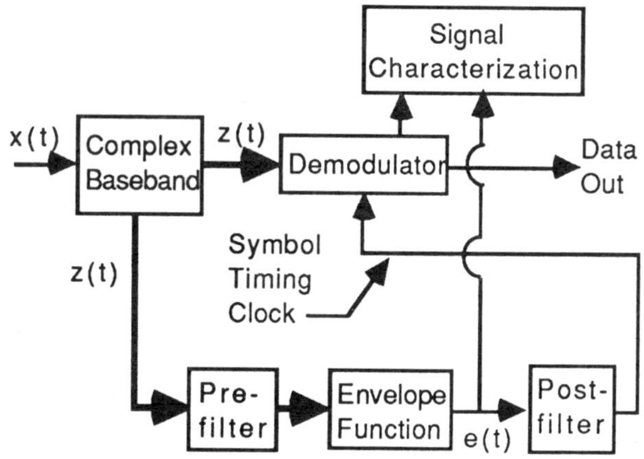

Figure 1. System configuration (dark lines depict complex data paths).

II. Problem Formulation

We start with the bandpass representation for the received signal:

$$x(t) = \mathrm{Re}\{\mathbf{X}(t)\, e^{j(\omega_0 t + \varnothing)}\}\,, \qquad (1)$$

where ω_0 and $\varnothing$ denote the carrier frequency and phase, respectively, and $\mathbf{X}(t)$ represents the bauded transmitted data stream. For PAM, $\mathbf{X}(t)$ is represented by:

$$\mathbf{X}(t) = \sum_{n=-\infty}^{\infty} a_n\, g(t - nT)\,, \qquad (2a)$$

Acknowledgment: The research described in this paper was carried out by the Jet Propulsion Laboratory, California Institute of Technology, under a contract with the National Aeronautics and Space Administration.

and for QAM:

$$\mathbf{X}(t) = \sum_{n=-\infty}^{\infty} a_n g(t - nT) + j\, b_n q(t - nT) . \quad (2b)$$

The signal pulse for PAM is denoted by g(t) and the corresponding pulses for QAM by g(t) and q(t) (g and q are assumed real), T is the baud interval and the data symbols (a_n, b_n) are independent, zero-mean discrete random variables. We assume that the complex baseband operation merely "de-spins" the carrier (without necessarily recovering carrier phase) to produce:

$$z(t) = \mathbf{X}(t)\, e^{j\phi} . \quad (3)$$

For practical considerations, we are interested in envelope functions of the data that are independent of the carrier phase. Here we focus primarily on the square envelope function:

$$e(t) = |z(t)|^2 = |\mathbf{X}(t)|^2 .$$

It is easy to analyze and illustrates the essential features of the characterization problem. For PAM, the mean-square envelope is given by:

$$E[e(t)] = E(a^2) \sum_{n=-\infty}^{\infty} \{g(t-nT)\}^2 , \quad (4a)$$

and for QAM:

$$E[e(t)] = E(a^2) \sum_{n=-\infty}^{\infty} \{g(t-nT)\}^2 + E(b^2) \sum_{n=-\infty}^{\infty} \{q(t-nT)\}^2 . \quad (4b)$$

Note that in either case of 1-D or 2-D data formats, the mean-square envelope function depends on the sum of squares of T-delayed versions of the signal pulse shape. Thus the mean-square envelope function is periodic in the baud interval (T). This allows us to utilize the Poisson sum formula to obtain:

$$E[e(t)] = \sum_{n=-\infty}^{\infty} C_n e^{2\pi j n/T} . \quad (5)$$

For PAM, the amplitudes, C_n, are given by:

$$C_n = E(a^2) \frac{1}{2\pi T} \int_{-\infty}^{\infty} G(\omega)\, G\!\left(\frac{2\pi n}{T} - \omega\right) d\omega , \quad (6a)$$

and for QAM:

$$C_n = E(a^2) \frac{1}{2\pi T} \int_{-\infty}^{\infty} G(\omega)\, G\!\left(\frac{2\pi n}{T} - \omega\right) d\omega$$

$$+ E(b^2) \frac{1}{2\pi T} \int_{-\infty}^{\infty} Q(\omega)\, Q\!\left(\frac{2\pi n}{T} - \omega\right) d\omega , \quad (6b)$$

where $G(\omega)$ and $Q(\omega)$ denote the Fourier transforms of g(t) and q(t), respectively, e.g.,

$$G(\omega) = \frac{1}{2\pi} \int_{-\infty}^{\infty} g(t)\, e^{-j\omega t}\, dt ,$$

and similarly for $Q(\omega)$.

Given the Fourier series expansion of E[e(t)] in (5) and (6), we immediately have the following lemmas for PAM and QAM modulation formats:

Lemma (PAM)

The mean-square envelope function does not contain a signal baud timing component (sinewave at frequency $2\pi/T$) if, and only if:

$$E(a^2) \int_{-\infty}^{\infty} G(\omega)\, G\!\left(\frac{2\pi}{T} - \omega\right) d\omega = 0 . \quad (7a)$$

Lemma (QAM)

The mean-square envelope function does not contain a signal baud timing component (sinewave at frequency $2\pi/T$) if, and only if:

$$E(a^2) \int_{-\infty}^{\infty} G(\omega)\, G\left(\frac{2\pi}{T} - \omega\right) d\omega = 0$$

$$+ \; E(b^2) \int_{-\infty}^{\infty} Q(\omega)\, Q\left(\frac{2\pi}{T} - \omega\right) d\omega = 0 \, . \qquad (7b)$$

We are now interested in determining what can be said about $G(\omega)$ and $Q(\omega)$ (or $g(t)$ and $q(t)$) when (7) is satisfied. Specifically, we wish to establish necessary and sufficient conditions on g and q that guarantee the presence of a signal baud timing component in the mean-square envelope function (or, equivalently, necessary and sufficient conditions that guarantee its absence).

III. Pulse Shape Design Considerations

Initially we consider 1-D PAM. From (7), we can immediately establish a necessary condition on g(t), viz., that g(t) must have bandwidth in excess of the Nyquist frequency, π/T (corresponding to sampling at the baud rate $\omega_S = 2\pi/T$). If not, a timing component will not be present in the mean-square envelope function. This necessary condition is well known having been established in the classic work of Gitlin [G&H], Mazo [Ma], Franks [Fr] and others. Practically, it states that if we compute the mean-square envelope function of an unknown, PAM signal and a sinewave at the baud rate is detected, then we know that the signal must have excess bandwidth.

What is not well known, however, is a sufficient condition(s) on g(t) that guarantees the existence of a signal baud timing component in E[e(t)]. Clearly, the requirement of excess bandwidth is not sufficient as can be seen in the case of time-limited pulses. For example, if g(t) is the rectangular pulse: $g(t) = 1$, $|t| \le T/2$ and $g(t) = 0$, otherwise; then $E[e(t)] = E[a^2]$ = constant and thus no timing component exists ($C_1 = 0$) and yet g(t) has bandwidth well in excess of π/T ($G(\omega)$ is a sinc function).

Even if we restrict g(t) to be the class of bandlimited pulses with bandwidth in excess of π/T, we can still easily construct examples that do not exhibit a timing component in the mean-square envelope. An interesting example is given by:

$$G(\omega) = \begin{cases} 0.5 + \sqrt{f(\omega) + 0.25} \, , & 0 \le \omega \le \pi/T \\ 0.5 - \sqrt{f\left(\frac{2\pi}{T} - \omega\right) + 0.25} \, , & \pi/T < \omega \le 2\pi/T \\ 0 \, , \quad 2\pi/T < \omega & \\ G(-\omega) \, , \quad \omega < 0 \, , & \end{cases} \qquad (8a)$$

where:

$$f(\omega) = a \sin(\omega T) - b\, \omega\, (\omega - \pi/T) - c\, \omega^3 (\omega - \pi/T) - \omega T/(4\pi) \, . \qquad (8b)$$

This particular example corresponds to a signal pulse with 100% excess bandwidth which obeys Nyquist's first pulse shaping criterion: zero intersymbol interference (ISI) ([LS&W; chp. 4], [B&D; chp. 5], [Sh; ch. 5], [L&M; chp. 6], [Pr; chp. 6]). Furthermore, if:

$$a = \frac{13\pi}{20\,(12 - \pi^2)} \; ; \; b = \frac{T^2}{\pi}\left(\frac{1}{4\pi} - a\right) ; \quad c = -\frac{1}{2}\left(\frac{T}{\pi}\right)^4 ,$$

then the first derivative of $G(\omega)$ is continuous and thus g(t) decays asymptotically (large $|t|$) as $|t|^{-2}$. A plot of $G(\omega)$, corresponding to this particular set of parameters (a,b,c), is presented in Figure 2.

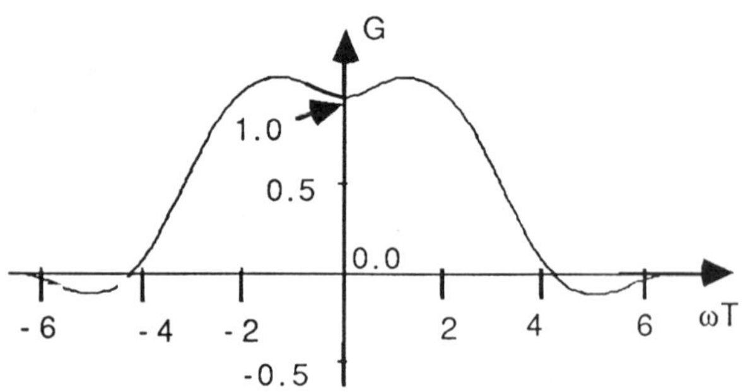

Figure 2. Fourier transform of excess bandwidth pulse with no timing component in mean-square envelope.

This is a well defined, bandwidth limited (efficient) pulse, satisfying Nyquist's zero ISI criterion, with realizable time-decay properties and 100% excess bandwidth and yet the mean-square envelope function does not contain a signal baud timing component. Specifically, this particular set of (a, b, c) parameters was chosen such that (7a) is satisfied.

It is not even sufficient that $G(\omega)$ be non-negative, e.g., we could make G non-zero over disjoint intervals: $G(\omega) = 1$ for $|\omega| \le \pi/T$ and $3\pi/T \le |\omega| \le 5\pi/T$ and $G(\omega) = 0$ otherwise. It is easy to see by inspection that again (7a) is satisfied for this particular example and yet this pulse contains excess bandwidth plus it satisfies the Nyquist zero ISI criterion.

Based on these results, we see that excess pulse bandwidth alone is not sufficient to guarantee a signal baud timing component in the mean-square envelope function for PAM. Thus the absence of a timing component does not necessarily imply that the PAM signal is devoid of excess bandwidth. The same conclusion also applies to QAM, i.e., excess bandwidth in the in-phase (g(t)) and/or quadrature (q(t)) signal pulses does not guarantee a signal baud timing component in E[e(t)].

However, for certain classes of signal pulse shapes, excess bandwidth is a sufficient (and necessary) condition. As a simple example, consider the following class of bandlimited signal pulse shapes, $\mathbf{g}_+$, with strictly positive, real Fourier transforms:

$\mathbf{g}_+ \equiv \{g(t) \mid \mathrm{Im}\,[G(\omega)] = 0$ and $G(\omega) > 0$ for $|\omega| \le \omega_{max}$ $(0 < \omega_{max} < \infty)$ and $G(\omega) = 0$ for $|\omega| > \omega_{max}\}$.

Then we have the following characterization theorem for PAM signals:

Characterization Theorem (PAM)

For $g(t) \in \mathbf{g}_+$, the mean-square envelope function of the PAM signal:

$$\sum_{n=-\infty}^{\infty} a_n\, g(t - nT)\,,$$

contains a signal baud timing component if, and only if, g(t) has excess bandwidth (i.e., $\omega_{max} > \pi/T$).

In proving this characterization theorem, we actually prove its equivalent inverse: For $g(t) \in \mathbf{g}_+$, the mean-square envelope function of the PAM signal does not contain a signal baud timing component if, and only if, g(t) does not have excess bandwidth. We already know that excess bandwidth is a necessary condition for the presence of a timing component. Vice versa, if there is no timing component, we know that (7a) is true by virtue of the PAM lemma. However, for $g(t) \in \mathbf{g}_+$ we have:

$$\int_{-\infty}^{\infty} G(\omega)\cdot G\!\left(\frac{2\pi}{T} - \omega\right) d\omega = \begin{cases} \displaystyle\int_{\frac{2\pi}{T}-\omega_{max}}^{\omega_{max}} G(\omega)\, G\!\left(\frac{2\pi}{T} - \omega\right) d\omega, & \omega_{max} > \pi/T \\ 0\,, & \omega_{max} \le \pi/T\,. \end{cases}$$

Thus, since $G(\omega) > 0$ $(|\omega| \le \omega_{max})$ we see that (7a) can only be satisfied if $\omega_{max} \le \pi/T$, i.e., if g(t) does not have excess bandwidth. This establishes the PAM characterization theorem. A similar theorem can be stated for QAM:

Characterization Theorem (QAM)

For $g(t) \in \mathbf{g}_+$ and $q(t) \in \mathbf{q}_+$, the mean-square envelope function of the 2-D QAM signal:

$$\sum_{n=-\infty}^{\infty} a_n\, g(t - nT) + j \sum_{n=-\infty}^{\infty} b_n\, q(t - nT)\,,$$

contains a signal baud timing component if, and only if, g(t) or q(t) has excess bandwidth.

Although restrictive, the above theorems cover a wide class of signal shapes, g(t) (and q(t)), including the important class of bandlimited pulse shapes with the raised cosine frequency characteristic [Sh], [LS&W], [B&D]. Furthermore, any $g(t) \in \mathbf{g}_+$ can be constructed from the auto-convolution of a real bandlimited function, x(t):

$$g(t) = x(t) * x(-t) \quad , \; x(t) \in \mathbf{b} \; , \qquad (9)$$

where: $\mathbf{b} \equiv \{x(t) \mid X(\omega) \neq 0 \text{ for } |\omega| \leq \omega_b \; (0 < \omega_b < \infty)$ and $X(\omega) = 0$ for $|\omega| > \omega_b\}$. Thus, (9) provides a simple procedure for generating bandlimited signal pulse shapes, $g(t) \in \mathbf{g}_+$, from the larger class of bandlimited functions, **b**.

An important question now arises: What is the largest possible class of pulse shapes ($\mathbf{g}_*$, say) for which excess bandwidth is a necessary and sufficient condition for the presence of a signal baud timing component in the mean-square envelope function ? Although we have not answered this question, we certainly know that $\mathbf{g}_*$ is not as large as the class of bandlimited functions. This is clear from the examples provided above. These same examples also demonstrate that $\mathbf{g}_*$ is not as large as the class of bandlimited pulse shapes satisfying the important Nyquist zero ISI criterion. These considerations lead us to conjecture that $\mathbf{g}_+$ contains all the practical pulse shapes in $\mathbf{g}_*$. Certainly, in designing self-synchronizing communication systems that incorporate envelope-square timing recovery, one must be careful to utilize appropriate excess bandwidth pulses (e.g., $g(t) \in \mathbf{g}_+$) that result in a strong signal baud timing component in the envelope function.

IV. Conclusions

In this paper, we have determined appropriate restrictions on the signal pulse shaping such that excess pulse bandwidth is a necessary and sufficient condition for the presence of a signal baud timing component in the mean-square envelope function. Alternatively, we can seek other baud timing-related properties of the mean-square envelope function for which excess pulse bandwidth is a necessary and sufficient condition. For example, is excess pulse bandwidth a necessary and sufficient condition that **all** the harmonics of the signal baud timing component (the C_n's in (5)) be present ? To the contrary, example (8) demonstrates that this is not true. In particular, for this example, $C_n = 0$ for $n \neq 0$ because: C_1 (and C_{-1}) vanishes as discussed above and the higher order C_n vanish by virtue of g(t) being bandlimited to $2\pi/T$. Nevertheless, the discovery of a baud timing-related property of the square-envelope function that is simple to check and is true if, and only if, there is excess pulse bandwidth would potentially be of significant practical importance.

References

[B&D] W. R. Bennett and J. R. Davey, **Data Transmission**, McGraw-Hill, New York, 1965.

[Fr] L. E. Franks, "Carrier and Bit Synchronization in Data Communication -- A Tutorial Review," **IEEE Trans. Comm.**, vol. COM-28, pp. 1107-1121, Aug. 1980.

[G&H] R. D. Gitlin and J. F. Hayes, "Timing Recovery and Scramblers in Data Transmission," **Bell Syst. Tech. J.**, vol. 54, pp. 569-593, March 1975.

[L&M] E. A. Lee and D. G. Messerschmitt, **Digital Communication**, Kluwer Academic Publishers, Boston 1988.

[LS&W] R. W. Lucky, J. Salz and E. J. Weldon, **Principles of Data Communication**, McGraw-Hill, New York, 1966.

[Ma] J. E. Mazo, "Jitter Comparison of Tones Generated by Squaring and by Fourth-Power Circuits," **Bell Syst. Tech. J.**, vol. 57, pp. 1489-1498, May-June 1978.

[Pr] J. G. Proakis, **Digital Communications**, 2nd Edition, McGraw-Hill, New York, 1989.

[Sh] K. S. Shanmugam, **Digital and Analog Communication Systems**, John Wiley & Sons, New York, 1979.

Adaptive Threshold Control Scheme in a Centralized Packet Radio Network with Common Direct-Sequence Spread-Spectrum Modulation *

In-Kyung Kim and Robert A. Scholtz
Communication Sciences Institute, Dept. of Electrical Engineering
University of Southern California, Los Angeles, CA 90089-2565

Abstract

This paper describes an automatic threshold control scheme for the detection of a header which is a frame synchronization word in an unslotted centralized packet radio network in which all communicators employ a common direct-sequence spread-spectrum modulation format. Adaptive threshold control is done by setting the header-detection threshold proportional to the maximum-likelihood estimate of the multiple-access noise level. The design value of false alarm probability determines the constant of proportionality under the assumption that the multiple-access noise is a complex Gaussian random process. Computer simulation is carried out to show that this Gaussian assumption is reasonably valid.

This scheme yields a false-alarm probability which is nearly invariant to the changes in the number of actual transmissions in the channel, and a detection efficiency essentially free of loss with respect to the ideal detection where the multiple access noise level is known to the receiver. Throughput bounds for an unslotted ALOHA system with this scheme are analyzed.

1 Common-Code SS Systems

The packet radio network described in this paper is a centralized packet radio network with star topology in which all radio terminals employ a common direct-sequence spread-spectrum (SS) modulation format to communicate with a single central receiver in an asynchronous random-access mode. The channel access scheme is unslotted ALOHA. As a modulation format, direct-sequence binary phase-shift keying (DS/BPSK) is employed in which each data bit is multiplied by a common spreading sequence and modulated onto a sinusoidal carrier for channel transmission. Each packet consists of a common header followed by a data sequence, in which the header sequence is used as a frame sync word.

At the central receiver, discrimination between two or more incoming packets must be made on the basis of time of arrival of the packets and offsets among the received version of the sequences. Therefore, the central receiver structure must possess a *multiple capture* capability which is provided by two strong properties of

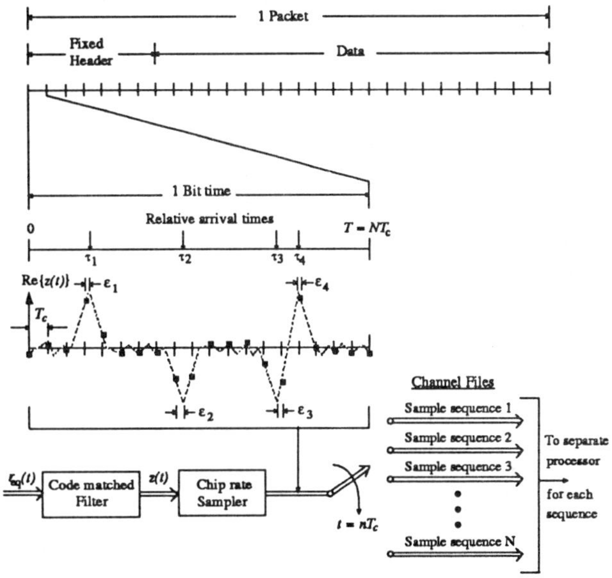

Figure 1: Multiple capture receiver

direct-sequence spread-spectrum signals, namely (1) very high time resolution, and (2) the similarity of nonsynchronously received signals to wide band noise. Here *multiple capture* refers to the nearly simultaneous reception of one or more packets in which each packet potentially can be received correctly if it is the sole occupant of a time-resolution cell, as determined by the spread-spectrum modulation. To avoid near-far problems, it is required that all radios can control their signal power to produce the same signal power at the central receiver.

Fig. 1 shows a multiple capture receiver structure in which received signals are passed through a common code-matched filter followed by a chip-rate sampler, which in turn feeds an array of channel files, one for each resolvable time-of-arrival [1]. We assume that there are N sample sequences, where N is the spreading ratio. Fig. 2 shows the header detector structure of the receiver where each sample sequence is correlated with the header sequence and the magnitude of the output is compared with the controlled threshold for detection of the presence of a packet. When a header is detected, the corresponding header detector outputs a capture signal which enables the data demodulator, and starts the packet counter that disables the threshold estimator feed switch until the packet has cleared

*This research was supported in part by the U.S. Army Research Office under Contract No. DAAL03-88-K-0059.

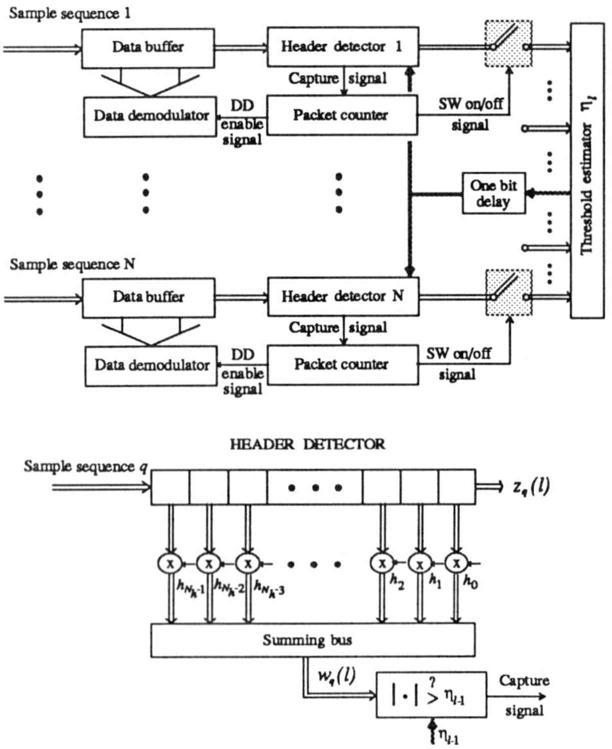

Figure 2: Header detector

the register. If the counter is not running, the capture signal makes the packet counter run and outputs a DD (data demodulation) enable signal and a SW off signal. When the counter completes the cycle, DD signal is disabled and SW on signal is enabled. Additional capture signal for the packet counter is ignored if the counter is already running.

2 Threshold Control Algorithms

The quality of the header detection process at the central receiver depends on the interference level in the receiver, which is a monotonic function of the number of active radios in the channel. As the number of active radios is increased in a fixed threshold receiver, the header detection probability and the false-alarm probability are both increased. In most cases, the system can not allow the false-alarm rate to increase beyond a certain level. Generally, it is preferable to use a strategy that maximizes the probability of detection for a given probability of false-alarm under the assumption that up to a certain false-alarm rate is tolerable. Because the number of active radios changes in time, the system requires a mechanism to try to keep the false-alarm probability constant. False-alarm probability regulation can be achieved by controlling the header-detection threshold adaptively in proportion to an estimate of the total noise and interference level in the receiver.

Let $z_j(l)$ denote the shift register output of the j-th header-detector at the l-th bit-time, and $\mathcal{U}_l$ denote the set of indices for which presence of a header or packet data bit at the l-th bit-time is not detected. Hence we assume that $z_j(l)$ for all $j \in \mathcal{U}_l$ are sample values of the accumulation of side lobe correlations of the received signals and are independent complex Gaussian random processes. Computer simulation shows that the Gaussian asumption is reasonably valid (see Results below). Then the maximum-likelihood estimate $\widehat{\sigma^2}(l)$ of the variance $\mathbf{E}\{|z_j(l)|^2\}$ for $j \in \mathcal{U}_l$ of the total noise $z_j(l)$ for $j \in \mathcal{U}_l$ at the l-th bit-time can be written as

$$\widehat{\sigma^2}(l) = \frac{1}{|\mathcal{U}_l|} \sum_{j \in \mathcal{U}_l} |z_j(l)|^2 \tag{1}$$

where $|\mathcal{U}_l|$ is the cardinality of $\mathcal{U}_l$. Theory says that if $|\mathcal{U}_l|$ is increased, then a more reliable multiple access noise power estimate is achievable.

Better performance is expected if the past history of noise estimate is combined. The controlled threshold value is set proportional to the estimate of the noise standard deviation. The proportionality constant is determined to give a specified false-alarm probability (system tolerance) at a given number of active radios (system design point).

An adaptive threshold control procedure for the detection of a header for the i-th data file at the l-th bit-time is the following: (i) Determine a tolerable false-alarm rate and the number of operating users. (ii) Determine how many previous values of the noise estimates to be combined in estimating the current noise level. Assume $L-1$ past values are combined (we will say L columns are used). (iii) Calculate proportionality constant K upon decision (i) and (ii) (see Section 3 below). (iv) Use the previous threshold η_{l-1} to detect existence of a header at the l-th bit-time. (v) Determine a new threshold η_l for the detection of a header at the $(l+1)$-st bit-time: Let $\hat{N}(l)$ denote the number of data files in which the presence of a header is not detected at the l-th bit-time, *i.e.*, $\hat{N}(l) = |\mathcal{U}_l|$. Then the new threshold value at the l-th bit-time is given by

$$\eta_l = K \sqrt{\frac{\sum_{j=0}^{L-1} \hat{N}(l-j)\widehat{\sigma^2}(l-j)}{\sum_{j=0}^{L-1} \hat{N}(l-j)}}. \tag{2}$$

Note procedure (iv) is based upon the fact that the channel traffic variation is slow compared to the bit-time duration, and provides time-adaptation of threshold control scheme in an unslotted-asynchronous network. If L is a small number, say 2 or 3, then (2) reduces to

$$\eta_l \simeq K \sqrt{\frac{\sum_{j=0}^{L-1} \widehat{\sigma^2}(l-j)}{L}} \tag{3}$$

because $\hat{N}(l) \simeq \hat{N}(l-1) \simeq \hat{N}(l-L+1)$. Equation (3) is used for numerical evaluations.

3 System Performance Analysis

In this section, we derive the false-alarm probability and the detection probability of a header in the adaptive threshold control system in the presence of m transmissions in the channel. Let $w_0(l)$ denote the summing bus output of the first header detector. False-alarm at the first header detector occurs if $|w_0(l)|$ exceeds the threshold η_{l-1} when all m transmitted signals arrive at outside $2T_c$ time interval from the sampling instant for detection, where T_c is the chip-time interval. Therefore the false-alarm probability for the first sample sequence at the l-th bit-time duration is given by

$$P_{fa}(l) = \Pr\{|w_0(l)| > \eta_{l-1} \mid \tau_k \notin \mathcal{I}_0, \forall k = 0, \cdots, m-1\} \quad (4)$$

where τ_k is the arrival time for the k-th signal and $\mathcal{I}_0 = [t_0 - T_c, t_0 + T_c]$, where t_0 is the actual sampling instant for the first sample sequence.

Let $R_c(p)$ and $R_h(p)$ denote the aperiodic partial correlations of the spreading sequence $\{c_j\}$ and the header sequence $\{h_j\}$ respectively. $R_c(p)$ is defined by (similar definition applies for $R_h(p)$)

$$R_c(p) \triangleq \begin{cases} \frac{1}{N}\sum_{j=0}^{N-1-p} c_j c_{j+p}, & \text{if } 0 \le p \le N-1 \\ 0, & \text{otherwise} \end{cases} . \quad (5)$$

Let $\sigma^2_{z,\mathrm{I}}$ denote $\mathbf{var}\{z_j(l)\}$ for $j \in \cup_{p=1}^{L}\mathcal{U}_{l-p}$ in the *abscence of receiver thermal noise* when *only one transmission* exists in the channel. Then $\sigma^2_{z,\mathrm{I}}$ can be written as [2]

$$\sigma^2_{z,\mathrm{I}} = \frac{2}{3(N-2)}\sum_{l=1}^{N-2}\left[R_c^2(l) + R_c^2(l+1) + R_c(l)R_c(l+1)\right]. \quad (6)$$

The samples $\{z_j(l-p) : j \in \mathcal{U}_{l-p}, p = 1, \cdots, L\}$ in the presence of m transmissions in the channel are assumed i.i.d. complex Gaussian random variables with mean zero and variance $\sigma^2_{z,m}$, where $\sigma^2_{z,m} = m\sigma^2_{z,\mathrm{I}} + \sigma^2_{z,\mathrm{T}}$, where $\sigma^2_{z,\mathrm{T}} = \frac{1}{2}N_0/PT_b$ is the variance of the receiver thermal noise component (additive white Gaussian process with one-sided power spectral density N_0), where P is the received signal power and T_b is the bit-time duration. Therefore η_{l-1} is Chi distributed random variable whose probability density function is given by

$$f_{\eta_{l-1}}(x) = \frac{2x^{2\hat{N}-1}\exp\left[-\left(\frac{x}{K\sigma_{z,m}/\sqrt{\hat{N}}}\right)^2\right]}{(\hat{N}-1)!(K\sigma_{z,m}/\sqrt{\hat{N}})^{2\hat{N}}} \quad (7)$$

where $\hat{N} = \hat{N}(l-1) + \hat{N}(l-2) + \cdots + \hat{N}(l-L) \simeq L\hat{N}(l)$. $w_0(l)$ is a complex Gaussian random variable with mean zero and variance $\sigma^2_{w,m} = \frac{1}{N_h}\sigma^2_{z,m}$ where N_h is the header sequence length. Therefore the false-alarm probability in (4) becomes

$$P_{fa}(l) = \int_0^\infty e^{-y^2/\sigma^2_{w,m}} f_{\eta_{l-1}}(y)dy \quad (8)$$

$$= \left[\frac{1}{1 + K^2N_h^2/\hat{N}}\right]^{\hat{N}}. \quad (9)$$

Observe that (9) is only a function of the proportionality constant K and $\hat{N}$, and that if $\hat{N}$ tends to infinity, (9) becomes $e^{-K^2N_h^2}$. Determination of the proportionality constant K is done with (9): If a tolerable false-alarm rate bound p_{fa}, the expected number $\bar{m}$ of transmitting radios, and L are provided, then K is determined by

$$K = \frac{1}{N_h}\sqrt{\tilde{N}(p_{fa}^{-1/\tilde{N}} - 1)} \quad (10)$$

where $\tilde{N} = L(N - \bar{m})$ when L is a small number.

Correct detection of a header at the first header-detector occurs when $|w_0(l)|$ exceeds η_{l-1}, given exactly one of the packet arrival times is in $\mathcal{I}_0$. The random variable $w_0(l)$ is complex Gaussian with mean $D(\epsilon)e^{j\phi}$ and variance $\sigma^2_{w,m-1}$, where $D(\epsilon) = 1 - \epsilon[1 - R_c(1) - R_h(1)R_c(N-1)]$, ϕ is a uniform random variable in $[0, 1/2)$, and $\sigma^2_{w,m-1} = [(m-1)\sigma^2_{z,\mathrm{I}} + \sigma^2_{z,\mathrm{T}}]/N_h$. Therefore $|w_0(l)|$ has a Rician density which is given by

$$f_{|w_0(l)|}(x) = \frac{2x}{\sigma^2_{w,m-1}}\exp\left(-\frac{x^2 + D^2(\epsilon)}{\sigma^2_{w,m-1}}\right)\mathrm{I}_0\left(\frac{2xD(\epsilon)}{\sigma^2_{w,m-1}}\right) \quad (11)$$

where $\mathrm{I}_0(\cdot)$ is the modified Bessel function of zero order. The header detection probability P_d when m transmissions exist is given by

$$P_d(\epsilon, m \mid \eta_{l-1}) = \mathrm{Q}\left(\frac{\sqrt{2}D(\epsilon)}{\sigma_{w,m-1}}, \frac{\sqrt{2}\eta_{l-1}}{\sigma_{w,m-1}}\right) \quad (12)$$

where $\mathrm{Q}(a, b)$ is Marcum's Q-function. By changing variables, $a = \frac{\sqrt{2}D(\epsilon)}{\sigma_{w,m-1}}$, $b = \frac{\sqrt{2}\eta_{l-1}}{\sigma_{w,m-1}}$, and $d^2 = \frac{2K^2N_h^2}{\hat{N}}(1 + \frac{\sigma^2_{z,\mathrm{I}}}{\sigma^2_{z,m-1}})$, the probability density function of the threshold in (7) is transformed into

$$f_b(x) = \frac{2}{d(\hat{N}-1)!}\left(\frac{x}{d}\right)^{2\hat{N}-1} e^{-(x/d)^2}. \quad (13)$$

Therefore the average header detection probability is given by

$$P_d(\epsilon, m) = \int_0^\infty \mathrm{Q}(a, x)f_b(x)dx \quad (14)$$

$$= 1 - \int_0^\infty xe^{-(a^2+x^2)/2}\mathrm{I}_0(ax)\left[\int_x^\infty f_b(y)dy\right]dx \quad (15)$$

$$= 1 - \frac{d^2e^{-\frac{a^2}{2}}}{2+d^2}\sum_{n=0}^{\hat{N}-1}\frac{1}{n!}\left(\frac{2}{2+d^2}\right)^n\int_0^\infty t^n e^{-t}\mathrm{I}_0(2\sqrt{\xi t})dt \quad (16)$$

where $\xi = \frac{a^2d^2}{2(2+d^2)}$ and $\int_0^\infty t^n e^{-t}\mathrm{I}_0(2\sqrt{\xi t})dt = n!\Phi(n+1, 1; \xi)$ where $\Phi(a, b; x)$ is the confluent hypergeometric function. Let define $\phi(n; x) \triangleq e^{-x}\Phi(n, 1; x)$. Then the header detection probability P_d can be finally written as

$$P_d(\epsilon, m) = 1 - \frac{d^2}{2+d^2}e^{-\frac{a^2}{2+d^2}}\sum_{n=0}^{\hat{N}-1}\left(\frac{2}{2+d^2}\right)^n\phi(n+1; \xi) \quad (17)$$

where $\phi(n, x)$ has the following recurrence relation:

$$n\phi(n+1; x) = (x + 2n - 1)\phi(n; x) + (1-n)\phi(n-1; x) \quad (18)$$

with $\phi(1;x) = 1$, and $\phi(2;x) = x+1$, so on. Evaluation of (17) requires the knowledge of $\hat{N}$, but if L is a small number, then $\hat{N}$ can be approximated by $L(N-m)$ for numerical evaluations.

4 Throughput Bounds

Lower and upper bounds of the throughput of the unslotted system with fixed packet length can be derived by considering the distributions of maximum and minimum numbers of interfering transmissions during the reception of a particular packet [3]. Packet arrival times are generated by a Poisson process with parameter μ and each packet comes with a new node (so the underlying population is infinity). The throughput measure is the maximum average number of captures in packet-length of time, constrained on the data bit error rate.

Let C denote the capture event of the packet and E (a missed-capture event) denote the complement of C. Then the probability of missed-capture $P(E)$ for an unslotted system is bounded by

$$P_{\mathrm{L}}(\mu) \leq P(E) \leq P_{\mathrm{U}}(\mu) \qquad (19)$$

where $P_{\mathrm{U}}(\mu) \triangleq 1 - \sum_{j=0}^{J} P_C(j) \Pr(N^* = j)$ and $P_{\mathrm{L}}(\mu) \triangleq \sum_{j=0}^{J} P_E(j) \Pr(N_* = j)$, where $P_C(j)$ is the probability of capture for a slotted system in the presence of j interfering transmissions, and $\Pr(N^* = j)$ and $\Pr(N_* = j)$ are the distributions of maximum and minimum numbers of interfering transmissions [3]. Note $P_C(j)$ can be expressed as the product of two conditional probabilities, namely, $P_C(j) = P_{\mathrm{h}}(j)P_{\mathrm{cf}}(j)$, where $P_{\mathrm{h}}(j)$ is the header detection probability of a packet given the packet is collision-free in the presence of j interfering transmissions, and $P_{\mathrm{cf}}(j)$ is the collision-free probability of the packet in the presence of j interfering transmissions. Note $P_{\mathrm{h}}(j) = P_{\mathrm{d}}(\epsilon, j+1)$ and $P_{\mathrm{cf}} = [1 - c/N]^j$ where c is a constant depending on the threshold value and the sampling rate. $c = 3$ is used for numerical evaluations.

Let p_{e} denote the constraint on the data bit error probability, and define

$$s_1(\mu) \triangleq \mu[1 - P_{\mathrm{U}}(\mu)] \qquad (20)$$

$$S_1(p_{\mathrm{e}}) \triangleq \max_{\mu}\{s_1(\mu)\ ;\ P_{\mathrm{Ub}}(\mu) \leq p_{\mathrm{e}}\} \qquad (21)$$

$$s_2(\mu) \triangleq \mu[1 - P_{\mathrm{L}}(\mu)] \qquad (22)$$

$$S_2(p_{\mathrm{e}}) \triangleq \max_{\mu}\{s_2(\mu)\ ;\ P_{\mathrm{Lb}}(\mu) \leq p_{\mathrm{e}}\} \qquad (23)$$

where $P_{\mathrm{Ub}}(\mu) \triangleq 1 - \sum_{j=0}^{J}(1 - P_{\mathrm{be}}(j)) \Pr(N^* = j)$ and $P_{\mathrm{Lb}}(\mu) \triangleq \sum_{j=0}^{J} P_{\mathrm{be}}(j) \Pr(N_* = j)$, where $P_{\mathrm{be}}(m)$ is data bit error probability of slotted system in the presence of $(m-1)$ interfering transmissions which is given by $P_{\mathrm{be}}(m) = \frac{1}{2}\sum_{d=\pm} Q\left(\frac{1-\epsilon(1-R_c(1)-dR_c(N-1))}{\sigma_{z,m-1}}\right)$ where $\sigma^2_{z,m-1} = (m-1)\sigma^2_{z,\mathrm{I}} + \sigma^2_{z,\mathrm{T}}$. Then the actual throughput $S^*(p_{\mathrm{e}})$ of unslotted system is bounded by

$$S_1(p_{\mathrm{e}}) \leq S^*(p_{\mathrm{e}}) \leq S_2(p_{\mathrm{e}}). \qquad (24)$$

5 Results

The auto-optimal phase m-sequence of period $N = 63$ with least side-lobe energy [4] is used as a spreading sequence. The header sequence consists of 16-bits of frame synchronization word with maximum side lobe partial correlation 2 [5]. For average performances P_{d} and P_{be} are averaged over ϵ. For worst performances P_{d} and P_{be} are evaluated at $\epsilon = 1/2$. Thermal noise components are ignored for numerical evaluations.

Fig. 3 compares the false-alarm probabilities for the threshold controlled system with the fixed threshold system designed at number of users 20 and false-alarm probability $1.6\text{x}10^{-8}$. The false-alarm probability for the threshold controlled system is essentially invariant to the changes in the number of actual transmissions in the channel. Fig. 4 shows how the header detection efficiency improves with increasing the number L of data columns for estimating interference noise level. The system is designed at desired false-alarm rate $1.6\text{x}10^{-7}$. In practice, $L = 2$ or 3 may be enough. Fig. 5 compares the header detection probabilities for the two systems designed at number of users 20 and false-alarm probability $1.6\text{x}10^{-16}$. On this figure, the intercept points with a vertical line would indicate the performance of the perfect threshold control system for a constant false-alarm rate. Note that *'s are closer to the vertical line on $p_{\mathrm{fa}} = 1.6\text{x}10^{-16}$ than x's and that *'s are always above the ideal ROC curves but x's are not. Fig. 6 shows the lower and upper bounds on throughput for the unslotted system designed at false-alarm probability $1.6\text{x}10^{-8}$, constrained on the data bit error rate. Worst and average performances are shown altogether.

Simulation results for Gaussian assumption validity are shown in Fig. 7. The histogram of the real part of the interference at the header detection correlator is compared with the ideal curve when the number m of actual transmitting radios in the channel is 30. Bin size for the histogram is one tenth of the sample standard deviation. The cases when $m = 10$ and $m = 5$ were also simulated and compared with the histogram of IMSL Gaussian random number generator. We conclude the Gaussian assumption is reasonably valid.

References

[1] D. I. Kim. *On the Performance of Common Spreading Code CDMA Packet Radio Systems with Multiple Capture Capability.* PhD thesis, University of Southern California, 1990.

[2] M. B. Pursley. Performance Evaluation for Phase-Coded Spread-Spectrum Multiple-Access Systems-Part I: System Analysis. *IEEE Trans. Commun.*, COM-25:795–799, August 1977.

[3] M. B. Pursley. Frequency-Hop Transmission for Satellite Packet Switching and Terrestrial Packet

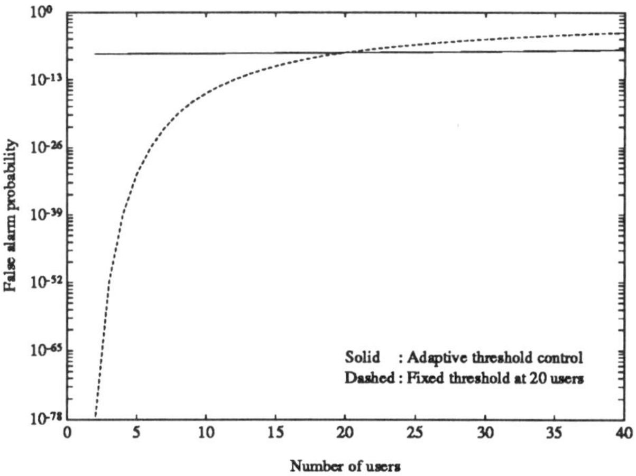

Figure 3: False-alarm sensitivities

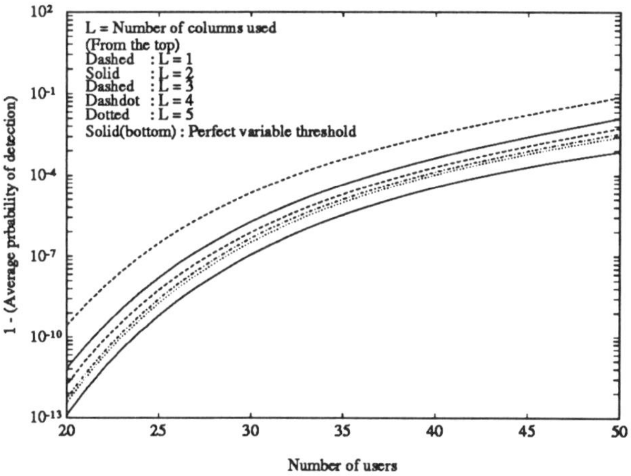

Figure 4: Header detection efficiency

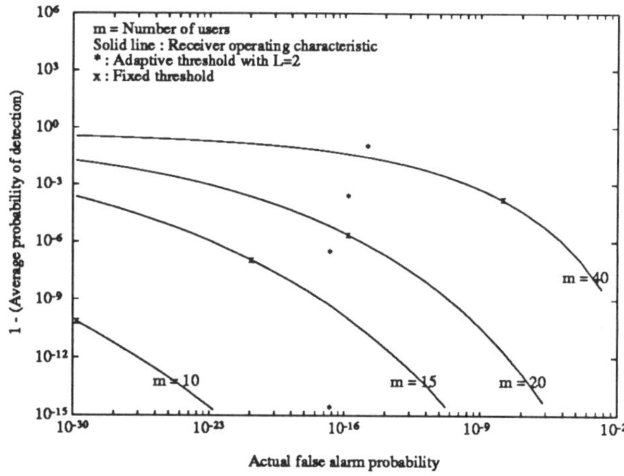

Figure 5: System robustness

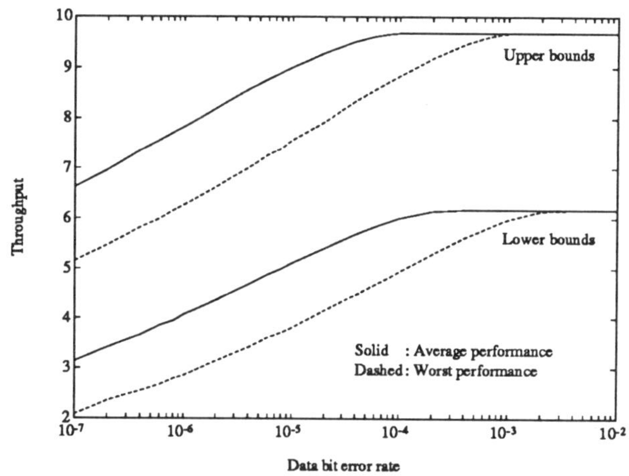

Figure 6: Throughput bounds for an unslotted system

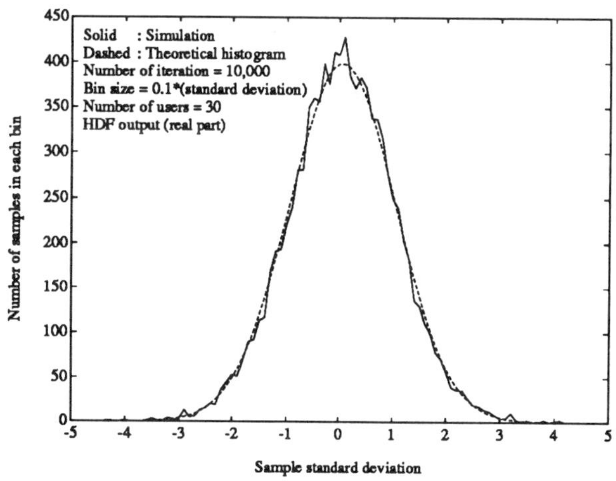

Figure 7: Real part of interference when number of user is 30

Radio Networks. *IEEE Trans. Inform. Theory*, COM-32:652–667, September 1986.

[4] M. B. Pursley and H. F. A. Roefs. Numerical Evaluation of Correlation Parameters for Optimal Phases of Binary Shift-Register Sequences. *IEEE Trans. Commun.*, COM-27:1597–1604, October 1979.

[5] R. Turyn. Sequences with Small Correlation. In H.B. Mann, editor, *Error Correcting Codes.* Wiley, New York, 1968.

[6] R. A. Scholtz. Frame Synchronization Techniques. *IEEE Trans. Commun.*, COM-28:1204–1213, August 1980.

[7] H. M. Finn and R. S. Johnson. Adaptive Detection Mode with Threshold Control as a Function of Spatially Sampled Clutter-Level Estimates. *RCA Review*, pages 415–464, september 1968.

Systematic Design of Digital Algorithms for Synchronization: from Theory to VLSI

Heinrich Meyr

Aachen University of Technology, ERT
Templergraben 55, W-5100 Aachen, Germany

Abstract

The methodology for the design and implementation of digital synchronization algorithms is discussed. Estimation theory is employed in the first step to systematically derive structures. Next the performance of design alternatives is evaluated taking quantization and discretization into account. In the third step the algorithms are mapped onto a suitable processor architecture. The necessity for integrated simulation tools ranging from the system level to the VLSI/DSP-level in this highly interactive design process is emphasized.

1. From analog to digital receivers

Communication theory is mostly concerned with coding/modulation. In most books on digital communications synchronization is addressed only superficially, if at all.

Until a few years ago synchronization was almost synonymous with phase-locked loop theory. All classical receivers could essentially be reduced to structures involving a phase-locked loop in one form or the other. Down conversion and phase recovery were performed simultaneously using circuits known as long loops. Clock recovery was done subsequently in base-band again using PLLs. The *structure* of the *synchronizer* was given, and the design of synchronization systems was concerned with the task of optimizing the performance (variance, cycle slip rate) by choosing suitable parameter values [1], [2], [3], [4], [5]. Despite the deceivingly simple appearance of a PLL model the task of analyzing and designing PLLs is far from simple as the large number of papers published in this area confirm. Probably even more convincing is to talk to hardware designers who actually build PLLs with a performance which allows coding-theorists to dream that perfect synchronization is not too far from reality.

With the advent of digital VLSI technology it became economically feasible to implement transmission schemes which operate far closer to the theoretical limits than was possible with analog technology. The introduction of VLSI technology was not a mere substitution of analog functional blocks but opened the gate for revolutionary new concepts and products. The traditional communications engineer was used to a two-dimensional design space involving *power* and *bandwidth*. Transmission schemes were either power limited or bandwidth limited. Today there is a new dimension added to the design space, namely signal *processing complexity*.

While any engineer involved in the design of products readily acknowledges the fundamental importance of the new dimension, many communication theorists still have many problems to truly accept it and properly take it into account in their work. Probably the main reason is based on the fact that there is no universal performance measure for complexity. It comes in different flavors depending on whether VLSI realization or programmable DSP solution is considered.

To be specific, the number of operations can be a meaningful performance measure if the receiver algorithm is implemented on a DSP. But it is meaningless if an algorithm is implemented as an ASIC. What counts, for example, in high data rate satellite receivers is whether the algorithm is amenable to parallel processing allowing the use of CMOS technology. Then the AT (area x cycle time) product is an appropriate measure. Ignoring the importance of an appropriate measure of complexity leads to meaningless conclusions when comparing algorithms.

2. Synthesis of synchronizer structures: from estimation theory to structures

The task of any receiver is to detect the symbol in the presence of various impairments transmitted over a channel. The design of a receiver belongs to either of two main classes. On the one hand there is the *ad-hoc approach* where structure is guessed on a-priori knowledge and optimized with respect to a given performance criterion. The alternative to an ad-hoc structure approach is a *derived structure approach* in which no a priori guesses are made. The design involves in a first step to establish a performance criterion and implement the algorithm that solves the problem on a suitable processor architecture.

For example, a maximum likelihood receiver requires maximization of the functional $p(\underline{r}|\underline{a})$ where $\underline{r}$ is a sufficient statistics of the received signal and $\underline{a}$ is the data sequence to be detected. Notice that in the probability density function $p(\underline{r}|\underline{a})$ there are no sync parameters. They are unwanted parameters which must be removed by averaging

$$p(\underline{r}|\underline{a}) = \int p(\underline{r}|\underline{a},\theta,\varepsilon)p(\varepsilon)p(\theta)d\varepsilon\, d\theta \qquad \theta: \ phase \qquad \varepsilon: \ clock \tag{1}$$

The truly optimal ML receiver *jointly* optimizes $p(\underline{r}|\underline{a},\theta,\varepsilon)$ with respect to all the trial parameters $(\underline{a},\theta,\varepsilon)$. There is no separate synchronization unit in the truly optimal receiver. Any synchronization unit is a suboptimal concession.

If we assume that the noise is weak then the ML-function $p(\underline{r}|\underline{a},\theta,\varepsilon)$ peaks around its true value (θ_0,ε_0). Neglecting a-priori knowledge on (ε,θ) yields

$$\begin{aligned} p(\underline{r}|\underline{a}) &\sim \max_{\theta,\varepsilon}\{p(\underline{r}|\underline{a},\theta,\varepsilon)p(\theta)p(\varepsilon)\} \\ \left(\hat{\theta},\hat{\varepsilon}\right) &\sim p\left(\underline{r}|\underline{a},\hat{\theta},\hat{\varepsilon}\right) \end{aligned} \tag{2}$$

The estimates are used as if they were true values in the subsequent detection process. This is the principle of synchronized detection as employed in any practical receiver.

For ML synchronization we must consider the data sequence $\underline{a}$ as an unwanted parameter. The ML-synchronizer then is an estimator for the joint estimation of the synchronization parameters $\left(\hat{\theta},\hat{\varepsilon}\right)$.

$$p(\underline{r}|\theta,\varepsilon) = \sum_{all\ sequences} p(\underline{a})p(\underline{r}|\underline{a},\theta,\varepsilon) \quad . \tag{3}$$

If the ML function $p(\underline{r}|\underline{a},\theta,\varepsilon)$ peaks at its maximum (weak noise assumption) then the sum can be approximated by just one term

$$p(\underline{r}|\theta,\varepsilon) \sim p(\underline{r}|\hat{\underline{a}}_{ML},\theta,\varepsilon) \tag{4}$$

Alternatively, assume that prior to data transmission a training sequence is transmitted which allows the acquisition of clock and phase. Because of the coding gain in the subsequent data transmission we obtain a very reliable estimate $\hat{\underline{a}}_{ML}$, such that $Pr(\hat{\underline{a}}_{ML}) \gg Pr(\underline{a})$ for all other sequences. Using this knowledge we again may neglect all but one term in the sum [eq. 3] to obtain [eq. 4]. (But note the different reasoning!) This is known as decision-directed synchronization.

It is important to see that the decision-directed synchronization is a special case of the averaging process [eq. 3], and not vice-versa, contrary to widespread thinking. Those not convinced asking for a formal 'proof' are encouraged to consider the classical Costas loop for BPSK. In this case the averaging of [eq. 3] can be performed analytically. The resulting nonlinearity tanh( ) in the loop can be approximated in the high SNR case by a hard limiter. The Costas loop then has exactly the form of a decision directed carrier synchronizer.

Returning to the general case, [eq.3] forms the basis to obtain practical synchronizer algorithms. The steps to achieve this goal are conceptually simple. Since averaging over the data sequence is analytically possible only in very special cases one has to resort to approximations. It can be shown [6] that virtually all known phase- and clock synchronizers are the result of different approximations to the averaging operation of [eq.3]. This includes algorithms such as the digital square and filter clock synchronizer (spectral analysis) [7], the Mueller + Müller [8] decision directed clock synchronizer, wave difference method, etc. which were hitherto not known to be related to ML-estimators.

Estimation theory is a very powerful tool for exploring the three-dimensional design space since it provides us with a general methodology to systematically derive and compare structures rather than rely on ad-hoc structures.

Just how powerful this tool is will be illustrated by two examples. Example 1 compares linear with nonlinear modulation schemes such as CPM. For linear modulation schemes a wide variety of practical algorithms with low complexity can be found. Only at great effort can ML synchronization algorithms be derived which, despite crude approximations, are far more complex than the ones employed for linear modulation. In addition, the algorithms are plagued by problems such as hang-up not present in digital algorithms for linear modulation.

Of course, a comparison must include the BER (bit error rate) over a realistic channel. The reader is referred to the interesting paper [9] where it is shown (surprisingly) that MPSK performs virtually as well as CPM over a nonlinear satellite channel, if compared properly.

Example 2 deals with coherent communications over fading channels which was long considered unfeasible. While this is true if PLL carrier recovery is employed it is definitely not true if systematically derived structures are employed [10], [11], [12], [13], [14], [15]. These examples clearly demonstrate the danger of relying on ad-hoc structures rather than systematically exploring the design space.

3. From structure to performance

Having found the structure, evaluation of performance is the next step in designing a digital receiver. This is traditionally the field where most papers on synchronization are published. But a sign of warning is necessary at this point. Many of these papers extrapolate implicit assumptions valid for analog signal processing to the digital world without questioning their applicability. Worse, they leave out critical parameters leading to erroneous conclusions.

Among the most useful (technology independent) theoretical results are bounds such as the Cramer-Rao bound. These bounds allow the implementation loss of realizable synchronizers to be assessed by computing the tracking error variance of synchronizers and comparing them with these bounds. The tracking error variance of synchronizers can be obtained analytically only if one neglects quantization (and to a large extent time discretization). Results thus obtained are once again bounds.

The only way to evaluate the performance of digital receiver implementation is by *computer simulation*. Computer simulation has become an indispensable tool for the evaluation and design. On the system level commercial available tools such as Boss®[1], SPW® and COSSAP®[2] allow the simulation of the entire transmission system comprising transmitter, channel and receiver. The simulation tools allow subsystems to be simulated *bit-level true* in order to assess the impact of time discretization and quantization. The proper choice of the sampling rate and quantization is of crucial importance since it largely determines the complexity of the VLSI implementation.

It is important that the overall receiver performance is optimized. The resulting transmission error rate (the ultimate performance measure) depends in a highly nonlinear and intricate way on many parameters such that a block-by-block optimization leads to highly increased complexity at negligible performance gain. A case study describing the entire design process can be found in [16].

Once a satisfactory bit true model exists on the system level a behavioral description suitable for VLSI realization must be created. There are high level hardware description languages such as VHDL and Verilog® which allow the incorporation of behavioral and structural information specific to hardware realization. The transfer of the behavioral/structural description used on the system level to the implementation level is crucial. This transfer amounts to linking two simulation worlds with different goals employing different simulation paradigms. A look at Fig. 1 displaying the different granularities of system description explains easily why it is impossible to reconcile the design task with *one* simulation tool and simulation paradigm. On the system level interest is primarily focussed on BER while on the implementation level the functional correctness is verified. While it is conceptionally possible to do system level simulation using VHDL the *computational efficiency* is one to two orders of magnitudes lower than using a system simulation paradigm [17] . Equally important is the *modeling efficiency*. By this term we mean the engineering hours required to obtain a simulation model. Here the system simulation tool outperforms VHDL description simply based on the fact

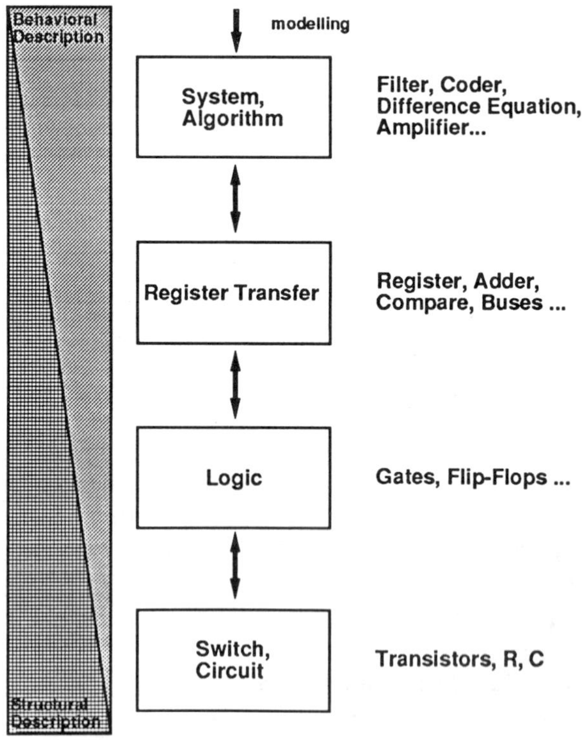

Figure 1 Different granularities of system description

1 Boss, SPW are registered trademarks of COMDISCO

2 COSSAP is a registered trademark of CADIS

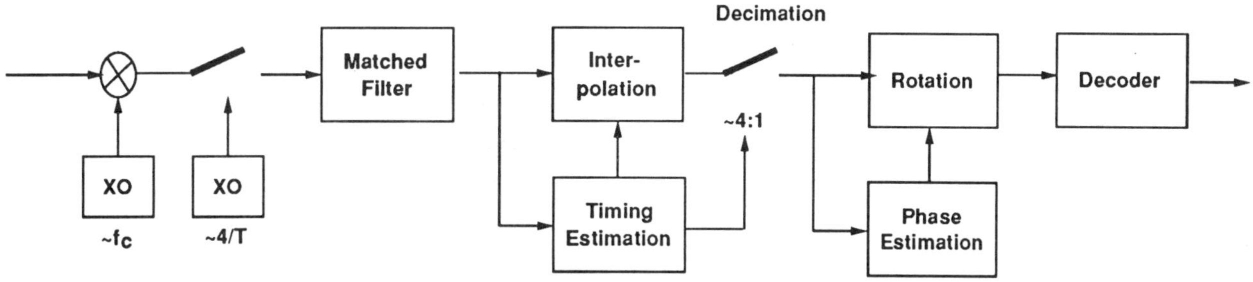

Figure 2 DIRECS block diagram

that information irrelevant on this level but required for implementation need not be specified. As a result, rapid prototyping is possible.

The design of an ASIC for communications most certainly is an iterative procedure involving interaction between the algorithmic and architectural level. Tools are needed to support rapid evaluation of architectural alternatives and to synthesize logic from a high-level description.

Tools for optimization and synthesis on the logic level using HDLs as input are commercially available now [eg. Synopsys Design Compiler, Racal Redac Silc. Syn, Mentor Autologic], while more advanced techniques like scheduling and allocation on register transfer level are currently moving from research (HYPER (UC Berkeley)) [18], [19], [20] to commercial exploitation.

4. Design Example: DIRECS

DIRECS is an acronym for **DI**gital**RE**ceiver **C**hip**S**et. The chip set represents a 100 Mbit//s all-digital receiver for trellis-codel 8–PSK modulation [21], [22], [23]. A prototype of the entire receiver has been manufactured and tested. It was put into operation in spring 1991. DIRECS consists of six application specific integrated circuits (ASICs), all of which are standard cell chips designed with a 2μm CMOS library. The overall complexity is in excess of 55000 equivalent gates. Parallel processing of two received symbols was applied to achive the envisaged throughput of 100 Mbit/s, although the receiver is clocked at 25 MHz. Figure 2 gives a structural outline of DIRECS.

After converting the RF signal to baseband and sampling the signal with fourfold oversampling, the complex samples are filtered with matched filters. Subsequently timing synchronization is performed by means of a timing estimator and a combined interpolation and decimation unit. The next step is phase synchronization accomplished with an estimator and a CORDIC processor for rotating the complex samples according to the phase estimate. Finally, a Viterbi decoder is used for the decoding of the trellis-coded signal.

Throughout the design of DIRECS extensive system simulations were used to explore the design alternatives and trade-offs on different levels of abstraction. It is mandatory to consider algorithm design and VLSI architecture as an entity [22], [24], [16]. Actually, the constraints of the VLSI implementation led to the necessity of modifying the algorithms. To evaluate the effects of finite wordlengths each component was modeled bittrue even where advanced implementation techniques such as redundant number representations were used. Table 1 gives an overview on area consumption and implementation losses at channel SNR of 10 dB in DIRECS.

function	active area	Impl. loss	comments
matched filters	70 mm^2	0.28 dB	16 parallel filters
timing synchro-nization	66 mm^2	0.24 dB	
phase synchro-nization	67 mm^2	0.14 dB	2 CORDIC processors (39 mm^2)
Viterbi decoder	47 mm^2	0.14 dB	
testing	13 mm^2	—	dedicated chip for system test [25]
fully digital receiver	263 mm^2	0.80 dB	

Table 1 Area versus Implementation Losses

Overall implementation loss of 0.8 dB is too large for a commercial product but the table shows that the area vs. performance trade-offs were well chosen. Improvement to a acceptable loss of 0,1 — 0,2 dB is possible as was demonstrated during the design process. Due to increased complexity a 1.0 μm technology is required which was not available (for cost reasons) at the time of chip fabrication.

Bibliography

[1] Lindsey, F.M. *Synchronization Systems in Communication and Control.* Prentice-Hall, Englewood Cliffs, NJ, 1972.

[2] Gardner, F.M. *Phaselock Techniques.* Wiley, New York, 1979.

[3] Meyr, H., Ascheid, G. *Synchronization in Digital Communications, Volume I.* John Wiley and Sons, Inc., New York London Sydney, 1989.

[4] Blanchard, A. *Phase-Locked Loops.* Wiley, New York, 1976.

[5] Viterbi, A.J. *Principles of Coherent Communications.* McGraw-Hill, New York, 1966.

[6] Meyr, H. *Synchronization in Digital Communications, Volume II.* John Wiley and Sons, Inc., New York London Sydney, in preparation.

[7] M. Oerder and H. Meyr. Digital filter and square timing recovery. *IEEE Trans. Commun., vol. COM-36,* (5):605–612, May 1988.

[8] Mueller, K.H., Müller, M. Timing recovery in digital synchronous data receivers. *IEEE Trans. Commun., vol. COM-24,* (5):516–531, May 1976.

[9] Ch. Rapp. Coded and uncoded M-PSK and CPM signals on nonlinear bandpass channels: An approach to a fair comparison. In *IEEE Globecom 1990 conference record,* 1990.

[10] Aghamohammadi, A. and Meyr, H. A New Method for Phase Synchronization and Automatic Gain Control of Linearly-Modulated Signals on Frequency-Flat Fading Channels. *IEEE Trans. Commun.,* 39:25–29, Jan. 1991.

[11] Fechtel, S.A. and Meyr, H. An Investigation of Channel Estimation and Equalization Techniques for Moderately Rapid Fading HF-Channels. In *Proc. ICC,* pages 768–772, Jun. 1991.

[12] Haeb, R. and Meyr, H. A Systematic Approach to Carrier Recovery and Detection of Digitally Phase-Modulated Signals on Fading Channels. *IEEE Trans. Commun.,* 37(7):748–754, Jul. 1989.

[13] Ten C.H. Kam, P.Y. An adaptive receiver with memory for slowly fading channels. *IEEE Trans. Commun., vol. COM-32,* (6):654–659, June 1984.

[14] Lodge J.H. Moher, M.L. TCMP- A Modulation and Coding Strategy for Recian Fading Channels. *IEEE J. Sel. Areas Commun., Vol. SAC-7,* (9):1347–1355, Dec. 1989.

[15] S. Stein. Fading Channel Issues in System Engineering. *IEEE J. Sel. Areas Commun., Vol. SAC-5,* (2):68–89, Feb 1987.

[16] Joeressen, O. , Oerder, M. , Serra, R. , Meyr, H. Direcs: System design of a 100Mbit/s digital receiver. In *submitted to IEE Proceedings, Part G,* 1991.

[17] P. Zepter and K. ten Hagen. Using VHDL with stream driven simulators for digital signal processing applications. In *EURO-VHDL'91 Proceedings,* Stockholm, Sweden, September 8-11 1991.

[18] R. A. Cottrell. Chip design using Silicon 1076. In *EURO-VHDL'91 Proceedings,* pages 306–312, Stockholm, Sweden, Sept. 1991. Institute for Microelectronics.

[19] J. Rabaey, C. Chu, P. Hoang, and M. Potkonjak. Fast prototyping of datapath-intensive architectures. *IEEE Design & Test of Computers,* June 1991.

[20] H. De Man, F. Catthoor, G. Goosens, J. Vanhoof, J. van Meerbergen, S. Note, and J. Huisken. Architecture-driven synthesis techniques for VLSI implementation of DSP algorithms. In *Proceedings of the IEEE,* pages 319–335. IEEE, February 1990.

[21] M. Oerder and H. Meyr. DIRECS: VLSI implementation of a 100MBit/s digital receiver. In *Proceedings of the Second International Workshop on Digital Signal Processing Techniques Applied to Space Communications, 1990,* Turino, Italy, 1990.

[22] M. Oerder and H. Meyr. VLSI implementation of synchronization algorithms in a 100 MBit/s digital receiver. In *IEEE Globecom 1990 conference record,* 1990.

[23] G. Fettweis and H. Meyr. New rotationally invariant trellis codes for 8PSK modulation. In *IEEE Int. Conf. Comm., ICC'89,* pages 319–335, Boston, Jun. 1989.

[24] G. Fettweis and H. Meyr. A 100MBit/s Viterbi decoder chip: Novel architecture and its realization. In *Int. Conf. on Comm., ICC'90,* (Atlanta, GA), April 1990.

[25] Serra, R. ,Oerder, M., Joeressen, O. ,Schotten, K., and Meyr, H. Digitaler 100mBit/s Empfänger in CMOS: Architektur- und Systemselbsttestkonzept. In *in 5th E.I.S. Workshop,* pages 33–342, April, 8.-9. 1991.

JOINT SYNCHRONIZATION AND DETECTION FROM SAMPLES TAKEN THROUGH INTEGRATE-AND-DUMP *

Ning Kong
Electrical Engineering Department
Texas A&M University
College Station, TX 77843

Costas N. Georghiades
Electrical Engineering Department
Texas A&M University
College Station, TX 77843

Abstract

We study the problem of joint synchronization and detection from samples taken by an integrate-and-dump filter, for both the optical Poisson and Gaussian channels. We show that the error probability due to timing error reduces to that with perfect synchronization as the sampling rate increases.

1 Introduction

The recent advent of high speed digital signal processing (DSP) chips has greatly motivated many studies on receiver algorithms which operate on sampled data instead of analog to produce the timing and symbol estimates [1,2,3,4,5,6,7,8]. In this paper, we study the problem of delay and sequence estimation for arbitrary baseband pulse shapes and arbitrary sampling rates as in [7] but in direct detection optical communications. We also study the same problem for the Gaussian channel with the samples taken not at the matched filter output but by an integrate-and-dump process.

The paper is organized as follows after the introduction. Section II has the derivation of the likelihood function for arbitrary baseband pulse shapes and sampling rates for optical Poisson systems and the same likelihood function using the integrate-and-dump filter for the Gaussian channel. The general results are then applied to several commonly used modulation schemes such as OOK and BPPM for the optical channel and NRZ signals for the Gaussian channel. Section III discusses the performance of the above described receivers both for optical and Gaussian channels. It contains the proof that the sequence error probability in the presence of timing error decreases monotonically to that with perfect synchronization as the sampling rate increases. Also, a lower bound on the error probability is derived for the Gaussian channel, and a comparison of this lower bound is made between the matched filter receiver and the integrate-and-dump receiver for NRZ signaling. Section IV shows simulation results on the performance of the receiver described above for OOK and BPPM, and Section V concludes.

*This work was partially supported by National Science Foundation grant NCR-9016329.

2 The Joint Likelihood Functions

2.1 The optical channel

In direct detection optical communications, for the popular modulation scheme PAM, an information sequence $\mathbf{x} = (x_0, x_1, \cdots, x_{N-1})$, where x_k has Q levels, is transmitted by modulating the intensity of a laser. The intensity $\lambda(t)$ of the corresponding received signal is

$$\lambda(t) = \sum_{k=0}^{N-1} x_k p(t - kT - \tau') + \lambda_n, \tag{1}$$

where λ_n is the noise intensity, τ' is a timing error, T is the symbol duration and $p(t)$ is an arbitrary baseband pulse of duration T seconds. The output of the photodetector is a doubly stochastic Poisson process N_t, with intensity (photons/s) given by (1).

The receiver samples N_t every $T_s = T/L$ seconds through integrate-and-dump to produce a sequence of samples $\mathbf{K}$,

$$\mathbf{K} = (K_{00}, K_{01}, \ldots, K_{0(L-1)}, K_{10}, K_{11}, \ldots, K_{1(L-1)}, \ldots, K_{(N-1)0}, K_{(N-1)1}, \ldots, K_{(N-1)(L-1)}),$$

which are further processed to extract timing and symbol estimates. Mathematically these samples are given by

$$K_{ij} = \int_{-\infty}^{\infty} g(t - iT - jT_s) dN_t, \tag{2}$$

where

$$g(t) = \begin{cases} 1, & 0 \le t \le T_s, \\ 0, & \text{otherwise}. \end{cases}$$

Because $\mathbf{K}$ is conditionally a Poisson random vector with independent components , the likelihood function for the joint estimation problem is

$$l(\mathbf{K}|\mathbf{x}, \tau') = \sum_{i=0}^{N-1} \sum_{j=0}^{L-1} [K_{ij} \ln \lambda_{ij}(\mathbf{x}, \tau') - \lambda_{ij}(\mathbf{x}, \tau')], \tag{3}$$

where

$$\lambda_{ij}(\mathbf{x}, \tau') = \lambda_n T_s + \lambda_s \sum_k x_k r_{ikj}, \tag{4}$$

1058-6393/91 $1.00 © 1991 IEEE

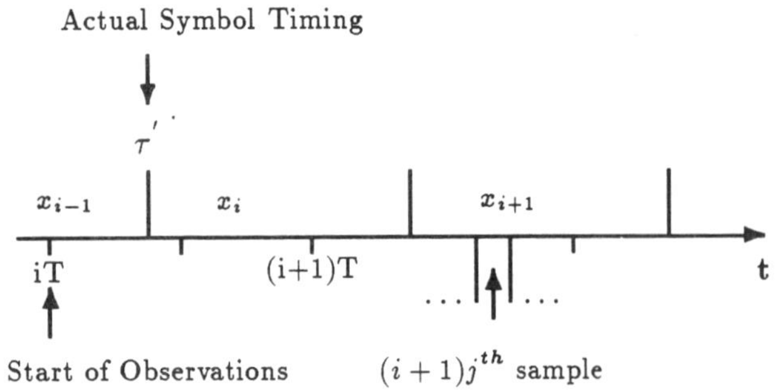

Figure 1: Timing Error of Unsynchronized Receiver

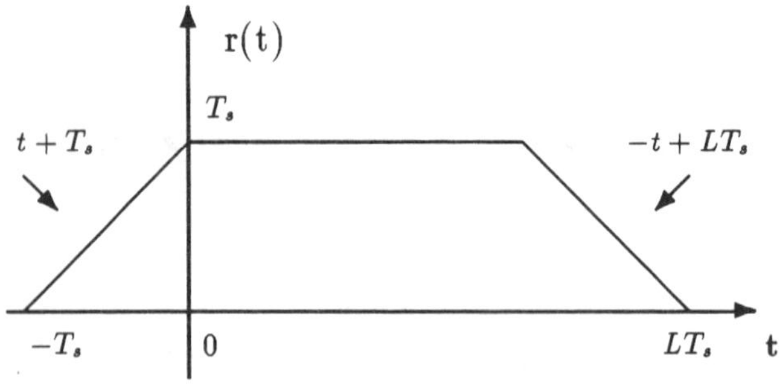

Figure 2: Correlation Function Between Rectangular Pulse and g(t)

$$r_{ikj} = r[(i-k)T + jT_s - \tau'], \tag{5}$$

and

$$r(t) = \int_{-\infty}^{\infty} g(\alpha)p(\alpha+t)d\alpha. \tag{6}$$

It can be shown further that

$$\lambda_{ij}(\mathbf{x}, \tau') = \begin{cases} \lambda_n T_s + \lambda_s x_{i-1} r_{i(i-1)j}, & j < n, \\ \lambda_n T_s + \lambda_s x_{i-1} r_{i(i-1)n} + \lambda_s x_i r_{iin}, & j = n, \\ \lambda_n T_s + \lambda_s x_i r_{iij}, & n < j \le L-1, \end{cases}$$

where $n = \lfloor \tau'/T_s \rfloor$. The above equation indicates that the received data in the ith transmission interval and j^{th} sampling instant has a mean controlled by not only the i^{th}symbol x_i but also the previous one, x_{i-1}; further more, we can see the intersymbol interference (ISI) is only in the $j = n^{th}$ chip of duration T_s. Apparently it will be reduced when T_s gets smaller, which is the intuitive idea of sampling more densely to get better performance. In Section III, we will prove theoretically that performance improves when the sampling rate increases.

If $p(t)$ is a rectangular pulse and $x_k \in \{0,1\}$, then we have the popular modulation scheme OOK. In this case, $r(t)$ is shown in Fig.2. For the special cases of one and two samples per symbol ($L = 1, 2$), when $0 \le \tau' \le T/2$, (3) reduces to

$$\begin{aligned} l(\mathbf{K}|\mathbf{x}, \tau) &= \sum_{i=0}^{N-1} \{K_{i0} \ln[\lambda_n + \lambda_s(x_i + (x_{i-1} - x_i)\tau)] \\ &\quad -\lambda_s T[x_i + (x_{i-1} - x_i)\tau]\}, \end{aligned} \tag{7}$$

for $L = 1$,and

$$\begin{aligned} l(\mathbf{K}|\mathbf{x}, \tau) = \sum_{i=0}^{N-1} \{ & K_{i0} \ln[\lambda_n + \lambda_s(x_i + (x_{i-1} - x_i)\tau)] \\ & + K_{i1} \ln(\lambda_n + \lambda_s x_i) - \lambda_s T[x_i + (x_{i-1} - x_i)\tau]\}, \end{aligned} \tag{8}$$

for $L = 2$, where $\tau = \tau'/T_s$, $T_s = T/L$. These two results are exactly the same as in [6] and [8], which are for $L = 1$ and $L = 2$ respectively.

For the pulse position modulation scheme, an information sequence $\mathbf{x} = (x_{0j}, x_{1j}, \cdots, x_{N-1,j})$, where $j = 0, 1, \cdots, Q-1$ and x_{kj} is one of the Q signals, is transmitted by modulating the intensity of the ith of Q slots in the T-second symbol interval. if the ith signal is transmitted, then $x_{ki} = 1$,and $x_{kj} = 0, j \ne i$. Each slot has duration T/Q. The resultant intensity of the received signal can be mathematically expressed as

$$\lambda(t) = \sum_{k=0}^{N-1}\sum_{j=0}^{Q-1} x_{kj} p(t - kT - j\frac{T}{Q} - \tau') + \lambda_n. \tag{9}$$

Then the demodulation is again to process the doubly stochastic Poisson process N_t by integrate-and-dump to obtain the photon counts or the samples $\mathbf{K}$ in time duration T_s seconds, $T_s = T/QL$, and QL is the sampling rate for PPM. (See Fig.3.) $\mathbf{K} = (K_{000}, K_{001}, \cdots, K_{00L-1}, K_{010}, K_{011}, \cdots, K_{01L-1}, \cdots, K_{N-1,Q-1,0}, K_{N-1,Q-1,1}, \cdots, K_{N-1,Q-1,L-1})$.

(10)

The likelihood function in this case is

$$l(\mathbf{K}|\mathbf{x}, \tau') = \sum_{i=0}^{N-1}\sum_{l=0}^{Q-1}\sum_{m=0}^{L-1} [K_{ilm} \ln \lambda_{ilm}(\mathbf{x}, \tau') - \lambda_{ilm}(\mathbf{x}, \tau')]$$

where

$$K_{ilm} = \int_{-\infty}^{\infty} g(t - iT - l\frac{T}{Q} - mT_s)dN_t, \tag{11}$$

$$\lambda_{ilm}(\mathbf{x}, \tau') = E\{K_{ilm}|\mathbf{x}, \tau'\} = \sum_{k=0}^{N-1}\sum_{j=0}^{Q-1} x_{kj} r_{ikljm},$$

$$r_{ikljm} = r[(i-k)T + (l-j-M)LT_s + (m-n)T_s - \Delta],$$

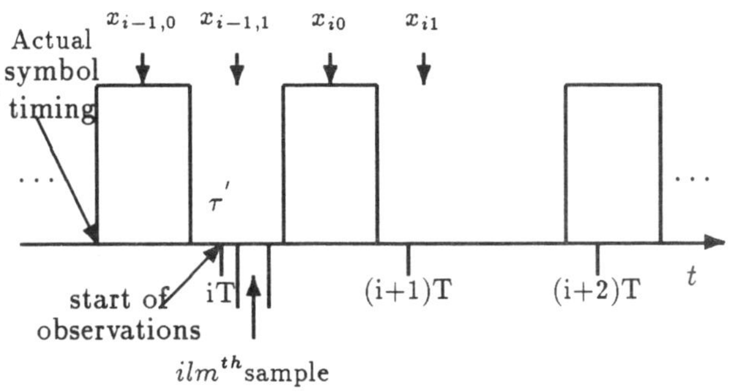

Figure 3: Timing Error for BPPM

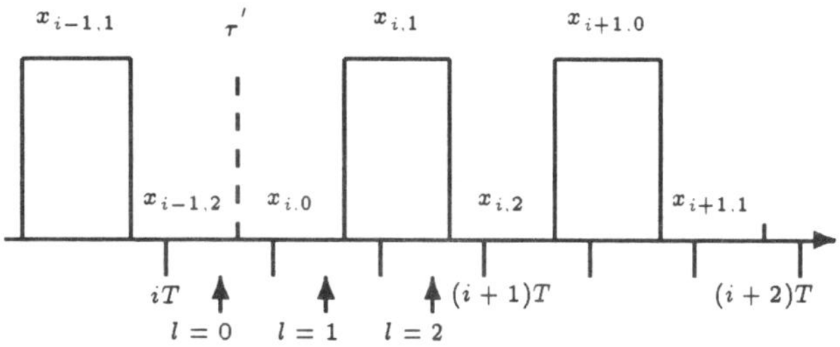

Figure 4: Q=3,M=0

and $\tau' = M\frac{T}{Q}+nT_s+\Delta$, $0 \le \Delta \le T_s$, $n = \lfloor\frac{\tau'-MT/Q}{T_s}\rfloor$ for $M = 0, 1, \cdots, Q-1$. The PPM scheme is more complicated than PAM in that it increases one more dimension in positioning during symbol interval T. The mean of ilm^{th} sample has three cases, depending on l and M. Which are:
When $l \ge M+1$, and $0 \le M \le Q-2$,

$$E\{K_{ilm}|\mathbf{x},\tau'\} =$$

$$\begin{cases} x_{i,l-M-1}r_{ii,l-M-1lm}, & m<n, \\ x_{i,l-M,}r_{ii,l-M,ln}+ & \\ x_{i,l-M-1}r_{ii,l-M-1,ln}, & m=n, \\ x_{i,l-M}r_{ii,l-M,lm}, & n<m<L. \end{cases}$$

When $l = M$, $0 \le M \le Q-1$,

$$E\{K_{ilm}|\mathbf{x},\tau'\} =$$

$$\begin{cases} x_{i-1,l-M+Q-1}r_{i,i-1,l-M+Q-1,lm}, & m<n, \\ x_{i,l-M}r_{ii,l-M,ln}+ & \\ x_{i-1,l-M+Q-1}r_{i,i-1,l-M+Q-1,ln}, & m=n, \\ x_{i,l-M}r_{ii,l-M,lm}, & n<m<L. \end{cases}$$

When $l \le M-1$, $1 \le M \le Q-1$,

$$E\{K_{ilm}|\mathbf{x},\tau'\} =$$

$$\begin{cases} x_{i-1,l-M+Q-1}r_{i,i-1,l-M+Q-1,lm}, & m<n, \\ x_{i-1,l-M+Q}r_{i,i-1,l-M+Q,lm}+ & \\ x_{i-1,l-M+Q-1}r_{i,i-1,l-M+Q-1,ln}, & m=n, \\ x_{i-1,l-M+Q}r_{i,i-1,l-M+Q,lm}, & n<m<L. \end{cases}$$

It is observed that for PPM, we not only have intersymbol interference but also interslot interference. For example, as shown in Fig.4 for $Q=3$ and $M=0$, when $l=0$,we have both intersymbol and interslot interference, when $l=1$, we have only interslot interference and so on. However, as T_s decreases or L increases, both intersymbol interference and interslot interference will reduce. If it is BPPM, $Q=2$ and assuming $\tau' \le \frac{T}{2}$ ($M=0$), then

$$\lambda_{i0m}(\mathbf{x},\tau') = \begin{cases} \lambda_s x_{i-1,1}T_s + \lambda_n T_s, \\ \lambda_s x_{i0}(T_s-\Delta) + \lambda_s x_{i-1,1}\Delta + \lambda_n T_s, \\ \lambda_s x_{i0}T_s + \lambda_n T_s, \end{cases}$$

and

$$\lambda_{i1m}(\mathbf{x},\tau') = \begin{cases} \lambda_s x_{i0}T_s + \lambda_n T_s, \\ \lambda_s x_{i1}(T_s-\Delta) + \lambda_s x_{i0}\Delta + \lambda_n T_s, \\ \lambda_s x_{i1}T_s + \lambda_n T_s, \end{cases}$$

which is the same as in [6].

2.2 The Gaussian channel

The output samples of the integrate-and-dump receiver for the Gaussian channel are:

$$\begin{aligned} R_{ij} &= \sum_k x_k \int_{-\infty}^{\infty} g(\alpha)p[\alpha+(i-k)T+jT_s-\tau']d\alpha \\ &+ N_{ij} = \sum_k x_k r_{ikj} + N_{ij}, \end{aligned} \qquad (12)$$

where r_{ikj} and $g(t)$ have the same definitions as for the optical channel, and

$$\begin{aligned} \mathbf{R} &= (R_{00}, R_{01}, \cdots, R_{0(L-1)}, \cdots, R_{(N-1)0}, R_{(N-1)1}, \\ &\cdots, R_{(N-1)(L-1)}) \end{aligned}$$

is a Gaussian random vector with the components independent of one another, with variance $N_0/2L$ and mean

$$\mu_{ij}(\mathbf{x},\tau') = \begin{cases} x_{i-1}r_{i(i-1)j}, & j \le n-1, \\ x_i r_{iin} + x_{i-1}r_{i(i-1)n}, & j = n, \\ x_i r_{iij}, & n < j \le L-1. \end{cases}$$

Then the likelihood function is

$$l_{\mathbf{x}}(\tau') = \sum_{i=0}^{N-1}\sum_{j=0}^{L-1}[R_{ij}\mu_{ij}(\mathbf{x},\tau') - \frac{1}{2}\mu_{ij}^2(\mathbf{x},\tau')]. \quad (13)$$

3 Performance

3.1 The optical channel

We first show that the error probability $Pr(\mathbf{x},\mathbf{y}|\tau')$ of deciding $\mathbf{y}$ while actually $\mathbf{x}$ is sent, using the receiver described in Section II, will theoretically reduce to the same without the timing error as the sampling rate L goes to infinity.

Let $l'_{\mathbf{z}}$ be the likelihood function without timing error ($\tau' = 0$) of sequence $\mathbf{z}$ with length N, and $l_{\mathbf{z}}$ the one with $\tau' \neq 0$; assume that $y_{-1} = x_{-1}$ and $y_N = x_N$. Then

$$\begin{aligned} Pr(\mathbf{x},\mathbf{y}|\tau') &= Pr(l_{\mathbf{y}} - l_{\mathbf{x}} \geq 0) \\ &= Pr\Bigg(\sum_{i=0}^{N-1}\sum_{j=0}^{L-1}\{K_{ij}\ln\frac{\lambda_{ij}(\mathbf{y},\tau')}{\lambda_{ij}(\mathbf{x},\tau')} \\ &\quad -[\lambda_{ij}(\mathbf{y},\tau') - \lambda_{ij}(\mathbf{x},\tau')]\} \geq 0\Bigg) \\ &= Pr\Bigg[\sum_{i=0}^{N-1}\sum_{j=0}^{L-1}K_{ij}\ln\frac{\lambda_{ij}(\mathbf{y},\tau')}{\lambda_{ij}(\mathbf{x},\tau')} \\ &\quad \geq \sum_{i=0}^{N-1}\sum_{j=0}^{L-1}\Delta\lambda_{ij}(\tau')\Bigg], \end{aligned} \tag{14}$$

while

$$\begin{aligned} Pr(\mathbf{x},\mathbf{y}|\tau' = 0) &= Pr(l'_{\mathbf{y}} - l'_{\mathbf{x}} \geq 0) \\ &= Pr\Bigg(\sum_{i=0}^{N-1}\sum_{j'=0}^{L-1}[K'_{ij'}\ln\frac{\lambda'_{ij'}(y_i)}{\lambda'_{ij'}(x_i)} \\ &\quad \geq \sum_{i=0}^{N-1}\sum_{j'=0}^{L-1}\Delta\lambda'_{ij'}]\Bigg). \end{aligned} \tag{15}$$

From Fig.1, note that

$$\begin{cases} K_{(i+1)j} &= K'_{ij'}, \\ \lambda_{(i+1)j} &= \lambda'_{ij'}, \end{cases} \quad j = 0,\cdots,n-1,$$

$$\begin{cases} K_{ij} &= K'_{ij'}, \\ \lambda_{ij} &= \lambda'_{ij'}, \end{cases} \quad j = n+1,\cdots,L-1.$$

After some simplification, we have

$$\begin{aligned} &Pr(\mathbf{x},\mathbf{y}|\tau') = \\ &Pr\{\sum_{i=0}^{N-1}\sum_{j'\neq n}[K'_{ij'}\ln\frac{\lambda'(y_i)}{\lambda'(x_i)} + K_{in}\ln\frac{\lambda_{in}(\mathbf{y},\tau')}{\lambda_{in}(\mathbf{x},\tau')}] \\ &\geq \sum_{i=0}^{N-1}[\sum_{j'\neq n}\Delta\lambda'_{ij'} + \Delta\lambda_{in}(\tau')]\}. \end{aligned} \tag{16}$$

As $T_s \rightarrow 0$ or equivalently $L \rightarrow \infty$, it can be easily shown that

$$\lim_{L\rightarrow\infty}\Delta\lambda_{in} = 0. \tag{17}$$

$$\lim_{L\rightarrow\infty}Pr(K_{in} = 0) = 1. \tag{18}$$

$$\begin{aligned} &\lim_{L\rightarrow\infty}Pr[\sum_{j'\neq n}K'_{ij'}\ln\frac{\lambda'_{ij'}(y_i)}{\lambda'_{ij'}(x_i)}] \\ &= Pr[\sum_{j'}K'_{ij'}\ln\frac{\lambda'_{ij'}(y_i)}{\lambda'_{ij'}(x_i)}]. \end{aligned}$$

$$\lim_{L\rightarrow\infty}\sum_{j'\neq n}\Delta\lambda'_{ij'} = \sum_{j'}\Delta\lambda'_{ij'}. \tag{19}$$

Hence

$$\lim_{L\rightarrow\infty}Pr(\mathbf{x},\mathbf{y}|\tau') = Pr(\mathbf{x},\mathbf{y}|\tau' = 0). \tag{20}$$

3.2 The Gaussian Channel

The performance in the Gaussian channel is much easier to analyze, because the difference Δl between $l_{\mathbf{x}}$ and $l_{\mathbf{y}}$ is also a Gaussian r.v. It can be shown that the lower bound on the error probability is

$$P_r(\mathbf{x},\mathbf{y}|\tau') \geq P_r(\Delta l_m \geq 0) = \frac{1}{2}\text{erfc}\left(\sqrt{\frac{LE_b\nu_m}{2N_0}}\right),$$

where ν_m is the mean of Gaussian r.v.$\Delta l_m = l_{\mathbf{x}} - l_{\mathbf{y}}$ with $\mathbf{x}$ and $\mathbf{y}$ differ only in the mth position.

If $p(t)$ is a rectangular (NRZ) pulse, the lower bound with arbitrary sampling rate L is

$$P_r(\mathbf{x},\mathbf{y}|\tau') \geq \frac{1}{2}\text{erfc}\left\{\sqrt{\frac{E_b}{N_0}[1 - \frac{2\tau'}{T} + 2L(\frac{\tau'}{T})^2]}\right\}, \tag{21}$$

which is the same lower bound from the output samples of the matched filter receiver in [7]. This implies that the integrate-and-dump receiver described above can potentially perform as well as the matched filter receiver (more complicated) which has correlated samples for NRZ signaling.

4 Simulation Results

In the simulation, the maximization of the likelihood function is implemented efficiently through a dynamic programming (Viterbi) algorithm τ' (M=16), using the following recursive formula:

$$l_i = l_{i-1} + \sum_{j=0}^{L-1}\left[K_{ij}\ln(1 + \frac{\mu_{ij}(\mathbf{x},\tau')}{\lambda_n T_s}) - \mu_{ij}(\mathbf{x},\tau')\right]. \tag{22}$$

The performance of the receiver with OOK signaling and BPPM is obtained by running 100,000 independent sequences of length $N = 20$ and is shown in Fig. 5. and Fig.6., which indicate that the bigger the sampling rate, the better the performance is, as expected.

5 Conclusion

The problem of sequence and timing error estimation for arbitrary baseband pulse shapes and arbitrary sampling rates in integrate-and-dump optical communications has been studied. It is shown that the error probability due to timing error reduces to the same with perfect synchronization as the sampling rate goes to infinity and a lower bound on the error probability is also derived. The general results are then applied to the popular cases of OOK and BPPM. We have also studied the same problem for the Gaussian channel with samples taken by an integrate-and-dump filter. A lower bound on the error probability is also derived, and it is shown to be the same as the one with samples taken from the matched filter for NRZ signaling.

References

[1] K. H. Mueller and M. Muller, "Timing Recovery in Digital Synchronous Data Receivers," *IEEE Transactions on Communications*, Vol.COM-24,No.5,pp. 516-531, May 1976.

[2] O. Agazzi, C. -P. J. Tzeng, D. G. Messerschmitt and D. A. Hodges, "Timing Recovery in Digital Subscriber Loops," *IEEE Transactions on Communications*, Vol. Com-33, No. 6, pp. 558-569, June 1985.

[3] F. M. Gardner, "A BPSK/QPSK Timing-Error Detector for Sampled Receivers," *IEEE Transactions on Communications*, Vol. COM-34, No. 5, May 1986.

[4] M. Oerder and H. Meyr, "Digital Filter and Square Timing Recovery," *IEEE Transactions on Communications*, Vol. COM-36, No. 5, May 1988.

[5] G. Ascheid, M. Oerder, J. Stahl and H. Meyr, "An All Digital Receiver Architecture for Bandwidth Efficient Transmission at High Data Rates," *IEEE Transactions on Communications*, Vol. COM-37, No. 8, August 1989.

[6] C. N. Georghiades, "Optimum Delay and Sequence Estimation From Incomplete Data," *IEEE Transanctions on Information Theory*, Vol. IT-36, No. 1, pp. 202-208, Jan. 1990.

[7] C. N. Georghiades and M. Moeneclaey, "Sequence Estimation and Synchronization From Nonsynchronized Samples ,"to appear in the *the IEEE Transanctions on Information Theory*, Nov. 1991.

[8] B. K. Kinard, "Sequence Decoding in the Presence of Timing Errors for NRZ Signaling," *Master's thesis*, Texas A&M University, August 1990.

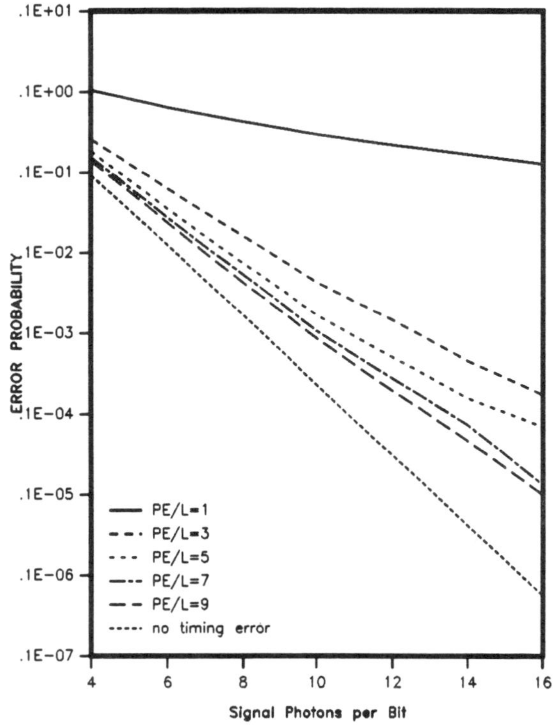

Figure 5: OOK,$\lambda_n T = 10^{-3}$ Photons/Bits

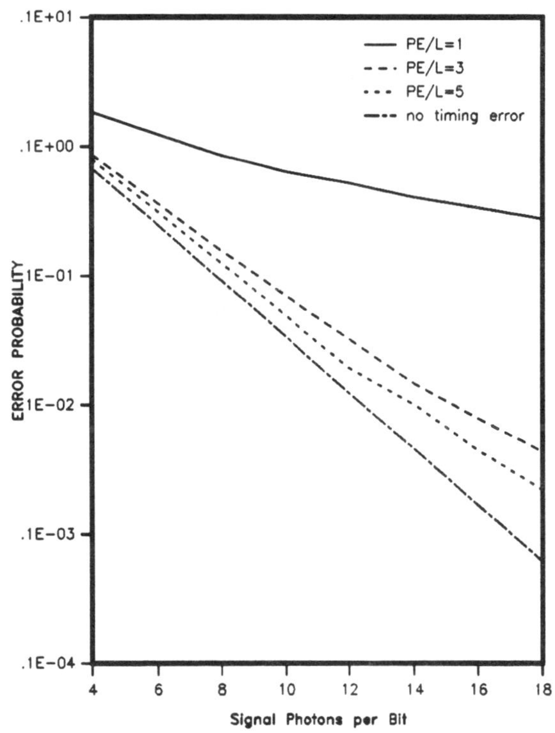

Figure 6: BPPM,$\lambda_n T' = 10^{-3}$ Photons/Bits

A Joint Source/Channel Coder Design*

Fuling Liu and Khalid Sayood
Dept. of Electrical Engineering
and the Center for Communication
& Information Science
Univ. of Nebraska, Lincoln, NE 68588-0511

Jerry D. Gibson†
Information Systems Laboratory
and the Telecommunications Program
Dept. of Electrical Engineering
Stanford Univ., Stanford, CA 94305

Abstract

Source coders and channel coders are generally designed separately without reference to each other. This approach is justified by a famous result of Shannons. However, there are many situations in practice in which the assumptions upon which this result is based are violated. Specifically, we examine the situation where there is residual redundancy at the source coder output. We have previously shown that this residual redundancy can be used to provide error correction using a Viterbi decoder. In this paper we present the second half of the design; the design of encoders for this situation. We show through simulation results that the proposed coders consistently outperform conventional source-channel coder pairs with gains of up to 12dB at high probability of error.

1 Introduction

One of Shannon's many fundamental contributions was his result that source coding and channel coding can be treated separately without any loss of performance for the overall system [1]. The basic design procedure is to select a source encoder which changes the source sequence into *iid* bits followed by a channel encoder which encodes the bits for reliable transmission over the channel. However, the separation argument no longer holds if either of the following two situations occur:

i. The input to the source decoder is different from the output of the source encoder, which happens when the link between the source encoder and source decoder is no longer error free, or

ii. The source coder output contains redundancy.

Case (i) occurs when the channel coder does not achieve zero error probability and case (ii) occurs when the source encoder is suboptimal. These two situations are common occurrences in practical systems where source or channel models are imperfectly known, complexity is a serious issue, or significant delay is not tolerable. Approaches developed for such situations are usually grouped under the general heading of joint source/channel coding.

Most joint source channel coding approaches can be classified in two main categories; (A) approaches which entail the modification of the source coder/decoder structure to reduce the effect of channel errors [2-10], and (B) approaches which examine the distribution of bits between the source and channel coders [11, 12]. The first set of approaches can be divided still further into two classes. One class of approaches examines the modification of the overall structure [2-5], while the other deals with the modification of the decoding procedure to take advantage of the redundancy in the source coder output [6-10].

In this paper we present an approach to joint source/channel coder design, which belongs to category A, and hence we explore a technique for designing joint source/channel coders, rather than ways of distributing bits between source coders and channel coders. We assume that the two nonideal situations referred to earlier are present. For a nonideal source coder, we use MAP arguments to design a decoder which takes advantage of redundancy in the source coder output to perform error correction. We then use the decoder structure to infer the encoder design.

2 The Design Criterion

For a discrete memoryless channel (DMC), let the channel input alphabet be denoted by $A =$

*This work was supported in part by NASA Lewis Research Center (NAG 3-806) and NASA Goddard Space Flight Center (NAG 5-916)

†On leave from Dept. of Electrical Engr. Texas A&M Univ.

1058-6393/91 $1.00 © 1991 IEEE

$\{a_0, a_1, \ldots, a_{M-1},\}$, and the channel input and output sequences by $Y = \{y_0, y_1, \ldots, y_{L-1}\}$ and $\hat{Y} = \{\hat{y}_0, \hat{y}_1, \ldots, \hat{y}_{L-1}\}$, respectively. If $\mathcal{A} = \{A_i\}$ is the set of sequences $A_i = \{\alpha_{i,0}, \alpha_{i,1}, \ldots, \alpha_{i,L-1}\}$, $\alpha_{i,k} \epsilon A$, then the optimum receiver (in the sense of maximizing the probability of making a correct decision) maximizes $P[C]$, where

$$P[C] = \sum_{A_i} P[C|\hat{Y}]P[\hat{Y}].$$

This in turn implies that the optimum receiver maximizes $P[C|\hat{Y}]$. When the receiver selects the output to be A_k, then $P[C|\hat{Y}] = P[Y = A_k|\hat{Y}]$. Thus, the optimum receiver selects the sequence A_k such that

$$P[Y = A_k|\hat{Y}] \geq P[Y = A_i|\hat{Y}] \ \forall_i.$$

Lemma 1

Let y_i be the input to a DMC. Given y_{i-1}, y_i is conditionally independent of $y_{n-k}, k > 1$. If $\hat{y}_0 = y_0$ then the optimum receiver selects a sequence A_i to maximize $\Pi_{i=1}^{L-1} P(y_i|y_{i-1}, \hat{Y}_i)$ where $\hat{Y}_k = \{\hat{y}_k, \hat{y}_{k+1}, \ldots, \hat{y}_{L-1}\}$.

The lemma addresses the situation in case (ii), i.e., the situation in which the source coder output (which is also the channel input sequence) contains redundancy. Using this lemma, we can design a decoder which will take advantage of dependence in the channel input sequence. The lemma provides the mathematical structure for the decoder. The physical structure can be easily obtained by examining the quantity to be maximized. The optimum decoder maximizes $P(Y|\hat{Y})$ or equivalently $\log P(Y|\hat{Y})$, but

$$\log P(Y|\hat{Y}) = \sum \log P(y_i|\hat{Y}_i, y_{i-1}) \qquad (1)$$

which is similar in form to the path metric of a convolutional decoder. Error correction using convolutional codes is made possible by explicitly limiting the possible codeword to codeword transitions, based on the previous code input and the coder structure. In this case, while there is no structure being imposed by the encoder, there is sufficient residual structure in the source coder output that can be used for error correction. This structure can be quantified in light of the Lemma. That is, the structure is reflected in the conditional probabilities, and can be utilized via the path metric in (1) in a decoder similar in structure to a convolutional decoder. However, to implement this decoder we need to be able to compute the path metric. Unfortunately the quantity $P(y_i|\hat{Y}_i, y_{i-1})$ is difficult to estimate. We have therefore used various approximations to this quantity with some success. In [8, 9] $P(y_i|\hat{Y}_i, y_{i-1})$ is approximated by $P(y_i|\hat{y}_i, y_{i-1})$ with excellent results. Other approximations can be found in [13].

In [9] we showed that the use of the decoder led to dramatic improvements under high error rate conditions. However at low error rates the performance improvement was from nonexistent to minimal. This is in contrast to standard error correcting approaches, in which the greatest performance improvements are at low error rates, with a rapid deterioration in performance at high error rates. In this work we combine the two approaches to develop a joint source channel codec which provides protection equal to the standard channel encoders at low error rates while providing significant error protection at high error rates.

3 Proposed Encoder Structure

In the conventional error protection approach we introduce structure in the transmitted bitstream. In the approach proposed in [9], we use the residual structure in the (generally nonbinary) source coder output sequence. To combine the two approaches, we need to introduce additional structure without disturbing the structure already present. Because of the nature of the decoding approach, a convolutional encoder would be most appropriate for introducing structure. However, a standard binary convolutional encoder will tend to destroy the structure in the source coder output. To preserve the residual structure while introducing additional structure we propose to use nonbinary convolutional encoders (NCE) whose input alphabet is the output alphabet of the source coder.

Let x_n, the input to the NCE, be selected from the alphabet $A = \{0, 1, 2, ..., N-1\}$, and let y_n, the output alphabet of the NCE, be selected from the alphabet $S = \{0, 1, 2, ..., M-1\}$. Then, two of the proposed NCEs can be described by the following mappings

1. $M = N^2$; $y_n = Nx_{n-1} + x_n$

The number of bits required to represent the output alphabet using a fixed length code is

$$\lceil \log_2(M) \rceil = \lceil \log_2(N^2) \rceil = \lceil 2\log_2(N) \rceil$$

Therefore in terms of rate, this coder is equivalent to a rate 1/2 convolutional encoder. The encoder memory in bits is $2\lceil \log_2(N) \rceil$ as each output value depends on two input values.

As an example, consider the situation when $N = 4$. Then $A = \{0, 1, 2, 3\}$ and $S = \{0, 1, 2, ..., 15\}$. Given the input sequence x_n : 0 1 3 0 2 1 1 0 3 3 and assuming the encoder is initialized with zeros, the output sequence will be y_n : 0 1 7 12 2 9 5 4 3 15.

The encoder memory is four bits. Notice that while the encoder output alphabet is of size N^2, at any given instant the encoder can only emit one of N different symbols as should be the case for a rate 1/2 convolutional encoder. For example if $y_{n-1} = 0$, then y_n will take on a value from $\{0, 1, 2, ..., (N-1)\}$. In general, given a value for y_{n-1}, y_n will take on a value from $\{\alpha N, \alpha N+1, \alpha N+2, ..., \alpha N+N-1\}$, where $\alpha = y_{n-1}(modN)$. This structure can be used by the decoder to provide error protection. The encoder is shown in Figure 1a.

2. $M = N^3$; $y_n = N^2 x_{2n-2} + N x_{2n-1} + x_{2n}$

The final encoder we consider is equivalent to a rate 2/3 convolutional coder. Notice that while the input output relationship looks similar to a rate 1/3 encoder, we generate one output for every two inputs. Thus, while the number of bits needed to represent one letter from the output alphabet is three times the bits needed to represent a letter from the input alphabet, because two input letters are represented by a single output letter, the rate is 2/3. Again, assuming a value of 4 for N, the output alphabet is of size 64, and for the input sequence used previously, the output sequence is y_n : 0 52 35 22 49 3.

The encoder memory is again 6 bits. A block diagram of the encoder is shown in Figure 1b. The rate of the encoder can also be inferred from the fact that while the encoder output alphabet is of size N^3, at any instant the encoder can transmit one of N^2 (instead of N) symbols. Given a value for y_{n-1}, y_n can take on a value from the alphabet $\{\gamma N^2, \gamma N^2+1, ..., \gamma N^2 + (N^2-1)\}$ where $\gamma = y_{n-1}(modN)$.

4 Binary Encoding of the NCE Output

We will make use of the residual structure in the source coder output (which is preserved in the NCE output) at the receiver. However, we can also make use of this structure in selecting binary codes for the NCE output. An intelligent assignment of binary codes can improve the error correcting performance of the system. Our strategy is to try to maximize the Hamming distance between codewords that are likely to be mistaken for one another.

First we obtain a partition of the alphabet based on the fact that given a particular value for y_{n-1}, y_n can only take on values from a subset of the full alphabet. To see this, consider the rate 1/2 NCE; then the alphabet S can be partitioned into the following sub-alphabets:

$$S_j = (jN, jN+1, ..., jN+N-1) \quad j = 0, 1, ...N-1$$

where the encoder will select letters from alphabet S_j at time n if $j = y_{n-1}(\mathrm{mod}N)$. Now for each sub-alphabet we have to pick N codewords out of M $(= N^2)$ possible choices. We first pick the sub-alphabet containing the most likely letter. The letters in the sub-alphabet are ordered according to their probability of occurrence. We assign a codeword a from the list of available codewords to the most probable symbol. Then, assign the complement of a to the next symbol on the list. Therefore the distance between the two most likely symbols in the list is $K = \lceil \log_2 M \rceil$ bits. We then pick a codeword b from the list which is at a Hamming distance of $K/2$ from a and assign it and its complement to the next two elements on the list. This process is continued with the selection of letters that are$K/2^k$ away from a at the kth step until all letters in the subalphabet have a codeword assigned to them. We then pick the sub-alphabet that contains the next most likely letter. It is assigned the available codeword at maximum distance from a. The procedure for assigning codewords within the sub-alphabet is then repeated. The assignment for a rate 1/2 with $N = 4$ code is shown in Table 1.

5 Simulation Results

The proposed approach was simulated using a two-bit DPCM system as the source coder, and the three NCE described in section 3. The source used were standard test images USC Girl, USC Couple and a 256x256 portion of Lena. The decoder structure used was that of a Viterbi decoder with branch metric log L

$$L = \frac{P(\hat{y}_i \mid y_i) P(y_i \mid y_{i-1}, y_{i-2})}{P(\hat{y}_i)}$$

where y_i denotes the NCE output and $\hat{y}_i$ denotes the corrupted channel output. The probabilities $P(y_i \mid y_{i-1}, y_{i-2})$ were estimated using a training sequence. This requires estimating MN^2 probabilities, which were estimated using the USC Girl image. The test images were the USC Couple and Lena images.

The proposed scheme was compared with a conventional source coder-convolutional coder combination. The source coder and source sequence were the same in both systems. The convolutional codes selected were the codes with maximal d_{free} and the same rate and memory characteristics as the proposed NCEs from [14]. The performance measure was the signal-to-noise-ratio (SNR) defined as

$$SNR = 10\log_{10}\frac{\sum u_i^2}{\sum (u_i - \hat{u}_i)^2}$$

where u_i is the input to the source encoder and $\hat{u}_i$ is the output of the source decoder.

The results show consistent improvement in performance for the proposed system. At low probabilities of error both systems perform very well. At high probabilities of error ($> 10^{-2}$), however, there is a substantial improvement in performance when the proposed system is used.

In Figures 2a and 2b we show the results of one of the simulations for the rate 1/2 codes. The binary assignment of Table 1 was used in the simulation. Notice the flatness of the performance curve for the proposed system. While the proposed system consistently outperforms the conventional system, it is at higher probabilities of error that the differences really become significant. At a probability of error of 10^{-1} there is almost a 6dB difference in the performance of the two systems! This "flattening out" of the performance curve makes the approach useful for a large variety of channel error conditions.

Similar performance improvements can be seen for the rate 2/3 system of the second mapping. The performance curves are shown in Figure 3. Notice that again the proposed system consistently outperforms the conventional system. In this case at a probability of error of 10^{-1} the performance improvement is more than 12dB! In fact, the proposed rate 2/3 system performs better than the conventional rate 1/2 system.

6 Conclusion

If the source and channel coder are designed in a "joint" manner, that is the design of each takes into account the overall conditions (source as well as channel statistics), we can obtain excellent performance over a wide range of channel conditions. In this paper we have presented one such design. The resulting performance improvement seems to validate this approach.

References

[1] C. E. Shannon. *Bell Syst. Tech. J.* 27:379–423, 623–656. 1948.

[2] E. Ayanoğlu and R. M. Gray. *IEEE Trans. Inform. Theory.* IT-33:855–865. Nov. 1987.

[3] T. C. Ancheta, Jr. Ph.D. dissertation, Dept. of Electrical Engr., Univ. of Notre Dame. Aug. 1977.

[4] K–Y Chang and R. W. Donaldson. *IEEE Trans. Commun.* COM-20:338–350. June 1972.

[5] D. J. Goodman and C. E. Sundberg. *Bell Syst. Tech. J.* 62:2017–203. Sept. 1983.

[6] R. C. Reininger and J. D. Gibson. *IEEE Trans. Commun.* COM-31:572–577. April 1983.

[7] R. Steele et al. *IEEE Trans. Commun.* COM-27:252–255. Jan. 1979.

[8] K. Sayood and J. C. Borkenhagen. *Proceedings IEEE ICC '86.* 1888–1892. June 1986.

[9] K. Sayood and J. C. Borkenhagen. *IEEE Transactions on Communications.* COM-39. June 1991.

[10] K. Sayood and J. D. Gibson. *Proc. 22nd Annual CISS*, Princeton, NJ. 380-385. Mar. 1988.

[11] J. W. Modestino et al. *IEEE Trans. Commun.* COM-29:1262–1274. Sept. 1981.

[12] D. Comstock and J. D. Gibson. *IEEE Trans. Commun.* COM-32:856–861. July 1984.

[13] K. Sayood, J. D. Gibson, and F. Liu. *Proc. 22nd Annual Asilomar Conference on Circuits, Systems, and Computers.* 102-106. Nov. 1988.

[14] S. Lin and D. J. Costello. *Error Control Coding.* Prentice Hall. 1983.

Table 1: Codeword Assignments

Symbol	Code	Symbol	Code
0	0000	8	1011
1	0011	9	0111
2	1100	10	0100
3	1111	11	1000
4	1110	12	0101
5	1101	13	1001
6	0001	14	1010
7	0010	15	0110

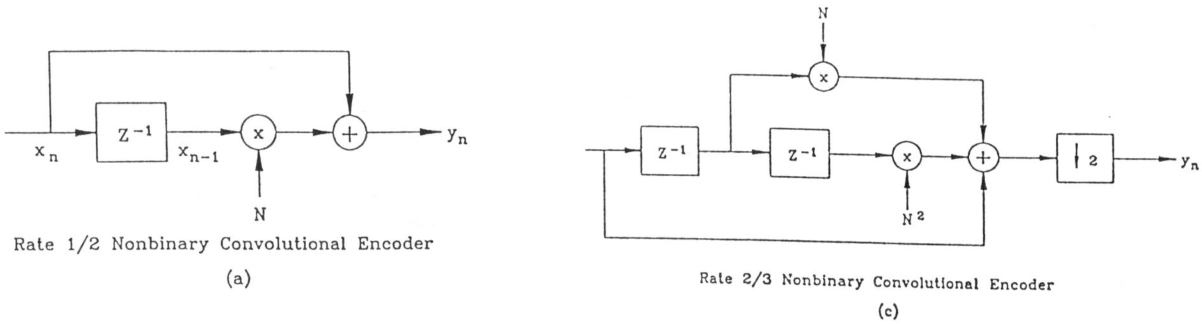

Figure 1. Proposed Nonbinary Convolutional Encoders.

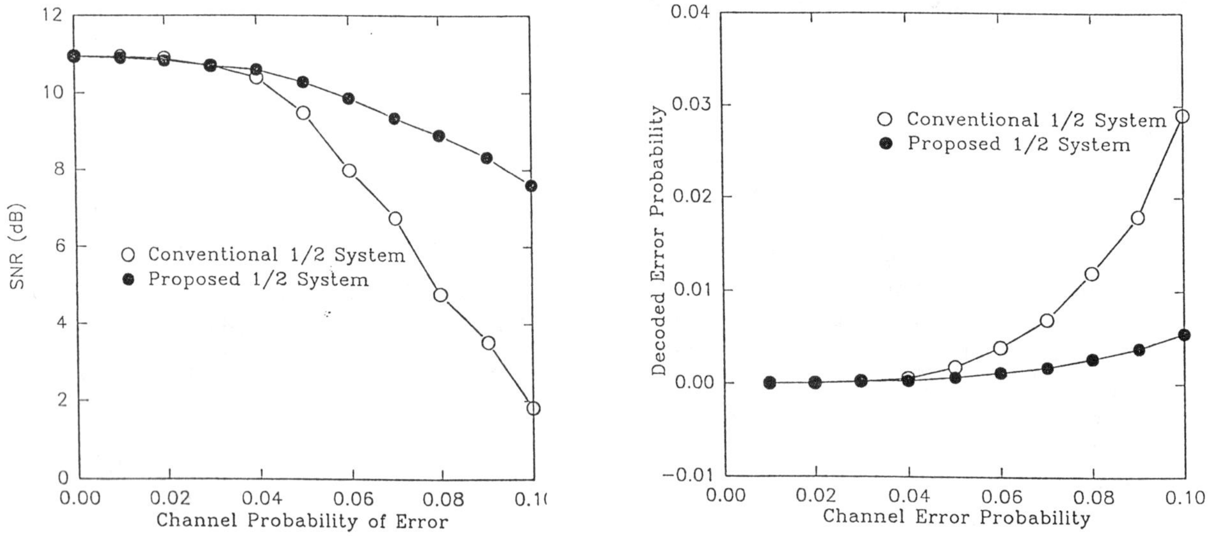

Figure 2. Performance Comparison of rate 1/2 Systems.

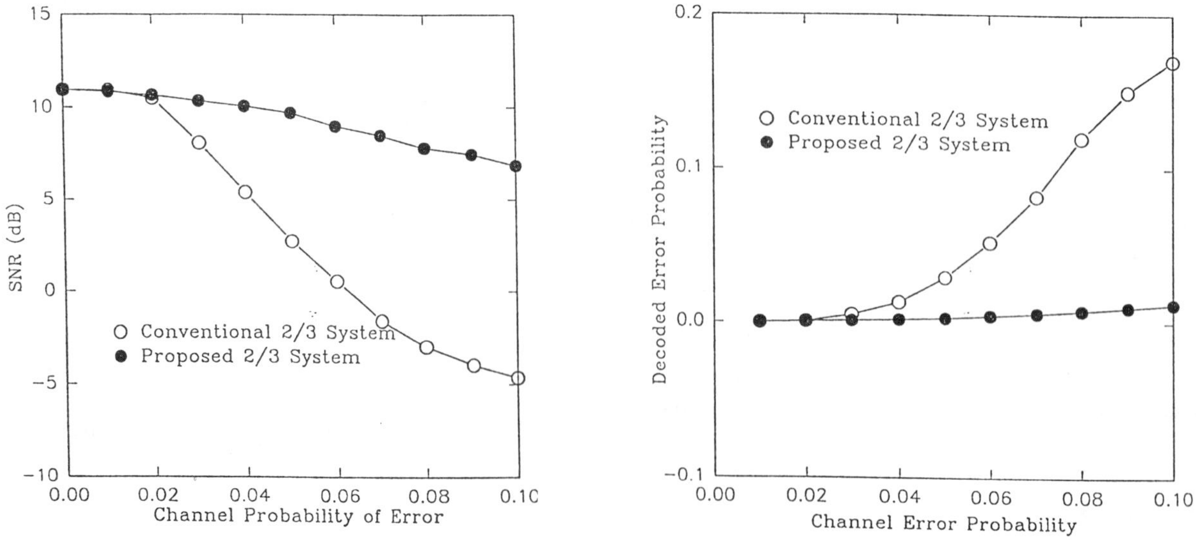

Figure 3. Performance Comparison of rate 2/3 Systems.

Robust Estimation of 3-D Motion Parameters in Presence of Correspondence Mismatches *

Subhasis Chaudhuri
Dept. of Electrical Engg.
Indian Institute of Technology
Powai, Maharashtra 400 076
INDIA

Shankar Chatterjee
Computer Laboratory for Analyzing Spatial Signals
Dept. of Elec. & Comp. Engg.
University of California, San Diego
La Jolla, Calif. 92093-0407

Abstract

Accuracy of a motion analysis scheme based on feature point correspondence is quite poor when there are mismatches between points. A small amount of mismatch (i.e., outlier data) may degrade the performance of the least-squares estimator significantly. Least median of squares (LMedS) estimator can provide the required robustness. This method works well even when almost half of the given data are outliers. A Monte-Carlo sampling technique is used to reduce the computational load. Subsequent use of the total least-squares (TLS) or the constrained least-squares (CLS) method on the trimmed data set enhances efficiency. Simulation experiments are carried out to evaluate the performance of the proposed method.

1 Introduction

The motion of a rigid object may be given by an affine transform

$$\overline{q}_i = \overline{p}_i \mathbf{R} + \overline{t}$$

where $\overline{p}_i$ is a row vector representing the coordinates of the ith point in the object, $\overline{q}_i$ is the coordinates of the same point after the motion, $\mathbf{R}$ is a 3×3 orthonormal rotation matrix, $\overline{t} = [t_x\ t_y\ t_z]$ is the translation of the object between two successive time frames. The motion understanding problem involves the estimation of the orthonormal matrix $\mathbf{R}$ and the vector $\overline{t}$ in presence of sensor noise, [1].Performances of the motion estimation methods depend on the accuracy with which the point correspondences have been established. Accuracy of these algorithms when there are mismatches in point correspondence has been investigated in [2], [3].

In range imaging, 3-D coordinates of feature points on the object are assumed to be available in an object centered coordinate system. The translation vector $\overline{t}$ can be easily estimated by noting the change in the centroid of the object in two successive frames, provided there is no occlusion. Thus, one needs to estimate only the rotation matrix $\mathbf{R}$, when the object is assumed to have undergone a rotation ψ about an axis, passing through the centroid of the object, with direction cosines (n_1, n_2, n_3). Given point correspondences, a constrained least squares estimate of $\mathbf{R}$ can be obtained [4], which imposes the orthonormality of the estimated $\mathbf{R}$ matrix. This implicitly assume the measurement errors confined to a single frame only. However, both frames are prone to measurement noise. To account for the noise in both time frames, the use of the total least squares (TLS) method has been suggested [5]. Although the TLS solution does not guarantee orthonormality, it is quite suitable to deal with an object undergoing simultaneous linear deformation (pseudo-rigid). Furthermore, the CLS method is not applicable in estimating motion parameters from visual images, but the TLS method can still be used.

Let T be a regression estimator on the data set $Z = \{z_i\}_{i=1}^{N}$, and Z' denote all possible corrupted samples obtained by replacing any m (outliers) of the original data. The maximum bias caused by such outliers is given by

$$bias_N(m; T, Z) = \sup_{Z'} ||T(Z') - T(Z)||.$$

The *breakdown point* of the estimator T is defined as,[6],

$$\epsilon_N^*(T) = \min_m \{\frac{m}{N};\ bias_N(m; T, Z) \text{ is infinite}\}.$$

and is independent of the probability distribution of the outlier. It may be noted that for the least squares method, a single bad outlier is sufficient to 'break down' the estimator ($i.e.$, $\epsilon^* = 1/N$). Numerical studies about the degradation of the performance of the LS estimators in presence of feature point mismatch have been reported in [2], [3].

*Partially supported by UCSD University Opportunity Fund #560-823-07427 and a grant from Naval Ocean Systems Center, San Diego, under contract #N66001-87-D-0136.

1058-6393/91 $1.00 © 1991 IEEE

Recently, Rousseeuw has suggested a new robust paradigm for linear regression, [6], [7], called the least median of squares (LMedS) estimator. It has been used in cluster analysis [8], in obtaining robust estimate of an autoregressive model [9], etc. The LMedS method has an asymptotic breakdown point $\epsilon^* = 0.5$, which is the maximum possible value that an estimator can achieve. Another advantage of this method is that one does not have to construct any influence function.

The organization of this paper is as follows. Least squares based methods are briefly presented in section 2. These LS methods will be used in conjunction with the LMedS estimator in section 3 to derive the motion analysis scheme. A brief sketch of the LMedS estimator, and the computational complexity of this method are presented in section 3. A Monte-Carlo sampling technique is used to speed up the computation. The results of the simulation experiments for range data as well as visual images are given in section 4. The merits and limitations of the proposed motion analysis scheme are discussed in section 5.

2 LS Estimation of Motion Parameters

2.1 Range Data

Let N be the number of feature points being considered. Assuming an object centered coordinate system, the motion equation can be given by $Q = P\mathbf{R} + \delta Q$, where Q and P are $N \times 3$ matrices representing the coordinates of the feature points in two successive frames, and δQ is the corresponding perturbations in the Q matrix. An LS estimate of the rotation matrix is given by $\hat{\mathbf{R}} = (P^TP)^{-1}P^TQ$. For a rigid object, the CLS estimate of the rotation matrix is given by

$$\hat{\mathbf{R}}_{CLS} = VU^T, \quad P^TQ = U\Sigma V^T \tag{1}$$

When the data in both frames are corrupted by sensor noise, we may use the total least squares (TLS) method proposed in [10], [11], [12]. The closed-form solution can be obtained as follows [13], [5] : obtain a singular value decomposition (SVD) of the augmented matrix $[P \mid Q]$

$$[P \mid Q] = U\Sigma V^T. \tag{2}$$

Here U and V are of dimensions $N \times 6$ and 6×6 respectively. For the noisy observation, a rank-3 approximation is considered. We may write V in the partitioned form

$$V = \begin{bmatrix} V_{11} & V_{12} \\ V_{21} & V_{22} \end{bmatrix}, \quad \text{each } V_{i,j} \text{ is } 3 \times 3$$

The TLS solution is given by

$$\hat{\mathbf{R}}_{TLS} = -V_{12}V_{22}^{-1}. \tag{3}$$

Constrained TLS is identical to the CLS under the assumption that the error covariance matrices in both frames have identical structure.

2.2 Perspective Projection

Let (x, y, z) and (x', y', z') represent the coordinates of the 3-D feature points, and (X, Y) and (X', Y') represent the corresponding points on the image plane in two successive time frames respectively. From geometry, we may obtain, [14]

$$[X' \; Y' \; 1] \, E \, [X \; Y \; 1]^T = 0 \tag{4}$$

where $E = G\mathbf{R}$, and

$$G = \begin{bmatrix} 0 & -t_z & t_y \\ t_z & 0 & -t_x \\ -t_y & t_x & 0 \end{bmatrix}.$$

Eq.(4) is linear and homogeneous in the nine unknowns in E matrix. Thus, the translation parameters can be estimated only up to a scale factor. Denoting $\alpha_j = \sqrt{(X_j^2 + Y_j^2 + 1)(X_j'^2 + Y_j'^2 + 1)}$, eq.(4) for each feature point, j, can be explicitly represented as

$$\frac{1}{\alpha_j} [X_jX_j' \; X_jY_j' \; X_j \; Y_jX_j' \; Y_jY_j' \; Y_j \; X_j' \; Y_j' \; 1] \cdot [e_1 \; e_2 \; e_3 \; e_4 \; e_5 \; e_6 \; e_7 \; e_8 \; e_9]^T = 0, \tag{5}$$

where e_i's are the elements of the E matrix.This can be written as $A\bar{e} = 0$, where A is an $N \times 9$ matrix. The LS solution involves perturbing the last column of the equation only so that it lies within the span of the data in first eight columns. The TLS solution acknowledges the possible presence of perturbations in all nine columns of the matrix A. If $A = U\Sigma V^T$, then $\hat{\bar{e}}_{TLS}$ is given by the right singular vector corresponding to the smallest singular value.

3 Least Median of Squares Estimate

Let $z_i = \{\bar{p}_i, \bar{q}_i\}$ be the measurements of the ith feature point on the object. If $\hat{\mathbf{R}}$ is an estimate of the rotation matrix, the residual for the ith feature point is defined as $r_i = |\bar{q}_i - \bar{p}_i\hat{\mathbf{R}}|$ (or $r_i = [X_i' \; Y_i' \; 1]\hat{E}[X_i \; Y_i \; 1]^T/\alpha_i$ for visual image data).

In LMedS methods, the parameter θ ($\mathbf{R}$ or E matrix) is estimated by solving

$$\phi(\theta) = \min_{\hat{\theta}} \; \text{med}_i \; r_i^2 \quad i = 1, 2, ..., N. \tag{6}$$

This estimator yields the smallest value of the median of the squared residuals for the entire data set. However, there is no analytic solution to eq. (6).

To solve the above, distinct k-tuples of data are chosen from the given data set, $\{z_i\}_{i=1}^N$, where $k \geq dim(\theta)$. There are $\binom{N}{k}$ different choices. A trial estimate of θ is computed for each. If $k = dim(\theta)$, then an exact solution for the parameter vector θ can be computed. Otherwise, the trial estimate θ may be computed in a least squares sense. Here we use CLS for the range data and TLS for the image data. The median of the squared residuals $r_i(\theta)$ for all the feature points is obtained. Then, for all possible k-tuples, one finds the smallest value of the median of residuals. The trial estimate that corresponds to the LMedS residual is the true LMedS estimate.

3.1 Nonuniquness of LMedS

It is interesting to note that, the LMedS estimator may *not* yield a unique solution, even in absence of any outlier or sensor noise. If *at least* half of the data points ($m \geq \lfloor N/2 \rfloor + 1$) lie in a hyperplane having dimension less than that of the parameter vector θ, then there is no unique estimate for θ. This is due to the fact that for each k-tuple chosen from such a subset of observations, there is no unique solution to the trial estimate as the observation matrix is rank deficient. This is analogous to the notion of sufficiently exciting (SE) signal in control theory. However, such a singular case in computer vision is unlikely.

3.2 Computational Issues

Computational complexity of the LMedS estimator is of the order of $O(N^{k+1})$. And the memory requirement is of the order of $(dim(\theta)+1)O(N^k)$. Space and time efficient algorithms for the LMedS estimator have been suggested in [15]. These algorithms deal with fitting a straight line through the given set of data.

The computation of the LMedS estimator for higher dimension, may be prohibitive. It increases exponentially with the increase in the size of k-tuple. In practice, one resorts to a Monte-Carlo sampling.

3.3 Sampling by Monte-Carlo Method

Here the probability of error P_e refers to the likelihood that one does not find the true estimate. One randomly choses a select number of samples from the given observations. If $\delta (< 0.5)$ is the fraction of data having correspondence mismatches, and M is the number of random sampling of k-tuples of data, then, [16],

$$P_c = 1 - P_e = 1 - [1 - \frac{\binom{N(1-\delta)}{k}}{\binom{N}{k}}]^M$$
$$\approx 1 - (1-(1-\delta)^k)^M, \text{for } k \ll N, \quad (7)$$

where P_c is the probability that at least one k-tuple chosen from M samples is free from outliers. The number of random sampling M required for a given amount of outlier contamination (δ) and for a particular choice of P_c is given in Tables 1 and 2. These tables indicate that one actually has to compute only a small number of trial solutions even for a very high value of P_c.

To improve the efficiency of the LMedS estimator [7], we select $N/2$ data points for which the residuals are the smallest using the LMedS estimate computed as above. The other half of the data may be considered to be outliers in the worst case (*i.e.*, $\delta = 0.5$). This requires the sorting ($O(N \log N)$) of the half of the residuals. The corresponding estimate may be called the *trimmed least median of squares* (TLMedS) estimate.

4 Simulation and Results

For the range data, an arbitrary shaped 3-D object consisting of a total of 50 feature points was generated using a random number generator. We assume that the points which could be placed anywhere on the object are visible at both instants of time (e.g., a wireframe, transparent object). A Gaussian random no. generator with zero mean and standard deviation 10 was used. The object was then rotated by an angle 30° around an axis passing through the origin with direction cosines (0.6040 0.7198 0.3421),[17]. We added independent Gaussian random noise with zero mean and varying standard deviations (σ) to the object points in both frames. A certain fraction (δ) of the given feature points were randomly permuted to obtain a mismatch in feature point correspondence. The CLS method as described in section 2 was used to estimate the motion parameters. The results for the non-robust CLS estimator for different values of σ are given in Table 3 (*CASE* A). Then we used a 4-tuple of observations at a time, and obtained the LMedS estimate (*CASE* B). Next, results in *CASE* B were chosen as the initial estimate, the residuals were sorted according to their magnitude, and only the smallest half of the residuals were retained. The CLS scheme was then used on the trimmed data set to obtain the corresponding TLMedS estimate (*CASE* C). To show that the LMedS estimates do not depend on the magnitude of the outliers, the coordinates of feature points in the first frame in *CASE* A, were multiplied by a factor of 10. The corresponding CLS estimates (*CASE* Ã) do not at all resemble the actual values of the parameters. However the corresponding LMedS and the TLMedS estimates remain unchanged.

For the results in Table 3, only 50 Monte-Carlo trials have been used, and the true parameters were recovered. The amount of mismatches in point correspondence was then increased to $\delta = 0.3$. The results are given in Table 4. Only 200 Monte-Carlo trials were performed, and the motion parameters were correctly recovered.

For perspective projection case, we used the same data set, but shifted along the z-direction so that the depth is always positive (in both frames). The object underwent a rotational as well as a translational motion. The results are given in Tables 5 and 6. P_e tolerated in Monte-Carlo sampling was less than 0.01%.

It may be seen from the results in Tables 5 and 6 that the non-robust estimate of the essential matrix is very poor in presence of correspondence mismatches. Although the correct parameters could be extracted using the proposed LMedS scheme, the estimates are quite sensitive to the sensor noise. The reason is that the elements of the estimated E matrix are not independent [17].

5 Conclusion

Since the pre-processing step of establishing the feature (point) correspondence between successive time frames is a very difficult task, one usually achieves such correspondence with only limited accuracy. By virtue of having a very high breakdown point, the LMedS estimator provides the much desired robustness against the correspondence mismatches. The applicability and the design procedure of a robust algorithm for the processing of a long sequence of motion data are currently under investigation.

References

[1] J. Weng, T.S. Huang and N.Ahuja, "Motion and Structure from Two Perspective Views: Algorithms, Error Analysis and Error Estimation", *IEEE Trans. Pattern Anal. Machine Intell.*, PAMI-11:451–476, May 1989.

[2] R. Haralick, H. Joo, C. Lee, X. Zhuang, V. Vaidya and M. Kim, "Pose Estimation from Corresponding Point Data", In H. Freeman, editor, *Machine Vision for Inspection and Measurement*, pages 1–84. Academic Press, Boston, 1989.

[3] S. Chaudhuri and S. Chatterjee, "On Analyzing the Performance of Least Squares Methods in Motion Estimation", In *Proc. 23rd Asilomar Conf. on Signals, Systems and Computer*, Monterey, CA, Nov., 1989.

[4] K.S. Arun, T.S. Huang and S.D. Blostein, "Least-Squares Fitting of two 3-D Point Sets", *IEEE Trans. Pattern Anal. Machine Intell.*, PAMI-9:698–700, September 1987.

[5] S. Chaudhuri and S. Chatterjee, "Estimation of Motion Parameters for a Deformable Object from Range Data", In *Proc. IEEE conf. on Computer Vision and Pattern Recognition*, San Diego, CA, 1989.

[6] P. Rousseeuw and A. Leroy, *Robust Regression and Outlier Detection*, John Wiley, New York, 1987.

[7] P. Rousseeuw, "Least Median of Squares Regression", *J. Am. Stat. Assn.*, 79:871–880, 1984.

[8] D. Kim, J. Kim, P. Meer, D. Mintz and A. Rozenfeld, "Robust Computer Vision: A Least Median of Squares Based Approach", In *Proc. DARPA Image Understanding Wksp*, pages 1117–1134, 1989.

[9] S. Blostein and H. Richardson, "Multidimensional Autoregressive Parameter Estimation Using Iteratively Reweighted Least Squares", In *Proc. IEEE Intl. Conf. on ASSP*, Albuquerque, NM, May 1990.

[10] G.H. Golub and C.F. Van Loan, "An Analysis of the Total Least Squares Problem", *SIAM J. Numer. Anal.*, 17:883–893, 1981.

[11] L.J. Gleser, "Estimation in a Multivariate Error in Variables Regression Model: Large Sample Results", *Ann. Statist.*, 9:22–44, 1981.

[12] S. van Huffel and J. Vandewalle, "The Total Least Squares Technique: Computation, Properties and Applications", In E.F. Deprettere, editor, *SVD and Signal Processing*, pages 189–208, North-Holland, Amsterdam, 1988.

[13] G.H. Golub and C.F. Van Loan, *Matrix Computations*, Johns Hopkins, Baltimore, 1987.

[14] T.S. Huang, "Determining 3-D Motion and Structure from Two Perspective Views", In T. Young and K.S. Fu, editors, *Handbook of Pattern Recognition and Image Processing*, pages 334–354, Academic Press, Orlando, 1986.

[15] H. Edelsbrunner and D. Souvaine, "Computing Least Median of Squares Regression Lines and Guided Topological Sweep", *J. Am. Stat. Assn.*, 85:115–119, 1990.

[16] S. Chaudhuri, *Feature Point Based Methods for the Estimation of 3-D Motion Parameters*, PhD thesis, University of California, San Diego, September, 1990.

[17] R.Y. Tsai and T.S. Huang, "Uniqueness and Estimation of Three-Dimensional Motion Parameters of Rigid Objects with Curved Surfaces", *IEEE Trans Pattern Anal. Machine Intell.*, PAMI-6:13–27, January 1984.

δ	k	M			
		$P_c = 96\%$	$P_c = 99\%$	$P_c = 99.9\%$	$P_c = 99.99\%$
0.12	4	4	6	8	11
	6	6	8	12	15
0.30	4	12	17	26	34
	6	26	37	56	74

Table 1: Number of random sampling required for the motion estimation from range data.

δ	k	M			
		$P_c = 96\%$	$P_c = 99\%$	$P_c = 99.9\%$	$P_c = 99.99\%$
0.12	10	10	15	22	29
	12	14	19	29	38
0.30	10	113	161	242	322
	12	231	331	496	661

Table 2: Number of random sampling required for the motion estimation from perspective projection data.

$M = 50$, $\delta = 0.12$, $k = 4$, CPU time=0.2s

θ_{orig}=(0.6040, 0.7198, 0.3421, 30°)

σ	CASE	n_1	n_2	n_3	ψ
0.0	A	0.6439	0.6605	0.3861	35.4
	Ã	0.4864	0.6285	-0.6070	24.9
	B	0.6040	0.7198	0.3421	30.0
	C	0.6040	0.7198	0.3421	30.0
0.3	A	0.6465	0.6568	0.3880	35.0
	Ã	0.4857	0.6287	-0.6073	24.6
	B	0.6017	0.7170	0.3519	29.8
	C	0.6033	0.7180	0.3471	29.6
1.0	A	0.6531	0.6476	0.3926	34.1
	Ã	0.4843	0.6292	-0.6079	24.1
	B	0.5960	0.7102	0.3747	29.2
	C	0.6016	0.7137	0.3588	28.6

Table 3: Estimates of the motion parameters for the range data in presence correspondence mismatches. Here, A) CLS, Ã) as in (A) but large outliers, B) LMedS, and C) TLMedS estimates.

$M = 200$, $\delta = 0.30$, $k = 4$, CPU time=0.9s

θ_{orig}=(0.6040, 0.7198, 0.3421, 30°)

σ	CASE	n_1	n_2	n_3	ψ
0.0	A	0.5576	0.7693	0.3120	37.1
	B	0.6040	0.7198	0.3421	30.0
	C	0.6040	0.7198	0.3421	30.0
0.3	A	0.5590	0.7673	0.3143	36.6
	B	0.5951	0.7268	0.3428	30.2
	C	0.6070	0.7143	0.3482	29.9
1.0	A	0.5629	0.7622	0.3197	35.4
	B	0.5739	0.7431	0.3442	30.9
	C	0.6139	0.7014	0.3621	29.6

Table 4: Estimates of the motion parameters for the range data when there are more outliers in the data.

$M = 100$, $\delta = 0.12$, $k = 10$, CPU time=1.8s

θ_{orig}=(0.1928, -0.6443, 0.0862, 0.4670, 0.1906, -0.3749, -0.3191, 0.1135, 0.1749)

σ	CASE	e_1	e_2	e_3	e_4	e_5	e_6	e_7	e_8	e_9
0.00	A	.0489	.5369	.4373	-.5946	-.0688	.1074	-.3410	.0759	.1619
	B	.1928	-.6443	.0862	.4670	.1906	-.3749	-.3191	.1135	.1749
	C	.1928	-.6443	.0862	.4670	.1906	-.3749	-.3191	.1135	.1749
0.03	A	.0493	.5366	.4366	-.5948	-.0685	.1075	-.3418	.0762	.1621
	B	.1863	-.6445	.0714	.4673	.1942	-.3859	-.3056	.1298	.1722
	C	.1866	-.6466	.0703	.4717	.1933	-.3815	-.3042	.1233	.1701
0.10	A	.0502	.5360	.4349	-.5954	-.0678	.1077	-.3435	.0768	.1626
	B	.1754	-.6619	.0340	.5099	.1961	-.3607	-.2665	.0956	.1457
	C	.1717	-.6656	.0260	.5169	.1951	-.3568	-.2576	.0904	.1406

Table 5: Estimates of the essential parameters for visual image data in presence of correspondence mismatches.

$M = 500$, $\delta = 0.30$, $k = 10$, CPU time=8.9s

θ_{orig}=(0.1928, -0.6443, 0.0862, 0.4670, 0.1906, -0.3749, -0.3191, 0.1135, 0.1749)

σ	CASE	e_1	e_2	e_3	e_4	e_5	e_6	e_7	e_8	e_9
0.00	A	-.0727	.6925	.3066	-.5831	-.1542	.2124	-.0980	-.0143	.0499
	B	.1928	-.6443	.0862	.4670	.1906	-.3749	-.3191	.1135	.1749
	C	.1928	-.6443	.0862	.4670	.1906	-.3749	-.3191	.1135	.1749
0.03	A	-.0721	.6926	.3064	-.5831	-.1542	.2121	-.0986	-.0145	.0501
	B	.1835	-.6535	.0522	.4856	.1932	-.3767	-.2868	.1160	.1607
	C	.1803	-.6554	.0392	.4913	.1950	-.3788	-.2740	.1194	.1553
0.10	A	-.0707	.6928	.3061	-.5831	-.1541	.2116	-.1000	-.0147	.0507
	B	.1524	-.6633	.0496	.5211	.2017	-.3911	-.1856	.1428	.1199
	C	.1579	-.6617	.0393	.5190	.2013	-.3898	-.1967	.1403	.1241

Table 6: Estimates of the essential parameters for visual image data when there are more outliers in the data.

Image Analysis by Information Decomposition and Synthesis*

James Shih-Jong Lee,
Center of Bio-Engr.
Mail Stop: FT-10
University of Washington
Seattle, WA 98195

Jenq-Neng Hwang, Eric W. Kelm
Information Processing Laboratory
Dept. of Electrical Engr., FT-10
University of Washington
Seattle, WA 98195

Abstract

Useful and irrelevant (or noise) information are often mixed together in an image and it is difficult for computer to remove the huge volume of irrelevant or noise data while retaining enough useful information. To provide the foundation for a good solution to this problem, "multiresolution decomposition and synthesis" image representation approaches have been developed. These approaches linearly or nonlinearly decompose the raw image data into a set of partial information channels, each channel reflects certain aspect (modality) of the information. The partial information can thus be processed independently or cooperatively. The processed decompositions can be selectively synthesized to recover the original image, to construct an improved image, or to highlight certain features. In this manner, useful information can be accessed/utilized/retained/modified and irrelevant information be ignored/rejected effectively. In this paper, we compare the image analysis capability using either the linear Gaussian pyramid and nonlinear morphological pyramid image decomposition/synthesis techniques, with emphasis on image enhancement.

1 Introduction

There are increasingly more medical, scientific, and engineering applications based on image processing and visualization of multidimensional data. In medical area, digital 3D imaging equipment is by now widely available in hospitals to provide non-invasive imaging of body structure and function. Clinical routine tasks drawn upon multidimensional imagery of human body include diagnostic evidedence gathering and decision making, surgical planning, radiation therapy planning, prognosis estimation and follow-up study assessment, etc. These tasks require complex human spatial reasoning abilities, such as mental representation of spatial morphology; mental recovery of spatial organ configurations and spatio-temporal anatomical and pathological events from images; reasoning and planning in space and time; mental imaging of spatial macroworld of the human body; comprehension of multidimensional structure-to-function causality; inference of spatial lesion loci from multidimensional images.

*This project was supported in part by a grant 1-F37-LM00006-02 from the National Library of Medicine.

In most image processing and analysis applications, data representation, which is often overlooked, is as critical as the algorithms applied to the data. Simple algorithms applied to proper representations could be more powerful than sophisticated algorithms applied to improper representations. Multiresolution decomposition and synthesis image representation approaches have thus been developed, which linearly/nonlinearly decompose the raw image data into a set of partial information channels, each channel reflects certain aspect (modality) of the information. The partial information can thus be processed independently or cooperatively. The processed decompositions can be selectively synthesized to recover the original image, to construct an improved image, or to highlight certain features. In this manner, useful information can be accessed/utilized/retained/modified and irrelevant information be ignored/rejected effectively.

There is an evidence that the early processing in human visual system uses a similar multiresolution representation [2]. By selectively decomposing the data and enhancing certain information, details which are not obvious can be highlighted and subsequently be extracted by relatively straightforward processes. Note that the decomposition is a complete representation of the data, namely, the original raw data can be exactly recovered from this representation.

The importance of analyzing images at many scales arises from the nature of images themselves. Since the world contains objects of many sizes, and these objects contain features of many sizes. In addition, objects can be at various distances from the viewer. As a result, any analysis procedure that is applied only at a single scale may miss information at other scales. The solution is to isolate features at different scales and to carry out analysis at all scales simultaneously. An object represented by a combination of multiscale data can be analyzed by selectively correlating these data. The multiscale representation provides a unique way to associate local information with global context by correlation between data of different scales. Thus, local features can be enhanced differently depending

1058-6393/91 $1.00 © 1991 IEEE

upon their global context encoded in the coarse representations.

In this paper, we studied the image enhancement capability for removing the structured/unstructured noise using both the linear Gaussian pyramid [1] and nonlinear morphological pyramid [3, 6] image decomposition techniques.

2 Multiresolution Image Decompositions

The multiresolution image representation technique is an iterative procedure designed to isolate image features at different scales and orientations, so as to support efficient scaled neighborhood operations through reduced image representation. The first image representation is a lowpass decomposition. In this decomposition, the original image $\mathbf{I} = \mathbf{L}_0$, which represents the zeroth level of the procedure, is (linearly or morphologically) lowpass filtered ($F(\cdot)$) and down-sampled ($D(\cdot)$, usually by a factor of two in each dimension) to obtain the next decomposition level, $\mathbf{L}_1$. $\mathbf{L}_1$ is then filtered and down-sampled in the same way to obtain $\mathbf{L}_2$. Further repetitions of the filter/down-sample steps generate the remaining lowpass decomposition levels $\{\mathbf{L}_i\}$ (see Figure 1):

$$\mathbf{L}_i = D(F(\mathbf{L}_{i-1})), \quad i \geq 1 \qquad (1)$$

The lowpass filter can be a linear convolution filters or nonlinear morphological filters such as dilation, erosion, opening, closing, etc. [4]. When the $F(\cdot)$ is a linear Gaussian convolution, the above technique is equivalent to the well known Gaussian pyramid data structure [1]. Nonlinear morphological filters are also gradually gaining their popularity in the multiresolution decomposition applications [3, 6].

The second image representation is a bandpass decomposition. In this decomposition, the bandpass component, $\mathbf{B}_i$, of each image can be generated by subtracting each lowpass decomposition level, $\mathbf{L}_i$, from the next lower level in the decomposition. Because these levels differ in their size, it is necessary to interpolate new sample values between those in a given level before that level is subtracted from the next lower level. The interpolation can be achieved by point replication ($U(\cdot)$) followed by a linear lowpass filtering ($P(\cdot)$) (see Figure 1).

$$\mathbf{B}_i = \mathbf{L}_i - P(U(\mathbf{L}_{i+1})), \quad i \geq 0 \qquad (2)$$

Decompositions provide insight into the data in different perspectives. As the bandpass decompositions reveal critical components of the image patterns at different scales for analysis, these components can be further decomposed into different orientations. In this way, the information of different directions can be isolated at multiple scales.

Given a multidimensional image, this decomposition approach decomposes the image into different bands in spatial frequency, orientation, or morphological pattern spectrum [5]. For example, the linear bandpass images, as with the Fourier transform, represent pattern components that are restricted in the spatial-frequency domain. But unlike the Fourier transform, these bandpass images are restricted to local volumes in the spatial domain. Spatial as well as spatial-frequency localization can be critical in the analysis of images that contain multiple objects.

3 Multiresolution Image Synthesis and Enhancement

An important property of these decompositions is that it is a complete image representation, i.e., the step used to construct the decomposition may be reversed to synthesize the original image exactly. To recover $\mathbf{L}_i$, the lowpass decomposition level $\mathbf{L}_{i+1}$ is interpolated and added to $\mathbf{B}_i$. This procedure can be repeated to recover $\mathbf{L}_{i-1}$, $\mathbf{L}_{i-2}$, and so on until the original image, $\mathbf{L}_0 = \mathbf{I}$, is recovered (see Figure 3).

$$\mathbf{L}_i = \mathbf{B}_i + P(U(\mathbf{L}_{i+1})), \quad i \geq 0 \qquad (3)$$

As the original image can be exactly recovered by image synthesis from the lowpass and bandpass decompositions, a modified image can be constructed by the synthesis of the modified lowpass and bandpass components. The modification can be performed to enhance features at desirable bands and directions and to reject undesirable artifact or to exclude certain undesirable information. As an example, highpass filtered images of different scales can be generated by selectively excluding low frequency components of the images. This can be accomplished by selective synthesis strategy. The highpass images of small scale can enhance local image features such as peaks, pits, ridges, valleys, and the boundaries of objects. As the scale of the highpass filtering increases, the objects themselves are enhanced.

3.1 Independent Enhancement Then Synthesis:

As shown in Figure 4, the modification can be done independently at each band and then synthesis. In this way, information can be modified differently at different scales. There is an evidence that the human visual system processes spatial-frequency (or pattern spectrum) information independently in adjacent bands. Application examples of the independent enhancement then synthesis strategy include multiband local contrast enhancement, multiband noise coring, multiband interpolation, and edge preserved averaging.

3.2 Between Band Correlation Then Synthesis:

As the bandpass decomposition provides both spatial and spatial-frequency (or pattern spectrum) localizations, spatial neighborhood operators can be applied to different spatial-frequency bands for detections and enhancements. An even more interesting possibility is to perform operations between bandpass channels of the same image (see Figure 5). The correlation strategies include a *weakening strategy* and an *enhancement strategy*.

A Weakening Strategy: Given a frequency band, a weakening strategy constrains the data intensity of a bandpass image by taking the minimum of its own pixel values and those of the (lower) adjacent bands to smooth out the bright noise. The adjacent bands can be preprocessed by neighborhood operations such as morphological dilations before the correlation. In the case of using $\mathbf{B}_{i+1}$ to correlate $\mathbf{B}_i$, we expand $\mathbf{B}_{i+1}$ and dilate it by an appropriate structuring element before performing the minimum operation with $\mathbf{B}_i$:

$$\mathbf{B}_i^w = \mathrm{MIN}(\mathbf{B}_i,\ \mathrm{DILATE}(U(\mathbf{B}_{i+1})) \qquad (4)$$

In this way, the value in $\mathbf{B}_i$ is constrained to the dilated value of the expanded $\mathbf{B}_{i+1}$.

Similarly, the correlation of $\mathbf{B}_i$ can also be performed using the adjacent higher frequency bands $\mathbf{B}_{i-1}$. More specifically, the weakening operations can be applied iteratively from the coarse (fine) resolutions to the fine (coarse) resolutions.

An Enhancement Strategy: Image features can be enhanced by band correlation in a similar fashion. Given a frequency band, an enhancement strategy increases the data intensities by a constrained dilation operation. The constraint is imposed by correlating its dilated image with the data from its adjacent bands by the following rule:

$$\mathbf{B}_i^e = \mathrm{MAX}(\mathbf{B}_i,\ \mathrm{DILATE}(U(\mathbf{B}_{i+1})) \qquad (5)$$

The enhancement strategy picks up the components in band $\mathbf{B}_{i+1}$ having intensity values higher than the intensity value in band $\mathbf{B}_i$. To further constrain the data for enhancement, only the data within the spatial dilation of band $\mathbf{B}_i$ are enhanced. The enhancement can be done by a maximum operation or by adding a fraction of the enhancement component to band $\mathbf{B}_i$. In addition, the enhancement correlation can be performed using the adjacent higher frequency band $\mathbf{B}_{i-1}$ as well. As the weakening operations, the enhancement operations can be applied iteratively from the coarse (fine) resolutions to the fine (coarse) resolutions.

In many applications, weakening and enhancement operations can be applied alternatively before image synthesis is performed to selectively detect and enhance desirable features and patterns in the image and suppress the undesirable information. Note that the strategies we describe here only represent a subset of the much richer possibilities in selective image enhancements.

4 Experimental Results in Image Enhancement

For our experiments, we used the weakening strategy of the "between band correlation then synthesis" approach to remove two types of noise: the unstructured noise (e.g., the random sparkle noise) and the structured noise (e.g., the high frequency imaging artifact) from the image which has been decomposed either by morphological pyramid or Gaussian pyramid techniques.

As shown in Figure 6(a), the original "Lenna" image was contaminated with white sparkle noise. The noise was successfully removed by the morphological pyramid band decomposition/synthesis approach based on opening operation with a flat 3×3 cross structure element (see Figure 6(b)). In contrast, the Gaussian pyramid approach with 5×5 Gaussian kernel was not entirely successful in removing the same noise (see Figure 6(c)). Similar result can also be seen in using both approaches for removing the structured noise. Figure 7(a) shows a 3D data of *guinea pig cochlea* contaminated with a high-frequency artifact (5 pixel wide stripe of noise). The noise was successfully removed by two consecutive morphological pyramid decomposition/synthesis procedures based on first opening with a flat 3×3 cross structure element then opening with another flat 5×5 cross structure element after synthesized from between band correlation (see Figure 7(b)). In contrast, the Gaussian pyramid approach with 5×5 Gaussian kernel was not entirely successful in removing the same noise (see Figure 7(c)).

References

[1] P. J. Burt and E. Adelson. The Laplacian pyramid as a compact image code. *IEEE Trans on Communication*, 31:532-540, 1983.

[2] F. W. Campbell, J. G. Robson. Application of Fourier analysis to the visibility of gratings. *J. of Physiol. (London)*, 197:551-558, 1968.

[3] R. M. Haralick, C. Lin, J. S. J. Lee, and X. Zhuang. Multiresolution morphology. *IEEE First Int'l Conf. on Computer Vision*, pp. 516-520, 1987.

[4] J. S. J. Lee, R. M. Haralick, and L. G. Shapiro. Morphologic edge detection. *IEEE Trans. Robotics and Automation*, 3(2):142-156, 1987.

[5] P. Maragos. Pattern spectrum and multiscale shape representation. *IEEE Trans. on Pattern Analysis and Machine Intelligence*, 11:701-716, July 1989.

[6] A. Toet. A morphological pyramidal image decomposition. *Pattern Recognition Letters*, 9:255-261, May 1989.

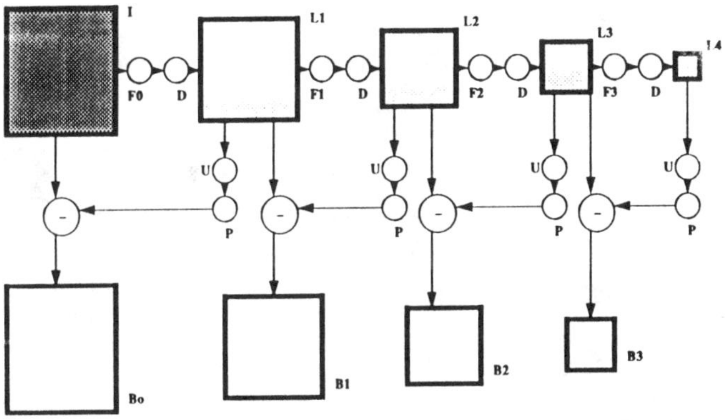

Figure 1: The construction procedures of the lowpass and bandpass decompositions.

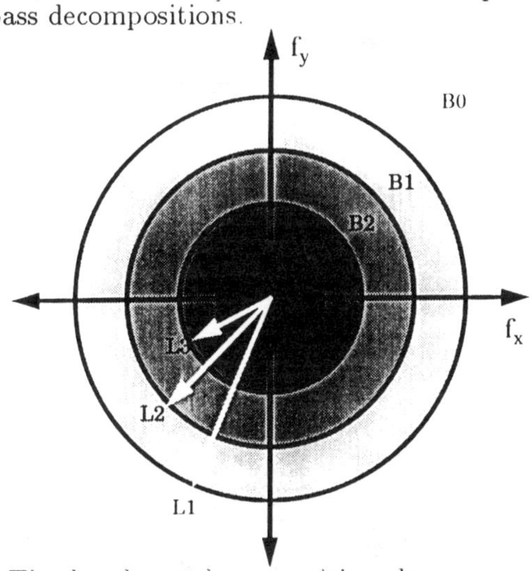

Figure 2: The bandpass decomposition decomposes the image into different bands in spatial-frequency or morphological-pattern scale.

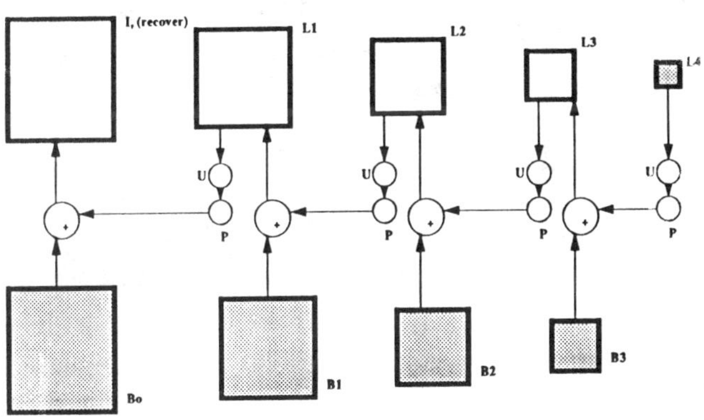

Figure 3: To recover $\mathbf{L}_i$, the lowpass decomposition level $\mathbf{L}_{i+1}$ is interpolated and added to $\mathbf{B}_i$. This procedure can be repeated to recover the original image.

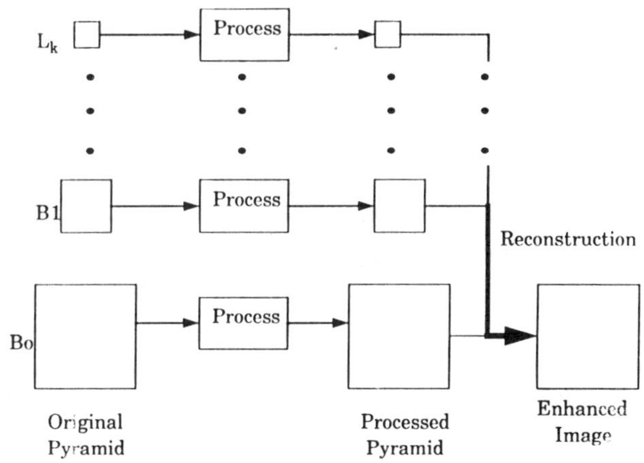

Figure 4: The modification can be done independently at each band and then synthesis.

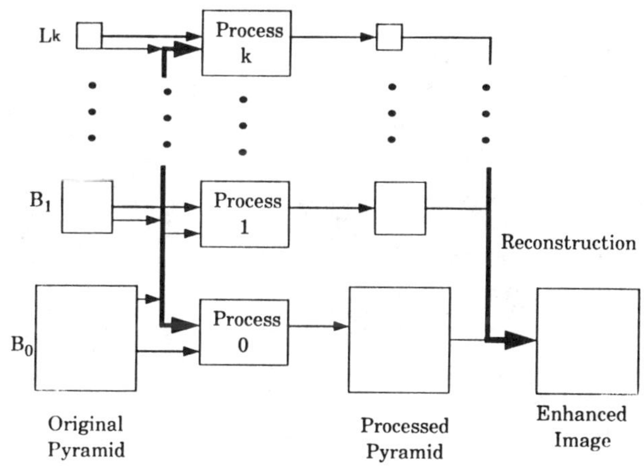

Figure 5: The correlation operations between bandpass channels of the same image are performed before synthesis.

(a) (b) (c)

Figure 6: Illustration of unstructured white noise being removed from "Lenna" using the decomposition and synthesis approach. (a) The original image contaminated with white sparkle noise. (b) The resulting image using morphological pyramid approach. (c) The resulting image using Gaussian pyramid approach.

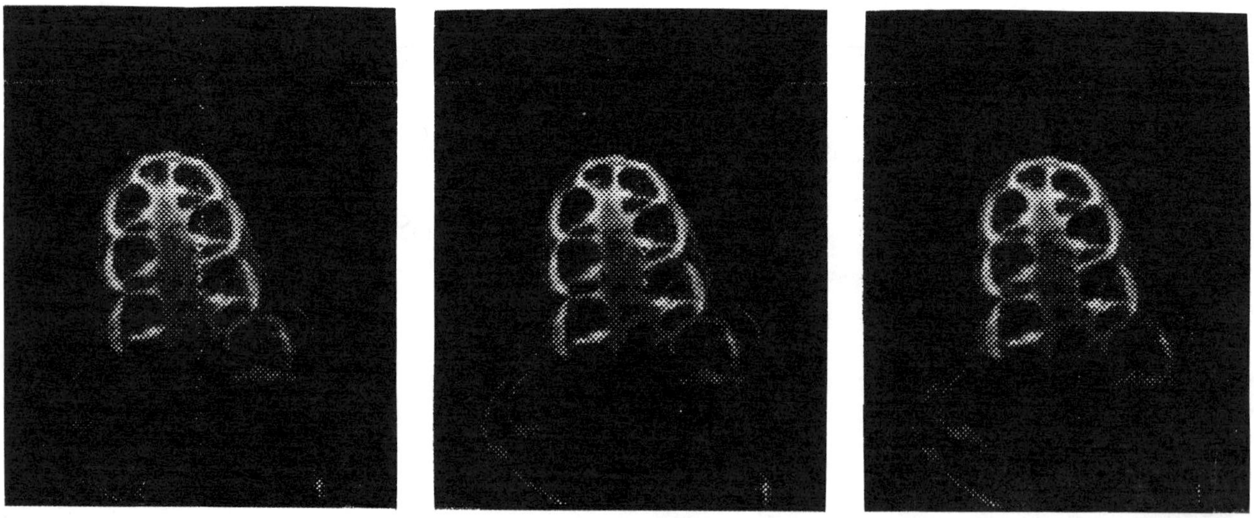

(a) (b) (c)

Figure 7: Illustration of structured high-frequency artifact being removed from a 3D image data of guinea pig cochlea using the decomposition and synthesis approach. (a) The original image contaminated with structured band noise. (b) The resulting image using morphological pyramid approach. (c) The resulting image using Gaussian pyramid approach.

Adapting Transform Coding to Texturally Segmented Images

S. SANEI R. I. KITNEY R. BENJAMIN

Electrical Eng. Department
Abadan Institute of Technology
Ahvaz, Iran

Biomedical Systems Group
Electrical Engineering Department
Imperial College, London SW7 2BT, England

Abstract

Non-stationarity in image textures requires implementation of variable size partitioning into stationary regions for coding purposes. The popular methods in regular decomposition however fail in accurately following the various region boundaries. In this paper the boundaries of different textures in the image are detected using a one-dimensional model of each row of the image. Auto-Regressive (AR) algorithm operates in 8 directions to enable an efficient detection of the boundaries. The edge map is then achieved by taking the result of edge detection operation in all directions into account. Transform coding and Run-length coding are applied to encode the textures and the boundaries respectively .

1 Introduction

Adaptive Transform Coding has been introduced as the best method for encoding natural multi-level images [1]. In traditional approaches different numbers of coefficients in transform domain are coded for transmission or storage according to the activity (level of detail) of the regions within the image frame. In these methods the sub-blocks are classified from low to high activity groups and less number of bits are allocated to the lowest whereas a large number of bits are used for encoding the highest activity regions [2]. With these techniques, the similarities between adjacent fixed size sub-blocks, and the long term stationarity of the data are ignored. In a new variable size coding criterion, the image frame is divided into variable size sub-blocks by regularly decomposing the image frame in such a way that smaller size blocks are in the area of higher activity and larger blocks contain less activity textures [3]&[4]. The decomposition representation normally has the form of a rooted tree. The root represents the entire image and the nodes on lower levels represent increasingly smaller regions of the picture. Variable size transform coding has been modified by introducing an efficient method of partitioning the image and also coding the corresponding tree with a minimum number of bits/pixel [4]. Following these criteria, the existing redundancy in coded data will be further reduced. Application of a hybrid Variable Size Transform Coding - Vector Quantization (VSTC-VQ) exploits the similarities in various block patterns [mainly *ac* coefficients] in order to reduce the overall number of bits per pixel [5]. However, application of regular decomposition criteria in partitioning the images does not necessarily follow the region boundaries or take the existing long-term stationarity of some patterns in the image into account. The above requirements may be fulfiled by accurate segmentation of images into uniform regions. Following sections describe the coding strategy based on detection of the boundaries and adapting transform coding to the arbitrary shape texturally segmented regions .

2 Coding Technique

In the method proposed here, the two components of the data, texture and boundaries, are formed by AR modelling of the data. The segmentation algorithm detects the boundaries between the two regions with a different set of AR coefficients. The boundaries are then coded using Run-length coding followed by allocation of variable length codewords to the data string. Transform coding will be adapted to the arbitrary shaped regions by an appropriate extrapolation of the regions to the proper size rectangles. The technique has been applied to the images which contain a number of arbitrary shaped regions of Brodatz textures [6]. The two major advantages of this technique over the recently proposed variable size transform coding methods are :

1058-6393/91 $1.00 © 1991 IEEE

- It removes the constraint on the shape of the blocks which has to be rectangular for transformation purposes,
- It exploits both short and long-term stationarity of various patterns in efficiently coding the data .

A transform coding method based on segmentation has also been proposed in which the image is segmented according to grey level variations [7]. This method however does not have the second advantage stated above. The overall coding procedure may be divided into 4 major parts ;

1 Segmentation based on texture variation
2 Restoration of the edge-map, thinning and linking the boundaries
3 Coding the boundaries
4 Coding the contents of the regions

2.1 Segmentation of the Image

Application of AR modelling in texture analysis, image classification, image segmentation and edge detection has been reported by some researchers [8]&[9]. Using the Deguchi and Horishita [9] method, texture features are derived from a 2-D linear prediction model. Assuming that the input image contains only two textures, an appropriate model is first determined for one of the textures. This model is then used to fit the entire image field and the fitting error at each pixel are calculated. The error statistics are then processed by a Bayesian classifier .

One of the problems with application of 2-D AR filters is that there is no absolute solution to the instability problem. Implementation of 1-D predictive filters is superior to 2-D ones for the above reason. There are many patterns which have certain direction (directional patterns). For these data, the best direction in which the rows are scanned is the one vertical to the pattern direction. However, only 8 scanning directions with 45^0 phase distance, are considered to deal with all existing patterns. The results are combined and/or compared in order to achieve the best representation of the boundaries .

In general, the current samples of a 1-D data of length N_w can be predicted from its previous samples as :

$$y(i)=\sum_{k=1}^{P} a_k y(i-k)+\epsilon(i) \quad \text{for } i=1, \ldots, N_w \qquad (1)$$

where, P is the model order, a_k is the kth prediction coefficient and $\{\epsilon(i)\}$ is a zero-mean uncorrelated process which has to be minimized for having the best set of coefficients. The model parameters $\{a_k\}$ are estimated using the autocorrelation sequences, $R_y(i)$, in each corresponding direction. The window size, N_w, and the prediction order P are the two important parameters which have to be determined for the corresponding patterns within the image .

In order to have an almost accurate prediction of a texture with a distinguishable fundamental frequency, f_0, N_w may be considered at least equal to $2/f_0$. In the case of having different fundamental frequencies along a row of the image, N_w will be equal to $2/f_{0min}$, where f_{0min} is the minimum fundamental frequency along the row. The fundamental frequencies can be measured by looking at the spectrum of the regions. In practice, N_w is chosen large enough to cover 2 or 3 cycles of the texture with highest fundamental period and small enough to be comparable with the length of the smallest uniform part of the related 1-D row .

Prediction order is estimated using the first Akaike [10] criterion called the Final Prediction Error (FPE) in which the order is selected to minimize the performance index ;

$$\mathrm{FPE}=\hat{\sigma}^2_{\epsilon w}\left(\frac{N_w+P+1}{N_w-P-1}\right) \qquad (2)$$

where $\hat{\sigma}^2_{\epsilon w}$ is the estimated variance of the linear prediction error. This performance index is calculated based on minimization of the mean-square error for a one-step predictor. The error is minimum when the set of prediction coefficients can well define the corresponding signal. At the boundaries the error jumps to a considerable value as there is no correlation among the adjacent samples. The error difference signal will then have a peak at each edge point. Figure 1 gives an example for the above concept .

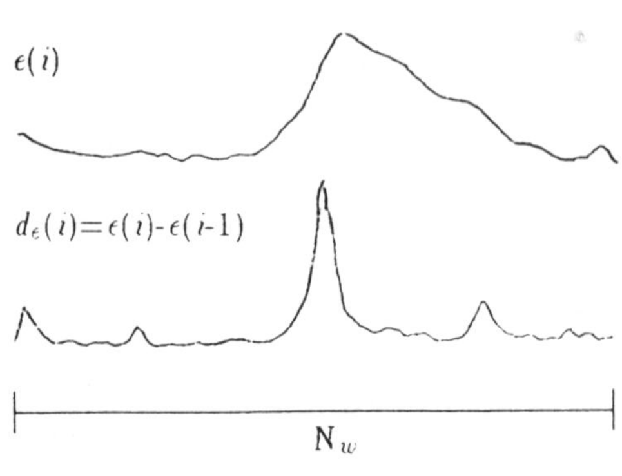

Figure 1 The error and the difference signal

This result is used to detect the texture boundaries in the proposed segmentation system simply by thresholding the difference signal. Figure 2 presents the result after applying the above AR-based system on an arbitrary combination of Brodatz textures. It is seen in Figure 2 that one more advantage of AR-based segmentation is its ability to detect the edges of two similar patterns with different phases .

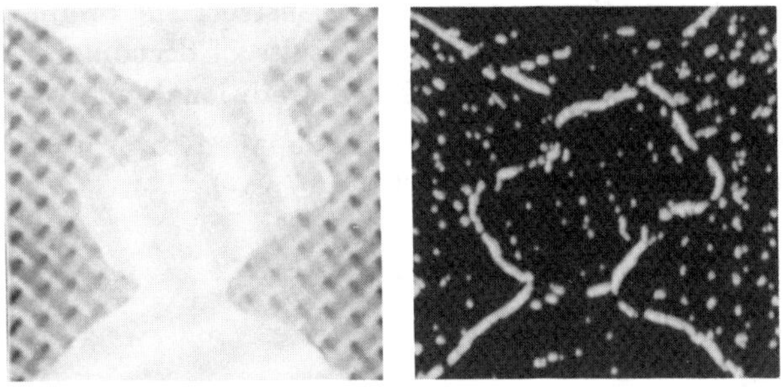

Figure 2 The result of 1-D AR-based edge detection

2.2 Thinning and linking the edges

The edge map achieved using the above method is affected by some artefacts because firstly, the patterns in the original image are not absolutely periodic; secondly, the size of the window, N_w, and the prediction order, P, do not match each particular segment well and are estimated from the entire frame length statistics, and finally, the prediction coefficients are estimated by minimization of the error signal and not necessarily by acurately matching the original signal and the predicted version. These artefacts may be reduced or removed by means of filtering and thresholding procedures. *Median filtering* and *out-range pixel smoothing* followed by thresholding reduce these artefacts to a very low level. Figure 3 shows the result of edge-map restoration .

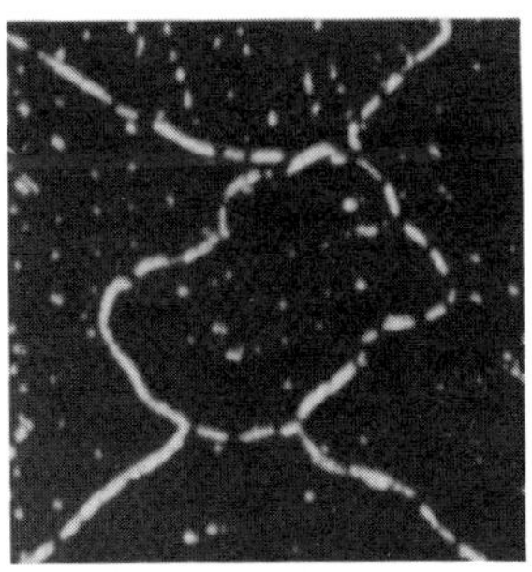

Figure 3 The result of edge restoration

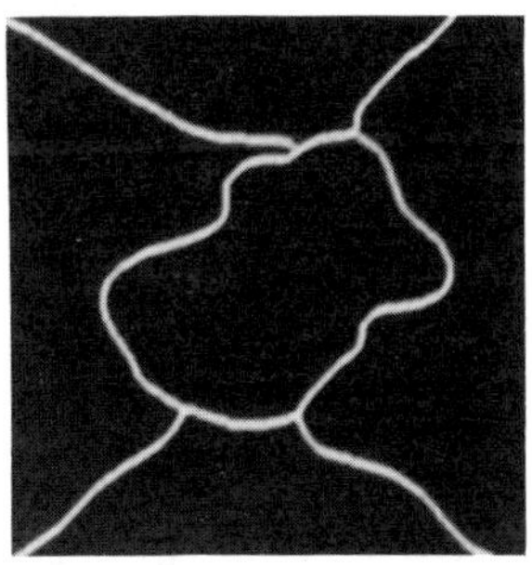

Figure 4 Linked edge-map

The edges are then thinned using the Safe-Point Thinning Algorithm (SPTA) [11]. At this stage the remaining artefacts can be removed by implementing a simple restoration method. The detected edges are then linked to provide a single-pixel-width connected contour. An appropriate linking strategy is still being studied. Here, the linking algorithm follows the following steps :

- sort the edge pieces according to their size
- measure the gradient of the edges at each end point
- connect the endpoints which
  a) are closest to each other,
  b) the different in gradient in these points is under a threshold level and
  c) belong to the longest edges among the non-connected pieces .

Figure 4 illustrates the resulting single-pixel-width linked edge-map .

2.3 Coding the Boundaries

Different coding techniques for binary images may be implemented to encode the bilevel contours. Here a 1-D Run-length coding has been applied. The lengths are defined as a number of blank points followed by an edge point. Variable length Huffman codes are then allocated to the lengths to further reduce the existing redundancy .

2.4 Application of Transform Coding to Arbitrary-Shape Image Segments

Transform coding was chosen for encoding the regions. With this method, each region is first extrapolated to a rectangle to facilitate the transformation. In addition, application of 2-D fast transformations requires the rectangle dimensions to be equal to 2^m, where m is an integer. Figure 5 shows one of the regions which has been extrapolated to a rectangle. Extrapolation of the regions demands that the whole texture remain stationary and that no extra information be added to the transformed data. This can be achieved by prediction of the blank pixels in the rectangle from the existing pixel values. In general, prediction has to be done in four directions to cover the whole area. For simplicity the prediction order was chosen to be fixed and equal to 3 .

Here DCT is applied in mapping the variable size rectangular patterns into the sets of transform coefficients. A certain number of coefficients are then selected and coded before transmission to the decoder. Zonal sampling is

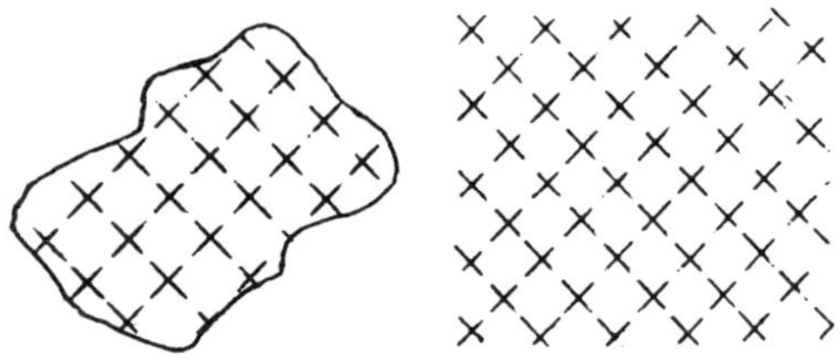

Figure 5 Extrapolation of an arbitrary shape region to a rectangle .

chosen for this purpose due to its simplicity. The simplest way of zonal coding is coding the coefficients selected by following a zig-zag path starting from the corner where the DC component is. On the other hand, with this new technique, each pattern has its corresponding data components in the spatial frequency domain which lies within a certain zone. An efficient coding of the transform parameters involves choosing the zone which contains the major dominant part of the components. The proposed algorithm moves the x-y coordinate within the image until the maximum energy of the AC components falls into the lower part of the coordinate. The position of the coordinate and the set of coefficients, which are selected by following a zig-zag path originating from the new coordinate origin, then need to be coded and transmitted to the decoder. To reduce the amount of overhead information, the new coordinate origin takes only a limited number of positions within the frame, say every 16 pixels .

The number of selected coefficients and the number of levels of quantization depend on the rate of compression required. For an acceptable result only the average of 12.5% of the coefficients quantized to 8 levels is sufficient. The DC level is quantized to 256 levels in order to avoid any visible change in grey level along the boundaries .

2.5 Decoding

The decoder requires the data together with the overhead information in order to reconstruct the original data. Assuming a constant rate coding process, the decoder may receive the data and the overhead information at a fixed rate of transmission. In this case there will be no buffering requirement in either coder or decoder. The reconstruction process a) decodes the boundaries by decoding the contours from the edge-map information, b) determines the positions and dimensions of the rectangles from the segments' edges by following the same criteria as in the encoder, c) decodes the transformation coefficients using the overhead information about the quantizer and position of the coordinates, d) takes the inverse transform of the coefficients to reconstruct the original rectangles, e) eliminates those pixels which fall outside the segments' boundaries by considering the already decoded edges and f) joins the segments to reconstruct the original data. Figure 6 compares the results of decoding the image shown in figure 2 for traditional adaptive DCT coding and when DCT is applied to arbitrary shape image segments for 0.6 bits/pixel .

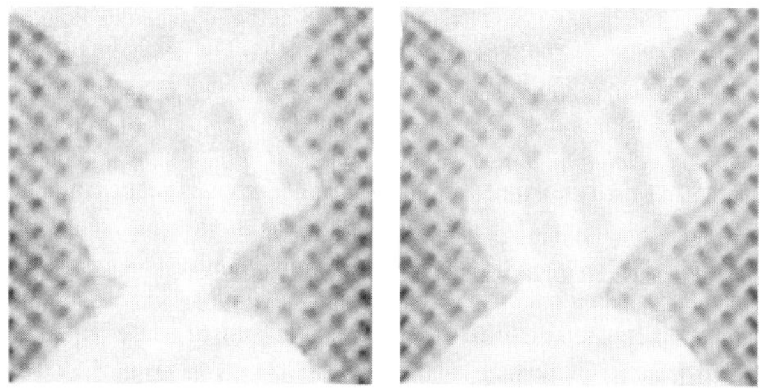

a b

Figure 6 0.6 bpp decoded image for a) traditional DCT coding and b) texturally segmented based DCT coding

3 Summary and Conclusions

In the coding technique presented in this paper the images are segmented by considering their textural content. DCT coding is then adapted to the arbitrary shaped regions to enable each uniform texture to be encoded separately. Despite this method giving an excellent performance for appropriately selected textures, its application to all natural images requires more work and modification. The developed methods of classification, segmentation and edge detection still have a long way to go to enable them to deal with arbitrary textures. The efficiency of the rest of the system also depends on the accuracy of the segmentation algorithm. The complexity of the system is high. However, since the Human Visual System (HVS) is more sensitive to the edges, the subjective quality of the reconstructed images is superior to those obtained using previous methods. In addition, application of a Vector Quantization - transform Coding (VQ-TC) may further reduce the bitrate by exploiting the similarities between certain patterns in different parts of the image .

References

[1] H. Yasuda, "Standardization Activities on Multimedia Coding in ISO", *Image Comm., vol.* 1.1, *Elsevier Science Publishers, pp.* 3-18, *June* 1989

[2] R. J. Clark, *Transform Coding of Images, Academic Press,* 1985

[3] D. J. Vaisey and A. Gersho, "Variable Block Size Image Coding", *IEEE, ICASSP Proc., pp.* 1854-7, 1989

[4] S. Sanei and R. I. Kitney, "Modified Variable Size Transform Coding of Images", *Proc. of ACSSC, Asilomar, USA, pp.* 322-5, 1990

[5] S. Sanei, "Compression Techniques for the Storage of Biomedical X-ray Images", *PhD Thesis, Imperial College, UK,* 1991

[6] P. Brodatz, *Textures, Phil Brodatz,* 1966

[7] Y. B. Yu, "Transform Coding of Images based on Shape Adaptive Models", *PhD Thesis, Imperial College, UK,* 1990

[8] R. L. Kashyap, "Characterization and Estimation of Two-dimensional ARMA models", *IEEE Trans. on Infor. Theory, vol.* 30, *pp.* 736-45, 1984

[9] K. Deguchi and I Morishita, "Texture Characterization and Texture-based Image Partitioning using 2-D Linear Estimation Techniques", *IEEE Trans. on Comput., vol. C-*27, *pp.* 739-45, 1978

[10] H. Akaike, "Statistical Predictor Identification", *Ann. Inst. Statist. Math., vol.* 22, *pp.* 203-217, 1970

[11] N. J. Naccache and R. Shinghal, "SPTA: A Proposed Algorithm for Thinning Binary Patterns", *IEEE Trans on Sys., Man., and Cybernetics, vol. SMC-*14, *No.* 3, *pp.* 409-418, 1984

Multilevel Segmentation and Analysis of Facsimile Images for Document Classification

Sally L. Wood
Electrical Engineering Department
Santa Clara University
Santa Clara, CA 95053

April Lee
Applied Signal Technology
160 Sobrante Way
Sunnyvale, CA 94086

Phillip L. Kelly
Applied Signal Technology
160 Sobrante Way
Sunnyvale, CA 94086

Abstract

Facsimile images are easily obtained in digital raster image form so that viewing, postprocessing, routing, and archiving can all be part of an integrated automated system which does not rely on paper copy. However, efficient operation of this type of system requires analysis of facsimile images so that text areas are converted from raster image representation to computer readable character strings, and documents are classified by content type for easier retrieval. Segmentation of the raster image can isolate text areas and classify the remaining non-text areas. Previous work demonstrated the effectiveness of both spatial frequency domain and raster image domain feature analysis using both the full image and smaller blocks. A more effective multilevel segmentation and classification system is presented. The first level indicates the major "visual" groupings of the information on the page and the second level explicitly identifies the number and type of internal elements. The medial axis transformation and contour shape parameters are used to analyze irregularly arranged areas and non-machine generated areas of a document. This approach is efficient because at each step the level of detail used in processing is appropriate for the decisions which are required. Examples of group 2, group 3 standard, and group 3 high resolution images demonstrate the segmentation and classification performance.

1 Introduction

Automatic processing of facsimile images which could classify and sort incoming transmissions and route them accordingly is a necessary component for an office automation system [1,2]. Combined with optical character recognition (OCR) [3] and other image compression techniques [4] page format analysis allows efficient processing of the image data for storage, selected retrieval, and merging with other data. Facsimile transmission allows an arbitrary mixture of printed text, drawings, and handwritten information to be sent in the form of a raster image. Most images are meant to be read by a human observer and are arranged so that the viewer can easily partition the image into a few significant segments. In most cases this is done visually by observing regions with white borders or with black form lines. However, each of these regions might also contain a mixture of text and nontext information, so additional partitioning might be needed.

Most currently used facsimile is either Group 2 or Group 3. The resolution of Group 2 is aproximately 100 dots per inch (dpi) horizontally and 100 lines per inch (lpi) vertically. Group 3 images offer a horizontal resolution of 200 dpi and a vertical resolution of either 100 lpi or 200 lpi. Although the 100 lpi resolution can pose a challenge for OCR, format analysis is effective at either resolution.

2 Segmentation by Prioritized Lines

The goal of the first level of segmentation is to divide the image into non-overlapping rectangular regions which correspond to visually coherent subsections of the document image. For example, in a document such as this one, the page would be divided into two columns and then boundaries around figures, tables, equations, and major sections would be located. In a letter format the subdivisions would typically correspond to the opening, the body, the closing, the date, and the letterhead. There is no uniquely correct solution to this type of segmentation, and resolution of issues such as whether or not individual paragraphs should be isolated depends on the requirements of later processing steps.

In order to focus on major divisions between image segments, a lower resolution auxilliary image is created. This grey scale image has an effective resolution of 12.5 pixels per inch both horizontally and vertically. While this resolution is insufficient to read most text, it is effective for localizing areas of black ink and bounding areas of white. Let $f(k,l)$ represent the original binary facsimile image. The low resolution grey level image, $g(m,n)$ is created by a local average low pass filter to achieve the desired resolution. The summation

$$g(m,n) = \sum_{k=mM}^{(m+1)M-1} \sum_{l=nN}^{(n+1)N-1} f(k,l) \qquad (1)$$

is followed by a scaling to maintain an 8 bit dynamic range. A second auxilliary image, $d(m,n)$, is created from $g(m,n)$ by selecting a low threshold to distinguish background from foreground and then performing a medial axis transformation on the thresholded image. This provides a local measure of the distance from each white or background pixel to the nearest dark pixel containing text or image information.

For both auxilliary images horizontal and vertical profiles of densities, parallel edges, and perpendicular edges are computed. For an image segment specified by the upper left corner (M_1, N_1) and the lower right corner (M_2, N_2) the row sums are given by

$$
\begin{aligned}
s_0(m) &= \sum_{n=N_1}^{N_2} g(m,n) \qquad (2)\\
s_{\|}(m) &= \sum_{n=N_1}^{N_2-1} |g(m,n+1) - g(m,n)|\\
s_{\perp}(m) &= \sum_{n=N_1}^{N_2} |g(m+1,n) - g(m,n)|
\end{aligned}
$$

and the column sums by

$$
\begin{aligned}
t_0(n) &= \sum_{m=M_1}^{M_2} g(m,n) \qquad (3)\\
t_{\|}(n) &= \sum_{m=M_1}^{M_2} |g(m+1,n) - g(m,n)|\\
t_{\perp}(n) &= \sum_{m=M_1}^{M_2-1} |g(m,n+1) - g(m,n)|
\end{aligned}
$$

Using both the horizontal and vertical profiles given above allows the identification of white and black lines in either orientation. A black line is indicated by a high value for the density sum and a high ratio of the density sum to the parallel difference sum. For example, in the profiles computed from $g(m,n)$ a horizontal black line would have a high value for $s_0(m)$ and, in comparison to $s_0(m)$, a low value of $s_{\|}(m)$ indicating a high average horizontal stroke width. The value of $s_{\perp}(m)$ will be high only at the top and bottom edges of the line. A white separating line would have a very low value for the density profile computed from $g(m,n)$ and a high value for the density profile computed from $d(m,n)$.

White and black lines detected in this manner can be given a priority based on the absolute profile values and the height and width of the lines. In general, black lines are given priority over white lines since black lines are often explicit form field boundaries. However, no priority is assigned based on orientation, because page orientation is not assumed to be known. Although for both white and black lines thresholds must be set to determine line locations and priorities, the values of these parameters are not critical for the major divisions which are determined by relative sizes and strengths. Sensitivity to these threshold values can be seen however in the limits for the low priority lines which cause segmentation to cease for lack of additional significant line separators.

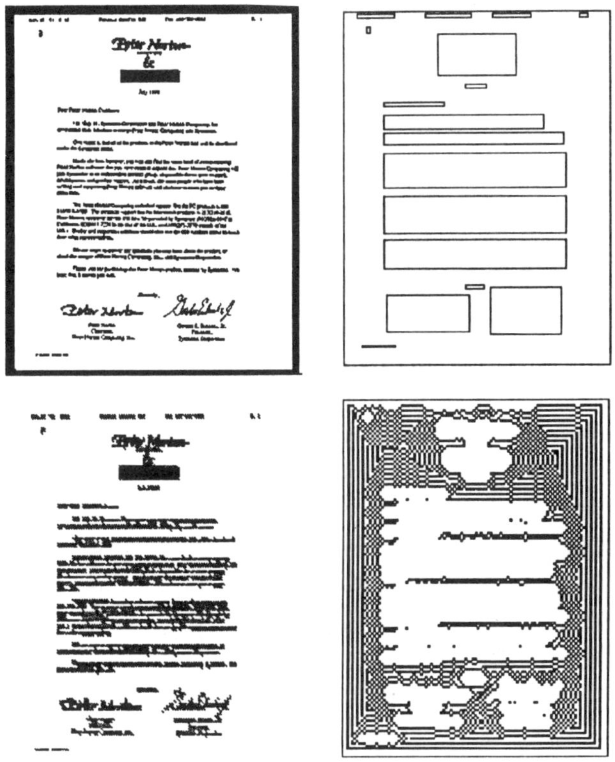

Figure 1: Fax Image 1: image, segmentation using recursive subdivision, reduced resolution image, and medial axis transform

Recursive segmentation of the complete low resolution image $g(m,n)$ proceeds until no reasonable separators remain. A data structure for each level 1 segment contains its bounding box, computed features based on profile values, the position of the highest priority black and white lines, and linkages to both the parent segment and to segments created from its subdivision. Initially the entire page is one segment. The higest priority line separator is identified, and then the segment is divided along that boundary. Two new segments are created and a new data structure for each is initialized which includes separator line priority information. The new segmentation line is also added to a separator list. The subdivided segment is made inactive, but remains linked to the new segments created from it and the segmentation line which divided it. This process continues with the identification of the highest priority separating line remaining in all of the active segments followed by a segment division until there are no significant lines available or until the number of segments created reaches a preset maximum limit.

A typical result from this first level of segmentation is shown in Figure 1. The fax image of a letter shown in the upper left was segmented as described above to generate the rectangular segments shown in the upper right. The two auxilliary images $g(m,n)$ and $d(m,n)$ are displayed in the middle row, and an analysis summary derived from segment characteristics is given

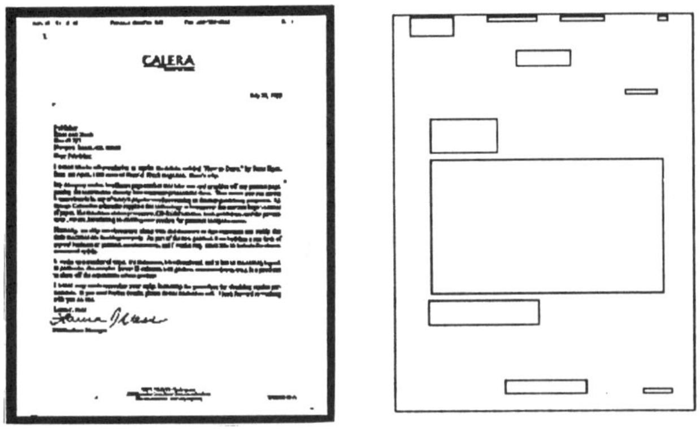

Figure 2: Fax Image 2: image and segmentation using recursive subdivision

at the bottom. The letterhead, date, opening, paragraphs of the body, closings, and information written by the fax system were all placed in separate segments.

SUMMARY

	NUM SEGS	PERCENT OF IMAGE
TEXT SEGS:	8	28.990000
FAX HDR SEGS:	6	1.160000
BLACK SEGS:	1	0.620000
DARK SEGS:	0	0.000000
HAND SEGS:	1	0.040000
TEXT/LINE SEGS:	1	3.640000
TEXT/HAND SEGS:	1	3.680000
OTHER SEGS:	0	0.000000

Up Direction for Page is ^

The automatic analysis of very general characteristics of the Figure 1 segments shown above indicates that the majority of the segments, both in number and area, were text. However, some segments were identified as combinations of text and handwriting or text and dark lines, and the information printed by the fax machine was identified. The "other" category, which is not used in this image includes halftone images and most line drawings. All of these categories are preliminary and based on the two auxilliary images. After the secondary segmentation the mixed segments would be divided into more homogeneous units. A page orientation is also predicted based on these results.

The fax image of a letter in Figure 2 was segmented in the same way as the letter of Figure 1. The results are similar except that the individual paragraphs were not separated. In both figures there is sufficient structural information to predict that the format is a letter format, and the desirability of subdividing to the paragraph level depends of further processing requirements. For automatic reading with OCR, either form would provide the OCR system with text only.

3 Problems with Line Segmentation

Although the method described is very effective for many page formats including multicolumn formats, there are some difficulties with this approach. The lack of a uniquely correct result of the segmentation process is evident in several respects. Paragraphs may be isolated from each other or remain combined in a larger body of text. Mixed segments which contain for example a signature and typed name or a letterhead symbol and mixed type sizes are visually treated as a single unit, but further analysis might require subdivision into homogeneous units. There are often cases where the visually coherent components can not be separated into regions with non overlapping rectangular boxes. Figure 3 shows an example of this in which the graphics for the soccer player can not be separated from the adjacent text by horizontal and vertical lines even though there is no place that the picture touches the text. The large unit classified as "other" included both the large text and the picture. This is an example of the general problem of mixed text and graphic information that also often occurs in large letterheads.

Complex formats and tabular layouts often haveno clear black or white boundary separating internal units. Even in simple formats skewed text makes blank areas between segments less detectable since in a horizontal projection the bottom of one segment edge can overlap the top of the opposite edge of the next segment.

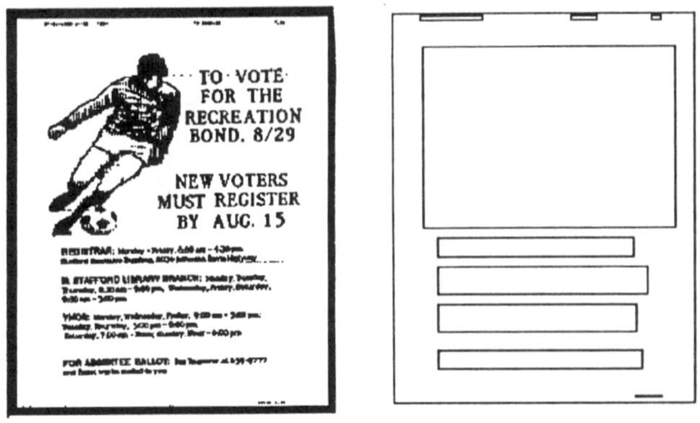

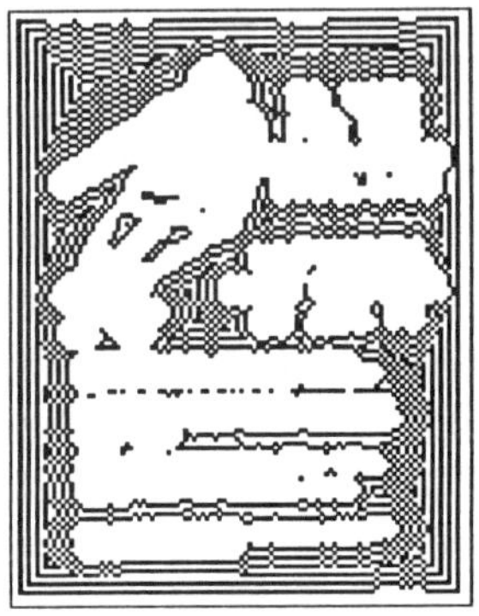

SUMMARY

	NUM SEGS	PERCENT OF IMAGE
TEXT SEGS:	4	17.080000
FAX HDR SEGS:	4	0.590000
BLACK SEGS:	0	0.000000
DARK SEGS:	0	0.000000
HAND SEGS:	0	0.000000
TEXT/LINE SEGS:	0	0.000000
TEXT/HAND SEGS:	0	0.000000
OTHER SEGS:	1	39.870000

Up Direction for Page is ^

Figure 3: Fax Image 3: image, segmentation using recursive subdivision, reduced resolution image, medial axis transform, and segmentation summary

4 Secondary Segmentation

Each of the segments defined by the recursive partitioning method was further analyzed using the original

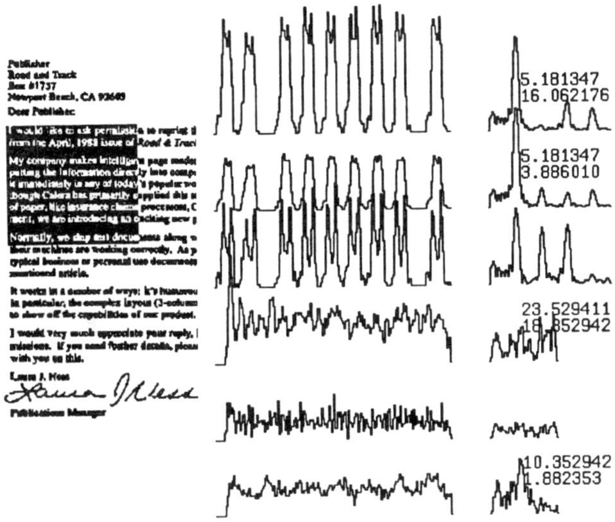

Figure 4: Profiles and Fourier transforms of segment of variable pitch text with regularly spaced lines

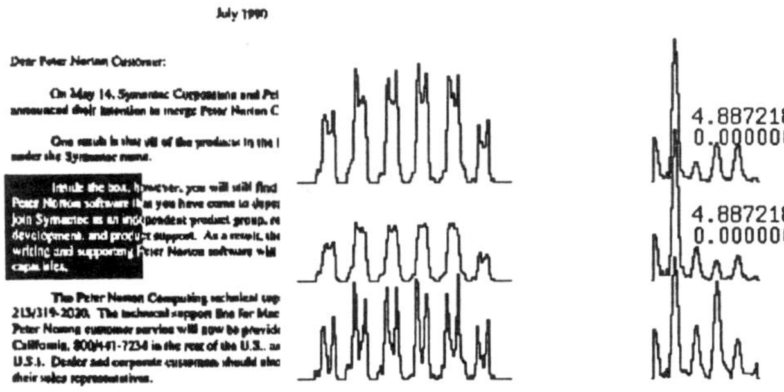

Figure 5: Profiles and Fourier transforms of narrow segment of skewed text

full resolution image for content based segmentation. The same profiles defined in equations 2 and 3 were used on the high resolution images to define and classify the content of individual segments. Figure 4 shows an example of this processing on a region of homogeneous text with regular spacing. In the first profile the individual lines of text are quite clear as are the two cases of additional paragraph spacing. These profiles can be analyzed to determine content based on both the characteristic pulse shape associated with lines of text and on the arrangement of these pulses. This has been shown to be a good discriminator for classifying text as composed of Roman characters, Chinese characters, and Arabic or handwritten characters [5]. Alternatively the frequency domain can be used to determine the strength of dominant periodicities or the absence of a periodicity.

4.1 Frequency Domain Analysis

A Discrete Fourier Transform (DFT) of profiles such as those shown in Figure 4 will have a sharp peak at the frequency corresponding to line spacing or pitch period for fixed pitch text. This peak is relatively insensitive to missing lines [5]. For variable pitch printed text there is a more uniform distribution of energy in the frequency domain. In the case of most non-text, low frequency components dominate. The characteristic peak associated with line spacing can be attenuated if paragraph spacing is not an integral multiple of the line spacing.

4.2 Cluster Analysis

The profiles determined from equations 2 and 3 can be analyzed by thresholding at a low value to separate the background or blank lines. In Figure 4 the steep slope of the profile for s_0 indicates that threshold value selection is not critical. After thresholding, each line or space is assigned a height and a type. As discussed in section 2, blank lines will have very small values for the density profile and thus will be separated by the thresholding. Text can be distinguished from lines based on average parallel stroke width. In addition, lines can be classified based on pulse shape as described in [5]. Initially all lines of the same type and same height are given a unique cluster identification. Then, clusters of the same type of line are successively merged based on the height information. All clusters must be separated by at least a minimum distance, and if remaining clusters are separated by more than a maximum distance, merging is stopped.

Using cluster information, further subdivision into homogeneous partitions is possible. Based on this approach, the segment in Figure 4 selected for illustration would be divided into three parts. Although all the text lines are assigned to the same cluster, there are two clusters of blank lines corresponding to the space between lines and the space between paragraphs. Thus a division along the paragraph boundaries would maintain homogeneous segments.

Classification can be affected by skewed text. In Figure 5 a narrow segment of skewed text taken from the letter in Figure 1 has properties similar to Figure 4. However, the natural segmentation of that text might lead to the segment shown in Figure 6. There the pulse waveform for the text has become more triangular due to the convolution of the text line pulse shape with a rectangular window of width equal to the line skew. A triangle is also the characteristic pulse shape for handwriting, as illustrated in Figure 7. This classification problem can be overcome by analyzing wide segments in several narrower parts.

Mixed segments will be broken into smaller segments to achieve homogeneity. The signature segment of Figure 7 for example would be distinguished from the three text lines by both type and by line height. Figure 7 also shows a mixed segment in which the two text sizes and the horizontal line would all be grouped separately to form three new segments.

5 Non-rectangular Segments

If the cluster analysis described above does not yield any further segmentation boundaries and there is no obvious periodic structure, then one further attempt is made to subdivide along boundaries of arbitrary shape. Using the complement of the medial axis transform image, objects can be identified and contours of objects which are not connected can be calculated.

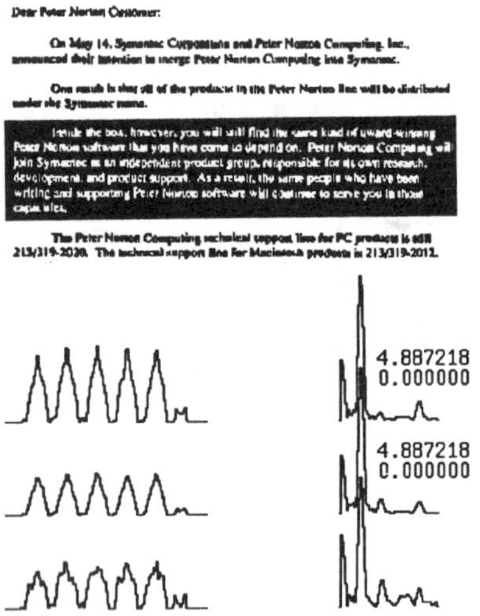

Figure 6: Profiles and Fourier transform of wider segment of skewed text

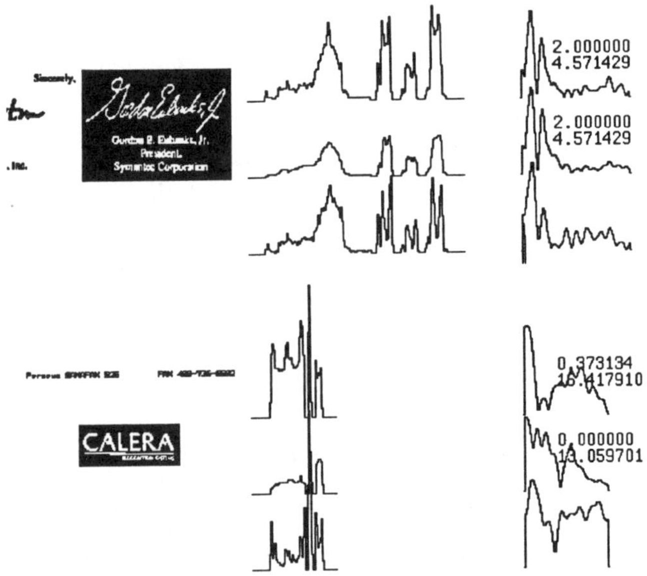

Figure 7: Horizontal profiles and Fourier transform of signature segment and segment with line and variable text size

Simple parameters of such contours such as the bounding octagon and the ratio of perimeter squared to area can be used to define further segmentation. For example, in the left half of Figure 8, the bottom three lines of the text near the graphic image from the large segment of Figure 3 are easily separated. After that is removed the top three lines are easily separated by a vertical line even though they appear to be touching in the low resolution image. The right half of Figure 8 shows a more complex letterhead and page layout.

6 Conclusions

A segmentation and classification method was tested for application to digitized facsimile images. The two level approach provided both a basic partition into components which are physically grouped together and a finer segmentation in which each segment is homogeneous. For the finer segmentation spatial domain techniques were superior to frequency domain techniques. These algorithms were tested on several group 2 and group 3 images and performed as expected. Since there is no uniquely correct result for a segmentation, quantitative measures of performance are not possible. In an application in which further processing places constraints on segmentation results, performance will be more readily evaluated.

References

[1] Wood, S. L. and A. Chekfa, "Format Analysis for a Page Reader for the Visually Impaired", **Proc. Ninth International Conf. IEEE EMBS**, pp2038-2039, Boston, Massachusetts, Nov. 13-16, 1987.

[2] Wood, S. L. "Image Processing Techniques For Sorting Facsimile Images", Technical Report **TN-055-88**, Applied Signal Technology, Sunnyvale California, 31 October 1988.

[3] "The OCReport", Vol. 1, No. 1, Calera Recognition Systems, Inc., Santa Clara, California, 1990.

[4] Brandt, J. W. and A.K. Jain, "A Medial Axis Transformation Algorithm for Compression and Vectorization of Document Images", **Proc. ICASSP-90**, Vol. 3, pp 1850-1853, Glasgow, Scotland, 23-26 May, 1989.

[5] Wood, S.L., B.A. Weisburn, and P.L.Kelly, "Image Processing for Segmentation and Classification of Facsimile Images", **Proceedings of the 24th Asilomar Conference on Signals, Systems, and Computers**, Pacific Grove, CA, November 5-7, 1990.

Figure 8: Subdivision into segments with non-rectangular boundaries

Segmentation and Object Tracking for the Microstructure Analysis of Soil

Gregory W. Donohoe, Michael F. Boccabella
EECE Dept., University of New Mexico
Albuquerque, NM 87131

John J. Gill, Maj., USAF
Phillips Laboaratory
Kirtland AFB, NM 87118

ABSTRACT

Civil engineering researchers are interested in the microstructural behavior of soils subjected to a crushing strain. This paper describes a machine vision system developed to help experimenters study this behavior. In laboratory experiments, a video camera views soil particles through a sapphire window in a containment vessel as they are subjected to compressive strain, or uniaxial loading, over a period of 5 to 20 minutes. The vision system segments the particles from the background, estimates the centroid and orientation of each, and tracks the particles throughout the experiment.

1 Introduction

Civil engineers use discrete element models to analyze the behavior of soil under strain. By understanding the microstructure of soils, they hope to be predict macroscopic behavior [2]. They represent granular soils with three-dimensional models, such as spheres and ellipsoids connected by friction bonds, and use computer simulations to study their behavior. Experimenters at the Phillips Laboratory subject real soil samples to compressive strain and observe their behavior in order to validate the computer models.

The experimenter places a sample of sand into a metal containment vessel fitted with sapphire windows with cross-shaped fiducial marks etched into them. He or she sets up video cameras to view the sand through the windows. Then he applies a vertical, uniaxial strain to the top of the vessel, compressing the soil, and records the video images onto an analog laser video disk.

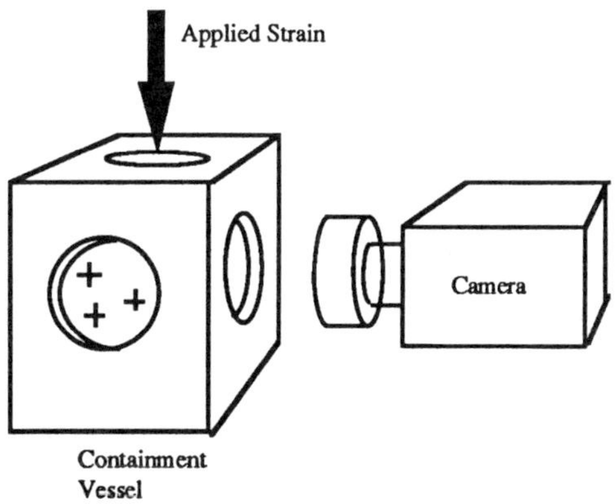

Fig. 1: Experimental apparatus

Finally, the experimenter examines the video sequence, finds visible soil particles, estimates their location and orientation, and tracks them from frame to frame. He sends the accumulated measurements to a discrete element analysis program to characterize the way the particles shift, rotate, and pack themselves together, and finally fracture. This report describes a system developed to automate the analysis of the video image sequences, relieving experimenters of a tedious

1058-6393/91 $1.00 © 1991 IEEE

and error-prone task.

2 Constraints

We treat the images as two-dimensional entities composed of essentially four regions (Fig. 2) These are (1) fiducial marks; (2) foreground particles; (3) background particles, partially obscured by another particle or a fiducial mark; and (4) voids between particles. A ring illuminator attached to the camera illuminates the scene.

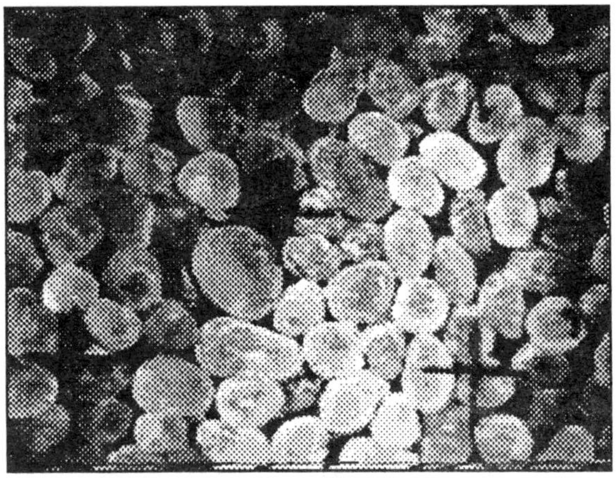

Fig. 2: Camera's view of experiment

Fiducial marks are necessary to register the images because the compression of the experimental apparatus causes the containment vessel, hence the image, to move. The laser-etched marks appear as black crosses to the human observer. Closer examination shows them to be rough, three-dimensional structures that exhibit a wide range of grey levels. The fixed illumination does not change throughout the experiment, so the appearance of the fiducial marks is nearly constant, although we can see the soil particles moving behind them.

The physical appearance of the particles depends on the choice of sand. The boundaries of the foreground particles are usually convex, closed curves. Contrast is good between the particles and the voids, but the boundaries between adjacent particles may not be detectable. Background particles consist of convex and concave curve segments. Voids are simply background particles that are too dimly lit to be detected. Uneven illumination produces illumination gradients across the image.

The images are recorded at the comfortable rate of about one image per second, then analyzed off-line. Thus, real-time processing constraints do not apply. The analysis package runs acceptably fast on a workstation. Still, the computational burden is significant and we prefer efficient algorithms.

We originally intended to constructed a fully automated system. We would present the first image to the system, which would find the foreground particles and the fiducial marks. It would then track particles through the sequence of frames. This autonomous artificial intelligence approach is difficult to implement and is prone to error. The prior probability that a pixel is near the center of a foreground particle is nearly impossible to determine without human intervention. If the system were to choose the wrong particles to track, this error would propagate through the entire experiment and ruin the analysis. So, we decided to simplify the problem by requiring the operator to designate the particles of interest and the fiducial marks with a mouse. This firmly establishes the prior probabilities, at least for the first frame. It does not pose an unreasonable burden on the experimenter, but gives him more control in guiding the analysis.

3 Fiducial Marks

Each fiducial mark is unique, and the marks show little variation between frames. This makes them good candidates for direct template matching. We define the template over a cross-shaped region selected by the user. This way the grains that peek out behind the fiducial marks don't appear in the template and do not interfere with the match. Fig. 3 below shows the portion of the image containing the fiducial mark, a cross-shape mask that excludes all but the region containing the fiducial mark , and the resulting template.

Fig 3: Fiducial mark (left), mask (center) and template (right)

We designate the region of the image belonging to the template as T, composed of template pixels $\{t(x,y)\}$. The image F is composed of grey level values $\{f(x,y)\}$. We considered the four common correlation measures listed below.

Squared Euclidean distance:

$$D_2(F,T) = \sum_{x',y' \in T} (f(x-x', y-y') - t(x', y'))^2 \qquad (1)$$

City block distance:

$$D_1(F,T) = \sum_{x',y' \in T} | f(x-x', y-y') - t(x', y') | \qquad (2)$$

Statistical cross correlation:

$$R_c(F,T) = \sum_{x',y' \in T} f(x-x', y-y')\, t(x',y') \qquad (3)$$

Morphological correlation

$$R_M(F,T) = \sum_{x',y' \in T} \max(f(x-x', y-y'), t(x',y')) \qquad (4)$$

Each of (1) - (4) produces a "match value" at each translation (x',y')across the image. We locate the fiducial marks by finding the translation that produces the best match value; this is a *minimum* for the distance measures and a *maximum* for the correlationmetrics. It can be shown that $R_c(F,T)$ is a special case of $D_2(F,T)$ when certain assumptions are made about image properties [1, p. 65]. Thus $D_2(F,T)$ is more dependable: it produces a unique match value in cases where $R_c(F,T)$ produces multiple matches. Similar arguments can be made about $D_1(F,T)$ and $R_M(F,T)$ [3].

$R_c(F,T)$ offers a computational advantage: we can take the Fast Fourier Transform of the image and the template and invoke the correlation theorem to implement (3). Unfortunately this advantage disappears when, as is this case, the template is not rectangular. For irregular templates like our cruciform fiducial marks, we can optimize direct computation to consider only those pixels under the nonzero part of the template. In the end we chose the city block distance of (2).

4 Finding the Particles

To locate the foreground particles in the images, we must segment them from the background and find their boundaries. We explored two segmentation methods: constrained region growing and edge detection with boundary completion. Edge detection and boundary completion proved to be much more reliable, so we discarded constrained region growing.

The user selects the particles of interest by clicking the mouse near the centroid of each. An excellent optimal edge detector developed by Shen and Castan is applied to the image [4], producing good but incomplete particle boundaries as shown in Fig. 4. We complete the boundaries by applying a set of heuristics described below.

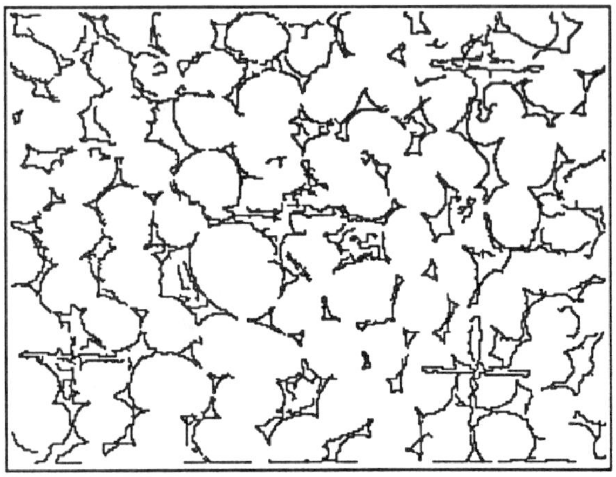

Fig. 4: Edge map

Each particle is roughly convex with respect to the center, and we know approximately where the center is for each particle. Using this center as a reference point, we map the edges into a modified polar coordinate system. We draw a ray at 0° from the center of the particle to the first edge pixel; the length of this ray represents a radius. A second ray is calculated at 180° to find the complementary radius. We add the two radii and save this diameter. Now we sweep the rays in a counterclockwise direction and repeat the process to produce a one-dimensional polar (diameter versus angle) representation of the boundary, as shown in Fig. 5. Using the diameter instead of the radius helps to compensate for errors in estimating the center of the particle.

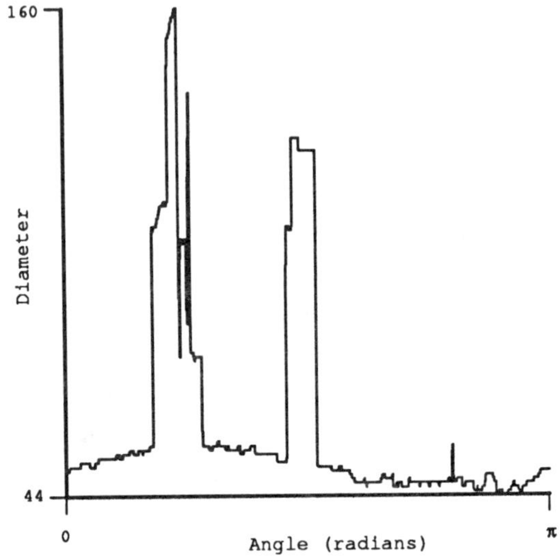

Fig. 5: Polar edge map with discontinuities

From the polar edge map we compute a histogram of diameters and choose the mode (the diameter that occurs most often) as a prototype. We find a point on the boundary that has this value and locate convex segments by radially searching along the edge until non-convex points are found. Non-convex points are presumed to represent gaps in the boundary. These gaps are filled with straight line segments, producing the smoothed polar plot of Fig. 6

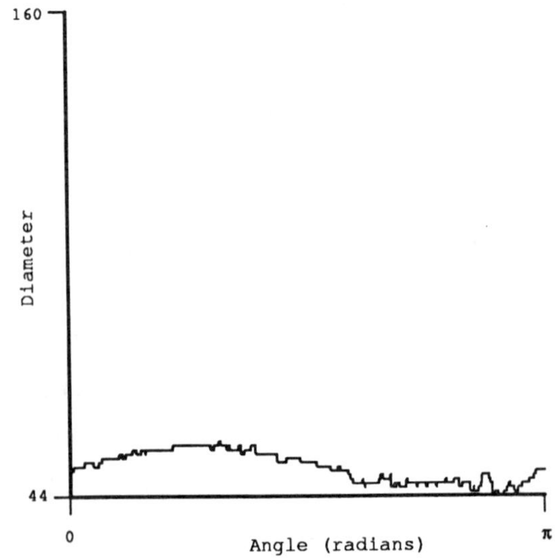

Fig. 6: Smoothed polar edge map

Once we have produced an entire closed boundary, it is easy to compute the central geometric moments up to order two. These provide an estimate of the area, and orientation for each particle [5].

The centroid thus acquired serves as the starting point for finding the particles in the next frame. Repeating the process outlined above, we track particles from frame to frame.

5 Conclusions

This work was motivated by the desire to relieve experimenters of the tedium of analyzing large image sequences by hand, and to reduce the human error introduced by such analysis. This software enables researchers to make better use of their time and money by performing more experiments. Once we have acquired the particle data, a reasonable next step is to compute statistical parameters and to highlight interesting behavior graphically on the video display. Examples of interesting behavior are clusters of particles that move together as a group, and "rogue" particles whose motion vector deviates from the norm. In the next phase of this project we will apply pattern recognition methods to define, detect and highlight these "interesting phenomena", then produce animated video sequences to display these phenomena to the civil engineering researchers.

References

[1] Ballard, D.H., and C.M. Brown, *Computer Vision*, Prentice-Hall, 1982.

[2] Gill, J.J., Ed., *Proc. Discrete Element Modeling/Soil Microstructure Working Group*, Shock Physics Division, Phillips Laboratory, New Mexico, March, 1991.

[3] Maragos, P., and R.W. Schafer, "Morphological Systems for Multidimensional Signal Processing", *Proceedings of the IEEE*, Vol. 78, No. 4, pp. 690-710, April 1990.

[4] Shen, J. and S. Castan, "An Optimal Linear Operator for Edge Detection," *Proc. Computer Vision and Pattern Recognition Conf.*, Miami, FL, 1986.

[5] Teague, M.R., "Image Analysis via the General Theory of Moments," *Journal of the Optical Society of America*, Vol. 70, No. 8, August 1980.

Optimal Boundary Smoothing For Curvature Estimation[1]

*Kwanghoon Sohn, Winser E. Alexander
**Jung H. Kim, Yonghoon Kim, Sung H. Yoon, Eui H. Park, C.A. Ntuen

* Department of ECE, NCSU, Raleigh, NC 29695-7910
** Image Processing & Vision System Laboratory
NC A&T State Univ., Greensboro, NC 27410

Abstract

Computing a curvature function on a digitized boundary is an ill-posed problem due to the discrete nature of the boundary. Thus, the boundary needs to be smoothed before computing the curvature function. We use a constrained regularization technique to obtain the optimal smooth boundary. The curvature function is computed from this optimal smooth boundary. This method solves a common critical problem of current curvature estimation methods in determining a unique smoothing factor. Using the above method, we derive a curvature function which is invariant under rotation, scale and translation.

1. Introduction

Since the information of the shape is concentrated at the points having high curvature [1], corner points are frequently used for a boundary representation. The *corner points* are usually detected in a curvature function space by capturing all the local extrema whose curvature values are above a certain threshold value. Hence, computing the curvature function properly is very important for unique corner point detection (or a unique boundary representation) and for reliable matching.

Despite the above importance of the curvature function in computer vision, the curvature function obtained naively from a digitized boundary is very ragged due to the discrete nature of a boundary data and noise, and it must be smooth to be used for further processing. The problem of curvature estimation is said to be *ill-posed* because the *existence, uniqueness* and *stability* of a solution are not guaranteed [2]. The computation of curvature involves the first and second derivatives and differentiating discrete and noisy data enhances the noise.

There are two methods to compute the curvature function on a digitized boundary curve in a wide sense: one is performed in a discrete domain and the other is performed in a continuous domain. The discrete domain has been used [3-6] to compute the curvature function directly using the concept of k-curvature to obtain the smooth curvature function rather than the ragged one. We have exploited the continuous domain by investigating several filtering techniques such as *average, median, lowpass* and *Gaussian filters* to reduce the effect of quantization and sampling to produce a smooth curvature function. However, a common critical problem of the existing methods is the difficulty in determining a unique smoothing factor.

We solve the above problem by using a *constrained regularization technique* which combines regularization and a noise equality constraint. The regularization technique transforms ill-posed problems into well-posed problems by introducing constraints such as a smoothness requirement. It results in a consistent, unique and stable solution. The degree of smoothing is determined by *a priori* noise information. That is, the noise equality constraint determines a unique smoothing factor. Since the curvature function computed from the optimal smooth boundary are invariant under rotation, scale and translation, we can detect corner points rotation, scale and translation invariantly, that is, we can represent a boundary invariantly.

This paper is structured as follows: In section 2, a new method for optimal boundary smoothing is intro-

[1] *Partially supported by the ARO grant no. DAAL03-90-0913.*

duced. A method of curvature estimation as well as problems of existing methods are discussed in section 3. Experimental results and invariance properties of curvature estimation are shown in section 4. Finally, concluding remarks are given in section 5.

2. Optimal Boundary Smoothing

We assume that an *original boundary* (f) is not known. It is only known that the function is smooth. The objective of the algorithm is to optimally estimate the original boundary f, given a measured boundary f_m and some knowledge about the noise f_n. Define

$$f_m = f + f_n \tag{1}$$

where f_m, f, and f_n are measured boundary, original boundary, and the noise vectors, respectively. Since we do not have any information about the original boundary, we find the optimally estimated smooth boundary (f_e) using constrained regularization. The following smoothness requirement is used as a *stabilizing functional* to obtain the smooth curvature estimate:

$$\min_{f_e} \int \left(f_e''(u)\right)^2 du \tag{2}$$

Using a discrete form, we summarize the problem as follows:

***Problem Statement*:** Find a boundary vector f_e to minimize the following stabilizing functional:

$$\min \left\{\sum \left[f_e(k+1) - 2f_e(k) + f_e(k-1)\right]^2\right\} \tag{3}$$

subject to the following noise equality constraint,

$$(f_m - f_e)^T (f_m - f_e) = \varepsilon = N\sigma_n^2 \tag{4}$$

where N is a number of data points, ε is a residual error and σ_n^2 is a noise variance.

■

We do not know the noise function, but we assume we know some of its statistical properties,

$$f_n^T f_n = \sum_k f_n^2(k) = \varepsilon$$
$$\text{if } E\{f_n(k)\} = 0, \text{ then } \varepsilon \approx N\sigma_n^2 \tag{5}$$

By combining Eq. 3 and Eq. 4, we obtain the following objective function,

$$J(f_e, \lambda) = f_e^T C_e^T C_e f_e + \lambda[(f_m - f_e)^T (f_m - f_e) - \varepsilon] \tag{6}$$

where λ is a Lagrange multiplier and C_e is a $N \times N$ *circulant matrix*,

$$C_e = \begin{bmatrix} -2 & 1 & 0 & \ldots & 0 & 0 & 1 \\ 1 & -2 & 1 & \ldots & 0 & 0 & 0 \\ & & & \vdots & & & \\ 0 & 0 & 0 & \ldots & 1 & -2 & 1 \\ 1 & 0 & 0 & \ldots & 0 & 1 & -2 \end{bmatrix} \tag{7}$$

Differentiating Eq. 6 with respect to f_e and λ yields

$$(f_m - f_e)^T (f_m - f_e) = \varepsilon = N\sigma_n^2$$
$$f_e = \left(I + \frac{1}{\lambda} C_e^T C_e\right)^{-1} f_m = \left(I + \gamma C_e^T C_e\right)^{-1} f_m \tag{8}$$

where $\gamma = 1/\lambda$ and I is an $N \times N$ identity matrix.

Since the size of the matrix used in Eq. 8 is $N \times N$ and N is usually large, it is clear that the inversion of the matrix is impractical. This difficulty can be avoided by making use of the properties of circulant matrices [7]. That is, we can estimate the boundary in a frequency domain using Discrete Fourier Transform (DFT) as follows:

$$F_e(u) = \frac{1}{1 + \gamma (D_c)_{ii}} F_m(u) \tag{9}$$

where $F_e(u)$ and $F_m(u)$ are DFTs of $f_e(k)$ and $f_m(k)$ and D_c is the diagonalized matrix of $C_e^T C_e$. Finally, we obtain the optimally estimated boundary vector, f_e as follows:

$$f_e = DFT^{-1}\{F_e(u)\} \tag{10}$$

The optimal value γ^* is iteratively obtained according to the constraint (Eq. 4). The constrained regularization technique using the noise equality constraint is a *data-driven* method. The level of noise of the boundary data determines the amount of smoothing.

3. Curvature Estimation

The curvature is generally defined as the derivative of the tangent angle to the curve. The curve can be parametrized by the parameter t (not necessarily the arclength parameter) and is represented by two single-valued functions as $x=X(t)$ and $y=Y(t)$. Then, the curvature function is obtained as:

$$\kappa(t) = \frac{d\theta}{ds} = \frac{\dot{X}(t)\ddot{Y}(t) - \ddot{X}(t)\dot{Y}(t)}{\left[\dot{X}^2(t) + \dot{Y}^2(t)\right]^{3/2}} \tag{11}$$

In a discrete image, the parametrized boundary is represented by an 8-neighbor Freeman chain code, i.e., it is represented by only 8 integer values (0-7), and is therefore severely affected by quantization. Since the formula to compute the curvature function (Eq. 11) involves the first and second derivatives and the data itself is noisy, the resulting curvature function computed on a digitized boundary is very ragged, and is difficult to use for further processing. Fig. 1 shows a boundary for the gun which was acquired through DT-2851 image acquisition system. Fig. 2(a) shows the resulting k-curvature function with k=1 computed by Groan-Verbeek's formula [6]. It shows the dominance of the quantization noise without smoothing, clearly too ragged to be used for further processing. As the k value is increased, it becomes less ragged while losing more of the local characteristics as shown in Fig. 2(b).

Filtering techniques such as *lowpass*, *Gaussian*, etc. filters may also be used to reduce the effect of quantization. Mokhtarian and Mackworth [8] use Gaussian smoothing to compute curvature at varying levels of detail. The Gaussian smoothed first and second derivatives of $X(t,\sigma)$ are obtained as:

$$\dot{X}(t,\sigma) = x(t) * \left(\frac{\partial g(t,\sigma)}{\partial t}\right)$$
$$\ddot{X}(t,\sigma) = x(t) * \left(\frac{\partial^2 g(t,\sigma)}{\partial t^2}\right) \quad (12)$$

Similarly, $\dot{Y}(t,\sigma)$ *and* $\ddot{Y}(t,\sigma)$ can be computed. Rewriting Eq. 11, the discrete curvature using Gaussian smoothing becomes:

$$k(t,\sigma) = \frac{\dot{X}(t,\sigma)\ddot{Y}(t,\sigma) - \ddot{X}(t,\sigma)\dot{Y}(t,\sigma)}{\left(\dot{X}^2(t,\sigma) + \dot{Y}^2(t,\sigma)\right)^{3/2}} \quad (13)$$

Fig. 3 shows the results after the Gaussian smoothing for a couple of values of σ. We have noisy estimate of curvature with σ=1, i.e., it is *undersmoothed* as a result of insufficient filtering. On the other hand, the filter with σ=8 *oversmooths* the curvature function. Thus, the curvature extrema are significantly attenuated, and poorly localized. However, we can get rid of the above problem by using the optimal smooth boundary ($x_e(t)$ and $y_e(t)$), which was obtained from the previous section, for curvature computation.

4. Experiments and Discussions

Fig. 4(a) shows a smooth boundary generated by the following parametric equation [9]:

$$x = s_f \times [50\sin(t+90)+20\sin(3t+50)] + x_0$$
$$y = s_f \times [60\sin(t+40)+20\sin(2t+70)] + y_0 \quad (14)$$

The continuous boundary was sampled at 360 points corresponding to t = 0,1, ... ,359 degrees. The curvature as a function of t of this smooth curve was obtained by Eq. 13 (Fig. 4-(b)). The original smooth boundary curve is quantized to test the performance of the algorithm. Fig. 5(a) shows the noisy curvature function of the quantized boundary computed by Eq. 13. The algorithm is applied to the x- and y-coordinates separately. The optimal values obtained by the above procedure for the quantized boundary are γ_x^* = 1069 and γ_y^* = 2900. Then the curvature function is computed using the smoothed x- and y-coordinate values according to Eq. 13. Fig. 5(b) shows the resulting curvature function about t. It shows the excellent results of the algorithm. However, each interval (t) has different arclength in general due to the effect of the Freeman chain code. For further processing we should make them uniform about the arclength. A *linear interpolation* was used for this purpose. Now, all the intervals have unique arclengths.

In order to see the effect of *scale* factor the boundary contour was generated with a scale factor, s_f=0.5 in Eq. 14, and quantized. The optimal values, γ_x^* = 1585 and γ_y^* = 5457, were obtained for this boundary. The resulting curvature function after the algorithm is applied is also very smooth and has the twice larger curvature value than the original one. We can see that the curvature functions are inversely proportional to the scale factor. However, the *scale* invariance is remedied by normalizing the curvature function with respect to the arclength as shown in Fig. 5(c). The original boundary is rotated by 150 degrees, translated by m_x = 20 and m_y = 20, and then quantized to see the rotation and translation invariance properties. The algorithm is applied to this boundary and the optimal values of γ are found as γ_x^* = 897 and γ_y^* = 1643. We can easily see the rotation, scale, and translation invariance properties.

5. Conclusion

Issues related to the optimal boundary smoothing for the invariant curvature estimation have been discussed in this paper. The current methods to compute curvature on a digitized boundary have common difficulty in determining a unique smoothing factor. We solved this problem by applying a constrained regularization technique to the digitized boundary. In addition, the algorithm preserves the useful invariance properties of the curvature function by normalizing it to the arclength. Thus, we can use the normalized curvature function with the valuable rotation, scale, and translation invariant properties which are necessary for the boundary representation and reliable matching.

References

[1] F. Attneave, "Some informational aspects of visual perception," *Psychol. Rev.*, vol 61, no 3, pp 183-193, 1954.

[2] J. Hadamard, *Lectures on the cauchy problem in linear partial differential equations*, Yale University Press, 1923.

[3] A. Rosenfeld and E. Johnston, "Angle detection on digital curves," *IEEE Trans. Comput.*, vol C-22, pp 875-878, Sep. 1973.

[4] A. Rosenfeld and J. Weszka, "An improved method of angle detection on digital curves," *IEEE Trans. Comput.*, vol C-24, pp 940-941, Sep. 1975.

[5] I. M. Anderson and J. C. Bezdek, "Curvature and tangential deflection of discrete arcs: A theory based on the commutator of scatter matrix pairs and its application to vertex detection in planar shape data," *IEEE Trans. Pattern Anal. Machine Intell.*, vol PAMI-6, pp 27-40, Jan. 1984.

[6] F. C. Groan and P. W. Verbeek, "Freeman-code probabilities of object boundary quantized contours," *Comput. Vision, Graphics, Image Processing*, vol 7, pp 391-402, 1978.

[7] B. Hunt, "The application of constrained least squares estimation of image restoration by digital computer," *IEEE Trans. Comput.*, vol C-22, no 9, pp 805-812, Sep. 1973.

[8] F. Mokhtarian and A. Mackworth, "Scale-based description and recognition of planar curves and two-dimensional shapes," *IEEE Trans. Pattern Anal. Machine Intell.*, vol PAMI-8, no 1, Jan. 1986.

[9] B. Shahraray and D. J. Anderson, "Optimal estimation of contour properties by cross-validated regularization," *IEEE Trans. Pattern Anal. Machine Intell.*, vol PAMI-11, no 6, pp 600-610, June, 1989.

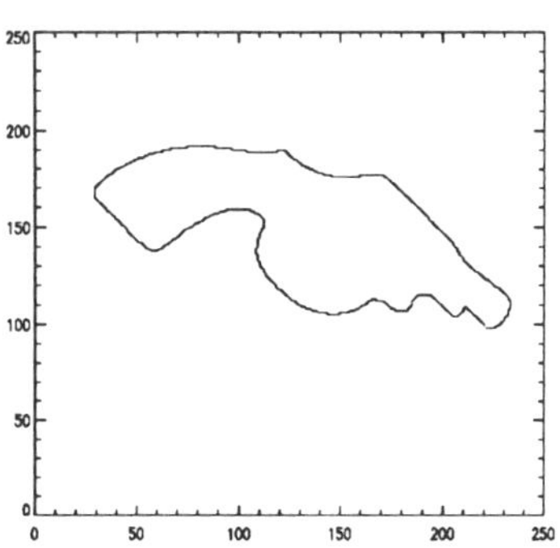

Fig. 1: Gun Boundary.

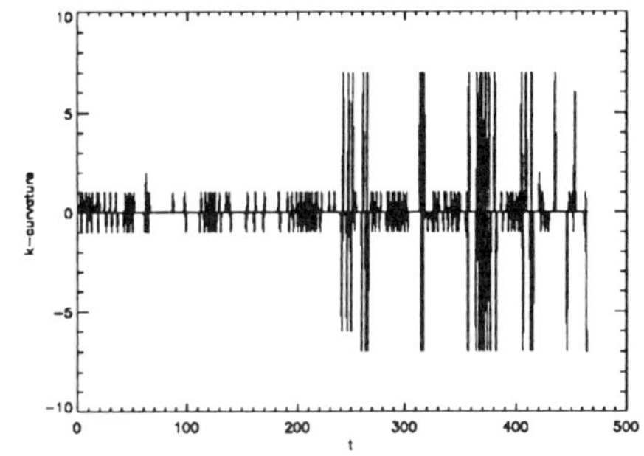

(a) k=1

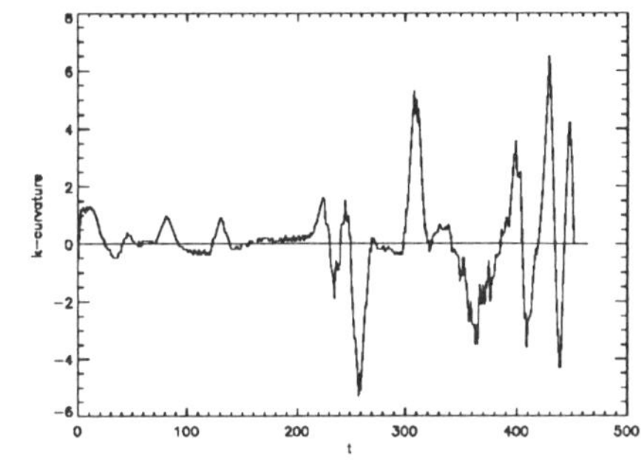

(b) k = 10

Fig 2.: k-curvature.

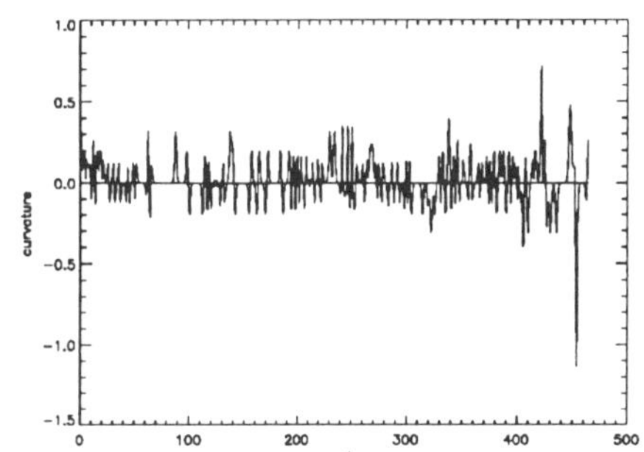

(a) σ = 1

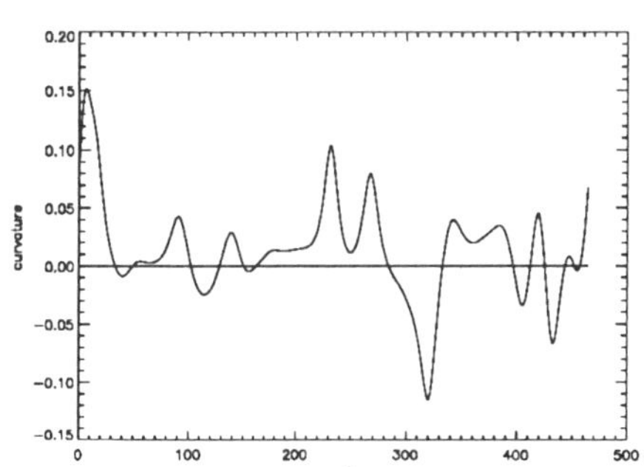

(b) σ = 8

Fig. 3: Gaussian Smoothed Curvature.

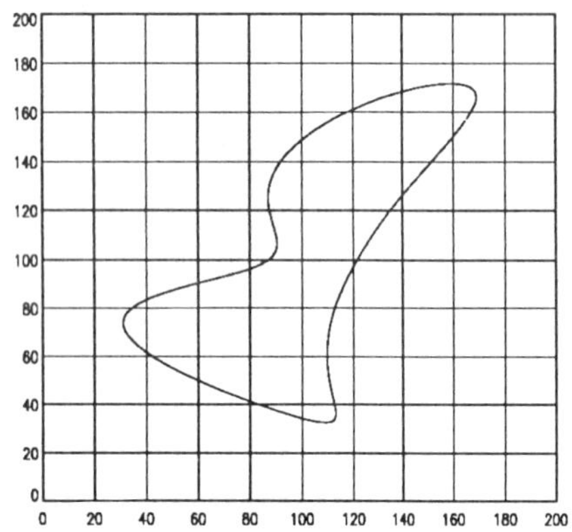

(a) Original Boundary

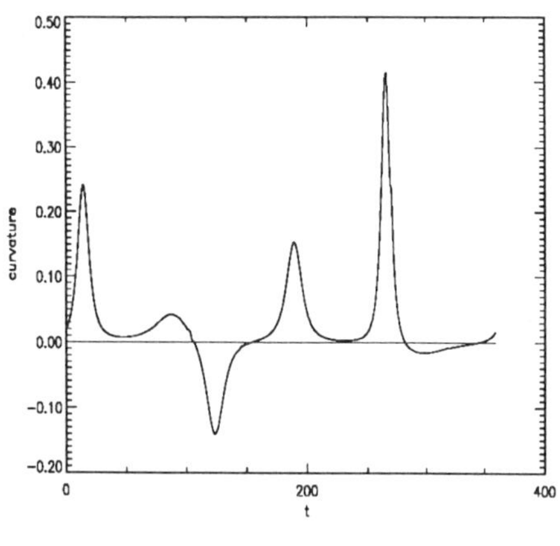

(b) Curvature (Original)

Fig. 4: Original Boundary and Curvature.

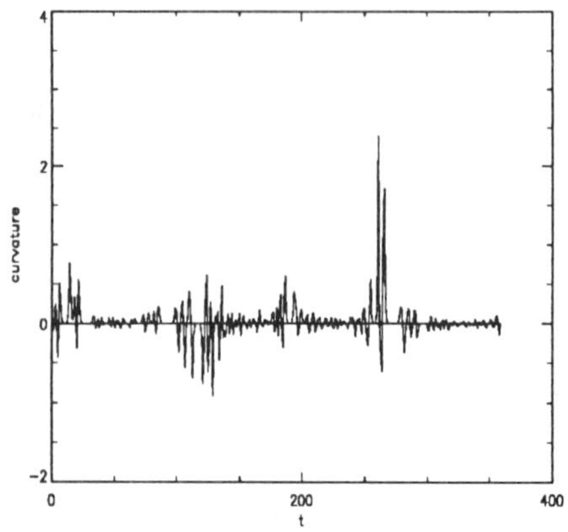

(a) Curvature (Quantized)

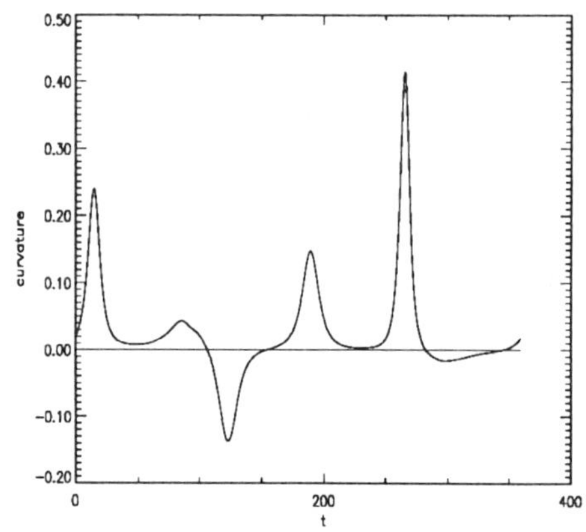

(b) Curvature (Regularized)

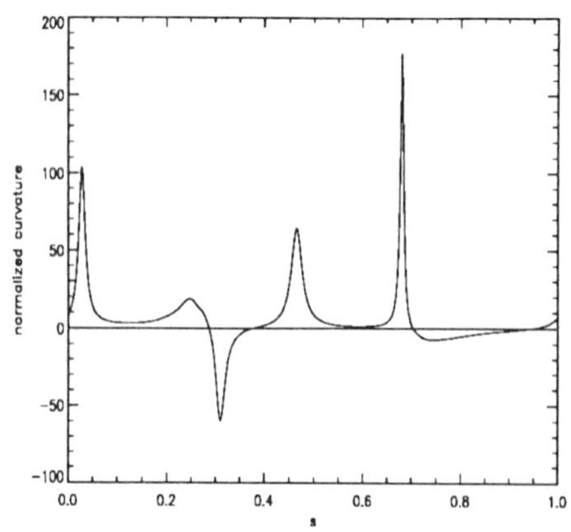

(c) Curvature (Normalized)

Fig. 5: Smoothed Curvature.

On A Consistency Issue in Multiresolution Stereo

Chienchung Chang
Qualcomm Inc.
10555 Sorrento Valley Road
San Diego, CA 92121, U.S.A.
Email:cchang@qualcomm.com

Shankar Chatterjee
Dept of Electrical and Computer Eng.
University of California, San Diego
La Jolla, CA 92093-0407, U.S.A.
Email:shankar@celece.ucsd.edu

Abstract

In this paper, we address an important problem in the multiresolution (pyramid) approach to the correspondence problem in computational vision: the *consistency* between the domain and range (disparity) pyramids. Consistency guarantees the validity of pyramidal implementation. We investigate some previous approaches in stereo vision and present a general framework in multiresolution stereo. Simulations on natural stereo pairs exhibit that results are superior with this consistency check.

1 Multiresolution Approach

When an object is located at a distance, it is hard for humans to discriminate detailed features. We can only guess at its rough shape and differentiate it from the background. As we approach the object, it becomes clearer and finer details become discernible. In the light of frequency analysis (*e.g.,* Fourier transform), an image is considered as a 2-D signal composed of different spatial frequency components. This implies that the information contained in the low-passed version is processed first by visual system. If and when needed, higher frequency components are added to depict details afterwards. Moreover, human visual system (HVS) is believed to have many different spatial frequency tuned binocular channels [8] [13] [9]. It processes various frequency components at different channels independently. The cross-channel information propagation disambiguates the object.

As shown in [1] [5], the stereo matching problem can be formulated as a constrained optimization problem. The appeal of an iterative relaxation algorithm is the feasibility of its local, parallel and distributed implementation. Nevertheless, given only the local influences of the site variables, global characteristics of the solution are arrived indirectly via the iterative propagation of local constraints across the grid lattices. Indirect propagation can lead to substantial computational inefficiency when the size of the grids tends to be extremely large. Convergence of the iterative process is often slow, and this severely erodes the power of massive parallelism. Local computations on a coarse grid over a given region are analogous to more global computations on a fine grid over the same region. This suggests the possibility of counteracting the sluggishness of global interactions by deploying local iterative algorithms over a multiresolution hierarchy of grids [17].

In *pyramidal relaxation methods*, the coarse levels trade spatial resolution. The inherent computational sluggishness of local iterative algorithms can also be interpreted from a spatial frequency perspective. Operating in the single-layer mode, the iterative relaxation algorithm usually yields rapid convergence for the high-frequency components of the solution [2] [3], which are obtainable merely from the local interactions of lattice variables. But the low-frequency components, representing more global properties, are, in general, hardly acquired. It is the slow (low-frequency) error components that resist elimination through the iterations and result in slow convergence. In the course of implementation, the majority of the iterations are simply to recover the low-frequency components of the desired solution. This dramatically neutralizes the advantages caused by local interactions. In fact, the order of the recovery of frequency components in single-resolution approach is opposite of that in the multiresolution method. It is this recovery order that endows the multiresolution approach with the merit of computation efficiency. When convergence is attained at the coarse level, that solution can be interpolated to the fine level. This interpolation constrains the low-frequency components of the desired solution at the fine level and introduces some high-frequency error which is easily reduced in just a few more iterations.

For efficient implementation of some vision algorithms, *e.g.* edge detection [18], stereo [1], shape from shading [17] [15] etc, an efficient data structure, called a multi-resolution *pyramid*, is often used to provide successively condensed representations of the information in the input image. For a given function $f(x,y)$, the pyramid is constructed as copies of $f(x,y)$ with both reduced spatial resolution and sample density. Let $f^0(x,y) \equiv f(x,y)$ be the original function (at the bottom level 0 of the pyramid). It is then iteratively transformed by an operator to generate a sequence of reduced versions of the original function. Usually the transformation operator T consists of two steps: low pass filtering (LPF) followed by subsampling (SS), *i.e.*

$$f^k = Tf^{k-1} = (LPF + SS)f^{k-1}$$
$$\triangleq [LPF(f^{k-1})]_{ss} = [f^{k-1}(x,y) \star \kappa(x,y)]_{ss} \qquad (1)$$

where $[f]_{ss}$ denotes the subsampling version of function f and f^k is the pyramidal representation at layer k. Typically, SS is regarded as a 2-D decimation process performed by a factor 2 along each axes. The low pass filtering is achieved using the convolution mask $\kappa(x,y)$, which is also called the *generating kernel*. The generating kernel is usually defined on a $(2W+1)\times(2W+1)$ rectangular window, which decides the bandwidth of the low-pass filter. When the low pass filter $\kappa(x,y)$ is a Gaussian function, the data structure is called *Gaussian pyramid* [4]. Pyramids are data structures that provide a sequence of condensed image representations. What is condensed may be image intensity, so that the constituent levels of the pyramid are reduced-resolution versions of the input image. But, it may also be descriptive information about features in the image, so that constituent levels represent increasingly coarser approximations of these features. For notational convenience, we call intensity, disparity and feature pyramids the corresponding multiscale representations of intensity, disparity and feature

1058-6393/91 $1.00 © 1991 IEEE

functions respectively. The bottom layer, level $k = 0$, is the original function $f^0(x,y) \equiv f(x,y)$ and the sequential higher level layers, $f^k(x,y)$, $k = 1, 2, ..., L-1$, are the low-pass filtered versions of $f^0(x,y)$ with size reduced accordingly. If the generating kernel is separable and symmetric [4], it can be stated more precisely as:

$$f^k(i,j) = \sum_{m=-W}^{W} \sum_{n=-W}^{W} \kappa(m,n) f^{k-1}(2i+m, 2j+n) \quad (2)$$

In multiresolution approach, the computation is first performed at the top layer (coarsest resolution) with relatively little cost. The result is then interpolated downward to the next finer layer as the initial solution. This process continues until the original grid in full-resolution is used and the final result is obtained. Among the merits of multiscale approach are: (i) a better initial condition provided from previous layer can significantly reduce computation at current layer and (ii) most iterations are concentrated on the coarse layers with computation proportional to the reduced domain size. Nevertheless, to ensure that the solution propagation down the pyramid provides a constructive effect, not an error propagation, it is required that the pyramid construction and solution interpolation be *consistent*.

The constraint propagation across the layers of pyramid structure is in the form of optimal solution interpolation. Since the interpolation of the previous level provides a better initial estimate of current level than the one randomly chosen, the search space is constrained and local search is possible (instead of the global search) at the current level. In the case of multiscale stochastic relaxation, the initial temperature $T_0^{(k)}$ at level k can be lower and the number of iterations required will be less as the process proceeds down the pyramid.

2 Consistency Across Resolution

In this section, we address an important problem in the multiresolution (pyramid) approach to the correspondence problem in computational vision: the *consistency* between the domain (*e.g.*, intensity, feature etc) and range (disparity) pyramids. Consistency validates pyramidal implementation. For the correspondence problem in stereo and motion, many existing algorithms [1] [6] [10] [12] tried to take advantage of the efficiency of pyramid structure while leaving the consistency unchecked. Since the major purpose of image analysis for machine vision is to reconstruct 3-D scenes from 2-D images, the mutual consistency here is in general referred to as the compliance between the pyramids built on 2-D image features and those based on the 3-D scene properties, in which we are interested. For example, in stereo matching, Barnard (1989) [1] employed intensity pyramid in the correspondence process. The interpolation, however, is performed on the disparity pyramid. The consistency is equivalent to there being a pair of transformation operators T, T' such that: if the disparity of a stereo pair $f_l(x,y)$, $f_r(x,y)$ at kth layer is $d^k(x,y)$, the following matching holds,

$$(Tf_l^k)(x,y) = (Tf_r^k)[x + T'd^k(x,y), y] \quad (3)$$

T and T' are called *conjugate pair* of operators such that the domain pyramid $\{f^k(x,y)\}_k$ is consistent with the range pyramid $\{d^k(x,y)\}_k$ and vice versa. That is, imposing the transformation T on the intensity pyramid implies a complementary transformation T' on disparity pyramid. Thus disparity at layer $(k+1)$, $d^{k+1}(x,y) = T'd^k(x,y)$ at layer $(k+1)$. In [1], intensity invariance is assumed, *i.e.*

$$f_l^k(x,y) = f_r^k[x + d^k(x,y), y], \; k \geq 0$$

Transformed by $T = LPF + SS$ on both sides, we have

$$Tf_l^k(x,y) = Tf_r^k[x + d^k(x,y), y]$$

$$\Rightarrow \left[\kappa(x,y) \star f_l^k(x,y)\right]_{ss} = \left[\kappa(x,y) \star f_r^k(x + d^k(x,y), y)\right]_{ss} \quad (4)$$

Hence,

$$[\kappa \star f_l^k](2x, 2y) = [\kappa \star f_r^k](2[x + \frac{d^k(2x,2y)}{2}], 2y)$$

$$\Rightarrow f_l^{k+1}(x,y) = f_r^{k+1}(x + \frac{d^k(2x,2y)}{2}, y) \triangleq f_r^{k+1}(x + d^{k+1}(x,y), y)$$

Thus, a multiscale representation of disparity consistent with the corresponding intensity pyramid is:

$$d^{k+1}(x,y) = \frac{d^k(2x,2y)}{2} \quad (5)$$

If $T = (LPF + SS)$, then the complementary operator is $T' = \frac{SS}{2}$. Namely, from the intensity invariance, it turns out that the intensity pyramid associated with T will give rise to a disparity pyramid formed by directly scaled subsampling, irrelevant of the low pass filter used. If the intensity is really invariant, the filter may be in any simple form as long as it can alleviate the aliasing effect due to the subsampling [14]. As shown in [7], if we impose on the generating kernel a constraint: $f^k(x,y)$ is to reconstruct $f^{k-1}(x,y)$ as close as possible at the reduced sampling rate, the optimum $\kappa(x,y)$ will be in the form of ideal low pass filter with Fourier transform pair defined as:

$$\kappa(x,y) = \frac{\sin x\frac{\pi}{2}}{x\pi} \frac{\sin y\frac{\pi}{2}}{y\pi} \quad (6)$$

$$K_d(u,v) = \begin{cases} 1 & 0 \leq |u|, |v| < \frac{\pi}{2} \\ 0 & \frac{\pi}{2} \leq |u|, |v| < \pi \end{cases} \quad (7)$$

It is capable of removing all components with frequencies beyond half the sampling frequency at $(k-1)$st level, which are major sources in causing aliasing effects during subsequent subsampling operation. It also keeps the low-pass information from distortion since the gain of the rectangular passband is equal to one. However, the *sinc* function (6) is persistent in spatial domain to an infinite extent, which makes the ideal low pass filter physically unrealizable. In general, alternatives, such as equiripple filters, are employed instead. The performance of the generating kernel is measured by their similarity to the ideal low pass filter in terms of two spectral indexes [14]:

- *Antialiasing index*:

$$\eta = \frac{\int_0^{\frac{\pi}{2}} \int_0^{\frac{\pi}{2}} |K(u,v)| du dv}{\int_0^{\pi} \int_0^{\pi} |K(u,v)| du dv} \quad (8)$$

- *Filtering index*:

$$\mu = \int_0^{\frac{\pi}{2}} \int_0^{\frac{\pi}{2}} \left| 1 - |K(u,v)| \right| du dv \quad (9)$$

The antialiasing index indicates the fraction of signal spectrum that will be folded back (aliased) into the spectral domain $(0, \frac{\pi}{2})$ at the next coarse level. Subsampling will be free from aliasing whenever the filter $K(u,v)$ makes $\eta = 1$. The filtering index reflects the degree of deviation from the ideal low pass filter $K_d(u,v)$. $\mu \geq 0$ with equality when $K(u,v) \equiv K_d(u,v)$. In fact, if the intensity is smooth and bandlimited, filtering may be omitted if we can sample the image beyond the Nyquist sampling rate.

On the other hand, if we want the solution in the disparity pyramid to be propagated in terms of a simple interpolation scheme similar to (4),

$$d^k(2x, 2y) = 2d^{k+1}(x,y),$$

the surface normal (p, q) at each layer will be identical:

$$p^{k+1}(x,y) = \frac{\partial d^{k+1}(x,y)}{\partial x} = \frac{\partial d^k(2x,2y)}{\partial(2x)} = p^k(2x,2y)$$

Similarly, $q^{k+1}(x,y) = q^k(2x,2y)$. In the usual model of image formation, the amount of light reflected by a surface patch is the reflectance of that patch multiplied by the incident light and the cosine of the difference between the angle of incidence and the angle of observer. If the surface is Lambertian, the intensity is proportional to the dot product of the surface normal and the incident illumination vector [11], *i.e.*,

$$f(x,y) \propto \frac{1 + p_s p + q_s q}{\sqrt{1 + p^2(x,y) + q^2(x,y)}}$$

where $(-p_s, -q_s, 1)$ and $(-p, -q, 1)$ denote the directions of light source and surface normal respectively. Note that the interpolation on the disparity pyramid according to (5) implies direct subsampling of the intensity,

$$f_s^{k+1}(x,y) = f_s^k(2x,2y), \quad s = l, r$$

and the intensity invariance for the stereo pairs at each resolution, *i.e.* $f_l^k(x,y) = f_r^k(x + d^k(x,y), y)$ for all k. Thus $T \Rightarrow T'$ if the surface is Lambertian. Unfortunately, most 3-D surfaces are not Lambertian (*e.g.*, specular) and intensities are far from invariant at different viewpoints. In this case, consistency between intensity and disparity pyramids derived above will become questionable.

To reliably reconstruct 3-D surfaces, robust features representing invariant properties can be derived from the image intensities. Feature invariance does *not* necessarily imply intensity invariance. Dense features, *e.g.* from the filtering of the intensity function, are usually the properties summarized over small neighborhoods of intensity surfaces. Many features, such as intensity gradients, zero crossings, edgels, Gabor features [5], characterize the relative image intensities and are thus more robust than the absolute intensities. If we process these features directly by the operator $T = LPF + SS$ to form a *feature pyramid* $\{h^k(\mathbf{x})\}_{k=0}^{L-1}$

$$h_s^k(\mathbf{x}) = T h_s^{k-1}(\mathbf{x}), \quad k \geq 1; s = l, r \tag{10}$$

$$h_s^0(\mathbf{x}) = \hat{T} f_s(\mathbf{x}), \quad s = l, r \tag{11}$$

where $\hat{T}$ is an operator used to extract feature function $h_s^0(\mathbf{x})$ from original intensity profiles $f_s(\mathbf{x})$. Following a similar derivation as in equations (4), (5), it can be easily shown that the feature invariance does give rise to a similar relationship between the feature and disparity pyramids, *i.e.*

$$h_l(\mathbf{x}) = h_r(\mathbf{x}+(d_{\mathbf{x}}, 0)) \text{ and } h_s^k = T h_s^{k-1} \Rightarrow d^k(\mathbf{x}) = \frac{d^{k-1}(2\mathbf{x})}{2} \tag{12}$$

However, if one constructs the intensity pyramid according to T and extract the features from the corresponding intensity profiles at each layer according to $\hat{T}$, it will, in general, violate the consistency criterion. Note that the two operators T and $\hat{T}$ usually do not commute:

$$T\hat{T}f \neq \hat{T}Tf \tag{13}$$

In summary, the feature pyramid, defined by (10) and (11), yields a consistent disparity pyramid defined by direction subsampling if the feature itself stands for invariant properties of intensity surfaces.

3 Matching Feature Vectors

A. Data Compatibility

Assume that a stereo pair of images $f_l(\mathbf{x})$, $f_r(\mathbf{x})$, defined over a 2-D regular $(N \times N)$ lattice $\Omega = \{\mathbf{x} = (m,n), 1 \leq m,n \leq N\}$, are used to generate two feature vector function $\mathbf{W}_l(\mathbf{x};\lambda)$, $\mathbf{W}_r(\mathbf{x};\lambda)$ in terms of Gabor wavelet transform [5]

$$\mathbf{W}_s(\mathbf{x};\lambda) = \{f_s(\mathbf{x}) \star g(\mathbf{x};\lambda,k,\alpha^j)\}_{k,j\in\mathbf{M}}, \quad s = l, r; \quad \mathbf{x} \in \Omega$$

where $g(\mathbf{x};\lambda,k,\alpha^j)$ denotes the discrete Gabor wavelet

$$g(m,n;\lambda,k,\alpha^j) =$$

$$\exp\left\{-\frac{\alpha^{2j}}{2}[\lambda m^2 + n^2] + i\pi\alpha^j[m\cos\frac{k\pi}{N} + n\sin\frac{k\pi}{N}]\right\} \tag{14}$$

and $m, n, k, j \in \mathbf{M}$ (the set of integers). Since corresponding image points are the projection of identical 3-D scene points, the feature vectors associated with them must be compatible. This entails compatibility constraint from observations, which can be characterized by a cost term in the performance functional

$$U_d(\mathbf{f}_l, \mathbf{f}_r|\mathbf{d}) = \sum_{\mathbf{x}\in\Omega} \left|\mathbf{W}_l(\mathbf{x};\lambda) - \mathbf{W}_r(\mathbf{x} + (d_{\mathbf{x}}, 0);\lambda)\right|^2 \tag{15}$$

where $\mathbf{d} = \{d_{\mathbf{x}}, \mathbf{x} \in \Omega\}$ denotes the disparity vector concatenated by all the nodal variables of the 2-D disparity function d(x,y) (disparity map) defined on Ω. $\mathbf{f}_l$, $\mathbf{f}_r$ are the vector representations of the stereo images. The performance term from observation is a metric of sum of squared errors (SSE).

B. Surface Smoothness

Some configurations of object surfaces in the real world are much more likely to occur than others. From observation, most existing 3-D surfaces are smooth almost everywhere except at depth or orientation discontinuities. The areas covered by these discontinuities are, nevertheless, negligible compared to whole image field. Thus it is at least experimentally feasible to impose the smoothness constraint over entire image plane. For a specific lattice point, smoothness may be encoded as the similarity of its disparity value with those of neighboring lattice points. Defining a homogeneous neighborhood structure associated with Ω as

$$\aleph_r = \{N_{\mathbf{x}}, \mathbf{x} \in \Omega\}, \quad N_{\mathbf{x}} = \{\mathbf{y}, |\mathbf{y} - \mathbf{x}|^2 \leq r\}$$

where $\aleph_r$ is called the *neighborhood* with radius r. The smoothness constraint can be described by a cost term

$$U_s(\mathbf{d}) = \sum_{\mathbf{x}\in\Omega} \sum_{\mathbf{y}\in N_{\mathbf{x}}} (d_{\mathbf{x}} - d_{\mathbf{y}})^2 \tag{16}$$

C. The Performance Functional

To ensure data compatibility as well as surface smoothness be satisfied at the same time, we can formulate the performance functional as a linear weighted sum of each individual cost:

$$U_p(\mathbf{d}|\mathbf{f}_l, \mathbf{f}_r) = U_d(\mathbf{f}_l, \mathbf{f}_r|\mathbf{d}) + \beta U_s(\mathbf{d}) + \gamma U_{god}(\mathbf{d}) \tag{17}$$

where $\beta \geq 0$, γ are the weighting coefficients of resultant performance function and U_{god} is the cost due to the gradient of disparity (GOD) constraint [5] [16], which is defined on a second neighborhood structure $\aleph'_r$ with radius r. The optimal disparity map is obtained as the function achieving the minimum of the overall performance functional U_p.

D. Consistency in Gabor Feature Pyramid

Gabor features bear biological evidence and are due to the oriented variations of intensity function, in effect, they represent invariant properties of intensity surface in a robust sense. In the pyramidal implementation of Gabor feature matching, using similar arguments as before, it can be easily shown that consistency does exist between the *Gabor feature pyramid* and the corresponding disparity pyramid. As stated in Section 2, if we let the feature pyramid in (10) and (11) be

$$\left\{h^k(\mathbf{x})\right\}_{k=0}^{K-1} \equiv \left\{\mathbf{W}^k(\mathbf{x};\lambda)\right\}_{k=0}^{K-1}$$

$$W_s^0(\mathbf{x};\lambda,\theta,\alpha) = \hat{T}_{\theta,\alpha} f_s(\mathbf{x}), \quad s = l, r \tag{18}$$

where $\hat{T}$ is an operator used to extract Gabor feature vector $\mathbf{W}_s(\mathbf{x})$ from intensity profiles $f_s(\mathbf{x})$. In the next subsection, we will show the fact experimentally: stereo matching using Gabor feature pyramid is superior to the one based on usual intensity pyramid.

4 Experimental Results and Discussion

The stereo pairs of images used in our algorithm include the Pentagon and Tree images. The bank of Gabor filters used in the simulation are with same aspect ratio $\lambda = 3$ and include 4 orientations (0, $\pi/4$, $\pi/2$ and $3\pi/4$ radians), 3 scales. Thus the feature vector at each lattice point consists of $K = 12$ complex components. The initial disparity map $\mathbf{d}_0$ is set as a zero vector in the experiments such that data compatibility constraint dominates the energy functional in the beginning of annealing. The initial temperatures are chosen to be 20.0 for Pentagon and Tree. For the sake of simplicity, the neighborhood structures used for smoothness and GOD constraints are assumed to be identical.

Images	Initial Temp	Ann Sched	Iter #	Neigh
Pentagon	$T_0 = 20.0$	1.0% off	700	$\aleph_2$
Tree	$T_0 = 20.0$	2.5% off	300	$\aleph_2$

Table 1: The computer simulation environments for single resolution case.

Tested	Level 2		Level 1		Level 0	
Images	$T_0^{(2)}$	Iter #	$T_0^{(1)}$	Iter #	$T_0^{(0)}$	Iter #
Tree	20.0	260	2.0	120	0.2	25
Pentagon	20.0	700	2.0	150	0.2	40

Table 2: The computer simulation environments in Gabor feature pyramid, in which the annealing schedules are identical to the single resolution case.

In our examples, we adopted neighborhood with radius 2, $\aleph_2$. For the single resolution case, the simulation environments are summarized in Table 1. For the multiresolution approach, a Gabor feature pyramid is constructed in accordance with a simple average followed by direct subsampling operations, which are imposed upon Gabor feature vectors. Due to the resolution of finite image size, 3-layer pyramid structure is used. The experimental results are shown in Figure 1 and 2 and the parameters of computer simulation in multiresolution approach are listed in Table 2. It can be easily seen that the performance based on Gabor feature pyramid is superior to the one based on Gaussian intensity pyramid under the identical simulation environments. It is also noted that the results from feature pyramid are closer to those in single resolution case, compared to intensity pyramid results. The fact that mutual consistency between domain (intensity or feature) and range (disparity) pyramids plays a critical role in multiscale implementation is verified experimentally.

References

[1] S.T. Barnard, Stochastic stereo matching over scale, *Intl. J. of Computer Vision*, vol 3:17–32, 1989.

[2] A. Brandt, Multi-level adaptive solutions to boundary-value problems, *Math. Comput.*, Vol.31:333–390, 1977.

[3] A. Brandt, Multi-level adaptive techniques (mlat) for partial differential equations: ideas and software, In J.R. Rice, editor, *Mathematical Software III*, pages 277–318. Academic Press, New York, NY, 1977.

[4] P.J. Burt, Fast filter transforms for image processing, *Comp. Vision, Graphic and Imag. Proc.*, Vol.16:20–51, 1981.

[5] C. Chang, *Area-Based Methods for Stereo Vision: the Computational Aspects and Applications*, PhD thesis, University of California, San Diego, 1991. Dept. of ECE.

[6] C. Chang and S. Chatterjee, Multiresolution stereo using simulated annealing, In *Proc. Intl. Joint Conf. on Neural Network*, pages II–885–890, San Diego, CA, June 1990.

[7] R.E. Crochiere and L.R. Rabiner, *Multirate Digital Signal Processing*, Prentice-Hall, Englewood Cliffs, NJ, 1983.

[8] F. Glazer, Multilevel relazation in low level computer vision, In A. Rosenfeld, editor, *Multiresolution Image Processing and Analysis*, pages 312–330. Springer-Verlag, New York, NY, 1984.

[9] C.S. Harris, editor, *Visual Coding and Adaptability*, Lawrence Erlbaum Assoc., Hillsdale, New Jersey, 1980.

[10] W. Hoff and N. Ahuja, Srufaces from stereo: integrating feature matching, disparity estimation, and contour detection, *IEEE Trans. on Patt. Anal. & Mach. Intel.*, PAMI-11(No.2):121–136, Feb 1989.

[11] B.K.P. Horn, *Robot Vision*, The MIT Press, Cambridge, MA, 1986.

[12] J. Konard and E. Dubois, Multigrid Bayesian estimation of image motion fields using stochastic relaxation, In *Proc. 2nd Intl. Conf. on Computer Vision*, pages 354–362, Tampa, Florida, Dec. 1988.

[13] D. Marr and T. Poggio, A computational theory of human stereo vision, *Royal Soc. of London*, B-204:301–328, 1979.

[14] P. Meer, E.S. Baugher, and A. Rosenfeld, Frequency domain analysis and synthesis of image pyramid generating kernels, *IEEE Trans. on Patt. Anal. & Mach. Intel.*, PAMI-9(No.4):512–522, July 1987.

[15] S. Peleg and G. Ron, Nonlinear multiresolution: a shape-from-shading example, *IEEE Trans. on Patt. Anal. & Mach. Intel.*, PAMI-12(No.12):1206–1210, Dec. 1990.

[16] S. Pollard, J. Mayhew, and J. Frisby, PMF: a stereo correspondence algorithm using disparity gradient limit, *Perception*, 14:449–470, 1985.

[17] D. Terzopoulos, Image analysis using multigrid relaxation methods, *IEEE Trans. on Patt. Anal. & Mach. Intel.*, PAMI-8(No.2):129–139, Mar. 1986.

[18] A. Witkin, Scale space filtering, In *Intl Joint Conf on Artificial Intelligence*, pages 1019–1021, Karlsruhe, France, 1983.

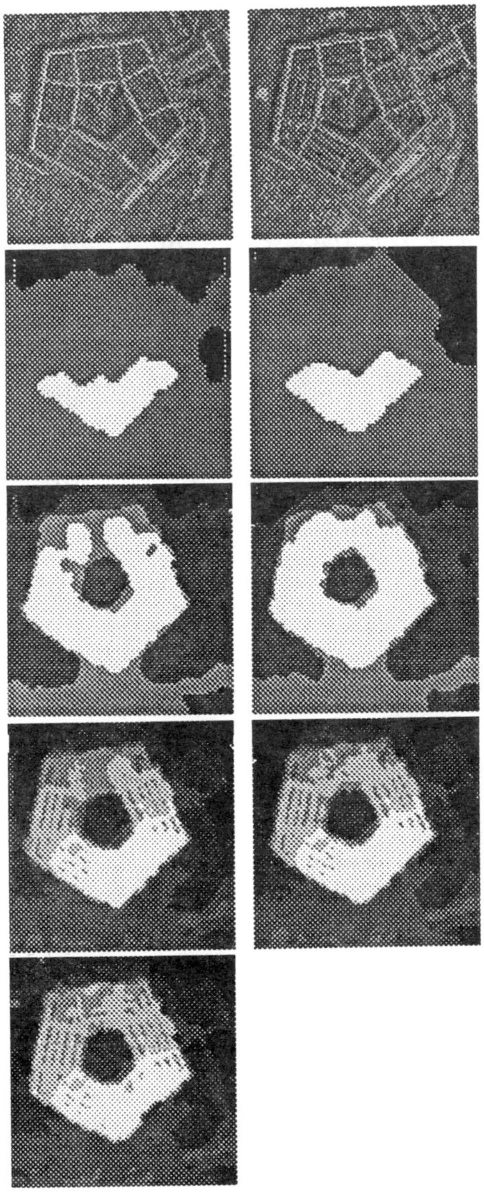

Figure 1: row 1: the original stereo images of Pentagon (256 x 256). row 2-4 (left column): the results due to multiresolution relaxation based on Gibbs sampler and Gabor intensity pyramid. The resolution is increased from row 2 to row 4. row 2-4 (right column): the results due to multiresolution relaxation based on Gibbs sampler and Gabor feature pyramid. It is readily perceived that the results based on feature pyramid are better than those from intensity pyramid, such as the *nutch* in the Pentagon (left of row 3). row 4: the result in single resolution case.

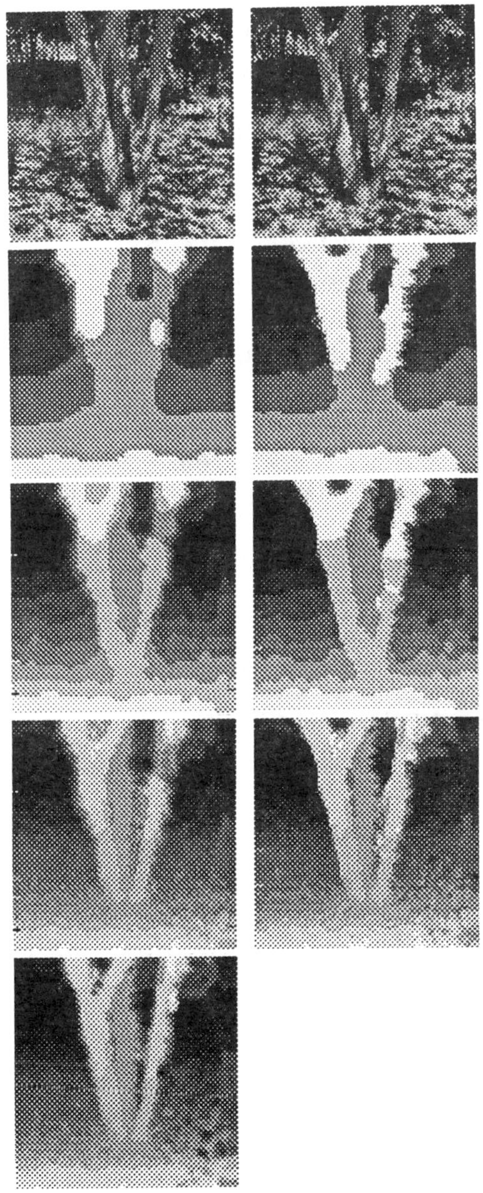

Figure 2: row 1: the original stereo images of Tree (256 x 256). row 2-4 (left column): the results due to multiresolution relaxation based on Gibbs sampler and Gabor intensity pyramid. The resolution is increased from row 2 to row 4. row 2-4 (right column): the results due to multiresolution relaxation based on Gibbs sampler and Gabor feature pyramid. It is readily perceived that the results based on feature pyramid are better than those from intensity pyramid, such as the trunk is *broken* (left of row 3 and 4). row 4: the result in single resolution case.

A Hybrid Approach Toward Model-Based Texture Segmentation*

Chienchung Chang
Qualcomm Inc.
10555 Sorrento Valley Road
San Diego, CA 92121, U.S.A.
Email:cchang@qualcomm.com

Shankar Chatterjee
Dept of Electrical and Computer Eng.
University of California, San Diego
La Jolla, CA 92093-0407, U.S.A.
Email:shankar@celece.ucsd.edu

Abstract

In this paper, we develop a hybrid texture segmentation algorithm, combining statistical (*maximum likelihood / maximum aposteriori*) and structural (*local consensus*) classification techniques. This algorithm exhibits several advantages over traditional algorithms based solely on statistical models: (i) homogeneity and integrity of each region in the texture mosaic can be included implicitly by a local voting scheme, in addition to explicit modelling through Gibbs random fields, (ii) modified maximum likelihood solution of hybrid segmentation algorithm provides a better initial estimate of region boundaries which can be used in a maximum a posteriori segmentation algorithm. With this initial estimate, an iterative, semi-deterministic relaxation algorithm, called *mean field annealing* (MFA), is used to locate the nearly global optimum solution efficiently.

1 Introduction

Most objects in the real world have textured surfaces. Since object boundaries are often important and informative components of the scene, it is critical to locate these boundaries before further processing of the image data in the areas such as remote sensing and analysis, computer vision etc. Texture statistics provide valuable information in locating the object boundaries which are intrinsically different from intensity edges insofaras they cannot be obtained directly from traditional edge detectors. In texture segmentation one deals with arbitrary texture types in regions of varying size and location. We assume that each constituent object has uniform pixel statistics as well as textural signatures, and the boundaries can be located completely by texture discriminants, *i.e.* these textured image regions can be segmented by merely localizing changes in their textural properties.

There are many existing approaches for texture segmentation. One powerful method is the model-based approach, mainly exploiting the embedded statistical properties obtained from the gray level image. In this paper, we present a hybrid texture segmentation algorithm, which integrates statistical and structural segmentation techniques constructively. Statistically, we model the intensities of each homogeneous region as a noncausal Gaussian Markov random field (GMRF) [1], where the correlation of pixel intensities can be well described. A second random variable (unobserved) is assigned to each site to account for scene constraints (*e.g.*, smoothness, homogeneity etc). The overall scene is then modelled as a realization of a Gibbs random fields (GRF) [17]. The scene constraints, formulated as penalized prior term, can aid in solving ill-posed problems in early vision modules and provide stable solutions. The statistical models, characterized by Markov random fields (MRF), renders image segmentation based on texture as a combinatorial optimization problem. The local property of MRF also facilitates the feasibility of parallel implementation of the relaxation algorithm.

To emphasize the structural integrity of the regions (and local dependence of textural attributes), a local voting (LV) [15] [7] [12] technique is introduced. This "post processing" attempts to resolve the ambiguities at the boundaries as well as within homogeneous regions of the textured scene. It also provides an interaction mechanism to enhance the intra-region correlation. The pixels are labelled in accordance with the scores from the competing texture candidates in a local support region. Counting the voting scores, measured by a function that minimizes the associated energy of MRF, from its neighbors in the support region, classification at the pixel site is accomplished based on a simple *winner-take-all* (WTA) rule.

Since such hybrid algorithm incorporates the natural constraints through LV, one obtains an improved *maximum likelihood* (ML) solution, over other segmentation algorithms which classify the pixels in a window, but retain the label only for the central pixel. It can be utilized to initialize an iterative, deterministic relaxation algorithm, called *mean field annealing* (MFA), to obtain the *maximum a posteriori* (MAP) estimates of pixel labels.

2 Image Model

In this section we briefly review the statistical formulation of both the observable intensity function and underlying texture classes in terms of GMRF and GRF respectively. We use a *doubly stochastic* model in the formulation, where two random variables are assigned to a pixel; one for characterizing the underlying texture (observed) and the other for class label (unobserved) [5].

2.1 GMRF Model for Texture

Let the intensity function, defined on a two dimensional plane Ω (size $M \times M$), of a certain texture class l be $y_l(s)$ with zero mean. In order to characterize the spatial interaction of the intensity field for a given texture region by a noncausal GMRF (*auto-normal*) model with a symmetric, homogeneous neighborhood system, we model the observation at each pixel of that region to be a weighted linear sum of contributions from its neighborhood intensities compensated by a zero mean, ad-

*This work was partially supported by a grant from California Space Institute, No 89-6387.

1058-6393/91 $1.00 © 1991 IEEE

ditively colored (correlated) Gaussian random variable [1]

$$y_l(\mathbf{s}) = \sum_{\mathbf{r}\in N} \theta_l(\mathbf{r}) y_l(\mathbf{s}+\mathbf{r}) + n_l(\mathbf{s}) \quad \forall \mathbf{s} \in \Omega \tag{1}$$

where $\theta_l(\mathbf{r})$ is used to indicate the degree of influence of its neighbors and N is the symmetric neighborhood whose size is determined by the order of GMRF [6], as shown in figure 1. The correlation matrix $\mathbf{\Psi}_l = [\Psi_l(\mathbf{s},\mathbf{r})]$ of the colored Gaussian process, $\{n_l(\mathbf{s})\}_{\mathbf{s}\in\Omega}$, associated with texture class l can be expressed as

$$\Psi_l(\mathbf{s},\mathbf{r}) \stackrel{\triangle}{=} E[n_l(\mathbf{s})n_l(\mathbf{r})] = -\theta_l(\mathbf{s}-\mathbf{r})\mu_l I_{N^*}(\mathbf{s}-\mathbf{r}) \tag{2}$$

where $I_{N^*}(\mathbf{r})$ is an indicator function, which equals to 1 when $\mathbf{r} \in N^* \stackrel{\triangle}{=} N \cup (0,0)$, and zero otherwise, and $\theta_l(\mathbf{r}) = \theta_l(-\mathbf{r})$, $\theta_l(\mathbf{0}) = -1$. While $\mathbf{s} = \mathbf{r}$, it can be easily seen that $\mu_l = E[n_l^2(\mathbf{s})]$ is the variance of Gaussian random variables $n_l(\mathbf{s})$, $\forall \mathbf{s} \in \Omega$. The formulation above will lead to a probability density function for $\{y(\mathbf{s})\}_{\mathbf{s}\in\Omega}$ with Markov property, *i.e.*

$$P(y(\mathbf{s})\,|y(\mathbf{r}),\mathbf{r}\neq\mathbf{s},\mathbf{r}\in\Omega) = P(y(\mathbf{s})|\,y(\mathbf{r}),\mathbf{r}\in N_{\mathbf{s}}), \quad \forall \mathbf{s}\in\Omega$$

where $N_{\mathbf{s}} = N + \mathbf{s}$ is the shifted version of the neighbor set N. The joint distribution of the site variables, $\{y_l(\mathbf{s})\}_{\mathbf{s}\in\Omega}$, is a multivariate normal,

$$P(\mathbf{y}_l) = \frac{|\mathbf{B}(\theta_l)|^{1/2}}{(2\pi\mu_l)^{1/2}} \exp[-\frac{1}{2\mu_l}\mathbf{y}_l^T\mathbf{B}(\theta_l)\mathbf{y}_l] \tag{3}$$

where $\mathbf{y}$ is the $M^2 \times 1$ column vector corresponding to the lexicographically ordered samples of a texture intensity function on image lattice Ω. $\mathbf{B}$ is a symmetric $M^2 \times M^2$ matrix whose diagonal elements are unity and off-diagonal elements $B_{m,n}$ are $-\theta_l(\mathbf{s}_m - \mathbf{s}_n)$. $\mathbf{s}_k$ is the two-dimensional coordinate corresponding to the kth component of vector $\mathbf{y}$. An $M \times M$ patch on 2-D image plane can be classified as one of L candidate textures, in ML sense, by minimizing the negative log likelihood [4],

$$-\log P(\mathbf{y}_l) \propto U(\mathbf{y}|l) \approx \frac{1}{\mu_l}\{C_l(0) - \theta_l^T \mathbf{C}_l\}$$

$$+\frac{1}{4\pi^2}\int_\Omega \log\frac{\mu_l}{1-\theta_l^T\phi(\mathbf{s})}d\mathbf{s}, \quad l = 1,\ldots,L \tag{4}$$

$$\phi(\mathbf{s}) = col.[\cos(\frac{2\pi\mathbf{s}^T\mathbf{r}}{M})], \mathbf{r}\in N; \quad \theta_l = col.[\theta_l(\mathbf{r})], \mathbf{r}\in N$$

$$\mathbf{C}_l = col.[C_l(\mathbf{r})], \mathbf{r}\in N; \quad C_l(\mathbf{r}) = \frac{1}{M^2}\sum_{\mathbf{s}\in\Omega} y(\mathbf{s})y(\mathbf{s}+\mathbf{r})$$

where $U(\mathbf{y}|l) \equiv U(\mathbf{y}|l_{\mathbf{u}} = l,\ \forall \mathbf{u}\in\Omega)$ and $l_{\mathbf{u}}$ is the label of texture class at location $\mathbf{u}$. That is, we assign all the pixels in the the $M \times M$ patch uniformly to a certain class $l \in L$ which maximizes the likelihood function $P(\mathbf{y}_l)$.

2.2 GRF Model for Texture Mosaic

To account for region homogeneity, boundary continuity, boundary closure etc, into a statistical formulation, we model the site labels for the whole image as a Gibbs random field, $\mathbf{l} = \{l_{\mathbf{u}}, \mathbf{u}\in\Omega\}$ [17], with joint distribution

$$P(\mathbf{l}) = \frac{1}{Z}\exp\{-U(\mathbf{l})/T\}, \text{ where } U(\mathbf{l}) = \sum_{c\in S_c} V_c(\mathbf{l}) \tag{5}$$

where Z is a normalization constant called the *partition function*, T is the *temperature* constant used to control the peak of Gibbs distribution and $V_c(\mathbf{l})$ is the clique potential associated with clique[1] c, which is an element of the clique set[2] S_c. $U(\mathbf{l})$ is the total energy for the configuration $\mathbf{l}$. Since the texture region is homogeneous, we can set the clique potential as

			6			
	5	4	3	4	5	
	4	2	1	2	4	
6	3	1	x	1	3	6
	4	2	1	2	4	
	5	4	3	4	5	
			6			

Figure 1: The neighborhood set contains the number of indicated by the number up to the order of GMRF order.

$$V_c(\mathbf{l}) = \begin{cases} \beta\cdot\delta(l_{\mathbf{u}} - l_{\mathbf{v}}) & \text{when } c = \{l_{\mathbf{u}}, l_{\mathbf{v}}\} \\ 0 & \text{otherwise} \end{cases} \tag{6}$$

where δ is the Kronecker delta function and β is the measure of homogeneity inside each region. A high β value encourages integrity of region clusters.

From Bayes' rule, we are able to formulate the posterior distribution of texture classes given the texture intensities as

$$P(\mathbf{l}|\mathbf{y}) = \frac{P(\mathbf{y}|\mathbf{l})\cdot P(\mathbf{l})}{P(\mathbf{y})} \propto P(\mathbf{y}|\mathbf{l})\cdot P(\mathbf{l}) \tag{7}$$

For the $M \times M$ patch, estimating the texture class by maximizing the a posteriori probability (MAP) is equivalent to maximizing the log likelihood penalized by a prior term, (assuming $P(\mathbf{y})$ to be constant, since it depends on observed data) *i.e.*

$$\arg\max_{\mathbf{l}} P(\mathbf{l}|\mathbf{y}) \equiv \arg\max_{\mathbf{l}} \left[\log P(\mathbf{y}|\mathbf{l}) + \log P(\mathbf{l})\right] \tag{8}$$

From (4) and (5), we know the MAP estimate can be obtained through the minimization of the posteriori energy function

$$U_p(\mathbf{l}|\mathbf{y}) = U(\mathbf{y}|\mathbf{l}) + U(\mathbf{l})/T \tag{9}$$

Texture segmentation is then formulated as a constrained optimization problem involving the minimization of the posterior energy function. The global optimum solution of the posterior energy function U_p is determined by the temperature constant T. When the system is in an excited state (high temperature), the solution is equivalent to an ML estimate. As the system is cooled down, the region homogeneity constraint plays an important role in locating the ground state of the energy well.

3 Relaxation Algorithms

3.1 Segmentation by MFA

Recently, one deterministic version of the *simulated annealing* (SA) algorithm, called *mean field annealing* (MFA) which minimizes the relative entropy of the system [2], was utilized to approximate the SA algorithm efficiently and successfully in a variety of applications in early vision modules, such as image restoration [10], surface reconstruction [8] etc. It is feasible to obtain a fast, iterative and parallel solution.

The state of a heat bath system with energy U_p is reflected in the system temperature T. As mentioned previously, if we

[1] *Clique* is the set of elements that every pair of elements within the clique are neighbors each other.

[2] The clique set is defined by the choice of neighborhood [1], [9].

start the annealing process from $T = \infty$, the ML solution will be an initial estimate. ML estimate provides much better initial configuration of region boundaries than the one randomly chosen in texture segmentation. This reasonably accurate initial estimate allows the MFA algorithm to closely approximate the ideal SA algorithm.

In the framework of MFA, one starts the annealing process from a relatively high temperature. At each fixed temperature, each site is visitd and the site variable is updated with a new value, which drives U_p to a lower energy state. During each updating, the mean effect of the neighbors are taken into account. The mean behavior of each site variable can be represented by its ensemble average, which is approximated by the average value through the evolution of annealing process. The updating is performed iteratively at each site until an equilibrium[3] is achieved. This equilibrium state is used as the initial configuration for another annealing process operating at a lower temperature, when the system is cooled down according to a prescribed schedule until a predefined threshold (temperature) is reached.

Here we describe a window based texture segmentation algorithm implemented by mean field annealing. It is then modified to yield a hybrid texture segmentation algorithm in the next subsections. With the assumption of the uniformity of texture attributes in a small window of the image except when the window spreads across the region boundaries, given a texture mosaic we use a fixed size window $\Omega_{\mathbf{s}}$ (usually square or circular in shape), centered at a pixel $\mathbf{s}$, and label all the pixels in this window uniformly with a single texture class. If we denote the classified pixel labels as vector $\hat{\mathbf{l}}_{\mathbf{s}}$, it will be a homogeneous vector with all of its components equal to $\hat{l}_{\mathbf{s}}$ (attribute of central pixel). The maximum a posteriori (MAP) classification is achieved by minimizing the posterior energy function, *i.e.*,

$$\hat{\mathbf{l}}_{\mathbf{s}} = \{\hat{l}_{\mathbf{u}} = \hat{l}_{\mathbf{s}}, \forall \mathbf{u} \in \Omega_{\mathbf{s}}\} = \arg\min_{\mathbf{l}_{\mathbf{s}}} U_p(\mathbf{l}_{\mathbf{s}}|\mathbf{y}_{\mathbf{s}}) \qquad (10)$$

where $\mathbf{y}_{\mathbf{s}}$, $\mathbf{l}_{\mathbf{s}}$, $\hat{\mathbf{l}}_{\mathbf{s}}$ are the column vectors associated with texture intensities, classes and estimated classes respectively for the window $\Omega_{\mathbf{s}}$. Moving this window, in arbitrary order, with the step size of one pixel, we label the central pixel according to the textural attribute of the corresponding window. This window-based classification technique was originally implemented, in [3] [5] with a fixed temperature and in [13] with deterministic relaxation, to obtain a pseudo-MAP solution. Due to the difficulty in estimating the appropriate temperature in the fixed temperature approach as well as the strong influence of initial estimates in a purely deterministic relaxation scheme, we employ MFA as a compromise.

3.2 Consensus Based on Local Evidence

In previous window-based segmentation algorithms, the image was classified pixelwise and the natural constraints (mainly region homogeneity) were imposed through parameters of statistical models for the occurrence of regions in the image. Even though the classification was performed patchwise, only the result (label) of the central pixel was taken into account in the segmentation. Too little homogeneity information was utilized and much of the information from the patch classification was wasted [16] [3] [13]. In this work, we use a second classification step, called local voting (LV) [7] [15], to implicitly model the interaction of texture attributes in the image as well as utilize the spatial coherence of homogeneous regions. It belongs to a class of methods based on local consensus, *e.g.* median/mode filtering etc. The effect of LV can be described as a distributed maximization of the total support for the unique winner. This idea is similar in spirit to the pixel labelling in [14].

The hybrid texture segmentation algorithm integrates the statistical window-based classification and structural local voting schemes cooperatively. That is, instead of labeling the pixel based only on the texture attribute of the window centered on it, we determine the pixel label by the voting scores of the possible texture classes from a local support region, *i.e.* we look for a texture class $l_{\mathbf{s}}$ at each location $\mathbf{s}$ in the image that maximizes the voting scores as follows:

$$\max_{l_{\mathbf{s}} \in L} \sum_{\mathbf{r} \in R_{\mathbf{s}}} w_{\mathbf{r}} \cdot H_{l_{\mathbf{s}}}(\hat{\mathbf{l}}_{\mathbf{r}}) \quad \text{and} \quad \hat{\mathbf{l}}_{\mathbf{r}} = \arg\min_{\mathbf{l}_{\mathbf{r}}} U_p(\mathbf{l}_{\mathbf{r}}|\mathbf{y}_{\mathbf{r}}) \qquad (11)$$

where H is the score measure used to combine the window-based classification results constructively. $R_{\mathbf{s}}$ is a local support region surrounding $\mathbf{s}$ and including $\mathbf{s}$, and $w_{\mathbf{r}} \geq 0$ is the weight for the vote by pixel at location $\mathbf{r} \in R_{\mathbf{s}}$. The above formulation implies that we collect the contributions of these windows centered on the pixels of support region $R_{\mathbf{s}}$ through the linear combination of score measures H. Thus voting is guided by the statistical classification results in $R_{\mathbf{s}}$.

Both voting and decision rules must be set to select the winner at each location in the image. Here, a "democratic voting" rule is used so that every member in the support region, including the central pixel, has the right to vote for the competing candidates of texture classes at the central pixel. However, the voting weight might be task- and site-dependent, *e.g.* inversely proportional to the Euclidean distance from the central pixel to indicate the decay of spatial correlation [15], or uniformly weighted for "real democracy" etc. As an example, for $w_{\mathbf{r}} = 0$, $\forall \mathbf{r} \neq \mathbf{s}$, it defaults to the usual window-based pixel classification [3]. The decision rule is based on the winner-take-all (WTA) strategy, *i.e.*, only one winner is allowed at each lattice site. The voting scores provide valuable information to suppress the competing candidates which may be potential winners in the event of a pixelwise window-based decision. One should note here, that if we do not use WTA, we can choose not to make "hard decisions", at every pixel [11] [15].

The prior knowledge of region homogeneity can be used to choose the "real democracy" weighting rule, which is $w_{\mathbf{r}} = w, \ \forall \mathbf{r} \in R_{\mathbf{s}}$. If the local support region coincides with the window and the score measure is set as

$$R_{\mathbf{s}} \equiv \Omega_{\mathbf{s}} \quad \text{and} \quad H_{l_{\mathbf{s}}}(\hat{\mathbf{l}}_{\mathbf{r}}) = \frac{1}{|R_{\mathbf{r}}|} \sum_{\mathbf{u} \in R_{\mathbf{r}}} I_{l_{\mathbf{s}}}(\hat{l}_{\mathbf{u}}) \qquad (12)$$

where $I_l(\hat{l})$ is an indicator function, which is equal to 1 only when $l = \hat{l}$ and zero otherwise, and $\mathbf{l}_{\mathbf{r}} = col.[l_{\mathbf{u}}, \ \mathbf{u} \in R_{\mathbf{r}}]$ is a column vector associated with line variables in $R_{\mathbf{r}}$, then the number of votes is employed as the classification criterion. We assume that there are L texture classes and, thus, L vote counters associated with each pixel, one for each texture class candidate. The hybrid segmentation algorithm, with MFA, can be summarized in the following:

Hybrid Texture Segmentation Algorithm

(0) Set the system state at a relative high value of temperature, $T = T_0$. (*Initialization phase*)

(1) Divide the whole image into small overlapped window-sized subimages and fit the GMRF model to each of these subimages by (10). (*Classification phase*)

(2) For each window subimage, label all the pixels in the block with the same texture class to indicate that they vote for

[3] The system will be in equilibrium if the total number of label changes between two successive iterations is very small.

the same class at all locations. *i.e.* the vote counters associated with the particular label at each pixel within the window are incremented by one. (*Voting phase*)

(3) A majority rule, WTA, is enforced on the label counters at each pixel. The label that gets the most votes is assigned to that pixel. (*Decision phase*)

(4) Reset all vote counters and repeat the classification, voting, and decision phases until the equilibrium state is reached at this temperature. (*Iteration phase*)

(5) Cool down the system temperature according to an annealing schedule, $T = g(T)$, and use the equilibrium state of the previous temperature to initiate the deterministic annealing process described in (2)-(4). (*Cooling phase*)

(6) Stop when the temperature is below a preset threshold T_h.

4 Experimental Results

In the experiments, natural texture mosaics are used as test examples for our algorithm. We use fourth order GMRF model to model each of these natural textural images, as for most textures, the correlation of pixel intensities decays dramatically as the spatial separation increases. We employ second order GRF as a region model to include the piecewise constant constraint except for the mosaic consisting of six textures, in which a fourth order GRF is assumed for the region model.

(a) Two-Class Mosaic

Type of Texture Mosaic	Two-Class Mosaic	Three-Class Mosaic	Six-Class Mosaic
GMRF Order	4	4	4
GRF Order	2	2	4
Window Size	9×9	7×7	7×7
Initial Temperature	100.0	100.0	100.0
Annealing Schedule	80% off	80% off	20% off
$V_c(\mathbf{l}) = \beta \cdot \delta(\cdot)$	$\beta = 1.0$	$\beta = 1.0$	$\beta = 1.0$
# of Iteration	4	5	12

Table 1: The summary of experimental environments.

The experimental environments are summarized in Table 1 and the results for this two-class texture mosaic are shown in figure 2. The initial temperature, $T_0 = 100.0$, is set high enough to ensure the likelihood term $U(\mathbf{y}|\mathbf{l})$ dominates over prior term $U(\mathbf{l})/T$. We use 9×9 window for Ω_S as well as R_S and assume that the conditions in (12) are satisfied. The ML and MAP region estimates derived by the hybrid model are shown in figure 2(d), 2(e). When they are compared with corresponding results obtained by the window-based method, as shown in figures 2(b) and 2(c), the efficiency of the hybrid model can be easily perceived. There exists a tradeoff between the quality of segmentation (such as the accuracy of region boundaries and intra-region homogeneity) and window size. The window size limits the resolvability of the window-based classification operator. The region whose size is much smaller than the window will be "invisible" to the segmentation window, since the majority of pixels inside the window will have different texture attributes whenever the window is centered at certain point in the region. A larger window yields more accurate classification within each homogeneous region, yet fuzzier texture boundaries.

(b) Three-Class Mosaic

The 3-class texture mosaic (128×128), consisting of wood, grass and wool, is shown in figure 3(a). The simulation parameters are identical to those in two-mosaic case. The original texture and segmented results are shown in figure 3.

(c) Six-Class Mosaic

Six textures, including wool, wood, pigskin, grass, calf leather and sand, are synthesized to form a chessboard like texture mosaic (256×256), as shown in figure 4(a). Under the GMRF formulation, the three member classes, *i.e.* wool, sand and pigskin, bear close resemblance in statistical characteristics such that there are high uncertainties in classifying these three close competitors with a "democratic" voting strategy. This phenomenon is obviously reflected by the abundant holes, interleaved mainly among those three textures, in the ML solutions, as shown in figures 4(b) and 4(d). In order to alleviate this uncertainty, we adopt a slower cooling schedule, $T_k = 0.8 \cdot T_{k-1}$, and a higher order of region model, 4th order, to increase the texture class dependence. The simulation results of the MAP region estimates after 12 iterations are shown in figures 4(c) and 4(e). The output quality of the hybrid model outstripes those without using LV technique in this complicated and difficult example.

5 Conclusion

Texture is often regarded as a second-order predicate for a group of pixels, implying that the region is integral as well as homogeneous. There may exist an uncertainty in the neighborhood of boundaries of "similar" textures. The hybrid segmentation algorithm focuses on the integrity and homogeneity through the use of GRF formulation of texture region and the LV technique. It was shown through experimental results that these two implicit and explicit models for texture region do cooperate consistently. Since any border pixel must group with certain local pixels to exhibit the textural property, uncertainty at the boundary can be removed to some extent if the contributions from its neighbors are considered.

Under a free boundary condition, which assumes that the image boundary pixels have less neighbors than the interior pixels, the pixels there will be more susceptible to misclassification in window-based segmentation algorithms due to insufficient data. To overcome this, a toroidal model (periodic boundary) is usually assumed. However, a toroidal model cannot resolve the problem completely since the texture properties on both ends of an image plane might be quite different. LV is capable of resolving this, since it allows the direct influence of the local neighbors in determining their own texture attributes.

References

[1] J. Besag, Spatial interaction and the statistical analysis of lattice system, *Journal of the Royal Statistical Soc.*, B-36:192–236, 1974.

[2] G.L. Bilbro, W.E. Snyder, and R.C. Mann, Mean-field approximation minimizes relative entropy, *Jour. of Optical Soci. America*, Vol-8(No.2):290–294, Feb. 1991.

[3] S. Chatterjee and R. Chellappa, Model based texture segmentation, In *Proc. 20th Asilomar Conf. on Signal System & Comp.*, Asilomar, CA, 1986.

[4] R. Chellappa and S. Chatterjee, Classification of textures using Gaussian Markov random fields, *IEEE Trans. on Acoust. Speech and Sig. Proc.*, ASSP-33:959–963, 1985.

[5] F.S. Cohen and D.B. Cooper, Simple parallel hierarchical and relaxation algorithms for segmentating noncausal Markovian fields, *IEEE Trans. on Patt. Anal. & Mach. Intel.*, PAMI-9(No.3):195–219, Mar. 1987.

[6] G.R. Cross and A.K. Jain, Markov random field texture models, In *IEEE Trans. on Pattern Anal. and Mach. Vision*, volume PAMI-5, pages 25–39, Jan. 1983.

[7] D. Drumheller and T. Poggio, On parallel stereo, In *Proc. IEEE Intl. Conf. on Robotics & Automation*, pages 1439–1448, Washington, D.C., 1986.

[8] D. Geiger and F. Girosi, Mean field theory for surface reconstruction, In *Proc. DARPA Image Understanding Workshop*, pages 617–630, 1989.

[9] S. Geman and D. Geman, Stochastic relaxation, Gibbs distributions and the Bayesian restoration of images, *IEEE Trans. on Patt. Anal. & Mach. Intel.*, PAMI-6, Nov. 1984.

[10] H.P. Hiriyannaiah, G.L. Bilbro, W.E. Snyder, and R.C. Mann, Restoration of piecewise-constant images by mean field annealing, *Jour. of Optical Soci. America*, Vol-6(No.12):1901–1912, Dec. 1989.

[11] J.T. Kent and K.V. Mardia, Spatial classification using fuzzy membership model, In *IEEE Trans. on Pattern Anal. and Mach. Vision*, volume PAMI-10:659–670, 1988.

[12] M. Leung and A.M. Peterson, Multiple channel neural network model for texture classification and segmentation, In *Proc. Intl. Conf. on Acoust., Speech & Sig. Proc.*, pages 2677–2680, Toronto, Canada, May 1991.

[13] B.S. Manjunath, T. Simchony, and R. Chellappa, Stochastic and deterministic networks for texture segmentation, *IEEE Trans. on Acoustics, Speech & Sig. Proc.*, Vol.38(No.6):1039–1049, June 1990.

[14] K.V. Mardia and T.J. Hainsworth, A spatial thresholding method for image segmentation, *IEEE Trans. on Pattern Anal. and Mach. Vision*, PAMI-10(No.6):919–927, 1988.

[15] K. Prazdny, Detection of binocular disparities, *Biological Cybernetics*, Vol.52:93–99, 1985.

[16] T. Simchony and R. Chellappa, Stochastic and deterministic algorithm for map texture segmentation, In *Proc. Intl. Conf. on Acoust., Speech & Sig. Proc.*, Apr. 1988.

[17] C.W. Therrien, A estimation-therotic approach to terrain image segmentation, *Comp. Vision, Graphic and Imag. Proc.*, Vol. 22:313–326, 1983.

Figure 2: upper left: the original 2-class texture mosaic (128 × 128): wool and pigskin. center left: the ML segmentation result by window-based algorithm. center right: the MAP segmentation result by window-based algorithm. bottom left: the ML segmentation result by hybrid algorithm. bottom right: the MAP segmentation result by hybrid algorith.

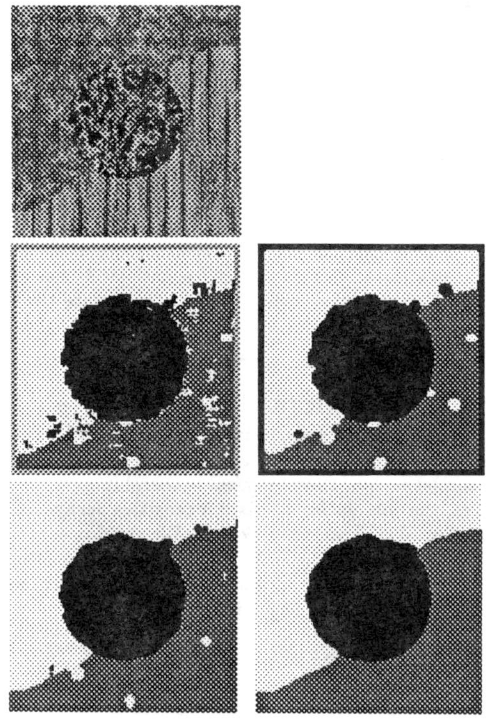

Figure 3: The segmentation results for 3-class texture mosaic (128 × 128): wood, wool and grass. The figure layout is the same as that in Figure 2.

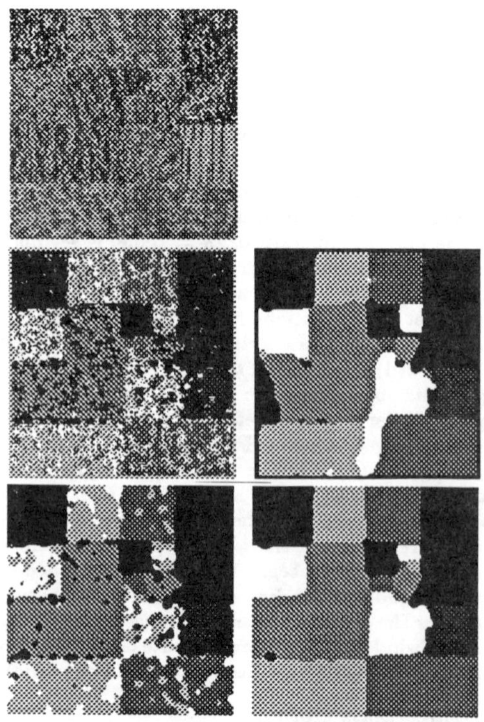

Figure 4: The segmentation results for 6-class texture mosaic (256 × 256): wool, wood, pigskin, grass, calf leather and sand. The figure layout is the same as that in Figure 2.

Computer Vision for Locating Buried Objects

Gregory A. Clark, Jose E. Hernandez, Nancy K. DelGrande, Robert J. Sherwood,
Shin-Yee Lu, Paul C. Schaich, and P.F. Durbin
Lawrence Livermore National Laboratory
Electrical Engineering
Livermore, CA, 94550

Abstract

Given two registered images of the earth, measured with aerial dual-band infrared (IR) sensors, we use advanced computer vision/automatic target recognition techniques to estimate the positions of buried land mines. The images are very difficult to interpret, because of large amounts of clutter. Conventional techniques use single-band imagery and simple correlations. They rely heavily on the judgment of the human doing the interpretation, and give unsatisfactory results with difficult data sets of the type we analyzed. Our automatic algorithms are able to eliminate most of the clutter and give greatly improved indications of regions in the image that could be interpreted as mines. The novelty of our approach lies in the following aspects: (1) A patented data fusion technique using two IR images and physical principles based on Planck's law, (2) A new region-based texture segmentation algorithm using Gabor Transform features and a clustering/thresholding algorithm based on a neural network (Self-Organizing Feature Map), (3) Prior knowledge of measured feasible temperatures and emissivities, and (4) Results with real data using buried surrogate mines.

1 Introduction

Given an image of interest, we use novel Computer Vision/Automatic Target Recognition techniques to automatically analyze the image and recognize objects contained in the image. In this application, we wish to recognize and classify buried objects in two-band infrared images of the earth.

A sandy-loam field covered by grass was selected for observation from a helicopter platform. See Fig. 1. For the layout we used nine buried surrogate mines, spaced 10 ft apart within a square grid that is 30 ft on a side. Surrogate mines were covered by 1-in., 2-in., and 4-in. layers of soil (going from west to east). They had 12-in. diameters (northernmost and southernmost rows) and 6-in. diameters (center row). There were sixteen 8-in. diameter aluminum objects exposed on the surface within the grid and an additional sixteen objects of varied sizes outside. A 2-in. wide aluminum metal band marked the grid perimeter. A novel technique for correcting for helicopter motion was created by R.J. Sherwood and used to improve the quality of the images [1]. Reference [1] describes the detailed physics of the problem. This paper emphasizes the image processing/vision aspects.

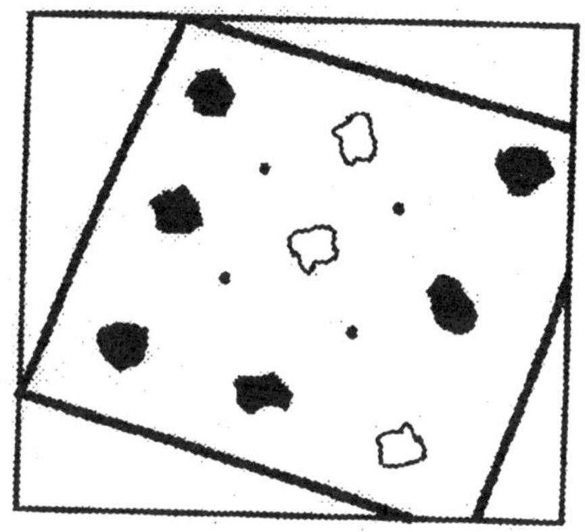

Figure 1: The locations of the buried surrogate mines (9 of them) are shown in this sketch. Note that a 2-in wide aluminum border surrounds the 30 ft. by 30 ft. area, and four aluminum fiducial markers are placed near the center of the layout.

2 Physical Principles for Data Fusion and Clutter Reduction

N.K. Del Grande created a novel (patented) technique, based on Planck's law, for using dual-band IR images [1]. She showed that the intensity of an IR image of the earth is proportional to the product of emissivity and temperature to a power involving the wavelength of the radiation [1]. See Fig. 2 and the Appendix. Because we wish to discriminate mines from soil by using the differing properties of temperature and emissivity, our goal is to separate the effects of emissivity and temperature from our measurements. We can do this by measuring infrared intensity at two different wavelengths (e.g. 5 microns and 10 microns) and then computing logarithms of the ratios of the two measurements to obtain emissivity alone and temperature alone (see Appendix). We call these results the emissivity map E and the temperature map T. As we shall see, this orthogonalization of the problem gives us a powerful technique for reducing clutter.

1058-6393/91 $1.00 © 1991 IEEE

We first compute the appropriate emissivity and temperature ratio images E and T, using the measured long wavelength and short wavelength images. See Fig. 2. The upper left-hand image is the long-wavelength (10 micron) raw image. The upper right-hand image is the short wavelength (5 micron) image. The lower left-hand image is the ratio E. The lower right-hand image is the temperature ratio T.

Now, we process the images along the two branches of the block diagram in Fig. 3. Following the lower branch of the diagram, we first apply rules based on prior physical knowledge of the experiment. For example, we know that the soil in the emisssivity ratio image (E) must have an average emissivity Eav that lies in a particular range. We discard from consideration pixels with values of $\ln(E/Eav)$ that lie outside the range -.05 to .05 (See Fig. 4b). We also discard from consideration pixels with values of T that lie outside the range σ to 2σ, where σ is the standard deviation of T (See Fig. 4a).

After applying these constraints to all of the pixels, the result is the set of pixels that can possibly represent soil sites above buried surrogate mines. These data were measured in the early afternoon, so the soil above a mine should be warmer than the average temperature measured over the course of a day. This helps eliminate image pixels that probably do not represent soil sites above buried surrogate mines obtained from early afternoon images which have warmer than average pixels (see Fig. 4c). Next, we use computer vision techniques to further improve the results shown in Fig. 4c.

3 Computer Vision Algorithms for Object Recognition

See the upper branch of the block diagram in Fig. 3.

- We first extract important features from the image, using Gabor Transforms, and use the features to do texture segmentation [1-3].

  The Gabor Transform of an image is the convolution of a Gabor filter with the image. The Gabor filter is a two-dimensional complex exponential multiplied by a Gaussian-shaped window function. It has been shown that Gabor filters provide a good model of biological vision. They model closely the response of striate cells in the visual cortex. We showed [2,3] that Gabor representations provide an automatic target recognition system with robustness to variations in the orientation, size, and translation of the objects in the image.

  Each pixel in the image has a Gabor feature vector associated with it. These vectors are rich in information, as each feature vector contains information about frequency content, orientation, and scale associated with the pixel value.

  In this particular application, we have used simple feature vectors composed of a single element, so our feature vectors are actually scalars. We use only one Gabor kernel at a single size corresponding roughly to the size of a mine.

- Texture segmentation is the process of using important image features to divide the image into regions that have common characteristics, referred to as textures. We use a special kind of neural network, called a self-organizing feature map (SOFM [5]) to cluster the Gabor feature vectors into classes, or texture types. The result of the clustering operation is a list showing which image pixels correspond to feature vectors in each of the texture classes in feature space. Given this list, we assign a color to each texture type, or class, and plot the appropriate colors for each of the pixels. This gives an image showing the clustering of the pixels into textures. This image is very valuable, because it helps us sort out which pixels might correspond to mines.

  Because we used simple scalar features in this application, we note that the general clustering operation is equivalent to a thresholding operation.

- The next step is called region growing. Here, we divide the pixels in the texture type list into spatial regions. We do this because the texture type list from the last step does not tell us which pixels in a given class (texture) lie adjacent to each other in space. Region growing provides this information. Region growing uses local spatial operations to identify regions (which in general are not contiguous) that belong to the same texture class. We assign a color to each region, and plot the appropriate colors for the pixels in the regions. This now gives us an image that shows regions that can be mines or other objects (see Fig. 4d).

  Refer now to the lower branch of the block diagram in Fig.3.

- We next apply physics rules to eliminate from consideration regions that cannot represent mines. We know that Fig. 4c represents pixels that are warm and can represent soil. To arrive at Fig. 4e, we take the logical intersection of Figures 4c and 4d, given that more than a chosen threshold (in this case 30 percent) of the pixels in the regions in Fig. 4d lie in Fig. 4c. This leaves us with pixels in regions that can reasonably be interpreted as mines (warm soil regions), and eliminates the rest.

- The results of region growing and validating pixels with physics rules often are good enough to greatly enhance our ability to identify mines by eye. However, the interpretation of some of the regions is still sometimes difficult, because of ambiguities in the shapes and sizes of the regions (see Fig. 4e). To deal with this, we apply rules based on shape and size information. We know that mines have a circular shape and we know their approximate sizes, so we use local operators to eliminate from consideration regions that have

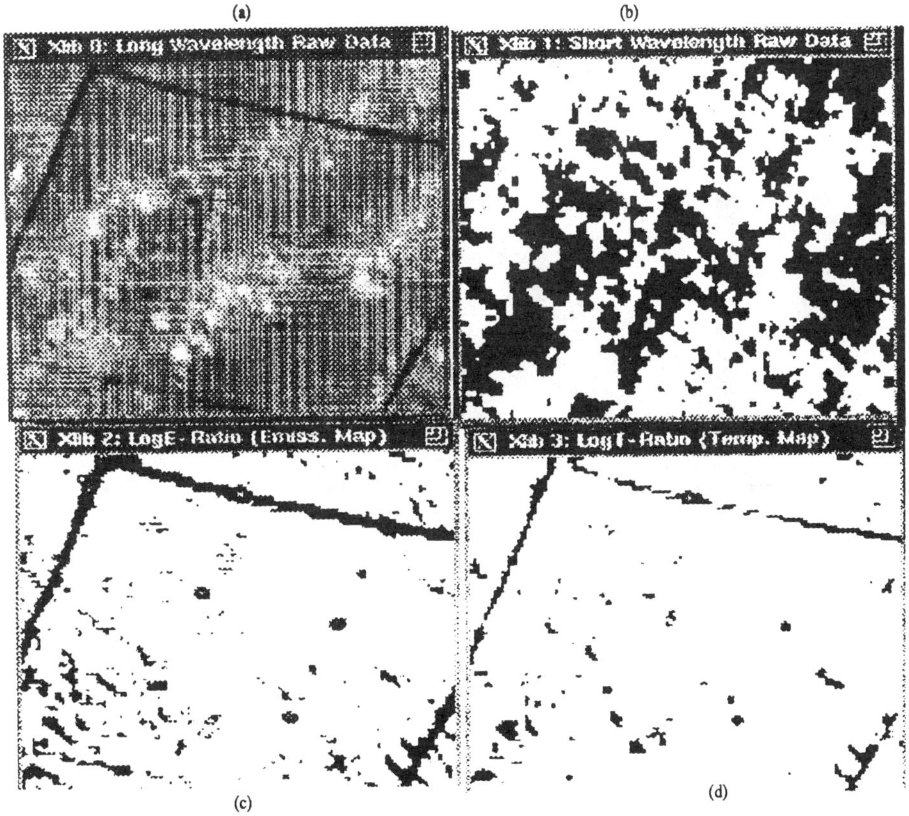

Figure 2: (a) The long wavelength (10 microns) infrared intensity image. (b) The short wavelength (5 microns) infrared intensity image. (c) The logarithm of the normalized emissivity ratio. (d) The logarithm of the normalized temperature ratio. (See appendix).

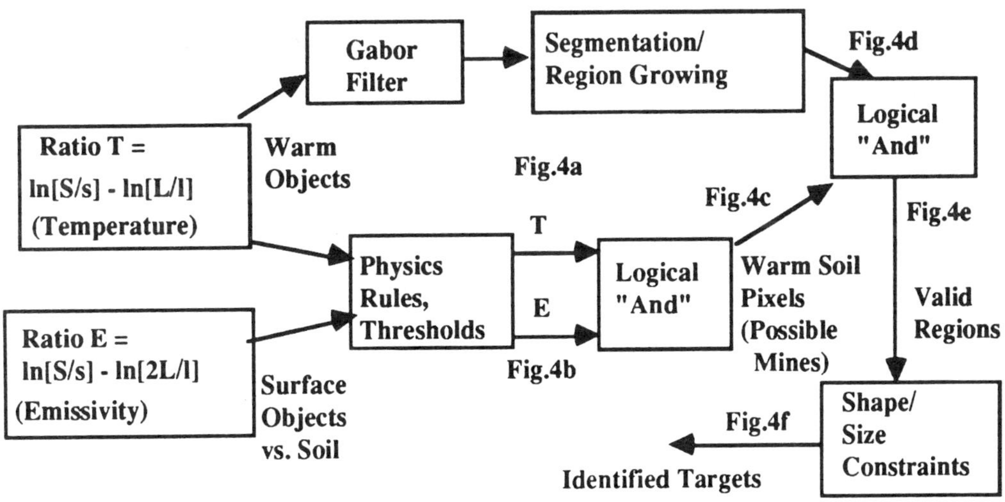

Figure 3: Block diagram showing the data processing steps taken to arrive at the results in Figs. 4a through 4f.

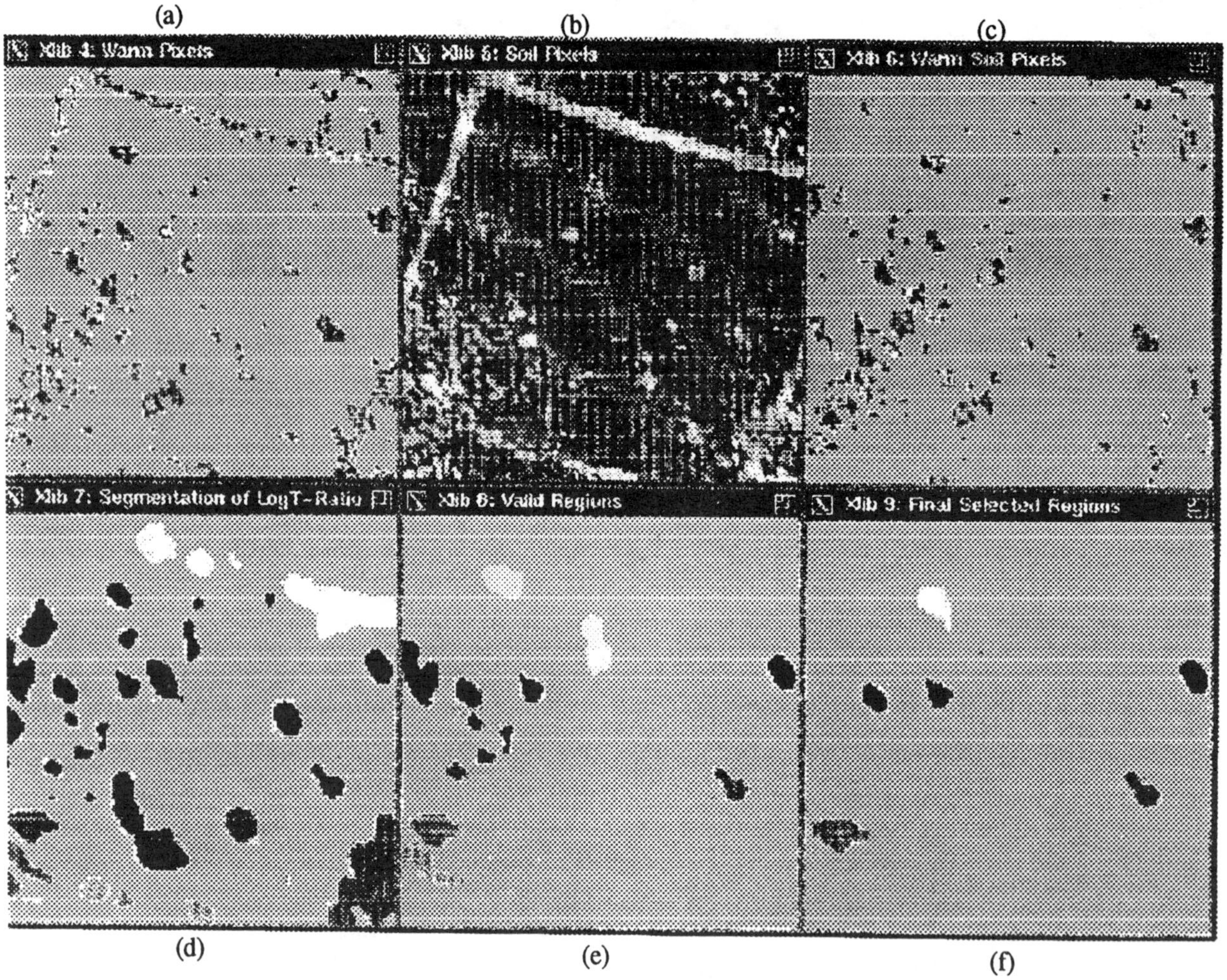

Figure 4: (a) Results of applying physics rules to threshold the values of the log of the normalized temperature ratio (See the Appendix). (b) Results of applying physics rules to threshold the values of the log of the normalized emissivity ratio (see the Appendix). (c) Logical intersection of Figs. 4a and 4b. (d) Log of the normalized temperature ratio after Gabor transformation, texture segmentation, and region growing. (e) Valid mine sites found by taking the logical intersection of Figs. 4c and 4c, given that more than 30 percent of the pixels in the regions in Fig. 4d also lie in Fig. 4c. (f) Final Selected Pixels, found by applying shape and size constraints to the pixels in Fig. 4e.

shapes and sizes that cannot logically be assigned to a mine. The result is shown in Fig. 4f (Final Selected Regions).

- Comparing Fig. 4f with Fig. 1 shows that the algorithms detected six of the nine buried mines, with one false alarm. This is very encouraging, but our current methods require that a human must examine an image to detect mine sites. Our long-term goal is to eliminate the need for human intervention. Not presented here are preliminary results we have obtained using a supervised back-propagation neural network to classify mine sites. Initial studies with very small training sets were successful, and we expect the technique to work well when large training sets are available.

4 Conclusion

We have shown initial results with a representative real data set. The data set is a very difficult one to interpret, because of large amounts of clutter and low signal-to-noise ratio.

Our overall approach of using physics, image processing, and a priori knowledge of the expected size and shape of the objects results in a greatly improved ability to interpret the mine images by the human eye The algorithms found six of the nine buried mines, while detecting one false alarm.

Our future plans include the use of more sophisticated feature extraction techniques, fusion of multiple sensors, more complex higher-order reasoning, and more thorough testing of a supervised neural network classifier to automatically classify objects in the image (eliminate the human from the process).

References

[1] N.K. Del Grande, G.A. Clark, P.F. Durbin, D.J.Fields, J.E. Hernandez, and R.J. Sherwood, Buried Object Remote Detection Technology for Law Enforcement, SPIE Orlando 91 Symposium, Orlando, Florida, April 1-5, 1991.

[2] G.A. Clark, J.E. Hernandez, B.S. Lawver, and R.J.Sherwood, Gabor Transforms and Neural Networks for Automatic Target Recognition, Workshop on Neural Networks, Auburn AL, Feb. 1991.

[3] G.A. Clark, J.E. Hernandez, and R.J. Sherwood, Image Invariance Properties of Gabor Transforms and Neural Networks for Automatic Target Recognition, Neural Networks for Decision, Estimation, and Control Workshop, Naval Underwater Systems Center, Newport, R. I., August7 and 8, 1990.

[4] S.-Y. Lu, J.E. Hernandez and G.A. Clark, Texture Segmentation by Clustering of Gabor Feature Vectors, submitted to the 1991 International Joint Conference on Neural Networks (IJCNN91), Seattle, WA, July 8-12, 1991.

[5] T. Kohonen, *Self-Organization an Associative Memory*, Springer-Verlag, 3rd edition, 1989.

5 Appendix

We use dual-band IR images to separate temperature and emissivity effects.

- Intensity at a given wavelength, I_λ is a multiplicative function of both emissivity ϵ and temperature T.

$$I_\lambda \sim \epsilon T^{50/\lambda} \ , \ \lambda = \text{wavelength}$$

- We can obtain temperature alone by computing the ratio

$$R = \frac{I_{10}}{I_5} = \frac{\epsilon_{10} T^{50/10}}{\epsilon_5 T^{50/5}} = \frac{\epsilon_{10}}{\epsilon_5} T^{-5}$$

For a gray body, $\epsilon_{10} \sim \epsilon_5$, so $R \sim T^{-5}$

This temperature ratio shows mostly pixels sensitive to temperature (buried objects).

- We can obtain emissivity alone by computing the ratio

$$\frac{I_{10}^2}{I_5} = \frac{\epsilon_{10}^2 (T^5)^2}{\epsilon_5 T^{10}} = \frac{\epsilon_{10}^2}{\epsilon_5} \sim \epsilon$$

This ratio is sensitive mostly to surface objects.

- For our processing, we compute normalized ratios

$$T = ln \frac{S/s}{L/\ell} \text{ and } E = ln \frac{S/s}{2L/\ell}$$

where

S is the short wavelength intensity (e.g. I_5)

s is the average value of the pixels in S

L is the long wavelength intensity (e.g. I_{10})

l is the average value of the pixels in L

Algebraic/Topological Combined Strategies for Sequential ATPG

E. Macii * A. R. Meo

Politecnico di Torino
Dip. di Automatica e Informatica
Corso Duca degli Abruzzi 24
Torino, ITALY 10129

Abstract

This paper describes an algorithm for Automatic Test Pattern Generation for sequential integrated circuits, based on a combination of Algebraic and Topological techniques. The algebraic concepts of Observability and Controllability Functions have been applied to an innovative version of the sequential D-Algorithm working on the combinational part of the network that has been partitioned in several Fanout-Free Regions. The ideas of Observability Order, Controllability Order, and Incompatibility Function have been introduced in order to improve the ATPG performance.

1 Introduction

The technological level reached in the implementation of integrated circuits, even allowing a very high concentration of transistors within a *single chip*, opens up a series of problems which become more and more complex with the increase of the scale of integration. Among them, a key role is covered by the device testing procedure, which is absolutely necessary to ensure a high level of reliability [1] [2] [3].

In a first approximation, an Automatic Test Pattern Generation procedure can be seen as a software package that, given a certain fault, provides the computation of the input stimuli vector (i.e. the *test pattern*) that must be applied to the circuit in order to detect the fault itself. Given the location of the fault, the procedure starts investigating all the possible paths on the network that allow its propagation until it can be observed on a primary output; then it computes the input sequence necessary to satisfy all the conditions imposed on the internal nodes of the network during the propagation phase [4].

Several innovative concepts, like *Circuit Partitioning* [5] [6], *Controllability Order* and *Observability Order* [7] [8] [9], *Incompatibility Function* have been added to this basic scheme in order to allow the management of very wide circuits and to improve the ATPG performance. A fault simulator is used to obtain the coverage of all the possible faults of the circuit [10].

*Currently dr. E. Macii is with the Department of Electrical and Computer Engineering, University of Colorado at Boulder.

The paper is organized as follows. Some theoretical definitions are presented in the next section. In Section 3, the strategy for test generation is outlined. In Section 4 an example of application is shown, while experimental results are reported in Section 5.

2 Theoretical Definitions

In this section we present the theoretical definitions of the most important elements of our test generation algorithm, that will be described in the next section.

2.1 Fanout-Free Region (FFR)

A FFR is a set of logic interconnected gates characterized by the property that each predecessor gate has only one successor; in other words, no fanout points are present in the connection scheme of the gates forming the FFR. A detailed description of FFRs theory is provided by [11].

2.2 Controllability Function (CF) of a FFR output node

The CF of a FFR output node denotes the set of conditions to be assumed by the input lines of the block to set (1-CF) or to reset (0-CF) the node itself.

2.3 Observability Function (OF) of a FFR input node

The OF of a FFR input node defines the set of values to be assumed by the others input lines to let the node be observable on the block output, that is to let the value of the output line dependent on the signal level of the input node under consideration.

2.4 Controllability Order (CO) of a node

The CO of a node is defined as a two-dimensional quantity: (α , β). The first field (α) represents the sequential order, that is the lower bound of the number of time frames required to excite the node; the second field (β) represents the combinational order, that is the lower bound of the number of combinational levels to be excited in addition to the complete time frame. The CO of a certain node, then, measures how difficult it is to set or to reset the node itself.

1058-6393/91 $1.00 © 1991 IEEE

2.5 Observability Order (OO) of a node

The OO of a node is defined as a two-dimensional quantity: $(\gamma\ ,\ \delta)$. The first field (γ) represents the sequential order, that is the lower bound of the number of time frames required to observe the node on a primary output; the second field (δ) represents the combinational order, that is the lower bound of the number of combinational levels to be observed in addition to the complete time frame. The OO of a certain node, then, measures how difficult it is to observe the node itself on a primary output.

3 Strategy for Test Generation

The procedure for test pattern generation that has been implemented can be subdivided in two parts; the first one consists of a set of pre-processing operations involving the topological description of the network; the second one performs the test generation for all the *single stuck-at* fault of the circuit, employing the results obtained in the first step.

3.1 Pre-Processing Operations

The following pre-processing operations are applied to the original circuit description in order to speed-up the test generation process.

1. Assuming that a *gate-level* description for the combinational portion of the network is available (i.e. the *Huffman Model* is employed), the first operation that must be performed is the circuit partitioning in FFRs, by collapsing the gates in logical blocks containing a number of elements as higher as possible and fanout-point free. The result of this operation is that the topological description of the network is now changed from *gate-level* to *block-level.*

2. The information stored during the partitioning phase, and referring to the physical interconnection of the gates, is used to compute the boolean equations of each FFR. However, a *minimization* step is required in order to reduce the complexity of the equations, i.e. the number of rows in the truth table of the function (in the following referred to as PLA). At this point, according to the definition given previously, the CFs of each node have been determined. It is important to observe that both 0-CF and 1-CF have been computed; therefore, no function inversion, that is transformation of a logic function from *sp* to *ps* form, is required during the COs determination.

3. From the logic equations of each FFR (i.e. the CFs), the OFs are computed. The PLAs obtained at the end of this step are then submitted to a minimization process to further reduce their complexity. As this operation has been completed, the *topological to algebraic translation* of the circuit is available. More specifically, the CFs and the OF, expressed in *sp* form, are available for each node of the network; in other words, 0-CF, 1-CF and OF PLAs for each circuit node are stored in the main memory.

4. The final step of the pre-processing phase is the COs and OOs computation. This operation is performed by simply applying, in an iterative way, the definitions given previously.

The data structure created by the pre-processing operations stays resident in the main memory of the computer until a test pattern for each fault is generated.

3.2 Test Generation Algorithm

The test generation algorithm can be divided in two phases:

- observability phase;
- controllability phase.

During the observability step, the fault is propagated in the circuit until it can be observed on a primary output; the propagation is obtained with the recursive expansion of the OF of each logic block; the choice of the best path that may be covered on the decision tree is driven by the OO of each node. It's clear that different time frames can be involved in the propagation of the fault; to each node of the search tree is then associated a *Timed Observability Function* of the type:

$$(x^*_{01}\ldots x^*_{0p})(x^*_{11}\ldots x^*_{1q})(x^*_{r1}\ldots x^*_{rs})\ldots$$

where x^*_{ij} denotes the value of the node j at the time frame i and the star denotes a possible complementation.

When a primary output is reached, the observability phase finishes and starts the controllability step of the research, which carries out the justification of all the variables that have been employed during the fault propagation. The Timed Observability Function becomes a *Timed Controllability Function.* The variables x^*_{ij} are excited in order from the largest times and the most external FFRs. When the Timed EF contains only variables associated to primary inputs, the same Timed EF describes the test pattern.

The procedure terminates succesfully (i.e. a test pattern is generated) if the justification operation requires a number of time frames which is smaller than a predefined threshold (i.e. the temporal length of the test pattern).

During the controllability phase a conflict on one or more variables can be detected. In this case it is necessary to execute some backtrack operations. Again, using the weights (i.e. the COs) previously determined, we can reduce the number of backtracks necessary to compute the test pattern.

3.3 Incompatibility Function (IF)

To improve the algorithm performance, the concept of *Incompatibility Function* has been introduced. The IF is the set of all product terms originating conflicts during the controllability phase. By means of the IF,

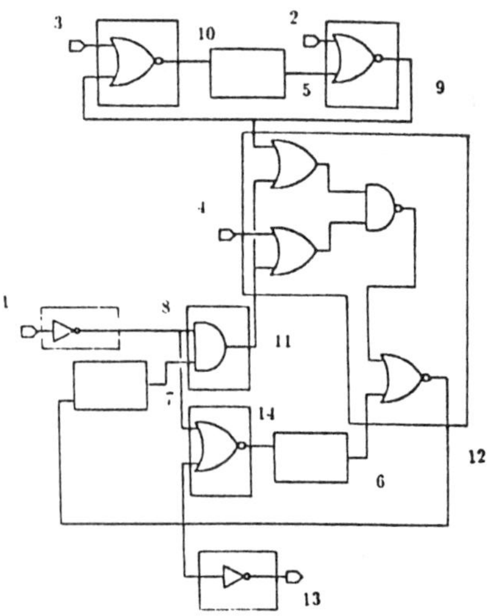

Figure 1: The *S27* circuit

the process can be substantially speeded up. In fact, before selecting a product term from the CFs the term is compared with the IF. If it is already covered by this function it implies that it is not excitable and therefore must be rejected immediately. Terms are added to the IF anytime they conflict with the present state. The IF is periodically minimized to make its manipulation easier and faster.

4 An Application Example

As an application example of the proposed algorithm, consider the circuit *S27*. It is the smallest circuit in the ISCAS'89 [14] set, and its schematic is represented in Fig.1. In that figure it is also shown the circuit partitioning in FFRs.

Assuming the FFR identification number is equal to its output node number, the boolean equations of each block are computed during the partitioning phase.

The *block-level* description of the network is then processed by means of the procedures for CFs and OFs computation and minimization, in order to determine Controllability and Observability conditions for each node. By applying COs and OOs computation algorithms, we obtain the Controllability and Observability Tables of the circuit, shown respectively in Tab.1 and Tab.2. In those tables, a variable is denoted by the number of the corresponding node, and appears like a positive number if its value is one and appears like a negative number if its value is zero.

Now we can perform the test pattern generation for a given fault of the circuit. Consider, for example, the node 8 stuck-at-1.

Node	*1-CO*	*0-CO*	*1-CF PTs*	*0-CF PTs*
1	(0,0)	(0,0)	none	none
2	(0,0)	(0,0)	none	none
3	(0,0)	(0,0)	none	none
4	(0,0)	(0,0)	none	none
5	(1,2)	(1,1)	none	none
6	(1,4)	(1,2)	none	none
7	(2,3)	(1,3)	none	none
8	(0,1)	(0,1)	-1	1
9	(1,2)	(0,1)	-2 -5	2, 5
10	(0,2)	(0,1)	-3 -9	3, 9
11	(2,4)	(0,2)	7 8	-8, -7
12	(1,3)	(0,3)	4 -6 9, -6 11	-9 -11, -4 -11, 6
13	(0,4)	(1,4)	-12	12
14	(0,4)	(0,2)	-8 -12	8, 12

Table 1: The *S27* Controllability Table

Node	*OO*	*Obs. node*	*OF PTs*
1	(0,4)	8	none
2	(0,3)	9	-5
3	(1,4)	10	-9
4	(0,2)	12	-6 9 -11
5	(0,3)	9	-2
6	(0,2)	12	11, 4 9
7	(0,3)	11	8
8	(0,3)	11	7
		14	-12
9	(0,2)	12	4 -6 -11
		10	-3
10	(1,3)	5	none
11	(0,2)	12	-4 -6, -6 -9
12	(0,1)	13	none
		7	none
		14	-8
13	(0,0)	none	none
14	(1,2)	6	none

Table 2: The *S27* Observability Table

Time frame # 0	Time frame # 1	Time frame # 2
1 -4	**-1 -2 4**	**-1 3**

Table 3: Test Pattern for node 8 stuck-at 1

Node	Obs. node	Timed OF
8	11	tf # 0: 7
11	12	tf # 0: -4 -6 7
12	13	tf # 0: -4 -6 7

Table 4: The Observability Stack

The Observability and the Controllability searching phases produce the test pattern shown in Tab.3.

The correctness of the solution found can be easily observed exploring the Observability and the Controllability Searching Stacks reported in Tab.4 and Tab.5.

5 Experimental Results

The ATPG package has been implemented in *C programming language*, on a DEC-VAX 9000, running VMS 5.4 operating system.

The *Hardware Description Language* (HDL) used for the description of the circuits, at gate-level, is the one included in the MOZART package [12], an integrated CAD environment for digital systems simulation and testing.

The minimizer used for the functions complexity reduction has been developed by the authors in a VAX-VMS environment and it is based on the same concepts employed by the ESPRESSO minimizer [13].

In Tab.6 and Tab.7 are shown the results obtained running the ATPG procedure on a large subset of the ISCAS'89 synchronous sequential circuits [14]. The description of the circuits has been translated to our system adding to each flip-flop a common clock signal, with no other modification. An asynchronous reset line has been added to the circuits whose names are followed by the * character in Tab.6 and Tab.7. For those circuits, our algorithm is not able to initialize all the memory elements.

Tab.6 contains circuit partitioning statistics and CPU time required for Functions and Orders computation.

Results concerning fault coverage and CPU time needed for test pattern generation are presented in

Nodes	Timed CF
contr. PT	tf # 0: -4 -6 7 -8
8	tf # 0: 1 -4 -6 7
7 -6	tf # 0: 1 -4 — tf # 1: 12 -14
-14	tf # 0: 1 -4 — tf # 1: 8 12
12	tf # 0: 1 -4 — tf # 1: 4 -6 8 9
9	tf # 0: 1 -4 — tf # 1: -2 4 -5 -6 8
8	tf # 0: 1 -4 — tf # 1: -1 -2 4 -5 -6
-6 -5	tf # 0: 1 -4 — tf # 1: -1 -2 4 — tf # 2: -10 -14
-14	tf # 0: 1 -4 — tf # 1: -1 -2 4 — tf # 2: 8 -10
-10	tf # 0: 1 -4 — tf # 1: -1 -2 4 — tf # 2: 3 8
8	tf # 0: 1 -4 — tf # 1: -1 -2 4 — tf # 2: -1 3

Table 5: The Controllability Stack

Circuit	Gates	Nodes	CPU time	
			CF & OF	CO & OO
S27	10	14	00:00:02	00:00:01
S208	96	52	00:01:05	00:00:04
S298	119	58	00:00:08	00:00:01
S344	160	73	00:00:06	00:00:01
S349	150	74	00:00:06	00:00:01
S382	158	85	00:00:06	00:00:02
S386	159	41	00:00:59	00:00:03
S400	148	89	00:00:06	00:00:02
S420	196	105	00:00:16	00:00:05
S444	181	101	00:00:07	00:00:02
S510 *	211	103	00:02:01	00:00:11
S526	194	86	00:01:56	00:00:05
S526n	193	86	00:01:56	00:00:05
S641	379	153	00:00:23	00:00:04
S713	393	176	00:00:20	00:00:03
S820	289	80	00:09:47	00:00:14
S832	287	80	00:10:05	00:00:16
S838	390	208	00:13:36	00:01:03
S953 *	395	238	00:01:13	00:00:18
S1196	529	206	00:00:22	00:00:39
S1238	508	219	00:04:16	00:00:37
S1423	657	288	00:00:45	00:00:09
S1488	653	130	00:08:48	00:00:57
S1494	647	130	00:08:27	00:00:59
S5378	2779	1185	00:43:57	00:06:41

Table 6: Statistics about some ISCAS'89 circuits

Tab.7. CPU time is expressed as HOUR:MIN:SEC.

Notice that the proposed algorithm generates test pattern only for faults located on the primary inputs or on the FFRs output nodes; the coverage for FFRs internal faults is obtained using a fault simulator.

6 Conclusions

A new approach to test generation for sequential circuits has been presented in this paper.

The circuit is first partitioned in several blocks, each having a single output; the boolean function implemented by the block is computed and from it the Controllability Functions of each output of the block with respect to the input lines and the Observability Function of each input of the block with respect to the output are extracted.

Examining the information contained in these functions, each fault is first propagated to a primary output and then justified up to primary inputs.

A backtracking procedure is adopted in order to scan the full set of possibilities: to reduce the otherwise enormous computational cost of finding a test pattern, Controllability Orders, Observability Orders and Incompatybility Function have been defined, orienting the search towards the choices having the highest success probability.

Experimental results are reported, showing the effectiveness of the method when applied to the standard set of synchronous sequential circuits [14].

The algorithm outlined in this paper is subject to further improvement. Our attention is focused on the management of the circuits having reconvergent fanout [15]. For that purpose, a nine-valued algebra

Circuit	Faults	F.C. #	F.C. %	CPU time
S27	32	29	90.6	00:00:04
S208	215	114	53.2	00:10:23
S298	179	133	74.3	00:02:39
S344	324	267	82.6	00:19:01
S349	332	272	81.9	00:27:39
S382	399	319	79.9	00:05:52
S386	384	246	64.1	00:07:53
S400	424	340	80.3	00:07:41
S420	430	121	28.1	13:06:43
S444	474	376	79.3	07:01:59
S510 *	564	508	90.1	00:38:18
S526	555	385	69.3	00:11:48
S526n	553	384	69.5	00:11:31
S641	465	339	72.9	00:09:16
S713	581	404	69.6	00:12:01
S820	850	661	77.8	13:17:09
S832	870	686	78.8	23:11:50
S838	857	191	22.3	24:31:13
S953 *	1079	940	87.1	31:01:40
S1196	1242	1110	89.4	05:22:41
S1238	1355	1059	78.1	10:44:00
S1423	1515	905	59.7	16:19:51
S1488	1486	1233	82.9	02:22:50
S1494	1506	1232	81.8	02:21:12
S5378	4603	2536	55.1	24:13:47

Table 7: Statistics about some ISCAS'89 circuits

has been developed [16] [17]. The results obtained with a preliminary implementation show that our efforts have been productive.

Acknowledgements

The authors are very grateful to prof. A. Lioy and prof. M. Sonza Reorda of the Politecnico di Torino for their valuable suggestions and their helpful advice and discussions. Moreover, E. Macii thanks dr. T. Estrada of the University of Colorado at Boulder for the review of this paper.

References

[1] M. A. Breuer, A. D. Friedman, *Diagnosis and Reliable Design of Digital Systems*, Computer Science Press, 1976.

[2] H. Fujiwara, *Logic Testing and Design for Testability*, The MIT Press, 1985.

[3] M. Abramovici, M. A. Breuer, A. D. Friedman, *Digital Systems Testing and Testable Design*, Computer Science Press, 1990.

[4] G. R. Putzolu, J. P. Roth, "A heuristic algorithm for the testing of asynchronous circuits", *IEEE Transactions on Computers*, Vol. C-20, No. 6, pp. 639-647, June 1971.

[5] M. J. Williams, J. B. Angell, "Enhancing Testability of Large Scale Integrated Circuits via Test Points and Additional Logic", *IEEE Transactions on Computers*, Vol. C-22, No. 1, pp. 46-60, January 1973.

[6] E. B. Eichelberger, T. W. Williams, "A Logic Design Structure for LSI Testability", *DAC-14: 14th IEEE Design Automation Conference*, pp. 462-468, June 1977.

[7] L. H. Goldstein, "Controllability / Observability Analysis of Digital Circuits", *IEEE Transactions on Circuits and Systems*, Vol. CAS-26, No. 9, pp. 685-693, September 1979.

[8] J. Grason, "TMEAS – A Testability Measurement Program", *DAC-16: 16th IEEE Design Automation Conference*, pp. 156-161, June 1979.

[9] J. E. Stephenson, J. Grason, "A Testability Measure for Register Transfer Level Digital Circuits", *FTCS-6: 6th IEEE International Symposium on Fault-Tolerant Computing*, pp. 101-107, June 1976.

[10] M. A. Breuer, "Functional Partitioning and Simulation of Digital Circuits", *IEEE Transactions on Computers*, Vol. C-19, No. 11, pp. 1038-1046, November 1970.

[11] S. Hong, "Fault Simulation Strategy for Combinational Logic Networks", *FTCS-8: 8th IEEE International Symposium on Fault-Tolerant Computing*, pp. 96-99, June 1978.

[12] S. Gai, P. Montessoro, F. Somenzi, "MOZART: a Concurrent Multi-Level Simulator", *IEEE Transactions on Computer Aided Design*, Vol. CAD-7, No. 9, pp. 1005-1016, September 1988.

[13] R. Brayton, G. Hachtel, C. McMullen, A. Sangiovanni, *Logic Minimization Algorithms for VLSI Synthesis*, Kluwer Academic Publishers, 1984.

[14] F. Brglez, D. Bryan, K. Kozminski, "Combinational Profiles of Sequential Benchmark Circuits", *ISCAS'89: IEEE International Symposium on Circuits and Systems*, pp. 1929-1934, May 1989.

[15] V. K. Agarwal, A. S. Fung, "Multiple Fault Testing of Large Circuits by Single Fault Test Sets", *IEEE Transactions on Computers*, Vol. C-30, No. 11, pp. 855-865, November 1981.

[16] P. Muth, "A Nine-Valued Circuit Model for Test Generation", *IEEE Transactions on Computers*, Vol. C-25, No. 6, pp. 630-636, June 1976.

[17] C. W. Cha, W. E. Donath, F. Ozguner, "9-V Algorithm for Test Pattern Generation of Combinational Digital Circuits", *IEEE Transactions on Computers*, Vol. C-27, No. 3, pp. 193-200, March 1978.

ON THE FAULT LOCATION IN COMBINATIONAL LOGIC CIRCUITS

R. Rajsuman and M. Saad
Department of Computer Engineering
Case Western Reserve University
Cleveland, OH 44106

B. Gupta
Department of Computer Science
Southern Illinois University
Carbondale, IL 62901

ABSTRACT

A method is presented to identify and locate a fault in the combinational logic circuits. A test pattern generator (called SPARK) has been developed, based upon a modified version of the D-algorithm. A fault dictionary is set-up during test generation. Through simulation, all activated lines under a test vector are identified. The test response is evaluated to identify possible paths which may have the fault. We do not stop testing as soon as a fault is detected. Based upon circuit topology, the circuit is further exercised using fan-in and fan-out points to identify a faulty line segment or a gate. The fault location algorithm and experimental results on some combinational circuits are given.

1. INTRODUCTION

The testing problem in VLSI circuits can be divided into two categories: (a) Fault detection and (b) Fault location A considerable work has been done in the past to detect a fault in MOS VLSI [1-4]. A significant body of literature exists which considers the manufacturing faults in VLSI systems. Several techniques such as exhaustive testing, syndrome testing, random testing are available, which detects manufacturing faults in the digital circuits. Techniques are available at the system level to locate/diagnose a fault [5-9]. Such techniques are also required for VLSI circuits. The testing techniques available today cannot locate a fault in VLSI circuits. For fault diagnosis in VLSI circuits, highly skilled engineers inspect the faulty circuit under microscopes to study the cause of failure.

To enhance the production yield and for process improvement, fault location is essential. Today, semiconductor manufacturers identify a faulty circuit by functional testing. To improve the production yield, process development engineers examine this faulty circuit under a microscope to identify the fault. The whole procedure is very complex, extremely time consuming and require specialized personnel as well as equipments. To overcome these problems, algorithmic techniques are urgently require to locate faults in a given MOS circuit.

The main problem of fault location attributes to the limited number of observation points in VLSI circuits. With the increasing chip complexity, both observability and controllability become extremely poor and some times, it become almost infeasible to evaluate the circuit response in the traditional designs. Few attempts have been made to use electron beam probes to access the internal points of a circuit, to identify a fault [10-11]. In this paper, a CAD approach is given to identify a fault. This technique drastically reduces the time which is spend today to determine the cause of failure. With this technique, fault can be analyzed faster and the cause of failure can be identified in a significantly small time. This will help in isolating the faults and reconfiguration of the system, and hence, significant improvement in reliability.

2. TEST GENERATION

In this section, we will briefly describe the test generation methodology implemented in our CAD system, called SPARK. The system considers a gate level model consists of primitive gates (NOT, NAND, NOR, AND and OR) to generate the test vectors. Consider a generalized combinational block with n primary inputs and m primary outputs i.e.:

$$Set\ of\ primary\ inputs\ \ X = [x_1, x_2, ..., x_n]$$

$$Set\ of\ primary\ outputs\ \ Y = [y_1, y_2, ..., y_m]$$

A path from the input x_i to the output y_j, which can be activated by a binary tuple applied at the inputs, will be denoted by P_{ij}. A test set which detects a fault in this combinational block, is given as:

$$T = [t_1, t_2, ..., t_t]$$

$$where\ \ t_i = [x_1, x_2, ..., x_n]\ ;\ x_i\ being\ '0'\ or\ '1'.$$

A test vector can detect the fault if the output response under that test is different than the response of the fault free circuit. To satisfy this requirement, for at least one vector, the output of the faulty circuit has to be complementary than the output of the fault free circuit. In a circuit, for a line stuck-at fault, a test can be derived using the D-algorithm.

Input to the SPARK is functional relationships between the primary inputs and the primary outputs. These boolean equations are fed to the system in the format of circuit topology. If an actual gate level schematics is available, it can be fed directly. If the boolean expressions are fed, then SPARK generates the equivalent gate level schematics of the circuit, consists of primitive gates (AND, OR, NOT, NOR and NAND). Each line in this gate level schematics is identified by a unique name and test vectors are generated for each line stuck-at fault. The test generation algorithm is given in Fig. 1. The associate procedures, Propagate and Implicate, are also given in Fig. 1. SPARK also calculates the expected response of a fault free circuit for each vector and creat a fault dictionary. All the lines through which the fault effect is propagated (activated lines) under the test are also identified. This step is particularly important for fault diagnosis as described in the next section.

```
/* Procedure name: Test generation
/* Parameters: Test_Cube, Init_Cube, T_Path, lines, Fault, Faulty_line,
/* Error, Skip, Impossible, Untraceable, <I_Qf, I_Qr>,
/* <P_Qf, P_Qr>, <B_Qf, B_Qr> are all global variables.
/* <I_Qf, I_Qr>, <P_Qf, P_Qr>, <B_Qf, B_Qr> are all queues.

For V=1 to lines Do
Begin
  For W=0 to 1 Do
  Begin
    Skip=false
    Error=false
    Impossible=false
    Untraceable=false
    Fault=W
    Faulty_line=V
    Initialize the values of T_Path to -1
    Primitive_Of_the_Fault (Test_Cube)
    <B_Qf, B_Qr> = nil
    While (no Out_Reached(Faulty_line) and Untraceable is false)
          or (Skip=true) Do
    Begin
      if (Skip=false) Then
        Init_Cube=Test_Cube
        <P_Qf, P_Qr>=nil
        Propagate(Faulty_Line)
      Otherwise Skip=false

      New_Faulty_line(Faulty_line, Fault)
      insert Faulty_line in T_Path
      <I_Qf, I_Qr>=nil
      Test_Cube_Stack=nil
      Implicate(Faulty_line)

      if (Impossible=true) Then
      Repeat
        if (propagation queue is not empty) Then
          delete Faulty_line from T_Path
          Skip=true
          Error=false
          Impossible=false
          try to propagate through another gate if any
        Otherwise
          if the backtracking queue is empty or there is only
             one element and <P_Qf, P_Qr> of that element is empty Then
             Untraceable=true
          Otherwise
             backtrack to the previous element in the backtracking
             queue and consider other propagation possibilities
      Until (Impossible=false) or (Untraceable=true)
    End
    Store_Test_Vector (Test_Cube)
  End
End

/* Procedure name: Propagate
/* Parameters: line:integer;

US=nil
All_Input_PCs(line)
While (stack US is not empty) Do
Begin
  remove the top two elements from US stack and intersect
  those primitive cubes. Then, insert the intersection
  of those cubes into the propagation queue
End
if (propagation queue is not empty) Then
   remove first element from the queue and intersect
   it with Test_Cube. Then
      copy the result of that intersection back into the
      Test_Cube only if the result of that intersection
      does not result in a conflict

/* Procedure name: Implicate
/* Parameters: line:integer;

if (line is an output of a gate) Then
Begin
  Step 1: find the indices R1 and R2 of those primitive cubes
          in the PCs table
  Step 2: insert R1 and R2 into the implication queue and push
          Test_Cube onto the stack of test cube Test_Cube_Stack
  Step 3: Repeat
            if (R1 has not been used) Then
               used1=true
               intersect PCs[R1].PC with Test_Cube
            Otherwise if (R2 has not been used) Then
               used2=true
               if (R2<>0) Then
                  Error=false
                  intersect PCs[R2].PC with Test_Cube
          Until Error is false or both indices R1 and R2
          have been used
  Step 4: if after step 3 Error is false Then
          Begin
            if there is one element in the implication cube
            and R2 has been used or R2=0 Then
               Impossible=true
            Otherwise
               remove one element from the end of implication
               queue and restore the old value of Test_Cube by
               making it equal to the top element in the
               Test_Cube_Stack
               if (R2 has not been used) Then
```

```
            used2=true
            if (R2<>0) Then
               Error=false
               intersect PCs[R2].PC with Test_Cube
      Repeat step 4 until Error=false or Impossible=true
      if Impossible=true Then stop
      Otherwise
         if (Error=false) Then
            update the value of Test_Cube
            go to the position, i, in Test_Cube
            whose value is not equal to -1
            implicate (i)
     End
     Otherwise
        if (Error=false) Then
           update the value of Test_Cube
           go to the position, i, in Test_Cube
           whose value is not equal to -1
           implicate (i)
End
Otherwise
  if (Error=false) Then
     go to the position, i, in Test_Cube whose
     value is not equal to -1
     implicate (i)
```

Fig. 1: Test generation algorithm as implemented in SPARK. The associated procedures are also given.

3. IDENTIFICATION OF A FAULT

In this section, the basic methodology is described to locate a fault in the given circuit. The procedure is best illustrated with the help of Fig. 2a. For simplicity, we assume that a fault is detected, i.e. for at least one test vector, the output response of the circuit is different than the calculated correct response. After detecting a fault, this faulty output line is traced backward, to obtain one or more paths connecting a primary input to the primary output. Any two paths in this list share a common fan-in point as shown in Fig. 2a. Names of all fan-in points are stored. In Fig. 2a, three different paths are shown, which share the fan-in points 'a' and 'b'. The name of paths, which might have the fault, are saved. For example, for Fig. 2a, we have three names:

$$T_{path} = [P_{pj}, P_{qj}, P_{rj}]$$

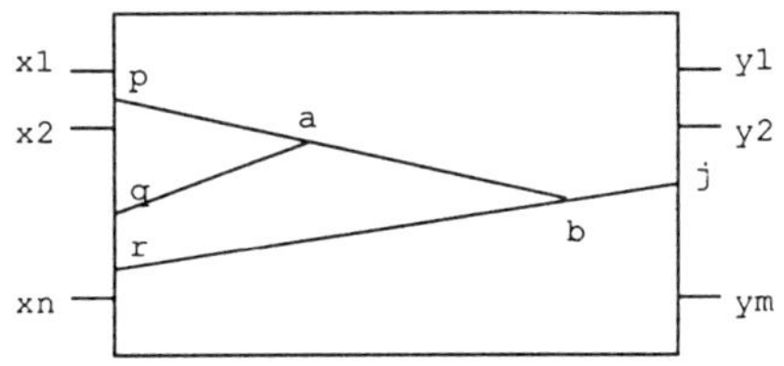

Fig. 2a: Illustration of paths connecting the primary inputs to the primary outputs, obtained after tracing the faulty response.

Also, using fan-in points, all paths are segmented and a segment list is created. For Fig. 2a, the segment list is:

$$L_{segment} = [S_{pa}, S_{ab}, S_{bj}, S_{qa}, S_{rb}]$$

It should be noted that every gate (except NOT gate) in the circuit, is a fan-in point. However, fault cannot be traced back through all the gates. The list of fan-in points contains the names of only those gates which are included in two or more paths in the path list. The gate level diagram for Fig. 2a, is shown in Fig. 2b. Different nodes are marked in Fig. 2b, to identify different paths. In Fig. 2b, some extra paths are also shown to represent the generalized block of Fig. 2a.

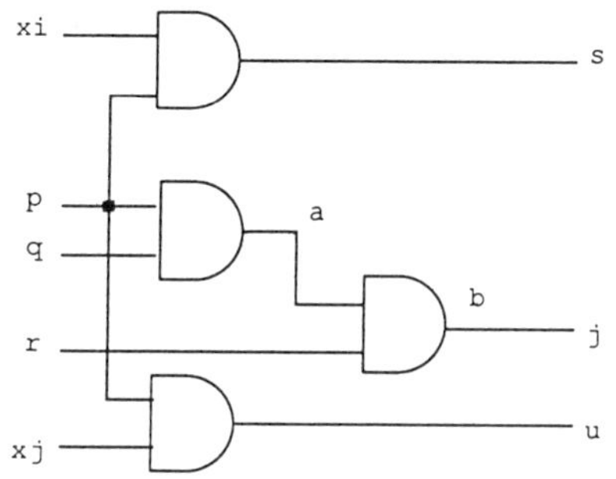

Fig. 2b: gate level diagram of the circuit used in fig. 2a.

After identification of a faulty segment, the segment can be further broken down into small parts, using fan-out points. This is shown in Fig. 3, with two fan-out points marked as 'c' and 'd'. The original list of path names is updated. The new list contains the names of paths which share the faulty segment of the original path. For example, for Fig. 3 we have three names:

$$T_{path} = [P_{ps}, P_{pj}, P_{pu}]$$

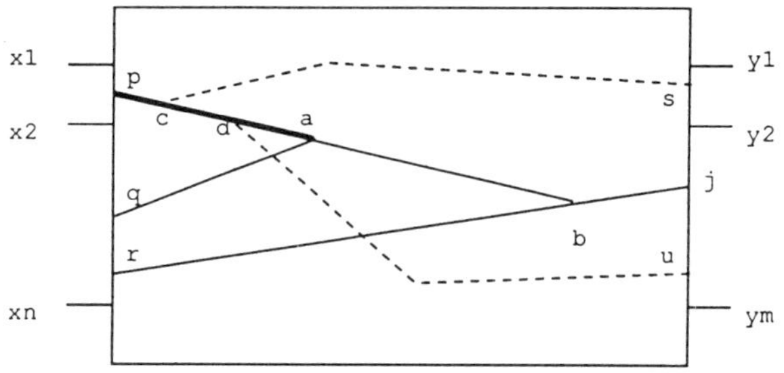

Fig. 3: Division of faulty segment using fan-out points.

So, if the faulty path is recognized as S_{pa}, we can further divide it and exercise each segment independently. This way, we can identify the atomic element which is faulty. The flow chart of the procedure which identifies a faulty segment is given in Fig. 4. The procedure is summarized as follows:

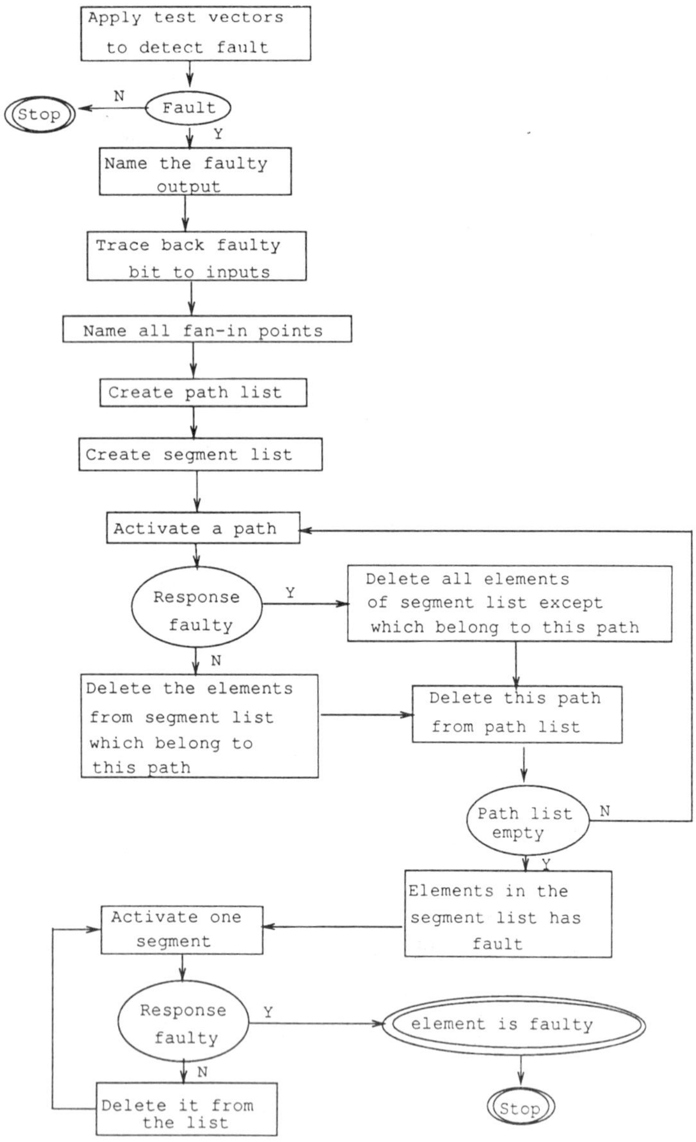

Fig. 4: Fault diagnosis algorithm.

The response of a circuit under a test set is evaluated to identify the faulty path from a primary input to a primary output. After identification of a faulty path, the path is segmented into small parts. These segments are named separately. The individual response of each segment is evaluated. The segments, whose response matches with the correct calculated response is deleted from the list. This identifies a single segment which has the fault. The faulty segment is further divided into small segments which are evaluated independently. This procedure is repeated until any further division is possible. The pseudo code of this procedure is given in Fig. 5.

```
/* Procedure name: Common_Part
/* Parameters: Loc_Path, Path:L_ar; flag:integer;

if all values of Loc_Path are equal to -1 Then
   Loc_Path=Path because Loc_Path is empty
Otherwise
Begin
  Step 1: find the index x in Loc_Path at which Path and
          Loc_Path start to have equal values
  Step 2: find the index y in Loc_Path at which Path and
          Loc_Path stop to have equal values
  Step 3: if there is an intersection between Loc_Path and Path Then
             go to step 4
             Otherwise stop
  Step 4: if flag=0 Then
             get the part of Loc_Path from 1 to x and y to
             Max_Lines (this is the intersection of the faulty Loc_Path
             and the nonfaulty Path)
          if flag=1 Then
             get the part of Loc_Path from x to y (this is the
             intersection of the faulty Loc_Path and the faulty Path)
End

/* Procedure name: Locate_The_Fault
/* Parameters: none

Scan the list of output lines and check whether any of
these lines is faulty
/* We refer to Table to decide whether a line is faulty or
/* not bychecking the flag value of the line
if output line is faulty Then
   Stop
Otherwise
   Step 1: initialize the values of Loc_Path to -1
   Step 2: Intersect Loc_Path with the paths of all faulty lines in
           the table, Table, and make Loc_Path be that intersection
   Step 3: Intersect Loc_Path with the paths of all nonfaulty lines
           in the Table, as long as those paths are subsets of Loc_Path
   Step 4: The fault is located at the line that is left after the
           above series of intersections
```

Fig. 5: Fault location algorithm as implemented in SPARK.

Example: Consider a simple combinational circuit as given in Fig. 6. The circuit has five inputs and three outputs and contains internal fan-out point. One line stuck-at fault is introduced in the circuit at a time after the circuit is described in SPARK. SPARK generates the test vectors, calculate the expected response of fault free circuit and the path which is activated by the test. This is given in Fig. 7 for the circuit shown in Fig. 6. SPARK applies the test vectors to the faulty

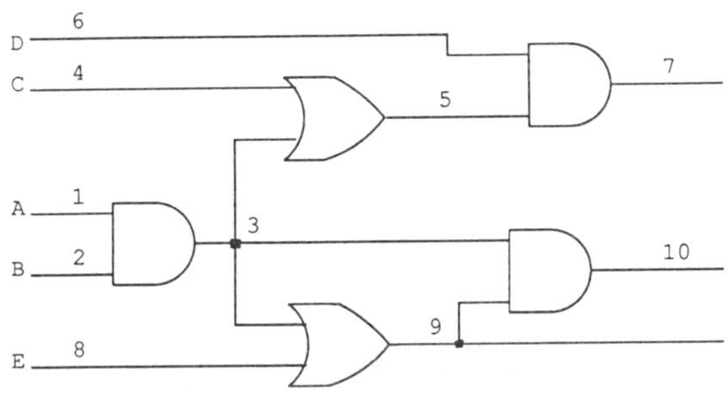

Fig 6: An example circuit to illustrate SPARK's operation.

Line #	Fault	Path	Inputs [abcde]	Outputs 9 10 7
1	s-a-1	1 3 10	1xx10	10x
1	s-a-0	1 3 10	1xx11	11x
2	s-a-1	2 3 10	1xx01	10x
2	s-a-0	2 3 10	1xx11	11x
3	s-a-1	3 10	1xx0x	10x
3	s-a-0	3 10	1xx11	11x
4	s-a-1	4 5 7	x100x	x00
4	s-a-0	4 5 7	x110x	x01
5	s-a-1	5 7	x100x	x00
5	s-a-0	5 7	x11xx	xx1
6	s-a-1	6 7	x01xx	xx0
6	s-a-0	6 7	x11xx	xx1
7	s-a-1	7	x0xxx	xx0
7	s-a-0	7	x11xx	xx1
8	s-a-1	8 9	0xx0x	00x
8	s-a-0	8 9	1xx0x	10x
9	s-a-1	9	0xx0x	00x
9	s-a-0	9	1xxxx	1xx
10	s-a-1	10	xxx0x	x0x
10	s-a-0	10	1xx11	11x

Fig. 7: Test set for the circuit given in fig. 6.

circuit and identifies the faulty path. It then keep intersecting the segments of faulty path with the segments of other paths. For example, when line #1 s-a-1 fault is introduced, SPARK finds the fault under test vector {abcde = 01xx1} and identifies path 1-3-10 as faulty. After intersecting with the fault free response of path 3-10 (test for line #3 s-at fault, test vector {abcde = x0xx1 and 11xx1}), it identifies fault on line #1.

When fault line #5 s-a-1 fault is introduced, SPARK finds the fault under test vector {abcde = x001x} and identifies path 4-5-7 as faulty. The fault is converged to path 5-7 and finally at line #5. □

4. DISCUSSION AND CONCLUSIONS

The desirable output of the above procedure is the identification of an atomic element (a single gate or a transmission line), which has the fault. In our system, we have used fan-in and fan-out points as the points by which a path can be divided into small segments. However, in the worst case, when a path cannot be divided into small segments, a single path from a primary input to the primary output is identified. Consider a simple combinational circuit as shown in Fig. 8. All the test vectors for this circuit are listed in Fig. 9. Let us assume a s-a-1 fault on line #5. This fault is detected under test vector {abcde = 111x0}. This fault is propagated through line #8. As line #5 and line #8 cannot be exercised independently using different vectors, SPARK identifies fault on line #8 instead of line #5.

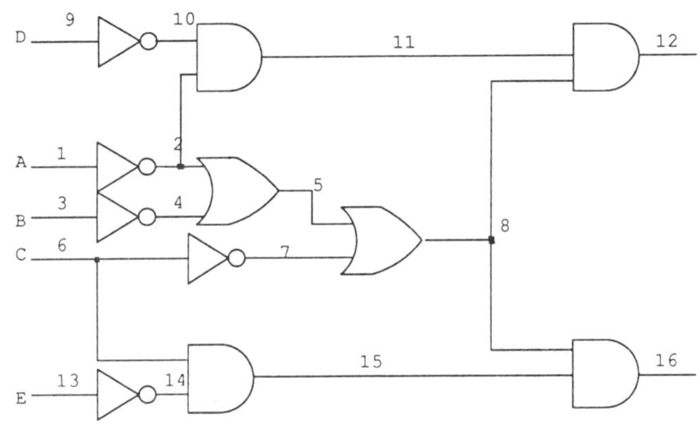

Fig. 8: An example circuit to show fault equivalence.

The situation as illustrated in Fig. 8, exists because of the fault equivalence. In such condition, a fault cannot be identified accurately. With the above procedure we can locate a fault accurately, only if we have sufficient number of fan-outs in the circuit. In the absence of fan-out points, or when the fault effect cannot be propagated through a given path, extra control points are required. Using some extra hardware, a circuit can be designed such that a test vector will uniquely activate a path. While generating test vectors for a given circuit, necessary points can be identified to enhance the circuit observability and controllability. For example, if the test vectors are generated by PODEM [2], the points encounter in backtracing can be made observation points. Using these extra observation points any path can be activated. If the observed response of a path is fault free, than the segments associated to that path are considered as fault free. After activating each path one can determine an atomic element which has the fault. In this paper we have not

Line #	Fault	Path	Inputs [edcba]	Outputs 16 12
1	s-a-1	1 2 11 12	x00x0	01
1	s-a-0	1 2 11 12	x00x1	00
2	s-a-1	2 11 12	x00x1	00
2	s-a-0	2 11 12	x00x0	01
3	s-a-1	3 4 5 8 16	0x101	10
3	s-a-0	3 4 5 8 16	0x111	00
4	s-a-1	4 5 8 16	0x111	00
4	s-a-0	4 5 8 16	0x101	10
5	s-a-1	5 8 16	0x111	00
5	s-a-0	5 8 16	0x10x	1x
6	s-a-1	6 15 16	0x0xx	0x
6	s-a-0	6 15 16	0x10x	1x
7	s-a-1	7 8 16	0x111	00
7	s-a-0	undetectable		
8	s-a-1	8 16	0x111	00
8	s-a-0	8 12	x00x0	01
9	s-a-1	9 10 11 12	x00x0	01
9	s-a-0	9 10 11 12	x10x0	00
10	s-a-1	10 11 12	x10x0	00
10	s-a-0	10 11 12	x00x0	01
11	s-a-1	11 12	xx0x1	00
11	s-a-0	11 12	x00x0	01
12	s-a-1	12	xxxx1	x0
12	s-a-0	12	x00x0	01
13	s-a-1	13 14 15 16	0x10x	1x
13	s-a-0	13 14 15 16	1x10x	0x
14	s-a-1	14 15 16	1x10x	0x
14	s-a-0	14 15 16	0x10x	1x
15	s-a-1	15 16	xx0xx	0x
15	s-a-0	15 16	0x10x	1x
16	s-a-1	16	xx0xx	0x
16	s-a-0	16	0x10x	1x

Fig. 9: Test set for the circuit given in fig. 8.

considered such modified designs. The basic rules to identify the minimum number of extra observation points as well as their locations are under investigation and will be addressed in future.

The fault model for the above procedure includes all single line stuck-at-1/0 faults only. All single transistor stuck-open/short faults and all single bridging faults with/without feedback effect have not been included in the SPARK at this time. It should be noted that in the case of a bridging fault between two lines, more than one path from the primary input to the primary output, shows the fault effect. However, after execution of fault location algorithm, a segment, common to all faulty paths is identified. This procedure is also able to detect and locate the faults which are masked in the presence of other faults during the propagation of fault effect. The main reason behind this, is the segmentation of a path and the evaluation of response for each segment independently. Hence, some undetected multiple faults are also detected as well as located by this technique. The detection and location of undetectable multiple faults is a very desirable feature for reliability analysis engineers and yield enhancement/process development engineers. The detection and diagnosis of multiple faults is under investigation.

References

1. A. D. Friedman and P. R. Menon, "Fault detection in digital circuits", Prentice Hall Inc., 1971.
2. P. Goel, "An implicit enumeration algorithm to generate tests for combinational logic circuits", IEEE Trans. Comp., vol. C-30, March 1980, pp. 215-222.
3. S. K. Jain and V. D. Agarwal, "Test generation for MOS circuits using D-algorithm", IEEE Design Auto. Conf., 1983, pp. 64-70.
4. R. Rajsuman, A. P. Jayasumana and Y. K. Malaiya, "Testing of complex gates", Electronics Letts., vol. 23, July 1987, pp. 813-814.
5. A. T. Dahbura and G. M. Masson, "An O(n2.5) fault identification algorithm for diagnosable systems", IEEE Trans. Comp., vol. C-23, April 1984, pp. 486-492.
6. H. Fujiwara and K. Kinoshita, "On the computational complexity of system diagnosis", IEEE Trans. Comp., vol. C-20, Oct. 1978, pp. 881-885.
7. G. G. L. Meyer, "A diagnosis algorithm for the BGM system level fault model", IEEE Trans. Comp., vol. C-33, Sep. 1984, pp. 756-758.
8. A. K. Somani, V. K. Agarwal and D. Avis, "A generalized theory for system level diagnosis", IEEE Trans. Comp., vol. C-36, May 1987, pp. 538-546.
9. E. T. Purcell, "Fault diagnosis assistant", IEEE Circuits and Devices magazine, pp. 47-59, Jan. 1988.
10. N. Kuji, T. Tamama and M. Nagatani, "FINDER: A CAD system based electron beam tester for fault diagnosis of VLSI circuits", IEEE Trans. CAD, vol. CAD-5, April 1986, pp. 313-319.
11. T. Tamama and N. Kuji, "Integrating an electron beam system into VLSI fault diagnoses", IEEE Design and Test, vol. 3, Aug. 1986, pp. 23-29.

Scheme for detecting CMOS stuck-open faults using single test patterns

Manjit S. Cheema
Department of Electrical Engineering
N.C. A&T State University,
Greensboro.N.C.27411.

P. K. Lala
Department of Electrical Engineering
N.C. A&T State University,
Greensboro.N.C.27411.

Abstract

It has been widely accepted that not all defects in CMOS logic can be represented by the stuck-at fault model. One example is the transistor stuck-open fault. The major problem with stuck-open faults is that it forces a combinational circuit to behave as a sequential circuit. It has been established that two pattern test technique may be inadequate to detect all stuck-open faults in the presence of stray ciruits delays and glitches. Furthermore 'robust tests' that can detect transistor stuck-open faults independent of glitches and circuit delays may not exist for all stuck-open faults. This paper presents a technique to modify CMOS circuits so that any transistor stuck-open fault in the circuit can be detected using only single test pattern. Two additional transistors and an additional input line are required for this purpose. It is also necessary to incorporate an inverter at the final output of the circuit.

1. INTRODUCTION

Complementry MOS (CMOS) technology has emerged as the the dominant technology for manufacturing VLSI digital circuits[1]. A CMOS circuit consists of a P-network connected between Vdd and the output node, and a N-network connected between GND and the output node.The circuit for P-network and N-network bear a dual relationship by DeMorgan's theorem. Fig.-1 shows the block diagram of a CMOS circuit. It is now generally accepted that not all faults in CMOS VLSI logic can be modelled by the stuck-at-0 and stuck-at-1 models used at the gate level. An example is the transistor stuck-open(s-open) fault.

The major problem with the stuck-open fault is that it forces a combinational circuit to behave as a sequential circuit and hence exhibit memory like property[2-4]. In other words, the circuit output depends not only on the current inputs but also on past

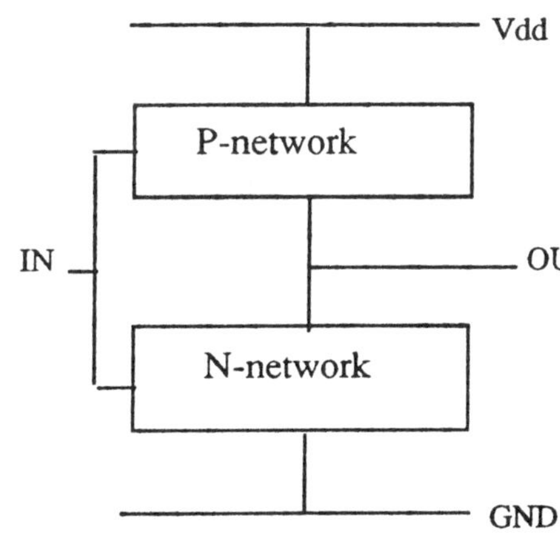

Fig.1 Block diagram of FCMOS circuit.

inputs. Testing for a stuck-open p-transistor (n-transistor) requires presetting the output node to logic 0 (logic 1) via an initializing input pattern and then applying a test pattern that establishes a conducting path to reverse the state of the output node to logic 1 (logic 0). In other words a two pattern strategy has to be used to detect stuck-open faults. The output node changes only if there is no stuck-open fault in the selected conducting path. It has been shown in [5,6] that two pattern tests may be invalidated in static CMOS circuit because of arbitrary circuit delays, glitches and timing skew in input changes. However these problems can be avoided by redesigning CMOS circuits[7-10]. In this paper a new design technique for testable CMOS circuits is presented. This technique makes the circuit fully testable for all single stuck-open faults using only single test pattern. The problem of test invalidation by glitches and circuit delay etc. is totally eliminated.

1058-6393/91 $1.00 © 1991 IEEE

2 Proposed approach for stuck-open fault detection

In order to detect stuck-open faults the the original CMOS circuit has been augmented by using two pass transistors(p1,n1), an inverter(p2,n2) and an external input line 'X'. The inverter acts as a buffer to boost the signal at the ouput. The block diagram implementation of the proposed scheme is shown in Fig.2. It should be noted that only a single inverter will be used to drive the final output of circuit under test, irrespective of the number of stages in the circuit.

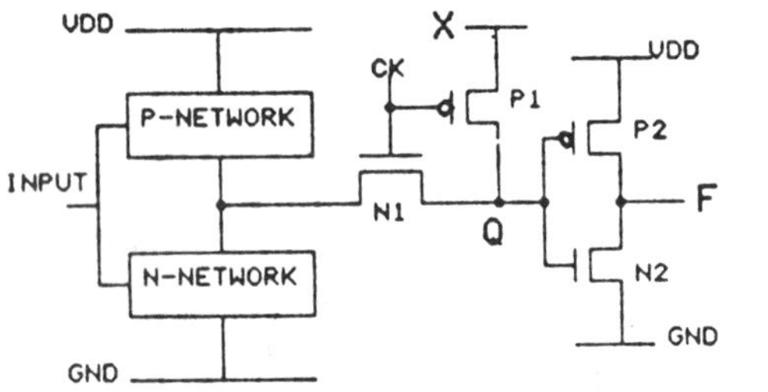

FIG.2 BLOCK DIAGRAM OF AUGMENTED CMOS CIRCUIT

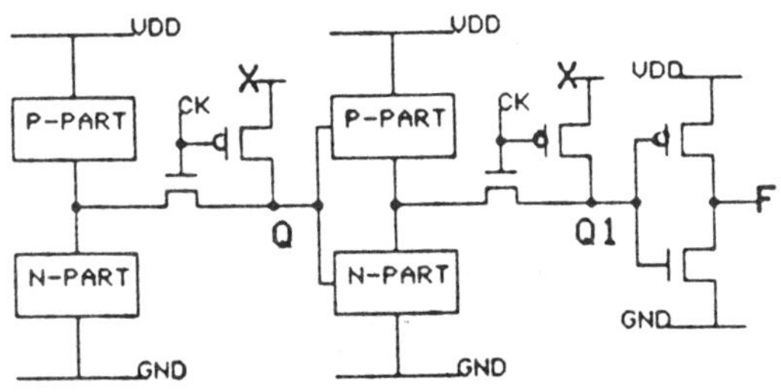

FIG.3 BLOCK DIAGRAM OF AUGMENTED TWO STAGE CMOS CIRCUIT

Hence if a circuit has several stages, the inverter is needed only to drive final output of the circuit, the intermediate stages can drive the following stages directly. This is illustrated in Fig.3. The first stage does not have any inverter at its output. It is assumed that the circuit gives the final output only after Ck makes a transition from 0 to 1. In theorem1 it is proved that any single stuck-open fault in the circuit can be detected by a single test pattern.

THEOREM1: All single stuck-open faults in the P-part or N-part of a CMOS circuit modified as shown in Fig.-2 or Fig.3 can be detected by using only a single test pattern .

Proof: During the test mode when the clock is low, the output of the circuit is always set to the opposite value to that expected, so that for any s-open fault in the selected conducting path the circuit output retains this value. Therefore when the P-network (N-network) is selected for testing and a selected test pattern is applied to the circuit, the output Q is set at logic 0(1). The transmission path from Vdd (GND) to the output node 'Q' will be activated by the test pattern if only if a stuck-open fault is not present in the selected path. Hence the output node Q will change from logic 0 (logic 1) to logic 1(logic 0) iff there is no s-open fault in the selected path in P-network (N-network). In other words if the output Q of the P-network (N-network) remains at logic 0 (logic 1) in response to a selected test pattern, a stuck-open fault has been detected. QED.

Let us apply the proposed technique to the CMOS circuit shown in Fig.4(a).The augmented version of this circuit is shown in Fig.4(b).

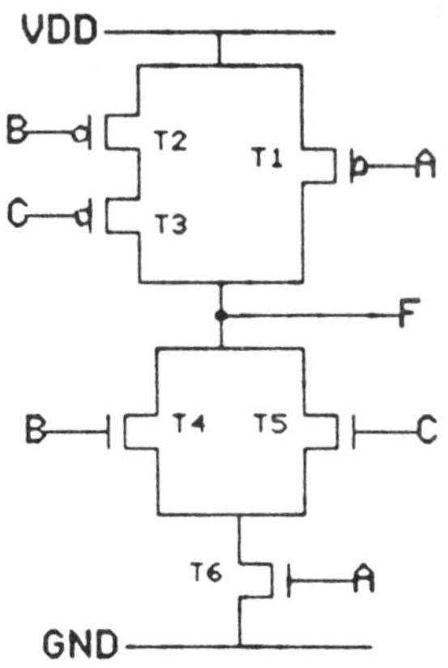

FIG.4(A) CMOS CIRCUIT IMPLEMENTING FUNCTION $F=\overline{A(B+C)}$

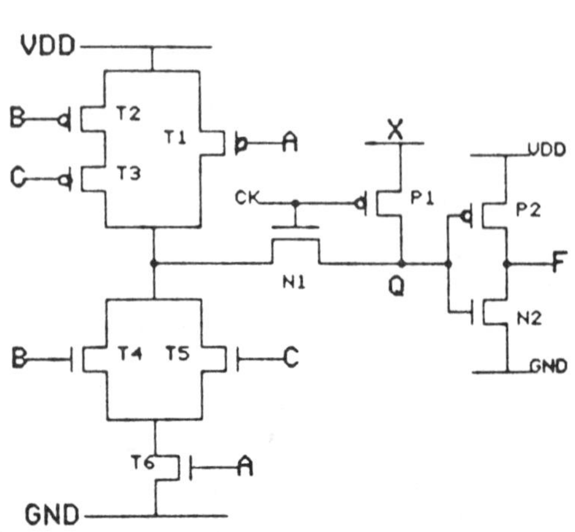

FIG.4(B) AUGMENTED CIRCUIT FOR STUCK-OPEN FAULT DETECTION

Let us suppose transistor T1 is stuck-open. The test pattern for this fault is : A=0, B=1, C=0, X=0. When the test pattern is applied, output node 'Q' will be set to logic '1' iff T1 is not stuck-open else it will be set to logic '0', hence the fault will be detected. A test pattern for each transistor stuck-open fault in the circuit of Fig.4(B) is given below:

Fault	Test pattern	Faulty(F)	Fault free(F)
T1 s-open	A=0, B=1, C=0, X=0	1	0
T2 or T3 s-open	A=1, B=0, C=0, X=0	1	0
T4 or T6 s-open	A=1, B=1, C=0, X=1	0	1
T5 or T6 s-open	A=1, B=0, C=1, X=1	0	1

Now let us illustrate the scheme for a multistage circuit shown in Fig.5 taken from [7]. It has been shown that no two pattern test seqence exist for certain stuck-open faults in this circuit[7]. However by using the technique suggested in this paper,the circuit can be tested for all single stuck-open faults using only single test patterns.

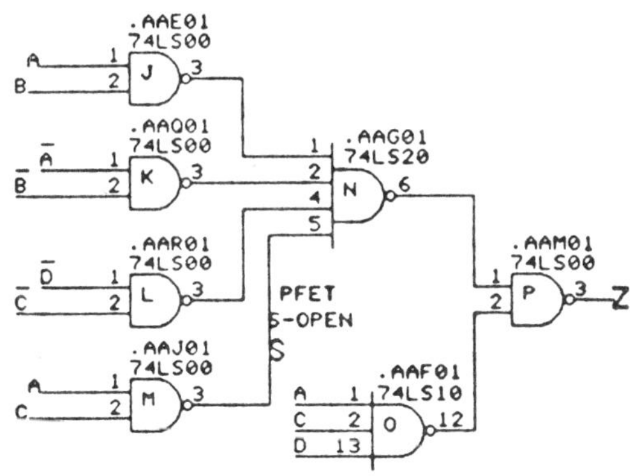

FIG.5 CMOS CIRCUIT TAKEN FROM [REF.7]

To illustrate let us derive a test for a P-transistor stuck-open fault connected to input A of the NAND gate marked 'J' in Fig5. The input pattern that will detect this fault is: A=0, B=1, C=0, D=1, X=0. When the pattern is applied, output node 'Z' will be discharged to '0' if and only if the above PFET stuck-open fault exists in the sensitized path else it will be charged to '1'. Now let us derive a test pattern for a stuck-open fault in a PFET driven by S in NAND gate marked 'N' in Fig.5 for which a three pattern test set has been suggested in [7]. This pattern is given as [ABCD]= [1001,1011,1010]. Using our approach this fault can be detected using a single test pattern :A=1, B=0, C=1, D=0, X=0. When this pattern is applied output 'Z' will be set to '0' in the absence of the stuck-open fault otherwise it will be set to '1'.

The total number of test patterns needed to detect all stuck-open faults can be derived using the following theorem.

THEOREM2: The maximum number of test patterns needed to test all stuck-open faults is equal to total number of independent paths from Vdd(GND) to output node.

Proof: If two or more transistors are in series in a path between Vdd or GND to output node, then the effect of one being stuck-open will be same as the effect of more than one stuck-open transistors connected in series. In other words, a single test pattern is sufficient to detect the presence of one or more than one stuck-open transistor in a path. Therefore, the total number of test patterns to detect all stuck-open faults cannot exceed the total number of independent paths from Vdd(GND) to the output. QED

To illustrate, let us consider circuit shown in Fig.4(b). It has four independent paths from Vdd(GND) to output node 'F'. As can be seen T2 or T3 stuck-open can be detected by just one test pattern. Similarly (T4&T6) and (T5&T6) can be tested using one test pattern for each path.

The effect of the additional transistors being faulty will be as follows:

N1 stuck-open: If N1 is stuck-open it will be easily detectable because the output of the circuit will get stuck at 0 (1) for X=0 (1). In other words the output will get stuck at the value assumed by X at that instant of time.

P1 stuck-open : If P1 is stuck open , it will not have any effect on normal operation of the circuit because P1 transfers external input to output only when CK is low during test mode. When CK is high the output is connected to Vdd or GND according to the input pattern applied to the circuit. However, if P1 gets s-open it can be detected by setting the output at 1(0) and applying X=0(1), while keeping the clock low.

P2 (N2) stuck-open : If P2(N2) is stuck open, it can be easily detected by setting the CK permanantly low, and then applying 0(1) at the additional input 'X'. The output 'F' will get stuck-at '0'(1) if P2(N2) is stuck-open.

3. Conclusion

A technique for detection of stuck-open faults in CMOS circuits has been presented. This approach needs only single test pattern for detecting such faults, and the circuit retains its combinational charateristic. The problem of test invalidation by circuit delays, timing skews etc. has been eliminated. Only two transistors and an additional input is needed to make any gate fully testable for all single stuck-open faults. Hence the area overhead is not significant. With a slight modification and little more increase in overhead the technique can be extended to detection of s-closed faults as well.

Acknowledgement

This work was supported by Air Force of Scientific Research under grant F-49620-89-C-0069.

References

1. Y.M. El-Ziq et al.,"Functional-level test generation for stuck faults in CMOS VLSI",Proc. Int'l test conf., pp 536-546, 1981.

2. R.Chandramouli,"On testing s-open faults,"Proc. Int'l Symp. fault-tolerant computing, pp 258-265, 1983.

3. J.P. Hayes,"Fault Modelling", IEEE Design and Test of computers, pp. 88-95, April 85.

4. R.Rajsuman, A.P. Jayasuman and Y.K. Malaya,"CMOS s-open fault testability",IEEE J. Solid-state circuits, vol.-24, no.1, pp 193-194, Feb. 89.

5. R.Rajsuman, A.P. Jayasuman and Y.K. Malaya,"CMOS open fault detection in the presence of glitches and timing skews",IEEE J. Solid-state circuits, vol.-24, no..4, Aug.89.

6. R.Rajsuman, A.P.Jayasuman and Y.K. Malaya,"CMOS open fault detection using single test patterns",IEEE J. Solid-state circuits, vol.26, no.1, Jan.1991.

7. S.M. Reddy,M.K. Reddy,"Testable realization for FET s-open faults in CMOS combinational logic circuits", IEEE Trans. computers, vol. c-35, pp 742-754, Aug. 86.

8. N.K. Jha and J.A. Abraham,"Design of testable CMOS logic circuits under arbitrary delays", Vol. CAD.-4, no.3, pp 264-469, July1985.

9. S.M. Reddy,M.K. Reddy and J.G. Kuhl," On testable design for CMOS logic circuits",Proc. 1983 Int. test conf., Philadelphia, PA, pp 435-445, Oct. 1983.

10. Sandip Kundu,"Design of multioutput CMOS combinational logic circuits for robust testability", IEEE Trans. on Computer-Aided-Design,Vol. 8, no. 11, pp. 1222-1225, Nov.89.

A Technique for Analog Fault Diagnosis using Element Modulation

Alvernon Walker
Department of ECE
Box 7911
North Carolina State Univ.
Raleigh, NC 27695

Parag K. Lala
Department of EE
North Carolina A&T State University
Greensboro, NC 27411

Winser E. Alexander
Department of ECE
Box 7911
North Carolina State Univ.
Raleigh, NC 27695

Abstract

This paper describes a new branch fault-diagnosis technique, which requires a single excitation source at one test frequency for solving of branch-diagnosis equations. Linearly independent branch-diagnosis equations are constructed by modulating selected network elements. A network element whose value can be varied in a controlled fashion is defined as a modulated element. It is assumed for simplicity that all modulated devices in the network under test have two states (a state is a predefined element value). A subset of the network elements are modulated. Necessary and sufficient conditions for the diagnosis of at least $q+1$ branch elements at faulty circuit nodes have been derived for a general network where q network elements are modulated. An approximate upper bound on this system of equations is also derived. This paper is concluded with an example of the proposed technique applied to a resistive circuit.

1 Introduction

Digital techniques and system modularity have been used for several decades to avoid problems with analog diagnostics [1]. These methods generally cannot be applied to current mixed VLSI circuits containing large analog subnetworks, because of the finite number of pins on integrated circuit packages, and performance degradation induced by parasitic capacitance, cross talk, etc. Interest in analog fault diagnosis has grown due to the limitations and costs of existing analog test procedures. These diagnostics techniques can be subdivided into two broad classes of algorithms; simulation-before-test (SBT) and simulation-after-test (SAT) methods [2]. In this paper we present a SAT algorithm for fault-diagnosis, which consists of a node fault location phase and a branch fault-diagnosis phase. The major contribution of this paper is the presentation of a new technique called *element modulation*, for implementing the branch diagnosis. The node location phase is similar to the node fault location method proposed by Biermacki [3],[4] and Sakla *et al* [5].

The node fault location phase is based on the consistency of a system of linear equations which are invariant on network element faults, and allows the detection of a bounded number of network node faults r. For a general network N with $n+1$ nodes, typically r is less than n; these networks are defined as r-node fault testable. The r-node fault testability of a network is related to the number of accessible network nodes (measurement nodes) and circuit topology [6]. This phase of the technique will not be discussed further in this paper, additional details can be found in reference 6. The branch-diagnosis equations for a network with modulated elements are derived in section 2 of this paper. Necessary and sufficient conditions for the solvability of a subnetwork with at least $q+1$ branches are also derived. In addition a first order approximate upper bound for this system of branch-diagnosis equations is obtained. An application of the technique is given in section 3.

2 Branch-diagnosis equations

In this section the branch-diagnosis equations for a network N containing modulated elements are derived. These equations are based on Tellegen's theorem [7]. Node, and branch constraints of the proposed technique are also developed. Henceforth, all modulated elements will be referred to as switches and their two parameter states will be defined as off and on. The low impedance state of the modulated elements is defined as on, and the high impedance state as off. The following notation will be used in this paper:

V_k branch voltage vector of network N,
I_k branch current vector of network N,
$\hat{V}_k$ branch voltage vector of adjoint network $\hat{N}$,
$\hat{I}_k$ branch current vector of adjoint network $\hat{N}$,
V_{pj} port voltage at port j,
I_{pj} port current at port j,
Y_k branch admittance vector of network N,
C^{nxm} Complex vector space which denote all n x

m matrices with complex entries.

Before the branch-diagnosis equations can be developed it is necessary to introduce a second circuit $\hat{N}$ with the same topology as the network N but not necessarily with the same branch characteristics [7]. $\hat{N}$ and N are adjoint networks of each other. All pairs of nodes used for network excitation and measurement are called *ports*. All voltages and currents in the following equations, with exception of port voltages and currents, are branch voltages and currents. The following equation is obtained from Tellegen's conservation of power theorem

$$\sum_{k=1}^{b} (\hat{V}_k I_k - V_k \hat{I}_k) = -\sum_{j=1}^{m} (\hat{V}_{pj} I_{pj} - V_{pj} \hat{I}_{pj}) \quad (1)$$

where b is the number of branches and m is the number of ports of network N (its adjoint network $\hat{N}$). Now if we assume that N is perturbed to $N+\Delta N$ (*i.e.* Y_k is perturbed to $Y_k+\Delta Y_k$) due to some branch fault(s), equation (1) becomes

$$\sum_{k=1}^{b} [\hat{V}_k (I_k+\Delta I_k) - (V_k+\Delta V_k)\hat{I}_k] = -\sum_{j=1}^{m} [\hat{V}_{pj}(I_{pj}+\Delta I_{pj}) - (V_{pj}+\Delta V_{pj})\hat{I}_{pj}] . \quad (2)$$

Substraction of (1) from (2) and the introduction of branch constraints [8] yield

$$\sum_{k=1}^{b} (V_k+\Delta V_k)\hat{V}_k \Delta Y_k = -\sum_{j=1}^{m} (\hat{V}_{pj}\Delta I_{pj} - \Delta V_{pj}\hat{I}_{pj}) . \quad (3)$$

The term(s) ΔY_k on the LHS of equation (3) is the deviation of branch k from its normal value. Since a single constant current excitation source is used, equation (3) reduces to

$$\sum_{k=1}^{b} (V_k+\Delta V_k)\hat{V}_k \Delta Y_k = \Delta V_{p1}\hat{I}_{p1} . \quad (4)$$

The switched (modulated) branch elements of N (and $\hat{N}$) represent two possible impedance states which both must appear in (4), if the on and off impedance states of the modulated branches deviate from their normal value. A network N with q modulated branches (where $q < b$) produces the following branch-diagnosis equations

$$\sum_{k=1}^{b-q} (V_k+\Delta V_k)\hat{V}_k \Delta Y_k + \sum_{k=b-q+1}^{b} (V_k+\Delta V_k)\hat{V}_k \Delta Y_{k(on)} + \sum_{k=b+1}^{b+q} (V_k+\Delta V_k)\hat{V}_k \Delta Y_{k(off)} = \Delta V_{p1}\hat{I}_{p1} . \quad (5)$$

The pattern of on and off switching branches will be discussed in terms of binary numbers of length q for clarity *i.e.* a network with q (=4) modulated branches configured as $Y_{1(on)}, Y_{2(off)}, Y_{3(on)}, Y_{4(off)}$ is represented by 0101. The network represented by (5) appears to contain $b+q$, branches which are more than the actual branches in N. The additional q branch elements in (5) are called *virtual branches*. The q coefficients associated with these branches and branches $(b\text{-}n+1,\ldots\ldots,b)$ are defined by the following mapping:

$$\begin{aligned}(V_{b+k}+\Delta V_{b+k})\hat{V}_{b+k} &= d_{k-1}[(V_{b-q+k}+\Delta V_{b-q+k})\hat{V}_{b-q+k}], \quad 1 \le k \le q,\\ (V_{b-q+k}+\Delta V_{b-q+k})\hat{V}_{b-q+k} &= \overline{d_{k-1}}[(V_{b-q+k}+\Delta V_{b-q+k})\hat{V}_{b-q+k}],\end{aligned} \quad (6)$$

where $d_0 \ldots . d_{q-1}$ are the bits of a binary number. There are 2^q combinations of branch switch settings for a network N with q modulated branch elements, which produce a system of 2^q equations similar to (5). A subnetwork of network N is branch diagnosable if it is possible to uniquely determine the admittance vector $\boldsymbol{Y}$ from the given measurements. This is possible if the rank of the system of 2^q equations is greater than or equal to the sum of the number of branches and virtual branches in the faulty subnetwork. The following theorems provide an explicit lower bound (Theorem 1) and an approximate upper bound (Theorem 2) on the rank of the system of 2^q linear equations; the proofs of these theorems have been omitted for brevity.

Theorem 1: For a network N with q modulated elements, at least $q+1$ of the 2^q equations are linearly independent if all branch and virtual branch voltages are nonzero.

Theorem 2: If a perturbed network $N+\Delta N$ which contains q modulated elements has the following properties:

i) number of node(s) faults less than or equal to r

ii) all branch and virtual branch error voltage $\Delta V_k \approx 0.0$

the first order approximate upper bound on the number of linearly independent equations is

$$rank\begin{bmatrix} a_{1,1} & a_{1,2} & a_{1,3} & \cdots & a_{1,b} & 0 & 0 & \cdots & 0 \\ \Phi_{1,1} & -a_{1,2} & \Phi_{3,1} & \cdots & \Phi_{b,1} & a_{1,2}+\Phi_{2,1} & 0 & \cdots & 0 \\ \Phi_{1,2} & \Phi_{2,2} & -a_{1,3} & \cdots & \Phi_{b,2} & 0 & a_{1,3}+\Phi_{3,2} & \cdots & 0 \\ \cdot & & & & & & & & \\ \cdot & & & & & & & & \\ 0 & -\Phi_{2,q} & -\Phi_{3,q-1} & \cdots & 0 & \Phi_{2,q} & \Phi_{3,q-1} & \cdots & 0 \\ 0 & -\Phi_{2,q} & -\Phi_{3,q-1} & \cdots & -\Phi_{b,1} & \Phi_{2,q} & \Phi_{3,q-1} & \cdots & \Phi_{b,1} \end{bmatrix} \qquad (7)$$

where,

$a_{1,k}$ correspond to $(V_k+\Delta V_k)\hat{V}_k$ terms in equation (5) for 2^q switch combinations,

$\Phi_{k,p}$ correspond to $(\partial a_{1,k}/\partial b_p)\Delta b_p$,

b_p is one of the q modulated branches, $1 \leq p \leq q$.

The number of unknowns in (5) for a given set of node faults can be determined using the following two-step procedure:

(1) determine the number l of nonzero columns in the incidence matrix **A** of a network N, for the set of faulty nodes [7],

(2) add 1 to l for each switch (virtual branches) incident at the faulty node(s).

The number of linearly independent equations generated when N is perturbed (*i.e.* $N+\Delta N$) are initially unknown. The number of branches and virtual branches, l, incident at a set of faulty nodes typically are larger than $q+1$. If the lower bound defined by Theorem 1 is used, the designer must add additional switches to N for branch fault diagnosability at faulty node sets with more than $q+1$ incident branches and virtual branches. The problem with this approach is that the additional switches may not be needed because the rank of the system of 2^q linear equations are typically larger than $q+1$. Currently there is no technique available to determine the number of linearly independent equations for a set of given faults; nevertheless it is possible to provide an upper bound on the number of linearly independent equations. The upper bound developed here is an approximate estimate for the perturbed network $N+\Delta N$ assuming $\Delta V_k \approx 0.0$.

The general system of branch-diagnosis equations can be developed for any subnetwork contained in network $N+\Delta N$, which meet node fault testability constraints. We will used 2^q equations in the system of equations. Generally, only a subset of these equations are required to implement the branch-diagnosis technique. Let equation (5) for a network N with q switches and b-branches yielding 2^q switch combinations, be defined as a system of equations

$$Qx = P, \qquad (8)$$

where,

$Q \in C^{u \times (b+q)}$, elements correspond to $(V_k+\Delta V_k)\hat{V}_k$ terms of (5), $1 \leq k \leq b+q$,

$x \in C^{(b+q)}$, with $x = [\Delta Y_1, \ldots, \Delta Y_{b+q}]^T$,

$P \in C^u$, and $P = [\Delta V_{pl[0]}\hat{I}_{pl[0]}, \ldots, \Delta V_{pl[u-1]}\hat{I}_{pl[u-1]}]^T$,

$u = 2^q$.

The set of nonzero rows of ΔY_n is identified during the node fault isolation phase of the technique [6]. Let S be a column-selector matrix such that $[(V_k+\Delta V_k)\hat{V}_k]S$ is a $2^q \times l$ matrix with $l<(b+q)$, and that $S^T[\Delta Y_k]$ retains all nonzero rows of ΔY_k. Therefore equation (8) becomes

$$QS \cdot S^T x = P. \qquad (9)$$

The admittance derivation of the faulty subnetwork can then be determined with linear techniques. Therefore, the nonzero rows of the column vector ΔY_k can be evaluated by

$$S^T x = (S^T \cdot Q^T \cdot QS)^{-1}(S^T \cdot Q^T)P. \qquad (10)$$

3 Example

Consider the 2-node fault testable network N with 9 switched branch elements as shown in Fig. 1. The accessible node 1,2,3,4 were selected based on the testability procedure proposed by Huang *et al* [6]. Let the normal element values of the unmodulated branches and on-admittance values of switched branch elements of the network in Fig. 1 be $y_i = 1.0\mho$, for $i = 1,2,\ldots,8,10,11,\ldots,14$, $y_9 = 2.0\mho$, and $y_{15} = 0.5\mho$. The non-faulty off-admittance of modulated branches are $y_i = 0.01\mho$, for $i = 2,3,5,8,10,\ldots,14$. The following calculations have been performed with routines developed at Argonne National Laboratories: LINPACK [9], and EISPACK [10]. The binary switch setting in the network is 0,0,0,0,0,0,0,0,0. Now let branches 3 and 7 be perturbed such that $y_{3(on)}+\Delta y_{3(on)} = 2.0\mho$, and $y_7+\Delta y_7 = 2.0\mho$ and all other branches remain the same. Let i_m be the terminal current vector which defines the current flowing through the accessible nodes 1,2,3,4 in Fig 1,

$$i_m = \begin{bmatrix} 1 \\ 0 \\ 0 \\ 0 \end{bmatrix}.$$

Nodes 5 and 6 were found to be faulty using the node fault diagnosis method which implement the first phase of this technique [6]. The number of nonzero columns in the incidence matrix **A** of Fig. 1 for nodes 5, and 6 is found to be 7. Three of the incident branches are switches, therefore the total number of branches and virtual branches in the subnetwork connected to these nodes is 10.

computed only once because they are a function of the adjoint network $\hat{N}$. The port voltage in (5) is also determined using Kirchhoff's voltage law and the node voltages found above. In this example only 64 switch combinations have been used to develop the system of branch-diagnosis equations. These voltages are used to compute the entries of the Q matrix and the column vector P in equation (8). The columns of the Q matrix which correspond to the branches and virtual branches incident at the faulty nodes (node 5 and 6) are extracted with the selector-matrix S. The $S^T x$ column vector is computed using the least square method. The final diagnosis is two branch faults:

$$\Delta y_{3(on)} = 1.0\mho \quad \Delta y_7 = 1.0\mho.$$

4 Conclusion

An analog fault diagnosis technique has been presented for the calculation of network element variations. The technique requires that a subset of the circuit nodes be accessible, a single excitation source at a single frequency, and q modulated branch elements. Given the node measurement voltages the branch element value can easily be computed by solving two sets of linear algebraic equations; node diagnosis equations, and branch-diagnosis

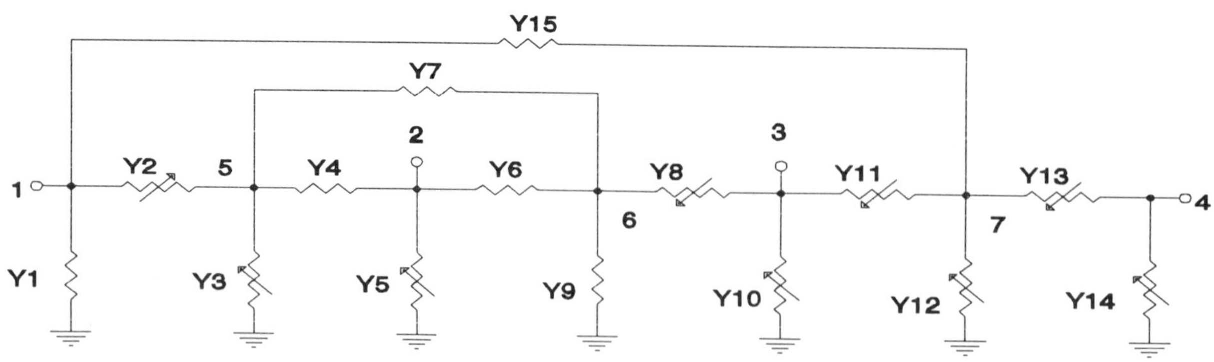

Figure 1 Resistive test network with modulated elements represented by variable resistors.

The branch and virtual branch voltages are computed using the node voltages of $N+\Delta N$ found during the node fault diagnosis phase. The branch voltages of the adjoint network $\hat{N}$ are computed from the node voltages of the adjoint network of N. These adjoint branch voltages are

equations. Furthermore, the technique proposed here allows one to determine the lower bound $q+1$ on the number of branches, and virtual branches that can be resolved for a given node fault set. Also an estimate of the highest possible rank of a system of branch-diagnosis equations is given. Further investigation is in progress to determine the accuracy of this technique in the presence of normal component variations and noisy node voltage measurements.

Acknowledgements

This work is partly supported by a Helen Carr Fellowship from the American Society for Engineering Education. Alvernon Walker expresses his gratitude and appreciation to Dr. Harold L. Martin, Dean-School of Engineering, North Carolina A&T State University, for his continued support and guidance, and Dr. Ethelbert N. Chukwu of North Carolina State University for his constructive suggestions in the rendering of theorem proofs.

References

[1] "Special Issue on Automatic Analog Fault Diagnosis," *IEEE Trans. Circuits Syst.*, Vol. CAS-26, July 1979.

[2] A.E. Salama, J.A. Starzyk, and J.W. Bandler,"A Unified Decomposition Approach for Fault Location in Large Analog Circuits," *IEEE Trans. Circuits Syst.*, Vol. CAS-31, pp. 609-622, July 1984.

[3] R.M. Biernacki, and J.W. Bandler," Fault Location of Analog Circuits," *IEEE 1980 Int. Symp. Circuits and Systems*, pp. 1078-1081.

[4] R.M. Biernacki, and J.W. Bandler," Multiple-fault Location of Analog Circuits," *IEEE Trans. Circuits Syst.*, Vol. CAS-28, pp. 361-366, May 1981.

[5] A.A. Sakla, E.I. El-Masry, and T.N. Trick," A Sensitivity Algorithm for Fault Detection in Analog Circuits," *IEEE 1980 Int. Symp. Circuits and Systems*, pp. 1075-1077.

[6] Z.F. Huang, C.S. Lin, and R.W. Liu," Node-Fault Diagnosis and a Design of Testability," *IEEE Trans. Circuits Syst.*, Vol. CAS-30, pp. 257-265, May 1983.

[7] L.O. Chua, and P.M. Lin, *Computer-Aided Analysis of Electronic Circuits: Algorithms and Computational Techniques*, Prentice-Hall, Englewood Cliffs, N.J., 1975.

[8] T.N. Trick, W. Mayeda, and A.A. Sakla," Calculation of Parameter Values from Node Voltage Measurements," *IEEE Trans. Circuits and Syst.*, vol. CAS-26, no. 7, July 1979.

[9] J.J. Dongarra, C.B. Moler, J.R. Bunch, and G.W. Stewart, *LINPACK User's Guide*, SIAM, Philadelphia, PA, 1979.

[10] B.T. Smith *et al.*, *Matrix Eigensystem Routine-EISPACK Guide*, Vol. 6, Spring-Verlag, Berlin, Heidelberg, New York, 1974.

Design of m-out-of-n Bit-Voters

Behrooz Parhami

Dept. of Electrical & Computer Engineering
University of California
Santa Barbara, CA 93106, USA

Abstract

Several synthesis methods for m-out-of-n bit-voting networks are described and the costs of the resultant designs are compared. In the special case of majority voting (with gate count used as the cost index), it is shown that a multiplexer-based design method yields the best realizations for small n, while designs based on selection networks are most efficient when n is large. Other design techniques are offered that might be more efficient with particular technologies or when certain standard building blocks are to be used in the hardware realization. Most of these designs can be extended to the general case of unequal vote weights (weighted voting).

0 Introduction

The idea of voting to obtain highly reliable results from multiple, less reliable, computation channels has been used extensively in the design of fault-tolerant computer systems. When voting is to be performed frequently and at a relatively low level, high-throughput hardware voters must be utilized. In this paper, we report some results from an ongoing research project dealing with general design paradigms for hardware bit- and word-voters as well as efficient algorithms for software voting on more complex data objects [2], [3], [4].

We consider the design of an m-out-of-n bit-voter (henceforth simply voter) defined as a circuit with n inputs x_i, $i = 1, 2, \ldots, n$, and a single output y, such that $y = 1$ iff at least m of the n inputs are 1. Note that our definition is asymmetric with respect to 0 and 1 in that an output of 0 does not imply and is not implied by the presence of at least m zeros at the inputs. Such asymmetry should not be worrisome and is in fact useful in the design of systems where one of the two output values is considered 'safe'. It is easy to modify the designs such that 0/1 symmetry is achieved and an extra output indicates when neither 0 nor 1 has the required quorum. Majority voting corresponds to the special case of $m = \lceil (n+1)/2 \rceil$ and has 0/1 symmetry for odd n.

1 Gate-Level Designs

A voter can be constructed as a two-level AND-OR (equivalently NAND-NAND) digital logic circuit with $g = n!/[m!\,(n-m)!]$ m-input AND gates and a single g-input OR gate for small values of the parameters m and n. Also, a two-level OR-AND realization, requiring $g' = n!/[(m-1)!\,(n-m+1)!]$ $(n-m+1)$-input OR gates and a single g'-input AND gate, is possible. In the first realization, all distinct subsets of m inputs are ANDed together and the voter output is 1 if at least one of the AND results is 1. In the second realization, all possible subsets of $n-m+1$ inputs are ORed together and the voter output is 0 if at least one of the OR results is 0.

The two-level AND-OR realization is 'simpler' than the two-level OR-AND version (in terms of both gate count and gate-input count) iff $m > (n+1)/2$. The complexities are equal for odd n if $m = (n+1)/2$. As an example, for a 2-out-of-5 voter, the two-level AND-OR design uses 10 two-input AND gates and a single 10-input OR gate while the OR-AND design is less complex with 5 four-input OR gates and one 5-input AND gate.

For large values of n, two-level designs are impractical. Assuming the use of f-input gates and ignoring the possibility of gate sharing, the total number of gates in the two-phase AND-OR and OR-AND realizations will change from $g+1$ and $g'+1$ to:

$$G = g\lceil (m-1)/(f-1)\rceil + \lceil (g-1)/(f-1)\rceil$$

$$G' = g'\lceil (n-m)/(f-1)\rceil + \lceil (g'-1)/(f-1)\rceil$$

These expressions are obtained by replacing each gate with a tree of f-input gates implementing the same function. The circuit delays will increase to $\lceil \log_f m\rceil + \lceil \log_f g\rceil$ and $\lceil \log_f (n-m+1)\rceil + \lceil \log_f g'\rceil$ gate levels.

With gate sharing, an exact general gate-count analysis becomes difficult. However bounds for the number of gates can be obtained that are close to actual values and show the excessive complexity of this approach for large values of n. It is thus imperative to explore more structured design techniques.

2 Decomposition-Based Designs

Hierarchical decomposition strategy (divide-&-conquer) can be used to facilitate the design. There are two ways to proceed with the decomposition approach:

1. Picking a partitioning scheme and then designing a suitable merging network.
2. Selecting a merging network and then designing the required partitioning algorithm.

With the first approach, we divide the inputs into disjoint subsets, enumerate the various combinations in which different subsets can contribute votes in such a way that the voting threshold is matched or exceeded, provide smaller voters to realize these contributions, and finally, design a logic network for combining the results. Because the subsets can be selected in many different ways, this approach does not lend itself to general analyses. We will thus limit our discussion to a simple example.

Consider the design of a 3-out-of-5 voter using the subsets $S_1 = \{x_1, x_2, x_3\}$ and $S_2 = \{x_4, x_5\}$. The combinations that match or exceed the threshold of 3 are:

3-of-3 in S_1 + (2-of-3 in S_1 and 1-of-2 in S_2) +
(1-of-3 in S_1 and 2-of-2 in S_2)

This yields the logical expression $x_1x_2x_3 + (x_1x_2 + x_2x_3 + x_3x_1)(x_4 + x_5) + (x_1 + x_2 + x_3)x_4x_5$ which directly translates into a 4-level logic circuit with 10 gates and 25 input lines.

We next explore the second decomposition strategy with multiplexers used as merging networks. Our interest in this approach arises from the availability of multiplexers as off-the-shelf universal components. The strategy is to select a subset of the inputs as control inputs to a multiplexer, determine the residual input functions, and then repeat the process for each function, if needed, until easily realizable functions are obtained.

For example, with a 2-input multiplexer in the first decomposition stage, the residual functions correspond to an m-out-of-(n–1) voter and an (m–1)-out-of-(n–1) voter. To design a 3-out-of-5 voter using 2-input multiplexers, we take x_1 as the first control variable. The residual functions corresponding to $x_1 = 0$ and $x_1 = 1$ are $x_2x_3x_4 + x_2x_3x_5 + x_2x_4x_5 + x_3x_4x_5$ and $x_2x_3 + x_2x_4 + x_2x_5 + x_3x_4 + x_3x_5 + x_4x_5$, yielding the result

$$[x'_1[x'_2(x_3x_4x_5)+x_2h] + x_1[x'_2h+x_2(x_3+x_4+x_5)]]$$

where $h = x_3x_4 + x_3x_5 + x_4x_5$ has the 2-input multiplexer realization $h = [x'_3(x_4x_5) + x_3(x_4+x_5)]$. The resulting circuit is shown in Figure 1a. Clearly, a 4-input multiplexer can replace the last two levels. With 8-input multiplexers, the expression becomes

$$[x'_1x'_2x_3h_1 + x'_1x_2x'_3h_1 + x'_1x_2x_3h_2 + x_1x'_2x'_3h_1 + x_1x'_2x_3h_2 + x_1x_2x'_3h_2 + x_1x_2x_3]$$

where $h_1 = (x_4x_5)$ and $h_2 = (x_4 + x_5)$. The resulting circuit is depicted in Figure 1b.

3 Arithmetic-Based Design

In the 'arithmetic' approach, the sign of $-m + \sum x_i$ is computed and the voting result is the complement of this sign. A parallel counter [5] to compute $\sum x_i$ followed by a circuit to compare the sum to m (or a single 'accumulative parallel counter') can easily compute the output. Designs for parallel counters are based on carry-save reduction techniques which are used extensively in reducing multiple-operand additions to two-operand additions. Cost and delay analyses for such reductions are available [5], [6]. One advantage of the arithmetic-based design is that it can be easily extended to the case of unequal vote weight as discussed in Section 6.

4 Design with Selection Networks

The design of an m-out-of-n voter is equivalent to selecting the mth largest value from among n input bits. Selection networks can be built from 2-sorter (comparator) cells. Knuth [1] defines three types of selection networks with n inputs:

1. Select the m largest values and move them to m outputs in no particular order.
2. Select the mth largest value and move it to a specified output line.
3. Select the m largest values and move them to m output lines in sorted order.

Denoting the number of 2-sorter or comparator cells by $U(m, n)$, $V(m, n)$, and $W(m, n)$ for type-1, type-2, and type-3 selectors above, we have:

$$U(m, n) \le V(m, n) \le W(m, n)$$

When dealing with bits, a two-sorter simply consists of a pair of 2-input gates: An OR to produce the larger and an AND to produce the smaller of the two values.

Type-3 selectors do more than what is required here. Type-2 selectors do exactly what we want. However, for most practical values of m and n, a type-1 selector augmented by an AND or OR circuit (that indicates whether all of the m largest values are 1s or whether all of the $n - m + 1$ smallest values are not all 0s) is both faster and more economical.

Consider the design of a 4-out-of-8 voter. The required type-1 selection network that selects the 4 largest bit values and moves them to the upper half of the output lines is given in Figure 2, where each heavy vertical line represents a comparator that moves the larger of the two input values to the upper and the smaller value to the lower line. This selector requires 14 comparators or 28 two-input gates with 4 gate levels of delay. A 4-input

AND gate connected to the upper 4 outputs completes the circuit. Note that a 5-out-of-8 majority voter results if we connect an OR gate to the lower 4 output lines.

5 Comparison of Various Designs

We will compare the designs only for simple majority voters (i.e., when $m = \lfloor n/2 \rfloor + 1$). Figure 3 shows the cost of majority voters designed based on 2-level logic expressions ('gate-level'), two-input multiplexer decomposition, the arithmetic-based approach, and selection networks, assuming maximum gate fan-in of 4. Figure 3 indicates that the gate-level or multiplexer-based approach is best for small values of n whereas selection networks offer the most economical solution for larger values of n. The theory of selection networks is well-developed and efficient designs are available [1].

A comparison of delays is much more difficult. If the designs are used with pipelining [2], the differences in latencies (number of gate levels) are not significant as far as throughput is concerned. However, the number of gate levels does affect the cost due to the requirement for latches between pipeline stages. A general analysis is impractical because the number of logic signals going from one pipeline stage to the next cannot be expressed as a simple function of the relevant parameters.

6 Weighted Bit-Voters

If each of the n bit-inputs x_i, $i = 1, 2, \ldots, n$, has an associated vote v_i such that the output 1 is produced iff the sum of votes associated with the inputs having the value 1 is no less than a threshold t, then we have a generalized t-out-of-Σv_i voter.

The techniques discussed in the preceding sections can be extended for such voters. The only exception concerns selection networks that can be used here only if the vote weights are small integers. The way to do this is to fan-out each input to v_i inputs of a (Σv_i)-input voter. However such designs tend to be highly inefficient.

The arithmetic-based approach is ideally suited to this case. As a simple example, consider the design of 6-input voter where the inputs have the associated votes of 2, 2, 2, 1, 1, 1 and the voting threshold is 5. The arithmetic expression to be evaluated is

$$-5 + 2x_1 + 2x_2 + 2x_3 + x_4 + x_5 + x_6$$

The reduction steps in our multiple-operand addition using full-adder and half-adder cells are shown in the following diagram. Here, 1011 is the 4-bit 2's-complement representation of the constant –5. Three of the inputs are written in the second column because they carry a weight of 2. If the weight of x_1 were 3, say, then x_1 would be written in both the first and the second column. Each vertical box encloses the inputs to a full or half adder and the associated horizontal box encloses the two outputs. The diagram shows level 1 with 2 full adders and level 2 with 1 full adder and 1 half adder. A 2-bit adder can be used to reduce the remaining y and z bits in the middle two columns, with the carry-out of this adder providing the final output. The leftmost 1 can be ignored since it causes only a complementation that cancels the complementation needed for obtaining the resultant output from the sign bit.

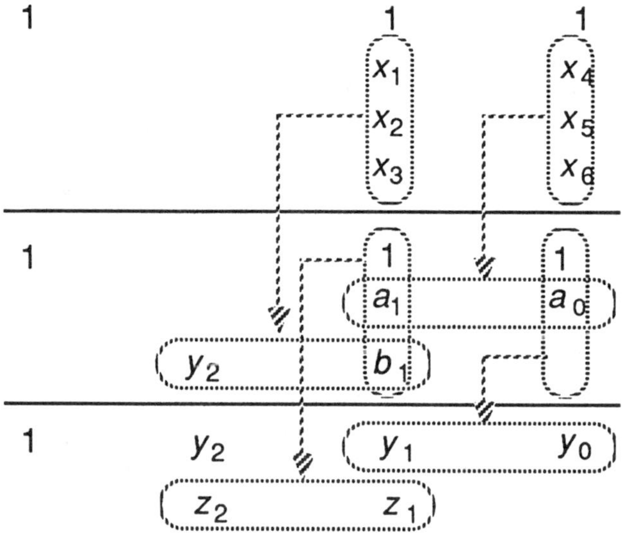

A particularly useful arithmetic-based design scheme is depicted in Figure 4. Here a pair of binary trees, composed of adders of appropriate lengths, tally the votes for 0 and 1 inputs. The 'compare and select' circuit at the right produces the final output. The number of adders required for an n-input voter is $2n - 2$ and the longest adder deals with operands of length $\log_2(\Sigma v_i)$.

The multiplexer-based approach works fine with weighted voting if proper care is taken to select the control variables in an optimal order.

7 Conclusions

A theoretical study of weighted voting [4] has shown that voting is at least as complex as sorting when the size of the input space is large. On the other hand with a small input space of size δ ($\delta = 2$ for bit-voting), linear-size, logarithmic-time voters are practically realizable. In this paper, we have explored and compared some useful design techniques for m-out-of-n and the more general t-out-of-Σv_i weighted bit-voters.

Despite the fact that the designs are quite practical, and in some cases asymptotically optimal, no claim is made as to their absolute efficiency or optimality. There may be other methods that yield better designs for a given set of requirements. The optimality question and several extensions to the design techniques discussed here are being investigated as part of an ongoing research project.

References

[1] Knuth, D.E., *The Art of Computer Programming — Vol. 3: Sorting and Searching*, Addison-Wesley, 1973, Section 5.3.4, pp. 220-246.

[2] Parhami, B., "High-Performance Parallel Pipelined Voting Networks", *Proc. of the International Parallel Processing Symp.*, Anaheim, CA, Apr./May 1991, pp. 491-494.

[3] Parhami, B., "Voting Networks", *IEEE Transactions on Reliability*, Vol. 40, pp. 380-394, Aug. 1991.

[4] Parhami, B., "The Parallel Complexity of Weighted Voting", *Proc. of the International Conf. on Parallel and Distributed Computing and Systems*, Washington, DC, Oct. 1991, pp. 382-385.

[5] Swartzlander, E.E., "Parallel Counters," *IEEE Transactions on Computers*, Vol. C-22, No. 11, Nov. 1973, pp. 1021-1024.

[6] Waser, S. and M.J. Flynn, *Introduction to Arithmetic for Digital System Designers*, Holt, Rinehart, & Winston, 1982.

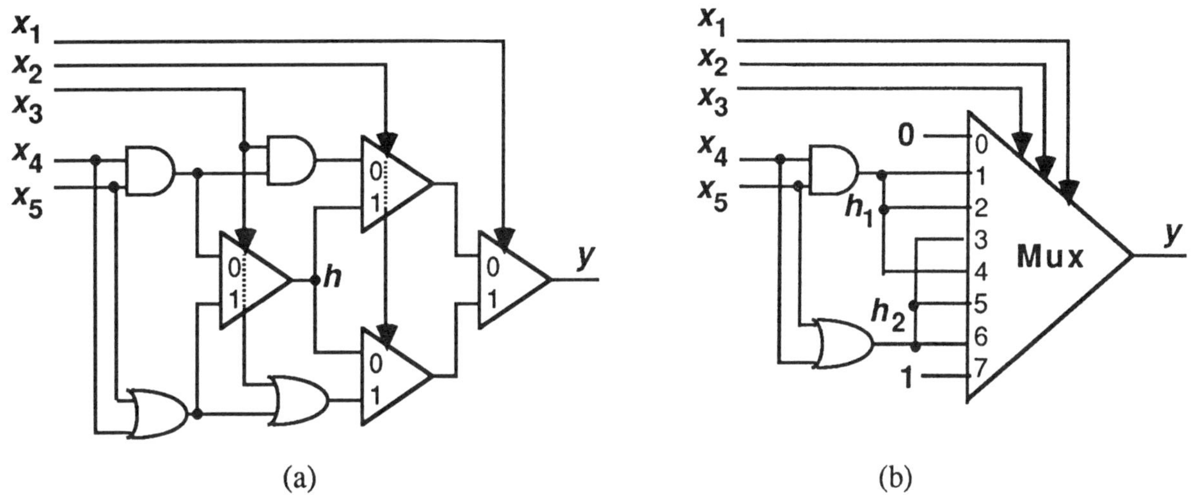

Figure 1. Multiplexer-based realizations for a 3-out-of-5 bit-voter.

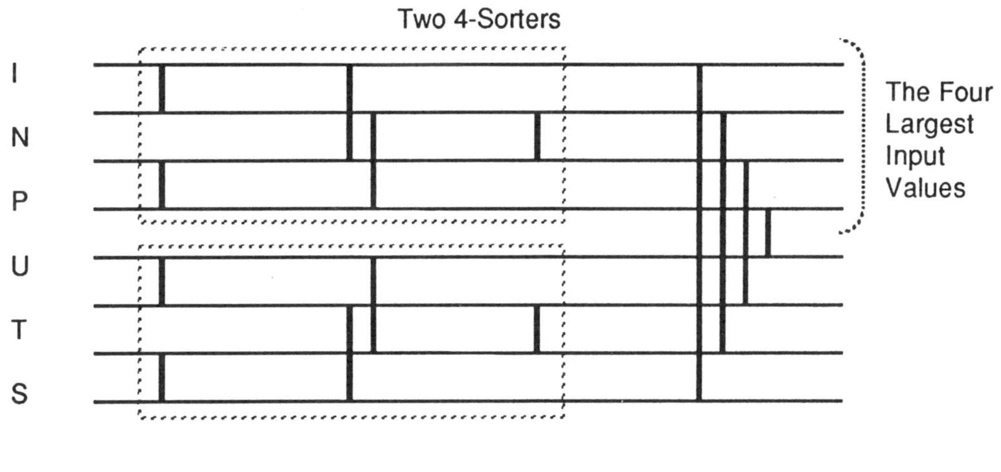

Figure 2. Network of comparators to select the 4 largest of 8 values.

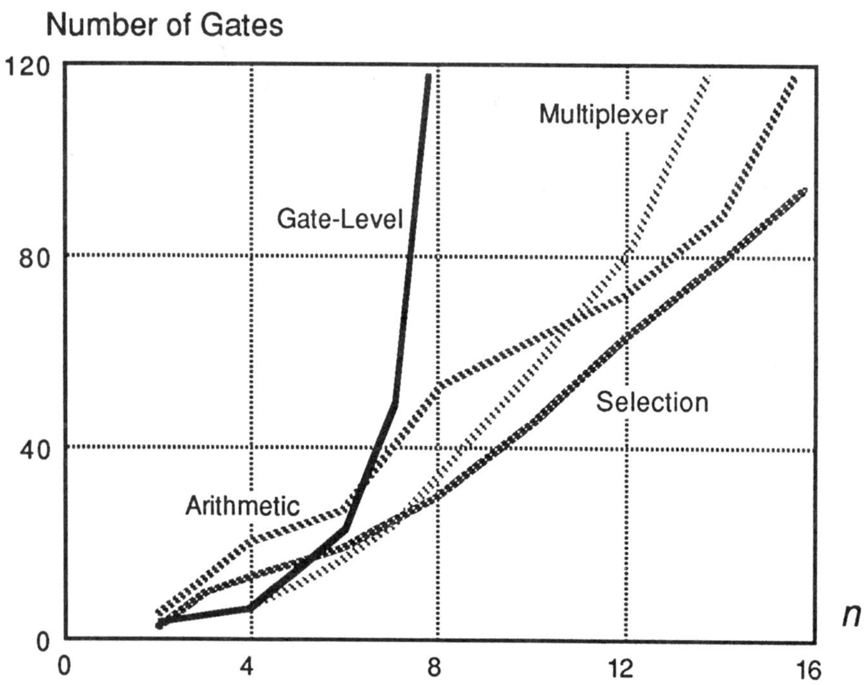

Figure 3. Cost of simple majority bit-voters with different designs as a function of input size *n*.

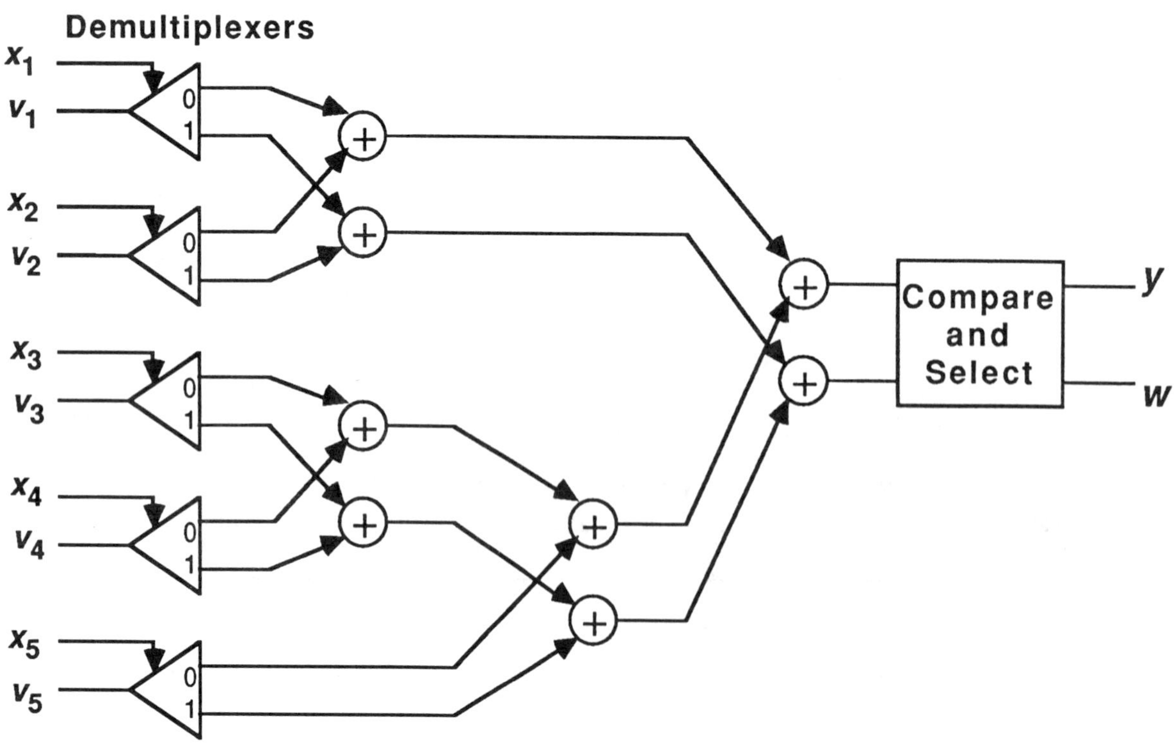

Figure 4. A 5-input weighted bit-voter constructed according to the 'vote-tallying' method.

Local Supervisor Model for Distributed Reconfiguration

L. S. DeBrunner
School of Electrical Engineering
and Computer Science
University of Oklahoma
Norman, Oklahoma 73019

F. G. Gray
Bradley Department of Electrical
Engineering
Virginia Polytechnic Institute and State University
Blacksburg, Virginia 24061

Abstract

The Local Supervisor Model is presented as a basis for comparing and designing distributed reconfiguration algorithms . Cells designated as local supervisors work together in parallel to change the global structure through a series of separate actions requiring only short communication paths between close neighbors. The use of the model in design and comparison is illustrated through an array application.

1 Introduction

To reach the high levels of reliability required by applications involving long missions and life critical situations, reconfiguration is necessary. In addition, hard core system components must be eliminated since they limit the reliability level a system can achieve.

Some method for system evaluation at early stages of development is needed to allow design decisions to be made early and with low cost [1]. Models also support the comparison of competing systems. The use of a model allows a precise, theoretical study of reconfiguration algorithms.

This paper proposes a model which is useful for comparing and designing distributed reconfiguration algorithms for highly reliable computing systems. The Local Supervisor Model (LSM) is a distributed model in which individual processors act in parallel as local supervisors to change the structure of the entire system. Only short communication paths are needed for communication between close neighbors; no global knowledge is required.

The Local Supervisor Model is related to the Distributed Recovery Strategy[7]. The Distributed Recovery Strategy (DRS) was used to develop reconfiguration algorithms for tree systems and loop systems. DRS does not provide an explicit method for developing a reconfiguration algorithm and does not allow multiple or near-coincident faults. LSM provides a method for designing reconfigurable systems which permits double and near-coincident faults.

This paper discusses the Local Supervisor Model and its uses. Section 2 gives background about DRS. Then, LSM is proposed in Section 3. Section 4 describes the use of LSM to design a reconfiguration algorithm for an array. Section 5 discusses how LSM was used to compare systems. Finally, Section 6 summarizes conclusions about LSM and its uses.

2 Distributed Recovery Strategy

In the Distributed Recovery Strategy, processors respond to errors in their neighborhoods. An error condition exists when a state in the basic graph (the minimum configuration required to perform the desired computation) is missing from the system's configuration. A faulty node is represented by the F state. A faulty link is represented by assigning the F state to the source node. In each case, the previous state of the node becomes "missing" from the system. Each state is assigned error conditions to which to respond. A state's error response set may be the empty set, which means that the state does not respond to any error condition. Based on the error response set assignments, the specified node reacts to a particular error condition by assuming local supervisor responsibilities and attempting to find a spare to assume the associated missing state. If no spare is found, the responding node assumes the missing state, and the iterative recovery process continues. If no spare can be found and the responding node cannot assume the missing state, then the recovery attempt fails. For the tasks of a faulty node to be reassigned, each node must periodically store backup information for the tasks associated with its error response assignments.

DRS requires several assumptions. A node must be able to determine whether each of its neighbors is faulty. Also, the error response sets associated with each pair of nodes must be disjoint. Finally, no more than one error may be present in the system at a given time [7]. These assumptions guarantee the reconfiguration process to be well-behaved.

DRS has been applied to tree systems and loop systems. Both of these architectures are based on graphs which have been well-studied and which have many useful graph theoretical properties. Although DRS provides a backbone of a reconfiguration algorithm, the actual design process is not specified. For example, DRS does not give any procedure for choosing error response sets, a critical issue in the design of an algorithm. Another shortcoming of DRS is the require-

1058-6393/91 $1.00 © 1991 IEEE

ment that only one error be present in the system at a given time. Thus, DRS does not allow multiple faults or near-coincident faults.

The Local Supervisor Model described below addresses these issues and provides a method for designing and comparing distributed reconfigurable systems.

3 Local Supervisor Model

This section first describes the reconfiguration strategy for the Local Supervisor Model in Section 3.1. The method for choosing error response sets is then outlined in Section 3.2. Finally, in Section 3.3, the use of spares as local supervisors is discussed.

3.1 Reconfiguration strategy for LSM

The reconfiguration strategy for LSM encompoasses that of the DRS. However, the LSM reconfiguration strategy has a method for handling multiple faults. Suppose a local supervisor responds to an error and attempts to find a spare with the required adjacencies, but none is available. The local supervisor then checks to see if it has the required adjacencies to take over itself. If it has the required adjacencies, it takes the missing state as in the DRS. If it does not have the required adjacencies, it looks for a spare in which all missing adjacencies are not in the error state's error response set. For each missing adjacency, if the spare has a neighbor whose previous state was the missing adjacency, then the local supervisor commands that spare to assume the error state.

The behavior of LSM is orderly. No node can respond to a perceived error if the missing state is present anywhere in the system. It has been proven that for this new approach, no two cells may have the same computation state[2]. Thus, the algorithm is well-behaved. The notation E(s) will be used to refer to the error condition in which state s is missing from the system.

3.2 Method for choosing error response sets

Another problem solved by LSM is the method for choosing error response sets. This section first gives representations for specifying error response sets. Then, selection constraints for error response sets are stated. Finally, heuristics for choosing error response sets are discussed.

Error response sets can be specified in several formats. A set representation in which each state has a set of errors assigned to it has been proposed [7]. This representation is not convenient for the development of algorithms. A more convenient format for larger systems is a collection of rules which describe the error response assignments[3]. The rule representation is typically useful in proofs. However, neither of these representations is very useful in choosing error response sets. To choose error response sets, an easily understood description of the error response set is needed.

We propose a directed graph (digraph) to specify the error response sets. This visual representation allows quick comprehension of the error response assignments. The tail of an arc is incident on the responding state, and the head of the arc is incident on the error state. The digraph will be superimposed on the basic graph, G_b. The set of all nodes adjacent to a given node for which the tail of the connecting arc is incident on the given node are the error conditions to which the given node responds.

Three absolute constraints must be met in assigning the error response sets.

1. The responding state must be adjacent to the error state in the basic graph.
2. Only one state responds to a particular error state.
3. Each error state is responded to by some other state.

These constraints relate to the digraph representation. In summary, there must exist an edge in G_b with endpoints at the responding state and the error state. Also, each node in the digraph must have exactly one head of an arc incident on it.

LSM provides several heuristics to be used in conjunction with these constraints to choose a good error response digraph.

1. It is desirable for the responding node to be adjacent in the redundant graph to all of the error node's neighbors in the basic graph for the initial case.
2. Choose assignments so that growth tends to go in the same direction.
3. Avoid short cycles.

The heuristics are needed for the design of a reconfiguration algorithm for an architecture in general.

3.3 Using spares as local supervisors

In various systems, using spares to respond to error conditions can produce a redundant system that has a more consistent reconfiguration response. However, when spares are used to respond to errors, the questions arises "What happens if the spare which responds to the error becomes faulty?" This problem is identified by considering how reconfiguration is carried out in general.

When spares can be used to respond to errors, a good method for making error response assignments is essential. Spare behavior must be defined carefully in situations in which spare cells display behavior dependent on their responsibilities.

LSM supports the use of spares as local supervisors as follows. First, for each spare that has error response duties, a different state should be added to the error response digraph. For the spare which responds to a normal computation error condition, an arc from that spare state to the state it responds to will be added to the error response digraph.

When a chain of spares is concerned and a spare responds to a normal computation error condition, the spare that responds after the spare responding to the normal computation error then takes on that spare's state. Arcs are also added to form a chain of these special states in the error response digraph to show the use of spares ad local supervisors. This part of the digraph indicates that the spares maintain the same function relative to the basic graph.

The constraints for error response assignments require that the in-degree for each vertex in the digraph be exactly 1. We propose that these constraints be relaxed for special spare states. We will allow vertices in the digraph corresponding to spare states to have an in-degree of 1 or 0. In summary, LSM allows spares to be used to respond to error conditions and gives heuristics for assigning spares to ensure proper behavior.

The Local Supervisor Model is more versatile and more practical than DRS because it allows multiple error conditions and it provides heuristics for choosing error response assignments. An algorithm designed using this strategy can be implemented as described above; however, the implementation could be approached differently. The LSM can be used to design the behavior of an algorithm. Then, the implementation may be streamlined to produce a faster, more efficient implementation than the direct implementation of LSM would yield.

4 Using LSM to design an algorithm

This section shows how LSM can be used to design an algorithm. The design process and simulation results are considered.

4.1 LSM design process

The problem addressed is the design of a reconfigurable array architecture. The basic graph is a 3×3 processor array with Von Neumann connections (North, East, South, West). The redundant graph has connections to the north, northeast, east, far east, southeast, south, southwest, far west, west and northwest neighbors. These redundant connections are proposed in [6]. The basic graph embedded in the redundant graph is shown in Figure 1. The error condition corresponding to the missing state i,j is indicated using the notation $E(S_{i,j})$. The redundant graph is a 7×7 array. This size is large enough to allow the investigation of all situations we want to consider. Both the basic graph and the redundant graph can easily be extended to larger sizes.

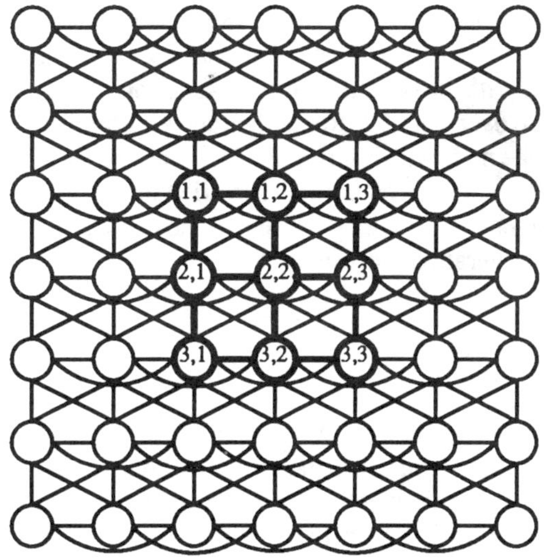

Figure 1. The Basic Graph Embedded in the Redundant Graph

The error response assignment selected is shown using digraph representation in Figure 2. This choice follows the first two heuristics stated previously. However, it doesn't meet the third guideline since it contains a short cycle which can result in an infinite looping error. If we do not allow the use of spares to act as local supervisors, a cycle must appear in the error response digraph. A larger cycle could be introduced instead by assigning state 3,1 to respond to error $E(S_{3,3})$, but that would violate the first guideline. Thus, the error response assignment chosen is fairly good. In a Section 5, we will consider the use of spares as local supervisors. The reconfiguration algorithm resulting from using this error response assignment with LSM for the specified redundant graph will be referred to as the Array Reconfiguration Algorithm (ARA).

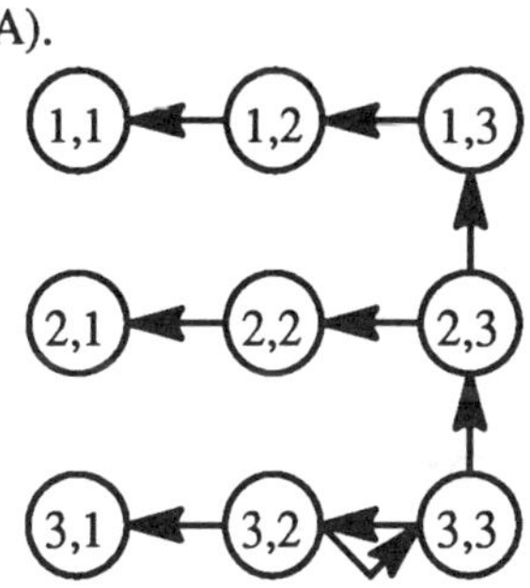

Figure 2. Error Response Assignment for the Array Reconfiguration Algorithm

The example in Figure 3 shows two local supervisors working cooperatively, one after another, in the Array Reconfiguration Algorithm. This example shows the reconfiguration for the error $E(S_{2,2})$. When the cell in state $S_{2,2}$ becomes faulty, the responding node in state 2,3 reacts by looking for an appropriate spare. It does not find a spare with the necessary connections to take state 2,2. But, the responding node has all necessary connections so it assumes state 2,2 creating condition $E(S_{2,3})$. The node in state 3,3 responds. It finds a spare with the needed adjacencies and commands that spare to take over state 2,3. Reconfiguration is complete. In each case, the node which acts as the local supervisor by responding to the error condition is specified in the error response assignment. This example shows the iterative nature of the reconfiguration process.

4.2 Simulation results

We used the Architecture Design and Assessment System (ADAS) [4] and custom C programs to verify ARA. The coverage and time requirements were compared to those of the White-Gray Local algorithm [6] which is based on the same redundant array structure.

ARA covers all single faults. The coverage of single faults was expected since the error response set was chosen so that the recovery strategy would not fail for any single fault. The White–Gray Local algorithm also covers all single faults. In each case, ARA requires the same or fewer steps for reconfiguration. However, ARA steps are more complex and thus much longer than those of the White–Gray Local algorithm.

ARA covers 72% of double faults compared to 75% for the White–Gray Local algorithm. The average number of steps required for reconfiguration with ARA is 1.7, while on average 2.4 steps are required for the White–Gray Local algorithm.

A double sequential fault is defined as "two single faults separated in time so that reconfiguration in response to the first fault has completed before the second fault occurs" [5]. ARA covers 83% of double sequential faults. The White–Gray Local algorithm covers 75% of double faults. ARA requires 2.7 steps on average, while the White–Gray Local algorithm requires 4.0 steps.

5 Using LSM to compare algorithms

LSM was used[2] to describe the behavior of the White-Gray Local Reconfiguration Algorithm. When the error response assignments are chosen as shown in Figure 4, the behavior of the modified ARA using these assignments is identical to the White-Gray Local Reconfiguration Algorithm. The value of LSM in comparing algorithms is that the the behavior of an algorithm can be specified in the LSM format. The difference in error response assignments indicates that the chief difference between ARA and the White-Gray Local Reconfiguration Algorithm is the use of spares to respond to errors. Also, the basic graph is embedded in the far northeast corner of the redundant graph for the White-Gray Local algorithm. ARA and the White-Gray local algorithm are show strong similarities because each exploits the structure of the redundant graph.

6 Conclusions

This paper proposes the Local Supervisor Model (LSM) as a tool for designing and comparing distributed reconfiguration algorithms. Local supervisors change the global structure of the system through a series of cooperative actions. Only short communication paths between close neighbors are required. The actions are iterative.

We used LSM to design the Array Reconfiguration Algorithm and to compare two reconfiguration algorithms for a given architecture.

ARA has been proven to reconfigure correctly for single faults in an $N \times M$ array embedded in an unbounded redundant graph[2]. An outline of this proof is given in [3].

In summary, we demonstrated that the Local Supervisor Model is a valuable tool for designing and comparing reconfigurable systems. This model can be used for any general reconfiguration algorithm.

References

[1] A. Avizienis, J. A. Abraham, W. C. Carter, J. Goldberg, D. A. Rennels, and R. Turn, *Fault Tolerance Analysis Task, Volume 2, Fault Tolerance and Security Technology Assessment,* March 1987, SDIO BM/C^3 Processor and Algorithm Working Group, RADC System Survivability Evaluation Contract, General Research Corporation, F04701-85-C-0136, TA0006.

[2] Linda S. DeBrunner, "Modeling Reconfiguration Algorithms for Regular Architectures," *Ph.D. Dissertation,* Virginia Polytechnic Institute and State University, 1991.

[3] Linda S. DeBrunner and F. Gail Gray, "A Distributed Reconfiguration Algorithm for an Array of Processors," *Proceedings of SOUTHEASTCON,* April 1990.

[4] G. A. Frank, D. L. Franke, and W. F. Ingogly, "An Architecture Design and Assessment System," *VLSI Design,* August 1985, pp. 30-50.

[5] F. Gail Gray, Linda S. DeBrunner, and Tennis S. White, "Topics Related to Methodology for Reliable Computer Design," *Final Report,* June 15, 1988.

[6] F. Gail Gray and Tennis S. White, "Summary of a Distributed Control Algorithm for a Dynamically Reconfigurable Array Architecture," *Proceedings of the International Conference on Wafer Scale Integration*, San Francisco, California, January 3–5, 1989, pp. 131–140.

[7] Raif M. Yanney and John P. Hayes, "Distributed Recovery in Fault-Tolerant Multiprocessor Networks," *IEEE Transactions on Computers,* Vol. C-35, No. 10, October 1986, pp. 871-879.

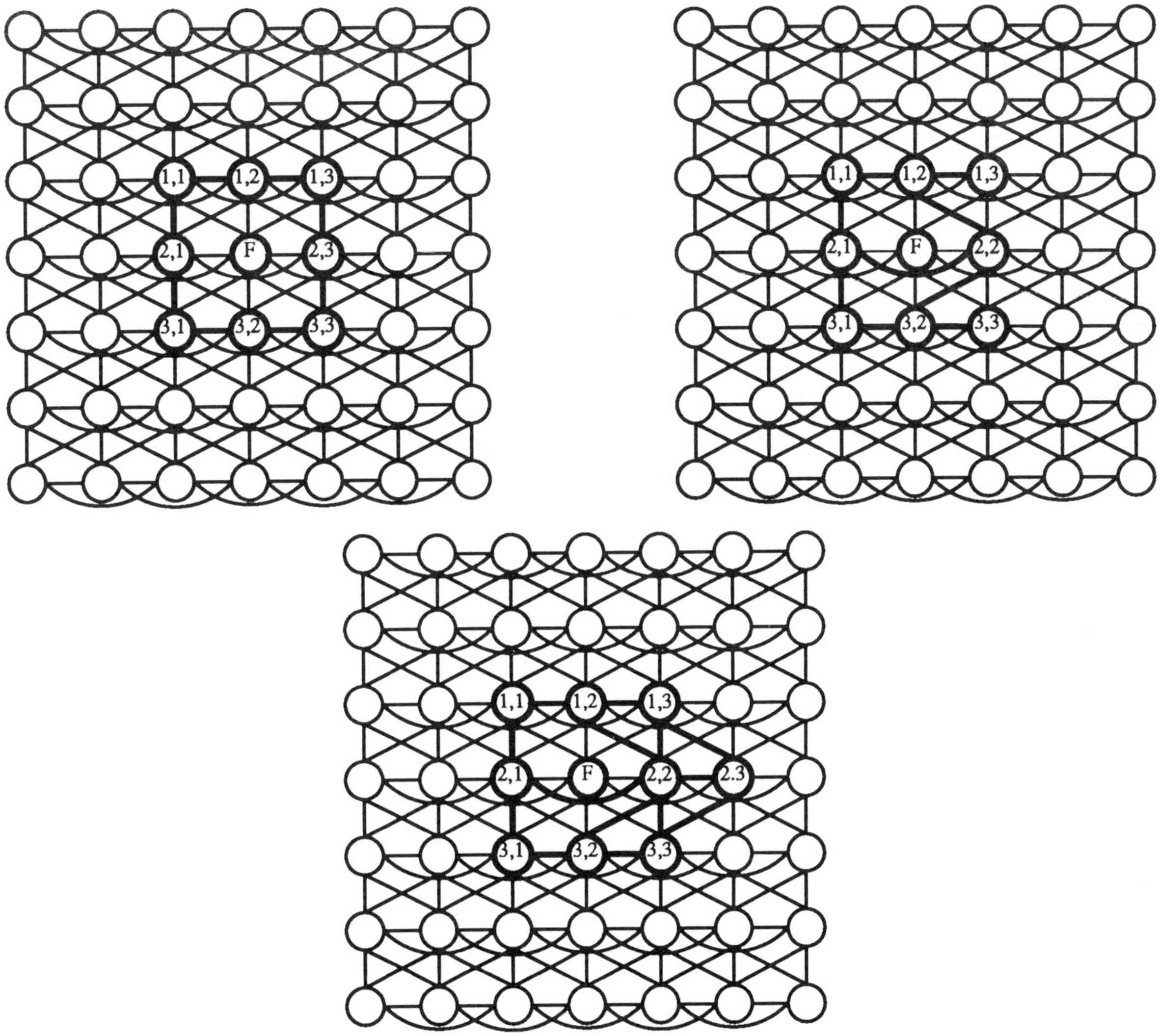

Figure 3. Reconfiguration for Error $E(S_{2,2})$

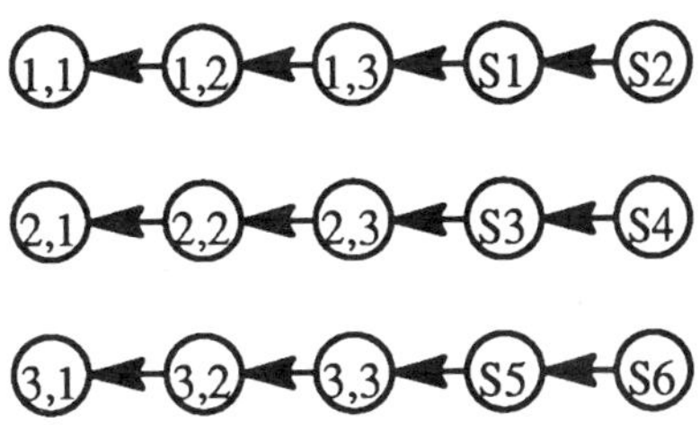

Figure 4. Error Response Digraph for LSM Model of White-Gray Local Algorithm

Author Index

Offices of the IEEE Computer Society

Headquarters Office
1730 Massachusetts Avenue, N.W.
Washington, DC 20036-1903
Phone: (202) 371-0101 — Fax: (202) 728-9614

Publications Office
P.O. Box 3014
10662 Los Vaqueros Circle
Los Alamitos, CA 90720-1264
Membership and General Information: (714) 821-8380
Publication Orders: (800) 272-6657 — Fax: (714) 821-4010

European Office
13, avenue de l'Aquilon
B-1200 Brussels, BELGIUM
Phone: 32-2-770-21-98 — Fax: 32-3-770-85-05

Asian Office
Ooshima Building
2-19-1 Minami-Aoyama, Minato-ku
Tokyo 107, JAPAN
Phone: 81-3-408-3118 — Fax: 81-3-408-3553

IEEE Computer Society

IEEE Computer Society Press Publications

Monographs: A monograph is an authored book consisting of 100-percent original material.
Tutorials: A tutorial is a collection of original materials prepared by the editors, and reprints of the best articles published in a subject area. Tutorials must contain at least five percent of original material (although we recommend 15 to 20 percent of original material).
Reprint collections: A reprint collection contains reprints (divided into sections) with a preface, table of contents, and section introductions discussing the reprints and why they were selected. Collections contain less than five percent of original material.
Technology series: Each technology series is a brief reprint collection — approximately 126-136 pages and containing 12 to 13 papers, each paper focusing on a subset of a specific discipline, such as networks, architecture, software, or robotics.
Submission of proposals: For guidelines on preparing CS Press books, write the Editorial Director, IEEE Computer Society Press, PO Box 3014, 10662 Los Vaqueros Circle, Los Alamitos, CA 90720-1264, or telephone (714) 821-8380.

Purpose

The IEEE Computer Society advances the theory and practice of computer science and engineering, promotes the exchange of technical information among 100,000 members worldwide, and provides a wide range of services to members and nonmembers.

Membership

All members receive the acclaimed monthly magazine *Computer*, discounts, and opportunities to serve (all activities are led by volunteer members). Membership is open to all IEEE members, affiliate society members, and others seriously interested in the computer field.

Publications and Activities

Computer **magazine:** An authoritative, easy-to-read magazine containing tutorials and in-depth articles on topics across the computer field, plus news, conference reports, book reviews, calendars, calls for papers, interviews, and new products.

Periodicals: The society publishes six magazines and five research transactions. For more details, refer to our membership application or request information as noted above.

Conference proceedings, tutorial texts, and standards documents: The IEEE Computer Society Press publishes more than 100 titles every year.

Standards working groups: Over 100 of these groups produce IEEE standards used throughout the industrial world.

Technical committees: Over 30 TCs publish newsletters, provide interaction with peers in specialty areas, and directly influence standards, conferences, and education.

Conferences/Education: The society holds about 100 conferences each year and sponsors many educational activities, including computing science accreditation.

Chapters: Regular and student chapters worldwide provide the opportunity to interact with colleagues, hear technical experts, and serve the local professional community.

Other IEEE Computer Society Press Titles

MONOGRAPHS

Analyzing Computer Architectures
Written by J.C. Huck and M.J. Flynn
(ISBN 0-8186-8857-2); 206 pages

Branch Strategy Taxonomy and Performance Models
Written by Harvey G. Cragon
(ISBN 0-8186-9111-5); 150 pages

Desktop Publishing for the Writer: Designing, Writing, and Developing
Written by Richard Ziegfeld and John Tarp
(ISBN 0-8186-8840-8); 380 pages

Digital Image Warping
Written by George Wolberg
(ISBN 0-8186-8944-7); 340 pages

Integrating Design and Test—CAE Tools for ATE Programming
Written by K.P. Parker
(ISBN 0-8186-8788-6); 160 pages

JSP and JSD—The Jackson Approach to Software Development (Second Edition)
Written by J.R. Cameron
(ISBN 0-8186-8858-0); 560 pages

National Computer Policies
Written by Ben G. Matley and Thomas A. McDannold
(ISBN 0-8186-8784-3); 192 pages

Physical Level Interfaces and Protocols
Written by Uyless Black
(ISBN 0-8186-8824-2); 240 pages

Protecting Your Proprietary Rights in Computer and High-Technology Industries
Written by Tobey B. Marzouk, Esq.
(ISBN 0-8186-8754-1); 224 pages

X.25 and Related Protocols
Written by Uyless Black
(ISBN 0-8186-8976-5); 304 pages

TUTORIALS

Advanced Computer Architecture
Edited by D.P. Agrawal
(ISBN 0-8186-0667-3); 400 pages

Advances in Distributed System Reliability
Edited by Suresh Rai and Dharma P. Agrawal
(ISBN 0-8186-8907-2); 352 pages

Autonomous Mobile Robots:
Perception, Mapping, and Navigation—Volume 1
Edited by S.S. Iyengar and A. Elfes
(ISBN 0-8186-9018-6); 552 pages

Autonomous Mobile Robots:
Control, Planning, and Architecture—Volume 2
Edited by S.S. Iyengar and A. Elfes
(ISBN 0-8186-9116-6); 536 pages

Broadband Switching: Architectures, Protocols, Design, and Analysis
Edited by C. Dhas, V.K. Konangi, and M. Sreetharan
(ISBN 0-8186-8926-9); 528 pages

Computer and Network Security
Edited by M.D. Abrams and H.J. Podell
(ISBN 0-8186-0756-4); 448 pages

Computer Architecture
Edited by D.D. Gajski, V.M. Milutinovic, H. Siegel, and B.P. Furht
(ISBN 0-8186-0704-1); 602 pages

Computer Arithmetic I
Edited by Earl E. Swartzlander, Jr.
(ISBN 0-8186-8931-5); 398 pages

Computer Arithmetic II
Edited by Earl E. Swartzlander, Jr.
(ISBN 0-8186-8945-5); 412 pages

Computer Communications: Architectures, Protocols and Standards (Second Edition)
Edited by William Stallings
(ISBN 0-8186-0790-4); 448 pages

Computer Graphics Hardware: Image Generation and Display
Edited by H.K. Reghbati and A.Y.C. Lee
(ISBN 0-8186-0753-X); 384 pages

Computer Graphics: Image Synthesis
Edited by Kenneth Joy, Nelson Max, Charles Grant, and Lansing Hatfield
(ISBN 0-8186-8854-8); 380 pages

Computer Vision: Principles
Edited by Rangachar Kasturi and Ramesh Jain
(ISBN 0-8186-9102-6); 700 pages

Computer Vision: Advances and Applications
Edited by Rangachar Kasturi and Ramesh Jain
(ISBN 0-8186-9103-4); 720 pages

Digital Image Processing (Second Edition)
Edited by Rama Chellappa
(ISBN 0-8186-2362-4); 400 pages

Digital Private Branch Exchanges (PBXs)
by Edwin Coover
(ISBN 0-8186-0829-3); 394 pages

Distributed Computing Network Reliability
Edited by Suresh Rai and Dharma Agrawal
(ISBN 0-8186-8908-0); 357 pages

Distributed Software Engineering
Edited by Sol Shatz and Jia-Ping Wang
(ISBN 0-8186-8856-4); 294 pages

Domain Analysis and Software Systems Modeling
Edited by Ruben Prieto-Diaz and Guillermo Arango
(ISBN 0-8186-8996-X); 312 pages

DSP-Based Testing of Analog and Mixed Signal Circuits
Edited by Matthew Mahoney
(ISBN 0-8186-0785-8); 272 pages

Fault Tolerant Computing
Edited by V. Nelson and B. Carroll
(ISBN 0-8186-8677-4); 428 pages

Formal Verification of Hardware Design
Edited by Michael Yoeli
(ISBN 0-8186-9017-8); 340 pages

Hard Real-Time Systems
Edited by J.A. Stankovic and K. Ramamritham
(ISBN 0-8186-0819-6); 624 pages

Integrated Services Digital Networks (ISDN) (Second Edition)
Edited by William Stallings
(ISBN 0-8186-0823-4); 406 pages

Local Network Technology (Third Edition)
Edited by William Stallings
(ISBN 0-8186-0825-0); 512 pages

Microprogramming and Firmware Engineering
Edited by V. Milutinovic
(ISBN 0-8186-0839-0); 416 pages

Modeling and Control of Automated Manufacturing Systems
Edited by A. A. Desrochers
(ISBN 0-8186-8916-1); 384 pages

Nearest Neighbor Pattern Classification Techniques
Edited by Belur V. Dasarathy
(ISBN 0-8186-8930-7); 464 pages

New Paradigms for Software Development
Edited by William Agresti
(ISBN 0-8186-0707-6); 304 pages

Object-Oriented Computing, Volume 1: Concepts
Edited by Gerald E. Petersen
(ISBN 0-8186-0821-8); 214 pages

Object-Oriented Computing, Volume 2: Implementations
Edited by Gerald E. Petersen
(ISBN 0-8186-0822-6); 324 pages

Parallel Architectures for Database Systems
Edited by A.R. Hurson, L.L. Miller, and S.H. Pakzad
(ISBN 0-8186-8838-6); 478 pages

Reduced Instruction Set Computers (RISC) (Second Edition)
Edited by William Stallings
(ISBN 0-8186-8943-9); 448 pages

Software Engineering Project Management
Edited by R. Thayer
(ISBN 0-8186-0751-3); 512 pages

Software Maintenance and Computers
Edited by D.H. Longstreet
(ISBN 0-8186-8898-X); 304 pages

Software Quality Assurance: A Practical Approach
Edited by T.S. Chow
(ISBN 0-8186-0569-3); 506 pages

Software Reusability
Edited by Peter Freeman
(ISBN 0-8186-0750-5); 304 pages

Software Reuse—Emerging Technology
Edited by Will Tracz
(ISBN 0-8186-0846-3); 400 pages

Software Risk Management
Edited by B.W. Boehm
(ISBN 0-8186-8906-4); 508 pages

Standards, Guidelines and Examples on System and Software Requirements Engineering
Edited by Merlin Dorfman and Richard H. Thayer
(ISBN 0-8186-8922-6); 626 pages

System and Software Requirements Engineering
Edited by Richard H. Thayer and Merlin Dorfman
(ISBN 0-8186-8921-8); 740 pages

Test Access Port and Boundary-Scan Architecture
Edited by C. M. Maunder and R. E. Tulloss
(ISBN 0-8186-9070-4); 400 pages

Test Generation for VLSI Chips
Edited by V.D. Agrawal and S.C. Seth
(ISBN 0-8186-8786-X); 416 pages

Visual Programming Environments: Paradigms and Systems
Edited by Ephraim Glinert
(ISBN 0-8186-8973-0); 680 pages

Visual Programming Environments: Applications and Issues
Edited by Ephraim Glinert
(ISBN 0-8186-8974-9); 704 pages

Visualization in Scientific Computing
Edited by G.M. Nielson, B. Shriver, and L. Rosenblum
(ISBN 0-8186-8979-X); 304 pages

VLSI Testing and Validation Techniques
Edited by H. Reghbati
(ISBN 0-8186-0668-1); 616 pages

Volume Visualization
Edited by Arie Kaufman
(ISBN 0-8186-9020-8); 494 pages

REPRINT COLLECTIONS

Distributed Computing Systems: Concepts and Structures
Edited by A. Ananda and B. Srinivasan
(ISBN 0-8186-8975-0); 416 pages

Expert Systems: A Software Methodology for Modern Applications
Edited by Peter Raeth
(ISBN 0-8186-8904-8); 476 pages

Milestones in Software Evolution
Edited by Paul W. Oman and Ted G. Lewis
(ISBN 0-8186-9033-X); 332 pages

Object-Oriented Databases
Edited by Ez Nahouraii and Fred Petry
(ISBN 0-8186-8929-3); 256 pages

Validating and Verifying Knowledge-Based Systems
Edited by Uma G. Gupta
(ISBN 0-8186-8995-1); 400 pages

ARTIFICIAL NEURAL NETWORKS TECHNOLOGY SERIES

Artificial Neural Networks—Concept Learning
Edited by Joachim Diederich
(ISBN 0-8186-2015-3); 160 pages

Artificial Neural Networks—Electronic Implementation
Edited by Nelson Morgan
(ISBN 0-8186-2029-3); 144 pages

Artificial Neural Networks—Theoretical Concepts
Edited by V. Vemuri
(ISBN 0-8186-0855-2); 160 pages

SOFTWARE TECHNOLOGY SERIES

Computer-Aided Software Engineering (CASE)
Edited by E.J. Chikofsky
(ISBN 0-8186-1917-1); 110 pages

Software Reliability Models: Theoretical Development, Evaluation and Applications
Edited by Y.K. Malaiya and P.K. Srimani
(ISBN 0-8186-2110-9); 136 pages

MATHEMATICS TECHNOLOGY SERIES

Computer Algorithms
Edited by Jun-ichi Aoe
(ISBN 0-8186-2123-0); 154 pages

Multiple-Valued Logic in VLSI Design
Edited by Jon T. Butler
(ISBN 0-8186-2127-3); 128 pages

COMMUNICATIONS TECHNOLOGY SERIES

Multicast Communication in Distributed Systems
Edited by Mustaque Ahamad
(ISBN 0-8186-1970-8); 110 pages

ROBOTICS TECHNOLOGY SERIES

Multirobot Systems
Edited by Rajiv Mehrotra and Murali R. Varanasi
(ISBN 0-8186-1977-5); 122 pages

NOTES